A TEXTBOOK OF

ORGANIC CHEMISTRY

(For BSc Students)

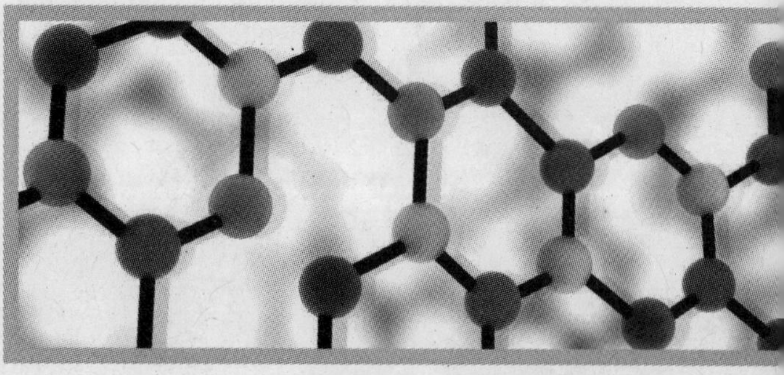

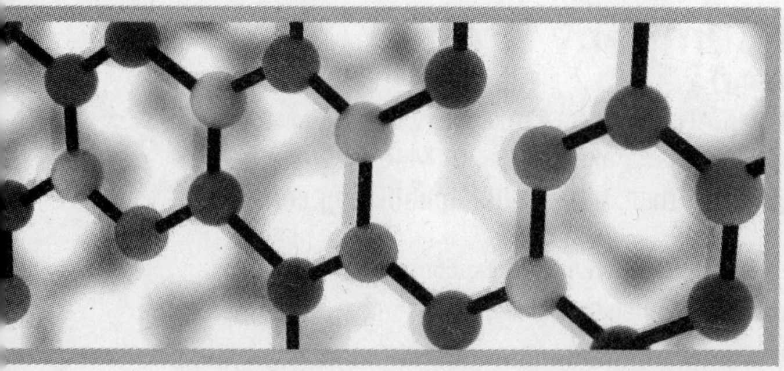

A TEXTBOOK OF

ORGANIC CHEMISTRY

(For BSc Students)

ARUN BAHL

MS (Boston), PhD (Edinburgh), CChem FRSC (London), FAIC (USA)
Formerly Professor of Organic Chemistry, Panjab University, Chandigarh

B S BAHL

MSc. Formerly Principal, and Head of the Postgraduate
Department of Chemistry, DAV College, Jalandhar

S. CHAND
PUBLISHING

S Chand And Company Limited

(ISO 9001 Certified Company)

RAM NAGAR, NEW DELHI - 110 055

S Chand And Company Limited
(ISO 9001 Certified Company)
Head Office: 7361, RAM NAGAR, QUTAB ROAD, NEW DELHI - 110 055
Phone: 23672080-81-82, 66672000 Fax: 91-11-23677446
www.**schandpublishing.com**; e-mail: **info@schandpublishing.com**

Branches:

Ahmedabad	:	Ph: 27541965, 27542369, ahmedabad@schandpublishing.com
Bengaluru	:	Ph: 22268048, 22354008, bangalore@schandpublishing.com
Bhopal	:	Ph: 4209587, bhopal@schandpublishing.com
Chandigarh	:	Ph: 2625356, 2625546, 4025418, chandigarh@schandpublishing.com
Chennai	:	Ph: 28410027, 28410058, chennai@schandpublishing.com
Cuttack	:	Ph: 2332580, 2332581, cuttack@schandpublishing.com
Dehradun	:	Ph: 2711101, 2710861, dehradun@schandpublishing.com
Guwahati	:	Ph: 2738811, 2735640, guwahati@schandpublishing.com
Hyderabad	:	Ph: 27550194, 27550195, hyderabad@schandpublishing.com
Jaipur	:	Ph; 2219175, 2219176, jaipur@schandpublishing.com
Jalandhar	:	Ph: 2401630, jalandhar@schandpublishing.com
Kochi	:	Ph: 2809208, 2808207, cochin@schandpublishing.com
Kolkata	:	Ph: 23353914, 23357458, kolkata@schandpublishing.com
Lucknow	:	Ph: 4065646, lucknow@schandpublishing.com
Mumbai	:	Ph: 22690881, 22610885, 22610886, mumbai@schandpublishing.com
Nagpur	:	Ph: 2720523, 2777666, nagpur@schandpublishing.com
Patna	:	Ph: 2300489, 2260011, patna@schandpublishing.com
Pune	:	Ph: 64017298, pune@schandpublishing.com
Raipur	:	Ph: 2443142, raipur@schandpublishing.com (Marketing Office)
Ranchi	:	Ph: 2361178, ranchi@schandpublishing.com
Sahibabad	:	Ph: 2771235, 2771238, delhibr-sahibabad@schandpublishing.com

First Edition 1948
Subsequent Editions and Reprints 1953, 56, 58, 59, 61, 63, 64, 65, 66, 69, 70, 71, 72, 73, 74 (Twice), 76 (Twice), 77, 79 (Twice), 80 (Twice), 82, 83 (Twice), 84, 85, 86, 87, 88, 89, 90, 91, 92, 93, 95, 96, 97 (Twice), 98 (Twice), 99, 2000, 2001, 2002, 2003, 2004, 2005, First Multicolour Illustrative Edition 2005, Reprints 2006, 2007 (Twice), 2008, 2009, 2010, 2011, 2012, 2013, 2014, 2015, Twenty-Second Edition 2016, Reprint 2017, 2018 (Twice) 2019

LPSPE 2019

ISBN: 978-93-5283-730-4 **Product Code:** H6OCH68CHEM10ENAV19L

PRINTED IN INDIA

By Nirja Publishers & Printers Pvt. Ltd., 54/3/2, Jindal Paddy Compound, Kashipur Road, Rudrapur-263153 Uttarakhand and published by S. Chand & Company Ltd.,7361, Ram Nagar, New Delhi -110 055.

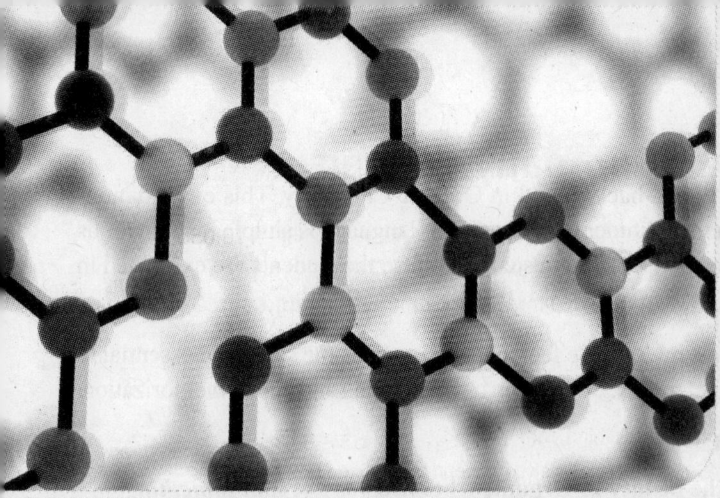

Preface

We feel great pleasure in presenting the twenty-second edition of this best-selling book for over 67 years to the students. It is written in a simple and easy-to-understand language. The text is supported by a number of illustrations and solved examples for better understanding of the concepts discussed. It caters to the needs of BSc (Organic Chemistry) students of Indian universities. It is used all over the country because no other book has our unique combination of three essentials: simplicity, brevity and accuracy.

We have received certain comments from teachers and students which are given below. These comments sum up book's high-quality and study-approach.

An Excellent Book for Organic Chemistry. This is an excellent book on Organic Chemistry. Anybody who wants to master the basic principles of organic chemistry must have this book. Authors lucidly explain each and every topic. I had used this book when I prepared for entrance examinations 15 years back...and now I recommend this to my students...

–Santosh Prasad

Absolutely Essential Textbook. This is a textbook that fits the way Organic Chemistry should be taught. It reads like a novel and teaches like a text. It has a superb focus, a fresh approach, and an easy readability. Emphasis is on reasoning rather than memorization. It is as helpful as a tutor. Five star book!!

–Chetan Garg

Approach Built for Today's Students. Absolutely God sent! As a student I find this text one of the best I have ever used. Its beautiful cover, clear color graphics and images create a pleasant aesthetic that transfers to the material and enhances the learning process.

–Annika Arora

Beautiful Visualization. This is the best textbook of organic chemistry you can find, hands down. Great illustrations, easy-to-understand text, step-by-step mechanisms and more than enough end-of-chapter problems. Visualization is so important for organic chemistry and beautifully presented here. Chapter and page layouts make sense and are easy to follow as now subjects are presented. If you are a teacher I would recommend you start using this textbook to teach organic chemistry course, your students will benefit from it.

–William Kamp

Special Features

Many BSc students do not have a good background in Organic Chemistry. This examination-oriented text is especially written with these students in mind. The language is simple, explanations clear, and presentation very systematic. Concepts that usually confuse the students are explained in a step-by-step manner. Our commitment to quality is total!

Concept-density per page has been kept low. We feel that this is a big time saver and essential to quick-learning and retention of the subject matter. Stress is on understanding and not on memorization.

Increased Focus on Fundamentals

"We very strongly believe that the beauty of Organic Chemistry is not that *something* occurs, but rather *why* it occurs. Understanding the "why" causes a subject to come alive. Hiding "why" hurts the subject and, in the long run, hurts the student."

In this new edition, we have retained all that was judged good in the previous edition. Time-tested functional group approach is retained. Chemistry of aliphatic and aromatic compounds is treated separately. Emphasis is on correlation of the structure of a functional group with its properties. A detailed molecular orbital and valence bond interpretation of the structure of each functional group is given. This enables the student to predict the properties of the functional group.

Fundamental principles of energetics, reaction rates, and stereochemistry are provided to lay a strong foundation. The subject matter is presented in terms of carbonium ions, carbanions, free radicals, and other reactive species. Application of spectroscopy in structure determination has been emphasized. Step-by-step reaction mechanisms are given throughout. All reaction mechanisms have been rewritten using the latest conventions. Fish-hook arrows are used to describe the free-radical mechanisms.

Solving Problems is an Essential Component of the Learning Process

The only way to do well in Organic Chemistry exams is by practice and repetition. End-of-chapter problems will help you achieve a deeper understanding of the material through constant reinforcement and practice.

This book contains more than 1650 latest university questions. It also contains more than 1700 multiple choice questions. By solving these examination problems, you can precisely know your own success-level. Complete answer or clue follows every question. This is the book which the examiners use.

Three special chapters have been included in this textbook to improve your problem-solving skills. They are: Predicting Trends (Acidities, Basicities, Boiling Points, and Solubilities) –Chapter 51, Multistep Synthesis –Chapter 52 and 500 New Multiple Choice Questions –Chapter 53. These three chapters contain about 850 problems, detailed guidelines are provided to help solve such problems effectively.

Excellent for Entrance Tests

This book is an important text for the Entrance Tests for Engineering and Medical Colleges. Your time is valuable, and when every moment counts, you need efficient study material. Your choice of a book can mean success or failure. Because today you need a book that can help you streak ahead of competition and succeed. No-one knows more about your needs than us. It is a tall claim, but it is true!

Value Addition

1. **Full-Color Presentation Facilitates the Learning Process**. One of the distinctive features of the text is its visual impact. This is the first Indian Organic Chemistry book to be completely done in Four-Color and on computer. All structures and reaction equations have been computer-generated using highly advanced Software. New color graphics, illustrations, and real pictures have been extensively used to highlight and reinforce the important points. Color has also been used to highlight change and concepts. All mechanisms have been redrawn using blue arrows to show the precise movement of electrons.
2. **Molecular Models.** We have developed more than 350 molecular models for incorporation into this text. The purpose of these models is to assist you to visualize organic molecules as three-dimensional objects.
3. **Expert Advice** on "How to Study Organic Chemistry" is given immediately after this preface.
4. **Guidelines** are provided to help you understand concepts that are considered difficult and catch careless mistakes before exams.
5. **Scientific Accuracy** has been checked and rechecked. Subject matter is modern and error-free.
6. **Extensive Glossary and Index** have been provided for quick cross-reference.

We Wish you Success!

Our target is your success and we are committed to make that happen. This book has everything you want to have in your Organic Chemistry text. We urge you to study regularly, and hope that this error-free book will make it easier for you to do so. If, after reading you feel you cannot answer a particular question, please feel free to write at:

Mail : 590, Sector 18-B, Chandigarh - 160 018
E-mail : arunbahl2000@gmail.com

We would be glad to receive suggestions and comments for further improvements.

Arun Bahl

B S Bahl

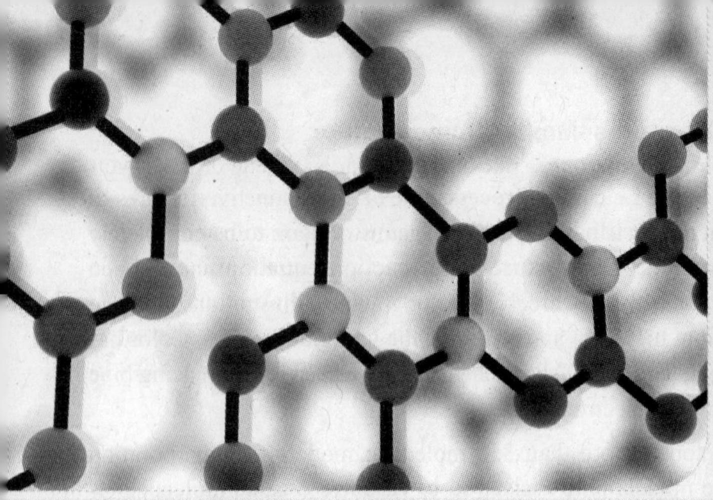

Study Tips

LEARN MORE–CRAM LESS

Each year thousands of students enter the Organic Chemistry classroom with fears about what is called by some, "The Impossible Subject". Indeed, Organic Chemistry can be very difficult, but it is manageable. It can also be very rewarding. Here we offer you a good, simple to follow plan that will allow you to get the kind of high grades you want in Organic Chemistry.

1. Have a Desire to Learn–Never Cram Organic Chemistry

Learning Organic Chemistry does not require any special talent. In order to succeed in your study of Organic Chemistry, you need only the desire to learn and the willingness to work hard.

Organic Chemistry is not a subject that can be crammed. Many students make the mistake of trying to memorize this subject. An understanding of the basic principles is more important than just memorization. It is true that facts must be learned, but you will be confused by them unless you understand them first.

> **"I have no idea where people get the idea that organic chemistry is memorization. I don't even know what you would even memorize. Anyone that tells you organic chemistry requires lots of memorization is seriously misinformed."**

2. Read Carefully–Never Skip Topics

First read each chapter carefully. Then close your book and try to reproduce the key facts you have studied. Be specific and write them down. This technique will rapidly reveal which items you know well and which require further study. Then spend more time on the items which caused trouble.

Organic Chemistry is cumulative. That is, what you learn today is a basis for new material tomorrow. It is important that you attempt to master each topic/concept before going to a new one. So, do not read selectively.

3. Never Get Behind–Set Deadlines

Some of us have poor study habits. A fatal sin in Organic Chemistry is to fall behind. Attempts at cramming a night or two before an examination bring the ghastly realization that the subject cannot be quickly learned. Also, the pace of the course does not allow much breathing space to catch up once one is left behind. So keep up!

Set a deadline for each chapter. Without setting deadlines, we do not get serious about getting something done. Studying organic chemistry for one hour every day is much better than studying seven hours on Sunday.

4. Learn the Basics of Organic Structures–Master Nomenclature

If you don't know how to name a simple compound, like an alcohol or an alkene or ether, you will fail every test. Guaranteed!

The whole language of Organic Chemistry is based on nomenclature and basic structures. The more you learn these basics, the easier it will be to understand how chemical reactions affect them. If you can easily pick out ketones, alcohols, carboxylic acids, ethers, amines and hydrocarbons (alkanes, alkenes, alkynes, cyclic hydrocarbons) and name them, you will be in a very good position to start understanding how and why reactions occur. Practice, Practice, Practice! Nomenclature is easy to understand once you have done it a few times.

5. Study Actively–Use Lots of Paper and Pencil

Write the structures, reactions, and mechanisms over and over again. Organic Chemistry is best assimilated by writing; and not by simply reading. There is a good reason for this. Organic structures, reactions, and mechanisms are complex. If you simply examine them, you may think you understand them thoroughly. This level of understanding comes to most of us through writing. Only by writing we pay attention to the details, such as, which atoms are connected to which atoms, which bonds break in a reaction and which bonds form. When we write reactions and mechanisms, connections are made in our brains that provide the long-term memory needed for success in Organic Chemistry. We can guarantee that your score in Organic Chemistry will be directly proportional to be the number of pages of paper that you will fill with your own writing in studying during this course.

6. Do a Lot of Practice Problems

Let's face it. The standard measuring stick for your performance is your performance on examinations. These are composed of problems that you must solve on your own, without reference to notes or textbook.

It is extremely important that you apply what you have learned. The ideas/concepts will not be fixed in your mind until you put them to use. To achieve this, do the problems given at the end of each chapter. Solve as many problems you can. Do them without notes, and don't peek at the answers. If you cannot understand why something occurs or does not occur, read the text again. Do problems all of the time, not just the day before an exam. This will ensure your success in examinations.

7. Stay Stress-Free

Keep yourself stress-free. This is because stress affects your memory and concentration. It makes you feel tired. Stress reduces your efficiency. Relieve yourself of extra strain by setting goals and keeping a diary.

The above guide should successfully lead you through the trials and tribulations of Organic Chemistry. Good Luck!! (and remember always – carbon only has four bonds!)

BE PREPARED–FOCUS ON UNDERSTANDING

WORK PROBLEMS–PRACTICE DAILY

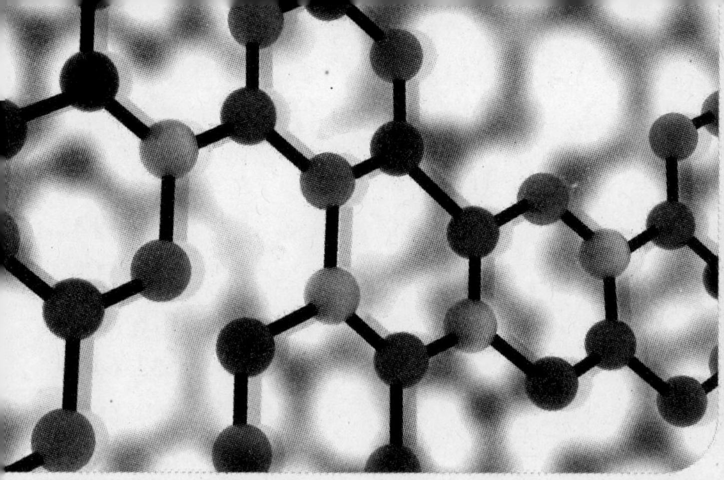

Brief Contents

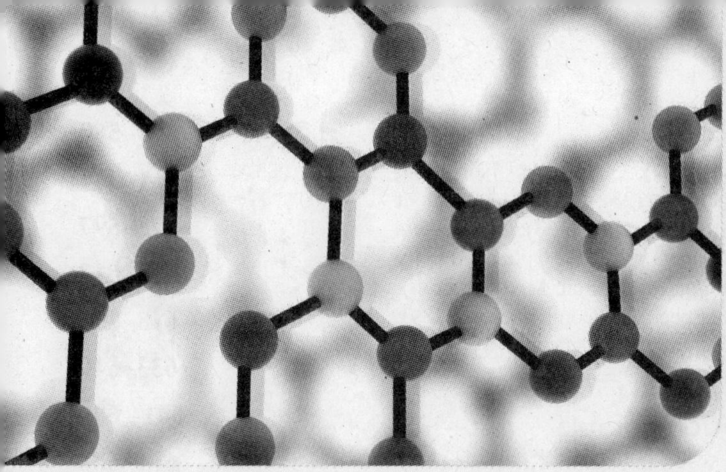

Detailed Contents

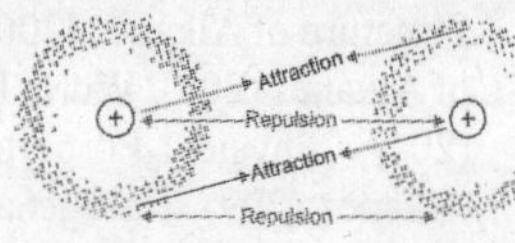

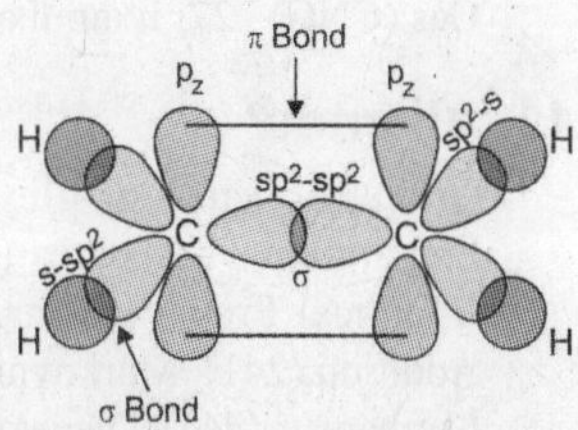

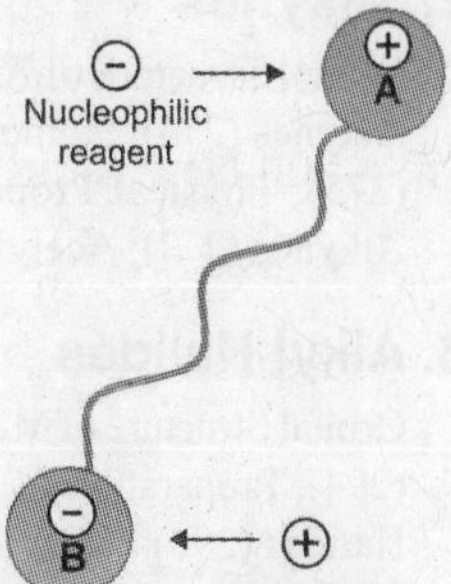

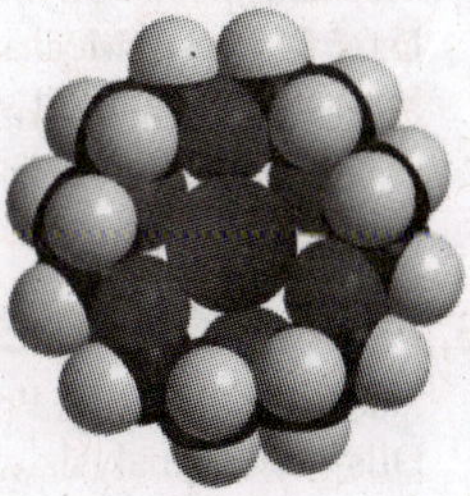

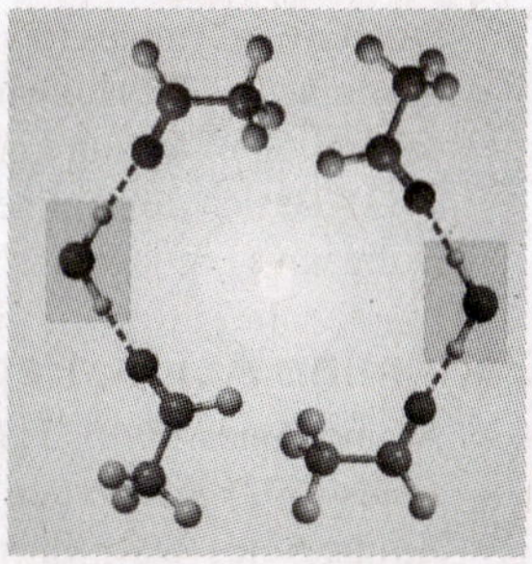

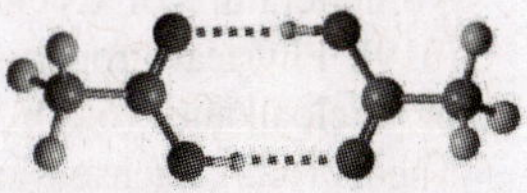

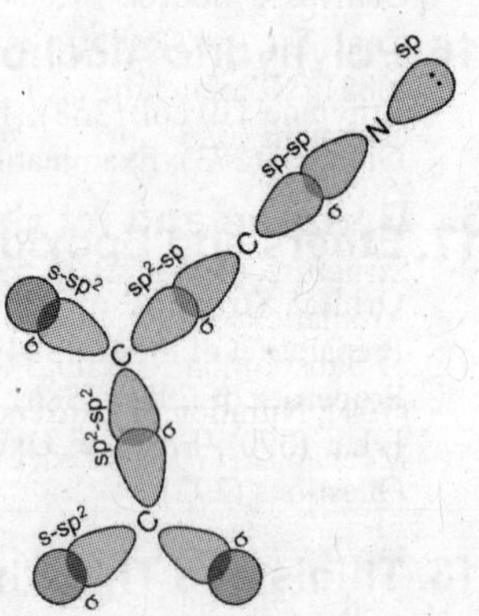

sp
N
sp-sp
σ
C
sp²-sp
s-sp²
σ
C
sp²-sp²
σ
C
s-sp²
σ

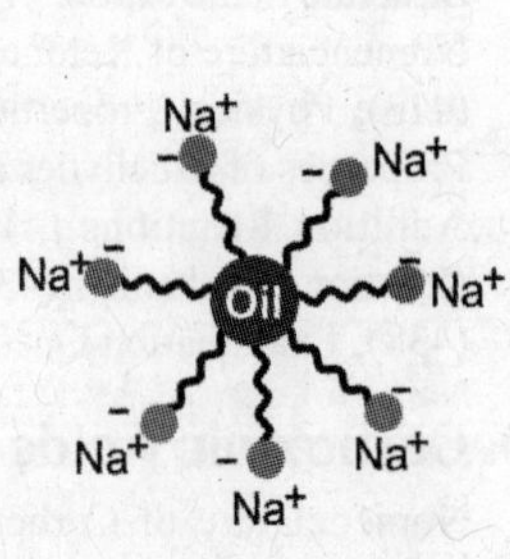

Na+
Oil

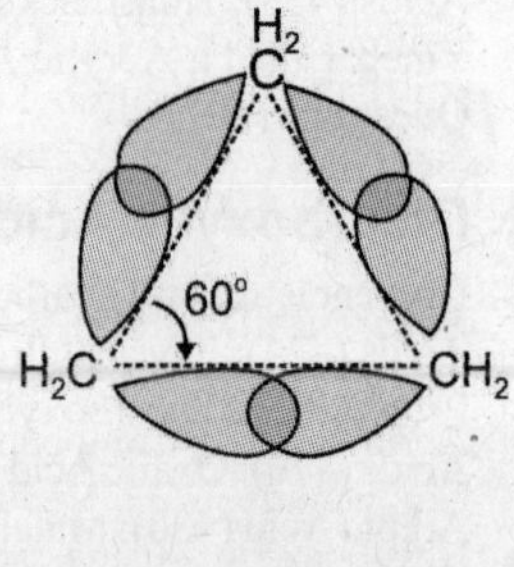

H2
C
60°
H2C
CH2

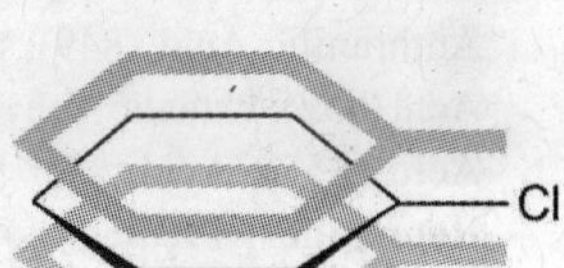
Cl

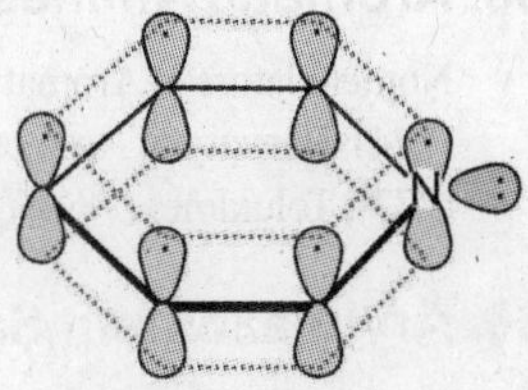
N

1 Getting Started

There are two main classes of chemical compounds, inorganic and organic. More than a century ago all substances then known were classified on the basis of source from which they were derived. The compounds which were of mineral origin were known as inorganic and those of vegetable or animal origin were called organic. For example, table salt, marble, and carbon dioxide were inorganic, whereas acetic acid (from vinegar), alcohol (from wine), tartaric acid (from grapes), citric acid (from lemons) and sucrose (from sugarcane) were organic.

It was assumed that the organic compounds could be produced only by living matter, for living matter was thought to possess a **Vital Force**. As long as this concept prevailed no effort was made to produce organic compounds in the laboratory, and the *Vital Force Theory* long went unchallenged.

In 1828, the German chemist Friedrich Wohler heated ammonium cyanate, derived from inorganic substances, and obtained the organic compound urea. The urea formed in this way proved to be identical in all respects with urea isolated from urine.

$$\underset{\text{Pot. cyanate}}{KOCN} + NH_4Cl \xrightarrow{\text{heat}} KCl + NH_4OCN$$

$$\underset{\text{Amm. cyanate}}{NH_4OCN} \xrightarrow{\text{heat}} \underset{\text{Urea}}{NH_2-\overset{\overset{\displaystyle O}{\|}}{C}-NH_2}$$

This synthesis of an organic compound from totally inorganic materials did not immediately break down the old concept. However, in the following years syntheses of many organic compounds were achieved. Thus by 1850, the *Vital Force Theory* was finally and decisively disproved. It had become evident, however, that most of the compounds formed by living cells (plants and animals) contained carbon. Hence, the emphasis was shifted from origin to composition, and organic compounds came to be considered as those containing carbon.

MODERN DEFINITION OF ORGANIC CHEMISTRY

Although the original meaning of the term organic chemistry no longer applies, it has been retained because of wide usage and also because so many of the substances involved in organic chemistry do come directly or indirectly from living matter. The modern definition may most simply be stated as : **Organic chemistry is defined as the study of carbon compounds.** However, there are several exceptions :

Carbon monoxide, CO	Carbonates, e.g., Na_2CO_3	Thiocyanates, e.g., NaSCN
Carbon dioxide, CO_2	Bicarbonates, e.g., $NaHCO_3$	Cyanates, e.g., KOCN
Carbon disulfide, CS_2	Cyanides, e.g., KCN	Carbides, e.g., CaC_2

These compounds are studied as part of inorganic chemistry because of their properties.

Points to Remember

1. **Organic chemistry is all around us.** Organic chemistry is that branch of chemistry that deals with the structure, properties, and reactions of compounds that contain carbon. It is a highly creative science. Chemists in general and organic chemists in particular can create new molecules never before proposed which, if carefully designed, may have important properties for the betterment of the human experience.
2. **All organic compounds contain the element carbon.** Most of them also contain hydrogen. Many organic compounds also contain other elements : oxygen, nitrogen, sulfur, phosphorus, and the halogens.

3. More than 13 million organic compounds are known.

WHY ORGANIC CHEMISTRY IS A SEPARATE DISCIPLINE ?

Organic compounds obey the same fundamental laws of chemistry that hold for inorganic compounds. However, they are studied as a separate discipline because of the following reasons :

(1) **Large Number of Compounds.** There are approximately 100,000 known inorganic compounds and this number is not rapidly increasing. On the other hand, there are over 35 million known organic compounds and several thousand new compounds are synthesized each year. If the study of 5 million organic compounds was included with that of carbon under inorganic chemistry, it would throw the subject out of balance.

(2) **Unique Chemical and Physical Properties.** There are marked differences between the composition, structure, and properties of organic and inorganic compounds (Table 1.1) which make their study as a separate branch more practical and useful.

(3) **Unique Character of Carbon.** Carbon has the ability to bond successively to other carbon atoms to form chains of varying lengths and shapes. This property of carbon is called **Catenation** and is responsible for the variety and large number of organic compounds.

Table 1.1 A COMPARISON OF PROPERTIES OF ORGANIC AND INORGANIC COMPOUNDS

Organic	Inorganic
1. Low melting points	**1.** High melting points
2. Low boiling points	**2.** High boiling points
3. Low solubility in water; high solubility in nonpolar solvents	**3.** High solubility in water; low solubility in nonpolar solvents
4. Flammable	**4.** Nonflammable
5. Solutions are nonconductors of electricity	**5.** Solutions are conductors of electricity
6. Exhibit covalent bonding	**6.** Exhibit ionic bonding
7. Exhibit isomerism	**7.** Isomers are limited to a few exceptions (e.g., the transition elements)
8. Chemical reactions are usually slow	**8.** Chemical reactions are fast

IMPORTANCE OF ORGANIC CHEMISTRY

No field of science is so closely related with our daily activities as is organic chemistry. The food we eat is mainly organic in nature. The changes which this food undergoes in our bodies, are organic chemical reactions. Metabolism, growth and maintenance of our bodily functions involves organic chemistry, as do the analogous changes taking place throughout the entire living world, plant and animal.

The clothes we wear, the dyes that color them, the soap and starch used to launder them, the leather in our shoes, as well as the dye and shoe polish, are products of organic chemical industry. Many of the structural materials in our houses and furniture, as well as paints and varnishes used for finishing them, are all organic. Many of the appointments in our motor cars, their fuels and lubricants, and the fuels which power our industrial plants are all organic in nature. The tremendous importance of organic compounds in everyday life is illustrated by the following list :

Food : Proteins, fats, carbohydrates
Clothing : Cotton, silk, wool, nylon, rayon, dacron
Shelter : Wood, paints, varnishes
Power and Transportation : Natural gas, petroleum products, coal
Medicines and Drugs : Penicillin G, streptomycin, LSD
Insecticides : DDT
Herbicides : Treflan, 2,4-D
Hormones and Steroids
Vitamins and Enzymes
Antiseptics and Anesthetics
Pigments and Dyes
Paper and inks
Photographic films and Developers
Perfumes and Flavors
Plastics, Rubber and Resins
Propellants and Explosives
Soaps and Detergents
Refrigerants

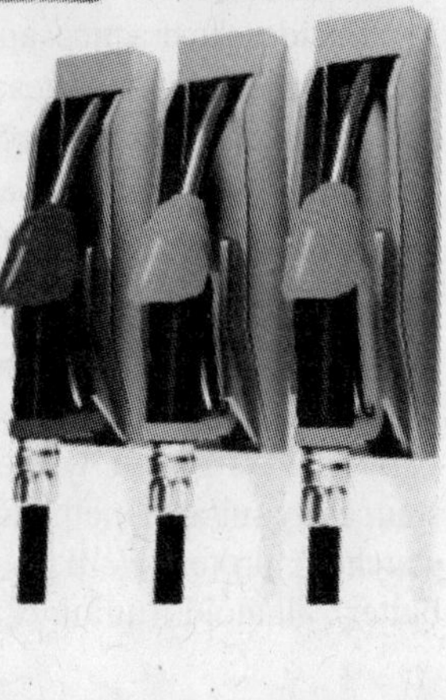

Frequent investigations based upon organic chemistry have profoundly influenced the lives of many people, sometimes even the course of civilization. At one time thousands of acres of land were required for the growing of plants such as Indigo and Turkey red. Organic chemical research showed how to synthesize these dyes from coal-tar, and thus released this land for other productive uses.

Importance of organic chemistry is enormous. It deals with many of the most important man-made artifacts of everyday life, such as gasoline, synthetic fibers, plastics, insecticides, moth-repellants, explosives, dyes, and medicines.

The course of civilization has been greatly affected by the advances in medicine. Much of the progress has been due to the introduction of such organic compounds as antiseptics, anesthetics, and antibiotics. Disease has been checked, the death rate reduced, and the life expectancy of mankind doubled within the last half century. Such profound alteration in human economy, of course, poses new problems for mankind. Possible over population, inadequate supplies of food, increased average age of population, and the like, now challenge the ingenuity of our leaders in the various fields of human endeavour.

Besides the practical applications, most biological problems are concerned with organic chemistry. The composition of tissues and of foods; their changes in metabolism, growth processes, physiology, disease, all involve the reactions of organic chemistry. An understanding of biological processes, the functioning of plants and animals, and many fundamental phases of agriculture, medicine, and the like, can be based only upon a knowledge of organic chemistry.

SOURCES OF ORGANIC COMPOUNDS

Historically, as has been mentioned, all organic compounds at one time came from plants or animals. Many of our most important substances are still derived from these sources directly or indirectly.

(1) **Plants and Animals.** Many organic compounds are obtained directly from plant or animal sources by suitable methods of isolation. A few familiar examples are carbohydrates (cellulose, sugars, starches), proteins (silk, wool, casein, food proteins), fats and oils (cottonseed, soybean oils, lard, butter), alkaloids (quinine, morphine, strychnine), hormones, vitamins, perfumes, favors, resins.

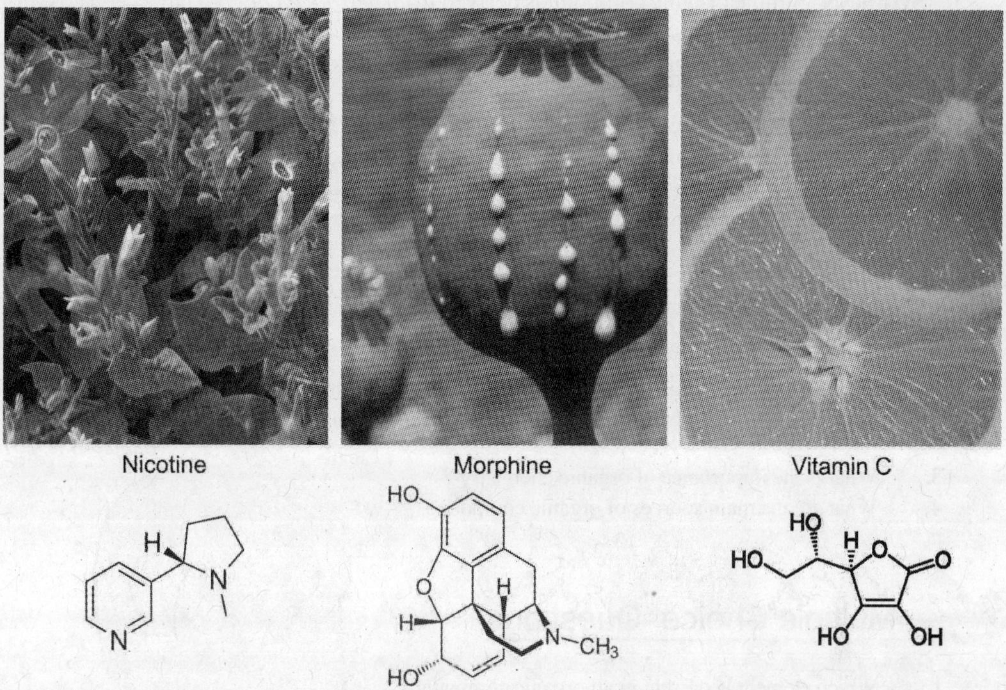

Three examples of organic compounds in living organisms. Tobacco contains nicotine, an addictive alkaloid. Rose hips contain vitamin C, essential for preventing scurvy. Opium poppies contain morphine.

(2) **Natural Gas and Petroleum.** Natural gas and petroleum are now the major sources of organic compounds. They are used as fuels and also, through synthetic organic reactions, for the production of hundreds of useful organic substances such as solvents, synthetic rubber, explosives, and plastics.

(3) **Coal.** Coal is another major source of organic compounds. It yields coke and coal-tar on pyrolysis or destructive distillation. More than 200 organic compounds have been directly isolated from coal-tar. These coal-tar products form the starting materials for the manufacture of thousands of useful aromatic compounds, including perfumes, drugs, dyes, photographic developers, and others.

Natural gas, Petroleum and Laboratory synthesis are the main sources of organic compounds today.

(4) **Fermentation.** Fermentation is defined as the production of chemicals by the action of microorganisms. By employing appropriate organisms hosts of useful substances are produced including alcohols, acids, vitamins, and antibiotics.

(5) **Synthesis.** Simple organic compounds derived from *petroleum or coal* have been converted into thousands of useful materials by synthetic methods. Many examples might be cited of synthetic organic compounds replacing those obtained from natural sources, such as dyes, rubber, fibers, plastics, drugs, vitamins. In many cases the synthetic materials are superior to the natural compounds replaced. For example, synthetic dyes are superior to those of natural origin. In other cases the synthetic materials are entirely unknown in nature and fill the requirements not satisfied from any other source. Examples are ether, glycol, aspirin, and the sulpha drugs. Synthetic organic chemistry touches almost every phase of life.

Examination Questions

1. What is organic chemistry ? Briefly discuss how organic and inorganic compounds differ.
2. What property of carbon accounts for the occurrence of so many carbon compounds ? (*Bhopal BSc, 2015*)
3. What is the importance of organic chemistry ?
4. What are the main sources of organic compounds ?

Multiple Choice Questions

1. Which element is present in all organic compounds ?

(*a*) Carbon (*b*) Nitrogen
(*c*) Oxygen (*d*) Phosphorus

Answer. (*a*)

2. Which property is generally characteristic of an organic compound ?

(*a*) low melting point (*b*) high melting point
(*c*) soluble in polar solvents (*d*) insoluble in nonpolar solvents

Answer. (*a*)

3. Which of the following is not a characteristic of organic compounds ?

(*a*) They usually have low melting points.
(*b*) They usually are only slightly soluble or insoluble in water.
(*c*) If water soluble they seldom conduct an electric current.
(*d*) Bonds which bind the atoms together are nearly always ionic.

Answer. (*d*)

4. Which statement explains why the element carbon forms so many compounds ?

(*a*) Carbon atoms combine readily with oxygen.
(*b*) Carbon atoms have very high electronegativity.
(*c*) Carbon readily forms ionic bonds with other carbon atoms.
(*d*) Carbon readily forms covalent bonds with other carbon atoms.

Answer. (*d*)

5. Catenation is a property of the carbon atom which describes its ability to

(*a*) bond with other carbon atoms (*b*) form double and triple bonds
(*c*) exist in plant and animal form (*d*) form bonds in its ground state

Answer. (*a*)

6. One of the major sources of organic compounds is

(*a*) Natural gas (*b*) Fermentation
(*c*) Sea water (*d*) Atmosphere

Answer. (*a*)

2 Purification of Organic Compounds

Organic compounds derived from natural sources are seldom pure. They are often mixed with other substances which also occur with them. Similarly, the compounds prepared in the laboratory are generally mixed with the products which may also have been formed during the course of reaction. In order to study its properties and to determine its structure, a given substance must be first of all obtained in a state of purity. The methods employed for purification depends upon the nature of the organic compound and the impurities present in it. The processes commonly used for the separation and purification of organic compounds are :

(1) Crystallization

(2) Sublimation

(3) Distillation

(4) Fractional distillation

(5) Distillation under reduced pressure

(6) Steam distillation

(7) Extraction with solvents

(8) Chromatography

CRYSTALLIZATION

This is the process most commonly employed for the purification of solid organic compounds. The impure solid is dissolved in the minimum volume of a suitable solvent. The soluble impurities pass into solution while the insoluble ones are left behind. The hot solution is then filtered and allowed to cool undisturbed till crystallization is complete. The crystals are then separated from the mother liquor by filtration and dried. The efficiency of the process of crystallization depends on the skill with which the following steps are carried out :

(1) **Choice of Solvent.** For a judicious choice of the solvent, about 50 mg of the solid is taken in

a test tube and shaken with a few drops of the solvent. The solvent is unsuitable if whole of the solid goes into solution at room temperature. If the solid dissolves upon heating and throws out maximum crystals on cooling, the solvent is suitable. This process is repeated with solvents such as water, benzene, ether, ethyl alcohol, acetone, etc., till the most satisfactory one is sorted out.

(2) **Preparation of Solution.** A suitable quantity of the solid substance is taken in a conical flask fitted with a reflux condenser. A small volume of the solvent selected in step (1) is also placed in the flask. The quantity of the solvent should be just enough to dissolve the whole solid on boiling. The heating may be done on a water bath or wire gauze according as the solvent is low boiling or high boiling.

(3) **Filtration of Hot Solution.** The hot solution obtained above is then filtered through a fluted filter paper placed in an ordinary glass funnel. If the quantity of the solution is large, it takes longer and the crystals may form in the funnel during filtration. To prevent this, a hot-water funnel may be used.

(4) **Crystallization.** The hot filtrate is then allowed to cool undisturbed in a beaker. The pure solid substance separates as crystals. When a rich crop of crystals has been obtained, crystallization is complete. Sometimes crystallization does not occur even after cooling the filtrate for a sufficiently long time. In such a case the crystallization is induced either by scratching the walls of the beaker with a glass rod or by adding a few crystals of the pure compounds to the solution.

(5) **Separation and Drying of Crystals.** The crystals are separated from the mother liquor by filtration. The filtration is generally effected under reduced pressure using a Buchner (or Hirsch) funnel as shown in Fig. 2.1. When the whole of the mother liquor has been drained into the filtration flask, the crystals are washed with small quantities of the pure cold solvent to remove adhering impurities. The crystals are then dried by pressing between pads of filter papers, in an oven, or in a vacuum desiccator (Fig. 2.2).

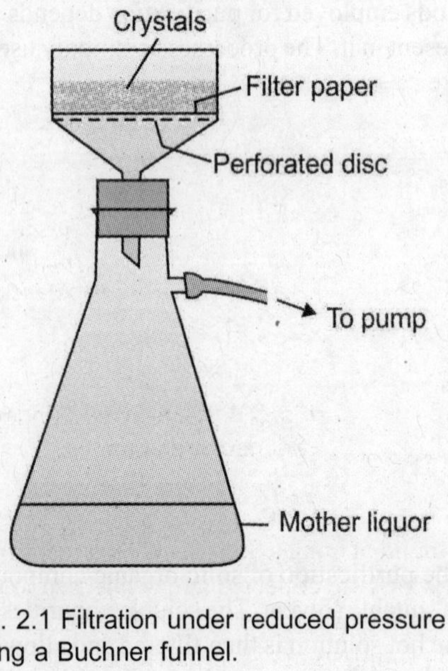

Fig. 2.1 Filtration under reduced pressure using a Buchner funnel.

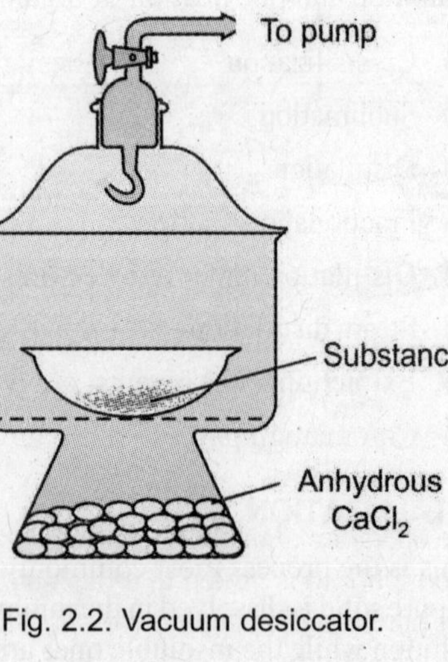

Fig. 2.2. Vacuum desiccator.

SUBLIMATION

Some substances when heated, pass directly from the solid to the vapor state without melting. The vapors when cooled give back the solid substance. This process known as sublimation is very helpful in separating volatile from nonvolatile solid. It is, however, of limited application as only a few substances like naphthalene, camphor, and benzoic acid can be purified by this process.

The impure substance is placed in a large beaker (Fig. 2.3). The beaker is covered with a watch glass. Heat is applied gently to the beaker, and the resulting vapors condense as crystals on the bottom surface of the watch glass. The watch glass can be removed after some time and the crystals collected.

Organic substances such as benzoic acid, naphthalene etc., which have high vapor pressure at temperatures below their melting points, can be sublimed relatively quickly. These can be conveniently purified by the method described above. Substances which have very small vapor pressure or tend to decompose upon heating, are purified by sublimation under reduced pressure.

The glass apparatus used for sublimation under reduced pressure is shown in Fig. 2.4. The chief features of this apparatus are a large heating and a large cooling surface with a small distance in between. This is necessary because the amount of the substance in the vapor phase is much too small in case of a substance with low vapor pressure.

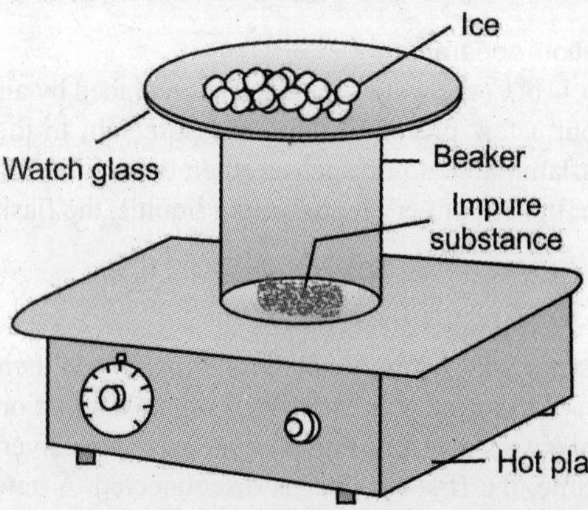

Fig. 2.3. Sublimation apparatus.

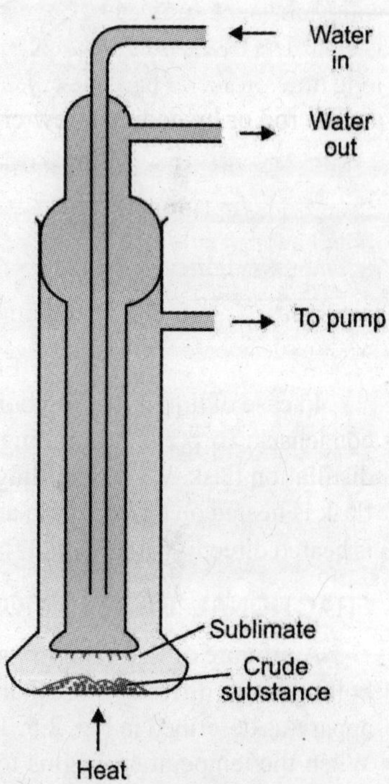

Fig. 2.4. Sublimation under reduced pressure.

DISTILLATION

The operation of distillation is employed for the purification of liquids from nonvolatile impurities. The impure liquid is boiled in a flask and the vapors so formed are collected and condensed to give back the pure liquid in another vessel. The nonvolatile impurities are left behind in the flask.

The apparatus used for distillation is shown in Fig. 2.5. It consists of a distillation-flask fitted with a thermometer in its neck and a condenser at the side-tube. The liquid to be purified is placed in the distillation-flask and the thermometer so adjusted that its bulb stands just below the opening of the side-tube. This ensures the correct recording of the temperature of the vapor passing over the condenser. A suitable vessel is attached to the lower end of the condenser to receive the condensed liquid.

On heating the distillation flask, the thermometer first records a rise in temperature which soon becomes constant. At this point, which is the boiling temperature of the pure liquid, most of the liquid passes over. Towards the end of the operation the temperature rises once again on account of the superheating of the vapor. The distillation is stopped at this stage and the receiver disconnected.

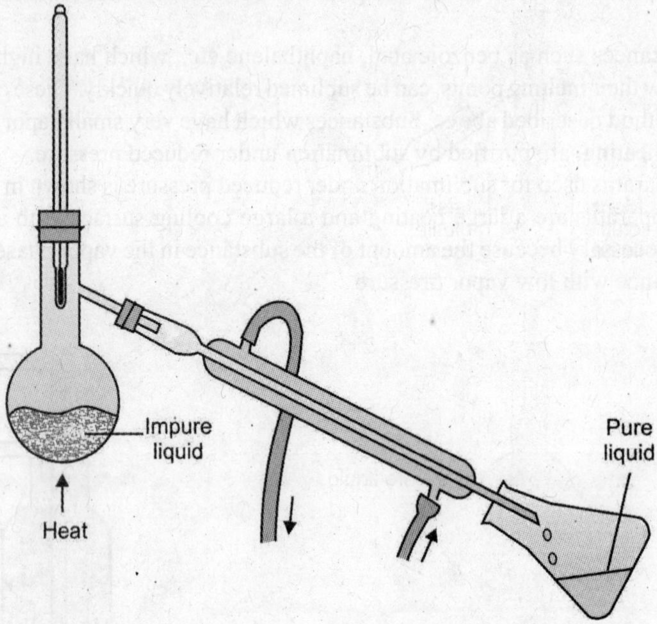

Fig. 2.5. Simple distillation apparatus.

In case of liquids having boiling points lower than 110°C, the water-condenser is replaced by air condenser. To prevent bumping, it is customary to put a few pieces of unglazed porcelain in the distillation flask. While distilling a very volatile and inflammable liquid such as ether, the distillation flask is heated on a water-bath and not on a wire gauze. In case of very high-boiling liquids, the flask is heated directly with a naked flame.

FRACTIONAL DISTILLATION

A mixture of two or more volatile liquids can be separated by fractional distillation. When their boiling points differ by more than 40°C, the operation can be carried with the help of simple distillation apparatus described in Fig. 2.5. The more volatile liquid passes over first and is collected in a receiver. When the temperature begins to rise for the second time, the first receiver is disconnected. A new receiver is attached as soon as the temperature becomes constant once again. Thus the distillate is collected in fractions and the process is termed *Fractional Distillation.*

When the liquids present in the mixture have their boiling points close to each other, the separation is best effected by fitting the distillation flask with a fractionating column which in turn is connected to the condenser (Fig. 2.6.). On heating, the vapors of the more volatile liquid A, along with a little of the vapors of less volatile liquid B, rise up and come in contact with the large cooling surface of the fractionating column. The vapors of B condense first and that of A pass on. The condensed liquid flowing down the column meets the fresh hot ascending vapor. It snatches more of B from the vapor mixture and gives up any dissolved vapor of A. This process is repeated at every bulb of the fractionating column, so that the vapor escaping at its top consists almost exclusively of A and the condensed liquid flowing back into the distillation flask is rich in B. If necessary, the process can be repeated with the distillate and the liquid left in the distillation flask.

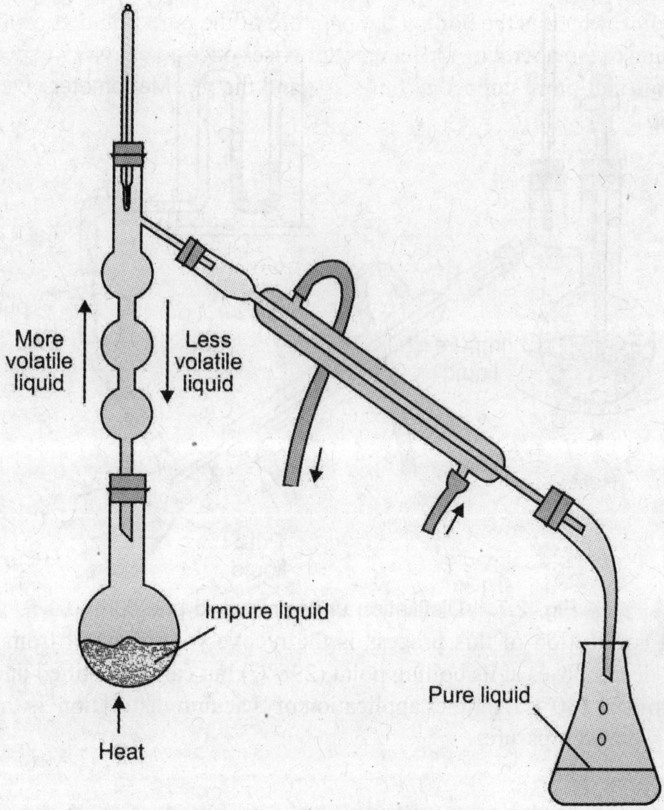

Fig. 2.6. Fractional distillation

The use of fractionating column has found a remarkable application in modern industry, especially, in the distillation of petroleum, coal-tar and crude alcohol.

DISTILLATION UNDER REDUCED PRESSURE

The 'straight' distillation is suitable only for liquids which boil without decomposition at atmospheric pressure. In case of organic liquids which decompose before their boiling point is reached, the distillation is carried under reduced pressure when the liquid boils at a lower temperature. A liquid boils when its vapor pressure is equal to the atmospheric pressure. When the pressure is reduced by suction, the liquid boils at a lower temperature. The apparatus used for distillation under reduced pressure is shown in Fig. 2.7.

(1) **Claisen flask,** having two necks. It is fitted with a long drawn jet, dipping in the liquid to be distilled. During the distillation, a stream of bubbles rises through the capillary of this jet and prevents bumping, which is so pronounced here than in ordinary distillation.

(2) **Condenser,** connected with the Claisen flask on the one hand and a filtration flask, serving as a receiver, on the other.

(3) **Manometer.** The receiver flask is connected to an exhaust pump through (*a*) a mercury manometer which tells the pressure under which the distillation is being carried, and (*b*) a trap, to eliminate any condensed liquid.

The pressure in the apparatus is reduced with the help of a water pump. Whenever a lower pressure is desired, the water pump is replaced by a mercury pump.

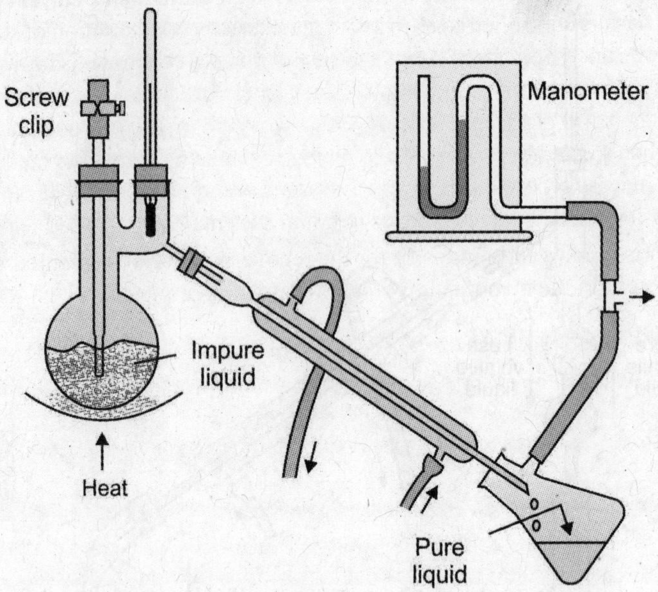

Fig. 2.7. Distillation under reduced pressure.

An important application of this process is the recovery of glycerol from spent-lye in soap industry. Glycerol decomposes at its boiling point (298°C) but can be distilled unchanged at 12 mm pressure when it boils at 180°C. Another application of 'vacuum distillation' is the concentration of sugar juice under reduced pressure.

STEAM DISTILLATION

Many substances that are insoluble in water and volatile in steam, can be purified by distillation in a current of steam (Steam Distillation). The nonvolatile impurities are left behind in the distillation flask.

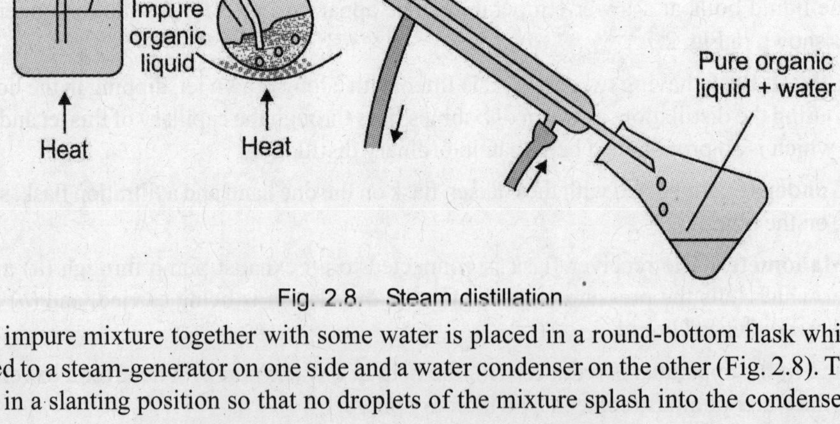

Fig. 2.8. Steam distillation

The impure mixture together with some water is placed in a round-bottom flask which is then connected to a steam-generator on one side and a water condenser on the other (Fig. 2.8). The flask is adjusted in a slanting position so that no droplets of the mixture splash into the condenser on brisk boiling and bubbling of steam. The mixture in the flask is heated and then a current of steam passed into it. Heating of the flask is controlled so as to avoid unnecessary condensation of steam in it. Steam picks up the volatile substance from the mixture and passes into the condenser. The distillate collected in the receiver consists of a mixture of water and the organic substance. The distillation is stopped when the droplets or the solid particles of the organic substance cease to appear in the condenser.

The distillate is then treated to recover the organic substance by a suitable method. In case it is a solid, the substance may be separated by simple filtration, and if it is a liquid, it can be removed by means of a separating funnel. The aqueous layer in both cases may be extracted with a solvent.

Steam distillation is employed in industry for the recovery of various essential oils from plants and flowers. It is also used in the manufacture of aniline and turpentine oil.

EXTRACTION WITH A SOLVENT

When an organic substance is present as solution in water, it can be recovered from the solution by the following steps :

(1) The aqueous solution is shaken with an immiscible organic solvent in which solute is more soluble.

(2) The solvent layer is separated by means of a separating funnel.

(3) The organic substance is then recovered from it by distilling off the solvent.

The process of removing a substance from its aqueous solution by shaking with a suitable organic solvent is termed **Extraction.**

Procedure. The aqueous solution is placed in a separating funnel. A small quantity of the organic solvent, say ether or chloroform, is then added to it. The organic solvent being immiscible with water, will form a separate layer. The mouth of the funnel is closed with a stopper or the palm of the hand and the contents shaken gently. The solute being more soluble in the organic solvent is transferred to it. The solvent layer is then separated by opening the tap and running out the lower layer. The organic substance dissolved in it is finally recovered by distilling off the solvent. It is always better to extract two or three times with smaller quantities of the solvent than once with the bulk of the solvent provided.

Soxhlet Extraction. When an organic substance is to be recovered from a solid, it is extracted by means of an organic solvent in which the impurities are insoluble. In actual practice the extraction from solids is often tedious and requires thorough contact and heating with the solvent. This is done in special apparatus, the *Soxhlet Extractor* (Fig. 2.10). It consists of a glass cylinder having a side tube T and siphon S. The cylinder carries a water condenser at the top and is fitted below into the neck of a boiling flask.

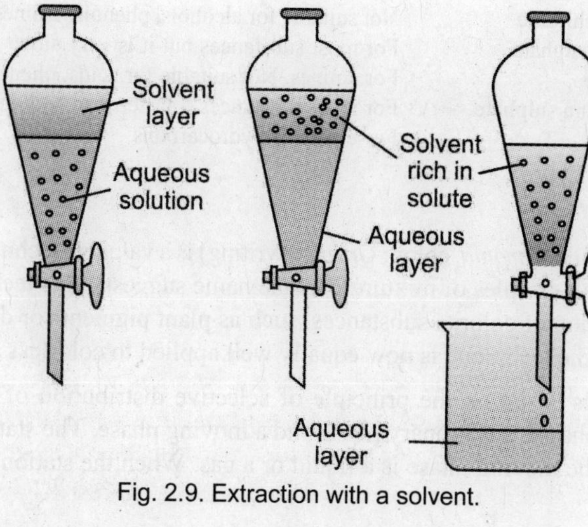

Fig. 2.9. Extraction with a solvent.

The powdered material is placed into the thimble made of stout filter paper and the apparatus is fitted up as shown in Fig. 2.10. The flask containing a suitable solvent is heated on a water-bath or sand bath. As the solvent boils, its vapors rise through the side tube T up into the water condenser. The condensed liquid drops on the solid in the thimble, dissolves the organic substance and filters out into the space between the thimble and the glass cylinder. As the level of liquid here rises, the solution flows through the siphon back into the boiling flask. The solvent is once again vaporized, leaving behind the extracted substance in the flask. In this way, a continuous stream of pure solvent drops on the solid material, extracts the soluble substance and returns to the flask. At the end of the operation the solvent in the boiling flask is distilled off, leaving the organic substance behind.

Soxhlet extractor is used with advantage for the extraction of oils and fats from flowers and seeds, and alkaloids from plants. The apparatus ensures maximum extraction with a limited quantity of the solvent.

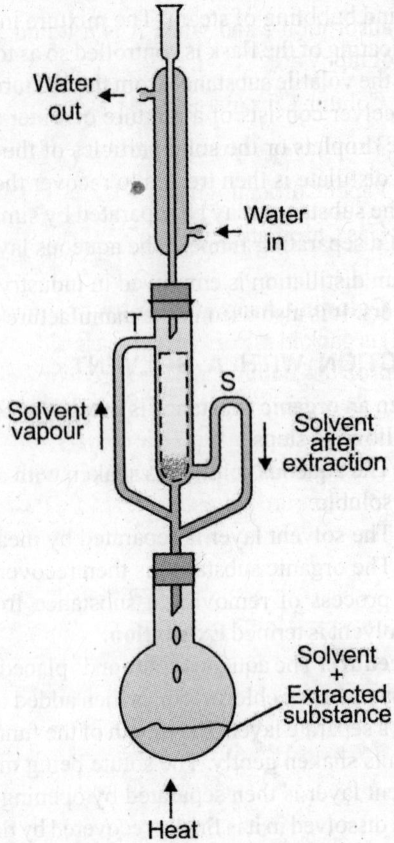

Fig. 2.10. Soxhlet extractor.

DRYING SOLUTIONS WITH DRYING AGENTS

There are various chemicals that can be used to remove the last traces of water from a pure liquid. These are summarized below.

Drying Agent	Use
Anhydrous calcium chloride	Not suitable for alcohols, phenols, amines and acidic compounds
Anhydrous sodium sulphate	For most substances but it is very slow
Potassium hydroxide	For amines. Not suitable for acids, phenols and esters.
Anhydrous magnesium sulphate	For most substances – faster than sodium sulphate
Sodium wire	For ethers or hydrocarbons

CHROMATOGRAPHY

Chromatography (Gk. *Chroma*, color ; *Graphe*, writing) is a valuable technique for the separation and purification of small samples of mixtures. As the name suggests, this technique was originally confined to the separation of colored substances, such as plant pigments or dyestuffs. The name is no longer accurate, as the technique is now equally well applied to colorless substances.

Chromatography is based on the principle of selective distribution of the components of a mixture between two phases, a stationary phase and a moving phase. The stationary phase can be a solid or liquid, while the moving phase is a liquid or a gas. When the stationary phase is solid, the basis is adsorption ; and when it is a liquid, the basis is partition. Some of the common types of chromatography are :

(1) Column Chromatography

(2) Thin Layer Chromatography

(3) Paper Chromatography

(4) Gas Chromatography

(1) **Column Chromatography.** In column chromatography the powdered adsorbent (alumina or silica gel) is packed into a vertical glass tube. A small quantity of the mixture to be separated is added to the top of the column where it remains adsorbed. A solvent (pentane, benzene, ether) is then allowed to flow down the tube. Separation of the components of the mixture takes place as a result of differential adsorption. If the components are visible, colored bands appear down the column. The flow of the solvent is continued until all the components are eluted from the bottom, one by one. The different fractions are collected and individual components recovered from the solution (Fig. 2.11).

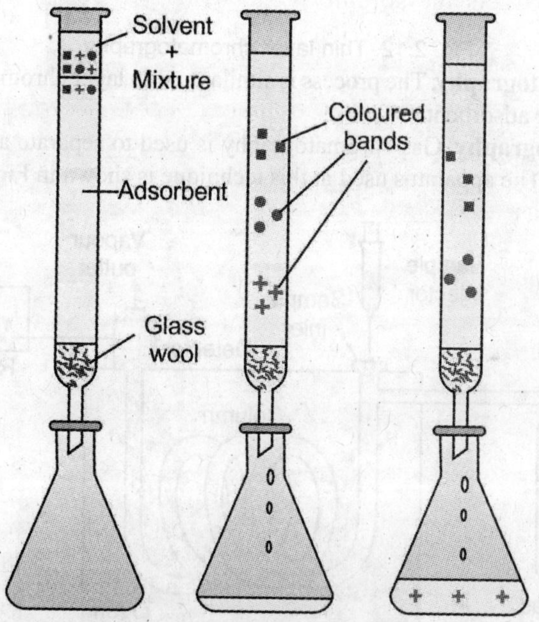

Fig. 2.11. Separation of a mixture by column chromatography. (a) mixture added to top of the column ; (b) appearance of bands ; (c) collection of fractions.

(2) **Thin Layer Chromatography (TLC).** With thin layer chromatography it is possible to separate even minute quantities of mixtures. A sheet of glass is coated with a thin layer of adsorbent (cellulose powder, alumina, silica). This is done by mixing the adsorbent with a suitable liquid, usually water, to form a slurry. This is applied to the glass by spreading or dipping. After drying the plate, a drop of the mixture is placed just above one edge, which is then dipped in a pool of solvent (Fig. 2.12). The solvent is drawn up the adsorbent layer by capillary action. The separation of the mixture occurs as in column chromatography.

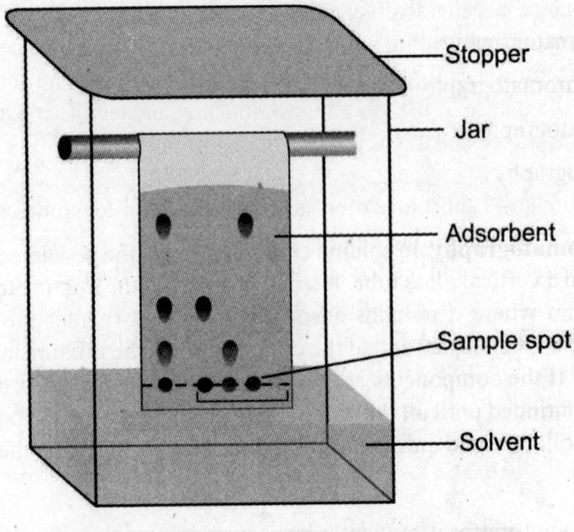

Fig. 2.12. Thin layer chromatography.

(3) **Paper Chromatography.** The process is similar to thin layer chromatography, except that a strip of paper acts as the adsorbent.

(4) **Gas Chromatography.** Gas chromatography is used to separate and analyze mixtures of volatile organic liquids. The apparatus used in this technique is shown in Fig. 2.13.

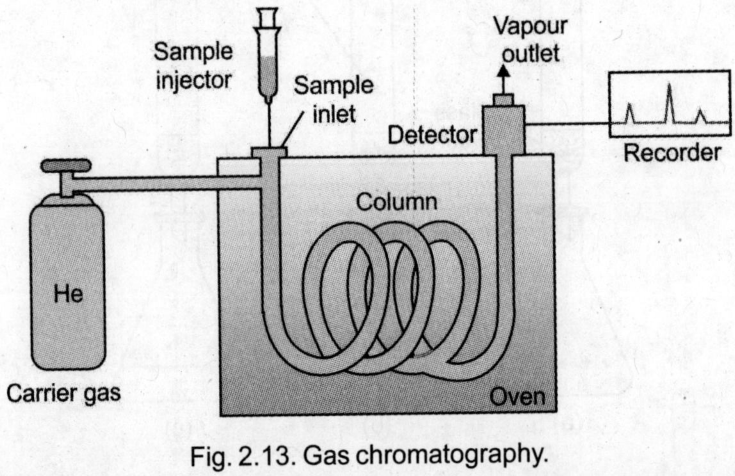

Fig. 2.13. Gas chromatography.

The column packing consists of porous, inert powder (crushed firebrick) impregnated with an oil (silicone or petroleum jelly). Helium is passed through the heated column, and a sample of the liquid mixture to be separated is injected into the helium stream. The mixture vaporizes and is carried through the column by the helium flow. The components of the mixture pass through the column at different rates, some quickly, some slowly. As each component reaches the detector, a peak is provided by the recorder. A gas chromatogram of a mixture of hydrocarbons is shown in Fig. 2.14.

The area under the peak is proportional to the concentration of that component in the sample. A component is identified by its retention time, that is, the length of time it remains in the column. By chilling the exit stream of helium (from vapor outlet) at appropriate times, each individual component can be condensed and collected.

TESTS OF PURITY

A pure organic substance has characteristic physical properties, crystalline form, refractive index, specific gravity, melting point and boiling point. If a given sample of substance shows the properties that the pure compound is known to possess, it may be considered pure. The numerical value of these properties changes with the nature of the amount of impurities present. In most laboratory work, the melting point of a solid substance and the boiling point of a liquid substance is considered a sufficient indication of its purity.

Melting Point. A pure solid substance melts sharply at a definite temperature, while an impure substance will have a lower and indefinite melting point.

Procedure. The apparatus employed for the determination of the melting point of a given solid substance is shown in Fig. 2.15. The crystals are powdered finely and charged into a capillary tube sealed at one end. The capillary tube should be 5 to 6 cm long and 1 mm in diameter. The substance should stand in the capillary 3-4 mm from the bottom when thoroughly packed. The capillary tube is wetted with the liquid in the bath and placed along side a thermometer fixed in an iron stand. The capillary remains sticking to the thermometer by itself and is so adjusted that the solid in it stands just opposite to the middle of the mercury bulb. The thermometer is now lowered in a beaker containing paraffin oil and the apparatus set up as shown in Fig. 2.15. The beaker is heated slowly and the temperature of the bath kept uniform by gentle but constant stirring with a ring stirrer. When the substance in the capillary just shows signs of melting, the burner is removed and the stirring continued. The temperature at which the substance just melts and becomes transparent is recorded. The experiment is repeated with a new capillary charged with fresh substance and the average of the two melting points thus determined is the correct melting point of the substance under examination.

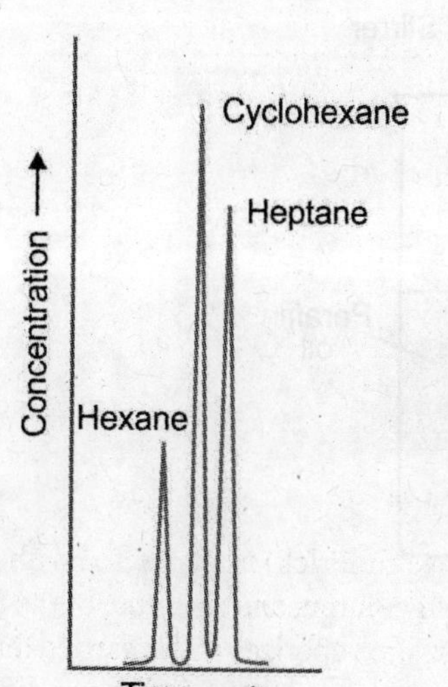

Fig. 2.14. A gas chromatogram of a mixture of hydrocarbons.

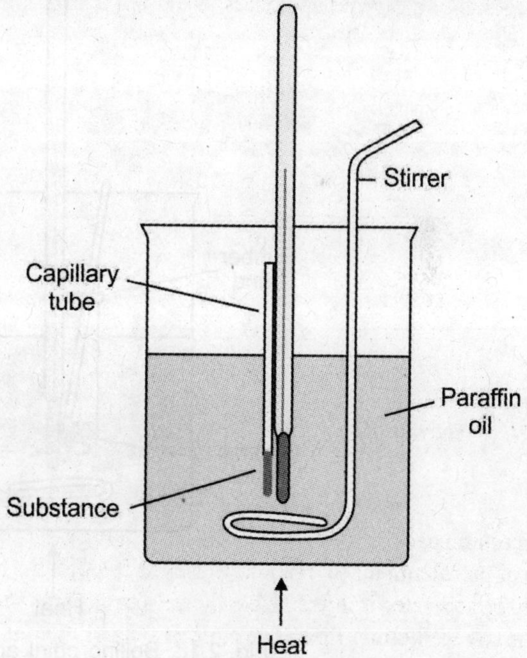

Fig. 2.15. Melting point apparatus.

Mixed Melting Point. The method of mixed melting point as a means of establishing the purity of a substance depends upon the fact that a pure organic compound possesses a sharp melting point while the presence of impurities usually lowers the melting point. A substance mixed with impurities does not melt sharply at a fixed temperature but does so over a range of temperatures. It will start melting at a certain temperature but will not become entirely liquid until a higher temperature is reached.

The substance whose purity is to be ascertained is mixed with a sample of the authentically pure substance. The melting point of the mixture is determined as usual. If this mixture melts sharply and the melting point comes out to be the same as that of the sample under experiment taken alone, it is proved that the two are indeed the samples of the same substance. In other words, the substance under trial is proved to be pure beyond doubt.

Boiling Point. A pure organic liquid boils at a fixed temperature which is characteristic of that substance.

(1) **Distillation Method**. If enough liquid is available, its boiling point is determined in an ordinary distillation apparatus. A pure liquid will distil at a constant temperature which is its boiling point. In case the liquid is impure, the boiling point will rise during distillation.

(2) **Capillary-tube Method**. When only a small quantity of the liquid is available, its boiling point is determined by the 'Capillary-tube Method'. A few drops of the liquid are placed in a thin-walled small test-tube (Fig. 2.16). A capillary tube sealed at about one centimeter from one end, is dropped into it. The glass tube containing the liquid and capillary, is then tied along side a thermometer so that the liquid stands just near the bulb. The thermometer is then lowered in a beaker containing paraffin oil or sulfuric acid (any liquid whose boiling point is higher than that of the substance under investigation).

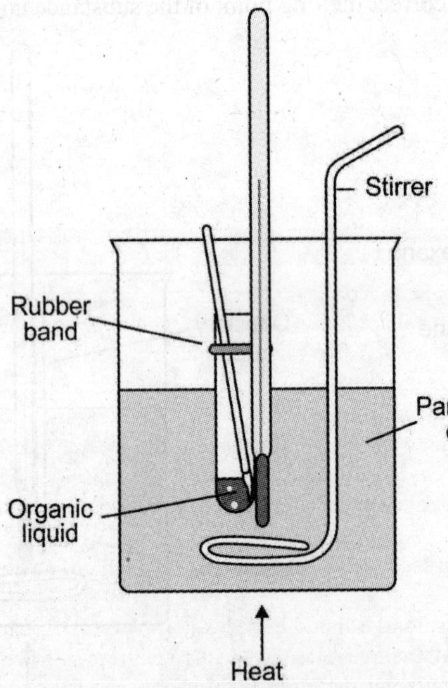

Fig. 2.16. Boiling point apparatus.

The beaker is heated and the bath liquid stirred continuously with a ring stirrer. When the boiling point is reached, bubbles issue in a rapid stream from the lower end of the capillary. The thermometer is read when the evolution of bubbles just stops. The experiment is repeated with a fresh liquid in a new capillary and the boiling point recorded as before. The mean of the two readings is taken to be the correct boiling point of the liquid under examination.

Examination Questions

1. How would you purify a solid organic compound ?
2. How would you purify a liquid organic compound ?
3. Write notes on :
(*a*) Fractional distillation
(*b*) Steam distillation
(*c*) Solvent extraction
4. How will you separate a mixture of naphthalene and sodium chloride ?

Answer. We can separate naphthalene from sodium chloride by sublimation technique. Naphthalene sublimes. Sodium chloride does not sublime. A substance is said to sublime if, on cooling, its vapor changes directly from vapor to solid without going through an intermediate state. Usually a substance which sublimes also changes from solid to vapor without melting to a liquid. If a mixture of naphthalene and sodium chloride is heated, the naphthalene turns directly to a vapor but the sodium chloride remains unchanged. When the vapor is cooled, solid naphthalene collects free from sodium chloride.

5. Which of the methods is best suited to separate a 500 ml sample of two miscible liquids whose boiling points differ by approximately 60° ?
(*a*) distillation (*b*) fractional distillation
(*c*) paper chromatography (*d*) use of a separatory funnel

Answer. (*a*)

6. For which of the following mixtures would Steam Distillation be the most appropriate method of separation ?
(*a*) Diethyl ether and water (*b*) Ethyl alcohol and water
(*c*) Aniline and sodium chloride (*d*) None of these

Answer. (*c*). Remember : Steam distillation is especially useful for separating a volatile liquid, which is virtually immiscible with water, from nonvolatile materials at a temperature below its boiling point.

7. Write a note on : Chromatography (*Madras BSc, 2015*)
8. Write a note on : Column Chromatography (*Delhi BSc, 2014*)
9. Write a note on : Thin-Layer Chromatography
10. What are the criteria by which the purity of an organic compound may be established ? (*Jiwaji BSc, 2013*)
11. Indicate whether the following statements are *true* or *false*.
(*a*) Animal charcoal is added to a solution during recrystallization to ensure even boiling.
(*b*) Sublimation is used for the separation and purification of compounds with low melting points.
(*c*) The boiling point of a liquid increases as the pressure is reduced.
(*d*) Paper chromatography is used only for the separation of colored compounds.
(*e*) An azeotrope is a liquid mixture of definite composition which boils at atmospheric pressure.

Answer. (*a*) *False.* Animal charcoal is added to the hot solution during recrystallization is to remove colored impurities.

(*b*) *False.* Sublimation applies to substances which do not melt on heating but which pass directly into the vapor state.

(*c*) *False.* The vapor pressure of a liquid increases with increasing temperature. A liquid boils when its vapor pressure is equal to the external pressure. A reduction in pressure lowers the boiling-point of the liquid.

(*d*) *False.* Paper chromatography applies to both colored and colorless compounds. Separation is more obvious with colored components, but the location of colorless substances on the paper may be shown by ultraviolet light in some cases or by treatment with suitable chemicals to form colored derivatives.

(*e*) *True*

3

Composition of Organic Compounds

DETECTION OF ELEMENTS

The first step in the analysis of an organic compound is the detection of elements present in it. Most of these compounds contain 2 to 5 different elements. The principal elements present are : carbon, hydrogen and oxygen. Often, in addition to these, they may contain nitrogen, sulfur and halogens. Phosphorus and metals are also present but only rarely. The order of abundance in which these elements are found in organic compounds is indicated below :

Carbon	–	Always present
Hydrogen	–	Nearly always present
Oxygen	–	Generally present
Nitrogen, Halogens and Sulfur	–	Less commonly present
Phosphorus and Metals	–	Rarely present

DETECTION OF CARBON AND HYDROGEN

If the compound under investigation is known to be organic, there is no need to test for carbon. This test is performed only to establish whether a given compound is organic or not. With the exception of few compounds e.g., carbon tetrachloride, all organic compounds also contain hydrogen. The presence of both these elements is confirmed by the following common test.

Copper Oxide Test. The organic substance is mixed intimately with about three times its weight of *dry* copper oxide. The mixture is then placed in a hard glass test-tube fitted with a bent delivery tube (Fig. 3.1), the other end of which is dipping into lime water in another test-tube. The mixture is heated strongly when the following reactions take place :

$$C + 2CuO \longrightarrow CO_2 + 2Cu$$

$$2H + CuO \longrightarrow H_2O + Cu$$

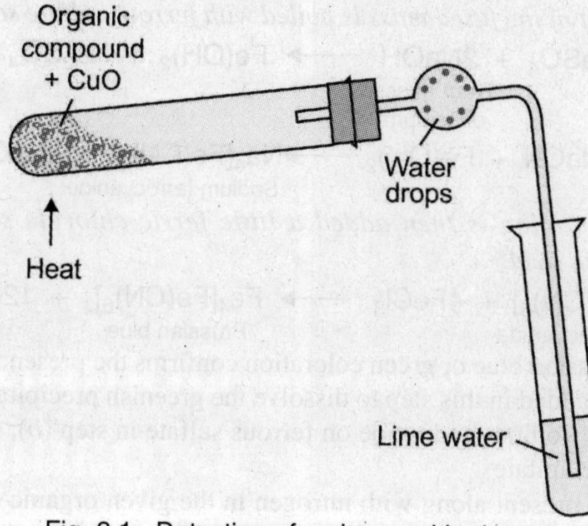

Fig. 3.1. Detection of carbon and hydrogen.

Thus if carbon is present, it is oxidized to carbon dioxide which turns lime water milky. If hydrogen is also present, it will be oxidized to water which condenses in small droplets on the cooler wall of the test-tube and inside the bulb. The formation of water is further confirmed by testing the condensed liquid with anhydrous copper sulfate (*white*) that is turned blue.

If the substance under investigation is a gas or a volatile liquid the above test is modified. The vapors of the substance are passed over heated copper oxide contained in a hard glass combustion tube. The issuing gases are tested for carbon dioxide and water vapor as described before.

While testing for hydrogen, it is essential that the apparatus and copper oxide used are absolutely dry. Cupric oxide being hygroscopic in nature is heated strongly just before use.

DETECTION OF OXYGEN

There is no conclusive test for oxygen, though its presence in organic compounds is often inferred by indirect methods.

(1) The substance is heated alone in a dry test-tube, preferably in an atmosphere of nitrogen. Formation of droplets of water on cooler parts of the tube obviously shows the presence of oxygen. A negative result, however, does not necessarily show the absence of oxygen.

(2) The second method is to test for the presence of various oxygen-containing groups such as *hydroxyl* (OH), *Carbonyl* (COOH), *aldehyde* (CHO), *nitro* (NO_2) etc. If any of these is detected, the presence of oxygen is confirmed.

(3) The sure test for oxygen depends on the determination of the percentage of all other elements present in the given compound. If the sum of these percentages falls short of hundred, the remainder gives the percentage of oxygen and thus confirms its presence.

DETECTION OF NITROGEN

The presence of nitrogen in an organic compound is shown by the following tests :

(1) **Soda-lime Test.** The given substance is mixed with double the amount of soda-lime (NaOH + CaO) and heated in a test tube. The vapors of ammonia evolved show the presence of nitrogen. A negative result, however, is not a proof of the absence of nitrogen. Many classes of nitrogenous compounds including *nitro* and *diazo* derivatives, do not respond to this test.

(2) **Sodium Test (*Lassaigne's Test*).** This is a good test for the detection of nitrogen in all classes of nitrogenous compounds. It involves the following steps :

(*a*) *The substance is heated strongly with sodium metal.*

$$\mathrm{Na} + \underbrace{\mathrm{C} + \mathrm{N}}_{\text{From Organic Compound}} \longrightarrow \mathrm{NaCN}$$

(*b*) *The water extract of the fused mass is boiled with ferrous sulfate solution.*

$$FeSO_4 + \underset{\text{(From excess of sodium)}}{2NaOH} \longrightarrow Fe(OH)_2 + Na_2SO_4$$

$$6NaCN + Fe(OH)_2 \longrightarrow \underset{\text{Sodium ferrocyanide}}{Na_4[Fe(CN)_6]} + 2NaOH$$

(*c*) *To the cooled solution is then added a little ferric chloride solution and excess of concentrated hydrochloric acid.*

$$\underset{\text{Sodium ferrocyanide}}{3Na_4[Fe(CN)_6]} + 4FeCl_3 \longrightarrow \underset{\text{Prussian blue}}{Fe_4[Fe(CN)_6]_3} + 12NaCl$$

The formation of Prussian blue or green coloration confirms the presence of nitrogen.

Hydrochloric acid is added in this step to dissolve the greenish precipitate of ferrous hydroxide produced by the excess of sodium hydroxide on ferrous sulfate in step (*b*), which would otherwise mark the Prussian blue precipitate.

In case sulfur is also present along with nitrogen in the given organic compound, a blood red coloration may appear while performing the above test. This is due to the formation of sodium sulfocyanide which again reacts with ferric chloride to produce blood red coloration :

$$Na + C + N + S \longrightarrow \underset{\text{Sodium sulfocyanide}}{NaCNS}$$

$$3NaCNS + FeCl_3 \longrightarrow \underset{\substack{\text{Ferric sulfocyanide} \\ \text{(Blood red)}}}{Fe(CNS)_3} + 3NaCl$$

Procedure of Sodium Test. Fix a fusion tube in an iron stand clamping it just near the upper end. Take a freshly cut piece of sodium of the size of a pea and dry it by pressing between the folds of a filter paper. Place the metal in a fusion tube and heat it from below. When it melts to a shining globule, put a pinch of the organic compound on it. Heat the tube with the tip of the flame till all reaction ceases and it becomes red hot. Now plunge it in about 50 ml of distilled water taken in a china dish and break the bottom of the tube by striking against the dish. Boil the contents of the dish for about ten minutes and filter. Label the filtrate as 'sodium extract' and proceed with it as follows.

Take a portion of the sodium extract in a test-tube and note if it is alkaline. If it is not, make it so by adding sodium hydroxide. Then add to it freshly prepared ferrous sulfate solution and heat it to boiling. Put 2-3 drops of ferric chloride solution, cool and acidify with conc. hydrochloric acid. A Prussian blue or green precipitate, or even coloration, confirms the presence of nitrogen.

DETECTION OF SULFUR

The presence of sulfur in organic substances is shown as described below :

(1) **Sodium Test.** Sulfur, if present, in the given organic compound, upon fusion with sodium reacts to form sodium sulfide.

$$2Na + S \longrightarrow \underset{\text{Sodium sulfide}}{Na_2S}$$

Thus, the 'Sodium extract' obtained from the fused mass may be tested as :

(*a*) To a portion, add freshly prepared sodium nitroprusside solution. A deep violet coloration indicates sulfur.

(*b*) Acidify a second portion of the extract with acetic acid and then add lead acetate solution. A black precipitate of lead sulfide confirms the presence of sulfur.

$$\underset{\text{Lead acetate}}{Pb(CH_3COO)_2} + Na_2S \longrightarrow \underset{\substack{\text{Lead sulfide} \\ \text{(Black ppt.)}}}{PbS} + 2CH_3COONa$$

(2) **Oxidation Test.** The organic substance is fused with a mixture of potassium nitrate and sodium carbonate. The sulfur, if present, is oxidized to sulfate.

$$Na_2CO_3 + S + 3O \longrightarrow Na_2SO_4 + CO_2$$

The fused mass is extracted with water, acidified with hydrochloric acid and then barium chloride solution is added to it. A white precipitate indicates the presence of sulfur.

$$BaCl_2 + Na_2SO_4 \longrightarrow \underset{\substack{\text{Barium sulfate} \\ \text{(White ppt.)}}}{BaSO_4} + 2NaCl$$

DETECTION OF HALOGENS

(1) **Sodium Test.** Upon fusion with sodium, the halogens in the organic compound are converted to the corresponding sodium halides. Thus,

$$Cl + Na \longrightarrow NaCl$$
$$Br + Na \longrightarrow NaBr$$
$$I + Na \longrightarrow NaI$$

Acidify a portion of 'sodium extract' with dilute nitric acid and add to it silver nitrate solution.
White precipitate soluble in ammonia indicates **Chlorine.**
Yellow precipitate sparingly soluble in ammonia indicates **Bromine.**
Yellow precipitate insoluble in ammonia indicates **Iodine.**

When nitrogen or sulfur is also present in the compound, the 'sodium extract' before testing for halogens is boiled with strong nitric acid to decompose the cyanide and the sulfide formed during the sodium fusion. If not removed, these radicals will form a white and black precipitate respectively on the addition of silver nitrate.

$$NaCN + HNO_3 \longrightarrow NaNO_3 + HCN \uparrow$$
$$Na_2S + 2HNO_2 \longrightarrow 2NaNO_2 + H_2S \uparrow$$

(2) **Copper Wire Test.** (*Beilstein's Test*). The copper wire flattened at one end is heated in an oxidizing Bunsen flame till it ceases to impart any green color to the flame. A small quantity of substance under investigation is now taken on the flattened end of the wire which is reinserted in the Bunsen flame. Upon heating for a while, the halogen present in the substance is converted to a volatile copper halide which imparts a blue or green color to the flame. This test though very sensitive, is not always reliable. A substance like urea which contains no halogen, also colors the flame green.

DETECTION OF PHOSPHORUS

The solid substance is heated strongly with an oxidizing agent such as conc. nitric acid or a mixture of sodium carbonate and potassium nitrate. The phosphorus present in the substance is thus oxidized to phosphate. The residue is extracted with water, boiled with some nitric acid, and then a hot solution of ammonium molybdate is added to it in excess. A yellow coloration or precipitate indicates the presence of phosphorus.

DETECTION OF METALS

The substance is strongly heated in a crucible, preferably of platinum, till all reaction ceases. An incombustible residue indicates the presence of a metal in the substance. The residue is extracted with dilute acid and the solution tested to the presence of metallic radical by the usual scheme employed for inorganic salts.

ESTIMATION OF ELEMENTS

Having known the elements present in a given organic compound, we proceed to determine their composition by weight. The estimation of carbon, hydrogen, nitrogen and other elements can be done accurately by methods described below. No dependable method is, however, available for the determination of oxygen and hence its amount is always determined by difference.

ESTIMATION OF CARBON AND HYDROGEN

Both carbon and hydrogen are estimated together in one operation. A known weight of the organic substance is burnt in excess of oxygen when the carbon and hydrogen present in it are oxidized to carbon dioxide and water respectively.

$$C_xH_y + \underset{(excess)}{O_2} \longrightarrow xCO_2 + y/2\ H_2O$$

The weights of carbon dioxide and water thus formed are determined and the amounts of *carbon* and *hydrogen* in the original substance calculated.

Apparatus. The apparatus employed for the purpose consists of three units :

(1) Oxygen supply

(2) Combustion tube

(3) Absorption apparatus

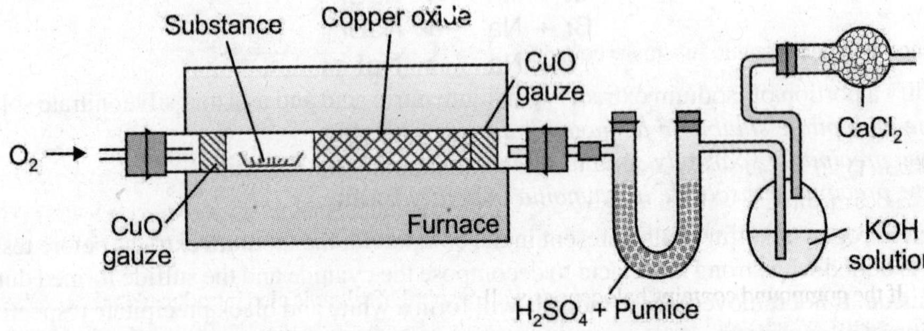

Fig. 3.2. Apparatus for the estimation of C and H.

Oxygen Supply. Oxygen from the aspirator is allowed to bubble through sulfuric acid contained in a Drechsel bottle and then passed through U-tubes charged with soda-lime. The oxygen gas thus freed from moisture and carbon dioxide enters the combustion tube.

Combustion Tube. A hard glass tube about 33 inches in length and 1/2 inch in diameter, and open at both ends is used for the combustion of the organic substance. It is filled as shown in Fig. 3.2 with (*i*) a roll of oxidized copper gauze to prevent the backward diffusion of the products of combustion ; (*ii*) a porcelain boat containing a known weight (about 0.2 g) of the organic substance; (*iii*) coarse copper oxide packed in about two-thirds of the entire length of the tube, and kept in position by loose asbestos plugs on either side ; and (*iv*) a roll of oxidized copper gauze placed towards the end of the combustion tube to prevent any vapors of the organic substance leaving the tube unoxidized. The combustion tube is enclosed in a furnace, heated by gas burners.

Absorption Apparatus. The products of combustion containing moisture and carbon dioxide are then passed on to the absorption apparatus which consists of : (*i*) a weighed U-tube packed with pumice soaked in concentrated sulfuric acid, to absorb water ; (*ii*) a set of bulbs containing strong solution of potassium hydroxide, to absorb carbon dioxide ; and finally (*iii*) a guard-tube filled with anhydrous calcium chloride to prevent the entry of moisture from atmosphere.

Procedure. To start with, before loading it with the boat, the combustion tube is detached from the absorption unit. The tube is heated strongly to dry its contents and carbon dioxide present in it is removed by passing a current of pure, dry oxygen through it. It is then cooled slightly and connected to the absorption apparatus. The other end of the combustion tube is opened for a while and the boat containing weighed organic substance introduced. The tube is again heated strongly till the whole of the substance in the boat has burnt away. This takes about two hours. Finally, a strong current of oxygen is passed through the combustion tube to sweep away any traces of carbon dioxide or moisture which may have been left in it. The U-tube and the potash bulbs are then detached and the increase in weight of each of them determined.

Calculations :

Let the weight of the substance taken = w g

Increase in weight of U-tube (H_2O) = x g

Increase in weight of potash bulbs (CO_2) = y g

Since 18 g of water contain hydrogen = 2 g

x g of water contain hydrogen = $\frac{2}{18} \times x g$

Also, 44 g of CO_2 contain carbon = 12 g

y g of CO_2 contain carbon = $\frac{12}{44} \times y g$

Hence the given organic substance contains

$$H = \frac{2x}{18} \times \frac{100}{w} \%$$

$$C = \frac{12y}{44} \times \frac{100}{w} \%$$

Notes: (1) If the organic substance under investigation also contains nitrogen, on combustion it will produce oxides of nitrogen which will also be absorbed in the potash bulbs. Hence a spiral of copper is introduced at the right extreme of the combustion tube, so that the oxides of nitrogen are reduced to free nitrogen which escapes unabsorbed.

(2) If the compound contains halogens as well, a spiral of silver is also introduced in the combustion tube. The free halogens, which would have been otherwise absorbed in the potash bulbs, are converted to silver halides and thus eliminated from the products of combustion.

(3) In case the substance also contains sulfur, the copper oxide in the combustion tube is replaced by lead chromate. The sulfur dioxide formed during combustion is thus converted to lead sulfate and prevented from passing into the absorption unit.

Example 1. *0.2475 g of an organic substance gave on combustion 0.4950 g of carbon dioxide and 0.2025 g of water. Calculate the percentage of carbon and hydrogen in it.*

Weight of the substance taken = 0.2475 g

Weight of CO_2 formed = 0.4950 g

Weight of H_2O formed = 0.2025 g

Since 18 g of H_2O contain hydrogen = 2 g

0.2025 g of H_2O contain hydrogen = $\frac{2 \times 0.2025}{18}$ g

Also, 44 g of CO_2 contain carbon = 12 g

0.4950 g of CO_2 contain carbon = $\frac{12 \times 0.4950}{44}$ g

Hence the given organic compound contains

$$C = \frac{12 \times 0.4950}{44} \times \frac{100}{0.2475} = 54.54\%$$

$$H = \frac{2 \times 0.2025}{18} \times \frac{100}{0.2475} = 9.09\%$$

Example 2. *0.2346 g of an organic compound containing carbon, hydrogen and oxygen only was analyzed by the combustion method. The increase in weight of the U-tube and the potash bulbs at the end of the operation was found to be 0.2754 g and 0.4488 g respectively. Determine the percentage composition of the compound.*

Wt of CO_2 (Increase in wt of potash bulbs) = 0.4488 g

Wt of H_2O (Increase in wt of U-tube) = 0.2754 g

Since 44 g of CO_2 contain carbon = 12 g

0.4488 g of CO_2 contain carbon = $\frac{12 \times 0.4488}{44}$ g

Also, 18 g of H_2O contain hydrogen $= 2\text{ g}$

0.2754 g of H_2O contain hydrogen $= \dfrac{2 \times 0.2754}{18}\text{ g}$

Wt of compound taken for analysis $= 0.2346\text{ g}$

$$\therefore \quad \text{Percentage of C} = \frac{12 \times 0.4488}{44} \times \frac{100}{0.2346} = 52.17$$

$$\text{Percentage of H} = \frac{2 \times 0.2754}{18} \times \frac{100}{0.2346} = 13.04$$

$$\text{Percentage of O} = 100 - (52.17 + 13.04) = 34.79$$

(By difference)

ESTIMATION OF NITROGEN

The two main methods for the estimation of nitrogen in an organic substance are :

(1) Dumas Method

(2) Kjeldahl's Method

Dumas Method. This method is based upon the fact that nitrogenous compound when heated with copper oxide in an atmosphere of carbon dioxide yield free nitrogen. Thus,

$$C_xH_yN_z + (CuO) \longrightarrow xCO_2 + y/2\, H_2O + z/2\, N_2 + (Cu)$$

Traces of oxides of nitrogen, which may be formed in some cases, are reduced to elemental nitrogen by passing over heated copper spiral.

Apparatus. The apparatus used in the Dumas method consists of :

(1) CO_2 Generator

(2) Combustion Tube

(3) Schiff's Nitrometer

CO_2 Generator. The carbon dioxide needed in this process is prepared by heating magnesite or sodium bicarbonate contained in a hard-glass test-tube, or by the action of dilute hydrochloric acid on marble in a Kipp's apparatus. The gas is passed into the combustion tube after being dried by bubbling through concentrated sulfuric acid contained in a Drechsel bottle.

Combustion Tube. The combustion tube of about 90 cm length, which is heated in a combustion furnace, is charged with :

(*i*) *a roll of oxidized copper gauze* to prevent the back diffusion of the products of combustion and to heat the organic substance mixed with copper oxide by radiation ; (*ii*) a weighed amount (about 0.2 g) of the *organic substance* mixed with excess of copper oxide ; (*iii*) *a layer of coarse copper oxide* packed in about two-thirds of the entire length of the tube, and kept in position by loose asbestos plugs on either side ; this oxidizes the organic vapors passing through it ; and (*iv*) *a reduced copper spiral* which reduces any oxides of nitrogen formed during combustion, to nitrogen.

Schiff's Nitrometer. The nitrogen gas obtained by the decomposition of the substance in the combustion tube is mixed with considerable excess of carbon dioxide. It is estimated by passing into Schiff's nitrometer where carbon dioxide is absorbed by caustic potash solution and the nitrogen collects in the upper part of the graduated tube.

Procedure. The apparatus is fitted up as shown in Fig. 3.3, and to start with, the tap of nitrometer is left open. Carbon dioxide is passed through the combustion tube to replace the air in it. When the gas bubbles rising through the potash solution fail to reach the top of it and are completely absorbed, it shows that only carbon dioxide is coming and that all the air has been displaced from the combustion tube. The nitrometer is then filled with potassium hydroxide solution by lowering the reservoir and the

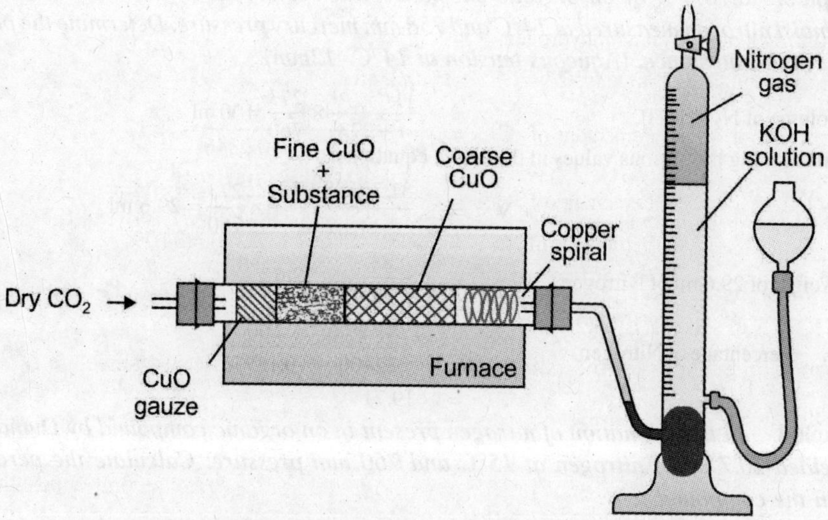

Fig. 3.3. Dumas method for estimation of nitrogen.

tap closed. The combustion tube is now heated in the furnace, raising the temperature gradually. The nitrogen set free from the compound collects in the nitrometer. When the combustion is complete, a strong current of carbon dioxide is sent through the apparatus in order to sweep the last traces of nitrogen from it. The volume of the gas collected is noted after adjusting the reservoir so that the solution in it and the graduated tube is the same. The atmospheric pressure and the temperature are also recorded.

Calculations :

Let the weight of the substance taken = w g

Volume of moist N_2 = v ml

Room temperature = t°C

= $(t + 273)$°A

Barometric pressure = P mm of Hg

Aqueous tension at room temp. = p mm of Hg

∴ Pressure of dry N_2 = $(P–p)$ mm of Hg

(*i*) **Calculation of Volume of N_2 at NTP :**

From the general Gas Equation, we have

$$\frac{P_1V_1}{T_1} = \frac{P_2V_2}{T_2}$$

where V_1 represents the volume of dry gas at pressure P_1 and temperature T_1 (in degrees absolute) and V_2 represents the volume of the gas at NTP (pressure P_2 = 760 mm of Hg; and temperature T_2 = 273°A)

Volume (V_2) of N_2 at NTP $= \dfrac{P_1V_1T_2}{T_1P_2}$

$$= \frac{(P - p)v}{t + 273} \times \frac{273}{760}$$

= V ml (Say)

(*ii*) **Calculation of Percentage of Nitrogen :**

Now 22,400 ml of N_2 at NTP weigh = 28 g

V ml of N_2 at NTP weigh $= \dfrac{28}{22{,}400} \times \text{V}g$

∴ Percentage of N $= \dfrac{28\text{V}}{22{,}400} \times \dfrac{100}{w}$

Example 3. *0.1877 g of an organic substance when analyzed by the Dumas method yield 31.7 ml of moist nitrogen measured at 14°C and 758 mm mercury pressure. Determine the percentage of nitrogen in the substance. (Aqueous tension at 14°C=12mm).*

Volume of N_2 at NTP $= \dfrac{v(P-p)}{t+273} \times \dfrac{273}{760} = \text{V ml}$

Substituting the various values in the above equations.

$$V = \frac{31.7\ (758-12)}{14+273} \times \frac{273}{760} = 29.6 \text{ ml}$$

Weight of 29.6 ml of Nitrogen $= \dfrac{28}{22400} \times 29.6$ g

∴ Percentage of Nitrogen $= \dfrac{28 \times 29.6}{22400} \times \dfrac{100}{0.1877}$

$= 19.71$

Example 4. *In the estimation of nitrogen present in an organic compound by Dumas method, 0.200g yielded 20.7 ml of nitrogen at 15°C and 760 mm pressure. Calculate the percentage of nitrogen in the compound.*

Volume of N_2 at NTP $= \dfrac{v \times P}{t+273} \times \dfrac{273}{760} = \text{V ml}$

Substituting the various values in the above equation,

$$V = \frac{20.7 \times 760}{288} \times \frac{273}{760} = 19.62 \text{ ml}$$

Weight of 19.62 ml of Nitrogen $= \dfrac{28}{22400} \times 19.62$ g

∴ Percentage of Nitrogen $= \dfrac{28}{22400} \times 19.62 \times \dfrac{100}{0.2}$

$= 12.26$

Kjeldahl's Method. This method is carried much more easily than the Dumas method. It is used largely in the analysis of foods and fertilizers.

Kjeldahl's method is based on the fact that when an organic compound containing nitrogen is heated with concentrated sulfuric acid, the nitrogen in it is quantitatively converted to ammonium sulfate. The resultant liquid is then treated with excess of alkali and the liberated ammonia gas absorbed in excess of standard acid. The amount of ammonia (and hence of nitrogen) is determined by finding the amount of acid neutralized by back titration with some standard alkali.

Procedure. A weighed quantity of the substance (0.3 to 0.5 g) is placed in a special long-necked '*Kjeldahl flask*' made of pyrex glass. About 25 ml of concentrated sulfuric acid together with a little potassium sulfate and copper sulfate is added to it. Potassium sulfate raises the boiling point and thus ensures complete reaction, while copper sulfate acts as a catalyst. The flask is loosely stoppered by a glass bulb and heated gently in an inclined position. The heating is continued till the brown color of the liquid first produced, disappears leaving the contents clear as before. At this point all the nitrogen in the substance is converted to ammonium sulfate. The Kjeldahl flask is then cooled and its contents diluted with some distilled water, and then carefully transferred into a one-litre round-bottomed flask. An excess of sodium hydroxide solution is poured down the side of the flask and it is fitted with a Kjeldahl trap and a water condenser, as shown in Fig. 3.5. The lower end of the condenser dips in a measured volume of excess of the N/10 H_2SO_4 solution. The liquid in the round bottomed flask is then heated and the liberated ammonia distilled into sulfuric acid. The Kjeldahl trap serves to retain any alkali splashed up on vigorous boiling. When no more ammonia passes over (test the distillate with red litmus), the receiver is removed. The excess of acid is then determined by titration with N/10 alkali, using phenolphthalein as the indicator.

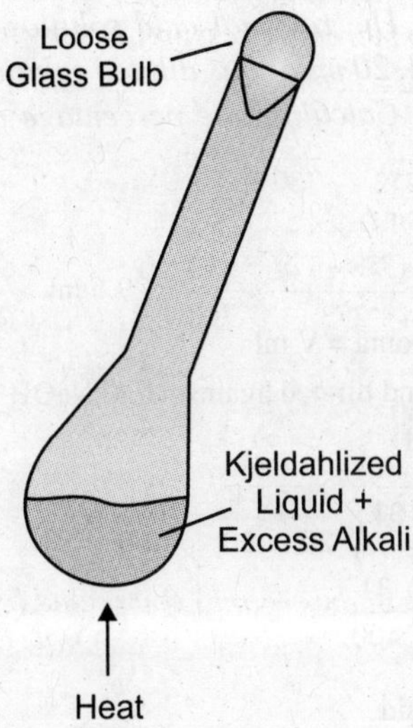

Fig. 3.4. Kjeldahl flask.

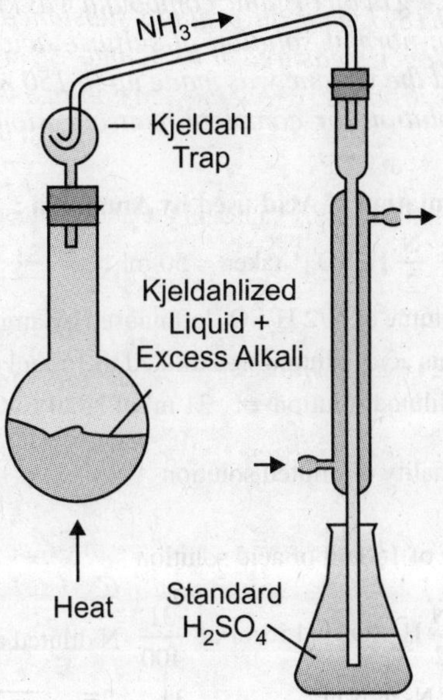

Fig. 3.5. Distillation of ammonia.

Calculations :

Let the weight of the organic substance be *w* g and V ml of N.HCl is required for complete neutralization of ammonia evolved.

$$\text{V ml of N.HCl} = \text{V ml of N.NH}_3$$

1000 ml of N.NH_3 contain 17 g of ammonia or 14 g of nitrogen

Amount of Nitrogen present in V ml of N.NH_3

$$= \frac{14}{1000} \times V \times N \text{ g}$$

$$\therefore \text{Percentage of Nitrogen} = \text{Wt of Nitrogen} \times \frac{100}{\text{Wt of substance}}$$

$$= \frac{14}{1000} \times V \times N \times \frac{100}{w}$$

$$= \frac{1.4\,VN}{w}$$

Where, N stands for normality of the acid used

V stands for volume of the acid used

w stands for weight of the substance kjeldahlized

Example 5. *0.257 g of an organic substance was heated with conc. sulfuric acid and then distilled with excess of strong alkali. The ammonia gas evolved was absorbed in 50 ml of N/10 hydrochloric acid which required 23.2 ml of N/10 NaOH for neutralization at the end of the operation. Determine the percentage of nitrogen in the substance.*

Volume of N/10 HCl neutralized by ammonia = 50 – 23.2 = 26.8 ml

Now, 26.8 ml of N/10 HCl ≡ 26.8 ml of N/10 NH_3
≡ 26.8 ml of N/10 N (*combined*)

26.8 ml of N/10 N solution contains $\frac{14}{10} \times \frac{26.8}{1000}$ g of nitrogen

This weight of nitrogen was originally present in 0.257 g of the organic substance

$$\therefore \quad \text{Percentage of Nitrogen} = \frac{14 \times 26.8}{10 \times 1000} \times \frac{100}{0.257} = 14.6$$

Example 6. *0.4 g of an organic compound was kjeldahlized and ammonia evolved was absorbed into 50 ml of semi-normal solution of sulfuric acid. The residual acid solution was diluted with distilled water and the volume was made up to 150 ml. 20 ml of this diluted solution required 31 ml of N/20 NaOH solution for complete neutralization. Calculate the percentage of nitrogen in the compound.*

(*i*) **Calculation of Acid used by Ammonia :**

Volume of $\frac{N}{2}$ H_2SO_4 taken = 50 ml

Let the volume of N/2 H_2SO_4 left unused by ammonia = V ml

V ml of this acid solution was diluted to 150 ml and titrated against N/20 NaOH

20 ml of diluted solution ≡ 31 ml of N/20 NaOH

∴ Normality of diluted solution $= 31 \times \frac{N}{20} \times \frac{1}{20} = \frac{31}{400} N$

Normality of 150 ml of acid solution $= \frac{31}{400} N$

i.e., V ml of $\frac{N}{2} H_2SO_4 \equiv 150 \text{ ml of } \frac{31}{400}$ N diluted acid

$$V = 150 \times \frac{31}{400} \times \frac{2}{1} = 23.25 \text{ ml}$$

i.e., Volume of $\frac{N}{2}$ H_2SO_4 left = 23.25 ml

∴ Volume of acid used up by ammonia = 50 – 23.25 = 26.75 ml

(*ii*) **Calculation of Percentage of N :**

26.75 ml of N/2 H_2SO_4 ≡ 26.75 ml of N/2 NH_3 solution

≡ 26.75 ml of N/2 N (*combined in solution*)

∴ Wt of nitrogen in 26.75 ml of N/2 solution

$$= \frac{14}{2} \times \frac{1}{1000} \times 26.75 \text{ g}$$

This wt of nitrogen was originally present in 0.4 of the substance

∴ Percentage of Nitrogen $= \frac{14}{2} \times \frac{1}{1000} \times 26.75 \times \frac{100}{0.4} = 46.81$

ESTIMATION OF HALOGENS

Carius Method. It involves oxidizing the organic substance with fuming nitric acid in the presence of silver nitrate. The halogen of the substance is thus converted to silver halide which is separated and weighed.

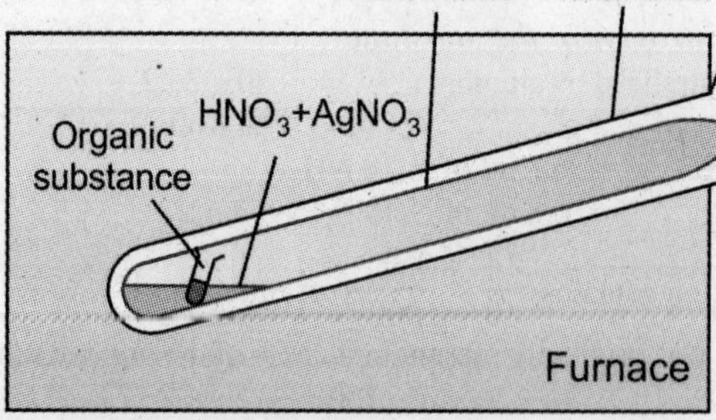

Fig. 3.6. Heating the bomb tube in a furnace.

The Carius method of estimating halogens involves the use of a stout hard glass tube (*Bomb tube*) of about 50 cm length and closed at one end. About 5 ml of fuming nitric acid together with 2 to 2.5 g of pure silver nitrate is placed in it. A narrow weighing tube, containing an accurately weighed quantity of the substance is then slipped into the bomb tube, taking care to avoid the mixing of the substance with nitric acid. The open end of the tube is then sealed off. It is now heated in an iron jacket of the furnace heated by burners from below. The temperature of the furnace is gradually raised to 300°C. After heating for about six hours, the tube is allowed to cool. The high pressure developed inside the tube is released by softening the sealed end with a small flame, when a hole is blown open through which the gases escape. The end of the tube is now cut off and the contents transferred into beaker. The halide formed is collected in a Gooch crucible, washed, dried and weighed.

Let w g of the organic substance give x g of the silver halide

$$\therefore \quad \text{Percentage of Halogen} = \frac{\text{At Wt of halogen}}{\text{M Wt of silver halide}} \times \frac{100\,x}{w}$$

Example 7. *0.197 g of an organic substance when heated with excess of strong nitric acid and silver nitrate gave 0.3525 g of silver iodide. Find the percentage of iodine in the compound.*

$$\text{Since,} \quad \underset{(108+127)}{AgI} \equiv \underset{127}{I}$$

In this case 0.197 g of the compound gave 0.3525 g of AgI

$$= \frac{0.3525 \times 127}{235} \text{ g iodine}$$

$$\therefore \quad \text{Percentage of Iodine} = \frac{0.3525 \times 127}{235} \times \frac{100}{0.197} = 96.70$$

Example 8. *0.2562 g of an organic substance when heated with excess of strong nitric acid and silver nitrate gave 0.3066 g of silver bromide. Find the percentage of bromine in the compound.*

$$\text{Since,} \quad \underset{(108+80)}{AgBr} \equiv \underset{80}{Br}$$

Here, 0.2562 g of the substance gave 0.3066 g of AgBr

$$= \frac{0.3066 \times 80}{188} \text{ g bromine}$$

$$\therefore \text{ Percentage of Bromine} = \frac{0.3066 \times 80}{188} \times \frac{100}{0.2562} = 50.92$$

ESTIMATION OF SULFUR AND PHOSPHORUS

Sulfur is estimated by a slight modification of the Carius method. The organic substance is heated with fuming nitric acid but no silver nitrate is added. The sulfur of the substance is oxidized to sulfuric acid which is then precipitated as barium sulfate by adding excess of barium chloride solution. The precipitate is filtered, washed, dried and weighed. From the weight of the barium sulfate so obtained, the percentage of sulfur can be calculated.

Phosphorus is also estimated like sulfur. The organic substance is heated with fuming nitric acid whereupon its phosphorus is oxidized to phosphoric acid. The phosphoric acid is precipitated as ammonium phosphomolybdate, $(NH_4)_3PO_4.12MoO_3$, by the addition of ammonia and ammonium molybdate solution, which is then separated, dried and weighed.

Example 9. *In an estimation of sulfur by the Carius method 0.2175 g of the substance gave 0.5825 g of barium sulfate. Calculate the percentage of sulfur in the substance.*

$$\text{We know,} \quad \underset{(137+32+64)}{BaSO_4} \equiv \underset{32}{S}$$

or, 233 g of $BaSO_4$ contain 32 g of sulfur

$$0.5825 \text{ g of } BaSO_4 \text{ contain } \frac{32}{233} \times 0.5825 \text{ g of sulfur}$$

This is the weight of sulfur present in 0.2175 g of the substance

$$\text{Hence, percentage of S} = \frac{32 \times 0.5825}{233} \times \frac{100}{0.2175} = 36.78$$

Example 10. *0.395 g of an organic compound by Carius method for the estimation of sulfur gave 0.582 g of $BaSO_4$. Find the percentage of sulfur in the compound.*

We know,

$$\underset{233}{BaSO_4} \equiv \underset{32}{S}$$

i.e. 233 g of $BaSO_4$ contains 32 g of sulfur

$$\therefore \quad 0.582 \quad " \quad " \quad " \quad \frac{32}{233} \times 0.582 \text{ g of sulfur}$$

This is the weight of sulfur present in 0.395 g of the substance

$$\text{Hence, percentage of S} = \frac{32}{233} \times 0.582 \times \frac{100}{0.395} = 20.24$$

ESTIMATION OF OXYGEN

Until recently, the percentage of oxygen in a given organic compound was found by difference. Any percentage of the total composition not accounted for by the foregoing determinations was taken to be oxygen. This rather faulty procedure of estimating oxygen has now been replaced by a direct experimental method described below.

Principle. The given organic sample is pyrolyzed (decomposed by heating) in a stream of nitrogen gas. The elemental oxygen along with other gaseous products is passed through hot carbon (1100°C). All the oxygen is thereby converted to carbon monoxide. This when passed through warm (175°C) iodine pentoxide (I_2O_5) is oxidized to carbon dioxide producing iodine.

$$\text{Org. Compound} \xrightarrow{\Delta} O_2 + \text{Gaseous Products}$$

$$O_2 + 2C \xrightarrow{1100°C} 2CO$$

$$5CO + I_2O_5 \xrightarrow{175°C} 5CO_2 + I_2$$

By determining the amount of CO_2 (or I_2) produced, the percentage of oxygen in the original organic compound can be calculated.

Procedure. The pyrolysis of the given organic compound is carried in a combustion apparatus similar to the one used for the estimation of carbon and hydrogen.

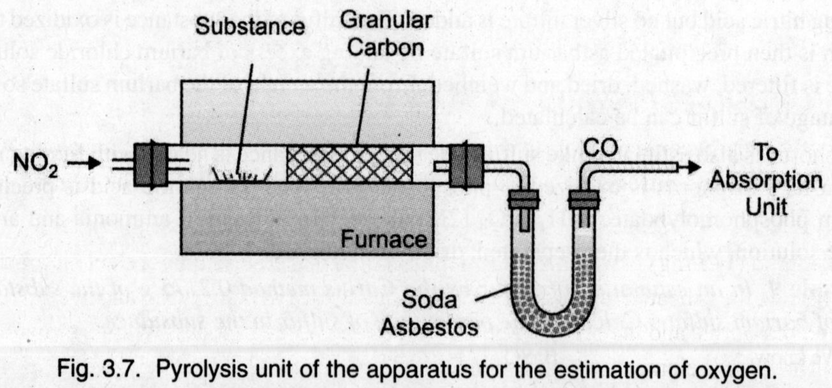

Fig. 3.7. Pyrolysis unit of the apparatus for the estimation of oxygen.

A weighed quantity of the sample is placed in a porcelain boat which is loaded into the combustion tube. The latter part of the combustion tube is packed with granular carbon. The temperature of the heating furnace jacket is kept at 1100°C. Pure nitrogen from a cylinder enters the tube from the left end and sweeps through it.

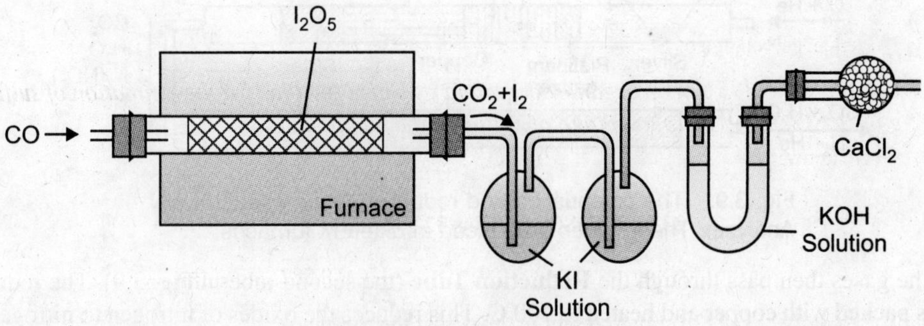

Fig. 3.8. Absorption unit.

The gases emerging from the combustion tube containing carbon monoxide are first passed through a U-tube containing soda-asbestos (asbestos impregnated with dry NaOH). This retains any acidic materials obtained from halogens, nitrogen or sulfur present in the organic sample. The gaseous mixture free from the acidic materials is then passed through a wide glass tube packed with iodine pentoxide kept at 175°C. It converts carbon monoxide to carbon dioxide and generates iodine.

The resulting gaseous mixture is now allowed to pass over a bed of potassium iodide (KI) laid in a wide glass tube. This removes iodine. The residual gases containing carbon dioxide are finally passed through a soda-asbestos U-tube where CO_2 is retained. The soda-asbestos tube is weighed before and after pyrolysis. The difference in weight gives the amount of CO_2 obtained from the given sample of the organic compound.

Calculations :

Let the weight of organic substance taken $= w$ g

Increase in the weight of KOH tube $= x$ g

Since $O \equiv CO \equiv CO_2$

32 g of oxygen present in the substance can form 44 g of CO_2

$\therefore$ 44 g of CO_2 are obtained from oxygen $=$ 32 g

x g of CO_2 are obtained from oxygen $= \dfrac{32}{44} \times x$ g

Hence the weight of oxygen in the weight of sample taken $= \dfrac{32}{44} \times x$ g

$\therefore$ the percentage of oxygen $= \dfrac{32}{44} \times \dfrac{x}{w} \times 100$

MODERN MICRO-ESTIMATION OF ELEMENTS

Recently it has become possible to determine elements automatically. The apparatus to determine carbon, hydrogen, and nitrogen in a compound is called **CHN Analyzer**. The method involves the following steps :

The sample (1–3 mg) is weighed in a platinum boat placed in stainless steel **Combustion Tube** (Fig. 3.9). The tube is heated in a furnace at 900°C in a stream of dry oxygen and helium (carrier gas). The compound burns, and the products pass along a tube. They first pass over three materials which are present to remove elements which would otherwise interfere with the analysis of C, H and N. Silver removes halogens as nonvolatile silver halides, and silver tungstate and silver vanadate remove sulfur and phosphorus with which they combine to form nonvolatile salts.

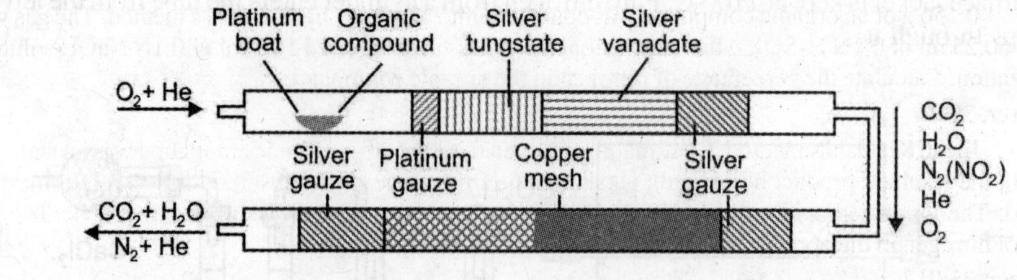

Fig. 3.9. The combustion and reduction tubes used in CHN Analyzer. These tubes are placed in different furnaces.

The gases then pass through the **Reduction Tube** (the second tube in Fig. 3.9). The reduction tube is packed with copper and heated to 700°C. This reduces the oxides of nitrogen to nitrogen and removes oxygen by formation of cupric oxide.

The gases which now contain CO_2, H_2O, N_2, and He, then pass through a gas chromatograph. The recorded chromatogram consists of three peaks (water, carbon dioxide, and nitrogen). The area under each peak is measure of the amount of each of the three components from which percentages can be calculated.

Examination Questions

1. How would you detect carbon and hydrogen in an organic compound ?
2. How would you detect nitrogen, halogens and sulfur in an organic compound ?
(Vishwa Bharti BSc, 2013; Annamalai BSc. 2015)
3. The presence of nitrogen in an organic compound can be detected by
   (*a*) Lassaigne's sodium test
   (*b*) Adding lead acetate solution to sodium extract
   (*c*) Adding $AgNO_3$ solution to sodium extract
   (*d*) Adding sodium nitroprusside to sodium extract

Answer. (*a*)

4. A compound (A) after fusion with sodium gives a precipitate when silver nitrate is added to a portion of the filtrate in acid solution. Compound (A) is likely to be
   (*a*) $CH_3CH_2CH_2Br$ (*b*) CH_3CH_2CHO
   (*c*) $CH_3CH_2CH_2OH$ (*d*) CH_3CH_2COOH

Answer. (*a*)

5. How are carbon and hydrogen estimated in an organic compound ?
6. 0.6 g of an organic compound gave on combustion 0.09 g of H_2O and 0.55 g of CO_2. Calculate the percentage composition of the compound if it contains only C, H, and oxygen.

Answer. C = 25% ; H = 1.66% ; and O = 73.34%.

7. 0.2 g of an organic compound containing carbon, hydrogen, and oxygen, yielded on combustion 0.14 g of carbon dioxide and 0.12 g of water. Calculate the percentage composition of the compound.

Answer. C = 19.90% ; H = 6.66% ; and O = 74.24%.

8. A compound which is believed to have the molecular formula C_4H_8O is submitted for elemental analysis. What percentage of carbon and hydrogen would you expect to be reported if C_4H_8O is the correct formula ?

Answer. C = 66.7% ; H – 11.1% ; and O – 22.2%.

9. How is nitrogen estimated in an organic compound ? *(Lucknow BSc, 2014)*
10. Describe Kjeldahl's method of estimation of nitrogen in an organic compound.
(Agra BSc, 2013 ; Berhampur BSc, 2014; Osmania BSc, 2015)

11. 0.156 g of an organic compound was converted into ammonia by Kjeldahl's method. The gas was absorbed in 60.25 ml of 0.1N H_2SO_4. After absorption, the excess acid required 16.5 ml of 0.1N NaOH solution for neutralization. Calculate the percentage of nitrogen in the organic compound.

Answer. 39.26

12. In the Kjeldahl's method of estimating nitrogen 0.4965 g of an organic compound was heated with conc. H_2SO_4, the resultant product boiled with alkali and the ammonia evolved absorbed in 150 ml decinormal sulfuric acid. The unreacted acid required 40 ml of 0.1N sodium carbonate for neutralization. Calculate the percentage of nitrogen in the organic compound.

Answer. 31.017

13. 0.4422 g of an organic compound was Kjeldahlized and ammonia evolved was absorbed in 100 ml of seminormal N/2 H_2SO_4. The residual acid required 131 ml of 0.25N NaOH solution. Determine the percentage of nitrogen in the compound.

Answer. 54.61

14. In Kjeldahl's estimation of nitrogen, ammonia evolved from 0.5 g of organic compound neutralized 10 ml of 1N acid. Hence the percentage of nitrogen in the organic compound is :

(*a*) 28 (*b*) 30 (*c*) 32 (*d*) 34

Answer. (*a*)

15. The ammonia evolved from 0.21 g of an organic compound by Kjeldahl's method neutralized 15 ml of N/20 H_2SO_4 solution. Calculate the percentage of nitrogen.

Answer. 5

16. 0.4 g of an organic compound on Kjeldahl's analysis gave enough ammonia to neutralize 40 ml of 0.1N H_2SO_4. Calculate the percentage of nitrogen in the compound.

Answer. 14

17. If ammonia obtained from 0.2 g of an organic compound by Kjeldahl's method neutralizes 25 ml of 0.1N acid, what is the percentage of nitrogen in the compound ?

Answer. 17.50

18. 0.35 g of an organic compound was Kjeldahlized and ammonia evolved was absorbed in 100 ml of N/5 H_2SO_4. The residual acid required 154 ml of N/10 NaOH for neutralization. Calculate the percentage of nitrogen in the compound. *(Shivaji BSc, 2015)*

Answer. 18.4

19. 1 g of an organic compound was analyzed by Kjeldahl's method. The ammonia produced was absorbed in 80 ml of N/5 H_2SO_4. The excess acid required 34 ml of N/10 NaOH for neutralization. Calculate the percentage of nitrogen in the compound. *(Berhampur BSc, 2014)*

Answer. 17.64

20. 0.788 g of an organic compound was Kjeldahlized and ammonia absorbed in 100 ml of 1N sulfuric acid. The excess acid required 73.7 ml of 1N sodium hydroxide solution for neutralization. Calculate the percentage of nitrogen in the compound. *(Osmania BSc, 2014)*

Answer. 46.72

21. Describe Dumas' method of estimation of nitrogen in an organic compound.

(Goa BSc, 2013 ; Merrut BSc, 2014)

22. In the Dumas method for the estimation of nitrogen, the sample is heated with copper oxide and the gas evolved is passed over :

(*a*) Copper oxide (*b*) Nickel (*c*) Copper gauze (*d*) None of these

Answer. (c)

23. 0.2 g of an organic compound gave 20.7 ml of nitrogen at 15°C and 760 mm pressure when analyzed by the Dumas method. The percentage of nitrogen in the given compound is

(*a*) 14.26 (*b*) 16.26 (*c*) 12.26 (*d*) 10.26

Answer. (*c*)

24. 1.00 g of an organic compound gave 18.60 ml of nitrogen at 17°C and 754 mm pressure in Dumas method. Determine the percentage of nitrogen in the compound. (Vapor pressure at 17°C = 14 mm).

(Goa BSc, 2014)

Answer. 2.17

25. In a Dumas experiment 0.2033 g of an organic compound gave 31.7 ml of moist nitrogen at 14°C and 758 mm pressure. If the aqueous vapor pressure at 14°C is 14 mm, calculate the percentage of nitrogen in the compound. *(Merrut BSc, 2013)*

Answer. 18.10

26. How are halogens estimated in organic compound ?

27. Describe the Carius method of estimation of chlorine in an organic compound. *(North Eastern Hill BSc, 2015)*

28. In the Carius method for the estimation of halogens, the sample is heated with

(*a*) Conc HNO_3 (*b*) Conc H_2SO_4 and Ag

(*c*) Conc HNO_3 and $AgNO_3$ (*d*) None of these

Answer. (*c*)

29. 0.2562 g of an organic compound gave 0.3066 g of silver bromide when heated with excess of nitric acid and silver nitrate. The percentage of bromine in the given compound is

(*a*) 20.92 (*b*) 30.92 (*c*) 40.92 (*d*) 50.92

Answer. (*d*)

30. 0.147 g of an organic compound containing chlorine was heated with nitric acid and silver nitrate and gave 0.287 g of silver chloride. Calculate the percentage of chlorine in the compound. *(North Eastern Hill BSc, 2015)*

Answer. 48.29

31. 0.207 g of an organic compound gave 0.282 g of silver bromide in Carius method. Determine the percentage of bromine in the compound.

Answer. 57.97

32. 0.394 g of an organic compound containing iodine on treatment by Carius method gave 0.705 g of silver iodide. Calculate the percentage of iodine in the compound.

Answer. 96.70

33. How is sulfur estimated in an organic compound ?

34. 0.2595 g of an organic compound when treated by Carius method gave 0.35 g of $BaSO_4$. Calculate the percentage of sulfur in the compound.

Answer. 18.52

35. 0.38 g of an organic compound gives 1.17 g of barium sulfate. Calculate the percentage of sulfur in the compound.

Answer. 42.28

4

Empirical and Molecular Formulae

(Determination of Molecular Weights)

EMPIRICAL FORMULA

The **Empirical Formula** of a compound is the *simplest formula which expresses its percentage composition*. Such a formula, while it tells us the ratio of the atoms of different elements present in the molecule, does not indicate the actual number of atoms of each element present in it.

CALCULATION OF EMPIRICAL FORMULA

Having determined the percentage of each element present in a compound as described in Chapter 3, its empirical formula is calculated by the following steps :

(1) *The percentage of each element is divided by its atomic weight.* This gives the relative number of different kinds of atoms in the molecule.

(2) *The numbers obtained in step* (1) *are then divided by the smallest of these numbers* (HCF), so as to give a simple atomic ratio.

(3) The figures expressing the simple atomic ratio, as obtained from step (2), give the number of atoms of each element in the molecule, minor fractions being neglected as due to experimental error. However, if necessary, they are further reduced to whole numbers by multiplying with a suitable factor.

Example 1. *An organic substance on analysis was found to contain 10.06 per cent carbon, 0.84 per cent hydrogen, and 89.10 per cent chlorine. Calculate its empirical formula.*

Step 1. Divide the percentage composition of each element by its respective atomic weight.

Carbon	*Hydrogen*	*Chlorine*
$\frac{10.06}{12} = 0.84$	$\frac{0.84}{1} = 0.84$	$\frac{89.10}{35.5} = 2.51$

Step 2. Divide these numbers by the smallest number (in this case 0.84).

$\frac{0.84}{0.84} = 1$	$\frac{0.84}{0.84} = 1$	$\frac{2.51}{0.84} = 3$

Hence, the empirical formula is $C_1H_1Cl_3$ or $CHCl_3$.

Example 2. *Percentage composition of an organic substance as determined by analysis was : Carbon 14.5 ; hydrogen 1.8 ; chlorine 64.46 ; oxygen 19.24. Calculate its empirical formula.*

Element	%age		At Wt		Atomic ratio		HCF		Atoms present
C	14.5	÷	12	=	1.21	÷	1.20	=	$1 \times 2 = 2$
H	1.8	÷	1	=	1.80	÷	1.20	=	$1.5 \times 2 = 3$
Cl	64.46	÷	35.5	=	1.81	÷	1.20	=	$1.5 \times 2 = 3$
O	19.24	÷	16	=	1.20	÷	1.20	=	$1 \times 2 = 2$

In this case, the final figures representing the number of atoms of each element are multiplied by 2 so as to get whole numbers. Hence, the empirical formula of the compound is $C_2H_3Cl_3O_2$.

Example 3. *When analyzed by the combustion method, 0.20 g of an organic substance yielded : (i) 0.147 g of carbon dioxide and 0.12 g of water and (ii) 74.6 ml of nitrogen gas at NTP. Find the simplest formula of the substance.*

Calculation of Percentage of Elements :

Since $CO_2 \equiv C$; $H_2O \equiv 2H$; 22,400 ml $N_2 = 28$ g

$$\text{Percentage of Carbon} = \frac{0.147 \times 12}{44} \times \frac{100}{0.2} = 20.04$$

$$\text{Percentage of Hydrogen} = \frac{0.12 \times 2}{18} \times \frac{100}{0.2} = 6.66$$

$$\text{Percentage of Nitrogen} = \frac{74.6 \times 28}{22400} \times \frac{100}{0.2} = 46.63$$

$$\text{Percentage of Oxygen (by difference)} = 26.67$$

Calculation of Empirical Formula :

Element	%age		At Wt		Atomic ratio		HCF		Atoms present
C	20.04	÷	12	=	1.67	÷	1.66	=	1
H	6.66	÷	1	=	6.66	÷	1.66	=	4
N	46.63	÷	14	=	3.33	÷	1.66	=	2
O	26.67	÷	16	=	1.66	÷	1.66	=	1

Hence, the empirical formula of the substance is CH_4N_2O.

Example 4. *0.21 g of an organic substance containing C, H, O and N only, gave on combustion 0.462 g carbon dioxide and 0.1215 g water. 0.104 g of it when distilled with caustic soda evolved ammonia which was neutralized by 15 ml of N/20 H_2SO_4. Calculate the empirical formula.*

Calculation of Percentage Composition :

Percentage of Nitrogen :

$$15 \text{ ml of } \frac{N}{20}\ H_2SO_4 \equiv 15 \text{ ml of } \frac{N}{20}\ NH_3 \equiv 15 \text{ ml of } \frac{N}{20}\ N \textit{ (Combined)}$$

$$\therefore \text{Wt of nitrogen in 0.104 g substance} = 15 \times \frac{14}{20} \times \frac{1}{1000} \text{ g} = 0.0105 \text{ g}$$

$$\text{\%age of Nitrogen} = \frac{0.0105}{0.104} \times 100 = 10.09$$

Percentage of C and H :

Since, $CO_2 \equiv C$; $H_2O \equiv 2H$

$$\text{\%age of C} = \frac{0.462 \times 12}{44} \times \frac{100}{0.21} = 60.0$$

$$\text{\%age of H} = \frac{0.1215 \times 2}{18} \times \frac{100}{0.21} = 6.43$$

Percentage of Oxygen (by difference) = 23.48

Calculation of Empirical Formula :

Element	%age		At Wt		Atomic ratio		HCF		Atoms present
C	60.00	÷	12	=	5.00	÷	0.72	=	7
H	6.43	÷	1	=	6.43	÷	0.72	=	9
N	10.09	÷	14	=	0.72	÷	0.72	=	1
O	23.48	÷	16	=	1.46	÷	0.72	=	2

Hence, the empirical formula is $C_7H_9NO_2$.

MOLECULAR FORMULA

The **Molecular Formula** of a substance expresses *the actual number of atoms of each element present in its molecule.* It may be the same as the empirical formula of the substance or an exact multiple of it.

Substance	Empirical Formula	Molecular Formula		
Methane	CH_4	CH_4	or	$(CH_4)_1$
Ethylene	CH_2	C_2H_4	or	$(CH_2)_2$
Acetic acid	CH_2O	$C_2H_4O_2$	or	$(CH_2O)_2$
Glucose	CH_2O	$C_6H_{12}O_6$	or	$(CH_2O)_6$

Thus we can write :

$$Molecular\ Formula = (Empirical\ Formula)_n$$

Where $n = 1, 2, 3$ etc. When $n = 1$, the molecular formula coincides with the empirical formula. It follows from the above relation that

$$M\ Wt = Emp\ Formula\ Wt \times n$$

$$\therefore\ n = \frac{M\ Wt}{Emp\ Formula\ Wt}$$

If the molecular weight is known, the value of n can be found with the help of this relation.

Example 1. *An organic compound has been found to possess the empirical formula CH_2O and molecular weight 90. Give its molecular formula. ($C = 12$; $H = 1$; $O = 16$).*

Molecular formula = (Empirical formula)$_n$

But $$n = \frac{\text{M Wt}}{\text{Emp F Wt}}$$

In this case, M Wt = 90

Emp F Wt = 12 + 2 + 16 = 30 $\qquad \therefore\ n = \frac{90}{30} = 3$

Hence, the molecular formula = $(CH_2O)_3$ or $C_3H_6O_3$

Example 2. *An organic liquid contains 12.8 per cent carbon, 2.1 per cent hydrogen and 85.1 per cent bromine. 0.188 g of it in a Victor Meyer experiment, displaced 24.2 ml of moist air measured at 14°C and 752 mm pressure. Find the molecular formula of the substance (Aq tension at 14°C = 12 mm).*

Calculation of Molecular Weight :

$$\text{Volume of dry air replaced at NTP} = \frac{24.2(752-12)}{287} \times \frac{273}{760} = 22.41\ \text{ml}$$

Now, the wt of 22.41 ml of vapor of substance = 0.188 g

$$\text{"} \quad \text{"} \quad 22400 \quad \text{"} \quad \text{"} \quad = \frac{0.188}{22.41} \times 22400 = 187.9\ g$$

∴ Molecular weight of the substance = 187.9

Calculation of Empirical Formula :

Element	%age		At Wt		Atomic ratio		HCF		Atoms present
C	12.8	÷	12	=	1.066	÷	1.064	=	1
H	2.1	÷	1	=	2.10	÷	1.064	=	2
Br	85.1	÷	80	=	1.064	÷	1.064	=	1

∴ Empirical formula = CH_2Br

Calculation of Molecular Formula :

Let the molecular formula be $(CH_2Br)_n$

But $$n = \frac{\text{M Wt}}{\text{Emp F Wt}} = \frac{187.9}{12+2+80} = 2$$

∴ Molecular formula = $C_2H_4Br_2$

Example 3. *An organic dibasic acid contains C = 17.39 ; H = 1.45 and Br = 57.97 per cent. The vapor density of its ethyl ester is 166. Determine the molecular formula of the acid.*

Calculation of Molecular Weight of Acid :

Let the dibasic acid be RH_2 and its ethyl ester $R(C_2H_5)_2$

Molecular weight of ethyl ester = 2 × VD = 2 × 166 = 332

Thus $R(C_2H_5)_2 = R + 2(24 + 5) = 332$ ∴ R = 274

∴ Molecular weight of the acid RH_2 = 274 + 2 = 276

Calculation of Empirical Formula :

Percentage of oxygen = 100 – (17.39 + 1.45 + 57.97) = 23.19

Element	%age		At Wt		Atomic ratio		HCF		Atoms present
C	17.39	÷	12	=	1.45	÷	0.72	=	2
H	1.45	÷	1	=	1.45	÷	0.72	=	2
Br	57.97	÷	80	=	0.72	÷	0.72	=	1
O	23.19	÷	16	=	1.45	÷	0.72	=	2

Hence, the empirical formula is $C_2H_2BrO_2$

Let the molecular formula of the acid be $(C_2H_2BrO_2)_n$.

$$\text{But } n = \frac{\text{M Wt}}{\text{Emp F Wt}} = \frac{276}{24+2+80+32} = \frac{276}{138} = 2$$

∴ Molecular formula is $C_4H_4Br_2O_4$.

Example 4. *0.2 g of anhydrous organic acid gave on combustion 0.04 g water and 0.195 g of carbon dioxide. The acid is found to be dibasic and 0.5 g of its silver salt leaves on ignition 0.355 g silver. What is the molecular formula of the acid ?*

Calculation of Molecular Weight :

$$\frac{\text{M Wt of silver salt}}{2 \times 108} = \frac{0.5}{0.355}$$

$$\text{M Wt of silver salt} = \frac{0.5}{0.355} \times 2 \times 108 = 304.2$$

If RH_2 be the dibasic acid, its silver salt is RAg_2

But $RH_2 = RAg_2 - 2Ag + 2H = 304.2 - 216 + 2 = 90.2$

Calculation of Empirical Formula :

$$\text{Percentage of C} = \frac{0.195 \times 12}{44} \times \frac{100}{0.2} = 26.59$$

Percentage of $H = \frac{0.040 \times 2}{18} \times \frac{100}{0.2} = 2.22$

Percentage of O (by difference) = 71.19

Element	%age		At Wt		Atomic ratio		HCF		Atoms present
C	26.59	÷	12	=	2.22	÷	2.22	=	1
H	2.22	÷	1	=	2.22	÷	2.22	=	1
O	71.19	÷	16	=	4.45	÷	2.22	=	2

∴ Empirical formula is CHO_2

Calculation of Molecular Formula :

Let it be $(CHO_2)_n$; $n = \frac{\text{M Wt}}{\text{Emp F Wt}} = \frac{90.2}{12+1+32} = 2$

Hence, molecular formula of the acid is $C_2H_2O_4$

Example 5. *An organic monobasic acid gave the following percentage composition : C = 70.59 ; H = 5.88 ; O = 23.53.*

0.272 g of the acid required 20 ml of N/10 NaOH for complete neutralization.

Deduce the empirical and the molecular formulae of the acid.

Calculation of Molecular Weight :

$$20 \text{ ml } \frac{N}{10} \text{ NaOH} \equiv 20 \text{ ml } \frac{N}{10} \text{ acid} \equiv \frac{20}{1000} \times \frac{N}{10} \text{ g acid}$$

Thus, $$\frac{20}{1000} \times \frac{N}{10} = 0.272$$

$$\therefore N = 500 \times 0.272 = 136$$

$$\text{Molecular Weight of acid} = N \times \text{Basicity} = 136 \times 1 = 136$$

Calculation of Empirical Formula :

Element	%age		At Wt		Atomic ratio		HCF		Atoms present
C	70.59	÷	12	=	5.88	÷	1.47	=	4
H	5.88	÷	1	=	5.88	÷	1.47	=	4
O	23.53	÷	16	=	1.47	÷	1.47	=	1

Hence, empirical formula is C_4H_4O

Calculation of Molecular Formula :

Let molecular formula be $(C_4H_4O)_n$

$$n = \frac{\text{M Wt}}{\text{Emp F Wt}} = \frac{136}{48+4+16} = \frac{136}{68} = 2$$

Hence, molecular formula is $C_8H_8O_2$.

Example 6. *A monoacid organic base gave the following results on analysis :*

(i) 0.100 g gave 0.2882 g CO_2 and 0.0756 g H_2O.

(ii) 0.200 g gave 21.8 ml nitrogen at 15°C and 760 mm.

(iii) 0.400 g of the platinichloride left on ignition, 0.125 g of Pt.

What is the molecular formula of the base ?

Calculation of Molecular Weight :

$$\text{M Wt of platinichloride} = 195 \times \frac{\text{Wt of salt taken}}{\text{Wt of Pt left}} = 195 \times \frac{0.4}{0.125} = 624$$

$$B = \frac{B_2H_2PtCl_6 - H_2PtCl_6}{2} = \frac{624-410}{2} = 107$$

∴ Molecular weight of the base is 107

Calculation of Empirical Formula :

Volume of nitrogen at NTP $= 21.8 \times \dfrac{273}{288} = 20.66$ ml

Percentage of $C = \dfrac{0.2882 \times 12}{44} \times \dfrac{100}{0.1} = 78.6$

Percentage of $H = \dfrac{0.0756 \times 2}{18} \times \dfrac{100}{0.1} = 8.4$

Percentage of $N = \dfrac{20.66 \times 28}{22400} \times \dfrac{100}{0.2} = 12.91$

Element	%age		At weight		At ratio		HCF		Atoms present
C	78.6	÷	12	=	6.55	÷	0.92	=	7
H	8.4	÷	1	=	8.40	÷	0.92	=	9
N	12.91	÷	14	=	0.92	÷	0.92	=	1

∴ Empirical Formula is C_7H_9N.

Calculation of Molecular Formula :

Let the molecular formula be $(C_7H_9N)_n$

But $n = \dfrac{\text{M Wt}}{\text{Emp F Wt}} = \dfrac{107}{84+9+14} = \dfrac{107}{107} = 1$

∴ Molecular Formula is $(C_7H_9N)_1$ or C_7H_9N.

MOLECULAR FORMULA OF GASEOUS HYDROCARBONS

The molecular formula of a gaseous hydrocarbon may be determined directly without a previous knowledge of its percentage composition. A known volume of the hydrocarbon is mixed with a measured oxygen and the mixture exploded in a eudiometer tube. The carbon and hydrogen of the substance is oxidized to form carbon dioxide and water respectively. The equation of explosion reaction can be written as

$$C_xH_y + (x+y/4)O_2 = xCO_2 + y/2\,H_2O$$

The following observations are recorded :

(1) *The volume of the gaseous products after explosion and cooling* (V_1). This corresponds to the volume of carbon dioxide and unreacted oxygen, the water vapor being condensed to liquid which occupies negligible volume.

(2) *The volume of the residual oxygen after introducing alkali* (V_2). When a little alkali is introduced in the eudiometer tube, it absorbs carbon dioxide leaving behind unreacted oxygen. The volume of carbon dioxide alone would thus be equal to (V_1–V_2).

The molecular formula of the hydrocarbon is then found by the application of Avogadro's Law.

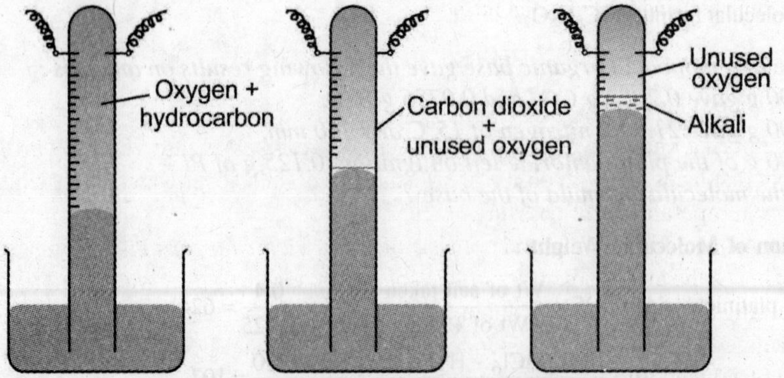

Fig. 4.1. Explosion of gaseous hydrocarbons in eudiometer tube.

Example 7. *10 ml of a gaseous hydrocarbon was mixed with 100 ml of oxygen and the mixture exploded in a eudiometer tube. The volume of the mixture after cooling was reduced to 90 ml which upon treatment with potash solution was further reduced to 70 ml. Determine the molecular formula of the hydrocarbon.*

Volume of oxygen used = 100 – 70 = 30 ml

Volume of carbon dioxide formed = 90 – 70 = 20 ml

If C_xH_y be the formula of the hydrocarbon,

$C_xH_y + (x + y/4)O_2 = xCO_2 + y/2\ H_2O$

Applying the converse of Avogadro's Law,

1 ml C_xH_y + $(x + y/4)$ ml O_2 = x ml CO_2 + $y/2$ ml H_2O

or 10 ml C_xH_y + $10(x + y/4)$ ml O_2 = $10x$ ml CO_2 + $10y/2$ ml H_2O

Thus $10(x + y/4) = 30$ and $10x = 20$

Hence $x = 2$ and $y = 4$

Therefore, the molecular formula of the hydrocarbon is C_2H_4.

Alternative Solution :

Total oxygen used (for CO_2 and H_2O) = 30 ml

$C + O_2 = CO_2$ $\therefore$ 1 vol $CO_2 \equiv$ 1 vol O_2

Hence for 20 ml CO_2 formed, volume of oxygen used = 20 ml

$\therefore$ Volume of oxygen used for water alone = 30 – 20 = 10 ml

$2H_2 + O_2 = 2H_2O$; 1 vol $O_2 \equiv$ 2 vol H_2O

Hence for 10 ml oxygen used, volume of water formed = 20 ml

Now, 10 ml hydrocarbon + 30 ml O_2 = 20 ml CO_2 + 20 ml H_2O

or 1 ml " + 3 ml O_2 = 2 ml CO_2 + 2 ml H_2O

Applying Avogadro's Law, 1 molecule of hydrocarbon yields two molecules of CO_2 (containing two C atoms) and two molecules of H_2O (containing four H atoms). Hence the molecular formula of the hydrocarbon is C_2H_4.

DETERMINATION OF MOLECULAR WEIGHT

The **Molecular Weight** of a compound is *the weight of one molecule of it as compared to the weight of an atom of hydrogen as 1.* The methods commonly employed for the determination of molecular weights of organic substances are :

Physical	Chemical
1. Victor Meyer method	1. Silver salt method for Acids
2. Dumas method	2. Platinichloride method for Bases
3. Hofmann method	3. Volumetric method for Acids and Bases
4. Freezing point and Boiling point method	

PHYSICAL METHODS

VICTOR MEYER METHOD FOR VOLATILE SUBSTANCES

In this method, a known weight of the substance is converted into vapor by dropping in a hot tube. The vapor displaces its own volume of air which is collected over water and its volume measured at the observed temperature and pressure.

The apparatus used in this method consists of : (*i*) a *Victor Meyer tube* of pyrex glass, having a side-tube leading to the arrangement for collection of displaced air over water ; (*ii*) an *outer jacket* of copper, containing a liquid boiling at about 30°C higher than the substance whose molecular weight is to be determined ; and (*iii*) a *tiny stoppered bottle* (Hofmann bottle), in which the substance is weighed and introduced into the heated Victor Meyer tube.

Procedure. The apparatus is fitted up as shown in Fig. 4.2. The liquid in the copper jacket is boiled by heating it with a Bunsen burner when the hot vapors of the liquid in turn heat the Victor Meyer tube. When no more air bubbles escape from the side-tube, a graduated tube filled with water is inverted over its end dipping in water. The Hofmann bottle containing a weighed quantity (0.1–0.2 g) of the substance and loosely stoppered, is then dropped into Victor Meyer tube, opening its cork for a while. The substance at once converts into vapors which blow out the stopper and displace an equal volume of air from the upper part of the Victor Meyer tube. The volume of air collected in the graduated tube is read after carefully levelling in a cylinder of water and the atmospheric pressure and temperature recorded. The volume is reduced to NTP and the weight of 22.4 liters of vapor calculated. This is numerically equal to the molecular weight of the substance.

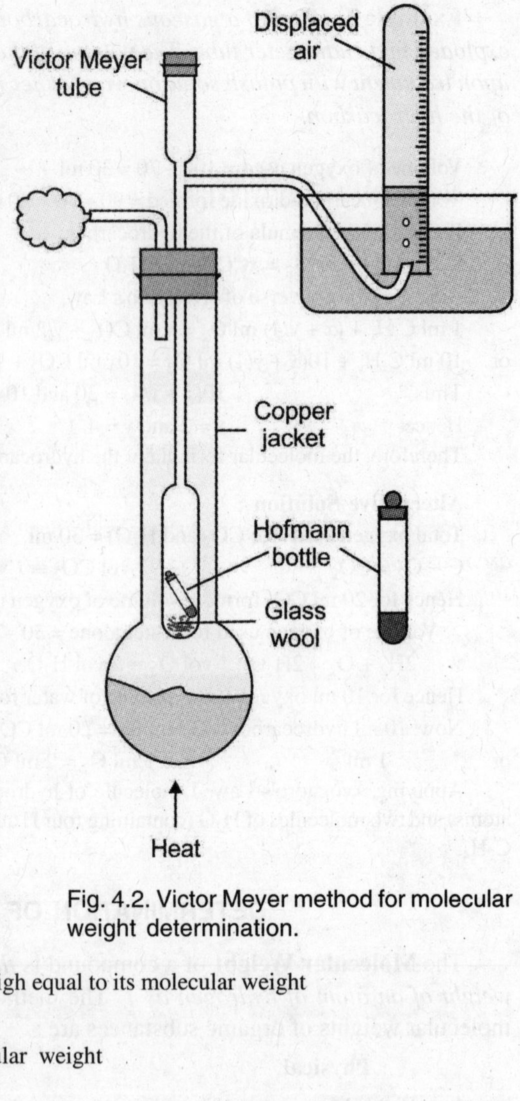

Fig. 4.2. Victor Meyer method for molecular weight determination.

Calculations :

Weight of the substance = w g

Let the volume of the air displaced be V_1 ml (Say) at temperature T_1 and pressure P_1. This volume is converted to NTP using the following relation :

$$\frac{P_1 \times V_1}{T_1} = \frac{P_2 \times V_2}{T_2} \quad \therefore \quad V_2 = \frac{P_1 \times V_1 \times T_2}{T_1 \times P_2}$$

V_2 is the volume of the air displaced at NTP. The volume of air displaced at NTP is equal to the volume of the vapors produced at NTP.

We know that 22.4 liters of a substance at NTP weigh equal to its molecular weight

V_2 ml at NTP weigh $= w$ g

22,400 ml at NTP weigh $= \frac{w}{V_2} \times 22,400 =$ Molecular weight

Example 1. *In a Victor Meyer determination of molecular weight 0.1 g of the substance displaced 27 ml of moist air measured at 15°C and 745 mm pressure. What is the molecular weight of substance ? (Vapor pressure of water at 15°C = 12.7 mm)*

Volume of displaced air, $V_{15} = 27$ ml

Atmospheric temperature $= 15°C$ or on absolute scale

$T_{15} = 273 + 15 = 288°A$

Pressure of dry air, $P_{15} = 745 - 12.7 = 732.3$ mm

$$\frac{P_N \times V_N}{T_N} = \frac{P_{15} \times V_{15}}{T_{15}} \quad \text{or} \quad V_N = \frac{P_{15} \times V_{15} \times T_N}{T_{15} \times P_N}$$

Hence, volume of displaced air at NTP $= \frac{732.3 \times 27 \times 273}{288 \times 760} = 24.66$ ml

Thus the weight of 24.66 ml of vapor of the substance = 0.1 g

Thus the weight of 22400 ml of vapor of the substance $= \frac{0.1 \times 22400}{24.66} = 90.84$ (*mol wt*)

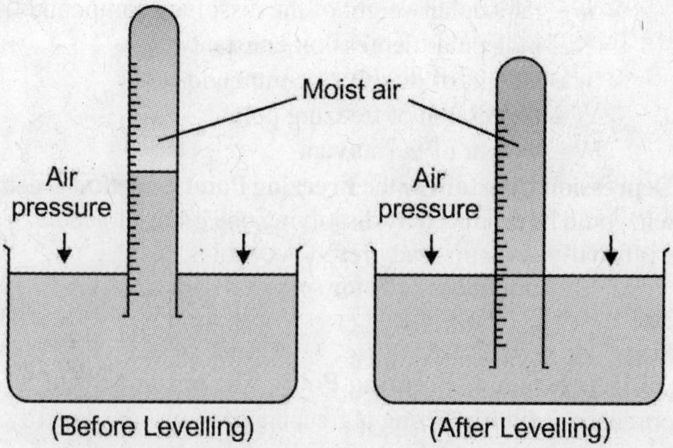

Fig 4.3. After levelling, the pressure of moist air collected is equal to the atmospheric pressure.

Example 2. *0.22 g of a substance when vaporized by Victor Meyer's method displaced 45.0 ml of air measured over water at 20°C and 755 mm pressure. Calculate the molecular weight of the substance. (Vapor pressure of water at 20°C is 17.4 mm)*

(*i*) **Calculation of Volume At NTP :**

Initial Conditions		*Final Conditions*	
Volume	$(V_1) = 45$ ml	Volume	$= V_2$
Pressure	$(P_1) = 755 - 17.4$	Pressure	$= 760$ mm
	$= 737.6$ mm	Temperature	$= 273°$A
Temperature (T_1)	$= 20 + 273$		
	$= 293°$A		

Applying the general Gas Equation

$$\frac{760 \times V_2}{273} = \frac{45 \times 737.6}{293}$$

$$\therefore V_2 = \frac{45 \times 737.6}{293} \times \frac{273}{760} = 40.70 \text{ ml.}$$

(*ii*) **Calculation of M Wt :**

Wt of 40.7 ml of vapor at NTP = 0.22 g

$\therefore$ Wt of 22400 ml of vapor at NTP $= \frac{0.22}{40.7} \times 22400 = 121.1$ g

Thus the Molecular Weight of the substance = 121.1

Dumas Method. This is based on the principle of weighing a known volume of the vapor of the substance in a glass bulb at an elevated temperature. The calculation of the molecular weight is done as in Victor Meyer method.

Hofmann Method. This method is suitable for determining molecular weight of substances at their boiling point under atmospheric pressure.

A known weight of the substance is vaporized above a mercury column in a barometric tube and the volume of the vapor formed noted. The molecular weight of the substance is then calculated as described under **Victor Meyer** method.

CRYOSCOPIC METHOD OR FREEZING POINT METHOD

The molecular weight of a nonvolatile organic compound can be easily determined by noting the depression of freezing point of solvent produced by dissolving a definite weight of the substance in a known weight of solvent.

$$m = \frac{100 \text{ K } w}{\Delta \text{T W}}$$

Where, m = molecular weight of the dissolved compound
K = molecular depression constant
w = weight of dissolved compound
ΔT = depression of freezing point
W = weight of the solvent

The **Molecular Depression Constant** or the **Freezing Point Constant** is defined as the depression of freezing point which would be produced by dissolving one gram-molecule of the solute in 100 g of the solvent. Its value is usually given in ready reference tables.

Values of K for some Solvents

Water	18.6°C	Acetic acid	38.2°C
Benzene	51.2°C	Phenol	73.0°C

Determination of Depression of Freezing Point. The apparatus which is almost exclusively employed for the determination of the lowering of freezing point was designed by Beckmann (Fig. 4.4). It consists of (*i*) the *inner freezing point tube* provided with a side-tube for introducing the solute and fitted with Beckmann thermometer reading up to 0.01°C and a platinum stirrer (*ii*) the *outer air-jacket* surrounding the inner tube which ensures a slower and more uniform rate of cooling of the liquid ; (*iii*) a stout *glass cylinder* which contains freezing mixture and is also provided with a stirrer.

In an actual determination 15-20 g of the solvent is taken in the inner freezing point tube and the apparatus set up as shown in the diagram. The solvent is gently stirred and the mercury thread of the thermometer watched carefully. The temperature first falls below the freezing point of the solvent but later as it begins to solidify, it rises rapidly owing to the latent heat set free, and finally remains steady at the true freezing point.

The freezing point of the solvent having been accurately determined, the solvent is remelted by removing the tube from the bath, and a weighed amount (0.1–0.2 g) of the solute is introduced through the side-tube. Now the freezing point of the solution is determined in the same way as that of the solvent alone. A further quantity of the solute may then be added and another reading taken. Knowing the depression of the freezing point produced, the molecular weight of the solute can be calculated.

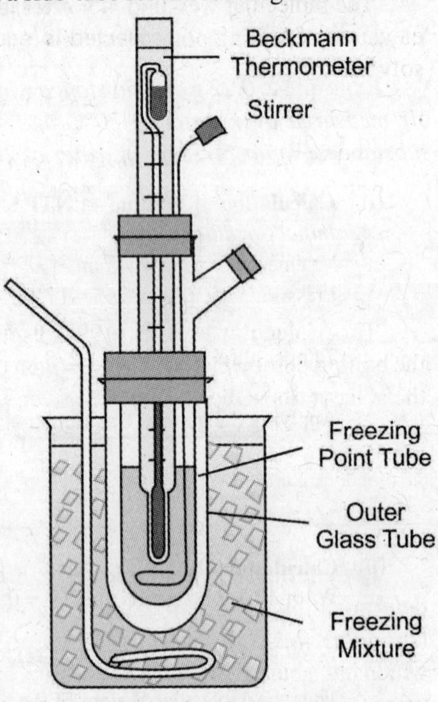

Fig. 4.4. Determination of depression of freezing point.

Example 1. *0.512 g of an organic substance when dissolved in 25 g of water lowered its freezing point by 0.19°C. Calculate the molecular weight of the substance. (Molecular depression constant for water is 18.5°C).*

Substituting the values in the expression

$$m = \frac{100 \text{ K } w}{\Delta \text{T W}}$$

$$= \frac{100 \times 18.5 \times 0.512}{0.19 \times 25} = 199.41$$

Example 2. *1.355 g of a substance dissolved in 55 g of acetic acid produced depression of 0.618°C in the freezing point. Calculate the molecular weight of the dissolved substance. The molecular depression of the freezing point of acetic acid (100 g) is 38.5°C.*

Depression of freezing point = 0.618°C
Weight of solute = 1.355 g
Weight of solvent = 55 g
Substituting the values in the expression

$$m = \frac{100\ \text{K}\ w}{\Delta\text{T W}}$$

$$= \frac{100 \times 38.5 \times 1.355}{0.618 \times 55} = 153.5$$

EBULLIOSCOPIC METHOD OR BOILING POINT METHOD

The molecular weight of a nonvolatile organic compound can also be determined by noting the elevation of boiling point produced by dissolving a definite weight of it in a known weight of the solvent.

$$m = \frac{100\ \text{K}\ w}{\Delta\text{T W}}$$

m = Molecular weight of the dissolved compound
K = Molecular elevation constant
w = Weight of dissolved substance
W = Weight of solvent
ΔT= Elevation of boiling point

The **Molecular Elevation Constant** or the **Boiling Point Constant** is defined as the elevation of the boiling point which would be produced by dissolving one gram-molecule of the solute in 100 g of the solvent. Its value is generally given in ready reference tables.

Values of K for some Solvents

Water	5.2°C	Benzene	27.7°C
Ethyl alcohol	11.5°C	Ether	21.0°C
Acetone	16.7°C	Acetic acid	25.3°C

Determination of Elevation of Boiling Point. The apparatus now commonly used for the determination of the elevation of boiling point was designed by Landsberger. It consists of : (*i*) *boiling flask* which sends vapors of the solvent into the inner tube ; (*ii*) an *inner boiling tube* which has a bulb with a hole in the side, and is graduated in ml. It is fitted with a thermometer reading up to 0.01°C and a glass tube having a rose-head at the lower end. The rose-head ensures uniform distribution of the vapors through the solvent. The bulb reflects back any particles of the solvent which happen to fly about when the boiling becomes brisk ; (*iii*) an *outer covering tube* which receives hot vapors from the inner tube through the hole H. This forms a protecting jacket round the boiling-tube and prevents the loss of heat due to radiation.

In an actual determination 10–15 g of solvent is placed in the inner tube and the vapors of the solvent passed through it. At the start, the solvent vapors condense and as more vapors are passed the liquid begins to boil. As soon as the temperature has become steady, the temperature is read on the thermometer. This is the boiling point of the pure solvent. Now the supply of solvent vapors is cut off for a while and a weighed quantity (0.2–0.5 g) of the substance whose molecular weight is to be determined is dropped into the solvent in the boiling tube. The boiling point of the solution is then determined as before. Immediately after reading the temperature, the thermometer and the rose-head are carefully raised out of the solution and the volume of the solution noted. The weight of the solvent can be found by multiplying its volume with density.

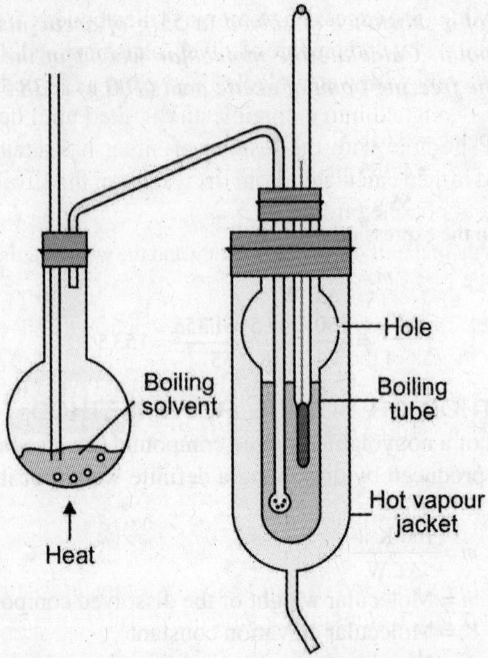

Fig. 4.5. Determination of elevation of boiling point by Landsberger method.

Example 1. *The boiling point of pure acetone is 56.38°C at normal pressure. A solution of 0.717 g of a compound in 10 g of acetone boiled at 56.88°C. What is the molecular weight of the compound ? The molecular elevation constant for acetone is 16.7°C.*

Elevation in boiling point = 56.88 – 56.38 = 0.50°C
Substituting the given values in the expression

$$m = \frac{100\ \text{K}\ w}{\Delta\text{T W}}$$

$$m = \frac{100 \times 16.7 \times 0.717}{0.50 \times 10} = 239.47$$

Example 2. *A solution made by dissolving 1.5126 g of a substance in 21.01 g of benzene has a boiling point 1.05°C higher than the boiling point of pure benzene. Find the molecular weight of the substance. Molecular elevation of boiling point of 100 g benzene is 25.7°C.*

Elevation of boiling point = 1.05°C
Substituting the given values in the expression

$$m = \frac{100\ \text{K}\ w}{\Delta\text{T W}}$$

$$m = \frac{100 \times 25.7 \times 1.5126}{1.05 \times 21.01} = 176.21$$

CHEMICAL METHODS

SILVER SALT METHOD FOR ACIDS

This method of determining molecular weights of organic acids is based on the fact that they form insoluble silver salts which upon heating decompose to leave a residue of metallic silver.

Procedure. A small quantity of the unknown acid is dissolved in water and treated with slight excess of ammonium hydroxide. The excess of ammonia is then boiled off. To this is added sufficient quantity of silver nitrate, when a white precipitate of silver salt is obtained. The precipitate is separated by filtration, washed successively with water, alcohol and ether and dried in the steam oven. About 0.2 g of the dry silver salt is weighed into a crucible and ignited until decomposition is complete. Ignition is repeated till the crucible with the residue of silver has attained constant weight. The molecular weight of the acid is then calculated from the weight of the silver salt taken and the weight of the residue of metallic silver obtained from it.

Calculations. Let the weight of the silver salt taken be x g and the weight of the residue of metallic silver be a g.

The weight of silver salt that would leave 108 g (equivalent weight of the silver) of residue $= \frac{x}{a} \times 108$ and this is the equivalent weight of the salt of the acid.

But Eq wt of acid = Eq Wt of the silver salt – Eq Wt of silver + Eq Wt of Hydrogen

$$= \frac{x}{a} \times 108 - 108 + 1 = \frac{x}{a} \times 108 - 107$$

Molecular weight of the acid

$$= \text{Eq Wt} \times \text{Basicity} = \left(\frac{x}{a} \times 108 - 107\right) \times n$$

Example 1. *0.759 g of the silver salt of a dibasic organic acid was ignited, when a residue of 0.463 g of metallic silver was left. Calculate the molecular weight of the acid.*

Weight of silver salt taken = 0.759 g

Weight of silver left as residue = 0.463 g

∴ The weight of silver salt that would leave 108 g of silver $= \frac{0.759}{0.463} \times 108 = 177.05$ and this is the equivalent weight of silver salt.

Now the equivalent weight of the acid = 177.05 – 108 + 1 = 70.05

∴ Molecular weight of the acid = 70.05 × 2 (basicity) = 140.1

Example 2. *When heated cautiously in a crucible, 0.3652 g of silver benzoate gave 0.1720 g silver. Find the molecular weight of benzoic acid.*

The weight of silver benzoate that would leave 108 g of silver

$$= \frac{0.3652}{0.1720} \times 108 = 229.3 \text{ g}$$

∴ Equivalent weight of benzoic acid = 229.3 – 108 + 1 = 122.3

Since benzoic acid is a monobasic acid, its molecular weight = 122.3 × 1 = 122.3

PLATINICHLORIDE METHOD FOR BASES

Most organic bases (amines) combine with hydrochloro-platinic acid, H_2PtCl_6, to form insoluble salts known as *platinichlorides*. These double salts may be represented by the general formula $B_2H_2PtCl_6$, where B stands for one equivalent of the base. Platinichlorides on ignition decompose to leave a residue of platinum.

$$B_2H_2PtCl_6 \longrightarrow Pt$$

Knowing the weights of the platinichloride taken and the platinum left behind, the molecular weight of the base can be calculated.

Let x g be the weight of the platinichloride taken and a g of the platinum residue after ignition.

(1) Since one molecule of the platinichloride of the base, $B_2H_2PtCl_6$, contains one atom of platinum (atomic weight = 195), 195 g of platinum will be left by 1 g molecular weight of the platinichloride.

But a g of platinum is left by x g platinichloride

$\therefore$ 195 g " " " $\frac{x}{a} \times 195$ g "

Hence Mol Wt of platinichloride $= \frac{x}{a} \times 195$

(2) The equivalent weight of the base, B

$$= \frac{B_2H_2PtCl_6 - H_2PtCl_6}{2}$$

$$= \frac{\text{M Wt of platinichloride} - 410}{2} = \frac{\frac{x}{a} \times 195 - 410}{2}$$

(3) If the acidity of the base is n, the Mol Wt of the base

$$= \left(\frac{x}{a} \times 195 - 410\right) \frac{n}{2}$$

Example 1. *0.6387 g of the platinichloride of a monoacid base on ignition gave 0.209 g of platinum. Find the molecular weight of the base. (Pt = 195)*

(*i*) 0.209 g platinum is left by 0.6387 g of platinichloride

$\therefore$ 195 g " " $\frac{0.6387}{0.209} \times 195$ g "

Hence, Mol Wt of the platinichloride $= \frac{0.6387}{0.209} \times 195 = 596.0$

(*i*) Eq Wt of base $= \frac{\text{M Wt of platinichloride} - 410}{2}$

$$= \frac{596 - 410}{2} = 93.0$$

(*ii*) Mol Wt of base = Eq Wt × Acidity = 93.0 × 1 = 93.0

Example 2. *0.98 g of the platinichloride of a diacid base left on ignition 0.3585 g of platinum. Find the molecular weight of the base.*

(*i*) 0.3585 g Pt is left by 0.98 g platinichloride

195 g " " $\frac{0.98}{0.3585} \times 195$ g "

$\therefore$ M Wt of platinichloride $= \frac{0.98}{0.3585} \times 195 = 533$

(*ii*) Eq Wt of base $= \frac{\text{M Wt of platinichloride} - 410}{2}$

$$= \frac{533 - 410}{2} = \frac{123}{2}$$

(*iii*) M Wt of base = Eq Wt × Acidity $= \frac{123}{2} \times 2 = 123$

VOLUMETRIC METHOD FOR ACIDS AND BASES

The molecular weight of an acid can be determined by dissolving a known weight of it (x g) in water and titrating the solution against standard alkali solution (say N/10) using phenolphthalein as indicator.

Suppose V ml of N/10 alkali neutralize x g acid

$\therefore$ 1000 ml " 1N " " $\frac{x}{V} \times 1000 \times 10$g acid

Since acids and bases neutralize each other in equivalent proportions, and 1000 ml of 1N alkali contain one gram equivalent of it, the Eq Wt of acid $= \frac{x}{V} \times 1000 \times 10$

and Mol Wt of acid = Eq Wt × Basicity

The molecular weight of a base may also be determined exactly as above by titrating a known weight of it against a standard acid and applying the relationship :

$$\text{Mol Wt of base} = \text{Eq Wt} \times \text{Acidity}$$

Example 1. *0.183 g of an aromatic monobasic acid required 15 ml of N/10 sodium hydroxide solution for exact neutralization. Calculate the molecular weight of the acid.*

15 ml N/10 NaOH solution $\equiv$ 0.183 g acid

1000 ml 1N NaOH solution $\equiv \frac{0.183}{15} \times 1000 \times 10$ g acid

But 1000 ml of 1N sodium hydroxide contain 1 g equivalent of it and will neutralize 1 g equivalent weight of acid.

$\therefore$ Eq Wt of acid $= \frac{0.183}{15} \times 1000 \times 10 = 122$

Mol Wt of acid = Eq Wt × Basicity = 122 × 1 = 122

Example 2. *0.25 g of a dibasic organic acid was dissolved in water and the volume made to 100 ml. 10 ml of this solution required 12.3 ml of N/30 sodium hydroxide solution for complete neutralization. Find the molecular weight of the acid.*

12.3 ml N/30 NaOH solution $\equiv$ 10 ml of acid

123 ml N/30 NaOH solution $\equiv$ 100 ml of acid

$\equiv$ 0.25 g acid

1000 ml 1N NaOH solution $\equiv \frac{0.25}{123} \times 1000 \times 30$ g

$\therefore$ Eq Wt of acid $= \frac{0.25}{123} \times 1000 \times 30 = 60.97$

Mol Wt of acid = Eq Wt × Basicity = 60.97 × 2 = 121.94

Example 3. *0.20 g of a diacid base required 25 ml of N/8 HCl for exact neutralization. What is the molecular weight of the base ?*

25 ml of N/8 HCl solution $\equiv$ 0.20 g of the base

1000 ml of 1N HCl solution $\equiv \frac{0.20}{25} \times 1000 \times 8$ g of base

But 1000 ml of 1N HCl contain 1 g equivalent of acid and will neutralize 1 g equivalent of the base.

$\therefore$ Eq Wt of base $= \frac{0.20}{25} \times 1000 \times 8 = 64$

Mol Wt of base = Eq Wt × Acidity = 64 × 2 = 128

Examination Questions

1. How will you determine molecular weight of an organic acid by silver salt method ?

(Calicut BSc, 2014)

2. Describe Platinichloride method of determination of molecular weight of an organic monoacidic base.

(Tezpur BSc, 2015)

3. 0.249 g of silver salt of a dibasic acid gave 0.162 g of silver on ignition. Calculate the molecular weight of the acid.

Answer. 118

4. 0.75 g of platinichloride of a monoacidic base on ignition gave 0.245 g of platinum. Find the molecular weight of the base. (At Wts : Pt = 195 ; Cl = 35.5)

Answer. 93.46

5. 0.177 g of a monobasic acid required 30 ml of N/10 NaOH solution for complete neutralization. Calculate the molecular weight of the acid.

Answer. 59

6. An organic substance of molecular weight 107 was found to contain C = 78.6%, H = 8.4%, and N = 12.92%. Determine the molecular formula of the substance. *(Utkal BSc, 2014)*

Answer. C_7H_9N

7. 0.2 g of an organic dibasic acid gave on combustion 0.04 g water and 0.195 g carbon dioxide. The molecular weight of the acid is 90. Assign molecular formula to the organic acid.

Answer. $C_2H_2O_4$

8. A dibasic acid gave the following results on analysis :

(*a*) 0.354 g of acid gave on combustion 0.5284 g of CO_2 and 0.162 g of H_2O.

(*b*) 0.5976 g of its silver salt left on ignition 0.3888 g of silver.

Determine the molecular formula of the acid.

Answer. $C_4H_6O_4$

9. An organic liquid contains C = 12.8%, H = 2.1%, and Br = 85.1%. 0.188 g of it in a Victor Meyer experiment displaced 24.2 ml of moist air measured at 14°C and 752 mm pressure. Find the molecular formula of the substance. (Aq. tension at 14°C is 12 mm). *(Assam BSc, 2015)*

Answer. $C_2H_4Br_2$

10. 5 ml of a gaseous hydrocarbon required 10 ml of oxygen for complete combustion and produced 5 ml of CO_2. Calculate the molecular formula of the hydrocarbon.

Answer. CH_4

11. Find the empirical and molecular formula of a monobasic acid containing C = 40%, H = 6.67%, and having equivalent weight 60.

Answer. Empirical formula = CH_2O

Molecular formula = $C_2H_4O_2$

12. A compound on analysis gave the following percentage composition by weight : H = 9.09%, C = 54.55%, and O = 36.26%. Its vapor density is 44. Find the molecular formula of the compound. (At Wts : C = 12, H = 1, O = 16).

Answer. $C_4H_8O_2$

13. A compound containing C, H, and O is found to contain C = 32% and H = 4%. Its molecular weight is 150. Find its molecular formula.

Answer. $C_4H_6O_6$

14. A monoacid base contains 53.3% C, 15.6% H, and 31.1% N. Its 0.1087 g platinichloride on ignition gave 0.0429 g of platinum. Find the molecular formula of the base.

Answer. C_2H_7N

15. Upon complete combustion, 0.29 g of an unknown organic compound containing carbon, oxygen, and hydrogen yielded 0.66 g of carbon dioxide and 0.27 g of water. What is the empirical formula ? From this data can you determine the molecular formula ?

Answer. The empirical formula is C_3H_6O. Molecular formula can not be determined because the molecular weight is not given.

Each of the following questions can be divided into two parts : (1) Calculation of empirical or molecular formulas on the basis of given data ; (2) Derivation of the structural formulas on the basis of given reactions or chemical tests. Attempt the following questions after studying the rest of the book so that you are well versed with the chemical properties of various compounds/families. Questions 24 to 40 involve reactions of aliphatic compounds. Questions 41 to 51 involve reactions of aromatic compounds.

16. Three isomeric organic compounds, with vapor density 28, gave the following results on analysis ; C = 85.71% and H = 14.29%. Each one of them decolorizes bromine water and alkaline $KMnO_4$. Give their structural formulas.

Answer. The structural formulas of the three isomers are :

$CH_3CH_2CH{=}CH_2$ (1-Butene) $CH_3CH{=}CHCH_3$ (2-Butene) $CH_3{-}C(CH_3){=}CH_2$ (2-Methylpropene)

17. An organic compound was found to contain C = 40.7%, H = 8.47%, N = 23.73% ; and VD = 29.5. It gives methyl cyanide on heating with phosphorus pentoxide and methylamine on treatment with bromine and KOH. Identify the compound and explain the reactions. *(Madras BSc, 2015)*

Answer. $CH_3-\overset{\overset{\large O}{\|}}{C}-NH_2$; Acetamide

18. Two organic compounds (A) and (B) have the same percentage composition : C = 88.82%, H = 11.18%. The VD of each compound is 27. Both of them decolorize bromine in CCl_4. Compound (A) gives a precipitate with ammoniacal $AgNO_3$ and oxidizes to propionic acid and CO_2. Compound (B) does not give a precipitate with ammoniacal $AgNO_3$ but oxidizes to oxalic acid and CO_2. Deduce the structural formulas of (A) and (B).

Answer. *(A)* = $CH_3CH_2C\equiv CH$; 1-Butyne

(B) = $CH_2=CH-CH=CH_2$; 1,3-Butadiene

19. Two organic compounds (A) and (B) have the same percentage composition : C = 40.00%; H = 6.67%. The VD of (A) is 15 while that of (B) is 30. Compound (A) reduces Fehling's solution but does not react with sodium carbonate. Compound (B) does not reduce Fehling's solution but produces effervescence with sodium carbonate solution. Deduce the structural formulas of (A) and (B).

Answer. *(A)* = HCHO ; Formaldehyde

(B) = CH_3COOH ; Acetic acid

20. An organic compound (A) containing C = 40.00% ; H = 6.67% reduced Fehling's solution. On treatment with NaOH, it gave an alcohol (B) and the sodium salt of an acid (C). Compound (B) can be obtained by reduction of (A). The vapor density of (A) is 15. Give structures of (A), (B), and (C) and explain all reactions with the help of equations. *(Kalyani BSc, 2013)*

Answer. *(A)* = $H-\overset{\overset{\large O}{\|}}{C}-H$; Formaldehyde

(B) = CH_3OH ; Methyl alcohol

(C) = HCOOH ; Formic acid

21. An alkyl halide (A) of molecular formula weight 109 was treated with Mg metal in ether solution. Dry CO_2 was passed through the reaction mixture. The reaction mixture was then hydrolysed with dilute H_2SO_4 to form a compound (B) which on treatment with PCl_5 followed by NH_3 gave compound (C). Compound (C) on boiling with Br_2 and KOH formed a compound (D) which contains C = 53.33%, H = 15.33%, and N = 31.11%. Deduce the structural formulas of (A), (B), (C) and (D) and write the reactions involved.

Answer. The compounds are :

(A) = CH_3CH_2Br ; Ethyl bromide

(B) = CH_3CH_2COOH ; Propanoic acid

(C) = $CH_3CH_2CONH_2$; Propanamide

(D) = $CH_3CH_2NH_2$; Ethylamine

22. An organic compound (A) gave the following results : C = 60.00% ; H = 13.30%. On oxidation (A) gave another product (B) which on oxidation gave acetic acid. What structural formulas would you assign to (A) and (B) ? Vapor density of (A) = 30. *(Rani Durgavati BSc, 2014)*

Answer. *(A)* = $CH_3-\overset{\overset{\large OH}{|}}{CH}-CH_3$; Isopropyl alcohol

(B) = $CH_3-\overset{\overset{\large O}{\|}}{C}-CH_3$; Acetone

23. An organic compound (A) containing C = 40.00% ; H = 6.67% reduced Fehling's solution. On treatment with NaOH, it gave an alcohol (B) and the sodium salt of an acid (C). Compound (B) can be obtained by the reduction of (A). The vapor density of (A) is 15. Give structures of (A), (B), and (C) and explain all reactions with the help of equations.

Answer. *(A)* = HCHO ; Formaldehyde

(B) = CH_3OH ; Methyl alcohol

(C) = HCOOH ; Formic acid

24. An organic compound contains C = 20%, H = 6.67%, N = 46.67% and the rest oxygen. Its molecular weight is 60. Find the molecular formula. The compound on slow heating gives ammonia gas and a solid residue, which gives violet color with alkaline copper sulphate solution. Identify the organic compound. *(Burdwan BSc, 2014)*

Answer. $NH_2-\overset{\overset{O}{\|}}{C}-NH_2$; Urea

25. An organic compound (A) containing 62.07% of carbon and 10.35% of hydrogen was obtained by the oxidation of a compound (B) containing 60% of carbon and 13.33% of hydrogen and vapor density 30. The compound (A) on treatment with PCl_5 gave a compound (C), the vapor density of which is 56.5. On oxidation, (A) gave an acid (D), silver salt of which contained 59.5% of silver. What are (A), (B), (C) and (D)?

Answer. (*A*) = CH_3CH_2CHO ; Propionaldehyde
(*B*) = $CH_3CH_2CH_2OH$; *n*-Propyl alcohol
(*C*) = $CH_3CH_2CHCl_2$; 1,1-Dichloropropane
(*D*) = CH_3CH_2COOH ; Propionic acid

26. An organic compound has the empirical formula CH_2O and VD = 30. It turns blue litmus red and reacts with sodium carbonate to give CO_2. Its sodium salt on heating with soda-lime gives methane. Give the structural formula of the compound and explain the reactions.

Answer. $CH_3-\overset{\overset{O}{\|}}{C}-OH$; Acetic acid

27. On combustion 0.324 g of a dibasic acid containing only C, H, and O, gave 0.3168 g CO_2 and 0.0648 g H_2O. 0.5 g of its silver salt gave on ignition 0.355 g of silver. The acid on treatment with H_2SO_4 gave CO_2 and CO. It could be readily oxidized by acidic $KMnO_4$. Calculate the molecular formula of the dibasic acid and give its structural formula.

Answer. HOOC—COOH ; Oxalic acid

28. An aliphatic organic compound (A) gave the following results : C = 61.01% ; H = 15.25% ; and N = 23.74%. On treatment with nitrous acid (A) gave another product (B) which contains : C = 60.00% ; H = 13.33% ; and no nitrogen. Compound (B) on careful oxidation gave (C) which forms iodoform on treatment with iodine and alkali and had molecular weight 58. What structural formulas would you assign to (A), (B), and (C). Explain the reactions with equations.

Answer. (*A*) = $CH_3-\underset{}{\overset{\overset{CH_3}{|}}{CH}}-NH_2$; Isopropylamine
(*B*) = $CH_3-\overset{\overset{CH_3}{|}}{CH}-OH$; Isopropyl alcohol
(*C*) = $CH_3-\overset{\overset{O}{\|}}{C}-CH_3$; Acetone

29. An organic compound (A), the vapor density of which is 53, contains C = 79.2%, H = 5.5%. On treatment with concentrated KOH solution it gave two compounds (B) and (C). (B) on oxidation gave (A) while (C) on distillation with sodalime gave benzene. Determine the structural formulas of (A), (B), and (C) and explain the reactions.

Answer. (*A*) = C_6H_5CHO ; Benzaldehyde
(*B*) = $C_6H_5CH_2OH$; Benzyl alcohol
(*C*) = C_6H_5COOH ; Benzoic acid

30. An organic compound (A) contains C = 76.6% and H = 6.38%. It is soluble in aqueous sodium hydroxide and gives violet color with $FeCl_3$. Its alkaline solution on being heated with CCl_4 gave an acid (B) of equivalent weight 138. Identify (A) and (B). *(Manipal BSc, 2015)*

Answer. (*A*) = Phenol
(*B*) = Salicylic acid

31. An organic compound having 79.25% of carbon and 5.66% of hydrogen had vapor density 53. It had a characteristic smell and on oxidation with $KMnO_4$, it gave an acid. The sodium salt of the above acid on distillation with sodalime gave benzene. What is the original compound ?

Answer. C_6H_5CHO ; Benzaldehyde

32. An organic liquid (A) containing C, H, and N on treatment with sodium nitrite and hydrochloric acid at 0–5°C gives another compound (B). Compound (B) has molecular weight 140.5 and it contains C = 51.24%, H = 3.55%, N = 19.99%, and Cl = 25.2%. Compound (B) on treatment with CuCN forms another compound (C). On reduction with $LiAlH_4$, (C) forms a primary amine which on treatment with nitrous acid gives an alcohol. The alcohol on oxidation forms benzoic acid. Identify (A), (B), and (C) and explain all the reactions.

Answer. *(A)* = $C_6H_5NH_2$; Aniline

(B) = $C_6H_5N_2^+Cl^-$; Benzenediazonium chloride

(C) = C_6H_5CN ; Benzonitrile

33. An organic compound (A) with vapor density 61.5 gave the following results on analysis: 0.123 g of (A) gave 0.264 g CO_2 and 0.045 g H_2O. 0.1845 g of (A) gave 16.8 ml of N_2 at NTP. When treated with Sn/HCl, it gave (B) containing C = 77.41%, H = 7.63%, and N = 15.05%. When (B) was diazotized and heated in the presence of water, compound (C) was obtained which on distillation with zinc dust gave benzene. Assign structural formulas to (A), (B), and (C).

Answer. *(A)* = $C_6H_5NO_2$; Nitrobenzene

(B) = $C_6H_5NH_2$; Aniline

(C) = C_6H_5OH ; Phenol

34. A compound (A) contains C = 66.4%, H = 5.53%, and Cl = 28.06%. When oxidized with $KMnO_4$ it yields an acid (B) containing C = 68.8%, and H = 4.9%. When (A) is treated with aq. NaOH, compound (C) is produced which contains C = 77.8%, and H = 7.4%. Give the structural formulas of (A), (B) and (C). Explain the reactions involved. *(Madurai BSc, 2014)*

Answer. *(A)* = $C_6H_5CH_2Cl$; Benzyl chloride

(B) = C_6H_5COOH ; Benzoic acid

(C) = $C_6H_5CH_2OH$; Benzyl alcohol

Multiple Choice Questions

1. A hydrocarbon is found to contain 81.80% carbon and 18.20% hydrogen. Its empirical formula will be

(a) C_2H_6 *(b)* C_3H_8 *(c)* C_4H_8 *(d)* C_3H_6

Answer. *(b)*

2. The percentage composition of an organic compound is C = 48.7, H = 13.6, and N = 37.8%. Its empirical formula will be

(a) $C_3H_{10}N_2$ *(b)* C_5H_7N *(c)* C_5H_9N *(d)* None of these

Answer. *(a)*

3. A compound (A) has the following percentage composition : carbon 52.17%, Hydrogen 13.04%, and oxygen. The number of hydrogen atoms in the empirical formula is

(a) 6 *(b)* 5 *(c)* 4 *(d)* 3

Answer. *(a)*

4. An organic compound has empirical formula CH_2O and molecular weight 90. Its molecular formula will be

(a) $C_3H_6O_3$ *(b)* $C_6H_{12}O_6$ *(c)* $C_2H_4O_2$ *(d)* None of these

Answer. *(a)*

5. An organic compound has 42.1% C and 6.4% H. Its molecular weight is 342. Its molecular formula will be

(a) $C_3H_4O_4$ *(b)* $C_4H_6O_4$ *(c)* $C_{12}H_{22}O_{11}$ *(d)* None of these

Answer. *(c)*

6. The empirical formula of a hydrocarbon was found to be CH_2. The molecular weight, determined by mass spectroscopy, is 84. The molecular formula is

(a) C_2H_4 *(b)* C_4H_8 *(c)* C_5H_{10} *(d)* C_6H_{12}

Answer. *(d)*

7. 0.177 g of a monobasic acid required 30 ml of N/10 NaOH solution for complete neutralization. Its molecular weight will be

(a) 49 *(b)* 59 *(c)* 69 *(d)* 79

Answer. *(b)*

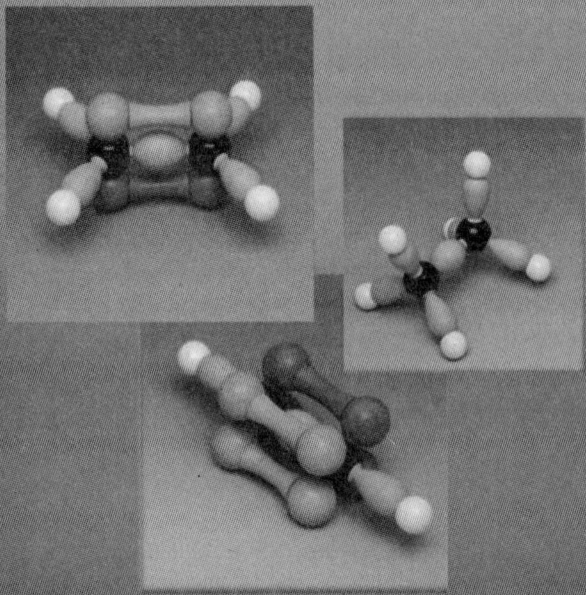

5

Structure of Organic Molecules

CLASSICAL CONCEPT OF BONDING

STRUCTURE OF ATOM

The views on the atomic structure which are accepted today have developed from the classical Rutherford-Bohr theory. According to this theory, the atom is made of a central positively charged *nucleus* containing positively charged particles called *Protons*, and neutral particles called *Neutrons*, both having unit mass. The nucleus is surrounded by negatively charged particles called *Electrons* which carry one unit negative charge and negligible weight.

The electrons are said to revolve around the nucleus in fixed orbits or energy levels. While the electron moves in such a level, it possesses a definite quantity of energy and it neither emits nor absorbs energy. The electrons are arranged in the orbits so that the maximum number of electrons in the various orbits starting from the one nearest the nucleus is 2, 8, 18, 32, 18, 8. The outermost orbit of electrons in different atoms (except those of inert gases), is incomplete and the electrons in it are known as the *Valence Electrons*.

WHY ATOMS COMBINE TO FORM MOLECULES ?

The classical concept of formation of molecules, proposed by Lewis and Kossel, is based upon the electronic structure of atoms. The atoms of inert gases have either 2 or 8 electrons in the outermost orbit. These gases do not enter into chemical combination and, therefore, are assumed to have complete or stable orbits. The atoms of all other elements have incomplete outermost orbits and tend to complete them by chemical combination with other atoms. G.N. Lewis proposed that it is the urge of atoms to complete their outermost orbits of electrons as in the inert gases, which is responsible for chemical combination. In other words, chemical combination between two atoms results from the redistribution of electrons between them so that both the atoms complete their outermost orbits or acquire stable electronic configuration possessed by the inert gases.

TYPES OF BONDS

There are three basic ways in which chemical combination occurs :

(1) Ionic or electrovalent bond
(2) Covalent bond
(3) Coordinate bond

Ionic or Electrovalent Bond. Ionic or electrovalent bonds are formed by transfer of valence electrons from one atom to another. This type of bond unites two atoms one of which has excess electrons than the stable number (2 or 8), and the other is short of electrons. Sodium chloride is a typical compound formed in this way.

Here the sodium atom (2, 8, 1) transfers its excess electron to chlorine atom (2, 8, 7), and thus both attain a stable inert gas type electron configuration. Sodium acquires the electron configuration of *neon* (2, 8) and becomes positively charged. Chlorine acquires the electron configuration of *argon* (2, 8, 8), and becomes negatively charged. These oppositely charged ions are held together by electrostatic force of attraction. This type of bond is commonly found in inorganic compounds.

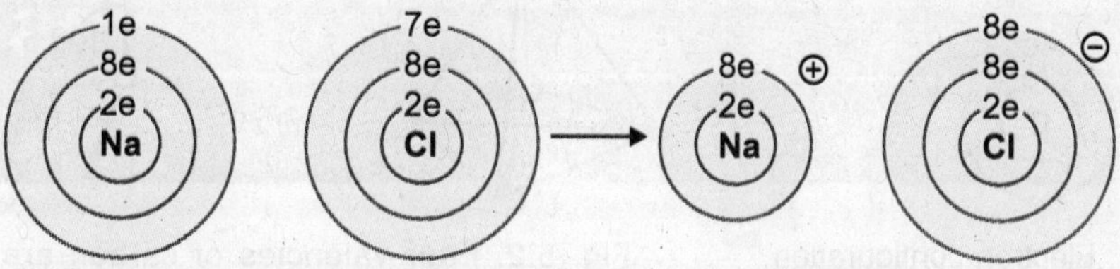

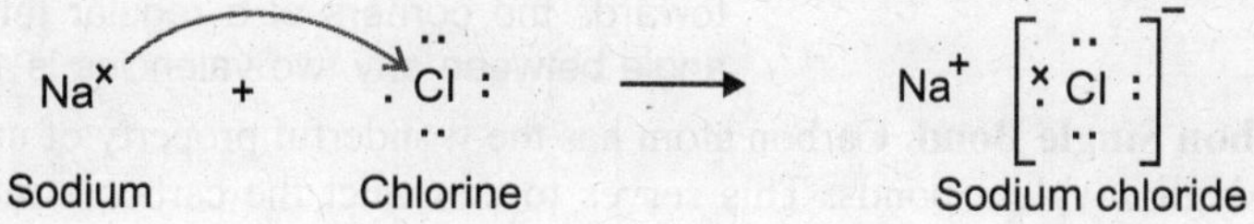

Ionic or electrovalent compounds are nonvolatile, soluble in water and possess high melting points. Their aqueous solutions conduct electric current.

Covalent Bond. Covalent bonds are formed by mutual sharing of electrons. This type of bond unites two atoms, both of which are short of electrons. The two atoms contribute one electron each and then share the resulting pair of electrons. Hydrogen is the simplest compound formed in this way.

$$\text{H}\cdot + {}^{\times}\text{H} \longrightarrow \text{H}\overset{\cdot}{\underset{\times}{}}\text{H} \quad \text{or} \quad \text{H—H}$$

Here the two electrons are shared, and give to each hydrogen atom the configuration of helium. This type of bond is termed covalent bond and is indicated by a line. Covalent bonds are commonly found in organic compounds.

Covalent compounds are volatile, generally insoluble in water but soluble in organic solvents. They possess low melting and boiling points. Their solutions do not conduct electric current.

Coordinate Bond. Coordinate bond is also formed by mutual sharing of electrons but in this case the two electrons that are shared come from the same atom. A coordinate bond unites two atoms, one of which has a spare pair of electrons and the other is short of a pair of electrons. The first atom (donor atom) contributes one pair (lone pair) of electrons and the second atom (acceptor atom) accepts it. After the formation of the bond, the lone pair of electrons is held in common. The coordinate bond is represented by an arrow, pointing away from the donor atom. An excellent illustration of the coordinate bond is found in the boron hydride-ammonia complex.

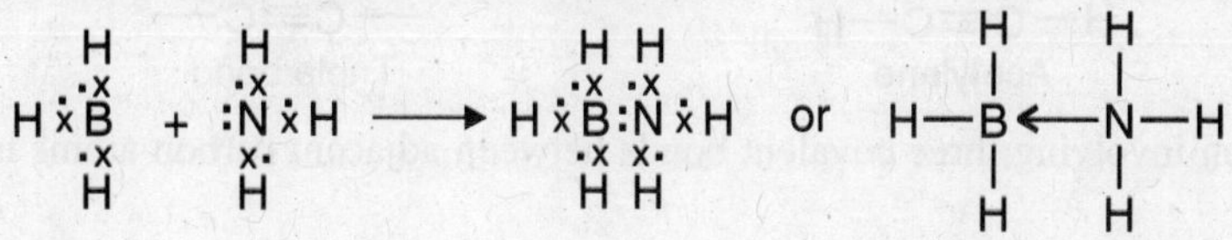

VALENCE OF CARBON

The atomic number of carbon is 6 and its atomic weight is 12. Its electron configuration is shown in Fig. 5.1. It has four electrons in the last orbit and tends to gain four more electrons by forming four covalent bonds with other hydrogen atoms. Thus, the structural formula of the simplest hydrocarbon methane (CH_4) can be written as :

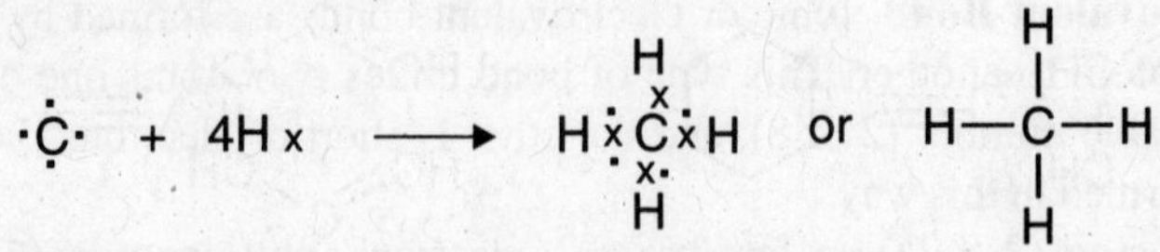

Similarly in all organic molecules carbon atom is tetravalent. That is, it has a valence of 4. According to Le Bel and van't Hoff the four valencies of carbon do not lie in one plane. They are directed towards the corners of a regular tetrahedron so that the angle between any two valencies is 109.5° (Fig. 5.2).

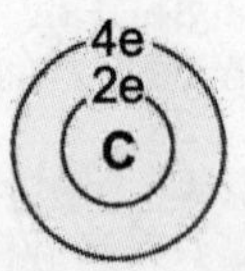

Fig. 5.1. Electron configuration of carbon.

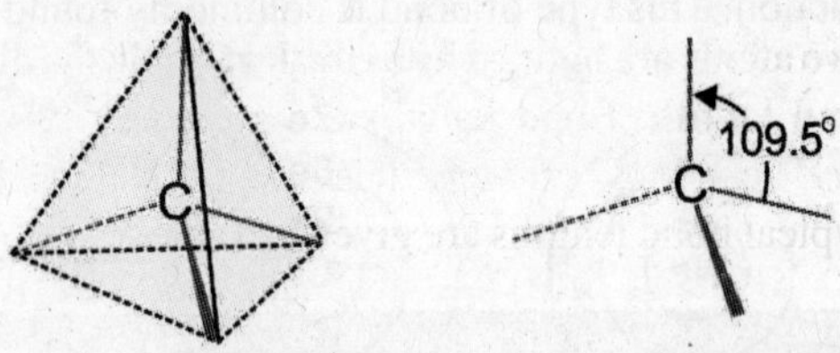

Fig. 5.2. Four valencies of carbon are directed towards the corners of a regular tetrahedron. The angle between any two valencies is 109.5°.

Carbon-Carbon Single Bond. Carbon atom has the wonderful property of uniting with other carbon atoms through covalent bonds. This serves to construct the carbon structure of organic molecules. Thus the molecules of hydrocarbons, ethane and propane contain two and three carbon atoms respectively linked by covalent bonds.

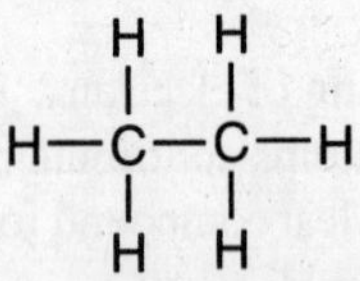

Ethane

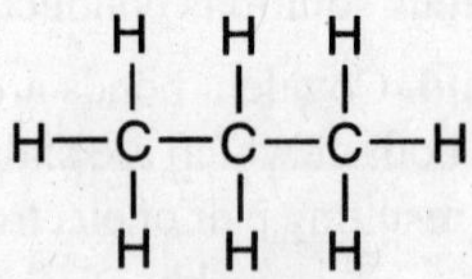

Propane

Carbon-Carbon Double Bond. In some compounds, two of the valencies of a carbon atom may be satisfied by union with the two valencies of another carbon atom. Thus in ethylene the two carbons are united by two covalent bonds.

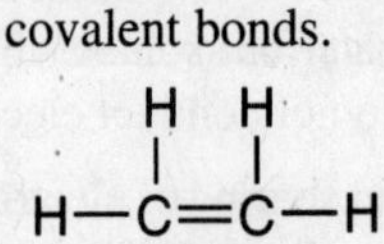

Ethylene

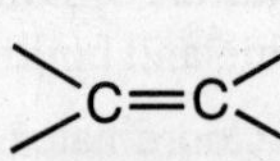

Carbon-Carbon double bond

Such a union involving two covalent bonds between adjacent carbon atoms is called **Double Bond.**

Carbon-Carbon Triple Bond. Sometimes two adjacent carbons are linked together by three covalent bonds. Thus acetylene molecule is represented as :

$$H-C\equiv C-H$$

Acetylene

$$-C\equiv C-$$

Triple bond

Such a union involving three covalent bonds between adjacent carbon atoms is called **Triple Bond.**

Cyclic Structures. We have given above some examples of substances where the molecules consist of carbon atoms joined together in chains that are free at both ends. There are numerous compounds known where carbon atoms join to form closed rings. These are called **Cyclic Compounds** or **Ring Compounds**.

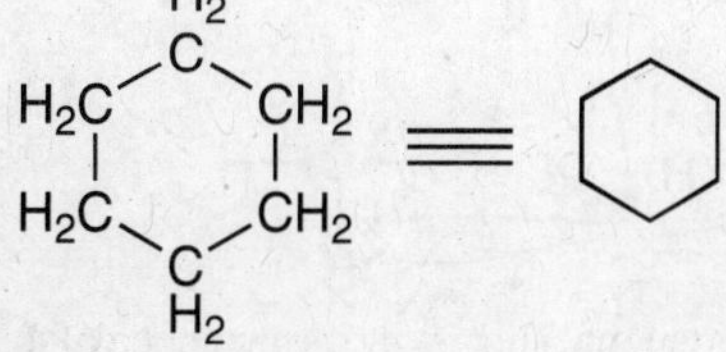

Cyclohexane

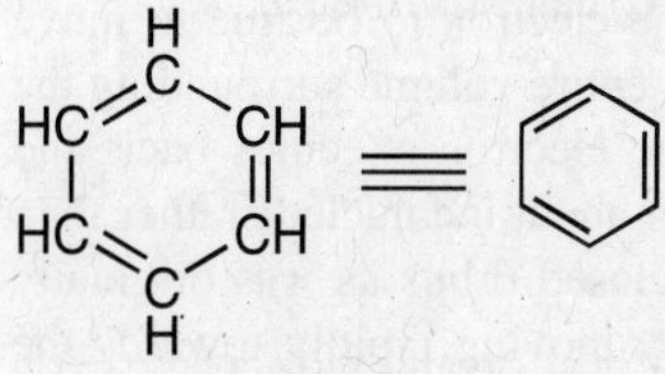

Benzene

BOND LENGTHS

When two atoms are bonded by a covalent bond, the distance between the centres of the two nuclei is called bond length. Bond lengths are measured by X-ray crystallography and by microwave spectroscopy. The unit of bond length is Angstrom (1Å=10^{-8} cm). For most bonds the values are 1 to 2 Å. Some typical bond lengths are given in Table 5.1.

Table 5.1 SOME BOND LENGTHS

Bond	Bond length (Å)	Bond	Bond length (Å)
C–H	1.09	C–O	1.43
C–C	1.54	C=O	1.21
C=C	1.34	C–N	1.47
C≡C	1.20	C–Cl	1.77
O–H	0.96	C–Br	1.91
N–H	1.02	C–I	2.12

BOND ENERGIES

Bond energy or bond strength is defined as the amount of energy required to break a bond in a molecule. Bond energies depend upon the type of bond as well as the structural environment in which the bond is situated. They are determined by quantitative measurements of heats of chemical reactions (calorimetry) and by spectroscopic methods. The unit of bond energy is kcal/mol. Some typical bond energies are given in Table 5.2.

Table 5.2 SOME BOND ENERGIES

Bond	Bond energy (kcal/mol)	Bond	Bond energy (kcal/mol)
C–H	99	C–O	86
C–C	83	C=O	179
C=C	146	C–N	73
C≡C	200	C–Cl	81
O–H	111	C–Br	68
N–H	93	C–I	51

MODERN CONCEPT OF BONDING

ORBITAL STRUCTURE OF ATOM

With the emergence of Wave Mechanics, the Bohr model of the atom suffered a serious set back and so did the principles of bonding that were based on it. The present day model of the atom considers the electron to be free to move about in the entire volume surrounding the nucleus. The electron executes back and forth motion about the nucleus rather than moving in closed orbits as was originally believed. By moving rapidly towards the nucleus and away from it in all possible directions, the electron virtually occupies the entire space around it. So that if the moving electron were visible to the eye, it would not appear as a particle but would give the appearance of a diffused spherical 'cloud' formed by its motion. Since the formation of this cloud is due to the motion of the electron, it is commonly referred to as *Electron Cloud*.

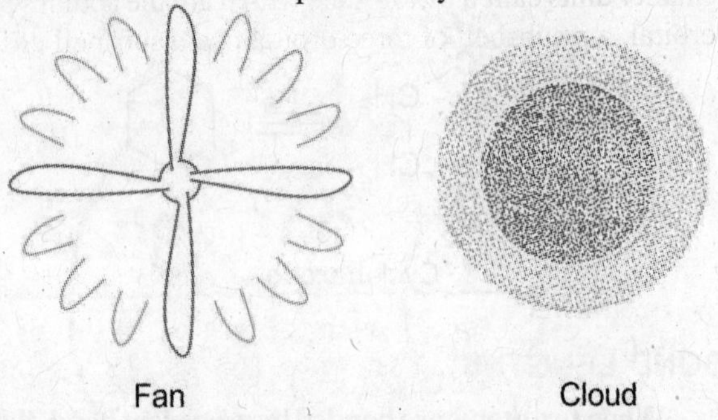

Fig. 5.3. The fan occupies only a portion of the circle, but it occupies effectively the whole circular space when it revolves rapidly. The same reasoning applies to the formation of a cloud when electron moves briskly in a shell.

The concept of electron cloud can be easily understood by considering the case of a revolving ceiling fan. When at rest or moving slowly, the four-blade fan occupies only a section of the circle but as the fan turns rapidly it appears to occupy the entire circular space. Similarly, the rapidly moving electron would occupy effectively the entire space around the nucleus but at any given time the electron would be found somewhere in three dimensional space best described by points of highest probability. **The shape and size of this region in space around the nucleus, where the probability of finding an electron is maximum, is called an Orbital**. An electron can be anywhere within an orbital at any given time, although it tends to occupy certain portions of its orbital to a greater extent than other portions.

The electrons in an atom may be identified by four **Quantum Numbers** *i.e.*, principal quantum number, azimuthal quantum number, magnetic quantum number and spin quantum number. The principal quantum number refers to the size of the orbital and also the energy of the electron in it, the azimuthal quantum number to the shape of the orbital, the magnetic quantum number, to the orientation of the orbital in space ; and the spin quantum number, to the direction of the spin of the electron about its own axis (Fig. 5.4). According to **Pauli's Exclusion Principle**, the two electrons in an orbital may have the same principal, azimuthal and magnetic quantum numbers but they must have different spin quantum numbers. Such pairs of electrons with opposite spin are referred to as 'spin-paired' electrons.

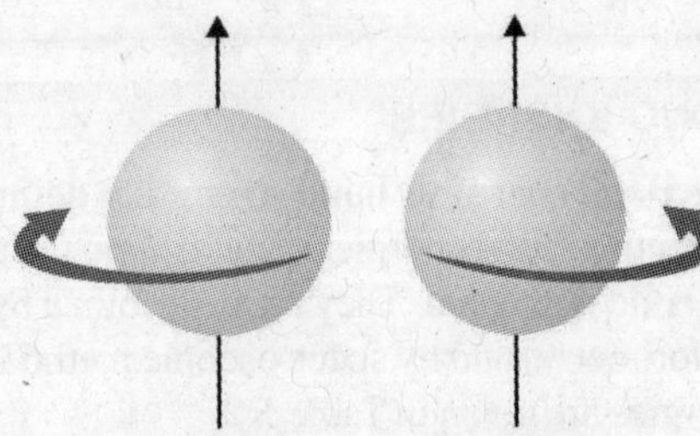

Fig. 5.4. An electron spins around its own axis in clockwise or anticlockwise direction.

Electrons having the same principal quantum number are said to occupy a **Shell** which is further subdivided into **Subshells**. Starting from the shell nearest to the nucleus, the shells enveloping each other are designated by numerals 1, 2, 3, 4, 5, 6 and 7 or by letters K, L, M, N, O, P and Q. The number of subshells in a shell is equal to the number of the principal shell. For example, the third shell contains three subshells and the fourth shell contains four subshells. So far only four subshells designated as *s*, *p*, *d* and *f* are known and their relative energies increase in the order mentioned above. The subshells of different shells are identified by giving the number of the principal shell before the type of the subshell *i.e., s, p, d* or *f*. The first shell has only one subshell designated as 1*s*; the second shell has two subshells designated as 2*s* and 2*p* ; the third shell has three subshells designated as 3*s*, 3*p*, 3*d* ; and the fourth shell has four subshells namely 4*s*, 4*p*, 4*d* and 4*f*. The subshells are further divided into smaller units called the orbitals which are the actual seat of electrons. An *s* subshell is made up of one orbital; a *p* subshell of three orbitals ; a *d* subshell of five orbitals and *f* subshell of seven orbitals.

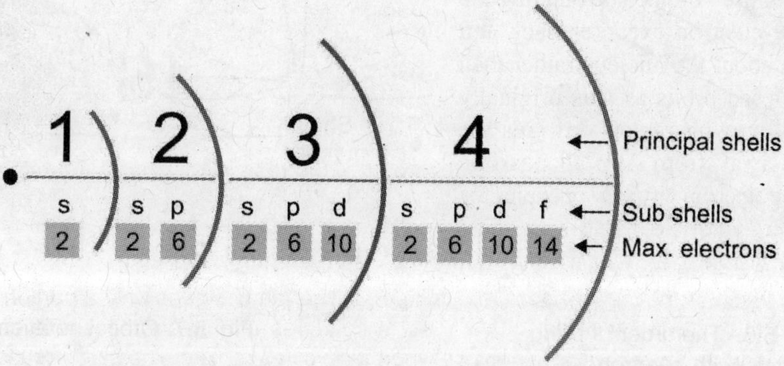

Fig. 5.5. Distribution of electrons in Shells and Subshells.

The outermost shell is known as the Valence Shell and the electrons that occupy this shell are called the Valence Electrons.

ELECTRON CONFIGURATION

The distribution of electrons in shells, subshells, and orbitals is governed by the following rules:

(1) *The total number of electrons that can be accommodated in a* **Shell** *is equal to* $2n^2$, *where n refers to the principal quantum number of the shell.* Thus the first shell can accommodate 2 electrons, the second 8, the third 18, and the fourth 32.

(2) *The total number of electrons that can be accommodated in a* **Subshell** *is equal to twice the number of orbitals it contains.* Thus an *s* subshell can have 2 electrons because it has only one orbital, *p* subshell can have 6 electrons due to three orbitals, *d* can have 10 electrons due to five orbitals, and *f* can have 14 electrons due to seven orbitals. Distribution of electron in shells and subshells is illustrated in Fig. 5.5.

(3) *The total number of electrons that can be accommodated in an orbital is 2 and these two electrons must have opposite spins* (Pauli's Exclusion Principle). Electrons with opposite spins are given the symbols ↑ and ↓ .

(4) *Electrons occupy the orbital of the lowest energy first and then the filling of the orbitals of higher energy starts.* Thus the electrons fill the 1*s* orbital before occupying 2*s* orbital. The energy of orbitals of the various subshells follows the sequence given below :

$$1s < 2s < 2p < 3s < 3p < 4s < 3d < 4p < 5s$$

This is also the order of filling various orbitals (Fig. 5.6).

(5) *In a subshell, all the available degenerate orbitals* (*i.e.,* the orbitals of same energy) *are occupied singly first and then pairing of electrons in each orbital occurs.* Thus, of the three degenerate orbitals of *p* subshell (p_x, p_y, and p_z), no one will have two electrons as long as any other is vacant. This is called **Hund's Rule**, or the **Principle of Maximum Multiplicity**.

The distribution of electrons in an atom, by convention, is always written in terms of electrons which each of the subshells occupies. In common practice, the number of electrons in the subshell are indicated as superscripts (Fig. 5.7).

The complete electron configuration of any element is written by listing the subshells in the order of increasing energy and the number of electrons occupying each subshell are indicated by superscripts. The electron configurations of the first ten elements are given in Table 5.3.

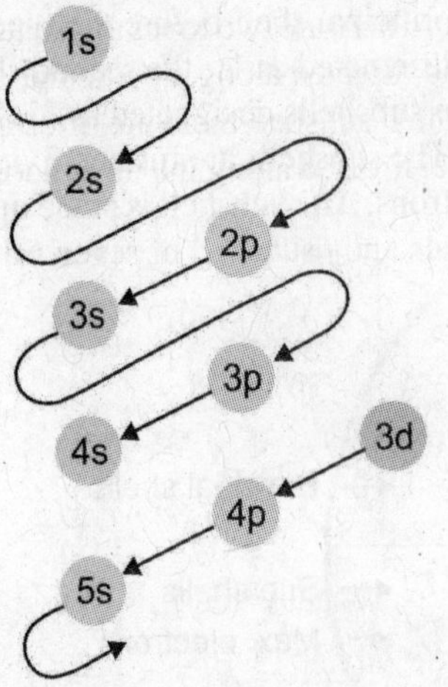

Fig. 5.6. The order of filling orbitals with electrons.

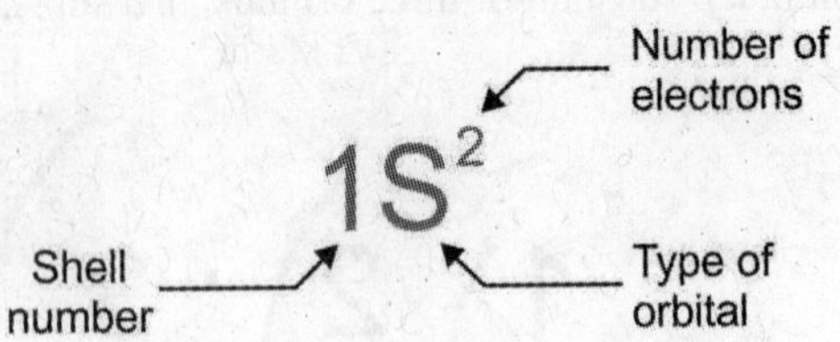

Fig. 5.7. Orbital notation.

Table 5.3 ELECTRON CONFIGURATION OF FIRST TEN ELEMENTS

Element	At. No.	Electron configuration			
H	1	$1s^1$			
He	2	$1s^2$			
Li	3	$1s^2$	$2s^1$		
Be	4	$1s^2$	$2s^2$		
B	5	$1s^2$	$2s^2$	$2p_x^1$	
C	6	$1s^2$	$2s^2$	$2p_x^1$	$2p_y^1$	
N	7	$1s^2$	$2s^2$	$2p_x^1$	$2p_y^1$	$2p_z^1$
O	8	$1s^2$	$2s^2$	$2p_x^2$	$2p_y^1$	$2p_z^1$
F	9	$1s^2$	$2s^2$	$2p_x^2$	$2p_y^2$	$2p_z^1$
Ne	10	$1s^2$	$2s^2$	$2p_x^2$	$2p_y^2$	$2p_z^2$

SHAPES AND ORIENTATION OF ORBITALS

An orbital represents a well-defined region in space in which the electron can be found. They have definite three-dimensional shapes. In organic chemistry we are mainly concerned with the *s* and *p* orbitals. The shapes and orientations of these orbitals are described below.

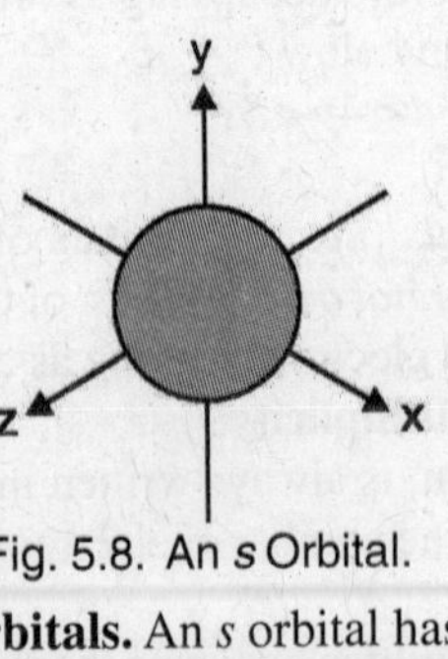

Fig. 5.8. An *s* Orbital.

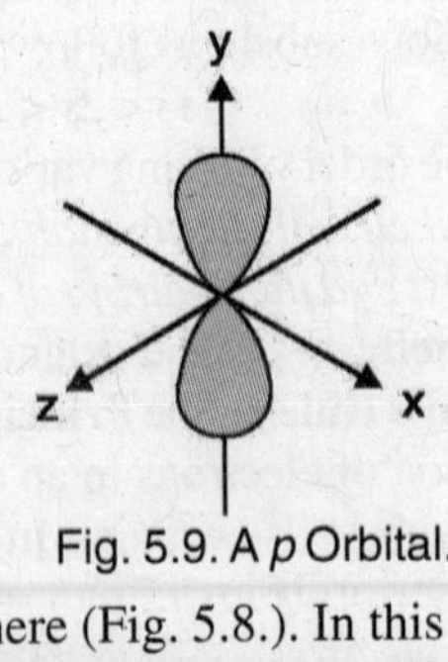

Fig. 5.9. A *p* Orbital.

Shape of *s* Orbitals. An *s* orbital has the shape of a sphere (Fig. 5.8.). In this orbital there is an equal probability of finding the electron in any direction away from the nucleus. All *s* orbitals have this shape. The difference between an electron in a 1*s* and one in 2*s* orbital is that, the electron in the 2*s* orbital is farther away from the nucleus and has greater energy.

Shape of *p* Orbitals. A *p* orbital is dumbbell shaped (Fig. 5.9). There are two lobes associated with this orbital. They are located on opposite sides of the nucleus and directed along a particular axis. In contrast to the *s* subshell for which there is only one orbital, in the *p* subshell there are three orbitals of equal energy for electrons to occupy. These three *p* orbitals are directed along the three coordinate axes as shown in Fig. 5.10.

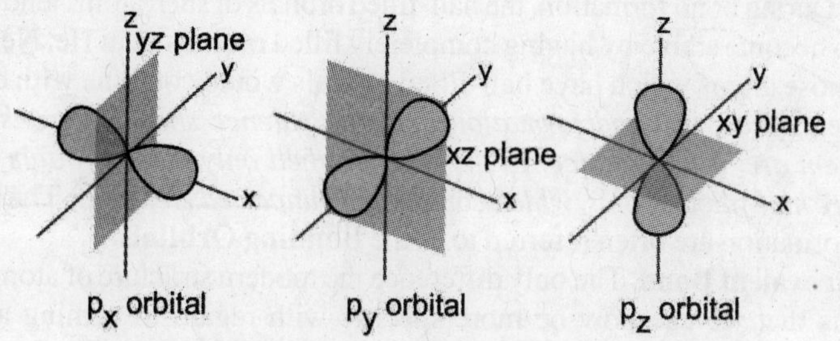

Fig. 5.10. Relative orientations of three *p* orbitals.

BONDING IN THE LIGHT OF MODERN ATOMIC STRUCTURE

Let us consider what happens when two atoms approach each other. When the atoms are far apart, there is no force between them. As the atoms move closer, the electron clouds overlap. There is an attraction between the electrons in one atom, and the nucleus of the other atom. There is also repulsion between the electrons from each atom and repulsion between the nuclei (Fig. 5.11). These repulsive forces become significant at small distances.

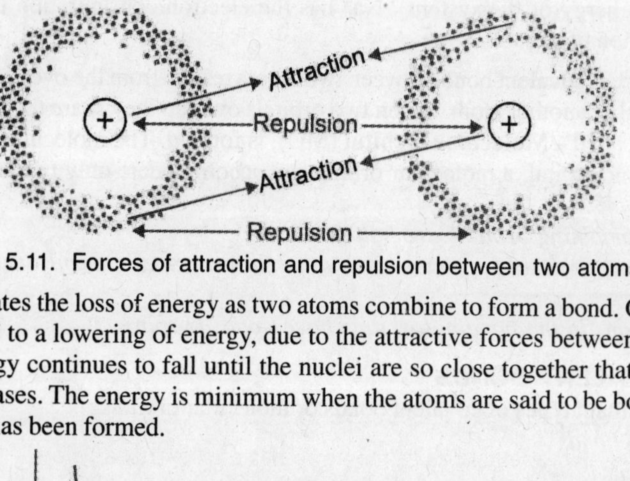

Fig. 5.11. Forces of attraction and repulsion between two atoms.

Fig. 5.12 illustrates the loss of energy as two atoms combine to form a bond. Overlapping of the atomic orbitals leads to a lowering of energy, due to the attractive forces between the electrons and the nuclei. The energy continues to fall until the nuclei are so close together that repulsion occurs, and the energy increases. The energy is minimum when the atoms are said to be bound together, that is, a chemical bond has been formed.

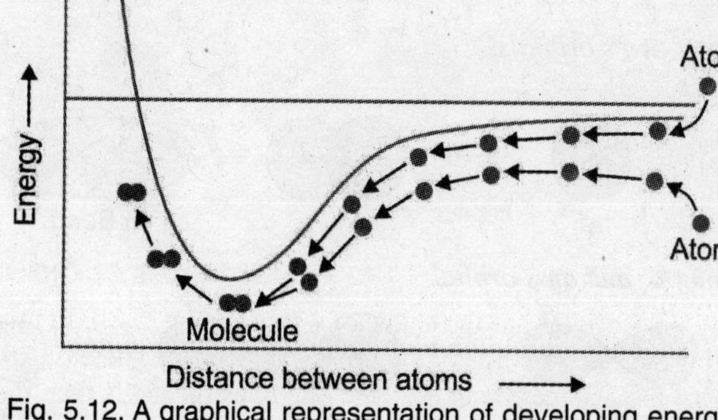

Fig. 5.12. A graphical representation of developing energy change as atoms combine to form a bond.

A chemical bond is formed if there is a decrease in the total energy of the system when the bonding atoms approach each other. Thus the new arrangement of electrons and nuclei obtained because of bond formation is less energetic (more stable) than the arrangement in isolated atoms. The electron configuration of chemically stable inert gases shows that all the orbitals in their atoms are completely filled. On the other hand, atoms of all other elements are known to be reactive and have half-filled orbitals. During bond formation, the half-filled orbitals of such atoms tend to acquire another electron in order to become stable by having completely filled orbitals as in He, Ne, Ar, etc. Thus it is evident that only those atoms which have half-filled orbitals would combine with each other to form molecules. In other words, *in bond formation only the valence shell electrons and the orbitals associated with them are of importance. In the valence shell only those orbitals take part in bond formation which are half-filled, that is, which contain only unpaired electrons.* The half-filled orbitals involved in bond formation are often referred to as the **Bonding Orbitals**.

Ionic or Electrovalent Bond. The only difference the modern structure of atom makes to the idea of ionic bonding is that we can now be more specific with regard to naming and indicating the positions of the electrons that are transferred. Thus in sodium chloride, it is the 3*s* electron of sodium that is transferred to the chlorine and it will occupy a 3*p* orbital in the chlorine atom.

$$\text{Na } 1s^2 2s^2 2p^6 3s^1 \longrightarrow \underset{\text{Neon Configuration}}{[\text{Na}]^+ \ 1s^2 2s^2 2p^6} + e^-$$

$$\text{Cl } 1s^2 2s^2 2p^6 3s^2 3p^5 + e^- \longrightarrow \underset{\text{Argon Configuration}}{[\text{Cl}]^- \ 1s^2 2s^2 2p^6 3s^2 3p^6}$$

Both Na^+ and Cl^- have attained a more stable electronic configuration of an inert gas (and thereby lowering the total energy of the system). Na^+ has the electron configuration of *neon*. Cl^- has the electron configuration of *argon*.

Covalent Bond. A covalent bond between two atoms results from the overlap of an orbital of one atom with an orbital of another atom. When two orbitals overlap they share the same region in space and a new orbital, called a **Molecular Orbital** (MO), is formed. The molecular orbital is a covalent bond. Like an atomic orbital, a molecular orbital can accommodate only two electrons. A covalent bond is formed only if :

(1) *the two combining orbitals are half-filled.*
(2) *the bonding orbitals approach each other in proper alignment needed for an effective overlap.*
(3) *the electrons in the bonding orbitals have opposite spins.*

TYPES OF COVALENT BONDS

There are two main types of covalent bonds or molecular orbitals :

(1) *Sigma* (σ) *bonds*
(2) *Pi* (π) *bonds*

Sigma Bond. A sigma bond is formed by the linear or end-to-end overlap of orbitals. They may be obtained :

(a) *By the overlap of two s orbitals,*

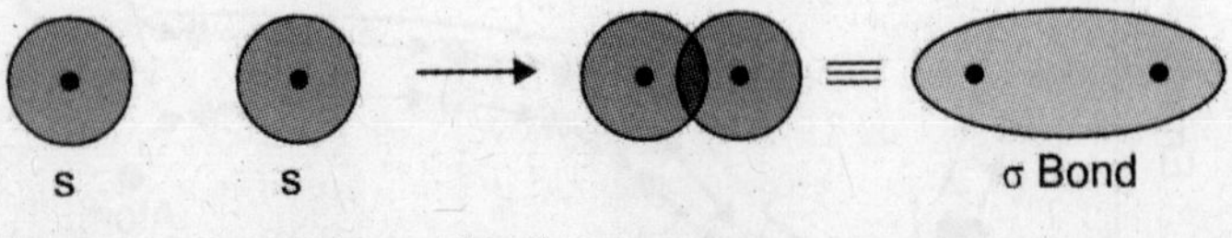

(b) *By the overlap of p_x and an s orbital,*

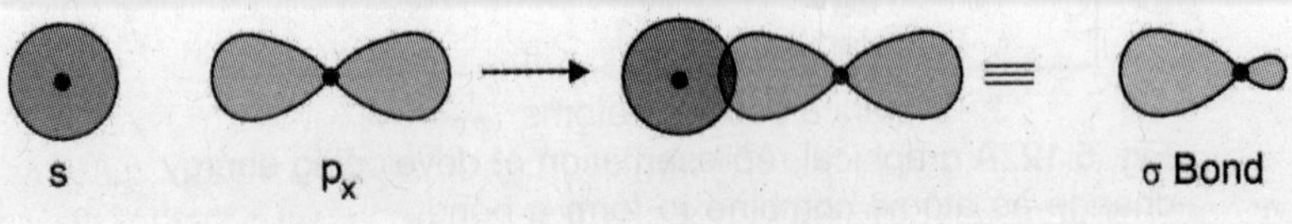

(c) *By the overlap of two p orbitals,*

Sigma bonds are symmetrical around the line drawn between the two nuclei. They are represented by a single line drawn between the atomic symbols. The electrons that occupy a sigma bond are called **Sigma (σ) Electrons.**

Pi Bond. A pi (π) bond is formed by parallel or side-to-side overlap of *p* orbitals. For example,

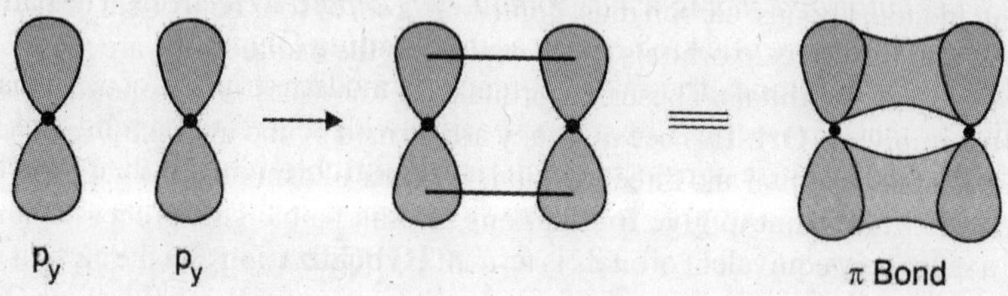

Like the *p* orbitals from which it is obtained, a π bond has two lobes. One half of the π bond lies above the plane containing the two nuclei and the other half lies below. The electrons that occupy π bond are known as the **Pi (π) Electrons.**

There are two important differences between sigma and pi bonds.

(1) Pi electrons are loosely held than a pair of electrons in a σ bond. They are of higher energy. As a result, π bonds are more easily broken and are more reactive than σ bonds.

(2) Rotation of atoms is not possible around a π bond. If any attempt is made to do so, the lobes of p orbitals will no longer be coplanar and will not overlap to form the π bond. This restriction in rotation around a π bond is responsible for *Cis* and *Trans* isomerism in alkenes.

HYBRIDIZATION

sp³ HYBRIDIZATION OF CARBON

Let us consider the electronic configuration of carbon in its ground or atomic state.

$$\text{C (Ground state)} = 1s^2\ \underbrace{2s^2 2p_x^{\ 1} 2p_y^{\ 1} 2p_z^{\ 0}}_{\text{Valence shell}}$$

In terms of energy level diagram, this electron configuration may be represented as in Fig. 5.13. Since there are only two unpaired electrons (half-filled orbitals), it might be expected that only two single covalent bonds will be formed.

On this basis, carbon would combine with two hydrogen atoms ($H = 1s^1$) to form a molecule CH_2. The two C–H bonds would be formed by overlap of the 2*p* orbitals (p_x and p_y) with the 1*s* orbital of each hydrogen atom. Since the angle separating the *p* orbitals is 90°, the C–H bonds would be at right angles to each other. But from chemical analysis we know that the simplest stable compound that carbon forms with hydrogen is methane (CH_4) and this compound contains four identical C–H bonds.

Now let us assume that one of the 2*s* electrons in the ground state is *moved* or *promoted* to the empty p_z orbital. Since $2p_z$ orbital is at a higher energy level than the 2*s* orbital, this promotion process would require input of energy. This energy is supplied in the form of heat or light. This new state of carbon is referred to as the **Excited State**. The electron configuration of the carbon atom in its excited state is :

$$\text{C (Excited state)} = 1s^2\ \underbrace{2s^1 2p_x^{\ 1} 2p_y^{\ 1} 2p_z^{\ 1}}_{\text{Valence shell}}$$

In terms of energy level diagram, this electron configuration may be represented as in Fig. 5.13. Since there are four unpaired electrons (half-filled orbitals) in the valence shell of the carbon atom in its excited state, it might be expected that four covalent bonds will be formed. On this basis, carbon would combine with four hydrogen atoms to form a molecule CH_4. The three C–H bonds would be formed by the overlap of three 2*p* orbitals (p_x, p_y and p_z) with the 1*s* orbital of each hydrogen atom. The fourth C–H bond will be formed by the overlap of the 2*s* orbital of carbon with the 1*s* orbital of a hydrogen atom. Since the angle separating the *p* orbitals in an atom is 90°, the three C–H bonds may be expected to be at right angles to each other. The fourth C–H bond involving the overlap of *s* orbitals will not have any directional characteristics because *s* orbitals are spherically symmetrical. This implies that two different types of C–H bonds are involved in the formation of methane molecule. This is not true. Experimentally, methane has been shown to contain four identical C–H bonds that are directed towards the corners of a regular tetrahedron.

To form four identical bonds, carbon must contribute a set of four equivalent orbitals. This can be achieved if the 2*s* and the three 2*p* orbitals (p_x, p_y and p_z) in the excited state are *mixed* or *hybridized* to give four new equivalent orbitals. These new orbitals are known as sp^3 (pronounced *ess-pee-three*) hybrid orbitals or simply **sp³ Orbitals** because they are formed by the interaction of one *s* and three *p* orbitals. Mixing of a pure *s* orbital and three *p* orbitals is rather like mixing of a gallon of pure red paint and three gallons of white paint to give four gallons of pink paint. This process of mixing of pure orbitals to give a set of new equivalent orbitals is termed **Hybridization** and the carbon is said to be in **Hybridized State**. The electron configuration of the carbon atom in its sp^3 hybridized state is :

$$\text{C (Hybridized state)} = 1s^2\ \underbrace{2(sp^3)^1 2(sp^3)^1 2(sp^3)^1 2(sp^3)^1}_{\text{Valence shell}}$$

In terms of energy level diagram, the above electron configuration may be represented as in Fig. 5.13.

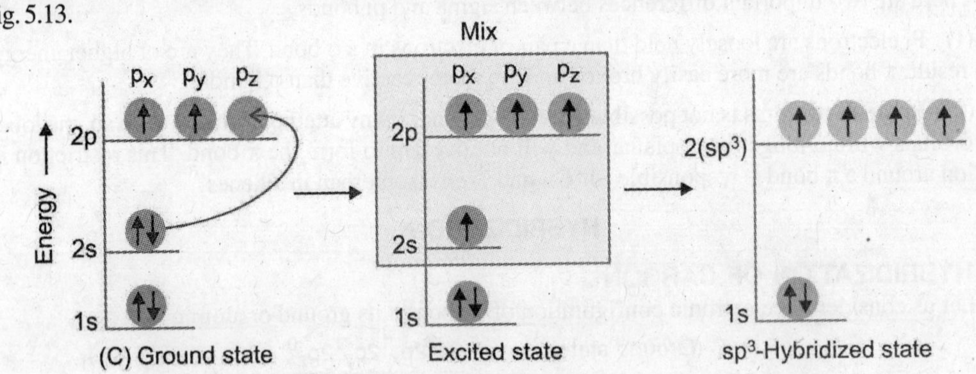

Fig. 5.13. Formation of four equivalent sp^3 hybrid orbitals of carbon.

The processes which lead to the formation of four equivalent sp^3 orbitals are summarized in Fig. 5.13.

Each sp^3 orbital contains one electron. Since each sp^3 orbital is obtained from one *s* and three *p* orbitals, it has 25% *s*-character and 75% *p*-character. As indicated, each sp^3 orbital has a large lobe and a small lobe (Fig. 5.14).

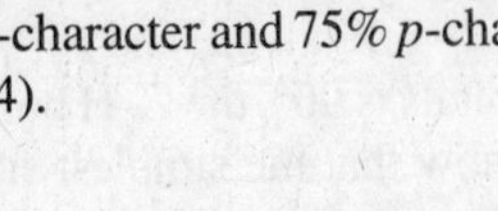

Fig. 5.14. Shape of an sp^3 hybrid orbital.

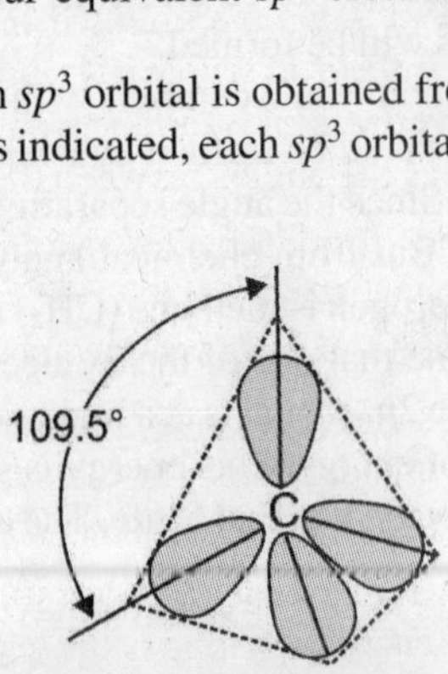

Fig. 5.15 Orientation of four sp^3 hybrid orbitals.

The four new sp^3 orbitals obtained above are identical (same energy and shape) but differ only in their orientation in space with respect to each other. The four sp^3 orbitals are arranged in such a way that their axes are directed towards the corners of a regular tetrahedron with carbon located at the centre. The angle between any two orbitals is therefore, 109.5°. The orientation of sp^3 orbitals is shown in Fig. 5.15. The smaller lobes are not indicated because they do not extend sufficiently far from nucleus to participate in bond formation.

The tetrahedral arrangement is favored because it allows the sp^3 orbitals to stay as far away from each other as possible and thereby reducing the electron-electron repulsion. This is in keeping with the fact that each sp^3 orbital contains an electron, and electrons stay as far apart as possible because they have the same charge.

At this stage, the student should clearly understand that carbon does not necessarily undergo bond formation in its ground state. The electron configuration may change prior to bond formation and the final arrangement that it acquires would depend upon the number of other atoms or groups to which it is attached.

Whenever carbon is bonded to four other atoms or groups (as in methane), it uses sp³ hybrid orbitals.

Bonding in Methane. In methane carbon forms single covalent bonds with four hydrogen atoms. Since the carbon atom is attached to four other atoms it uses sp^3 orbitals to form these bonds. Fig. 5.16 shows how the bonds in methane are formed.

Each C–H covalent bond is the result of the overlap of an sp^3 orbital from carbon and 1*s* orbital from hydrogen.

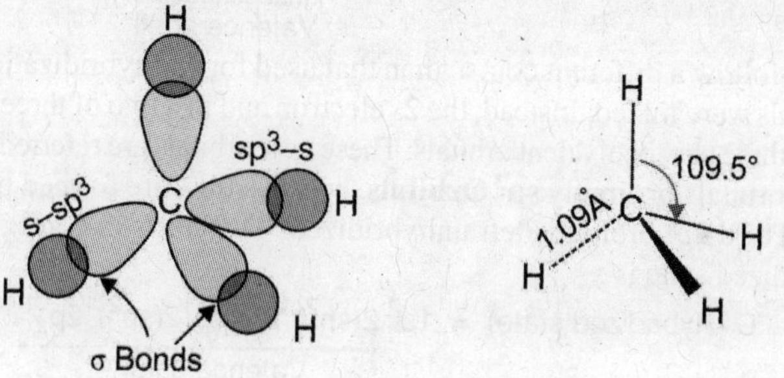

Fig. 5.16. Bonding in methane.

Since the four sp^3 orbitals are oriented in such a way that their axes are directed towards the corners of a regular tetrahedron with carbon located at the centre, the resulting C–H bonds are also directed towards the vertices of a tetrahedron with carbon at the centre. Thus, the bond angles in methane are the same as the angles between the axes of the sp^3 orbitals, that is 109.5°.

The covalent bonds formed by the overlap of sp^3 orbitals and *s* orbitals are sigma (σ) bonds because the electron density in each bond is symmetrical about the line joining the centres of two bonded atoms. Thus, all C–H bonds in methane are sigma bonds.

Electron diffraction and spectroscopic studies have shown that methane has a tetrahedral structure and all the C–H bonds are identical. They are the same length (1.09 Å). The energy required to break any of the four bonds is the same (102 kcal). The angle between any pair of bonds is 109.5°.

Bonding in Ethane. In ethane there are six C–H covalent bonds and one C–C covalent bond. As in the case of methane, each C–H bond is the result of the overlap of an sp^3 hybrid orbital from carbon and 1*s* orbital from hydrogen. The C–C bond arises from the overlap of sp^3 orbitals, one from each carbon atom. Fig. 5.17 shows how the bonds in ethane are formed.

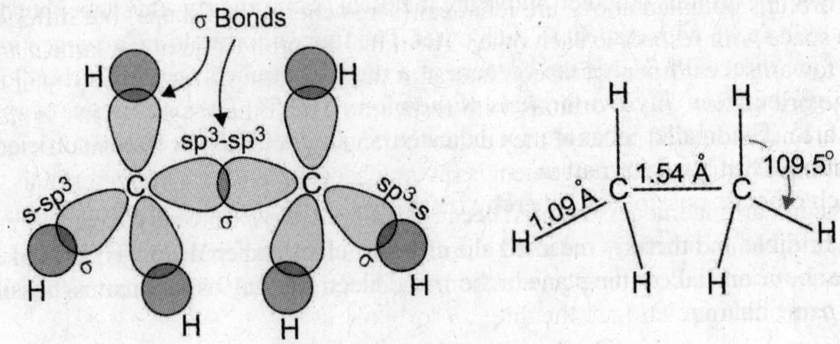

Fig. 5.17 Bonding in Ethane.

All C–H bonds and the C–C bond are sigma (σ) bonds.

Electron diffraction and spectroscopic studies have given the following measurements for the molecule of ethane : H–C–H and H–C–C bond angles are 109.5°, C–H bond length is 1.09Å and C–C bond length is 1.54 Å.

sp^2 HYBRIDIZATION OF CARBON

As in the case of sp^3 hybridization, we start with the carbon atom in its ground state.

$$\text{C (Ground state)} = 1s^2 \underbrace{2s^2 2p_x^1 2p_y^1 2p_z^0}_{\text{Valence shell}}$$

The promotion of one of the two 2*s* electrons to the empty p_z orbital gives the excited state.

$$\text{C (Excited state)} = 1s^2 \underbrace{2s^1 2p_x^1 2p_y^1 2p_z^1}_{\text{Valence shell}}$$

At this point we follow a different course than that used for sp^3 hybridization, in which case one 2*s* and three 2*p* orbitals were mixed. Instead, the 2*s* electron and just two of three 2*p* orbitals are mixed or hybridized to give three new equivalent orbitals. These new orbitals are referred to as sp^2 (pronounced *ess-pee-two*) hybrid orbitals or simply **sp^2 Orbitals**, because they are formed by interaction of one *s* and two *p* orbitals. Third $2p_z$ orbital is left unhybridized. The electron configuration of the carbon atom in its sp^2 hybridized state is :

$$\text{C (Hybridized state)} = 1s^2 \underbrace{2(sp^2)^1 2(sp^2)^1 2(sp^2)^1 2p_z^1}_{\text{Valence shell}}$$

The processes which lead to the formation of three equivalent sp^2 hybrid orbitals are summarized in Fig. 5.18.

Each sp^2 orbital contains an unpaired electron. The shape of an sp^2 orbital is similar to that of an sp^3 orbital.

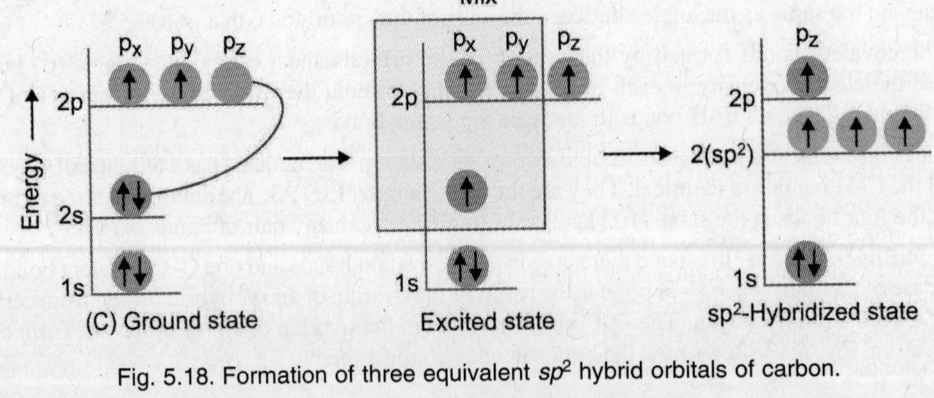

Fig. 5.18. Formation of three equivalent sp^2 hybrid orbitals of carbon.

The sp^2 orbitals obtained above are identical (same energy and shape) but differ only in their orientation in space with respect to each other. The three sp^2 orbitals lie in the same plane with their axes directed towards the corners of an equilateral triangle. The angle between any pair of orbitals is thus 120°. The orientation of sp^2 orbitals with respect to each other is shown in Fig. 5.19 (*a*). The smaller lobes are not indicated because they do not extend sufficiently far from nucleus to participate in bond formation. The trigonal arrangement is favored because it allows the sp^2 orbitals to stay as far apart from each other as possible and thereby reducing the electron-electron repulsions.

The unhybridized p_z orbital is oriented along an axis perpendicular to the plane of sp^2 orbitals, with each lobe above and below the plane of the sp^2 orbitals. Fig. 5.19 (*c*) indicates the position of the unhybridized p_z orbital in relation to the three sp^2 orbitals.

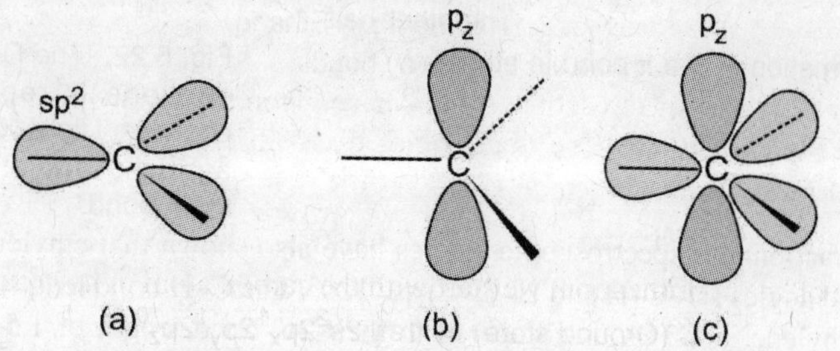

Fig. 5.19. (*a*) Orientation of three sp^2 orbitals. (*b*) The unhybridized p_z orbital. (*c*) Orientation of the p_z orbital in relation to the sp^2 orbitals.

It should be remembered that whenever carbon is bonded to three other atoms or groups (as in ethylene) it always uses sp^2 hybrid orbitals and a p_z orbital to form its bonds.

Bonding in Ethylene. Each carbon atom in ethylene ($H_2C{=}CH_2$), is attached to two hydrogen atoms by single covalent bonds and to another carbon atom by a double bond. Since each carbon is attached to three other atoms, it uses sp^2 hybrid orbitals plus an unhybridized p_z orbital to form its bonds.

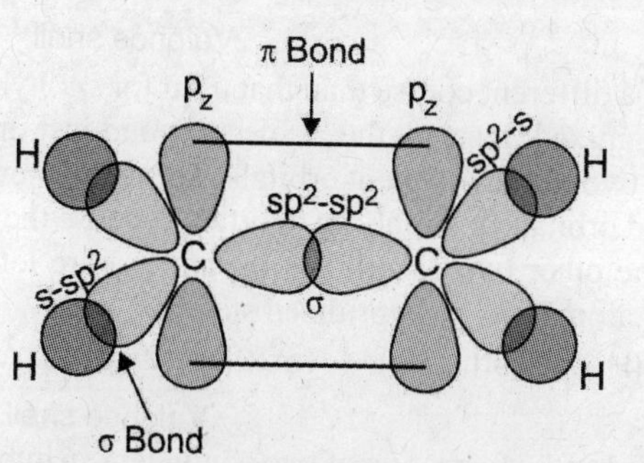

Fig. 5.20. Bonding in ethylene.

In ethylene there are four C–H single covalent bonds and one C–C double bond. Each C–H bond is a sigma (σ) bond and results from the overlap of an sp^2 orbital from carbon and 1*s* orbital from hydrogen.

One of the two bonds in the double bond is also a σ bond and results from end-to-end (linear) overlap of the sp^2 orbitals, one from each carbon atom.

The second bond in the double bond is a pi (π) bond and results from side-to-side (lateral) overlap of two unhybridized p_z orbitals, one from each carbon atom. It should be noted that p_z orbitals can overlap only when all six atoms lie in the same plane, that is, the plane of the σ bonds. Like the *p* orbitals from which it is formed, a π bond consists of two equal parts. One part lies above the plane of the carbons and hydrogens and the other part lies below this plane. These two parts together make up one (π) bond.

Although the C–C double bond in ethylene is represented by two equivalent lines, remember that one line represents a σ bond and the other the π bond (Fig. 5.22).

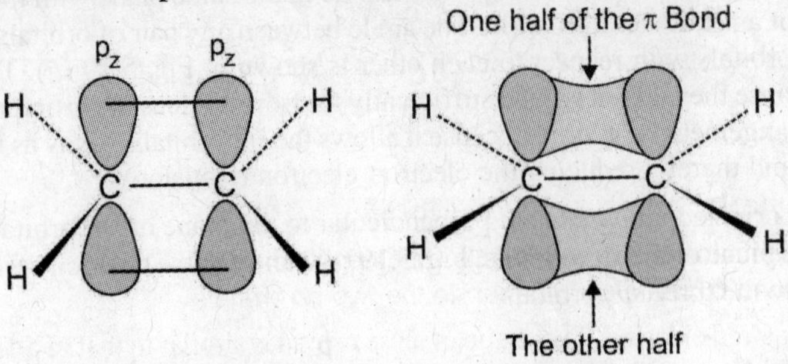

Fig. 5.21. Formation of the π bond in ethylene.

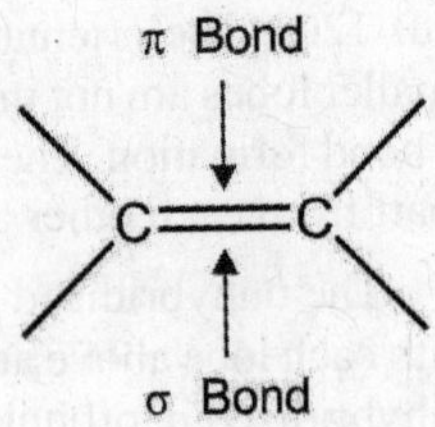

Fig. 5.22. The C–C double bond in ethylene is represented by two equivalent lines. One line represents a σ bond. The other line represents the π bond.

Electron diffraction and spectroscopic studies have also shown that ethylene is a planar (flat) molecule. All bond angles (H–C–H and H–C–C) are 120°. The C–H bond length is 1.09Å. The C–C bond length in ethylene is 1.34Å as compared to the C–C bond length of 1.54Å in ethane. This shortening is to be expected on the basis of additional orbital overlap associated with the two bonds.

sp HYBRIDIZATION OF CARBON

As in the case of sp^2 hybridization, we again start with the carbon atom in its ground state.

$$\text{C (Ground state)} = 1s^2\ \underbrace{2s^2 2p_x^{\,1} 2p_y^{\,1} 2p_z^{\,0}}_{\text{Valence shell}}$$

The promotion of one of the two 2*s* electrons to the empty p_z orbital gives the excited state.

$$\text{C (Excited state)} = 1s^2\ \underbrace{2s^1 2p_x^{\,1} 2p_y^{\,1} 2p_z^{\,1}}_{\text{Valence shell}}$$

At this point we follow a different course than that used for sp^2 hybridization, in which case one 2*s* and two 2*p* orbitals were mixed. Instead, the 2*s* electron and just one of the three 2*p* orbitals are *mixed* or *hybridized* to give two new equivalent orbitals. These two new orbitals are referred to as *sp* (pronounced *ess-pee*) hybrid orbitals or simply **sp orbitals** because they are formed by interaction of one *s* and one *p* orbital. The other two 2*p* orbitals (p_y and p_z) are left unhybridized. The electron configuration of the carbon atom in its *sp* hybridized state is :

$$\text{C (Hybridized state)} = 1s^2\ \underbrace{2(sp)^1 2(sp)^1 2p_y^{\,1} 2p_z^{\,1}}_{\text{Valence shell}}$$

The processes which lead to the formation of two equivalent *sp* hybrid orbitals are summarized in Fig. 5.23.

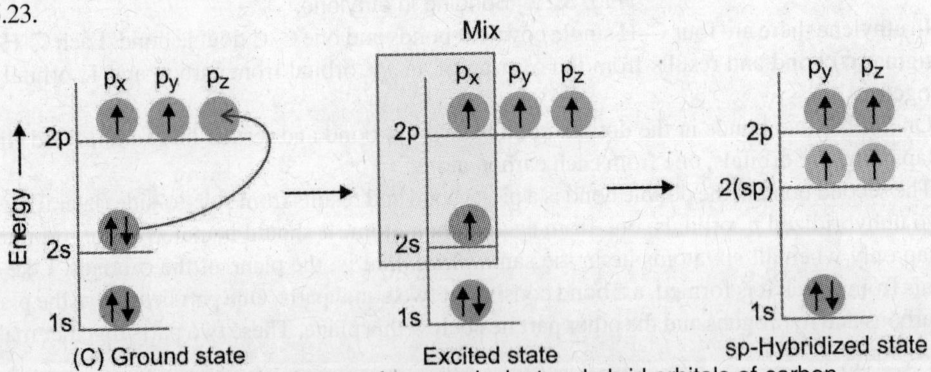

Fig. 5.23. Formation of two equivalent *sp* hybrid orbitals of carbon.

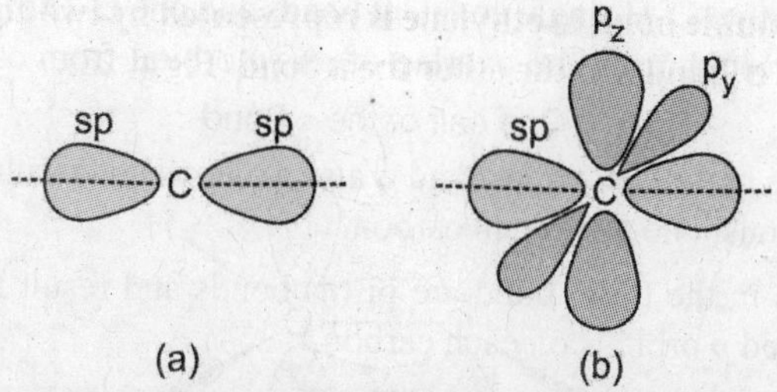

Fig. 5.24. (*a*) Orientation of two *sp* orbitals (*b*) Orientation of the unhybridized p_y and p_z orbitals in relation to the two *sp* orbitals.

Each *sp* orbital contains an unpaired electron. The shape of an *sp* orbital is similar to that of an sp^2 orbital. The *sp* orbitals obtained above are identical (same energy and shape) but differ only in their orientation in space with respect to each other. They lie in a straight line. The angle between the two *sp* orbitals is thus 180° (Fig. 5.24*a*). The linear arrangement is favored because it allows the *sp* orbitals to stay as far apart from each other as possible and thereby reducing the electron-electron repulsions. The unhybridized p_y and p_z orbitals are at right angles to the line of the *sp* hybrid orbitals. Fig. 5.24(*b*) indicates the positions of the p_y and p_z orbitals in relation to the two *sp* orbitals.

It should be remembered that whenever carbon is bonded to two other atoms or groups (as in acetylene) it always uses sp hybrid orbitals and two 2p (p_y and p_z) orbitals to form its bonds.

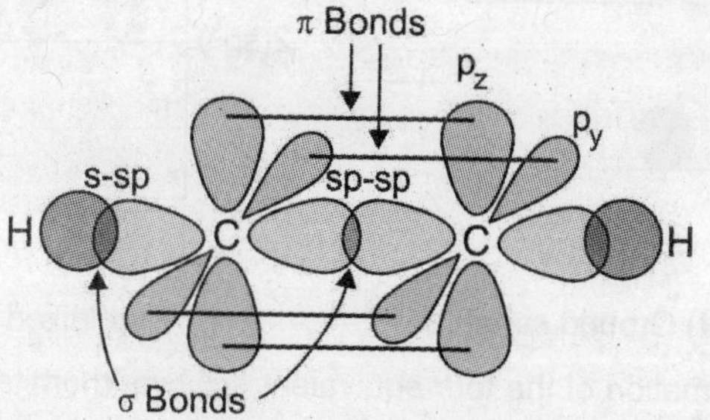

Fig. 5.25. Bonding in Acetylene.

Bonding in Acetylene. Each carbon atom in acetylene (H–C≡C–H) is attached to one hydrogen atom by a single covalent bond and to another carbon atom by a triple bond. Since each carbon is attached to two other atoms it uses *sp* hybrid orbitals plus two unhybridized 2*p* orbitals (p_y and p_z) to form its bonds.

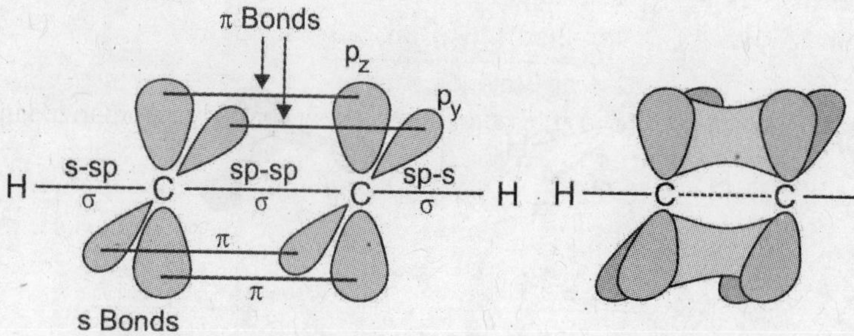

Fig. 5.26. Formation of the two π bonds in Acetylene.

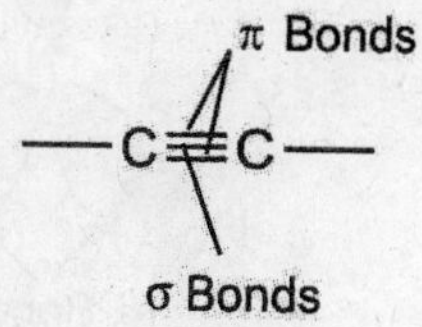

Fig. 5.27. The C—C triple bond in acetylene is representated by three equivalent lines. Two lines represent the π bonds. One line represents the σ bond.

In acetylene there are two C–H single covalent bonds and one C–C triple bond. Each C–H bond is a sigma (σ) bond and results from the overlap of an *sp* orbital from carbon and 1*s* orbital from hydrogen.

One of the three bonds in the C–C triple bond is also a σ bond and results from end-to-end (linear) overlap of the two *sp* orbitals, one from each carbon.

The other two bonds in the triple bond are pi (π) bonds and result from side-to-side (lateral) overlap of the unhybridized *p* orbitals on each carbon.

Although the C–C triple bond is represented by three equivalent lines, remember that one line represents a σ bond and the other two the π bonds.

Electron diffraction studies have also shown that acetylene is a linear molecule. The H–C–C bond angle is 180°. The C–H bond length is 1.09 Å. The C–C bond length is 1.20 Å.

HYBRIDIZATIONS OF NITROGEN

Atomic number of nitrogen is seven. It has five electrons in the valence shell (outermost shell). Like carbon, the nitrogen atom also undergoes sp^3, sp^2 and *sp* hybridizations. Fig. 5.28 illustrates the formation of sp^3 hybrid orbitals of nitrogen.

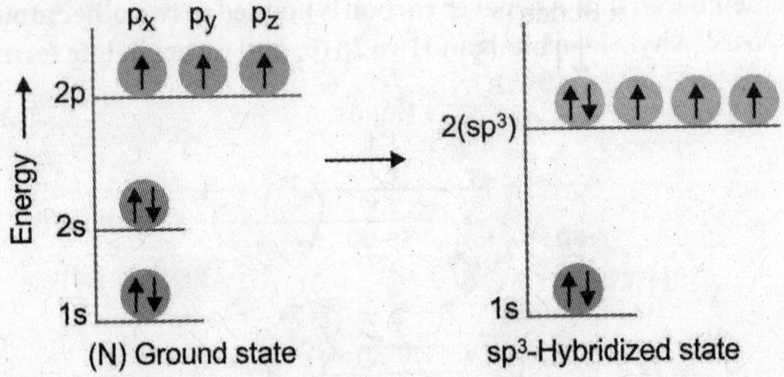

Fig 5.28. Formation of the four equivalent sp^3 hybrid orbitals of nitrogen.

The four sp^3 orbitals of nitrogen are directed towards the corners of a regular tetrahedron, just as the sp^3 orbitals of carbon do. Nitrogen has one more electron than carbon. As a result, one of the sp^3 hybrid orbitals of nitrogen is completely filled and cannot take part in bond formation. Whenever nitrogen is attached to *three* other atoms or groups (as in ammonia or methylamine), it always uses sp^3 hybrid orbitals to form its bonds. Fig. 5.29 illustrates the formation of bonds in ammonia.

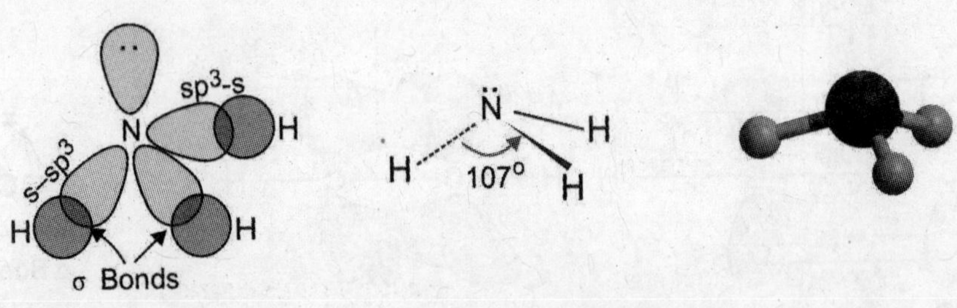

Fig. 5.29. Bonding in Ammonia.

The three N–H bonds in ammonia are sigma (σ) bonds. Each of these bonds is formed by the overlap of an sp^3 orbital from nitrogen and 1*s* orbital from hydrogen. The H–N–H bond angle is 107°. This angle is slightly less than the normal tetrahedral angle of 109.5°. This is because the sp^3 orbital containing the unshared pair of electrons is somewhat larger than those containing bonding pairs. Thus the angles between the σ bonds on the other side of the molecule are compressed slightly.

Also, nitrogen can form compounds using sp^2 and sp hybrid orbitals. Whenever nitrogen is attached to two other atoms or groups (as in imines or azo compounds), it uses two sp^2 hybrid orbitals *plus* the unhybridized p_z orbital to form its bonds (Fig. 5.30).

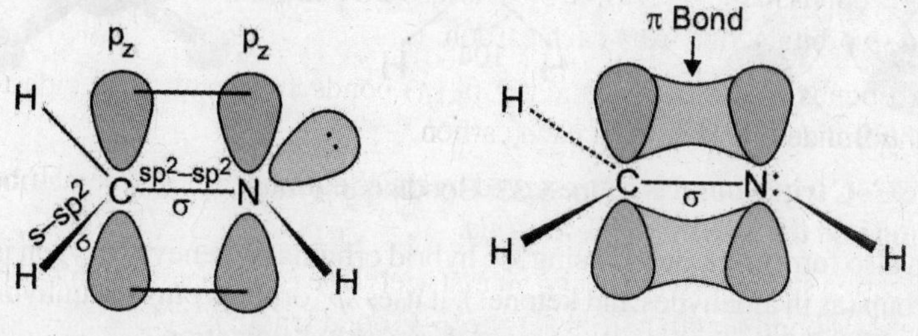

Fig.5.30. Bonding in CH_2=NH. Both carbon and nitrogen are sp^2 hybridized.

Whenever nitrogen is attached to only one other atom or group of atoms (as in hydrogen cyanide, methyl cyanide, or nitrogen), it uses an sp hybrid orbital *plus* the two unhybridized $2p$ orbitals (p_y and p_z) to form its bonds (Fig. 5.31).

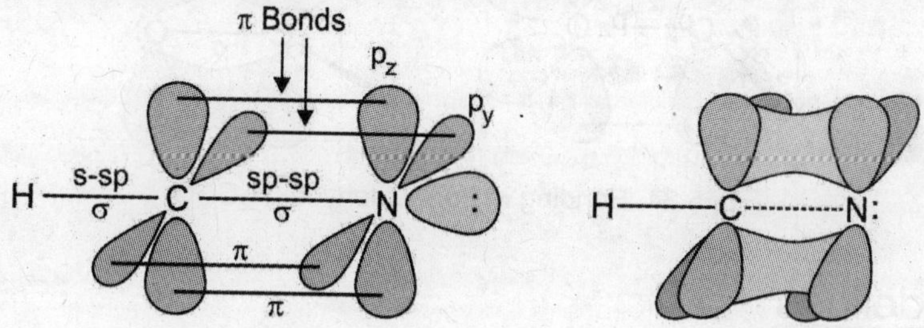

Fig. 5.31. Bonding in Hydrogen cyanide, H–C≡N.

HYBRIDIZATIONS OF OXYGEN

Atomic number of oxygen is eight. It has six electrons in the valence shell (outermost shell). Like carbon or nitrogen, the oxygen atom undergoes sp^3 and sp^2 hybridizations. Fig. 5.32. illustrates the formation of sp^3 hybrid orbitals of oxygen.

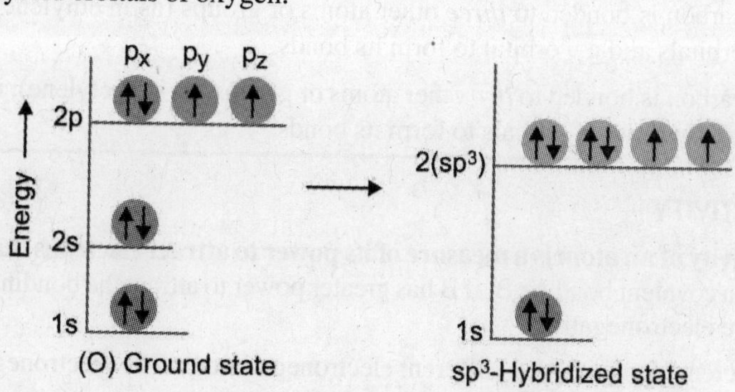

Fig. 5.32. Formation of the four equivalent sp^3 hybrid orbitals of oxygen.

The four sp^3 hybrid orbitals of oxygen, are directed towards the corners of a regular tetrahedron, just as the sp^3 orbitals of carbon or nitrogen. Oxygen has one more electron than nitrogen. As a result, two of the sp^3 orbitals of oxygen are completely filled and cannot take part in bond formation. Whenever oxygen is attached to two other atoms or groups (as in water, ethers, or alcohols), it always uses sp^3 hybrid orbitals to form its bonds. Fig. 5.33 illustrates the formation of bonds in water.

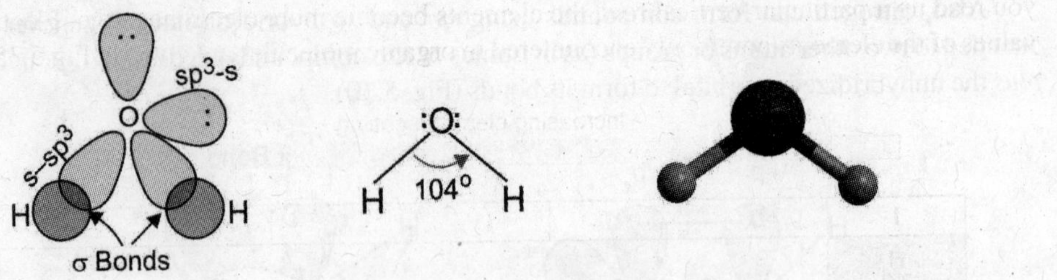

Fig. 5.33. Bonding in Water.

Oxygen can also form compounds using sp^2 hybrid orbitals. Whenever oxygen is attached to one other atom or group (as in aldehydes and ketones), it uses sp^2 orbitals *plus* the unhybridized p_z orbital to form its bonds (Fig. 5.34). Oxygen does not undergo *sp* hybridization.

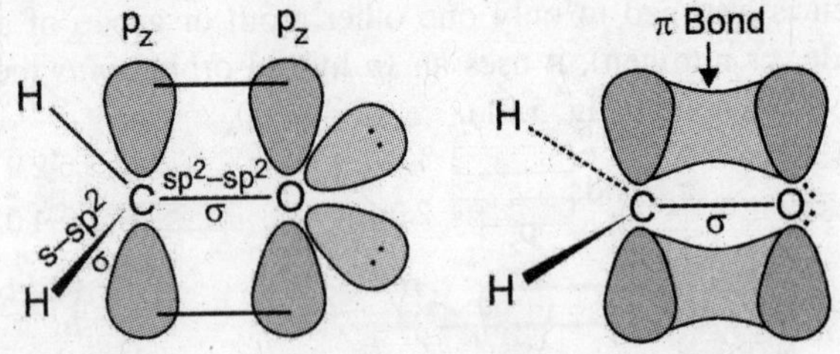

Fig. 5.34. Bonding in Formaldehyde, $H_2C{=}O$.

Guidelines

Drawing Molecular Orbital Structure of Organic Molecules

When you are asked to draw orbital structure of an organic molecule, always remember :

(1) Whenever carbon is bonded to *four* other atoms or groups (as in methane or ethane), it always uses sp^3 hybrid orbitals to form its bonds.

(2) Whenever carbon is bonded to *three* other atoms or groups (as in ethylene), it always uses sp^2 hybrid orbitals and a *p* orbital to form its bonds.

(3) Whenever carbon is bonded to *two* other atoms or groups (as in acetylene), it always uses *sp* hybrid orbitals and two *p* orbitals to form its bonds.

ELECTRONEGATIVITY

Electronegativity of an atom is a measure of its power to attract electrons that it is sharing in a covalent bond. In a covalent bond A : B, if B has greater power to attract the bonding pair of electrons, it is said to be more electronegative.

Atoms of different elements have different electronegativities. The electronegativity of an atom depends upon two factors :

(1) **Atomic Number.** That is, the magnitude of the positive charge on the nucleus.

(2) **Atomic Radii.** That is, the distance of the valence electrons from the nucleus.

Atoms with higher atomic number and less atomic radius have higher electronegativities. The trend of electronegativities in the Periodic Table is rather regular. As you read to the right in a particular horizontal row of Periodic Table, the elements become more electronegative. Likewise, as you read up a particular vertical row, the elements become more electronegative. Electronegativity values of the elements commonly encountered in organic molecules are given in Fig.5.35.

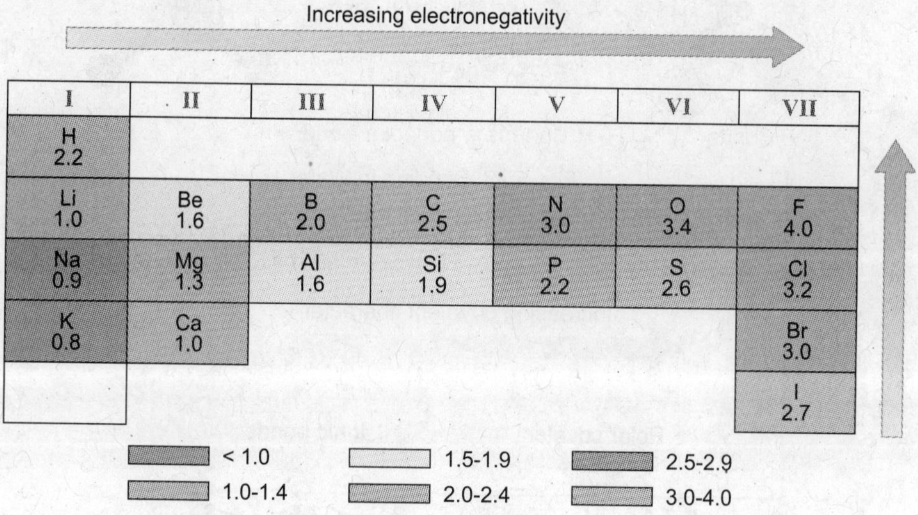

Fig.5.35. Electronegativities of some common elements.

The concept of electronegativity has proved very useful in predicting the nature of bonds as also the properties of compounds.

POLAR AND NONPOLAR BONDS

When a covalent bond is formed between two atoms with different electronegativities, the electrons involved in the bond are not shared. The atom with higher electronegativity pulls the bonding electrons closer to it. In other words, the electron density in the molecular orbital would be greater around the atom with higher electronegativity. The result of this displacement of the molecular orbital toward the more electronegative atom will be that the more electronegative atom will acquire a small negative and the less electronegative atom will acquire a small positive charge.

Let us now consider the C–Cl covalent bond. Because chlorine is more electronegative than carbon, the electron density in the molecular orbital would be higher around chlorine atom than around the carbon atom. Thus the chlorine atom acquires a small negative charge and the carbon atom acquires a small positive charge. This may be indicated in the following way.

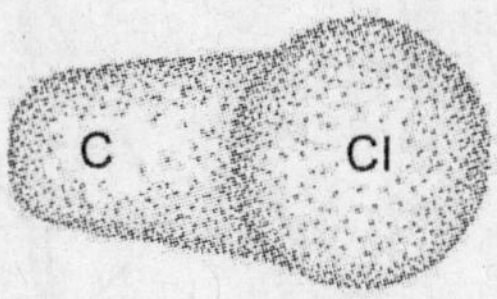

or $\overset{\delta+}{C}—\overset{\delta-}{Cl}$

C–Cl Bond is a polar bond.

The Greek letter delta, δ, stands for *small*. The symbol δ_+ means a small positive charge, and δ_- means a small negative charge. Such a bond which appears to have a positive end and a negative end, is said to be a **Polar Bond**.

When a covalent bond is formed between two atoms with same electronegativities (C–C, H–H, and F–F), the electrons involved in the bond are shared equally. Electron density in the molecular orbital binding the two atoms together is same around each atom. There is no positive and negative end. Such a bond is said to be a **Nonpolar Bond**.

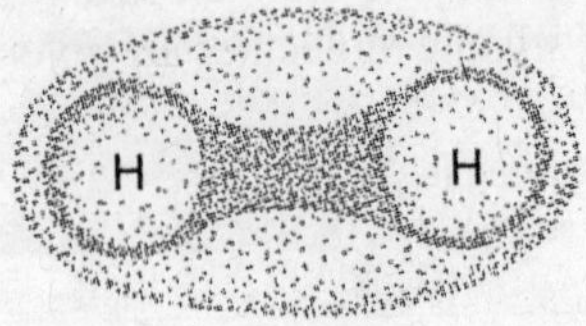

or H—H

H–H Bond is a nonpolar bond.

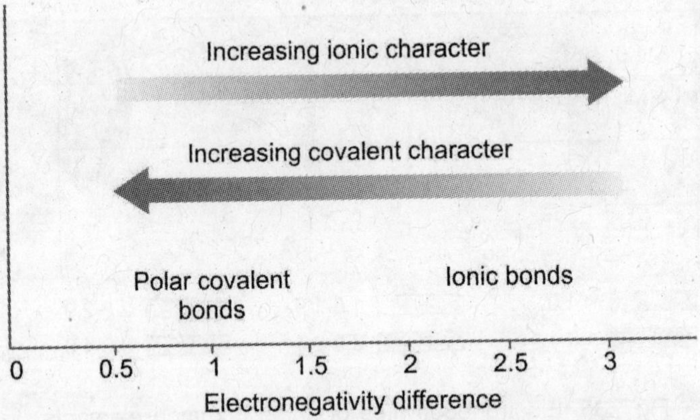

Fig. 5.36. Relationship between electronegativity difference and bond polarity. If the difference in electronegativity between bonded atoms is less than 0.5, it is a nonpolar bond. If the difference is between 0.5 to 1.9, it is a polar bond. If the difference is greater than 1.9, it is an ionic bond.

POLAR AND NONPOLAR MOLECULES

A molecule is said to be a polar molecule if it satisfies the following two conditions :

(1) **The molecule must contain one or more polar bonds.**

(2) **The polar bonds must be so directed that there are separate centres of positive and negative charges in the molecule.**

Carbon dioxide has two polar carbon-oxygen bonds, but the molecule is not polar. The carbon dioxide molecule is linear, and the centre of the two negative charges is at the same place as the centre of the positive charge - the carbon atom (Fig. 5.37).

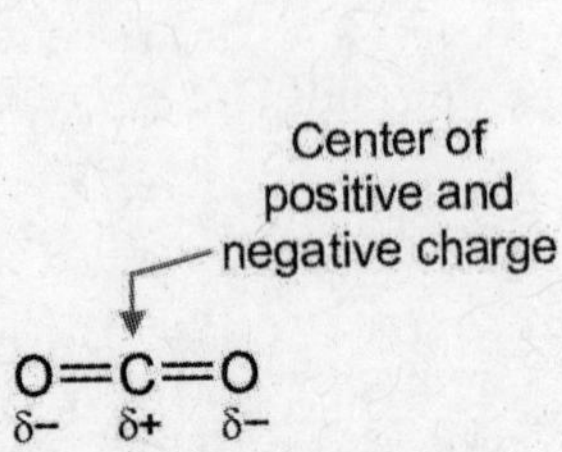

Fig. 5.37. Carbon dioxide is a nonpolar molecule.

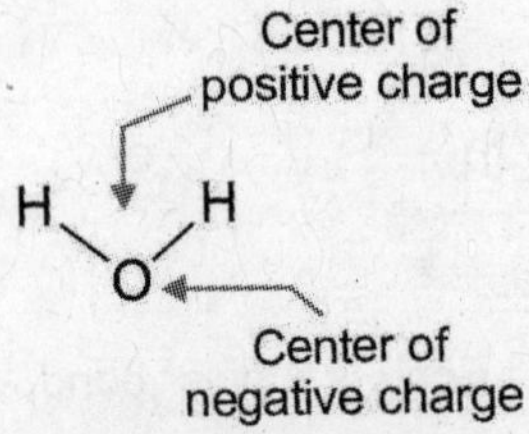

Fig. 5.38. Water is a polar molecule.

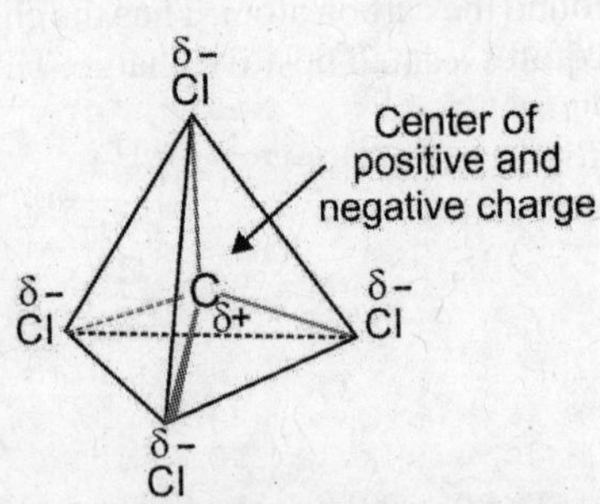

Fig. 5.39. Carbon tetrachloride is a nonpolar molecule.

Water molecule has two polar oxygen-hydrogen bonds. The H–O–H bond angle is 104°. The centre of the positive charge is between the hydrogen atoms (Fig. 5.38). As the centre of positive charge and the centre of negative charge do not coincide, water molecule is polar.

Carbon tetrachloride has four polar carbon-chlorine bonds, but the molecule is not polar. The carbon tetrachloride is tetrahedral (all Cl–C–Cl angles are 109.5°), and the centre of four negative charges is at the same place as the centre of the positive charge - the carbon atom (Fig. 5.39).

Polar molecules behave like small magnets and possess a dipole moment. The dipole moment (μ) is the product of the magnitude of the charges and the distance between the charges.

$$\text{Dipole Moment} = \text{Charge} \times \text{Distance}$$

Molecules that do not have a dipole moment are called **Nonpolar Molecules**.

Polar molecules have a great attraction for each other as also for ions, and thus acquire characteristic physical and chemical properties.

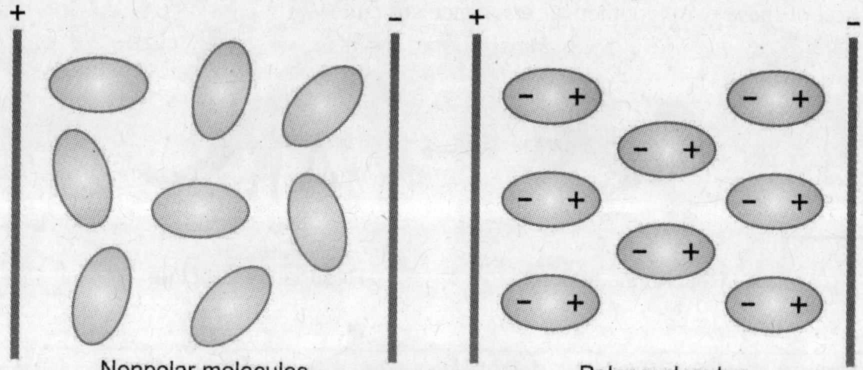

Nonpolar molecules   Polar molecules

The effect of an electric field on the orientation of nonpolar molecules and polar molecules. Nonpolar molecules orient themselves randomly, whereas polar molecules orient themselves so that their negative ends point toward the positive plate and their positive ends point toward the negative plate.

RESONANCE CONCEPT

A number of organic compounds cannot be accurately represented by one structure. For example, benzene is ordinarily represented as :

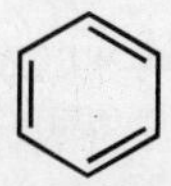

This structure has three carbon-carbon single bonds and three carbon-carbon double bonds. However, it has been determined experimentally that all carbon-carbon bonds in benzene are identical and have the same bond length (1.39Å). Furthermore, the carbon-carbon bond length of (1.39 Å) is intermediate between the normal carbon-carbon double-bond length (1.34 Å) and the normal carbon-carbon single-bond length (1.54 Å). Actually two alternative structures (1 and 2) can be written for benzene :

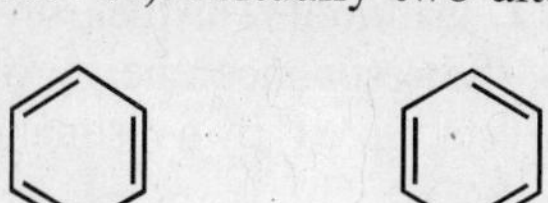

1   2

These two structures differ only in the position of electrons. Neither (1) nor (2) is a correct representation of benzene. The actual structure of benzene lies somewhere between these two structures. This phenomenon in which two or more structures can be written for a compound which involve identical positions of atoms is called **Resonance**.

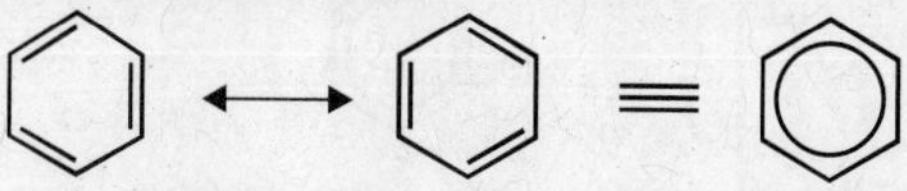

1   2   3

Benzene   Benzene

The actual structure of the molecule is said to be a **Resonance Hybrid** of various possible alternative structures. The alternative structures are referred to as the **Resonance Structures** or **Canonical Forms**. A double headed arrow (↔) between the resonance structures is used to represent the resonance hybrid. Thus in the case of benzene, (1) and (2) represent the resonance structures. Actual structure of the molecule may be represented as hybrid of these two resonance structures, or by the single structural formula (3).

It should be clearly understood that the resonance structures (1) and (2) are not actual structures of the benzene molecule. They exist only in theory. None of these structures adequately represents the molecule. In resonance theory, we view the benzene molecule (which is of course a real entity) as being hybrid of these two hypothetical resonance structures.

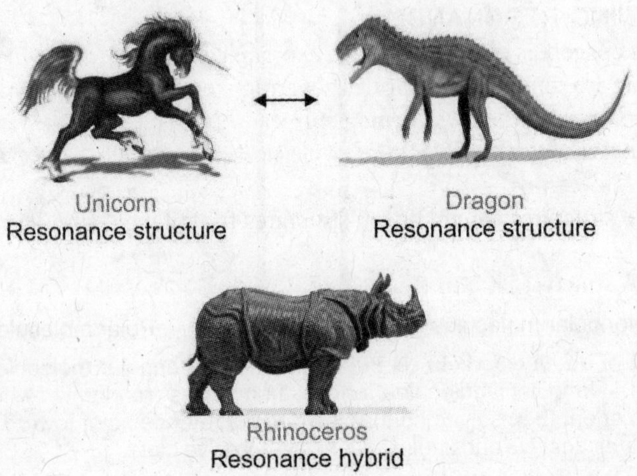

Analogy between a resonance hybrid and a rhinoceros. The resonance structures, like the unicorn and the dragon, are not real. They are mythical creatures. Only the hybrid, like the rhinoceros, describes the real structure.

A green mark produced by overprinting blue with yellow is a good visual analogy to the resonance concept. Yellow and blue correspond to the resonance structures (canonical forms), and the green corresponds to the hybrid structure.

The resonance hybrid is more stable than any one of the various resonance structures. The difference in energy between the hybrid and the most stable resonance structure is known as the Resonance Energy. Resonance energy can be determined by the difference between the calculated and experimental heats of combustion (energy given off as heat when one mole of compound is burned) of the compound. For example, it has been calculated that the hypothetical structure (1) or (2) would have a heat of combustion of 797 kcal/mol. The measured value for the heat of combustion of benzene is 759 kcal/mol. Therefore, the resonance energy of benzene is (797–759) kcal/mol or 38 kcal/mol. The benzene is said to be "stabilized" by a resonance energy of 38 kcal/mol (Fig. 5.40).

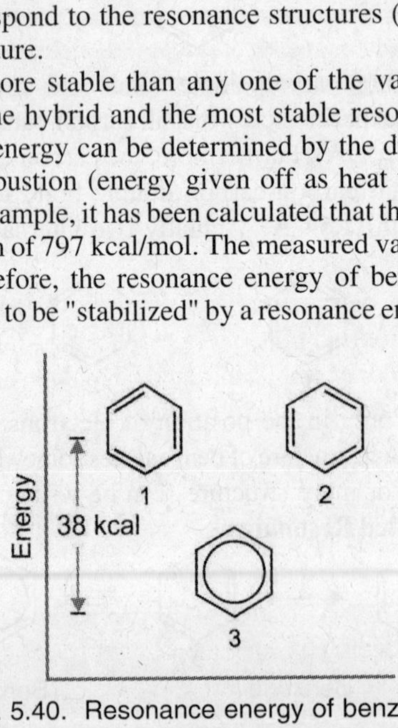

Fig. 5.40. Resonance energy of benzene.

Another species that is not correctly represented by a single structure is the acetate ion. As in the case of benzene, acetate ion is a hybrid of two resonance structures. Both carbon-oxygen bonds in the acetate ion are identical and have the same bond length (1.26 Å). The carbon-oxygen bond length of 1.26 Å is intermediate between the normal carbon-oxygen double-bond length (1.20 Å) and the normal carbon-oxygen single-bond length (1.43 Å).

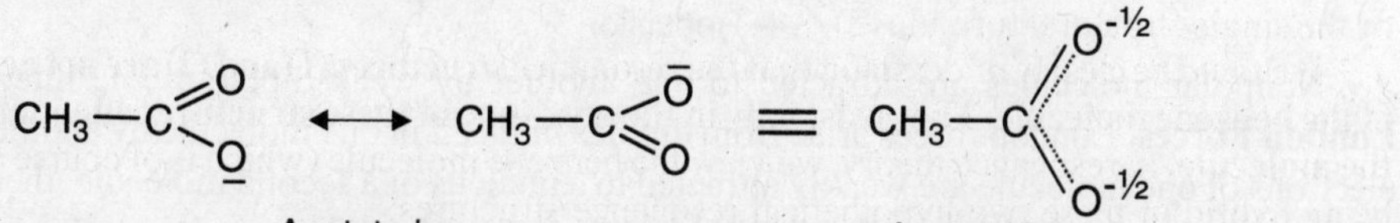

Acetate ion Acetate ion

RULES GOVERNING RESONANCE

The resonance concept is extremely useful in organic chemistry, and will come up a number of times. The following are some important rules governing resonance.

(1) Resonance occurs whenever a molecule can be represented by two or more structures differing only in the arrangement of electrons, without shifting any atoms. **Resonance only involves the delocalization of electrons.**

(2) Resonance structures are not actual structures for the molecule. They are nonexistent and hypothetical.

(3) Resonance structures are interconvertible by one or a series of short electron-shifts. For example,

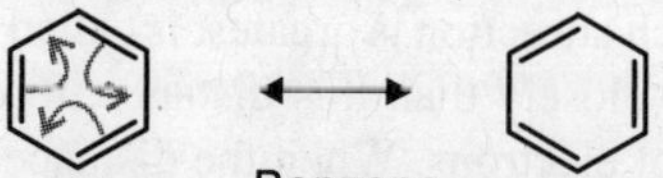

Benzene

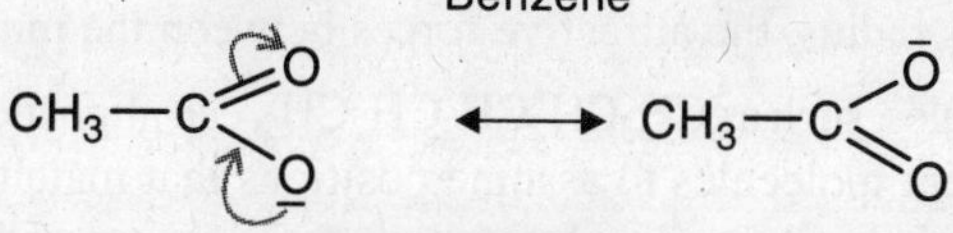

Acetate ion

(4) Resonance hybrid represents the actual structure of the molecule. The structure of the resonance hybrid is intermediate between the various resonance structures and is not a mixture of them.

(5) Resonance hybrid is represented by a double headed arrow (⟷). This should not be confused with the two arrows (⇌) used to denote equilibrium between two different compounds.

(6) Resonance hybrid is more stable than any of its contributing forms (resonance structures).

(7) Resonance always increases the stability of a molecule and lessens its reactivity.

ATTRACTIONS BETWEEN MOLECULES

DIPOLE-DIPOLE INTERACTIONS

Molecules attract and repel one another, except in a dispersed gas. These attractions and repulsions arise primarily from dipole-dipole interactions: attractions between opposite charges and repulsions between like charges. For example, in liquid CH_3Cl the partially negative chlorine atom in one molecule is attracted to the partially positive carbon in another molecule.

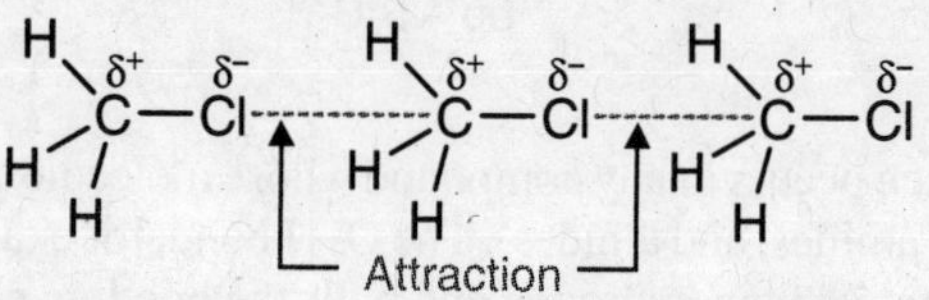

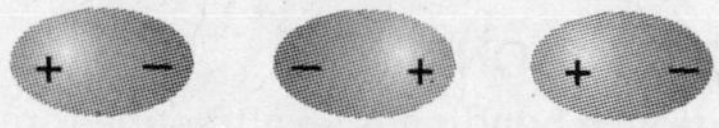

When the chlorine ends of two CH_3Cl molecules approach closely, the two molecules repel each other.

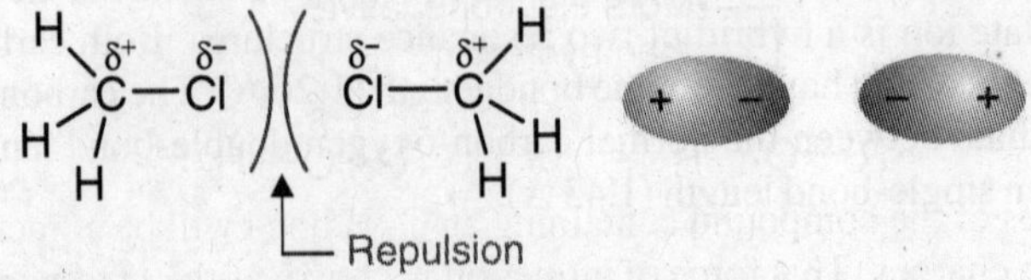

Nonpolar molecules are attracted to one another by weak dipole-dipole interactions called **London Forces.** London forces arise from dipoles *induced* in one molecule by another. In this case, electrons of one molecule are weakly attracted to a nucleus of a second molecule; then the electrons of the second molecule are repelled by the electrons of the first. The result is an uneven distribution of electrons and an induced dipole. Fig. 5.41 depicts how an induced dipole can arise when two molecules approach each other.

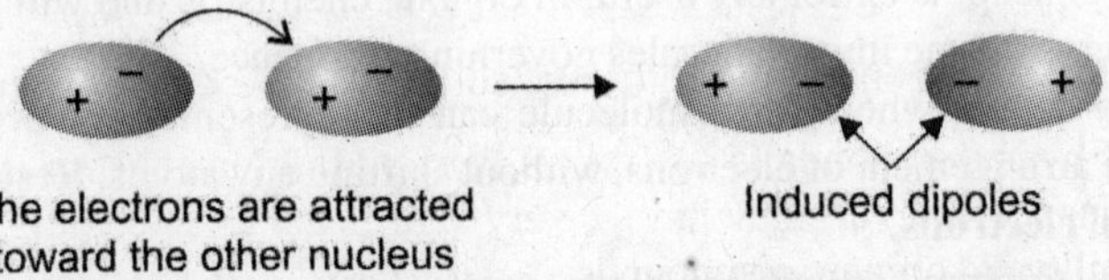

Fig. 5.41. Nonpolar molecules can induce dipoles in each other.

The various dipole-dipole interactions (attractive and repulsive) are collectively called **van der Waals Forces.** The distance between molecules has an important effect on the strength of van der Waals forces. The distance at which attraction is greatest is called the **van der Waals Radius.** If two atoms approach each other more closely than this distance, repulsions develop between the two nuclei and between the two sets of electrons. When the distance between two molecules becomes larger than the van der Waals radius, the attractive forces between the molecules decrease.

Continuous-chain molecules, such as $CH_3CH_2CH_2CH_2CH_3$, can align themselves in zigzag chains, enabling the atoms of different molecules to assume positions that match the van der Waals radii. Maximal van der Waals attractions can develop between such long-chain molecules. Branched molecules cannot approach one another closely enough for all the atoms to assume optimal van der Waals distances. Because more energy is necessary to overcome van der Waals attractions and to free molecules from the liquid state, continuous-chain compounds have higher boiling points than branched compounds of the same molecular weight and otherwise similar structures.

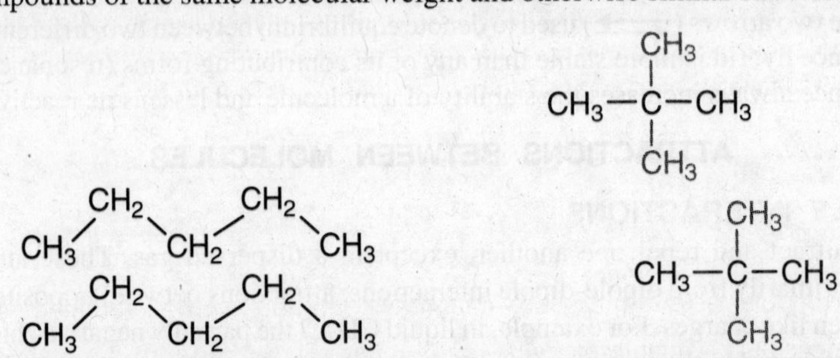

Continuous chain; higher boiling
bp 36°C

Branched chain;lower boiling
bp 4.5°C

HYDROGEN BONDING

Hydrogen bonding is an attractive force which occurs in any compound whose molecules contain O–II or N–II bonds (as in water, alcohols, acids, amines, and amides). Thc O–II bond, for cxamplc, is a highly polar bond. Oxygen is more electronegative than hydrogen and pulls the bonding electrons closer to it. As a result of this displacement, the oxygen atom acquires a small negative charge (δ_-) and the hydrogen atom a small positive charge (δ_+).

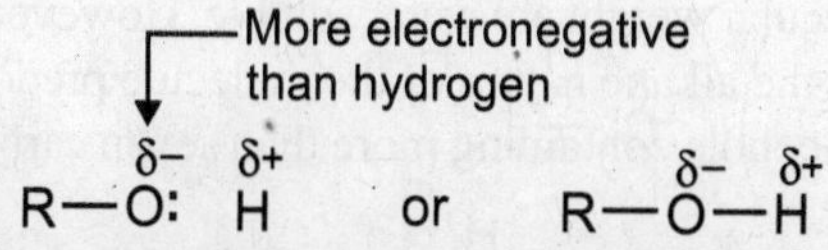

Adjacent molecules of the compound containing an O–H bond will be attracted to each other by virtue of these opposite charges. This force of attraction is known as the **Hydrogen Bond**. Usually a hydrogen bond is represented by a dotted line (Fig. 5.42).

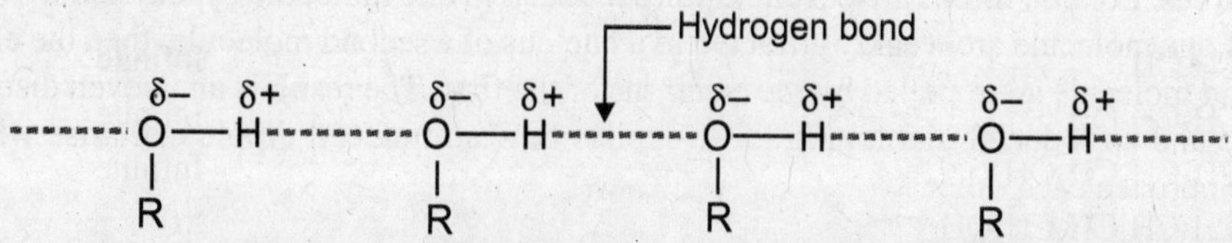

Fig. 5.42. Hydrogen bonds (shown by dotted lines) between alcohol molecules.

The strengths of hydrogen bonds (5 to 10 kcal per bond) are much less than the strengths of ordinary covalent bonds. However, they have a very significant effect on the physical properties (boiling points, solubility) of organic compounds.

Effect on Boiling Points. It is understandable that substances having nearly the same molecular weights, have the same boiling point. The boiling points of alkanes and ethers of comparable molecular weights are not far apart, but the boiling points of alcohols having almost equal molecular weights are considerably higher.

$CH_3—CH_2—CH_3$	$CH_3—O—CH_3$	$CH_3—CH_2—OH$
Propane	Dimethyl ether	Ethanol
(MW 44 ; *bp* –45°C)	(MW 46 ; *bp* –25°C)	(MW 46 ; *bp* +78°C)

This can be explained on the basis of hydrogen bonding. Ethanol forms hydrogen bonds. Extra energy in the form of heat is required to break the hydrogen bonds holding the molecules together before it can be volatilized. Propane and dimethyl ether do not form hydrogen bonds and, therefore, have low boiling points.

Effect on Water-Solubility. A hydrogen-bonded substance is usually soluble in another hydrogen-bonded substance. For example, alcohols are soluble in water but alkanes are not. This is because a nonpolar alkane molecule cannot break into the hydrogen-bonded sequence in water (Fig. 5.43). It cannot replace the hydrogen bonds that would have to be broken to let it in.

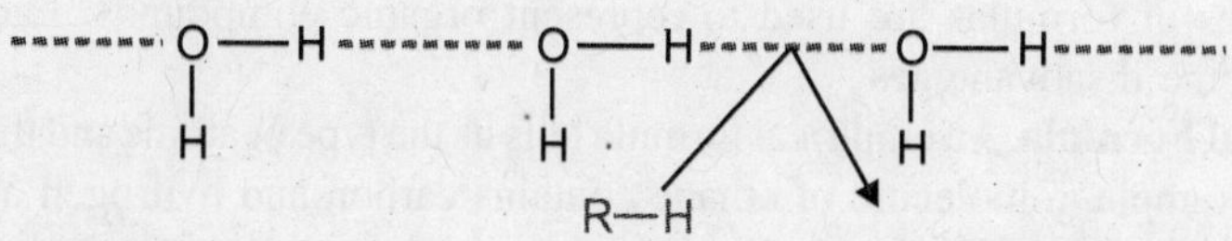

Fig. 5.43. Nonpolar alkane molecules cannot slip into the hydrogen-bonded sequence.

An alcohol molecule is capable of hydrogen bonding. It can slip into the hydrogen bonded sequence in water. It can replace the hydrogen bonds that must be broken to let it in.

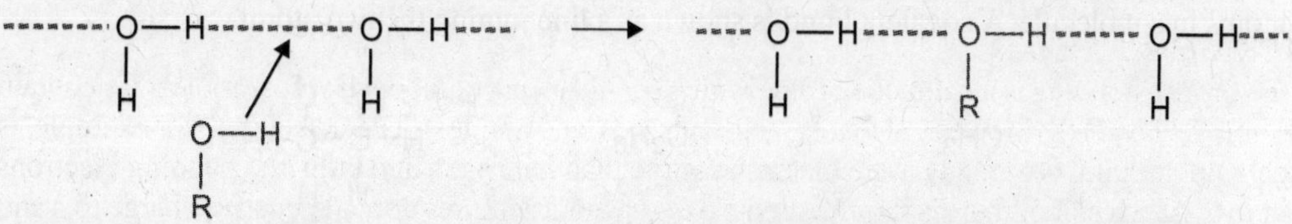

Fig. 5.44. Polar alcohol molecules can slip into the hydrogen-bonded sequence in water.

Thus alcohols of low molecular weight are water soluble. However, when the alkyl group (R–) is four or more carbons in length the alkane nature of the molecule predominates, and water solubility falls off sharply (Table 5.4). Alcohols containing more than seven carbons are insoluble in water.

Table 5.4 SOLUBILITY OF ALCOHOLS IN WATER

Alcohol	Solubility ($g/100\ ml$ of H_2O)
CH_3OH	Infinite
CH_3CH_2OH	Infinite
$CH_3CH_2CH_2OH$	Infinite
$CH_3CH_2CH_2CH_2OH$	8.0
$CH_3CH_2CH_2CH_2CH_2OH$	2.7
$CH_3CH_2CH_2CH_2CH_2CH_2OH$	0.6

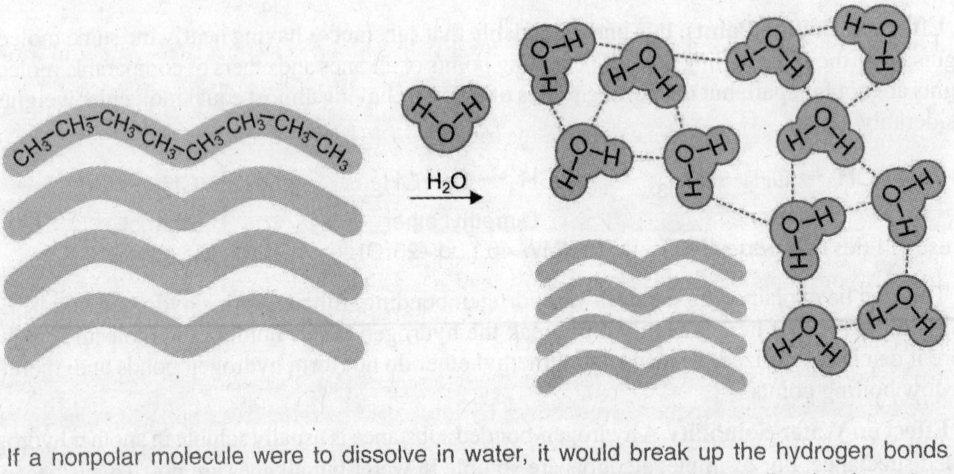

If a nonpolar molecule were to dissolve in water, it would break up the hydrogen bonds between the water molecules. Therefore, nonpolar substances do not dissolve in water.

REPRESENTATION OF ORGANIC COMPOUNDS WITH PAPER AND PENCIL

Different types of formulas are used to represent organic compounds. Each type has some advantages and some disadvantages.

(1) **Empirical Formula.** An empirical formula tells us the type of atoms and their relative ratio in a molecule. For example, a molecule of ethane contains carbon and hydrogen atoms in a ratio of 1 : 3 ; the empirical formula is CH_3.

(2) **Molecular Formula.** A molecular formula tells us the actual number of atoms in a molecule. For example, the molecular formula of ethane is C_2H_6.

(3) **Structural Formula.** A structural formula is more informative. It clearly shows the arrangement of atoms in a molecule. A covalent bond is shown as a line joining the two atoms.

CH_3	C_2H_6	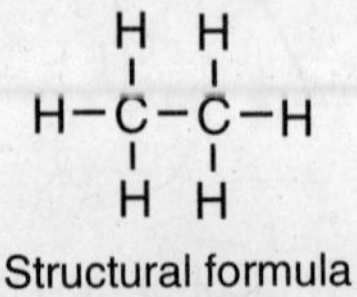
Empirical formula for ethane	Molecular formula for ethane	Structural formula for ethane

(4) **Condensed Structural Formula.** As the name implies, it is a short hand method of representing the structural formula. The convention is to omit most of the covalent bonds that connect the atoms within the molecule.

CH_3CH_3 is the condensed structural formula for 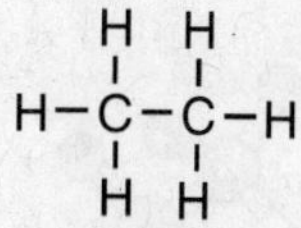

CH_3CH_2OH is the condensed structural formula for 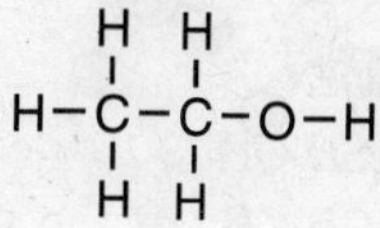

Structural formulas may be condensed even further if a molecule has two or more identical groups of atoms. In these cases, brackets are used to include a repetitive group of atoms.

$(CH_3)_2CHOH$ is the same as 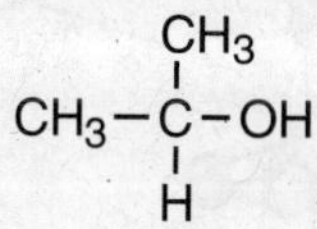

For sake of clarity, double or triple bonds are usually shown in a condensed structural formula.

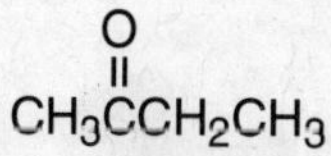 is the same as 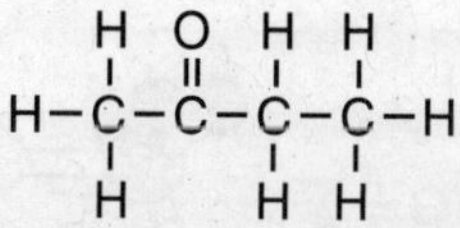

(5) **Skeletal Formula (Line-angle Formulas).** Perhaps the ultimate abbreviation of structures is the use of lines to represent the carbon framework. Drawing **skeletal formulas** is easy:

(*a*) Carbon atoms usually aren't shown. Instead, a carbon atom is assumed to be at the intersection of two lines (bonds) and at the end of each line. Occasionally, a carbon atom might be indicated for emphasis or clarity.

(*b*) Hydrogen atoms bonded to carbon aren't shown. Because carbon always has a valence of four, we mentally supply the correct number of hydrogen atoms for each carbon.

(*c*) All atoms other than carbon and hydrogen are shown. (Hydrogen atoms attached to all noncarbon atoms must also be shown.)

Examples of skeletal formulas:

$CH_3(CH_2)_3CH_3$	$(CH_3)_2CHCH_2CH_3$	$(CH_3)_4C$
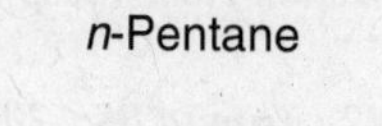	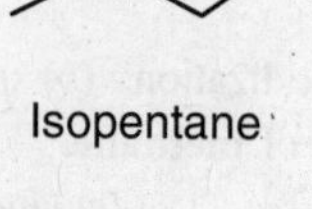	
n-Pentane	Isopentane	Neopentane

$CH_3CH{=}CHCH_2CH_3$ 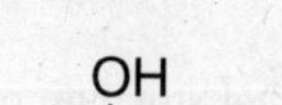

2-Pentene

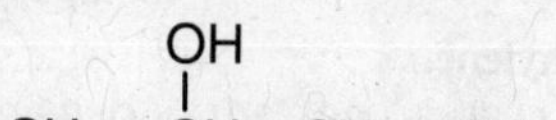

2-Butanol

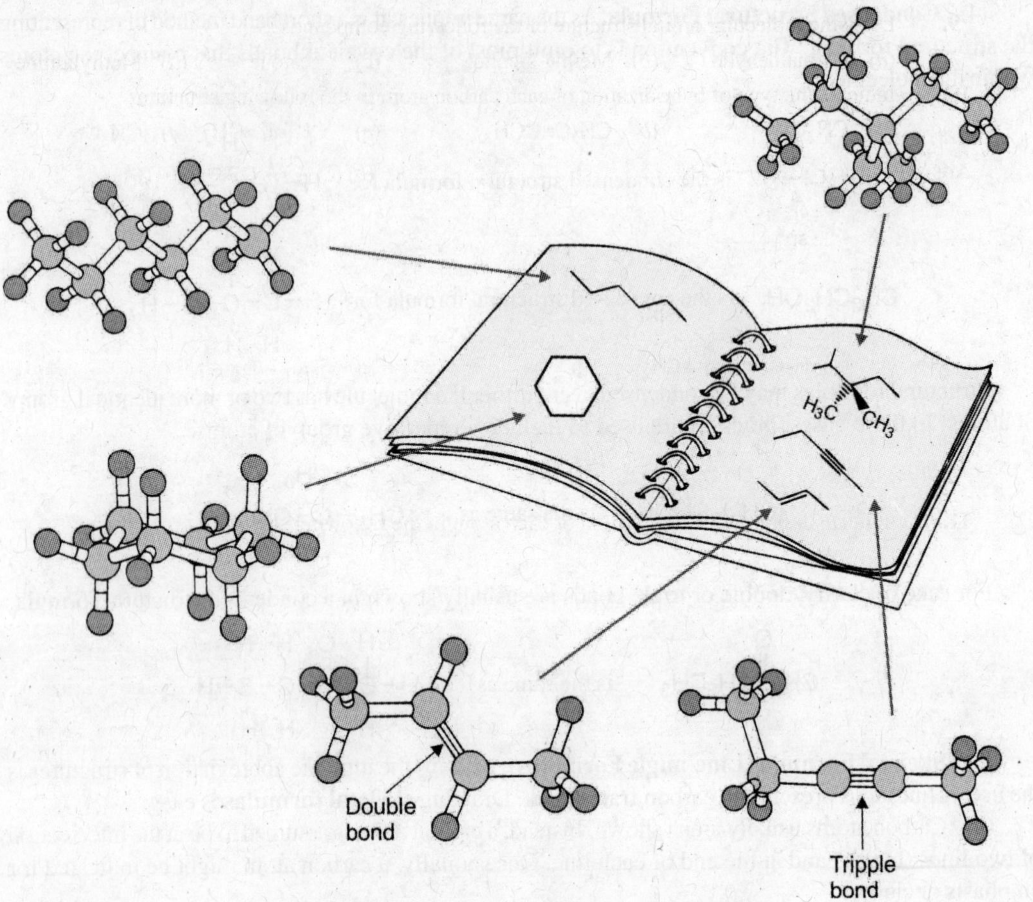

In skeletal formulas (Line-angle formulas) lines are used to represent the carbon framework.

Examination Questions

1. Define or explain the following terms or concepts :
   - (*a*) Covalent bond
   - (*b*) Ionic bond
   - (*c*) Sigma bond
   - (*d*) Pi bond
   - (*e*) Hybridization
   - (*f*) Polar molecule
   - (*g*) Resonance
   - (*h*) Hydrogen bonding
   - (*i*) Resonance energy
2. What is hybridization ? What are the distinguishing features of sp^3 and sp^2 orbitals ? Explain with suitable examples.
3. Write notes on : (*a*) sp^3 hybridization ; (*b*) sp^2 hybridization ; and (*c*) *sp* hybridization.
4. Discuss the orbital structure of **methane.**

   (*Jammu BSc, 2012 ; Tripura BSc, 2013 ; Agra BSc, 2015*)
5. Explain why four covalent bonds in methane are equivalent.
6. Discuss the orbital structure of **ethane.**

   (*Solapur BSc, 2013 ; Kerala BSc, 2014*)
7. Discuss the orbital structure of **ethylene.**

   (*Burdwan BSc, 2013 ; Lucknow BSc, 2013 ; Delhi BSc, 2014 ; Manipal BSc, 2015*)
8. Discuss the orbital structure of **acetylene.**

   (*Delhi BSc, 2013 ; Goa BSc, 2013 ; Kerala BSc, 2014 ; Tamil Nadu BSc, 2015*)

9. Draw the molecular orbital structure of the following compounds :

(*a*) Formaldehyde (*b*) Methyl alcohol (*c*) Acetone (*d*) Methylamine

10. Indicate the type of hybridization of each carbon atom in the following structures :

(*a*) CH_3CH_3 (*b*) $CH_3C{\equiv}CCH_3$ (*c*) $CH_2{=}C{=}CH_2$ (*d*) $CH_3C{\equiv}N$

Answer. (*a*) 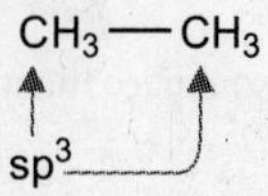

(*b*) 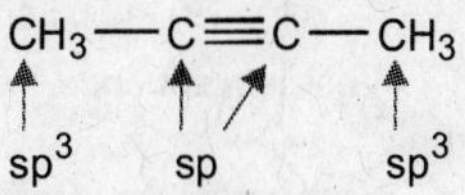

(*c*) 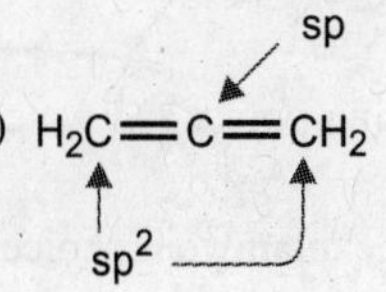

(*d*) 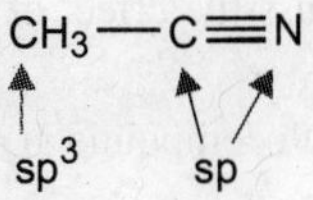

11. Indicate the type of hybridization of each atom in the following structures :

(*a*) H—C(=O)—H (*b*) CH_3—O—CH_3 (*c*) H—C≡N (*d*) CH_3NH_2

Answer. (*a*) 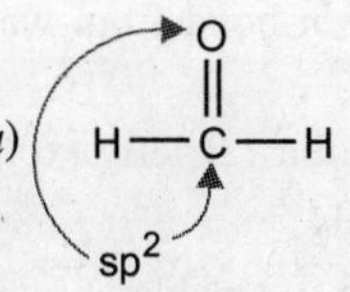

(*c*) 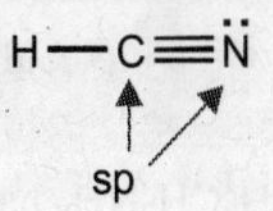

(*b*) 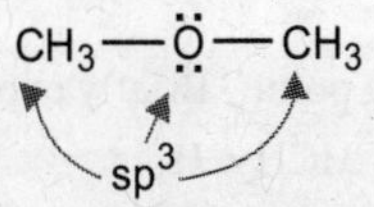

(*d*) 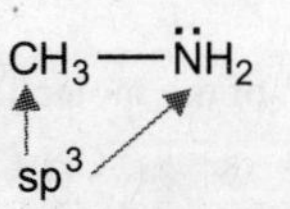

12. Indicate the type of hybrid orbitals used by carbon in each of the following structures :

(*a*) H—C(=O)—H (*b*) H—C(=O)—OH (*c*) CH_3Cl (*d*) CH_3OH

Answer. (*a*) sp^2 ; (*b*) sp^2 ; (*c*) sp^3 ; (*d*) sp^3

13. What two hybrid atomic orbitals overlap to form the C–C σ bond in acetaldehyde, CH_3CHO ?

Answer. Csp^3 and Csp^2

14. What two hybrid atomic orbitals overlap to form the C–C σ bond in acetonitrile, $CH_3C{\equiv}N$?

Answer. Csp^3 and Csp

15. What two hybrid atomic orbitals overlap to form the C–C σ bond in allene, $H_2C{=}C{=}CH_2$?

Answer. Csp and Csp^2

16. What is the approximate value of any HCC bond angle in $H_2C{=}CHCCl_3$?

Answer. 120°

17. What is the approximate value of the CCC bond angle in $CH_3CH_2CH_2OH$?

Answer. 109.5°

18. What is the approximate value of the CCC bond angle in $CH_3C{\equiv}CCH_3$?

Answer. 180°

19. What are polar and nonpolar bonds ? Give examples.

20. What are polar and nonpolar molecules ? Give examples.

21. Which of the following compounds are polar, and which are nonpolar ?

(*a*) CO_2 (*b*) CH_3Cl (*c*) CCl_4 (*d*) CH_3OH

(*e*) $CHCl_3$ (*f*) CH_3COCH_3

Answer. (*a*) and (*c*) are Nonpolar.

(*b*), (*d*), (*e*), and (*f*) are Polar.

22. What is resonance ? Give the conditions necessary for resonance. Illustrate your answer with a suitable example. (*Nagpur BSc, 2014*)

23. Write a note on : Resonance.

(*Madras BSc, 2013 ; Pune BSc, 2013 ; Panjab BSc, 2014 ; Baroda BSc, 2015*)

24. Write a note on : Hydrogen bonding.

25. Discuss the effect of hydrogen bonding on the boiling points and water-solubility of organic compounds.

26. Which compound is more soluble in water ? Briefly explain your choice.

$(CH_3)_2NH$ or $CH_3CH_2CH_3$

Answer. $(CH_3)_2NH$ is more soluble in water since it can hydrogen bond with water. Alkanes are not capable of hydrogen bonding with water.

27. Which compound is more soluble in water ? Briefly explain your choice.

CH_3OCH_3 or CH_3CH_2OH

Answer. CH_3CH_2OH is more soluble in water since it can donate a hydrogen bond to water and accept a hydrogen bond from water. CH_3OCH_3 can only accept a hydrogen bond from water; it does not have hydrogen which can hydrogen bond to water.

28. Which of the molecules below has the higher boiling point ? Briefly explain your choice.

$(CH_3)_3N$ or $CH_3CH_2CH_2NH_2$

Answer. $CH_3CH_2CH_2NH_2$ has the higher boiling point since it is capable of intermolecular hydrogen bonding.

29. Which of the molecules below has the higher boiling point ? Briefly explain your choice.

$CH_3CH_2CH_2CH_2CH_3$ or $(CH_3)_2CHCH_2CH_3$

Answer. $CH_3CH_2CH_2CH_2CH_3$ has the higher boiling point. As the degree of branching increases, the surface area of the molecule decreases and the potential for intermolecular attraction via London forces decreases. Greater surface area leads to a more intermolecular attraction which results in a higher boiling point.

30. Which of the molecules below has the higher boiling point ? Briefly explain your choice.

$CH_3CH_2CH_2OH$ or $CH_3CH_2OCH_3$

Answer. $CH_3CH_2CH_2OH$ has the higher boiling point since it is capable of intermolecular hydrogen bonding.

31. Draw the line-angle formula (skeletal formula) for $(CH_3)_2CHCH_2CHO$.

Answer. 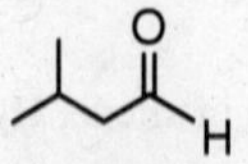

32. Draw the line-angle formula for $(CH_3)_2CHCH_2CH_2NH_2$.

Answer. 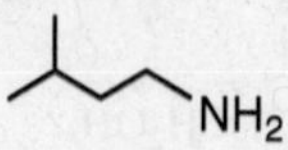

33. Draw the line-angle formula (skeletal formula) for $(CH_3CH_2)_2C{=}O$.

Answer. 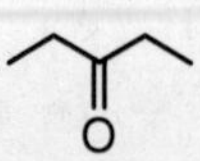

Multiple Choice Questions

1. In stable organic compounds, carbon will always form

(a) 2 bonds (b) 4 bonds (c) 3 bonds (d) 5 bonds

Answer. (b)

2. Carbon-carbon double bonds consist of :

(a) one σ bond, one π bond (b) two σ bonds, one π bond

(c) one σ bond, two π bonds (d) two σ bonds, two π bonds

Answer. (a)

3. Acetylene has a total of :

(a) one σ bond, two π bonds (b) two σ bonds, four π bonds

(c) three σ bonds, two π bonds (d) one σ bond, four π bonds

Answer. (c)

4. In propene there are

(a) eight σ bonds and one π bond (b) seven σ bonds and two π bonds

(c) six σ bonds and three π bonds (d) nine σ bonds

Answer. (a)

5. In propyne there are

(a) six σ bonds and two π bonds (b) seven σ bonds and one π bond

(c) six σ bonds and one π bond (d) eight σ bonds

Answer. (a)

6. 1-Buten-3-yne has

(a) six σ and four π bonds (b) seven σ and three π bonds

(c) eight σ and two π bonds (d) nine σ and one π bond

Answer. (b)

7. Compound in which carbons use only sp^3 hybrid orbitals for bond formation is

(a) $CH_3CH_2CH_3$ (b) $CH_3C{\equiv}CH$

(c) $CH_3CH{=}CH_2$ (d) $CH_2{=}CH{-}CH{=}CH_2$

Answer. (a)

8. Compound in which carbon uses sp^3 hybrid orbitals for bond formation is

(a) 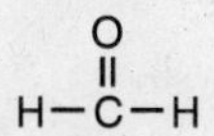 (b) 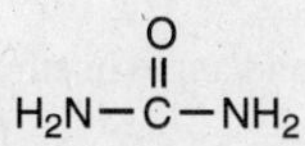

(c) 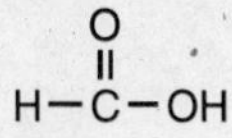 (d) $(CH_3)_3COH$

Answer. (d)

9. When the carbon atom is sp^2 hybridized in a compound, it is bonded to

(a) 2 other atoms (b) 4 other atoms (c) 3 other atoms (d) 5 other atoms

Answer. (c)

10. Compound in which carbons use only sp^2 hybrid orbitals for bond formation is

(a)  (b) 

(c) $CH_2{=}CH{-}CH{=}CH_2$ (d) $CH_3CH{=}C{=}CH_2$

Answer. (c)

11. Which of the following bonding patterns of carbon is not allowed in the formation of an organic compound ?

(*a*)  (*b*) 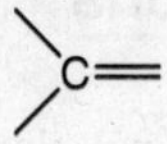 (*c*)  (*d*) 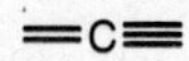

Answer. (*d*)

12. Choose the correct hybridization for the atom indicated in the molecule below.

$CH_3CH_2CH_2CH_3$
↑

(*a*) *sp* (*b*) sp^2 (*c*) sp^3 (*d*) none of these

Answer. (*c*)

13. Choose the correct hybridization for the atom indicated in the molecule below.

$CH_3CH_2CH_2CH_3$
↑

(*a*) *sp* (*b*) sp^2 (*c*) sp^3 (*d*) none of these

Answer. (*c*)

14. Acrylonitrile (CH_2=CHCN) contains ________ σ bonds and ________ π bonds.

(*a*) 4, 1 (*b*) 6, 3 (*c*) 4, 3 (*d*) 6, 1

Answer. (*b*)

15. How many carbon-carbon σ bonds are present in the molecule shown ?

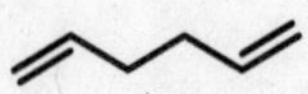

(*a*) 2 (*b*) 3 (*c*) 4 (*d*) 5

Answer. (*d*)

16. How many π bonds are present in the molecule shown?

(*a*) 1 (*b*) 2 (*c*) 4 (*d*) 6

Answer. (*b*)

17. Which of the following is a planar molecule ?

(*a*) Formaldehyde (*b*) Acetone (*c*) Formic acid (*d*) Acetic acid

Answer. (*a*)

18. What is bond angle between the hybrid orbitals in methane ?

(*a*) 180° (*b*) 120° (*c*) 109.5° (*d*) 115.5°

Answer. (*c*)

19. The H–C–C bond angle in ethane is

(*a*) 60° (*b*) 109.5° (*c*) 120° (*d*) 118°28'

Answer. (*b*)

20. The H–C–H bond angle in ethylene is

(*a*) 60° (*b*) 120° (*c*) 90° (*d*) 180°

Answer. (*b*)

21. What is the bond length of a carbon-carbon double bond ?

(*a*) 1.20 Å (*b*) 1.34 Å (*c*) 1.54 Å (*d*) 1.68 Å

Answer. (*b*)

22. The carbon-carbon bond lengths in rank of increasing bond length is :

(*a*) triple, double, single (*b*) single, double, triple
(*c*) single, triple, double (*d*) triple, single, double

Answer. (*a*)

23. Which of the following is the correct order of bond lengths :

(*a*) C–C < C=C < C≡C

(*b*) C–C > C≡C > C=C

(*c*) C≡C > C–C > C=C

(*d*) C≡C < C–C > C=C

Answer. (*a*)

24. Which of the following hydrocarbons has the shortest C–C bond length ?

(*a*) $CH_2{=}CH_2$ (*b*) CH_3CH_3 (*c*) HC≡CH (*d*) 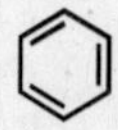

Answer. (*c*)

25. The carbon-carbon bond length is maximum in

(*a*) $CH_2{=}CH_2$ (*b*) CH_3CH_3 (*c*) HC≡CH (*d*) 

Answer. (*b*)

26. Which statement is true ?

(*a*) Resonance hybrids are inherently unstable.

(*b*) Resonance hybrids are more stable than any individual resonance form.

(*c*) Resonance hybrids are averages of all resonance forms resembling the less stable forms.

(*d*) Resonance hybrids are averages of all resonance forms resembling the more stable forms.

Answer. (*d*)

27. Resonance structures of a molecule have

(*a*) same arrangement of atoms

(*b*) different arrangement of atoms

(*c*) same number of paired electrons

(*d*) different number of paired electrons

Answer. (*a*) and (*c*)

28. Which of the following compounds have planar molecules ?

(*a*) ethyl alcohol (*b*) formaldehyde (*c*) diethyl ether (*d*) 1,3-butadiene

Answer. (*b*) and (*d*)

29. Which of the following compounds will show dipole moment ?

(*a*) *cis*-1,2-dichloroethylene

(*b*) *o*-dichlorobenzene

(*c*) *trans*-1,2-dichloroethylene

(*d*) *p*-dichlorobenzene

Answer. (*a*) and (*b*)

30. Which molecule has a nonzero dipole moment ?

(*a*) Cl_2 (*b*) CO_2 (*c*) CCl_4 (*d*) $CHCl_3$

Answer. (*d*)

31. Which of the following compounds have lowest dipole moment ?

(*a*) carbon tetrachloride

(*b*) chloromethane

(*c*) dichloromethane

(*d*) chloroform

Answer. (*a*)

32. Which molecule has the greatest dipole moment

(*a*) CH_3Cl (*b*) CH_3Br (*c*) CH_3F (*d*) CH_3I

Answer. (*c*)

33. Which of the following compounds have highest dipole moment ?

(*a*) dichloromethane (*b*) chloroform

(*c*) chloromethane (*d*) carbon tetrachloride

Answer. (*a*)

34. Which of the following molecules contains a polar covalent bond ?

(*a*) H_2 (*b*) F_2 (*c*) CH_3Cl (*d*) NaCl

Answer. (*c*)

35. Which one of the molecules shown below has *no* net molecular dipole moment ?

(*a*) CH_3Cl (*b*) $H_2C{=}CH_2$ (*c*) CH_2O (*d*) CH_3OH

Answer. (*b*)

36. Which one of the molecules shown below has a net molecular dipole moment ?

(*a*) CCl_4 (*b*) CO_2 (*c*) CH_3CCl_3 (*d*) $BeCl_2$

Answer. (*c*)

37. Which of the molecules below can hydrogen bond to another of the same compound ?

(*a*) $CH_3CH_2OCH_2CH_3$

(*b*) $CH_3CH_2COOCH_3$

(*c*) $(CH_3CH_2)_2CHOH$

(*d*) $CH_3CH_2COCH_2CH_3$

Answer. (*c*)

38. Which of the following has the highest boiling point ?

(*a*) 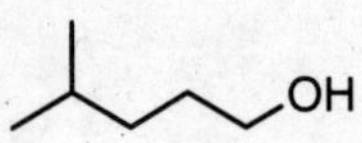

(*b*) 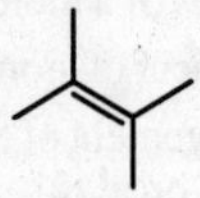

(*c*) 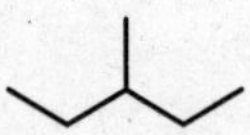

(*d*) 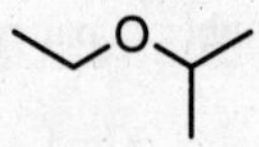

Answer. (*a*)

6

Organic Reactions and their Mechanism

In a chemical reaction, the reactant molecule undergoing attack is referred to as the **Substrate** and the general term used to describe the attacking species is the **Reagent**. The substrate and the reagent interact to yield the products of the reaction.

Substrate + Reagent → Products

The carbon bonds in the substrate molecule are broken (or cleaved) to give fragments which are very reactive and constitute transitory intermediates. At once they may react with other species present in the environment to form new bonds to give the products.

The steps of an organic reaction showing the breaking and making of new bonds of carbon atoms in the substrate leading to the formation of the final products through transitory intermediates, are often referred to as its **Mechanism**.

Substrate → Intermediate (transitory) → Products

Most of the attacking reagents carry either a positive or a negative charge. The positively charged reagents attack the regions of high electron density in the substrate molecule. On the other hand, the negatively charged reagents will attack the regions of low electron density in the substrate molecule.

FACTORS WHICH INFLUENCE A REACTION

A reaction may occur or may not occur depending upon the density of electrons at the site of reaction in the substrate. The factors which influence the electron density in the substrate are:

(1) Inductive Effect
(2) Mesomeric Effect
(3) Electromeric Effect

Inductive Effect. It involves σ electrons. The σ electrons which form a covalent bond are seldom shared equally between the two atoms. This is because different atoms have different electronegativity values, *i.e.*, different powers of attracting the electrons in the bond. Consequently, electrons are displaced towards the more electronegative atom. This introduces a certain degree of polarity in the bond. The more electronegative atom acquires a small negative charge (δ–). The less electronegative atom acquires a small positive charge (δ+).

Consider the carbon-chlorine bond. As chlorine is more electronegative, it will become negatively charged with respect to the carbon atom.

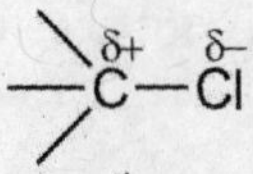
1

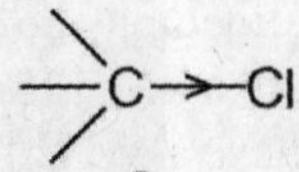
2

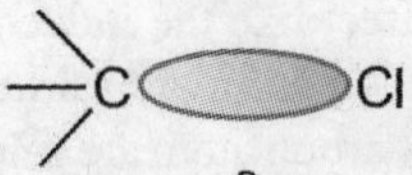
3

Structure (1) indicates the relative charges on the two atoms. In (2), the arrow head placed in the middle of the bond indicates the direction in which the electrons are drawn. In (3), the more heavily shaded part shows the region in which the electron density is greatest.

The inductive effect (I Effect) refers to the polarity produced in a molecule as a result of higher electronegativity of one atom compared to another.

The carbon-hydrogen bond is used as a standard. Zero effect is assumed in this case. Atoms or groups which lose electrons toward a carbon atom are said to have a **+I Effect**. Such groups will be referred to in this book as electron-releasing. Those atoms or groups which draw electrons away from a carbon atom are said to have a **–I Effect**. Such groups will be referred to as electron-attracting.

Some common atoms or groups which cause +I or –I Effects are shown below :

(*a*) **–I Effect Groups (Electron-attracting) :**

NO_2, F, Cl, Br, I, OH, $C_6H_5^-$

(*b*) **+I Effect Groups (Electron-releasing) :**

$(CH_3)_3C^-$, $(CH_3)_2CH^-$, $CH_3CH_2^-$, CH_3^-

Tertiary alkyl groups exert greater +I effect than secondary which in turn exert a greater effect than primary.

An inductive effect is not confined to the polarization of one bond. It is transmitted along a chain of carbon atoms, although it tends to be insignificant beyond the second carbon.

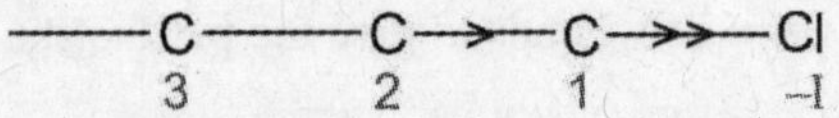

The inductive effect of C_1 upon C_2 is significantly less than the effect of the chlorine atom on C_1.

The inductive effect results in a permanent state of the molecule and can be observed practically in the form of dipole moments. The effect does not depend upon the presence of a reagent.

Mesomeric Effect. It involves π electrons of double and triple bonds.

The mesomeric effect (M effect) refers to the polarity produced in a molecule as a result of interaction between two π bonds or a π bond and lone pair of electrons. The effect is transmitted along a chain in a similar way as are inductive effects.

The mesomeric effect is of great importance in conjugated compounds. (Conjugated compounds are those in which the carbon atoms are linked alternately by single and double bonds). In such systems, the π electrons get delocalized as a consequence of mesomeric effect, giving a number of resonance structures of the molecule.

Consider a carbonyl group (>C=O). The oxygen atom is more electronegative than the carbon atom. As a result, the π electrons of the carbon-oxygen double bond get displaced toward the oxygen atom. This gives the following resonance structures :

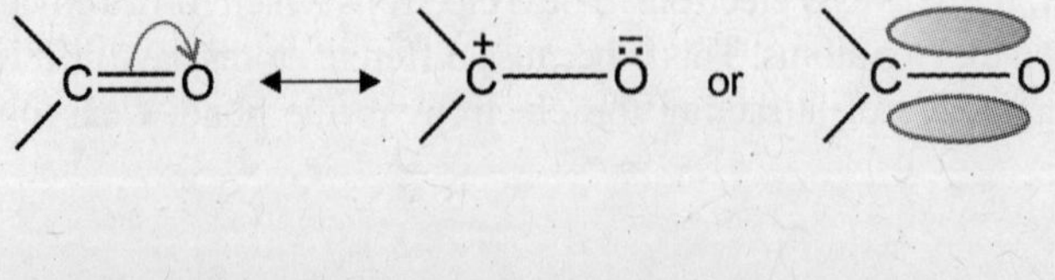

The mesomeric effect is represented by a curved arrow. The head of the arrow indicates the movement of a pair of π electrons. If the carbonyl group is conjugated with a carbon-carbon double bond, the above polarization will be transmitted further via the π electrons.

The mesomeric effect like the inductive effect may be positive or negative. Atoms which lose electrons toward a carbon atom are said to have a **+M Effect**. Those atoms or groups which draw electrons away from a carbon atom are said to have a **–M Effect**.

Some common atoms or groups which cause +M or –M effects are listed below :

(*a*) **+M Effect Groups :**

Cl, Br, I, NH_2, NR_2, OH, OCH_3

(*b*) **–M Effect Groups :**

NO_2, CN, >C=O

The +M effect of the bromine atom is shown below :

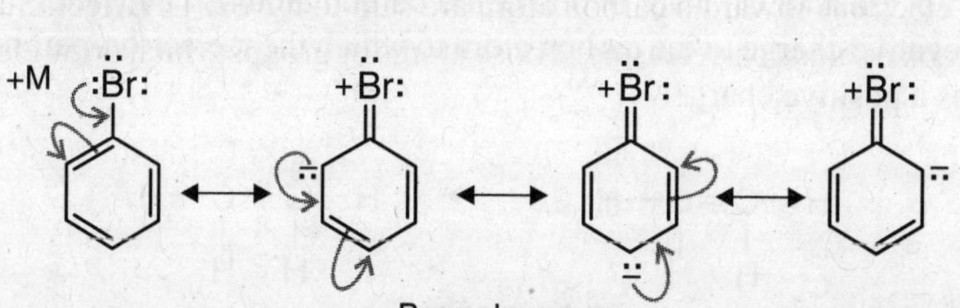

Bromobenzene

The –M effect of the nitro group is shown below :

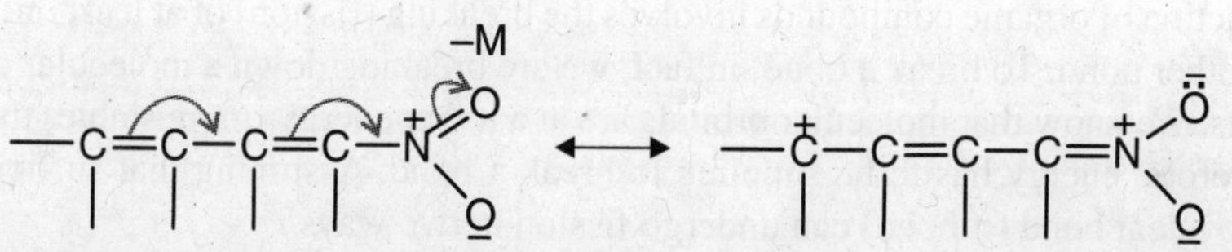

The mesomeric effect like the inductive effect results in a permanent state of the molecule. It does not depend upon the presence of a reagent.

The inductive and mesomeric effects indicate the charge distribution in a molecule. Thus they provide an effective way of determining the point of attack of electrophiles and nucleophiles on the molecule.

Electromeric Effect. Like the mesomeric effect, it also involves the π electrons.

The electromeric effect (E Effect) refers to the polarity produced in a multiple bonded compound as it is approached by a reagent.

When a double or a triple bond is exposed to an attack by an electrophile E^+ (a reagent), the two π electrons which form the π bond are completely transferred to one atom or the other. The electromeric effect is represented as :

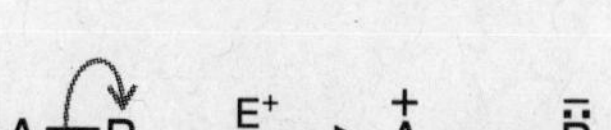

The curved arrow shows the displacement of the electron pair. The atom A has lost its share in the electron pair and B has gained this share. As result, A acquires a positive charge and B a negative charge. Notice that the arrow points away from the centre of the bond and towards the atom that gains the electron pair.

Consider the example where an electrophile (E^+) attacks a carbon-carbon double bond in the ethylene molecule. We know that the double bond is made up of one σ bond and one π bond. The electrons in the π bond are quite exposed. Under the influence of the electric field of the positively charged electrophile, the symmetry of the molecular orbital is disturbed entirely in favor of one carbon atom as shown in Fig. 6.1.

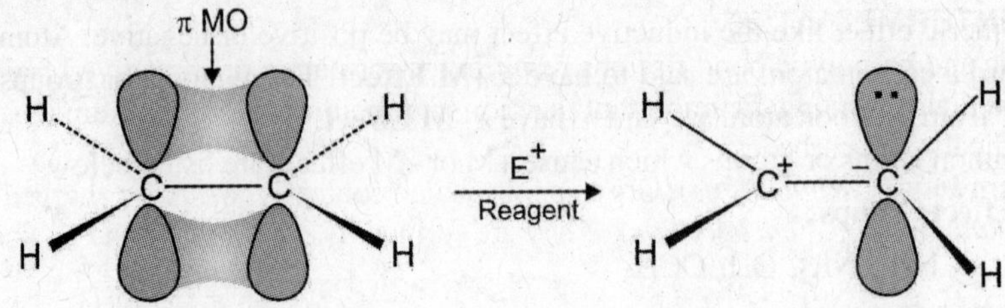

Ethylene

Fig. 6.1. The presence of an electrophile disturbs the symmetry of the π MO in favor of one of the carbon atoms in ethylene.

This gives a negative charge to the carbon atom to which the π electron-pair migrates, while the other atom acquires a positive charge.

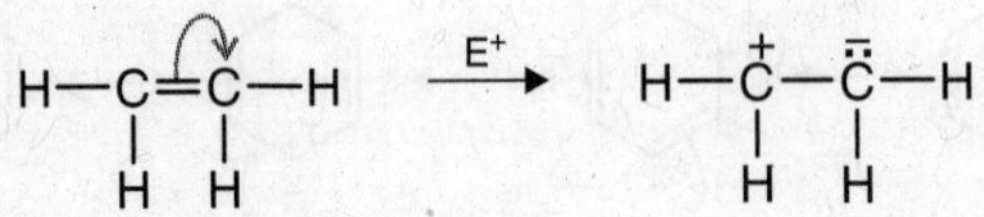

The electromeric effect is a temporary effect. It takes place only in the presence of a reagent.

HOMOLYTIC AND HETEROLYTIC FISSION

Every reaction of organic compounds involves the breaking (fission) of at least one bond and the making of another bond. To break a bond, in fact, we are breaking down a molecular orbital to give atomic orbitals. We know that molecular orbitals are at a lower energy (more stable) than the atomic orbitals. Therefore, energy has to be supplied to break a bond. Assuming that sufficient energy is available, a covalent bond (σ bond) can undergo fission in two ways :

(1) By Homolytic Fission or Homolysis
(2) By Heterolytic Fission or Heterolysis

Homolytic Fission. In this process each of the atoms acquires one of the bonding electrons.

$$A{-}B \text{ or } A:B \longrightarrow A\cdot + \cdot B$$

The products, A· and ·B, are called **Free Radicals**. They are electrically neutral and have one unpaired (odd) electron associated with them. Free radicals are extremely reactive because of the tendency of this electron to become paired at the earliest opportunity. Consequently, reactions which proceed *via* the intermediate formation of free radicals often take place very rapidly. Homolytic fission is the most common mode of fission in the vapor phase.

Homolytic reactions are usually initiated by heat, light or organic peroxides.

Heterolytic Fission. In this process one of the atoms acquires both of the bonding electrons when the bond is broken.

$$A{-}B \longrightarrow \overset{+}{A} + :\bar{B} \quad (1)$$

$$A{-}B \longrightarrow \bar{A}: + \overset{+}{B} \quad (2)$$

In example (1), *B* is more electronegative than *A* which thereby acquires both the bonding electrons and becomes negatively charged. The arrow indicates that the sigma electrons that form the *A–B* bond are leaving *A* and becoming the exclusive property of *B*. In example (2), *A* is shown to be more electronegative than *B*.

The products of heterolytic fission are ions. Reactions which involve heterolytic fission take place at measurable rates. Heterolytic fission occurs most readily with polar compounds in polar solvents.

REACTION INTERMEDIATES

Heterolytic and homolytic bond fissions result in the formation of short-lived fragments called reaction intermediates. Among the important reaction intermediates are carbonium ions, carbanions, carbon free radicals, and carbenes.

Carbonium Ions *(Carbocations)*. Organic ions which contain a positively charged carbon atom are called *carbonium ions* or *carbocations*. They are formed by heterolytic bond fission.

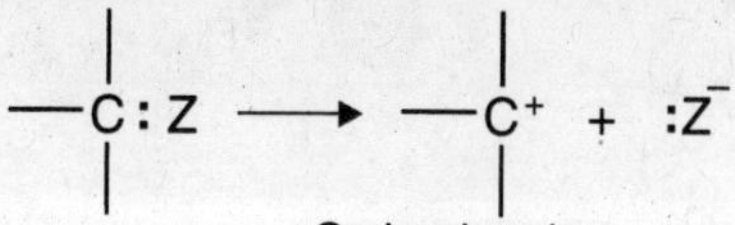

Carbonium ion

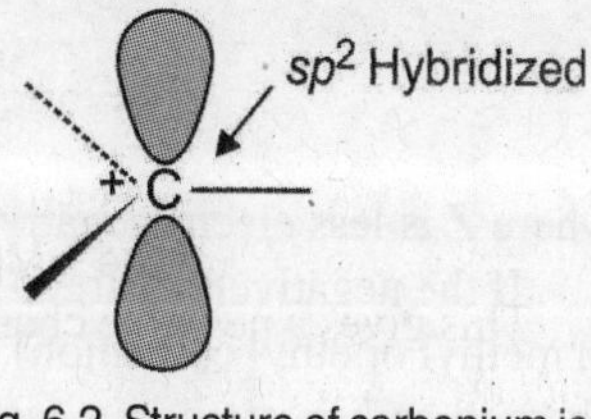

Fig. 6.2. Structure of carbonium ion.

where Z is more electronegative than carbon.

The positively charged carbon atom in a carbonium ion uses sp^2 hybrid orbitals to form three σ bonds (Fig. 6.2). An empty *p* orbital extends above and below the plane of the σ bonds. This empty *p* orbital makes the carbon atom electron-deficient and gives it a positive charge. Thus a carbonium ion will combine with any substance (*e.g.*, nucleophiles) which can donate a pair of electrons.

Carbonium ions are named after the parent alkyl group and adding the words *carbonium ion*. For example,

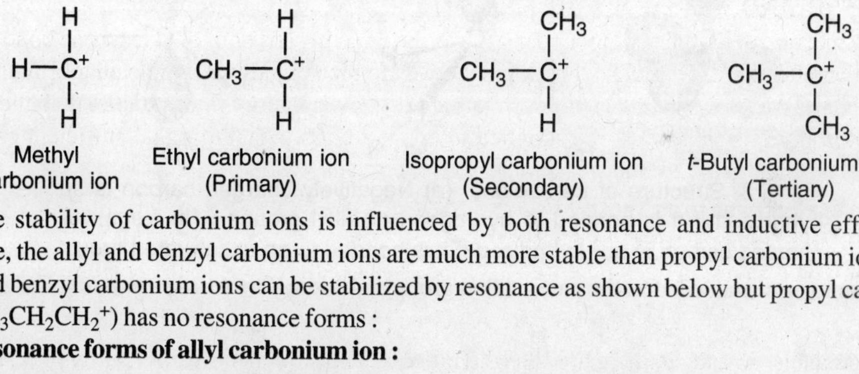

Methyl carbonium ion | Ethyl carbonium ion (Primary) | Isopropyl carbonium ion (Secondary) | *t*-Butyl carbonium ion (Tertiary)

The stability of carbonium ions is influenced by both resonance and inductive effects. For example, the allyl and benzyl carbonium ions are much more stable than propyl carbonium ions. Both allyl and benzyl carbonium ions can be stabilized by resonance as shown below but propyl carbonium ion ($CH_3CH_2CH_2^+$) has no resonance forms :

Resonance forms of allyl carbonium ion :

$$CH_2{=}CH{-}\overset{+}{C}H_2 \longleftrightarrow \overset{+}{C}H_2{-}CH{=}CH_2$$

Resonance forms of benzyl carbonium ion :

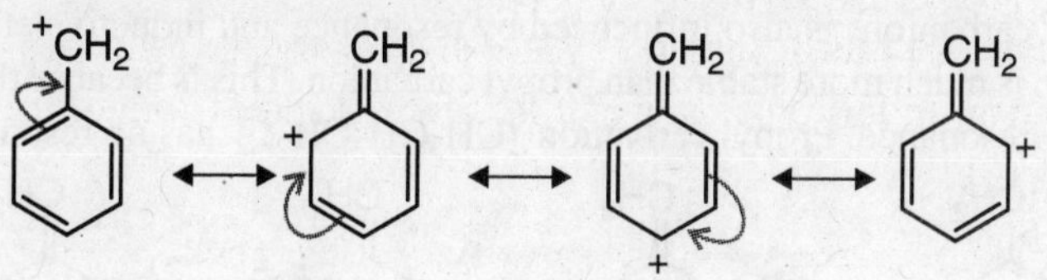

Electron-releasing groups (+I groups) also stabilize carbonium ions by partial neutralization of the positive charge on carbon.

Thus, a tertiary carbonium ion is more stable than a secondary, which in turn in more stable than a primary because of the +I effect associated with alkyl groups.

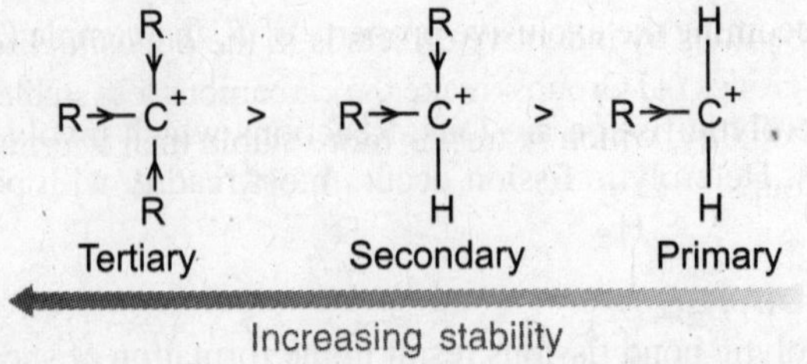

Electron-attracting groups (–I Groups) like $–NO_2$, –Br will make a carbonium ion less stable.

Carbanions. Organic ions which contain a negatively charged carbon atom are called *carbanions*. They are also formed by heterolytic bond fission.

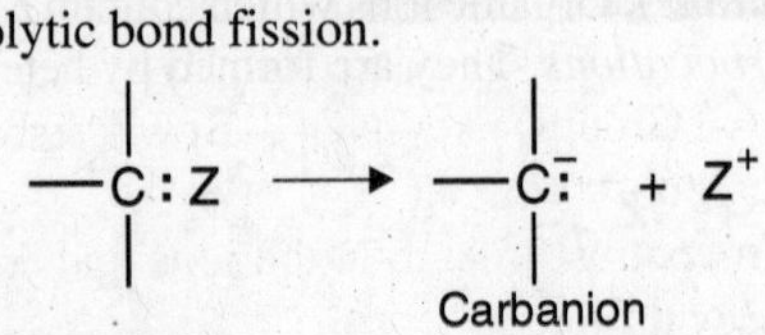

where Z is less electronegative than carbon.

If the negatively charged carbon atom in a carbanion is bonded to hydrogens or alkyl groups (as in methyl or ethyl carbanion), it uses sp^3 hybrid orbitals to form the three σ bonds. If the negatively charged carbon atom in a carbanion is bonded to an unsaturated group (as in benzyl carbanion), it uses sp^2 hybrid orbitals to form the three σ bonds. The completely filled sp^3 or p orbital makes the carbon electron-rich and gives it a negative charge. Thus a carbanion will combine with any substance (*e.g.*, electrophiles) which can accept a pair of electrons.

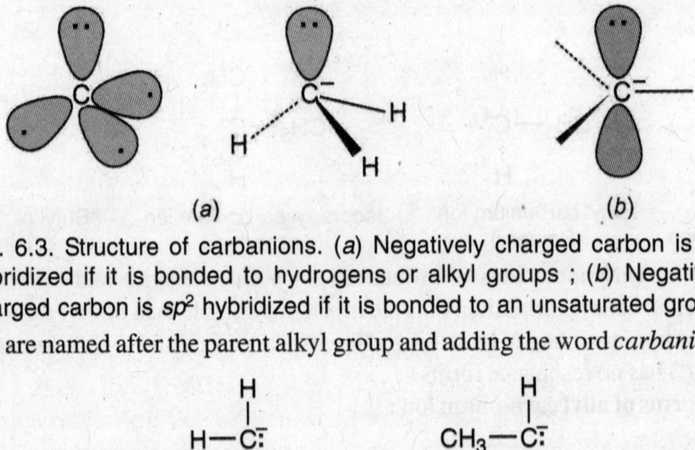

Fig. 6.3. Structure of carbanions. (*a*) Negatively charged carbon is sp^3 hybridized if it is bonded to hydrogens or alkyl groups ; (*b*) Negatively charged carbon is sp^2 hybridized if it is bonded to an unsaturated group.

Carbanions are named after the parent alkyl group and adding the word *carbanion*. For example,

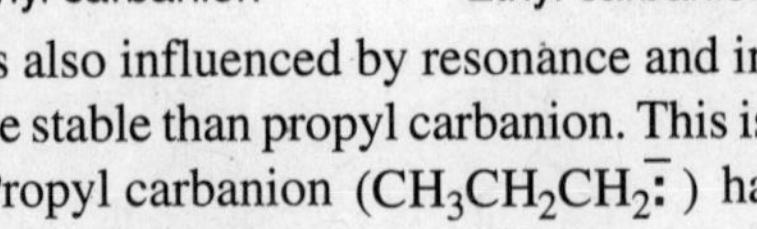

Methyl carbanion   Ethyl carbanion

The stability of carbanions is also influenced by resonance and inductive effects. For example, the benzyl carbanion is much more stable than propyl carbanion. This is because the benzyl carbanion can be stabilized by resonance. Propyl carbanion ($CH_3CH_2\bar{C}H_2{:}$) has no resonance forms.

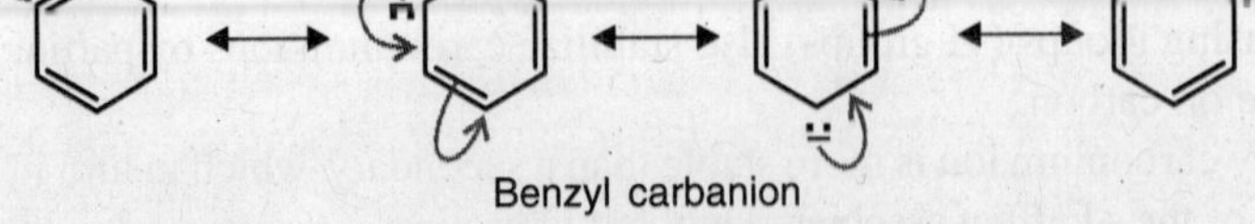

Benzyl carbanion

Stabilization of carbanions by inductive effects is in the opposite direction from the carbonium ions. Electron-releasing groups (+I Groups) make the carbanions less stable. Thus a primary carbanion is more stable than a secondary, which in turn is more stable than a tertiary because of the +I effect associated with alkyl groups.

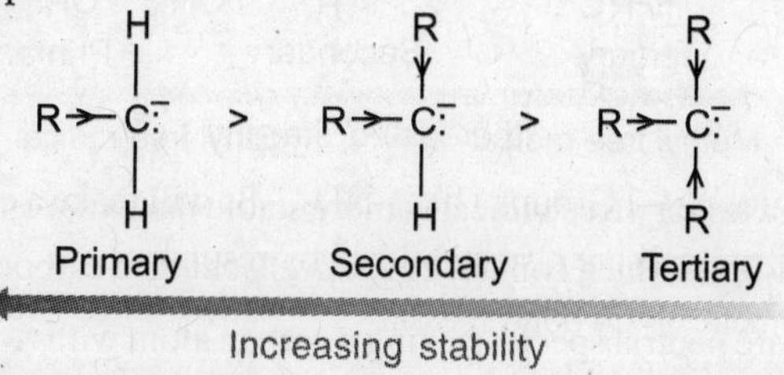

Electron-attracting groups (–I Groups) like $-NO_2$, –Br will stabilize carbanion by partial removal of the negative charge on the carbon.

Carbon Free Radicals. In contrast to carbonium ions and carbanions, these have no charge. They are formed by homolytic bond fission.

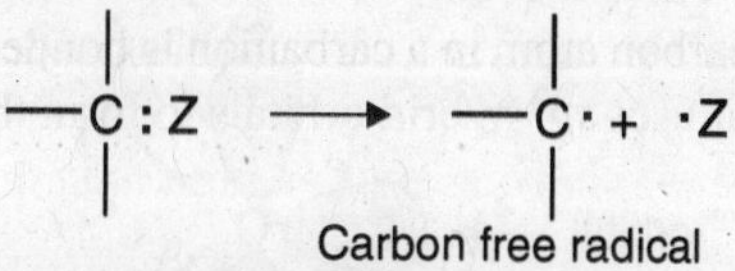

Here Z and carbon atom have similar electronegativities.

What do you want to be when you grow up ?
Free radical

The carbon atom in a carbon free radical uses sp^2 hybrid orbitals to form three σ bonds (Fig. 6.4). A half-filled *p* orbital extends above and below the plane of σ bonds. Thus carbon free radicals are electrically neutral and have one unpaired (odd) electron associated with them.

They are extremely reactive because of the tendency of this electron to become paired at the earliest opportunity. Free radicals combine with other free radicals or with other molecules to produce larger free radicals.

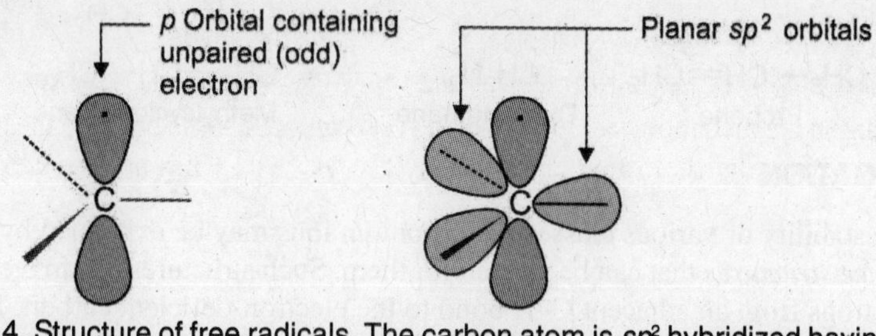

Fig. 6.4. Structure of free radicals. The carbon atom is sp^2 hybridized having three sp^2 orbitals and *p* orbital containing the unpaired (odd) electron. The sp^2 orbitals are trigonal planar and the *p* orbital is perpendicular to the plane of sp^2 orbitals.

Carbon free radicals are named after the parent alkyl group and adding the word *free radical*. For Example,

$$H-\underset{H}{\overset{H}{\underset{|}{\overset{|}{C}}}}\cdot \qquad\qquad CH_3-\underset{H}{\overset{H}{\underset{|}{\overset{|}{C}}}}\cdot$$

Methyl free radical　　　　Ethyl free radical

Like carbonium ions, a tertiary free radical is more stable than a secondary, which in turn is more stable than a primary. Free radicals are also stabilized by resonance.

Carbenes. Carbenes are neutral species having a carbon atom with two bonds and two electrons.

$-\ddot{C}-$　　　*For example* : Methylene (H_2C:)

Carbenes are highly reactive. They act as strong electrophiles because they can accept a pair of electrons to complete their outer shell.

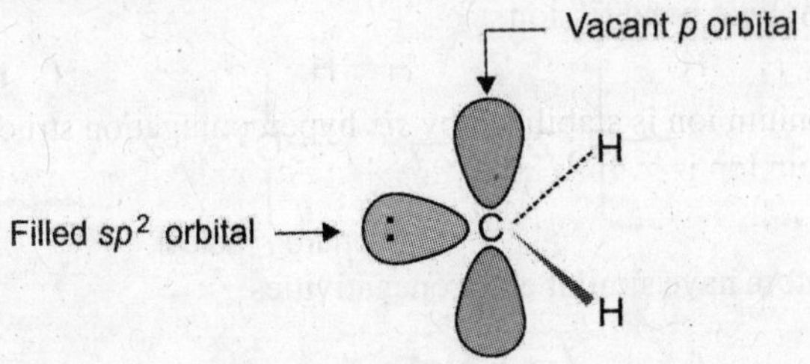

Fig. 6.5. Structures of methylene (H_2C:). The carbon atom is sp^2 hybridized.

Carbenes may be generated in a number of ways :

(1) By reaction of chloroform in the presence of a strong alkali (For example alkoxides).

$$CHCl_3 + R\bar{O}\text{:} \longrightarrow ROH + Cl_3\bar{C}\text{:}$$

$$Cl_3\bar{C}\text{:} \longrightarrow Cl_2C\text{:} + \bar{Cl}\text{:}$$

(2) By decomposition of diazomethane or ketene in the presence of UV light.

$$\underset{\text{Diazomethane}}{H_2C=\overset{+}{N}=\overset{-}{N}} \xrightarrow{UV} H_2C\text{:} + N_2$$

$$\underset{\text{Ketene}}{H_2C=C=O} \xrightarrow{UV} H_2C\text{:} + CO$$

An important reaction of carbenes is their addition to a carbon-carbon double bond to form a cyclopropane derivative.

$$\underset{\text{Propene}}{CH_3-CH=CH_2} + \underset{\text{Diazomethane}}{CH_2N_2} \xrightarrow{UV} \underset{\text{Methylcyclopropane}}{CH_3-\overset{CH_2}{\overset{/\ \backslash}{CH-CH_2}}}$$

HYPERCONJUGATION

The relative stability of various classes of carbonium ions may be explained by the number of *no-bond resonance structures* that can be written for them. Such structures are arrived at by shifting the bonding electrons from an adjacent C–H bond to the electron-deficient carbon. In this way, the positive charge originally on carbon is dispersed to the hydrogen. This manner of electron release by assuming *no-bond* character in the adjacent C–H bond is called **Hyperconjugation** or **No-Bond Resonance.**

The more hyperconjugation structures (no-bond resonance structures) that can be written for a species, the more stable is the species. For example,

(1) Ethyl carbonium ion is stabilized by *three* hyperconjugation structures :

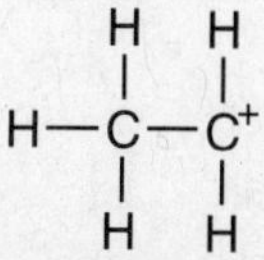

Ethyl carbonium ion

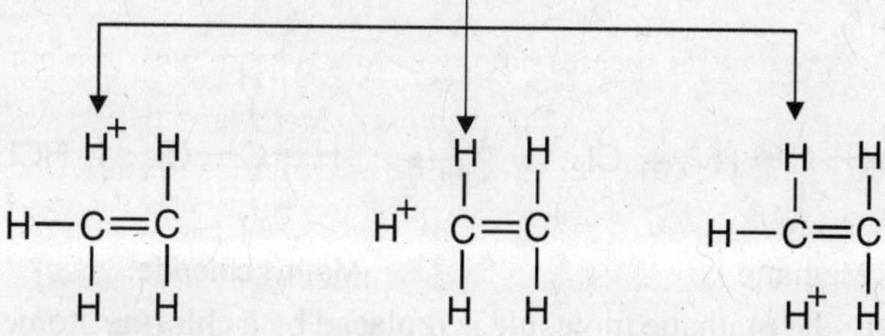

(2) Isopropyl carbonium ion is stabilized by *six* hyperconjugation structures.

(3) *t*-Butyl carbonium ion is stabilized by *nine* hyperconjugation structures.

Thus, the following order of stability holds :

$$(CH_3)_3\overset{+}{C} > (CH_3)_2\overset{+}{C}H > CH_3\overset{+}{C}H_2 > \overset{+}{C}H_3$$

In general, resonance effects (mesomeric effects) are more important than hyperconjugation effects. The allyl and benzyl carbonium ions are more stable than most alkyl carbonium ions because the former are stabilized by resonance while the latter are stabilized by hyperconjugation.

CLASSIFICATION OF REAGENTS

Organic reagents fall into two main groups :

(1) Electrophiles or Electrophilic Reagents

(2) Nucleophiles or Nucleophilic Reagents

Electrophiles. *A reagent which can accept an electron pair in a reaction is called an electrophile*. The name electrophile means "electron-loving" and indicates that it attacks regions of high electron density (negative centres) in the substrate molecule. Electrophiles are electron-deficient. They may be positive ions (including carbonium ions) or neutral molecules with electron-deficient centres. An electrophile can be represented by the general symbol E^+.

Nucleophiles. *A reagent which can donate an electron pair in a reaction is called a nucleophile*. The name nucleophile means "nucleus-loving" and indicates that it attacks regions of low electron density (Positive centres) in the substrate molecule. Nucleophiles are electron-rich. They may be negative ions (including carbanions) or neutral molecules with free electron pairs. A nucleophile can be represented by a general symbol $Nu:^-$.

Examples of electrophiles and nucleophiles are given below :

Electrophiles	Nucleophiles
H^+	Cl^-, Br^-, I^-
Cl^+, Br^+, I^+	OH^-
$\overset{+}{N}O_2$, R_3C^+	CN^-
$^+SO_3H$	RCH_2^-
$AlCl_3$, BF_3	NH_3, RNH_2
	H_2O, ROH

TYPES OF ORGANIC REACTIONS

The reactions of organic compounds can be classified into four main types :

(1) Substitution Reactions

(2) Addition Reactions

(3) Elimination Reactions

(4) Rearrangement Reactions

Substitution Reactions. Substitution reactions are those reactions in which an atom or group of atoms directly attached to a carbon in the substrate molecule is replaced by another atom or group of atoms. For example,

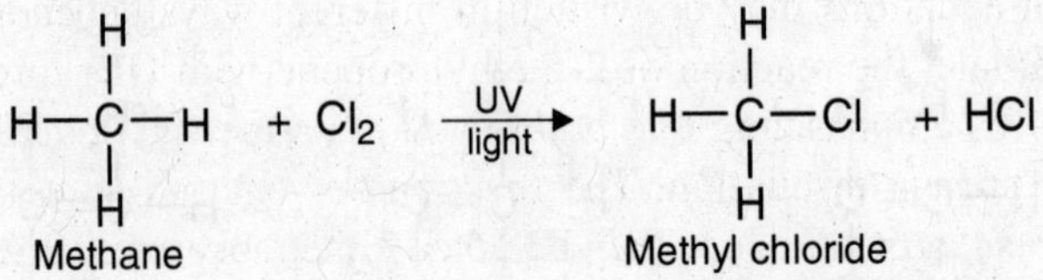

A hydrogen atom of the methane molecule is replaced by a chlorine atom.

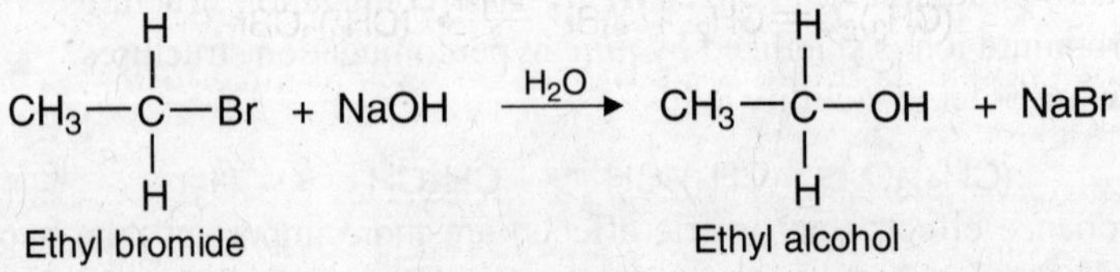

In the above reaction, the bromine atom of ethyl bromide is substituted by a hydroxyl group.

Addition Reactions. Addition reactions are those in which atoms or groups of atoms are simply added to a double or triple bond without the elimination of any atom or other molecules. In these reactions, at least one π bond is lost while two new σ bonds are formed. Double bonds become saturated, and triple bonds are converted into double bonds or may become saturated by further addition. For example,

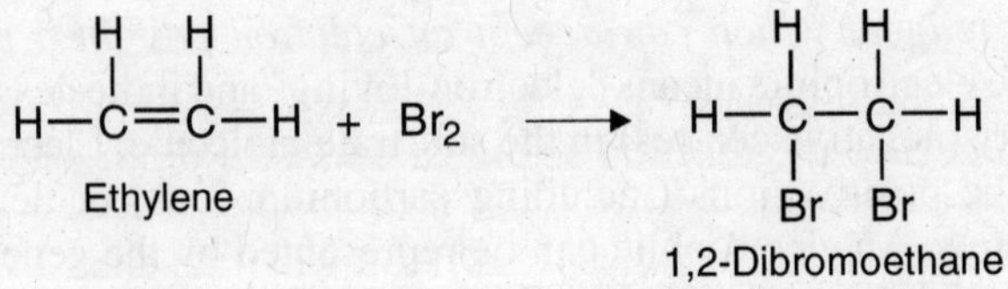

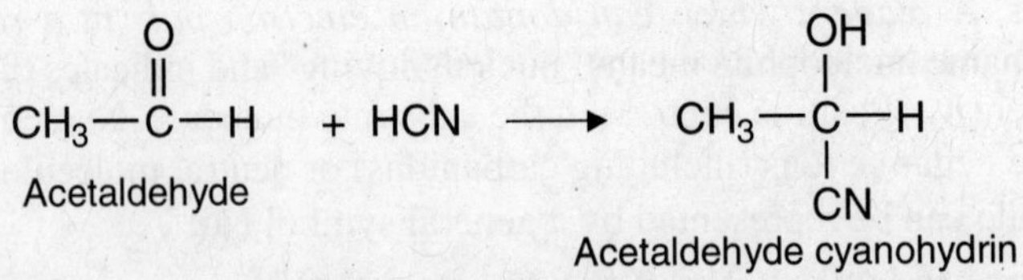

Elimination Reactions. Elimination reactions are those which involve the removal of atoms or groups of atoms from two adjacent atoms in the substrate molecule to form a multiple bond. Elimination reactions may be regarded as reverse of addition reactions. In these reactions, two σ bonds are lost and a new π bond is formed. Saturated compounds become unsaturated. For example,

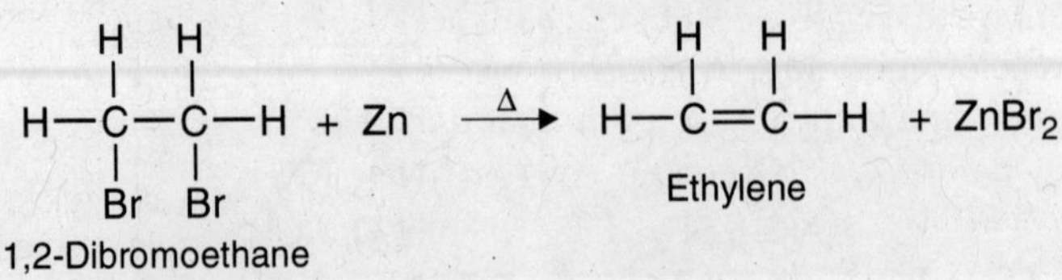

Rearrangement Reactions. Rearrangement reactions involve the migration of an atom or group of atoms from one site to another within the same molecule. The product is always the structural isomer of the original compound. For example,

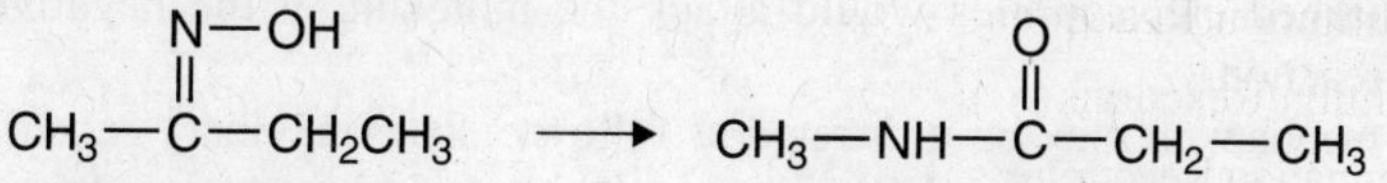

Methyl ethyl ketoxime → *N*-Methylpropionamide

Points to Remember

Organic reaction equations may be written in different ways depending on the emphasis desired. For example, the reaction of 2-methylpropene with HBr might be written in the format A + B → C, emphasizing that *both* reaction partners are equally important for the purposes of the present discussion. The reaction solvent and notes about other reaction conditions such as temperature can then be noted either above or below the reaction arrow. For example,

$$(CH_3)_2C{=}CH_2 + HBr \xrightarrow{\text{ether}} (CH_3)_3CBr$$

Alternatively, we might choose to write the same reaction in the format

$$A \xrightarrow{B} C$$

emphasizing that reagent A is the starting material whose chemistry is of greater interest. Reagent B is then placed above the reaction arrow, together with notes about solvent and reaction conditions. For example,

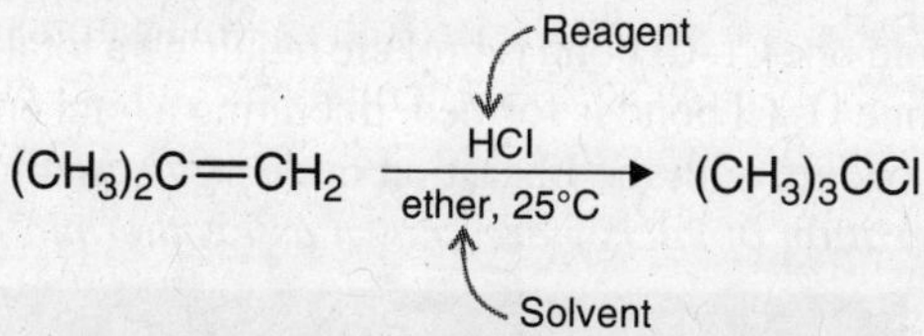

Both reaction formats are used in this book, and it is important that the different roles of chemicals shown next to the reaction arrow be understood.

TYPES OF REACTIONS MECHANISMS

Reaction mechanisms are divided into two main classes :

(1) Polar or Ionic Mechanism

(2) Free Radical Mechanism

Polar or Ionic Mechanism. This type of mechanism applies to organic reactions in which heterolytic bond fission takes place. Here the substrate molecule develops polarity (positive and negative charged centres) by the displacement of electrons due to inductive, mesomeric, or electromeric effects. These positive and negative centres can be attacked by electrophiles or nucleophiles. *The mechanism of reactions involving the attack of electrophilic or nucleophilic reagents on the polar substrate molecules is referred to as polar or ionic mechanism.*

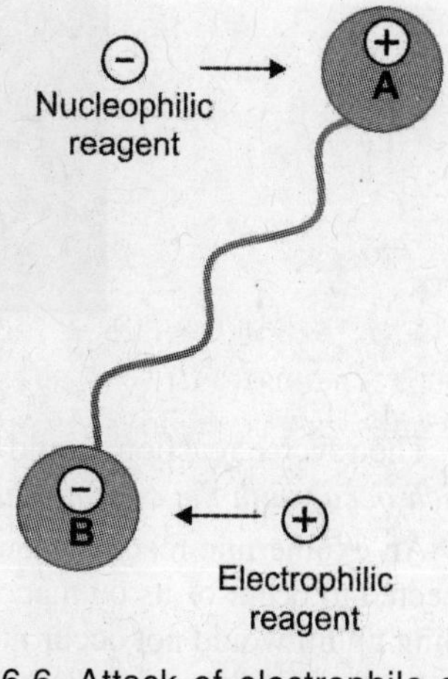

Fig. 6.6. Attack of electrophile and nucleophile on the polar AB molecule.

The cardinal principle of electrostatic that *the unlike charges attract while like charges repel each other* forms the basis of all polar mechanisms. The negatively charged nucleophiles attack the substrate molecule at the positive centre causing a nucleophilic reaction. On the other hand, the positively charged electrophiles would attack the molecule at the negative centre causing an electrophilic reaction.

A polar mechanism may be illustrated as follows : The imaginary substrate molecule AB has positive end A and a negative end B (Fig. 6.6). This would be attacked by nucleophiles at A and electrophiles at B.

Free Radical Mechanism. This type of mechanism applies to organic reactions in which homolytic bond fission takes place. The first step is the formation of free radicals (R·) from a reagent by homolytic fission. These free radicals can then attack the substrate to give the product.

$$R\cdot + A{-}B \longrightarrow R{-}A + B\cdot$$

The free radicals carry no charge. They can attack any part of the substrate molecule regardless of the electronic distribution of the substrate. The mechanism of reactions involving the attack of free radicals on the substrate molecule is referred to as a *free radical mechanism.*

ENERGY REQUIREMENTS OF ORGANIC REACTIONS

Exothermic and Endothermic Reactions. In an organic reaction, some bonds break while new ones are formed. Energy is required for breaking of bonds and also it is liberated in the formation of bonds. Consider the reaction between methane and chlorine to give chloromethane (CH_3Cl) and hydrogen chloride.

$$H_3C{-}H + Cl{-}Cl \longrightarrow H_3C{-}Cl + H{-}Cl$$

Bond Energy (*kcal/mol*): C–H 99; Cl–Cl 58; C–Cl 81; H–Cl 103

Here one C–H bond and one Cl–Cl bond is broken, requiring a total energy of 157 kcal/mol. At the same time one C–Cl and one H–Cl bond is formed, liberating a total energy of 184 kcal/mol. The net result is that the reaction occurs with the liberation of energy equal to 27 kcal/mol. *Such reactions which occur with the liberation of energy are known as exothermic reactions.*

Reaction of glycerol with $KMnO_4$ is example of an exothermic reaction

The reverse reaction would require the consumption of 27 kcal/mol of energy. *Such reactions which occur with the consumption of energy are known as endothermic reactions.*

An exothermic reaction could be considered like a ball rolling down a hill side and would be expected to occur of its own accord. An endothermic reaction which may be considered like a ball rolling uphill would not occur normally, unless energy is supplied to the system. The total chemical energy stored in the molecules before and after the reaction is described as the **potential energy** of the system.

Activation Energy. A reaction always proceeds by breaking of bonds in the reacting molecules. Therefore, whether a reaction is exothermic or endothermic, to start with energy must come from some source to break the bonds.

We know from our knowledge of Physical Chemistry that molecules of the reactants are in a state of rapid motion and possess kinetic energy. The reaction occurs when the reacting molecules approach in proper alignment and collide. On such collisions, the kinetic energy possessed by the molecules is transformed into potential energy of the system. Thus to start a reaction, the required energy is supplied by the collisions of the reacting molecules, where rate could be enhanced if necessary, say by heating.

Not all collisions between the reacting molecules are fruitful. It is only the collisions of molecules which possess a certain minimum energy that can bring about the reaction. The molecules that have come to possess higher potential energy through collisions are said to have been activated to enter into reaction.

The minimum amount of potential energy that must be provided by collisions of the reacting molecules for the reaction to occur is known as the **Activation Energy** (E_a).

It is true that for an exothermic reaction, the collisions of the reactant molecules readily supply the required initial energy and the reaction takes place spontaneously. On the other hand, for an endothermic reaction the molecules need to be activated by supplying energy, say in the form of heat to make the reaction go.

Imagine some balls attempting to climb over a hill from one valley to another (Fig. 6.7). Only those balls will cross over which possess a certain minimum of energy to reach the hilltop from where they would get over the barrier. Similarly, only such reacting molecules that possess a certain minimum energy could change into products but before doing so, these would have to cross over the *activation energy* barrier.

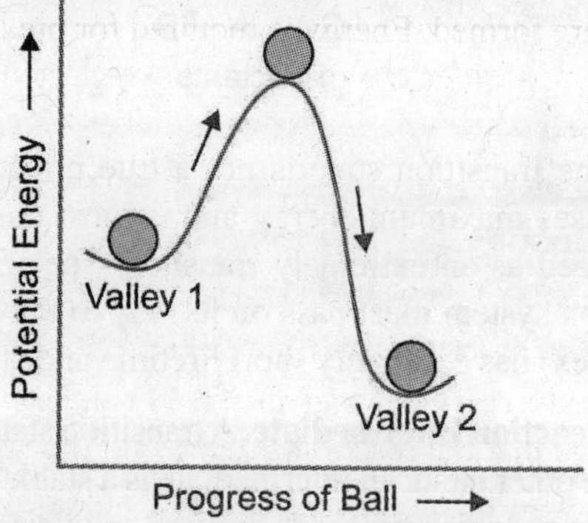

Fig. 6.7. Activation energy anology.

Consider the energy changes during the course of the reaction,

$$C + A{-}B \longrightarrow C{-}A + B$$

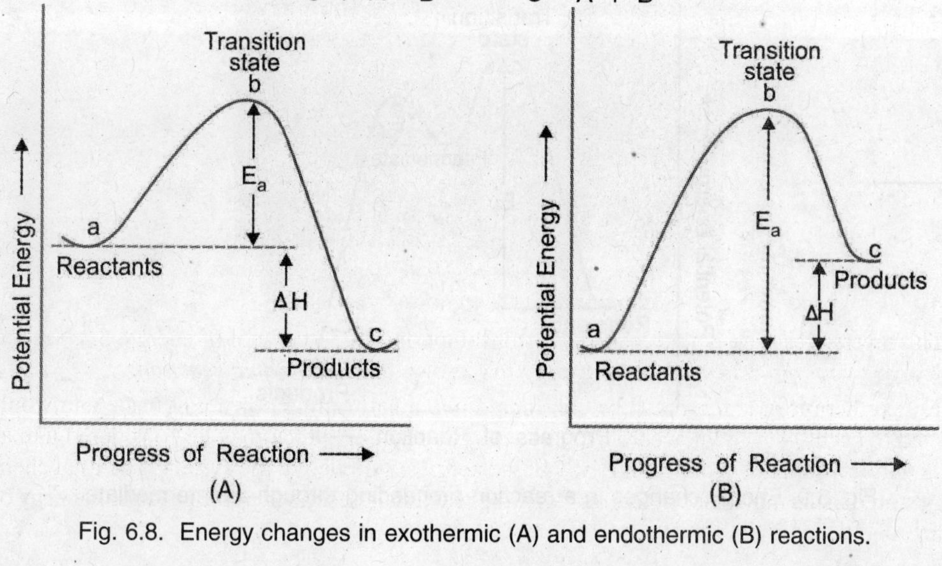

Fig. 6.8. Energy changes in exothermic (A) and endothermic (B) reactions.

In the beginning both C and A–B possess certain potential energy indicated by the point (*a*) on the curve (Fig. 6.8). These reacting molecules also possess kinetic energy which on collisions is transformed into potential energy. This results in the increase of potential energy and the system moves up along the curve till the cliff (*b*) is reached. The energy of cliff state is a sort of temporary phase and leads to products C–A + B (*c*) when the potential energy of the system is again changed into kinetic energy and then into heat or any other form of energy.

Fig. 6.8 shows that for an exothermic reaction, the system originally possesses more potential energy than the products and the excess energy (ΔH) is liberated as heat. For an endothermic reaction, the system to start with has less potential energy than the products (–ΔH) and, therefore, it absorbs heat from the surroundings.

Transition State. The reaction C + A—B → C—A + B could be visualized to take place by the following steps. The molecule C approaches A—B from a direction remote from B (proper alignment). While C draws nearer to A, B starts being repelled from A until a stage is reached when C and B are rather loosely attached to A and are approximately equidistant from it. This is the least stable arrangement and is called the *transition state (TS) or activated complex.* The sequence of events may be represented as :

$$\underset{\text{Reactants}}{C + A{-}B} \longrightarrow \underset{\text{Transition state}}{C\cdots A\cdots B} \longrightarrow \underset{\text{Products}}{C{-}A + B}$$

The transition state is not a true molecule, the bonds being partial. In this state the system possesses maximum energy and is most unstable. Hence the transition state of a system could be described as an extremely transitory specific arrangement of atoms and groups through which a reaction system must pass on its way to the products. In other words, the transition state (activated complex) has extremely short lifetime and at once decomposes to give the products.

Reaction Intermediate. A transition state refers to an imaginary molecule and cannot be isolated. On the other hand, an intermediate is a stable entity and can be isolated under appropriate conditions. A reaction which proceeds through an intermediate has to surmount two energy barriers. One for the conversion of the reactants to the intermediate (E_a) and the other for the conversion of the intermediate into products ($E_{a'}$) as shown in Fig. 6.9.

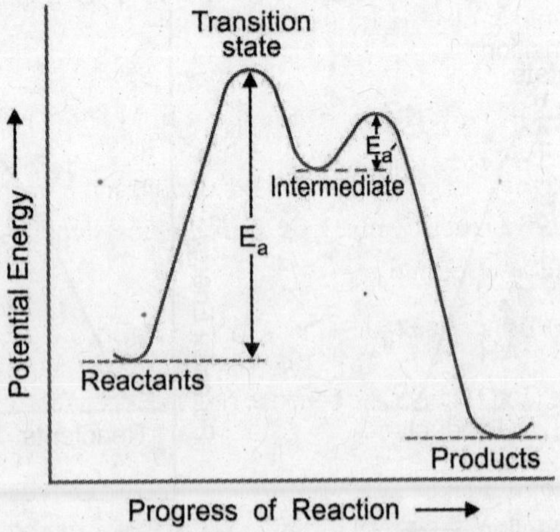

Fig. 6.9. Energy changes in a reaction proceeding through an intermediate.

SOME IMPORTANT MECHANISMS

MECHANISMS OF SUBSTITUTION REACTIONS

Reactions which involve the replacement or substitution of one or more atoms or groups of a compound by other atoms or groups are known as *substitution reactions*. These reactions may be initiated by a nucleophile, electrophile, or free radical.

Free-Radical Substitution Reactions. These reactions are initiated by free radicals. The chlorination of methane in the presence of ultraviolet light is an example of free-radical substitution.

$$\underset{\text{Methane}}{CH_4} + Cl_2 \xrightarrow[\text{light}]{\text{UV}} \underset{\text{Methyl chloride}}{CH_3Cl} + HCl$$

The mechanism of the above reaction involves the following steps :

(1) **Initiation Step.** A chlorine molecule undergoes homolytic fission in the presence of ultraviolet light to give chlorine free radicals.

$$Cl{-}Cl \xrightarrow[\text{light}]{\text{UV}} Cl\cdot + Cl\cdot$$

(2) **Propagation Steps.** A chlorine atom attacks the methane molecule to give methyl free radicals and hydrogen chloride.

$$Cl\cdot + CH_4 \longrightarrow \underset{\text{Methyl free radical}}{\cdot CH_3} + HCl$$

The methyl free radical attacks a chlorine molecule to yield methyl chloride and chlorine atom.

$$CH_3\cdot + Cl{-}Cl \longrightarrow CH_3{-}Cl + Cl\cdot$$

These propagation steps are repeated again and again.

(3) **Termination Steps.** These involve the formation of stable molecules by combination of free radicals.

$$Cl\cdot + \cdot Cl \longrightarrow Cl{-}Cl$$
$$CH_3\cdot + \cdot Cl \longrightarrow CH_3{-}Cl$$
$$CH_3\cdot + \cdot CH_3 \longrightarrow CH_3{-}CH_3$$

Further details of this reaction are given in Chapter *10*.

Electrophilic Substitution Reactions. When a substitution reaction involves the attack by an electrophile, the reaction is referred to as electrophilic substitution. The bromination of benzene in the presence of $FeBr_3$ is an example of electrophilic substitution.

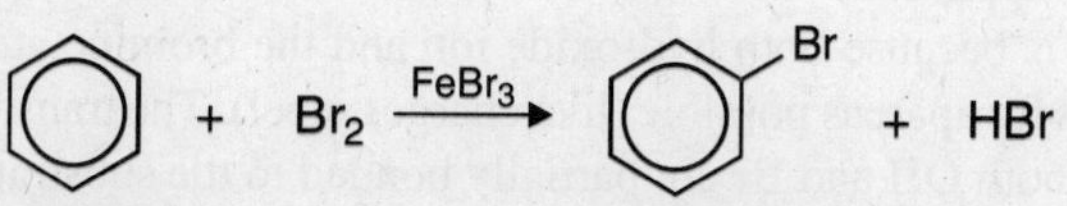

Benzene Bromobenzene

The mechanism of the above reaction involves the following steps :

Step 1. Formation of the electrophile.

$$Br{-}Br + FeBr_3 \longrightarrow \underset{\text{Electrophile}}{Br^+} + FeBr_4^-$$

Step 2. The electrophile (Br^+) attacks the benzene ring to form a resonance stabilized carbonium ion.

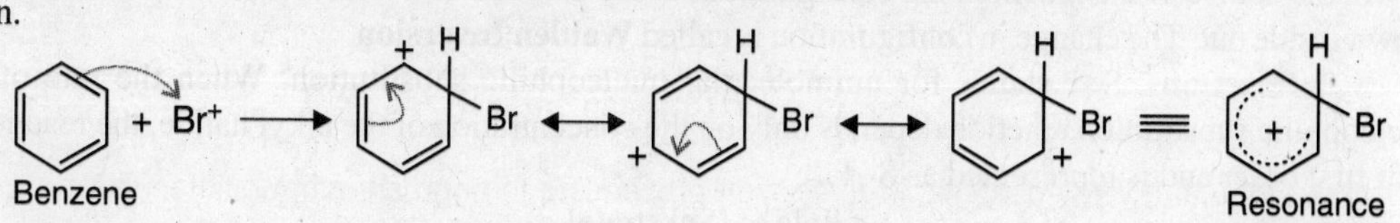

Benzene Resonance hybrid

Step 3. Elimination of proton to give the substitution product.

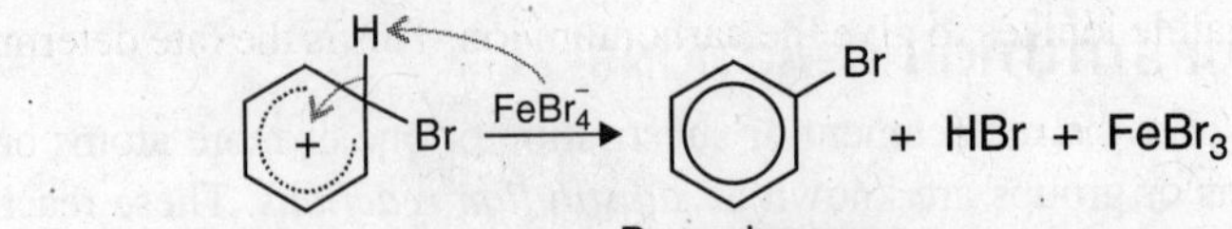

Bromobenzene

More examples of electrophilic substitution are discussed in Chapter 33.

Nucleophilic Substitution Reactions. When a substitution reaction involves the attack by a nucleophile, the reaction is referred to as S_N (S stands for substitution and N for a nucleophile). The hydrolysis of alkyl halides by aqueous NaOH is an example of nucleophilic substitution.

$$\text{R—X} + \underset{\text{Nucleophile}}{\bar{\text{O}}\text{H}} \longrightarrow \text{R—OH} + \underset{\text{Leaving group}}{\bar{\text{X}}}$$

The nucleophilic substitution reactions are divided into two classes :

(1) S_N2 Reactions

(2) S_N1 Reactions

S_N2 Reactions. S_N2 stands for bimolecular nucleophilic substitution. When the rate of a nucleophilic substitution reaction depends on the concentration of both the substrate and the nucleophile, the reaction is of second order and is represented as S_N2.

$$\text{Rate} \propto [\text{Substrate}]\,[\text{Nucleophile}]$$

Evidently, the rate-determining step involves the participation of both the substrate and the nucleophile.

Consider the hydrolysis of methyl bromide by aqueous NaOH. The reaction and the transition state are represented in Fig. 6.10.

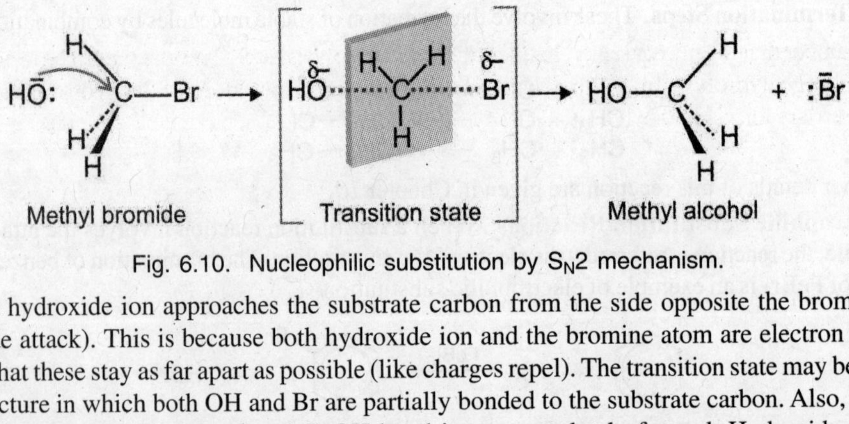

Methyl bromide Transition state Methyl alcohol

Fig. 6.10. Nucleophilic substitution by S_N2 mechanism.

The hydroxide ion approaches the substrate carbon from the side opposite the bromine atom (backside attack). This is because both hydroxide ion and the bromine atom are electron rich. It is natural that these stay as far apart as possible (like charges repel). The transition state may be pictured as a structure in which both OH and Br are partially bonded to the substrate carbon. Also, the C–Br bond is not completely cleaved and C–OH bond is not completely formed. Hydroxide ion has a diminished negative charge because it has started to share its electrons with the substrate carbon. The bromine atom also carries a negative charge because it has started removing its shared pair of electrons from the carbon atom. In the transition state, the three C–H bonds lie in one plane. The C–OH and C –Br bonds are perpendicular to the plane of C–H bonds. The energy needed for the cleavage of C–Br bond is partially provided by the energy liberated by the C–OH bond formation.

In the course of the reaction, the configuration of the carbon is inverted, rather like an umbrella blown inside out. The change in configuration is called **Walden Inversion**.

S_N1 Reactions. S_N1 stands for unimolecular nucleophilic substitution. When the rate of a nucleophilic substitution reaction depends only on the concentration of the alkyl halide, the reaction is of first order and is represented as S_N1.

$$\text{Rate} \propto [\text{Substrate}]$$

Consider the hydrolysis of *t*-butyl bromide. The reaction consists of two steps :

Step 1. The alkyl halide ionizes to give the carbonium ion. This is the rate determining step.

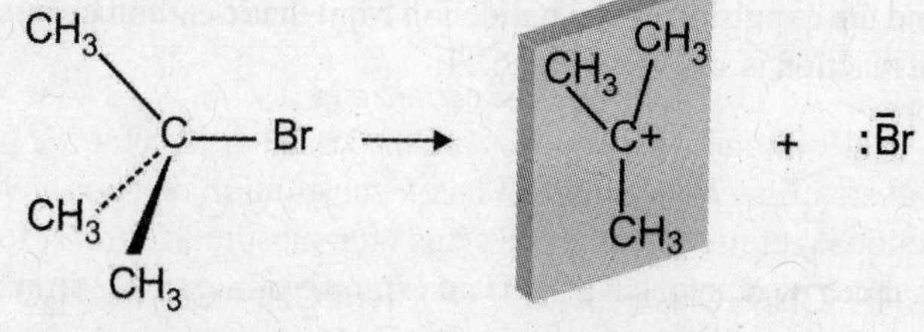

tert-Butyl bromide　　　Carbonium ion

The carbonium ion is planar. This is because the central positively charged carbon atom is sp^2 hybridized.

Step 2. The nucleophile can attack the planar carbonium ion from either side to give *t*-butyl alcohol.

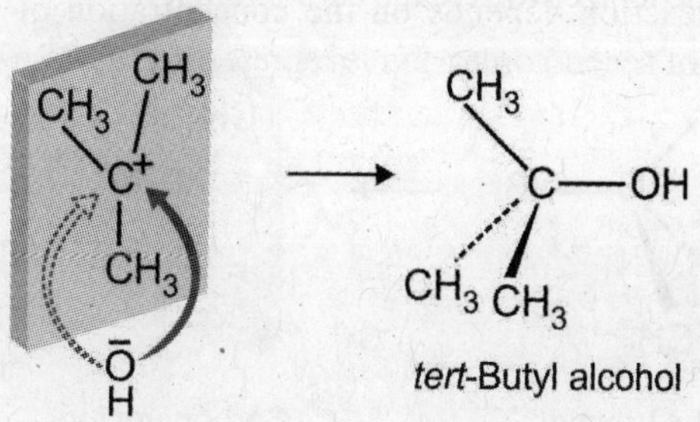

tert-Butyl alcohol

Remember that the primary alkyl halides undergo hydrolysis by S_N2 mechanism. The tertiary alkyl halides undergo hydrolysis by S_N1 mechanism. This is because the attack of the hydroxide ion on the crowded tertiary alkyl halides is quite difficult.

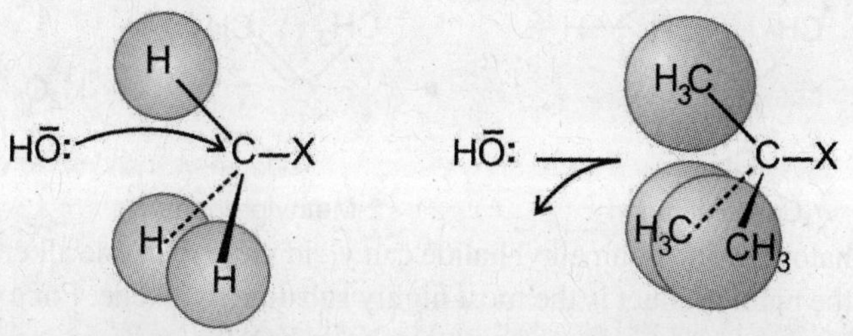

Easy attack (Primary halide)　　　Difficult attack (Tertiary halide)

Secondary alkyl halides may undergo hydrolysis by either S_N1 or S_N2 mechanism.

MECHANISMS OF ELIMINATION REACTIONS

Elimination reactions are reverse of addition reactions. Here two or four atoms or groups attached to the adjacent carbon atoms in the substrate molecule are eliminated to form a multiple bond. The dehydrohalogenation of alkyl halides with alcoholic alkalis is an example of elimination.

$$R-CH_2-CH_2-X \xrightarrow{\bar{O}H} R-CH=CH_2 + H_2O + \bar{X}$$

These reactions like substitution reactions are also divided into two classes :

(1) E2 Reactions

(2) E1 Reactions

E2 Reactions. E2 stands for bimolecular elimination. When the rate of an elimination reaction depends upon the concentration of a substrate and the nucleophile, the reaction is of second order and is represented as E2. E2 like S_N2 is also a one step process in which the abstraction of the proton from the β-carbon and the expulsion of the halide ion from the α-carbon occurs simultaneously. The mechanism of such a reaction is shown in Fig. 6.11.

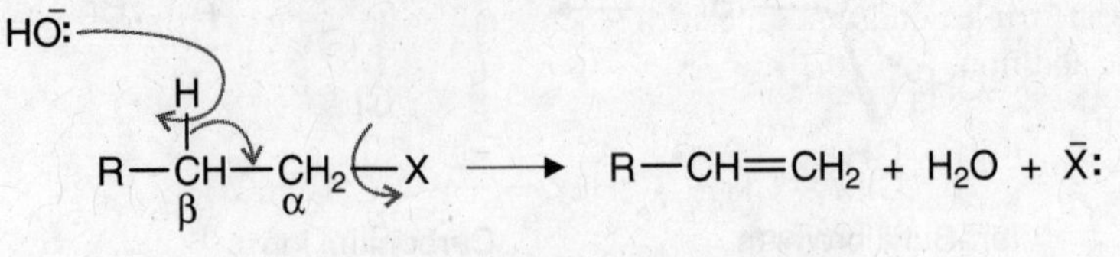

Fig. 6.11. Elimination by E2 mechanism.

E1 Reaction. E1 stands for unimolecular elimination. When the rate of an elimination reaction depends only on the concentration of the substrate, the reaction is of first order and is designated as E1. E1 like S_N1 reactions are also two step processes.

Step 1. The alkyl halide ionizes to give the carbonium ion.

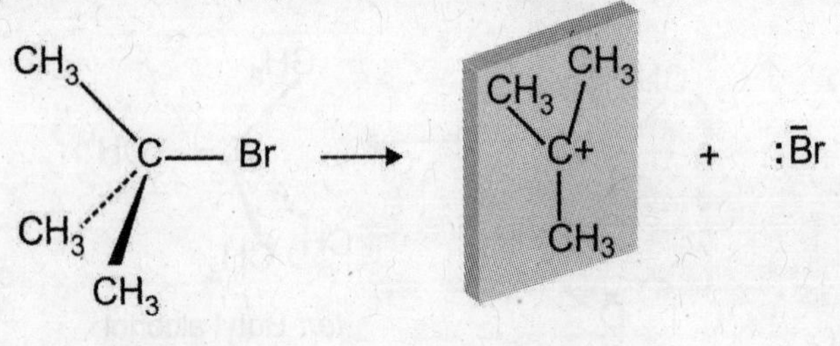

tert-Butyl bromide — Carbonium ion

Step 2. A proton is abstracted by the base from the adjacent β-carbon atom to give the alkene.

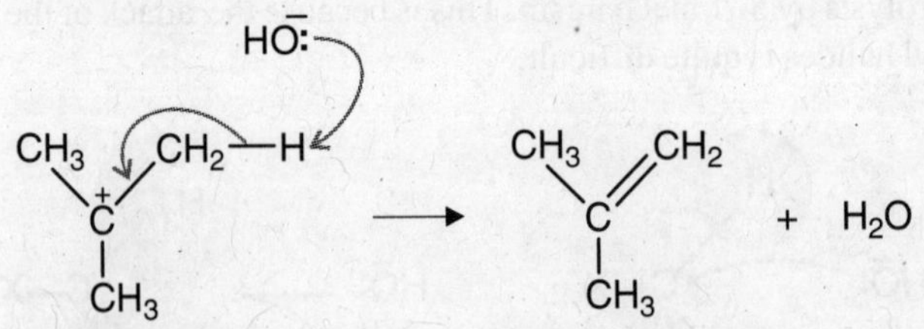

Carbonium ion — 2-Methylpropene

If the dehydrohalogenation of an alkyl halide can yield more than one alkene, then according to the **Saytzeff Rule**, the main product is the most highly substituted alkene. For example, two alkenes are possible when 2-bromobutane is heated with alcoholic KOH.

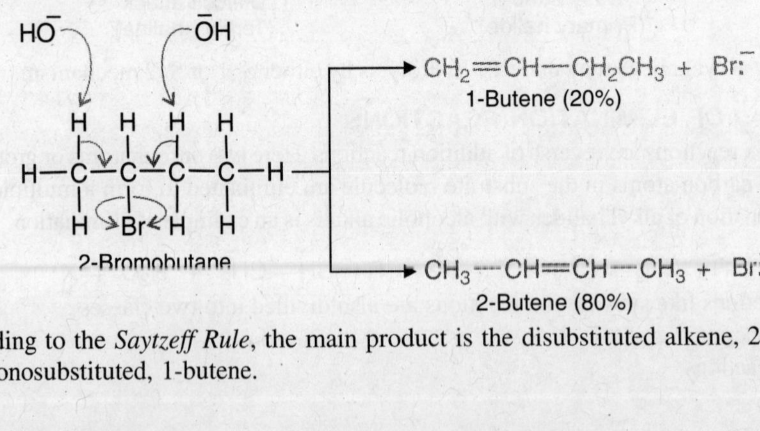

2-Bromobutane → 1-Butene (20%) + 2-Butene (80%)

According to the *Saytzeff Rule*, the main product is the disubstituted alkene, 2-butene, rather than the monosubstituted, 1-butene.

MECHANISMS OF ADDITION REACTIONS

Addition reactions are those in which atoms or groups of atoms are simply added to a double or triple bond without the elimination of any atom or other molecules. These reactions may be initiated by electrophiles, nucleophiles, or free radicals.

Electrophilic Addition Reactions. When an addition reaction involves the initial attack by an electrophile, the reaction is referred to as electrophilic addition. Compounds containing carbon-carbon double and triple bonds undergo such reactions. The addition of HBr to ethylene is an example of electrophilic addition.

$$\underset{\text{Ethylene}}{CH_2{=}CH_2} + HBr \longrightarrow \underset{\text{Ethyl bromide}}{\overset{H\quad\;\; Br}{CH_2{-}CH_2}}$$

The mechanism of the above reaction involves the following steps :

Step 1. Hydrogen bromide gives a proton (H^+) and bromide ion ($Br{:}^-$).

$$H{-}Br \longrightarrow \underset{\text{Electrophile}}{\overset{+}{H}} + \underset{\text{Nucleophile}}{Br\bar{:}}$$

Step 2. The proton (electrophile) attacks the π bond of ethylene to give a carbonium ion.

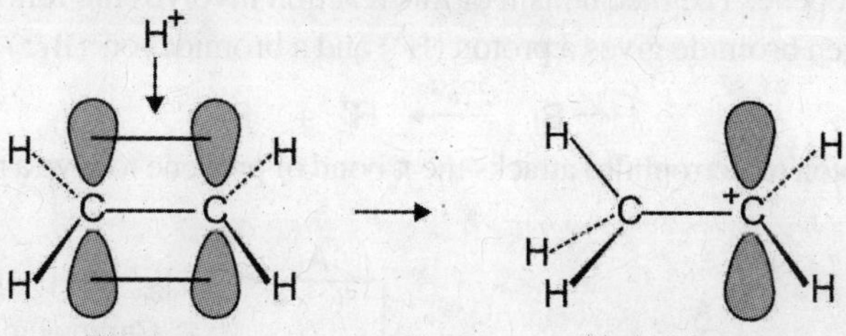

Ethylene   Carbonium ion

This step can also be written as :

$$\overset{+}{H} + CH_2{=}CH_2 \longrightarrow \underset{\text{Carbonium ion}}{CH_3{-}\overset{+}{C}H_2}$$

Step 3. The bromide ion (nucleophile) attacks the carbonium ion to give the addition product.

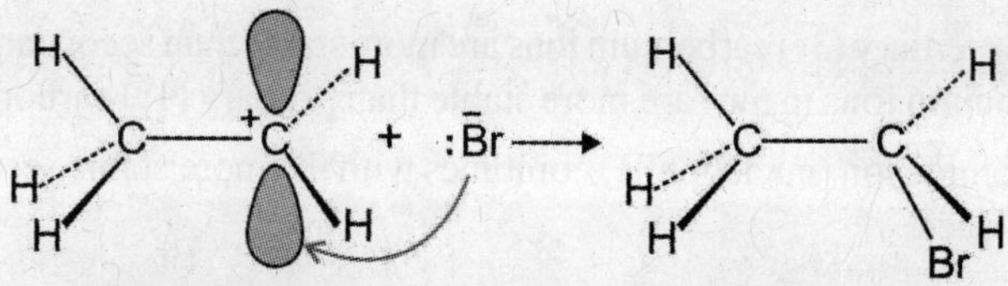

Carbonium ion   Ethyl bromide

This step can also be written as :

$$\underset{\text{Carbonium ion}}{CH_3{-}\overset{+}{C}H_2} + Br\bar{:} \longrightarrow \underset{\text{Ethyl bromide}}{CH_3{-}CH_2{-}Br}$$

Other reagents like HCl, HOCl, H_2SO_4, H_2O, Br_2, Cl_2, etc., add to alkenes similarly.

When an alkene is symmetrical about the double bond, as ethylene is, the product formed in addition reaction is the same no matter which way the reagent becomes attached to the alkene.

$$CH_2{=}CH_2 + HBr \longrightarrow CH_3{-}CH_2Br$$

$$CH_2{=}CH_2 + HBr \longrightarrow BrCH_2{-}CH_3$$

Identical

If, however, both the alkene and the adding reagent are unsymmetrical, two alternatives are possible :

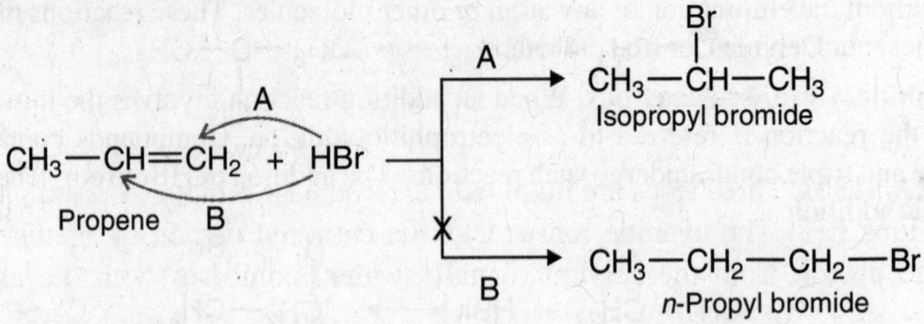

Experimentally it has been found that isopropyl bromide is the major product.

After studying a large number of such addition reactions, the Russian chemist Valdimir Markovnikov put forward the following rule : **When an unsymmetrical reagent adds to an unsymmetrical double bond, the positive part of the reagent becomes attached to the double-bonded carbon atom which bears the greatest number of hydrogen atoms.**

Markovnikov Rule can be rationalized in terms of modern mechanistic theory. Consider the addition of HBr to propene. The mechanism of this reaction involves the following steps :

Step 1. Hydrogen bromide gives a proton (H^+) and a bromide ion (Br^-)

$$H{-}Br \longrightarrow H^+ + Br^-$$

Step 2. The proton (electrophile) attacks the π bond of propene to give a more stable carbonium ion.

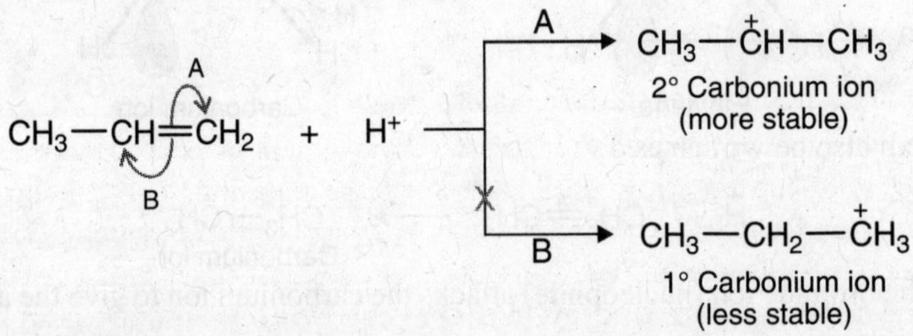

Remember : The tertiary (3°) carbonium ions are more stable than secondary (2°) carbonium ions. Secondary (2°) carbonium ions in turn are more stable than primary (1°) carbonium ions.

Step 3. The bromide ion (nucleophile) combines with the more stable secondary carbonium ion to give the major product.

$$\underset{\text{2° carbonium ion}}{CH_3{-}\overset{+}{C}H{-}CH_3} + Br^- \longrightarrow \underset{\text{Isopropyl bromide}}{CH_3{-}\underset{}{\overset{Br}{\overset{|}{C}H}}{-}CH_3}$$

Markovnikov rule may now be restated : **Addition of an unsymmetrical reagent to an unsymmetrical double bond proceeds in such a way as to involve the most stable carbonium ion.**

Nucleophilic Addition Reactions. When an addition reaction involves the initial attack by a nucleophile, the reaction is referred to as nucleophilic addition. Aldehydes and ketones which contain carbon-oxygen double bonds undergo such reactions.

The carbon-oxygen double bond (carbonyl group), as we have recently learned, is highly polar in character. This is because of higher electronegativity of oxygen as compared to carbon. The carbonyl group may be represented as shown below :

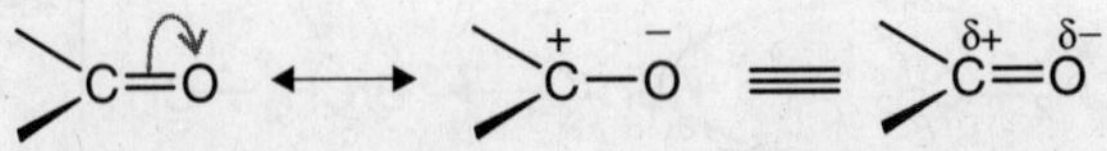

The addition of HCN to acetone is an example of nucleophilic addition.

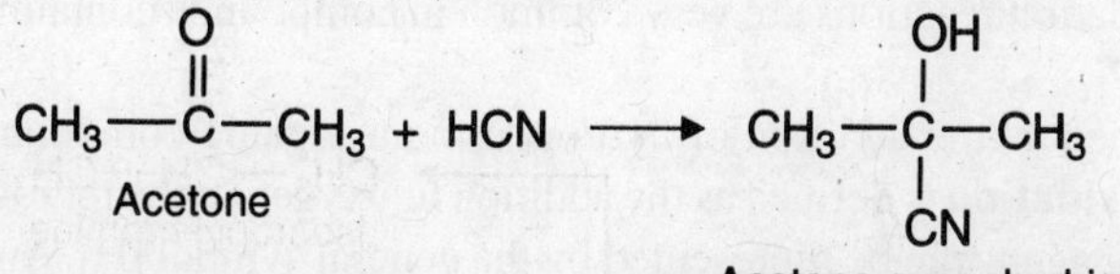

MECHANISM. Three steps are involved : (*1*) The base removes a proton from HCN to give cyanide ions ; (*2*) The cyanide ion attacks the carbonyl carbon of acetone to form an anion; (*3*) The proton from the solvent (usually water) combines with the anion to give cyanohydrin.

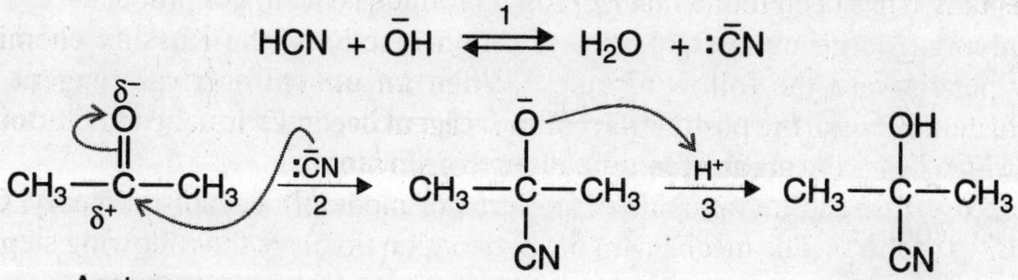

Points to Remember

Writing Organic Reaction Mechanisms

(1) A mechanism must explain the formation of product(s).

(2) A curved arrow (↷) is used to indicate the movement of electrons. This convention is often used carelessly, but the arrow does have a precise meaning. The tail shows where the electron is initially, and the head points to where it will be when the step is complete.

Warning : Many students tend to represent the movement of electrons by arrows that start nowhere and end nowhere. Many students start arrows where there are no electrons. The arrow must start at a point where there are electrons and go to a location to which the electrons will move.

(3) Carbonium ions are ions which contain a positively charged carbon atom. The order of carbonium ion stability is :

Tertiary > Secondary > Primary

(4) Carbanions are ions which contain a negatively charged carbon atom. The order of carbanion stability is:

Primary > Secondary > Tertiary

(5) Free radicals are electrically neutral. They have no charge. The order of free radical stability is:

Tertiary > Secondary > Primary

(6) Electrophiles are electron-poor. They attack negative centres in molecules.

(7) Nucleophiles are electron-rich. They attack positive centres in molecules.

(8) Carbonyl carbon atoms, C=O, are electropositive ; carbonyl oxygen atoms are electronegative.

(9) Never move atoms when changing from one resonance structure to another. *Only move electrons* !

(10) **Never mix ionic and free radical mechanisms.** If you wish to explain a particular reaction by ionic mechanism, do not bring in free radicals in any step.

REDUCTION AND OXIDATION REACTIONS

Reduction and oxidation reactions are very common in compounds containing double and triple bonds.

Reduction is defined as the addition of hydrogen to an organic compound or replacement of oxygen by hydrogen. **Oxidation** is defined as the addition of oxygen or the removal of hydrogen from a compound. A reducing agent can be represented by the general symbol [H]. An oxidizing agent can be represented by the symbol [O].

Reduction Reactions :

$$\underset{\text{Ethylene}}{CH_2{=}CH_2} \xrightarrow{[H]} \underset{\text{Ethane}}{CH_3-CH_3}$$

$$\underset{\text{Acetaldehyde}}{CH_3-\overset{\overset{O}{\|}}{C}-H} \xrightarrow{[H]} \underset{\text{Ethanol}}{CH_3-CH_2-OH}$$

Oxidation Reactions :

$$\underset{\text{Acetaldehyde}}{CH_3-\overset{\overset{O}{\|}}{C}-H} \xrightarrow{[O]} \underset{\text{Acetic acid}}{CH_3-\overset{\overset{O}{\|}}{C}-OH}$$

$$\underset{\text{Isopropyl alcohol}}{CH_3-\overset{\overset{OH}{|}}{CH}-CH_3} \xrightarrow{[O]} \underset{\text{Acetone}}{CH_3-\overset{\overset{O}{\|}}{C}-CH_3}$$

The result of reduction or oxidation depends upon the nature of the original compound and that of the reagent used. While reduction seldom causes a rupture of the carbon chain, oxidation does it very often.

SOME TYPICAL REDUCTION REACTIONS

The following reduction reactions are frequently used in Organic Chemistry :

(1) **Catalytic Hydrogenation.** This involves the use of hydrogen gas in the presence of a catalyst (Ni, Pt or Pd). Catalytic hydrogenation is commonly used for alkenes and aromatic compounds. Aldehydes, ketones and nitriles may also be reduced by this method.

$$\underset{\text{Ethylene}}{CH_2{=}CH_2} + H_2 \xrightarrow[\Delta]{Ni} \underset{\text{Ethane}}{CH_3-CH_3}$$

$$\underset{\text{Acetylene}}{HC{\equiv}CH} + 2H_2 \xrightarrow[\Delta]{Ni} \underset{\text{Ethane}}{CH_3-CH_3}$$

$$\underset{\text{Acetaldehyde}}{CH_3-\overset{\overset{O}{\|}}{C}-H} + H_2 \xrightarrow[\Delta]{Ni} \underset{\text{Ethanol}}{CH_3-\overset{\overset{OH}{|}}{CH_2}}$$

$$\underset{\text{Benzonitrile}}{C_6H_5-C{\equiv}N} + 2H_2 \xrightarrow[\Delta]{Ni} \underset{\text{Benzylamine}}{C_6H_5-CH_2-NH_2}$$

In each of the above reactions, two hydrogen atoms add across each *pi* bond. The sp^2 hybridized sigma bonds of atoms involved become sp^3 sigma bonds, whether the pi bond is a part of carbon-carbon double bond or a carbon-oxygen double bond (Fig. 6.12). In the case of triple bonds, *sp* bonds become sp^3 bonds. In each case an unsaturated compound becomes saturated.

(2) **Reduction with Lithium Aluminium Hydride.** Lithium aluminium hydride is a general reducing agent for aldehydes, ketones and nitriles. In this method the compound is first treated with lithium aluminium hydride and then with water.

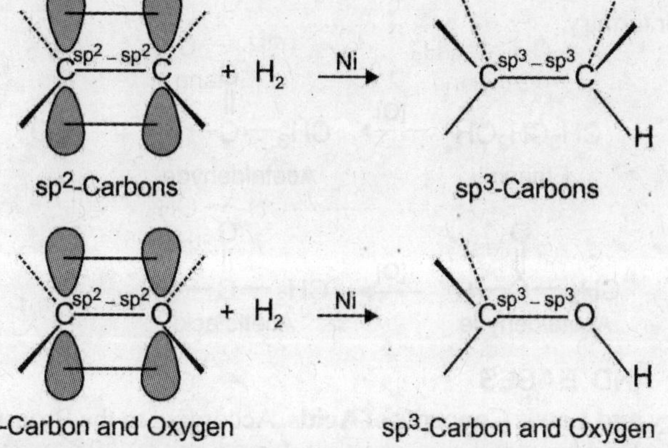

Fig. 6.12. Addition of hydrogen to carbon-carbon and carbon-oxygen double bonds.

$$\underset{\text{Acetone}}{CH_3-\overset{\overset{\large O}{\|}}{C}-CH_3} \xrightarrow[(2)\ H_2O]{(1)\ LiAlH_4} \underset{\text{Isopropyl alcohol}}{CH_3-\overset{\overset{\large OH}{|}}{C}H-CH_3}$$

$$\underset{\text{Ethyl cyanide}}{CH_3CH_2C\equiv N} \xrightarrow[(2)\ H_2O]{(1)\ LiAlH_4} \underset{\text{Propylamine}}{CH_3CH_2CH_2NH_2}$$

Sodium borohydride reacts with carbonyl compounds by a similar mechanism.

Lithium aluminium hydride also reduces carboxylic acids and esters.

(3) **Reduction with Metals.** Some metals have found use in the reduction of certain functional groups. For example, treatment of nitrobenzene with iron filings in hydrochloric acid yields aniline.

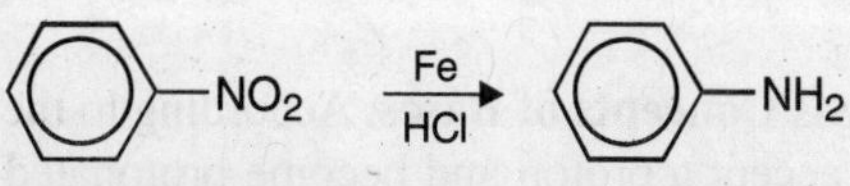

Nitrobenzene Aniline

Sodium metal in ethanol is used to reduce ketones and esters to alcohols.

SOME TYPICAL OXIDATION REACTIONS

The most complete oxidation of an organic compound is its conversion to carbon dioxide. This can be effected by burning the compound in air. However, controlled oxidations can be carried out. Some carbon-hydrogen bonds and carbon-carbon bonds are more susceptible to oxidation than others. A π bond, for example, is the point of attack in oxidation of alkenes.

(1) Alkenes are oxidized to glycols with cold dilute potassium permanganate.

$$\underset{\text{Ethylene}}{CH_2=CH_2} \xrightarrow[H_2O]{[O]} \underset{\text{Ethylene glycol}}{\underset{OH\quad\ \ OH}{\underset{|\qquad\quad |}{CH_2-CH_2}}}$$

(2) Treatment of alkenes with hot alkaline potassium permanganate results in the cleavage of the molecule at the π bond.

$$\underset{\text{2-Butene}}{CH_3-CH=CH-CH_3} \xrightarrow{[O]} \underset{\text{Acetic acid}}{2CH_3-\overset{\overset{\large O}{\|}}{C}-OH}$$

(3) Alcohols and aldehydes when oxidized with a mixture of potassium dichromate and sulfuric acid, ultimately yield carboxylic acids. The nature of the products depends on whether the alcohol is primary, secondary or tertiary.

$$\underset{\text{Ethanol}}{CH_3CH_2OH} \xrightarrow{[O]} \underset{\text{Acetaldehyde}}{CH_3-\overset{\overset{\large O}{\|}}{C}-H} + H_2O$$

$$\underset{\text{Acetaldehyde}}{CH_3-\overset{\overset{\large O}{\|}}{C}-H} \xrightarrow{[O]} \underset{\text{Acetic acid}}{CH_3-\overset{\overset{\large O}{\|}}{C}-OH}$$

ORGANIC ACIDS AND BASES

Bronsted-Lowry and Lewis Concepts of Acids. According to the Bronsted-Lowry concept, an acid is any substance that can donate a proton and dissociate in aqueous solution.

$$\underset{\text{Acid}}{HX} + H_2O \rightleftharpoons H_3\overset{+}{O} + X\bar{:}$$

For strong acids, such as HCl or HNO_3, the equilibrium lies far to the right. For weak acids, the equilibrium lies far to the left. Most of the organic acids are weak acids. Their strength is measured by the dissociation constant K_a.

$$K_a = \frac{\left[H_3O^+\right]\left[X^-\right]}{\left[HX\right]} \qquad [\ ]\text{ indicates molar concentration}$$

The stronger the acid, the larger is K_a. The K_a values are usually very small, and are often quoted in logarithmic form as $pK_a = -\log K_a$. Most pK_a values fall within the range 1–14, and the stronger the acid, the smaller is its pK_a.

According to the Lewis concept, an acid is any substance that can accept a pair of electrons. Example: $ZnCl_2$, $AlCl_3$ and $FeCl_3$.

Bronsted-Lowry and Lewis Concepts of Bases. According to the Bronsted-Lowry concept, a base is any substance that can accept a proton and become protonated in aqueous solution.

$$\underset{\text{Base}}{B\bar{:}} + H_2O \rightleftharpoons BH^+ + OH^-$$

For strong bases, such as KOH or NaOH, the equilibrium lies far to the right. For weak bases, the equilibrium lies far to the left. Most of the organic bases are weak bases. Their relative strength is measured by its K_b.

$$K_b = \frac{\left[BH^+\right]\left[OH^-\right]}{\left[B:^-\right]} \qquad [\ ]\text{ indicates molar concentration}$$

The stronger the base, the larger is K_b. The K_b values are usually very small, and are often quoted in logarithmic form as $pK_b = -\log K_b$. The stronger the base, smaller is its pK_b value.

According to the Lewis concept, a base is any substance that can donate a pair of electrons. Example: NH_3, CH_3NH_2, and pyridine.

CONJUGATE ACIDS AND BASES

Bronsted defined acids and bases in terms of a hydrogen ion (proton, H^+) being lost or gained. An acid is a proton donor and a base is a proton acceptor. The concept of conjugate pairs is a helpful tool in describing acid-base reactions. The species left behind after a Bronsted acid donates a proton is the conjugate base of that acid. Likewise, the base, once it has gained a proton in the acid-base exchange, becomes the conjugate acid. All these terms are illustrated in the generalized reaction below. Notice that the species being transferred is H^+ (a hydrogen cation and not a neutral H or an anion H^-). The equations must be balanced electrically and atomically.

Acid		Base		Conjugate acid		Conjugate Base
HX	+	Y	⟶	HY^+	+	X^-
HCl	+	H_2O	⟶	H_3O^+	+	Cl^-

Examination Questions

1. Define or explain the following terms or concepts :
(*a*) Homolytic bond fission (*b*) Heterolytic bond fission
(*c*) Inductive effect (*d*) Mesomeric effect
(*e*) Carbonium ions (*f*) Carbanions
(*g*) Free radicals (*h*) Carbenes
(*i*) Exothermic reaction (*j*) Endothermic reaction
(*k*) Electrophilic substitution (*l*) Nucleophilic substitution
(*m*) Electrophilic addition (*n*) Nucleophilic addition

2. Describe homolytic and heterolytic fission of a covalent bond. How carbonium ions, carbanions and free radicals are obtained ?

3. Explain the following terms with suitable examples :
(*a*) Carbonium ions and Carbanions
(*b*) Electrophiles and Nucleophiles
(*c*) Lewis acids and Lewis bases

4. What is Inductive effect ? Give one example of a system where this effect is operative. (*Kalyani BSc, 2015*)

5. What is Mesomeric effect ? Give one example of a system where this effect is operative. (*Bhopal BSc, 2014*)

6. What is the essential difference between Inductive effect and Mesomeric effect ? (*Jiwaji BSc, 2014*)

7. Write notes on :
(*a*) Inductive effect (*Tamil Nadu BSc, 2015*)
(*b*) Mesomeric effect (*Kerala BSc, 2014*)
(*c*) Hyperconjugation
(*d*) Carbonium ions (*Madras BSc, 2014*)
(*e*) Carbanions (*Delhi BSc, 2015*)
(*f*) Free radicals (*Nagpur BSc, 2014*)
(*g*) Carbenes

8. Write notes on :
(*a*) Stability of Carbonium ions (*Baroda BSc, 2015*)
(*b*) Stability of Carbanions (*Osmania BSc, 2015*)

9. Explain the relative stability of primary, secondary, and tertiary carbonium ions. (*Anna BSc, 2014*)

10. Explain the relative stability of primary, secondary, and tertiary carbanions. (*Bundelkhand BSc, 2014*)

11. Explain why benzyl carbonium ion is more stable than ethyl carbonium ion.

(Panjab BSc, 2015)

12. Explain why benzyl carbanion is more stable than ethyl carbanion.

13. What are carbonium ions ? Arrange the following according to their increasing stability. Explain your answer.

(*a*) $CH_3CH_2CH_2\overset{+}{C}H_2$ (*b*) $(CH_3)_3\overset{+}{C}$ (*c*) $CH_3CH_2\overset{+}{C}HCH_3$

Answer. (*b*) > (*c*) > (*a*)

14. Select the most stable carbonium ion from the following

(*a*) $\overset{+}{C}H_3$ (*b*) $(CH_3)_2\overset{+}{C}H$ (*c*) $CH_3\overset{+}{C}H_2$ (*d*) $(CH_3)_3\overset{+}{C}$

Answer. (*d*) is most stable.

15. What are Electrophiles ? Give two examples. *(Magadh BSc, 2014)*

16. What are Nucleophiles ? Give two examples. *(Sambalpur BSc, 2015)*

17. Which of the following molecules and ions are electrophilic ?

(A)	(B)	(C)	(D)	(E)	(F)	(G)	(H)
CH_4	Br^+	H_2O	$H_2C{=}CH_2$	BF_3	NH_3^+	NO_2^+	Br^-

Answer. The electrophiles are : (B), (E), and (G).

18. Which of the following molecules and ions are nucleophilic ?

(A)	(B)	(C)	(D)	(E)	(F)	(G)	(H)
CH_4	H_2O	BH_3	$H_2C{=}CH_2$	$O\overline{H}$	Br^+	NH_3	Br^-

19. Classify the following into electrophilic and nucleophilic reagents :

(*a*) H^+ (*b*) Cl^- (*c*) NO_2^+ (*d*) HO^-

Answer. (*a*) and (*c*) are electrophiles

(*b*) and (*d*) are nucleophiles

20. Classify the following as electrophiles and nucleophiles :

(*a*) CN^x (*b*) H_2O (*c*) Br^+

(*d*) NH_3 (*e*) NO_2^+ (*f*) ROH

Answer. (*c*) and (*e*) are electrophiles

(*a*), (*b*), (*d*), and (*f*) are nucleophiles

21. Define electrophiles and nucleophiles, and classify the following accordingly :

(*a*) $CH_3CH_3\overset{+}{C}H_2$ (*b*) $CH_3\overset{-}{C}HCH_3$ (*c*) $N\overset{+}{O_2}$ (*d*) NH_3

Answer. (*a*) and (*c*) are electrophiles

(*b*) and (*d*) are nucleophiles

22. What are different types of organic reactions ? Explain them with suitable examples.

23. What is the essential difference between a free radical reaction and an ionic reaction ?

24. Classify the following reagents as either Lewis acids or Lewis bases :

(*a*) $AlCl_3$ (*b*) $(CH_3)_3N$:

(*c*) HF (*d*) $CH_3-\ddot{\underset{..}{S}}-CH_3$

Answer. (*a*) and (*c*) are Lewis acids.

(*b*) and (*d*) are Lewis bases.

25. Which of the following compounds are Lewis acids ?

(*a*) $AlCl_3$ (*b*) CH_3OH (*c*) $FeBr_3$ (*d*) SO_3

Answer. Compounds (*a*), (*c*), and (*d*) are Lewis acids.

Multiple Choice Questions

1. An inductive effect refers to the movement of electrons through

(*a*) π bonds (*b*) *p* orbitals

(*c*) nonpolar bonds (*d*) sigma bonds

Answer. (*d*)

2. Electrophiles are

(*a*) electron rich (*b*) electron poor

(*c*) electrophobic (*d*) carbanions

Answer. (*b*)

3. Which of the following species are electrophiles ?

$$\underset{\text{I}}{CH_3—\overset{+}{C}H_2—CH_3} \qquad \underset{\text{II}}{H\ddot{\underset{\cdot\cdot}{O}}^{-}} \qquad \underset{\text{III}}{\ddot{N}H_3} \qquad \underset{\text{IV}}{H^+}$$

(*a*) II, III (*b*) II, IV (*c*) I, IV (*d*) I, III

Answer. (*c*)

4. Which of the following is *not* an electrophile ?

(*a*) BH_3 (*b*) H^+ (*c*) $AlCl_3$ (*d*) $H_2C=CH_2$

Answer. (*d*)

5. Which of the following is *not* an electrophile ?

(*a*) $CH_3CH_2OCH_2CH_3$ (*b*) BF_3

(*c*) $(CH_3)_3C^+$ (*d*) HOCl

Answer. (*a*)

6. Nucleophiles are

(*a*) electron rich (*b*) electron poor

(*c*) nucleophobic (*d*) carbocations

Answer. (*a*)

7. In order for a reagent to behave as a nucleophile it must have

(*a*) an overall positive charge. (*b*) an overall negative charge.

(*c*) a non-bonding electron pair. (*d*) a nitrogen or sulfur atom.

Answer. (*c*)

8. Nucleophiles have

(*a*) only bonding electron pairs (*b*) nonbonding electron pairs

(*c*) no nonbonding electron pairs (*d*) a positive charge

Answer. (*b*)

9. Which reagent is a good nucleophile ?

(*a*) NH_3 (*b*) HBr (*c*) Br_2 (*d*) BH_3

Answer. (*a*)

10. Which of the following is *not* a nucleophile ?

(*a*) NH_3 (*b*) H_2O (*c*) H^+ (*d*) $H_2C=CH_2$

Answer. (*c*)

11. Which of the following is **not** a nucleophile ?

(*a*) H_2O (*b*) CH_3NH_2

(*c*) C_2H_5SH (*d*) C_6H_{12} (cyclohexane)

Answer. (*d*)

12. Which of the following is *not* a nucleophile ?
(*a*) Hydroxide, HO^- (*b*) Water, H_2O
(*c*) Methane, CH_4 (*d*) Chloride, Cl^-

Answer. (*c*)

13. Carbocations are
(*a*) neutral species (*b*) electrophilic
(*c*) nucleophilic (*d*) stable molecules

Answer. (*b*)

14. Carbocations are
(*a*) carbon atoms with a negative charge
(*b*) carbon atoms with a positive charge
(*c*) carbon atoms with no charge
(*d*) any atom with a positive charge

Answer. (*b*)

15. What is the hybridization of the C atom in the methyl cation ?
(*a*) C atom is sp^3 hybridized. (*b*) C atom is not hybridized.
(*c*) C atom is sp^2 hybridized. (*d*) C atom is *sp* hybridized.

Answer. (*c*)

16. A carbocation is an example of a
(*a*) transition state (*b*) reactive intermediate
(*c*) negatively charged species (*d*) neutral species

Answer. (*b*)

17. A tertiary carbocation (carbonium ion) is more stable than either a secondary or primary carbocation because
(*a*) it carries three positive charges
(*b*) it has a pyramidal configuration
(*c*) it has a trigonal planar configuration
(*d*) it possesses three electron-donating substituent groups

Answer. (*d*)

18. The order of stability of carbonium ions is
(*a*) tertiary > secondary > primary (*b*) secondary > tertiary > primary
(*c*) primary > secondary > tertiary (*d*) primary > tertiary > secondary

Answer. (*a*)

19. The order of stability of carbanions is
(*a*) primary > secondary > tertiary (*b*) secondary > tertiary > primary
(*c*) tertiary > secondary > primary (*d*) tertiary > primary > secondary

Answer. (*a*)

20. Carbocations undergo rearrangement reactions
(*a*) to give more stable carbocations (*b*) by hydride shifts
(*c*) by methyl shifts (*d*) for all of these reasons

Answer. (*d*)

21. Arrange the following carbocations in order of *increasing* stability (least stable).

$$(CH_3)_2CH\overset{+}{C}H_2 \quad (CH_3)_3\overset{+}{C} \quad CH_3CH_2\overset{+}{C}HCH_3$$

1 2 3

(*a*) 2 < 1 < 3 (*b*) 3 < 1 < 2 (*c*) 1 < 2 < 3 (*d*) 1 < 3 < 2

Answer. (*d*)

22. Free radicals

(*a*) contain no electrons (*b*) contain unpaired electrons

(*c*) contain all paired electrons (*d*) none of these

Answer. (*b*)

23. In what kind of atomic orbital is the unpaired electron of the methyl radical?

(*a*) a *p* orbital (*b*) an sp^2 orbital

(*c*) an *sp* orbital (*d*) an sp^3 orbital

Answer. (*a*)

24. Homolytic cleavage means a bond

(*a*) breaks to form two radicals (*b*) breaks to form ionic species

(*c*) is formed from two radicals (*d*) is formed from two ions

Answer. (*a*)

25. Homolytic bond cleavage gives

(*a*) cations (*b*) anions

(*c*) radicals (*d*) cations and anions

Answer. (*c*)

26. Kinetics is the study of

(*a*) how fast reactions occur

(*b*) energy and its transformation in equilibration reactions

(*c*) the interaction of molecules with light

(*d*) none of these

Answer. (*a*)

27. A concerted reaction is one where ?

(*a*) Multiple reagents react in a series of steps.

(*b*) The reaction occurs in a single step.

(*c*) There is a large activation energy barrier to the reaction.

(*d*) There is no activation energy barrier to the reaction.

Answer. (*b*)

28. What is the classification of the following reaction ?

$$CH_2{=}CHCH_2CH_2Br \; + \; CH_3ONa \longrightarrow CH_2{=}CHCH_2CH_2OCH_3$$

(*a*) Elimination (*b*) Substitution

(*c*) Rearrangement (*d*) Addition

Answer. (*b*)

29. What do curved arrows show in a reaction mechanism ?

(*a*) The movement of electrons.

(*b*) The movement of protons.

(*c*) The collisions of reacting molecules.

(*d*) The motion of starting materials and reagents.

Answer. (*a*)

30. The curved arrow shows electron movement

(*a*) toward a nucleophile (*b*) toward an electrophile

(*c*) away from an electrophile (*d*) in either direction

Answer. (*b*)

31. A curved arrow shows

(*a*) an equilibrium (*b*) a resonance structure

(*c*) electron pair movement (*d*) none of these

Answer. (*c*)

32. Which statement about curved arrows is *not* correct ?

(*a*) Curved arrows always start at an atom.

(*b*) Curved arrows are drawn to indicate the movements of electrons.

(*c*) The arrow is drawn in the direction of the electron flow.

(*d*) Arrows are drawn away from negative charges and/or toward positive charges.

Answer. (*a*)

33. An acid is a material that

(*a*) accepts a proton (*b*) donates a proton

(*c*) donates an electron (*d*) accepts a neutron

Answer. (*b*)

34. Brønsted-Lowry acids are defined as

(*a*) proton donors (*b*) proton acceptors

(*c*) electron pair donors (*d*) electron pair acceptors

Answer. (*a*)

35. Lewis acids

(*a*) donate electrons (*b*) accept electrons

(*c*) are electron rich (*d*) are proton donors

Answer. (*b*)

36. A conjugate base is

(*a*) a species that forms when a conjugate acid donates a proton to a base

(*b*) a species that forms when an acid donates a proton to a base

(*c*) a complex between two bases

(*d*) a base with its valence electrons removed

Answer. (*b*)

37. A conjugate base results from

(*a*) an acid losing a proton (*b*) a base accepting a proton

(*c*) an acid gaining a proton (*d*) a base accepting an electron pair

Answer. (*a*)

38. Acids with large K_a values have

(*a*) large pK_a values (*b*) pK_a values equal to their K_a values

(*c*) small pK_a values (*d*) no relationship to a pK_a value

Answer. (*c*)

7

Isomerism

In the study of organic chemistry we come across many cases when two or more compounds consist of equal number of like atoms. These compounds have the same molecular formula but differ from each other in physical or chemical properties, and are called **Isomers** and the phenomenon is called **Isomerism**. Since isomers have the same molecular formula, the difference in their properties must be due to different modes of combination or arrangement of atoms within the molecule. There are two main types of isomerism :

(1) Constitutional Isomerism

(2) Stereoisomerism

Constitutional Isomerism. When the isomerism is due to difference in the arrangement of atoms within the molecule, without any reference to space, the phenomenon is called *Constitutional Isomerism*. In other words, **Constitutional isomers are compounds that have the same molecular formula but different structural formulas.** Constitutional isomerism is of five types :

(*a*) Chain Isomerism

(*b*) Position Isomerism

(*c*) Functional Isomerism

(*d*) Metamerism

(*e*) Tautomerism

Stereoisomerism. When isomerism is caused by the different arrangements of atoms or groups in space, the phenomenon is called *Stereoisomerism*. **The stereoisomers have the same structural formulas but differ in arrangement of atoms in space.** In other words, stereoisomerism is exhibited by such compounds which have the same structural formula but differ in *configuration*. (The term configuration refers to the three-dimensional arrangement of atoms that characterizes a particular compound). Stereoisomerism is of two types :

(*a*) Geometrical or Cis-Trans Isomerism

(*b*) Optical Isomerism

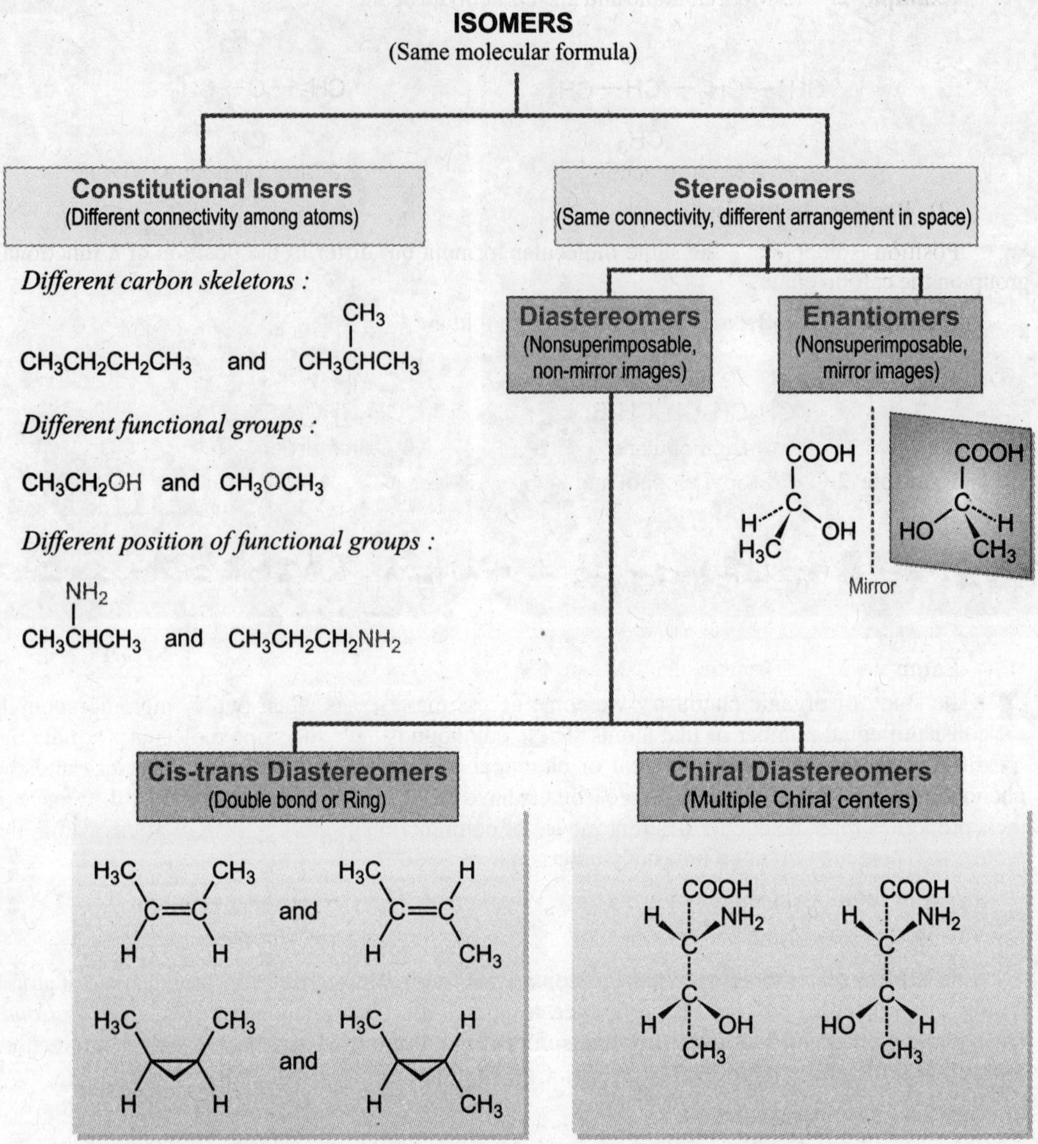

CONSTITUTIONAL ISOMERISM (STRUCTURAL ISOMERISM)

In constitutional isomerism the isomers have the same molecular formula but differ in constitutional formula, that is, in the order in which the different atoms are linked in the molecule. Constitutional isomerism is of five types :

(1) **Chain Isomerism**

Chain isomers (also called skeletal isomers) have the same molecular formula but differ in the order in which the carbon atoms are bonded to each other.

Example 1. *n*-Butane and Isobutane

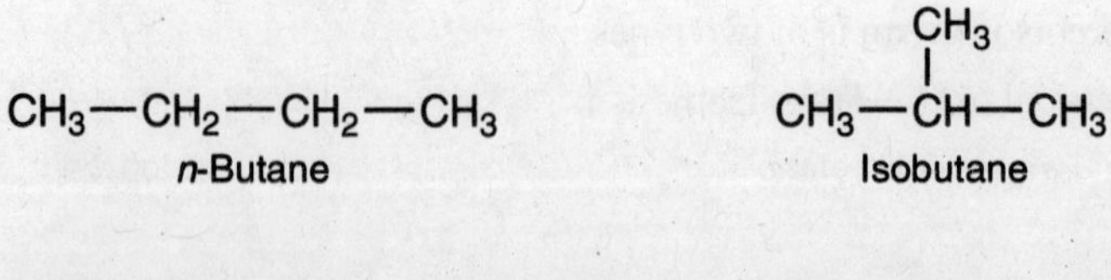

Example 2. 2-Methylbutane and 2,2-Dimethylpropane

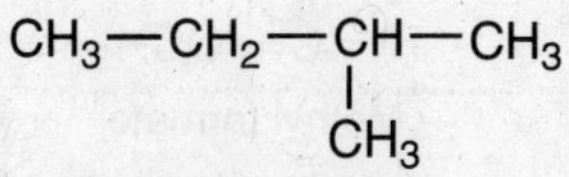

2-Methylbutane

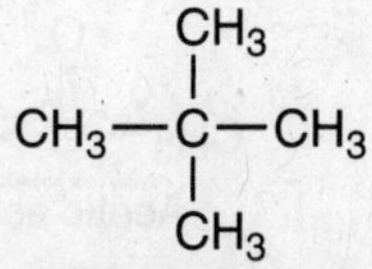

2,2-Dimethylpropane

(2) **Position Isomerism**

Position isomers have the same molecular formula but differ in the position of a functional group on the carbon chain.

Example 1. 1-Bromobutane and 2-Bromobutane

$CH_3CH_2CH_2CH_2Br$

1-Bromobutane

2-Bromobutane

Example 2. *n*-Propyl alcohol and Isopropyl alcohol

$CH_3—CH_2—CH_2OH$

n-Propyl alcohol

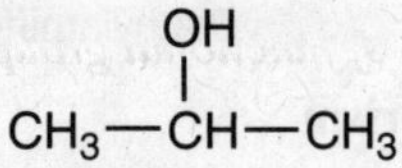

Isopropyl alcohol

Example 3. 1-Butene and 2-Butene

$CH_3CH_2CH{=}CH_2$

1-Butene

$CH_3CH{=}CHCH_3$

2-Butene

Example 4. 2-Pentanone and 3-Pentanone

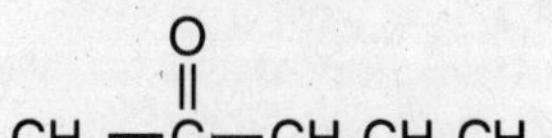

2-Pentanone

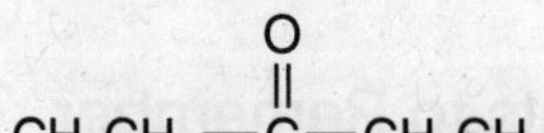

3-Pentanone

Example 5. *o*-Dichlorobenzene and *m*-Dichlorobenzene

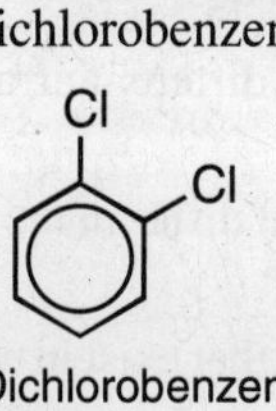

o-Dichlorobenzene

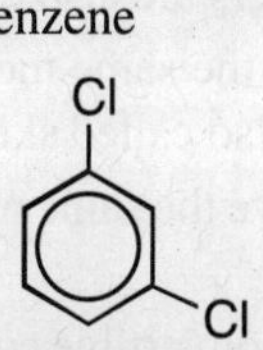

m-Dichlorobenzene

(3) **Functional Isomerism**

Functional isomers have the same molecular formula but different functional groups.

Example 1. Ethyl alcohol and Dimethyl ether

CH_3CH_2OH

Ethyl alcohol

$CH_3—O—CH_3$

Dimethyl ether

Example 2. Acetone and Propionaldehyde

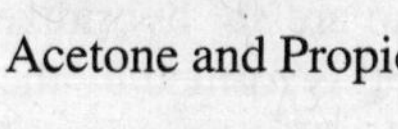

Acetone

Propionaldehyde

Example 3. Acetic acid and Methyl formate

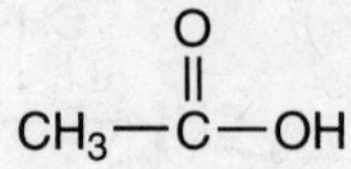

Acetic acid

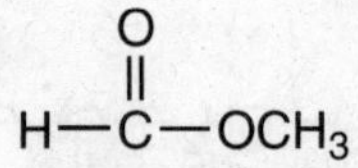

Methyl formate

(4) **Metamerism**

This type of isomerism is due to the unequal distribution of carbon atoms on either side of the functional group. Members belong to the same homologous series.

Example 1. Diethyl ether and Methyl propyl ether

$CH_3CH_2—O—CH_2CH_3$ — Diethyl ether

$CH_3—O—CH_2CH_2CH_3$ — Methyl propyl ether

Example 2. Diethylamine and Methylpropylamine

$CH_3CH_2—NH—CH_2CH_3$ — Diethylamine

$CH_3CH_2CH_2—NH—CH_3$ — Methylpropylamine

(5) **Tautomerism**

It is a special type of functional isomerism in which the isomers are in dynamic equilibrium with each other. For example, ethyl acetoacetate is an equilibrium mixture of the following two forms. At room temperature, the mixture contains 93% of keto-form plus 6% of the enol-form.

Keto-Form ⇌ Enol-Form

Points to Remember

1. Constitutional isomers are also called structural isomers.
2. Chain isomers have the same molecular formula but different arrangement of carbon atoms. Chain isomers are also called skeletal isomers.
3. Position isomers have the same molecular formula but differ in the position of a substituent or functional group.
4. Tautomers are interconvertible constitutional isomers that exist in equilibrium with each other.

GEOMETRICAL ISOMERISM

Geometrical isomerism (also called *cis-trans* isomerism) results from a restriction in rotation about double bonds, or about single bonds in cyclic compounds.

(1) **Geometrical Isomerism in Alkenes**

The carbon atoms of the carbon-carbon double bond are sp^2 hybridized. The carbon-carbon double bond consists of a σ bond and a π bond. The σ bond is formed by the overlap of sp^2 hybrid orbitals. The π bond is formed by the overlap of *p* orbitals. The presence of the π bond locks the molecule in one position. The *two* carbon atoms of the C=C bond and the *four* atoms that are attached to them lie in one plane and their positions in space are fixed. Rotation around the C=C bond is not possible because rotation would break the π bond (Fig. 7.1).

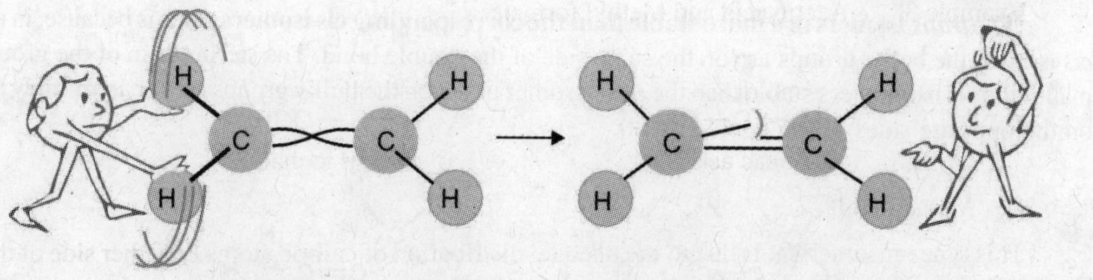

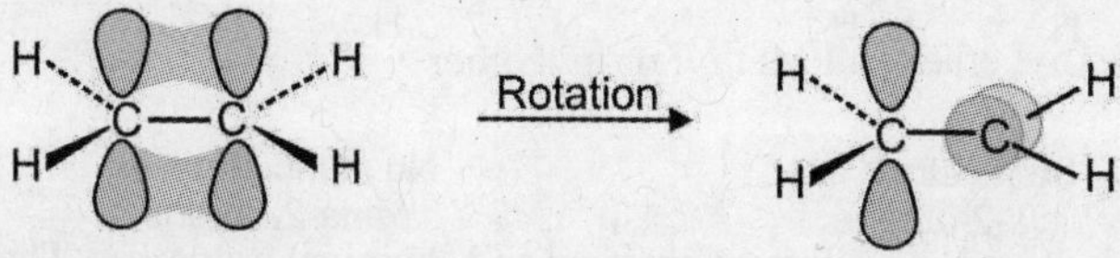

Fig.7.1 Rotation about π bond is not possible because it would break the π bond.

This restriction of rotation about the carbon-carbon double bond is responsible for the geometrical isomerism in alkenes. A popular analogy for this situation is based upon two boards and two nails. Driving one nail through two boards will not prevent free rotation of the two boards. But once a second nail is used, the boards cannot be freely rotated.

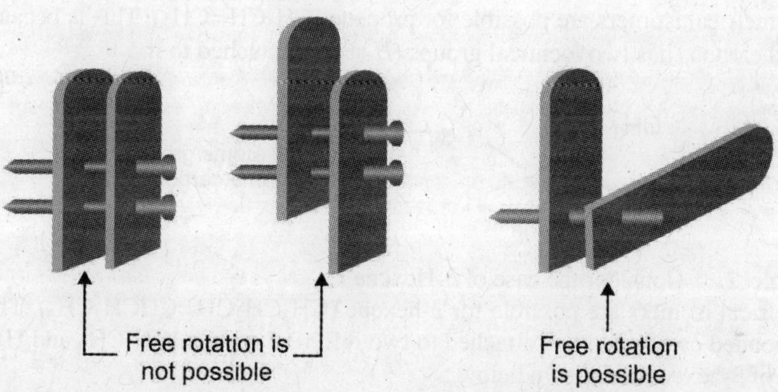

Consider the case of 2-butene. It exists in two spacial arrangements :

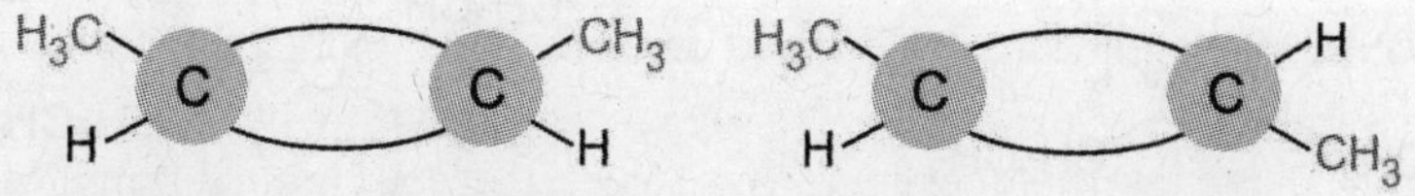

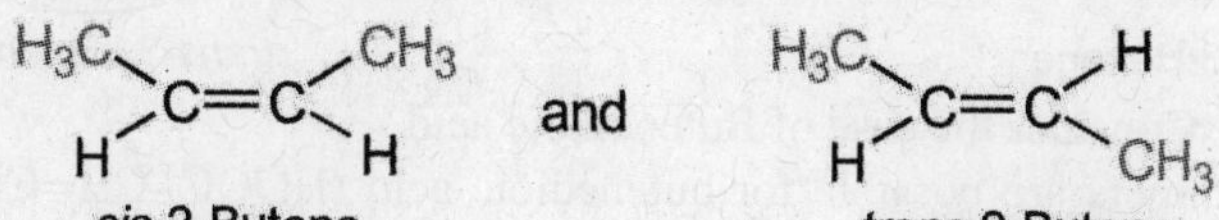

cis-2-Butene (bp 4°C) *trans*-2-Butene (bp 1°C)

These two compounds are referred to as geometrical isomers and are distinguished from each other by the terms *cis* and *trans*. **The *cis* isomer is one in which two similar groups are on the same side of the double bond. The *trans* isomer is that in which two similar groups are on the opposite sides of the double bond.** Consequently, this type of isomerism is often called *cis-trans* isomerism. Geometrical isomers are stereoisomers, because they have the same structural formula but different spacial arrangement of atoms.

The *trans* isomers are more stable than the corresponding cis isomers. This is because, in the *cis* isomer, the bulky groups are on the same side of the double bond. The steric strain of the groups makes the *cis* isomer less stable than the *trans* isomer in which the bulky groups are far apart (they are on the opposite sides of the double bond).

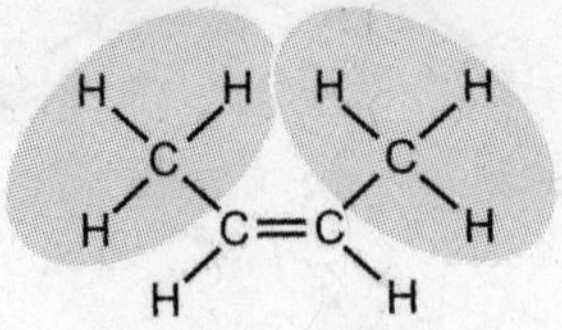

Steric strain in *cis*-2-butene

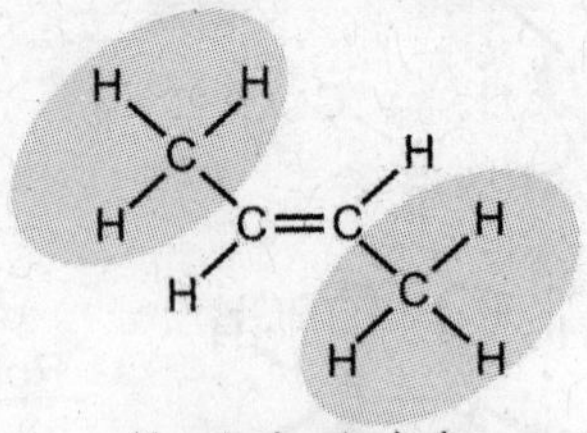

No steric strain in *trans*-2-butene

The geometrical isomers have different physical and chemical properties. They can be separated by conventional techniques like fractional distillation, gas chromatography etc.

All alkenes do not show geometrical isomerism. Geometrical isomerism is possible only when each double bonded carbon atom is attached to two different atoms or groups. The following examples illustrate this condition for the existence of geometrical isomers.

Example 1. Consider the case of Propene

No geometrical isomers are possible for propene ($CH_3CH{=}CH_2$). This is because one of the double bonded carbons has two identical groups (*H* atoms) attached to it.

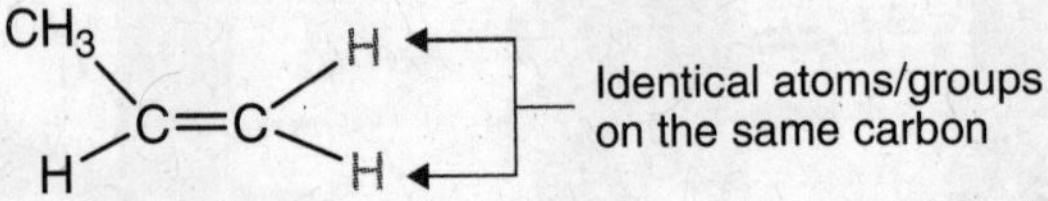

Propene

Example 2. Consider the case of 3-Hexene

Geometrical isomers are possible for 3-hexene ($CH_3CH_2CH{=}CHCH_2CH_3$). This is because each double bonded carbon atom is attached to two different groups (CH_3CH_2 and H). The *cis* and *trans* isomers of 3-hexene are shown below :

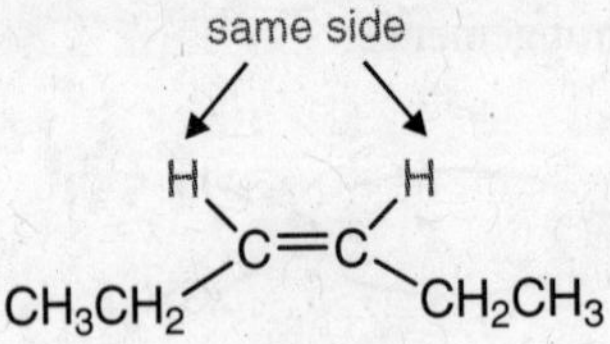

cis-3-Hexene

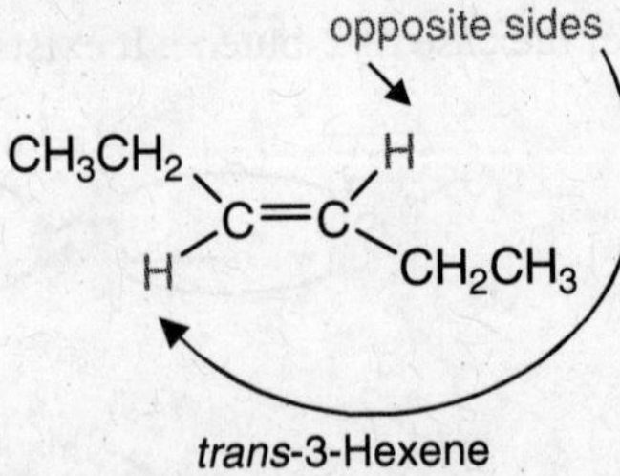

trans-3-Hexene

Example 3. Consider the case of Butenedioic acid

Geometrical isomers are possible for butenedioic acid (HOOC–CH=CH–COOH). This is because each double bonded carbon atoms has two different groups attached to it (H and COOH).

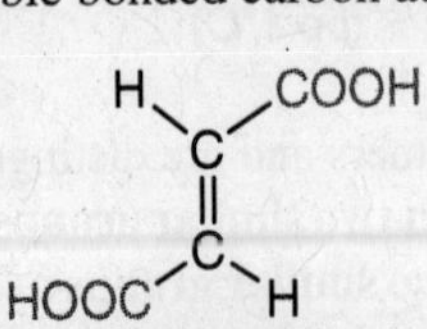

trans-Butenedioic acid
(Fumaric acid)
mp 130°C

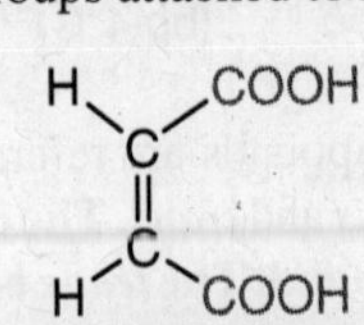

cis-Butenedioic acid
(Maleic acid)
mp 288°C

These isomers differ in physical properties such as melting point, dipole moment, and acidities. The *trans*-isomer has no dipole, since it is symmetrical, unlike the *cis*-isomer. The *cis*-isomer is stronger acid than the *trans*-isomer. Even chemical properties differ in the butenedioic acids. For example, the *cis*-acid undergoes dehydration at 100°C.

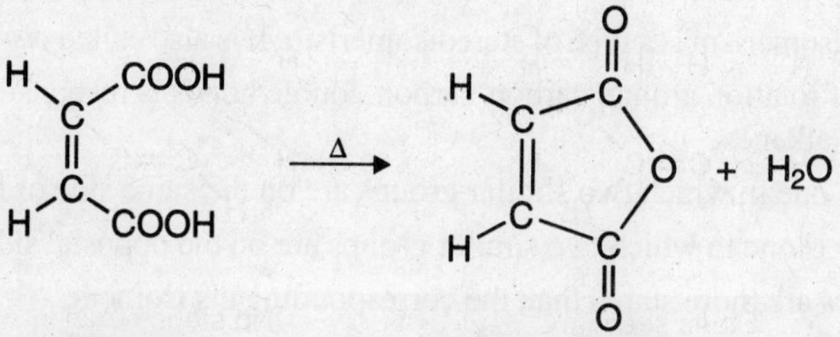

Maleic acid Maleic anhydride

However, the *trans*-acid does not form an anhydride until heated to 240°C. At that temperature there is sufficient thermal energy to break the double bond, so that the *cis*-acid and the anhydride are formed in succession.

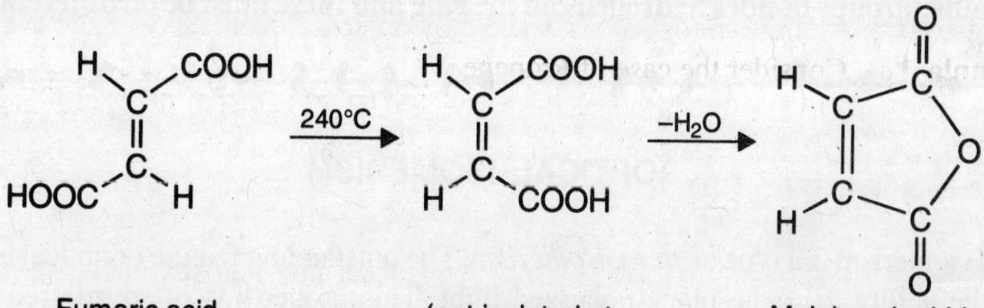

Fumaric acid (not isolated at this temperature) Maleic anhydride

(2) Geometrical Isomerism in Cyclic Compounds

Geometrical isomerism is also possible in cyclic compounds. There can be no rotation about carbon-carbon single bonds forming a ring because rotation would break the bonds and break the ring. For example, 1,2-dimethylcyclopropane exists in two isomeric forms.

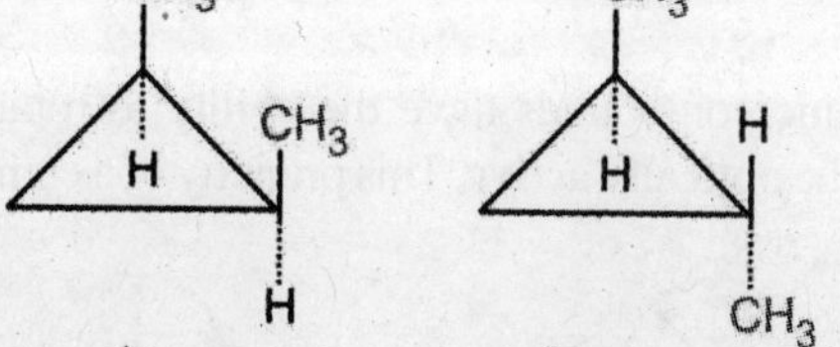

cis *trans*

In *cis*-1,2-dimethylcyclopropane, the two methyl groups are on the same side of the ring. In *trans*-1,2-dimethylcyclopropane, they are on opposite sides. **A requirement for geometrical isomerism in cyclic compounds is that there must be at least two other groups besides hydrogens on the ring and these must be on different ring carbon atoms.** For example, no geometrical isomers are possible for 1,1-dimethylcyclopropane.

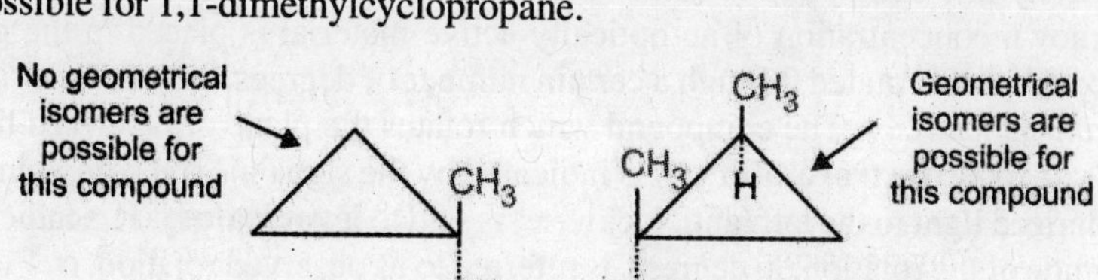

Points to Remember

1. Stereoisomers have the same structural formulas but differ in arrangement of atoms in space.
2. Geometrical isomerism is a type of stereoisomerism. It is also called *cis-trans* isomerism.
3. Restriction of rotation around carbon-carbon double bonds is responsible for the geometrical isomerism in alkenes.
4. *Cis* isomer is one in which two similar groups are on the same side of the double bond.
5. *Trans* isomer is one in which two similar groups are on the opposite sides of the double bond.
6. *Trans* isomers are more stable than the corresponding cis isomers.
7. **All alkenes do not show geometrical isomerism. Geometrical isomerism is possible only when each double bonded carbon atom is attached to two different atoms or groups.**
8. Geometrical isomerism is also possible in cyclic compounds.
9. **A requirement for geometrical isomerism in cyclic compounds is that there must be at least two other groups besides hydrogens on the ring and these must be on different ring carbon atoms.**

OPTICAL ISOMERISM

Optical isomerism is a type of *stereoisomerism*. The outstanding feature of optical isomers is that they have the ability to rotate plane-polarized light. This property is often referred to as optical activity and requires a brief discussion.

OPTICAL ACTIVITY - PLANE-POLARIZED LIGHT

Light from ordinary electric lamp is composed of waves vibrating in many different planes. When it is passed through Nicol prism (made of calcite, $CaCO_3$) or polaroid lense, light is found to vibrate in only one plane, and is said to be **plane-polarized** or simply polarized. The diagrams illustrate the vibrations of ordinary and polarized light from a beam propagated perpendicularly to the plane of paper.

Solutions of some organic compounds have the ability to rotate the plane of polarized light. These compounds are said to be **optically active**. This property of a compound is called **optical activity**.

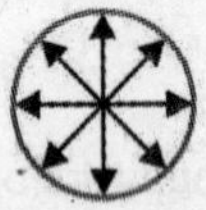

Ordinary Light

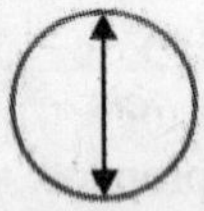

Plane-Polarized Light (Polarized Light)

Optical activity in a compound is detected and measured by means of a **polarimeter**. When a solution of a known concentration of an optically active material is placed in the polarimeter, the beam of polarized light is rotated through a certain number of degrees, either to the right (*clockwise*) or to the left (*anti-clockwise*). The compound which rotates the plane of polarized light to the right (clockwise) is said to be **dextrorotatory**. It is indicated by the sign (+). The compound which rotates the plane of polarized light to the left (anticlockwise) is said to **levorotatory**. It is indicated by the sign (–). The magnitude of the rotation, in degrees, is referred to as observed rotation, α. Fig. 7.2 shows the parts of a polarimeter.

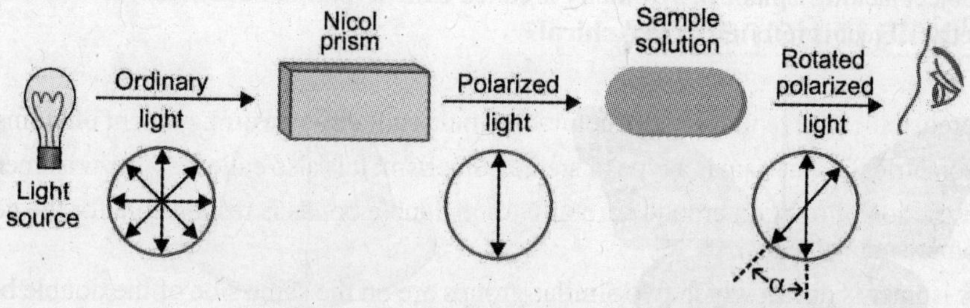

Fig.7.2. A simple polarimeter in operation.

SPECIFIC ROTATION

Optically active compounds rotate the plane of polarized light. The degree of rotation depends upon the number of molecules of the compound encountered by light along its path. It is, therefore, necessary to introduce some standard by which the rotating powers of different compounds may be compared.

Specific rotation is defined as the rotation produced by a solution of length 10 centimeters and unit concentration (1 g/ml) for the given wavelength of light at the given temperature. Consider a system in which the plane of polarized light is rotated through α degrees at a temperature of t°C and using the sodium D-line as the source. The specific rotation $[\alpha]_D^t$ is given by $[\alpha]_D^t = \frac{100\alpha}{lc}$ where α is the observed rotation ; *l* is the length of the tube in decimeters ; and *c* is the number of grams per 100 ml of solution. The specific rotation varies with both the wavelength of light and the temperature.

Problem. A 1.20 g sample of cocaine, $[\alpha]_D = -16°$, was dissolved in 7.50 ml of chloroform and placed in a sample tube having a path length of 5.0 cm. What was the observed rotation ? Is cocaine dextrorotatory or levorotatory ?

Solution. Use the following relationship :

$$[\alpha]_D^t = \frac{\alpha}{\ell \times c}$$

Where, α = observed rotation in degrees

ℓ = length of tube (dm) = path length

c = concentration of sample (g/ml)

The given values are :

$[\alpha]_D^t = -16°$

$\alpha = ?$

ℓ = 5.0 cm = 0.50 dm (1 dm = 10 cm)

c = 1.20 g/7.50 ml = 0.160 g/ml

$$\text{Thus,} -16° = \frac{\alpha}{0.50 \times 0.160}$$

$$\alpha = -16° \times 0.50 \times 0.160$$

$$= -1.3°$$

The observed rotation is –1.3°.

Cocaine is levorotatory.

PLANE OF SYMMETRY

A plane which divides an object into two symmetrical halves, is said to be *plane of symmetry*. For example, a person or a hat has a plane of symmetry (Fig. 7.3). A person's hand or gloves lack a plane of symmetry.

An object lacking a plane of symmetry is called **Chiral** (pronounced as *Ki-ral*) or **Dissymmetric**. A symmetric object is referred to as **Achiral.**

Fig. 7.3. Planes of symmetry.

A chiral object cannot be superimposed on its mirror image. A left hand, for example, does not possess a plane of symmetry, and its mirror image is not another left hand but a right hand (Fig. 7.4). The two are not identical, because they cannot be superimposed. If we were to lay one hand on top of the other, the fingers and the thumbs would clash.

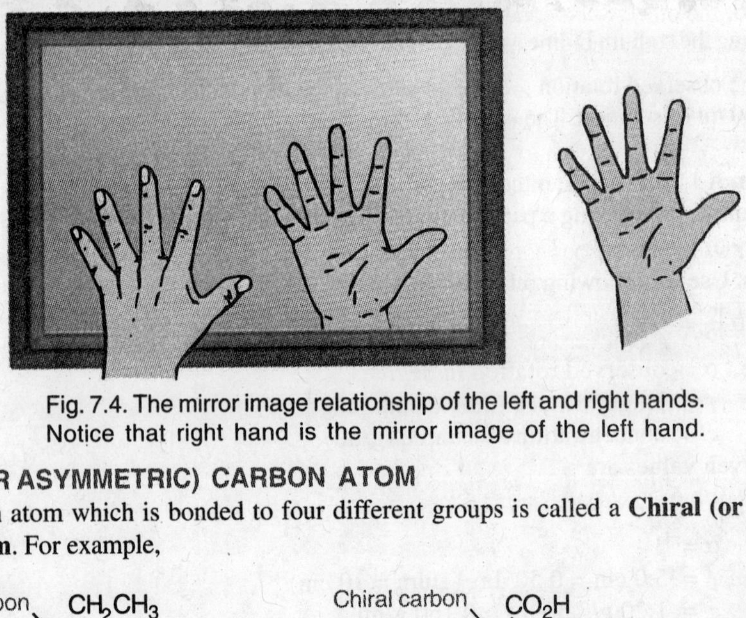

Fig. 7.4. The mirror image relationship of the left and right hands. Notice that right hand is the mirror image of the left hand.

CHIRAL (OR ASYMMETRIC) CARBON ATOM

A carbon atom which is bonded to four different groups is called a **Chiral (or Asymmetric) Carbon Atom**. For example,

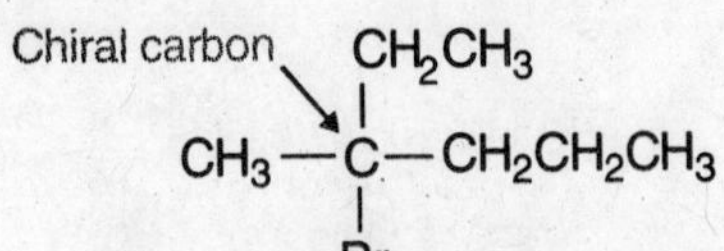

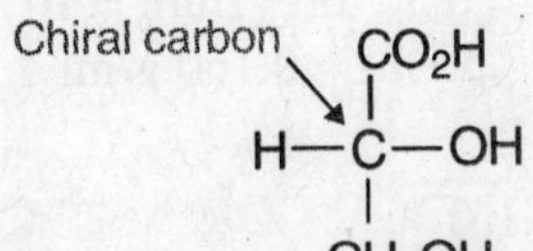

The term **chiral** (Greek **handedness**; pronounced **kairal**) carbon atom means that a carbon atom is bonded to four different groups and that a molecule of this type lacks a plane of symmetry. Such a molecule is also called **Asymmetric** or **Dissymmetric.**

OPTICAL ISOMERISM

A substance which has the ability to rotate the plane of polarized light is said to be *optically active* and exhibits *optical isomerism.* **Optical isomers are isomers that are identical in all properties except for their effect on plane of polarized light.**

The isomer which rotates the plane of polarized light to the right (clockwise direction) is known as **Dextrorotatory Isomer** or (+)-isomer. The isomer which rotates the plane of polarized light to the left (anticlockwise direction) is known as the **Levorotatory Isomer** or (–)-isomer. The optical rotatory powers of two isomers are equal in magnitude but opposite in sign. An equimolar mixture of the two isomers, therefore, will not rotate the plane of polarized light at all and is said to be **Racemic Mixture.**

Optical isomers have the same physical properties : melting point, boiling point, density, etc. They have the same specific rotation but with opposite signs.

OPTICAL ISOMERISM OF LACTIC ACID

Lactic acid (2-Hydroxypropanoic acid) is an example of a compound which shows optical isomerism. It contains one chiral carbon atom.

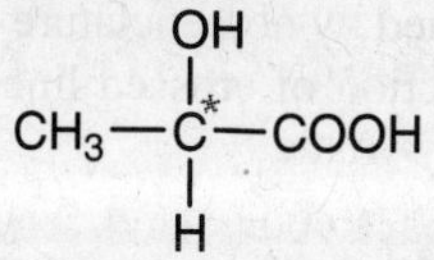

Lactic acid. The chiral carbon is shown by an asterisk.

Two three dimensional structures are possible for Lactic acid (Fig. 7.5).

These structures are not identical because they cannot be superimposed on each other. One is the mirror image of the other. Such nonsuperimposable mirror image forms are optical isomers and are called *enantiomers*. Thus, three forms of lactic acid are known. Two are optically active and the third is optically inactive.

1. (+)-**Lactic Acid**. It rotates the plane of polarized light to the right (clockwise direction) and is called dextrorotatory.

2. (–)-**Lactic Acid**. It rotates the plane of polarized light to the left (anticlockwise direction) and is called levorotatory. (–)-Lactic acid is the mirror image of (+)-lactic acid and *vice versa*.

3. (±)-**Lactic Acid**. It does not rotate the plane of polarized light. That is, it is optically inactive. It is an equimolar mixture of (+)-and (–)-forms (racemic mixture).

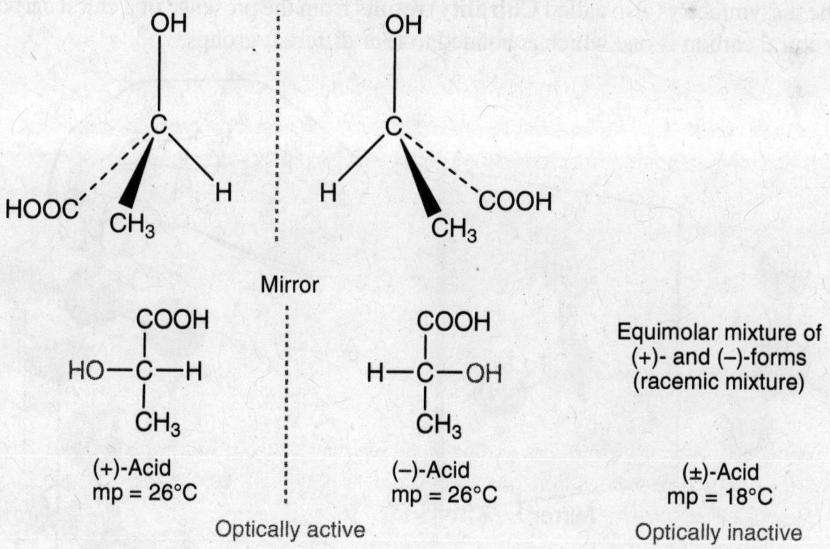

Fig. 7.5. Isomers of Lactic acid. In the upper line two three-dimensional structures are shown. In the lower line a commonly used Fischer projection is given. The vertical lines represent bonds going away from the observer/reader and horizontal lines represent bonds coming toward the observer.

FISCHER PROJECTIONS

Fischer projections provide an easy way to draw three-dimensional molecules on paper. All atoms are projected onto one plane. The Fischer projection of one of the enantiomers of lactic acid is shown below :

Fischer projection of (+)-Lactic acid.

The **Fischer rules** for showing the arrangement around a chiral carbon are as follows:

1. The carbon chain of the compound is arranged vertically, with the most oxidized carbon at the top. Or, put the carbon number *1* (as defined by nomenclature rules) at the top.
2. Represent the chiral carbon(s) at the intersection of crossed lines. Some teachers like to show the chiral carbon also.

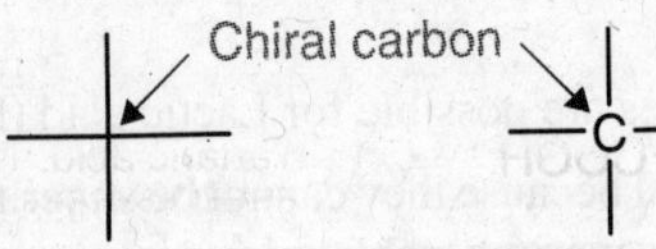

3. Groups attached to the vertical lines are understood to be going back behind the plane of the paper. That is, the vertical lines represent bonds going away from the observer. The chiral carbon is in the paper plane.
4. Groups attached to the horizontal lines are understood to be coming forward out of the paper plane. That is, the horizontal lines represent bonds coming toward the observer.

CONDITION FOR OPTICAL ISOMERISM

The necessary condition for a molecule to exhibit optical isomerism is that it should be *dissymmetric*. That is, the molecule should not be superimposable on its mirror image. In simple molecules, the dissymmetry (also called **Chirality**) results from the presence of a chiral carbon atom. Recall that a chiral carbon is one which is bonded to four different groups.

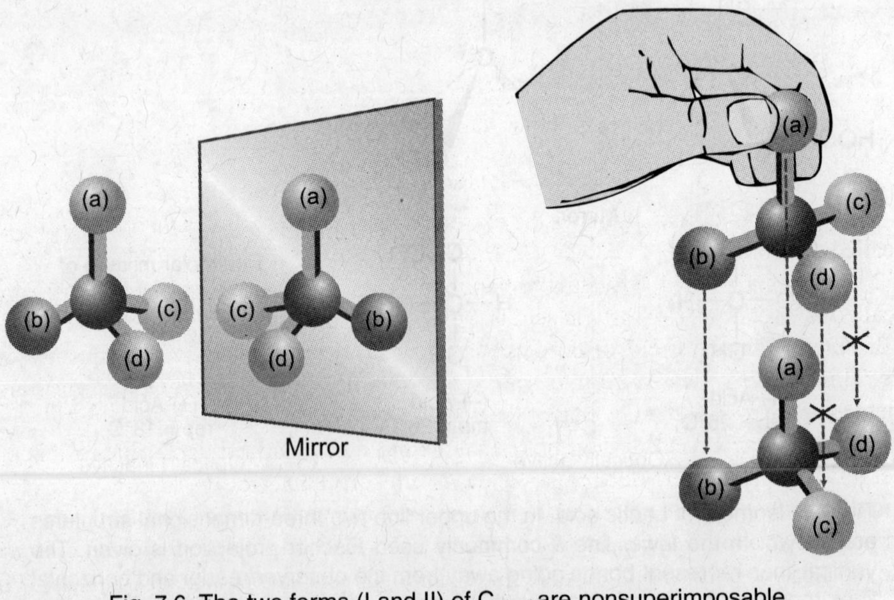

Fig. 7.6. The two forms (I and II) of C_{abcd} are nonsuperimposable.

Consider a chiral molecule C_{abcd}. Fig. 7.6 shows two three dimensional models for this molecule. The form (II) is the mirror image of (I) and the two cannot be superimposed. If you try to superimpose the four groups attached to a chiral carbon atom in form (I) and form (II), only two of the four groups coincide, while the other two are in conflict. This shows that the two forms represent different compounds.

The nonsuperimposable mirror image forms of a chiral molecule are called **Enantiomers** (Greek, *enantio* = opposite). They represent two optical isomers : (+) and (–). Their opposite rotatory powers are due to the opposite arrangements of groups around the asymmetric carbon atom.

It is true that most of the compounds which contain chiral carbon atoms show optical isomerism. But it is not always so. There are some compounds (such as *meso*-tartaric acid) which have asymmetric carbons but being *achiral*, do not show optical isomerism. On the other hand, there are certain compounds (such as substituted allenes and biphenyls) which have no chiral carbons but being dissymmetric, show optical isomerism.

OPTICAL ISOMERISM OF TARTARIC ACID

Tartaric acid (2,3-Dihydroxybutanedioic acid) contains two chiral carbon atoms.

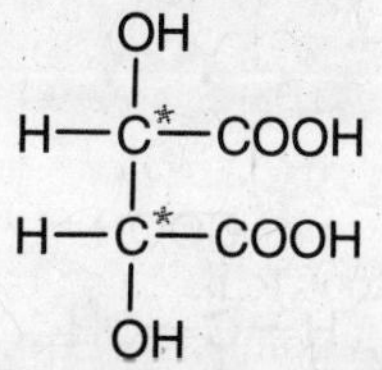

Tartaric acid. The two chiral carbons are shown by asterisks.

Four forms of tartaric acid are known (Fig. 7.7). Two of them are optically active and two are optically inactive. The optically active forms are related to each other as an object to its mirror image. That is, they are *enantiomers*.

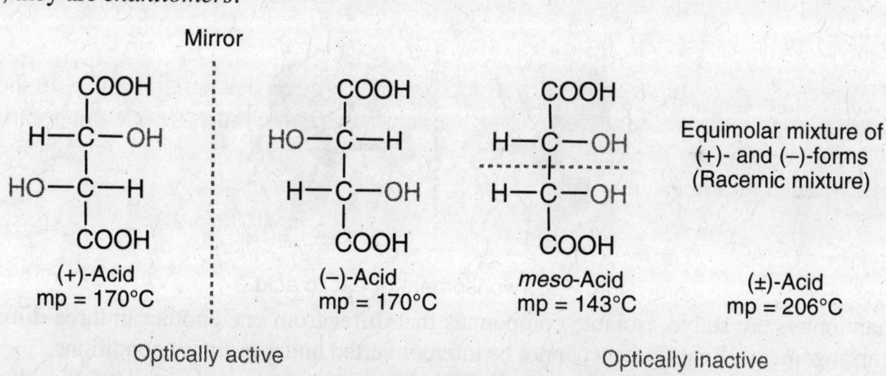

Fig. 7.7. Isomers of tartaric acid.

1. (+)-**Tartaric Acid**. It rotates the plane of polarized light to the right (clockwise direction) and is called dextrorotatory.
2. (–)-**Tartaric Acid**. It rotates the plane of polarized light to the left (anticlockwise direction) and is called levorotatory. (–)-Tartaric acid is the mirror image of (+)-tartaric acid and *vice versa*.
3. ***meso*-Tartaric Acid**. It possesses a plane of symmetry and is consequently optically inactive. This optically inactive form is said to be *internally compensated* (Optical rotation of one asymmetric carbon is cancelled by that of the other).
4. (±)-**Tartaric Acid**. It does not rotate the plane of polarized light. That is, it is optically inactive. It is an equimolar mixture of (+)- and (–)-forms (racemic mixture).

A molecule containing *n* chiral centres can give rise to a maximum of 2^n optical isomers, unless the compound has a *meso* form, in which case it is $2^n - 1$. For example, *four optical* isomers are possible for 2-bromo-2-butanol. It has *two* chiral carbon atoms but no plane of symmetry (Fig. 7.8).

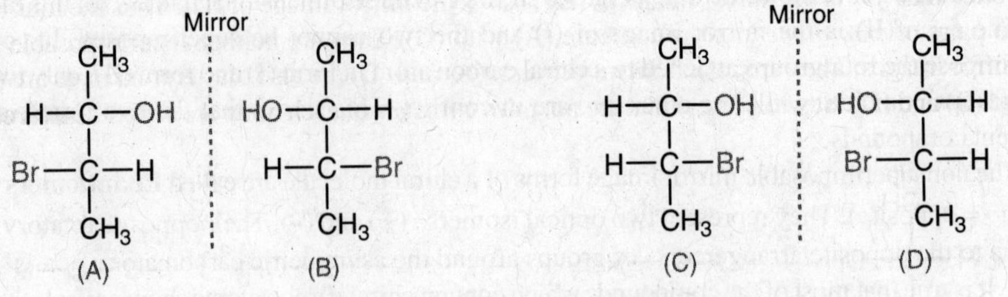

Fig. 7.8. Four isomers of 3-Bromo-2-butanol.

Optical isomers which are not enantiomers (mirror images) such as *(A)* and *(C)* or *(B)* and *(D)* of the above example, are known as **Diastereomers**.

PROPERTIES OF ENANTIOMERS

Optical isomers that are mirror images are called **Enantiomers**. These always exist as discrete pairs. For example, there are two optical isomers of lactic acid (Fig. 7.9). Notice that (A) is the mirror image of (B). They are a pair of enantiomers.

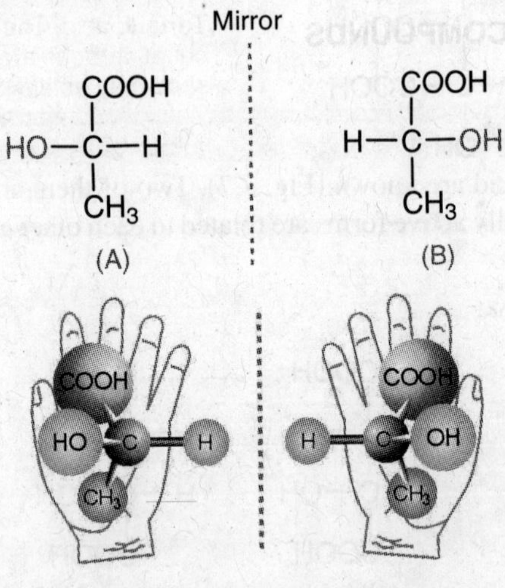

Fig. 7.9. Two isomers of Lactic acid.

Enantiomers are stable, isolable compounds that differ from one another in three-dimensional spatial arrangements. Enantiomers cannot be interconverted under ordinary conditions.

Enantiomers have identical properties in all respects except in their interaction with plane of polarized light. Enantiomers have the same melting point, density, solubility, color, and reactivity toward acids and bases. They differ, however, in the direction in which they rotate the plane of polarized light. Both rotate the plane of polarized light to exactly the same extent (same angle) but one rotates the plane to the right (clockwise ; called *dextrorotatory*), while the other rotates the plane to the left (anticlockwise ; called *levorotatory*).

A mixture of equal amounts of two enantiomers is called a **Racemic Mixture**. Such a mixture is optically inactive (does not rotate the plane of polarized light) because the two components rotate the plane of polarized light equally in opposite directions and cancel one another.

PROPERTIES OF DIASTEREOMERS

In general, each chiral carbon atom in a molecule doubles the number of theoretically possible isomers. Hence, molecule with *n* chiral carbon atoms should have 2^n stereoisomers. Fig. 7.10 shows the four isomers of 3-bromo-2-butanol, which has *two* chiral carbon atoms.

Notice that (A) is the mirror image of (B) ; (C) is the mirror image of (D). Thus the four isomers are two pairs of *enantiomers*. Now compare (A) with (C). They are neither superimposable nor are they mirror images. They are called *diastereomers*. (A) and (D) are also diastereomers, as are (B) and (C), and (B) and (D). Stereoisomers that are not mirror images of each other are called **Diastereomers**.

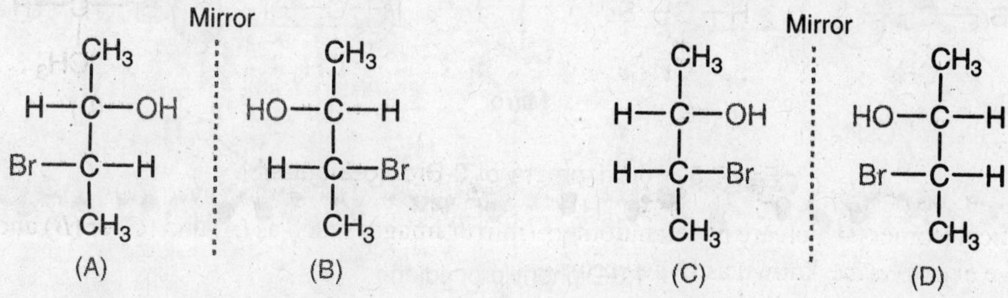

Fig. 7.10. Four isomers of 3-Bromo-2-butanol.

Diastereomers have different properties. Two diastereomers will have different melting points, boiling points, and solubilities. They will have different chemical reactivities toward most reagents.

PROPERTIES OF MESO COMPOUNDS

A compound with two or more chiral carbon atoms but also having a plane of symmetry (a mirror plane) is called the **Meso Compound**. Fig. 7.11 shows two meso compounds. These molecules have planes of symmetry dividing them midway between the two chiral carbons in each. Notice that one-half of the molecule is the mirror image of the other. Both molecules are optically inactive, even though each has two chiral centres. Neither will rotate the plane polarized light.

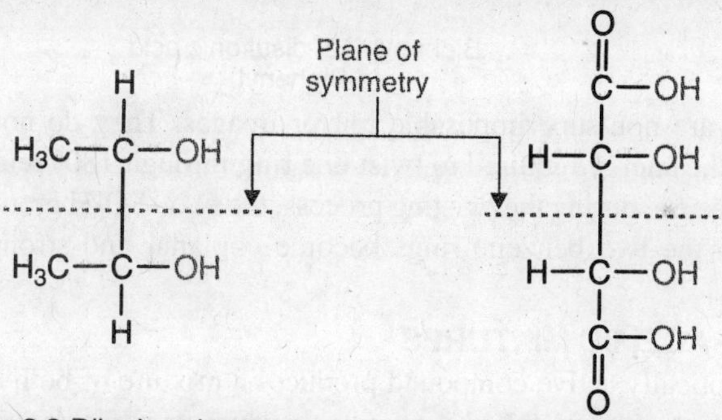

meso-2,3-Dihydroxybutane *meso*-Tartaric acid

Fig. 7.11. Examples of meso compounds.

OPTICAL ACTIVITY WITHOUT CHIRAL CARBONS

Compounds containing a chiral carbon can exist in optically active forms. However, compounds which do not possess a chiral carbon atom can also exist in optically active forms provided that the molecule is dissymmetric. Example are :

(1) **Allene Derivatives**

Some derivatives of allene ($CH_2{=}C{=}CH_2$) exhibit optical isomerism. Example is 1,3-diphenylpropadiene. In allenes, the central carbon is *sp* hybridized, and the terminal carbons are sp^2 hybridized. The central carbon forms two sp-sp^2 σ bonds. The central carbon also has two *p* orbitals which are mutually perpendicular. These form π bonds with the *p* orbitals on the other carbon atoms. As a result, the substituents at one end of the molecule are in a plane which is perpendicular to that of the substituents at the other end, so that the compound exists in two forms which are non-superimposable mirror images and are optically active.

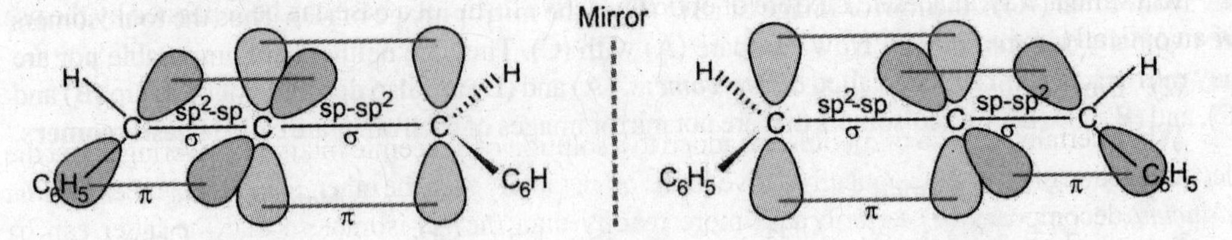

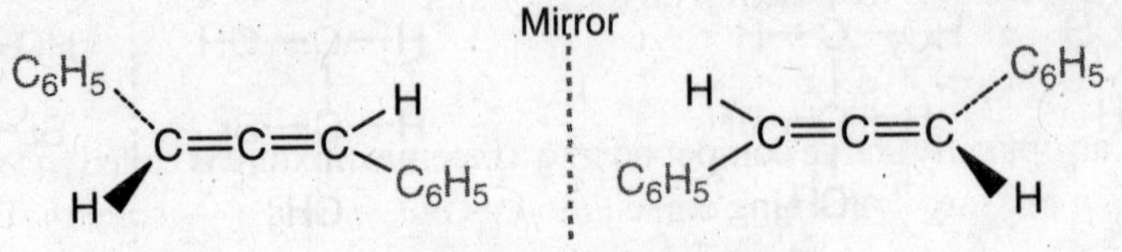

1,3-Diphenylpropadiene

(2) **Biphenyl Derivatives**

Substituted biphenyls show optical isomerism when substituents in the 2-positions are large enough to prevent rotation about the bond joining the two benzene rings. For example, biphenyl 2,2'-disulfonic acid exists in two forms.

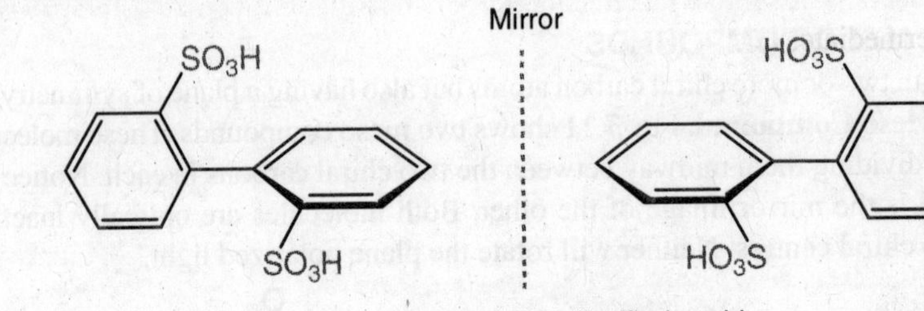

Biphenyl-2,2'-disulfonic acid
(a Biphenyl)

These two forms are non-superimposable mirror images. They do not interconvert at room temperature because the energy required to twist one ring through 180° relative to the other is too high. This in turn is because, during the twisting process, the two $-SO_3H$ groups must come into very close proximity when the two benzene rings become coplanar and strong repulsive forces are introduced.

RESOLUTION OF RACEMIC MIXTURES

Synthesis of an optically active compound produces a mixture of both (+)- and (–)-isomers in equal amounts. Such a mixture is called a *racemic mixture* or a *racemate*. **The separation of a racemic mixture into its two optically active components (+ and – isomers) is known as Resolution.** Following methods are used :

(1) **Chemical Resolution**

Optically active isomers of the same compound resemble one another so closely in all properties except optical that it is not possible to separate them by ordinary laboratory methods. If, however, the racemic mixture is made to combine with another optically active compound, differences are set up which can be exploited for separation. Suppose for example, that racemic lactic acid is allowed to combine with the optically active base, (–)-strychnine. The salts formed will be :

(–)-strychnine (+)-lactate and
(–)-strychnine (–)-lactate

These salts are not true enantiomers. They differ considerably in properties and, in particular, in solubility in various solvents. By fractional crystallization from a suitable solvent, they can be separated. Then treatment with dilute mineral acid removes the optically active base and leaves the two separate samples of (+)-lactic acid and (–)-lactic acid.

In a similar way, a racemic mixture of optically active forms of a base can be separated by the use of an optically active acid.

(2) **Biochemical Resolution**

When certain bacteria or moulds are added to a solution of a racemic mixture they bring about the decomposition of one of the optically active forms more rapidly than the other. For example, *Penicillium glaucum* decomposes (+)-tartaric acid more readily than the (–)-isomer so that the latter can be obtained from the residue after treatment with the mould.

RACEMIZATION

Conversion of an optically active compound into a racemic mixture is called *racemization*. Recall : Racemic mixture is a mixture containing equal amounts of (+)- and (–)-isomers. Thus,

$$2(+)\text{-A} \longrightarrow (+)\text{-A} + (-)\text{-A}$$

or

$$2(-)\text{-A} \longrightarrow (+)\text{-A} + (-)\text{-A}$$

Racemization can be accomplished by means of heat, light, or by conversion of the isomer into an optically inactive intermediate which reverts to the racemic mixture. The conversion of either of the optically active lactic acids into a racemic mixture by heating its aqueous solution may proceed through an *enol* intermediate.

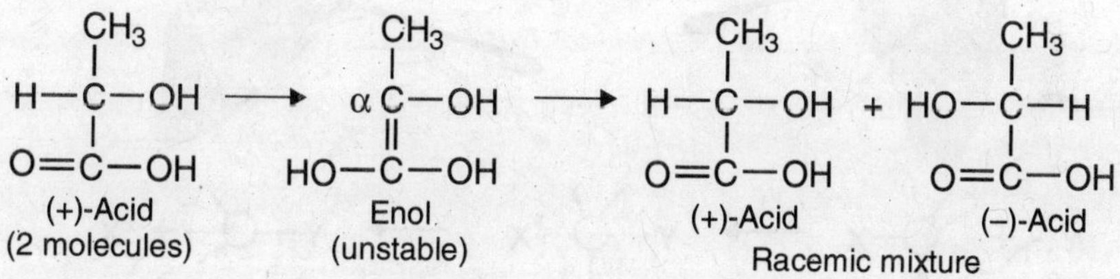

When the proton migrates back to the carbon atom, the process can involve the opening of either of the two linkages of the double bond and can produce either the original configuration or the opposite of it.

CHIRAL SYNTHESIS

When a compound containing a chiral carbon atom is synthesized by ordinary laboratory methods from a symmetric compound, the product is a racemic mixture. If, however, such a synthesis is carried under the *asymmetric influence* of a suitable optically active reagent, only one of the optically active isomers (+ or –) is formed. This process in which a chiral compound is synthesized from a symmetric compound to yield the (+)-isomer or (–)-isomer directly is termed **Chiral Synthesis**. For example, the reduction of pyruvic acid, $CH_3COCOOH$, in the laboratory leads to (±)-lactic acid (Racemic mixture). On the other hand, pyruvic acid is reduced by *yeast* to (–)-lactic acid only.

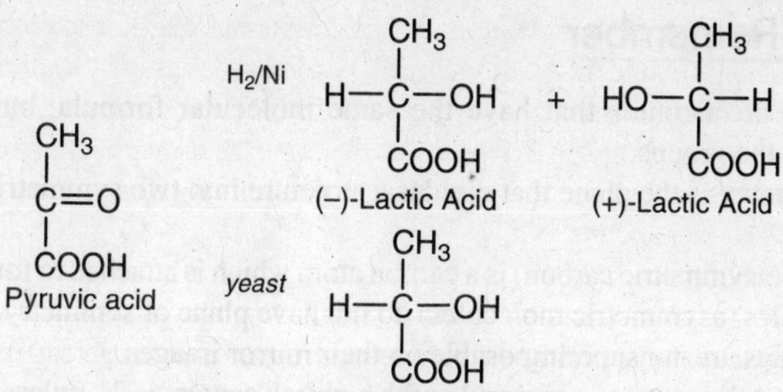

WALDEN INVERSION

When a group attached to a chiral carbon atom is replaced, the configuration of the new compound may be opposite to that of the original. That is, the product is the enantiomer (mirror image) of the expected substance. This phenomenon is known as **Walden Inversion.** Factors which determine whether or not an inversion will occur are the nature of the reagent, the nature of the solvent, the temperature, and the nature of the substance being acted upon.

The **mechanism** of Walden inversion involves three steps :

(1) Approach of the attacking group from the side of the molecule opposite to the position of the group which is later replaced.

(2) Formation of an intermediate transition state in which three groups and the chiral carbon atom lie in one plane.

(3) Separation of the group being replaced. This change is analogous to an umbrella being blown inside out (Fig. 7.12).

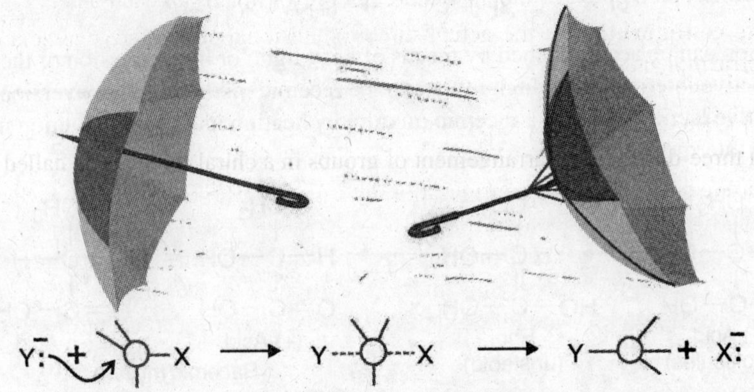

Fig. 7.12. Representation of Walden inversion.

The conversion of (+)-chlorosuccinic acid into (–)-malic acid is an example of Walden inversion.

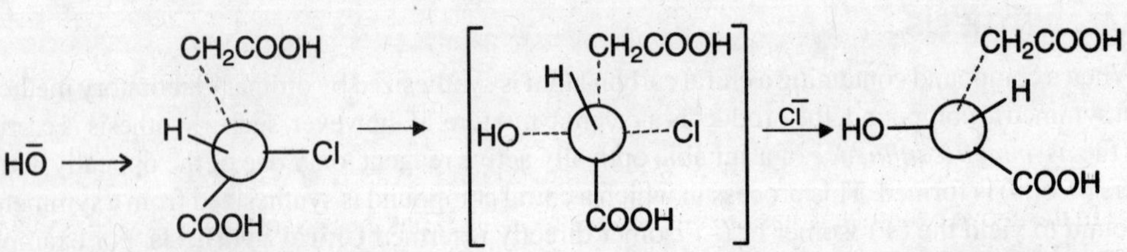

(+)-Chlorosuccinic acid (–)-Malic acid

Points to Remember

1. **Stereoisomers** are isomers that have the same molecular formula, but different spatial arrangement of the groups.
2. **Plane of symmetry** is the plane that divides a structure into two symmetrical halves that are mirror images.
3. **Chiral carbon** (asymmetric carbon) is a carbon atom which is attached to four different groups.
4. **Chiral molecules** (asymmetric molecules) do not have plane of symmetry.
5. **Chiral molecules** are nonsuperimposable on their mirror images.
6. **Number of stereoisomers** of a molecule with n chiral centers is 2^n, unless the compound has a meso form, in which case it is $2^n - 1$.

7. **Optical isomers** are isomers that are identical in properties except for their effect on the plane of polarized light. They contain at least one chiral carbon atom.
8. **Optically active molecule** is a molecule which has the ability to rotate the plane of polarized light.
9. **Enantiomers** are optical isomers that are mirror images.
10. **Dextrorotatory** molecules rotate the plane of polarized light to the right (clockwise direction). They are designated by the prefix (+) or *d*.
11. **Levorotatory** molecules rotate the plane of polarized light to the left (anticlockwise direction). They are designated by the prefix (–) or *l*.
12. **Racemic mixture** is a mixture that contains equal amounts of two enantiomers. It is designated by the prefix (±).
13. **Meso compounds** have two or more chiral centres but have a plane of symmetry between them. They are optically inactive.
14. **Diastereomers** are optical isomers that are not mirror images of each other.
15. **Absolute configuration** is the actual three-dimensional spatial arrangement of groups in asymmetric molecules.

R,S SYSTEM OF SPECIFYING CONFIGURATIONS

The actual three-dimensional arrangement of groups in a chiral molecule is called its **Absolute Configuration**. We can specify the configuration by using the **R,S System**. The R,S system is also called the **Cahn-Ingold-Prelog System** (named after its inventors).

In the R,S system, we arrange the four groups attached to the chiral carbon in decreasing order of priority (1, 2, 3, 4) by applying the rules given below. We then view the chiral center with the lowest priority group (4) on the side opposite the observer (group priorities are shown in parentheses) :

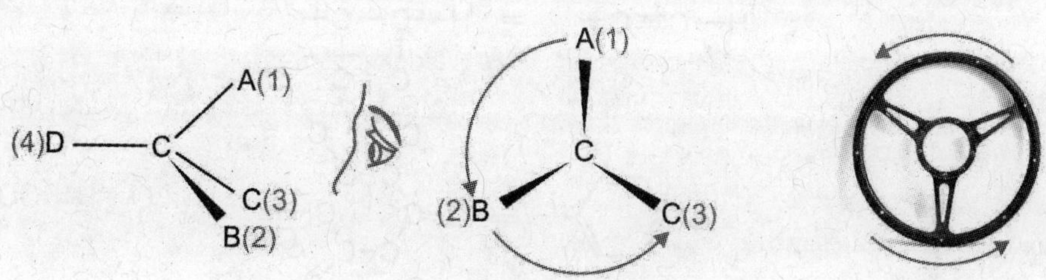

Anticlockwise = S Left turn = S

As seen by the observer

In the above example, **A** has the highest priority (1) ; **B** has the next highest (2) ; **C** has the next (3) ; and **D** has the lowest priority (4). We now go from high to low priority (1→2→3). To do so we move in an *anticlockwise* direction, and therefore we assign to this structure the **S Configuration** (S = *sinister* = left). In the structure given below, we go from highest to lowest priority by moving in a *clockwise* direction. We therefore assign to this structure the **R Configuration** (R = *rectus* = right).

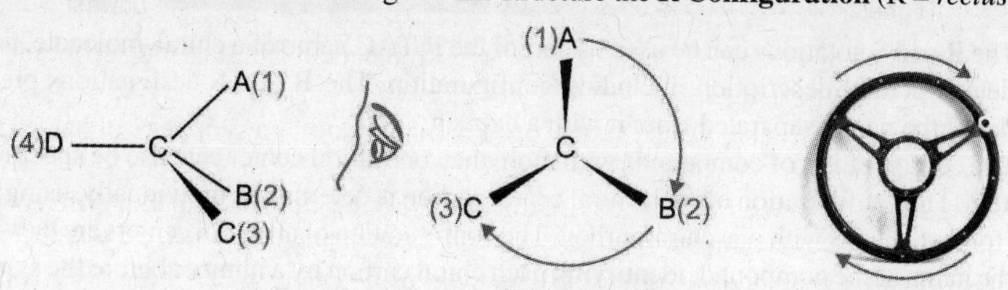

Clockwise = R Right turn = R

As seen by the observer

To establish the *group priorities* we use the following **Sequence Rules** :

Rule 1. Of the atoms attached directly to the chiral carbon atom, the one with the highest atomic number has the highest priority. For example,

I > Br > Cl > F > O > N > C > H

Highest ⟶ Lowest

Rule 2. If the atoms attached to the chiral carbon atom are the same, we determine priority by going to the next atom away from the chiral carbon atom. For example,

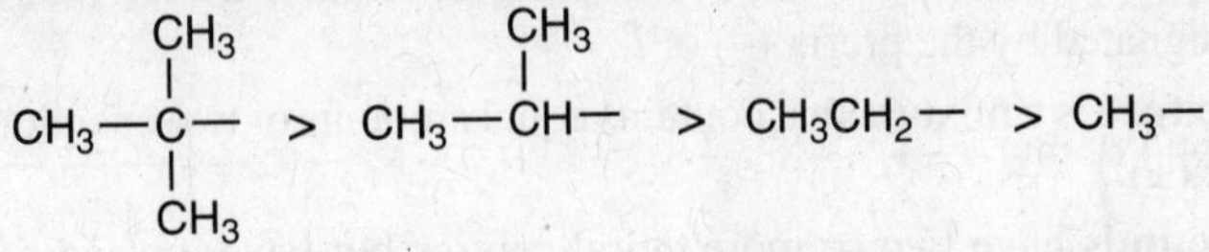

Highest ⟶ Lowest

Ethyl has a higher priority than methyl because the ethyl group has (CHH attached to the first carbon, whereas the methyl carbon has only hydrogens (HHH), and C has priority over H. Isopropyl is of higher priority than ethyl because it has two carbons attached to the first carbon and ethyl has only one. If there is no difference at the second atom in the chain, we go to the next atom and so forth.

Rule 3. A double bond is treated as though each atom of the double bond were bonded to two atoms:

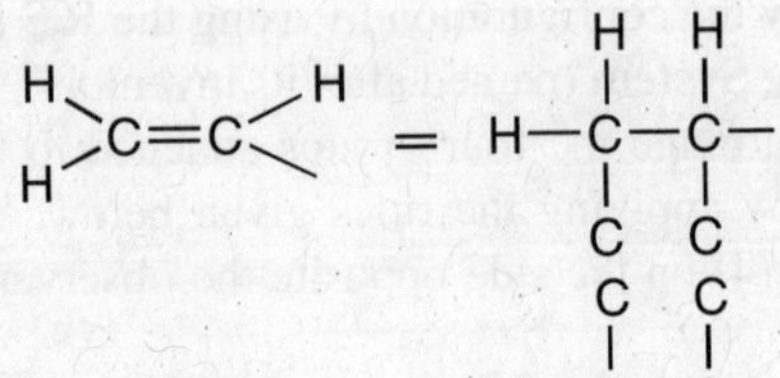

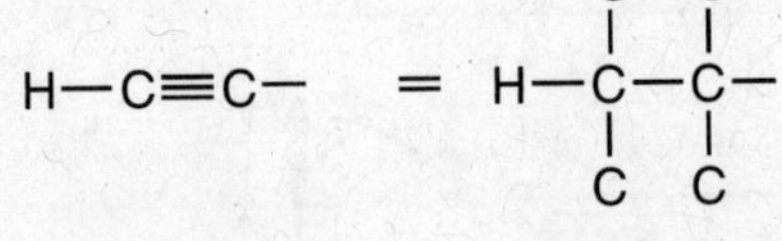

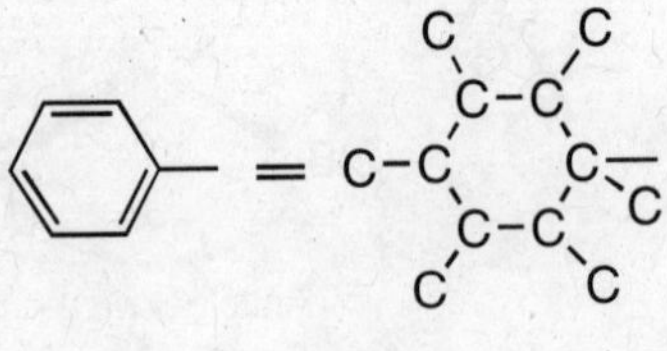

For example,

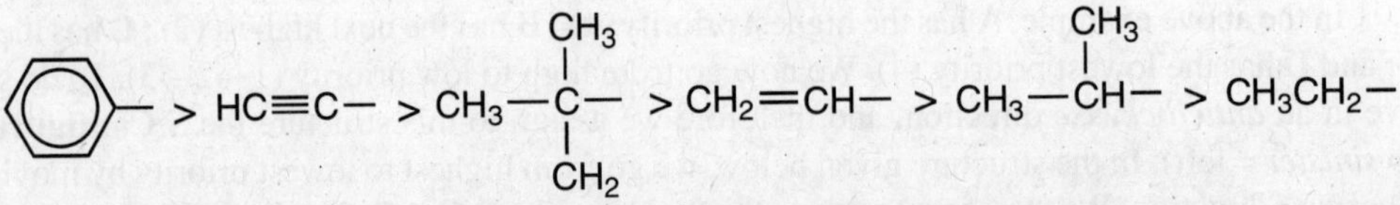

Highest ⟶ Lowest

The **R** and **S** notations can be used as part of the IUPAC name of a chiral molecule, to provide a complete structural description, including configuration. The **R** and **S** designations precede the remainder of the name, separated from it with a hyphen.

The configuration of compounds with more than one chiral center can also be specified by the R,S system. The configuration of each chiral center carbon is determined individually, using the same rules as for compounds with one chiral carbon. The configuration of all chiral centers are then specified before the name of the compound, identifying each chiral carbon by a number before the symbol **R** or **S**. For example, the compound 2R,3S-3-chloro-2-pentanol has the **R** configuration at carbon 2 and the **S** configuration at carbon 3.

Points to Remember

1. Most errors in designating the configurations of molecules by R,S system come from a failure to point the lowest priority group away from the observer. You can easily assign the wrong configuration to a molecule if you do not point the lowest priority group away from yourself.
2. The priority sequence for the most common atoms and groups is as follows :

(High Priority) —I, —Br, —Cl, —SH, —F, —O—C(=O)—R, —OR, —OH, $—NO_2$, —NH—C(=O)—R, $—NR_2$, —NHR, $—NH_2$, $—CCl_3$, —C(=O)—Cl, —C(=O)—OR, —C(=O)—OH, —C(=O)—NH_2, —C(=O)—R, —C(=O)—H, —CN, $—CH_2OH$, C_6H_5—, $—CR_3$, $—CHR_2$, $—CH_2R$, $—CH_3$, —H **(Low Priority)**

3. The Fischer projection formula is commonly used to show configuration without going to the trouble of drawing three-dimensional structure. In this convention the chiral center and the bonds to it are represented by crossed vertical and horizontal lines. It is understood that **groups attached by vertical bonds point away from the observer ; those bonded horizontally point toward the observer.** Some chemists show chiral carbon also.

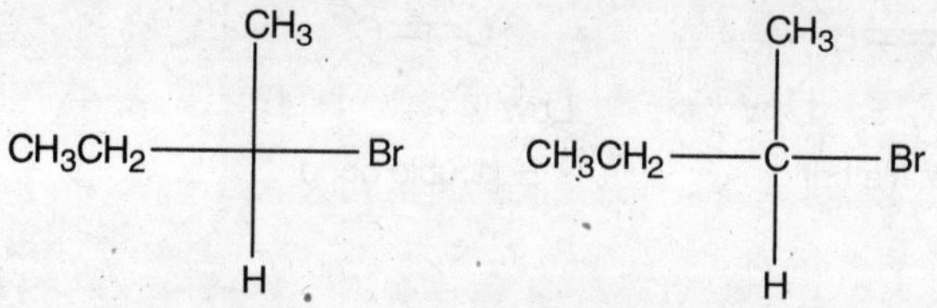

Fischer Projection Formula

4. In exam you are ordinarily given a Fischer projection formula and asked to specify configuration using the symbols **R** and **S**. It is easy to do so if you follow the following *step-by-step* method. Consider the following example :

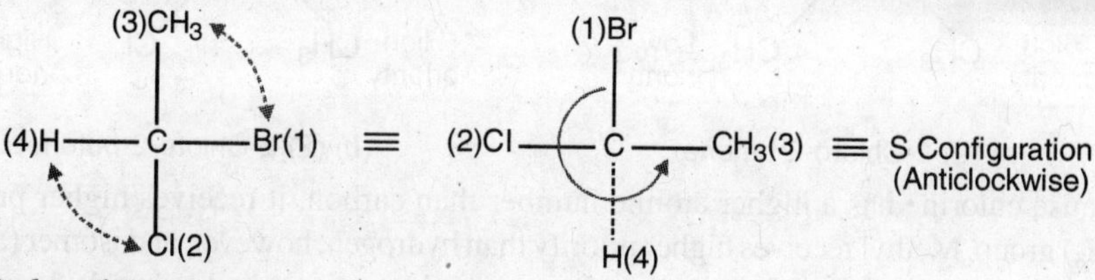

Order of Priority

(1) Br (atomic number 79.9) (*Highest*)

(2) Cl (atomic number 35.5)

(3) CH_3 (C, atomic number 12)

(4) H (atomic number 1) (*Lowest*)

Step 1. *Determine the priority of groups attached to the chiral carbon atom.* Apply the sequence rules carefully. In the above example, the order of priority is :

$$Br > Cl > CH_3 > H$$

Step 2. Position the lowest priority group (in this case H) down and away from the observer. This is done by interchanging groups bonded to the asymmetric carbon. *Remember that interchanging any pair of groups in a Fischer projection inverts the configuration.*

Interchanging two pairs of groups retains the original configuration. Thus, the interchange operation must always be done in pairs to avoid a configuration change.

In the above example, a dotted line has been used to highlight a bond between the chiral carbon and the lowest priority group (H). The interchange operations have been shown by dotted double-headed curved arrows.

Step 3. Specify the direction of decreasing priority of the three groups (1→2→3). Ignore the lowest priority group. *If the groups 1, 2, and 3 are arranged in clockwise fashion, the configuration is R. If the groups occur in anticlockwise fashion, the configuration is S.*

In the above example, the configuration is **S** because the groups 1, 2, and 3 are arranged in anticlockwise fashion (shown by circular arc).

E, Z SYSTEM OF SPECIFYING CONFIGURATIONS

The configuration of geometrical isomers can be designated as *cis* or *trans*. If the two identical groups are on the same side of the double bond, the isomer is *cis* ; and if they occur on opposite sides, the isomer is *trans*. This method does not work when there are no identical groups, as in BrClC=CFI (1-Bromo-1-chloro-2-fluoro-2-iodoethene). The configuration of these and all other geometrical isomers can be specified by using the notations ***Z*** and ***E***. To do this, first determine the group of highest priority on each carbon. If the two high-priority groups are together on the same side, the configuration is ***Z*** (German, *zusammen* = together). If they are on opposite sides, the configuration is ***E*** (German, *entgegen* = opposite).

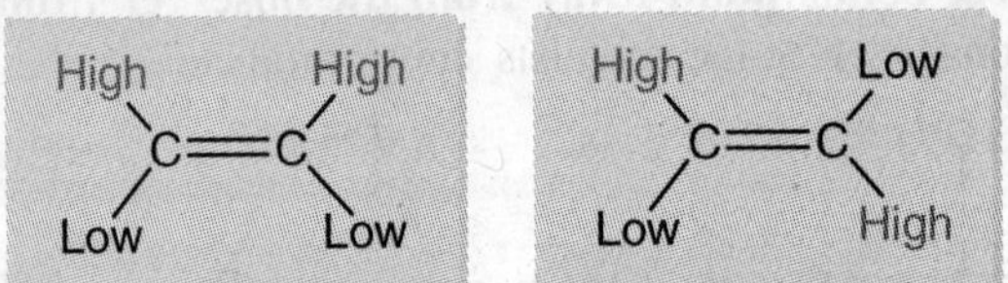

Z double bond   *E* double bond

For example,

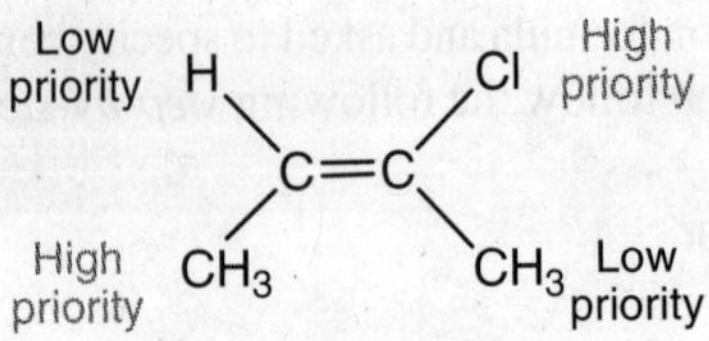

(*a*) (*E*)-2-Chloro-2-butene

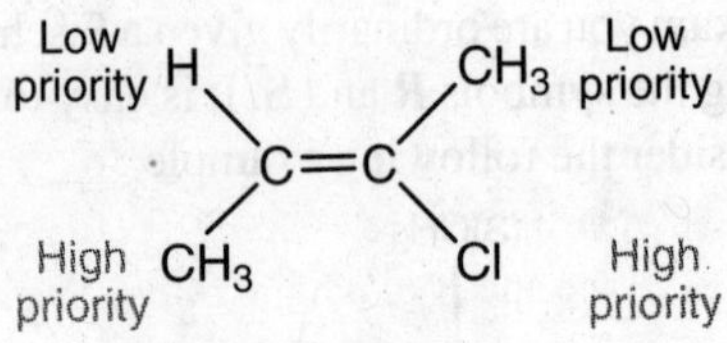

(*b*) (*Z*)-2-Chloro-2-butene

Because chlorine has a higher atomic number than carbon, it receives higher priority than a methyl (CH_3) group. Methyl receives higher priority than hydrogen, however, and isomer (a) is therefore assigned *E* geometry (high-priority groups on opposite sides of the double bond). Isomer (b) has *Z* geometry (high-priority groups on "zed zame zide" of the double bond).

Examination Questions

1. Define or explain the following terms :

(*a*) Structural isomerism
(*b*) Stereoisomerism
(*c*) Geometrical isomerism
(*d*) Optical isomerism
(*e*) Chiral carbon
(*f*) Chiral molecule
(*g*) Enantiomers
(*h*) Diastereomers
(*i*) Meso compounds
(*j*) Racemic mixture

2. Write a note on : Constitutional isomerism
3. Write notes on :
   (*a*) Chain Isomerism
   (*b*) Position isomerism
   (*c*) Functional isomerism
   (*d*) Metamerism
4. Write a note on : Geometrical isomerism.

(Ranchi BSc, 2014 ; Baroda BSc, 2014 ; Jadavpur BSc, 2015)

5. State the necessary conditions for a compound to show Geometrical isomerism. Illustrate your answer with examples.

(Dibrugarh BSc, 2014)

6. Describe the isomerism exhibited by maleic and fumaric acids.
7. Which of the following compounds show geometrical isomerism :
   (*a*) 2-Butene (*b*) 2-Methyl-2-butene
   (*c*) 2-Pentene (*d*) 1,2-Dichloropropane

Answer. (*a*) and (*c*)

8. Which of the following compounds show geometrical isomerism :
   (*a*) $CH_3-CH=CH_2$ (*b*) $CH_3-\underset{}{\overset{Br}{C}}=\overset{Br}{C}-CH_2CH_3$ (*c*) $CH_3CH_2CH_2CH=CHCH_3$

Answer. (*b*) and (*c*)

9. Which is more stable, cis-2-methyl-3-hexene or trans-2-methyl-3-hexene ? Explain

Answer. The trans isomer is more stable. The cis isomer has two large substituents on the same side of the double bond, which cause steric strain.

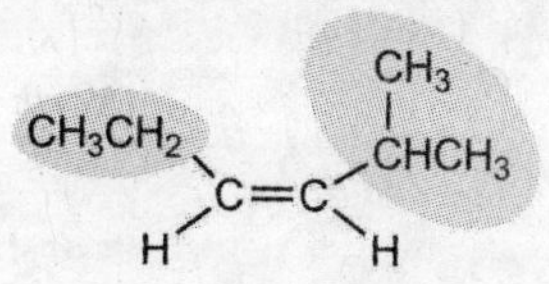

cis-2-Methyl-3-hexene

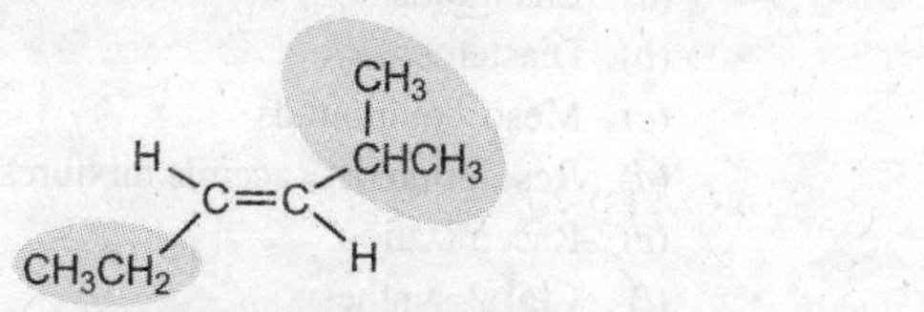

trans-2-Methyl-3-hexene

10. What is optical activity ? How is it measured ?
11. What is specific rotation ?
12. If the specific rotation of one enantiomer of 2-butanol is +13.5° what is the specific rotation of the other enantiomer ? Explain.

Answer. The magnitude of the specific rotation is the same for both enantiomers. This parameter differs only in sign. If one enantiomer of 2-butanol has a specific rotation of +13.5°, the other enantiomer must have a specific rotation of –13.5°.

13. Which of the following solvents is not suitable for determining the specific rotation of an asymmetric unknown ? Explain
   (*a*) Water (*b*) Methyl alcohol
   (*c*) 2-Butanol (*d*) Diethyl ether

Answer. The solvent (*c*) (2-butanol) is unsuitable. It has an asymmetric center. It would rotate plane-polarized light in the polarimeter and, possibly, obscure or interfere with the rotation of the unknown asymmetric compound.

14. Write a note on : Optical isomerism.

(Madras BSc, 2014 ; Vikram BSc, 2014 ; Sambalpur BSc, 2015 ; West Bengal BSc, 2015)

15. State the necessary conditions for a compound to show optical isomerism.

(Gujarat BSc, 2014)

16. Which of the following compounds are chiral ? Label all chiral centres.

(*a*) 2, 4-Dimethylheptane (*b*) 5-Ethyl-3,3-dimethylheptane

(*c*) 2-Chloropentane (*d*) 3-Methyl-1-pentene

Answer. Chiral centers (asymmetric carbons) are labeled with an asterisk :

(*a*) 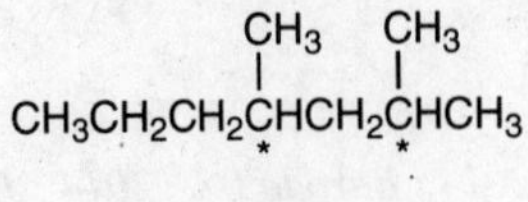

2,4-Dimethylheptane
(chiral)

(*b*) 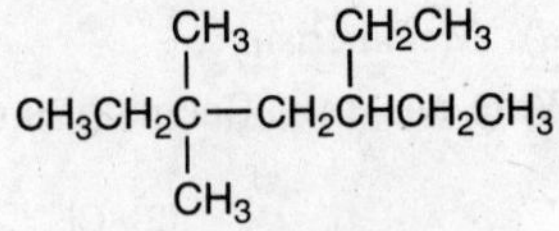

5-Ethyl-3,3-dimethylheptane
(achiral)

(*c*) 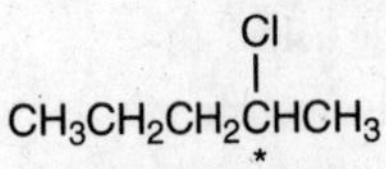

2-Chloropentane
(chiral)

(*d*) 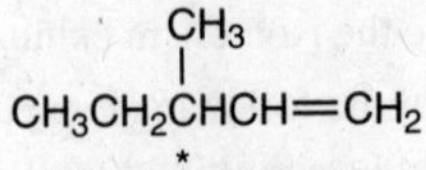

3-Methyl-1-pentene
(chiral)

17. Write a note on : Optical Isomerism of Lactic acid. (*Solapur BSc, 2014 ; Agra BSc, 2015*)

18. Write a note on : Optical Isomerism of Tartaric acid. (*Indore BSc, 2014 ; Anna BSc, 2015*)

19. Which of the following compounds show optical isomerism :

(*a*) 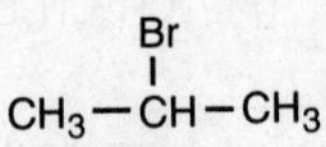 (*b*) 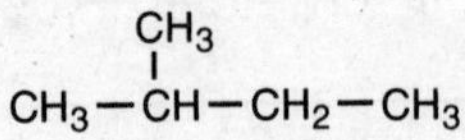 (*c*) 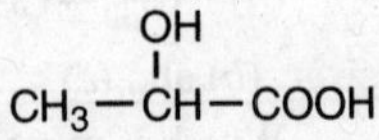

Answer. (*c*)

20. Write notes on :

(*a*) Enantiomers (*Baroda BSc, 2015*)

(*b*) Diastereomers (*Tezpur BSc, 2014*)

(*c*) Meso compounds (*North Eastern Hill BSc, 2014*)

(*d*) Resolution of Racemic mixtures

(*e*) Racemization

(*f*) Chiral synthesis

(*g*) Walden inversion (*Annamalai BSc, 2015*)

21. Describe two methods used for resolving racemic mixtures into optically active forms.

22. Assign priorities to the substituents in each of the following sets :

(*a*) –H, –Br, $-CH_2CH_3$, $-CH_2CH_2OH$

(*b*) –COOH, $-COOCH_3$, $-CH_2OH$, –OH

(*c*) –Br, $-CH_2Br$, –Cl, $-CH_2Cl$

Answer. (*a*) –Br, $-CH_2CH_2OH$, $-CH_2CH_3Cl$, –H

Highest ⟶ Lowest

(*b*) –OH, $-COOCH_3$, –COOH, $-CH_2OH$

Highest ⟶ Lowest

(*c*) –Br, –Cl, $-CH_2Br$, $-CH_2Cl$

Highest ⟶ Lowest

23. Describe **R**, **S** notations used in fixing the positions of groups in an optically active compound, with an example.

24. Assign **R** or **S** configuration to each of the following compounds : (*Kakatiya BSc, 2015*)

(*a*) 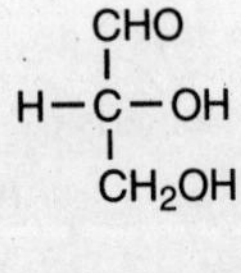 (*b*) 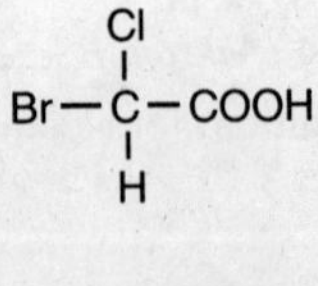

Answer. (*a*) Order of Priority is : OH > CHO > CH_2OH > H

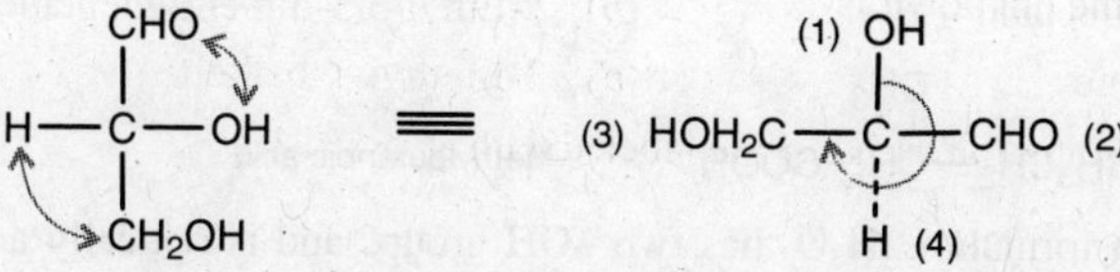

R configuration (clockwise)

(*b*) Order of Priority is : Br > Cl > COOH > H

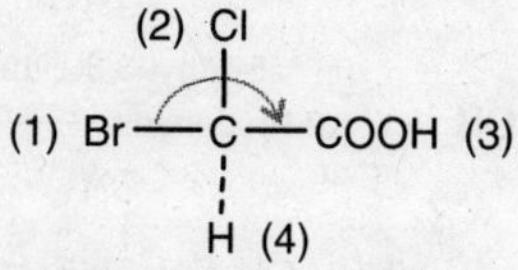

R configuration (clockwise)

25. Assign **R** or **S** configuration to each of the following compounds :

(*a*) 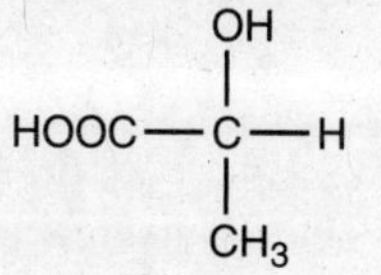 (*b*) 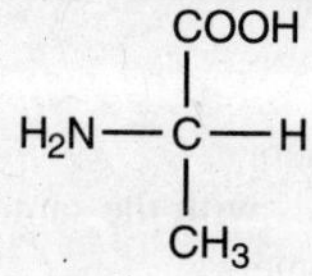

Answer. (*a*) Order of Priority is : HO > COOH > CH_3 > H

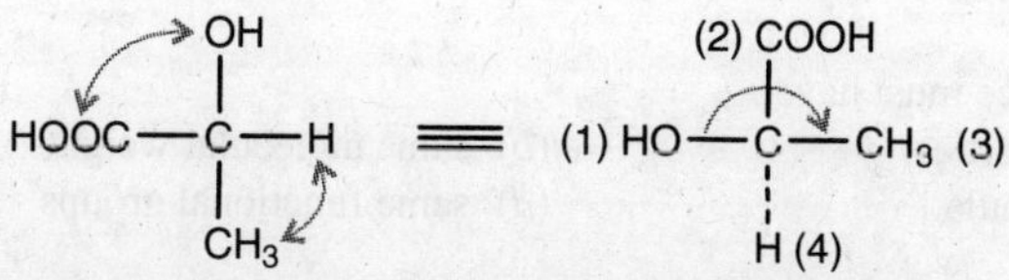

R configuration (clockwise)

(*b*) Order of Priority is : NH_2 > COOH > CH_3 > H

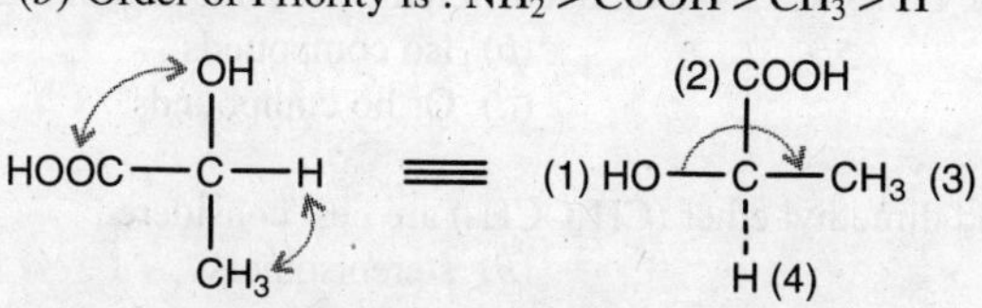

R configuration (clockwise)

26. What do you understand by the **E**, **Z** notations ?

27. Assign *E* or *Z* configuration to the following alkenes : (*Delhi BSc, 2015*)

(*a*) 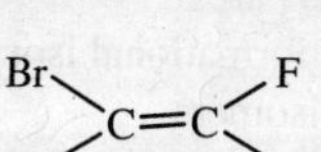 (*b*) 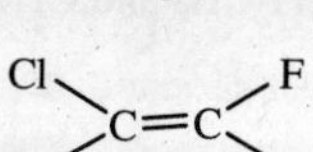 (*c*) 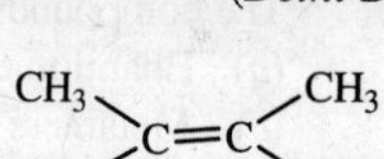

Answer. (*a*) 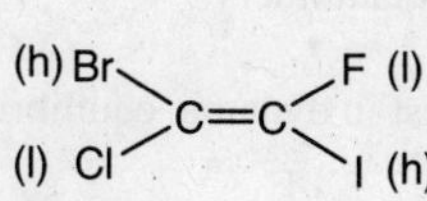 *E* configuration

(*b*) 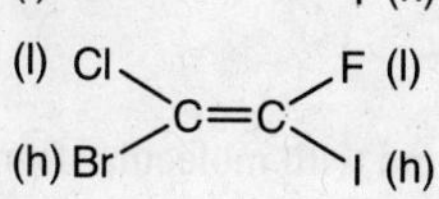 *Z* configuration

(*c*) 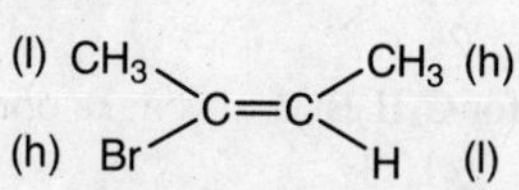 *E* configuration

28. A primary alcohol of formula $C_5H_{12}O$ is optically active. What is its structure ?

Answer. The structure of the unknown is :

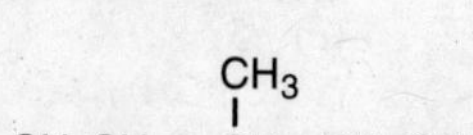

2-Methyl-1-butanol

29. An acid of formula $C_5H_{10}O_2$ is optically active. What is its structure ? *(Madurai BSc, 2015)*

Answer. The structure of the unknown is

$$CH_3CH_2-\underset{}{\overset{CH_3}{\overset{|}{C}H}}-COOH \quad \text{2-Methylbutanoic acid}$$

30. A compound with formula $C_3H_8O_2$ has two –OH groups and is optically active. What is its structure ?

Answer. The structure of the unknown is :

$$CH_3-\overset{OH}{\overset{|}{C}H}-CH_2OH \quad \text{1,2-Propanediol}$$

Multiple Choice Questions

1. Stereoisomers are
(*a*) constitutional isomers
(*b*) different compounds with the same connectivity
(*c*) conformational isomers
(*d*) identical isomers

Answer. (*b*)

2. The isomers of a substance must have
(*a*) same chemical properties (*b*) same molecular weight
(*c*) same structural formula (*d*) same functional groups

Answer. (*b*)

3. Compounds with the same molecular formula but different structural formulas are called
(*a*) Alkoxides (*b*) Iso compounds
(*c*) Isomers (*d*) Ortho compounds

Answer. (*c*)

4. Ethanol (CH_3CH_2OH) and dimethyl ether (CH_3OCH_3) are best considered :
(*a*) structural isomers (*b*) stereoisomers
(*c*) enantiomers (*d*) diastereomers

Answer. (*a*)

5. The compounds $CH_3CH_2OCH_2CH_3$ and $CH_3OCH_2CH_2CH_3$ are
(*a*) Enantiomers (*b*) Conformational isomers
(*c*) Metamers (*d*) Optical isomers

Answer. (*c*)

6. Which of the following statements is false about tautomers ?
(*a*) Tautomers are structural isomers
(*b*) Tautomers are structural isomers which exist in dynamic equilibrium
(*c*) Tautomerism involves movement of atoms
(*d*) Tautomers have independent existence

Answer. (*d*)

7. How many isomers are possible for the compound with molecular formula C_4H_8 ?
(*a*) 2 (*b*) 4 (*c*) 6 (*d*) 8

Answer. (*c*)

8. How many structural isomers are possible for C_4H_9Br ?
(*a*) 2 (*b*) 3 (*c*) 4 (*d*) 5

Answer. (c)

9. How many isomeric aromatic hydrocarbons are possible for C_8H_{10} ?
(*a*) 3 (*b*) 4 (*c*) 5 (*d*) 6

Answer. (*b*)

10. Which compound is not an isomer of the other three ?

(a) n-Pentane (b) 2,2-Dimethylpropane
(c) 2-Methylbutane (d) 2,3-Dimethylbutane

Answer. (d)

11. Alkenes show geometrical isomerism due to :
(a) Asymmetry (b) Rotation around a single bond
(c) Resonance (d) Restricted rotation around a double bond

Answer.(d)

12. Which of the following compounds exhibit geometrical isomerism ?
(a) 1-Pentene (b) 2-Methyl-2-pentene
(c) 2-Pentene (d) 2-Methyl-2-butene

Answer. (c)

13. Which of the following compounds may exist as *cis-trans* isomers ?
(a) 1-Butene (b) 2-Butene
(c) Cyclopropane (d) Acetone

Answer. (b)

14. Geometrical isomerism is shown by
(a) Lactic acid (b) Maleic acid
(c) 1-Butene (d) 1,1-Dichloroethylene

Answer. (b)

15. Which of the following compounds show geometrical isomerism
(a) $(CH_3)_3N$ (b) $CH_3CH{=}CH_2$
(c) $(CH_3)_2NH$ (d) $CH_3CH{=}CHCH_3$

Answer. (d)

16. Which of the following compounds will show geometrical isomerism ?
(a) $CH_2{=}CHCl_2$ (b) $ClCH{=}CHBr$
(c) $CH_2{=}CHCl$ (d) $Cl_2C{=}CBr_2$

Answer. (b)

17. Which of the following compounds will not show geometrical isomerism ?
(a) BrCH=CHBr (b) BrCH=CHCl
(c) $CH_3-\underset{}{\overset{Br}{\overset{|}{C}}}=CHBr$ (d) $CH_3-\overset{CH_3}{\overset{|}{C}}=CHCH_3$

Answer. (d)

18. A molecule is said to be chiral
(a) if it contains plane of symmetry
(b) if it contains center of symmetry
(c) if it cannot be superimposed on its mirror image
(d) if it can be superimposed on its mirror image

Answer. (c)

19. Which of the statements is false regarding chiral compounds ?
(a) rotate the plane of polarized light (b) have cis and trans isomers
(c) exist as enantiomers (d) can be detected with a polarimeter

Answer. (b)

20. An optically active compound
(a) must contain atleast four carbons
(b) when in solution rotate the plane of polarized light
(c) must always contain an asymmetric carbon atom
(d) in solution always give a negative reading in polarimeter.

Answer. (b)

21. Plane-polarized light is affected by
(a) Identical molecules (b) All polymers
(c) Chiral molecules (d) All biomolecules

Answer. (c)

22. It is possible to distinguish between optical isomers

(*a*) by using chemical tests (*b*) by mass spectrometry
(*c*) by IR spectroscopy (*d*) by polarimetry

Answer. (*d*)

23. Optical isomers that are mirror images are called :
(*a*) Tautomers (*b*) Diastereomers
(*c*) Enantiomers (*d*) Metamers

Answer. (*c*)

24. Optical isomers that are not mirror images are called
(*a*) Diastereomers (*b*) Enantiomers
(*c*) Metamers (*d*) Meso compounds

Answer. (*a*)

25. Enantiomers have which of the following characteristics ?
(*a*) rotate ordinary light
(*b*) have the same melting point
(*c*) are superimposable mirror images
(*d*) react with optically active molecules at the same rate

Answer. (*b*)

26. Which of the following statements is false about enantiomers ?
(*a*) rotate plane-polarized light
(*b*) are superimposable mirror images
(*c*) are nonsuperimposable mirror images
(*d*) have the same melting point

Answer. (*b*)

27. A meso compound :
(*a*) is an achiral molecule which contains chiral carbons
(*b*) contains a plane of symmetry or a center of symmetry
(*c*) is optically inactive
(*d*) is characterized by all of the above

Answer. (*d*)

28. What is the possible number of optical isomers for a compound containing *n* dissimilar asymmetric carbon atoms ?
(*a*) n^2 (*b*) 2^n (*c*) $n + 1$ (*d*) $n + 2$

Answer. (*b*)

29. What is the possible number of optical isomers for a compound containing 2 dissimilar asymmetric carbon atoms ?
(*a*) 2 (*b*) 4 (*c*) 6 (*d*) 8

Answer. (*b*)

30. *meso*-Tartaric acid is
(*a*) sometimes optically active (*b*) always optically active
(*c*) sometimes optically inactive (*d*) always optically inactive

Answer. (*d*)

31. Which of the following compounds will be optically active ?
(*a*) Propanoic acid (*b*) 3-Chloropropanoic acid
(*c*) 2-Chloropropanoic acid (*d*) 3-Chloropropene

Answer. (*c*)

32. Which of the following compounds will be optically active ?
(*a*) Succinic acid (*b*) *meso*-Tartaric acid
(*c*) Lactic acid (*d*) Chloroacetic acid

Answer. (*c*)

33. Which of the following isomeric compounds show optical isomerism :
(*a*) 1-Aminopentane (*b*) 2 Aminopentane
(*c*) 3-Aminopentane (*d*) 2,2-Dimethylpropylamine

Answer. (*b*)

34. 2-Butanol is optically active because it contains :

(*a*) an asymmetric carbon
(*b*) a plane of symmetry
(*c*) a hydroxyl group
(*d*) a center of symmetry

Answer. (*a*)

35. Optical isomerism is shown by
(*a*) *n*-Butyl chloride
(*b*) *sec*-Butyl chloride
(*c*) *tert*-Butyl chloride
(*d*) Isobutyl chloride

Answer. (*b*)

36. Which of the following compounds is an optically active compound ?
(*a*) $CH_3-\underset{\displaystyle |}{\overset{\displaystyle OH}{CH}}-COOH$
(*b*) $CHCl_3$
(*c*) CH_3CH_2COOH
(*d*) CH_3CH_2OH

Answer. (*a*)

37. Lactic acid, $CH_3-\overset{\displaystyle OH}{\overset{|}{CH}}-COOH$, is a molecule which shows
(*a*) Geometrical isomerism
(*b*) Tautomerism
(*c*) Optical isomerism
(*d*) Metamerism

Answer. (*c*)

38. How many optical isomers are possible for lactic acid ?
(*a*) 2 (*b*) 4 (*c*) 6 (*d*) 8

Answer. (*a*)

39. Which of the following represents a racemic mixture ?
(*a*) 75% (R)-2-butanol, 25% (S)-2-butanol
(*b*) 25% (R)-2-butanol, 75% (S)-2-butanol
(*c*) 50% (R)-2-butanol, 50% (S)-2-butanol
(*d*) none of the above

Answer. (*c*)

40. Consider (R)- and (S)-2-butanol. Which physical property distinguishes the two compounds ?
(*a*) melting point
(*b*) solubility in common solvents
(*c*) Rotation of plane-polarized light
(*d*) Infrared spectrum

Answer. (*c*)

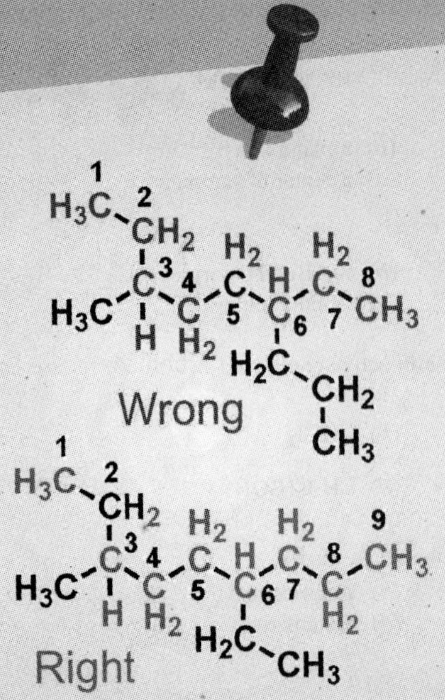

8 Nomenclature of Organic Compounds

You must learn to translate structures into names and names into structures

TYPES OF ORGANIC COMPOUNDS

There are 6 million organic compounds. In dealing with such a large number of compounds some classification is essential. There are five *main* types of organic compounds.

(1) **Aliphatic Compounds.** Compounds which consist of open-chain of carbon atoms are called aliphatic compounds. There is no limit to the number of atoms involved. Examples are :

Ethane	$CH_3—CH_3$
Propane	$CH_3—CH_2—CH_3$
Ethyl alcohol	$CH_3—CH_2—OH$
Acetic acid	$CH_3—COOH$
n-Butylamine	$CH_3—CH_2—CH_2—CH_2—NH_2$

(2) **Saturated and Unsaturated Compounds.** Compounds which contain only carbon and hydrogen are called *hydrocarbons*. A hydrocarbon is said to be *saturated* if it contains only C–C single bonds. A hydrocarbon is said to be *unsaturated* if it contains C=C or C≡C multiple bonds. Examples are :

Ethane	$CH_3—CH_3$	**It is saturated**
Ethylene	$CH_2{=}CH_2$	**It is unsaturated**
Acetylene	$HC{\equiv}CH$	**It is unsaturated**

The term unsaturation is also sometimes used to describe a compound containing multiple bonds between other pairs of atoms, *e.g.* C=O, C≡N.

(3) **Aromatic Compounds.** Benzene and all compounds that have structures and chemical properties resembling benzene are called aromatic compounds. Examples are :

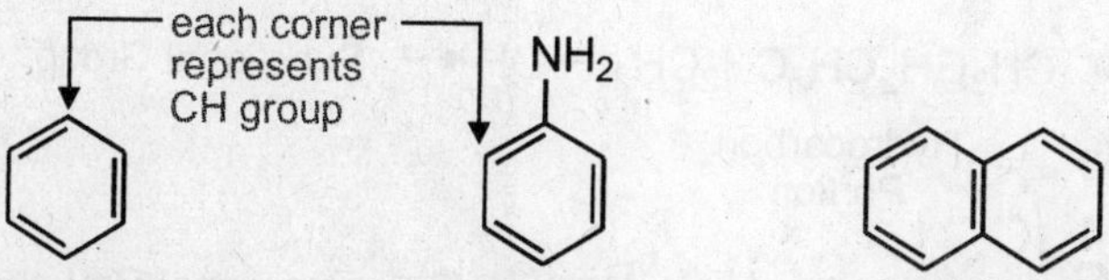

Benzene Aniline Naphthalene

(4) **Alicyclic Compounds.** Cyclic compounds which consist only of carbon atoms are called alicyclic or carbocyclic compounds. Examples are :

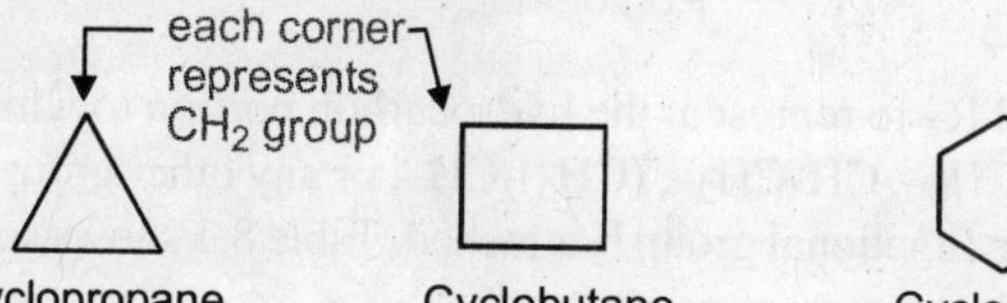

Cyclopropane Cyclobutane Cyclohexane

(5) **Heterocyclic Compounds.** Cyclic compounds in which the ring atoms are of carbon and some other element (For example, *N*, *S* or *O*) are called heterocyclic compounds. Examples are :

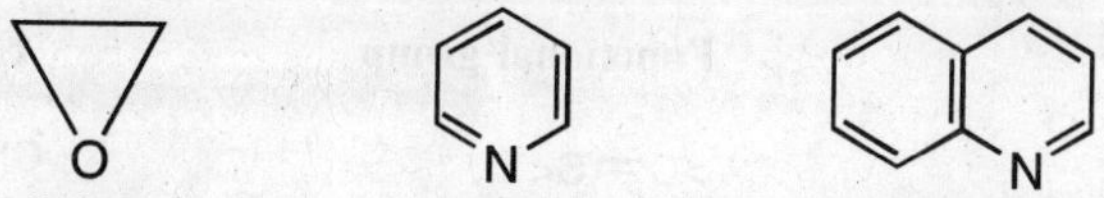

Oxirane (ethylene oxide) Pyridine Quinoline

FUNCTIONAL GROUPS

The basic idea of the functional group is at the heart of much of the organic chemistry we shall study. We will carry out many transformations of organic molecules. In most cases the change will occur at one "spot" in the original reacting molecule. That spot is the *functional group.*

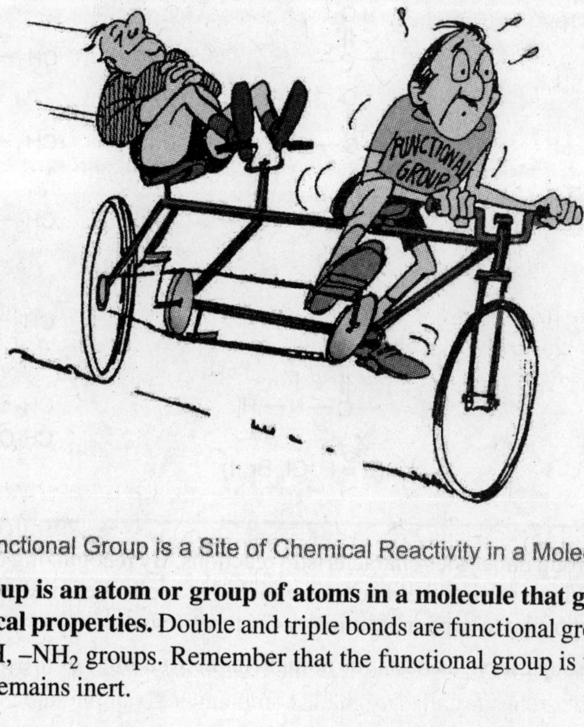

A Functional Group is a Site of Chemical Reactivity in a Molecule.

A functional group is an atom or group of atoms in a molecule that gives the molecule its characteristic chemical properties. Double and triple bonds are functional groups. Other examples include –Cl, –Br, –OH, $–NH_2$ groups. Remember that the functional group is the action group. The hydrocarbon portion remains inert.

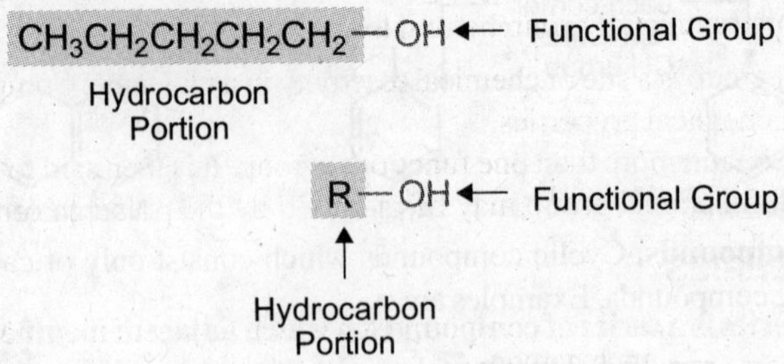

We often use the symbol R– to represent the hydrocarbon portion to which the functional group is attached. Thus R– can be CH_3–, CH_3CH_2–, $(CH_3)_2CH$–, or any other group of *C* and *H* atoms with one free valence by which the functional group is attached. Table 8.1 shows some common functional groups and the corresponding classes of compounds.

Table 8.1 CLASSES AND FUNCTIONAL GROUPS OF ORGANIC COMPOUNDS

Class	Functional group	Example
Alkene	>C=C<	$CH_2=CH_2$
Alkyne	—C≡C—	$CH \equiv CH$
Alcohol	—OH	CH_3CH_2-OH
Ether	$-\overset{\mid}{\underset{\mid}{C}}-O-\overset{\mid}{\underset{\mid}{C}}-$	CH_3-O-CH_3
Aldehyde	$-\overset{O}{\overset{\parallel}{C}}-H$	$CH_3-\overset{O}{\overset{\parallel}{C}}-H$
Ketone	$-\overset{O}{\overset{\parallel}{C}}-$	$CH_3-\overset{O}{\overset{\parallel}{C}}-CH_3$
Carboxylic acid	$-\overset{O}{\overset{\parallel}{C}}-OH$	$CH_3-\overset{O}{\overset{\parallel}{C}}-OH$
Ester	$-\overset{O}{\overset{\parallel}{C}}-O-\overset{\mid}{\underset{\mid}{C}}-$	$CH_3-\overset{O}{\overset{\parallel}{C}}-OCH_3$
Amine	$-\overset{H}{\overset{\mid}{N}}-H$	CH_3-NH_2
Amide	$-\overset{O}{\overset{\parallel}{C}}-\overset{H}{\overset{\mid}{N}}-H$	$CH_3-\overset{O}{\overset{\parallel}{C}}-NH_2$
Halide	–X (X = F, Cl, Br, I)	CH_3CH_2-Br

Each functional group undergoes characteristic reactions. By recognizing the functional group in a molecule, it is possible to predict the reactions which that molecule will undergo. The concept of functional group is important to organic chemistry for three reasons :

1. Functional groups serve as basis for nomenclature (naming) of organic compounds.

2. Functional groups serve to classify organic compounds into classes (families). All compounds with the same functional group belong to the same class.
3. A functional group is a site of chemical reactivity in a molecule. Compounds in the same class have similar chemical properties.

A molecule can contain more than one functional group. It is then said to be **Polyfunctional**, and the properties of each functional group may be modified by the presence of the others.

HOMOLOGOUS SERIES

A homologous series is a series of compounds in which adjacent members differ by a CH_2 unit. The individual members are called **Homologs**. For example, the homologous series of alcohols can be represented as :

General formula is ROH or $C_nH_{2n+1}OH$

n	R	Formula
1	CH_3-	CH_3OH
2	CH_3CH_2-	CH_3CH_2OH
3	$CH_3CH_2CH_2-$	$CH_3CH_2CH_2OH$
4	$CH_3CH_2CH_2CH_2-$	$CH_3CH_2CH_2CH_2OH$

The *general characteristics* of a homologous series are :

1. All compounds in the series contain the same elements and the functional group.
2. All compounds in the series can be represented by a general formula.
3. The molecular formula of each homolog differs from one *above* and one *below* it by a CH_2 unit.
4. All compounds in the series can be prepared by similar methods.
5. All compounds in the series have similar chemical properties.
6. There is a gradual variation in physical properties with increasing molecular weight.

PRINCIPLES OF NOMENCLATURE

In early days of organic chemistry, each new compound was given an individual name. Such a name was based on the source, some property, or some other trivial reason. Thus formic acid, HCOOH, was so named as it was obtained by distillation of red ants (Latin, *formica* = ants). A group of sedatives was named barbiturates after the name of a woman Barbara. The structure of many of these compounds was not known at that time. An ordinary name given to a compound without reference to its structure is called a **Common Name** or **Trivial name**. The common names are like nicknames.

IUPAC SYSTEM OF NOMENCLATURE

With the rapid growth of organic chemistry, the number of compounds increased fantastically (now about 6 million). It became impossible to give common names to such a large number of compounds.

In 1957, the *International Union of Pure and Applied Chemistry* evolved a scheme for giving systematic names to organic compounds on the basis of structure. This is known as the **IUPAC System**. (IUPAC is pronounced as *eye-you-pack*). This system has set rules for naming organic molecules from their structures.

The systematic name of a compound derived from its structural formula by applying IUPAC rules, is referred to as its **IUPAC Name**. One organic compound can have only one IUPAC name. It is superior to a common name as it gives an insight into the structure of the molecule. Knowing the IUPAC name of a compound, we can at once write its structural formula. *Remember that the common names identify compounds while the IUPAC names represent structures.*

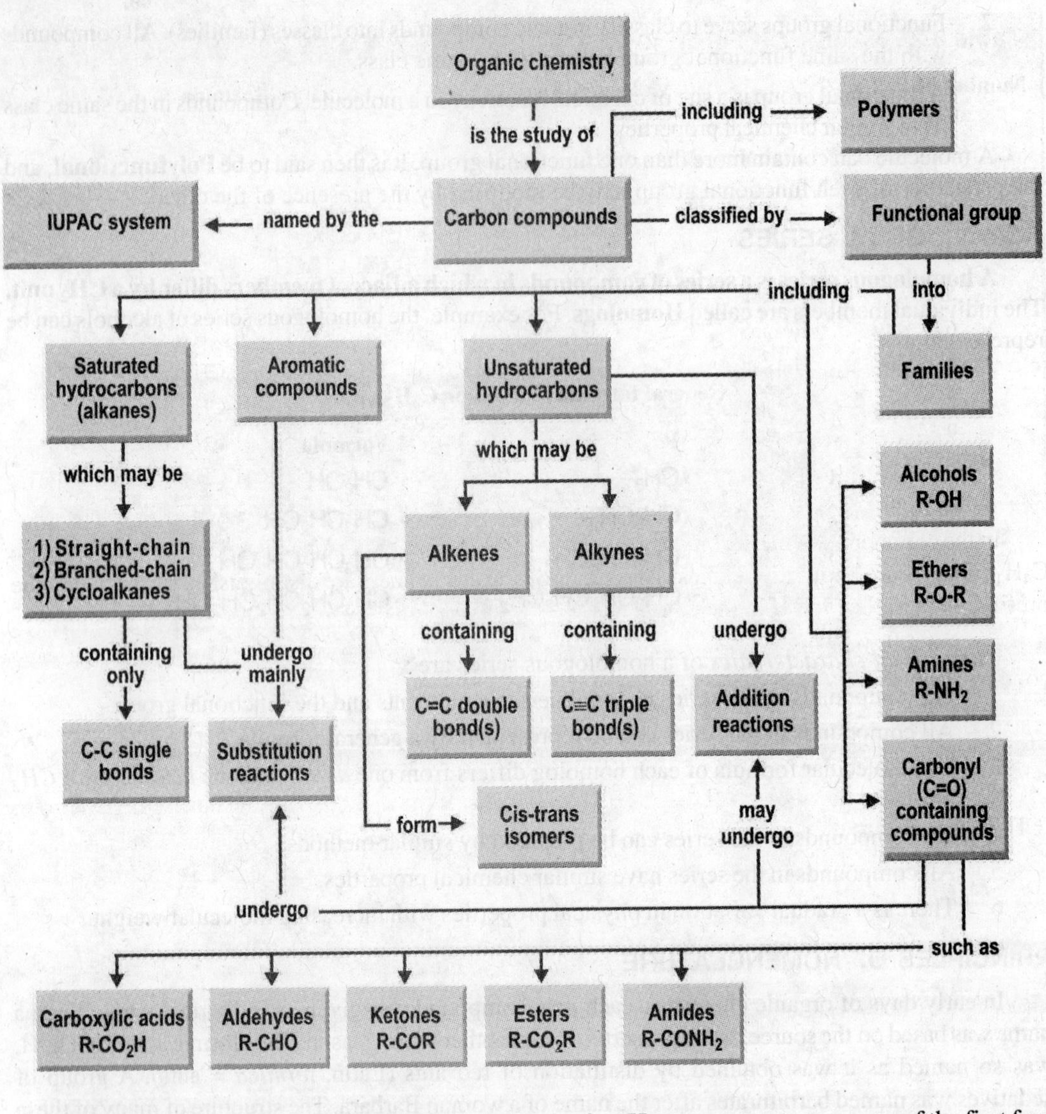

IUPAC names are increasingly used in current practice. However, common names of the first few members of a homologous series have been retained in the IUPAC system. In this book both IUPAC and common names will be used.

Naming organic compounds is an essential skill. You must be able to give correct name for a structure or derive a correct structure from a name. The IUPAC system may appear difficult. In reality, learning systematic nomenclature is probably the easiest part of organic chemistry. All that is required is a good deal of practice. Study the following sections very carefully. To help you master organic nomenclature, we have included many helpful ideas, tips, and suggestions .

ALKANES

Alkanes are hydrocarbons that contain only single bonds. (Recall : Hydrocarbons are compounds that contain only carbon and hydrogen). The first four members of the series are known by their common names : *Methane, Ethane, Propane*, and *Butane*. The names of larger alkanes are derived from the Greek prefixes that indicate the number of carbon atoms in the molecule. Thus *pentane* has 5 carbons, *hexane* has 6, and so on.

Table 8.2 NAMES OF FIRST 10 STRAIGHT CHAIN ALKANES

Number of carbon atoms	Name	Molecular formula	Structural formula
1	Methane	CH_4	CH_4
2	Ethane	C_2H_6	CH_3CH_3
3	Propane	C_3H_8	$CH_3CH_2CH_3$
4	Butane	C_4H_{10}	$CH_3CH_2CH_2CH_3$
5	Pentane	C_5H_{12}	$CH_3CH_2CH_2CH_2CH_3$
6	Hexane	C_6H_{14}	$CH_3CH_2CH_2CH_2CH_2CH_3$
7	Heptane	C_7H_{16}	$CH_3CH_2CH_2CH_2CH_2CH_2CH_3$
8	Octane	C_8H_{18}	$CH_3CH_2CH_2CH_2CH_2CH_2CH_2CH_3$
9	Nonane	C_9H_{20}	$CH_3CH_2CH_2CH_2CH_2CH_2CH_2CH_2CH_3$
10	Decane	$C_{10}H_{22}$	$CH_3CH_2CH_2CH_2CH_2CH_2CH_2CH_2CH_2CH_3$

In the common system all isomeric alkanes have the same parent name. For example, two isomeric C_4H_{10} alkanes are known as butanes. The names of various isomers are distinguished by prefixes. The prefix indicates the type of branching present in the molecule.

(1) **Meaning of the Prefix *n*-.** Prefix *n*- is used for those alkanes in which all carbons are in one continuous chain. The prefix *n*- stands for normal or straight–chain.

$CH_3CH_2CH_2CH_3$
n-Butane

$CH_3CH_2CH_2CH_2CH_3$
n-Pentane

(2) **Meaning of the Prefix Iso-.** Prefix *iso*- is used for those alkanes which have a methyl group (CH_3–) attached to the second last carbon atom of the continuous chain.

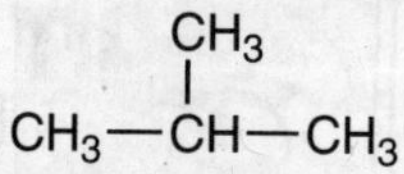
Isobutane

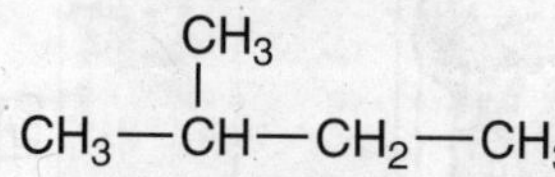
Isopentane

(3) **Meaning of the Prefix Neo-.** Prefix *neo*- is used for those alkanes which have two methyl groups attached to the second last carbon atom of the continuous chain.

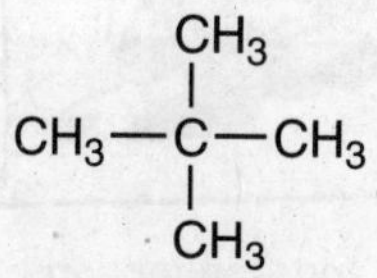
Neopentane

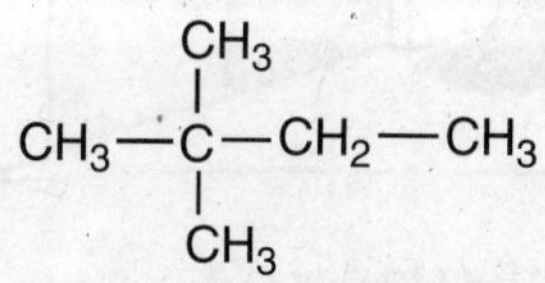
Neohexane

Guidelines

In the common system of naming alkanes, the following prefixes are encountered very often. Memorize their meaning.

Prefix	Meaning of the Prefix
n-	Alkane is unbranched.
iso-	Alkane contains $(CH_3)_2CH$– and no other branches.
neo-	Alkane contains $(CH_3)_3C$– and no other branches.

Day-to-day Life : Follow Rules and Conventions
IUPAC Nomenclature : Follow Rules and Conventions

Classification of Carbon Atoms

The structural formulas of alkanes contain four types of carbons :

(1) **Primary Carbon.** A carbon atom attached to one other (or no other) carbon atom is called primary carbon (1° carbon ; 1° = primary).

(2) **Secondary Carbon.** A carbon atom attached to two other carbon atoms is called secondary carbon (2° carbon ; 2° = secondary).

(3) **Tertiary Carbon.** A carbon atom attached to three other carbon atoms is called tertiary carbon (3° carbon ; 3° = tertiary).

(4) **Quaternary Carbon.** A carbon atom attached to four other carbon atoms is called quaternary carbon (4° carbon ; 4° = quaternary)

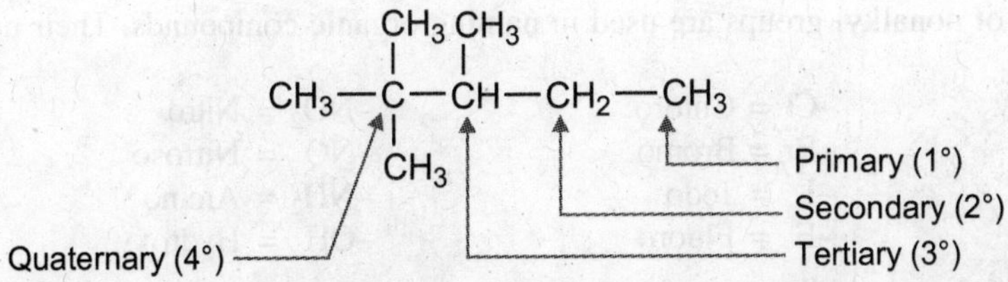

Hydrogen atoms attached to 1°, 2°, 3° carbon atoms are often referred to as primary, secondary and tertiary hydrogen atoms. *You must learn to recognize primary, secondary, tertiary, and quaternary carbons.*

Alkyl Groups

An alkyl group is formed by removing one hydrogen atom from an alkane.

$$\underset{\text{Alkane}}{RH} \xrightarrow{-H} \underset{\text{Alkyl group}}{R-}$$

$$\underset{\text{Methane}}{CH_4} \xrightarrow{-H} \underset{\text{Methyl group}}{CH_3-}$$

$$\underset{\text{Ethane}}{CH_3CH_3} \xrightarrow{-H} \underset{\text{Ethyl group}}{CH_3CH_2-}$$

The symbol R– is often used to represent an alkyl group. The grouping R– (e.g.CH_3CH_2–) is not a compound and must be bonded to another atom or group of atoms. Alkyl groups are named by dropping *-ane* from the name of the corresponding alkane, and adding the ending-*yl*.

Alkane – ane + yl = Alkyl

Table 8.3 lists the structures and names of some common alkyl groups.

Table 8.3 NAMES OF COMMON ALKYL GROUPS

Parent alkane	Structure of parent alkane	Structure of alkyl group	Name of alkyl group
Methane	CH_4	CH_3-	methyl
Ethane	CH_3CH_3	CH_3CH_2-	ethyl
Propane	$CH_3CH_2CH_3$	$CH_3CH_2CH_2-$	*n*-propyl
		$CH_3\overset{\vert}{C}HCH_3$	isopropyl
n-Butane	$CH_3CH_2CH_2CH_3$	$CH_3CH_2CH_2CH_2-$	
		$CH_3CH_2\overset{\vert}{C}HCH_3$	
Isobutane	$CH_3\overset{CH_3}{\overset{\vert}{C}}HCH_3$	$CH_3\overset{CH_3}{\overset{\vert}{C}}HCH_2-$	isobutyl
		$CH_3\overset{CH_3}{\overset{\vert}{\underset{\vert}{C}}}CH_3$	*t*-butyl (tertiary-butyl)

Nonalkyl Groups

A number of nonalkyl groups are used in naming organic compounds. Their names are given below :

–Cl = Chloro	$-NO_2$ = Nitro
–Br = Bromo	–NO = Nitroso
–I = Iodo	$-NH_2$ = Amino
–F = Fluoro	–OH = Hydroxy

IUPAC Rules for Naming Alkanes

The IUPAC system is much the same for all classes of organic compounds. The IUPAC rules for naming alkanes are given below. *Master them* !

Rule 1. **Select the longest continuous carbon chain.** Remember that this chain does not have to be that portion of the molecule that is written horizontally. This is a mistake made by many beginning students.

Rule 2. **Name the longest chain.** The longest carbon chain is chosen as the basis for the name. The names are those given in Table 8.2. These are the *Parent Names.*

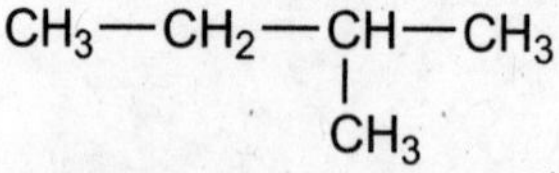

The longest continuous chain has four carbon atoms. Thus, the compound is named as a butane.

Rule 3. **Number the longest chain.** The carbon atoms in the longest chain are numbered. The numbering is started from that end which will give numbers having the lowest value to carbons carrying substituents.

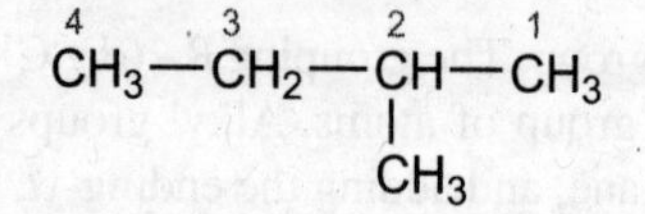

and not

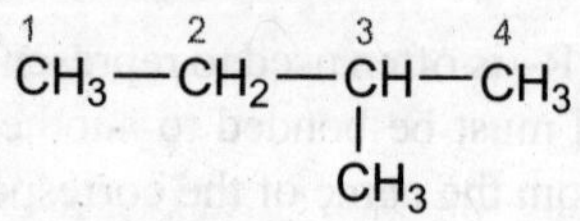

Rule 4. **Identify the substituent.** Name the substituent. Indicate its position by the number of the carbon atom to which it is attached.

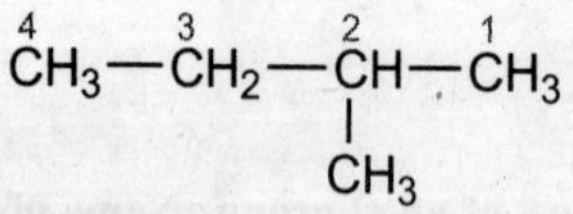

The attached group is located on carbon 2 of the chain, and it is a methyl group.

Rule 5. **Prefix the position number and name of the substituent onto the parent name. The whole name is written as one-word.** Notice that the number and name of the substituent are separated by a hyphen.

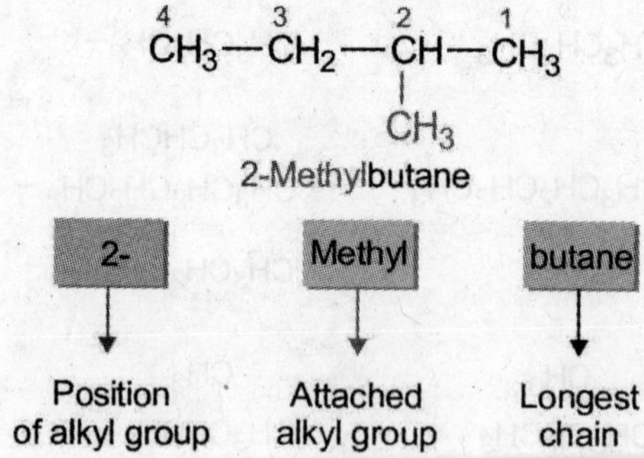

The following additional rules are used when more than one substituent is attached to the longest continuous carbon chain.

Rule 6. Identify the substituents by names and position numbers. When the same substituent is present two or more times in the molecule, prefixes *di,tri-, tetra-, penta-*, etc. are used. Position of each substituent is indicated by a separate number. These position numbers, separated by commas, are put just before the name of the substituent, with the hyphen before and after the numbers when necessary.

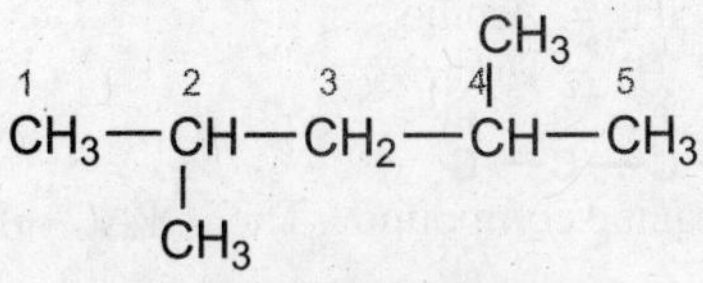

2,4-Dimethylpentane

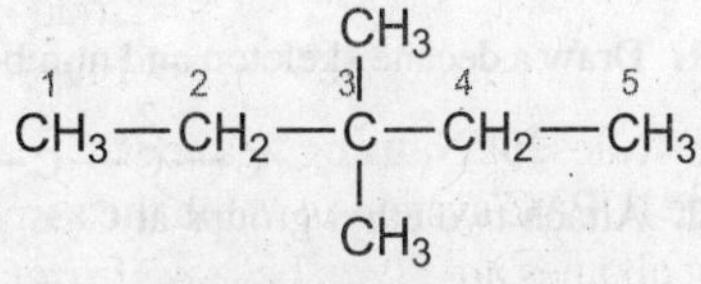

3,3-Dimethylpentane

Rule 7. When two or more different substituents are present, their names are arranged in alphabetic order and added to the name of the parent alkane, again as one word.

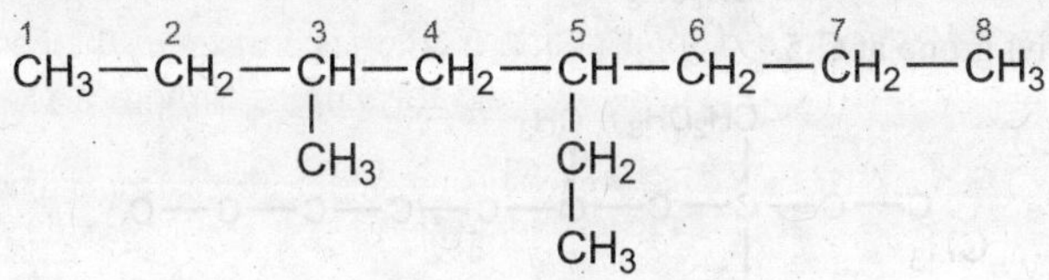

5-Ethyl-3-methyloctane

Guidelines

Use the following step-by-step procedure to write the IUPAC names from structural formulas

Consider the following structural formula :

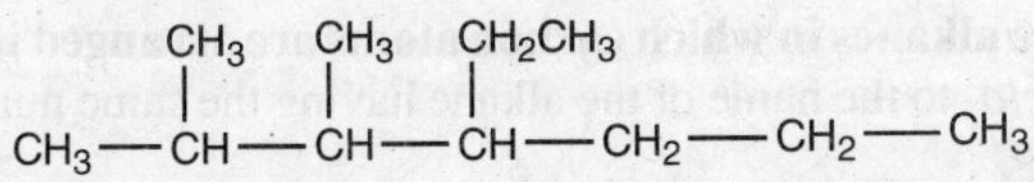

Step 1. Identify the longest chain. Many beginning students find it helpful to box the longest chain. In this example, the longest chain contains seven carbons. The seven carbon chain is *heptane*.

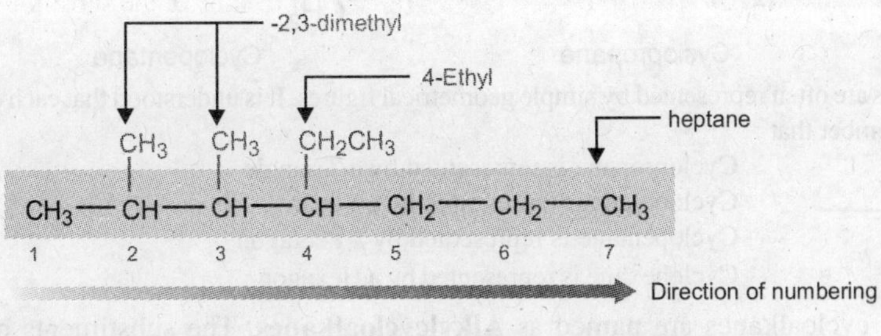

Step 2. Number of chain. The chain is numbered from left to right. This gives the lowest numbers to the attached alkyl groups.

Step 3. Identify the alkyl groups. There are two methyl groups at C-2 and C-3. There is one ethyl group at C-4.

Step 4. Write the IUPAC name. In this case the IUPAC name is *4-Ethyl-2,3-dimethylheptane*. Always remember : (*a*) Numbers are separated from each other by commas ; (*b*) Numbers are separated from names by hyphens ; (*c*) Prefixes *di-, tri-* are not taken into account in alphabetizing substituent names.

Guidelines

Use the following step-by-step procedure to draw structural formulas from the IUPAC names

Consider the following IUPAC name :

3,3-Diethyl-5-methyldecane

Step 1. Draw a decane skeleton and number it.

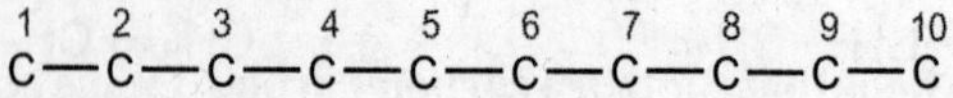

Step 2. Attach two ethyl groups at C-3.

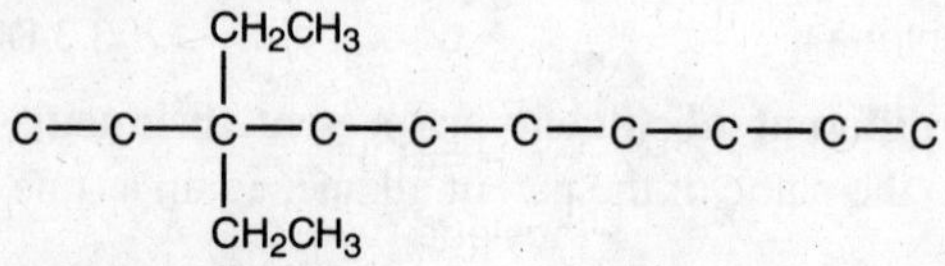

Step 3. Attach a methyl group at C-5.

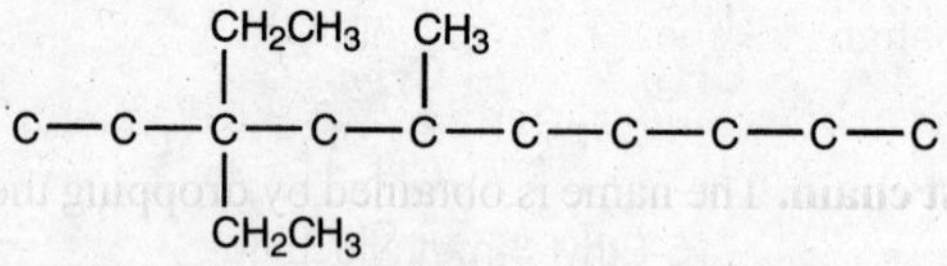

Step 4. Supply hydrogen atoms so that each carbon atom has four bonds.

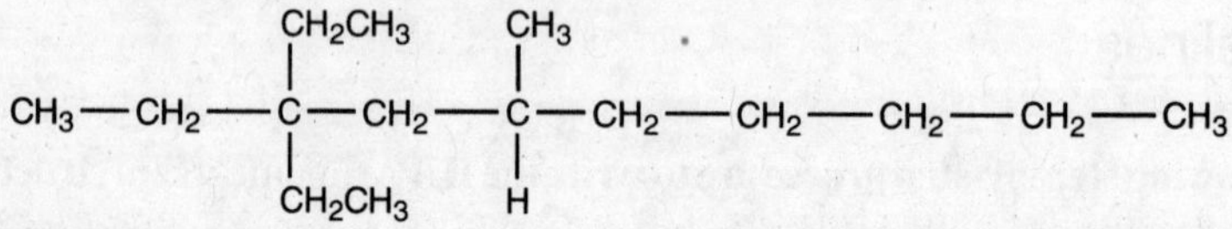

CYCLOALKANES

Cycloalkanes are alkanes in which carbon atoms are arranged in a ring. They are named by attaching the prefix *cyclo-* to the name of the alkane having the same number of carbons as in the ring

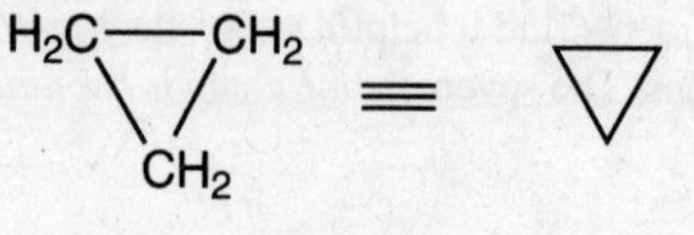

Cyclopropane

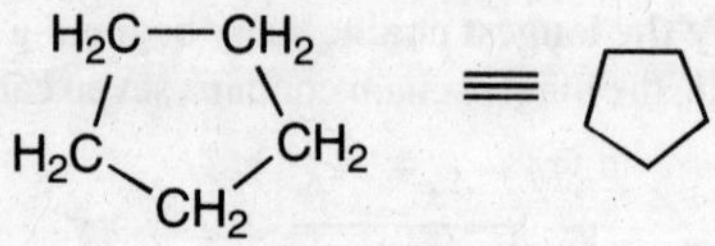

Cyclopentane

Cycloalkanes are often represented by simple geometrical figures. It is understood that each corner represents CH_2 group. Remember that :

Cyclopropane is represented by a Triangle
Cyclobutane is represented by a Square
Cyclopentane is represented by a Pentagon
Cyclohexane is represented by a Hexagon

Substituted cycloalkanes are named as **Alkylcycloalkanes.** The substituents on the ring are named, and their positions are indicated by numbers. The ring is numbered so that the substituents are given the lowest possible numbers.

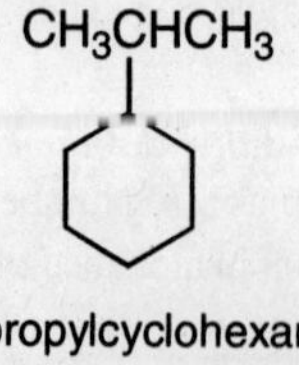

Isopropylcyclohexane

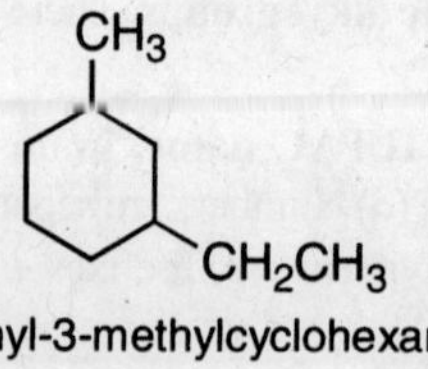

1-Ethyl-3-methylcyclohexane

The ring is designated as a substituent if the alkane chain contains a greater number of carbon atoms than the ring.

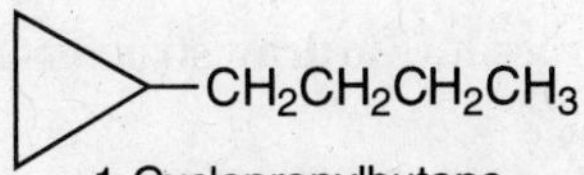

1-Cyclopropylbutane
(*not* Butylcyclopropane)

ALKENES

Alkenes are hydrocarbons that contain a carbon-carbon double bond. Their common names are obtained by changing the ending *-ane* of the corresponding alkane to *-ylene* (Alkane – ane + ylene).

$CH_2{=}CH_2$	$CH_3{-}CH{=}CH_2$	$CH_3{-}C(CH_3){=}CH_2$
Ethylene	Propylene	Isobutylene

IUPAC Rules for Naming Alkenes

The IUPAC rules for naming alkenes are given below :

Rule 1. Select the longest carbon chain containing the double bond. This is the parent chain.

Rule 2. Name the longest chain. The name is obtained by dropping the final *-ane* from the name of the corresponding alkane, and adding the ending *-ene*.

Alkane –ane + ene = Alkene

Rule 3. Number the chain from the end closer to the double bond.

Rule 4. Indicate the position of the double bond by the number of the first (lowest numbered) carbon atom involved in the double bond.

Rule 5. Alkyl groups and other substituents are numbered, named, and placed as prefixes in alphabetic order. For example,

$CH_2{=}CH_2$	$CH_3{-}CH{=}CH_2$	$CH_3{-}C(CH_3){=}CH_2$
Ethene	Propene	2-Methylpropene

$$\overset{4}{C}H_3{-}\overset{3}{C}H_2{-}\overset{2}{C}H{=}\overset{1}{C}H_2$$

1-Butene

$$\overset{1}{C}H_3{-}\overset{2}{C}(CH_3){=}\overset{3}{C}H{-}\overset{4}{C}H_2{-}\overset{5}{C}H_3$$

2-Methyl-2-pentene

Rule 6. Alkenes containing two double bonds are named as **Alkadienes.**

$$\overset{1}{C}H_2{=}\overset{2}{C}{=}\overset{3}{C}H_2$$

1,2-Propadiene

$$\overset{1}{C}H_2{=}\overset{2}{C}H{-}\overset{3}{C}H{=}\overset{4}{C}H_2$$

1,3-Butadiene

Rule 7. In naming **cycloalkenes**, number the ring to give the double bonded carbons the numbers 1 and 2. Choose the direction of numbering so that the substituents get the lowest numbers. The position of the double bond is not indicated because it is known to be between C-1 and C-2.

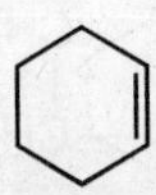

Cyclohexene

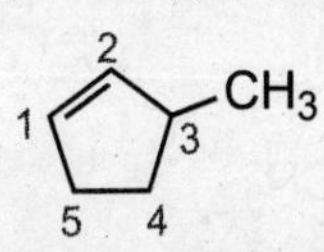

3-Methylcyclopentene

ALKYNES

Alkynes are hydrocarbons that contain a carbon-carbon triple bond. The simplest alkyne, HC≡CH, is commonly called acetylene.

IUPAC Rules for Naming Alkynes

The IUPAC rules for naming alkynes are analogous to those for alkenes.

Rule 1. Select the longest carbon chain containing the triple bond. This is the parent chain.

Rule 2. Name the longest chain. The name is obtained by dropping the final *-ane* from the name of the corresponding alkane, and adding the ending *-yne*.

Alkane – ane + yne = Alkyne

Rule 3. Number the chain from the end closer to the triple bond.

Rule 4. Indicate the position of the triple bond by the number of the first (lowest numbered) carbon atom involved in the triple bond.

Rule 5. Alkyl groups and other substituents are numbered, named, and placed as prefixes in alphabetic order. For example,

HC≡CH — Ethyne

CH_3—C≡CH — Propyne

$\overset{4}{C}H_3\overset{3}{C}≡\overset{2}{C}\overset{1}{C}H_3$ — 2-Butyne

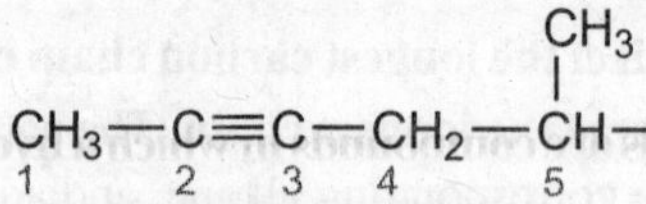

5-Methyl-2-hexyne
(*not* 2-Methyl-4-hexyne)

Rule 6. Alkynes containing two triple bonds are named as **Alkadiynes.**

$\underset{1}{HC}≡\underset{2}{C}—\underset{3}{C}≡\underset{4}{CH}$ — 1,3-Butadiyne

ALKYL HALIDES

Alkyl halides are compounds which contain carbon-halogen bonds. They are classified as primary, secondary, or tertiary, depending upon whether the halogen atom is bonded to a primary, secondary, or tertiary carbon atom. Abbreviations for these terms are 1°, 2°, and 3°, respectively. The common names of alkyl halides are obtained by naming the alkyl group attached to the halogen and adding the name of the halide as a separate word.

CH_3—Cl — Methyl chloride

CH_3CH_2—Br — Ethyl bromide

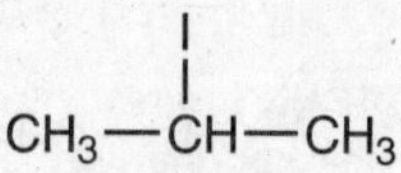

Isopropyl iodide

IUPAC Rules for Naming Alkyl Halides

The IUPAC names of alkyl halides are obtained by using the following rules :

Rule 1. Select the longest chain to which the halogen is attached and give it the name of the corresponding alkane.

Rule 2. Prefix the name of the alkane by chloro, bromo, iodo, or fluoro.

Rule 3. Number the chain so as to give the carbon carrying the halogen atom the lowest possible number.

Rule 4. Other substituents are numbered, named and placed as prefixes in alphabetic order. If there are two or more identical halogen substituents, the prefixes *di-*, *tri-*, *tetra-*, etc. are used. For example.

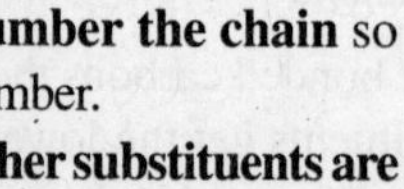

2-Bromopropane

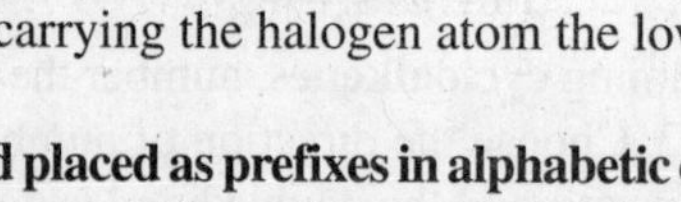

1-Chloro-2-methylbutane

$$\overset{5}{CH_3}-\overset{4}{CH_2}-\overset{3}{C}(Cl)_2-\overset{2}{CH_2}-\overset{1}{CH_2}-Br$$

1-Bromo-3,3-dichloropentane

ORGANOMETALLIC COMPOUNDS

Organometallic compounds are compounds containing a carbon-metal bond. They are named as *Alkylmetals.*

CH_3CH_2Li	$(CH_3CH_2)_4Pb$
Ethyllithium	Tetraethyllead

If the metal is bonded to an inorganic anion as well as a carbon atom, the compound is named as a derivative of the inorganic salt.

CH_3CH_2MgBr Ethylmagnesium bromide

ALCOHOLS

Alcohols are compounds in which a hydroxyl group (–OH) is bonded to a saturated carbon. They are classified as primary, secondary, or tertiary, depending upon whether the –OH group is attached to a primary, secondary, or tertiary carbon atom. Abbreviations for these terms are 1°, 2°, and 3°, respectively. The common names of alcohols are obtained by naming the alkyl group attached to the –OH group and adding alcohol as a separate word.

CH_3CH_2OH	$CH_3CH_2CH_2OH$	$CH_3-CH(OH)-CH_3$
Ethyl alcohol	*n*-Propyl alcohol	Isopropyl alcohol

IUPAC Rules for Naming Alcohols

In the IUPAC system, alcohols are named as *Alkanols*. Following rules are used :

Rule 1. Select the longest chain to which the hydroxyl group is attached.

Rule 2. Name the longest chain. The name is obtained by dropping the final *-e* from the name of the alkane that contains the same number of carbon atoms, and adding the ending *-ol*.

Alkane – e + ol = Alkanol

Rule 3. Number the chain to give the lowest number to carbon attached to the hydroxyl group.

Rule 4. Indicate the position of the hydroxyl group by the number of the carbon attached to the hydroxyl group.

Rule 5. Other substituents are numbered, named, and placed as prefixes in alphabetic order. For example,

CH_3OH	CH_3CH_2OH	$CH_3-CH(OH)-CH_3$	$\overset{4}{CH_3}-\overset{3}{CH}(CH_3)-\overset{2}{CH_2}-\overset{1}{CH_2}-OH$
Methanol	Ethanol	2-Propanol	3-Methyl-1-butanol

Rule 6. Alcohols containing two or three –OH groups are named as **Alkanediols** and **Alkanetriols** respectively. Notice that *-e* of the corresponding alkane name is retained.

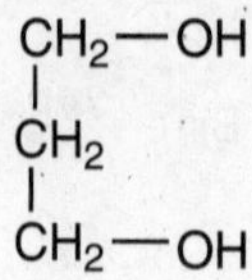

1,3-Propanediol

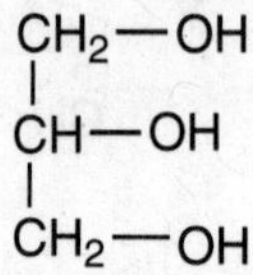

1,2,3-Propanetriol

ETHERS

Ethers are compounds in which an oxygen atom is bonded to two organic groups (R–O–R). In the common system, the two alkyl groups attached to the oxygen atom are named in alphabetic order and the word *ether* is added. If the groups are same, the prefix *di-* is used.

CH_3OCH_3	Dimethyl ether
$CH_3CH_2OCH_3$	Ethyl methyl ether

In the IUPAC system, ethers are named as **Alkoxyalkanes**. The smaller alkyl group *plus* the oxygen atom is called an *alkoxy* substituent.

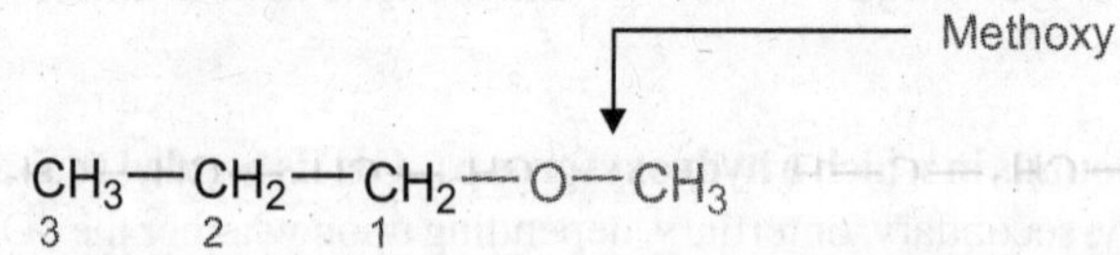

1-Methoxypropane

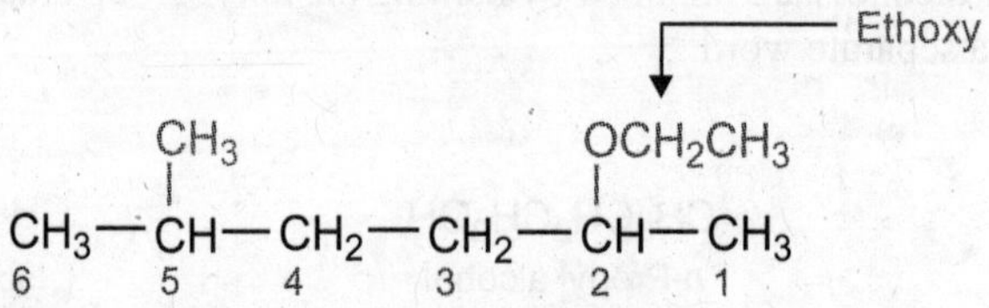

2-Ethoxy-5-methylhexane

OXIRANES (EPOXIDES)

Oxiranes are cyclic ethers in which the ether oxygen is part of the three membered ring. Oxiranes are also called *Epoxides*. Because they are readily prepared from alkenes, they are commonly known as *Alkene oxides*. In the IUPAC system, they are named as *Alkyloxiranes*. Substituents on the oxirane ring require a numbering system. The oxygen atom is given the number *1*.

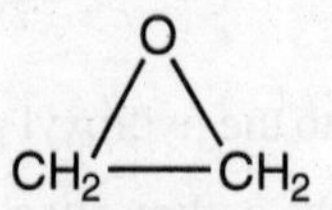

Oxirane (Ethylene oxide)

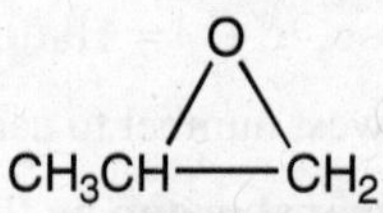

2-Methyloxirane (Propylene oxide)

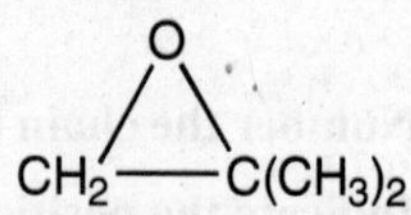

2,2-Dimethyloxirane (Isobutylene oxide)

ALDEHYDES

Aldehydes are compounds in which the carbonyl group (C=O) is bonded to an organic group and hydrogen (or two hydrogens). Their common names are related to those of carboxylic acids.

Formula	Common Name	IUPAC Name
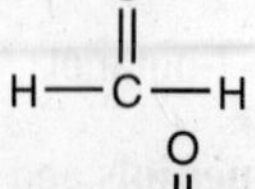	Formaldehyde	Methanal
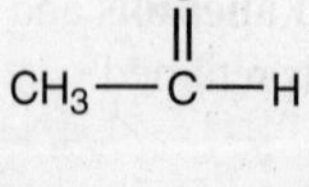	Acetaldehyde	Ethanal

$CH_3CH_2-C(=O)-H$	Propionaldehyde	Propanal
$CH_3CH_2CH_2-C(=O)-H$	Butyraldehyde	Butanal

IUPAC Rules for Naming Aldehydes

In the IUPAC system, aldehydes are named as *Alkanals*. Following rules are used :

Rule 1. Select the longest chain containing the aldehyde group.

Rule 2. Name the longest chain. The name is obtained by dropping the final *-e* from the name of the corresponding alkane, and adding the ending *-al*.

Alkane – e + al = Alkanal

Rule 3. Number the chain by assigning the number *1* to the aldehyde carbon. The number *1* is not used to indicate the position of the carbonyl carbon, since it is always located at the end of the chain.

Rule 4. Other substituents are numbered, named, and placed as prefixes in alphabetic order.

$$\underset{4}{CH_3}-\underset{3}{CH(CH_3)}-\underset{2}{CH_2}-\underset{1}{C(=O)}-H$$

3-Methylbutanal

$$\underset{5}{CH_3}-\underset{4}{CH_2}-\underset{3}{CH(Br)}-\underset{2}{CH_2}-\underset{1}{C(=O)}-H$$

3-Bromopentanal

Rule 5. When there are two aldehyde groups in a molecule, it is named as ***Alkanedial***. Notice that *-e* of the corresponding alkane name is retained.

$$H-\underset{6}{C(=O)}-\underset{5}{CH_2}-\underset{4}{CH_2}-\underset{3}{CH_2}-\underset{2}{CH_2}-\underset{1}{C(=O)}-H$$

1,6-Hexanedial

Note : Many students are very sloppy in writing *a*'s and *o*'s so that an *a* looks like an *o*. Remember that the ending for an aldehyde is *-al* and for an alcohol *-ol*. Be sure to write clearly so that *a* looks like an *a*.

KETONES

Ketones are compounds in which the carbonyl group (C=O) is bonded to two organic groups. Their common names are obtained by naming the alkyl groups attached to the carbonyl group and adding the word *ketone*.

Formula	Common Name	IUPAC Name
$CH_3-C(=O)-CH_3$	Dimethyl ketone (Acetone)	Propanone
$CH_3-C(=O)-CH_2CH_3$	Ethyl methyl ketone	Butanone

IUPAC Rules for Naming Ketones

In the IUPAC system, ketones are named as *Alkanones*. Following rules are used :

Rule 1. Select the longest chain containing the carbonyl carbon atom.

Rule 2. Name the longest chain. The name is obtained by dropping the final *-e* from the name of the corresponding alkane, and adding the ending *-one*.

Alkane – e + one = Alkanone

Rule 3. Number the chain to give the lowest number to the carbonyl carbon.

Rule 4. Other substituents are numbered, named, and placed as prefixes in alphabetic order. For example,

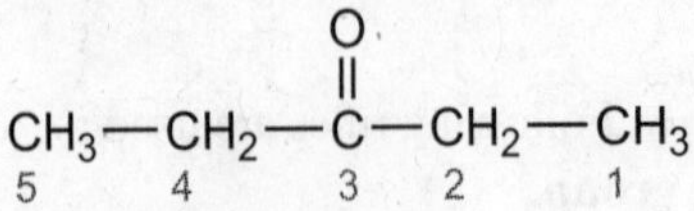

3-Pentanone

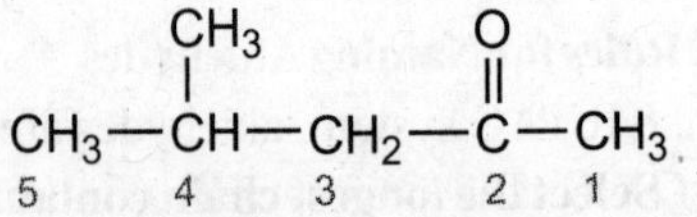

4-Methyl-2-pentanone

Rule 5. When there are two carbonyl groups in a molecule, it is named as **Alkanedione**. Notice that *-e* of the corresponding alkane name is retained.

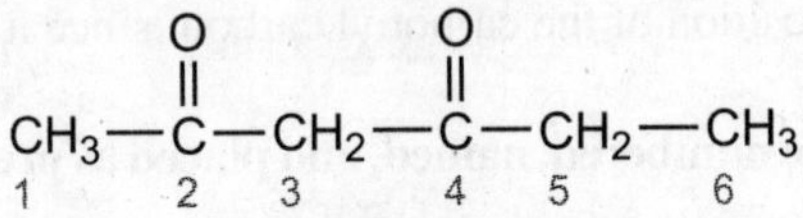

2,4-Hexanedione

CARBOXYLIC ACIDS

Carboxylic acids are compounds which contain the carboxyl group (–COOH). Their common names end in *-ic acid.*

Formula	Common Name	IUPAC Name
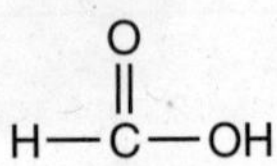	Formic acid	Methanoic acid
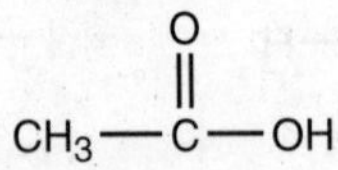	Acetic acid	Ethanoic acid
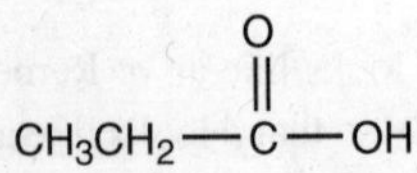	Propionic acid	Propanoic acid
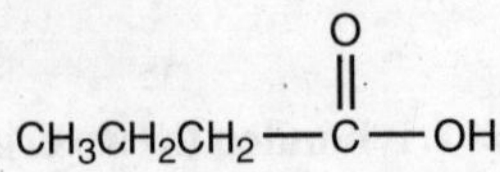	Butyric acid	Butanoic acid

IUPAC Rules for Naming Carboxylic Acids

In the IUPAC system, carboxylic acids are named as *Alkanoic acids*. Following rules are used :

Rule 1. Select the longest chain containing the carboxyl group.

Rule 2. Name the longest chain. The name is obtained by dropping the final *-e* from the name of the corresponding alkane, and adding the ending *-oic acid.*

Alkane – e + oic acid = Alkanoic acid

Rule 3. Number the chain starting with carboxyl carbon as number *1*. The number *1* is not used to indicate the position of the carboxyl carbon, since it is always located at the end of the chain.

Rule 4. Other substituents are numbered, named, and placed as prefixes in alphabetic order. For example,

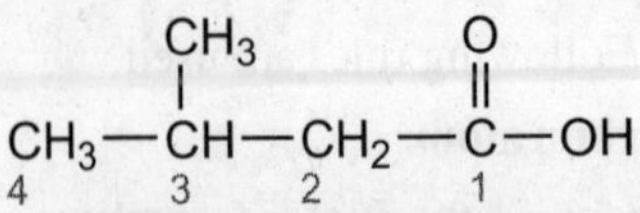

3-Methylbutanoic acid

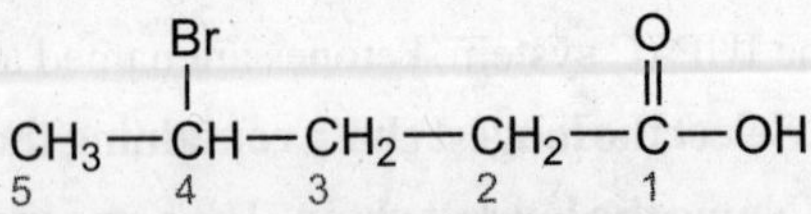

4-Bromopentanoic acid

Rule 5. When there are two carboxyl groups in a molecule, it is named as **Alkanedioic acid**. Notice that *-e* of the corresponding alkane name is retained.

$$\underset{1}{\overset{O}{\overset{\|}{HO-C}}}-\underset{2}{CH_2}-\underset{3}{CH_2}-\underset{4}{CH_2}-\underset{5}{CH_2}-\underset{6}{\overset{O}{\overset{\|}{C}}}-OH$$

1,6-Hexanedioic acid

CARBOXYLIC ACID DERIVATIVES

Carboxylic acid derivatives are compounds in which the hydroxyl part of the carboxyl group is replaced by various other groups.

(*a*) **Acid Chlorides.** Acid chlorides are derived from acids by replacing *OH* group by *Cl* atom. They are named by changing the ending *-ic acid* of the corresponding carboxylic acid to *-yl chloride.*

Formula	Common Name	IUPAC Name
$CH_3-\overset{O}{\overset{\|}{C}}-Cl$	Acetyl chloride	Ethanoyl chloride
$CH_3CH_2-\overset{O}{\overset{\|}{C}}-Cl$	Propionyl chloride	Propanoyl chloride

(*b*) **Esters.** Esters are derived from acids by replacing the *OH* group by *OR* group. Their names consist of two words. The first word is the name of the alkyl group attached to the oxygen atom. The second word is derived from the carboxylic acid name with *-ic acid* changed to *-ate*.

Formula	Common Name	IUPAC Name
$H-\overset{O}{\overset{\|}{C}}-OCH_3$	Methyl formate	Methyl methanoate
$CH_3-\overset{O}{\overset{\|}{C}}-OC_2H_5$	Ethyl acetate	Ethyl ethanoate

(*c*) **Anhydrides.** Anhydrides are derived from acids by replacing *OH* by *OCOR*. They are named by changing the suffix *acid* of the parent acid to *anhydride*.

Formula	Common Name	IUPAC Name
$CH_3-\overset{O}{\overset{\|}{C}}-O-\overset{O}{\overset{\|}{C}}-CH_3$	Acetic anhydride	Ethanoic anhydride
$C_2H_5-\overset{O}{\overset{\|}{C}}-O-\overset{O}{\overset{\|}{C}}-C_2H_5$	Propionic anhydride	Propanoic anhydride

(*d*) **Amides.** Amides are derived from acids by replacing *OH* by NH_2. They are named by replacing the ending *-oic acid* (of the IUPAC name) or *-ic acid* (of the common name) of the corresponding carboxylic acid with the ending *-amide*.

Formula	Common Name	IUPAC Name
$H-\overset{O}{\overset{\|}{C}}-NH_2$	Formamide	Methanamide
$CH_3-\overset{O}{\overset{\|}{C}}-NH_2$	Acetamide	Ethanamide

SULFONIC ACIDS

Sulfonic acids are compounds which contain the sulfo group ($-SO_3H$). Their IUPAC names are obtained by adding the suffix -sulfonic acid to the alkane name.

$CH_3CH_2SO_3H$

Ethanesulfonic acid

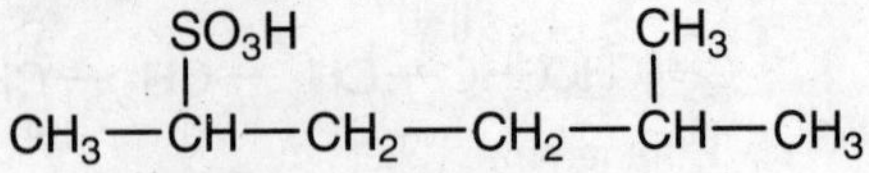

5-Methyl-2-hexanesulfonic acid

NITRO COMPOUNDS

Nitro compounds are compounds which contain the nitro group ($-NO_2$). Their IUPAC names are obtained by adding the prefix *nitro-* to the alkane name.

CH_3NO_2

Nitromethane

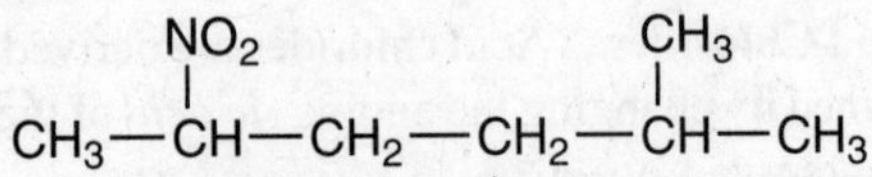

5-Methyl-2-nitrohexane

CYANIDES OR NITRILES

Cyanides are compounds which contain the nitrile group ($-CN$). In the common system, they are named as alkyl cyanides. In the IUPAC system, they are named as **Alkanenitriles**. Numbering includes the nitrile carbon atom.

$\overset{2}{C}H_3-\overset{1}{C}N$

Ethanenitrile
(Methyl cyanide)

$\overset{5}{C}H_3\overset{4}{C}H_2\overset{3}{C}H_2\overset{2}{C}H_2-\overset{1}{C}N$

Pentanenitrile
(*n*-Butyl cyanide)

The prefix *cyano* is used for the nitrile group when it occurs as a substituent. In that case, the carbon of the $-CN$ group is not numbered.

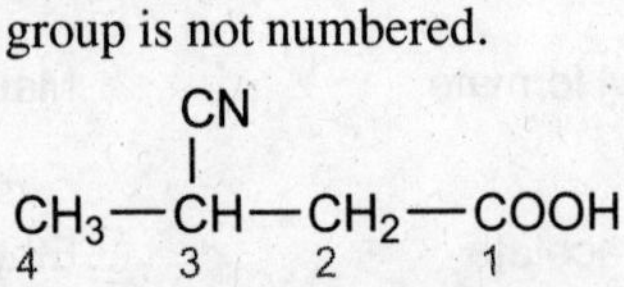

3-Cyanobutanoic acid

AMINES

Amines are derivatives of ammonia (NH_3) in which one or more hydrogen atoms have been replaced by alkyl or aryl groups. They are classified as primary, secondary, or tertiary depending on the number of groups attached to the nitrogen atom. Abbreviations for these terms are 1°,2° and 3°, respectively.

$R-NH_2$ **Primary amines** have one alkyl group attached to the nitrogen atom.
R_2NH **Secondary amines** have two alkyl groups attached to the nitrogen atom.
R_3N **Tertiary amines** have three alkyl groups attached to the nitrogen atom.

The *R* groups in secondary and tertiary amines may be same or different.

In the common system, amines are named by adding the suffix *-amine* to the name of the alkyl group or groups. The entire name is written *as one word*. The names of the alkyl groups are arranged in alphabetic order.

CH_3NH_2

Methylamine

$CH_3-NH-CH_3$

Dimethylamine

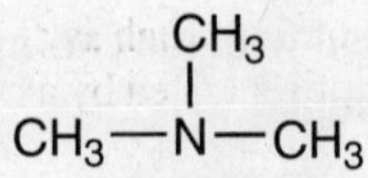

Trimethylamine

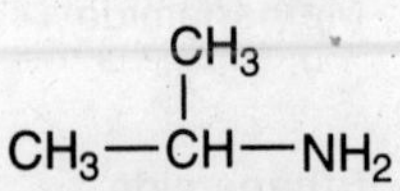

Isopropylamine

$CH_3-NH-CH_2CH_3$

Ethylmethylamine

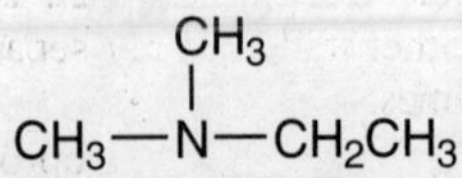

Ethyldimethylamine

In the IUPAC system, primary amines are named by replacing the final *-e* of the parent alkane by *-amine*. If necessary, a number is added to indicate the position of $-NH_2$ group.

$CH_3CH_2NH_2$ — Ethanamine

$CH_3-CH(CH_3)-NH_2$ — 2-Propanamine

Secondary or tertiary amines are named as *N*-substituted derivatives of primary amines. The largest of the groups attached to nitrogen is chosen as the organic group of the primary amine. The remaining alkyl groups are named as substituents by using the prefix *N*- to indicate that they are attached to nitrogen.

$CH_3CH_2-NH-CH_3$ — *N*-Methylethanamine

$CH_3CH_2-N(CH_3)-CH_2CH_2CH_2CH_3$ — *N*-Ethyl-*N*-methylbutanamine

$CH_3-N(CH_3)-\overset{1}{C}H_2-\overset{2}{C}H(CH_3)-\overset{3}{C}H_2-\overset{4}{C}H_3$

2-Methyl-*N*,*N*-dimethylbutanamine

For diamines, the final *-e* of the hydrocarbon name is retained.

$H_2NCH_2CH_2CH_2CH_2CH_2CH_2NH_2$ — 1,6-Hexanediamine

When it is necessary to name $-NH_2$ as a substituent, it is called the *amino* group.

$H_2NCH_2CH_2CH_2CH_2OH$ — 4-Amino-1-butanol

Guidelines

Spellings and Punctuation of IUPAC Names

(1) IUPAC names are almost always written as single words, with notable exception of organic salts and compounds named as acids and acid derivatives.

Ethanol	Ethanoic acid
2-Pentanone	Ethanoyl chloride

(2) Commas are used between two adjacent numbers or letter symbols, and hyphens are used to separate numbers and letter symbols in names.

3,4-Dimethyl-3-heptene   *N*,*N*-Dimethylethanamide

(3) Structural prefixes such as *cis*-, *trans*-, *meso*-, *sec*-, (*or s*-), *tert*- (*or t*-) are *italicized* and joined to the name by a hyphen. These prefixes are ignored in alphabetizing compound names or in capitalising names at the beginning of a sentence.

trans-1,2-Dimethylcyclohexane was isolated from.........

cis-2-Butene is less stable than..........

When these words are not part of a chemical name, they are not italicized.

a cis isomer

a meso compound

(4) Structural prefixes such as di-, tri-, and tetra- are treated as part of the basic name and therefore are neither italicized nor separated by a hyphen. These prefixes are not taken into account in alphabetizing compound names.

4-Ethyl-2,4-dimethylhexane   3-Ethyl-4,7-dimethylnonane

(5) Structural prefixes such as cyclo-, iso-, and neo- are treated as part of the basic name and therefore are neither italicized nor separated by a hyphen. These prefixes are ignored when alphabetizing compound names.

Isopropylamine	Cyclohexane
Ethylisopropylamine	Cyclohexylamine

Next time you make a new compound, just write me a memo with its IUPAC name on it. I will understand. Do not run around with models like an idiot.

NOMENCLATURE OF POLYFUNCTIONAL COMPOUNDS

The functional group present in a molecule determines its class. If it contains an –OH group, the molecule is an alcohol. If it contains a –COOH group, it is a carboxylic acid. Now let us consider a structure which contains both –OH and –COOH functional groups. It can be named as an alcohol or a carboxylic acid. The IUPAC system gives higher priority to –COOH over –OH, and the structure is named as a carboxylic acid. The –OH group is considered as a substituent.

When a compound contains two or more different types of functional groups (Polyfunctional compound), the functional group which specifies its class is called the **Principal Functional Group.** The other functional groups are referred to as **Substituents**. For example, in the following structure :

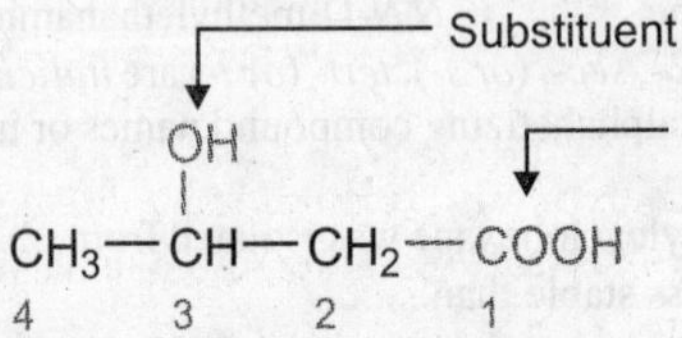

the –COOH group is the principal functional group, while –OH group is a substituent. The principal functional group specifies the class of structure as carboxylic acid. The IUPAC name is 3-Hydroxybutanoic acid.

Selection of the Principal Functional Group

The IUPAC system has laid down the priority of functional groups for determining the class of a polyfunctional compound. Table 8.4 gives a list of functional groups in decreasing order of priority for citation as the principal functional group. That is, the functional group which occurs higher up in the priority table is the principal functional group and specifies the class. Therefore, by having a look at the priority table, you can at once know the class of polyfunctional structure.

Table 8.4 NOMENCLATURE PRIORITY FOR DETERMINING THE PRINCIPAL FUNCTIONAL GROUP. HIGHEST PRIORITY GROUP IS AT THE TOP

Class	Functional group	Suffix used
Carboxylic acid	$-\overset{O}{\overset{\|\|}{C}}-OH$	–oic acid
Sulfonic acid	$-SO_3H$	–sulfonic acid
Ester	$-\overset{O}{\overset{\|\|}{C}}-O-$	alkyl –oate
Acid halide	$-\overset{O}{\overset{\|\|}{C}}-X$	–oyl halide
Amide	$-\overset{O}{\overset{\|\|}{C}}-NH_2$	–amide
Nitrile	$-CN$	–nitrile
Aldehyde	$-\overset{O}{\overset{\|\|}{C}}-H$	–al
Ketone	$-\overset{O}{\overset{\|\|}{C}}-$	–one
Alcohol	$-OH$	–ol
Amine	$-\overset{\|}{N}-$	–amine
Ethers	$-O-$	(ether)
Alkene	$-\overset{\|}{C}=\overset{\|}{C}-$	–ene
Alkyne	$-C\equiv C-$	–yne

Table 8.5 PREFIXES USED FOR FUNCTIONAL GROUPS

–Br	Bromo	–R	Alkyl	$-\overset{O}{\overset{\|\|}{C}}-$	Oxo		
–Cl	Chloro	–OR	Alkoxy	$-NO_2$	Nitro		
–F	Fluoro	–OH	Hydroxy	–NO	Nitroso		
–I	Iodo	$-NH_2$	Amino	–CN	Cyano		

IUPAC Rules for Naming Polyfunctional compounds

The IUPAC names of polyfunctional compounds are obtained by using the following rules:

Rule 1. Identify the principal functional group. This gives the class name of the structure. For example in the following structure :

$$CH_3-\underset{}{\overset{OH}{\overset{|}{CH}}}-CH_2-\overset{CH_3}{\overset{|}{CH}}-COOH$$

The two functional groups in the above structure are –COOH and –OH. Since –COOH ranks higher in the priority table, the structure is named as a carboxylic acid.

Rule 2. Number the longest chain containing the principal functional group from the end closer to it.

$$\underset{5}{CH_3}-\underset{4}{\overset{OH}{\overset{|}{CH}}}-\underset{3}{CH_2}-\underset{2}{\overset{CH_3}{\overset{|}{CH}}}-\underset{1}{COOH}$$

Rule 3. Write the parent name corresponding to the number of carbons in the longest chain.

Longest chain = 5 carbons

Parent name = pentanoic acid

Rule 4. Arrange the substituent names with position numbers in alphabetic order.

4-Hydroxy-2-methyl

Rule 5. Prefix substituent names with the parent name.

4-Hydroxy-2-methylpentanoic acid

Rule 6. The following functional groups are always named as substituents. Their names are prefixed with the parent name.

—Cl	Chloro	—R	Alkyl
—Br	Bromo	—OR	Alkoxy
—I	Iodo	$—NH_2$	Amine
—F	Fluoro	$—NO_2$	Nitro
—CN	Cyano	—NO	Nitroso

Rule 7. Carbon-Carbon double or triple bonds are usually indicated by integrating *-en-* or *-yn-* into the *suffix*, as in *-enoic acid* and *-ynol.*

Rule 8. Compounds containing a double bond (C=C) and a triple bond (C≡C) in the main chain are named as **alkenynes.** The position number of the double bond is inserted before *-alken-* and that of triple bond before *-yne*.

The following examples illustrate the use of these IUPAC rules :

Example 1

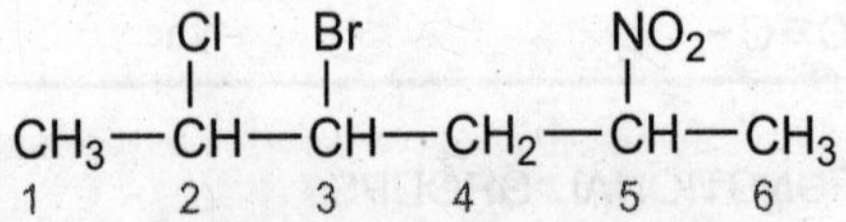

3-Bromo-2-chloro-5-nitrohexane

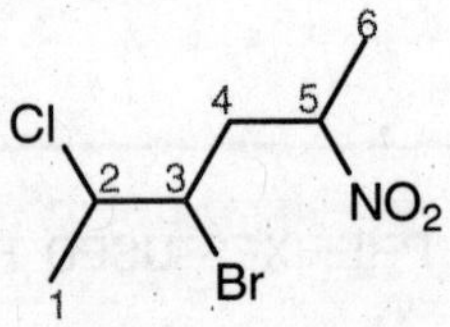

Comments : The functional groups are —Cl, —Br, and $—NO_2$. These functional groups are always named as substituents. None of them is the principal functional group. The names of substituents are arranged in alphabetic order.

Example 2

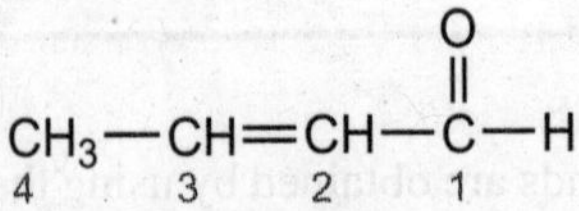

2-Butenal

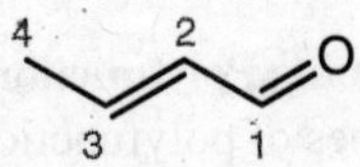

Comments : The functional groups are C=C and —CHO. The —CHO group is the principal functional group as it ranks higher in the priority table (Table 8.4). The compound is named as an aldehyde. The position number of —CHO is not indicated before *-al* as it is always *1*.

Example 3

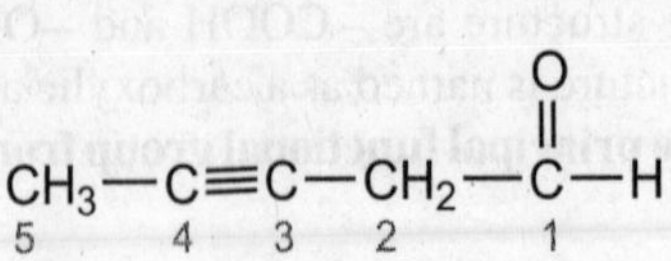

3-Pentynal

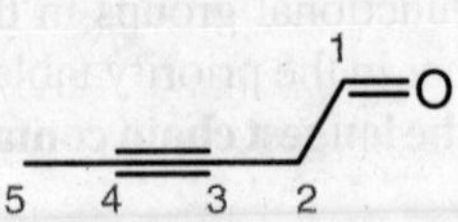

Comments : The functional groups are C≡C and —CHO. The —CHO group is the principal functional group as it ranks higher in the priority table. The compound is named as an aldehyde.

Example 4

$CH_3-CH_2-\overset{O}{\overset{\|}{C}}-CH_2-\overset{O}{\overset{\|}{C}}-H$

5 4 3 2 1

3-Oxopentanal

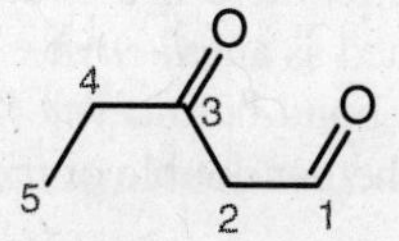

Comments : The functional groups are C=O and –CHO. The –CHO group is the principal functional group as it ranks higher in the priority table. The compound is named as an aldehyde. Both carbonyl carbons form part of the longest chain. The ketonic carbonyl group (C=O) is indicated by the prefix *Oxo*.

Example 5

$CH_3-\overset{O}{\overset{\|}{C}}-CH_2-CH_2-OH$

1 2 3 4

4-Hydroxy-2-butanone

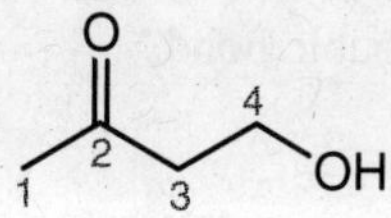

Comments : The functional groups are C=O and –OH. The C=O group is the principal functional group as it ranks higher in the priority table. The compound is named as a ketone. The –OH group is indicated by the prefix *Hydroxy*.

Example 6

$CH_3-CH=CH-\overset{O}{\overset{\|}{C}}-CH_2-CH_3$

6 5 4 3 2 1

4-Hexen-3-one

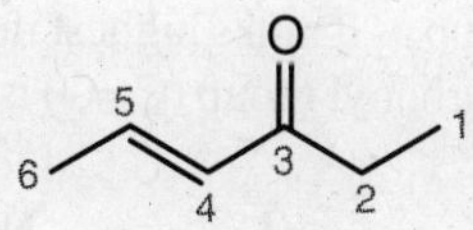

Comments : The functional groups are C=C and C=O. The C=Ogroup is the principal functional group as it ranks higher in the priority table. The compound is named as a ketone.

Example 7

$CH_2=CH-CH_2-\overset{OH}{\overset{|}{C}H}-CH_3$

5 4 3 2 1

4-Penten-2-ol

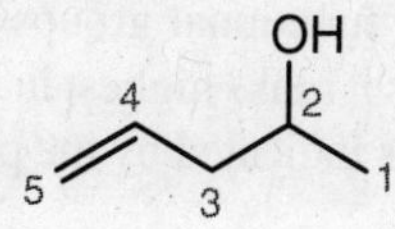

Comments : The functional groups are C=Cand –OH. The –OH group is the principal functional group as it ranks higher in the priority table. The compound is named as an alcohol.

Example 8

$CH_3-CH_2-\overset{OCH_3}{\overset{|}{C}H}-CH_2-OH$

4 3 2 1

2-Methoxy-1-butanol

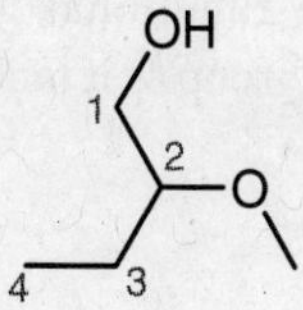

Comments : The functional groups are –OCH_3 and –OH. The compound is named as an alcohol. The –OCH_3 group is always treated as a substituent. It is indicated by the prefix *Methoxy*.

Example 9

$CH_3-CH=CH-C\equiv CH$

5 4 3 2 1

3-Penten-1-yne

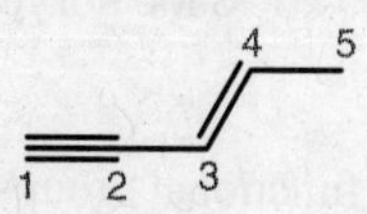

Comments : When there is a double bond (C=C) and a triple bond (C≡C) in the main chain the compound is named as an *alkenyne*. The position number of the double bond is inserted before *alken* and that of triple bond before *yne*. The main carbon chain is numbered from the end closer to the multiple bond, whether double or triple bond.

Example 10

$$\underset{5}{HC}\equiv\underset{4}{C}-\underset{3}{CH_2}-\underset{2}{CH}=\underset{1}{CH_2}$$

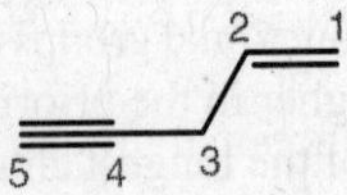

1-Penten-4-yne

Comments : The compound contains a double bond and a triple bond. It is named as an *alkenyne*. Remember that when either bond can be given the same position number, the chain is numbered from the end closer to double bond.

Example 11

$$\underset{5}{CH_3}-\underset{4}{\overset{\overset{\displaystyle O}{\|}}{C}}-\underset{3}{CH_2}-\underset{2}{\overset{\overset{\displaystyle CH_3}{|}}{CH}}-\underset{1}{COOH}$$

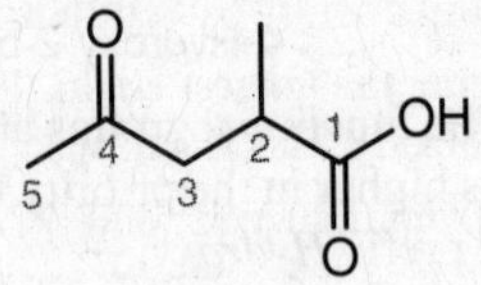

2-Methyl-4-oxopentanoic acid

Comments : The functional groups are C=O and –COOH. The –COOH group is the principal functional group as it ranks highest in the priority table. The compound is named as a carboxylic acid. The ketonic carbonyl group (C=O) is indicated by the prefix *Oxo*.

Example 12

$$\underset{6}{CH_3}-\underset{5}{\overset{\overset{\displaystyle CH_3}{|}}{CH}}-\underset{4}{CH_2}-\underset{3}{\overset{\overset{\displaystyle NH_2}{|}}{CH}}-\underset{2}{CH_2}-\underset{1}{COOH}$$

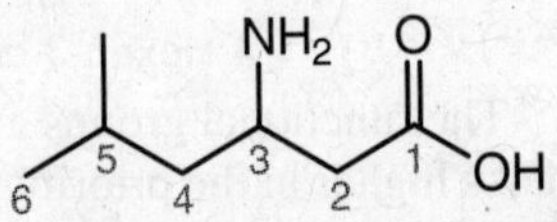

3-Amino-5-methylhexanoic acid

Comments : The functional groups are $-NH_2$ and –COOH. The –COOH group is the principal functional group as it ranks highest in the priority table. The compound is named as a carboxylic acid. The $-NH_2$ group is indicated by the prefix *Amino*.

Example 13

$$\underset{4}{CH_3}-\underset{3}{\overset{\overset{\displaystyle OH}{|}}{CH}}-\underset{2}{CH_2}-\underset{1}{\overset{\overset{\displaystyle O}{\|}}{C}}-OCH_2CH_3$$

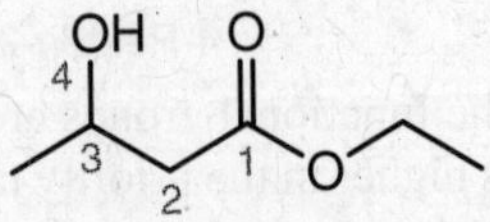

Ethyl 3-hydroxybutanoate

Comments : The functional groups are $-\overset{\overset{\displaystyle O}{\|}}{C}-O-$ and –OH. The –COO– group is the principal functional group as it ranks higher in the priority table. The compound is named as an ester.

Example 14

$$\underset{5}{CH_3}-\underset{4}{CH_2}-\underset{3}{\overset{\overset{\displaystyle OCH_3}{|}}{CH}}-\underset{2}{CH_2}-\underset{1}{\overset{\overset{\displaystyle O}{\|}}{C}}-Cl$$

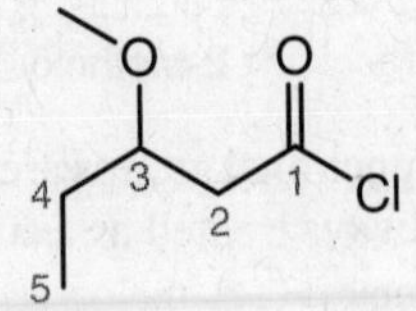

3-Methoxypentanoyl chloride

Comments : The functional groups are $-\overset{\overset{\displaystyle O}{\|}}{C}-Cl$ and $-OCH_3$. The compound is named as an acid halide. The group $-OCH_3$ is always treated as a substituent. It is indicated by the prefix *Methoxy*.

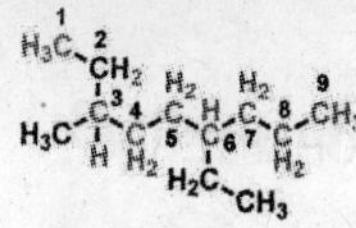

Example 15

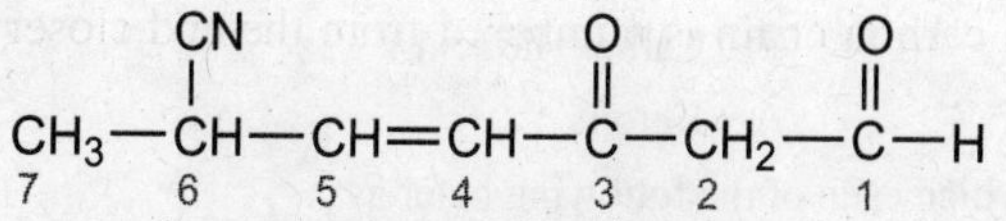

6-Cyano-3-oxo-4-heptenal

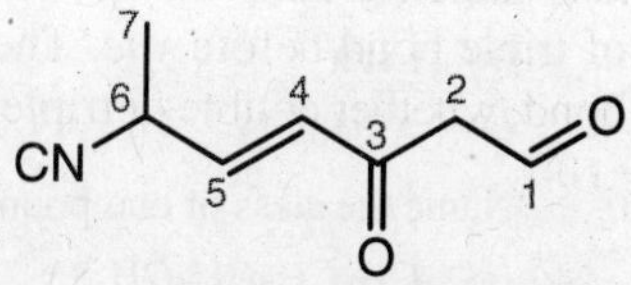

Comments : The functional groups are —CN, C=C, —CO—, and —CHO. The —CHO group ranks higher in the priority table. The compound is named as an aldehyde. The ketonic carbonyl group (C=O) is indicated by the prefix *Oxo*. The —CN group is indicated by the prefix *Cyano*.

Guidelines

Common Mistakes Made by Students in Using IUPAC Rules

(1) **Not recognizing the longest chain.** The longest chain does not have to be that portion of the molecule written horizontally. This is a mistake made by many beginning students.

(2) **Numbering from the wrong end.** When in doubt how to number the chain, do it both ways to pick the name with the lowest substituent group number.

(3) **Failure to write the same number twice and/or failure to the prefix di-, tri-, etc., when identical substituents are attached to the same carbon.**

(4) **Failure to arrange the substituent names in alphabetic order.** Remember that the structural prefixes such as di-, tri-, tetra-, iso-, and neo- are not taken into account when alphabetizing.

(5) **Failure to write IUPAC names as single words.** Remember that most of the IUPAC names are written as single words. Exceptions are organic salts and compounds named as acids and acid derivatives.

(6) Many students have a tendency to write names which are derived partially from the IUPAC system and partially from the common/trivial system. For example, 2-propanol is a correct IUPAC name for isopropyl alcohol, the name isopropanol is an objectionable combination of the IUPAC and common names.

(7) Forgetting to put cyclo- into the names of ring compounds.

(8) **Wrong selection of the principal functional group in naming polyfunctional compounds**. This is a mistake made by many students. Always follow the priority order given in Table 8.4.

In this chapter we have discussed the nomenclature of aliphatic compounds only. The nomenclature of aromatic compounds is given in Chapter 32.

Examination Questions

1. Define or explain the following terms :

(*a*) Aliphatic compounds (*b*) Aromatic compounds
(*c*) Alicyclic compounds (*d*) Heterocyclic compounds
(*e*) Carbocyclic compounds (*f*) Functional group
(*g*) Alkyl group (*h*) Homologous series

2. What do you understand by the term Functional group ? Give the structures and names of five important functional groups.

3. Write a note on : Homologous Series.

4. The following two compounds belong to a homologous series. Fill in the structures of the missing members between these two compounds.

$$CH_3CH_2Cl \ldots\ldots\ldots\ldots\ldots\ldots CH_3(CH_2)_7CH_2Cl$$

5. Write a condensed structural formula for any :

(*a*) Alcohol (*b*) Alkyl halide
(*c*) Ester (*d*) Carboxylic acid
(*e*) Aldehyde (*f*) Ketone

6. Name the class of compounds to which each of the following belongs :

(a) $CH_3CH_2CH_2CH_3$ (b) CH_3OH
(c) $CH_3CH_2OCH_3$ (d) $CH_3CH_2NH_2$
(e) HCOOH (f) $CH_3C{\equiv}CH$

7. Name the following alkyl groups :

(*a*) CH_3– (*b*) CH_3CH_2–
(*c*) $CH_3CH_2CH_2$– (*d*) $(CH_3)_2CH$–

8. What do you understand by the term principal functional group ? When the following groups are present in a molecule what will be its class name :

(*a*) –OH, –COOH, $–NO_2$ (*b*) $–NO_2$, $–NH_2$, $–CONH_2$
(*c*) –Br, $–CH_3$, –CHO

Answer. (*a*) Carboxylic acid ; (*b*) Amide ; (*c*) Aldehyde

A. Writing Structural Formulas From Names

9. Write a structure for each of the following compounds. Explain why the given name is wrong and give a correct name in each case.

(*a*) 1-Methylpentane (*b*) 2-Ethylbutane
(*c*) 2,3-Dichloropropane (*d*) 1,1,3-Trimethylpropane
(*e*) 3-Bromo-2-methylpropane

Answer. To conserve space, the structural formulas are not written, but you should write them out.

(*a*) The longest consecutive chain has six carbon atoms ; the correct name is *Hexane*.
(*b*) The longest chain was not selected ; the correct name is *3-Methylpentane*.
(*c*) The numbering was started at the wrong end ; the correct name is *1,2-Dichloropropane*.
(*d*) The longest chain was not selected ; the correct name is *2-Methylpentane*.
(*e*) The numbering was started at the wrong end ; the correct name is *1-Bromo-2-methylpropane*.

10. Draw structures corresponding to the following IUPAC names :

(*a*) 2-Methyl-1,5-hexadiene
(*b*) 3-Ethyl-2,2-dimethyl-3-heptene
(*c*) 3,5-Dimethyl-4-hexen-1-yne
(*d*) 1-Methyl-1,3-cyclopentadiene
(*e*) 1,5-Heptadiyne

Answer. The structural formula for (*c*) is :

$$CH_3-\underset{}{\overset{CH_3}{\overset{|}{C}}}=CH-\overset{CH_3}{\overset{|}{CH}}-C\equiv CH$$

11. Write the structural formulas for each of the following compounds :

(*a*) 1-Ethoxy-1-propanol (*b*) 1,2,3-Propanetriol

Answer. The structural formula for (*a*) is :

$$CH_3-CH_2-\overset{OH}{\overset{|}{CH}}-OCH_2CH_3$$

12. Write the structural formulas for each of the following compounds :

(*a*) 2-Chloro-3,5-dimethylhexane (*b*) 1-Methylcyclohexane
(*c*) 1,4-Pentadiene (*d*) 2-Methyl-4-nitro-2-pentanol
(*e*) 4-Methyl-2-pentyne

Answer. (*c*) $CH_2{=}CH{-}CH_2{-}CH{=}CH_2$

13. Write the structural formula of 2,2,4-trimethylpentane.

Answer. The structural formula of 2,2,4-trimethylpentane is

$$\begin{array}{c} \quad CH_3 \qquad\quad\;\; CH_3 \\ \quad | \qquad\qquad\quad | \\ CH_3-C-CH_2-CH-CH_3 \\ | \qquad\qquad\qquad \\ CH_3 \qquad\qquad\quad \end{array}$$

14. Write the structural formulas of the following compounds :

(*a*) Cyclopentane (*b*) 3-Methyl-1-butanol

(*c*) 2-Chloro-5-methylhexanal

Answer. The structural formula for (*c*) is :

$$\begin{array}{c} CH_3 \qquad\qquad\qquad\qquad\quad O \\ | \qquad\qquad\qquad\qquad\quad \| \\ CH_3-CH-CH_2-CH_2-CH-C-H \\ \qquad\qquad\qquad\qquad\quad | \qquad \\ \qquad\qquad\qquad\qquad\quad Cl \qquad \end{array}$$

15. Give the structural formulas for each of the following compounds :

(*a*) 2-Chloro-2-methylpropane (*b*) 2,2,4-Trimethylpentane

(*c*) 2-Propenal

Answer. The structural formula for (*c*) is :

$$\begin{array}{c} \qquad\qquad\quad O \\ \qquad\qquad\quad \| \\ CH_2=CH-C-H \end{array}$$

B. Writing Names From Structural Formulas

16. Write the IUPAC names for each of the following compounds :

(*a*) $CH_3CH=CH\overset{O}{\overset{\|}{C}}H$ (*b*) $CH_3CH=CHCH_2OH$

(*c*) $(CH_3)_2C=CHCH_2CH_2\overset{O}{\overset{\|}{C}}CH_3$ (*d*) $(CH_3)_2C=CHCH_2CH_2\overset{OH}{\overset{|}{C}}HCH_3$

Answer. (*a*) 2-Butenal (*b*) 2-Buten-1-ol

(*c*) 6-Methyl-5-hepten-2-one (*d*) 6-Methyl-5-hepten-2-ol

17. Give the IUPAC names for the following compounds :

(*a*) $CH_2=CH-CH_2-\overset{O}{\overset{\|}{C}}-CH_3$ (*b*) $CH_3-C\equiv C-\overset{CH_3}{\overset{|}{C}}H-CH_3$

(*c*) $CH_3-\overset{OH}{\overset{|}{C}}H-CH_2-\overset{O}{\overset{\|}{C}}-OH$

Answer. (*a*) 4-Penten-2-one (*b*) 4-Methyl-2-pentyne

(*c*) 3-Hydroxybutanoic acid

18. Write the IUPAC names for each of the following compounds :

(*a*) $CH_3-\overset{OH}{\overset{|}{C}}H-CH_2-CH_2-CH_3$ (*b*) $CH_3-CH_2-\overset{O}{\overset{\|}{C}}-CH_2-CH_3$

(*c*) $CH_3-CH_2-CH_2COOH$ (*d*) $CH_3-\overset{O}{\overset{\|}{C}}-CH_2-\overset{O}{\overset{\|}{C}}-CH_3$

Answer. (*d*) 2,4-Pentanedione

19. Write the IUPAC names for each of the following compounds :

(*a*) $CH_3-CH_2-CH_2-\overset{O}{\overset{\|}{C}}-CH_2Cl$ (*b*) $CH_3-\overset{CH_3}{\overset{|}{C}}H-C\equiv CH$

(*c*) $HO-CH_2-CH_2-COOH$

Answer. (*a*) 1-Chloro-2-pentanone (*b*) 3-Methyl-1-butyne

(*c*) 3-Hydroxypropanoic acid

20. Give the IUPAC name for :

$$CH_3CH_2-\overset{\overset{\displaystyle O}{\|}}{C}-CH_2CH_2-\overset{\overset{\displaystyle O}{\|}}{C}-OCH_3$$

Answer. Methyl 4-oxohexanoate

21. Give IUPAC names for the following compounds :

(*a*) $H_2C{=}CHCH_2CH(CH_3)_2$ (*b*) $CH_3CH_2CH{=}CHCH(CH_3)_2$

(*c*) $CH_3CH{=}CH\overset{\overset{\displaystyle CH_3}{|}}{C}HCH_2C{\equiv}CH$ (*d*) $CH_3C{\equiv}CCH_2CH_2C{\equiv}CH$

Answer. (*a*) 4-Methyl-1-pentene (*b*) 2-Methyl-3-hexene
(*c*) 4-Methyl-5-hepten-1-yne (*d*) 1,5-Heptadiyne

22. Give IUPAC names for the following structures :

(*a*) $(CH_3)_2CHCH_2CH_2COCl$ (*b*) $CH_3CH_2CH(CH_3)CN$

(*c*) $H_2C{=}CHCH_2CH_2CONH_2$ (*d*) $(CH_3CH_2)_2CHCN$

Answer. (*a*) 4-Methylpentanoyl chloride (*b*) 2-Methylbutanenitrile
(*c*) 4-Pentenamide (*d*) 2-Ethylbutanenitrile

Multiple Choice Questions

1. The correct IUPAC name of the following compound is

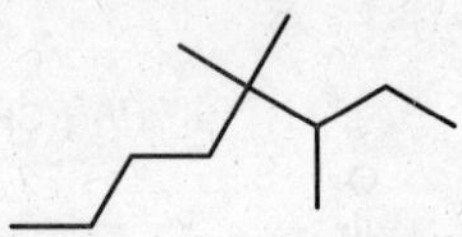

(*a*) 2-Ethyl-3,5-dimethylheptane. (*b*) 6-Ethyl-5,5-dimethylheptane.
(*c*) 3,4,4-Trimethyloctane. (*d*) 5,5,6-Trimethyloctane.

Answer. (*c*)

2. What is the IUPAC name of the following ?

$$CH_3CH_2CH_2CH_2\underset{\underset{\displaystyle CH_2CH_3}{|}}{\overset{\overset{\displaystyle CH_2CH_3}{|}}{C}}HCHCH_3$$

(*a*) 5,6-Diethylhexane (*b*) 2,3-Diethylhexane
(*c*) 5-Ethyl-6-methylheptane (*d*) 4-Ethyl-3-methylheptane

Answer. (*d*)

3. What is the IUPAC name of the following ?

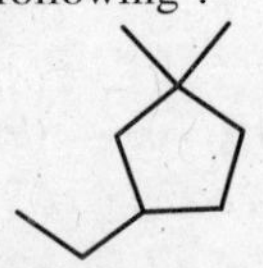

(*a*) 1-Ethyl-4.4-dimethylcyclopentane (*b*) 1-Ethyl-3,3-dimethylcyclopentane
(*c*) 3-Ethyl-1,1-dimethylcyclopentane (*d*) 4-Ethyl-1,1-dimethylcyclopentane

Answer. (*c*)

4. What is the IUPAC name of the compound below ?

$$CH_3\underset{\underset{\displaystyle OH}{|}}{C}HCH_2CH_2CH_2C(CH_3)_3$$

(*a*) 5,5-Dimethyl-2-hexanol (*b*) 2,2-Dimethyl-5-hexanol
(*c*) 5,5-Dimethyl-2-pentanol (*d*) 2,2-Dimethyl-5-pentanol

Answer. (*a*)

5. What is the IUPAC name of the following compound ?

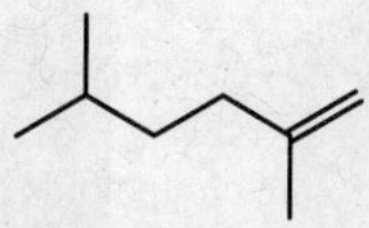

(*a*) 2,5-Dimethyl-1-hexene (*b*) 1,4-Dimethyl-1-hexene
(*c*) 2,5-Dimethyl-2-hexene (*d*) 2,5-Dimethyl-5-hexene

Answer. (*a*)

6. What is the IUPAC name of the following compound ?

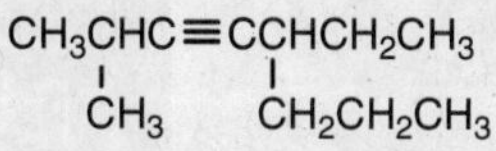

(*a*) 2-Methyl-5-propyl-3-heptyne (*b*) 1-Isopropyl-3-propyl-1-pentyne
(*c*) 5-Ethyl-2-methyl-3-octyne (*d*) 4-Ethyl-7-methyl-5-octyne

Answer. (*c*)

7. What is the IUPAC name of the following diene ?

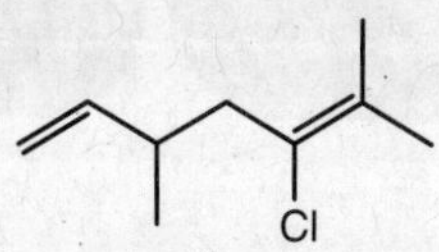

(*a*) 3-Chloro-2,5-dimethyl-2,6-heptadiene
(*b*) 3-Chloro-2,5-dimethyl-1,5-heptadiene
(*c*) 5-Chloro-3,5-dimethyl-1,6-heptadiene
(*d*) 5-Chloro-3,6-dimethyl-1,5-heptadiene

Answer. (*d*)

8. What is the name of the following carboxylic acid ?

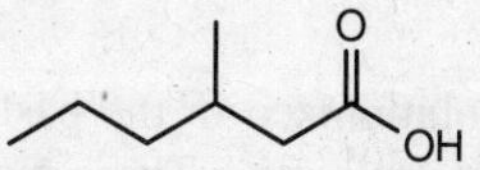

(*a*) 3-Propyllbutanoic acid (*b*) 3-Methyl-3-propylpropanoic acid
(*c*) 3-Methylpentanoic acid (*d*) 3-Methylhexanoic acid

Answer. (*d*)

9. What is the name of the following ester ?

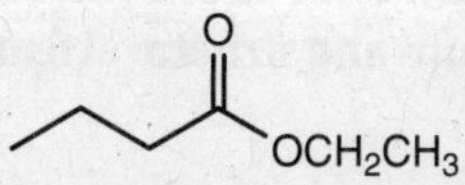

(*a*) Propyl ethanoate (*b*) Ethyl propanoate
(*c*) Ethoxy butanoate (*d*) Ethyl butanoate

Answer. (*d*)

10. What is the name of the following compound ?

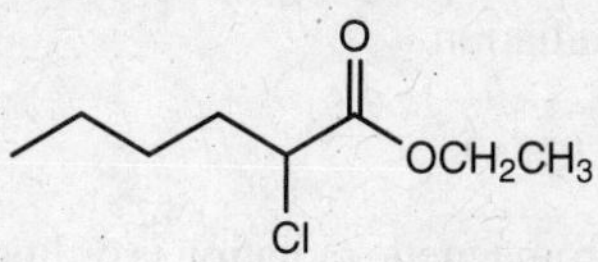

(*a*) 2-Chlorohexyl ethanoate (*b*) 1-Chlorohexyl ethanoate
(*c*) Ethyl-2-chlorohexanoate (*d*) Ethyl-1-chlorohexanoate

Answer. (*c*)

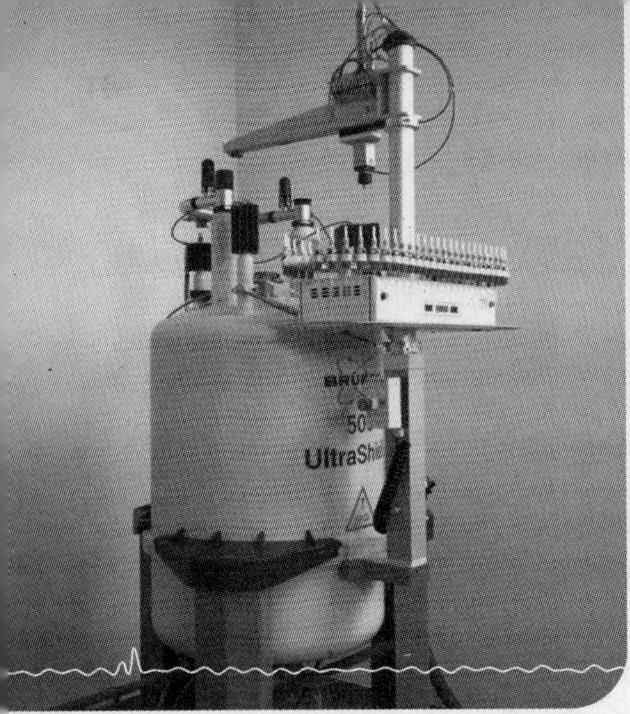

9

Spectroscopy

Until about 1945, the structure of a newly discovered organic compound was based almost entirely on its molecular formula and chemical reactions. The compound was converted into a known compound by a series of specific reactions, or it was synthesized from smaller known compounds by specific reactions. This way of finding the structure of an organic compound was time-consuming and not always 100% accurate. The conclusions were as reliable as the understanding of the reactions that were carried out.

Organic chemistry has been revolutionized in the past 50 years by the use of **Spectroscopic Methods** for analysis and structure determination. These methods have three major advantages over most chemical methods :

(1) **Spectroscopic methods are easier and faster to do than most chemical tests or reactions.**

(2) **Spectroscopic methods provide far more information about molecular structure. Practically all functional groups and structural features can be detected with very small amounts of sample.**

(3) **Spectroscopic methods are nondestructive and, if necessary, the entire sample can be recovered.**

There are four spectroscopic methods which are very widely used in Organic Chemistry. They are : ultraviolet-visible, infrared, nuclear magnetic resonance, and mass spectroscopy. The basis of these methods is electromagnetic radiation.

ELECTROMAGNETIC RADIATION

The most familiar form of electromagnetic radiation is ordinary white light, which can be dispersed by a prism into a spectrum of colors. This visible light represents only a small part of the entire **Electromagnetic Spectrum,** which extends from high-energy cosmic rays to low-energy radio or radar waves. Electromagnetic radiation can be described as a stream of energetic particles called **Photons** or as a **Wave Motion**.

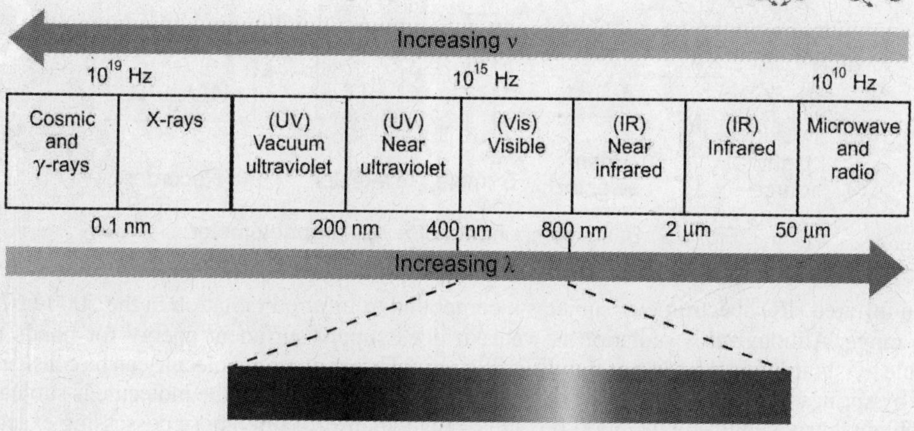

Fig. 9.1. Electromagnetic spectrum.

The fundamental equations of electromagnetic radiation are :

$$E = hv \quad \text{and} \quad v\lambda = c$$

where E is the energy of radiation, h is Planck's constant, v (pronounced as *nu*) is the frequency of radiation, λ (pronounced as *lambda*) is the wavelength of the radiation, and c is the velocity of light which is constant. Since v and λ are related by a constant (c), the energy of radiation can be described by either frequency or the wavelength.

Fig. 9.1 shows the various components of the electromagnetic spectrum. In organic chemistry, the important wavelength regions are the ultraviolet, visible, and infrared.

The ultraviolet and visible regions are adjacent, and the boundary is somewhat arbitrary since it depends upon human perception of light. The commonly used unit for wavelength in this region is the **Nanometer** (1 nm = 10^{-9} meter = 0.1Å, angstrom). The infrared region is at longer wavelengths (lower energy, since $E = hc/\lambda$). For this region, the scale may be expressed in wavelength (microns, μ) or **Wave Number**. The latter is generally preferred. The wave number is the reciprocal of the wavelength in centimeters.

PRINCIPLE OF SPECTROSCOPY

All organic compounds interact with electromagnetic radiation, that is, they absorb energy. When a molecule absorbs energy, a transformation occurs that may be either temporary or permanent. Lower-energy radiation may cause a molecular rotation or a bond vibration. Higher-energy radiation may cause the promotion of electrons to higher energy levels or bond cleavage.

Whether the transformation involves molecular rotation, bond vibration, or electronic transition, the molecule absorbs only the wavelength of radiation with exactly the energy necessary for the transition. The absorption of wavelengths of radiation is selective for a particular transition which depends on the structure of the molecule. By measuring the absorption spectra of known compounds, we can correlate the wavelengths of energy absorbed with characteristic structural features. This information is then used to determine the structure of unknown compounds.

SPECTROPHOTOMETER

The instrument that is used to measure the amount of electromagnetic radiation absorbed by an organic molecule is called **Spectrophotometer** or **Spectrometer**. It consists of a light source of radiation, with a prism that can select the desired wavelengths, which are passed through a sample of the compound being investigated (Fig. 9.2). The radiation that is absorbed by the sample can be detected, analyzed, and recorded. The recording is called the **Spectrum** of the compound.

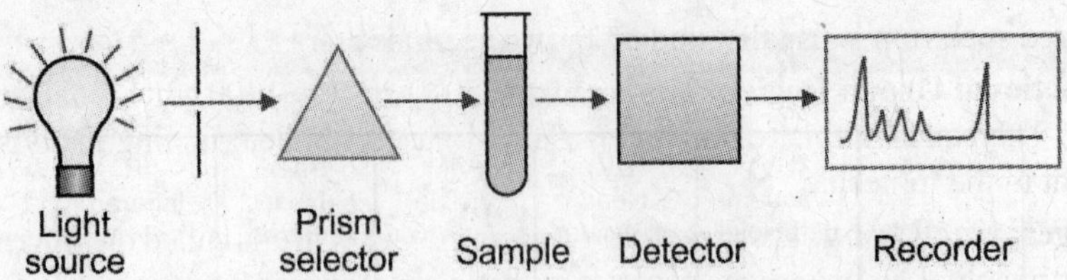

Fig. 9.2. Basic components of a spectrophotometer.

INFRARED SPECTROSCOPY

An infrared (IR) spectrometer subjects a compound to infrared radiation in the 5000-667 cm^{-1} (2µm) range. Although this radiation is weak, it does supply sufficient energy for bonds in the molecule to vibrate by **Stretching** or **Bending** (Fig. 9.3). The atoms of a molecule can be considered as linked by springs that are set in motion by the application of energy. As the molecule is subjected to the individual wavelengths in the 5000-667 cm^{-1} range, it absorbs only those possessing exactly the energy required to cause a particular vibration. Energy absorptions are recorded as bands (peaks) on chart paper.

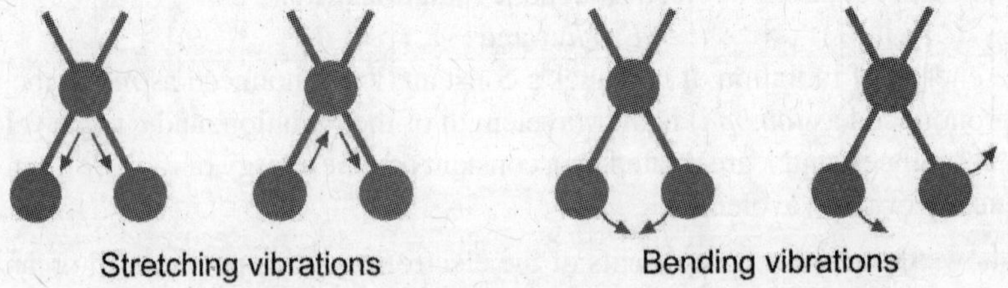

Fig. 9.3. Molecular vibrations caused by infrared radiation.

Since different bonds and functional groups absorb at different wavelengths, an infrared spectrum is used to determine the structure of organic molecules. For example, carbon-carbon triple-bond is stronger than a carbon-carbon double bond and requires a shorter wavelength (greater energy) to stretch. The same considerations apply to carbon-oxygen and carbon-nitrogen bonds.

C≡C	C=O	C≡N
2100–2200 cm^{-1}	1690–1750 cm^{-1}	2210–2260 cm^{-1}
C=C	C—O	C—N
1620–1680 cm^{-1}	1050–1400 cm^{-1}	1050–1400 cm^{-1}

Thus, from the position of an absorption peak, one can identify the group that caused it. Fig. 9.4 shows the general areas in which various bonds absorb in the infrared.

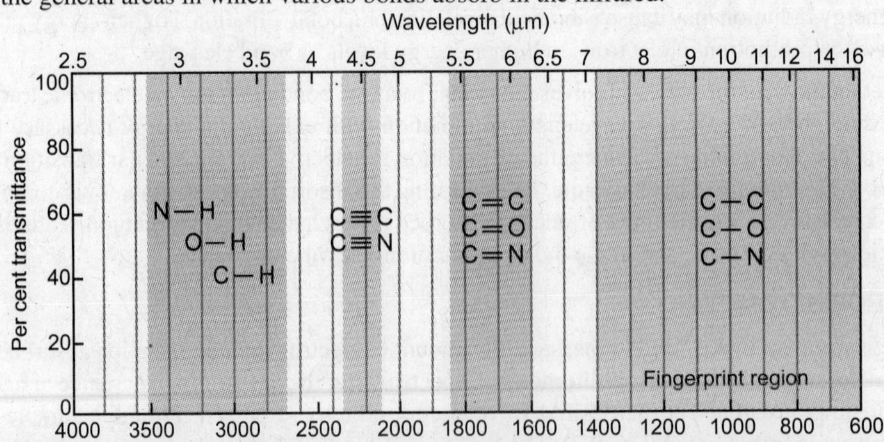

Fig. 9.4. Area of absorption for various bonds in the infrared.

An infrared spectrum is usually studied in two sections :

(1) Functional Group Region. The area from 5000 cm^{-1} to 1300 cm^{-1} is called the functional group region. The bands in this region are particularly useful in determining the type of functional groups present in the molecule.

(2) Fingerprint Region. The area from 1300 cm^{-1} to 667 cm^{-1} is called the fingerprint region. A peak-by-peak match of an unknown spectrum with the spectrum of the suspected compound in this region can be used, much like a fingerprint, to confirm its identity. Table 9.1 shows some characteristic infrared absorption bands. Fig. 9.5 shows some examples of infrared spectra.

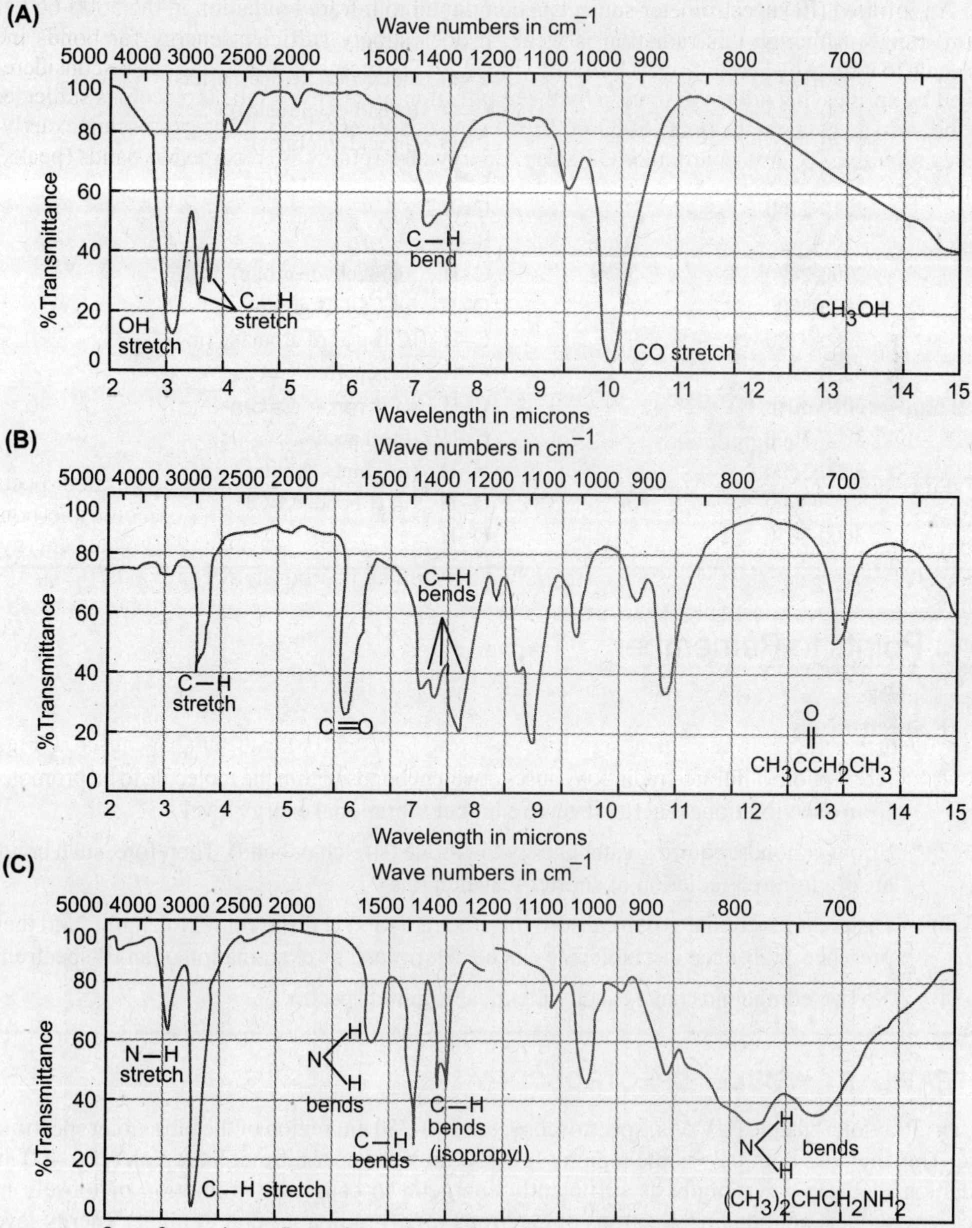

Fig. 9.5. Some examples of IR spectra. (A) Methanol. (B) Butanone. (C) Isobutylamine

Table 9.1 SOME CHARACTERISTIC IR ABSORPTION BANDS

Range in cm^{-1}	Bond (Remarks)
1050-1400	C–O (in ethers, alcohols, esters)
1050-1400	C–N (in amines)
1315-1475	C–H (in alkanes)
1340-1500	NO_2 (two peaks)
1450-1600	C=C (in aromatic rings ; several peaks)
1620-1680	C=C (in alkenes)
1630-1690	C=O (in amides)
1690-1750	C=O aldehydes, ketones, esters)
1700-1725	C=O (in carboxylic acids)
1770-1820	C=O (in acid chlorides)
2100-2200	C≡C
2210-2260	C≡N
2500	S–H
2700-2800	C–H (of aldehyde group)
2500-3000	O–H (of COOH group)
3000-3100	C–H (C is part of aromatic ring)
3330	C–H (C is part of C≡C)
3020-3080	C–H (C is part of C=C)
2800-3000	C–H (in alkanes)
3300-3500	N–H (in amines, amides)
3200-3600	O–H (in H-bonded ROH)
3600-3650	O–H

Points to Remember

IR Spectroscopy

(1) Absorption of infrared radiation causes covalent bonds within the molecule to be promoted from one vibrational energy level to a higher vibrational energy level.

(2) Stronger bonds require greater energy to vibrate (stretch or bend). Therefore, such bonds absorb infrared radiation of shorter wavelengths.

(3) Different functional groups absorb infrared radiation at different wavelengths, and their presence or absence in a molecule can be determined by examination of an IR spectrum.

(4) No two compounds have exactly identical infrared spectra.

ULTRAVIOLET-VISIBLE SPECTROSCOPY

In ultraviolet-visible (UV-Vis) spectroscopy, the 200-750 nm region of the ultraviolet spectrum is used. This includes both the visible region (400-750 nm) and near ultraviolet region (200-400 nm). Radiation of these wavelengths is sufficiently energetic to cause the promotion of loosely held electrons, such as nonbonding electrons or electrons involved in a π-bond to higher energy levels. For absorption in this particular region of ultraviolet spectrum, the molecule must contain conjugated double bonds. If the conjugation is extensive, the molecule will absorb in the visible region.

The ultraviolet-visible spectrum is composed of only a few broad bands of absorption (Fig. 9.6). The wavelength of maximum absorbance is referred to as λ_{max}. The following points should be kept in mind while interpreting a UV-Vis spectrum.

(1) Nonconjugated alkenes show an intense absorption below 200 nm and is therefore inaccessible to most commonly used UV spectrometers. For example, ethylene has $\lambda_{max} = 171$ nm. This absorption comes from the light-induced promotion of a π-electron to the next higher energy level.

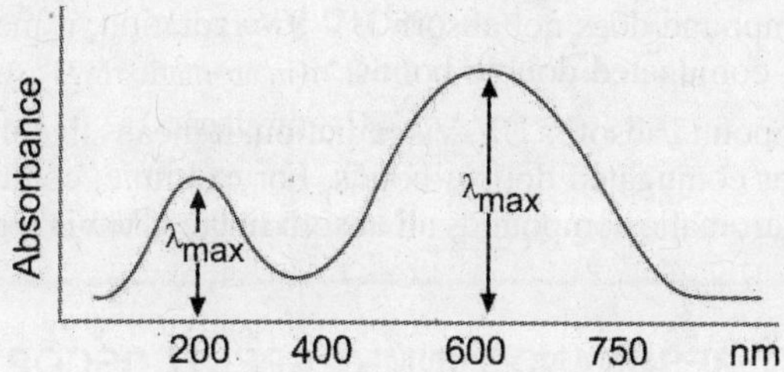

Fig. 9.6. A sample UV-Vis spectrum.

(2) Nonconjugated carbonyl compounds have a very weak absorption band in the 200-300 nm region. This band arises from excitation of one of the nonbonding electrons (from an unshared pair) to the next higher energy level. For example,

$CH_3-\overset{\overset{O}{\|}}{C}-CH_3$

Acetone
279 nm

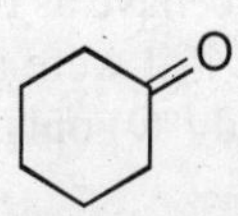

Cyclohexanone
291 nm

(3) When a molecule contains two or more nonconjugated carbon-carbon double bonds, the UV spectrum is that expected of a simple alkene. However, when the double bonds are conjugated, λ_{max} is shifted to longer wavelengths. For example,

$CH_2{=}CH-CH_2-CH_2-CH{=}CH_2$
1,5-Hexadiene
178 nm

$CH_3CH{=}CH-CH{=}CHCH_3$
2,4-Hexadiene
227 nm

(4) Conjugation of a carbon-carbon double bond and a carbonyl group shifts the λ_{max} of both groups to longer wavelengths. For example,

$(CH_3)_2C{=}CH-CH_2CH_3$ 180 nm

$(CH_3)_2CH-CH_2-\overset{\overset{O}{\|}}{C}-CH_3$ 283 nm

$(CH_3)_2C{=}CH-\overset{\overset{O}{\|}}{C}-CH_3$ 230 nm (for C=C)
327 nm (for C=O)

(5) As the number of double bonds in conjugation increases, λ_{max} also increases. For example,

$CH_2{=}CH-CH{=}CH_2$
1,3-Butadiene
215 nm

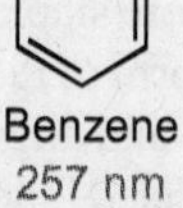

Benzene
257 nm

Points to Remember

UV-Vis Spectroscopy

(1) Absorption of ultraviolet-visible radiation (200-700 nm) causes electrons within molecules to be promoted from one energy level to a higher electronic energy level.

(2) If an organic compound does not absorb UV-Vis radiation, it means that the compound does not contain conjugated double bonds.

(3) If an organic compound absorbs UV-Vis radiation, it means that the compound contains a carbonyl group or conjugated double bonds. For example, conjugated dienes, carbonyl compounds, and aromatic compounds all absorb in the UV-Vis region.

NUCLEAR MAGNETIC RESONANCE (NMR) SPECTROSCOPY

Nuclear magnetic resonance spectroscopy involves absorption of electromagnetic radiation in the radio frequency region. Absorption of radiowaves in the presence of magnetic field is accompanied by a special type of nuclear transition, and for this reason we call this type of spectroscopy nuclear magnetic resonance (NMR) spectroscopy.

The nuclei of certain atoms behave as if they are spinning charges. Any spinning charge creates a magnetic field and behaves as if it were a tiny bar magnet. Of the three nuclei most common in organic compounds (^{1}H, ^{12}C, and ^{16}O) only the hydrogen nucleus proton behaves in this manner.

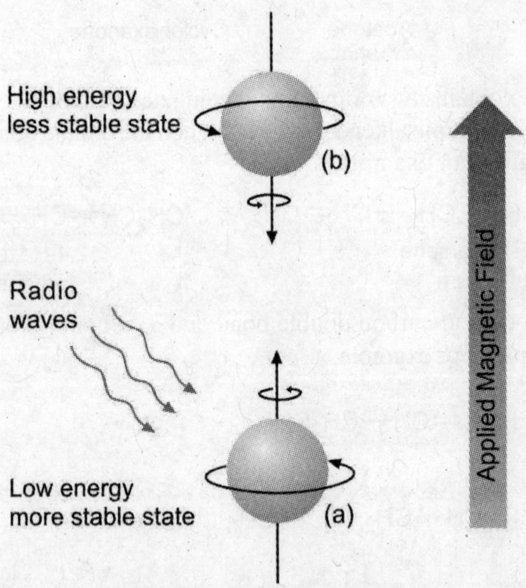

Fig. 9.7. Orientation of a spinning hydrogen nucleus in an applied magnetic field. (*a*) shows the nucleus in more stable low-energy state. (*b*) shows the nucleus in less stable high-energy state. Applying radiowaves *flips* the nucleus from the more stable 'aligned' state (*a*) to less stable 'nonaligned' state (*b*).

When a proton in an organic molecule is placed in a strong magnetic field, it can align with the field or against it. In the more stable low-energy state, it is aligned *with* the magnetic field (Fig. 9.7). If energy is supplied in the form of radiowaves of exactly the right frequency, radiation will be absorbed and the nucleus will "flip" and align *against* the applied magnetic field in the higher energy state. For example, a proton in an applied field of 14000 gauss requires a frequency of 60 million hertz (cycles per second) for this particular transition. When this exact frequency is applied, it is absorbed, and the absorption recorded on a chart paper. In practice, either the magnetic field can be held constant and the radio frequency varied, or more commonly, the radio frequency can be held constant and the magnetic field varied. Fig. 9.8 shows the various components of a NMR spectrometer.

Position of Signals ; Chemical Shifts. The main purpose of NMR is not to detect the presence of protons in a molecule. It can distinguish between protons in different chemical environments within the molecule. Protons on benzene ring, or on a carbon bearing a chlorine, or on a carbon adjacent to a carbonyl group absorb radio frequency energies at different applied magnetic fields, and appear at different locations (chemical shifts) on the recording paper. Also, the position of absorption is relatively constant for protons in a particular chemical or structural environment. Hence, *the number of signals recorded on the NMR chart paper indicates the number of different types of protons in a molecule.* The position of the peak can give information about the molecular structure in the vicinity of the proton.

To understand the application of NMR, we should know what are **Equivalent** and **Non-equivalent protons**. Equivalent protons are protons in the same molecule that are chemically indistinguishable. In other words, equivalent protons are positioned in structurally and chemically equivalent areas in the molecule. As an example, let us consider the following compounds.

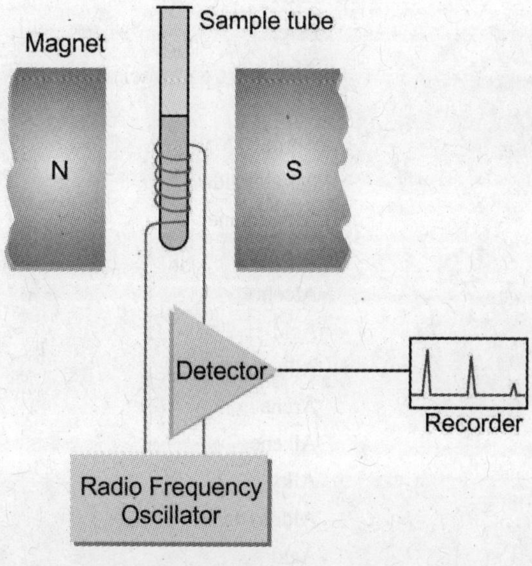

Fig. 9.8. Schematic diagrams of a NMR spectrometer.

$\underset{a}{CH_3}—\underset{b}{CH_2}—Cl$
Ethyl chloride
(2 Signals)

$\underset{a}{CH_3}—\underset{b}{CH_2}—\underset{c}{CH_2}—Cl$
n-Propyl chloride
(3 Signals)

$\underset{a}{CH_3}—\underset{b}{CH}(Cl)—\underset{c}{CH_3}$
Isopropyl chloride
(2 Signals)

Equivalent protons are indicated by the same letter, non-equivalent protons by different letters. Notice that *n*-propyl chloride can be easily distinguished from its isomer isopropyl chloride, by simply examining the NMR spectra of the two compounds. *n*-propyl chloride gives three NMR signals and has three non-equivalent sets of protons, whereas isopropyl chloride gives two NMR signals and has two sets of equivalent protons and one non-equivalent proton.

Delta and Tau Scales. Nuclear magnetic resonance chart paper is rectangular with a linear scale of δ (*delta*) units across the bottom. Most chart papers have scales from δ = 0 to δ = 8 or 9 ppm (parts per million). To every sample a small amount of **Tetramethylsilane** (TMS), $(CH_3)_4Si$, is added as a reference (the TMS signal is at δ = 0). All other signals are relative to TMS. Sometimes τ (*tau*) scale is used where the TMS signal is at 10.0 ppm. Hence τ = 10 – δ.

Table 9.2 lists some characteristic chemical shifts (from TMS) of different types of protons.

Table 9.2 SOME TYPICAL NMR CHEMICAL SHIFTS OF PROTONS

Type of Proton	Type of Compound	δ Value
$R–CH_3$	Alkane	0.8–1.2
$Ar–CH_3$	Arene	2.2–2.5
$C=C–CH_3$	Alkene	1.6–1.9
$R–O–CH_3$	Ether	3.2–3.3
$Ar–O–CH_3$	Ether	3.7–4.0
$R–CO–CH_3$	Ketone	2.1–2.4
$Ar–CO–CH_3$	Ketone	2.4–2.6
$CH_3–COOR$	Ester	1.9–2.2
$R–COOCH_3$	Ester	3.0–3.9
$CH_3–N<$	Amine	2.2–2.6
$R–CH_2–R$	Alkane	1.1–1.5
$R–CH_2–Ar$	Arene	2.5–2.9
$R–CH_2–X$	Alkyl halide	2.7–4.5
$R–CH_2–NO_2$	Nitroalkane	4.4–4.6
$R–CH_2–OH$	Alcohol	3.4–4.0
$Ar–CH_2–OH$	Alcohol	4.0–5.0
$R–CH_2–COOH$	Acid	2.0–2.6
$R_3C–H$	Alkane	1.4–1.6
Ar–H	Arene	6.0–7.5
C=C–H	Alkene	4.5–6.0
C≡C–H	Alkyne	1.7–1.9
R–CHO	Aldehyde	9.0–10.0
R–COOH	Acid	10–12
R–OH	Alcohol	2–6
Ar–OH	Phenol	6–8
$R–NH_2$	Amine	1–5

As an example, let us consider the NMR spectrum of benzyl alcohol (Fig. 9.9 A). Notice that there are four signals in this spectrum.

(1) The signal on the far right at δ = 0 is that of the TMS reference.
(2) The signal at δ = 7.3 is a single sharp peak due to the five chemically equivalent phenyl protons.
(3) The signal at δ = 4.6 is a single sharp peak due to the two chemically equivalent methylene ($–CH_2–$) protons.
(4) The signal at δ = 2.4 is a single sharp peak due to the hydroxy (–OH) proton.

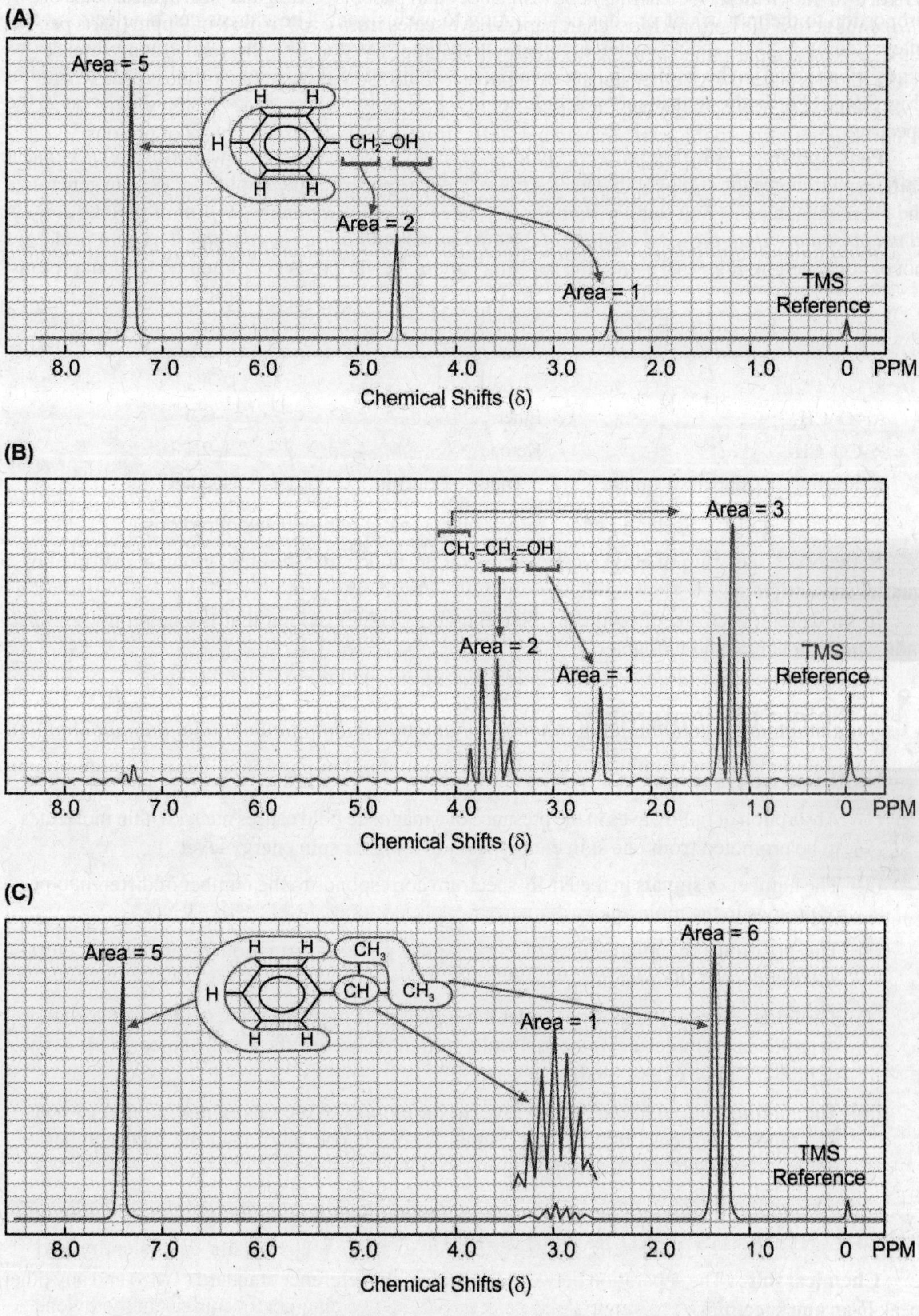

Fig. 9.9. Some examples of NMR spectra. (A) Benzyl alcohol. (B) Ethyl alcohol. (C) Isopropylbenzene

Peak Area ; Integration. The relative areas under the various peaks of a NMR spectrum are in proportion to the number of protons contributing to each signal. These areas can be electronically integrated by a NMR spectrometer. Comparison of the areas provides the ratio among the various kinds of protons in the molecule. For example, consider the NMR spectrum of benzyl alcohol (Fig. 9.9 A). The hydrogens in the molecule are in a 1 : 2 : 5 ratio, like the corresponding areas in the spectrum.

Peak Splitting ; Splitting Pattern. Hydrogens on adjacent carbons, each with a different chemical shift, can influence the signal of the other. This influence appears as peak splitting. We can generalize the phenomenon by saying that *the number of peaks into which a particular proton's signal is split, equals one more than the total number of protons on directly adjacent carbons*. The split peaks are not of equal height. Fig. 9.10 shows the splitting pattern for various types of non-equivalent protons.

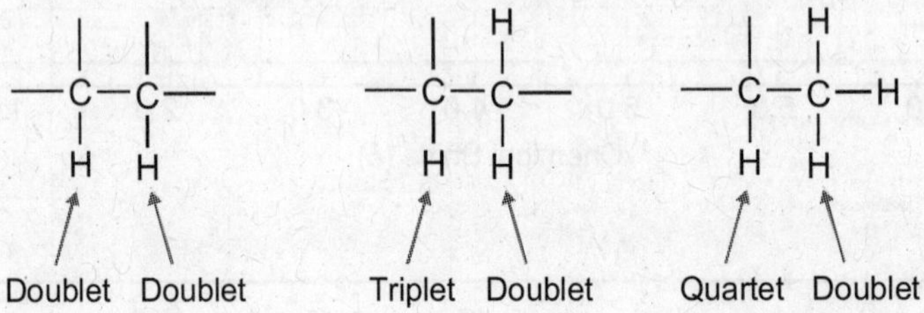

Fig. 9.10. Splitting pattern for various types of non-equivalent protons.

In the NMR spectrum of ethyl alcohol (Fig. 9.9 B), notice that the ethyl group is indicated by a quartet and a triplet.

In the NMR spectrum of isopropyl benzene (Fig. 9.9 C), notice that the isopropyl group is indicated by a heptet and a doublet.

Points to Remember

NMR Spectroscopy

(1) Absorption of radiowaves in the presence of a magnetic field causes nuclei within molecules to be promoted from one spin energy level to a higher spin energy level.

(2) The number of signals in the NMR spectrum corresponds to the number of different types of protons in the molecule.

(3) The position (chemical shift) of each signal gives information about the structural environment of the protons.

(4) The relative areas (Integration) under the signals give the ratio of the numbers of each type of proton in the molecule. If the molecular formula is known, the actual number of each type of proton can be determined.

(5) The splitting pattern of each signal gives us the number of protons on neighboring carbons. The number of peaks into which a signal is split is one more than the total number of protons on directly adjacent carbons.

IMPORTANT TERMS USED IN NMR SPECTROSCOPY

Chemical shift : The separation between the peak of the reference standard (TMS) and any other peak in an nmr spectrum.

δ scale : An nmr scale in which the TMS peak is set at zero and peaks to the left of it (downfield) have increasingly positive values, in parts per million (ppm).

TMS : Abbreviation for tetramethylsilane, $(CH_3)_4Si$, the reference standard in nmr spectrometry.

Downfield shift : The protons of most organic compounds are less shielded than those of TMS and therefore require less energy for resonance to take place. The peaks appear therefore at lower fields, or *downfield* of the TMS reference peak.

Spin-spin coupling : The splitting of a signal into two or more peaks because of magnetic interactions between neighboring protons.

$(n + 1)$ rule : A rule that allows us to predict into how many smaller peaks a given signal is split by neighboring protons. The rule states that a signal is split into $(n + 1)$ peaks, where n represents the number of equivalent *next-door protons*.

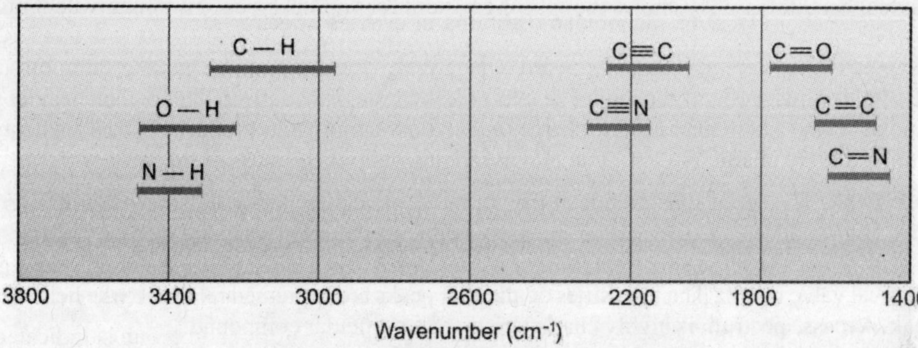

Group Assignments in Infrared Spectra

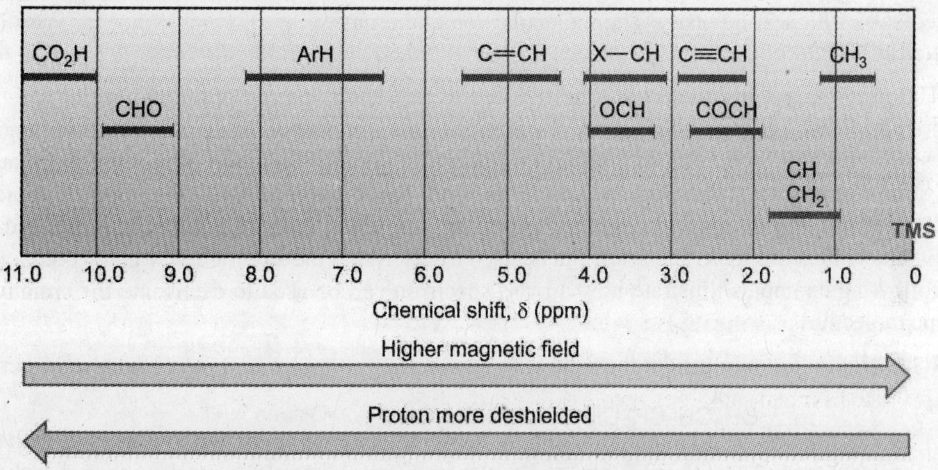

Group Assignments in NMR Spectra

MASS SPECTROSCOPY

In a mass spectrometer, organic molecules are vaporized and bombarded with a beam of very high-energy electrons (Fig. 9.11). The resulting collisions impart considerable energy to the molecule, which in turn emit electrons to produce positively charged ions. These ions possess so much energy that they often fragment through various bond cleavages to produce new positively charged ions.

$$ABC \xrightarrow[\text{beam}]{\text{electron}} \begin{array}{l} ABC^+ \quad \text{(Molecular ion)} \\ AB^+ \quad C^+ \\ A^+ \quad BC^+ \end{array}$$

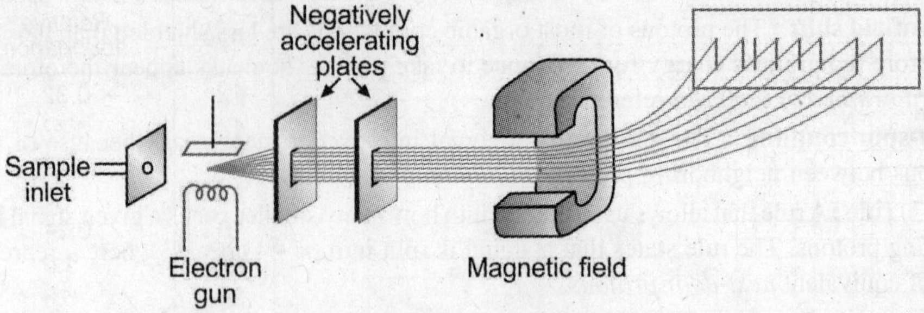

Fig. 9.11. Schematic diagrams of a mass spectrometer.

The positive ions are accelerated toward a negatively charged plate by passing them through a magnetic field, which deflects the ions. The ions of lighter mass are deflected more than heavier ions. Each kind of ion has a particular mass to charge (*m/e*) ratio. For most ions produced in the fragmentation of the molecule, the charge is +1 so, that *m/e* usually represents the mass of the ion.

Base Peak. The set of ions produced from a molecule can be analyzed since each has its own *m/e* ratio, and produces a signal whose intensity is due to the relative abundance of that ion. *The largest peak found in a mass spectrum (that of highest intensity) is called the base peak,* and is given the numerical value of 100. The intensities of all other peaks are expressed relative to the height of the base peak. A mass spectrum is highly characteristic of a particular compound.

Molecular Ion. *The ion formed by removing one electron from the parent molecule is called molecular ion or parent ion.* The molecular ion peak is usually represented as M^+. It may or may not be the peak of highest intensity. The molecular ion is the most important ion, since its mass is the **Molecular Weight** of the parent molecule.

The use of mass spectroscopy is not limited to the determination of the molecular weight of compounds. It also has great utility in the elucidation of the structure of a molecule. The fragmentation processes of organic molecules follow certain patterns in that the fragments broken off are related to bond strengths and the stability of the species formed. Since certain structural features in molecule produce definite characteristic fragmentation patterns, the identification of these fragments and their relative intensities in a mass spectrum can be used to determine the structure of organic compounds. The following examples illustrate how a mass spectrum can be used to determine the structure of organic molecules :

Example 1. Let us consider the following isomers.

$CH_3CH_2CH_2CH_2CH_2CH_2CH_2CH_3$ *n*-Octane

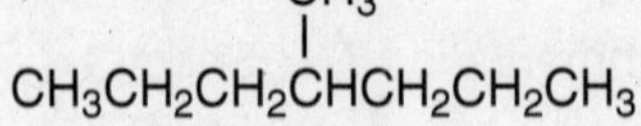

4-Methylheptane

n-Octane can be easily distinguished from 4-methylheptane, since the methyl group branched at C-4 in the latter compound, is easily broken off as CH_3^+ fragment and will give a peak of large intensity at $m/e = 15$

Example 2. Consider the mass spectrum shown in Fig. 9.12. The fragmentation pattern and relative abundance of peaks is as follows. A molecular weight of 32, which corresponds to O_2, is ruled out, since the only other possible fragment would be O with a peak of relatively *high* intensity at $m/e = 16$. An *m/e* peak is present at a value of 16, but has a very *low* relative intensity, 0.21. The peak at $m/e = 15$ is due to a CH_3^+ fragment, and the overall spectrum indicates an elemental composition of CH_4O. Consideration of other peaks present leads to the conclusion that the molecule producing this spectrum is methyl alcohol (CH_3OH).

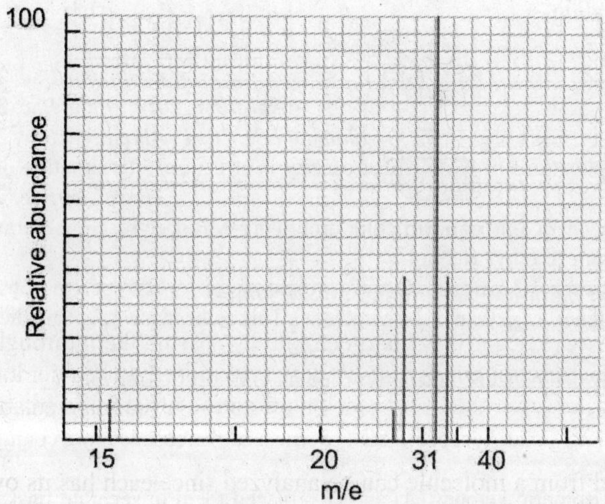

m/e	Relative abundance
12	0.32
13	0.72
14	2.4
15	1.3
16	0.21
17	1.0
28	6.3
29	64
30	3.8
31	100
32	66
33	0.98
34	0.14

Fig. 9.12. Mass spectrum of Methyl alcohol, CH_3OH.

Points to Remember

Mass Spectroscopy

(1) Mass spectroscopy involves organic molecules being bombarded by a very high-energy electron beam.

(2) The peak of highest intensity in a mass spectrum is referred to as the base peak.

(3) Fragmentation processes can produce numerous fragments, from which the structures of organic molecules can be deduced.

(4) When one electron is removed from a molecule, a molecular ion is produced. The *m/e* value of the molecular ion peak is the molecular weight of the compound being investigated. The molecular weights obtained by mass spectroscopy are extremely accurate.

Examination Questions

1. (*a*) Explain why spectroscopic methods are increasingly used for structure determination of organic compounds.

(*b*) What type of information one can obtain from : (*i*) IR spectrum ; (*ii*) UV spectrum ; and (*iii*) NMR spectrum.

2. What type of molecular transitions are brought about by absorption of :
(*a*) Ultraviolet-visible radiation
(*b*) Infrared radiation

3. Write notes on :
(*a*) UV spectroscopy *(Jadavpur BSc, 2014)*
(*b*) IR spectroscopy *(Agra BSc, 2014 ; Pune BSc, 2015)*
(*c*) NMR spectroscopy *(Jammu BSc, 2014 ; Allahabad BSc, 2015)*

4. Discuss the principle and applications of NMR spectroscopy in structure determination of organic compounds. *(Burdwan BSc, 2015)*

5. What is meant by the term *Chemical Shift* of a particular proton in NMR spectroscopy ? Draw the splitting patterns of protons in NMR spectrum of ethyl alcohol.

6. How will you distinguish between 1,3-pentadiene and 1,4-pentadiene by UV spectroscopy ?
Answer. 1,3-Pentadiene will have higher λ_{max} than 1,4-pentadiene. This is because 1,3-Pentadiene is conjugated ; whereas 1,4-pentadiene is not.

7. How will you distinguish between benzene and anthracene by UV spectroscopy ? *(Mysore BSc, 2014)*
Answer. Anthracene will have higher λ_{max} than benzene. This is because anthracene is more conjugated than benzene.

8. A chemist prepared the two isomeric ketones (*A*) and (*B*), placed them in separate flasks, but forgot to label them. How could you differentiate the two by UV spectroscopy

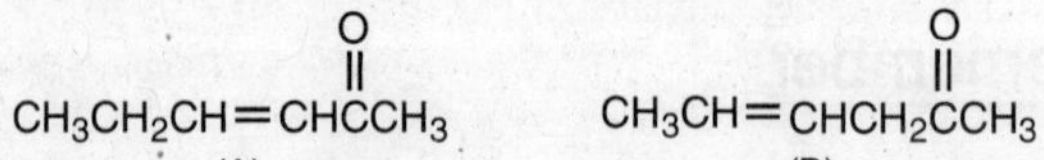

(A) (B)

Answer. In the UV spectra, (*A*), a conjugated system, would show λ_{max} at higher wavelengths than (*B*), a nonconjugated compound.

9. Two isomeric dienes (*A*) and (*B*) having the molecular formula C_5H_8 absorb at λ_{max} 223 nm and λ_{max} 178 nm respectively. Write the structures of the two isomers.
Answer.
(A) is $CH_2{=}CH{-}CH{=}CH{-}CH_3$; 1,3-Pentadiene
(B) is $CH_2{=}CH{-}CH_2{-}CH{=}CH_2$; 1,4-Pentadiene

10. Arrange the following compounds in the increasing order of their UV absorption maxima :
(*a*) Ethylene (*b*) Naphthalene (*c*) Anthracene (*d*) 1,3-Butadiene
Answer. (*a*) < (*d*) < (*c*) < (*b*)

11. How do you identify the following functional groups by IR spectroscopy :
(*a*) –OH (*b*) $-NH_2$ (*c*) –CN (*d*) >C=O

12. Discuss the principle of IR spectroscopy. How will you distinguish between the following pairs of compounds on the basis of IR spectroscopy :
(*a*) Ethyl alcohol *and* diethyl ether
(*b*) Acetic acid *and* ethyl acetate
Answer. (*a*) Ethyl alcohol shows a strong/broad absorption at 3200-3500 cm^{-1} (due to O–H bond) while diethyl ether does not ; and (*b*) Acetic acid shows a strong/broad absorption at 2500–3500 cm^{-1} (due to O–H bond) while ethyl acetate does not.

13. How would you distinguish between the following pairs of compounds by IR spectroscopy ?
(*a*) $CH_3CH_2CH_2N(CH_3)_2$ *and* $CH_3CH_2CH_2NH_2$
(*b*) $CH_3CH_2CH_2COOH$ *and* $CH_3CH_2COOCH_3$
(*c*) $CH_3CH_2COCH_3$ *and* $CH_3CH_2COOCH_3$
Answer. (*a*) $CH_3CH_2CH_2NH_2$ shows N–H absorption at 3300-3500 cm^{-1} while $CH_3CH_2CH_2N(CH_3)_2$ does not.
(*b*) $CH_3CH_2CH_2COOH$ shows O–H absorption at 2500-3500 cm^{-1} while $CH_3CH_2CH_2COOCH_3$ does not.
(*c*) $CH_3CH_2COOCH_3$ shows C–O absorption at 1050-1350 cm^{-1} while $CH_3CH_2COCH_3$ does not.

14. Two compounds, (*A*) and (*B*), have the same molecular formula C_2H_6O. They have different IR spectra. Compound (*A*) shows a strong/broad absorption at 3400 cm^{-1}, while compound (*B*) does not. Suggest formulas for (*A*) and (*B*) which account for the difference.

Answer. *(A)* is ethyl alcohol (CH_3CH_2OH) ; *(B)* is dimethyl ether (CH_3OCH_3). The strong absorption at 3400 cm^{-1} in *(A)* is due to the O–H bond.

15. Draw the structure of a compound with each of the following molecular formulas that will show only one peak in its NMR spectrum :

(*a*) $C_3H_6Cl_2$ (*b*) C_5H_{12} (*c*) C_2H_6O (*d*) C_4H_6

Answer. Possible structures are :

(*a*) 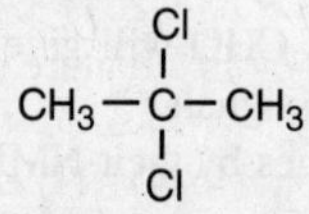 (*b*) 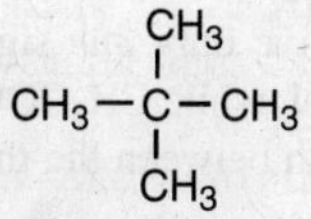

(*c*) CH_3-O-CH_3 (*d*) $CH_3C{\equiv}CCH_3$

16. How many NMR signals do you expect from each of the following compounds. Indicate also the splitting pattern of the various signals.

(*a*) CH_3OCH_3 (*b*) $CH_3OCH_2CH_3$ (*c*) CH_3CH_2OH

Answer. (*a*) One signal (singlet) ; (*b*) Three signals (singlet, quartet, triplet) ; and (*c*) Two signals (singlet, quartet).

17. How many signals (ignoring the splitting patterns) would you see in the NMR spectra of the following compounds ?

(*a*) Butanone (*b*) Cyclobutane (*c*) *p*-xylene (*d*) 2-Propanol

Answer. (*a*) Three signals ; (*b*) One signal ; (*c*) Two signals ; and (*d*) Three signals.

18. Predict the signal pattern of the CH_3 protons in the NMR spectra of the following compounds.

(*a*) CH_3CHBr_2 (*b*) 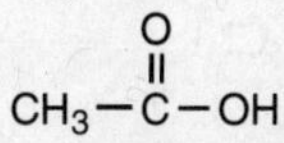

(*c*) CH_3CH_2Br (*d*) 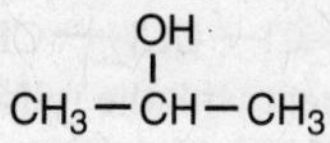

Answer. Apply the (n + 1) rule. (*a*) The CH_3 group has one neighboring proton and would therefore be a *doublet* ; (*b*) The CH_3 group has no neighboring proton, and would therefore be a *singlet* ; (*c*) The CH_3 group has two neighboring equivalent protons and would therefore be a *triplet* ; and (*d*) The two CH_3 groups are equivalent and have one neighboring proton. The CH_3 signal would therefore be a *doublet.*

19. The NMR spectrum of compound (*A*) C_2H_6O, shows one signal only, a *singlet.* Deduce the structure of (*A*).

Answer. (*A*) is CH_3–O–CH_3

20. The NMR spectrum of compound (*A*), C_5H_{12}, gives only one signal, a *singlet.* What is the structure of (*A*)? (*Delhi BSc, 2015*)

Answer. The molecular formula, C_5H_{12}, indicates that (*A*) is an alkane. The *single* signal in the NMR spectrum indicates that all hydrogens are equivalent. The only possible C_5H_{12} isomer that fits the data is $(CH_3)_4C$ (2,2-Dimethylpropane).

21. The UV spectrum for a compound with formula C_3H_6O shows a weak absorption band at 280 nm. The NMR spectrum shows only one signal, a *singlet*. What is the structure of the compound ?

Answer. The weak UV absorption at 280 nm is characteristic of a carbonyl group (C=O), leaving C_2H_6 for the remainder of the original formula. This suggests two CH_3 groups. The structure of acetone (CH_3COCH_3) fits all the data.

22. The NMR spectrum of a compound C_4H_9Br consists of a single sharp line. What is the structure of the compound ? (*Punjabi BSc, 2014*)

Answer. The compound is *tert*-Butyl bromide, $(CH_3)_3$ CBr.

23. Suggest the structure of a compound each with the following NMR spectral features :

(*a*) An alcohol with two NMR peaks

(*b*) A compound C_5H_{10} with a single NMR peak

(*c*) A compound C_4H_6 with a single NMR peak.

Answer. (*a*) Methyl alcohol ; (*b*) Cyclopentane ; (*c*) 2-Butyne

24. How would you distinguish between the following pairs of compounds by NMR spectroscopy ?

(*a*) $CH_3-\overset{O}{\overset{\|}{C}}-CH_3$ and $CH_3CH_2-\overset{O}{\overset{\|}{C}}-H$

(*b*) $CH_3-\overset{O}{\overset{\|}{C}}-CH_3$ and $CH_3-\overset{O}{\overset{\|}{C}}-OCH_3$

Answer. (*a*) CH_3COCH_3 will give only *one* signal, while CH_3CH_2CHO will give *three* signals; and (*b*) CH_3COCH_3 will give only *one* signal, while CH_3COOCH_3 will give *two* signals.

25. How will you distinguish between the three dibromobenzenes by their NMR spectra.

(*Annamalai BSc, 2015*)

Answer. *o*-Dibromobenzene will show *two* peaks ; *m*-dibromobenzene will show *three* peaks ; and *p*-dibromobenzene will show only *one* peak.

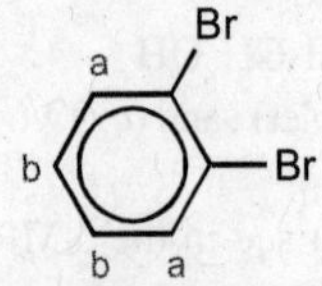

o-Dibromobenzene
(two peaks)
1:1

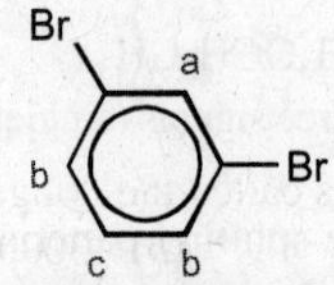

m-Dibromobenzene
(three peaks)
1:2:1

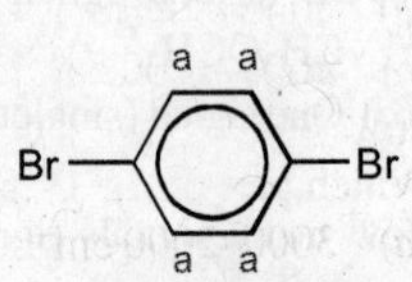

p-Dibromobenzene
(one peak)

26. Give a structure consistent with the following NMR data :
Molecular formula = $C_3H_5Cl_3$

(*a*) Singlet at δ 2.20, *3H* (*b*) Singlet at δ 4.02, *2H*

Answer. The compound is $\overset{b}{CH_2Cl}-CCl_2-\overset{a}{CH_3}$

27. Give a structure consistent with the following NMR data : Molecular formula = $C_{10}H_{14}$

(*a*) Singlet at δ 1.30, *9H* (*b*) Singlet at δ 7.28, *5H*

Answer. The compound is :

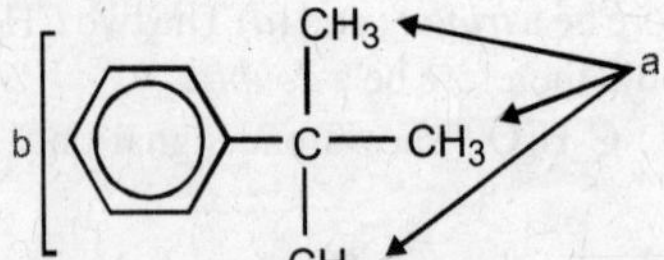

tert-Butylbenzene

28. Suggest a structure consistent with the following NMR data : Molecular formula = C_9H_{12}

(*a*) Singlet at δ 6.78, 3*H* (*b*) Singlet at δ 2.25, 9*H*

Answer. The compound is :

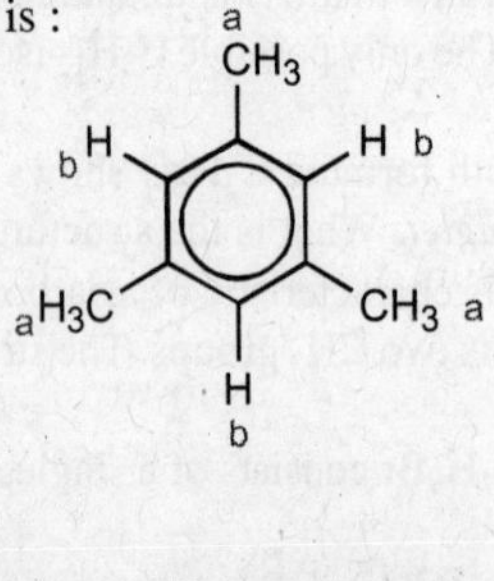

Mesitylene

Multiple Choice Questions

1. Infrared spectroscopy provides valuable information about

(*a*) Molecular weight (*b*) Melting point
(*c*) Conjugation (*d*) Functional groups

Answer. (*d*)

2. Which of the following processes leads to the absorption of infrared radiation by a molecule ?

(*a*) Nuclear spin flip (*b*) Bond vibration
(*c*) Bond breaking (*d*) Electron excitation

Answer. (*b*)

3. Define "wavenumber."

(*a*) wavelength ×speed of light (*b*) 1/wavelength
(*c*) wavelength/speed of light (*d*) speed of light/wavelength

Answer. (*b*)

4. The frequency of peaks in infrared spectra is expressed in wavenumbers. Which of the following is the unit of wavenumber ?

(*a*) cm (*b*) cm^3 (*c*) cm^{-1} (*d*) $J\ mol^{-1}$

Answer. (*c*)

5. What part of an IR spectrum is called the functional group region ?

(*a*) 4000–3000 cm^{-1} (*b*) 4000–2000 cm^{-1}
(*c*) 4000–1000 cm^{-1} (*d*) 3000–1000 cm^{-1}

Answer. (*c*)

6. Which part of the IR spectrum is called the "fingerprint region" ?

(*a*) 3000–2000 cm^{-1} (*b*) 3000–1000 cm^{-1}
(*c*) 2000–1000 cm^{-1} (*d*) 1000–600 cm^{-1}

Answer. (*d*)

7. A strong signal at 3400 cm^{-1} in an IR spectrum indicates the presence of a(n)

(*a*) Alcohol (*b*) Ether (*c*) Carbonyl (*d*) Amine

Answer. (*a*)

8. A strong signal at 1700 cm^{-1} in an IR spectrum indicates the presence of a(n)

(*a*) Alcohol (*b*) Ether (*c*) Carbonyl (*d*) Amine

Answer. (*c*)

9. Which of the following compounds gives an infrared spectrum with a peak at approximately 1720 cm^{-1} (strong peak) ?

(*a*)

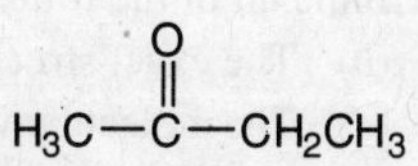
(*b*)

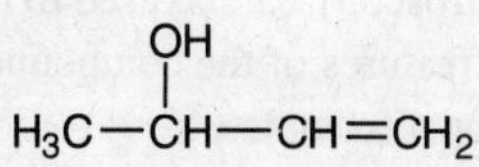
(*c*)

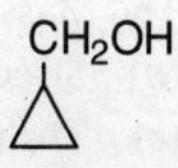
(*d*)

Answer. (*b*)

10. NMR stands for

(*a*) new molecular reactions (*b*) nuclear material resonance
(*c*) nuclear magnetic resonance (*d*) nuclear magnetic resistance

Answer. (*c*)

11. What type of radiation is used in nuclear magnetic resonance (NMR) spectroscopy ?

(*a*) Radio waves (*b*) Ultraviolet light
(*c*) Infrared radiation (*d*) Microwaves

Answer. (*a*)

12. Which feature in the ^1H NMR spectrum provides information about the number of types of chemically distinct protons in a molecule ?

(*a*) Integral (*b*) Number of signals
(*c*) Chemical shift (*d*) Multiplicity

Answer. (*b*)

13. Which of the following compounds is frequently used as an internal reference in ^1H NMR spectroscopy ?

(*a*) TNS (*b*) DMS (*c*) TMS (*d*) DMSO

Answer. (*c*)

14. Integrating a ^1H NMR spectrum gives the

(*a*) number of peaks (*b*) number of absorptions
(*c*) absolute number of hydrogen atoms (*d*) relative number of hydrogen atoms

Answer. (*d*)

15. Downfield chemical shifts result from
(*a*) reducing the number of hydrogen atoms
(*b*) increasing the number of hydrogen atoms
(*c*) adjacent electron-donating groups
(*d*) adjacent electron-withdrawing groups

Answer. (*d*)

16. How many kinds of non-equivalent protons are there in 3-methylpentane ?
(*a*) 4 (*b*) 3 (*c*) 6 (*d*) 5

Answer. (*a*)

17. Which of the following describes the ^{1}H NMR spectrum of 1,2-dichloroethane ?
(*a*) one singlet (*b*) two doublets (*c*) two triplets (*d*) one doublet

Answer. (*a*)

18. Which of the following combinations of peaks appears in the ^{1}H NMR spectrum of 2,3-dimethylbutane ?
(*a*) two doublets (*b*) a doublet and a septet
(*c*) a doublet and an octet (*d*) a singlet and two doublets

Answer. (*b*)

19. Which of the following combinations of peaks appears in the 1H NMR spectrum of diethyl ether, $CH_3CH_2OCH_2CH_3$?
(*a*) a triplet and a quartet (*b*) two singlets
(*c*) a quartet and a sextet (*d*) a triplet and a doublet

Answer. (*a*)

20. In what region of the ^{1}H NMR spectrum would you find the signal for the acidic hydrogen of a carboxylic acid ?
(*a*) 0–2.5 ppm (*b*) 2.5–5 ppm (*c*) 7.5–10 ppm (*d*) 10–12 ppm

Answer. (*d*)

21. The Mass Spectroscopy can be used to determine all of the following except ?
(*a*) Structural features of the compound (*b*) The exact structure of the compound
(*c*) The molecular formula (*d*) The molecular weight

Answer. (*b*)

22. Which of the following corresponds to the molecular ion in a mass spectrum ?
(*a*) Any ion formed by fragmentation of the molecule
(*b*) The ion formed by removal of an electron from the molecule
(*c*) The most abundant ion
(*d*) The ion with the smallest *m/z*

Answer. (*b*)

23. The molecular ion peak gives information about
(*a*) the height of the peak (*b*) the molecular weight
(*c*) fragmentation patterns (*d*) IR radiation absorption

Answer. (*b*)

24. What is plotted along the x-axis of a mass spectrum ?
(*a*) mass/energy (*b*) mass × charge
(*c*) mass/charge (*d*) mass

Answer. (*c*)

25. Which of the following is the definition of the base peak in a mass spectrum ?
(*a*) The peak corresponding to the molecular ion.
(*b*) The peak corresponding to the most abundant ion.
(*c*) The peak corresponding to the ion with lowest *m/z*.
(*d*) The peak corresponding to any ion formed by fragmentation of the molecule.

Answer. (*b*)

10

Alkanes and Petroleum

Methane flare from a lake in Alaska.

Alkanes are the simplest organic compounds made of carbon and hydrogen only. They have the general formula C_nH_{2n+2} where $n = 1, 2, 3$ etc. The first three members of this class can be represented as :

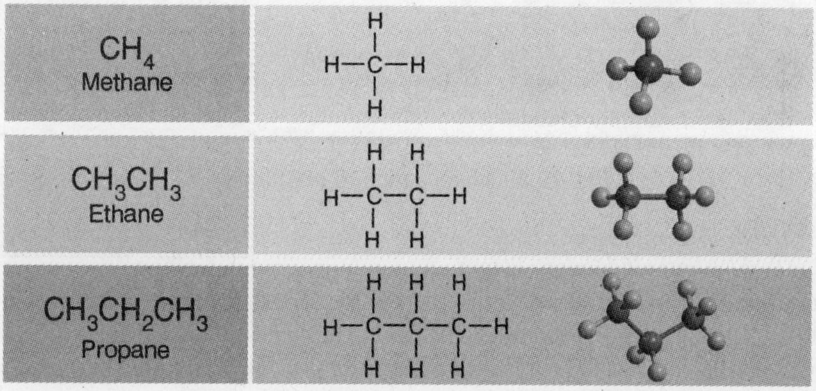

The carbon atoms in their molecules are bonded to each other by single covalent bonds. Each carbon is again bonded to enough hydrogen atoms to give maximum covalence of 4. Since the carbon skeleton of alkanes is fully 'saturated' with hydrogens, they are also called **Saturated Hydrocarbons**.

Alkanes contain strong C–C and C–H covalent bonds. Therefore, this class of hydrocarbons are relatively chemically inert. Hence they are sometimes referred to as **Paraffins** (Latin, *parum affinis* = little affinity).

STRUCTURE

Let us consider methane (CH_4) and ethane (CH_3–CH_3) for illustrating the orbital make up of alkanes. In methane, carbon forms single bonds with four hydrogen atoms. Since the carbon atom is attached to four other atoms, it uses sp^3 hybrid orbitals to form these bonds. Each C–H bond is the result of the overlap of an sp^3 orbital from carbon and an *s* orbital from hydrogen (Fig. 10.1). All C–H bonds are σ bonds.

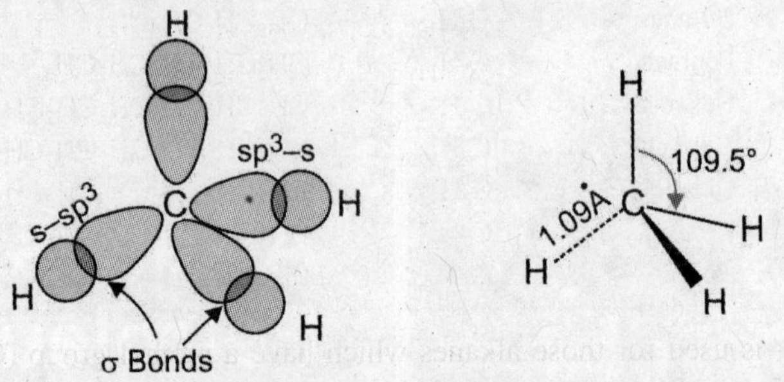

Fig. 10.1. Orbital structure of Methane.

In ethane, there are six C–H covalent bonds and one C–C covalent bond. As in the case of methane, each C–H bond is the result of overlap of an sp^3 hybrid orbital from carbon and an *s* orbital from hydrogen. The C–C bond arises from the overlap of the sp^3 orbitals, one from each carbon (Fig. 10.2). All C–H bonds and the C–C bond are σ bonds.

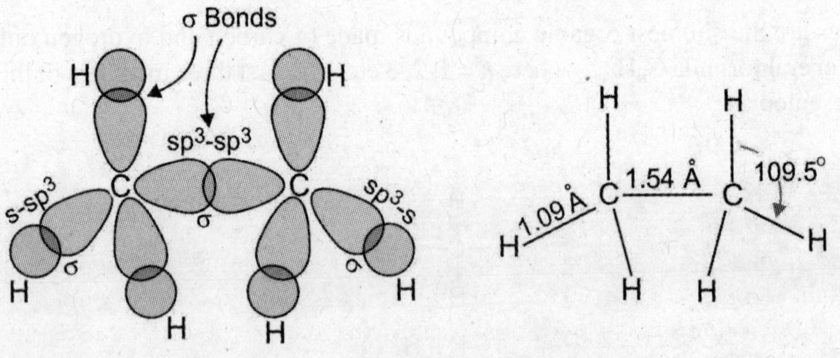

Fig. 10.2. Orbital structure of Ethane.

NOMENCLATURE

There are two systems of naming alkanes :

(1) **Common System**. The first four members of the series are called by their common names (trivial names) : *methane*, *ethane*, *propane*, and *butane*. The names of larger alkanes are derived from the Greek prefixes that indicate the number of carbon atoms in the molecule. Thus, *pentane* has *five* carbons, *hexane* has *six*, and so forth (Table 10.1).

In the common system all isomeric alkanes have the same parent name. For example, two isomeric C_4H_{10} alkanes are known as butanes. The names of various isomers are distinguished by prefixes. The prefix indicates the type of branching present in the molecule.

(1) Prefix *n*- is used for those alkanes in which all carbons are in one continuous chain. The prefix *n*- stands for normal.

$CH_3CH_2CH_2CH_3$ — *n*-Butane

$CH_3CH_2CH_2CH_2CH_3$ — *n*-Pentane

Table 10.1 NAMES OF ALKANES

Number of carbon atoms	Name	Molecular formula	Structure of the normal isomer
1	Methane	CH_4	CH_4
2	Ethane	C_2H_6	CH_3CH_3
3	Propane	C_3H_8	$CH_3CH_2CH_3$
4	Butane	C_4H_{10}	$CH_3CH_2CH_2CH_3$
5	Pentane	C_5H_{12}	$CH_3CH_2CH_2CH_2CH_3$
6	Hexane	C_6H_{14}	$CH_3CH_2CH_2CH_2CH_2CH_3$
7	Heptane	C_7H_{16}	$CH_3CH_2CH_2CH_2CH_2CH_2CH_3$
8	Octane	C_8H_{18}	$CH_3CH_2CH_2CH_2CH_2CH_2CH_2CH_3$
9	Nonane	C_9H_{20}	$CH_3CH_2CH_2CH_2CH_2CH_2CH_2CH_2CH_3$
10	Decane	$C_{10}H_{22}$	$CH_3CH_2CH_2CH_2CH_2CH_2CH_2CH_2CH_2CH_3$

(2) Prefix *iso-* is used for those alkanes which have a methyl group (CH_3–) attached to the second last carbon atom of the continuous chain.

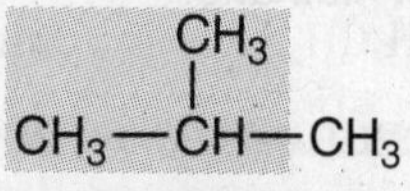

Isobutane

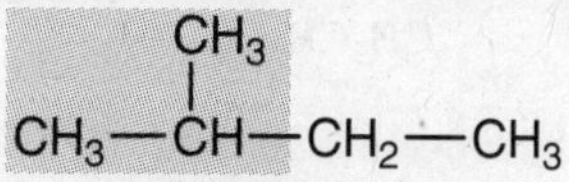

Isopentane

(3) Prefix *neo-* is used for those alkanes which have two methyl groups attached to the second last carbon atom of the continuous chain.

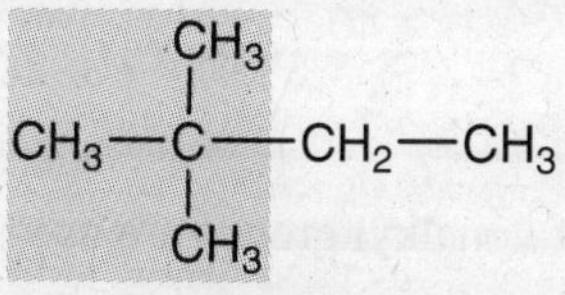

Neohexane

Primary, Secondary, and Tertiary Carbons. The structural formulas of alkanes contain four types of carbon atoms :

(1) A carbon atom attached to one other (or no other) carbon is called **primary carbon** (*1° carbon*; *1° = primary*).

(2) A carbon atom attached to two other carbon atoms is called **secondary carbon** (*2° carbon*; *2° = secondary*).

(3) A carbon atom attached to three other carbon atoms is called **tertiary carbon** (*3° carbon*; *3° = tertiary*).

(4) A carbon atom attached to four other carbon atoms is called **quaternary carbon** (*4° carbon*; *4° = quaternary*).

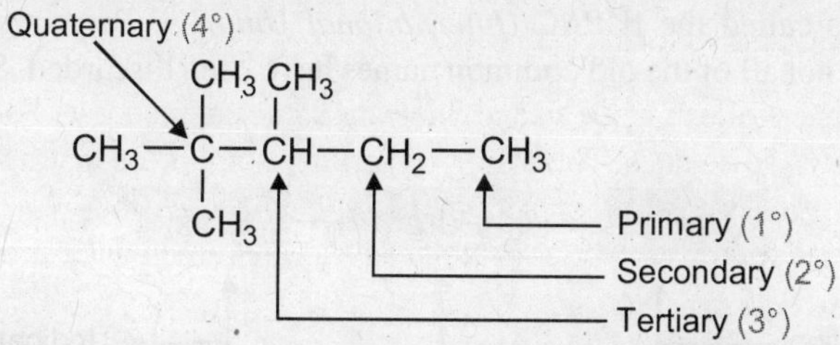

Hydrogen atoms attached to 1°, 2°, 3° carbon atoms are often referred to as primary, secondary, and tertiary hydrogen atoms.

Alkyl Groups. An alkyl group is formed by removing one hydrogen from an alkane. They are named simply by dropping *-ane* from the name of the corresponding alkane and replacing it by *-yl* (**alkane - ane + yl = alkyl**). For example,

$$\underset{\text{Alkane}}{RH} \xrightarrow{-H} \underset{\text{Alkyl group}}{R-}$$

$$\underset{\text{Methane}}{CH_4} \xrightarrow{-H} \underset{\text{Methyl group}}{CH_3-} \qquad \underset{\text{Ethane}}{CH_3CH_3} \xrightarrow{-H} \underset{\text{Ethyl group}}{CH_3CH_2-}$$

Table 10.2 shows the names of some important alkyl groups.

Table 10.2 NAMES OF COMMON ALKYL GROUPS

Parent alkane	Structure	Alkyl group	Name of alkyl group
Methane	CH_4	CH_3-	methyl
Ethane	CH_3CH_3	CH_3CH_2-	ethyl
Propane	$CH_3CH_2CH_3$	$CH_3CH_2CH_2-$	*n*-propyl
		$CH_3\overset{\vert}{C}HCH_3$	isopropyl
n-Butane	$CH_3CH_2CH_2CH_3$	$CH_3CH_2CH_2CH_2-$	*n*-butyl
		$CH_3CH_2\overset{\vert}{C}HCH_3$	*sec*-butyl (secondary-butyl)
Isobutane	$CH_3\overset{CH_3}{\overset{\vert}{C}}HCH_3$	$CH_3\overset{CH_3}{\overset{\vert}{C}}HCH_2-$	isobutyl
		$CH_3\underset{\vert}{\overset{CH_3}{\overset{\vert}{C}}}CH_3$	*t*-butyl or *tert*-butyl (tertiary-butyl)

Nonalkyl Groups. A number of nonalkyl groups are used in naming organic compounds. For example,

Group	Name	Group	Name
–F	fluoro	$-NO_2$	nitro
–Cl	chloro	$-NH_2$	amino
–Br	bromo	–OH	hydroxy
–I	iodo		

(2) **IUPAC System**. The common system has some limitations. Pentane has three isomers ; hexane has five. The more complicated the alkane, the greater the number of special prefixes needed to name the isomers. In order to device a system of nomenclature that could be used for even the most complicated compounds, committees of chemists have met periodically since 1892. The system resulting from these meetings is called the IUPAC (*International Union of Pure and Applied Chemistry*) system. In this system, not all of the old common names have been discarded. Some common names have been retained.

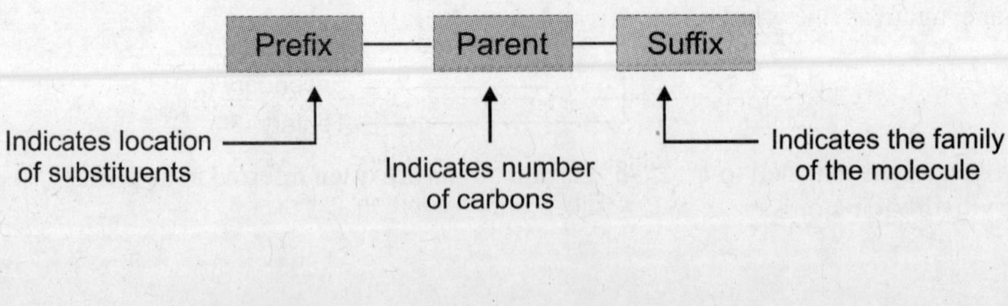

The IUPAC system is much the same for all families of organic compounds. The steps for alkanes are as follows :

Step 1. Name the longest chain. The longest continuous carbon chain is chosen as the basis for the name. The names are those given in Table 10.1. These are *parent names*.

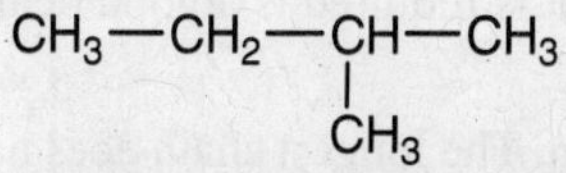

The longest continuous chain has four carbon atoms. Thus, the compound is named as a butane.

Step 2. Number the longest chain. The carbon atoms in the longest chain are numbered. The numbering is started from that end which will give numbers having the lowest value to carbons carrying substituents.

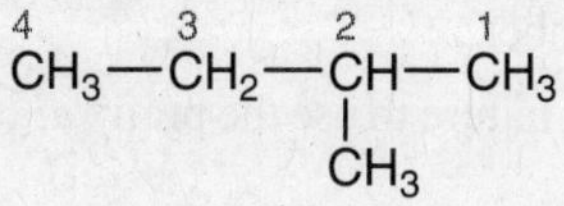

and not

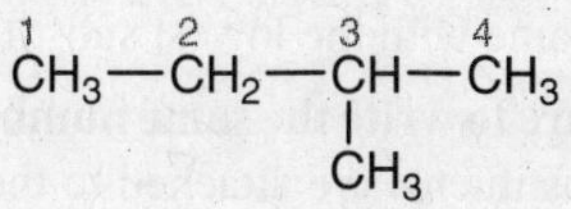

Step 3. Locate and name the substituents. Each substituent is named, and the position of each substituent is indicated by the number of the carbon atom to which it is attached.

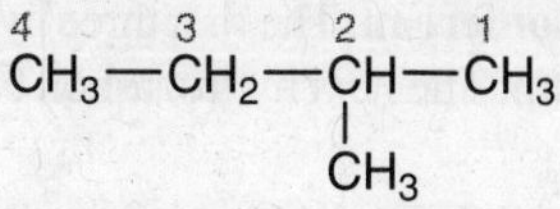

The attached group is located on carbon 2 of the chain, and it is a methyl group.

Step 4. Combine the longest chain and substituents into the name. The position and the name of the substituent are added to the name of the longest chain and written as one word.

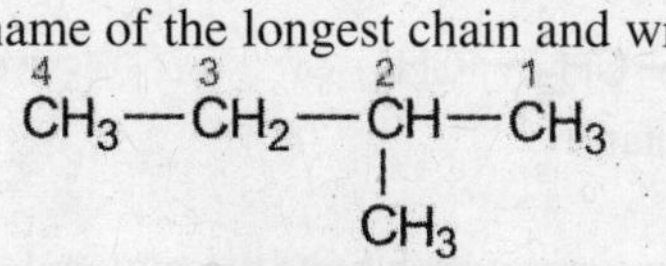

2-Methylbutane

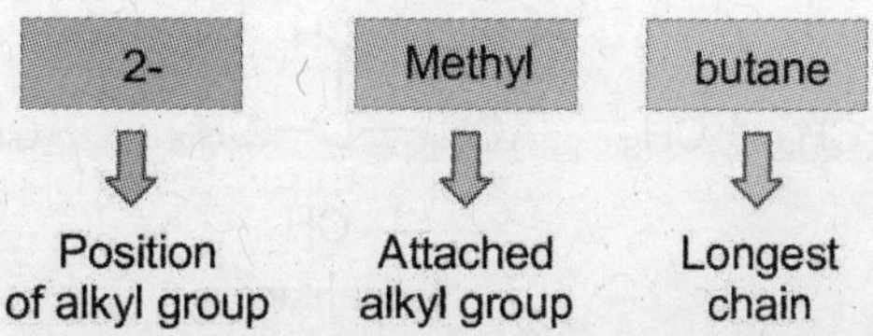

Additional steps are needed when more than one substituent is attached to the longest chain.

Step 5. Indicate the number and position of substituents. If the same substituent is present two or more times in the molecule, the number of this substituent is indicated by a prefix di-, tri-, tetra-, penta-, etc., and the location of each is indicated by a separate number. These position numbers, separated by commas, are put just before the name of the substituent, with hyphen before and after the numbers when necessary.

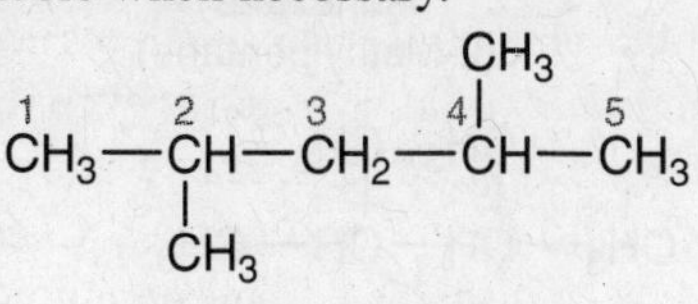

2,4-Dimethylpentane

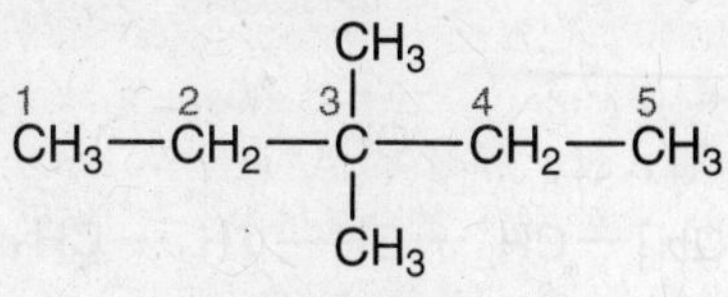

3,3-Dimethylpentane

If two or more substituents are present, their names are alphabetized and added to the name of the parent alkane, again as one word.

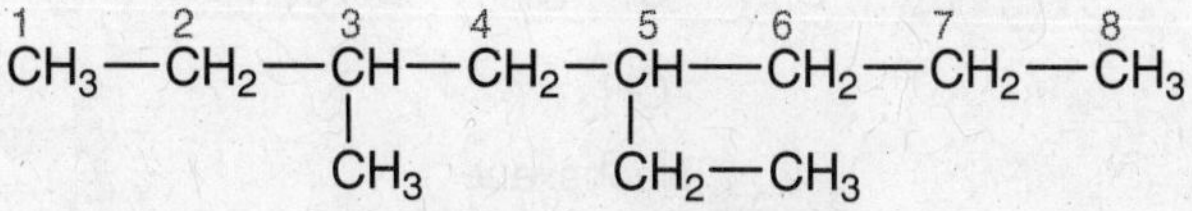

5-Ethyl-3-methyloctane

Points to Remember

The IUPAC system may appear to be difficult. In reality, learning the systematic nomenclature is probably the easiest part of organic chemistry. All that is required is a good deal of practice. Avoid the following pitfalls :

(1) **Failure to identify the longest continuous chain**. The longest chain does not have to be that portion of the molecule written horizontally. This is a mistake made by many beginning students.

(2) **Numbering from the wrong end**. When in doubt how to number the chain, do it both ways to pick the name with the lowest substituent group number.

(3) **Failure to write the same number twice** and/or failure to use the prefix *di-*, *tri-* etc., when identical substituents are attached to the same carbon.

ISOMERISM IN ALKANES

Alkanes offer for study an excellent example of **Chain Isomerism**. The first three hydrocarbons of the series (methane, ethane, propane) do not exhibit isomerism. The next hydrocarbon butane (C_4H_{10}) exists in two isomeric forms.

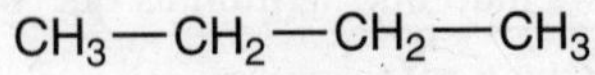

n-Butane

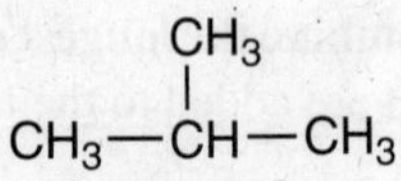

Isobutane
(2-Methylpropane)

Similarly, there are three pentanes which have the molecular formula C_5H_{12}.

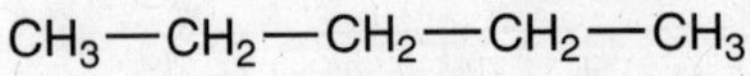

n-Pentane

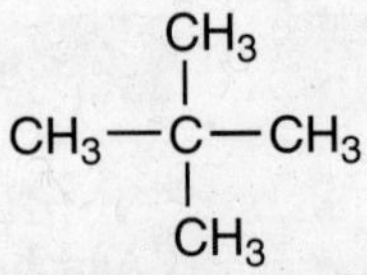

Neopentane
(2,2-Dimethylpropane)

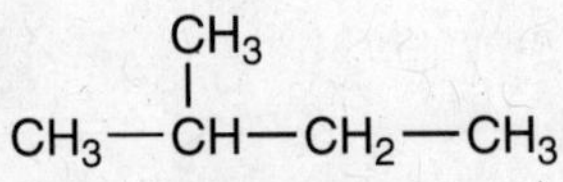

Isopentane

Hexane (C_6H_{14}) exists in five isomeric forms.

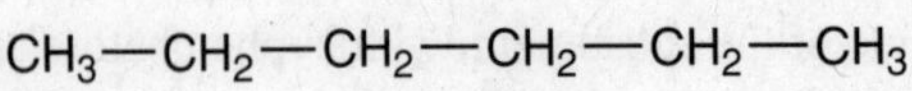

n-Hexane

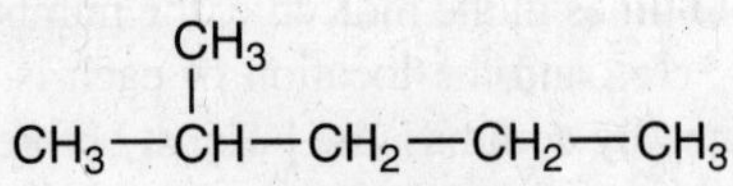

Isohexane
(2-Methylpentane)

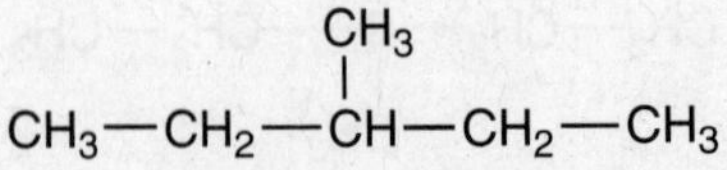

3-Methylpentane

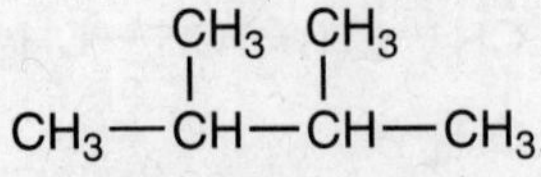

2,3-Dimethylbutane

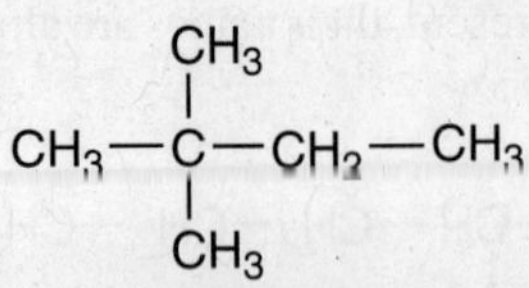

Neohexane
(2,2-Dimethylbutane)

As we go higher in the series C_nH_{2n+2}, with the increase in the number of carbons, the number of isomers increases very rapidly. For example, there are 9 heptanes, 18 octanes, and 75 decanes.

NATURAL SOURCES OF ALKANES

The two main sources of alkanes are *natural gas* and *petroleum*. Both of these substances are frequently found together in underground deposits. Natural gas contains about 80% methane and 10% ethane, the remaining 10% being a mixture of higher members. Petroleum is a chief source of alkanes containing up to 40 carbons.

METHODS OF PREPARATION

Alkanes are prepared by the following methods :

(1) **Hydrogenation of Alkenes or Alkynes**. Alkenes or alkynes react with hydrogen in the presence of nickel catalyst at 200-300°C to form alkanes. Other catalysts which can be used are platinum and palladium.

(*a*)
$$\underset{\text{Alkene}}{R-CH{=}CH_2} + H_2 \xrightarrow[\Delta]{\text{Ni}} \underset{\text{Alkane}}{R-CH_2-CH_3}$$

$$\underset{\text{Ethylene}}{CH_2{=}CH_2} + H_2 \xrightarrow[\Delta]{\text{Ni}} \underset{\text{Ethane}}{CH_3-CH_3}$$

$$\underset{\text{Propene}}{CH_3-CH{=}CH_2} + H_2 \xrightarrow[\Delta]{\text{Pt}} \underset{\text{Propane}}{CH_3-CH_2-CH_3}$$

(*b*)
$$\underset{\text{Alkyne}}{R-C{\equiv}CH} + 2H_2 \xrightarrow[\Delta]{\text{Ni}} \underset{\text{Alkane}}{R-CH_2-CH_3}$$

$$\underset{\text{Acetylene}}{CH{\equiv}CH} + 2H_2 \xrightarrow[\Delta]{\text{Ni}} \underset{\text{Ethane}}{CH_3-CH_3}$$

(2) **Reduction of Alkyl Halides**. Alkyl halides undergo reduction with nascent hydrogen to form alkanes.

$$\underset{\text{Alkyl halide}}{R-X} + 2[H] \longrightarrow \underset{\text{Alkane}}{R-H} + HX$$

$$\underset{\text{Methyl iodide}}{CH_3-I} + 2[H] \longrightarrow \underset{\text{Methane}}{CH_4} + HI$$

$$\underset{\text{Ethyl bromide}}{CH_3CH_2-Br} + 2[H] \longrightarrow \underset{\text{Ethane}}{CH_3-CH_3} + HBr$$

The hydrogen for reduction may be obtained by using any of the following reducing agents : Zn + HCl ; Zn + CH_3COOH ; Zn–Cu couple in ethanol ; or $LiAlH_4$; Hydrogen gas and Ni or Pt catalyst may also be used.

(3) **Decarboxylation of Carboxylic Acids**. When the sodium salt of a carboxylic acid is heated strongly with sodalime (NaOH + CaO), a molecule of carbon dioxide is split off as carbonate and an alkane is formed. Notice that the alkane so produced contains one carbon less than the original acid.

$$R-COONa + NaOH \xrightarrow{\Delta} \underset{\text{Alkane}}{RH} + Na_2CO_3$$

$$\underset{\text{Sodium acetate}}{CH_3-COONa} + NaOH \xrightarrow{\Delta} \underset{\text{Methane}}{CH_4} + Na_2CO_3$$

(4) **Hydrolysis of Grignard Reagents**. Alkyl magnesium halides (Grignard reagents) are obtained by treating alkyl halides with magnesium in anhydrous ether. These on treatment with water give alkanes.

$$RX + Mg \xrightarrow{\text{ether}} \underset{\text{Alkylmagnesium halide}}{RMgX}$$

$$RMgX + HOH \longrightarrow \underset{\text{Alkane}}{RH} + MgX(OH)$$

$$\underset{\text{Methylmagnesium iodide}}{CH_3MgI} + HOH \longrightarrow \underset{\text{Methane}}{CH_4} + MgI(OH)$$

(5) **Wurtz Synthesis**. Higher alkanes are produced by heating an alkyl halide (RX) with sodium metal in dry ether solution. Two molecules of the alkyl halide lose their halogen atoms as NaX. The net result is the joining of two alkyl groups to yield a symmetrical alkane (R—R type) having an even number of carbon atoms.

$$R—X + 2Na + X—R \xrightarrow{\text{ether}} \underset{\text{symmetrical alkane}}{R—R} + 2NaX$$

$$\underset{\text{Methyl bromide}}{CH_3—Br} + 2Na + Br—CH_3 \xrightarrow{\text{ether}} \underset{\text{Ethane}}{CH_3—CH_3} + 2NaBr$$

The mechanism involves the formation of an extremely reactive organosodium intermediate:

$$RX + 2Na \longrightarrow [RNa] + NaX$$

$$RX + [RNa] \longrightarrow R—R + NaX$$

The reaction between two different alkyl halides and sodium gives a mixture of products that are difficult to separate. For example, methyl chloride and ethyl chloride when treated with sodium give three products : propane from the combination of methyl chloride and ethyl chloride; ethane from the combination of two molecules of methyl chloride ; and *n*-butane from two molecules of ethyl chloride.

$$\underset{\text{Ethyl chloride}}{CH_3CH_2—Cl} + 2Na + \underset{\text{Methyl chloride}}{Cl—CH_3} \xrightarrow{\text{ether}} \underset{\text{Propane}}{CH_3CH_2—CH_3} + 2NaCl$$

$$\underset{\text{Methyl chloride}}{CH_3—Cl} + 2Na + Cl—CH_3 \xrightarrow{\text{ether}} \underset{\text{Ethane}}{CH_3—CH_3} + 2NaCl$$

$$\underset{\text{Ethyl chloride}}{CH_3CH_2—Cl} + 2Na + Cl—CH_2CH_3 \xrightarrow{\text{ether}} \underset{n\text{-Butane}}{CH_3CH_2—CH_2CH_3} + 2NaCl$$

LIMITATION. As shown above, the use of two different alkyl halides in Wurtz reaction always leads to a mixture of alkanes. The separation of these alkanes is not always easy because of little difference in their boiling points. The problem of isolating the desired product limits the usefulness of this method to the preparation of only the symmetrical alkanes.

(6) **Corey-House Alkane Synthesis**. This method was developed in late 1960s. An alkyl halide is first converted to lithium dialkyl cuperate, LiR_2Cu. This is then treated with an alkyl halide to give an alkane.

$$\underset{\text{Methyl bromide}}{CH_3Br} + 2Li \xrightarrow{\text{ether}} \underset{\text{Methyllithium}}{CH_3Li} + LiBr$$

$$2CH_3Li + CuI \xrightarrow{\text{ether}} \underset{\text{Lithium dimethyl cuperate}}{Li(CH_3)_2Cu} + LiI$$

$$\underset{\text{Ethyl bromide}}{CH_3CH_2Br} + Li(CH_3)_2Cu \longrightarrow \underset{\text{Propane}}{CH_3CH_2{-}CH_3} + CH_3Cu + LiBr$$

This method is particularly suitable for the preparation of unsymmetrical alkanes, *i.e.*, those of the type R—R′.

(7) **Kolbe's Synthesis**. When a concentrated solution of sodium salt of a carboxylic acid is electrolyzed, an alkane is formed.

$$\underset{\text{Sodium carboxylate}}{2RCO\bar{O}\overset{+}{Na}} + 2H_2O \longrightarrow \underset{\text{at anode}}{R{-}R + 2CO_2} + \underset{\text{at cathode}}{2NaOH + H_2}$$

$$\underset{\text{Sodium acetate}}{2CH_3CO\bar{O}\overset{+}{Na}} + 2H_2O \longrightarrow \underset{\text{at anode}}{CH_3{-}CH_3 + 2CO_2} + \underset{\text{at cathode}}{2NaOH + H_2}$$

This reaction is only suitable for the preparation of symmetrical alkanes, *i.e.*, those of the type, R–R.

PHYSICAL PROPERTIES

(1) First four alkanes *methane*, *ethane*, *propane* and *butane* are gases. Next thirteen members (C_5 to C_{17}) are colorless liquids. Higher alkanes are waxlike solids.

PHYSICAL STATES OF IMPORTANT ALKANES

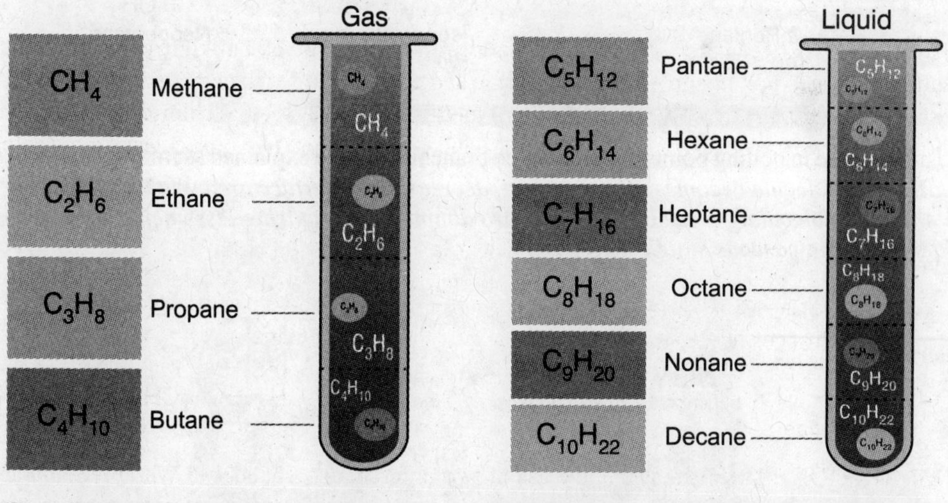

(2) Alkanes are nonpolar compounds. Their solubility characteristic may be predicted by what is commonly known as the 'like dissolves like' rule. What this rule means is that nonpolar compounds are soluble in other nonpolar solvents and that polar compounds are generally soluble in other polar solvents. Thus, alkanes are soluble in the nonpolar solvents like carbon tetrachloride and benzene, but they are insoluble in polar solvents such as water.

(3) In general, the specific gravities of alkanes increase with increasing molecular weight.

Table 10.3 MELTING AND BOILING POINTS OF ALKANES

Name	Formula	bp °C	mp °C
Methane	CH_4	– 161.5	– 182.5
Ethane	CH_3CH_3	– 88.6	– 183.3
Propane	$CH_3CH_2CH_3$	– 42.1	– 187.7
Butane	$CH_3(CH_2)_2CH_3$	0	– 138.4
Pentane	$CH_3(CH_2)_3CH_3$	36.0	– 129.7
Hexane	$CH_3(CH_2)_4CH_3$	68.7	– 95.3

(4) When a substance boils, it changes from a liquid to a gas, a process requiring energy. The boiling points of normal hydrocarbons increase with increasing molecular weight. One explanation for this trend is that as molecules become larger there are more forces of attraction between them, and more energy is needed to go from the liquid to the gaseous state. Except for the very small alkanes, the boiling point rises 20-30°C for each addition of a carbon atom to the chain (Fig. 10.3).

Among isomeric alkanes, the straight-chain compound has the highest boiling point. For the other isomers, the greater the number of branches, the lower the boiling point. For example,

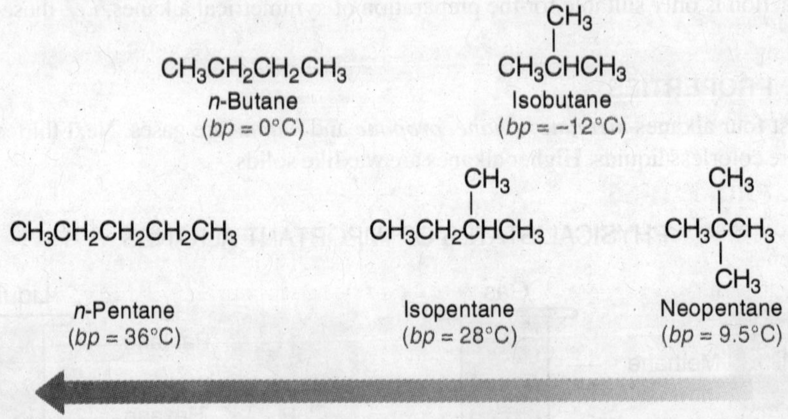

The decrease in boiling points with increased branching can be explained as follows: *as branching increases, the molecule becomes more compact, decreasing its surface area. With a smaller surface area, the intermolecular forces of attraction are diminished, requiring less energy to go from the liquid state to the gaseous state.*

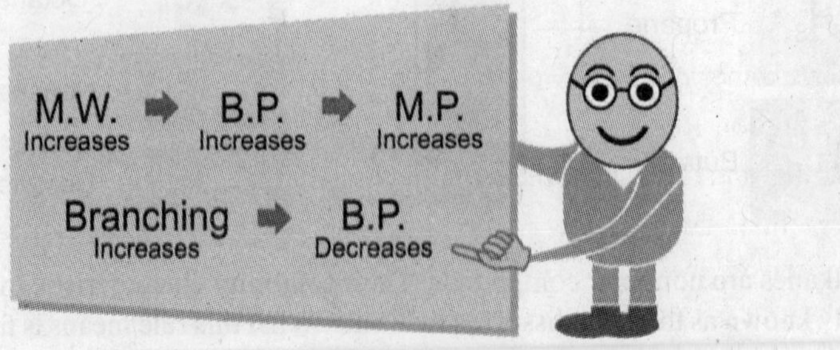

(5) The melting points of alkanes also increase with increasing molecular weight. However, unlike the change in boiling point, there is no regularity in the change in melting point with the number of carbon atoms in a molecule. Also, differences in melting points between straight-chain and branched-chain compounds follow no regular pattern.

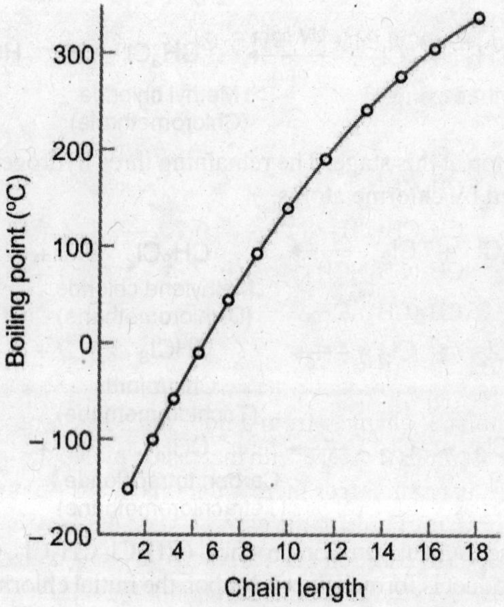

Fig. 10.3. Boiling points of normal alkanes.

(6) ***IR spectrum***. The infrared spectra of alkanes show absorptions corresponding to the C–H stretching frequencies at 2850-3000 cm^{-1}. The C–C stretching absorption have variable frequencies and are usually weak. Methyl (CH_3–) and methylene (–CH_2–) groups normally have characteristic C–H bending vibrations at 1400-1470 cm^{-1}. Methyl groups also show a weaker band near 1380 cm^{-1}.

CHEMICAL PROPERTIES

As already observed, alkanes are relatively stable to common reagents such as acids, alkalis, oxidizing agents, at room temperature. This is due to the fact that the electronegativities of carbon (2.60) and hydrogen (2.1) do not differ appreciably. Thus the bond electrons in C–H are practically equally shared between them and the bond is almost nonpolar. The C–C bond is completely nonpolar. Therefore, *polar reagents find no reaction sites on alkane molecules*. Furthermore, the C–H and C–C bonds are strong bonds. This explains why alkanes are stable to acids, alkalis, oxidizing agents etc., at room temperature.

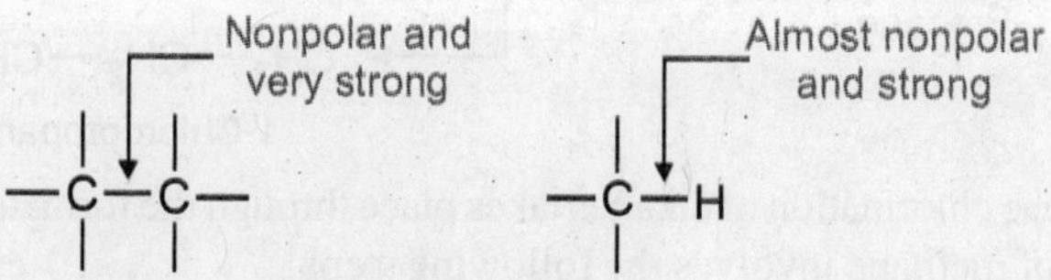

However, alkanes undergo two types of reactions :

(*a*) Substitution Reactions

(*b*) Thermal and Catalytic Reactions

These reactions take place at high temperatures or on absorption of light energy through the formation of highly reactive *free radicals*.

Some of the important reactions of alkanes are described below :

(1) **Halogenation**. This involves the substitution of hydrogen atoms of alkanes with halogen atoms.

(*a*) **Chlorination**. Alkanes react with chlorine in the presence of ultraviolet light, or diffused sunlight, or at a temperature of 300–400°C, yielding a mixture of products. For example, methane reacts with chlorine to give methyl chloride and HCl.

$$\underset{\text{Methane}}{CH_4} + Cl_2 \xrightarrow[\text{or } \Delta]{\text{UV light}} \underset{\substack{\text{Methyl chloride}\\ \text{(Chloromethane)}}}{CH_3Cl} + HCl$$

The reaction does not stop at this stage. The remaining three hydrogen atoms of methyl chloride can be successively replaced by chlorine atoms.

$$CH_3Cl + Cl_2 \longrightarrow \underset{\substack{\text{Methylene chloride}\\ \text{(Dichloromethane)}}}{CH_2Cl_2} + HCl$$

$$CH_2Cl_2 + Cl_2 \longrightarrow \underset{\substack{\text{Chloroform}\\ \text{(Trichloromethane)}}}{CHCl_3} + HCl$$

$$CHCl_3 + Cl_2 \longrightarrow \underset{\substack{\text{Carbon tetrachloride}\\ \text{(Tetrachloromethane)}}}{CCl_4} + HCl$$

In actual practice, all the four substitution products (CH_3Cl, CH_2Cl_2, $CHCl_3$, CCl_4) are obtained. The extent to which each product is formed depends upon the initial chlorine to methane ratio. Carbon tetrachloride is the major product if an excess of chlorine is used. Methyl chloride is the major product if an excess of methane is used.

Ethane and higher alkanes react with chlorine in a similar way and all possible substitution products are obtained. For example, ethane reacts with chlorine to give chloroethane as the monosubstitution product.

$$\underset{\text{Ethane}}{CH_3CH_3} + Cl_2 \xrightarrow[\text{heat}]{\text{UV}} \underset{\text{Chloroethane}}{CH_3CH_2Cl} + HCl$$

Propane contains two types (primary and secondary) of hydrogens. Therefore, it gives two monosubstitution products. Generally speaking, a tertiary hydrogen is more readily replaced than a secondary hydrogen. A secondary hydrogen is more readily replaced than a primary hydrogen.

$$\underset{\text{Propane}}{CH_3-CH_2-CH_3} + Cl_2 \longrightarrow \underset{\text{2-Chloropropane}}{CH_3-\underset{|}{\overset{Cl}{C}}H-CH_3} \quad \text{and} \quad \underset{\text{1-Chloropropane}}{CH_3-CH_2-CH_2Cl}$$

MECHANISM. The chlorination of alkanes takes place through the formation of free radicals. For example, chlorination of methane involves the following steps.

(1) **Chain-Initiation Step.** Chlorine molecule undergoes homolytic fission to give chlorine free radicals.

$$Cl : Cl \xrightarrow[\text{or } \Delta]{\text{UV light}} Cl\cdot + Cl\cdot$$

(2) **Chain-Propagation Steps.** (*a*) Chlorine free radical attacks methane to produce HCl and methyl free radical. *Notice that fish-hook arrows are used to indicate the movement of single electrons.*

$$Cl\cdot + \underset{\text{Methane}}{H : CH_3} \longrightarrow H{:}Cl + \underset{\text{Methyl radical}}{\cdot CH_3}$$

(*b*) Methyl free radical attacks chlorine molecule to give methyl chloride and chlorine free radical.

$$H_3C\cdot + Cl:Cl \longrightarrow \underset{\text{Methyl chloride}}{H_3C:Cl} + Cl\cdot$$

Steps (*a*) and (*b*) are repeated over and over again.

(3) **Chain-Termination Steps**. The above chain reaction comes to halt when any two free radicals combine. For example,

$$Cl\cdot + \cdot Cl \longrightarrow Cl—Cl$$
$$CH_3\cdot + \cdot Cl \longrightarrow CH_3—Cl$$
$$CH_3\cdot + \cdot CH_3 \longrightarrow CH_3—CH_3$$

A chlorine free radical can also attack methyl chloride to form chloromethyl free radical. This free radical participates further in the chain reaction to yield methylene chloride (dichloromethane).

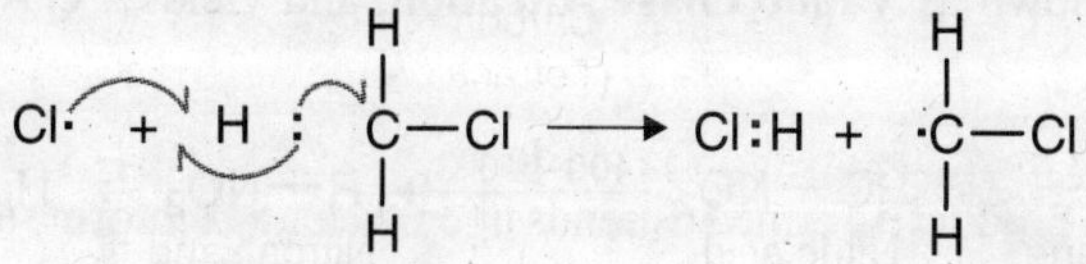

Methyl chloride   Chloromethyl radical

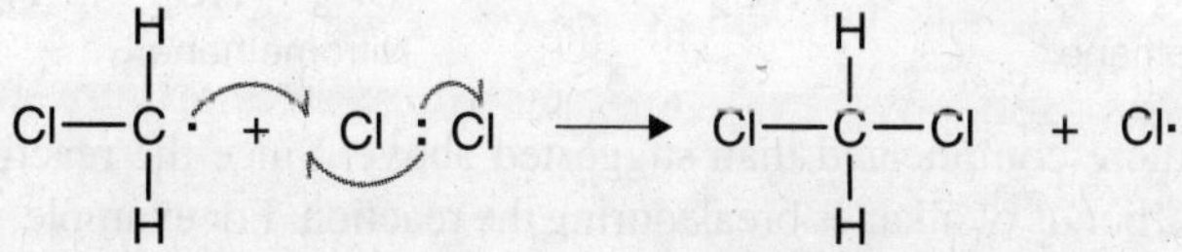

Dichloromethane

Similarly, chloroform (trichloromethane) and carbon tetrachloride (tetrachloromethane) are obtained by further chain reaction.

PROOF OF FREE RADICAL MECHANISM. If the reaction actually follows the above mechanism, then the addition of substances that are sources of free radicals should initiate the reaction even in the dark at room temperature. This is actually so. For example, methane reacts with chlorine in the dark at room temperature in the presence of catalytic amounts of dibenzoyl peroxide.

$$\underset{\text{Dibenzoyl peroxide}}{C_6H_5—\overset{O}{\overset{\|}{C}}—O—O—\overset{O}{\overset{\|}{C}}—C_6H_5} \xrightarrow[25°C]{\text{dark}} \underset{\text{Benzoyl free radical}}{2C_6H_5—\overset{O}{\overset{\|}{C}}—O\cdot}$$

$$C_6H_5—\overset{O}{\overset{\|}{C}}—O\cdot \longrightarrow CO_2 + \underset{\text{Phenyl free radical}}{C_6H_5\cdot}$$

Phenyl free radical then reacts with chlorine molecule to produce a chlorine free radical.

$$C_6H_5\cdot + Cl:Cl \longrightarrow \underset{\text{Chlorobenzene}}{C_6H_5:Cl} + \underset{\text{Chlorine free radical}}{Cl\cdot}$$

Once the chlorine free radicals are produced, the reaction can proceed in the manner described above.

(*b*) **Bromination.** Bromine reacts with alkanes in a similar manner but less vigorously.

(*c*) **Iodination.** Iodine reacts with alkanes reversibly. The hydrogen iodide formed as the by-product is a powerful reducing agent and is capable of reducing the iodoalkane to the alkane.

$$CH_4 + I_2 \rightleftharpoons \underset{\text{Methyl iodide}}{CH_3I} + HI$$

However, alkanes can be iodinated in the presence of an oxidizing agent such as HIO_3 (or HNO_3) which removes the hydriodic acid (HI) as it is formed.

$$5HI + HIO_3 \longrightarrow 3I_2 + 3H_2O$$

(*d*) **Fluorination**. Fluorine is most reactive. It reacts with alkanes explosively under most conditions. Fluoroalkanes can, however, be obtained from alkanes by the action of fluorine diluted with nitrogen.

(2) **Nitration**. This involves the substitution of a hydrogen atom of alkane with $-NO_2$ group. At ordinary temperatures, alkanes do not react with nitric acid. However, when a mixture of an alkane and nitric acid vapors is heated at 400-500°C, one hydrogen atom on the alkane is replaced by a nitro group. The process is known as **Vapor Phase Nitration**, and yields a class of compounds called *nitroalkanes*.

$$\underset{\text{Alkane}}{R{-}H} + \underset{\text{Nitric acid}}{HO{-}NO_2} \xrightarrow{400\text{–}500°C} \underset{\text{Nitroalkane}}{R{-}NO_2} + H_2O$$

$$\underset{\text{Methane}}{CH_3{-}H} + HO{-}NO_2 \xrightarrow{450°C} \underset{\text{Nitromethane}}{CH_3{-}NO_2} + H_2O$$

The reaction is more complicated than suggested above. Since the reaction is carried at high temperatures, the C–C bonds of alkanes break during the reaction. For example, ethane actually gives a mixture of nitroethane and nitromethane.

$$\underset{\text{Ethane}}{CH_3CH_3} + HONO_2 \xrightarrow{450°C} \underset{\text{Nitroethane}}{CH_3CH_2NO_2} + \underset{\text{Nitromethane}}{CH_3NO_2} + H_2O$$

Higher alkanes yield even more complex mixtures.

(3) **Sulfonation**. This involves the substitution of a hydrogen atom of alkane with $-SO_3H$ group. At ordinary temperatures neither concentrated nor fuming sulfuric acid reacts with alkanes. However, when alkanes are subjected to a prolonged reaction with fuming H_2SO_4, one hydrogen atom on the alkane is replaced by the sulfonic acid group ($-SO_3H$).

$$\underset{\text{Alkane}}{R{-}H} + \underset{(fuming)}{HO{-}SO_3H} \xrightarrow{\Delta} \underset{\text{Alkanesulfonic acid}}{R{-}SO_3H} + H_2O$$

where R = $C_6H_{13}-$ or larger alkyl group. Lower alkanes like methane and ethane do not give this reaction.

(4) **Oxidation ; Combustion**. When ignited in the presence of excess oxygen, alkanes burn to form carbon dioxide and water. Large quantity of heat (the heat of combustion) is also evolved.

$$C_nH_{2n+2} + \frac{3n+1}{2} O_2 \xrightarrow{\text{flame}} nCO_2 + (n+1)H_2O + \text{Heat}$$

$$\underset{\text{Methane}}{CH_4} + 2O_2 \longrightarrow CO_2 + 2H_2O + 212.8 \text{ kcal/mol}$$

$$\underset{\text{Ethane}}{2CH_3CH_3} + 7O_2 \longrightarrow 4CO_2 + 6H_2O + 745.6 \text{ kcal/mol}$$

The combustion of alkanes is an important reaction because it forms the basis for the extensive use of these hydrocarbons as source of heat and power.

When burnt in insufficient supply of oxygen, alkanes form carbon monoxide and carbon (carbon black). For example,

$$2CH_4 + 3O_2 \xrightarrow{\text{flame}} 2CO + 4H_2O$$
$$CH_4 + O_2 \xrightarrow{\text{flame}} C + 2H_2O$$

(5) **Isomerization**. Normal alkanes are converted to their branched-chain isomers in the presence of aluminium chloride and HCl at 25°C. For example,

$$\underset{n\text{-Butane}}{CH_3-CH_2-CH_2-CH_3} \xrightarrow[\text{HCl}]{\text{AlCl}_3} \underset{\substack{\text{Isobutane}\\\text{(2-Methylpropane)}}}{CH_3-\underset{}{\overset{CH_3}{\overset{|}{C}H}}-CH_3}$$

(6) **Pyrolysis (Cracking)**. The decomposition of a compound by heat is called pyrolysis (Greek : *pyro*, fire + *lysis*, loosening). This process when applied to alkanes is known as *cracking*. When alkanes are heated to a high temperature in the absence of air, a thermal decomposition occurs. Large alkane molecules are broken down into a mixture of smaller, lower molecular weight *alkanes*, *alkenes*, and *hydrogen*. Pyrolysis generally requires temperatures in the range 500-800°C. However, in the presence of a catalyst (finely divided silica-alumina) reactions can be carried at less high temperatures. This is called **Catalytic Cracking**.

Ethane when heated to 500°C in the *absence of air*, gives a mixture of methane, ethylene, and hydrogen.

$$\underset{\text{Ethane}}{CH_3CH_3} \xrightarrow[\text{absence of air}]{500°C} \underset{\text{Ethylene}}{CH_2{=}CH_2} + \underset{\text{Methane}}{CH_4} + H_2$$

MECHANISM. The mechanism of the above reaction involves the following steps :

(1) **Chain-Initiation Step**. Ethane molecule decomposes to give two methyl free radicals.

$$H_3C : CH_3 \xrightarrow{\Delta} \underset{\text{Methyl free radical}}{2H_3C\cdot}$$

(2) **Chain-Propagation Steps**. (*a*) Methyl free radical attacks ethane to produce methane and ethyl free radical.

$$H_3C\cdot + \underset{\text{Ethane}}{H{:}CH_2CH_3} \longrightarrow \underset{\text{Methane}}{CH_4} + \underset{\text{Ethyl radical}}{\cdot CH_2CH_3}$$

(*b*) Ethyl free radical gives ethylene and a hydrogen atom.

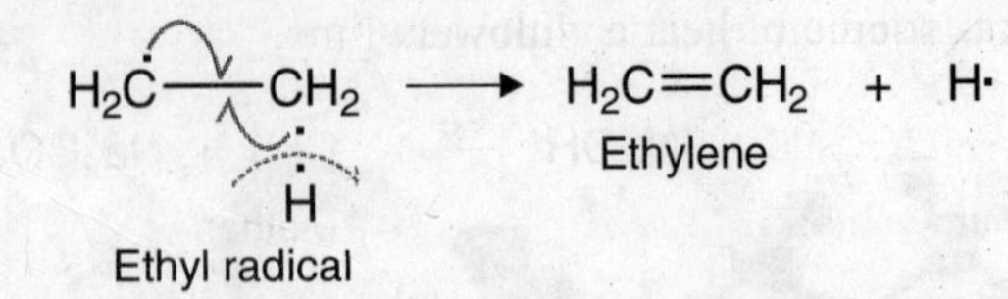

(*c*) Hydrogen atom attacks ethane to give ethyl free radical and hydrogen molecule.

Steps (*b*) and (*c*) are repeated over and over again.

(3) **Chain-Termination Step**. Unreacted ethyl free radicals combine with hydrogen atoms to regenerate ethane which can further take part in the reaction.

$$\underset{\text{Ethyl radical}}{CH_3CH_2\cdot} + \cdot H \longrightarrow \underset{\text{Ethane}}{CH_3CH_3}$$

Propane when heated to 600°C in the absence of air, gives a complex mixture of propylene, ethylene, methane, and hydrogen.

$$\underset{\text{Propane}}{CH_3CH_2CH_3} \xrightarrow[\text{absence of air}]{600°C} \underset{\text{Propylene}}{CH_3CH{=}CH_2} + \underset{\text{Ethylene}}{CH_2{=}CH_2} + \underset{\text{Methane}}{CH_4} + H_2$$

Large molecules yield even more complex mixtures because the initial products can also undergo pyrolysis. The pyrolysis (or cracking) is of great importance to the petroleum industry because it provides means of making smaller, more volatile molecules from larger, less volatile hydrocarbon molecules.

(7) **Aromatization**. Alkanes containing 6 to 10 carbon atoms are converted into benzene and its homologues at high temperatures and in the presence of a catalyst. For example, when *n*-hexane is passed over Cr_2O_3 supported over alumina at 600°C, benzene is produced. Platinum supported over alumina can also be used as a catalyst. The reaction involves simultaneous cyclization and dehydrogenation.

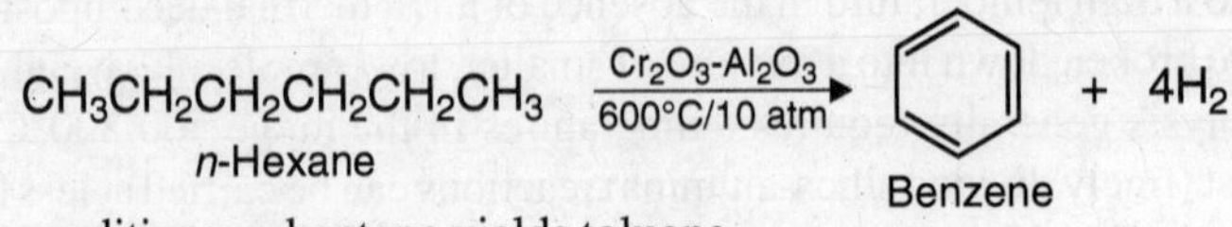

Under similar conditions, *n*-heptane yields toluene.

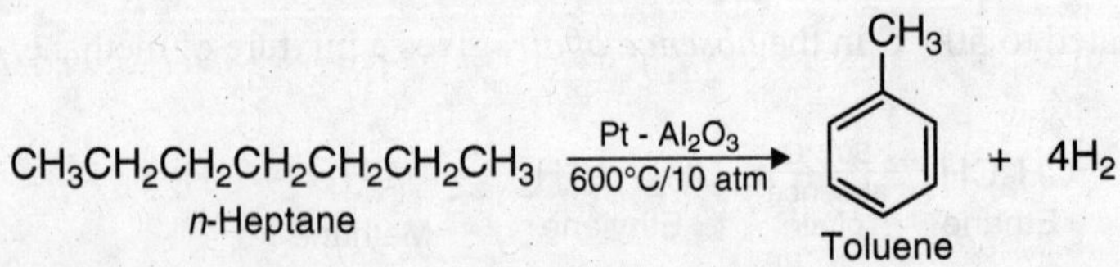

This reaction is called **Aromatization Reaction**. It provides an excellent method of passing from *Aliphatic* to *Aromatic series*.

METHANE, CH_4

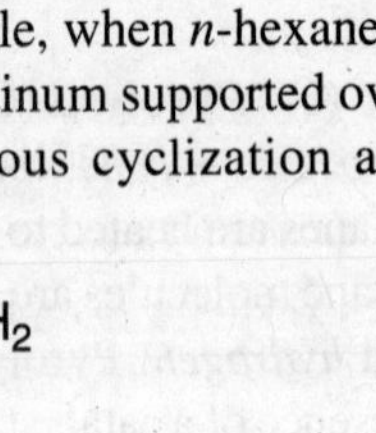

Methane is the first member of the alkane series. It is also called **Marsh Gas** because it is found bubbling from marshy places where it is produced as a result of bacterial decomposition of dead vegetable matter. It also occurs under pressure in many coal seams. The coal miners call it **Fire Damp** because it forms explosive mixtures with air. The biggest source of methane is *Natural Gas* issuing from the petroleum wells and it contains 80-90 per cent methane. It also occurs in *Coal Gas* up to the extent of 35 per cent.

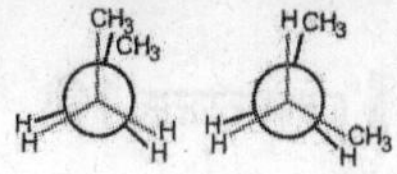

Preparation. Methane can be obtained :

(1) By heating anhydrous sodium acetate with soda-lime.

$$\underset{\text{Sodium acetate}}{CH_3\text{—}COONa} + NaOH \xrightarrow{\Delta} \underset{\text{Methane}}{CH_4} + Na_2CO_3$$

The heating can be carried in a pyrex test-tube and the gas collected over water.

(2) By the reduction of methyl iodide, using nascent hydrogen, formed by the action of ethanol on zinc-copper couple.

$$\underset{\text{Methyl iodide}}{CH_3I} + 2[H]\ (\textit{from couple}) \longrightarrow \underset{\text{Methane}}{CH_4} + HI$$

A zinc-copper couple is prepared by allowing zinc to stand in a solution of copper sulphate. The couple is then placed in a round-bottomed flask and a mixture of methyl iodide and 95 per cent ethanol is added from the tap funnel. Methane produced is collected over water.

(3) By treating aluminium carbide with water or dilute acid.

$$\underset{\text{Aluminium carbide}}{Al_4C_3} + 12HCl \longrightarrow \underset{\text{Methane}}{3CH_4} + 4AlCl_3$$

(4) By hydrolysis of methylmagnesium iodide.

$$\underset{\text{Methylmagnesium iodide}}{CH_3MgI} + H_2O \longrightarrow \underset{\text{Methane}}{CH_4} + Mg(OH)I$$

The above methods are used for the preparation of methane on small-scale. Larger quantities of methane are obtained from natural gas, from crude oil, or as by products in the refining of petroleum products.

Properties. Methane is a colorless gas, insoluble in water but soluble in organic solvents like acetone, ethanol, ether, and benzene. Methane gives all the general reactions of alkanes. Other characteristic reactions are :

(1) In the presence of UV light or on ignition, chlorine reacts with methane with an explosion to form carbon and hydrogen chloride.

$$CH_4 + 2Cl_2 \xrightarrow{\text{UV light}} C + 4HCl$$

(2) When methane is treated with steam in the presence of a nickel catalyst, carbon monoxide and hydrogen are formed. This mixture can be passed under pressure over heated ZnO + Cr_2O_3 catalyst to yield methanol.

$$\underset{\text{Methane}}{CH_4} + H_2O\ (\textit{steam}) \xrightarrow[\Delta]{Ni} CO + 3H_2$$

$$CO + 2H_2 \xrightarrow[Cr_2O_3]{ZnO} \underset{\text{Methanol}}{CH_3OH}$$

(3) When methane is heated in the *absence of air* or oxygen at 1500°C, acetylene is formed.

$$2CH_4 \xrightarrow{1500°C} \underset{\text{Acetylene}}{2HC{\equiv}CH} + 3H_2$$

(4) When methane is burnt in a limited supply of air, carbon black is produced.

$$CH_4 + O_2 \xrightarrow{\text{flame}} 2H_2O + \underset{\text{Carbon black}}{C}$$

Uses. Methane is used : (1) as a domestic fuel in the form of natural gas ; (2) in the manufacture of methanol ; (3) in the manufacture of carbon black which is used in printing inks, shoe polishes, gramophone records, and rubber tyres.

ETHANE, $CH_3–CH_3$

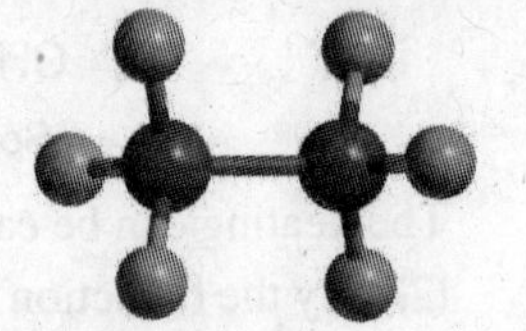

Ethane occurs to the extent of 10-20 per cent along with methane in *Natural Gas*.

Preparation. Ethane can be prepared :

(1) By heating sodium propionate with soda-lime in a pyrex test-tube.

$$\underset{\text{Sodium propionate}}{CH_3CH_2COONa} + NaOH \xrightarrow{\Delta} \underset{\text{Ethane}}{CH_3CH_3} + Na_2CO_3$$

(2) By the reduction of ethyl iodide, using nascent hydrogen, formed by the action of ethanol on zinc-copper couple.

$$\underset{\text{Ethyl iodide}}{CH_3CH_2I} + 2[H]\ (\textit{from couple}) \longrightarrow \underset{\text{Ethane}}{CH_3CH_3} + HI$$

(3) By the action of sodium metal on methyl iodide in dry ether solution. (*Wurtz* Reaction)

$$\underset{\text{Methyl iodide}}{CH_3I} + 2Na + ICH_3 \xrightarrow{\text{ether}} \underset{\text{Ethane}}{CH_3CH_3} + 2NaI$$

Thin slices of sodium metal are suspended in dry ether contained in a small flask fitted with a reflux condenser and a tap-funnel. Methyl iodide is slowly added from the tap-funnel and the ethane evolved is collected over water.

(4) By passing a mixture of ethylene and hydrogen over heated nickel catalyst.

$$\underset{\text{Ethylene}}{CH_2{=}CH_2} + H_2 \xrightarrow[\Delta]{Ni} \underset{\text{Ethane}}{CH_3CH_3}$$

(5) By hydrolysis of ethylmagnesium bromide.

$$\underset{\text{Ethylmagnesium bromide}}{CH_3CH_2MgBr} + H_2O \longrightarrow \underset{\text{Ethane}}{CH_3CH_3} + Mg(OH)Br$$

Properties. Ethane is a colorless gas. It is sparingly soluble in water but dissolves in organic solvents like ethanol, ether, and benzene. It gives all the general reactions of alkanes.

CONFORMATIONS OF ALKANES

Free rotation is possible around C–C single bonds (s bonds). **The different spatial arrangements of a molecule that can be obtained by rotation around C–C single bonds are called Conformations or Conformation Isomers.**.

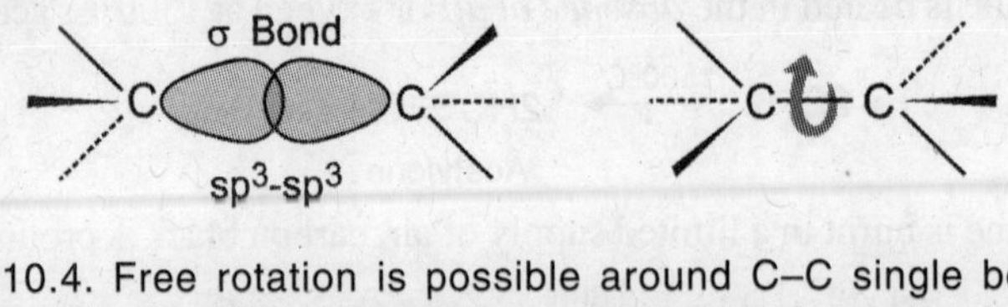

Fig.10.4. Free rotation is possible around C–C single bonds (σ bonds). The different spatial arrangements of a molecule that result from rotation about a single bond are called conformations.

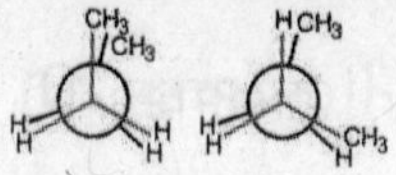

Conformations of Ethane. Let us consider ethane, $H_3C–CH_3$, as an example. The two tetrahedral methyl groups can rotate about the carbon-carbon bond axis yielding several arrangements.

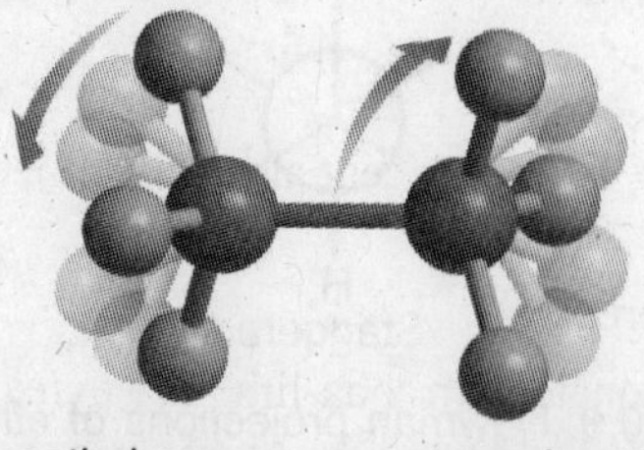

Fig.10.5. In ethane, the two methyl groups can rotate about the C–C σ bonds. As a result, an infinite number of spatial arrangements (conformations) of ethane are possible.

The two extreme arrangements of ethane represented by ball-and-stick models are shown in Fig. 10.6.

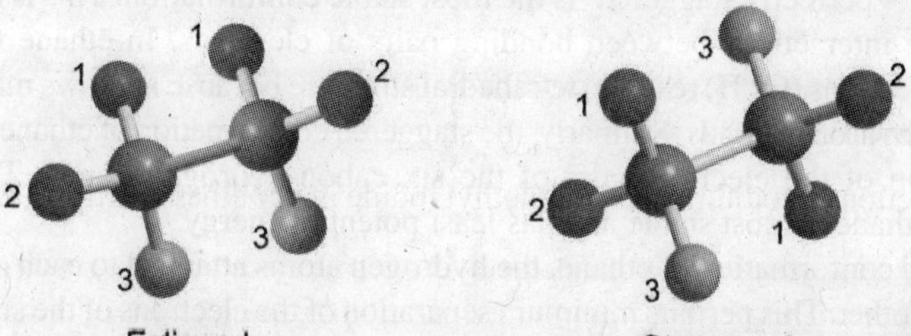

Fig. 10.6. Ball-and-Stick models of eclipsed and staggered conformations of ethane.

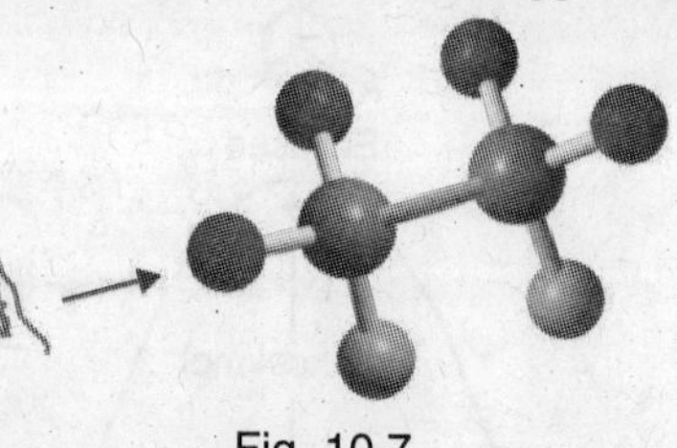

Fig. 10.7

When viewed from one end of the model along C–C axis (Fig. 10.7), the models shown above represent the following conformational forms.

(1) The **Eclipsed Form** in which the rear methyl group is completely eclipsed and only the one nearer the eye is visible.

(2) The **Staggered Form** in which the rear methyl group has been rotated upside down (Fig. 10.6) and all the six H atoms of the two methyl groups are staggered symmetrically.

(3) Evidently there could be several other arrangements or forms possible in between the eclipsed form and the staggered form. Such an arrangement lying anywhere between the two extreme forms is called a **Skew Form**.

As suggested by Melvin Newman, in ethane the front carbon atom is represented by the intersection of bonds from it, while the rear carbon appears as a circle (Fig. 10.8).

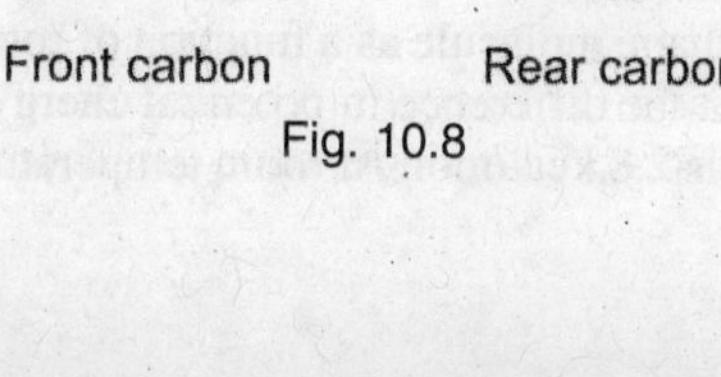

Fig. 10.8

The Newman projection formulas for the conformations of ethane could be depicted as in Fig. 10.9.

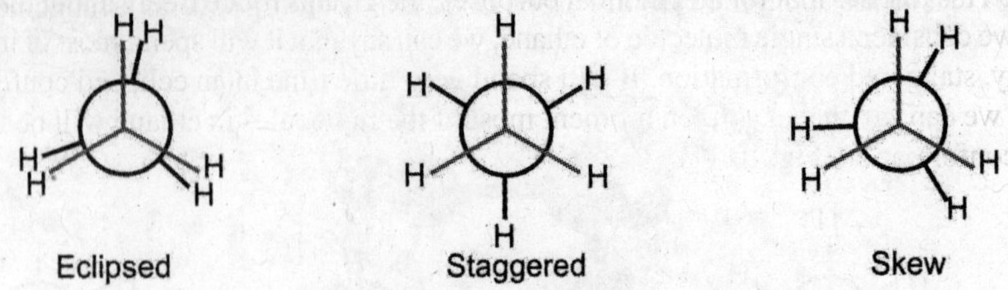

Fig. 10.9. Newman projections of ethane.

It may be noticed that the bonds of the rear carbon in the eclipsed form are shown slightly displaced simply to make them visible.

The different conformations of ethane are not of equal stability. The staggered form in which the hydrogen atoms are 'perfectly staggered' is the most stable conformation. This is easily explained in terms of repulsive interactions between bonding pairs of electrons. In ethane the interactions of bonding pairs of electrons (C : H) result in tetrahedral structure because it allows maximum separation of the four carbon-hydrogen bonds. Similarly, the staggered conformation of ethane permits maximum possible separation of the electron pairs of the six cabon-hydrogen bonds. Thus the staggered conformation of ethane is most stable and has least potential energy.

In the eclipsed conformation of ethane, the hydrogen atoms attached to each carbon are in direct opposition to each other. This permits minimum separation of the electrons of the six carbon-hydrogen

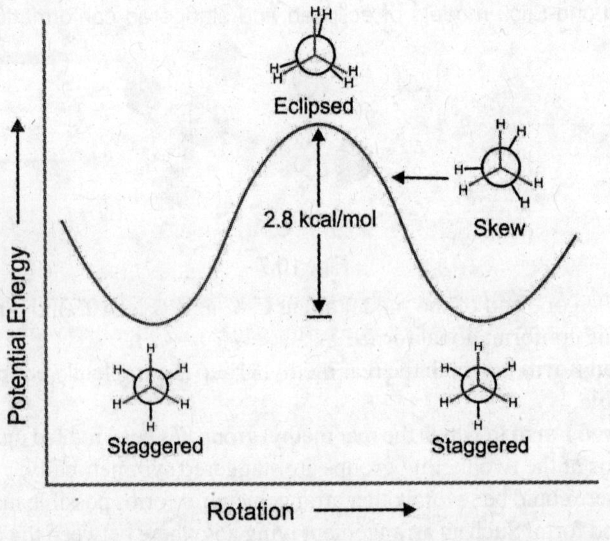

Fig. 10.10. Potential energy of ethane as function of degree of rotation about the C–C bond.

bonds. The eclipsed conformation is, therefore, of highest energy and has the lowest stability. Evidently, all the skew conformations of ethane will have stabilities in between that of the staggered conformation and that of eclipsed conformation. Hence the relative stabilities of the various conformations of ethane are :

Staggered > Skew > Eclipsed

A plot of potential energy of an ethane molecule as a function of rotation about carbon-carbon bond is shown in Fig. 10.10. Notice that the difference in potential energy between the staggered and the eclipsed conformations of ethane is 2.8 kcal/mol. At room temperature, however, molecules have an average of 15-20 kcal/mol. Thus ethane molecules have more than enough energy to surmount this small barrier. This implies that for all practical purposes, the groups rotate freely about the C–C bond. However, if we consider a single molecule of ethane, we can say that it will spend most of its time in the lowest energy, staggered conformation. It will spend very little time in an eclipsed conformation. In other words, we can say that at a given moment most of the molecules in ethane will be in staggered or in a skew conformation (Fig. 10.11).

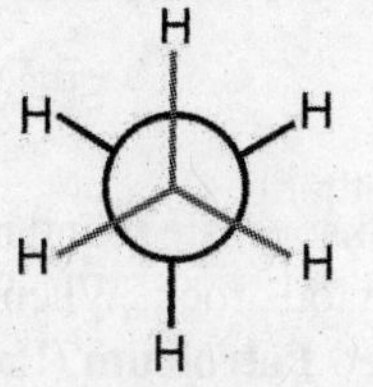

Fig. 10.11. Ordinarily ethane is mostly staggered form.

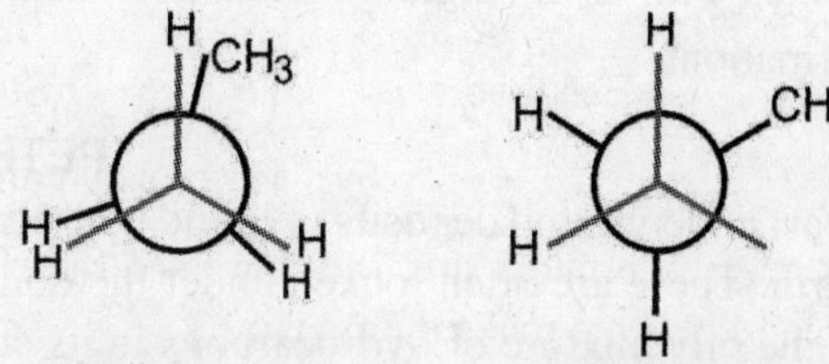

Eclipsed Staggered

Fig. 10.12. Conformations of propane.

Conformations of Propane. The conformations of propane are similar to those of ethane except that one of the hydrogen atoms is replaced by a methyl group. The two extreme conformations (eclipsed and staggered) of propane are shown in Fig. 10.12.

Conformations of Butane. They are rather complicated. All eclipsed conformations are equivalent in case of ethane and propane. But in the case of butane, two distinct conformations are possible ; one in which a methyl group is eclipsed by a methyl group and the other in which the methyl groups are eclipsed by hydrogens (Fig. 10.13).

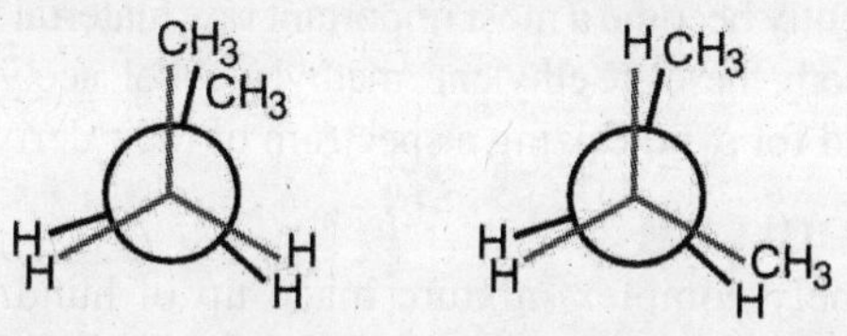

Fig. 10.13. Eclipsed conformations of butane.

Of these two eclipsed conformations, the one having larger methyl groups eclipsing (1), will naturally experience more repulsive forces than the one where a methyl group is faced with a hydrogen atom (2). Thus the conformation (2) is favored from stability standpoint.

Likewise, it will be observed that for butane two different staggered conformations are possible (Fig. 10.14).

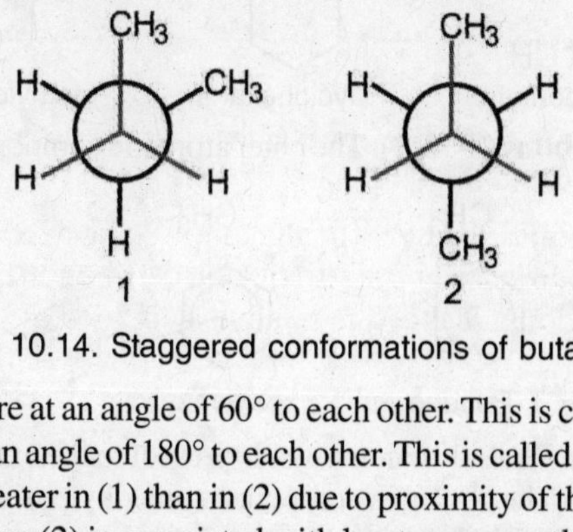

1 2

Fig. 10.14. Staggered conformations of butane.

In (1) the methyl groups are at an angle of 60° to each other. This is called **Gauche Conformation**. In (2) the methyl groups are at an angle of 180° to each other. This is called **Anti** or **Trans Conformation**. The repulsive forces will be greater in (1) than in (2) due to proximity of the methyl groups in the former. Consequently, the conformation (2) is associated with least energy and maximum stability.

In larger alkanes too, the most stable conformations are generally staggered, with the largest groups *anti* to each other. The *gauche* conformations are only slightly less stable than the *anti* ones, and are also present. Thus, while the structural formulas of alkanes are usually written in a straight line for convenience, it is important to keep in mind the zig-zag nature of hydrocarbon chains.

Because there is continuous rotation about all the C–C single bonds in a molecule, organic molecules with C–C single bonds are not static balls and sticks–they have many interconvertible conformations.

PETROLEUM

Vast underground deposits of liquid hydrocarbons, mainly alkanes, are found in many regions of the earth. These are often locked under the dome-shaped impervious rock. When drilled out of the mine, the oily mixture of hydrocarbons in its crude form is called **Petroleum** (Latin, *Petra* = rock, *oleum* = oil), or **Crude Oil**. Accompanying the reservoirs of petroleum is **Natural Gas**, consisting of gaseous hydrocarbons, largely methane.

COMPOSITION OF NATURAL GAS

The gas found above the petroleum deposits in mines or by itself, is referred to as *Natural Gas*. It is a mixture of low molecular weight alkanes. It is composed primarily of methane (80%), ethane (13%), propane (3%), butane (1%), C_5 to C_8 alkanes (0.5%) and nitrogen (1.3%).

The natural gas is used as domestic fuel for heating and cooking. For this purpose, the hydrocarbons from C_3 to C_8 are removed as liquids by compression or 'solvent absorption'. The remaining gas consisting of methane and ethane (also nitrogen) is compressed in steel cylinders and sold as **Bottled Gas**. In large towns, compressed natural gas is distributed by pipelines.

The natural gas has currently become a most important raw material for several industrial organic chemicals. The prominent among these are ethylene, methyl alcohol, acetylene, and ammonia. Ethylene and acetylene in turn are used for synthesizing a spectrum of their derivatives.

COMPOSITION OF PETROLEUM

Petroleum is an extremely complex mixture made up of hundreds of compounds, mostly hydrocarbons. It consists of alkanes, cycloalkanes, and aromatic hydrocarbons. Besides, there are present small amounts of sulphur and nitrogen compounds. The ratio of the various components much depends on the origin of the oil.

(1) **Alkanes** (30-70%). Petroleum contains straight-chain alkanes (*n*-alkanes) of well over 30 carbon atoms. Also present are many branched isomers (methyl *n*-alkanes) in the range C_6–C_8.

(2) **Cycloalkanes** (16-64%). The cycloalkanes commonly present in petroleum are :

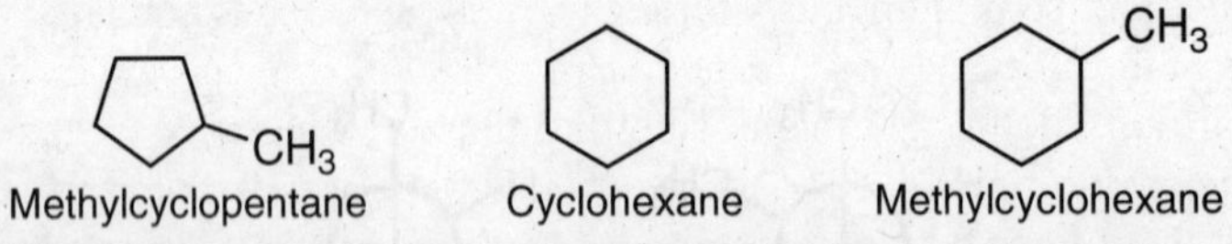

(3) **Aromatic Hydrocarbons** (8-15%). The chief aromatic components of petroleum are :

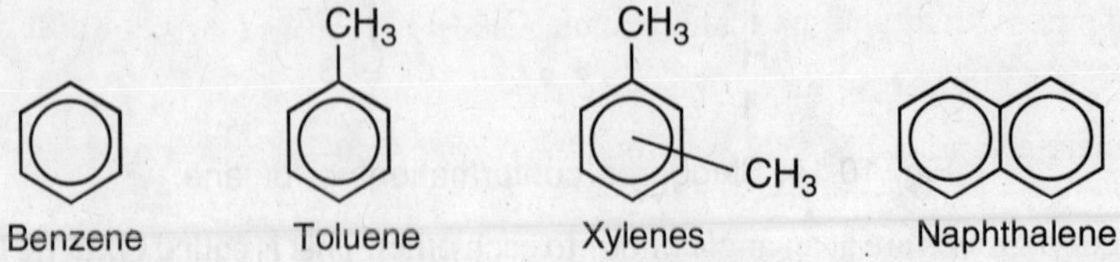

(4) **S, N, O Compounds**. The sulfur compounds present to the extent of 6% include mercaptans (R–SH) and sulfides (R–S–R). The Nitrogen compounds are alkylpyridines, quinolines and pyrroles. The oxygen compounds found include alcohols, phenols and resins.

PETROLEUM REFINING

Petroleum as it comes from the ground is a rather viscous and highly colored liquid. It often possesses unpleasant odor which is largely due to the presence of sulphur compounds in it. Technically it is called **Crude Oil**.

Crude oil consists chiefly of mixture of alkanes boiling between a wide range, say 0°C and 400°C, and as such it is unsuitable for most technical purposes. If, however, the oil is subjected to fractional distillation, it can be separated into a number of fractions each of which has a technical name and finds use in industry. The process of dividing petroleum into fractions with different boiling ranges and free from undesirable impurities, is termed **Refining**.

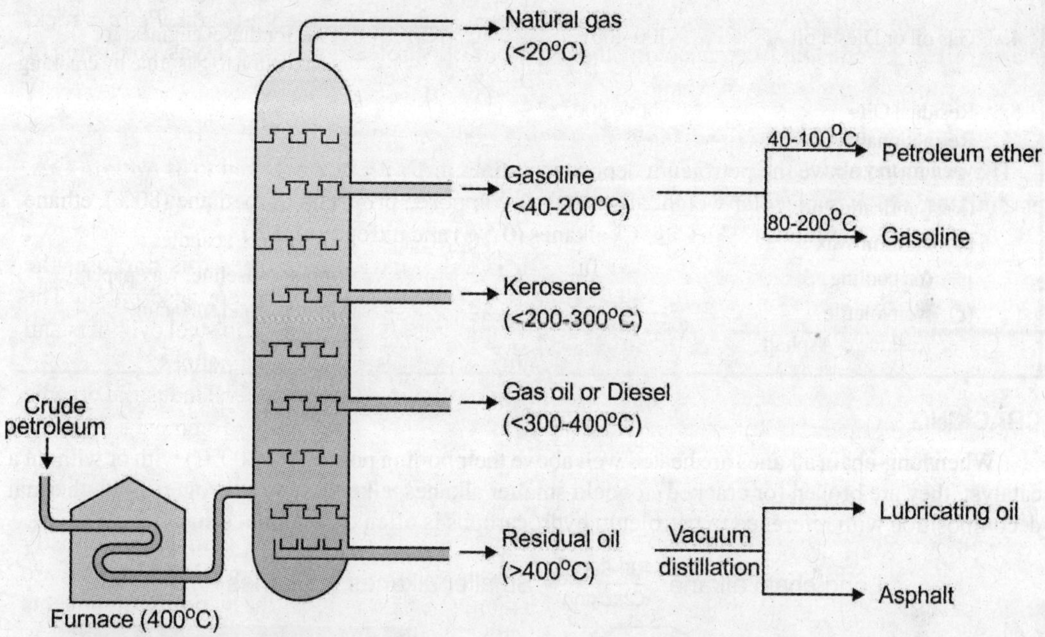

Fig. 10.15. Fractional distillation of petroleum (crude oil).

The distillation of petroleum is carried in a tubular furnace with a tall steel fractionating column (bulb). The crude oil is pumped continuously through heated pipes and flashed into the fractionating column. The vapors of the oil as they rise up the fractionating column, become cooler and condense at the shelves at various heights. The highest boiling fraction condenses, at the bottom and the lowest boiling fraction at the top (Fig. 10.15). The uncondensed gases escape along with the gasoline vapors at the head of the column. Outlets are provided in the side of the column at suitable heights to withdraw a number of fractions, the actual number of fractions and their boiling point ranges depending on the source of the petroleum and the trade demand. Table 10.4 shows the composition and uses of the main petroleum fractions.

In order to secure a yield of petrol or gasoline, the higher boiling straight run 'gas oil' (Diesel) or 'Residual oil' are cracked and simultaneously fractionally distilled. The *Catalytic Cracking* is more modern and efficient. The process is used in preference to the simple *Thermal Cracking*. The stock is charged into the 'Reactor' containing the catalyst at elevated temperature. The vapors of the cracked stock then pass to the fractionating unit and the products are : (*i*) *Cracked gasoline* ; and (*ii*) *Cracked fuel oil* (Fig. 10.16).

Table 10.4 COMPOSITION AND USES OF THE MAIN PETROLEUM FRACTIONS

	Name	Boiling Range°C	Composition	Uses
1.	Natural gas	Up to room temperature	C_4–C_5	Fuel gas
2.	Straight run Gasoline– Refractionated to yield :	40-200	C_5–C_{11}	
	(*a*) Petroleum ether	40-80	—	Solvent
	(*b*) Gasoline or Petrol	80-200	—	Motor fuel ; solvent ; dry-cleaning
3.	Kerosene	200-300	C_{11}–C_{16}	Illuminant ; fuel for stoves ; for making oil gas
4.	Gas oil or Diesel oil	300-400	C_{16}–C_{18}	Fuel for diesel engines; for conversion to gasoline by cracking
5.	Residual Oil– Refractionated under vacuum to give :	above 400	—	
	(*a*) Lubricating oil	—	C_{18}–C_{20}	Lubrication
	(*b*) Paraffin Wax (on cooling)	—	C_{20}–C_{30}	Ointments ; candles ; toilets; vaseline; wax paper
	(*c*) Nonvolatile Residue ; Asphalt	—	—	Paints ; road surfacing

CRACKING

When long-chain alkanes are heated well above their boiling point (500-800°C) with or without a catalyst, they are broken (or cracked) to yield smaller alkanes, alkenes, and hydrogen. Such thermal decomposition with reference to petroleum hydrocarbons is often called **Cracking**.

$$\text{Long-chain alkane} \xrightarrow[\text{Cracking}]{400\text{-}800°C} \text{Smaller alkanes + alkenes + } H_2$$

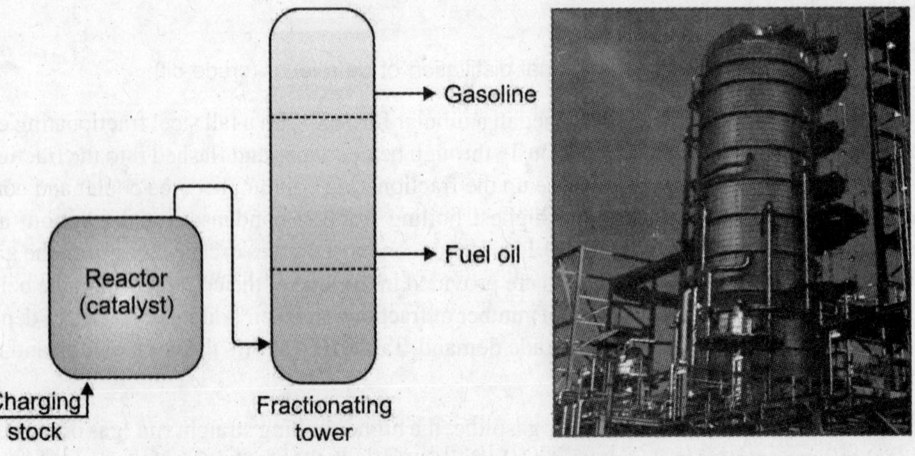

Fig. 10.16. Catalytic Cracking.

In the process of cracking, C–C and C–H bonds may rupture at various places. Thus by breaking of a single alkane, the number of products is quite large. For example,

$$CH_3CH_2CH_2CH_3 \xrightarrow{600^\circ C} \begin{cases} CH_3CH_2CH{=}CH_2 + H_2 & \text{1-Butene} \\ CH_3CH{=}CHCH_3 + H_2 & \text{2-Butene} \\ CH_3CH{=}CH_2 + CH_4 & \text{Propene} \\ CH_2{=}CH_2 + CH_3CH_3 & \text{Ethylene} \end{cases}$$

n-Butane

In *Catalytic Cracking*, silica or alumina is used as a catalyst. The process of cracking is of tremendous importance in petroleum industry.

(1) **Gasoline Production**. Crude Oil contains about 18 per cent of the gasoline fraction. To meet the growing demand of gasoline (*motor fuel*), the higher-boiling fractions of crude oil are converted to lower-boiling hydrocarbons of gasoline range. Thus, kerosene could be changed to gasoline by catalytic cracking.

$$\underset{(br\ 200\text{-}300^\circ C)}{\text{Kerosene}\ (C_{11}\text{—}C_{16})} \xrightarrow[50^\circ C]{Al_2O_3\text{–}SiO_2} \underset{(br\ 80\text{-}200^\circ C)}{\text{Gasoline}\ (C_5\text{—}C_{11})}$$

Catalytic cracking of higher-boiling fractions of petroleum increases the proportion of gasoline and also produces alkenes and branched-chain hydrocarbons by isomerization.

$$\underset{n\text{-Hexane}}{CH_3CH_2CH_2CH_2CH_2CH_3} \xrightarrow[\Delta]{\text{catalyst}} \underset{\text{2,3-Dimethylbutane}}{CH_3\text{—}\overset{\overset{\large CH_3}{|}}{CH}\text{—}\overset{\overset{\large CH_3}{|}}{CH}\text{—}CH_3}$$

Thus the product is high quality motor fuel, its octane number being greatly improved.

(2) **Production of Petrochemicals**. The thermal cracking of ethane and propane is done to produce ethylene, propylene, and other alkenes.

$$\underset{\text{Ethane}}{CH_3\text{—}CH_3} \xrightarrow[\text{pressure}]{800^\circ C} \underset{\text{Ethylene}}{CH_2{=}CH_2} + H_2$$

$$\underset{\text{Propane}}{CH_3\text{—}CH_2\text{—}CH_3} \xrightarrow[\text{pressure}]{800^\circ C} \underset{\text{Propene}}{CH_2{=}CH\text{—}CH_3} + H_2$$

Ethylene and propene are raw materials for numerous organic compounds (See Petrochemicals).

(3) **Manufacture of Fuel Gas**. Higher-boiling fractions could be converted to gaseous fuel by cracking. By employing very high temperature, the liquid hydrocarbons are converted to what we call **Oil Gas** for laboratory use. The oil gas so produced is a mixture of methane, ethane, ethylene, and hydrogen. The oil gas is actually prepared by dropping kerosene into a red-hot iron retort and removing the tarry matter.

SYNTHETIC PETROL

Fortunately, India has built up its natural petroleum sources during the recent years. However, the supply of the motor fuel *petrol* or *gasoline* from all sources put together falls far short of the actual needs of a developing country as our's. Therefore, the production of petrol by synthetic methods is engaging the attention of our organic chemists. The 'synthetic petrol' has been produced in Germany from coal, lignite, and other carbonaceous material. The chief methods used for preparing petrol from coal are :

(1) **Bergius Process**. In this method, finely powdered coal and a suitable catalyst is made into a paste with *Heavy Oil* produced in the process. The coal paste is then preheated and pumped to the *Converter*. Here the paste is heated to 450°C and hydrogen bubbled through it at 250 atm pressure. The coal, which is really a mixture of high molecular complex organic compounds deficient in hydrogen, and elementary carbon, combines with hydrogen to form higher saturated compounds. These as a result of cracking and simultaneous hydrogenation yield mixtures of alkanes. Thus the vapors leaving the *Converter* upon condensation yield 'Synthetic Petroleum' or *Crude Oil*. This crude oil upon fractional distillation produces : (*i*) **Petrol ;** (*ii*) **Middle oil** ; and (*iii*) **Heavy oil.** The Middle oil fraction is again hydrogenated in the vapor phase in contact with a solid catalyst to give more petrol. Actually, the second process yields four times the petrol obtained by the primary hydrogenation of coal.

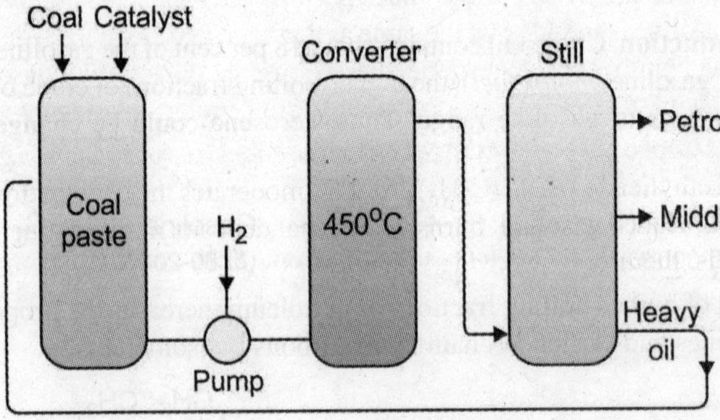

Fig. 10.17. Manufacture of Synthetic Petrol from Coal.

The heavy oil obtained in the process is recirculated for making fresh coal paste. One ton of soft coal treated as above yields 140 gallons of petrol. The *Bergius Process* is at present the most promising of all methods so far invented for the production of synthetic motor fuel (Fig. 10.17).

(2) **Fischer-Tropsch Process (*Reduction of Carbon monoxide*).** This method was developed in 1923 by two German chemists, Franz Fischer and Hans Tropsch. Carbon monoxide needed in this process is made by passing steam over red-hot 'coke'. The water gas thus obtained is mixed with hydrogen and passed at a pressure of 5-10 atmospheres into a chamber containing a 'cobalt catalyst' at 200°C. The product is a mixture consisting mainly of liquid hydrocarbons.

$$nCO + (2n+1)H_2 \longrightarrow \underset{\text{Alkane}}{C_nH_{2n+2}} + nH_2O$$

The 'artificial petroleum' so obtained is separated by fractional distillation as described under the *Bergius Process*. The various fractions separated are petrol, kerosene oil, lubricating oil, diesel oil and paraffin wax.

OCTANE NUMBER

Gasoline (or petrol) is a complex mixture of C_4 to C_{10} hydrocarbons. The branched-chain alkanes give a relatively smooth performance when burnt in the internal combustion engines of motor cars, and are excellent fuels. Normal alkanes, on the other hand, detonate and cause knocking of the engine, and are bad fuels. Different gasolines vary in respect of their knock property, depending on their composition.

The knock property of various gasolines is stated in terms of what we call the **Octane Number** or **Octane Rating**. 2,2,4-Trimethylpentane (Isooctane), with no tendency to knock, is assigned the octane number 100, and heptane that causes great knocking is assigned the octane number 0.

$CH_3CH_2CH_2CH_2CH_2CH_2CH_3$

Heptane
(Octane No = 0)

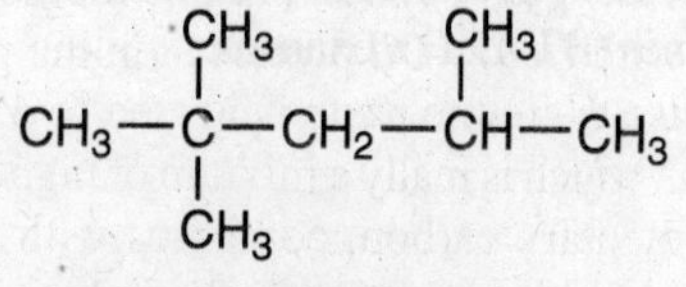

2,2,4-Trimethylpentane
(Isooctane)
(Octane No = 100)

The Octane Number of various gasolines was determined by matching their knock performance with that of mixtures of isooctane and heptane in an experimental engine.

The Octane Number of a particular gasoline is that per cent of isooctane in a mixture of isooctane and heptane which gives equivalent knock performance. For example, the knock performance of a given sample of gasoline is found to be the same as that of a mixture of 42% of isooctane and 58% of heptane. The Octane Number of the gasoline, therefore, is 42. Similarly, a gasoline with the same knock performance as a mixture of 90% isooctane and 10% heptane, is designated as *90 Octane*.

Octane Number is improved by addition of TEL. The knock performance of a gasoline is improved by the addition of tetraethyllead (*TEL*), $(C_2H_5)_4Pb$. *TEL* moderates the combustion of the fuel so that it burns smoothly. As a 'leaded gasoline' burns, lead metal deposits in the engine. Ethylene bromide (1,2-Dibromoethane) is, therefore, also added to react with this lead to form lead bromide which is volatile at engine temperature.

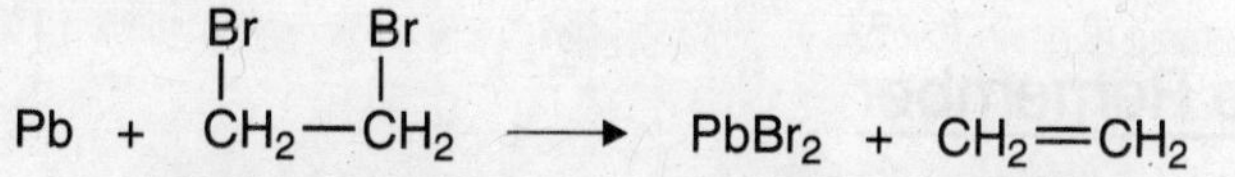

Ethylene bromide (*volatile*) Ethylene

Since lead is toxic, the lead bromide that is exhausted from the engine pollutes the atmosphere. Therefore, 'no-lead' or 'low-lead' gasolines are currently used.

No-lead Gasoline with improved Octane Number. 'No-lead' gasolines have been recently obtained by enhancing the percentage of branched-chain alkanes, alkenes, and aromatics (benzene, toluene, xylenes). The following processes are used :

(1) **Isomerization (*Reforming*).** The middle fractions of petroleum, containing predominantly straight-chain alkanes, are isomerized or *reformed* by passing over $AlCl_3$ catalyst at 200°C.

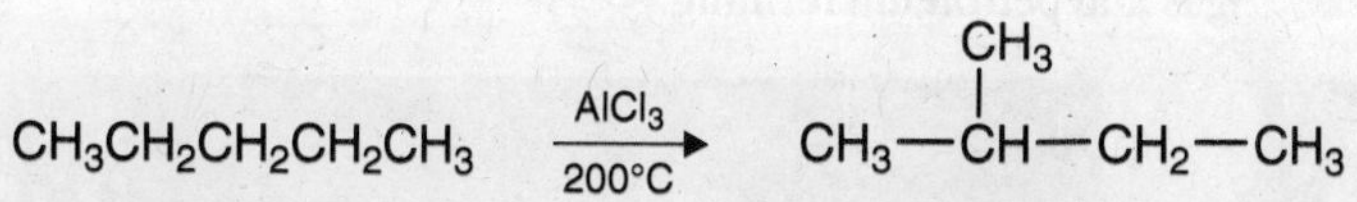

Pentane (Octane No = 62) → Isopentane (Octane No = 90)

(2) **Alkylation.** Isobutylene produced as a by-product of cracking of petroleum, alkylates isobutane to form isooctane in the presence of conc. H_2SO_4.

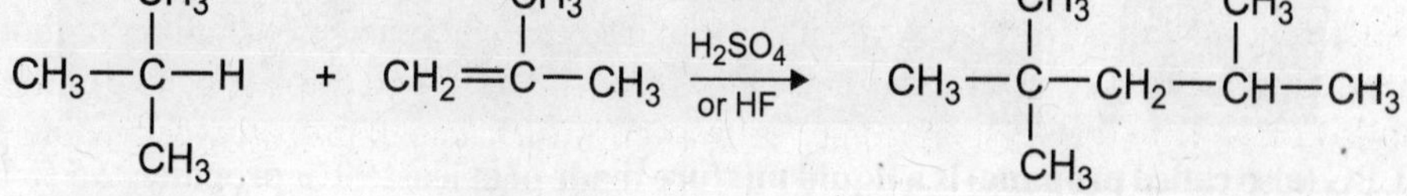

Isobutane Isobutylene 2,2,4-Trimethylpentane (Octane No = 100)

(3) **Aromatization (*Platforming*).** The octane number is improved significantly by converting low-octane alkanes into high-octane components. The C_6–C_8 straight-chain alkanes (petroleum

naphtha), in the presence of platinum catalyst (Pt–Al_2O_3) at 500°C, yield a mixture of benzene, toluene and xylenes (BTX). For example,

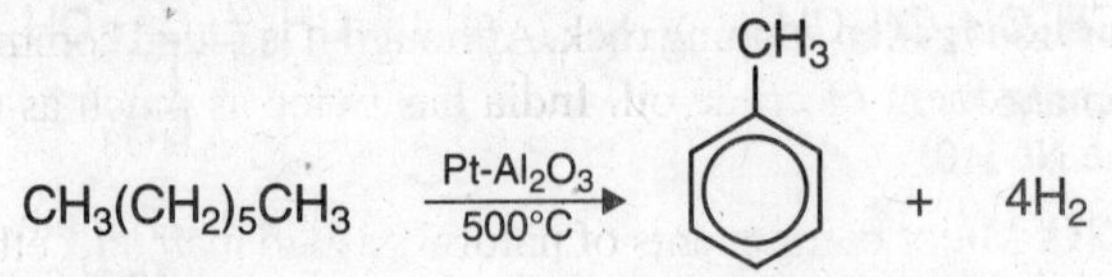

$$CH_3(CH_2)_5CH_3 \xrightarrow[500°C]{Pt\text{-}Al_2O_3} \text{Toluene} + 4H_2$$

Heptane (Octane No = 0) — Toluene (Octane No = 104)

The gasolines prepared by isomerization and alkylation, when blended with BTX are about 90 Octane. This is the fuel that is now sold as no-lead gasoline.

CETANE NUMBER

A number given to diesel fuel sample which is a measure of its performance in a compression engine. **It is the volume percent of hexadecane (Cetane) in a blend with 1-methylnaphthalene that gives the same performance in a standard compression engine as the fuel sample under the same operating conditions.**

Hexadecane is given an arbitrary cetane number of 100 and 1-methylnaphthalene a number of 0. The diesel engine charges air into a chamber and compresses it during intake. Liquid fuel is then injected into the compressed air, where combustion occurs, generating the energy required for locomotion. Fuels with cetane numbers greater than 45 are required for good performance in a compression engine.

Points to Remember

Alternative Fuels

1. An alternative fuel is anything that is not gasoline or diesel.
2. The following organic compounds have been tried as alternate fuels :
   - **Biodiesel** - made from vegetable oil
   - **Ethanol** - made from corn or sugarcane
   - **Methanol** - made from natural gas and wood chips.
   - **Natural Gas** - also called CNG (compressed natural gas). It is a fossil fuel found in earth's crust.
   - **Propane** - also called LPG (liquified petroleum gas). It is a liquid by-product of natural gas and petroleum refining.

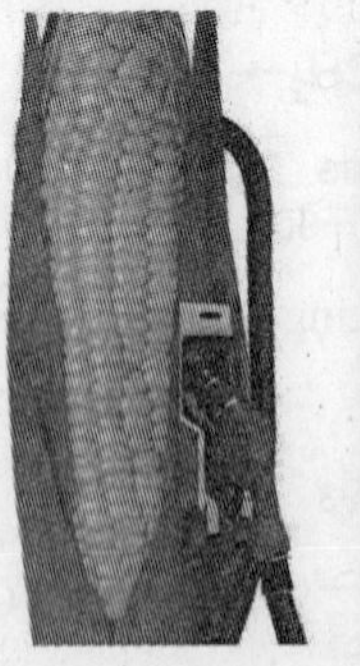

3. **LPG (also called propane) is a liquid mixture made of at least 90% propane, 2.5 % butane and higher hydrocarbons, with a remaining balance of ethane and propene.**
4. CNG and propane are successful vehicle fuels. CNG is being projected and promoted as transportation fuel of the 21st century.

COMPRESSED NATURAL GAS (CNG)

Natural gas is a fossil fuel found in earth's crust. It is commonly found trapped between liquid petroleum and hard impervious called capping rock. Although it is found commonly with or near crude oil deposits, it is not a component of crude oil. India has twice as much as natural gas reserves as crude oil reserves.

Composition of CNG. Major components of natural gas are methane, ethane, and propane. The proportions, of these gases vary according to the region where the gas was recovered. Other components are carbon dioxide, nitrogen and sulphur, which may occur in trace amounts. Typical composition of natural gas found in India is :

Methane	85-90 %	C_4 and higher	85-90 %
Ethane	4-5 %	C_6 and higher	0.2-0.3 %
Propane	1.7-2 %	$CO_2 + N_2$	3-9 %
		O_2	0.5-0.6 %
		Others	0.1-0.2 %

Advantages of CNG. Due to several advantages, compressed natural gas (CNG) is considered the most promising vehicle fuel and should be promoted as the fuel of tomorrow. The benefits offered by CNG compared to conventional fuels (gasoline, diesel) are summarized below :

1. Natural gas exists naturally and it is abundantly available.
2. CNG provides more energy per kg indicating that vehicle operation with this fuel is more efficient.
3. Octane rating is more therefore thermal efficiency is high.
4. It exhibits very low particulate emissions.
5. Life of engine components is increased, as deposits are less.
6. Natural gas is a safe fuel. It has vapor density nearly three times lighter than air, hence it rises and dissipates quickly when released.

Examination Questions

Part 1. Alkanes

1. Write the structural formulas and give IUPAC names for all isomeric alkanes of the molecular formula C_5H_{12}.

Clue. There are *3* alkanes of formula C_5H_{12}.

2. Write the structural formulas and give IUPAC names for all isomeric alkanes of formula C_6H_{14}.

Clue. There are *5* alkanes of formula C_6H_{14}.

3. How are alkanes prepared ? Describe their important reactions.

4. Explain why alkanes are relatively unreactive.

5. Write equations for the preparation of *n*-butane from :

(*a*) *n*-Butyl bromide

(b) Ethyl bromide

(c) 2-Butene

6. What happens when ethyl chloride is treated with sodium metal ?

Answer. *n*-Butane is formed. This is Wurtz reaction. For reaction equation see text.

7. What happens when methane is treated with iodine in the presence of an oxidizing agent, HIO_3 ?

(*Roorkee BSc, 2015*)

Answer. Methyl iodide is formed. For reaction equation see text.

8. What happens when ethane is treated with concentrated nitric acid at 450°C ?

(*Pune BSc, 2014*)

Answer. A mixture of nitroethane and nitromethane is formed. For reaction equation see text.

9. What happens when *n*-hexane is treated with Cr_2O_3 supported over alumina at 600°C ?

Answer. Benzene is formed. For reaction equation see text.

10. Discuss the mechanism of chlorination of methane.

(*Anna BSc, 2014 ; Rohilikhand BSc, 2014 ; Kakatiya BSc, 2015 ; North Eastern Hill BSc, 2015*)

11. Arrange each series of compounds in order of increasing boiling point.

(*a*) *n*-Pentane, *n*-hexane, *n*-butane

(*b*) *n*-Hexane, *n*-pentane, 2-methylpentane

(*c*) *n*-Octane, 2,2,3-trimethylpentane, 2-methylheptane

Answer. (*a*) *n*-Butane, *n*-pentane, *n*-hexane. For this series of compounds, the order of increasing boiling points follows the order of increasing molecular weights.

(*b*) *n*-Pentane, 2-methylpentane, *n*-hexane. Notice that *n*-pentane has the lowest molecular weight (72), and therefore the lowest boiling point. 2-Methylpentane and *n*-hexane are isomers and have the same molecular weight (86). The more branched isomer has fewer intermolecular attractive forces, and therefore 2-methylpentane has a lower boiling point than *n*-hexane.

(*c*) 2,2,3-Trimethylpentane, 2-methylheptane, *n*-octane. All three compounds have the same molecular weight (114). The most branched isomer has the lowest boiling point, and the least branched the highest boiling point.

12. Arrange the following isomeric alkanes in order of decreasing boiling points :

(*a*) Pentane (*b*) 2,2-Dimethylpropane

(c) 2-Methylbutane (*Panjab BSc, 2015*)

Answer. The structures of (*a*), (*b*), and (*c*) are given below :

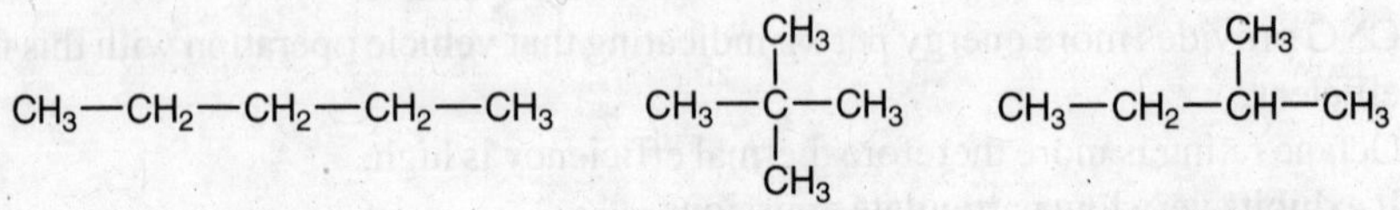

Pentane 2,2-Dimethylpropane 2-Methylbutane

We see that pentane has the least branching of these three isomers, 2,2-dimethylpropane has the most, and 2-methylbutane is in between. Since the boiling points should decrease as the amount of branching increases, the sequence of boiling points should be

Pentane > 2-Methylbutane > 2,2-Dimethylpropane

In fact, the boiling points of these three compounds are 36°C, 28°C, and 10°C, respectively.

13. Give evidence to show that chlorination of methane involves a free-radical mechanism. (*Kerala BSc, 2014*)

14. Write a note on : Halogenation of alkanes. (*Lucknow BSc, 2014*)

15. Write a note on : Wurtz reaction.

16. Write a note on : Conformations of alkanes. (*Delhi BSc, 2014 ; Madras BSc, 2015*)

17. Discuss the conformations of ethane.

18. Discuss the conformations of *n*-butane.

19. What are conformational isomers ? Discuss the conformational isomers of ethane. (*Bhopal BSc, 2014*)

20. Draw the Newman projection for both the least and the most stable conformations of 1,2-dibromoethane. Explain your answer. (*Osmania BSc, 2015*)

Answer. The least and the most stable conformations of 1,2-dibromoethane are :

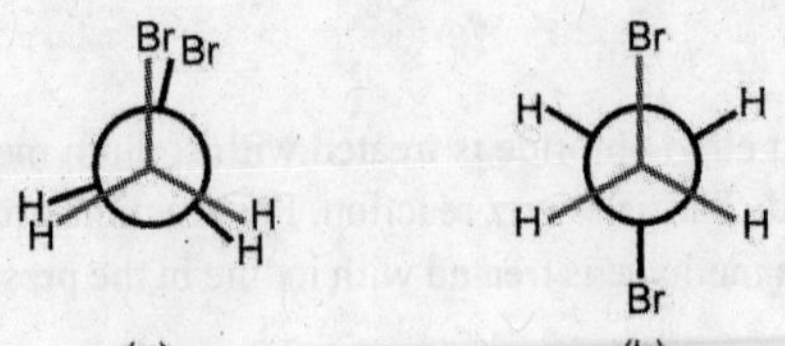

(a) Eclipsed (Least stable) (b) Staggered (Most stable)

(*a*) is least stable, owing to the large size and high electron density of bromine atoms (maximum repulsion). (*b*) is most stable because there is maximum separation between the two bromine atoms (minimum repulsion).

21. How will you synthesize butane from ethyl bromide ?

Clue. Use Wurtz reaction.

22. How will you synthesize ethane from methane ?

Answer. Following steps are involved :

$$CH_4 \xrightarrow[UV]{Cl_2} CH_3Cl \xrightarrow[ether]{Na} CH_3—CH_3$$

23. How will you synthesize methane from acetic acid ?

Answer. Following steps are involved

$$CH_3-\overset{O}{\overset{\|}{C}}-OH \xrightarrow{NaOH} CH_3-\overset{O}{\overset{\|}{C}}-O\overset{-}{}\overset{+}{Na} \xrightarrow[\Delta]{Sodalime} CH_4$$

24. How will you synthesize propane from methane ?

Answer. Do not use Wurtz reaction in this case. Remember that Wurtz reaction is suitable only for the preparation of symmetrical alkanes. Use Corey-House synthesis. Following steps are involved :

Step 1. Formation of ethyl chloride from methane.

$$CH_4 \xrightarrow[UV]{Cl_2} CH_3Cl \xrightarrow{Na} CH_3CH_3 \xrightarrow[UV]{Cl_2} CH_3CH_2Cl$$

Step 2. Formation of propane from ethyl chloride *via* Corey-House synthesis.

$$CH_3CH_2Cl \xrightarrow{Li} \underset{\text{(2 molecules)}}{CH_3CH_2Li} \xrightarrow{CuI} (CH_3CH_2)_2CuLi \xrightarrow{CH_3Cl} CH_3CH_2CH_3$$

25. How will you synthesize 2-methylhexane from isopropyl bromide ?

Answer. Use Corey-House synthesis :

$$\underset{\text{Isopropyl bromide}}{CH_3\overset{Br}{\overset{|}{C}}HCH_3 \equiv CH_3\overset{CH_3}{\overset{|}{C}}HBr} \xrightarrow{Li} \underset{\text{(2 molecules)}}{CH_3\overset{CH_3}{\overset{|}{C}}HLi} \xrightarrow{CuI} (CH_3\overset{CH_3}{\overset{|}{C}}H)_2CuLi$$

$$\xrightarrow{CH_3CH_2CH_2CH_2Br} \underset{\text{2-Methylhexane}}{CH_3\overset{CH_3}{\overset{|}{C}}HCH_2CH_2CH_2CH_3}$$

26. How will you synthesize 2-methylhexane from *n*-propyl bromide ? (*Utkal BSc, 2015*)

Answer. Use Corey-House synthesis :

$$\underset{n\text{-Propyl bromide}}{CH_3CH_2CH_2Br} \xrightarrow{Li} \underset{\text{(2 molecules)}}{CH_3CH_2CH_2Li} \xrightarrow{CuI} (CH_3CH_2CH_2)_2CuLi$$

$$\xrightarrow{CH_3—\overset{CH_3}{\overset{|}{C}}H—CH_2Br} \underset{\text{2-Methylhexane}}{CH_3-\overset{CH_3}{\overset{|}{C}}H-CH_2CH_2CH_2CH_3}$$

27. An alkane with molecular weight 72 formed only one substitution product. Suggest a structure for the alkane.

Answer. The molecular weight of the alkane is 72. This corresponds to the molecular formula C_5H_{12} ($n = 5$ in C_nH_{2n+2}). Three isomers are possible for C_5H_{12}.

$$\underset{n\text{-Pentane}}{CH_3CH_2CH_2CH_2CH_3} \qquad \underset{\text{2-Methylbutane}}{CH_3—\overset{CH_3}{\overset{|}{C}}H—CH_2—CH_3} \qquad \underset{\text{2,2-Dimethylpropane}}{CH_3—\overset{CH_3}{\overset{|}{\underset{CH_3}{\underset{|}{C}}}}—CH_3}$$

Since the alkane forms only one monosubstitution product, it must have only one type of hydrogens. Therefore, the unknown alkane is 2,2-dimethylpropane.

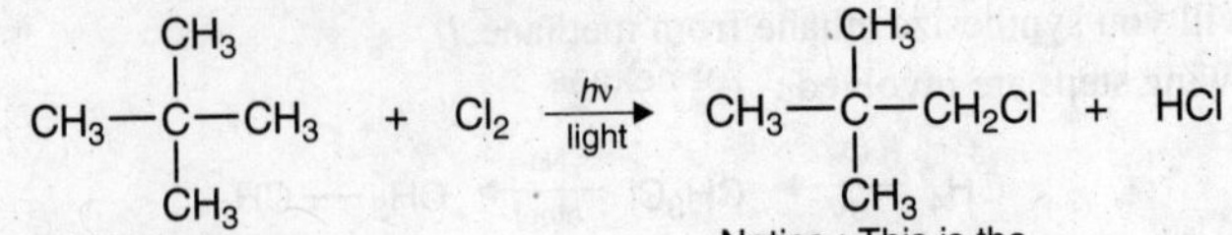

2,2-Dimethylpropane

Notice : This is the only possible substitution product.

28. A hydrocarbon C_4H_{10} (A) on monochlorination gives a compound C_4H_9Cl (B). Compound (B) on treatment with sodium metal gives 2,2,3,3-tetramethylbutane. What are the structural formulas of (A) and (B) ?

Answer. (*A*) is 2-Methylpropane ; (*B*) is 2-Chloro-2-methylpropane.

Reaction equations are :

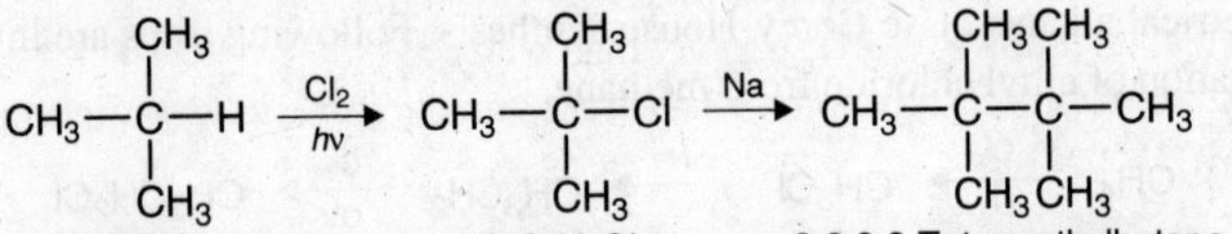

(*A*) C_4H_{10} (*B*) C_4H_9Cl 2,2,3,3-Tetramethylbutane

Part 2. Petroleum and Natural Gas

29. Describe giving details the distillation of crude petroleum into different fractions. *(Indore BSc, 2014)*

30. Write a note on : Petroleum refining. *(Karnataka BSc, 2015)*

31. Write a note on : Octane number.

32. Write a note on : Cetane number.

33. Write a note on : Catalytic cracking.

34. Write a note on : Synthetic petrol.

35. Write a note on : Fischer-Tropsch process.

36. Write a note on : CNG.

Multiple Choice Questions

1. Hydrocarbons are

(*a*) Composed of carbon and hydrogen
(*b*) Composed of carbon, hydrogen, and oxygen
(*c*) Composed of carbon and oxygen
(*d*) Composed of carbon and nitrogen

Answer. (*a*)

2. Which of following statements is false about propane ?

(*a*) all bond angles are 109.5°
(*b*) each carbon is sp^3 hybridized
(*c*) the compound is combustible
(*d*) the compound undergoes polymerization to give polypropylene.

Answer. (*d*)

3. Which of the following is a correct name according to the IUPAC rules ?

(*a*) 2-Methylcyclohexane (*b*) 2-Ethyl-2-methylpentane
(*c*) 3,4-Dimethylpentane (*d*) 3-Ethyl-2-methylpentane

Answer. (*d*)

4. A tertiary carbon is bonded directly to :

(*a*) 2 hydrogens (*b*) 3 carbons
(*c*) 2 carbons (*d*) 4 carbons

Answer. (*b*)

5. How many isomers are possible for pentane ?

(*a*) 2 (*b*) 3 (*c*) 4 (*d*) 5

Answer.(*b*).

6. How many isomers are possible for hexane ?

(*a*) 4 (*b*) 5 (*c*) 6 (*d*) 7

Answer. (*b*)

7. How many isomers are possible for heptane ?

(*a*) 9 (*b*) 10 (*c*) 11 (*d*) 12

Answer. (*a*)

8. Marsh gas mainly contains

(*a*) $CH_2{=}CH_2$ (*b*) CH_4 (*c*) H_2S (*d*) $CHCl_3$

Answer. (*b*)

9. Which statement is false ?

(*a*) Many alkanes are soluble in water.

(*b*) All alkanes have a lower density than water.

(*c*) At room temperature some alkanes are liquids, some solids, some gases.

(*d*) All alkanes burn.

Answer. (*a*)

10. As the molecular weight of alkanes increases, how do the boiling point and melting point change?

(*a*) boiling point increases ; melting point increases.

(*b*) boiling point decreases ; melting point decreases

(*c*) boiling point increases ; melting point decreases

(*d*) boiling point increases ; melting point increases sequentially for alkanes over four carbons.

Answer. (*d*)

11. The branching of alkanes that produces symmetrical structures :

(*a*) raises the boiling point ; raises the melting point

(*b*) raises the boiling point ; lowers the melting point

(*c*) lowers the boiling point ; lowers the melting point

(*d*) lowers the boiling point ; raises the melting point

Answer. (*a*)

12. How does the melting point of propane and ethane compare ?

(*a*) the melting point of propane is greater.

(*b*) the melting point of ethane is greater.

(*c*) the melting points are within 5°C of one another.

(*d*) no relationship exists among these alkanes and their melting points.

Answer. (*b*)

13. How do the boiling points of butane, propane, and ethane compare ?

(*a*) the boiling point of ethane is greatest.

(*b*) the boiling point of butane is greatest.

(*c*) the boiling point of propane is greatest.

(*d*) there is no relationship among these three alkanes and their boiling points.

Answer. (*b*)

14. Which of the following alkanes will have the *highest* boiling point ?

(*a*) *n*-Octane (*b*) Isopentane (*c*) *n*-Butane (*d*) Neopentane

Answer. (*a*)

15. Which of the following alkanes will have the *lowest* boiling point ?

(*a*) *n*-Heptane (*b*) Isopentane (*c*) *n*-Hexane (*d*) Neopentane

Answer. (*d*)

16. Methane can be prepared by :

(*a*) the reaction of iodomethane with sodium in dry ether

(*b*) the reaction of methanol with concentrated H_2SO_4

(*c*) the reaction of sodium methanoate with soda-lime

(*d*) the reaction of sodium ethanoate with soda-lime

Answer. (*d*)

17. Which of the following reactions can be used to prepare alkanes ?
(*a*) Corey-House synthesis (*b*) Williamson synthesis
(*c*) Friedel-Crafts reactions (*d*) None of these

Answer. (*a*)

18. Which of the following reactions can be used to prepare alkanes ?
(*a*) Wurtz reaction (*b*) Wolf-kishner reduction
(*c*) Kolbe's electrolysis (*d*) All of these

Answer. (*d*)

19. Which of the following compounds does not dissolve in concentrated H_2SO_4 on warming ?
(*a*) *n*-Hexane (*b*) Diethyl ether
(*c*) 1-Butene (*d*) Aniline

Answer. (*a*)

20. 2,3-Dimethyl-2-butene undergoes catalytic hydrogenation to give
(*a*) 2,3-Dimethylbutane (*b*) 2-Methylpentane
(*c*) 2,2-Dimethylbutane (*d*) 3-Methylpentane

Answer. (*a*)

21. Why is the halogenation of alkanes considered a chain reaction ?
(*a*) it occurs quickly.
(*b*) it occurs without the generation of intermediates.
(*c*) each step generates the reactive intermediate that causes the next step to occur.
(*d*) the reaction allows long chains of halogenated alkanes to be formed.

Answer. (*c*)

22. The chlorination of methane to give CCl_4 is an example of
(*a*) an electrophilic addition (*b*) a free-radical substitution
(*c*) a nucleophilic addition (*d*) an electrophilic substitution

Answer. (*b*)

23. In the chlorination of alkanes, the first step in which chlorine free radicals are produced is called :
(*a*) initiation (*b*) activation
(*c*) propagation (*d*) deactivation

Answer. (*a*)

24. Which halogen *does not* appreciably react with methane in a free-radical substitution reaction ?
(*a*) chlorine (*b*) bromine
(*c*) iodine (*d*) fluorine

Answer. (*c*)

25. What product is formed in the free-radical bromination of methane ?
(*a*) bromomethane (*b*) dibromomethane
(c) tribromomethane (*d*) all of these

Answer. (*d*)

26. Chlorination of an alkane as compared to bromination proceeds
(*a*) at a slower rate
(*b*) at a faster rate
(*c*) with equal rates
(*d*) with equal or different rate depending upon the source of alkane.

Answer. (*b*)

27. How many monochlorinated isomers would result from the reaction of chlorine with *n*-butane in the presence of UV light ?
(*a*) 2 (*b*) 4 (*c*) 6 (*d*) 8

Answer. (*a*)

28. The combustion of pentane produces :
(*a*) Pentene (*b*) $HCl + H_2O$
(*c*) Pentyne (*d*) $CO_2 + H_2O$

Answer. (*d*)

29. The combustion of one mole of propane, C_3H_8, produces how many moles of H_2O ?
(*a*) 2 (*b*) 3 (*c*) 4 (*d*) 5

Answer. (*c*)

30. The major carbon compound formed from the incomplete combustion of a hydrocarbon in air is

(*a*) carbon dioxide (*b*) carbon monoxide
(*c*) water (*d*) alkyl chains

Answer. (*b*)

31. The thermal decomposition of alkanes in the absence of air is called :

(*a*) Combustion (*b*) Oxidation
(*c*) Cracking (*d*) Hydrogenation

Answer. (*c*)

32. LPG (Household cooking gas) is mainly a mixture of

(*a*) Methane + Ethane (*b*) Acetylene + O_2
(*c*) Butane + Isobutane (*d*) Acetylene + H_2

Answer. (*c*)

33. Octane number is related to

(*a*) Gasoline (*b*) Kerosene oil
(*c*) Diesel oil (*d*) Lubricating oil

Answer. (*a*)

34. The octane rating of gasoline provides information on

(*a*) its antiknock properties (*b*) its ignition properties
(*c*) its percentage proportion of octane (*d*) its percentage content of isooctane

Answer. (*a*)

35. In any fuel, the percentage by volume of isooctane in a mixture of isooctane and *n*-heptane which will knock under same conditions as the fuel being tested, is called :

(*a*) Cracking (*b*) Iodine number
(*c*) Aromatization (*d*) Octane number

Answer. (*d*)

36. Which of the following compounds is assigned the Octane Number of zero :

(*a*) *n*-Octane (*b*) 2,3,3-Trimethylpentane
(*c*) *n*-Heptane (*d*) 2,2,4-Trimethylpentane

Answer. (*c*)

37. Which of the following compounds is assigned the Octane Number of 100 :

(*a*) *n*-Heptane (*b*) 2,3,3-Trimethylpentane
(*c*) *n*-Octane (*d*) 2,2,4-Trimethylpentane

Answer. (*d*)

38. Gasoline with an octane number of 90 is equivalent in knocking characteristics to a mixture of heptane and isooctane of the following composition :

(*a*) 20% heptane + 80% isooctane (*b*) 90% heptane + 10% isooctane
(*c*) 80% heptane + 20% isooctane (*d*) 10% heptane + 90% isooctane

Answer. (*d*)

39. A fuel with octane number 90 means it is as good as a mixture of :

(*a*) 90% Isooctane + 10% *n*-heptane
(*b*) 90% *n*-Heptane + 10% isooctane
(*c*) 2 Litre of 90% isooctane + 2 litre of benzene
(*d*) 2 Litre of 90% *n*-heptane + 2 litre of benzene

Answer. (*a*)

40. Which of the following compounds is used as an antiknock compound ?

(*a*) Ethyllithium (*b*) Tetraethyllead
(*c*) Ethyl acetate (*d*) Lead acetate

Answer. (*b*)

41. Which of the following methods can be used to increase the octane rating of gasoline ?

(*a*) Adding branched-chain alkanes (*b*) Adding tetraethyllead
(*c*) Adding aromatic hydrocarbons (*d*) All of these

Answer. (*d*)

11

Alkenes

Ethylene is produced by most fruits and vegetables. It speed up ripening.
Life of the produce is doubled, if ethylene is removed by absorption.
Alumina impregnated on potassium permanganate is used for ethylene absorption.

Alkenes are hydrocarbons that contain a carbon-carbon double bond (C=C) in their molecule. They have the general formula C_nH_{2n} (n = number of carbon atoms) in comparison to C_nH_{2n+2} of alkanes. Alkenes contain two hydrogen atoms less than alkanes and are thus designated unsaturated hydrocarbons. As a class, alkenes are commonly known as **Olefines** (Latin, *Oleum* = oil; *ficare* = to make) because the lower members form oily products on treatment with chlorine or bromine.

Ethylene is the first member of the series. Propylene is the second member of the series. These are represented as :

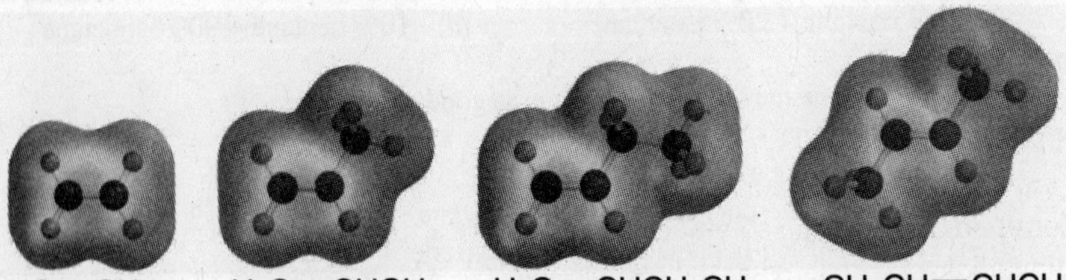

$H_2C{=}CH_2$	$H_2C{=}CHCH_3$	$H_2C{=}CHCH_2CH_3$	$CH_3CH{=}CHCH_3$
Ethylene (Ethene)	Propylene (Propene)	1-Butene	2-Butene

Alkenes seldom occur free in nature. They are, however, produced in large amounts by cracking of petroleum.

STRUCTURE

Let us consider ($H_2C{=}CH_2$) for illustrating the orbital make up of alkenes.

In ethylene the carbon atoms are sp^2 hybridized. They are attached to each other by a σ bond and a π bond. The σ bond results from the overlap of two sp^2 hybrid orbitals (*i.e.*, one from each carbon). The π bond is formed from overlap of the unhybridized p orbitals (Fig. 11.1).

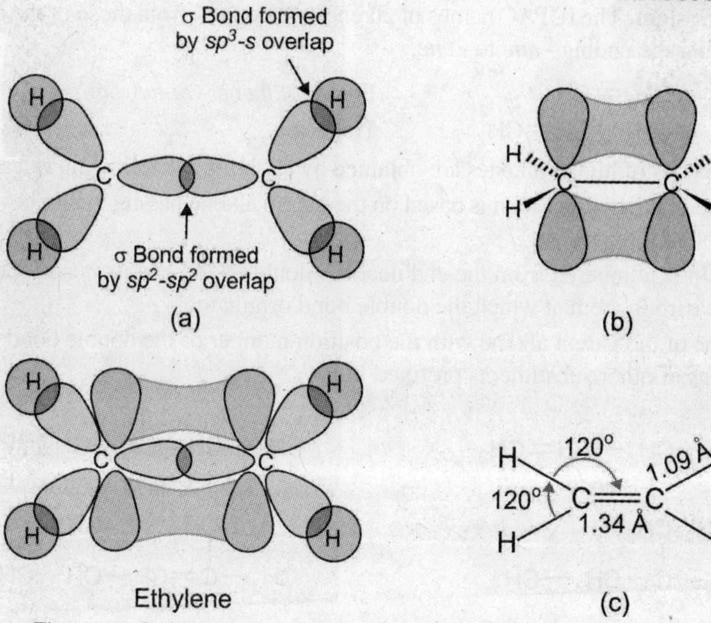

Fig. 11.1. Orbital structure of ethylene. (*a*) Formation of σ bonds; (*b*) Formation of the π bond ; (*c*) Geometry of the molecule.

The remaining sp^2 orbitals form σ bonds with either carbon or hydrogen atoms. Ethylene is a planar molecule.

REMEMBER

(1) The carbon-carbon double bond in alkenes is made up of one σ bond and one π bond.

(2) Alkenes are more reactive than alkanes. This is due to the availability of the more exposed π electrons.

(3) Higher alkenes are related to ethylene by replacement of hydrogen substituents by alkyl groups. There are two different types of carbon-carbon bonds in propene, $CH_3CH=CH_2$. The double bond is the σ + π type, and the bond to the methyl group is a σ bond formed by sp^3–sp^2 overlap.

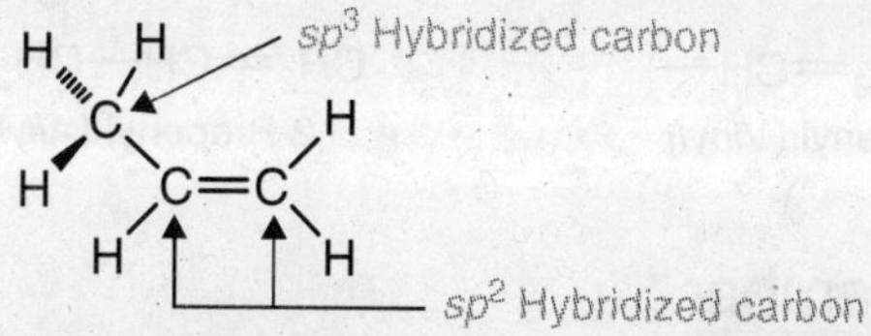

NOMENCLATURE

There are two ways of naming alkenes :

(1) **Common System**. The common names of the first four members are derived from those of the corresponding alkanes by changing the ending *–ane* to *–ylene*. Greek letters are used to distinguish isomers having double bond at the first (α) or the second (β) carbon of the chain.

$CH_2{=}CH_2$	Ethylene (*Ethane – ane + ylene*)
$CH_2{=}CH{-}CH_3$	Propylene
$CH_3{-}\overset{\overset{\large CH_3}{\vert}}{C}{=}CH_2$	Isobutylene
$CH_2{=}CH{-}CH_2{-}CH_3$	α-Butylene
$CH_3{-}CH{=}CH{-}CH_3$	β-Butylene

(2) **IUPAC System**. The IUPAC names of alkenes are derived from those of the corresponding alkanes by changing the ending *–ane* to *–ene*.

$CH_2{=}CH_2$ Ethane (*Ethane – ane + ene*)

$CH_3{-}CH{=}CH_2$ Propene

The IUPAC names of higher alkenes are obtained by applying the following rules :

(1) The name of the hydrocarbon is based on the parent alkene having the longest carbon chain of which double bond is a part.

(2) This chain is numbered from the end near the double bond and its position is indicated by the number of the carbon atom at which the double bond originates.

(3) The name of the parent alkene with the position number of the double bond is written first and then the names of other substituents prefixed to it.

$\overset{4}{C}H_3{-}\overset{3}{C}H_2{-}\overset{2}{C}H{=}\overset{1}{C}H_2$

1-Butene

$\overset{1}{C}H_3{-}\overset{2}{C}H{=}\overset{3}{C}H{-}\overset{4}{C}H_3$

2-Butene

$\underset{1}{C}H_2{=}\underset{2}{\overset{\overset{CH_3}{|}}{C}}{-}\underset{3}{C}H_2{-}\underset{4}{C}H_3$

2-Methyl-1-butene

$\underset{1}{C}H_3{-}\underset{2}{\overset{\overset{CH_3}{|}}{C}}{=}\underset{3}{C}H{-}\underset{4}{\overset{\overset{CH_3}{|}}{C}H}{-}\underset{5}{C}H_3$

2,4-Dimethyl-2-pentene

(4) When there are two or three double bonds in a molecule, the ending *–ane* of the corresponding alkane is replaced by *–adiene* or *–atriene* to get the name of the hydrocarbon.

$CH_2{=}C{=}CH_2$

Propadiene

$\overset{1}{C}H_2{=}\overset{2}{C}H{-}\overset{3}{C}H{=}\overset{4}{C}H_2$

1,3-Butadiene

Alkenyl Groups. The monovalent groups obtained by the removal of one hydrogen atom from alkenes are called *alkenyl groups*. The IUPAC names of these groups are derived by replacing the *–e* of the parent alkene by *–yl* (*e.g., ethene –e + yl*). The carbon atoms constituting the group are numbered so that the carbon with free valence is always number 1. In the following examples, the common names are given in brackets.

$CH_2{=}CH{-}$

Ethenyl (*Vinyl*)

$\overset{3}{C}H_2{=}\overset{2}{C}H{-}\overset{1}{C}H_2{-}$

2-Propenyl (*Allyl*)

Points to Remember

Line formulas provide a very convenient representation for many open-chain structures. In line formulas, the line represents a single bond. An intersection or endpoint represents a carbon atom with its complement of hydrogens. Double and triple bonds are shown as double and triple lines. All functional groups are drawn out.

 means $CH_3(CH_2)_3CH_3$

 means $CH_2{=}CH_2$

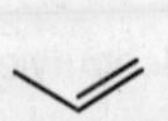 means $CH_3CH{=}CH_2$

 means $CH_2{=}CHCH{=}CH_2$

DEGREE OF UNSATURATION

It is not always possible to write isomeric structures from the given molecular formula. However, we can interpret molecular formulas using the following formula. It will help you to determine the number of double bonds or rings in the molecule.

$$n = \mathrm{C} + 1 - \frac{1}{2}\,(\mathrm{H} - \mathrm{N} + \mathrm{X})$$

Where n is the degree of unsaturation, and C, H, N, and X are the numbers of carbon, hydrogen, nitrogen, and halogen atoms respectively in the molecule. Thus by pulling the values of C, H, N and X you can find the values of the degree of unsaturation.

Remember :

If $n = 0$ Molecule is saturated.

$n = 1$ Molecule can have one double bond or one ring.

$n = 2$ Molecule can have two double bonds or one triple bond or one double bond *plus* one ring.

$n = 4$ Molecule may be aromatic (One ring *plus* three double bonds).

The above formula is very useful. The more you use it, the more you will appreciate its application. However, additional information is generally required to arrive at the exact structural formula of the given unknown.

Example 1. Consider the case of C_4H_6.

Apply the following formula :

$$n = \mathrm{C} + 1 - \frac{1}{2}\,(\mathrm{H} - \mathrm{N} + \mathrm{X})$$

$$= 4 + 1 - \frac{1}{2}(6)$$

$$= 5 - 3$$

$$= 2$$

The value of $n = 2$. This means that the molecule may have two double bonds, or one triple bond or one double bond *plus* one ring. Thus we can write the following possible structures for C_4H_6:

$CH_2{=}CH{-}CH{=}CH_2$ (1) $CH_2{=}C{=}CH{-}CH_3$ (2) $CH_3CH_2{-}C{\equiv}CH$ (3)

$CH_3{-}C{\equiv}C{-}CH_3$ (4)

(5)

Example 2. Consider the case of C_5H_{10}

Apply the following formula :

$$n = \mathrm{C} + 1 - \frac{1}{2}\,(\mathrm{H} - \mathrm{N} + \mathrm{X})$$

$$= 5 + 1 - \frac{1}{2}(10)$$

$$= 6 - 5$$

$$= 1$$

The value of $n = 1$. This means that the molecule may have one double bond or one ring. Thus we can write the following possible structures for C_5H_{10} :

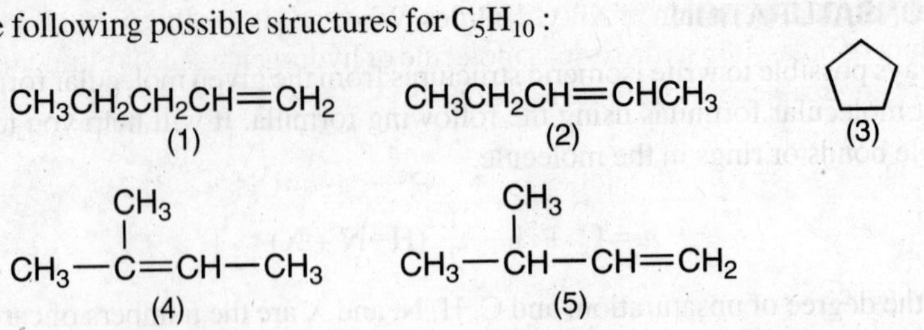

METHODS OF PREPARATION

Most of the methods of preparation of alkenes involve elimination of atoms or groups from two adjacent carbon atoms.

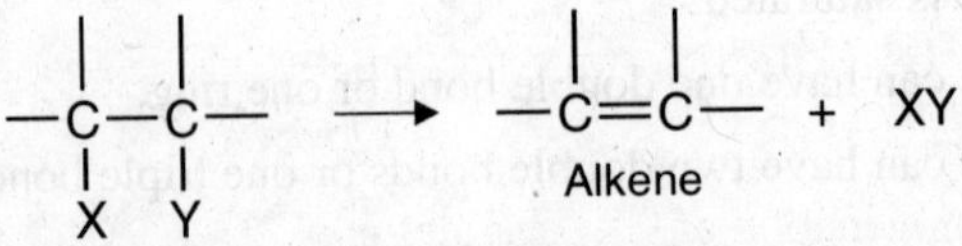

(1) **By Dehydration of Alcohols.** When an alcohol is heated in the presence of sulfuric acid, a molecule of water is eliminated and an alkene is formed.

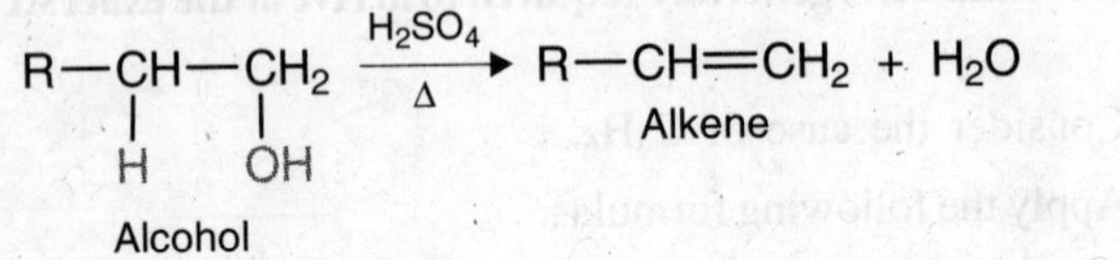

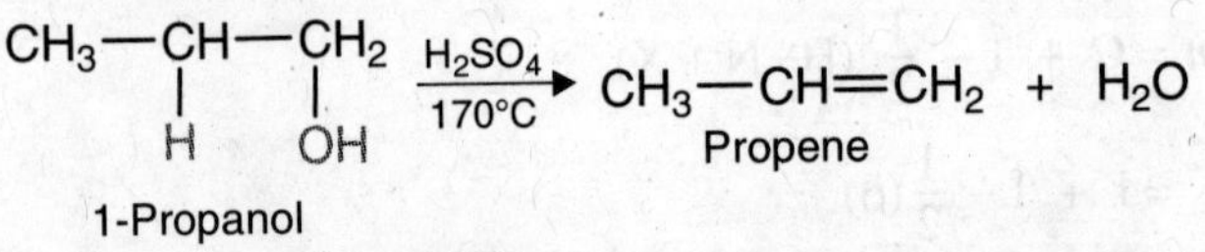

$$CH_3-CH_2-OH \xrightarrow[170°C]{H_2SO_4} CH_2=CH_2 + H_2O$$

Ethyl alcohol → Ethylene

The ease of dehydration of alcohols is : 3° alcohol > 2° alcohol > 1° alcohol.

With unsymmetrical secondary or tertiary alcohols elimination can proceed in two ways and a mixture of alkenes is formed. For example, dehydration of 2-butanol gives 2-butene (major product) and 1-butene (minor product).

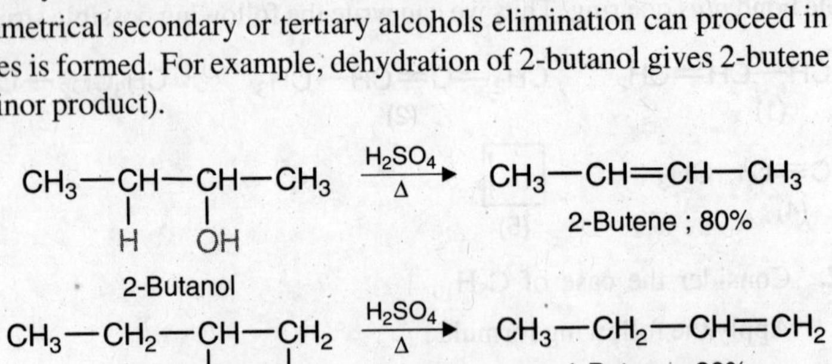

This is in accordance with **Saytzeff Rule**. The rule states that *when such alternatives exist, hydrogen is preferentially eliminated from the carbon atom with fewer number of hydrogen atoms.*

Dehydration of an alcohol may also be brought by passing the alcohol vapors over heated alumina (Al_2O_3) at 400°C. Other dehydrating agents which may be used are phosphorus pentoxide (P_2O_5) and phosphoric acid (H_3PO_4).

(2) **By Dehydrohalogenation of Alkyl Halides**. When an alkyl halide is heated with an alcoholic solution of sodium or potassium hydroxide, a molecule of hydrogen halide is eliminated and an alkene is formed. Where applicable, Saytzeff rule is followed.

$$\underset{\text{Alkyl halide}}{R-\underset{H}{CH}-\underset{X}{CH_2}} + KOH \xrightarrow[\Delta]{\text{alcohol}} \underset{\text{Alkene}}{R-CH=CH_2} + KX + H_2O$$

$$\underset{\text{1-Bromopropane}}{CH_3-\underset{H}{CH}-\underset{Br}{CH_2}} + KOH \xrightarrow[\Delta]{\text{alcohol}} \underset{\text{Propene}}{CH_3-CH=CH_2} + KBr + H_2O$$

$$\underset{\text{Ethyl bromide}}{CH_3-CH_2-Br} + NaOH \xrightarrow[\Delta]{\text{alcohol}} \underset{\text{Ethylene}}{CH_2=CH_2} + NaBr + H_2O$$

The ease of dehydrohalogenation of alkyl halides is :

$$3° \text{ alkyl halide} > 2° \text{ alkyl halide} > 1° \text{ alkyl halide}$$

(3) **By Dehalogenation of vicinal Dihalides**. A compound having two halogen atoms on adjacent carbon atoms is called vicinal dihalide (or *vic*-dihalide). Alkenes are formed when *vic*-dihalides are heated with zinc dust in ethyl alcohol.

$$\underset{\textit{vic}\text{-Dihalide}}{R-\underset{Br}{CH}-\underset{Br}{CH_2}} + Zn \xrightarrow[\Delta]{\text{alcohol}} \underset{\text{Alkene}}{R-CH=CH_2} + ZnBr_2$$

$$\underset{\text{1,2-Dibromopropane}}{CH_3-\underset{Br}{CH}-\underset{Br}{CH_2}} + Zn \xrightarrow[\Delta]{\text{alcohol}} \underset{\text{Propene}}{CH_3-CH=CH_2} + ZnBr_2$$

(4) **By Controlled Hydrogenation of Alkynes**. Alkynes react with hydrogen in the presence of *Lindlar's catalyst* to give alkenes. Lindlar's catalyst is Pd poisoned with $CaCO_3$ plus quinoline.

$$\underset{\text{Alkyne}}{R-C\equiv C-H} + H_2 \xrightarrow[\text{quinoline}]{\text{Pd-CaCO}_3} \underset{\text{Alkene}}{R-CH=CH_2}$$

$$\underset{\text{Propyne}}{CH_3-C\equiv C-H} + H_2 \xrightarrow[\text{quinoline}]{\text{Pd-CaCO}_3} \underset{\text{Propene}}{CH_3-CH=CH_2}$$

$$\underset{\text{Acetylene}}{H-C\equiv C-H} + H_2 \xrightarrow[\text{quinoline}]{\text{Pd-CaCO}_3} \underset{\text{Ethylene}}{CH_2=CH_2}$$

(5) **By Cracking of Alkanes**. Alkanes when heated at 500-800°C in the absence of air decompose to yield lower molecular weight alkenes, alkanes, and hydrogen.

$$\underset{\text{Ethane}}{CH_3-CH_3} \xrightarrow{600°C} \underset{\text{Ethylene}}{CH_2=CH_2} + CH_4 + H_2$$

$$\underset{\text{Propane}}{CH_3-CH_2-CH_3} \xrightarrow{600°C} \underset{\text{Propene}}{CH_3-CH=CH_2} + \underset{\text{Ethylene}}{CH_2=CH_2} + CH_4 + H_2$$

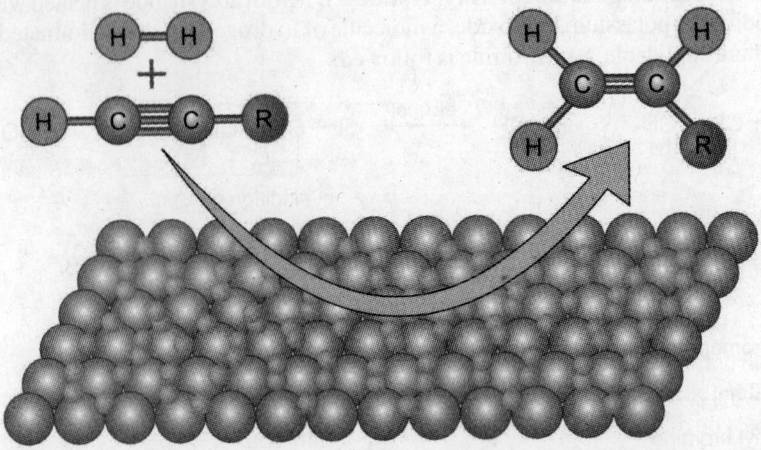

Alkynes react with hydrogen in the presence of Lindlar catalyst (Pd - $BaCO_3$ or $CaCO_3$) to form alkenes. Notice that the reaction stops at the alkene stage. Alkanes are not formed.

PHYSICAL PROPERTIES

(1) The first three alkenes (ethene, propene, and butenes) are gases at ordinary temperature; the next fourteen members are liquids; and the alkenes with more than 18 carbon atoms in the molecule are solids. They are all colorless and odorless, except ethene, which has a rather pleasant odor.

(2) Alkenes are only slightly soluble in water but dissolve freely in organic solvents.

(3) Their boiling points, melting points, and specific gravities, in general, rise with increase of molecular weight in the homologous series. The melting points and boiling points of some 1-alkenes are given in the Table 11.1.

Table 11.1 MELTING AND BOILING POINTS OF 1-ALKENES

Name	Formula	mp °C	bp °C
Ethylene	$CH_2{=}CH_2$	–169	–104
Propene	$CH_2{=}CHCH_3$	–185	–47
1-Butene	$CH_2{=}CHCH_2CH_3$	–130	–6.5
1-Pentene	$CH_2{=}CH(CH_2)_2CH_3$	–138	32
1-Hexene	$CH_2{=}CH(CH_2)_3CH_3$	–141	64

(4) ***IR Spectrum.*** The IR spectra of alkenes show C–H stretching absorptions at 3000-3100 cm^{-1}. Carbon-carbon double bonds give stretching absorption peaks at 1620-1680 cm^{-1}. Absorptions arising from C–H bending vibrations are at 600-1000 cm^{-1}.

CHEMICAL PROPERTIES

Alkenes are more reactive than alkanes. This is because :

(1) The π electrons of a double bond are located much farther from the carbon nuclei and are thus less firmly bound to them.

(2) The overlap of atomic orbitals in forming a π bond is not as effective as that in σ bonds. Thus a π bond is weaker than a σ bond and more easily broken.

The most important reactions of alkenes are the addition reactions. In these reactions the π bond is broken and is replaced by two new σ bonds. For example, when an alkene reacts with E–Nu, the carbon-carbon π bond is broken and a new C–E σ bond and C–Nu σ bond are formed.

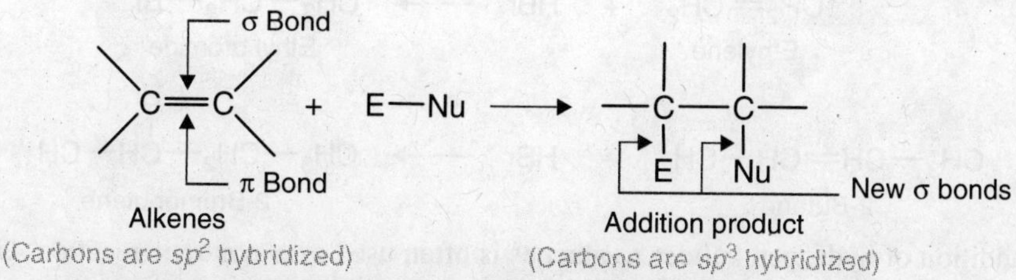

Mechanism of Addition to Carbon-Carbon Double Bond. As mentioned above, the π electrons of a double bond are less firmly bound to the carbon nuclei. They are more exposed. A carbon-carbon double bond is, in effect, electron rich. It can act as supplier of electrons to electrophilic reagents. Recall that electrophilic reagents are electron deficient species. They may be positive ions capable of forming a covalent bond with carbon.

Let us now consider the addition of a reagent E–Nu to a carbon-carbon double bond. This involves the following steps :

Step 1. Attack of the π bond on the electrophile forms a carbocation.

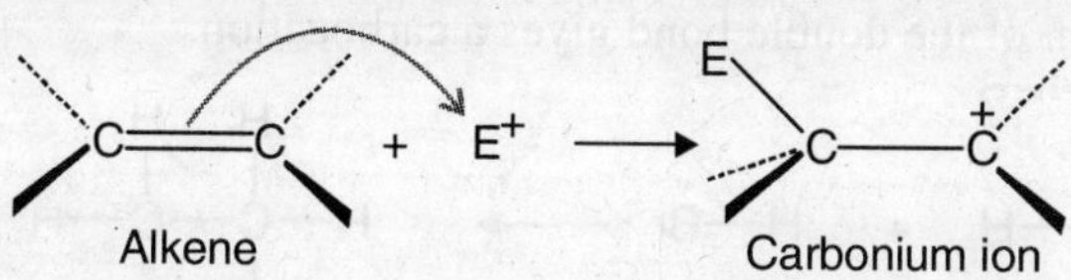

Step 2. Attack by a nucleophile gives the addition product.

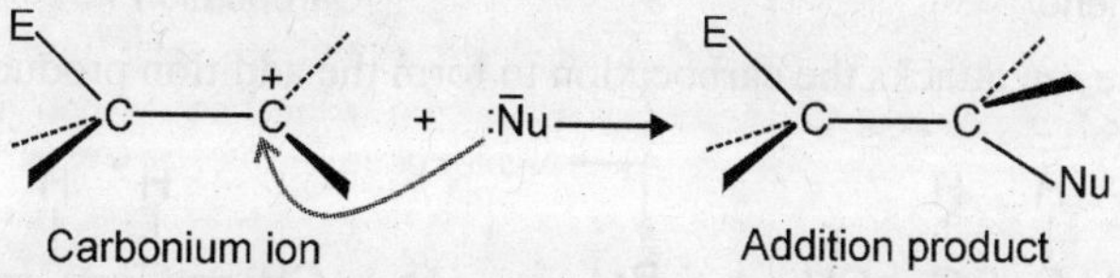

These steps are further explained in Fig. 11.2. Because the addition reactions are initiated by electrophiles, they are referred to as **electrophilic addition reactions**. In these reactions.

E^+ is usually H^+, Br^+, or Cl^+.
Nu:⁻ is usually Br^-, Cl^-, HO^-, or $\hat{O}SO_3H$.

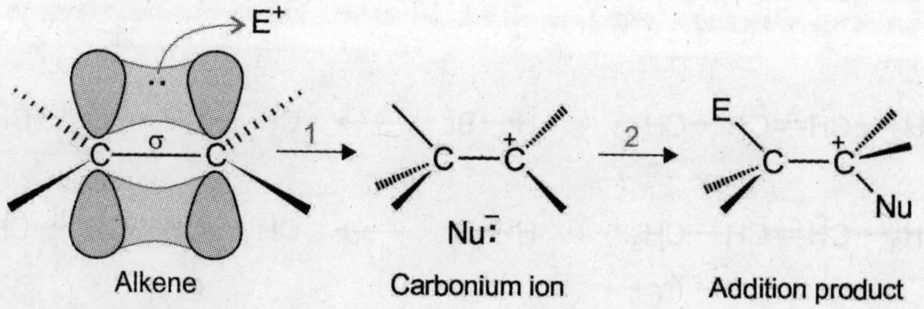

Fig. 11.2. In alkenes, the electrons in the π bond are spread farther from the carbon nuclei and are loosely held than the σ electrons. The π bond acts as a nucleophile. A strong electrophile attracts the π electrons to form a new σ bond, forming a carbocation. The curved arrow shows the movement of electrons, from the electron-rich π bond to the electron-poor electrophile. The carbocation can then be attacked by a nucleophile to form the addition product.

(1) **Addition of Hydrogen Halides**. Alkenes react with hydrogen halides (HCl, HBr, or HI) to form alkyl halides.

$$CH_2{=}CH_2 + HBr \longrightarrow CH_3{-}CH_2{-}Br$$

Ethylene → Ethyl bromide

$$CH_3{-}CH{=}CH{-}CH_3 + HBr \longrightarrow CH_3{-}CH_2{-}CH(Br){-}CH_3$$

2-Butene → 2-Bromobutane

The addition of hydrogen halides to alkenes is often used as a synthetic reaction. Usually, the gaseous HX is bubbled through a solution of the alkene. (Concentrated aqueous solutions of hydrogen halides give mixtures of products because water can also add to double bonds.) The relative reactivity of HX in this reaction is HI > HBr > HCl > HF. The strongest acid (HI) is the most reactive toward alkenes, while the weakest acid (HF) is the least reactive.

MECHANISM. In the above reaction HBr adds to ethylene to form bromoethane. The proton in HBr is electrophilic; it reacts with the alkene to form a carbocation. Bromide ion reacts rapidly with the carbocation to give a product in which elements of HBr have added to the ends of the double bond.

Step 1. Protonation of the double bond gives a carbocation.

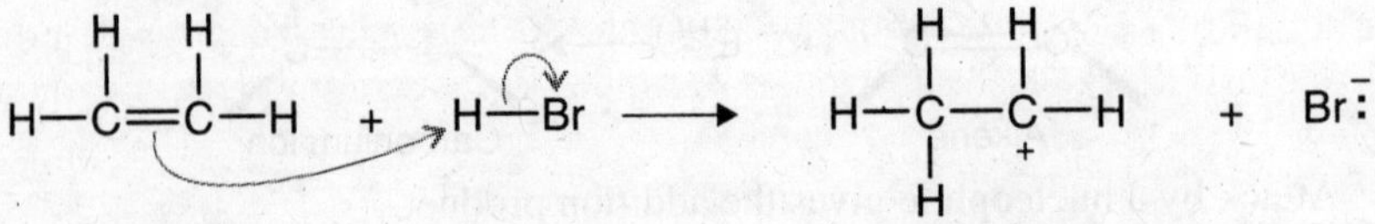

Ethylene   Carbocation

Step 2. Bromide ion attacks the carbocation to form the addition product.

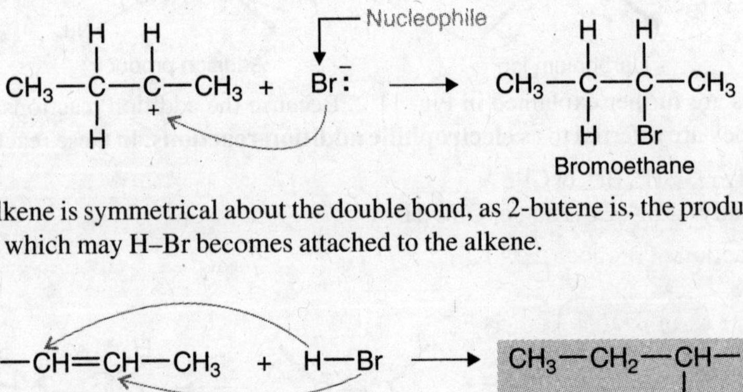

Bromoethane

When an alkene is symmetrical about the double bond, as 2-butene is, the product formed is the same no matter which may H–Br becomes attached to the alkene.

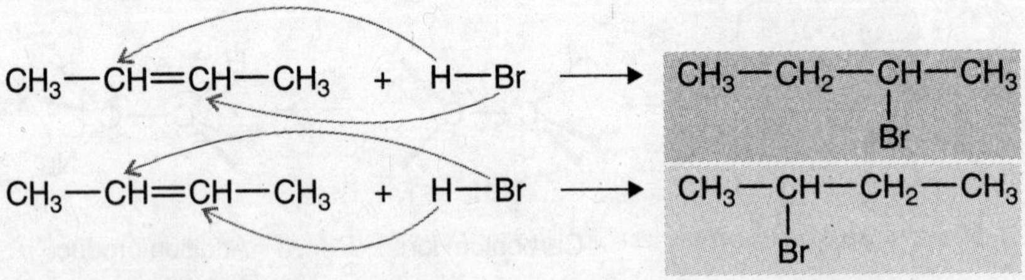

1-Butene   Identical (2-Bromobutane)

Markovnikov Rule. If an alkene is *unsymmetrical* (that is, if the groups attached to the two sp^2-hybridized carbons differ), there is the possibility of two different products from HX addition.

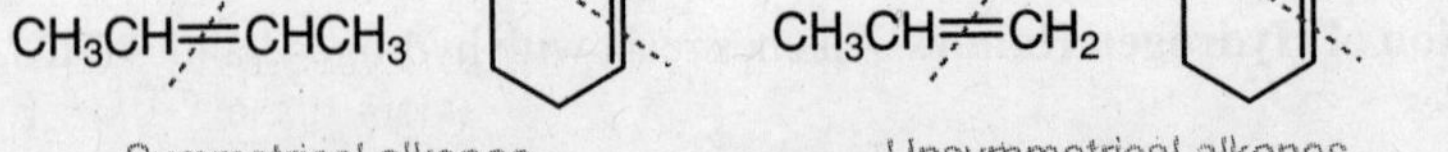

Symmetrical alkenes   Unsymmetrical alkenes

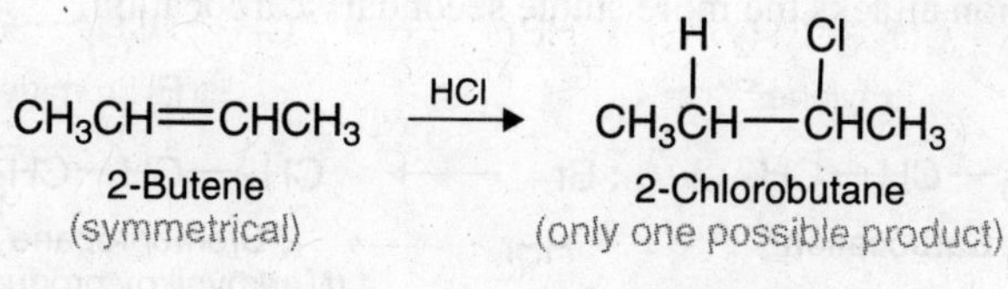

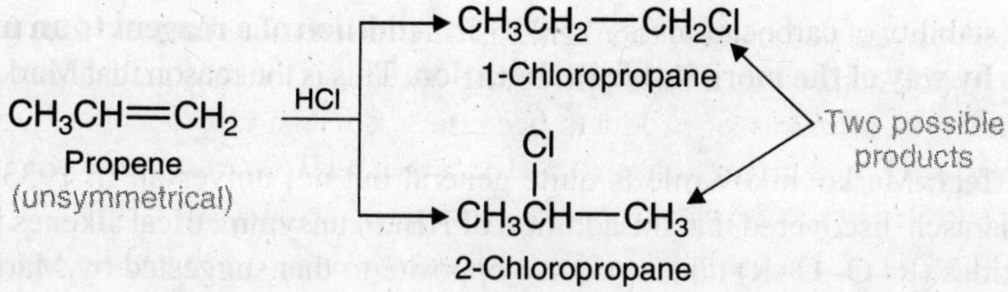

In an electrophilic addition that can lead to two products, one product usually predominates over the other. In 1869, the Russian Chemist Vladimir Markovnikov formulated the following empirical rule. **In additions of HX to unsymmetrical alkenes, the H^+ of HX goes to the double-bonded carbon that already has the greatest number of hydrogens.**

By Markovnikov's rule, we would predict that the reaction of HCl with propene yields 2-chloropropane (and not the 1-chloro isomer). Examples of reactions that obey Markovnikov's rule are given below :

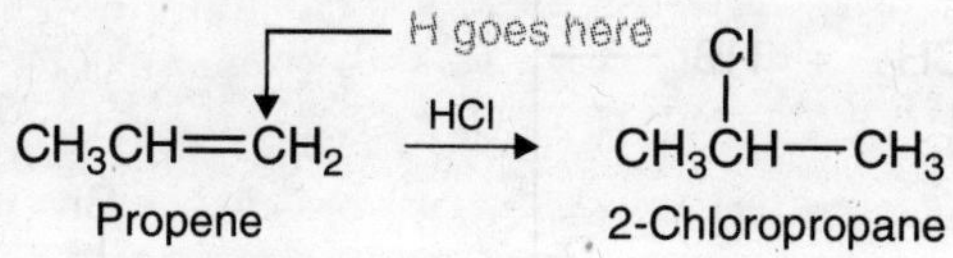

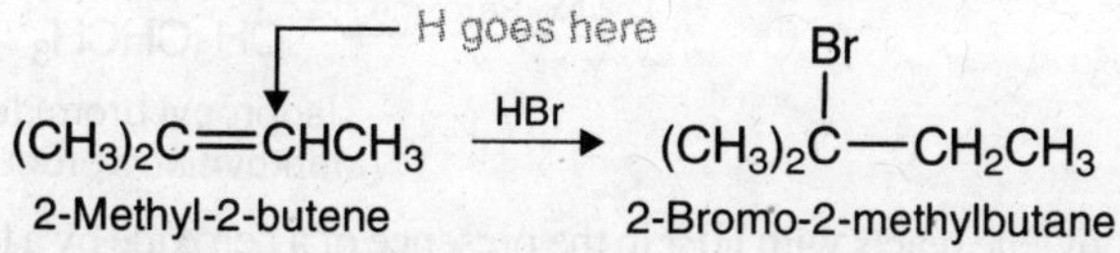

EXPLANATION. Markovnikov formulated this empirical rule because of experimental observations. We can explain this rule on the basis of mechanism of HX addition. Let us again consider the reaction of propene with HBr.

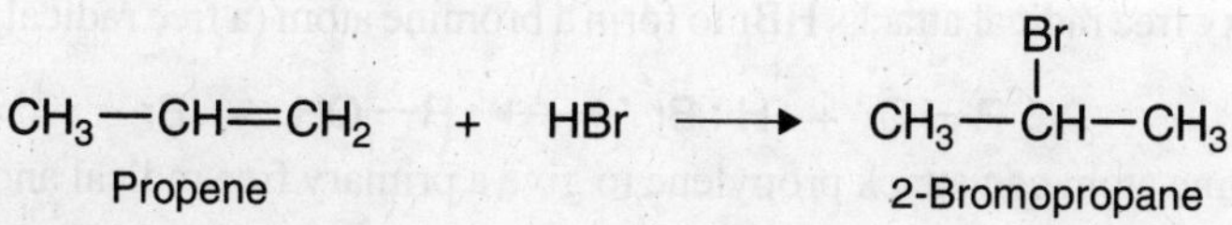

The above reaction takes place by the following steps :

Step 1. Formation of a carbocation. Notice that for propene, two carbocations are possible.

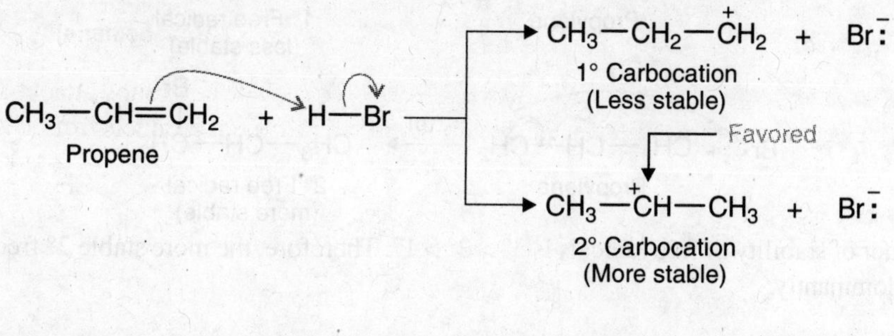

Step 2. Bromide ion attacks the more stable secondary carbocation.

$$\underset{\text{2° Carbocation}}{CH_3-\overset{+}{C}H-CH_3} + \overset{-}{:Br} \longrightarrow \underset{\substack{\text{2-Bromopropane}\\ \text{(Markovnikov product)}}}{CH_3-\underset{}{\overset{Br}{\overset{|}{C}H}}-CH_3}$$

The order of stability of carbocations is 3° > 2° > 1°. **Addition of a reagent to an unsymmetrical alkene proceeds by way of the more stable carbocation.** This is the reason that Markovnikov's rule is followed.

Peroxide Effect. Markovnikov rule is quite general but not universal. In 1933 the American chemist M.S. Kharasch discovered that the addition of HBr to unsymmetrical alkenes in the presence of organic peroxides (R–O–O–R) takes a course opposite to that suggested by Markovnikov rule. This phenomenon of *anti*-Markovnikov addition of HBr caused by the presence of peroxides is known as *Peroxide Effect*. For example, when propylene reacts with HBr in the presence of a peroxide, the major product is *n*-propyl bromide, whereas in the absence of a peroxide the major product is isopropyl bromide.

$$\underset{\text{Propylene}}{CH_3-CH=CH_2} + HBr \xrightarrow{\text{Peroxide}} \underset{\substack{n\text{-Propyl bromide}\\ \text{(anti-Markovnikov product)}}}{CH_3CH_2CH_2Br}$$

$$\underset{\text{Propylene}}{CH_3-CH=CH_2} + HBr \xrightarrow{\text{No Peroxide}} \underset{\substack{\text{Isopropyl bromide}\\ \text{(Markovnikov product)}}}{CH_3\overset{Br}{\overset{|}{C}H}CH_3}$$

MECHANISM. Propylene reacts with HBr in the presence of a peroxide by a free radical mechanism. Following steps are involved :

Step 1. Peroxide dissociates to give alkoxy free radicals.

$$R-O-O-R \xrightarrow{\Delta} 2R-O\cdot$$

Step 2. Alkoxy free radical attacks HBr to form a bromine atom (a free radical).

$$R-O\cdot + H:Br \longrightarrow R-OH + Br\cdot$$

Step 3. Bromine atom can attack propylene to give a primary free radical and a secondary free radical.

$$Br\cdot + \underset{\text{Propylene}}{CH_3-CH\mathrel{\ddot{=}}CH_2} \xrightarrow{(a)} \underset{\substack{\text{1° Free radical}\\ \text{(less stable)}}}{CH_3-\overset{Br}{\overset{|}{C}H}-\dot{C}H_3}$$

$$Br\cdot + \underset{\text{Propylene}}{CH_3-CH\mathrel{\ddot{=}}CH_2} \xrightarrow{(b)} \underset{\substack{\text{2° Free radical}\\ \text{(more stable)}}}{CH_3-\dot{C}H-\overset{Br}{\overset{|}{C}H_2}}$$

The order of stability of free radicals is 3° > 2° > 1°. Therefore, the more stable 2° free radical is formed predominantly.

Step 4. More stable 2° free radical attacks the H—Br molecule to form *anti*-Markovnikov product and a bromine atom. The bromine atom goes back to step 3.

$$CH_3-\dot{C}H-CH_2Br + H:Br \longrightarrow \underset{n\text{-Propyl bromide}}{CH_3-CH_2-CH_2Br} + Br\cdot$$

Remember that HCl and HI do not give *anti*-Markovnikov products in the presence of peroxides. This is because :

(*a*) The H–Cl bond (103 kcal/mol) is stronger than H–Br bond (87 kcal/mol). It is not broken by the alkoxy free radicals obtained from peroxides.

(*b*) The H–I bond (71 kcal/mol) is weaker than H–Br bond (87 kcal/mol). It is broken by the alkoxy free radicals obtained from peroxides. But the iodine atoms so formed readily combine with each other to yield iodine molecules rather than attack the double bond in alkenes.

(2) **Addition of Hypohalous Acids**. Alkenes react with hypohalous acids (HOX) to give halohydrins. Markovnikov rule is followed in case of unsymmetrical alkenes. In these acids, halogen is the positive part and –OH group is the negative part.

$$\underset{\text{Ethylene}}{CH_2=CH_2} + H\overset{-}{O}\overset{+}{Cl} \longrightarrow \underset{\text{Ethylene chlorohydrin}}{\underset{|}{\overset{OH}{}}CH_2-\underset{}{\overset{Cl}{|}}CH_2}$$

$$\underset{\text{Propylene}}{CH_3-CH=CH_2} + H\overset{-}{O}\overset{+}{Cl} \longrightarrow \underset{\substack{\text{Propylene chlorohydrin}\\ \text{(Markovnikov product)}}}{CH_3-\overset{\overset{OH}{|}}{C}H-\overset{\overset{Cl}{|}}{C}H_2}$$

(3) **Addition of H_2SO_4.** Sulfuric acid undergoes addition to an alkene just as hydrogen halide does. The product is an alkyl hydrogen sulfate. Markovnikov rule is followed.

$$\underset{\text{Propene}}{CH_3-CH=CH_2} + H-OSO_3H \longrightarrow \underset{\substack{\text{1-Methylethyl}\\ \text{hydrogen sulfate}}}{CH_3-\overset{\overset{OSO_3H}{|}}{C}H-CH_3}$$

MECHANISM. Following steps are involved :

Step 1. Protonation of the double bond gives a more stable carbocation.

$$\underset{\text{Propene}}{CH_3-CH=CH_2} + H^+ \longrightarrow \underset{\text{2° Carbocation}}{CH_3-\overset{+}{C}H-CH_3}$$

Step 2. Nucleophilic attack by $\bar{O}SO_3H$.

$$CH_3-\overset{+}{C}H-CH_3 + \bar{O}SO_3H \longrightarrow \underset{\substack{\text{1-Methylethyl}\\ \text{hydrogen sulfate}}}{CH_3-\underset{\underset{OSO_3H}{|}}{C}H-CH_3}$$

When alkyl hydrogen sulfates are heated with water, alcohols are formed.

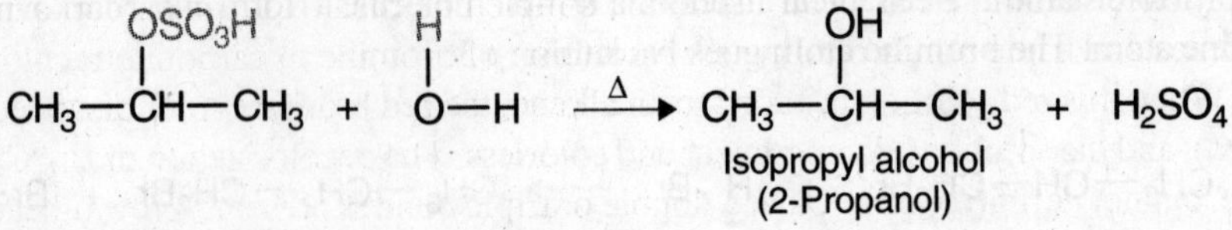

The overall result of the above reactions appears to be Markovnikov addition of H_2O to the double bond.

(4) **Addition of Water (Hydration).** Alkenes react with water in the presence of a strong acidic catalyst (dilute H_2SO_4) to form an alcohol. Markovnikov rule is followed.

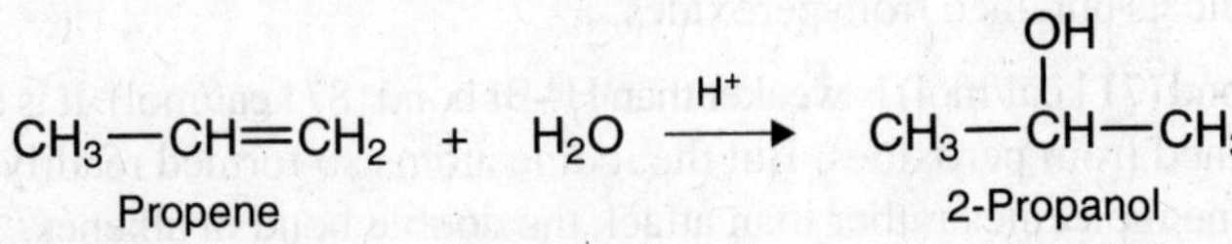

MECHANISM. Following steps are involved :

Step 1. Protonation of the double bond forms a more stable carbocation.

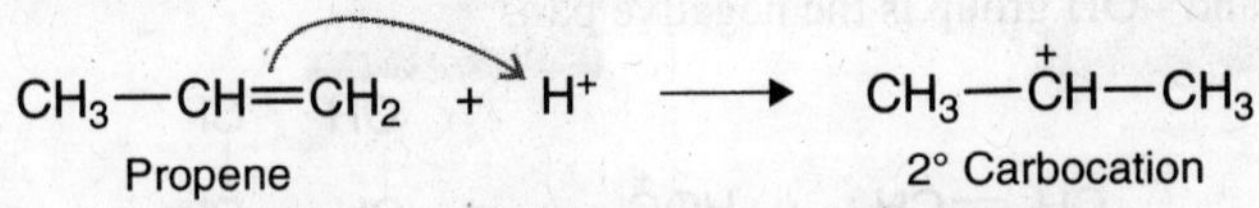

Step 2. Nucleophilic attack by water gives a protonated alcohol.

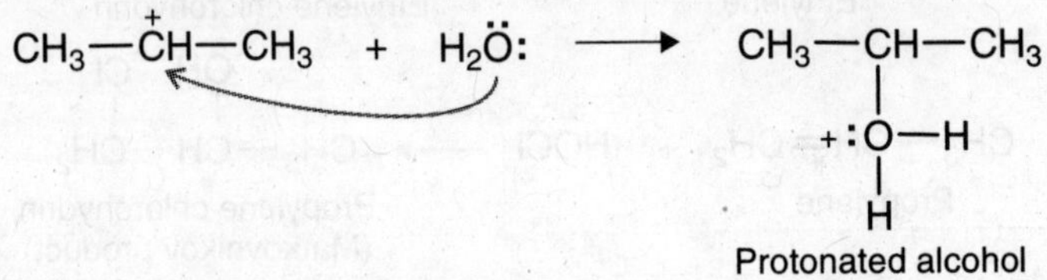

Step 3. Abstraction of proton gives alcohol.

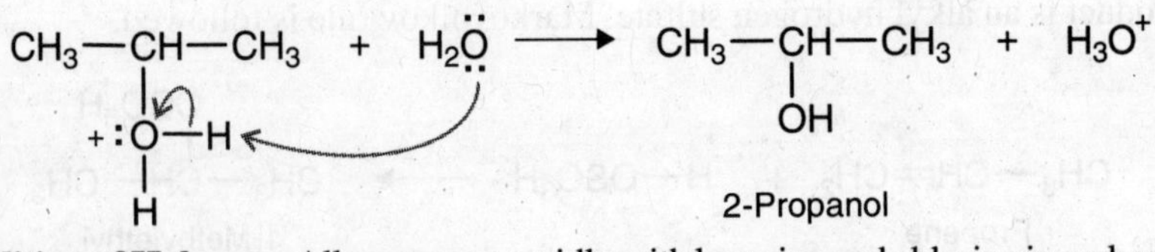

(5) **Addition of Halogens.** Alkenes react rapidly with bromine and chlorine in carbon tetrachloride (inert solvent) solvent to form ***vic*-dihalides.**

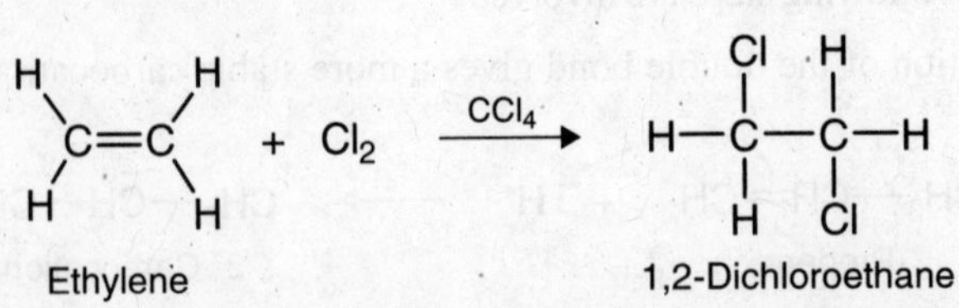

The addition of halogens is an *anti addition* to the carbon-carbon double bond.

Neither F_2 nor I_2 is a useful reagent for alkene addition reactions. Fluorine undergoes an explosive free-radical reaction with organic compounds. Iodine adds to a double bond, but the reaction is reversible and the equilibrium lies on the alkene-iodine side.

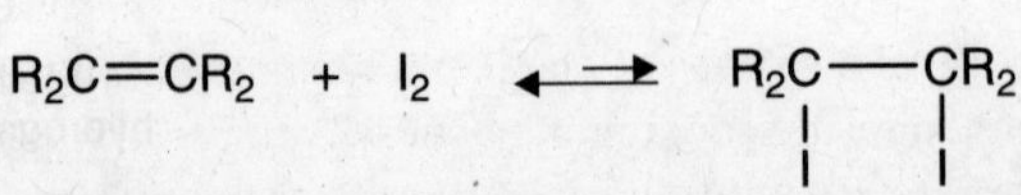

The addition of bromine has been used as a simple chemical test for carbon-carbon double bonds of alkenes and triple bonds of alkynes. A solution of bromine in carbon tetrachloride is deep red in color. When this red solution is added to an alkene, the red bromine color disappear (we say it is decolorized), and the solution becomes clear and colorless. The decolorization of Br_2/CCl_4 solution is suggestive, but not definitive proof, that a double or triple bond is present. A few other compounds such as phenols, aldehydes, and ketones, also decolorize Br_2/CCl_4 solution. Alkanes do not react with Br_2 in CCl_4 at room temperature and in the dark.

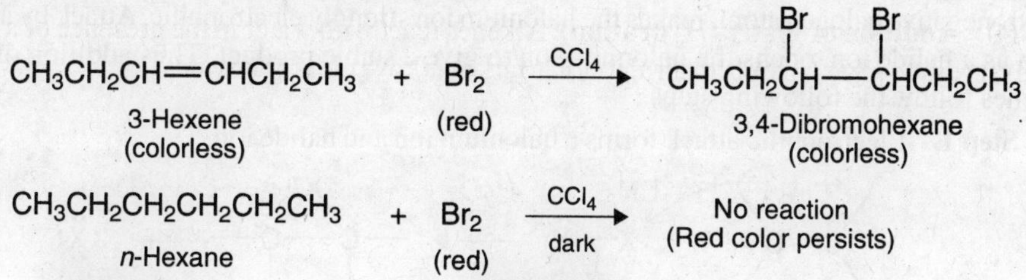

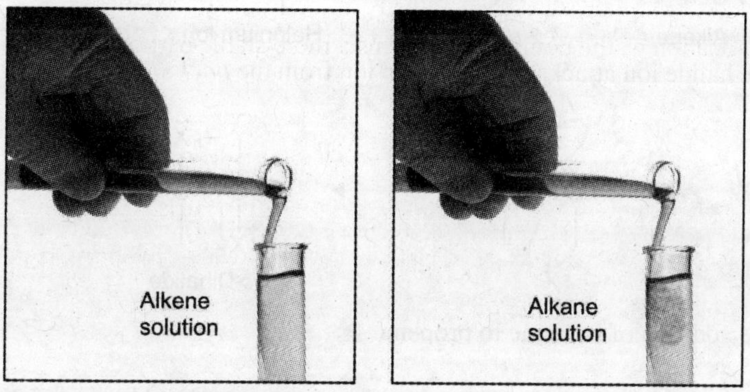

11.3. When a solution of bromine (red) is added to 3-hexene, the bromine color quickly disappears because bromine adds across the double bond. When bromine is added to *n*-hexane (at right), the color persists.

Mechanism of addition of halogens to alkenes. Halogens add to alkenes to form *vicinal* dihalides.

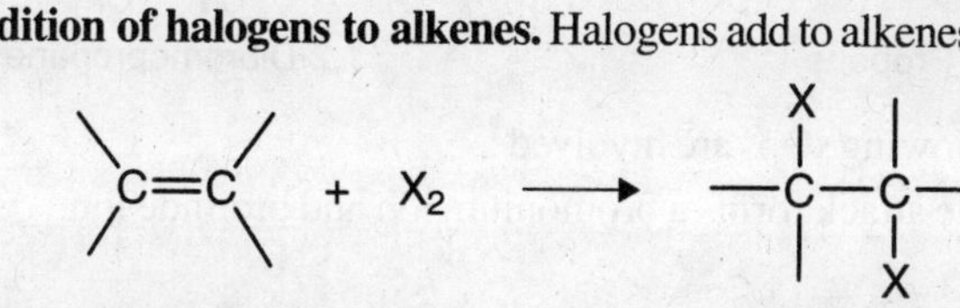

Alkene ($X_2 = Cl_2$ or Br_2) *anti*-Addition

A halogen molecule (Cl_2 or Br_2) is electrophilic. A nucleophile can react with a halogen, displacing the halide ion.

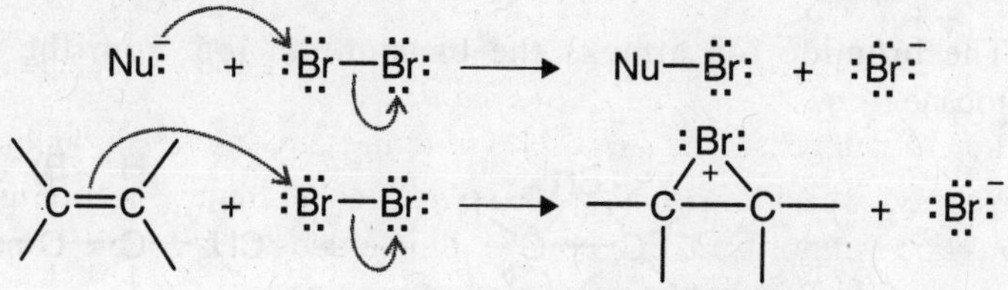

Bromonium ion

In the above reaction, the pi electrons of an alkene attack the bromine molecule, expelling the bromide ion. A bromonium ion results, containing a three-membered ring with a positive charge on the bromine atom. A similar reaction with chlorine forms chloronium ion.

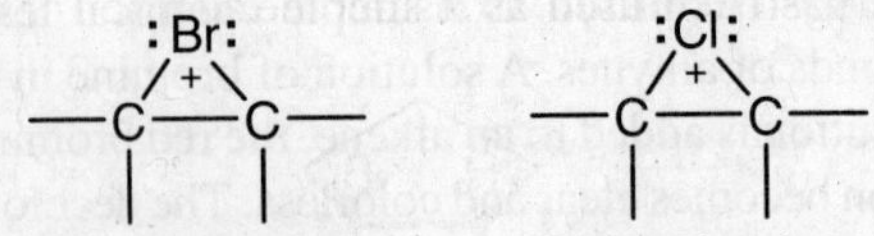

Bromonium ion Chloronium ion

Unlike a normal carbocation, all the atoms in a halonium ion have filled octets. The three-membered ring has considerable ring strain, however, which, combined with a positive charge on an electronegative halogen atom, makes the halonium ion strongly electrophilic. Attack by a nucleophile, such as a halide ion, opens the halonium ion to give a stable product. Thus addition of halogens to alkenes follow the following steps :

Step 1. Electrophilic attack forms a halonium ion and halide ion.

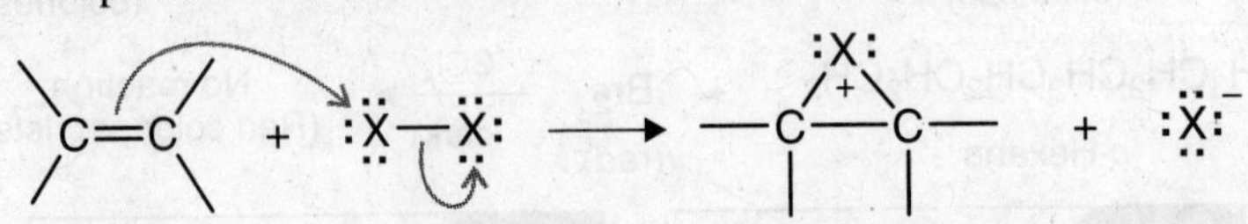

Alkene Halonium ion

Step 2. The halide ion attacks the halonium ion from the *back side* to form *vic*-dihalide.

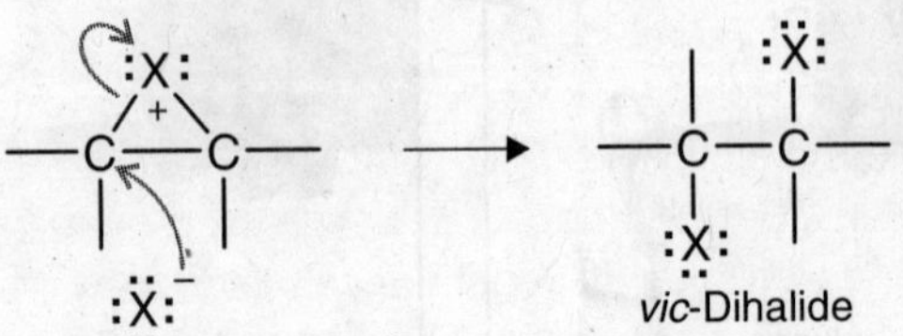

vic-Dihalide

Consider the addition of bromine to propene.

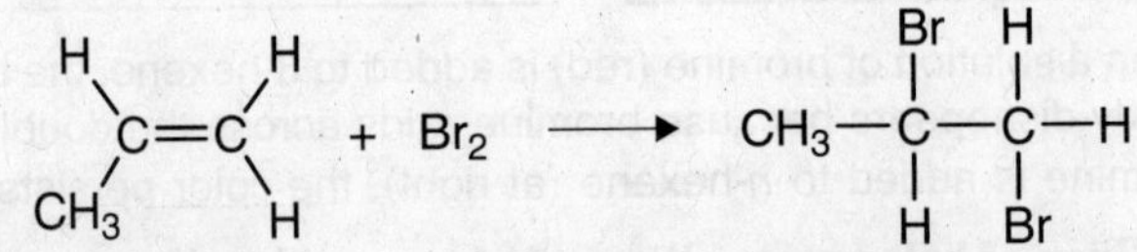

Propene 1,2-Dibromopropane

MECHANISM. Following steps are involved :

Step 1. Electrophilic attack forms a bromonium ion and bromide ion.

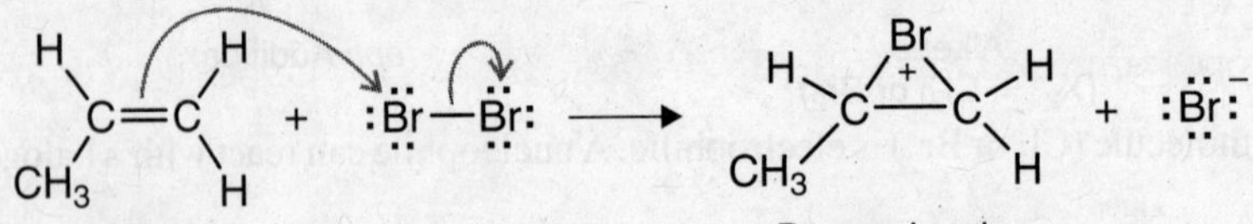

Propene Bromonium ion

Step 2. The bromide ion attacks the bromonium ion from the *back side* to form 1,2-dibromopropane.

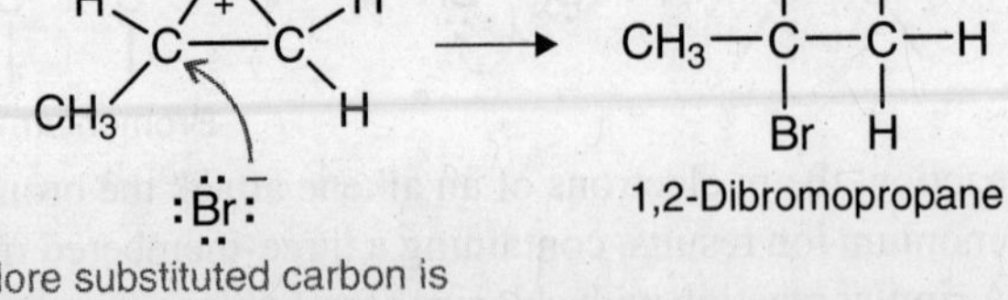

1,2-Dibromopropane

More substituted carbon is attacked by the halide ion

Stereochemistry of Halogen Addition. Consider the addition of bromine to cyclopentene.

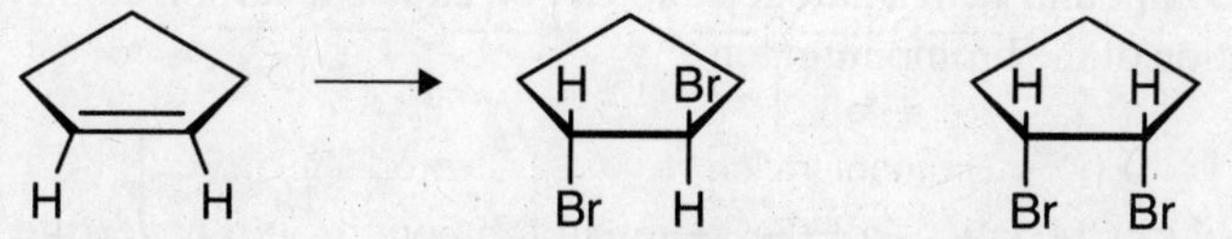

Cyclopentene — *trans*-1,2-Dibromocyclopentane — *cis*-1,2-Dibromocyclopentane (not formed)

The addition of bromine to cyclopentene is a stereospecific *anti addition* reaction. When a nucleophile attacks a halonium ion, it must do so from the *back side*. This back side attack results in *anti* stereochemistry of addition.

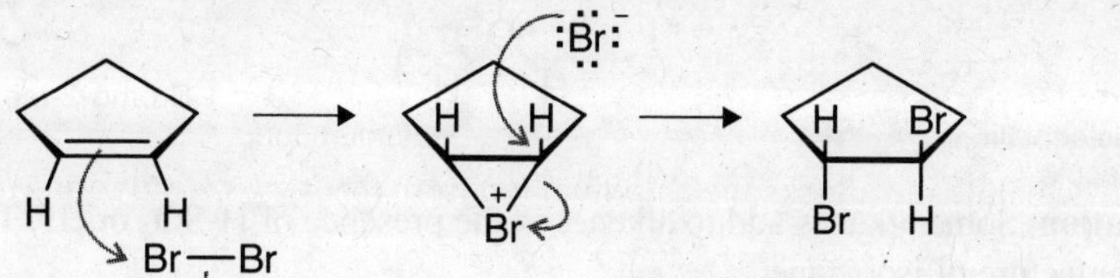

trans-1,2-Dibromocyclopentane

Mixed Addition. Bromination reactions of alkenes proceed by way of a bromonium-ion intermediate, followed by attack by a bromide ion to yield the dibromide. Is the second step limited to attack only by bromide ion ? Can other nucleophiles compete with bromide ion in the second step to yield other products ? Consider the case of a bromination reaction carried out with Br_2 in a solution containing Cl^- (from, say, NaCl). In this reaction, two nucleophiles (Br^- and Cl^-) are present. In such a case, *mixed dihalide products* are observed – along with the dibromoalkane, we find some bromochloroalkane.

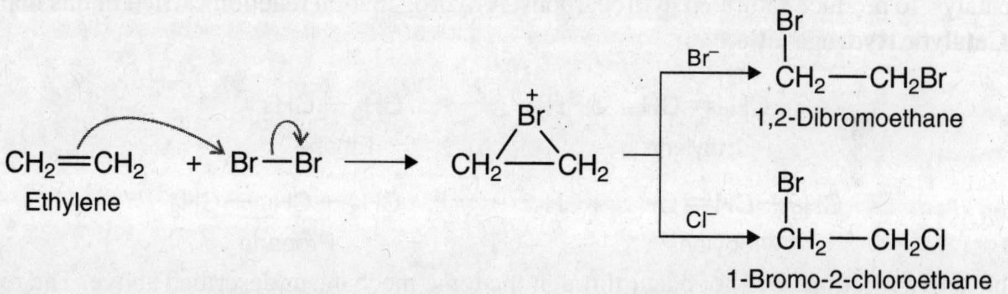

When propene is treated with Br_2 + Cl^- (from NaCl), only one bromochloropropane is isolated as a product.

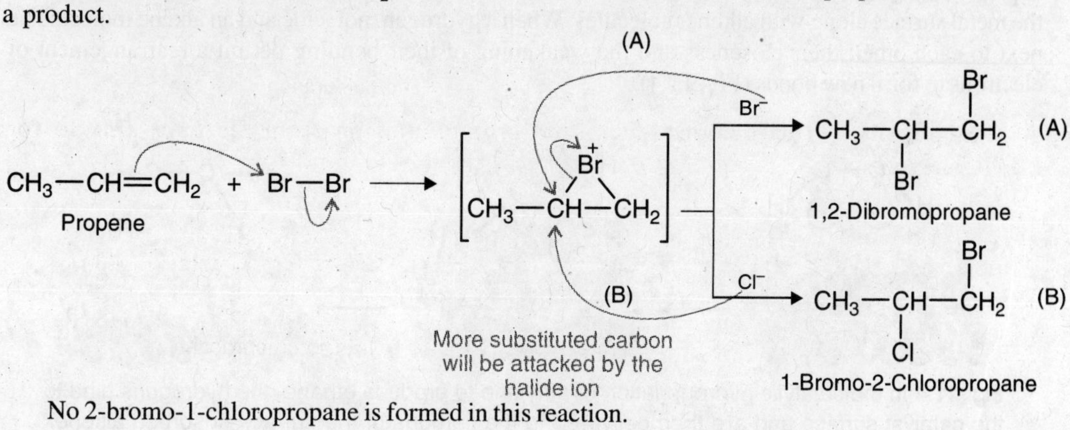

No 2-bromo-1-chloropropane is formed in this reaction.

Addition of Halogens and Water. When an alkene is treated with a mixture of Cl_2 or Br_2 in water, a 1,2-halohydrin (a compound with a halogen and OH on adjacent carbon atoms) is formed.

Step 1. Formation of the bromonium ion.

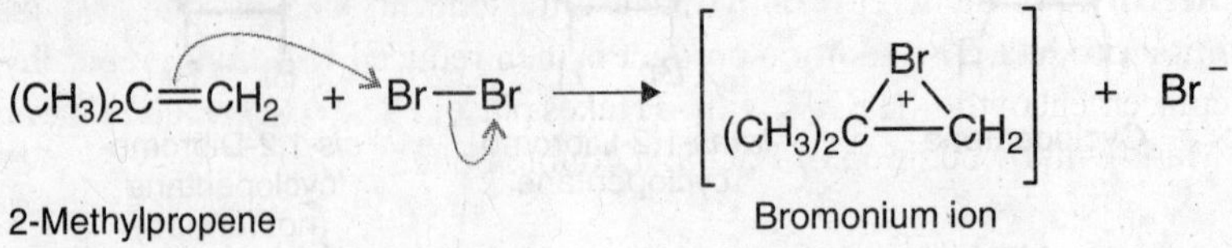

Step 2. Water attacks the bromonium ion.

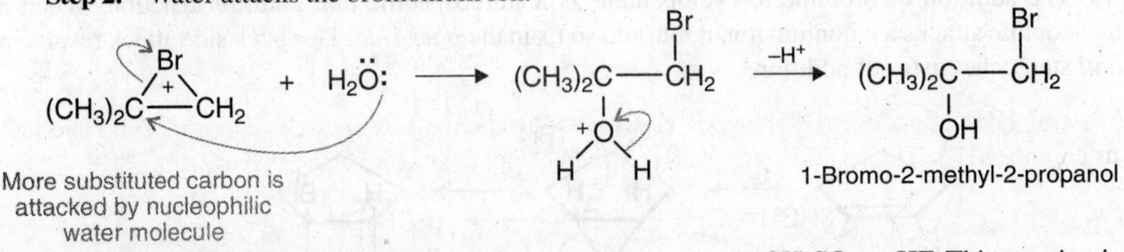

(6) **Alkylation**. Some alkanes add to alkenes in the presence of H_2SO_4 or HF. This reaction is used in the manufacture of isooctane.

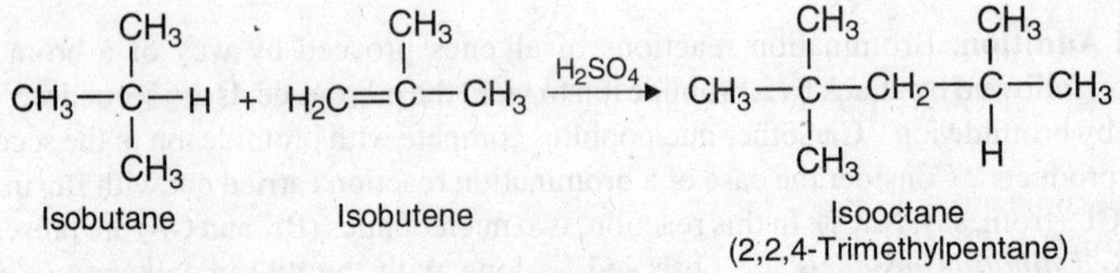

(7) **Addition of Hydrogen.** Alkenes add hydrogen under pressure and in the presence of Ni, Pt, or Pd catalyst to produce saturated hydrocarbons. A hydrogenation reaction carried in this manner is called **Catalytic Hydrogenation.**

$$\underset{\text{Ethylene}}{CH_2{=}CH_2} + H_2 \xrightarrow[\Delta]{Ni} \underset{\text{Ethane}}{CH_3{-}CH_3}$$

$$\underset{\text{Propylene}}{CH_3{-}CH{=}CH_2} + H_2 \xrightarrow[\Delta]{Ni} \underset{\text{Propane}}{CH_3{-}CH_2{-}CH_3}$$

The above reaction does not occur through the ionic mechanism described above. The reaction depends on high affinity of hydrogen gas for certain metals - Ni, Pt, Pd. The hydrogen is adsorbed on the metal surface along with alkene molecules. When a hydrogen molecule and an alkene molecule lie next to each other, their closeness and the weakening of their bonding permit a rearrangement of electrons to form new bonds (Fig. 11.4).

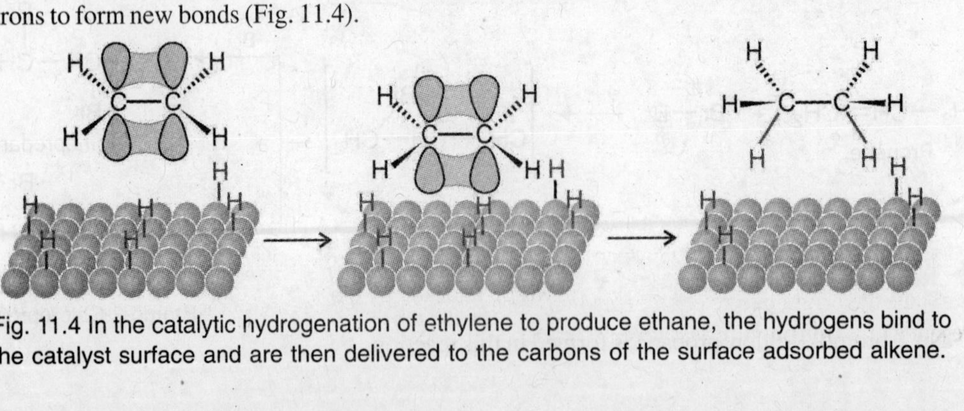

Fig. 11.4 In the catalytic hydrogenation of ethylene to produce ethane, the hydrogens bind to the catalyst surface and are then delivered to the carbons of the surface adsorbed alkene.

These hydrogenation reactions usually result in addition of two hydrogens to the same side of the carbon-carbon double bond (*Cis-Addition*). This is to be expected because the H_2 and alkene molecules lie side by side on the catalyst surface at the time of reaction.

(8) **Oxymercuration-demercuration**. Alkene is treated with aqueous solution of mercuric acetate to give an addition product. The addition product is then reduced with sodium borohydride to form the alcohol. Replacement of the –HgOAc with –H takes place. The overall result of the above reactions appears to be Markovnikov addition of H_2O to a double bond.

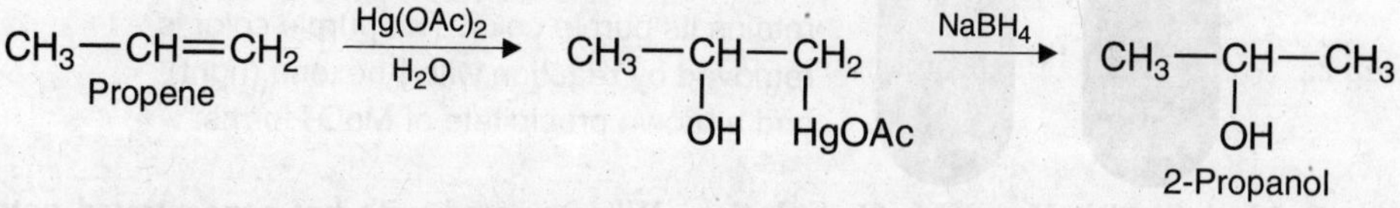

(9) **Hydroboration**. Diborane (B_2H_6) reacts with alkenes to form trialkylboranes. Diborane adds as borane (BH_3). The positive part of BH_3 is the boron ; the negative part is hydrogen.

$$\underset{\text{Alkene}}{3RCH{=}CH_2} + BH_3 \longrightarrow \underset{\text{Trialkylborane}}{(RCH_2CH_2)_3B}$$

Trialkylboranes are used for the synthesis of primary alcohols by reaction with alkaline aqueous solution of hydrogen peroxide.

$$(RCH_2CH_2)_3B + H_2O_2 \xrightarrow[\bar{O}H]{H_2O} \underset{1°\text{Alcohol}}{3RCH_2CH_2OH} + H_3BO_3$$

The overall result of the above reactions appears to be *anti*-Markovnikov addition of H_2O to a double bond. Thus,

$$\underset{\text{Propene}}{CH_3{-}CH{=}CH_2} \xrightarrow[2.\ H_2O_2/\bar{O}H]{1.\ B_2H_6} \underset{\text{1-Propanol}}{CH_3{-}CH_2{-}CH_2{-}OH}$$

(10) **Oxidation with Cold KMnO$_4$ Solution**. Alkenes react with cold dilute potassium permanganate solution to form glycols. Glycols are compounds whose molecules contain two –OH groups on adjacent carbons. The reaction looks like the addition of two –OH groups of hydrogen peroxide, HO–OH, to a double bond.

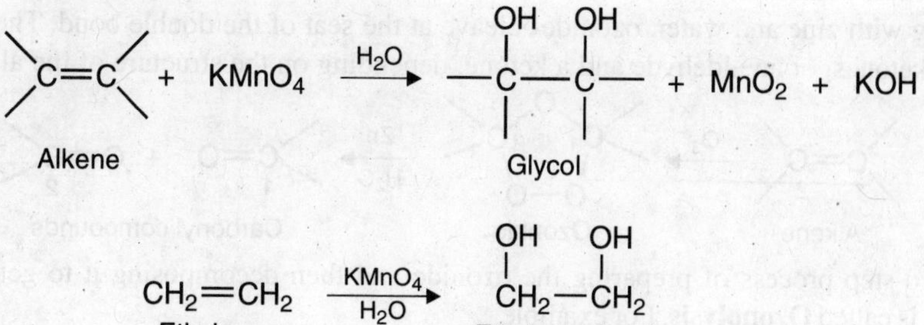

Since the bright purple color of $KMnO_4$ disappears during the reaction ; it is used as a **Test for the Presence of a Double Bond (Baeyer's Test)**.

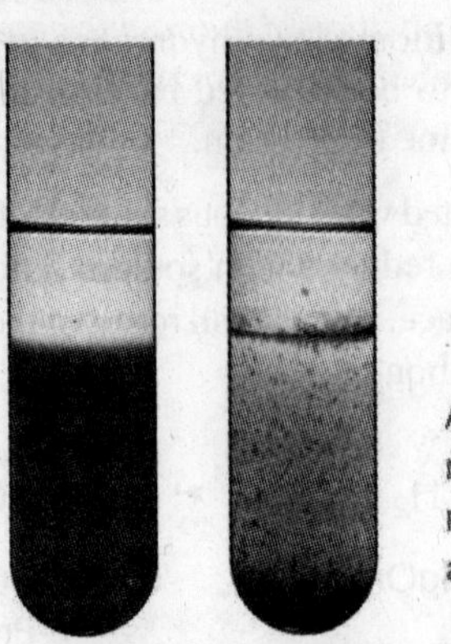

Aqueous $KMnO_4$ shaken with hexane (left) retains its purple color. The purple color is removed by reaction with l-hexene (right) and a brown precipitate of MnO_2 forms.

(11) **Oxidation with Hot $KMnO_4$ Solution**. When treated with hot concentrated potassium permanganate solution, alkenes split at the double bond to form ketones and/or acids.

$$\underset{\text{Propylene}}{CH_3-CH=CH_2} \xrightarrow[\text{hot } KMnO_4]{[O]} \underset{\text{Acetic acid}}{CH_3-\overset{\overset{\large O}{\|}}{C}-OH} + CO_2 + H_2O$$

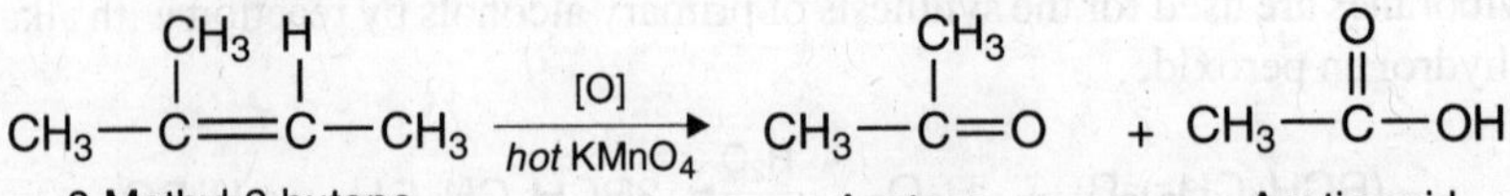

2-Methyl-2-butene → Acetone + Acetic acid

(12) **Catalytic Oxidation**. Alkenes react with oxygen in the presence of silver catalyst at 250-400°C to form **Epoxides**.

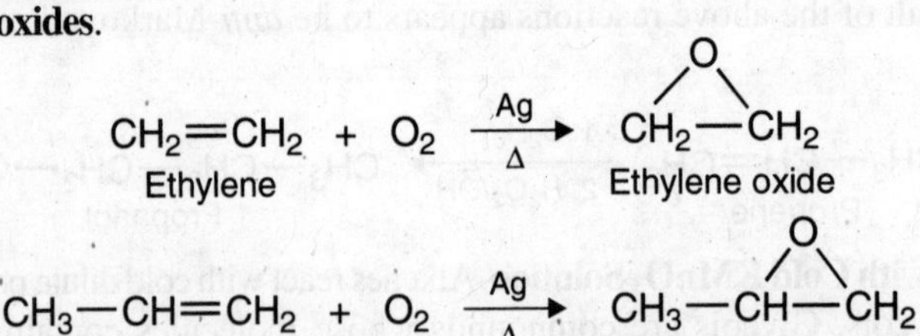

Ethylene → Ethylene oxide; Propylene → Propylene oxide

(13) **Oxidation with Ozone**. When ozone is passed through an alkene in an inert solvent, it adds across the double bond to form an **Ozonide**. Ozonides are explosive compounds. They are not isolated. On warming with zinc and water, ozonides cleave at the seat of the double bond. The products are aldehydes, ketones, or an aldehyde and a ketone, depending on the structure of the alkene.

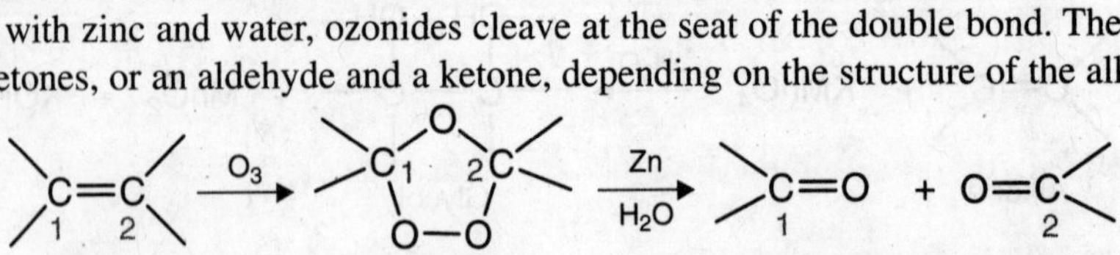

Alkene → Ozonide → Carbonyl compounds

The two-step process of preparing the ozonide and then decomposing it to get the carbonyl compounds is called **Ozonolysis**. For example,

$$\underset{\text{Ethylene}}{CH_2=CH_2} \xrightarrow[(2)\ Zn/H_2O]{(1)\ O_3} \underset{\text{Formaldehyde (2 molecules)}}{CH_2=O + O=CH_2}$$

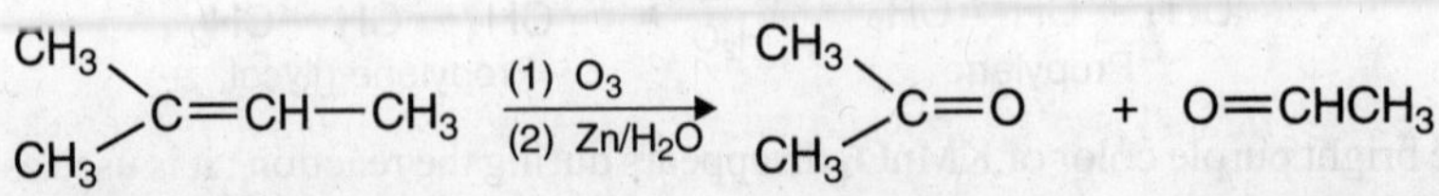

2-Methyl-2-butene → Acetone + Acetaldehyde

Ozonolysis is probably the best method for *locating the position of the double bonds in unknown alkenes.* The oxygenated carbons in carbonyl compounds obtained by ozonolysis are the ones that were joined by a double bond in the original alkene. Suppose an alkene on ozonolysis gives the carbonyl compounds.

$$CH_3-\overset{\displaystyle CH_3}{\underset{*}{C}}{=}O \quad \text{and} \quad O{=}\overset{\displaystyle H}{\underset{*}{C}}-CH_3$$

Acetone Acetaldehyde

Joining the oxygenated carbons (marked by asterisk) by a double bond, we get the following structure of the unknown alkene.

$$CH_3-\overset{\displaystyle CH_3}{C}{=}\overset{\displaystyle H}{C}-CH_3$$

2-Methyl-2-butene

(14) **Combustion**. When burnt in air, alkenes are oxidized to CO_2 and water.

$$CH_2{=}CH_2 + 3O_2 \longrightarrow 2CO_2 + 2H_2O + \text{Heat}$$

(15) **Polymerization**. The process by which simple molecules join together to form large molecules is known as *polymerization*. Simple alkenes polymerize to form long chain addition polymers. For example, ethylene gives polyethylene. Propylene gives polypropylene.

$$nCH_2{=}CH_2 \xrightarrow{\text{polymerization}} \left[-CH_2-CH_2-\right]_n$$

Ethylene Polyethylene

$$nCH_3CH{=}CH_2 \xrightarrow{\text{polymerization}} \left[-\overset{\displaystyle CH_3}{CH}-CH_2-\right]_n$$

Propylene Polypropylene

These reactions are catalyzed by HF, H_2SO_4, or organic peroxides. High temperature and pressure are generally required. For more details and discussion of the mechanism see Chapter 49.

(16) **Substitution of Alkenes by Halogens (Allylic Substitution).** When alkenes are treated with Cl_2 or Br_2 at high temperatures, one of their allylic hydrogens is replaced by halogen atom. Allylic position is the carbon adjacent to one of the unsaturated carbon atoms.

$$CH_3-CH{=}CH_2 + Cl_2 \xrightarrow{500°C} ClCH_2-CH{=}CH_2 + HCl$$

Propene Allyl chloride (3-Chloro-1-propene)

MECHANISM. Allylic substitution takes place by a free-radical mechanism. Following steps are involved :

Step 1. Chain-Initiation Step. Chlorine molecule undergoes homolytic fission to give chlorine free radicals.

$$Cl{:}Cl \xrightarrow{\Delta} Cl\cdot + \cdot Cl$$

Step 2. Chain-Propagation Steps. (*a*) Chlorine free radical attacks propene to give HCl and an allyl free radical.

$$Cl\cdot + H{:}CH_2-CH{=}CH_2 \longrightarrow H{:}Cl + \cdot CH_2-CH{=}CH_2$$

Allyl radical

(*b*) The allyl free radical attacks chlorine molecule to produce allyl chloride and a chlorine free radical.

$$\dot{C}H_2-CH=CH_2 + Cl:Cl \longrightarrow Cl:CH_2-CH=CH_2 + Cl\cdot$$

Allyl radical　　　　　　　　Allyl chloride

Steps (*a*) and (*b*) are repeated over and over again.

Step 3. Chain-Termination Steps. The above chain reaction comes to halt when any two free radicals combine. For Example,

$$Cl\cdot + \cdot Cl \longrightarrow Cl-Cl$$

$$Cl\cdot + \cdot CH_2CH=CH_2 \longrightarrow Cl-CH_2CH=CH_2$$

$$2CH_2=CH-CH_2\cdot \longrightarrow CH_2=CH-CH_2-CH_2-CH=CH_2$$

ETHYLENE, *Ethene*, $CH_2=CH_2$

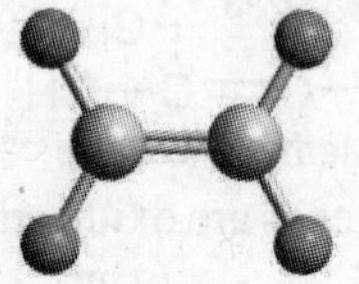

Ethylene is the first member of the alkene series. It is produced in large amounts by the cracking of petroleum.

Preparation. Ethylene can be obtained :

(1) By heating ethanol with excess concentrated sulfuric acid at 170°C. (*Laboratory Method*)

$$\underset{\text{Ethanol}}{CH_3CH_2OH} \xrightarrow[170°C]{\text{conc } H_2SO_4} \underset{\text{Ethylene}}{CH_2=CH_2} + H_2O$$

(2) By heating ethyl iodide with alcoholic KOH.

$$\underset{\text{Ethyl iodide}}{CH_3-CH_2-I} + KOH \xrightarrow[\Delta]{\text{ethanol}} \underset{\text{Ethylene}}{CH_2=CH_2} + KI + H_2O$$

(3) By heating 1,2-dibromoethane with zinc dust in ethanol.

$$\underset{\text{1,2-Dibromoethane}}{\begin{matrix} CH_2-Br \\ | \\ CH_2-Br \end{matrix}} + Zn \xrightarrow[\Delta]{\text{ethanol}} \underset{\text{Ethylene}}{\begin{matrix} CH_2 \\ \| \\ CH_2 \end{matrix}} + ZnBr_2$$

(4) By controlled addition of hydrogen to acetylene in the presence of Lindlar's catalyst.

$$\underset{\text{Acetylene}}{CH\equiv CH} + H_2 \xrightarrow[\text{quinoline}]{Pd/CaCO_3} \underset{\text{Ethylene}}{CH_2=CH_2}$$

(5) By passing ethanol vapors over alumina catalyst at 350°C.

$$\underset{\text{Ethanol}}{CH_3CH_2OH} \xrightarrow[350°C]{Al_2O_3} \underset{\text{Ethylene}}{CH_2=CH_2} + H_2O$$

(6) By cracking of natural gas and petroleum.

$$\underset{\text{Ethane}}{CH_3-CH_3} \xrightarrow{600°C} \underset{\text{Ethylene}}{CH_2=CH_2} + CH_4 + H_2$$

Properties (Physical). Ethylene is a colorless gas with a sweet odor. It is slightly soluble in water but dissolves readily in organic solvents such as ethanol and ether.

(Chemical). Some of the important reactions of ethylene are listed in Table 11.2.

Table 11.2 CHEMICAL REACTIONS OF ETHYLENE

$CH_2{=}CH_2$ (Ethylene)

Reagent	Product	Name
H_2; Ni, Δ	$CH_3{-}CH_3$	Ethane
Br_2	$CH_2Br{-}CH_2Br$	1,2-Dibromoethane
HBr	$CH_3{-}CH_2Br$	Bromoethane
HOCl	$CH_2(OH){-}CH_2Cl$	Ethylene chlorohydrin
1. H_2SO_4; 2. H_2O, Δ	$CH_3{-}CH_2OH$	Ethyl alcohol
$KMnO_4$; H_2O, Cold	$CH_2(OH){-}CH_2(OH)$	Ethylene glycol
O_2; Ag, Δ	$CH_2{-}CH_2$ bridged by O (epoxide)	Ethylene oxide
1. O_3; 2. Zn, H_2O	$2H{-}C({=}O){-}H$	Formaldehyde
Polymerize	$\left[CH_2{-}CH_2\right]_n$	Polyethylene

Uses. Ethylene is used in the manufacture of many industrially important polymers (*e.g.*, PVC, EPR, Sarans, polyethylene). These polymers are used in the manufacture of floor tiles, shoe soles, synthetic fibres, raincoats, purses, saran wrap, and pipes.

PROPENE, *Propylene*, $CH_3{-}CH{=}CH_2$

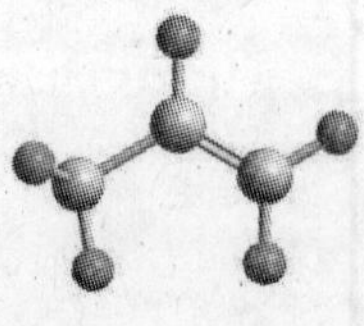

It is produced on large scale along with ethylene in the cracking of petroleum.

Preparation. Propene may be obtained :

(1) By dehydration of 1-propanol or 2-propanol.

$$\underset{\text{2-Propanol (Isopropyl alcohol)}}{CH_3{-}\overset{OH}{\overset{|}{C}H}{-}\overset{H}{\overset{|}{C}H_2}} \xrightarrow[\Delta]{H_2SO_4} \underset{\text{Propene}}{CH_3{-}CH{=}CH_2} + H_2O$$

(2) By dehydrohalogenation of 1-bromopropane or 2-bromopropane.

$$\underset{\text{2-Bromopropane (Isopropyl bromide)}}{CH_3{-}\overset{Br}{\overset{|}{C}H}{-}\overset{H}{\overset{|}{C}H_2}} + KOH \xrightarrow{\text{alcohol}} \underset{\text{Propene}}{CH_3{-}CH{=}CH_2} + H_2O + KBr$$

Properties. Propene is a colorless gas, bp –48°C. The chemical properties are listed in Table 11.3. Propene is an unsymmetrical alkene. Markovnikov rule is applied except in the addition of HBr in the presence of peroxides.

Uses. Propene is used in the manufacture of ethylene-propylene rubbers (EPR), acetone, isopropyl alcohol, *n*-propyl alcohol, and allyl bromide.

Table 11.3 CHEMICAL REACTIONS OF PROPENE (PROPYLENE). MARKOVNIKOV RULE IS FOLLOWED.

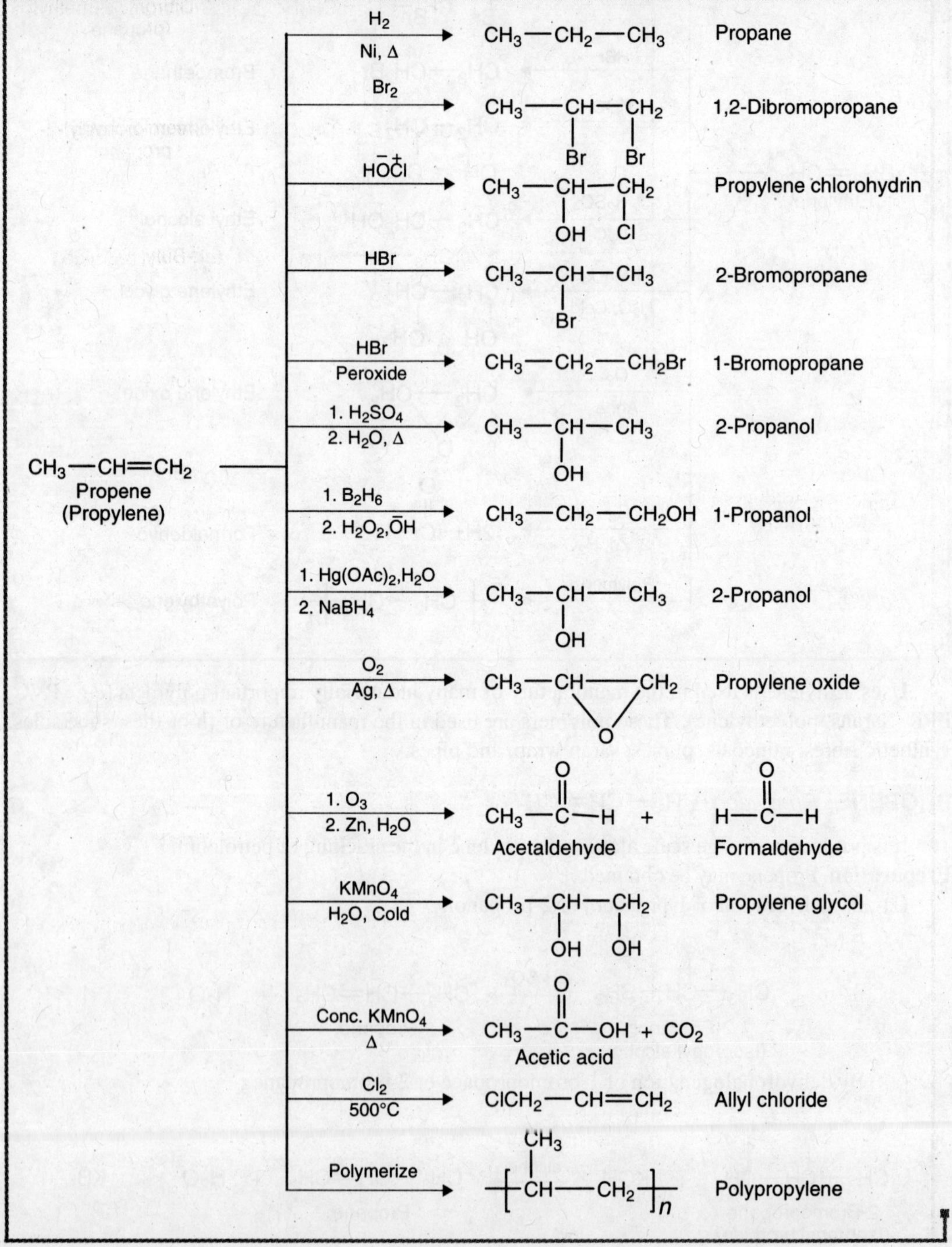

Table 11.4 CHEMICAL REACTIONS OF 2-METHYL-1-PROPENE (ISOBUTYLENE). MARKOVNIKOV RULE IS FOLLOWED.

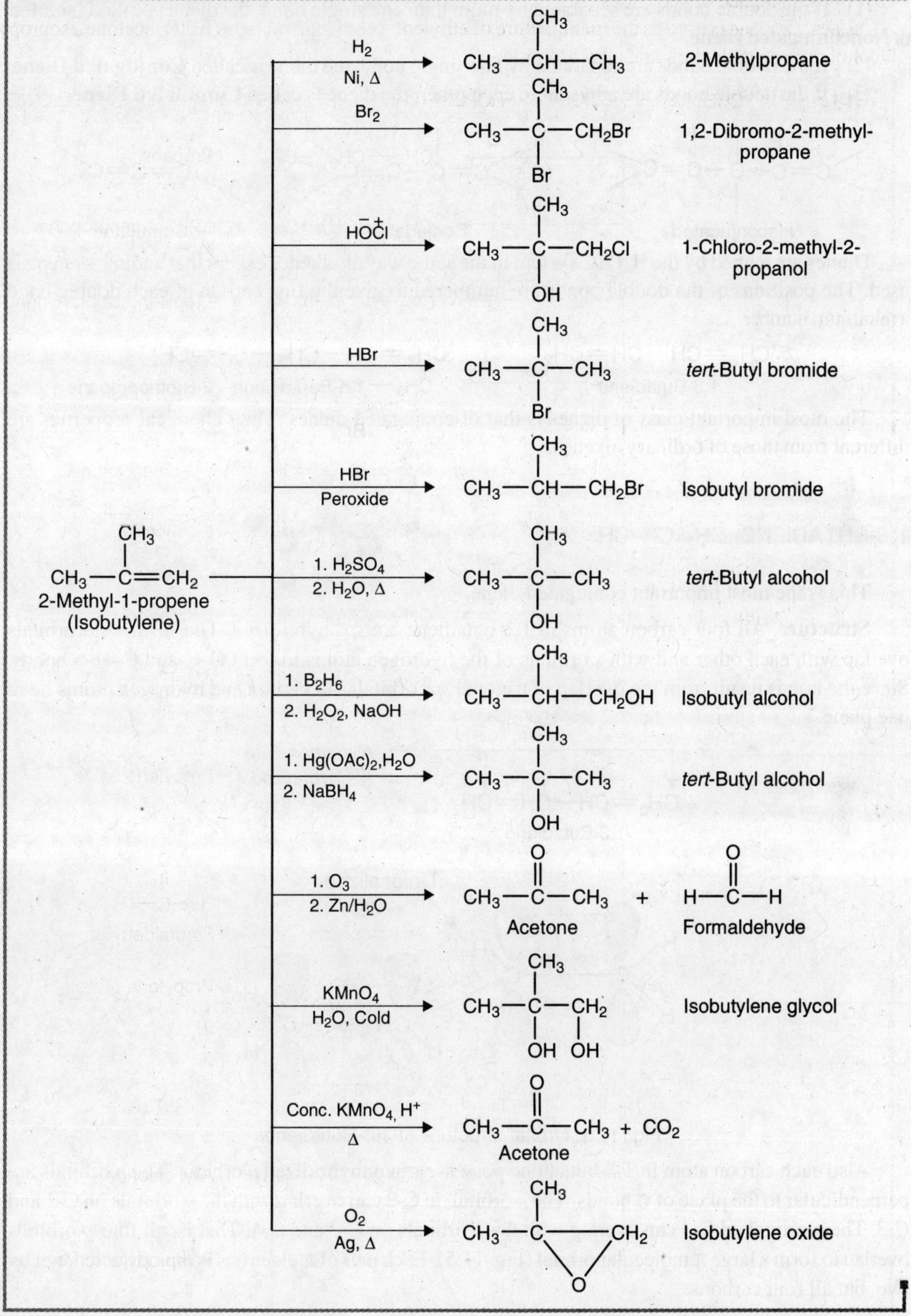

DIENES

Alkenes containing two carbon-carbon double bonds are called dienes or alkadienes.

(1) If the double bonds are separated by more than one single bond, the diene is called **Isolated** or **Nonconjugated Diene.**

(2) If the double bonds are separated by one single bond, the diene is called **Conjugated Diene.**

(3) If the double bonds are adjacent to each other, the diene is called **Cumulated Diene.**

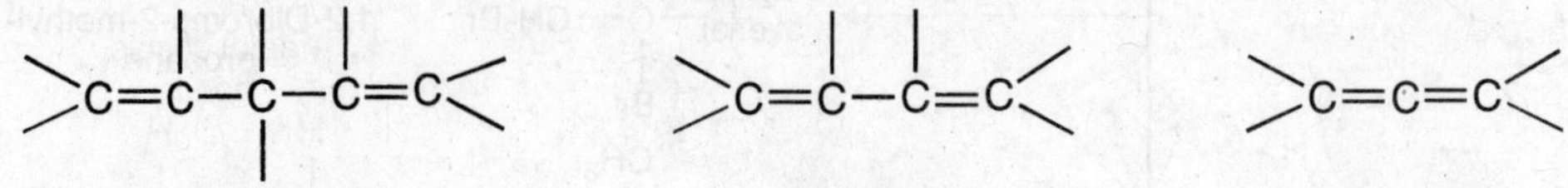

Nonconjugated Conjugated Cumulated

Dienes are named by the IUPAC system in the same way as alkenes except that ending *–adiene* is used. The positions of the double bonds are numbered to give the first carbon of each double bond a minimum number.

$CH_2{=}CH{-}CH{=}CH_2$ 1,3-Butadiene

$CH_2{=}CH{-}CH_2{-}CH{=}CH_2$ 1,4-Pentadiene

The most important class of dienes is that of conjugated dienes. Their chemical properties are different from those of ordinary alkenes.

1, 3-BUTADIENE, $CH_2{=}CH{-}CH{=}CH_2$

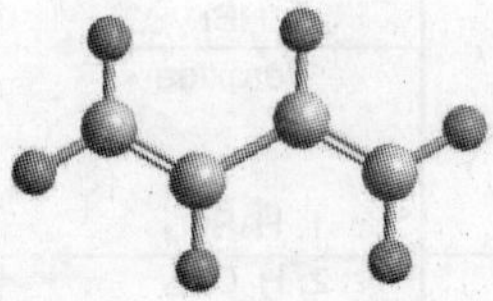

This is the most important conjugated diene.

Structure. All four carbon atoms in 1,3-butadiene are sp^2 hybridized. The sp^2 hybrid orbitals overlap with each other and with *s* orbitals of the hydrogen atoms to form C–C and C–H σ bonds. Since the bonds result from the overlap of trigonal sp^2 orbitals, all carbon and hydrogen atoms lie in one plane.

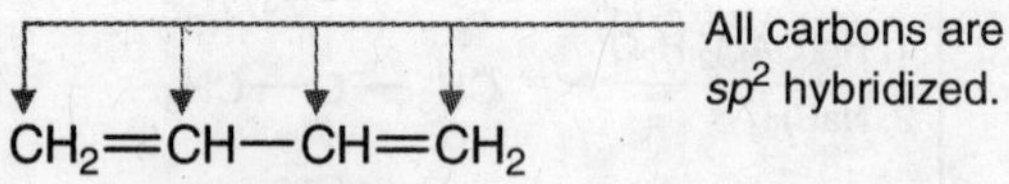

1,3-Butadiene

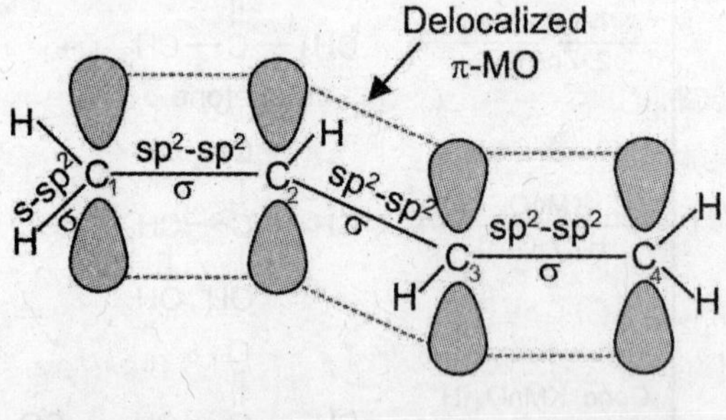

Fig. 11.5. Orbital structure of 1,3-Butadiene.

Also each carbon atom in 1,3-butadiene possesses an unhybridized *p* orbital . The *p* orbitals are perpendicular to the plane of σ bonds. The *p* orbital on C-2 can overlap with the *p* orbitals on C-1 and C-3. The *p* orbital on C-3 can overlap with the *p* orbitals on C-2 and C-4. That is, all four *p* orbitals overlap to form a large π molecular orbital (Fig. 11.5). Each pair of π electrons is thus attracted, not by two, but all four carbons.

The overlap of p orbitals of C-2 and C-3 in both directions, which allows the π electrons to be spread over a larger area, is referred to as **Delocalization**. This delocalization of π electrons is responsible for greater stability of 1,3-butadiene.

The C–C single bonds in 1,3-butadiene are shorter (1.48 Å) than normal carbon-carbon single bonds. The C–C double bonds are longer (1.37 Å) than normal isolated carbon-carbon double bonds.

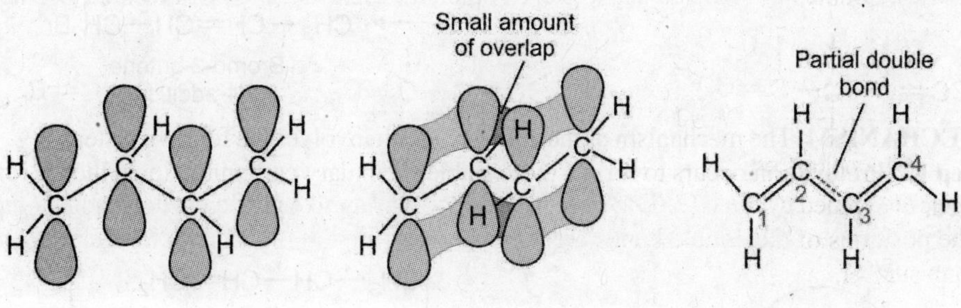

Structure of 1,3-Butadiene. All carbon atoms are sp^2 hybridized. All atoms lie in one plane. The delocalized molecular orbital contains four electrons. The central carbon-carbon single bond is shorter than the 1.54 Å bonds typical of alkanes because of its partial double bond character.

Preparation. 1,3-Butadiene can be prepared :

(1) **From Acetylene**

$$\underset{\text{Acetylene}}{2HC{\equiv}CH} \xrightarrow[NH_4Cl]{CuCl} \underset{\text{Vinylacetylene}}{HC{\equiv}C{-}CH{=}CH_2} \xrightarrow[Pd/BaSO_4]{H_2} \underset{\text{1,3-Butadiene}}{CH_2{=}CH{-}CH{=}CH_2}$$

$Pd/BaSO_4$ is Lindlar's catalyst.

(2) **From 1-Butene**

$$\underset{\text{1-Butene}}{CH_3CH_2CH{=}CH_2} \xrightarrow{Br_2/600°C} CH_3\overset{Br}{\overset{|}{C}}HCH{=}CH_2 \xrightarrow[\Delta]{\text{alc KOH}} \underset{\text{1,3-Butadiene}}{CH_2{=}CH{-}CH{=}CH_2}$$

(3) **From 1, 4-Butanediol** by acid-catalyzed dehydration.

$$\underset{\text{1,4-Butanediol}}{\overset{OH}{\overset{|}{C}}H_2CH_2CH_2\overset{OH}{\overset{|}{C}}H_2} \xrightarrow[\Delta]{H^+} \underset{\text{1,3-Butadiene}}{CH_2{=}CH{-}CH{=}CH_2} + 2H_2O$$

(4) **From *n*-Butane** by catalytic dehydrogenation.

$$\underset{n\text{-Butane}}{CH_3CH_2CH_2CH_3} \xrightarrow[600°C]{Cr_2O_3/Al_2O_3} \underset{\text{1,3-Butadiene}}{CH_2{=}CH{-}CH{=}CH_2}$$

Properties. 1,3-Butadiene is a colorless gas, bp –4.5°C. It gives the following reactions :

(1) **Addition of Halogen Acids**. 1,3-Butadiene reacts with halogen acids (HBr or HCl) to yield a mixture of two compounds. They are 3-bromo-1-butene and 1-bromo-2-butene. The first product results from **1,2-addition** to one of the double bonds. The second product results from addition to terminal (1,4) positions with the formation of a new double bond between C-2 and C-3. This latter process is known as **1,4-addition**. At low temperatures the 1,2-addition is preferred, whereas at high temperatures 1,4-addition predominates.

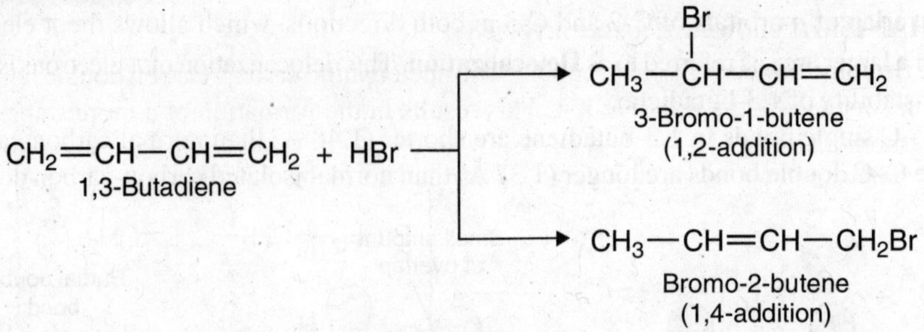

MECHANISM. The mechanism of the above reaction involves the following steps :

Step 1. Protonation occurs to form a more stable secondary carbonium ion which is resonance stabilized.

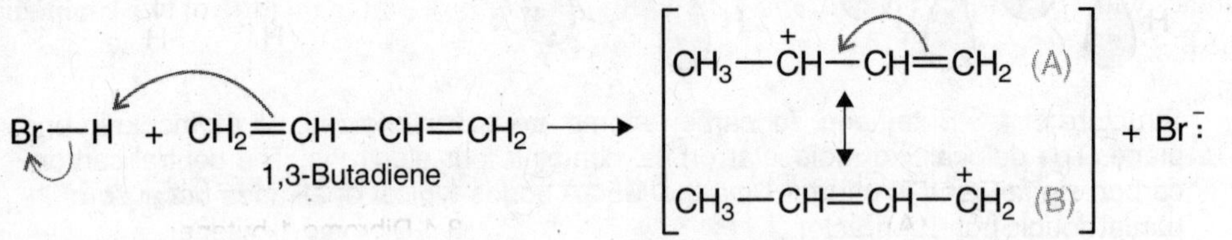

Step 2. Bromide ion, :$\bar{\text{Br}}$, combines with (A) to give 3-bromo-1-butene (1,2-addition). It combines with (B) to give 1-bromo-2-butene (1,4-addition). Thus a mixture of two compounds is obtained.

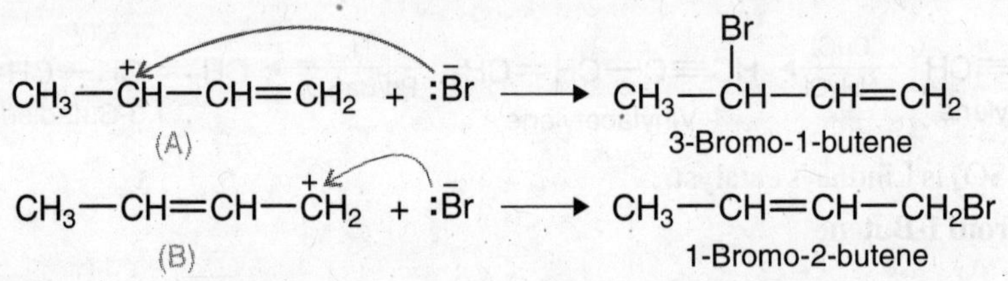

In the reaction of 1,3-butadiene with an equimolar quantity of HBr, the ratio of 1,2-addition to 1,4-addition varies with the temperature at which the reaction is carried out. At –80°C, the 1,2-addition product predominates. At 40°C the 1,4-addition product predominates.

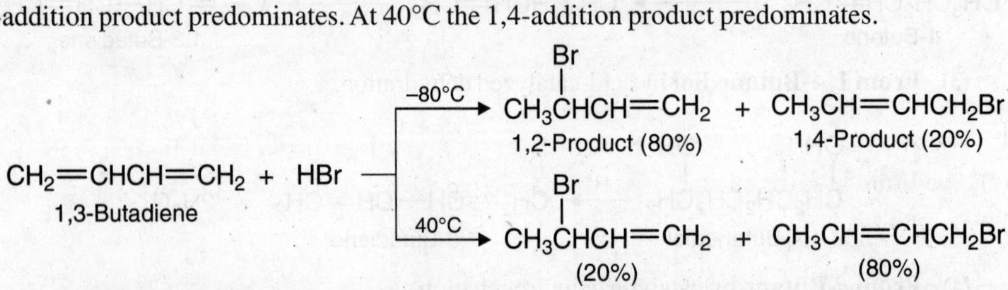

(2) **Addition of Halogens.** 1,3-Butadiene reacts with halogens (Cl_2 or Br_2) in the presence of an inert solvent (CCl_4) to give a mixture of two dibromo compounds.

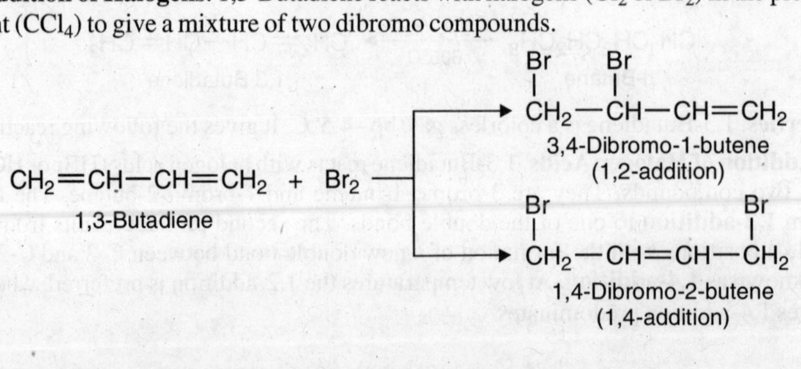

MECHANISM. Following steps are involved :

Step 1. Bromine molecule is electrophilic. The nucleophilic carbon-carbon double bond can react with bromine, displacing the bromide ion. This results in the formation of a more stable resonance stabilized carbonium ion.

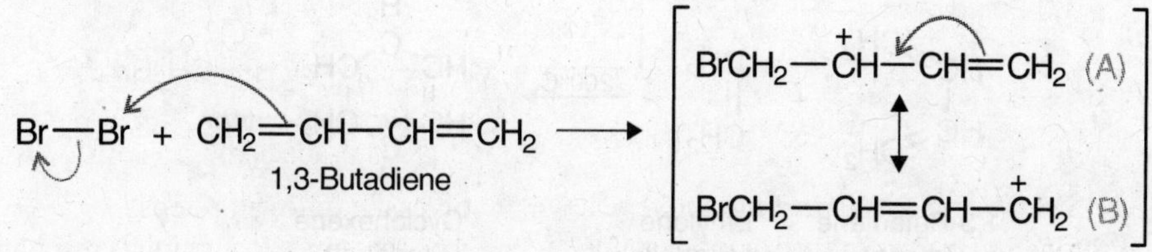

Step 2. The bromide ion combines with (*A*) to form 3,4-dibromo-1-butene (*1,2-addition*). It combines with (*B*) to form 1,4-dibromo-2-butene (*1,4-addition*). Thus a mixture of two compounds is formed.

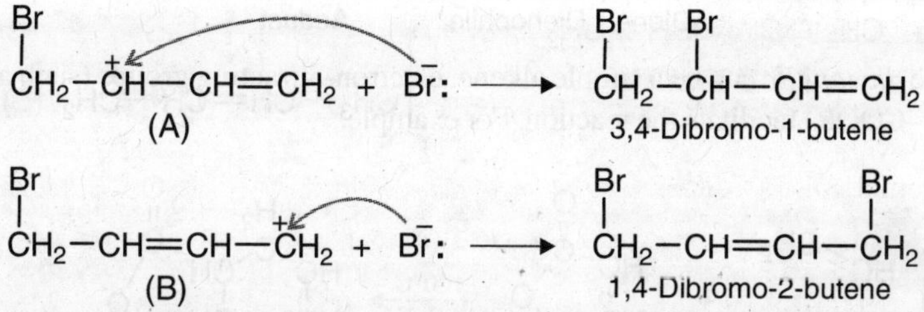

(3) **Addition of Water.** 1,3-Butadiene reacts with water in the presence of H_2SO_4 to give a mixture of 1,2- and 1,4-addition products.

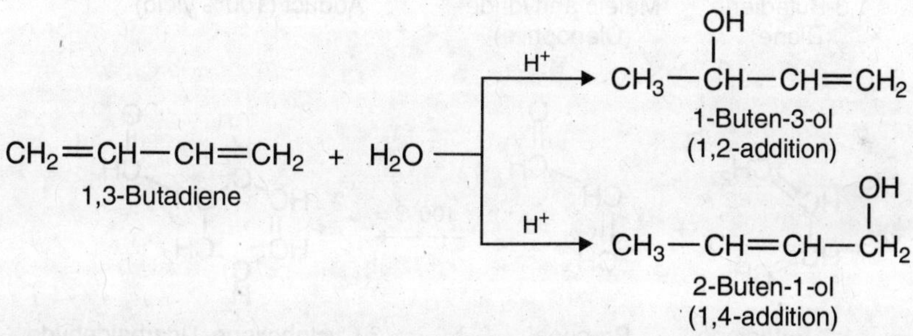

(4) **Addition of Hydrogen**. 1,3-Butadiene reacts with H_2 in the presence of a catalyst to give a mixture of 1-butene and 2-butene.

$$\underset{\text{1,3-Butadiene}}{CH_2{=}CH{-}CH{=}CH_2} \xrightarrow[Pt]{H_2} \underset{\substack{\text{1-Butene}\\ \text{(1,2-addition)}}}{CH_3CH_2CH{=}CH_2} + \underset{\substack{\text{2-Butene}\\ \text{(1,4-addition)}}}{CH_3CH{=}CHCH_3}$$

(5) **Polymerization**. 1,3-Butadiene polymerizes in the presence of peroxides to give polybutadiene (*Buna Rubber*).

$$n\,\underset{\text{1,3-Butadiene}}{CH_2{=}CH{-}CH{=}CH_2} \xrightarrow{R_2O_2} \underset{\substack{\text{Polybutadiene}\\ (\textit{Buna Rubber})}}{\left[\!-CH_2{-}CH{=}CH{-}CH_2-\!\right]_n}$$

The structure of the polymer suggests that 1,4-addition predominates.

(6) **Diels-Alder Reaction**. This involves the treatment of 1,3-butadiene (or any other conjugated diene) with an alkene or an alkyne. No catalyst is required.

The alkene or alkyne used in Diels-Alder reaction is referred to as **dienophile** (diene-lover). The product of Diels-Alder reaction is called the **adduct**. The net result is the formation of two new σ bonds and one new π bond at the expense of three original π bonds. The simplest example of a Diels-Alder reaction is the reaction of 1,3-butadiene with ethylene.

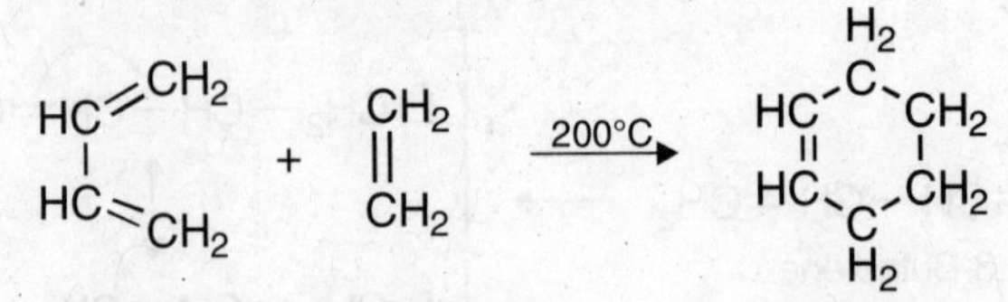

1,3-Butadiene (diene) | Ethylene (dienophile) | Cyclohexene (adduct)

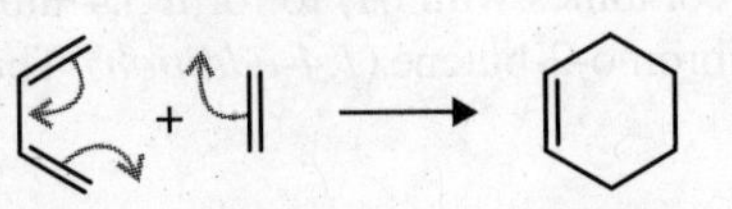

Diene | Dienophile | Adduct

Although the *dienophile* can be a simple alkene, electron-attracting groups (such as –CN, –CHO, –COR, –COOH, –COOR) facilitate the reaction. For example,

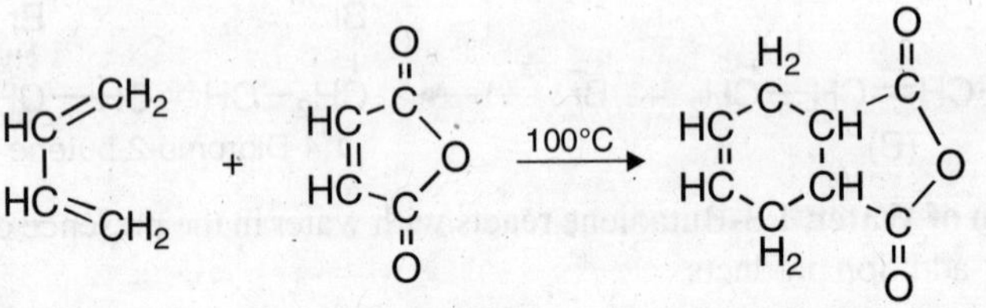

1,3-Butadiene (Diene) | Maleic anhydride (Dienophile) | Adduct (100% yield)

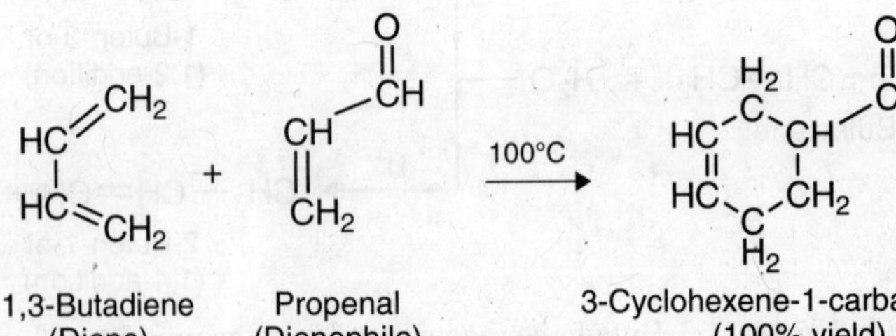

1,3-Butadiene (Diene) | Propenal (Dienophile) | 3-Cyclohexene-1-carbaldehyde (100% yield)

The Diels-Alder reaction is widely used in the synthesis of *six membered* ring compounds.

MECHANISM. The Diels-Alder reactions have a one-step concerted mechanism. A diene reacts with an electron-poor alkene to form a new cyclohexene ring. For example,

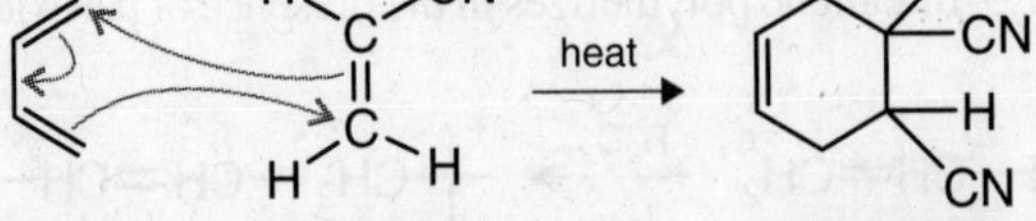

Diene | Dienophile | Cyclohexane ring (Adduct)

The Diels-Alder reactions have a concerted mechanism, with all the bond breaking and bond making occurring in one single step. Three pairs of electrons move simultaneously, requiring a transition state with overlap between the end *p* orbitals of the diene and those of the dienophile.

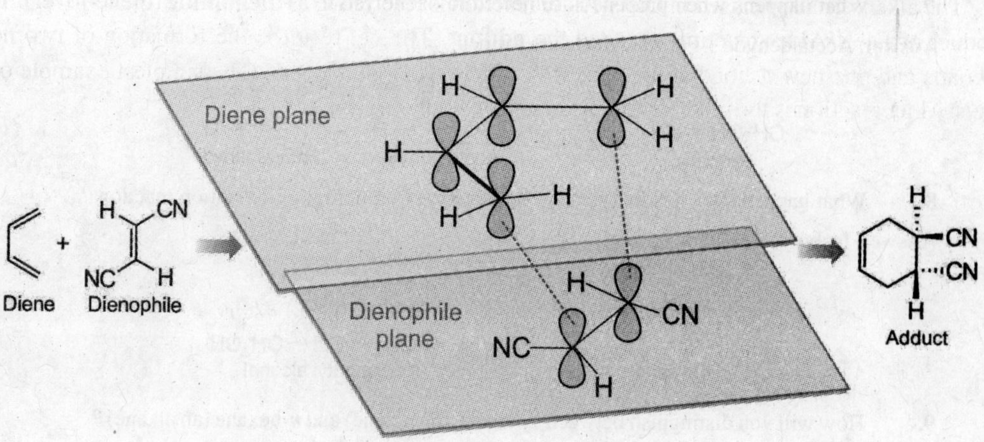

Points to Remember

1. The Diels-Alder reaction involves the reaction of a conjugated diene with an alkene or alkyne.
2. The alkene or alkyne is called *dienophile*. The product of Diels-Alder reaction is called *adduct*.
3. A six-membered ring is always formed in a Diels-Alder reaction.

Examination Questions

Part 1. Alkenes

1. Write the structural formulas and give IUPAC names for all isomeric alkenes of the molecular formula C_5H_{10}.

Clue. There are 5 alkenes of formula C_5H_{10}.

2. How are alkenes prepared ? Describe their important reactions.

3. How is ethylene prepared in the laboratory ? What happens when ethylene is treated with :

(*a*) Br_2/CCl_4 (*b*) Dilute H_2SO_4
(*c*) Concentrated H_2SO_4 (*d*) Dilute cold $KMnO_4$
(*e*) HBr (*f*) O_3 *then* Zn/H_2O

4. Give two methods for the preparation of propene. What happens when propene is treated with :

(*a*) Br_2/CCl_4 (*b*) HBr/peroxide
(*c*) HBr (*d*) HOCl
(*e*) $Cl_2/500°C$ (*f*) Cold dil. $KMnO_4$

5. What happens when propene is treated with hydrogen bromide ? (*Kanpur BSc, 2014*)

Answer. 2-Bromopropane (Isopropyl bromide) is formed.

$$\underset{\text{Propene}}{CH_3-CH=CH_2} + HBr \longrightarrow \underset{\text{2-Bromopropane}}{CH_3-\overset{\overset{Br}{|}}{CH}-CH_3}$$

6. What happens when propene is treated with HBr in the presence of a peroxide ?
(*Bhopal BSc, 2014 ; Shivaji BSc, 2015*)

Answer. 1-Bromopropane (*n*-Propyl bromide) is formed.

$$\underset{\text{Propene}}{CH_3-CH=CH_2} + HBr \xrightarrow{R_2O_2} \underset{\text{1-Bromopropane}}{CH_3CH_2CH_2Br}$$

7. What happens when propene is subjected to ozonolysis ?

Answer. Acetaldehyde + Formaldehyde are formed.

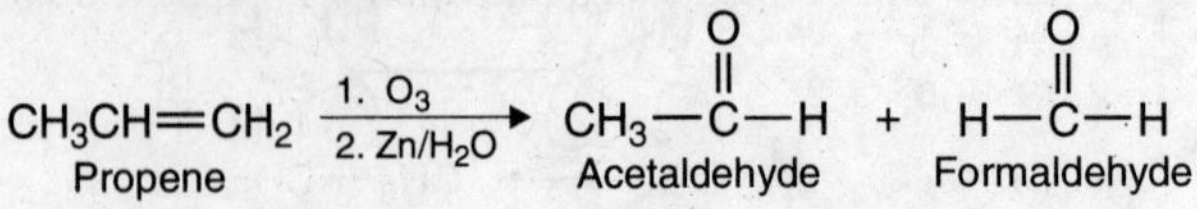

8. What happens when isobutylene is subjected to hydroboration-oxidation reaction ?

Answer. Isobutyl alcohol is formed.

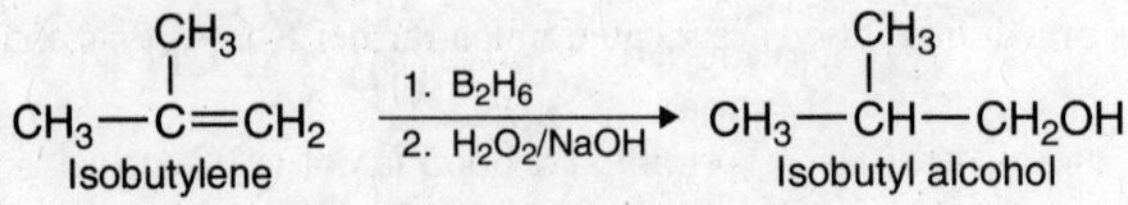

9. How will you distinguish between 1-hexene (an alkene) and *n*-hexane (an alkane) ?

(*Annamali BSc, 2014 ; Jiwaji BSc, 2014 ; Tamil Nadu BSc, 2015*)

Note. In such questions, you are expected to give only visual chemical tests. Indicate clearly what you would do and observe. Always give the reaction equation for the positive test.

Answer. There are two tests :

(*a*) **Reaction with cold dilute $KMnO_4$ solution** (*Baeyer's Test*). 1-Hexene reacts with cold dilute $KMnO_4$ solution ; *n*-hexane does not. The observed change will be disappearance of the purple color of $KMnO_4$, and appearance of brown precipitate of MnO_2.

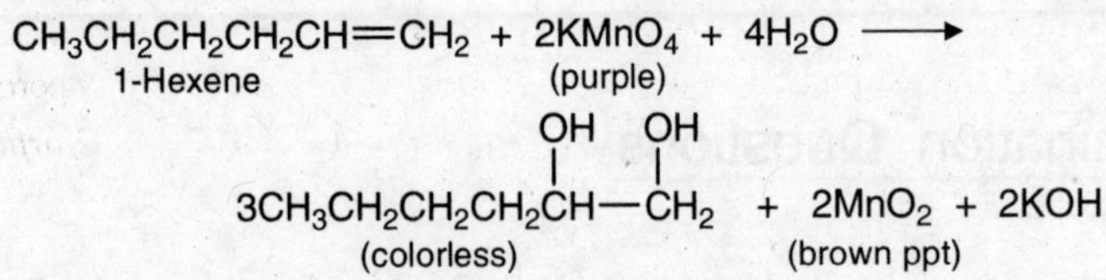

(*b*) **Reaction with Br_2 in CCl_4.** 1-Hexene will decolorize the red color of the solution of Br_2 in CCl_4 immediately.

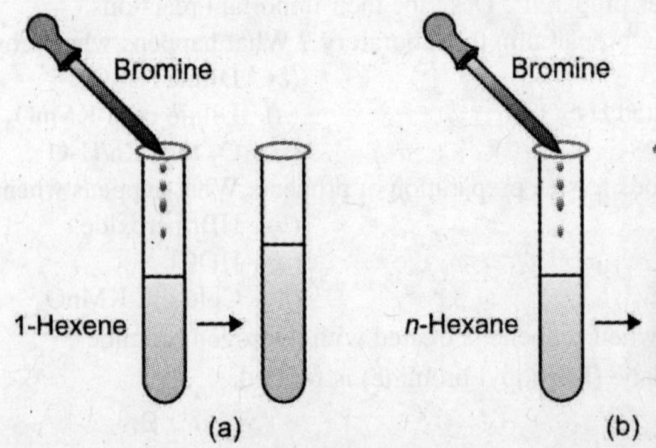

1-Hexene decolorizes red color of bromine in carbon tetrachloride solution. *n*-Hexane does not react with this reagent at room temperature.

n-Hexane will react only when the mixture is subjected to high temperature or UV light.

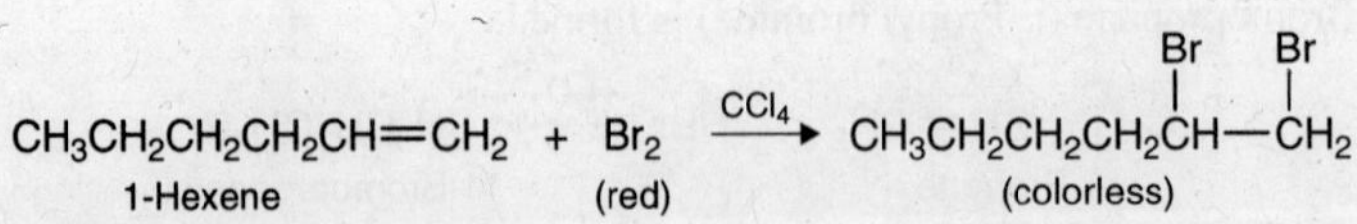

10. How will you distinguish between propene and propane ?

Answer. Both propene and propane are gases. Pass them through dilute cold $KMnO_4$ solution (*purple*) or Br_2 in CCl_4 solution (*red*). Propene will decolorize both the solutions ; propane does not react.

$$\underset{\text{Propene}}{CH_3CH{=}CH_2} + \underset{\text{(red)}}{Br_2} \xrightarrow{CCl_4} \underset{\text{(colorless)}}{CH_3{-}\overset{\overset{\large Br}{|}}{C}H{-}\overset{\overset{\large Br}{|}}{C}H_2}$$

11. Explain why alkenes are more reactive than alkanes.

Answer. Alkenes are more reactive than alkanes. This is because :

(*a*) The π electrons of a double bond are located much farther from the carbon nuclei and are thus less firmly bound to them.

(*b*) The overlap of atomic orbitals in forming a π bond is not as effective as that in σ bonds. Thus a π bond is weaker than a σ bond and more easily broken.

12. Write a note on : Markovnikov rule. (*Panjab BSc, 2015*)

13. Write a note on : Peroxide effect. (*Delhi BSc, 2015*)

14. Write a note on : Ozonolysis. (*Agra BSc, 2014*)

15. How would you establish the position of the double bond in an alkene ? Give an example to prove your method.

16. What is hydroboration ? Give an example.

17. Give the general mechanism of electrophilic addition reactions.

18. Give the mechanism of addition of bromine to ethylene. (*Kerala BSc, 2015*)

19. Give the mechanism of addition of HBr to ethylene. (*Mysore BSc, 2014*)

20. Give the mechanism of addition of HBr to propene. (*Utkal BSc, 2015*)

21. Give the mechanism of addition of HBr to propene in the presence of a peroxide.

(*Burdwan BSc, 2014 ; North Bengal BSc, 2015*)

22. Identify (A), (B), and (C) :

$$\underset{\text{Propane}}{CH_3CH_2CH_3} \xrightarrow[\text{uv}]{Cl_2} (A) \xrightarrow[\text{alcohol}/\Delta]{KOH} (B) \xrightarrow[\text{peroxide}]{HBr} (C)$$

Answer. (*A*) = $CH_3\overset{\overset{\large Cl}{|}}{C}HCH_3 + CH_3CH_2CH_2Cl$; 2- and 1-Chloropropane

(*B*) = $CH_3CH{=}CH_2$; Propene

(*C*) = $CH_3CH_2CH_2Br$; 1-Bromopropane

23. Identify (A), (B), and (C) :

$$\underset{\text{Propane}}{CH_3CH_2CH_3} \xrightarrow[\text{uv}]{Cl_2} (A) \xrightarrow[\text{alcohol}/\Delta]{KOH} (B) \xrightarrow[500°C]{Cl_2} (C)$$

Answer. (*A*) = $CH_3\overset{\overset{\large Cl}{|}}{C}HCH_3 + CH_3CH_2CH_2Cl$; 2- and 1-Chloropropane

(B) = $CH_3CH{=}CH_2$; Propene

(C) = $ClCH_2CH{=}CH_2$; Allyl chloride

24. How will you synthesize ethylene from ethane ?

Answer. Following steps are involved :

$$\underset{\text{Ethane}}{CH_3CH_3} \xrightarrow[\text{uv}]{Cl_2} CH_3CH_2Cl \xrightarrow[\text{alcohol}/\Delta]{KOH} \underset{\text{Ethylene}}{CH_2{=}CH_2}$$

25. How will you synthesize 2,3-dimethylbutane from propene ?

Answer. Use Wurtz reaction :

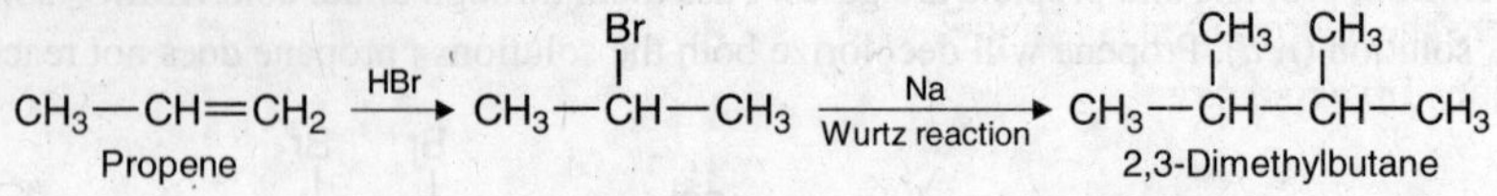

26. How will you synthesize propane from isopropyl alcohol ? *(Sambalpur BSc, 2014)*

Answer. Following steps are involved :

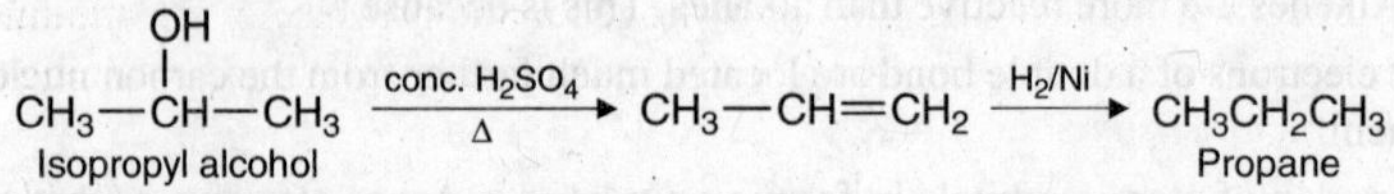

27. How will you synthesize isopropyl alcohol from *n*-propyl alcohol ? *(Cochin BSc, 2014)*

Answer. Following steps are involved :

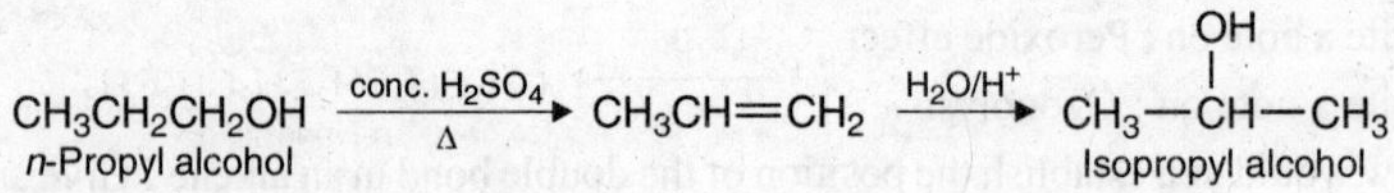

28. How will you synthesize *n*-propyl alcohol from isopropyl alcohol ? *(Dibrugarh BSc, 2015)*

Answer. Following steps are involved :

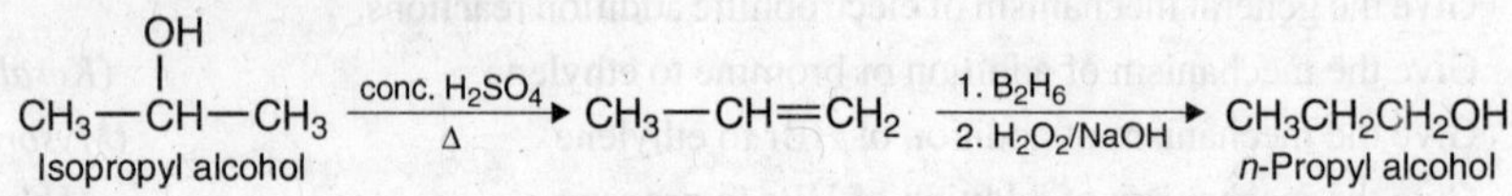

29. How will you synthesize isopropyl alcohol from *n*-propyl bromide ? *(Dibrugarh BSc, 2015)*

Answer. Following steps are involved :

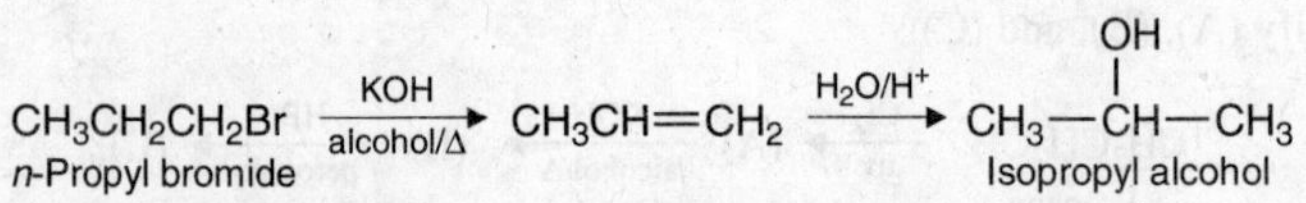

30. How will you synthesize 1-bromopropane from 2-bromopropane ?

Answer. Following steps are involved :

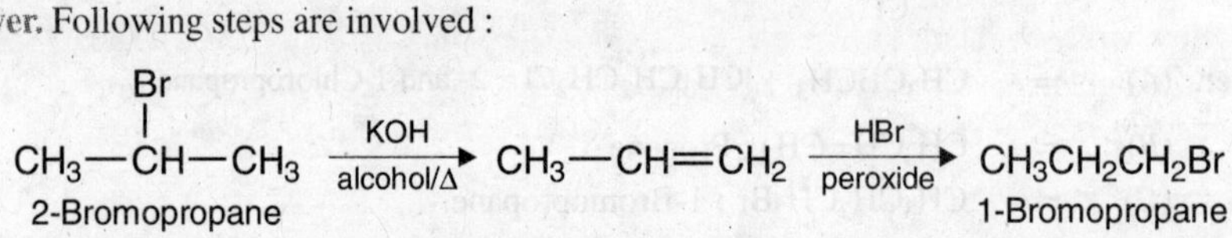

31. How will you synthesize 2-butene from 1-butene ? *(Rajasthan BSc, 2014)*

Answer. Following steps are involved :

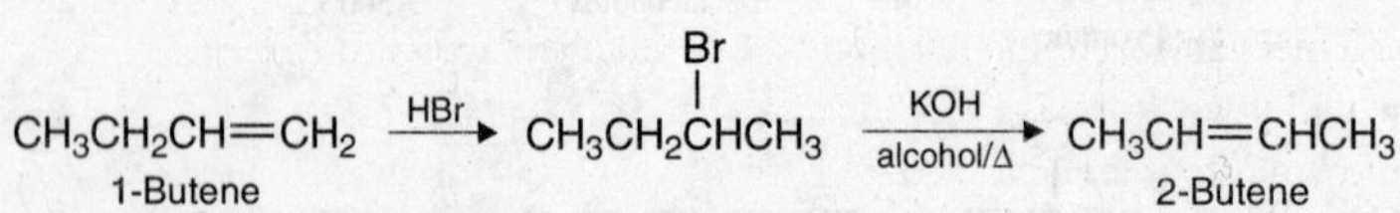

32. How will you synthesize allyl chloride from propane ? *(Andhra BSc, 2015)*

Answer. Following steps are involved :

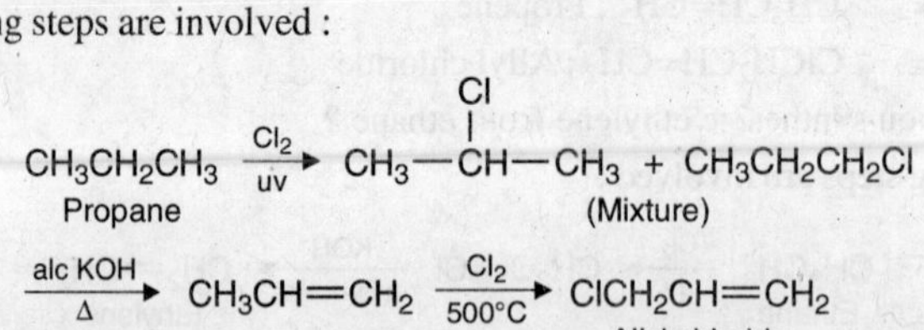

33. A hydrocarbon reacts with conc. H_2SO_4 to form an alkyl hydrogen sulphate which on hydrolysis yields *tert*-Butyl alcohol. What is the structural formula of the hydrocarbon ?

Answer. The hydrocarbon is *Isobutylene*.

Reactions Involved Are :

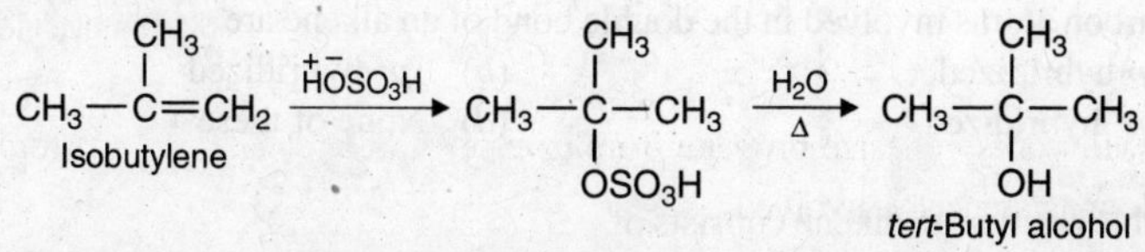

34. A hydrocarbon, A, adds one mole of hydrogen in the presence of a platinum catalyst to form -hexane. When A is oxidized with hot concentrated $KMnO_4$, a single carboxylic acid containing three carbon atoms is isolated. Give the structure and name of A.

Answer. The carbon skeleton must be the same as that of *n*-hexane ; there is one double bond, which must be located in the centre of the chain (three carbons on each side). Therefore, *(A)* is *3-Hexene*.

Reactions Involved Are :

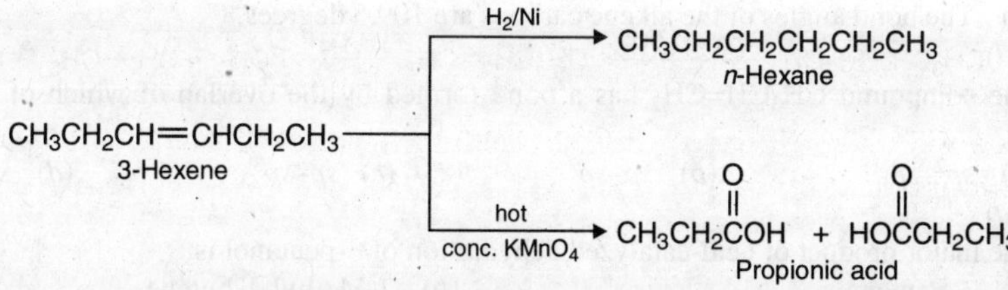

Part 2. Dienes

35. Discuss the molecular orbital structure of 1,3-butadiene. *(Delhi BSc, 2015)*

36. How is 1,3-butadiene prepared ? Describe its important reactions.

37. What happens when 1,3-butadiene is treated with bromine in carbon tetrachloride ?

Answer. A mixture of 3,4-dibromo-1-butene *and* 1,4-dibromo-2-butene is formed. For reaction equation see text.

38. What happens when 1,3-butadiene is treated with HBr ?

Answer. A mixture of 3-bromo-1-butene *and* 1-bromo-2-butene is formed. For reaction equation see text.

39. Write a note on : Stability of conjugated dienes.

40. Write a note on : Diels-Alder reaction. *(Madras BSc, 2015)*

41. How will you distinguish between 1,3-butadiene and *n*-butane. *(Saurashtra BSc, 2014)*

Answer. 1,3-Butadiene immediately decolorizes the red color of Br_2 in CCl_4 solution. *n*-Butane does not react with this reagent.

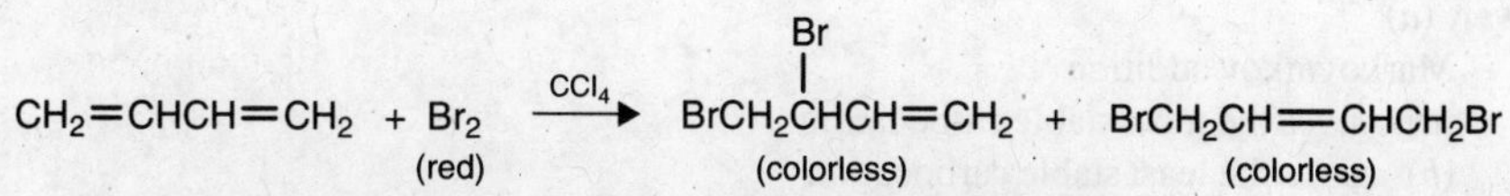

42. Give a mechanism for the reaction of 1,3-butadiene with HBr. *(Baroda BSc, 2014)*

43. Give a mechanism for the reaction of 1,3-butadiene with bromine. *(Cuttack BSc, 2014)*

44. How will you synthesize 1,3-butadiene from 1-butene ? *(Allahabad BSc, 2015)*

Answer. Following steps are involved :

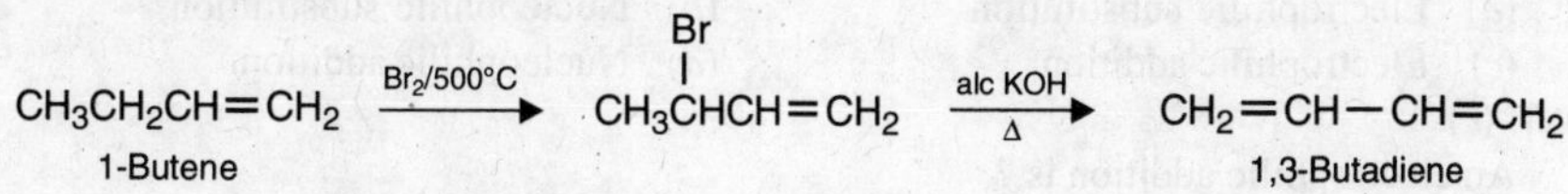

45. How will you synthesize allene from propene ?

Answer. Following steps are involved :

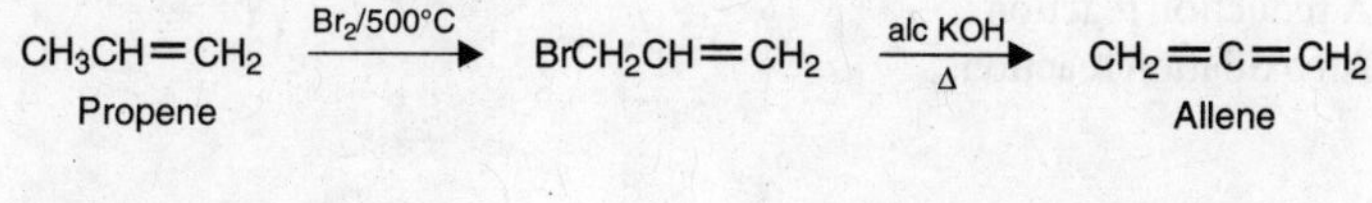

Multiple Choice Questions

1. The carbon atoms involved in the double bond of an alkene are
(*a*) *sp* hybridized (*b*) sp^2 hybridized
(*c*) sp^3 hybridized (*d*) None of these

Answer. (*b*)

2. A double bond in an alkene consists of
(*a*) two *sigma* bonds (*b*) two *pi* bonds
(*c*) one *sigma* bond and one *pi* bond (*d*) one *sigma* bond and two *pi* bonds

Answer. (*c*)

3. Which of the following statements is *not* true ?
(*a*) All atoms attached to the alkene carbons are in the same plane.
(*b*) The carbons of an alkene are sp^2 hybridized.
(*c*) The *p* orbitals of the *pi* bond are perpendicular to the plane of the atoms attached to the alkene.
(*d*) The bond angles of the alkene carbons are 109.5 degrees.

Answer. (*d*)

4. The compound $CH_3CH{=}CH_2$ has a bond formed by the overlap of which of the following hybrid orbitals ?
(*a*) sp^2–sp^3 (*b*) sp–sp^2 (*c*) sp–sp^3 (*d*) sp^3–sp^3

Answer. (*a*)

5. The major product of acid-catalyzed dehydration of 3-pentanol is :
(*a*) 1-Pentene (*b*) 2-Methyl-1-butene
(*c*) 2-Pentene (*d*) 3-Methyl-1-butene

Answer. (*c*)

6. The dehydrohalogenation of 2-bromobutane with alcoholic KOH gives mainly
(*a*) 2-Butene (*b*) 2-Butyne (*c*) 1-Butene (*d*) 1-Butyne

Answer. (*a*)

7. Markovnikov's rule states ?
(*a*) Hydrogen halides add so that the proton adds to the end of the double bond with the most hydrogen atoms on the carbon.
(*b*) Hydrogen halides add so that the proton adds to the end of the double bond with the least hydrogen atoms on the carbon.
(*c*) Hydrogen halides add so that the halide adds to the end of the double bond with the most hydrogen atoms on the carbon.
(*d*) Hydrogen halides additions occur by the halide atom adding first.

Answer. (*a*)

8. Markovnikov addition
(*a*) gives the most stable carbocation
(*b*) gives the least stable carbocation
(*c*) is addition to a carbon atom containing the least hydrogen atoms
(*d*) none of these

Answer. (*a*)

9. The most typical reaction of simple alkenes is
(*a*) Electrophilic substitution (*b*) Nucleophilic substitution
(*c*) Electrophilic addition (*d*) Nucleophilic addition

Answer. (*c*)

10. An electrophilic addition is ?
(*a*) An addition to an alkene initiated by a nucleophilic attack.
(*b*) An addition to an alkene initiated by an electrophilic attack.
(*c*) A reduction reaction.
(*d*) An oxidation reaction.

Answer. (*b*)

11. Why does a H^+ ion attacking a carbon-carbon double bond add to the carbon with the least number of substituents ?

(*a*) the reaction is resonance stabilized.
(*b*) the hybrid geometry favors this process.
(*c*) nucleophile tends to attack stable centers of negative charge.
(*d*) a more stable carbonium ion is generated.

Answer. (*d*)

12. Propene reacts with bromine to form 1,2-dibromopropane. This is an example of

(*a*) Nucleophilic addition (*b*) Electrophilic addition
(*c*) Nucleophilic substitution (*d*) Electrophilic substitution

Answer. (*b*)

13. Baeyer's reagent is

(*a*) dilute $KMnO_4$ (*b*) $HCl + ZnCl_2$ (*c*) Br_2 in CCl_4 (*d*) NH_2NH_2

Answer. (*a*)

14. In the addition of HX to a double bond, the hydrogen goes to the carbon that already has more hydrogens is a statement of

(*a*) Markovnikov's rule (*b*) Hund's rule (*c*) Huckel rule (*d*) Saytzeff rule

Answer. (*a*)

15. Markovnikov's addition of HBr is not applicable to

(*a*) Propene (*b*) 1-butene (*c*) 1-pentene (*d*) 2-butene

Answer. (*d*)

16. In the reaction of $CH_3CH_2CH{=}CH_2$ with HCl, the H of the HCl will become attached to which carbon ?

(*a*) C-1 (*b*) C-2 (*c*) C-3 (*d*) C-4

Answer. (*a*)

17. The disappearance of the purple color of $KMnO_4$ in its reaction with alkene is known as

(*a*) Markovnikov test (*b*) Grignard test (*c*) Baeyer test (*d*) Wurtz test

Answer. (*c*)

18. Which of the following will give a negative test when treated with bromine in carbon tetrachloride?

(*a*) Butane (*b*) 2-Butene (*c*) 1,3-Butadiene (*d*) 2-Butyne

Answer. (*a*)

19. Ethylene reacts with HI to give

(*a*) Iodoethane (*b*) 2,2-Diiodoethane
(*c*) 1,1-Diiodoethane (*d*) None of these

Answer. (*a*)

20. Which of the following reagents will react with propene ?

(*a*) Hot alkaline $KMnO_4$ (*b*) Sodium metal
(*c*) Cold dilute HNO_3 (*d*) $LiAlH_4$

Answer. (*a*)

21. Which of the following compounds will have zero dipole moment ?

(*a*) *cis*-1,2-dibromoethylene (*b*) 1,1-dibromoethylene
(*c*) *trans*-1,2-dibromoethylene (*d*) all of these

Answer. (*c*)

22. 2-Methylpropene reacts with HBr to give

(*a*) *tert*-Butyl bromide (*b*) Isobutane
(*c*) *n*-Butyl bromide (*d*) None of these

Answer. (*a*)

23. 2-Butene reacts with HBr to give

(*a*) 1-Bromobutane (*b*) 2,3-Dibromobutane
(*c*) 2-Bromobutane (*d*) 2,2-Dibromobutane

Answer. (*c*)

24. Which of the following alkenes reacts with HBr in the presence of a peroxide to give *anti*-Markovnikov's product ?

(*a*) 1-Butene (*b*) 2,3-Dimethyl-2-butene
(*c*) 2-Butene (*d*) 3-Hexene

Answer. (*a*)

25. Propene reacts with HBr in the presence of a peroxide to give

(*a*) *n*-Propyl bromide (*b*) Allyl bromide
(*c*) Isopropyl bromide (*d*) Vinyl bromide

Answer. (*a*)

26. 2-Methylpropene reacts with HBr in the presence of peroxide to give

(*a*) A primary alkyl bromide (*b*) A secondary alkyl bromide
(*c*) A tertiary alkyl bromide (*d*) A vicinal dibromide

Answer. (*a*)

27. Propene is treated with dilute potassium permanganate. What is the major product ?

(*a*) 1,2-Ethanediol (*b*) 1,2-Propanediol
(*c*) Propanal (*d*) Ethanal

Answer. (*b*)

28. Propene reacts with Cl_2 in H_2O to give

(*a*) 1-Chloro-2-propanol (*b*) 2-Chloro-2-propanol
(*c*) 1-Chloro-1-propanol (*d*) 2-Chloro-1-propanol

Answer. (*a*)

29. 1-Butene reacts with Cl_2 in H_2O to give

(*a*) 1-Chloro-2-butanol (*b*) 2-Chloro-2-butanol
(*c*) 1-Chloro-1-butanol (*d*) 2-Chloro-1-butanol

Answer. (*a*)

30. Catalytic hydrogenation of 3-methyl-1-butene gives :

(*a*) Isobutane (*b*) 2,2-Dimethylbutane
(*c*) 2-Methylbutane (*d*) 2,3-Dimethylbutane

Answer. (*c*)

31. Which of the following alkenes gives only acetic acid on oxidation with hot concentrated $KMnO_4$.

(*a*) Ethylene (*b*) 1-Butene (*c*) Propene (*d*) 2-Butene

Answer. (*d*)

32. Which of the following compounds will give only acetaldehyde on ozonolysis ?

(*a*) 1-Butene (*b*) Acetylene (*c*) 2-Butene (*d*) Ethylene

Answer. (*c*)

33. Which of the following alkenes will give a mixture of acetone and formaldehyde on ozonolysis ?

(*a*) 2-butene (*b*) 2-methyl-2-butene (*c*) 1-butene (*d*) 2-methylpropene

Answer. (*d*)

34. Which of the following alkenes will give a mixture of acetone and acetaldehyde an ozonolysis ?

(*a*) 1-butene (*b*) 2-methyl-2-butene
(*c*) 2-butene (*d*) 2-methylpropene

Answer. (*b*)

35. A hydrocarbon, C_6H_{12}, on ozonolysis gives only one product which does not give silver mirror with Tollens' reagent. The hydrocarbon is

(*a*) 2,3-Dimethyl-2-butene (*b*) 2-Hexene
(*c*) 2-Methyl-2-pentene (*d*) 3-Hexene

Answer. (*a*)

36. Combustion of an alkene with sufficient oxygen will produce

(*a*) carbon dioxide and water (*b*) carbon monoxide and water
(*c*) only carbon dioxide (*d*) only carbon monoxide

Answer. (*a*)

37. Conjugated dienes contain

(*a*) only two double bonds
(*b*) two double bonds connected to the same carbon atom

(c) alternating single and double bonds
(d) two double bonds separated by two single bonds

Answer. (c)

38. 1,2-Butadiene has
(a) only *sp* hybridized carbon atoms
(b) only sp^2 hybridized carbon atoms
(c) only sp^3 hybridized carbon atoms
(d) sp, sp^2, sp^3 hybridized carbon atoms

Answer. (d)

39. How many σ (sigma) bonds are there in CH_2=CH—CH=CH_2 ?
(a) 3 (b) 6
(c) 9 (d) 12

Answer. (c)

40. Propadiene, CH_2=C=CH_2, is
(a) a planar compound (b) a cumulated diene
(c) an isolated diene (d) a conjugated diene

Answer. (b)

41. Which of the following compounds have planar molecules ?
(a) 1,3-Butadiene (b) Dimethyl ether
(c) 1-Butene (d) Allene

Answer. (a)

42. Which of the following has a bond formed by the overlap of an sp^2 hybrid orbital with an *sp* hybrid orbital ?
(a) CH_2=CH_2 (b) CH_2=$CHCH_3$
(c) CH_2=C=CH_2 (d) $CH_3C{\equiv}CCH_2CH_3$

Answer. (c)

43. Conjugated alkenes are characterized by ?
(a) alternating π and σ bonds.
(b) adjacent π bonds.
(c) π bonds separated by two or more σ bonds.
(d) a cyclic system containing 6π electrons.

Answer. (a)

44. What kind of reactions do dienes most commonly undergo ?
(a) nucleophilic addition reactions
(b) electrophilic addition reactions
(c) nucleophilic substitution reactions
(d) electrophilic substitution reactions

Answer. (b)

45. 1,4 Additions are ?
(a) Additions of 1 or 4 atoms to the diene.
(b) The addition of a reagent to atoms 1 and 4 of the alkene.
(c) The addition of reagents to atoms 1 and 4 of a cumulated diene.
(d) The addition of a reagent to atoms 1 and 4 of the conjugated diene.

Answer. (d)

46. 1,3-Butadiene reacts with bromine to mainly give
(a) 3,4-Dibromo-1-butene (b) 4-Bromo-1-butene
(c) 1,4-Dibromo-2-butene (d) 1-Bromo-2-butene

Answer. (c)

47. A Diels-Alder reaction is a method of making
(a) cyclohexenes (b) cyclobutanes
(c) cyclohexanes (d) hexenes

Answer. (a)

12 Alkynes

The combustion of acetylene provides tremendous amounts of heat, sufficient to cut metal using an oxyacetylene torch.

So far we have considered only two types of hydrocarbons : (1) the Alkanes which are fully saturated ; and (2) the Alkenes which contain two H-atoms less than the corresponding alkanes and are characterized by the presence of a double bond in the molecule. There is yet another class of hydrocarbons known as **Alkynes**, which are more highly unsaturated than even the alkenes. Alkynes contain four H–atoms less than the corresponding alkanes and are characterized by the presence of a triple bond in the molecule. The first and the most important member of this series of hydrocarbons is acetylene, HC≡CH, and hence these are also called the **Acetylenes**, and the triple bond is often referred to as the **Acetylenic Linkage**.

STRUCTURE

Let us consider acetylene (H–C≡C–H) for illustrating the orbital make up of alkynes. Each carbon atom in acetylene is attached to one hydrogen atom by a single covalent bond and to another carbon atom by a triple bond. Since each carbon is attached to two other atoms it uses *sp* hybrid orbitals plus two unhybridized 2*p* orbitals (p_y and p_z) to form its bonds.

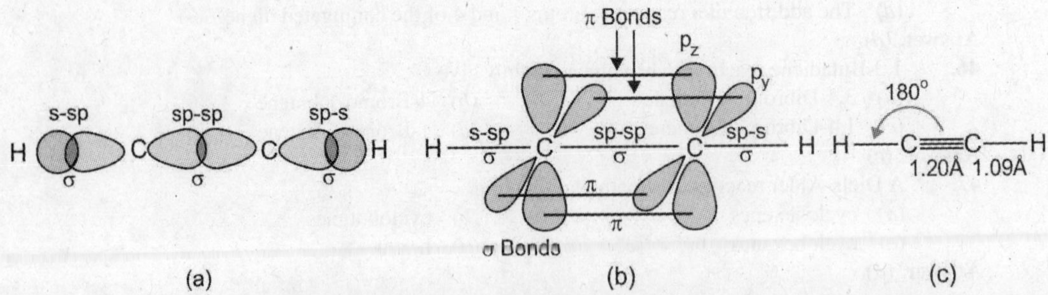

Fig. 12.1. Orbital structure of acetylene. (a) Formation of σ bonds; (b) Formation of π bonds ; (c) Geometry of the molecule.

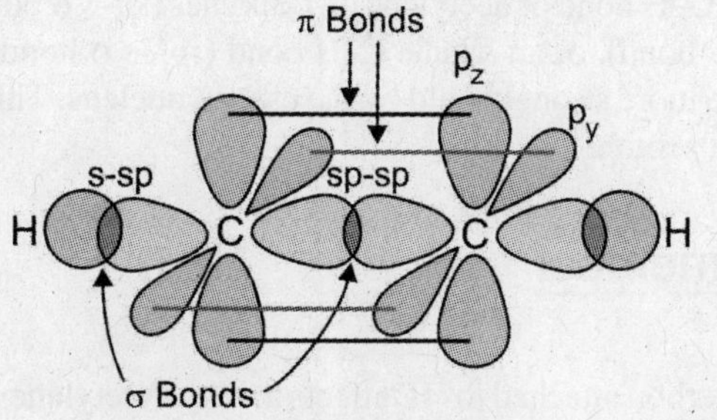

Fig. 12.2. Bonding in acetylene.

In acetylene there are two C–H single covalent bonds and one C–C triple bond. Each C–H bond is a sigma (σ) bond and results from the overlap of an *sp* orbital from carbon and 1*s* orbital from hydrogen. One of the three bonds in the C–C triple bond is also a σ bond and results from end-to-end (linear) overlap of the two *sp* orbitals, one from each carbon. The other two bonds in the triple bond are pi (π) bonds and result from side-to-side (lateral) overlap of the unhybridized *p* orbitals on each carbon. **Although the C–C triple bond is represented by three equivalent lines, remember that one line represents a σ bond and the other two the π bonds.**

Electron diffraction studies have also shown that acetylene is a linear molecule. The H–C–C bond angle is 180°. The C–H bond length is 1.09 Å. The C–C bond length is 1.20 Å.

Remember. (1) The carbon-carbon triple bond in alkynes is made up of one σ bond and two π bonds.

(2) Like alkenes, alkynes undergo addition reactions. These reactions are due to the availability of more exposed π electrons.

ACIDITY OF ALKYNES

Acetylene and 1-alkynes are acidic in nature because they readily donate protons to strong bases.

$$R-C\equiv C{:}H \ + \ \text{Base} \longrightarrow R-C\equiv \bar{C}{:} + H-\text{Base}$$

Moreover, acetylene is a stronger acid than ethylene, which in turn is a stronger acid than ethane. The acidity of acetylene or 1-alkynes can be explained by the atomic orbital description of the C—H bonds. We know that *s* electrons are closer to the nucleus than are *p* electrons. Therefore, **the more *s* character there is in a hybrid atomic orbital, the closer the electrons of the bond are to the nucleus.** The amount of *s* character in various hybrid orbitals is as follows :

Hybrid Orbital	% s Character
sp	50
sp^2	33.3
sp^3	25

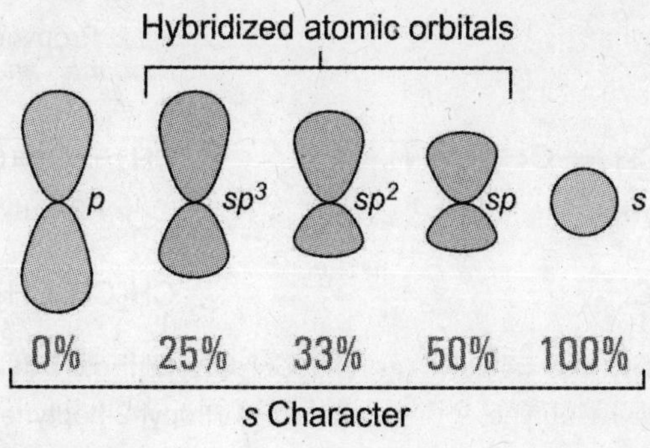

The *s* character of the C–H bond of acetylene or 1-alkynes (*sp*–*s* σ bond) is greater than that of an alkene C–H bond (sp^2–*s* σ bond), or an alkane C–H bond (sp^3–*s* σ bond). Thus the electrons of the C–H bond in 1-alkynes are more strongly held by the carbon nucleus. This makes it easier to remove the hydrogen (≡C–H) as a proton.

Points to Remember

1. Hybridization of the carbon attached to 'H' affects acidity. Acetylene is more acidic than ethylene or ethane.

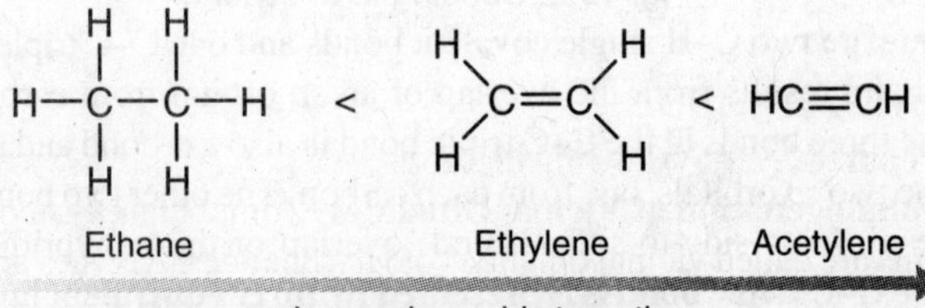

Ethane   Ethylene   Acetylene

Increasing acid strength

2. Acetylene and terminal alkynes (1-alkynes) are acidic. They react with strong bases like $NaNH_2$, ammoniacal silver nitrate (Tollens' reagent), and ammoniacal cuprous chloride to form alkynides.

NOMENCLATURE

There are three systems of naming alkynes :

(1) **Common System.** Acetylene is the common name for the first member of the series.

$$H{-}C{\equiv}C{-}H \qquad \text{Acetylene}$$

(2) **Derived System**. In this system, the higher alkynes are regarded as alkyl derivatives of acetylene.

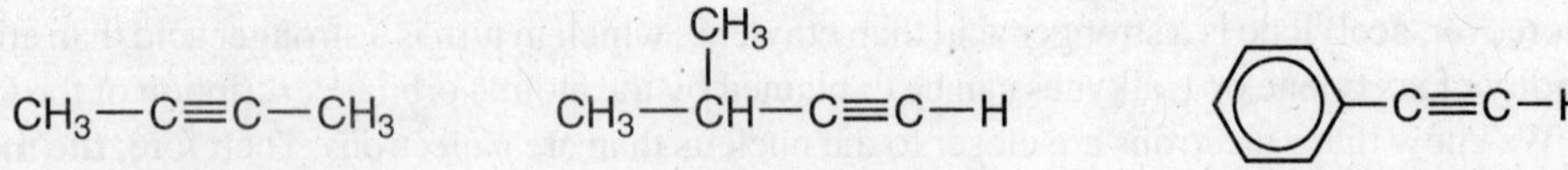

Dimethylacetylene   Isopropylacetylene   Phenylacetylene

(3) **IUPAC System**. The IUPAC names of alkynes are obtained by dropping the ending *-ane* of the parent alkane and adding the suffix *-yne*. If necessary the carbon chain including the triple bond is numbered from the end nearest this bond. The position of the triple bond is indicated by prefixing the number of the carbon preceding it to the name of the alkyne.

$$H{-}C{\equiv}C{-}H$$

Ethyne
(*Ethane –ane + yne*)

$$CH_3{-}C{\equiv}C{-}H$$

Propyne
(*Propane –ane + yne*)

$$\overset{4}{CH_3}{-}\overset{3}{CH_2}{-}\overset{2}{C}{\equiv}\overset{1}{C}{-}H$$

1-Butyne

$$\overset{1}{CH_3}{-}\overset{2}{C}{\equiv}\overset{3}{C}{-}\overset{4}{CH_3}$$

2-Butyne

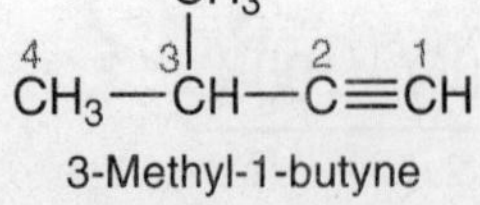

3-Methyl-1-butyne

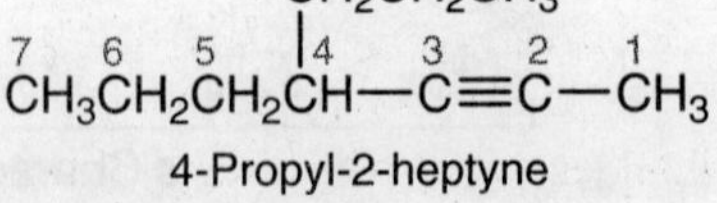

4-Propyl-2-heptyne

When both a double and triple bond are present, the hydrocarbon is named an **alkenyne** with numbers as low as possible given to the multiple bonds. In case of a choice, the *double bond gets the lower number*.

$CH_3CH{=}CHC{\equiv}CH$ — 3-Penten-1-yne (*not* 2-penten-4-yne)

$HC{\equiv}CCH_2CH{=}CH_2$ — 1-Penten-4-yne (*not* 4-penten-1-yne)

The monovalent groups derived from alkynes are named by deleting the final '*e*', of the parent hydrocarbon and adding the suffix *-yl*.

$H{-}C{\equiv}C{-}$ Ethynyl (*Ethyne −e + yl*)

$CH_3{-}C{\equiv}C{-}$ Propynyl

METHODS OF PREPARATION

Alkynes are prepared by the following methods :

(1) **Dehydrohalogenation of Vicinal Dihalides**. Compounds that contain halogen atoms on adjacent carbon atoms are called vicinal dihalides or *vic*-dihalides. Alkynes are obtained by treatment of vicinal dihalides with alcoholic KOH followed by sodium amide ($NaNH_2$). This is a useful method since the vicinal dihalides are readily prepared from alkenes by the addition of halogens.

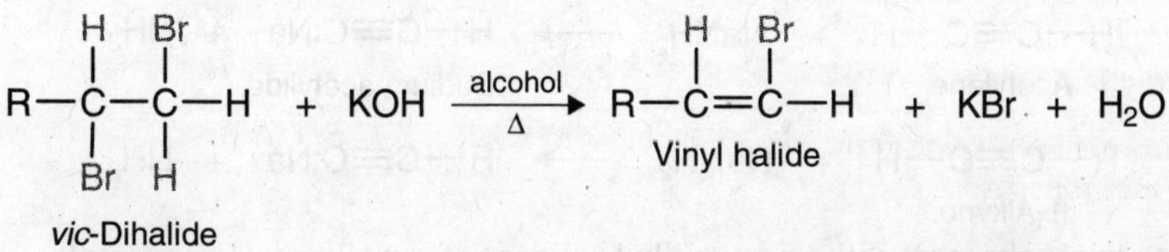

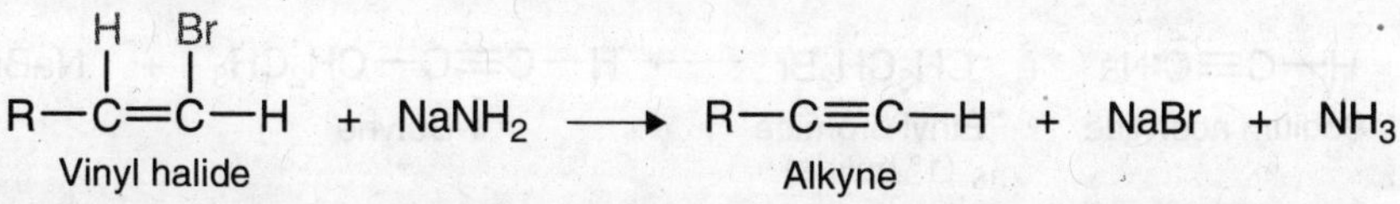

Notice that the product of the first HBr elimination is a substituted vinyl bromide. The vinyl halides are unreactive and a stronger base ($NaNH_2$) is used to remove the second HBr molecule. Acetylene and propyne can be prepared by this method as follows :

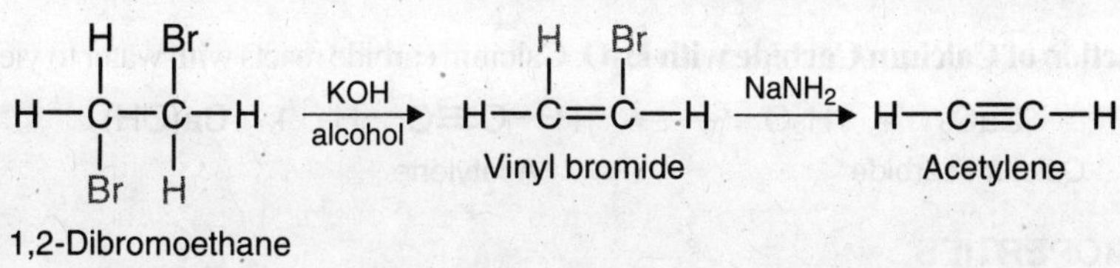

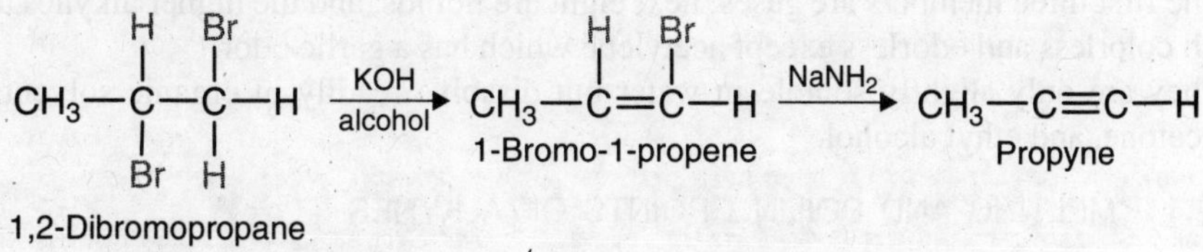

The *vic*-dihalides may be directly treated with $NaNH_2$ to give alkynes.

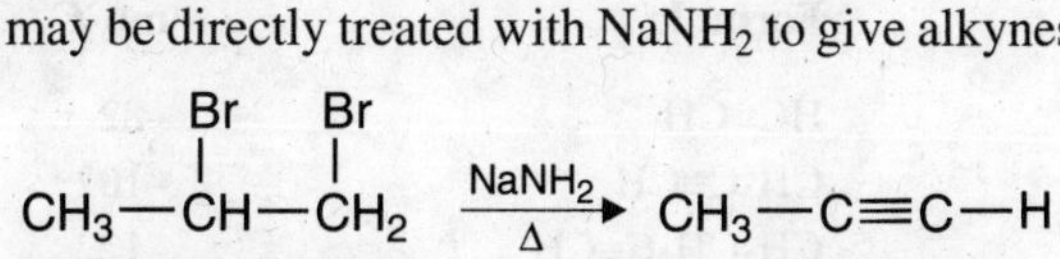

The *gem*-dihalides can also be directly treated with $NaNH_2$ to give alkynes.

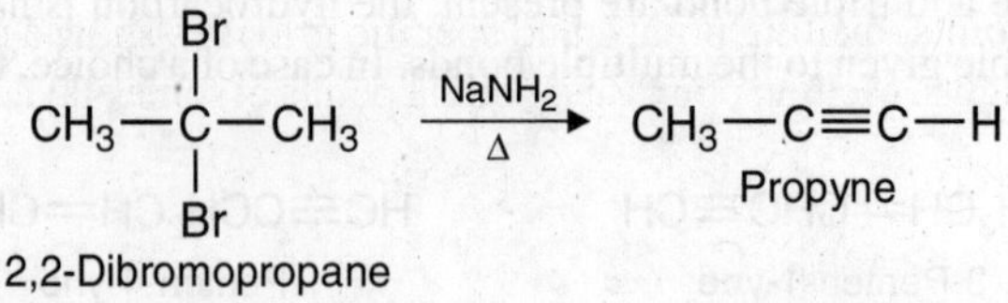

(2) **Dehalogenation of Tetrahalides**. When 1,1,2,2-tetrahalides are heated with zinc dust in alcohol, alkynes are formed.

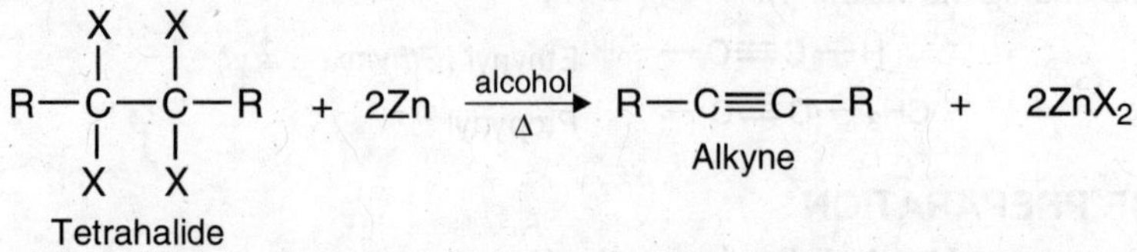

This reaction is not of great utility for preparing alkynes, because the tetrahalides are themselves generally made by addition of halogens to alkynes.

(3) **Reaction of Sodium Acetylides with 1° Alkyl Halides**. Sodium acetylides are prepared by treatment of acetylene or 1-alkynes with $NaNH_2$ in liquid ammonia.

$$\underset{\text{Acetylene}}{H-C\equiv C-H} + NaNH_2 \longrightarrow \underset{\text{Sodium acetylide}}{H-C\equiv \bar{C}{:}\overset{+}{Na}} + NH_3$$

$$\underset{\text{1-Alkyne}}{R-C\equiv C-H} + NaNH_2 \longrightarrow R-C\equiv \bar{C}{:}\overset{+}{Na} + NH_3$$

These sodium salts react with primary alkyl halides to form alkynes. For example,

$$\underset{\text{Sodium acetylide}}{H-C\equiv \bar{C}{:}\overset{+}{Na}} + \underset{\substack{\text{Ethyl bromide}\\ \text{(1° halide)}}}{CH_3CH_2Br} \longrightarrow \underset{\text{1-Butyne}}{H-C\equiv C-CH_2CH_3} + NaBr$$

$$CH_3CH_2C\equiv \bar{C}{:}\overset{+}{Na} + CH_3CH_2Br \longrightarrow \underset{\text{3-Hexyne}}{CH_3CH_2C\equiv CCH_2CH_3} + NaBr$$

The main advantage of this method is that it can be used to convert lower alkynes into higher alkynes.

(4) **Reaction of Calcium Carbide with H_2O**. Calcium carbide reacts with water to yield acetylene.

$$\underset{\text{Calcium carbide}}{CaC_2} + H_2O \longrightarrow \underset{\text{Acetylene}}{H-C\equiv C-H} + Ca(OH)_2$$

PHYSICAL PROPERTIES

(1) The first three members are gases, next eight are liquids, and the higher alkynes are solids. They are all colorless and odorless except acetylene which has a garlic odor.

(2) They are only slightly soluble in water but dissolve readily in organic solvents such as benzene, acetone, and ethyl alcohol.

Table 12.1 MELTING AND BOILING POINTS OF ALKYNES

Name	Formula	mp°C	bp°C
Acetylene	$HC\equiv CH$	–82	–75
Propyne	$CH_3C\equiv CH$	–102	–23
1-Butyne	$CH_3CH_2C\equiv CH$	–122	9
1-Pentyne	$CH_3CH_2CH_2C\equiv CH$	–98	40

(3) Their boiling points, melting points, and specific gravities show a regular increase with the increase in molecular weights. Melting points and boiling points of some alkynes are given in Table 12.1.

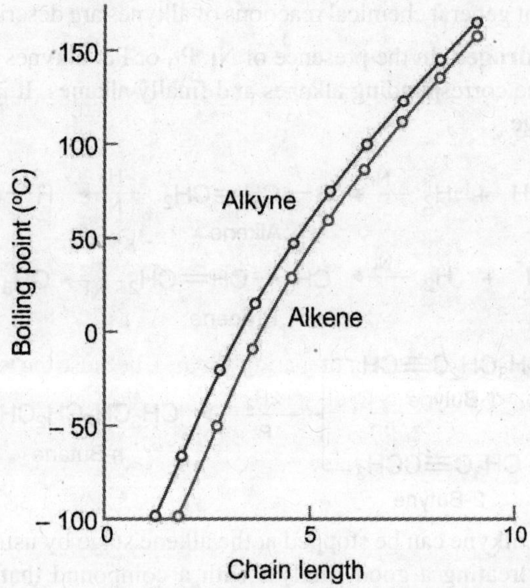

Fig. 12.3. The boiling points of alkynes are generally a little higher than those of the alkenes.

(4) Alkynes generally have slightly higher boiling points than the corresponding alkenes (Fig. 12.3). This observation is in accord with our discussions about London forces. Because the alkyne is linear (180°) and the alkene has a bent shape (120°), the alkynes are easier to stack next to one another. This is similar to the discussion in chapter 9 on the lower boiling points of the branched hydrocarbons.

(5) ***IR Spectrum.*** The IR spectra of alkynes show C–H stretching absorptions at ~3300 cm^{-1}. Carbon-carbon triple bond gives stretching absorption peaks at 2100–2260 cm^{-1}. These absorptions are weak and will be absent if the triple bond is symmetrically substituted.

CHEMICAL PROPERTIES

Alkynes give the same kind of addition reactions as do alkenes. However, with alkynes the addition may take place in *one* step or *two* steps, depending upon conditions.

(*i*) Addition of one molecule of the reagent.

$$-C\equiv C- \;+\; X-Y \longrightarrow -\underset{X}{\underset{|}{C}}=\underset{Y}{\underset{|}{C}}-$$

(*ii*) Addition of second molecule of the reagent.

$$-\underset{X}{\underset{|}{C}}=\underset{Y}{\underset{|}{C}}- \;+\; X-Y \longrightarrow -\overset{X}{\overset{|}{\underset{X}{\underset{|}{C}}}}-\overset{Y}{\overset{|}{\underset{Y}{\underset{|}{C}}}}- \quad \text{or} \quad -\overset{Y}{\overset{|}{\underset{X}{\underset{|}{C}}}}-\overset{X}{\overset{|}{\underset{Y}{\underset{|}{C}}}}-$$

Addition reactions occur with carbon-carbon triple bonds for the same reason that they occur with the carbon-carbon double bond, namely, the availability of the loosely held π electrons.

Other reactions of alkynes result from acidic hydrogen atoms in acetylene and 1-alkynes. In the presence of strong bases the following reaction takes place.

$$—C{\equiv}C—H \;+\; :\underset{\text{base}}{B} \longrightarrow —C{\equiv}C: \;+\; H—B$$

Some of the important general chemical reactions of alkynes are described below :

(1) **Addition of Hydrogen**. In the presence of Ni, Pt, or Pd, alkynes add up two molecules of hydrogen first forming the corresponding alkenes and finally alkanes. It is not possible to stop the reaction at the alkene stage.

$$\underset{\text{Alkyne}}{R—C{\equiv}CH} + H_2 \xrightarrow{\text{Ni}} \underset{\text{Alkene}}{R—CH{=}CH_2} \xrightarrow[\text{Ni}]{H_2} \underset{\text{Alkane}}{R—CH_2—CH_3}$$

$$\underset{\text{Propyne}}{CH_3—C{\equiv}CH} + H_2 \xrightarrow{\text{Ni}} \underset{\text{Propene}}{CH_3—CH{=}CH_2} \xrightarrow[\text{Ni}]{H_2} \underset{\text{Propane}}{CH_3—CH_2—CH_3}$$

$$\left.\begin{array}{c} \underset{\text{1-Butyne}}{CH_3CH_2C{\equiv}CH} \\ \\ \underset{\text{2-Butyne}}{CH_3C{\equiv}CCH_3} \end{array}\right] \xrightarrow[\text{Pt}]{2H_2} \underset{n\text{-Butane}}{CH_3CH_2CH_2CH_3}$$

Hydrogenation of an alkyne can be stopped at the alkene stage by using a 'poisoned' (partially deactivated) catalyst by treating a good catalyst with a compound that makes the catalyst less effective. **Lindlar's catalyst** is a poisoned palladium catalyst, composed of powdered barium sulfate mixed with palladium, poisoned with quinoline. Nickel boride (Ni_2B) is a newer alternative to Lindlar's catalyst that is more easily made and often give better yields.

In an internal alkyne, *syn* addition gives *cis* product. For example, when 2-hexyne is hydrogenated using Lindlar catalyst, the product is *cis*-hexene.

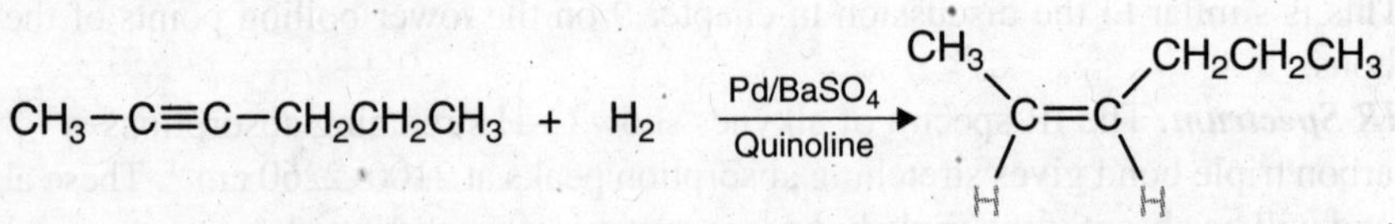

2-Hexyne *cis*-2-Hexene

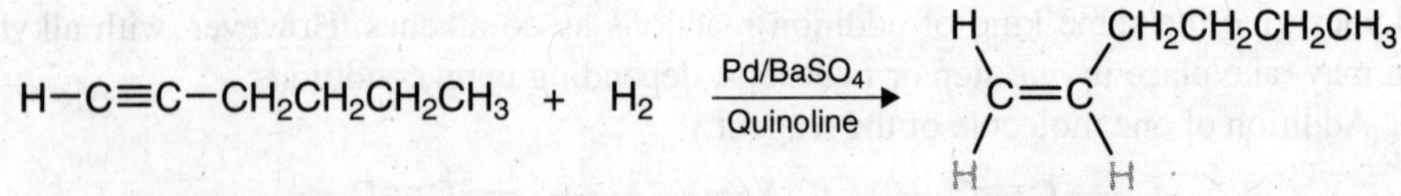

1-Hexyne *cis*-2-Hexene

Quinoline is a basic heterocyclic compound.

(2) **Metal-Ammonia Reduction to *trans* Alkenes.** To form a *trans* alkene, two hydrogens must be added to the alkyne with *anti* stereochemistry. Sodium metal in liquid ammonia reduces alkynes with *anti* stereochemistry, so this reduction is used to convert alkynes to *trans* alkenes.

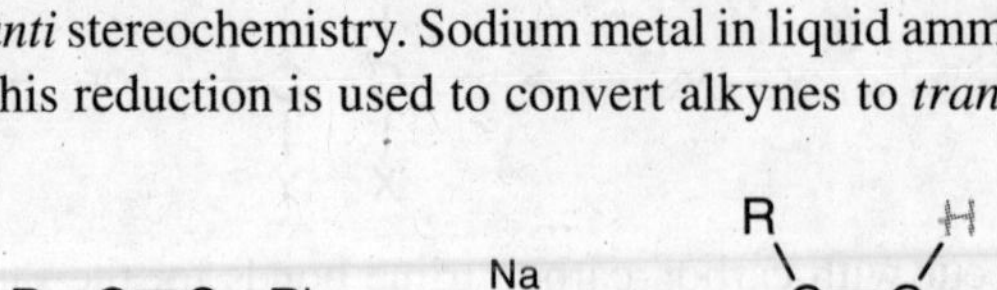

Alkyne *trans*-Alkene

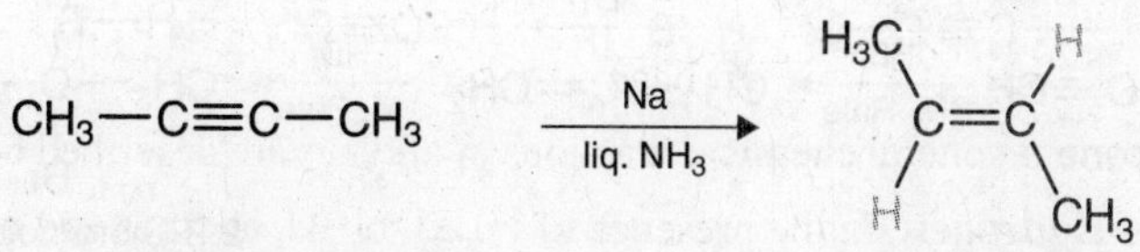

2-Butyne trans-2-Butene

(3) **Addition of Halogens.** Halogens add to alkynes in two steps forming a dihalide and then a tetrahalide.

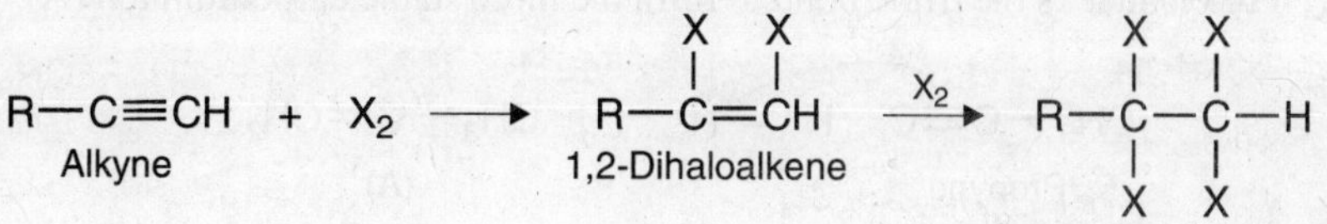

Alkyne 1,2-Dihaloalkene 1,1,2,2-Tetrahaloalkane

$$CH_3-C\equiv CH + Br_2 \longrightarrow CH_3-CBr=CHBr \xrightarrow{Br_2} CH_3-CBr_2-CHBr_2$$

Propyne 1,2-Dibromopropene 1,1,2,2-Tetrabromopropane

(4) **Addition of Halogen Acids.** Halogen acids also add to symmetrical alkynes in two stages. After one molecule of the acid has been added to a symmetrical alkyne, the product is an unsymmetrical derivative of alkene so that the addition of a second molecule of the acid takes place in accordance with the **Markovnikov rule**. Thus both the halogens become attached to the same carbon atom.

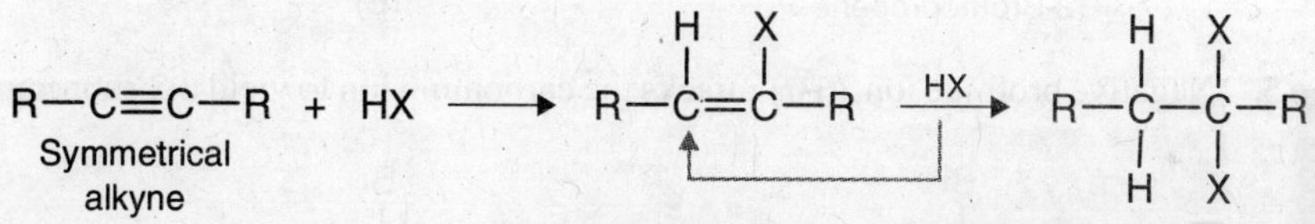

Symmetrical alkyne

Apply Markovnikov rule. Positive part of the adding reagent goes (in this case hydrogen) to the carbon with more hydrogens.

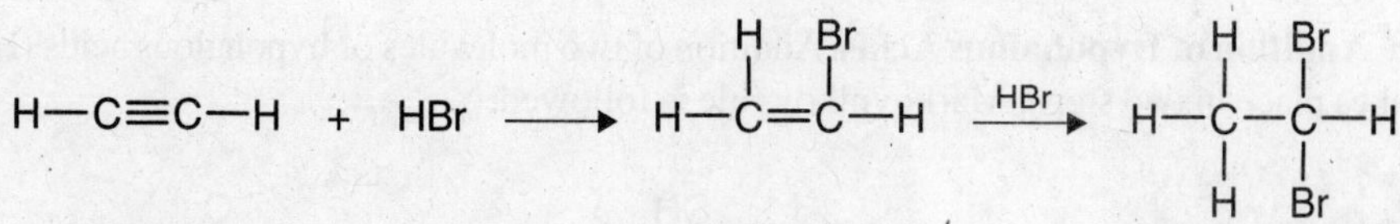

Acetylene 1-Bromoethene 1,1-Dibromoethane

The addition of halogen acid to unsymmetrical alkynes follows Markovnikov rule in the first as well as in the second step.

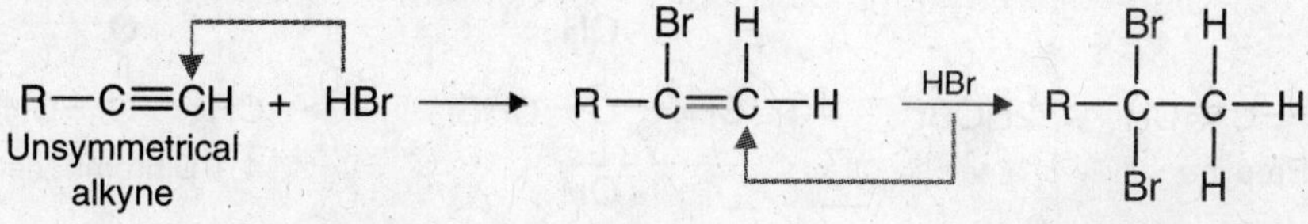

Unsymmetrical alkyne

Apply Markovnikov rule in both steps. Positive part of the adding reagent (in this case hydrogen) goes to the carbon atom with more hydrogens.

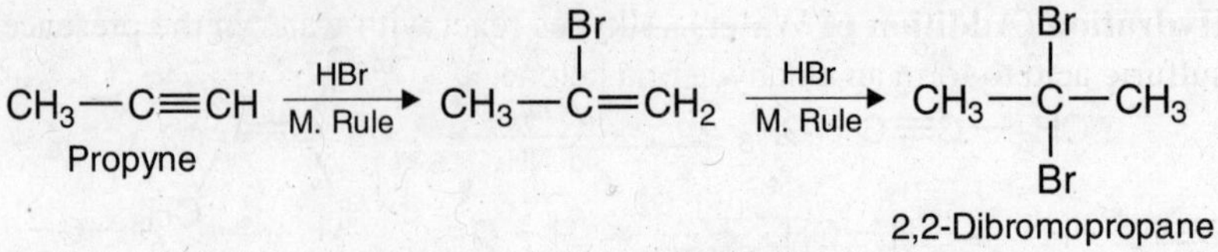

The mechanism of the above reaction involves the following steps :

Step 1. HBr ionizes to give a proton (electrophile) and a bromide ion (nucleophile).

$$HBr \longrightarrow \underset{\text{Proton}}{H^+} + \underset{\text{Bromide ion}}{:\bar{B}r}$$

Step 2. Proton attacks the triple bond to form the more stable carbonium ion (A).

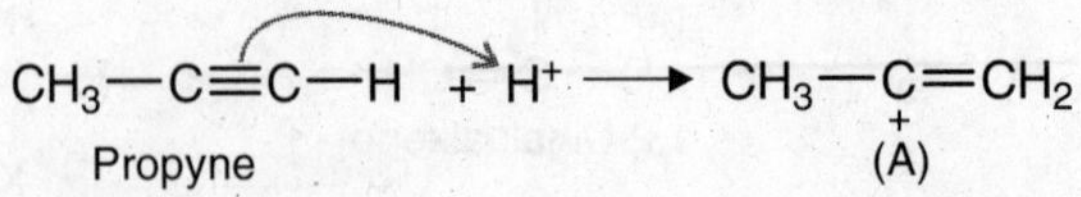

Step 3. Negative bromide ion $(:\bar{B}r)$ attacks the carbonium ion to yield 2-bromopropene.

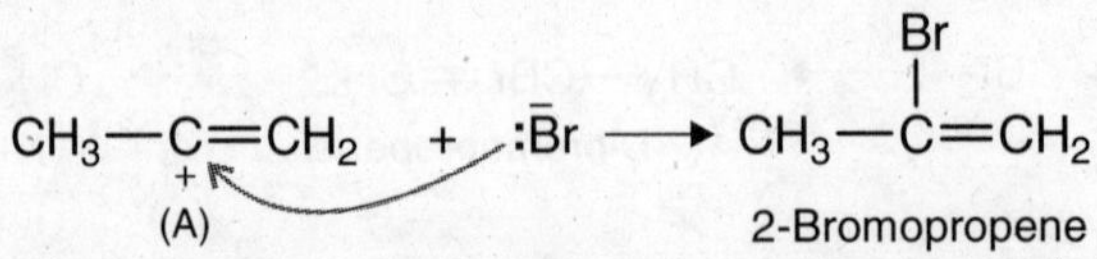

Step 4. Proton attacks 2-bromopropene to form a stable carbonium ion (B).

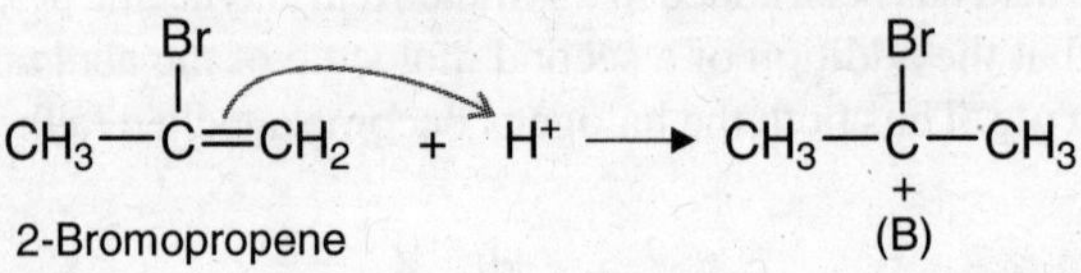

Step 5. Negative bromide ion $(:\bar{B}r)$ attacks the carbonium ion to yield 2,2-dibromopropane.

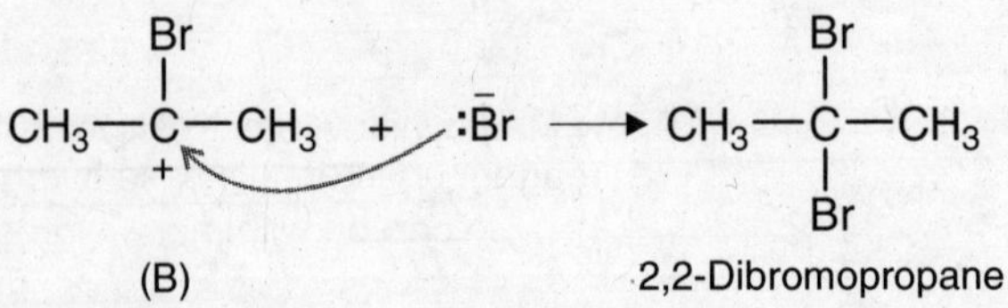

(5) **Addition of Hypohalous Acids**. Addition of two molecules of hypohalous acids (HOX) to alkynes takes place in two steps. Markovnikov rule is followed.

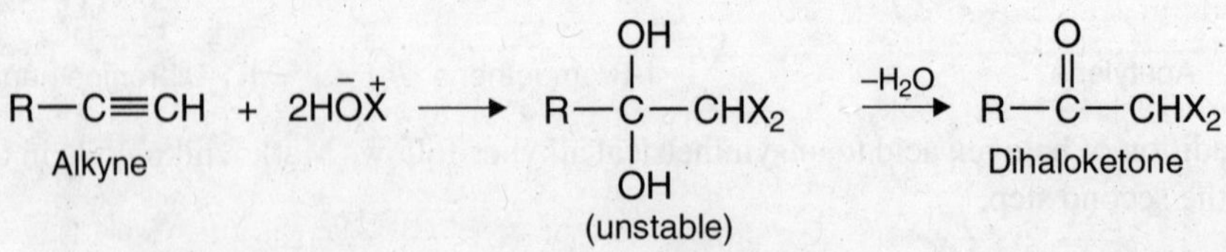

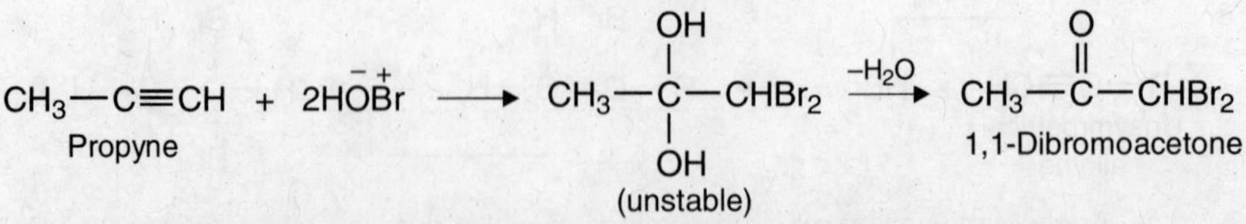

Acetylene under the same conditions gives dibromoacetaldehyde.

(6) **Hydration (Addition of Water)**. Alkynes react with water in the presence of mercuric sulfate and sulfuric acid to form an aldehyde or a ketone.

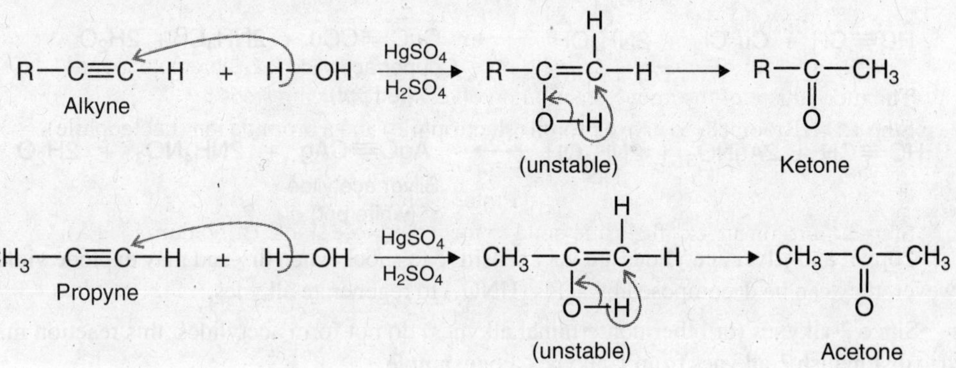

Acetylene under the same conditions gives acetaldehyde.

(7) **Hydroboration-oxidation of Alkynes.** Alkynes react with diborane followed by treatment with alkaline H_2O_2 to form carbonyl compounds. Terminal alkynes always yield aldehydes. Internal alkynes always yield ketones.

$$\underset{\text{Propyne}}{CH_3-C\equiv C-H} \xrightarrow[\text{(2) } H_2O_2/NaOH]{\text{(1) } B_2H_6} \underset{\text{Propanal}}{CH_3CH_2-\overset{\overset{\displaystyle O}{\|}}{C}-H}$$

$$\underset{\text{2-Butyne}}{CH_3-C\equiv C-CH_3} \xrightarrow[\text{(2) } H_2O_2/NaOH]{\text{(1) } B_2H_6} \underset{\text{Butanone}}{CH_3CH_2-\overset{\overset{\displaystyle O}{\|}}{C}-CH_3}$$

(8) **Addition of Hydrogen Cyanide**. Alkynes react with HCN in the presence of barium cyanide catalyst. Thus acetylene yields vinyl cyanide or acrylonitrile from which the synthetic fiber 'Orlon' is made.

$$\underset{\text{Acetylene}}{H-C\equiv C-H} + HCN \xrightarrow[\text{pressure}]{Ba(CN)_2} \underset{\substack{\text{Vinyl cyanide}\\ \text{(Acrylonitrile)}}}{CH_2=CH-CN}$$

(9) **Salt Formation**. We have already pointed out that hydrogens in acetylene or 1-alkynes ($\equiv$C–H) are acidic. They can be replaced by metals to form salts known as **Acetylides**.

(*a*) **Formation of Sodium Acetylides**. Acetylene or 1-alkynes react with sodium amide ($NaNH_2$) to form sodium acetylides.

$$\underset{\text{Acetylene}}{H-C\equiv C-H} + NaNH_2 \xrightarrow[NH_3]{\text{liquid}} \underset{\text{Sod acetylide}}{H-C\equiv \overset{-}{C}\!:\!\overset{+}{Na}} + NH_3$$

$$\underset{\text{1-Alkyne}}{R-C\equiv C-H} + NaNH_2 \xrightarrow[NH_3]{\text{liquid}} R-C\equiv \overset{-}{C}\!:\!\overset{+}{Na} + NH_3$$

Sodium acetylides react with primary alkyl halides to yield higher alkynes.

$$R-C\equiv \overset{-}{C}\!:\!\overset{+}{Na} + \underset{1^\circ\text{ halide}}{R'X} \longrightarrow \underset{\text{Alkyne}}{R-C\equiv C-R'} + NaX$$

(*b*) **Formation of Copper and Silver Acetylides.** Acetylene and 1-alkynes react with ammoniacal solutions of cuprous chloride and silver nitrate to form acetylides of these metals.

$$HC{\equiv}CH + Cu_2Cl_2 + 2NH_4OH \longrightarrow \underset{\text{Copper acetylide (red ppt)}}{\overset{+}{Cu}\overset{-}{C}{\equiv}\overset{-}{C}\overset{+}{Cu}} + 2NH_4Cl + 2H_2O$$

$$HC{\equiv}CH + 2AgNO_3 + 2NH_4OH \longrightarrow \underset{\text{Silver acetylide (white ppt)}}{\overset{+}{Ag}\overset{-}{C}{\equiv}\overset{-}{C}\overset{+}{Ag}} + 2NH_4NO_3 + 2H_2O$$

Copper and silver acetylides are very sensitive to shock when dry and may explode violently. However, they can be decomposed by acids (HNO_3) to regenerate alkynes.

Since 2-alkynes (or other nonterminal alkynes) do not form acetylides, this reaction may be used to distinguish 2-alkynes from 1-alkynes. For example,

$$\underset{\text{1-Butyne}}{CH_3CH_2C{\equiv}CH} + AgNO_3 \xrightarrow{NH_4OH} \underset{\text{(White ppt)}}{CH_3CH_2{-}C{\equiv}\overset{-}{C}\overset{+}{Ag}}$$

$$\underset{\text{2-Butyne}}{CH_3C{\equiv}CCH_3} + AgNO_3 \xrightarrow{NH_4OH} \text{No Reaction}$$

(10) **Oxidation with dilute $KMnO_4$.** Alkynes react with neutral aqueous $KMnO_4$ to form α-diketones. Purple color of $KMnO_4$ disappears during the reaction.

$$\underset{\text{Alkyne}}{R{-}C{\equiv}C{-}R'} + KMnO_4 \xrightarrow[{[O]}]{H_2O} \underset{\alpha\text{-Diketone}}{R{-}\overset{\overset{O}{\|}}{C}{-}\overset{\overset{O}{\|}}{C}{-}R'} + MnO_2 + KOH$$

(11) **Oxidation with $KMnO_4$.** The oxidation of alkynes with alkaline potassium permanganate cleaves the molecule at the site of the triple bond to form carboxylic acids and CO_2.

$$\underset{\text{1-Alkyne}}{R{-}C{\equiv}C{-}H} + 4[O] \longrightarrow R{-}COOH + CO_2$$

$$\underset{\text{2-Alkyne}}{R'{-}C{\equiv}C{-}R} + 4[O] \longrightarrow R'{-}COOH + R{-}COOH$$

$$\underset{\text{Propyne}}{CH_3{-}C{\equiv}C{-}H} + 4[O] \longrightarrow \underset{\text{Acetic acid}}{CH_3{-}COOH} + CO_2$$

Acetylene under these conditions yields oxalic acid. This degradation reaction of alkynes is useful in determining the structure of alkynes.

(12) **Ozonolysis.** Alkynes react with ozone to give ozonides. These ozonides on decomposition with H_2O yield diketones, which are oxidized to acids by hydrogen peroxide produced in the reaction.

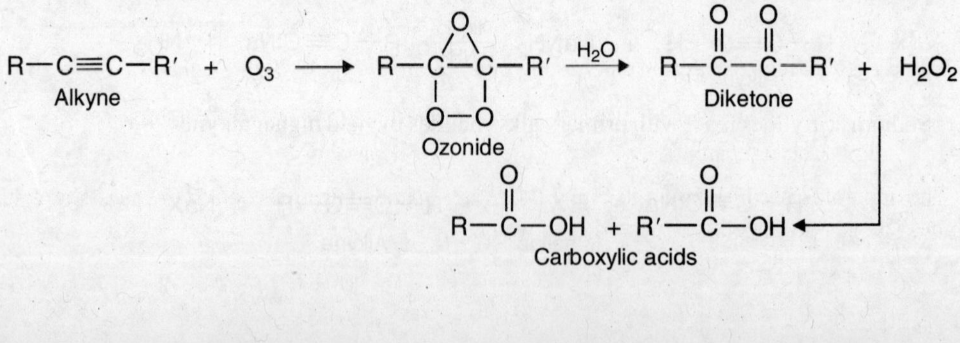

Propyne under these conditions gives a mixture of acetic acid and formic acid. Acetylene yields both glyoxal and formic acid.

(13) **Polymerization**. Alkynes when passed through a red hot iron tube polymerize to form aromatic hydrocarbons. The reaction provides a method for passing from aliphatic to aromatic compounds.

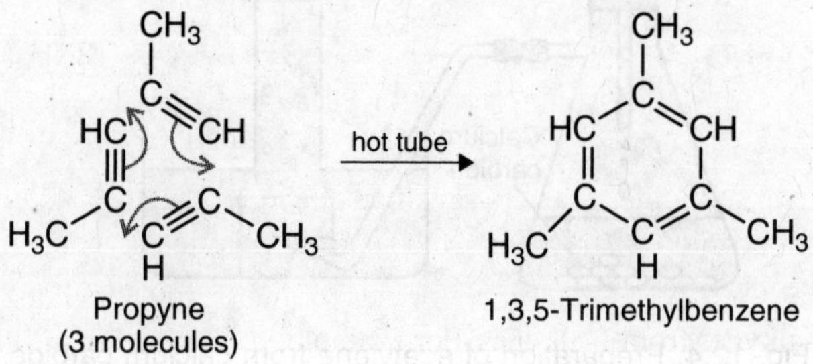

Acetylene under these conditions yields benzene.

ACETYLENE, *Ethyne*, H—C≡C—H 

Acetylene is the most important alkyne. It is a link between inorganic and organic chemistry. It is used as a starting material for a very large number of industrially important organic compounds.

Preparation. Acetylene can be obtained :

(1) By treatment of 1,2-dibromoethane with alcoholic KOH followed by $NaNH_2$.

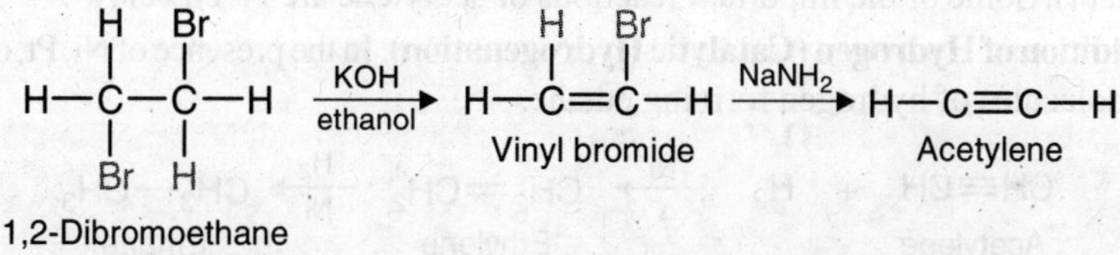

(2) By heating 1,1,2,2-tetrabromoethane with zinc dust in alcohol.

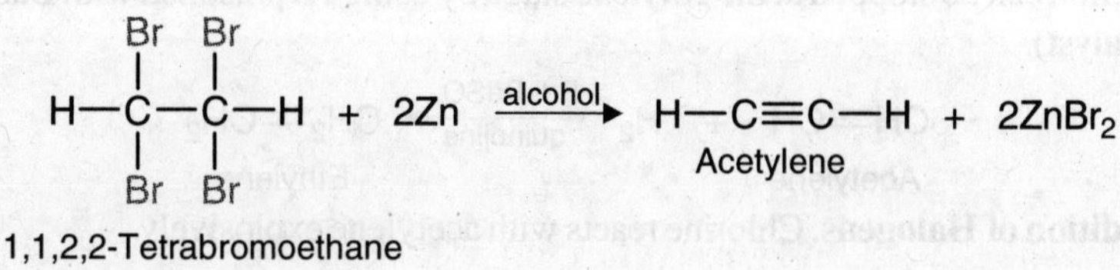

(3) By heating iodoform with silver powder.

$$\underset{\text{Iodoform}}{CHI_3} + 6Ag + \underset{\text{Iodoform}}{I_3CH} \xrightarrow{\Delta} \underset{\text{Acetylene}}{HC{\equiv}CH} + 6AgI$$

(4) By the action of water on calcium carbide. (*Laboratory Method*)

$$\underset{\text{Calcium carbide}}{CaC_2} + 2H_2O \longrightarrow \underset{\text{Acetylene}}{H{-}C{\equiv}C{-}H} + Ca(OH)_2$$

The apparatus used is shown in Fig. 12.4. Water is added dropwise on to the pieces of calcium carbide. The gas produced is collected over water.

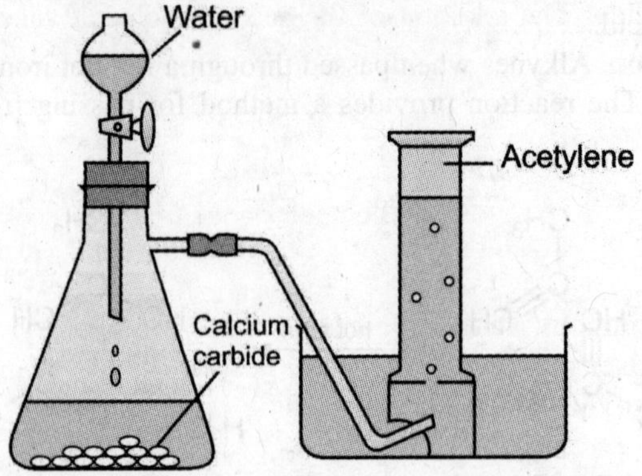

Fig. 12.4. Preparation of acetylene from calcium carbide.

(5) Acetylene is manufactured by heating methane or natural gas at 1500°C in an electric arc.

$$\underset{\text{Methane}}{2CH_4} \xrightarrow[\text{arc}]{\text{Electric}} \underset{\text{Acetylene}}{H{-}C{\equiv}C{-}H} + 3H_2$$

Properties (*Physical*). Acetylene is a colorless gas, bp –75°C. It is practically insoluble in water, but very soluble in acetone.

(*Chemical*). Some of the important reactions of acetylene are given below :

Properties (*Physical*). Acetylene is a colorless gas, bp –75°C. It is practically insoluble in water, but very soluble in acetone.

(*Chemical*). Some of the important reactions of acetylene are given below :

(1) **Addition of Hydrogen (Catalytic Hydrogenation)**. In the presence of Ni, Pt, or Pd, acetylene adds up two molecules of hydrogen forming ethane.

$$\underset{\text{Acetylene}}{CH{\equiv}CH} + H_2 \xrightarrow[\Delta]{Ni} \underset{\text{Ethylene}}{CH_2{=}CH_2} \xrightarrow[Ni]{H_2} \underset{\text{Ethane}}{CH_3{-}CH_3}$$

The reaction can be stopped at the ethylene stage by using Pd poisoned with $BaSO_4$ + quinoline (**Lindlar's Catalyst**).

$$\underset{\text{Acetylene}}{CH{\equiv}CH} + H_2 \xrightarrow[\text{quinoline}]{Pd{-}BaSO_4} \underset{\text{Ethylene}}{CH_2{=}CH_2}$$

(2) **Addition of Halogens**. Chlorine reacts with acetylene explosively.

$$\underset{\text{Acetylene}}{H{-}C{\equiv}C{-}H} + Cl_2 \longrightarrow 2C + 2HCl$$

Acetylene and chlorine can be made to react smoothly by using solutions of reagents.

$$\underset{\text{Acetylene}}{H{-}C{\equiv}C{-}H} + Cl_2 \longrightarrow \underset{\text{1,2-Dichloroethylene}}{H{-}\overset{Cl}{\overset{|}{C}}{=}\overset{Cl}{\overset{|}{C}}{-}H} \xrightarrow{Cl_2} \underset{\text{1,1,2,2-Tetrachloroethane}}{H{-}\underset{Cl}{\underset{|}{\overset{Cl}{\overset{|}{C}}}}{-}\underset{Cl}{\underset{|}{\overset{Cl}{\overset{|}{C}}}}{-}H}$$

Acetylene reacts with bromine in carbon tetrachloride in a similar manner. The red color of bromine is discharged.

(3) **Addition of Halogen Acids**. Halogen acids add to acetylene in two steps forming a monohalide and then a dihalide. The addition of the second molecule takes place according to Markovnikov rule.

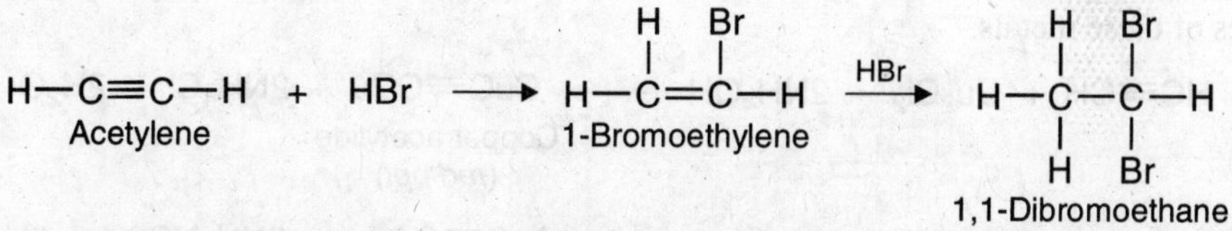

(4) **Addition of Hypohalous Acids**. Hypohalous acids add to acetylene in two steps. The addition of second molecule takes place according to *Markovnikov rule*. The product is unstable and undergoes dehydration to yield dihaloacetaldehyde.

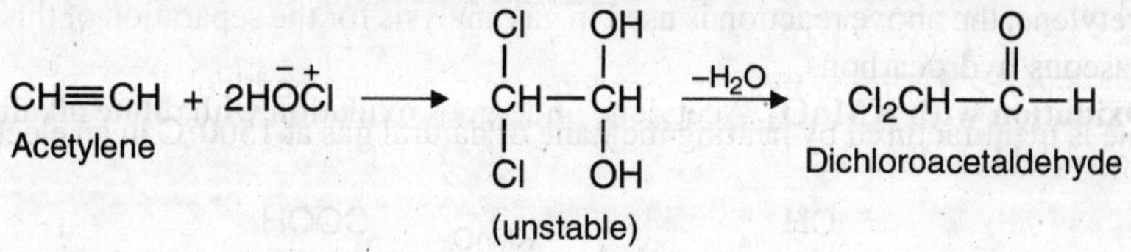

(5) **Addition of Water (Hydration)**. Acetylene reacts with water in the presence of mercuric sulfate and sulfuric acid to give vinyl alcohol. Vinyl alcohol is unstable and rearranges to yield acetaldehyde.

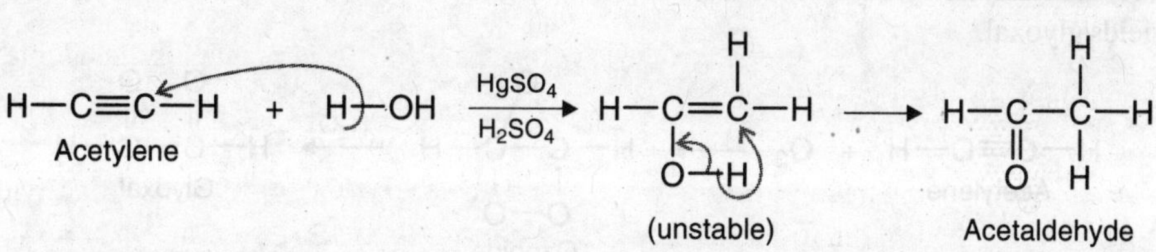

(6) **Addition of HCN**. Acetylene reacts with hydrogen cyanide in the presence of barium cyanide to give acrylonitrile (vinyl cyanide). Acrylonitrile if polymerized gives **Orlon**.

$$\underset{\text{Acetylene}}{H—C{\equiv}C—H} + H—CN \xrightarrow{Ba(CN)_2} \underset{\text{Acrylonitrile}}{CH_2{=}CH—CN}$$

(7) **Addition of Alcohols**. Acetylene reacts with alcohols in the presence of KOH at high temperatures and pressures to form vinyl ethers.

$$CH{\equiv}CH + \underset{\text{Methanol}}{CH_3O—H} \xrightarrow[\Delta]{KOH} \underset{\text{Methyl vinyl ether}}{CH_3O—CH{=}CH_2}$$

$$CH{\equiv}CH + \underset{\text{Ethanol}}{CH_3CH_2O—H} \xrightarrow[\Delta]{KOH} \underset{\text{Ethyl vinyl ether}}{CH_3CH_2O—CH{=}CH_2}$$

(8) **Addition of Acetic Acid**. Acetylene reacts with acetic acid in the presence of mercuric sulfate to give vinyl acetate. Vinyl acetate if polymerized gives polyvinyl acetate plastics.

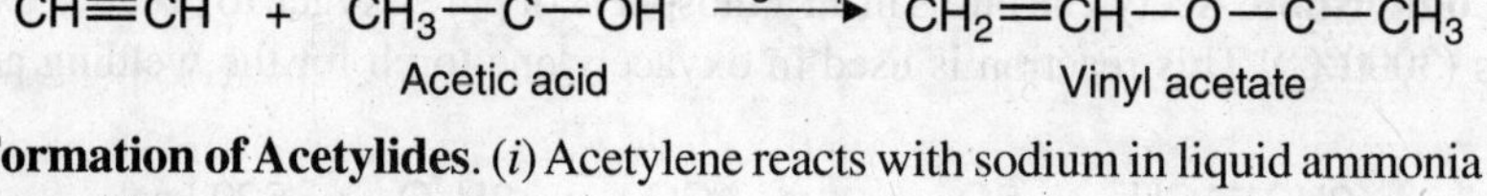

(9) **Formation of Acetylides**. (*i*) Acetylene reacts with sodium in liquid ammonia or $NaNH_2$ to form sodium acetylide.

$$\underset{\text{Acetylene}}{2H-C\equiv C-H} + 2Na \xrightarrow[NH_3]{\text{liquid}} \underset{\text{Sodium acetylide}}{2H-C\equiv C:\overset{-}{\overset{+}{N}}a} + H_2\uparrow$$

(*ii*) Acetylene reacts with ammoniacal solutions of cuprous chloride and silver nitrate to give acetylides of these metals.

$$HC\equiv CH + Cu_2Cl_2 + 2NH_4OH \longrightarrow \underset{\substack{\text{Copper acetylide} \\ (\textit{red ppt})}}{CuC\equiv CCu} + 2NH_4Cl + 2H_2O$$

$$HC\equiv CH + 2AgNO_3 + 2NH_4OH \longrightarrow \underset{\substack{\text{Silver acetylide} \\ (\textit{white ppt})}}{AgC\equiv CAg} + 2NH_4NO_3 + 2H_2O$$

These reactions serve as tests for acetylene. Since acetylides are decomposed by acids to regenerate acetylene, the above reaction is used in gas analysis for the separation of this hydrocarbon from other gaseous hydrocarbons.

(10) **Oxidation with $KMnO_4$.** Acetylene undergoes oxidation with dilute alkaline potassium permanganate to yield oxalic acid.

$$\underset{\text{Acetylene}}{\begin{matrix} CH \\ ||| \\ CH \end{matrix}} + 4[O] \xrightarrow[\overline{O}H]{KMnO_4} \underset{\text{Oxalic acid}}{\begin{matrix} COOH \\ | \\ COOH \end{matrix}}$$

(11) **Ozonolysis**. Acetylene reacts with ozone to give an ozonide which on decomposition with water yields glyoxal.

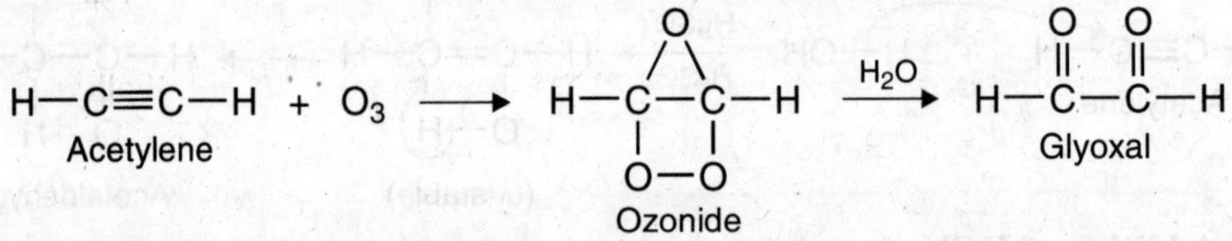

(12) **Dimerization**. Acetylene can add to itself to form a *dimer* in the presence of cuprous chloride and ammonium chloride in HCl.

$$HC\equiv CH + H-C\equiv CH \xrightarrow[NH_4Cl]{Cu_2Cl_2} \underset{\text{Vinyl acetylene}}{CH_2=CH-C\equiv CH}$$

Vinyl acetylene reacts with HCl to give **Chloroprene** (2-Chloro-1,3-butadiene).

$$\underset{\text{Vinyl acetylene}}{CH_2=CH-C\equiv CH} + HCl \longrightarrow \underset{\text{Chloroprene}}{CH_2=CH-\overset{\overset{Cl}{|}}{C}=CH_2}$$

Chloroprene if polymerized gives the synthetic rubber **Neoprene**.

(13) **Polymerization**. Acetylene polymerizes to give benzene when passed through a heated iron tube (400°C) or a tube containing a complex organo-nickel catalyst at 70°C.

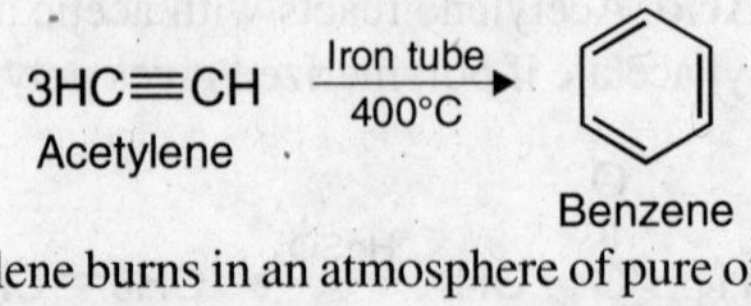

(14) **Combustion.** Acetylene burns in an atmosphere of pure oxygen to produce extremely high temperatures (3000°C). This reaction is used in oxyacetylene torch for the welding and cutting of metals.

$$2HC\equiv CH + 5O_2 \longrightarrow 4CO_2 + 2H_2O + 620 \text{ kcal}$$

Uses. Acetylene is used : (*1*) in oxyacetylene torch used for welding and cutting metals ; (*2*) for the preparation of acetaldehyde, acetic acid, and ethyl alcohol ; (*3*) for the manufacture of PVC, polyvinyl acetate, polyvinyl ethers, orlon, and neoprene rubber.

PROPYNE, *Methylacetylene,* $CH_3-C\equiv C-H$

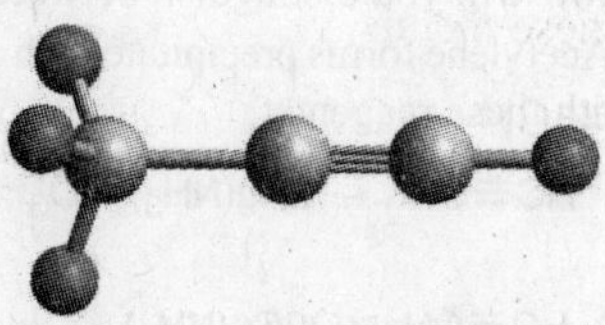

Preparation. Propyne can be prepared : (1) By treatment of 1,2-dibromopropane with alcoholic KOH followed by $NaNH_2$.

$$\underset{\text{1,2-Dibromopropane}}{CH_3-\overset{H}{\underset{Br}{C}}-\overset{Br}{\underset{H}{C}}-H} \xrightarrow[-HBr]{\text{KOH, alcohol}} \underset{\text{1-Bromo-1-propene}}{CH_3-\overset{H}{C}=\overset{Br}{C}-H} \xrightarrow[-HBr]{NaNH_2} \underset{\text{Propyne}}{CH_3-C\equiv C-H}$$

(2) By the action of methyl iodide on sodium acetylide.

$$\underset{\text{Acetylene}}{H-C\equiv C-H} + NaNH_2 \xrightarrow[NH_3]{\text{liquid}} \underset{\text{Sod acetylide}}{H-C\equiv \bar{C}{:}\overset{+}{Na}} + NH_3$$

$$H-C\equiv \bar{C}{:}\overset{+}{Na} + CH_3I \longrightarrow \underset{\text{Propyne}}{H-C\equiv C-CH_3} + NaI$$

Properties. Propyne is a colorless gas, *bp* 23°C. It gives all the reactions of alkynes.

Examination Questions

1. Discuss the molecular orbital structure of acetylene.
(*Delhi BSc, 2013 ; Goa BSc, 2013 ; Kerala BSc, 2014 ; Tamil Nadu BSc, 2015*)
2. How are alkynes prepared ?
3. How is acetylene prepared in the laboratory ?
4. What are the uses of acetylene ?
5. How does acetylene react with the following reagents ?
(*a*) H_2/Pd (*b*) $H_2/Pd/BaSO_4$ (*c*) HBr (*d*) $H_2O/H_2SO_4/HgSO_4$
(*e*) $NaNH_2$ (*f*) $AgNO_3/NH_4OH$ (*g*) $HCN/Ba(CN)_2$ (*h*) Cu_2Cl_2/NH_4OH
6. How does propyne react with the following reagents ?
(*a*) $NaNH_2$ (*b*) $AgNO_3/NH_4OH$ (*c*) O_3 *then* Zn/H_2O (d) $H_2O/H_2SO_4/HgSO_4$

Answer. The products are :
(*a*) $CH_3C\equiv CNa$ (*b*) $CH_3C\equiv CAg$ (*c*) Acetone (*d*) Acetic acid + Formic acid

7. What happens when calcium carbide is treated with water ?
8. What happens when acetylene is passed through red hot iron tube ?
9. What happens when acetylene is passed through ammoniacal silver nitrate solution ?
(*Lucknow BSc, 2015*)
10. What happens when acetylene is passed through ammoniacal cuprous chloride solution ?
(*Pondichery BSc, 2013*)
11. What happens when acetylene is passed through dilute sulfuric acid in the presence of mercuric sulfate ?

Answer. Acetaldehyde is formed. For reaction equation see text.

12. Write a note on : Acidity of Acetylene. *(Kota BSc, 2015)*

13. Write a note on : Acidity of 1-Alkynes. *(Madurai BSc, 2014)*

14. Explain why acetylene forms a silver acetylide but ethylene does not.

15. How will you distinguish between acetylene and ethylene ?

Answer. Acetylene forms precipitate with ammoniacal silver nitrate or cuprous chloride solution. Ethylene does not react with these reagents.

$$HC{\equiv}CH + 2Ag(NH_3)_2NO_3 \longrightarrow \underset{\text{white ppt}}{AgC{\equiv}CAg} + 2NH_4NO_3 + 2NH_3$$

$$HC{\equiv}CH + 2Cu(NH_3)_2Cl \longrightarrow \underset{\text{red ppt}}{CuC{\equiv}CCu} + 2NH_4Cl + 2NH_3$$

16. How will you distinguish between 1-butyne and 2-butyne ? *(Goa BSc, 2015)*

Answer. 1-Butyne forms precipitate with ammoniacal solution of cuprous chloride or silver nitrate. 2-Butyne does not react with these reagents because it contains no acidic hydrogens.

$$C_2H_5C{\equiv}CH + Cu(NH_3)_2Cl \longrightarrow \underset{\text{red ppt}}{C_2H_5C{\equiv}CCu} + NH_4Cl + NH_3$$

$$C_2H_5C{\equiv}CH + Ag(NH_3)_2NO_3 \longrightarrow \underset{\text{white ppt}}{C_2H_5C{\equiv}CAg} + NH_4NO_3 + NH_3$$

17. How will you synthesize acetaldehyde from acetylene ?

18. How will you synthesize vinyl acetate from acetylene ?

19. How will you synthesize vinyl methyl ether from acetylene ?

20. How will you synthesize acetylene from ethylene ?

Answer. Following steps are involved :

$$\underset{\text{Ethylene}}{CH_2{=}CH_2} \xrightarrow{Br_2} BrCH_2CH_2Br \xrightarrow{NaNH_2} \underset{\text{Acetylene}}{HC{\equiv}CH}$$

21. How will you synthesize 1-butyne from acetylene ? *(West Bengal BSc, 2014)*

Answer. Following steps are involved :

$$\underset{\text{Acetylene}}{HC{\equiv}CH} \xrightarrow{NaNH_2} HC{\equiv}\overset{-}{C}\overset{+}{Na} \xrightarrow{CH_3CH_2Br} \underset{\text{1-Butyne}}{HC{\equiv}CCH_2CH_3}$$

22. How will you synthesize acetone from propene ? *(Kalyani BSc, 2015)*

Answer. Following steps are involved :

$$\underset{\text{Propene}}{CH_3CH{=}CH_2} \xrightarrow{Br_2} CH_3\underset{|}{\overset{Br}{C}}H{-}\overset{Br}{C}H_2 \xrightarrow{NaNH_2} \underset{\text{Propyne}}{CH_3C{\equiv}CH} \xrightarrow[H_2SO_4/HgSO_4]{H_2O} \underset{\text{Acetone}}{CH_3{-}\overset{O}{\overset{\|}{C}}{-}CH_3}$$

23. How will you synthesize Neoprene from acetylene ? *(Kota BSc, 2014)*

Answer. Following steps are involved :

$$\underset{\text{Acetylene}}{HC{\equiv}CH} + \underset{\text{Acetylene}}{H{-}C{\equiv}CH} \xrightarrow[NH_4Cl]{Cu_2Cl_2} \underset{\text{Vinylacetylene}}{CH_2{=}CH{-}C{\equiv}CH}$$

$$\xrightarrow[\text{(M rule)}]{HCl} \underset{\text{Chloroprene}}{CH_2{=}CH{-}\overset{Cl}{\overset{|}{C}}{=}CH_2} \xrightarrow[R_2O_2]{\text{Polymerization}} \underset{\text{Neoprene}}{\left[{-}CH_2{-}CH{=}\overset{Cl}{\overset{|}{C}}{-}CH_2{-} \right]_n}$$

24. An alkyne (*A*) with molecular formula C_5H_8 reacts with neither sodamide nor with ammoniacal cuprous chloride. Assign structural formula to (*A*).

Answer. (*A*) = $CH_3CH_2C{\equiv}CCH_3$ (2-Pentyne).

25. A compound has a molecular formula, C_4H_6. When it is treated with excess hydrogen in the presence of nickel catalyst, a new compound, C_4H_{10}, is formed. When it is treated with ammoniacal silver nitrate, a white precipitate is formed. What is the structural formula of the given compound ?

Answer. The given compound is 1-Butyne.

26. An unknown compound (*A*) has the molecular formula, C_5H_8. Suggest a possible structure for (*A*) from the following data :

(*a*) One mole of (*A*) takes up two moles of hydrogen.
(*b*) (*A*) forms a precipitate with ammoniacal solution of cuprous chloride or silver nitrate.
(*c*) Treatment of (*A*) with H_2O in the presence of dilute sulfuric acid and mercuric sulfate yields 2-pentanone as the main product.

Answer. (*A*) = $CH_3CH_2CH_2C{\equiv}CH$ (1-Pentyne)

27. An unknown compound (*B*) has the molecular formula, C_5H_8. Suggest a possible structure for (*B*) from the following data :

(*a*) One mole of (*B*) takes up two moles of chlorine.
(*b*) Compound (*B*) decolorizes a solution of Br_2 in CCl_4.
(*c*) Compound (*B*) forms no precipitate with ammoniacal $AgNO_3$.
(*d*) Treatment of (*B*) with H_2O in the presence of dilute sulfuric acid and mercuric sulfate catalyst yields an equimolar mixture of 2-pentanone and 3-pentanone.

Answer. (*B*) = $CH_3CH_2C{\equiv}CCH_3$ (2-Pentyne)

28. An unknown compound (*C*) has the molecular formula C_5H_8. Suggest a possible structure for (*C*) from the following data :

(*a*) One mole of (*C*) takes up two moles of bromine.
(*b*) (*C*) does not react with ammoniacal $AgNO_3$.
(*c*) Ozonolysis of one mole of (*C*) yields two moles of formaldehyde and one mole of 2-oxopropanal.

Answer. (*C*) = $CH_2{=}C(CH_3){-}CH{=}CH_2$ (2-Methyl-1,3-butadiene)

29. Compound (*A*), C_6H_{12}, is treated with chlorine in the presence of carbon tetrachloride to give (*B*), $C_6H_{12}Cl_2$. Compound (*B*) is treated with alcoholic KOH, followed by $NaNH_2$, resulting in the formation of (*C*). Compound (*C*) is treated with hydrogen in the presence of nickel catalyst to give 2-methylpentane. Compound (*C*) does not react with sodamide or an ammoniacal solution of silver nitrate. Ozonolysis of (*A*) gives two aldehydes, (*D*) and (*E*). Compounds (*E*) is identified as acetaldehyde. From this information, deduce the structural formulas of (*A*), (*B*), (*C*), and (*D*). Write equations for all reactions.

Answer. (*A*) = $CH_3{-}CH{=}CH{-}CH(CH_3){-}CH_3$ (*B*) = $CH_3{-}CH(Cl){-}CH(Cl){-}CH(CH_3){-}CH_3$

(*C*) = $CH_3{-}C{\equiv}C{-}CH(CH_3){-}CH_3$ (*D*) = $CH_3{-}CH(CH_3){-}C(=O){-}H$

Multiple Choice Questions

1. A triple bond consists of

(*a*) 2 *sigma* bonds and 1 *pi* bond (*b*) 3 *sigma* bonds
(*c*) 1 *sigma* bond and 2 *pi* bonds (*d*) 3 *pi* bonds

Answer. (*c*)

2. The bond angles associated with the hybrid orbitals of a carbon involved in a triple bond is

(*a*) 180° (*b*) 120°
(*c*) 109° (*d*) 45°

Answer. (*a*)

3. How many distinct alkynes exist with a molecular formula of C_4H_6 ?

(*a*) 0 (*b*) 1 (*c*) 2 (*d*) 3

Answer. (*c*)

4. How many distinct terminal alkynes exist with a molecular formula of C_5H_8 ?
(*a*) 1 (*b*) 2 (*c*) 3 (*d*) 4

Answer. (*b*)

5. Which of the following alkynes has the lowest boiling point ?
(*a*) 3,3-Dimethyl-1-butyne (*b*) 1-Hexyne
(*c*) 2-Hexyne (*d*) 3-Hexyne

Answer. (*a*)

6. The pi bond of an alkyne is ________ and ________ than the *pi* bond of an alkene.
(*a*) shorter, stronger (*b*) shorter, weaker
(*c*) longer, stronger (*d*) longer, weaker

Answer. (*b*)

7. What is the approximate C-C-C bond angle in propyne, $CH_3C{\equiv}CH$?
(*a*) 180° (*b*) 120° (*c*) 109.5° (*d*) 90°

Answer. (*a*)

8. The length of a triple bond is
(*a*) equal to the length of a single bond (*b*) less than the length of a single bond
(*c*) greater than the length of a single bond (*d*) greater than the length of a double bond

Answer. (*b*)

9. Which of the following is *not* a true statement ?
(*a*) Butyne is more acidic than butane.
(*b*) An electron in an *sp* orbital is (on average) closer to the nucleus than an electron in an sp^2 orbital.
(*c*) A triple bond consists of two *sigma* bonds and one *pi* bond.
(*d*) A triple bond is shorter than a double bond.

Answer. (*c*)

10. In the following hydrogenation reactions ;

$$HC{\equiv}CH \xrightarrow{H_2/Ni} CH_2{=}CH_2 \xrightarrow{H_2/Ni} CH_3CH_3$$

The hybrid state of the carbon atom changes from
(*a*) $sp^3 \rightarrow sp^2 \rightarrow sp$ (*b*) $sp \rightarrow sp^2 \rightarrow sp^3$
(*c*) $sp^2 \rightarrow sp^3 \rightarrow sp$ (*d*) $sp^3 \rightarrow sp \rightarrow sp^2$

Answer. (*b*)

11. Which of the following hydrocarbons has acidic hydrogens ?
(*a*) 1-Butene (*b*) 1-Butyne (*c*) 2-Butene (*d*) 2-Butyne

Answer. (*b*)

12. Lindlar's catalyst is
(*a*) $LiAlH_4$ (*b*) $Pd/BaSO_4$ in Quinoline
(*c*) NH_2NH_2 (*d*) $HCl/ZnCl_2$

Answer. (*b*)

13. The higher reactivity of an alkene or alkyne, as compared to an alkane, is due to
(*a*) *sigma* bonds (*b*) *pi* bonds
(*c*) hydrogen bonds (*d*) None of these

Answer. (*b*)

14. Which of the following statements about alkenes and alkynes is correct ?
(*a*) Alkynes are reduced *more* readily than alkenes.
(*b*) Alkynes are reduced *less* readily than alkenes.
(*c*) Alkynes and alkenes are reduced with *equal* speed.
(*d*) Alkynes and alkenes can not be reduced.

Answer. (*a*)

15. Which of the following sequences regarding acid-strength is correct ?
(*a*) $HC{\equiv}CH > CH_3CH_2OH > CH_3COOH$
(*b*) $HC{\equiv}CH > CH_3COOH > CH_3CH_2OH$
(*c*) $CH_3COOH > HC{\equiv}CH > CH_3CH_2OH$
(*d*) $CH_3COOH > CH_3CH_2OH > HC{\equiv}CH$

Answer. (*d*)

16. 1,2-Dichloroethane reacts with excess of $NaNH_2$ to form

(*a*) Vinyl chloride (*b*) Ethylene
(*c*) Ethyl chloride (*d*) Acetylene

Answer. (*d*)

17. Propyne is formed by

(*a*) Polymerization of acetylene
(*b*) Reaction of acetylene with methane
(*c*) Reaction of acetylene with methyl chloride
(*d*) Reaction of sodium acetylide with methyl chloride

Answer. (*d*)

18. Which of the following compounds on hydrolysis gives acetylene ?

(*a*) CaC_2 (*b*) Mg_2C_3 (*c*) Al_4C_3 (*d*) Cu_2Cl_2

Answer. (*a*)

19. Which one of the following compounds will (*a*) decolorizes dilute cold $KMnO_4$; (*b*) decolorizes bromine water ; and (*c*) gives a white precipitate with ammoniacal $AgNO_3$ solution.

(*a*) 1-Hexene (*b*) 1-Hexyne (*c*) 2-Hexene (*d*) 2-Hexyne

Answer. (*b*)

20. Ethylene and acetylene can be distinguished by using

(*a*) Bromine in CCl_4 (*b*) Tollens' reagent
(*b*) Baeyer's reagent (*d*) Phenylhydrazine

Answer. (*b*)

21. 1-Butyne can be distinguished from 2-butyne by using

(*a*) potassium permanganate (*b*) bromine in CCl_4
(*c*) Tollens' reagent (*d*) chlorine in CCl_4

Answer. (*c*)

22. 1-Butyne reacts with

(*a*) $NaNH_2$ (*b*) Dil H_2SO_4 and $HgSO_4$
(*c*) HBr (*d*) All of these

Answer. (*d*)

23. Propyne can be converted to propene by using

(*a*) H_2 + Lindlar's catalyst (*b*) NH_2NH_2
(*c*) H_2 + Pt catalyst (*d*) NH_2NH_2 + KOH

Answer. (*a*)

24. 2-Butyne undergoes catalytic hydrogenation in the presence of Lindlar's catalyst to give

(*a*) 2-Butene (*b*) Butane
(*c*) 1-Butene (*d*) 2-Methylpropene

Answer. (*a*)

25. Which of the following describes an unsymmetrical addition reaction ?

(*a*) Propyne with 1 mole H_2, Ni and heat (*b*) Propyne with 2 moles Cl_2 in CCl_4
(*c*) Propyne with 1 mole Br_2 in CCl_4 (*d*) Propyne with 1 mole HBr

Answer. (*d*)

26. What is the major organic product that results when 1-heptyne is treated with 2 equivalents of HBr ?

(*a*) 2,3-Dibromo-1-heptene (*b*) 2,3-Dibromo-2-heptene
(*c*) 1,2-Dibromoheptane (*d*) 2,2-Dibromoheptane

Answer. (*d*)

27. The strength of a triple bond is

(*a*) equal to the strength of a double bond
(*b*) less than the strength of a double bond
(*c*) greater than the strength of a double bond
(*d*) less than the strength of a single bond

Answer. (*c*)

28. Which of the following is the *most* acidic ?

(*a*) 1-Butyne (*b*) 2-Butyne
(*c*) 1-Butene (*d*) Butane

Answer. (*a*)

29. Indicate the correct relative acidities.
(*a*) ethane > ethyne > ethene (*b*) ethyne > ethane > ethane
(*c*) ethyne > ethene > ethane (*d*) ethane > ethane > ethyne

Answer. (*c*)

30. Addition of two moles of HCl to propyne gives :
(*a*) 2,2-Dichloropropane (*b*) 1,3-Dichloropropane
(*c*) 1,2-Dichloropropane (*d*) None of these

Answer. (*a*)

31. Propyne reacts with aqueous H_2SO_4 in the presence of $HgSO_4$ to form
(*a*) acetone (*b*) 1-propanol
(*c*) acetaldehyde (*d*) 2-propanol

Answer. (*a*)

32. Acetylene reacts with water in the presence of sulfuric acid and mercuric sulfate to give
(*a*) Acetone (*b*) Acetic acid
(*c*) Formaldehyde (*d*) Acetaldehyde

Answer. (*d*)

33. Which of the following compounds will react with ammoniacal silver nitrate ?
(*a*) 1-Butene (*b*) 1-Butyne
(*c*) 2-Butene (*d*) 2-Butyne

Answer. (*b*)

34. Which of the following compounds does not react with ammoniacal $AgNO_3$ solution ?
(*a*) Acetylene (*b*) 1-Butyne
(*c*) Propyne (*d*) 2-Butyne

Answer. (*d*)

35. Which of the following compounds reacts with ammoniacal cuprous chloride to give a precipitate ?
(*a*) 1-Butene (*b*) 1-Butyne
(*c*) 2-Butene (*d*) 2-Butyne

Answer. (*b*)

36. Ozonolysis of 2-butyne gives
(*a*) Formic acid (*b*) Propanoic acid
(*c*) Acetic acid (*d*) Butanoic acid

Answer. (*c*)

37. Which alkyne yields propanoic acid as the only product upon treatment with ozone followed by hydrolysis ?
(*a*) 1-Butyne (*b*) 2-Hexyne
(*c*) 1-Pentyne (*d*) 3-Hexyne

Answer. (*d*)

38. When acetylene is passed through hot iron tube at 400°C, it gives
(*a*) Benzene (*b*) Toluene
(*c*) *o*-Xylene (*d*) Mesitylene

Answer. (*a*)

39. When propyne is passed through hot iron tube at 400°C, it gives
(*a*) Benzene (*b*) Toluene
(*c*) *m*-Xylene (*d*) Mesitylene

Answer. (*d*)

40. The monomer for Neoprene is
(*a*) Isoprene (*b*) Acrylonitrile
(*c*) Chloroprene (*d*) 1,3-Butadiene

Answer. (*c*)

41. A compound, C_4H_6, reacts with bromine and forms a white precipitate with ammoniacal silver nitrate solution. It reacts with dilute H_2SO_4 in the presence of mercuric sulfate to form 2-butanone. The compound could be
(*a*) 1-Butyne (*b*) 1-Butene
(*c*) 2-Butyne (*d*) 2-Butene

Answer. (*a*)

13

Alkyl Halides

Ethyl chloride spray, a skin "refrigerant," is used as a topical anesthetic.

Alkyl halides are compounds in which a halogen atom is attached to carbon. For example,

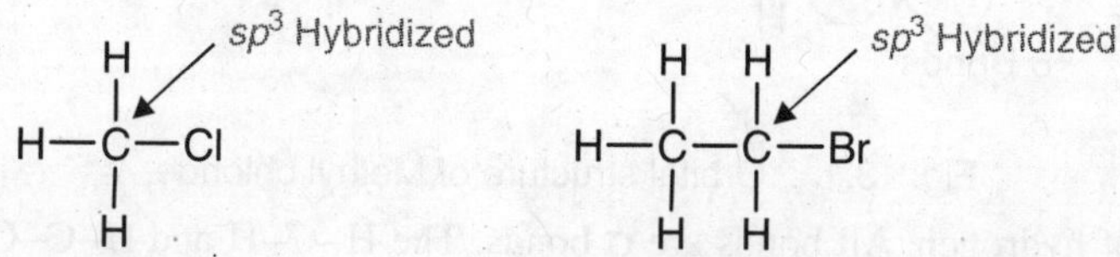

Methyl chloride　　　Ethyl bromide

They have the general formula :

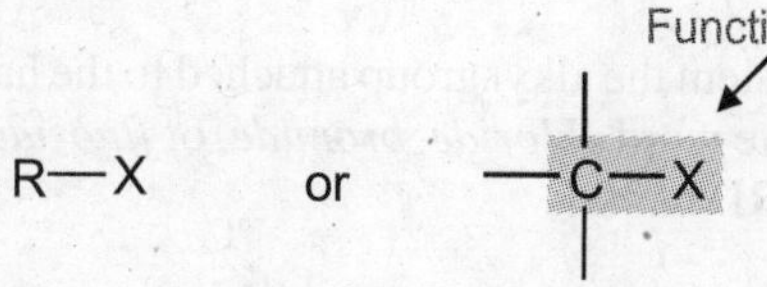

Where R = alkyl group ; X = Cl, Br, I, or F. The halogen atom bonded to carbon is the functional group of alkyl halides.

Alkyl halides are classified as *Primary* (1°), *Secondary* (2°), or *Tertiary* (3°), depending upon whether the X atom is attached to a primary, secondary, or a tertiary carbon.

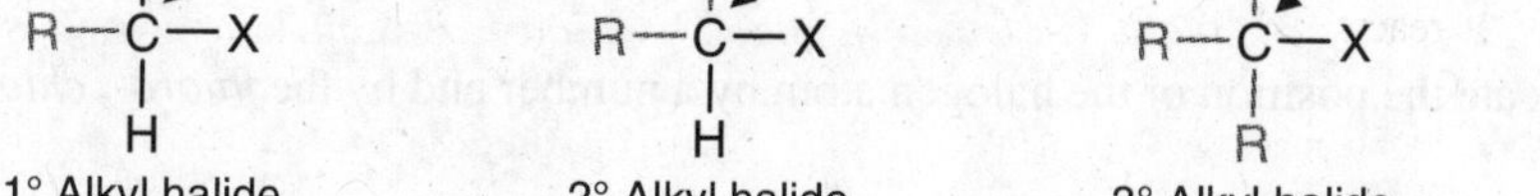

1° Alkyl halide　　　2° Alkyl halide　　　3° Alkyl halide

Alkyl halides are among the most useful organic compounds. They are frequently used to introduce alkyl groups into other molecules.

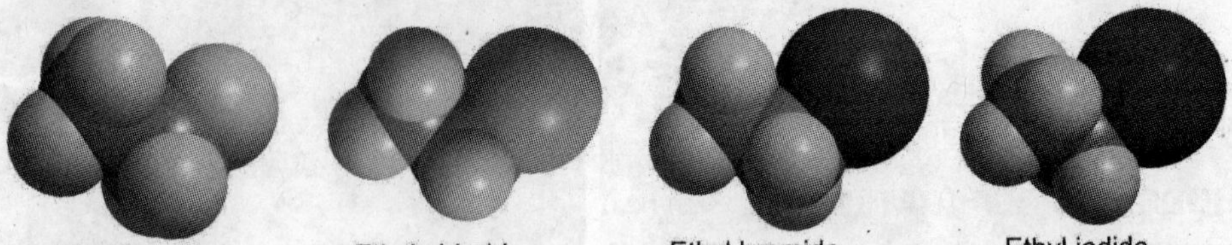

Ethyl fluoride	Ethyl chloride	Ethyl bromide	Ethyl iodide
bp −38°C	bp 12°C	bp 38°C	bp 72°C

Space-filling models of the Ethyl halides. The heavier halogens are larger, with much greater surface areas.

STRUCTURE

Let us consider methyl chloride (CH_3Cl) for illustrating the orbital make up of alkyl halides. In methyl chloride, the carbon atom is sp^3 hybridized. The chlorine atom has a half-filled *p* orbital in its valence shell. The C–Cl bond is formed by the overlap of an sp^3 orbital of carbon and the half-filled *p* orbital of chlorine atom (Fig. 13.1). Each C–H bond is formed by the overlap of an sp^3 orbital of carbon

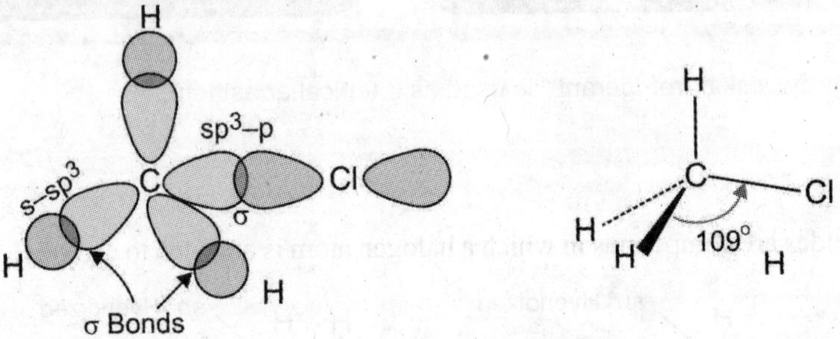

Fig. 13.1. Orbital structure of Methyl chloride.

and the *s* orbital of hydrogen. All bonds are σ bonds. The H–C–H and H–C–Cl bond angles are approximately tetrahedral.

NOMENCLATURE

Alkyl halides are named in two ways :

(1) **Common System.** In this system the alkyl group attached to the halogen atom is named first. This is then followed by an appropriate word *chloride*, *bromide*, or *fluoride*. Notice that the common names of alkyl halides are TWO-WORD names.

$CH_3—Br$	$CH_3CH_2—Cl$	$CH_3—CH(Br)—CH_3$
Methyl bromide	Ethyl chloride	Isopropyl bromide

(2) **IUPAC System.** The IUPAC names of alkyl halides are obtained by using the following rules:

(*a*) Select the longest carbon chain containing the halogen atom and name the alkyl halide as a derivative of the corresponding hydrocarbon.

(*b*) Number the chain so as to give the carbon carrying the halogen atom the *lowest* possible number.

(*c*) Indicate the position of the halogen atom by a number and by the *fluoro-*, *chloro-*, *bromo-*, or *iodo-*.

(*d*) Name other substituents and indicate their positions by numbers.

The examples given below show how these rules are applied. Notice that the IUPAC names of alkyl halides are ONE-WORD names.

$CH_3CH_2—Br$	$CH_3CH_2CH_2—I$	$\overset{1}{CH_3}—\overset{2}{CH}(Cl)—\overset{3}{CH_3}$	$\overset{4}{CH_3}—\overset{3}{CH}(CH_3)—\overset{2}{CH_2}—\overset{1}{CH_2Br}$
Bromoethane	1-Iodopropane	2-Chloropropane	1-Bromo-3-methylbutane

METHODS OF PREPARATION

Alkyl halides can be prepared by the following methods :

(1) **Halogenation of Alkanes.** Alkanes react with Cl_2 or Br_2 in the presence of UV light or at high temperature (400°C) to give alkyl halides along with polyhalogen derivatives.

$$\underset{\text{Methane}}{CH_4} \xrightarrow[\text{UV light}]{Cl_2} CH_3Cl + CH_2Cl_2 + CHCl_3 + CCl_4$$

This method is not used in the laboratory because of the difficulty of separating the products.

(2) **Addition of Halogen Acids to Alkenes.** Halogen acids (HCl, HBr, HI) add to alkenes to yield alkyl halides. The mode of addition follows *Markovnikov rule*, except for the addition of HBr in the presence of organic peroxides (R–O–O–R).

$$\underset{\text{2-Alkene}}{R—CH{=}CH—R} + HX \longrightarrow \underset{\text{Alkyl halide}}{R—CH_2—CH(X)—R}$$

$$\underset{\text{Ethylene}}{CH_2{=}CH_2} + HI \longrightarrow \underset{\text{Ethyl iodide}}{CH_3—CH_2—I}$$

$$\underset{\text{1-Alkene}}{R—CH{=}CH_2} + HBr \longrightarrow R—CH(Br)—CH_3$$

$$\underset{\text{Propene}}{CH_3—CH{=}CH_2} + HBr \longrightarrow \underset{\substack{\text{2-Bromopropane}\\(\textit{Markovnikov product})}}{CH_3—CH(Br)—CH_3}$$

$$\underset{\text{Propene}}{CH_3—CH{=}CH_2} + HBr \xrightarrow{\text{peroxide}} \underset{\substack{\text{1-Bromopropane}\\(\textit{anti}\text{-Markovnikov product})}}{CH_3—CH_2—CH_2Br}$$

(3) **Action of Halogen Acids on Alcohols.** Alcohols react with HBr or HI to produce alkyl bromides or alkyl iodides. Alkyl chlorides are produced by the action of *dry* HCl in the presence of zinc chloride catalyst.

$$\underset{\text{Alcohol}}{R—OH} + H—X \longrightarrow \underset{\text{Alkyl halide}}{R—X} + H_2O$$

$$\underset{\text{Ethyl alcohol}}{CH_3CH_2OH} + HCl \xrightarrow{ZnCl_2} \underset{\text{Ethyl chloride}}{CH_3CH_2Cl} + H_2O$$

$$\underset{n\text{-Propyl alcohol}}{CH_3CH_2CH_2OH} + HBr \longrightarrow \underset{n\text{-Propyl bromide}}{CH_3CH_2CH_2Br} + H_2O$$

(4) **Action of Phosphorus Halides on Alcohols.** Alcohols react with phosphorus halides (PX_5 or PX_3) to form alkyl halides.

$$R{-}OH + PX_5 \text{ (or } PX_3) \longrightarrow R{-}X$$
Alcohol → Alkyl halide

$$2CH_3CH_2OH + PCl_5 \longrightarrow 2CH_3CH_2Cl + POCl_3 + H_2O$$
Ethyl alcohol → Ethyl chloride

$$3CH_3CH_2OH + PBr_3 \longrightarrow 3CH_3CH_2Br + H_3PO_3$$
Ethyl bromide

$$3CH_3OH + PI_3 \longrightarrow 3CH_3I + H_3PO_3$$
Methyl alcohol → Methyl iodide

PBr_3 or PI_3 are produced *in situ* by the addition of Br_2 and I_2 to red phosphorus.

$$2P + 3Br_2 \longrightarrow 2PBr_3$$
$$2P + 3I_2 \longrightarrow 2PI_3$$

(5) **Action of Thionyl Chloride on Alcohols.** Alcohols react with thionyl chloride ($SOCl_2$) in the presence of pyridine to produce alkyl chlorides. Pyridine (C_5H_5N) absorbs hydrogen chloride as it is formed.

$$R{-}OH + SOCl_2 \xrightarrow{\text{pyridine}} R{-}Cl + SO_2 + HCl$$
Alcohol → Alkyl chloride

$$CH_3CH_2OH + SOCl_2 \xrightarrow{\text{pyridine}} CH_3CH_2Cl + SO_2 + HCl$$
Ethyl alcohol → Ethyl chloride

(6) **Halogen Exchange Reaction.** This reaction is particularly suitable for preparing alkyl iodides. The alkyl bromide or chloride is heated with a concentrated solution of sodium iodide in acetone.

$$CH_3CH_2{-}Br + NaI \xrightarrow[\Delta]{\text{acetone}} CH_3CH_2{-}I + NaBr$$
Ethyl bromide → Ethyl iodide

Alkyl fluorides are also prepared by treating an alkyl chloride or bromide with inorganic fluorides.

$$2CH_3{-}Cl + Hg_2F_2 \longrightarrow 2CH_3{-}F + Hg_2Cl_2$$
Methyl chloride → Methyl fluoride

PHYSICAL PROPERTIES

(1) CH_3Cl, CH_3Br, CH_3F, and CH_3CH_2Cl are gases at room temperature. Other alkyl halides up to C_{18} are colorless liquids. Those beyond C_{18} are colorless solids.

Table 13.1 BOILING POINTS OF SOME ALKYL HALIDES

Name	Structure	bp°C
Methyl chloride	CH_3Cl	–24 (*gas*)
Methyl bromide	CH_3Br	5 (*gas*)
Methyl iodide	CH_3I	43 (*liquid*)
Ethyl chloride	CH_3CH_2Cl	13 (*gas*)
Ethyl bromide	CH_3CH_2Br	38 (*liquid*)
Ethyl iodide	CH_3CH_2I	72 (*liquid*)

(2) Alkyl halides are insoluble in water but soluble in organic solvents. The insolubility in water is due to their inability to form hydrogen bonds with water.

(3) Alkyl bromides and iodides are denser than water. Alkyl chlorides and fluorides are lighter than water.

(4) Alkyl halides have higher boiling points than alkanes of comparable molecular weight. For a given halogen atom, the boiling points of alkyl halides increase with the increase in the size of the alkyl group. For a given alkyl group, the boiling points of alkyl halides follow the order RI > RBr > RCl > RF (See Table 13.1).

(5) ***IR Spectrum.*** Alkyl halides show strong absorptions in the *infrared* region arising from the stretching vibrations of the carbon-halogen bond. Characteristic absorptions are observed at 950–1350 cm^{-1} for C–F, at 510–775 cm^{-1} for C–Cl at 490–650 cm^{-1} for C–Br, and at 465–600 cm^{-1} for C–I.

CHEMICAL PROPERTIES

Alkyl halides are very reactive compounds. They undergo *substitution*, *elimination*, and *reduction* reactions. Alkyl halides also react with metals to form organometallic compounds.

A. SUBSTITUTION REACTIONS

The carbon-halogen bond in alkyl halides is *polar* because of the high electronegativity of the halogen atom relative to carbon. The carbon atom is therefore a good target for attack by nucleophiles (electron rich species). In fact, the nucleophilic substitution reactions are the most common reactions of alkyl halides. They may be represented as :

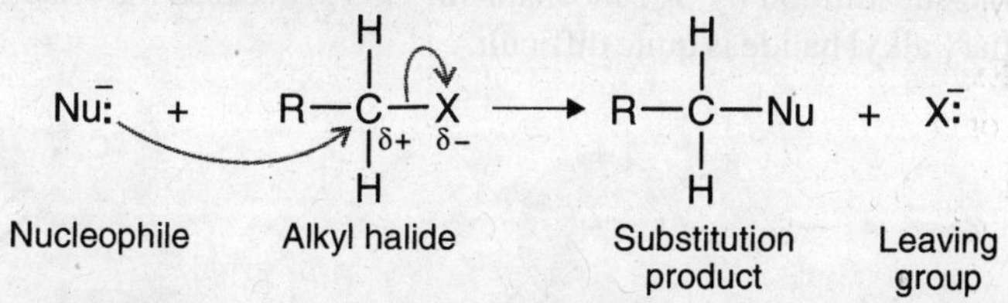

As you will recall from our earlier discussion (Chapter 6) alkyl halides can undergo nucleophilic substitution by two mechanisms :

S_N2 MECHANISM. S_N2 stands for bimolecular nucleophilic substitution. In this mechanism, the attack of the nucleophile and the ejection of the halide ion takes place simultaneously. The reactants on their way to products pass through a transition state where the C–OH bond is *half-formed* and the C–X bond is *half-broken*. Consider the example of the hydrolysis of methyl bromide by aqueous NaOH. The reaction and the transition state are shown in Fig. 13.2.

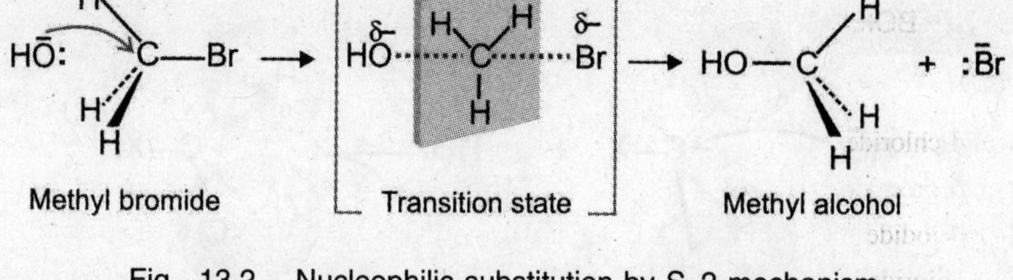

Fig. 13.2. Nucleophilic substitution by S_N2 mechanism.

S_N1 MECHANISM. S_N1 stands for unimolecular nucleophilic substitution. This mechanism involves two steps. Consider the hydrolysis of *tert*-butyl bromide with aqueous NaOH.

Step 1. The alkyl halide ionizes to give a planar carbonium ion. The carbonium ion is planar because the central positively charged carbon atom is sp^2 hybridized.

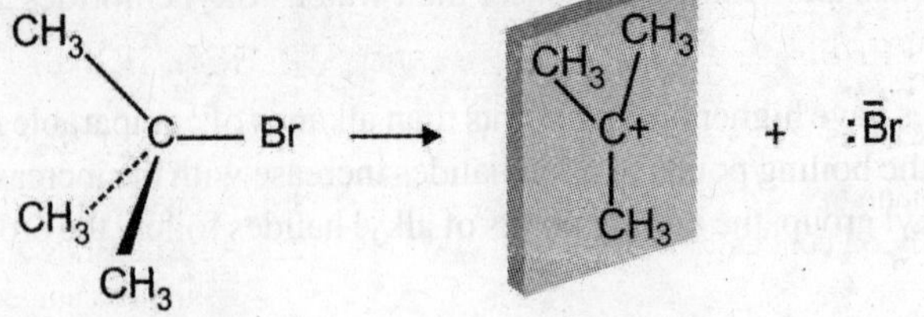

tert-Butyl bromide   Carbonium ion

Step 2. The nucleophile can attack the planar carbonium ion from either side to give the product.

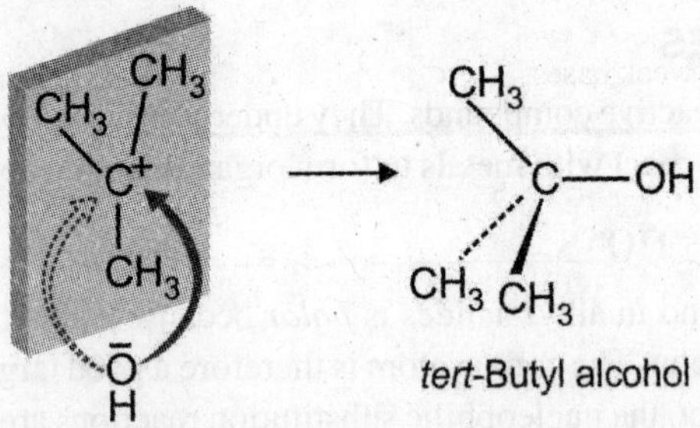

tert-Butyl alcohol

Remember that the primary alkyl halides undergo substitution by S_N2 mechanism. The tertiary alkyl halides undergo substitution by S_N1 mechanism. This is because the attack of the nucleophile on the crowded tertiary alkyl halide is quite difficult.

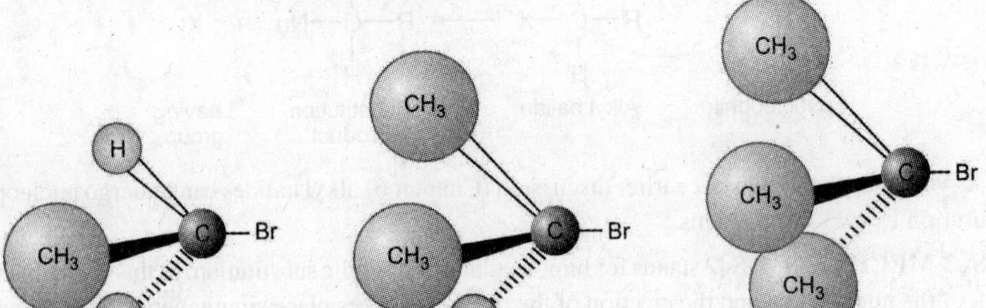

Steric crowding in ethyl bromide, isopropyl bromide, and *t*-butyl bromide.

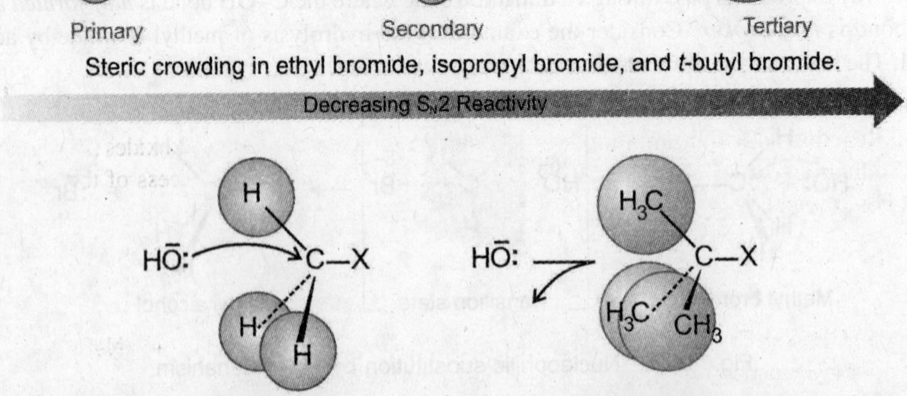

Easy attack (Primary halide)   Difficult attack (Teritiary halide)

The secondary alkyl halides may undergo nucleophilic substitution by either S_N1 or S_N2 mechanism depending on the solvent. S_N1 mechanism will predominate if an ionizing solvent is present. S_N2 mechanism will predominate if ionizing solvent is absent.

Table 13.2 DIFFERENCE BETWEEN S_N1 and S_N2 REACTIONS

S_N1 Reactions	S_N2 Reactions
1. A two-step mechanism.	1. A one-step mechanism.
2. Rate depends only on concentration of alkyl halide.	2. Rate depends on concentration of alkyl halide and nucleophile.
3. Is unimolecular.	3. Is bimolecular.
4. Rate depends on structure of halide : $3° > 2° > 1° > CH_3X$	4. Rate depends on structure of halide : $CH_3X > 1° > 2° > 3°$
5. Reaction is favored by polar solvents.	5. Reaction is favored by nonpolar solvents.
6. Reaction can occur with weak bases.	6. Reaction requires strong bases.

Some of the important nucleophilic substitution reactions of alkyl halides are described below :

(1) **Reaction with Aqueous KOH.** Alkyl halides react with aqueous potassium hydroxide to form alcohols. The halogen atom is substituted by –OH group.

$$\underset{\text{Methyl iodide}}{CH_3I} + KOH \xrightarrow[\Delta]{H_2O} \underset{\text{Methyl alcohol}}{CH_3OH} + KI$$

$$\underset{\text{Ethyl bromide}}{CH_3CH_2Br} + KOH \xrightarrow[\Delta]{H_2O} \underset{\text{Ethyl alcohol}}{CH_3CH_2OH} + KBr$$

MECHANISM. In the above reaction $H\bar{O}\!:$ is the nucleophile.

$$H\bar{O}\!: + CH_3\overset{\delta+}{C}H_2\!-\!\overset{\delta-}{Br} \xrightarrow{S_N2} \underset{\text{Ethyl alcohol}}{CH_3CH_2\!-\!OH} + :\bar{Br}$$

(2) **Reaction with Moist Silver Oxide.** Alkyl halides on treatment with a suspension of silver oxide in moist ether produce alcohols. Halogen atom is substituted by –OH group.

$$Ag_2O + H_2O \longrightarrow 2AgOH$$

$$\underset{\text{Ethyl bromide}}{CH_3CH_2Br} + AgOH \xrightarrow{\Delta} \underset{\text{Ethyl alcohol}}{CH_3CH_2OH} + AgBr$$

(3) **Reaction with Sodium Alkoxides.** Alkyl halides react with sodium alkoxides (RONa) to form ethers. Sodium alkoxides are prepared by dissolving metallic sodium in excess of the appropriate alcohol. For example,

$$CH_3CH_2OH + Na \longrightarrow \underset{\text{Sod ethoxide}}{CH_3CH_2\bar{O}\overset{+}{Na}} + H_2$$

$$\underset{\text{Ethyl bromide}}{CH_3CH_2Br} + \overset{+}{Na}\bar{O}CH_2CH_3 \longrightarrow \underset{\text{Diethyl ether}}{CH_3CH_2OCH_2CH_3} + NaBr$$

This method of making ethers is called **Williamson Ether Synthesis.**

MECHANISM. In the above reaction $CH_3CH_2\bar{O}:$ is the nucleophile.

$$CH_3CH_2\bar{O}: + CH_3\overset{\delta+}{C}H_2-\overset{\delta-}{Br} \xrightarrow{S_N2} \underset{\text{Diethyl ether}}{CH_3CH_2OCH_2CH_3} + :\bar{Br}$$

Ethers can also be produced by heating an alkyl halide with dry silver oxide.

$$\underset{\text{Methyl iodide}}{2CH_3I} + Ag_2O \xrightarrow{\Delta} \underset{\text{Diethyl ether}}{CH_3-O-CH_3} + 2AgI$$

(4) **Reaction with Ammonia.** When an alkyl halide is heated with an alcoholic solution of ammonia in a sealed tube, alkylation of ammonia takes place. A mixture of different classes of amines results.

$$\underset{\text{Ethyl bromide}}{CH_3CH_2Br} + \underset{\text{(in ethanol)}}{NH_3} \xrightarrow{\Delta} \underset{\text{Ethylamine (1°)}}{CH_3CH_2NH_2} + HBr$$

$$CH_3CH_2NH_2 + CH_3CH_2Br \longrightarrow \underset{\text{Diethylamine (2°)}}{(CH_3CH_2)_2NH} + HBr$$

$$(CH_3CH_2)_2NH + CH_3CH_2Br \longrightarrow \underset{\text{Triethylamine (3°)}}{(CH_3CH_2)_3N} + HBr$$

$$(CH_3CH_2)_3N + CH_3CH_2Br \longrightarrow \underset{\text{Tetraethylammonium bromide (4°)}}{(CH_3CH_2)_4\overset{+}{N}\overset{-}{Br}}$$

Each amine formed exists in equilibrium with its salt. For example,

$$\underset{\text{Ethylamine}}{CH_3CH_2NH_2} + HBr \rightleftharpoons \underset{\text{Ethylammonium bromide}}{CH_3CH_2\overset{+}{N}H_3\overset{-}{Br}}$$

(5) **Reaction with Sodium Cyanide.** Alkyl halides react with sodium cyanide in a suitable solvent (generally aqueous ethanol) to form alkyl cyanides or nitriles. Halogen atom is replaced by –CN group.

$$\underset{\text{Ethyl bromide}}{CH_3CH_2-Br} + NaCN \xrightarrow[\text{ethanol, }\Delta]{\text{aqueous}} \underset{\text{Ethyl cyanide}}{CH_3CH_2-CN} + NaBr$$

MECHANISM. Cyanide ion is an excellent nucleophile. It attacks ethyl bromide by an S_N2 mechanism to form ethyl cyanide.

$$NaCN \xrightarrow{\text{solvent}} \overset{+}{Na} + :\bar{C}N$$

$$:\bar{C}N + CH_3\overset{\delta+}{C}H_2-\overset{\delta-}{Br} \xrightarrow{S_N2} \underset{\text{Ethyl cyanide}}{CH_3CH_2-CN} + :\bar{Br}$$

Alkyl cyanides are useful synthetic reagents. They can be easily converted into carboxylic acids and 1° amines.

$$R-C\equiv N + 2H_2O \xrightarrow[\text{hydrolysis}]{H^+} \underset{\text{Carboxylic acid}}{R-\overset{O}{\overset{\|}{C}}-OH}$$

$$R-C\equiv N + 4[H] \xrightarrow[\text{reduction}]{LiAlH_4} \underset{\text{1° Amine}}{R-CH_2-NH_2}$$

Alkyl halides react with silver cyanide to form isocyanides.

$$\underset{\text{Ethyl bromide}}{CH_3CH_2Br} + AgCN \longrightarrow \underset{\text{Ethyl isocyanide}}{CH_3CH_2NC} + AgBr$$

The explanation for this lies in the structure of silver cyanide which is thought to exist in the form of a chain :

$$Ag{-}C{\equiv}N{:}Ag{-}C{\equiv}N{:}Ag{-}C{\equiv}N \textit{ etc.}$$

Thus the silver atom is linked to both a nitrogen and a carbon atom. Accordingly both isomers are possible.

(6) **Reaction with RCOOAg.** When an alkyl halide is heated with an alcoholic solution of the silver salt of a carboxylic acid, an ester is formed.

$$\underset{\text{Silver acetate}}{CH_3{-}\overset{\overset{\displaystyle O}{\|}}{C}{-}\overset{-}{O}\overset{+}{Ag}} + \underset{\text{Ethyl bromide}}{BrCH_2CH_3} \xrightarrow[\Delta]{\text{ethanol}} \underset{\text{Ethyl acetate}}{CH_3{-}\overset{\overset{\displaystyle O}{\|}}{C}{-}OCH_2CH_3} + AgBr$$

(7) **Reaction with Acetylides.** Alkyl halides react with sodium acetylides to form higher alkynes.

$$CH_3Br + \underset{\text{Sodium acetylide}}{\overset{+}{Na}\overset{-}{C}{\equiv}CH} \longrightarrow \underset{\text{Propyne}}{CH_3{-}C{\equiv}CH} + NaBr$$

$$CH_3Br + \underset{\text{Sodium propynide}}{\overset{+}{Na}\overset{-}{C}{\equiv}CCH_3} \longrightarrow \underset{\text{2-Butyne}}{CH_3C{\equiv}CCH_3} + NaBr$$

(8) **Reaction with KSH.** Alkyl halides react with alcoholic potassium hydrosulfide to form thiols. Halogen atom is substituted by –SH group.

$$\underset{\text{Ethyl iodide}}{CH_3CH_2{-}I} + KSH \xrightarrow[\Delta]{\text{ethanol}} \underset{\text{Ethanethiol}}{CH_3CH_2{-}SH} + KI$$

(9) **Reaction with K_2S.** Alkyl halides react with potassium sulfide to form dialkyl sulfides.

$$\underset{\text{Ethyl iodide}}{2CH_3CH_2{-}I} + K_2S \longrightarrow \underset{\text{Diethyl sulphide}}{CH_3CH_2{-}S{-}CH_2CH_3} + 2KI$$

(10) **Reaction with $AgNO_2$.** Alkyl iodides react with silver nitrite to form nitroalkanes.

$$\underset{\text{Ethyl iodide}}{CH_3CH_2I} + AgNO_2 \longrightarrow \underset{\text{Nitroethane}}{CH_3CH_2NO_2} + AgI$$

B. ELIMINATION REACTIONS

(11) **Reaction with Alcoholic KOH.** Alkyl halides react with alcoholic potassium hydroxide to form alkenes. The reaction involves the elimination of HX from the alkyl halides and is called *dehydrohalogenation reaction*.

$$\underset{\text{Ethyl bromide}}{\underset{}{\overset{\overset{\displaystyle H}{|}}{C}H_2{-}\overset{\overset{\displaystyle Br}{|}}{C}H_2}} + KOH \xrightarrow[\Delta]{\text{ethanol}} \underset{\text{Ethylene}}{CH_2{=}CH_2} + KBr + H_2O$$

$$\underset{\text{1-Bromopropane}}{CH_3\overset{\overset{\displaystyle H}{|}}{C}H{-}\overset{\overset{\displaystyle Br}{|}}{C}H_2} + KOH \xrightarrow[\Delta]{\text{ethanol}} \underset{\text{Propene}}{CH_3CH{=}CH_2} + KBr + H_2O$$

MECHANISM. In ethanol an equilibrium occurs between the solvent and potassium hydroxide to produce potassium ethoxide.

$$\underset{\text{Ethanol}}{CH_3CH_2OH} + KOH \rightleftharpoons \underset{\text{Pot ethoxide}}{CH_3CH_2\overset{-}{O}\overset{+}{K}} + H_2O$$

Potassium ethoxide is a strong base. It favours elimination and substitution reactions. There is always a competition between elimination and substitution reactions. For example, ethyl bromide on treatment with alcoholic KOH can give either ethylene or diethyl ether. The attacking nucleophile is $CH_3CH_2O{:}^-$.

Elimination :

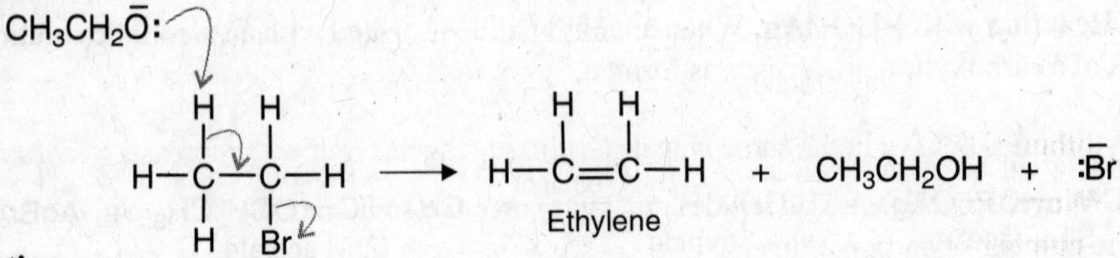

Substitution :

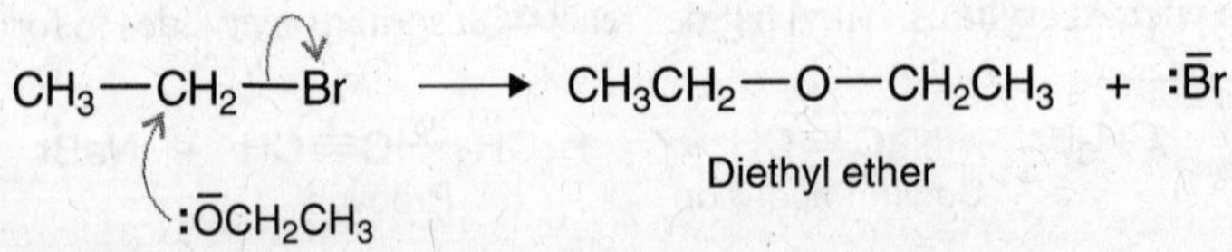

The ratio of the elimination to substitution product depends on the structure of the alkyl halide and experimental conditions. Primary and secondary alkyl halides undergo dehydrohalogenation by E2 mechanism. Tertiary alkyl halides do so by E1 mechanism (*See Chapter* 6).

Saytzeff Rule. If the dehydrohalogenation of an alkyl halide can yield more than one alkene, then according to the *Saytzeff rule*, the main product is the most highly substituted alkene. For example, two alkenes are possible when 2-bromobutane is heated with alcoholic KOH.

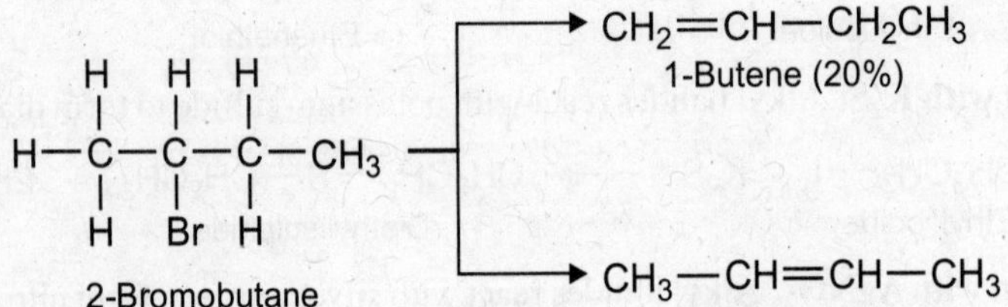

Notice that the major product is 2-butene, a disubstituted alkene.

C. MISCELLANEOUS REACTIONS

(12) **Reduction.** Alkyl halides are reduced to alkanes by (*i*) Zn + HCl ; (*ii*) $LiAlH_4$; and (*iii*) H_2 in the presence of Ni or Pd.

$$\underset{\text{Ethyl bromide}}{CH_3CH_2Br + H_2} \xrightarrow[\textit{or } Pd]{Ni} \underset{\text{Ethane}}{CH_3CH_3} + HBr$$

$$\underset{\text{1-Iodopropane}}{CH_3CH_2CH_2I} + 2[H] \xrightarrow{Zn/HCl} \underset{\text{Propane}}{CH_3CH_2CH_3} + HI$$

$$\underset{\text{2-Bromopropane}}{CH_3-\overset{\overset{\displaystyle Br}{|}}{C}H-CH_3} + 2[H] \xrightarrow[\text{ether}]{LiAlH_4} \underset{\text{Propane}}{CH_3CH_2CH_3} + HBr$$

(13) **Reaction with Mg.** Alkyl halides react with magnesium metal in dry ether to form Grignard reagents.

$$\underset{\text{Methyl iodide}}{CH_3I} + Mg \xrightarrow{\text{ether}} \underset{\text{Methylmagnesium iodide}}{CH_3MgI}$$

$$\underset{\text{Ethyl bromide}}{CH_3CH_2Br} + Mg \xrightarrow{\text{ether}} \underset{\text{Ethylmagnesium bromide}}{CH_3CH_2MgBr}$$

(14) **Reaction with Lithium.** Alkyl halides react with lithium in dry ether to form alkyllithiums.

$$\underset{\text{Ethyl bromide}}{CH_3CH_2Br} + 2Li \xrightarrow{\text{ether}} \underset{\text{Ethyllithium}}{CH_3CH_2Li} + LiBr$$

Alkyllithiums behave in the same way as Grignard reagents, but with increased reactivity.

(15) **Wurtz Reaction.** Alkyl halides react with metallic sodium in dry ether to give alkanes with double the number of carbon atoms.

$$\underset{\text{Ethyl bromide}}{CH_3CH_2Br} + 2Na + BrCH_2CH_3 \xrightarrow{\text{ether}} \underset{n\text{-Butane}}{CH_3CH_2CH_2CH_3} + 2NaBr$$

(16) **Halogenation.** Alkyl halides react with Cl_2 or Br_2 in the presence of UV light or at high temperature to form polyhalogenation derivatives. For example, methyl chloride reacts with chlorine to yield a mixture of methylene dichloride, chloroform, and carbon tetrachloride.

$$CH_3Cl \xrightarrow[\text{UV light}]{Cl_2} CH_2Cl_2 \xrightarrow{Cl_2} CHCl_3 \xrightarrow{Cl_2} CCl_4$$

(17) **Friedel-Crafts Alkylation.** Alkyl halides react with benzene in the presence of anhydrous $AlCl_3$ to form alkylbenzenes.

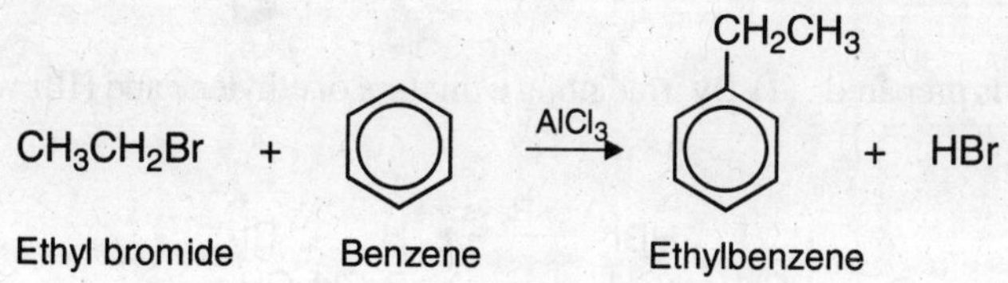

METHYL IODIDE, *Iodomethane*, CH_3I

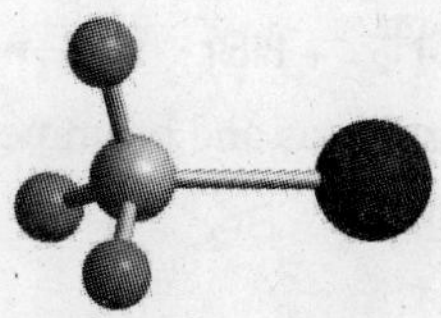

Preparation. It can be prepared : (1) By warming a mixture of methyl alcohol, red phosphorus and iodine.

$$\underset{\text{Methyl alcohol}}{6CH_3OH} + 2P + 3I_2 \xrightarrow{\text{warm}} \underset{\text{Methyl iodide}}{6CH_3I} + 2H_3PO_3$$

(2) By the action of dimethyl sulfate on potassium iodide solution in the presence of calcium carbonate.

$$\underset{\text{Dimethyl sulphate}}{(CH_3)_2SO_4} + KI \longrightarrow \underset{\text{Methyl iodide}}{CH_3I} + K(CH_3)SO_4$$

Properties. Methyl iodide is a sweet-smelling liquid, *bp* 42.5°C. When freshly prepared, it is colorless but becomes brown on standing, which is due to its decomposition to iodine. Since it is a liquid, it is easier to handle than methyl chloride (a gas) and is used a great deal as a **Methylating Agent** in organic synthesis.

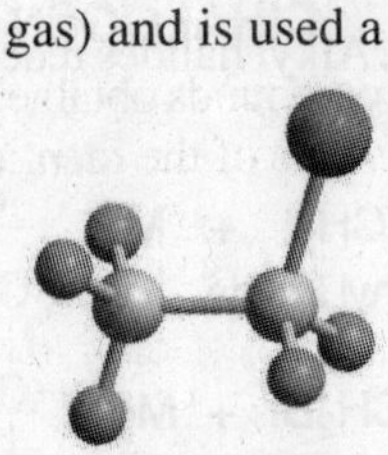

ETHYL CHLORIDE, *Chloroethane*, CH_3CH_2Cl

Preparation. It can be prepared : (1) By passing dry HCl into ethyl alcohol in the presence of zinc chloride.

$$\underset{\text{Ethyl alcohol}}{CH_3CH_2OH} + HCl \xrightarrow{ZnCl_2} \underset{\text{Ethyl chloride}}{CH_3CH_2Cl} + H_2O$$

(2) By the chlorination of ethane at 400°C.

$$\underset{\text{Ethane}}{CH_3CH_3} + Cl_2 \xrightarrow{400°C} CH_3CH_2Cl + \text{Polyhalogen Derivatives}$$

Ethyl chloride is separated by fractional distillation.

(3) By the addition of HCl to ethylene in the presence of aluminium chloride.

$$\underset{\text{Ethylene}}{CH_2{=}CH_2} + HCl \xrightarrow{AlCl_3} \underset{\text{Ethyl chloride}}{CH_3CH_2Cl}$$

Properties. Ethyl chloride is a gas at room temperature, *bp* 12.5°C. It is used as a refrigerating agent and in the manufacture of tetraethyllead.

ETHYL BROMIDE, *Bromoethane*, CH_3CH_2Br

Preparation. It is prepared : (1) By irradiating a mixture of ethylene and HBr with γ-radiation from a cobalt-60 source.

$$HBr \xrightarrow{\gamma\text{-Rays}} H\cdot + Br\cdot$$
$$Br\cdot + CH_2{=}CH_2 \longrightarrow BrCH_2CH_2\cdot$$
$$BrCH_2CH_2\cdot + HBr \longrightarrow BrCH_2CH_3 + Br\cdot$$

(2) By the action of red phosphorus and bromine on ethyl alcohol.

$$\underset{\text{Ethyl alcohol}}{6CH_3CH_2OH} + 2P + Br_2 \longrightarrow \underset{\text{Ethyl bromide}}{6CH_3CH_2Br} + {}_2H_3PO_3$$

(3) By the action of hydrogen bromide on ethyl alcohol.

$$\underset{\text{Ethyl alcohol}}{CH_3CH_2OH} + HBr \longrightarrow CH_3CH_2Br + H_2O$$

Hydrogen bromide used in the above reaction is prepared *in situ* by the action of concentrated sulfuric acid on potassium bromide.

$$2KBr + H_2SO_4 \longrightarrow 2HBr + K_2SO_4$$

Properties. Ethyl bromide is a colorless liquid, *bp* 38.3°C. It is insoluble in water but dissolves in ethanol and ether. Ethyl bromide gives all the typical reactions of alkyl halides. It is used as an **Ethylating Agent** in organic synthesis.

DIHALOGEN DERIVATIVES

Dihalogen derivatives are compounds obtained by replacing two hydrogen atoms of a hydrocarbon by two halogen atoms. The presence of the identical halogen atoms is indicated by the prefix *di-* and position numbers. For example,

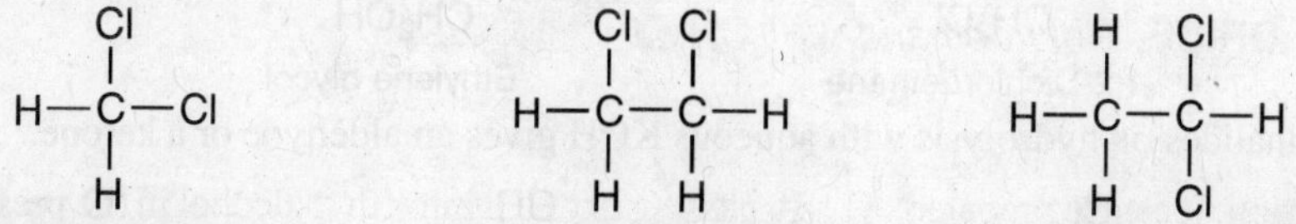

Dichloromethane (*Methylene dichloride*) — 1,2-Dichloroethane (*Ethylene dichloride*) — 1,1-Dichloroethane (*Ethylene dichloride*)

If two halogen atoms are attached to *adjacent* carbons, the compound is referred to as a **vicinal (*vic*-) Dihalide**. If two halogen atoms are attached to the *same* carbon, then it is known as a **geminal (*gem*-) Dihalide**. Notice that in the above examples, 1,2-dichloroethane is a *vic*-dihalide ; 1,1-dichloroethane is a *gem*-dihalide.

METHODS OF PREPARATION

***gem*-Dihalides** are prepared : (1) By the action of phosphorus pentahalides on aldehydes and ketones.

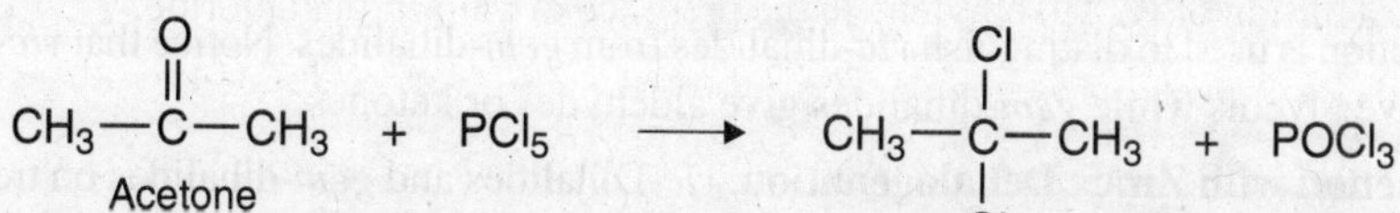

Acetone → 2,2-Dichloropropane

(2) By the addition of hydrogen halides to alkynes. Markovnikov rule is followed.

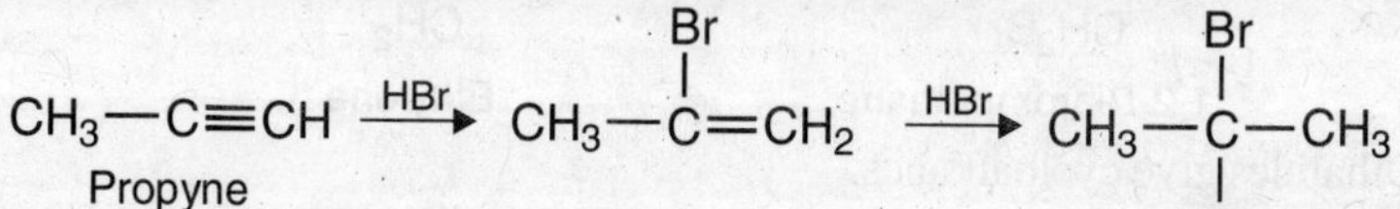

Propyne → 2,2-Dibromopropane

***vic*-Dihalides** are prepared : (1) By the addition of halogens to alkenes.

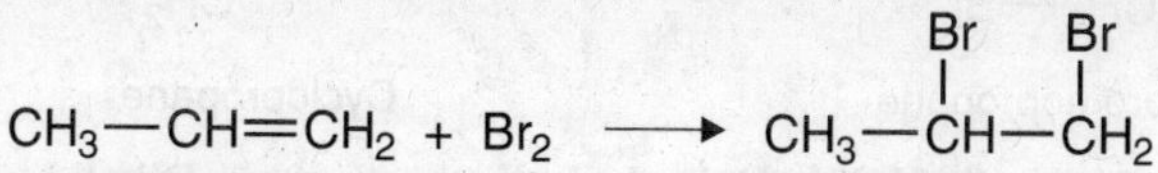

Propene → 1,2-Dibromopropane

(2) By the action of phosphorus halides (or hydrogen halides) on glycols.

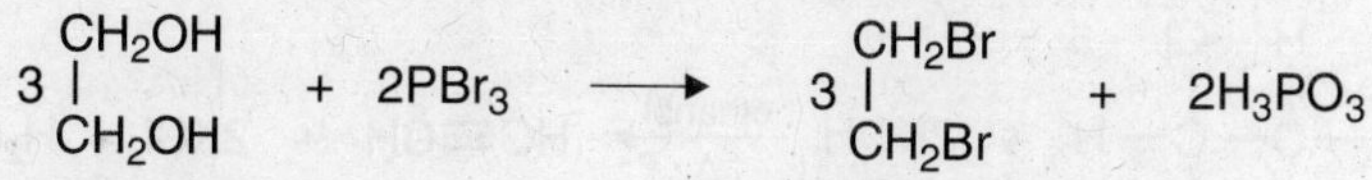

Ethylene glycol → 1,2-Dibromoethane (Ethylene dibromide)

PHYSICAL PROPERTIES

(1) Dihalides are sweet-smelling, colorless liquids. They have relatively high boiling points.

(2) They are insoluble in water and soluble in organic solvents.

(3) They are heavier than water. Diiodomethane (CH_2I_2) is the densest organic liquid. Its density is 3.325.

CHEMICAL PROPERTIES

The chemical properties of dihalides are very similar to those of alkyl halides. Thus they undergo both substitution and elimination reactions.

(1) **Hydrolysis with Aqueous NaOH or KOH.** *vic*-Dihalides on heating with aqueous sodium hydroxide give glycols.

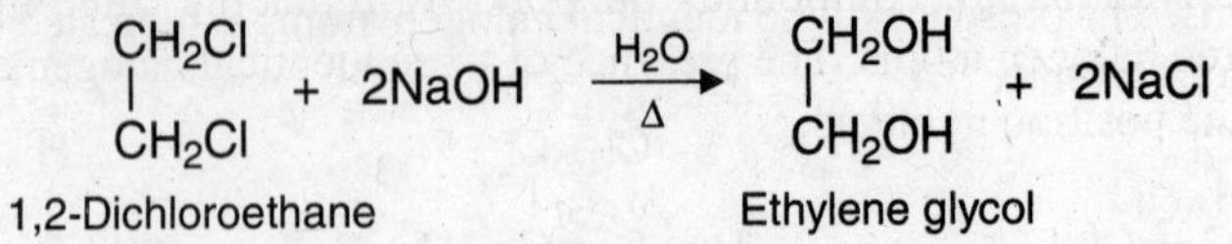

gem-Dihalides on hydrolysis with aqueous KOH gives an aldehyde or a ketone.

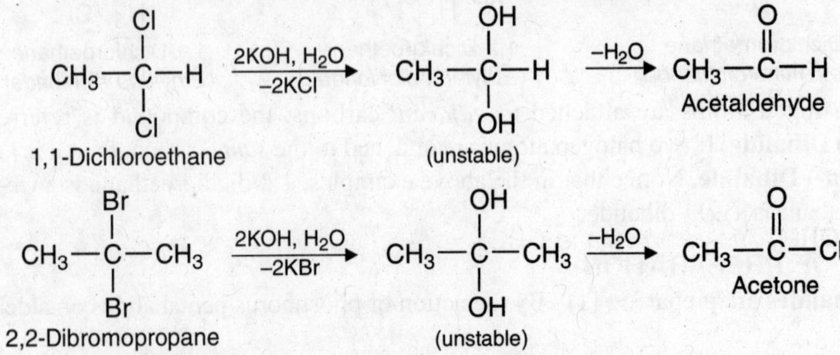

This reaction is used to distinguish *vic*-dihalides from *gem*-dihalides. Notice that *vic*-dihalides on hydrolysis give glycols while *gem*-dihalides give aldehydes or ketones.

(2) **Reaction with Zinc ; Dehalogenation.** *vic*-Dihalides and *gem*-dihalides on treatment with zinc dust in methanol give alkenes.

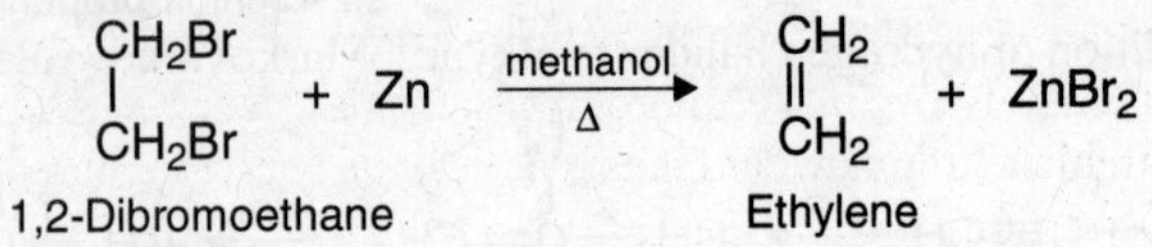

1,3- to 1,6-Dihalides give cycloalkanes.

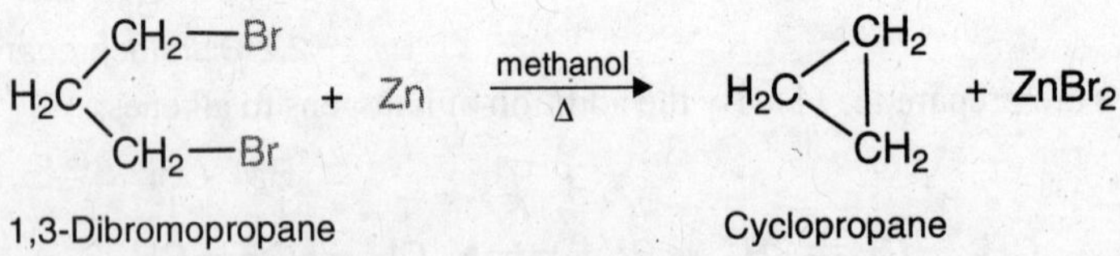

(3) **Reaction with Alcoholic KOH ; Dehydrohalogenation.** *vic*-Dihalides and *gem*-dihalides on treatment with alcoholic potassium hydroxide give alkynes.

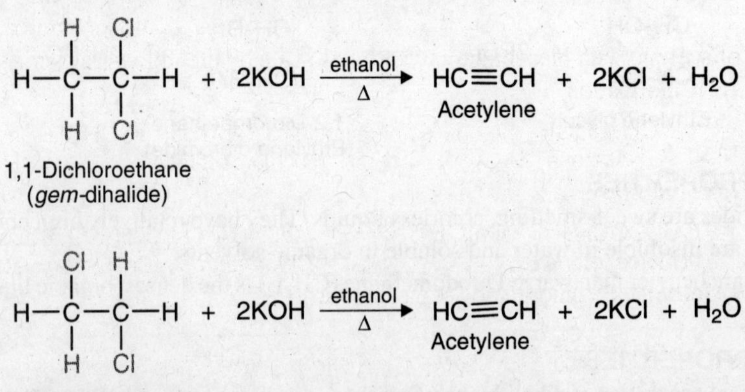

TRIHALOGEN DERIVATIVES

Trihalogen derivatives are compounds obtained by replacing three hydrogen atoms of a hydrocarbon by three halogen atoms. The presence of three identical halogen atoms is indicated by the prefix *tri-* and the position numbers.

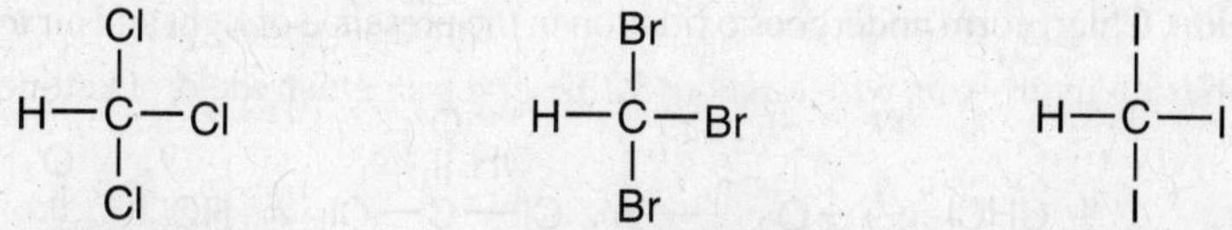

Trichloromethane (*Chloroform*) Tribromomethane (*Bromoform*) Triiodomethane (*Iodoform*)

CHLOROFORM, *Trichloromethane*, $CHCl_3$

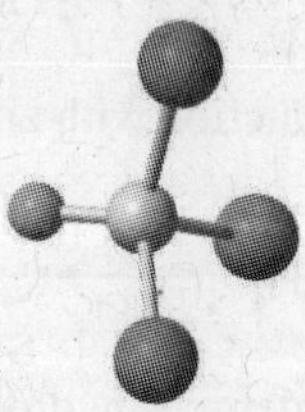

Chloroform is an important trihalogen derivative of methane. In the past chloroform was extensively used as a general anesthetic for surgery but it is rarely used for this purpose now because it causes extensive liver damage.

Preparation. Chloroform is prepared :

(1) **From Ethyl Alcohol (or Acetone) and Bleaching Powder.** By heating ethyl alcohol or acetone with bleaching powder, $Ca(OCl)_2$. The bleaching powder acts as source of chlorine and calcium hydroxide. This method is used to make chloroform in the laboratory and on commercial scale. Reaction of ethyl alcohol with bleaching powder takes place by the following three steps.

Step 1. Oxidation

$$\underset{\text{Ethyl alcohol}}{CH_3{-}CH_2OH} + Cl_2 \longrightarrow \underset{\text{Acetaldehyde}}{CH_3{-}CHO} + 2HCl$$

Step 2. Chlorination

$$\underset{\text{Acetaldehyde}}{CH_3{-}CHO} + 3Cl_2 \longrightarrow \underset{\substack{\text{Chloral}\\ \text{(Trichloroacetaldehyde)}}}{CCl_3{-}CHO} + 3HCl$$

Step 3. Hydrolysis

$$\underset{\text{Chloral}}{2CCl_3{-}CHO} + Ca(OH)_2 \longrightarrow \underset{\text{Chloroform}}{2CHCl_3} + \underset{\text{Calcium formate}}{(HCOO)_2Ca}$$

Reaction of acetone with bleaching powder takes place by the following two steps.

Step 1. Chlorination

$$\underset{\text{Acetone}}{CH_3COCH_3} + 3Cl_2 \longrightarrow \underset{\text{Trichloroacetone}}{CCl_3COCH_3} + 3HCl$$

Step 2. Hydrolysis

$$\underset{\text{Trichloroacetone}}{CCl_3COCH_3} + Ca(OH)_2 \longrightarrow \underset{\text{Chloroform}}{2CHCl_3} + \underset{\text{Calcium acetate}}{(CH_3COO)_2Ca}$$

(2) **From Methane.** By chlorination of methane at 400°C.

$$CH_4 \xrightarrow[400°]{Cl_2} CH_3Cl + CH_2Cl_2 + CHCl_3 + CCl_4$$

Chloroform is separated from the products by fractional distillation. Today most of the chloroform is manufactured by this method.

Properties (*Physical*). Chloroform is a colorless liquid with a characteristic sickly smell, bp 61°C. It is almost insoluble in water but is soluble in most organic solvents. Chloroform vapors when inhaled cause temporary unconsciousness.

(*Chemical*). The chemical properties of chloroform are as follows :

(1) **Oxidation.** Chloroform undergoes oxidation in the presence of light and air to form phosgene (carbonyl chloride).

$$\underset{\text{Chloroform}}{CHCl_3} + \tfrac{1}{2}O_2 \longrightarrow \underset{\text{Phosgene}}{Cl-\overset{\overset{\displaystyle O}{\|}}{C}-Cl} + HCl$$

Chloroform is stored in dark brown bottles to prevent the formation of phosgene, as it is highly poisonous.

(2) **Reduction.** It undergoes reduction with zinc and hydrochloric acid in the presence of ethyl alcohol to form dichloromethane.

$$\underset{\text{Chloroform}}{CHCl_3} + 2[H] \xrightarrow[HCl]{Zn} \underset{\text{Dichloromethane}}{CH_2Cl_2} + HCl$$

(3) **Hydrolysis.** Chloroform undergoes hydrolysis with hot aqueous sodium hydroxide to give sodium formate.

$$\underset{\text{Chloroform}}{CHCl_3} + 4NaOH \longrightarrow \underset{\text{Sodium formate}}{HCO\overset{-}{O}\overset{+}{Na}} + 3NaCl + 2H_2O$$

(4) **Chlorination.** Chloroform reacts with chlorine in the presence of diffused sunlight or UV light to form carbon tetrachloride.

$$\underset{\text{Chloroform}}{CHCl_3} + Cl_2 \xrightarrow{\text{UV light}} \underset{\text{Carbon tetrachloride}}{CCl_4} + HCl$$

(5) **Nitration.** Chloroform reacts with nitric acid to form *chloropicrin* or nitrochloroform. Chloropicrin is used as an insecticide.

$$Cl_3C-H + HO-NO_2 \longrightarrow \underset{\text{Chloropicrin}}{Cl_3C-NO_2} + H_2O$$

(6) **Reaction with Silver.** Chloroform reacts with silver powder to form acetylene.

$$\underset{\text{Chloroform}}{CHCl_3} + 6Ag + Cl_3CH \longrightarrow \underset{\text{Acetylene}}{HC\equiv CH} + 6AgCl$$

(7) **Reaction with Acetone.** Chloroform undergoes condensation with acetone in the presence of alkali to form *chloretone*. Chloretone is used as a drug.

$$\underset{\text{Acetone}}{CH_3-\overset{\overset{\displaystyle O}{\|}}{C}-CH_3} + \underset{\text{Chloroform}}{H-CCl_3} \longrightarrow \underset{\text{Chloretone}}{CH_3-\overset{\overset{\displaystyle OH}{|}}{\underset{\underset{\displaystyle CCl_3}{|}}{C}}-CH_3}$$

(8) **Reimer-Tiemann Reaction.** Chloroform reacts with phenol in sodium hydroxide to form salicylaldehyde.

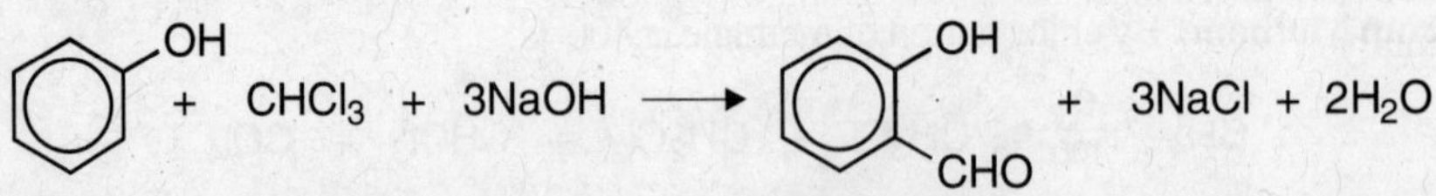

Phenol Salicylaldehyde

(9) **Isocyanide Reaction**. Chloroform reacts with primary amines in the presence of alcoholic potassium hydroxide to form an isocyanide or isonitrile.

$$\underset{\text{Chloroform}}{CHCl_3} + 3KOH + \underset{\text{Amine}}{R{-}NH_2} \xrightarrow{\Delta} \underset{\text{Isocyanide}}{R{-}\overset{-}{N}{\equiv}\overset{+}{C}} + 3KCl + 3H_2O$$

Isocyanides have strong disagreeable odors. Their formation is used as a **Test for Primary Amines**.

Uses. Chloroform is used : (1) as a solvent for fats, waxes, and rubber ; and (2) in the preparation of chloropicrin and chloretone.

IODOFORM, *Triiodomethane*, CHI_3

Iodoform resembles chloroform in the methods of preparation and properties.

Preparation. Iodoform is prepared :

(1) **From Ethanol or Acetone**. By the action of iodine and alkali on ethanol or acetone.

$$\underset{\text{Ethanol}}{CH_3{-}CH_2OH} + I_2 \longrightarrow \underset{\text{Acetaldehyde}}{CH_3{-}CHO} + 2HI$$

$$CH_3{-}CHO + 3I_2 \longrightarrow \underset{\text{Iodal}}{I_3C{-}CHO} + 3HI$$

$$I_3C{-}CHO + KOH \longrightarrow \underset{\text{Iodoform}}{CHI_3} + HCOOK$$

Sodium carbonate can also be used in place of KOH.

(2) **Electrolytic Method**. By the electrolysis of a solution of ethanol, potassium iodide, and sodium carbonate. The solution is maintained at 60–70°C. During electrolysis, iodine is set free at the anode and sodium hydroxide at the cathode which is subsequently converted to sodium carbonate by passing a current of carbon dioxide. These substances then react with ethanol to form iodoform. Being a cheaper method it is used for the industrial preparation of iodoform.

Properties. Iodoform is a yellow crystalline solid, mp 119°C. It is insoluble in water but dissolves in ethyl alcohol or ether. Iodoform undergoes hydrolysis with hot aqueous sodium hydroxide to form sodium formate.

$$\underset{\text{Iodoform}}{CHI_3} + 4NaOH \xrightarrow{\Delta} \underset{\text{Sodium formate}}{HCOONa} + 3NaI + 2H_2O$$

Iodoform is used as an antiseptic.

CARBON TETRACHLORIDE, *Tetrachloromethane*, CCl_4

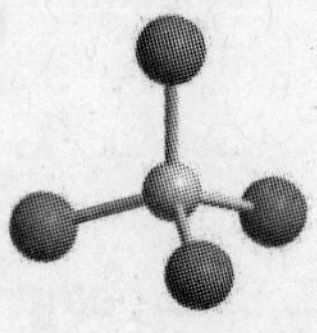

Carbon tetrachloride is the most important tetrahalogen derivative of methane.

Preparation. Carbon tetrachloride is manufactured :

(1) **From Carbon Disulfide**. By the action of chlorine on carbon disulfide. Aluminium chloride is used as a catalyst.

$$CS_2 + 3Cl_2 \xrightarrow{AlCl_3} \underset{\text{Carbon tetrachloride}}{CCl_4} + \underset{\text{Sulphur monochloride}}{S_2Cl_2}$$

Carbon tetrachloride is removed from the mixture by fractional distillation.

(2) **From Methane**. By chlorination of methane with excess of chlorine at 400°C.

$$CH_4 + 4Cl_2 \xrightarrow{400°C} CCl_4 + 4HCl$$

Properties (*Physical*). Carbon tetrachloride is a colorless liquid, *bp* 77°C. It is insoluble in water but soluble in ethanol and ether. It is an excellent solvent for oils, waxes, fats, and greases. Carbon tetrachloride is used as a fire-extinguisher under the name **Pyrene** because it is non-inflammable. When a spray of carbon tetrachloride is directed at fire, its dense vapors prevent the oxygen of air from reaching the burning articles. However, carbon tetrachloride also forms phosgene. Hence after using *Pyrene* to extinguish a fire, the room should be well ventilated.

(*Chemical*). Carbon tetrachloride is inert to most organic reagents. It gives the following reactions :

(1) **Reaction with Steam**. Carbon tetrachloride vapors react with steam at 500°C to give phosgene.

$$\underset{}{CCl_4} + \underset{\text{steam}}{H_2O} \xrightarrow{500°C} \underset{\text{Phosgene}}{Cl-\overset{\overset{\displaystyle O}{\|}}{C}-Cl} + 2HCl$$

(2) **Reduction**. Carbon tetrachloride can be reduced by moist iron filings to give chloroform.

$$CCl_4 + 2[H] \xrightarrow{Fe/H_2O} \underset{\text{Chloroform}}{CHCl_3} + HCl$$

(3) **Reaction with HF**. Carbon tetrachloride reacts with hydrogen fluoride in the presence of antimony pentafluoride to form dichlorodifluoromethane (*Freon-12*).

$$CCl_4 + 2HF \xrightarrow{SbF_5} \underset{\text{Freon-12}}{CCl_2F_2} + 2HCl$$

Freon-12 (*bp*–29°C) is widely used as a refrigerant and propellant in aerosol sprays of all kinds.

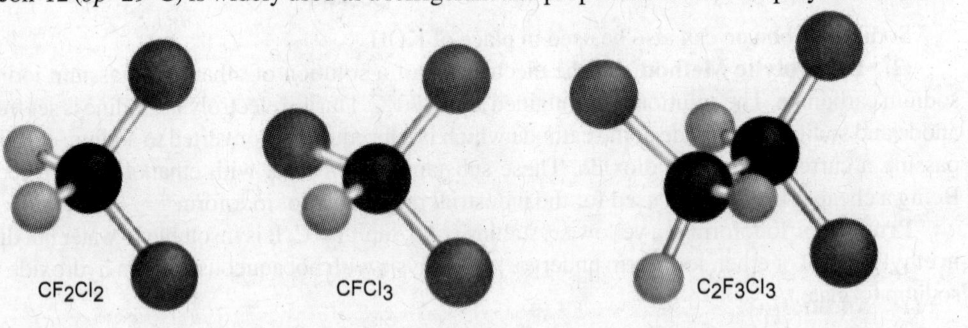

CF_2Cl_2 is Freon-12. Other CFCs include $CFCl_3$ (Freon-11) and $C_2F_3Cl_3$ (Freon-113).

UNSATURATED HALIDES

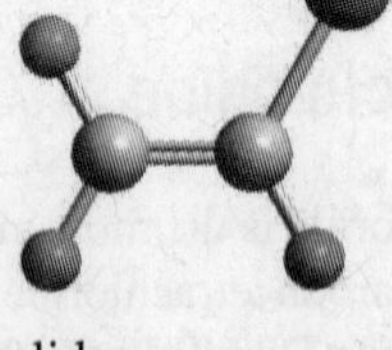

VINYL CHLORIDE, *Chloroethene*, $CH_2=CH-Cl$

Vinyl chloride is the most important of the unsaturated halides.

Preparation. Vinyl chloride is obtained :

(1) By the controlled addition of hydrogen chloride to acetylene. $HgCl_2$ is used as a catalyst.

$$\underset{\text{Acetylene}}{CH\equiv CH} + HCl \xrightarrow[150°C]{HgCl_2} \underset{\text{Vinyl chloride}}{CH_2=CH-Cl}$$

(2) By the action of chlorine on ethylene at 500°C.

$$\underset{\text{Ethylene}}{CH_2{=}CH_2} + Cl_2 \xrightarrow{500°C} \underset{\text{Vinyl chloride}}{CH_2{=}CH{-}Cl}$$

(3) By heating 1,2-dichloroethane with alcoholic KOH.

$$\underset{\text{1,2-Dichloroethane}}{\overset{\displaystyle Cl\quad\;\; H}{\overset{|\qquad\;\; |}{CH_2{-}CH{-}Cl}}} + KOH \xrightarrow[\Delta]{\text{ethanol}} \underset{\text{Vinyl chloride}}{CH_2{=}CH{-}Cl} + KCl + H_2O$$

Properties (*Physical*). Vinyl chloride is a colorless gas at room temperature, *bp* –13°C.

(*Chemical*). Vinyl chloride contains a carbon-carbon double bond and a halogen atom. The halogen atom is quite inert. It does not react with nucleophilic reagents the way other alkyl halides do. For example, it is not hydrolysed by sodium hydroxide.

$$CH_2{=}CH{-}Cl + NaOH \longrightarrow \text{No Reaction}$$

$$(\textit{Cf.}\ CH_3CH_2{-}Cl + NaOH \xrightarrow{H_2O} CH_3CH_2OH + NaCl)$$

Why Halogen of Vinyl Chloride is Inert ? The reason for this is that the C–Cl bond in vinyl chloride is stronger than the C–Cl bond in an alkyl halide. This is so because a *p* orbital on chlorine interacts with the *p* orbital on the adjacent carbon atom (Fig. 13.3). The new delocalized orbital gives partial double bond character to the C–Cl bond in vinyl chloride. As a result the chlorine atom becomes firmly bound to carbon and cannot be replaced easily.

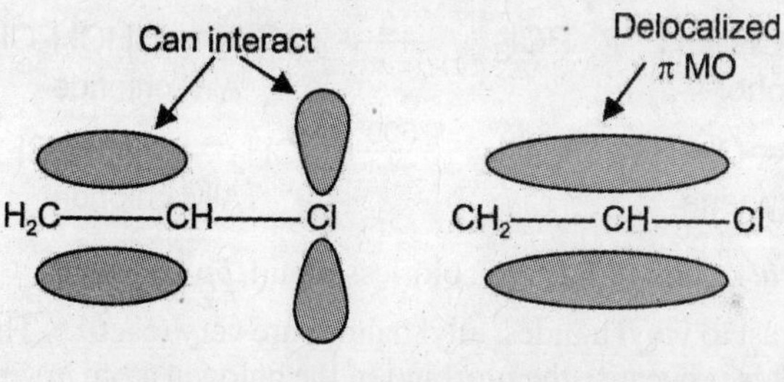

Fig. 13.3. *p*-orbital overlap in vinyl chloride.

However, the carbon-carbon double bond of vinyl chloride shows the usual addition reactions.

(1) **Addition of Cl_2.** Chlorine adds to vinyl chloride to form 1,1,2-trichloroethane.

$$\underset{\text{Vinyl chloride}}{CH_2{=}CH{-}Cl} + Cl_2 \longrightarrow \underset{\text{1,1,2-Trichloroethane}}{\overset{\displaystyle Cl\quad\;\; Cl}{\overset{|\qquad\;\; |}{CH_2{-}CH{-}Cl}}}$$

(2) **Addition of HBr**. Hydrogen bromide adds to vinyl chloride to form 1-bromo-1-chloroethane. Markovnikov rule is followed.

$$CH_2{=}CH{-}Cl + HBr \longrightarrow \underset{\text{1-Bromo-1-chloroethane}}{\overset{\displaystyle Br}{\overset{|}{CH_3{-}CH{-}Cl}}}$$

(3) **Polymerization**. Vinyl chloride undergoes polymerization in the presence of benzoyl peroxide ($C_6H_5CO{-}O{-}O{-}COC_6H_5$) to form polyvinyl chloride (*PVC*).

$$\underset{\text{Vinyl chloride}}{n\overset{\displaystyle \quad\;\; Cl}{\overset{\qquad |}{CH_2{=}CH}}} \xrightarrow{\text{peroxide}} \underset{\text{PVC}}{\left[\overset{\displaystyle \qquad\quad Cl}{\overset{\qquad\quad |}{CH_2{-}CH}}\right]_n}$$

PVC is used in the manufacture of synthetic leather goods, rain coats, plastic curtains, plastic pipes, floor tiles, and packaging materials.

ALLYL IODIDE, *3-Iodo-1-propene*, $CH_2{=}CHCH_2I$

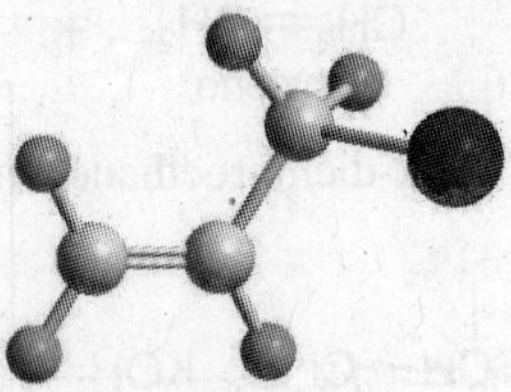

Allyl iodide is widely used in organic synthesis.

Preparation. It is prepared : (1) By heating glycerol with hydriodic acid.

$$\underset{\text{Glycerol}}{\begin{matrix} CH_2OH \\ | \\ CHOH \\ | \\ CH_2OH \end{matrix}} + 3HI \xrightarrow[-H_2O]{\Delta} \begin{matrix} CH_2I \\ | \\ CHI \\ | \\ CH_2I \end{matrix} \xrightarrow[-I_2]{\Delta} \underset{\text{Allyl iodide}}{\begin{matrix} CH_2I \\ | \\ CH \\ \| \\ CH_2 \end{matrix}}$$

(2) By heating allyl chloride with sodium iodide in acetone.

$$\underset{\text{Allyl chloride}}{CH_2{=}CHCH_2Cl} + NaI \xrightarrow[\Delta]{\text{acetone}} \underset{\text{Allyl iodide}}{CH_2{=}CHCH_2I} + NaCl$$

This halogen-exchange reaction is called **Finkelstein Reaction**. Allyl chloride used in the reaction may be obtained by heating allyl alcohol with PCl_3 or chlorination of propene at 500°C.

$$\underset{\text{Allyl alcohol}}{3CH_2{=}CHCH_2OH} + PCl_3 \xrightarrow{\Delta} \underset{\text{Allyl chloride}}{3CH_2{=}CHCH_2Cl} + H_3PO_3$$

$$\underset{\text{Propene}}{CH_2{=}CHCH_3} + Cl_2 \xrightarrow{500^\circ} \underset{\text{Allyl chloride}}{CH_2{=}CHCH_2Cl} + HCl$$

Properties. (*Physical*). Allyl iodide is a colorless liquid, *bp* 103.1°C.

(***Chemical***) In contrast to vinyl halides, allyl halides are very reactive. They undergo nucleophilic substitution reactions. This is because the *p*-orbital of the halogen atom in vinyl halides interacts with the adjacent π MO. The delocalization gives partial double bond character to the C–X bond. This means that the halogen atom is firmly bound to the carbon atom and cannot be replaced easily. In allyl halides, this type of interaction is not possible because the p-orbital of the halogen atom is separated from the π MO of the double bond by a saturated sp^3-hybridized carbon atom (Fig. 13.4). As a result the halogen atom in allyl halides can be easily replaced. Thus the reactions of allyl halides are analogous to those of alkyl halides.

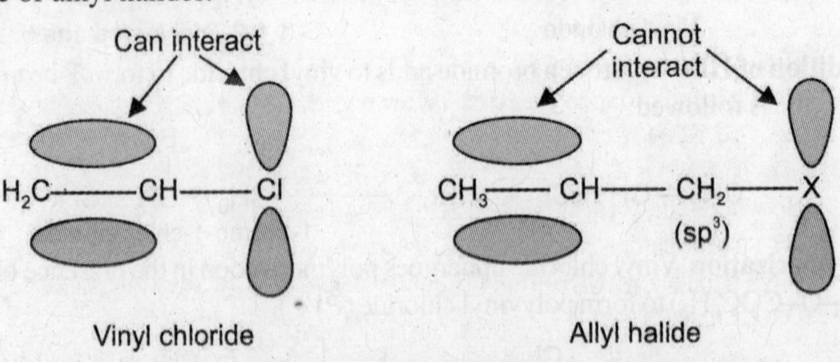

Fig. 13.4. In allyl halides, the *p*-orbital of the halogen atom cannot interact with the π MO of the C–C double bond. This is because they are separated by a saturated sp^3 hybridized carbon atom.

(1) **Substitution Reactions**. Following are some of the important nucleophilic substitution reactions of allyl iodide.

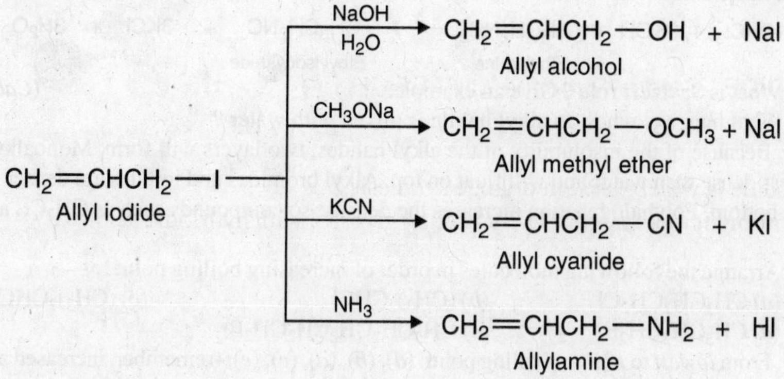

(2) **Addition Reactions**. The carbon-carbon double bond in allyl iodide shows the usual electrophilic addition reactions. Markovnikov rule is followed.

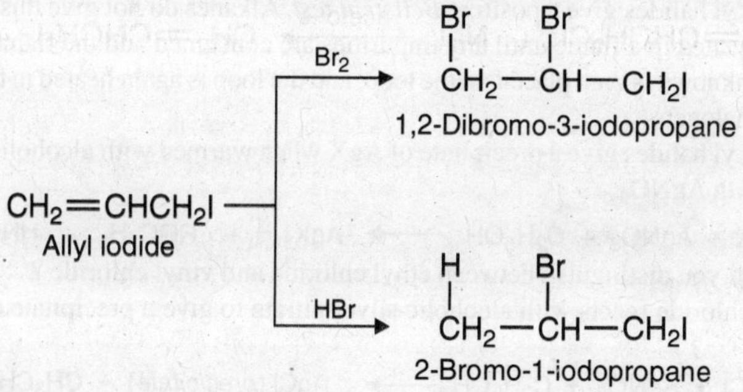

Examination Questions

1. Write structural formulas for alkyl bromides of the molecular formula C_4H_9Br. Name each according to the IUPAC system.

Clue. There are 4 isomers of formula C_4H_9Br.

2. Write structural formulas for all alkyl bromides of molecular formula $C_5H_{11}Br$.

Clue. There are 7 isomers of formula $C_5H_{11}Br$.

3. How are alkyl halides prepared ? Describe their synthetic utility.

(Bhopal BSc, 2014 ; Patna BSc, 2015)

4. How is Freon-12 prepared ? *(Kota BSc, 2014)*

5. How is ethyl bromide prepared ? How does it react with the following reagents :

(*a*) Aq KOH (*b*) Na (*c*) *Alc* KOH

(*d*) NaCN (*e*) CH_3CH_2ONa (*f*) NaI / Acetone

6. How does ethyl iodide react with the following reagents :

(*a*) Acetic acid (*b*) Mg (*c*) *Alc* KOH (*d*) Na

7. What happens when *n*-propyl bromide is treated with alcoholic KOH ?

Answer. Propene is formed :

$$\underset{n\text{-Propyl bromide}}{CH_3CH_2CH_2Br} + Alc\ KOH \longrightarrow \underset{\text{Propene}}{CH_3CH{=}CH_2} + KBr + H_2O$$

8. What happens when ethyl alcohol is treated with bleaching powder ?

Answer. Chloroform is formed. For reaction equations see text.

9. What happens when chloroform is heated with ethylamine in the presence of alcoholic KOH ? *(Lucknow BSc, 2014)*

Answer. Ethyl isocyanide is formed.

$$\underset{\text{Ethylamine}}{CHCl_3 + 3KOH + CH_3CH_2NH_2} \xrightarrow{\Delta} \underset{\text{Ethyl isocyanide}}{CH_3CH_2NC} + 3KCl + 3H_2O$$

10. What is Saytzeff rule ? Give an example. *(Calicut BSc, 2015)*

11. What happens when an alkyl halide is mixed with water ?

Answer. Because of the insolubility of the alkyl halides, two layers will form. Monoalkylfluorides and chlorides are less dense than water and will float on top. Alkyl bromides and iodides are denser than water and will sink to the bottom. Polyhalogenation increases the density, so compounds such as CH_2Cl_2 and $CHCl_3$ sink to the bottom.

12. Arrange the following molecules in order of increasing boiling point :
(*a*) $CH_3CH_2CH_2Cl$ (*b*) $(CH_3)_2CHCl$ (*c*) $(CH_3)_2CHCH_2Br$
(*d*) $CH_3CH_2CH_2F$ (*e*) $CH_3CH_2CH_2CH_2CH_2Br$

Answer. From *lowest* to *highest* boiling point: (*d*), (*b*), (*a*), (*c*), (*e*). (remember, increased size = increased boiling point, increased halogen weight/size = increased boiling point, and increased branching = lower boiling point).

13. How will you distinguish between an alkyl halide and an alkene ? *(Delhi BSc, 2015)*

Answer. (*a*) Alkyl halides give a positive *Beilstein test.* Alkanes do not give this test. In this test a small loop of copper wire is heated in a flame until any impurities are consumed and the flame has become colorless. A drop or piece of the unknown is then placed on the loop, and the loop is again heated in the flame. A green flame shows the presence of halogen.

(*b*) Alkyl halides give a precipitate of AgX when warmed with alcoholic silver nitrate solution. Alkanes do not react with $AgNO_3$.

$$RX + AgNO_3 + C_2H_5OH \longrightarrow AgX\downarrow + ROC_2H_5 + HNO_3$$

14. How will you distinguish between ethyl chloride and vinyl chloride ? *(Andhra BSc, 2014)*

Answer. Ethyl chloride reacts with alcoholic silver nitrate to give a precipitate of AgCl. Vinyl chloride does not give this test.

$$\underset{\text{Ethyl chloride}}{CH_3CH_2Cl} + AgNO_3 + C_2H_5OH \longrightarrow AgCl\ (\textit{precipitate}) + CH_3CH_2OC_2H_5 + HNO_3$$

15. How will you distinguish between chloroform and carbon tetrachloride ?

Answer. When chloroform is heated with aniline and alcoholic KOH, offensive smell of isocyanide (or carbylamine) is produced. Carbon tetrachloride does not give this test.

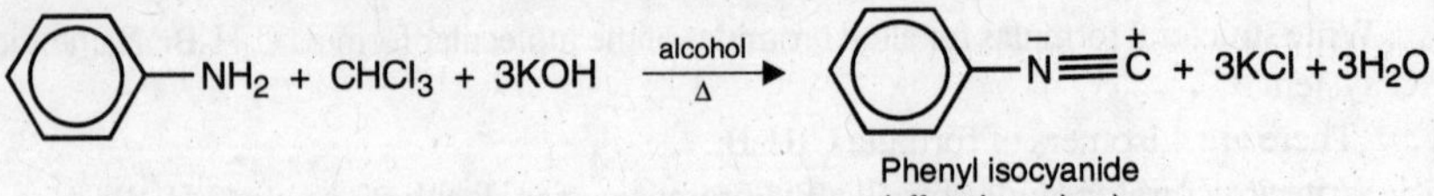

Phenyl isocyanide
(offensive odour)

16. Explain why vinyl chloride is less reactive than ethyl chloride.
(Madras BSc, 2014 ; Jiwaji BSc, 2014 ; Panjab BSc, 2015)

17. Explain why allyl chloride is more reactive than vinyl chloride.
(Kalyani BSc, 2014 ; West Bengal BSc, 2015)

18. Discuss the mechanism of nucleophilic substitution reactions of alkyl halides.
(North Eastern Hill BSc, 2014)

19. Discuss the mechanism of S_N1 and S_N2 reactions of alkyl halides.
(Devi Ahilya BSc, 2014 ; Anna BSc, 2015)

20. What is the easiest way to summarize the competition between S_NI and S_N2 ?

Answer. A good way to summarize the competition is to remember that S_N2 reactions are controlled by steric factors and S_N1 reactions are controlled by electronic factors.

21. Give the mechanism for the reaction of methyl bromide with aqueous NaOH to form methyl alcohol.

22. Give a mechanism for the reaction of *tert*-butyl bromide with aqueous NaOH to form *tert*-butyl alcohol. (*Jadavpur BSc, 2015*)

23. Discuss the mechanism of elimination reactions of alkyl halides.

24. Discuss the mechanism of E2 and E1 reactions of alkyl halides. (*Mysore BSc, 2014*)

25. How will you convert ethyl bromide into the following compounds :

(*a*) *n*-Butane (*b*) Ethyl cyanide (*c*) Ethyl alcohol (*d*) Ethyl acetate

26. How will you synthesize isopropyl bromide from *n*-propyl bromide ? (*Pune BSc, 2014*)

Answer. Following steps are involved :

$$\underset{n\text{-Propyl bromide}}{CH_3CH_2CH_2Br} \xrightarrow[\text{alcohol}/\Delta]{KOH} CH_3-CH=CH_2 \xrightarrow[\text{Markovnikov addition}]{HBr} \underset{\text{Isopropyl bromide}}{CH_3-\overset{Br}{\overset{|}{C}H}-CH_3}$$

27. How will you synthesize *n*-propyl bromide form isopropyl bromide ? (*Shivaji BSc, 2015*)

Answer. Following steps are involved :

$$\underset{\text{Isopropyl bromide}}{CH_3-\overset{Br}{\overset{|}{C}H}-CH_3} \xrightarrow[\text{alcohol}/\Delta]{KOH} CH_3-CH=CH_2 \xrightarrow[\text{peroxide}]{HBr} \underset{n\text{-Propyl bromide}}{CH_3CH_2CH_2Br}$$

28. How will you synthesize propionic acid from ethyl bromide ?

Answer. Following steps are involved :

$$\underset{\text{Ethyl bromide}}{CH_3CH_2Br} \xrightarrow{NaCN} CH_3CH_2CN \xrightarrow{H_2O/H^+} \underset{\text{Propanoic acid}}{CH_3CH_2-\overset{O}{\overset{||}{C}}-OH}$$

29. How will you synthesize vinyl bromide from ethyl alcohol ? (*Aligarh BSc, 2014*)

Answer. Following steps are involved :

$$CH_3CH_2OH \xrightarrow[\Delta]{H_2SO_4} CH_2=CH_2 \xrightarrow[CCl_4]{Br_2} BrCH_2-CH_2Br \xrightarrow{\text{alc. KOH}} CH_2=CHBr \text{ (Vinyl bromide)}$$

30. How will you synthesize allyl chloride from propane ?

(*Baroda BSc, 2014 ; Cochin BSc, 2014 ; Tamil Nadu BSc, 2015*)

Answer. Following steps are involved :

$$\underset{\text{Propane}}{CH_3CH_2CH_3} \xrightarrow[uv]{Cl_2} CH_3\overset{Cl}{\overset{|}{C}H}CH_3 + CH_3CH_2CH_2Cl \xrightarrow{\text{alc. KOH}} CH_3CH=CH_2 \xrightarrow[500^\circ]{Cl_2} \underset{\text{Allyl chloride}}{ClCH_2CH=CH_2}$$

31. A primary alkyl halide (*A*), C_4H_9Br, reacted with hot alcoholic KOH to give compound (*B*). Compound (*B*) reacted with HBr to give an isomer of (*A*), (*C*). When (*A*) was reacted with sodium it gave a compound (*D*), C_8H_{18}, which was different than the compound when *n*-butyl bromide was reacted with sodium. Give the structural formula of (*A*) and write equations for all the reactions.

Answer. (*A*) = 1-Bromo-2-methylpropane

(*B*) = 2-Methylpropene

(*C*) = 2-Bromo-2-methylpropane

(*D*) = 2,5-Dimethylhexane

32. Treatment of 2-bromobutane with hot alcoholic KOH gives a mixture of three isomeric butenes (*A*), (*B*), and (*C*). Ozonolysis of the minor product, (*A*), gives formaldehyde and another aldehyde in equimolar amounts. Both (*B*) and (*C*) give the same single product, (*D*), on ozonolysis. What are the structural formulas of (*A*), (*B*), (*C*), and (*D*) ? Give equations for all the reactions.

Answer. (*A*) is 1-Butene ; (*B*) is *cis*-2-Butene ; (*C*) is *trans*-2-Butene ; and (*D*) is Acetaldehyde.

Multiple Choice Questions

1. Which statement best describes a carbon-chlorine bond?
   (*a*) The bond is covalent and polar, with chlorine carrying a partial negative charge.
   (*b*) The bond is ionic and polar, with chlorine carrying a full negative charge.
   (*c*) The bond is covalent and polar, with chlorine carrying a partial positive charge.
   (*d*) The bond is ionic and polar, with chlorine carrying a full positive charge.

Answer. (*a*)

2. What is the Cl–Cl–Cl bond angle of CCl_4 ?
   (*a*) 120° (*b*) 60° (*c*) 109.5° (*d*) 90°

Answer. (*c*)

3. Saytzeff rule states the ______ is formed most readily.
   (*a*) least substituted alkane (*b*) most substituted alkane
   (*c*) least substituted alkene (*d*) most substituted alkene

Answer. (*d*)

4. Allylic radicals are stabilized by
   (*a*) resonance (*b*) interaction with light
   (*c*) electron-withdrawing atoms (*d*) oxygen atoms

Answer. (*a*)

5. Which of the following poisonous gas is formed when chloroform is exposed to light and air ?
   (*a*) Mustard gas (*b*) Carbon monoxide
   (*c*) Phosgene (*d*) Chlorine

Answer. (*c*)

6. Freon-12, CCl_2F_2, is used as a
   (*a*) Local anesthetic (*b*) Dry-cleaning agent
   (*c*) Refrigerant (*d*) Disinfectant

Answer. (*c*)

7. Which of the following compounds has been suggested as causing depletion of the ozone layer in the upper stratosphere ?
   (*a*) CH_4 (*b*) CCl_2F_2
   (*c*) CF_4 (*d*) CH_2Cl_2

Answer. (*b*)

8. Which of the following reagents cannot be used to prepare an alkyl chloride from an alcohol ?
   (*a*) $HCl + ZnCl_2$ (*b*) $SOCl_2$ (*c*) NaCl (*d*) PCl_5

Answer. (*c*)

9. 2-Propanol reacts with KBr and concentrated H_2SO_4 to give
   (*a*) 1-Bromopropane (*b*) 1,2-Dibromopropane
   (*c*) 2-Bromopropane (*d*) 2,2-Dibromopropane

Answer. (*c*)

10. The best way to prepare 1,2-Dibromoethane is to treat :
    (*a*) Ethylene with Br_2
    (*b*) Acetylene with excess of HBr
    (*c*) Ethylene with excess of HBr
    (*d*) 1,2-Dichloroethane with Br_2

Answer. (*a*)

11. Alkyl halides undergo
    (*a*) Electrophilic substitution reactions
    (*b*) Electrophilic addition reactions
    (*c*) Nucleophilic substitution reactions
    (*d*) Nucleophilic addition reactions

Answer. (*c*)

12. *n*-Propyl iodide reacts with sodium ethoxide to give :

(*a*) $CH_3CH_2OCH_2CH_3$ (*b*) $CH_3CH_2OCH_2CH_2CH_3$

(*c*) $CH_3CH_2OCH_3$ (*d*) $CH_3OCH_2CH_2CH_3$

Answer. (*b*)

13. 1-Bromobutane reacts with alcoholic KOH to mainly give

(*a*) 1-Butene (*b*) 2-Butene

(*c*) 1-Butanol (*d*) 2-Butanol

Answer. (*b*)

14. 2-Bromobutane reacts alcoholic KOH to mainly give

(*a*) 1-Butene (*b*) 2-Butene

(*c*) 1-Butanol (*d*) 2-Butanol

Answer. (*b*)

15. Isopropyl bromide reacts with alcoholic KOH to give

(*a*) Propene (*b*) Isopropyl alcohol

(*c*) Propane (*d*) *n*-Propyl alcohol

Answer. (*a*)

16. 2,2-Dichloropropane reacts with aqueous KOH to give

(*a*) 2,2-Propanediol (*b*) Propanal

(*c*) Acetone (*d*) Propene

Answer. (*c*)

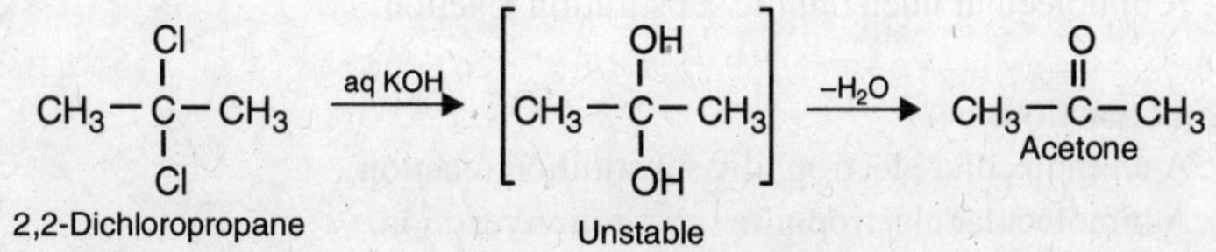

17. 1,1-Dichloropropane reacts with aqueous KOH to give

(*a*) 1,1-Propanediol (*b*) Acetaldehyde

(*c*) Propanone (*d*) Propyne

Answer. (*b*)

18. Which of the following compounds will not give a white precipitate with alcoholic $AgNO_3$.

(*a*) Bromobenzene (*b*) 1-Bromopropane

(*c*) Ethyl bromide (*d*) 2-Bromopropane

Answer. (*a*)

19. Which of the following compounds will give a white precipitate with alcoholic $AgNO_3$?

(*a*) Vinylbenzene (*b*) Chlorobenzene

(*c*) Vinyl chloride (*d*) Allyl chloride

Answer. (*d*)

20. Carbylamine test involves heating a mixture of

(*a*) alcoholic KOH, methyl iodide, and sodium metal

(*b*) alcoholic KOH, methyl iodide, and primary amine

(*c*) alcoholic KOH, chloroform, and primary amine

(*d*) alcoholic KOH, methyl alcohol, and primary amine

Answer. (*c*)

21. When chloroform is heated with aqueous NaOH, it gives

(*a*) Formic acid (*b*) Sodium formate

(*c*) Acetic acid (*d*) Sodium acetate

Answer. (*b*)

22. Which alkyl halides react most readily by nucleophilic substitution ?

(*a*) CH_3CH_2Cl (*b*) CH_3CH_2I

(*c*) CH_3CH_2Br (*d*) CH_3CH_2F

Answer. (*b*)

23. Which of the following factors influence whether a reaction will proceed by an S_N1, S_N2, E1, or E2 mechanism ?

(*a*) Structure of the alkyl halide
(*b*) Solvent
(*c*) Concentration of reagents
(*d*) Nature of the nucleophile
(*e*) All of these

Answer. (*e*)

24. Which compound reacts most rapidly by an S_N1 mechanism ?

(*a*) Methyl chloride (*b*) Isopropyl chloride
(*c*) Ethyl chloride (*d*) *tert*-Butyl chloride

Answer. (*d*)

25. Which of the following compounds would react most rapidly in an S_N2 reaction ?

(*a*) CH_3CH_2I (*b*) $CH_2{=}CH{-}I$
(*c*) $(CH_3)_2CHI$ (*d*) $(CH_3)_3CI$

Answer. (*a*)

26. The term S_N2 stands for ?

(*a*) An electrophilic addition reaction.
(*b*) A nucleophilic reaction.
(*c*) A unimolecular substituent reaction.
(*d*) A bimolecular nucleophilic substitution reaction.

Answer. (*d*)

27. An S_N2 reactions is ?

(*a*) A unimolecular electrophilic substitution reaction.
(*b*) A bimolecular electrophilic substitution reaction.
(*c*) A unimolecular nucleophilic substitution reaction.
(*d*) A bimolecular nucleophilic substitution reaction.

Answer. (*d*)

28. The S_N2 reaction is known to occur with

(*a*) racemization (*b*) partial inversion
(*c*) almost complete inversion (*d*) mutarotation

Answer. (*c*)

29. Which statement best describes the mechanism of an S_N2 reaction ?

(*a*) front side attack with retention of configuration
(*b*) back side attack with inversion of configuration
(*c*) front side and back side attack with racemization
(*d*) front side attack with inversion of configuration

Answer. (*b*)

14

Organometallic Compounds

Organometallic compounds are those compounds which contain at least one carbon-metal bond.

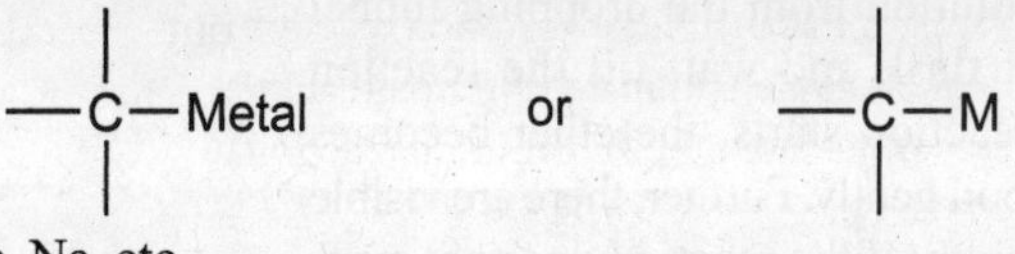

Where M = Mg, Li, Pb, Zn, Na, etc.

Organometallic compounds are named by simply adding the name of the metal to that of the organic group bonded to the metal. For example,

CH_3—Mg—I Methylmagnesium iodide	CH_3CH_2—Mg—Br Ethylmagnesium bromide
CH_3—Li Methyllithium	$(CH_3CH_2)_4Pb$ Tetraethyllead

GRIGNARD REAGENTS, RMgX

Organomagnesium halides are called **Grignard Reagents**. They are named after Victor Grignard (Nobel Prize, 1912), who discovered them and developed their use as synthetic reagents. The Grignard reagents are highly reactive and used in the synthesis of alkanes, alkenes, alcohols, aldehydes, ketones, and carboxylic acids. Their general formula can be written as :

RMgX or R—MgX

Where R = Alkyl or aryl group
X = Cl, Br, or I

Examples are :

CH_3—MgI Methylmagnesium iodide	CH_3CH_2—MgBr Ethylmagnesium bromide

Preparation. Grignard reagents are prepared in the laboratory by the action of alkyl halides on magnesium metal in the presence of *dry ether*. Dry ether is pure anhydrous diethyl ether, $C_2H_5OC_2H_5$.

$$\underset{\text{Alkyl halide}}{R{-}X} + Mg \xrightarrow[\text{reflux}]{\text{ether}} \underset{\substack{\text{Alkylmagnesium halide}\\(\textit{Grignard reagent})}}{R{-}MgX}$$

$$CH_3{-}I + Mg \xrightarrow[\text{reflux}]{\text{ether}} \underset{\substack{\text{Methylmagnesium}\\\text{iodide}}}{CH_3{-}MgI}$$

$$CH_3CH_2{-}Br + Mg \xrightarrow[\text{reflux}]{\text{ether}} \underset{\text{Ethylmagnesium bromide}}{CH_3CH_2{-}MgBr}$$

Alkyl iodides are more reactive than the bromides, which are in turn more reactive than chlorides. Alkyl bromides are most suitable for the preparation of Grignard reagents.

Laboratory Procedure. The apparatus used in the preparation of Grignard reagents is shown in Fig. 14.1. The round-bottomed flask is fitted with a reflux condenser and a dropping funnel. Magnesium ribbon cut into small pieces (5 mm) is suspended in dry ether placed in the round-bottomed flask. Approximately 1 : 1 mixture of the alkyl halide and dry ether is placed in the dropping funnel. If methylmagnesium iodide is to be prepared, use methyl iodide, CH_3I. If ethylmagnesium bromide is to be prepared, use ethyl bromide, C_2H_5Br.

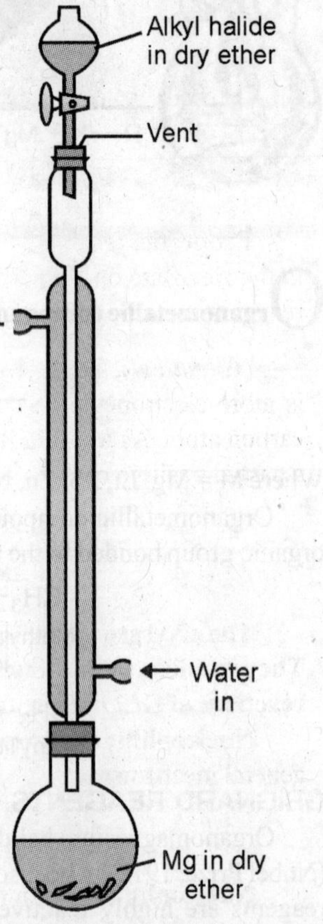

Fig. 14.1. Preparation of Grignard reagent.

Add 2-3 ml of the solution from the dropping funnel into the round-bottomed flask and wait till the reaction commences. When the reaction starts, the ether becomes cloudy and also begins to boil gently. Further, there are visible signs of attack on magnesium. If the reaction does not start, drop one or two crystals of iodine into the flask. Once the reaction starts, the addition of solution from the dropping funnel is maintained at such a rate that the ether refluxes gently. When the reaction is complete, a clear solution of the Grignard reagent in ether is obtained. This is then treated *in situ* with various reagents to get the desired synthetic products.

Role of Ether. Grignard reagents are usually prepared in diethyl ether ($CH_3CH_2OCH_2CH_3$) as solvent. It is thought that two ether oxygen atoms complex with the magnesium atom, stabilizing the reagent.

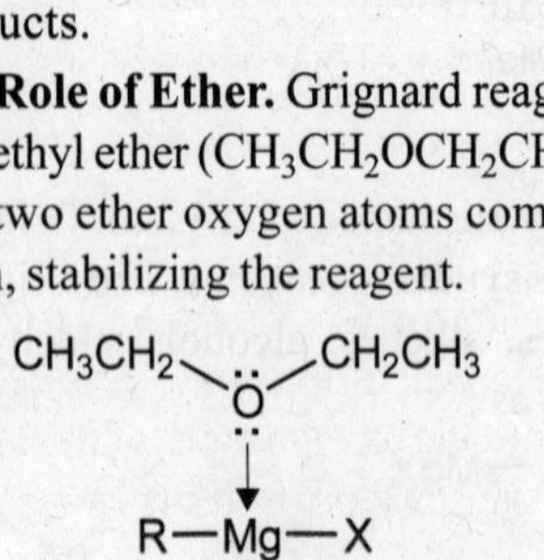

Two molecules of diethyl ether complex with the Mg atom of the Grignard reagent.

Essential Precautions. (*1*) All reagents should be anhydrous (moisture-free) and apparatus oven-dried. Moisture reacts with the Grignard reagent and prevents its formation.

$$RMgX + H_2O \rightarrow RH + MgX(OH)$$

(2) There should be no naked flames anywhere near the apparatus. Diethyl ether catches fire very quickly.

I need a good software for my son.
One that will correct his organic chemistry and opinions !

Properties (*Physical*). Grignard reagents are nonvolatile, colorless solids. They are seldom isolated in the free state on account of their explosive nature. Therefore, for synthetic purposes the Grignard reagents are always prepared and used in ether solution.

(*Chemical*). The C–Mg bond in Grignard reagents is covalent but highly polar. The carbon atom is more electronegative than magnesium. The electrons of the C–Mg bond are drawn towards the carbon atom. As a result, the carbon atom has a partial negative charge and the magnesium atom has a partial positive charge.

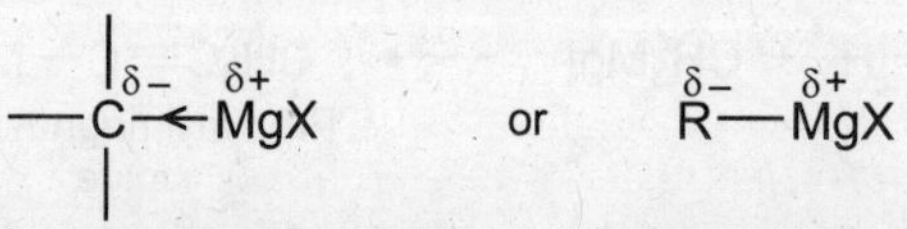

The alkyl groups in Grignard reagents being electron-rich can act as carbanions or nucleophiles. They would attack polarized molecules at points of low electron density. Thus the characteristic reactions of Grignard reagents are nucleophilic substitution and nucleophilic addition reactions.

Nucleophilic Substitution. The Grignard reagents undergo substitution reactions by the following general mechanism :

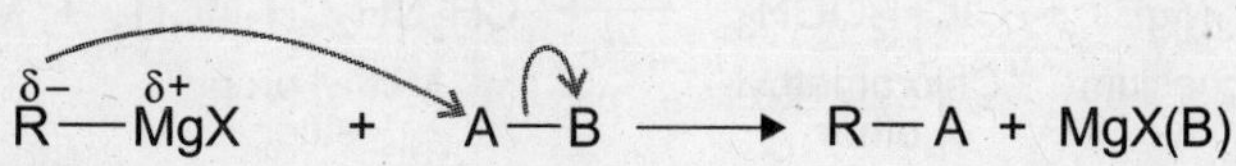

Nucleophilic Addition. Carbonyl compounds are attacked by Grignard reagents to form addition products which on acid-hydrolysis yield alcohols, acids, esters, etc. Addition always proceeds such that the positive part of the Grignard reagent ($MgX^{\delta+}$) combines with the oxygen atom of the carbonyl group. The negative part ($R^{\delta-}$) of the Grignard reagent goes to the carbon atom of the carbonyl group.

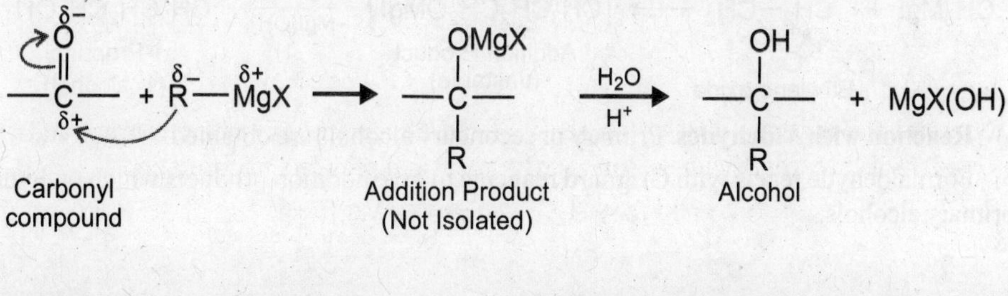

Grignard reagents react with a variety of compounds yielding almost the entire range of organic substances. The following reactions illustrate the synthetic importance of Grignard reagents :

(1) **Reaction with Compounds Containing Active Hydrogens.** By active hydrogen we mean a hydrogen that is more acidic than that in alkanes. Thus compounds like water, alcohols, and amines which contain active hydrogens react with Grignard reagents to form alkanes.

$$\underset{\text{Methylmagnesium iodide}}{CH_3MgI} + \underset{\text{Water}}{HO{-}H} \longrightarrow \underset{\text{Methane}}{CH_4} + MgI(OH)$$

$$CH_3MgI + \underset{\text{Ethanol}}{C_2H_5O{-}H} \longrightarrow CH_4 + MgI(OC_2H_5)$$

$$CH_3MgI + \underset{\text{Ethylamine}}{C_2H_5NH{-}H} \longrightarrow CH_4 + MgI(NHC_2H_5)$$

Since methane evolved in the above reactions can be measured, it gives a method for the estimation of –OH and –NH_2 groups in organic compounds.

(2) **Reaction with Alkyl Halides.** Grignard reagents react with saturated alkyl halides to form higher alkanes or alkenes.

$$\underset{\text{Methylmagnesium iodide}}{\overset{\delta- \;\; \delta+}{CH_3MgI}} + \overset{\delta+ \;\; \delta-}{CH_3CH_2{-}Br} \longrightarrow \underset{\text{Propane}}{CH_3CH_2CH_3} + MgI(Br)$$

$$CH_3MgI + \underset{\text{Allyl bromide}}{CH_2{=}CHCH_2Br} \longrightarrow \underset{\text{1-Butene}}{CH_2{=}CHCH_2CH_3} + MgI(Br)$$

(3) **Reaction with Alkynes.** Terminal alkynes (1-alkynes) react with Grignard reagents to form alkynylmagnesium halides which on subsequent treatment with alkyl halides form higher alkynes.

$$\underset{\text{Propyne}}{CH_3C{\equiv}C{-}H} + CH_3MgI \longrightarrow \underset{\text{Propynylmagnesium iodide}}{CH_3C{\equiv}C{-}MgI} + CH_4$$

$$CH_3C{\equiv}C{-}MgI + CH_3I \longrightarrow \underset{\text{2-Butyne}}{CH_3C{\equiv}C{-}CH_3} + MgI_2$$

(4) **Reaction with Lower Halogenated Ethers.** Grignard reagents react with lower halogenated ethers to form higher ethers.

$$\underset{\text{Ethylmagnesium bromide}}{CH_3CH_2MgBr} + \underset{\text{Chloromethyl ether}}{ClCH_2OCH_3} \longrightarrow \underset{\text{Methyl propyl ether}}{CH_3CH_2CH_2OCH_3} + MgBr(Cl)$$

(5) **Reaction with Ethylene Oxide.** Grignard reagents react with ethylene oxide to give an addition product which on hydrolysis forms primary alcohols.

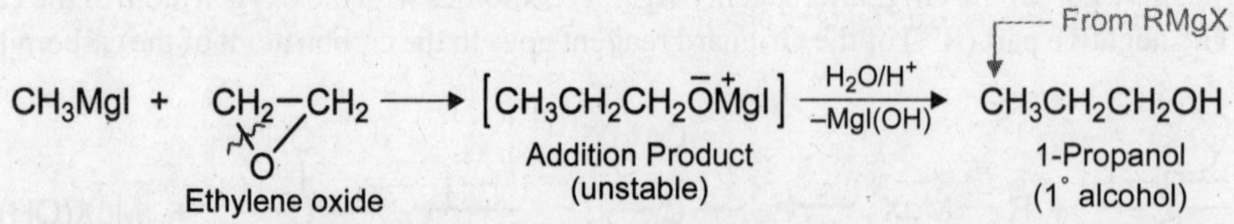

(6) **Reaction with Aldehydes.** Primary or secondary alcohols are obtained.

(*a*) Formaldehyde reacts with Grignard reagents to give addition products which on hydrolysis yield primary alcohols.

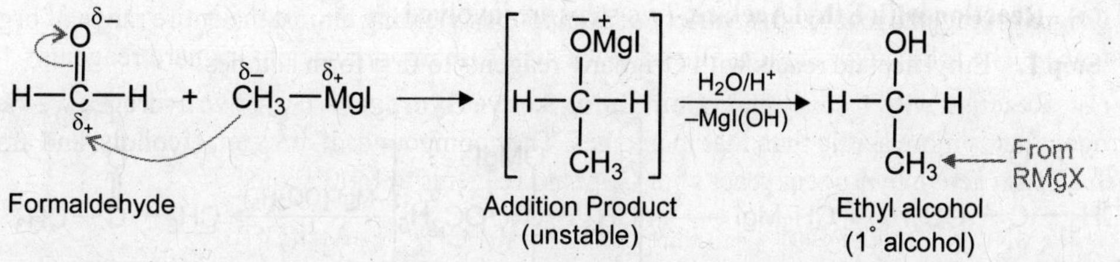

(*b*) Other aldehydes react with Grignard reagents to give addition products which on hydrolysis yield secondary alcohols.

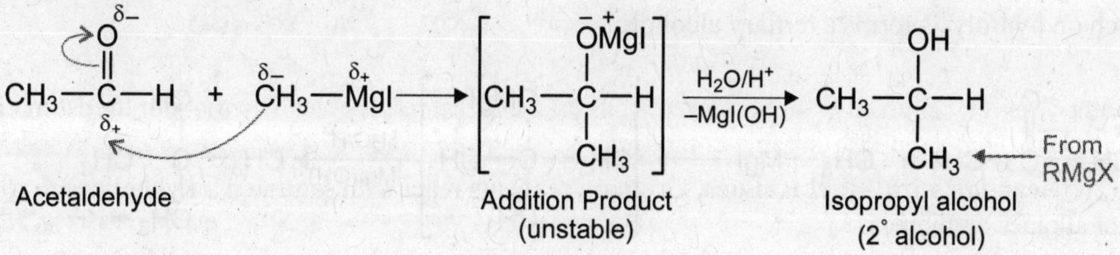

(7) **Reaction with Ketones**. Grignard reagents react with ketones to give addition products which on hydrolysis form tertiary alcohols.

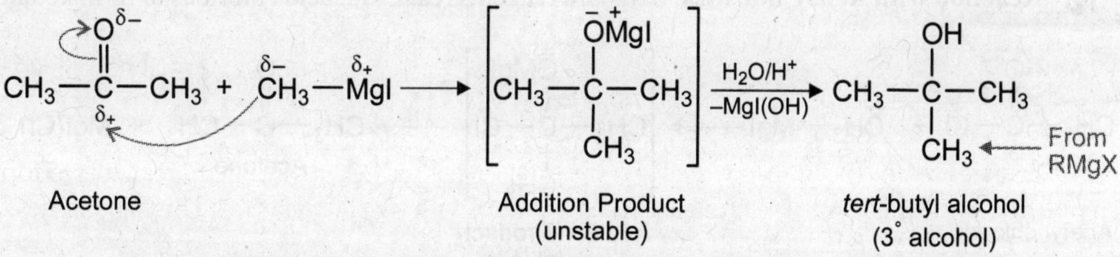

(8) **Reaction with Esters**. Grignard reagents react with formic esters to form secondary alcohols, while other esters yield tertiary alcohols.

(*a*) **Reaction with Ethyl Formate**. Two steps are involved :

Step 1. Ethyl formate reacts with Grignard reagents to first form aldehydes.

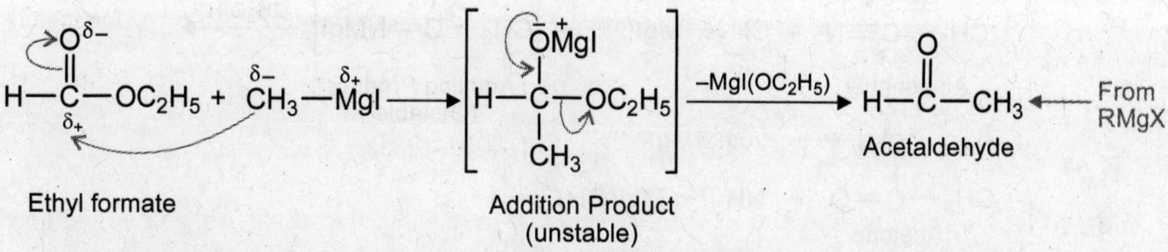

Step 2. The aldehyde then reacts with more of the Grignard reagent to give an addition product which on hydrolysis forms a secondary alcohol.

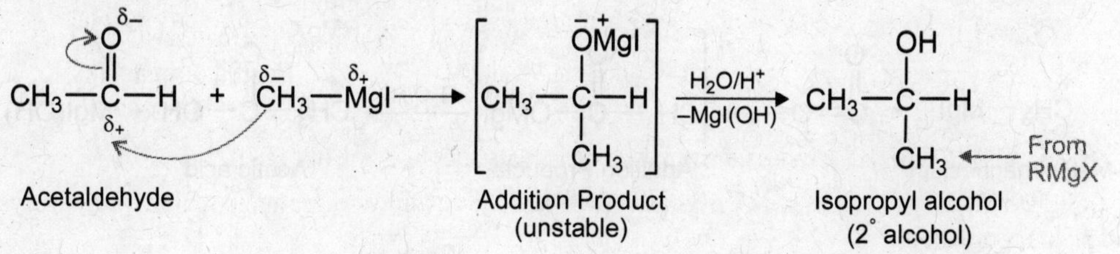

(*b*) **Reaction with Ethyl Acetate.** Two steps are involved :

Step 1. Ethyl acetate reacts with Grignard reagents to first form ketones.

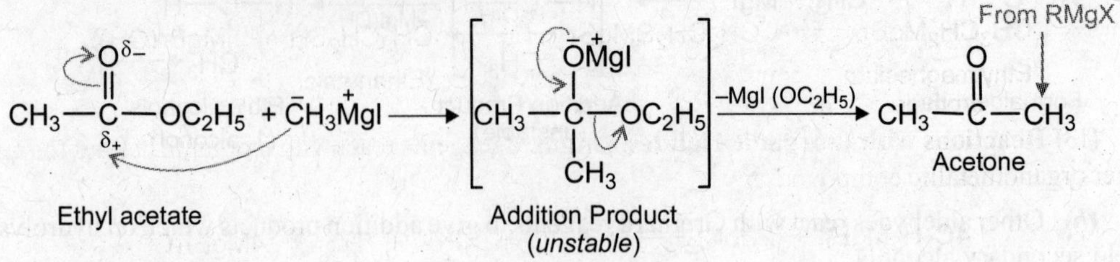

Step 2. The ketone then reacts with more of the Grignard reagent to give an addition product which on hydrolysis forms a tertiary alcohol.

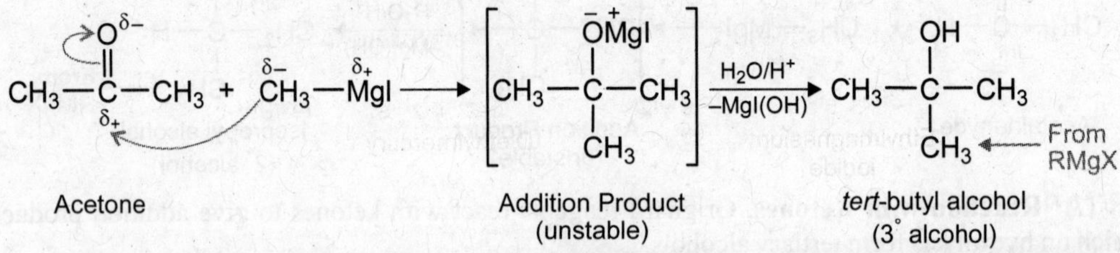

(9) **Reaction with Acid Chlorides**. Grignard reagents react with acid chlorides to form ketones.

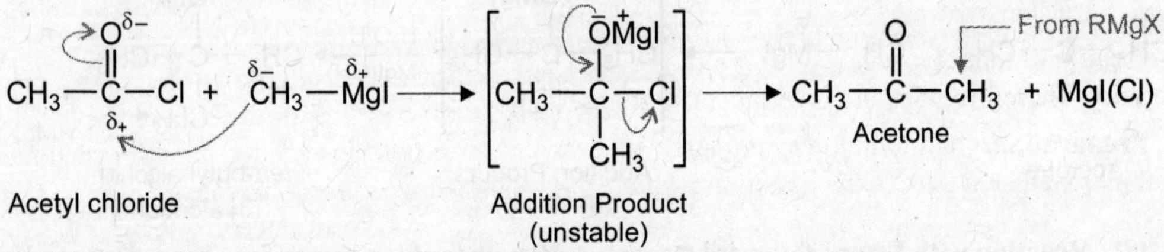

(10) **Reaction with Cyanides**. Grignard reagents react with cyanides to give addition products which on hydrolysis with dilute HCl yield ketones.

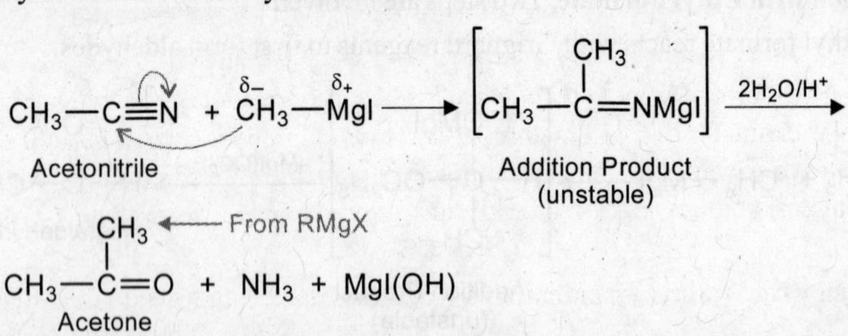

(11) **Reaction with Carbon Dioxide**. Grignard reagents react with carbon dioxide (O=C=O) to give addition products which on hydrolysis yield carboxylic acids.

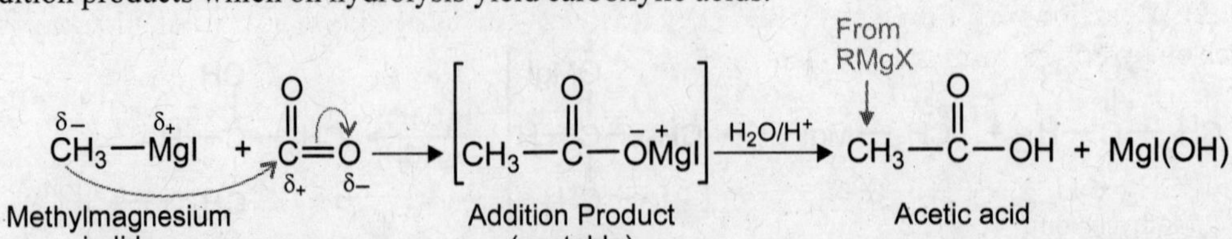

(12) **Reaction with Sulfur**. Sulfur reacts with a Grignard reagent to give a product which on hydrolysis gives the corresponding thioalcohol.

$$\underset{\text{Ethylmagnesium bromide}}{CH_3CH_2MgBr} \xrightarrow{S} CH_3CH_2SMgBr \xrightarrow{H_2O/H^+} \underset{\text{Ethanethiol}}{CH_3CH_2SH} + MgBr(OH)$$

(13) **Reactions with Inorganic Halides**. Grignard reagents react with inorganic halides to form other organometallic compounds.

$$\underset{\text{Ethylmagnesium bromide}}{4C_2H_5MgBr} + 2PbCl_2 \longrightarrow \underset{\text{Tetraethyllead}}{(C_2H_5)_4Pb} + Pb + 4MgBr(Cl)$$

$$\underset{\text{Methylmagnesium iodide}}{4CH_3MgI} + SiCl_4 \longrightarrow \underset{\text{Tetramethylsilane}}{(CH_3)_4Si} + 4MgI(Cl)$$

$$\underset{\text{Ethylmagnesium iodide}}{2C_2H_5MgI} + HgCl_2 \longrightarrow \underset{\text{Diethylmercury}}{(C_2H_5)_2Hg} + 2MgI(Cl)$$

ORGANOLITHIUM COMPOUNDS, RLi

Organolithium compounds are those compounds which contain C—Li bond. For example,

$$\underset{\text{Methyllithium}}{CH_3—Li} \qquad \underset{\text{Ethyllithium}}{CH_3CH_2—Li} \qquad \underset{n\text{-Propyllithium}}{CH_3CH_2CH_2—Li}$$

Organolithium compounds behave in the same way as Grignard reagents, but with enhanced reactivity. Therefore, they are now increasingly used in organic synthesis.

Preparation. Organolithium compounds are obtained by the reaction of alkyl halides with lithium metal in dry ether at –10°C. The reaction is carried in an inert atmosphere of nitrogen.

$$\underset{\text{Methyl iodide}}{CH_3—I} + Li \xrightarrow[-10^\circ C]{\text{ether}} \underset{\text{Methyllithium}}{CH_3—Li} + LiI$$

$$\underset{\text{Ethyl bromide}}{CH_3CH_2—Br} + Li \xrightarrow[-10^\circ C]{\text{ether}} \underset{\text{Ethyllithium}}{CH_3CH_2—Li} + LiBr$$

Properties. The increased reactivity of organolithium compounds, over Grignard reagents is due to the greater polar character of C–Li in comparison to C–Mg bond. Organolithium compounds are particularly sensitive towards air and moisture. They react with compounds containing active hydrogens, aldehydes, ketones, carbon dioxide, and ethylene oxide in the same way as do Grignard reagents.

(1) **Reaction with Water**. Organolithium compounds react with water to form alkanes.

$$\underset{\text{Methyllithium}}{CH_3—Li} + H—OH \longrightarrow \underset{\text{Methane}}{CH_4} + LiOH$$

(2) **Reaction with Ethylene Oxide.** They react with ethylene oxide to give an addition product which on hydrolysis form primary alcohols.

$$\underset{\text{Methyllithium}}{\overset{\delta-}{CH_3}—\overset{\delta+}{Li}} + \underset{O}{CH_2—CH_2} \longrightarrow \underset{\text{(unstable)}}{[CH_3CH_2CH_2\overset{-\,+}{OLi}]} \xrightarrow[(-LiOH)]{H_2O/H^+} \underset{\text{1-Propanol (1° alcohol)}}{CH_3CH_2CH_2OH}$$

From RLi

(3) **Reaction with Aldehydes.** Primary or secondary alcohols are obtained.

(*a*) Formaldehyde reacts with organolithium compounds to give addition products which on hydrolysis yield primary alcohols.

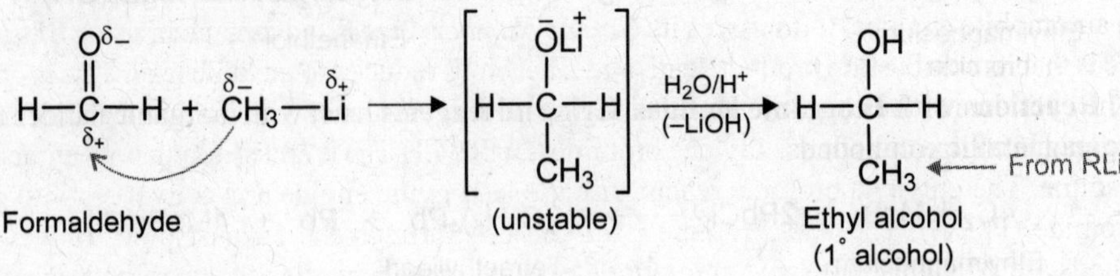

(*b*) Other aldehydes react with organolithium compounds to give addition products which on hydrolysis yield secondary alcohols.

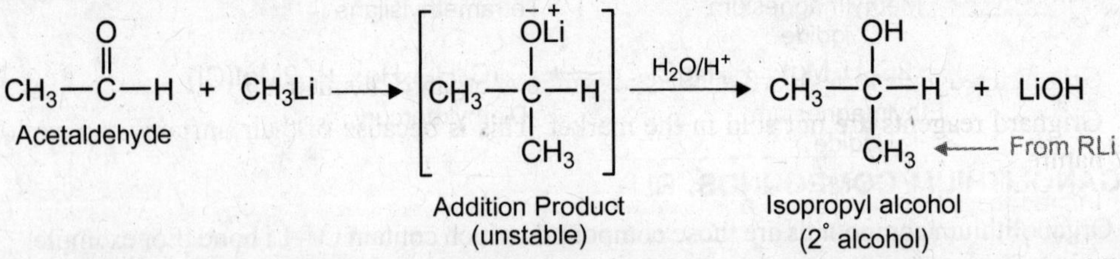

(4) **Reaction with Ketones.** Organolithium compounds react with ketones to give addition products which on hydrolysis form tertiary alcohols.

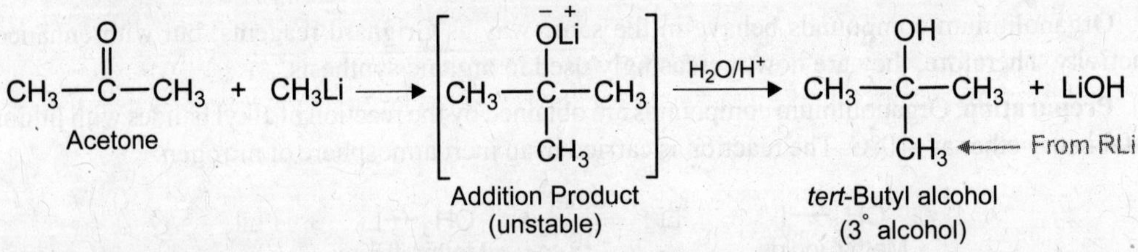

(5) **Reaction with Carbon Dioxide.** Organolithium compounds react with carbon dioxide (O=C=O) to give addition products which on hydrolysis yield carboxylic acids.

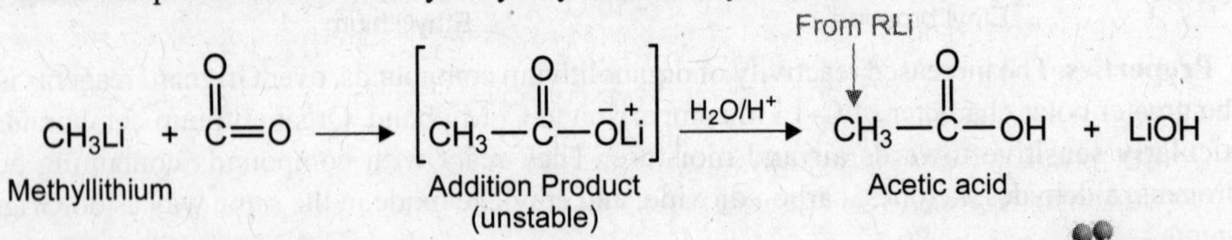

TETRAETHYLLEAD, $(CH_3CH_2)_4$ Pb

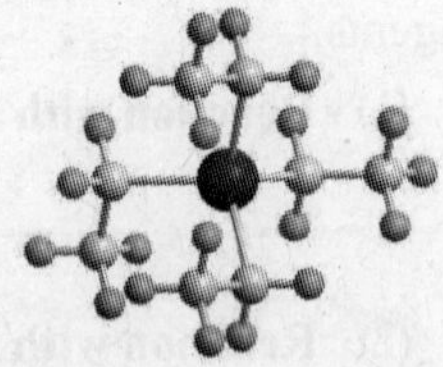

Tetraethyllead is an important organometallic compound. It finds wide use in industry under the name **TEL**.

Preparation. Tetraethyllead is prepared :

(1) By the action of ethylmagnesium iodide on lead chloride.

$$\underset{\text{Ethylmagnesium iodide}}{4C_2H_5MgI} + 2PbCl_2 \longrightarrow \underset{\text{Tetraethyllead}}{(C_2H_5)_4Pb} + Pb + 4MgICl$$

(2) By heating ethyl chloride with a sodium-lead alloy.

$$\underset{\text{Ethyl chloride}}{4C_2H_5Cl} + 4Na + Pb \longrightarrow \underset{\text{Tetraethyllead}}{(C_2H_5)_4Pb} + 4NaCl$$

Properties. Tetraethyllead is a colorless liquid, bp 202°C. Like other lead compounds, it is a highly toxic substance. It is soluble in organic liquids like gasoline but is insoluble in water. Being largely covalent in nature, tetraethyllead has low reactivity.

Use as an Anti-Knock. Tetraethyllead is an important *Anti-Knock* additive in gasoline (petrol) used in automobile engines. It improves its *Octane Number*. One disadvantage in using this gasoline additive is that its combustion product, lead oxide, becomes reduced to metallic lead. This is deposited inside the cylinders of the engine, causing clogging of the cylinder valves. To avoid accumulation of metallic lead in engine cylinders, ethylene bromide ($BrCH_2CH_2Br$) is added along with tetraethyllead into gasoline. The ethylene bromide reacts with the lead in the engine and converts it into gaseous lead bromide ($PbBr_2$). The lead bromide is discharged through the exhaust pipe. However, lead compounds are very poisonous and automobile exhaust poses a big health hazard in large cities.

Points to Remember

1. Grignard reagents are important tools of synthetic organic chemistry.
2. Grignard reagents are not sold in the market. This is because of their unstable and explosive nature.
3. In practice, a Grignard reagent is produced by dropping a solution of the alkyl halide in dry ether into the reaction flask containing magnesium ribbon suspended in dry ether or tetrahydrofuran (THF). The ether solution of the Grignard reagent is used immediately in the flask in which it is prepared.
4. Grignard reagents react as if they are carbanions. They are powerful nucleophiles and bases.
5. Grignard reagents cannot be prepared from halides which contain acidic groups in their molecules. For example, Grignard reagents containing the following groups cannot be prepared.

$$—OH,\quad —NH_2,\quad —COOH,\quad —C\equiv C—H$$

$$—\overset{O}{\overset{\|}{C}}—H,\quad —\overset{O}{\overset{\|}{C}}—R,\quad —\overset{O}{\overset{\|}{C}}—OR,\quad —C\equiv N$$

6. Grignard reagents will not react with compounds which contain acidic hydrogens. (For example, H_2O, ROH, RNH_2 and RCOOH). The acidic hydrogens will first react with the Grignard reagent and form an alkane.
7. Grignard reagents react
   with formaldehyde to form primary alcohols.
   with ethylene oxide to form primary alcohols.
   with other aldehydes to form secondary alcohols.
   with ketones to form tertiary alcohols.
   with carbon dioxide to form carboxylic acids.

Examination Questions

1. What are organometallic compounds ? How are they named ?

2. What are Grignard reagents ? Describe their preparation and synthetic uses.

(*Mangalore BSc, 2014 ; Shivaji BSc, 2014 ; Andhra BSc, 2015*)

3. How is methylmagnesium iodide prepared ? Using methylmagnesium iodide how will you prepare a primary, secondary and tertiary alcohol.

4. What happens when methylmagnesium iodide is treated with the following reagents and the product hydrolysed ?

(*a*) Formaldehyde (*b*) Acetaldehyde (*c*) Acetone

(*d*) Ethylene oxide (*e*) Carbon dioxide (*f*) 2-Butanone

Answer. (*f*) 2-Methyl-2-butanol is formed.

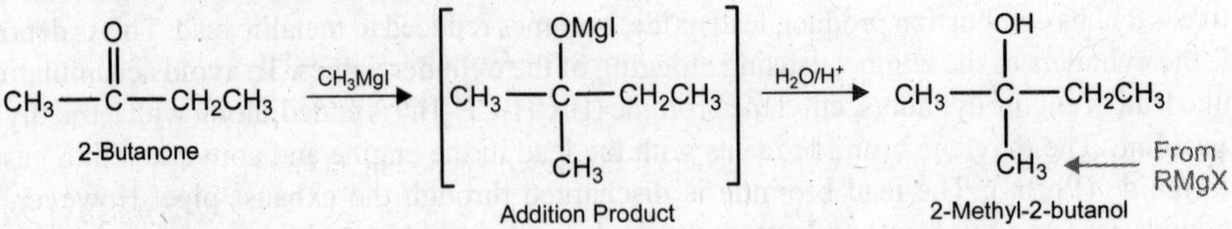

5. What happens when ethylmagnesium bromide is treated with the following reagents and the product hydrolysed ?

(*a*) Formaldehyde (*b*) Acetaldehyde (*c*) Acetone

(*d*) Ethylene oxide (*e*) Carbon dioxide (*f*) 2-Butanone

(*g*) Acetonitrile

Answer. (*g*) 2-Butanone is formed.

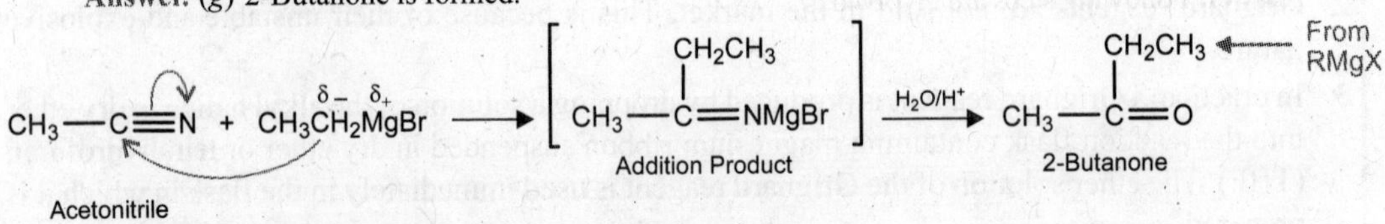

6. Give the use of Grignard reagents in preparing :

(*a*) an alkane (*b*) a primary alcohol

(*c*) a secondary alcohol (*d*) a tertiary alcohol

(*e*) a carboxylic acid

7. Write a note on : Synthetic applications of Grignard reagents.

(*Delhi BSc, 2014 ; Kurukshetra BSc, 2014 ; Panjab BSc, 2015*)

8. Write a note on : Synthetic applications of Organolithium compounds.

9. Write a mechanism for the reaction of formaldehyde with methylmagnesium iodide to form ethyl alcohol.

10. Identify (*A*), (*B*), and (*C*) :

Answer. (*A*) = CH_3CH_2OH ; Ethyl alcohol

(*B*) = CH_3CH_2Cl ; Ethyl chloride

(*C*) = $CH_3CH_2NH_2$; Ethylamine

11. Identify (*A*), (*B*), and (*C*) :

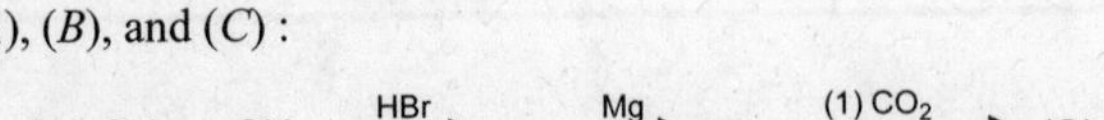

Answer. (*A*) = $CH_3CH_2CH_2Br$; *n*-Propyl bromide

(*B*) = $CH_3CH_2CH_2MgBr$; *n*-Propylmagnesium bromide

(*C*) = $CH_3CH_2CH_2COOH$; Butanoic acid

12. A student maintains that a Grignard reagent undergoes reaction with water to yield an alcohol ? What is wrong with this statement ?

Answer. The carbon of the Grignard reagent is partially *negative* and undergoes reaction with a *positive* group (such as H^+), not with a negative group (such as HO^-).

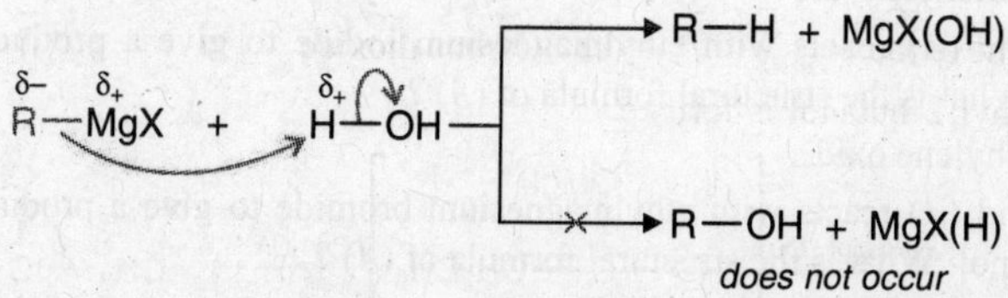

13. How will you synthesize ethyl alcohol from methylmagnesium iodide ? *(Rajasthan BSc, 2014)*

14. How will you synthesize *n*-propyl alcohol from methylmagnesium iodide ? *(Kalyani BSc, 2014 ; Rohilkhand BSc, 2015)*

15. How will you synthesize isopropyl alcohol from methylmagnesium iodide ? *(Madras BSc, 2015)*

16. How will you synthesize 2-butanol from acetylene ? *(Magadh BSc, 2014)*

Answer. Following steps are involved :

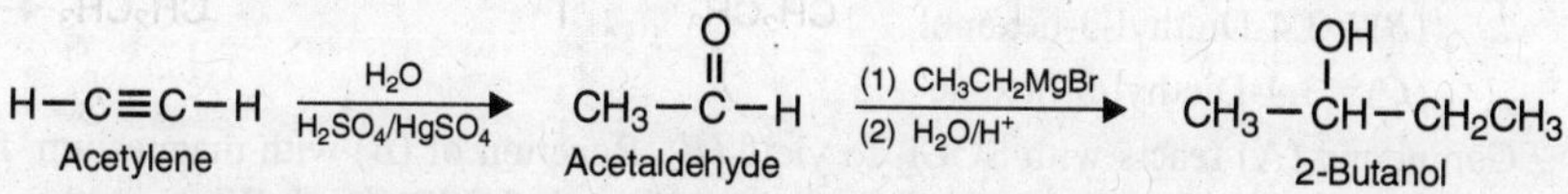

17. How will you synthesize *tert*-butyl alcohol from propyne ? *(Dibrugarh BSc, 2014 ; Meerut BSc, 2015)*

Answer. Following steps are involved :

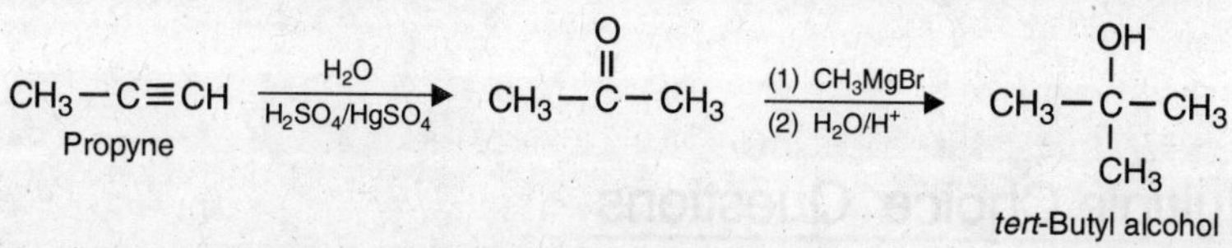

18. How will you synthesize acetic acid from methane ? *(Gulbarga BSc, 2014)*

Clue. Steps are : (*i*) Cl_2/UV ; (*ii*) Mg ; (*iii*) CO_2 ; and (*iv*) H_2O/H^+.

19. How will you synthesize propionic acid from methylmagnesium iodide ? *(Anna BSc, 2014 ; Tamil Nadu BSc, 2015)*

Answer. Following steps are involved :

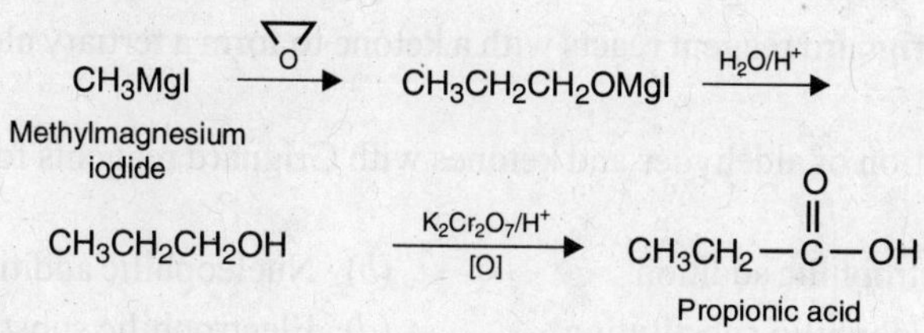

20. How will you synthesize acetone from methyl iodide ? *(Agra BSc, 2014)*

21. How will you synthesize 2-butanone from ethyl bromide ? *(Utkal BSc, 2014)*

Answer. Following steps are involved :

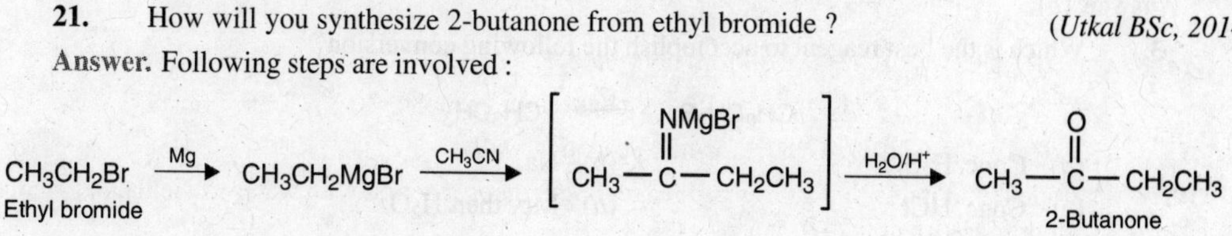

22. Compound (*A*) reacts with ethylmagnesium bromide to give a product which on hydrolysis yields *n*-propyl alcohol. What is the structural formula of (*A*) ?

Answer. (*A*) is Formaldehyde.

23. Compound (*A*) reacts with ethylmagnesium iodide to give a product which on hydrolysis yields *n*-butyl alcohol. What is the structural formula of (*A*) ?

Answer. (*A*) is Ethylene oxide.

24. Compound (*A*) reacts with ethylmagnesium bromide to give a product which on hydrolysis yields 3-methyl-3-pentanol. What is the structural formula of (*A*) ?

Answer. (*A*) is 2-Butanone

25. Compound (*A*), which is an alkyl cyanide reacts with ethylmagnesium iodide to give compound (*B*). Compound (*B*) reacts with dilute HCl to yield 3-pentanone. What is the structural formula of (*A*) ?

Answer. (*A*) is Ethyl cyanide.

26. A ketone $C_8H_{16}O$ (*A*) reacts with ethylmagnesium bromide to give a product which on hydrolysis yields an alcohol (*B*). Compound (*B*) undergoes dehydration with concentrated sulfuric acid to form (*C*) which on ozonolysis gives 3-pentanone as the only product. What are (*A*), (*B*), and (*C*) ? Write equations for all reactions.

Answer. (*A*) = 4-Ethyl-3-hexanone
(*B*) = 3,4-Diethyl-3-hexanol
(*C*) = 3,4-Diethyl-3-hexene

27. Compound (A) reacts with $SOCl_2$ to yield (*B*). Reaction of (*B*) with magnesium, followed by reaction with acetaldehyde and then aqueous acid, results in 5-methyl-2-heptanol. What are the structures of compounds (*A*) and (*B*) ?

Answer. (A) = $CH_3CH_2-\underset{}{\overset{CH_3}{\overset{|}{C}H}}-CH_2CH_2OH$ (B) $CH_3CH_2-\overset{CH_3}{\overset{|}{C}H}-CH_2CH_2Cl$

Multiple Choice Questions

1. Which of the following is not a true statement ?

(*a*) A Grignard reagent reacts with carbon dioxide to form a carboxylic acid.
(*b*) A Grignard reagent reacts with an aldehyde to form a secondary alcohol.
(*c*) A Grignard reagent reacts with an acyl halide to form a secondary alcohol.
(*d*) A Grignard reagent reacts with a ketone to form a tertiary alcohol.

Answer. (*c*)

2. The reaction of aldehydes and ketones with Grignard reagents followed by hydrolysis is an example of

(*a*) Electrophilic addition (*b*) Nucleophilic addition
(*c*) Nucleophilic substitution (*d*) Electrophilic substitution

Answer. (*b*)

3. Which is the best reagent to accomplish the following conversion ?

$$CH_3CH_2Br \xrightarrow{(?)} CH_3CH_3$$

(*a*) Conc. H_2SO_4 (*b*) Na
(*c*) Conc. HCl (*d*) Mg, then H_2O

Answer. (*d*)

4. What is the major product of the following reaction ?

$$CH_3CH_2-\overset{\overset{\Large O}{\|}}{C}-H \;+\; CH_3MgBr \longrightarrow \xrightarrow{H_2O/H^+}$$

(*a*) 1-Butanol (*b*) Butanal

(*c*) 2-Butanol (*d*) Butanone

Answer. (*c*)

5. Which of the following gives a tertiary alcohol when treated with Grignard reagents ?

(*a*) $H-\overset{\overset{\Large O}{\|}}{C}-H$ (*b*) $CH_3-\overset{\overset{\Large O}{\|}}{C}-H$

(*c*) $CH_3-\overset{\overset{\Large O}{\|}}{C}-CH_3$ (*d*) None of these

Answer. (*c*)

6. Which of the following compounds will react with methylmagnesium bromide to give *tert*-butyl alcohol ?

(*a*) Acetyl chloride (*b*) Acetone

(*c*) Isopropyl alcohol (*d*) Acetaldehyde

Answer. (*b*)

7. Lithium reacts with *n*-butyl chloride to give

(*a*) *n*-Butane (*b*) *n*-Butyllithium

(*c*) Chlorine (*d*) Hydrogen

Answer. (*b*)

8. Organometallic compounds of lithium are stable in

(*a*) Water (*b*) Air

(*c*) Ethyl alcohol (*d*) Ether

Answer. (*d*)

9. Which of the following statements about organometallic compounds is *false* ?

(*a*) Alkyllithium reagents (RLi) add to the carbonyl group of aldehydes and ketones.

(*b*) Grignard reagents (RMgBr) add to the carbonyl group of aldehydes and ketones.

(*c*) Alkyllithium and Grignard reagents do not add to esters.

(*d*) Grignard reagents are prepared in ether or tetrahydrofuran (THF).

Answer. (*c*)

10. Grignard reagents react with carbonyl groups to produce ?

(*a*) Alkenes (*b*) Alkynes

(*c*) Alcohols (*d*) Aldehydes and Ketones

Answer. (*c*)

11. Formaldehyde (HCHO) reacts first with CH_3MgCl and then with dilute acid to give a

(*a*) primary alcohol (*b*) secondary alcohol

(*c*) tertiary alcohol (*d*) quaternary alcohol

Answer. (*a*)

12. The product of the reaction of a Grignard reagent with an aldehyde is ?

(*a*) 1° Alcohol (*b*) 2° Alcohol

(*c*) 3° Alcohol (*d*) Ketone

Answer. (*b*)

13. Which of the following compounds gives a secondary alcohol upon reaction with methylmagnesium bromide ? Assume the usual acid workup.

(*a*) Butyl formate (*b*) 3-Pentanone
(*c*) Pentanal (*d*) Methyl butanoate

Answer. (*c*)

14. Tertiary alcohols may be prepared by the action of a Grignard reagent on

(*a*) a ketone (*b*) a secondary alcohol
(*c*) an ester (*d*) an aldehyde

Answer. (*a*)

15. The reaction of a Grignard reagent with carbon dioxide forms

(*a*) a ketone with one more carbon than the Grignard reagent.
(*b*) a primary alcohol.
(*c*) a carboxylic acid with one more carbon than the Grignard reagent.
(*d*) a secondary alcohol.

Answer. (*c*)

16. CH_3CHO reacts first with ethyllithium (CH_3CH_2Li) and then with dilute acid to give

(*a*) Ethane (*b*) 2-Butanol
(*c*) Butane (*d*) No reaction

Answer. (*b*)

17. Grignard reagents react with ethylene oxide to give:

(*a*) Ethers
(*b*) Esters
(*c*) 1° alcohols with two carbons added to the Grignard alkyl group
(*d*) 2° alcohols with two carbons added to the Grignard alkyl group

Answer. (*c*)

18. The reaction of the methylmagnesium bromide with ethylene oxide gives, after addition of aqueous acid, a

(*a*) primary alcohol (*b*) secondary alcohol
(*c*) tertiary alcohol (*d*) ketone

Answer. (*a*)

19. The final product of the reaction of a nitrile and a Grignard reagent is ?

(*a*) an acid (*b*) an aldehyde
(*c*) an amine (*d*) a ketone

Answer. (*d*)

20. The reaction of an ester with a Grignard reagent followed by hydrolysis yields:

(*a*) A 1° alcohol (*b*) A 2° alcohol
(*c*) A 3° alcohol (*d*) An ether

Answer. (*c*)

15

Alcohols

Ethanol is considered as the fuel of the future.

Alcohols are compounds in which a hydroxyl (–OH) group is attached to saturated carbon atom.

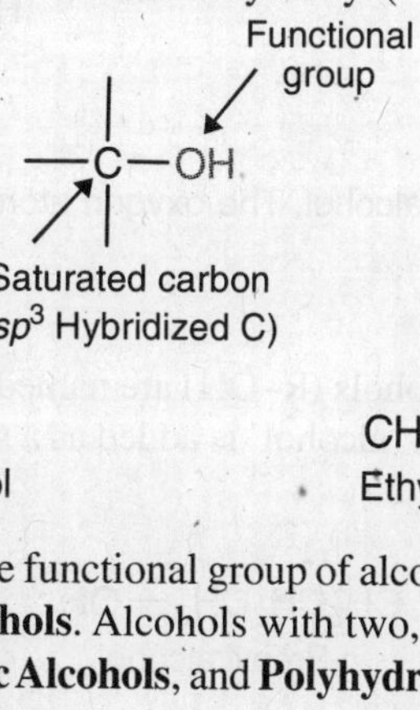

R—OH
General formula

Examples are :

CH_3OH — Methyl alcohol

CH_3CH_2OH — Ethyl alcohol

The hydroxyl group is the functional group of alcohols. Alcohols containing one hydroxyl group are called **Monohydric Alcohols**. Alcohols with two, three, or more hydroxyl groups are known as **Dihydric Alcohols, Trihydric Alcohols**, and **Polyhydric Alcohols** respectively. For example,

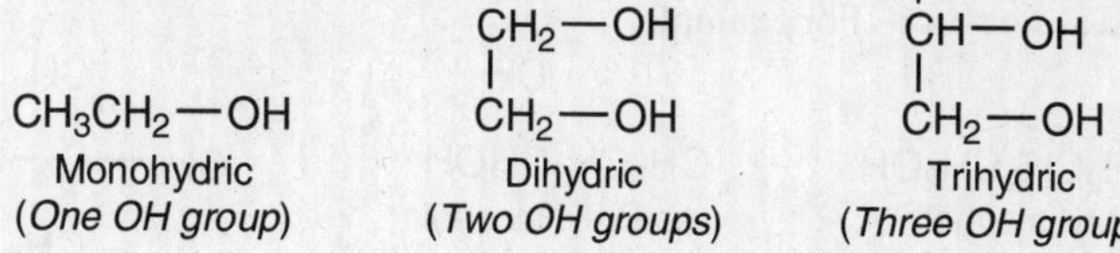

Monohydric (*One OH group*) — Dihydric (*Two OH groups*) — Trihydric (*Three OH groups*)

MONOHYDRIC ALCOHOLS

Monohydric alcohols contain one –OH group attached to a saturated carbon. They may be represented as (R–OH).

PRIMARY (1°), SECONDARY (2°), AND TERTIARY (3°) ALCOHOLS

Monohydric alcohols are classified as primary (1°), secondary (2°), or tertiary (3°), depending upon whether the –OH group is attached to a primary, a secondary, or a tertiary carbon.

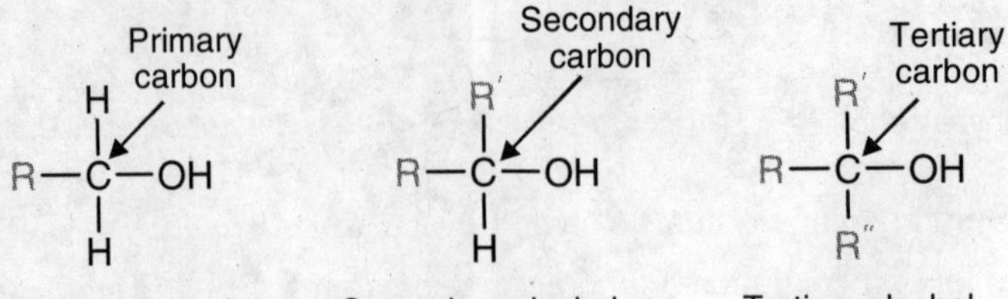

Primary alcohol Secondary alcohol Tertiary alcohol

STRUCTURE

Let us consider methyl alcohol (CH_3OH) for illustrating the orbital make up of alcohols. In methyl alcohol both oxygen and carbon are sp^3 hybridized. Two of the sp^3 orbitals of oxygen are completely filled and cannot take part in bond formation (Fig. 15.1). The C–O bond in methyl alcohol is formed by overlap of an sp^3 orbital of carbon and an sp^3 orbital of oxygen. The O–H bond is formed by overlap of an sp^3 orbital of oxygen and *s* orbital of hydrogen. The C–O–H bond angle is 105°. It is less than the normal tetrahedral angle. This is because the two completely filled sp^3 orbitals of oxygen repel each other. This results in reduction of the bond angle.

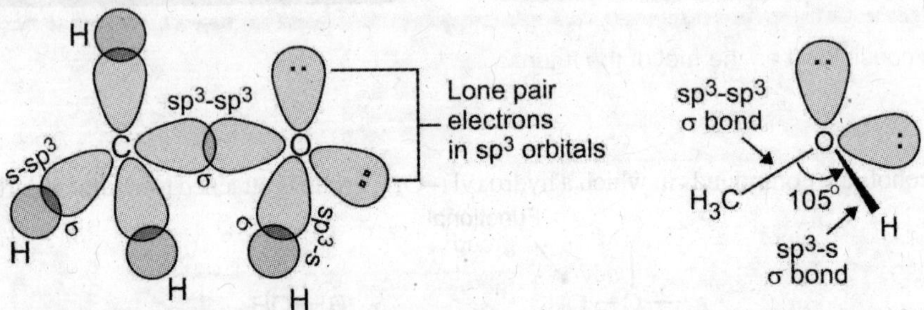

Fig.15.1. Orbital structure of methyl alcohol. The oxygen atom is sp^3 hybridized.

NOMENCLATURE

Alcohols are named by three systems :

(1) **Common System**. In this system alcohols (R–OH) are named as *Alkyl Alcohols*. The alkyl group attached to the –OH group is named and 'alcohol' is added as a separate word. For example,

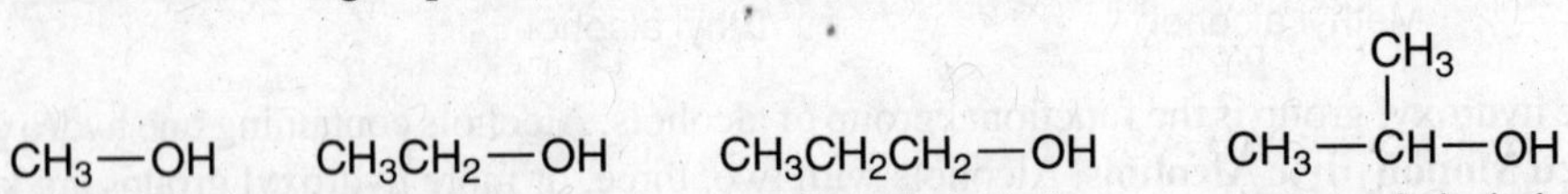

Methyl alcohol Ethyl alcohol *n*-Propyl alcohol Isopropyl alcohol

As we go higher in the series, it becomes necessary to indicate whether a particular alcohol is primary (1°), secondary (2°), or tertiary (3°). The prefix secondary is abbreviated as *sec*-. The prefix tertiary is abbreviated as *tert*- or *t*-. For example,

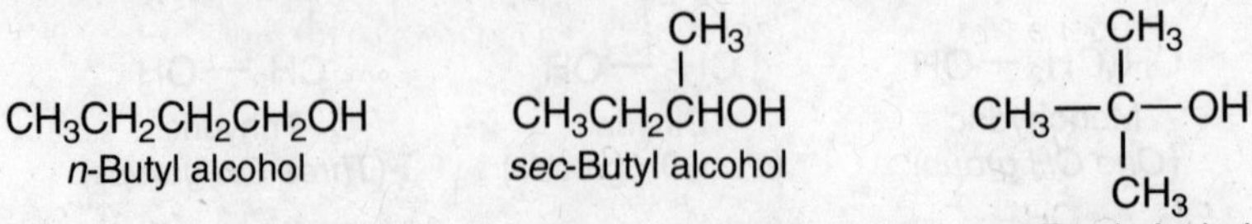

n-Butyl alcohol *sec*-Butyl alcohol *tert*-Butyl alcohol

(2) **Carbinol System**. In this system alcohols are considered as derivatives of methyl alcohol which is called *Carbinol*. The alkyl group attached to the carbon carrying the –OH group are named in alphabetic order. Then the suffix *–carbinol* is added. For example,

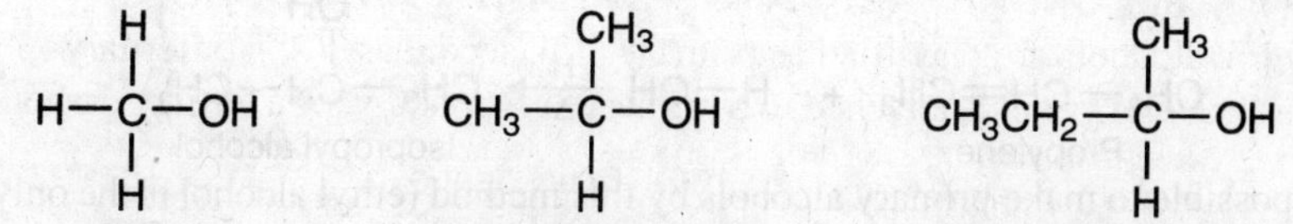

Carbinol (Methyl alcohol) — Dimethylcarbinol (Isopropyl alcohol) — Ethylmethylcarbinol (*sec*-Butyl alcohol)

(3) **IUPAC System**. In this system alcohols are named as *Alkanols*. The IUPAC rules are :

(*a*) Select the longest continuous carbon chain containing the –OH group.

(*b*) Change the name of the alkane corresponding to this chain by dropping the ending *-e* and adding the suffix *-ol*.

(*c*) Number the chain so as to give the carbon carrying the –OH group, the lowest possible number. The position of the –OH group is indicated by this number.

(*d*) Indicate the positions of other substituents, or multiple bonds by numbers.

The examples given below show how these rules are used :

$CH_3—OH$
Methanol
(*Methane – e + ol*)

$CH_3CH_2—OH$
Ethanol
(*Ethane – e + ol*)

$\overset{3}{C}H_3\overset{2}{C}H_2\overset{1}{C}H_2OH$
1-Propanol

$Cl\overset{3}{C}H_2\overset{2}{C}H_2\overset{1}{C}H_2OH$
3-Chloro-1-propanol

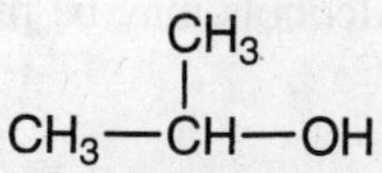

2-Propanol

$\overset{3}{C}H_2{=}\overset{2}{C}H—\overset{1}{C}H_2—OH$
2-Propen-1-ol
(Allyl alcohol)

METHODS OF PREPARATION

Alcohols are prepared by the following methods :

(1) **Hydrolysis of Alkyl Halides**. Alkyl halides react with aqueous sodium hydroxide to form alcohols.

$$\underset{\text{Alkyl halide}}{R—X} + NaOH \xrightarrow[\Delta]{H_2O} \underset{\text{Alcohol}}{R—OH} + NaX$$

$$\underset{\text{Ethyl bromide}}{CH_3CH_2—Br} + NaOH \xrightarrow[\Delta]{H_2O} \underset{\text{Ethyl alcohol}}{CH_3CH_2—OH} + NaBr$$

(2) **Hydration of Alkenes**. Alkenes react with sulfuric acid to produce alkyl hydrogen sulfates (Markovnikov rule is followed). Alkyl hydrogen sulfates on hydrolysis give alcohols. For example,

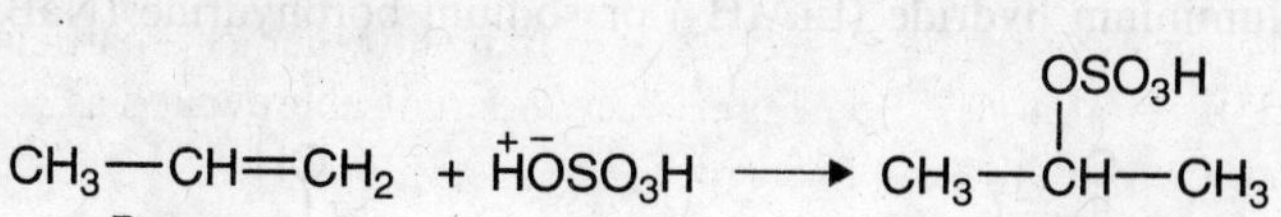

Propene — Isopropyl hydrogen sulfate

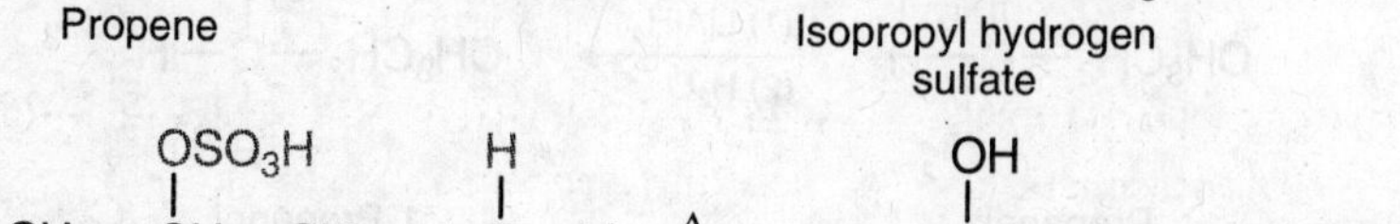

Isopropyl alcohol

The overall result of the above reactions appears to be Markovnikov addition of H_2O (Hydration) to a double bond.

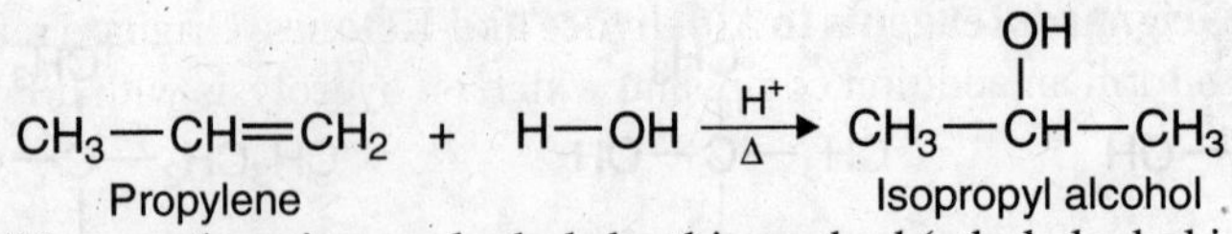

It is not possible to make primary alcohols by this method (ethyl alcohol is the only exception).

(3) **Hydroboration-Oxidation of Alkenes**. Alkenes react with diborane, B_2H_6 to form trialkylboranes. Diborane adds as borane, BH_3. The positive part of BH_3 is the boron, the negative part is hydrogen. For example,

$$\underset{\text{Propylene}}{3CH_3-CH=CH_2} + BH_3 \longrightarrow \underset{\text{Tripropylborane}}{(CH_3CH_2CH_2)_3B}$$

Trialkylboranes are used for making primary alcohols by reaction with alkaline aqueous solution of hydrogen peroxide.

$$(CH_3CH_2CH_2)_3B + H_2O_2 \xrightarrow{\bar{O}H} \underset{\substack{n\text{-Propyl alcohol} \\ \text{(1-Propanol)}}}{3CH_3CH_2CH_2OH} + H_3BO_3$$

The overall result of the above reactions appears to be *anti*-Markovnikov addition of H_2O to a double bond.

$$\underset{\text{Propylene}}{CH_3-CH=CH_2} \xrightarrow[\text{(2) } H_2O/\bar{O}H]{\text{(1) } B_2H_6} \underset{n\text{-Propyl alcohol}}{CH_3CH_2CH_2OH}$$

(4) **Hydrolysis of Esters**. Alcohols may be prepared by base or acid-catalyzed hydrolysis of esters.

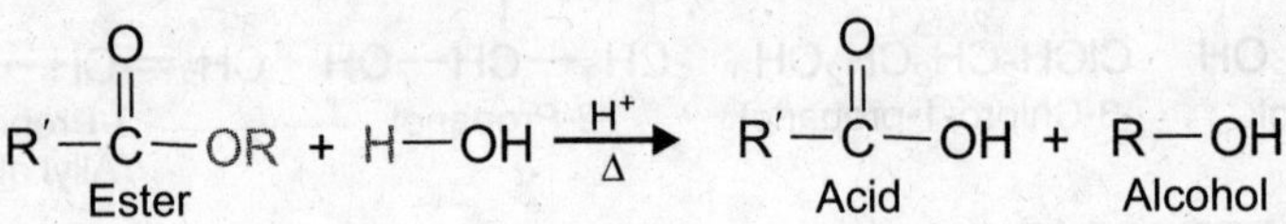

(5) **Catalytic Hydrogenation of Aldehydes and Ketones.** Catalytic hydrogenation also reduces aldehydes and ketones to 1° and 2° alcohols, respectively, using H_2 and Pd (or another metal catalyst). H_2 adds to the C=O in much the same way that it adds to the C=C of an alkene.

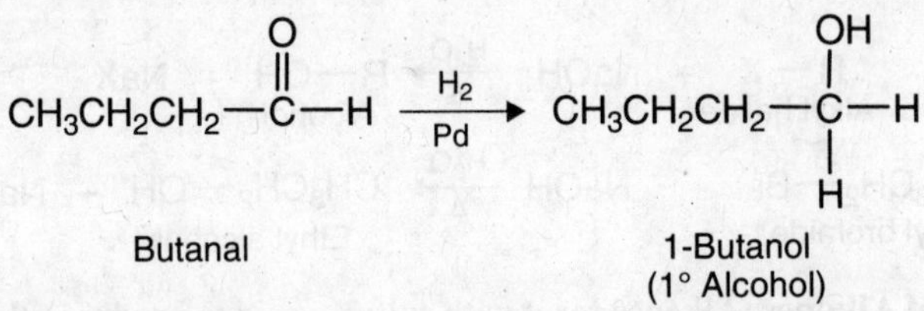

(6) **Reduction of Carbonyl Compounds with Metal Hydrides.** Aldehydes and ketones undergo reduction with lithium aluminium hydride ($LiAlH_4$) or sodium borohydride ($NaBH_4$) to form alcohols.

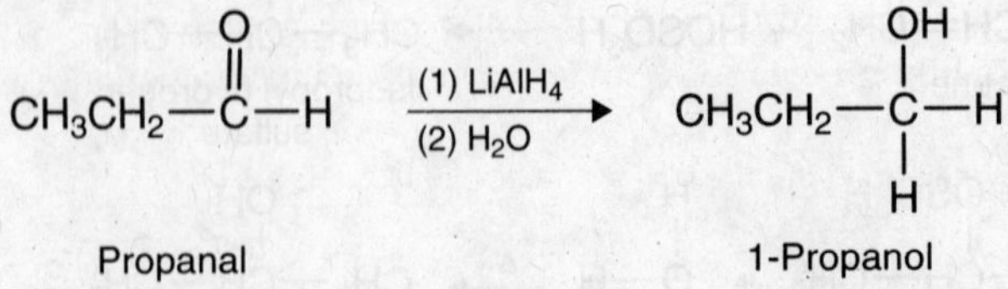

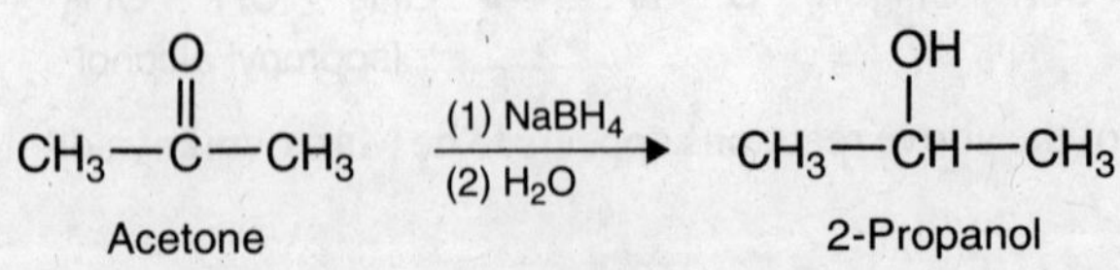

(7) **Addition of Grignard Reagents to Aldehydes and Ketones**. Grignard reagents react with aldehydes or ketones to form an addition compound which on hydrolysis with dilute acid gives the corresponding alcohols.

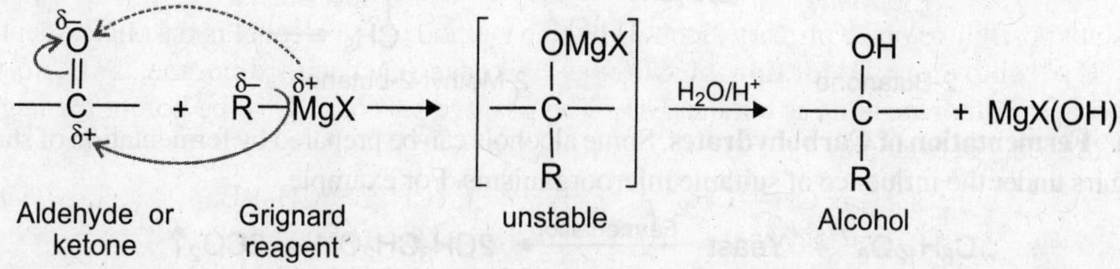

Primary alcohols are obtained by treating a Grignard reagent with formaldehyde or ethylene oxide.

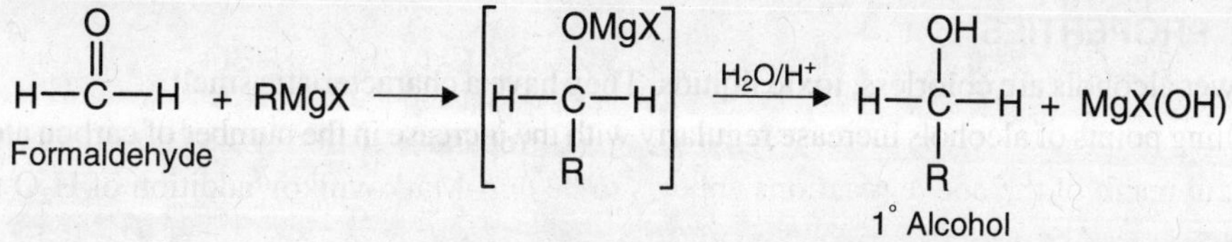

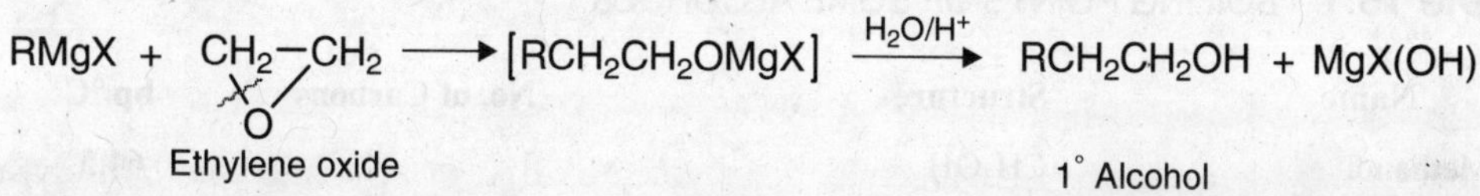

Secondary alcohols are obtained by treating a Grignard reagent with aldehydes other than formaldehyde.

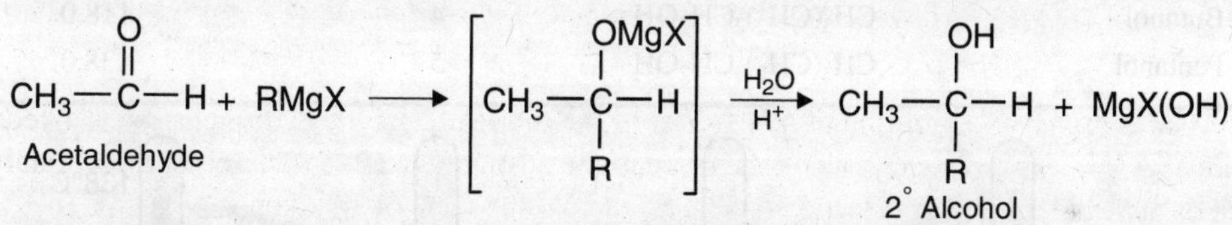

Tertiary alcohols are obtained by treating Grignard reagents with ketones.

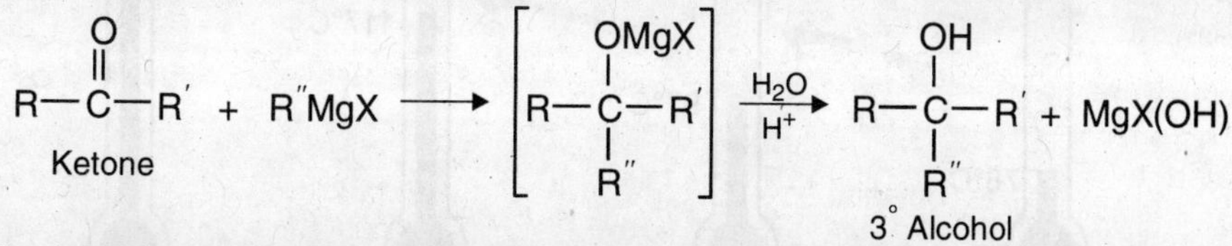

(8) **Addition of Lithium Reagents to Aldehydes and Ketones.** Lithium reagents (RLi) are closely related to Grignard reagents. These reagents are prepared similar to Grignard reagents, by the reaction of lithium metal with an organic halide. They undergo the same types of reactions as Grignard reagents do. Because Li is more electropositive than Mg, lithium reagents are more reactive than Grignard reagents. Therefore, lithium reagents can be used for the synthesis of even very hindered alcohols. Also, lithium reactions can be run at very low temperatures.

$$RX + 2\,Li \longrightarrow RLi + LiX$$

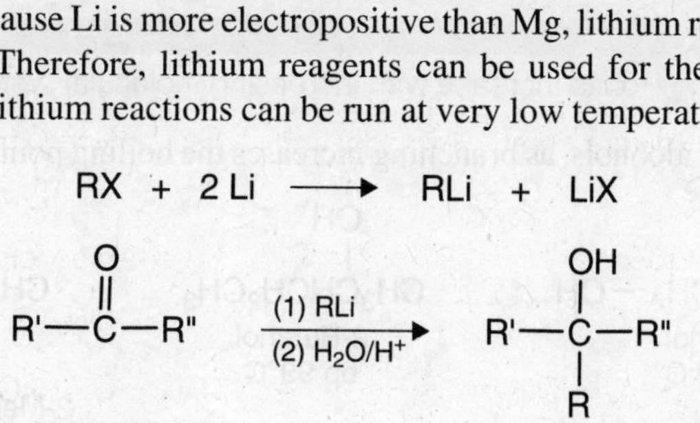

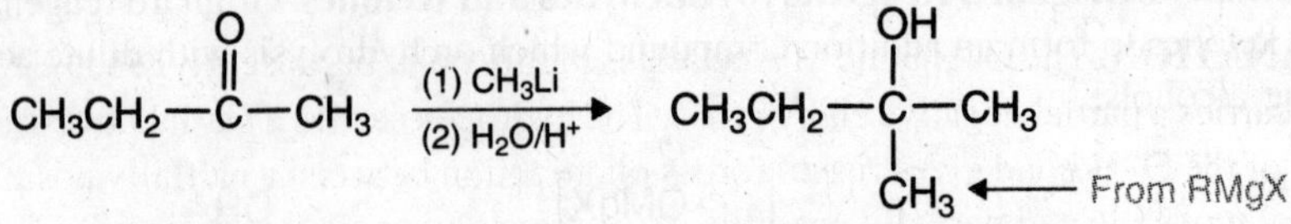

2-Butanone 2-Methyl-2-butanol

(9) **Fermentation of Carbohydrates**. Some alcohols can be prepared by fermentation of starches and sugars under the influence of suitable microorganisms. For example,

$$\underset{\text{Glucose}}{C_6H_{12}O_6} + \text{Yeast} \xrightarrow{\text{Fermentation}} \underset{\text{Ethyl alcohol}}{2CH_3CH_2OH} + 2CO_2\uparrow$$

PHYSICAL PROPERTIES

(1) Lower alcohols are colorless, toxic liquids. They have a characteristic smell.

(2) Boiling points of alcohols increase regularly with the increase in the number of carbon atoms (Table 15.1).

Table 15.1 BOILING POINTS OF SOME ALCOHOLS

Name	Structure	No. of Carbons	bp °C
Methanol	CH_3OH	1	64.5
Ethanol	CH_3CH_2OH	2	78.3
1-Propanol	$CH_3CH_2CH_2OH$	3	97.0
1-Butanol	$CH_3(CH_2)_2CH_2OH$	4	118.0
1-Pentanol	$CH_3(CH_2)_3CH_2OH$	5	138.0

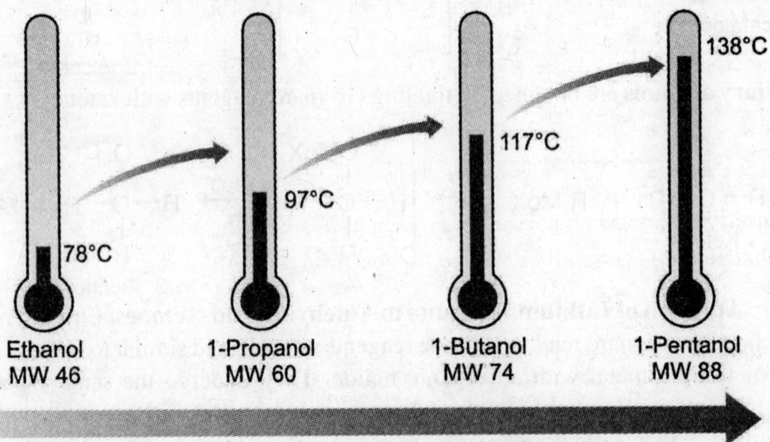

Boiling Points Increase with increasing molecular weight.

(3) Among isomeric alcohols, as branching increases the boiling points decrease. For example,

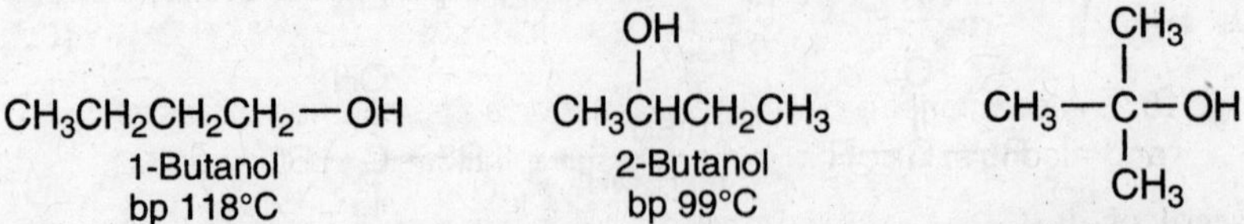

1-Butanol bp 118°C 2-Butanol bp 99°C 2-Methyl-2-propanol bp 83°C

(4) Boiling points of alcohols are much higher than those of the corresponding alkanes.

EXPLANATION. The O–H bond of alcohols is highly polar because oxygen is electronegative. The oxygen carries a partial negative charge (δ–). The hydrogen carries a partial positive charge (δ+). The polarity of the O–H bond gives rise to forces of attraction between a partially positive hydrogen in one alcohol molecule and partially negative oxygen in another alcohol molecule. These forces of attraction are referred to as **Hydrogen Bonding**. The reason that alcohols have higher boiling points is that a great deal of energy (in the form of heat) is required to overcome these attractive forces.

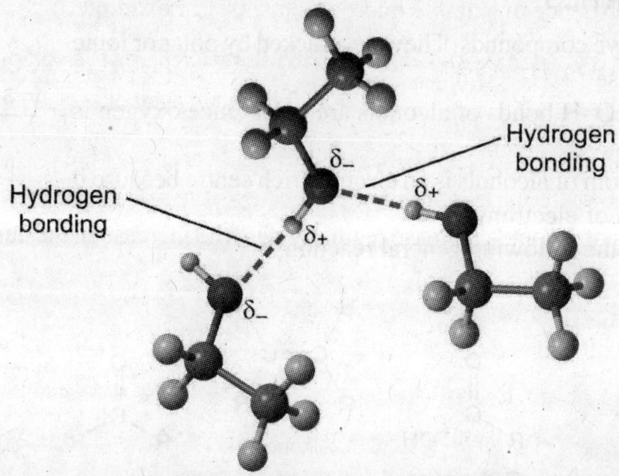

Hydrogen bonding in alcohols. A weak attraction between a positively polarized OH hydrogen and a negatively polarized oxygen holds molecules together.

Alkanes, which have no –OH group, do not undergo hydrogen bonding. Therefore, their boiling points are low.

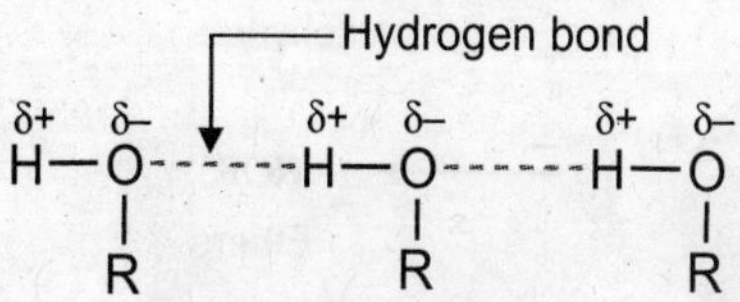

Hydrogen bonding in alcohol molecules

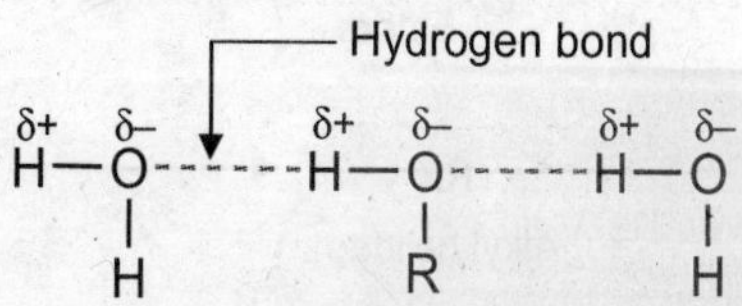

Hydrogen bonding between alcohol and water molecules

Figure 15.2 shows comparison of boiling points of some alkanes, chloroalkanes, and alcohols. Notice that alcohols generally have the higher boiling points.

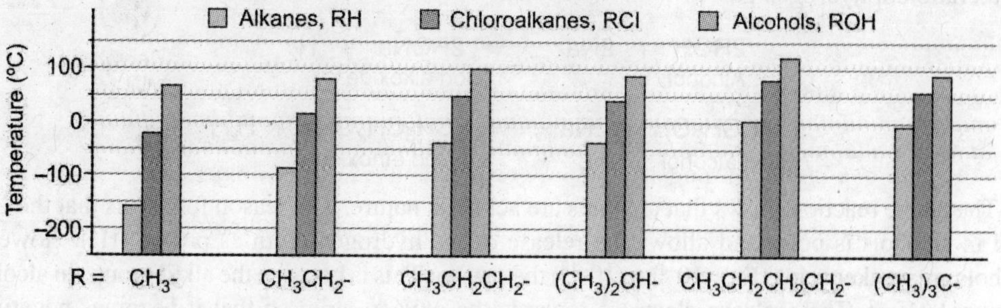

Fig. 15.2. A comparison of boiling points for some alkanes, chloroalkanes, and alcohols. Alcohols generally have the higher boiling points.

(5) Lower alcohols (C_1 to C_3) are completely soluble in water. As we go higher in the series, the water solubility falls rapidly.

EXPLANATION. A compound that forms hydrogen bonds between its own molecules can also form hydrogen bonds with water. These alcohol-water hydrogen bonds are the cause of high solubility of lower alcohols in water. In higher alcohols, the nonpolar alkyl group becomes more important. They have less tendency to form hydrogen bonds with water and hence are less soluble.

(6) ***IR Spectrum.*** Alcohols show a strong broad band in the 3200-3600 cm^{-1} region due to O–H stretching. They show a broad band in the 1000-1200 cm^{-1} region due to C–O stretching.

CHEMICAL PROPERTIES

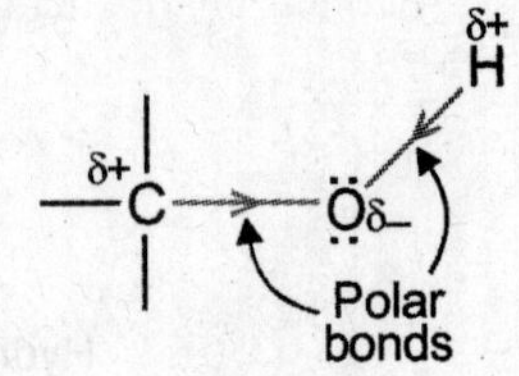

Alcohols are reactive compounds. They are attacked by polar or ionic reagents. This is because :

(1) The C–O and O–H bonds of alcohols are polar since oxygen is highly electronegative.

(2) The oxygen atom of alcohols is an electron-rich centre because it has two unshared pairs of electrons.

Alcohols undergo the following general reactions :

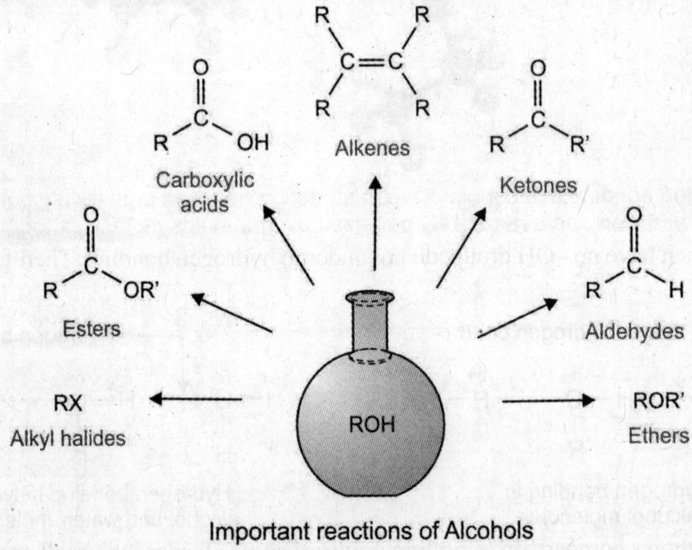

Important reactions of Alcohols

(1) **Reaction with Active Metals.** Alcohols react with sodium or potassium to form alkoxides with the liberation of hydrogen gas.

$$\underset{\text{Alcohol}}{2ROH} + 2Na \longrightarrow \underset{\text{Sod alkoxide}}{2R\overset{-}{O}\overset{+}{Na}} + H_2$$

$$\underset{\text{Ethyl alcohol}}{2CH_3CH_2OH} + 2Na \longrightarrow \underset{\text{Sod ethoxide}}{2CH_3CH_2\overset{-}{O}\overset{+}{Na}} + H_2\uparrow$$

The above reaction shows that alcohols are acidic in nature. The reason for this is that the O–H bond in alcohols is polar and allows the release of the hydrogen atom as proton (H^+). However, alcohols are weaker acids ($K_a = 10^{-16}$ to 10^{-18}) than water. This is because the alkyl groups in alcohols have a +I effect. They release electrons towards the oxygen atom so that it becomes negatively charged. This negative charge on oxygen makes the release of the positive proton more difficult.

Tertiary alcohols are less acidic than secondary alcohols. The secondary alcohols are less acidic than primary alcohols. This is because the +I effect would be maximum in tertiary alcohols, as they contain three alkyl groups attached to the carbon bearing the –OH group.

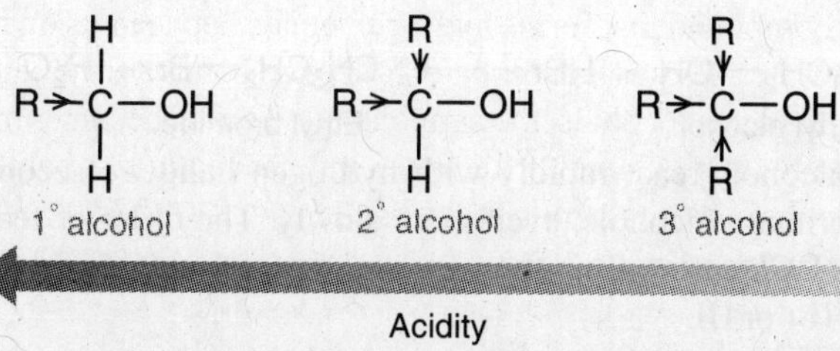

Alcohols are not acidic enough to react with aqueous NaOH or KOH.

$$\text{R—OH} + \text{NaOH} \longrightarrow \text{No Reaction}$$

Sodium metal reacts vigorously with alcohols such as methyl alcohol or ethyl alcohol with the evolution of hydrogen gas (Seen as bubbles)

(2) **Reaction with Phosphorus Halides.** Alcohols react with phosphorus pentahalides (PX_5) and phosphorus trihalides (PX_3) to form alkyl halides.

(*i*)

$$\underset{\text{Alcohol}}{\text{R—OH}} + PX_5 \longrightarrow \underset{\text{Alkyl halide}}{\text{R—X}} + POX_3 + HX$$

$$\underset{\text{Ethyl alcohol}}{CH_3CH_2\text{—OH}} + PBr_5 \longrightarrow \underset{\text{Ethyl bromide}}{CH_3CH_2\text{—Br}} + POBr_3 + HBr$$

(*ii*)

$$\underset{\text{Alcohol}}{3\text{R—OH}} + PX_3 \longrightarrow \underset{\text{Alkyl halide}}{3\text{R—X}} + H_3PO_3$$

$$\underset{\text{Methyl alcohol}}{3CH_3\text{—OH}} + PI_3 \longrightarrow \underset{\text{Methyl iodide}}{3CH_3\text{—I}} + H_3PO_3$$

(3) **Reaction with Thionyl Chloride.** Alcohols react with thionyl chloride ($SOCl_2$) to form alkyl chlorides.

$$\underset{\text{Alcohol}}{\text{R—OH}} + SOCl_2 \longrightarrow \underset{\text{Alkyl chloride}}{\text{R—Cl}} + SO_2\uparrow + HCl\uparrow$$

$$\underset{\text{Ethyl alcohol}}{CH_3CH_2\text{—OH}} + SOCl_2 \longrightarrow \underset{\text{Ethyl chloride}}{CH_3CH_2\text{—Cl}} + SO_2\uparrow + HCl\uparrow$$

(4) **Reaction with Hydrogen Halides.** Alcohols react with hydrogen halides (HX) to form the corresponding alkyl halides.

$$\underset{\text{Alcohol}}{\text{R—OH}} + HX \longrightarrow \underset{\text{Alkyl halide}}{\text{R—X}} + H_2O$$

where HX = HI, HBr, HCl

$$\underset{\text{Ethyl alcohol}}{CH_3CH_2{-}OH} + HBr \longrightarrow \underset{\text{Ethyl bromide}}{CH_3CH_2{-}Br} + H_2O$$

In general, tertiary alcohols react rapidly with hydrogen halides ; secondary alcohols react somewhat slower ; and primary alcohols, even more slowly. The order of reactivity of hydrogen halides is HI > HBr > HCl. HCl reacts only in the presence of a catalyst (anhydrous $ZnCl_2$). No catalyst is required in the case of HBr or HI.

MECHANISM. Primary alcohols react with hydrogen halides by an S_N1 mechanism. The mechanism of the reaction between ethyl alcohol (1° alcohol) and hydrogen bromide is described below :

Step 1. Protonation of ethyl alcohol.

$$\underset{\text{Ethyl alcohol}}{CH_3{-}CH_2{-}\ddot{O}{-}H} + H^+ \longrightarrow \underset{\text{Protonated alcohol}}{CH_3{-}CH_2{-}\overset{+}{\ddot{O}}(H){-}H}$$

Step 2. Nucleophile (Br^-) attacks the carbon holding the protonated hydroxyl group to form ethyl bromide.

$$Br^- + \underset{\text{Protonated alcohol}}{CH_3{-}CH_2{-}\overset{+}{\ddot{O}}(H){-}H} \xrightarrow{S_N2} \underset{\text{Ethyl bromide}}{CH_3{-}CH_2{-}Br} + H_2O$$

Tertiary and secondary alcohols react with hydrogen halides by an S_N2 mechanism.

(5) **Reaction with Nitric Acid**. Alcohols react with nitric acid to form alkyl nitrates. For example,

$$\underset{\text{Ethyl alcohol}}{CH_3CH_2{-}OH} + HO{-}NO_2 \longrightarrow \underset{\text{Ethyl nitrate}}{CH_3CH_2{-}O{-}NO_2} + H_2O$$

(6) **Reaction with Sulfuric Acid**. The reaction of alcohols with sulfuric acid is very sensitive to reaction conditions. For example,

(*i*) When ethyl alcohol is treated with concentrated sulfuric acid at room temperature, ethyl hydrogen sulfate is produced.

$$\underset{\text{Ethyl alcohol}}{C_2H_5{-}OH} + HO{-}SO_2OH \xrightarrow[\text{temperature}]{\text{room}} \underset{\text{Ethyl hydrogen sulfate}}{C_2H_5{-}O{-}SO_2OH} + H_2O$$

(*ii*) **Dehydration of Alcohols to Alkenes**. When ethyl alcohol is treated with concentrated sulfuric acid at 170°C, ethylene is formed. Notice that only one alcohol molecule is involved in the reaction.

$$\underset{\text{Ethyl alcohol}}{CH_3{-}CH_2{-}OH} \xrightarrow[170°C]{\text{Conc } H_2SO_4} \underset{\text{Ethylene}}{CH_2{=}CH_2} + H_2O$$

MECHANISM. The mechanism of the above reaction involves the following steps :

Step 1. Protonation of the alcohol molecule.

$$\underset{\text{Ethyl alcohol}}{CH_3{-}CH_2{-}\ddot{O}{-}H} + H^+ \longrightarrow \underset{\text{Protonated alcohol}}{CH_3{-}CH_2{-}\overset{+}{\ddot{O}}(H){-}H}$$

Step 2. Elimination of water molecule from the protonated alcohol gives a stable carbonium ion.

$$CH_3{-}CH_2{-}\overset{+}{\ddot{O}}(H){-}H \longrightarrow \underset{\text{Ethyl carbonium ion}}{CH_3{-}\overset{+}{C}H_2} + H_2O$$

Step 3. Carbonium ion loses a proton to give the alkene.

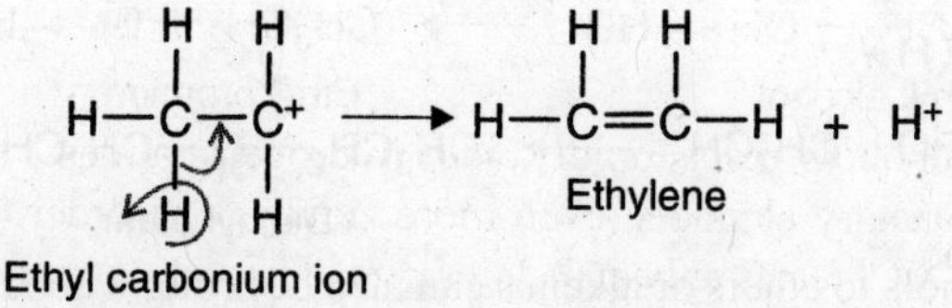

The ease of dehydration of alcohols follows the order 3° > 2° > 1° which is also the order of stability of the carbonium ions.

Dehydration of secondary and tertiary alcohols containing four or more carbon atoms gives a mixture of two alkenes. For example,

$$\underset{\text{2-Butanol}}{CH_3CH_2\overset{\overset{OH}{|}}{C}HCH_3} \xrightarrow{\text{Conc } H_2SO_4} \underset{\substack{\text{1-Butene} \\ (\textit{Minor})}}{CH_3CH_2CH{=}CH_2} + \underset{\substack{\text{2-Butene} \\ (\textit{Major})}}{CH_3CH{=}CHCH_3}$$

The alkene produced in greater abundance is indicated by Saytzeff Rule. It states that the alkene formed preferentially is the one containing the higher number of alkyl groups. Therefore, in the above example, 2-butene is the major product.

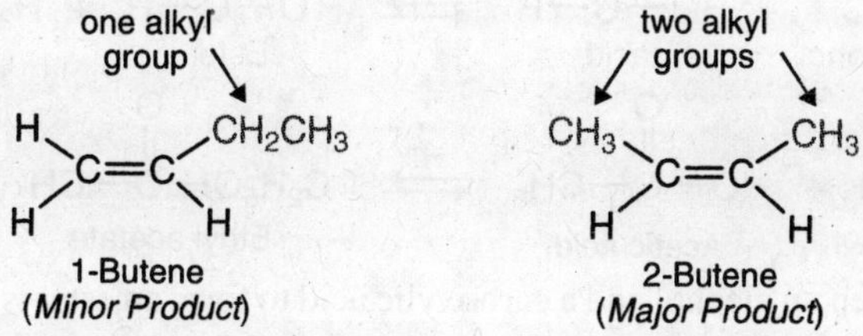

(*iii*) **Dehydration of Alcohols to Ethers.** When excess of ethyl alcohol is treated with concentrated sulfuric acid at 140°C, diethyl ether is formed. Notice that two alcohol molecules are involved in the reaction.

$$\underset{\text{Ethyl alcohol}}{CH_3CH_2{-}OH} + HO{-}CH_2CH_3 \xrightarrow[140°C]{\text{Conc } H_2SO_4} \underset{\text{Diethyl ether}}{CH_3CH_2{-}O{-}CH_2CH_3} + H_2O$$

MECHANISM. The mechanism of the above reaction involves the following steps :

Step 1. Protonation of the alcohol moelcule.

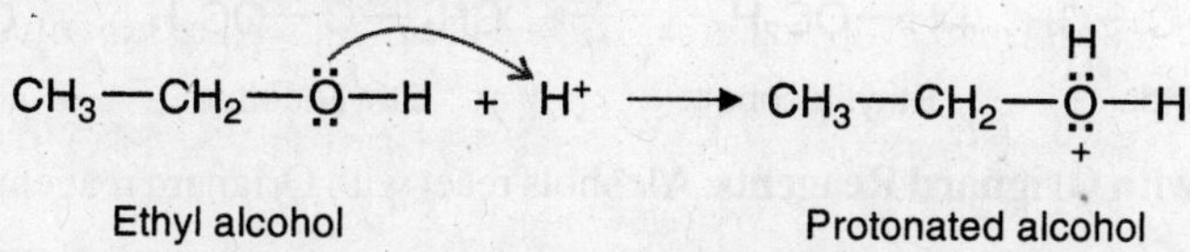

Step 2. Elimination of water molecule from the protonated alcohol gives a stable carbonium ion.

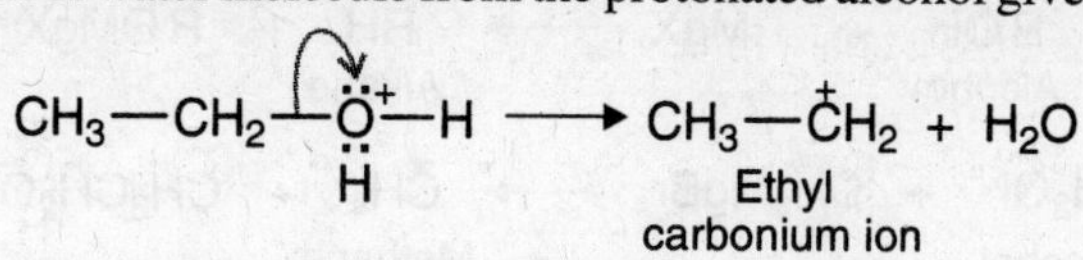

Step 3. Carbonium ion attacks another alcohol molecule to form an oxonium ion.

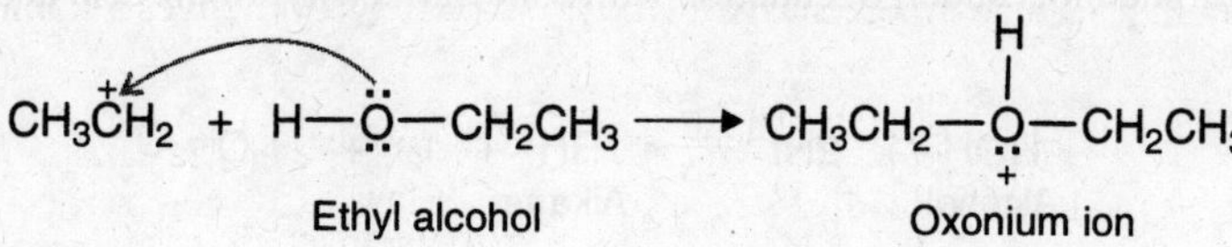

Step 4. Oxonium ion loses a proton to form an ether.

$$CH_3CH_2-\overset{+}{\underset{|}{O}}(H)-CH_2CH_3 \longrightarrow \underset{\text{Diethyl ether}}{CH_3CH_2-\ddot{O}-CH_2CH_3} + H^+$$

Dehydration of alcohols to ethers or alkenes can also be brought about by passing the vapors of the alcohol over heated alumina catalyst. For example,

$$2CH_3CH_2OH \xrightarrow[200°C]{Al_2O_3} \underset{\text{Diethyl ether}}{CH_3CH_2-O-CH_2CH_3} + H_2O$$

$$CH_3CH_2OH \xrightarrow[350°C]{Al_2O_3} \underset{\text{Ethylene}}{CH_2=CH_2} + H_2O$$

(7) **Reaction with Carboxylic Acids**. Alcohols react with carboxylic acids to form esters. Concentrated sulfuric acid is used as a catalyst. The reaction is reversible and can be shifted in the forward direction by removing water as soon as it is formed.

$$\underset{\text{Alcohol}}{R'OH} + \underset{\text{acid}}{HO-\overset{O}{\overset{\|}{C}}-R} \underset{}{\overset{H^+}{\rightleftharpoons}} \underset{\text{Ester}}{R'O-\overset{O}{\overset{\|}{C}}-R} + H_2O$$

$$\underset{\text{Ethyl alcohol}}{C_2H_5OH} + \underset{\text{Acetic acid}}{HO-\overset{O}{\overset{\|}{C}}-CH_3} \overset{H^+}{\rightleftharpoons} \underset{\text{Ethyl acetate}}{C_2H_5O-\overset{O}{\overset{\|}{C}}-CH_3} + H_2O$$

The reaction between an alcohol and a carboxylic acid to form an ester is called **Esterification**.

(8) **Reaction with Acid Halides and Acid Anhydrides**. Alcohols react with acid halides and acid anhydrides to form esters.

$$\underset{\text{Acetyl chloride}}{CH_3-\overset{O}{\overset{\|}{C}}-Cl} + \underset{\text{Ethyl alcohol}}{H-OC_2H_5} \xrightarrow{\Delta} \underset{\text{Ethyl acetate}}{CH_3-\overset{O}{\overset{\|}{C}}-OC_2H_5} + HCl$$

$$\underset{\text{Acetic anhydride}}{CH_3-\overset{O}{\overset{\|}{C}}-O-\overset{O}{\overset{\|}{C}}-CH_3} + \underset{\text{Ethyl alcohol}}{H-OC_2H_5} \xrightarrow{\Delta} \underset{\text{Ethyl acetate}}{CH_3-\overset{O}{\overset{\|}{C}}-OC_2H_5} + \underset{\text{Acetic acid}}{CH_3-\overset{O}{\overset{\|}{C}}-OH}$$

(9) **Reaction with Grignard Reagents**. Alcohols react with Grignard reagents (RMgX) to form alkanes.

$$\underset{\text{Alcohol}}{R'OH} + RMgX \longrightarrow \underset{\text{Alkane}}{RH} + R'OMgX$$

$$\underset{\text{Ethyl alcohol}}{CH_3CH_2OH} + CH_3MgBr \longrightarrow \underset{\text{Methane}}{CH_4} + CH_3CH_2OMgBr$$

(10) **Reduction**. Alcohols undergo reduction with concentrated hydriodic acid and red phosphorus to produce alkanes.

$$\underset{\text{Alcohol}}{ROH} + 2HI \xrightarrow[\Delta]{P} \underset{\text{Alkane}}{RH} + I_2 + H_2O$$

$$\underset{\text{Ethyl alcohol}}{CH_3CH_2OH} + 2HI \xrightarrow[\Delta]{P} \underset{\text{Ethane}}{CH_3CH_3} + I_2 + H_2O$$

(11) **Oxidation**. Alcohols can be oxidized. The nature of the product depends on the type of alcohol and the conditions of the reaction. Most widely used oxidizing agents are $KMnO_4 + H_2SO_4$ or $Na_2Cr_2O_7 + H_2SO_4$. **Oxidation of alcohols can be used to distinguish between primary, secondary, and tertiary alcohols.**

Primary alcohols are first oxidized to aldehydes and then to acids.

$$\underset{\text{Ethyl alcohol}}{CH_3CH_2OH} \xrightarrow[Na_2Cr_2O_7/H^+]{[O]} \underset{\text{Acetaldehyde}}{CH_3-\overset{\overset{\displaystyle O}{\|}}{C}-H} \xrightarrow[Na_2Cr_2O_7/H^+]{[O]} \underset{\text{Acetic acid}}{CH_3-\overset{\overset{\displaystyle O}{\|}}{C}-OH}$$

Notice that the acid and the alcohol contain the same number of carbon atoms. The reaction can be stopped at the aldehyde stage by removing it from the oxidizing medium as it is formed (*e.g.*, by distillation).

Secondary alcohols are oxidized to the corresponding ketones.

$$\underset{\text{Isopropyl alcohol}}{CH_3-\overset{\overset{\displaystyle OH}{|}}{C}H-CH_3} \xrightarrow[Na_2Cr_2O_7/H^+]{[O]} \underset{\text{Acetone}}{CH_3-\overset{\overset{\displaystyle O}{\|}}{C}-CH_3} + H_2O$$

Further oxidation under very drastic conditions breaks up the ketone molecule, producing carboxylic acids containing fewer carbon atoms per molecule.

Tertiary alcohols are stable to oxidation under normal conditions.

$$\underset{\textit{tert}\text{-Butyl alcohol}}{CH_3-\overset{\overset{\displaystyle CH_3}{|}}{\underset{\underset{\displaystyle CH_3}{|}}{C}}-OH} \xrightarrow[Na_2Cr_2O_7/H^+]{[O]} \text{No Reaction}$$

Other reagents used for these oxidation reactions are :

(*a*) Alkaline potassium permanganate : $KMnO_4/OH$

(*b*) Chromic acid : $H_2Cr_2O_7$

This reagent is prepared *in situ* by the reaction of CrO_3 with dilute sulfuric acid (Jones' reagent).

(*c*) Pyridine chlorochromate (PCC), $C_5H_6NCrO_3Cl$, in dichloromethane solvent.

PCC in dichloromethane is a special oxidizing agent. It is used to convert primary alcohols to aldehydes. This reagent does not oxidize aldehydes to carboxylic acids.

$$\underset{\text{1-Butanol}}{CH_3CH_2CH_2CH_2OH} \xrightarrow[CH_2Cl_2]{PCC} \underset{\text{Butanal}}{CH_3CH_2CH_2-\overset{\overset{\displaystyle O}{\|}}{C}-H}$$

(12) **Reaction with the Hot Copper ; Dehydrogenation.** Different types of alcohols give different products when their vapors are passed over copper gauze at 300°C.

Primary alcohols lose hydrogen and give an aldehyde.

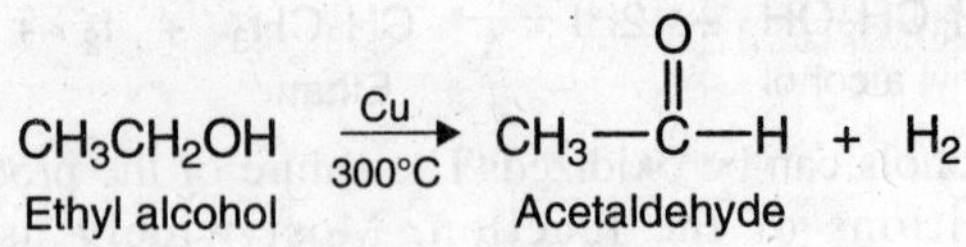

Ethyl alcohol — Acetaldehyde

Secondary alcohols lose hydrogen and yield a ketone.

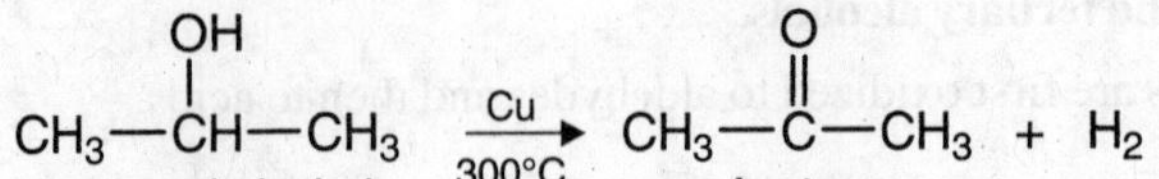

Isopropyl alcohol (2-Propanol) — Acetone

Tertiary alcohols are not dehydrogenated but lose a molecule of water to give alkenes.

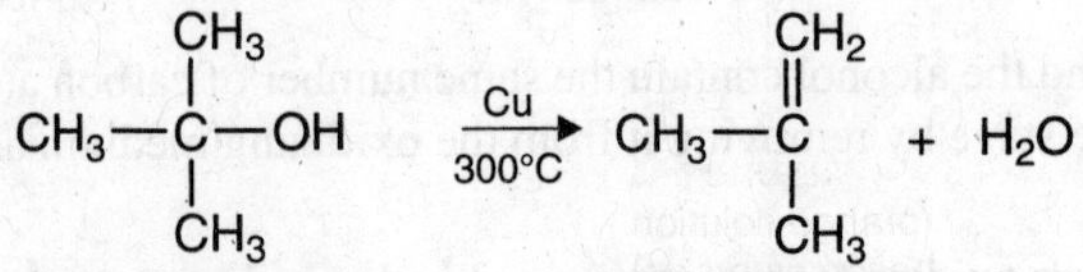

tert-Butyl alcohol — 2-Methylpropene

TESTS FOR AN –OH GROUP

The following three tests may be used to detect the presence of an –OH group in organic compounds. For these tests, take the liquid compound or a solution of the solid compound in an inert solvent such as *dry* ether or benzene.

(1) Add to it small pieces of *Sodium metal.* If bubbles of hydrogen gas are given off, the compound contains an –OH group.

$$2R{-}OH + 2Na \longrightarrow 2R{-}ONa + H_2\uparrow$$

(2) Add to it *Phosphorus pentachloride*. If the mixture becomes warm with evolution of HCl gas, the given compound contains an –OH group.

$$R{-}OH + PCl_5 \longrightarrow R{-}Cl + POCl_3 + HCl\uparrow$$

(3) Add *Acetyl chloride* or *Benzoyl chloride* to the substance. The separation of an oily layer of the ester and evolution of HCl gas indicates the presence of an –OH group. It may be noted that if the compound was taken in an organic solvent, the oily ester may dissolve in it and thus limiting the test to the evolution of HCl gas only.

HOW TO DISTINGUISH BETWEEN 1°, 2°, AND 3° ALCOHOLS ?

The following tests are used to distinguish between primary, secondary, and tertiary alcohols.

(1) **Lucas Test**. In this test, alcohols are treated with a solution of HCl and zinc chloride (*Lucas reagent*) to form alkyl halides. Zinc chloride serves as a catalyst.

$$\underset{\text{Alcohol}}{R{-}OH} + HCl \xrightarrow{ZnCl_2} \underset{\text{Alkyl halide}}{R{-}Cl} + H_2O$$

The three types of alcohols undergo this reaction at different rates. Tertiary alcohols react with *Lucas reagent* very rapidly. Secondary alcohols react somewhat slower. Primary alcohols react with *Lucas reagent* even more slowly.

In practice, the *Lucas test* is carried out as follows : An alcohol is mixed, at room temperature, with concentrated HCl and $ZnCl_2$. The alkyl chloride, which is formed, is insoluble in the medium. It causes the solution to become cloudy before it separates as a distinct layer.

(a) With **Tertiary alcohols** cloudiness appears immediately.

(b) With **Secondary alcohols** cloudiness appears in 5 minutes.

(c) With **Primary alcohols** the solution remains clear. This is because primary alcohols do not react with *Lucas reagent* at room temperature. High temperatures are needed.

Table 15.2 DICHROMATE TEST

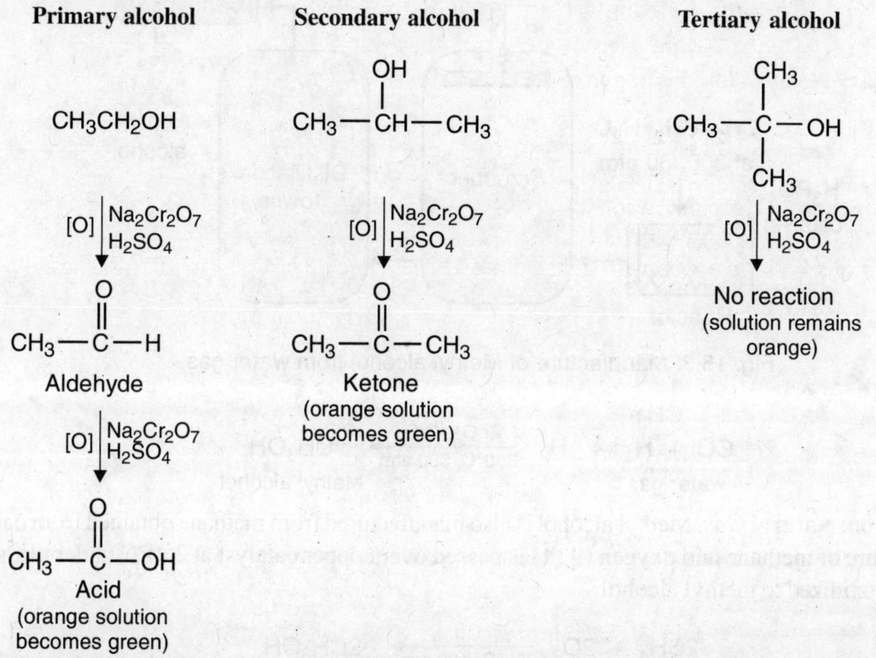

(2) **Dichromate Test**. This test is based on the fact that different types of alcohols give different products on oxidation. The alcohol is treated at room temperature with sodium dichromate in sulfuric acid (orange solution). Identification of the products gives us information regarding the type of the alcohol.

(*a*) **Primary alcohols** give a carboxylic acid containing the same number of carbons. There will be a change in color of the solution from orange to green.

(*b*) **Secondary alcohols** give a ketone containing the same number of carbons. There will be a change in color of the solution from orange to green.

(c) **Tertiary alcohols** do not react under these conditions. Solution will remain orange. (Table 15.2)

METHYL ALCOHOL, *Methanol*, CH_3OH

Methyl alcohol is also known as **Wood Alcohol** because it was formerly obtained by the destructive distillation of wood.

Manufacture. Methyl alcohol may be prepared by any of the general methods described before. For large scale production, the following three methods can be used :

(1) **From Water Gas**. Today, methanol is mostly manufactured by this method. Steam is passed through red-hot coke to form water gas. Water gas is a mixture of carbon monoxide and hydrogen.

$$C + H_2O\ (\textit{steam}) \xrightarrow{1300°C} \underbrace{CO + H_2}_{\text{water gas}}$$

Water gas is mixed with half its volume of hydrogen. The mixture is compressed to 300 atmospheres. It is then passed over zinc oxide-chromium oxide catalyst at 300°C. Methyl alcohol vapors are produced which are condensed (Fig. 15.3).

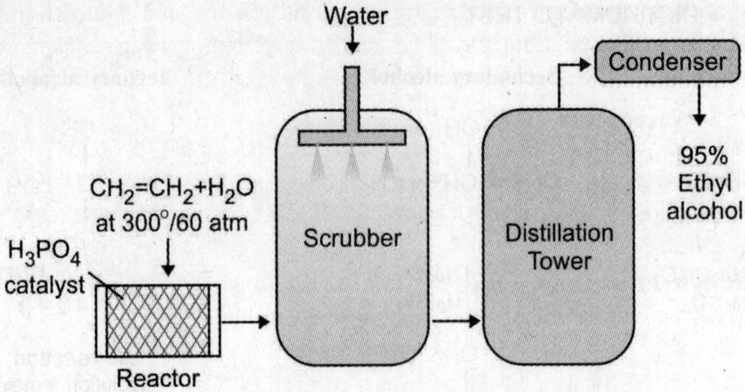

Fig. 15.3. Manufacture of Methyl alcohol from water gas.

$$\underbrace{CO + H_2}_{\text{water gas}} + H_2 \xrightarrow[300°C,\ 300\ atm]{ZnO\ +\ CrO_3} \underset{\text{Methyl alcohol}}{CH_3OH}$$

(2) **From Natural Gas**. Methyl alcohol is also manufactured from methane obtained from natural gas. A mixture of methane and oxygen (9 : 1) is passed over copper catalyst at 250°C under pressure. Methane is oxidized to methyl alcohol.

$$\underset{\text{Methane}}{2CH_4} + 2O_2 \xrightarrow[250°C,\ 100\ atm]{Cu} \underset{\text{Methyl alcohol}}{2CH_3OH}$$

Properties. Methyl alcohol is a colorless liquid, bp 64.5°C. It is miscible with water in all proportions. When taken internally, methyl alcohol is a poison. As little as 30 ml can cause death, and lesser amounts can cause irreversible blindness. A mixture of methyl alcohol vapor and air is easily ignited by even a small spark. Chemically it behaves like a typical monohydric alcohol. It gives most of the general reactions mentioned before.

Uses. Methyl alcohol is used : (*1*) as a solvent for paints and varnishes ; (*2*) as an antifreeze for automobile radiators ; (*3*) to denature ethyl alcohol ; (*4*) as a motor fuel. A 20% mixture of methyl alcohol and gasoline makes a good motor fuel ; and (*5*) in the manufacture of formaldehyde.

ETHYL ALCOHOL, *Ethanol*, CH_3CH_2OH

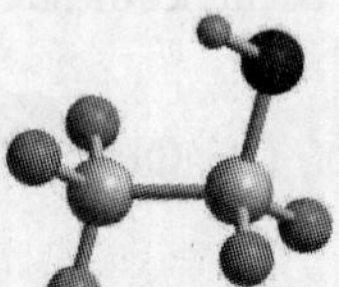

Ethyl alcohol is the alcohol of wine, beer, whiskey, and similar beverages. It is often referred to as simply 'alcohol'. Ethyl alcohol is also known as **Grain Alcohol**, since it can be prepared from starchy grains.

Manufacture. Ethyl alcohol is manufactured by the following methods :

(1) **Fermentation of Sugars**. Ethyl alcohol has been made for centuries by the fermentation of sugars in the presence of yeast.

$$\underset{\text{Sucrose}}{C_{12}H_{22}O_{11}} + H_2O \xrightarrow{\text{Invertase}} \underset{\text{Glucose}}{C_6H_{12}O_6} + \underset{\text{Fructose}}{C_6H_{12}O_6}$$

$$C_6H_{12}O_6 \xrightarrow{\text{Zymase}} \underset{\text{Ethyl alcohol}}{2CH_3CH_2OH} + 2CO_2$$

Molasses is a cheap source of glucose, fructose, and sucrose and it forms an excellent raw material for making ethyl alcohol. Ethyl alcohol is also produced from cheap starchy materials such as potatoes, maize and barley. The starchy material is first converted into sugar which is then fermented by yeast.

Just prior to World War II, the fermentation of molasses accounted for about 75% of total world ethanol production. Today less than 10% of ethanol is made by this method. Over 90% is synthesized from ethylene gas which is a by-product of petroleum industry. In India, molasses is still the major source of ethyl alcohol.

(2) **Sulfuric Acid Process**. This involves indirect hydration of ethylene. Ethylene is treated with concentrated sulfuric acid at 75°C to form ethyl hydrogen sulfate.

$$\underset{\text{Ethylene}}{CH_2{=}CH_2} + H_2SO_4 \xrightarrow{75°C} \underset{\text{Ethyl hydrogen sulfate}}{CH_3{-}CH_2{-}OSO_3H}$$

Ethyl hydrogen sulfate is then hydrolysed to form ethyl alcohol. The yield of ethyl alcohol is 90 per cent.

$$\underset{\text{Ethyl hydrogen sulfate}}{CH_3{-}CH_2{-}OSO_3H} + H_2O \longrightarrow \underset{\text{Ethyl alcohol}}{CH_3{-}CH_2{-}OH} + H_2SO_4$$

(3) **Phosphoric Acid Process**. This involves direct hydration of ethylene. Phosphoric acid is used as a catalyst.

$$\underset{\text{Ethylene}}{CH_2{=}CH_2} + H_2O \xrightarrow[300°C,\ 60\ atm]{H_3PO_4} \underset{\text{Ethyl alcohol}}{CH_3{-}CH_2{-}OH}$$

Phosphoric acid process was first used commercially by Shell (USA) in 1947. Ethylene and water are preheated to 300°C and then passed under high pressure into a reactor. The reactor is a stainless steel vessel containing solid phosphoric acid catalyst. The reaction products are then cooled and passed into the scrubber where ethyl alcohol is dissolved in water. The aqueous solution of alcohol is sent to the distillation tower to get pure alcohol (Fig. 15.4). The yield of ethanol is 95 per cent.

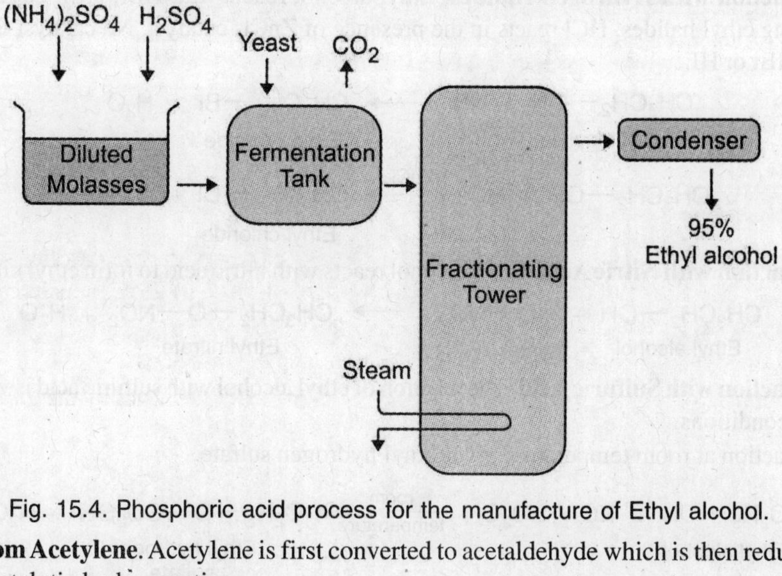

Fig. 15.4. Phosphoric acid process for the manufacture of Ethyl alcohol.

(4) **From Acetylene**. Acetylene is first converted to acetaldehyde which is then reduced to ethyl alcohol by catalytic hydrogenation.

$$\underset{\text{Acetylene}}{H{-}C{\equiv}C{-}H} + H_2O \xrightarrow[HgSO_4]{H_2SO_4} \underset{\text{Acetaldehyde}}{CH_3{-}\overset{O}{\overset{\|}{C}}{-}H}$$

$$\underset{\text{Acetaldehyde}}{CH_3{-}\overset{O}{\overset{\|}{C}}{-}H} + H_2 \xrightarrow[\Delta]{Ni} \underset{\text{Ethyl alcohol}}{CH_3{-}CH_2{-}OH}$$

Properties. (***Physical***). Ethyl alcohol is a colorless, volatile liquid with a characteristic pleasant odor, bp 78°C. It mixes with water in all proportions. Thus, water is effective in extinguishing the fires of ethyl alcohol. Water dilutes ethyl alcohol to the point that it can no longer produce an adequate volume of flammable vapor.

(***Chemical***). Ethyl alcohol gives all the general reactions of the class described before.

(1) **Reaction with Sodium or Potassium.** Ethyl alcohol reacts with sodium or potassium to form the corresponding ethoxides.

$$\underset{\text{Ethyl alcohol}}{2CH_3CH_2OH} + 2Na \longrightarrow \underset{\text{Sodium ethoxide}}{2CH_3CH_2\bar{O}\overset{+}{Na}} + H_2$$

(2) **Reaction with Phosphorus Halides.** Ethyl alcohol reacts with phosphorus halides to give ethyl halides.

$$\underset{\text{Ethyl alcohol}}{CH_3CH_2{-}OH} + PCl_5 \longrightarrow \underset{\text{Ethyl chloride}}{CH_3CH_2{-}Cl} + POCl_3 + HCl$$

$$3CH_3CH_2{-}OH + PBr_3 \longrightarrow \underset{\text{Ethyl bromide}}{3CH_3CH_2{-}Br} + H_3PO_3$$

(3) **Reaction with Thionyl Chloride.** Ethyl alcohol reacts with thionyl chloride ($SOCl_2$) to give ethyl chloride.

$$\underset{\text{Ethyl alcohol}}{CH_3CH_2{-}OH} + SOCl_2 \longrightarrow \underset{\text{Ethyl chloride}}{CH_3CH_2{-}Cl} + SO_2 + HCl$$

(4) **Reaction with Hydrogen Halides.** Ethyl alcohol reacts with HBr, HCl, or HI to give the corresponding ethyl halides. HCl reacts in the presence of $ZnCl_2$ catalyst. No catalyst is required in the case of HBr or HI.

$$\underset{\text{Ethyl alcohol}}{CH_3CH_2{-}OH} + HBr \longrightarrow \underset{\text{Ethyl bromide}}{CH_3CH_2{-}Br} + H_2O$$

$$CH_3CH_2{-}OH + HCl \xrightarrow{ZnCl_2} \underset{\text{Ethyl chloride}}{CH_3CH_2{-}Cl} + H_2O$$

(5) **Reaction with Nitric Acid.** Ethyl alcohol reacts with nitric acid to form ethyl nitrate.

$$\underset{\text{Ethyl alcohol}}{CH_3CH_2{-}OH} + HO{-}NO_2 \longrightarrow \underset{\text{Ethyl nitrate}}{CH_3CH_2{-}O{-}NO_2} + H_2O$$

(6) **Reaction with Sulfuric Acid.** The reaction of ethyl alcohol with sulfuric acid is very sensitive to reaction conditions.

(*a*) Reaction at room temperature gives ethyl hydrogen sulfate.

$$\underset{\text{Ethyl alcohol}}{C_2H_5{-}OH} + HO{-}SO_2OH \xrightarrow[\text{temperature}]{\text{room}} \underset{\text{Ethyl hydrogen sulfate}}{C_2H_5{-}O{-}SO_2OH} + H_2O$$

(*b*) Reaction at 170°C gives ethylene.

$$\underset{\text{Ethyl alcohol}}{CH_3CH_2OH} \xrightarrow[170°C]{H_2SO_4} \underset{\text{Ethylene}}{CH_2{=}CH_2} + H_2O$$

(*c*) Reaction at 140°C gives diethyl ether.

$$\underset{\text{Ethyl alcohol}}{CH_3CH_2{-}OH + HO{-}CH_2CH_3} \xrightarrow[140°C]{\text{Conc } H_2SO_4} \underset{\text{Diethyl ether}}{CH_3CH_2{-}O{-}CH_2CH_3} + H_2O$$

(7) **Reaction with Alumina**. Ethyl alcohol undergoes dehydration when its vapors are passed over heated alumina (Al_2O_3) catalyst.

$$\underset{\text{Ethyl alcohol}}{CH_3CH_2OH} \xrightarrow[350°C]{Al_2O_3} \underset{\text{Ethylene}}{CH_2{=}CH_2} + H_2O$$

$$2CH_3CH_2OH \xrightarrow[200°C]{Al_2O_3} \underset{\text{Diethyl ether}}{CH_3CH_2OCH_2CH_3} + H_2O$$

(8) **Reaction with Carboxylic Acids**. Ethyl alcohol reacts with carboxylic acids in the presence of sulfuric acid to form esters.

$$C_2H_5OH + \underset{\text{Acetic acid}}{HO{-}\overset{\overset{O}{\|}}{C}{-}CH_3} \xrightleftharpoons{H^+} \underset{\text{Ethyl acetate}}{C_2H_5O{-}\overset{\overset{O}{\|}}{C}{-}CH_3} + H_2O$$

(9) **Reaction with Acid Halides and Acid Anhydrides.** Ethyl alcohol reacts with acid halides and acid anhydrides to form esters.

$$\underset{\text{Acetyl chloride}}{CH_3{-}\overset{\overset{O}{\|}}{C}{-}Cl} + \underset{\text{Ethyl alcohol}}{H{-}OC_2H_5} \xrightarrow{\Delta} \underset{\text{Ethyl acetate}}{CH_3{-}\overset{\overset{O}{\|}}{C}{-}OC_2H_5} + HCl$$

$$\underset{\text{Acetic anhydride}}{CH_3{-}\overset{\overset{O}{\|}}{C}{-}O{-}\overset{\overset{O}{\|}}{C}{-}CH_3} + \underset{\text{Ethyl alcohol}}{H{-}OC_2H_5} \xrightarrow{\Delta} \underset{\text{Ethyl acetate}}{CH_3{-}\overset{\overset{O}{\|}}{C}{-}OC_2H_5} + \underset{\text{Acetic acid}}{CH_3{-}\overset{\overset{O}{\|}}{C}{-}OH}$$

(10) **Reaction with Grignard Reagents**. Ethyl alcohol reacts with Grignard reagents to form alkanes.

$$CH_3CH_2OH + \underset{\text{Methylmagnesium bromide}}{CH_3MgBr} \xrightarrow{\Delta} \underset{\text{Methane}}{CH_4} + CH_3CH_2OMgBr$$

(11) **Reduction**. Ethyl alcohol undergoes reduction with concentrated hydriodic acid and red phosphorus to produce ethane.

$$CH_3CH_2OH + 2HI \xrightarrow[\Delta]{P} \underset{\text{Ethane}}{CH_3CH_3} + I_2 + H_2O$$

(12) **Oxidation.** Ethyl alcohol undergoes oxidation with $Na_2Cr_2O_7 + H_2SO_4$ or $KMnO_4 + H_2SO_4$ to form acetaldehyde or acetic acid.

$$CH_3CH_2OH \xrightarrow[Na_2Cr_2O_7/H^+]{[O]} \underset{\text{Acetaldehyde}}{CH_3{-}\overset{\overset{O}{\|}}{C}{-}H} \xrightarrow[Na_2Cr_2O_7/H^+]{[O]} \underset{\text{Acetic acid}}{CH_3{-}\overset{\overset{O}{\|}}{C}{-}OH}$$

ROADSIDE BREATH TEST. A novel application of the oxidation of ethyl alcohol is found in the roadside breath test, a screening device intended to give a rough measure of the quantity of alcohol in the blood. It is based on the premise that the ethanol absorbed into the blood from the stomach and intestine is continuously transferred into the lungs and exhaled. A volume of exhaled breath is said to contain 1/2100 of the amount of ethanol in an identical volume of blood. It is an offense for a motor vehicle driver to have a blood alcohol level in excess of 150 mg/100 ml of blood (0.15%).

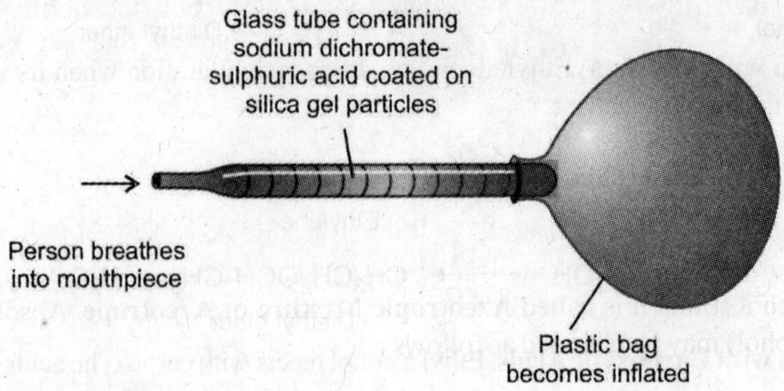

Fig. 15.5. Apparatus for Roadside Breath Test.

The operation of the **Breathalyzer** used in this test (Fig. 15.5) is based on the potassium dichromate-sulfuric acid oxidation of ethanol:

$$\underset{\text{Orange}}{3C_2H_5OH + 2K_2Cr_2O_7 + 8H_2SO_4} \longrightarrow 3CH_3COOH + \underset{\text{Green}}{2Cr_2(SO_4)_3} + 2K_2SO_4 + 11H_2O$$

The particles of silica gel in a sealed gas ampule are impregnated with the reagents. Before the ampule is used, the ends are broken off; then one end is fitted with a plastic mouthpiece, and the other is attached to the neck of a flattened bag. When air containing ethanol is blown through the tube, a chemical reaction takes place (the temperature of the tube increases), and the chromium sulfate produced is shown by the formation of a green color in place of the original orange reagent. If the green color extends beyond the yellow centre mark of the tube, the motorist is considered drunk and taken to the police station for further tests.

(13) **Reaction with Hot Copper**. When ethyl alcohol vapors are passed over copper gauze at 300°C, acetaldehyde is formed.

$$CH_3CH_2OH \xrightarrow[300^\circ C]{Cu} \underset{\text{Acetaldehyde}}{CH_3-\overset{\overset{\displaystyle O}{\|}}{C}-H} + H_2$$

(14) **Formation of Iodoform.** When ethyl alcohol is warmed with aqueous KOH and iodine, iodoform is formed.

$$CH_3CH_2OH + 4I_2 + 6KOH \xrightarrow{\text{warm}} \underset{\text{Iodoform}}{CHI_3} + HCO\overset{-+}{OK} + 5KI$$

Uses. Ethyl alcohol is used : (*1*) in the manufacture of alcoholic beverages ; (2) as an industrial solvent ; (*3*) in the manufacture of drugs, flavoring extracts, and perfumes ; (*4*) in hospitals as an antiseptic ; (*5*) as a low freezing (fp –114.1°C) and mobile liquid in scientific apparatus such as thermometers and spirit levels ; (*6*) as an antifreeze in automobile radiators ; (*7*) in lacquers and varnishes ; (*8*) as a preservative for biological specimens ; (*9*) in the manufacture of acetaldehyde, acetic acid, ethyl chloride, and ethyl acetate ; and (*10*) in the manufacture of synthetic rubber.

Ethyl alcohol is the *alcohol* of wine, beer and whiskey. Ethyl alcohol may be the first organic chemical routinely manufactured by humans.

Points to Remember

Absolute Alcohol. Absolute ethyl alcohol is 100% pure ethyl alcohol. Commercial alcohol (Rectified Spirit) is a mixture of 95% ethyl alcohol and 5% water. This mixture cannot be separated by further distillation. Such a solution is called **Azeotropic Mixture** or **Azeotrope**. Absolute ethyl alcohol (Water-free alcohol) may be obtained as follows :

(*a*) *Laboratory Method*. Quicklime is added to the commercial alcohol. The mixture is refluxed for 8 hours. It is then distilled to give absolute ethyl alcohol.

$$\underset{95\%}{CH_3CH_2OH} + \underset{5\%}{H_2O} + CaO \xrightarrow{\Delta} \underset{100\%}{CH_3CH_2OH} + Ca(OH)_2\downarrow$$

(*b*) *Industrial Method*. Benzene is added to the commercial alcohol. Distillation of the mixture yields three fractions :

At 65°C, a constant boiling mixture of ethyl alcohol, benzene, and water (a ternary azeotrope).
At 68°C, a constant boiling mixture of ethyl alcohol and benzene (a binary azeotrope).
At 78°C, absolute alcohol is obtained.

Denatured Alcohol. Denatured alcohol is commercial ethyl alcohol to which small amounts of very poisonous substances have been added. The manufacture and sale of ethyl alcohol is under strict government control. Heavy excise duty is levied on sale of alcoholic beverages. For industrial purposes, ethyl alcohol is duty-free. Therefore, the commercial alcohol is denatured by addition of poisonous substances like methyl alcohol or pyridine.

Alcoholic Beverages. The largest use of ethyl alcohol is as a beverage. Wines contain about 12% ethyl alcohol. Beers contain about 4%. Whiskey and brandy contain 40-50% ethyl alcohol. The alcoholic content of a beverage is indicated by a measure known as **Proof Spirit**. For some obscure reason, the proof value is twice the alcoholic content by volume. Thus whiskey, which is 50% alcohol, is said to be 100 PROOF.

Effect of concentration of ethyl alcohol in blood on an average adult.

When you drink alcoholic beverages, the ethyl alcohol finds its way in your blood. Some of the effects of various concentrations of ethyl alcohol in blood. Prolonged, excessive intake of ethyl alcohol causes permanent liver damage. When ethyl alcohol is taken along with barbiturates or tranquilizers, it produces coma and may cause death.

PROPYL ALCOHOLS, *Propanols*, C_3H_7OH

Two isomeric propyl alcohols are known :

$$\underset{\substack{\text{1-Propanol} \\ (n\text{-Propyl alcohol})}}{CH_3CH_2CH_2OH} \qquad \underset{\substack{\text{2-Propanol} \\ \text{(Isopropyl alcohol)}}}{CH_3-\overset{\overset{\large OH}{|}}{C}H-CH_3}$$

***n*-Propyl alcohol, 1-Propanol, $CH_3CH_2CH_2OH$.** It was originally obtained from fusel oil, but is now obtained by hydrogenation of carbon monoxide. A more recent method is by the catalytic reduction of propargyl alcohol.

$$\underset{\text{Propargyl alcohol}}{HC{\equiv}CCH_2OH} + 2H_2 \xrightarrow{\text{Ni or Pt}} \underset{\text{1-Propanol}}{CH_3CH_2CH_2OH}$$

1-Propanol, bp 97.4°C, is miscible with water, ethanol and ether. It is used in the preparation of propanoic acid, toilet preparations such as lotions, etc.

Isopropyl alcohol, 2-Propanol, $(CH_3)_2CHOH$. It is prepared industrially by passing propene from cracked petroleum into concentrated sulfuric acid, then diluting with water, and distilling off the isopropanol. Isopropyl ether is obtained as a by-product (*cf.* ethanol) :

$$CH_3CH{=}CH_2 + (HO)_2SO_2 \longrightarrow (CH_3)_2CHOSO_2OH$$

$$(CH_3)_2CHOSO_2OH + CH_3CH{=}CH_2 \longrightarrow [(CH_3)_2CHO]_2SO_2$$

$$(CH_3)_2CHOSO_2OH + H_2O \longrightarrow (CH_3)_2CHOH + H_2SO_4$$

$$[(CH_3)_2CHO]SO_2 + 2H_2O \longrightarrow 2(CH_3)_2CHOH + H_2SO_4$$

$$[(CH_3)_2CHO]SO_2 + (CH_3)_2CHOH \longrightarrow [(CH_3)_2CH]_2O + (CH_3)_2CHOSO_2OH$$

Isopropanol is also prepared by *direct* hydration by passing a mixture of propene and steam at 220-250°C, and under pressure (220 atm), over a catalyst of tungsten oxide plus zinc oxide, on a silica carrier.

Isopropanol, bp 82.4°C, is soluble in water, ethanol and ether. It is used for preparing esters, acetone, keten, as a solvent, and for high-octane fuel.

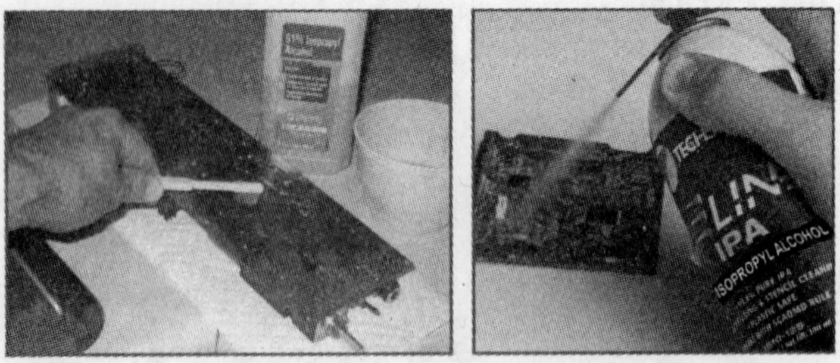

Isopropyl alcohol dissolves a wide range of nonpolar compounds. It is also relatively non-toxic and evaporates quickly. It is used widely as a solvent and as a cleaning fluid. Examples of this use include cleaning electronic devices such as the lenses of lasers in optical disc drives (e.g. CD, DVD). It is also used to clean LCD and glass computer monitor screens. Typically, 70% solution of isopropanol in water is used for this purpose (universal cleaning solution).

BUTYL ALCOHOLS, *Butanols*, C_4H_9OH

There are 4 isomeric butyl alcohols :

$CH_3CH_2CH_2OH$	$CH_3CH_2-\underset{}{\overset{OH}{\overset{\vert}{C}H}}-CH_3$	$CH_3-\overset{CH_3}{\overset{\vert}{C}H}-CH_2OH$	$CH_3-\underset{CH_3}{\underset{\vert}{\overset{CH_3}{\overset{\vert}{C}}}}-OH$
n-Butyl alcohol (1-Butanol) bp 117°C	*sec*-Butyl alcohol (2-Butanol) bp 99°C	Isobutyl alcohol (2-Methylpropanol) bp 107°C	*tert*-Butyl alcohol (2-Methyl-2-propanol) bp 83°C

n-Butyl alcohol, 1-butanol, bp 117.4°C, is prepared industrially from propene by the Oxo process. It is widely used as a solvent.

Isobutyl alcohol, 2-methyl-1-propanol , bp 108°C, is also obtained by the Oxo process. It behaves as a primary alcohol, but it readily rearranges due to the presence of a branched chain near the COH group, e.g., when it is treated with hydrochloric acid, isobutyl chloride and *tert*-butyl chloride are obtained.

$$(CH_3)_2CHCH_2OH \xrightarrow{HCl} (CH_3)_2CHCH_2Cl + (CH_3)_3CCl$$

sec-Butyl alcohol, 2-butanol, bp 100°C, is prepared industrially by the hydration of 1- or 2-butene (from cracked petroleum) by means of concentrated sulfuric acid (*cf.* ethanol and isopropanol). It is used for the preparation of butanone, esters, and as a lacquer solvent.

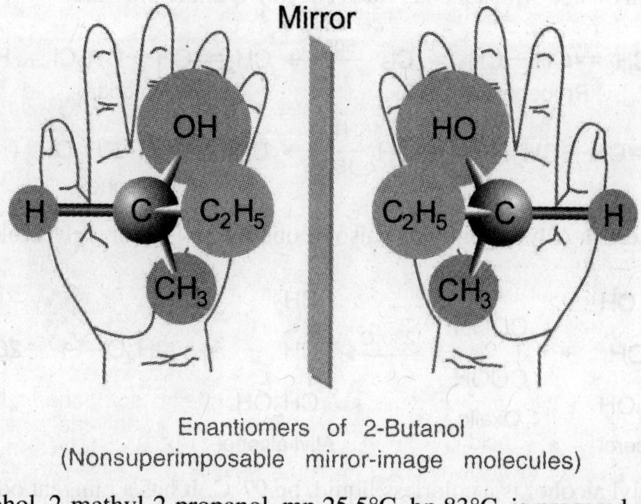

Enantiomers of 2-Butanol
(Nonsuperimposable mirror-image molecules)

tert-Butyl alcohol, 2-methyl-2-propanol, mp 25.5°C, bp 83°C, is prepared synthetically by the hydration of isobutene (from cracked petroleum). It is mainly used as an alkylating agent.

UNSATURATED ALCOHOLS

Unsaturated monohydric alcohols are of two types according as the OH is attached to a carbon which is singly or doubly-linked to another carbon. The simplest representatives of these alcohols are :

$$\underset{\text{Vinyl alcohol}}{CH_2{=}CH{-}OH} \qquad \underset{\text{Allyl alcohol}}{CH_2{=}CH{-}CH_2{-}OH}$$

In general, the first type of compounds resemble aromatic phenols while the second type resemble aliphatic alcohols.

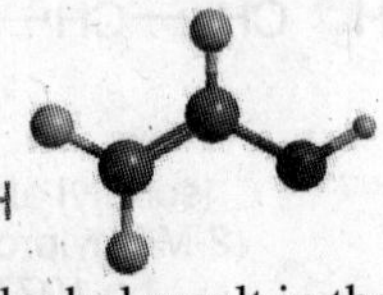

VINYL ALCOHOL, $CH_2{=}CH{-}OH$

All attempts to prepare vinyl alcohol result in the formation of acetaldehyde. This is due to the fact that this compound shows keto-enol tautomerism. The vinyl alcohol as soon as it is produced, ketonizes to form acetaldehyde.

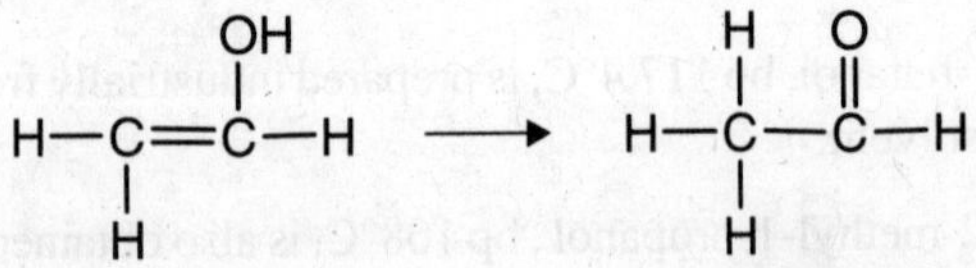

Vinyl alcohol — Acetaldehyde

However, derivatives of vinyl alcohol such as vinyl bromide and vinyl acetate are quite stable.

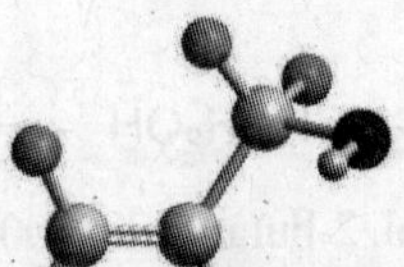

ALLYL ALCOHOL, *2-Propen-1-ol*, $CH_2{=}CH{-}CH_2OH$

Allyl alcohol is an important unsaturated alcohol.

Preparation. Allyl alcohol is prepared :

(1) **From Propene.** Propene is first converted into allyl chloride by high temperature chlorination. Allyl chloride is then treated with aqueous NaOH at 150°C under pressure.

$$\underset{\text{Propene}}{CH_2{=}CH{-}CH_3} + Cl_2 \xrightarrow{500°C} \underset{\text{Allyl chloride}}{CH_2{=}CH{-}CH_2Cl} + HCl$$

$$CH_2{=}CH{-}CH_2Cl + NaOH \xrightarrow[150°C]{H_2O} \underset{\text{Allyl alcohol}}{CH_2{=}CH{-}CH_2OH} + NaCl$$

(2) **From Glycerol.** Allyl alcohol may also be obtained by heating glycerol with oxalic acid at 230°C.

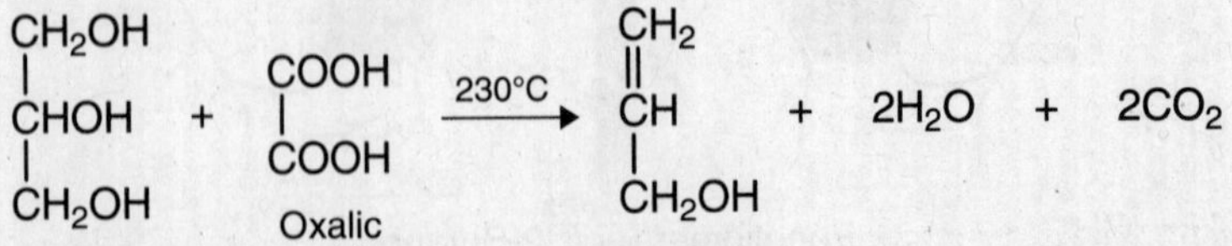

Glycerol — Oxalic acid — Allyl alcohol

Properties. Allyl alcohol is a colorless liquid, bp 97°C. It has a pungent odor. Allyl alcohol is completely miscible with water.

Allyl alcohol contains a carbon-carbon double bond and a primary –OH group. It shows reactions characteristic of these groups (Fig. 15.6).

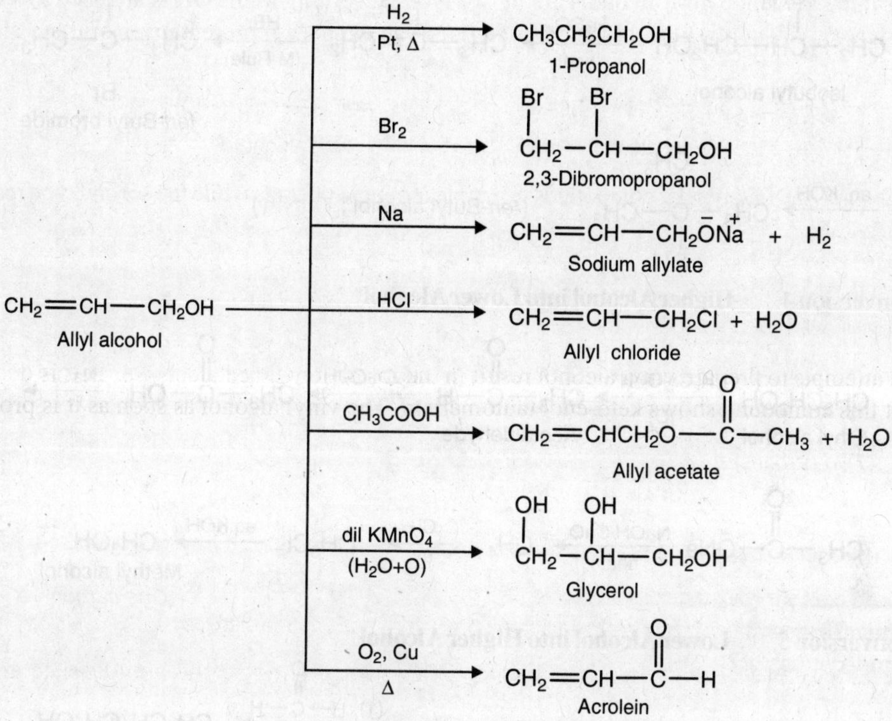

Fig. 15.6. Reactions of Allyl alcohol.

Uses. Allyl alcohol is used : (*1*) in the preparation of glycerol ; and (*2*) in the manufacture of allyl esters which are used in perfumes, pharmaceuticals, and food flavors.

FIVE IMPORTANT INTERCONVERSIONS

Conversion 1. Primary Alcohol into Secondary Alcohol

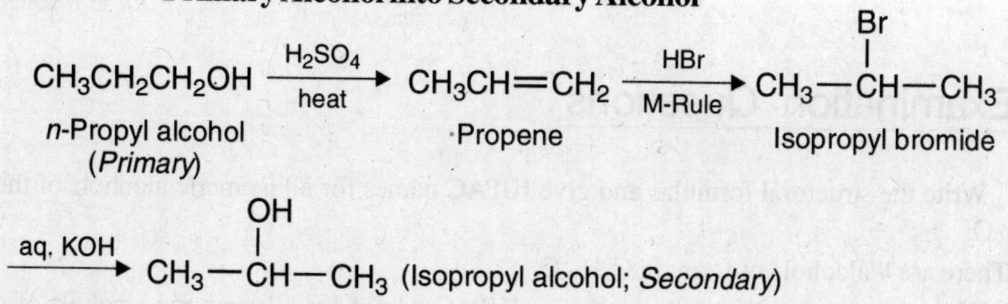

Conversion 2. Secondary Alcohol into Tertiary Alcohol

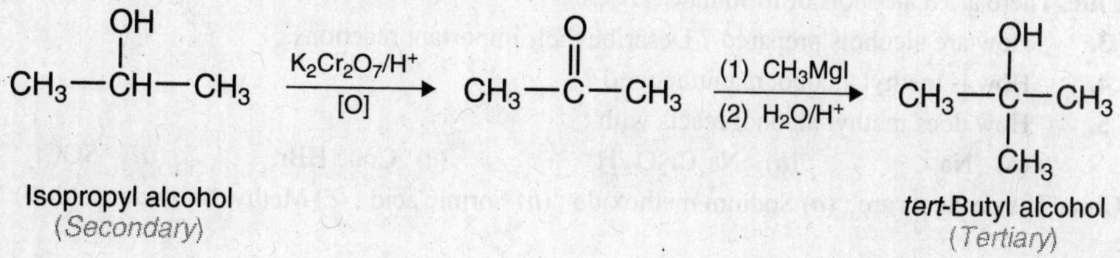

Conversion 3. Primary Alcohol into Tertiary Alcohol

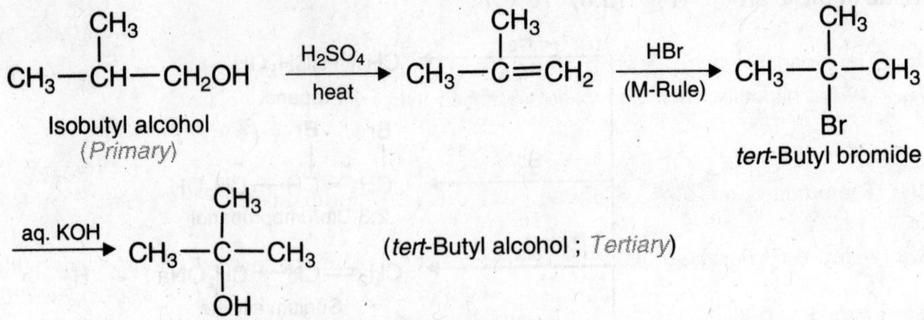

Conversion 4. Higher Alcohol into Lower Alcohol

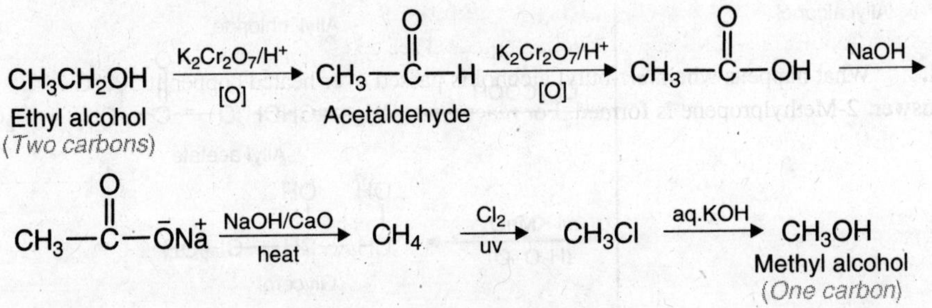

Conversion 5. Lower Alcohol into Higher Alcohol

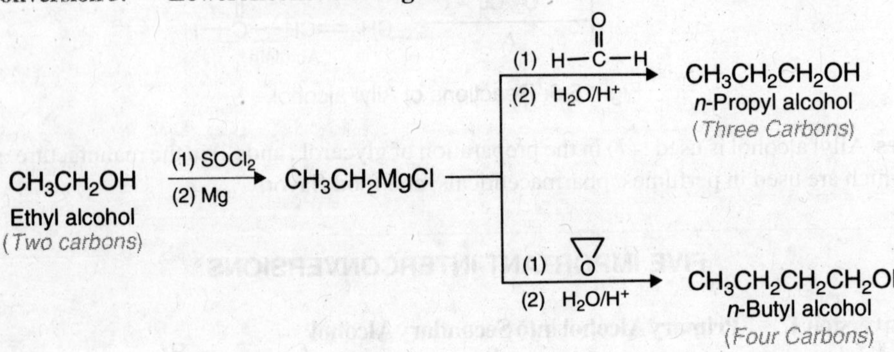

Examination Questions

1. Write the structural formulas and give IUPAC names for all isomeric alcohols of the molecular formula $C_4H_{10}O$.

Clue. There are *4* alcohols of formula $C_4H_{10}O$.

2. Write the structural formulas and give IUPAC names for all isomeric alcohols of the formula $C_5H_{12}O$.

Clue. There are *8* alcohols of formula $C_5H_{12}O$.

3. How are alcohols prepared ? Describe their important reactions.

4. How is methyl alcohol manufactured ?

5. How does methyl alcohol reacts with :

(*a*) Na (*b*) $Na_2Cr_2O_7/H^+$ (*c*) Conc HBr (*d*) $SOCl_2$

Clue. The products are : (*a*) Sodium methoxide ; (*b*) Formic acid ; (*c*) Methyl bromide ; and (*d*) Methyl chloride.

6. How does ethyl alcohol reacts with :

(*a*) Na (*b*) Conc H_2SO_4/170°C (*c*) PCl_5 (*d*) $K_2Cr_2O_7/H^+$

(*Kashmir BSc, 2014*)

Clue. The products are : (*a*) Sodium ethoxide ; (*b*) Ethylene ; (*c*) Ethyl chloride ; and (*d*) Acetic acid.

7. What happens when 1-propanol is treated with :

(*a*) PCl_5 (*b*) $SOCl_2$ (*c*) Conc H_2SO_4/170°C (*d*) Na

(*Bilaspur BSc, 2015*)

Clue. The products are : (*a*) 1-Chloropropane ; (*b*) 1-Chloropropane ; (*c*) Propene ; and (*d*) Sodium propoxide.

8. How do primary, secondary, and tertiary alcohols differ in their behavior towards oxidation ?

(*Rohilkhand BSc, 2014*)

9. What happens when ethyl alcohol is treated with concentrated sulfuric acid under different conditions ?

(*Goa BSc, 2014*)

10. What happens when ethyl alcohol is treated with alkaline solution of iodine ?

Answer. Iodoform is formed. For reaction equation see text.

11. What happens when *tert*-butyl alcohol is passed over heated copper at 300°C ?

Answer. 2-Methylpropene is formed. For reaction equation see text.

12. How will you distinguish among primary, secondary, and tertiary alcohols ?

(*Madras BSc, 2014 ; Gulbarga BSc, 2014 ; Periyar BSc, 2015*)

13. How will you distinguish between 1-propanol and 2-propanol ? (*Delhi BSc, 2015*)

Answer. Lucas test is used. 2-Propanol (2° alcohol) produces cloudiness within 5 minutes with Lucas reagent ($HCl/ZnCl_2$). Cloudiness is due to the formation of immiscible alkyl halide. 1-Propanol (1° alcohol) does not react with Lucas reagent at room temperature.

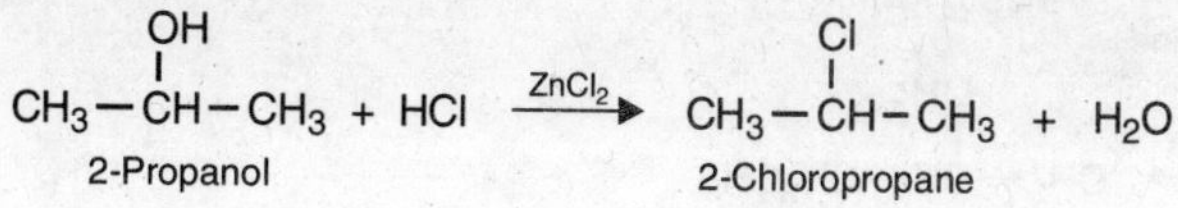

14. How will you distinguish between methyl alcohol and ethyl alcohol ?

(*Bihar BSc, 2014 ; Madurai BSc, 2015*)

Answer. Iodoform test is used. Ethyl alcohol reacts with alkaline solution of iodine (I_2/NaOH) to form yellow precipitate of iodoform. The alkaline solution of iodine first oxidizes ethyl alcohol to acetaldehyde which further reacts with this solution to form iodoform. Methyl alcohol does not give this test.

$$\underset{\text{Ethyl alcohol}}{CH_3CH_2OH} + 4I_2 + 6NaOH \longrightarrow \underset{\text{Iodoform}}{CHI_3\downarrow} + HCOONa + 5NaI + 5H_2O$$

15. How will you distinguish between *n*-butyl alcohol and diethyl ether ?

(*Tezpur BSc, 2014 ; Shivaji BSc, 2015*)

Answer. *n*-Butyl alcohol reacts with sodium to give hydrogen (seen as bubbles). Diethyl ether does not give this test.

$$\underset{n\text{-Butyl alcohol}}{CH_3CH_2CH_2CH_2OH} + Na \longrightarrow \underset{\text{Sod. butoxide}}{CH_3CH_2CH_2CH_2\bar{O}\overset{+}{Na}} + H_2\uparrow$$

16. Explain why alcohols have higher boiling points than alkanes of comparable molecular weights.

(*Delhi BSc, 2014 ; Baroda BSc, 2014 ; Gorkhpur BSc, 2015 ; Nagpur BSc, 2015*)

17. Explain why ethyl alcohol has higher boiling point than diethyl ether. (*Andhra BSc, 2014*)

18. Rank the following compounds in increasing order of acidity and explain your answer.

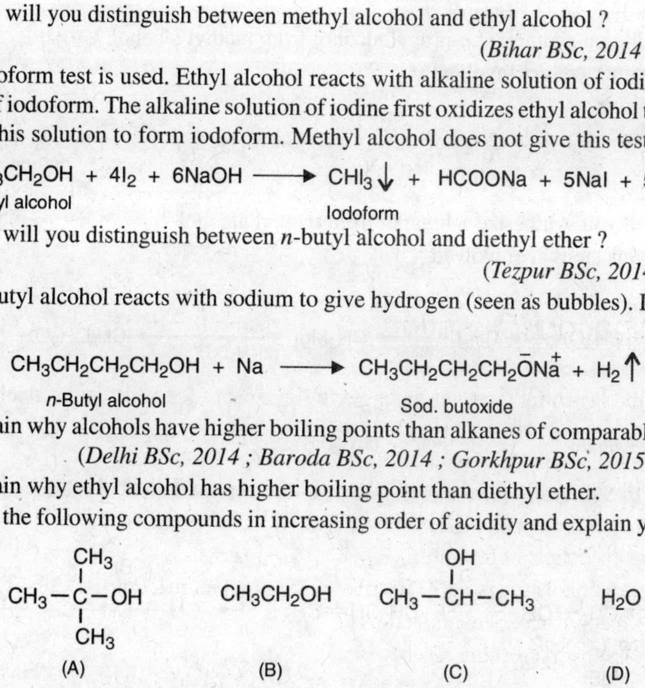

Answer. The increasing order of acidity is (lowest ⟶ highest) :

(A) < (C) < (B) < (D)

As the number of alkyl groups attached to the carbon bearing the hydroxyl group increases, the acidity decreases. Alkyl groups are electron-releasing and they tend to stabilize the alcohol, making it more difficult to lose a proton. In addition, the presence of more alkyl groups destabilizes the negatively charged conjugate base.

19. Explain the meaning of the following terms : (*a*) Absolute alcohol ; (*b*) Denatured alcohol ; and (*c*) Proof spirit.

20. Discuss the mechanism of acid–catalyzed dehydration of ethyl alcohol. *(Bangalore BSc, 2014 ; Panjab BSc, 2015)*

21. How will you synthesize the following compounds from the indicated starting materials ?

(*a*) Isopropyl alcohol *from* *n*-propyl alcohol
(*b*) *tert*-Butyl alcohol *from* isopropyl alcohol *(Agra BSc, 2014)*
(*c*) Methyl alcohol *from* ethyl alcohol *(Osmania BSc, 2015)*
(*d*) *n*-Propyl alcohol *from* ethyl alcohol *(Ranchi BSc, 2014)*
(*e*) Allyl alcohol *from* propene

22. How will you synthesize ethyl alcohol from acetylene ? *(Punjabi BSc, 2015)*

Answer. Following steps are involved :

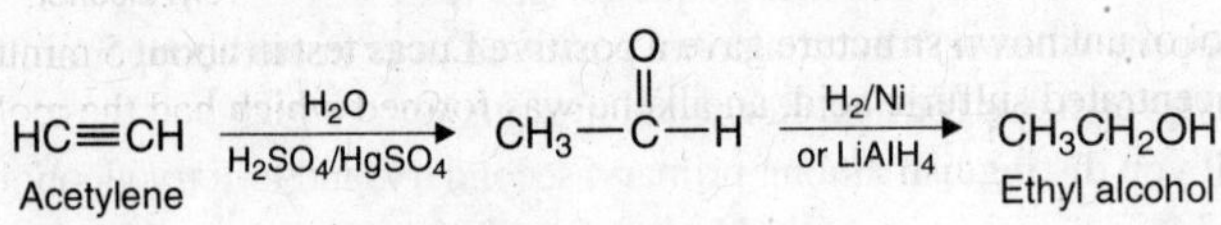

23. How will you synthesize 2-butanol from acetylene ? *(Kalyani BSc, 2015)*

Answer. Following steps are involved :

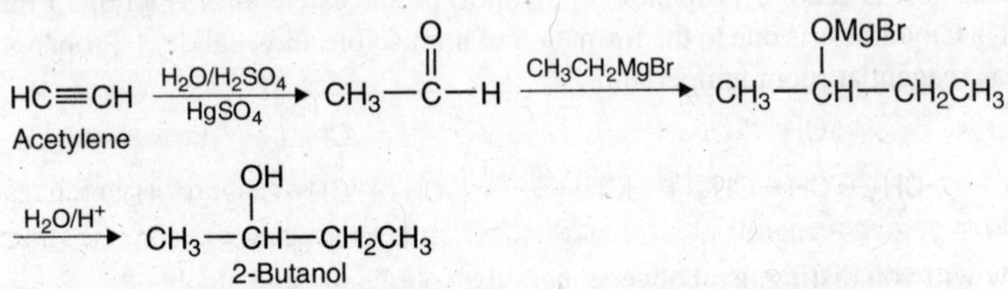

24. How will you synthesize *n*-propyl alcohol from methyl alcohol ?

Answer. Following steps are involved :

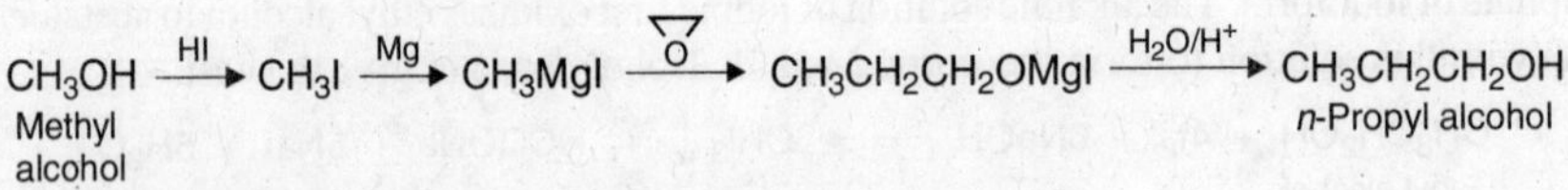

25. How will you synthesize 1-butanol from methyl alcohol ?

Answer. Following steps are involved :

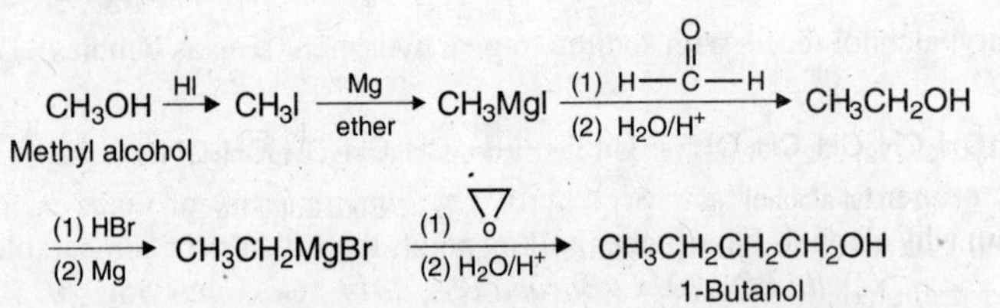

26. How will you synthesize isobutyl alcohol from *n*-propyl alcohol ?

Answer. Following steps are involved :

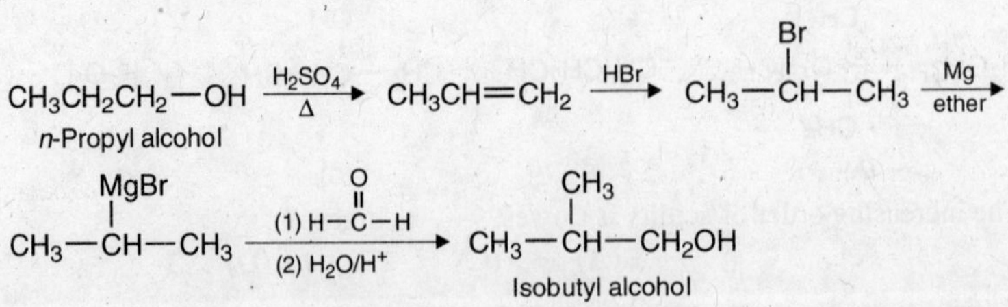

27. How will you synthesize 1-propanol from 2-propanol ? (*Manipur BSc, 2015*)

Answer. Following steps are involved :

$$\underset{\text{2-Propanol}}{CH_3-\overset{\overset{\large OH}{|}}{C}H-CH_3} \xrightarrow[\Delta]{H_2SO_4} CH_3CH=CH_2 \xrightarrow[\text{peroxide}]{HBr} CH_3CH_2CH_2Br$$

$$\xrightarrow{\text{aq KOH}} \underset{\text{1-Propanol}}{CH_3CH_2CH_2OH}$$

28. How will you synthesize allyl alcohol from propene ?

Answer. Following steps are involved :

$$\underset{\text{Propene}}{CH_3CH=CH_2} \xrightarrow[500°C]{Cl_2} ClCH_2CH=CH_2 \xrightarrow{\text{aq. KOH}} \underset{\text{Allyl alcohol}}{HOCH_2CH=CH_2}$$

29. An alcohol of unknown structure gave a positive Lucas test in about 5 minutes. When the alcohol was treated with hot concentrated sulfuric acid, an alkene was formed which had the molecular formula C_4H_8. Ozonolysis of this compound gave a single product C_2H_4O. What was the structure of the alcohol ?

Answer. The unknown alcohol was *2-Butanol*.

Reactions Involved Are :

$$\underset{\text{2-Butanol}}{CH_3\overset{\overset{\large OH}{|}}{C}HCH_2CH_3} \xrightarrow[170°C]{H_2SO_4} CH_3CH=CHCH_3 \xrightarrow[(2)\ Zn/H_2O]{(1)\ O_3} \underset{\text{Acetaldehyde}}{2CH_3-\overset{\overset{\large O}{||}}{C}-H}$$

30. A hydrocarbon (A) reacts with hydrogen iodide to form compound (B) which reacts with aqueous potassium hydroxide to give compound (C). Oxidation of (C) gives Acetone. What are the structural formulas of (A), (B), and (C) ? Give equations for all reactions.

Answer. (*A*) is Propylene ; (*B*) is Isopropyl iodide ; (*C*) is Isopropyl alcohol.

Reactions Involved Are :

$$\underset{(A)}{CH_3-CH=CH_2} \xrightarrow[\text{(M.Rule)}]{HI} \underset{(B)}{CH_3-\overset{\overset{\large I}{|}}{C}H-CH_3} \xrightarrow{\text{aq. NaOH}}$$

$$\underset{(C)}{CH_3-\overset{\overset{\large OH}{|}}{C}H-CH_3} \xrightarrow[[O]]{K_2Cr_2O_7/H^+} \underset{\text{Acetone}}{CH_3-\overset{\overset{\large O}{||}}{C}-CH_3}$$

31. A Compound (A) reacts with thionyl chloride to give compound (B). (B) reacts with magnesium to form a Grignard reagent which is treated with acetone, and the product is hydrolysed to give 2-methyl-2-butanol. What are the structural formulas of (A) and (B) ?

Answer. (*A*) is Ethyl alcohol ; (*B*) is Ethyl chloride.

Reactions Involved Are :

$$\underset{(A)}{CH_3CH_2OH} \xrightarrow{SOCl_2} \underset{(B)}{CH_3CH_2Cl} \xrightarrow{Mg} CH_3CH_2MgCl \xrightarrow[(2)\ H_2O/H^+]{(1)\ (CH_3)_2C=O} \underset{\text{2-Methyl-2-butanol}}{CH_3CH_2-\overset{\overset{\large OH}{|}}{\underset{\underset{\large CH_3}{|}}{C}}-CH_3}$$

Multiple Choice Questions

1. The number of structural isomers of alcohols with molecular formula C_3H_7OH is

(*a*) 5 (*b*) 4 (*c*) 3 (*d*) 2

Answer. (*d*)

2. The number of structural isomers of alcohols with molecular formula C_4H_9OH is

(*a*) 5 (*b*) 4 (*c*) 3 (*d*) 6

Answer. (*b*)

3. Why do alcohols have boiling points much higher than hydrocarbons of similar molecular weight ?

(*a*) Alcohols have greater van der Waals attraction forces.

(*b*) Alcohol molecules have greater molecular symmetry.

(*c*) Hydrogen bonds must be broken in the process of volatilization.

(*d*) Alcohols must overcome greater ionic forces in the process of volatilization.

Answer. (*c*)

4. The high boiling points of alcohols, as compared to the corresponding alkanes, are due to

(*a*) Hydrogen bonding (*b*) Heavy oxygen atom

(*c*) Water solubility (*d*) None of these

Answer. (*a*)

5. Which of the following has the highest boiling point ?

(*a*) diethyl ether (*b*) *n*-Butyraldehyde

(*c*) *n*-propyl chloride (*d*) *n*-Butyl alcohol

Answer. (*d*)

6. Which of the following has the highest boiling point ?

(*a*) $CH_3CH_2CH_2CH_2Cl$ (*b*) $HOCH_2CH_2CH_2CH_2OH$

(*c*) $CH_3CH_2CH_2CH_2OH$ (*d*) $CH_3CH_2OCH_2CH_3$

Answer. (*b*)

7. Which of the following compounds has the highest boiling point ?

(*a*) $CH_3CH_2-\overset{\overset{\displaystyle O}{\|}}{C}-H$ (*b*) $CH_3CH_2CH_2OH$

(*c*) $CH_3-\overset{\overset{\displaystyle O}{\|}}{C}-CH_3$ (*d*) $CH_3CH_2OCH_3$

Answer. (*b*)

8. Which of the following has the highest boiling point ?

(*a*) CH_3CH_2OH (*b*) $CH_3CH_2CH_2CH_2OH$

(*c*) CH_3OH (*d*) $CH_3CH_2CH_2OH$

Answer. (*b*)

9. Rank the following substances in order of increasing boiling point (lowest $\rightarrow$ highest):

$CH_3CH_2CH_2CH_2OH$	$(CH_3)_2CHOCH_3$	$(CH_3)_3COH$
(1)	(2)	(3)

(*a*) (1) < (2) < (3)

(*b*) (2) < (1) < (3)

(*c*) (2) < (3) < (1)

(*d*) (3) < (2) < (1)

Answer. (*c*)

10. Which of the following compounds is the least soluble in water ?

(*a*) $HOCH_2CH_2OH$ (*b*) $CH_3CH_2CH_2OH$

(*c*) CH_3CH_2OH (*d*) $CH_3CH_2CH_2CH_2OH$

Answer. (*d*)

11. Methanol is known as :

(*a*) Rubbing alcohol (*b*) Grain alcohol

(*c*) Wood alcohol (*d*) Denatured alcohol

Answer. (*c*)

12. Rectified spirit is

(*a*) 100% Ethanol (*b*) 90% Ethanol (*c*) 100% Methanol (*d*) 95% Ethanol

Answer. (*d*)

13. Grain alcohol is another name for

(*a*) Methyl alcohol (*b*) Isopropyl alcohol

(*c*) Ethyl alcohol (*d*) *n*-Propyl alcohol

Answer. (*c*)

14. Lucas reagents is

(*a*) $HCl/NaNO_2$ (*b*) H_2/Pd

(*c*) $HCl/ZnCl_2$ (*d*) $H_2/Pd/BaSO_4$

Answer. (*c*)

15. Lucas test is used to determine the type of

(*a*) alcohols (*b*) acids

(*c*) amines (*d*) carbohydrates

Answer. (*a*)

16. Which of the following gives positive Iodoform test :

(*a*) 1-Propanol (*b*) 2-Propanol

(*c*) 3-Propanol (*d*) None of these

Answer. (*b*)

17. Which of the following reagents will replace –OH group by a halogen atom ?

(*a*) HOCl (*b*) Br_2 (*c*) $SOCl_2$ (*d*) I_2

Answer. (*c*)

18. The compound which reacts most readily with Lucas reagent is

(*a*) CH_3CH_2Cl (*b*) $(CH_3)_2CHOH$ (*c*) CH_3CH_2OH (*d*) $(CH_3)_3COH$

Answer. (*d*)

19. Which of the following compounds reacts slowest with Lucas reagent at room temperature ?

(*a*) 1-butanol (*b*) 2-propanol

(*c*) 2-butanol (*d*) 2-methyl-2-propanol

Answer. (*a*)

20. Which of the following compounds will react fastest with Lucas reagent ?

(*a*) 1-propanol (*b*) 2-methyl-1-propanol

(*c*) 2-propanol (*d*) 2-methyl-2-propanol

Answer. (*d*)

21. Which of the following compounds reacts fastest with HBr ?

(*a*) 1-propanol (*b*) 2-methyl-1-propanol

(*c*) 2-propanol (*d*) 2-methyl-2-propanol

Answer. (*d*)

22. When ethanol is treated with sodium metal :

(*a*) Sodium ethoxide is formed (*b*) The sodium is oxidized
(*c*) The acidic hydrogen in reduced (*d*) All these occur

Answer. (*d*)

23. Which of the following alcohols will give a yellow precipitate of iodoform with iodine and dilute NaOH solution ?

(*a*) 1-Propanol (*b*) 2-Propanol
(*c*) 1-Butanol (*d*) 2-Methyl-2-propanol

Answer. (*b*)

24. Ethyl alcohol can react with concentrated H_2SO_4 to give

(*a*) Ethylene (*b*) Diethyl ether
(*c*) Ethyl hydrogen sulfate (*d*) All of these

Answer. (*d*)

25. Ethanol on heating with concentrated H_2SO_4 at 170°C gives

(*a*) ethylene (*b*) ethyl hydrogen sulfate
(*c*) diethyl ether (*d*) diethyl sulfate

Answer. (*a*)

26. The major product of acid-catalyzed dehydration of 2-butanol is :

(*a*) 2-Butene (*b*) 2-Butyne
(*c*) 1-Butene (*d*) 1-Butyne

Answer. (*a*)

27. Which is the best reagent for carrying out the following conversion ?

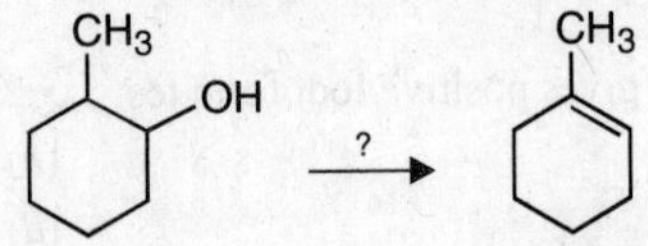

(*a*) $LiAlH_4$ (*b*) Conc H_2SO_4
(*c*) H_2/Ni (*d*) NaOH

Answer. (*b*)

28. Which of the following reaction mixtures will give an organic product which is ionic ?

(*a*) $CH_3CH_2OH + PCl_5$ (*b*) $CH_3CH_2OH + KBr + H_2SO_4$
(*c*) $CH_3CH_2OH + Na$ (*d*) $CH_3CH_2OH + SOCl_2$

Answer. (*c*)

29. Hydroboration-oxidation of propene gives :

(*a*) Isopropyl alcohol (*b*) *n*-Propyl alcohol
(*c*) Isobutyl alcohol (*d*) *tert*-Butyl alcohol

Answer. (*b*)

30. Hydroboration-oxidation of 2-Methylpropene gives

(*a*) 2-Methyl-2-propanol (*b*) 1,2,3-Propanetriol
(*c*) 2-Methyl-1-propanol (*d*) 1,2-Propanediol

Answer. (*c*)

31. Which of the following compounds will not be *easily* oxidized ?

(*a*) Primary alcohol (*b*) Secondary alcohol
(*c*) Tertiary alcohol (*d*) Aldehyde

Answer. (*c*)

32. Which of the following is most resistant to oxidation ?

(*a*) CH_3CH_2OH (*b*) $(CH_3)_2CHOH$

(*c*) $HOCH_2CH_2OH$ (*d*) $(CH_3)_3COH$

Answer. (*d*)

33. Oxidation of a primary alcohol with $KMnO_4/\bar{O}H$ produces

(*a*) a carboxylic acid (*b*) an ether

(*c*) a ketone (*d*) an ester

Answer. (*a*)

34. Oxidation of a secondary alcohol with $K_2Cr_2O_7/H^+$ produces

(*a*) a carboxylic acid (*b*) a ketone

(*c*) an aldehyde (*d*) an ester

Answer. (*b*)

35. Isopropyl alcohol reacts with acidic sodium dichromate to give

(*a*) Acetaldehyde (*b*) Acetic acid

(*c*) Propionic acid (*d*) Acetone

Answer. (*d*)

36. Isopropyl alcohol can be converted to acetone by treatment with

(*a*) $HCl/ZnCl_2$ (*b*) $Na_2Cr_2O_7/H_2SO_4$

(*c*) NaOH (*d*) $LiAlH_4$

Answer. (*b*)

37. Which of the bonds in Ethyl alcohol (CH_3CH_2OH) will undergo heterolytic bond fission most readily ?

(*a*) O–H (*b*) C–H

(*c*) C–O (*d*) C–C

Answer. (*a*)

38. The mechanism of dehydration of an alcohol to give an alkene involves formation of :

(*a*) Carbonium ions (*b*) Carbanions

(*c*) Free radicals (*d*) Carbenes

Answer. (*a*)

39. The mechanism of dehydration of an alcohol to give an ether involves formation of :

(*a*) Carbonium ions (*b*) Carbanions

(*c*) Free radicals (*d*) Carbenes

Answer. (*a*)

40. The acid-catalyzed dehydration mechanism for alcohols is best described as a(n) :

(*a*) E1 (*b*) E2

(*c*) S_N1 (*d*) S_N2

Answer. (*a*)

41. When ethanol is heated with concentrated H_2SO_4, a gas is produced. Which of the following compounds is formed when this gas is treated with bromine in CCl_4 ?

(*a*) Bromomethane (*b*) 1,2-Dibromoethane

(*c*) Bromoethane (*d*) 1,1,2,2-Tetrabromoethane

Answer. (*b*)

16

Polyhydric Alcohols

Compounds which contain more than one –OH group are called **Polyhydric Alcohols** or **Polyols**. They are further classified according to the number of –OH groups present in the molecule.

DIHYDRIC ALCOHOLS (DIOLS)

Compounds which contain two –OH groups on different carbons are called **Dihydric Alcohols** or **Diols**. Compounds with two –OH groups on one carbon are seldom encountered. This is because they are unstable and undergo spontaneous decomposition to give the corresponding carbonyl compound and water. Diols were commonly known as **Glycols**, since they have a sweet taste (Greek, *glycys* = sweet).

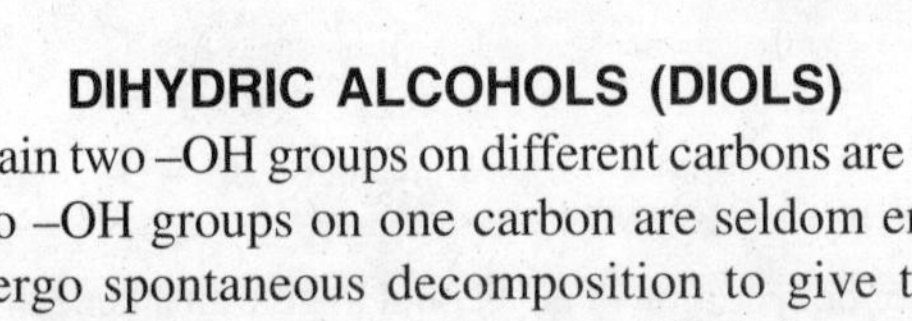

NOMENCLATURE

Diols have both common and IUPAC names. The IUPAC names are obtained by adding the suffix *diol* to the name of the parent alkane. Numbers are used to indicate the positions of the two –OH groups. For example,

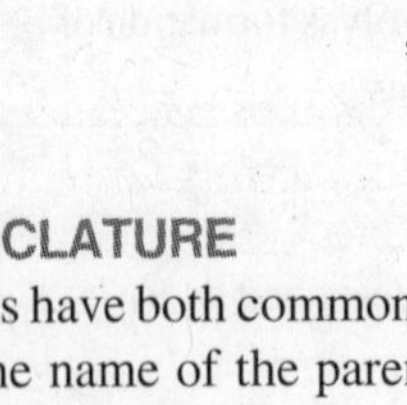

1,2-Ethanediol
(Ethylene glycol)

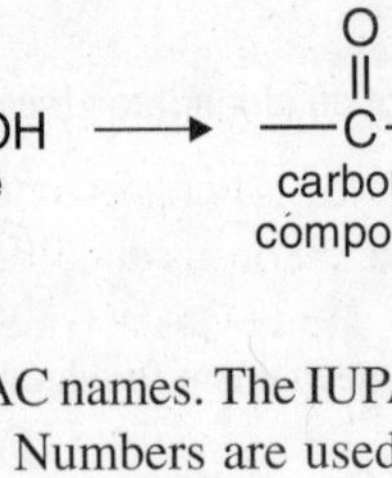

1,2-Propanediol
(Propylene glycol)

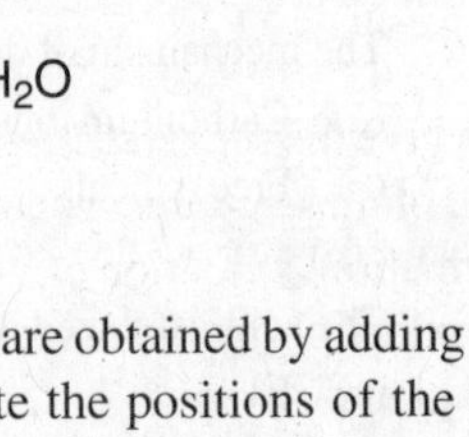

1,3-Propanediol
(Trimethylene glycol)

Points to Remember

1. Compounds which contain two hydroxy groups are called Diols.
2. Compounds which contain two hydroxy groups on adjacent carbons are called Glycols, *vic*-Diols, or 1,2-Diols.
3. Compounds which contain three hydroxy groups are called Triols.
4. The common name of a *vic*-Diol is obtained by naming the alkene from which the diol could be formed, *plus* the word glycol.
5. The IUPAC names of polyhydric alcohols are obtained by adding the suffix *diol*, *triol*, to the name of the corresponding alkane. Numbers are used to indicate the positions of hydroxy groups.

PROPERTIES

Lower diols are colorless, viscous liquids. They are soluble in water. Their boiling points are very much higher than those of monohydric alcohols of similar molecular weight. For example, the boiling point of ethylene glycol (MW = 62) is 197°C while that of 1-propanol (MW = 60) is 97°C. This is because the presence of two –OH groups in the molecule gives rise to extensive hydrogen bonding. This also explains their high viscosity, because the molecules of liquid diol being associated by hydrogen bonds, cannot move freely relative to each other. The high solubility of diols is due to their ability to form hydrogen bonds at both –OH groups with water molecules.

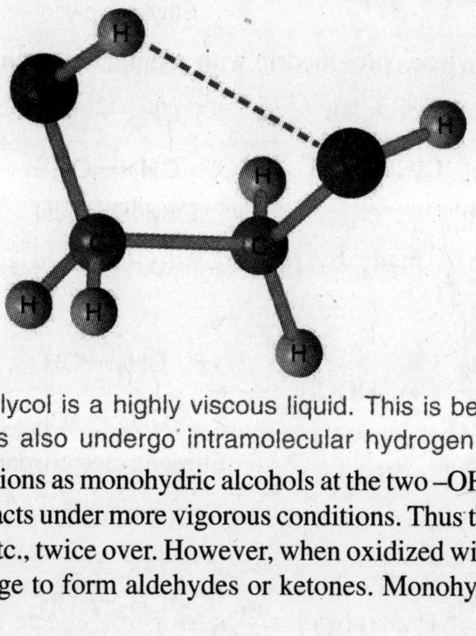

Ethylene glycol is a highly viscous liquid. This is because the molecules also undergo intramolecular hydrogen bonding

Diols give the same reactions as monohydric alcohols at the two –OH groups separately, although the second of these groups reacts under more vigorous conditions. Thus they undergo ester formation, halide formation, oxidation etc., twice over. However, when oxidized with periodic acid (HIO_4) diols undergo oxidation by cleavage to form aldehydes or ketones. Monohydric alcohols do not display such oxidative cleavage.

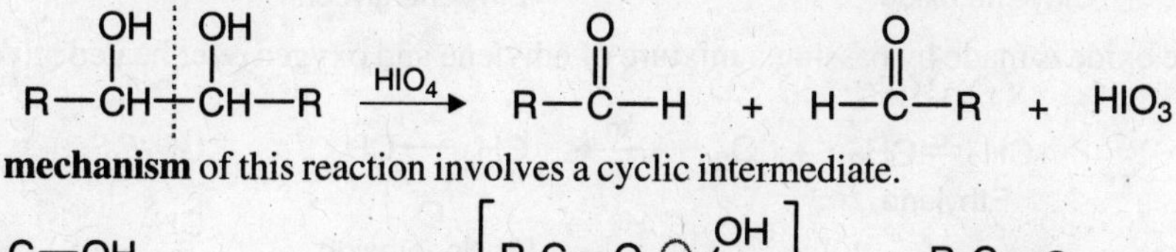

The **mechanism** of this reaction involves a cyclic intermediate.

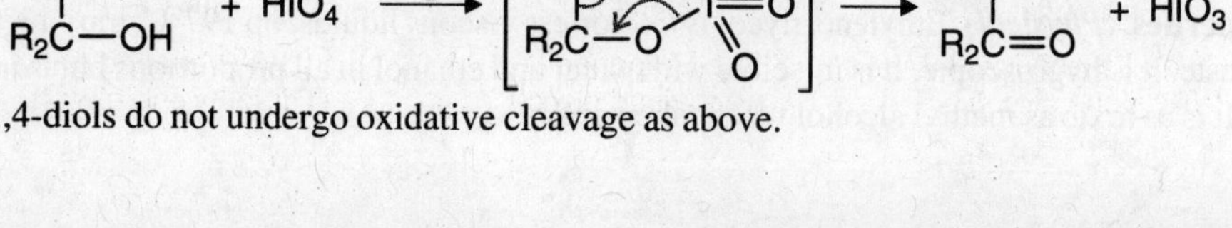

1,3- or 1,4-diols do not undergo oxidative cleavage as above.

ETHYLENE GLYCOL, *1,2-Ethanediol*, HO–CH$_2$–CH$_2$–OH

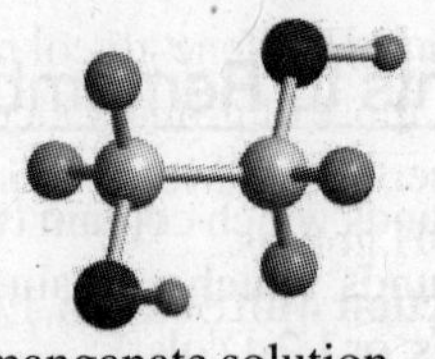

Ethylene glycol is the most important member of the class. It is often referred to as *Glycol*.

Preparation. It is prepared as follows :

(1) **By oxidation of ethylene with cold dilute potassium permanganate solution.**

$$\underset{\text{Ethylene}}{\begin{matrix} CH_2 \\ \| \\ CH_2 \end{matrix}} + (H_2O + O) \xrightarrow[KMnO_4]{Cold} \underset{\text{Ethylene glycol}}{\begin{matrix} CH_2-OH \\ | \\ CH_2-OH \end{matrix}}$$

(2) **By oxidation of ethylene with osmium tetroxide in the presence of sodium bisulfide.**

$$\underset{\text{Ethylene}}{\begin{matrix} CH_2 \\ \| \\ CH_2 \end{matrix}} + (H_2O + O) \xrightarrow[NaHSO_3]{OsO_4} \underset{\text{Ethylene glycol}}{\begin{matrix} CH_2-OH \\ | \\ CH_2-OH \end{matrix}}$$

(3) **By hydrolysis of 1,2-dibromoethane with aqueous sodium carbonate solution.**

$$\underset{\text{1,2-Dibromoethane}}{\begin{matrix} CH_2Br \\ | \\ CH_2Br \end{matrix}} + 2Na_2CO_3 + 2H_2O \longrightarrow \underset{\text{Ethylene glycol}}{\begin{matrix} CH_2OH \\ | \\ CH_2OH \end{matrix}} + 2NaHCO_3 + 2NaBr$$

(4) **By hydrolysis of ethylene chlorohydrin with sodium bicarbonate.**

$$2\underset{\text{Ethylene chlorohydrin}}{\begin{matrix} CH_2-OH \\ | \\ CH_2-Cl \end{matrix}} + NaHCO_3 \longrightarrow 2\underset{\text{Ethylene glycol}}{\begin{matrix} CH_2-OH \\ | \\ CH_2-OH \end{matrix}} + NaCl + CO_2$$

Ethylene chlorohydrin is made by passing ethylene into aqueous hypochlorous acid ($H_2O + Cl_2 \longrightarrow HOCl + HCl$).

$$\underset{\text{Ethylene}}{\begin{matrix} CH_2 \\ \| \\ CH_2 \end{matrix}} + HOCl \longrightarrow \underset{\text{Ethylene chlorohydrin}}{\begin{matrix} CH_2-OH \\ | \\ CH_2-Cl \end{matrix}}$$

(5) **By hydrolysis of ethylene oxide with H_2O at 200°C under pressure or with dilute H_2SO_4 at 60°C.**

$$\underset{\text{Ethylene oxide}}{CH_2-CH_2 \text{ (bridged by O)}} + H_2O \xrightarrow[60°C]{H^+} \underset{\text{Ethylene glycol}}{\begin{matrix} CH_2-CH_2 \\ | \quad\;\; | \\ OH \quad OH \end{matrix}}$$

Ethylene oxide is made by passing a mixture of ethylene and oxygen over heated silver catalyst.

$$\underset{\text{Ethylene}}{CH_2=CH_2} + O_2 \xrightarrow[\Delta]{Ag} \underset{\text{Ethylene oxide}}{CH_2-CH_2 \text{ (bridged by O)}}$$

Properties. (*Physical*). Ethylene glycol is a colorless viscous liquids, bp 197°C, mp –13°C. It has a sweet taste. It is hygroscopic. It is miscible with water and ethanol in all proportions but is insoluble in ether. It is as toxic as methyl alcohol when taken orally.

(*Chemical*). Ethylene glycol molecule contains two primary –OH groups. Its chemical reactions are, therefore, those of primary alcohols twice over. Generally, one –OH group is attacked completely before the other reacts. More vigorous conditions are sometimes needed for reactions of the second of the two –OH groups.

(1) **Reaction with Sodium.** Ethylene glycol reacts with sodium at 50°C to form monoalkoxide and dialkoxide when temperature is raised.

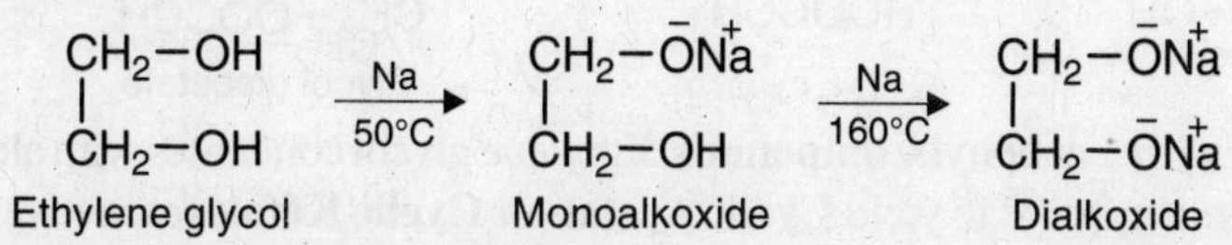

(2) **Reaction with Phosphorus Pentahalides**. Ethylene glycol reacts with PCl_5 first to form ethylene chlorohydrin and then 1,2-dichloroethane. PBr_5 reacts in the same way.

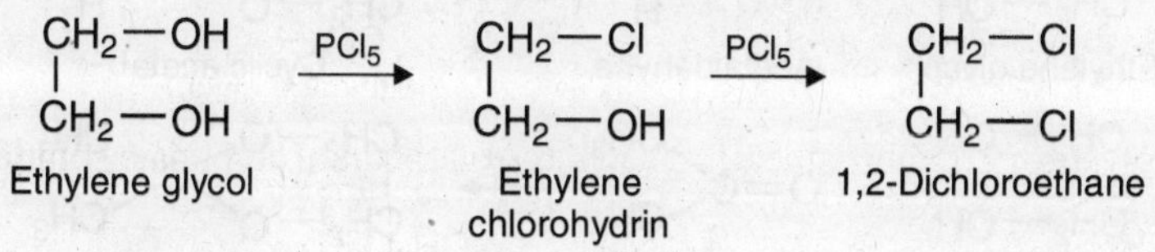

(3) **Reaction with Phosphorus Trihalides.** Ethylene glycol reacts with PBr_3 or PCl_3 to form the corresponding dihalides.

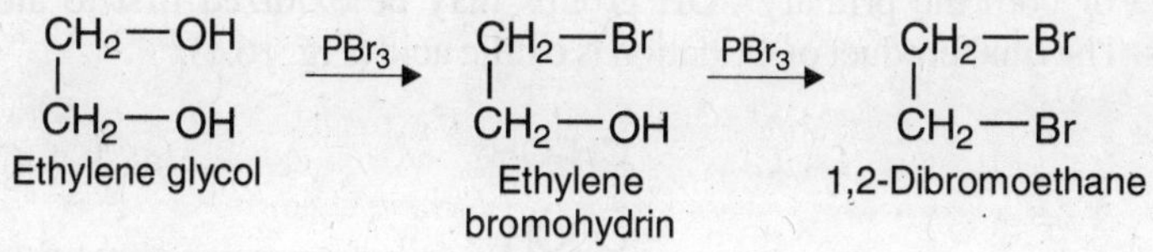

Ethylene glycol reacts with phosphorus triiodide to form 1,2-diiodoethane which is unstable and decomposes to form ethylene.

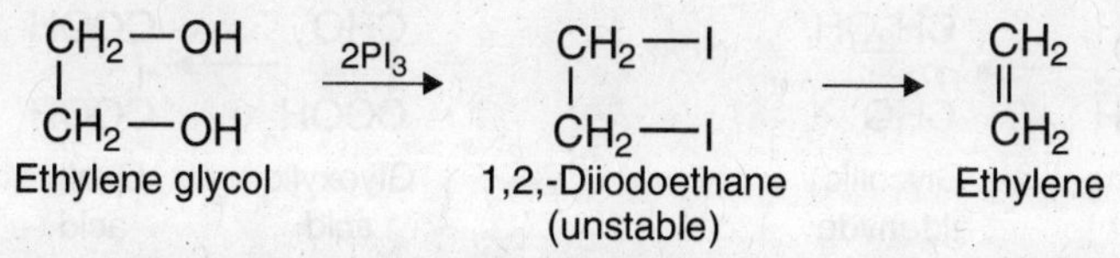

(4) **Reaction with Hydrogen Chloride.** Ethylene glycol reacts with HCl in two steps first forming ethylene chlorohydrin at 160°C and then at 200°C to yield 1,2-dichloroethane.

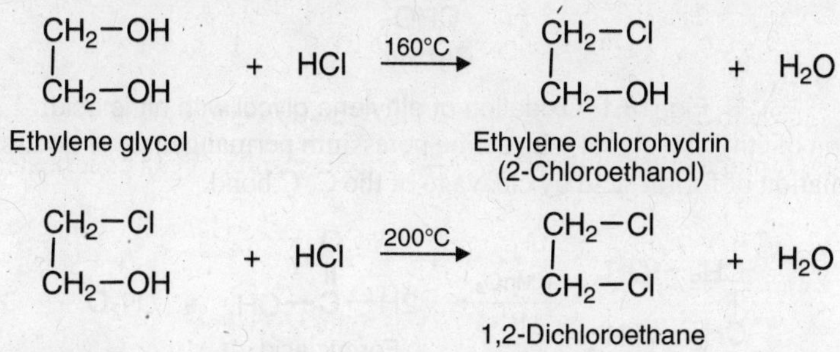

(5) **Reaction with Carboxylic acids.** Ethylene glycol reacts with carboxylic acids in the presence of mineral acid to form the corresponding esters or diesters.

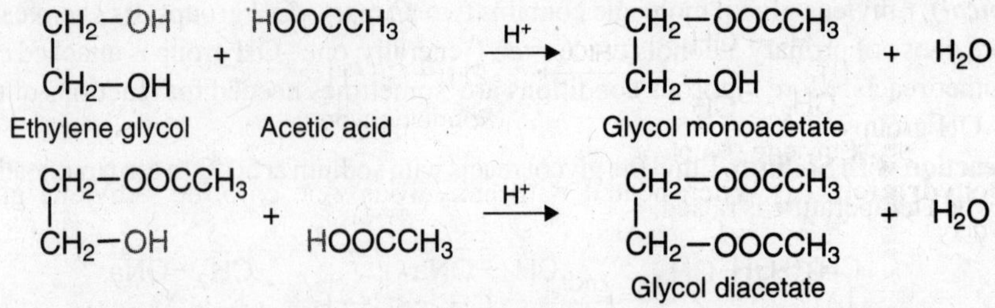

(6) **Reaction with Carbonyl Compounds.** Ethylene glycol condenses with aldehydes or ketones in the presence of mineral acid to yield **Cyclic Acetals** or **Cyclic Ketals** respectively.

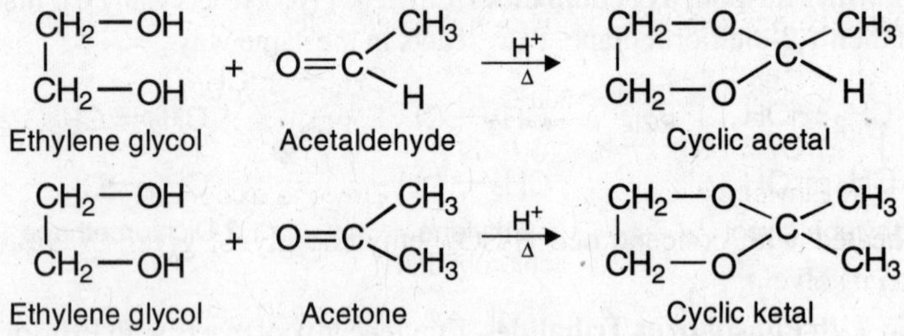

(7) **Oxidation.** (*a*) The oxidation of ethylene glycol with nitric acid yields a number of substances as one or both the primary –OH groups may be oxidized first to aldehyde and then carboxylic groups. The final product of oxidation is oxalic acid (Fig. 16.1).

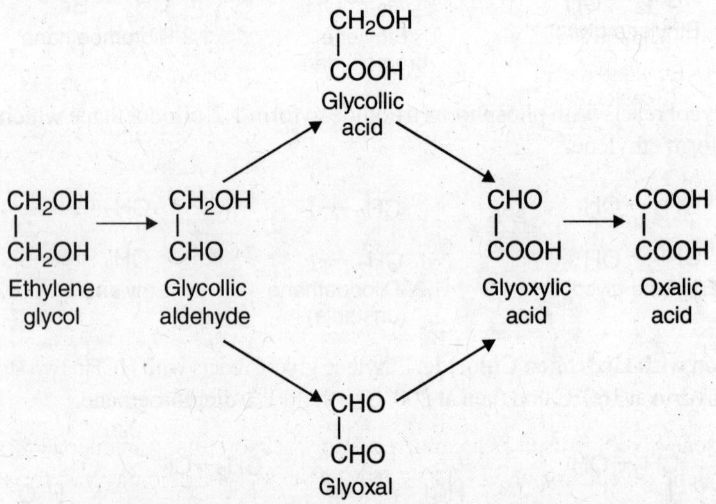

Fig. 16.1. Oxidation of ethylene glycol with nitric acid.

(*b*) Oxidation of ethylene glycol with acidic potassium permanganate or potassium dichromate results in the formation of formic acid by cleavage of the C–C bond.

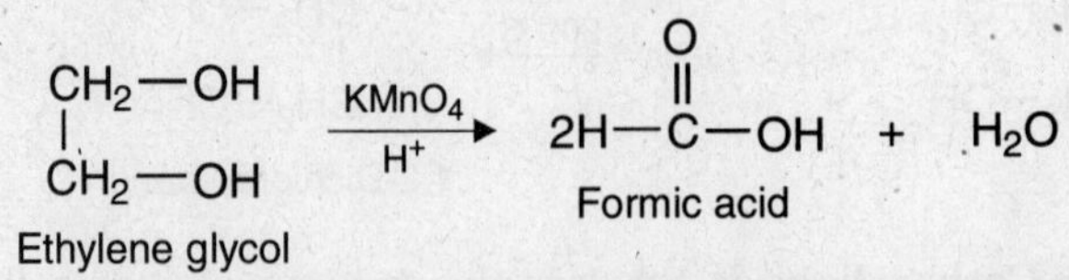

(*c*) When oxidized with periodic acid or lead tetraacetate, ethylene glycol gives formaldehyde.

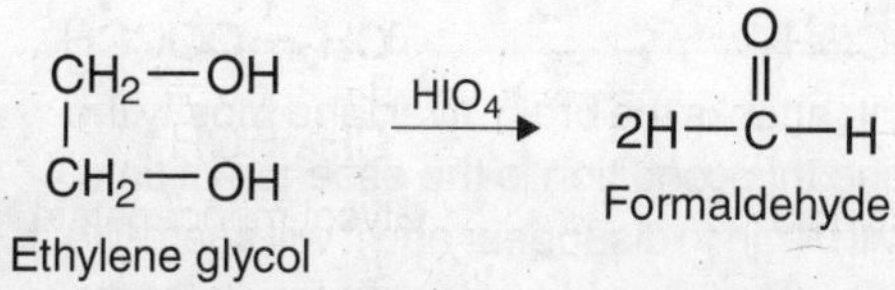

(8) **Dehydration.** (*a*) When heated with anhydrous zinc chloride, ethylene glycol gives acetaldehyde.

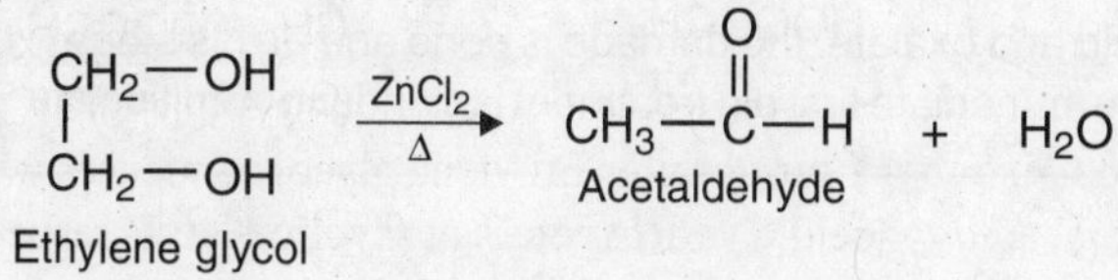

(*b*) When heated alone at 500°C, it gives ethylene oxide.

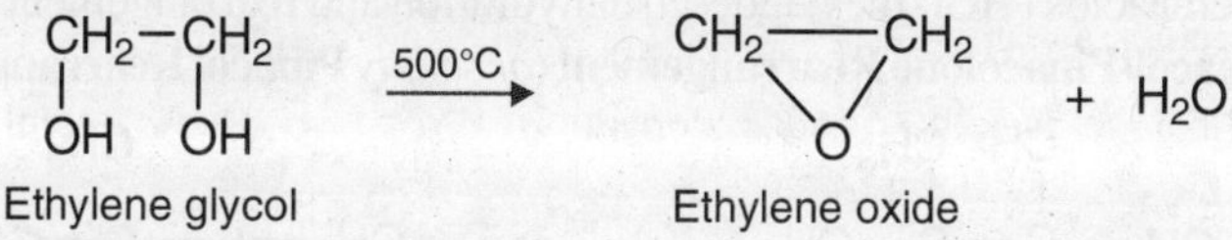

(*c*) When heated with concentrated H_2SO_4, ethylene glycol gives **Dioxane.** Dioxane is an important industrial solvent.

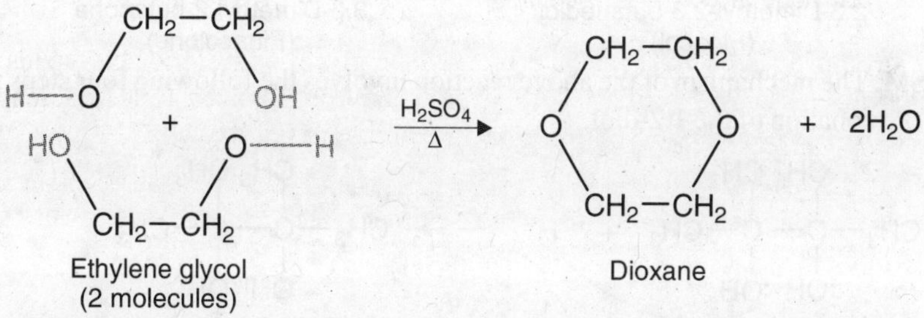

(*d*) When heated with concentrated phosphoric acid, it gives diethylene glycol.

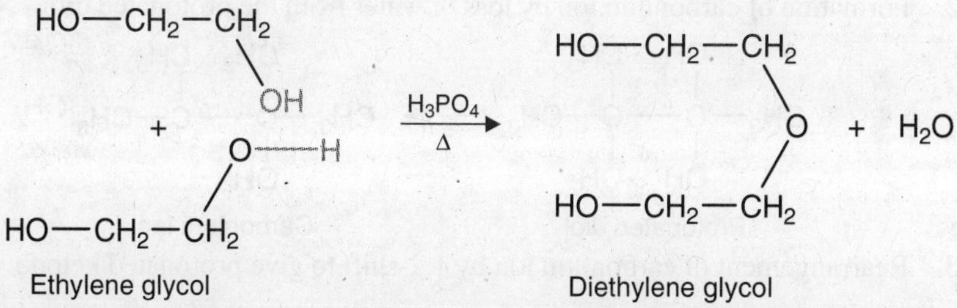

Uses. Ethylene glycol is used as an antifreeze for car radiators. An aqueous mixture containing 50 per cent ethylene glycol freezes at –34°C. It is also used as a deicing fluid for aeroplane wings. Ethylene glycol is used in manufacture of Dacron, dioxane, and ethylene oxide.

Ethylene is used for deicing aeroplanes. It is also used as coolant for car radiators.

Ethylene Glycol is a toxic poison. It attacks the brain, liver, and mostly the kidneys. The biggest hurdle in treating this condition is the essence of time. After about 6 hours, the symptoms will begin to disappear and it will seem like the problem is over...but it is not over. That's just how long it takes for the chemical to do its damage to the kidneys. Once the ethylene glycol is converted to formic acid and oxalate, the damage is done and at this stage it is rarely reversible. The immediate first aid treatment is to induce vomiting.

PINACOLS

Completely substituted *1,2-diols,* such as 2,3-dimethyl-2,3-butanediol, are called **Pinacols.** On treatment with mineral acids (HCl) they undergo dehydration and rearrangement to form ketones. This reaction is called **Pinacol-Pinacolone Rearrangement** (*or simply* **Pinacol Rearrangement).** For example,

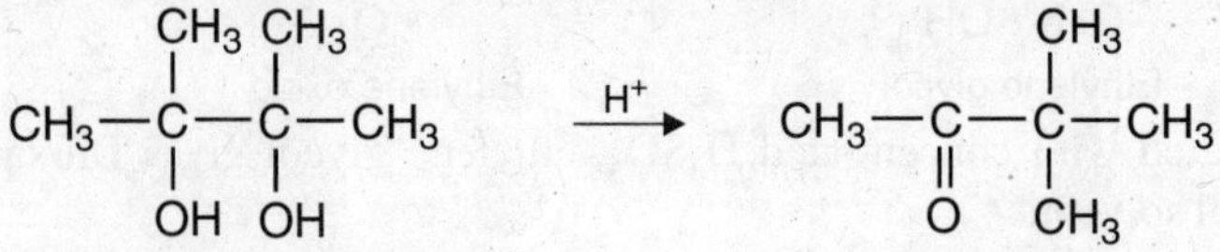

2,3-Dimethyl-2,3-butanediol (Pinacol) → 3,3-Dimethyl-2-butanone (Pinacolone)

MECHANISM. The mechanism of the above reaction involves the following four steps :

Step 1. Protonation of the 1,2-diol.

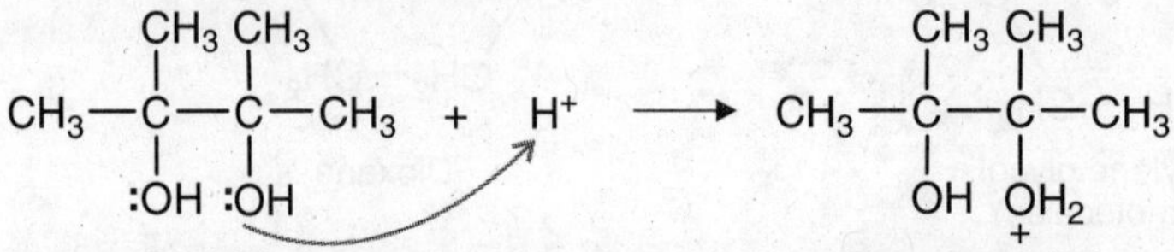

Pinacol → Protonated diol

Step 2. Formation of carbonium ion by loss of water from the protonated diol.

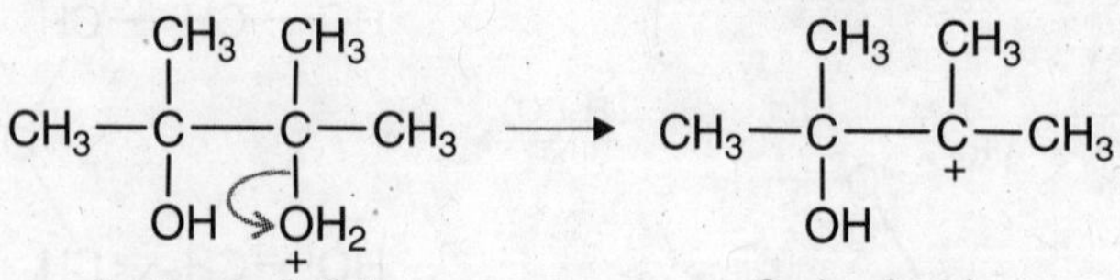

Protonated diol → Carbonium ion

Step 3. Rearrangement of carbonium ion by 1,2-shift to give protonated ketone.

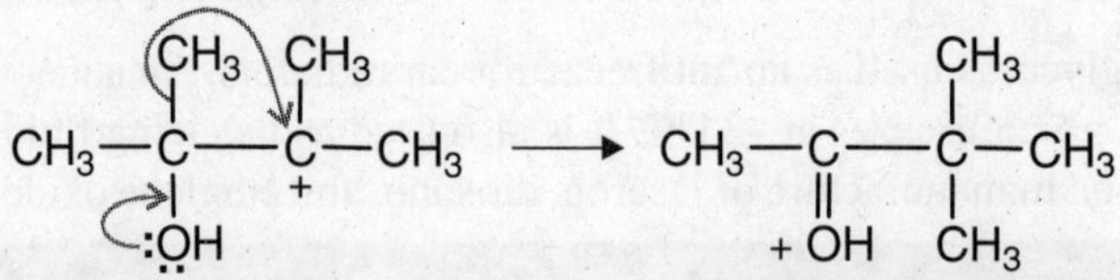

Carbonium ion → Protonated ketone

Step 4. Formation of ketone by loss of proton from the protonated ketone.

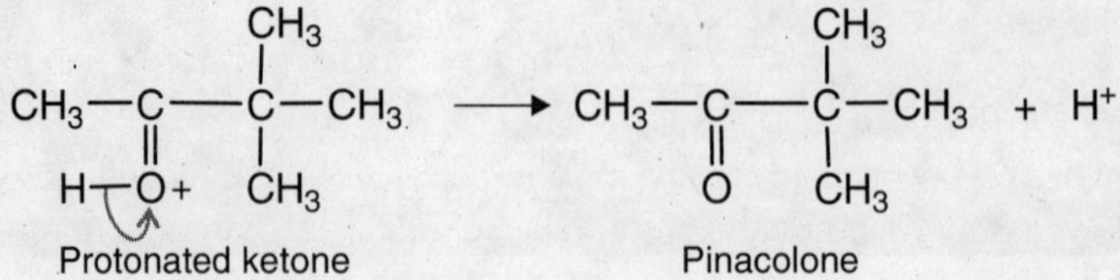

Protonated ketone → Pinacolone

TRIHYDRIC ALCOHOLS (TRIOLS)

Compounds which contain three –OH groups are called **Trihydric Alcohols** or **Triols**. The introduction of the third –OH group in a diol molecule raises the boiling point by about 100°C, increasing the viscosity. An increase in the number of –OH groups also enhances the hydrogen bonding ability and association, thereby raising the boiling point. For example,

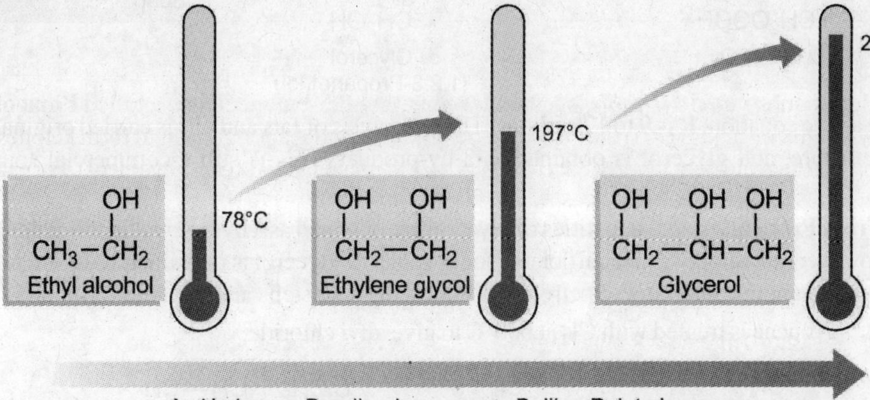

As Hydrogen Bonding Increases – Boiling Points Increase.

Glycerol contains three –OH groups and undergoes intramolecular hydrogen bonding, thereby increasing the boiling point and viscosity.

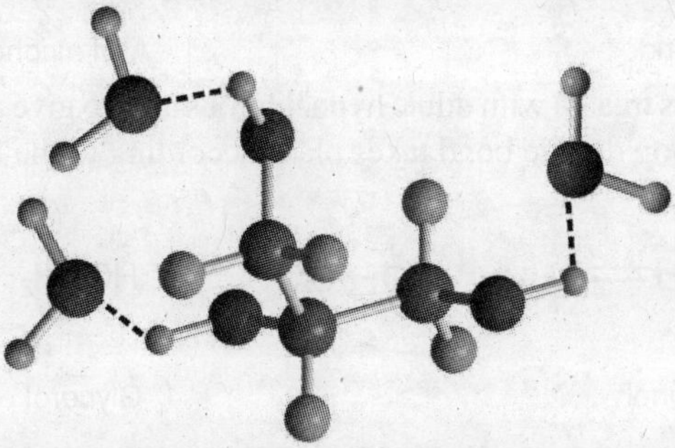

Glycerol is highly soluble in water. This is because it also forms hydrogen bonds with water molecules (Intermolecular hydrogen bonding).

Triols give many reactions of the primary and secondary –OH groups which they contain. They also give the simple reactions of –OH groups in triplicate. The only important member of the class is glycerol or 1,2,3-propanetriol.

GLYCEROL, *1,2,3-Propanetriol*, $HOCH_2–CH(OH)–CH_2OH$

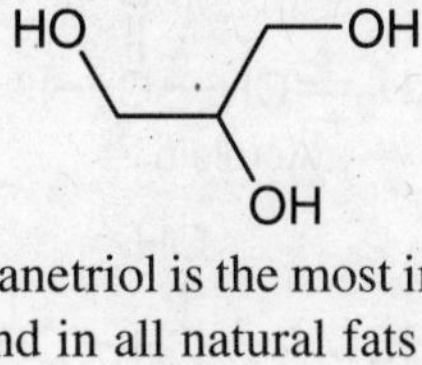

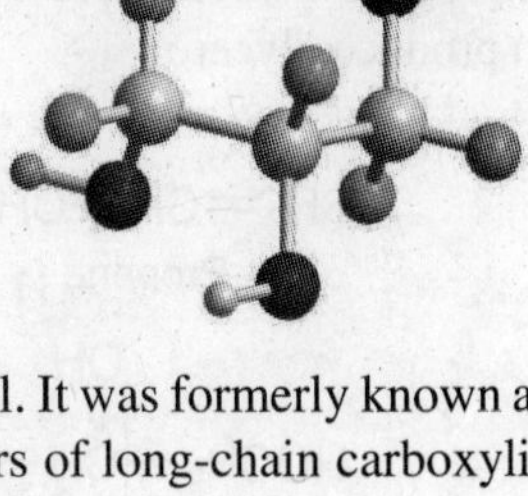

Glycerol or 1,2,3-propanetriol is the most important trihydric alcohol. It was formerly known as *Glycerine*. Glycerol is found in all natural fats and oils as glyceryl esters of long-chain carboxylic acids.

Manufacture. Glycerol is manufactured :

(1) **From Fats and Oils.** Natural oils and fats are triesters of glycerol and long-chain carboxylic acids (mainly palmitic, stearic, and oleic acids). On hydrolysis with alkali, the fats and oils produce glycerol and the salts of the long-chain acids which are called soaps.

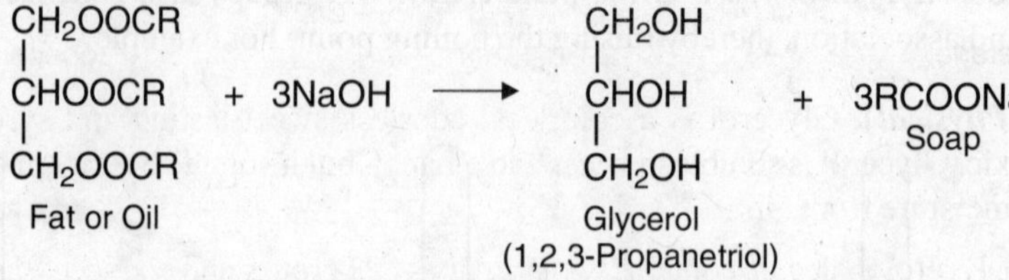

In the above equation, R is 9 to 17 carbons. The hydrolysis of fats and oils is carried originally for soap manufacture, and glycerol is obtained as a by-product. This is still a commercial source of glycerol.

(2) **From Propene.** Large quantities of glycerol are obtained as a by-product in the manufacture of soap. However, this supply is not sufficient. Today much of glycerol is manufactured from propene obtained by the catalytic cracking of petroleum. Following four steps are involved :

Step 1. Propene is treated with Cl_2 at 600°C to give allyl chloride.

$$\underset{\text{Propene}}{CH_3{-}CH{=}CH_2} + Cl_2 \xrightarrow{600°C} \underset{\text{Allyl chloride}}{ClCH_2{-}CH{=}CH_2}$$

Step 2. Allyl chloride is treated with dil. NaOH to give allyl alcohol.

$$\underset{\text{Allyl chloride}}{ClCH_2{-}CH{=}CH_2} + NaOH \longrightarrow \underset{\text{Allyl alcohol}}{HOCH_2{-}CH{=}CH_2}$$

Step 3. Allyl alcohol is treated with dilute hypochlorous acid to give a chlorohydrin. The addition of HOCl to the carbon-carbon double bond takes place according to the Markovnikov rule.

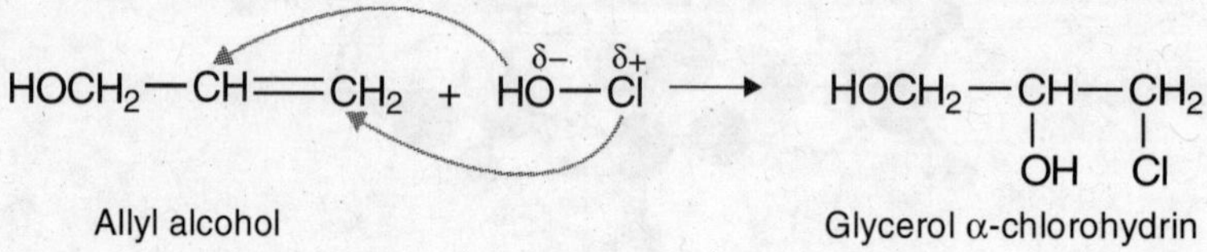

Step 4. The chlorohydrin is treated with dil. NaOH to yield glycerol.

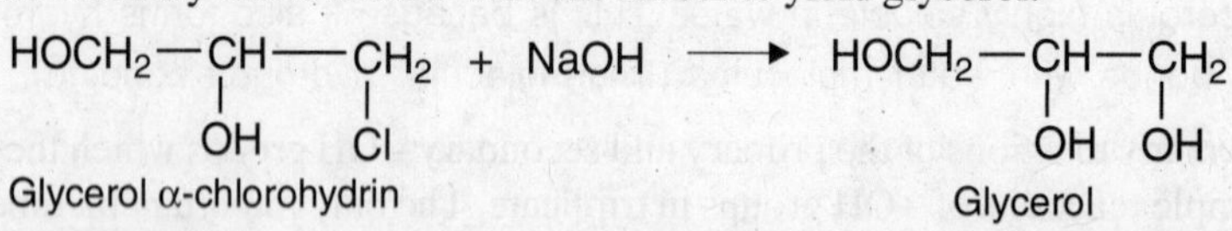

Alternative Method. Propene may be oxidized with oxygen (air) in the presence of CuO at 350°C to give acrolein. Acrolein is then treated with hydrogen peroxide (*Hydroxylation*) followed by reduction to produce glycerol.

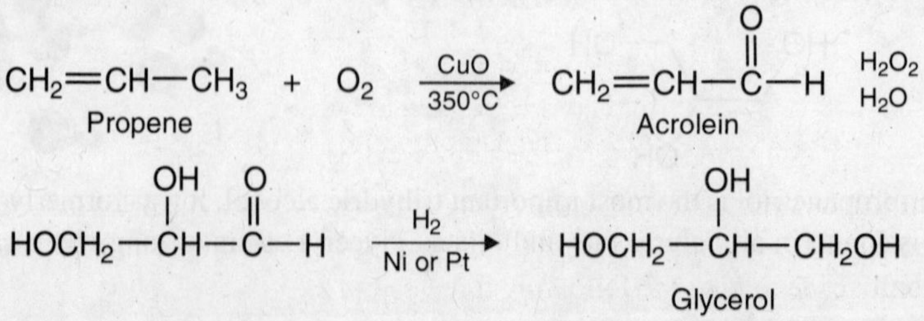

In a country like India where petroleum is not found in abundance, propene required for the above synthesis may also be obtained from acetone of the wood distillation industry.

$$\underset{\text{Acetone}}{CH_3-\overset{O}{\overset{\|}{C}}-CH_3} \xrightarrow[\text{Pt or Ni}]{H_2} \underset{\text{2-Propanol}}{CH_3-\overset{OH}{\overset{|}{C}H}-CH_3} \xrightarrow[250°C]{Al_2O_3} \underset{\text{Propene}}{CH_3-CH{=}CH_2}$$

Properties (*Physical*). Glycerol is a colorless, odorless sweet-tasting and syrupy liquid, bp 290°C. It is nontoxic. Glycerol is soluble in water and ethanol, but insoluble in ether. It is hygroscopic, that is, it absorbs moisture from air.

(*Chemical*). Glycerol molecule contains two primary –OH groups and one secondary –OH group. It undergoes many of the reactions to be expected of these types of alcohols. The carbon atoms in glycerol are indicated as α, β, and α′.

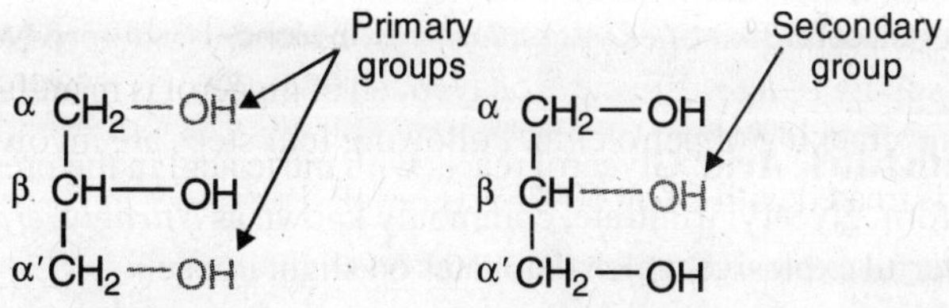

Glycerol Glycerol

In general, the two primary –OH groups are more reactive than the secondary –OH group. Some of the reactions that are characteristic of monohydroxy compounds are modified to a certain extent by the presence of three –OH groups.

(1) **Reaction with Sodium.** When glycerol is treated with sodium at room temperature, one of the primary –OH groups is attacked to form monosodium glycerolate. At higher temperatures, the second primary –OH group is also attacked to give disodium glycerolate. The secondary –OH group does not react at all.

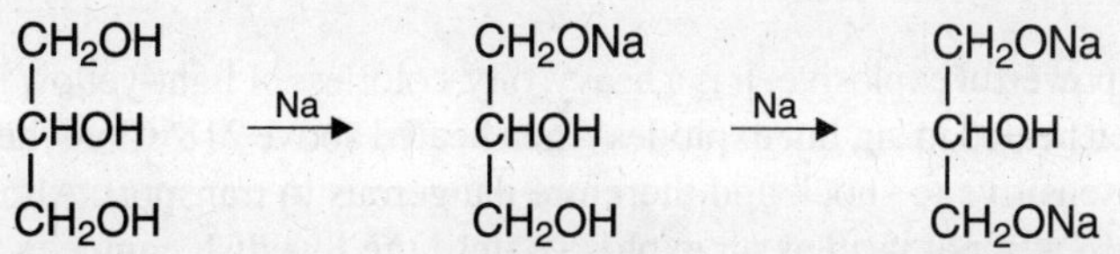

Glycerol Monosodium glycerolate Disodium glycerolate

(2) **Reaction with PCl_5.** Glycerol reacts with PCl_5 to form glyceryl trichloride. All these –OH groups are replaced by Cl atoms.

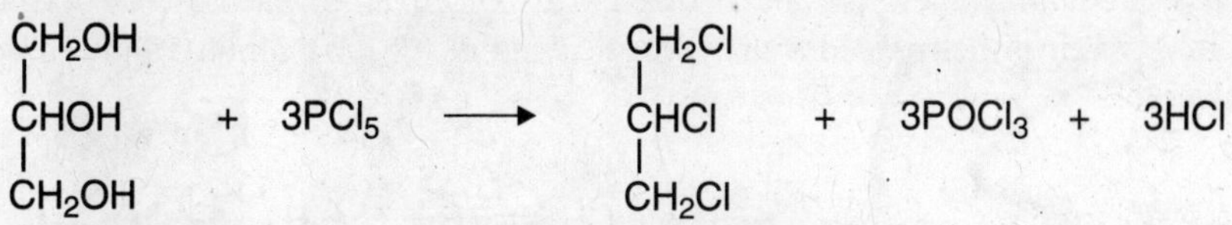

Glycerol Glyceryl trichloride (1,2,3-Trichloropropane)

(3) **Reaction with Carboxylic Acids.** Glycerol reacts with monocarboxylic acids to form mono-, di-, and triesters depending on the amount of acid used. For example, glycerol reacts with a mixture of acetic acid and acetic anhydride to give the following three esters :

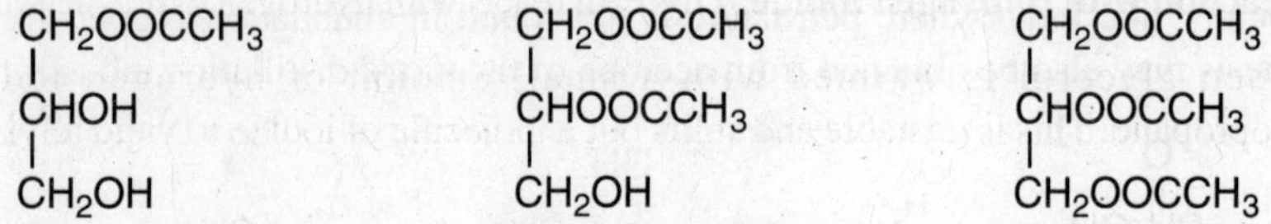

Glyceryl monoacetate   Glyceryl diacetate   Glyceryl triacetate

(4) **Reaction with Hydrogen Chloride.** When HCl is passed into glycerol at 110°C, both α- and β-glycerol monochlorohydrins are formed. If the reaction is carried for a long time, glycerol α,α'-dichlorohydrin and glycerol α,β-dichlorohydrin are formed.

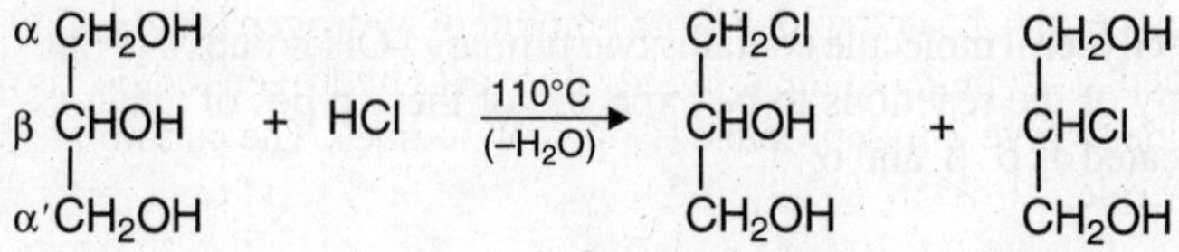

Glycerol   α-Mono-chlorohydrin   β-Mono-chlorohydrin

(5) **Reaction with Nitric Acid.** Glycerol reacts with nitric acid in the presence of sulfuric acid (catalyst), at 25°C to form glyceryl trinitrate, commonly known as *Nitroglycerine*. It is a pale yellow oily liquid. It is a powerful explosive which detonates on slight impact.

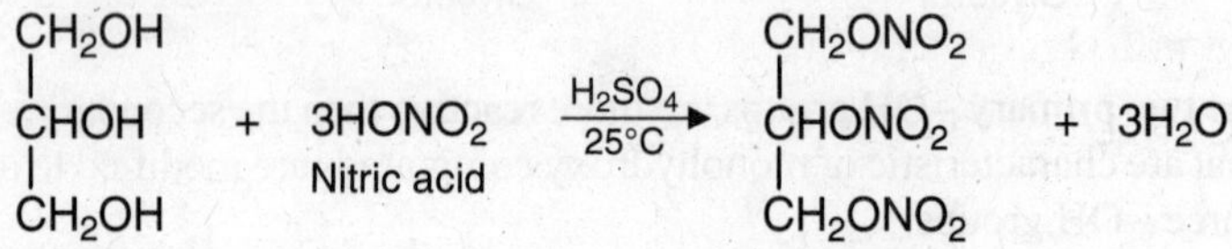

Glycerol   Glyceryl trinitrate (Nitroglycerine)

Points to Remember

Nitroglycerin is a powerful explosive. It is a heavy, oily, colorless or light-yellow liquid. Nitroglycerin burns quietly when heated in air, but explodes when heated above 218°C or when heated in a closed vessel. It is very sensitive to shock and therefore dangerous to transport. Although discovered in 1846, nitroglycerin was not used as an explosive until the Swedish engineer and inventor Alfred Nobel used it in making dynamite in 1866. Nitroglycerin is a common explosive today and is usually mixed with an inert, porous material such as sawdust. When detonated, it produces about 10,000 times its own volume of gas. Nitroglycerin is also used medically, in doses of 0.2 to 0.6 mg, as a heart medication under the trade names Nitrostat and Nitrospan.

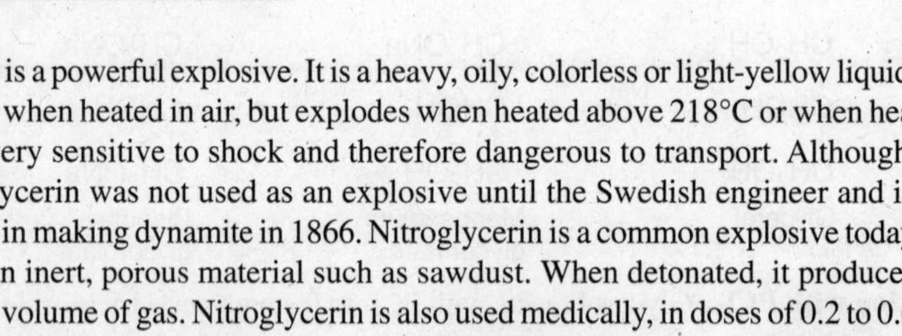

(6) **Reaction with Hydrogen Iodide.** Glycerol reacts with hydrogen iodide in two ways:

(*a*) When glycerol is warmed with a small amount of hydrogen iodide, it gives 1,2,3-triiodopropane. This is unstable and splits out a molecule of iodine to yield allyl iodide.

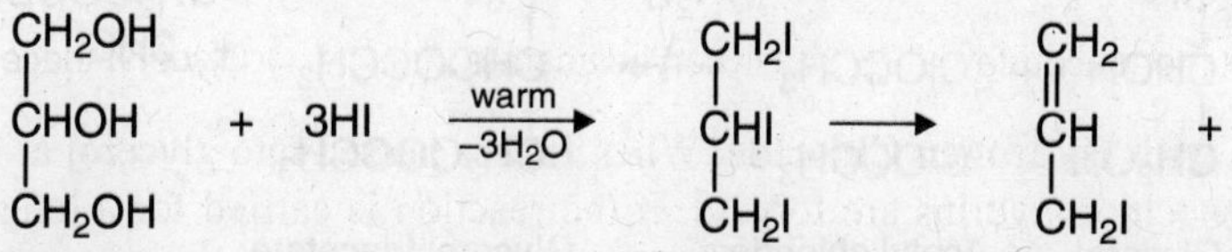

Glycerol (unstable) Allyl iodide

(*b*) When glycerol is heated with a large amount of hydrogen iodide, allyl iodide first produced as above is reduced to propene. In the presence of excess hydrogen iodide, propene adds a molecule of hydrogen iodide to give 2-iodopropane (isopropyl iodide). The addition takes place according to the Markovnikov rule.

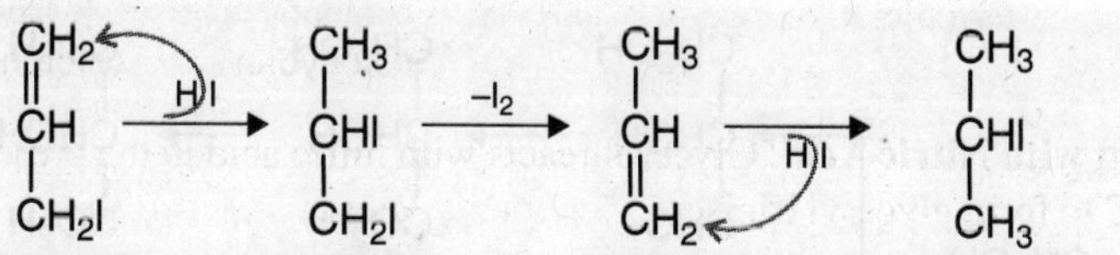

Allyl iodide (unstable) Propene 2-Iodopropane

(7) **Reaction with Oxalic Acid.** Glycerol reacts with oxalic acid in two ways :

(*a*) At 110°C glycerol reacts with oxalic acid to form glycerol monoformate.

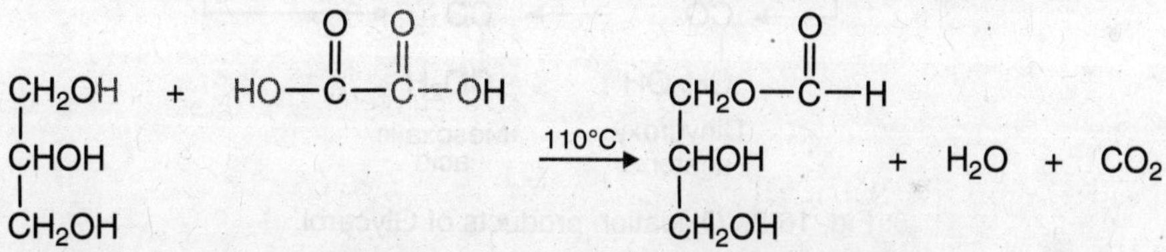

Glycerol Oxalic acid Glycerol monoformate

Glycerol monoformate on hydrolysis gives formic acid and glycerol is regenerated.

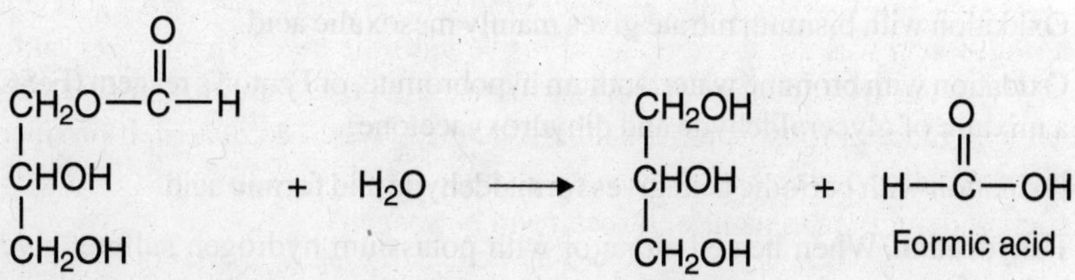

Glycerol monoformate Glycerol

(*b*) At 260°C glycerol reacts with oxalic acid to form allyl alcohol.

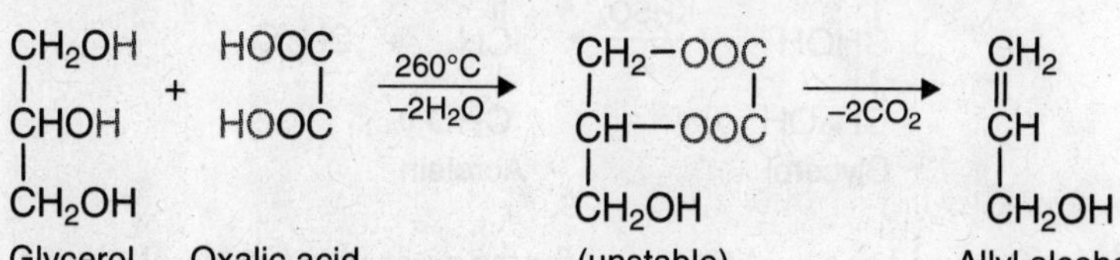

Glycerol Oxalic acid (unstable) Allyl alcohol

(8) **Reaction with Acetyl Chloride.** Glycerol reacts with acetyl chloride (or acetic anhydride) to form glyceryl triacetate.

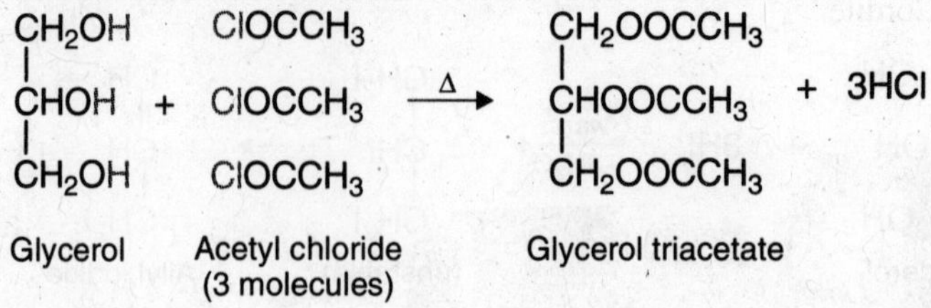

(9) **Oxidation.** The two primary alcohol groups in glycerol are capable of being oxidized to the aldehyde and then the carboxyl group. The secondary alcohol group can be oxidized to the carbonyl group. Thus glycerol can give rise to a variety of oxidation products depending on the nature of the oxidizing agent (Fig. 16.2).

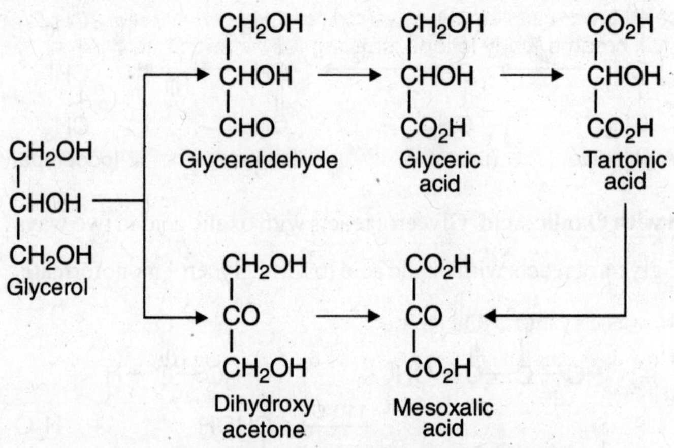

Fig. 16.2. Oxidation products of Glycerol.

(*a*) Oxidation with dilute HNO_3 gives glyceric acid and tartonic acid.

(*b*) Oxidation with concentrated HNO_3 gives mainly glyceric acid.

(*c*) Oxidation with bismuth nitrate gives mainly mesoxalic acid.

(*d*) Oxidation with bromine water, sodium hypobromite, or Fenton's reagent ($FeSO_4$+H_2O_2), gives a mixture of glyceraldehyde and dihydroxyacetone.

(*e*) Oxidation with periodic acid gives formaldehyde and formic acid.

(10) **Dehydration.** When heated alone or with potassium hydrogen sulfate, glycerol eliminates two molecules of water to form acrolein.

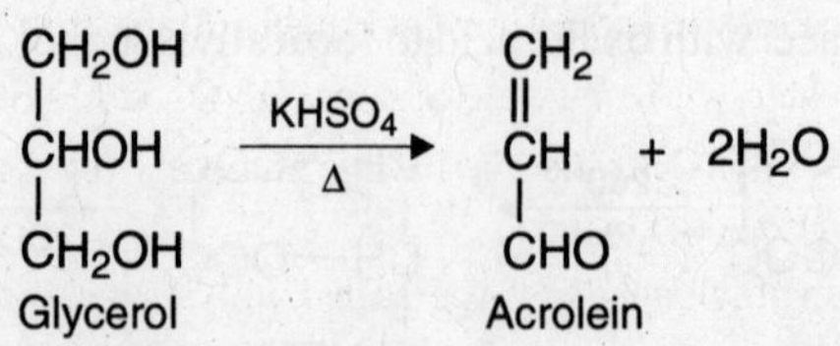

This reaction is often used as a qualitative test for the presence of glycerol, since acrolein is easily detected on account of its peculiar odor.

Uses. Glycerol is nontoxic and an excellent *humectant* (moisture-retaining agent). It is used in bakery products, hand lotions, vanishing creams, shaving soaps and tooth pastes. Glycerol is used in production of plastics, synthetic fibers, and surface coatings. It is used in the manufacture of explosives like Dynamite and Cordite.

Glycerol is an excellent moisture-retaining agent. It is used in vanishing creams, body lotions, shaving foams and tooth pastes.

Examination Questions

1. (*a*) How is ethylene glycol prepared ?

(*b*) How does ethylene glycol react with the following reagents :

(*i*) PCl_5 (*ii*) PI_3 (*iii*) Sodium (*iv*) HIO_4

(*Devi Ahiliya BSc, 2015*)

2. (*a*) How is glycerol prepared ?

(*b*) How does glycerol react with the following reagents ?

(*i*) $KHSO_4$/heat (*ii*) HI (*iii*) PCl_5 (*iv*) Sodium

(*v*) HNO_3/H_2SO_4 (*vi*) Oxalic acid (*vii*) Acetyl chloride

3. How is glycerol prepared from oils ? Starting from glycerol how are the following compounds prepared : (*i*) Nitroglycerine ; (*ii*) Allyl alcohol ; (*iii*) Isopropyl iodide.

4. Give the synthesis and uses of ethylene glycol.

5. Give the synthesis and uses of glycerol. (*Jiwaji BSc, 2014*)

6. What happens when

(*a*) Ethylene glycol is heated with conc. H_2SO_4.

(*b*) Glycerol is treated with oxalic acid at 260°C.

(*c*) Glycerol is heated with potassium hydrogen sulfate.

(*d*) Glycerol is treated with a mixture of conc. HNO_3 and H_2SO_4.

Answer. The products are : (*a*) Dioxane ; (*b*) Allyl alcohol ; (*c*) Acrolein ; (*d*) Glyceryl trinitrate (Nitroglycerine). For reaction equations see text.

7. What happens when ethylene glycol is distilled with a small amount of phosphoric acid ?

Answer. Diethylene glycol is formed. Reaction equation is given in the text.

8. Write a note on : Pinacol rearrangement.

(*Delhi BSc, 2013 ; Gorakhpur BSc, 2013 ; Sambalpur BSc, 2014*)

9. Discuss the mechanism of Pinacol-Pinacolone rearrangement.

(Mysore BSc, 2013 ; Panjab BSc, 2015)

10. An organic compound of formula $C_3H_8O_3$ on acetylation with acetyl chloride (*or* acetic anhydride) gives a compound of formula $C_9H_{14}O_6$. How many hydroxyl groups are present in the compound ? What is its possible structure ?

Answer. Alcohols react with acetyl chloride (*or* acetic anhydride) to form esters.

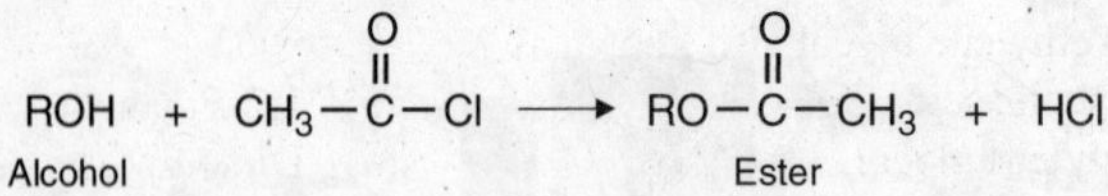

Notice that for the OH group present in the alcohol, C_2H_2O is added. In the given problem, the increase is $C_6H_6O_3$ ($3 \times C_2H_2O$). Thus there are three OH groups in the molecule. The compound could be glycerol.

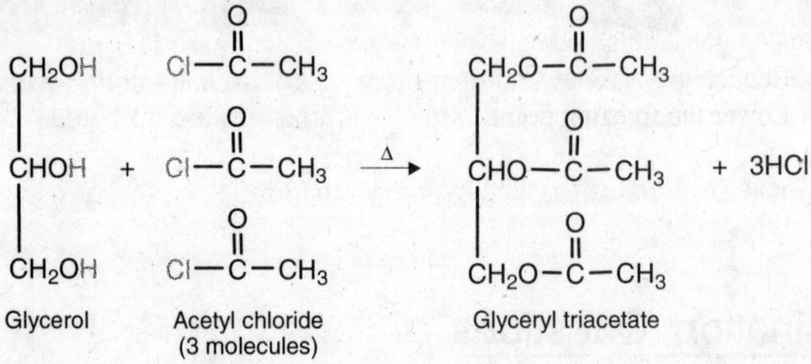

11. How will you synthesize the following compounds from ethylene glycol ?

(*a*) Oxalic acid (*b*) Acetylene (*c*) Dioxane

Answer. Step are : (*a*) heat with HNO_3 ; (*b*) (*i*) PCl_5, (*ii*) $NaNH_2$; (*c*) heat with conc. H_2SO_4.

12. How will you synthesize succinic acid from ethylene glycol ? *(Kakatiya BSc, 2015)*

Answer. Following steps are involved :

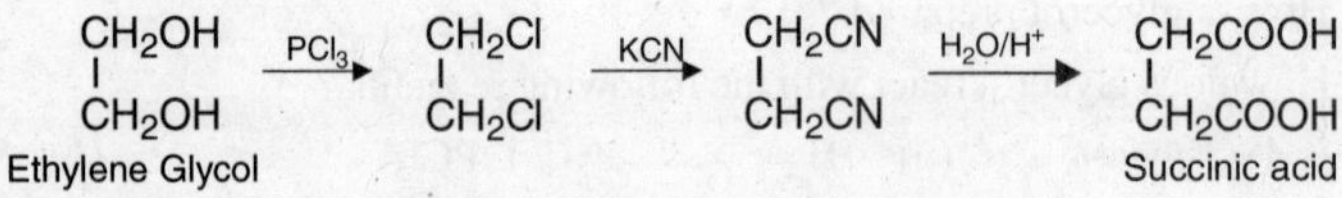

13. How will you synthesize 1,2-butanediol from 1-butanol ?

Answer. Following steps are involved :

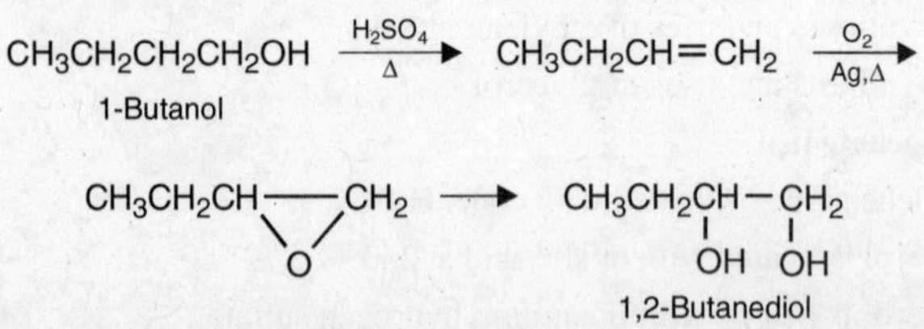

14. How will you synthesize the following compounds from glycerol ?

(*a*) Allyl alcohol (*b*) Nitroglycerine (*c*) Acrolein

Answer. Steps are : (*a*) heat with oxalic acid at 230°C; (*b*) heat with a mixture of conc. HNO_3 and H_2SO_4; (*c*) heat with $KHSO_4$. For reaction equations see text.

15. How will you synthesize glycerol (1,2,3-propanetriol) from propene ?

(Madras BSc, 2013 ; Patna BSc, 2014 ; Delhi BSc, 2015)

16. How will you synthesize glycerol from isopropyl alcohol ?

17. How will you synthesize glycerol from acetone ? (*Kanpur BSc, 2013; Kerala BSc, 2014*)

Multiple Choice Questions

1. What is ethylene glycol also known as ?
(*a*) Glycolic acid (*b*) 1,2-Ethanediol, MEG, glycol
(*c*) Butylene glycol (*d*) Dimethyl glycol

Answer. (*b*)

2. Ethylene glycol is produced from ___________, *via* the intermediate ethylene oxide.
(*a*) Chloroform (*b*) Alkene
(*c*) Ethylene (*d*) Alkane

Answer. (*c*)

3. In cold countries ethylene glycol is added to water in car radiators. This helps to
(*a*) Reduce the viscosity (*b*) Make water a better lubricant
(*c*) Lower the freezing point (*d*) Lower the boiling point

Answer. (*c*)

4. Ethylene oxide undergoes acid-hydrolysis to form
(*a*) Ethylene glycol (*b*) Formic acid
(*c*) Ethyl alcohol (*d*) Acetic acid

Answer. (*a*)

5. Ethylene glycol undergoes oxidation with hot acidic $KMnO_4$ to form :
(*a*) Formic acid (*b*) Formaldehyde
(*c*) Acetic acid (*d*) Acetaldehyde

Answer. (*a*)

6. Ethylene glycol reacts with excess of PCl_5 to give
(*a*) Chloroethane (*b*) 1,2-Dichloroethane
(*c*) Hexachloroethane (*d*) 1,3-Dichloroethane

Answer. (*b*)

7. 1,2-Ethanediol reacts with anhydrous zinc chloride to form
(*a*) Ethylene (*b*) Acetaldehyde
(*c*) Acetylene (*d*) Ethyl chloride

Answer. (*b*)

8. When ethylene glycol is heated with concentrated HNO_3, it forms
(*a*) Oxalic acid (*b*) Ethylene oxide
(*c*) Dioxane (*d*) Diethylene glycol

Answer. (*a*)

9. Which of the following is used as an antifreeze ?
(*a*) Ethylene glycol (*b*) Glycerol
(*c*) Diethyl ether (*d*) Picric acid

Answer. (*a*)

10. Arrange the compounds in the order of increasing solubility in water (*lowest first*).
I. CH_3OH II. $CH_3CH_2CH_2CH_2CH_2OH$
III. $CH_3CH_2CH_2OH$ IV. $HOCH_2CH_2CH_2CH_2CH_2OH$
(*a*) IV, III, II, I (*b*) II, III, IV, I
(*c*) III, IV, II, I (*d*) II, IV, III, I

Answer. (*b*)

11. Which of the following has the *highest* boiling point ?
(*a*) $CH_3OCH_2CH_2CH_2OH$ (*b*) $CH_3OCH_2CH_2OCH_3$
(*c*) $HOCH_2CH_2CH_2OH$ (*d*) $CH_3OCH_2CH_2OH$

Answer. (*c*)

12. Periodic acid (HIO_4) is used to ?
(*a*) oxidize aldehydes (*b*) oxidize geminal diols
(*c*) oxidize vicinal diols (*d*) oxidize ketones

Answer. (*c*)

13. What is glycerol also known as ?
(*a*) 2,3-Benzodiazine (*b*) Cupric bromide
(*c*) 1,2,3-Propanetriol (*d*) 1,2-Butanediol

Answer. (*c*)

14. As a sugar substitute, glycerol is 60 percent as sweet as __________.
(*a*) Glucose (*b*) Fructose
(*c*) Starch (*d*) Sucrose

Answer. (*d*)

15. When glycerol is heated with oxalic acid at 260°C, it gives
(*a*) 1,2-Propanediol (*b*) Vinyl alcohol
(*c*) 1,3-Propanediol (*d*) Allyl alcohol

Answer. (*d*)

16. When glycerol is heated with potassium hydrogen sulfate ($KHSO_4$), it forms
(*a*) Acrolein (*b*) Acetic acid
(*c*) Allyl alcohol (*d*) Propionic acid

Answer. (*a*)

17. Glycerol on warming with a small amount of hydriodic acid gets converted to
(*a*) Propene (*b*) 3-Iodopropene
(*c*) 1,3-Diiodopropane (*d*) 2-Iodopropane

Answer. (*b*)

18. When glycerol is treated with a mixture of concentrated $HNO_3 + H_2SO_4$, it forms
(*a*) Nitroethane (*b*) 1-Nitropropane
(*c*) Nitroglycerine (*d*) 2-Nitropropane

Answer. (*c*)

19. In the manufacture of dynamite, one of the chemicals used is
(*a*) Glycerol (*b*) Glycerol triacetate
(*c*) Glycerol trinitrate (*d*) Glycerol triiodide

Answer. (*c*)

20. Which of the following compounds is least soluble in water ?
(*a*) Glycerol (*b*) Ethyl alcohol
(*c*) Ethylene glycol (*d*) Ethyl chloride

Answer. (*d*)

21. Compound (A) reacts with sodium metal to form one mole of H_2. The compound (A) can be
(*a*) $CH_3CH_2CH{=}CH_2$ (*b*) $HOCH_2CH_2CH_2CH_2OH$
(*c*) $CH_2{=}CH{-}CH{=}CH_2$ (*d*) $CH_3CH_2CH_2CH_2OH$

Answer. (*b*)

22. What is the chemical name of Nitroglycerin (IUPAC) ?
(*a*) 1,2,3-Trinitroxypropane (*b*) 1,2,3-Tribromopropane
(*c*) 5-Benzene-1,2,3-triol chloride (*d*) 6-Amino-4-cyclohex-4-ene-1,2,3-triol

Answer. (*a*)

23. Glycerol is used to produce nitroglycerin, or glycerol-trinitrate (GTN), which is an essential ingredient of smokeless gunpowder and various explosives such as dynamite, gelignite and propellants like __________.
(*a*) Ballistite (*b*) Smokeless powder
(*c*) Cordite (*d*) Internal ballistite

Answer. (*c*)

24. Dynamites containing ____________, which increase the viscosity of the mix, are commonly known as "gelatins".
(*a*) Cordite (*b*) Nitrocellulose
(*c*) Nitroglycerin (*d*) Smokeless powder

Answer. (*b*)

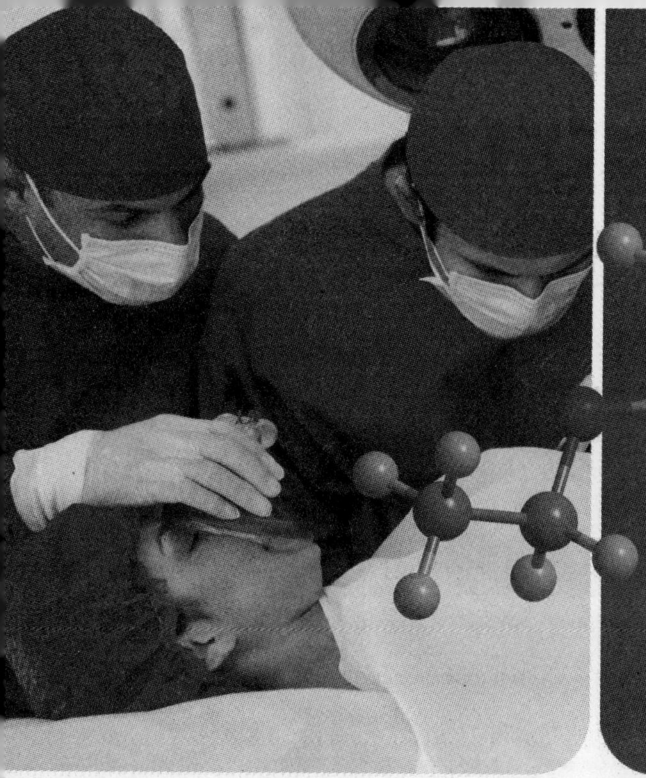

17

Ethers and Epoxides

Ethers are a class of compounds which contain an oxygen atom bonded to two alkyl groups. They are further designated as **Symmetrical Ethers** or **Unsymmetrical Ethers** according as the two alkyl groups attached to the oxygen atom are same or different.

R—Ö—R	R—Ö—R′
Symmetrical ether	Unsymmetrical ether

For example,

CH_3CH_2—Ö—CH_2CH_3	CH_3—Ö—CH_2CH_3
Symmetrical ether	Unsymmetrical ether
R groups are **same**.	R groups are **different**

The functional group of ethers is C–O–C

Ethers may be considered as derivatives of water in which both hydrogen atoms have been replaced by alkyl groups. They may also be considered to be derivatives of alcohols in which the hydroxyl hydrogen has been replaced by an alkyl group.

H—O—H	R—O—R′	R—O—H
Water	Ether	Alcohol

STRUCTURE

In ethers, the oxygen atom is sp^3 hybridized. The two C–O σ bonds are formed by sp^3–sp^3 overlap. The two unshared pairs of electrons on oxygen occupy sp^3 hybrid orbitals. The C–O–C angle is very close to 110° (Fig. 17.1)

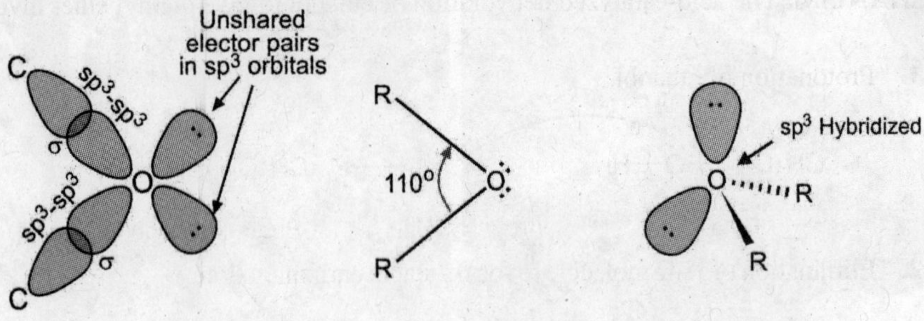

Fig. 17.1. Orbital structure of an ether molecule.

NOMENCLATURE

(1) **Common System**. The two alkyl groups attached to oxygen are named in alphabetic order and the word ether is added. If the two alkyl groups are same (R–O–R), the prefix *di-* is used. Although the prefix *di-* is often omitted, it is better to include it to avoid confusion.

$$CH_3CH_2—O—CH_2CH_3 \qquad CH_3—O—CH_2CH_3$$

Diethyl ether (Ethyl ether) — Ethyl methyl ether

(2) **IUPAC System.** In this system, ethers are named as **Alkoxyalkanes**. The larger of the alkyl groups is considered to be the alkane. The name of the alkane is prefixed by the name of the alkoxy group and position number.

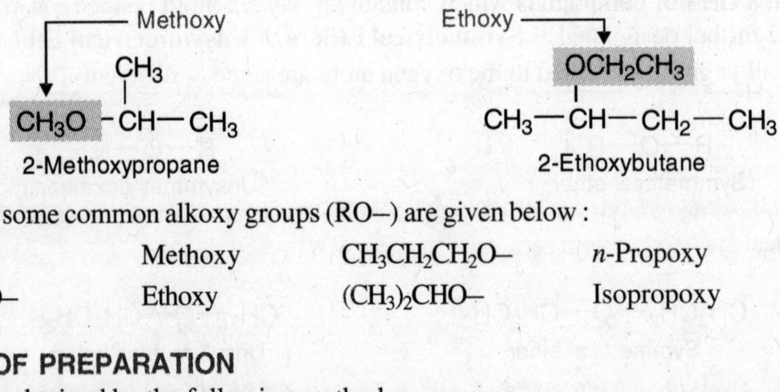

2-Methoxypropane — 2-Ethoxybutane

Names of some common alkoxy groups (RO–) are given below :

$CH_3O–$	Methoxy	$CH_3CH_2CH_2O–$	*n*-Propoxy
$CH_3CH_2O–$	Ethoxy	$(CH_3)_2CHO–$	Isopropoxy

METHODS OF PREPARATION

Ethers are obtained by the following methods :

(1) **Dehydration of Alcohols**. Symmetrical ethers are prepared by heating an excess of alcohol with concentrated H_2SO_4 at 140°C.

$$\underset{\text{Alcohol (2 molecules)}}{R—OH + H—OR} \xrightarrow[140°C]{H_2SO_4} \underset{\text{Ether}}{R—O—R} + H_2O$$

$$\underset{\text{Ethanol}}{C_2H_5—OH} + \underset{\text{Ethanol}}{H—OC_2H_5} \xrightarrow[140°C]{H_2SO_4} \underset{\text{Diethyl ether}}{C_2H_5—O—C_2H_5} + H_2O$$

The starting alcohol in the above reaction must be primary and the reaction temperature must be kept at 140°C. This is because alkene formation is favored at higher temperatures and with secondary or tertiary alcohols.

MECHANISM. The acid-catalyzed dehydration of ethanol to give diethyl ether involves four steps :

Step 1. Protonation of ethanol.

$$\underset{\text{Ethanol}}{CH_3CH_2{-}\ddot{O}{-}H} + \underset{\text{(from } H_2SO_4)}{H^+} \longrightarrow \underset{\text{Protonated alcohol}}{CH_3CH_2{-}\overset{+}{\ddot{O}}(H){-}H}$$

Step 2. Elimination of H_2O molecule to form a stable carbonium ion.

$$CH_3CH_2{-}\overset{+}{\ddot{O}}(H){-}H \longrightarrow \underset{\text{Carbonium ion}}{CH_3\overset{+}{C}H_2} + H_2O$$

Step 3. Carbonium ion attacks another molecule to give an oxonium ion.

$$CH_3\overset{+}{C}H_2 + \underset{\text{Ethanol}}{CH_3CH_2{-}\ddot{O}{-}H} \longrightarrow \underset{\text{Oxonium ion}}{CH_3CH_2{-}\overset{+}{O}(H){-}CH_2CH_3}$$

Step 4. Oxonium ion loses a proton to yield diethyl ether.

$$CH_3CH_2{-}\overset{+}{O}(H){-}CH_2CH_3 \longrightarrow \underset{\text{Diethyl ether}}{CH_3CH_2{-}\ddot{O}{-}CH_2CH_3} + H^+$$

(2) **Passing Alcohol Vapors over Al_2O_3.** Ethers may be obtained by passing alcohol vapors over hot activated Al_2O_3 at 250°C.

$$\underset{\text{Ethanol}}{C_2H_5{-}OH} + \underset{\text{Ethanol}}{H{-}OC_2H_5} \xrightarrow[250°C]{Al_2O_3} \underset{\text{Diethyl ether}}{C_2H_5{-}O{-}C_2H_5} + H_2O$$

(3) **Williamson Ether Synthesis**. This involves the treatment of a sodium alkoxide with an alkyl halide. Sodium alkoxides are obtained by treating alcohols with sodium metal ($ROH + Na \rightarrow RONa + H_2$). Both symmetrical and unsymmetrical ethers can be made by this method.

$$\underset{\text{Sod alkoxide}}{R{-}\bar{O}\overset{+}{Na}} + \underset{\text{Alkyl halide}}{R'{-}X} \longrightarrow \underset{\text{Ether}}{R{-}O{-}R'} + NaX$$

$$\underset{\text{Sod methoxide}}{CH_3{-}\bar{O}\overset{+}{Na}} + \underset{\text{Bromoethane}}{C_2H_5{-}Br} \longrightarrow \underset{\text{Ethyl methyl ether}}{CH_3{-}O{-}C_2H_5} + NaBr$$

$$\underset{\text{Sod ethoxide}}{C_2H_5{-}\bar{O}\overset{+}{Na}} + \underset{\text{Bromoethane}}{C_2H_5{-}Br} \longrightarrow \underset{\text{Diethyl ether}}{C_2H_5{-}O{-}C_2H_5} + NaBr$$

MECHANISM. The Williamson synthesis involves a displacement of halide ion from an alkyl halide by an alkoxide. It follows S_N2 mechanism.

$$\underset{\text{Sodium ethoxide}}{CH_3CH_2\bar{O}\,\overset{+}{Na}} + \underset{\text{Bromoethane}}{CH_3CH_2{-}Br} \xrightarrow{S_N2} \underset{\text{Diethyl ether}}{CH_3CH_2{-}O{-}CH_2CH_3} + NaBr$$

The alkyl halide used in this reaction must be primary. If secondary or tertiary alkyl halides are used, the reaction leads mainly to alkene formation.

(4) **Heating Alkyl Halides with Dry Silver Oxide**. This method is useful only for symmetrical ethers.

$$\begin{matrix} CH_3CH_2-Cl \\ CH_3CH_2-Cl \end{matrix} + Ag_2O \xrightarrow{\Delta} \underset{\text{Diethyl ether}}{CH_3CH_2-O-CH_2CH_3} + 2AgCl$$

Ethyl chloride
(2 molecules)

(5) **Action of Diazomethane on Alcohols**. Methyl ethers can be obtained by treatment of primary or secondary alcohols with diazomethane. Fluoroboric acid (HBF_4) is used as a catalyst.

$$R-O-H + CH_2N_2 \xrightarrow{HBF_4} R-O-CH_3 + N_2$$

$$\underset{\text{Ethyl alcohol}}{CH_3CH_2-O-H} + CH_2N_2 \xrightarrow{HBF_4} \underset{\text{Ethyl methyl ether}}{CH_3CH_2-O-CH_3} + N_2$$

PHYSICAL PROPERTIES

(1) Dimethyl ether and ethyl methyl ether are gases. All others are colorless liquids with pleasant odors.

(2) Lower ethers are highly volatile and very flammable.

(3) Boiling points of ethers show a gradual increase with the increase in molecular weight.

Table 17.1 BOILING POINTS OF SOME ETHERS

Compound	Formula	bp (°C)
Dimethyl ether	CH_3OCH_3	–25
Ethyl methyl ether	$CH_3OC_2H_5$	7.6
Diethyl ether	$C_2H_5OC_2H_5$	35
Diisopropyl ether	$(CH_3)_2CHOCH(CH_3)_2$	91

(4) Ethers have lower boiling points than isomeric alcohols. This is because ether molecules cannot form hydrogen bonds with each other as they have no –OH groups. The boiling points of ethers are close to the boiling points of alkanes. For example, diethyl ether (MW=74) boils at 35°C ; *n*-pentane (MW=72) boils at 36°C.

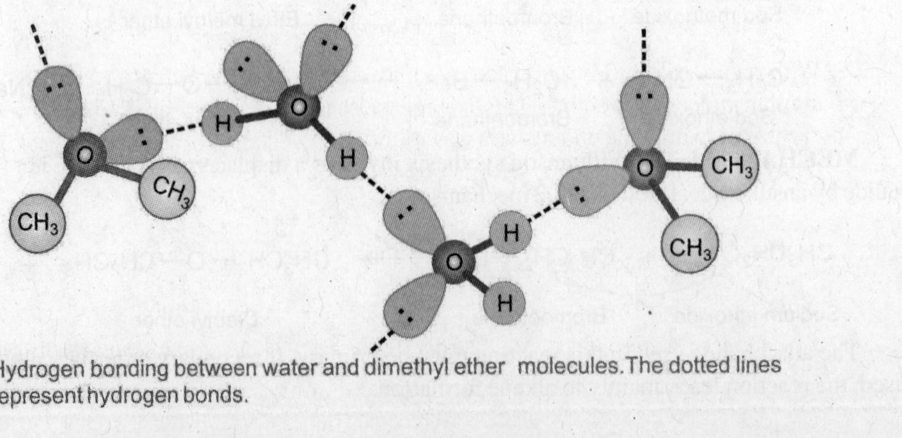

Hydrogen bonding between water and dimethyl ether molecules. The dotted lines represent hydrogen bonds.

(5) Ethers are slightly soluble in water. This is because they can form hydrogen bonds with water.

(6) ***IR Spectrum.*** Ethers show a strong, broad band due to C–O stretching in the 1060-1300 cm^{-1} range.

Points to Remember

Comparison of Boiling Points and Solubilities of Alkanes, Ethers, and Alcohols

1. The boiling points and solubility characteristic of ethers may be understood from their structures. Ethers, unlike water or alcohols, have no hydrogens directly attached to the oxygen atom. As a result there is no hydrogen bonding between one ether molecule and another, as there is between alcohol molecules. We can therefore predict that the boiling points of ethers will be considerably lower than those of alcohols, and about the same as those of alkanes, of comparable molecular weights.

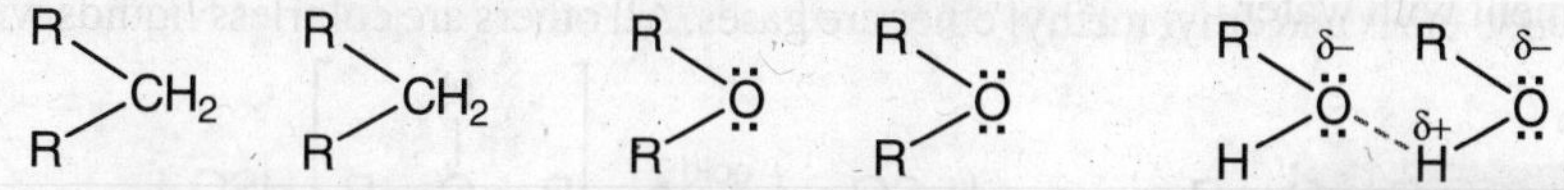

Alkanes : No hydrogen bonding between molecules; low boiling points

Ethers : No hydrogen bonding between molecules; low boiling points

Alcohols : Hydrogen bonding between molecules; high boiling points

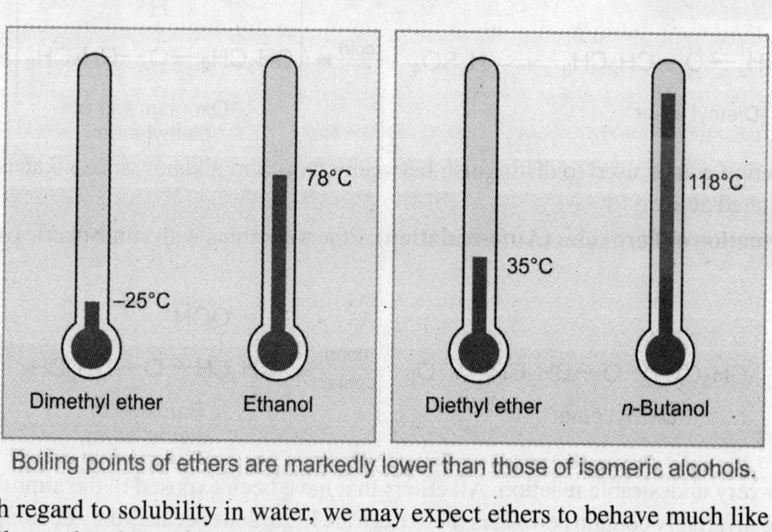

Boiling points of ethers are markedly lower than those of isomeric alcohols.

2. With regard to solubility in water, we may expect ethers to behave much like alcohols of similar molecular weights. This is because the oxygen atom in ethers, as in alcohols, is capable of hydrogen bonding with water molecules. Alkanes, of course, are insoluble because no hydrogen bonding with water is possible.

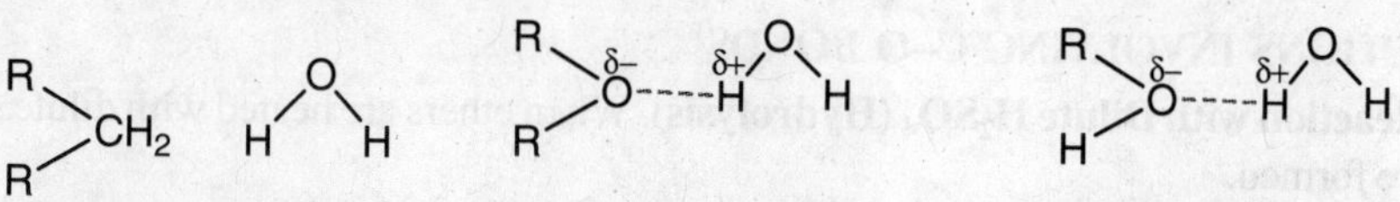

Alkanes : No hydrogen bonding with water ; insoluble

Ethers : Hydrogen bonding with water; soluble

Alcohols : Hydrogen bonding with water; soluble

CHEMICAL PROPERTIES

Ethers are quite stable compounds. They do not react with bases, reducing agents, oxidizing agents, and active metals.

A. REACTION INVOLVING ALKYL GROUPS

(1) **Halogenation**. When ethers are treated with Cl_2 or Br_2 in the dark, substitution products are obtained. The extent of substitution depends upon the reaction conditions. Halogenation preferentially takes place at α-carbon atoms.

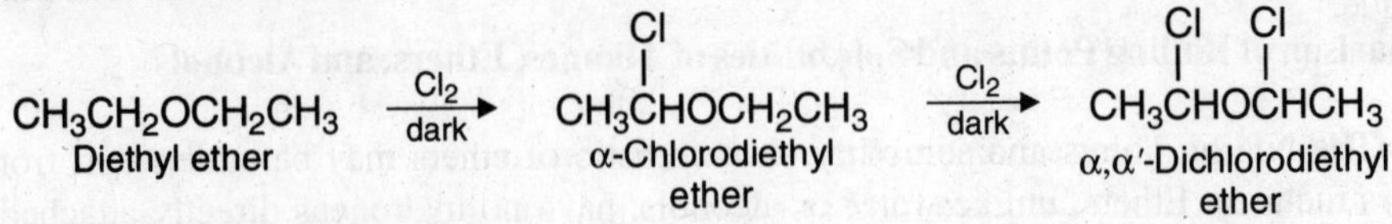

B. REACTIONS AT ETHER OXYGEN

(2) **Formation of Oxonium Salts**. Ethers react with cold concentrated H_2SO_4 or HCl to give oxonium salts. Ethers donate a pair of electrons to a hydrogen ion to form these salts. The oxonium salts once formed are soluble in the acid solution. The ethers can be regenerated from the oxonium salts by treatment with water.

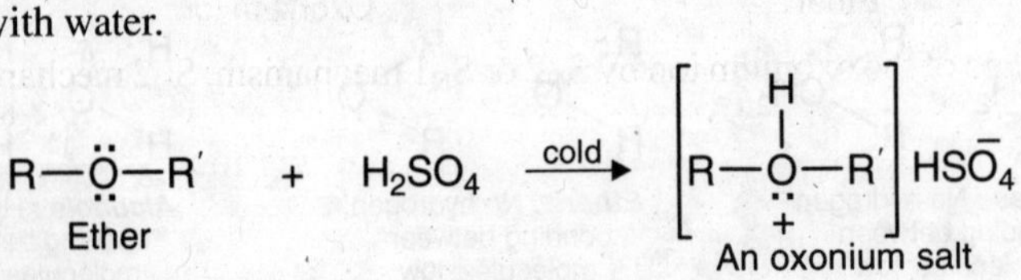

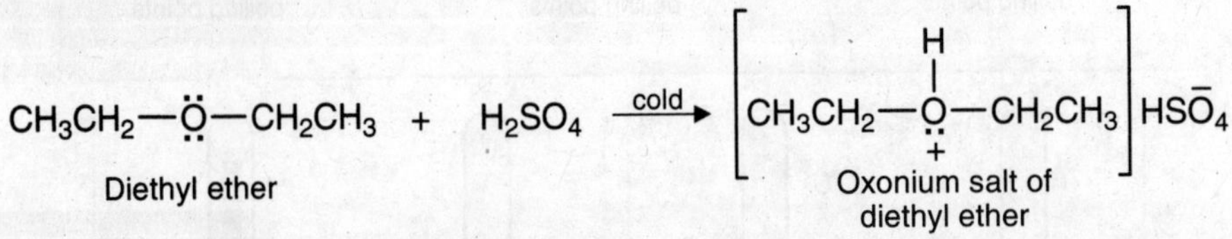

This reaction can be used to distinguish between ethers and alkanes since alkanes do not react with concentrated acids.

(3) **Formation of Peroxides (Autooxidation)**. Ethers combine with atmospheric oxygen to form peroxides.

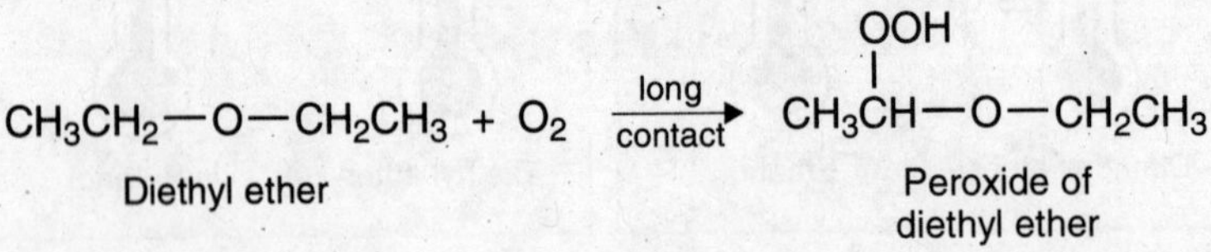

This is a very undesirable reaction. All ethers that have been exposed to the atmosphere for any length of time invariably contain peroxides. Peroxides are hazardous because they decompose violently at high temperatures. When ether that contains peroxides is distilled, the peroxides concentrate in the residue. At the end of distillation, the temperature increases, and the residual peroxides may explode. For this reason, ethers should never be evaporated to dryness.

C. REACTIONS INVOLVING C–O BONDS

(4) **Reaction with Dilute H_2SO_4 (Hydrolysis)**. When ethers are heated with dilute sulfuric acid, alcohols are formed.

$$\underset{\text{Diethyl ether}}{CH_3CH_2{-}O{-}CH_2CH_3} + H_2O \xrightarrow[\Delta]{H^+} \underset{\text{Ethanol}}{2CH_3CH_2OH}$$

(5) **Reaction with PCl_5.** Ethers react with phosphorus pentachloride to give alkyl chlorides.

$$R-O-R' + PCl_5 \xrightarrow{\Delta} R-Cl + R'-Cl + POCl_3$$

$$\underset{\text{Diethyl ether}}{CH_3CH_2-O-CH_2CH_3} + PCl_5 \xrightarrow{\Delta} \underset{\text{Ethyl chloride}}{2CH_3CH_2Cl} + POCl_3$$

(6) **Cleavage by Ethers.** Ethers react with hot concentrated HI or HBr to give an alcohol and an alkyl halide.

$$\underset{\text{Ether}}{R-O-R'} + HX \xrightarrow{\Delta} \underset{\text{Alcohol}}{R-OH} + \underset{\text{Alkyl halide}}{R'X}$$

MECHANISM. The mechanism of the above reaction involves the following steps :

Step 1. Protonation of ether to form an oxonium ion.

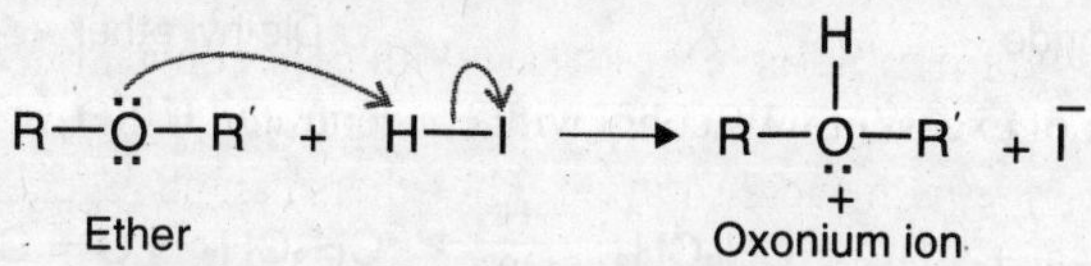

Step 2. Cleavage of the oxonium ion by S_N2 or S_N1 mechanism. S_N2 mechanism is favored if the alkyl halide is primary.

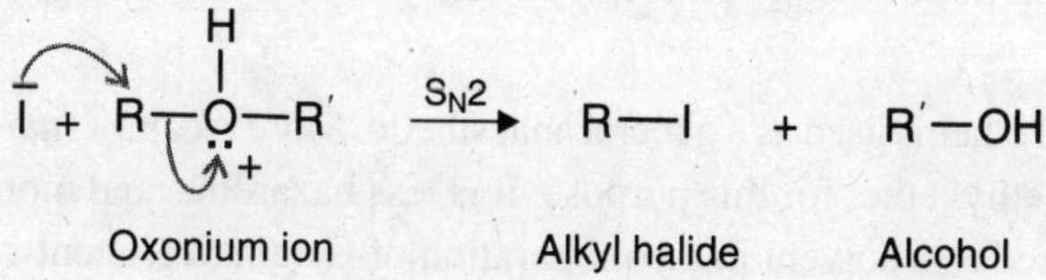

S_N1 mechanism is favored if the alkyl groups are secondary or tertiary.

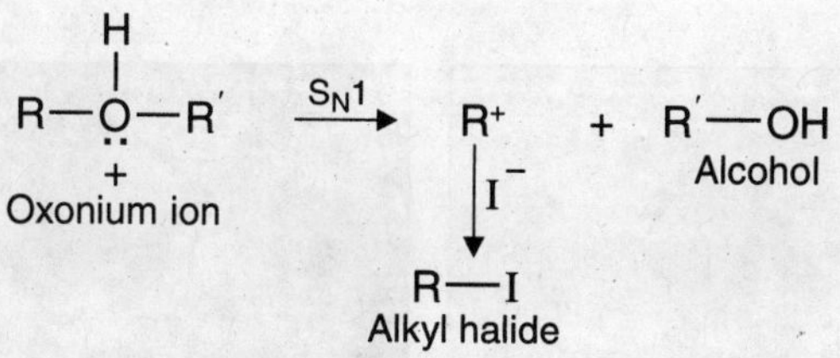

If an excess of acid is used, the alcohol initially produced is also converted to an alkyl halide by the reaction :

$$R'-OH + \underset{(excess)}{H-X} \longrightarrow R'-X + H_2O$$

For example,

$$\underset{\text{Diethyl ether}}{CH_3CH_2-O-CH_2CH_3} + HI \xrightarrow{\Delta} \underset{\text{Ethyl iodide}}{CH_3CH_2I} + \underset{\text{Ethanol}}{CH_3CH_2OH}$$

$$CH_3CH_2-O-CH_2CH_3 + \underset{(excess)}{2HI} \xrightarrow{\Delta} \underset{\text{Ethyl iodide}}{2CH_3CH_2I} + H_2O$$

(7) **Reaction with Acetyl Chloride.** Ethers react with acetyl chloride in the presence of anhydrous zinc chloride to form ethyl chloride and ethyl acetate.

$$\underset{\text{Diethyl ether}}{C_2H_5OC_2H_5} + \underset{\text{Acetyl chloride}}{CH_3-\overset{O}{\overset{\|}{C}}-Cl} \xrightarrow[\Delta]{ZnCl_2} \underset{\text{Ethyl chloride}}{C_2H_5Cl} + \underset{\text{Ethyl acetate}}{CH_3-\overset{O}{\overset{\|}{C}}-OC_2H_5}$$

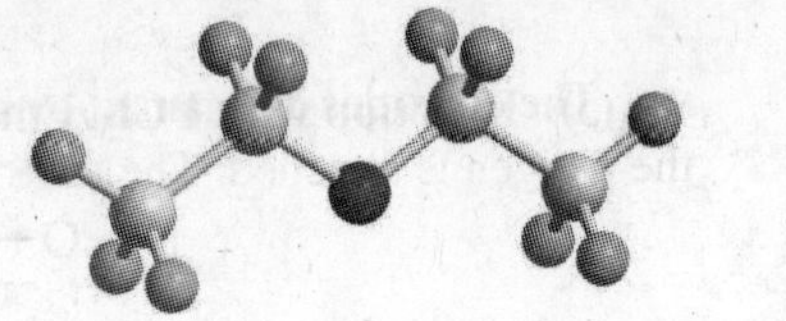

DIETHYL ETHER, *Ethyl ether*, $CH_3CH_2–O–CH_2CH_3$

Diethyl ether is the most important member of this class. It is often referred to as simply *Ether*.

Preparation. Diethyl ether may be obtained :

(1) By treating ethyl bromide with sodium ethoxide. (*Williamson Synthesis*)

$$\underset{\text{Sod ethoxide}}{CH_3CH_2\overset{-}{O}\overset{+}{Na}} + \underset{\text{Ethyl bromide}}{CH_3CH_2Br} \longrightarrow \underset{\text{Diethyl ether}}{CH_3CH_2—O—CH_2CH_3} + NaBr$$

(2) By heating ethyl chloride with dry silver oxide (Ag_2O).

$$\underset{\text{Ethyl chloride}}{2CH_3CH_2—Cl} + Ag_2O \longrightarrow \underset{\text{Diethyl ether}}{CH_3CH_2—O—CH_2CH_3} + 2AgCl$$

(3) By dehydration of excess ethyl alcohol with concentrated H_2SO_4 at 140°C.

$$\underset{\text{Ethyl alcohol (2 molecules)}}{CH_3CH_2—OH + HO—CH_2CH_3} \xrightarrow[140°C]{H^+} \underset{\text{Diethyl ether}}{CH_3CH_2—O—CH_2CH_3} + H_2O$$

The chemical properties of diethyl ether have already been discussed under the general reactions of ethers.

Uses. (1) Diethyl ether is used as a general anaesthetic. More recently, *halothane* (CF_3–CHClBr) has largely replaced diethyl ether for this purpose. It is less hazardous and more rapid in action ; and (2) Diethyl ether is used as a solvent in the preparation of Grignard reagents. It is also an excellent solvent for fats, waxes, oils, plastics, and lacquers.

An ether inhaler used by William Morton in the first public demonstration of ether anesthesia on October 16, 1846, at Massachusetts General Hospital.

EPOXIDES (OXIRANES)

Epoxides (or oxiranes) are cyclic ethers in which the ethereal oxygen is part of a three-membered ring.

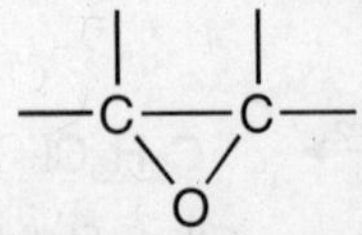

An epoxide

The epoxides are named by numbering the alkyl chain and indicating the two carbons to which the oxygen is attached. The prefix "epoxy" denotes this functional group.

$$\underset{\text{(1,2-Epoxyethane)}}{\underset{\text{Ethylene oxide}}{CH_2\!-\!CH_2 \text{ (bridged by O)}}} \qquad \underset{\text{(1,2-Epoxypropane)}}{\underset{\text{Propylene oxide}}{CH_3\!-\!CH\!-\!CH_2 \text{ (CH and } CH_2 \text{ bridged by O)}}}$$

ETHYLENE OXIDE, *Epoxyethane*, $(CH_2)_2O$

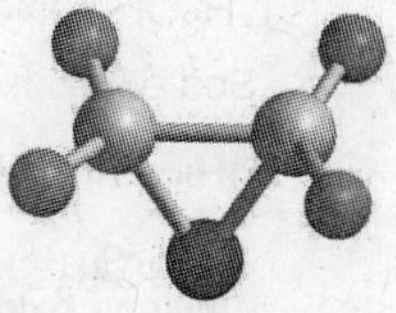

Ethylene oxide is the most important member of this class.

Preparation. Ethylene oxide may be obtained :

(1) By oxidation of ethylene with oxygen in the presence of silver catalyst.

$$\underset{\text{Ethylene}}{2CH_2{=}CH_2} + O_2 \xrightarrow[250°C]{Ag} \underset{\text{Ethylene oxide}}{2CH_2\!-\!CH_2 \text{ (bridged by O)}}$$

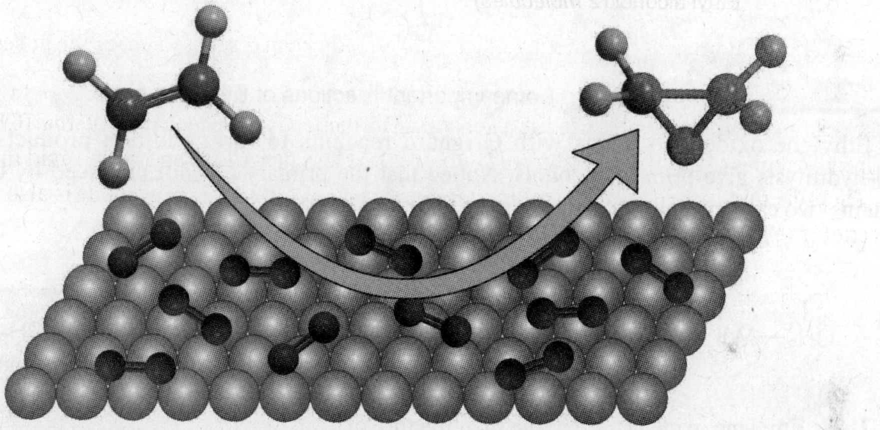

Ethylene reacts with oxygen in the presence of heated silver catalyst to form Ethylene oxide.

(2) By oxidation of ethylene with peracetic acid.

$$\underset{\text{Ethylene}}{CH_2{=}CH_2} + CH_3CO_3H \longrightarrow \underset{\text{Ethylene oxide}}{CH_2\!-\!CH_2 \text{ (bridged by O)}}$$

Properties. Ethylene oxide is a toxic, flammable gas, bp 14°C. It is a very reactive compound. Recall that a similar three-membered ring compound, cyclopropane, was highly reactive because of its strained configuration. Ethylene oxide, by the same reason, has a tendency to open up its ring. Thus oxygen atom combines with the reactive hydrogen atom of various compounds to form a hydroxy group, –OH, as shown below :

$$\underset{\text{Ethylene oxide}}{CH_2\!-\!CH_2 \text{ (bridged by O)}} + H\!-\!A \longrightarrow \underset{\text{Addition product}}{\underset{\displaystyle OH}{CH_2}\!-\!CH_2\!-\!A}$$

HA may be water (H–OH), alcohol (H–ÖR), ammonia (H–NH_2), or a halogen acid (H–Br). Some of the important reactions of ethylene oxide are shown in Fig. 17.2.

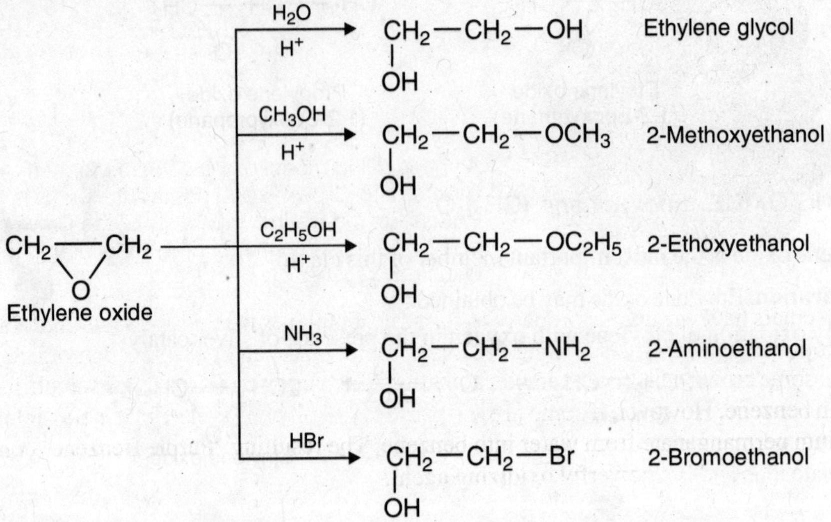

Fig. 17.2. Some important reactions of Ethylene oxide.

Ethylene oxide also reacts with Grignard reagents to form addition products which on acid-hydrolysis give primary alcohols. Notice that the primary alcohol produced by this method contains two carbons more than the starting Grignard reagent.

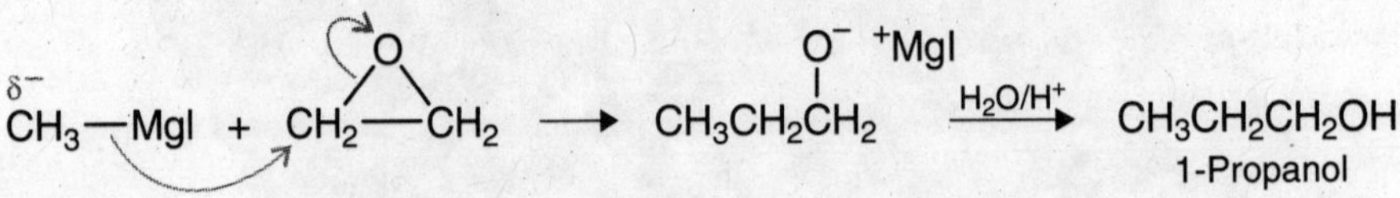

Uses. Ethylene oxide is used in the manufacture of :

(1) Nonionic detergents ; (2) Dacron ; (3) Ethylene glycol which is used as a permanent antifreeze; and (4) Ethyl cellosolve (2-ethoxyethanol), which is used as a solvent for quick-drying varnishes and enamels.

CROWN ETHERS

Crown ethers are cyclic polyethers. Examples are,

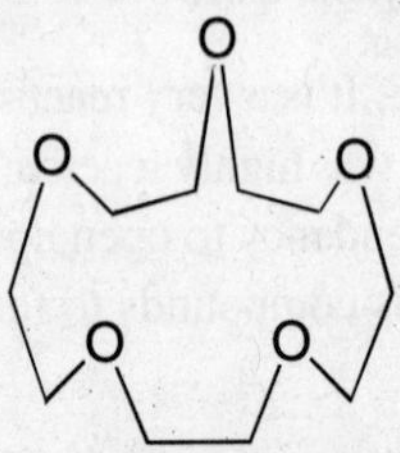

[15]Crown-5

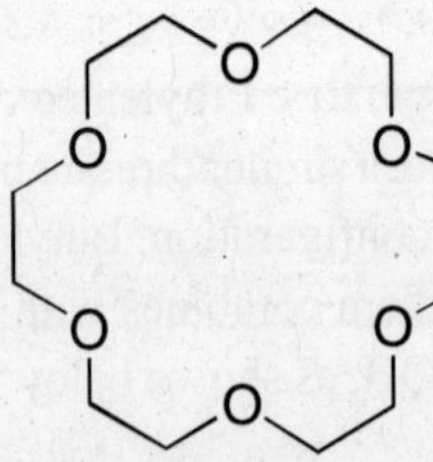

[18]Crown-6

These compounds are called **Crown Ethers** because of their crown-like shape. The first number in their common names indicates the ring size and the second number indicates the number of oxygen atoms.

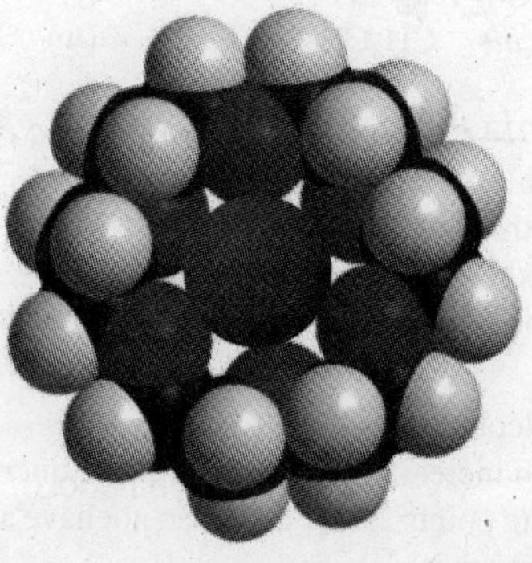

A space-filling model of the complex formed between 18-Crown-6 and potassium ion (K^+). Potassium ion fits into the cavity of the Crown ether where it is bound by acid-base interaction with the oxygens.

Crown ethers have the unique property of forming complexes with positive ions like Na^+, K^+, and Li^+. This complexing ability is so strong that ionic compounds can be dissolved in organic solvents that contain some crown ether. For example, potassium permanganate ($KMnO_4$) is soluble in water but insoluble in benzene. However, if some [18]Crown-6 is dissolved in benzene, it is possible to extract the potassium permanganate from water into benzene. The resulting "Purple Benzene" contains free permanganate ions and is a powerful oxidizing agent.

Points to Remember

1. **Alkoxy group** is an alkyl group attached to oxygen (RO–).
2. **Crown ether** is a large cyclic polyether.
3. **Cyclic ether** is an ether in which the oxygen atom is part of a heterocyclic ring.
4. **Epoxidation** is the conversion of an alkene into an epoxide/oxirane using a peracid.
5. **Epoxide** is a three-membered cyclic ether, also called an oxirane.
6. **Ether** is a family of compounds in which an oxygen atom is attached to two carbon atoms.
7. **Ether cleavage** is the cleaving of an ether into an alcohol and an alkyl halide (usually iodide) using hydrogen iodide.
8. **Oxirane** is a three-membered cyclic ether, also called an epoxide.
9. **Peracid** is a compound of general formula RCO_3H formed from a carboxylic acid by reaction with hydrogen peroxide.
10. **Polyether** is a compound containing many ether groups.
11. **Williamson synthesis** is a synthesis of ethers involving an S_N2 nucleophilic attack by an alkoxy group on an alkyl halide.

Examination Questions

1. Write the structural formulas of all the isomeric compounds that can be represented by the molecular formula $C_4H_{10}O$. Write their IUPAC names.

2. (*a*) How is diethyl ether prepared ?

(*b*) How does diethyl ether react with the following reagents ?

(*i*) O_2/long contact (*ii*) Cold conc. H_2SO_4 (*iii*) PCl_5 (*iv*) Hot conc. HI

3. (*a*) How is ethylene oxide prepared ?
(*b*) How does ethylene oxide react with the following reagents ?
(*i*) H_2O/H^+ (*ii*) HBr (*iii*) CH_3CH_2OH/H^+ (*iv*) NH_3

4. Write a note on : Williamson's Synthesis.
(*Magadh BSc, 2014 ; Tripura BSc, 2014 ; Banaras BSc, 2015 ; Orissa BSc, 2015*)

5. Write a note on : Cleavage of ethers by acids.

6. What happens when sodium ethoxide is heated with ethyl bromide ?

7. What happens when diethyl ether is heated with excess of concentrated HI ?

8. Write a step-by-step mechanism for the reaction of diethyl ether with hot concentrated HI.
(*Bhopal BSc, 2014 ; Kota BSc, 2015*)

9. Why do ethers boil so much lower than isomeric alcohols ?

Answer. Alcohols are capable of hydrogen bonding, which increases the strength of the intermolecular forces between the alcohol molecules, and hence raises their boiling points. Since ethers do not have a hydrogen bonded to their oxygen atom, they cannot hydrogen bond to one another.

10. Explain why the boiling point of ethyl alcohol is higher than that of diethyl ether though they have the same molecular weight. (*Anna BSc, 2014 ; Rohilikhand BSc, 2015*)

11. *n*-Butyl ether (bp 118°C) has a much higher boiling point than its isomer diethyl ether (bp 35°C). Explain (*Shivaji BSc, 2014 ; Madras BSc, 2015*)

Answer. Boiling point is raised by hydrogen bonding between like molecules of a compound. *n*-Butyl alcohol has a higher boiling point because its molecules can form hydrogen bonds to each other. Diethyl ether has got low boiling point because its molecules can not form hydrogen bonds to each other.

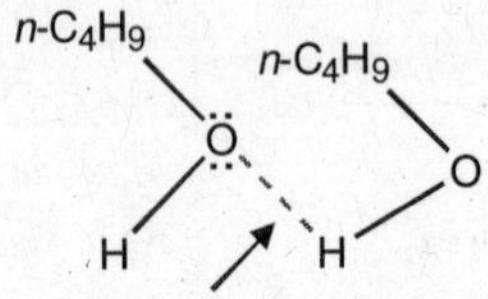

n-Butyl alcohol has a higher boiling point because of hydrogen bonding between its molecules.

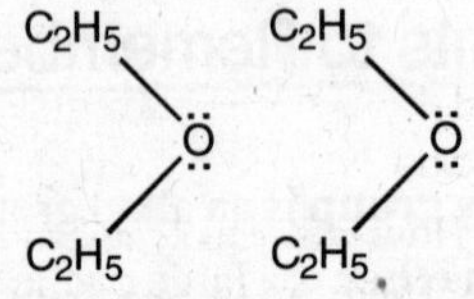

Diethyl ether has a low boiling point because it can not form hydrogen bonds between its molecules.

12. Arrange the following compounds in order of increasing boiling point :
(*a*) $HOCH_2CH_2OH$ (*b*) $CH_3OCH_2CH_2OH$
(*c*) $CH_3CH_2CH_2CH_3$ (*d*) $CH_3OCH_2CH_2OCH_3$

Answer. Lowest boiling point ⟶ Highest boiling point
(*c*) < (*d*) < (*b*) < (*a*)

13. Arrange the following molecules in order of their solubility in water.
(*a*) $CH_3OCH_2CH_2OCH_3$ (b) $CH_3CH_2CH_2CH_3$
(*c*) $CH_3CH_2OCH_2CH_3$ (d) $CH_3CH_2CH_2CH_2OCH_3$

Answer. From *lowest* to *highest* solubility: (*b*), (*c*), (*d*), (*a*)

14. Explain why epoxides are more reactive than ethers.

Answer. The ring atoms (**two** carbons and **one** oxygen) are sp^3 hybridized. The atoms are bonded to each other via σ bonds. The two sp^3 hybrid orbitals of oxygen are completely filled and contain the nonbonded electron pairs. The C–O–C bond angle of an epoxide must be 60°, a considerable deviation from the tetrahedral bond angle. For this reason, epoxides have angle strain, making them much more reactive than other ethers.

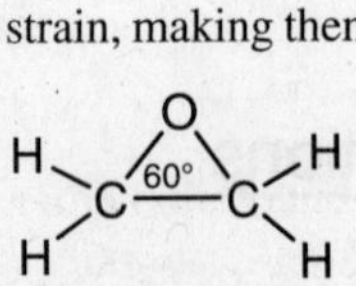

A strained, three-membered ring

15. How will you distinguish between diethyl ether and *n*-pentane ?
(*Baroda BSc, 2013 ; Dibrugarh BSc, 2014 ; Delhi BSc, 2015*)

Answer. Diethyl ether reacts with concentrated sulfuric acid to give a homogeneous solution containing the oxonium salt. *n*-Pentane does not react with sulfuric acid. In this case two distinct layers are formed.

$$\underset{\text{Diethyl ether}}{CH_3CH_2-\ddot{O}-CH_2CH_3} + H_2SO_4 \xrightarrow{\text{cold}} \underset{\substack{\text{Oxonium salt of diethyl ether}\\ \text{(soluble in } H_2SO_4)}}{\left[CH_3CH_2-\overset{H}{\underset{+}{O}}-CH_2CH_3\right] H\bar{S}O_4}$$

16. How will you distinguish between diethyl ether and 1-pentene ?

Answer. 1-Pentene decolorizes purple color of cold, dilute $KMnO_4$ solution. Diethyl ether does not react with this reagent.

$$\underset{\text{1-Pentene}}{CH_3CH_2CH_2CH{=}CH_2} + \underset{\text{(purple)}}{KMnO_4} \xrightarrow{H_2O} \underset{\text{1,2-Pentanediol}}{CH_3CH_2CH_2\overset{OH}{CH}-\overset{OH}{CH_2}} + MnO_2 + KOH$$

17. How will you distinguish between diethyl ether and ethyl alcohol ?

(Bihar BSc, 2013 ; Ranchi BSc, 2014 ; Kalyani BSc, 2015)

Answer. Ethyl alcohol reacts with sodium to produce hydrogen gas (seen as bubbles). Diethyl ether does not react with sodium.

$$\underset{\text{Ethyl alcohol}}{CH_3CH_2OH} + Na \longrightarrow \underset{\text{Sodium ethoxide}}{CH_3CH_2\bar{O}\overset{+}{Na}} + H_2\uparrow$$

18. How will you synthesize diethyl ether from ethyl iodide ?

Answer. Following steps are involved :

$$\underset{\text{Ethyl iodide}}{CH_3CH_2I} \xrightarrow{\text{aq. NaOH}} CH_3CH_2OH \xrightarrow{Na} CH_3CH_2\bar{O}\overset{+}{Na} \xrightarrow{CH_3CH_2I} \underset{\text{Diethyl ether}}{CH_3CH_2OCH_2CH_3}$$

19. How will you synthesize diethyl ether from ethyl alcohol ? *(Tamil Nadu BSc, 2014)*

Answer. There are two methods :

Method 1. *Via* acid-catalyzed dehydration of ethyl alcohol at 140°C.

$$2CH_3CH_2OH \xrightarrow[140°C]{H_2SO_4} CH_3CH_2-O-CH_2CH_3 + H_2O$$

Method 2. *Via* Williamson's synthesis.

$$\underset{\text{Ethyl alcohol}}{CH_3CH_2OH} \xrightarrow{Na} CH_3CH_2\bar{O}\overset{+}{Na}$$

$$CH_3CH_2OH \xrightarrow{HBr} CH_3CH_2Br$$

$$CH_3CH_2\bar{O}\overset{+}{Na} + CH_3CH_2Br \longrightarrow \underset{\text{Diethyl ether}}{CH_3CH_2OCH_2CH_3}$$

20. How will you synthesize *n*-butyl alcohol from ethylene ? *(Berhampur BSc, 2015)*

Answer. Following steps are involved :

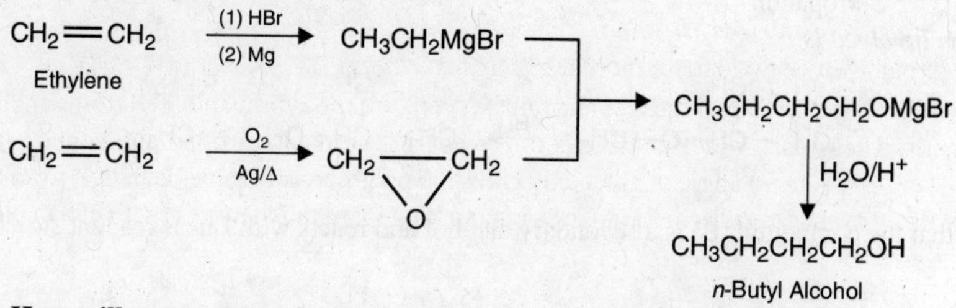

21. How will you synthesize *t*-butyl ethyl ether from *t*-butyl alcohol ?

Answer. Following steps are involved :

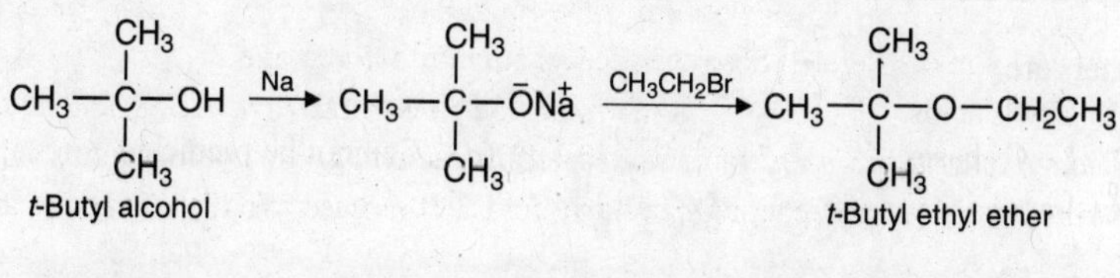

22. How will you synthesize 2-cyclohexylethanol from cyclohexanol ?

Answer. Following steps are involved :

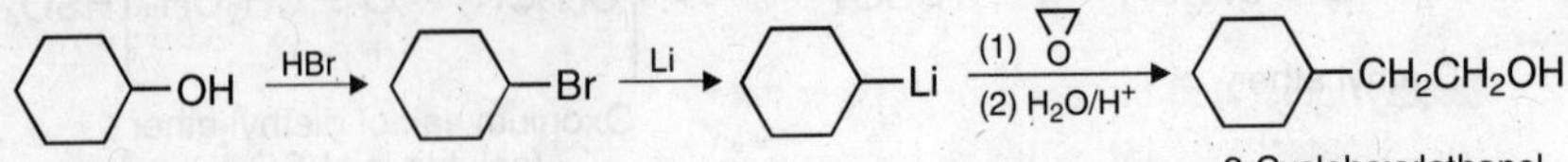

Cyclohexanol 2-Cyclohexylethanol

23. Compound (A), $C_4H_{10}O$, is found to be soluble in sulfuric acid. (A) does not react with sodium or potassium permanganate. When (A) is heated with excess of hydriodic acid, it is converted into a single alkyl iodide. What is the structural formula of (A) ?

Answer. (*A*) is Diethyl ether. (*A*) can not be an alcohol, since alcohols react with sodium and potassium permanganate.

Reactions Involved Are :

$$CH_3CH_2-O-CH_2CH_3 \xrightarrow{H_2SO_4} \left[CH_3CH_2-\overset{+}{\underset{H}{\ddot{O}}}-CH_2CH_3\right]\overline{HSO_4}$$

Soluble oxonium salt

(A) $C_4H_{10}O$

$$CH_3CH_2-O-CH_2CH_3 \xrightarrow{2HI/\Delta} 2CH_3CH_2I$$

24. A compound of molecular formula $C_4H_{10}O_3$ exhibited chemical behavior typical of both an alcohol and an ether. Its reaction with excess hot hydrogen iodide produced 1,2-diiodoethane as the only organic product. What is its structure ?

Answer. The structure of the unknown is : $HOCH_2CH_2OCH_2CH_2OH$

Reaction Involved Is :

$$HOCH_2CH_2OCH_2CH_2OH + \text{excess } HI \xrightarrow{\Delta} 2\ ICH_2CH_2I$$

1,2-Diiodoethane

Notice that both –OH groups of the unknown react with concentrated HI. The ether linkage also undergoes cleavage with excess concentrated HI.

25. Compound (A) ($C_4H_{10}O$) was insoluble in water and did not react with sodium metal. It did react with HI and produced methyl iodide and compound (B), which reacted with sodium metal. Compound (B) reacted with Lucas reagent in 5 minutes. What are the structures of (A) and (B) ?

Answer. The unknown compounds are :

(A) = 2-Methoxypropane
(B) = 2-Propanol

Reaction Involved Is :

$$\underset{(A)}{CH_3-\overset{CH_3}{\overset{|}{CH}}-O-CH_3} \xrightarrow{HI} \underset{(B)}{CH_3-\overset{CH_3}{\overset{|}{CH}}-OH} + CH_3I$$

Notice that the compound (B) is a secondary alcohol and reacts with Lucas reagent slowly.

Multiple Choice Questions

1. Ethers are

(*a*) Lewis acids (*b*) Neutral
(*c*) Lewis bases (*d*) Can not be predicted

Answer. (*c*)

2. Ethanol is isomeric with
(*a*) Dimethyl ether (*b*) Ethanal
(*c*) Diethyl ether (*d*) Propanone

Answer. (*a*)

3. Diethyl ether and methyl propyl ether are
(*a*) Conformational isomers (*b*) Metamers
(*c*) Geometrical isomers (*d*) Enantiomers

Answer. (*b*)

4. The reaction of a sodium alkoxide with an alkyl halide is called
(*a*) Wurtz-Fittig reaction (*b*) Perkin reaction
(*c*) Williamson's synthesis (*d*) Aldol condensation

Answer. (*c*)

5. Ethyl bromide reacts with sodium methoxide to form
(*a*) Diethyl ether (*b*) Ethyl methyl ether
(*c*) Dimethyl ether (*d*) *n*-Propyl alcohol

Answer. (*b*)

6. Ethyl alcohol reacts with concentrated H_2SO_4 at 140°C to form
(*a*) Acetone (*b*) Ethylene
(*c*) Diethyl ether (*d*) Acetic acid

Answer. (*c*)

7. Ethers are kept in brown bottles because
(*a*) Brown bottles are cheaper than colorless clear bottles
(*b*) Ethers absorb moisture
(*c*) Ethers evaporate readily
(*d*) Ethers are oxidized to explosive peroxides

Answer. (*d*)

8. Ethers are stored in brown bottles. This is because on exposure to air and light ethers are converted to
(*a*) Peroxides (*b*) Oxonium ions
(*c*) Ozonides (*d*) Electrophiles

Answer. (*a*)

9. Ethers can be freed from peroxides by treatment with
(*a*) Ferrous salt (*b*) Sodium carbonate
(*c*) Ferric salt (*d*) Sodium bicarbonate

Answer. (*c*)

10. Which of the following statements is false ?
(*a*) Diethyl ether has been used as a general anesthetic
(*b*) Ethyl alcohol is present in all alcoholic beverages
(*c*) Methyl alcohol is produced by fermentation of sugars
(*d*) Ethylene glycol is a common antifreeze for automobiles

Answer. (*c*)

11. Ethers react with cold concentrated H_2SO_4 to form
(*a*) Oxonium salts (*b*) Alkenes
(*c*) Alkoxides (*d*) Zwitterions

Answer. (*a*)

12. Which of the following reagents readily react with ethyl methyl ether ?
(*a*) NaOH (*b*) Conc HI (*c*) $KMnO_4$ (*d*) H_2O

Answer. (*b*)

13. Diethyl ether reacts with sodium metal to give
(*a*) Ethanol (*b*) Sodium ethoxide
(*c*) Ethanal (*d*) Nothing happens

Answer. (*d*)

14. Diethyl ether on heating with excess concentrated HI gives
(*a*) Methyl iodide (*b*) Isopropyl iodide
(*c*) Ethyl iodide (*d*) *n*-Propyl iodide

Answer. (*c*)

15. Diethyl ether reacts with excess of hot concentrated HI to form
(*a*) $CH_3CH_2I + CH_3CH_2OH$ (*b*) Only CH_3CH_2OH
(*c*) $CH_3CH_2OH + CH_2{=}CH_2$ (*d*) Only CH_3CH_2I

Answer. (*d*)

16. Which is the best reagent to accomplish the following conversion ?

$$CH_3CH_2OCH_2CH_3 \xrightarrow{?} CH_3CH_2Br$$

(*a*) Br_2 in CCl_4 (*b*) NaBr
(*c*) Br_2 in H_2O (*d*) Conc HBr

Answer. (*d*)

17. Which of the following pairs of compounds will not form hydrogen bonds with each other ?
(*a*) CH_3OH and CH_3CH_2OH
(*b*) CH_3SH and CH_3CH_2SH
(*c*) CH_3OCH_3 and $CH_3CH_2OCH_2CH_3$
(*d*) CH_3COOH and H_2O

Answer. (*c*)

18. The compound with the lowest boiling point is
(*a*) H_2O (*b*) CH_3CH_2OH
(*c*) CH_3OCH_3 (*d*) $CH_3CH_2CH_3$

Answer. (*d*)

19. Which of the following functional group series is ranked according to increasing boiling points ?
(*a*) diethyl ether, ethane, ethanol, ethanethiol
(*b*) ethane, ethanol, diethyl ether, ethanethiol
(*c*) ethane, diethyl ether, ethanethiol, ethanol
(*d*) diethyl ether, ethane, ethanethiol, ethanol

Answer. (*c*)

20. The following compounds have identical molecular weights. Which would have the lowest boiling point ?
(*a*) 1-Methoxypropane (*b*) 1-Butanol
(*c*) 1,1-Dimethylethanol (*d*) 2-Butanol

Answer. (*a*)

21. Cyclic ethers with three-membered ring are called
(*a*) Lactones (*b*) Oxiranes (*c*) Alkoxides (*d*) Epoxy resins

Answer. (*b*)

22. Ethylene reacts with oxygen in the presence of silver catalyst at 300°C to form
(*a*) Diethyl ether (*b*) Ethylene glycol
(*c*) Ethyl alcohol (*d*) Ethylene oxide

Answer. (*d*)

23. Ethylene oxide reacts with HBr to give
(*a*) 1-Bromoethanol (*b*) Ethyl bromide
(*c*) 2-Bromoethanol (*c*) Ethylene glycol

Answer. (*c*)

24. Acid-catalyzed hydrolysis of ethylene oxide yields
(*a*) CH_3CH_2OH (*b*) $HOCH_2CH_2OH$
(*c*) $CH_3CH_2CH_2OH$ (*d*) $HOCH_2CH_2CH_2OH$

Answer. (*b*)

25. The product of the reaction of ethylene oxide with acidic methanol is
(*a*) $CH_3OCH_2CH_2OH$ (*b*) $CH_3CH_2CH_2CH_2OH$
(*c*) $HOCH_2CH_2OH$ (*d*) $CH_3CH_2OCH_2CH_3$

Answer. (*a*)

26. Ethylene oxide reacts with ammonia to give
(*a*) 1-Aminoethanol (*b*) Ethylamine
(*c*) 2-Aminoethanol (*d*) Acetamide

Answer. (*c*)

18

Thiols and Thioethers

(Mercaptans and Sulfides)

Thiols and thioethers are responsible for the pungent odors of onions and garlic.

Since sulfur is directly below oxygen in Group VI of the periodic table, there are sulfur counterparts of alcohols and ethers. The sulfur analogs of alcohols are called *Thiols, Mercaptans, or Alkyl Hydrogen Sulfides.* The sulfur analogs of ethers are called *Thioethers or Sulfides.*

R—OH	R—SH	R—O—R	R—S—R
Alcohol	Thiol	Ether	Thioether

THIOLS OR MERCAPTANS

Just as alcohols are regarded as alkyl derivatives of water, thiols could be considered as derived from hydrogen sulfide (H_2S).

$$\text{H—O—H} \xrightarrow[+\text{R}]{-\text{H}} \underset{\text{Alcohol}}{\text{R—O—H}} \qquad \Big\| \qquad \text{H—S—H} \xrightarrow[+\text{R}]{-\text{H}} \underset{\text{Thiol}}{\text{R—S—H}}$$

The –SH group is the functional group in thiols. It is called **Sulfhydryl, Mercapto,** or **Thiol group**. Like H_2S, thiols are weakly acidic. They react with mercuric ions to form insoluble salts. Therefore, thiols were given the name **Mercaptans** (Latin, *mercurium captans* = mercury catching).

Points to Remember

1. The letter *thia* or *thio* in any name imply the presence of sulfur.
2. Thiols and thioethers are responsible for the pungent odors of onions and garlic. Several thiols give the skunk its signature odor. This property of thiols is used in a positive way to make natural gas and LPG safe for household use. The hydrocarbons (such as methane and propane) in these gases are odorless and extremely flammable. Because the presence of these hydrocarbons cannot be detected by the human nose, suppliers add trace amounts of low molecular weight thiols to their gas products. Leaks can then be detected by smell before there is a concentration large enough to ignite and cause a fire or explosion.

STRUCTURE OF THIOLS

Let us consider methanethiol (CH_3SH) for illustrating the orbital make up of thiols. In methanethiol both sulfur and carbon are sp^3 hybridized. Two of the sp^3 orbitals of sulfur are completely filled and cannot take part in bond formation (Fig.18.1).

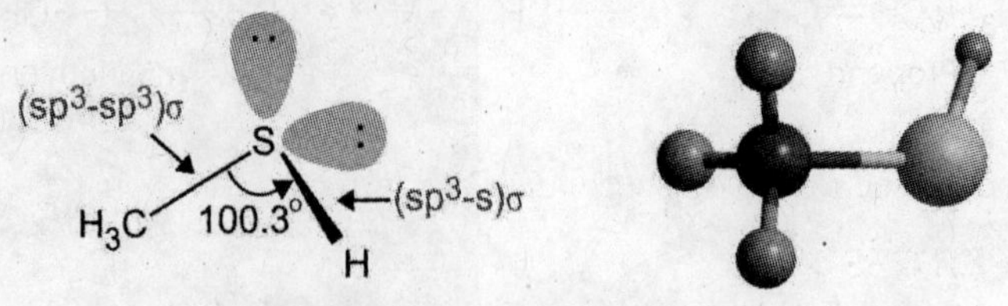

Fig.18.1 Structure of Methanethiol.

The C–S bond in methanethiol is formed by overlap of an sp^3 orbital of carbon and an sp^3 orbital of sulfur. The S–H bond is formed by overlap of an sp^3 orbital of sulfur and *s* orbital of hydrogen. The C–S–H bond angle is 100.3°. It is less than the normal tetrahedral angle. This is because the two completely filled sp^3 orbitals of sulfur repel each other. This results in reduction of the bond angle.

NOMENCLATURE

The members of this class are commonly named as *Alkyl mercaptans*. Their IUPAC names are obtained by adding the suffix *thiol* to the name of the corresponding alkane. For example,

	Common Name	**IUPAC Name**
CH_3SH	Methyl mercaptan	Methanethiol
CH_3CH_2SH	Ethyl mercaptan	Ethanethiol
$CH_3CH_2CH_2SH$	*n*-Propyl mercaptan	1-Propanethiol

METHODS OF PREPARATION

Thiols are obtained :

(1) By heating alkyl halides with potassium hydrosulfide (KSH).

$$\underset{\text{Ethyl iodide}}{CH_3CH_2I} + KSH \xrightarrow{\Delta} \underset{\text{Ethanethiol}}{CH_3CH_2SH} + KI$$

(2) By heating alkyl halides with thiourea followed by basic-hydrolysis.

$$\underset{\text{Ethyl bromide}}{CH_3CH_2Br} + \underset{\text{Thiourea}}{S{=}C(NH_2){-}NH_2} \xrightarrow{\Delta} \underset{\text{Ethylisothiourea Salt}}{\left[CH_3CH_2{-}\overset{+}{S}{=}C(NH_2){-}NH_2\right]Br^-}$$

$$\xrightarrow[NaOH]{H_2O} \underset{\text{Ethanethiol}}{CH_3CH_2SH} + \underset{\text{Urea}}{O{=}C(NH_2){-}NH_2}$$

(3) By reaction of Grignard reagents with sulfur followed by acidic-hydrolysis.

$$\underset{\text{Ethylmagnesium bromide}}{CH_3CH_2MgBr} \xrightarrow{S} CH_3CH_2SMgBr \xrightarrow{H_2O/H^+} \underset{\text{Ethanethiol}}{CH_3CH_2SH}$$

(4) By passing a mixture of alcohol vapor and hydrogen sulfide over a thoria catalyst at 400°C.

$$\underset{\text{Ethanol}}{CH_3CH_2OH} + H_2S \xrightarrow[\Delta]{ThO_2} \underset{\text{Ethanethiol}}{CH_3CH_2SH} + H_2O$$

(5) By addition of hydrogen sulfide to alkenes in the presence of sulfuric acid catalyst. Markovnikov rule is followed.

$$\underset{\text{Propene}}{CH_3-CH{=}CH_2} + H-SH \xrightarrow[\Delta]{H_2SO_4} \underset{\substack{\text{2-Propanethiol}\\ (2^\circ\ \text{Thiol})}}{CH_3-\overset{\overset{\displaystyle SH}{|}}{CH}-CH_3}$$

This method is used for making secondary and tertiary thiols.

PHYSICAL PROPERTIES

(1) Methanethiol is a gas. Ethanethiol and higher members are colorless volatile liquids.

(2) Lower members have strong repulsive odors. For example, methanethiol and ethanethiol are added to natural gas in minute amounts to make gas leaks detectable by smell.

(3) Boiling points of thiols are much lower than those of the corresponding alcohols. This is because of the absence of hydrogen bonding in thiols. For Example,

Thiol	bp°C	Alcohol	bp°C
CH_3SH	6	CH_3OH	56
CH_3CH_2SH	35	CH_3CH_2OH	78
$CH_3CH_2CH_2SH$	68	$CH_3CH_2CH_2OH$	98

(4) Thiols are insoluble in water. However, they are soluble in ethanol and ether.

(5) IR SPECTRUM. Thiols show a weak S—H stretching band at 2550-2590 cm^{-1}.

CHEMICAL PROPERTIES

Thiols resemble alcohols in most of their chemical properties ; the main difference is their behavior towards oxidizing agents.

(1) **Reaction with Sodium**. Thiols react with sodium to form salts known as mercaptides.

$$\underset{\text{Ethanethiol}}{2CH_3CH_2SH} + Na \longrightarrow \underset{\text{Sodium ethyl mercaptide}}{2CH_3CH_2\overset{-}{S}\overset{+}{Na}} + H_2\uparrow$$

(2) **Reaction with Acids and Acid Halides**. Thiols react with acids and acid halides to form thioesters.

$$\underset{\text{Acetic acid}}{CH_3-\overset{\overset{\displaystyle O}{\|}}{C}-OH} + \underset{\text{Ethanethiol}}{HSC_2H_5} \longrightarrow \underset{\text{Ethyl thioacetate}}{CH_3-\overset{\overset{\displaystyle O}{\|}}{C}-SC_2H_5} + H_2O$$

$$\underset{\text{Acetyl chloride}}{CH_3-\overset{\overset{\displaystyle O}{\|}}{C}-Cl} + HSC_2H_5 \longrightarrow \underset{\text{Ethyl thioacetate}}{CH_3-\overset{\overset{\displaystyle O}{\|}}{C}-SC_2H_5} + HCl$$

(3) **Reaction with Aldehydes and Ketones**. Thiols react with aldehydes and ketones in the presence of HCl to form mercaptals (thioacetals) and mercaptols (thioketals) respectively.

$$\underset{\text{Acetaldehyde}}{CH_3-\overset{\overset{\displaystyle O}{\|}}{C}-H} + 2C_2H_5SH \xrightarrow{HCl} \underset{\text{Diethyl methyl mercaptal}}{CH_3-\overset{\overset{\displaystyle SC_2H_5}{|}}{\underset{\underset{\displaystyle SC_2H_5}{|}}{C}}-H} + H_2O$$

$$\underset{\text{Acetone}}{CH_3-\overset{\overset{\displaystyle O}{\|}}{C}-CH_3} + 2C_2H_5SH \xrightarrow{HCl} \underset{\text{Diethyl dimethyl mercaptol}}{CH_3-\overset{\overset{\displaystyle SC_2H_5}{|}}{\underset{\underset{\displaystyle SC_2H_5}{|}}{C}}-CH_3} + H_2O$$

Diethyl dimethyl mercaptol on oxidation with potassium permanganate gives **Sulfonal** which is an important hypnotic.

$$\underset{\text{Diethyl dimethyl mercaptol}}{CH_3-\overset{\overset{\displaystyle SC_2H_5}{|}}{\underset{\underset{\displaystyle SC_2H_5}{|}}{C}}-CH_3} + 4[O] \longrightarrow \underset{\text{Sulfonal}}{CH_3-\overset{\overset{\displaystyle SO_2C_2H_5}{|}}{\underset{\underset{\displaystyle SO_2C_2H_5}{|}}{C}}-CH_3}$$

(4) **Reaction with Bases**. Thiols are more acidic than alcohols. They react with bases to form salts.

$$\underset{\text{Ethanethiol}}{CH_3CH_2SH} + NaOH \longrightarrow \underset{\text{Sodium ethyl mercaptide}}{CH_3CH_2\overset{-}{S}\overset{+}{N}a} + H_2O$$

(5) **Reaction with Metallic Salts and Oxides**. Thiols react with metallic salts and oxides to form salts.

$$\underset{\text{Ethanethiol}}{2C_2H_5SH} + \underset{\text{Lead acetate}}{(CH_3COO)_2Pb} \longrightarrow \underset{\text{Lead diethyl mercaptide}}{(C_2H_5S)_2Pb} + 2CH_3COOH$$

$$\underset{\text{Ethanethiol}}{2C_2H_5SH} + HgO \longrightarrow \underset{\text{Mercury diethyl mercaptide}}{(C_2H_5S)_2Hg} + H_2O$$

(6) **Oxidation**. The most significant difference in the chemical behavior of thiols and alcohols is the ease with which thiols are oxidized.

(*a*) Mild oxidizing agents such as halogens or hydrogen peroxide convert thiols to disulfides.

$$\underset{\text{Ethanethiol}}{2C_2H_5SH} + I_2 \longrightarrow \underset{\text{Diethyl disulfide}}{C_2H_5-S-S-C_2H_5} + 2HI$$

$$2C_2H_5SH + H_2O_2 \longrightarrow \underset{\text{Diethyl disulfide}}{C_2H_5-S-S-C_2H_5} + 2H_2O$$

The thiol-disulfide interconversion is extremely important in biochemistry, where disulfide bridges form the cross-links between protein chains that help stabilize the three–dimensional conformations of proteins.

$$\text{Protein}-SH + HS-\text{Protein} \longrightarrow \underset{\text{A cross-linked protein}}{\text{Protein}-S-S-\text{Protein}}$$

(*b*) Strong oxidizing agents such as potassium permanganate or conc. nitric acid convert thiols into sulfonic acids.

$$\underset{\text{Ethanethiol}}{C_2H_5SH} + 3[O] \xrightarrow[\text{or } HNO_3]{KMnO_4} \underset{\text{Ethanesulfonic acid}}{C_2H_5-SO_3H}$$

THIOETHERS OR SULFIDES

Thioethers are sulfur analogs of ethers. Just as ethers are regarded as derivatives of water, thioethers could be considered as derived from hydrogen sulfide (H_2S).

$$\underset{\text{Water}}{H{-}O{-}H} \xrightarrow[+2R]{-2H} \underset{\text{Ether}}{R{-}O{-}R} \quad \Big\| \quad \underset{\text{Hydrogen sulfide}}{H{-}S{-}H} \xrightarrow[+2R]{-2H} \underset{\text{Thioether } (Sulfide)}{R{-}S{-}R}$$

The **Sulfide Group** (–S–) is the functional group in thioethers.

NOMENCLATURE

(1) **Common System**. The common names of thioethers are obtained by writing the names of alkyl groups followed by the word *sulfide*.

$$\underset{\text{Dimethyl sulfide}}{CH_3{-}S{-}CH_3} \qquad \underset{\text{Ethyl methyl sulfide}}{CH_3{-}S{-}CH_2CH_3}$$

(2) **IUPAC System**. In this system thioethers are named as *Alkylthioalkanes*. Smaller alkyl group bonded to the sulfur atom is regarded as the substituent.

$$\underset{\text{1-Methylthiopropane}}{CH_3{-}S{-}CH_2CH_2CH_3} \qquad \underset{\text{2-Methylthiobutane}}{CH_3{-}\overset{\overset{SCH_3}{|}}{C}H{-}CH_2CH_3}$$

STRUCTURE OF THIOETHERS

Let us consider dimethyl sulfide (CH_3SCH_3) for illustrating the orbital make up of thioethers. In dimethyl sulfide both sulfur and carbon atoms are sp^3 hybridized. Two of the sp^3 orbitals of sulfur are completely filled and cannot take part in bond formation (Fig.18.2). The C–S bonds in dimethyl sulfide are formed by overlap of an sp^3 orbitals of carbons and an sp^3 orbitals of sulfur. The C–S–C bond angle is 99°. It is less than the normal tetrahedral angle. This is because the two completely filled sp^3 orbitals of sulfur repel each other. This results in reduction of bond angle.

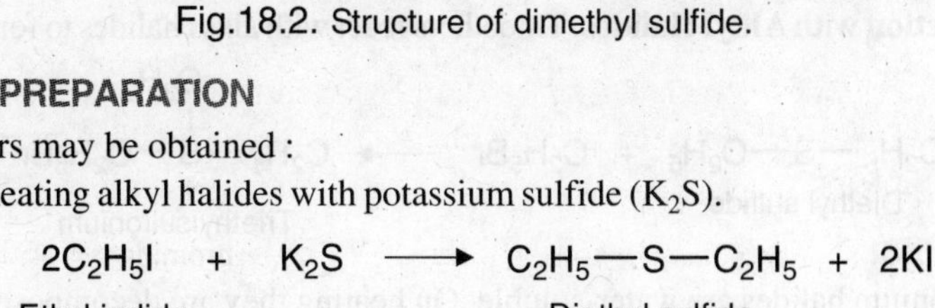

Fig.18.2. Structure of dimethyl sulfide.

METHODS OF PREPARATION

Thioethers may be obtained :

(1) By heating alkyl halides with potassium sulfide (K_2S).

$$\underset{\text{Ethyl iodide}}{2C_2H_5I} + K_2S \longrightarrow \underset{\text{Diethyl sulfide}}{C_2H_5{-}S{-}C_2H_5} + 2KI$$

(2) By heating alkyl halides with sodium mercaptides (*Cf. Williamson's ether synthesis*).

$$\underset{\text{Sodium ethyl mercaptide}}{C_2H_5\overset{-}{S}\overset{+}{Na}} + CH_3I \longrightarrow \underset{\text{Ethyl methyl sulfide}}{C_2H_5{-}S{-}CH_3} + NaI$$

(3) By heating ethers with phosphorus pentasulfide (P_2S_5).

$$\underset{\text{Diethyl ether}}{5C_2H_5{-}O{-}C_2H_5} + P_2S_5 \longrightarrow \underset{\text{Diethyl sulfide}}{5C_2H_5{-}S{-}C_2H_5} + P_2O_5$$

(4) By passing thiols over a mixture of alumina and zinc sulfide at 300°C.

$$\underset{\text{Ethanethiol}}{2C_2H_5SH} \xrightarrow[300^\circ C]{Al_2O_3 + ZnS} \underset{\text{Diethyl sulfide}}{C_2H_5{-}S{-}C_2H_5} + H_2S$$

PHYSICAL PROPERTIES

(1) Thioethers are unpleasant-smelling, colorless oily liquids. Their boiling points are higher than those of corresponding ethers.

Thioether	bp°C	Ether	bp°C
CH_3SCH_3	37	CH_3OCH_3	24
$C_2H_5SC_2H_5$	92	$C_2H_5OC_2H_5$	35

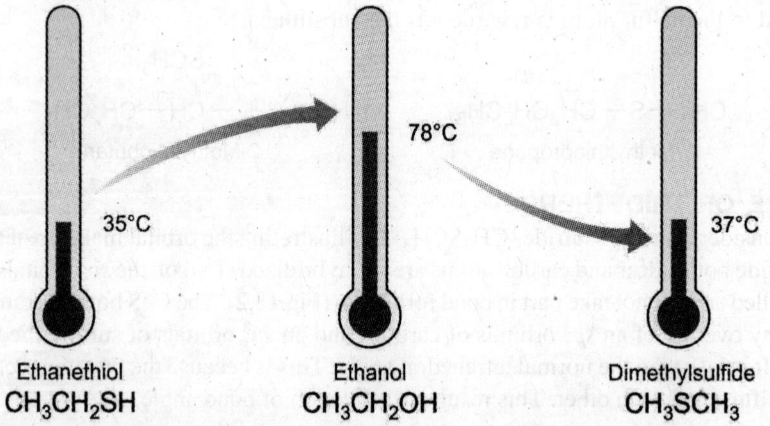

Boiling points of Ethanethiol, Ethanol and Dimethyl sulfide

(2) Thioethers are insoluble in water but soluble in ethanol and ether.

(3) **IR Spectrum.** Thioethers show a weak C—S stretching band at 600–800 cm^{-1}.

CHEMICAL PROPERTIES

Some of the important reactions of thioethers are described below :

(1) **Reaction with Alkyl Halides.** Thioethers react with alkyl halides to form sulfonium salts.

$$\underset{\text{Diethyl sulfide}}{C_2H_5{-}S{-}C_2H_5} + C_2H_5Br \longrightarrow \underset{\text{Triethylsulfonium bromide}}{C_2H_5{-}\overset{\overset{C_2H_5}{|}}{\underset{+}{S}}{-}C_2H_5Br^-}$$

The sulfonium halides are water-soluble. On heating they are decomposed to thioether and alkyl halide. These salts react with moist silver oxide (AgOH) to form strongly basic sulfonium hydroxides.

$$(C_2H_5)_3\overset{+}{S}\overset{-}{Br} + AgOH \longrightarrow \underset{\text{Triethylsulfonium hydroxide}}{(C_2H_5)_3\overset{+}{S}\overset{-}{O}H} + AgBr$$

(2) **Reaction with Halogens**. Thioethers react with halogens to give dihalides.

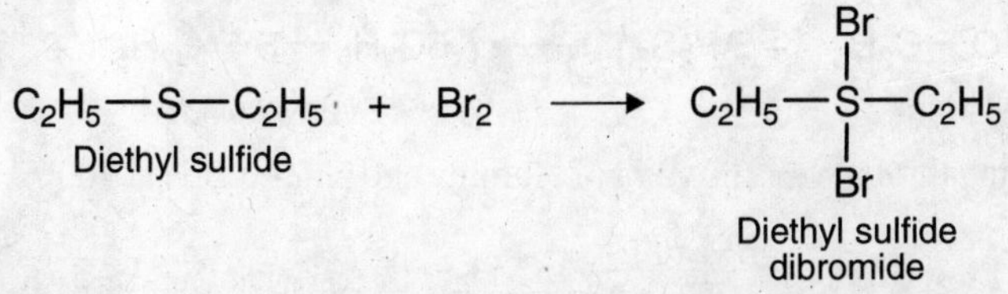

(3) **Hydrolysis**. Thioethers undergo hydrolysis with aqueous NaOH or KOH to form H_2S and alcohols.

$$\underset{\text{Diethyl sulfide}}{C_2H_5—S—C_2H_5} + 2H_2O \xrightarrow{\bar{O}H} \underset{\text{Ethanol}}{2C_2H_5OH} + H_2S$$

(4) **Oxidation**. Thioethers are easily oxidized.

(*a*) Mild oxidizing agents such as dilute nitric acid or hydrogen peroxide at 25°C (room temperature) convert thioethers into sulfoxides.

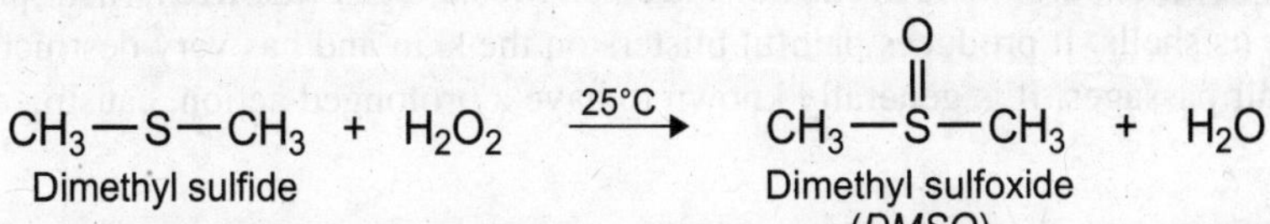

Dimethyl sulfoxide (*mp* 10°C ; *bp* 189°C) is a valuable solvent for both polar and nonpolar organic compounds. It has remarkable ability to penetrate the skin, carrying along with it whatever impurities are present. It must be handled very cautiously.

(*b*) Strong oxidizing agents such as potassium permanganate or hydrogen peroxide at 100ºC convert thioethers into sulfones.

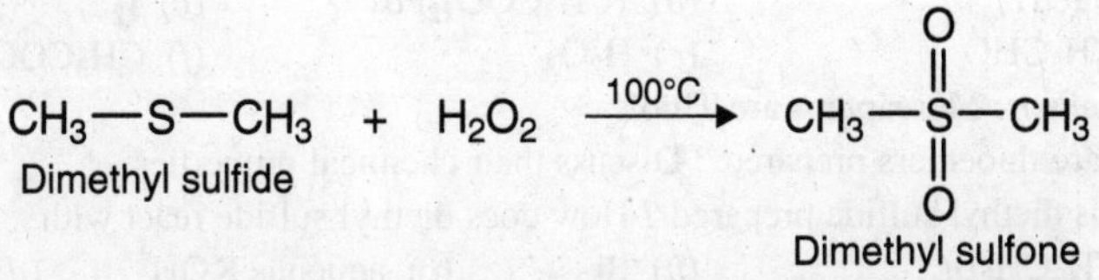

The sulfones are crystalline compounds and are often used for the identification of a liquid thioether.

MUSTARD GAS, *2,2'-Dichlorodiethyl Sulfide*, $(ClCH_2CH_2)_2S$

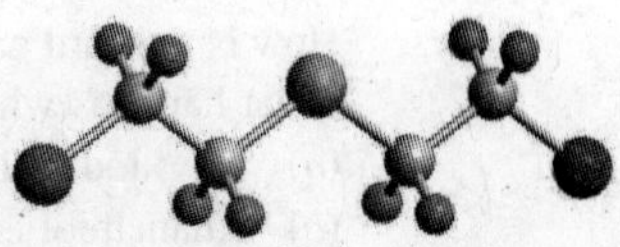

It was used as a poison gas in World War I.

Preparation. Mustard gas may be obtained :

(1) From ethylene by the following steps.

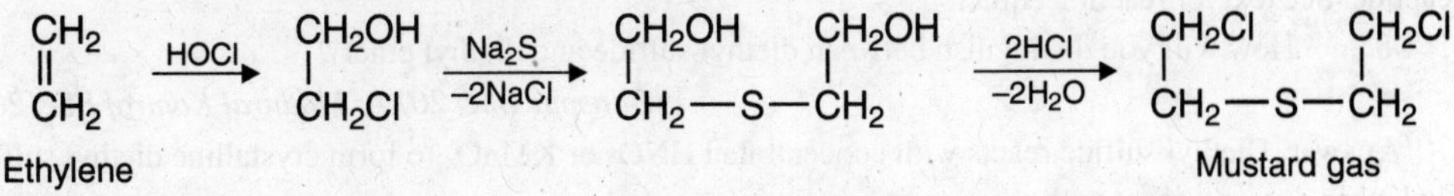

(2) By treating ethylene with sulfur monochloride.

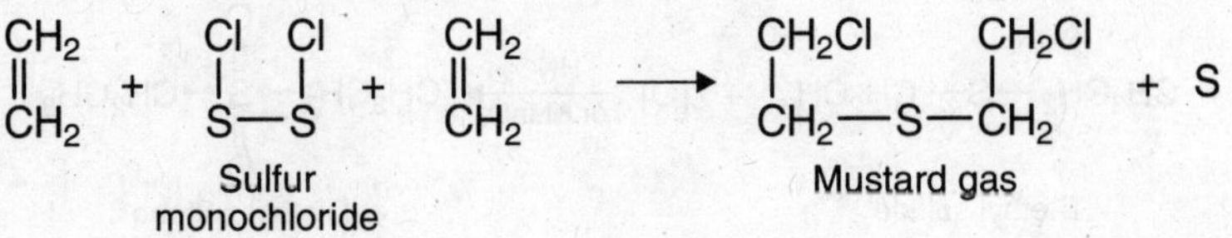

Mustard gas is poisonous. It has been used as a weapon of mass destruction.

Properties. Mustard gas is not a gas but an oily liquid with mustard-like smell, *bp* 215°C. It is insoluble in water but soluble in ethanol and ether. Mustard gas vaporizes when sprinkled by means of bursting its shells. It produces painful blisters on the skin and has very destructive effect on the lungs and air passages. It is generally known to have a prolonged action, causing death after about four days.

Examination Questions

1. How are thiols prepared ? Discuss their chemical properties.
2. How is ethyl mercaptan prepared ? How does ethyl mercaptan react with
   (*a*) NaOH (*b*) $(CH_3COO)_2Pb$ (*c*) I_2
   (*d*) CH_3CHO (*e*) H_2O_2 (*f*) CH_3COCl
   Remember : Mercaptans are Thiols.
3. How are thioethers prepared ? Discuss their chemical properties.
4. How is diethyl sulfide prepared ? How does diethyl sulfide react with
   (*a*) CH_3CH_2Br (*b*) Br_2 (*c*) aqueous KOH (*d*) H_2O_2
   Remember : Diethyl sulfide is a Thioether.
5. How is sulfonal prepared ?
6. How is mustard gas prepared ? (*Manipur BSc, 2014*)
7. What happens when
   (*a*) ethyl iodide is treated with KSH
   (*b*) ethanethiol is treated with KOH
   (*c*) ethanethiol is added to lead acetate solution

Answer. The products are (*a*) Ethanethiol ; (*b*) Potassium ethyl mercaptide ; and (*c*) Lead diethyl mercaptide. See text for reaction equations.

8. How will you distinguish between diethyl sulfide and diethyl ether ? (*Bilaspur BSc, 2014 ; Madurai Kamraj BSc, 2015*)

Answer. Diethyl sulfide reacts with concentrated HNO_3 or $KMnO_4$ to form crystalline diethyl sulfone. Diethyl ether does not react with these reagents.

$$\underset{\text{Diethyl sulfide}}{CH_3CH_2-S-CH_2CH_3} + [O] \xrightarrow[\text{or } KMnO_4]{HNO_3} \underset{\text{Diethyl sulfone}}{CH_3CH_2-\overset{O}{\overset{\|}{\underset{\underset{O}{\|}}{S}}}-CH_2CH_3}$$

9. How will you distinguish between ethyl mercaptan (ethanethiol) and ethyl alcohol ? *(Delhi BSc, 2015)*

Answer. Ethyl mercaptan dissolves in dilute NaOH forming a mercaptide *or* gives a precipitate with $HgCl_2$. Ethyl alcohol does not react with NaOH *or* $HgCl_2$.

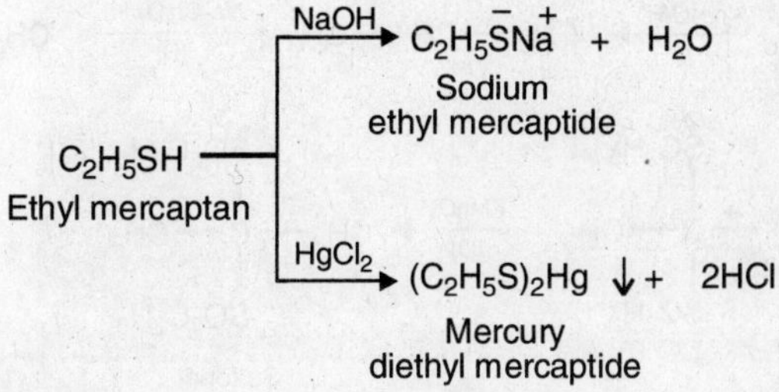

10. As thiols increase in the number of carbons, their odor decreases. Explain.

Answer. Because as the number of carbons increases, the volatility of the compound drops, and so does the percentage of sulfur.

11. Explain why n-propyl alcohol is more soluble in water than 1-propanethiol. *(Agra BSc, 2014)*

Answer. Sulfur is less electronegative than oxygen, so as an S–H bond is less polarized than an O–H bond. Hydrogen bonding is therefore weaker with thiols than alcohols, so that alcohol will be more able to interact with water molecules through hydrogen bonding. Thus, n-propyl alcohol ($CH_3CH_2CH_2OH$) is more soluble in water than 1-propanethiol ($CH_3CH_2CH_2SH$).

12. Explain why the boiling point of 1-butanol (*bp* 117°C) is higher than 1-butanethiol (*bp* 98.5°C). *(Osmania BSc, 2015)*

Answer. Sulfur is less electronegative than oxygen, so an S–H bond is less polarized than an O–H bond. Therefore, thiols do not make hydrogen bonds that are as strong as the hydrogen bonds made by alcohols. The stronger the hydrogen bonds, the higher the boiling point since hydrogen bonds in the liquid state must be broken upon boiling. Therefore, 1-butanol, with the stronger hydrogen bonds, has ($CH_3CH_2CH_2CH_2OH$) higher boiling point than 1-butanethiol ($CH_3CH_2CH_2CH_2SH$).

13. Explain why the boiling points of thioalcohols are lower than those of alcohols containing same number of carbon atoms.

14. Explain why dimethyl sulfoxide (*bp* 189°C) has a much higher boiling point than dimethyl sulfide (*bp* 37°C) *(North Bengal BSc, 2015)*

Answer. Dimethyl sulfoxide and dimethyl sulfide have the following structures :

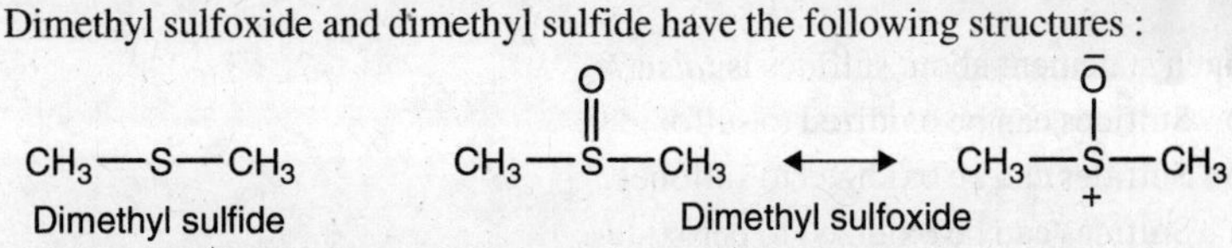

The sulfur-oxygen bond in dimethyl sulfoxide is highly polar, with a *positive* charge on sulfur and a *negative* charge on the oxygen. This polarity causes the molecules of dimethyl sulfoxide to associate. Energy (heat) is required to break up this molecular association. Therefore, the boiling point of dimethyl sulfoxide is much higher than that of dimethyl sulfide.

15. How will you synthesize 1-propanethiol from propene ? *(Jadavpur BSc, 2014)*

Answer. Following two methods can be used :

Method 1. *Via* Grignard reaction.

$$\underset{\text{Propene}}{CH_3CH{=}CH_2} \xrightarrow[\text{peroxide}]{\text{HBr}} CH_3CH_2CH_2Br \xrightarrow[\text{ether}]{\text{Mg}} CH_3CH_2CH_2MgBr \xrightarrow{\text{S}}$$

$$CH_3CH_2CH_2SMgBr \xrightarrow{H_2O/H^+} \underset{\text{1-Propanethiol}}{CH_3CH_3CH_2SH}$$

Method 2. *Via* reaction of alkyl halide with sodium hydrosulfide

$$\underset{\text{Propene}}{CH_3CH{=}CH_2} \xrightarrow[\text{peroxide}]{\text{HBr}} CH_3CH_2CH_2Br \xrightarrow{\text{NaSH}} \underset{\text{1-Propanethiol}}{CH_3CH_2CH_2SH}$$

16. How will you synthesize sulfonal from propene ? *(Baroda BSc, 2015)*

Answer. Following steps are involved :

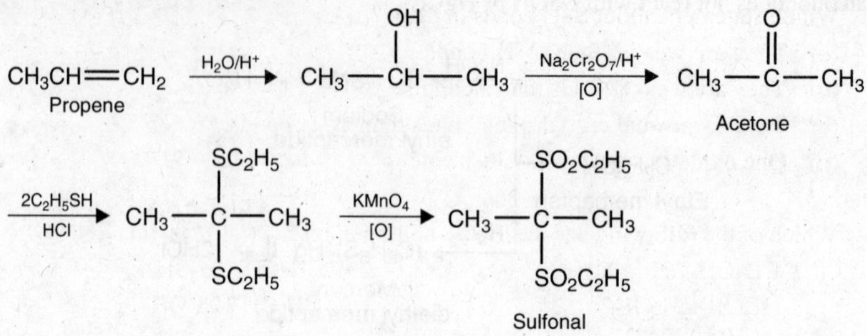

17. *n*-Propyl bromide reacts with excess of KSH to form (*A*), C_3H_8S. Compound (*A*) reacts with concentrated nitric acid to form (*B*), $C_3H_8O_3S$. Compound (*B*) reacts with PCl_5 and gives a chloride (*C*). Identify (*A*), (*B*), and (*C*).

Answer. (*A*) = $CH_3CH_2CH_2SH$ (1-Propanethiol)
(*B*) = $CH_3CH_2CH_2SO_3H$ (*n*-Propylsulfonic acid)
(*C*) = $CH_3CH_2CH_2SO_2Cl$ (*n*-Propylsulfonyl chloride)

Multiple Choice Questions

1. The common name for CH_3SCH_3 is
(*a*) Methane methane sulfide (*b*) Dimethyl sulfide
(*c*) Dimethyl sulfoxide (*d*) Dimethyl sulfone

Answer. (*b*)

2. Which of the following is *not* a property of thiols (RSH) ?
(*a*) They are all solids (*b*) They can be oxidized to disulfides
(*c*) They have foul odors (*d*) They are weak acids

Answer. (*a*)

3. Which statement about sulfides is *false* ?
(*a*) Sulfides can be oxidized to sulfoxides.
(*b*) Sulfides can be oxidized to sulfones.
(*c*) Sulfides can be oxidized to peroxides.
(*d*) Sulfides can be alkylated to sulfonium salts.

Answer. (*c*)

4. Thiols are prepared by reacting alkyl halides with ?
(*a*) KSH (*b*) SO_2 (*c*) SO_3 (*d*) H_2SO_3

Answer. (*a*)

5. Classify the following reaction.
$CH_3CH_2S{-}SCH_2CH_3 \rightarrow 2CH_3CH_2SH$
(*a*) oxidation (*b*) reduction (*c*) neither (*d*) both

Answer. (*b*)

6. Thiols are alcohol analogs in which the oxygen has been replaced by sulphur (e.g., CH_3SH). Given the fact that the S–H bond is less polar than the O–H bond, which of the following statements comparing thiols and alcohols is correct ?
(*a*) Hydrogen bonding forces are weaker in thiols.
(*b*) Hydrogen bonding forces are stronger in thiols.

(*c*) Hydrogen bonding forces would be the same.

(*d*) No comparison can be made without additional information.

Answer. (*a*)

7. Which statement about S–H bonds in *false* ?

(*a*) They are weaker than O–H bonds.

(*b*) They are weaker acids than alcohols.

(*c*) They form weaker hydrogen bonds than O–H bonds.

(*d*) One oxidative product is a disulfide.

Answer. (*b*)

8. Which of the following has the lowest boiling point ?

(*a*) CH_3CH_2OH (*b*) $CH_3CH_2CH_2SH$

(*c*) $HOCH_2CH_2OH$ (*d*) $CH_3CH_2CH_2OH$

Answer. (*b*)

9. Which of the following is least soluble in water ?

(*a*) CH_3OH (*b*) CH_3CH_2OH

(*c*) CH_3SH (*d*) $HOCH_2CH_2OH$

Answer. (*c*)

10. *n*-Butyl bromide reacts with NaSH to give

(*a*) $CH_3CH_2CH_2SH$ (*b*) CH_3SCH_3

(*c*) $CH_3CH_2CH_2CH_2SH$ (*d*) $CH_3CH_2SCH_2CH_3$

Answer. (*c*)

11. Arrange the compounds in the order of increasing boiling point (*least first*).

I. $H_3C–O–CH_3$ II. H_2O

III. CH_3CH_2OH IV. CH_3CH_2SH

(*a*) II, I, III, IV (*b*) I, IV, II, III

(*c*) IV, I, III, II (*d*) I, IV, III, II

Answer. (*d*)

12. Arrange the compounds in the order of increasing boiling point (*least first*).

I. $H_3C–O–CH_3$ II. H_2O

III. CH_3CH_2OH IV. CH_3CH_2SH

(*a*) II, I, III, IV (*b*) I, IV, II, III

(*c*) IV, I, III, II (*d*) I, IV, III, II

Answer. (*d*)

19

Aldehydes and Ketones

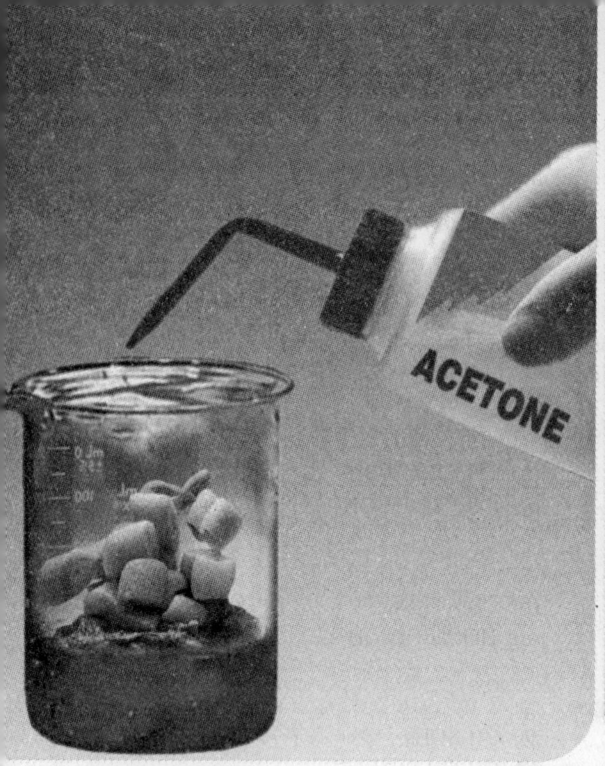

Styrofoam dissolves in acetone.

Both aldehydes and ketones contain a carbon-oxygen double bond (>C=O). This unit is referred to as the **Carbonyl Group**.

The carbonyl group

Aldehydes and ketones are collectively called **Carbonyl Compounds**. In aldehydes, the carbonyl carbon is bonded to one hydrogen and one alkyl group. Formaldehyde, HCHO, in which the carbonyl carbon is bonded to two hydrogen atoms is an exception. In ketones, the carbonyl carbon is always bonded to two alkyl groups. These alkyl groups may be same or different.

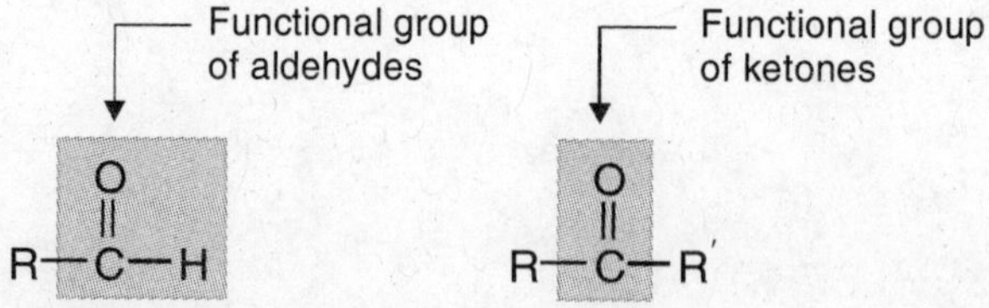

STRUCTURE OF THE CARBONYL GROUP

The carbonyl group, like the carbon-carbon double bond of alkenes, is composed of one σ bond and one π bond.

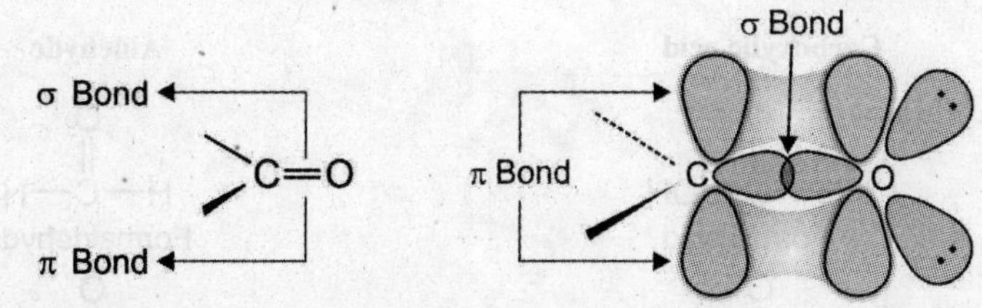

Both the carbon and oxygen are sp^2 hybridized. The σ bond is formed by the overlap of an sp^2 orbital of carbon and an sp^2 orbital of oxygen. The π bond is formed by the overlap of unhybridized *p* orbitals of the two atoms. The two unshared electron pairs of oxygen occupy the sp^2 hybrid orbitals of oxygen. Because the carbonyl carbon is sp^2 hybridized, the three atoms attached to it lie in the same plane. The bond angles between the attached atoms are approximately 120° (Fig. 19.1).

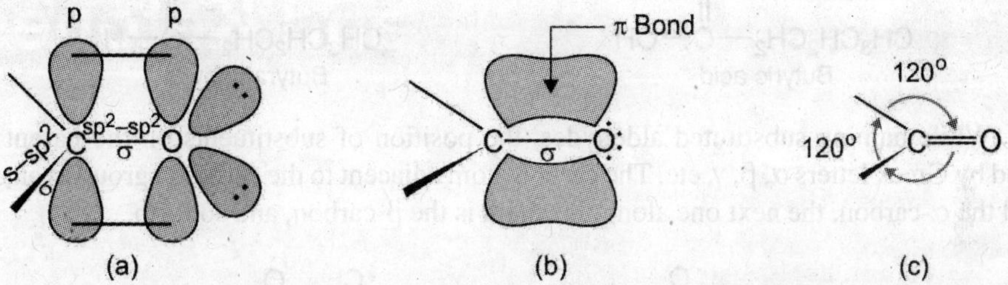

Fig. 19.1. The carbon-oxygen double bond in aldehydes and ketones is composed of a σ bond and a π bond. (*a*) The σ bond is formed by the overlap of an sp^2 orbital of carbon and an sp^2 orbital of oxygen ; (*b*) The π bond is formed by the overlap of the unhybridized *p* orbitals ; (*c*) Geometry of the carbonyl group.

The electrons in the π bond of the carbonyl group are *not* equally shared. In fact, they are pulled more toward the more electronegative oxygen atom. As a result, the bond is polarized, with the oxygen atom being slightly negative (δ_-) and the carbon atom being slightly positive (δ_+). This is often indicated as :

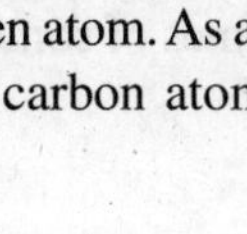

Alternatively, the polar nature of the carbonyl group can also be indicated by the following resonance structures :

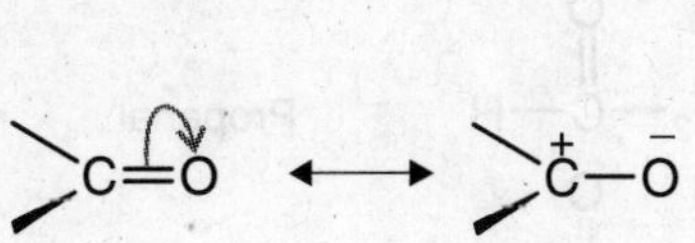

NOMENCLATURE OF ALDEHYDES

There are two systems of naming aldehydes.

(1) **Common System**. Common names for aldehydes are obtained from the names of the corresponding carboxylic acids. The aldehyde name is obtained by replacing the ending *-ic acid* of the acid name with *-aldehyde*.

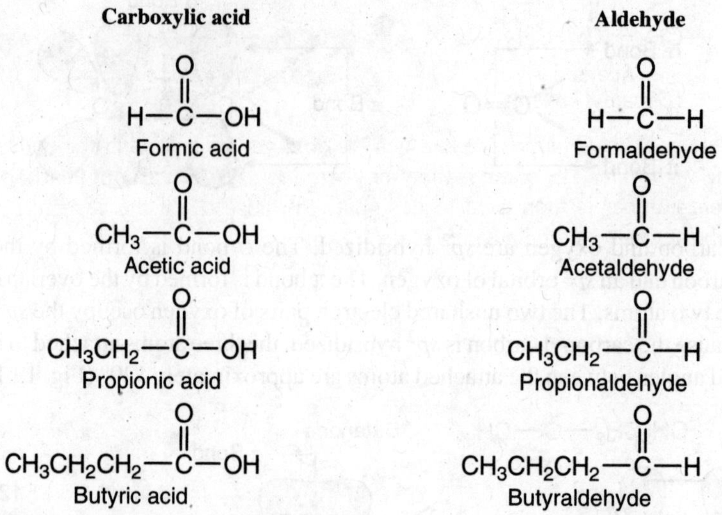

While naming substituted aldehydes, the position of substituents on the parent chain is indicated by Greek letters α, β, γ, etc. The carbon atom adjacent to the carbonyl group in an aldehyde is called the α-carbon, the next one along the chain is the β-carbon, and so forth.

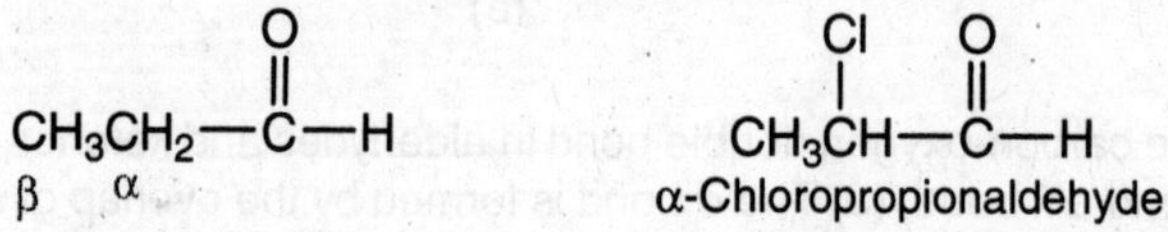

(2) **IUPAC System**. IUPAC names for aldehydes are obtained by replacing the ending *-e* of the corresponding alkane with *-al*. Since the aldehyde functional group is always at the end of the chain, there is no need to specify its position. However, when other substituents are present, the carbonyl carbon is assigned number 1. The IUPAC names of some aldehydes are given below :

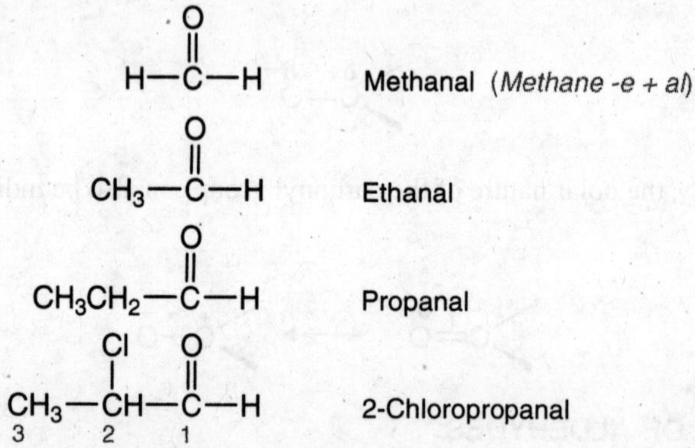

NOMENCLATURE OF KETONES

There are two systems of naming ketones.

(1) **Common System**. Common names of ketones are obtained by simply naming the two alkyl groups attached to the carbonyl group and adding the word *ketone*. The simplest ketone, dimethyl ketone, is usually referred to as *acetone*.

$$CH_3-\overset{O}{\overset{\|}{C}}-CH_3 \qquad CH_3CH_2-\overset{O}{\overset{\|}{C}}-CH_3$$

Acetone (*Dimethyl ketone*) — Ethyl methyl ketone

(2) **IUPAC System**. IUPAC names for ketones are obtained by replacing the ending *-e* of the corresponding alkane with *-one*. The chain is then numbered to give the carbonyl carbon the *lowest* possible number. This number is then used to designate the position of the carbonyl group. The IUPAC names of some ketones are given below :

$$CH_3-\overset{O}{\overset{\|}{C}}-CH_3 \quad \text{Propanone } (\textit{Propane} - e + \textit{one})$$

$$CH_3CH_2-\overset{O}{\overset{\|}{C}}-CH_3 \quad \text{Butanone}$$

$$CH_3CH_2CH_2-\overset{O}{\overset{\|}{C}}-CH_3 \quad \text{2-Pentanone}$$

METHODS OF PREPARATION

Aldehydes and ketones may be prepared by the following methods :

(1) **Oxidation of Alcohols.** Aldehydes and ketones can be prepared by the controlled oxidation of primary and secondary alcohols using an acidified solution of potassium dichromate or permanganate. Primary alcohols produce aldehydes.

$$\underset{\text{1° Alcohol}}{RCH_2OH} + [O] \xrightarrow[H^+]{K_2Cr_2O_7} \underset{\text{Aldehyde}}{R-\overset{O}{\overset{\|}{C}}-H}$$

$$\underset{\text{Ethyl alcohol}}{CH_3CH_2OH} + [O] \xrightarrow[H^+]{K_2Cr_2O_7} \underset{\text{Acetaldehyde}}{CH_3-\overset{O}{\overset{\|}{C}}-H} + H_2O$$

The aldehydes formed in the above reaction are very easily oxidized to carboxylic acids if allowed to remain in the reaction mixture.

$$\underset{\text{Aldehyde}}{R-\overset{H}{\overset{|}{C}}=O} + [O] \xrightarrow[H^+]{K_2Cr_2O_7} \underset{\text{Carboxylic acid}}{R-\overset{OH}{\overset{|}{C}}=O} + H_2O$$

Ketones are prepared by the oxidation of secondary alcohols.

$$\underset{\text{2° Alcohol}}{R-\overset{OH}{\overset{|}{CH}}-R'} + [O] \xrightarrow[H^+]{K_2Cr_2O_7} \underset{\text{Ketone}}{R-\overset{O}{\overset{\|}{C}}-R'} + H_2O$$

$$\underset{\substack{\text{2-Propanol}\\ \text{(Isopropyl alcohol)}}}{CH_3-\overset{OH}{\overset{|}{CH}}-CH_3} + [O] \xrightarrow[H^+]{K_2Cr_2O_7} \underset{\text{Acetone}}{CH_3-\overset{O}{\overset{\|}{C}}-CH_3} + H_2O$$

Ketones are not easily oxidized further and can be obtained in high yield by this method. Alternatively, ketones can be obtained from secondary alcohols by **Oppenauer Oxidation**. In this process, the appropriate secondary alcohol is refluxed with excess of acetone in the presence of aluminium tertiary-butoxide, $AlOC(CH_3)_3$, catalyst.

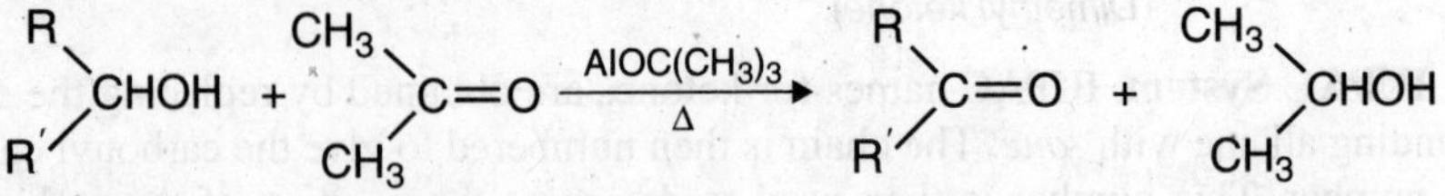

2° Alcohol Acetone Ketone 2-Propanol

Other agents used for these oxidation reactions are :

(*a*) Alkaline potassium permanganate : $KMnO_4/\bar{O}H$

(*b*) Chromic acid : $H_2Cr_2O_7$

This reagent is prepared *in situ* from CrO_3 or $Na_2Cr_2O_7$ with dilute sulfuric acid.

(*c*) Pyridine chlorochromate (PCC), $C_5H_6NCrO_3Cl$, in dichloromethane solvent.

PCC in dichloromethane is a special oxidizing agent. It is used to convert primary alcohols to aldehydes. This reagent does not oxidize aldehydes to carboxylic acids.

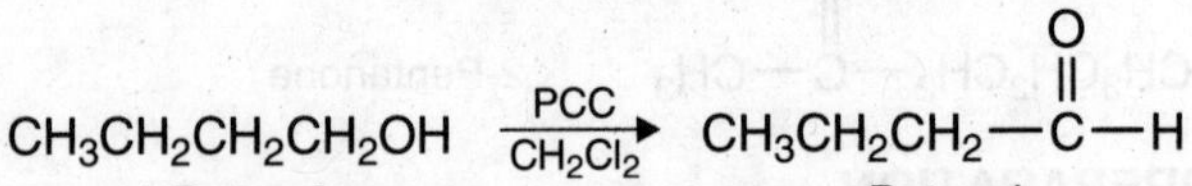

1-Butanol Butanal

(2) **Catalytic Dehydrogenation of Alcohols**. Aldehydes may be prepared by passing the vapors of primary alcohols over a copper catalyst heated to about 300°C.

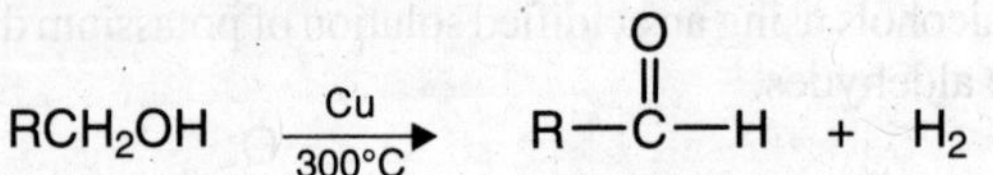

1° Alcohol Aldehyde

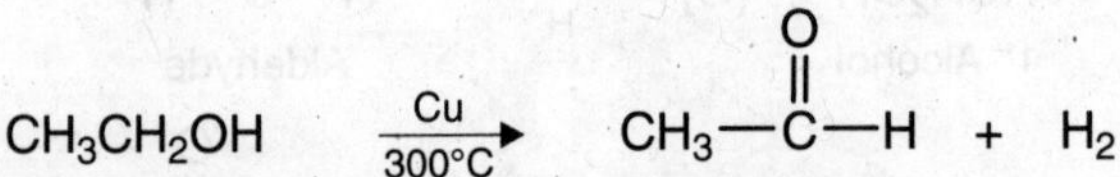

Ethyl alcohol Acetaldehyde

Similarly ketones are produced from secondary alcohols.

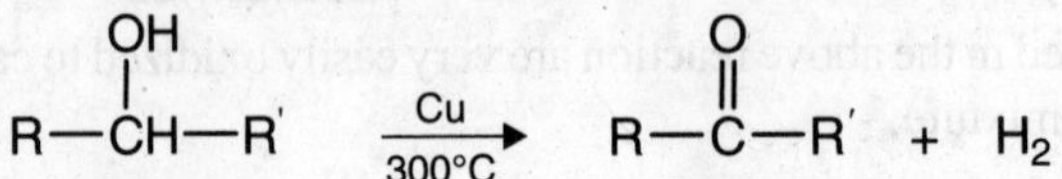

2° Alcohol Ketone

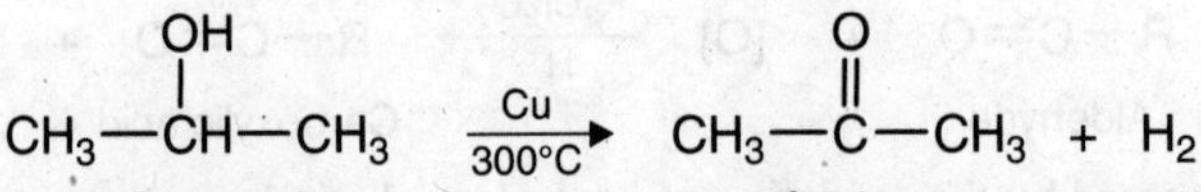

2-Propanol (Isopropyl alcohol) Acetone

(3) **Oxidation of Alkenes (Ozonolysis)**. Aldehydes and ketones can be obtained by ozonolysis of alkenes. This involves the treatment of the alkenes with ozone to give ozonides. The ozonides are not isolated because they are often explosive in dry state. They are decomposed with $Zn+H_2O$ to form aldehydes and ketones.

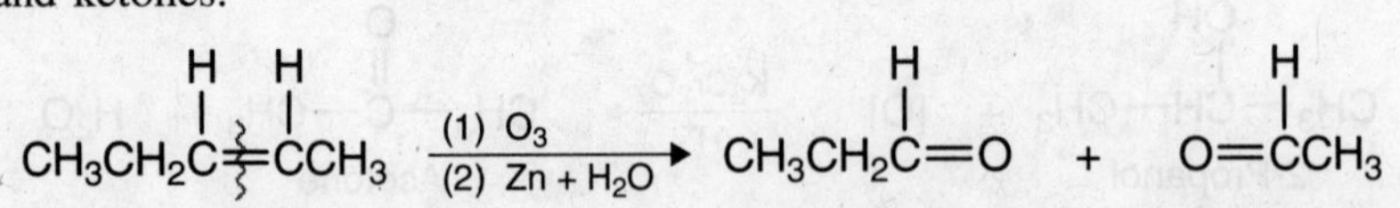

2-Pentene Propanal Acetaldehyde

Ozonolysis of alkenes is not a good preparative method for aldehydes and ketones. This is because a mixture of carbonyl compounds is often produced. However, if the starting alkene is symmetrical, only one carbonyl compound will be obtained. For example,

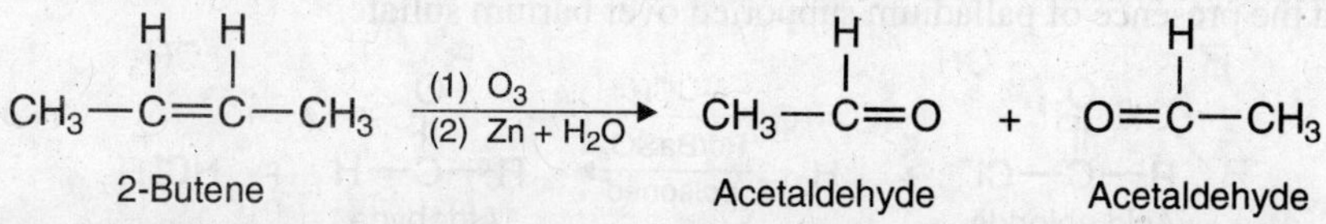

(4) **Hydration of Alkynes**. Hydration of acetylene yields acetaldehyde. Water adds to acetylene in the presence of mercuric sulfate and sulfuric acid to form an unstable enol-intermediate. This intermediate rearranges to give acetaldehyde.

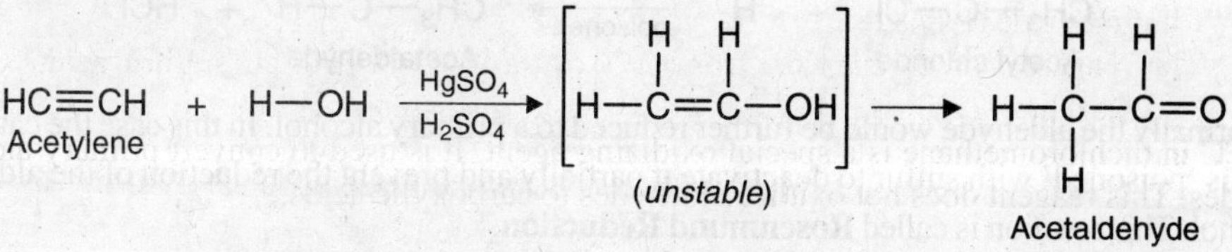

Hydration of alkynes, other than acetylene, gives ketones. Water adds according to the *Markovnikov rule* to give an unstable enol-intermediate. This intermediate rearranges to form ketones.

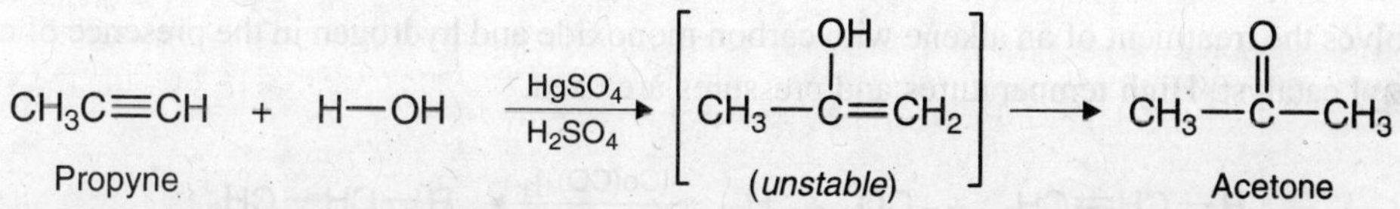

(5) **Hydrolysis of gem-Dihalides**. A *gem*-dihalide is a compound that has two halogen atoms attached to the same carbon. Aldehydes are prepared by alkaline hydrolysis of those *gem*-dihalides in which two halogen atoms are attached to the terminal carbon atom. For example,

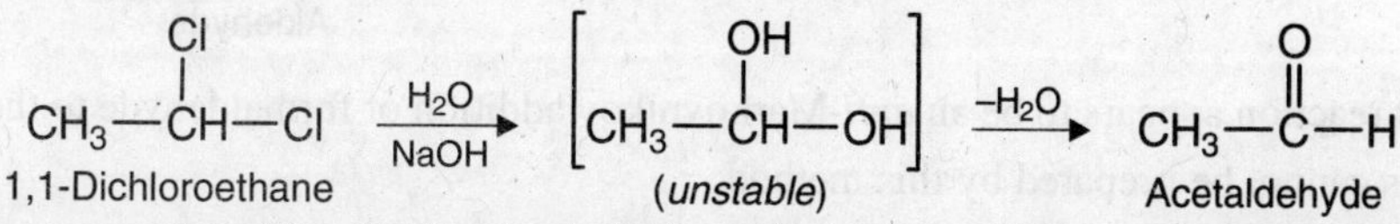

Ketones are produced by alkaline hydrolysis of those *gem*-dihalides in which the two halogen atoms are attached to a non-terminal carbon atom. For example,

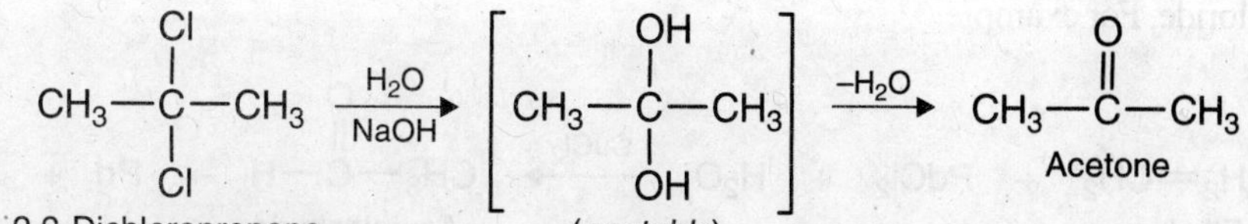

(6) **Pyrolysis of Calcium Salts of Acids**. Symmetrical ketones may be prepared by heating calcium salts of acids at 400°C. For example,

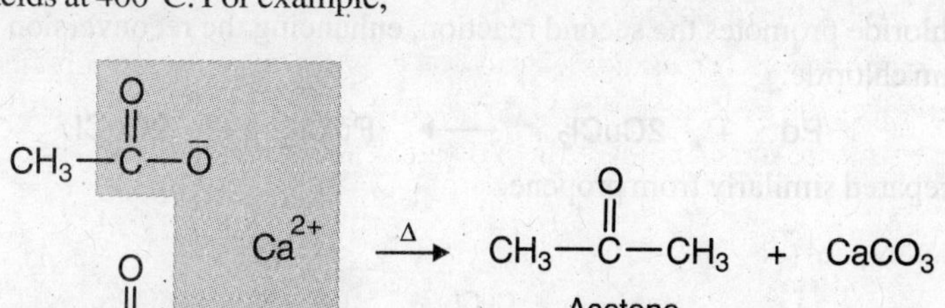

The calcium salts of acids are obtained by heating carboxylic acids with calcium oxide. Aldehydes cannot be prepared by this method.

(7) **Reduction of Acid Chlorides**. Aldehydes can be prepared by the hydrogenation of acid chlorides in the presence of palladium supported over barium sulfate.

$$\underset{\text{Acid chloride}}{R{-}\overset{\overset{\displaystyle O}{\|}}{C}{-}Cl} + H_2 \xrightarrow[\text{poisoned}]{Pd/BaSO_4} \underset{\text{Aldehyde}}{R{-}\overset{\overset{\displaystyle O}{\|}}{C}{-}H} + HCl$$

$$\underset{\text{Acetyl chloride}}{CH_3{-}\overset{\overset{\displaystyle O}{\|}}{C}{-}Cl} + H_2 \xrightarrow[\text{poisoned}]{Pd/BaSO_4} \underset{\text{Acetaldehyde}}{CH_3{-}\overset{\overset{\displaystyle O}{\|}}{C}{-}H} + HCl$$

Normally the aldehyde would be further reduced to a primary alcohol. In this case the catalyst Pd/$BaSO_4$ is 'poisoned' with sulfur to deactivate it partially and prevent the reduction of the aldehyde to an alcohol. This reaction is called **Rosenmund Reduction**.

Ketones cannot be prepared by this method.

(8) **Oxo Process**. The Oxo process is an industrially important method for producing aldehydes. It involves the treatment of an alkene with carbon monoxide and hydrogen in the presence of cobalt carbonyl catalyst. High temperatures and pressures are used.

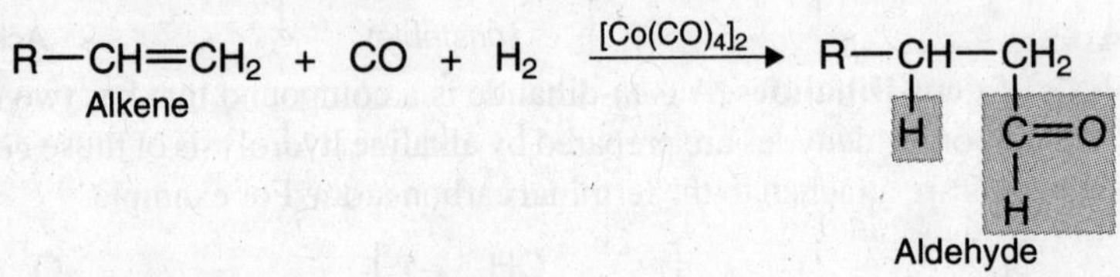

The net reaction appears to be an *anti*-Markovnikov addition of formaldehyde to the alkene.

Ketones cannot be prepared by this method.

(9) **Wacker Process**. Both aldehydes and ketones can be prepared by this method. This process involves the treatment of an alkene with an acidified aqueous solution of palladium chloride and cupric chloride. For example,

$$\underset{\text{Ethylene}}{CH_2{=}CH_2} + PdCl_2 + H_2O \xrightarrow{CuCl_2} \underset{\text{Acetaldehyde}}{CH_3{-}\overset{\overset{\displaystyle O}{\|}}{C}{-}H} + Pd + 2HCl$$

$$Pd + HCl \longrightarrow PdCl_2$$

The cupric chloride promotes the second reaction, enhancing the reconversion of the palladium back into palladium chloride.

$$Pd + 2CuCl_2 \longrightarrow PdCl_2 + 2CuCl$$

Acetone is prepared similarly from propene.

$$\underset{\text{Propene}}{CH_3CH{=}CH_2} + PdCl_2 + H_2O \xrightarrow{CuCl_2} \underset{\text{Acetone}}{CH_3{-}\overset{\overset{\displaystyle O}{\|}}{C}{-}CH_3} + Pd + 2HCl$$

PHYSICAL PROPERTIES

(1) Formaldehyde is a gas at room temperature. Acetaldehyde boils at 20°C. Other lower aldehydes and ketones are colorless liquids.

(2) Lower aldehydes possess rather unpleasant, pungent smells whereas the ketones have pleasant, sweet odors.

(3) Density of aldehydes and ketones is less than that of water.

Table 19.1 BOILING POINTS OF SOME ALDEHYDES AND KETONES

Name	Formula	bp°C
Aldehydes		
Formaldehyde	HCHO	–21
Acetaldehyde	CH_3CHO	20
Propionaldehyde	CH_3CH_2CHO	49
n-Butyraldehyde	$CH_3CH_2CH_2CHO$	76
Ketones		
Acetone	CH_3COCH_3	56
Butanone	$CH_3COCH_2CH_3$	80
2-Pentanone	$CH_3COCH_2CH_2CH_3$	102
3-Pentanone	$CH_3CH_2COCH_2CH_3$	101

(4) **Boiling Points of Aldehydes and Ketones**. Because of the polarity of the carbonyl group, aldehydes and ketones are polar compounds. The polar character of the molecules gives rise to intermolecular attractions. These attractive forces, called **dipole-dipole attractions**, occur between the partial negative charge on the carbonyl oxygen of one molecule and the partial positive charge on the carbonyl carbon of another molecule.

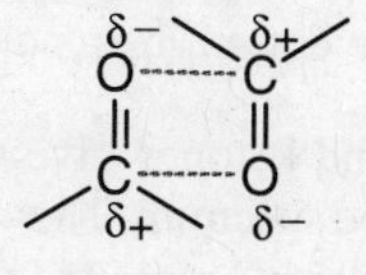

Dipole-dipole attractions among carbonyl compounds

Dipole-dipole attractions, although important, are not as strong as interactions due to hydrogen bonding. As a result, the boiling points of aldehydes and ketones are higher than those of nonpolar alkanes, but lower than those of alcohols, of comparable molecular weights. For example,

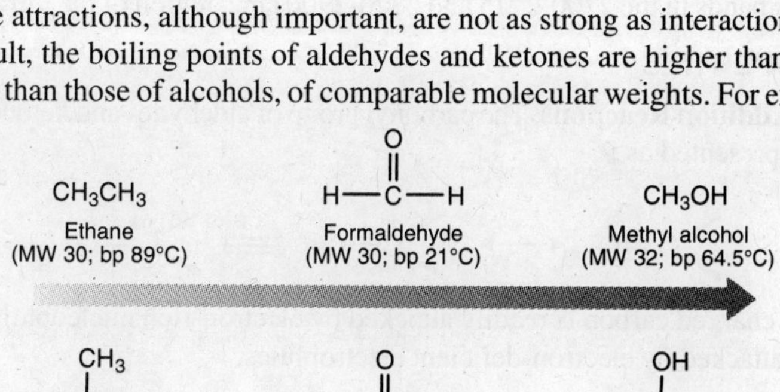

(5) **Solubility in Water**. The lower aldehydes and ketones are soluble in water because aldehydes and ketones form hydrogen bonds with water, even though they are incapable of intermolecular hydrogen bonding with themselves.

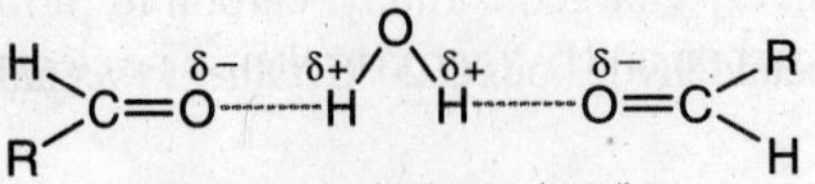

Intermolecular hydrogen bonding between water and aldehydes

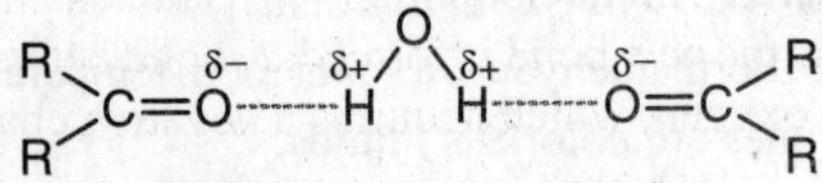

Intermolecular hydrogen bonding between water and ketones

As the hydrocarbon portion of the molecule increases, the solubility in water decreases rapidly. Aldehydes and ketones with more than six carbons are essentially insoluble in water. However, the higher as well as the lower aldehydes and ketones are soluble in organic solvents such as benzene, ether, and carbon tetrachloride.

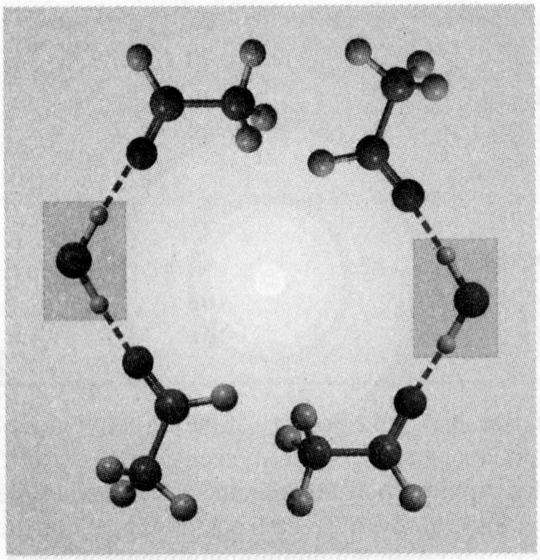

Acetaldehyde dissolved in water.
The dotted lines indicate hydrogen bonds.

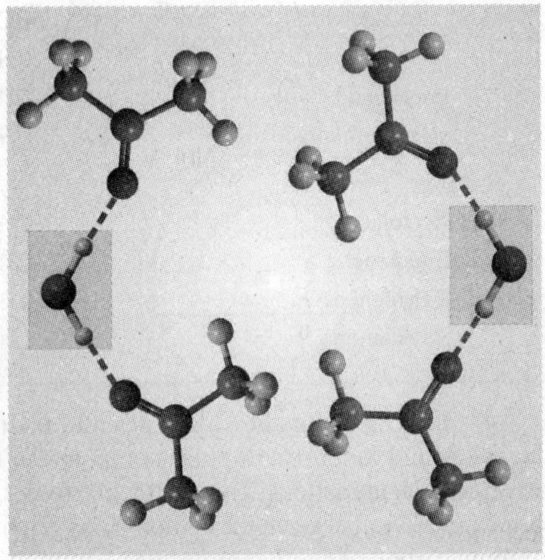

Acetone dissolved in water.
The dotted lines indicate hydrogen bonds.

(6) **IR spectrum.** Carbonyl groups of aldehydes and ketones give rise to very strong C=O stretching bands in the 1665-1780 cm^{-1} region of the infrared spectrum. The –CHO group of aldehydes also gives two weak bands in the 2700-2775 and 2820-2900 cm^{-1} region of the infrared spectrum.

CHEMICAL PROPERTIES

Nucleophilic Addition Reactions. The carbonyl group of aldehydes and ketones is a highly polar group. It may be represented as :

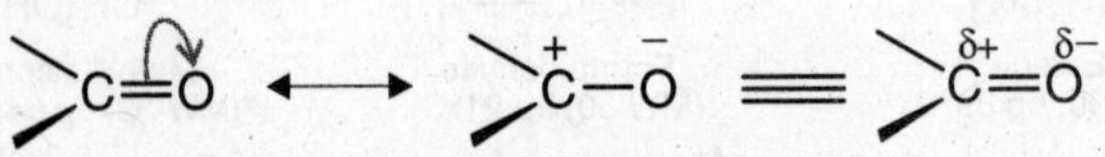

The positively charged carbon is readily attacked by electron-rich nucleophiles. The negatively charged oxygen is attacked by electron-deficient electrophiles.

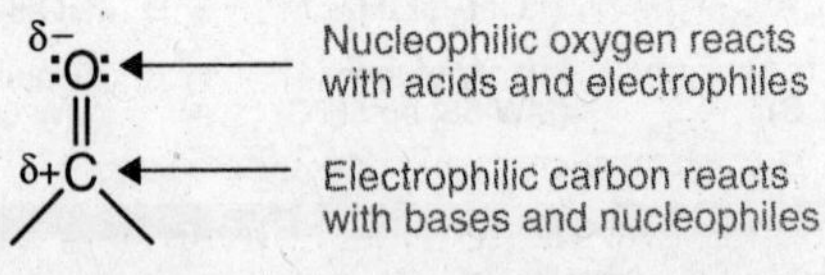

Aldehydes and ketones undergo nucleophilic addition reactions by the following general mechanism.

Step 1. The nucleophile ($Nu:^-$) attacks the positively charged carbonyl carbon to form a new bond. As the new bond is formed, π-bond between the carbon and oxygen is broken. The electron pair goes to oxygen, which acquires a negative charge.

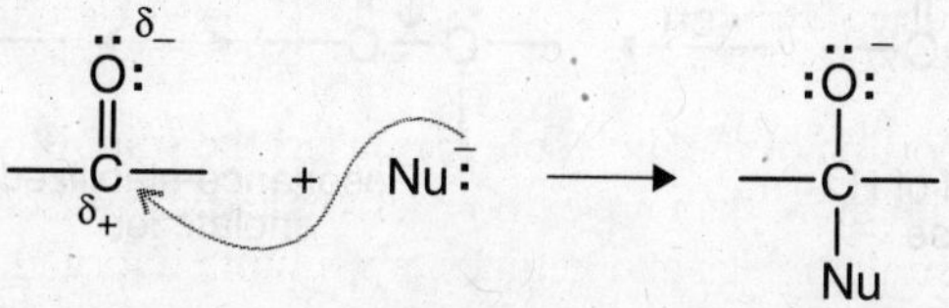

Step 2. The electrophile (*e.g.*, H^+) attacks the negatively charged oxygen to form the addition product.

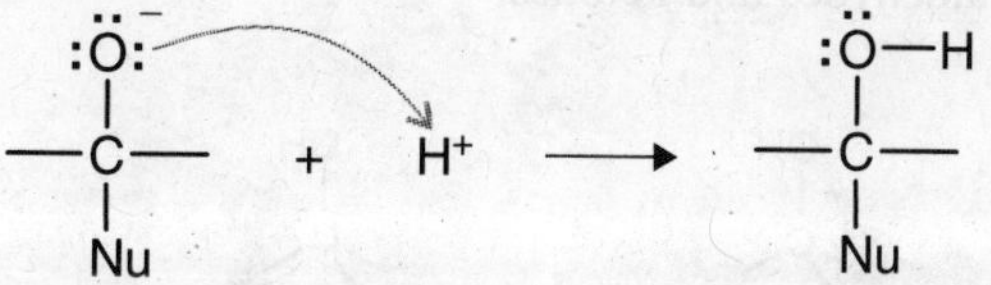

Addition product

The nucleophilic addition reactions of carbonyl compounds may be catalyzed by acids or bases.

Base-catalyzed addition. Bases convert a weak neutral nucleophile to a strong one by removing a proton. The strong nucleophile then adds to the carbonyl group as shown above.

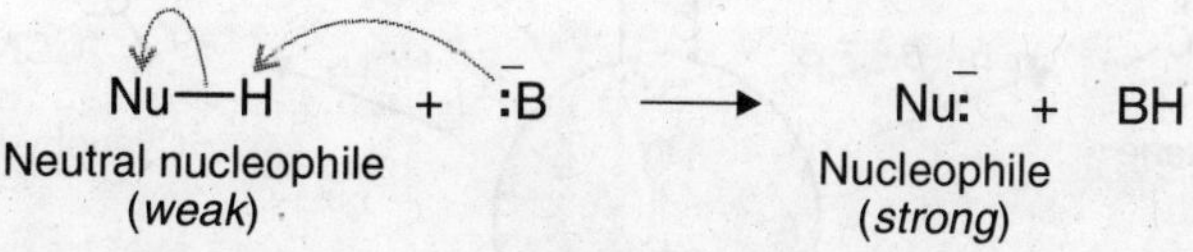

Acid-catalyzed addition. The acid-catalyzed nucleophilic addition occurs by the following mechanism :

Step 1. The hydrogen ion from the acid attacks the negatively charged carbonyl oxygen to give protonated carbonyl group. The protonated carbonyl group is resonance stabilized.

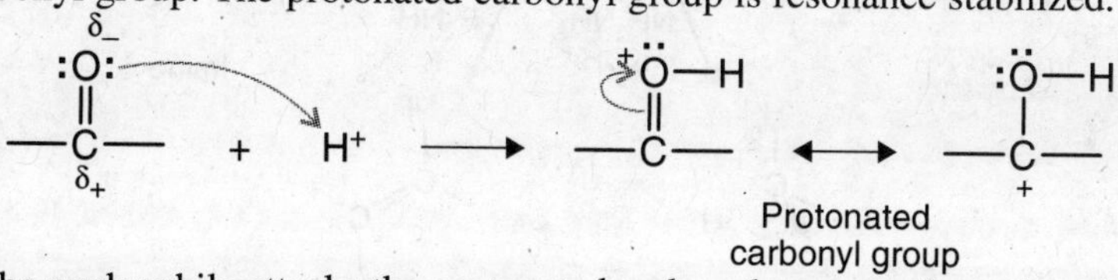

Step 2. The nucleophile attacks the protonated carbonyl group to form the addition product.

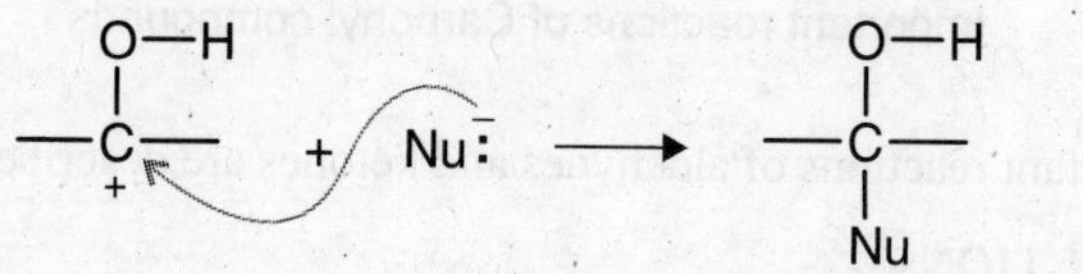

Addition product

Notice that the addition product is the same whether the reaction is acid-catalyzed or base-catalyzed. The nucleophile always adds to the carbonyl carbon and the proton (electrophile), to the oxygen. Generally ketones are less reactive than aldehydes in nucleophilic addition reactions.

Acidity of α-Hydrogens. A carbon atom next to the carbonyl group is called an α-carbon. A hydrogen attached to an α-carbon is referred to as an α-hydrogen. The α-hydrogens of aldehydes and ketones are acidic in nature. The acidity is due to the fact that the anion, which results from the removal of an α-hydrogen by a base $B:^-$, is stabilized by resonance. The resonance-stabilized anion is called **Enolate Ion**.

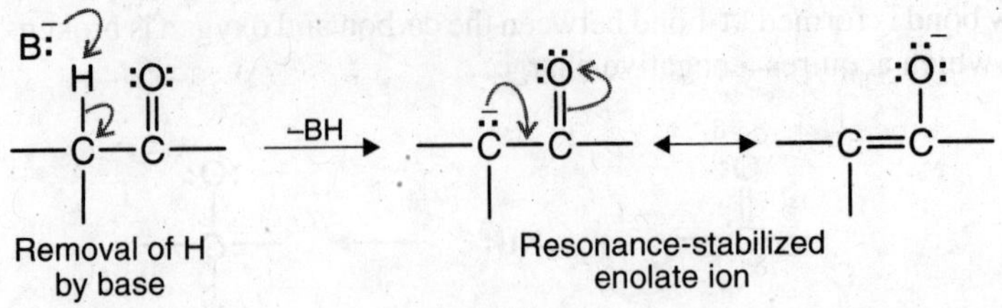

The α-carbon of the enolate ion is negatively charged. It can act as a nucleophile. The formation of the enolate ion followed by its addition to a carbonyl group is the process involved in all the condensation reactions of aldehydes and ketones.

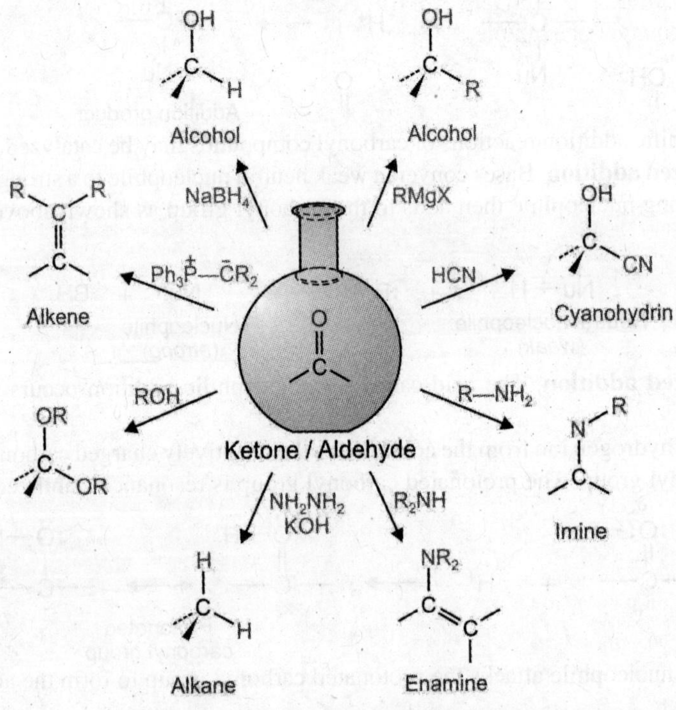

Important reactions of Carbonyl compounds

Some of the important reactions of aldehydes and ketones are described below :

A. ADDITION REACTIONS

(1) **Addition of Sodium Bisulfite**. Aldehydes and methyl ketones react with a saturated aqueous solution of sodium bisulfite ($NaHSO_3$) to form solid addition compounds.

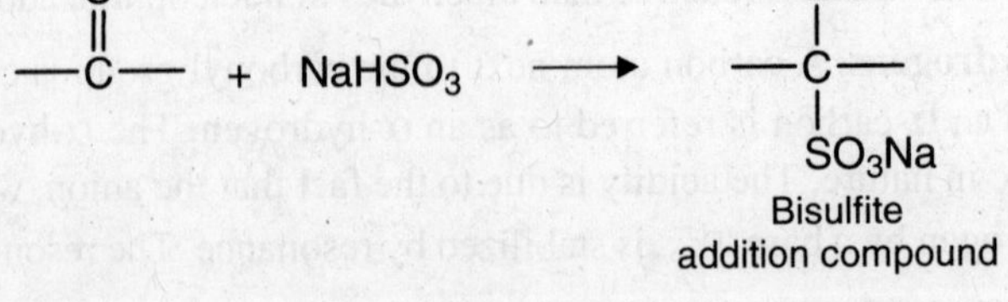

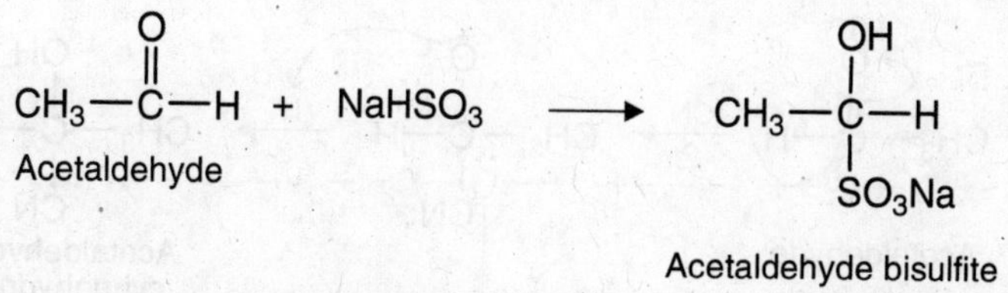

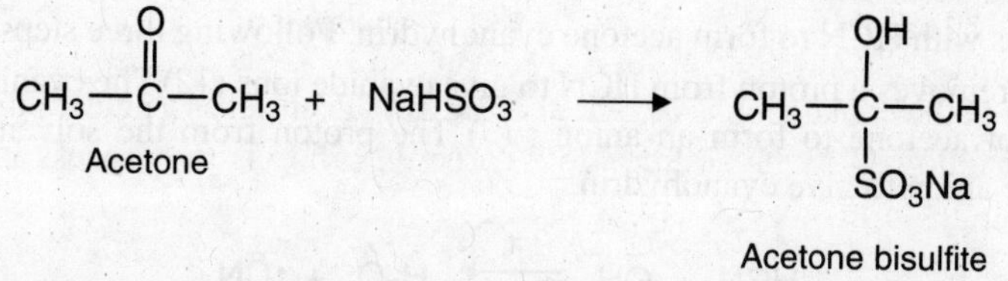

The bisulfite addition compounds can be decomposed with dilute acids or bases to regenerate the carbonyl compound. For example,

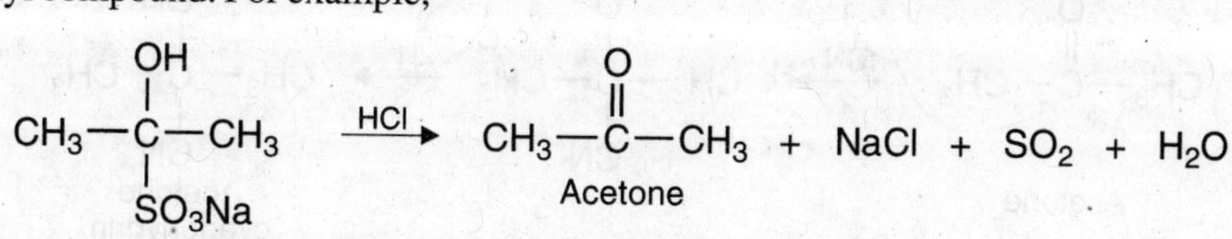

Therefore, the formation and decomposition of bisulfite addition compounds is used for the purification and separation of carbonyl compounds from mixtures.

(2) **Addition of Hydrogen Cyanide**. Aldehydes and ketones react with hydrogen cyanide to form **Cyanohydrins**. The reaction is carried in the presence of a basic catalyst.

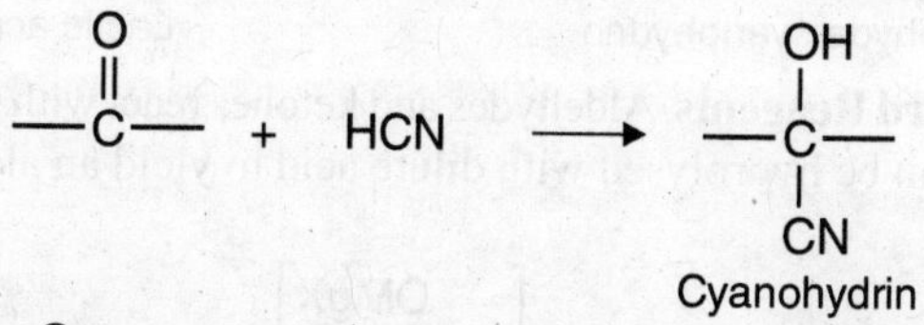

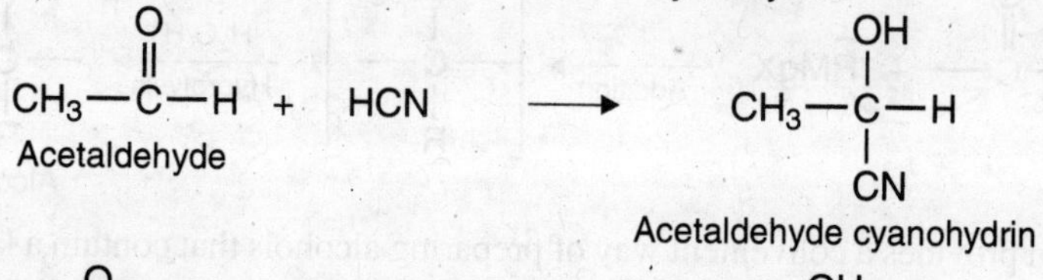

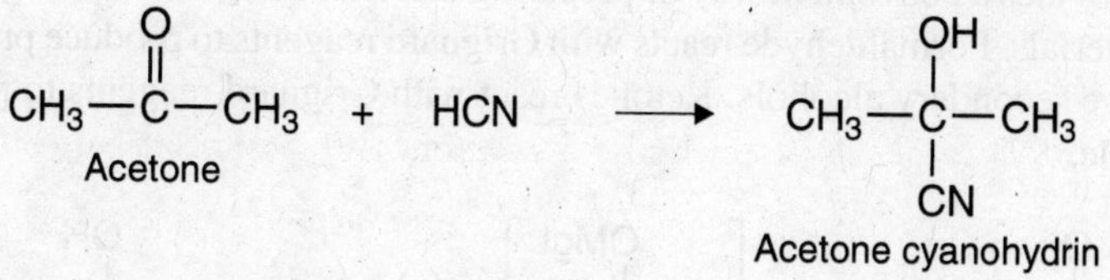

HCN is a very poisonous gas. It is produced *in situ* by the action of dilute sulfuric acid on potassium cyanide.

MECHANISM. Acetalidehyde reacts with HCN to form acetaldehyde cyandhydrin. Following three steps are involved : (*1*) The base removes a proton from HCN to give cyanide ions ; (2) The cyanide ion attacks the carbonyl carbon of acetaldehyde to form an anion ; (*3*) The proton from the solvent (usually water) combines with the anion to give cyanohydrin.

$$HCN + \bar{O}H \overset{1}{\rightleftharpoons} H_2O + :\bar{C}N$$

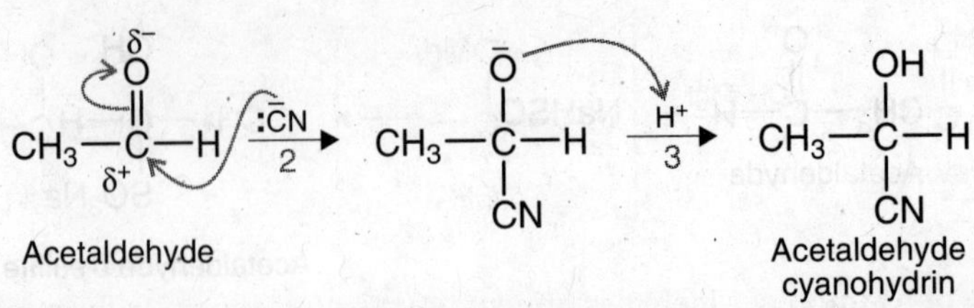

Acetone reacts with HCN to form acetone cyanohydrin. Following three steps are involved :

(*1*) The base removes a proton from HCN to give cyanide ions ; (2) The cyanide ion attacks the carbonyl carbon of acetone to form an anion ; (*3*) The proton from the solvent (usually water) combines with the anion to give cyanohydrin.

$$HCN + \bar{O}H \overset{1}{\rightleftharpoons} H_2O + :\bar{C}N$$

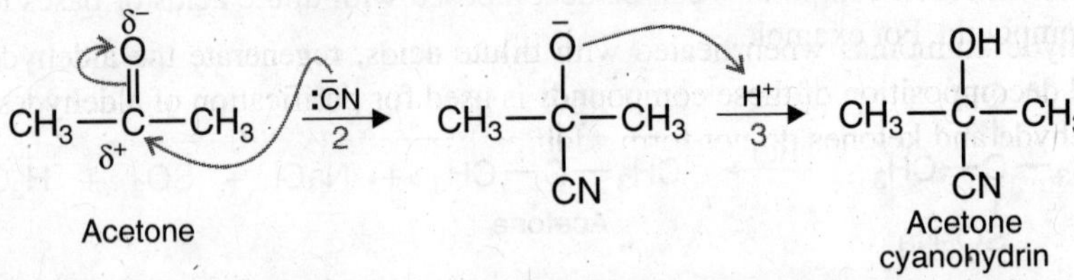

Like other nitriles, cyanohydrins can be hydrolysed to give α-hydroxy carboxylic acids. For example,

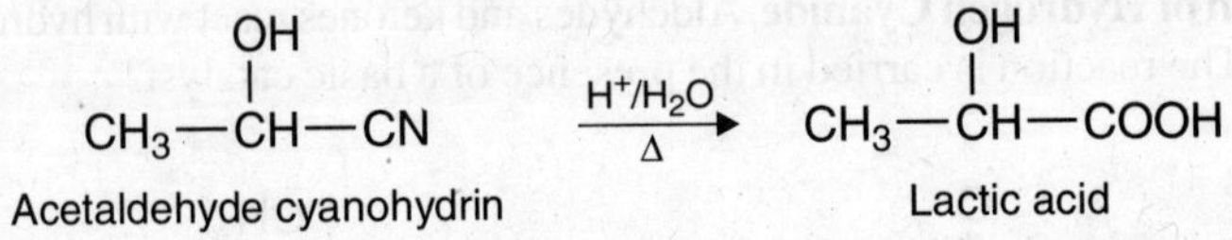

(3) **Addition of Grignard Reagents.** Aldehydes and ketones react with Grignard reagents to give an addition product which can be hydrolysed with dilute acid to yield an alcohol.

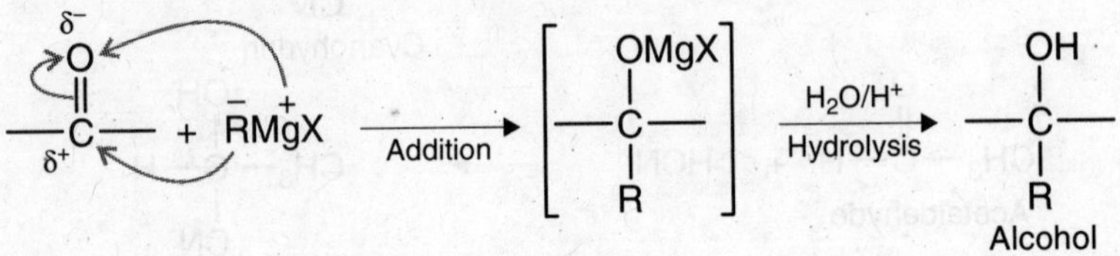

The reaction provides a convenient way of preparing alcohols that contain a larger carbon chain than the starting materials. Formaldehyde reacts with Grignard reagents to produce primary alcohols. Other aldehydes give secondary alcohols. Ketones react with Grignard reagents to produce tertiary alcohols. For example,

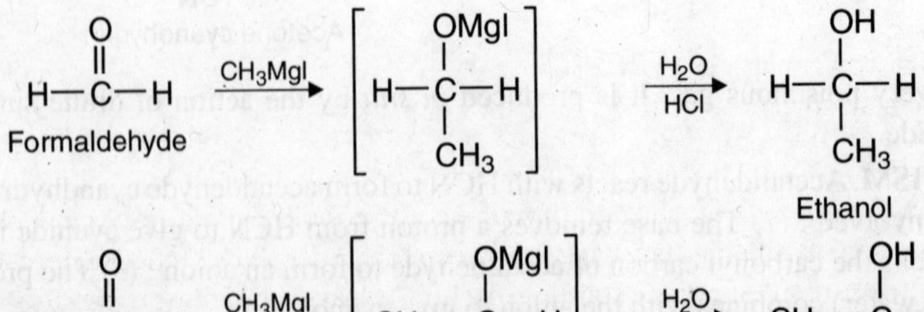

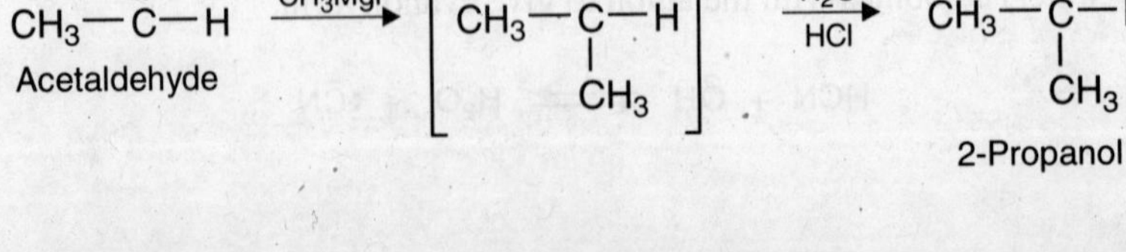

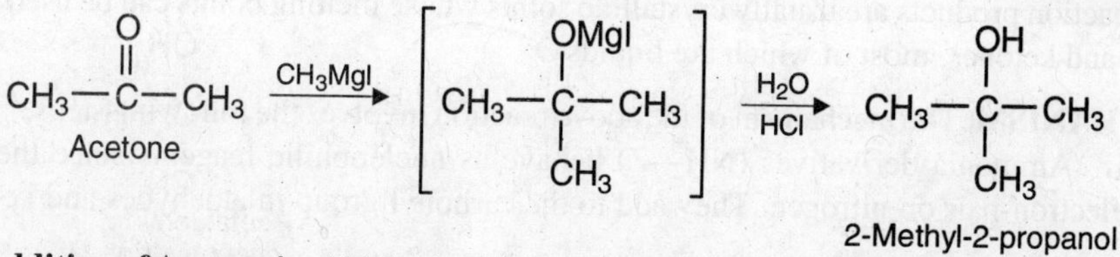

(4) **Addition of Ammonia.** Aldehydes (except formaldehyde) react with ammonia to form solid aldehyde ammonias. For example,

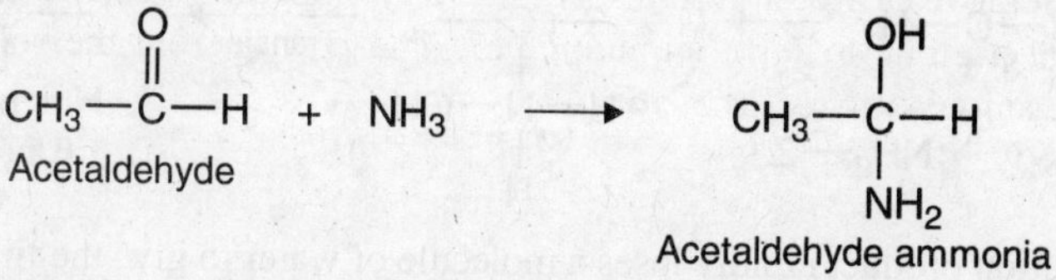

The aldehyde ammonias when heated with dilute acids, regenerate the aldehydes. Thus the formation and decomposition of these compounds is used for purification of aldehydes.

Formaldehyde and ketones do not form addition compounds with ammonia, but instead yield complex condensation products. For example,

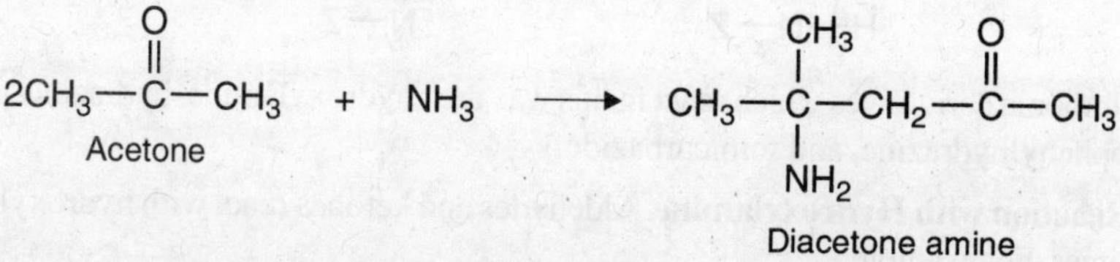

B. ADDITION REACTIONS FOLLOWED BY LOSS OF WATER

(5) **Addition of Alcohols.** Alcohols react with aldehydes in the presence of anhydrous HCl to form unstable addition products known as **Hemiacetals**. These hemiacetals react further with alcohol to form stable compounds known as **Acetals**. Notice that the acetals are *gem*-diethers.

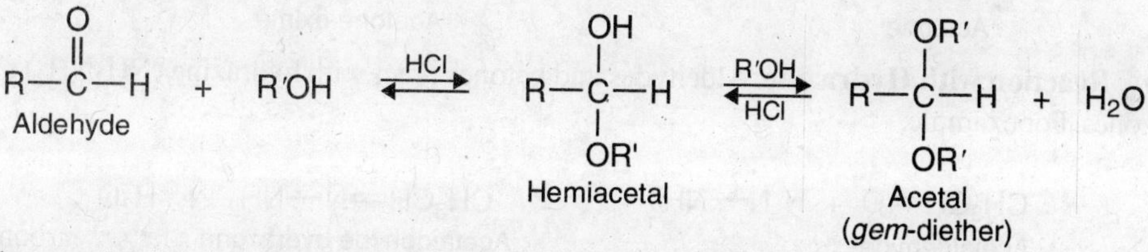

The reaction is reversible. A large excess of alcohol is used to shift the equilibrium in favor of acetal formation. The reaction of acetaldehyde with methyl alcohol results in the formation of acetaldehyde dimethyl acetal.

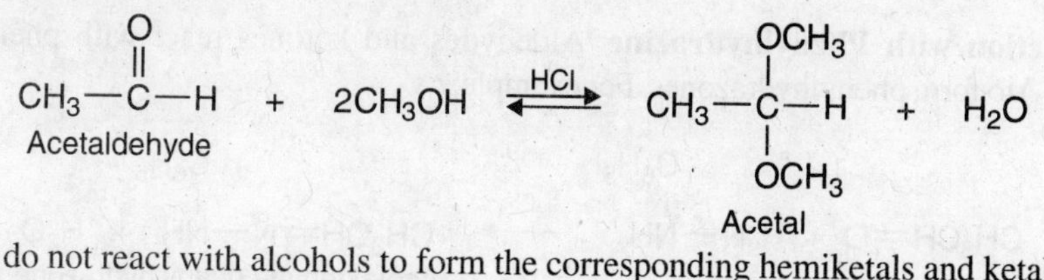

Ketones do not react with alcohols to form the corresponding hemiketals and ketals.

(6) **Reaction with Ammonia Derivatives.** Some ammonia derivatives (NH_2–Z) react with aldehydes and ketones to form compounds containing carbon-nitrogen double bonds, together with the elimination of a water molecule.

$$>C=O \; + \; H_2N-Z \longrightarrow \; >C=N-Z \; + \; H_2O$$

The reaction products are usually crystalline solids whose melting points can be used to identify aldehydes and ketones, most of which are liquids.

MECHANISM. The mechanism of the above reaction involves the following steps :

Step 1. Ammonia derivatives (NH_2–Z) behave as nucleophilic reagents since they have an unshared electron-pair on nitrogen. They add to the carbonyl group in aldehydes and ketones.

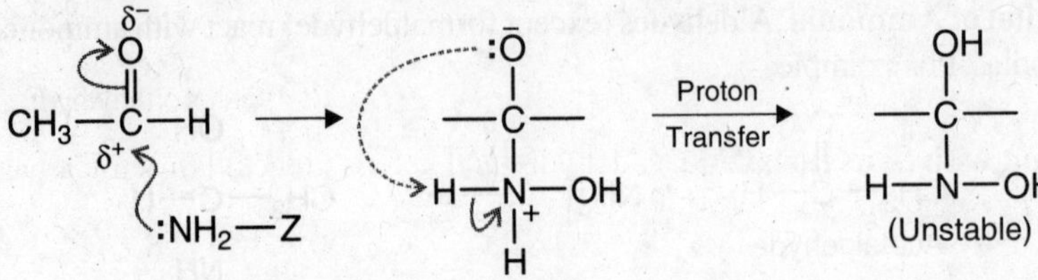

Step 2. Addition product rapidly loses a molecule of water to give the final product.

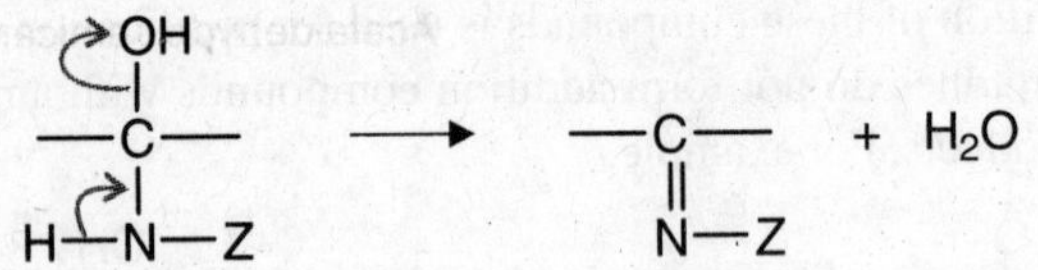

The ammonia derivatives which react in this way are : hydroxylamine, hydrazine, phenylhydrazine, 2,4-dinitrophenylhydrazine, and semicarbazide.

(*a*) **Reaction with Hydroxylamine**. Aldehydes and ketones react with hydroxylamine (NH_2OH) to form oximes. For example,

$$\underset{\text{Acetaldehyde}}{CH_3CH{=}O} + H_2N{-}OH \longrightarrow \underset{\text{Acetaldehyde oxime}}{CH_3CH{=}N{-}OH} + H_2O$$

$$\underset{\text{Acetone}}{(CH_3)_2C{=}O} + H_2N{-}OH \longrightarrow \underset{\text{Acetone oxime}}{(CH_3)_2C{=}N{-}OH} + H_2O$$

(*b*) **Reaction with Hydrazine**. Aldehydes and ketones react with hydrazine (NH_2NH_2) to form hydrazones. For example,

$$\underset{\text{Acetaldehyde}}{CH_3CH{=}O} + H_2N{-}NH_2 \longrightarrow \underset{\text{Acetaldehyde hydrazone}}{CH_3CH{=}N{-}NH_2} + H_2O$$

$$\underset{\text{Acetone}}{(CH_3)_2C{=}O} + H_2N{-}NH_2 \longrightarrow \underset{\text{Acetone hydrazone}}{(CH_3)_2C{=}N{-}NH_2} + H_2O$$

(*c*) **Reaction with Phenylhydrazine**. Aldehydes and ketones react with phenylhydrazine ($NH_2NHC_6H_5$) to form phenylhydrazones. For example,

$$\underset{\text{Acetaldehyde}}{CH_3CH{=}O} + H_2N{-}\overset{\overset{C_6H_5}{|}}{NH} \longrightarrow \underset{\text{Acetaldehyde phenylhydrazone}}{CH_3CH{=}N{-}\overset{\overset{C_6H_5}{|}}{NH}} + H_2O$$

$$\underset{\text{Acetone}}{(CH_3)_2C{=}O} + H_2N{-}\overset{\overset{C_6H_5}{|}}{NH} \longrightarrow \underset{\text{Acetone phenylhydrazone}}{(CH_3)_2C{=}N{-}\overset{\overset{C_6H_5}{|}}{NH}} + H_2O$$

(*d*) **Reaction with 2,4-Dinitrophenylhydrazine**. Aldehydes and ketones react with 2,4-dinitrophenylhydrazine to form 2,4-dinitrophenylhydrazones. For example,

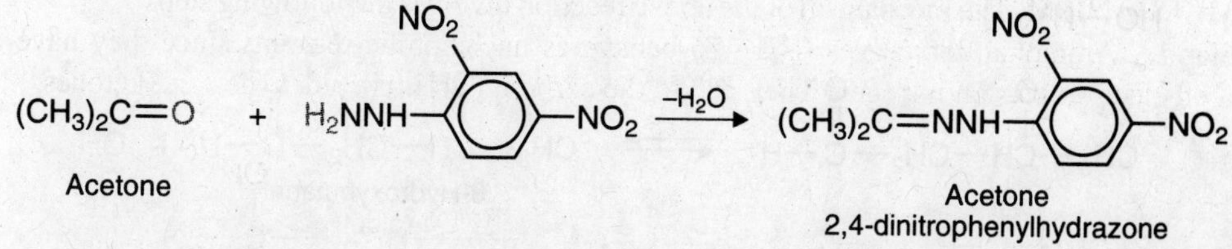

(*e*) **Reaction with Semicarbazide**. Aldehydes and ketones react with semicarbazides to form semicarbazones. For example,

$$\underset{\text{Acetaldehyde}}{CH_3CH{=}O} + H_2N{-}NH\overset{O}{\overset{\|}{C}}NH_2 \longrightarrow \underset{\text{Acetaldehyde semicarbazone}}{CH_3CH{=}N{-}NH\overset{O}{\overset{\|}{C}}NH_2} + H_2O$$

$$\underset{\text{Acetone}}{(CH_3)_2C{=}O} + H_2N{-}NH\overset{O}{\overset{\|}{C}}NH_2 \longrightarrow \underset{\text{Acetone semicarbazone}}{(CH_3)_2C{=}N{-}NH\overset{O}{\overset{\|}{C}}NH_2} + H_2O$$

C. REACTIONS INVOLVING ALKYL GROUPS

(7) **Aldol Condensation**. Aldehydes containing α-hydrogens undergo self-addition in the presence of a base to form products called **Aldols**. The reaction is called *Aldol Condensation*. The term aldol is derived from the combination of the words Aldehyde and Alcohol, the two functional groups present in the product. For example, two molecules of acetaldehyde combine with each other in the presence of dilute NaOH to form 3-hydroxybutanal.

$$\underset{\text{Acetaldehyde}}{CH_3{-}\overset{O}{\overset{\|}{C}}{-}H} + \underset{\text{Acetaldehyde}}{\overset{H}{\overset{|}{C}H_2}{-}\overset{O}{\overset{\|}{C}}{-}H} \xrightarrow{\bar{O}H} \underset{\text{3-Hydroxybutanal (aldol)}}{CH_3{-}\overset{OH}{\overset{|}{C}H}{-}CH_2{-}\overset{O}{\overset{\|}{C}}{-}H}$$

MECHANISM. The reaction is reversible and involves the following steps :

Step 1. The enolate ion is formed

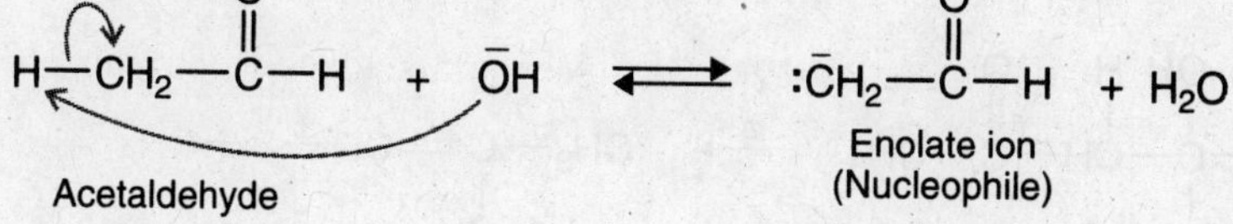

Step 2. The enolate ion attacks the carbonyl carbon of another unionized aldehyde molecule.

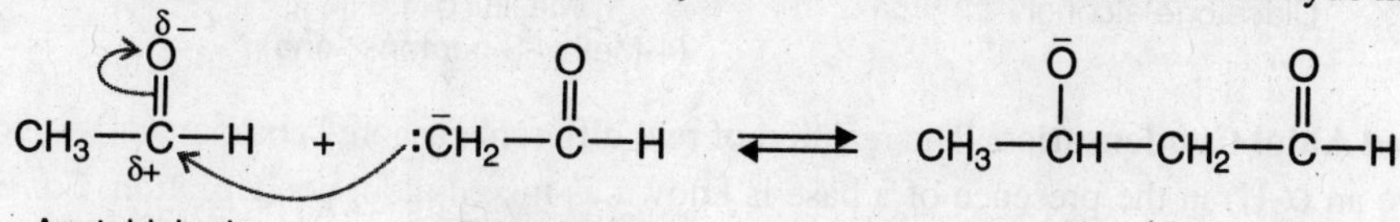

Step 3. The negative oxygen in the product accepts a proton from water to give aldol.

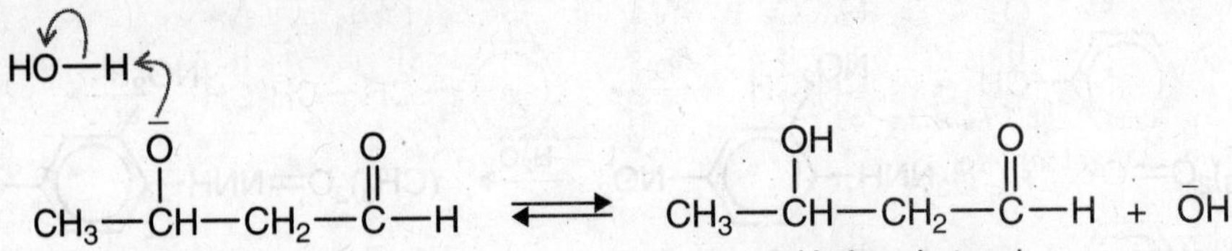

3-Hydroxybutanal (aldol)

The product in all aldol condensations has several common structural characteristics :

1. The product is a larger molecule that contains a newly formed carbon-carbon bond.
2. The newly formed C–C bond occurs between the α carbon of one aldehyde molecule and what was originally the carbonyl carbon of the second aldehyde molecule.
3. The product contains two functional groups: one carbonyl group and one hydroxyl group.
4. The hydroxyl group is always attached to the beta carbon (β carbon), which is two carbons away from the carbonyl group.

Aldols are easily dehydrated either by heating or by treatment with dilute acid to form α, β-unsaturated aldehydes. For example,

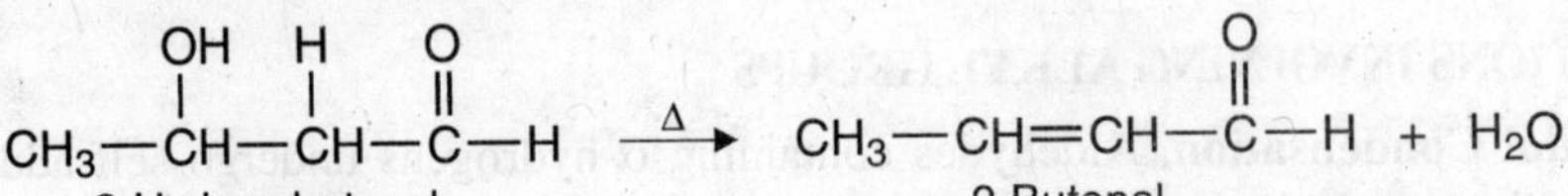

3-Hydroxybutanal → 2-Butenal (Crotonaldehyde)

Ketones containing α-hydrogens also undergo aldol condensation to form **Ketols**. For example, two molecules of acetone combine with each other in the presence of barium hydroxide to form 4-hydroxy-4-methyl-2-pentanone (diacetone alcohol).

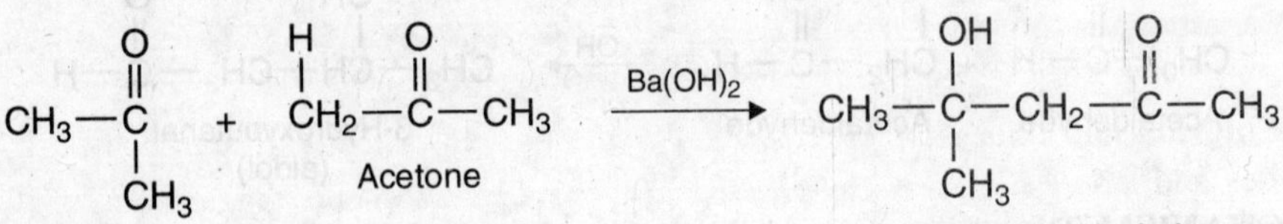

Acetone + Acetone → Diacetone alcohol

Ketols are also easily dehydrated by heating or by treatment with dilute acid to form α,β-unsaturated ketones. For example,

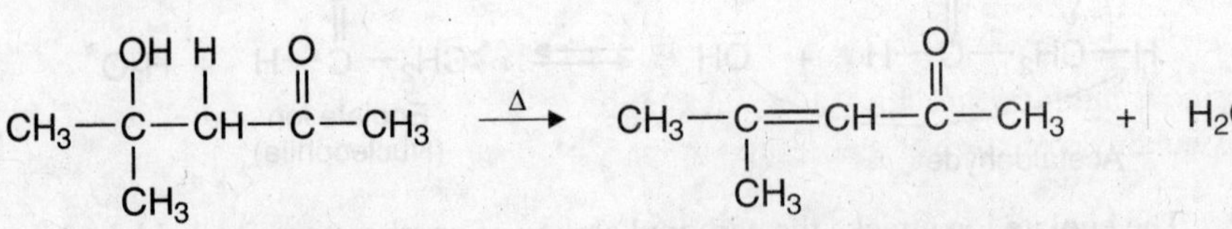

Diacetone alcohol → Mesityl oxide (4-Methyl-3-penten-2-one)

Mixed Aldol Condensation. The reaction of two different carbonyl compounds (one of which must have an α-H) in the presence of a base is known as mixed aldol condensation. For example, acetaldehyde reacts with benzaldehyde (which has no α-H) in the presence of a base to form cinnamaldehyde.

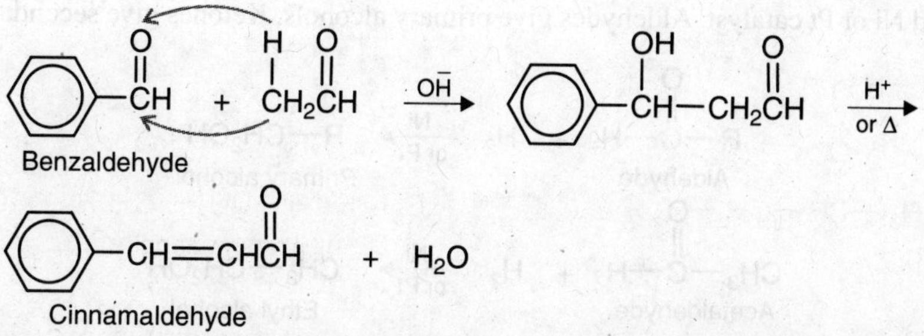

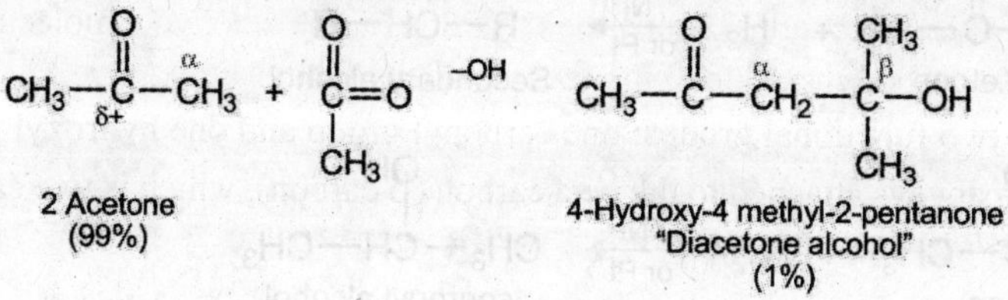

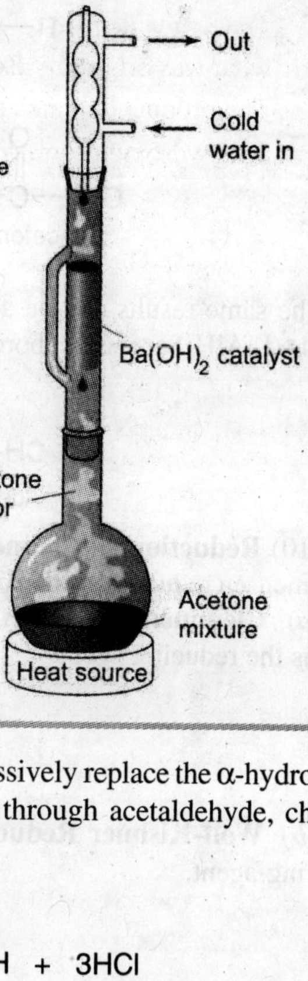

The Aldol condensation of acetone gives only 1% product at equilibrium, yet a clever technique gives a good yield. Acetone refluxes onto a basic catalyst such as $Ba(OH)_2$. The nonvolatile diacetone alcohol does not reflux, so its equilibrium concentration gradually increases until all the acetone is converted to diacetone alcohol.

(8) **Chlorination**. Under suitable conditions, chlorine will successively replace the α-hydrogens in aldehydes and ketones. For example, when chlorine is bubbled through acetaldehyde, chloral (trichloroethanal) is obtained.

$$\underset{\text{Acetaldehyde}}{CH_3-\overset{O}{\overset{\|}{C}}-H} + 3Cl_2 \longrightarrow \underset{\text{Chloral}}{CCl_3-\overset{O}{\overset{\|}{C}}-H} + 3HCl$$

If chlorine is bubbled through warm acetone, successive replacement of the methyl hydrogens takes place, yielding a mixture of chloropropanones.

$$\underset{\text{Acetone}}{CH_3-\overset{O}{\overset{\|}{C}}-CH_3} \xrightarrow{Cl_2} CH_3-\overset{O}{\overset{\|}{C}}-CH_2Cl + CH_3-\overset{O}{\overset{\|}{C}}-CHCl_2\text{, etc.}$$

D. REDUCTION REACTIONS

Aldehydes and ketones can be reduced to alcohols or alkenes.

(9) **Reduction to Alcohols.** Aldehydes and ketones can be reduced to alcohols by treatment with hydrogen and Ni or Pt catalyst. Aldehydes give primary alcohols. Ketones give secondary alcohols.

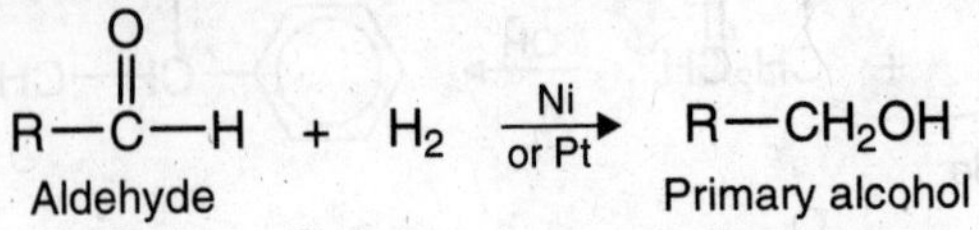

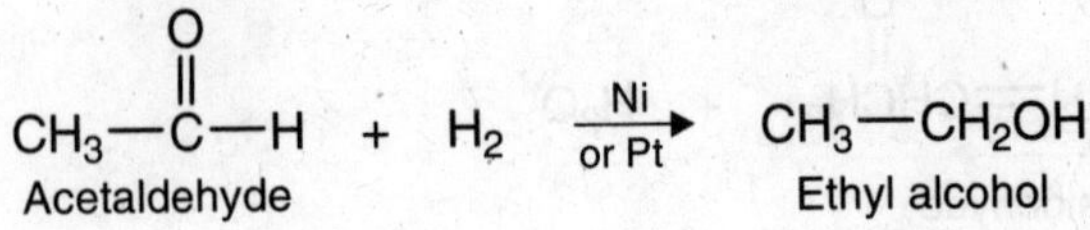

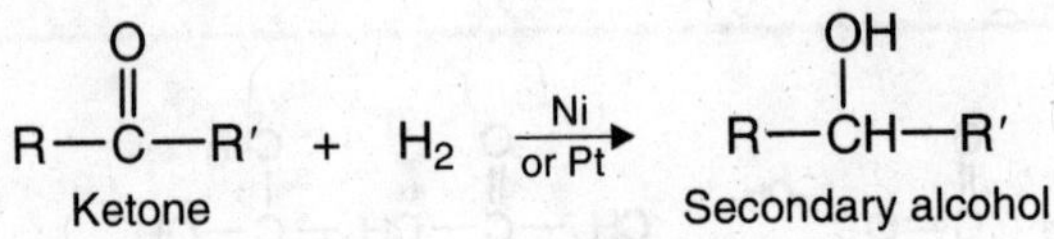

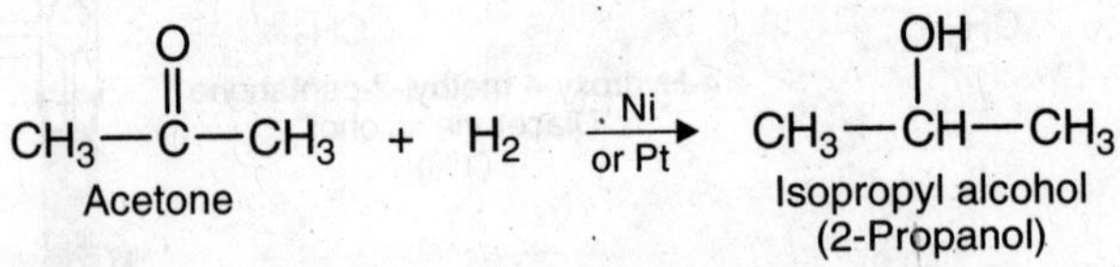

The same results can be achieved with chemical reducing agents such as lithium aluminium hydride ($LiAlH_4$) or sodium borohydride ($NaBH_4$). For example,

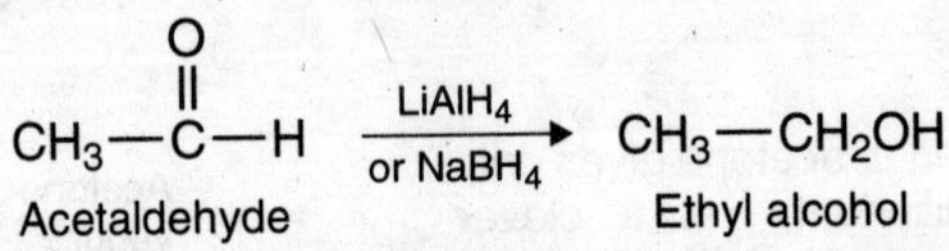

(10) **Reduction to Alkanes.** Aldehydes and ketones can be reduced to alkanes by either the Clemmensen reduction or the Wolf-Kishner reduction.

(*a*) **Clemmensen Reduction.** This involves the use of zinc-mercury amalgam in hydrochloric acid as the reducing agent.

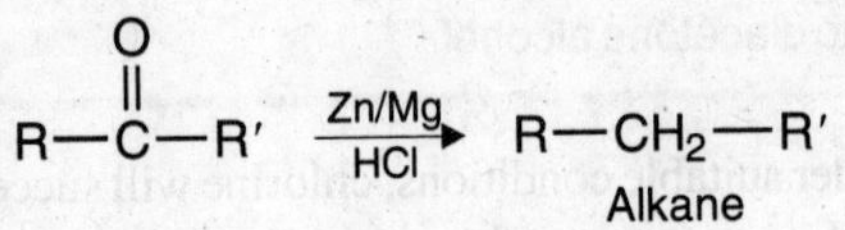

(*b*) **Wolf-Kishner Reduction.** This involves the use of a basic solution of hydrazine as the reducing agent.

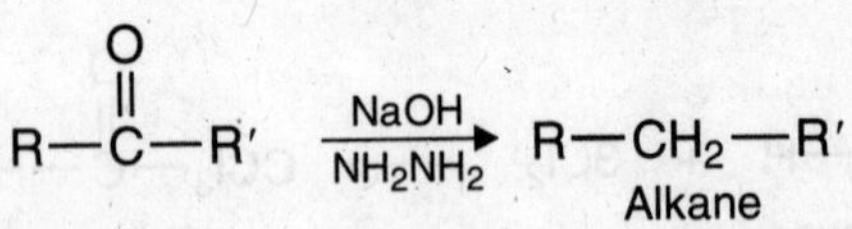

(11) **Reduction to Pinacols.** Ketones when reduced in neutral or alkaline medium, form pinacols (symmetrical 1,2-diols). For example, acetone undergoes reduction with magnesium amalgam to form 2,3-dimethylbutane-2,3-diol.

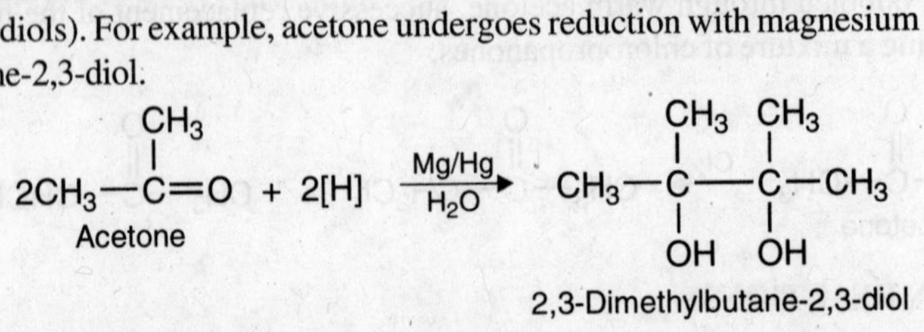

Aldehydes do not give this reaction.

E. OXIDATION REACTIONS

Although the chemical reactivity of aldehydes and ketones is very similar, their behavior toward oxidizing agents is quite different. Aldehydes are easily oxidized. Ketones are oxidized only under drastic conditions.

(12) **Oxidation of Aldehydes**. Aldehydes can be oxidized with sodium (or potassium) dichromate in acidic medium to form carboxylic acids containing the same number of carbon atoms.

$$\underset{\text{Aldehyde}}{R-\overset{O}{\overset{\|}{C}}-H} \xrightarrow[H_2SO_4]{Na_2Cr_2O_7} \underset{\text{Carboxylic acid}}{R-\overset{O}{\overset{\|}{C}}-OH} \qquad \underset{\text{Acetaldehyde}}{CH_3-\overset{O}{\overset{\|}{C}}-H} \xrightarrow[H_2SO_4]{Na_2Cr_2O_7} \underset{\text{Acetic acid}}{CH_3-\overset{O}{\overset{\|}{C}}-OH}$$

Potassium permanganate can also be used in place of sodium dichromate. Aldehydes can be oxidized by much milder oxidizing agents such as *Tollens' reagent, Fehling's solution* and *Benedict's solution*.

Tollens' Reagent. Tollen's reagent is an ammoniacal solution of silver oxide. It is obtained by adding ammonia to a precipitate of silver oxide present in a solution of silver nitrate and sodium hydroxide. When Tollens' reagent is used to oxidize an aldehyde, the silver ion is reduced to metallic form and if the reaction is carried out in a clean test tube, deposits as a *mirror*. The silver mirror formed indicates the presence of an aldehyde group in a molecule.

$$\underset{\text{Aldehyde}}{R-\overset{O}{\overset{\|}{C}}-H} + 2Ag(NH_3)_2OH \longrightarrow R-\overset{O}{\overset{\|}{C}}-\overset{-}{O}\overset{+}{N}H_4 + \underset{\text{Silver mirror}}{2Ag\downarrow} + H_2O + 3NH_3$$

The Tollens' test detects the presence of readily oxidised functional groups in organic molecules, such as aldehydes, by the rapid deposition of a silver mirror on the glass walls of the flask.

Fehling's Solution. Fehling's solution is an alkaline solution of cupric ion complexed with sodium potassium tartrate ions. When Fehling's solution is used to oxidize an aldehyde, the complexed cupric ion (*deep blue*) is reduced to cuprous oxide (*red*). The presence of the red precipitate of cuprous oxide serves as an indication of an aldehyde group in a molecule.

The presence of an aldehyde group in a molecule is shown by the formation of a brick-red precipitate of copper (I) oxide in the Fehling's test.

$$\underset{\text{Aldehyde}}{R-\overset{O}{\overset{\|}{C}}-H} + 2Cu(OH)_2 + NaOH \longrightarrow R-\overset{O}{\overset{\|}{C}}-\overset{-}{O}\overset{+}{Na} + \underset{\text{red}}{Cu_2O\downarrow} + 3H_2O$$

Benedict's solution is an alkaline solution of cupric ion complexed with citrate ions. It reacts in the same way as Fehling's solution.

(13) **Oxidation of Ketones**. Ketones can be oxidized by strong oxidizing agents such as alkaline $KMnO_4$ or hot concentrated HNO_3 to form two carboxylic acids with fewer carbon atoms than the original ketone. This is because the ketone is broken into two fragments by attack on either side of the carbonyl group.

$$R-\overset{O}{\overset{\|}{C}}-R' \xrightarrow[\Delta]{HNO_3} R-\overset{O}{\overset{\|}{C}}-OH + R'-\overset{O}{\overset{\|}{C}}-OH$$

Either of these bonds can break   Carboxylic acids

$$\underset{\text{2-Butanone}}{CH_3CH_2-\overset{O}{\overset{\|}{C}}-CH_3} \xrightarrow[\Delta]{HNO_3} \underset{\text{Propanoic acid}}{CH_3CH_2-\overset{O}{\overset{\|}{C}}-OH} + \underset{\text{Acetic acid}}{CH_3-\overset{O}{\overset{\|}{C}}-OH}$$

Ketones do not react with Tollens' reagent, Fehling's solution, or Benedict's solution.

F. SOME OTHER IMPORTANT REACTIONS

(14) **Haloform Reaction**. Acetaldehyde and methyl ketones react rapidly with halogen (Cl_2, Br_2, or I_2) in the presence of alkali to form haloform. For example,

$$\underset{\text{Methyl ketone}}{R-\overset{O}{\overset{\|}{C}}-CH_3} + 3Br_2 + 4NaOH \xrightarrow{\Delta} R-\overset{O}{\overset{\|}{C}}-\overset{-}{O}\overset{+}{Na} + \underset{\text{Bromoform}}{CHBr_3} + 3H_2O + 3NaBr$$

This reaction, called the *Haloform reaction*, takes place in two steps.

Step 1. Three hydrogen atoms on the α-carbon are successively replaced by halogen atoms.

$$\underset{\text{Methyl ketone}}{R-\overset{\overset{O}{\|}}{C}-CH_3} + 3Br_2 + 3NaOH \longrightarrow \underset{\alpha,\alpha,\alpha\text{-Tribromoketone}}{R-\overset{\overset{O}{\|}}{C}-CBr_3} + 3H_2O + 3NaBr$$

Step 2. The α,α,α-trihaloketone molecule is cleaved to give a trihalomethane.

$$R-\overset{\overset{O}{\|}}{C}-CBr_3 + NaOH \longrightarrow R-\overset{\overset{O}{\|}}{C}-\overset{-}{O}\overset{+}{Na} + \underset{\text{Bromoform}}{CHBr_3}$$

The haloform reaction is used as a diagnostic test for the presence of $-COCH_3$ group. For this purpose, a solution of iodine is added to an unknown compound in an aqueous alkali solution. A positive test will yield a bright yellow precipitate of iodoform (Chloroform and bromoform are liquids). **This is known as Iodoform test and is given by methyl ketones and acetaldehyde. Ethyl alcohol and secondary alcohols also give a positive iodoform test**. This is because ethyl alcohol is converted to acetaldehyde and secondary alcohols are converted to methyl ketones under the conditions used for the test.

The haloform reaction can also be used to distinguish methyl ketones from other ketones. For example,

$$\underset{\text{2-Pentanone}}{CH_3CH_2CH_2-\overset{\overset{O}{\|}}{C}-CH_3} \xrightarrow[NaOH]{I_2} \underset{\text{Iodoform}}{CHI_3\downarrow}$$

$$\underset{\text{3-Pentanone}}{CH_3CH_2-\overset{\overset{O}{\|}}{C}-CH_2CH_3} \xrightarrow[NaOH]{I_2} \text{No Precipitate Formed}$$

(15) **Cannizzaro Reaction**. Aldehydes which lack an α-hydrogen, when heated with concentrated NaOH, undergo a disproportionation reaction. One half of the aldehyde molecules are oxidized to a carboxylic acid and one half are reduced to an alcohol. This reaction is known as *Cannizzaro Reaction*. (Remember : Aldehydes with α-hydrogens do not undergo this reaction. Under these conditions, they undergo an Aldol condensation reaction). For example,

$$\underset{\substack{\text{Formaldehyde}\\(\text{No }\alpha\text{-hydrogens})}}{2H-\overset{\overset{O}{\|}}{C}-H} + NaOH \xrightarrow{\Delta} \underset{\text{Methanol}}{CH_3OH} + \underset{\text{Sodium formate}}{H-\overset{\overset{O}{\|}}{C}-\overset{-}{O}\overset{+}{Na}}$$

MECHANISM. Two steps are involved :

Step 1. Attack of the HO^- on the carbonyl group.

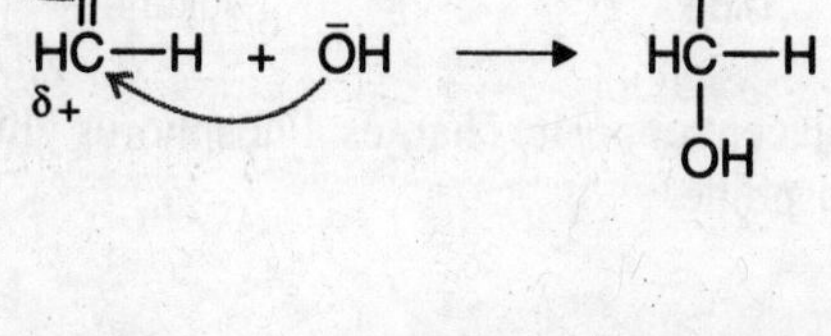

Step 2. Hydride transfer.

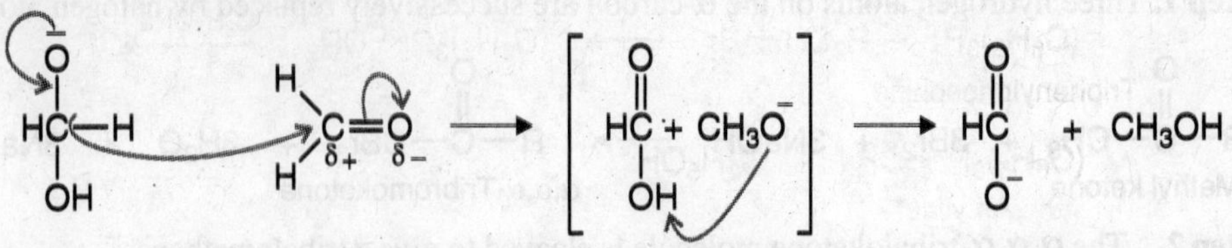

Ketones do not give this reaction.

(16) **Reformatsky Reaction**. This involves the treatment of an aldehyde or ketone with α-bromo ester in the presence of zinc. The product (after acid-hydrolysis) is a β-hydroxy ester. For example,

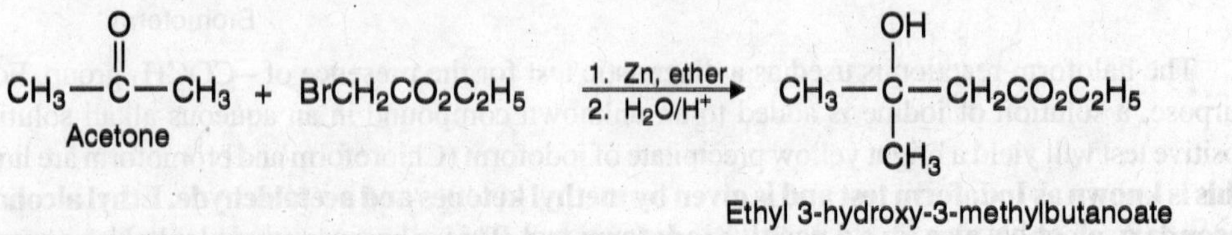

MECHANISM. Three steps are involved :

Step 1. Formation of zinc salt of enol of the ester.

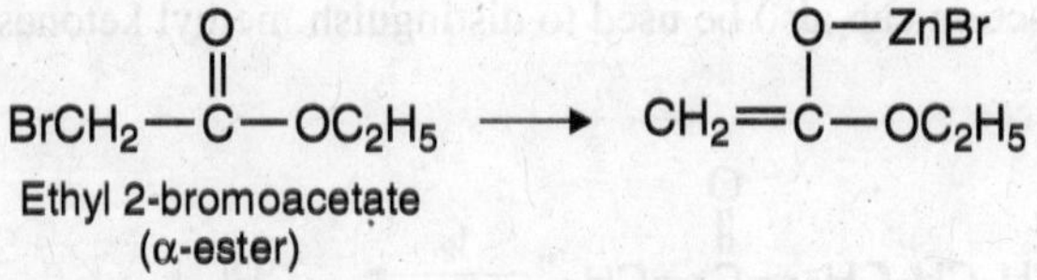

Step 2. The zinc salt reacts with the carbonyl compound.

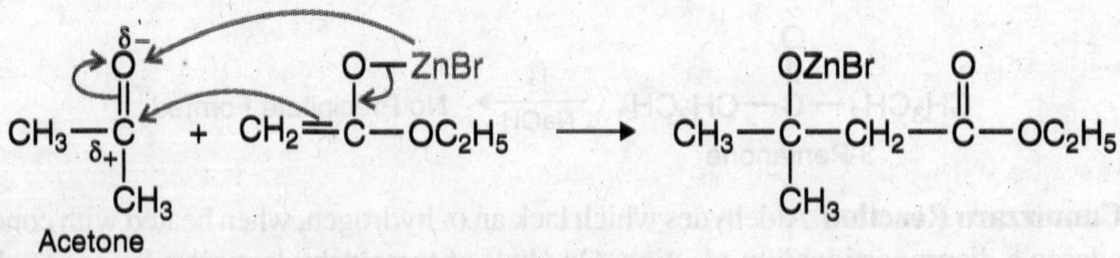

Step 3. Acid-hydrolysis gives the β-hydroxy ester.

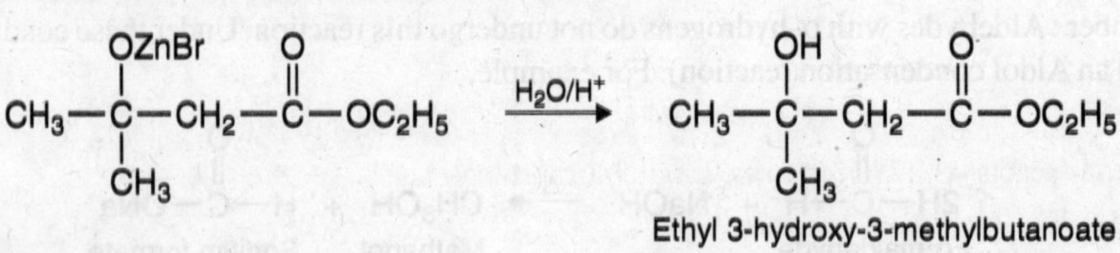

(17) **Wittig Reaction**. This involves the treatment of aldehydes and ketones with phosphorus ylides to form alkenes.

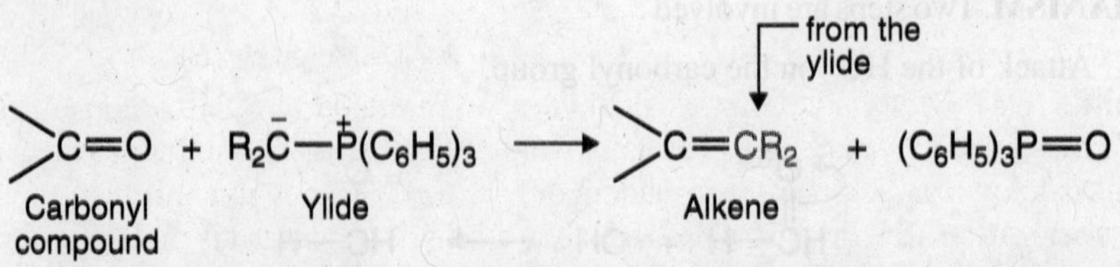

An ylide is a molecule with adjacent opposite charges. Phosphorus ylides are prepared from primary alkyl halides and triphenylphosphine.

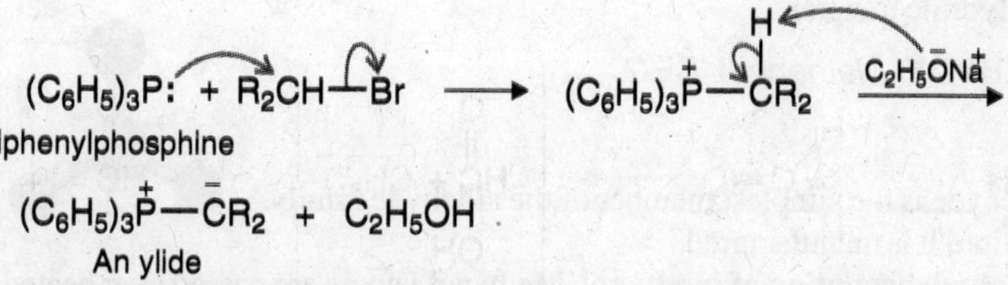

MECHANISM. Two steps are involved :

Step 1. The negative carbon of the ylide attacks the carbonyl carbon to form a *betaine*. A betaine is a molecule having nonadjacent opposite charges.

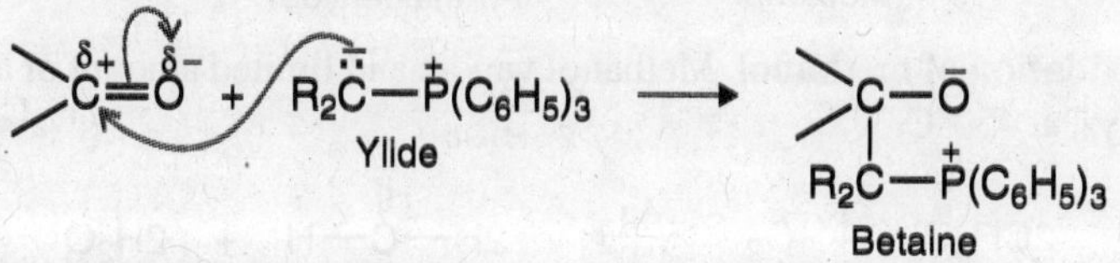

Step 2. The betaine undergoes elimination of triphenylphosphine oxide to give the alkene.

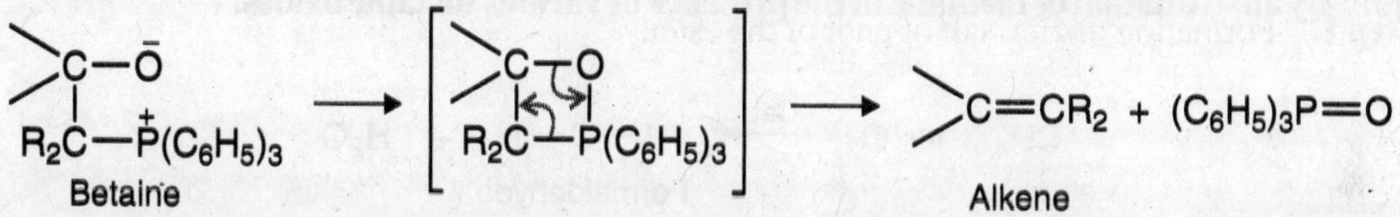

The Wittig reaction is an excellent method of making alkenes from aldehydes and ketones. For example,

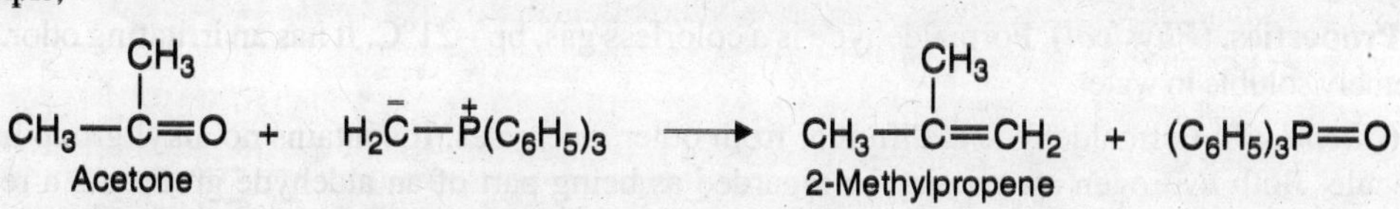

(18) **Reaction with Phosphorus Pentachloride.** Both aldehydes and ketones react with PCl_5 to form *gem*-dihalides. For example, acetaldehyde gives 1,1-dichloroethane.

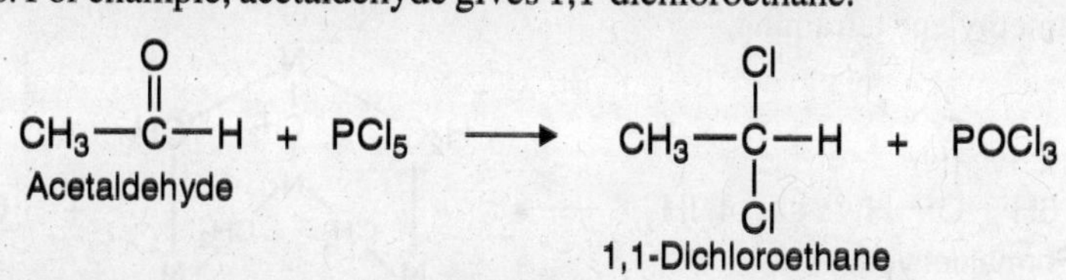

Acetone reacts with PCl_5 to form 2,2-dichloropropane.

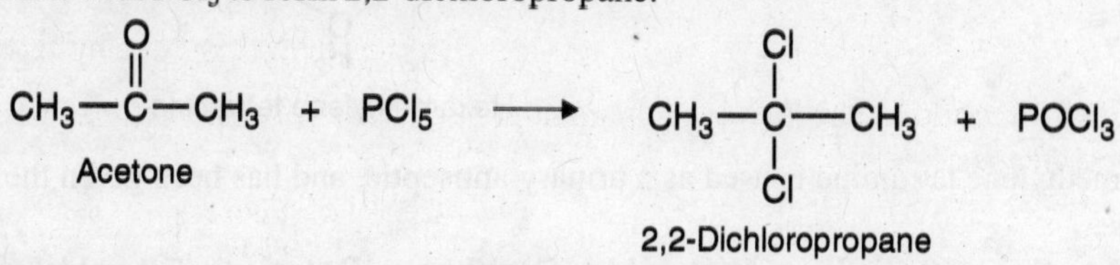

(19) **Schiff's Test.** Dilute solutions of aldehydes when added to Schiff's reagent (Rosaniline hydrochloride dissolved in water and its red color decolorized by passing sulfur dioxide) restores its red color. This is known as Schiff's test for aldehydes. Ketones do not give this test.

(20) **Polymerization.** The first few members of the aldehyde series readily undergo polymerization to give a variety of products (for details see formaldehyde and acetaldehyde). Ketones, however, are not much susceptible to polymerization.

FORMALDEHYDE, *Methanal*, HCHO

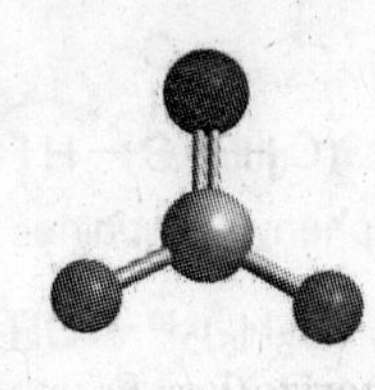

Formaldehyde is the simplest member of the aldehyde family.

Preparation. It is manufactured :

(1) **By dehydrogenation of methanol.** Methanol vapors are passed over heated silver catalyst at 300°C.

$$\underset{\text{Methanol}}{CH_3OH} \xrightarrow[300^\circ]{Ag} \underset{\text{Formaldehyde}}{H-\overset{\overset{O}{\|}}{C}-H} + H_2$$

(2) **By air-oxidation of methanol.** Methanol vapors and limited amount of air are passed over heated silver catalyst at 450°C.

$$\underset{\text{Methanol}}{2CH_3OH} + O_2 \xrightarrow[450^\circ]{Ag} \underset{\text{Formaldehyde}}{2H-\overset{\overset{O}{\|}}{C}-H} + 2H_2O$$

(3) **By air-oxidation of methane in the presence of various metallic oxides.**

$$CH_4 + O_2 \xrightarrow[\Delta]{Cat.} \underset{\text{Formaldehyde}}{H-\overset{\overset{O}{\|}}{C}-H} + H_2O$$

Since formaldehyde is a gas, the product is marketed as 40% aqueous solution under the name **Formalin.**

Properties. (*Physical*). Formaldehyde is a colorless gas, bp –21°C. It has an irritating odor. It is extremely soluble in water.

(*Chemical*). Formaldehyde is different from other aldehydes. It contains no alkyl group in the molecule. Both hydrogen atoms may be regarded as being part of an aldehyde group. As a result, several reactions of formaldehyde are different from those of other aldehydes.

(1) **Reaction with Ammonia**. When treated with ammonia, it does not give an aldehyde-ammonia, but forms hexamethylene tetramine.

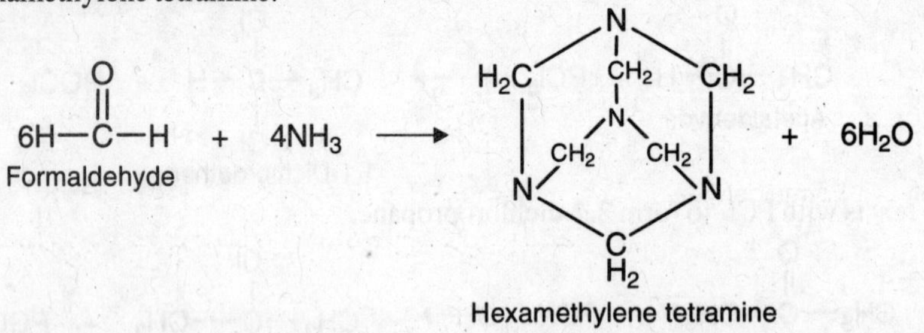

Hexamethylene tetramine is used as a urinary antiseptic, and has been given the trade name of **Urotropine.**

(2) **Reaction with Sodium Hydroxide (Cannizzaro Reaction)**. Formaldehyde reacts with concentrated NaOH solution to form methanol and sodium formate.

$$\underset{\text{Formaldehyde}}{2HCHO} + NaOH \longrightarrow \underset{\text{Methanol}}{CH_3OH} + \underset{\text{Sodium formate}}{HCOONa}$$

(3) **Reaction with Alcohols**. Like other aldehydes, it reacts with alcohols to form acetals. For example, formaldehyde reacts with methanol in the presence of HCl to form **Methylal.**

$$\mathrm{H{-}\overset{\overset{\Large O}{\|}}{C}{-}H + 2CH_3OH \xrightleftharpoons{HCl} H{-}\underset{\underset{\Large OCH_3}{|}}{\overset{\overset{\Large OCH_3}{|}}{C}}{-}H + H_2O}$$

Methylal

(4) **Polymerization**. Formaldehyde polymerizes as follows :

(*a*) When aqueous solution of formaldehyde (*Formalin*) is evaporated to dryness, **Paraformaldehyde** is formed.

$$\mathrm{HOCH_2(CH_2O)_nCH_2OH}$$

Paraformaldehyde

$(n = 6 - 50)$

Paraformaldehyde is a crystalline solid, mp 121-123°C. It regenerates formaldehyde on heating. Because paraformaldehyde is readily reconverted to formaldehyde upon gentle heating, it serves as a convenient storage form of formaldehyde.

(*b*) Formaldehyde on treatment with concentrated H_2SO_4 gives **Polyoxymethylene**, $(CH_2O)_n2H_2O$, where $n > 100$. Polyoxymethlyenes are insoluble white solids which regenerate formaldehyde on heating.

(*c*) When formaldehyde gas is allowed to stand at room temperature, it slowly undergoes polymerization and forms a white solid called **Metaformaldehyde** or **Trioxane**.

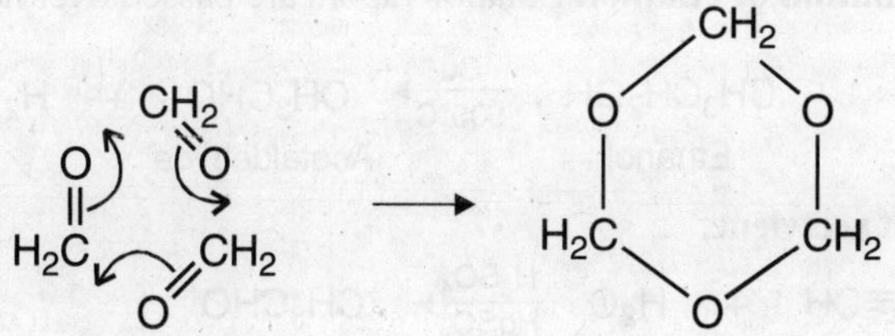

Formaldehyde (3 molecules) — Metaformaldehyde (mp 61–62°C)

(5) **Condensation with Phenol**. Formaldehyde condenses with phenol to give a synthetic plastic

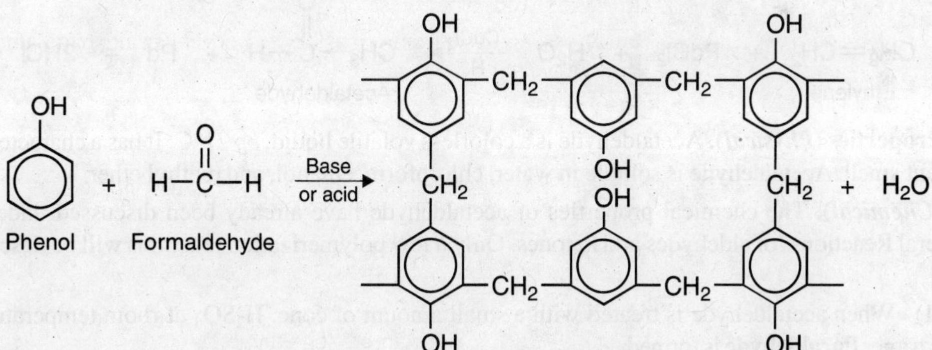

Bakelite

Bakelite. Phenol is refluxed with formalin and 0.88 ammonia (catalyst) when an oil separates. The oily liquid is transferred to an open vessel and heated until a test sample, on cooling in water, is found to be hard and brittle. It is then left to cool to give bakelite.

Uses. Formaldehyde is sold as 40% aqueous solution under the name FORMALIN and is used in this form for most purposes. Formalin is used : (*1*) as a general antiseptic ; (*2*) in the manufacture of urinary antiseptic *Urotropine* ; (*3*) in the preservation of biological specimens ; (*4*) in the manufacture of synthetic dyes like Pararosaniline and Indigo ; and (*5*) in the manufacture of synthetic plastics such as *Bakelite* and *Melmac*.

ACETALDEHYDE, *Ethanal*, CH_3CHO

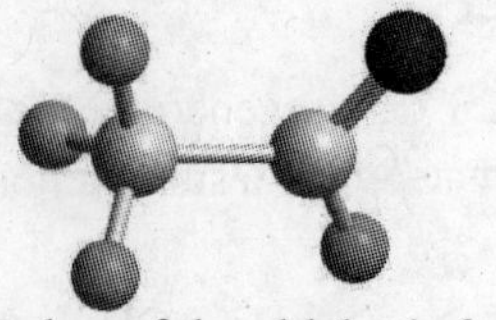

Acetaldehyde is the most important member of the aldehyde family.

Laboratory Preparation. Acetaldehyde is prepared in the laboratory by oxidation of ethanol with acidified sodium dichromate solution.

$$\underset{\text{Ethanol}}{CH_3CH_2OH} + [O] \xrightarrow[H^+]{Na_2Cr_2O_7} \underset{\text{Acetaldehyde}}{CH_3-\overset{\overset{\Large O}{\|}}{C}-H} + H_2O$$

Manufacture. Acetaldehyde is manufactured :

(1) **By air-oxidation of ethanol.** Ethanol vapors and limited amount of air are passed over heated silver catalyst at 500°C.

$$\underset{\text{Ethanol}}{CH_3CH_2OH} + O_2 \xrightarrow[500°C]{Ag} \underset{\text{Acetaldehyde}}{CH_3CHO} + 2H_2O$$

(2) **By dehydrogenation of ethanol.** Ethanol vapors are passed over heated copper catalyst at 500°C.

$$\underset{\text{Ethanol}}{CH_3CH_2OH} \xrightarrow[500°C]{Cu} \underset{\text{Acetaldehyde}}{CH_3CHO} + H_2$$

(3) **By hydration of acetylene.**

$$\underset{\text{Acetylene}}{HC\equiv CH} + H_2O \xrightarrow[HgSO_4]{H_2SO_4} \underset{\text{Acetaldehyde}}{CH_3CHO}$$

(4) **By Wacker Process.** This involves the treatment of ethylene with an acidified aqueous solution of palladium chloride and cupric chloride.

$$\underset{\text{Ethylene}}{CH_2=CH_2} + PdCl_2 + H_2O \xrightarrow[H^+]{CuCl_2} \underset{\text{Acetaldehyde}}{CH_3-\overset{\overset{\Large O}{\|}}{C}-H} + Pd + 2HCl$$

Properties (*Physical*). Acetaldehyde is a colorless volatile liquid, *bp* 21°C. It has a characteristic pungent smell. Acetaldehyde is soluble in water, chloroform, ethanol, and diethyl ether.

(*Chemical*). The chemical properties of acetaldehyde have already been discussed under the 'General Reactions' of aldehydes and ketones. Only a few polymerization reactions will be described here.

(1) When acetaldehyde is treated with a small amount of conc. H_2SO_4 at room temperature, a cyclic trimer **Paraldehyde** is formed.

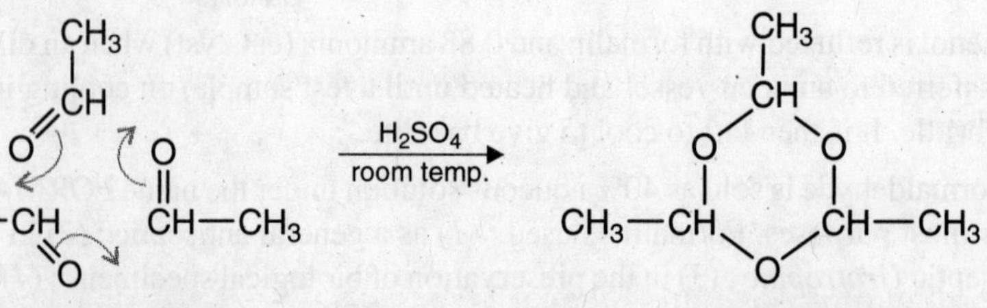

Paraldehyde is a liquid, *bp* 128°C. It regenerates acetaldehyde on distilling with conc. H_2SO_4.

(2) When acetaldehyde is treated with a small amount of conc. H_2SO_4 at 0°C, a cyclic tetramer **Metaldehyde** is formed.

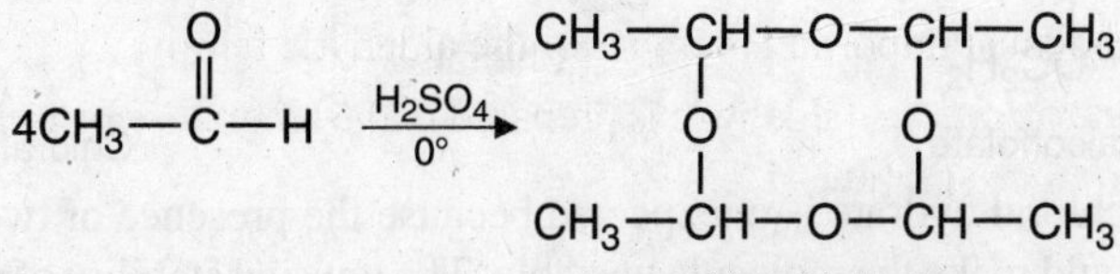

Metaldehyde

Metaldehyde is a solid, *mp* 246°C. This tetramer is used for killing slugs and snails.

Uses. Acetaldehyde is used in the manufacture of : (*1*) Acetic acid ; (*2*) Acetic anhydride ; (*3*) Chloral ; (*4*) 1,3-Butadiene (used in rubbers) ; and (*5*) Dyes.

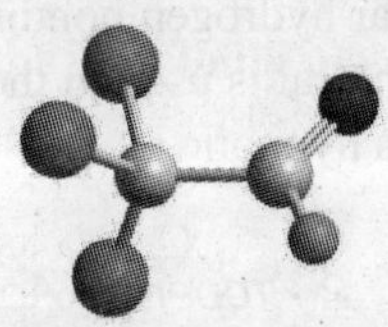

CHLORAL, *Trichloroethanal*, CCl_3CHO

Preparation. Chloral is prepared on large scale by the action of chlorine on ethanol. Chlorine is first passed into cold ethanol and then at 60°C till no further absorption of chlorine takes place. The final product is chloral alcoholate which separates as a crystalline solid. This on distillation with conc. H_2SO_4 gives chloral. The reactions involved may be shown as :

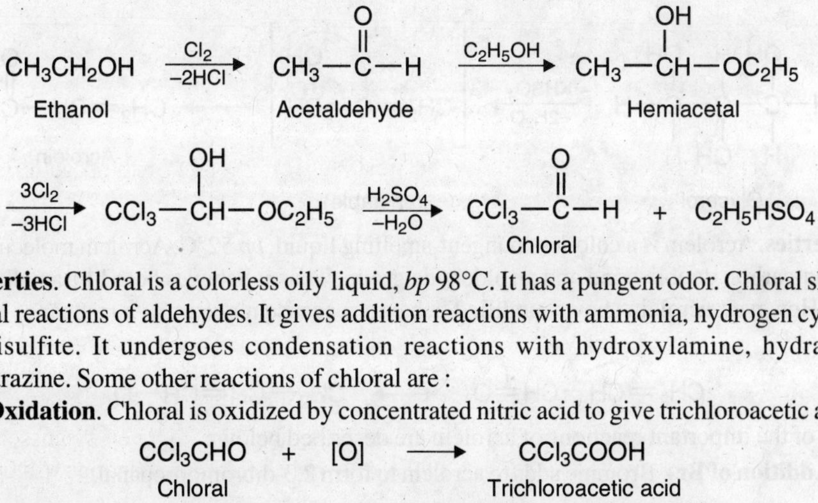

Properties. Chloral is a colorless oily liquid, *bp* 98°C. It has a pungent odor. Chloral shows most of the usual reactions of aldehydes. It gives addition reactions with ammonia, hydrogen cyanide, and sodium bisulfite. It undergoes condensation reactions with hydroxylamine, hydrazine, and phenylhydrazine. Some other reactions of chloral are :

(1) **Oxidation**. Chloral is oxidized by concentrated nitric acid to give trichloroacetic acid.

$$\underset{\text{Chloral}}{CCl_3CHO} + [O] \longrightarrow \underset{\text{Trichloroacetic acid}}{CCl_3COOH}$$

(2) **Reduction**. Chloral is reduced by aluminium ethoxide to give trichloroethanol.

$$\underset{\text{Chloral}}{CCl_3CHO} \xrightarrow{Al(OC_2H_5)_3} \underset{\text{Trichloroethanol}}{CCl_3CH_2OH}$$

(3) **Reaction with NaOH**. When heated with concentrated sodium hydroxide solution, it yields pure chloroform.

$$\underset{\text{Chloral}}{CCl_3CHO} + NaOH \xrightarrow{\Delta} \underset{\text{Chloroform}}{CHCl_3} + \underset{\text{Sodium formate}}{HCOONa}$$

(4) **Reaction with Water and Ethanol**. When chloral is treated with water or ethanol, it forms chloral hydrate (*mp* 57°C) or chloral alcoholate (*mp* 46°C), respectively. These compounds are stable, and the water or ethanol can only be removed by treatment with concentrated sulfuric acid.

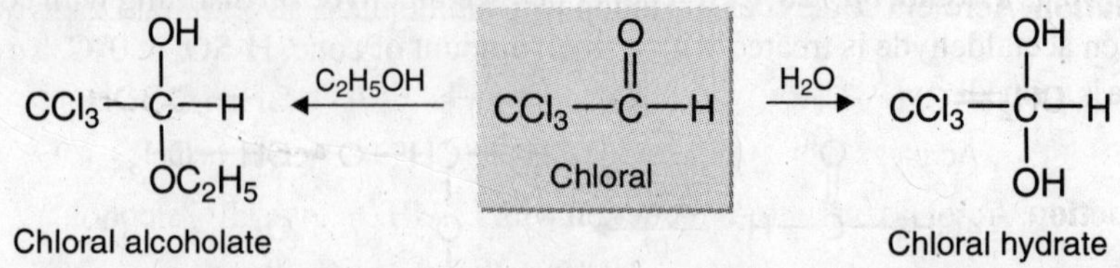

The stability of chloral hydrate is unexpected because the presence of two –OH groups on the same carbon atom should make the molecule unstable. The unusual stability of chloral hydrate can be explained as follows : The strong *–I Effect* of the three chlorine atoms gives the carbon atom of the carbonyl group a small positive charge which prevents the release of a hydroxyl ion. Thus the elimination of water molecule becomes difficult. Also, the IR spectrum of chloral hydrate shows intramolecular hydrogen bonding between the chlorine and hydrogen atoms.

Uses. Chloral is used in the manufacture of *DDT*, an important insecticide. It has been used in medicine as a hypnotic.

ACROLEIN, *2-Propenal, Acraldehyde,* CH_2=CHCHO

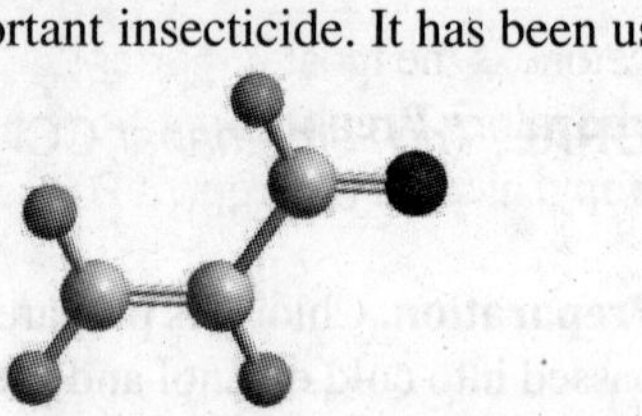

Acrolein is the simplest unsaturated aldehyde.

Preparation. Acrolein is prepared by the dehydration of glycerol with potassium hydrogen sulfate.

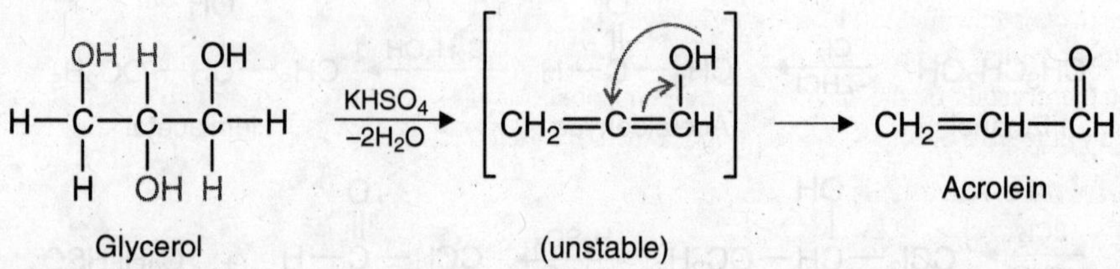

Properties. Acrolein is a colorless pungent-smelling liquid, *bp* 52°C. Acrolein molecule contains both carbon-carbon double bond and an aldehyde group. Thus it behaves both like an alkene and an aldehyde. However, acrolein shows modified behavior because it occurs as a resonance hybrid.

Some of the important reactions of acrolein are described below :

(1) **Addition of Br_2**. Bromine adds to acrolein to form 2,3-dibromopropanal.

$$\underset{\text{Acrolein}}{CH_2{=}CH{-}CHO} + Br_2 \longrightarrow \underset{\text{2,3-Dibromopropanal}}{\underset{Br}{CH_2}{-}\underset{Br}{CH}{-}CHO}$$

(2) **Addition of HCl**. Hydrogen chloride adds to acrolein to form 3-chloropropanal.

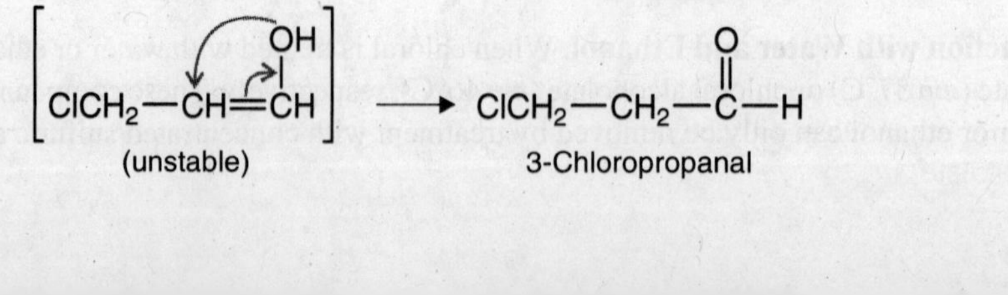

(3) **Oxidation**. Acrolein undergoes oxidation with ammoniacal silver nitrate to give acrylic acid.

$$\underset{\text{Acrolein}}{CH_2{=}CH{-}CHO} + [O] \xrightarrow{AgNO_3} \underset{\text{Acrylic acid}}{CH_2{=}CH{-}COOH}$$

(4) **Reduction**. Acrolein undergoes reduction with $LiAlH_4$ to give allyl alcohol.

$$\underset{\text{Acrolein}}{CH_2{=}CH{-}CHO} \xrightarrow{LiAlH_4} \underset{\text{Allyl alcohol}}{CH_2{=}CH{-}CH_2OH}$$

Uses. Acrolein is used : (*1*) as a tear gas ; (*2*) in the manufacture of insecticides ; and (*3*) in the preparation of acrolein-urea-formaldehyde resins.

ACETONE, *Propanone, Dimethyl ketone*, $CH_3{-}CO{-}CH_3$

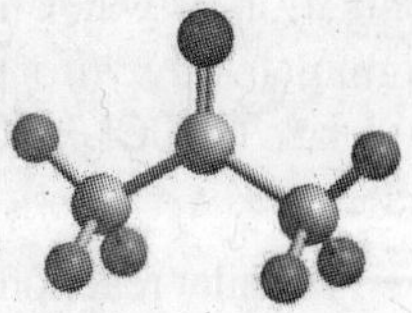

Acetone is the most important member of the ketone family.

Laboratory Preparation. Acetone is prepared in the laboratory by oxidation of isopropyl alcohol (2-Propanol) with acidified sodium dichromate solution.

$$\underset{\text{Isopropyl alcohol}}{CH_3{-}\overset{\overset{\large OH}{|}}{C}H{-}CH_3} + [O] \xrightarrow[H^+]{Na_2Cr_2O_7} \underset{\text{Acetone}}{CH_3{-}\overset{\overset{\large O}{\|}}{C}{-}CH_3} + H_2O$$

A mixture of isopropyl alcohol and sodium dichromate solution is run into boiling dilute H_2SO_4. As the liquid comes in contact with the acid, a vigorous reaction takes place and the orange dichromate is reduced to green chromic sulfate. Impure acetone distills over. The impure acetone is purified by adding sodium bisulfite solution, filtering off the crystals of the bisulfite compound obtained, and regenerating acetone by decomposing the crystals with dilute acid or alkali. The acetone is then distilled (*bp* 56°C).

Manufacture. Acetone is manufactured :

(1) **By air-oxidation of isopropyl alcohol at 500°C.** No catalyst is required

$$\underset{\substack{\text{Isopropyl alcohol}\\ \text{(2-Propanol)}}}{CH_3{-}\overset{\overset{\large OH}{|}}{C}H{-}CH_3} + O_2 \xrightarrow{500°C} \underset{\text{Acetone}}{CH_3{-}\overset{\overset{\large O}{\|}}{C}{-}CH_3} + H_2O$$

(2) **By dehydrogenation of isopropyl alcohol.** Isopropyl alcohol vapors are passed over heated copper catalyst at 300°C.

$$\underset{\text{Isopropyl alcohol}}{CH_3{-}\overset{\overset{\large OH}{|}}{C}H{-}CH_3} \xrightarrow[300°C]{Cu} \underset{\text{Acetone}}{CH_3{-}\overset{\overset{\large O}{\|}}{C}{-}CH_3} + H_2$$

(3) **By Wacker Process.** This involves the treatment of propene with an acidified aqueous solution of palladium chloride and cupric chloride.

$$\underset{\text{Propene}}{CH_3CH{=}CH_2} + PdCl_2 + H_2O \xrightarrow{CuCl_2} \underset{\text{Acetone}}{CH_3{-}\overset{\overset{\large O}{\|}}{C}{-}CH_3} + Pd + 2HCl$$

Acetone is also obtained as a by-product of *Cumene Process* for the manufacture of phenol. The processes for the preparation of acetone by fermentation of carbohydrates and also from calcium acetate, are obsolete.

Properties (*Physical*). Acetone is a colorless, volatile, highly flammable liquid, *bp* 56°C. It has an ethereal odor. It is miscible with water, ethanol, and ether in all proportions.

(*Chemical*). The chemical properties of acetone have already been discussed under 'General Reactions' of aldehydes and ketones.

(1) **Haloform Reaction.** When acetone is treated with bleaching powder ($CaOCl_2 = CaO + Cl_2$), which supplies both chlorine and alkali, it first forms trichloroacetone which on hydrolysis gives chloroform.

$$\underset{\text{Acetone}}{CH_3-\overset{\overset{O}{\|}}{C}-CH_3} + 3Cl_2 \longrightarrow \underset{\text{Trichloroacetone}}{CCl_3-\overset{\overset{O}{\|}}{C}-CH_3} + 3HCl$$

$$CCl_3-\overset{\overset{O}{\|}}{C}-CH_3 + Ca(OH)_2 \longrightarrow \underset{\text{Chloroform}}{CHCl_3} + CH_3COOH$$

A similar reaction with iodine and alkali gives iodoform.

(2) **Condensation Reactions.** Acetone undergoes the following condensation reactions :

(*a*) When treated with dry HCl, acetone first gives mesityl oxide and then phorone.

Formation of Mesityl Oxide :

$$\underset{\text{Acetone}}{(CH_3)_2C{=}O} + \underset{\text{Acetone}}{H_3C-\overset{\overset{O}{\|}}{C}-CH_3} \xrightarrow[-H_2O]{HCl} \underset{\text{Mesityl oxide}}{(CH_3)_2C{=}CH-\overset{\overset{O}{\|}}{C}-CH_3}$$

Formation of Phorone :

$$\underset{\text{Mesityl oxide}}{(CH_3)_2C{=}CH-\overset{\overset{O}{\|}}{C}-CH_3} + \underset{\text{Acetone}}{O{=}C(CH_3)_2} \xrightarrow[-H_2O]{HCl} \underset{\text{Phorone}}{(CH_3)_2C{=}CH-\overset{\overset{O}{\|}}{C}-CH{=}C(CH_3)_2}$$

(*b*) When treated with H_2SO_4, acetone gives mesitylene.

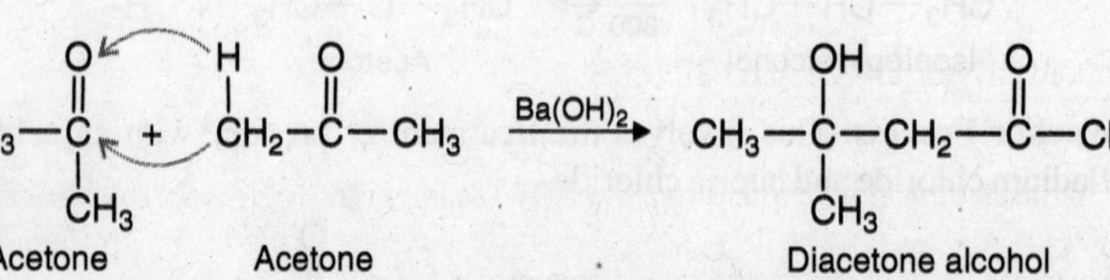

(*c*) When treated with $Ba(OH)_2$, acetone undergoes aldol condensation to form diacetone alcohol.

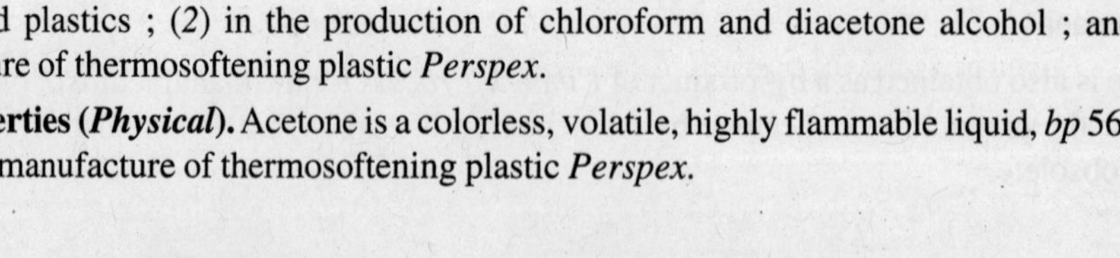

Uses. Acetone is used : (*1*) as a solvent for acetylene, cellulose derivatives, varnishes, lacquers, resins, and plastics ; (*2*) in the production of chloroform and diacetone alcohol ; and (*3*) in the manufacture of thermosoftening plastic *Perspex*.

Properties (*Physical*). Acetone is a colorless, volatile, highly flammable liquid, *bp* 56°C. It has an etherea7e manufacture of thermosoftening plastic *Perspex*.

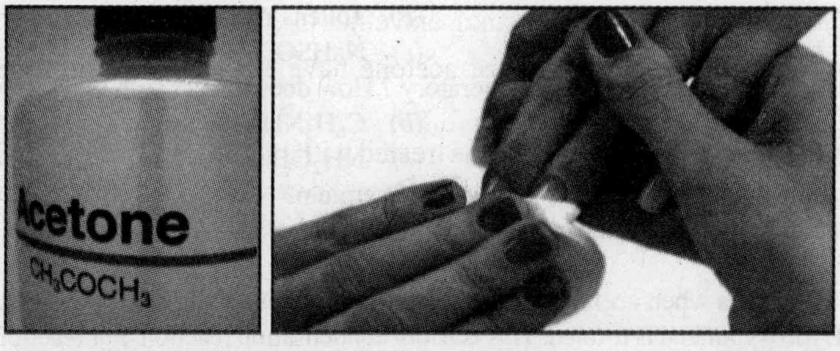

Propanone, commonly known as acetone, is one of the most widely used solvents in industry, the laboratory, and the home. Acetone is used as a nail polish remover.

METHYL ETHYL KETONE, *2-Butanone*, $CH_3COCH_2CH_3$

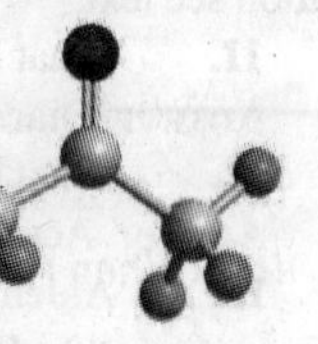

It is technically known by the initials *MEK*.

Preparation. 2-Butanone is obtained :

(1) **By treatment of 2-butanol with $Na_2Cr_2O_7/H_2SO_4$ or Cu at 300°C.**

$$\underset{\text{2-Butanol}}{CH_3-\underset{OH}{\underset{|}{CH}}-CH_2CH_3} \xrightarrow[\Delta]{Na_2Cr_2O_7/H^+} \underset{\text{2-Butanone}}{CH_3-\overset{O}{\overset{\|}{C}}-CH_2CH_3}$$

(2) **By Wacker Process.** This involves the treatment of 1-butene with an acidified aqueous solution of palladium chloride and cupric chloride.

$$\underset{\text{1-Butene}}{CH_3CH_2CH{=}CH_2} + PdCl_2 + H_2O \xrightarrow{CuCl_2} \underset{\text{2-Butanone}}{CH_3CH_2\overset{O}{\overset{\|}{C}}CH_3} + Pd + 2HCl$$

Methyl ethyl ketone is a pleasant-smelling liquid, *bp* 80°C. It is soluble in water. It gives all the characteristic reactions of ketones.

Methyl ethyl ketone is used in petroleum industry in dewaxing of lubricating oils. It is an excellent solvent for *fingernail polish* and is used as a solvent for vinyl resins and synthetic rubber.

Examination Questions

1. Write the structural formulas and give IUPAC names for all aldehydes and ketones of the molecular formula $C_5H_{10}O$. *(Annamalai BSc, 2015)*

Clue. There are four aldehydes and three ketones of formula $C_5H_{10}O$.

2. How are aldehydes prepared ? Describe their important reactions.

3. How are ketones prepared ? Describe their important reactions.

4. How is acetaldehyde prepared in the laboratory ? How does it react with the following reagents:

(*a*) $NaHSO_3$ (*b*) HCN (*c*) CH_3MgI
(*d*) NH_2OH (*e*) Dilute NaOH (*f*) $Br_2/NaOH$

5. How is formaldehyde prepared in the laboratory ? How does it react with the following reagents:
(*a*) NH_3 (*b*) Tollens' reagent (*c*) HCN
(*d*) Hydroxylamine (*e*) $NaHSO_3$

6. How is acetone prepared in the laboratory ? How does it react with the following reagents :
(*a*) $NaHSO_3$ (*b*) $C_6H_5NHNH_2$ (*c*) CH_3MgI
(*d*) Dilute NaOH (*e*) NH_2OH (*f*) $I_2/NaOH$

7. Describe the preparation and properties of acrolein.

8. What is the structure of the carbonyl group ? How does it react with ; (*i*) HCN ; (*ii*) $NaHSO_3$; and (*iii*) NH_2OH.

9. What happens when acetaldehyde is treated with dilute NaOH ? (*Kanpur BSc, 2014*)

Answer. 3-Hydroxybutanal is formed. This is Aldol condensation reaction. For reaction equation see text.

10. What happens when formaldehyde is treated with concentrated NaOH solution ? (*Shivaji BSc, 2015*)

Answer. Methyl alcohol and sodium formate are formed. This is Cannizzaro reaction. For reaction equation see text.

11. What happens when acetone is heated with solid barium hydroxide ?

Answer. Diacetone alcohol is formed. This is Aldol condensation reaction. For reaction equation see text.

12. What happens when calcium acetate is heated ?

Answer. Acetone is formed. For reaction equation see text.

13. Aldehydes and ketones form hydrocarbons by :
(*a*) Rosenmund's reduction (*b*) Cannizzaro reaction
(*c*) Clemmensen reduction (*d*) Stephen's reduction

Answer. (*c*)

14. How will you distinguish between acetaldehyde and acetone ?
(*Dibrugarh BSc, 2014 ; North Eastern Hill BSc, 2014 ; Delhi BSc, 2015*)

Answer. Following tests are used :

(*a*) **Tollens' test.** Acetaldehyde forms silver mirror with Tollens' reagent. Tollens' reagent is a solution of silver nitrate in ammonium hydroxide [actually $Ag(NH_3)_2OH$]. Acetone does not give this test.

$$\underset{\text{Acetaldehyde}}{CH_3-\overset{\overset{O}{\|}}{C}-H} + 2Ag(NH_3)_2OH \xrightarrow{\Delta} CH_3-\overset{\overset{O}{\|}}{C}-ONH_4 + \underset{\text{silver ppt.}}{2Ag} + H_2O + 3NH_3$$

(*b*) **Fehling's test.** Acetaldehyde forms red precipitate with Fehling's solution. Fehling's solution is an alkaline solution of cupric ion complexed with sodium potassium tartrate. Acetone does not give this test.

$$\underset{\text{Acetaldehyde}}{CH_3-\overset{\overset{O}{\|}}{C}-H} + 2Cu(OH)_2 + 2NaOH \xrightarrow{\Delta} CH_3-\overset{\overset{O}{\|}}{C}-ONa + \underset{\text{red ppt.}}{Cu_2O} + 3H_2O$$

15. How will you distinguish between formaldehyde and acetaldehyde ? (*Madras BSc, 2015*)

Answer. Acetaldehyde forms yellow precipitate of iodoform with an alkaline solution of iodine (Iodoform test). Formaldehyde does not give this test.

$$\underset{\text{Acetaldehyde}}{H-\overset{\overset{O}{\|}}{C}-CH_3} + 4NaOH + 3I_2 \longrightarrow H-\overset{\overset{O}{\|}}{C}-ONa + \underset{\text{yellow ppt.}}{CHI_3} + 3H_2O + 3NaI$$

16. How will you distinguish between 2-pentanone and 3-pentanone ?

Answer. 2-Pentanone forms yellow precipitate of iodoform with an alkaline solution of iodine (Iodoform test). 3-Pentanone does not give this test.

$$\underset{\text{2-Pentanone}}{CH_3CH_2CH_2\overset{\overset{O}{\|}}{C}CH_3} + 4NaOH + I_2 \longrightarrow CH_3CH_2CH_2\overset{\overset{O}{\|}}{C}ONa + \underset{\text{yellow ppt.}}{CHI_3} + 3H_2O + 3NaI$$

17. Give the general mechanism of nucleophilic addition reactions of carbonyl compounds ?

18. Give the mechanism of addition of HCN to acetaldehyde.

19. Give the mechanism of addition of HCN to acetone.

20. Discuss the mechanism of Aldol condensation.
(Baroda BSc, 2014 ; Ranchi BSc, 2014 ; West Bengal BSc, 2015)

21. Discuss the mechanism of Cannizzaro reaction.

22. Write a note on : Aldol condensation. *(Agra BSc, 2014)*

23. Write a note on : Cannizzaro reaction. *(Kurukshetra BSc, 2014)*

24. Write a note on : Wittig reaction. *(Jadavpur BSc, 2014)*

25. Write a note on : Reformatsky reaction. *(Panjab BSc, 2015)*

26. Identify (A), (B) and (C) :

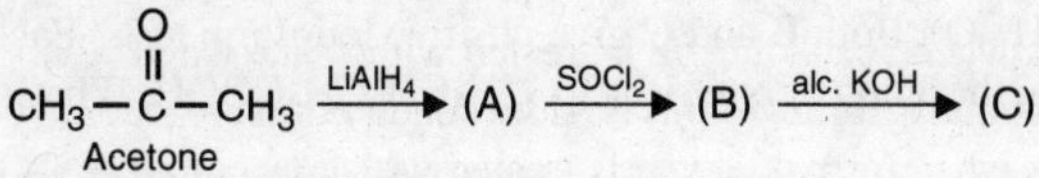

Answer. (*A*) = $CH_3-CH(OH)-CH_3$; Isopropyl alcohol

(*B*) = $CH_3-CH(Cl)-CH_3$; Isopropyl chloride

(*C*) = $CH_3-CH{=}CH_2$; Propene

27. How will you synthesize acetaldehyde from formaldehyde ? *(Aligarh BSc, 2014)*

Answer. Following steps are involved :

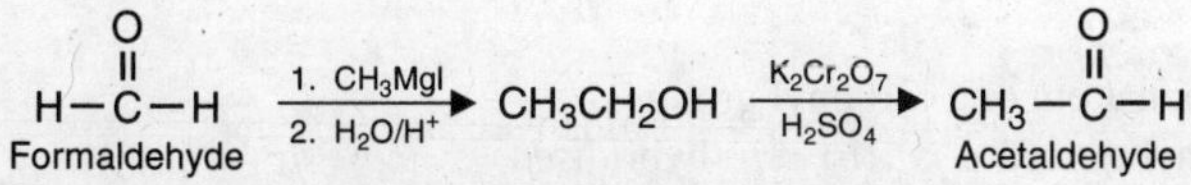

28. How will you synthesize acetone from acetaldehyde ? *(Kalyani BSc, 2015)*

Answer. Following steps are involved :

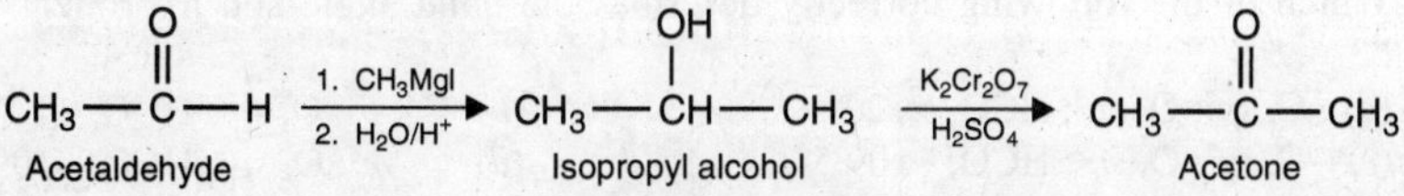

29. How will you synthesize lactic acid from acetylene ? *(Indore BSc, 2014)*

Clue. Steps are : (*i*) $H_2O/H_2SO_4/HgSO_4$; (*ii*) HCN ; and (*iii*) H_2O/H^+

30. How will you synthesize 2-butanone from ethyl alcohol ?

Clue. Steps are : (*i*) $SOCl_2$; (*ii*) Mg/ether ; (*iii*) CH_3CHO ; (*iv*) H_2O/H^+ ; and (*v*) $Na_2Cr_2O_7/H_2SO_4$.

31. An alkene, C_6H_{12}, after ozonolysis yielded two products. One of these gave a positive iodoform reaction but a negative Tollens' test. The other product gave a positive Tollens' test, but negative iodoform reaction. What is the structure and IUPAC name of the alkene ?

Answer. The alkene is 2-Methyl-2-pentene.

32. Compound (*A*), $C_5H_{12}O$, does not give a precipitate with phenylhydrazine. Oxidation of (*A*) with potassium dichromate gives (*B*), $C_5H_{10}O$. Compound (*B*) reacts readily with phenylhydrazine but does not give Tollens' test. The original compound (*A*) can be dehydrated with sulfuric acid to give a hydrocarbon (*C*), C_5H_{10}. Ozonolysis of the hydrocarbon (*C*) gives acetone and acetaldehyde. Deduce the structural formulas of (*A*), (*B*), and (*C*).

Answer. (*A*) = 2-Methyl-2-butanol (*B*) = 3-Methyl-2-butanone (*C*) = 2-Methyl-2-butene

33. An unsaturated hydrocarbon reacts with HCl to give a product that is identical with the product obtained by treating acetaldehyde with PCl_5. What is the structural formula of the hydrocarbon ?

Answer. The unsaturated hydrocarbon is Acetylene.

34. Compound (*A*), $C_4H_{10}O$, undergoes oxidation to give (*B*), C_4H_8O. (*B*) forms an oxime but does not give Tollens' test. (*B*) reacts with iodine and potassium hydroxide to give iodoform. Deduce the structures of (*A*) and (*B*) and explain the reactions.

Answer. (*A*) = 2-Butanol (*B*) = 2-Butanone

35. Compound (*A*), $C_5H_{11}Br$, forms a Grignard reagent to form *n*-pentane. When (*A*) is treated with aqueous potassium hydroxide, compound (*B*), $C_5H_{12}O$, is formed. Compound (*B*) gives a positive Lucas test. When (*B*) is treated with potassium dichromate in sulfuric acid, (*C*) is formed. Compound (*C*) gives a negative iodoform test. Deduce the structural formulas of (*A*), (*B*), and (*C*).

Answer. (*A*) = 3-Bromopentane
(*B*) = 3-Pentanol
(*C*) = 3-Pentanone

36. Unknown **A** (C_5H_{10}) rapidly decolorizes a solution of bromine in carbon tetrachloride. When **A** is dissolved in cold concentrated sulfuric acid and then heated with water, **B** is formed. **B** ($C_5H_{12}O$) reacts with chromic acid to give **C** ($C_5H_{10}O$). Both **B** and **C** give positive iodoform tests. Each iodoform reaction mixture contains, in addition to CHI_3, the anion of butyric acid, $CH_3CH_2CH_2COOH$. What are the structures of **A**, **B**, and **C** ?

Answer. (*A*) = 1-Pentene (*B*) = 2-Pentanol (*C*) = 2-Pentanone

37. Deduce the structure of a compound that is consistent with the following data.
Tollens' test : positive
2,4-Dinitrophenylhydrazine: yellow-orange product with the formula $C_9H_{10}N_4O_4$.

Answer. The unknown is Propanal.

Multiple Choice Questions

1. The carbon atom of a carbonyl group is
(*a*) *sp* hybridised (*b*) sp^2 hybridised (*c*) sp^3 hybridised (*d*) None of these

Answer. (*b*). Whenever carbon is bonded to three other atoms or groups, it uses sp^2 hybrid orbitals to form its bonds.

2. Which of the following correctly describes the bond angle and hybridizations present in formaldehyde ?
(*a*) C, sp^2; O, sp^3; HCO, ~120° (*b*) C, sp^2; O, sp^2; HCO, ~120°
(*c*) C, sp^2; O, sp^2; HCO, ~109.5° (*d*) C, sp^3; O, sp^2; HCO, ~109.5°

Answer. (*b*)

3. The positively polarized carbon atom of a carbonyl group acts as:
(*a*) an electrophile and a Lewis base. (*b*) a nucleophile and a Lewis base.
(*c*) an electrophile and a Lewis acid. (*d*) a nucleophile and a Lewis acid.

Answer. (*c*)

4. The double bond between carbon and oxygen is similar to an alkene C=C, except that C=O is:
(*a*) shorter and weaker. (*b*) shorter and stronger.
(*c*) longer and weaker. (*d*) longer and stronger.

Answer. (*b*)

5. Which statement about the carbonyl group is NOT true ?
(*a*) The carbonyl carbon is sp^2 hybridized.
(*b*) The bond angles among the three atoms attached to the carbonyl carbon are 120°.
(*c*) The three atoms attached to the carbonyl carbon form a nonplanar geometry.
(*d*) The carbonyl group forms resonance structures.

Answer. (*c*)

6. Acetone contains
(*a*) nine σ bonds plus one π bond (*b*) ten σ bonds
(*c*) eight σ bonds plus two π bonds (*d*) nine π bonds plus one σ bond

Answer. (*a*)

7. Formalin is
(*a*) 10% solution of formaldehyde in water (*b*) 20% solution of formaldehyde in water
(*c*) 40% solution of formaldehyde in water (*d*) 80% solution of formaldehyde in water

Answer. (*c*)

8. Which of the following will have the highest boiling point ?

(*a*) Propanone (*b*) 2-Pentanone (*c*) Butanone (*d*) 2-Hexanone

Answer. (*d*). Boiling points increase with molecular weight. The higher the molecular weight, the higher is the boiling point. All of the given compounds are ketones. 2-Hexanone has the highest molecular weight.

9. Which of the following will have the highest boiling point ?

(*a*) Methanal (*b*) Ethanal (*c*) Propanal (*d*) Butanal

Answer. (*d*). The boiling points increase with molecular weight. All of the given compounds are aldehydes. Butanal has the highest molecular weight.

10. What property of low-molecular weight aldehydes and ketones accounts for the magnitude of their boiling points ?

(*a*) The ability to form strong H-bonds between their molecules.

(*b*) The ability of the carbonyl oxygen to form H-bonds with other carbonyl groups.

(*c*) The ability of the polar carbonyl group to attract other polar molecules.

(*d*) The ability of the carbonyl group to attract electrophiles and form bonds.

Answer. (*c*)

11. Primary alcohols have boiling points that are ____________ the corresponding aldehydes.

(*a*) Lower than (*b*) Higher than (*c*) About the same

Answer. (*b*). For compounds of similar molecular weights, boiling points increase with the degree of hydrogen bonding. *Remember* : Alcohols form strong hydrogen bonds and will boil at a higher temperature than the corresponding aldehydes.

12. Which of the following represents the correct ranking in terms of increasing boiling point ?

(*a*) *n*-butane < 1-butanol < diethyl ether < 2-butanone

(*b*) *n*-butane < 2-butanone < diethyl ether < 1-butanol

(*c*) *n*-butane < diethyl ether < 1-butanol < 2-butanone

(*d*) *n*-butane < diethyl ether < 2-butanone < 1-butanol

Answer. (*d*)

13. The melting points of aldehydes and ketones tend to :

(*a*) decrease with increasing molecular weight.

(*b*) increase with increasing molecular weight.

(*c*) remain unchanged with increasing molecular weight.

(*d*) be unpredictable due to resonance.

Answer. (*b*)

14. Which action best accounts for the solubility of aldehydes and ketones in water ?

(*a*) Polar interactions between solute molecules.

(*b*) H-bonding between solute molecules.

(*c*) Van der Waals forces

(*d*) H-bonding between solute and solvent molecules

Answer. (*d*)

15. Ketones are prepared by the oxidation of

(*a*) Primary alcohol (*b*) Secondary alcohol (*c*) Tertiary alcohol (*d*) None of these

Answer. (*b*)

16. Cyclopentanol undergoes oxidation to give :

(*a*) Cyclopentene (*b*) Cyclopentanone (*c*) Cyclopentane (*d*) Cyclopentanal

Answer. (*b*)

17. Which of the following compounds reacts with sodium bisulfite and ammoniacal silver nitrate solution ?

(*a*) $CH_3CH_2{-}C({=}O){-}OH$ (*b*) $CH_3CH_2{-}C({=}O){-}H$

(*c*) $CH_3CH_2CH{=}CH_2$ (*d*) $CH_3C{\equiv}CCH_3$

Answer. (*b*)

18. Acetone reacts with HCN to form a cyanohydrin. It is an example of

(*a*) Electrophilic addition (*b*) Electrophilic substitution
(*c*) Nucleophilic addition (*d*) Nucleophilic substitution

Answer. (*c*)

19. Which of the following reagents will react readily with both aldehydes and ketones ?

(*a*) Grignard reagent (*b*) Fehling's reagent
(*c*) Tollens' reagent (*d*) Schiff's reagent

Answer. (*a*)

20. Boiling acetaldehyde (CH_3CHO) reacts with chlorine gas to give

(*a*) $CH_3—\overset{O}{\overset{\|}{C}}—Cl$ (*b*) $CCl_3—\overset{O}{\overset{\|}{C}}—H$
(*c*) CH_3CHCl_2 (*d*) $CHCl_3$

Answer. (*b*)

21. The reaction of ethanal with one equivalent of methanol and a trace of an acid will give

(*a*) Acetal (*b*) Hemiacetal (*c*) Ketal (*d*) Hemiketal

Answer. (*b*)

22. Acetone undergoes reduction with hydrazine in the presence of NaOH to form propane. This reaction is known as

(*a*) Clemmensen reduction (*b*) Wolf-Kishner reduction
(*c*) Rosenmund reduction (*d*) Reformatsky reaction

Answer. (*b*)

23. Acetone undergoes reduction with zinc amalgam in HCl to form propane. This reaction is known as

(*a*) Clemmensen reduction (*b*) Wolf-Kishner reduction
(*c*) Rosenmund reduction (*d*) Aldol condesation

Answer. (*a*)

24. Cannizzaro reaction is not given by

(*a*) Formaldehyde (*b*) Trimethylacetaldehyde
(*c*) Acetaldehyde (*d*) Benzaldehyde

Answer. (*c*)

25. When formaldehyde is treated with 50% NaOH solution, it undergoes

(*a*) Cannizzaro reaction (*b*) Wurtz reaction
(*c*) Aldol condensation (*d*) Hydrolysis

Answer. (*a*)

26. The reduction of a ketone

(*a*) always gives a primary alcohol (*b*) always gives a secondary alcohol
(*c*) always gives a carboxylic acid (*d*) always gives a ketal

Answer. (*b*)

27. Reduction of acetaldehyde with H_2/Ni gives

(*a*) Ethyl alcohol (*b*) Acetic acid (*c*) Ethylene (*d*) Ethane

Answer. (*a*)

28. On reduction with $LiAlH_4$, which of the following compounds could yield an optically active compound ?

(*a*) Propanal (*b*) Propanone (*c*) Butanal (*d*) Butanone

Answer. (*d*)

29. The reduction of cyclohexanone with $LiAlH_4$ (or $NaBH_4$) will give

(*a*) an alcohol (*b*) an organic acid (*c*) an aldehyde (*d*) a hemiketal

Answer. (*a*)

30. Aldehydes undergo oxidation with $KMnO_4/H^+$ to give

(*a*) Alcohols (*b*) Acetals (*c*) Ketones (*d*) Acids

Answer. (*d*)

31. Oxidation of acetaldehyde with $Na_2Cr_2O_7/H^+$ gives

(*a*) Acetone (*b*) Acetic acid (*c*) Ethanol (*d*) Ethylene glycol

Answer. (*b*)

32. A hydrazone will result from the reaction of hydrazine with

(*a*) a phenol (*b*) an aldehyde (*c*) an alcohol (*d*) an acid

Answer. (*b*)

33. Aldehydes can be distinguished from ketones by using

(*a*) Ammoniacal $AgNO_3$ solution (*b*) Phenylhydrazine

(*c*) Saturated $NaHSO_3$ solution (*d*) Thionyl chloride

Answer. (*a*)

34. The appearance of a silver mirror in Tollens' test indicates the presence of :

(*a*) an aldehyde (*b*) a ketone (*c*) an alcohol (*d*) an alkene

Answer. (*a*)

35. Acetaldehyde on treatment with Tollens' reagent gives a precipitate of

(*a*) Ag (*b*) $AgNO_3$ (*c*) Cu_2O (*d*) None of these

Answer. (*a*)

36. Which of the following compounds will give a positive test with Fehling's solution ?

(*a*) Formaldehyde (*b*) Acetone (*c*) Ethyl acetate (*d*) Acetic acid

Answer. (*a*)

37. Acetaldehyde on treatment with Fehling's solution gives a precipitate of

(*a*) Cu (*b*) Cu_2O (*c*) CuO (*d*) None of these

Answer. (*b*)

38. Which of the following will react with Fehling's solution giving a red precipitate ?

(*a*) CH_3CH_2CHO (*b*) $CH_3CH_2OCH_2CH_3$

(*c*) $CH_3CH_2CH_2OH$ (*d*) $CH_3-\overset{\overset{\displaystyle O}{\|}}{C}-CH_3$

Answer. (*a*)

39. Which of the following compounds will give an iodoform test ?

(*a*) Benzoic acid (*b*) Ethanol (*c*) Benzyl chloride (*d*) Methanol

Answer. (*b*)

40. Which of the following compounds does not give a positive iodoform test :

(*a*) Ethanol (*b*) Ethanal (*c*) Methanol (*d*) Propanone

Answer. (*c*)

41. Which of the following compounds does not give iodoform test on treatment with I_2/NaOH ?

(*a*) Ethanol (*b*) Propanone (*c*) 2-Propanol (*d*) Butanone

Answer. (*d*)

42. Which of the following compounds will give a positive iodoform test ?

(*a*) Benzaldehyde (*b*) 2-Pentanone (*c*) 3-Hexanone (*d*) 3-Pentanone

Answer. (*b*)

43. An organic compound (A), C_3H_8O, on oxidation gives (B), C_3H_6O. The compound (A) could be

(*a*) an aldehyde (*b*) a ketone (*c*) an alcohol (*d*) an ester

Answer. (*c*)

44. An organic compound (A), C_3H_8O, on oxidation gives (B), C_3H_6O. The compound (B) could be

(*a*) a carbonyl compound (*b*) an alcohol (*c*) a carboxylic acid (*d*) an ether

Answer. (*a*)

45. An unknown compound gave a positive Tollens' test. Treatment of the unknown with I_2/NaOH gave a solid which was identified as iodoform. The unknown was ?

(*a*) Ethanal (*b*) Acetone (*c*) Propanal (*d*) Acetophenone

Answer. (*a*)

46. Organic compound (A), C_4H_8O, does not react with sodium or PCl_5. However, it reduces an alkaline solution of a copper (II) salt on heating. Compound (A) could be

(*a*) an aldehyde (*b*) a primary alcohol

(*c*) a ketone (*d*) a secondary alcohol

Answer. (*a*)

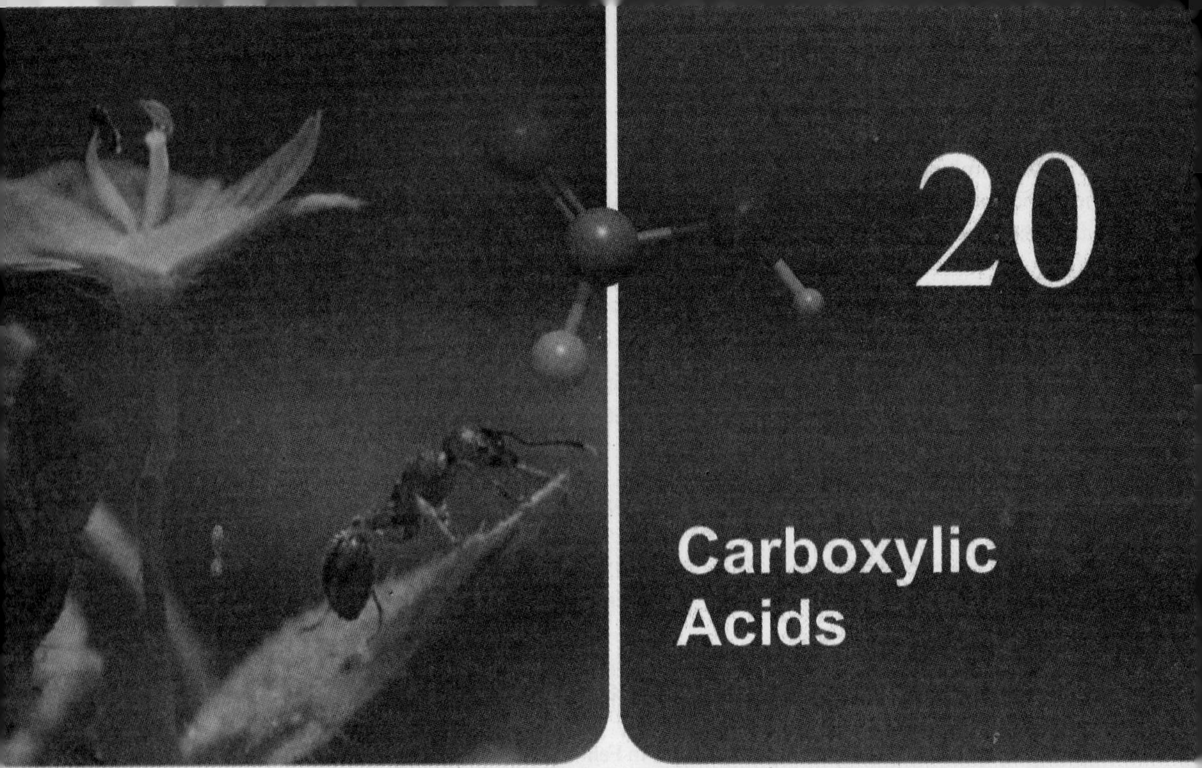

20

Carboxylic Acids

Formic acid derives its name from the fact that it was first obtained by distillation of ants (Latin, *formica* = ant).

Organic compounds which contain the carboxyl functional group (–COOH) are called the **Carboxylic Acids**. Their general formula is :

$$R-\overset{\overset{\displaystyle O}{\|}}{C}-OH \quad \text{or} \quad R-COOH \quad \text{or} \quad R-CO_2H$$

where R is H or an alkyl group.

The name *carboxyl* is derived from CARBonyl (C=O) and hydrOXYL (–OH) because in the carboxyl group these two groups are directly bonded to each other. The properties of the carboxyl group are not simply those of carbonyl and hydroxyl groups combined; the two groups interact to give carboxylic acids their own distinctive properties.

Carboxylic acids are further classified as monocarboxylic acids, dicarboxylic acids, tricarboxylic acids etc., according as the number of –COOH groups present in the molecule is 1, 2, 3 or more. The long-chain monocarboxylic acids are commonly called **Fatty Acids** because many of them are obtained by the hydrolysis of animal fats or vegetable oils.

MONOCARBOXYLIC ACIDS

NOMENCLATURE

There are three ways of naming carboxylic acids :

(1) **Common System**. The simple carboxylic acids are better known by their common names. The common names are usually derived from the Latin or Greek word that indicates the original source of the acid. They do not follow any rule except that all common names of acids end in *–ic acid.* For example,

$HCOOH$	Formic acid
CH_3COOH	Acetic acid
CH_3CH_2COOH	Propionic acid
$CH_3CH_2CH_2COOH$	Butyric acid
$CH_3CH_2CH_2CH_2COOH$	Valeric acid

In the common system, Greek letters (α, β, γ, δ, etc.) are used to indicate the position of substituents. The carbon atom adjacent to the carboxyl carbon is assigned the letter α, the next carbon on the chain β, the next one γ, and so on. For example,

$$\overset{\delta}{C}-\overset{\gamma}{C}-\overset{\beta}{C}-\overset{\alpha}{C}-COOH$$

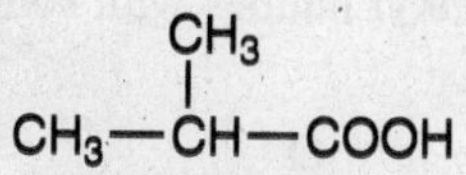

α-Methylpropionic acid
(Isobutyric acid)

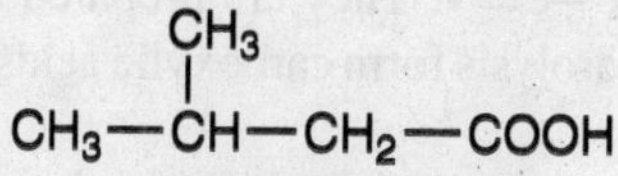

β-Methylbutyric acid
(Isovaleric acid)

(2) **Derived System.** Some carboxylic acids may be named as substituted acetic acids. For example,

$(CH_3)_3CCOOH$	Trimethylacetic acid
$C_6H_5CH_2COOH$	Phenylacetic acid

(3) **IUPAC System.** In the IUPAC system, the carboxylic acids are named as **Alkanoic acids (Alkane – e + oic acid)**. The systematic name of the individual acid is obtained by replacing the ending '*e*' of the corresponding alkane by *–oic acid*. For example,

$HCOOH$	Methanoic acid (*Methane –e + oic acid*)
CH_3COOH	Ethanoic acid
CH_3CH_2COOH	Propanoic acid
$CH_3CH_2CH_2COOH$	Butanoic acid

For naming higher members, the longest continuous chain containing the carboxyl group (–COOH) is selected. The number of carbon atoms in this chain gives the name of the parent alkane. The position of substituents is indicated by numbers. The carboxyl carbon is always given number 1, the carbon adjacent to it is given the number 2, and so on. For example,

$$\overset{5}{C}-\overset{4}{C}-\overset{3}{C}-\overset{2}{C}-\overset{1}{C}OOH$$

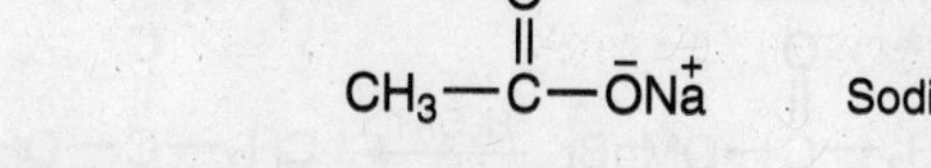

3,4-Dimethylhexanoic acid

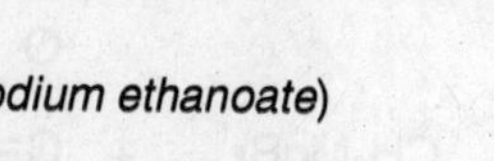

2-Bromopropanoic acid

Salts of carboxylic acids are named by replacing the ending *–ic acid* with *–ate*. This applies to common as well as IUPAC names.

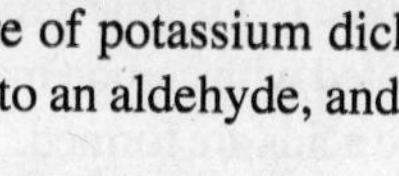

Sodium acetate (*Sodium ethanoate*)

METHODS OF PREPARATION

Carboxylic acids are prepared by the following methods :

(1) **Oxidation of Primary Alcohols or Aldehydes.** Primary alcohols or aldehydes undergo oxidation with a mixture of potassium dichromate and sulfuric acid to form carboxylic acids. The alcohol is first oxidized to an aldehyde, and then to a carboxylic acid.

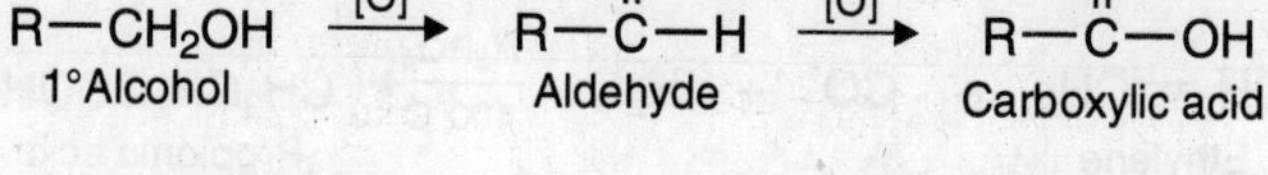

1°Alcohol — Aldehyde — Carboxylic acid

$$\underset{\text{Ethyl alcohol}}{CH_3CH_2OH} \xrightarrow[H_2SO_4]{K_2Cr_2O_7} \underset{\text{Acetaldehyde}}{CH_3CHO} \xrightarrow[H_2SO_4]{K_2Cr_2O_7} \underset{\text{Acetic acid}}{CH_3COOH}$$

Potassium permanganate can also be used in place of potassium dichromate.

(2) **Hydrolysis of Nitriles (Alkyl Cyanides)**. Nitriles or alkyl cyanides are compounds with the general formula R—C≡N. They are prepared by treating alkyl halides with sodium cyanide. The nitriles on acid-hydrolysis form carboxylic acids.

$$\underset{\text{Alkyl halide}}{R{-}X} \xrightarrow[-NaX]{NaCN} \underset{\substack{\text{Alkyl cyanide}\\ \text{(nitrile)}}}{R{-}C{\equiv}N} \xrightarrow{H_2O/H^+} \underset{\text{Carboxylic acid}}{R{-}\overset{\overset{O}{\|}}{C}{-}OH} + NH_3$$

$$\underset{\substack{\text{Methyl}\\ \text{bromide}}}{CH_3Br} \xrightarrow[-NaBr]{NaCN} CH_3{-}C{\equiv}N \xrightarrow{H_2O/H^+} \underset{\text{Acetic acid}}{CH_3{-}\overset{\overset{O}{\|}}{C}{-}OH} + NH_3$$

Notice that the acid produced by this method contains one more carbon atom than the original alkyl halide.

(3) **Hydrolysis of Esters**. When an ester is boiled with concentrated aqueous NaOH, sodium salt of the acid is formed. This on treatment with dilute HCl gives the corresponding carboxylic acid. For example,

$$\underset{\text{Ethyl acetate}}{CH_3{-}\overset{\overset{O}{\|}}{C}{-}OC_2H_5} + NaOH \xrightarrow{\Delta} \underset{\text{Sodium acetate}}{CH_3{-}\overset{\overset{O}{\|}}{C}{-}\overset{-}{O}\overset{+}{Na}} + \underset{\text{Ethanol}}{C_2H_5OH}$$

$$CH_3{-}\overset{\overset{O}{\|}}{C}{-}\overset{-}{O}\overset{+}{Na} + HCl \longrightarrow \underset{\text{Acetic acid}}{CH_3{-}\overset{\overset{O}{\|}}{C}{-}OH} + NaCl$$

(4) **Reaction of Grignard Reagents with CO_2**. Grignard reagents (RMgX) react with carbon dioxide to form addition products that can be hydrolysed to carboxylic acids.

$$R{-}X + Mg \xrightarrow{\text{ether}} RMgX$$

$$RMgX + \overset{\overset{O}{\|}}{C}{=}O \longrightarrow \underset{\text{(Not isolated)}}{R{-}\overset{\overset{O}{\|}}{C}{-}OMgX} \xrightarrow{H_2O/H^+} \underset{\text{Carboxylic acid}}{R{-}\overset{\overset{O}{\|}}{C}{-}OH}$$

$$\underset{\substack{\text{Methylmagnesium}\\ \text{bromide}}}{CH_3MgBr} + \overset{\overset{O}{\|}}{C}{=}O \longrightarrow \underset{\text{(Not isolated)}}{CH_3{-}\overset{\overset{O}{\|}}{C}{-}OMgBr} \xrightarrow{H_2O/H^+} \underset{\text{Acetic acid}}{CH_3{-}\overset{\overset{O}{\|}}{C}{-}OH}$$

Notice that the acid formed by this method contains one more carbon than the original Grignard reagent. Organolithium compounds can also be used in place of Grignard reagents.

(5) **Carboxylation of Alkenes**. When an alkene is heated with carbon monoxide and steam under pressure with phosphoric acid (H_3PO_4) at 400°C, carboxylic acids are formed. This is a recent industrial method of making carboxylic acids and is called **Koch Reaction**.

$$\underset{\text{Ethylene}}{CH_2{=}CH_2} + CO + H_2O \xrightarrow[400^\circ C]{H_3PO_4} \underset{\text{Propionic acid}}{CH_3CH_2COOH}$$

$$\underset{\text{Propene}}{CH_3-CH=CH_2} + CO + H_2O \xrightarrow[400°C]{H_3PO_4} \underset{\text{Isobutyric acid}}{CH_3-\overset{\overset{CH_3}{|}}{C}H-COOH}$$

(6) **Malonic Ester Synthesis**. This involves the reaction of an alkyl halide with the sodium derivative of diethyl malonate to give a substituted malonic ester. This can be hydrolysed and decarboxylated to yield an acid.

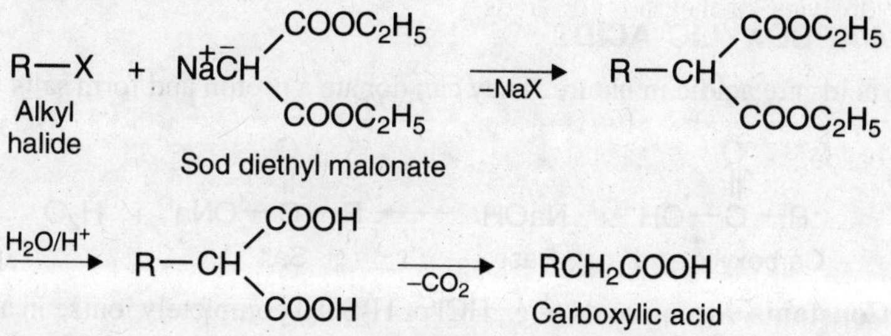

Notice that the acid produced by this method contains two carbon atoms more than the starting alkyl halide.

PHYSICAL PROPERTIES

(1) Lower carboxylic acids (up to C_{10}) are liquids with sharp or disagreeable odors. Higher members are waxlike solids and almost odorless.

(2) Boiling points of carboxylic acids increase regularly with increase in molecular weight.

(3) Melting points of carboxylic acids increase irregularly with increase of molecular weight. The 'even' members have markedly higher melting points than 'odd' members (Table 20.1).

Table 20.1 MELTING AND BOILING POINTS OF SOME CARBOXYLIC ACIDS

Name	Formula	mp °C	bp °C
Formic acid	HCOOH	8	101
Acetic acid	CH_3COOH	16	118
Propionic acid	CH_3CH_2COOH	–21	141
Butyric acid	$CH_3(CH_2)_2COOH$	–7	164
Valeric acid	$CH_3(CH_2)_3COOH$	–35	186

(4) Boiling points of carboxylic acids are higher than those of alcohols of same molecular weight. For example,

	Formula	MW	bp °C
Acetic acid	CH_3COOH	60	118
1-Propanol	$CH_3CH_2CH_2OH$	60	97

This is attributed to hydrogen-bonding between acid molecules.

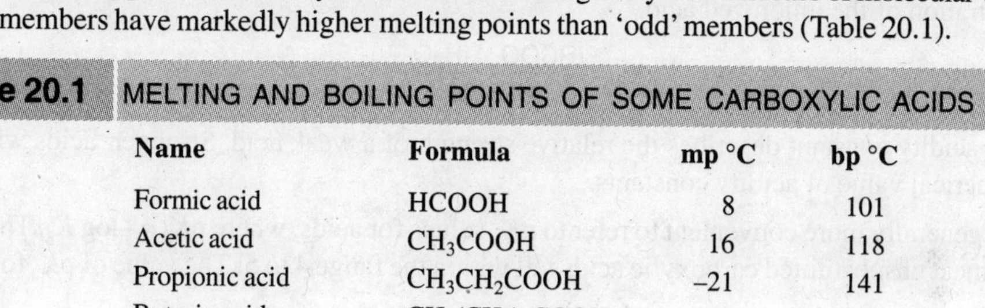

Acetic acid dimer

(5) First four members are completely soluble in water. This is due to the ability of the carboxyl group to form hydrogen bonds with water molecules. *As the hydrocarbon chain in the acid lengthenes, the solubility of the acid in water falls off rapidly*. Carboxylic acids are readily soluble in ethanol, ether, and benzene.

(6) ***IR spectrum.*** Carboxylic acids show a broad O—H stretching vibrations at 2500-3300 cm^{-1}, which overlaps the normally sharp C—H stretching band. They also show carbonyl (C=O) stretching frequency at 1710-1720 cm^{-1}.

ACIDITY OF CARBOXYLIC ACIDS

Carboxylic acids are acidic in nature. They can donate a proton and form salts with bases.

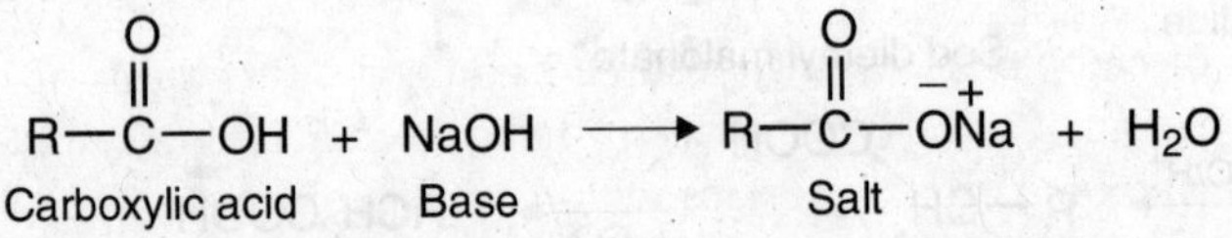

Acidity Constants. Strong acids (*e.g.*, HCl or H_2SO_4) completely ionize in aqueous solution. Carboxylic acids are weak acids. They are only partially ionized in aqueous solution and an equilibrium exists between the ionized and unionized forms.

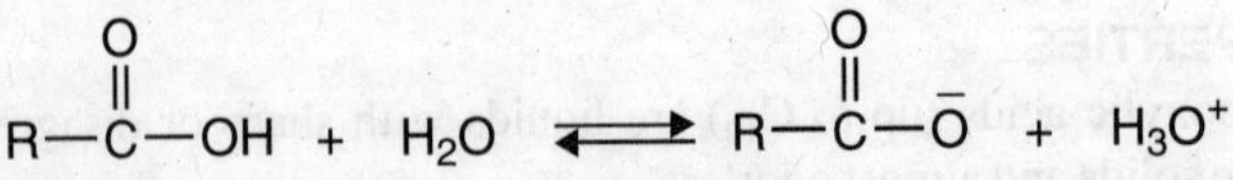

The extent of ionization is described by an equilibrium constant K_a, which is known as **Acidity Constant**. It is defined as the concentration of the products of ionization in moles per litre divided by the concentration of the unionized acid.

$$K_a = \frac{[RCOO^-][H_3O^+]}{[RCOOH]}$$

The acidity constant describes the relative strength of a weak acid. Stronger acids will have higher numerical value of acidity constants.

It is generally more convenient to refer to pK_a values for acids, where $pK_a = -\log K_a$. Thus, pK_a values for most unsubstituted carboxylic acids fall within the range 4 to 5. The value of pK_a for acetic acid is 4.76.

Points to Remember

1. Acidity constants can be described in terms of K_a values or pK_a values. The pK_a is the negative logarithm of the K_a. That is, $K_a = 10^{-pK_a}$. The smaller the pK_a, the stronger the acid.
2. Stronger acids will have larger K_a and smaller pK_a. Weak acids will have smaller K_a and larger pK_a.

Explanation of Acidity. Carboxylic acids are acidic and lose a proton readily because the carboxylate ion formed by ionization or reaction with a base is stabilized by resonance.

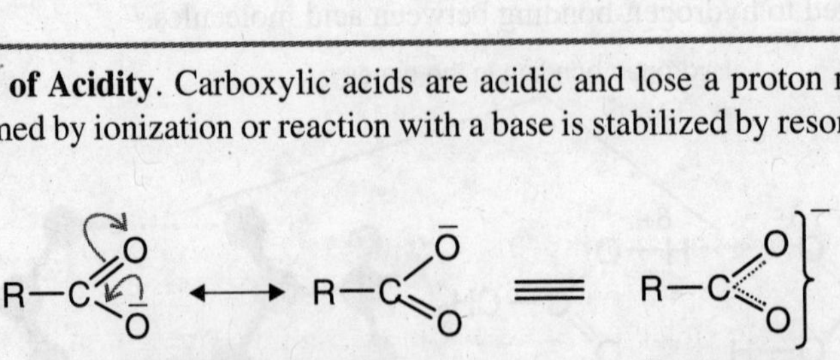

Resonance forms of carboxylate ion — Resonance hybrid

X-Ray studies support the fact that carboxylate ion exists as a resonance hybrid. For example, in formic acid the carbon-oxygen bonds have different lengths, whereas in sodium formate the two carbon-oxygen bond lengths are identical and intermediate in length between those of normal double and single carbon-oxygen bonds.

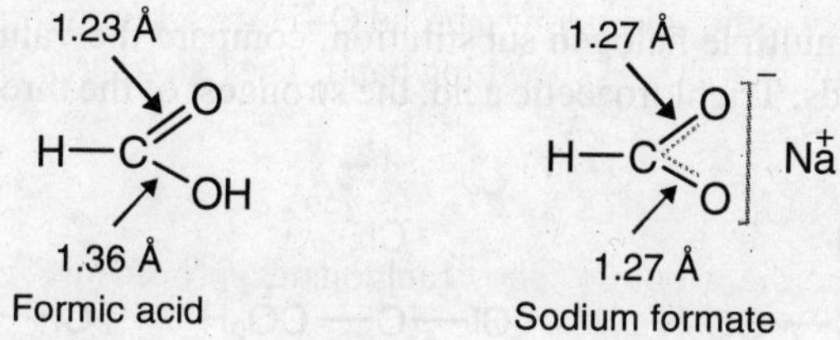

The stability of carboxylate ion can also be explained on the basis of its molecular orbital structure. The carbon atom of carboxylate ion is sp^2-hybridized. It is bound to each oxygen atom by a σ bond. The unused carbon *p* orbital overlaps with *p* orbitals of both oxygen atoms to form stable delocalized molecular orbital (Fig. 20.1).

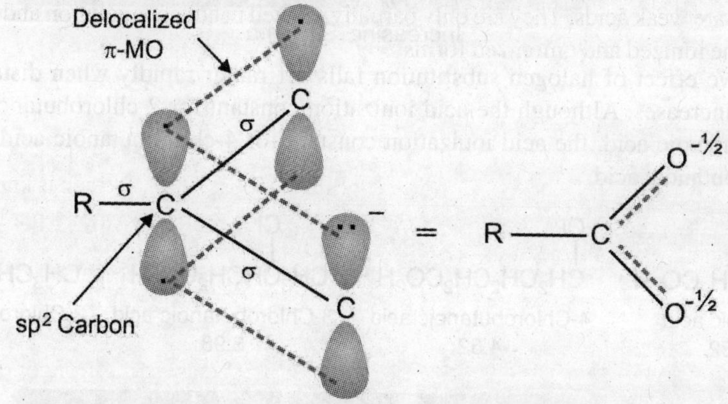

Fig. 20.1. Orbital structure of carboxylate ion.

Notice that the four electrons are bound to three atoms (1 *carbon* + 2 *oxygens*). This delocalization of π electrons is responsible for the extra-stability of the carboxylate ion.

Effect of Substituents on Acidity. The most important factor affecting the acidity is the *Inductive Effect* of substituents on the α-carbon atom.

(1) **Electron-releasing alkyl groups decrease the acidity.** This is because the electron-releasing groups increase the negative charge on the carboxylate ion and destabilize it. The loss of proton becomes more difficult. Also, as the size of the alkyl group increases, acidity decreases. For example,

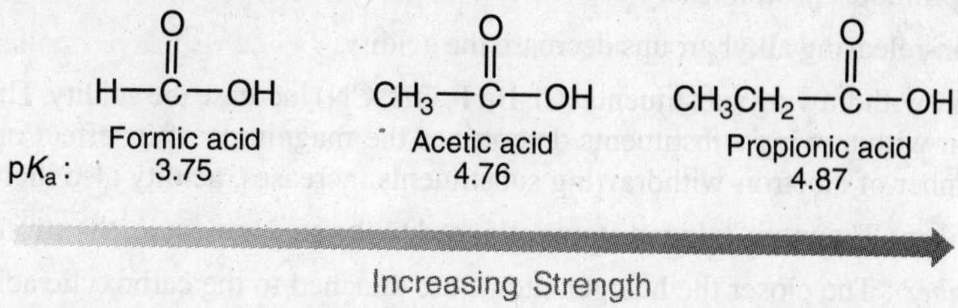

(2) **Electron-withdrawing substituents (Cl, Br, F, OH, CN) increase the acidity.** Compare, for example, the acidities of acetic acid and the halogen-substituted acetic acids. As the electronegativity of the halogen increases, its inductive effect increases, and the strength of the halogen-substituted acid increases. Fluoroacetic acid is the strongest of the monohalogenated acetic acids.

	CH_3CO_2H	ICH_2CO_2H	$BrCH_2CO_2H$	$ClCH_2CO_2H$	FCH_2CO_2H
	Acetic acid	Iodoacetic acid	Bromoacetic acid	Chloroacetic acid	Fluoroacetic acid
pK_a:	4.76	3.18	2.90	2.86	2.59

Increasing Strength →

To see the effects of multiple halogen substitution, compare the values of pK_a for the mono-, di-, and trichloroacetic acids. Trichloroacetic acid, the strongest of the three acids, is a stronger acid than H_3PO_4.

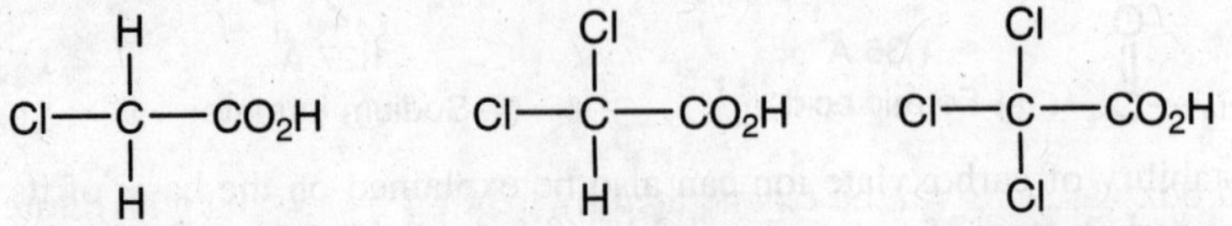

	Chloroacetic acid	Dichloroacetic acid	Trichloroacetic acid
pK_a :	2.86	1.48	0.70

Increasing Strength →

The inductive effect of halogen substitution falls off rather rapidly when distance from the carboxyl group increases. Although the acid ionization constant for 2-chlorobutanoic acid is 100 times that for butanoic acid, the acid ionization constant for 4-chlorobutanoic acid is only about 2 times that for butanoic acid.

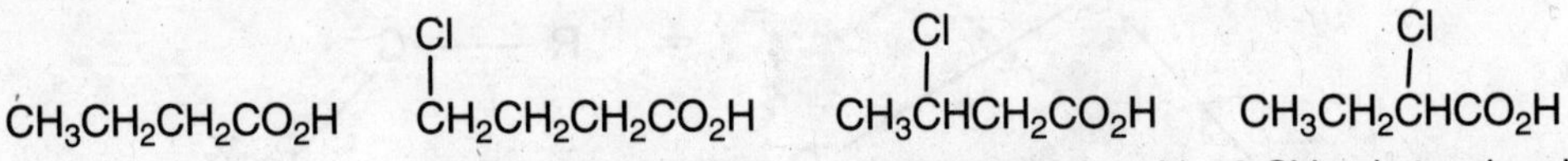

	Butanoic acid	4-Chlorobutanoic acid	3-Chlorobutanoic acid	2-Chlorobutanoic acid
pK_a :	4.82	4.52	3.98	2.83

Increasing Strength →

Points to Remember

Acidity of Carboxylic Acids

1. Strength of an acid is determined by the readiness with which they will donate a proton.
2. Acid strength will be increased by any factor which increases stability of the anion or which promotes proton loss.
3. Electron-releasing alkyl groups decrease the acidity.
4. Electron-withdrawing substituents (Cl, Br, F, OH, CN) increase the acidity. The strength of electron-withdrawing substituents determines the magnitude of its effect on acidity. As the number of electron-withdrawing substituents increases, acidity also increases.
5. *Remember* : The more halogen atoms attached to the acid molecule, the stronger the acid.
6. *Remember* : The closer the halogen atoms are attached to the carboxylic acid functional group, the stronger the acid.

CHEMICAL PROPERTIES

The main chemical reactions of carboxylic acids are described below :

(1) **Salt Formation**. Carboxylic acids react with hydroxides, carbonates, and bicarbonates to form the corresponding salts. For example,

$$\underset{\text{Acetic acid}}{CH_3-\overset{O}{\overset{\|}{C}}-OH} + NaOH \longrightarrow \underset{\text{Sodium acetate}}{CH_3-\overset{O}{\overset{\|}{C}}-\bar{O}\overset{+}{Na}} + H_2O$$

$$2CH_3-\overset{O}{\overset{\|}{C}}-OH + Na_2CO_3 \longrightarrow 2CH_3-\overset{O}{\overset{\|}{C}}-\bar{O}\overset{+}{Na} + H_2O + CO_2$$

$$CH_3-\overset{O}{\overset{\|}{C}}-OH + NaHCO_3 \longrightarrow CH_3-\overset{O}{\overset{\|}{C}}-\bar{O}\overset{+}{Na} + H_2O + CO_2$$

The carboxylic acids can be regenerated by treating these salts with dilute mineral acids. For example,

$$\underset{\text{Sod acetate}}{CH_3-\overset{O}{\overset{\|}{C}}-\bar{O}\overset{+}{Na}} + HCl \longrightarrow \underset{\text{Acetic acid}}{CH_3-\overset{O}{\overset{\|}{C}}-OH} + NaCl$$

(2) **Formation of Acid Halides**. Carboxylic acids react with phosphorus halides or thionyl chloride ($SOCl_2$), to form acid halides. For example,

$$\underset{\text{Acetic acid}}{CH_3-\overset{O}{\overset{\|}{C}}-OH} + PCl_5 \longrightarrow \underset{\text{Acetyl chloride}}{CH_3-\overset{O}{\overset{\|}{C}}-Cl} + POCl_3 + HCl$$

$$3CH_3-\overset{O}{\overset{\|}{C}}-OH + PCl_3 \longrightarrow 3CH_3-\overset{O}{\overset{\|}{C}}-Cl + H_3PO_3$$

$$CH_3-\overset{O}{\overset{\|}{C}}-OH + SOCl_2 \longrightarrow CH_3-\overset{O}{\overset{\|}{C}}-Cl + SO_2 + HCl$$

MECHANISM. Following steps are involved :

Step 1. The carboxylic acid reacts with thionyl chloride ($SOCl_2$) to give an acyl chlorosulfite intermediate.

$$\underset{\text{Carboxylic acid}}{R-\overset{O}{\overset{\|}{C}}-OH} + \underset{\text{Thionyl chloride}}{SOCl_2} \longrightarrow \underset{\text{Acyl chlorosulfite intermediate}}{R-\overset{O}{\overset{\|}{C}}-O-\overset{O}{\overset{\|}{S}}-Cl}$$

Step 2. If nucleophile Cl^- ion attacks the carbonyl carbon of the acyl chlorosulfite intermediate to form a tetrahedral intermediate.

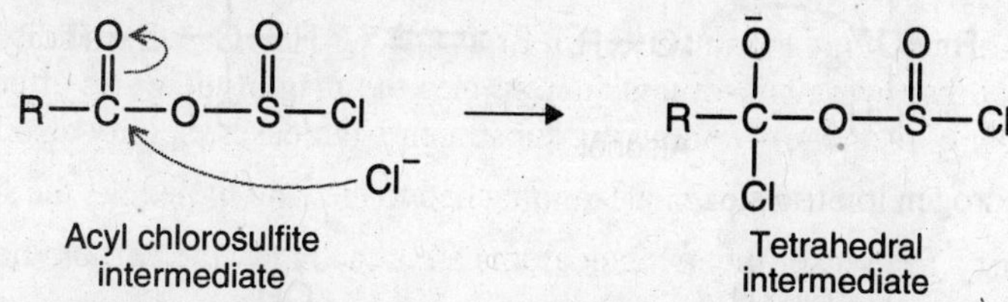

Step 3. Oxygen in the tetrahedral intermediate donates a pair of electrons to restore the carbon oxygen double bond and at the same time expels the chlorosulfite ion, thereby forming the acyl chloride. The chlorosulfite group (OSOCl) decomposes to SO_2 and chloride ion.

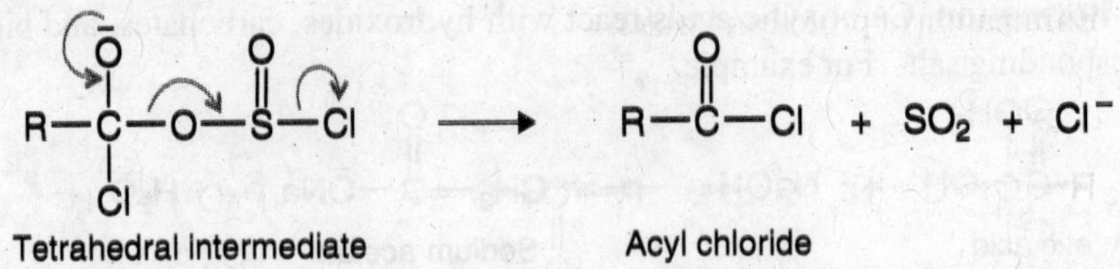

(3) **Formation of Amides**. Carboxylic acids react with ammonia to give salts, which on heating yield amides.

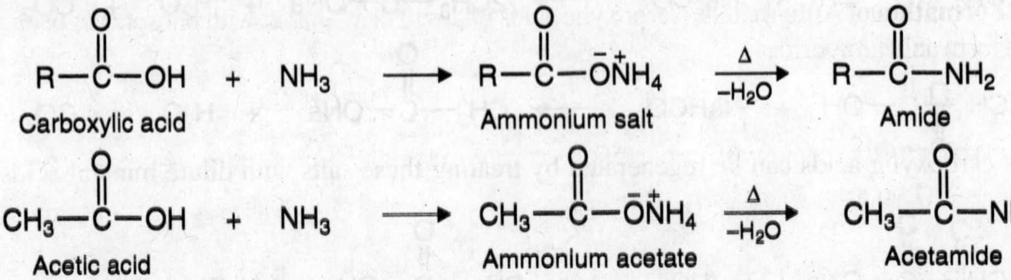

(4) **Formation of Esters**. Carboxylic acids react with alcohols in the presence of a strong acid catalyst like H_2SO_4 or HCl to form esters. The reaction is reversible and is called **Esterification.**

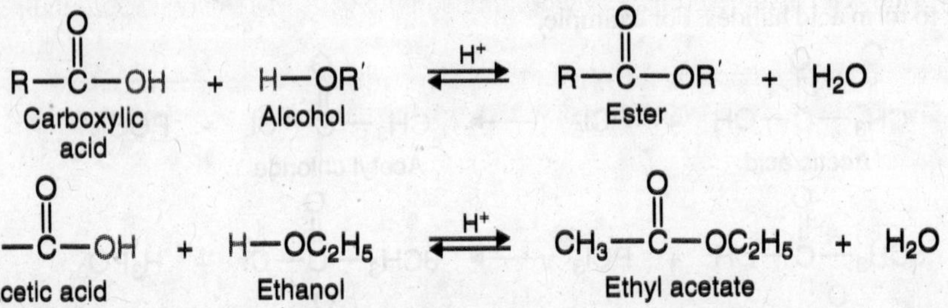

The equilibrium can be shifted to the right by using excess of alcohol or removal of water by distillation.

MECHANISM. The mechanism of esterification involves the following steps :

Step 1. Protonation of carboxylic acid.

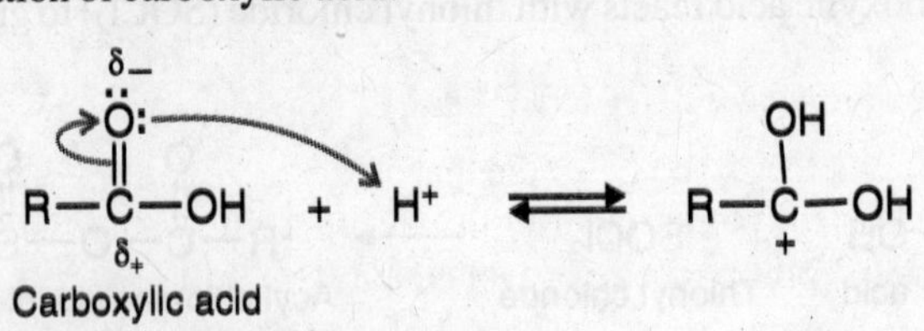

Step 2. Attack by nucleophilic R'OH.

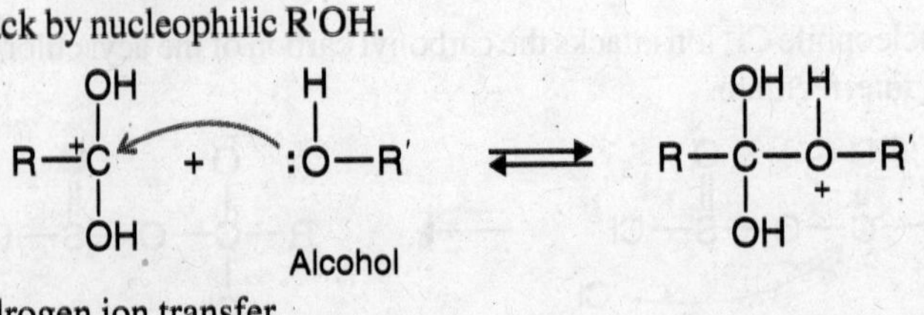

Step 3. Hydrogen ion transfer.

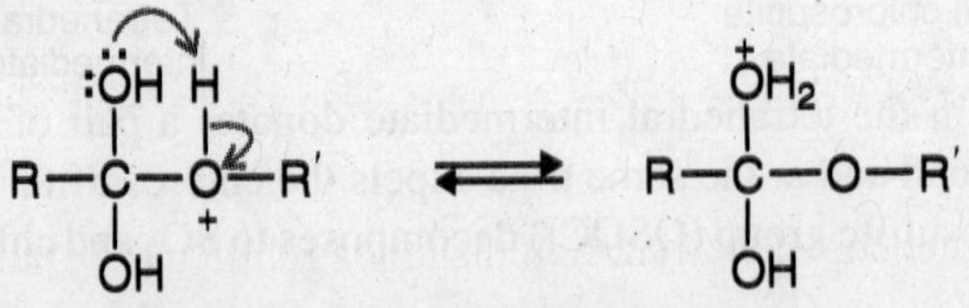

Step 4. Elimination of proton and water.

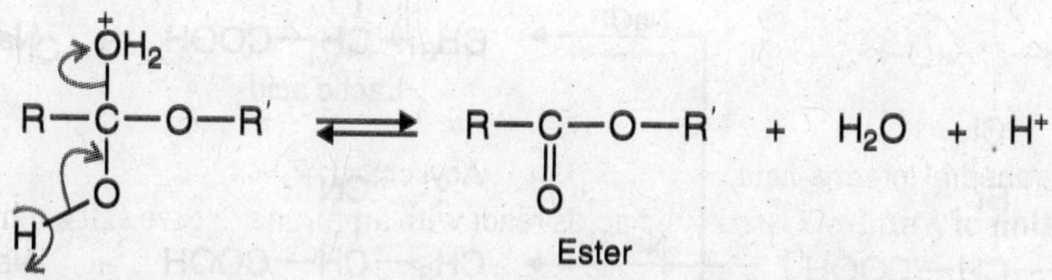

(5) **Formation of Anhydrides**. Carboxylic acids undergo dehydration with phosphorus pentoxide (P_2O_5) to form acid anhydrides.

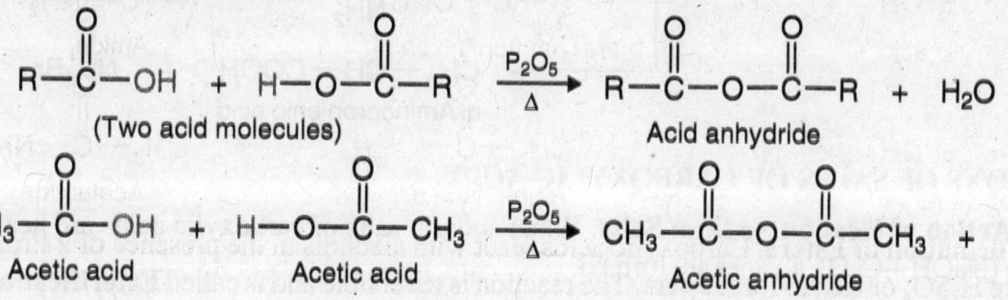

Anhydrides can also be prepared by treating sodium salts of acids with acid halides.

(6) **Reduction**. Carboxylic acids undergo reduction with lithium aluminium hydride to give primary alcohols.

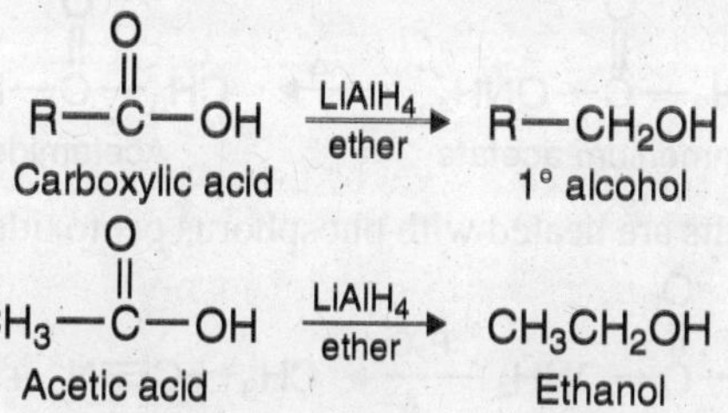

Carboxylic acids cannot be reduced with H_2/Ni or Na + C_2H_5OH.

(7) α-**Halogenation**. When a carboxylic acid that contains α-hydrogens is treated with Cl_2 or Br_2 in the presence of phosphorus, the α-hydrogen atoms are replaced by chlorine or bromine atoms. This reaction is known as the **Hell-Volhard-Zelinsky** (or HVZ) **Reaction**.

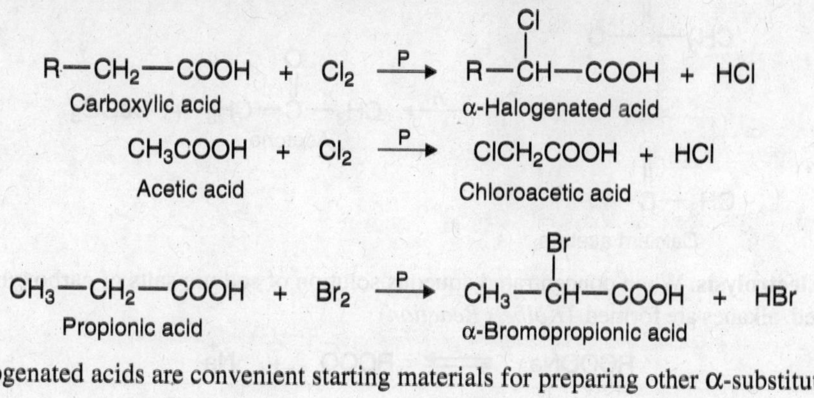

α-Halogenated acids are convenient starting materials for preparing other α-substituted acids.

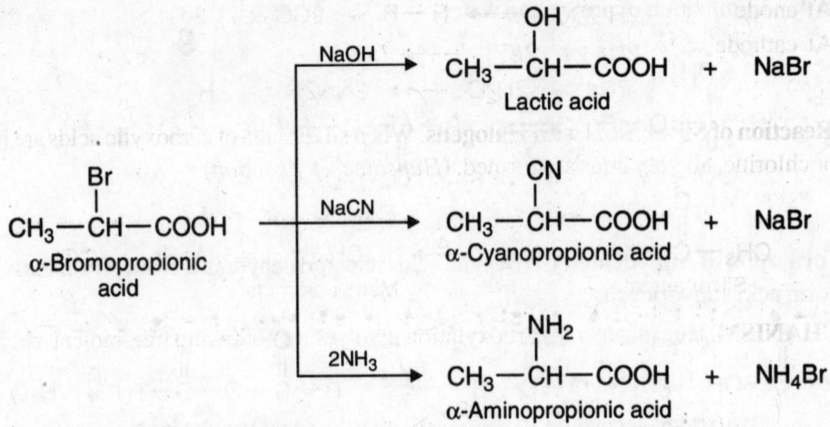

REACTIONS OF SALTS OF CARBOXYLIC ACIDS

(1) **Action of Heat on Sodium Salts.** When sodium salts of carboxylic acids are heated with soda-lime (NaOH + CaO), alkanes are formed.

$$\underset{\text{Sodium acetate}}{CH_3-\overset{O}{\overset{\|}{C}}-\overset{-}{O}\overset{+}{Na}} + NaOH \xrightarrow[\Delta]{CaO} \underset{\text{Methane}}{CH_4} + Na_2CO_3$$

(2) **Action of Heat on Ammonium Salts.** When ammonium salts are heated alone, amides are formed.

$$\underset{\text{Ammonium acetate}}{CH_3-\overset{O}{\overset{\|}{C}}-\overset{-}{O}\overset{+}{N}H_4} \xrightarrow{\Delta} \underset{\text{Acetamide}}{CH_3-\overset{O}{\overset{\|}{C}}-NH_2}$$

When the ammonium salts are heated with phosphorus pentoxide (P_2O_5), cyanides are formed.

$$\underset{\text{Ammonium acetate}}{CH_3-\overset{O}{\overset{\|}{C}}-\overset{-}{O}\overset{+}{N}H_4} \xrightarrow[\Delta]{P_2O_5} \underset{\text{Methyl cyanide}}{CH_3-C\equiv N} + 2H_2O$$

(3) **Action of Heat on Calcium Salts.** When calcium or barium salts of carboxylic acids are heated, ketones are formed.

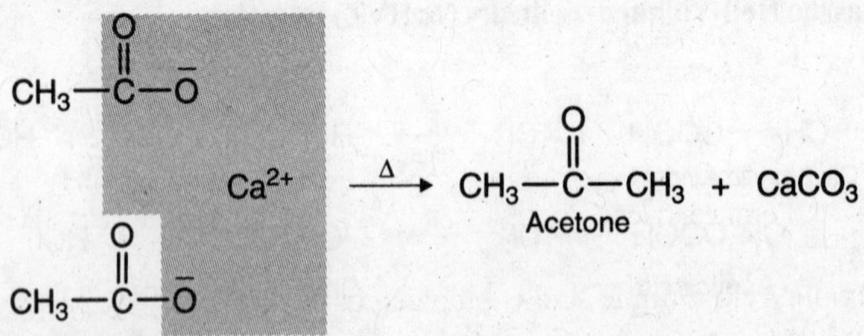

Calcium acetate

(4) **Electrolysis.** When concentrated aqueous solution of sodium salts of carboxylic acids are electrolyzed, alkanes are formed. (*Kolbe's Reaction*)

$$RCOONa \rightleftharpoons RCO\overset{-}{O} + \overset{+}{Na}$$

At anode : $2RCO\overset{-}{O} \longrightarrow R-R + 2CO_2 + 2e^-$

At cathode : $2\overset{+}{Na} + 2e^- \longrightarrow 2Na$

$$2Na + 2H_2O \longrightarrow 2NaOH + H_2$$

(5) **Reaction of Silver Salts with Halogens**. When silver salts of carboxylic acids are heated with bromine or chlorine, alkyl halides are formed. (*Hunsdiecker Reaction*)

$$\underset{\text{Silver acetate}}{CH_3-\overset{\overset{O}{\|}}{C}-\overset{-}{O}\overset{+}{Ag}} + Br_2 \xrightarrow{\Delta} \underset{\text{Methyl bromide}}{CH_3Br} + AgBr + CO_2$$

MECHANISM. Hunsdiecker decarboxylation involves the following free-radical mechanism :

$$\underset{\text{Silver acetate}}{CH_3-\overset{\overset{O}{\|}}{C}-\overset{-}{O}\overset{+}{Ag}} \xrightarrow[CCl_4]{Br_2} CH_3-\overset{\overset{O}{\|}}{C}-OBr + AgBr$$

$$CH_3-\overset{\overset{O}{\|}}{C}-O:Br \xrightarrow[\text{(or UV)}]{\Delta} CH_3-\overset{\overset{O}{\|}}{C}-O\cdot + \cdot Br$$

$$CH_3-\overset{\overset{O}{\|}}{C}-O\cdot \longrightarrow CH_3\cdot + CO_2$$

$$CH_3\cdot + Br:O-\overset{\overset{O}{\|}}{C}-CH_3 \longrightarrow \underset{\text{Methyl bromide}}{CH_3Br} + CH_3-\overset{\overset{O}{\|}}{C}-O\cdot$$

FORMIC ACID, *Methanoic Acid*, HCOOH

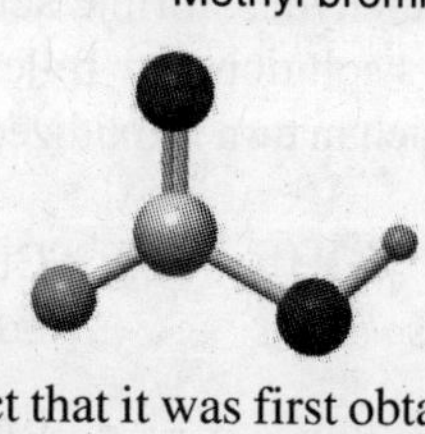

Formic acid derives its name from the fact that it was first obtained by distillation of ants (Latin, *formica* = ant).

Preparation. Formic acid may be obtained as follows :

(1) **From Methanol**. By the oxidation of methanol with acidified sodium dichromate solution.

$$\underset{\text{Methanol}}{CH_3OH} + [O] \xrightarrow{Na_2Cr_2O_7/H^+} \underset{\text{Formaldehyde}}{H-\overset{\overset{O}{\|}}{C}-H}$$

$$\underset{\text{Formaldehyde}}{H-\overset{\overset{O}{\|}}{C}-H} + [O] \xrightarrow{Na_2Cr_2O_7/H^+} \underset{\text{Formic acid}}{H-\overset{\overset{O}{\|}}{C}-OH}$$

(2) **From Oxalic Acid**. Formic acid is prepared in the laboratory by heating oxalic acid with glycerol catalyst at 110°C.

$$\underset{\text{Oxalic acid}}{\begin{matrix}COOH\\|\\COOH\end{matrix}} \xrightarrow[\Delta]{\text{glycerol}} \underset{\text{Formic acid}}{\begin{matrix}H\\|\\COOH\end{matrix}} + CO_2$$

The above decarboxylation reaction was discussed in Chapter *16*.

(3) **From Carbon Monoxide**. Formic acid is prepared industrially by heating carbon monoxide with sodium hydroxide at 150°C and under pressure of 8 atmospheres.

$$CO + NaOH \xrightarrow[\text{pressure}]{\Delta} \underset{\text{Sodium formate}}{H-\overset{\overset{\displaystyle O}{\|}}{C}-\overset{-\ +}{ONa}}$$

The free acid is liberated from its sodium salt by reaction with sulfuric acid.

$$\underset{\text{Sodium formate}}{H-\overset{\overset{\displaystyle O}{\|}}{C}-\overset{-\ +}{ONa}} + H_2SO_4 \longrightarrow \underset{\text{Formic acid}}{H-\overset{\overset{\displaystyle O}{\|}}{C}-OH} + NaHSO_4$$

Properties (*Physical*). Formic acid is a colorless, sharp smelling, irritating liquid, *bp* 100.8°C. It is miscible with water, ethanol, and ether. It is very corrosive and causes burns and blisters on the skin.

(*Chemical*). Formic acid is unique because it contains both an aldehyde group and a carboxyl group.

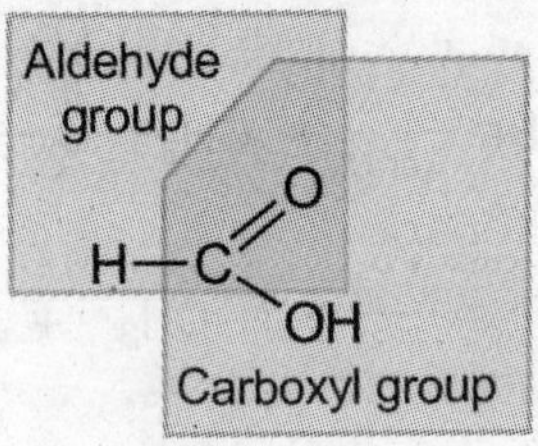

The reactions of formic acid are in keeping with its above structural feature.

(1) **Reducing Action**. Like aldehydes, formic acid is easily oxidized and is therefore a strong reducing agent. It reduces Fehling's solution and Tollens' reagent. Both these reagents first convert the formic acid into formate ion which in turn is oxidized.

$$HCO\bar{O} + \underset{(\textit{Fehling's solution})}{2\overset{2+}{Cu}} + 5\bar{O}H \longrightarrow \underset{\text{red ppt}}{Cu_2O\downarrow} + \overset{2-}{CO_3} + 3H_2O$$

$$HCO\bar{O} + \underset{(\textit{Tollens' reagent})}{2\overset{+}{Ag}} + 3\bar{O}H \longrightarrow \underset{\text{silver mirror}}{2Ag\downarrow} + \overset{2-}{CO_3} + 2H_2O$$

Formic acid decolorizes acidified potassium permanganate solution.

$$5HCO\bar{O} + 2Mn\bar{O}_4 + 6\overset{+}{H} \longrightarrow 2\overset{2+}{Mn} + 8H_2O + 5CO_2\uparrow$$

Formic acid also reduces mercuric chloride to mercurous chloride.

$$HCOOH + \underset{\substack{\text{Mercuric}\\\text{chloride}}}{2HgCl_2} \longrightarrow \underset{\substack{\text{Mercurous}\\\text{chloride}}}{Hg_2Cl_2\downarrow} + 2HCl + CO_2\uparrow$$

(2) **Salt Formation**. Formic acid is the strongest acid among all the members of the homologous series. It forms salts with alkalis, carbonates, and bicarbonates.

$$HCOOH + NaOH \longrightarrow \underset{\text{Sodium formate}}{HCO\overset{-\ +}{ONa}} + H_2O$$

$$HCOOH + NaHCO_3 \longrightarrow HCO\overset{-\ +}{ONa} + CO_2\uparrow + H_2O$$

(3) **Esterification**. Formic acid reacts with alcohols to form esters. It is not necessary to add a mineral acid to catalyze the reaction since the formic acid acts as its own catalyst.

$$HCOOH + CH_3OH \underset{}{\overset{\Delta}{\rightleftharpoons}} \underset{\text{Methyl formate}}{HCOOCH_3} + H_2O$$

(4) **Reaction with PCl_5.** Formic acid reacts with PCl_5 to give formyl chloride which at once decomposes into hydrogen chloride and carbon monoxide.

$$HCOOH + PCl_5 \longrightarrow \underset{\substack{\text{Formyl chloride} \\ \text{(unstable)}}}{HCOOCl} + POCl_3 + HCl$$

$$HCOOCl \longrightarrow HCl + CO$$

(5) **Action of Heat**. When formic acid is heated above 160°C under pressure, it decomposes to give carbon dioxide and hydrogen.

$$HCOOH \xrightarrow[\text{pressure}]{\Delta} CO_2 + H_2$$

(6) **Reaction with Sulfuric Acid**. When formic acid is warmed with concentrated sulfuric acid, it decomposes to give carbon monoxide and water. This reaction is used to prepare carbon monoxide in the laboratory.

$$HCOOH \xrightarrow[\Delta]{H_2SO_4} H_2O + CO$$

(7) **Action of Heat on Formates**. Salts of formic acid are called formates.

(*a*) When *Sodium formate* is heated to about 360°C, it decomposes to give sodium oxalate and hydrogen. Sodium oxalate on treatment with dilute H_2SO_4 gives oxalic acid.

$$\underset{\substack{\text{Sodium formate} \\ \text{(2 molecules)}}}{\begin{matrix} HCOONa \\ HCOONa \end{matrix}} \xrightarrow{\Delta} \underset{\text{Sodium oxalate}}{\begin{matrix} COONa \\ | \\ COONa \end{matrix}} + H_2$$

(*b*) When *calcium formate* is heated, it gives formaldehyde.

$$\underset{\text{Calcium formate}}{(HCOO)_2Ca^{2+}} \xrightarrow{\Delta} \underset{\text{Formaldehyde}}{HCHO} + CaCO_3$$

(*c*) When *ammonium formate* is heated, it gives formamide.

$$H-\overset{\overset{\displaystyle O}{\|}}{C}-OH + NH_3 \longrightarrow \underset{\text{Ammonium formate}}{H-\overset{\overset{\displaystyle O}{\|}}{C}-O\overset{+}{N}H_4} \xrightarrow[-H_2O]{\Delta} \underset{\text{Formamide}}{H-\overset{\overset{\displaystyle O}{\|}}{C}-NH_2}$$

Uses. Formic acid is used : (*1*) in textile dyeing and finishing ; (*2*) as coagulating agent for rubber latex ; (*3*) in leather tanning ; (*4*) in the preservation of fruits ; (*5*) as an antiseptic ; (*6*) in the treatment of gout ; and (*7*) in the laboratory preparation of carbon monoxide.

ACETIC ACID, *Ethanoic Acid*, CH_3COOH

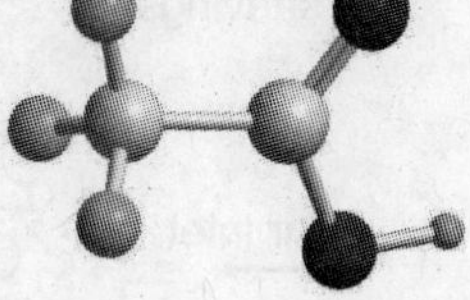

Acetic acid occurs in fruit juices which have become sour as a result of fermentation. It is a chief constituent of vinegar (Latin, *acetum* = vinegar).

Manufacture. Acetic acid can be prepared by any of the general methods described before. The following methods are used for the large-scale preparation of acetic acid.

(1) **By Air-Oxidation of Acetaldehyde**. Now-a-days acetic acid is mainly manufactured by the air-oxidation of acetaldehyde in the presence of manganous acetate.

$$\underset{\text{Acetaldehyde}}{2CH_3-\overset{\overset{\displaystyle O}{\|}}{C}-H} + O_2 \xrightarrow[50\text{–}70^\circ C]{(CH_3COO)_2Mn} \underset{\text{Acetic acid}}{2CH_3-\overset{\overset{\displaystyle O}{\|}}{C}-OH}$$

Acetaldehyde used in the above process may be obtained from acetylene.

$$\underset{\text{Acetylene}}{HC\equiv CH} + H_2O \xrightarrow[HgSO_4]{H_2SO_4} \underset{\text{Acetaldehyde}}{CH_3-\overset{\overset{\displaystyle O}{\|}}{C}-H}$$

(2) **By Air-Oxidation of Butane**. Acetic acid is also manufactured by the air-oxidation of butane (obtainable from petroleum) in the presence of cobalt acetate catalyst.

$$\underset{\text{Butane}}{CH_3CH_2CH_2CH_3} + O_2 \xrightarrow[200^\circ C]{(CH_3COO)_2Co} \underset{\text{Acetic acid}}{CH_3COOH}$$

(3) **Quick Vinegar Process**. This process is limited to the production of vinegar for food industry. Vinegar is 4 to 6 per cent acetic acid and is obtained from fermented liquors containing 12 to 15 per cent ethanol. The aqueous solution of ethanol in contact with air and under the influence of *Bacterium aceti* produces acetic acid.

$$\underset{\text{Ethanol}}{2CH_3CH_2OH} + O_2 \xrightarrow{B.\ aceti} \underset{\text{Acetaldehyde}}{3CH_3CHO} + 2H_2O$$

$$2CH_3CHO + O_2 \xrightarrow{B.\ aceti} \underset{\text{Acetic acid}}{2CH_3COOH}$$

In the *Quick Vinegar Process*, large wooden vats filled with wood shavings impregnated with old vinegar (*Bacterium aceti*) are used in series. The vats are fitted with a perforated cover and the ethanol solution is introduced at the top of the first vat and allowed to trickle down the shavings. Air is drawn in through the holes in the lower walls of the vessel and passes up in the opposite direction to the trickling liquid (Fig. 20.2). The heat generated in the oxidation of ethanol to acetic acid maintains the temperature at 35°C which is favorable to the growth and activity of bacteria. The liquor leaving the bottom of each vat is introduced at the top of the next vat. The maximum concentration of acetic acid obtained by this method is 10 per cent. The process requires 8 to 10 days for completion.

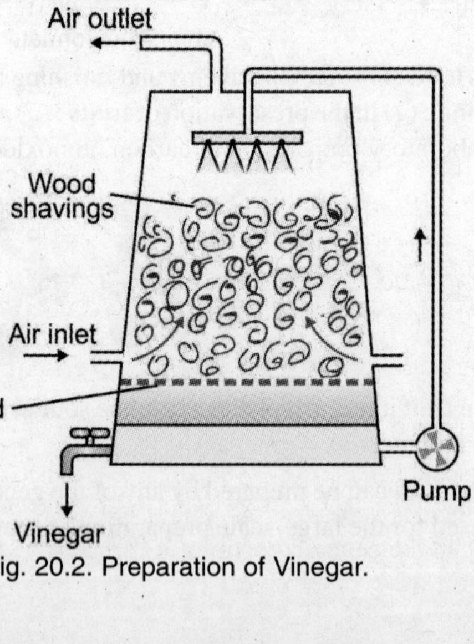

Fig. 20.2. Preparation of Vinegar.

Properties. Acetic acid is a colorless, corrosive liquid with vinegar odor and sour taste, bp 118°C, mp 16.6°C. The vapors are suffocating and cause damage to the lungs. Pure acetic acid is called **Glacial Acetic Acid** because it forms an ice-like solid when cooled. It is miscible with water, ether, and ethanol. Acetic acid is a typical carboxylic acid and gives all the general reactions of the class.

Uses. Acetic acid is used in the manufacture of : (*1*) white vinegar ; (*2*) cellulose acetate and polyvinyl acetate ; and (*3*) acetone and esters used in perfumes, dyes, plastics, and pharmaceuticals.

Table 20.2 COMPARISON OF FORMIC ACID AND ACETIC ACID

Formic Acid, HCOOH	**Acetic Acid, CH_3COOH**
Points of Resemblance	
(1) Forms salts with bases, carbonates, and bicarbonates.	(1) Forms salts with bases, carbonates, and bicarbonates.
(2) Forms esters when treated with alcohols in the presence of conc. H_2SO_4.	(2) Forms esters when treated with alcohols in the presence of conc. H_2SO_4.
(3) Reacts with PCl_5 to give formyl chloride which being unstable decomposes to give CO + HCl.	(3) Reacts with PCl_5 to give acetyl chloride.
(4) Forms an amide (formamide) when its ammonium salt is heated.	(4) Forms an amide (acetamide) when its ammonium salt is heated.
Points of Difference	
(1) When formic acid is heated above 160°C under pressure, it gives $CO_2 + H_2$.	(1) Acetic acid is stable to heat.
(2) When formic acid is heated with conc. H_2SO_4, it gives $CO + H_2O$	(2) Acetic acid does not react with conc. H_2SO_4.
(3) Formic acid does not react with Cl_2 in the presence of red P.	(3) Acetic acid reacts with Cl_2 in the presence of P to form mono-, di-, or trichloroacetic acid.
(4) Gives formaldehyde when its calcium salt is heated.	(4) Gives acetone when its calcium salt is heated.
(5) When its sodium salt is heated with soda-lime, hydrogen is formed.	(5) When its sodium salt is heated with soda-lime, methane is formed.
(6) When its sodium salt is heated to 360°C, sodium oxalate is formed.	(6) Its sodium salt is stable at 360°C.
(7) Electrolysis of aqueous solution of sodium salt gives hydrogen.	(7) Electrolysis of aqueous solution of sodium salt gives ethane.
(8) Formic acid reduces Fehling's solution and Tollens' reagent.	(8) Acetic acid does not react with Fehling's solution or Tollens' reagent.

PROPIONIC ACID, *Propanoic acid*, CH_3CH_2COOH

Preparation. Propionic acid is obtained by the oxidation of *n*-propyl alcohol with sodium dichromate and sulfuric acid.

$$\underset{n\text{-Propyl alcohol}}{CH_3CH_2CH_2OH} + 2[O] \xrightarrow[H_2SO_4]{Na_2Cr_2O_7} \underset{\text{Propionic acid}}{CH_3CH_2COOH} + H_2O$$

Properties. Propionic acid is a colorless oily liquid, *mp* –22°C, *bp* 141°C. It is miscible with water, ethanol, and ether. It gives all the general reactions of carboxylic acids.

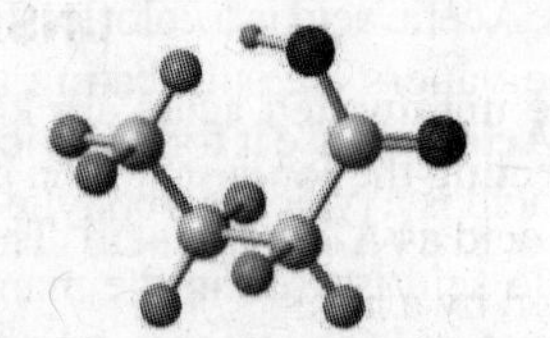

n-BUTYRIC ACID, *Butanoic acid*, $CH_3CH_2CH_2COOH$

Preparation. *n*-Butyric acid is obtained by the oxidation of *n*-butyl alcohol with sodium dichromate and sulfuric acid.

$$\underset{n\text{-Butyl alcohol}}{CH_3CH_2CH_2CH_2OH} + 2[O] \xrightarrow[H_2SO_4]{Na_2Cr_2O_7} \underset{n\text{-Butyric acid}}{CH_3CH_2CH_2COOH} + H_2O$$

Properties. *n*-Butyric acid is a rancid-smelling viscous liquid, *mp* –4.7°C, *bp* 164°C. It is miscible with water, ethanol and ether. It gives all the general reactions of carboxylic acids. It is used in the tanning industry for deliming of hides.

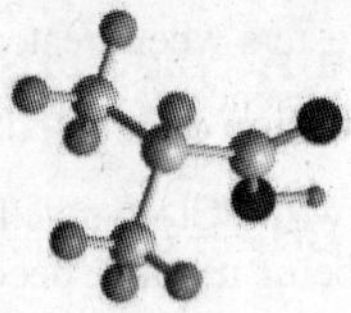

ISOBUTYRIC ACID, *2-Methylpropanoic acid*, $(CH_3)_2CHCOOH$

Preparation. Isobutyric acid is obtained :

(1) By the oxidation of isobutyl alcohol with sodium dichromate and sulfuric acid.

$$\underset{\text{Isobutyl alcohol}}{CH_3-\overset{\displaystyle CH_3}{\overset{|}{C}H}-CH_2OH} + 2[O] \xrightarrow[H_2SO_4]{Na_2Cr_2O_7} \underset{\text{Isobutyric acid}}{CH_3-\overset{\displaystyle CH_3}{\overset{|}{C}H}-COOH} + H_2O$$

(2) From isopropyl alcohol by the following steps :

$$\underset{\text{Isopropyl alcohol}}{CH_3-\overset{\displaystyle CH_3}{\overset{|}{C}H}-OH} \xrightarrow{SOCl_2} CH_3-\overset{\displaystyle CH_3}{\overset{|}{C}H}-Cl \xrightarrow{NaCN} CH_3-\overset{\displaystyle CH_3}{\overset{|}{C}H}-CN \xrightarrow{H_2O/H^+} \underset{\text{Isobutyric acid}}{CH_3-\overset{\displaystyle CH_3}{\overset{|}{C}H}-COOH}$$

Properties. Isobutyric acid is a colorless oily liquid, *bp* 154°C. It is sparingly soluble in water. It gives all the typical reactions of carboxylic acids.

PALMITIC ACID, $CH_3(CH_2)_{14}COOH$; STEARIC ACID, $CH_3(CH_2)_{16}COOH$

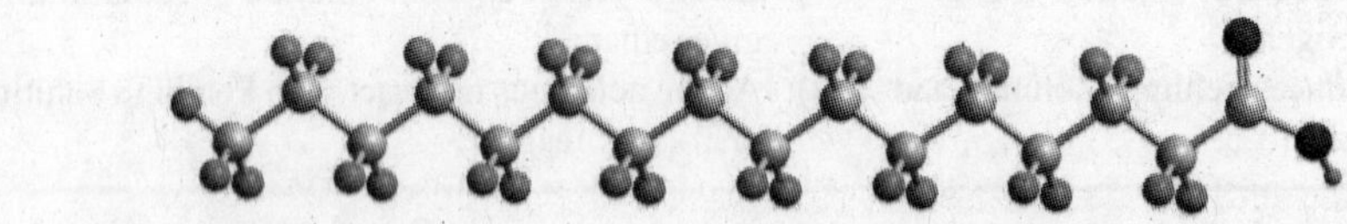

Palmitic Acid

Stearic Acid

These acids occur as glyceryl esters in natural fats from which they are prepared by hydrolysis. Palmitic acid takes its name from *Palm Oil* and stearic acid from *Stear*, meaning tallow.

Palmitic acid and stearic acid are colorless waxy solids which melt at 64°C and 72°C respectively. They are insoluble in water and soluble in ethanol and ether. Their sodium and potassium salts are the constituents of ordinary soaps. These acids are also used in the manufacture of candles.

UNSATURATED ACIDS

Most of the unsaturated acids are known by their common names. Their IUPAC names are obtained by selecting the longest carbon chain containing the double bond and the carboxyl group, and naming the acid as **Alkenoic acid**. The position of the double bond with respect to the carboxyl group is indicated by a number.

$$\underset{\text{Propenoic acid (Acrylic acid)}}{CH_2{=}CH{-}COOH} \qquad \underset{\text{2-Butenoic acid (Crotonic acid)}}{\overset{4}{C}H_3\overset{3}{C}H{=}\overset{2}{C}H{-}\overset{1}{C}OOH}$$

ACRYLIC ACID, *Propenoic acid*, $CH_2{=}CH{-}COOH$

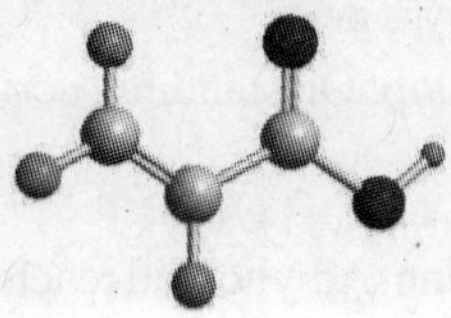

Preparation. Acrylic acid is obtained :

(1) By oxidation of acrolein with ammoniacal silver nitrate.

$$\underset{\text{Acrolein}}{CH_2{=}CH{-}CHO} + [O] \xrightarrow[NH_4OH]{AgNO_3} \underset{\text{Acrylic acid}}{CH_2{=}CH{-}COOH}$$

(2) From allyl alcohol by the following steps :

$$\underset{\text{Allyl alcohol}}{CH_2{=}CH{-}CH_2OH} \xrightarrow{Br_2} CH_2Br{-}CHBr{-}CH_2OH \xrightarrow[{[O]}]{HNO_3} CH_2Br{-}CHBr{-}COOH \xrightarrow[(-ZnBr_2)]{Zn/\Delta} \underset{\text{Acrylic acid}}{CH_2{=}CH{-}COOH}$$

(3) From propionic acid by the following steps. This method involves the use of *Hell-Volhard-Zelinsky* (HVZ) *reaction*.

$$\underset{\text{Propionic acid}}{CH_3CH_2COOH} \xrightarrow[HVZ]{Br_2/P} CH_3\overset{Br}{\overset{|}{C}}HCOOH \xrightarrow{alc\ KOH} CH_2{=}CHCOO\overset{-}{}\overset{+}{K} \xrightarrow{H_2O/H^+} \underset{\text{Acrylic acid}}{CH_2{=}CH{-}COOH}$$

(4) By carbonylation of acetylene in the presence of water and nickel carbonyl catalyst.

$$\underset{\text{Acetylene}}{HC{\equiv}CH} + CO + H_2O \xrightarrow{Ni(CO)_4} \underset{\text{Acrylic acid}}{CH_2{=}CH{-}COOH}$$

Properties. (*Physical*). Acrylic acid is a colorless pungent-smelling liquid, *bp* 141°C. It is miscible with water, alcohol, and ether.

(*Chemical*). Acrylic acid behaves both as an alkene and a carboxylic acid.

A. REACTIONS OF THE DOUBLE BOND

(1) **Reduction**. Acrylic acid undergoes reduction with sodium and ethanol to give propionic acid.

$$\underset{\text{Acrylic acid}}{CH_2{=}CHCOOH} + 2[H] \xrightarrow[C_2H_5OH]{Na} \underset{\text{Propionic acid}}{CH_3CH_2COOH}$$

(2) **Addition of Halogens**. Acrylic acid reacts with Cl_2 or Br_2 to form addition products.

$$\underset{\text{Acrylic acid}}{CH_2{=}CHCOOH} + Br_2 \longrightarrow \underset{\alpha,\beta\text{-Dibromopropionic acid}}{\overset{Br}{\overset{|}{C}}H_2{-}\overset{Br}{\overset{|}{C}}HCOOH}$$

(3) **Addition of Halogen Acids**. Acrylic acid reacts with halogen acids to give addition products. Markovnikov rule is not followed.

$$\underset{\text{Acrylic acid}}{CH_2{=}CHCOOH} + HBr \longrightarrow \underset{\beta\text{-Bromopropionic acid}}{BrCH_2CH_2COOH}$$

(4) **Reaction with Dil. $KMnO_4$**. Acrylic acid reacts with dilute cold $KMnO_4$ solution to give glyceric acid.

$$\underset{\text{Acrylic acid}}{CH_2{=}CHCOOH} + H_2O + [O] \xrightarrow[KMnO_4]{\text{cold}} \underset{\text{Glyceric acid}}{CH_2(OH){-}CH(OH)COOH}$$

(5) **Polymerization**. On standing, acrylic acid slowly polymerizes to a glassy solid.

B. REACTIONS OF –COOH GROUP

(6) **Salt Formation**. Acrylic acid reacts with sodium hydroxide or sodium carbonate to form salts.

$$\underset{\text{Acrylic acid}}{CH_2{=}CH{-}\overset{O}{\overset{\|}{C}}{-}OH} + NaOH \longrightarrow \underset{\text{Sodium acrylate}}{CH_2{=}CH{-}\overset{O}{\overset{\|}{C}}{-}\bar{O}\overset{+}{Na}} + H_2O$$

(7) **Reaction with PCl_5**. Acrylic acid reacts with phosphorus pentachloride to give acryl chloride.

$$\underset{\text{Acrylic acid}}{CH_2{=}CH{-}\overset{O}{\overset{\|}{C}}{-}OH} + PCl_5 \longrightarrow \underset{\text{Acryl chloride}}{CH_2{=}CH{-}\overset{O}{\overset{\|}{C}}{-}Cl} + POCl_3 + HCl$$

(8) **Esterification**. Acrylic acid reacts with ethanol in the presence of acid catalyst to give ethyl acrylate.

$$\underset{\text{Acrylic acid}}{CH_2{=}CH{-}\overset{O}{\overset{\|}{C}}{-}OH} + C_2H_5OH \xrightarrow{H^+} \underset{\text{Ethyl acrylate}}{CH_2{=}CH{-}\overset{O}{\overset{\|}{C}}{-}OC_2H_5} + H_2O$$

Esters of acrylic acid are used to produce colorless plastics such as *Lucite* and *Plexiglass*.

CROTONIC ACID, *trans-2-Butenoic acid*, $CH_3CH{=}CHCOOH$

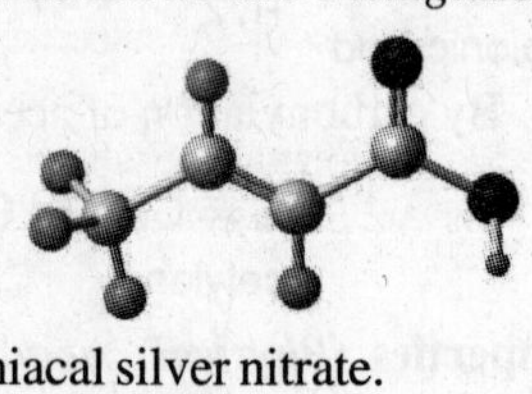

Preparation. (1) By the oxidation of crotonaldehyde with ammoniacal silver nitrate.

$$\underset{\text{Crotonaldehyde}}{CH_3CH{=}CHCHO} + [O] \xrightarrow[NH_4OH]{AgNO_3} \underset{\text{Crotonic acid}}{CH_3CH{=}CHCOOH}$$

(2) By heating acetaldehyde and diethyl malonate in the presence of pyridine (*Knoevenagel Reaction*) followed by the hydrolysis and decarboxylation.

$$\underset{\text{Acetaldehyde}}{CH_3CH{=}O} + \underset{\text{Diethyl malonate}}{H_2C(COOC_2H_5)_2} \xrightarrow[\Delta,(-H_2O)]{\text{pyridine}} CH_3CH{=}C(COOC_2H_5)_2$$

$$\xrightarrow[\Delta]{\text{aq. KOH}} CH_3CH{=}C(CO\bar{O}\overset{+}{K})_2 \xrightarrow[(-CO_2)]{H^+/\Delta} \underset{\text{Crotonic acid}}{CH_3CH{=}CHCOOH}$$

Properties. Crotonic acid is a colorless solid, *mp* 72°C. Like acrylic acid, it gives reactions of an alkene and a carboxylic acid. It undergoes bromination with *N*-bromosuccinimide to form γ-bromocrotonic acid, $BrCH_2CH{=}CHCOOH$.

OLEIC ACID, *cis-9-Octadecenoic acid*, $CH_3(CH_2)_7CH{=}CH(CH_2)_7COOH$

Oleic acid is found in oils and fats in the form of glyceryl esters. It is obtained by basic-hydrolysis of *Olive Oil*.

Properties. Oleic acid is a colorless liquid, *mp* 16°C, *bp* 286°C. It is insoluble in water but dissolves in ethanol and ether. It gives the reactions of alkenes and carboxylic acids. Oleic acid reacts with ozone to form the corresponding ozonide. The ozonide on treatment with zinc and water decomposes to yield two aldehydes.

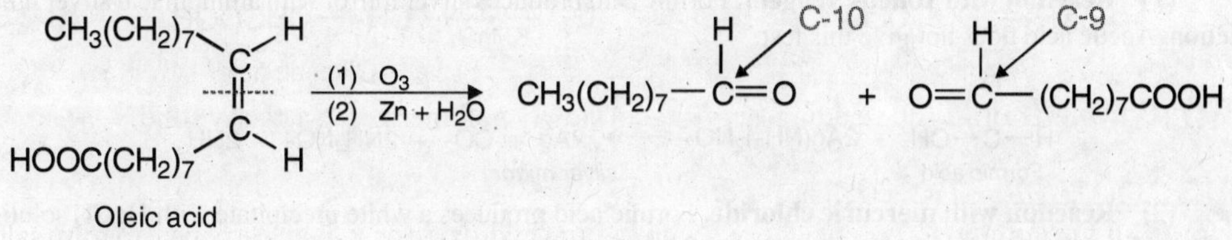

This reaction shows that the position of the double bond in oleic acid is between C-9 and C-10.

Uses. Oleic acid is used in the manufacture of soaps, lubricants, and synthetic detergents.

Examination Questions

1. How are aliphatic carboxylic acids prepared ? Describe their important reactions.
2. How is acetic acid manufactured ? How does it react with :

   (*a*) PCl_5 (*b*) $SOCl_2$ (*c*) CH_3CH_2OH/H^+
   (*d*) P_2O_5 (*e*) $LiAlH_4$ (*f*) Aqueous NaOH

Answer. The products are : (*a*) Acetyl chloride ; (*b*) Acetyl chloride ; (*c*) Ethyl acetate ; (*d*) Acetic anhydride ; (*e*) Ethyl alcohol ; and (*f*) Sodium acetate. For reaction equations see text.

3. What is glacial acetic acid ?
4. How is acrylic acid prepared ? How does it react with :
   (*a*) sodium and ethyl alcohol
   (*b*) dilute cold potassium permanganate solution
   (*c*) phosphorus pentachloride
5. What happens when acetic acid is treated with sodium bicarbonate ? (*Agra BSc, 2014*)

Answer. Sodium acetate is formed.

$$\underset{\text{Acetic acid}}{CH_3{-}\overset{O}{\overset{\|}{C}}{-}OH} + NaHCO_3 \longrightarrow \underset{\text{Sodium acetate}}{CH_3{-}\overset{O}{\overset{\|}{C}}{-}\bar{O}\overset{+}{Na}} + H_2O + CO_2$$

6. What happens when acetic acid is treated with phosphorus pentachloride ?

Answer. Acetyl chloride is formed.

$$\underset{\text{Acetic acid}}{CH_3{-}\overset{O}{\overset{\|}{C}}{-}OH} + PCl_5 \longrightarrow \underset{\text{Acetyl chloride}}{CH_3{-}\overset{O}{\overset{\|}{C}}{-}Cl} + POCl_3 + HCl$$

7. What happens when propanoic acid is treated with bromine in the presence of red phosphorus? (*Madras BSc, 2015*)

Answer. 2-Bromopropanoic acid is formed.

$$\underset{\text{Propanoic acid}}{CH_3{-}CH_2{-}COOH} + Br_2 \xrightarrow{P} \underset{\text{2-Bromopropanoic acid}}{CH_3{-}\overset{Br}{\overset{|}{CH}}{-}COOH} + HBr$$

8. What happens when acetic acid is heated with phosphorus pentoxide ?

Answer. Acetic anhydride is formed.

$$\underset{\text{Acetic acid}}{CH_3-\overset{\overset{\displaystyle O}{\|}}{C}-OH} + \underset{\text{Acetic acid}}{H-O-\overset{\overset{\displaystyle O}{\|}}{C}-CH_3} \xrightarrow[\Delta]{P_2O_5} \underset{\text{Acetic anhydride}}{CH_3-\overset{\overset{\displaystyle O}{\|}}{C}-O-\overset{\overset{\displaystyle O}{\|}}{C}-CH_3} + H_2O$$

9. How will you distinguish between formic acid and acetic acid ? *(Mangalore BSc, 2015)*

Answer. Following tests are used :

(1) **Reaction with Tollens' reagent**. Formic acid produces silver mirror with ammoniacal silver nitrate solution. Acetic acid does not give this test.

$$\underset{\text{Formic acid}}{H-\overset{\overset{\displaystyle O}{\|}}{C}-OH} + 2Ag(NH_3)_2NO_3 \longrightarrow \underset{\textit{silver mirror}}{2Ag} + CO_2 + 2NH_4NO_3 + 2NH_3$$

(2) **Reaction with mercuric chloride**. Formic acid produces a white precipitate with $HgCl_2$ solution. Acetic acid does not give this test.

$$H-\overset{\overset{\displaystyle O}{\|}}{C}-OH + 2HgCl_2 \longrightarrow \underset{\text{Mercurous chloride}}{Hg_2Cl_2} + CO_2 + 2HCl$$

10. How will you distinguish between propionic acid and crotonic acid ?

Answer. Crotonic acid is an unsaturated acid ; it decolorizes red color of Br_2/CCl_4 solution. Propionic acid does not give this test.

$$CH_3CH{=}CHCOOH + \underset{(\textit{red})}{Br_2} \longrightarrow \underset{(\textit{colorless})}{CH_3\overset{\overset{\displaystyle Br}{|}}{C}H-\overset{\overset{\displaystyle Br}{|}}{C}HCOOH}$$

11. Write a note on : Hell-Volhard-Zelinsky reaction. *(Delhi BSc, 2015)*

12. Write a note on : Acidity of carboxylic acids.

13. Discuss the effect of substituents on the acidity of monocarboxylic acids.

(Baroda BSc, 2014 ; Indore BSc, 2014 ; Nagpur BSc, 2015)

14. Explain why formic acid is a stronger acid than acetic acid.

15. Explain why acetic acid is a stronger acid than propionic acid.

16. Explain why chloroacetic acid is a stronger acid than acetic acid.

(Patna BSc, 2014 ; Tamil Nadu BSc, 2015)

17. Explain : Acetic acid is a stronger acid than ethyl alcohol. *(Panjab BSc, 2015)*

Answer. Acetic acid is a stronger acid than ethyl alcohol because the acetate ion formed by ionization is stabilized by resonance. The negative charge is delocalized over two oxygen atoms. In the ethoxide ion, on the contrary, the negative charge is localized on the single oxygen atom. There is no resonance. Ethyl alcohol is, therefore, much weaker acid than acetic acid.

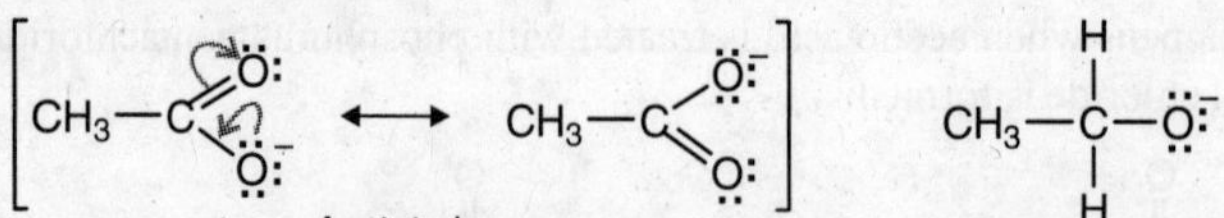

Acetate ion (Resonance structures)

Ethoxide ion (No resonance structures)

18. Arrange the following compounds in order of increasing acidity :

$$ClCH_2COOH,\ Cl_2CHCOOH,\ Cl_3CCOOH$$

Answer. Trichloroacetic acid is the strongest. The order is :

$$Cl_3CCOOH > Cl_2CHCOOH > ClCH_2COOH$$

19. Arrange the following compounds in order of increasing acidity :

$$HCOOH,\ ClCH_2COOH,\ CH_3COOH$$

Answer. Chloroacetic acid is the strongest. The order is :

$$ClCH_2COOH > HCOOH > CH_3COOH$$

20. Explain why boiling points of carboxylic acids are higher than the corresponding alcohols. (*Mysore BSc, 2015*)

21. Discuss the mechanism of esterification of carboxylic acids. (*Delhi BSc, 2014*)

22. Identify (A), (B), and (C) in the following reaction sequence :

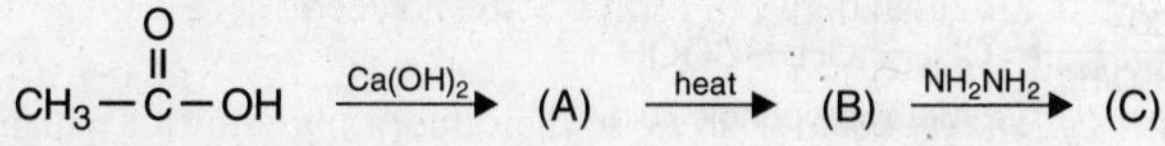

Answer. (A) = $(CH_3COO)_2Ca$; Calcium acetate

(B) = $(CH_3)_2C{=}O$; Acetone

(C) = $(CH_3)_2C{=}NNH_2$; Acetone hydrazone

23. How will you synthesize acetic acid from acetylene ?

Answer. Following steps are involved :

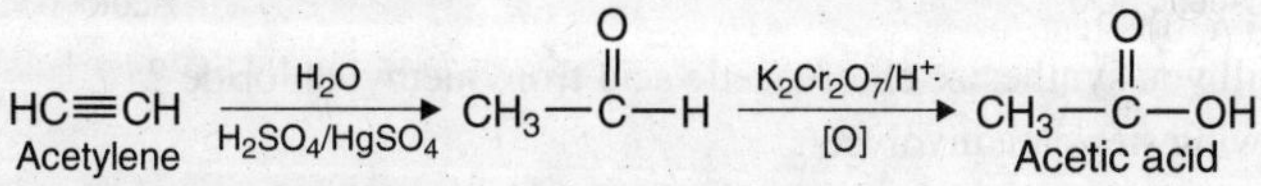

24. How will you synthesize acetic acid from propionic acid ? (*Madras BSc, 2015*)

Answer. Following steps are involved :

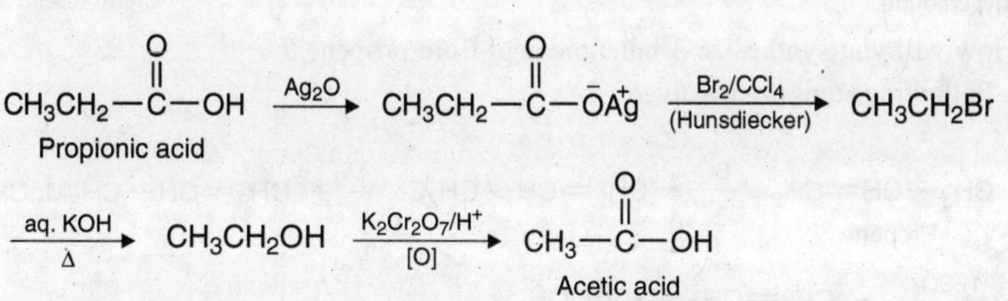

25. How will you synthesize propionic acid from acetic acid ? (*Tamil Nadu BSc, 2014*)

Answer. Following steps are involved :

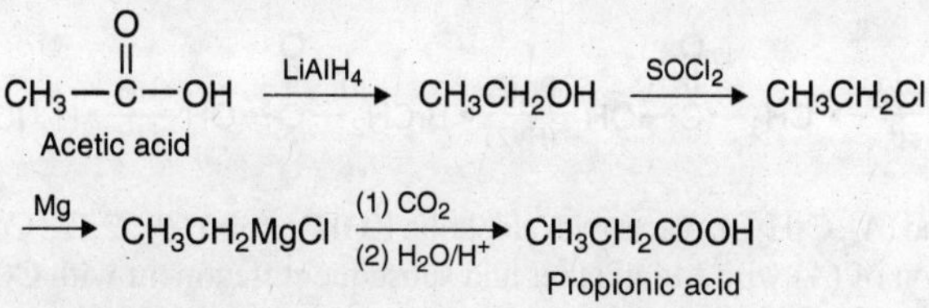

26. How will you synthesize propanoic acid from ethyl alcohol ? (*Agra BSc, 2015*)

Answer. Following steps are involved :

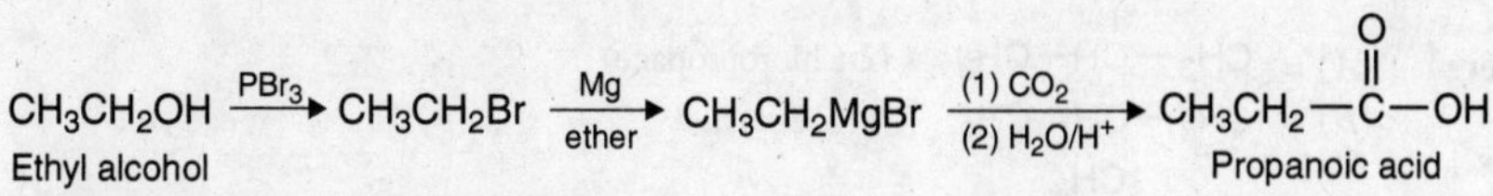

27. How will you synthesize propionic acid from ethylene ?

Answer. Following steps are involved :

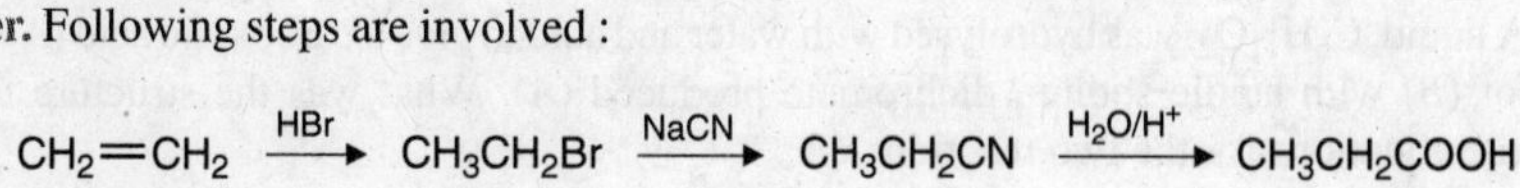

28. How will you synthesize 2-methylpropanoic acid from propene ? *(Baroda BSc, 2013)*

Answer. Following steps are involved :

$$\underset{\text{Propene}}{CH_3-CH=CH_2} \xrightarrow{HBr} CH_3-\overset{\overset{CH_3}{|}}{CH}-Br \xrightarrow[\text{ether}]{Mg} CH_3-\overset{\overset{CH_3}{|}}{CH}-MgBr$$

$$\xrightarrow[(2)\ H_2O/H^+]{(1)\ CO_2} \underset{\text{2-Methylpropanoic acid}}{CH_3-\overset{\overset{CH_3}{|}}{CH}-COOH}$$

29. How will you synthesize acetone from acetic acid ? *(Anna BSc, 2014)*

Answer. Following steps are involved :

$$\underset{\text{Acetic acid}}{2CH_3-\overset{\overset{O}{\|}}{C}-OH} \xrightarrow{Ca(OH)_2} (CH_3COO)_2Ca \xrightarrow[-CaCO_3]{\Delta} \underset{\text{Acetone}}{CH_3-\overset{\overset{O}{\|}}{C}-CH_3}$$

30. How will you synthesize chloroacetic acid from methyl chloride ?

Answer. Following steps are involved :

$$\underset{\text{Methyl chloride}}{CH_3Cl} \xrightarrow{Mg} CH_3MgCl \xrightarrow[(2)\ H_2O/H^+]{(1)\ CO_2} CH_3-\overset{\overset{O}{\|}}{C}-OH \xrightarrow[(HVZ)]{P/Cl_2} \underset{\text{Chloroacetic acid}}{ClCH_2-\overset{\overset{O}{\|}}{C}-OH}$$

31. How will you synthesize 3-butenoic acid from propene ?

Answer. Following steps are involved :

$$\underset{\text{Propene}}{CH_3-CH=CH_2} \xrightarrow[500^\circ C]{Cl_2} CH_2=CH-CH_2Cl \xrightarrow{Mg} CH_2=CH-CH_2MgCl$$

$$\xrightarrow[(2)\ H_2O/H^+]{(1)\ CO_2} \underset{\text{3-Butenoic acid}}{CH_2=CH-CH_2COOH}$$

32. How will you synthesize glycine from ethyl alcohol ? *(Sagar BSc, 2015)*

Answer. Following steps are involved :

$$\underset{\text{Ethyl alcohol}}{CH_3CH_2OH} \xrightarrow[H^+]{K_2Cr_2O_7} CH_3-\overset{\overset{O}{\|}}{C}-OH \xrightarrow[(HVZ)]{Br_2/P} BrCH_2-\overset{\overset{O}{\|}}{C}-OH \xrightarrow{NH_3} \underset{\text{Glycine}}{H_2NCH_2-\overset{\overset{O}{\|}}{C}-OH}$$

33. Compound *(A)*, C_3H_7Cl, reacts with alcoholic KOH to form *(B)*, C_3H_6. Compound *(B)* decolourises Br_2/CCl_4 solution. Reaction of *(A)* with Mg in ether and subsequent treatment with CO_2 and dilute acid gives a compound *(C)*, whose molecular formula is $C_4H_8O_2$. When we add compound *(C)* to aqueous $NaHCO_3$ solution, bubbles are evolved. Give the structural formulas of *(A)*, *(B)*, and *(C)*. Write equations for all the reactions involved.

Answer. $(A) = CH_3-\overset{\overset{Cl}{|}}{CH}-CH_3$ (2-Chloropropane)

$(B) = CH_3-CH=CH_2$ (Propene)

$(C) = CH_3-\overset{\overset{CH_3}{|}}{CH}-COOH$ (2-Methylpropanoic acid)

34. A liquid, $C_6H_{12}O_2$, was hydrolysed with water and acid to give a carboxylic acid *(A)* and an alcohol *(B)*. Oxidation of *(B)* with acidic sodium dichromate produced *(A)*. What was the structure of the original compound ? Write equations for the two reactions.

Answer. The unknown is $CH_3CH_2-\overset{\overset{O}{\|}}{C}-OCH_2CH_2CH_3$ (*n*-Propyl propionate)

35. An unknown ester, $C_5H_{10}O_2$, was hydrolysed with water and acid to give a carboxylic acid (*A*) and an alcohol (*B*). Treatment of (*B*) with PBr_3 gave an alkyl bromide (*C*). When (*C*) was treated with KCN, a product (*D*) was formed which, on hydrolysis with water and acid, gave a carboxylic acid (*A*). Give the structure and name of the original ester. Identify (*A*) through (*D*) and write equations for the reactions described.

Answer. The unknown ester is $CH_3CH_2-\overset{O}{\overset{\|}{C}}-OCH_2CH_3$ (Ethyl propionate). The other unknowns are:

(*A*) = CH_3CH_2COOH (Propionic acid) (*B*) = CH_3CH_2OH (Ethyl alcohol)
(*C*) = CH_3CH_2Br (Ethyl bromide) (*D*) = CH_3CH_2CN (Propionitrile)

Multiple Choice Questions

1. Which of the following is *not* true regarding carboxylic acids?
(*a*) Carboxylic acids with five or fewer carbon atoms quite water-soluble.
(*b*) Carboxylic acids are almost completed dissociated in water.
(*c*) The carbon-oxygen bonds of the formate anion ($HCOO^-$) are the same length.
(*d*) The carbon atom of formic acid, HCOOH, is sp^2 hybridized.

Answer. (*b*)

2. What statement regarding carboxylate anions is *not* true ?
(*a*) The carbon-oxygen double bond is shorter than the carbon-oxygen single bond.
(*b*) The carboxylate is more stable than an alkoxide ion.
(*c*) The negative charge is delocalized over both oxygens.
(*d*) They can be formed by the reaction of carboxylic acids and aqueous sodium bicarbonate.

Answer. (*a*)

3. Vinegar is a
(*a*) 5% solution of acetic acid in water (*b*) 25% solution of acetic acid in water
(*c*) 50% solution of acetic acid in water (*d*) 40% solution of formic acid in water

Answer. (*a*)

4. The complete hydrolysis of a nitrile gives
(*a*) an acid (*b*) an ester
(*c*) an anhydride (*d*) an acid halide

Answer. (*a*)

5. The reaction of a Grignard reagent with carbon dioxide forms
(*a*) a ketone with one more carbon than the Grignard reagent.
(*b*) a primary alcohol.
(*c*) a carboxylic acid with one more carbon than the Grignard reagent.
(*d*) a secondary alcohol.

Answer. (*c*)

6. What Grignard reagent is needed to prepare pentanoic acid from the reaction of a Grignard reagent with carbon dioxide?
(*a*) Ethylmagnesium bromide (*b*) Propylmagnesium bromide
(*c*) Butylmagnesium bromide (*d*) Methylmagnesium bromide

Answer. (*c*)

7. The interaction that has the greatest effect on the physical properties (boiling point, melting point, solubility) of carboxylic acids is ?
(*a*) van der Waals interactions (*b*) non-polar interactions
(*c*) large dipole interactions (*d*) hydrogen bonding

Answer. (*d*)

8. Which of the following has the *highest* boiling point ?
(*a*) Pentane (*b*) 2-Pentanol
(*c*) 1-Chloropentane (*d*) Pentanoic acid

Answer. (*d*)

9. Which compound has the highest boiling point ?
(a) CH_3CH_3 (b) CH_3OCH_3 (c) CH_3CH_2OH (d) CH_3COOH

Answer. (d)

10. The greater acidity of carboxylic acids compared to alcohols arises primarily from :
(a) the electron-donating effect of the hydroxyl group
(b) the electron-withdrawing effect of the carboxyl oxygen
(c) the acidity of α-hydrogens of carboxylic acids
(d) the resonance stability associated with the carboxylate ion

Answer. (d)

11. Which of the following statements is false about the acid-strength of acetic acid ?
(a) Acetic acid is a stronger acid than monochloroacetic acid.
(b) Acetic acid is a stronger acid than propionic acid.
(c) Acetic acid is a weaker acid than trichloroacetic acid.
(d) Acetic acid is a weaker acid than formic acid.

Answer. (a)

12. Which of the following compounds is the strongest acid ?
(a) CH_3COOH (b) CH_3CH_2COOH
(c) CF_3COOH (d) $CH_3CH_2CH_2COOH$

Answer. (c)

13. Which of the following is the strongest acid ?
(a) Formic acid (b) Trichloroacetic acid
(c) Acetic acid (d) Trifluoroacetic acid

Answer. (d)

14. Which of the following is the strongest acid ?
(a) Butanoic acid (b) 2-Chlorobutanoic acid
(c) 3-Chlorobutanoic acid (d) 4-Chlorobutanoic acid

Answer. (b)

15. Which of the following is the strongest acid in aqueous solution ?
(a) CH_3COOH (b) $ClCH_2COOH$
(c) CH_3CH_2COOH (d) $Cl_2CHCOOH$

Answer. (d)

16. Which is the strongest acid ?
(a) CH_3COOH (b) $Cl_2CHCOOH$
(c) $ClCH_2COOH$ (d) Cl_3CCOOH

Answer. (d)

17. Which of the following compounds is most acidic ?
(a) CH_3COOH (b) $ClCH_2COOH$
(c) CH_3CH_2COOH (d) FCH_2COOH

Answer. (d)

18. Which of the following compounds is least acidic ?
(a) CH_3CH_2COOH (b) $BrCH_2CH_2COOH$
(c) $CH_3-CH(Br)-COOH$ (d) $CH_3-CH(F)-COOH$

Answer. (a)

19. Which of the following will give acetic acid on acid-hydrolysis ?
(a) Ethyl acetate (b) Acetone
(c) Methyl propionate (d) Lactic acid

Answer. (a)

20. Propanenitrile undergoes acid-hydrolysis to give
(*a*) Formic acid (*b*) Propionic acid
(*c*) Acetic acid (*d*) Butyric acid
Answer. (*b*)
21. The characteristic reaction of carboxylic acids is :
(*a*) electrophilic addition (*b*) electrophilic substitution
(*c*) nucleophilic addition (*d*) nucleophilic substitution
Answer. (*d*)
22. Which of the following compounds will react with Tollens' reagent to give metallic silver ?
(*a*) Formic acid (*b*) Ethyl alcohol
(*c*) Acetic acid (*d*) Acetone
Answer. (*a*)
23. Which of the following compounds on treatment with $NaHCO_3$ will liberate CO_2 ?
(*a*) Acetic acid (b) Ethylamine
(*c*) Acetone (d) Ethyl alcohol
Answer. (*a*)
24. Which of the following reagents will convert acetic acid into acetyl chloride ?
(*a*) NaCl (*b*) $HCl/ZnCl_2$
(*c*) $SOCl_2$ (*d*) HCl
Answer. (*c*)
25. Butyric acid reacts with PCl_5 to give
(*a*) Benzoyl chloride (*b*) 1-Chlorobutane
(*c*) Butyryl chloride (*d*) 1-Chloropropane
Answer. (*c*)
26. Acetic acid undergoes reduction with $LiAlH_4$ to give
(*a*) Ethanol (*b*) Ethane
(*c*) Ethanal (*d*) Ethyne
Answer. (*a*)
27. Acetic acid reacts with methyl alcohol in the presence of an acid catalyst to give
(*a*) Methyl formate (*b*) Ethyl formate
(*c*) Methyl acetate (*d*) Ethyl acetate
Answer. (*c*)
28. Calcium acetate on strong heating gives
(*a*) Methane + $CaCO_3$ (*b*) Ethane + $CaCO_3$
(*c*) Acetone + $CaCO_3$ (*d*) Ethane + CaO
Answer. (*c*)
29. $CH_3(CH_2)_4\,COO^-Na^+$ is the sodium salt of hexanoic acid. The water solubility of this salt is :
(*a*) higher than that of hexanoic acid (*b*) lower than that of hexanoic acid
(*c*) completely insoluble (*d*) not predictable
Answer. (*a*)
30. Ammonium acetate on strong heating gives
(*a*) Urea (*b*) Formamide
(*c*) Uric acid (*d*) Acetamide
Answer. (*d*)
31. Which reaction does not yield an ester as one of the products ?
(*a*) A carboxylic acid is heated with an alcohol
(*b*) A Grignard reagent is added to a carboxylic acid
(*c*) An acid halide is treated with an alcohol
(*d*) An alkyl halide is heated with the salt of a carboxylic acid
Answer. (*b*)
32. Silver acetate reacts with Br_2 to form methyl bromide, carbon dioxide, and AgBr. This is an example of
(*a*) Wurtz reaction (*b*) Etard's reaction
(*c*) Hunsdiecker reaction (*d*) Perkin reaction
Answer. (*c*)

21

Dicarboxylic Acids

Exterior wood tends to turn dark grey after being exposed and oxalic acid solutions are used to lighten and brighten dingy grey wood.

They contain two carboxyl groups, one at each end of a saturated hydrocarbon chain. Their general formula is

$$HO-\overset{\displaystyle O}{\overset{\|}{C}}-(CH_2)_n-\overset{\displaystyle O}{\overset{\|}{C}}-OH \quad \text{or} \quad HOOC(CH_2)_nCOOH$$

where $n = 0,1,2,3$, etc.

Dicarboxylic acids are named as **Alkanedioic Acids** in the IUPAC system. Most simple dicarboxylic acids have common names and these are the names that we will use. The formulas and names of the first nine members of the series are listed below.

Table 21.1 NAMES OF DICARBOXYLIC ACIDS

Formula	Common Name	IUPAC Name
HOOC–COOH	Oxalic acid	Ethanedioic acid
HOOC–CH_2–COOH	Malonic acid	Propanedioic acid
HOOC–$(CH_2)_2$–COOH	Succinic acid	Butanedioic acid
HOOC–$(CH_2)_3$–COOH	Glutaric acid	Pentanedioic acid
HOOC–$(CH_2)_4$–COOH	Adipic acid	Hexanedioic acid
HOOC–$(CH_2)_5$–COOH	Pimelic acid	Heptanedioic acid
HOOC–$(CH_2)_6$–COOH	Suberic acid	Octanedioic acid
HOOC–$(CH_2)_7$–COOH	Azelaic acid	Nonanedioic acid
HOOC–$(CH_2)_8$–COOH	Sebacic acid	Decanedioic acid

METHODS OF PREPARATION

The methods of preparation of dicarboxylic acids are analogous to those used for monocarboxylic acids, only we have to develop two COOH groups instead of one.

(1) **By Cyanide Synthesis**. The starting material is a α-halogenated acid or polymethylene dibromide.

$$\underset{\text{Chloroacetic acid}}{Cl-CH_2-COOH} \xrightarrow{NaCN} \underset{\text{Cyanoacetic acid}}{CN-CH_2-COOH} \xrightarrow{H_2O/H^+} \underset{\text{Malonic acid}}{HOOC-CH_2-COOH}$$

$$\underset{\substack{\text{1,2-Dibromoethane}\\ \text{(Ethylene dibromide)}}}{\begin{matrix} CH_2Br \\ | \\ CH_2Br \end{matrix}} \xrightarrow{2NaCN} \begin{matrix} CH_2CN \\ | \\ CH_2CN \end{matrix} \xrightarrow{H_2O/H^+} \underset{\text{Succinic acid}}{\begin{matrix} CH_2COOH \\ | \\ CH_2COOH \end{matrix}}$$

(2) **By Oxidation of Cyclic Ketones**

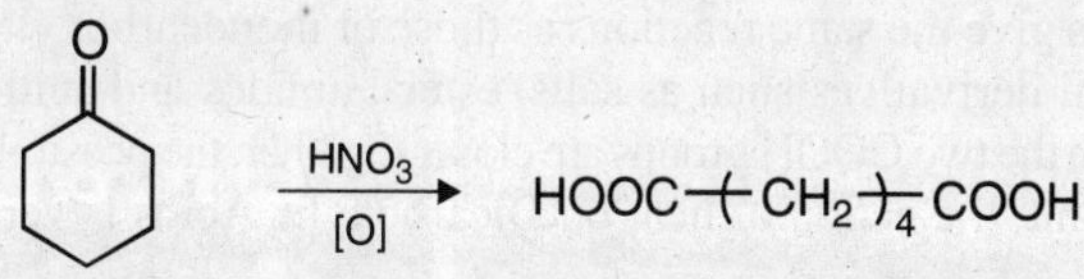

Cyclohexanone Adipic acid

(3) **By Malonic ester and Acetoacetic ester Synthesis**. For discussion see *Chapter 25* Adipic acid.

PHYSICAL PROPERTIES

(1) Dicarboxylic acids are crystalline solids, their melting points being much higher than the corresponding monocarboxylic acid.

(2) The lower members dissolve readily in water but the solubility falls with the increase of molecular weight.

(3) The melting points of dicarboxylic acids show 'alternation' or 'oscillation' from one member to the other. The melting point of an 'even acid' (having even number of carbon atoms) is always higher than that of the 'odd acid' lying immediately above and below it.

Table 21.2 MELTING POINTS ; VALUES OF K_1 AND K_2

Acid	mp°C	$K_1 \times 10^{-5}$	$K_2 \times 10^{-5}$
Oxalic acid	187	5400	5.2
Malonic acid	135	170	0.23
Succinic acid	185	6.8	0.24
Glutaric acid	97.5	4.6	0.28
Adipic acid	151	3.8	0.23
Pimelic acid	1.05	3.5	0.27

(4) **Acidity**. The dicarboxylic acids ionize in two steps :

$$\underset{\text{Oxalic acid}}{HOOC-COOH} \underset{}{\overset{K_1}{\rightleftharpoons}} HOOC-CO\bar{O} + H^+$$

$$HOOC-CO\bar{O} \overset{K_2}{\rightleftharpoons} \bar{O}OC-CO\bar{O} + H^+$$

The electron-withdrawing effect of the second COOH group enhances the acidity of the first COOH. Therefore, the value of the first dissociation constant K_1 is higher than that of acetic acid (K_a=1.5×10^{-5}). The inductive effect becomes weaker as the two COOH groups are further separated by CH_2 groups and K_1 decreases as we ascend the series.

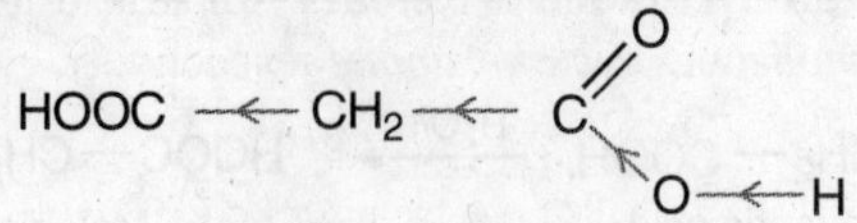

Inductive effect of the second COOH on the first COOH

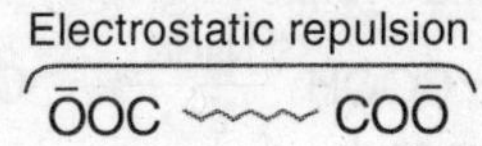

As evident from the Table 21.2, the value of K_2 is smaller than K_a of acetic acid (with the exception of oxalic acid) because the removal of proton from the second COOH is more difficult. This is so because the electrostatic repulsion due to negative charges in the dianion, $\bar{O}OC—CO\bar{O}$, destabilizes the dianion. That is why the value of K_2 is much less than K_1.

CHEMICAL PROPERTIES

Dicarboxylic acids give the same reactions as those of monocarboxylic acids in duplicate. Thus they give mono- and di-derivatives such as salts, esters, amides and acid halides. However, in the lower members in which the two COOH groups are closer together, the possibilities of mutual interactions increase. From this point, the action of heat on Dicarboxylic Acids is very interesting and needs a special mention.

ACTION OF HEAT ON DICARBOXYLIC ACIDS

The reactions that occur when dicarboxylic acids are heated, are unique indeed. These depend on the carbon chain length separating the two COOH groups.

(1) **Acids with 1-C or no-C chain undergo Decarboxylation**. Oxalic acid and malonic acid decarboxylate, when heated, to form *monocarboxylic acids*.

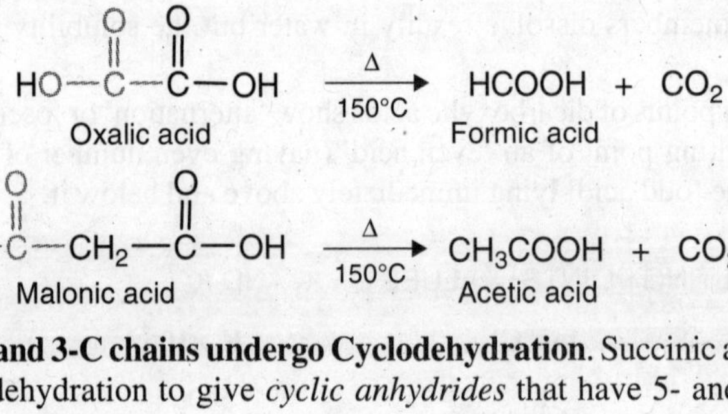

Oxalic acid — Formic acid

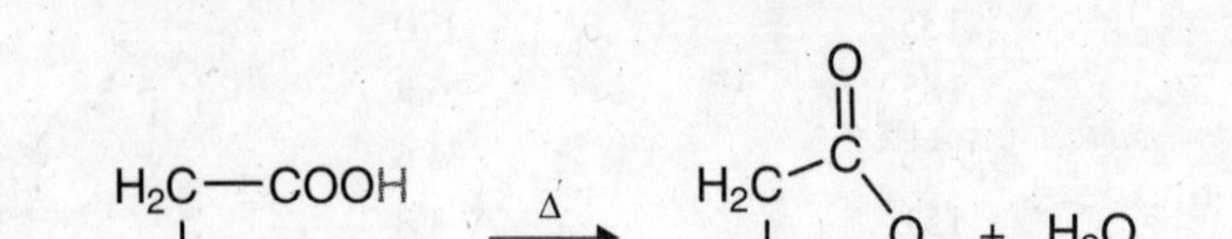

Malonic acid — Acetic acid

(2) **Acids with 2-C and 3-C chains undergo Cyclodehydration**. Succinic acid and glutaric acid react by intramolecular dehydration to give *cyclic anhydrides* that have 5- and 6-membered rings respectively.

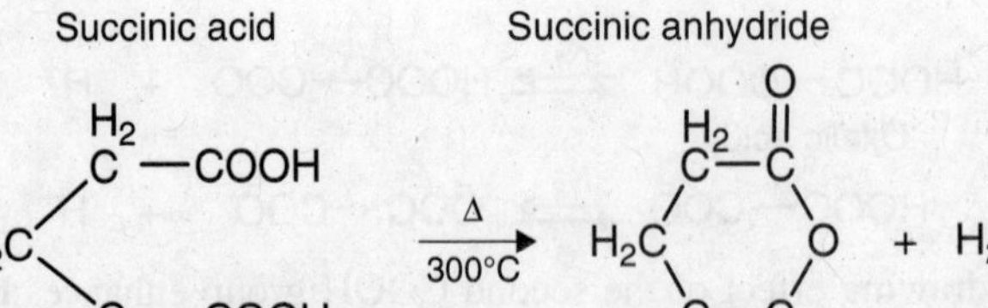

Succinic acid — Succinic anhydride

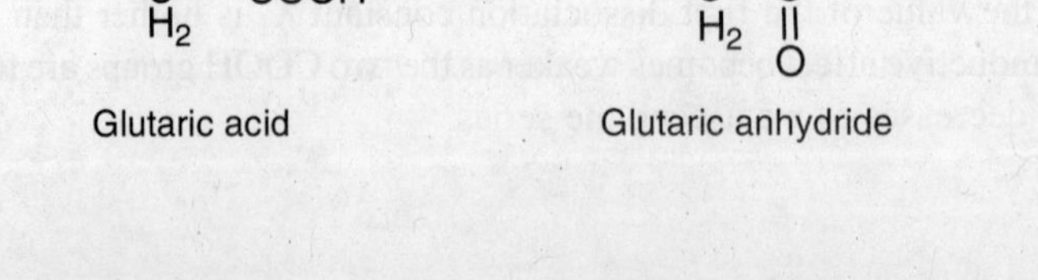

Glutaric acid — Glutaric anhydride

(3) **Acids with 4-C and 5-C chains undergo Cyclization to Ketones.** Adipic acid and pimelic acid decarboxylate and cyclize to give *cyclopentanone* and *cyclohexanone* respectively.

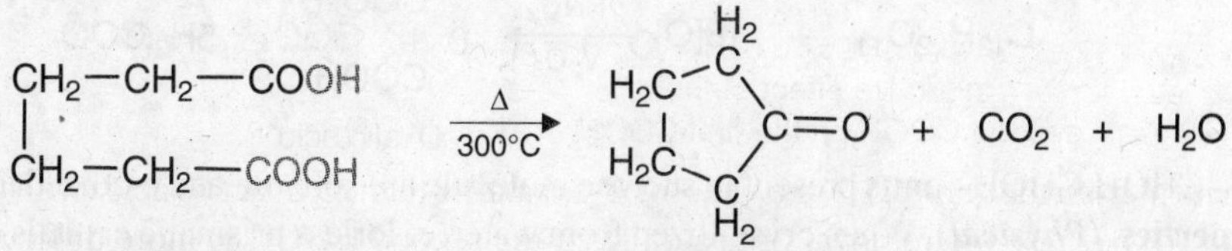

Adipic acid Cyclopentanone

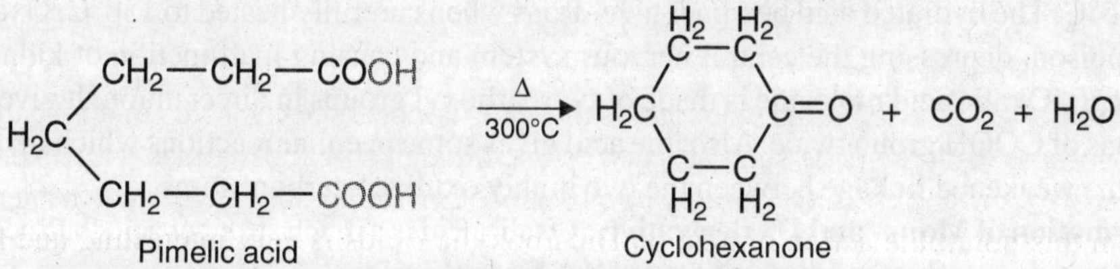

Pimelic acid Cyclohexanone

(4) **Acids with 6-C or higher chains undergo Intermolecular Dehydration to form Linear Polymers.** The dicarboxylic acids having six-carbon chain react by intermolecular dehydration when a linear polymer results.

$$\underset{\text{Suberic acid (2 molecules)}}{HO\overset{O}{\overset{\|}{C}}(CH_2)_6\overset{O}{\overset{\|}{C}}OH + HO\overset{O}{\overset{\|}{C}}(CH_2)_6\overset{O}{\overset{\|}{C}}OH} \xrightarrow[-H_2O]{\Delta} \underset{\text{A dimer}}{HO\overset{O}{\overset{\|}{C}}(CH_2)_6\overset{O}{\overset{\|}{C}}O\overset{O}{\overset{\|}{C}}(CH_2)_6\overset{O}{\overset{\|}{C}}OH}$$

The reaction continues with *n* molecules to yield a *linear polymer*.

OXALIC ACID, *Ethanedioic acid*, HOOC-COOH

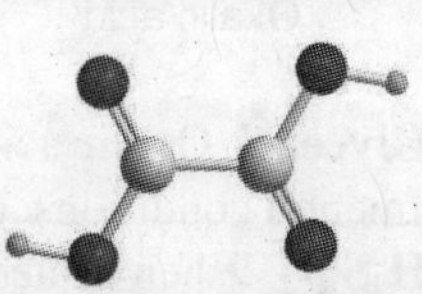

It occurs as potassium hydrogen oxalate in the wood *sorrel*, rhubarb, and tomatoes. The insoluble calcium oxalate is found in some stony deposits in kidneys and bladder in human body.

Preparation. (1) **Manufacture.** Oxalic acid is made industrially by heating sodium formate to 400°C.

$$\underset{\text{Sodium formate}}{2HCO\bar{O}\overset{+}{Na}} \xrightarrow{400°C} \underset{\text{Sodium oxalate}}{\overset{+}{Na}\bar{O}OC-CO\bar{O}\overset{+}{Na}} + H_2$$

The sodium oxalate thus formed is dissolved in water and calcium hydroxide added to precipitate calcium oxalate. The solution is filtered and the filtrate treated with calculated quantity of dilute sulfuric acid to liberate the oxalic acid.

$$\underset{\text{Sodium oxalate}}{\begin{matrix} CO\bar{O}\overset{+}{Na} \\ | \\ CO\bar{O}\overset{+}{Na} \end{matrix}} + Ca(OH)_2 \longrightarrow \underset{\text{Calcium oxalate}}{\begin{matrix} CO\bar{O} \\ | \\ CO\bar{O} \end{matrix}\; Ca^{2+}\downarrow} + 2NaOH$$

$$\underset{\text{Calcium oxalate}}{\begin{matrix} CO\bar{O} \\ | \\ CO\bar{O} \end{matrix}\; Ca^{2+}} + H_2SO_4 \longrightarrow \underset{\text{Oxalic acid}}{\begin{matrix} COOH \\ | \\ COOH \end{matrix}} + CaSO_4\downarrow$$

Calcium sulfate precipitates, and oxalic acid is crystallized from the filtrate as the hydrate, $(COOH)_2.2H_2O$.

(2) **Laboratory Method.** In the laboratory, oxalic acid is made by oxidation of sucrose or molasses with concentrated nitric acid in the presence of vanadium pentoxide as catalyst.

$$C_{12}H_{22}O_{11} + 18[O] \xrightarrow[V_2O_5]{HNO_3} 6 \begin{array}{c} COOH \\ | \\ COOH \end{array} + 5H_2O$$

Oxalic acid

The –CHOH.CHOH– units present in sucrose molecule are split out and oxidized to oxalic acid.

Properties. (*Physical*). When crystallized from water, colorless prismatic crystals of oxalic acid dihydrate, $(COOH)_2.2H_2O$, are obtained. The dihydrate melts at 101.5°C, while the anhydrous acid melts at 189.5°C. The hydrated acid becomes anhydrous when carefully heated to 150°C. Oxalic acid is an active poison, depressing the central nervous system and causing malfunction of kidneys.

(*Chemical*). Oxalic acid molecule is made of two carboxyl groups in direct union. It gives all the usual reactions of COOH group twice. Also, the acid gives some peculiar reactions which involve the cleavage of the weakened linkage between the two highly oxidized carbon atoms.

(1) **Formation of Mono- and Di- derivatives**. Oxalic acid is a much stronger acid than acetic acid and readily forms two series of salts, esters, acid halides and amides.

$$\begin{array}{c} \overset{-\ +}{COONa} \\ | \\ \overset{-\ +}{COONa} \end{array} \qquad \begin{array}{c} COOC_2H_5 \\ | \\ COOC_2H_5 \end{array} \qquad \begin{array}{c} COCl \\ | \\ COCl \end{array} \qquad \begin{array}{c} CONH_2 \\ | \\ CONH_2 \end{array}$$

Disodium oxalate | Diethyl oxalate | Oxalyl chloride | Oxamide

(2) **Action of Heat**. When heated at 150°C, it decarboxylates to give formic acid.

$$\underset{\text{Oxalic acid}}{HOOC-COOH} \xrightarrow{150°C} \underset{\text{Formic acid}}{HCOOH} + CO_2$$

(3) **Action with Glycerol**. Oxalic acid reacts with glycerol to form formic acid or allyl alcohol, depending upon experimental conditions. (*See Glycerol*)

(4) **Action with H_2SO_4**. When heated with concentrated sulfuric acid, it is decomposed to give carbon dioxide, carbon monoxide and water.

$$\underset{\text{Oxalic acid}}{HOOC-COOH} \xrightarrow[\Delta]{\text{Conc } H_2SO_4} CO_2 + CO + H_2O$$

(5) **Oxidation**. It is readily oxidized, for example with acidified potassium permanganate.

$$\underset{\text{Oxalic acid}}{HOOC-COOH} + [O] \xrightarrow[H_2SO_4]{KMnO_4} 2CO_2 + H_2O$$

Uses. Oxalic acid is used : (*1*) For removing ink stains and for bleaching straw for hats, since it reduces brown ferric compounds to soluble and almost colorless ferrous salts ; (*2*) As a mordant in dyeing and calico printing ; (*3*) In manufacture of inks and metal polishes ; (*4*) For preparing allyl alcohol and formic acid in the laboratory ; and (*5*) In *Redox Titrations*.

MALONIC ACID, *Propanedioic acid*, $HOOC-CH_2-COOH$

This acid occurs as calcium salt in sugar beet.

Preparation. Malonic acid is prepared from chloroacetic acid by treating with aqueous sodium cyanide to give cyanoacetic acid, and subsequent hydrolysis with dilute hydrochloric acid.

$$\underset{\text{Chloroacetic acid}}{Cl-CH_2-COOH} \xrightarrow{NaCN} \underset{\text{Cyanoacetic acid}}{CN-CH_2-COOH} \xrightarrow{H_2O/H^+} \underset{\text{Malonic acid}}{HOOC-CH_2-COOH}$$

Properties. Malonic acid is a white crystalline solid, mp 135°C. It is soluble in water and ethanol, but sparingly soluble in ether. It gives all the usual reactions of a dicarboxylic acid.

(1) **Action of Heat**. Acetic acid is formed.

$$\underset{\text{Malonic acid}}{HOOC{-}CH_2{-}COOH} \xrightarrow{150°C} \underset{\text{Acetic acid}}{CH_3COOH} + CO_2$$

(2) **Action with P_2O_5**. On heating with P_2O_5, malonic acid gives carbon suboxide.

$$\underset{\text{Malonic acid}}{O{=}C(OH){-}CH_2{-}C(OH){=}O} \xrightarrow[\Delta]{P_2O_5} \underset{\text{Carbon suboxide}}{O{=}C{=}C{=}C{=}O} + 2H_2O$$

(3) **Action with Aldehydes**. When a pyridine solution of malonic acid and an aldehyde is heated, α,β-unsaturated acid is produced.

$$RCH{=}O + \underset{\text{Malonic acid}}{H_2C(COOH)_2} \xrightarrow[\Delta]{\text{Pyridine}} \underset{\alpha,\beta\text{-Unsaturated acid}}{RCH{=}CHCOOH} + H_2O + CO_2$$

Uses. Its diethyl ester is a valuable synthetic reagent for the preparation of a variety of carboxylic acids.

SUCCINIC ACID, *Butanedioic acid*, $HOOCCH_2CH_2COOH$

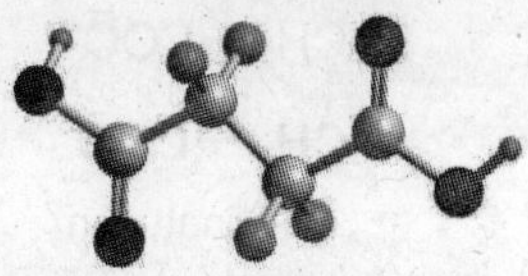

Succinic acid was first obtained as a result of distillation of *amber* and hence its name (Latin, *succinum* = amber). It is also produced during the alcoholic fermentation of sugar.

Preparation. (1) Succinic acid is prepared from ethylene bromide by treating with sodium cyanide and subsequent hydrolysis of ethylene cyanide.

$$\underset{\text{Ethylene bromide}}{\begin{array}{c}CH_2Br\\|\\CH_2Br\end{array}} \xrightarrow{2NaCN} \underset{\text{Ethylene cyanide}}{\begin{array}{c}CH_2CN\\|\\CH_2CN\end{array}} \xrightarrow{H_2O/H^+} \underset{\text{Succinic acid}}{\begin{array}{c}CH_2COOH\\|\\CH_2COOH\end{array}}$$

(2) It is obtained industrially by catalytic reduction of maleic acid.

$$\underset{\text{Maleic acid}}{\begin{array}{c}CHCOOH\\||\\CHCOOH\end{array}} + H_2 \xrightarrow[\Delta]{Ni} \underset{\text{Succinic acid}}{\begin{array}{c}CH_2COOH\\|\\CH_2COOH\end{array}}$$

Succinic acid can also be synthesized with the help of *Malonic Ester Synthesis*.

Properties. (*Physical*). Succinic acid forms white monoclinic prisms, mp 188°C. It is soluble in ethyl alcohol and ether, but moderately soluble in water.

(*Chemical*). Succinic acid gives all the normal reactions of a dicarboxylic acid.

(1) **Action of Heat.** On heating to 300°C, it loses a molecule of water to form succinic anhydride.

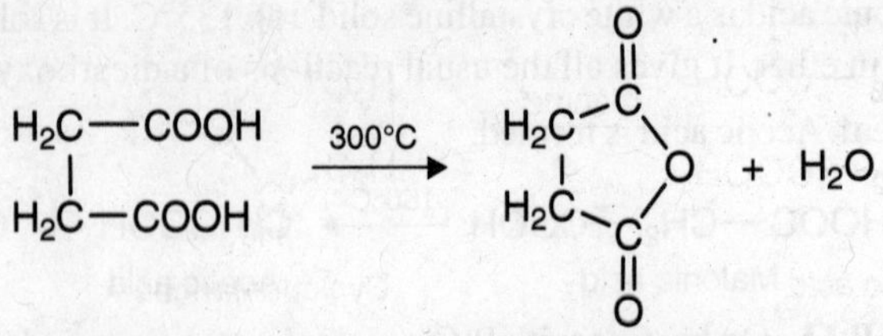

Succinic acid Succinic anhydride

(2) **Reaction with Ammonia**. With ammonia it forms ammonium succinate which upon heating loses a molecule of ammonia to yield succinimide.

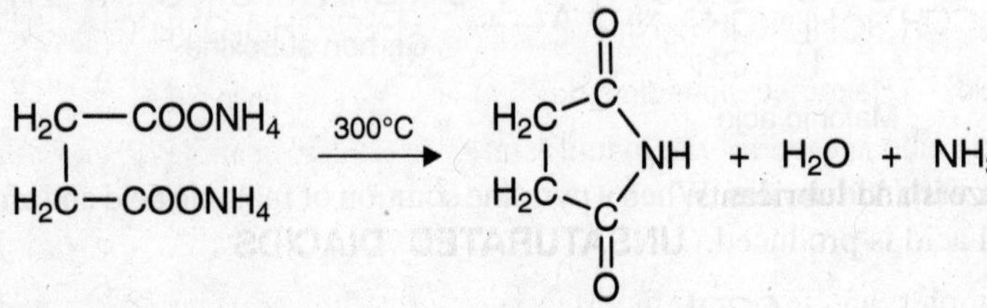

Ammonium succinate Succinimide

When succinimide is treated with an alkaline solution of Br_2 at 0°C, it forms *N*-bromosuccinimide (NBS) which is a valuable reagent for allylic bromination.

(3) **Electrolysis of Potassium Salt**. The electrolysis of a strong solution of potassium succinate gives ethylene.

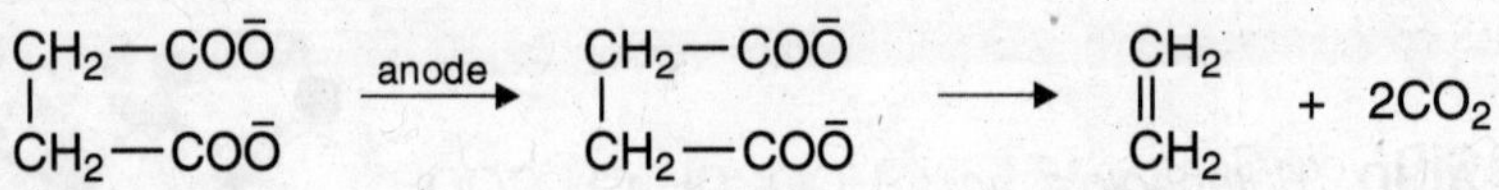

Succinate ion (*unstable*) Ethylene

Uses. Succinic acid is used : (*1*) in the manufacture of lacquers and dyes, (*2*) in volumetric analysis as a standard substance for acid-base titrations.

ADIPIC ACID, *Hexanedioic acid*, $HOOC(CH_2)_4COOH$

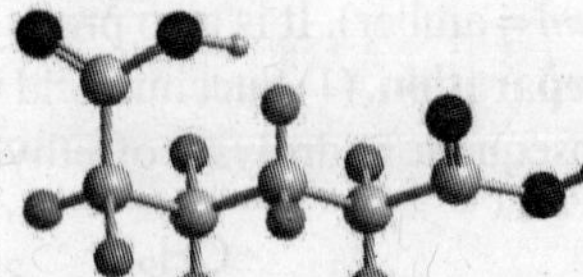

Adipic acid is prepared industrially by : (1) Oxidation of cyclohexane with air in the presence of boric acid catalyst to give cyclohexanol ; (2) Oxidation of cyclohexanol with concentrated nitric acid to form cyclohexanone and then adipic acid.

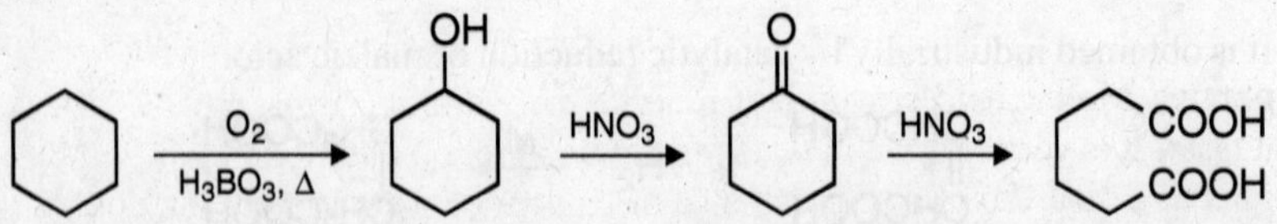

Cyclohexane Cyclohexanol Cyclohexanone Adipic acid

The cyclohexane needed for the above process is obtained by the catalytic hydrogenation of benzene.

Properties. Adipic acid is a white crystalline solid, mp 153°C. It is slightly soluble in water, but soluble in alcohol and ether. It gives all the normal reactions of a dicarboxylic acid.

When heated to 300°C, it yields cyclopentanone.

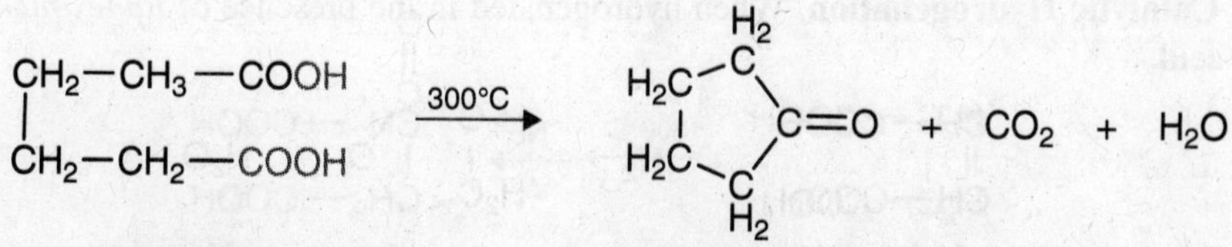

Adipic acid Cyclopentanone

When adipic acid is heated with hexamethylenediamine, a polyamide Nylon-6,6 is formed by splitting out water molecules.

$$\mathrm{HO\overset{O}{\overset{\|}{C}}(CH_2)_4\overset{O}{\overset{\|}{C}}OH} + \mathrm{H_2N(CH_2)_6NH_2} \xrightarrow{\Delta} \left[\mathrm{\overset{O}{\overset{\|}{C}}(CH_2)_4\overset{O}{\overset{\|}{C}}NH(CH_2)_6NH}\right]_n + 2n\mathrm{H_2O}$$

Adipic acid Hexamethylenediamine Nylon-6,6

Uses. Adipic acid is used for the manufacture of : (*1*) *Nylon* and *Polyurethane foams* ; (*2*) esters used as plasticizers and lubricants.

UNSATURATED DIACIDS

The two simplest and the most important unsaturated dicarboxylic acids are :

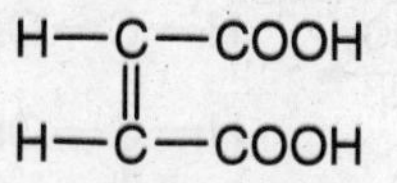

Maleic acid
cis-Butenedioic acid

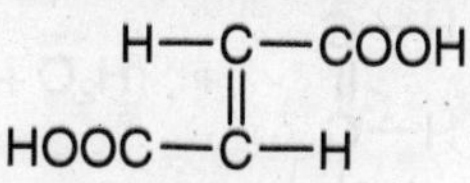

Fumaric acid
trans-Butenedioic acid

These afford the best known example of geometrical isomerism.

MALEIC ACID, *cis-Butenedioic acid*

It is prepared commercially by catalytic oxidation of 2-butene (from petroleum) or benzene.

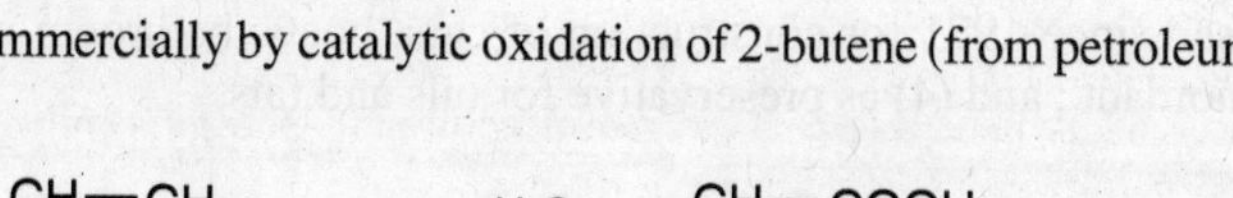

2-Butene Maleic acid

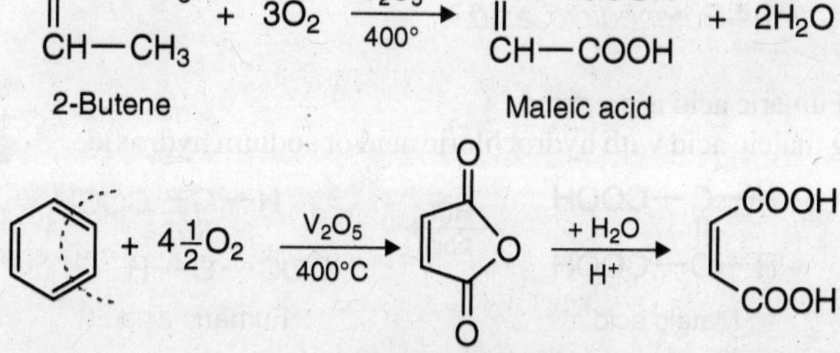

Maleic anhydride Maleic acid

Properties. Maleic acid crystallizes in large monoclinic prisms, mp 130.5°C, having repulsive astringent taste. It is very soluble in water and ethyl alcohol, but moderately soluble in ether. It gives the reactions of a dicarboxylic acid and also of an activated carbon-carbon double bond.

(1) **Action of Heat**. On heating to 150°C, it dehydrates to form maleic anhydride.

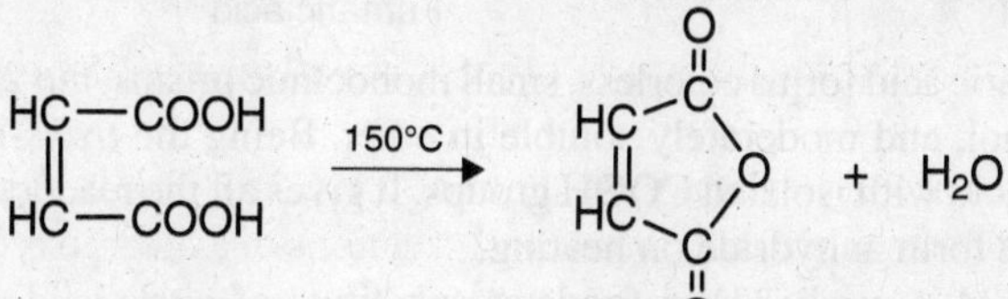

Maleic acid Maleic anhydride

(2) **Catalytic Hydrogenation**. When hydrogenated in the presence of *Raney nickel*, it forms succinic acid.

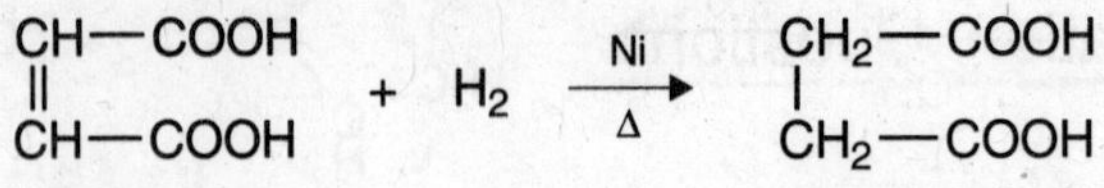

Maleic acid Succinic acid

(3) **Hydration**. It adds a molecule of water across the double bond in the presence of a mineral acid, yielding maleic acid.

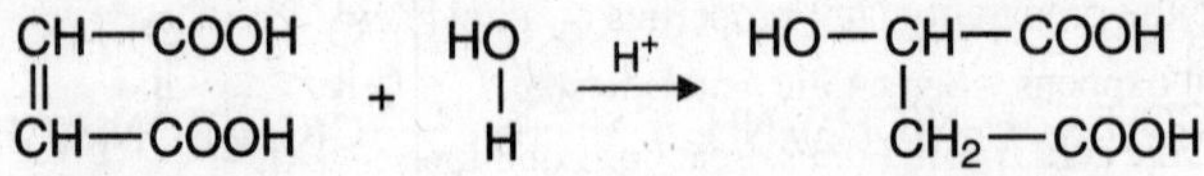

Maleic acid Malic acid

(4) **Reaction with Alkaline $KMnO_4$.** When treated with alkaline $KMnO_4$ solution, it undergoes hydroxylation to form *meso*-tartaric acid.

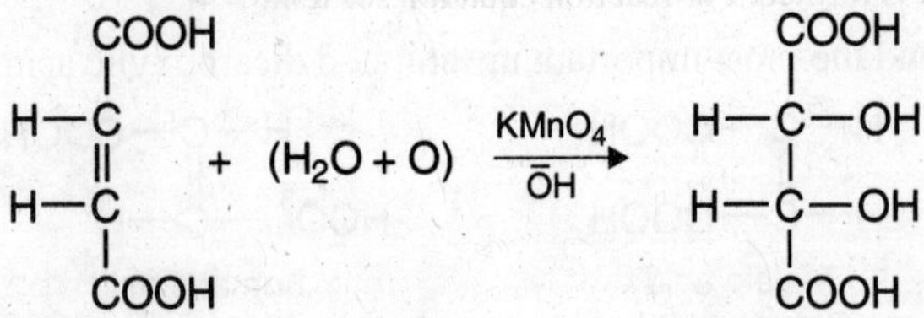

Maleic acid *meso*-Tartaric acid

Uses. Maleic acid is used : (1) for preparing succinic acid, malic acid, tartaric acid, fumaric acid, and several other derivatives ; (2) as maleic anhydride for polyester resins needed in fibrous glass material, widely used since 1971 for construction of cars, boats, bathroom units, gasoline storage tanks, etc ; (3) as mordant ; and (4) as preservative for oils and fats.

FUMARIC ACID, *trans-Butenedioic acid*

Preparation. Fumaric acid is prepared :

(1) By boiling maleic acid with hydrochloric acid or sodium hydroxide.

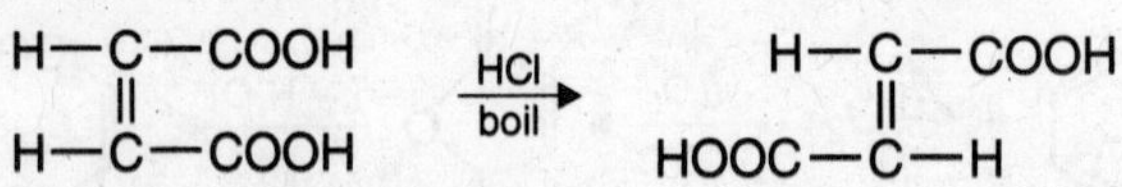

Maleic acid Fumaric acid

(2) By oxidation of furfural with sodium chlorate.

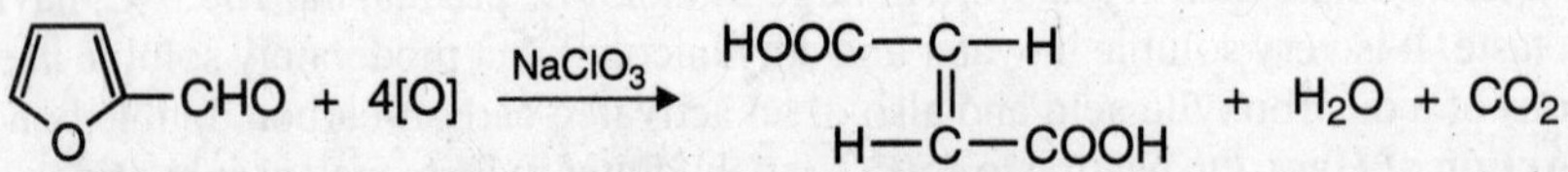

Furfural Fumaric acid

Properties. Fumaric acid forms colorless, small monoclinic prisms, mp 287°C. It is very soluble in water and ethyl alcohol, and moderately soluble in ether. Being the *trans*-isomer, it behaves like a normal dicarboxylic acid with isolated COOH groups. It gives all the reactions shown by maleic acid, except that it does not form anhydride on heating.

Uses. Fumaric acid is used : (1) in foods as substitute of citric acid and tartaric acid ; (2) as mordant ; (3) in making inks ; and (4) as modifier for polyesters.

Examination Questions

1. Give the preparation and properties of oxalic acid.
2. Give the preparation and properties of malonic acid.
3. Give the preparation and properties of succinic acid.
4. Give the preparation and properties of adipic acid.
5. What happens when oxalic acid is heated ?

Answer. Formic acid is formed. For reaction equation see text.

6. What happens when oxalic acid is oxidized with acidic $KMnO_4$?

Answer. Carbon dioxide and water are formed. For reaction equation see text.

7. What happens when malonic acid is heated ?

Answer. Acetic acid is formed. For reaction equation see text.

8. What happens when succinic acid is heated ?

Answer. Succinic anhydride is formed. For reaction equation see text.

9. What happens when maleic acid is heated ?

Answer. Maleic anhydride is formed. For reaction equation see text.

10. How will you distinguish between oxalic acid and acetic acid ? (*Shivaji BSc, 2015*)

Answer. Oxalic acid gives carbon dioxide on heating, which forms white precipitate with lime-water. Acetic acid does not give this test.

$$\underset{\text{Oxalic acid}}{HOOC-COOH} \xrightarrow{\Delta} \underset{\text{Formic acid}}{HCOOH} + CO_2$$

11. How will you distinguish between oxalic acid and succinic acid ? (*Kalyani BSc, 2015*)

Answer. Oxalic acid gives carbon dioxide on heating, which forms white precipitate with lime-water. No carbon dioxide is evolved when succinic acid is heated. (When succinic acid is heated, succinic anhydride is formed).

$$\underset{\text{Oxalic acid}}{HOOC-COOH} \xrightarrow{\Delta} \underset{\text{Formic acid}}{HCOOH} + CO_2$$

$$CO_2 + Ca(OH)_2 \longrightarrow \underset{\text{white ppt.}}{CaCO_3} + H_2O$$

12. How will you distinguish between maleic acid and fumaric acid ? (*Kerala BSc, 2014*)

Answer. Maleic acid on heating gives maleic anhydride (mp 250°C). No anhydride is formed when fumaric acid is heated.

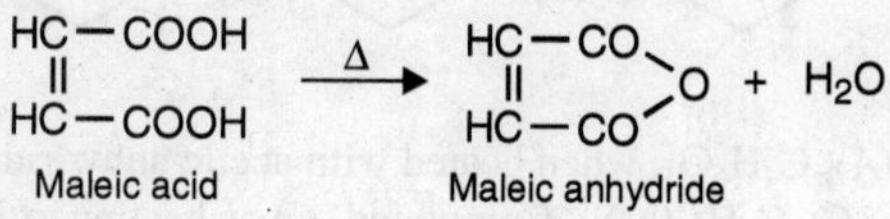

13. How will you synthesize oxalic acid from formic acid ? (*Bihar BSc, 2014*)

Answer. Following steps are involved :

$$\underset{\text{Formic acid}}{H-\overset{O}{\overset{\|}{C}}-OH} \xrightarrow{NaOH} \underset{\text{(2 molecules)}}{H-\overset{O}{\overset{\|}{C}}-ONa} \xrightarrow[300°]{NaOH} \begin{matrix} COONa \\ | \\ COONa \end{matrix} \xrightarrow{H_2SO_4} \underset{\text{Oxalic acid}}{\begin{matrix} COOH \\ | \\ COOH \end{matrix}}$$

14. How will you synthesize oxalic acid from ethylene ? *(Calicut BSc, 2014)*

Answer. Following steps are involved :

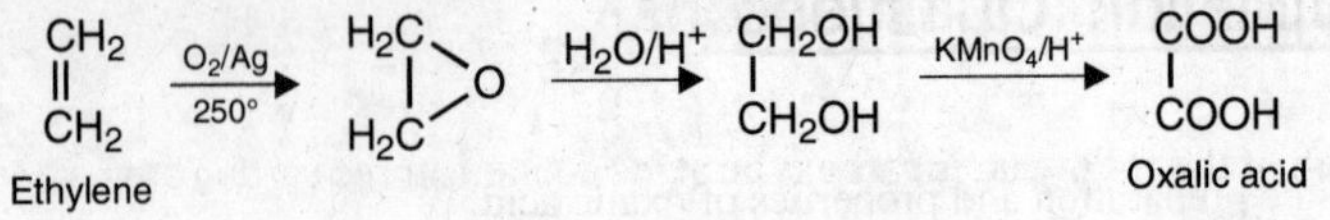

15. How will you synthesize malonic acid from acetic acid ? *(Sri Venkateshwara BSc, 2015)*

Answer. Following steps are involved :

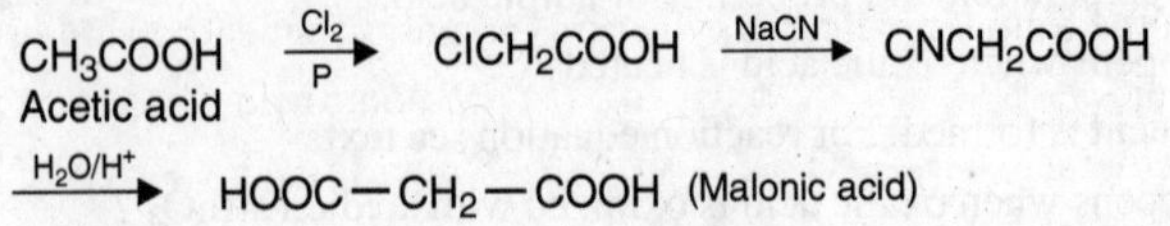

16. How will you synthesize succinic acid from ethylene ? *(Anna BSc, 2014)*

Answer. Following steps are involved :

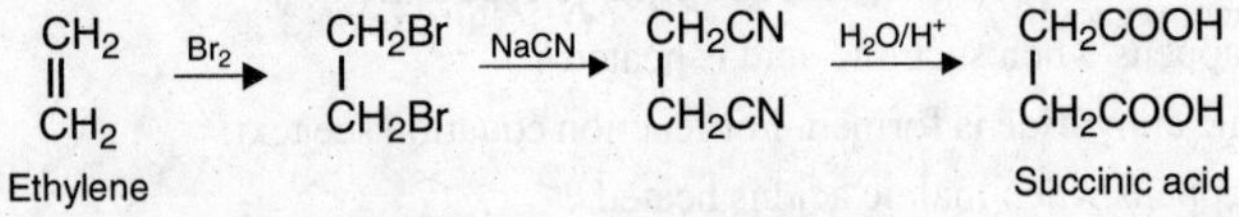

17. How will you synthesize maleic acid from succinic acid ?

Answer. Following steps are involved :

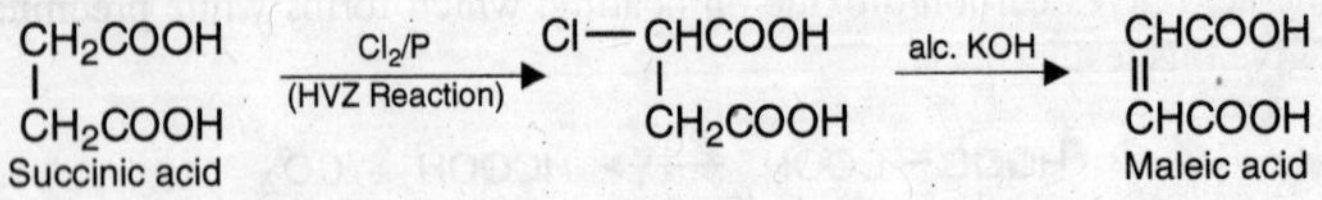

18. How will you synthesize *N*-bromosuccinimide from succinic acid ?

Answer. Following steps are involved :

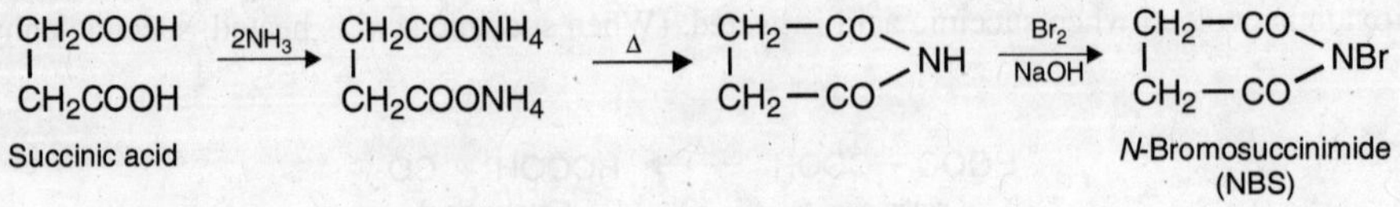

19. How will you synthesize adipic acid from benzene ? *(Jiwaji BSc, 2015)*

Answer. Following steps are involved :

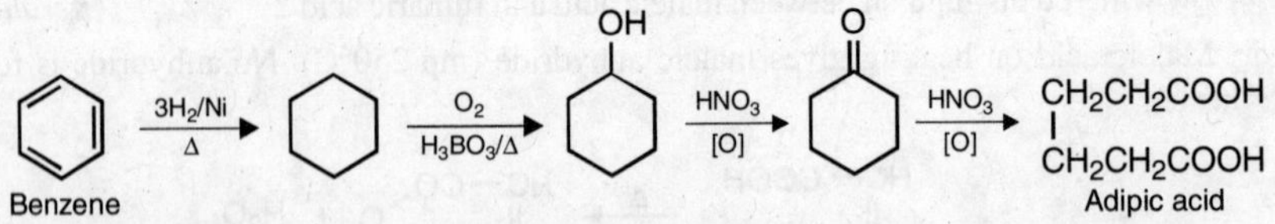

20. Compound (*A*), $C_4H_6O_4$, when heated with acetic anhydride gives (*B*), $C_4H_4O_3$. Compound (*B*) reacts with ammonia to give (*C*), $C_4H_{12}O_4N_2$. Compound (*C*) on heating to 300°C forms (*D*), $C_4H_5O_2N$. Both (*B*) and (*C*) can be hydrolysed to (*A*). Identify (*A*), (*B*), (*C*), and (*D*).

Answer. (*A*) = Succinic acid

(*B*) = Succinic anhydride

(*C*) = Ammonium succinate

(*D*) = Succinimide

Multiple Choice Questions

1. Which of the following dicarboxylic acids contains three carbon atoms ?
   (*a*) Adipic acid (*b*) Malonic acid
   (*c*) Succinic acid (*d*) Glutaric acid

Answer. (*b*)

2. Which of the following dicarboxylic acids contains four carbon atoms ?
   (*a*) Adipic acid (*b*) Malonic acid
   (*c*) Succinic acid (*d*) Glutaric acid

Answer. (*c*)

3. Which of the following dicarboxylic acids contains six carbon atoms ?
   (*a*) Adipic acid (*b*) Malonic acid
   (*c*) Succinic acid (*d*) Glutaric acid

Answer. (*a*)

4. Which of the following terms describes a useful method of carrying out the following reaction ?

$$NCCH_2CH_2CN \longrightarrow HOOCCH_2CH_2COOH$$

   (*a*) Reduction (*b*) Acylation
   (*c*) Hydrolysis (*d*) Alkylation

Answer. (*c*)

5. Chromic acid oxidation of $HOCH_2CH_2CHO$ gives
   (*a*) $HOCH_2CH_2CO_2H$ (*b*) HO_2CCH_2CHO
   (*c*) $HO_2CCH_2CO_2H$ (*d*) $HOCH_2CH_2CHO$

Answer. (*c*)

6. Decarboxylation of malonic acid gives
   (*a*) acetaldehyde (*b*) acetone
   (*c*) ethyl acetate (*d*) acetic acid

Answer. (*d*)

7. Ethylmalonic acid is decarboxylated by heating to give
   (*a*) diethyl ethylmalonate (*b*) ethylmalonic acid
   (*c*) disodium ethylmalonate (*d*) butanoic acid

Answer. (*d*)

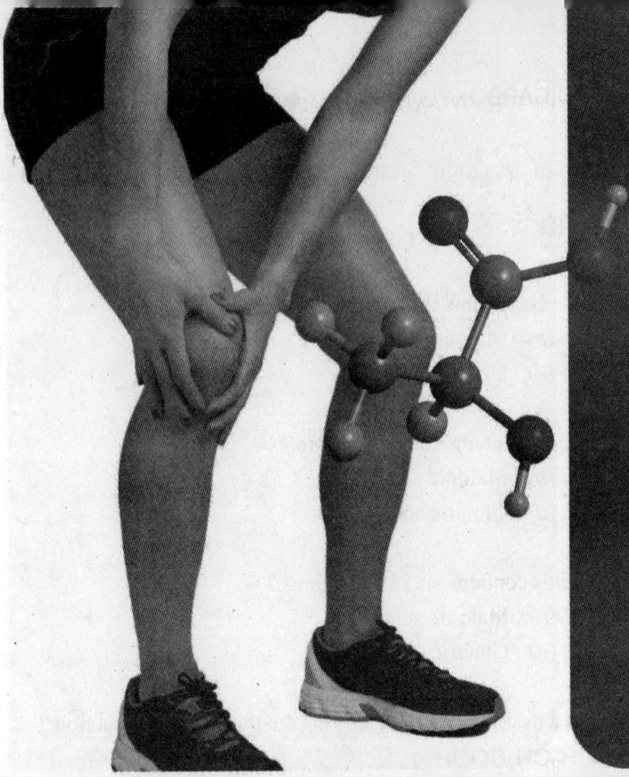

22 Hydroxy and Halo Acids

Lactic acid, a product of anaerobic glycolysis, causes painful muscle contractions. Upon crossing the finish line, the winners show the combined effects of oxygen debt and lactic acid buildup in muscles.

The carboxylic acids in which one or more hydrogen atoms of the hydrocarbon group have been replaced by atoms or groups such as OH, X (halogen) or NH_2, are referred to as **Substituted Acids**. They are designated as *hydroxy acids, halo acids,* or *amino acids* depending on the substituent present. For example,

$HO-CH_2-COOH$	$Br-CH_2-COOH$	H_2N-CH_2-COOH
Hydroxyacetic acid	Bromoacetic acid	Aminoacetic acid

Some of the hydroxy acids are very common naturally occurring substances and are usually referred to by common names. The halo acids, on the other hand, are mostly synthetic and are assigned IUPAC names. The positions of the hydroxy and halogen substituents on the carbon chain are indicated by Greek letters or by numbers.

$$\begin{array}{cccccc} \epsilon & \delta & \gamma & \beta & \alpha & \\ 6 & 5 & 4 & 3 & 2 & 1 \\ C- & C- & C- & C- & C- & COOH \end{array}$$

For example,

$$\underset{\text{OH}}{\overset{|}{}}\quad CH_3-\underset{|}{\overset{OH}{CH}}-COOH \qquad\qquad CH_3-\overset{Br}{\overset{|}{CH}}-CH_2-COOH$$

α-Hydroxypropionic acid	β-Bromobutyric acid
2-Hydroxypropanoic acid	3-Bromobutanoic acid

Here we will discuss the hydroxy acids and the halo acids only, while the amino acids will be taken up along with *Proteins*.

HYDROXY ACIDS

They are the derivatives of carboxylic acids in which one or more H atoms of the hydrocarbon group are replaced by as many OH groups. They are referred to as α, β, γ, δ etc., hydroxy acids according as the OH is bonded to α, β, γ or δ carbon of the chain. Some of the important hydroxy acids are :

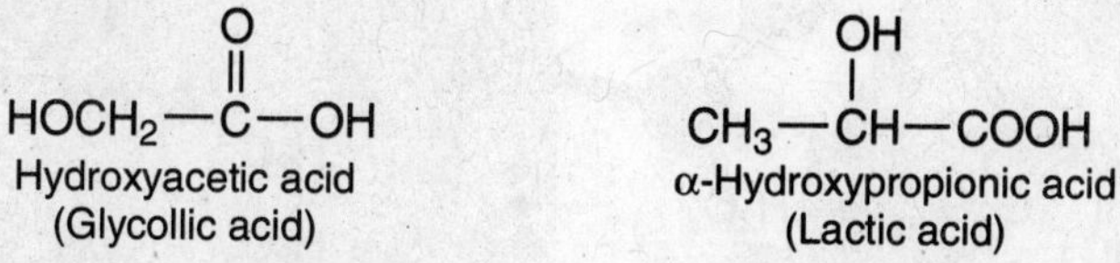

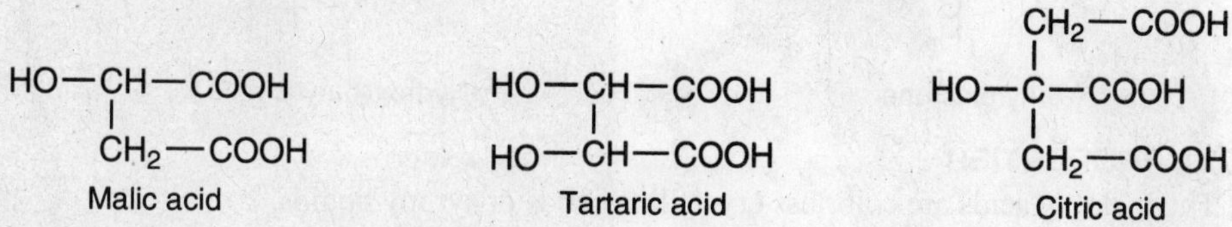

METHODS OF PREPARATION

FOR α-HYDROXY ACIDS

(1) **By hydrolysis of cyanohydrins** derived from aldehydes or ketones.

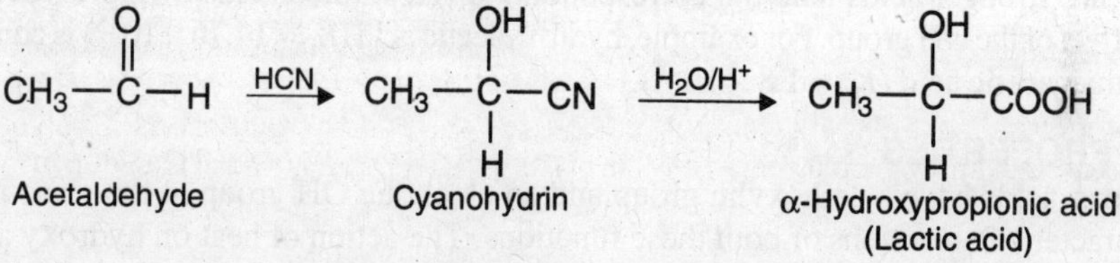

(2) **By hydrolysis of halo acids with dilute aqueous alkali.** The desired halo acid may be obtained by HVZ reaction from the corresponding acid.

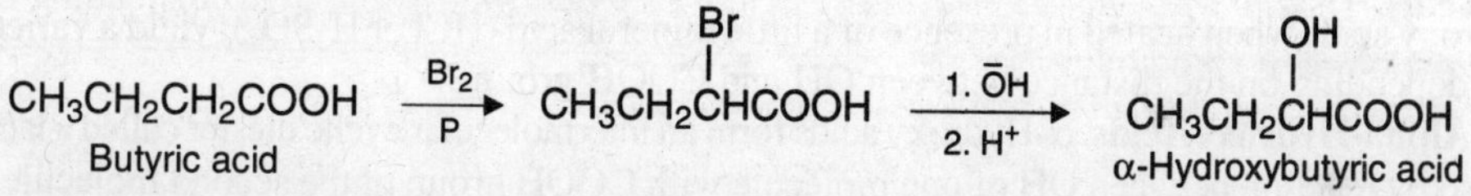

FOR β-HYDROXY ACIDS

(3) **By reduction of β-keto acids or esters** with $LiAlH_4$ in the presence of pyridine.

$$CH_3COCH_2COOC_2H_5 \xrightarrow[\text{pyridine}]{LiAlH_4} CH_3CH(OH)CH_2COOC_2H_5 \xrightarrow[H^+]{H_2O} CH_3CH(OH)CH_2COOH$$

Ethyl acetoacetate (a β-keto ester) → β-Hydroxybutyric acid

(4) **By Reformatsky reaction** *e.g.*,

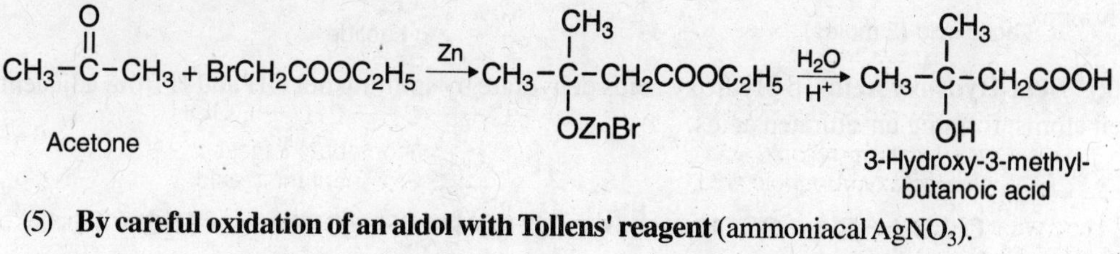

(5) **By careful oxidation of an aldol with Tollens' reagent** (ammoniacal $AgNO_3$).

$$CH_3CH(OH)CH_2CHO \xrightarrow{[O]} CH_3CH(OH)CH_2COOH$$

Acetaldehyde aldol → β-Hydroxybutyric acid

FOR γ, δ, AND LONGER HYDROXY ACIDS

(6) **By hydrolysis of the corresponding lactone.**

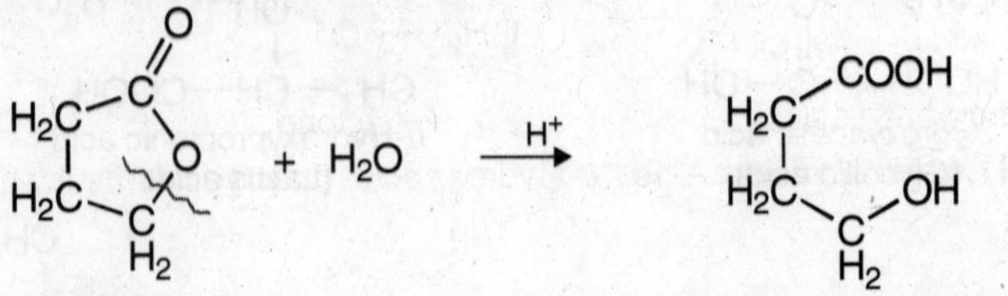

γ-Butyrolactone γ-Hydroxybutyric acid

PHYSICAL PROPERTIES

(1) The hydroxy acids are colorless crystalline solids or syrupy liquids.

(2) Their melting and boiling points are much higher than those of the parent unsubstituted acids.

(3) They are more soluble in water than either the parent carboxylic acid or the alcohol, since both the functional groups (OH and COOH) in them can form hydrogen bonds with water. They are less soluble in alcohol and ether.

(4) They are stronger acids than the corresponding unsubstituted acids owing to the electron-withdrawing effect of the OH group. For example, hydroxyacetic acid ($K_a = 14.76 \times 10^{-5}$) is considerably stronger acid than acetic acid ($K_a = 1.8 \times 10^{-5}$).

CHEMICAL PROPERTIES

The hydroxy acids have a carboxylic group and an alcoholic OH group in the same molecule. They give characteristic reactions of both these functions. The action of heat on hydroxy acids is of special interest.

ACTION OF HEAT ON HYDROXY ACIDS

Hydroxy acids when heated in presence of a little mineral acid (HCl or H_2SO_4), yield a variety of products depending on the distance between OH and COOH groups.

(1) **Alpha-Hydroxy Acids.** α-Hydroxy acids form an intermolecular cyclic diester called a lactide. They do so by reaction between OH of one molecule with COOH group of the second molecule.

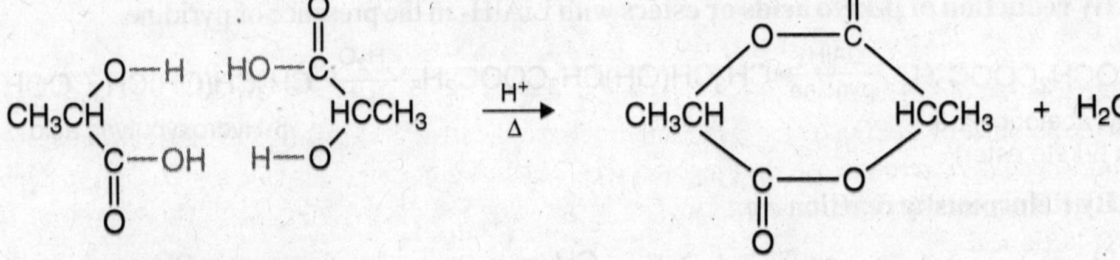

Lactic acid (2 moles) Lactide

(2) **Beta-Hydroxy Acids.** β-Hydroxy acids dehydrate by splitting out OH and H from adjacent carbon atoms, forming unsaturated acids.

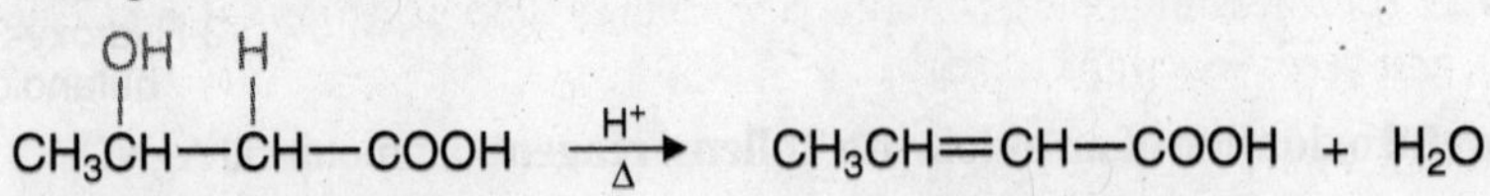

β-Hydroxybutyric acid Crotonic acid

(3) **Gamma and Delta Hydroxy Acids.** γ- and δ-Hydroxy acids form internal esters known as lactones. The COOH at one end of the molecule reacts with OH on the other, to form five- and six-membered cyclic esters *i.e.*, γ- and δ-lactones.

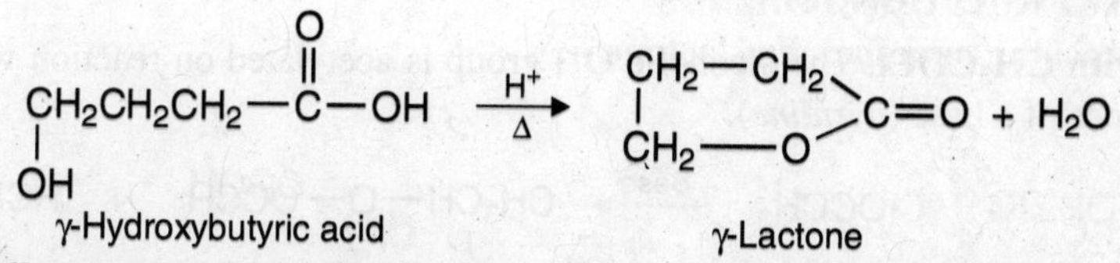

(4) **Epsilon and Longer Hydroxy Acids.** ε-Hydroxy acids undergo intermolecular esterification to form linear polymers.

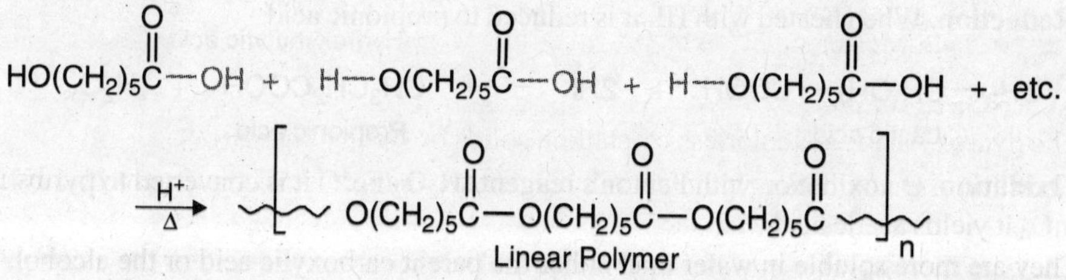

LACTIC ACID, *α-Hydroxypropionic acid, 2-Hydroxypropanoic acid*

Lactic acid is the main constituent of milk that has gone sour and hence its name (L. *Lactis* = milk).

Preparation. (1) By the hydrolysis of acetaldehyde cyanohydrin.

$$CH_3-\overset{H}{\overset{|}{C}}=O \xrightarrow{HCN} CH_3-\overset{OH}{\overset{|}{C}H}-CN \xrightarrow{H_2O/H^+} CH_3-\overset{OH}{\overset{|}{C}H}-COOH$$

Acetaldehyde — Cyanohydrin — Lactic acid

(2) **From Molasses.** Lactic acid is manufactured by fermentation of molasses (or *milk whey* containing lactose) by the microorganism *Bacillus acidi lactiti* (BAL).

$$\underset{\text{(from molasses)}}{C_6H_{12}O_6} \xrightarrow{BAL} \underset{\text{Lactic acid}}{2CH_3-CHOH-COOH}$$

A dilute solution of molasses (or *whey*) is treated with BAL (sour milk). The fermentation is carried at 35-40°C in the presence of $CaCO_3$. As the lactic acid is produced, it reacts with $CaCO_3$ to form calcium lactate. The calcium lactate is filtered off and decomposed with calculated quantity of dilute H_2SO_4. The insoluble calcium sulfate is removed and the lactic acid set free in the solution is recovered by distillation in *vacuo*. The product is D-lactic acid.

Properties. (*Physical*). Lactic acid is a colorless, crystalline solid, mp 53°C, and has a sour taste. The acid obtained from molasses is D-isomer. The synthetic product is racemic lactic acid, mp 18°C, specific rotation [α] = +3.82. Lactic acid is soluble in water, alcohol, and ether.

(*Chemical*). Lactic acid molecule contains a secondary alcohol group (>CHOH) and a carboxyl group (COOH), and gives reactions of both.

REACTION INVOLVING COOH GROUP

(1) **Formation of Salts.** It reacts with alkalis to form salts.

$$\underset{\text{Lactic acid}}{CH_3CH(OH)COOH} + NaOH \longrightarrow \underset{\text{Sodium lactate}}{CH_3CH(OH)CO\overset{-}{O}\overset{+}{Na}} + H_2O$$

REACTIONS INVOLVING OH GROUP

(2) **Reaction with CH_3COCl.** The alcoholic OH group is acetylated on reaction with acetyl chloride in the presence of a base (*pyridine*).

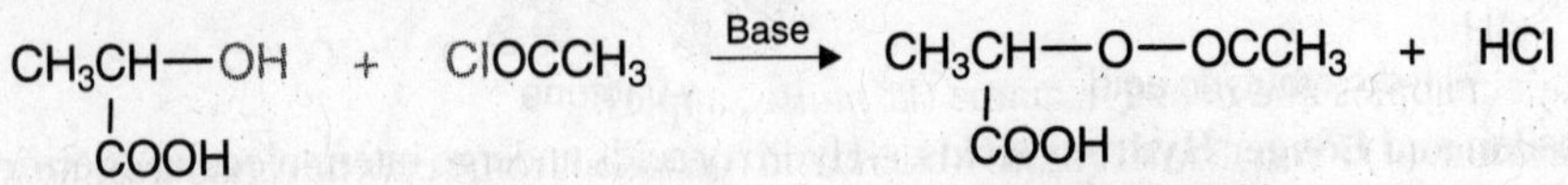

Lactic acid — Acetyllactic acid

(3) **Reduction**. When heated with HI, it is reduced to propionic acid.

$$CH_3-CH(OH)-COOH + 2HI \xrightarrow{\Delta} CH_3CH_2COOH + H_2O$$

Lactic acid — Propionic acid

(4) **Oxidation**. On oxidation with Fenton's reagent (H_2O_2/Fe^{3+}) it is converted to pyruvic acid. With $KMnO_4$ it yields acetic acid.

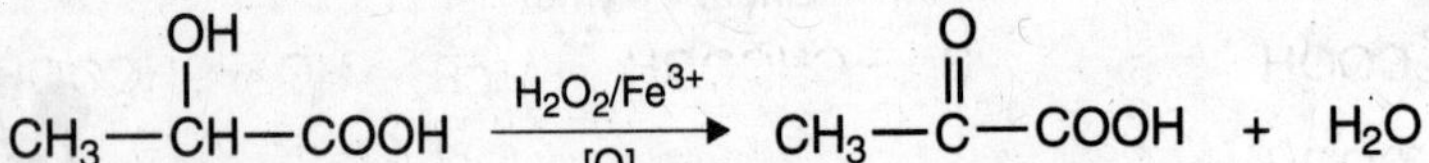

Lactic acid — Pyruvic acid

REACTIONS INVOLVING BOTH OH AND COOH

(5) **Reaction with PCl_5.** Both the OH and COOH groups are attacked by phosphorus pentachloride to form lactyl chloride.

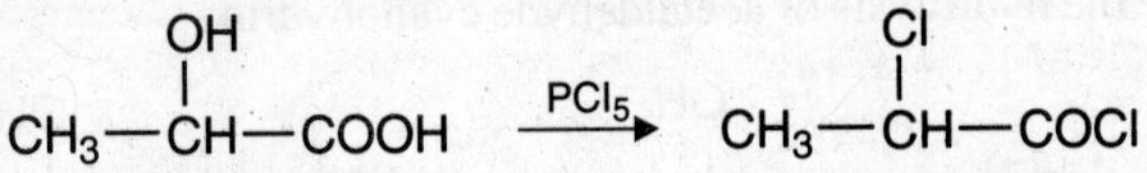

Lactic acid — Lactyl chloride

(6) **Action of Heat**. When heated alone, it forms an intermolecular cyclic diester called a **Lactide**.

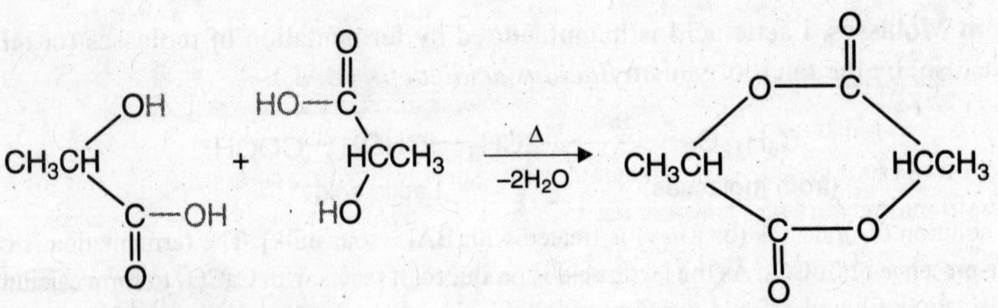

Lactic acid (2 moles) — Lactide

(7) **Reaction with H_2SO_4.** When heated with dilute sulfuric acid at 130°C, lactic acid is decomposed to yield acetaldehyde and formic acid.

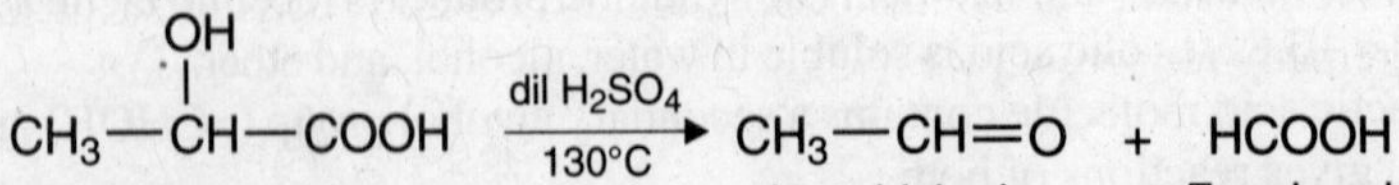

Lactic acid — Acetaldehyde — Formic acid

With concentrated H_2SO_4, formic acid is further decomposed to CO and H_2O. (*Test*)

Uses. Lactic acid is used : (1) in dairy products ; (2) as acidulant in beverages and candies ; (3) for deliming of hides ; (4) as a mordant ; (5) Ethyl and butyl lactates are used as plasticizers ; (6) Calcium and iron lactates are used in medicine.

MALIC ACID, *Monohydroxysuccinic acid*

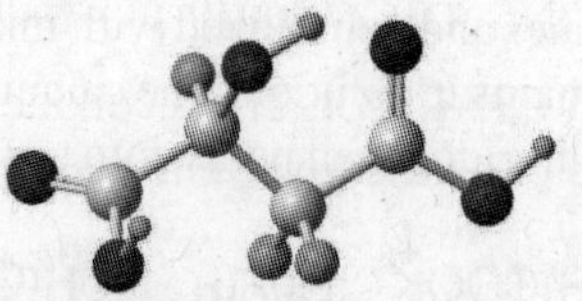

It occurs in apples and hence its name (L. *malus*, apple).

Preparation. (1) **From Maleic acid**. Maleic acid (obtained from benzene) upon treatment with hot dilute sulfuric acid under pressure gives malic acid.

$$\underset{\text{Maleic acid}}{\begin{array}{c} CHCOOH \\ \| \\ CHCOOH \end{array}} + HO—H \xrightarrow[\Delta,\ \text{pressure}]{\text{dil } H_2SO_4} \underset{\text{DL-Malic acid}}{\begin{array}{l} CH(OH)COOH \\ | \\ CH_2COOH \end{array}}$$

(2) **From Succinic acid**. Careful monobromination of succinic acid and subsequent treatment with moist silver oxide forms malic acid.

$$\underset{\text{Succinic acid}}{\begin{array}{l} CH_2COOH \\ | \\ CH_2COOH \end{array}} \xrightarrow[-HBr]{Br_2} \underset{\text{Monobromosuccinic acid}}{\begin{array}{l} Br—CHCOOH \\ \quad\ | \\ \quad\ CH_2COOH \end{array}} \xrightarrow[-AgBr]{AgOH} \underset{\text{Malic acid}}{\begin{array}{l} HO—CHCOOH \\ \quad\ | \\ \quad\ CH_2COOH \end{array}}$$

Properties. (*Physical*). Malic acid forms colorless crystals, mp 100°C, having sour taste. It is very soluble in water and alcohol ; slightly soluble in ether. It contains an asymmetric carbon atom and shows optical isomerism, the natural product is levorotatory.

$$HOOC—\underset{\text{2° Alcohol group}}{\overset{OH}{\overset{|}{CH}}}—CH_2—COOH$$

(***Chemical***). Malic acid is a secondary alcohol and a dicarboxylic acid, and gives reactions accordingly.

(1) **Reduction**. When reduced with HI, it yields succinic acid.

$$\underset{\text{Malic acid}}{\begin{array}{l} CH(OH)—COOH \\ | \\ CH_2COOH \end{array}} + 2HI \xrightarrow{\Delta} \underset{\text{Succinic acid}}{\begin{array}{l} CH_2COOH \\ | \\ CH_2COOH \end{array}} + H_2O + I_2$$

(2) **Action of Heat**. Malic acid, when strongly heated, splits out a molecule of water from adjacent carbons to yield maleic acid and fumaric acid.

$$\underset{\text{Malic acid}}{\begin{array}{l} HO—CH—COOH \\ \quad\quad | \\ \quad\quad CH_2—COOH \end{array}} \xrightarrow[-2H_2O]{\Delta} \underset{\text{Maleic acid}}{\begin{array}{l} H—C—COOH \\ \quad\ \| \\ H—C—COOH \end{array}} + \underset{\text{Fumaric acid}}{\begin{array}{l} \quad\quad H—C—COOH \\ \quad\quad\quad\quad \| \\ HOOC—C—H \end{array}}$$

At 300°C, fumaric acid rearranges to give maleic acid which dehydrates to maleic anhydride.

Uses. Maleic acid is used : (1) as acidulant in soft drinks and foodstuffs ; (2) in medicine as purgative and as a remedy for sore throat.

TARTARIC ACID, *2,3-Dihydroxysuccinic acid*

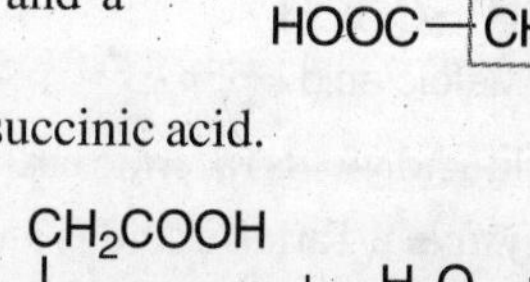

Tartaric acid occurs free and as potassium hydrogen tartarate in grape juice.

Preparation. (1) **From Grape Juice**. Argol which is potassium hydrogen tartarate (*Cream of tartar*), is formed as a sludgy precipitate during the fermentation of grape juice in wine making. It is dissolved in boiling water and neutralized with milk-of-lime. Calcium tartarate is precipitated, while potassium tartarate remains in solution. The calcium tartarate is separated by filtration and the filtrate treated with calcium chloride when potassium tartarate is also precipitated as calcium tartarate.

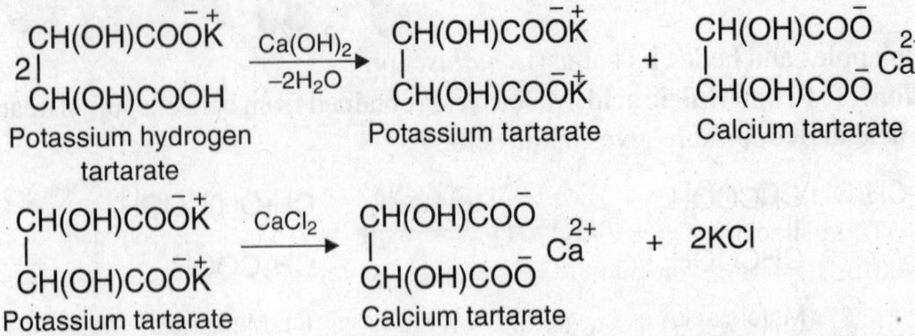

The calcium tartarate obtained in the two steps is treated with a calculated quantity of dilute H_2SO_4 to liberate tartaric acid. The calcium sulfate precipitate that is formed is removed by filtration, and the filtrate concentrated by evaporation to get the crystals of tartaric acid.

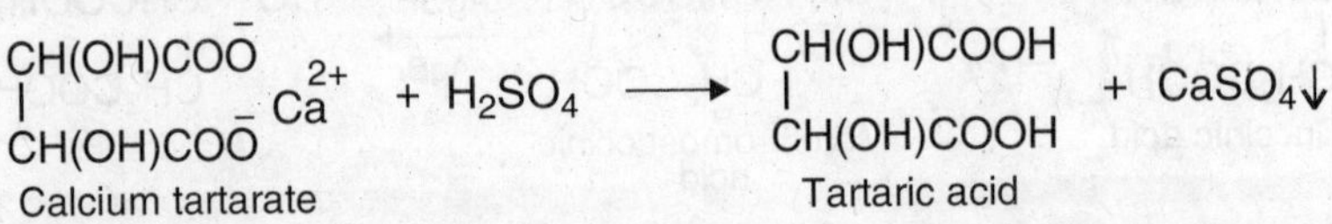

(2) **From Maleic Acid**. Maleic acid produced industrially by oxidation of cyclohexane is treated with alkaline $KMnO_4$ to get *meso*-tartaric acid.

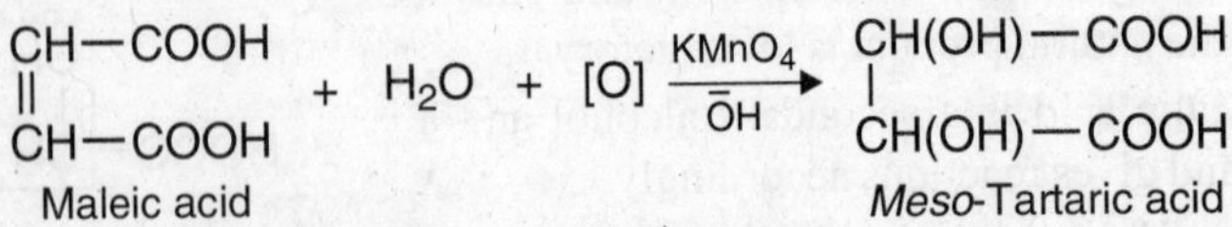

Similarly, fumaric acid yields racemic tartaric acid.

Properties. (*Physical*). Tartaric acid is a white crystalline substance, mp 171°C, having acidic taste. It is soluble in water and alcohol, but sparingly soluble in ether. Since it has two asymmetric carbon atoms, tartaric acid shows optical isomerism. The natural tartaric acid is the *dextro* variety, $\alpha = +12.7$.

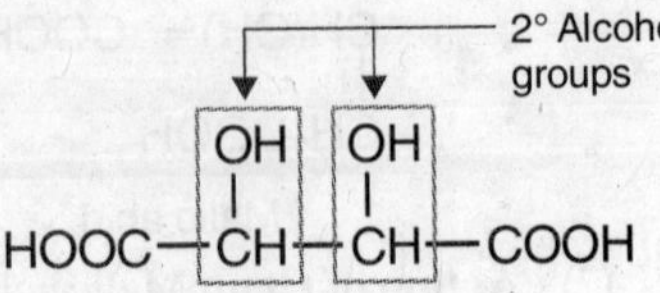

(*Chemical*). Tartaric acid molecule has two secondary alcohol groups (>CHOH) and two –COOH groups, and behaves accordingly.

(1) **Formation of Salts.** It forms two series of salts when treated with alkalis.

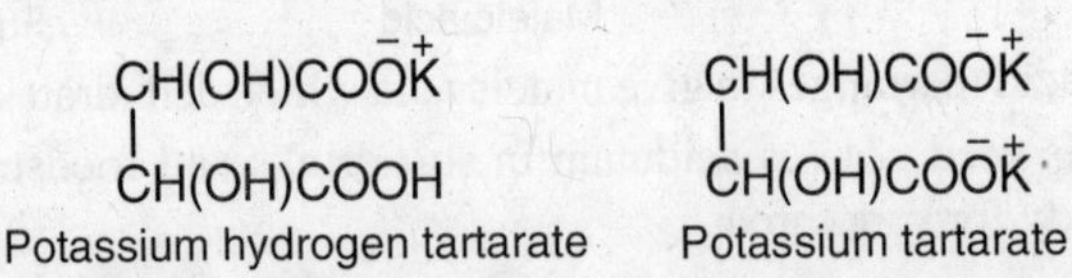

(2) **Action of Heat**. When heated, tartaric acid is converted to pyruvic acid.

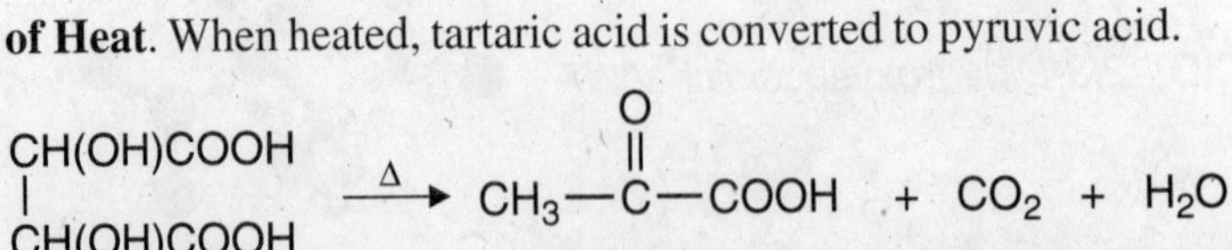

However, on strong heating, tartaric acid chars, giving smell of burnt sugar. (*Test*)

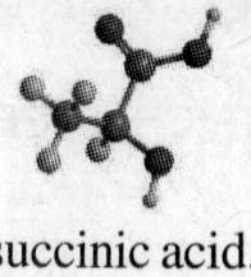

(3) **Reaction with HBr**. It reacts with hydrobromic acid to form 2,3-dibromosuccinic acid.

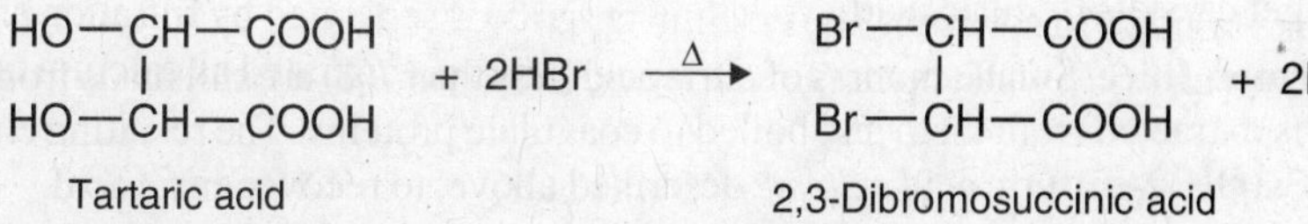

(4) **Reduction**. When heated with HI, tartaric acid is reduced to malic acid and then succinic acid.

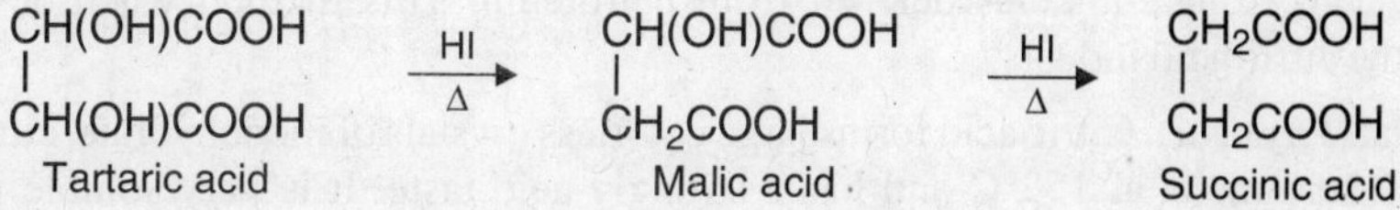

(5) **Oxidation**. (*i*) On oxidation with HNO_3, it is converted to tartonic acid and then oxalic acid.

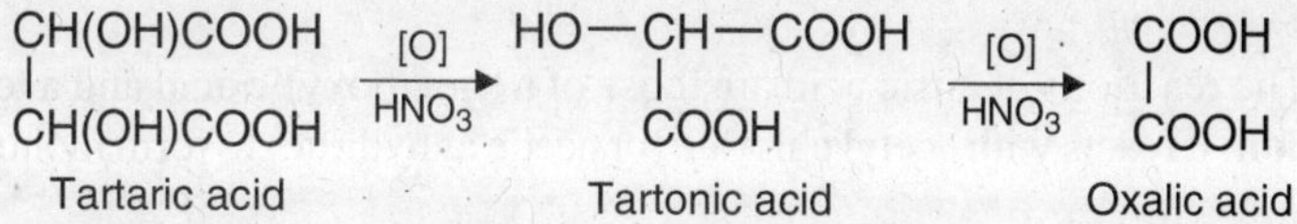

(*ii*) When warmed with ammoniacal silver nitrate, tartaric acid is oxidized to tartonic acid and silver mirror is formed. (*Test*)

(*iii*) **Fenton's reagent** (Fe^{3+}/H_2O_2) oxidizes tartaric acid to dihydroxymaleic acid.

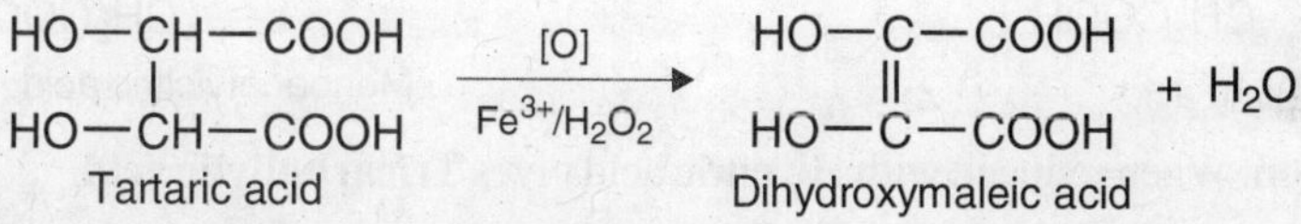

(6) **Formation of Complexes**. In Fehling solution ($CuSO_4$ + Sodium potassium tartarate + NaOH), the insoluble copper hydroxide first formed dissolves due to the formation of a soluble complex,

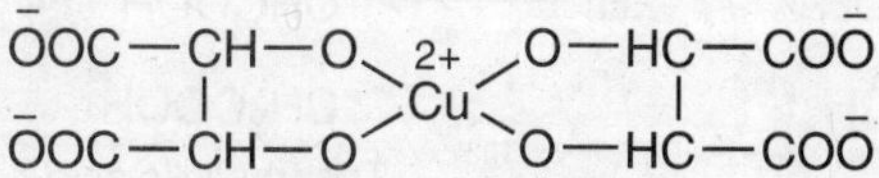

Sodium cupritartarate

Uses. Tartaric acid is used : (1) in carbonated beverages and effervescent tablets ; (2) in baking powder ; (3) as mordant ; (4) in silvering of mirrors ; (5) in tanning ; (6) for preparing tartar emetic, cream of tartar, Rochelle salt etc.

CITRIC ACID, *2-Hydroxypropane-1,2,3-tricarboxylic acid*

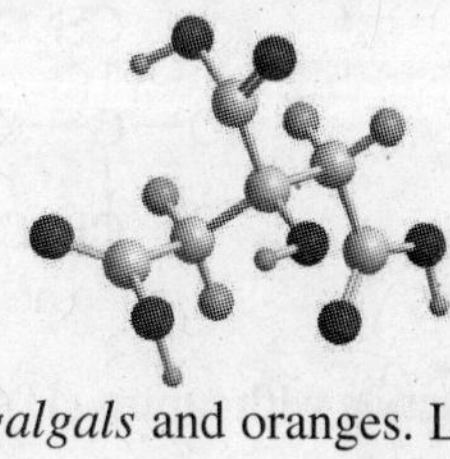

It occurs in the juice of citrus fruits such as lemons, limes, *galgals* and oranges. Lemon juice contains 7-10 percent citric acid.

Preparation. (1) **From Molasses**. Molasses containing sucrose is diluted with water and subjected to fermentation with a microorganism *Aspergillus niger*.

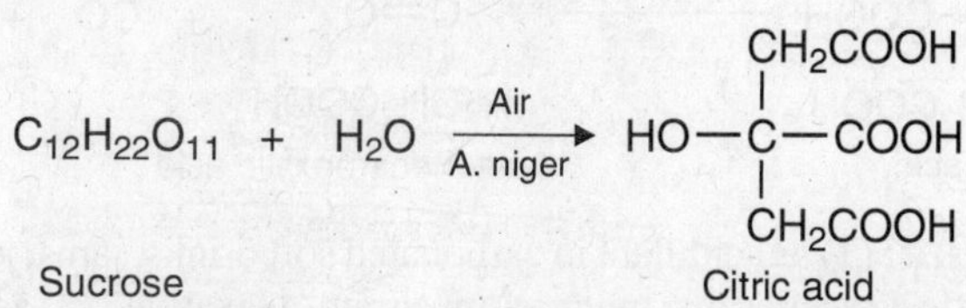

The fermentation process is carried for 7 to 10 days at 26–28°C. The resulting solution of citric acid is neutralized with $Ca(OH)_2$ to form insoluble calcium citrate. This is washed with water and

decomposed with dilute sulfuric acid. The calcium sulfate is filtered off and the solution concentrated under vacuum to get crystals of citric acid.

(2) **From Lemon Juice**. Small amounts of citric acid (less than 7%) are still made from citrus fruit wastes. The juice is extracted from them and boiled to coagulate proteins. The resulting clear solution is worked by the Ca(OH)2-sulfuric acid *scheme* described above, to recover citric acid.

(3) **From Petroleum**. Recently it has been shown that certain strains of *Candida* (a yeast) can produce citric acid from *n*-alkanes derived from petroleum. This method when developed will revolutionize the citric acid industry.

Properties. (*Physical*). Citric acid forms large colorless crystals of monohydrate, citric acid-H_2O. The anhydrous acid melts at 152°C and has a strongly acid taste. It is very soluble in water and alcohol, but is sparingly soluble in ether. It is not optically active as it contains no asymmetric carbon atom. It is nontoxic.

(*Chemical*). The reactions of citric acid are those of a tricarboxylic acid and a tertiary alcohol.

(1) **Acetylation**. It reacts with acetyl chloride (or acetic anhydride) to form monoacetylcitric acid.

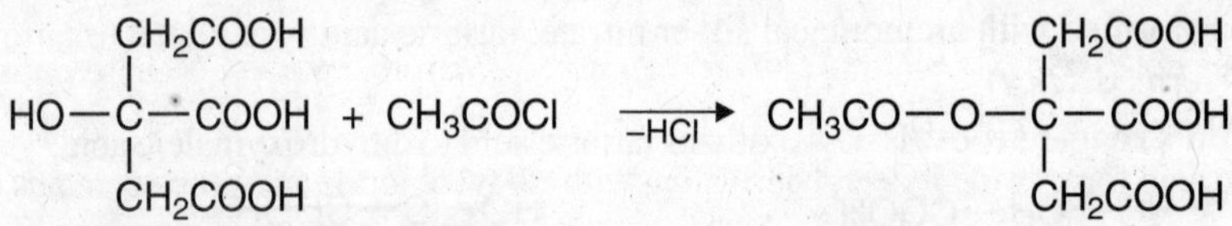

Citric acid   Monoacetylcitric acid

(2) **Reduction**. When reduced with HI, citric acid gives **Tricarballylic acid**.

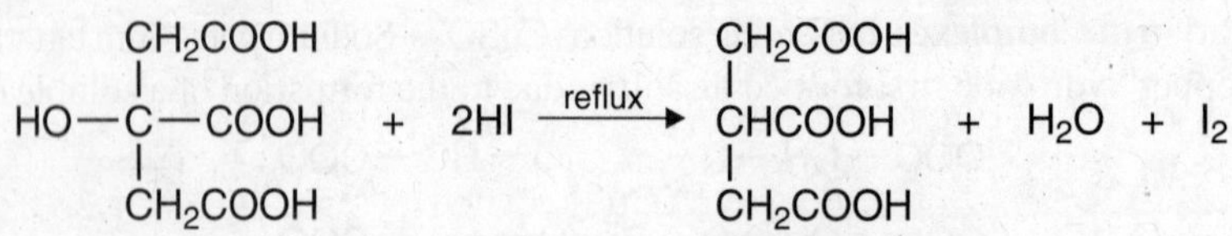

Citric acid   Tricarballylic acid

(3) **Action of Heat**. When heated at 150°C, citric acid gives a molecule of water to form **Aconitic acid**.

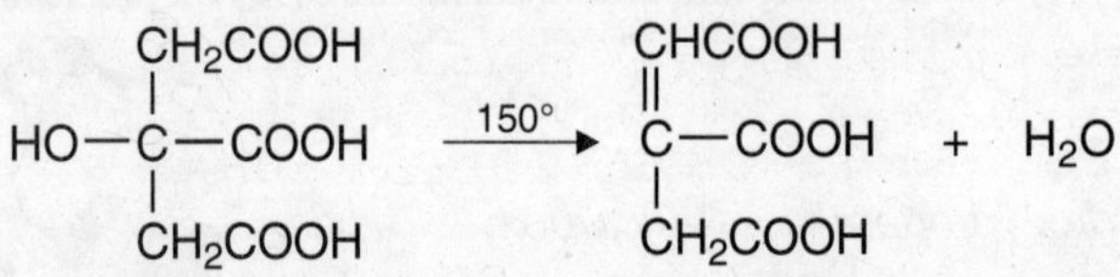

Citric acid   Aconitic acid

(4) **Reaction with fuming H_2SO_4.** On heating with fuming sulfuric acid, it gives acetonecarboxylic acid.

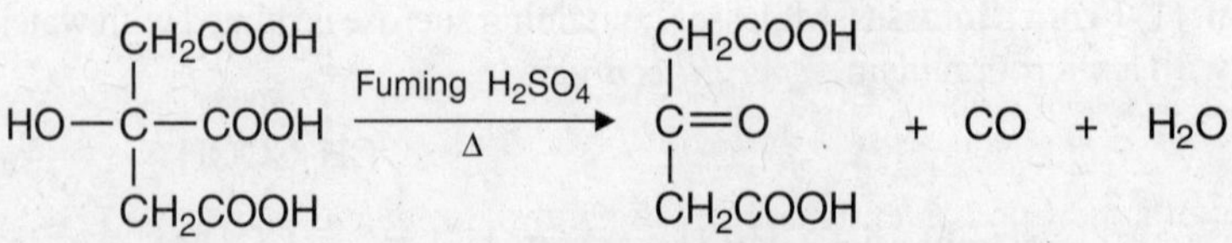

Citric acid   Acetonecarboxylic acid

Uses. Citric acid is used : (1) as acidulant in carbonated soft drinks, jams, jellies, candies, etc., (2) as medicinal in form of effervescent salts, magnesium citrate, a laxative ; (3) as mordant ; (4) as ferric ammonium citrate in the preparation of blueprint papers ; and (5) as esters (*e.g.*, tributyl citrate) that are good plasticizers for lacquers and varnishes.

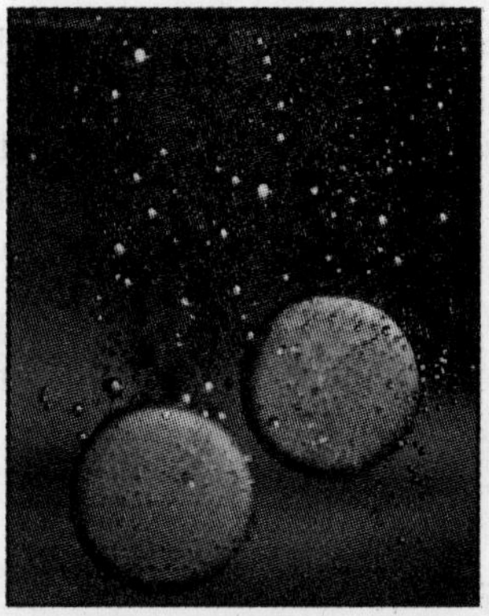

A commercial remedy for excess stomach acid. The bubbles are carbon dioxide, CO_2 from reaction between citric acid $C_6H_8O_7$, and sodium bicarbonate, $NaHCO_3$.

Structure. (1) As determined by elemental analysis and molecular weight determination, the molecular formula of citric acid is $C_6H_8O_7$.

(2) It gives three series of salts and esters, showing thereby the presence of three COOH groups in the molecule.

(3) It does not give a molecule of CO_2 on heating. Therefore, no two COOH groups are attached to the same carbon.

(4) Citric acid forms a monoacetyl derivative with acetyl chloride and hence contains one alcoholic OH group.

(5) From the above considerations, citric acid can have one of the following two possible structures.

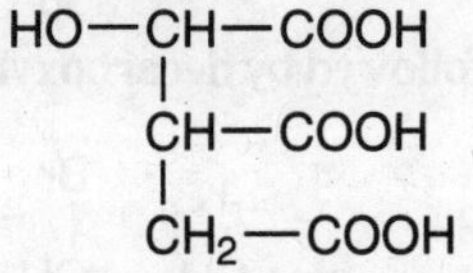

I

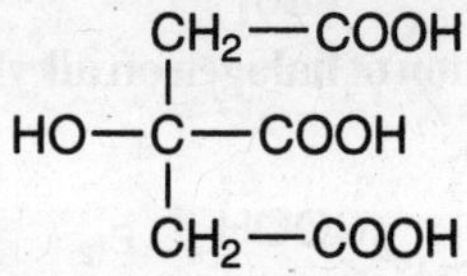

II

(6) **The Synthesis of Citric acid** from glycerol establishes that it has the structure II and not I.

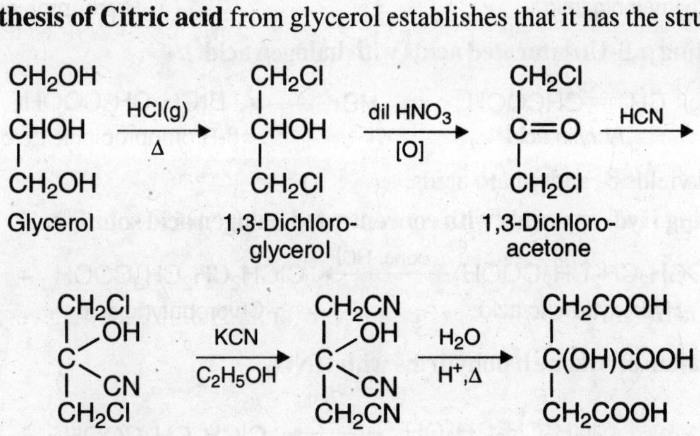

The product of synthesis is identical with the natural citric acid.

HALO ACIDS

Halo acids or halogen-substituted acids are those in which one or more H atoms of the hydrocarbon chain are replaced by halogen atoms (F, Cl, Br, or I). They are designated as α, β, γ etc., according as the halogen is bonded to α, β or γ-carbon of the chain with respect to COOH group. Thus we have,

$ClCH_2COOH$	$\overset{\beta}{}\ \ \overset{\alpha}{}$ $BrCH_2CH_2COOH$	$\overset{\gamma}{}\ \ \overset{\beta}{}\ \ \overset{\alpha}{}$ $ICH_2CH_2CH_2COOH$
Chloroacetic acid	β-Bromopropionic acid (3-Bromopropanoic acid)	γ-Iodobutyric acid (4-Iodobutanoic acid)

METHODS OF PREPARATION

(1) **By Hell-Volhard-Zelinsky Reaction.** This is the usual method of preparing α-halo acids. The corresponding aliphatic carboxylic acid is reacted with bromine or chlorine in the presence of red phosphorus (or a phosphorus halide) to give an α-halo acid.

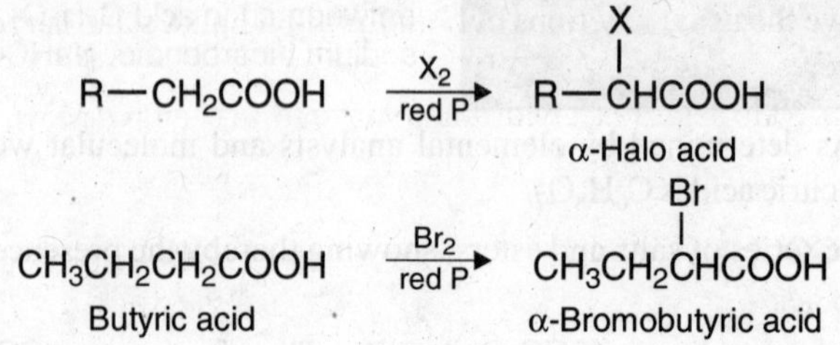

If excess of bromine or chlorine is used, the products obtained are α,α-dihalo acids or α,α,α-trihalo acids.

$$\underset{\text{Monochloroacetic acid}}{ClCH_2COOH} \xrightarrow[P]{Cl_2} \underset{\text{Dichloroacetic acid}}{Cl_2CHCOOH} \xrightarrow[P]{Cl_2} \underset{\text{Trichloroacetic acid}}{Cl_3CCOOH}$$

(2) **By the action of halogen on alkylmalonic acid followed by decarboxylation.**

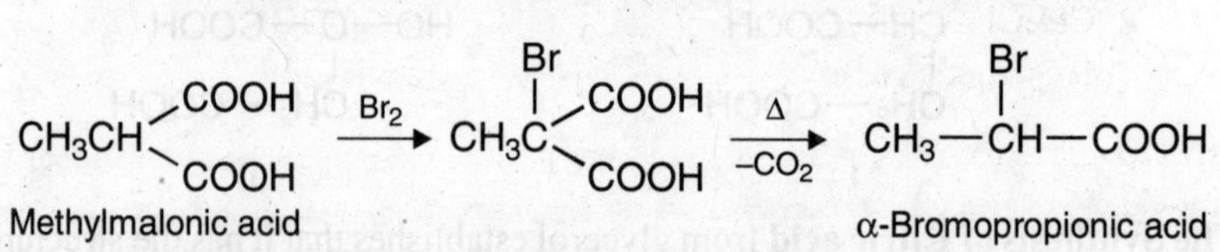

(3) **By treating α,β-Unsaturated acids with halogen acid.**

$$\underset{\text{Acrylic acid}}{CH_2{=}CHCOOH} + HBr \longrightarrow \underset{\beta\text{-Bromopropionic acid}}{BrCH_2CH_2COOH}$$

This method yields β- and γ-halo acids.

(4) **By heating Hydroxy acids with concentrated halogen acid solution.**

$$\underset{\gamma\text{-Hydroxybutyric acid}}{HOCH_2CH_2CH_2COOH} \xrightarrow{\text{conc. HCl}} \underset{\gamma\text{-Chlorobutyric acid}}{ClCH_2CH_2CH_2COOH} + H_2O$$

(5) **By oxidation of Alkene halohydrins with HNO_3.**

$$\underset{\text{Propylene chlorohydrin}}{ClCH_2CH_2CH_2OH} \xrightarrow[HNO_3]{[O]} \underset{\beta\text{-Chloropropionic acid}}{ClCH_2CH_2COOH}$$

PHYSICAL PROPERTIES

(1) Halo acids are colorless liquids or solids at room temperature.

(2) They dissolve in water giving strongly acidic solutions.

(3) **Relative Acid Strength.** They are stronger acids than the parent unsubstituted acids. This is due to the electron-withdrawing inductive effect of the halogen atom.

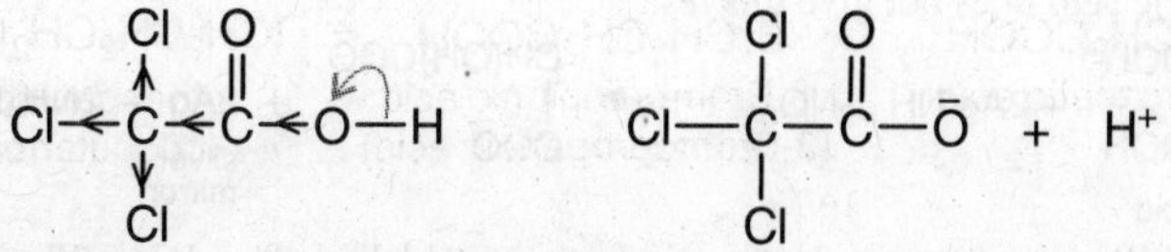

The acid strength depends on the number of halogen atoms present and their position on the carbon chain. The acidity decreases as the halogen is removed farther away from the COOH groups due to the inductive effect.

CHEMICAL PROPERTIES

The halo acids give the usual reactions of both the alkyl halides and carboxylic acids. Two of its important reactions are :

(1) **Amination.** α-Halo acids react with ammonia to form α-amino acids.

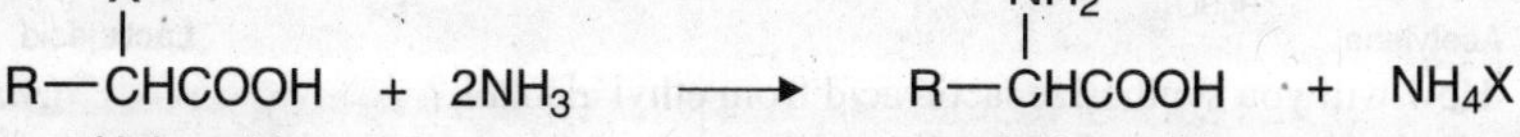

(2) **Hydrolysis with Alkalis.** α-Halo acids on treatment with aqueous alkali (NaOH or KOH) give α-hydroxy acids.

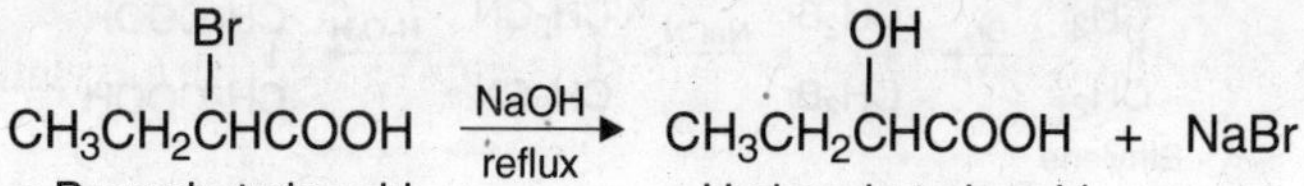

Similarly β-, γ- or δ-halo acids are hydrolysed to give hydroxy acids which give α,β-unsaturated acids and lactones (*Cf.* Action of Heat on Hydroxy acids).

Examination Questions

1. What are hydroxy acids ? How are they prepared ?
2. What is the effect of heat on hydroxy acids ? (*Orissa BSc, 2014 ; West Bengal BSc, 2015*)
3. Give the preparation and properties of lactic acid. (*Rohilkhand BSc, 2015*)
4. Give the preparation and properties of tartaric acid.
5. Give the preparation and properties of citric acid.
6. How is the structure of citric acid established ?
7. What happens when lactic acid is heated ?

Answer. A lactide is formed. For reaction equation see text.

8. What happens when tartaric acid is treated with hydrogen iodide ?

Answer. Succinic acid is formed. For reaction equation see text.

9. What happens when tartaric acid is oxidized with nitric acid ?

Answer. Oxalic acid is formed. For reaction equation see text.

10. How will you distinguish between citric acid and tartaric acid ?

Answer. When tartaric acid is treated with ammoniacal silver nitrate solution (Tollens' reagent), silver mirror is formed. Citric acid does not give this test.

$$\underset{\text{Tartaric acid}}{\begin{array}{l}CH(OH)COOH\\ |\\ CH(OH)COOH\end{array}} + 2Ag(NH_3)_2NO_3 \longrightarrow \begin{array}{l}CH(OH)CO\bar{O}\\ |\\ CO\bar{O}\end{array} + \underset{\text{silver mirror}}{2Ag} + 2NH_4NO_3 + 2NH_3$$

11. How will you synthesize lactic acid from acetaldehyde ? *(Goa BSc, 2014 ; Delhi BSc, 2015)*

Answer. Following steps are involved :

$$\underset{\text{Acetaldehyde}}{CH_3-\overset{\overset{O}{\|}}{C}-H} \xrightarrow{HCN} CH_3-\overset{\overset{OH}{|}}{CH}-CN \xrightarrow{H_2O/H^+} \underset{\text{Lactic acid}}{CH_3-\overset{\overset{OH}{|}}{CH}-COOH}$$

12. How will you synthesize lactic acid from acetylene ? *(Burdwan BSc, 2014)*

Answer. Following steps are involved :

$$\underset{\text{Acetylene}}{HC\equiv CH} \xrightarrow[HgSO_4]{H_2O/H_2SO_4} CH_3\overset{\overset{O}{\|}}{C}H \xrightarrow{HCN} CH_3\overset{\overset{OH}{|}}{C}HCN \xrightarrow{H_2O/H^+} \underset{\text{Lactic acid}}{CH_3\overset{\overset{OH}{|}}{C}HCOOH}$$

13. How will you synthesize lactic acid from ethyl alcohol ? *(Aligarh BSc, 2014)*

Answer. Following steps are involved :

$$\underset{\text{Ethyl alcohol}}{CH_3CH_2OH} \xrightarrow{K_2Cr_2O_7/H^+} CH_3\overset{\overset{O}{\|}}{C}H \xrightarrow{HCN} CH_3\overset{\overset{OH}{|}}{C}HCN \xrightarrow{H_2O/H^+} \underset{\text{Lactic acid}}{CH_3\overset{\overset{OH}{|}}{C}HCOOH}$$

14. How will you synthesize tartaric acid from ethylene ? *(Kalyani BSc, 2014)*

Answer. Following steps are involved :

$$\underset{\text{Ethylene}}{\begin{array}{l}CH_2\\ \|\\ CH_2\end{array}} \xrightarrow{Br_2} \begin{array}{l}CH_2Br\\ |\\ CH_2Br\end{array} \xrightarrow{NaCN} \begin{array}{l}CH_2CN\\ |\\ CH_2CN\end{array} \xrightarrow{H_2O/H^+} \begin{array}{l}CH_2COOH\\ |\\ CH_2COOH\end{array}$$

$$\xrightarrow[\text{HVZ reaction}]{Br_2/P} \begin{array}{l}BrCHCOOH\\ |\\ BrCHCOOH\end{array} \xrightarrow[(AgOH)]{\text{moist } Ag_2O} \underset{\text{Tartaric acid}}{\begin{array}{l}HO-CH-COOH\\ |\\ HO-CH-COOH\end{array}}$$

15. How will you synthesize citric acid from glycerol ?

Multiple Choice Questions

1. Lactams are ?
   (*a*) cyclic acids (*b*) cyclic esters
   (*c*) cyclic anhydrides (*d*) cyclic amides

Answer. (*d*)

2. A cyclic ester is called a
   (*a*) Lactone (*b*) Semicarbazone
   (*c*) Lactose (*d*) Dioxane

Answer. (*a*)

3. α-Bromobutyric acid is also called
   (*a*) 1-Bromobutanoic acid (*b*) 2-Bromobutanoic acid
   (*c*) 2-Bromobutanoate acid (*d*) 3-Bromobutanoic acid

Answer. (*b*)

4. The self-condensation reaction of one molecule of 5-hydroxyhexanoic acid gives
   (*a*) an anhydride (*b*) a lactone
   (*c*) a ketone (*d*) a lactam

Answer. (*b*)

23 Carboxylic Acid Derivatives

The carboxylic acid derivatives may be regarded as compounds in which the –OH group of a carboxylic acid has been replaced by another functional group. Their general formula is :

$$R-\overset{\overset{\large O}{||}}{C}-X \quad (X = Cl, Br, NH_2, OR, O-\overset{\overset{\large O}{||}}{C}-R)$$

All carboxylic acid derivatives contain the **acyl group**. For this reason carboxylic acid derivatives are often called **acyl compounds**.

$$R-\overset{\overset{\large O}{||}}{C}- \qquad \text{Acyl group}$$

Acyl halides	Anhydrides	Esters	Amides
$R-\overset{\overset{O}{\|\|}}{C}-X$ X = F, Cl, Br or I	$R-\overset{\overset{O}{\|\|}}{C}-O-\overset{\overset{O}{\|\|}}{C}-R$	$R-\overset{\overset{O}{\|\|}}{C}-OR'$	$R-\overset{\overset{O}{\|\|}}{C}-NR'_2$ R' = H or alkyl
$CH_3-\overset{\overset{O}{\|\|}}{C}-Cl$ Acetyl chloride (Ethanoyl chloride)	$CH_3-\overset{\overset{O}{\|\|}}{C}-O-\overset{\overset{O}{\|\|}}{C}-CH_3$ Acetic anhydride (Ethanoic anhydride)	$CH_3-\overset{\overset{O}{\|\|}}{C}-OC_2H_5$ Ethyl acetate (Ethyl ethanoate)	$CH_3-\overset{\overset{O}{\|\|}}{C}-NH_2$ Acetamide (Ethanamide)

All these compounds have the common structural feature of the carbon-oxygen double bond.

ACID HALIDES

Acid halides are the derivatives of acids in which the –OH function of carboxyl group has been replaced by a halogen atom.

$$\mathrm{R{-}\overset{\overset{\displaystyle O}{\|}}{C}{-}OH \xrightarrow[+\,X]{-\,OH} R{-}\overset{\overset{\displaystyle O}{\|}}{C}{-}X}$$

The acid halides are also referred to as **Acyl halides** ; the acyl group is R–CO–.

The acid chlorides are by far the most important of all the acid halides and will be discussed in detail. Their type formula may be represented as,

Functional group

$$\mathrm{R{-}\boxed{\overset{\overset{\displaystyle O}{\|}}{C}{-}Cl} \qquad R{-}CO{-}Cl \quad or \quad RCOCl}$$

They are named by dropping the ending '*–ic acid*' from either the common name or the IUPAC name and adding the suffix '*–yl chloride*'

Formula	**Common Name**	**IUPAC Name**
CH_3COCl	Acetyl chloride	Ethanoyl chloride
CH_3CH_2COCl	Propionyl chloride	Propanoyl chloride

METHODS OF PREPARATION

Acid chlorides are prepared from appropriate carboxylic acids :

(1) **By reaction with Thionyl Chloride ($SOCl_2$)**

$$\mathrm{\underset{Acid}{R{-}\overset{\overset{\displaystyle O}{\|}}{C}{-}OH} + SOCl_2 \longrightarrow \underset{Acid\ chloride}{R{-}\overset{\overset{\displaystyle O}{\|}}{C}{-}Cl} + HCl + SO_2}$$

$$\mathrm{\underset{Acetic\ acid}{CH_3{-}\overset{\overset{\displaystyle O}{\|}}{C}{-}OH} + SOCl_2 \longrightarrow \underset{Acetyl\ chloride}{CH_3{-}\overset{\overset{\displaystyle O}{\|}}{C}{-}Cl} + HCl + SO_2}$$

This method is superior to others as the by-products being gases escape leaving the acid chloride in the pure state.

(2) **By reaction with Phosphorus Trichloride (PCl_3)**

$$\mathrm{R{-}\overset{\overset{\displaystyle O}{\|}}{C}{-}OH + PCl_3 \longrightarrow R{-}\overset{\overset{\displaystyle O}{\|}}{C}{-}Cl + H_3PO_3}$$

$$\mathrm{\underset{Acetic\ acid}{CH_3{-}\overset{\overset{\displaystyle O}{\|}}{C}{-}OH} + PCl_3 \longrightarrow \underset{Acetyl\ chloride}{CH_3{-}\overset{\overset{\displaystyle O}{\|}}{C}{-}Cl} + H_3PO_3}$$

(3) **By reaction with Phosphorus Pentachloride (PCl_5)**

$$R-\overset{O}{\overset{\|}{C}}-OH + PCl_5 \longrightarrow R-\overset{O}{\overset{\|}{C}}-Cl + POCl_3 + HCl$$

$$\underset{\text{Propionic acid}}{CH_3CH_2-\overset{O}{\overset{\|}{C}}-OH} + PCl_5 \longrightarrow \underset{\text{Propionyl chloride}}{CH_3CH_2-\overset{O}{\overset{\|}{C}}-Cl} + POCl_3 + HCl$$

PHYSICAL PROPERTIES

(1) All simple acid chlorides are colorless liquids of lower boiling point than the corresponding acid.

(2) They have sharp pungent smell, and are lachrymators (tear-producing).

(3) They emit pale fumes of hydrogen chloride when exposed in air on account of their reaction with water vapor.

(4) They are insoluble in water but slowly dissolve in it because of hydrolysis.

(5) The C=O stretching band in *IR Spectrum* occurs at frequency range 1780-1850 cm^{-1}.

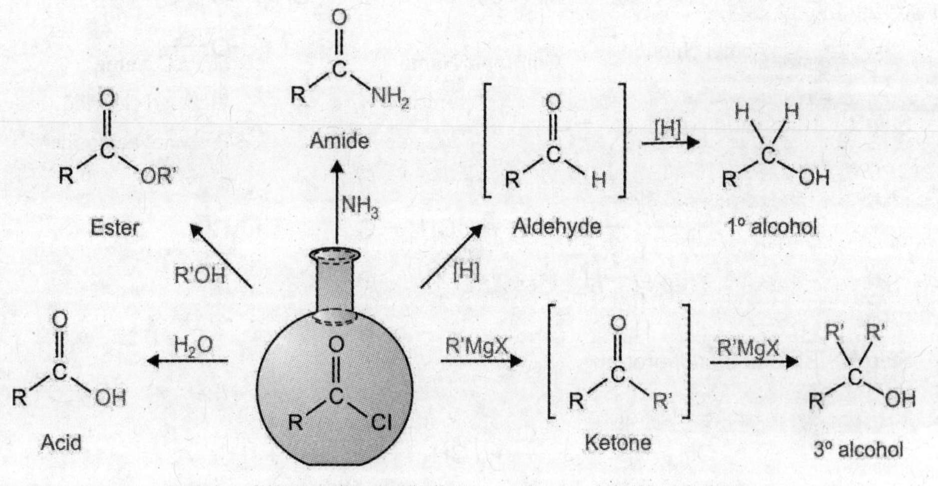

Nucleophilic acyl substitution reactions of Acid chlorides.

CHEMICAL PROPERTIES

In RCOCl we have a carbonyl group attached to a highly electronegative chlorine atom. Due to the electron-withdrawing effect of the Cl atom, the electron density on the carbonyl carbon is diminished further. As a result, acid chlorides are readily attacked by nucleophiles, whereby the Cl atom is replaced by a nucleophile.

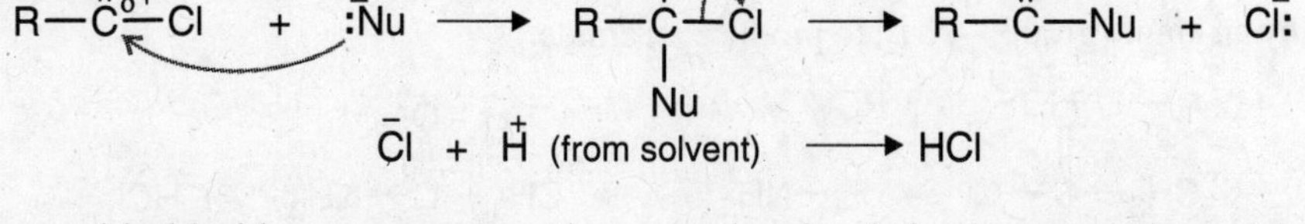

Thus acid chlorides are very reactive compounds, their characteristic reactions being nucleophilic substitution reactions.

$$CH_3-\overset{\overset{\large O}{||}}{C}-Cl + \underset{\text{(Nucleophile)}}{NuH} \longrightarrow CH_3-\overset{\overset{\large O}{||}}{C}-Nu + HCl$$

They react with weak nucleophiles as water, alcohols, ammonia and amines to produce the corresponding acid, ester, amide or substituted amides. Some of their important reactions are listed below.

(1) **Hydrolysis.** They react with water when the chlorine atom is displaced by OH group to reform the original organic acid.

$$\underset{\text{Acetyl chloride}}{CH_3-\overset{\overset{\large O}{||}}{C}-Cl} + HOH \longrightarrow \underset{\text{Acetic acid}}{CH_3-\overset{\overset{\large O}{||}}{C}-OH} + HCl$$

MECHANISM. It is an addition-elimination reaction. Following steps are involved :

Step 1. Addition of nucleophile.

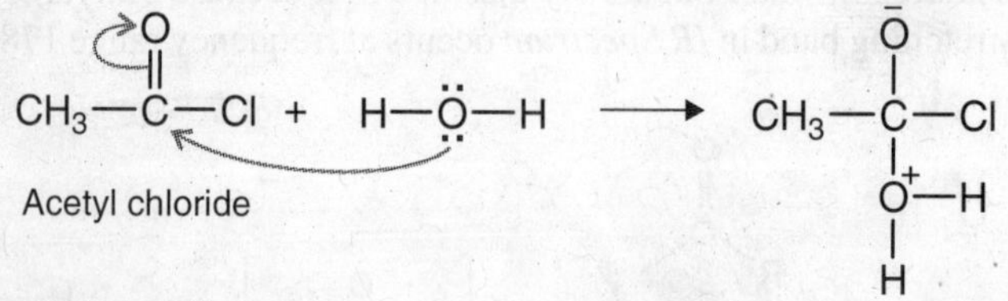

Step 2. Elimination of Cl^-.

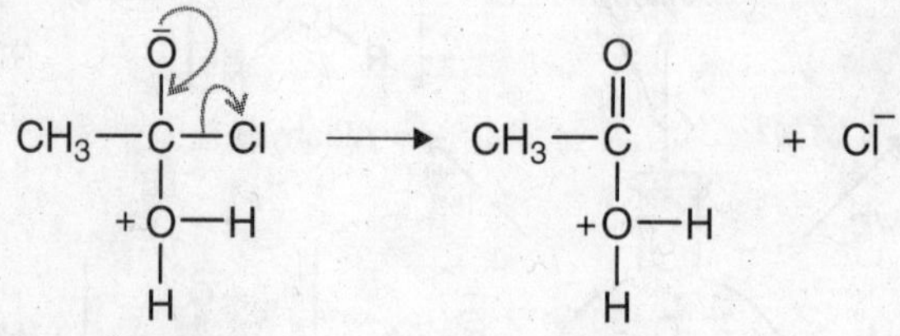

Step 3. Elimination of proton.

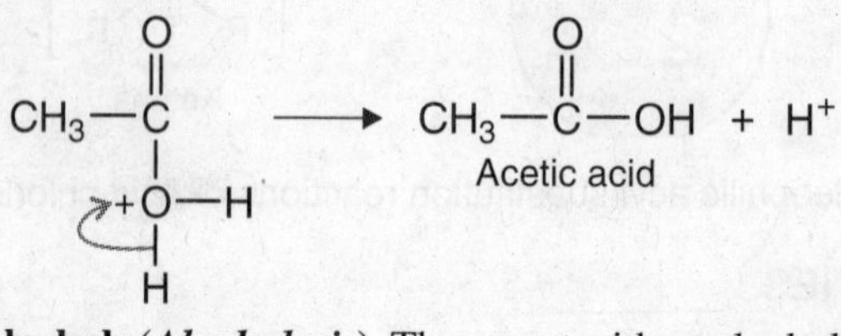

(2) **Reaction with Alcohols (*Alcoholysis*).** They react with an alcohol when the chlorine atom is displaced by alkoxy group, –OR′, to produce an ester.

$$\underset{\text{Acetyl chloride}}{CH_3-\overset{\overset{\large O}{||}}{C}-Cl} + \underset{\text{Ethyl alcohol}}{HOC_2H_5} \longrightarrow \underset{\text{Ethyl acetate}}{CH_3-\overset{\overset{\large O}{||}}{C}-OC_2H_5} + HCl$$

(3) **Reaction with Ammonia (*Ammonolysis*).** They react with ammonia when the Cl atom is displaced by an amino group, $-NH_2$, to produce an amide.

$$\underset{\text{Acetyl chloride}}{CH_3-\overset{\overset{\large O}{||}}{C}-Cl} + \underset{\substack{\text{Ammonia}\\\text{(Excess)}}}{H-NH_2} \longrightarrow \underset{\text{Acetamide}}{CH_3-\overset{\overset{\large O}{||}}{C}-NH_2} + HCl$$

$$HCl + NH_3 \longrightarrow NH_4Cl$$

MECHANISM. Following steps are involved :

Step 1. Addition of nucleophile.

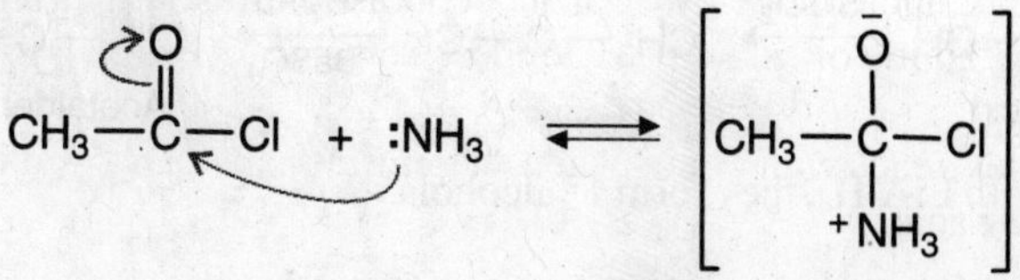

Tetrahedral intermediate (Unstable)

Step 2. Elimination of Cl^-.

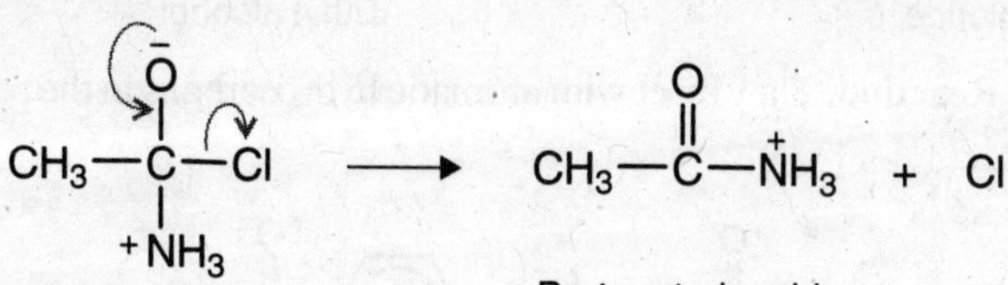

Protonated amide

Step 3. Protonated amide reacts with another ammonia molecule to form acetamide.

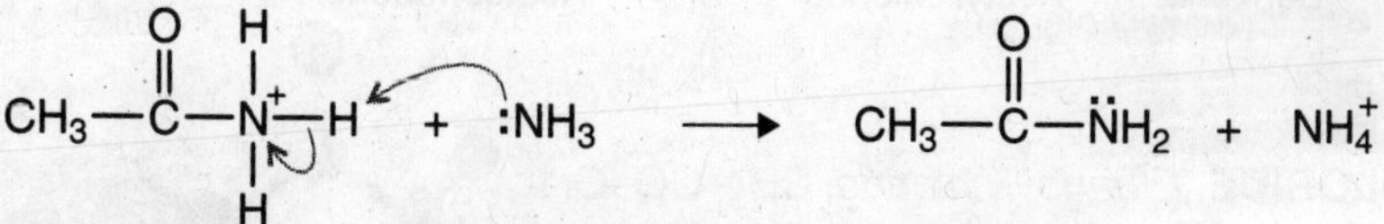

(Acts as base) Acetamide

(4) **Reaction with 1° and 2° Amines.** Acid halides react with 1° and 2° amines to produce substituted amides.

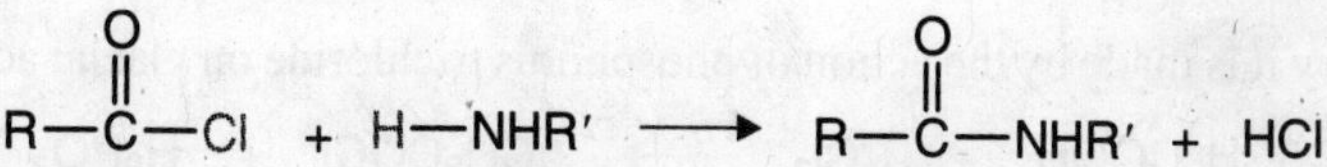

1°amine *N*-Alkylamide

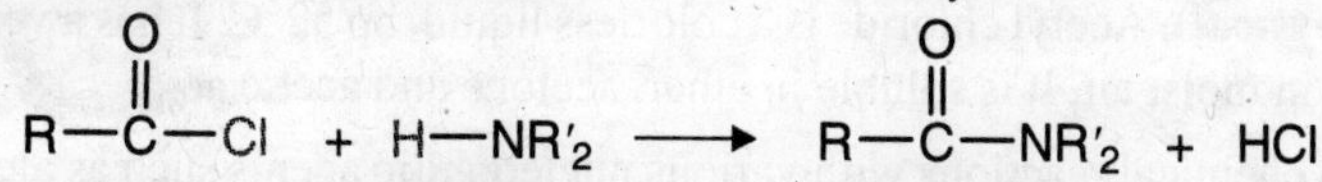

2°amine *N,N*-Dialkylamide

(5) **Reaction with Salts of Carboxylic Acids.** They react with sodium salts of carboxylic acids to form anhydrides.

Acetyl chloride Acetic anhydride

(6) **Reduction.** (*a*) When reduced with hydrogen in the presence of 'poisoned' palladium catalyst, they form aldehydes.

Acetyl chloride Acetaldehyde

This reaction is called **Rosenmund Reduction**. Normally the aldehyde would be further reduced to a primary alcohol. In this case the catalyst is 'poisoned' with barium sulfate to deactivate it partially and prevent the reduction of the aldehyde to an alcohol. Rosenmund reduction is used to make aldehydes from acids. For example,

$$\underset{\text{Acetic acid}}{CH_3-\overset{O}{\overset{\|}{C}}-OH} \xrightarrow{SOCl_2} CH_3-\overset{O}{\overset{\|}{C}}-Cl \xrightarrow[BaSO_4]{H_2/Pd} \underset{\text{Acetaldehyde}}{CH_3-\overset{O}{\overset{\|}{C}}-H}$$

(*b*) When reduced with $LiAlH_4$, they form 1° alcohols.

$$\underset{\text{Acetyl chloride}}{CH_3-\overset{O}{\overset{\|}{C}}-Cl} + 4[H] \xrightarrow{LiAlH_4} \underset{\text{Ethyl alcohol}}{CH_3-CH_2-OH} + HCl$$

(7) **Friedel-Crafts Reaction.** They react with aromatic hydrocarbons in the presence of anhydrous $AlCl_3$ to form ketones.

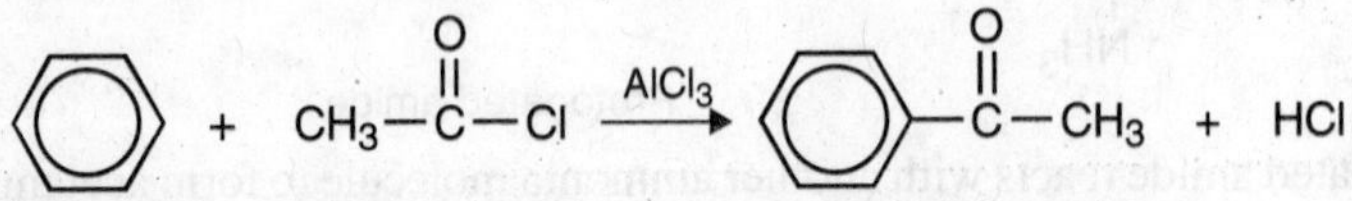

Benzene Acetyl chloride Acetophenone

ACETYL CHLORIDE, *Ethanoyl chloride*, CH_3–CO–Cl

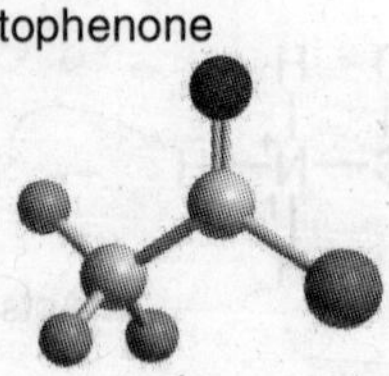

According to the accepted abbreviation of acetyl group, CH_3CO–, as *Ac*, the formula of acetyl chloride may be written as **AcCl**.

In the laboratory it is made by the action of phosphorus trichloride on glacial acetic acid in cold.

$$3CH_3COOH + PCl_3 \longrightarrow 3CH_3COCl + H_3PO_3$$

Properties (*Physical*). Acetyl chloride is a colorless liquid, bp 52°C. It has a very pungent smell and fumes strongly in moist air. It is soluble in ether, acetone and acetic acid.

(*Chemical*). Its chemical reactions with various nucleophilic agents such as alcohols, ammonia, and amines have been discussed under the general treatment of acid halides. In these reactions a H atom of the functional group OH or NH_2 is replaced by the acetyl group (CH_3CO–). This process known as **Acetylation** *i.e., introduction of acetyl group*, is very useful in synthetic organic chemistry.

$$\underset{\text{Acetyl chloride}}{CH_3-\overset{O}{\overset{\|}{C}}-Cl} + \underset{\text{Alcohol}}{H-OR} \longrightarrow \underset{\text{Ester}}{CH_3-\overset{O}{\overset{\|}{C}}-OR} + HCl$$

$$\underset{\text{Acetyl chloride}}{CH_3-\overset{O}{\overset{\|}{C}}-Cl} + \underset{\text{Amine}}{H_2N-R} \longrightarrow \underset{\textit{N-}\text{Alkylacetamide}}{CH_3-\overset{O}{\overset{\|}{C}}-NHR} + HCl$$

Uses. The chief use of acetyl chloride is as *Acetylating Agent* in the preparation of several derivatives, dyestuffs, and pharmaceuticals. It is also used to determine the number of OH groups in organic molecules.

ANHYDRIDES

The compounds derived by loss of a water molecule between two molecules of a carboxylic acid are called **Acid Anhydrides** or simply **Anhydrides.**

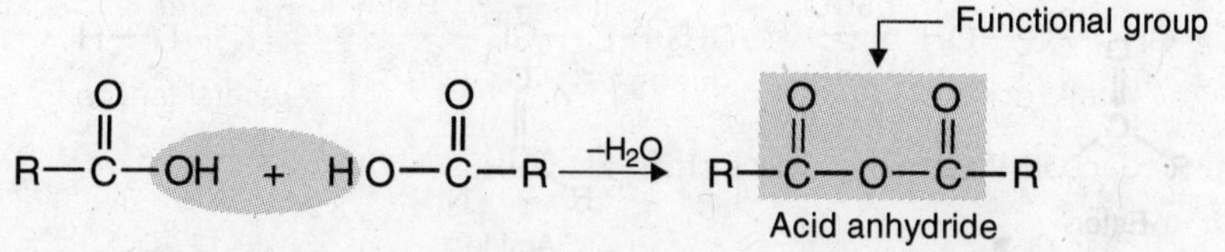

The functional group of this series of acid derivatives is –CO–O–CO–.

Anhydrides are named by replacing the word 'acid' in the name of the parent acid by 'anhydride'.

Formula	Common Name	IUPAC Name
$CH_3-CO-O-CO-CH_3$	Acetic anhydride	Ethanoic anhydride
$CH_3CH_2-CO-O-CO-CH_2CH_3$	Propionic anhydride	Propanoic anhydride

Anhydrides derived from two molecules of different acids are called **Mixed Anhydrides.**

METHODS OF PREPARATION

Acid anhydrides may be prepared :

(1) **By reaction of Acid halide with a Carboxylic acid**

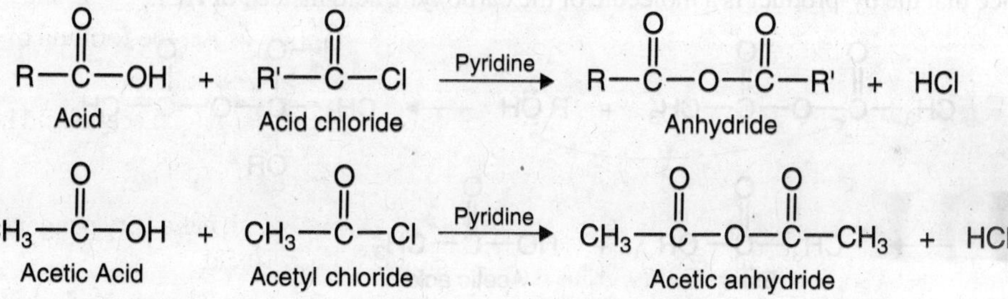

(2) **By reaction of Acid halide with Salt of a Carboxylic acid**

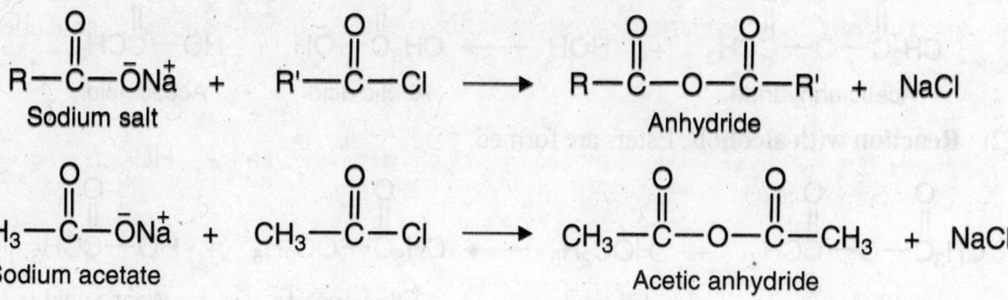

PHYSICAL PROPERTIES

(1) The lower simple anhydrides are colorless liquids with a sharp pungent smell.

(2) They are insoluble in water for lack of hydrogen bonding. But they dissolve after a while due to conversion to the soluble acid by hydrolysis.

(3) They have higher boiling points than the acid from which they are derived. This is so because of their large size and enhanced van der Waal's interactions.

(4) They are characterized by the C=O stretching band in *IR spectrum* at frequency range 1800-1850 and 1740-1790 cm^{-1}.

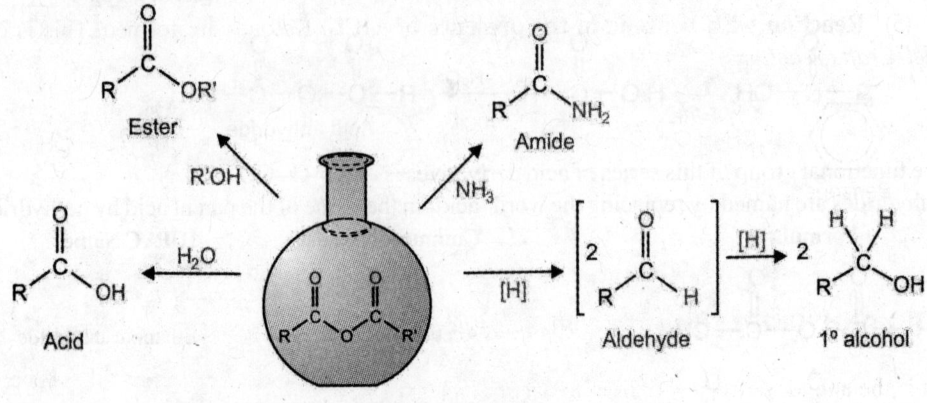

Important reactions of Acid anhydrides.

CHEMICAL PROPERTIES

Anhydrides react with nucleophiles in the same way as acid chlorides. But since –OCOR group is not as electronegative as a Cl atom, anhydrides are less reactive than acid chlorides. Thus anhydrides give all the nucleophilic substitution reactions shown by acid chlorides but with less vigor. The mechanism of the nucleophilic substitution here is the same as stated for acid chlorides with the difference that the by-product is a molecule of the carboxylic acid instead of HCl.

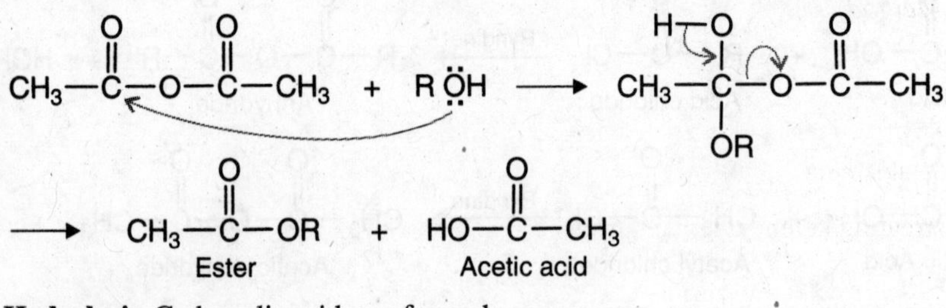

(1) **Hydrolysis**. Carboxylic acids are formed.

$$CH_3COOCOCH_3 + HOH \longrightarrow CH_3COOH + HOCOCH_3$$

Acetic anhydride → Acetic acid + Acetic acid

(2) **Reaction with alcohols**. Esters are formed.

$$CH_3CO{-}O{-}COCH_3 + HOC_2H_5 \longrightarrow CH_3CO{-}OC_2H_5 + HO{-}COCH_3$$

Ethanol → Ethyl acetate + Acetic acid

(3) **Reaction with ammonia**. Amides are formed.

$$CH_3CO{-}O{-}COCH_3 + H{-}NH_2 \longrightarrow CH_3CO{-}NH_2 + HO{-}COCH_3$$

Ammonia → Acetamide + Acetic acid

(4) **Reaction with amines**. Substituted amides are formed.

$$\underset{}{CH_3-\overset{O}{\overset{\|}{C}}-O-\overset{O}{\overset{\|}{C}}-CH_3} + \underset{\text{1°amine}}{H-NHR} \longrightarrow \underset{N\text{-Alkylacetamide}}{CH_3-\overset{O}{\overset{\|}{C}}-NHR} + CH_3-\overset{O}{\overset{\|}{C}}-OH$$

(5) **Reaction with benzene in the presence of $AlCl_3$**. Ketones are formed. This is called *Friedel-Crafts reaction.*

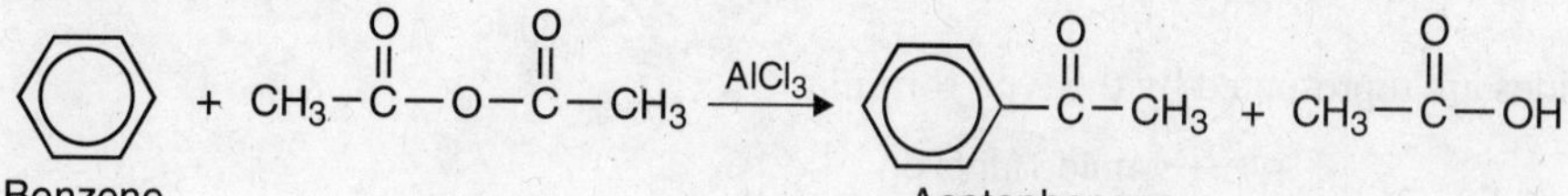

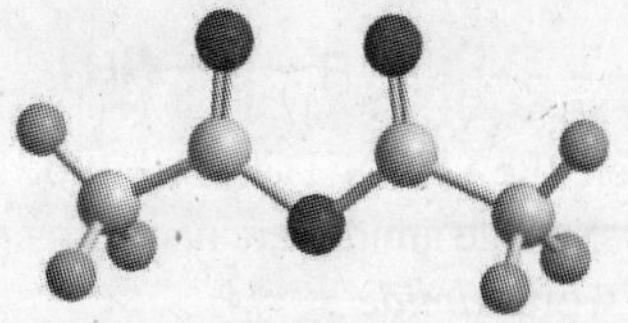

ACETIC ANHYDRIDE, *Ethanoic anhydride*, $(CH_3CO)_2O$

It is the anhydride corresponding to acetic acid.

Preparation. It may be prepared :

(1) By distilling a mixture of anhydrous sodium acetate and acetyl chloride. (*Laboratory Method*)

$$\underset{\text{Sodium acetate}}{CH_3-\overset{O}{\overset{\|}{C}}-\overset{-}{O}\overset{+}{Na}} + CH_3-\overset{O}{\overset{\|}{C}}-Cl \longrightarrow \underset{\text{Acetic anhydride}}{CH_3-\overset{O}{\overset{\|}{C}}-O-\overset{O}{\overset{\|}{C}}-CH_3} + NaCl$$

(2) By air-oxidation of acetaldehyde in the presence of Co-Cu-Acetates at 50-70°C under pressure. (*Industrial Method*)

$$\underset{\text{Acetaldehyde}}{2CH_3-\overset{O}{\overset{\|}{C}}-H} + O_2 \xrightarrow[\text{50–70°C}]{\text{Co–Cu Acetates}} \underset{\text{Acetic anhydride}}{CH_3-\overset{O}{\overset{\|}{C}}-O-\overset{O}{\overset{\|}{C}}-CH_3} + H_2O$$

(3) By treating ketene with acetic acid.

$$\underset{\text{Acetic acid}}{CH_3-\overset{O}{\overset{\|}{C}}-OH} + \underset{\text{Ketene}}{CH_2=C=O} \longrightarrow \underset{\text{Acetic anhydride}}{CH_3-\overset{O}{\overset{\|}{C}}-O-\overset{O}{\overset{\|}{C}}-CH_3}$$

The ketene is produced by dehydration of acetic acid using triethyl phosphate as catalyst at 700°C under reduced pressure.

$$CH_3-\overset{O}{\overset{\|}{C}}-OH \longrightarrow CH_2=C=O + H_2O$$

Properties. Acetic anhydride is a colorless liquid, 139.9°C, sp gr 1.083. It has a pungent sharp odor. It is sparingly soluble in cold water but miscible with alcohol, ether, and acetic acid. It gives reactions parallel to those of acetyl chloride, although less vigorously. These have been discussed under the general treatment.

Uses. Acetic anhydride is usually a more satisfactory *Acetylating Agent* than acetyl chloride. As such it is used in the preparation of medicines (*aspirin*) and plastics (*cellulose acetate, vinyl acetate*).

AMIDES

The derivatives of carboxylic acids in which the –OH function of the –COOH group has been replaced by an amino group, $-NH_2$, are called Amides or Acid Amides.

$$R-\overset{O}{\overset{\|}{C}}-OH \xrightarrow[+\,NH_2]{-\,OH} \underset{\text{Amide}}{R-\overset{O}{\overset{\|}{C}}-NH_2}$$

Amides are represented by the type formula

Amide function

$$R-\overset{O}{\overset{\|}{C}}-NH_2 \quad \text{or} \quad R-CO-NH_2 \quad \text{or} \quad RCONH_2$$

where the $-CONH_2$ is the Amide Function or the Amide Group.

Simple amides are named by replacing the ending '*–ic acid*' (Common) or '*–oic acid*' (IUPAC) by the word '*amide*'.

Formula	Common Name	IUPAC Name
$H-\overset{O}{\overset{\|}{C}}-NH_2$	Formamide	Methanamide
$CH_3-\overset{O}{\overset{\|}{C}}-NH_2$	Acetamide	Ethanamide

METHODS OF PREPARATION

Amides are obtained by treatment of ammonia with carboxylic acids, acid chlorides, anhydrides, or esters.

(1) **From carboxylic acids**

$$\underset{\text{Acetic acid}}{CH_3-\overset{O}{\overset{\|}{C}}-OH} \xrightarrow{NH_3} CH_3-\overset{O}{\overset{\|}{C}}-\overset{-\,+}{ONH_4} \xrightarrow{\Delta} \underset{\text{Acetamide}}{CH_3-\overset{O}{\overset{\|}{C}}-NH_2} + H_2O$$

(2) **From acid chlorides**

$$\underset{\text{Acetyl chloride}}{CH_3-\overset{O}{\overset{\|}{C}}-Cl} + 2NH_3 \longrightarrow \underset{\text{Acetamide}}{CH_3-\overset{O}{\overset{\|}{C}}-NH_2} + NH_4Cl$$

(3) **From acid anhydrides**

$$\underset{\text{Acetic anhydride}}{CH_3-\overset{O}{\overset{\|}{C}}-O-\overset{O}{\overset{\|}{C}}-CH_3} + NH_3 \longrightarrow \underset{\text{Acetamide}}{CH_3-\overset{O}{\overset{\|}{C}}-NH_2} + \underset{\text{Acetic acid}}{CH_3-\overset{O}{\overset{\|}{C}}-OH}$$

(4) **From esters**

$$\underset{\text{Ethyl acetate}}{CH_3-\overset{O}{\overset{\|}{C}}-OC_2H_5} + NH_3 \longrightarrow \underset{\text{Acetamide}}{CH_3-\overset{O}{\overset{\|}{C}}-NH_2} + C_2H_5OH$$

PHYSICAL PROPERTIES

(1) Almost all amides are white, odorless (when pure), crystalline solids.

(2) They have sharp melting points that are higher than those of corresponding acids, owing to intermolecular hydrogen bonding.

(3) Simple amides upto six-carbons are very soluble in water but the water-solubility declines as we ascend the series.

(4) The characteristic C=O stretch band in the *IR spectrum* occurs at 1680 cm^{-1}.

CHEMICAL PROPERTIES

Amides are least reactive of the acid derivatives. They owe their chemical stability to resonance.

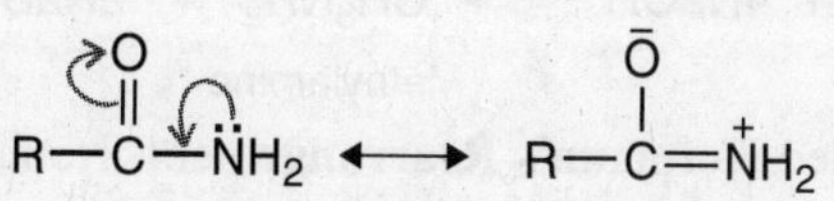

Since the electron-pair on N atom is not readily available to an acid, amides are much less basic than amines. In fact, they are neutral compounds and do not form salts with acids. Some of their general reactions are listed below :

(1) **Hydrolysis**. The amides can be hydrolysed in acid or in alkaline solution on prolonged heating.

$$\underset{\text{Acetamide}}{CH_3-\overset{O}{\overset{\|}{C}}-NH_2} + H_2O + HCl \longrightarrow \underset{\text{Acetic acid}}{CH_3-\overset{O}{\overset{\|}{C}}-OH} + NH_4Cl$$

$$CH_3-\overset{O}{\overset{\|}{C}}-NH_2 + NaOH \longrightarrow \underset{\text{Sodium acetate}}{CH_3-\overset{O}{\overset{\|}{C}}-\bar{O}\overset{+}{Na}} + NH_3$$

(2) **Reaction with Nitrous Acid ($NaNO_2$ + HCl).** Amides on treatment with nitrous acid form carboxylic acids. (*Cf* Amines)

$$\underset{\text{Amide}}{R-\overset{O}{\overset{\|}{C}}-NH_2} + \underset{\text{Nitrous acid}}{O=N-OH} \longrightarrow \underset{\text{Acid}}{R-\overset{O}{\overset{\|}{C}}-OH} + N_2 + H_2O$$

$$\underset{\text{Propionamide}}{CH_3CH_2-\overset{O}{\overset{\|}{C}}-NH_2} + O=N-OH \longrightarrow \underset{\text{Propionic acid}}{CH_3CH_2-\overset{O}{\overset{\|}{C}}-OH} + N_2 + H_2O$$

(3) **Reduction**. Lithium aluminium hydride ($LiAlH_4$) reduces amides to primary amines.

$$\underset{\text{Amide}}{R-\overset{O}{\overset{\|}{C}}-NH_2} \xrightarrow{LiAlH_4} \underset{1^\circ\text{ amine}}{R-CH_2-NH_2}$$

(4) **Dehydration with P_2O_5**. When heated with a strong dehydrating agent such as phosphorus pentoxide, they form nitriles.

$$\underset{\text{Acetamide}}{CH_3-\overset{O}{\overset{\|}{C}}-NH_2} \xrightarrow[-H_2O]{P_2O_5} \underset{\substack{\text{Acetonitrile}\\ \text{(Methyl cyanide)}}}{CH_3C\equiv N}$$

(5) **Hofmann's Degradation of Amides**. Treatment of an amide with bromine and sodium hydroxide solution produces a primary amine.

$$\underset{\text{Amide}}{R-\overset{\overset{\displaystyle O}{\|}}{C}-NH_2} + Br_2 + 4NaOH \longrightarrow \underset{1^\circ \text{ amine}}{RNH_2} + 2NaBr + Na_2CO_3 + 2H_2O$$

Specific Example :

$$\underset{\text{Acetamide}}{CH_3-\overset{\overset{\displaystyle O}{\|}}{C}-NH_2} + Br_2 + 4NaOH \longrightarrow \underset{\text{Methylamine}}{CH_3NH_2} + 2NaBr + Na_2CO_3 + 2H_2O$$

This reaction is also called **Hofmann's Rearrangement**. Notice that the overall result is the removal of the carbonyl group from the amide. The product contains one carbon less than the original amide. The reaction provides a useful method for the descent of a homologous series. The mechanism of this reaction is given in Chapter 27 (Amines).

ACETAMIDE, *Ethanamide*, $CH_3–CO–NH_2$

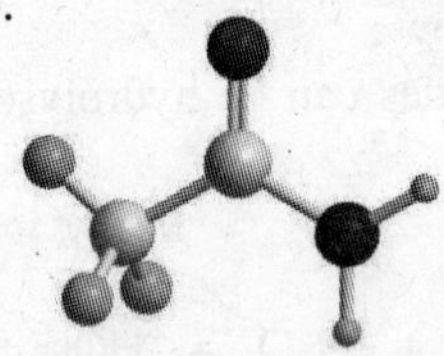

It is the most important amide.

Preparation. (1) It is made by treatment of ethyl acetate with a concentrated aqueous solution of ammonia.

$$\underset{\text{Ethyl acetate}}{CH_3-\overset{\overset{\displaystyle O}{\|}}{C}-OC_2H_5} + NH_3 \longrightarrow \underset{\text{Acetamide}}{CH_3-\overset{\overset{\displaystyle O}{\|}}{C}-NH_2} + C_2H_5OH$$

(2) In the laboratory it is prepared by heating ammonium acetate.

$$\underset{\text{Ammonium acetate}}{CH_3-\overset{\overset{\displaystyle O}{\|}}{C}-\overset{-}{O}\overset{+}{N}H_4} \longrightarrow \underset{\text{Acetamide}}{CH_3-\overset{\overset{\displaystyle O}{\|}}{C}-NH_2} + H_2O$$

In actual practice, ammonium carbonate is first heated with excess of glacial acetic acid. The ammonium acetate thereby produced dissociates at elevated temperatures to give acetamide.

Properties. Acetamide forms small, colorless, needle-like crystals, mp 82°C, bp 223°C. It has mousy odor but this is due to impurity and can be removed by recrystallization from benzene. It is very soluble in water and alcohol ; only slightly soluble in ether. Chemically it behaves as a typical amide and gives all the general reactions discussed before.

ESTERS

Esters are derivatives of carboxylic acids in which the –OH group has been replaced by –OR (alkoxy) group.

$$\underset{\text{Carboxylic acid}}{R-\overset{\overset{\displaystyle O}{\|}}{C}-OH} \xrightarrow[+\,OR']{-OH} \underset{\text{Ester}}{R-\overset{\overset{\displaystyle O}{\|}}{C}-OR'}$$

The esters of carboxylic acids are often referred to as **Carboxylic Esters**. The type formula of this class of acid derivatives is :

Ester Function

$$R-\overset{O}{\overset{\|}{C}}-OR' \quad \text{or} \quad R-CO-OR \quad \text{or} \quad RCOOR'$$

Ester Linkage

The ester functional group (or function) may be represented as –CO– OR′ or –COOR′.

Esters are by far the most important class of acid derivatives. A large number of them occur in flowers and fruits which owe their fragrance to these compounds. They are used in many synthetic products such as perfumes (for essences), pesticides, fibers, solvents and plasticizers.

NOMENCLATURE

The name of an ester is of two words. The name of the alkyl group attached to oxygen is the first word. The second word is the name of the parent acid with the ending '*–ic acid*' changed to '*–ate*.'

Formula	Common Name	IUPAC Name
$H-\overset{O}{\overset{\|}{C}}-OCH_3$	Methyl formate	Methyl methanoate
$CH_3-\overset{O}{\overset{\|}{C}}-OCH_2CH_3$	Ethyl acetate	Ethyl ethanoate
$CH_3(CH_2)_2-\overset{O}{\overset{\|}{C}}-OCH_3$	Methyl butyrate	Methyl butanoate

METHODS OF PREPARATION

(1) **By Fischer Esterification**. This method consists of heating a mixture of a carboxylic acid and an alcohol in the presence of concentrated H_2SO_4 or dry HCl as catalyst.

$$\underset{\text{Carboxylic acid}}{R-\overset{O}{\overset{\|}{C}}-OH} + \underset{\text{Alcohol}}{HO-R'} \xrightleftharpoons{H_2SO_4} \underset{\text{Ester}}{R-\overset{O}{\overset{\|}{C}}-OR'} + H_2O$$

$$\underset{\text{Acetic acid}}{CH_3COOH} + \underset{\text{Ethanol}}{HOC_2H_5} \xrightleftharpoons{H_2SO_4} \underset{\text{Ethyl acetate}}{CH_3COOC_2H_5} + H_2O$$

The reaction is carried to completion by using an excess of a reactant or by removing the water or ester as they are formed. The mechanism of this reaction is given in Chapter *20*.

(2) **By reaction of Acid chlorides with Alcohols.** Pyridine is added to remove HCl as it is formed.

$$\underset{\text{Acid chloride}}{R-\overset{O}{\overset{\|}{C}}-Cl} + HOR' \xrightarrow{\text{Pyridine}} \underset{\text{Ester}}{R-\overset{O}{\overset{\|}{C}}-OR'} + HCl$$

(3) **By reaction of Acid anhydrides with Alcohols**

$$\underset{\text{Acid anhydride}}{R-\overset{O}{\overset{\|}{C}}-O-\overset{O}{\overset{\|}{C}}-R} + HOR' \xrightarrow{H^+} \underset{\text{Ester}}{R-\overset{O}{\overset{\|}{C}}-OR'} + R-\overset{O}{\overset{\|}{C}}-OH$$

(4) **By reaction of Carboxylate salts and Alkyl halides**

$$\underset{\text{Sodium carboxylate}}{R-\overset{\overset{O}{\|}}{C}-\overset{-}{O}\overset{+}{Na}} + \underset{\text{Alkyl halide}}{XR'} \longrightarrow \underset{\text{Ester}}{R-\overset{\overset{O}{\|}}{C}-OR'} + NaX$$

(5) **By Transesterification (Ester Interchange)**

$$R-\overset{\overset{O}{\|}}{C}-OR' + R''OH \xrightarrow[\text{or } OH^-]{H^+} \underset{\text{Ester}}{R-\overset{\overset{O}{\|}}{C}-OR''} + R'OH$$

(6) **By reaction of a Carboxylic acid with Diazomethane.** This method yields methyl esters only.

$$R-\overset{\overset{O}{\|}}{C}-OH + \underset{\text{Diazomethane}}{CH_2N_2} \longrightarrow R-\overset{\overset{O}{\|}}{C}-OCH_3 + N_2$$

PHYSICAL PROPERTIES

(1) Lower carboxylic esters are neutral colorless liquids possessing characteristic fruity odors.

(2) The boiling points of methyl and ethyl esters are much lower than their parent acid, since the former are unassociated liquids. Thus ethyl acetate boils at 77°C, while acetic acid boils at 118°C.

(3) They are generally insoluble in water and soluble in most organic solvents.

(4) Many esters are excellent solvents for other organic substance *e.g.*, ethyl acetate and amyl acetate.

(5) In the *IR Spectrum*, esters show characteristic C=O stretching frequency at 1735 cm^{-1}.

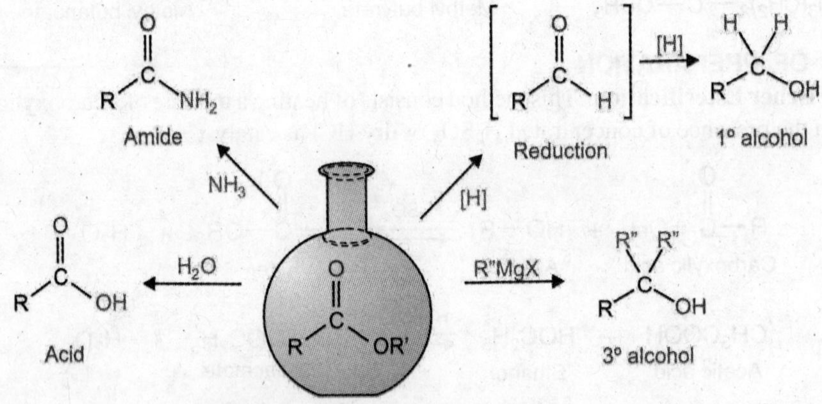

Important reactions of Esters.

CHEMICAL PROPERTIES

Esters gives nucleophilic substitution reactions similar to the acid halides and anhydrides. The carbonyl carbon in their molecule is electron-deficient owing to the electron-withdrawing effect of —OR' and carries a partial positive charge.

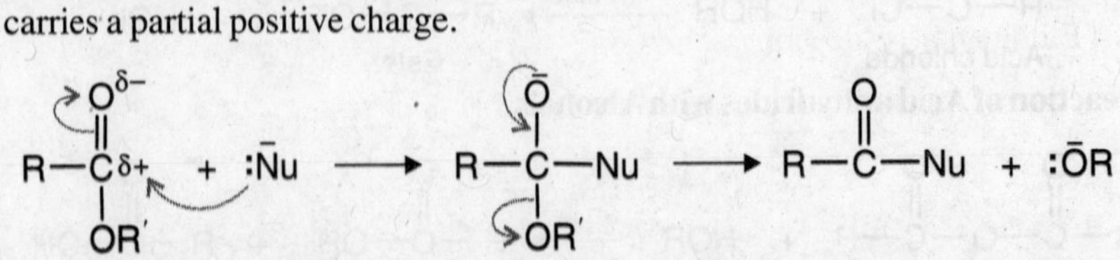

However, esters undergo nucleophilic reactions less readily than do acid halides or anhydrides. This is attributed to the resonance effect.

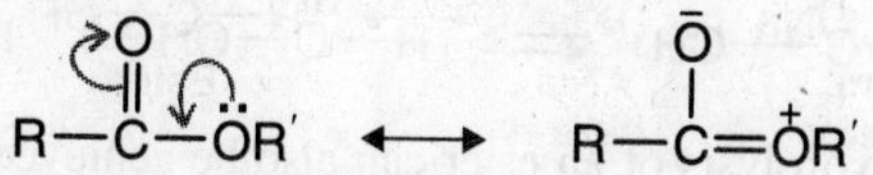

Thus the delocalization of unshared electron pair on $-\ddot{O}R$, decreases the electron-deficiency of carbonyl carbon which becomes less prone to nucleophilic attack.

The general reactions of esters are listed below :

(1) **Acid Hydrolysis.** Esters when heated with water in the presence of an acid catalyst (H_2SO_4 or HCl) are hydrolysed to give the parent carboxylic acid and alcohol. The reaction is reversible.

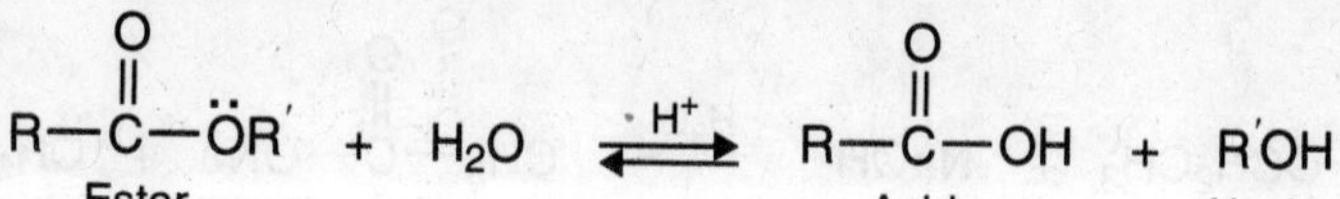

For example,

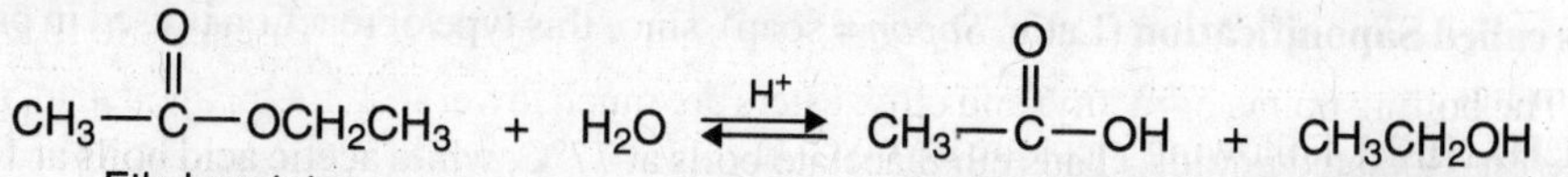

MECHANISM. Following steps are involved in the acid-hydrolysis of an ester :

Step 1. Attack by H^+ on oxygen atom of the carbonyl group.

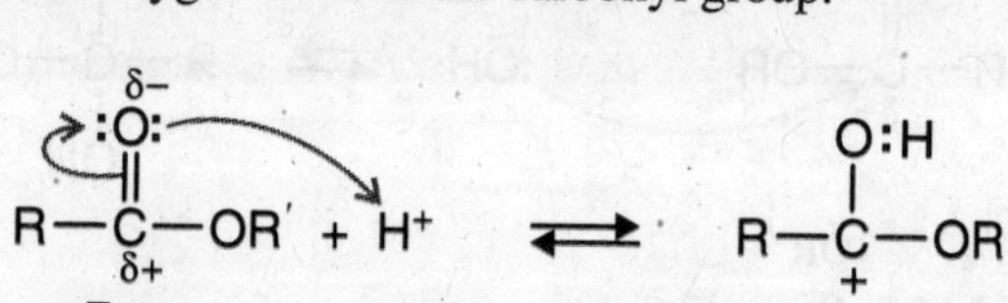

Step 2. Addition of water to electron-deficient carbon.

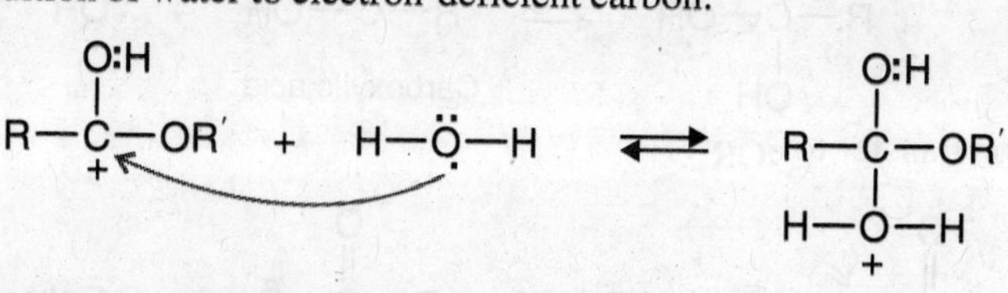

Step 3. Hydrogen ion transfer.

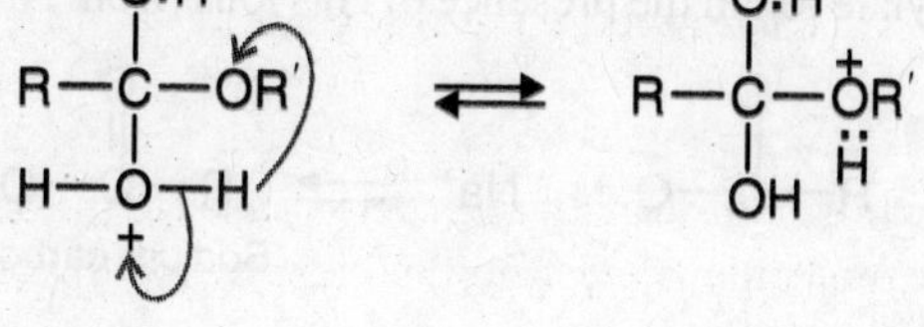

Step 4. Elimination of alcohol.

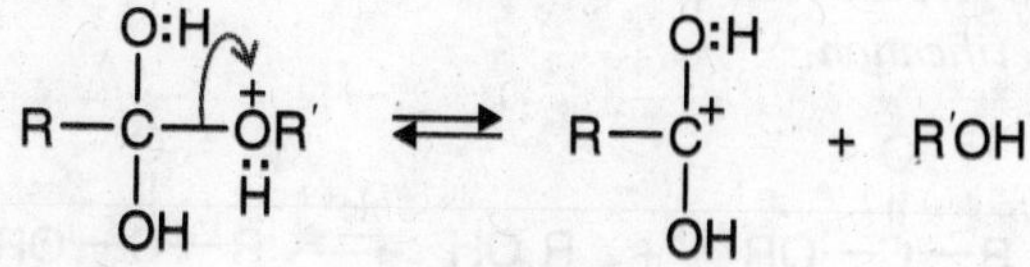

Step 5. Regeneration of the catalyst and formation of carboxylic acid.

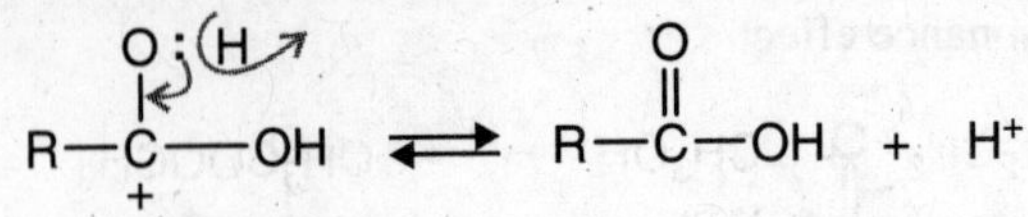

Basic-Hydrolysis. The hydrolysis of an ester can also be achieved by refluxing with aqueous solution of a strong base such as sodium hydroxide.

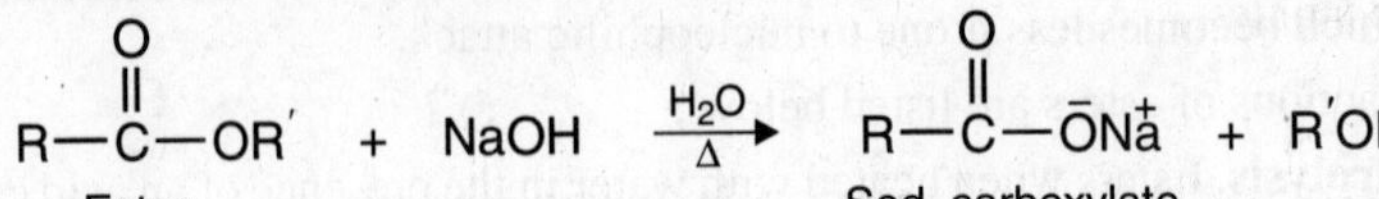

Ester — Sod. carboxylate

For example,

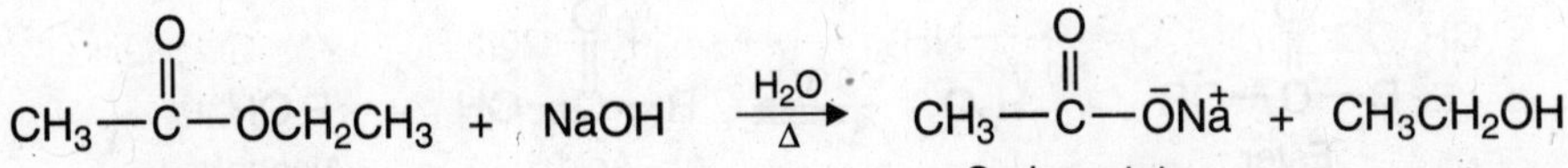

Ethyl acetate — Sod. acetate

The alkaline hydrolysis (basic-hydrolysis) of an ester to form sodium salt of the parent acid and alcohol is called Saponification (Latin, *Sapon* = soap), since this type of reaction is used in preparing soaps from fats.

MECHANISM. Following steps are involved :

Step 1. Attack by HO^- on electron-deficient carbonyl carbon.

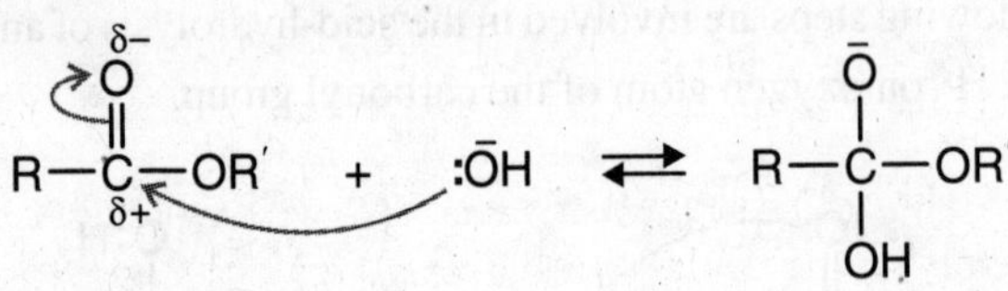

Step 2. Elimination of $:\bar{O}R'$

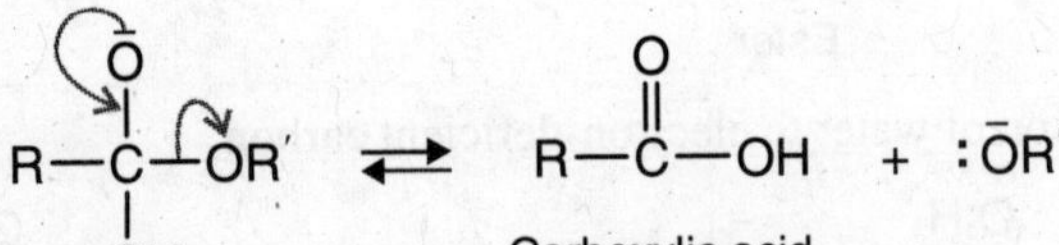

Carboxylic acid

Step 3. Proton transfer to $:\bar{O}R'$.

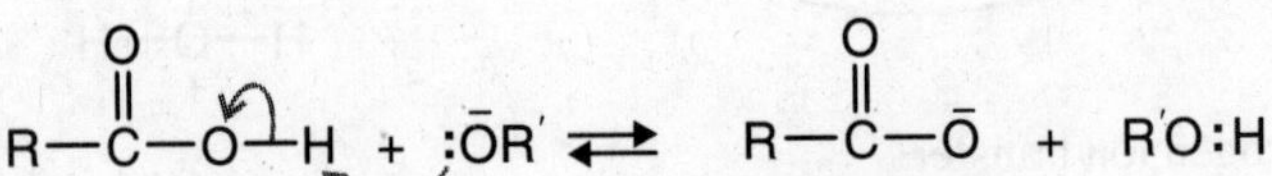

Carboxylate ion

Step 4. Carboxylate ion in the presence of Na^+ ions (from NaOH) gives sodium carboxylate.

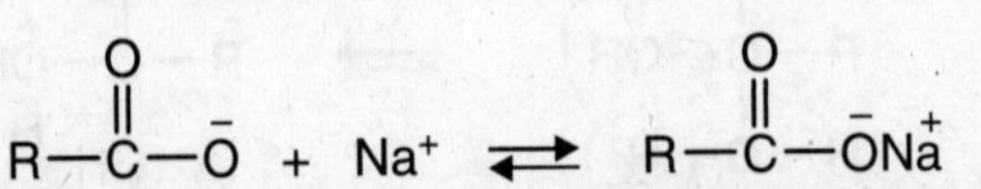

Sodium carboxylate

(2) **Transesterification.** Ester of an alcohol can react with another lower alcohol in the presence of a mineral acid, to give the ester of second alcohol. *The interchange of alcohol portions of the esters is termed transesterification.*

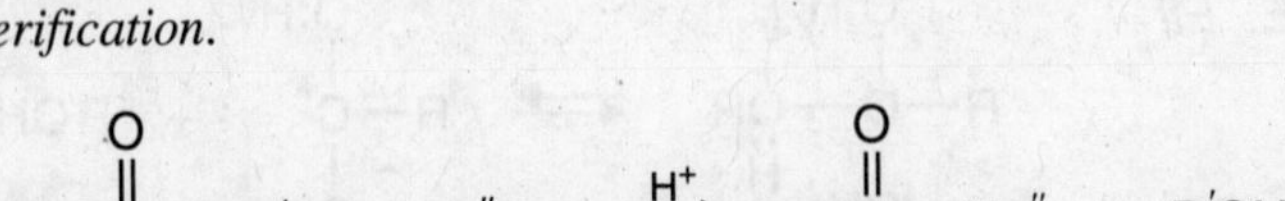

The lower alcohol being more volatile escapes and the reaction proceeds to completion.

For example,

$$\underset{\text{Ethyl acetate}}{CH_3COOCH_2CH_3} + \underset{\text{Methanol}}{CH_3OH} \xrightarrow{H^+} \underset{\text{Methyl acetate}}{CH_3COOCH_3} + \underset{\text{Ethanol}}{C_2H_5OH}$$

(3) **Reaction with Ammonia (*Ammonolysis*).** Esters react with ammonia, often in a solvent such as ethanol, to give amides.

$$\underset{\text{Ester}}{R{-}\overset{O}{\overset{\|}{C}}{-}OR'} + \underset{\text{Ammonia}}{H{-}NH_2} \longrightarrow \underset{\text{Amide}}{R{-}\overset{O}{\overset{\|}{C}}{-}NH_2} + \underset{\text{Alcohol}}{R'OH}$$

$$\underset{\text{Ethyl acetate}}{CH_3COOC_2H_5} + H{-}NH_2 \longrightarrow \underset{\text{Acetamide}}{CH_3{-}CONH_2} + \underset{\text{Ethanol}}{C_2H_5OH}$$

(4) **Reduction to Alcohols**. Esters can be reduced to the primary alcohol corresponding to the acid from which it was derived by lithium aluminium hydride, $LiAlH_4$, or by sodium and alcohol.

$$\underset{\text{Ester}}{R{-}\overset{O}{\overset{\|}{C}}{-}OR'} + 4[H] \xrightarrow[\text{ether}]{LiAlH_4} \underset{1^\circ\text{ alcohol}}{R{-}CH_2OH} + R'OH$$

$$\underset{\text{Methyl acetate}}{CH_3{-}CO{-}OCH_3} + 4[H] \xrightarrow[\text{ether}]{LiAlH_4} \underset{\text{Ethanol}}{CH_3{-}CH_2OH} + \underset{\text{Methanol}}{CH_3OH}$$

(5) **Reaction with Grignard Reagents**. Esters react with Grignard reagents to form ketones which at once react with another molecule of the Grignard reagent to yield a tertiary alcohol.

$$\underset{\text{Ester}}{R{-}\overset{O}{\overset{\|}{C}}{-}OR'} \xrightarrow{R''MgX} R{-}\underset{R''}{\underset{|}{\overset{OMgX}{\overset{|}{C}}}}{-}OR' \longrightarrow \underset{\text{Ketone}}{R{-}\overset{O}{\overset{\|}{C}}{-}R''} + XMgOR'$$

$$R{-}\overset{O}{\overset{\|}{C}}{-}R'' \xrightarrow{R''MgX} R{-}\underset{R''}{\underset{|}{\overset{OMgX}{\overset{|}{C}}}}{-}R'' \xrightarrow{H_2O} \underset{3^\circ\text{ alcohol}}{R{-}\underset{R''}{\underset{|}{\overset{OH}{\overset{|}{C}}}}{-}R''}$$

(6) **Claisen Condensation**. Esters that have hydrogen atoms on the α carbon atom undergo condensation reactions to form β-keto esters, in the presence of a strong base (C_2H_5ONa).

$$\underbrace{CH_3\overset{O}{\overset{\|}{C}}{-}OC_2H_5 + HCH_2\overset{O}{\overset{\|}{C}}OC_2H_5}_{\text{Ethyl acetate (2 molecules)}} \xrightarrow[-C_2H_5OH]{C_2H_5\bar{O}\overset{+}{Na}} \underset{\text{Ethyl acetoacetate}}{CH_3\overset{O}{\overset{\|}{C}}CH_2\overset{O}{\overset{\|}{C}}OC_2H_5}$$

ETHYL ACETATE, *Ethyl ethanoate*, $CH_3COOCH_2CH_3$

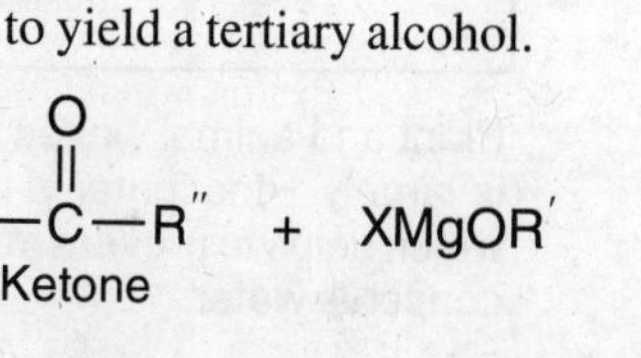

Ethyl acetate is a typical example of an ester of an organic acid.

Preparation. It is prepared in industry and in the laboratory by heating ethanol with glacial acetic acid in the presence of concentrated sulfuric acid.

$$\underset{\text{Acetic acid}}{CH_3COOH} + \underset{\text{Ethanol}}{C_2H_5OH} \xrightleftharpoons{H_2SO_4} \underset{\text{Ethyl acetate}}{CH_3COOC_2H_5} + H_2O$$

Properties. Ethyl acetate is a colorless liquid, bp 77°C, having a pleasant fruity odor. It is sparingly soluble in water but freely soluble in chloroform, alcohol and ether. It is a neutral nontoxic liquid. It gives all the general reactions of esters described before.

Uses. Ethyl acetate is used : (1) as a solvent in industry, notably for lacquers and resins; (2) in artificial fruit flavors ; and (3) in organic synthesis *e.g.*, for making ethyl acetoacetate.

Plant and animal waxes are high-molecular-weight esters. Beeswax, for example, is largely 1-triacontanyl palmitate. The leaves of most plants are coated with wax, which helps to prevent microorganisms from attacking them and also allows them to conserve water.

WHAT ARE WAXES ?

They are esters of long-chain fatty acids and long-chain monohydric alcohols, each containing 16 to 34 carbon atoms. They are produced by plants and animals, and melt between 35–100°C. They have a satiny 'waxy' feel and are insoluble in water, but soluble in organic solvents. In general they are found in nature as mixtures of esters and besides they also contain some free acids and alcohols.

Beeswax, obtained from the honeycomb of the bee, is largely *Myricyl Palmitate*, $C_{15}H_{13}COOC_{30}H_{61}$. It melts between 62-65°C and is used in preparing boot polishes, cosmetic creams, lipsticks, floor and furniture polishes, and candles.

Carnauba Wax, derived from the leaves of Carnauba Palm native in Brazil, is mainly *Myricyl Cerotate*, $C_{15}H_{51}COOC_{30}H_{61}$. It melts between 80–87°C and is used in polishes, as a coating on mimeograph stencils, and carbon papers.

Spermaceti Wax, obtained from head-cavity of the sperm whale, consists mainly of *Cetyl Palmitate*, $C_{15}H_{31}COOC_{16}H_{33}$. It melts between 42–50°C and is primarily used as emollient (skin-soother) in ointments and cosmetics. It is also used as laundry wax for lustering linen, making soap and candles.

UREA, *Carbamide*, $H_2N–CO–NH_2$

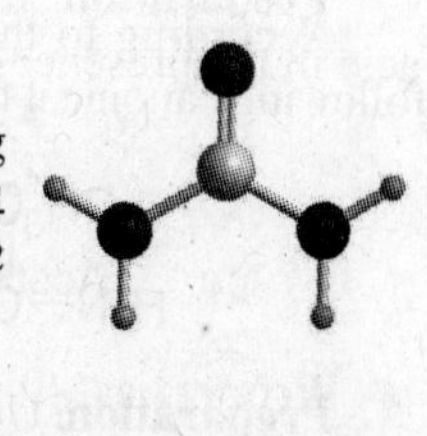

It is the normal end product of human metabolism of nitrogen-containing foods (proteins). An adult person excretes about 30 grams of urea in urine in 24 hours. Urea is the diamide of carbonic acid. The carbon atom in urea is sp^2 hybridized.

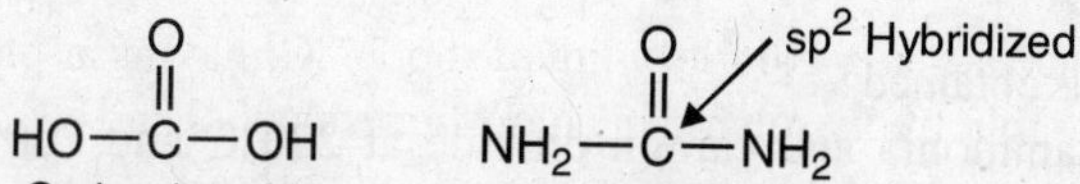

Carbonic acid   Urea (diamide)

Structure. In urea the two nitrogen atoms, the oxygen atom, and the carbon atom are all sp^2 hybridized. These sp^2 hybrid orbitals overlap with each other and with *s* orbitals of the four hydrogen atoms forming C—O, C—N, and N—H σ bonds as shown in Fig.23.1 These σ bonds lie in one plane and are approximately 120° apart.

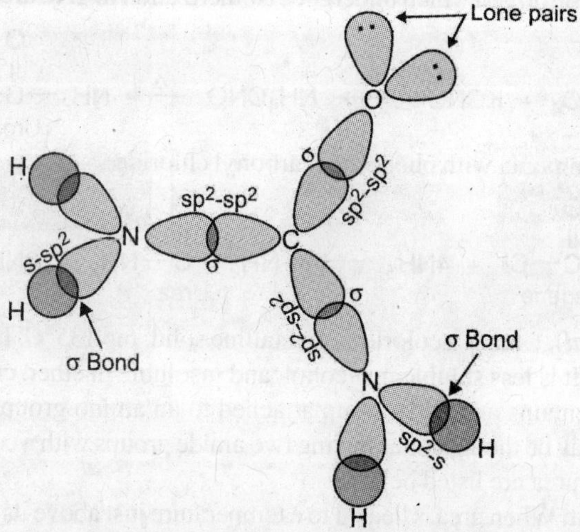

Fig. 23.1. Formation of σ bonds in urea.

In addition to the sp^2 orbitals, each atom in urea also possesses an unhybridized *p* orbital which is perpendicular to the plane of σ bonds. The *p* orbital on each nitrogen atom contains a pair of electrons, the *p* orbital on carbon contains one electron, and the *p* orbital on oxygen also contains one electron. These *p* orbitals laterally overlap forming a delocalized π molecular orbital above and below the plane of the σ bonds (Fig.23.2).

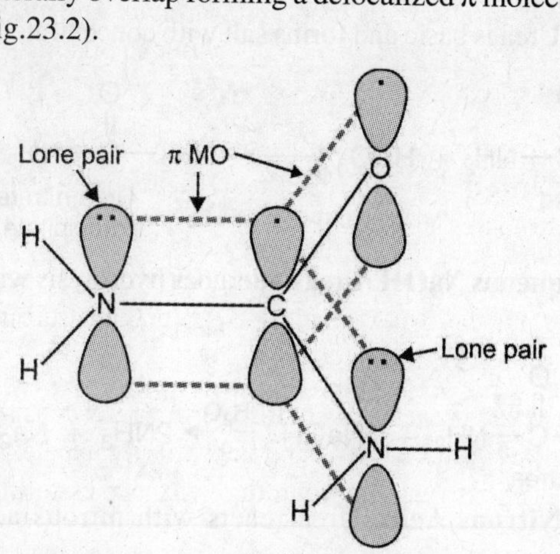

Fig. 23.2. Formation of delocalized π molecular orbital in urea molecule.

According to the Valence Bond Theory, urea is considered to be a *resonance hybrid* of the following canonical forms.

$$H_2N-\overset{\overset{O}{\|}}{C}-NH_2 \longleftrightarrow H_2\overset{+}{N}=\overset{\overset{\bar{O}}{|}}{C}-NH_2 \longleftrightarrow H_2N-\overset{\overset{\bar{O}}{|}}{C}=\overset{+}{N}H_2$$

Preparation. Urea is obtained :

(1) By heating ammonia and carbon dioxide at 200°C and 200 atmospheres pressure. (*Commercial Method*)

$$2NH_3 + CO_2 \xrightarrow[200\ atm.]{200°C} \underset{Urea}{NH_2-\overset{\overset{O}{\|}}{C}-NH_2} + H_2O$$

(2) By evaporating an aqueous solution containing ammonium sulfate and potassium cyanate. Ammonium cyanate is first formed which undergoes isomerization to give urea. (*Laboratory Method*)

$$(NH_3)_2SO_4 + KCNO \longrightarrow NH_4CNO \xrightarrow{\Delta} \underset{Urea}{NH_2-\overset{\overset{O}{\|}}{C}-NH_2}$$

(3) By treating ammonia with phosgene (carbonyl chloride).

$$\underset{Phosgene}{Cl-\overset{\overset{O}{\|}}{C}-Cl} + 4NH_3 \longrightarrow \underset{Urea}{NH_2-\overset{\overset{O}{\|}}{C}-NH_2} + 2NH_4Cl$$

Properties. (*Physical*). Urea is a colorless, crystalline solid, mp 135°C. It is very soluble in water giving neutral solutions. It is less soluble in alcohol, and insoluble in ether, chloroform and benzene.

(*Chemical*). Urea contains an amide group attached to an amino group, and gives reactions of both these functions. It may be thought of as having two amide groups with a common –CO– function. Some of the reactions of urea are listed below.

(1) **Effect of Heat.** When urea is heated to a temperature just above its melting point (132°C) it loses ammonia and biuret is formed. This reaction is the basis of the well-known **Biuret Test.**

$$\underset{Urea}{NH_2-\overset{\overset{O}{\|}}{C}-NH-H} + \underset{Urea}{NH_2-\overset{\overset{O}{\|}}{C}-NH_2} \xrightarrow{\Delta} \underset{Biuret}{NH_2-\overset{\overset{O}{\|}}{C}-NH-\overset{\overset{O}{\|}}{C}-NH_2} + NH_3$$

(2) **Salt Formation.** Urea is basic and forms salt with concentrated nitric acid.

$$\underset{Urea}{NH_2-\overset{\overset{O}{\|}}{C}-NH_2} + HNO_3 \longrightarrow \underset{\substack{Urea\ nitrate \\ (precipitate)}}{NH_2-\overset{\overset{O}{\|}}{C}-NH_2.HNO_3}$$

(3) **Reaction with Aqueous NaOH.** Urea undergoes hydrolysis with aqueous NaOH solution to give ammonia.

$$\underset{Urea}{NH_2-\overset{\overset{O}{\|}}{C}-NH_2} + 2NaOH \xrightarrow[\Delta]{H_2O} 2NH_3 + Na_2CO_3$$

(4) **Reaction with Nitrous Acid.** Urea reacts with nitrous acid to form nitrogen and carbon dioxide.

$$\underset{\text{Urea}}{NH_2-\overset{O}{\overset{\|}{C}}-NH_2} + 2HNO_2 \longrightarrow 2N_2 + CO_2 + 3H_2O$$

(5) **Reaction with Alkaline Hypohalites.** Urea reacts with an alkaline solution of sodium hypobromite ($Br_2 + NaOH$) to give nitrogen. This reaction is the basis of the clinical estimation of urea in urine.

$$\underset{\text{Urea}}{NH_2-\overset{O}{\overset{\|}{C}}-NH_2} + 3NaOBr + 2NaOH \longrightarrow 3NaBr + Na_2CO_3 + N_2 + 3H_2O$$

(6) **Reaction with Hydrazine.** Urea reacts with hydrazine at 100°C to form semicarbazide.

$$\underset{\text{Urea}}{NH_2-\overset{O}{\overset{\|}{C}}-NH_2} + H-\overset{H}{\overset{|}{N}}-NH_2 \xrightarrow{\Delta} \underset{\text{Semicarbazide}}{NH_2-\overset{O}{\overset{\|}{C}}-\overset{H}{\overset{|}{N}}-NH_2} + NH_3$$

(7) **Reaction with Malonic Ester.** Urea reacts with malonic ester in the presence of sodium ethoxide to give barbituric acid.

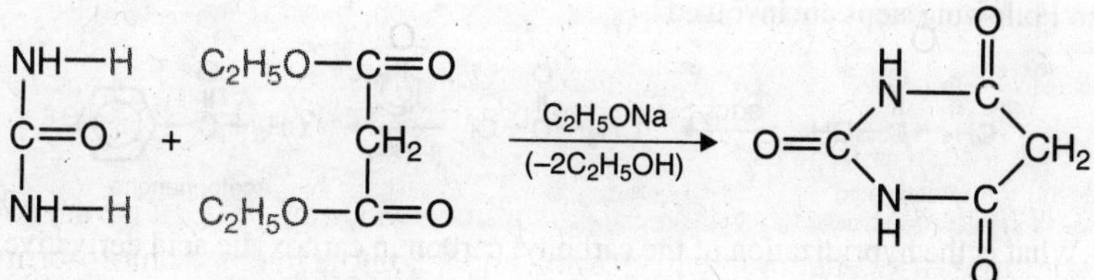

Urea Malonic ester Barbituric acid

(8) **Reaction with Formaldehyde.** Urea reacts with formaldehyde in the presence of an acid or base to form urea-formaldehyde resins.

$$\underset{\text{Urea}}{NH_2-\overset{O}{\overset{\|}{C}}-NH_2} + n\underset{\text{Formaldehyde}}{H-\overset{O}{\overset{\|}{C}}-H} \longrightarrow \underset{\substack{\text{Urea-formaldehyde resin}\\ \text{(Condensation polymer)}}}{\left[NH-\overset{O}{\overset{\|}{C}}-NH-CH_2\right]_n} + nH_2O$$

(9) **Dehydration of Urea.** Urea undergoes dehydration to yield melamine.

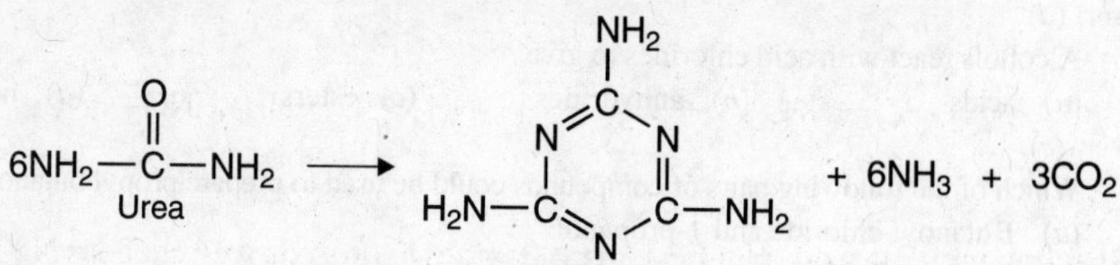

Melamine

Melamine is primarily used in the production of melamine-formaldehyde resins which have much greater hardness than urea-formaldehyde resins. Both melamine-formaldehyde and urea-formaldehyde resins have very varied uses including adhesives, laminates, moulding compounds, coatings, and textile finishes.

Uses. Urea is used : (1) chiefly as fertilizer ; (2) as animal feed ; (3) for making barbiturates; (4) for manufacture of urea-formaldehyde resins ; and (5) as stabilizer for explosives (nitrocellulose).

Examination Questions

I. ACID HALIDES

1. How are acid chlorides prepared ? Give their important reactions.
2. What happens when acetyl chloride is treated with ammonia ?

Answer. Acetamide is formed. For reaction equation see text.

3. Write a note on : Rosenmund reduction. *(Delhi BSc, 2015)*
4. Write a mechanism for the reaction of acetyl chloride with water. *(Nehu BSc, 2014 ; Indore BSc, 2014 ; Baroda BSc, 2015)*
5. How will you synthesize acetyl chloride from acetic acid ?
6. How will you synthesize acetyl chloride from methyl chloride ?

Answer. Following steps are involved :

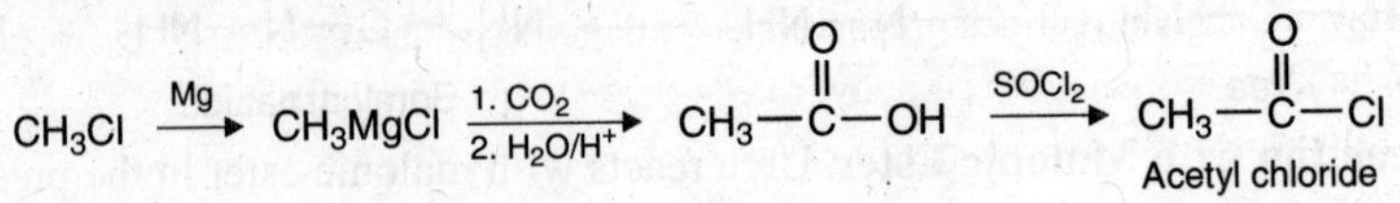

7. How will you synthesize acetophenone from acetic acid ?

Answer. Following steps are involved :

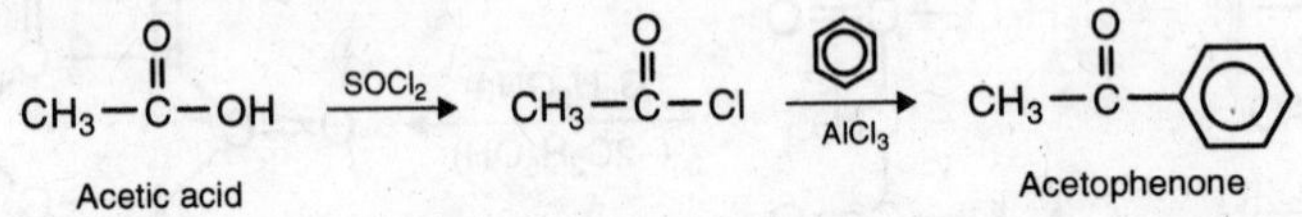

8. What is the hybridization of the carbonyl carbon in carboxylic acid derivatives ?
   (*a*) sp^3 (*b*) sp (*c*) sp^2 (*d*) p

Answer. (*c*)

9. Which of the following statements is False about acetyl chloride ? *(Agra BSc, 2014)*
   (*a*) undergoes hydrolysis to give formic acid
   (*b*) reacts with NH_3 to give acetamide
   (*c*) reacts with alcohols to form esters
   (*d*) can be prepared by treatment of acetic acid by PCl_5

Answer. (*a*)

10. Acid chlorides react with
    (*a*) alcohols (*b*) amines
    (*c*) carboxylic acids (*d*) all of these

Answer. (*d*)

11. Alcohols react with acid chlorides to give
    (*a*) acids (*b*) anhydrides (*c*) esters (*d*) none of these

Answer. (*c*)

12. Which of the following pairs of compounds could be used to prepare propyl butanoate ?
    (*a*) Butanoyl chloride and 1-propanol
    (*b*) Propanoyl chloride and 1-butanol
    (*c*) Butanoyl chloride and 1-pentanol
    (*d*) Butanoyl chloride and phenol

Answer. (*a*)

13. Acid chlorides are reduced with lithium aluminium hydride to give
    (*a*) carboxylic acids (*b*) alcohols
    (*c*) ketones (*d*) aldehydes

Answer. (*b*)

14. Amides are commonly prepared by reacting acyl halides, esters, or anhydrides with ?

(*a*) ammonium hydroxide (*b*) ammonium chloride

(*c*) ammonia (*d*) tertiary amines

Answer. (*c*)

15. An amine reacts with an acid chloride to give

(*a*) an acid (*b*) an ester (*c*) a lactone (*d*) an amide

Answer. (*d*)

16. Acid chlorides undergo reduction to the corresponding aldehydes on treatment with hydrogen in the presence of $Pd/BaSO_4$. This reaction is called :

(*a*) Clemmensen reduction (*b*) Rosenmund reduction

(*c*) Wolf-Kishner reduction (*d*) None of these

Answer. (*b*)

II. ACID ANHYDRIDES

1. How are acid anhydrides prepared ? Give their important reactions.

2. What happens when acetic anhydride is treated with ethyl alcohol ?

Answer. Ethyl acetate and acetic acid are formed. For reaction equation see text.

3. How will you synthesize acetic anhydride from ethyl alcohol ? (*Jiwaji BSc, 2015*)

Answer. Following steps are involved :

Step 1. Conversion of ethyl alcohol into acetic acid.

$$CH_3CH_2OH + [O] \xrightarrow{K_2Cr_2O_7/H^+} CH_3-\overset{O}{\overset{\|}{C}}-OH$$

Step 2. Conversion of acetic acid into acetic anhydride.

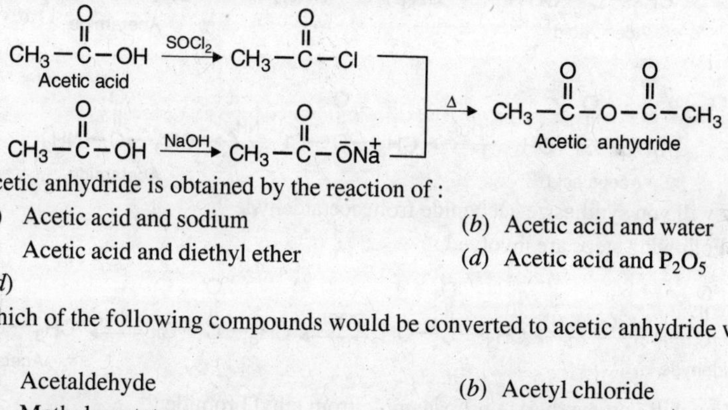

4. Acetic anhydride is obtained by the reaction of :

(*a*) Acetic acid and sodium (*b*) Acetic acid and water

(*c*) Acetic acid and diethyl ether (*d*) Acetic acid and P_2O_5

Answer. (*d*)

5. Which of the following compounds would be converted to acetic anhydride when treated with sodium acetate ?

(*a*) Acetaldehyde (*b*) Acetyl chloride

(*c*) Methyl acetate (*d*) Acetamide

Answer. (*b*)

6. Acetic acid reacts with propionic acid to give :

(*a*) Acetic anhydride (*b*) Propionic anhydride

(*c*) Acetic propionic anhydride (*d*) All of these

Answer. (*d*)

7. Which reagents will convert propanoic acid to propanoic anhydride ?

(*a*) H_2O, H_3O^+ (*b*) $SOCl_2$, then CH_3CH_2COONa

(*c*) $CH_3CH_2CH_2MgBr$, then H_3O^+ (*d*) All of the above

Answer. (*b*)

8. Anhydrides react with water to give

(*a*) acids (*b*) esters (*c*) amides (*d*) none of these

Answer. (*a*)

III. Amides

1. How are amides prepared ? Give their important reactions.
2. What happens when acetamide is treated with phosphorus pentoxide ?

(Delhi BSc, 2013 ; Indore BSc, 2015)

Answer. Acetonitrile is formed.

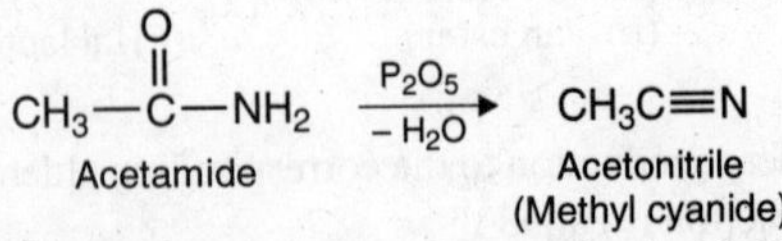

3. What happens when acetamide is treated with bromine in NaOH solution ? *(Kerala BSc, 2015)*

Answer. Methylamine is formed.

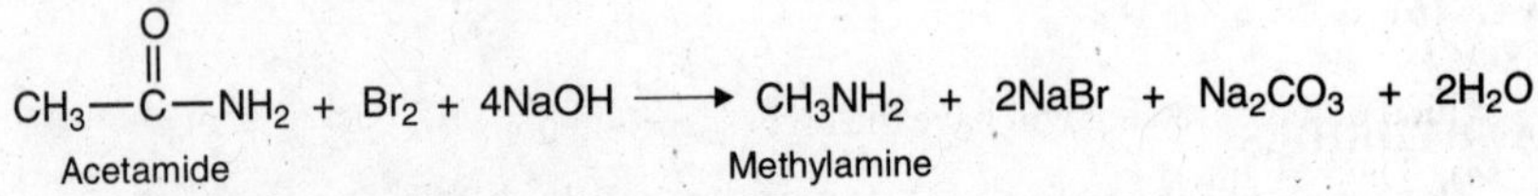

4. Write resonance structures of acetamide.
5. Explain why amides are much weaker bases than aliphatic amines. *(Pune BSc, 2014)*
6. What is Hofmann's degradation of amides ?
7. How will you synthesize methylamine from acetamide ?
8. How will you synthesize acetamide from acetic acid ? *(Bundelkhand BSc, 2015)*

Answer. There are two methods :

Method 1

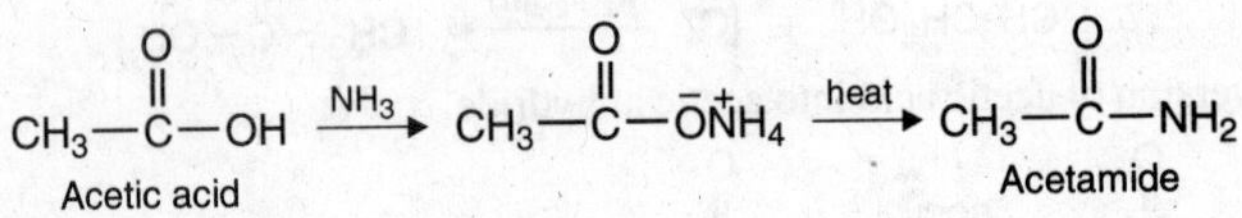

Method 2

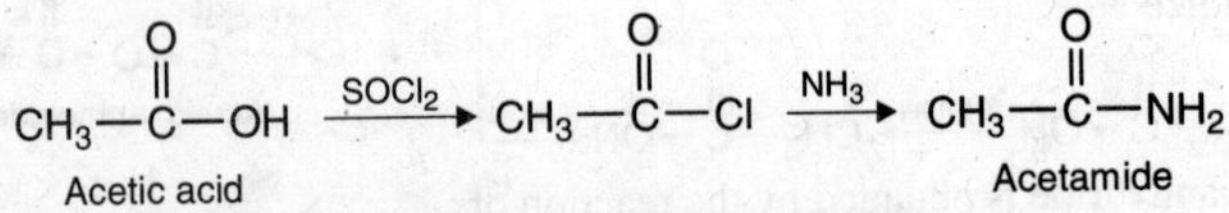

9. How will you synthesize acetamide from acetaldehyde ?

Answer. Following steps are involved :

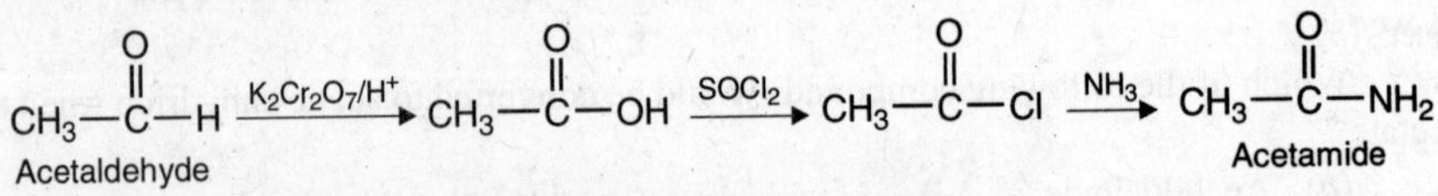

10. How will you synthesize propanamide from ethyl bromide ?

Clue. Steps are : *(i)* NaCN ; *(ii)* H_2O/H^+ ; *(iii)* $SOCl_2$; and *(iv)* NH_3.

11. Acetamide on heating with P_2O_5 gives :

(a) Methylamine *(b)* Ethylamine
(c) Methyl cyanide *(d)* None of these

Answer. *(c)*

12. Acetamide is a much weaker base than ammonia. This is because

(a) the electron withdrawing effect of the C=O group makes the lone pair of electrons on nitrogen atom less available for protonation.
(b) the electron withdrawing effect of the C=O group makes the lone pair of electrons on nitrogen atom more available for protonation.
(c) the presence of CH_3 group sterically hinders the protonation of the NH_2 group.
(d) an H atom from the nitrogen is less easily lost from CH_3CONH_2 than from NH_3.

Answer. *(a)*

13. Which of the following statements is *False* about acetamide ?

(*a*) reacts with $Br_2/NaOH$ to form tertiary amine
(*b*) undergoes acid-hydrolysis to form acetic acid
(*c*) undergoes dehydration with P_2O_5 to form a nitrile
(*d*) is a white crystalline solid, soluble in water

Answer. (*a*)

14. The C—C—O bond angle in $CH_3-\overset{\overset{\large O}{\|}}{C}-NH_2$ is approximately

(*a*) 60° (*b*) 109.5°
(*c*) 90° (*d*) 120°

Answer. (*d*)

15. Which of the following will have the highest boiling point ?

(*a*) Diethyl ether (*b*) *n*-Butane
(*c*) Acetamide (*d*) *n*-Propylamine

Answer. (*c*)

16. Acetamide can be converted into methylamine by

(*a*) P_2O_5 (*b*) $Br_2/NaOH$
(*c*) PCl_5 (*d*) $LiAlH_4$

Answer. (*b*)

17. Acetamide on reduction with $LiAlH_4$ gives

(*a*) Acetaldehyde (*b*) Ethylamine
(*c*) Acetone (*d*) Methylamine

Answer. (*b*)

18. By which of the following methods can CH_3CONH_2 be converted to CH_3CN ?

(*a*) Oxidize with $Na_2Cr_2O_7/H^+$ (*b*) Heat with concentrated sulfuric acid
(*c*) Heat to 200°C (*d*) Warm with P_2O_5

Answer. (*d*)

19. Which of the following reagents will react with acetamide to form methylamine ?

(*a*) $LiAlH_4$ (*b*) $Br_2/NaOH$
(*c*) PCl_5 (*d*) H_2/Ni

Answer. (*b*)

IV. Esters

1. How are esters prepared ? Give their important reactions.

2. What happens when ethyl acetate is treated with ammonia ?

Answer. Acetamide is formed.

$$\underset{\text{Ethyl acetate}}{CH_3COOC_2H_5} + H{-}NH_2 \longrightarrow \underset{\text{Acetamide}}{CH_3{-}CONH_2} + \underset{\text{Ethanol}}{C_2H_5OH}$$

3. Explain why the boiling point of ethyl acetate is lower than that of butyric acid even though the molecular weights of the two compounds are the same. (*Utkal BSc, 2015*)

Answer. Butyric acid has a much higher boiling point than ethyl acetate because in the acid there are strong intermolecular forces of attraction (Hydrogen bonds).

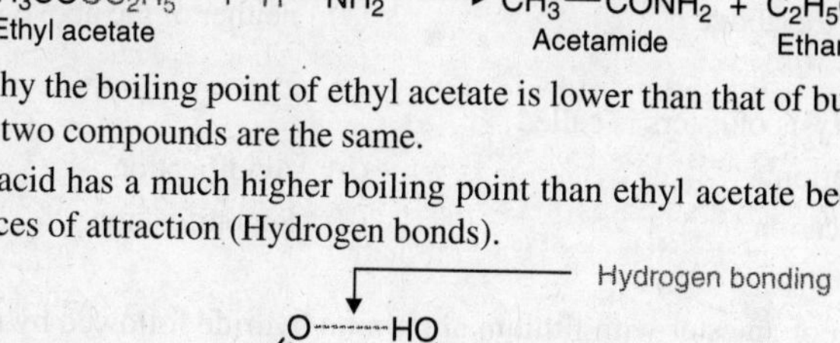

No hydrogen bonds are possible among molecules of the ester.

4. How will you synthesize ethyl acetate from ethyl alcohol ? *(Calicut BSc, 2014)*

Answer. Following steps are involved :

$$\underset{\text{Ethyl alcohol}}{CH_3CH_2OH} \xrightarrow{K_2Cr_2O_7/H^+} CH_3-\overset{\overset{\displaystyle O}{\|}}{C}-OH \xrightarrow{CH_3CH_2OH/H^+} \underset{\text{Ethyl acetate}}{CH_3-\overset{\overset{\displaystyle O}{\|}}{C}-OCH_2CH_3}$$

5. How will you synthesize ethyl acetate from acetic acid ? *(Baroda BSc, 2015)*

Answer. Following steps are involved :

$$\underset{\text{Acetic acid}}{CH_3-\overset{\overset{\displaystyle O}{\|}}{C}-OH} \xrightarrow{LiAlH_4} CH_3CH_2OH \xrightarrow{CH_3COOH/H^+} \underset{\text{Ethyl acetate}}{CH_3-\overset{\overset{\displaystyle O}{\|}}{C}-OCH_2CH_3}$$

6. How will you synthesize ethyl acetate from ethyl bromide ?

Clue. Steps are : (*i*) Aqueous KOH ; (*ii*) $K_2Cr_2O_7/H^+$; and (*iii*) CH_3CH_2OH/H^+.

7. Discuss the mechanism of Esterification. *(Madras BSc, 2014)*
8. What is Saponification ? Give its mechanism.
9. Discuss the mechanism of basic hydrolysis of an ester. *(Utkal BSc, 2015)*
10. Esters can be formed by nucleophilic substitution reactions of :
    (*a*) carboxylic acids and alcohols ; acid-catalyzed
    (*b*) acid chlorides and alcohols
    (*c*) both of the above
    (*d*) neither of the above

Answer. (*c*)

11. The conversion $CH_3-\overset{\overset{\displaystyle O}{\|}}{C}-OCH_3 \longrightarrow CH_3CH_2OH$ is called :
    (*a*) Oxidation (*b*) Dehydration
    (*c*) Reduction (*d*) Hydrolysis

Answer. (*c*)

12. The following reaction is called :

$$CH_3CH_2-\overset{\overset{\displaystyle O}{\|}}{C}-OCH_2CH_3 \xrightarrow[H_2O]{NaOH} CH_3CH_2-\overset{\overset{\displaystyle O}{\|}}{C}-\overset{-}{O}\overset{+}{Na} + CH_2CH_2OH$$

(*a*) Saponification (*b*) Condensation
(*c*) Elimination (*d*) Esterification

Answer. (*a*)

13. Hydrolysis of an ester can be accomplished by :
    (*a*) base-promoted hydrolysis (*b*) acid-catalyzed hydrolysis
    (*c*) both of the above (*d*) neither of the above

Answer. (*c*)

14. Basic-hydrolysis of esters is called
    (*a*) Acetylation (*b*) Acidification
    (*c*) Esterification (*d*) Saponification

Answer. (*d*)

15. The reaction of an ester with lithium aluminum hydride followed by hydrolysis yields:
    (*a*) A 1° alcohol (*b*) A 2° alcohol
    (*c*) A 3° alcohol (*d*) An ether

Answer. (*a*)

16. Ethyl acetate is hydrolyzed by water to give

(*a*) a lactone (*b*) an ester

(*c*) an anhydride (*d*) a carboxylic acid and an alcohol

Answer. (*d*)

17. The reaction of an ester with a Grignard reagent followed by hydrolysis yields:

(*a*) A 1° alcohol (*b*) A 2° alcohol

(*c*) A 3° alcohol (*d*) An ether

Answer. (*c*)

18. What is the best choice of reagent(s) to achieve the following conversion ?

$$CH_3CH_2CH_2{-}\overset{\overset{\displaystyle O}{\|}}{C}{-}OCH_3 \xrightarrow{?} CH_3CH_2CH_2CH_2{-}OH$$

(*a*) $LiAlH_4$, ether; followed by H_3O^+ (*b*) NaOH, H_2O

(*c*) H_2O (*d*) H_2SO4

Answer. (*a*)

19. Which products result from the reaction $CH_3CH_2COOCH_3$ with H_2O, NaOH ?

(*a*) $CH_3CH_2COOH + HOCH_3$

(*b*) $CH_3CH_2COO^-Na^+ + HOCH_3$

(*c*) $CH_3CH_2CH_2OH + HOCH_3$

(*d*) Propanoic anhydride + $NaOCH_3$

Answer. (*b*). $CH_3CH_2COO^-Na^+ + HOCH_3$. This is saponification, also called alkaline ester hydrolysis. The carboxylate is often shown without the sodium cation: $CH_3CH_2COO^-$.

20. An unknown ester, $C_5H_{10}O_2$, was hydrolysed with dilute HCl to produce an acid (A) and an alcohol (B). Treatment of (B) with PBr_3 gave an alkyl bromide (C). When (C) was treated with KCN, a product (D) was formed that, on hydrolysis with dilute HCl, gave acid (A). Give the structure and name of the original ester. Identify (A) through (D) and write equations for the reactions involved.

Answer. The unknown ester is *Ethyl propionate*.

Reactions Involved Are :

$$\underset{\text{Ethyl propionate}}{CH_3CH_2{-}\overset{\overset{\displaystyle O}{\|}}{C}{-}OCH_2CH_3} \xrightarrow{H_2O/H^+} \underset{\substack{\text{Propionic acid}\\ \text{(A)}}}{CH_3CH_2{-}\overset{\overset{\displaystyle O}{\|}}{C}{-}OH} + \underset{\substack{\text{Ethyl alcohol}\\ \text{(B)}}}{CH_3CH_2OH}$$

$$\underset{\substack{\text{Ethyl alcohol}\\ \text{(B)}}}{CH_3CH_2OH} \xrightarrow{PBr_3} \underset{\substack{\text{Ethyl bromide}\\ \text{(C)}}}{CH_3CH_2Br} \xrightarrow{KCN} \underset{\substack{\text{Propionitrile}\\ \text{(D)}}}{CH_3CH_2CN} \xrightarrow{H_2O/H^+} \underset{\substack{\text{Propionic acid}\\ \text{(A)}}}{CH_3CH_2COOH}$$

V. Urea

1. How is urea prepared ? Describe its important reactions.

2. Discuss the molecular orbital structure of urea. (*Madurai BSc, 2015*)

3. What happens when urea is

(*a*) heated alone (*b*) treated with nitrous acid

(*c*) treated with hydrazine

Answer. (*a*) Biuret is formed ; (*b*) nitrogen and carbon dioxide are formed ; and (*c*) semicarbazide is formed. For reaction equations see text.

4. Explain why urea is basic. (*Baroda BSc, 2014 ; Anna BSc, 2015*)

Answer. Urea is a hybrid of the following resonance structures :

$$H_2N{-}\overset{\overset{\displaystyle O}{\|}}{C}{-}NH_2 \longleftrightarrow H_2\overset{+}{N}{=}\overset{\overset{\displaystyle \bar{O}}{|}}{C}{-}NH_2 \longleftrightarrow H_2N{-}\overset{\overset{\displaystyle \bar{O}}{|}}{C}{=}\overset{+}{N}H_2$$

Urea is basic. This is because the oxygen atom carries a negative charge (See the resonance structures) and can combine with a proton to form a new O–H bond. The positive ion formed when oxygen takes up a proton is also stabilized by resonance.

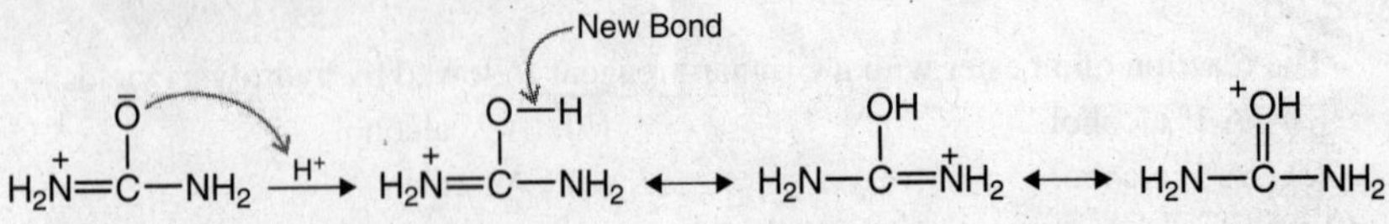

5. Write a note on : Biuret Test *(Kalyani BSc, 2014)*

Answer. This test is used to identify urea in the laboratory.

Laboratory Procedure. Warm a few crystals of urea gently in a dry test-tube. When the crystals melt, continue warming until the liquid solidifies owing to much higher melting point of Biuret. The ammonia given off during the heating can be readily detected by its smell. Allow the tube to cool, and dissolve the remaining solid by warming it with a little dilute NaOH solution. Now add to the tube one drop of very dilute copper sulfate solution. The liquid turns pink.

Reaction Equation

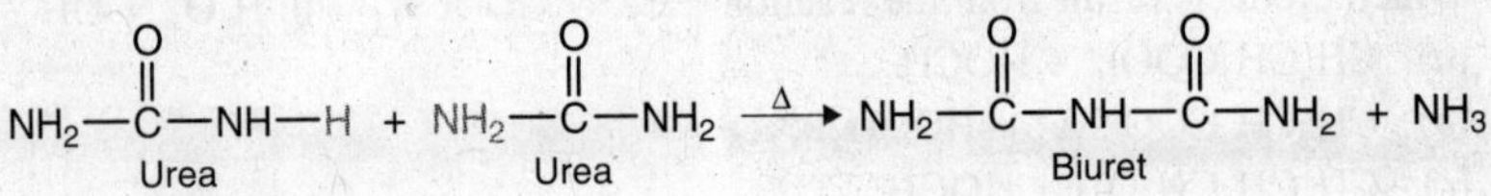

Comments. Biuret test is also given by proteins and peptides which contain –NH–CO–groups.

6. The structure of urea is shown below. The carbon atom in urea is :

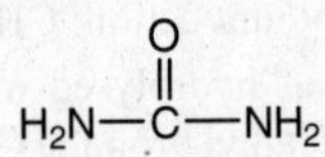

(*a*) *sp*^{3} hybridized (*b*) *sp*^{2} hybridized

(*c*) *sp* hybridized (*d*) not hybridized

Answer. (*b*)

7. Carbonyl chloride reacts with ammonia to give

(*a*) Urea (*b*) Acetone

(*c*) Acetamide (*d*) Chloroform

Answer. (*a*)

8. Which of the following is not a constituent of normal urine ?

(*a*) Albumin (*b*) Urea

(*c*) Sodium chloride (*d*) Uric acid

Answer. (*a*)

24

Nitriles, Isonitriles, Nitroalkanes and Diazomethane

Water-proof awning fabric is made from polyacrylonitrile which in-turn is obtained from acrylonitrile monomers.

NITRILES (CYANIDES)

Nitriles have the general formula:

$$R{-}C{\equiv}N \quad \text{or} \quad RCN$$

Nitriles are named as follows:

(1) **Cyanide System.** In this system, they are named as **Alkyl cyanides.**

CH_3CN	CH_3CH_2CN	$CH_3{-}\underset{}{\overset{CH_3}{\overset{\vert}{C}}}H{-}CN$
Methyl cyanide	Ethyl cyanide	Isopropyl cyanide

(2) **Common System.** Their common names are derived from the *common names* of the acids to which they are hydrolyzed. The ending *–ic acid* of the corresponding acid is dropped and the suffix *–onitrile* is added. For example, CH_3CN upon hydrolysis yields acetic acid,

$$CH_3{-}C{\equiv}N + H_2O \xrightarrow{H^+} \underset{\text{Acetic acid}}{CH_3{-}COOH} + NH_3$$

and is, therefore, named as acetonitrile.

Acetic acid – ic acid + onitrile = Acetonitrile

(3) **IUPAC System.** In this system, the longest carbon chain containing the cyan group (CN) determines the parent name. The IUPAC name of an individual nitrile is derived by adding the name of the parent hydrocarbon on the suffix *-nitrile*. The name is written as *one word.* Thus CH_3CN is named as ethanenitrile, the parent hydrocarbon being ethane. While naming substituted nitriles, the carbon atom triply bonded to the nitrogen is assigned number 1.

Structure	Cyanide Name	Common Name	IUPAC Name
CH_3-CN	Methyl cyanide	Acetonitrile	Ethanenitrile
CH_3CH_2-CN	Ethyl cyanide	Propiononitrile	Propanenitrile
$CH_3CH_2CH_2-CN$	Propyl cyanide	Butyronitrile	Butanenitrile
$CH_3-CH(CH_3)-CN$	Isopropyl cyanide	Isobutyronitrile	2-Methylpropanenitrile

STRUCTURE

Let us consider acetonitrile. (CH_3CN) to illustrate the molecular orbitals structure of nitriles. In acetonitrile the methyl carbon is sp^3 hybridized. The carbon and the nitrogen atoms of the carbon-nitrogen triple bond are *sp* hybridized.

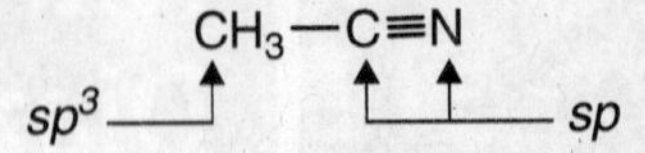

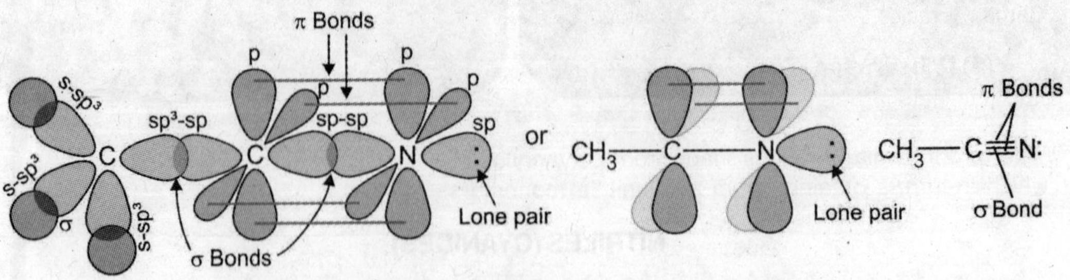

Fig. 24.1. Molecular orbital structure of Acetonitrile.

The σ bonds result from s–sp^3, sp^3–sp, and sp–sp overlaps. The π bonds result by the overlap of *p* orbitals from the *sp* hybridized atoms of the carbon-nitrogen triple bond (as in alkynes). The carbon-nitrogen triple bond is made up of one σ bonds and two π bonds (Fig.24.1).

The H-C-H and H-C-C bond angles are 109.5°. The C-C-N angle is 180°. The lone-pair of electrons on nitrogen resides in the *sp* hybrid orbital of nitrogen.

METHODS OF PREPARATION

Nitriles are prepared by:

(1) **Action of Metal cyanides on Alkyl halides.** Nitriles can be prepared conveniently by heating an alkyl halide with sodium or potassium cyanide in aqueous-ethanolic solution.

$$\underset{\text{Alkyl halide}}{R-X} + NaCN \longrightarrow \underset{\text{Nitrile}}{R-CN} + NaX$$

$$\underset{\text{Ethyl bromide}}{CH_3CH_2-Br} + NaCN \longrightarrow \underset{\text{Propanenitrile}}{CH_3CH_2-CN} + NaBr$$

When nitriles are prepared from alkyl halides by this method, an extra carbon atom is introduced. This is the key step for ascending the series of carboxylic acids (or 1° alcohols)

$$RCOOH \xrightarrow[H^+]{CH_3OH} RCOOCH_3 \xrightarrow{LiAlH_4} RCH_2OH \xrightarrow{PBr_3}$$

$$RCH_2Br \xrightarrow{KCN} RCH_2CN \xrightarrow[H^+]{H_2O} RCH_2COOH$$

(2) **Dehydration of Amides or Ammonium salts of Carboxylic acids.** Simple amides on drastic dehydration by heating with phosphorus pentoxide yield nitriles.

$$\underset{\text{Amide}}{R-\overset{O}{\overset{\|}{C}}-NH_2} \xrightarrow[\Delta]{P_2O_5} \underset{\text{Nitrile}}{R-C\equiv N} + H_2O$$

$$\underset{\text{Propanamide}}{CH_3CH_2-\overset{O}{\overset{\|}{C}}-NH_2} \xrightarrow[\Delta]{P_2O_5} \underset{\text{Propanenitrile}}{CH_3CH_2-CN} + H_2O$$

Ammonium salts of carboxylic acids dehydrate first to form amide, which on loss of a molecule of water yields nitrile. Thus when carboxylic acid vapors mixed with ammonia are passed over heated alumina at 500°C, a nitrile results. (*Industrial method*)

$$\underset{\text{Carboxylic acid}}{R-\overset{O}{\overset{\|}{C}}-OH} + NH_3 \longrightarrow \underset{\text{Ammonium salt}}{R-\overset{O}{\overset{\|}{C}}-\overset{-}{O}\overset{+}{N}H_4} \xrightarrow[-H_2O]{Al_2O_3, 500°C} \underset{\text{Amide}}{R-\overset{O}{\overset{\|}{C}}-NH_2} \xrightarrow[\Delta]{P_2O_5} \underset{\text{Nitrile}}{R-C\equiv N}$$

(3) **Dehydration of Aldoximes.** Aldoximes on dehydration with acetic anhydride or phosphorus pentoxide yield nitriles.

$$\underset{\text{Aldoxime}}{R-\overset{H}{\overset{|}{C}}=N-OH} \xrightarrow[\text{OR } P_2O_5]{(CH_3CO)_2O} \underset{\text{Nitrile}}{R-C\equiv N} + H_2O$$

$$\underset{\text{Acetaldoxime}}{CH_3-\overset{H}{\overset{|}{C}}=N-OH} \xrightarrow[\text{OR } P_2O_5]{(CH_3CO)_2O} \underset{\substack{\text{Acetonitrile}\\\text{(Ethanenitrile)}}}{CH_3-C\equiv N} + H_2O$$

PHYSICAL PROPERTIES

(1) The lower nitriles are colorless volatile liquids (CH_3CN *bp* 82°C, CH_3CH_2CN *bp* 97°C).
(2) They are readily soluble in water as also in organic solvents (alcohol, ether, benzene).
(3) A nitrile, like a carbon-carbon triple bond, absorbs in the *IR Spectrum* at about 2250 cm^{-1}.

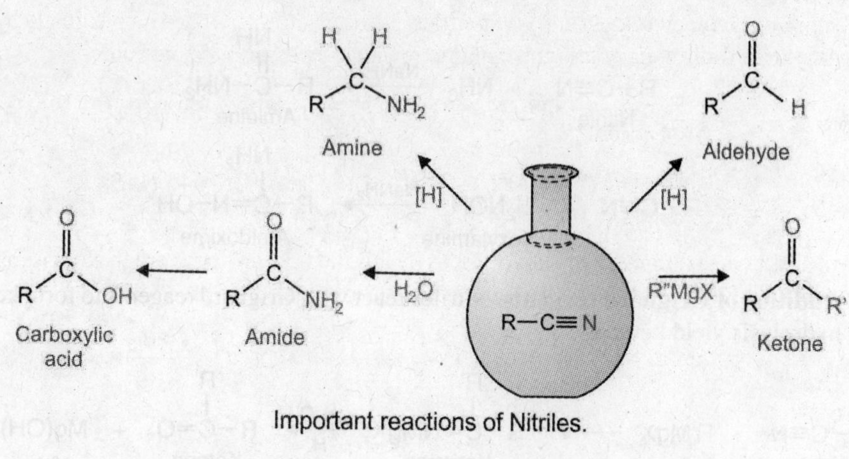

Important reactions of Nitriles.

CHEMICAL PROPERTIES

The important reactions of nitriles are given below :

(1) **Hydrolysis.** On boiling with aqueous mineral acid or aqueous alkali, nitriles are hydrolyzed to a carboxylic acid and ammonia. This reaction takes place through the formation of an amide, which is then hydrolyzed to a carboxylic acid.

$$\underset{\text{Nitrile}}{R-C\equiv N} \xrightarrow[H_2O]{H^+ \text{ or } \bar{O}H} \underset{\text{Amide}}{R-\overset{\overset{O}{\|}}{C}-NH_2} \xrightarrow[H_2O]{H^+ \text{ or } \bar{O}H} \underset{\text{Carboxylic acid}}{R-\overset{\overset{O}{\|}}{C}-OH} + NH_3$$

$$\underset{\text{Acetonitrile}}{CH_3-C\equiv N} \xrightarrow[H_2O]{H^+ \text{ or } \bar{O}H} \underset{\text{Acetamide}}{CH_3-\overset{\overset{O}{\|}}{C}-NH_2} \xrightarrow[H_2O]{H^+ \text{ or } \bar{O}H} \underset{\text{Acetic acid}}{CH_3-\overset{\overset{O}{\|}}{C}-OH} + NH_3$$

When a mineral acid (say HCl) is used for hydrolysis, the product is a carboxylic acid and ammonia. On the other hand, if an alkali (say NaOH) is used, the hydrolysis results in the formation of sodium salt of the carboxylic acid.

$$R-C\equiv N + 2H_2O + HCl \longrightarrow RCOOH + \overset{+}{N}H_4\overset{-}{C}l$$

$$R-C\equiv N + H_2O + NaOH \longrightarrow RCO\overset{-}{O}\overset{+}{N}a + NH_3$$

(2) **Alcoholysis.** Nitriles on boiling with excess of alcohol in the presence of concentrated sulfuric acid or hydrogen chloride, form esters.

$$\underset{\text{Nitrile}}{R-C\equiv N} + \underset{\text{Alcohol}}{R'-OH} + H_2O \xrightarrow{H^+} \underset{\text{Ester}}{R-\overset{\overset{O}{\|}}{C}-OR'} + NH_4^+$$

$$\underset{\text{Propanenitrile}}{CH_3CH_2-C\equiv N} + \underset{\text{Methanol}}{CH_3OH} \xrightarrow{H^+} \underset{\text{Methyl propionate}}{CH_3CH_2-\overset{\overset{O}{\|}}{C}-OCH_3} + NH_4^+$$

(3) **Addition of Ammonia and Hydroxylamine.** Nitriles when treated with ammonia or hydroxylamine in the presence of $NaNH_2$ as catalyst, form amidines and amidoximes respectively.

$$\underset{\text{Nitrile}}{R-C\equiv N} + NH_3 \xrightarrow{NaNH_2} \underset{\text{Amidine}}{R-\overset{\overset{NH}{\|}}{C}-NH_2}$$

$$R-C\equiv N + \underset{\text{Hydroxylamine}}{H_2NOH} \xrightarrow{NaNH_2} \underset{\text{Amidoxime}}{R-\overset{\overset{NH_2}{|}}{C}=N-OH}$$

(4) **Addition of Grignard reagents.** Nitriles react with Grignard reagents to form ketimines, which on hydrolysis yield ketones.

$$\underset{\text{Nitrile}}{R-C\equiv N} + R'MgX \longrightarrow \underset{\text{Ketimine}}{R-\overset{\overset{R'}{|}}{C}=NMgX} \xrightarrow[H^+]{H_2O} \underset{\text{Ketone}}{R-\overset{\overset{R'}{|}}{C}=O} + Mg(OH)X$$

$$\underset{\text{Acetonitrile}}{CH_3-C\equiv N} + \underset{\text{Methylmagnesium bromide}}{CH_3MgBr} \longrightarrow CH_3-\overset{\overset{CH_3}{|}}{C}=NMgBr \xrightarrow[H^+]{H_2O} \underset{\text{Acetone}}{CH_3-\overset{\overset{CH_3}{|}}{C}=O} + Mg(OH)Br$$

(5) **Reduction.** On treatment with lithium aluminium hydride, $LiAlH_4$, nitriles undergo complete reduction to yield primary amines.

$$\underset{\text{Nitrile}}{R-C\equiv N} + 4[H] \xrightarrow[\text{ether}]{LiAlH_4} \underset{\text{Primary amine}}{R-CH_2-NH_2}$$

This reduction may also be carried with sodium and ethanol (*Mendius Reaction*), or with hydrogen and metal catalyst (Raney nickel).

$$\underset{\text{Acetonitrile}}{CH_3-C\equiv N} + 2H_2 \xrightarrow{Ni} \underset{\text{Ethylamine}}{CH_3-CH_2-NH_2}$$

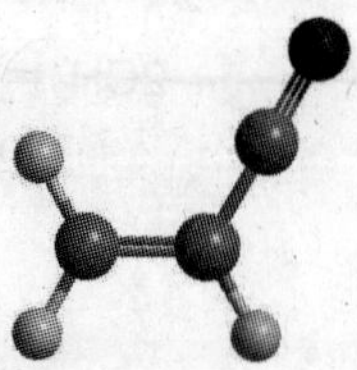

ACRYLONITRILE

It is commercially the most important nitrile. It is also called *vinyl cyanide*.

Structure. Molecular orbital structure of acrylonitrile is shown in Fig.24.2.

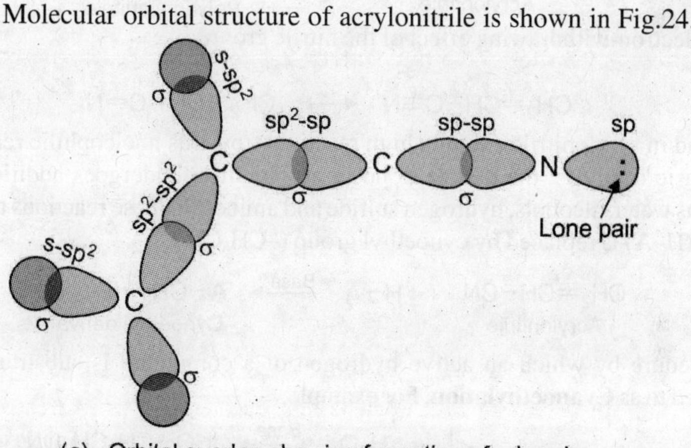

(*a*) Orbital overlap showing formation of σ bonds.

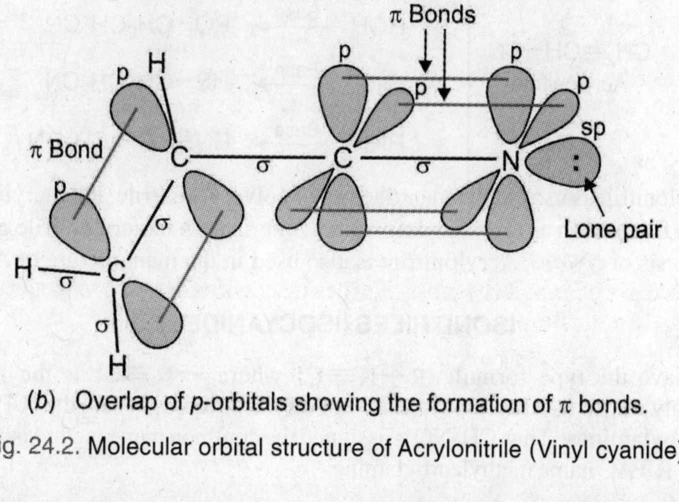

(*b*) Overlap of *p*-orbitals showing the formation of π bonds.

Fig. 24.2. Molecular orbital structure of Acrylonitrile (Vinyl cyanide).

Preparation. (1) Acrylonitrile is chiefly made by the catalytic addition of hydrogen cyanide to acetylene.

$$\underset{\text{Acetylene}}{H-C\equiv C-H} + H-C\equiv N \xrightarrow[NH_4Cl, 90°C]{Cu_2Cl_2} \underset{\text{Acrylonitrile}}{H-\overset{H}{\overset{|}{C}}=\overset{C\equiv N}{\overset{|}{C}}-H} \text{ or } CH_2=CH-CN$$

(2) It is also obtained from ethylene and propylene derived from petroleum sources.

$$\underset{\text{Ethylene}}{CH_2=CH_2} \xrightarrow[Ag,\ 200°C]{O_2} \underset{\text{Ethylene oxide}}{CH_2\overset{O}{-}CH_2} \xrightarrow{HCN} \overset{OH}{\overset{|}{CH_2}}-CH_2-CN \xrightarrow{-H_2O} \underset{\text{Acrylonitrile}}{CH_2=CH-CN}$$

Properties. Acrylonitrile is a colorless liquid bp 77.3°C. It polymerizes readily in the presence of a catalyst (organic peroxides and other oxidizing agents) to form polyacrylonitrile.

$$\underset{\text{Propene}}{2CH_2=CH-CH_3} + 2NH_3 + 3O_2 \xrightarrow[450°C]{\text{catalyst}} \underset{\text{Acrylonitrile}}{2CH_2=CH-CN} + 6H_2O$$

$$\underset{\text{Acrylonitrile}}{n\,CH_2=CH-CN} \xrightarrow{R_2O_2} \underset{\text{Polyacrylonitrile}}{\left[-CH_2-\overset{CN}{\overset{|}{CH}}-\right]_n}$$

Owing to the electron-withdrawing effect of the nitrile group,

$$CH_2=CH-C\equiv N \longleftrightarrow \overset{+}{C}H_2-CH=C=\overset{-}{N}:$$

the double bond in acrylonitrile exhibits high reactivity towards nucleophilic reagents. Thus in the presence of basic catalysts, the double bond in acrylonitrile undergoes addition of a number of reagents such as water, alcohols, hydrogen sulfide and amines. In these reactions the active hydrogen of the reagent (H–A) is replaced by cynoethyl group ($-CH_2CH_2CN$).

$$\underset{\text{Acrylonitrile}}{CH_2=CH-CN} + H-A \xrightarrow{\text{Base}} \underset{\text{Cynoethyl derivative}}{A-CH_2-CH_2-CN}$$

This procedure by which an active hydrogen of a compound is substituted by cyanoethyl group is referred to as **Cyanoethylation**. For example,

$$\underset{\text{Acrylonitrile}}{CH_2=CH-CN} \begin{cases} + H_2O \xrightarrow{\text{Base}} HO-CH_2CH_2CN \\ + ROH \xrightarrow{\text{Base}} RO-CH_2CH_2CN \\ + H_2S \xrightarrow{\text{Base}} HS-CH_2CH_2CN \\ + RNH_2 \xrightarrow{\text{Base}} RNH-CH_2CH_2CN \end{cases}$$

Uses. Acrylonitrile is used in the manufacture of polyacrylonitrile. Polyacrylonitrile is used for making synthetic fibers such as *Orlon* and *Acrilan*. Dimerization of acrylonitrile gives adiponitrile, used in the synthesis of *Nylons*. Acrylonitrile is also used in the manufacture of *ABS* plastic.

ISONITRILES (ISOCYANIDES)

Isonitriles have the type formula $R-\overset{+}{N}\equiv\overset{-}{C}:$ where $-\overset{+}{N}\equiv\overset{-}{C}:$ is the functional group. They are commonly named as **Alkyl isocyanides** or **Alkyl isonitriles**, where the IUPAC system names them as **Alkylcarbylamines**. Thus CH_3NC is assigned the common name methyl isocyanide or methyl isonitrile, and the IUPAC name methylcarbylamine.

METHODS OF PREPARATION

(1) By heating an alkyl iodide with silver cyanide in aqueous ethanolic solution.

$$\underset{\text{Alkyl iodide}}{R{-}I} + \underset{\text{Silver cyanide}}{Ag{-}C{\equiv}N} \xrightarrow{\Delta} \underset{\text{Isonitrile}}{R{-}\overset{+}{N}{\equiv}\overset{-}{C}{:}} + AgI$$

$$CH_3{-}I + Ag{-}C{\equiv}N \xrightarrow{\Delta} \underset{\substack{\text{Methyl isonitrile}\\ \text{(Methyl isocyanide)}}}{CH_3{-}\overset{+}{N}{\equiv}\overset{-}{C}{:}} + AgI$$

(2) By heating a primary amine and chloroform ($CHCl_3$) with ethanolic potassium hydroxide solution. (**Carbylamine Reaction**)

$$\underset{1^\circ\text{ amine}}{RNH_2} + CHCl_3 + 3KOH \xrightarrow{\Delta} \underset{\text{Isonitrile}}{RNC} + 3KCl + 3H_2O$$

$$CH_3NH_2 + CHCl_3 + 3KOH \xrightarrow{\Delta} \underset{\text{Methyl isonitrile}}{CH_3NC} + 3KCl + 3H_2O$$

PROPERTIES

(Physical). Isonitriles are highly unpleasant smelling liquids, insoluble in water. They have lower boiling points than the isomeric nitriles ; CH_3CN *bp* 81°C, CH_3NC *bp* 68°C. In the *IR Spectrum*, they show the $\overset{+}{N}{\equiv}\overset{-}{C}{:}$ stretch absorption at 2185-2120 cm^{-1}.

(Chemical). (1) **Hydrolysis**. They are hydrolyzed by dilute acids to form a primary amine and formic acid, but are not hydrolyzed by alkalis.

$$\underset{\text{Alkyl isonitrile}}{R{-}\overset{+}{N}{\equiv}\overset{-}{C}{:}} + 2H_2O \xrightarrow{\text{Acid}} \underset{1^\circ\text{ Amine}}{R{-}NH_2} + \underset{\text{Formic acid}}{HCOOH}$$

$$\underset{\text{Alkyl isonitrile}}{R{-}\overset{+}{N}{\equiv}\overset{-}{C}{:}} + H_2O \xrightarrow{\text{Alkali}} \text{No reaction}$$

In contrast, nitriles (RCN) undergo both acid and base hydrolysis to form the corresponding carboxylic acids (RCOOH). This reaction brings out clearly that in isonitriles the alkyl group R is bonded to nitrogen, while in nitriles it is bonded to the terminal carbon.

$$\underset{\text{Methyl isonitrile}}{CH_3{-}\overset{+}{N}{\equiv}\overset{-}{C}{:}} + 2H_2O \xrightarrow{H^+} \underset{\text{Methylamine}}{CH_3{-}NH_2} + HCOOH \quad (i)$$

$$\underset{\text{Acetonitrile}}{CH_3{-}C{\equiv}N} + 2H_2O \xrightarrow{H^+} \underset{\text{Acetic acid}}{CH_3{-}\overset{O}{\overset{\|}{C}}{-}OH} + NH_3 \quad (ii)$$

In the reaction (*i*) the original $H_3C{-}N$ bond remains intact, and in reaction (*ii*) the original $H_3C{-}C$ bond remains intact.

(2) **Reduction.** Isonitriles are reduced to secondary amines either by sodium and ethanol, or by catalytic hydrogenation (Pt, H_2).

$$\underset{\text{Alkyl isonitrile}}{R{-}\overset{+}{N}{\equiv}\overset{-}{C}{:}} + 2H_2 \xrightarrow{Na/C_2H_5OH} \underset{2^\circ\text{ Amine}}{R{-}NH{-}CH_3}$$

$$\underset{\text{Propane isonitrile}}{CH_3CH_2{-}\overset{+}{N}{\equiv}\overset{-}{C}{:}} + 2H_2 \xrightarrow{Pt} \underset{\text{Ethylmethylamine}}{CH_3CH_2{-}NH{-}CH_3}$$

ALKYL NITRITES

Alkyl nitrites are the esters of nitrous acid. They have the type formula R–O–N=O. They are isomeric with nitroalkanes. Of the esters of nitrous acid, ethyl nitrite is by far the most important and typical member.

ETHYL NITRITE, CH_3CH_2–O–N=O

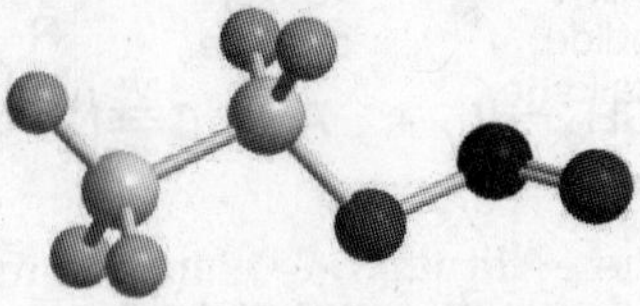

It is isomeric with nitroethane, $C_2H_5NO_2$.

Preparation. It may be prepared :

(1) By adding dilute sulfuric acid to a chilled mixture of ethyl alcohol and sodium nitrite ($NaNO_2$).

$$\underset{\text{Ethyl alcohol}}{CH_3CH_2OH} + \underset{\text{Nitrous acid}}{HONO} \xrightarrow[0°C]{H_2SO_4} \underset{\text{Ethyl nitrite}}{CH_3CH_2—O—N{=}O} + H_2O$$

(2) By the reaction of ethyl iodide with potassium nitrite.

$$\underset{\text{Ethyl iodide}}{CH_3CH_2I} + KNO_2 \xrightarrow{0°C} \underset{\text{Ethyl nitrite}}{CH_3CH_2—O—N{=}O} + KI$$

Properties. (*Physical*). It is a yellowish, volatile liquid, bp 16.4°C, having a characteristic smell of apples. It is soluble in alcohol and ether ; decomposes in water.

(*Chemical*). (1) **Hydrolysis**. Ethyl nitrite is readily hydrolysed by water, dilute alkali, or dilute acid, regenerating nitrous acid.

$$\underset{\text{Ethyl nitrite}}{CH_3CH_2—O—N{=}O} + H—OH \xrightarrow[\text{or } \bar{O}H]{H^+} \underset{\text{Nitrous acid}}{HON{=}O} + C_2H_5OH$$

(2) **Reduction**. When reduced with tin and hydrochloric acid, it yields ethyl alcohol and ammonia.

$$\underset{\text{Ethyl nitrite}}{CH_3CH_2—O—N{=}O} + 6[H] \xrightarrow{Sn/HCl} \underset{\text{Ethyl alcohol}}{CH_3CH_2OH} + NH_3 + H_2O$$

Uses. Ethyl nitrite is used : (1) in organic reactions as a source of nitrous acid in the absence of water ; and (2) in medicine for treatment of asthma and heart diseases.

NITROALKANES

Nitroalkanes may be considered as the *nitro* derivatives of alkanes. They have the type formula R–NO_2, where the –NO_2 functional group is bonded directly to the carbon atom of the alkyl group.

NOMENCLATURE

They are named by prefixing '*nitro*' to the name of the parent alkane. The total IUPAC name emerges as one-word, the position of the –NO_2 group on the hydrocarbon chain being indicated by a number.

$$\underset{\text{Nitromethane}}{CH_3—NO_2} \qquad \underset{\text{Nitroethane}}{CH_3CH_2—NO_2} \qquad \underset{\text{2-Nitropropane}}{CH_3—\overset{\overset{NO_2}{|}}{CH}—CH_3}$$

Like alcohols, nitroalkanes are designated as **Primary** (1°), **Secondary** (2°), or **Tertiary** (3°), according as the –NO_2 group is attached to a primary, a secondary or a tertiary carbon atom.

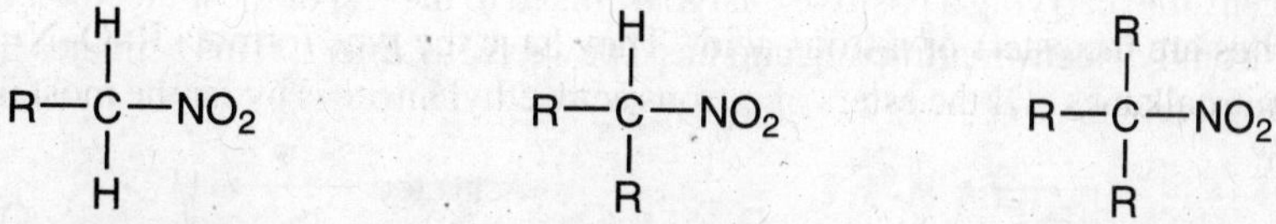

1° nitroalkane 2° nitroalkane 3° nitroalkane

METHODS OF PREPARATION

(1) **By vapor-phase Nitration of Alkanes.** Nitroalkanes are now produced commercially by passing a gaseous mixture of an alkane and nitric acid through a metal tube at 400°C. With alkanes other than methane, a mixture of nitroalkanes is obtained which can be separated by fractional distillation.

$$\underset{\text{Methane}}{CH_4} + \underset{\text{Nitric acid}}{HONO_2} \xrightarrow{400°C} \underset{\text{Nitromethane}}{CH_3NO_2} + H_2O$$

$$\underset{\text{Ethane}}{CH_3CH_3} + HONO_2 \xrightarrow{400°C} \underset{\text{Nitroethane}}{CH_3CH_2NO_2} + H_2O$$

(2) **By reaction of Silver nitrite with Alkyl halides**

$$\underset{\text{Ethyl bromide}}{CH_3CH_2Br} + AgNO_2 \longrightarrow \underset{\text{Nitroethane}}{CH_3CH_2NO_2} + AgBr$$

(3) **By reaction of Sodium nitrite with α-Halo carboxylic acids**

$$\underset{\text{Chloroacetic acid}}{ClCH_2COOH} + NaNO_2 \xrightarrow{\Delta} \underset{\text{Nitroacetic acid}}{O_2NCH_2COOH} + NaCl$$

$$O_2NCH_2COOH \xrightarrow[-CO_2]{\Delta} \underset{\text{Nitromethane}}{O_2NCH_3}$$

(4) **By oxidation of Primary amines**

$$\underset{\text{1-Aminobutane}}{CH_3(CH_2)_3NH_2} + 3[O] \xrightarrow{\text{Peracetic acid}} \underset{\text{1-Nitrobutane}}{CH_3(CH_2)_3NO_2} + H_2O$$

PHYSICAL PROPERTIES

(1) The lower nitroalkanes are colorless, pleasant-smelling liquids.

(2) They are sparingly soluble in water giving neutral solutions.

(3) Since they are polar, nitroalkanes are excellent solvents for polar and ionic compounds.

(4) Due to appreciable electrostatic attraction between their polar molecules, the boiling points of nitroalkanes are abnormally high. CH_3NO_2 boils at 101°C, and $C_2H_5NO_2$ at 115°C.

(5) Nitroalkanes have much higher boiling points than the less polar isomeric alkyl nitrites *e.g.*, C_2H_5ONO has *bp* 17°C in contrast to nitroethane, 115°C.

(6) They show strong infrared bands at 1580 and 1375 cm^{-1}.

CHEMICAL PROPERTIES

The nitro group ($-NO_2$) can be represented as hybrid of two resonance forms.

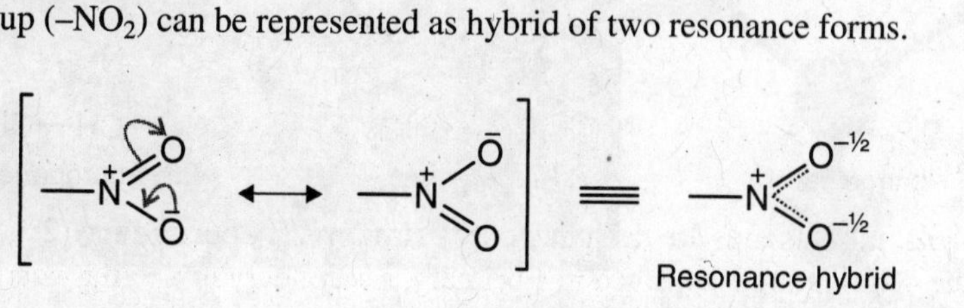

Resonance hybrid

The nitrogen atom carrying a positive charge is linked to the rest of the molecule. Therefore, the nitro group is strongly electron-attracting and has a powerful *–I Effect*. This activates the hydrogens attached to the α-carbon.

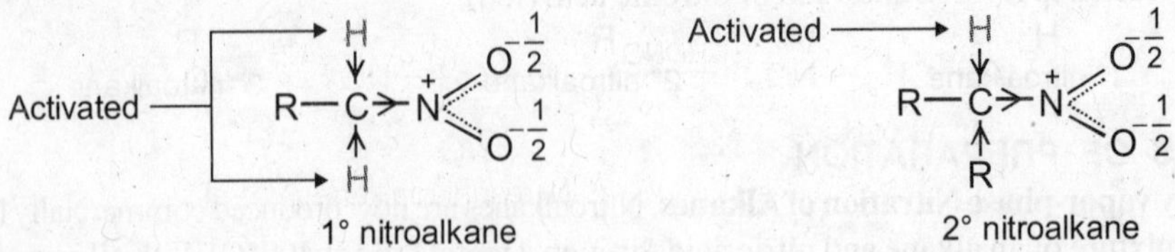

Hence, the most reactions of nitroalkanes are those involving the α-hydrogens of primary and secondary nitroalkanes.

(1) **Acidity ; Nitro-Aci Tautomerism**. Nitroalkanes owe their acidity to the formation of a resonance- stabilized anion.

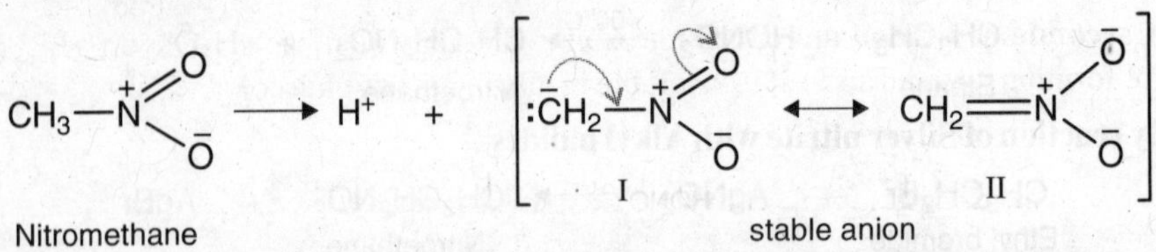

1° and 2° nitroalkanes are known to exist as a tautomeric mixture of **Nitro form** and **Aci form** (a strong acid). These are thought of as derived by protonation of the resonance forms I and II of the anion.

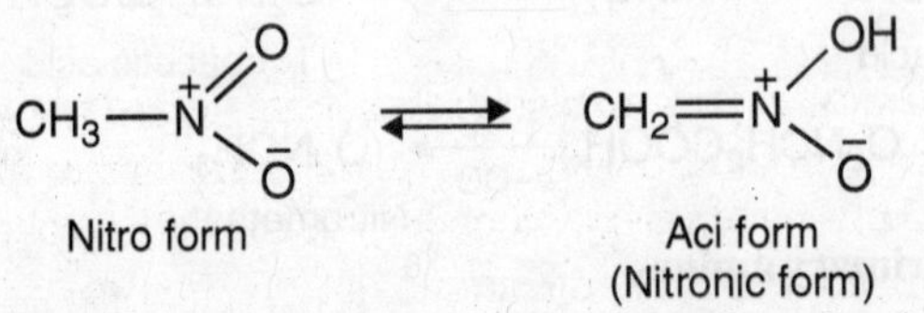

The equilibrium normally lies almost entirely to the *nitro* form but in the presence of an alkali (NaOH) the *aci* form exists. Thus 1° and 2° nitroalkanes dissolve in alkalies to produce the salts of the *aci* form. These on acidification regenerate the *aci* form which reverts to the stable *nitro* form.

(2) **Reaction with Halogens.** 1° and 2° nitroalkanes are readily halogenated in the α-position by treatment with chlorine or bromine in alkaline solution.

$$CH_3{-}NO_2 \xrightarrow[NaOH]{Cl_2} \underset{\text{Chloropicrin}}{CCl_3{-}NO_2}$$

Chloropicrin is an important insecticide.

Chloropicrin is trichloronitromethane (CCl_3NO_2). It is used in insecticides and fungicides.

(3) **Reaction with Nitrous Acid**. 1°, 2° and 3° nitroalkanes having two, one or no α-hydrogen atoms react differently with nitrous acid.

(*a*) **A primary nitroalkane** reacts with nitrous acid to form nitroso-nitroalkane which dissolves in NaOH solution to form sodium salt of nitronic acid (red).

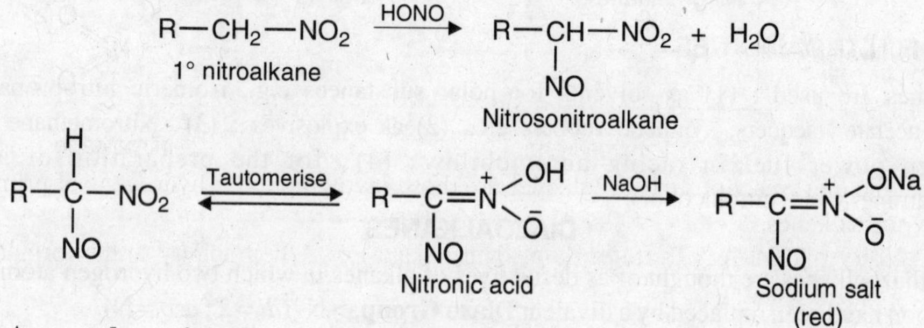

(*b*) **A secondary nitroalkane** gives nitrosonitroalkane which has no α-hydrogen, and hence incapable of forming nitronic acid. It is insoluble in alkali and has blue color.

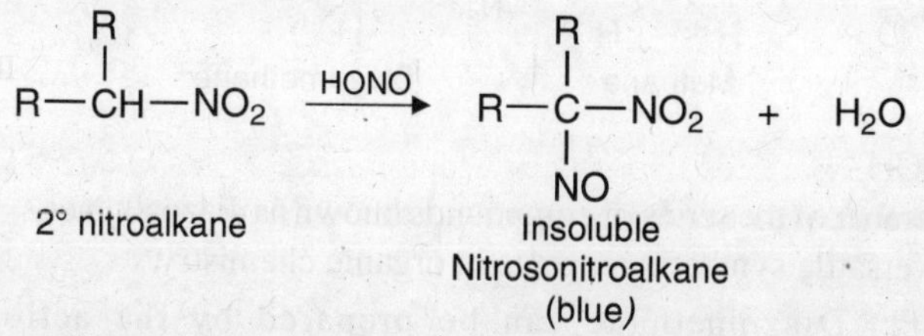

(*c*) **A tertiary nitroalkane** having no α-hydrogen atom fails to react with nitrous acid.

The different behavior of nitrous acid with 1°, 2°, and 3° nitroalkanes forms the basis of the **Victor Meyer Test** for distinction between primary, secondary and tertiary alcohols.

(4) **Condensation with Aldehydes and Ketones**. 1° and 2° nitroalkanes in the presence of dilute alkali undergo aldol-addition or condensation with aldehydes and ketones to give **Nitroalcohols**.

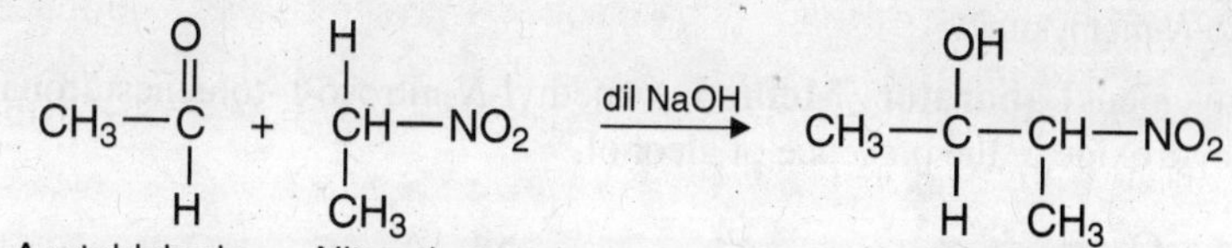

(5) **Reduction**. Nitroalkanes can be reduced with tin or iron, and hydrochloric acid to form primary **Amines**.

$$\underset{\text{Nitroethane}}{C_2H_5NO_2} + 6[H] \xrightarrow{\text{Sn/HCl}} \underset{\text{Ethylamine}}{C_2H_5{-}NH_2} + 2H_2O$$

When reduced in neutral solution with zinc dust and aqueous ammonium chloride, nitroalkanes yield **Alkylhydroxylamines**.

$$\underset{\text{Nitroethane}}{C_2H_5NO_2} + 4[H] \xrightarrow[\text{warm}]{\text{Zn/NH}_4\text{Cl}} \underset{\text{Ethylhydroxylamine}}{C_2H_5{-}NHOH} + H_2O$$

(6) **Hydrolysis to Hydroxylamine**. Nitroalkanes when boiled with concentrated hydrochloric acid or sulfuric acid, are hydrolysed to form a carboxylic acid and hydroxylamine.

$$\underset{\text{Nitroethane}}{CH_3CH_2{-}NO_2} + H_2O \xrightarrow[\Delta]{H^+} \underset{\text{Acetic acid}}{CH_3COOH} + \underset{\text{Hydroxylamine}}{NH_2OH}$$

(7) **Thermal Decomposition**. Upon rapid heating nitroalkanes decompose with great violence. Advantage is taken of this reaction in the commercial use of nitroalkanes as explosives.

$$\underset{\text{Nitromethane}}{CH_3NO_2} \xrightarrow{\Delta} \tfrac{1}{2}N_2 + CO_2 + \tfrac{3}{2}H_2$$

USES OF NITROALKANES

Nitroalkanes are used : (1) as solvents for polar substances e.g., isomeric nitropropanes for cellulose acetate, lacquers, synthetic rubbers etc., (2) as explosives ; (3) Nitromethane is used as a high power fuel in racing automobiles ; (4) for the preparation of amines, hydroxylamines, and nitroalcohols.

DIAZOALKANES

The diazoalkanes are thought of as derivatives of alkanes in which two hydrogen atoms on the same carbon have been replaced by a divalent **Diazo Group**, $>N_2$ (*di* = 2 ; *azo* = N).

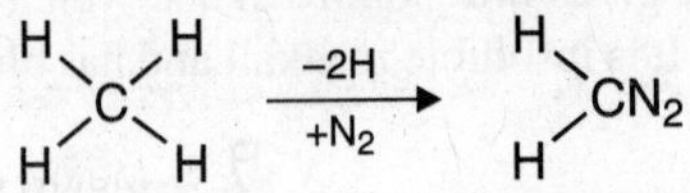

Methane Diazomethane

DIAZOMETHANE, CH_2N_2

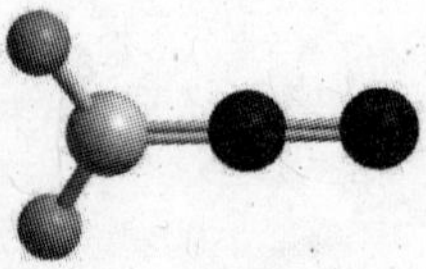

It is the parent member of the series of compounds known as diazoalkanes. It is one of the most versatile synthetic reagents in organic chemistry.

Preparation. (1) Diazomethane can be prepared by the action of a base on a *N*-nitroso-*N*-methylamide.

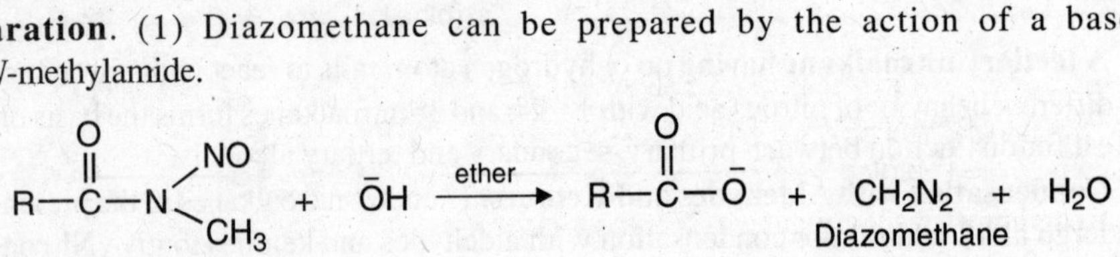

Diazomethane

N-Nitroso-*N*-methylamide

(2) In a convenient **Laboratory Method**, *N*-methyl-*N*-nitroso-*p*-toluenesulfonamide is treated with potassium hydroxide in the presence of alcohol.

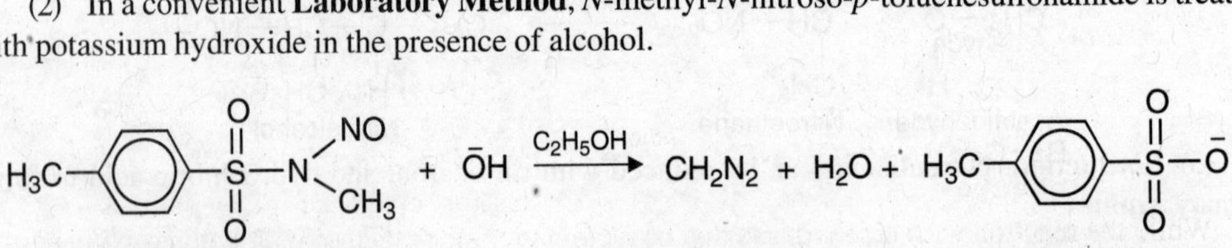

N-Methyl-*N*-nitroso-*p*-toluenesulphonamide Diazomethane *p*-Toluenesulphonate ion

The nitroso amide is available from direct nitrosation of *N*-methyl-*p*-toluenesulfonamide.

Properties. (*Physical*). Diazomethane is an intensely yellow gas at room temperature, bp –23°C. It is highly toxic and shock sensitive. It is dangerously explosive when pure. Since it is safe to handle the gas in ether solution, it is prepared and used for various synthetic reactions in ether.

(*Chemical*). Diazomethane can be represented as a resonance hybrid of the following forms :

$$\underset{\text{I}}{CH_2{=}\overset{+}{N}{=}\overset{-}{N}{:}} \longleftrightarrow \underset{\text{II}}{\overset{-}{\ddot{C}}H_2{-}\overset{+}{N}{\equiv}N{:}}$$

Most of the reactions of diazomethane can be readily explained in terms of the nucleophilic attack on form II followed by loss of nitrogen, N_2.

(1) Decomposition to Methylene by light (Photolysis) or heat (Pyrolysis)

$$\underset{\text{Diazomethane}}{H_2\ddot{C}-\overset{+}{N}\equiv N} \xrightarrow[\text{or } \Delta]{h\nu} \underset{\text{Methylene}}{H_2\ddot{C}} + N_2$$

Methylene is a very reactive species and gives many interesting reactions by insertion at single bonds such as C–H, C–C, and C–Halogen.

(2) Reaction with Mineral Acids. Hydrochloric acid and sulfuric acid convert diazomethane to diazonium salts which at once decompose to form methyl chloride or methyl sulfate, and nitrogen.

$$\underset{\text{Diazomethane}}{CH_2=\overset{+}{N}=\overset{-}{N}} + HCl \longrightarrow \underset{\text{(unstable)}}{CH_3-\overset{+}{N}\equiv N\overset{-}{Cl}} \longrightarrow \underset{\text{Methyl chloride}}{CH_3Cl} + N_2$$

(3) Methylation reactions with Carboxylic acids, Alcohols, Phenols and Amines. Diazomethane reacts with many compounds containing reactive hydrogen atoms under appropriate conditions, when –H is converted to $-CH_3$ group.

$$\underset{\text{Carboxylic acid}}{R-\overset{O}{\overset{\|}{C}}-OH} + CH_2N_2 \xrightarrow{\text{ether}} \underset{\text{Methyl ester}}{R-\overset{O}{\overset{\|}{C}}-OCH_3} + N_2$$

$$\underset{\text{Phenol}}{C_6H_5-O-H} + CH_2N_2 \xrightarrow{\text{ether}} \underset{\text{Anisole}}{C_6H_5-O-CH_3} + N_2$$

$$\underset{\text{Alcohol}}{R-O-H} + CH_2N_2 \xrightarrow[\text{ether}]{BF_3} \underset{\text{Methyl ether}}{R-O-CH_3} + N_2$$

$$\underset{1°\text{ amine}}{R-\overset{H}{\overset{|}{N}}-H} + CH_2N_2 \xrightarrow[\text{ether}]{BF_3} \underset{2°\text{ Methylamine}}{R-\overset{H}{\overset{|}{N}}-CH_3}$$

(4) Reaction with Aldehydes and Ketones

$$\underset{\text{Aldehyde}}{R-\overset{O}{\overset{\|}{C}}-H} + CH_2N_2 \xrightarrow{\text{ether}} \underset{\text{Methyl ketone}}{R-\overset{O}{\overset{\|}{C}}-CH_3} + N_2$$

$$\underset{\text{Ketone}}{R-\overset{O}{\overset{\|}{C}}-R'} + CH_2N_2 \xrightarrow{\text{ether}} \underset{\text{Higher homologue}}{R-\overset{O}{\overset{\|}{C}}-CH_2-R'} + N_2$$

While the reaction with aldehydes is that of methylation, the reaction with ketones proceeds by insertion of $:CH_2$ in C–C bond.

(5) Reaction with Acid halides to give Carboxylic acids with one more carbon atom. Acid chlorides react with excess of diazomethane to give a diazoketone. This is decomposed by water in the presence of silver oxide (colloidal silver) to yield a carboxylic acid with one more carbon atom.

$$\underset{\text{Acid chloride}}{R-\overset{O}{\overset{\|}{C}}-Cl} + CH_2N_2 \longrightarrow \underset{\text{Diazoketone}}{R-\overset{O}{\overset{\|}{C}}-CHN_2} + HCl$$

$$R-\overset{O}{\overset{\|}{C}}-CHN_2 + H_2O \xrightarrow[\Delta]{Ag_2O} \underset{\text{Carboxylic acid}}{R-CH_2-\overset{O}{\overset{\|}{C}}-OH} + N_2$$

The HCl produced in the first step is removed by reaction with excess of diazomethane as in reaction (2).

The overall reaction, $RCOCl \longrightarrow RCH_2COOH$ is called **Arndt-Eistert Reaction.** This is one of the most important reactions of converting a carboxylic acid into the next higher homologue.

$$\underset{(n \text{ Carbons})}{RCOOH} \xrightarrow{SOCl_2} RCOCl \xrightarrow{CH_2N_2} RCOCHN_2 \xrightarrow[Ag_2O]{H_2O} \underset{(n+1 \text{ Carbons})}{RCH_2COOH}$$

MECHANISM. Following 3 steps are involved:

Step 1. Acid chlorides react with diazomethane to form diazoketones.

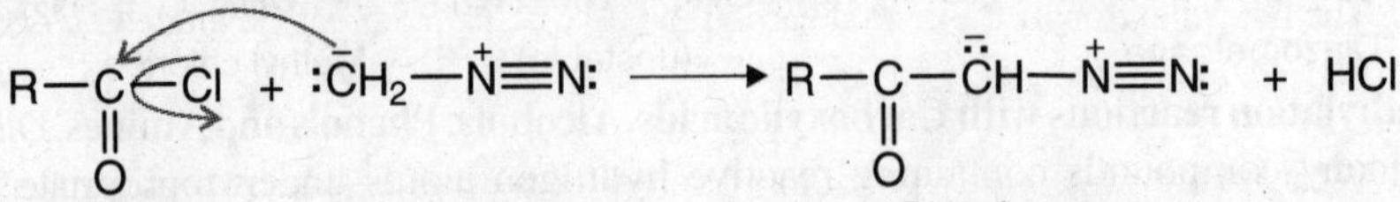

Acid halide Diazomethane Diazoketone

Step 2. Diazoketones undergo loss of nitrogen in the presence of colloidal silver (Ag_2O) to form α-ketocarbenes.

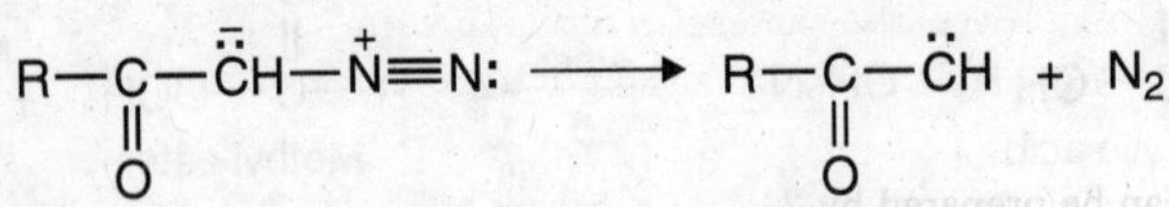

α-Ketocarbene

Step 3. Ketocarbenes undergo rearrangement by 1,2-*shift* of the alkyl group *R* to give ketenes. Ketenes on hydrolysis yield carboxylic acids.

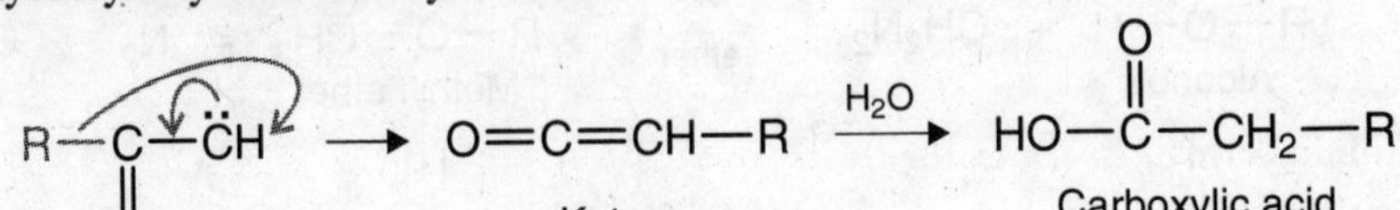

Ketene Carboxylic acid

(6) **Reaction with Alkanes to give higher homologues.** Here a CH_2 group is inserted in C–C bond

$$\underset{\text{Propane}}{CH_3CH_2CH_3} + CH_2N_2 \xrightarrow[-N_2]{hv} \underset{n\text{-Butane}}{CH_3CH_2CH_2CH_3} + \underset{\text{Isobutane}}{CH_3\overset{\overset{CH_3}{|}}{C}HCH_3}$$

(7) **Addition to Alkenes to give Cyclopropane derivatives**

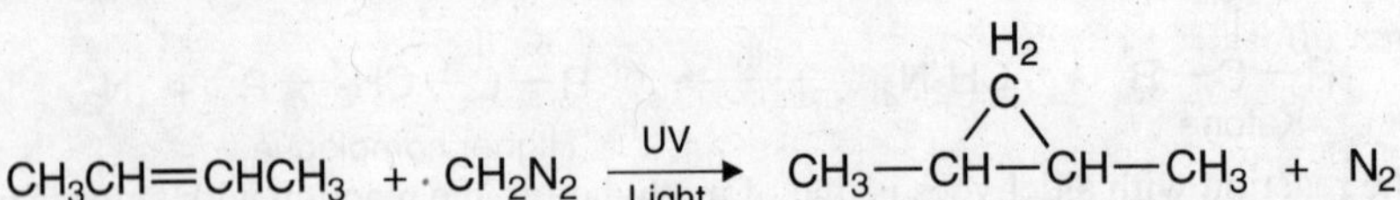

2-Butene 1,2-Dimethylcyclopropane

Examination Questions

1. Discuss the structure of acetonitrile. (*Osmania BSc,2014*)
2. How are nitriles prepared ? Give their important reactions.
3. What happens when acetonitrile is hydrolysed with dilute HCl. (*Bhopal BSc,2015*)

Answer. Acetic acid is formed.

$$\underset{\text{Acetonitrile}}{CH_3CN} \xrightarrow{H_2O/H^+} \underset{\text{Acetic acid}}{CH_3-\overset{\overset{O}{||}}{C}-OH}$$

4. How will you synthesize acetonitrile from acetic acid ? *(Delhi BSc,2014 ; Panjab BSc,2015)*

Answer. Following steps are involved :

$$\underset{\text{Acetic acid}}{CH_3-\overset{O}{\overset{\|}{C}}-OH} \xrightarrow{SOCl_2} CH_3-\overset{O}{\overset{\|}{C}}-Cl \xrightarrow{NH_3} CH_3-\overset{O}{\overset{\|}{C}}-NH_2 \xrightarrow[\Delta]{P_2O_5} \underset{\text{Acetonitrile}}{CH_3CN}$$

5. What is the hybridization of the N atom in HCN ?

(*a*) *p* (*b*) sp^2 (*c*) sp^3 (*d*) *sp*

Answer. (*d*)

6. The hybridization of the central carbon in $CH_3C{\equiv}CN$ and the bond angle CCN are

(*a*) sp^2, 180° (*b*) *sp*, 180° (*c*) sp^2, 120° (*d*) sp^3, 109°

Answer. (*b*)

7. What atomic orbitals are used for formation of the sigma bond of a carbon–nitrogen triple bond ?

(*a*) an sp^2 orbital of carbon and a $2sp^2$ orbital of nitrogen

(*b*) an *sp* orbital of carbon and an sp^2 orbital of nitrogen

(*c*) an *sp* orbital of carbon and an *sp* orbital of nitrogen

(*d*) an sp^2 orbital of carbon and an *sp* orbital of nitrogen

Answer. (*c*)

8. Nitriles can be prepared by ?

(*a*) the hydration of amines (*b*) the dehydration of acids

(*c*) the reduction of acids (*d*) the dehydration of amides

Answer. (*d*)

9. The conversion of a nitrile to an acid is achieved by ?

(*a*) an oxidation (*b*) a reduction

(*c*) a dehydration (*d*) a hydrolysis

Answer. (*d*)

10. The complete hydrolysis of a nitrile gives

(*a*) an acid (*b*) an ester (*c*) an anhydride (*d*) an acid halide

Answer. (*a*)

11. Nitriles are reduced to

(*a*) alcohols (*b*) carboxylic acids (*c*) amines (*d*) amides

Answer. (*c*)

12. The final product of the reaction of a nitrile and a Grignard reagent is ?

(*a*) an acid (*b*) an aldehyde (*c*) an amine (*d*) a ketone

Answer. (*d*)

13. Discuss the molecular orbital structure of acrylonitrile.

14. Describe the preparation, properties and uses of acrylonitrile.

15. How will you synthesize methyl isonitrile from methylamine ? *(Pune BSc,2014)*

Answer. Carbylamine reaction is used.

$$\underset{\text{Methylamine}}{CH_3NH_2} + CHCl_3 + 3KOH \xrightarrow{\Delta} \underset{\text{Methyl isonitrile}}{CH_3NC} + 3KCl + 3H_2O$$

16. How will you distinguish between ethyl cyanide and ethyl isocyanide ?

Answer. Ethyl isocyanide on hydrolysis with dilute HCl gives ethylamine + formic acid. The presence of formic acid in the product can be tested with ammoniacal silver nitrate solution (Tollens' reagent). Recall that formic acid produces silver precipitate (silver mirror) with Tollens' reagent.

$$\underset{\text{Ethyl isocyanide}}{CH_3CH_2NC} + 2H_2O \xrightarrow{H^+} CH_3CH_2NH_2 + \underset{\text{Formic acid}}{HCOOH}$$

$$HCOOH + 2Ag(NH_3)_2NO_2 \longrightarrow \underset{\text{Silver ppt.}}{2Ag} + CO_2 + 2NH_4NO_3 + 2NH_3$$

Ethyl cyanide on hydrolysis with dilute acid gives propionic acid. Propionic acid does not react with Tollens' reagent.

$$\underset{\text{Ethyl cyanide}}{CH_3CH_2CN} \xrightarrow{H_2O/H^+} \underset{\text{Propionic acid}}{CH_3CH_2COOH} + NH_3$$

$$CH_3CH_2COOH + 2Ag(NH_3)_2NO_3 \longrightarrow \text{No Reaction}$$

17. How is ethyl nitrite prepared ? Give its important reactions.

18. How are nitroalkanes prepared ? Give their important reactions.

19. What happens when nitroethane is reduced with Sn/HCl ?

Answer. Ethylamine is formed.

$$\underset{\text{Nitroethane}}{C_2H_5NO_2} + 6[H] \xrightarrow{Sn/HCl} \underset{\text{Ethylamine}}{C_2H_5{-}NH_2} + 2H_2O$$

20. How will you distinguish between nitroethane and ethyl nitrite ?

Answer. Nitroethane (1° nitroalkane) reacts with nitrous acid to form crystalline nitrosonitroethane which dissolves in NaOH solution to form sodium salt of nitronic acid (*red*). Ethyl nitrite does not give this test.

$$\underset{\text{Nitroethane}}{CH_3CH_2NO_2} \xrightarrow{HONO} \underset{\text{Nitrosonitroalkane}}{CH_3{-}CH(NO){-}NO_2} + H_2O$$

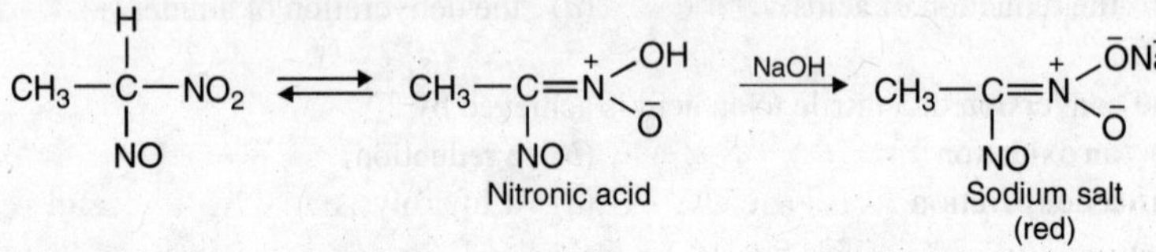

21. How is diazomethane prepared ? Describe its synthetic applications.

22. What happens when 2-butene is treated with diazomethane? (*Calicut BSc,2015*)

Answer. 1,2-Dimethylcyclopropane is formed.

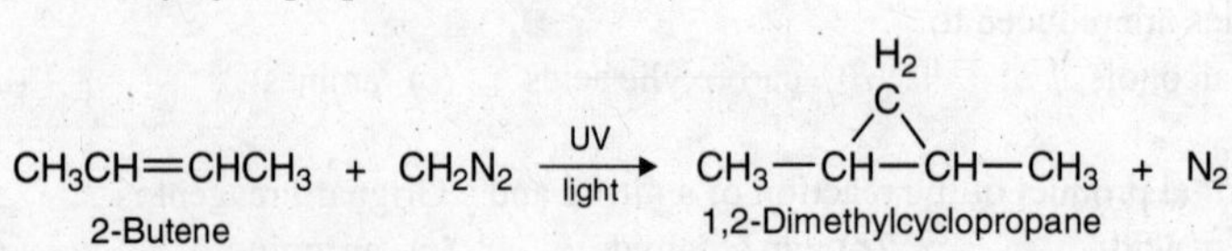

23. How will you synthesize propionic acid from acetic acid ? (*Indore BSc,2014*)

Answer. Use Arndt-Eistert reaction.

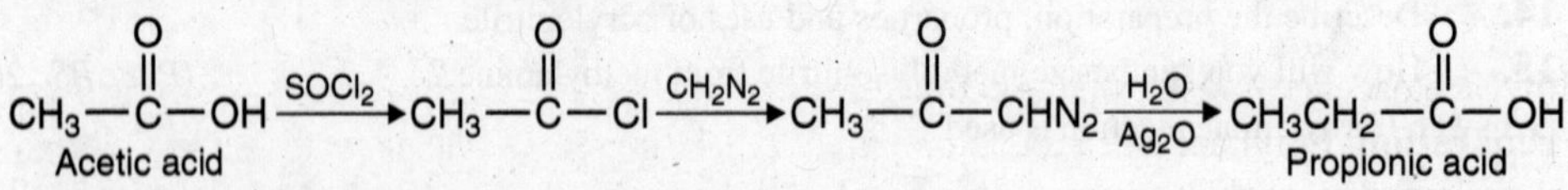

24. Discuss the mechanism of Arndt-Eistert reaction.

25. Write a note on: Arndt-Eistert reaction.

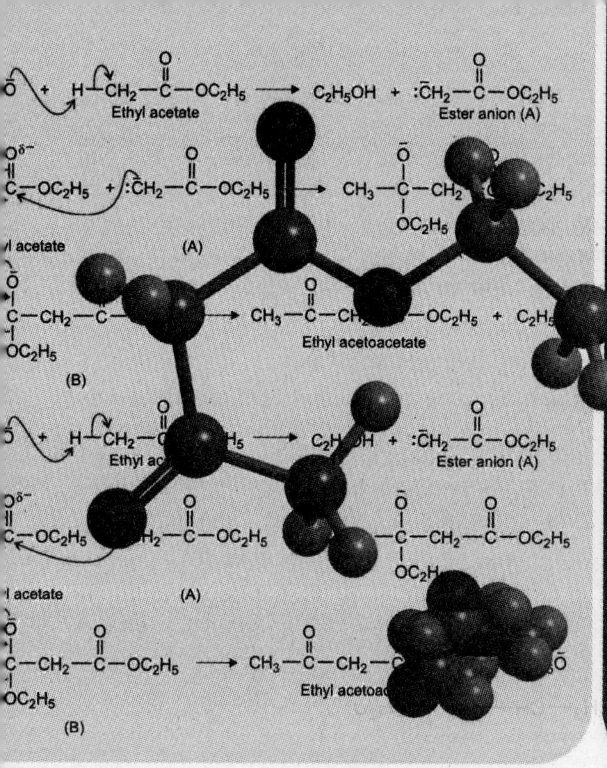

25

Acetoacetic Ester and Malonic Ester

The class of compounds which contain a methylene group (–CH_2–) directly bonded to two electron withdrawing groups such as –$COCH_3$, –$COOC_2H_5$, –CN, are called **Active Methylene Compounds**. This is so because the –CH_2– group in them is acidic and reactive. Ethyl acetoacetate (*Acetoacetic ester*) and Diethyl malonate (*Malonic ester*) belong to this class.

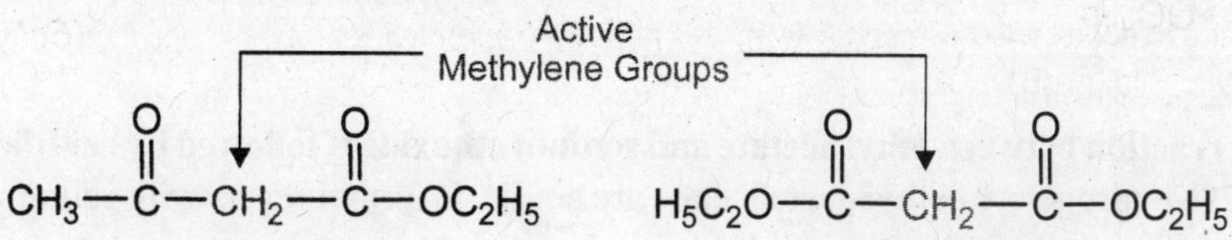

Ethyl acetoacetate (Acetoacetic ester)   Diethyl malonate (Malonic ester)

ETHYL ACETOACETATE, $CH_3COCH_2COOC_2H_5$

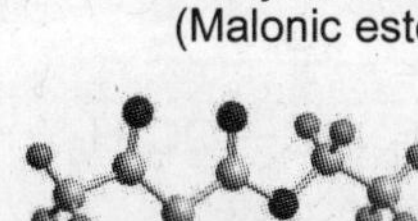

Ethyl acetoacetate is also called **Acetoacetic ester**. Its IUPAC name is **Ethyl 3-oxobutanoate**.

Preparation. Ethyl acetoacetate is prepared by heating ethyl acetate with sodium ethoxide in ethanol, followed by acidification.

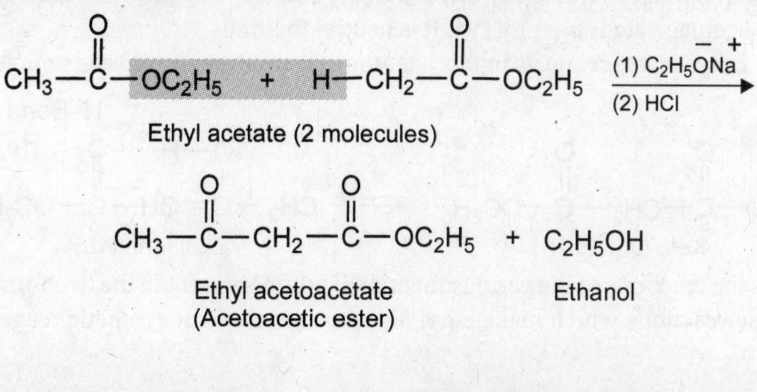

Laboratory Method. A mixture of ethyl acetate and sodium ethoxide ($C_2H_5OH + Na \rightarrow C_2H_5ONa$) is heated at 78°C for 8 hours. The mixture is then cooled and HCl or acetic acid is slowly added. The oily layer is separated, dried, and distilled. The fraction passing between 178-181°C is ethyl acetoacetate.

This reaction is an example of a general reaction known as **Claisen Condensation**. It involves base-catalyzed condensation of two ester molecules to form an alcohol and a β-keto ester. Ethyl acetoacetate is a β-keto ester.

MECHANISM. The mechanism of the above reaction involves three steps :

Step 1. Ethoxide ion (from $C_2H_5O^-Na^+$) attacks ethyl acetate to give ethyl alcohol and the ester anion.

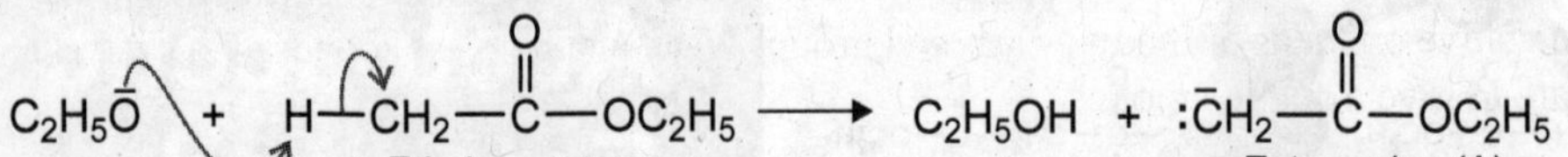

Ethyl acetate — Ester anion (A)

Step 2. Ester anion attacks the carbonyl group of a second molecule of ethyl acetate.

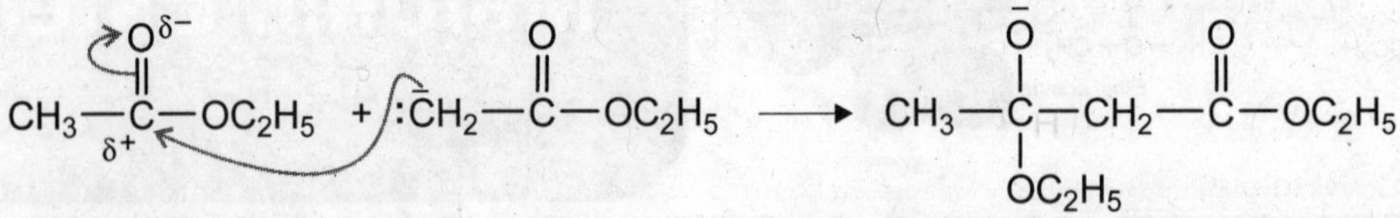

Ethyl acetate (A) (B)

Step 3. Ethoxide ion is eliminated.

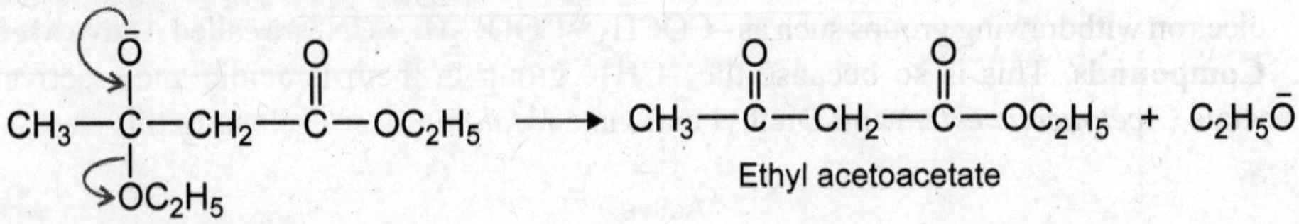

(B) — Ethyl acetoacetate

Why the reaction between ethyl acetate and sodium ethoxide is followed by acidification ? This is because the α-hydrogens of ethyl acetoacetate are acidic. It further reacts with sodium ethoxide to form a salt. Acidification of the salt gives back ethyl acetoacetate.

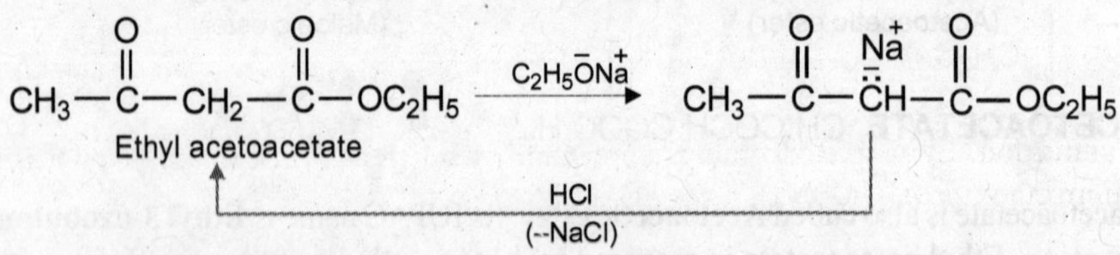

Properties (*Physical*). Ethyl acetoacetate is a colorless pleasant-smelling liquid, bp 180.4°C, density 1.0282. It is sparingly soluble in water but freely soluble in organic solvents. Refractive index of ordinary ethyl acetoacetate is $n_D = 1.4194$. It is neutral to litmus.

(*Chemical*). Ethyl acetoacetate, is in fact a tautomeric mixture of *keto* and *enol* forms.

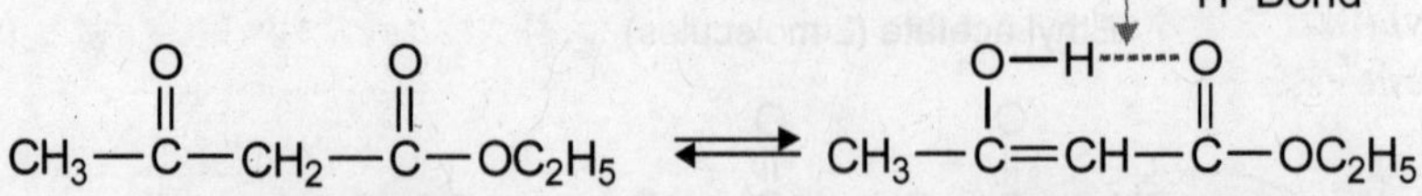

Keto form (90%) — Enol form (10%)

Thus it gives the reactions of the various functional groups present in the two forms. Here we will discuss only those reactions which make ethyl acetoacetate a useful synthetic reagent.

(1) **Acidity of Methylene hydrogens ; Formation of salts**. Ethyl acetoacetate contains a methylene group (–CH_2–) flanked by two carbonyl groups. The C–H bond in CH_2 group is readily ionizable because the proton removal forms a very stable carbanion.

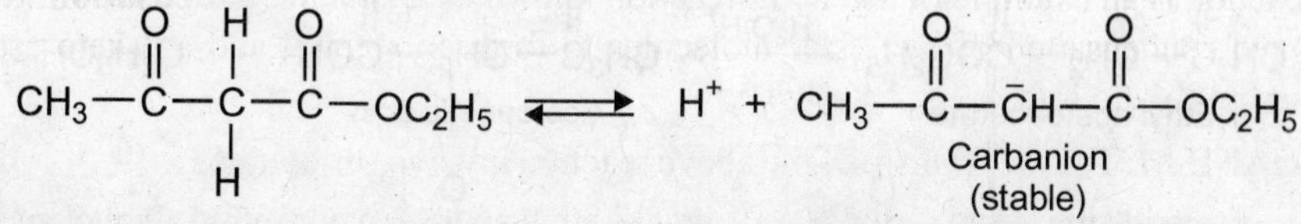

The acidity of the C—H bond of methylene group is attributed to two factors :

(*a*) **Inductive Effect**. The inductive effect caused by the electron attracting power of the electronegative oxygens of the two carbonyl groups weakens the C—H bonds. Thus the H atom can dissociate to give a stable anion.

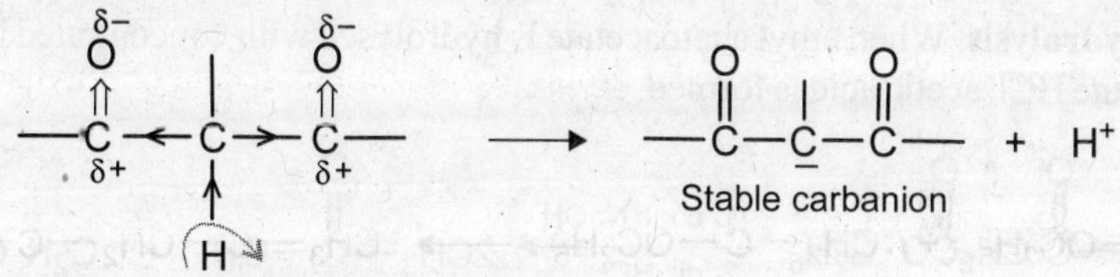

(*b*) **Resonance-Stabilization of Carbanion**. The acidity of the C–H bond is greatly enhanced because the negative charge in the carbanion is delocalized into the two carbonyl groups by resonance. The highly resonance-stabilized carbanion may be represented as :

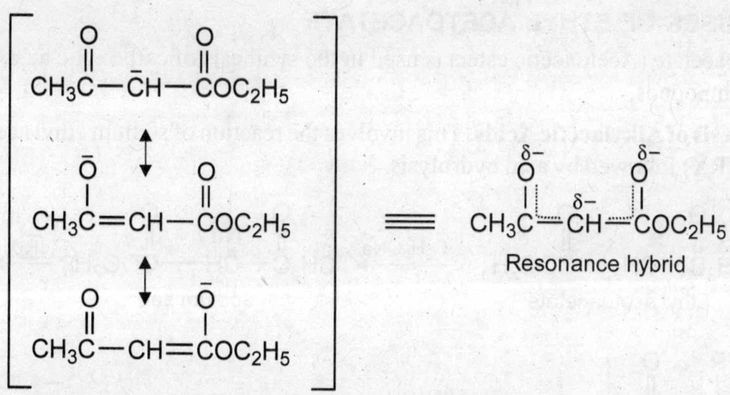

Salt Formation. Ethyl acetoacetate is appreciably acidic and when treated with a strong base such as sodium ethoxide (C_2H_5ONa), forms its sodium salt.

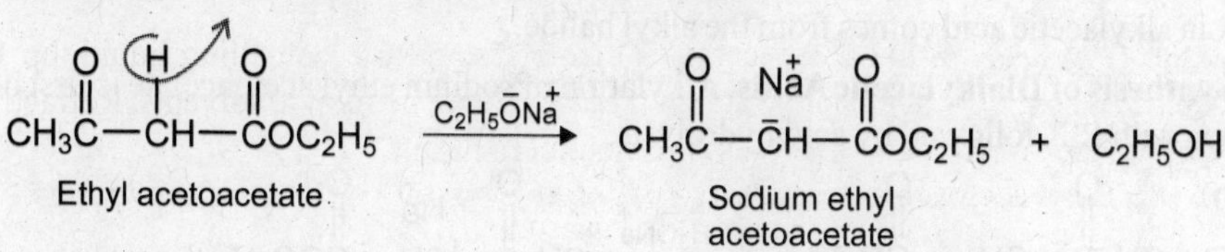

(2) **Alkylation**. Ethyl acetoacetate anion is nucleophilic and reacts with alkyl halides to give alkyl acetoacetic ester.

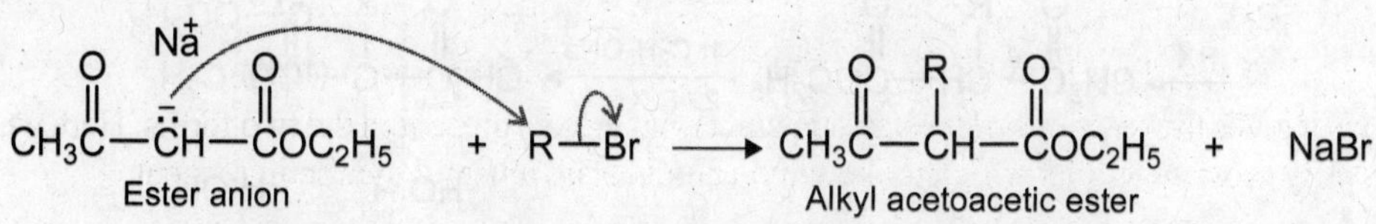

(3) **Ketonic Hydrolysis**. When ethyl acetoacetate is hydrolysed with dilute HCl, acetoacetic acid is formed. Acetoacetic acid undergoes decarboxylation on heating.

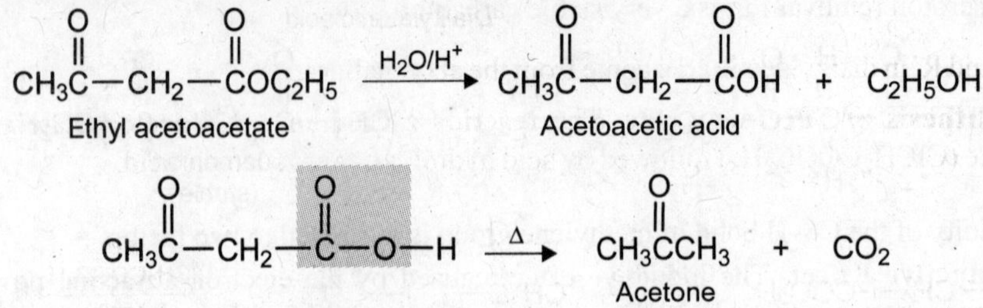

This type of hydrolysis of ethyl acetoacetate (or its alkyl derivatives) to give a ketone is called *ketonic hydrolysis*.

(4) **Acid Hydrolysis**. When ethyl acetoacetate is hydrolysed with concentrated NaOH and then acidified with dilute HCl, acetic acid is formed.

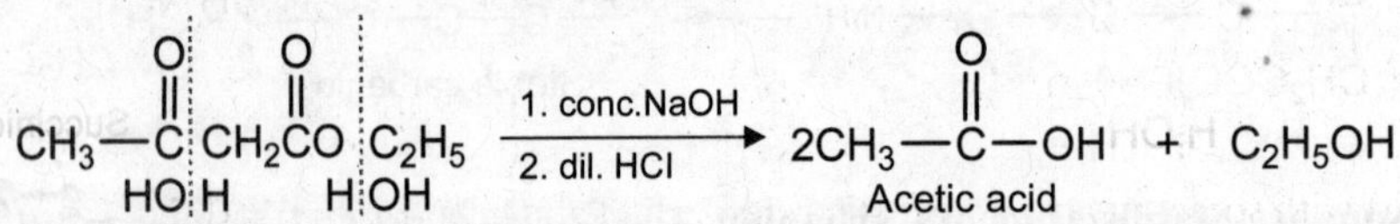

This type of hydrolysis of ethyl acetoacetate (or its alkyl derivatives) to give a carboxylic acid is called *acid hydrolysis*.

SYNTHETIC USES OF ETHYL ACETOACETATE

Ethyl acetoacetate (Acetoacetic ester) is used in the synthesis of carboxylic acids, ketones, and heterocyclic compounds.

(1) **Synthesis of Alkylacetic Acids**. This involves the reaction of sodium ethyl acetoacetate with an alkyl halide (RX) followed by acid hydrolysis.

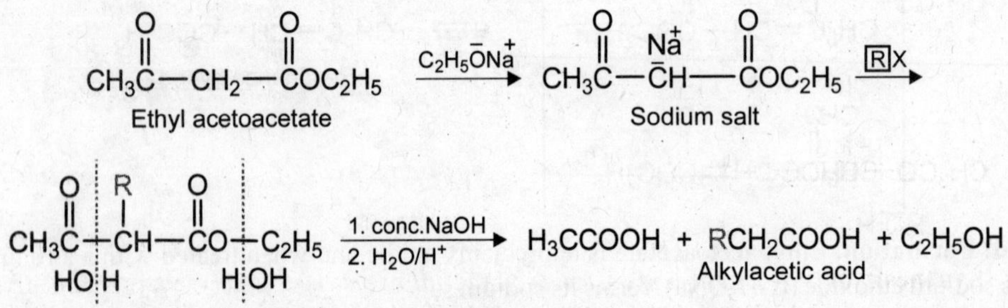

The R in alkylacetic acid comes from the alkyl halide.

(2) **Synthesis of Dialkylacetic Acids**. Alkylation of sodium ethyl acetoacetate is first done with RX and then with R′X followed by acid hydrolysis.

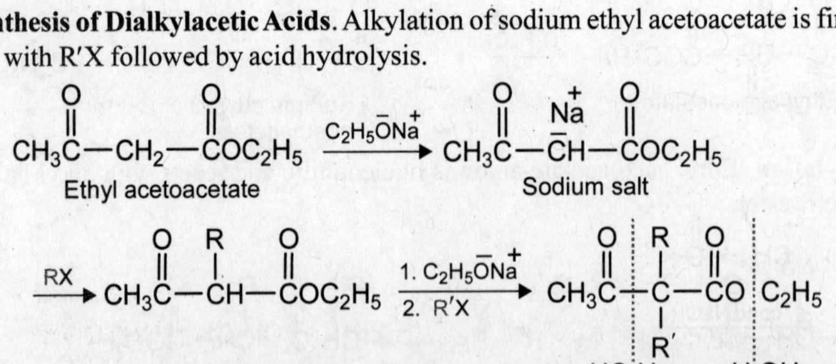

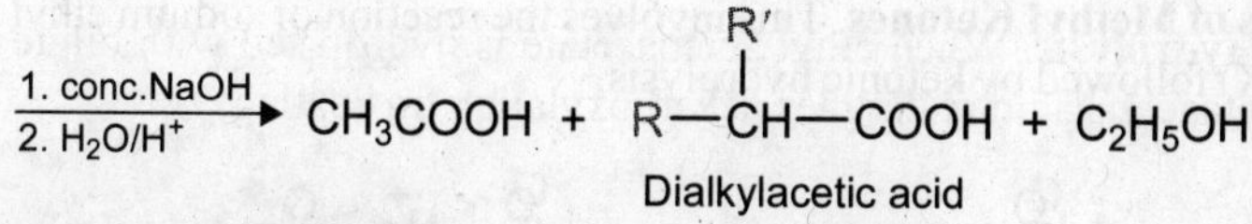

The R and R′ in dialkylacetic acid come from the alkyl halides.

(3) **Synthesis of Succinic Acids**. The reaction of sodium ethyl acetoacetate with ethyl chloroacetate ($ClCH_2COOC_2H_5$) followed by acid hydrolysis gives succinic acid.

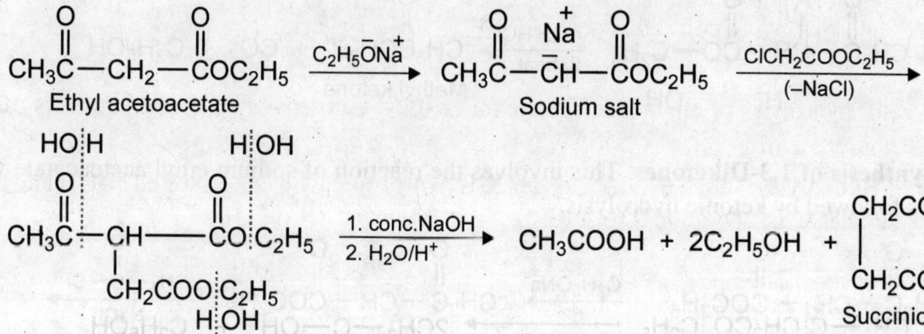

If we start with sodium alkylacetoacetic esters, alkylsuccinic acids are formed.

(4) **Synthesis of Higher Normal Diacids**. The reaction of sodium ethyl acetoacetate (2 molecules) with an alkylene diiodide followed by acid hydrolysis gives a normal dicarboxylic acid.

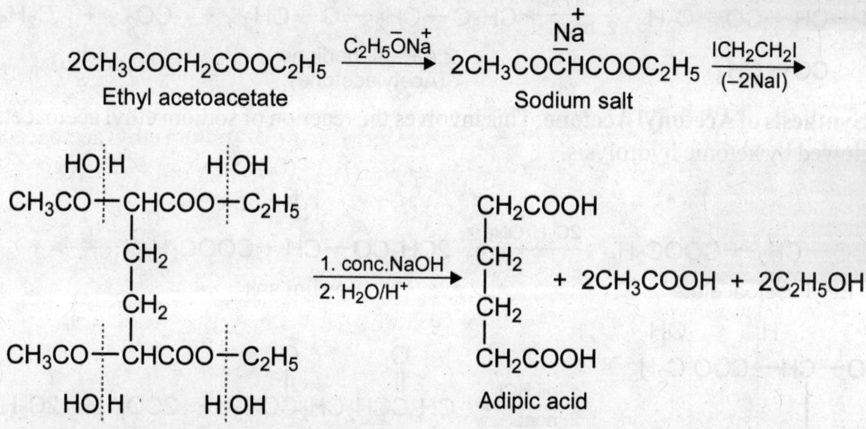

(5) **Synthesis of α,β-Unsaturated Acids**. This involves base catalyzed reaction of ethyl acetoacetate with an aldehyde or a ketone followed by acid hydrolysis.

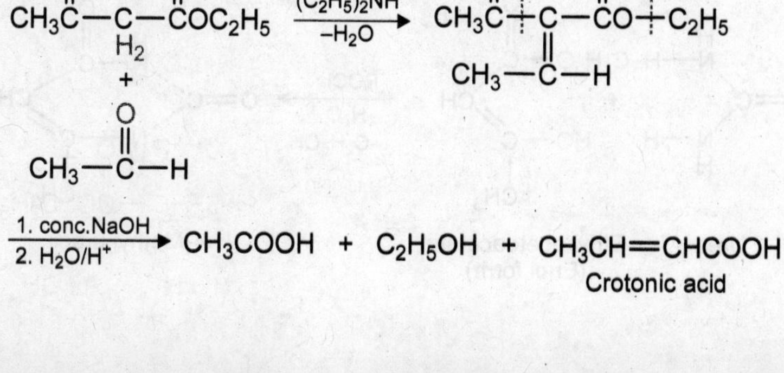

(6) **Synthesis of Methyl Ketones**. This involves the reaction of sodium ethyl acetoacetate with an alkyl halide (RX) followed by ketonic hydrolysis.

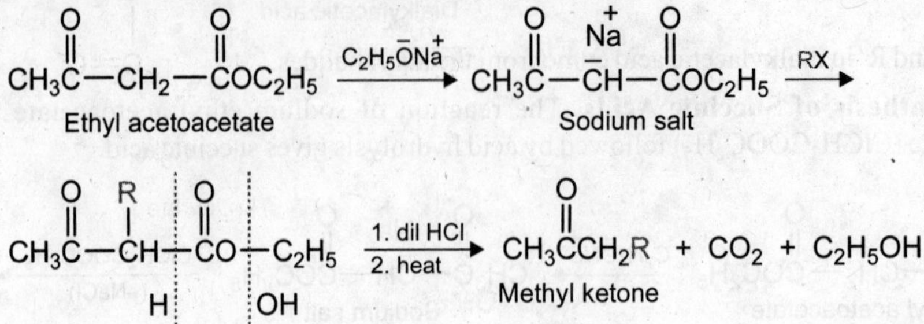

(7) **Synthesis of 1,3-Diketones**. This involves the reaction of sodium ethyl acetoacetate with acid halides followed by ketonic hydrolysis.

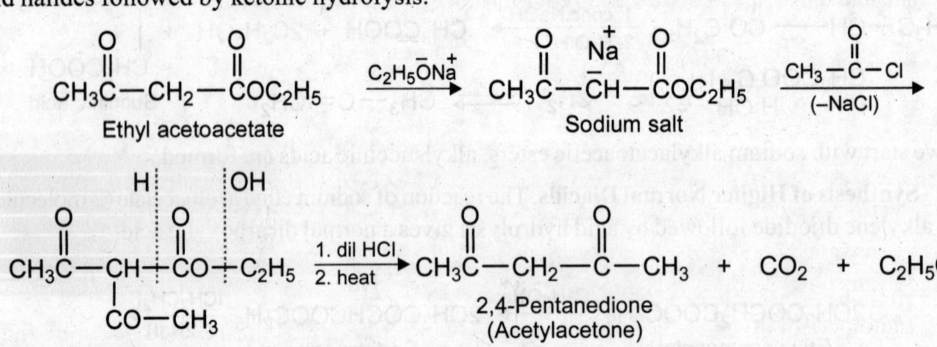

(8) **Synthesis of Acetonyl Acetone**. This involves the reaction of sodium ethyl acetoacetate with iodine followed by ketonic hydrolysis.

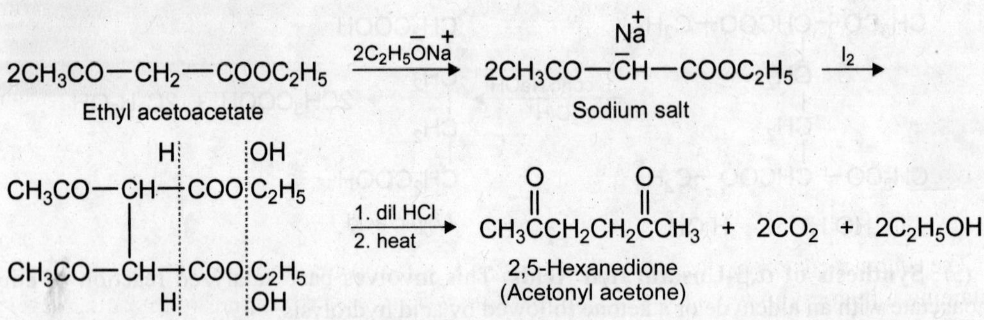

(9) **Synthesis of 4-Methyl Uracil**. Ethyl acetoacetate (in its *enol* form) reacts with urea in the presence of phosphoryl chloride to give 4-methyl uracil.

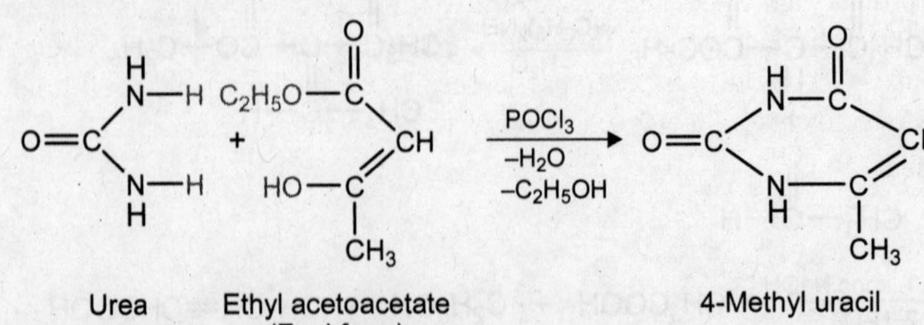

(10) **Synthesis of Antipyrine**. Ethyl acetoacetate reacts with phenylhydrazine to give antipyrine.

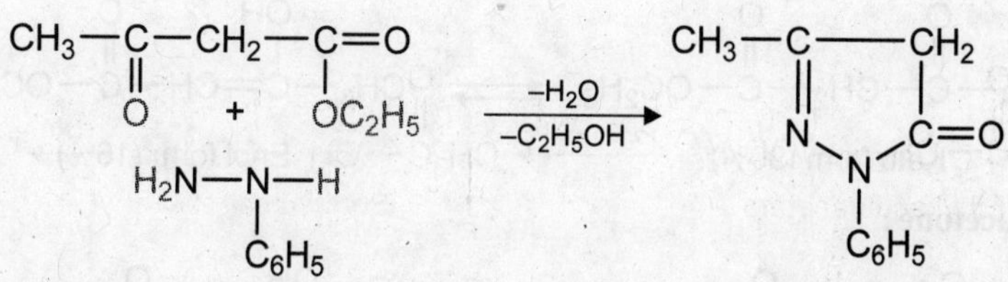

Antipyrine

TAUTOMERISM

When two structural isomers are mutually interconvertible and exist in dynamic equilibrium, they are called Tautomers and the phenomenon is termed Tautomerism. For example, acetone exhibits tautomerism and may be represented as an equilibrium mixture of two isomers.

Keto form Enol form

Here the *keto* form changes to the *enol* form by migration of a proton to carbonyl oxygen. Then a pair of electrons shifts from the C–H bond to C–C bond.

How Tautomerism differs from Resonance ? (1) The tautomers are distinct molecules while resonance forms have no real independent existence.

(2) Tautomerism involves the movement of both electrons and atoms (a hydrogen atom in the above case) whereas resonance structures differ in respect of positions of electrons only.

Keto-Enol Tautomerism. Aldehydes, ketones, and other carbonyl compounds (*e.g.*, esters) exhibit this special type of tautomerism. It involves migration of a proton (H^+) from α-carbon to carbonyl oxygen by the following mechanism.

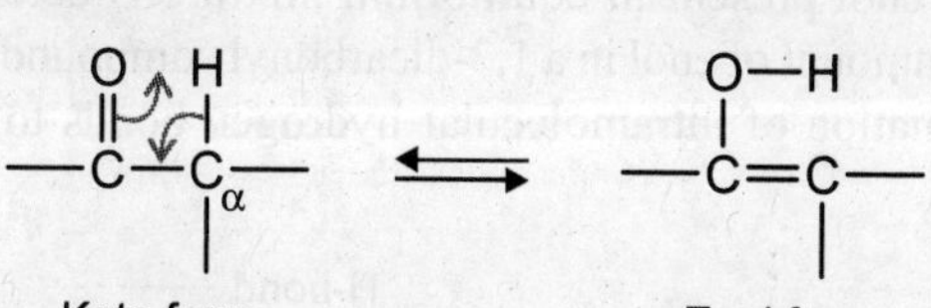

Keto form Enol form

The tautomer containing the carbonyl group (C=O) is designated as the **Keto form**. The other one containing a hydroxy group attached to a doubly bonded carbon is referred to as the **Enol form** (alkENe + alcohOL). This kind of tautomerism is called the **Keto-Enol Tautomerism**.

We can easily determine the percentage of the keto-enol equilibrium mixtures by NMR spectroscopy. In ordinary aldehydes and ketones with just one carbonyl group, the enol form comprises less than 1% of the equilibrium mixture. However, compounds with two carbonyl groups separated by a methylene group ($-CH_2-$) can exist to a great degree in the enol form under equilibrium conditions. This will be illustrated by the following examples of keto-enol tautomerism.

(*a*) **Acetone :**

Keto form Enol form (.00015%)

(*b*) **Ethyl acetoacetate :**

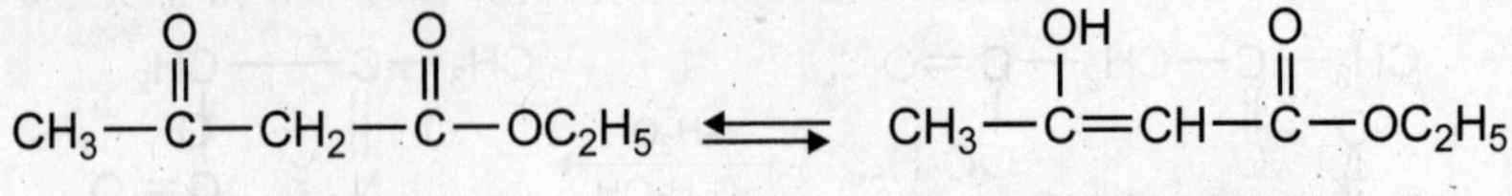

Keto form (90%) Enol form (16%)

(*c*) **Acetylacetone :**

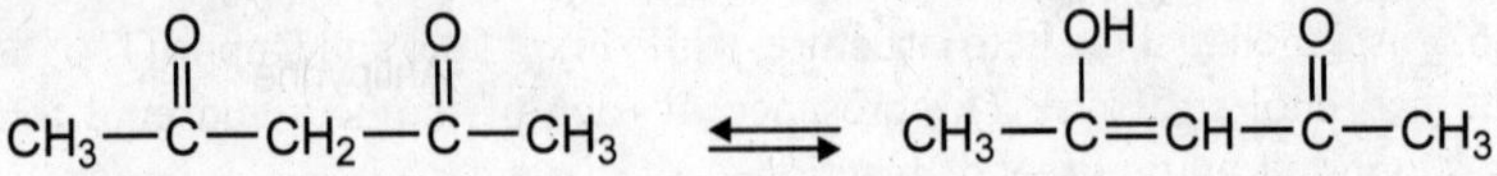

Keto form (24.0%) Enol form (76.0%)

EXPLANATION. Owing to the electron-withdrawing capacity of the carbonyl group, α-hydrogen of the keto compound ionizes to form a carbanion (enolate ion). The carbanion is resonance-stabilized and may be represented as

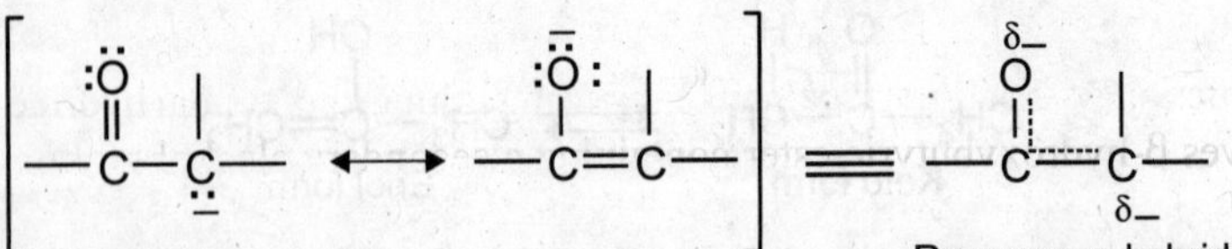

Resonance hybrid

On acidification, the protonation of the cation at oxygen produces the *enol* form, while protonation at carbon yields the *keto* form.

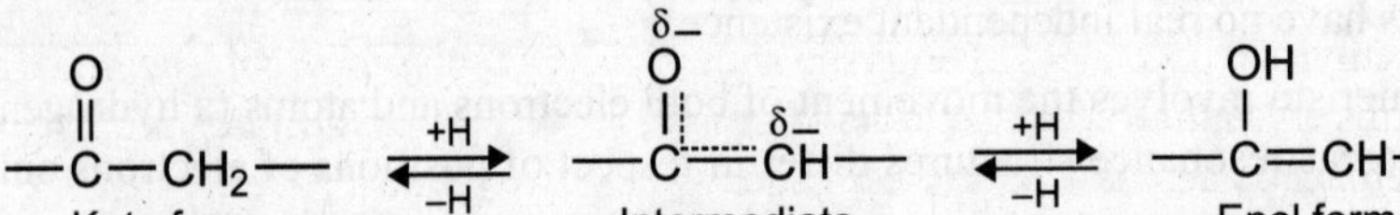

Keto form Intermediate Enol form

Both of these reactions are reversible and the net result is keto-enol equilibrium.

The proportion of the enol present in equilibrium mixture is determined by its stability. The existence of an appreciable amount of enol in a 1,3-dicarbonyl compound *e.g.*, acetylacetone (72%) is on two accounts : (1) Formation of intramolecular hydrogen bonds to give a six-membered ring, lending stability to the enol.

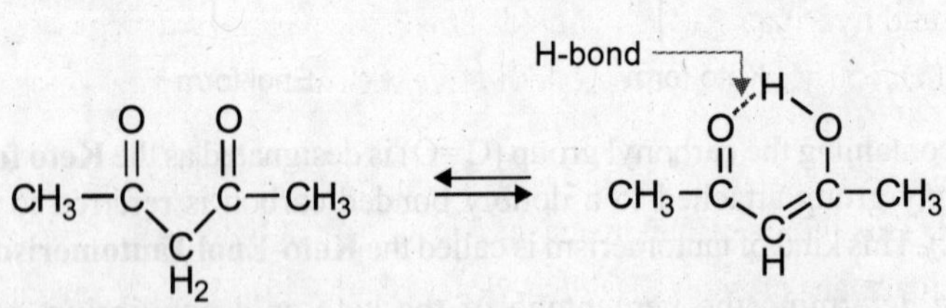

Keto form Enol form

(2) The enol form is resonance-stabilized, while the keto form is not.

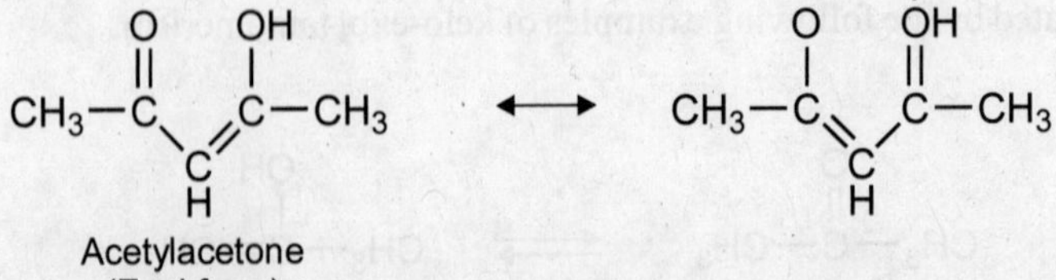

Acetylacetone
(Enol form)

KETO-ENOL TAUTOMERISM OF ETHYL ACETOACETATE

Ethyl acetoacetate offers a classical and most thoroughly investigated example of keto-enol tautomerism.

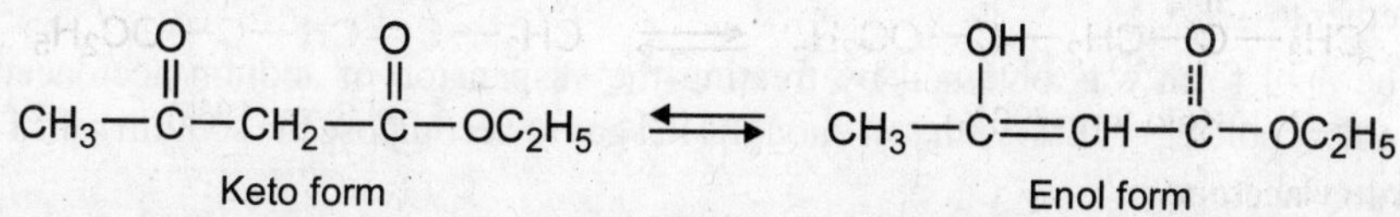

Geuther (1863) assigned to it the keto structure, while Frankland and Duppa (1865) showed that ethyl acetoacetate had enol structure. The presence of each of the keto and enol forms in ethyl acetoacetate was supported by two sets of reactions.

Reactions Supporting the Keto form. (1) Ethyl acetoacetate forms a bisulfite compound with sodium hydrogen sulfite.

(2) Ethyl acetoacetate forms a cyanohydrin with hydrogen cyanide.

(3) Ethyl acetoacetate forms an oxime with hydroxylamine, and a phenylhydrazone with phenylhydrazine.

(4) On reduction with sodium amalgam or by using lithium aluminium hydride in pyridine, ethyl acetoacetate gives β-hydroxybutyric ester containing a secondary alcohol group.

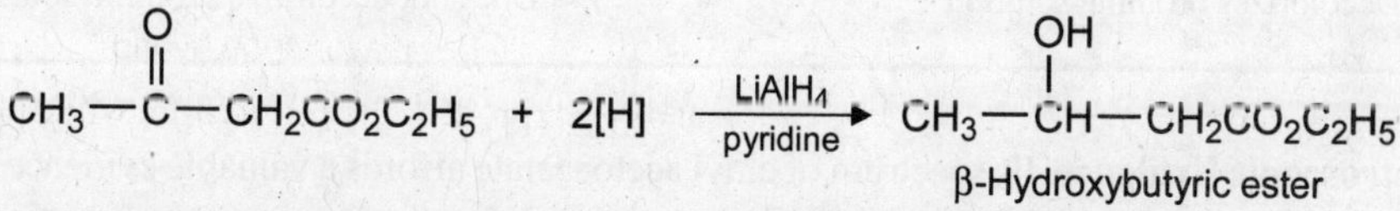

(5) When ethyl acetoacetate is hydrolysed with dil. NaOH solution in cold, acidified and extracted with ether, it forms crystalline acetoacetic acid (Kruger, 1952). This on heating yields acetone showing thereby that acetoacetic acid is a β-keto acid and ethyl acetoacetate a β-keto ester.

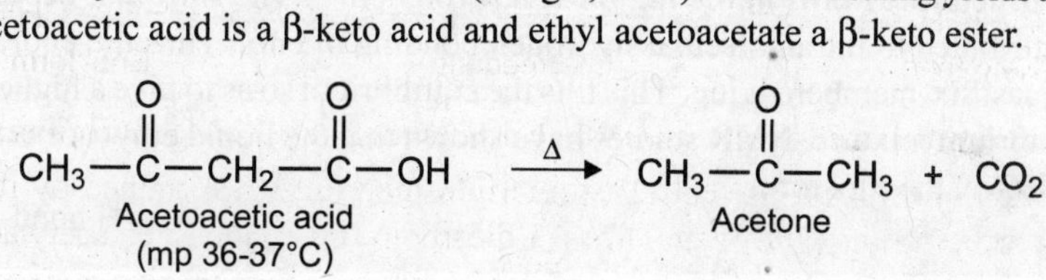

Reactions Supporting the Enol form. (1) Ethyl acetoacetate reacts with sodium metal to form the sodium derivative and hydrogen gas is evolved. This indicates the presence of OH group.

(2) Ethyl acetoacetate forms an acetyl derivative with acetyl chloride which shows the presence of OH group.

(3) When ethyl acetoacetate is treated with alcoholic bromine solution, the brown color of the latter is discharged. This indicates the presence of an alkene double bond (C=C).

(4) Ethyl acetoacetate produces reddish-violet color with $FeCl_3$, a reaction characteristic of phenolic OH. This points to the presence of C=C–OH structural unit in ethyl acetoacetate as in phenols.

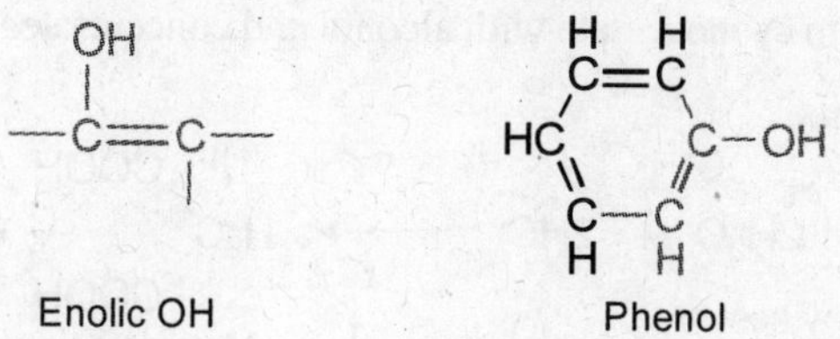

Separation of Keto and Enol forms. Ludwing Knorr (1911) isolated the two tautomeric forms of ethyl acetoacetate.

(1) Keto form was obtained as crystals by cooling a solution of ordinary ethyl acetoacetate in petroleum ether to –78°C.

(2) The enol form was obtained by treating the suspension of sodium acetoacetic ester in petroleum ether cooled to –78°C with just enough HCl gas to decompose the sodium salt. The product was an oily liquid.

The keto and enol forms isolated as above were found to have distinctive properties conforming to their structure.

Table 25.1 PROPERTIES OF KETO AND ENOL FORMS

Keto form	Enol form
1. Long colorless needles, mp –39°C	1. Colorless oily liquid, at –78°C
2. Refractive Index, $n_D^{20} = 1.4171$	2. $n_D^{20} = 1.4432$
3. Density 1.0368^{10}	3. 1.0119^{10}
4. No coloration with $FeCl_3$	4. Instense coloration (red-violet)
5. Decolorizes bromine solution	5. Does not decolorize bromine solution.

Spectroscopic Evidence. IR spectrum of ethyl acetoacetate affords a valuable evidence regarding its structure. It shows an absorption band at 1718 cm^{-1} indicating the presence of a ketonic carbonyl; and another band at 1650 cm^{-1} due to unsaturated carbonyl (C=C–C=O). But enol of ethyl acetoacetate containing this structure absorbs in the region 1685–1665 cm^{-1}. The shift that occurs by keto-enol equilibrium is attributed to intramolecular hydrogen bonding of enol. Thus the enol is stabilized by formation of a quasi six-membered ring. This tilts the equilibrium so as to give a higher percentage of enol in the equilibrium mixture. NMR studies have shown that the liquid ethyl acetoacetate contains approximately 75% of the enol form.

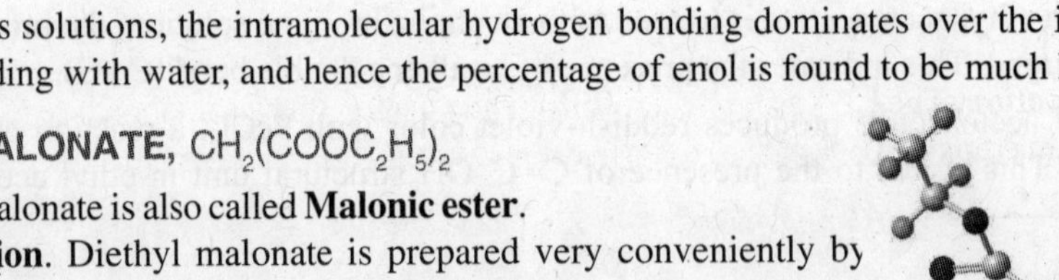

Keto form (25%) Enol form (75%)

In aqueous solutions, the intramolecular hydrogen bonding dominates over the intermolecular hydrogen bonding with water, and hence the percentage of enol is found to be much less, about 10.

DIETHYL MALONATE, $CH_2(COOC_2H_5)_2$

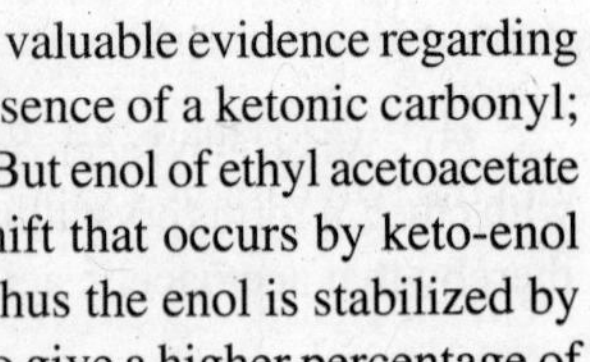

Diethyl malonate is also called **Malonic ester**.

Preparation. Diethyl malonate is prepared very conveniently by boiling sodium or potassium cyanoacetate with alcohol and concentrated hydrochloric acid.

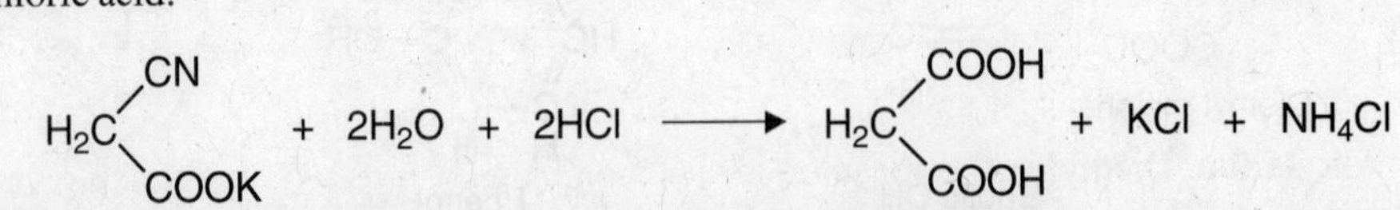

Pot cyanoacetate Malonic acid

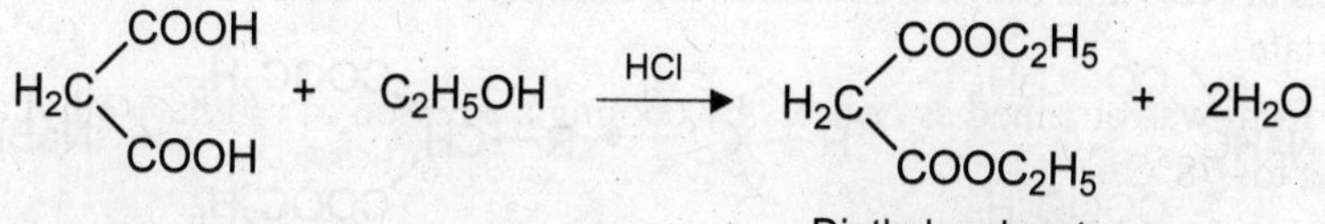

Diethyl malonate
(Malonic ester)

The cyanoacetate required for the process is obtained from acetic acid by the following steps :

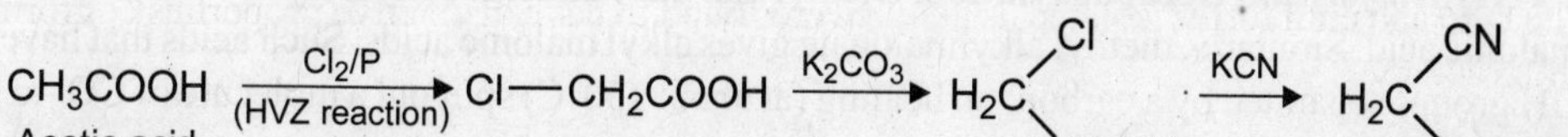

Acetic acid   Pot. cyanoacetate

Properties (*Physical*). Diethyl malonate is a colorless, pleasant-smelling liquid, bp 199.2°C. It is sparingly soluble in water, freely soluble in alcohol and ether.

(*Chemical*). (1) **Acidity of CH_2 group ; Formation of Salts**. Like ethyl acetoacetate, diethyl malonate contains a methylene group joined to two carbonyl groups. The H atoms of the CH_2 group are acidic. This is attributed to two factors : (*a*) the electron-attracting power of the electronegative oxygen of the carbonyl group (*inductive effect*) ; and (*b*) the resonance-stabilization of the resultant cation (diethyl malonate anion). Thus,

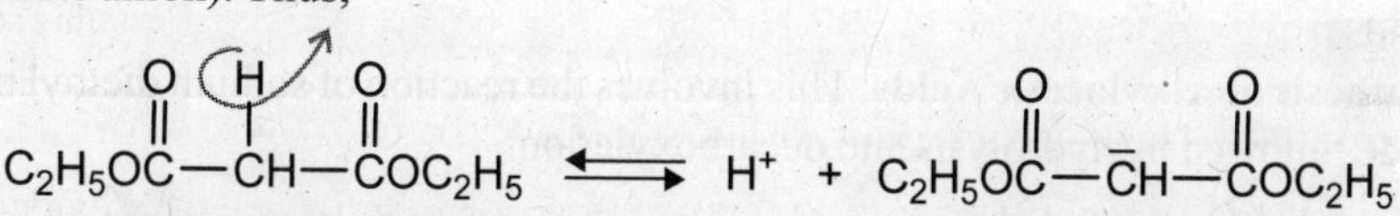

Diethyl malonate   Diethyl malonate anion

The diethyl malonate anion is highly resonance-stabilized so that its negative charge is delocalized into the two carbonyl groups.

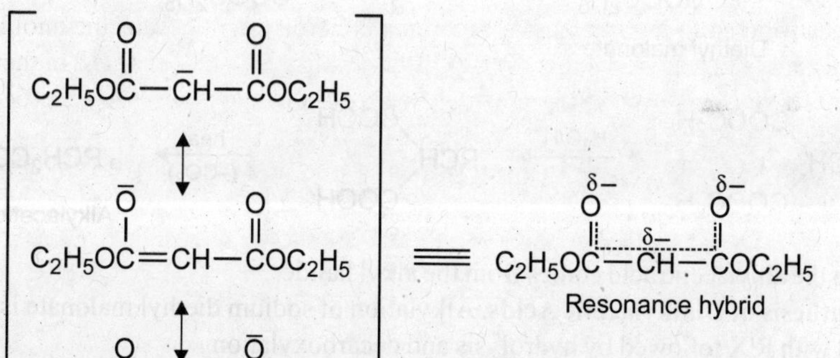

Resonance hybrid

Salt Formation. The CH_2 group being sufficiently acidic, diethyl malonate reacts with a strong base like sodium ethoxide (C_2H_5ONa) to form the sodium salt.

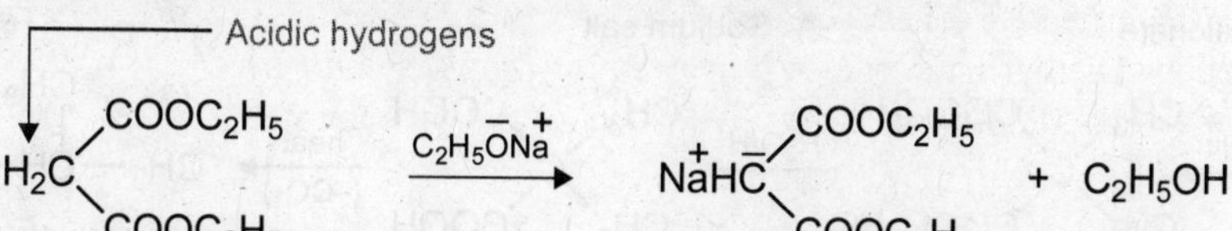

Diethyl malonate   Sodium diethyl malonate

(2) **Alkylation**. Diethyl malonate anion is nucleophilic and reacts with halides to give diethyl alkylmalonate.

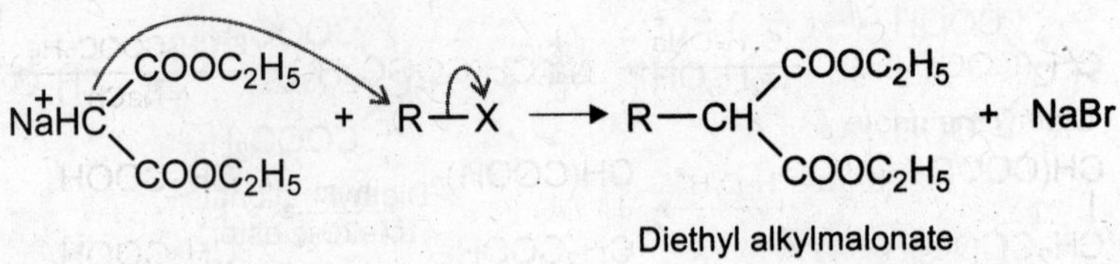

(3) **Hydrolysis and Decarboxylation.** Diethyl malonate undergoes hydrolysis with dilute HCl to give malonic acid. Similarly, diethyl alkylmalonate gives alkyl malonic acids. Such acids that have two –COOH groups separated by a carbon, on heating (at about 150°C) split out a molecule of CO_2 to give the monocarboxylic acid.

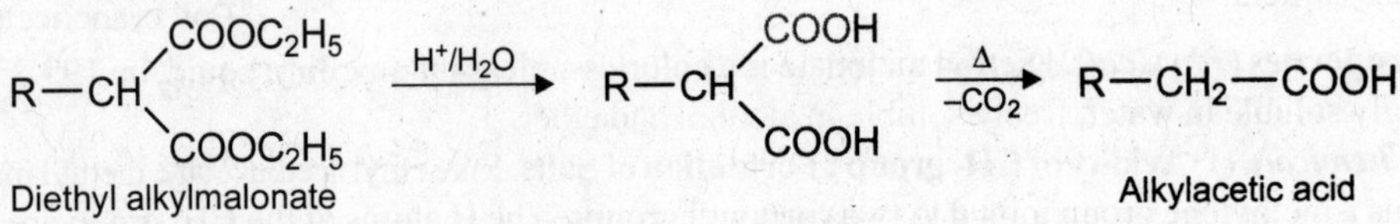

SYNTHETIC USES OF DIETHYL MALONATE

Diethyl malonate is used in the synthesis of carboxylic acids, keto acids, α-amino acids, and barbituric acid.

(1) **Synthesis of Alkylacetic Acids.** This involves the reaction of sodium diethyl malonate with an alkyl halide followed by hydrolysis and decarboxylation.

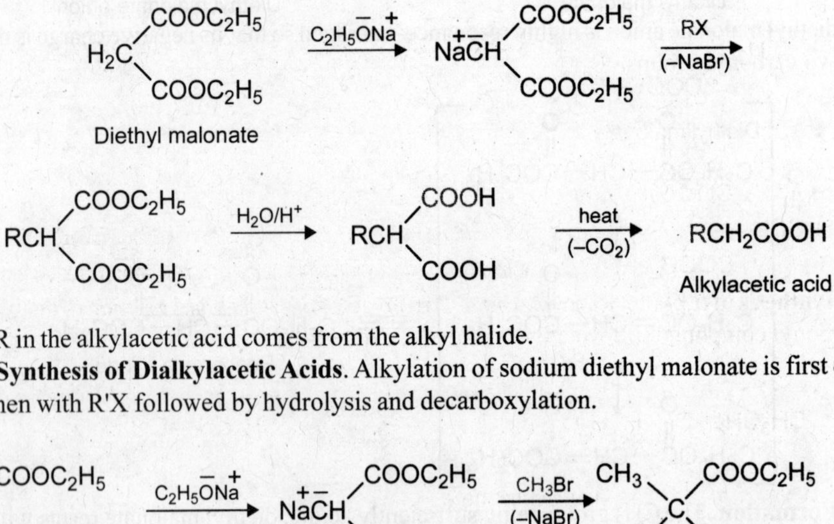

The R in the alkylacetic acid comes from the alkyl halide.

(2) **Synthesis of Dialkylacetic Acids.** Alkylation of sodium diethyl malonate is first done with RX and then with R'X followed by hydrolysis and decarboxylation.

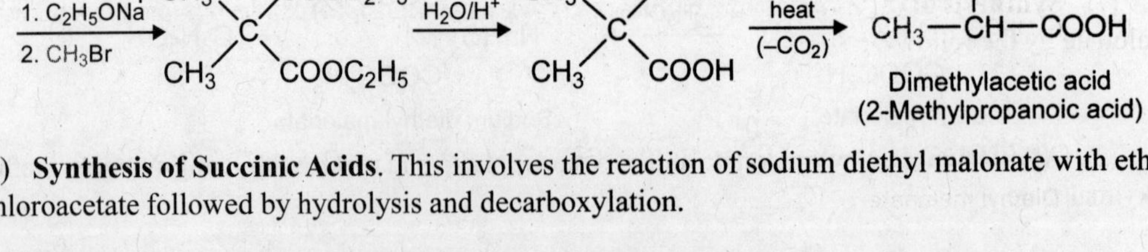

(3) **Synthesis of Succinic Acids.** This involves the reaction of sodium diethyl malonate with ethyl chloroacetate followed by hydrolysis and decarboxylation.

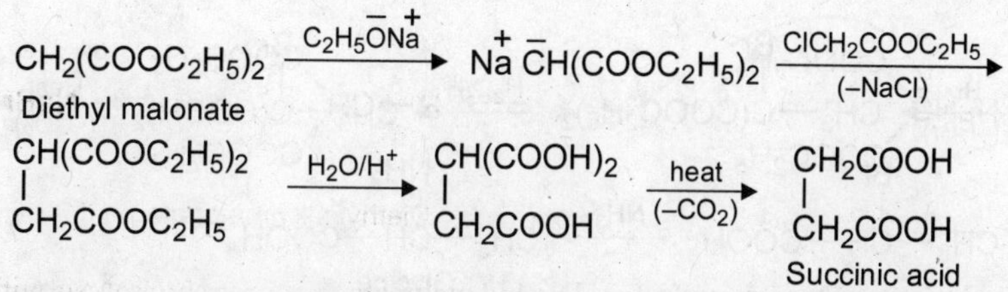

If we start with sodium alkylmalonic esters, alkylsuccinic acids are formed.

(4) **Synthesis of Higher Normal Diacids**. This involves the reaction of sodium diethyl malonate (2 molecules) with an alkylene diiodide followed by hydrolysis and decarboxylation gives a normal dicarboxylic acid.

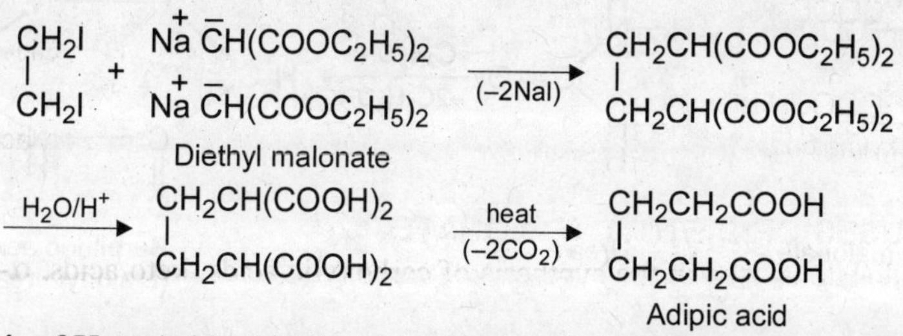

(5) **Synthesis of Keto Acids**. This involves the reaction of sodium diethyl malonate with acid halides followed by hydrolysis and decarboxylation.

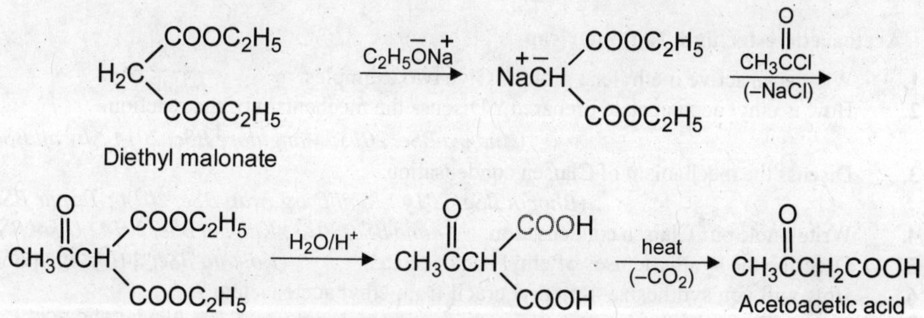

(6) **Synthesis of α,β-Unsaturated Acids**. This involves base catalyzed reaction of diethyl malonate with a carbonyl compound followed by hydrolysis and decarboxylation.

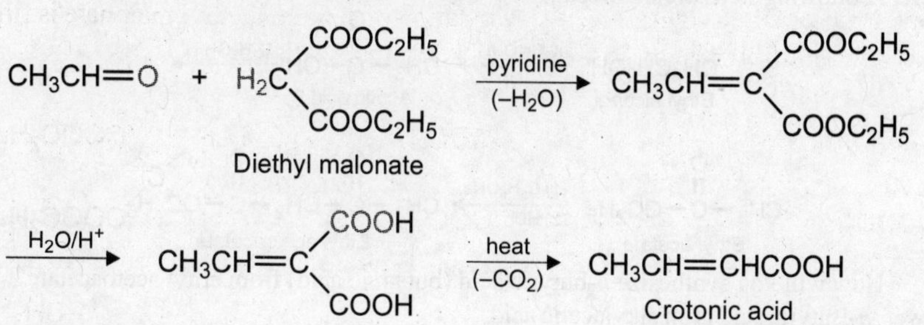

(7) **Synthesis of α-Amino acids**. Glycine (a typical α-amino acid) can be obtained from diethyl malonate by the following steps :

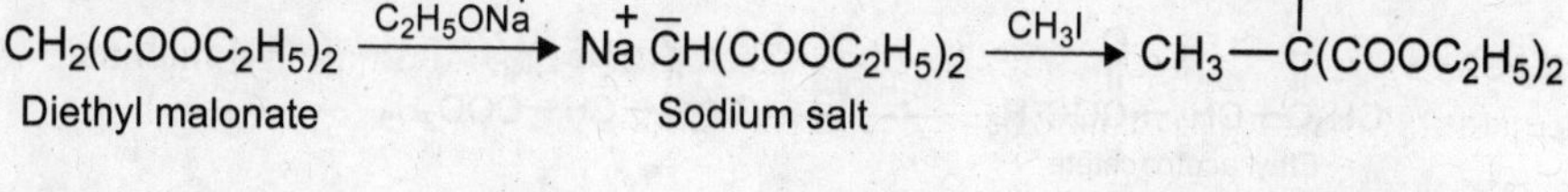

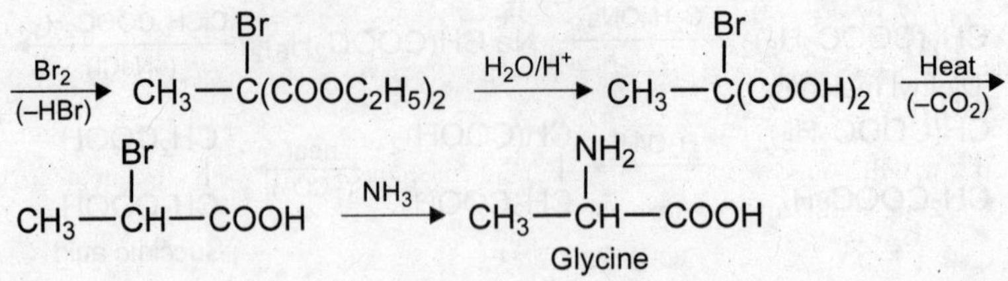

(8) **Synthesis of Barbituric Acid**. Diethyl malonate reacts with urea to give barbituric acid.

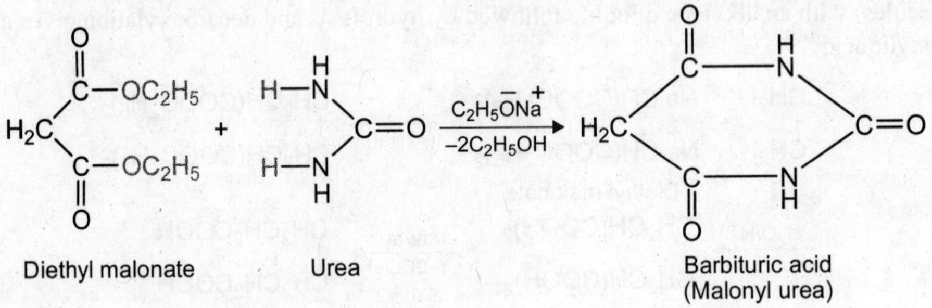

Examination Questions

Part 1. Acetoacetic ester and Tautomerism

1. What is an active methylene group ? Give two examples.
2. How is ethyl acetoacetate prepared ? Discuss the mechanism of the reaction.
   (Andhra BSc, 2013; Mangalore BSc, 2014 ; Jiwaji BSc, 2015)
3. Discuss the mechanism of Claisen condensation.
   (Bhopal BSc, 2014 ; Rani Durgawati BSc, 2014 ; Tezpur BSc, 2015)
4. Write a note on : Claisen condensation. *(Anna BSc, 2013 ; Kalyani BSc, 2014 ; Utkal BSc, 2015)*
5. Describe the synthetic uses of ethyl acetoacetate. *(Kolkata BSc, 2013; Agra BSc, 2014)*
6. How will you synthesize 4-methyl uracil from ethyl acetoacetate ?
7. How will you synthesize Antipyrine from ethyl acetoacetate ?
8. How will you synthesize ethyl acetoacetate from ethyl alcohol ?

Answer. Following steps are involved :

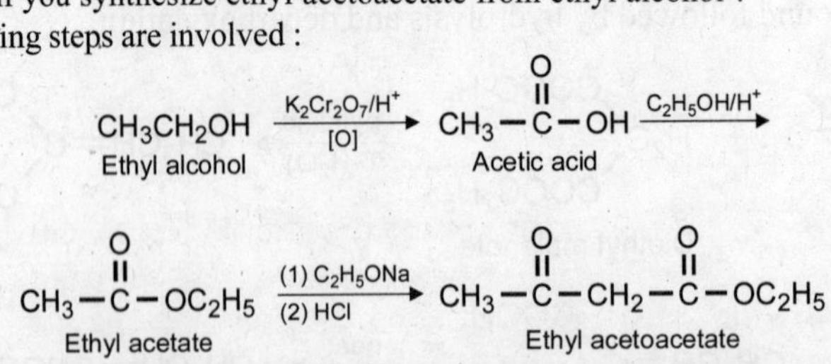

9. How will you synthesize *n*-butyric acid (butanoic acid) from ethyl acetoacetate ?

Answer. *n*-Butyric acid is an alkylacetic acid.

RCH_2COOH	$CH_3CH_2CH_2COOH$
Alkylacetic acid	n-butyric acid

Step involved are :

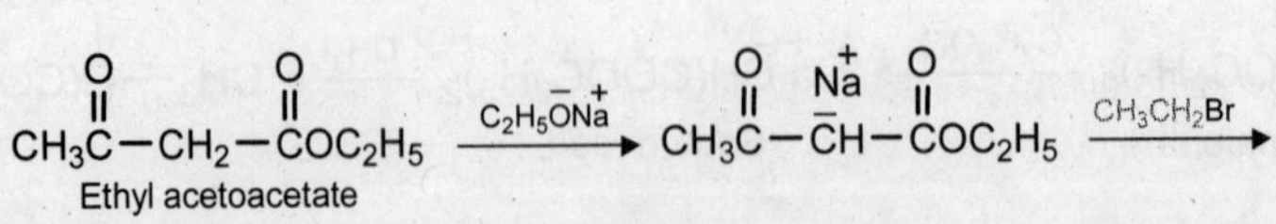

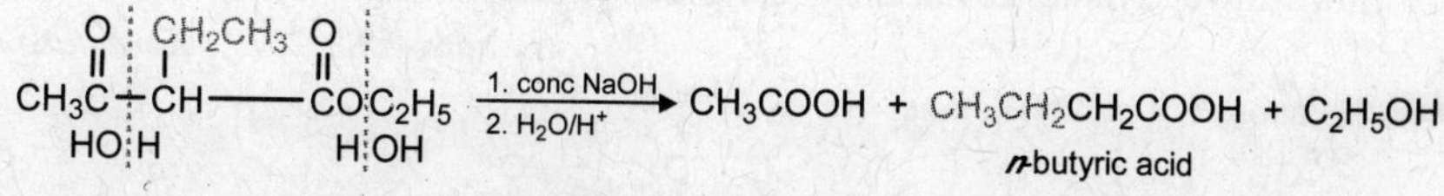

10. How will you synthesize propionic acid (propanoic acid) from ethyl acetoacetate ?

Clue. Propionic acid is an alkylacetic acid.

RCH_2COOH (Alkylacetic acid) CH_3CH_2COOH (Propionic acid)

Steps involved are : (*i*) C_2H_5ONa ; (*ii*) CH_3Br ; (*iii*) conc. NaOH followed by dil. HCl (acid hydrolysis).

11. How will you synthesize 2-methylbutanoic acid from ethyl acetoacetate ?

Clue. 2-Methylbutanoic acid is a dialkylacetic acid.

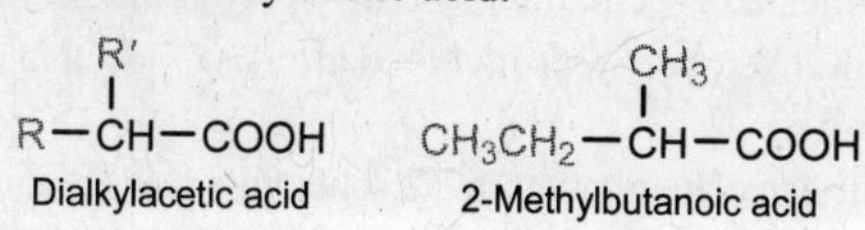

Steps involved are : (*i*) C_2H_5ONa ; (*ii*) CH_3Br ; (*iii*) C_2H_5ONa ; (*iv*) CH_3CH_2Br ; (*v*) conc. NaOH followed by dilute HCl (acid hydrolysis).

12. How will you synthesize succinic acid from ethyl acetoacetate ?

(*Orissa BSc, 2014 ; Awadh BSc, 2015*)

13. How will you synthesize methylsuccinic acid from ethyl acetoacetate ?

Clue. Steps are : (*i*) C_2H_5ONa ; (*ii*) $CH_3CH(Cl)COOC_2H_5$ = ethyl 2-chloropropanoate ; (*iii*) conc. NaOH followed by dilute HCl (acid-hydrolysis).

14. How will you synthesize 2,3-dimethylbutanoic acid from ethyl acetoacetate ?

Clue. 2,3-Dimethylbutanoic acid is a dialkylacetic acid. (*Rohilkhand BSc, 2014*)

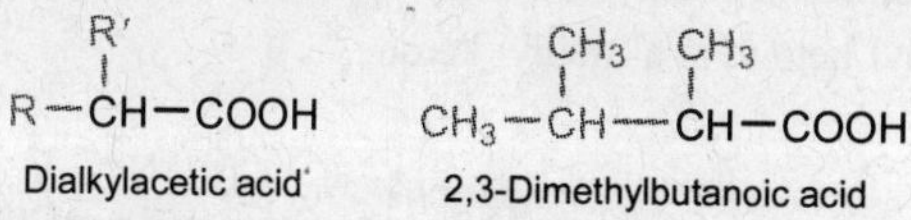

Steps involved are : (*i*) C_2H_5ONa ; (*ii*) CH_3Br ; (*iii*) C_2H_5ONa ; (*iv*) Isopropyl bromide = $(CH_3)_2CHBr$; (*v*) conc. NaOH followed by dilute HCl (acid hydrolysis).

15. How will you synthesize isobutyric acid (2-methylpropanoic acid) from ethyl acetoacetate ?

Answer. Isobutyric acid is a dialkylacetic acid. (*Andhra BSc, 2015*)

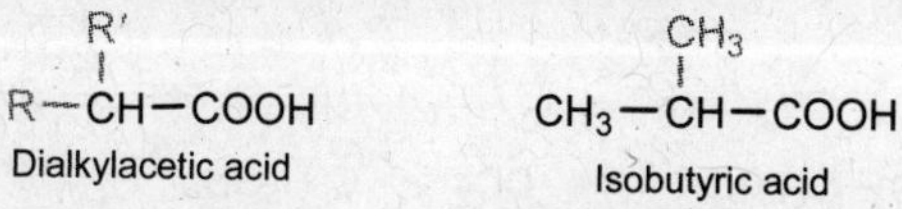

Steps involved are :

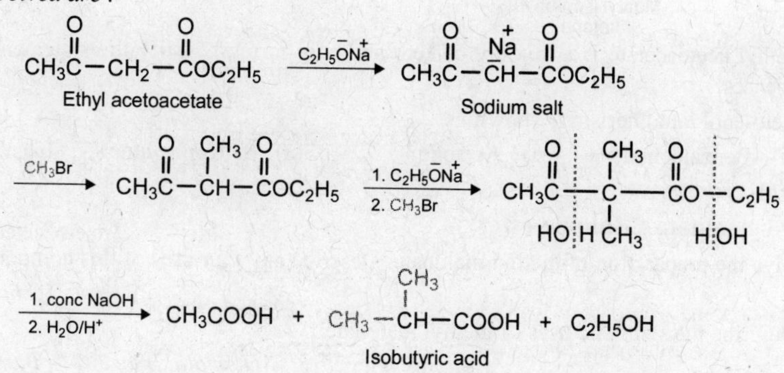

16. How will you synthesize adipic acid from ethyl acetoacetate ? (*Madurai BSc, 2015*)

17. How will you synthesize crotonic acid from ethyl acetoacetate ?

18. How will you synthesize butanone (ethyl methyl ketone) from ethyl acetoacetate ?

(Banaras BSc, 2013 ; Karnataka BSc, 2015)

Answer. Butanone is a methyl ketone.

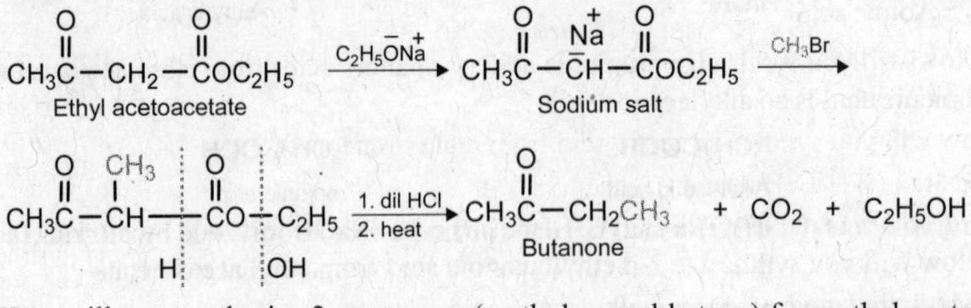

19. How will you synthesize 2-pentanone (methyl propyl ketone) from ethyl acetoacetate ?

Clue. 2-Pentanone is a methyl ketone. Steps involved are : (*i*) C_2H_5ONa ; (*ii*) CH_3CH_2Br ; (*iii*) dilute HCl followed by heat (ketonic hydrolysis).

20. How will you synthesize acetylacetone (2,4-pentanedione) from ethyl acetoacetate ?

(Sholapur BSc, 2014)

21. How will you synthesize acetonyl acetone (2,5-hexanedione) from ethyl acetoacetate ?

(Lucknow BSc, 2015)

22. What is tautomerism ? How does it differ from resonance ?

23. Write a note on : Keto-enol tautomerism.

(Utkal BSc, 2013; Delhi BSc, 2014 ; Burdwan BSc, 2015)

24. How it can be established that ethyl acetoacetate is a mixture of keto and enol forms ? How can the two forms be separated ?

25. How will you synthesize methyl isopropyl ketone from ethyl acetoacetate ?

Answer. Methyl isopropyl ketone is a methyl ketone.

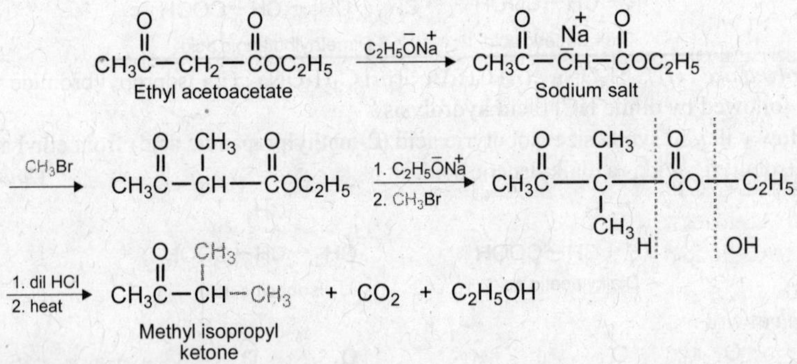

26. "Ethyl acetoacetate is a mixture of keto and enol forms." Justify the statement with suitable experimental evidences.

27. Keto-enol tautomerism is shown by

(*a*) Benzaldehyde (*b*) Acetone (*c*) Benzophenone (*d*) Acetic acid

Answer. (*b*)

Part 2. Diethyl malonate (Malonic ester)

28. Give the preparation of diethyl malonate. Describe any four uses of this compound.

(Shivaji BSc, 2014)

29. Describe the synthetic uses of diethyl malonate.

(Madras BSc, 2013 ; Annamalai BSc, 2014 ; Baroda BSc, 2015)

30. What happens when malonic ester is heated with urea ?

Answer. Barbituric acid is formed. For reaction equation see text.

31. How will you synthesize barbituric acid (malonyl urea) from diethyl malonate ?

32. How will you synthesize the following compounds from diethyl malonate ?

(*a*) Dimethylacetic acid (*Ranchi BSc, 2014*)

(*b*) Succinic acid (*Sagar BSc, 2015*)

(*c*) Adipic acid

(*d*) Crotonic acid

(*e*) Glycine (*Delhi BSc, 2013*)

33. How will you synthesize acetic acid from diethyl malonate ?

Clue. Steps are : (*i*) H_2O/H^+ ; and (*ii*) heat

34. How will you synthesize *n*-butyric acid (butanoic acid) from diethyl malonate ?

(*Jiwaji BSc, 2013 ; Sambalpur BSc, 2014 ; Cuttack BSc, 2015*)

Answer. *n*-Butyric acid is an alkylacetic acid.

RCH_2COOH — Alkylacetic acid $CH_3CH_2CH_2COOH$ — *n*-Butyric acid

Steps involved are :

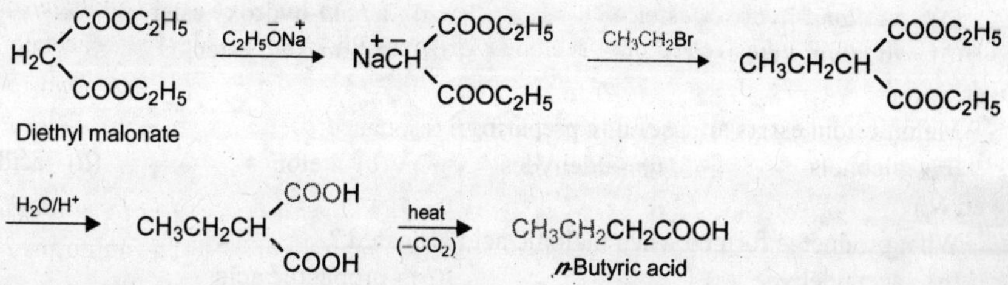

Multiple Choice Questions

1. Base-catalyzed condensation of two ester molecules to form an alcohol and β-keto ester is called

(*a*) Claisen condensation (*b*) Corey-House reaction

(*c*) Aldol condensation (*d*) Transesterification

Answer. (*a*)

2. Ethyl acetoacetate undergoes *acid-hydrolysis* with dilute HCl to form

(*a*) Acetoacetic acid (*b*) Succinic acid

(*c*) Acetic acid (*d*) Adipic acid

Answer. (*a*)

3. The ethyl derivative of acetoacetic ester on basic-hydrolysis gives

(*a*) Acetic acid (*b*) Acetic acid and propionic acid

(*c*) Propionic acid (*d*) Acetic acid and *n*-butyric acid

Answer. (*a*)

4. Ethyl acetate on heating with sodium ethoxide gives

(*a*) Ethyl acetoacetate (*b*) Sodium acetate

(*c*) Ethyl alcohol (*d*) Diethyl ether

Answer. (*a*)

5. When ethyl acetoacetate is subjected to ketonic hydrolysis, the ketone obtained is

(*a*) Dimethyl ketone (*b*) Methyl ethyl ketone

(*c*) Diethyl ketone (*d*) Methyl *n*-propyl ketone

Answer. (*c*)

6. Ethyl acetoacetate reacts with phenylhydrazine to give

(*a*) Antipyrine (*b*) Aspirin

(*c*) 4-Methyl uracil (*d*) DDT

Answer. (*a*)

7. The acetoacetic ester synthesis is used to prepare
   (*a*) α,β-unsaturated carbonyl compounds
   (*b*) methyl ketones
   (*c*) enamines
   (*d*) benzoic acid

Answer. (*b*)

8. Acid hydrolysis and decarboxylation of acetoacetic ester gives
   (*a*) acetaldehyde (*b*) acetone
   (*c*) ethyl acetate (*d*) ethyl formate

Answer. (*b*)

9. What product is formed when 1-bromobutane is used in an acetoacetic ester synthesis ?
   (*a*) 2-hexanone (*b*) 2-pentanone
   (*c*) 2-heptanone (*d*) hexanoic acid

Answer. (*c*)

10. The product of a Claisen condensation between two esters is usually
    (*a*) an *alpha*-hydroxy ester (*b*) a *beta*-hydroxy ester
    (*c*) an *alpha*-keto ester (*d*) a *beta*-keto ester

Answer. (*d*)

11. Malonic acid esters are useful in preparing ?
    (*a*) alcohols (*b*) aldehydes (*c*) ketones (*d*) acids

Answer. (*d*)

12. What product is formed when malonic acid is heated ?
    (*a*) acetaldehyde (*b*) propanoic acid
    (*c*) formic acid (*d*) acetic acid

Answer. (*d*)

13. What would be the product of the malonic ester synthesis if methyl bromide were used as the alkyl halide ?
    (*a*) methanoic acid (*b*) methyl malonate
    (*c*) propanone (*d*) propanoic acid

Answer. (*d*)

14. Acid-catalyzed hydrolysis of malonic ester gives
    (*a*) a monoacid (*b*) a diester
    (*c*) a diacid (*d*) a carboxylate salt

Answer. (*c*)

15. The reaction of malonic ester with excess ethoxide base and excess ethyl bromide gives
    (*a*) no reaction (*b*) a monosubstituted malonic ester
    (*c*) a disubstituted malonic ester (*d*) a trisubstituted malonic ester

Answer. (*c*)

16. Tautomers are
    (*a*) structural isomers (*b*) conformational isomers
    (*c*) geometric isomers (*d*) stereo isomers

Answer. (*a*)

17. Tautomers are
    (*a*) resonance forms (*b*) identical molecules
    (*c*) constitutional isomers (*d*) none of these

Answer. (*c*)

18. Diethyl malonate reacts with urea to give
    (*a*) Butyric acid (*b*) Barbituric acid (*c*) Glutaric acid (*d*) Mandelic acid

Answer. (*b*)

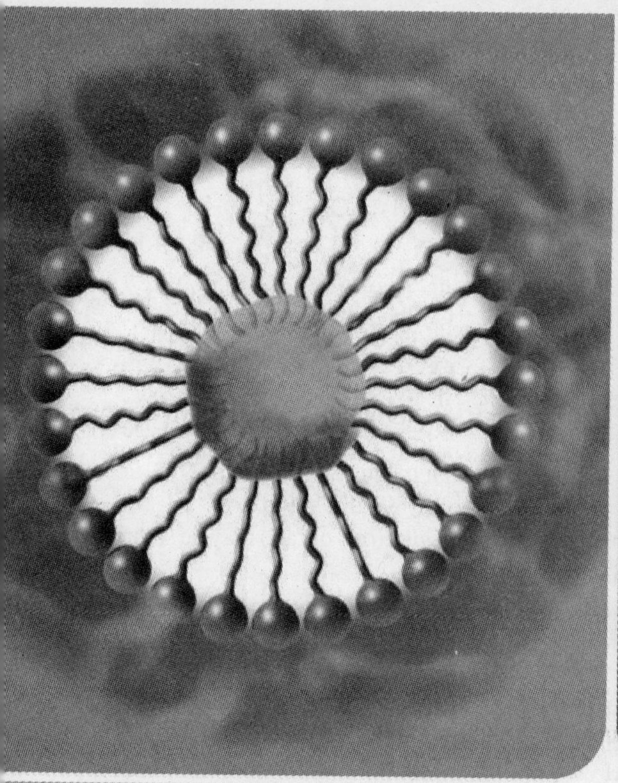

26

Fats and Oils; Soaps and Detergents

Fats and oils belong to the naturally occurring group of compounds called lipids. Lipids are constituents of plants and animals that are insoluble in water but soluble in organic solvents (ether, benzene). Fats and oils are those lipids which are saponifiable, while others such as steroids are non-saponifiable.

WHAT ARE FATS AND OILS ?

Natural fats and oils are the triesters of glycerol with long-chain carboxylic acids (12 to 20 carbons). These are known as **Triglycerides**.

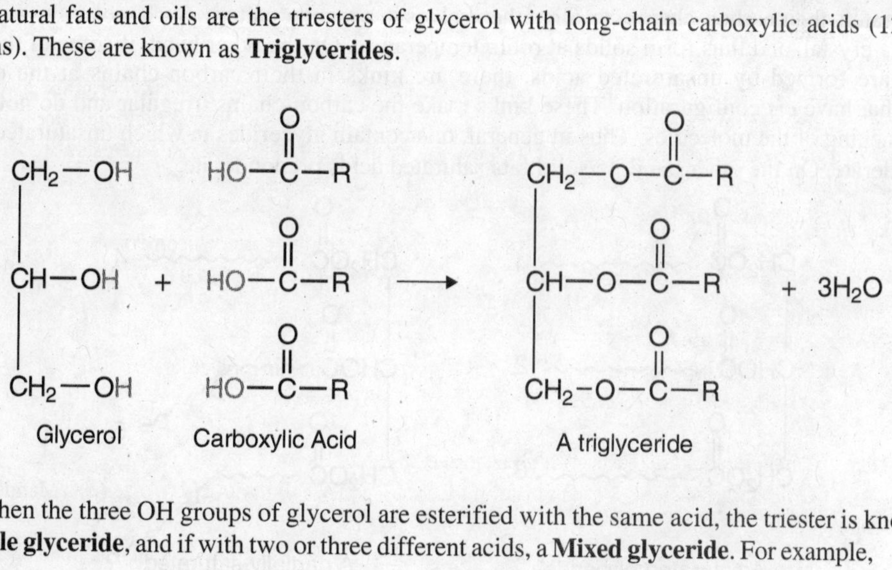

Glycerol   Carboxylic Acid   A triglyceride

When the three OH groups of glycerol are esterified with the same acid, the triester is known as a **Simple glyceride**, and if with two or three different acids, a **Mixed glyceride**. For example,

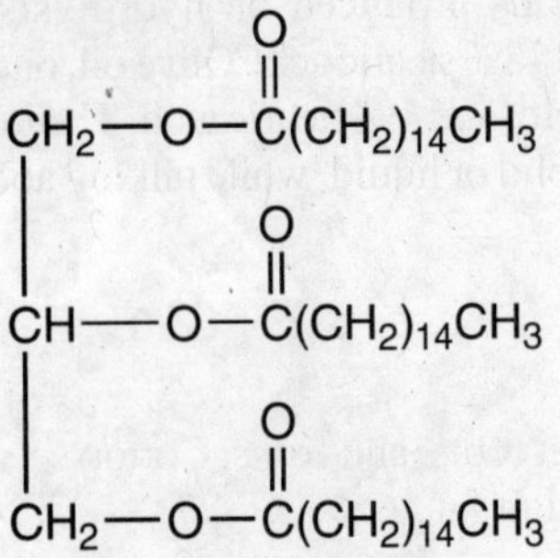

Glyceryl tripalmitate
(a simple glyceride)

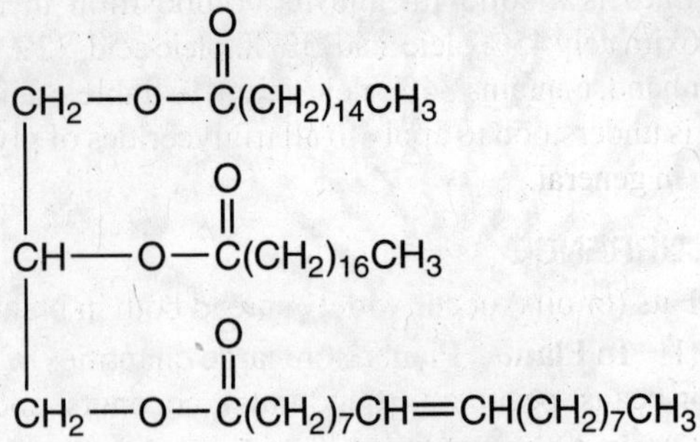

Glyceryl palmitostearooleate
(a mixed glyceride)

The carboxylic acids that form ester chains in natural triglycerides (oils or fats) may be saturated or unsaturated. Some of these are listed in the table below.

Table 26.1 SOME COMMON FATTY ACIDS

Name	Formula	mp°C
Myristic acid	$CH_3(CH_2)_{12}COOH$	58
Palmitic acid	$CH_3(CH_2)_{14}COOH$	63
Stearic acid	$CH_3(CH_2)_{16}COOH$	70
Oleic acid	$CH_3(CH_2)_7$ CH=CH$(CH_2)_7$COOH	4
	10 9 1	
Linoleic acid	$CH_3(CH_2)_4$ CH=CHCH$_2$CH=CH$(CH_2)_7$COOH	–5
	13 12 10 9	

DISTINCTION BETWEEN FATS AND OILS

The distinction between fats and oils is primarily based upon the difference in melting points. **Fats are composed of solid triesters of glycerol, while oils are composed of liquid triesters of glycerol at room temperature.** However, this distinction is not well founded as the physical state depends on climate and weather. The same glyceride may be solid at a hill station and liquid in plains.

The melting point of a fat or oil depends on its structure. If all the fatty acids forming a triglyceride are saturated, their carbon-chains can align themselves in a regular pattern. Such molecules can pack well in a crystal, and thus form solids at room temperature. On the other hand, if some of the ester chains are formed by unsaturated acids, there are kinks in their carbon-chains at the double bonds that have *cis* configuration. These kinks make the carbon-chains irregular and do not allow close-packing of the molecules. Thus in general, oils contain glycerides in which unsaturated acids preponderate. On the other hand, in solid fats saturated acids preponderate.

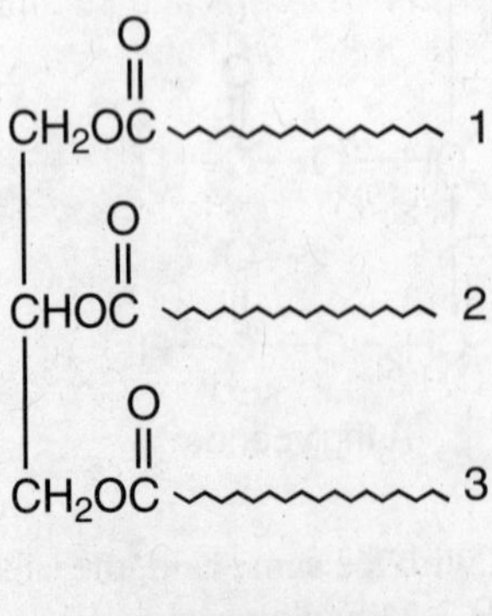

A saturated glyceride
(solid)

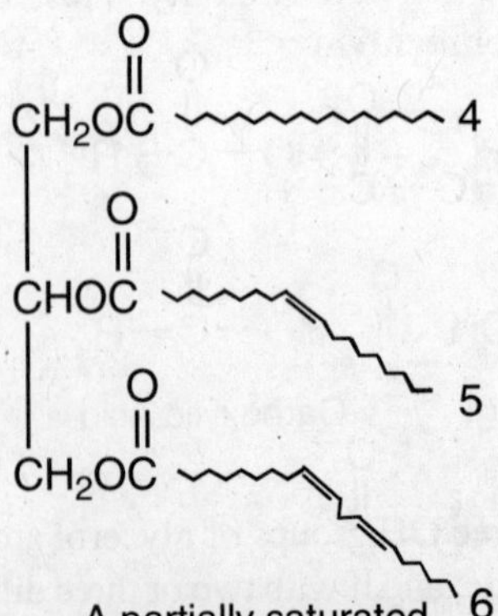

A partially saturated
triglyceride (liquid)

Lard is a solid fat and its composition in terms of fatty acids produced on hydrolysis is approximately 45% oleic acid, 5% linoleic acid, 32% palmitic acid, and 18% stearic acid. Olive oil, on the other hand, contains 84% oleic acid, 4% linoleic acid, 9% palmitic acid, and 3% stearic acid. The term 'fat' is understood to apply to all triglycerides of glycerol, whether solid or liquid, while talking about them in general.

OCCURRENCE

Fats (or oils) occur widely spread both in plants and animals.

(1) **In Plants.** Plants store large quantities of fats in their seeds, roots, and fruits. Cotton seeds, castor beans, peanuts (ground nuts), coconuts, and olives have a high fat content.

(2) **In Animals.** In animals, the fat deposits are to be found mostly under the skin and around the intestines and kidneys. Lard (*fat of hogs*) and tallow (*fat of kettle, sheep or horses*) are used industrially for soap-making.

EXTRACTION

Fats and oils are extracted from the natural products by three operations.

(1) **Rendering.** The animal tissues containing the fat are chopped off and heated dry or with water until the fat melts and can be removed.

(2) **Pressing.** Oils are obtained from seeds by crushing between steel rollers and then pressed in a *hydraulic press*.

(3) **Solvent Extraction.** It is often applied to the residue after pressing or rendering for complete removal of oil or fat. The solvents used include petroleum ether and benzene.

PROPERTIES

(Physical). (1) Oils and fats are liquids or solids having a greasy feel. When pure, they are colorless, odorless, and tasteless.

(2) They are insoluble in water but soluble in organic solvents such as ether, chloroform, benzene, and hot alcohol.

(3) They have a lower specific gravity than water and consequently will float on the surface when mixed with it.

(4) They readily form emulsions when agitated with water in the presence of soap, gelatin or other emulsifiers.

(Chemical). Fats and oils are triesters of glycerol with saturated and unsaturated fatty acids. Their reactions are those of ester groups in triplicate and carbon-carbon double bonds.

(1) **Hydrolysis.** They are readily hydrolysed by heating with acids or alkalies or superheated steam. When boiled with sodium or potassium hydroxide solution, the hydrolysis products are sodium or potassium salts of long-chain fatty acids. The latter are called **Soaps** and alkaline hydrolysis is referred to as **Saponification**.

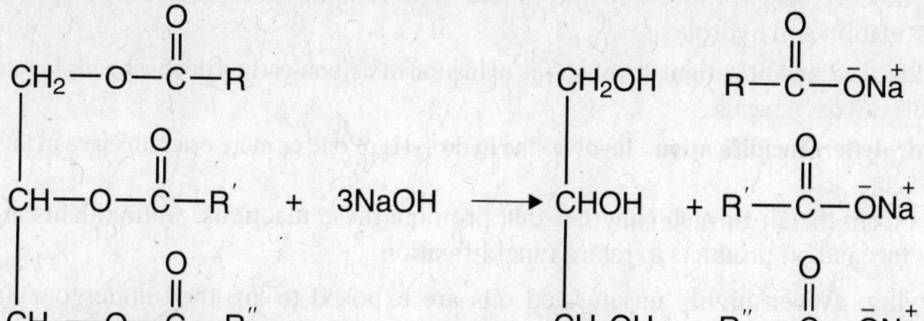

Oil or Fat (triglyceride) — Glycerol — Soap

(2) **Hydrogenation.** Vegetable oils are triglycerides of unsaturated fatty acids such as oleic acid and linoleic acid. On catalytic hydrogenation at low pressures, hydrogen adds across the carbon-carbon double bonds of the acid components of the triglycerides. This results in the formation of saturated glycerides which are solid fats at room temperature. This hydrogenation process is called **Hardening**. For example,

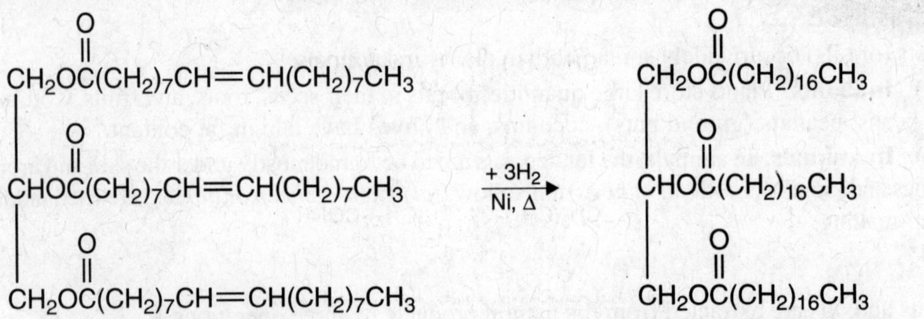

Glyceryl trioleate (mp –17°C) (A liquid vegetable oil) — Glyceryl tristearate (mp 55°C) (A solid fat)

Partial hydrogenation of vegetable oils is used for the manufacture of **Vegetable Ghee** *(Dalda)*. Complete hydrogenation according to the above equation would produce a hard brittle fat.

(3) **Hydrogenolysis.** Like a simple ester, the ester groups of triglycerides *(fats or oils)* can be reduced with hydrogen in the presence of copper chromite catalyst, $CuCr_2O_4$, at high pressure and temperature. This reaction, called *hydrogenolysis* (hydrogenation involving cleavage) gives in addition to glycerol, long-chain primary alcohols corresponding to the acid portion.

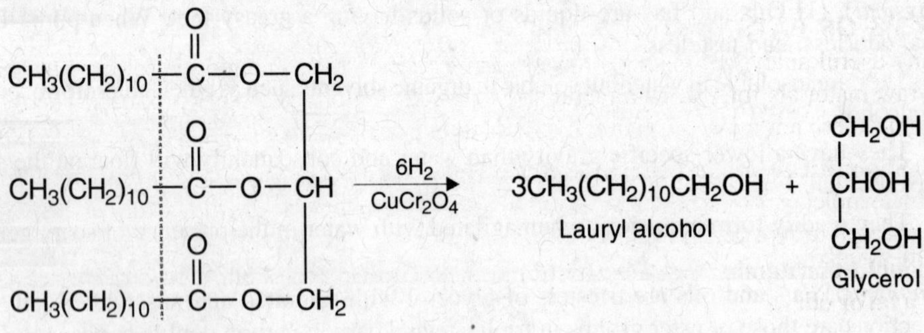

Glyceryl trilaurate — Lauryl alcohol — Glycerol

The long-chain alcohols so obtained are used for manufacturing *Detergents.*

(4) **Rancidification.** When fats and oils are left exposed to moist air, they develop foul-smell and sour taste. They are said to have become rancid. The rancidification is caused by two types of reactions : oxidation and hydrolysis.

Oxidative rancidification. Involves the oxidation of carbon-carbon double bonds in fats and oils to produce volatile carboxylic acids.

Hydrolytic rancidification. Involves the hydrolysis of one or more ester linkages in fats and oils to produce the original acid.

Bacteria from the air furnish enzymes that promote these reactions. Antioxidants are added to many edible fats and oil products to retard rancidification.

(5) **Drying.** When highly unsaturated oils are exposed to air, they undergo oxidation and polymerization to form a thin waterproof film. Such oils are called **Drying Oils** and the reaction is referred to as **Drying**. Linseed oil, which is rich in linolenic acid, is a common drying oil used in oil-based paints. Nondrying oils are either saturated or only moderately unsaturated.

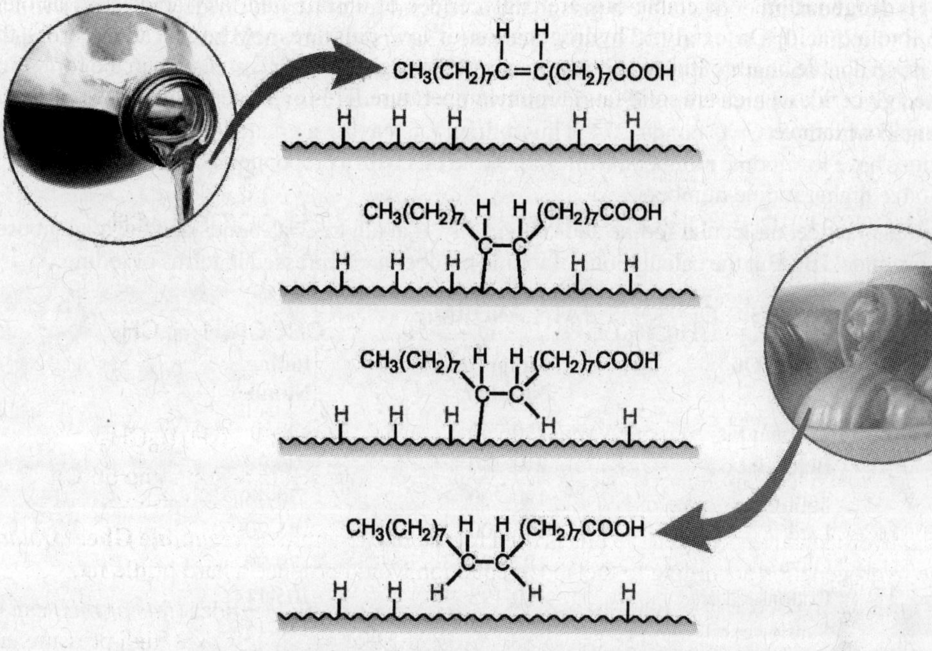

The hydrogenation of an oil to a fat is called Hardening, because a liquid is converted to a solid.

ANALYSIS OF FATS AND OILS

Many useful analytical methods have been devised to have quality control over fats and oils used as raw materials for the manufacture of soaps, drying oils and other products. The two most important of these are the determination of the *Saponification Number* and the *Iodine Number*.

(1) **Saponification Number (*Saponification Value*).** The saponification number tells the approximate molecular weight of a fat or oil, as also whether lower or higher fatty acids preponderate in the formation of its ester groups.

Saponification number is defined as the number of milligrams of KOH required to saponify one gram of a fat or oil.

In the saponification reaction, one mole of a fat reacts with three moles of KOH since the former has three ester groups. If M be the *molecular weight* of the fat, M grams of it require $3 \times 56 = 168$ grams or 168,000 milligrams of KOH for saponification. Therefore,

$$\text{Saponification Number of Fat} = \frac{168000}{M}$$

Knowing the saponification number as determined experimentally, M can be calculated.

As shown by the above expression, the saponification number of a fat is inversely proportional to the molecular weight. A lower value of saponification number, therefore, shows the preponderance of high-molecular-weight fatty acid residues (C_{18}) in the given fat. A higher saponification number, on the other hand, indicates the preponderance of low-molecular-weight fatty acid residues. The saponification number is characteristic of a particular fat or oil and serves for its identification.

(2) **Iodine Number (*Iodine Value*).** The degree of unsaturation of a fat or oil is measured by its *Iodine Number*.

Iodine number is the number of grams of iodine that would add to carbon-carbon double bonds (C=C) present in 100 grams of the fat or oil.

The greater the number of double bonds in the acid residues of a triglyceride, the greater will be the amount of iodine that adds to 100 grams. Because a saturated fatty acid has no double bonds, it takes up no iodine, so that its iodine number is zero. Oleic acid, which has one C=C bond, has an iodine number of 90, whereas linoleic acid with two C=C bonds, has an iodine number of 181, and linolenic acid with three C=C bonds, 274. Thus animal fats having a greater proportion of saturated acid residues have low iodine numbers, while vegetable oils with a preponderance of unsaturated acid residues have higher iodine numbers.

In actual practice, molecular iodine adds sluggishly, if at all, to C=C bonds and the reagent used is iodine bromide, IBr. But the calculations of iodine number are expressed in terms of iodine.

Table 26.2 ANALYSIS OF SOME FATS AND OILS

Fat or Oil	Saponification Number	Iodine Number
Coconut fat	250-260	8-10
Butter fat	210-230	26-28
Tallow	190-200	30-48
Lard	193-200	46-70
Olive oil	187-196	79-90
Cottonseed oil	190-198	105-114
Sunflower oil	188-194	140-156
Linseed oil	187-195	170-185

The above Table shows the saponification and iodine numbers of several fats and oils. Notice the high saponification number of butter fat, which is an indication that it contains some short carbon-chain acid residues. Notice also the low iodine number of butter fat, which has few double bonds and is consequently a solid at room temperature. Linseed oil, on the contrary, has a higher iodine number which shows that it is highly unsaturated.

MANUFACTURE OF VEGETABLE GHEE

Liquid vegetable oils are largely triglycerides of unsaturated fatty acids (oleic, linoleic). On hydrogenation in the presence of nickel and at low pressures, some of the carbon-carbon double bonds (C=C) are saturated to give a solid fat.

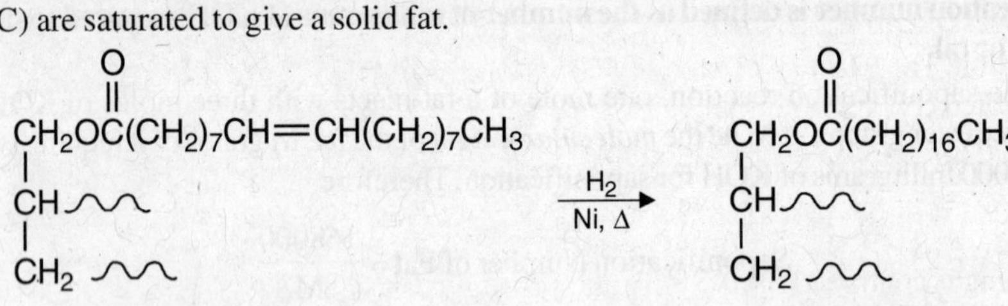

Vegetable oil Vegetable fat

In actual practice, the degree of hardening is carefully controlled to produce a fat of the desired consistency. The partially hydrogenated vegetable oil thus formed is deodorized by treatment with steam in vacuum and decolorized by agitation with Fullers earth and carbon. The product is mixed with vitamins and antioxidants and is sold as **Vanaspati** or **Vegetable ghee** for kitchen use. Vegetable ghee is preferred over oils for convenience, better appearance and taste, as also its stability to turn rancid.

Margarine is a substitute of butter which is prepared by hydrogenation of *groundnut oil* (peanut oil) or *cottonseed oil*. The resulting product is often churned with milk and artificially colored to give it a flavor and consistency resembling that of butter.

In India, the main oil stock used for making vegetable ghee is *groundnut oil*, or *cottonseed oil* for better quality product.

WHAT ARE SOAPS ?

Soaps are sodium or potassium salts of fatty acids, chiefly oleic, stearic, palmitic, lauric and myristic acids. Ordinary soaps are the products of hydrolysis of oils and fats with sodium hydroxide. The oils and fats are mixed triglycerides, and as such soaps are mixtures of salts of saturated and unsaturated long chain carboxylic acids containing 12 to 18 carbon atoms.

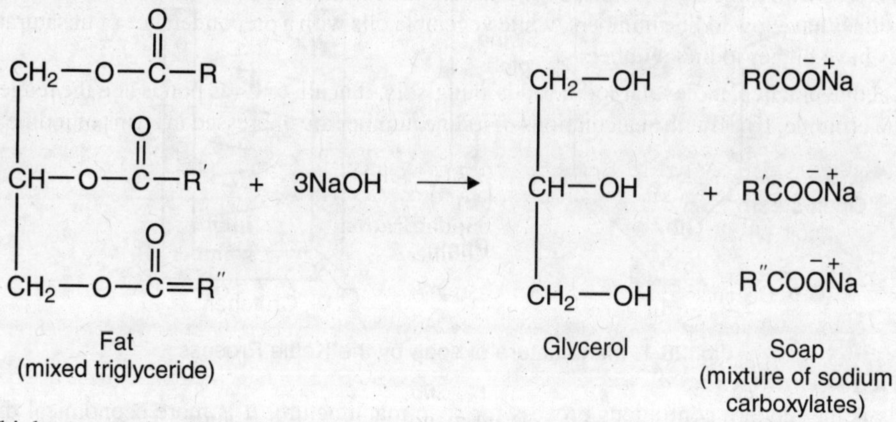

A higher proportion of salts of saturated acids (palmitic, stearic) gives **Hard Soaps,** while a higher proportion of salts of unsaturated acids (oleic acid) yields **Soft Soaps**.

The potassium soaps, produced by the saponification of fats with KOH, are usually softer and more soluble than the sodium soaps. Therefore, potassium soaps are used mainly in shaving creams and liquid soaps *(Shampoo)*.

MANUFACTURE OF SOAP

There are generally two methods used for soap manufacture :

(*a*) Kettle Process (*b*) Hydrolyser Process

The *Kettle Process* is the oldest and is now used in small factories or for special production on a limited scale. The *Hydrolyser Process* is the modern continuous process which is far more economical and gives the purest stuff.

KETTLE PROCESS

This process is carried in a steel tank or Kettle from which it gets the name. The various steps are :

(1) **Boiling.** The fat and NaOH solution are charged into the kettle and boiled with steam entering from a perforated coil at the bottom.

$$\underset{(Fat)}{C_3H_5(COOC_{17}H_{35})_3} + \underset{(Lye)}{3NaOH} \xrightarrow[\text{steam}]{\Delta} \underset{\substack{\text{Sodium stearate}\\(Soap)}}{3C_{17}H_{35}CO\overset{-}{O}\overset{+}{Na}} + \underset{\text{Glycerol}}{C_3H_5(OH)_3}$$

The boiling is continued till saponification is complete (about 80%).

(2) **Salting Out.** Salt (NaCl) is then added and boiling resumed till soap has separated *(common ion effect)*. The soap being lighter floats to the surface as curdy mass. The lower layer containing glycerol and salt is drawn off, leaving soap plus unreacted fat behind.

(3) **Addition of Fresh Lye.** The soap from step (2) contains 20% fat which escaped esterification earlier. It is boiled with a fresh supply of NaOH solution. More soap is formed. The lower layer is drawn off. The soap layer left in the kettle is boiled with water to wash away excess of lye.

(4) **Finishing.** The 'neat soap' while still molten, is pumped into the *crutcher* by means of swing pipe. The *crutcher* is a steam jacketed tank fitted with a blade-stirrer. Here the soap is mixed with color, perfumes, germicides, etc., till it becomes a homogeneous mass. The crutched soap is poured into frames and on solidification cut into cakes. For making toilet soap, the neat crutched soap is shredded, dried, and stamped into cakes.

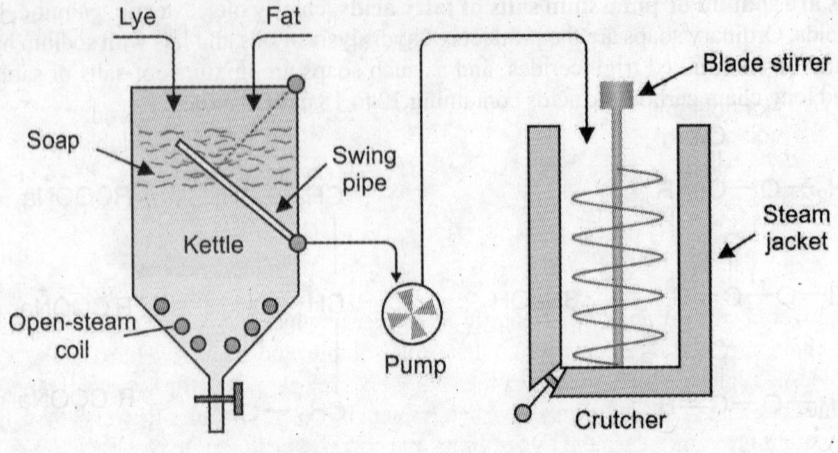

Fig. 26.1. Manufacture of soap by the Kettle Process.

HYDROLYSER PROCESS

This is the modern continuous process for soap manufacture. It is more economical than the *Kettle Process* and gives better quality soap.

The *Hydrolyser Process* consists of (*a*) Hydrolysis of fat with water in the presence of zinc oxide at high temperature and pressure ; (*b*) Distillation of fatty acids under vacuum ; and (*c*) Neutralization of the condensed fatty acids with alkali.

$$\underset{\text{Fat}}{C_3H_5(COOR)_3} + 3H_2O \xrightarrow[\Delta,\ \text{pressure}]{ZnO} \underset{\text{Fatty acids}}{3RCOOH} + \underset{\text{Glycerol}}{C_3H_5(OH)_3}$$

$$RCOOH + NaOH \longrightarrow \underset{\text{Soap}}{RCO\bar{O}\overset{+}{Na}} + H_2O$$

The flowsheet of the *Hydrolyser Process* is shown in Fig. 26.2. The fat mixed with zinc oxide catalyst, and water are heated to 230-250°C under 40-45 atm pressure in the *Hydrolyser*. The fatty acids mixed with water are discharged at the top into a steam flash-tank (not shown). The water vaporizes,

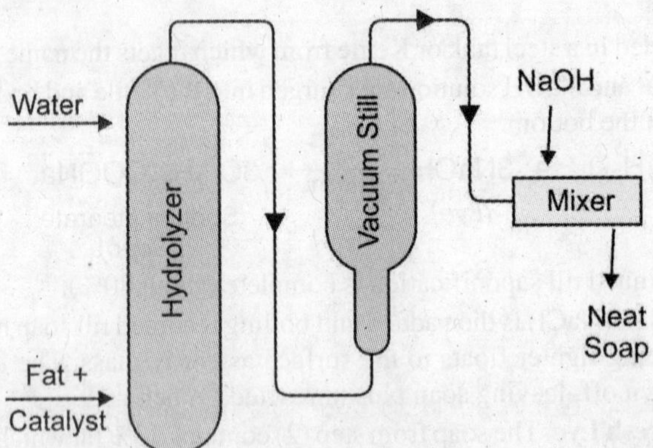

Fig. 26.2. Flowsheet of Soap Manufacture by Hydrolyser Process.

cooling the fatty acids which are then vacuum-distilled in the *Vacuum Still*. The vapors of fatty acids are passed through water condenser (not shown). The condensed fatty acids are finally neutralized in the Mixer and the product is neat soap.

CLEANSING ACTION OF SOAP

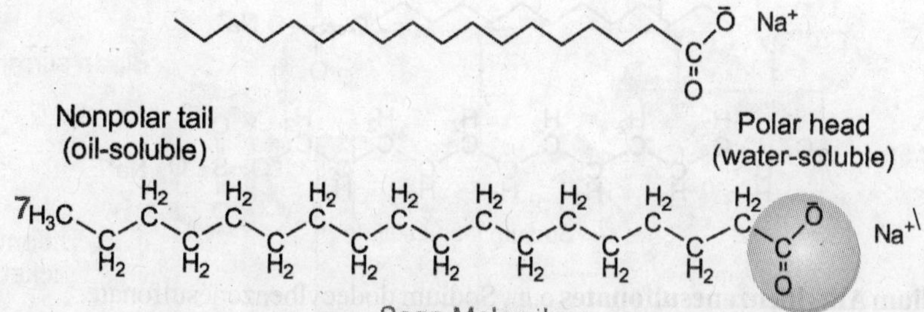

Soap Molecule

Skin is covered with a thin layer of natural oil or grease which catches dirt. The soap molecules have a long-chain nonpolar hydrocarbon tail that is oil-soluble, and a negative head (COO^-) which is water-soluble. As soap water is poured over the skin (or a dirty garment), the hydrocarbon tails of the soap molecules peg into it, while the negative head is held in water. The grease layer is then dislodged from skin by rubbing or from garment by tumbling and stirring. Each grease globule thus separated is pincushioned by hydrocarbon tails with negative heads outwards in water. The negative globules keep apart by mutual repulsions, and are said to have been emulsified. The emulsified grease globules bearing dirt can be readily washed with water (Fig. 26.3).

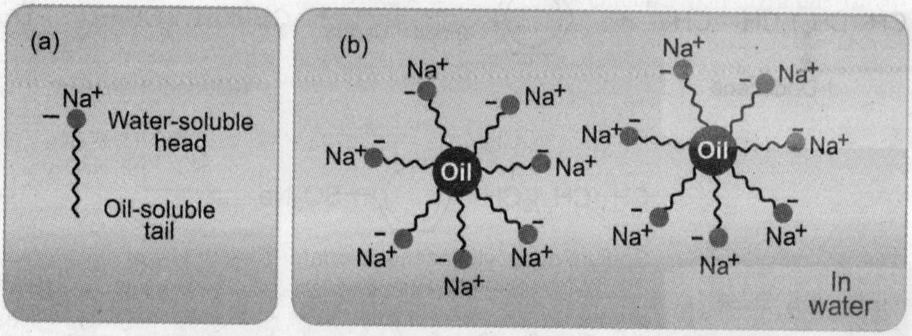

Fig. 26.3. (*a*) A soap molecule (*b*) Emulsified grease globule.

SYNTHETIC DETERGENTS (SYNDETS)

The synthetic detergents (also called Syndets) are substitutes of soaps. Their development is a recent innovation but syndets have already replaced about 80% of the world's demand for soap. The two chief types of detergents are :

(1) **Sodium Alkyl Sulfates** e.g., Sodium lauryl sulfate.

$$CH_3(CH_2)_{10}CH_2-O-\overset{\overset{O}{||}}{\underset{\underset{O}{||}}{S}}-\bar{O}\overset{+}{Na}$$

Sodium lauryl sulfate, one of the earliest synthetic detergents, is made from lauryl alcohol obtained by the hydrogenolysis of fats.

$$\underset{\text{Lauryl alcohol}}{CH_3(CH_2)_{10}CH_2OH} + HOSO_3H \xrightarrow{-H_2O} \underset{\text{Lauryl hydrogen sulfate}}{CH_3(CH_2)_{10}CH_2OSO_3H}$$

$$\downarrow NaOH$$

$$\underset{\text{Sodium lauryl sulfate}}{CH_3(CH_2)_{10}CH_2OS\bar{O}_3\overset{+}{Na}}$$

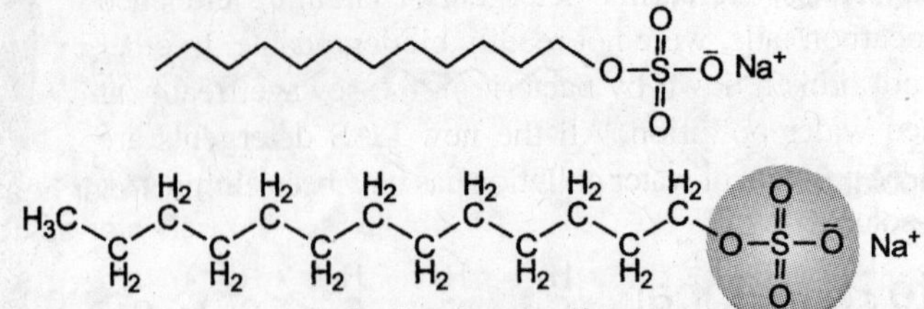

Sodium lauryl sulfate

(2) **Sodium Alkylbenzenesulfonates** e.g., Sodium dodecylbenzenesulfonate.

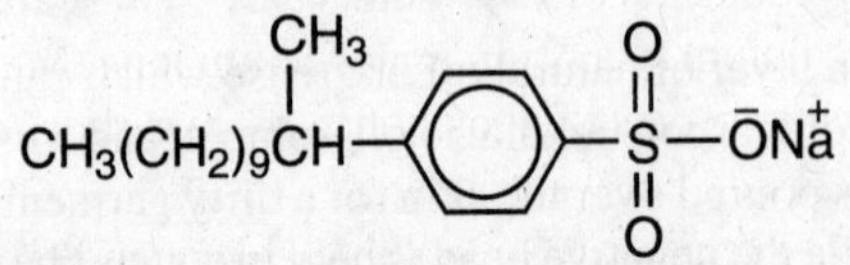

The long-chain alkylbenzenesulfonates (**LAS**) as shown above are most widely used synthetic detergents since 1975. Sodium dodecylbenzenesulfonate is produced from 1-dodecene (available from petroleum refinery) by Friedel-Crafts reaction.

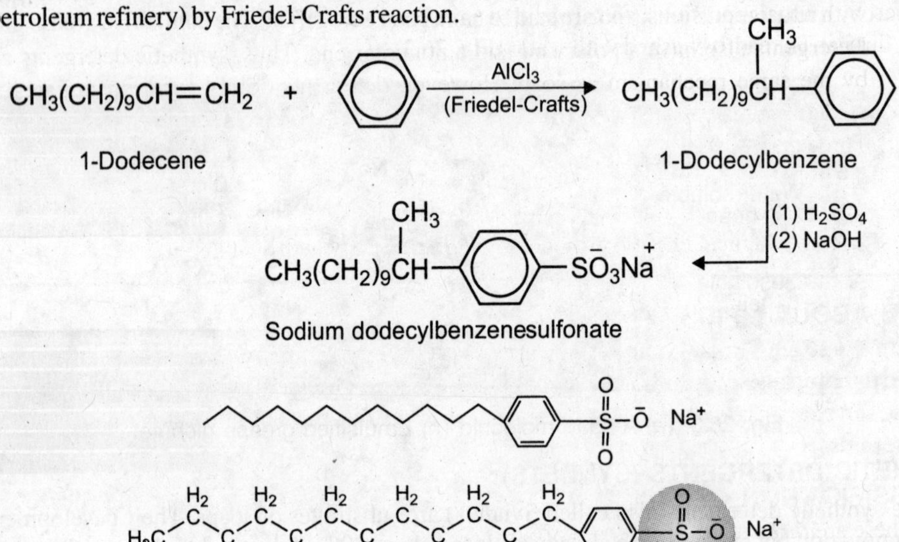

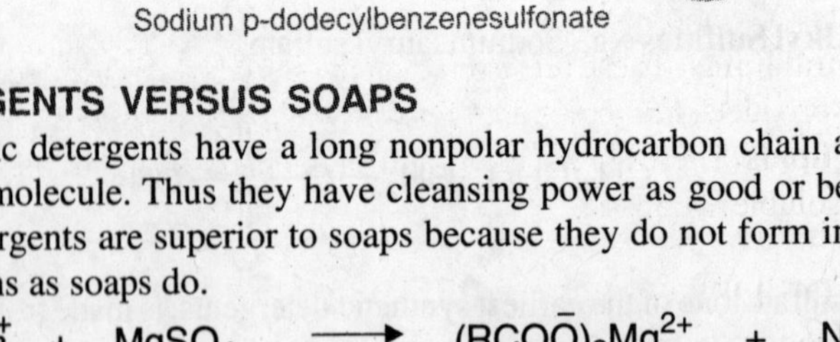

Sodium p-dodecylbenzenesulfonate

SYNTHETIC DETERGENTS VERSUS SOAPS

Like soaps, synthetic detergents have a long nonpolar hydrocarbon chain and a highly polar group at the end of the molecule. Thus they have cleansing power as good or better than ordinary soap. The synthetic detergents are superior to soaps because they do not form insoluble salts with Ca^{2+}, Mg^{2+}, and Fe^{3+} ions as soaps do.

$$\underset{\text{Soap}}{2RCO\bar{O}\overset{+}{Na}} + MgSO_4 \longrightarrow \underset{\text{(Insoluble)}}{(RCOO)_2Mg^{2+}} + Na_2SO_4$$

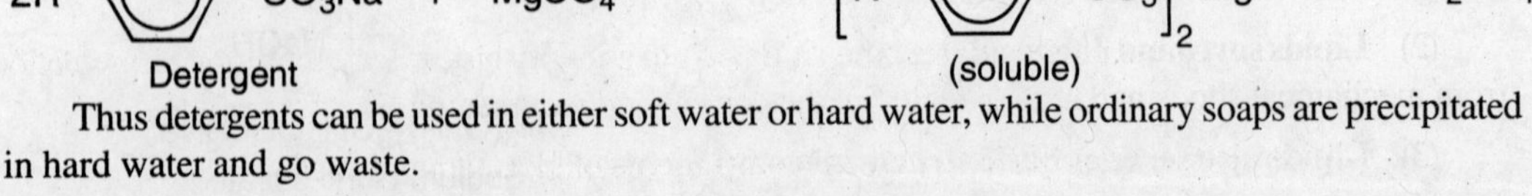

Thus detergents can be used in either soft water or hard water, while ordinary soaps are precipitated in hard water and go waste.

Unlike ordinary soaps, the earlier detergents containing branched chains in the hydrocarbon tails, were not readily biodegradable. In other words, they were not broken down by bacteria in the sewage treatment plants. They caused water pollution. All the new LAS detergents are biodegradable, and the problem of water pollution has now been eliminated.

Points to Remember

1. The basic-hydrolysis of fats and oils is called saponification.
2. The sodium or potassium salts of fatty acids are called soaps.
3. Water that contains calcium (Ca^{2+}) and magnesium ions (Mg^{2+}) is said to be hard water.
4. When boiled, hard water leaves a deposit of calcium carbonate, $CaCO_3$. This scale builds up in tea kettles and inside hot water heaters.
5. When soap is added to hard water, insoluble compounds form which appear as sticky scum. This scum leaves a deposit on clothes, skin, and hair.
6. The development of synthetic detergents by chemists was a great advantage for people with relatively hard tap water in their homes.
7. Detergents also have a polar end and a nonpolar end. Thus, synthetic detergents clean by the same mechanism as soap. However, detergents do not form precipitates with calcium or magnesium ions and can be used freely in hard water.
8. Three types of energy are needed for good cleaning results :
   (a) chemical energy, provided by the soap or detergent.
   (b) thermal energy, provided by warm or hot water.
   (c) mechanical energy, provided by a machine or hands.

MORE ABOUT LIPIDS

Lipids are a group of compounds with widely different compositions and structures. **They are defined as naturally occurring organic compounds that are insoluble in water, but soluble in nonpolar organic solvents such as ether, chloroform, or carbon tetrachloride.** Examples are : Oils, Fats, Terpenes, and Steroids.

CLASSIFICATION OF LIPIDS

There are *three* main classes of lipids :

(1) **Simple lipids.** These are esters which on hydrolysis yield only fatty acids and alcohols. Examples are : Fats, Oils, and Waxes.

(2) **Compound lipids.** These are esters which on hydrolysis produce fatty acids, alcohols, and one or more other products. Examples are : Phospholipids, Spingolipids, and Glycolipids.

(3) **Derived lipids.** These are not esters and do not undergo hydrolysis. Examples are : Terpenes, Steroids, and Fat-soluble Vitamins.

IMPORTANCE OF LIPIDS

Lipids are important to humans for *three* reasons :

(1) **Lipids serve as source of energy.** Excess carbohydrates and other energy giving foods are converted and stored in the body as lipids and fats. These fat reservoirs constitute one of the main storage forms of chemical energy and carbon in the body.

(2) **Lipids surround and insulate some vital body organs.** In this capacity, they provide protection from mechanical shock and help to maintain the correct body temperature.

(3) **Lipids also serve as basic structural components of all cell membranes.**

Examination Questions

1. What are Oils and Fats and how they differ from each other. *(Lucknow BSc, 2014)*
2. What is Soap ? How is Soap manufactured ?
3. What are Detergents ? How do they differ from Soaps ? *(Devi Ahilya BSc, 2015)*
4. (*a*) What are Lipids ?
   (*b*) How are Lipids classified ?
   (*c*) Why are Lipids important to humans ?
5. Write notes on :
   (*a*) Hydrogenation of oils
   (*b*) Saponification value
   (*c*) Iodine value
   (*d*) Cleansing action of soap *(Pondicherry BSc, 2014 ; Panjab BSc, 2015)*
   (*e*) Lipids
6. Explain why the melting points of unsaturated fatty acids are lower than those of saturated fatty acids.

Answer. Intermolecular dispersion forces involving the long hydrocarbon chains hold fatty acid molecules together in the solid state. Stronger intermolecular forces involve dipole-dipole attractions and hydrogen bonding between the carboxylic acid groups. Saturated fatty acids pack close together, maximizing the attractive forces. A *cis* carbon-carbon double bond in the long hydrocarbon chain introduces a kink in the carbon chain of unsaturated fatty acids, forcing non-ideal molecular packing thus diminishing the effects of intermolecular attractive forces. Therefore, there are weaker attractive forces between unsaturated fatty acid molecules than in their fully saturated counterparts, and as a result, unsaturated fatty acids have lower melting points.

7. Characterize the structural features necessary to make a good synthetic detergent.

Answer. A good synthetic detergent should have a long hydrocarbon chain with a very polar head group at one end of the molecule. This combination will allow for the formation of micelle structures in aqueous solution that will dissolve hydrophobic substances such as dirt, grease, and oil. The polar head group should not form insoluble salts with the ions normally found in hard water, such as Ca^{2+}, Mg^{2+}, and Fe^{2+}.

8. Identify some of the detergents used in shampoos and dish-washing liquids. Are they primarily anionic, neutral, or cationic detergents ?

Answer. The detergents used as shampoos and dish-washing liquids are alkylbenzenesulfonates and have the general formula :

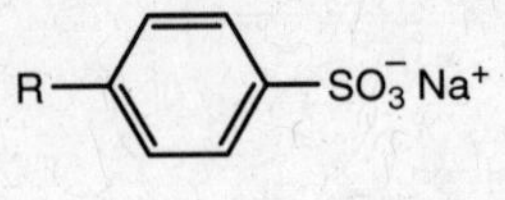

Where R represents a long hydrocarbon chain. They are primarily anionic detergents.

Multiple Choice Questions

1. Fatty acids are
   (*a*) Unsaturated dicarboxylic acids
   (*b*) Long-chain alkanoic acids
   (*c*) Aromatic carboxylic acids
   (*d*) Aromatic dicarboxylic acids

Answer. (*b*)

2. Sodium or potassium salts of fatty acids are called
   (*a*) Proteins (*b*) Terpenes (*c*) Carbohydrates (*d*) Soaps

Answer. (*d*)

3. Which of the following molecular formulas corresponds to that of an unsaturated fatty acids ?
(*a*) $C_{16}H_{33}COOH$ (*b*) $C_{19}H_{39}COOH$
(*c*) $C_{18}H_{35}COOH$ (*d*) $C_{17}H_{35}COOH$

Answer. (*c*)

4. Soap is
(*a*) a mixture of salts of fatty acids
(*b*) a salt of glycerol
(*c*) a mixture of ethers
(*d*) a mixture of aromatic ethers

Answer. (*a*)

5. Which of the following is a soap ?
(*a*) $C_{18}H_{37}COONa$ (*b*) CH_3COONa
(*c*) $C_{17}H_{35}C_6H_4SO_3Na$ (*d*) $C_3H_8O_3$

Answer. (*a*)

6. Lye is used in soap-making. Lye is a concentrated solution of which ionic compound ?
(*a*) $CuSO_4$ (*b*) NaCl (*c*) K_2CO_3 (*d*) NaOH

Answer. (*d*)

7. Fats and oils are
(*a*) monoesters of glycerol
(*b*) diesters of glycerol
(*c*) triesters of glycerol
(*d*) diesters of glycol

Answer. (*c*). Fats are solid triesters of glycerol. Oils are liquid triesters of glycerol.

8. Oleic acid is a fatty acid containing
(*a*) 12 carbons (*b*) 14 carbons
(*c*) 16 carbons (*d*) 18 carbons

Answer. (*d*)

9. Both stearic acid and linoleic acid have 18 carbons. Linoleic acid is unsaturated, while stearic acid is saturated. The melting point of stearic acid :
(*a*) is higher than linoleic acid
(*b*) is lower than linoleic acid
(*c*) is same as linoleic acid
(*d*) can not predict, insufficient information

Answer. (*a*)

10. Liquid oils can be converted to solid fats by
(*a*) Hydrogenation (*b*) Saponification
(*c*) Hydrolysis (*d*) Oxidation of double bonds

Answer. (*a*)

11. Partial hydrogenation of vegetable oils in the presence of Ni catalyst at 200°C gives
(*a*) Vanaspati ghee (*b*) Margarine
(*c*) Both (*a*) and (*b*) (*d*) None of these

Answer. (*a*)

12. Alkaline hydrolysis of oils (or fats) is called :
(*a*) Saponification (*b*) Fermentation
(*c*) Diazotization (*d*) Rancidification

Answer. (*a*)

13. Saponification of a fat
(*a*) always results in the formation of insoluble soaps
(*b*) produces glycerol and soap
(*c*) is used in the production of detergents
(*d*) is used in the production of lactic acid

Answer. (*b*)

14. Which of the following compounds will not be classified as lipids ?

(*a*) Fats (*b*) Waxes
(*c*) Soaps (*d*) Oils

Answer. (*b*)

15. Synthetic detergents can be represented by the following general formula

(*a*) RONa (*b*) $ROSO_3Na$ (*c*) RCOONa (*d*) RCOOH

Answer. (*b*)

16. The degree of unsaturation of a fat can be determined by means of its

(*a*) Iodine number (*b*) Octane number
(*c*) Saponification number (*d*) Melting point

Answer. (*a*)

17. Which of the following occurs when a greasy plate is placed in soapy water and scrubbed ?

(*a*) The hydrophilic head of the soap bonds to grease.
(*b*) The hydrophilic tail of the soap bonds to grease.
(*c*) The hydrophilic tail of the soap bonds to water.
(*d*) The soap froths but does not actually bond with the grease.

Answer. (*b*)

18. When oil and water are pured into the same vessel, two layers are observed with the less dense oil on top. These two liquids are said to be .....

(*a*) hydrophilic (*b*) immiscible
(*c*) hydrophobic (*d*) insoluble

Answer. (*b*)

19. Soaps cannot be used in hard water because they ....

(*a*) contain sodium. (*b*) create environmental problems.
(*c*) are too greasy. (*d*) form precipitates.

Answer. (*d*)

20. When sodium stearate, $CH_3(CH_2)_{16}COONa$, is added to hard water, the formula of one possible precipitate is ....

(*a*) $CH_3(CH_2)_{16}COOK$ (*b*) $(CH_3(CH_2)_{16}COO)_2Cd$
(*c*) $(CH_3(CH_2)_{16}COO)_2Ca$ (*d*) $(CH_3(CH_2)_{16}COO)_2Ba$

Answer. (*c*)

21. An advantage of detergents over soaps is ....

(*a*) detergents do not precipitate in hard water.
(*b*) detergents are cheaper to make.
(*c*) detergents are biodegradable.
(*d*) detergents always contain anionic heads.

Answer. (*a*)

27 Amines

Amines are responsible for the foul smell of dead fish.

Amines are derivatives of ammonia (NH_3) in which one or more hydrogen atoms have been replaced by alkyl groups.

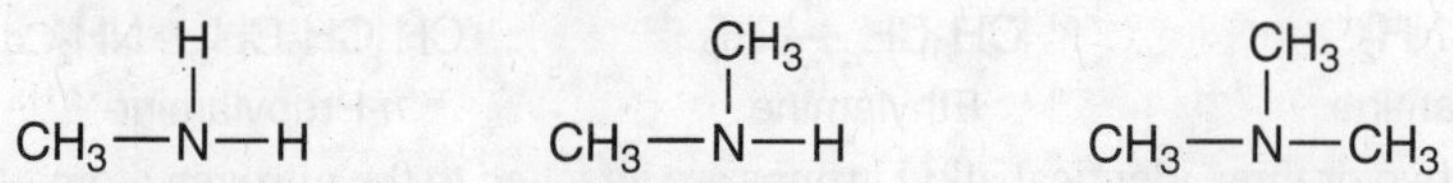

Methylamine | Dimethylamine | Trimethylamine

Amines are classified as **Primary** (1°), **Secondary** (2°), or **Tertiary** (3°), according to the number of alkyl groups attached to the nitrogen atom.

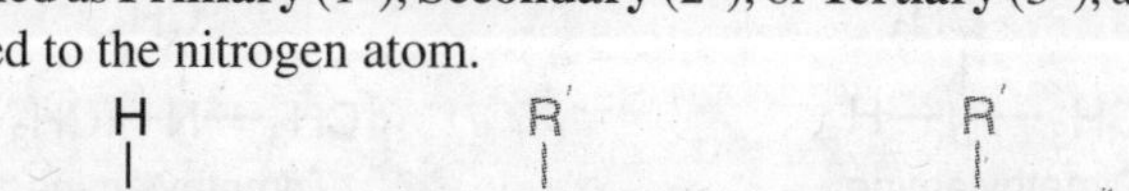

Primary (1°) | Secondary (2°) | Tertiary (3°)

A primary amine has only one alkyl group directly attached to the nitrogen. A secondary amine has two alkyl groups directly attached to the nitrogen. A tertiary amine has three alkyl groups directly attached to the nitrogen.

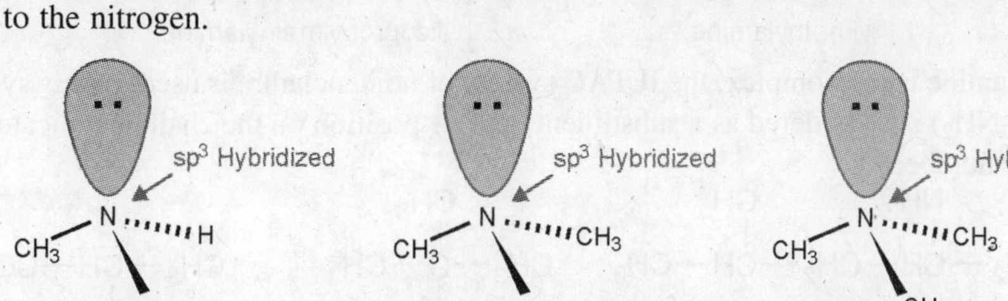

Methylamine (a primary amine) | Dimethylamine (a secondary amine) | Trimethylamine (a tertiary amine)

Orbital structure of Amines. The nitrogen atom of amines is sp^3 hybridized.

This classification is different from that of alkyl halides or alcohols. Their classification is based on the number of groups attached to the carbon that has the halide or hydroxyl group.

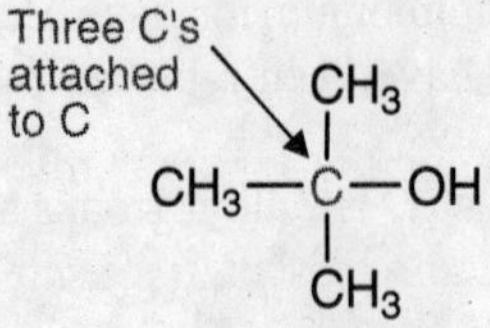

tert-Butyl alcohol
(3° alcohol)

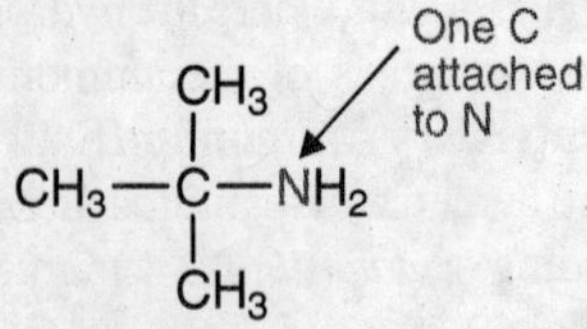

tert-Butylamine
(1° amine)

STRUCTURE

Let us consider methylamine (CH_3NH_2) for illustrating the orbital make up of amines. In methylamine both nitrogen and carbon are sp^3 hybridized. One of the sp^3 orbitals of nitrogen is completely filled and cannot take part in bond formation. The C–N σ bond in methylamine is formed by overlap of an sp^3 orbital of carbon and an sp^3 orbital of nitrogen. Each N–H σ bond is formed by the overlap of an sp^3 orbital of nitrogen and *s* orbital of hydrogen.

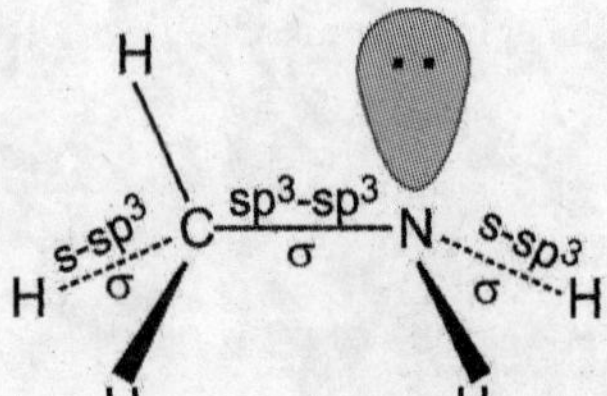

Fig. 27.1 Structure of Methylamine

Each C–H σ bond is formed by overlap of an sp^3 orbital of carbon and *s* orbital of hydrogen. All bond angles are approximately tetrahedral.

NOMENCLATURE

The following rules are applied in naming amines :

(1) Amines are named by naming the alkyl groups attached to the nitrogen atom followed by the ending *-amine*. Notice that the names are written as one word.

$CH_3—NH_2$	$CH_3CH_2—NH_2$	$CH_3CH_2CH_2—NH_2$
Methylamine	Ethylamine	*n*-Propylamine

(2) When two or three identical alkyl groups are attached to the nitrogen atom, the prefix *di-* or *tri-* is added to the name of the amine.

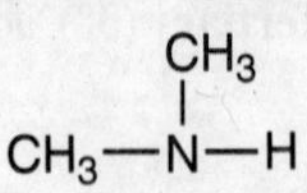

Dimethylamine

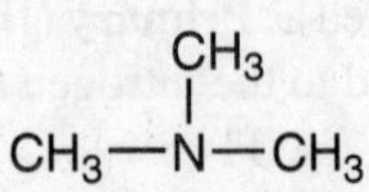

Trimethylamine

(3) When two or three different alkyl groups are attached to the nitrogen atom, they are named in alphabetical order.

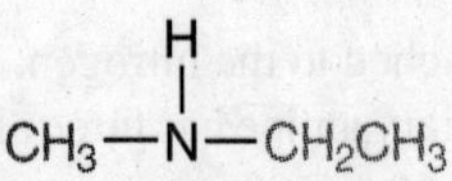

Ethylmethylamine

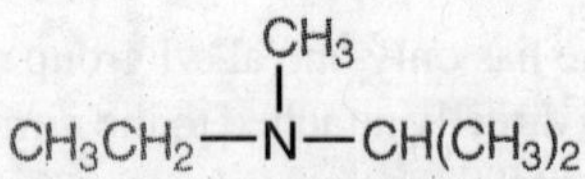

Ethylisopropylmethylamine

(4) If the amine is too complex, the IUPAC system of nomenclature is used. In this system the amino group (–NH_2) is considered as a substituent, and its position on the chain is indicated by the lowest possible number.

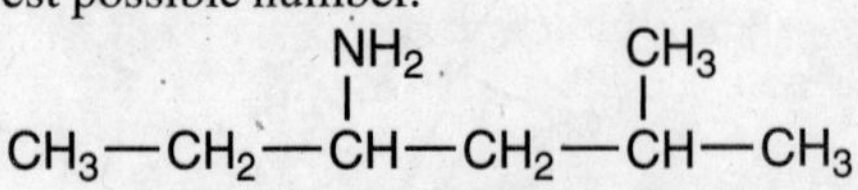

5-Methyl-3-aminohexane

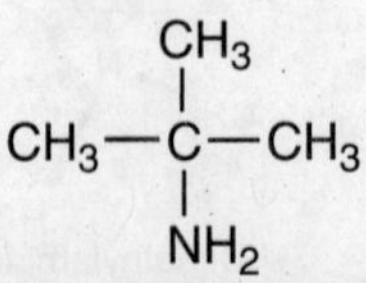

2-Amino-2-methylpropane
(*tert*-Butylamine)

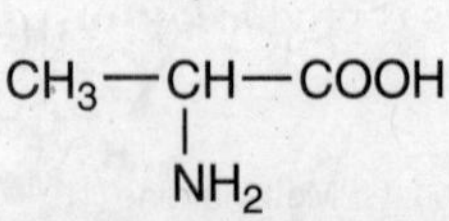

2-Aminopropanoic acid

SALTS OF AMINES

All amines are basic. They react with acids to form salts. The salts can be considered as being related to ammonium (NH_4^+) salts where the hydrogens of ammonium ion have been replaced by alkyl groups. When all four hydrogens of the ammonium ion have been replaced by alkyl groups, the compound is called **Quaternary Ammonium Salt.**

The amine salts are named as substituted ammonium salts. The alkyl groups are named first, and are followed by the ending *-ammonium*. The name of the anion present in the salt is then written as a second word. For example.

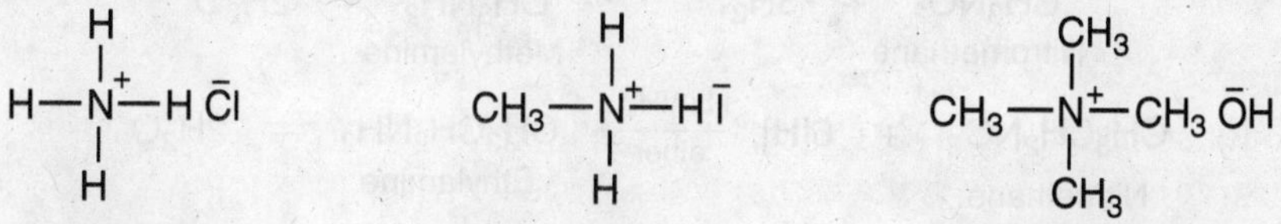

Ammonium chloride — Methylammonium iodide — Tetramethylammonium hydroxide (*Quaternary Salt*)

METHODS OF PREPARATION

The general methods of preparation of amines may be classified as follows :

I. METHODS YIELDING MIXTURE OF AMINES

(1) **Reaction of Alkyl Halides with Ammonia**. Amines can be prepared by heating an alkyl halide with alcoholic ammonia in a sealed tube. A mixture of primary, secondary, tertiary amines, and quaternary ammonium salt is obtained.

$$\underset{\text{Methyl iodide}}{CH_3I} + H{-}NH_2 \longrightarrow \underset{\text{Methylamine}}{CH_3NH_2} + HI$$

$$CH_3NH_2 + ICH_3 \longrightarrow \underset{\text{Dimethylamine}}{(CH_3)_2NH} + HI$$

$$(CH_3)_2NH + ICH_3 \longrightarrow \underset{\text{Trimethylamine}}{(CH_3)_3N} + HI$$

$$(CH_3)_3N + CH_3I \longrightarrow \underset{\text{Tetramethylammonium iodide}}{(CH_3)_4\overset{+}{N}\overset{-}{I}}$$

The amines, however, at once react with the hydrogen halide produced in the reaction to form salts:

$$CH_3\overset{+}{N}H_3\overset{-}{I}\ ;\ (CH_3)_2\overset{+}{N}H_2\overset{-}{I}\ ;\ \text{and}\ (CH_3)_3\overset{+}{N}H\overset{-}{I}$$

Thus in actual practice, the resulting mixture consists of different amine salts and the quaternary ammonium salt. This method is not suitable for laboratory synthesis because the mixture is often difficult to separate. On industrial scale, the mixture is separated by adding alkali to liberate the free amines and fractionally distilling. The primary, secondary, and tertiary amines pass over, while the quaternary ammonium salt is left behind.

(2) **Reaction of Alcohols with Ammonia**. Amines can be prepared by passing the vapors of alcohol and ammonia over heated alumina at 400°C. A mixture of primary, secondary, and tertiary amines is produced.

$$\underset{\text{Methanol}}{CH_3OH} + HNH_2 \xrightarrow[\Delta]{Al_2O_3} \underset{\text{Methylamine}}{CH_3NH_2} + H_2O$$

$$CH_3NH_2 + CH_3OH \longrightarrow \underset{\text{Dimethylamine}}{(CH_3)_2NH} + H_2O$$

$$(CH_3)_2NH + CH_3OH \longrightarrow \underset{\text{Trimethylamine}}{(CH_3)_3N} + H_2O$$

The primary amines may be obtained as the main product by using a large excess of ammonia.

II. METHODS FOR PRIMARY AMINES

(1) **Reduction of Nitroalkanes**. Primary amines can be obtained by reduction of nitroalkanes with H_2 + Pt (or Ni) or lithium aluminium hydride.

$$\underset{\text{Nitromethane}}{CH_3NO_2} + 3H_2 \xrightarrow{Pt} \underset{\text{Methylamine}}{CH_3NH_2} + 2H_2O$$

$$\underset{\text{Nitroethane}}{CH_3CH_2NO_2} + 6[H] \xrightarrow[\text{ether}]{LiAlH_4} \underset{\text{Ethylamine}}{CH_3CH_2NH_2} + 2H_2O$$

(2) **Reduction of Nitriles**. Primary amines can be prepared by reduction of nitriles (alkyl cyanides) with H_2 + Ni or lithium aluminium hydride.

$$\underset{\substack{\text{Methyl cyanide}\\\text{(Acetonitrile)}}}{CH_3C{\equiv}N} + 4[H] \xrightarrow[\text{ether}]{LiAlH_4} \underset{\text{Ethylamine}}{CH_3CH_2NH_2}$$

$$\underset{\substack{\text{Ethyl cyanide}\\\text{(Propanonitrile)}}}{CH_3CH_2C{\equiv}N} + 2H_2 \xrightarrow{Ni} \underset{n\text{-Propylamine}}{CH_3CH_2CH_2NH_2}$$

(3) **Reduction of Oximes**. Primary amines may be obtained by reduction of oximes of aldehydes and ketones with lithium aluminium hydride. Other reducing agents which can be used in place of lithium aluminium hydride are : H_2 + Ni and Na + C_2H_5OH.

$$\underset{\text{Acetaldoxime}}{CH_3CH{=}N{-}OH} + 4[H] \xrightarrow[\text{ether}]{LiAlH_4} \underset{\text{Ethylamine}}{CH_3CH_2NH_2} + H_2O$$

(4) **Reduction of Amides**. Primary amines can be obtained by reduction of simple amides with lithium aluminium hydride. Notice that the product contains the same number of carbons as the original amide.

$$\underset{\text{Acetamide}}{CH_3{-}\overset{\overset{\displaystyle O}{\|}}{C}{-}NH_2} + 4[H] \xrightarrow[\text{ether}]{LiAlH_4} \underset{\text{Ethylamine}}{CH_3CH_2NH_2} + H_2O$$

(5) **Hofmann's Degradation of Amides**. This is a good laboratory method for the conversion of an amide to a pure primary amine. The amide is warmed with bromine and concentrated aqueous NaOH solution

$$\underset{\text{Amide}}{R{-}\overset{\overset{\displaystyle O}{\|}}{C}{-}NH_2} + Br_2 + 4NaOH \longrightarrow \underset{1^\circ\text{ Amine}}{R{-}NH_2} + 2NaBr + Na_2CO_3 + 2H_2O$$

This reaction is also called *Hofmann's Rearrangement*. Notice that the overall result is the removal of the carbonyl group from the amide. The product contains one carbon less than the original amide. The method provides a useful technique for the *descent* of a homologous series.

MECHANISM. The Hofmann's degradation of amides proceeds through the following steps:

$$2NaOH + Br_2 \longrightarrow \overset{+}{Na}\overset{-}{O}Br + \overset{+}{Na}\overset{-}{Br} + H_2O$$

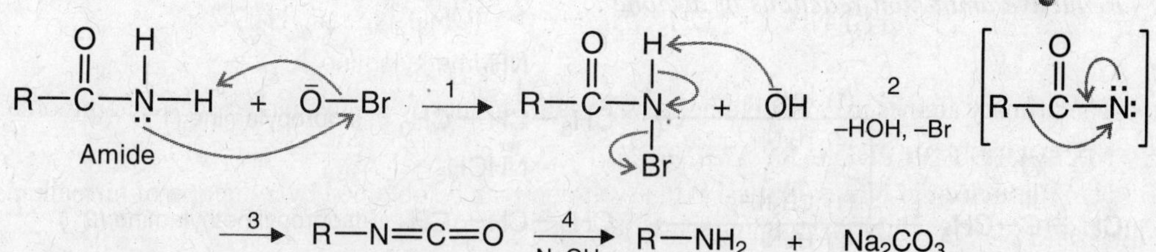

(6) **Gabriel Phthalimide Method**. This involves the treatment of phthalimide with potassium hydroxide to form the potassium salt. The salt is then heated with an alkyl halide to give *N*-alkylphthalimide, which in turn reacts with potassium hydroxide to form a potassium phthalate salt and a pure primary amine.

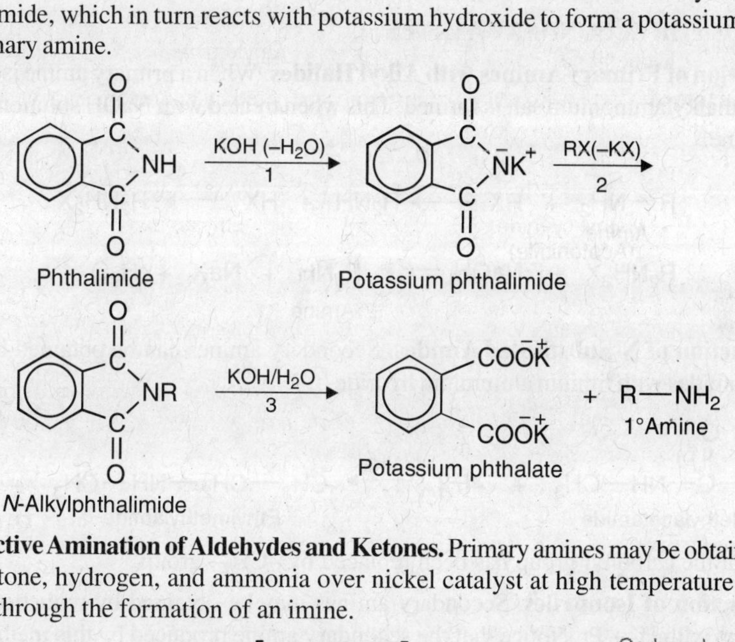

(7) **Reductive Amination of Aldehydes and Ketones.** Primary amines may be obtained by passing aldehyde or ketone, hydrogen, and ammonia over nickel catalyst at high temperature. The reaction probably goes through the formation of an imine.

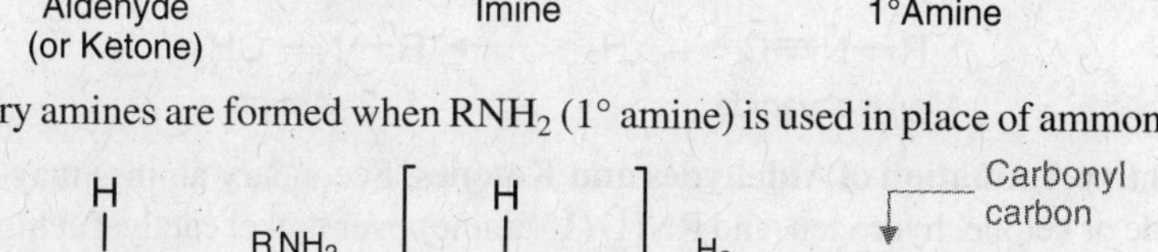

Secondary amines are formed when RNH_2 (1° amine) is used in place of ammonia.

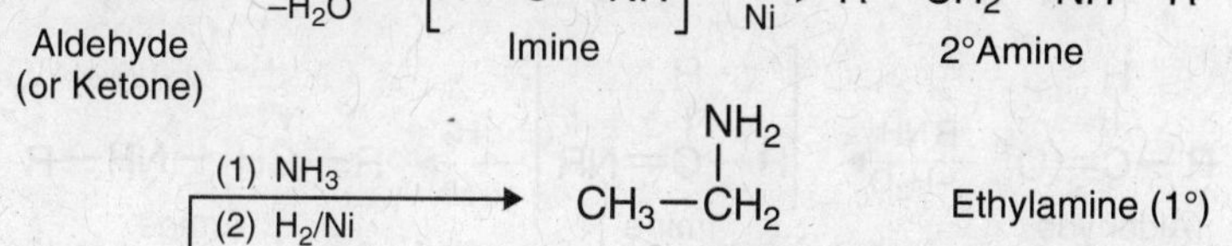

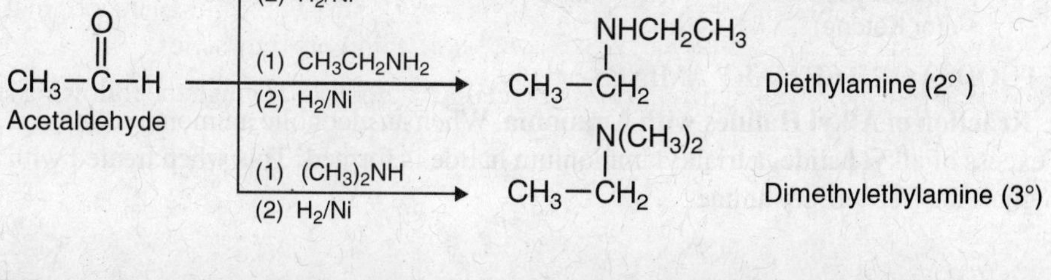

Reductive amination reactions of acetone :

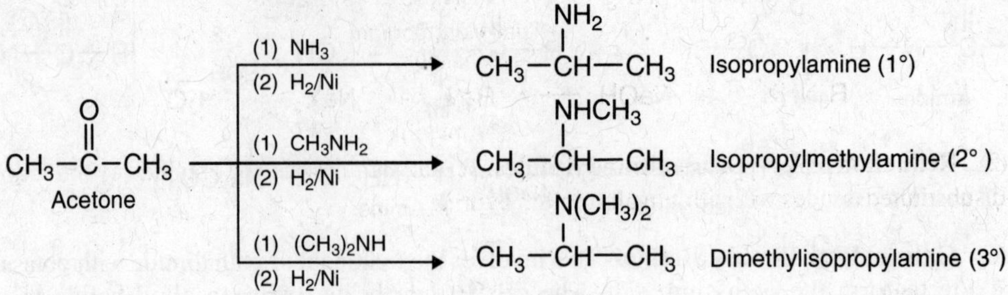

III. METHODS FOR SECONDARY AMINES

(1) **Reaction of Primary Amines with Alkyl Halides.** When a primary amine is heated with an alkyl halide, a dialkylammonium salt is formed. This when treated with NaOH solution yields the free secondary amine.

$$\underset{\text{1°Amine}}{R{-}NH_2} + RX \longrightarrow R_2NH + HX \longrightarrow R_2\overset{+}{N}H_2\overset{-}{X}$$

$$R_2\overset{+}{N}H_2\overset{-}{X} + NaOH \longrightarrow \underset{\text{2°Amine}}{R_2NH} + NaX + H_2O$$

(2) **Reduction of N-Substituted Amides.** Secondary amines can be obtained by reduction of *N*-substituted amides with lithium aluminium hydride.

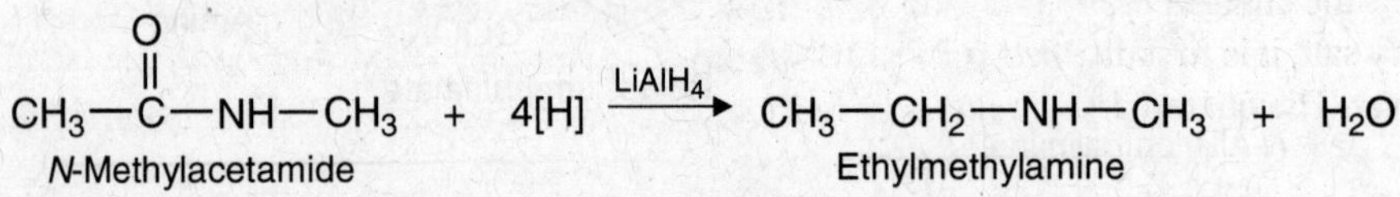

Notice that the carbonyl group has been replaced by $-CH_2-$ group.

(3) **Reduction of Isonitriles.** Secondary amines may be obtained by reduction of isonitriles (or isocyanides) with H_2 + Pt. Notice that the secondary amine produced by this method will always have one $-CH_3$ group.

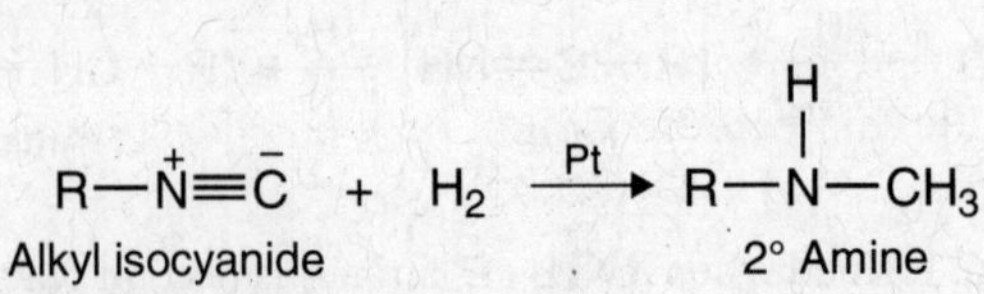

(4) **Reductive Amination of Aldehydes and Ketones.** Secondary amines may be obtained by passing aldehyde or ketone, hydrogen, and RNH_2 (1° amine) over nickel catalyst at high temperature.

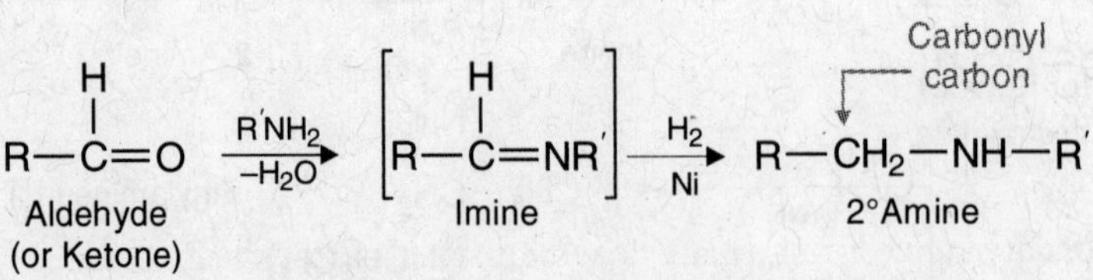

IV. METHODS FOR TERTIARY AMINES

(1) **Reaction of Alkyl Halides with Ammonia.** When an alcoholic ammonia solution is heated with an excess of alkyl halide, a trialkylammonium halide is formed. This when treated with NaOH solution gives the free tertiary amine.

$$3RX + NH_3 \longrightarrow \underset{\text{Trialkylammonium halide}}{R_3\overset{+}{N}H\overset{-}{X}} + 2HX$$

$$R_3\overset{+}{N}H\overset{-}{X} + NaOH \longrightarrow \underset{3^\circ\text{ Amine}}{R_3N} + NaX + H_2O$$

(2) **Reduction of *N,N*-Disubstituted Amides**. Tertiary amines may be obtained by reduction of *N,N*-disubstituted amides with lithium aluminium hydride.

$$\underset{N,N\text{-Dimethylacetamide}}{CH_3-\overset{\overset{\large O}{\|}}{C}-N(CH_3)_2} + 4[H] \xrightarrow{LiAlH_4} \underset{\text{Dimethylethylamine}}{CH_3CH_2N(CH_3)_2} + H_2O$$

(3) **Decomposition of Tetraalkylammonium Hydroxides**. Tertiary amines may be obtained by strongly heating tetraalkylammonium hydroxides. The hydroxides are prepared by treating the corresponding tetraalkylammonium iodide with moist silver oxide.

$$(CH_3)_4\overset{+-}{NI} + Ag_2O + H_2O \longrightarrow \underset{\text{Tetramethylammonium hydroxide}}{(CH_3)_4\overset{+-}{NOH}} + AgI$$

$$(CH_3)_4\overset{+-}{NOH} \xrightarrow{\Delta} \underset{\text{Trimethylamine}}{(CH_3)_3N} + CH_3OH$$

SEPARATION OF MIXTURE OF AMINES

When the mixture contains salts of primary, secondary, and tertiary amines along with the quaternary salt, it is first distilled with KOH solution. The mixture of the three amines (1°, 2° and 3°) distills over. The quaternary salt does not react with KOH and being nonvolatile is left behind.

$$\underset{1^\circ\text{Amine salt}}{R-\overset{+}{N}H_3\overset{-}{Cl}} + KOH \xrightarrow{\Delta} \underset{1^\circ\text{Amine}}{R-NH_2} + KCl + H_2O$$

2° and 3° Amine salts also react in this way.

The distillate contains the mixture of primary, secondary, and tertiary amines. It may be separated by the following methods :

(1) **Fractional Distillation**. The mixture of primary, secondary, and tertiary amines may be separated by fractional distillation because their boiling points are quite different. This method is extensively used in industry.

(2) **Hofmann Method**. This involves the treatment of the mixture with diethyl oxalate.

(*i*) The primary amine forms a dialkyloxamide, which is a solid.

$$\underset{\text{Ethyl oxalate}}{\begin{matrix} CO-OC_2H_5 \\ | \\ CO-OC_2H_5 \end{matrix}} + \underset{1^\circ\text{ Amine}}{\begin{matrix} H-NHR \\ \\ H-NHR \end{matrix}} \longrightarrow \underset{\text{Oxamide } (solid)}{\begin{matrix} CO-NHR \\ | \\ CO-NHR \end{matrix}} + 2C_2H_5OH$$

(*ii*) The secondary amine forms a dialkyloxamic ester, which is an oily liquid.

$$\underset{\text{Ethyl oxalate}}{\begin{matrix} CO-OC_2H_5 \\ | \\ CO-OC_2H_5 \end{matrix}} + \underset{2^\circ\text{ Amine}}{H-NR_2} \longrightarrow \underset{\text{Oxamic ester } (Liquid)}{\begin{matrix} CO-NR_2 \\ | \\ CO-OC_2H_5 \end{matrix}} + C_2H_5OH$$

(*iii*) The tertiary amine does not react at all.

All right, who took my lucky test tube ?

The reaction mixture is now fractionally distilled. The tertiary amine distills over and forms the first fraction. This is followed by the oxamic ester which forms the second fraction. The oxamide remains behind in the distillation flask.

The oxamide and the oxamic ester separated as above are hydrolysed with KOH to give back the amines which are distilled off.

$$\underset{\text{Oxamide}}{\begin{array}{l}CO-NHR\\ |\\ CO-NHR\end{array}} + \begin{array}{l}HOK\\ \\ HOK\end{array} \longrightarrow \underset{\text{Potassium oxalate}}{\begin{array}{l}CO-OK\\ |\\ CO-OK\end{array}} + \underset{1^\circ\ \text{Amine}}{2RNH_2}$$

$$\underset{\text{Oxamic ester}}{\begin{array}{l}CO-NR_2\\ |\\ CO-OC_2H_5\end{array}} + \begin{array}{l}HOK\\ \\ HOK\end{array} \longrightarrow \underset{\text{Potassium oxalate}}{\begin{array}{l}CO-OK\\ |\\ CO-OK\end{array}} + \underset{2^\circ\ \text{Amine}}{R_2NH} + C_2H_5OH$$

(3) **Hinsberg Method**. This involves the treatment of the mixture with benzenesulfonyl chloride (*Hinsberg reagent*). The solution is then made alkaline with aqueous NaOH.

(*i*) The primary amine gives N-alkylbenzenesulfonamide. This forms salt with NaOH, which is soluble in water.

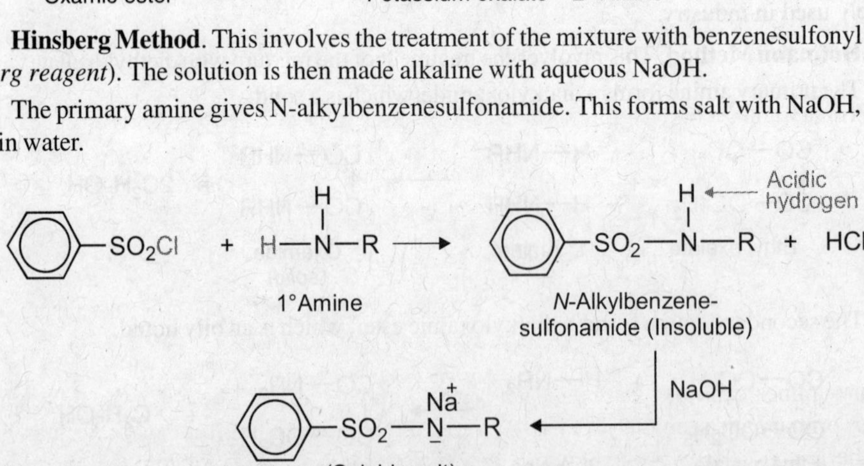

(*ii*) The secondary amine gives N,N-dialkylbenzenesulfonamide. This does not form salt with NaOH (No acidic hydrogens) and is insoluble in alkali solution.

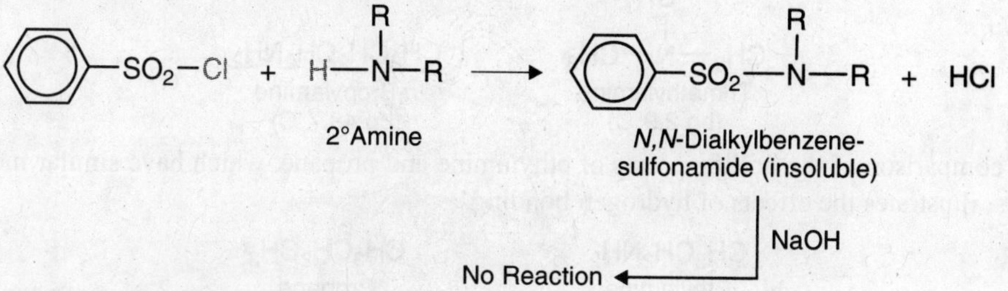

(*iii*) The tertiary amine does not react.

The resulting alkaline solution is distilled when the tertiary amine passes over and the remaining mixture is filtered. The filtrate on acidification gives the sulfonamide of the primary amine, while the solid residue is the sulfonamide of the secondary amine. The two sulfonamides thus isolated are hydrolysed with conc. IICl and distilled over NaOH to yield the respective amines.

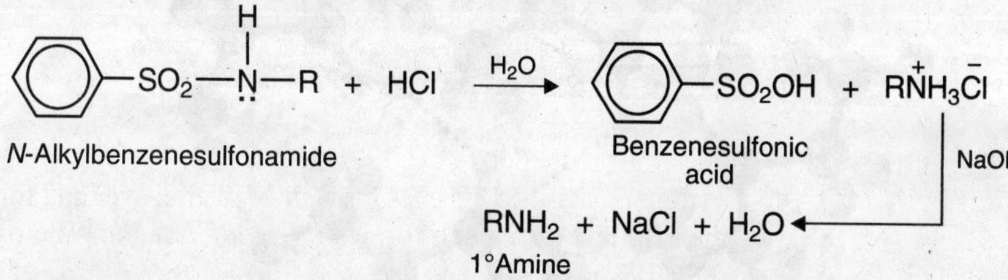

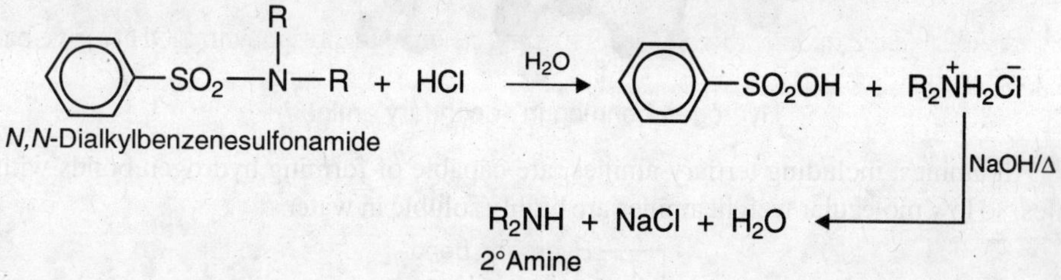

Nowadays benzenesulfonyl chloride has been replaced by *p*-toluenesulfonyl chloride, $CH_3–C_6H_4–SO_2Cl$, since the substituted sulfonamides thus formed are stable solids which can be easily recrystallized.

PHYSICAL PROPERTIES

(1) Lower amines are gases or low-boiling liquids and possess a characteristic ammonia-like smell.

(2) Primary and secondary amines are capable of forming hydrogen bonds with each other.

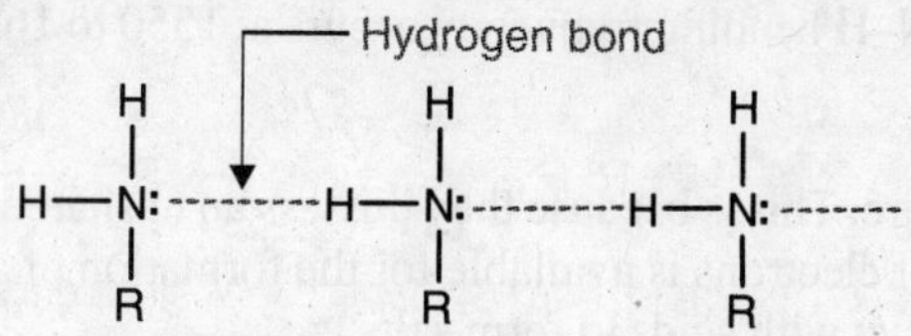

Hydrogen bonding in amine molecules

Tertiary amines can not form hydrogen bonds with each other. The ability to form hydrogen bonds gives primary and secondary amines relatively high boiling points for their molecular weights, although the H—N- - -H hydrogen bond is not as strong as H—O- - -H hydrogen bond.

A comparison of the boiling points of trimethylamine and *n*-propylamine, which have the same molecular weights, illustrates the effects of hydrogen bonding.

$(CH_3)_3N$	$CH_3CH_2CH_2NH_2$
Trimethylamine	*n*-Propylamine
(bp 2.9°C)	(bp 48.7°C)

A comparison of the boiling points of ethylamine and propane, which have similar molecular weights, illustrates the effects of hydrogen bonding.

$CH_3CH_2NH_2$	$CH_3CH_2CH_3$
Ethylamine	Propane
(bp +17°C)	(bp −42°C)

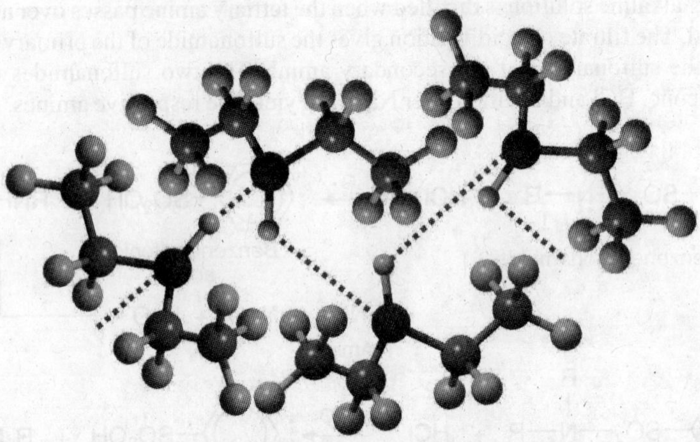

Hydrogen bonding in secondary amines

(3) All amines, including tertiary amines, are capable of forming hydrogen bonds with water molecules, so low molecular weight amines are highly soluble in water.

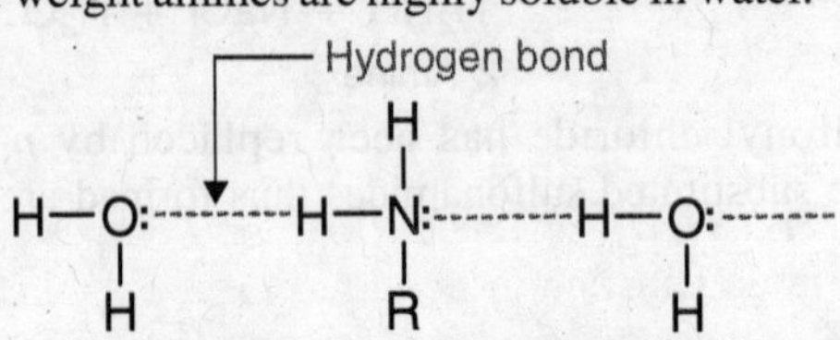

Hydrogen bonding between amine and water molecules

(4) ***IR Spectrum.*** The infrared spectrum of amines show N–H stretching at 3330 to 3570 cm^{-1}. Primary amines have two sharp peaks in this region, secondary amines only one, and tertiary amines none (no N–H bonds). The N–H bending frequency occurs at 1550 to 1650 cm^{-1}.

BASICITY OF AMINES

Amines are basic in nature. This is because they possess an unshared pair (lone pair) of electrons on nitrogen. This lone pair of electrons is available for the formation of a new bond with a proton or Lewis acids. Thus amines react with acids to form salts.

$$\underset{\text{Amine}}{R{-}\ddot{N}H_2} + H{-}Cl \longrightarrow \underset{\text{Alkylammonium chloride}}{R{-}\overset{+}{N}H_3\ \overset{-}{Cl}}$$

Basicity Constants. Strong bases (*e.g.*, NaOH or KOH) completely ionize in aqueous solution. Amines are weak bases. They are only partially ionized in aqueous solution and an equilibrium exists between the ionized and unionized forms.

$$R{-}NH_2 + H_2O \rightleftharpoons R{-}\overset{+}{N}H_3 + \overset{-}{O}H$$

The extent of ionization is described by an equilibrium constant K_b, which is known as **Basicity Constant**. It is defined as the concentration of the products of ionization in moles per liter divided by the concentration of the unionized base.

$$K_b = \frac{[RNH_3^+][OH^-]}{[RNH_2]} \qquad pK_b = -\log K_b$$

The basicity constant describes the relative strength of a weak base. **Stronger bases will have higher numerical values of K_b (Lower pK_b values). Weaker bases will have lower numerical values of K_b (Higher pK_b values).**

Relative Basicities. Aliphatic amines are stronger bases than ammonia. This is because the alkyl groups are electron-releasing. They increase the electron density around the nitrogen, thereby increasing the availability of the lone pair of electrons.

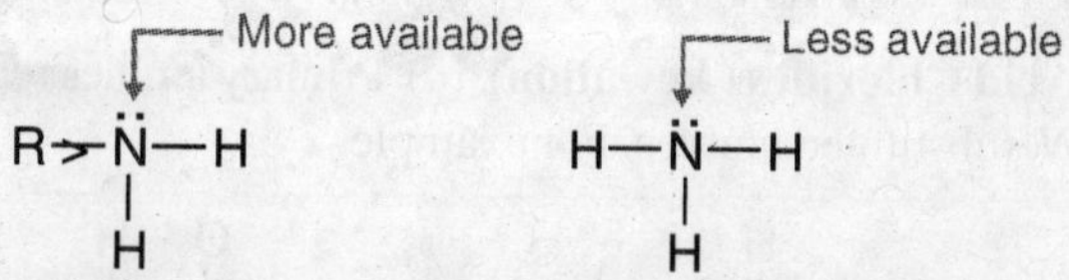

Amine　　Ammonia

Fig. 27.2. Amines are more basic than ammonia because the electron release by alkyl groups increases lone-pair availability.

The greater the number of electron-releasing alkyl groups, the greater the availability of nitrogen's lone pair and stronger the base.

	NH_3	$CH_3{-}NH_2$	$(CH_3)_2NH$	$(CH_3)_3N$
	Ammonia	Methylamine	Dimethylamine	Trimethylamine
pK_b	4.784	3.36	3.27	4.19

Notice that the dimethylamine is a stronger base than methylamine. However, trimethylamine is a weaker base than both dimethylamine and methylamine. Although in trimethylamine the electron-density is further increased, *the steric crowding of three methyl groups makes the approach and bonding by a proton relatively difficult.* The electrons are there but the path is blocked.

Points to Remember

1. Amines are basic.
2. Basicity of amines is due to the presence of the lone pair (unshared pair) of electrons on nitrogen.
3. Electron-releasing groups (+I groups) increase basicity of amines.
4. Electron-withdrawing groups (–I groups) decrease basicity of amines.
5. Amines are more basic than ammonia.
6. Any factor which reduces the availability of the lone pair of electrons on nitrogen reduces the basicity of the amine.
   Lone pair *more* available – Amine *more* basic.
   Lone pair *less* available – Amine *less* basic.

CHEMICAL PROPERTIES

The main reactions of amines are due to the presence of the lone pair of electrons on nitrogen. This lone pair of electrons is available for donation to electron-seeking reagents. Amines are nucleophilic reagents.

(1) **Salt Formation**. Amines are basic compounds. They react with mineral acids to form salts. For example,

$$\underset{\text{Ethylamine}}{CH_3CH_2{-}NH_2} + HCl \longrightarrow \underset{\text{Ethylammonium chloride}}{CH_3CH_2{-}\overset{+}{N}H_3\overset{-}{Cl}}$$

When these salts are treated with NaOH, parent amines are liberated.

$$CH_3CH_2{-}\overset{+}{N}H_3\overset{-}{Cl} + NaOH \longrightarrow CH_3CH_2NH_2 + NaCl + H_2O$$

(2) **Reaction with Alkyl Halides**. When an amine is treated with an alkyl halide, hydrogen atoms of the *N* atom are successively replaced by alkyl groups. In the final step, the tertiary amine adds one molecule of the alkyl halide to yield a quaternary ammonium salt.

$$\underset{1°\text{Amine}}{RNH_2} \xrightarrow[-HX]{RX} \underset{2°\text{Amine}}{R_2NH} \xrightarrow[-HX]{RX} \underset{3°\text{Amine}}{R_3N} \xrightarrow[-HX]{RX} \underset{\text{Quaternary salts}}{R_4\overset{+}{N}X}$$

(3) **Reaction with Acid Chlorides (Acylation)**. (*a*) Primary amines react with acid chlorides or acid anhydrides to form *N*-substituted amides. For example,

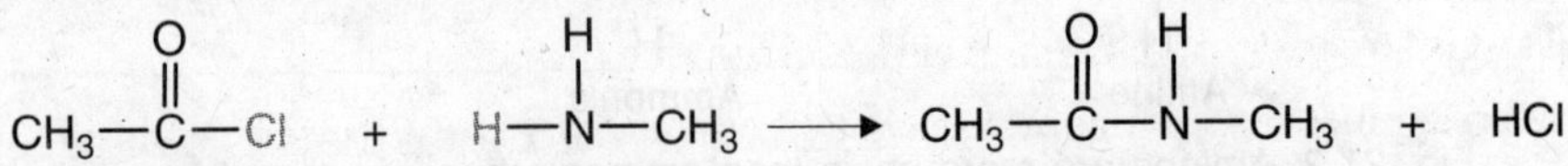

Acetyl chloride + Methylamine → *N*-Methylacetamide + HCl

MECHANISM. The mechanism involves the nucleophilic attack by the amine on the electrophilic carbon.

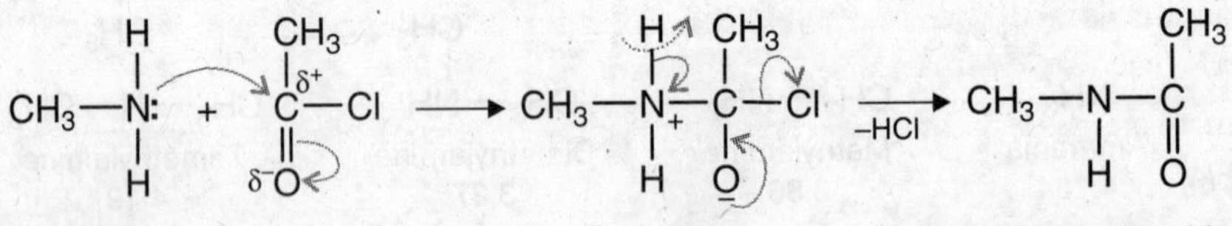

Methylamine (Unstable) *N*-Methylacetamide

(*b*) Secondary amines react with acid chlorides to form *N*,*N*-disubstituted amides.

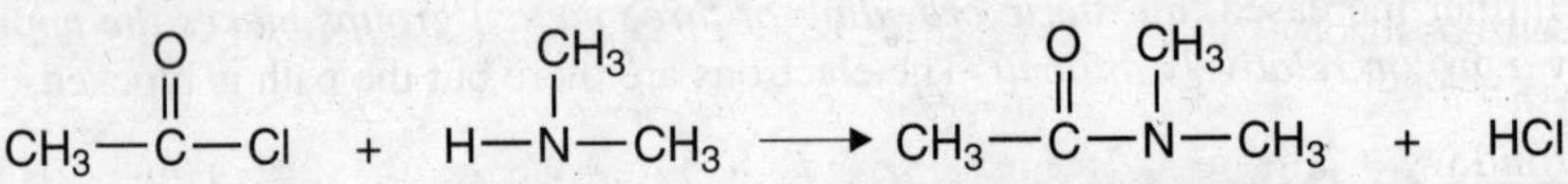

Acetyl chloride + Dimethylamine → *N*,*N*-Dimethylacetamide + HCl

(*c*) Tertiary amines do not react since they do not have a replaceable hydrogen on the nitrogen.

(4) **Reaction with Nitrous Acid**. Nitrous acid (HONO) is an unstable substance and is therefore prepared *in situ* by the reaction of sodium nitrite with dilute HCl at 0°C.

$$NaNO_2 + HCl \longrightarrow HONO + NaCl$$

(*a*) Primary amines react with nitrous acid to form alcohols and nitrogen gas.

$$\underset{1°\text{Amine}}{R{-}NH_2} + HONO \longrightarrow \underset{\text{Alcohol}}{ROH} + N_2\uparrow + H_2O$$

$$\underset{\text{Ethylamine}}{C_2H_5{-}NH_2} + HONO \longrightarrow \underset{\text{Ethanol}}{C_2H_5OH} + N_2\uparrow + H_2O$$

(*b*) Secondary amines react with nitrous acid to form *N*-nitrosoamines which are water-insoluble yellow oils.

$$\underset{\text{2°Amine}}{R_2N{-}H} + HONO \longrightarrow \underset{\substack{N\text{-Nitrosoamine}\\(\textit{yellow oil})}}{R_2N{-}N{=}O} + H_2O$$

$$\underset{\text{Diethylamine}}{(C_2H_5)_2N{-}H} + HONO \longrightarrow \underset{\substack{N\text{-Nitrosodiethylamine}\\(\textit{yellow oil})}}{(C_2H_5)_2N{-}N{=}O} + H_2O$$

(*c*) Tertiary amines react with nitrous acid to form trialkylammonium nitrite salts which are soluble in water.

$$\underset{\text{3°Amine}}{R_3N} + HONO \longrightarrow \underset{\substack{\text{Trialkylammonium nitrite}\\(\textit{soluble})}}{R_3\overset{+}{N}H\overset{-}{NO_2}}$$

$$\underset{\text{Triethylamine}}{(C_2H_5)_3N} + HONO \longrightarrow \underset{\substack{\text{Triethylammonium nitrite}\\(\textit{soluble})}}{(C_2H_5)_3\overset{+}{N}H\overset{-}{NO_2}}$$

This reaction is used as the basis of a test to distinguish between primary, secondary, and tertiary amines. The test is known as the **Nitrous Acid Test**. To summarize this test :

(*a*) Primary amines react with nitrous acid to produce nitrogen gas (seen as bubbles).

(*b*) Secondary amines react with nitrous acid to produce a yellow oily layer.

(*c*) Tertiary amines react with nitrous acid to form soluble nitrite salts. There is no visible sign of reaction.

(5) **Reaction with Benzenesulfonyl Chloride (Sulfonylation)**. Benzenesulfonyl chloride is obtained by the reaction of benzenesulfonic acid with PCl_5 or $SOCl_2$.

(*a*) Primary amines react with benzenesulfonyl chloride to form *N*-alkylbenzenesulfonamide.

Benzenesulfonyl chloride + 1°Amine ⟶ *N*-Alkylsulfonamide (Precipitate) + HCl

(*b*) Secondary amines react with benzenesulfonyl chloride to form *N*,*N*-dialkylbenzenesulfonamide.

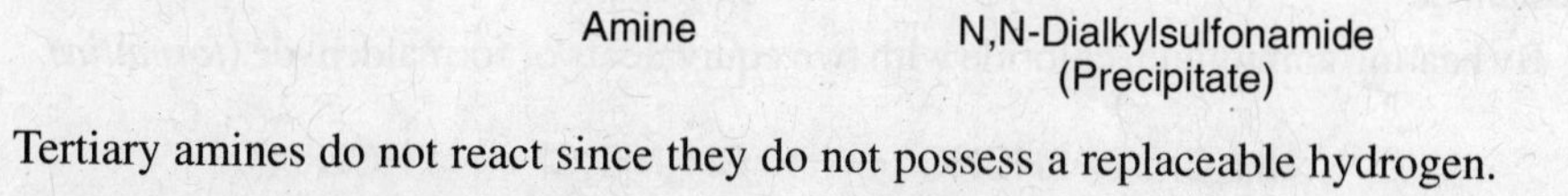

Amine ⟶ N,N-Dialkylsulfonamide (Precipitate)

(*c*) Tertiary amines do not react since they do not possess a replaceable hydrogen.

The sulfonamides from primary amines still have a replaceable hydrogen on the nitrogen. This hydrogen is acidic. Thus sulfonamides from primary amines *dissolve in bases* to form soluble salts.

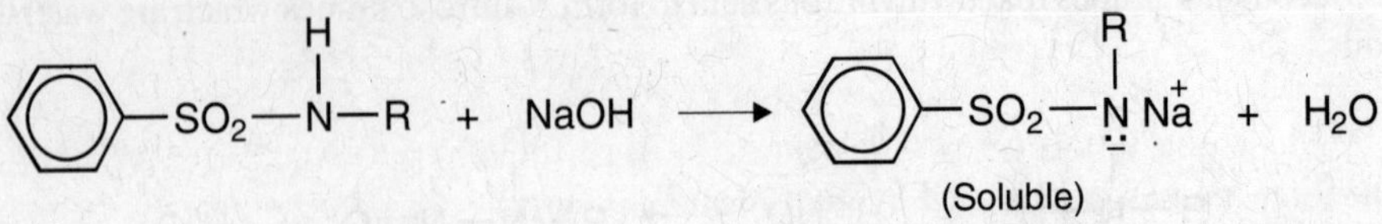

The sulfonamides from secondary amines do not have a replaceable hydrogen and so *do not dissolve in bases.*

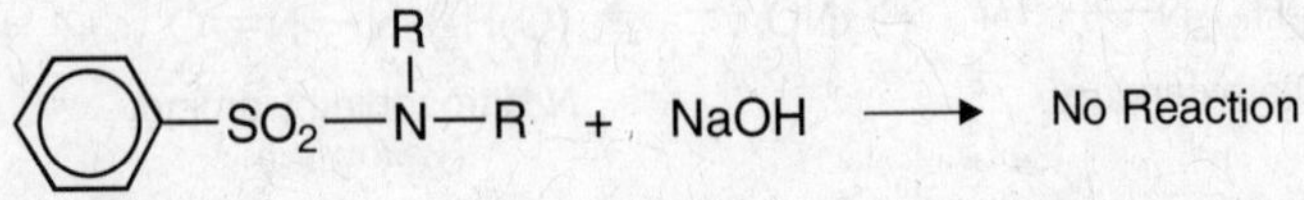

These reactions are used as the basis of a test to distinguish between primary, secondary, and tertiary amines. The test is known as the **Hinsberg Test**. To summarize this test :

(*a*) Primary amines react with benzenesulfonyl chloride to form a precipitate that is *soluble* in NaOH solution.

(*b*) Secondary amines react with benzenesulfonyl chloride to give a precipitate that is *insoluble* in NaOH solution.

(*c*) Tertiary amines *do not react* with benzenesulfonyl chloride.

(6) **Carbylamine Reaction**. Primary amines react with chloroform and a solution of KOH in ethanol to form isocyanides (isonitriles or carbylamines). Secondary and tertiary amines do not give this reaction.

$$\underset{\text{1° Amine}}{R{-}NH_2} + CHCl_3 + 3KOH \longrightarrow \underset{\text{Isocyanide}}{R{-}\overset{+}{N}{\equiv}\overset{-}{C}} + 3KCl + 3H_2O$$

The isocyanides or carbylamines are evil-smelling and can be easily detected. This reaction is used to distinguish primary amines from secondary and tertiary amines.

(7) **Reaction with Carbon Disulfide**. Primary amines react with carbon disulfide to form a dithiocarbamic acid. Dithiocarbamic acids undergo decomposition with mercuric chloride to give the corresponding alkyl isothiocyanate.

$$\underset{\text{1° Amine}}{R{-}NH_2} + CS_2 \xrightarrow{\Delta} \underset{\text{Dithiocarbamic acid}}{RNH{-}\overset{\overset{\large S}{\|}}{C}{-}SH} \xrightarrow{HgCl_2} \underset{\text{Alkyl isothiocyanate}}{RNCS} + HgS + 2HCl$$

METHYLAMINE, CH_3NH_2

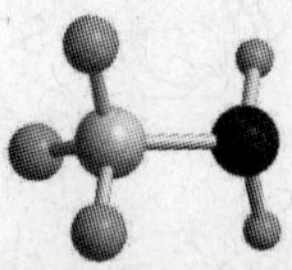

Methylamine is the simplest primary amine. It is prepared :

(1) By Hofmann degradation of acetamide. (*Laboratory Method*)

$$\underset{\text{Acetamide}}{CH_3{-}\overset{\overset{\large O}{\|}}{C}{-}NH_2} + Br_2 + 4NaOH \longrightarrow \underset{\text{Methylamine}}{CH_3NH_2} + 2NaBr + Na_2CO_3 + 2H_2O$$

(2) By heating ammonium chloride with two equivalents of formaldehyde (*formalin*).

$$NH_4Cl + \underset{\text{Formaldehyde}}{2HCHO} \xrightarrow{\Delta} \underset{\text{Methylammonium chloride}}{CH_3\overset{+}{N}H_3\overset{-}{Cl}} + HCOOH$$

$$CH_3\overset{+}{N}H_3\overset{-}{Cl} + NaOH \longrightarrow CH_3NH_2 + NaCl + H_2O$$

(3) By passing a mixture of ammonia and methanol over alumina at 450°C.

$$CH_3{-}OH + H{-}NH_2 \xrightarrow[450°C]{Al_2O_3} CH_3{-}NH_2 + H_2O$$

Methylamine further reacts with methanol to form dimethylamine and trimethylamine. The mixture of the three amines is separated by fractional distillation.

Methylamine is a gas, bp –7.6°C. It is soluble in water. It gives all the reactions of a primary amine. Methylamine is used as a refrigerant.

ETHYLAMINE, $CH_3CH_2NH_2$

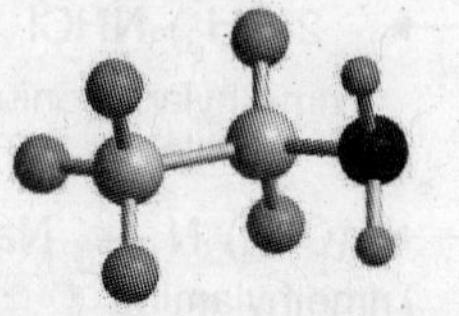

Ethylamine is prepared :

(1) By Hofmann degradation of propionamide. (*Laboratory Method*)

$$\underset{\text{Propionamide}}{C_2H_5\overset{\overset{\displaystyle O}{\|}}{C}NH_2} + Br_2 + 4NaOH \longrightarrow \underset{\text{Ethylamine}}{C_2H_5NH_2} + 2NaBr + Na_2CO_3 + 2H_2O$$

(2) By passing a mixture of ethylene and ammonia over a cobalt catalyst at 450°C.

$$CH_2{=}CH_2 + NH_3 \xrightarrow[450°C]{Co} CH_3CH_2NH_2$$

Ethylamine is a colorless liquid, bp 19°C. It has ammonia-like smell. It is soluble in water. Ethylamine gives all the reactions of a primary amine.

DIMETHYLAMINE, $(CH_3)_2NH$

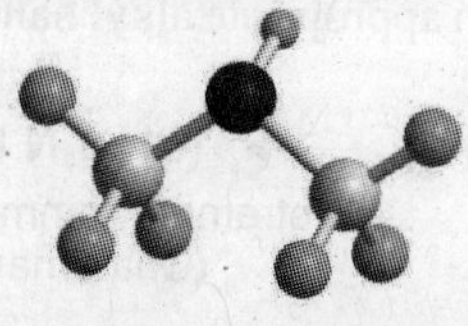

Dimethylamine is the simplest secondary amine. It is prepared :

(1) By heating ammonium chloride with four equivalents of formaldehyde (*formalin*).

$$NH_4Cl + \underset{\text{Formaldehyde}}{4HCHO} \longrightarrow \underset{\text{Dimethylammonium chloride}}{(CH_3)_2\overset{+}{N}H_2\overset{-}{Cl}} + 2HCOOH$$

$$(CH_3)_2\overset{+}{N}H_2\overset{-}{Cl} + NaOH \longrightarrow \underset{\text{Dimethylamine}}{(CH_3)_2NH} + NaCl + H_2O$$

(2) By hydrolysis of *p*-nitroso-*N*,*N*-dimethylaniline with NaOH solution.

p-Nitroso-*N*,*N*-Dimethylamine Dimethylamine

Dimethylamine is a colorless gas, bp 7°C. It has a strong ammoniacal odor. It is soluble in water. Dimethylamine gives all the reactions of a secondary amine. Dimethylamine is used for dehairing of hides. It is also used in the manufacture of synthetic fibre solvents, and rocket fuels.

TRIMETHYLAMINE, $(CH_3)_3N$

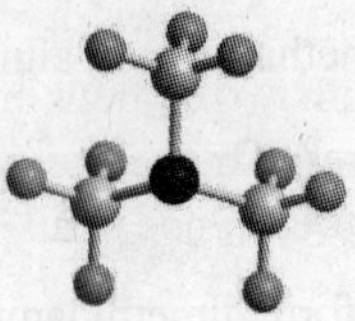

Trimethylamine is the simplest tertiary amine. It is obtained :

(1) By heating a solid mixture of ammonium chloride and paraformaldehyde.

$$2NH_4Cl + 9HCHO \longrightarrow \underset{\text{Trimethylammonium chloride}}{2(CH_3)_3\overset{+}{N}H\overset{-}{Cl}} + 3CO_2 + 3H_2O$$

$$(CH_3)_3\overset{+}{N}H\overset{-}{Cl} + NaOH \longrightarrow \underset{\text{Trimethylamine}}{(CH_3)_3N} + NaCl + H_2O$$

(2) By the thermal decomposition of tetramethylammonium hydroxide.

$$(CH_3)_4\overset{+}{N}\,\overset{-}{O}H \xrightarrow{\Delta} (CH_3)_3N + CH_3OH$$

Trimethylamine is a gas, bp 3.5°C. It is soluble in water. Trimethylamine is used as a source of methyl chloride.

$$(CH_3)_3N + 4HCl \xrightarrow{\Delta} \underset{\text{Methyl chloride}}{3CH_3Cl} + NH_4Cl$$

QUATERNARY AMMONIUM SALTS

Tetraalkylammonium Halides. These are obtained by treating ammonia with excess of an alkyl halide or by heating a tertiary amine with an appropriate alkyl halide. For example,

$$\underset{\text{Trimethylamine}}{(CH_3)_3N} + CH_3I \xrightarrow{\Delta} \underset{\substack{\text{Tetramethylammonium iodide} \\ \text{(Quaternary salt)}}}{(CH_3)_4\overset{+}{N}\,\overset{-}{I}}$$

Tetraalkylammonium halides have properties very similar to the ionic inorganic salts. They are crystalline solids. Their aqueous solutions conduct an electric current.

Tetraalkylammonium Hydroxides. These are obtained by heating tetraalkylammonium halides with aqueous solution of silver oxide.

$$(CH_3)_4\overset{+}{N}\,\overset{-}{I} + Ag_2O + H_2O \longrightarrow \underset{\text{Tetramethylammonium hydroxide}}{(CH_3)_4\overset{+}{N}\,\overset{-}{O}H} + AgI$$

Tetraalkylammonium hydroxides are deliquescent crystalline solids which are soluble in water. They are strong bases comparable in strength to NaOH or KOH. On heating above 125°C, they decompose to form tertiary amines. This reaction is called **Hofmann's Elimination**. Examples, are,

$$\underset{\text{4° Hydroxide}}{(CH_3)_4\overset{+}{N}\,\overset{-}{O}H} \xrightarrow{\Delta} \underset{\text{Trimethylamine}}{(CH_3)_3N} + CH_3OH$$

$$R{-}\overset{+}{N}(CH_3)_3\,\overset{-}{O}H \xrightarrow{\Delta} \underset{\text{Trimethylamine}}{(CH_3)_3N} + \textit{Alkene} + H_2O$$

In the second equation above, *R* is an alkyl group with two carbons or more. **In Hofmann's elimination, the least substituted alkene is the favored product.** If more than one alkyl group can be eliminated as alkene, ethyl group is eliminated (as *ethylene*) in preference to all other groups. For example,

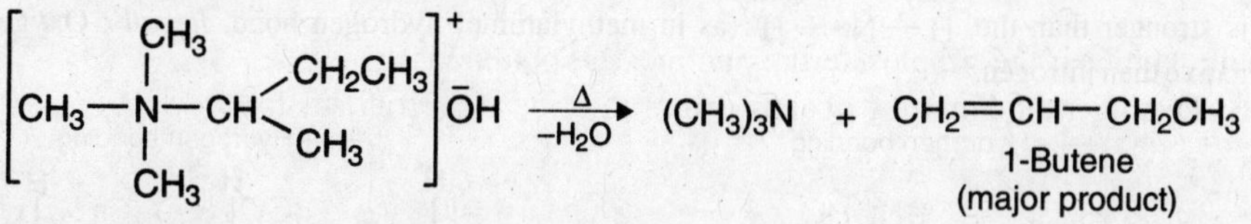

MECHANISM. The Hofmann's elimination of quaternary ammonium hydroxides takes place by E2 mechanism as follows :

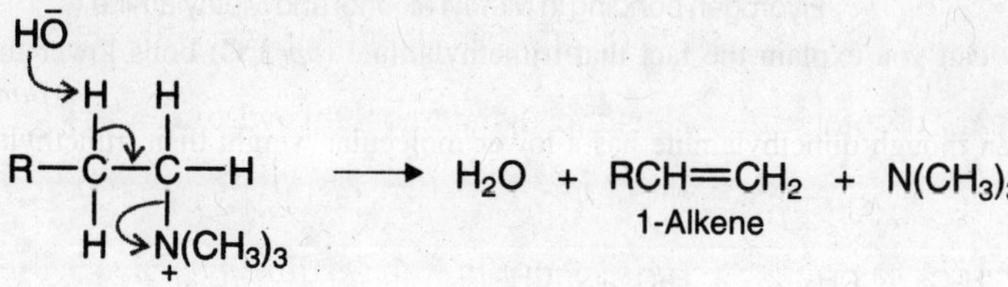

STEPPING UP AND DOWN THE AMINE SERIES

Reduction of nitriles and Hofmann degradation of amides are the two key reactions used for stepping *up* and *down* the amine homologous series.

(1) **Conversion of Methylamine into Ethylamine**

$$\underset{\text{Methylamine}}{CH_3NH_2} \xrightarrow{HONO} CH_3OH \xrightarrow{SOCl_2} CH_3Cl \xrightarrow{NaCN} CH_3CN \xrightarrow[4[H]]{LiAlH_4} \underset{\text{Ethylamine}}{CH_3CH_2NH_2}$$

(2) **Conversion of Ethylamine into Methylamine**

$$\underset{\text{Ethylamine}}{CH_3CH_2NH_2} \xrightarrow{HONO} CH_3CH_2OH \xrightarrow[[O]]{K_2Cr_2O_7/H^+} CH_3-\overset{O}{\overset{\|}{C}}-OH$$

$$\xrightarrow{SOCl_2} CH_3-\overset{O}{\overset{\|}{C}}-Cl \xrightarrow{NH_3} CH_3-\overset{O}{\overset{\|}{C}}-NH_2 \xrightarrow[NaOH]{Br_2} \underset{\text{Methylamine}}{CH_3NH_2}$$

Examination Questions

1. Write the names and structures of *four* isomeric amines having the molecular formula C_3H_9N.
2. Write the structures of *eight* isomeric amines having the molecular formula $C_4H_{11}N$.
3. What are amines ? How are they prepared ? Discuss the action of nitrous acid on primary, secondary, and tertiary amines.
4. Give one method each for the preparation of a primary, a secondary, and a tertiary amine.
5. What happens when ethylamine is treated with nitrous acid ?
6. Discuss the action of nitrous acid on primary, secondary, and tertiary amines.

(*Kota BSc, 2013 ; Sambalpur BSc, 2014 ; Baroda BSc, 2015*)

7. Explain why the boiling point of methyl alcohol (bp 64.5°C) is higher than that of methylamine (bp –6.5°C). *(Patna BSc, 2014 ; Andhra BSc, 2015)*

Answer. The boiling point of methyl alcohol is considerably higher than that of methylamine, although both compounds have similar molecular weights. This is because H—O- - -H hydrogen bond (as in methyl alcohol) is stronger than the H—N- - -H (as in methylamine) hydrogen bond. *Recall* : Oxygen is more electronegative than nitrogen.

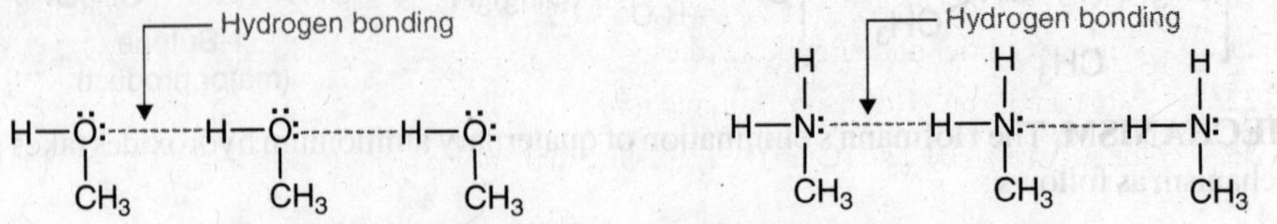

Hydrogen bonding in Methyl alcohol and Methylamine

8. How can you explain the fact that trimethylamine (*bp* 3°C) boils lower than dimethylamine (*bp* 7°C) ? *(Bangalore BSc, 2014)*

Answer. Even though dimethylamine has a lower molecular weight than trimethylamine, it boils at a higher temperature.

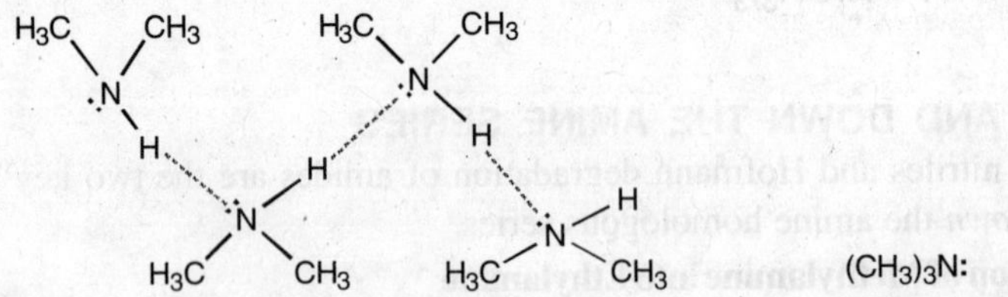

Dimethylamine　　　Trimethylamine

Liquid dimethylamine forms hydrogen bonds that must be broken in the boiling process. Since extra energy must be added to break these bonds, dimethylamine has a higher boiling point than trimethylamine which does not form hydrogen bonds.

9. How will you distinguish between primary, secondary, and tertiary amines ? *(Delhi BSc, 2015)*

10. How will you distinguish between ethylamine and dimethylamine ? *(Panjab BSc, 2014 ; Agra BSc, 2015)*

11. How will you distinguish between diethylamine and triethylamine ?

12. How will you separate a mixture of primary, secondary, and tertiary amines ? *(Berhampur BSc, 2013 ; Tripura BSc, 2014)*

13. Discuss the basicity of amines.

14. Explain why ethylamine is a stronger base than ammonia. *(Manipal BSc, 2014)*

15. Explain why dimethylamine is more basic than methylamine.

16. Explain why dimethylamine is a stronger base than trimethylamine. *(Goa BSc, 2014)*

17. Ethylamine and acetamide both contain an amino group. Explain why ethylamine is basic while acetamide is not. *(Madras BSc, 2015)*

Answer. Ethylamine is basic because the lone pair of electrons on nitrogen is available for the formation of a new bond with proton. Acetamide is not basic and does not form salts with acids. This is because lone pair of electrons on nitrogen is delocalized by resonance with the carbonyl group, which makes it less available for protonation.

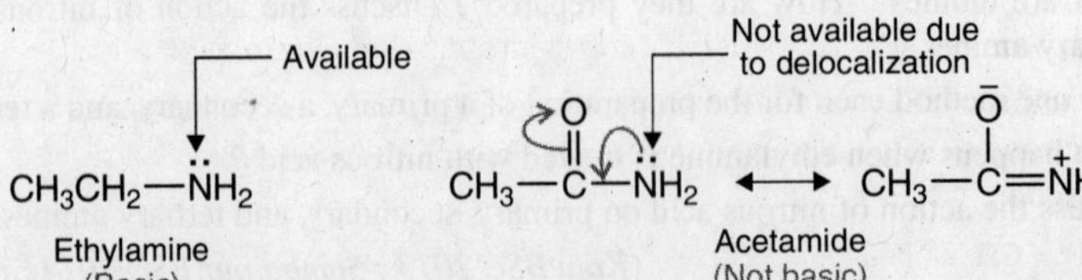

Ethylamine (Basic)　　　Acetamide (Not basic)

18. Arrange the following compounds in decreasing order of basicity : Ammonia, Dimethylamine, Ethylamine, and Benzamide.

Answer. The order is :

(*Strongest Base*) Dimethylamine > Ethylamine > Ammonia > Benzamide (*Weakest Base*)

19. Write a note on : Hofmann's degradation of amides. (*Jammu BSc, 2014*)

20. Discuss the mechanism of Hofmann's degradation of amides. (*Panjab BSc, 2015*)

21. What is Gabriel phthalimide synthesis ? (*Osmania BSc, 2014*)

22. Write a note on : Hofmann's elimination.

23. Identify (A), (B), and (C) in the following reaction sequence :

$$CH_3CH_2OH \xrightarrow{SOCl_2} (A) \xrightarrow{NaCN} (B) \xrightarrow{LiAlH_4} (C)$$

Answer. (A) = CH_3CH_2Cl

(B) = CH_3CH_2CN

(C) = $CH_3CH_2CH_2NH_2$

24. How will you synthesize *n*-butylamine from *n*-propyl chloride ?

Answer. Following steps are involved :

$$\underset{n\text{-Propyl chloride}}{CH_3CH_2CH_2Cl} \xrightarrow{NaCN} CH_3CH_2CH_2CN \xrightarrow{LiAlH_4} \underset{n\text{-Butylamine}}{CH_3CH_2CH_2CH_2NH_2}$$

25. How will you synthesize *n*-propylamine from *n*-propyl chloride ? (*Baroda BSc, 2015*)

Answer. Following steps are involved :

$$\underset{n\text{-Propyl chloride}}{CH_3CH_2CH_2Cl} \xrightarrow{aq.KOH} CH_3CH_2CH_2OH \xrightarrow[\text{or PCC in } CH_2Cl_2]{Cu/300°C}$$

$$CH_3CH_2-\overset{O}{\overset{\|}{C}}-H \xrightarrow[H_2/Ni]{NH_3} \underset{n\text{-Propylamine}}{CH_3CH_2CH_2NH_2}$$

26. How will you synthesize ethylamine from *n*-propyl chloride ?

Answer. Following steps are involved :

$$\underset{n\text{-Propyl chloride}}{CH_3CH_2CH_2Cl} \xrightarrow{aq.KOH} CH_3CH_2CH_2OH \xrightarrow[{[O]}]{K_2Cr_2O_7/H^+} CH_3CH_2-\overset{O}{\overset{\|}{C}}-OH$$

$$\xrightarrow[(2)\ NH_3]{(1)\ SOCl_2} CH_3CH_2-\overset{O}{\overset{\|}{C}}-NH_2 \xrightarrow[NaOH]{Br_2} \underset{\text{Ethylamine}}{CH_3CH_2NH_2}$$

27. How will you synthesize 1,4-diaminobutane from ethylene ?

Answer. Following steps are involved :

$$\underset{\text{Ethylene}}{\begin{matrix} CH_2 \\ \| \\ CH_2 \end{matrix}} \xrightarrow[CCl_4]{Br_2} \begin{matrix} CH_2Br \\ | \\ CH_2Br \end{matrix} \xrightarrow{NaCN} \begin{matrix} CH_2CN \\ | \\ CH_2CN \end{matrix} \xrightarrow{LiAlH_4} \underset{\text{1,4-Diaminobutane}}{\begin{matrix} CH_2CH_2NH_2 \\ | \\ CH_2CH_2NH_2 \end{matrix}}$$

28. How will you synthesize ethylamine from methylamine and *vice-versa* ? (*Orissa BSc, 2015*)

29. An amine (*A*) has the molecular formula C_3H_9N. When (*A*) was treated with nitrous acid at 0–5°C, a reaction took place and an oily layer separated from the reaction mixture. Write a structure for (*A*) and give an equation for the reaction.

Answer. (*A*) = $CH_3—NH—CH_2CH_3$ (Ethylmethylamine)

30. Compound (*A*), C_3H_7NO, reacts with bromine in the presence of NaOH to give (*B*). Compound (*B*) reacts with nitrous acid to form ethyl alcohol and nitrogen. Identify (*A*) and (*B*) and show the reactions involved.

Answer. (*A*) = $CH_3CH_2-\overset{\overset{\large O}{\|}}{C}-NH_2$ (Propanamide)

(*B*) = $CH_3CH_2NH_2$ (Ethylamine)

31. Compound (*A*) C_4H_9NO, reacts with bromine and sodium hydroxide to yield (*B*), C_3H_9N. Compound (*C*), $C_4H_7O_2Na$, reacts with P_2O_5 to give (*D*), C_4H_7N. (*D*) undergoes hydrolysis with dilute HCl to give (*E*), $C_4H_8O_2$, of which (*C*) is the sodium salt. (*B*) can be converted into an alcohol which on oxidation gives a compound identical to that obtained by distillation of dry salt of acetic acid. Identify (*A*), (*B*), (*C*), (*D*), and (*E*).

Answer. (*A*) = $(CH_3)_2CH-\overset{\overset{\large O}{\|}}{C}-NH_2$ (*D*) = $(CH_3)_3CHCN$

(*B*) = $(CH_3)_2CHNH_2$ (*E*) = $(CH_3)_3CHCOOH$

(*C*) = $(CH_3)_2CH-\overset{\overset{\large O}{\|}}{C}-ONa$

32. An optically active amide (*A*), $C_5H_{11}NO$, on acid-hydrolysis gives an acid and ammonia. When (*A*) is treated with bromine and alkali, a compound is obtained which on treatment with nitrous acid yields an optically active alcohol and nitrogen. The alcohol gives a positive iodoform test. What is the structure of (*A*) and show the reactions involved.

Answer. (*A*) = is 2-Methylbutanamide

33. Compound (*A*), $C_3H_6O_2$, when treated with excess of ammonia and then heated formed (*B*), C_3H_7NO. (*B*) when heated with P_2O_5 formed (*C*), C_3H_5N. (*C*) on complete acid-hydrolysis gave (*A*). (*C*) on reduction with lithium aluminium hydride formed a basic nitrogenous compound (*D*), C_3H_9N. (*D*) on treatment with nitrous acid formed an alcohol (*E*), C_3H_8O. (E) on oxidation formed compound (*A*). Give the structures of (*A*), (*B*), (*C*), (*D*), and (*E*) and show the reactions involved.

Answer. The unknown compounds are :

(*A*) = Propionic acid (*B*) = Propionamide

(*C*) = Ethyl cyanide (*D*) = *n*-Propylamine

(*E*) = *n*-Propyl alcohol

Multiple Choice Questions

1. Which of the following statements is *false* about primary amines ?

(*a*) They can be prepared by reduction of nitriles with $LiAlH_4$.
(*b*) They do not form salts with acids.
(*c*) The react with ice-cold nitrous acid to form nitrogen gas.
(*d*) They are basic and soluble in water.

Answer. (*b*)

2. Which of the following statements is *false* about secondary amines ?

(*a*) They react with chloroform and NaOH to form isocyanides.
(*b*) They can be prepared by reduction of isocyanides.
(*c*) They react with acid chlorides to give substituted amides.
(*d*) They react with ice-cold nitrous acid to give nitrosamine (yellow oil).

Answer. (*a*)

3. The hybridization of nitrogen in an amine is

(*a*) sp (b) sp^3 (*c*) sp^2 (d) sp^4

Answer. (*b*)

4. Which of the following compounds is a secondary amine ?

(*a*) $CH_3-\overset{NH_2}{\overset{|}{C}H}-CH_3$ (*b*) $H_2NCH_2CH_2NH_2$

(*c*) $CH_3CH_2NHCH_3$ (*d*) $(CH_3)_3N$

Answer. (*c*)

5. A sample of pure amine molecules is found to possess no intermolecular H-bonding. This sample is most likely :

(*a*) 1° amine (*b*) 2° amine (*c*) 3° amine (*d*) all of the these

Answer. (*c*)

6. Which of the following is least soluble in water ?

(*a*) Methylamine (*b*) Trimethylamine (*c*) Dimethylamine (*d*) Aniline

Answer. (*d*)

7. Which of the following is most soluble in water ?

(*a*) $CH_3CH_2CH_3$ (*b*) $CH_3-\overset{I}{\overset{|}{C}H}-CH_3$

(*c*) $CH_3CH_2CH_2Cl$ (*d*) $CH_3CH_2CH_2NH_2$

Answer. (*d*)

8. Which amine is *not* soluble in water ?

(*a*) Methylamine (*b*) Dimethylamine

(*c*) Trimethylamine (*d*) All are water-soluble

Answer. (*d*). Amines form H-bonds with water. Therefore, all low-molecular weight amines are soluble in water.

9. Consider a 1°, 2° and 3° amine, all of equivalent molecular weight. Which amine is most likely to have the lowest boiling point ?

(*a*) 1° amine (*b*) 2° amine

(*c*) 3° amine (*d*) Not enough information to determine

Answer. (*c*). The 3° amines do not form hydrogen bonds to one another. The 1° and 2° amines can form hydrogen bonds to themselves and therefore have higher boiling points than 3° amines.

10. Which of the following will have the highest boiling point ?

(*a*) Methylamine (*b*) Diethylamine

(*c*) Ethylamine (*d*) Triethylamine

Answer. (*d*)

11. Which compound has the highest boiling point ?

$CH_3CH_2CH_3$	CH_3CH_2OH	$CH_3CH_2NH_2$
1	2	3

(*a*) 1 (*b*) 2

(*c*) 3 (*d*) 2 and 3 have similar boiling points

Answer. (*b*). Amines are fairly polar, but they boil at temperatures lower than those of alcohols of similar chain length and structure. Amines do have higher boiling points than alkanes because they possess H-bond donors and acceptors.

12. Which of the following has the highest boiling point ?

(*a*) $CH_3CH_2CH_2CH_3$ (*b*) $CH_3CH_2CH_2CH_2Cl$

(*c*) $CH_3CH_2CH_2CH_2NH_2$ (*d*) $CH_3CH_2CH_2CH_2F$

Answer. (*c*)

13. Amines are generally classified as

(*a*) Weak acids (b) Strong acids

(*c*) Weak bases (d) Strong bases

Answer. (*c*)

14. Which of the following is most basic ?

(*a*) Ammonia (*b*) Methylamine

(*c*) Dimethylamine (*d*) Trimethylamine

Answer. (*c*)

15. Acetamide reacts with Br_2/NaOH to give

(*a*) Methylamine (*b*) Urea

(*c*) Ethylamine (*d*) Acetyl bromide

Answer. (*a*)

16. Acetonitrile undergoes reduction with $LiAlH_4$ to form :

(*a*) Methylamine (*b*) Dimethylamine

(*c*) Ethylamine (*d*) Trimethylamine

Answer. (*c*)

17. Reduction of an imine will give an

(*a*) Acid (*b*) Amide

(*c*) Amine (*d*) Alcohol

Answer. (*c*)

18. Alkyl halides react with ammonia in the presence of base to form :

(*a*) primary amines (*b*) nitrated alkyl halides

(*c*) amides (*d*) quaternary ammonium salts

Answer. (*a*)

19. The major product of the following reaction is :

$$CH_3CH_2Cl + (CH_3)_2NH \longrightarrow ?$$

(*a*) 1° amine (*b*) 2° amine

(*c*) 3° amines (*d*) amide

Answer. (*c*)

20. Aliphatic primary amines react with cold nitrous acid to form

(*a*) Alcohols (*b*) Diazonium salts

(*c*) Nitriles (*d*) Nitroalkanes

Answer. (*a*)

21. Which type of amine produces N_2 when treated with HONO ?

(*a*) Primary (*b*) Secondary

(*c*) Tertiary (*d*) Quaternary

Answer. (*a*)

22. Methylamine reacts with nitrous acid to form

(*a*) CH_3CH_3 (*b*) CH_3OH

(*c*) CH_3NO_2 (*d*) CH_3CH_2OH

Answer. (*b*)

23. Which of the following compounds react with nitrous acid to form an alcohol ?

(*a*) $CH_3CH_2NH_2$ (*b*) $(CH_3CH_2)_2NH$

(*c*) $(CH_3CH_2)_3N$ (*d*) 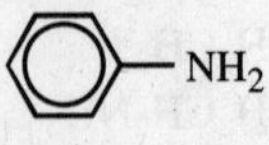

Answer. (*a*)

24. Which of the following amines react with nitrous acid to form nitrosoamine ?

(*a*) Ethylamine (*b*) Triethylamine

(*c*) Diethylamine (*d*) Isopropylamine

Answer. (*c*)

25. Which of the following does not react with acyl chlorides to form amides :

(*a*) ammonia (*b*) 1° amine

(*c*) 2° amine (*d*) 3° amine

Answer. (d)

26. Hinsberg's reagent is

(*a*) Pd + $BaSO_4$ (*b*) *p*-Toluenesulfonic acid

(*c*) NH_2NH_2 + KOH (*d*) Benzenesulfonic acid

Answer. (*d*)

27. Which of the following compounds will dissolve in aqueous NaOH after undergoing reaction with Hinsberg reagent ?

(*a*) $CH_3CH_2NH_2$ (*b*) $(CH_3CH_2)_2NH$

(*c*) $(CH_3CH_2)_3N$ (*d*) $(CH_3)_3N$

Answer. (*a*)

28. Which of the following compounds does not react with acetyl chloride ?

(*a*) Methylamine (*b*) Dimethylamine

(*c*) Aniline (*d*) Trimethylamine

Answer. (*d*)

29. A primary amine can be identified by using

(*a*) HCl (*b*) $CHCl_3$

(*c*) NaOH (*d*) $CHCl_3$ + KOH

Answer. (*d*)

30. Carbylamine test is given by

(*a*) Primary amines (*b*) Secondary amines

(*c*) Tertiary amines (*d*) None of these

Answer. (*a*)

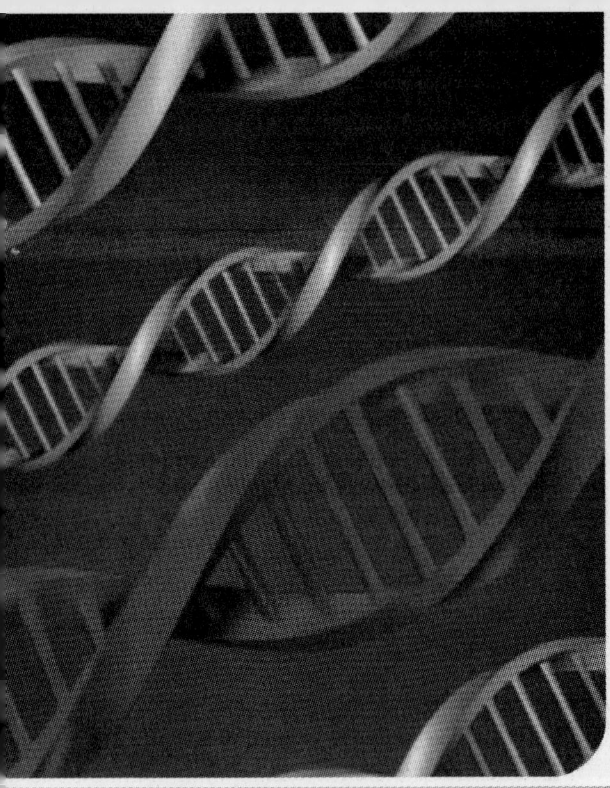

28

Amino Acids, Peptides and Proteins

Amino acids are the building blocks of the molecular structure of the important and very complex class of compounds known as proteins. The proteins on hydrolysis yield mixtures of the component amino acids. Therefore, to understand the structure and chemistry of proteins, we have to first undertake the study of amino acids.

AMINO ACIDS

Amino acids are compounds that contain at least one carboxyl group (–COOH) and one amino group ($-NH_2$). In α-amino acids, the amino group is located on the alpha (α) carbon (the carbon next to the carboxyl group). Their general formula is :

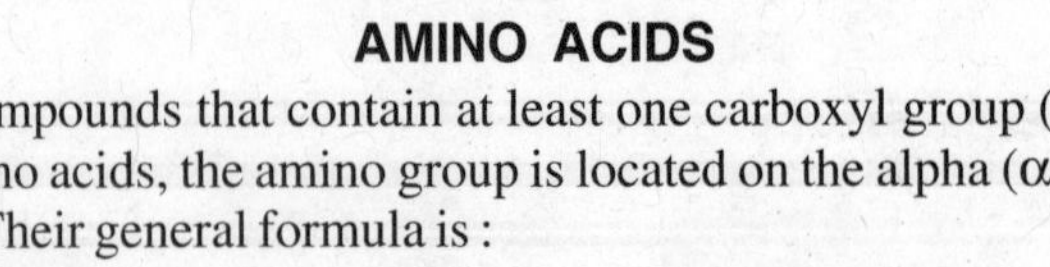

α-Amino acid

where *R* may be *H*, alkyl group, or may be quite complex containing other functional groups.

Points to Remember

1. About 20 amino acids are commonly found in proteins, but hundreds occur naturally.
2. All amino acids found in proteins are α-amino acids.
3. All amino acids found in proteins (except glycine) are optically active.
4. Names, structures, and 3-letter symbols of common amino acids are given in Table 28.1. Study them carefully.
5. **Essential amino acids** are those which cannot be produced in sufficient quantity by the human body. They must be included in the diet. Examples are : valine, phenylalanine, and tryptophan.
6. **Non-essential amino acids** are those which can be synthesized from other compounds by the body. Examples are : glycine, alanine, and serine.

ZWITTERION - DIPOLAR NATURE OF AMINO ACIDS

In solution, amino acids exist as charged molecules. This is because they contain both an acidic –COOH group and a basic –NH_2 group in the same molecule. Transfer of a proton takes place in a kind of internal acid-base reaction.

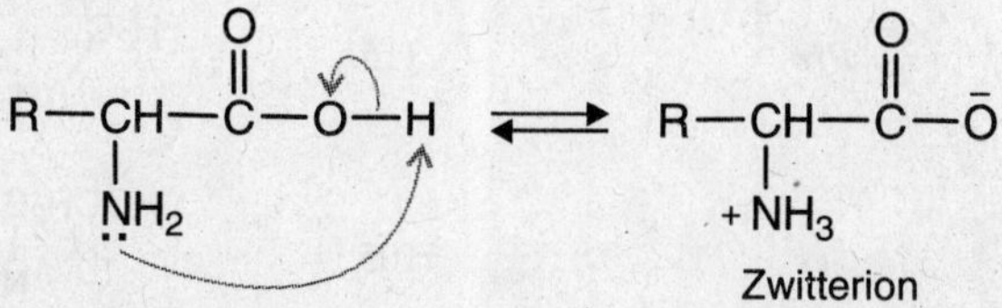

Zwitterion

TABLE 28.1 STRUCTURES OF THE 20 COMMON AMINO ACIDS FOUND IN PROTEINS.

Name	Abbreviations	Molecular weight	Structure
NEUTRAL AMINO ACIDS			
Alanine	*Ala*	89	$CH_3CHCOOH$ | NH_2
Asparagine	*Asn*	132	O ‖ $H_2N—CCH_2CHCOOH$ | NH_2
Cysteine	*Cys*	121	$HSCH_2CHCOOH$ | NH_2
Glutamine	*Gln*	146	O ‖ $H_2N—C—CH_2CH_2CHCOOH$ | NH_2
Glycine	*Gly*	75	CH_2COOH | NH_2
Isoleucine	*Ile*	131	$CH_3CH_2CH(CH_3)CHCOOH$ | NH_2
Leucine	*Leu*	131	$(CH_3)_2CHCH_2CHCOOH$ | NH_2
Methionine	*Met*	149	$CH_3SCH_2CH_2CHCOOH$ | NH_2
Phenylalanine	*Phe*	165	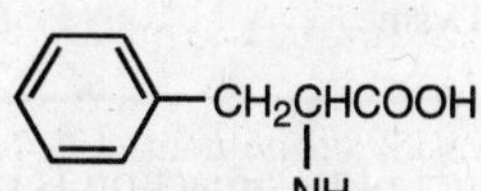

Name	Abbreviations	Molecular weight	Structure
Proline	*Pro*	115	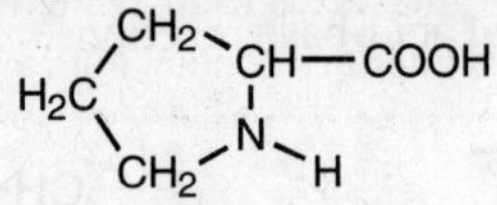
Serine	*Ser*	105	$HOCH_2CH(NH_2)COOH$
Threonine	*Thr*	119	$CH_3CH(OH)CH(NH_2)COOH$
Tryptophan	*Trp*	204	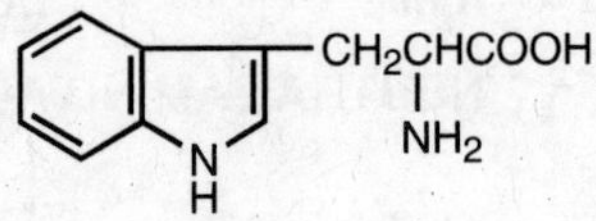
Tyrosine	*Tyr*	181	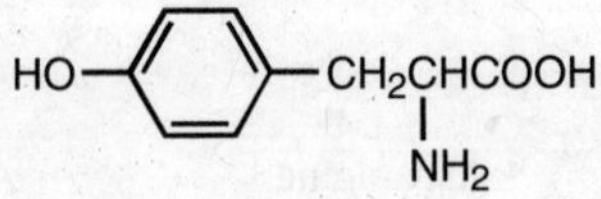
Valine	*Val*	117	$(CH_3)_2CHCH(NH_2)COOH$
ACIDIC AMINO ACIDS			
Aspartic acid	Asp	133	$HOOCCH_2CH(NH_2)COOH$
Glutamic acid	*Glu*	147	$HOOCCH_2CH_2CH(NH_2)COOH$
BASIC AMINO ACIDS			
Arginine	*Arg*	174	$H_2N{-}C({=}NH){-}NHCH_2CH_2CH_2CH(NH_2)COOH$
Histidine	*His*	155	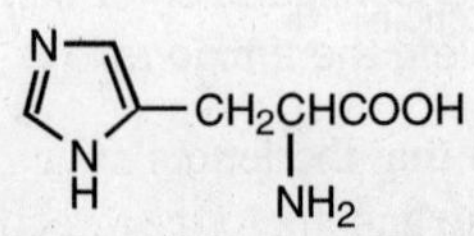
Lysine	*Lys*	146	$H_2NCH_2CH_2CH_2CH_2CH(NH_2)COOH$

The product of this reaction is called a *zwitterion* (or *dipolar ion*). **A zwitterion is an amino acid molecule containing both a positive and a negative charge**. The zwitterion is the more common form in which amino acids exist in aqueous solution.

Points to Remember

1. Amino acids form zwitterions by transferring a proton internally.
2. Zwitterions contain both a positive and a negative charge. They are neutral and have no net charge.

OPTICAL ACTIVITY OF AMINO ACIDS

Glycine does not have an asymmetric carbon atom and, therefore, does not have optical isomers. Alanine and all other amino acids have an asymmetric carbon at position 2, the α-carbon atom. For this reason they are all optically active and exist in D and L forms, which are nonsuperimposable mirror image isomers. If the carboxyl group is written at the top, the D-form refers to the isomer with the $-NH_2$ group on the right ; the L-form is the isomer with the $-NH_2$ group on the left. The reference compound for this assignment is glyceraldehyde.

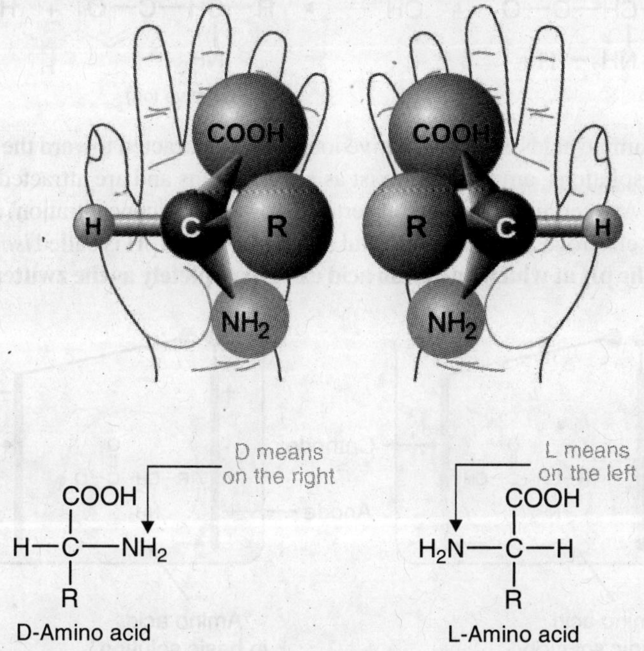

Fig. 28.1. Optical isomers of an amino acid. If the amino group on the asymmetric carbon atom projects to the right, the amino acid is said to be of D-configuration. If the amino group on the asymmetric carbon projects to the left, the amino acid is said to be of L-configuration.

Recall that the letters D and L refer only to the relative configuration around the asymmetric carbon atom and not to the direction of optical rotation. The direction of optical rotation is indicated by a (+) sign for a dextrorotatory amino acid and a (–) sign for a levorotatory one.

Most of the naturally occurring amino acids have the L-configuration. The products of laboratory synthesis are generally optically inactive. This is due to the formation of racemates containing equal amounts of D- and L-forms. The optical rotation of amino acids depends upon the *p*H of the solution. In neutral solution, most of the amino acids are dextrorotatory.

ISOELECTRIC POINT

Amino acids in the zwitterion form are amphoteric. That is, they react readily with acids or bases. The reaction with bases converts the ammonium group ($-NH_3^+$) into an amino group ($-NH_2$). The reaction with acids converts the carboxylate group $-COO^-$) into a carboxyl group ($-COOH$).

Amino acid in Acidic solution :

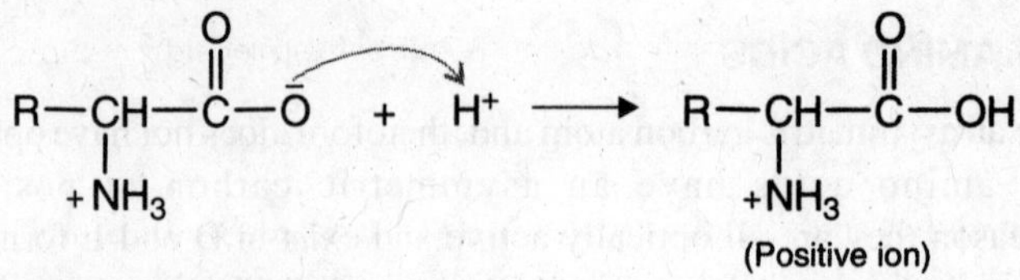

Amino acid in Basic solution :

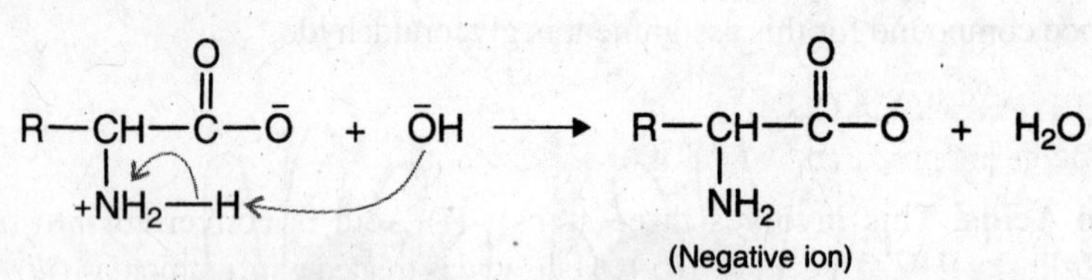

In acidic solutions, amino acids exist as positive ions and are attracted toward the *cathode* (negative electrode). In basic solutions, amino acids exist as negative ions and are attracted toward the *anode* (positive electrode). At a certain pH (that is, at certain hydrogen ion concentration) amino acids would not migrate to either electrode and exist as neutral zwitterions. This pH is called *isoelectric point*. **The isoelectric point is the pH at which an amino acid exists completely as the zwitterion.**

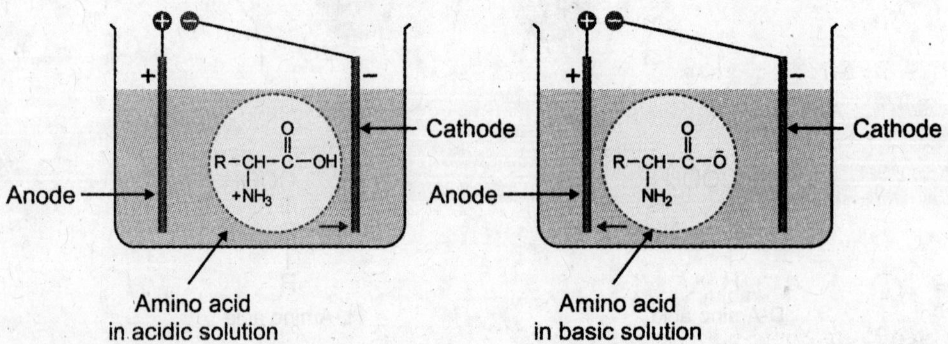

Each amino acid has a characteristic isoelectric point. Proteins, which are composed of amino acids, also have characteristic isoelectric points. The migration (or non-migration) under varying conditions of pH and electric field (*Electrophoresis*), is used to identify and separate amino acids from mixtures.

Points to Remember

1. Isoelectric point is the pH at which an amino acid completely exists as zwitterion.
2. Different amino acids (and proteins) have different isoelectric points.

NOMENCLATURE

They are named as amino-substituted aliphatic carboxylic acids and are designated α, β, γ, etc., depending on the position of NH_2 group on the carbon chain. For example,

Formula	Common Name
H_2N-CH_2-COOH	Aminoacetic acid (*Glycine*)
$CH_3-\underset{}{CH}(NH_2)-COOH$	α-Aminopropionic acid (*Alanine*)
$H_2N-\overset{\beta}{C}H_2-\overset{\alpha}{C}H_2-COOH$	β-Aminopropionic acid
$H_2N-\overset{\gamma}{C}H_2-\overset{\beta}{C}H_2-\overset{\alpha}{C}H_2-COOH$	γ-Aminobutyric acid

In IUPAC name of an amino acid, the position of NH_2 is indicated by the number of carbon on which it lies, taking C of COOH as '1'. Thus α-aminopropionic acid is named as 2-aminopropanoic acid.

For our present study α-amino acids are the most important because they are the final products of hydrolysis of proteins.

METHODS OF PREPARATION

α-Amino acids are prepared by the following methods :

(1) **From Acids**. This involves three steps : (*i*) Acid is converted into α-halo acid by Hell-Volhard-Zelinsky (HVZ) reaction ; (*ii*) α-Halo acid is treated with ammonia ; (*iii*) Ammonium salt is hydrolysed.

$$\underset{\text{Acetic acid}}{CH_3COOH} \xrightarrow{Cl_2/P} HCl + Cl-CH_2COOH \xrightarrow{3NH_3,\ -NH_4Cl}$$

$$H_2N-CH_2CO\overset{-}{O}\overset{+}{N}H_4 \xrightarrow{H^+/H_2O} \underset{\substack{\text{Glycine}\\ \text{(Aminoacetic acid)}}}{H_2N-CH_2COOH}$$

(2) **By Strecker Synthesis**. This involves two steps :

Step 1. An aldehyde is treated with a mixture of HCN and ammonia.

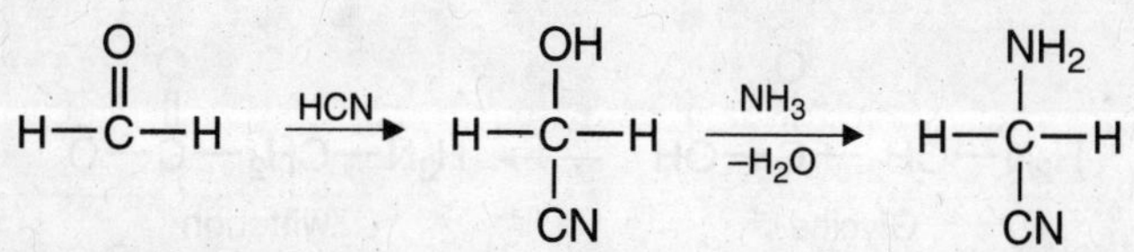

Formaldehyde α-Amino nitrile

Step 2. α-Amino nitrile is hydrolysed.

$$\underset{\alpha\text{-Amino nitrile}}{H_2N-CH_2-CN} \xrightarrow{H^+/H_2O} \underset{\text{Glycine}}{H_2N-CH_2-COOH}$$

(3) **By Gabriel Phthalimide Synthesis**. This involves the treatment of potassium phthalimide with an ester of a α-halo acid followed by hydrolysis.

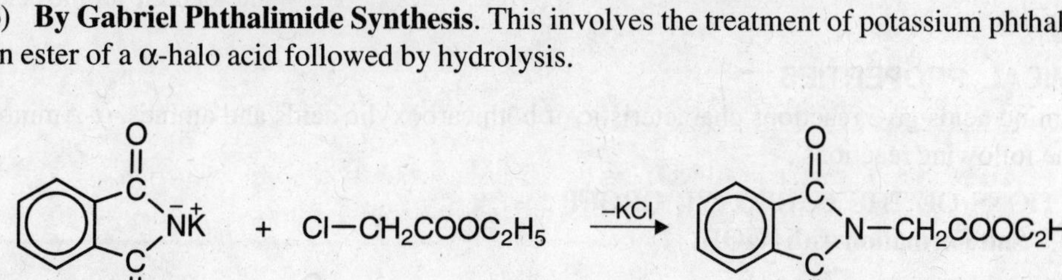

Potassium phthalimide

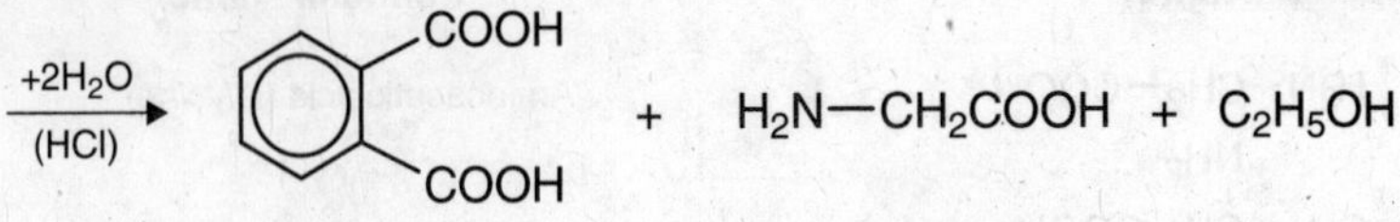

Phthalic acid Glycine

(4) **Koop Synthesis**. The treatment of α-keto acids with ammonia forms an imine which on catalytic reduction gives amino acid.

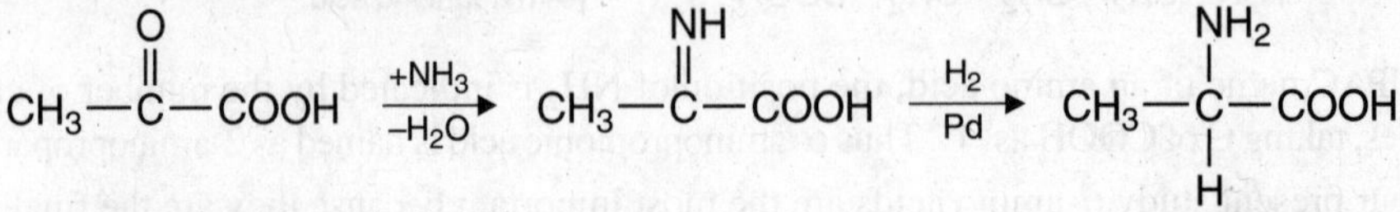

α-Ketopropionic acid α-Aminopropionic acid

PHYSICAL PROPERTIES

(1) Amino acids are colorless, crystalline substances having sweet taste. They melt with decomposition at high temperatures.

(2) They are soluble in polar solvents like water and insoluble in organic solvents such as petroleum ether, ethanol and benzene.

(3) Excepting glycine, all α-amino acids possess an asymmetric carbon atom and are optically active.

Glycine (optically inactive) Alanine (optically active)

(4) Their water solutions are neutral to litmus owing to the formation of their internal salt or Zwitterion.

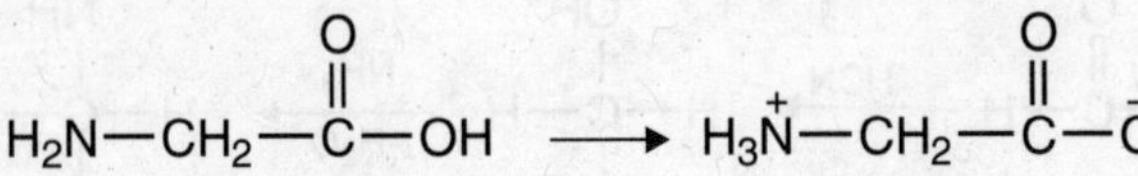

Glycine Zwitterion

(5) **From Natural Proteins**. Natural proteins are polymers of different amino acid units joined to each other by amide linkages. Thus proteins when boiled with dilute hydrochloric acid yield mixtures of amino acids. The mixture of amino acids is esterified and the individual esters separated by fractional distillation. They are then hydrolysed to get the free amino acids. Physical techniques such as fractional precipitation, chromatography, and electrophoresis are now widely used for the recovery of amino acids.

CHEMICAL PROPERTIES

Amino acids give reactions characteristic of both carboxylic acids and amines. α-Amino acids give the following reactions :

REACTIONS OF THE CARBOXYL GROUP

(1) **Salt Formation with NaOH**

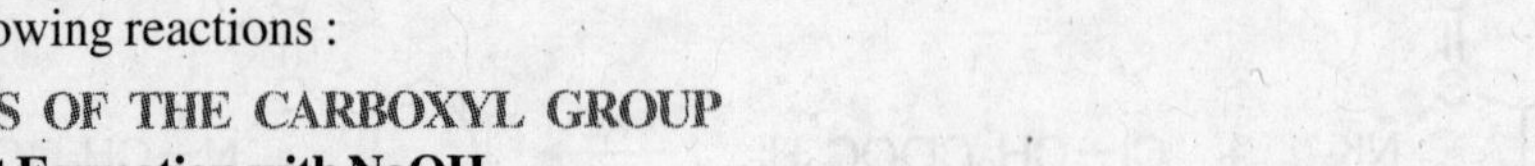

Glycine Sodium aminoacetate

(2) **Reduction with $LiAlH_4$**

$$\underset{\text{Glycine}}{H_2NCH_2-\overset{O}{\overset{\|}{C}}-OH} \xrightarrow[(+4H)]{LiAlH_4} \underset{\text{2-Aminoethanol}}{H_2NCH_2CH_2OH}$$

(3) **Esterification**. α-Amino acids react with alcohols in the presence of HCl. The acid not only acts as a catalyst but also reacts with the amino group. The free ester can be obtained by treatment of the hydrochloride with silver hydroxide (Moist Ag_2O).

$$\underset{\text{Glycine}}{H_2NCH_2-\overset{O}{\overset{\|}{C}}-OH} + C_2H_5OH \xrightarrow{HCl} \overset{-}{Cl}\ H_3\overset{+}{N}CH_2-\overset{O}{\overset{\|}{C}}-OC_2H_5 + H_2O$$

$$\overset{-}{Cl}\ H_3\overset{+}{N}CH_2-\overset{O}{\overset{\|}{C}}-OC_2H_5 \xrightarrow[(-AgCl\ ;\ -H_2O)]{AgOH} \underset{\text{Ethyl α-Aminoacetate}}{H_2NCH_2-\overset{O}{\overset{\|}{C}}-OC_2H_5}$$

REACTIONS OF THE AMINO GROUP

(4) **Salt Formation with Mineral Acids**

$$HCl + \underset{\text{Glycine}}{H_2N-CH_2COOH} \longrightarrow \underset{\text{Glycine hydrochloride}}{\overset{-}{Cl}\ H_3\overset{+}{N}-CH_2COOH}$$

(5) **Acylation**. α-Amino acids react with acid chlorides or anhydrides to form N-acyl amino acids.

$$\underset{\text{Acetyl chloride}}{CH_3-\overset{O}{\overset{\|}{C}}-Cl} + \underset{\text{Glycine}}{H-NHCH_2CO_2H} \longrightarrow \underset{\textit{N}\text{-Acetylglycine}}{CH_3-\overset{O}{\overset{\|}{C}}-NHCH_2CO_2H} + HCl$$

$$\underset{\text{Acetic anhydride}}{CH_3-\overset{O}{\overset{\|}{C}}-O-\overset{O}{\overset{\|}{C}}-CH_3} + \underset{\text{Glycine}}{H-NHCH_2CO_2H} \longrightarrow \underset{\textit{N}\text{-Acetylglycine}}{CH_3-\overset{O}{\overset{\|}{C}}-NHCH_2CO_2H} + CH_3CO_2H$$

(6) **Reaction with Nitrous Acid ($NaNO_2$ + HCl)**

$$HONO + \underset{\text{Glycine}}{H_2NCH_2COOH} \longrightarrow \underset{\text{Glycollic acid}}{HOCH_2COOH} + N_2 + H_2O$$

(7) **Reaction with Formaldehyde**

$$H_2C{=}O + \underset{\text{Glycine}}{H_2NCH_2COOH} \longrightarrow \underset{\textit{N}\text{-Methylene glycine}}{H_2C{=}NCH_2COOH} + H_2O$$

(8) **Reaction with 2,4-Dinitrofluorobenzene (DNFB).** (*Memorise :2,4-Dinitrofluorobenzene is also called Sangar's reagent*). Amino acids react with this reagent to produce yellow colored dinitrophenylamino acids (DNP-amino acids).

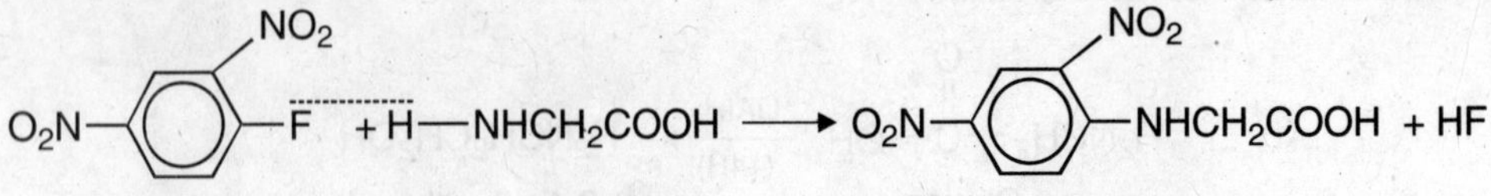

(DNFB) Glycine DNP-Glycine

OTHER REACTIONS

(9) **Effect of Heat**. α-Amino acids undergo dehydration (at 200°C) to give diketopiperazine.

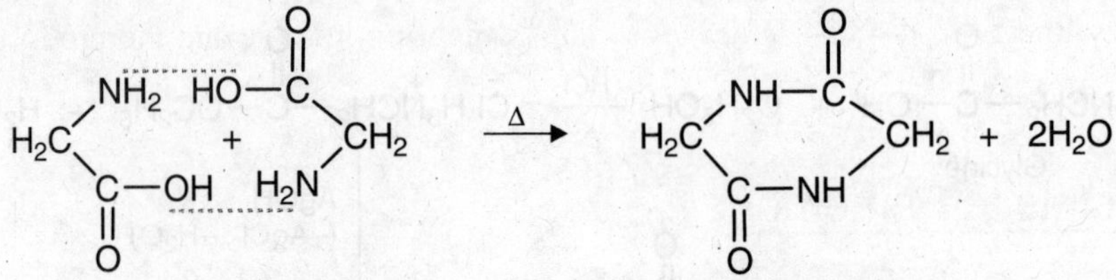

Glycine (2 molecules) Diketopiperazine

GLYCINE, *Aminoacetic acid*, H_2N-CH_2-COOH

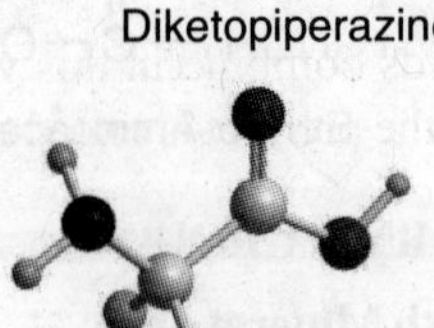

It is so named because of its sweet taste (Greek, *glycys = sweet*).

Preparation. Glycine may be readily obtained in the laboratory by treating chloroacetic acid with concentrated ammonium hydroxide solution.

$$Cl-CH_2COOH + NH_4OH \longrightarrow H_2N-CH_2COOH + NH_4Cl$$

A sample of pure glycine can also be prepared by *Gabriel Phthalimide Synthesis*.

Properties. Glycine is white crystalline substance, mp 235°C with decomposition. It is very soluble in water but insoluble in ethanol and ether. Its aqueous solutions are neutral owing to the formation of *Zwitterion*.

$$H_2N-CH_2-COOH \longrightarrow \underset{\text{Zwitterion}}{H_3\overset{+}{N}-CH_2-CO\overset{-}{O}}$$

It gives all the general reactions of α-amino acids already discussed.

Uses. Glycine is used : (1) in medicine for certain muscular disorders ; (2) for the manufacture of resins and varnishes.

PEPTIDES

WHAT ARE PEPTIDES - PEPTIDE LINKAGE

Two amino acids can combine with the elimination of water. Notice that the reaction is taking place between the carbonyl group of one amino acid and the amino group of another.

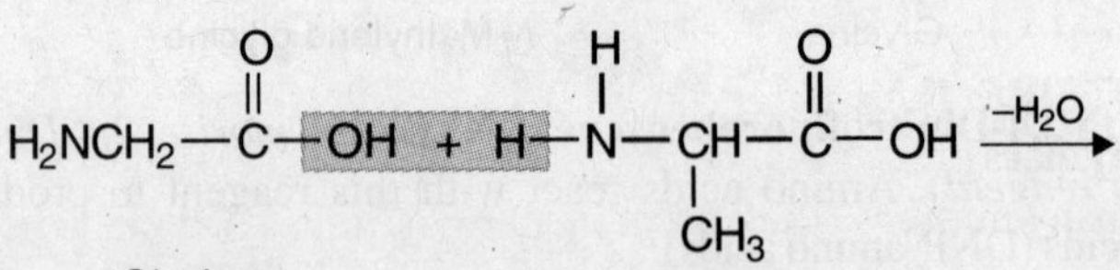

Glycine Alanine

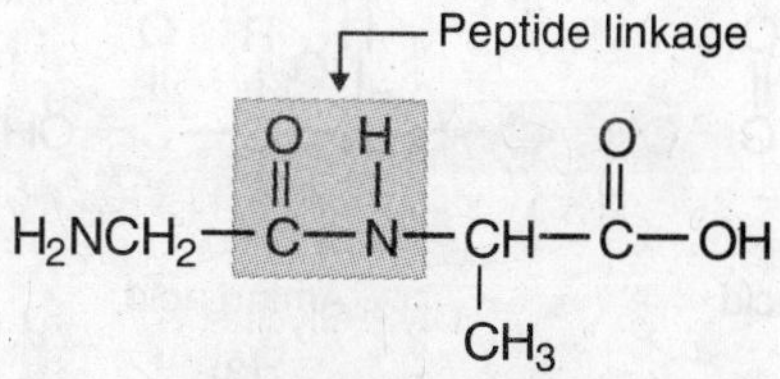

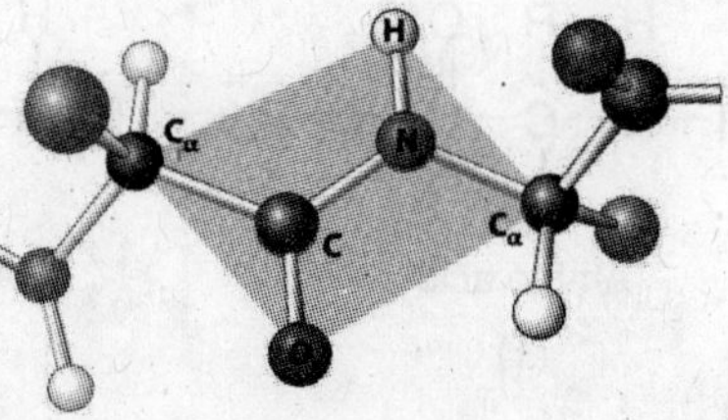

Glycylalanine (Dipeptide)

The resulting $-\overset{\overset{\displaystyle O}{\|}}{C}-N-$ linkage is called a **peptide linkage** and the product is called a **dipeptide**.

Points to Remember

1. When three amino acids combine in this way, the product is called a **tripeptide**. When four amino acids combine, the product is called a **tetrapeptide**. When many amino acids combine in this way, the product is called a **polypeptide**. Proteins are polypeptides containing more than 100 amino acids.
2. In a peptide, the amino acid that contains the free amino group is called the ***N*-terminal residue**. It is always written on the *left-hand side* of the chain. Similarly, the amino acid that contains the free carboxyl group is called the ***C*-terminal residue**. It is always written on the *right-hand side* of the chain. For example, a tripeptide from glycine, alanine, and phenylalanine should be written as :

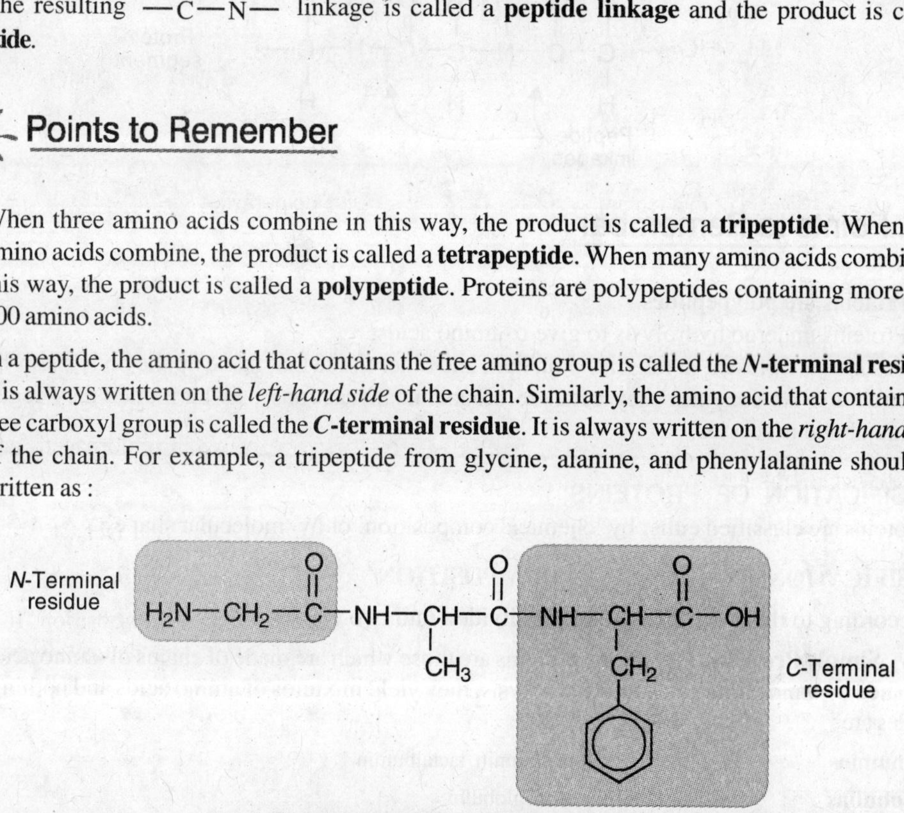

Glycylalaninephenylamine

Glycine is the *N*-terminal amino acid and phenylalanine is the *C*-terminal amino acid.

3. Peptides are named by listing the amino acids present in order they occur starting from the *N*-terminal amino acid. The typical amino acid suffix '-ine' is replaced by the '-yl' for all amino acids except the *C*-terminal amino acid.
4. Peptides can also be represented by standard **3-letter symbols** (Table 28.1). For example, glycylalanylphenylalanine can be written as *Gly-Ala-Phe*.

PROTEINS

WHAT ARE PROTEINS ?

Proteins are polymers of α-amino acids bonded by peptide linkages. Their molecular weights range from 5000 to many millions. They occur in all living cells. Without proteins life would not be possible.

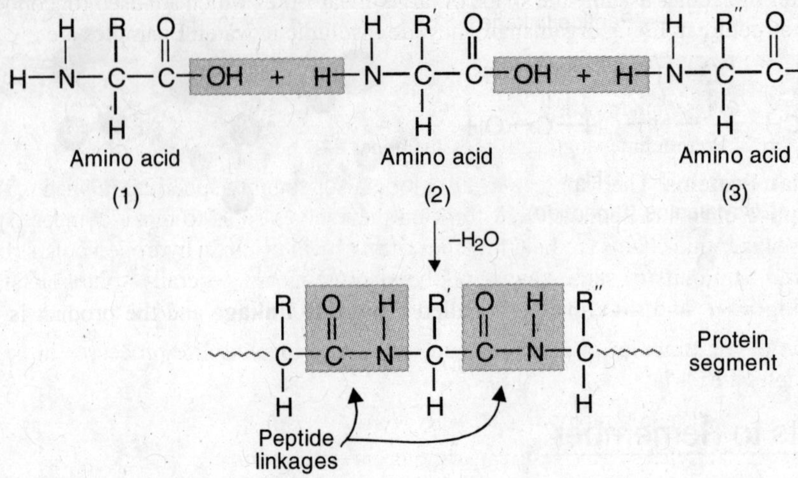

Points to Remember

1. Proteins are polypeptides.
2. Proteins undergo hydrolysis to give α-amino acids.
3. Proteins are present in muscle, skin, hair, and other tissue that make up the body's non-bony structure. As enzymes they catalyze biochemical reactions ; as harmones they regulate metabolic processes.

CLASSIFICATION OF PROTEINS

Proteins are classified either by 'chemical composition' or by 'molecular shape'.

CLASSIFICATION BY CHEMICAL COMPOSITION

According to this system, proteins are divided into two classes.

(1) **Simple Proteins**. The simple proteins are those which are made of chains of amino acid units only, joined by amide linkages. Upon hydrolysis, they yield mixtures of amino acids and nothing else. Examples are :

Albumins : Egg albumin, serum albumin, lactalbumin.

Globulins : Tissue globulin, serum globulin.

Glutenins : Glutenin in wheat, coryzenin in rice.

(2) **Conjugated Proteins**. These molecules are made of a *simple protein* united covalently or non-covalently with a non-protein factor. The non-protein portion is called a **Prosthetic group** (Gr. *prosthesis* = an addition) or **Cofactor**. Examples are :

Glycoproteins : Mucin in saliva ; prosthetic group, carbohydrate.

Phosphoproteins : Casein in milk ; prosthetic group, phosphoric acid.

Chromoproteins : Hemoglobin in red blood cells ; prosthetic group, colored iron pigment.

CLASSIFICATION BY MOLECULAR SHAPE

This shape-classification divides proteins into two types depending on their molecular shape: (*i*) *Fibrous proteins*, which are fibre-like ; (*ii*) *Globular proteins*, which are globe-like or spherical.

(1) **Fibrous Proteins**. In these proteins, the polypeptide chains coil about one another and are held together by strong inter-chain hydrogen bonds (between H and O of amino acid units of different

chains). Thus their molecules assume the shape of large linear fibres which are used for connections, support, and the structure in living organisms. They are insoluble in water. Examples are :

Keratin : Present in skin, hair and nails.

Myosin : Present in muscles.

Collagen : Present in tendons, cartilages and bones.

(2) **Globular Proteins**. Their molecules are globular or roughly spherical in shape. They are made of polypeptide chains coiled back and forth on themselves so as to make compact spheroid molecules. The polypeptide chains are held together firmly by inter-chain hydrogen bonds (between H and O of amino acid units of same chain). Globular proteins are generally soluble in water, or solutions of acids, bases, and salts.

The function of the globular proteins is to maintain and regulate life processes in the living organisms. Examples are :

Enzymes : Pepsin from stomach ; helps digestion.

Insulin : Secreted by pancreas ; regulates glucose metabolism.

Hemoglobin : Present in blood ; transports O_2 from lungs to all cellular tissues.

Antibodies : Protect the body from outside infection *e.g.*, gamma globulin in blood.

Cytochromes : Present in blood ; act as electron carriers.

ISOLATION OF PROTEINS

Proteins invariably occur as mixtures in the animal (or vegetable) source material. It is treated with a solvent and sodium chloride, sodium carbonate or sodium hydroxide in very low concentrations. From the solution so obtained, proteins are precipitated by a suitable reagent like ethanol, ammonium or magnesium sulfate or phosphotungstic acid. The mixtures of proteins are then separated by column chromatography. Careful control of pH and cataphoresis are amongst the modern techniques used for securing individual proteins from mixtures.

STRUCTURE OF PROTEINS

Proteins have a complex 3-dimensional structure. It is studied step-by-step at four different levels, rather than as a whole.

(1) **Primary Structure**. *The primary structure of a protein refers to the sequence of amino acids*. The amino acid sequence in many proteins have been determined and following observations have been made :

(*a*) A protein may contain more than one amino acid chain.

(*b*) Sequence of amino acids in the protein chains is essentially random.

(*c*) Small variations in the sequence of amino acids have pronounced effects on the properties of protein.

(2) **Secondary Structure**. *The secondary structure of a protein refers to the shape in which long amino acid chains exist*. Many proteins consist of amino acid chain coiled into a spiral known as an α-helix. The spiral is held together by hydrogen bonds between N–H and C=O groups vertically adjacent to one another in the helix. See Fig. 28.2.

(3) **Tertiary Structure**. An α-helix may be considered to be a piece of rope which is free to bend, twist and fold. *The tertiary structure of a protein refers to the final 3-dimensional shape that results from twisting, bending, and folding of protein helix.*

(4) **Quaternary Structure**. Complex proteins are often formed from two or more amino acid chains rather than a single amino acid chain. Each chain is a complete protein with a characteristic primary, secondary, and tertiary structure. *The quaternary structure refers to the way in which these amino acid chains of a complex protein are associated with each other*.

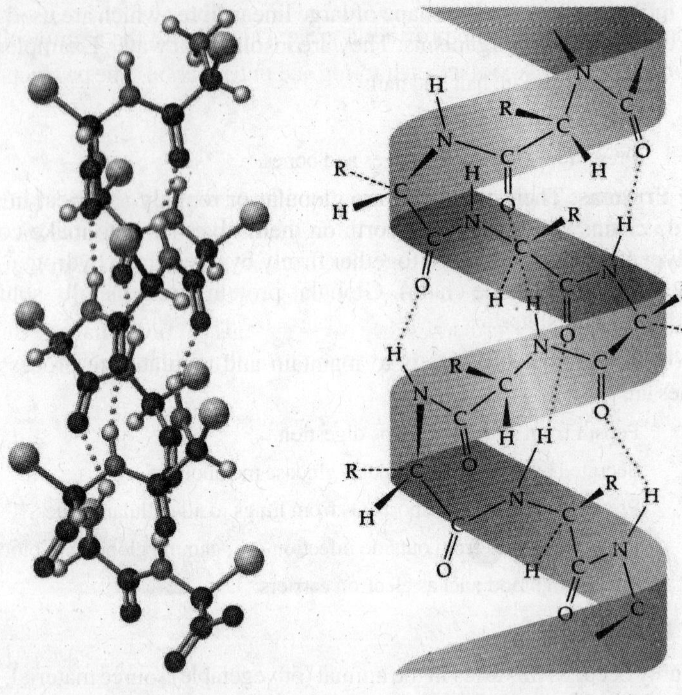

Fig. 28.2. Hydrogen bonding in the α-helix structure of a protein molecule.

GENERAL PROPERTIES OF PROTEINS

The important general properties of proteins are given below :

(1) Most proteins are colorless amorphous solids. They have no definite melting points.

(2) Proteins, like amino acids, exist as Zwitterions.

(3) Most proteins are optically active.

(4) Proteins, like amino acids, have isoelectric points.

(5) Proteins form colloidal dispersions in water. They can pass through a filter paper but not through a membrane.

(6) **Precipitation**. Proteins are easily precipitated (or coagulated) :

(*a*) By heat	(*d*) By silver nitrate
(*b*) By ethyl alcohol	(*e*) By picric acid or tannic acid
(*c*) By concentrated inorganic acids	(*f*) By ultraviolet rays and X-rays

(7) **Hydrolysis**. Simple proteins can be hydrolysed with acids (HCl), alkalis (NaOH), or enzymes to give component amino acids.

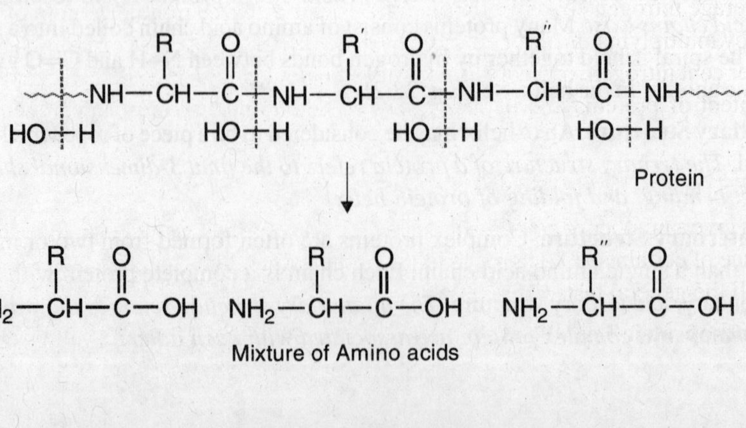

Mixture of Amino acids

Conjugated proteins also yield the prosthetic groups along with the amino acids.

(8) **Oxidation**. Proteins are oxidized on burning and putrification. The products include nitrogen, CO_2, and water.

DENATURATION OF PROTEINS

The physical properties of proteins can often be altered by heating, or by adding salts, dilute acids, or other chemicals. Proteins changed in this way are said to be **denatured** and the process is called **denaturation**. A simple example is the heating of white of an egg which is coagulated to solid mass (denatured protein). There are two important effects of denaturation :

(1) Denaturation does not alter the primary structures of proteins. **It only changes the secondary and tertiary structures of proteins**. The highly ordered natural protein structures are converted to more random conformations.

(2) Denaturation results in **loss of biological activity** of proteins.

When an egg is cooked, protein denaturation is irreversible. The egg white does not become clear and runny again when it is cooled.

IMPORTANCE OF PROTEINS

Proteins are indispensable components of all living things, including the simplest forms of life, such as bacteria, algae, and other microorganisms. They are required in the diets of all animals, including humans, to synthesize tissues, enzymes certain harmones, and some blood components, In addition, they are used in the maintenance and repair of existing tissues and as source of energy for the body.

DETERMINATION OF THE PROTEIN CONTENT OF FOODS

The analysis of protein content of foods (or biological material) is based on its nitrogen content. Since the average nitrogen content of proteins is 16 per cent, the protein content of a substance can be obtained by multiplying its nitrogen value by a factor $100/16 = 6.25$. For example, if a certain food contains 2 per cent nitrogen, on analysis its protein content would be $2 \times 6.25 = 12.5$ per cent. The total nitrogen content of proteins and peptides may be determined by the Dumas method or Kjeldahl method.

COLOR TESTS OF PROTEINS

Proteins give a number of color reactions which serve as their test. In fact, these reactions are due to the presence of certain specific groups in the protein structure. Therefore, a given protein may not respond to all the general tests listed below.

(1) **Xanthoproteic Test**. When a protein is warmed with conc. HNO_3, a yellow color is produced.

Protein + Conc. HNO_3 = Yellow Color

This test is given by a protein that consists of amino acids containing a benzene ring, such as tyrosine and tryptophan. The aromatic rings undergo nitration to give yellow products.

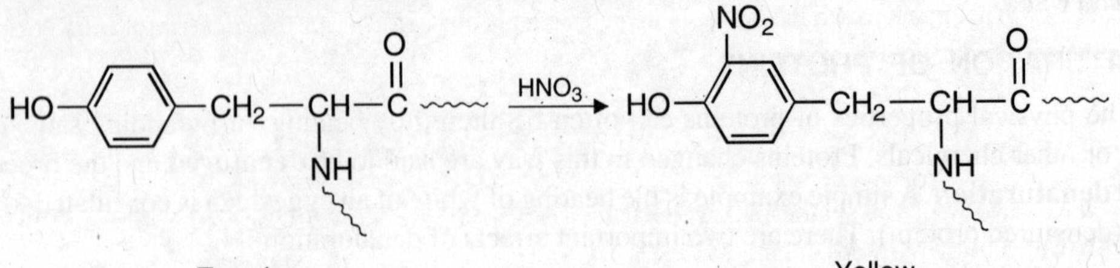

(2) **Biuret Test**. A small amount of NaOH solution is added to the protein. After mixing the two reagents, copper sulfate solution is added. Violet color is produced.

Protein + NaOH + Copper sulfate = Violet Color

This test is given by all proteins that contain two or more peptide linkages.

(3) **Ninhydrin Test**. When proteins are boiled with a dilute aqueous solution of ninhydrin, a violet color is produced.

Protein + Ninhydrin = Violet Color

This test is given by α-amino acids, proteins, and dipeptides.

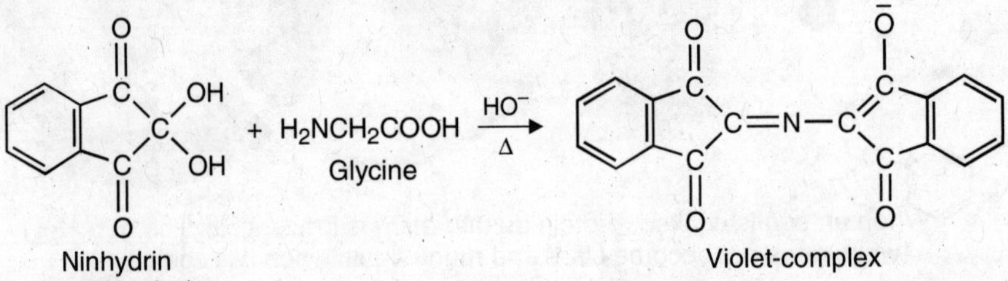

(4) **Millon's Test**. The Millon's reagent consists of mercury dissolved in nitric acid (forming a mixture of mercuric and mercurous nitrates and nitrites). When Millon's reagent is added to a protein, a white precipitate is formed.

Protein + Millon's Reagent = White Precipitate

This test is given by proteins which yield tyrosine on hydrolysis.

(5) **Hopkins-Cole Test**. Concentrated sulfuric acid is added down the side of a test-tube containing a solution of protein and glyoxalic acid, to form a layer. A violet ring appears between the two layers.

Protein + Gloxalic acid + conc. H_2SO_4 = Violet Ring

This test is given by proteins which have the amino acid tryptophan in its structure.

Examination Questions

1. What are α-amino acids ? How are they related to proteins ?
2. Amino acids act as proton donors as well as proton acceptors. Comment.
3. How are α-amino acids prepared ? Describe the important reactions of α-amino acids.
   (Mysore BSc, 2014 ; Panjab BSc, 2015)
4. Write a note on : Zwitterion.
5. Write a note on : Isoelectric Point of Amino Acids. *(Baroda BSc, 2014 ; Kakatiya BSc, 2015)*

6. What happens when glycine is heated ?

Answer. Diketopiperazine is formed. For reaction equation see text.

7. What happens when glycine is treated with nitrous acid ?

Answer. Glycollic acid is formed. For reaction equation see text.

8. What happens when alanine is treated with nitrous acid ? *(North Eastern Hills BSc, 2014)*

Answer. Lactic acid is formed

$$\underset{\text{Alanine}}{CH_3-\underset{NH_2}{\underset{|}{CH}}-COOH} + HONO \longrightarrow \underset{\text{Lactic acid}}{CH_3-\underset{OH}{\underset{|}{CH}}-COOH} + N_2 + H_2O$$

9. How will you synthesize glycine from acetic acid ? *(Jiwaji BSc, 2013 ; Baroda BSc, 2015)*

Answer. Following steps are involved :

$$\underset{\text{Acetic acid}}{CH_3COOH} \xrightarrow{Cl_2/P} ClCH_2COOH \xrightarrow{NH_3} H_2NCH_2COONH_4 \xrightarrow{H^+/H_2O} \underset{\text{Glycine}}{H_2NCH_2COOH}$$

10. How will you synthesize alanine from ethyl chloride ? *(Mysore BSc, 2013 ; Agra BSc, 2014)*

Answer. Following steps are involved :

$$\underset{\text{Ethyl chloride}}{CH_3CH_2Cl} \xrightarrow{NaCN} CH_3CH_2CN \xrightarrow{H^+/H_2O} CH_3CH_2COOH \xrightarrow[P]{Br_2}$$

$$CH_3\overset{\overset{Br}{|}}{C}HCOOH \xrightarrow{NH_3} CH_3\overset{\overset{NH_2}{|}}{C}HCOONH_4 \xrightarrow{H^+/H_2O} \underset{\text{Alanine}}{CH_3\overset{\overset{NH_2}{|}}{C}HCOOH}$$

11. How will you distinguish between glycine and acetamide ? *(Bundelkhand BSc, 2015)*

Answer. When glycine is boiled with dilute alkaline solution of ninhydrin, a violet color is produced. Acetamide does not give this test.

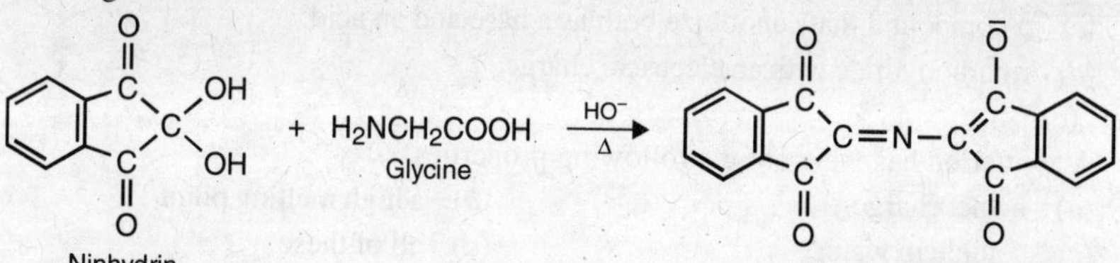

Ninhydrin (2 molecules) + Glycine → Violet-complex

12. What are proteins ?

13. What is the importance of proteins ?

14. How are proteins classified ?

15. Discuss the structure of proteins.

16. Describe the general properties of proteins.

17. Most naturally occurring amino acids are L-amino acids. Does this mean they are levorotatory? Explain.

Answer. No, this does not mean they are levorotatory. The "L" designation arises because the amino group is on the left in the Fischer projection and thus correlates to L-glyceraldehyde. The direction of optical rotation must be determined experimentally.

18. What is the major force responsible for the formation of an á-helix in protein secondary structure?

Answer. Hydrogen bonding

19. Which type of protein, globular or fibrous, tends to function primarily as structural parts of an organism ?

Answer. Fibrous

20. The nonprotein part of a conjugated protein is called a ________ group.

Answer. Prosthetic

21. Write a note on : Color Tests of Proteins.

22. Write a note on : Denaturation of Proteins. *(Madras BSc, 2013 ; Nehu BSc, 2015)*

Multiple Choice Questions

1. Amino acids have:
   (*a*) high melting points and low solubility in water.
   (*b*) large dipole moments and are more acidic than most carboxylic acids.
   (*c*) high melting points and large dipole moments.
   (*d*) low solubility in water and small dipole moments.

Answer. (*c*)

2. How many standard amino acids are there ?
   (*a*) 4 (*b*) 12 (*c*) 20 (*d*) 30

Answer. (*c*)

3. Proteins which provide all the essential amino acids in about the right proportions are called:
   (*a*) globular proteins. (*b*) complete proteins.
   (*c*) glycoproteins. (*d*) nucleoproteins.

Answer. (*b*)

4. Which of the following amino acids are classified as essential amino acids ? (More than one answer is possible.)
   (*a*) isoleucine (*b*) proline (*c*) tryptophan (*d*) histidine

Answer. (*a*), (*c*), (*d*)

5. A zwitterion is
   (*a*) an ion that is positively charged in solution
   (*b*) an ion that is negatively charged in solution
   (*c*) a compound that can ionize both as a base and an acid.
   (*d*) a carbohydrate with an electrical charge.

Answer. (*c*)

6. A zwitterion has which of the following properties
   (*a*) no net charge (*b*) a high melting point
   (*c*) soluble in water (*d*) all of these

Answer. (*d*)

7. An aqueous solution of glycine is neutral because of the formation of :
   (*a*) Carbanion (*b*) Zwitterion
   (*c*) Carbonium ions (*d*) Free radicals

Answer. (*b*)

8. Which one of the following compounds form zwitterions ?
   (*a*) carbonyl compounds (*b*) amino acids
   (*c*) phenols (*d*) heterocyclic compounds

Answer. (*b*)

9. The pH at which the amino acid shows no tendency to migrate when placed in an electric field is known as its :
   (*a*) Isoelectric point (*b*) Dipole moment
   (*c*) Iodine number (*d*) Wavelength

Answer. (*a*)

10. The isoelectric point of a protein is
    (*a*) the pH at which the protein molecule has no charges on its surface.
    (*b*) the pH at which a protein in solution has an equal number of positive and negative charges.
    (*c*) the electric charge under isothermal conditions
    (*d*) None of these.

Answer. (*b*)

11. The isoelectric point is important in:
(*a*) electrophoresis.
(*b*) determination of the C-terminal amino acid.
(*c*) determination of the N-terminal amino acid.
(*d*) the ninhydrin test.

Answer. (*a*)

12. Which of the following reactions is suitable for the preparation of α-amino acids ?
(*a*) Schmidt reaction (*b*) Hofmann's degradation of amides
(*c*) Strecker's synthesis (*b*) Reduction of nitro compounds

Answer. (*c*)

13. Glycine is
(*a*) NH_2CH_2COOH (*b*) $NH_2CH_2CH_2CH_2CH_2NH_2$
(*c*) $NO_2CH_2CH_2COOH$ (*b*) $BrCH_2COOH$

Answer. (*a*)

14. Glycine reacts with nitrous acid to form :
(*a*) Glycollic acid (*b*) Diketopiperazine
(*c*) Methylamine (*d*) Ethyl alcohol

Answer. (a)

15. When glycine is heated, it forms
(*a*) Diketopiperazine (*b*) Acrylic acid
(*c*) Butyric acid (*d*) Butyrlactam

Answer. (*a*)

16. Glycine is a unique amino acid because it
(*a*) has no chiral carbon (*b*) has a sulfur containing *R* group
(*c*) cannot form a peptide bond (*d*) is an essential amino acid

Answer.(*a*)

17. Which of the following organic ions results when glycine is treated with concentrated HCl ?
(*a*) $\overset{+}{N}H_3CH_2COOH$ (*b*) $NH_2CH_2CO\bar{O}$
(*c*) $\overset{+}{N}H_3CH_2CO\bar{O}$ (*d*) $HOCH_2CO\bar{O}$

Answer. (*a*)

18. Proteins are
(*a*) polyamides (*b*) Polymers of ethylene
(*c*) α-Aminocarboxylic acids (*d*) Polymers of propylene

Answer. (*a*)

19. The five elements present in most naturally occurring proteins are :
(*a*) C, H, O, P, and S (*b*) N, C, H, O, and I
(*c*) N, S, C, H, and O (*d*) C, H, O, S, and I

Answer. (*c*)

20. The nitrogen content of proteins can be quantitatively determined by
(*a*) Carius method (*b*) Kjeldahl's method
(*c*) Victor Meyer's method (*d*) Rast method

Answer. (*b*)

21. Which of the following is the main structural feature of proteins ?
(*a*) Peptide linkage (*b*) Ester linkage
(*c*) Ether linkage (*d*) α,β-Linkage

Answer. (*a*)

22. Peptide bonds are:
(a) ester linkages (b) imido linkages
(c) ether linkages (d) amide linkages

Answer. (d)

23. Which of the following is not an example of a protein ?
(a) hemoglobin (b) keratin
(c) insulin (d) cholesterol

Answer. (d)

24. The linear arrangement of amino acid units in proteins is called :
(a) primary structure (b) secondary structure
(c) tertiary structure (d) quaternary structure

Answer. (a)

25. The primary structure of a protein refers to :
(a) whether the protein is fibrous or globular
(b) the amino acid sequence in the polypeptide chain
(c) the orientation of the amino acid side chains in space
(d) the presence or absence of an α-helix.

Answer. (b)

26. The α-Helix is a common form of
(a) Primary structure (b) Tertiary structure
(c) Secondary structure (d) None of these

Answer. (c)

27. The α-Helix is held in a coiled conformation partially because of :
(a) Optical activity (b) Hydrogen bonding
(c) Resonance (d) Delocalization

Answer. (b)

28. The double helical structure of DNA is held together by
(a) sulfur-sulfur linkages (b) peptide bonding
(c) hydrogen bonding (d) glycosidic bonds

Answer. (c)

29. The primary structure of a protein refers to:
(a) the sequence of its amino acids.
(b) the orientation of á-helices.
(c) the orientation of pleated sheets.
(d) the orientation of peptide subunits within a complex protein.

Answer. (a)

30. Which of the following arrangements is usually *not* found in the secondary structure of proteins ?
(a) α-helix (b) double helix
(c) random coil (d) pleated sheet

Answer. (b)

31. Upon hydrolysis, proteins give
(a) Amino acids (b) Hydroxy acids
(c) Fatty acids (d) Alcohols

Answer. (a)

32. Complete hydrolysis of proteins produces :
(a) Ammonia and carbon dioxide (b) Urea and uric acid
(c) A mixture of amino acids (d) Glycogen and a fatty acid

Answer. (c)

33. Digestion of proteins involves

(*a*) changes in secondary structure only

(*b*) cleavage of peptide linkages

(*c*) removal of all carboxyl groups in the form of CO_2.

(*d*) removal of all NH_2 groups in the form of NH_3

Answer. (*b*)

34. Irreversible precipitation of proteins caused by heating is called :

(*a*) Polymerization (*b*) Denaturation

(*c*) Electrophoresis (*d*) Inversion

Answer. (*b*)

35. Precipitation or coagulation of proteins may be caused by

(*a*) Heat (*b*) Changes in pH

(*c*) Heavy metal salts (*d*) All of these

Answer. (*d*)

36. Ninhydrin test is given by

(*a*) Carbohydrates (*b*) Proteins

(*c*) Alkanes (*d*) Alkenes

Answer. (*b*)

37. Which of the following tests is not used for testing proteins ?

(*a*) Ninhydrin test (*b*) Biuret test

(*c*) Xanthoproteic test (*d*) Tollens' test

Answer. (*d*)

38. A protein solution on warming with concentrated HNO_3 may turn yellow. This test is called :

(*a*) Xanthoproteic test (*b*) Ninhydrin test

(*c*) Biuret test (*d*) Million's test

Answer. (*a*)

39. A compound gives a positive Tollens' test but negative Ninhydrin test. It is

(*a*) a protein (*b*) an amino acid

(*c*) a monosaccharide (*d*) pyridine

Answer. (*c*)

29

Cycloalkanes

People at play with the Boat and Chair forms of cyclohexane.

Cycloalkanes or cycloparaffins are saturated hydrocarbons in which the carbon atoms are joined by single covalent bonds to form a ring. They are also called Alicyclic Compounds. The prefix *ali-* is added because of their similarity to aliphatic compounds. The unsubstituted cycloalkanes form a homologous series with the general formula C_nH_{2n}. The first member of the series is cyclopropane, C_3H_6.

NOMENCLATURE

The IUPAC rules for naming cycloalkanes are as follows :

(1) The name of an unsubstituted cycloalkane is obtained by attaching the prefix *cyclo-* to the name of the corresponding normal alkane having the same number of carbon atoms as in the ring.

(2) Substituents on the ring are named, and their positions are indicated by numbers. *The ring is numbered so that the carbons bearing the substituents will have the lowest numbers.*

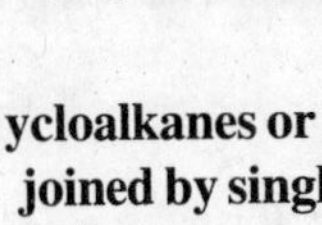

Cyclopropane

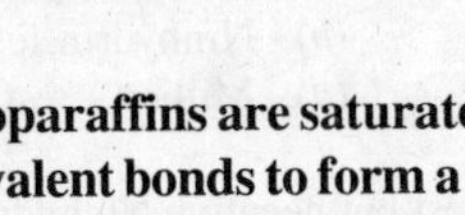

Bromocyclobutane

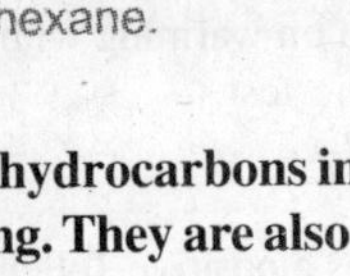

Cyclopentane

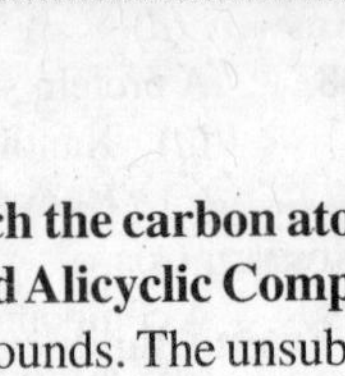

Cyclohexane

SYMBOLS OF CYCLOALKANES

For convenience and simplicity, cycloalkanes are often represented by simple geometric figures: a *triangle* for cyclopropane, a *square* for cyclobutane, a *pentagon* for cyclopentane, a *hexagon* for cyclohexane and so on. *It is understood that each corner represents one carbon and two hydrogens.*

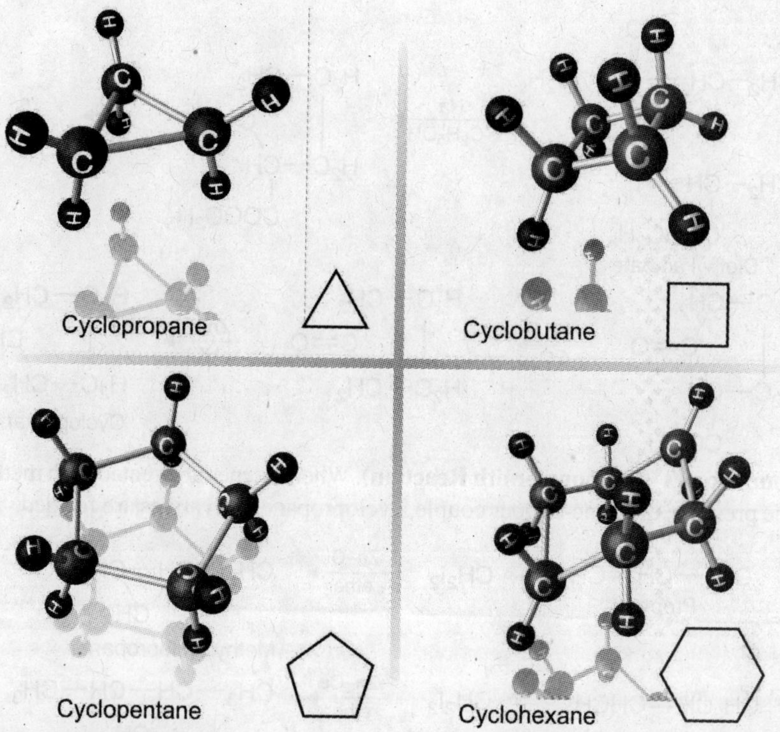

METHODS OF PREPARATION

Cycloalkanes are obtained :

(1) **From Dihalides.** Terminal dihalides when treated with sodium or zinc form cycloalkanes. This reaction is an extension of Wurtz reaction and is useful for the preparation of 3- to 6-membered rings.

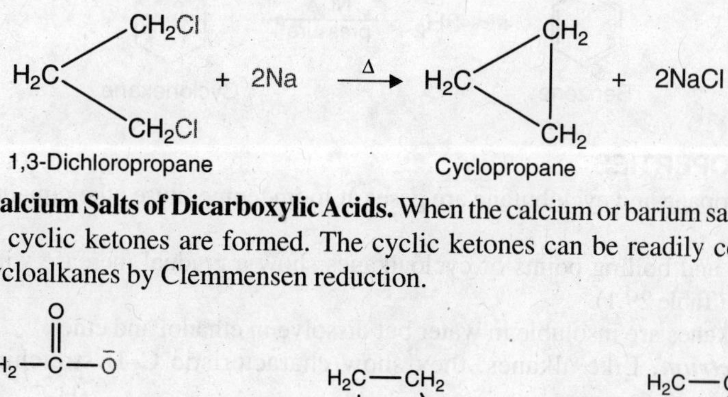

1,3-Dichloropropane Cyclopropane

(2) **From Calcium Salts of Dicarboxylic Acids.** When the calcium or barium salts of dicarboxylic acids are heated, cyclic ketones are formed. The cyclic ketones can be readily converted into the corresponding cycloalkanes by Clemmensen reduction.

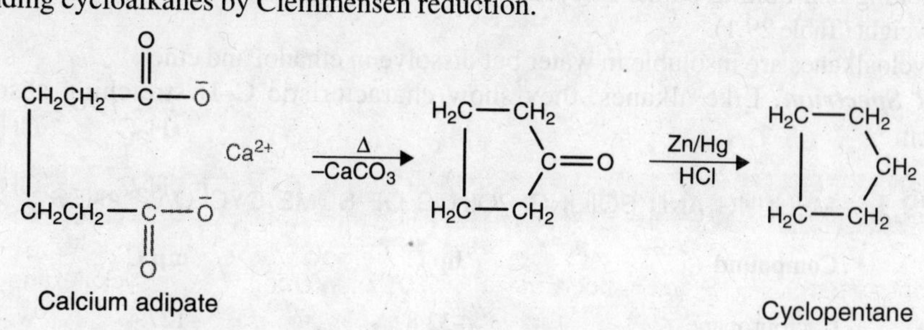

Calcium adipate Cyclopentane

(3) **From Esters of Dicarboxylic Acids (Dieckmann Reaction).** Esters of dicarboxylic acids when treated with sodium undergo intramolecular Acetoacetic Ester condensation and a β-ketoester is formed. The β-ketoesters on hydrolysis give cyclic ketones. These on reduction yield the corresponding cycloalkanes.

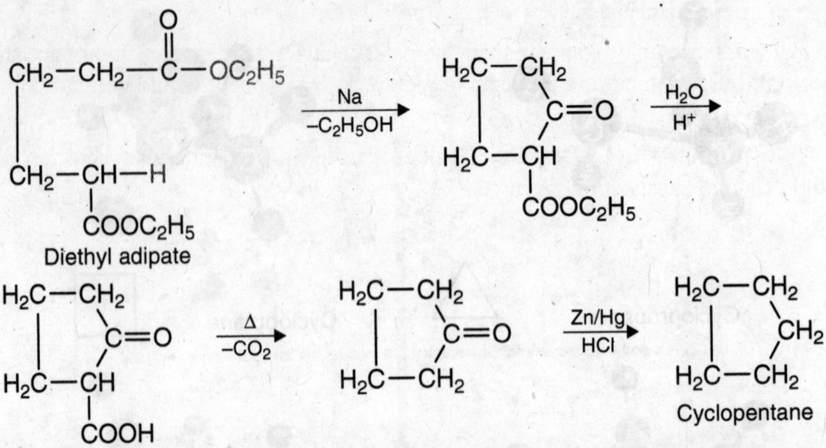

(4) **From Alkenes (Simmons-Smith Reaction).** When alkenes are treated with methylene iodide (CH_2I_2) in the presence of a zinc-copper couple, cyclopropane derivatives are formed.

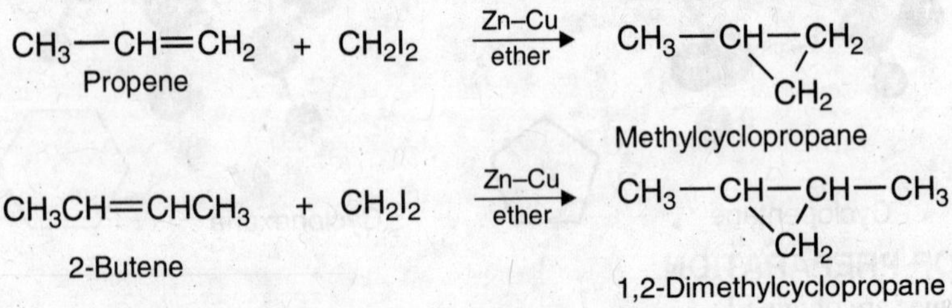

(5) **From Aromatic Hydrocarbons.** Six-membered cycloalkanes can be prepared by the catalytic reduction of benzene and its derivatives.

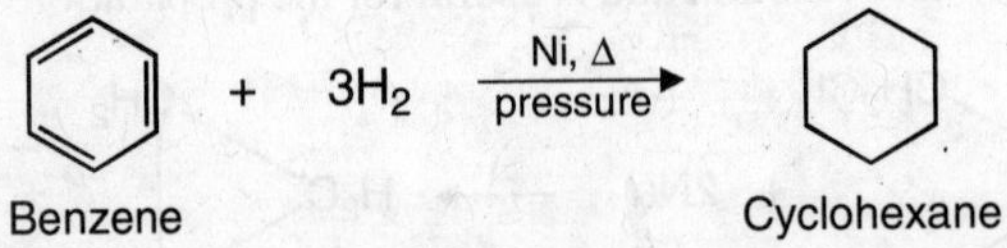

PHYSICAL PROPERTIES

(1) Cyclopropane and cyclobutane are gases at room temperature ; the remaining cycloalkanes are liquids.

(2) Melting and boiling points of cycloalkanes show a gradual increase with the increase in molecular weight (Table 29.1).

(3) Cycloalkanes are insoluble in water but dissolve in ethanol and ether.

(4) ***IR Spectrum.*** Like alkanes, they show characteristic C–H stretching absorption at 2850-3050 cm^{-1}.

Table 29.1 MELTING AND BOILING POINTS OF SOME CYCLOALKANES

Compound	bp°C	mp°C
Cyclopropane	–32.8	–127.4
Cyclobutane	12.5	–90.7
Cyclopentane	49.3	–13.9
Cyclohexane	80.7	6.6

CHEMICAL PROPERTIES

Cycloalkanes resemble alkanes in their chemical behavior. However, cyclopropane and cyclobutane are the exceptions. With certain reagents they undergo ring-opening and give addition products.

A. SUBSTITUTION REACTIONS

(1) **Substitution with Cl_2 and Br_2.** Cycloalkanes react with chlorine and bromine in the presence of UV light to give substitution products.

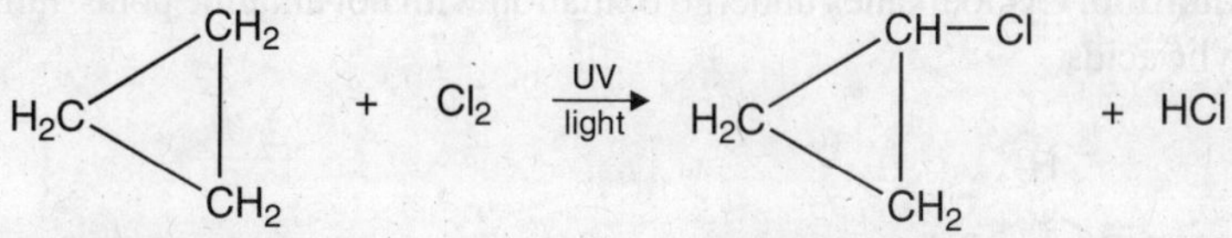

Cyclopropane Chlorocyclopropane

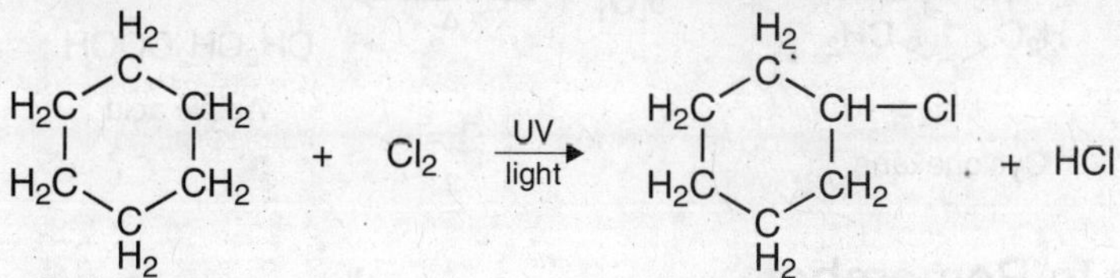

Cyclohexane Chlorocyclohexane

B. RING-OPENING REACTIONS

(2) **Addition of Cl_2 and Br_2.** Cyclopropane reacts with chlorine and bromine in the *dark* to form addition products. CCl_4 is used as a solvent.

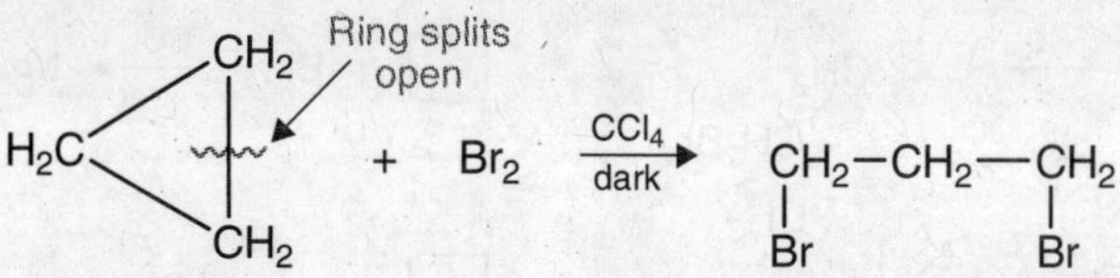

Cyclopropane 1,3-Dibromopropane

Cyclobutane and higher members do not give this reaction.

(3) **Addition of HBr and HI.** Cyclopropane reacts with concentrated HBr and HI to yield 1-bromopropane and 1-iodopropane respectively.

1-Bromopropane

Cyclopropane

Cyclobutane and higher members do not give this reaction.

(4) **Addition of Hydrogen.** Cyclopropane and cyclobutane react with hydrogen in the presence of Ni catalyst to give propane and *n*-butane respectively. Notice that higher temperature is required for cyclobutane.

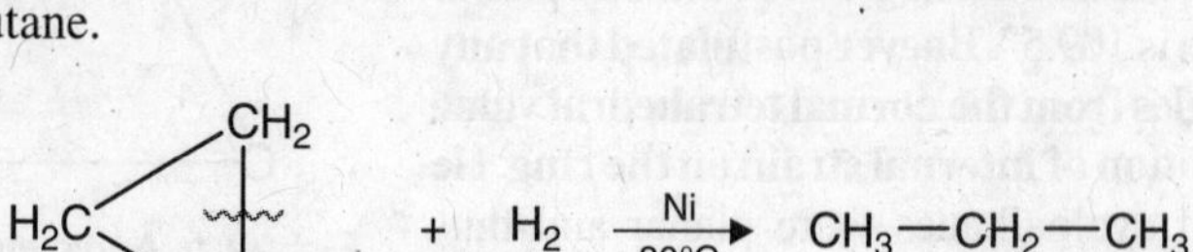

Cyclopropane

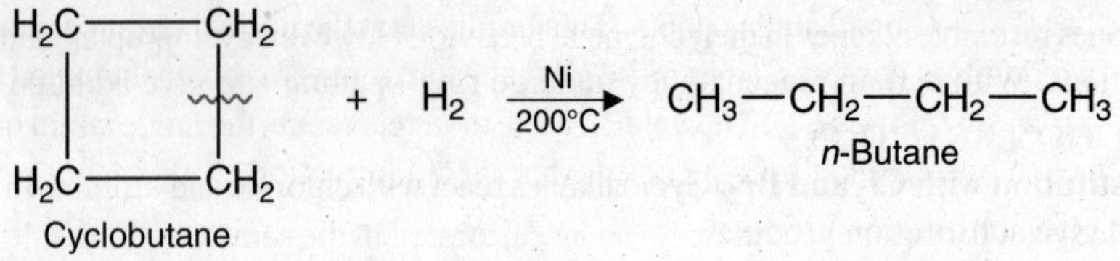

(5) **Oxidation.** Cycloalkanes undergo oxidation with hot alkaline potassium permanganate to form dicarboxylic acids.

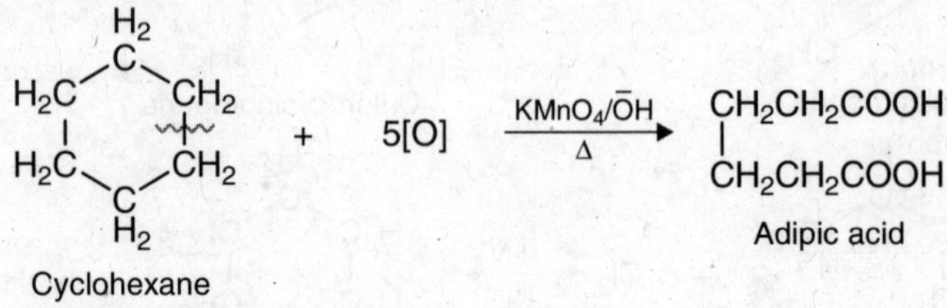

Points to Remember

Comparison of addition reactions of cyclopropane and cyclobutane.

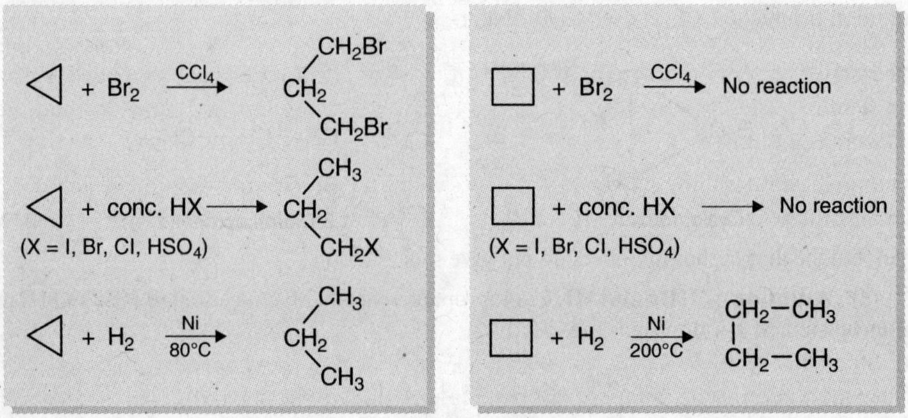

STABILITY OF CYCLOALKANES

BAEYER STRAIN THEORY

In 1885 Adolf Baeyer proposed a theory to explain the relative stability of the first few cycloalkanes. He based his theory on the fact that the normal angle between any pair of bonds of a carbon atom is 109.5°. **Baeyer postulated that any deviation of bond angles from the normal tetrahedral value would impose a condition of internal strain on the ring.** He also assumed that all cycloalkanes were planar and thus calculated the angles through which each of the valency bond was deflected from the normal direction in the formation of the various rings. This he called **Angle Strain**, which determined the stability of the ring.

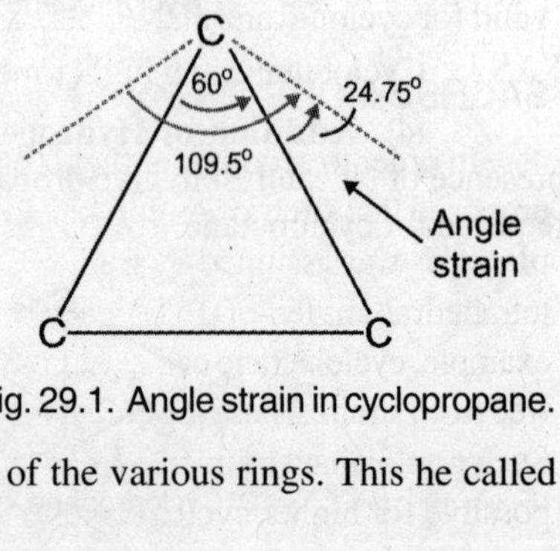

Fig. 29.1. Angle strain in cyclopropane.

In cyclopropane, the three carbon atoms occupy the corners of an equilateral triangle. Thus cyclopropane has C–C–C bond angles of 60°. This implies that the normal tetrahedral angle of 109.5° between any two bonds is compressed to 60°, and that each of the two bonds involved is pulled in by $\frac{1}{2}(109.5°–60°) = 24.75°$ (Fig. 29.1). The value 24.75° then represents the angle strain or the deviation through which each bond bends from the normal tetrahedral direction.

The angle strain for other cycloalkanes can be calculated in the same way. The values are given in Table 29.2. Whether the angle strain is positive or negative, its magnitude determines the extent of strain in the ring.

Table 29.2 ANGLE STRAIN IN CYCLOALKANES

Compound	Bond angle	Angle strain
Cyclopropane	60°	24.75°
Cyclobutane	90°	9.75°
Cyclopentane	108°	0.75°
Cyclohexane	120°	–5.27°

The angle strain is maximum in the case of cyclopropane. Thus according to the Baeyer Strain theory, cyclopropane should be highly strained molecule and consequently most unstable. The cyclopropane ring should, therefore, be expected to open-up on slightest provocation and thus releasing the strain within it. This is actually so. Cyclopropane undergoes ring-opening reactions with Br_2, HBr, and H_2(Ni) to give open-chain addition products.

The angle strain in the case of cyclobutane is less than that in case of cyclopropane. This signifies greater stability. Thus as expected, cyclobutane undergoes ring opening reactions but only under more drastic conditions.

The angle strain is minimum in the case of cyclopentane. This implies that cyclopentane is under least strain and should be most stable. Thus it is not surprising that cyclopentane does not undergo ring-opening reactions.

The angle strain in the case of cyclohexane is higher than that in case of cyclopentane. This strain increases continuously with increase in the number of carbon atoms in the ring. According to the Baeyer Strain theory, cyclohexane and the higher cycloalkanes should become increasingly unstable and hence more reactive. Contrary to this prediction, cyclohexane and the higher members are found to be quite stable. They do not undergo ring-opening reactions. Instead they resemble open-chain alkanes in reactivity, that is, they react by substitution. Thus, the Baeyer Strain theory satisfactorily accounts for the exceptional reactivity of cyclopropane, cyclobutane, and cyclopentane, but it is not valid for cyclohexane and higher members.

SACHSE-MOHR THEORY

In order to account for the stability of cyclohexane and higher members, Sachse and Mohr (1918) proposed that **such rings can become free from strain if all the ring carbons are not forced into one plane,** as was assumed by Baeyer. If the ring assumed a 'folded' or 'puckered' condition, the normal tetrahedral angles of 109.5° are retained and as a result, the strain within the ring is relieved. For example, cyclohexane can exist in two non-planar puckered conformations both of which are completely free from strain. These are called the **Chair Form** and the **Boat Form** because of their shape (Fig. 29.2). Such non-planar strain-free rings in which the ring carbons can have normal tetrahedral angles are also possible for higher cycloalkanes.

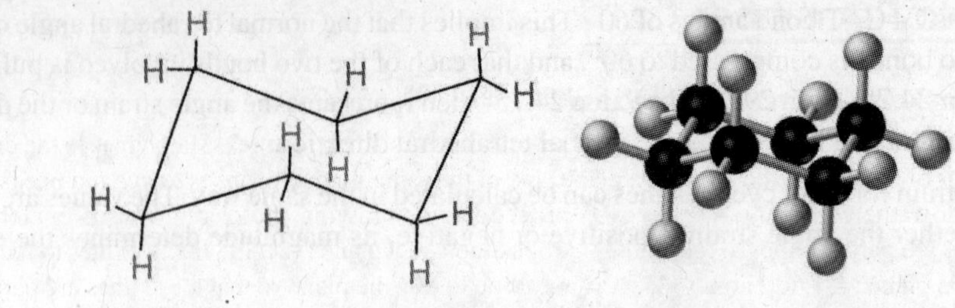

Chair form of cyclohexane

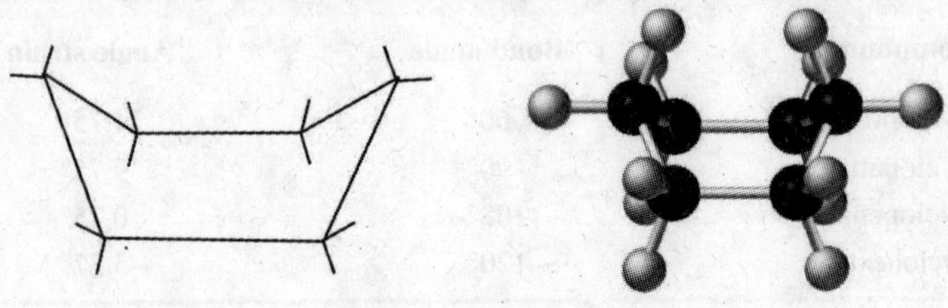

Boat form of cyclohexane

Fig. 29.2. The chair and boat forms of cyclohexane.

The chair form of cyclohexane is more stable than the boat form. Under ordinary conditions, cyclohexane molecules will mostly exist in the chair form.

Examination of the chair form of cyclohexane reveals that the hydrogen atoms can be divided into two categories. Six of the bonds to hydrogen atoms point straight up or down almost perpendicular to the plane of the molecule. These are called **Axial Hydrogens**. The other six hydrogens lie slightly above or slightly below the plane of the cyclohexane ring, and are called **Equatorial Hydrogens** (Fig. 29.3).

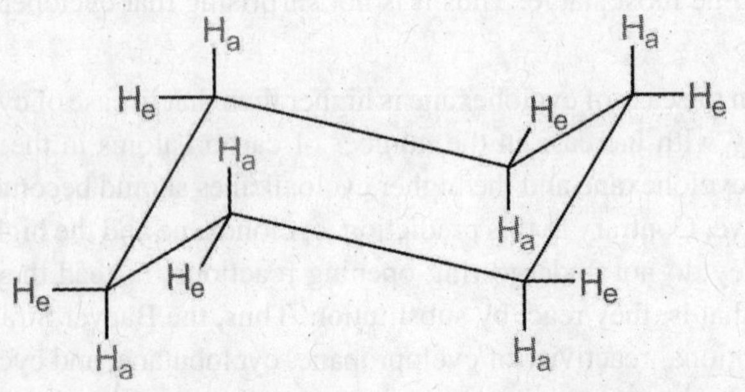

Fig. 29.3. Axial and equatorial hydrogens in the chair form. Axial hydrogens are shown as H_a. Equatorial hydrogens are shown as H_e.

MOLECULAR ORBITAL THEORY

We can explain the relative stability of cycloalkanes in terms of MO theory as follows :

We know that a covalent bond between two atoms is formed by the overlap of orbitals of the atoms involved. The greater the extent of overlap the stronger is the bond formed. The atomic orbitals overlap to the maximum extent if they overlap along their axes. As the axes of sp^3 orbitals are at angles of 109.5° to each other, the C–C bonds will have their maximum strength if the C–C–C bond angles have a value of 109.5°.

Points to Remember

Drawing Chair-Form of Cyclohexane Rings

The chair conformation of a cyclohexane ring has several distinguishing features. Many lines in the drawing are parallel. Four of the carbon atoms lie in a plane, and of the remaining two, one is *above* and one is *below* this plane. Follow the following step-by-step procedure for drawing the chair-form of cyclohexane rings : (*1*) Draw two parallel lines to locate four carbon atoms in a plane ; (*2*) Add the two lines to locate the carbon above the plane ; (*3*) Add lines to locate the carbon below the plane so that these lines are parallel to those in step 2 ; (*4*) Draw the six axial bonds *up* from the carbons that are up and *down* from carbons that are down ; and (*5*) Draw the equatorial bonds parallel to the existing carbon-carbon bonds. These steps are illustrated below :

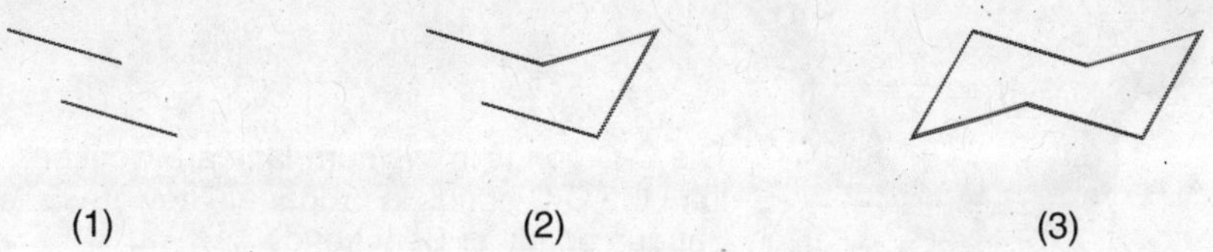

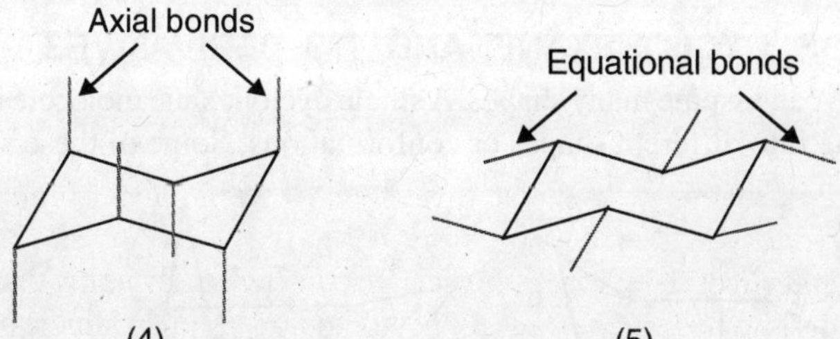

Cyclopropane has C–C–C bond angles of 60°. Cyclobutane has C–C–C bond angles have a value of 90°. The higher cycloalkanes and alkanes have C–C–C bond angles of 109.5°. As shown in Fig. 29.4, the small bond angles of cyclopropane indicate that the overlap of sp^3 orbitals of carbon is less than the overlap of sp^3 orbitals of carbon in alkanes (e.g., propane).

The bond angles of cyclopropane are less than the bond angles of cyclobutane, which in turn are less than the bond angles of higher cycloalkanes of *n*-alkanes. Therefore, the overlap of orbitals in cyclopropane is less than that in cyclobutane, which in turn in less than that in higher cycloalkanes or *n*-alkanes. The overlap of sp^3 orbitals of carbons in cyclopentane, higher cycloalkanes or *n*-alkanes is

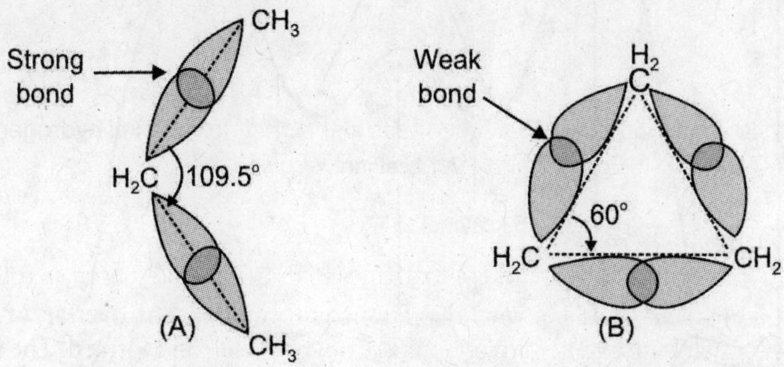

Fig. 29.4. Overlap between sp^3 orbitals in (*a*) propane and (*b*) cyclopropane. Maximum overlap occurs in propane.

maximum because in these cases it is possible for the sp^3 orbitals to overlap along their axes, the bond angles being approximately equal to 109.5°. This implies that the C–C bonds in cyclopropane are weaker than the C–C bonds of cyclobutane, which in turn are weaker than the C–C bonds in higher cycloalkanes and *n*-alkanes. Thus, it is not surprising that cyclopropane undergoes ring-opening reactions very readily and more drastic conditions are required to bring about the cleavage of the cyclobutane ring. Cyclopentane and higher members, as expected, do not undergo ring opening reactions and behave very much like the alkanes.

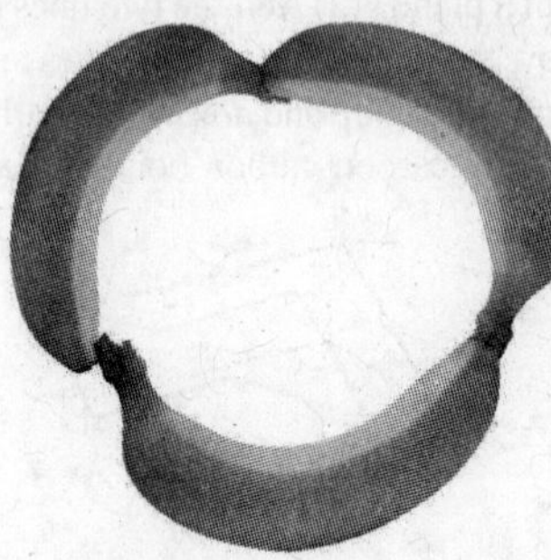

The C-C bonds in cyclopropane are weaker than the C-C bonds in propane. They are called Banana bonds or Bent bonds.

CONFORMATIONS OF CYCLOHEXANE AND ITS DERIVATIVES

The cyclohexane ring can assume many shapes. A single cyclohexane molecule is in a continuous state of flexing or flipping into different shapes or conformations. Some of these shapes are shown below :

These conformations arise due to rotation around carbon-carbon bonds. The *chair form* and the *boat form* are the two extreme cases.

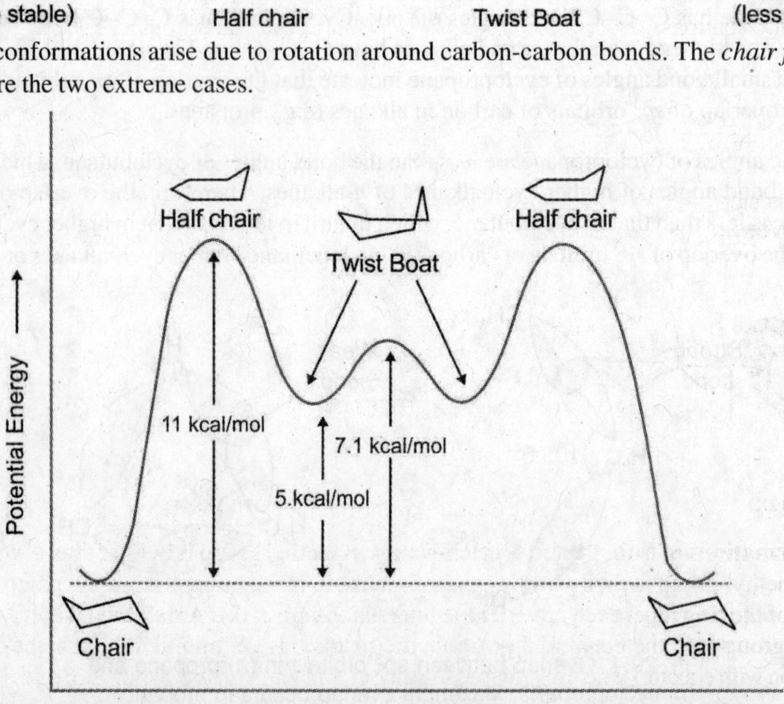

Fig. 29.5. Relative energies of the conformations of cyclohexane molecule.

Figure 29.5 shows the energy requirements for the interconversion of the different conformations of cyclohexane. Notice that the chair form has the lowest energy, while the half-chair has the highest energy. At any given time, we would expect most of the cyclohexane molecules to be in the chair form. Indeed, it has been calculated that about 99.9% of cyclohexane molecules are in the chair form at any one time.

Equatorial and Axial Hydrogens. The carbon atoms of the chair form of cyclohexane roughly lie in one plane, and an axis can be drawn perpendicular to this plane (Figure 29.6).

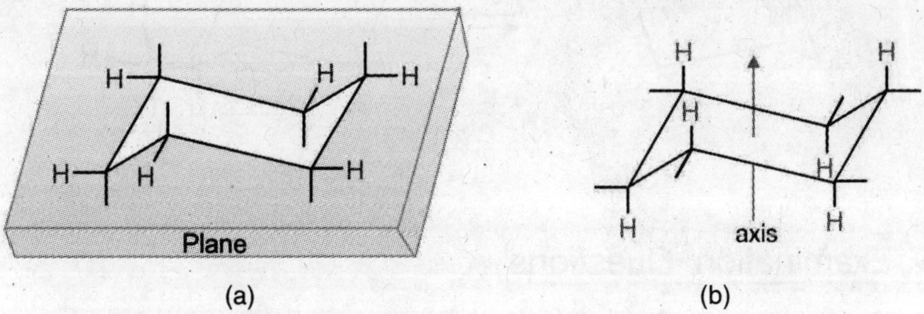

Fig. 29.6. The equatorial and axial hydrogens of cyclohexane. (*a*) Equatorial hydrogens lie in the plane of the ring carbons. (*b*) Axial hydrogens lie (up or down) parallel to the perpendicular axis.

Each carbon atom of cyclohexane is bonded to two hydrogens. The bond to one of these hydrogens lies in the rough plane of the ring ; this hydrogen is called **Equatorial Hydrogen**. The bond to the other hydrogen atom is parallel to the axis ; this hydrogen atom is called **Axial Hydrogen.**

Each of the six carbon atoms of cyclohexane has one equatorial and one axial hydrogen atom. Remember that there are six equatorial hydrogens, and six axial hydrogens. In the flipping and reflipping between conformations, axial becomes equatorial, while equatorial becomes axial.

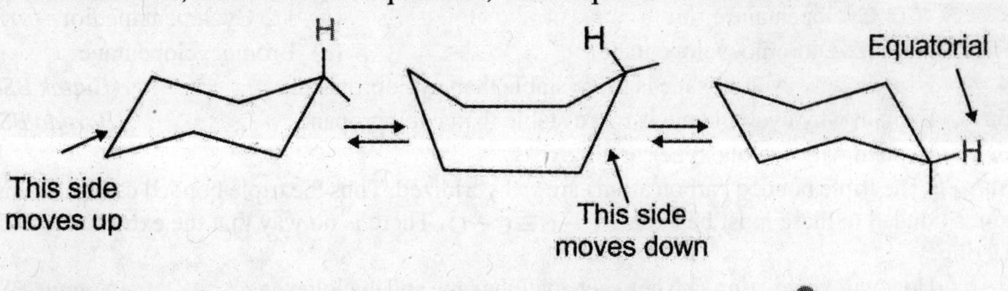

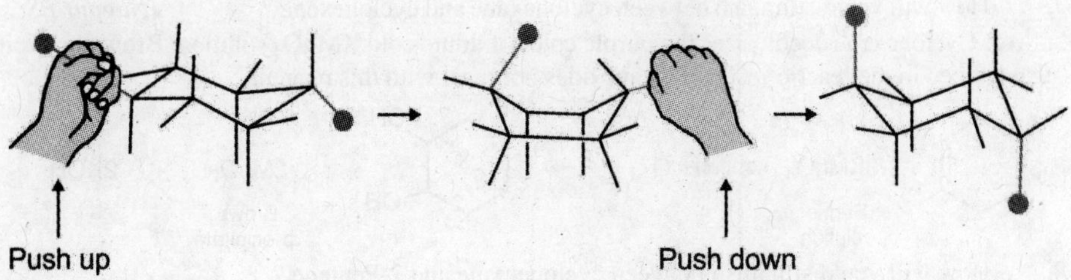

Conformations of Substituted Cyclohexanes. A methyl group is bulkier than a hydrogen atom. When the methyl group in methylcyclohexane is in the axial position, the methyl group and the axial hydrogens of the ring repel each other. These interactions are called **Axial-Axial Interactions.** When the methyl group is in the equatorial position, the repulsions are minimum. Thus, the energy of the conformation with equatorial methyl group is lower. At room temperature, about 95% methylcyclohexane molecules are in the conformation in which the methyl group is equatorial.

The bulkier the group, the greater is the energy difference between equatorial and axial conformations. In other words, a cyclohexane ring with a bulky substituent (e.g., *t*-Butyl group) is more likely to have that group in the equatorial position.

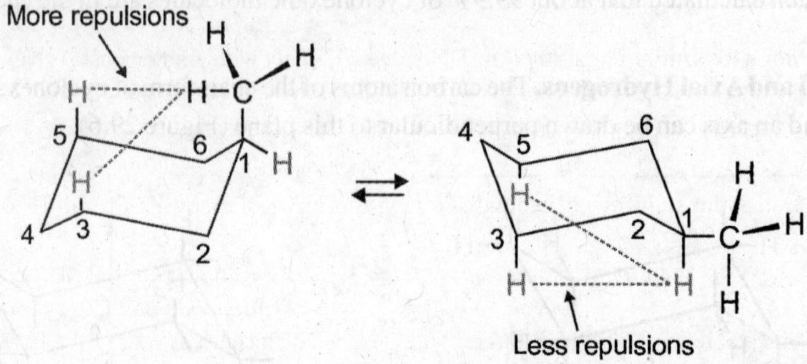

Examination Questions

1. What are cycloalkanes ? How are they prepared ? Describe important reactions of cycloalkanes.

2. (*a*) How can cyclopropane be prepared ?
   (*b*) How does cyclopropane react with :
   (*i*) Conc. HBr (*ii*) H_2/Ni at 80°C
   (*iii*) Cl_2/UV light (*iv*) Br_2/CCl_4 in dark
   (*c*) The C–C bonds in cyclopropane are weaker than the C–C bonds in propane. Explain this fact on the basis of molecular orbital theory. (*Panjab BSc, 2015*)

3. How does cyclopentane react with :
   (*a*) H_2/Ni (*b*) dilute cold $KMnO_4$ (*c*) Br_2/CCl_4 (*d*) HBr

Answer. The products are :
(*a*) Cyclopentane (*b*) 1,2-Cyclopentanediol
(*c*) 1,2-Dibromocyclopentane (*d*) Bromocyclopentane

4. Explain why cyclohexane is more stable than cyclopropane. (*Cuttak BSc, 2014*)

5. Explain why cyclobutane is more stable than cyclopropane. (*Baroda BSc, 2015*)

6. Explain why cyclohexyne cannot exist.

Answer. The triple bonded carbon atoms are *sp* hybridized. Thus the triple bonded carbon atoms and the carbon atoms bonded to these must be linear, $C-C\equiv C-C$. There is no way that the extra two carbon atoms can form a ring.

7. How will you distinguish between cyclohexane and cyclohexene ? (*Nagpur BSc, 2014*)

Answer. Cyclohexene decolorizes the purple color of dilute cold $KMnO_4$ solution. Brown precipitate of MnO_2 is also formed in the reaction. Cyclohexane does not react with this reagent.

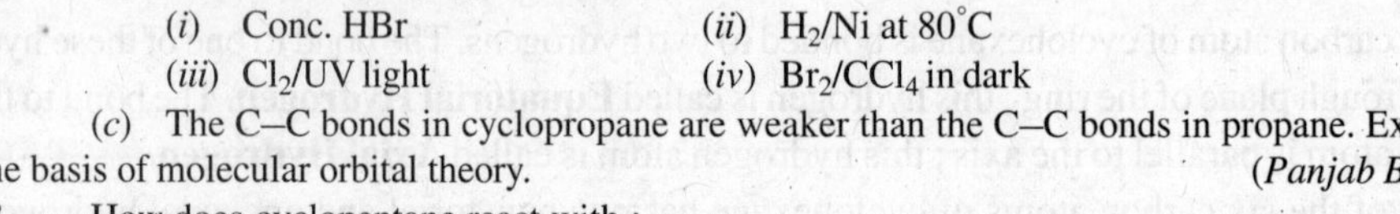

8. How will you distinguish between cyclopentane and 2-Pentene ?

Clue. Use dilute cold $KMnO_4$ or bromine in carbon tetrachloride.

9. The labels from bottles of cyclohexene, *n*-hexane and 1-hexyne have come off and fallen to the floor. Describe simple tests that would enable you to correctly relabel the bottles.

Answer. Perform the following two tests :

(1) **Reaction with Br_2 in carbon tetrachloride solution :** Cyclohexene and 1-hexyne will decolorize the red color of the solution of bromine in carbon tetrachloride. n-Hexane does not react. This test will help us to relabel the bottle of *n*-hexane.

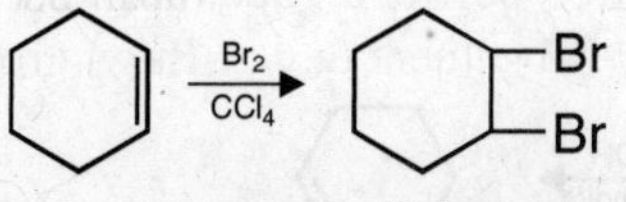

(Colourless)

$$CH_3(CH_2)_3C{\equiv}CH \xrightarrow[CCl_4]{Br_2} CH_3(CH_2)_3-\underset{Br}{\overset{Br}{C}}-\underset{Br}{\overset{Br}{C}}-H$$

(Colourless)

(2) **Reaction with ammoniacal $AgNO_3$.** 1-Hexyne will react with ammoniacal $AgNO_3$ to form a white precipitate, cyclohexene does not react. This test will help us to relabel the bottles of cyclohexane and 1-hexyne.

$$CH_3(CH_2)_3C{\equiv}CH + 2Ag(NH_3)_2NO_3 \longrightarrow \underset{\text{(White ppt)}}{CH_3(CH_2)_3C{\equiv}CAg} + 2NH_4NO_3 + 2NH_3$$

10. Discuss the stability of cycloalkanes on the basis of Baeyer's Strain Theory and its later developments.

11. Discuss the stability of cycloalkanes on the basis of Molecular Orbital Theory.
(Allahabad BSc, 2014 ; Madras BSc, 2015)

12. Write a note on : Conformations of cyclohexane and its derivatives. *(Delhi BSc, 2014)*

13. Write notes on :

(*a*) Baeyer's Strain Theory
(*b*) Sachse-Mohr's Theory of Strainless Rings
(*c*) Stability of Cycloalkanes *(Vikram BSc, 2014; Anna BSc, 2015)*
(*d*) Dieckmann Reaction

14. Discuss the mechanism for the reaction of cyclohexane with chlorine in the presence of ultraviolet light. *(Kerala BSc, 2013)*

Answer. Cyclohexane reacts with chlorine in the presence of ultraviolet light to form chlorocyclohexane.

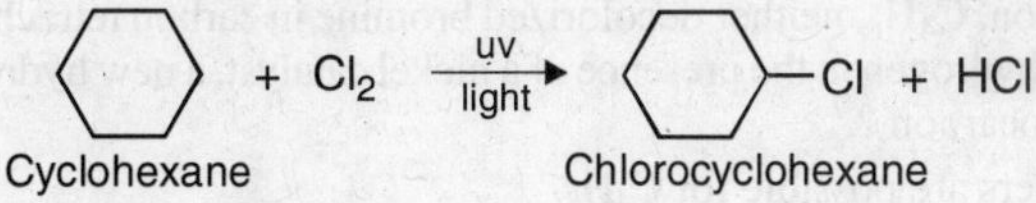

Mechanism. Following steps are involved :

Step 1. Initiation. Chlorine undergoes homolytic fission to form chlorine free radicals.

$$Cl:Cl \xrightarrow[\text{light}]{\text{uv}} Cl^{\cdot} + {}^{\cdot}Cl$$

Step 2. Chain-Propagation Steps. (*a*) Chlorine free radical attacks the cyclohexane molecule to give HCl and cyclohexyl free radical. *Notice that fish-hook arrows are used to indicate the movement of single electrons.*

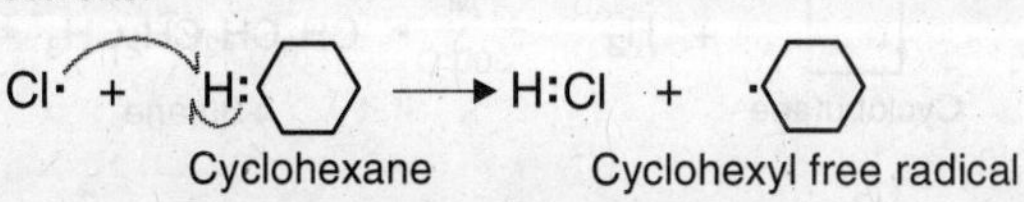

Steps (*a*) and (*b*) are repeated over and over again.

Step 3. Chain-Termination Steps. The above chain reaction comes to halt when any two free radicals combine. For example,

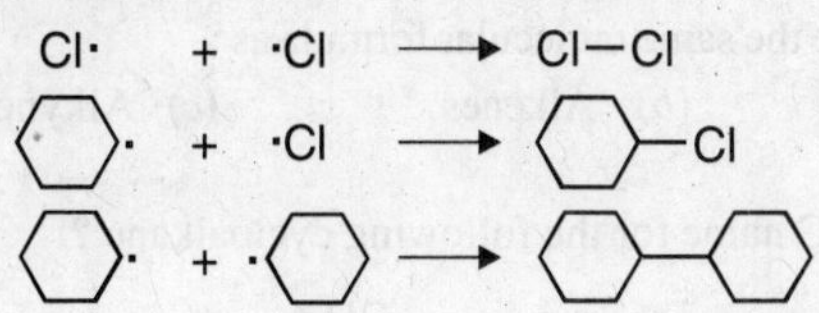

15. How will you synthesize 1-Bromopropane from 1,3-Dibromopropane ?

Clue. Steps are : (*i*) Zn/heat ; and (*ii*) Conc. HBr

16. How will you synthesize Butanoic acid from Cyclopropane ?

Answer. Following steps are involved :

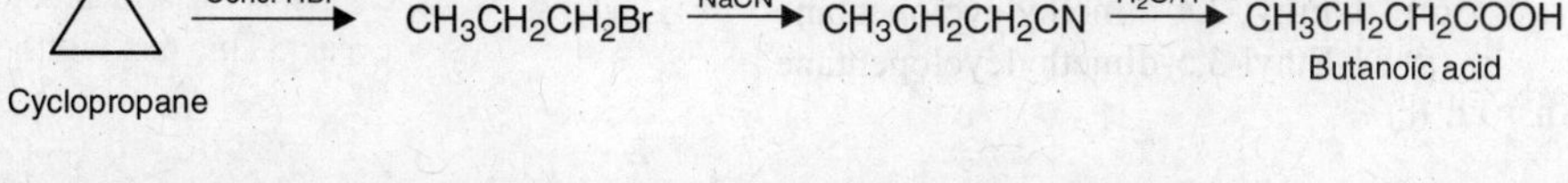

17. How will you synthesize cyclohexene from cyclohexane ? *(Bihar BSc, 2015)*

Answer. Following steps are involved :

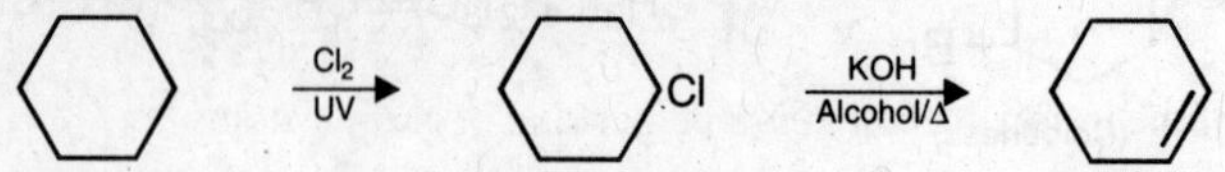

Cyclohexane — Chlorocyclohexane — Cyclohexene

18. How will you synthesize cyclohexane from Phenol ?

Answer. Following steps are involved :

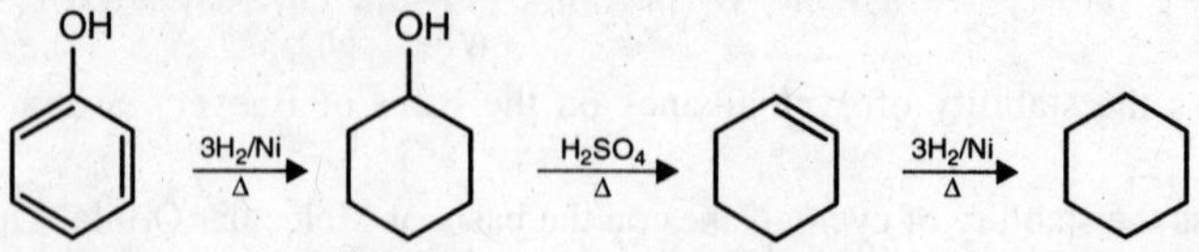

Phenol — Cyclohexanol — Cyclohexene — Cyclohexane

19. How will you synthesize Cyclopentane from adipic acid ?

Answer. Following steps are involved :

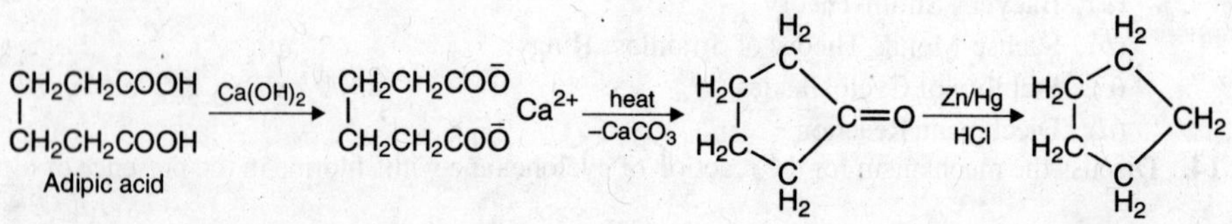

Cyclopentane

20. A hydrocarbon, C_4H_8, neither decolorized bromine in carbon tetrachloride nor reacted with HBr. When heated to 200°C with hydrogen in the presence of a nickel catalyst, a new hydrocarbon, C_4H_{10}, was formed. What was the original hydrocarbon ?

Answer. Three isomers are possible for C_4H_8.

$CH_3CH_2CH{=}CH_2$ (1-Butene) $\quad$ $CH_3CH{=}CHCH_3$ (2-Butene) $\quad$  Cyclobutane

The given compound is cyclobutane because 1-butene and 2-butene both decolorize bromine in carbon tetrachloride and react with HBr. Cyclobutane does not react with Br_2/CCl_4 or HBr.

 $+ H_2 \xrightarrow[200°C]{Ni} CH_3CH_2CH_2CH_3$

Cyclobutane — *n*-Butane

Multiple Choice Questions

1. Cycloalkanes have the same molecular formula as :

(*a*) Alkanes (*b*) Alkenes (*c*) Alkynes (*d*) Cycloalkenes

Answer. (*b*)

2. What is the IUPAC name for the following cycloalkane ?

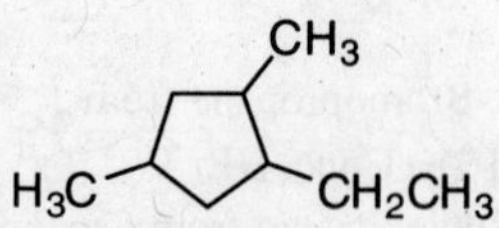

(*a*) 2,4-Dimethyl-1-ethylcyclopentane
(*b*) 1,3-Dimethyl-5-ethylcyclopentane
(*c*) 1-Ethyl-2,4-dimethylcyclopentane
(*d*) 1-Ethyl-3,5-dimethylcyclopentane

Answer. (*c*)

3. The C–C–C bond angle in cyclopropane is closest to :

(*a*) 120° (*b*) 109.5° (*c*) 60° (*d*) 90°

Answer. (*c*)

4. Which of the following contains a planar ring of carbon atoms ?

(*a*) cyclohexane (*b*) cyclobutane
(*c*) cyclopentane (*d*) cyclopropane

Answer. (*d*)

5. What are the hybridizations of carbons (1) and (2) respectively in the following structure ?

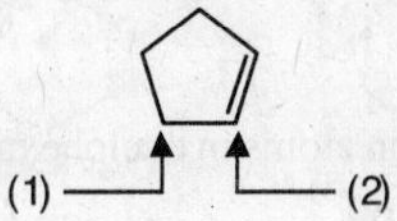

(*a*) sp^3 and sp^2 (*b*) sp^2 and sp^3
(*c*) sp^3 and sp (*d*) sp^2 and sp^2

Answer. (*a*)

6. What is the best description for the indicated bond in the following compound ?

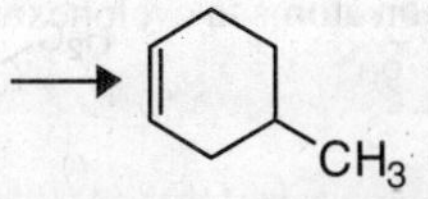

(*a*) $\sigma\ sp^3-sp^3$ (*b*) $\sigma\ sp^3-sp^3$ and $\pi\ 2p-2p$
(*c*) $\sigma\ sp^2-sp^2$ and $\pi\ 2p-2p$ (*d*) $\sigma\ sp-sp$ and $\sigma\ sp-sp$

Answer. (*c*)

7. The greater the orbital overlap on adjacent atoms,

(*a*) the stronger the bond
(*b*) the weaker the bond
(*c*) makes no difference in bond strength
(*d*) orbitals never overlap

Answer. (*a*)

8. Which of the following compounds will give cyclopropane on treatment with sodium in dry ether?

(*a*) 1,3-Dibromopropane (*b*) 1,1-Dibromopropane
(*c*) 1,2-Dibromopropane (*d*) 2,2-Dibromopropane

Answer. (*a*)

9. Which of the following cycloalkanes is most reactive ?

(*a*) Cyclopropane (*b*) Cyclohexane (*c*) Cyclobutane (*d*) Cycloheptane

Answer. (*a*)

10. Which of the following compounds will react most readily with concentrated sulfuric acid ?

(*a*) Ethane (*b*) Cyclohexane (*c*) Propane (*d*) Cyclohexene

Answer. (*d*)

11. A compound of formula C_6H_{12} does not react with concentrated sulfuric acid. The compound could be

(*a*) Alkane (*b*) Cycloalkane (*c*) Alkene (*d*) Cycloalkene

Answer. (*b*)

12. Which of the following molecules will decolorize bromine in carbon tetrachloride most readily ?

(*a*) 1,2-Dimethylcyclopropane (*b*) Cyclopentane
(*c*) 1,2-Dimethylcyclobutane (*d*) Cyclohexane

Answer. (*a*)

13. Cyclopropane reacts with concentrated HBr to give.

(*a*) 1-Bromopropane (*b*) Bromocyclopropane
(*c*) 2-Bromopropane (*d*) 1,2-Dibromopropane

Answer. (*a*)

14. Cyclobutane reacts with hydrogen in the presence of nickel catalyst at 200°C to give

(*a*) Butane (*b*) 1-Butene (*c*) 2-Butene (*d*) None of these

Answer. (*a*)

15. Cyclohexanol can be converted into cyclohexene by heating with

(*a*) Zn(Hg) and HCl (*b*) Concentrated H_2SO_4
(*c*) $SOCl_2$ (*d*) H_2 and Ni

Answer. (*b*)

16. The bond angle between carbon atoms in cyclohexane is

(*a*) 109.5° (*b*) 60° (*c*) 90° (*d*) 120°

Answer. (*a*)

17. Which of the cycloalkanes is not expected to have ring strain ?

(*a*) Cyclopropane (*b*) Cyclobutane (*c*) Cycloheptane (*d*) None of these

Answer. (*c*)

18. The bond angle between carbon atoms in cyclohexane is

(*a*) 60° (*b*) 90° (*c*) 109.5° (*d*) 120°

Answer. (*c*)

19. Which of the following statements is false about cyclohexane ?

(*a*) It is a saturated cyclic hydrocarbon
(*b*) All C–C–C bond angles are 109°28′
(*c*) It is very unstable, strained compound
(*d*) It can exist in two conformations which are designated as the boat-form and the chair-form.

Answer. (*c*)

20. The most stable conformation of cyclohexane is the :

(*a*) Haworth form (*b*) Boat form
(*c*) Newman form (*d*) Chair form

Answer. (*d*)

21. What percentage of cyclohexane molecule is estimated to be in the boat form at any given time ?

(*a*) over 99% (*b*) between 90% and 99%
(*c*) approximately 50% (*d*) less than 1%

Answer. (*d*)

30

Ureides and Purines

UREIDES

Acetyl derivatives of urea are called **Ureides**. The ureides prepared by the action of an acyl chloride or acid anhydride of a monocarboxylic acid are referred to as **Simple Ureides**.

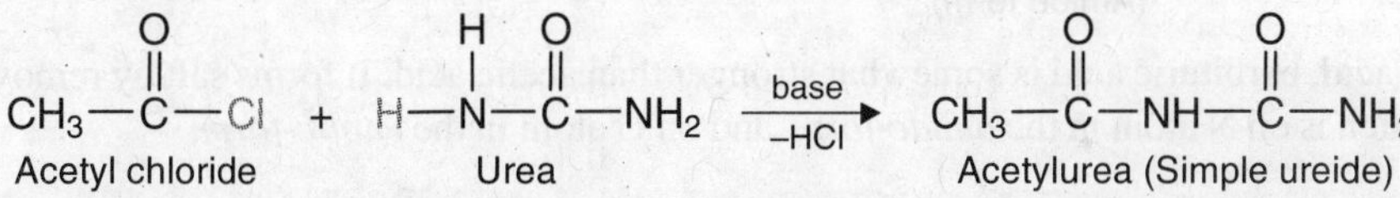

Oxalic acid and malonic acid, or their derivatives, react with the two NH_2 groups of urea to form five and six-membered ring compounds. These are termed **Cyclic Ureides.**

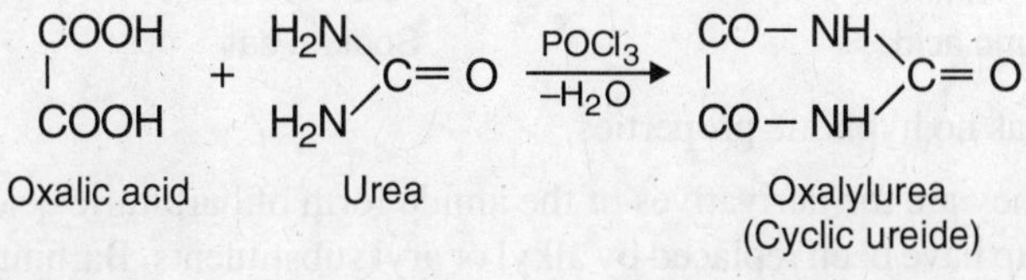

Similarly, diethylmalonate reacts with urea to form a six-membered cyclic ureide, *Malonylurea.*

Ureides are crystalline compounds. Like amides, they are hydrolysed by alkalis to form the parent acid and urea. The cyclic ureides are acidic due to enolization and hence form metallic salts. Many of them are excellent drugs.

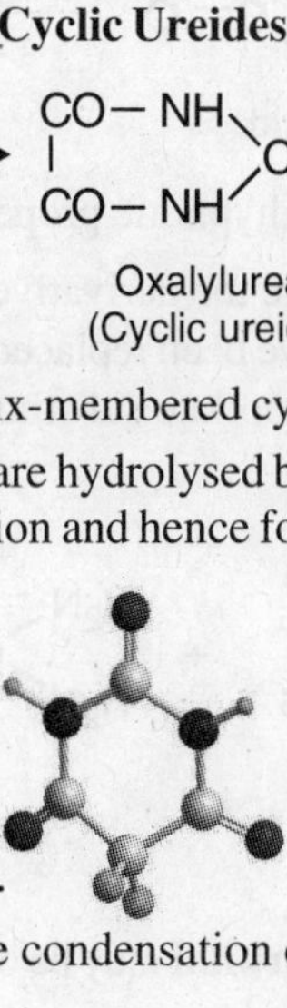

BARBITURIC ACID, *Malonylurea*

It is an important 6-membered cyclic ureide or diimide.

Preparation. Barbituric acid is best prepared by the condensation of dimethyl malonate with urea in the presence of sodium ethoxide.

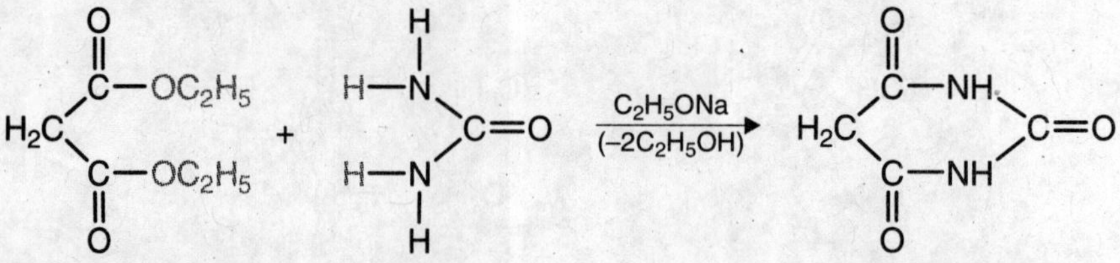

Diethyl malonate Urea Barbituric acid

Properties. Barbituric acid is a colorless crystalline solid, mp 248°C. On warming with alkali solution, it is hydrolysed to malonic acid and urea.

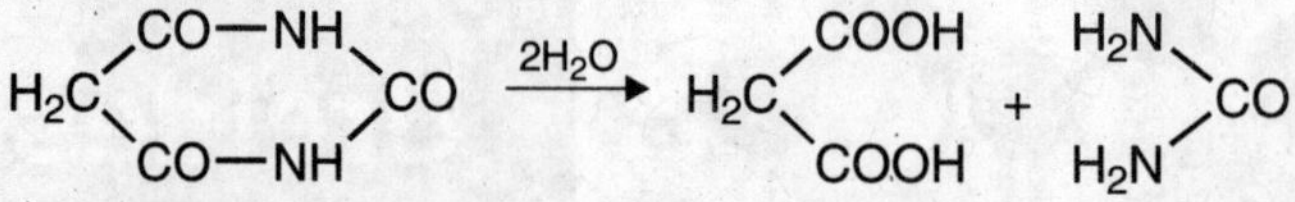

Barbituric acid Malonic acid Urea

Barbituric acid dissolves in water giving strongly acidic solutions. Its acid character is attributed to the presence of the enol form along with the amide form.

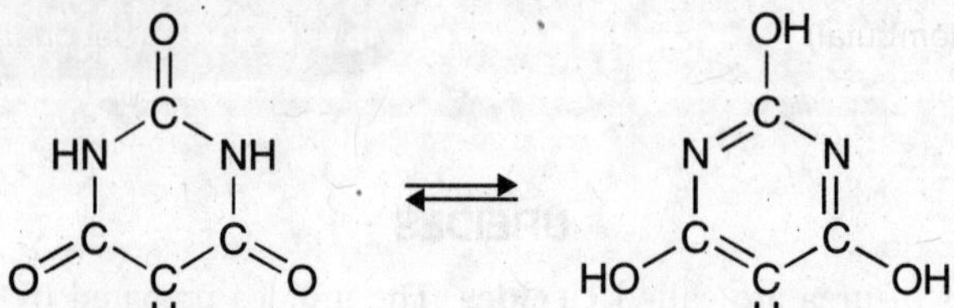

Barbituric Acid (Amide form) (Enol form)

As an acid, barbituric acid is somewhat stronger than acetic acid. It forms salt by removal of the H atom which is on N atom in the *amide-form* and on O atom in the *enolic-form.*

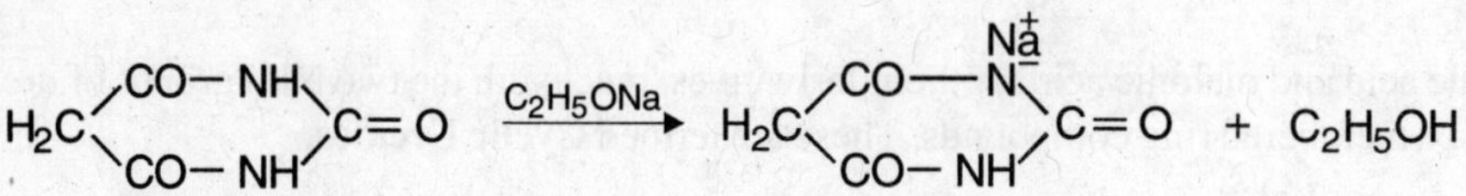

Barbituric acid Sodium salt

Barbituric acid has no hypnotic properties.

Barbiturates. They are the derivatives of the amide form of barbituric acid in which the two H atoms of the CH_2 group have been replaced by alkyl or aryl substituents. Barbiturates are prepared by condensing urea with appropriate substituted malonic esters.

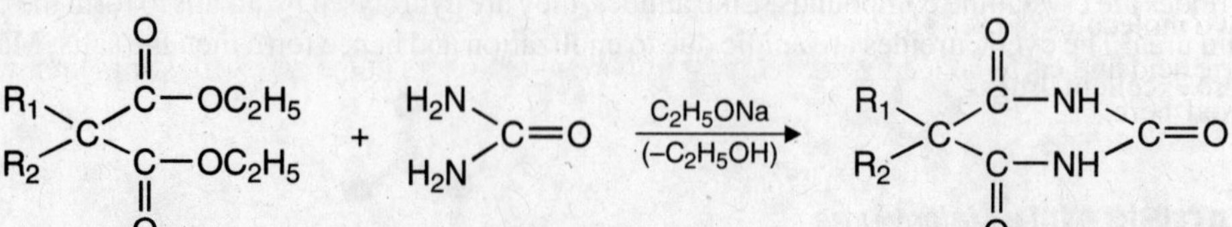

Substituted malonic ester Urea a Barbiturate

For example, **Phenobarbital** may be obtained by condensing urea with ethylphenylmalonic ester. Some of the popular barbiturates are :

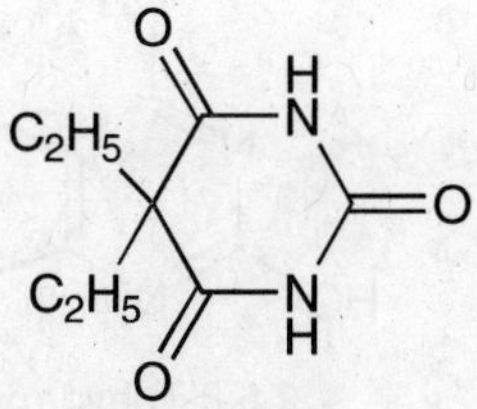

Barbital
(Veronal)

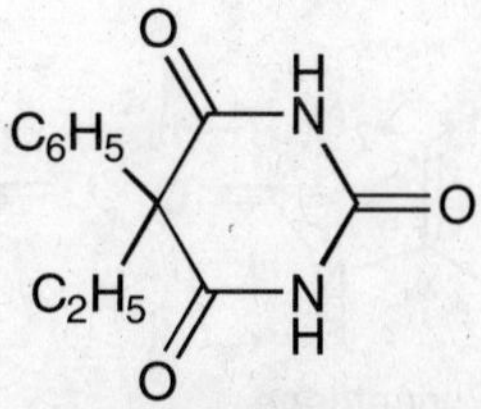

Phenobarbital
(Luminal)

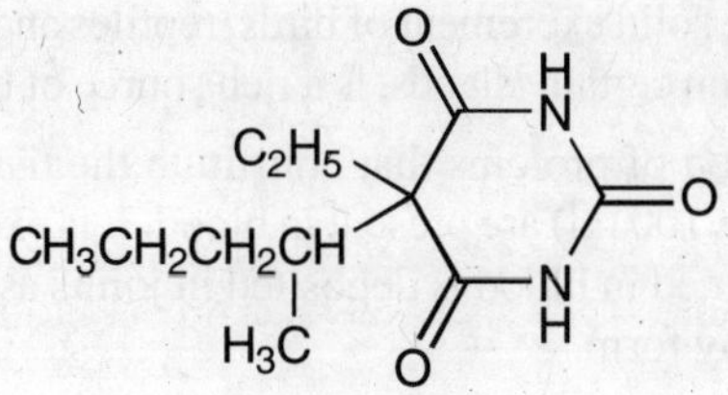

Pentobarbital
(Nembutal)

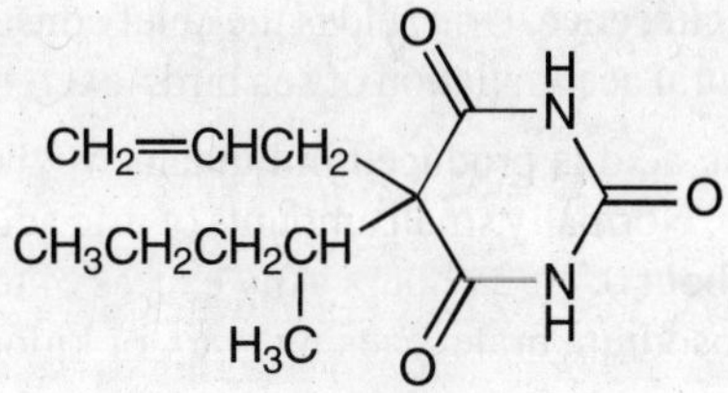

Secobarbital
(Seconal)

Barbiturates have a depressing effect on the central nervous system and as a result produce sleep. They are widely used as sedatives and soporifics. An overdose of barbiturate sleeping pills slows breathing and heart rate so as to cause death. Phenobarbital is used as sedative and to prevent epileptic fits.

PURINES

Purine is a bicyclic compound (mp 216-217°C) having the structure

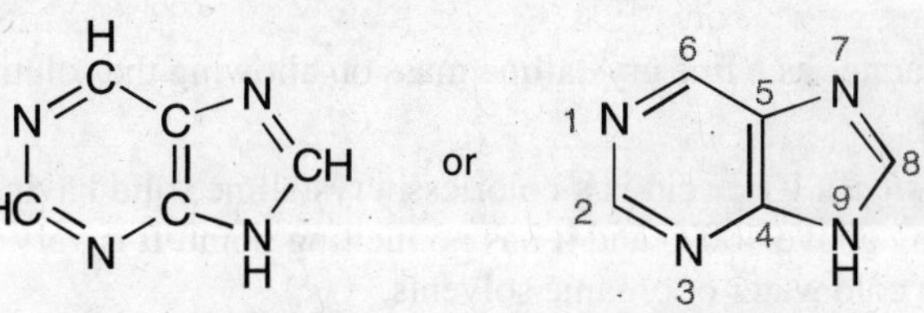

Numbering of purine ring system

It is made of a five and a six-membered heterocyclic ring containing two nitrogen atoms. These are fused together along the carbon atoms 4 and 5.

The class of compounds such as uric acid, caffeine, theobromine, theophylline, and xanthine which could be regarded as derivatives of purine, are designated as **Purines**. At one time, uric acid and other purines were represented by rectangular formulas since they could be thought of as formed from two molecules of urea and a dicarboxylic acid (*Diureides*).

Uric acid and caffeine are by far the most important members of the class of purines and will be discussed here.

URIC ACID, *2,6,8-Purinetrione*

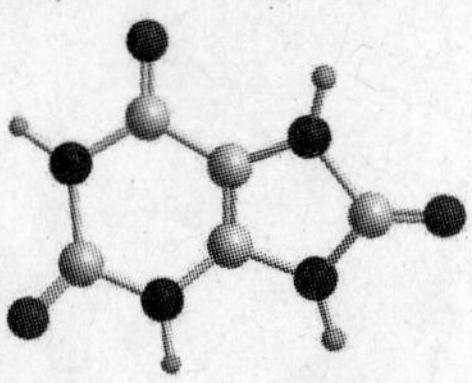

This most important derivative of purine is, in fact, a mixture of two tautomeric forms.

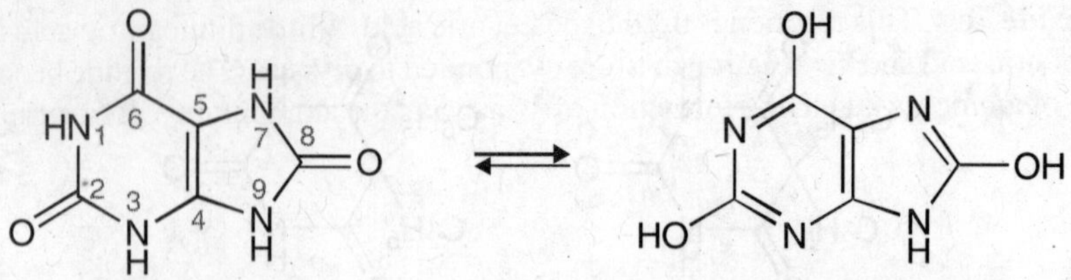

2,6,8-Purinetrione (Keto form) 2,6,8-Trihydroxypurine (Enol form)

Absorption spectra studies have shown that the *keto form* dominates.

Occurrence. Uric acid is the chief constituent of the solid excrement of birds, reptiles and insects. The natural accumulation of sea birds' excrement found on certain islands, is a rich source of uric acid.

Uric acid is produced in humans by the degradation of proteins that constitute the tissues and muscles. Normally small amounts of uric acid (upto 0.3g/100 ml) are present in blood. It is excreted in urine, about 0.7 g/24 hours. Any excess of uric acid formed in blood is deposited in joints as crystals of monosodium urate, causing *gout*, or kidney stone may form.

Isolation from Urine. Human urine contains small amounts of uric acid and its salts (Na or NH_4–urates). It is evaporated to get a concentrated solution and then strong hydrochloric acid is added. Thus the urates are decomposed to form free uric acid which precipitates on cooling. The total solid mass is recrystallized from hot water to give pure uric acid.

Extraction from Guano. Uric acid is produced on a large scale from Guano (the sea birds' excrement) which is largely ammonium urate. This is powdered and boiled with sodium hydroxide solution, until the evolution of ammonia ceases. The hot solution of sodium urate so obtained is filtered and poured into hydrochloric acid.

$$\text{Amm—Urate} \xrightarrow[-NH_3]{NaOH} \text{Na—Urate} \xrightarrow[\text{cool}]{HCl} \text{Uric Acid}$$

The uric acid separates as a fine crystalline mass on allowing the solution to stand in cold. It is filtered and dried in air.

Properties. (*Physical*). Uric acid is a colorless, crystalline solid having no taste or smell. It is decomposed on heating above 400°C and it has no melting point. It is very sparingly soluble in hot water and insoluble in cold water or organic solvents.

(*Chemical*). (1) **Formation of Salts.** Uric acid is a weak dibasic acid. Thus it reacts with sodium, ammonium, or lithium hydroxide to form two series of salts e.g., monosodium and disodium urate. Which of the two acidic OH groups in 2,6,8-trihydroxypurine give the disodium salt, is not yet certain.

While the sodium and ammonium urates are sparingly soluble in water, lithium urate is freely soluble. That is why *'lithiated water'* is often used as a remedy to secure the elimination of uric acid from the system of patients suffering from gout.

(2) **Reaction with $POCl_3$.** Uric acid reacts with phosphoryl chloride to from 2,6,8-trichloropurine which shows the presence of three OH groups in its molecule.

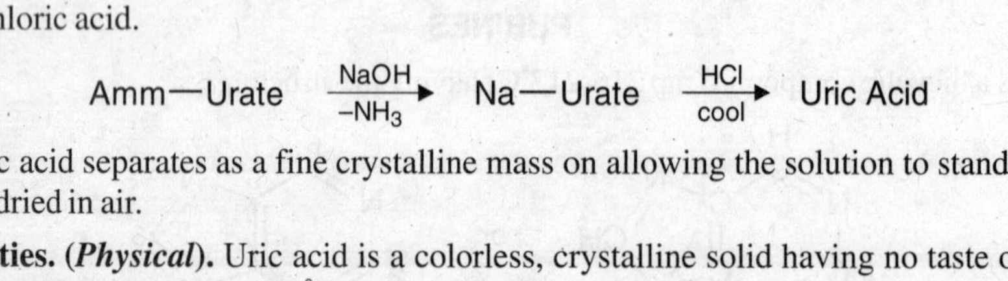

2,6,8-Trihydroxypurine (Uric acid) 2,6,8-Trichloropurine

(3) **Murexide Test.** This reaction is used to detect uric acid. A little dilute nitric acid is added to a few crystals of uric acid and the reaction mixture evaporated to dryness. The residue become purple on the addition of ammonia due to the formation of *Murexide*, the ammonium salt of purpuric acid.

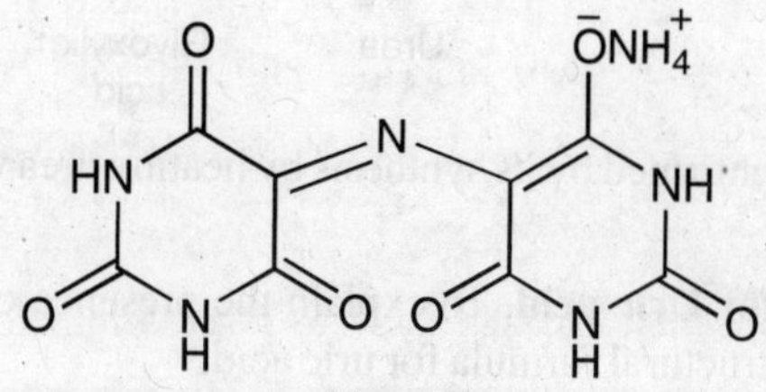

Murexide

STRUCTURE OF URIC ACID

(1) Uric acid has the molecular formula $C_5H_4N_4O_3$.

(2) **Presence of 4 N—H Groups.** Direct methylation of uric acid forms its *tetramethyl derivative* which upon hydrolysis yields four molecules of *methylamine*. This indicates the presence of four imine (—NH—) groups.

$$\underset{\text{Uric acid}}{C_5O_3(N{-}H)_4} \xrightarrow{CH_3I} \underset{\text{Tetramethyluric acid}}{C_5O_3(N{-}CH_3)_4} \xrightarrow{H_2O} \underset{\text{Methylamine}}{4CH_3NH_2}$$

(3) **Presence of Alloxan Moiety**. Uric acid on oxidation with dilute nitric acid gives *alloxan and urea.*

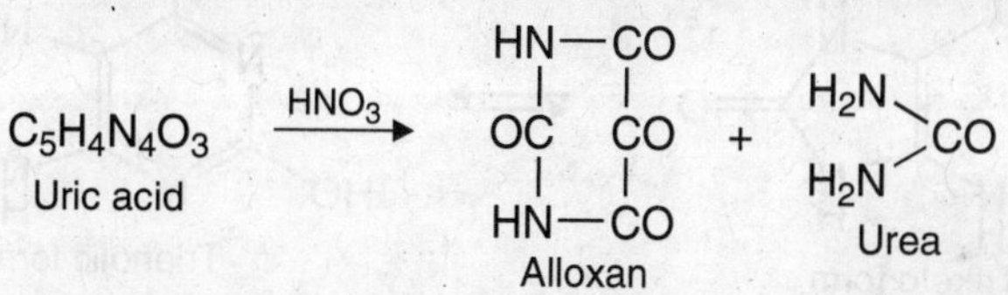

That alloxan is a diureide having the above structure, is shown by its hydrolysis.

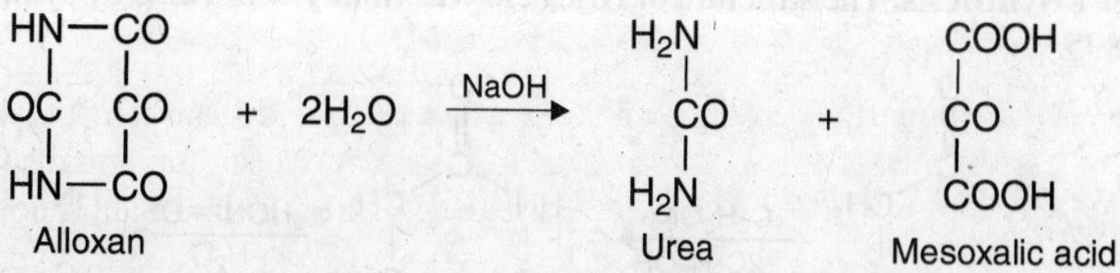

The structure of alloxan was confirmed by its synthesis from urea and mesoxalic acid (reverse of above reaction).

(4) **Presence of Allantoin Unit.** On oxidation with alkaline $KMnO_4$, uric acid forms allantoin.

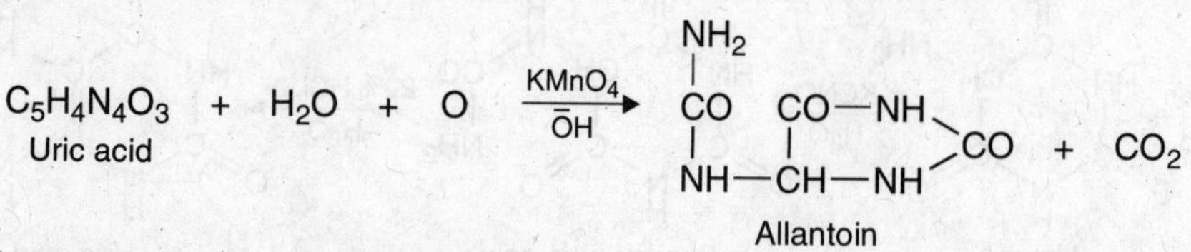

That allantoin is a diureide having the above structure, is proved by its hydrolysis to give two molecules of urea and one of glyoxylic acid.

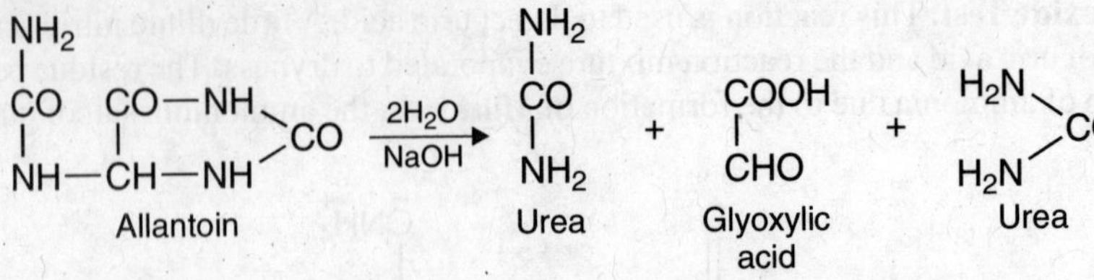

Allantoin structure was confirmed by its synthesis by heating urea with glyoxylic acid (reverse of above reaction).

(5) **Medicus Formula for Uric acid.** To explain the presence of alloxan and allantoin units, Medicus gave the following structural formula for uric acid.

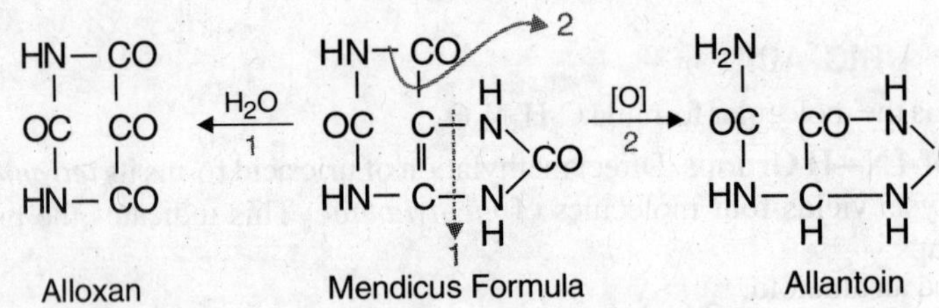

(6) **Tautomeric Mixture.** When reacted with phosphoryl chloride ($POCl_3$), uric acid gives 2,6,8-trichloropurine indicating the presence of three OH groups. Thus uric acid is a tautomeric mixture of the triketo and the trienol forms.

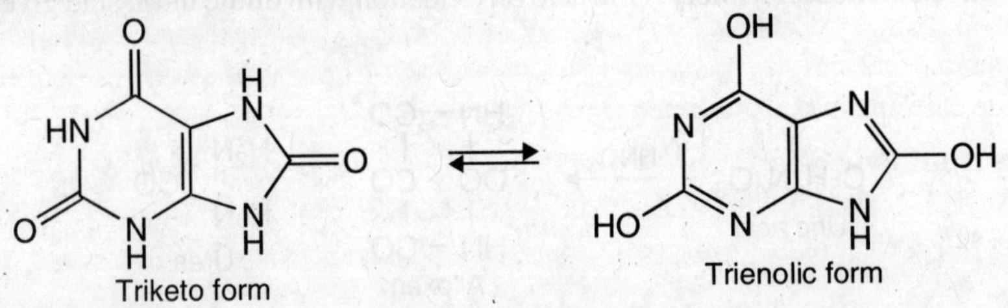

It has been shown by IR spectra that the *keto* form dominates in the equilibrium mixture.

(7) **Fischer's Synthesis.** The structure of Uric acid was finally confirmed by synthesis completed by Fischer in 1895.

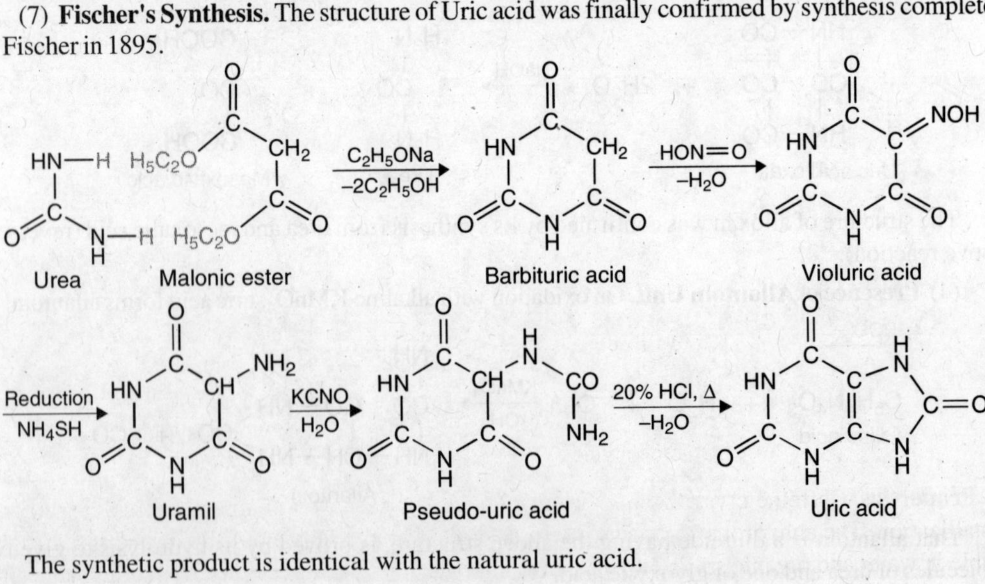

The synthetic product is identical with the natural uric acid.

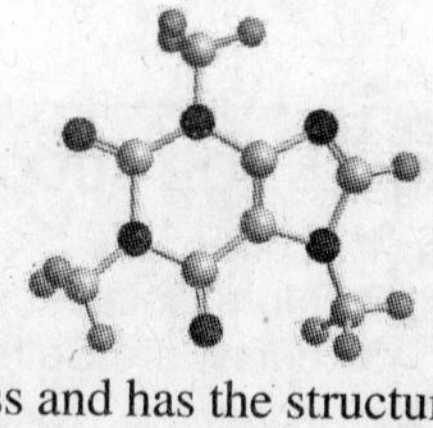

CAFFEINE, *Theine*, $C_8H_{10}O_2N_4$

It is an important member of the class and has the structure

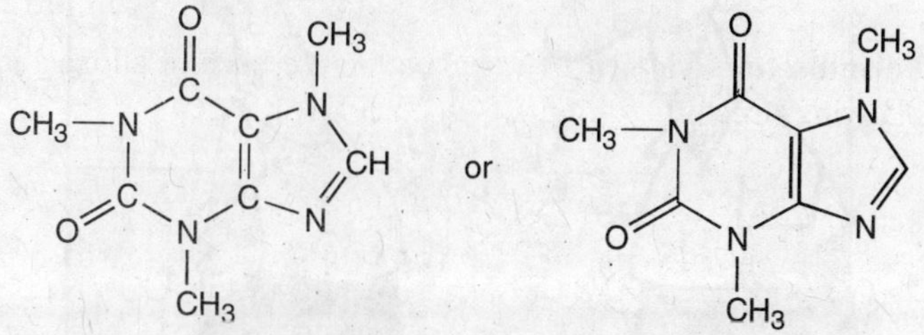

Caffeine occurs in dried tea leaves (up to 5%), coffee beans (up to 2%), and in cola leaves (up to 2.5%). It is a mild stimulant to the central nervous system and a nitrogen heterocyclic compound. Thus, it is now termed an alkaloid.

Isolation from Waste Tea. The damaged leaves and stem of the tea plant are boiled with water and filtered. The filtrate containing caffeine, proteins, and tannins is treated with basic lead acetate. The proteins and tannins get precipitated and are filtered off. To the filtrate is then added dilute sulfuric acid to remove any excess of lead acetate as insoluble lead sulfate. The resulting solution is decolorized with animal charcoal and caffeine extracted from it with chloroform. Chloroform is distilled off and the residue of crude caffeine is recrystallized from water to get the pure product.

Synthesis from Uric acid (Fischer, 1899). Uric acid is treated with methyl iodide in alkaline solution to give 1,3,7-trimethyluric acid. This on heating with phosphoryl chloride ($POCl_3$), forms chlorocaffeine which on reduction with hydrogen iodide yields caffeine.

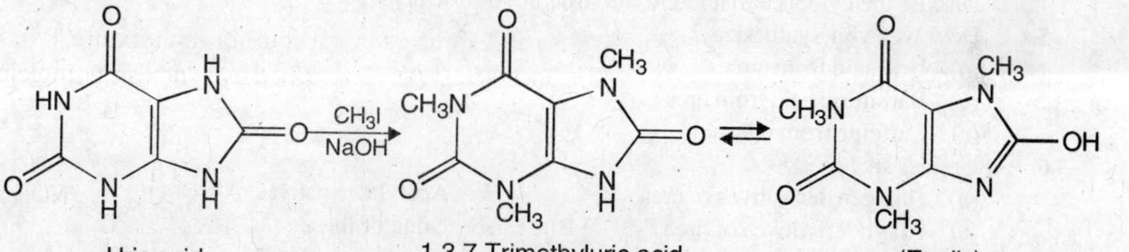

Uric acid 1,3,7-Trimethyluric acid (Enolic)

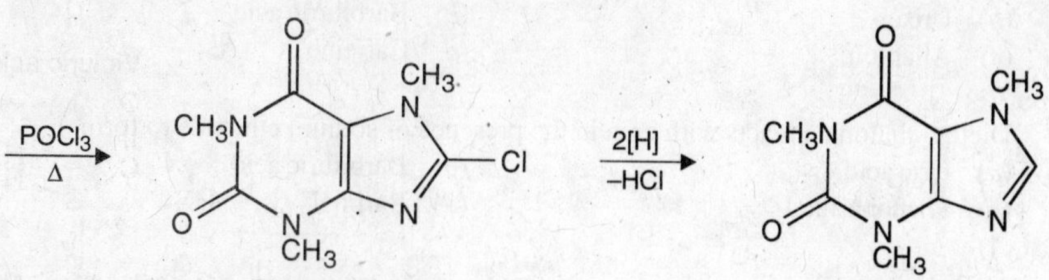

Chlorocaffeine Caffeine

Properties. Caffeine crystallizes in beautiful colorless needles with one molecule of water of crystallization. The anhydrous compound melts at 238°C. It has a feeble bitter taste and is sparingly soluble in water and organic solvents, being insoluble in ether. It is a weak base and forms salts with strong acids.

Can you think of any others?

Caffeine is a mild stimulant to the central nervous system, the heart and the respiratory system. That is why we feel refreshed when we take tea, coffee, or cola soft drinks. Caffeine is used in non-prescription stimulants and sleep-preventing pills. It is a diuretic.

Examination Questions

1. What are ureides ? Give two examples.
2. What are barbiturates ? Give two examples.
3. What are purines ? Give two examples.
4. Discuss the evidence that led to the structure of uric acid.
5. How will you synthesize
   (*a*) Uric acid from urea
   (*b*) Barbituric acid from urea
   (*c*) Caffeine from Uric acid
6. Ureides are
   (*a*) Halogen derivatives of urea (*b*) Acetyl derivatives of urea
   (*c*) Alkyl derivatives of urea (*d*) None of these

Answer. (*b*)

7. Uric acid on oxidation with alkaline $KMnO_4$ forms :
   (*a*) Urea (*b*) Barbituric acid
   (*c*) Allantoin (*d*) Caffeine

Answer. (*c*)

8. Diethyl malonate reacts with urea in the presence of sodium ethoxide to form
   (*a*) Uric acid (*b*) Barbituric acid
   (*c*) Phenobarbital (*d*) Barbital

Answer. (*b*)

31

Carbohydrates

The carbohydrates are an important class of naturally occurring organic compounds. These include glucose (grape sugar), fructose (honey), sucrose (cane sugar), starch (potatoes), and cellulose (wood). They are all composed of C, H and O. In general, carbohydrates can be represented by the formula $C_m(H_2O)_n$. Thus glucose, $C_6H_{12}O_6$, can be written as $C_6(H_2O)_6$. It was a French who gave them the class name **Carbohydrates** (Carbon-Hydrates). Today we know the structure of carbohydrates. They do not contain water molecules as such. The class name carbohydrates is evidently erroneous. But it persists because it is firmly rooted in the chemical nomenclature.

Carbohydrates are often referred to as **Saccharides** (Latin, *Saccharum* = sugar) because of the sweet taste of the simpler members of the class, the *sugars*. Carbohydrates invariably have common names which are characterized by the ending *–ose*.

WHAT ARE CARBOHYDRATES ?

The carbohydrates are polyfunctional compounds. They contain the following functional groups:

(1) Alcoholic hydroxy groups, –OH

(2) Aldehyde group –CHO

(3) Ketone group, >C = O

A precise definition of the term 'Carbohydrate' can be given as : **Polyhydroxyaldehydes or polyhydroxyketones, and large molecules that produce these compounds on hydrolysis**. This may be illustrated by citing the example of the two simplest carbohydrates containing three carbon atoms.

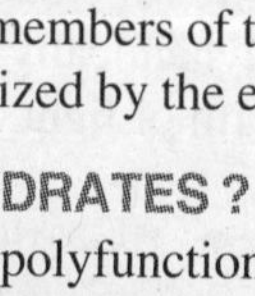

Glyceraldehyde
(*Polyhydroxyaldehyde*)

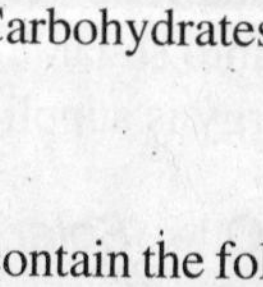

Dihydroxyacetone
(*Polyhydroxyketone*)

The above definition of 'carbohydrate' is not entirely satisfactory. It envisages the presence of free carbonyl group (>C=O) in the simple carbohydrate molecule which is not true.

We know that a carbonyl compound (aldehyde or ketone) reacts with an alcohol to form a hemiacetal.

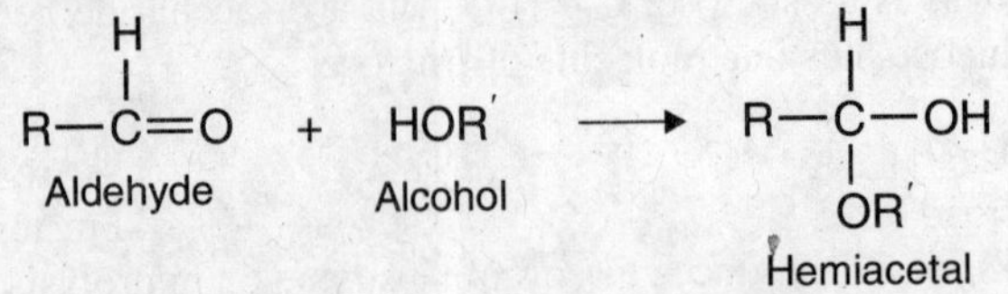

In carbohydrates, we have an aldehyde group which combines with an alcoholic OH of the same molecule to form an internal hemiacetal. We will also notice later that it is by elimination of H_2O between the hemiacetal OH group of two sugar molecules that larger carbohydrate molecules are formed.

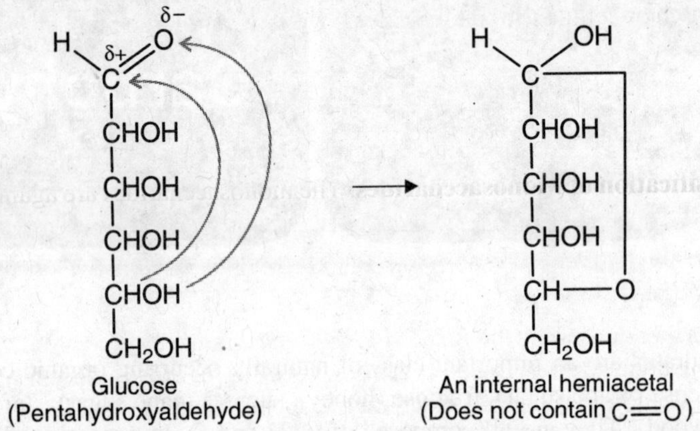

In light of the above, the definition of carbohydrate may be improved as : **A polyhydroxy compound that has an aldehyde or a ketone function present, either free or as hemiacetal or acetal.**

FORMATION OF CARBOHYDRATES *(Photosynthesis)*

The formation of carbohydrates in nature occurs in green plants by a process called *photosynthesis*. Plants contain the green pigment *chlorophyll*, which catalyses the conversion of carbon dioxide and water into sugar. The reaction is thermodynamically unfavorable, but proceeds because the necessary energy is supplied by the sun in the form of light.

$$6CO_2 + 6H_2O + \textit{Solar Energy} \xrightarrow{\textit{Chlorophyll}} \underset{\text{Sucrose}}{C_6(H_2O)_6} + 6O_2\uparrow$$

While plants build up carbohydrates from carbon dioxide and water, animals degrade carbohydrates by eating plants and combine the carbohydrates with oxygen from the air to carry out the reverse of photosynthesis.

CLASSIFICATION OF CARBOHYDRATES

The carbohydrates are divided into three major classes depending on the number of simple sugar units present in their molecule. In other words, the basis of classification of carbohydrates will be the number of simple sugar molecules produced on hydrolysis. The molecules so obtained may be of the same or different sugars.

(1) **Monosaccharides (*Simple Sugars*).** These are single unit carbohydrates (polyhydroxyaldehydes or polyhydroxyketones) that cannot be broken into simpler carbohydrates upon hydrolysis. Glucose and fructose are the examples.

$$\underset{\text{Glucose or Fructose}}{C_6H_{12}O_6} + H_2O \xrightarrow{H^+} \text{No reaction}$$

(2) **Oligosaccharides**. They are made of 2 to 10 units of monosaccharides or simple sugars. The oligosaccharides containing two monosaccharide units are called **Disaccharides**, and those containing three units **Trisaccharides**. Thus sucrose, $C_{12}H_{22}O_{11}$, is a disaccharide because on hydrolysis it gives one molecule of glucose plus one molecule of fructose.

$$\underset{\text{Sucrose}}{C_{12}H_{22}O_{11}} + H_2O \xrightarrow{H^+} \underset{\text{Glucose}}{C_6H_{12}O_6} + \underset{\text{Fructose}}{C_6H_{12}O_6}$$

On the other hand, *raffinose*, produces three simple sugars on hydrolysis and is designated as a trisaccharide.

$$\underset{\text{Raffinose}}{C_{18}H_{32}O_{16}} + 2H_2O \xrightarrow{H^+} \underset{\text{Glucose}}{C_6H_{12}O_6} + \underset{\text{Fructose}}{C_6H_{12}O_6} + \underset{\text{Galactose}}{C_6H_{12}O_6}$$

(3) **Polysaccharides**. They contain more than ten monosaccharide units in the molecule. Thus one molecule of starch or cellulose upon hydrolysis yields a very large number (n) of glucose units.

$$\underset{\text{Starch or Cellulose}}{(C_6H_{10}O_5)_n} + H_2O \xrightarrow{H^+} \underset{\text{Glucose}}{nC_6H_{12}O_6}$$

Further Classification of Monosaccharides. The monosaccharides are again classified on the basis of two factors :

(1) **By the Carbonyl Function.** Those containing the aldehyde function, –CHO, are called **Aldoses**. Others containing the keto group, –CO–, are called **Ketoses**.

(2) **By the Number of Carbon Atoms (3 to 8) in the Molecule.** The monosaccharides containing 3, 4, 5, 6 etc., carbon atoms are designated as **trioses, tetroses, pentoses, hexoses,** and so on.

The subclasses of monosaccharides based on the above factors are listed below :

No. of carbons in monosaccharide	Subclass	
	Aldose	**Ketose**
3	Aldotriose	Ketotriose
4	Aldotetrose	Ketotetrose
5	Aldopentose	Ketopentose
6	Aldohexose	Ketohexose
7	Aldoheptose	Ketoheptose

For example, *erythrose*, a four-carbon sugar with aldehyde function, is a **Aldotetrose**. Fructose, a six-carbon sugar having ketone function, is a **Ketohexose**.

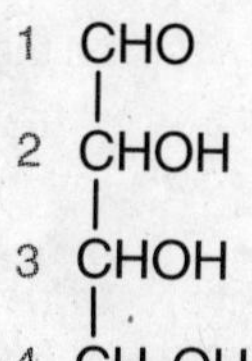

an aldotetrose
(Erythrose)

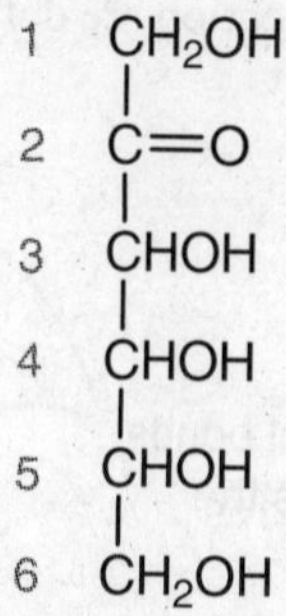

a ketohexose
(Fructose)

Sugars and Non-Sugars. The monosaccharides and oligosaccharides are soluble crystalline substances having a sweet taste. They are collectively known as *Sugars*. Polysaccharides on the other hand, are insoluble amorphous substances and are called *Non-Sugars*.

D AND L DESIGNATIONS

The simplest of all carbohydrates that fit the definition we have given for carbohydrates are the trioses, glyceraldehyde and dihydroxyacetone. Glyceraldehyde is an aldotriose, and dihydroxyacetone is a ketotriose.

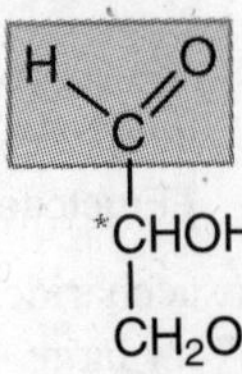

Glyceraldehyde
(An aldotriose)

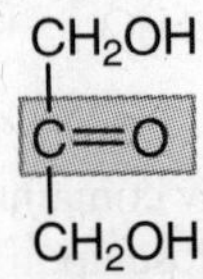

Dihydroxyacetone
(A ketotriose)

Glyceraldehyde contains one asymmetric (chiral) carbon atom (marked by an asterisk) and can thus exist in two optically active forms, called the D-form and the L-form. The following figure contains the ball and stick models of the D-and L-forms of glyceraldehyde and shows the spatial relationship of the groups attached to the asymmetric carbon atom in each case. Clearly, the two forms are mirror images that cannot be superimposed, that is, **they are enantiomers**.

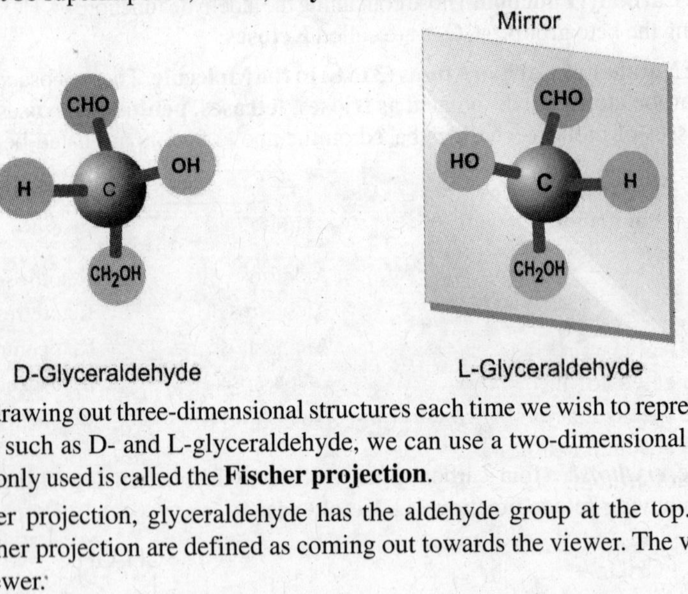

D-Glyceraldehyde L-Glyceraldehyde

Rather than drawing out three-dimensional structures each time we wish to represent an optically active compound such as D- and L-glyceraldehyde, we can use a two-dimensional projection. One projection commonly used is called the **Fischer projection**.

In the Fischer projection, glyceraldehyde has the aldehyde group at the top. The horizontal bonds in the Fischer projection are defined as coming out towards the viewer. The vertical bonds go away from the viewer.

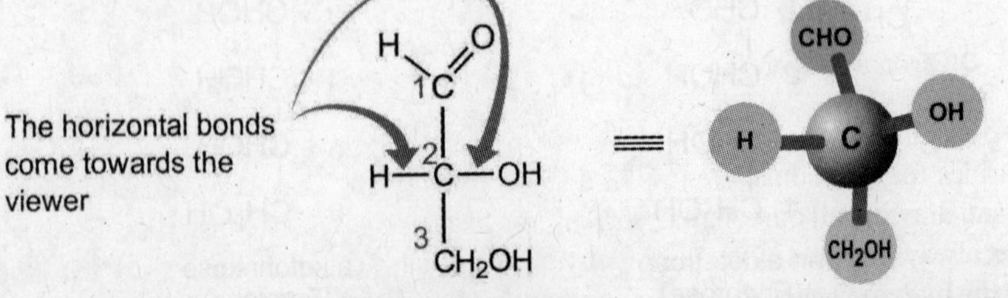

Fischer Projection D-Glyceraldehyde

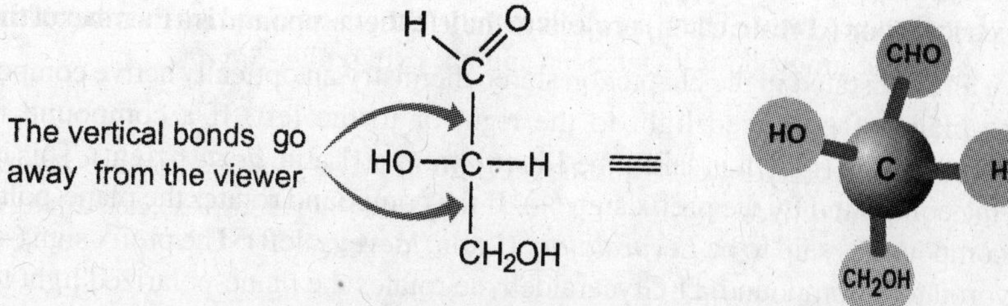

Fischer Projection L-Glyceraldehyde

A simpler way of drawing Fischer projections is to omit the asymmetric carbon atoms. Thus, D-glyceraldehyde may be represented as follows.

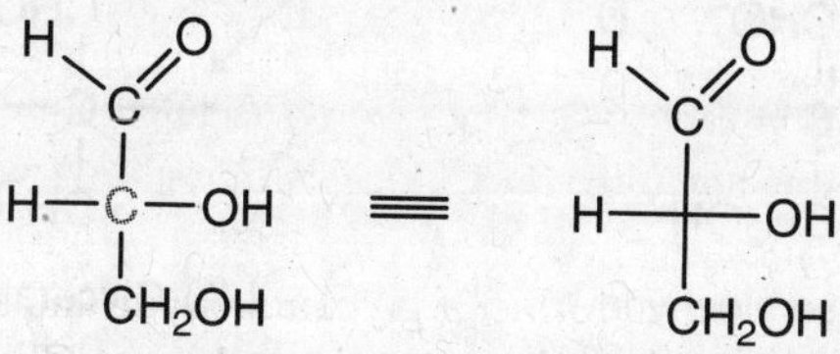

The same rules apply for drawing Fischer projections of more complex molecules. For example, the Fischer projection of an aldotetrose molecule may be drawn as indicated below.

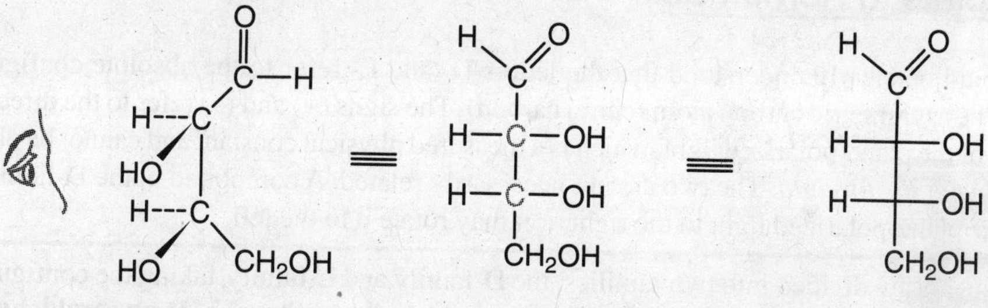

In the name D-glyceraldehyde, the prefix letter D means that the hydroxyl group on the carbon number 2 is projected to the *right* in the Fischer projection. The prefix letter L in the name L-glyceraldehyde means that the hydroxyl group is projected to the *left*. The prefixes D- and L- are used to tell us the absolute configuration, that is, the exact spatial relationship of the groups around the asymmetric carbon atom.

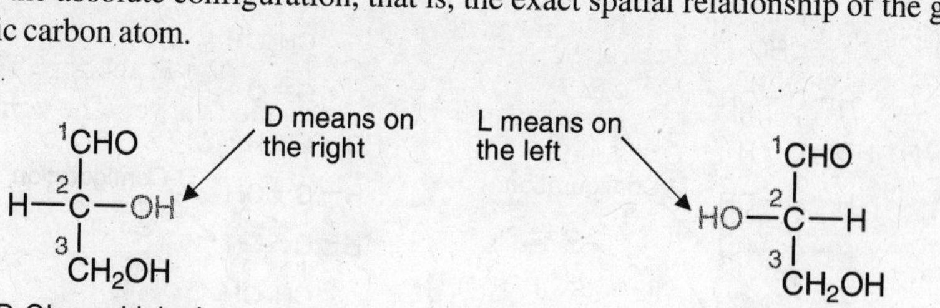

D-Glyceraldehyde L-Glyceraldehyde

The two forms of glyceraldehyde are especially important because the more complex monosaccharides may be considered to be derived from them. They serve as a reference point for designating and drawing all other monosaccharides. In carbohydrate chemistry, the Fischer projection formulas are always written with the aldehyde or ketone groups at the top of the structure. By definition, **if the hydroxyl group on the asymmetric carbon atom farthest from aldehyde or ketone**

group projects to the right, the compound is a member of the D-family. If the hydroxyl group on the farthest asymmetric carbon (chiral carbon) projects to the left, the compound is a member of the L-family.

As we have already stated in the chapter on stereochemistry an optically active compound is one that rotates the plane of polarized light to the right or to the left. If a compound rotates the plane-polarized light to the right, it is said to be *Dextrorotatory* (Latin, *dexter*, right). This is indicated in the name of the compound by the prefix sign (+). If the compound rotates the plane-polarized light to the left, the compound is said to be *Levorotatory* (Latin, *laeveus*, left). The prefix sign (–) is used to designate a levorotatory compound. D-Glyceraldehyde rotates the plane-polarized light to the right. This can be indicated in the name by calling this compound D(+)-glyceraldehyde (pronounced, *dee* plus glyceraldehyde). Similarly, L-glyceraldehyde rotates the plane-polarized light to the left. This can be indicated in the name by calling this compound L(–)-glyceraldehyde (pronounced, *ell* minus glyceraldehyde).

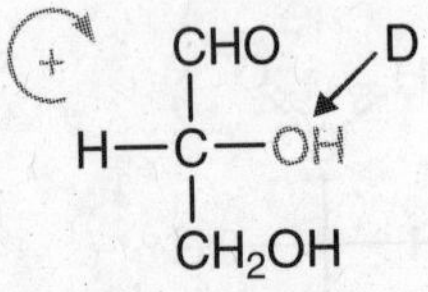

D(+)-Glyceraldehyde
rotates plane-polarized light to the right

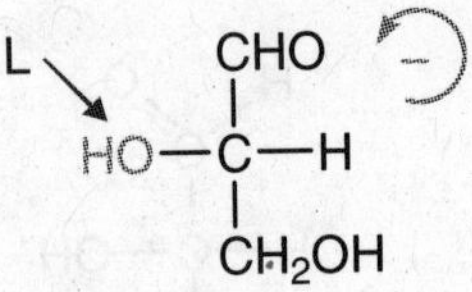

L(–)-Glyceraldehyde
rotates plane-polarized light to the left

Points to Remember

It should be clearly understood that the letters D- and L- refer to the absolute configuration around the asymmetric carbon atom (chiral carbon). The signs (+) and (–) refer to the direction of rotation of the plane-polarized light, which is a measured physical constant and cannot be obtained *by looking at the formula*. The two are not necessarily related. A compound in the D-family may rotate the plane-polarized light to the right or it may rotate it to the left.

The sugars are divided into two families, the D-family and L-family, taking the configuration of glyceraldehyde as standard. **The sugars having the same configuration as of D-glyceraldehyde at the asymmetric carbon most distant from the carbonyl group are designated as D-Sugars. Those with opposite configuration (that of L-glyceraldehyde) are called L-Sugars.** Thus natural glucose is D(+)-glucose and fructose is D(–)-fructose.

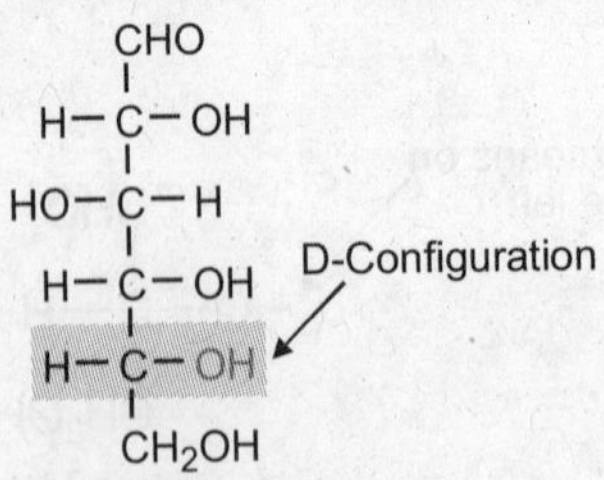

D(+)-Glucose

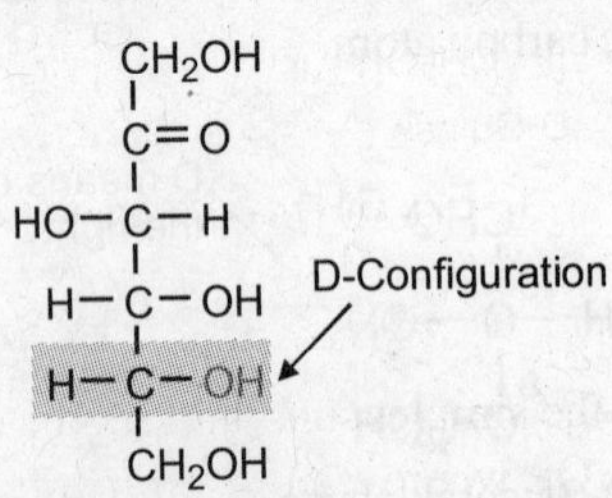

D(–)-Fructose

If there are *n* asymmetric carbon atoms in a molecule, the number of optical isomers is 2^n. Thus we have 2 aldotrioses, 4 aldotetroses, 8 aldopentoses, and 16 aldohexoses. Half of these compounds belong to the L-family. Although the naturally occurring sugars generally belong to the D-family, there are an equal number of compounds that have the L-configuration. The configurations of the D-sugars up to hexoses as determined by experiment and reasoning are given in Fig. 31.1.

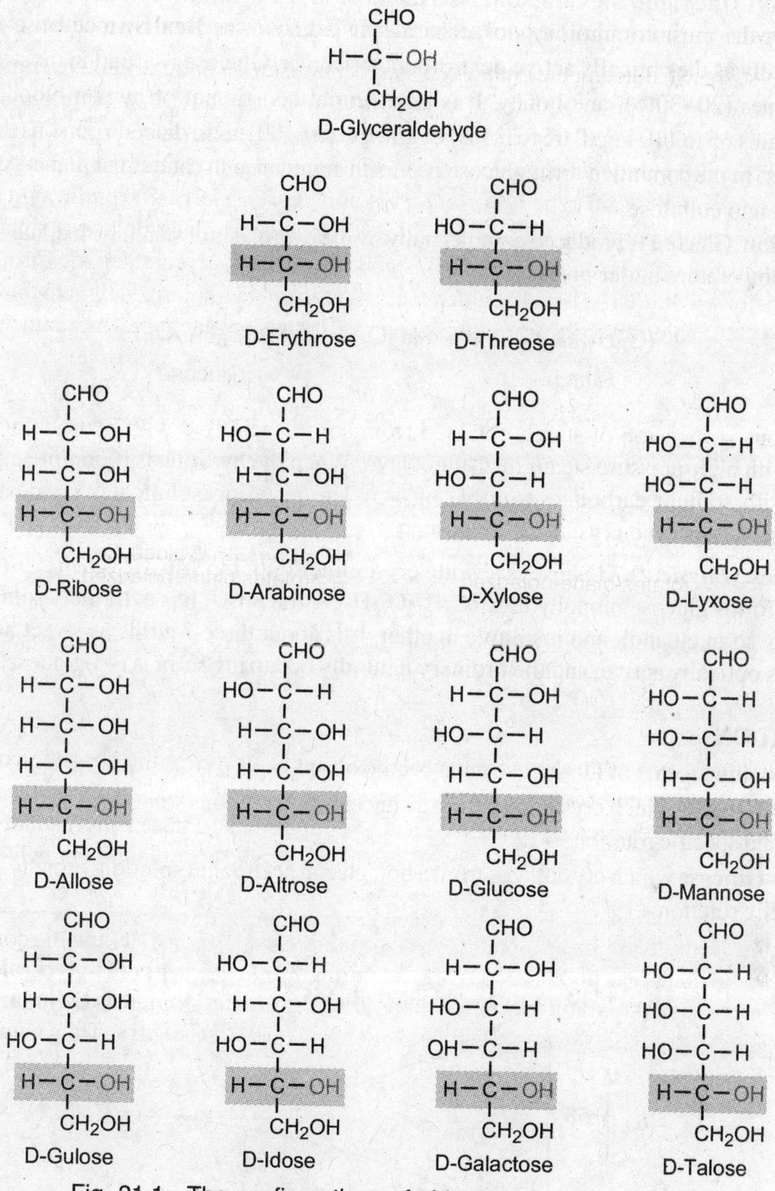

Fig. 31.1. The configurations of aldoses up to six carbons.

MONOSACCHARIDES

They are the simplest one-unit sugars (*mono* = one; *saccharide* = sugar). They have the general formula $C_nH_{2n}O_n$, where *n* varies from three to eight. Thus we have,

$C_3H_6O_3$	Trioses	$C_6H_{12}O_6$	Hexoses
$C_4H_8O_4$	Tetroses	$C_7H_{14}O_7$	Heptoses
$C_5H_{10}O_5$	Pentoses	$C_8H_{16}O_8$	Octoses

The most important naturally occurring monosaccharides are *pentoses* and *hexoses*. Of these glucose and fructose are undoubtedly the most important and typical.

GLUCOSE, *Dextrose*, $C_6H_{12}O_6$

Glucose is the most common monosaccharide. It is known as **Dextrose** because it occurs in nature principally as the optically active dextrorotatory isomer. Glucose is found in most sweet fruits, especially grapes (20–30%), and honey. It is an essential constituent of human blood. The blood normally contains 65 to 110 mg (0.06 to 0.1%) of glucose per 100 ml. In diabetic persons the level may be much higher. In the combined form glucose occurs in abundance in cane sugar and polysaccharides such as starch and cellulose.

Preparation. Glucose is produced commercially by the hydrolysis of starch with dilute hydrochloric acid at high temperature under pressure.

$$\underset{\text{Starch}}{(C_6H_{10}O_5)_n} + nH_2O \xrightarrow{HCl} \underset{\text{Glucose}}{nC_6H_{12}O_6}$$

An aqueous suspension of starch obtained from corn is acidified with hydrochloric acid. It is then heated with high-pressure steam in an autoclave. When the hydrolysis is complete, the liquid is neutralized with sodium carbonate to a pH of 4-5. The resulting solution is concentrated under reduced pressure to get the crystals of glucose.

Properties. (*Physical*). Glucose is a white crystalline solid, mp 146°C. When crystallized from cold water, it forms glucose monohydrate ($C_6H_{12}O_6.H_2O$), mp 86°C. It is extremely soluble in water, only sparingly so in ethanol, and insoluble in ether. It is about three-fourths as sweet as cane-sugar (sucrose). It is optically active, and the ordinary naturally occurring form is (+)-glucose.

MUTAROTATION

Two crystalline forms of D-glucose have been isolated.

(1) **α-D-Glucose** which crystallizes from a concentrated aqueous solution at 30°C. It has melting point 146°C and specific rotation +112°.

(2) **β-D-Glucose** which crystallizes from a hot, glacial acetic acid solution. It melts at 148-150°C and has specific rotation +19°.

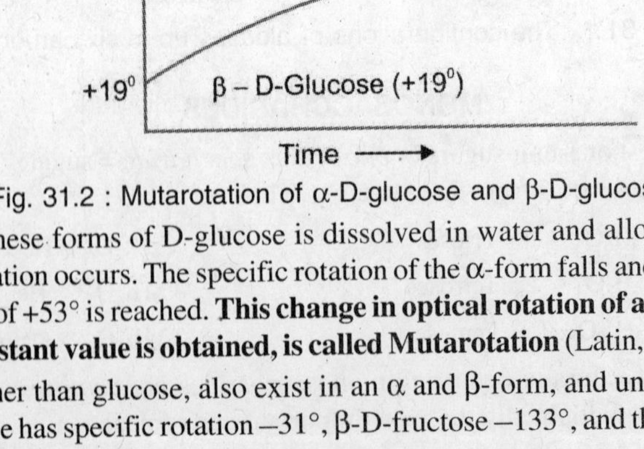

Fig. 31.2 : Mutarotation of α-D-glucose and β-D-glucose.

When either of these forms of D-glucose is dissolved in water and allowed to stand, a gradual change in specific rotation occurs. The specific rotation of the α-form falls and that of the β-form rises until a constant value of +53° is reached. **This change in optical rotation of a solution of either form of glucose until a constant value is obtained, is called Mutarotation** (Latin, *muto* = to change).

Many sugars, other than glucose, also exist in an α and β-form, and undergo mutarotation. For example, α-D-fructose has specific rotation –31°, β-D-fructose –133°, and the constant value –92°.

EXPLANATION. Glucose forms a stable cyclic hemiacetal between the CHO group and the OH group on the fifth carbon. In this process the first carbon becomes asymmetric, giving two isomers (I and II) which differ only in the configuration of the asymmetric carbon.

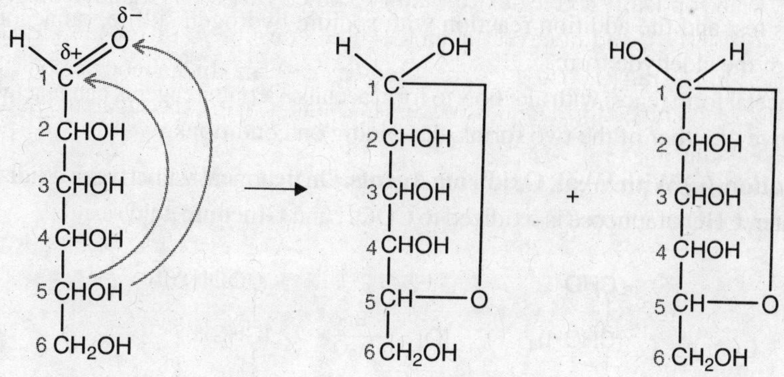

I — II

D-Glucose — α-D-Glucose — β-D-Glucose

These are called **Anomers** and the new asymmetric carbon is referred to as **Anomeric Carbon**. The anomer I with OH to the right is designated as the α-D-glucose and the other with OH to the left β-D-glucose.

The mutarotation occurs because of the slow interconversion of α-D-glucose and β-D-glucose *via* the open-chain form until equilibrium is established giving a constant specific rotation +53°.

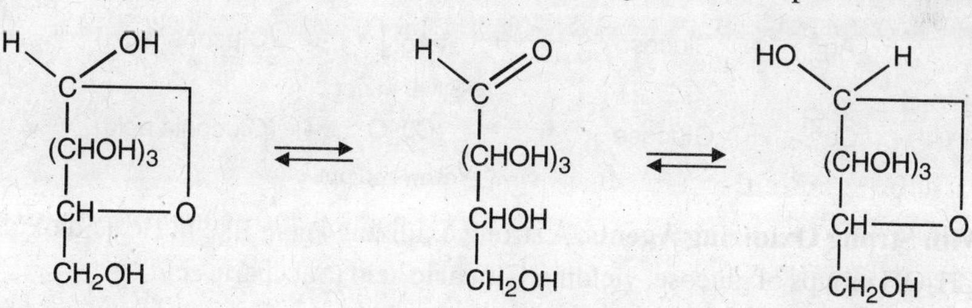

I
α-D-Glucose
specific rotation + 112°

D-Glucose
(Open-chain form)

II
β-D-Glucose
specific rotation + 19°

The concentrations of the open-chain form and the cyclic forms I and II at equilibrium are 0.01%, 36% and 64%, respectively.

(*Chemical*). We have seen that D-glucose is an equilibrium mixture of a straight chain form and a cyclic hemiacetal form.

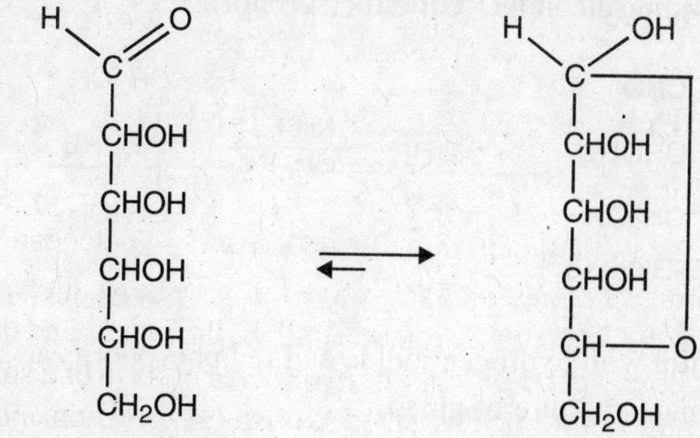

Open-chain form
(0.01%)

Cyclic hemiacetal form
(99.99%)

Even though it is mainly the hemiacetal form, ordinary D-glucose gives the characteristic reactions of simple aldehydes. A trace of open-chain form enters into reaction with the reagent and then the equilibrium shifts to the left so that the entire sample behaves as aldehyde. However, D-glucose fails to give Schiff's test and the addition reaction with sodium hydrogen sulfite, on account of the low concentration of the aldehyde form.

The hemiacetal form reacts with alcohols to form acetals. Certain reagents can react with D-glucose to give derivative of either of the two forms, depending on conditions.

(1) **Oxidation.** (*a*) **With Weak Oxidizing Agents.** On treatment with a weak oxidizing agent such as bromine water, CHO of glucose is oxidized to COOH and **Gluconic acid** results.

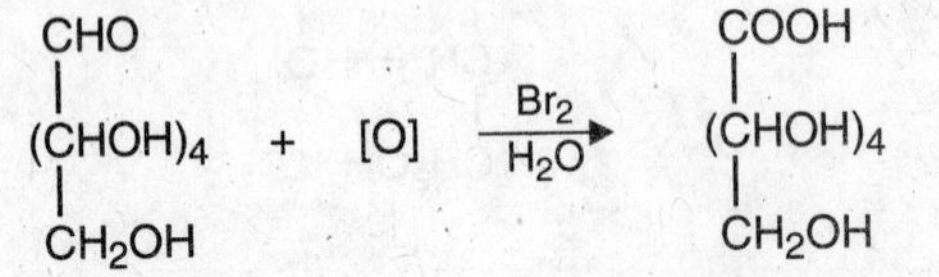

D-Glucose — Gluconic acid

For the same reason, glucose reduces *Tollens' Reagent*, Ag^+ (ammonia complex), and *Fehling's solution*, Cu^{2+} (tartarate complex), to give metallic silver and cuprous oxide, Cu_2O, respectively. These reactions are used as tests for the presence of glucose.

$$Ag^+ + \text{Glucose} \longrightarrow \underset{\text{Silver mirror}}{Ag\downarrow} + \text{(Gluconic acid)}$$

$$Cu^{2+} + \text{Glucose} \longrightarrow \underset{\text{Yellow-red ppt}}{Cu_2O} + \text{(Gluconic acid)}$$

(*b*) **With Strong Oxidizing Agents.** A strong oxidizing agent like nitric acid oxidizes both the CHO and CH_2OH groups of glucose, yielding **Glucaric acid** (Saccharic acid).

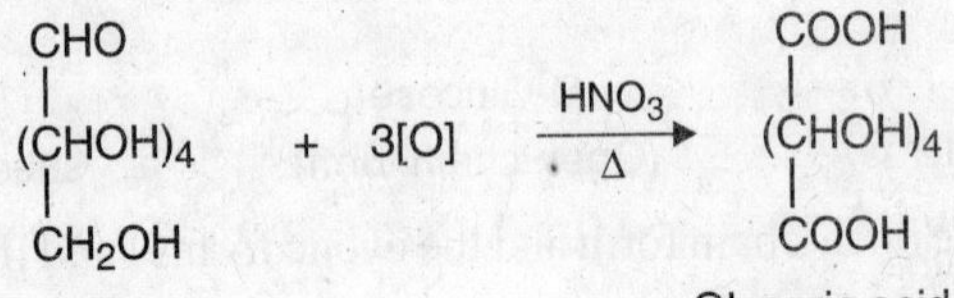

D-Glucose — Glucaric acid (Saccharic acid)

(2) **Reduction**. Glucose on reduction with sodium borohydride, $NaBH_4$, or catalytic reduction (H_2/Pd), gives the corresponding alcohol D-**Glucitol** (sorbitol).

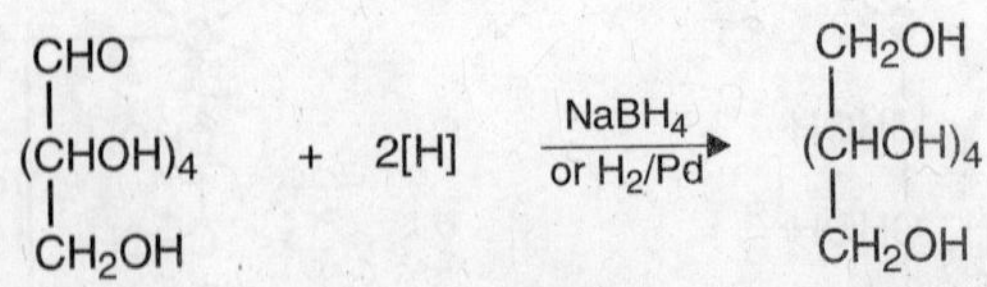

D-Glucose — D-Glucitol (Sorbitol)

The reduction of glucitol with hydrogen iodide and red phosphorus at 100°C, gives *n*-hexane. This proves the straight-chain structure of glucose.

(3) **Acetylation**. Glucose reacts with acetic anhydride in the presence of anhydrous zinc chloride, to form **Pentaacetyl glucose**.

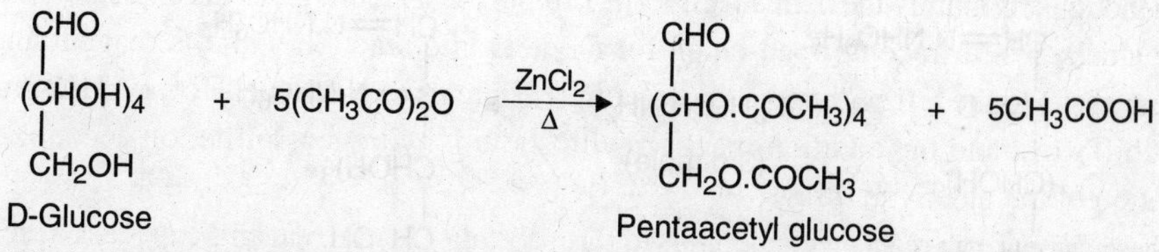

This gives a clue to the presence of five OH groups in glucose molecule.

(4) **Addition of HCN**. Like other aldehydes, hydrogen cyanide reacts by addition at the aldehydic carbonyl group. Since this creates a new asymmetric carbon, two isomeric cyanohydrins differing only in configuration at the aldehydic carbon are obtained.

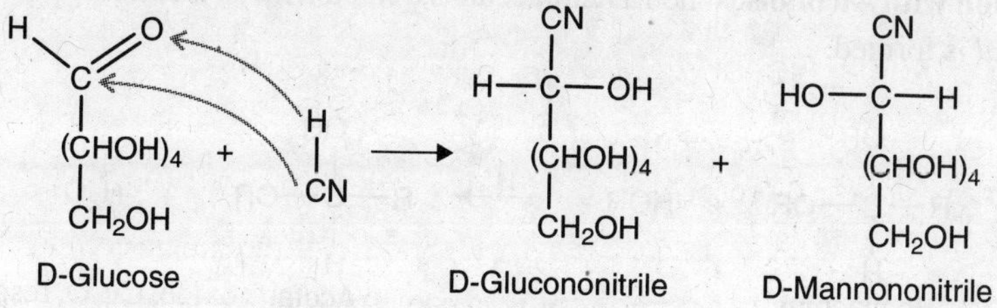

This reaction is important while synthesizing a higher aldose from a lower member.

(5) **Reaction with Hydroxylamine**. Glucose condenses with hydroxylamine, NH_2OH, to form **Glucose oxime**.

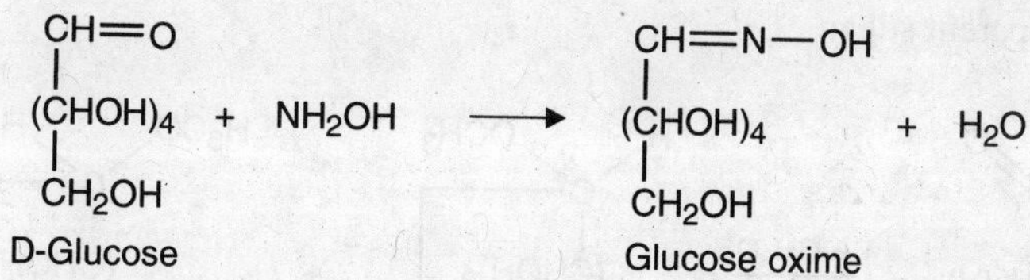

(6) **Reaction with Phenylhydrazine**. When warmed with excess phenylhydrazine, glucose first forms phenylhydrazone by condensation with CHO group. The reaction does not stop at this stage. The adjacent CHOH group is then oxidized by a second molecule of phenylhydrazine which itself is reduced to aniline and ammonia. The resulting carbonyl group reacts with a third molecule of the reagent to yield the final product **Glucosazone**.

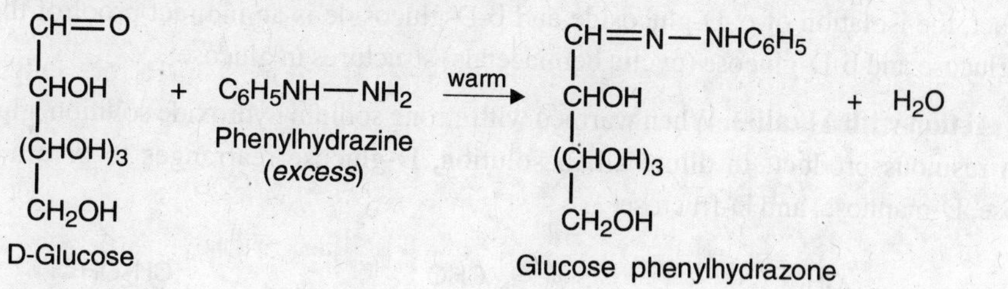

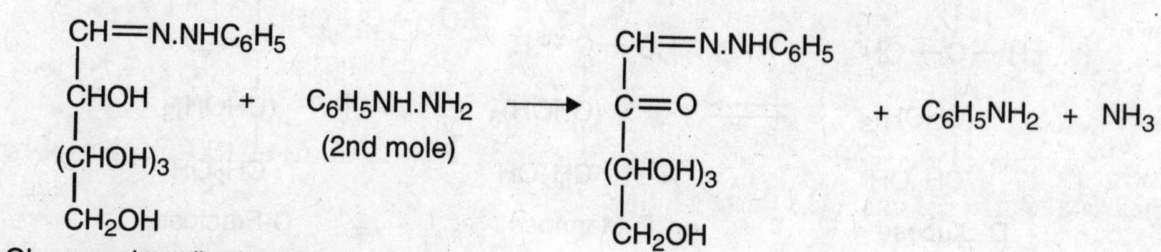

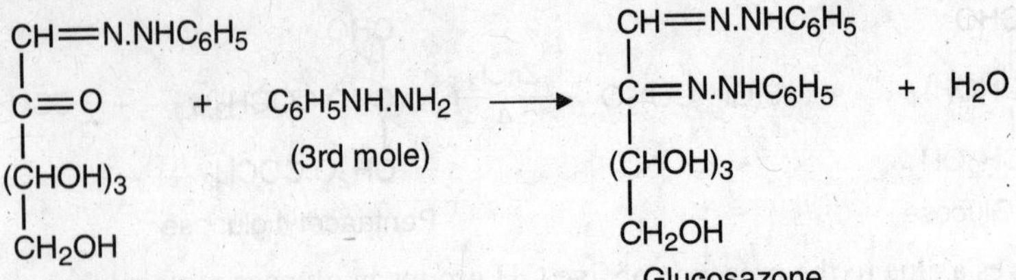

Glucose phenylhydrazone Glucosazone

Osazones are well-defined crystalline substances formed by aldoses and ketoses. They are often used for the identification of monosaccharides.

(7) **Reaction with Alcohols**. When a hemiacetal is treated with an alcohol in presence of a trace of acid, an *acetal* is formed.

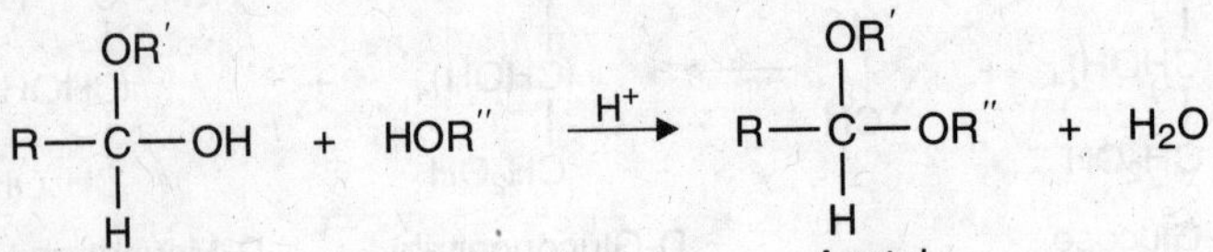

Hemiacetal Acetal

We have seen previously that glucose is an equilibrium mixture of two hemiacetals (α-glucose and β-glucose). Thus when glucose is treated with methanol in presence of dry HCl, these hemiacetal structures are converted to the corresponding acetal known as **Glycosides**. The individual glycosides are named after the parent sugar.

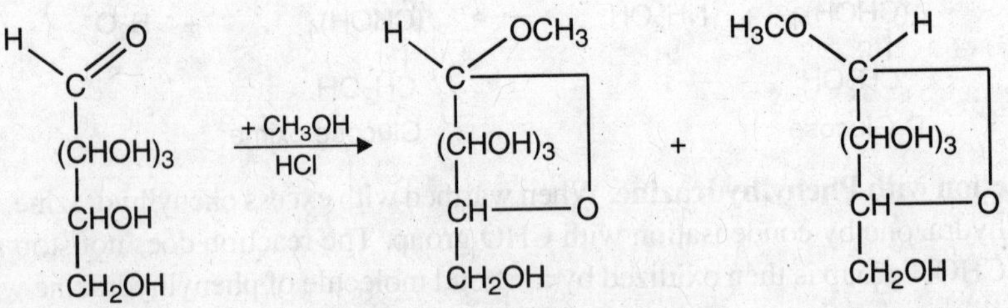

D-Glucose α-D-Glucoside β-D-Glucoside

In fact, the isolation of α-D-glucoside and β-D-glucoside is an indirect proof of the existence of α-D-glucose and β-D-glucose (cyclic hemiacetals) structures in glucose.

(8) **Action with Alkalies.** When warmed with strong sodium hydroxide solution, glucose forms a brown resinous product. In dilute alkali solution, D-glucose rearranges to give a mixture of D-glucose, D-mannose, and D-fructose.

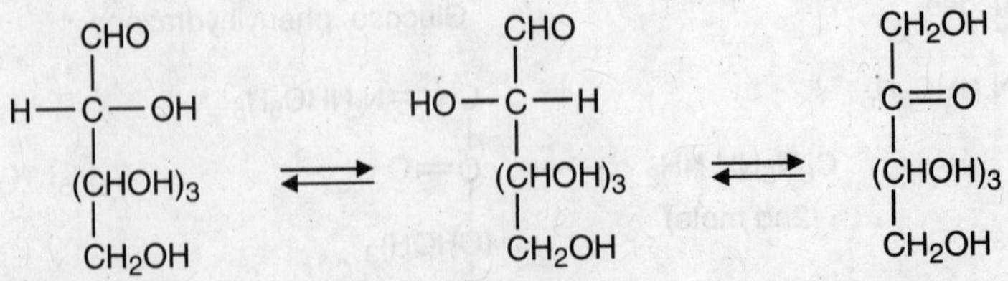

D-Glucose D-Mannose D-Fructose

The above equilibrium is established *via* the enediol starting from any of these three hexoses.

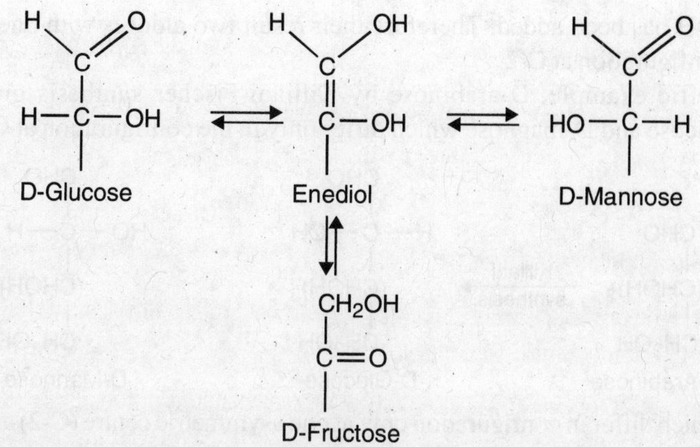

That is why D-fructose, although it has a ketonic C=O group, reduces Fehling's solution or Tollens' reagent. The rearrangement reaction of a monosaccharide in weakly alkaline solutions to give a mixture of isomeric sugars, is named as **Lobry de Bruyn Van Ekenstein rearrangement**.

(9) **Fermentation**. A solution of D-glucose is readily fermented by the enzyme *zymase* present in yeast, in the absence of air, to form **ethanol** and carbon dioxide.

$$\underset{\text{Glucose}}{C_6H_{12}O_6} \xrightarrow{\text{Zymase}} \underset{\text{Ethanol}}{2C_2H_5OH} + 2CO_2\uparrow$$

This reaction called alcoholic fermentation is the basis of manufacture of wines and alcohol.

(10) **Dehydration**. When heated strongly, or when treated with warm concentrated sulfuric acid, glucose is dehydrated to give a black mass of **Sugar Carbon**.

Uses. Glucose is used : (1) as a sweetening agent in syrups and confectionery ; (2) as food for infants ; (3) as a reducing agent in silvering of mirrors and to convert *indigo blue* to *indigo white* in vat dyeing ; and (4) as a raw material for wine and alcohol manufacture.

SYNTHESIS OF MONOSACCHARIDES

A number of interesting methods have been devised for the synthesis and interconversions of monosaccharides. They have proved to be very valuable not only in deducing the configuration of glucose and its isomers, but also in synthesizing and characterizing many new sugars.

A. Chain Lengthening of Aldoses ; Killiani-Fischer Synthesis

The aldose chains may be lengthened by one carbon atom by a procedure known as the **Killiani-Fischer Synthesis**. Thus an aldose can be converted to the next higher member by the following steps: (1) Formation of cyanohydrin ; (2) Hydrolysis of –CN to –COOH, giving aldonic acid; (3) Conversion of aldonic acid to lactone by heating ; (4) Reduction of lactone with sodium borohydride, $NaBH_4$, to get the higher aldose. For illustration, the overall change is the creation of an asymmetric centre at C-2

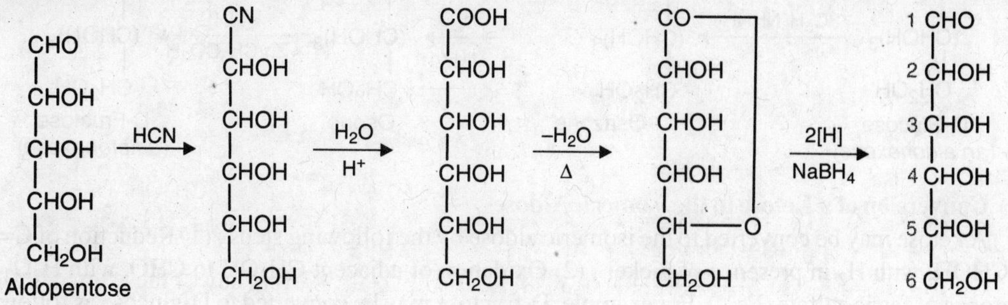

where a new CHOH has been added. Therefore their result two aldoses with one carbon more and differing only in configuration at C-2.

Taking a specific example, D-arabinose by Killiani-Fischer synthesis gives two isomeric aldohexoses, D-glucose and D-mannose which differ only in the configuration at C-2.

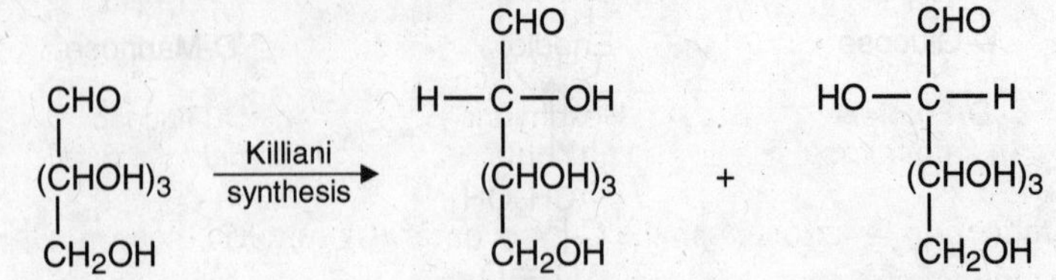

D-Arabinose D-Glucose D-Mannose

Such sugars which differ in configuration only at one asymmetric centre (C-2) are called **Epimers**.

B. Chain Shortening of Aldoses ; Ruff Degradation

Aldose chain may be shortened by one carbon from the aldehyde end by the reaction sequence known as **Ruff Degradation**. Thus an aldose can be converted to the next lower aldose in two steps; (1) Oxidation of the aldose to aldonic acid using bromine water ; (2) Oxidative decarboxylation of aldonic acid by treating the calcium salt with hydrogen peroxide in the presence of ferric sulfate catalyst. For illustration, an aldohexose can be stepped down to aldopentose as follows :

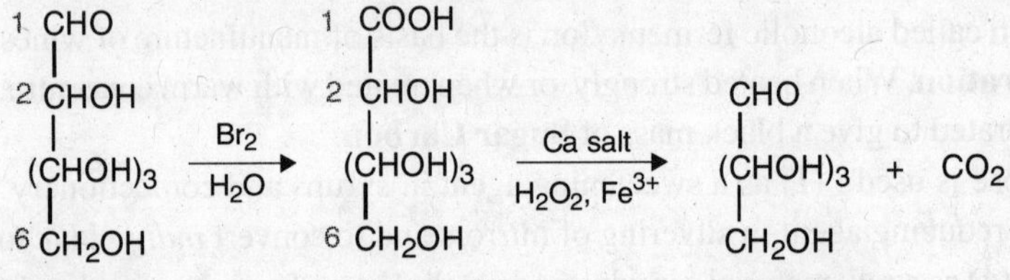

Aldohexose (D-Glucose) Aldonic acid Aldopentose (D-Arabinose)

Since asymmetry is lost at C-2, two aldohexoses that differ only in configuration at this carbon yield the same aldopentose.

C. Conversion of an Aldose to the Isomeric Ketose

An aldose may be converted to the isomeric ketose by the following steps : (1) Formation of *Osazone* by treating with excess phenylhydrazine ; (2) Hydrolysis of osazone to *Osone* with dilute HCl ; (3) Reduction of osone to ketose with zinc and acetic acid, when –CHO is reduced in preference to >C=O. For example, D-glucose may be converted to D-fructose as follows :

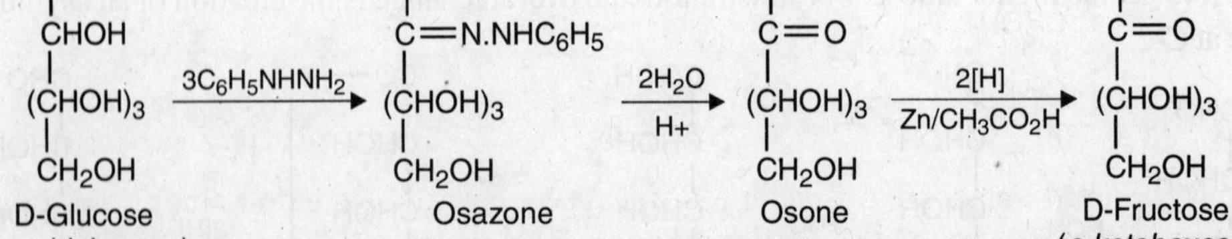

D-Glucose (*an aldohexose*) Osazone Osone D-Fructose (*a ketohexose*)

D. Conversion of a Ketose to the Isomeric Aldose

A ketose may be converted to the isomeric aldose by the following steps : (1) Reduction of C=O to CHOH, with H_2 in presence of nickel ; (2) Oxidation of adjacent CH_2OH to CHO, with H_2O_2 in presence of ferric sulfate (Fe^{3+}). For example, D-fructose may be converted to D-glucose as follows :

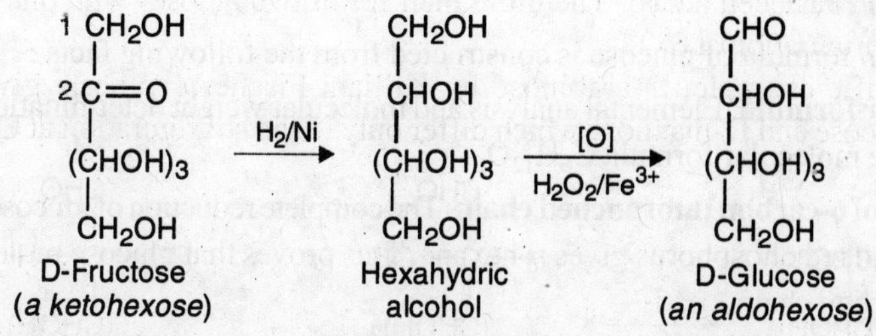

Because on reduction of ketohexose the C-2 becomes asymmetric theoretically two polyhydric alcohols and two aldohexoses may be formed.

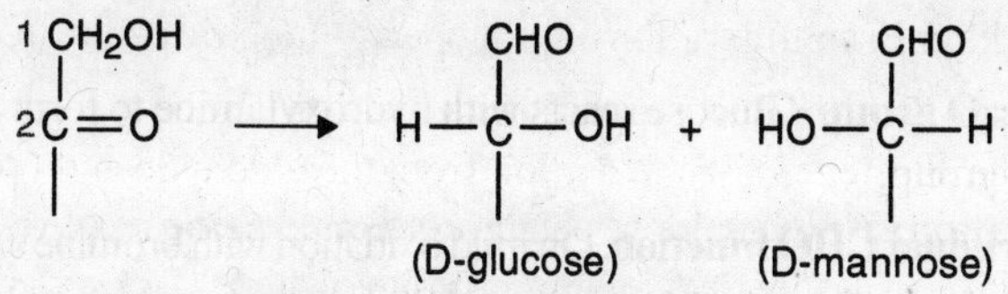

But in actual practice, one *i.e.*, D-glucose predominates.

E Conversion of an Aldose to its Epimer ; Epimerization

An aldose can be converted to its epimer with opposite configuration at C-2 by the following steps : (1) Oxidation of aldose with bromine water to give the aldonic acid ; (2) Heating the aldonic acid with pyridine, C_5H_5N, to give an equilibrium mixture of original and its epimeric acid ; (3) Separation of the epimeric acid and reduction of its lactone to yield the epimer aldose. For example, D-glucose may be converted to its epimer D-mannose as follows :

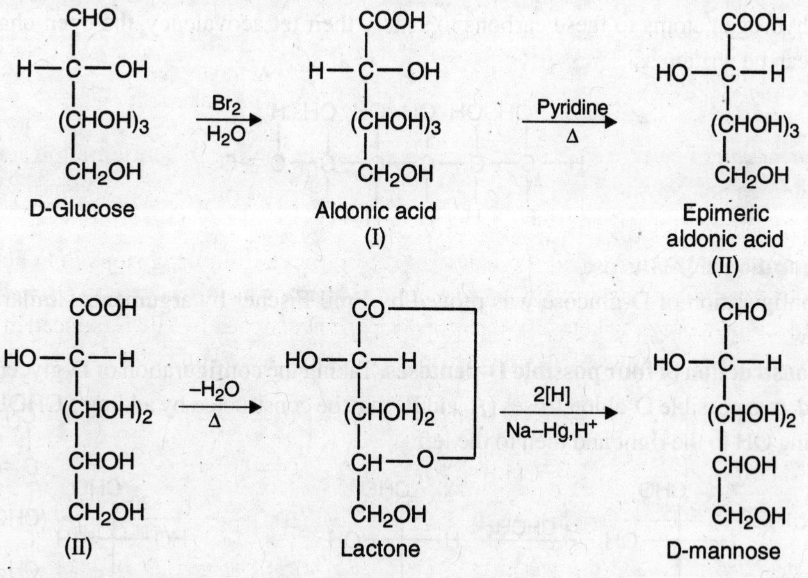

It may be noted that the configuration at C-2 has been changed, and the remaining part is identical in the two sugars. The change in configuration of one asymmetric carbon (C-2) in a compound containing two or more asymmetric carbon atoms is known as **Epimerization**.

STRUCTURE OF GLUCOSE

The structure of glucose may be discussed under the following heads : (*a*) Open chain formula ; (*b*) Configuration ; (*c*) Cyclic structure ; (*d*) Haworth representation.

A. Open-chain Formula

The open-chain formula of glucose is constructed from the following facts :

(1) **Molecular formula**. Elemental analysis and molecular weight determination have established that glucose has the molecular formula $C_6H_{12}O_6$.

(2) **Presence of 6-carbon unbranched chain**. The complete reduction of glucose with concentrated hydrogen iodide and red phosphorus gives *n*-hexane. This proves that glucose molecule is made of an unbranched six-carbon chain.

(3) **Presence of 5 OH groups**. Glucose reacts with acetic anhydride to form a pentaacetyl derivative. This shows the presence of five hydroxy groups. Since glucose is a stable compound, no two OH groups are attached to the same carbon. In other words, the five OH groups are on different carbons.

(4) **Presence of C=O group**. Glucose reacts with hydroxylamine to form an oxime. It suggests the presence of a carbonyl group.

(5) **Presence of terminal CHO function**. On mild oxidation with bromine water, glucose is converted to gluconic acid which when reduced with excess of HI yields *n*-hexanoic acid.

$$\underset{\text{Glucose}}{C_5H_{11}O_5.CHO} \xrightarrow{[O]} \underset{\text{Gluconic acid}}{C_5H_{11}O_5.COOH} \xrightarrow{HI} \underset{n\text{-Hexanoic acid}}{CH_3(CH_2)_4COOH}$$

This shows that glucose contains a six carbon straight chain with CHO at one end, which has been oxidized to COOH.

(6) **Construction of open-chain formula**. Knowing that glucose has a straight 6-carbon chain with a terminal CHO, the five OH groups can be placed one each on the remaining five carbons. Supplying hydrogen atoms to these carbons to satisfy their tetracovalency, the open-chain structure of glucose can be written as :

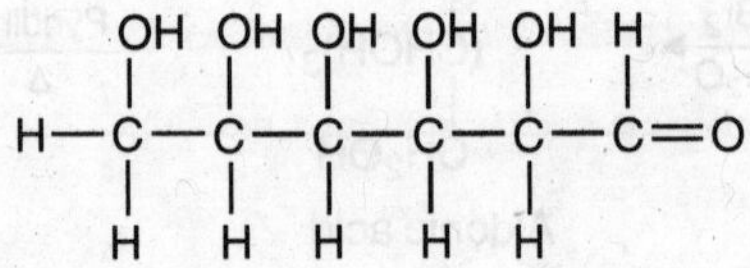

B. Configuration of D-Glucose

The configuration of D-glucose was proved by Emil Fischer by arguments similar to the ones stated below.

(1) **Construction of four possible D-pentoses.** Taking the configuration of D-glyceraldehyde as the standard, two possible D-aldotetroses (A and B) may be constructed by adding a CHOH just below CHO, placing OH to the right and then to the left.

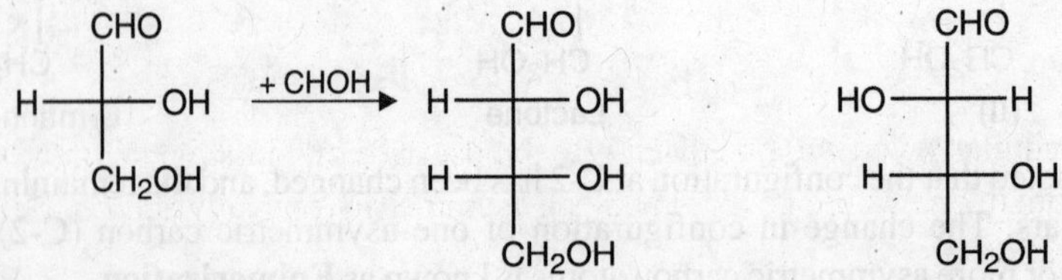

D-glyceraldehyde A B

Similarly, each of the two D-tetroses (A and B) gives two D-aldopentoses. Thus four possible D-aldopentoses are :

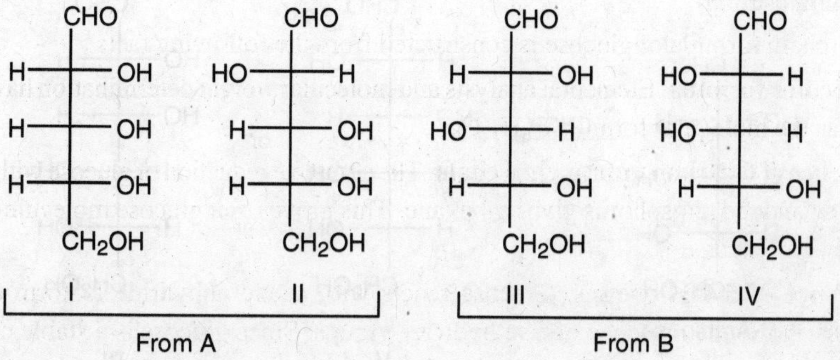

(2) **D-Arabinose has configuration II or IV.** Oxidation of D-arabinose with nitric acid oxidizes the terminal CHO and CH_2OH groups yielding two optically active (asymmetric) dicarboxylic acids. The forms II and IV can form two optically active (asymmetric) diacids, while I and III can give *meso* acids only that have a plane of symmetry. Therefore, D-arabinose is either II or IV.

(3) **Configuration II confirmed for D-arabinose.** D-Arabinose by Killiani-Fischer synthesis yields two epimeric aldohexoses, D-glucose and D-mannose. These on oxidation with nitric acid form two optically active (asymmetric) dicarboxylic acids. This is theoretically possible only if D-arabinose has the configuration II and not IV.

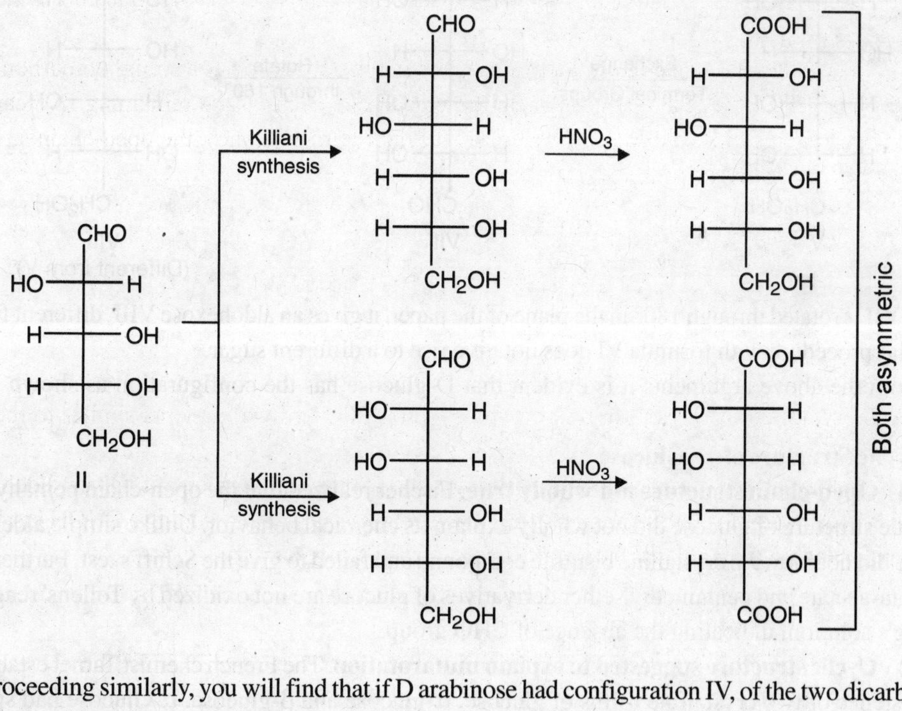

Proceeding similarly, you will find that if D arabinose had configuration IV, of the two dicarboxylic acids derived from it, one would be *meso* and one asymmetric (optically active). Hence D-arabinose has the configuration II.

(4) **Ruff degradation of D-glucose and D-mannose produces D-arabinose in each case.** In Ruff degradation the CHOH below CHO is destroyed. Therefore, the configuration of the two aldohexoses, D-glucose and D-mannose, can be derived by adding a new CHOH below CHO in form II of D-arabinose.

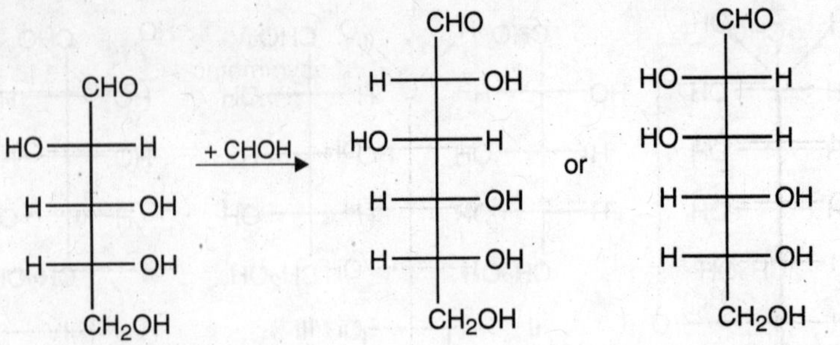

D-Arabinose II   V   VI

Hence D-glucose has configuration V or VI.

(5) **D-Glucose and L-glucose yield the same dicarboxylic acid**. This means that these two sugars differ only in respect of the position of the terminal groups (CHO and CH_2OH). Therefore, the exchange of the terminal groups in D-glucose should be able to give a different aldohexose (L-glucose). Let us now examine configurational formulas V and VI (One of which is D-glucose) from this angle.

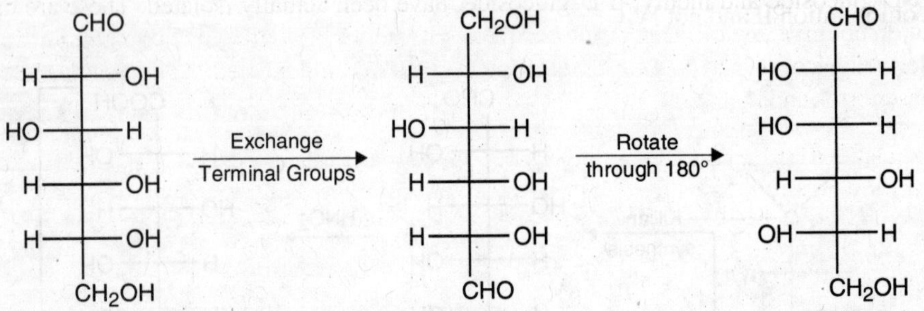

V   VII   VIII (Different from V)

If VII is rotated through 180° in the plane of the paper, it gives an aldohexose VIII, different from V. A similar procedure with formula VI does not give rise to a different sugar.

From the above arguments it is evident that D-glucose has the configuration as shown by the form V.

C. Cyclic Structure of D-Glucose

(1) **Open-chain structure not wholly true**. Fischer realized that the open-chain pentahydroxy aldehyde structure of glucose did not wholly explain its chemical behavior. Unlike simple aldehydes, glucose did not form the crystalline bisulfite compound and failed to give the Schiff's test. Furthermore, the penta-acetate and pentamethyl-ether derivatives of glucose are not oxidized by Tollens' reagent or Fehling's solution, indicating the absence of CHO group.

(2) **Cyclic structure suggested to explain mutarotation**. The French chemist Tarnet established the existence of two crystalline forms of glucose, α-glucose and β-glucose. α-Glucose had specific rotation +112°, while β-glucose +19°. The optical rotation of each of these forms changed gradually with time till finally a constant value of +53° was reached. To explain this phenomenon of mutarotation, it was visualized that the α and β glucose were in reality the cyclic hemiacetal forms of glucose which were interconvertible *via* the open-chain form. The constant value of +19° represented the state of equilibrium between α-D-glucose and β-D-glucose.

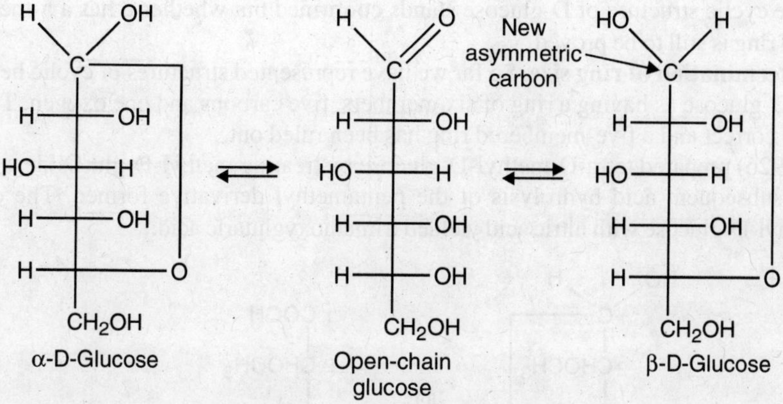

The concentrations of α-D-glucose, β-D-glucose, and the open-chain glucose at equilibrium are 36%, 64%, and less than 0.01% respectively. This explains why D-glucose can react both as an aldehyde and a cyclic hemiacetal in which CHO is absent.

(3) **Glycoside formation confirms cyclic structure.** Glucose when treated with methanol in the presence of dry HCl, gives two isomeric acetals or glycosides. These crystalline glycosides, methyl-α-D-glucoside and methyl-β-D-glucoside, have been actually isolated. These are optically active but do not give any of the reactions of free CHO group. Evidently, the two glycosides are the methyl derivatives of α- and β-D-glucose, formed as a result of the reaction between the hemiacetal OH of these forms and methanol.

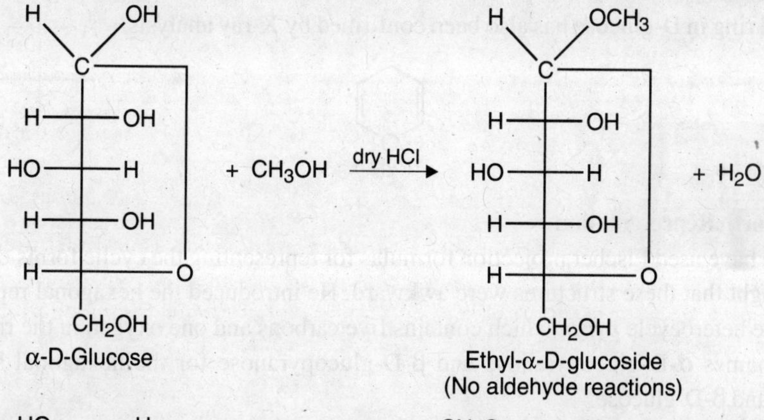

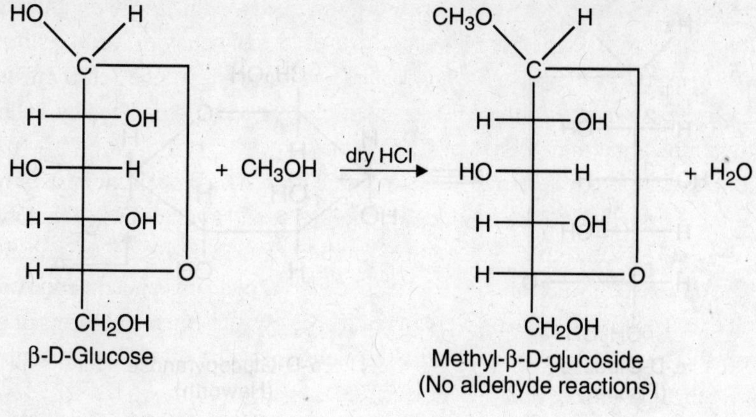

Thus the cyclic structure of D-glucose stands confirmed but whether it has a 5-membered or a 6-membered ring is still to be proved.

(4) **Determination of ring size.** So far we have represented structures of cyclic hemiacetals or anomers of D-glucose as having a ring of six members, five carbons and one oxygen. This has been proved to be correct and a five-membered ring has been ruled out.

Hirst (1926) prepared tetra-O-methyl-D-glucose by treating methyl-D-glucoside with dimethyl sulfate and subsequent acid hydrolysis of the pentamethyl derivative formed. The oxidation of tetra-O-methyl-D-glucose with nitric acid yielded trimethoxyglutaric acid.

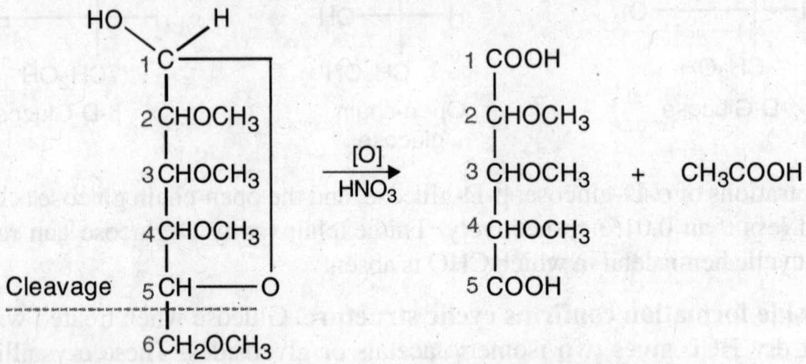

Tetra-O-methyl-D-glucose 2,3,4-Trimethoxyglutaric acid

Obviously, the two carboxylic carbons (1,5) of the trimethoxyglutaric acid are the one's originally involved in ring formation. Hence, there must have existed an oxide ring between C-1 and C-5. Tracing back the reaction sequence, it stands proved that D-glucose has a six-membered ring. The presence of a 6-membered ring in D-glucose has also been confirmed by X-ray analysis.

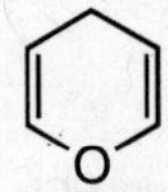

Pyran

D. The Haworth Representation

So far we have used Fischer projection formulas for representing the cyclic forms of D-glucose. Haworth thought that these structures were awkward. He introduced the hexagonal representations resembling the heterocycle *pyran* which contains five carbons and one oxygen in the ring. Thus, he claimed the names α-D-glucopyranose and β-D-glucopyranose for the hexagonal structures of α-D-glucose and β-D-glucose.

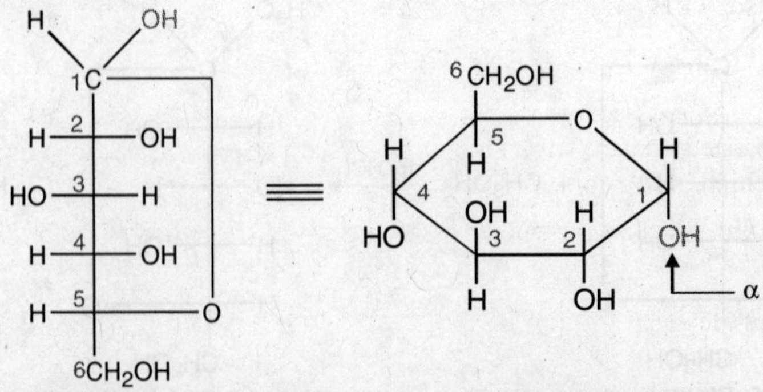

α-D-Glucose (Fischer) α-D-Glucopyranose (Haworth)

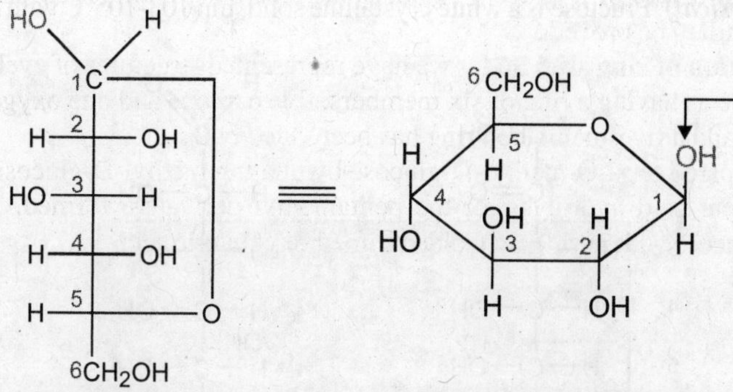

β-D-Glucose (Fischer) β-D-Glucopyranose (Haworth)

It may be noted that in Haworth formula, all the OH groups on the right in Fischer formula are directed below the plane of the ring, while those on the left go above the plane. The terminal CH_2OH projects above the plane of the ring.

FRUCTOSE, *Levulose*, $C_6H_{12}O_6$

This important ketohexose is also called *levulose* because the naturally occurring form of fructose is levorotary. Fructose is found along with glucose in the juices of ripe fruits and in honey. In the combined state, it occurs in sucrose, and in the polysaccharide **Inulin** obtained from plants such as *dahlia, artichoke* and *dandelion*.

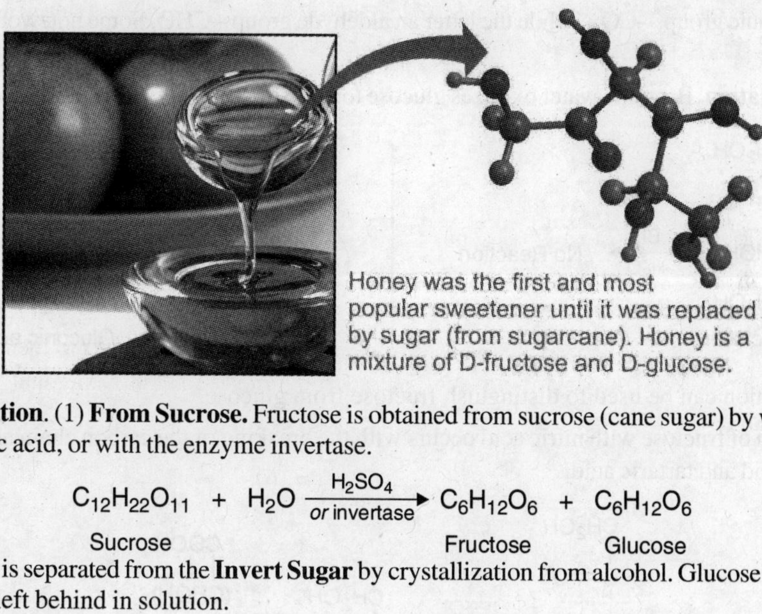

Honey was the first and most popular sweetener until it was replaced by sugar (from sugarcane). Honey is a mixture of D-fructose and D-glucose.

Preparation. (1) **From Sucrose.** Fructose is obtained from sucrose (cane sugar) by warming with dilute sulfuric acid, or with the enzyme invertase.

$$\underset{\text{Sucrose}}{C_{12}H_{22}O_{11}} + H_2O \xrightarrow[\text{or invertase}]{H_2SO_4} \underset{\text{Fructose}}{C_6H_{12}O_6} + \underset{\text{Glucose}}{C_6H_{12}O_6}$$

Fructose is separated from the **Invert Sugar** by crystallization from alcohol. Glucose being soluble in alcohol is left behind in solution.

(2) **From Inulin**. Fructose is prepared commercially by the hydrolysis of *inulin* by heating with dilute sulfuric acid.

$$\underset{\text{Inulin}}{(C_6H_{10}O_5)_n} + nH_2O \xrightarrow[\Delta]{H_2SO_4} \underset{\text{Fructose}}{nC_6H_{12}O_6}$$

The resulting solution is neutralized with barium hydroxide and the precipitated barium sulfate removed by filtration. The filtrate is concentrated under reduced pressure to yield the crystals of pure fructose.

Properties. (***Physical***). Fructose is a white crystalline solid, mp 103-105°C with decomposition. It is the sweetest of all sugars and very soluble in water. It is sparingly soluble in ethanol and insoluble in ether. D(+)-Fructose, the natural sugar, exhibits mutarotation and the final value of specific rotation is –92°.

(***Chemical***). Fructose is a pentahydroxy ketone and as compared to that of glucose its open-chain structure may be written as :

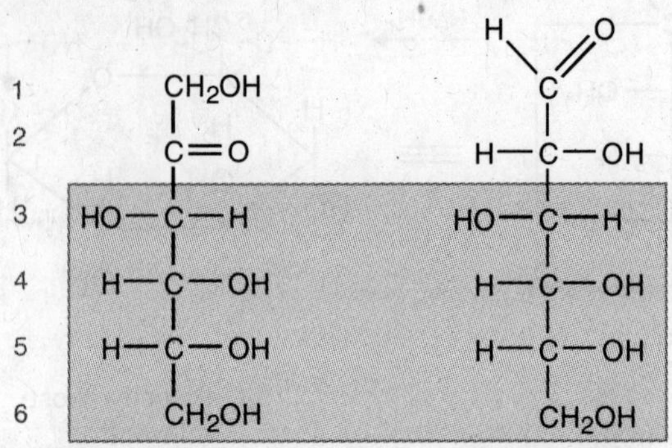

D-Fructose D-Glucose

In the two sugars, the configuration of the three asymmetric carbons in the box is identical. Thus fructose gives reactions similar to glucose, any difference being due to the fact that the former contains a ketonic group, –CO–, while the latter an aldehyde group –CHO. Some noteworthy reactions of fructose are :

(1) **Oxidation.** Bromine water oxidizes glucose to gluconic acid, but does not react with fructose.

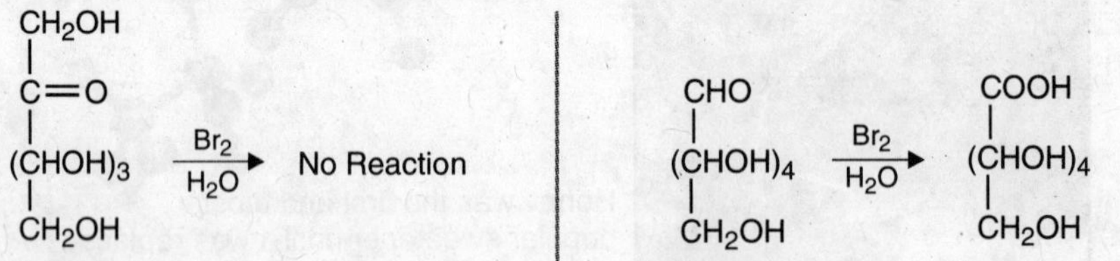

Fructose Glucose Gluconic acid

This reaction can be used to distinguish fructose from glucose.

Oxidation of fructose with nitric acid occurs with the breaking of the carbon chain and formation of glycollic acid and tartaric acid.

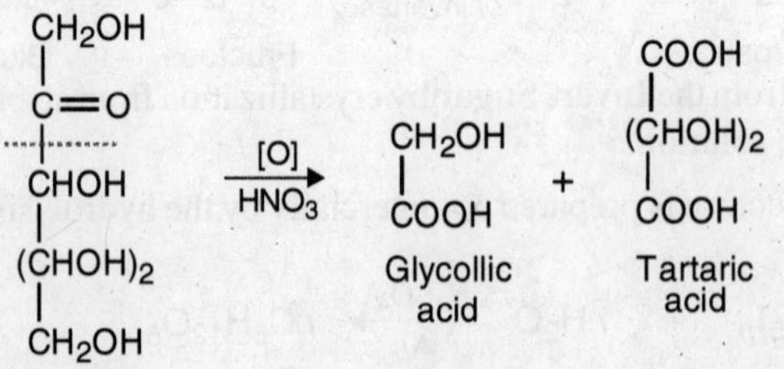

Fructose

(2) **Reduction.** Reduction of fructose gives two epimeric hexahydric alcohols as C-2 now becomes asymmetric.

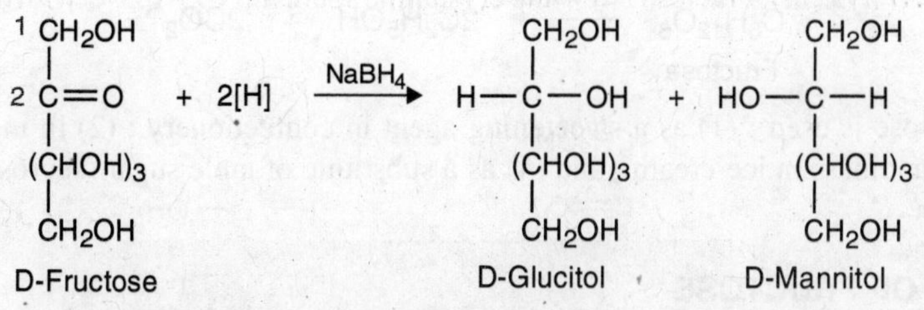

(3) **Action with Fehling's solution and Tollens' reagent**. Unlike a simple ketone, fructose reduces Fehling's solution and Tollens' reagent. This is because the alkaline reagents rearrange fructose to give D-glucose and D-mannose which are strong reducing agents.

(4) **Action with Phenylhydrazine**. When treated with excess phenylhydrazine, $C_6H_5NHNH_2$, fructose first forms phenylhydrazone. Then CH_2OH involving C-1 is oxidized to –CHO by a second molecule of phenylhydrazine. The –CHO finally reacts with a third molecule of phenylhydrazine to give the *osazone*.

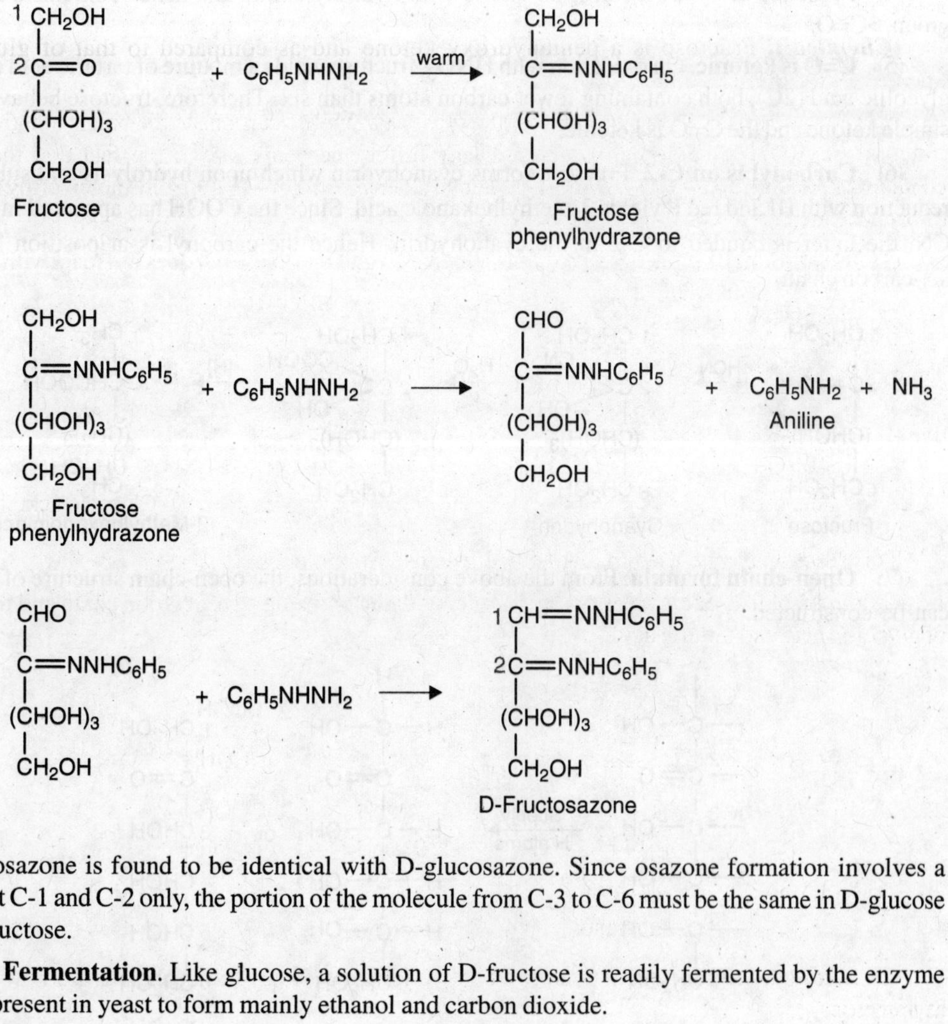

D-Fructosazone is found to be identical with D-glucosazone. Since osazone formation involves a change at C-1 and C-2 only, the portion of the molecule from C-3 to C-6 must be the same in D-glucose and D-fructose.

(5) **Fermentation**. Like glucose, a solution of D-fructose is readily fermented by the enzyme *zymase* present in yeast to form mainly ethanol and carbon dioxide.

$$\underset{\text{Fructose}}{C_6H_{12}O_6} \xrightarrow{\text{zymase}} 2C_2H_5OH + 2CO_2\uparrow$$

Uses. Fructose is used : (1) as a sweetening agent in confectionery ; (2) in medicinal syrups; (3) to prevent sandiness in ice-cream ; and (4) as a substitute of table sugar (sucrose) for the obese and the diabetic.

STRUCTURE OF FRUCTOSE

A. Open-chain Formula

(1) **Molecular formula.** As established by elemental analysis and molecular weight determination, the molecular formula of fructose is $C_6H_{12}O_6$.

(2) **Presence of 6-C chain**. Fructose on complete reduction with HI and red P forms *n*-Hexane. Therefore, it contains an unbranched 6-carbon chain.

(3) **Presence of 5 OH groups**. Fructose reacts with acetic anhydride to give pentaacetyl derivative. It indicates the presence of five hydroxy groups, each attached to a separate carbon.

(4) **Presence of C=O**. It forms an oxime with hydroxylamine and hence contains a carbonyl group, >C=O.

(5) **C=O is ketonic**. On oxidation with HNO_3, fructose yields a mixture of tartaric acid (4C) and glycollic acid (2C), both containing fewer carbon atoms than six. Therefore, fructose behaves like a simple ketone and the C=O is ketonic.

(6) **Carbonyl is on C-2**. Fructose forms cyanohydrin which upon hydrolysis and subsequent reduction with HI and red P yields 2-methylhexanoic acid. Since the COOH has appeared in place of CN, the latter is bonded to C-2 in the cyanohydrin. Hence the carbonyl is at position 2 on the six-carbon chain.

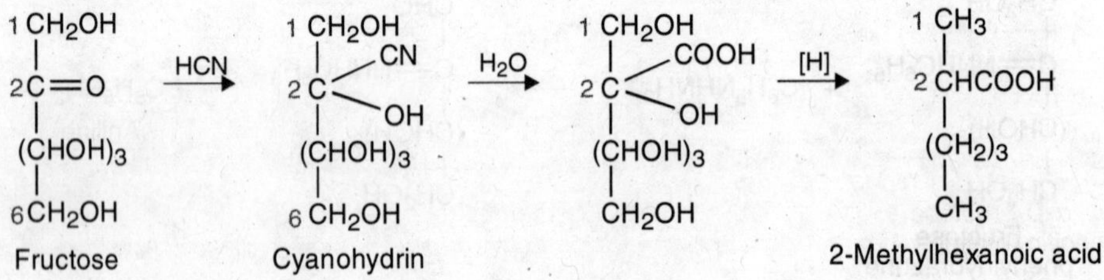

(7) **Open-chain formula**. From the above considerations, the open-chain structure of fructose can be constructed.

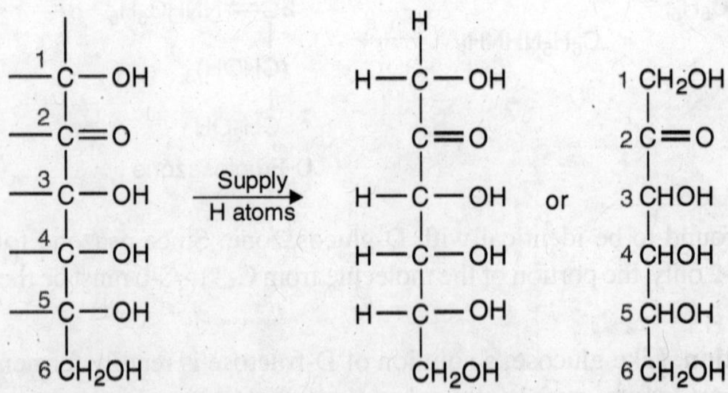

B. Configuration

Formation of Glucosazone. Fructose reacts with excess phenylhydrazine, $C_6H_5NHNH_2$, to form glucosazone. Or, that the osazone formed by fructose is identical with that obtained from D-glucose. This shows that the configuration of asymmetric carbon atoms C-3, C-4, and C-5 in D-fructose is the same as in D-glucose. Thus,

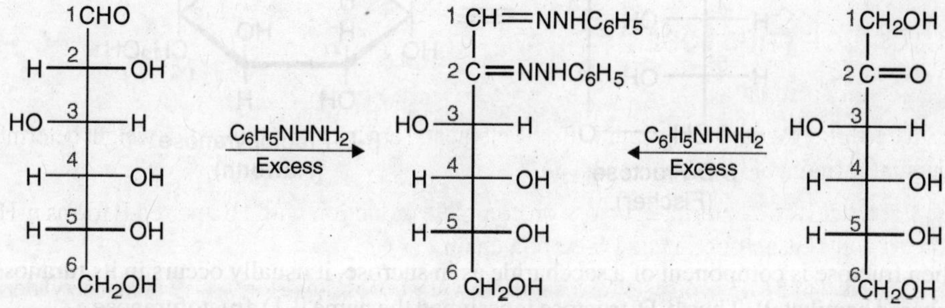

D-Glucose Glucosazone D-Fructose

Hence the configuration of D-fructose is proved as shown above.

C. Cyclic Structure of D-Fructose

By analogy to D-glucose and to explain its dual behavior, a six-membered hemiketal ring structure was proposed for D-fructose. Since D-Fructose also exhibits mutarotation, the α and β forms in equilibrium *via* the open-chain form may be represented as :

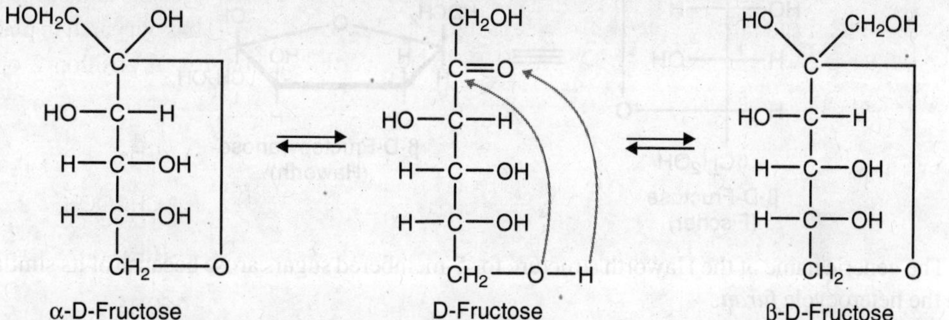

α-D-Fructose specific rotation −21° D-Fructose (open-chain) β-D-Fructose specific rotation −133°

Ordinary fructose is β-D-fructose.

D. The Haworth Representation

The Fischer projection formulas α-D-fructose and β-D-fructose have been represented by Haworth as follows.

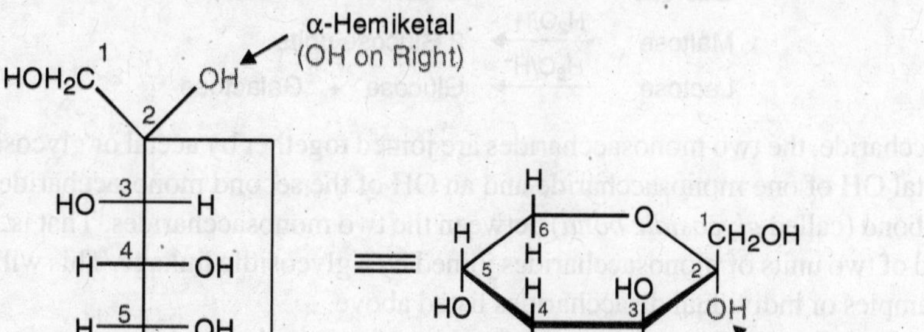

α-D-Fructose α-D-Fructopyranose

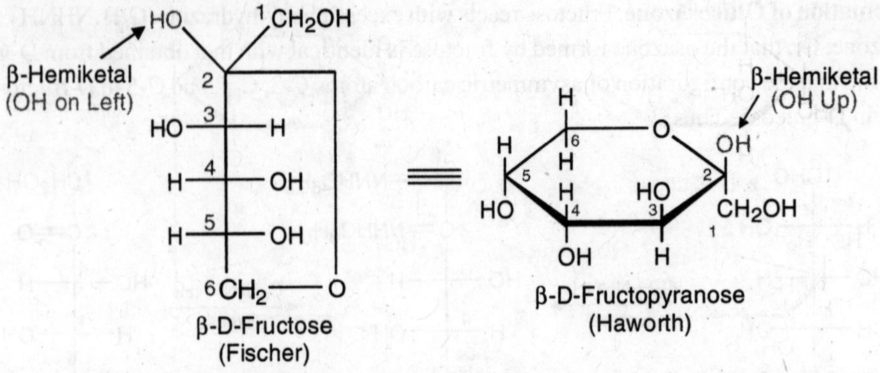

When fructose is component of a saccharide as in sucrose, it usually occurs in its furanose form (5-membered hemiketal). Thus β-D-fructose is assigned the name β-D-fructofuranose.

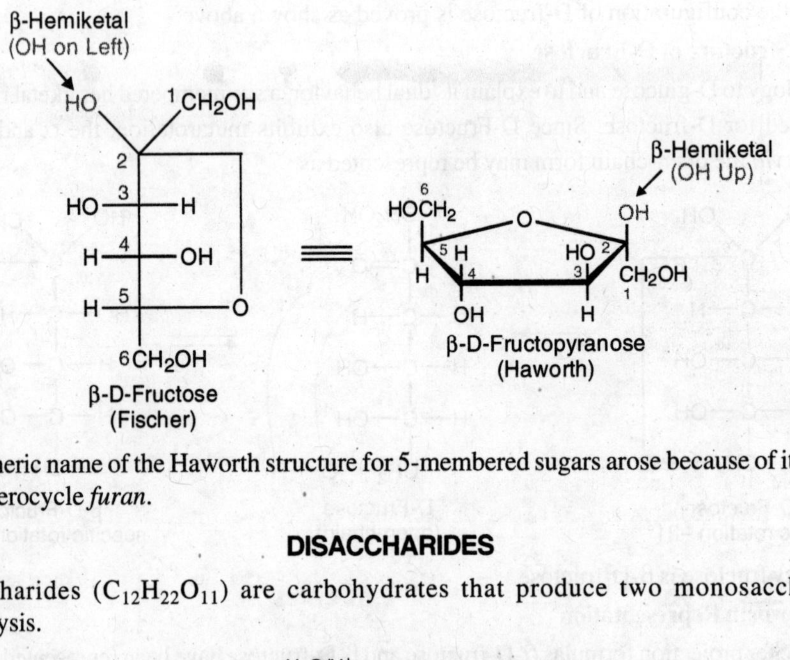

The generic name of the Haworth structure for 5-membered sugars arose because of its similarity with the heterocycle *furan*.

DISACCHARIDES

Disaccharides ($C_{12}H_{22}O_{11}$) are carbohydrates that produce two monosaccharides on acid-hydrolysis.

$$\text{Sucrose} \xrightarrow{H_2O/H^+} \text{Glucose} + \text{Fructose}$$

$$\text{Maltose} \xrightarrow{H_2O/H^+} \text{2 Glucose units}$$

$$\text{Lactose} \xrightarrow{H_2O/H^+} \text{Glucose} + \text{Galactose}$$

In a disaccharide, the two monosaccharides are joined together by acetal or glycoside formation. The hemiacetal OH of one monosaccharide and an OH of the second monosaccharide, dehydrate to establish the bond (called *glycosidic bond*) between the two monosaccharides. That is, disaccharides are composed of two units of monosaccharides joined by a glycosidic linkage. This will be illustrated by taking examples of individual disaccharides listed above.

Sucrose (*Cane Sugar*). Sucrose is ordinary table sugar. It is obtained from cane sugar. Sucrose is composed of α-D-glucose unit and β-D-fructose unit. These units are joined by α,β-glycosidic linkage between C-1 of the glucose unit and C-2 of the fructose unit.

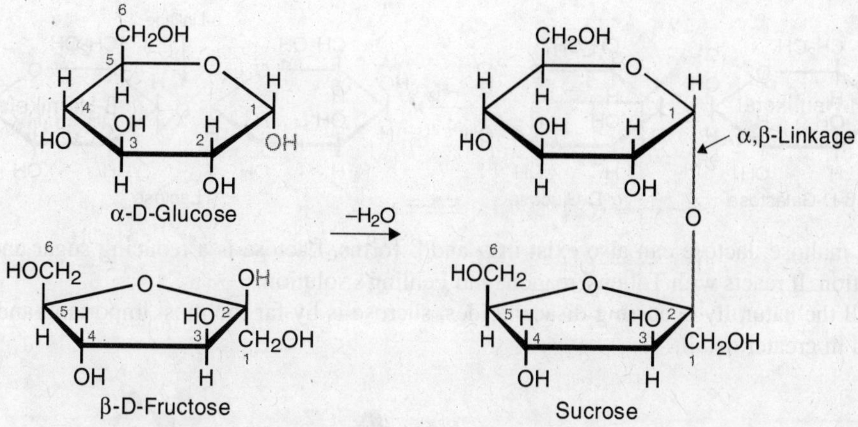

Notice that in the above structure of sucrose, hemiacetal structure is missing. That is why sucrose : (*a*) does not form an osazone with phenylhydrazine ; (*b*) does not reduce Tollens' reagent or Fehling's solution (Sucrose is a non-reducing sugar) ; (*c*) does not exhibit mutarotation.

In everyday life, when we say the word 'sugar' we usually refer to sucrose, the most widely occurring disaccharide in nature. Sucrose is prepared commercially in pure form in greater quantities than any other chemical substance.

Maltose. It is obtained from starch. It is composed of two α-D-glucose units joined by an α-glycosidic linkage between C-1 of one unit and C-4 of the other unit.

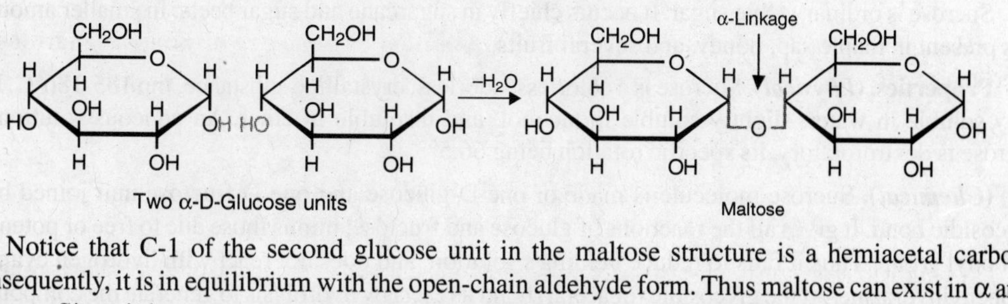

Notice that C-1 of the second glucose unit in the maltose structure is a hemiacetal carbon. Consequently, it is in equilibrium with the open-chain aldehyde form. Thus maltose can exist in α and β forms. Since it has a potential aldehyde group, maltose shows mutarotation, forms osazone, and reduces Fehling's solution (Maltose is a reducing sugar).

Lactose (*Milk Sugar*). It is found in the milk of all animals. Cow's milk contains 4-5% and human milk 6-7% lactose. Lactose is composed of β-D-galactose unit and α-D-glucose unit joined by β-D-glycosidic linkage between C-1 of the galactose and C-4 of the glucose unit.

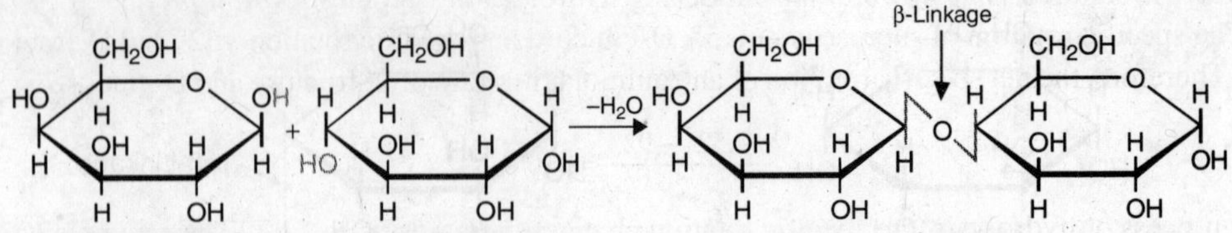

β-D-Galactose α-D-Glucose Lactose

Like maltose, lactose can also exist in α and β forms. Lactose is a reducing sugar and shows mutarotation. It reacts with Tollens' reagent and Fehling's solution.

Of all the naturally occurring disaccharides, sucrose is by far the most important, and will be discussed in greater detail.

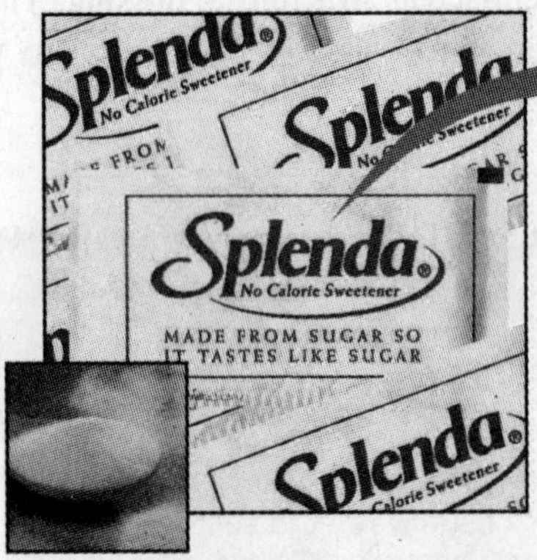

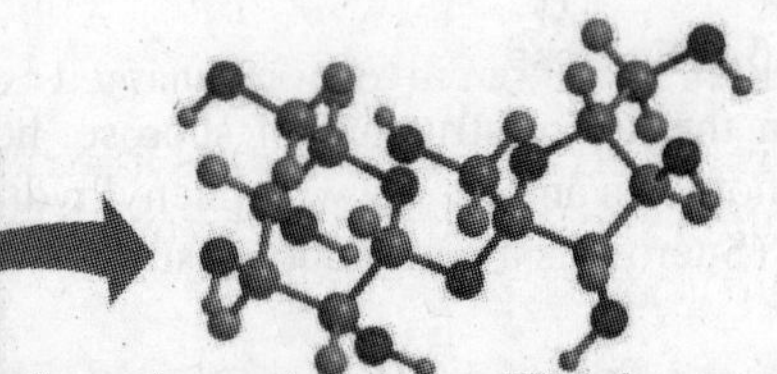

Sucralose is an artificial sweetener. Sucralose is not broken down by the body, so it is noncaloric. It is about 320 to 1,000 times as sweet as sucrose (table sugar). It is stable under heat and over a broad range of pH conditions. Therefore, it can be used in baking. Common brand name of sucralose-based sweetener is Splenda.

SUCROSE, *Cane Sugar*, $C_{12}H_{22}O_{11}$

Sucrose is ordinary table sugar. It occurs chiefly in sugarcane and sugar beets. In smaller amounts it is present in maple sap, honey, and several fruits.

Properties. (*Physical*). Sucrose is a colorless, odorless, crystalline substance, mp 185-186°C. It is very soluble in water, slightly soluble in alcohol, and insoluble in ether. An aqueous solution of sucrose is dextrorotatory, its specific rotation being 66.5°.

(*Chemical*). Sucrose molecule is made of one D-glucose and one D-fructose unit joined by a glycosidic bond. It gives all the reactions of glucose and fructose, minus those due to free or potential carbonyl group. Thus it fails to reduce Fehling's solution, and does not react with hydrogen cyanide or phenylhydrazine. At the glycosidic (or acetal) bond it reacts by hydrolysis to generate the component units, the D-glucose and D-fructose.

Hydrolysis or Inversion of Sucrose. The hydrolysis of sucrose by boiling with a mineral acid, or by the enzyme *invertase*, produces a mixture of equal molecules of D-fructose and D-glucose.

$$C_{12}H_{22}O_{11} + H_2O \xrightarrow[\text{or Invertase}]{H^+} \underset{\text{D-Glucose}}{C_6H_{12}O_6} + \underset{\text{D-Fructose}}{C_6H_{12}O_6}$$

Sucrose solution is *dextrorotatory* but during hydrolysis it becomes *levorotatory*.

The specific rotation of sucrose is +66.5°. D-Glucose has specific rotation +52° and D-fructose –92°. Therefore, the net specific rotation of an equimolar mixture of D-fructose and D-glucose is.

$$\frac{+52^\circ - 92^\circ}{2} = -20^\circ$$

In the process of hydrolysis, the specific rotation changes from +66.5° to –20°. The sign of specific rotation changes from (+) to (–), or is said to 'invert'. Hence the hydrolysis of sucrose to D-glucose and D-fructose, is termed **Inversion** and the hydrolysis mixture is called **Invert Sugar**.

$$\underset{\text{specific rotation } +66.5^\circ}{\text{Sucrose}} \xrightarrow{\text{Inversion}} \underbrace{\text{D–Glucose} + \text{D–Fructose}}_{\substack{\text{Invert Sugar} \\ \text{specific rotation } -20^\circ}}$$

The enzyme that brings about inversion is named as **Invertase**.

Uses. Sucrose is used : (1) as a food and as ingredient of jams, jellies, confections, syrups etc., (2) in the manufacture of sucrose octaacetate required to denature alcohol, to render paper transparent, and to make anhydrous adhesives.

WHAT ARE REDUCING AND NON-REDUCING SUGARS ?

Those saccharides which reduce Fehling's reagent and Tollens' reagent are called **Reducing Sugars**, and those which do not, **Non-Reducing Sugars**. The ability to reduce or not to reduce Fehling's and Tollens' reagents depends on certain characteristic structural features.

Structural Features of Reducing Sugars : (1) They contain α-hydroxyaldehyde or α-hydroxyketone functions.

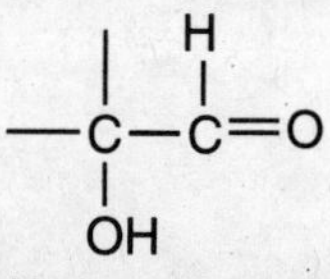

α-Hydroxyaldehyde

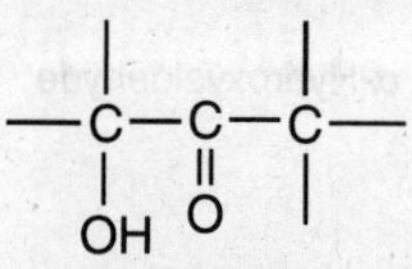

α-Hydroxyketone

(2) They contain cyclic hemiacetal or hemiketal structures in equilibrium with the open-chain form having a free CHO or C=O group as the case may be. Such sugars, therefore, reduce Fehling's and Tollens' reagents. **Reducing sugars also exhibit mutarotation**.

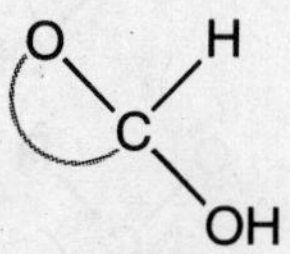

Hemiacetal

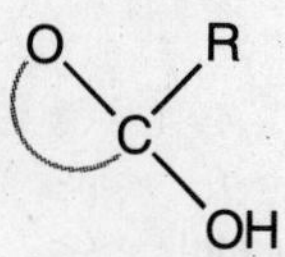

Hemiketal

Structural Features of Non-Reducing sugars. (1) They do not contain free aldehyde or ketone groups with OH on the carbon adjacent to the carbonyl group.

(2) They contain acetal or ketal structure that is stable. Their cyclic structures cannot be opened into an open-chain form having a free carbonyl group. Thus they are unable to reduce Fehling's or Tollens' reagent. For the same reason, **non-reducing sugars do not exhibit mutarotation** (Table 31.1).

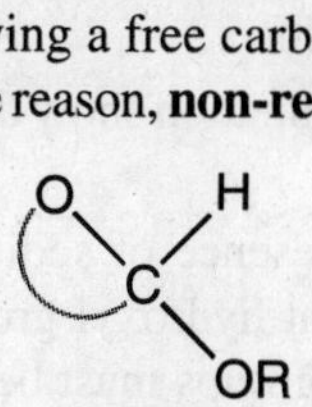

Acetal

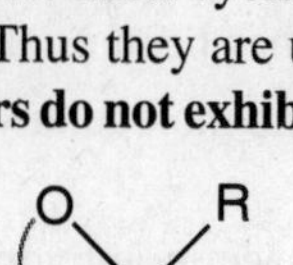

Ketal

Examples. (1) All monosaccharides contain free or potential CHO or C=O functions and are reducing sugars *e.g.*, D-glucose and D-fructose.

(2) Of the disaccharides, sucrose is non-reducing because the anomeric carbons of both monosaccharides are involved in glycoside or acetal formation. Maltose and lactose are reducing sugars, because in one of the monosaccharide units there is a hemiacetal group that can be opened to give free CHO function.

Table 31.1 PROPERTIES OF REDUCING AND NONREDUCING SUGARS

Reducing Sugars		Non-reducing Sugars
(1) Give positive *Tollens' test*		(1) Give negative *Tollens' test*
Ag^+ (ammonia complex) →	$Ag°\downarrow$	
Clear solution	Silver mirror	
Benedict's test		*Benedict's test*
Cu^{++} (citrate complex) →	$Cu_2O\downarrow$	
Clear blue solution	Yellow-red precipitate	
Fehling's test		*Fehling's test*
Cu^{++} (tartrate complex) →	$Cu_2O\downarrow$	
Clear blue solution	Yellow-red precipitate	
(2) Mutarotate		(2) Do not mutarotate
(3) Functional units present		(3) Functional units present
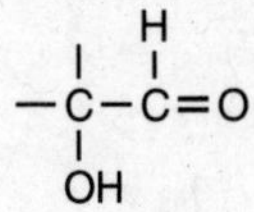 α-Hydroxyaldehyde		
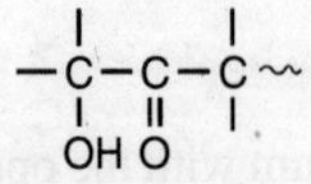 α-Hydroxyketone		
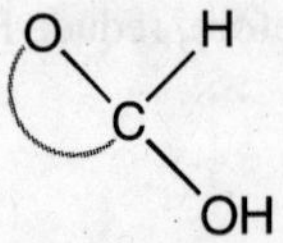 Hemiacetal		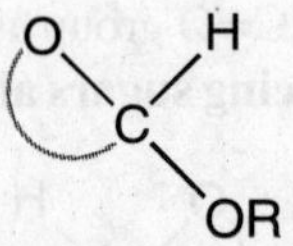 Acetal
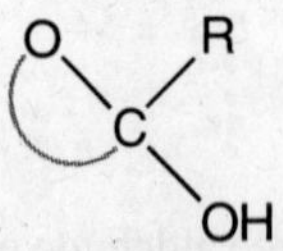 Hemiketal		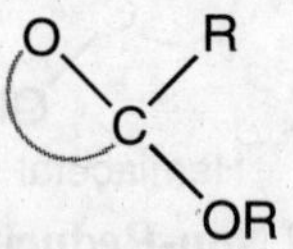 Ketal

STRUCTURE OF SUCROSE

The structure of sucrose has been established as follows :

(1) Elemental analysis and molecular weight determination show that the molecular formula of sucrose is $C_{12}H_{22}O_{11}$.

(2) Sucrose reacts with acetic anhydride in the presence of sodium acetate to form sucrose octaacetate. This reaction indicates the presence of eight hydroxyl groups in a sucrose molecule. Since sucrose is a stable compound, the eight hydroxyl groups must be present on separate carbon atoms.

(3) Hydrolysis of sucrose with dilute acids yields an equimolecular mixture of D-glucose and D-fructose. This indicates that the sucrose molecule is made up one unit of these monosaccharides.

(4) Sucrose does not reduce Tollens' reagent or Fehling's solution ; does not form an osazone (except on prolonged boiling), when glucosazone is formed (due to hydrolysis of sucrose); does not form methyl glucosides ; and does not undergo mutarotation. All these observations indicate that the cyclic forms of glucose and fructose are joined together through glycosidic linkage at points where the carbonyl groups would otherwise become available, that is, C-1 in glucose and C-2 in fructose.

(5) Sucrose reacts with dimethyl sulfate in an alkaline solution to form octamethylsucrose, which on hydrolysis yields a mixture of 2,3,4,6-tetramethyl-D-glucopyranose and 1,3,4,6-tetramethyl-D-Fructofuranose. The formation of these compounds indicates that the glucose unit in sucrose has a pyranose form (6-membered ring), and the fructose unit in the furanose form (5-membered ring).

$$C_{12}H_{14}O_3(OH)_8 + (CH_3)_2SO_4 \xrightarrow{NaOH} \underset{\text{Octamethylsucrose}}{C_{12}H_{14}O_3(OCH_3)_8} \xrightarrow[H^+]{H_2O}$$

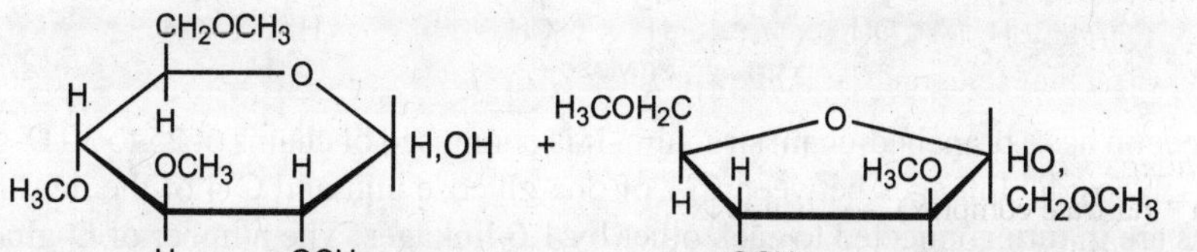

2,3,4,6-Tetramethyl-D-glucopyranose (α- and β-forms)

1,3,4,6-Tetramethyl-D-fructofuranose (α- and β-forms)

(6) Sucrose is hydrolysed by *maltase*, an enzyme that hydrolyses only α-glycosides. It is also hydrolysed by *invertase*, an enzyme that hydrolyses β- but not α-fructofuranosides. These observations indicate that sucrose is both an α-glucoside and a β-fructoside.

The above evidence clearly indicates that sucrose has the following structure :

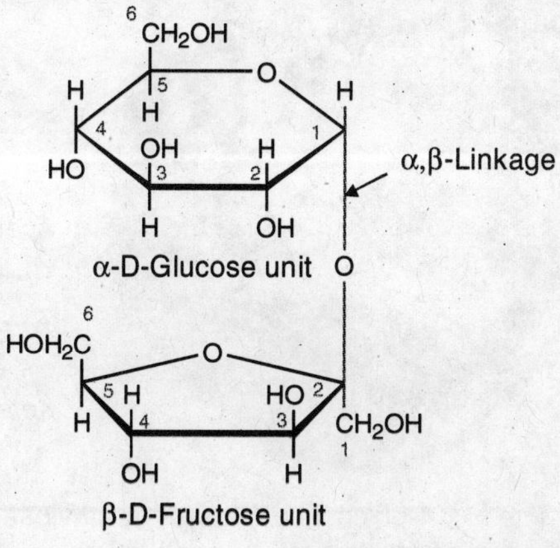

Structure of Sucrose

POLYSACCHARIDES

Polysaccharides are made of many (10 or more) monosaccharide units joined together by glycosidic linkages. By far the most important naturally occurring polysaccharides are *Starch* and *Cellulose*.

STARCH, $(C_6H_{10}O_5)_n$

Starch is the main contributor of carbohydrates in our diet. It exists exclusively in plants, stored in the seeds, roots, and fibres as food reserve. The chief sources of starch are cereals, potatoes, corn, and rice.

Manufacture. Starch occurs naturally as small globules enclosed in cellular tissue. These are soaked in water and crushed. The pulp thus obtained is carried by a stream of water on to a sieve. The starch is separated from the liquid passing through the sieve by decantation, and dried.

Structure. Starch is actually a mixture of two structurally different polysaccharides, **Amylose** (20%) and **Amylopectin** (80%). When starch is heated with hot water, it can be separated into these components. The part that is soluble in water is *amylose* and remaining fraction is *amylopectin*. Both amylose and amylopectin are composed of D-glucose units. The amylose molecule is made up of D-glucose units joined by α-glycosidic linkages between C-1 of one glucose unit and C-4 of the next glucose unit. The number of D-glucose units in amylose ranges from 60-300.

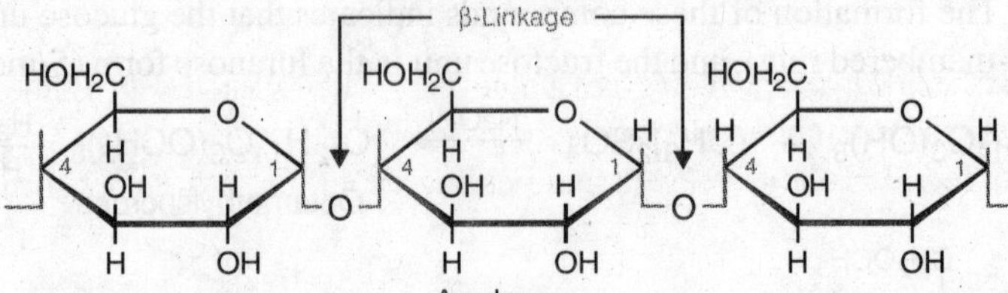

Amylose

Amylopectin has a branched-chain structure. It is composed of chains of 25 to 30 D-glucose units joined by α-glycosidic linkages between C-1 of one glucose unit and C-4 of the next glucose unit. These chains are in turn connected to each other by 1,6-linkages. The number of D-glucose units in amylopectin ranges from 300 to 6000.

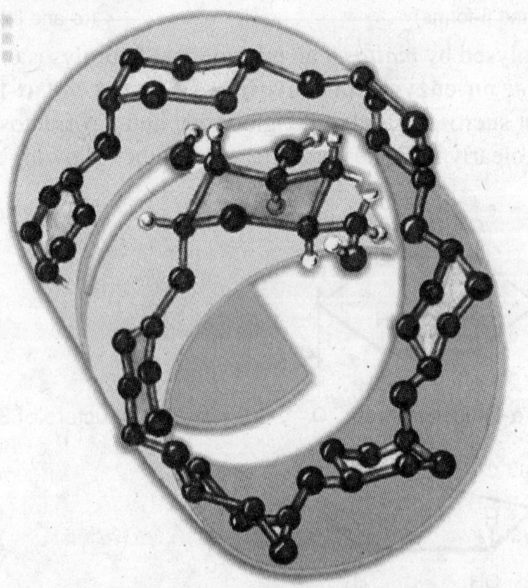

Structure of Amylose. The α-1,4-glycosidic linkages in amylose cause this carbohydrate to form a left-handed helix.

Properties. Starch is a colorless amorphous, powder having no definite melting point. It gives colloidal solutions in water.

(1) **Hydrolysis**. The hydrolysis of starch with hot dilute acids yields a syrupy mixture of glucose, maltose, and high-molecular weight saccharides. The mixture is called **Dextrin** and marketed as **Corn Syrup**. The digestive enzymes in humans convert starch ultimately to glucose which is the source of energy. In the presence of the enzyme **Amylase** (*diastase*) starch is converted to maltose, an important step in the manufacture of alcohol from starch.

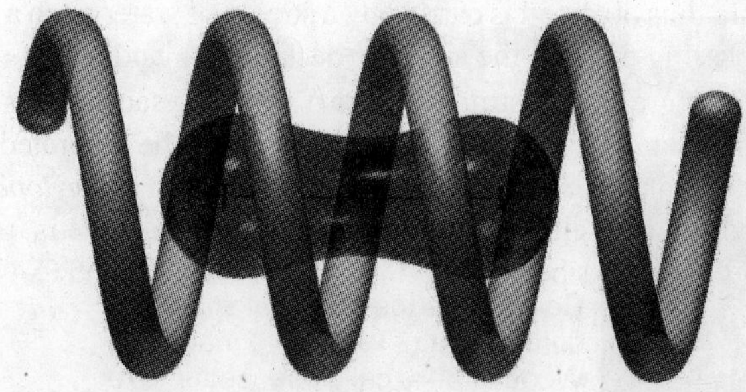

The amylose helix forms a blue charge-transfer complex with molecular iodine.

(2) **Iodine Test**. It is *amylose* which gives a blue color when treated with iodine. It is believed that amylose forms a helical structure and iodine is lodged in it.

Starch does not reduce Tollens' reagent and Fehling's solution. It does not form an osazone with phenylhydrazine.

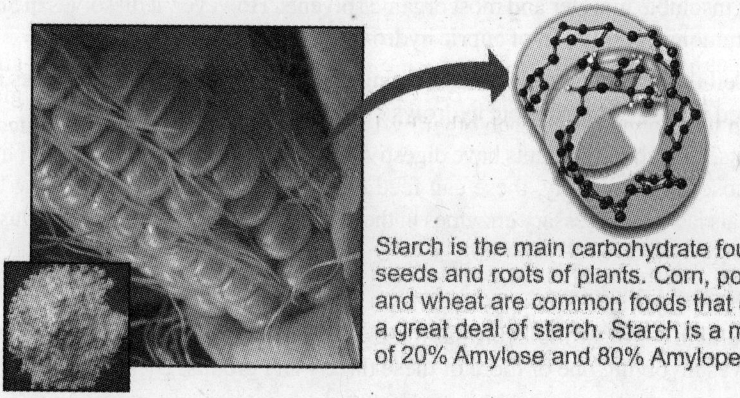

Starch is the main carbohydrate found in seeds and roots of plants. Corn, potatoes, and wheat are common foods that contain a great deal of starch. Starch is a mixture of 20% Amylose and 80% Amylopectin.

CELLULOSE, $(C_6H_{10}O_5)_n$

Cellulose is the main structural material of trees and other plants. Wood is 50% cellulose, while cotton wool is almost pure cellulose. Other sources of cellulose are straw, corncobs, bagasse, and similar agricultural wastes.

Manufacture. Cotton wool is about 97 per cent cellulose. It is ready for use after washing away the waxes and fats associated with it. The cellulose required for making paper is obtained from wood. *Lignin* and resinous substances present along with cellulose, are removed by digesting the wood chips under pressure with a solution of calcium hydrogen sulfite, $Ca(HSO_3)_2$. The cellulose separates as insoluble fibres which are washed with water, bleached and dried.

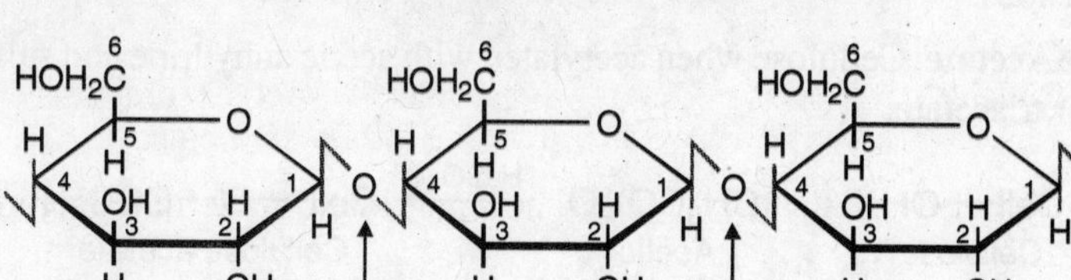

Cellulose

Structure. Cellulose is a straight-chain polysaccharide composed of D-glucose units. These units are joined by β-glycosidic linkages between C-1 of one glucose unit and C-4 of the next glucose unit. The number of D-glucose units in cellulose ranges from 300-2500.

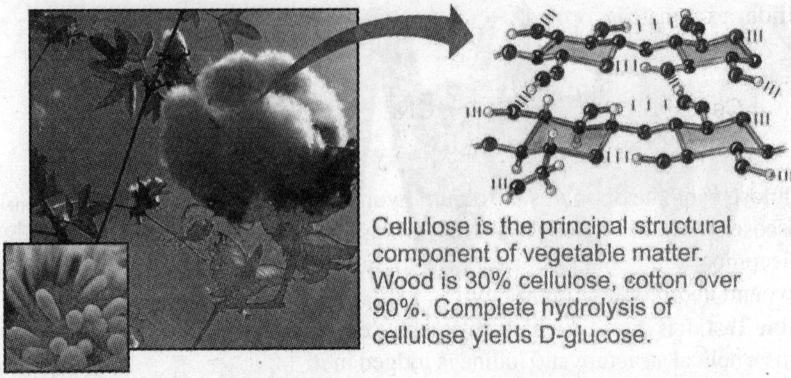

Cellulose is the principal structural component of vegetable matter. Wood is 30% cellulose, cotton over 90%. Complete hydrolysis of cellulose yields D-glucose.

Properties. Cellulose is a colorless amorphous solid having no melting point. It decomposes on strong heating. It is insoluble in water and most organic solvents. However, it dissolves in *Schweitzer's reagent* which is ammoniacal solution of cupric hydroxide.

Hydrolysis. Cellulose when hydrolysed by heating with dilute HCl (or H_2SO_4), gives D-glucose. Cellobiose is formed in case of incomplete hydrolysis.

The cattle, goats, and other ruminants have digestive enzymes (**Cellulases**) capable of hydrolysing cellulose into glucose. Consequently, these can feed directly on cellulose (grass, straw). Man and many other mammals lack the necessary enzymes in their digestive tract, and they cannot use cellulose as foodstuff.

Important Derivatives of Cellulose

Each glucose unit in cellulose molecule has three –OH groups. Thus cellulose gives industrially important derivatives involving one or more of these three –OH groups.

(1) **Nitrated Celluloses**. Cellulose reacts with nitric acid in the presence of sulfuric acid to form an ester *Cellulose Nitrate*, called **Gun Cotton** or **Cordite**.

$$\underset{\text{Cellulose}}{\text{Cell—OH}} + HNO_3 \xrightarrow{H_2SO_4} \underset{\text{Cellulose nitrate}}{\text{Cell—}ONO_2}$$

Gun cotton, which looks like ordinary cotton, is highly explosive, It is used in the manufacture of smokeless *gunpowder* or *propelling powder*.

By partial nitration, cellulose gives cellulose dinitrate, known as **Pyroxylin**. When dissolved in ether and alcohol, it forms a transparent film called **Collodion**. Collodion is used as a covering over cuts and skin abrasions.

(2) **Cellulose Acetate**. Cellulose when acetylated with acetic anhydride and sulfuric acid, forms cellulose acetate or triacetate.

$$\underset{\text{Cellulose}}{\text{Cell—OH}} + \underset{\text{Acetic anhydride}}{(CH_3CO)_2O} \xrightarrow{H_2SO_4} \underset{\text{Cellulose acetate}}{\text{Cell—O—}COCH_3}$$

The cellulose acetate (or triacetate) in acetone solution is passed through a *spinneret* into a stream of warm air to evaporate the acetone, a product called **Acetate Rayon** is obtained. The acetate fabrics accept color well and are wrinkle resistant. Cellulose acetate is also used in making motion picture film.

(3) **Cellulose Xanthate**. When cellulose is treated with sodium hydroxide solution and then with carbon disulfide, a xanthate is formed.

$$\underset{\text{Cellulose}}{\text{Cell—OH}} \xrightarrow{\text{NaOH}} \text{Cell—}\bar{\text{O}}\overset{+}{\text{Na}} \xrightarrow{\text{CS}_2} \underset{\substack{\text{Cellulose xanthate} \\ (\textit{soluble})}}{\text{Cell—O—}\overset{\overset{\text{S}}{\|}}{\text{S}}\text{—}\bar{\text{S}}\overset{+}{\text{Na}}}$$

The cellulose xanthate dissolves in sodium hydroxide solution to give a viscous solution and is named as **Viscose**. When the viscose is passed through a spinneret (fine holes) into a dilute acid bath, xanthate is decomposed, the regenerate appearing in the form of threads of **Viscose Rayon**. These threads are wound and processed into fabric.

$$\text{Cell—O—}\overset{\overset{\text{S}}{\|}}{\text{S}}\text{—}\bar{\text{S}}\overset{+}{\text{Na}} \xrightarrow{\text{H}^+} \underset{\substack{\text{Cellulose} \\ (\textit{insoluble})}}{\text{Cell—OH}} + \text{CS}_2$$

Examination Questions

1. What are carbohydrates ? How are they classified ?
2. Establish the structure of glucose ?
3. Discuss the cyclic structure of D-glucose. *(Jharkhand BSc, 2015)*
4. Establish the structure of fructose.
5. Discuss the cyclic structure of D-fructose. *(Patna BSc, 2014)*
6. Discuss the structure of sucrose ?
7. What happens when glucose is treated with the following reagents :
   (*a*) Br_2/H_2O (*b*) Conc HNO_3 (*c*) HI/P (*d*) Phenylhydrazine
8. Most monosaccharides are highly water soluble. Explain.

Answer. The presence of a large number of hydroxyl groups on a relatively small organic molecule results in high water solubility. Favorable hydrogen bonding between monosaccharides and water occurs readily.

9. When pure á-D-glucopyranose is dissolved in water, the optical rotation of the resulting solution changes over a period of time. What is the name of this phenomenon and why does it occur ?

Answer. The phenomenon is called *mutarotation*. In solution, á-D-glucopyranose is converted to the equilibrium mixture of á- and â-anomers *via* the open-chain form.

10. What are the important industrial applications of cellulose ?
11. How will you distinguish between glucose and fructose ?
(Madras BSc, 2013 ; Kerala BSc, 2014 ; Indore BSc, 2015)

Answer. There are two ways : (*a*) Glucose rotates plane polarized light to the right (It is dextrorotatory). Fructose rotates the plane of polarized light to the left (It is levorotatory).

(*b*) Glucose decolorizes red color of bromine water. It is oxidized to gluconic acid. Fructose does not react with bromine water.

$$\underset{\text{Glucose}}{\begin{array}{c}\text{CHO}\\|\\(\text{CHOH})_4\\|\\\text{CH}_2\text{OH}\end{array}} + \underset{\text{red}}{\text{Br}_2 + \text{H}_2\text{O}} \longrightarrow \underset{\substack{\text{Gluconic acid}\\\text{colourless}}}{\begin{array}{c}\text{COOH}\\|\\(\text{CHOH})_4\\|\\\text{CH}_2\text{OH}\end{array}} + 2\text{HBr}$$

12. How will you distinguish between glucose and sucrose (cane sugar) ?
(Andhra BSc, 2014 ; Delhi BSc, 2015)

Answer. Glucose is a reducing sugar. It reacts with ammoniacal solution of silver oxide (or silver nitrate) to form silver mirror. Sucrose is a non-reducing sugar. It does not react with Tollens' reagent.

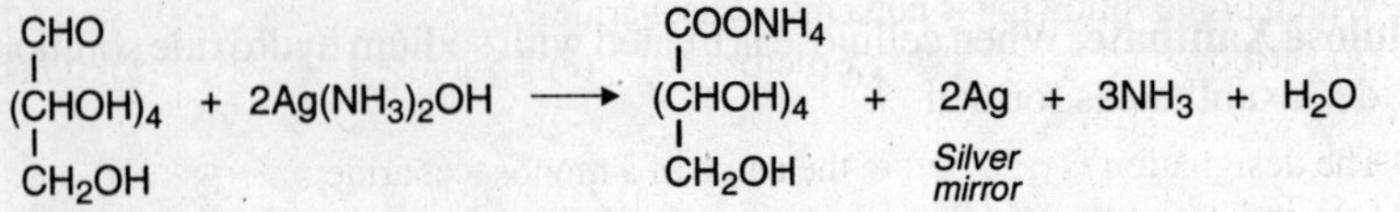

Glucose

13. Write a note on : Mutarotation. *(North Eastern Hill BSc, 2013 ; Orissa BSc, 2015)*

14. Write a note on : Inversion of cane sugar.

15. Write a note on : Ruff degradation. *(Jiwaji BSc, 2013)*

16. Write a note on : Killiani Fischer synthesis. *(Manipur BSc, 2015)*

17. Explain why D-glucose and D-fructose form the same osazone.

18. Explain why the optical rotation of the aqueous solution of D(+)-glucose changes with time and finally becomes static.

19. Explain why sucrose is not a reducing sugar. *(Ranchi BSc, 2014)*

20. Starting from glucose, how will you prepare : *(a)* Sorbitol ; and *(b)* *n*-Hexane.

21. Starting from glucose how will you prepare : *(a)* Glucosazone ; and *(b)* Glucaric acid.

22. How will you convert glucose into fructose. *(Delhi BSc, 2015)*

23. How will you convert fructose into glucose ? *(Banaras BSc, 2014)*

24. How will you convert glucose into arabinose and *vice versa* ?

25. How will you convert :

(a) An aldohexose into an aldopentose

(b) An aldopentose into an aldohexose

26. Compound (A) ($C_4H_8O_4$) gives a positive test with Tollen's reagent. Treatment of (A) with bromine water produces an optically active compound (B). Oxidation of (A) with nitric acid gives an optically inactive compound (C). Compound (A) can be prepared from L-glyceraldehyde by a Kiliani-Fischer synthesis. What are the structures of (A), (B), and (C) ?

Answer. The compounds are :

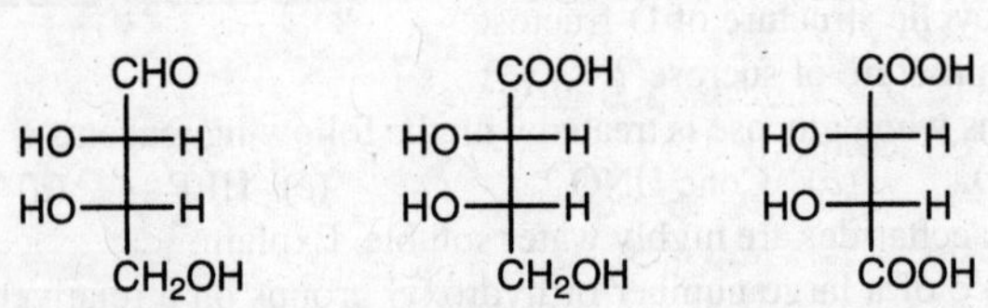

(A) (B) (C)

Multiple Choice Questions

1. How many molecules of carbon dioxide and water are needed to make a molecule of glucose in photosynthesis ?

(a) $5CO_2$ and $5H_2O$ *(b)* $5CO_2$ and $6H_2O$

(c) $6CO_2$ and $6H_2O$ *(d)* $6CO_2$ and $2H_2O$

Answer. *(c)*

2. Monosaccharides are classified according to :

(a) the number of carbon atoms in the molecule.

(b) whether they contain an aldehyde or a ketone group.

(c) their configurational relationship to glyceraldehyde.

(d) all of the above.

Answer. *(d)*

3. Which is a monosaccharide ?

(a) Sucrose *(b)* Maltose *(c)* Galactose *(d)* Cellulose

Answer. *(c)*

4. Which is a disaccharide ?

(a) Glucose *(b)* Maltose *(c)* Fructose *(d)* Cellulose

Answer. *(b)*

5. Which of the following is not a monosaccharide ?
(*a*) Ribose (*b*) Fructose (*c*) Sucrose (*d*) Glucose

Answer. (*c*)

6. The designation D or L before the name of a monosaccharide
(*a*) indicates the direction of rotation of polarized light.
(*b*) indicates the length of the carbon chain in the carbohydrate.
(*c*) indicates the position of the OH group on the carbon next to the primary alcohol group.
(*d*) indicates the position of the asymmetric carbon atoms in the carbohydrate.

Answer. (*c*)

7. The principal sugar in blood is
(*a*) Fructose (*b*) Glucose (*c*) Sucrose (*d*) Galactose

Answer. (*b*)

8. Glucose cannot be classified as
(*a*) a hexose (*b*) an oligosaccharide
(*c*) an aldose (*d*) a monosaccharide

Answer. (*b*)

9. Which of the following statements is false about glucose ?
(*a*) it is a reducing sugar.
(*b*) it is a disaccharide.
(*c*) it has a pyranose form.
(*d*) it is a polyalcohol.

Answer. (*b*)

10. Which of the following statements is false about α-D-glucose ?
(*a*) it has a pyranose ring.
(*b*) it is a hemiacetal.
(*c*) it shows mutarotation.
(*d*) it is the purest form of table sugar.

Answer. (*d*)

11. α-D-Glucose is different from β-D-glucose
(*a*) in the configuration at C-1
(*b*) because they are mirror images of each other
(*c*) because they are enantiomers
(*d*) because they are geometrical isomers

Answer. (*a*)

12. α-D-Glucopyranose is a(n) :
(*a*) hemiacetal (*b*) hemiketal (*c*) acetal (*d*) ketal

Answer. (*a*)

13. The number of asymmetric carbon atoms in the α-D-glucopyranose molecule is :
(*a*) 2 (*b*) 3 (*c*) 4 (*d*) 5

Answer. (*d*)

14. Which of the following statements is false about an aldohexose ?
(*a*) It is a monosaccharide.
(*b*) It contains a potential aldehyde group.
(*c*) α-D-Glucopyranose is an aldohexose.
(*d*) Fructose is an aldohexose.

Answer. (*d*)

15. All of the following monosaccharides give the same osazone except
(*a*) Galactose (*b*) Glucose (*c*) Fructose (*d*) Mannose

Answer. (*a*)

16. Mutarotation is a term related to
(*a*) Interconversion of anomers
(*b*) Relationship of D- and L- families
(*c*) Hydrolysis of sucrose
(*d*) Number of monsaccharides in a carbohydrate

Answer. (*a*)

17. The mutarotation of glucose is characterized by :

(*a*) a change from an aldehyde to ketone structure.
(*b*) a change of specific rotation from a (+) to a (–) value.
(*c*) the presence of an intramolecular bridge structure.
(*d*) the irreversible change from α-D to the β-D form.

Answer. (*c*)

18. Which of the following statements is false about glyceraldehyde ?

(*a*) Its IUPAC name is 1,2-dihydroxypropanal
(*b*) It is isomeric with 1,3-dihydroxypropanone
(*c*) It is optically active
(*d*) It shows mutarotation

Answer. (*d*)

19. Common table sugar is

(*a*) Glucose (*b*) Sucrose (*c*) Fructose (*d*) Maltose

Answer. (*b*)

20. Which of the following carbohydrates is sweeter than sucrose ?

(*a*) Glucose (*b*) Fructose (*c*) Lactose (*d*) None of these

Answer. (*b*)

21. By approximately what factor is the sweetness of saccharin greater than that of sugar ?

(*a*) 5 (*b*) 50 (*c*) 500 (*d*) 1000

Answer. (*c*)

22. Which of the following statements is false about sucrose ?

(*a*) It is also called table sugar.
(*b*) It may be fermented by yeast to produce alcohol.
(*c*) It reduces Fehling's solution.
(*d*) It does not reduce Tollens' reagent.

Answer. (*c*)

23. The sugar that yields only glucose on hydrolysis is

(*a*) Lactose (*b*) Sucrose (*c*) Maltose (*d*) Fructose

Answer. (*c*)

24. A reducing sugar will

(*a*) react with Fehling's test
(*b*) not react with Fehling's test
(*c*) have fewer calories
(*d*) always be a ketone

Answer. (*a*)

25. Which of the following carbohydrates is not a reducing sugar ?

(*a*) Glucose (*b*) Sucrose (*c*) Fructose (*d*) Lactose

Answer. (*b*)

26. Which of the following carbohydrates will not give a red precipitate of Cu_2O when heated with Benedict's solution ?

(*a*) Maltose (*b*) Glucose (*c*) Sucrose (*d*) Fructose

Answer. (*c*)

27. Which of the following compounds reduces Tollens' reagent ?

(*a*) Glucose (*b*) Sucrose (*c*) Methanol (*d*) Acetic acid

Answer. (*a*)

28. The reagent that can be used to differentiate an aldose and a ketose is :

(*a*) Bromine water (*b*) Fehling's solution
(*c*) Tollens' reagent (*d*) None of these

Answer. (*a*)

29. In the Fischer projection of D-(+)-glyceraldehyde, the hydroxyl group on the asymmetric carbon center is:

(*a*) on the bottom. (*b*) on the top.
(*c*) at the left. (*d*) at the right.

Answer. (*d*)

30. All chiral D-sugars rotate plane-polarized light:

(*a*) clockwise.
(*b*) counterclockwise.
(*c*) in a direction that cannot be predicted but must be determined experimentally.
(*d*) because they are optically inactive.

Answer. (*c*)

31. In a Haworth projection of a five-membered cyclic hemiacetal,

(*a*) the C6 carbon is drawn on the right.
(*b*) the C2 carbon is drawn on the left.
(*c*) the ring oxygen is drawn in the front.
(*d*) the ring oxygen is drawn in the back.

Answer. (*d*)

32. In solution, glucose exists as:

(*a*) the open-chain form only.
(*b*) the cyclic hemiacetal form only.
(*c*) an equilibrium mixture of the open-chain form and cyclic acetal forms.
(*d*) an equilibrium mixture of the open-chain form and cyclic hemiacetal forms.

Answer. (*d*)

33. Fructose forms:

(*a*) a five-membered cyclic hemiacetal.
(*b*) a six-membered cyclic hemiacetal.
(*c*) a six-membered lactone.
(*d*) a five-membered lactone.

Answer. (*a*)

34. A pyranose with the hydroxyl group on the anomeric carbon pointing up in the Haworth structure is designated:

(*a*) α (*b*) β (*c*) (*d*) δ

Answer. (*b*)

35. The á and â anomers of a D-glucopyranose are related as:

(*a*) enantiomers. (*b*) meso compounds.
(*c*) structural isomers. (*d*) epimers.

Answer. (*d*)

36. Which of the following statements is *true* ?

(*a*) All monosaccharides in either the ketose of aldose family are reducing sugars.
(*b*). Only aldoses, but not ketoses, are reducing sugars.
(*c*) Only ketoses, but not aldoses, are reducing sugars.
(*d*) All disaccharides, not monosaccharides, are reducing sugars.

Answer. (*a*)

37. Which of the following would give a positive Tollen's test ?

(*a*) á-D-glucopyranose (*b*) methyl â-D-glucopyranoside
(*c*) sucrose (*d*) methyl á-D-ribofuranoside

Answer. (*a*)

38. Which of the following is a reducing sugar ?

(*a*) á-D-fructofuranose (*b*) the methyl glycoside of cellobiose
(*c*) ethyl â-D-glucopyranoside (*d*) methyl á-D-fructofuranoside

Answer. (*a*)

39. Treatment of a monosaccharide with silver oxide and excess methyl iodide will:

(*a*) methylate all hydroxyl groups present.
(*b*) cleave the sugar between C5 and C6.
(*c*) cleave the sugar between C1 and C6.
(*d*) methylate only the hydroxyl at C1.

Answer. (*a*)

40. Two molecules of phenylhydrazine condense with each molecule of a given aldohexose to give an osazone in which:

(*a*) a phenylhydrazone derivative is formed at C1 only.
(*b*) a phenylhydrazone is formed at C2 only.
(*c*) both C1 and C2 are converted to phenylhydrazones.
(*d*) C1 and C2 are connected by a carbon-carbon double bond.

Answer. (*c*)

41. What derivative results when an aldopentose is treated with excess phenylhydrazine and catalytic acid ?

(*a*) pyranose (*b*) furanose (*c*) aldonic acid (*d*) osazone

Answer. (*d*)

42. What is the result of reaction involving D-glyceraldehyde and $NaBH_4$?

(*a*) ethanol and CO_2 (*b*) dihydoxyacetone
(*c*) glyceric acid (*d*) glycerol

Answer. (*d*)

43. Sucrose is hydrolyzed to what two monosaccharides ?

(*a*) glucose and fructose (*b*) glucose and galactose
(*c*) maltose and fructose (*d*) galactose and lactose

Answer. (*a*)

44. A linkage of two anomeric carbons is found in:

(*a*) cellobiose. (*b*) sucrose. (*c*) gentiobiose. (*d*) amylose.

Answer. (*b*)

45. Cellobiose is composed of D-glucose units linked by:

(*a*) â-1,6' glycosidic bonds. (*b*) á-1,6' glycosidic bonds.
(*c*) á-1,4' glycosidic bonds. (*d*) â-1,4' glycosidic bonds.

Answer. (*d*)

46. Cellulose and amylose are:

(*a*) 1,4' linked polymers of D-glucose which differ in the stereochemistry of this linkage.
(*b*) 1,6' linked polymers of D-glucose which differ in the stereochemistry of this linkage.
(*c*) polymers of D-glucose which are structurally isomeric.
(*d*) polymers of D-galactose which are structurally isomeric.

Answer. (*a*)

47. Starch

(*a*) is a trisaccharide.
(*b*) is also called amylose.
(*c*) is also called amylopectin.
(*d*) is a mixture of amylose + amylopectin.

Answer. (*d*)

48. The monosaccharide obtained by hydrolysis of starch is :

(*a*) D-Glucose (*b*) Maltose (*c*) D-Galactose (*d*) D-Ribose

Answer. (*a*)

49. Which of the following statements is false about cellulose ?

(*a*) It is a polymer of glucose molecules joined in β-1,4 linkages.
(*b*) It is a major component of cotton.
(*c*) It is used in the manufacture of Dacron fibres.
(*d*) It is used in the manufacture of Rayon fibres

Answer. (*c*)

50. Which of the following products is not derived from cellulose ?

(*a*) Rayon (*b*) Insulin (*c*) Gun cotton (*d*) Paper

Answer. (*b*)

32

Aromatic Compounds

> **There is a story told that one evening Kekule fell asleep while sitting in front of a fire. He dreamed about chains of atoms having the forms of twisting snakes. Suddenly one of the snakes caught hold of its own tail, forming a whirling ring. Kekule awoke, freshly inspired, and spent the remainder of the night working on his now-famous hypothesis. He is said to have written, "Let us learn to dream, gentlemen, and then perhaps we shall learn the truth."**

WHAT ARE AROMATIC COMPOUNDS ?

The term 'aromatic compounds' was first used by Kekule to classify benzene and its derivatives, many of which possessed fragrant odor or 'aroma'. This he did to distinguish them from the already known aliphatic compounds which lacked such odor. As new facts came to light, the term 'aromatic' came to be associated with 'chemical stability' rather than any aroma. With the development of structural theory, the stable character of aromatic compounds was attributed to the presence of a ring structure specific of benzene. So that, now **the term aromatic compounds stands for the whole series of compounds which contain one or more benzene rings in their molecule**. With the introduction of the new name of 'Arenes' for all aromatic hydrocarbons (Benzene, Naphthalene, Anthracene, etc.), a precise definition of aromatic compounds may be given as **Arenes and their Derivatives**.

Like the aliphatic compounds, aromatic compounds include the hydrocarbons, hydroxy derivatives (phenols), ethers, aldehydes, ketones, carboxylic acids, amines etc. We will discuss these in the following chapters. Besides, a number of relatively complex compounds, dyes, drugs, explosives, alkaloids and terpenes will also form a subject of our study.

COAL AS A SOURCE OF AROMATIC COMPOUNDS

Till about 1950, the chief source of benzene and other aromatic compounds was Coal. With the development of the Petroleum Industry, this source is no longer that important.

Coal is formed by the thermal decomposition of trees buried under earth's crust long time back. It has a sieve carbon structure in which complex aromatic molecules are entrapped.

Carbonization of Coal. When coal is heated in steel retorts at high temperatures in the absence of air, it *carbonizes* to coke. The process called **Carbonization** or **Coking** is accompanied by evolution of gaseous products. On condensation, these products yield a black viscous oil, the **Coal Tar**, aqueous ammonia (ammonia liquor), and the coal gas that escapes uncondensed.

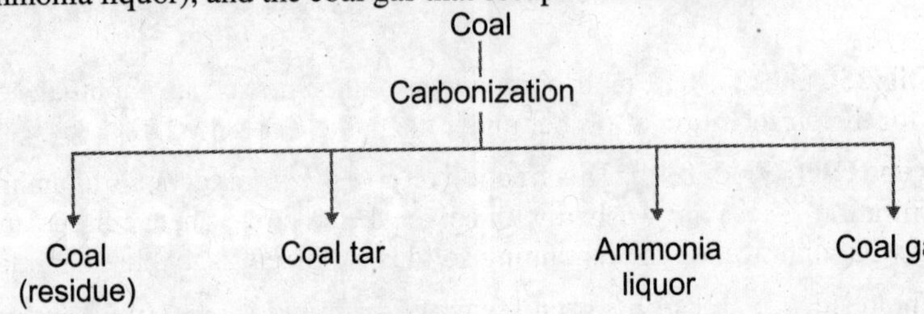

Until 1955, the carbonization of coal was done for the manufacture of coal gas needed as domestic fuel. With the availability of *natural gas* or *oil gas* from petroleum refineries, the carbonization of coal for gas manufacture has ceased. So that, the carbonization of coal is now exclusively carried to meet the demand of *coke* and *coal gas* for the steel industry. The process is operated at temperature in the range 1200-1350°C. The coal tar that is obtained contains a high proportion of aromatic compounds.

DISTILLATION OF COAL TAR

Coal tar is a black viscous distillate obtained in the coking of coal by the steel industry. It is a mixture of a large number of aromatic compounds which are isolated by distillation processes. In an ordinary continuous distillation process, coal tar yields the following fractions.

Table 32.1 FRACTIONS OF COAL TAR

	Fraction	Temperature Range	Chief Components
I.	LIGHT OIL	up to 200°C	Benzene, Toluene, Xylene
II.	MIDDLE OIL	200 — 250°C	Naphthalene, phenol, cresols
III.	HEAVY OIL	250 — 300°C	Naphthalene, cresols
IV.	ANTHRACENE OIL	300 — 350°C	Anthracene
V.	PITCH	Residue	

The various fractions listed above are worked up for their aromatic components.

I. Light Oil (up to 200°C). Besides benzene, toluene and xylenes, the light oil fraction contains small amounts of acidic substances, phenol and cresols ; and basic substances such as *or* pyridine. It is treated with concentrated sulfuric acid at low temperature to remove bases. Then it is neutralized with sodium hydroxide which removes phenols and excess acid. The oil thus purified on redistillation yields benzene (bp 80°C), toluene (bp 110°C), and xylenes (bp 135-145°C).

$$\text{Light oil} \xrightarrow[\text{Cold}]{H_2SO_4} \begin{matrix}\text{Bases}\\\text{removed}\end{matrix} \xrightarrow{\text{NaOH}} \begin{matrix}\text{Phenols}\\\text{removed}\end{matrix} \xrightarrow{\text{Distil}} \begin{matrix}\text{Benzene}\\\text{Toluene}\\\text{Xylenes}\end{matrix}$$

II. Middle Oil (200–250°C). It contains naphthalene, phenol and cresols. The naphthalene settles out upon cooling, is separated by centrifuging, and is purified by sublimation. After the naphthalene is removed, phenol and cresols are recovered from the remainder oil by treating with sodium hydroxide solution and then saturating with carbon dioxide. The mixture of phenols so liberated is fractionally distilled to yield phenol (bp 183°C) and a mixture of isomeric cresols (bp 191—201°C).

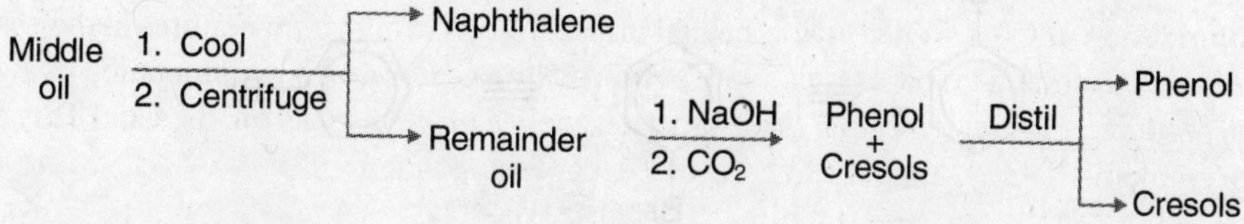

III. Heavy Oil (250–300°C). The main components of this fraction are naphthalene and cresols. It is used as such for the preservation of timber under the name **Creosote Oil**.

IV. Anthracene Oil (300–350°C). This fraction on cooling forms crystals of anthracene which are removed by filtration. These are washed with solvents to remove phenanthrene and carbazole, also present along with anthracene. The remaining solid is sublimed to get pure anthracene.

V. Pitch. The residue in the still is used for water proofing of roofs, road surfacing, and for making acid-resistant stoneware.

PETROLEUM AS A SOURCE OF AROMATIC COMPOUNDS

Until about 1940, coal tar was the only source of aromatics. In the past sixty years, the demand for aromatic compounds has far outstripped the amount available from coal tar. Consequently, petroleum has now become a major source of aromatics.

C_6–C_8 Fraction of petroleum naphtha is heated over Cr_2O_3–Al_2O_3 at 500°C at 15 atm pressure (Aromatization) or Pt–Al_2O_3 at 500°C at 15 atm pressure (Platforming). This process converts alkanes and cycloalkanes to arenes. The cycloalkanes undergo dehydrogenation to form arenes directly. The alkanes first undergo dehydrogenation to form cycloalkanes (Cyclization) which then undergo further dehydrogenation to give arenes (Aromatization).

Example 1. Conversion of *n*-hexane into benzene

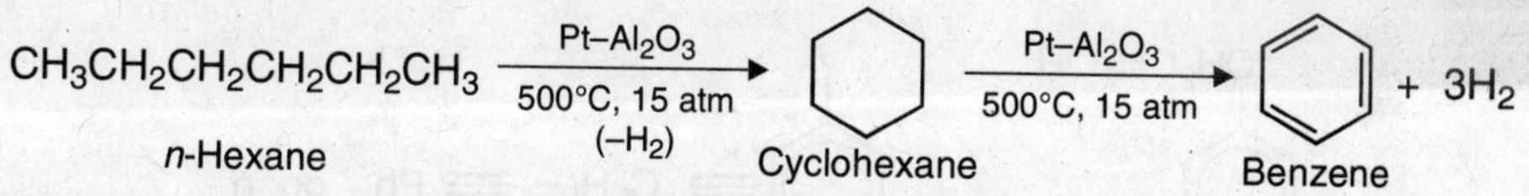

Example 2. Conversion of *n*-heptane into toluene

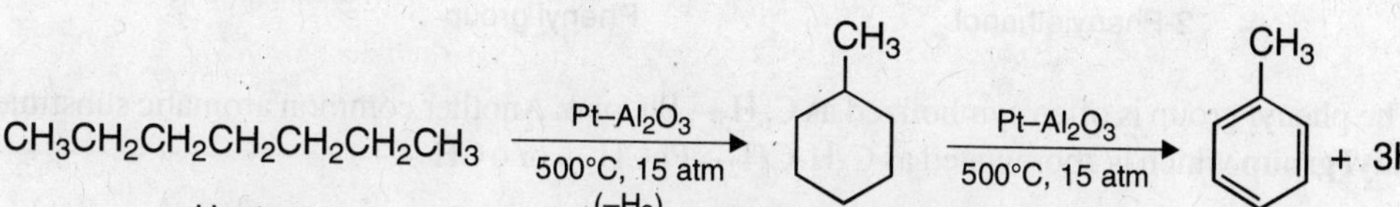

Similarly, xylenes are formed in the same process from dimethylcyclohexanes. The mixture of arenes that results is fractionated to give the individual arenes. Today, most of the benzene and almost all the toluene and xylenes, are derived from petroleum. These in turn are the basic raw materials for most other aromatic compounds.

NOMENCLATURE OF AROMATIC COMPOUNDS

As with aliphatic compounds, we encounter a blend of *common* and IUPAC names, while naming aromatic compounds. As a matter of fact, the use of common names is particularly prevalent.

A. Naming of Monosubstituted Benzenes. Benzene is represented by a regular hexagon with a circle inscribed in it. The monosubstituted derivatives of benzene are those in which a hydrogen of the ring is replaced by another substituent. These are named by prefixing the name of the substituent to the word 'benzene'. For example, benzene in which one of the hydrogens is replaced by bromine (Br), is named as bromobenzene. Here, the bromine atom may occupy any of the six positions on the benzene ring.

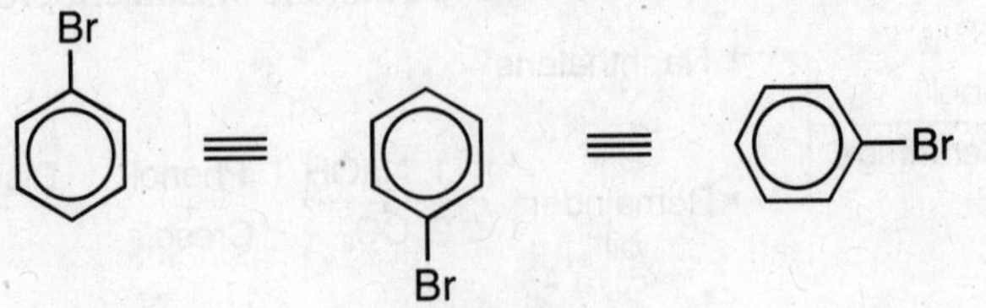

Bromobenzene

Other monosubstituted benzenes that can be named in the same manner are :

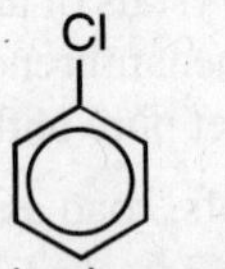

Chlorobenzene

Iodobenzene

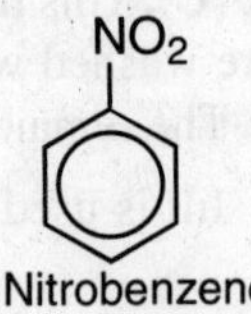

Nitrobenzene

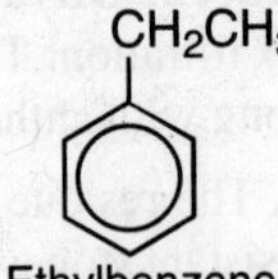

Ethylbenzene

Some benzene derivatives have *common names* which are currently accepted. For example,

Toluene

Phenol

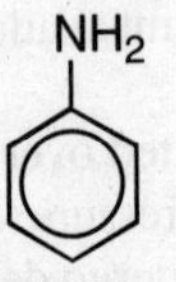

Aniline

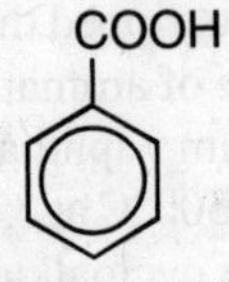

Benzoic acid

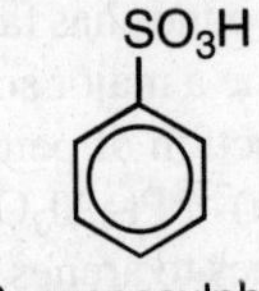

Benzenesulphonic acid

Sometimes, it is more convenient to name the benzene ring as a substituent, the **Phenyl group.**

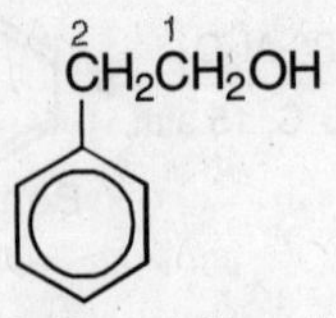

2-Phenylethanol

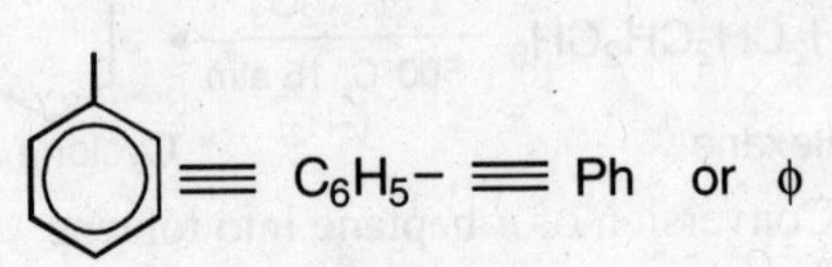

Phenyl group

The phenyl group is often symbolized as C_6H_5–, Ph, or ϕ. Another common aromatic substituent is **Benzyl group** which is abbreviated as $C_6H_5CH_2$–, $PhCH_2$–, or ϕCH_2–.

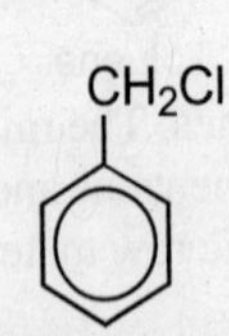

Benzyl chloride

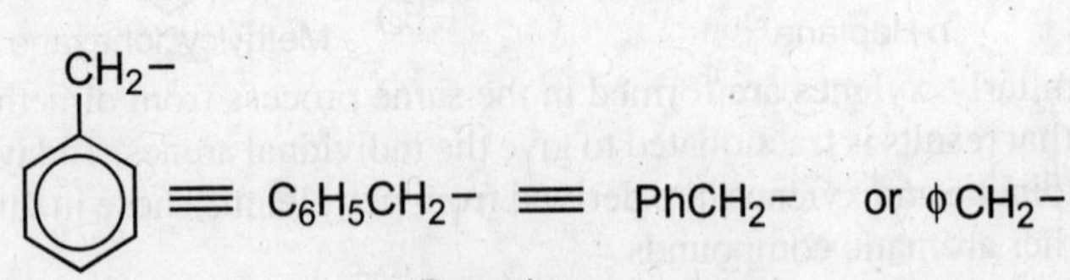

Benzyl group

B. Naming of Disubstituted Benzenes. When there are two substituents on the ring, three positional isomers are possible. Their relative positions are indicated by prefixes **Ortho** (*o*-), **Meta** (*m*-), and **Para** (*p*-), or by numbers. For example, the three isomers of dibromobenzene are named as :

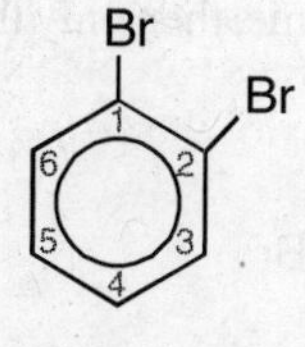

o-Dibromobenzene
(1,2–)

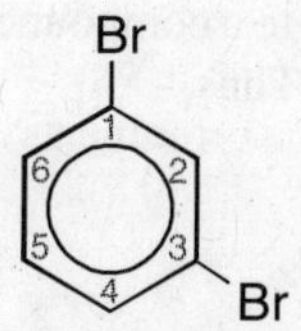

m-Dibromobenzene
(1,3–)

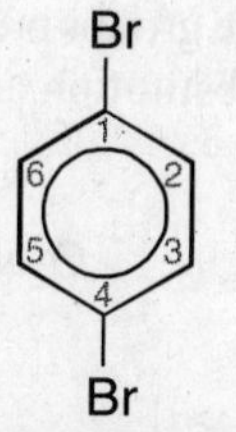

p-Dibromobenzene
(1,4–)

When a common name is used, the substituent which is responsible for the name (CH_3 in toluene ; OH in phenol) is considered on Carbon 1 (top position).

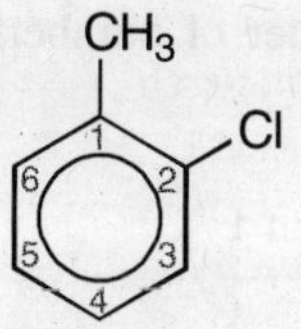

o-Chlorotoluene
2-Chlorotoluene

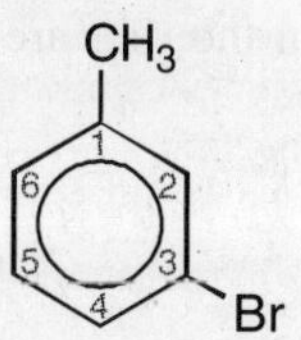

m-Bromotoluene
3-Bromotoluene

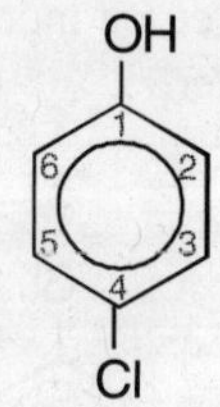

p-Chlorophenol
4-Chlorophenol

When two substituents are different, they are usually put in alphabetical order. Thus,

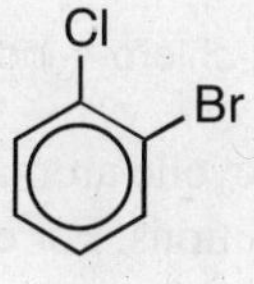

1-Bromo-2-chlorobenzene

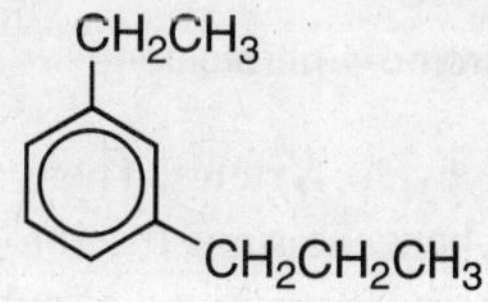

1-Ethyl-3-propylbenzene

C. Naming of Polysubstituted Benzenes. When three or more substituents are placed on the benzene ring, numbers must be used to designate their positions. One of the substituents is put on the top of the hexagon and numbered 1. The numbering of the remaining positions is done clockwise or anti-clockwise, whichever gives lowest possible group of numbers to the other substituents. For example,

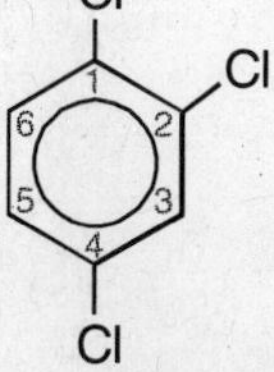

(Numbering clockwise)
1,2,4-Trichlorobenzene
right

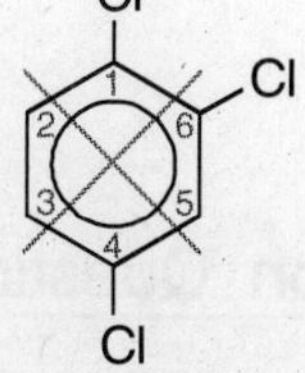

(Numbering anticlockwise)
1,4,6-Trichlorobenzene
wrong

The three possible isomeric trichlorobenzenes bear the following IUPAC names.

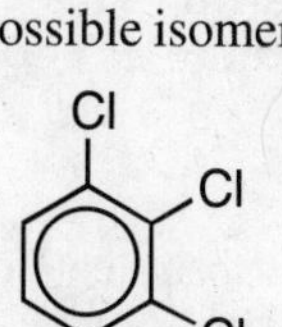

1,2,3-Trichlorobenzene

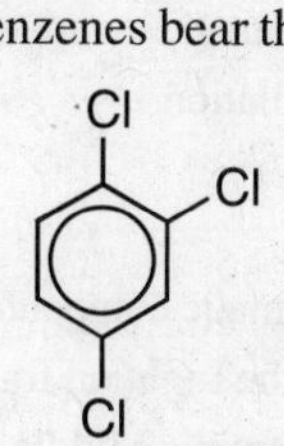

1,2,4-Trichlorobenzene

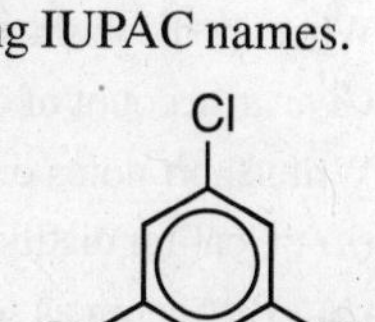

1,3,5-Trichlorobenzene

If one of the groups present gives rise to a compound with a special name, then only the remaining positions of substitution are mentioned. Thus,

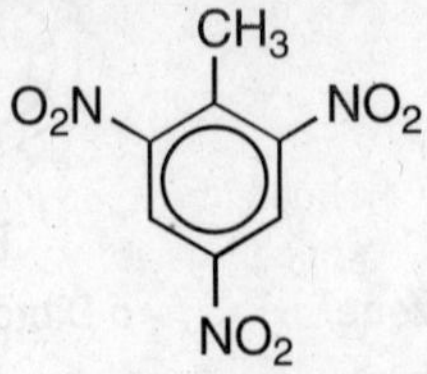

2,4,6-Trinitrotoluene

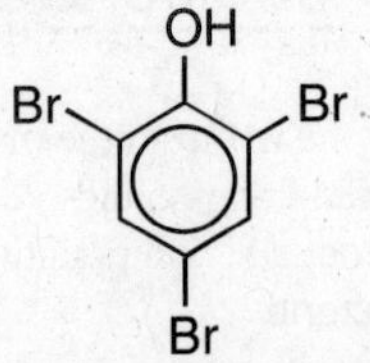

2,4,6-Tribromophenol

When three or more substituents on the ring are different, the order of numbering is in the alphabetic order.

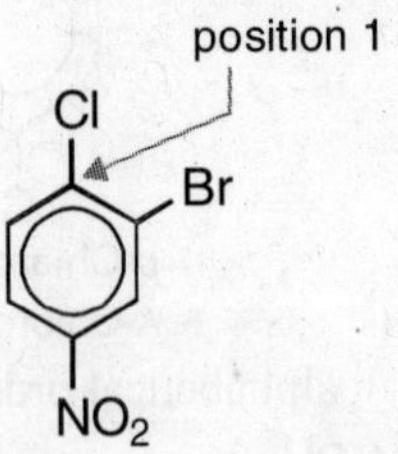

2-Bromo-4-nitrotoluene

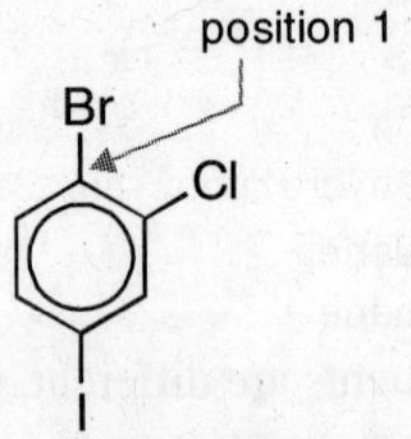

1-Bromo-2-chloro-4-iodobenzene

D. Fused Polycyclic Arenes. There are numerous polycyclic aromatic hydrocarbons (arenes) having one or more benzene rings fused together in *ortho* positions. For each an IUPAC numbering system is used to indicate positions of substituents.

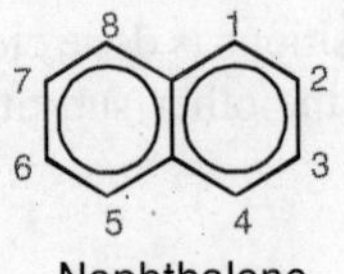

Naphthalene

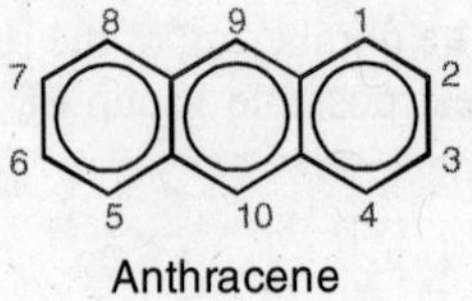

Anthracene

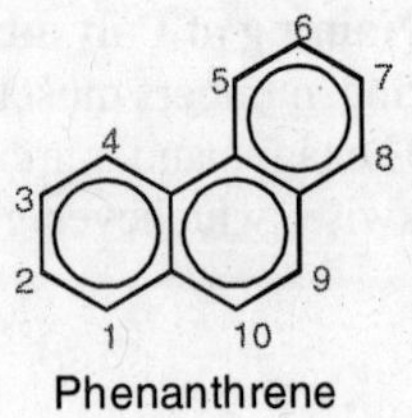

Phenanthrene

Examination Questions

1. What are aromatic hydrocarbons ? What are the main differences between aromatic and aliphatic compounds ?
2. What are the main sources of aromatic compounds ?
3. Give an account of coal-tar distillation.
4. Write short notes on :
   (*a*) Coal-tar distillation (*Manipal BSc, 2014*)
   (*b*) Petroleum as source of aromatic compounds
5. How is benzene obtained from the Light oil fraction of coal-tar distillation ?
6. How is naphthalene obtained from the Middle oil fraction of coal-tar distillation ?

Multiple Choice Questions

1. Coal-tar is the main source of :

(*a*) Aromatic compounds (*b*) Aliphatic compounds
(*c*) Heterocyclic compounds (*d*) None of these

Answer. (*a*)

2. Which of the following fractions of coal-tar distillation contains benzene ?

(*a*) Light oil (*b*) Middle oil
(*c*) Heavy oil (*d*) Anthracene oil

Answer. (*a*)

3. Which of the following fractions of coal-tar distillation contains naphthalene ?

(*a*) Light oil (*b*) Middle oil
(*c*) Heavy oil (*d*) Anthracene oil

Answer. (*b*)

4. Aromatic hydrocarbons are also called :

(*a*) Arenes (*b*) Huckel's compounds
(*c*) Trienes (*d*) Alkoxy compounds

Answer. (*a*)

5. Which of the following is an aromatic compound ?

(*a*) $CH_3CH_2CH_2Br$ (*b*) $CH_3-\underset{}{CH}(CH_3)-CH_3$

(*c*) 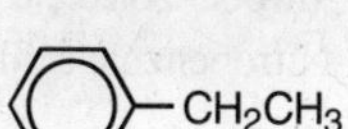

(*d*) 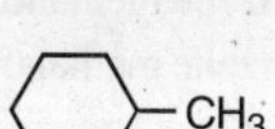

Answer. (*c*)

6. Which of the following compounds is *not* an aromatic compound ?

(*a*) 
(*b*) 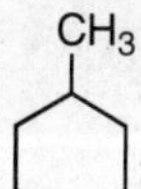

(*c*) 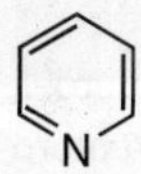

(*d*) 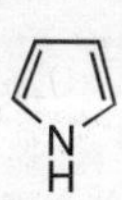

Answer. (*b*)

7. Which of the following compounds is known as aniline ?

(*a*) 

(*b*) 

(*c*) 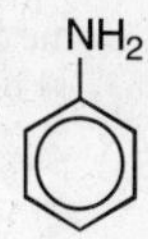

(*d*) 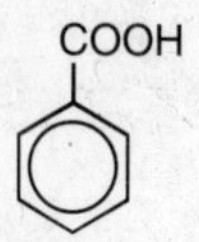

Answer. (*c*)

8. Which of the following structures represent *m*-dichlorobenzene ?

(*a*) 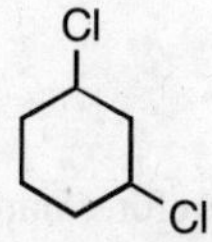

(*b*) 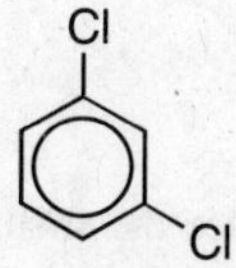

(*c*) 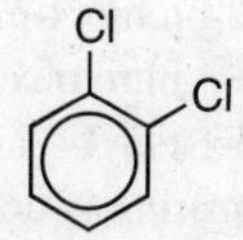

(*d*) 

Answer. (*b*)

9. Which of the following is also an acceptable name for 3-nitrophenol ?
(*a*) *o*-nitrophenol (*b*) *m*-nitrophenol
(*c*) *p*-nitrophenol (*d*) 3-cresol

Answer. (*b*)

10. Which of the following is another accepted name for methyl phenyl ketone ?
(*a*) anisole (*b*) cresol
(*c*) acetophenone (*d*) benzophenone

Answer. (*c*)

11. Which of the following is also an acceptable name for 1,3,5-trimethylbenzene ?
(*a*) Styrene (*b*) Mesitylene
(*c*) Cumene (*d*) 1,3,5-Xylene

Answer. (*b*)

12. The best name for the following compound is :

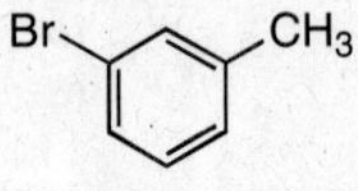

(*a*) 3-Methyl-bromobenzene (*b*) 3-Bromoaniline
(*c*) 3-Methylbromobenzene (*d*) 3-Bromotoluene

Answer. (*d*)

13. The correct IUPAC name for the following compound is :

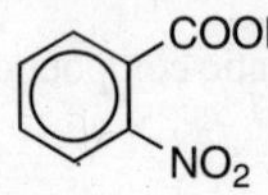

(*a*) *p*-Nitrobenzene methanoic acid (*b*) *o*-Nitrobenzoic acid
(*c*) *o*-Nitrobenzene methanoic acid (*d*) *m*-Nitrobenzoic acid

Answer. (*b*)

14. The correct IUPAC name for the following compound is :

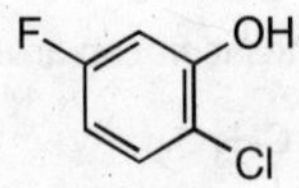

(*a*) 2-Chloro-5-fluoro-1-hydroxybenzene
(*b*) 1-Hydroxy-2-chloro-5-fluorobenzene
(*c*) 1-Fluoro-3-hydroxy-4-chlorobenzene
(*d*) None of these

Answer. (*d*). The name should be 2-Chloro-5-fluorophenol.

15. The correct IUPAC name for the following compound is :

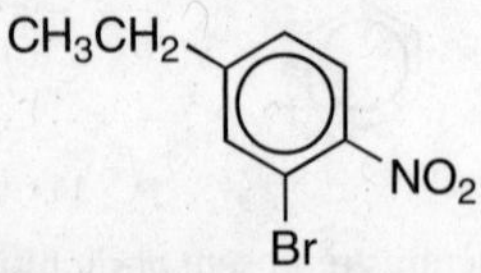

(*a*) 3-Bromo-4-ethyl-1-nitrobenzene
(*b*) 1-Ethyl-2-bromo-4-nitrobenzene
(*c*) 2-Bromo-4-nitrotoluene
(*d*) 3-Bromo-1-ethyl-4-nitrobenzene

Answer. (*d*).The numbering is done to produce the smallest total. Therefore, number from the ethyl group toward the nitro group. List the substituents in alphabetic order.

33

Benzene and its Homologues

Benzene and its homologues constitute the most important class of *Arenes* (aromatic hydrocarbons). Their discussion forms the basis for our study of all other aromatic compounds.

NOMENCLATURE

Benzene and its homologues are frequently called by the *Common Names* which are accepted by the IUPAC system. The homologues of benzene having a single alkyl group are named as **Alkylbenzenes**. These are made of a benzene ring to which is attached the alkyl group. The alkyl group bonded to a ring carbon is designated as the **Side-Chain**.

Benzene

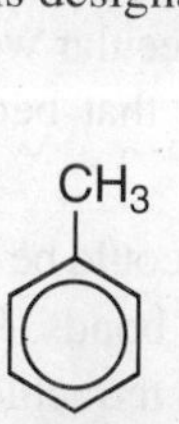

Toluene (Methylbenzene)

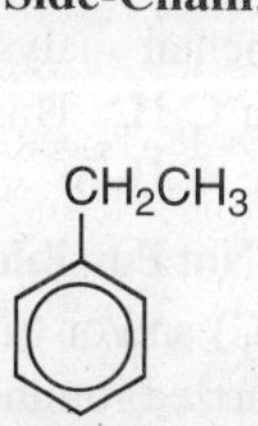

Ethylbenzene

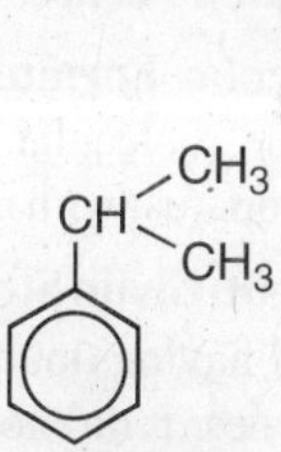

Cumene (Isopropylbenzene)

Dimethylbenzenes have the common name **Xylenes**. The three isomeric xylenes are :

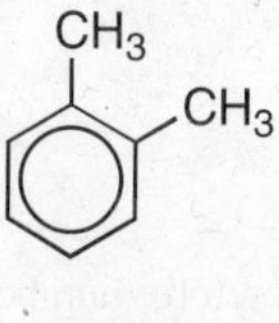

o-Xylene

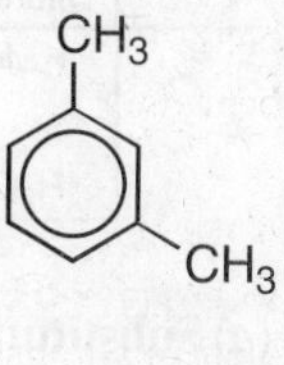

m-Xylene

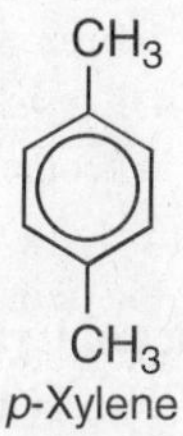

p-Xylene

Trimethylbenzenes are invariably assigned systematic names, the positions of CH_3–groups being shown by numbers. The three isomeric trimethylbenzenes are named as follows :

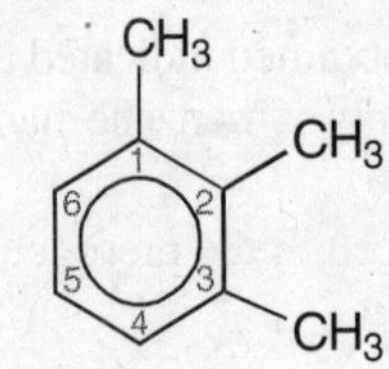

1,2,3-Trimethylbenzene

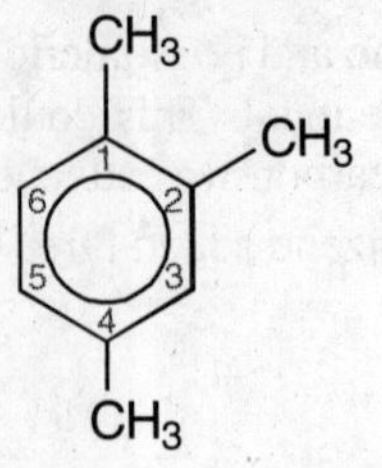

1,2,4-Trimethylbenzene

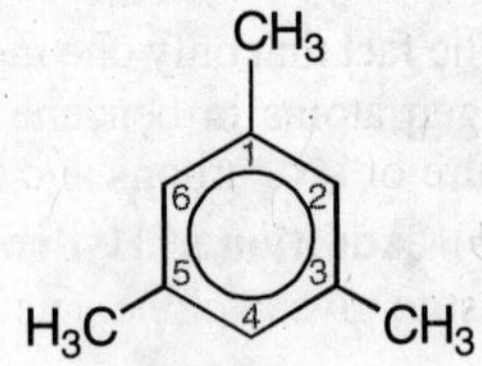

1,3,5-Trimethylbenzene

Aryl Groups. The groups resulting from arenes by dropping a H atom of the benzene ring are called *aryl groups* (Ar–).

Phenyl (Ph)

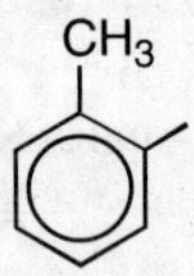

o-Tolyl

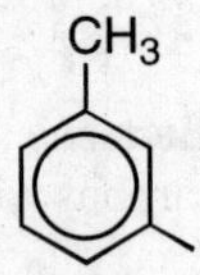

m-Tolyl

p-Tolyl

Aralkyl Groups. The groups derived by removal of H atom from the side-chain are called *aralkyl groups*. The groups furnished by toluene are

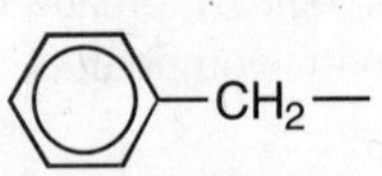

Benzyl

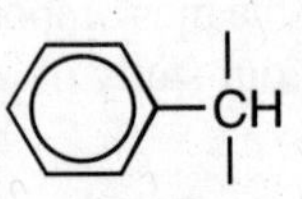

Benzal

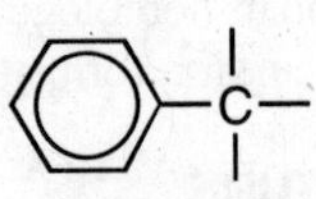

Benzo

STRUCTURE OF BENZENE

The structure of benzene has been derived as follows :

(1) **Molecular Formula**. Elemental analysis and molecular weight determination showed that benzene had the molecular formula C_6H_6. This indicated that benzene was a highly unsaturated compound (compare it with *n*-hexane, C_6H_{14}).

(2) **Straight-Chain Structure Not Possible**. Benzene could be constructed as a straight-chain or ring compound having double (C=C) and/or triple (C≡C) bonds. But benzene did not behave like alkenes or alkynes. It did not decolorize bromine in carbon tetrachloride or cold aqueous potassium permanganate. It did not add water in the presence of acids.

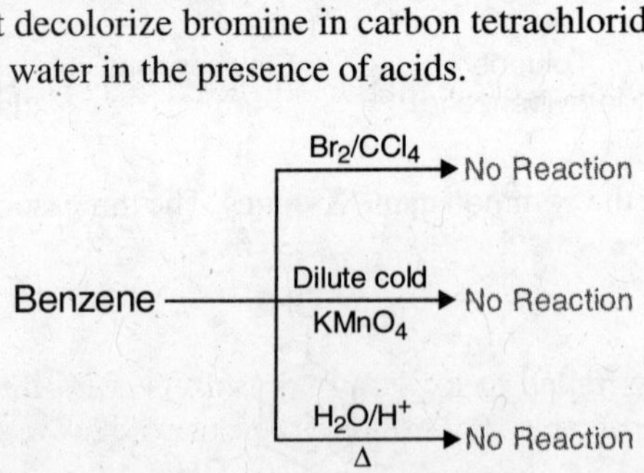

(3) **Evidence of Cyclic Structure**. (*a*) **Substitution of Benzene.** Benzene reacted with bromine in the presence of $FeBr_3$ (catalyst) to form monobromobenzene.

$$C_6H_6 + Br_2 \xrightarrow{FeBr_3} C_6H_5Br + HBr$$

Benzene Bromobenzene

The fact that only one monobromo and no isomeric products were obtained indicated that all six hydrogen atoms in benzene were identical. This could be possible only if benzene had a cyclic structure of six carbons and to each carbon was attached one hydrogen.

(*b*) **Addition of Hydrogen.** Benzene added three moles of hydrogen in the presence of nickel catalyst to give cyclohexane.

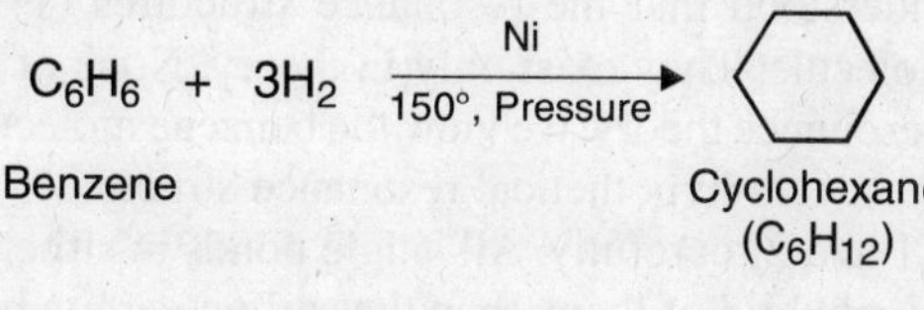

This confirmed the cyclic structure of benzene and also showed the presence of three carbon-carbon double bonds.

(4) **Kekule's Structure for Benzene**. In 1865, August Kekule suggested that benzene consisted of a cyclic planar structure of six carbons with alternate double and single bonds. To each carbon was attached one hydrogen. Benzene according to this proposal, was simple, 1,3,5-cyclohexatriene.

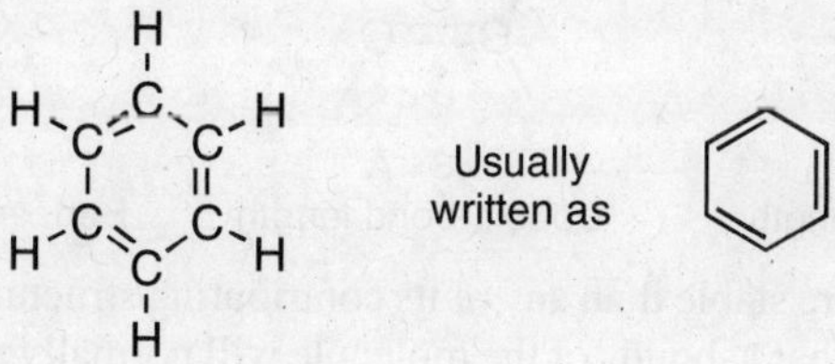

There were two objections : (*i*) If the Kekule's structure was correct, there should exist two *ortho* isomers of dibromobenzene. In one isomer, the two bromine atoms should be on carbons that are connected by a double bond, as shown in structure (*a*). In the other isomer, the bromines should be on carbons connected by a single bond, as shown in structure (*b*). In fact, only one *ortho*-dibromobenzene could be prepared.

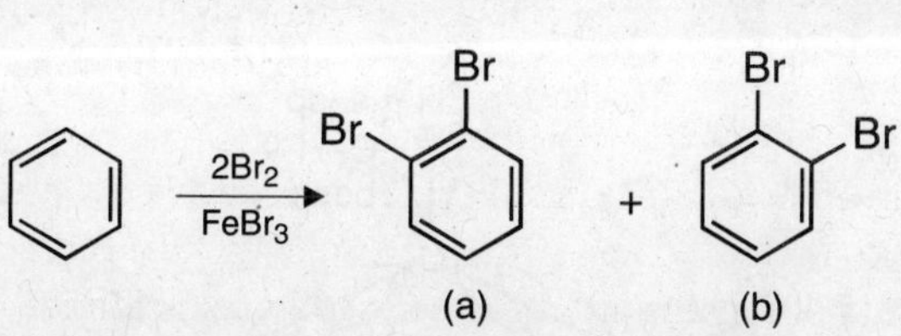

To overcome this objection, Kekule further suggested that benzene was a mixture of two forms (*1* and *2*) in rapid equilibrium.

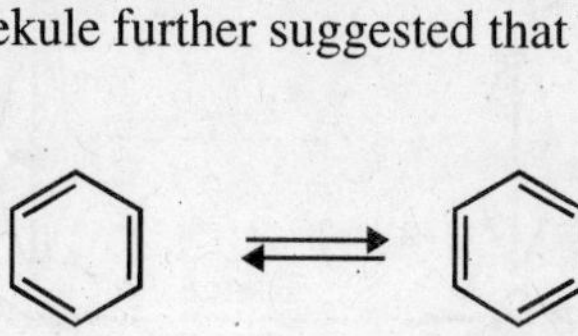

1 2

(*ii*) Kekule's structures failed to explain why benzene with three double bonds did not give addition reactions like other alkenes. For example, benzene did not react with HBr or Br_2 in CCl_4.

(5) **Resonance Description of Benzene**. The phenomenon in which two or more structures can be written for a substance which involve identical positions of atoms is called **Resonance**. The actual structure of the molecule is said to be **Resonance Hybrid** of various possible alternative structures. The alternative structures are referred to as the **Resonance Structures** or **Contributing Forms**. A

double headed arrow (↔) between the resonance structures is used to represent the resonance hybrid. Thus in the case of benzene, Kekule's structures (1) and (2) represent the resonance structures. Actual structure of the molecule may be represented as hybrid of these two resonance structures, or by the single structural formula (3).

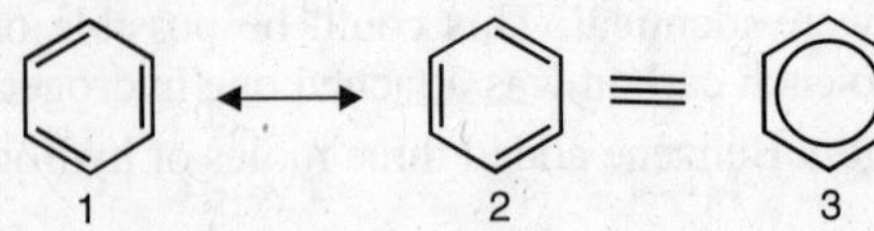

It should be clearly understood that the resonance structures (1) and (2) are not the actual structures of the benzene molecule. They exist only in theory. None of these structures adequately represents the molecule. In resonance theory, we view the benzene molecule (which of course is a real entity) as being hybrid of these two hypothetical resonance structures.

Look at the structures (1 and 2) carefully. All single bonds in structure (1) are double bonds in structure (2), that is consider a hybrid of them, then the carbon-carbon bonds in benzene are neither single bonds nor double bonds. Rather, they are something halfway between. This is exactly what we find experimentally. Spectroscopic measurements show that benzene is planar and that all of its carbon-carbon bonds are of equal length, 1.40 Å. This value lies in between the carbon-carbon single bond length (1.54 Å) and the carbon-carbon double bond length 1.34 Å).

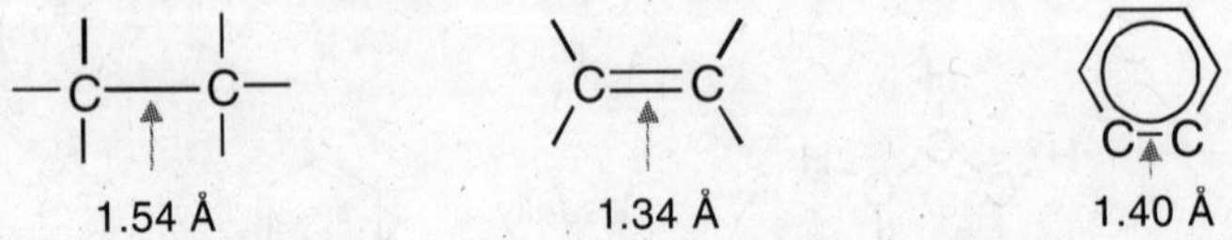

Resonance hybrid is more stable than any of its contributing structures. For benzene, the stability due to resonance is so great that π-bonds of the molecule will normally resist breaking. This explains lack of reactivity of benzene towards addition.

(6) **Molecular Orbital Structure of Benzene**. The structure of benzene is best described in terms of the modern molecular orbital theory. All six carbon atoms in benzene are sp^2 hybridized. The sp^2 hybrid orbitals overlap with each other and with *s* orbitals of the six hydrogen atoms forming C–C and C–H σ bonds (Fig. 33.1).

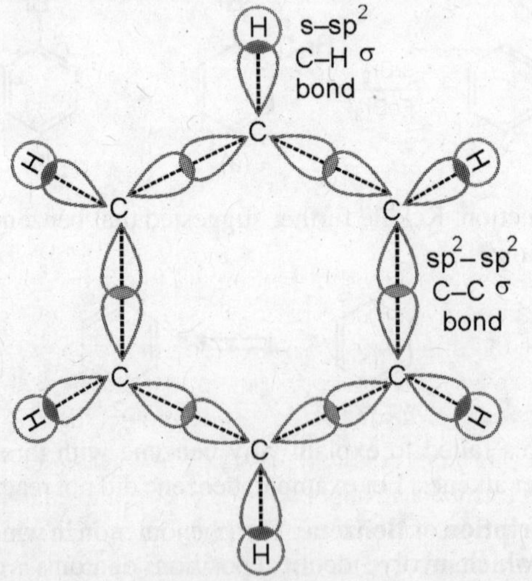

Fig. 33.1. Formation of σ-bonds in benzene.

Since the σ-bonds result from the overlap of planar sp^2 orbitals, all carbon and hydrogen atoms in benzene lie in the same plane. All σ-bonds in benzene lie in one plane and all bond angles are 120° (Fig. 33.2).

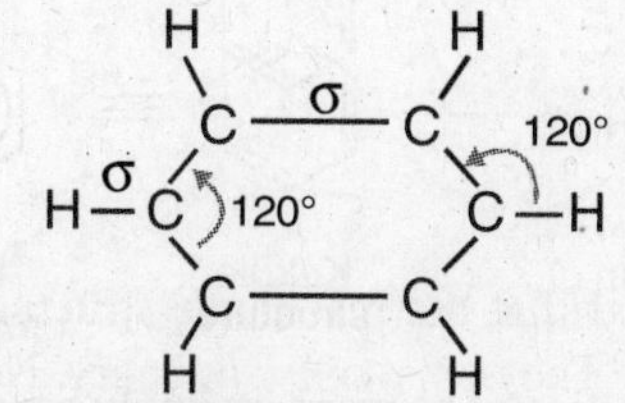

Fig. 33.2. All σ-bonds in benzene lie in one plane. All bond angles are 120°.

Also each carbon atom in benzene possesses an unhybridized *p*-orbital containing one electron. These *p*-orbitals are perpendicular to the plane of σ-bonds. The lateral overlap of these *p*-orbitals produces a π molecular orbital containing six electrons (Fig. 33.3). One half of this π molecular orbital lies above and the other half lies below the plane of the σ-bonds.

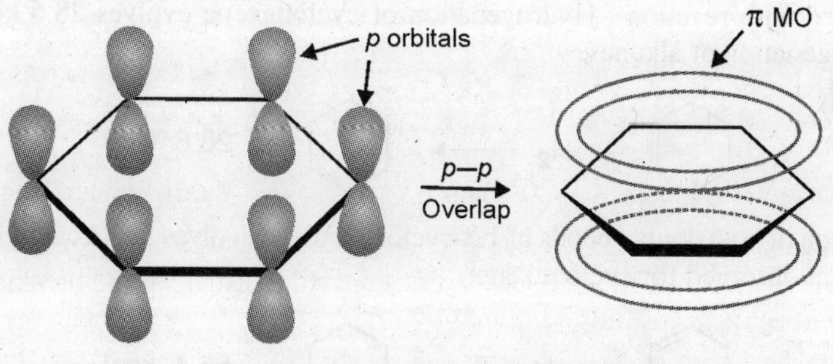

Fig. 33.3. Formation of π molecular orbital in benzene.

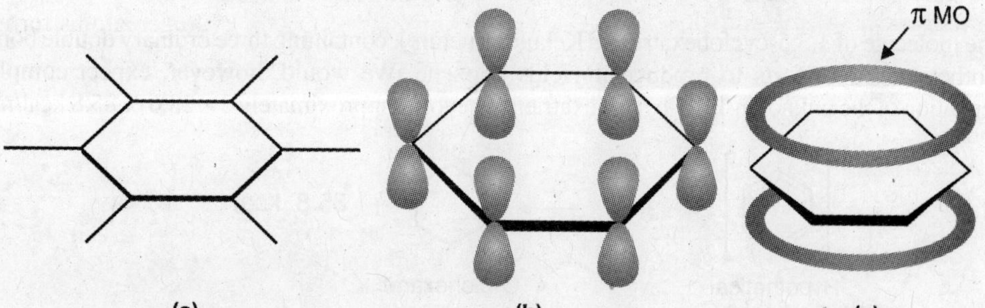

Orbital structure of benzene. All carbons atoms in benzene are sp^2 hybridized. (*a*) The carbon-carbon and carbon-hydrogen σ bonds are formed by sp^2-sp^2 and sp^2-*s* overlap. All σ bonds in benzene lie in one plane. (*b*) All carbon atoms in benzene have a *p* orbital on them. (*c*) The delocalized π molecular orbital is formed by the overlap of the six *p* orbitals (*c*) These *p* orbitals overlap to form a delocalized π molecular orbital.

The six electrons of the *p* orbitals cover all the six carbon atoms, and are said to be delocalized. As a result of delocalization is formed a stronger π-bond and a more stable molecule. Thermochemical data has actually shown that the stabilization energy of a delocalized π molecular orbital, as in benzene, is 36 kcal/mole compared to the *p*-orbitals forming three ordinary π-bonds as in 1,3,5-cyclohexatriene (Kekule structure). Thus benzene gives substitution reactions in which the stability of the benzene ring is retained.

There are three ways in which benzene can be represented :

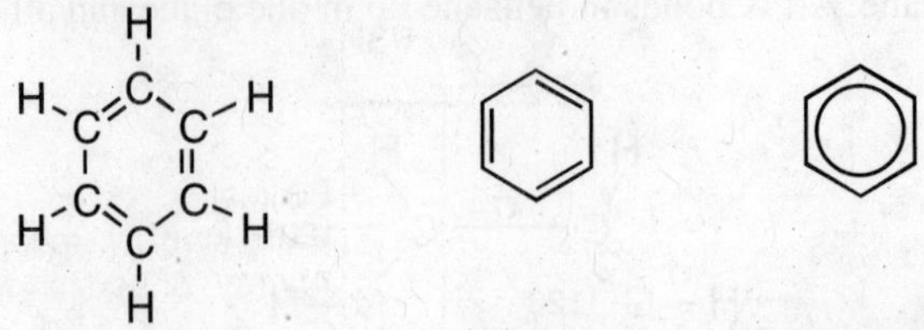

Expanded form | Kekule structure | Short-hand representation

A common short-hand representation of benzene is simply hexagon with a circle inside. The circle represents the π molecular orbital. In this book short-hand representation has been used ordinarily. Kekule structure has been used only in describing the reaction mechanisms where we want to keep track of the bonding electrons.

RESONANCE ENERGY OF BENZENE

Benzene's special stability is due to the formation of the delocalized π molecular orbital. The magnitude of this extra stability can be estimated by measuring the changes in heat of hydrogenations that are associated with reactions. Hydrogenation of cyclohexene evolves 28.6 kcal/mol, a value typical for hydrogenation of alkenes.

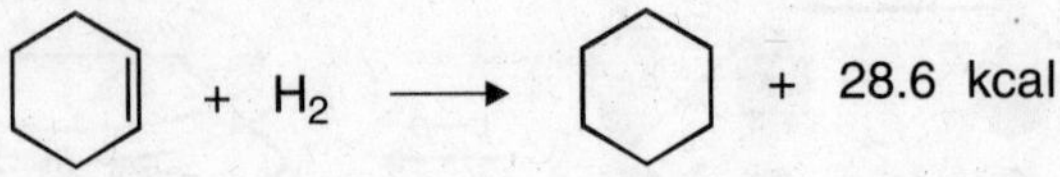

Hydrogenation of both double bonds of 1,3-cyclohexadiene evolves 54.4 kcal/mol, approximately double the amount observed for cyclohexene.

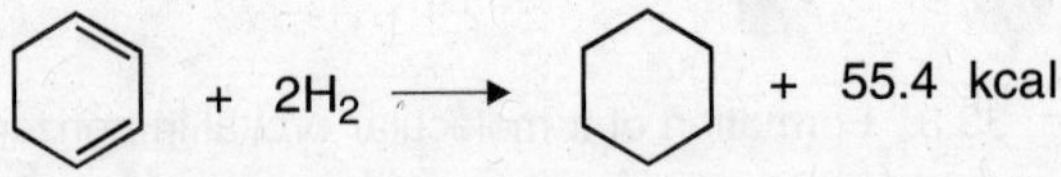

The molecule of 1,3,5-cyclohexatriene (Kekule structure), containing three ordinary double bonds, is hypothetical–any efforts to produce it yields benzene. We would, however, expect complete hydrogenation of the unknown 1,3,5-cyclohexatriene to evolve approximately 3×28.6 or 85.8 kcal/mol.

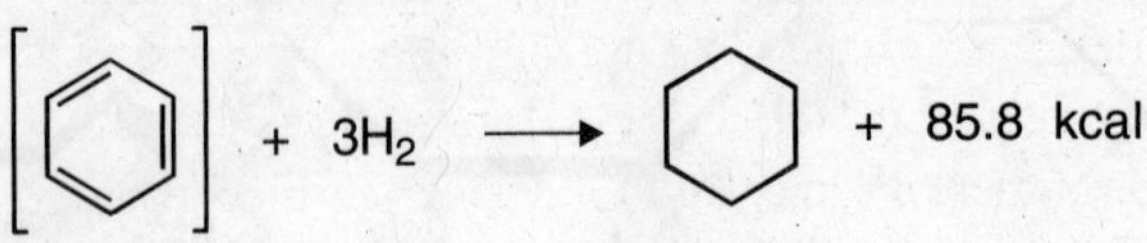

Hypothetical 1,3,5-Cyclohexatriene | Cyclohexane

Hydrogenation of benzene gives only 49.8 kcal/mole.

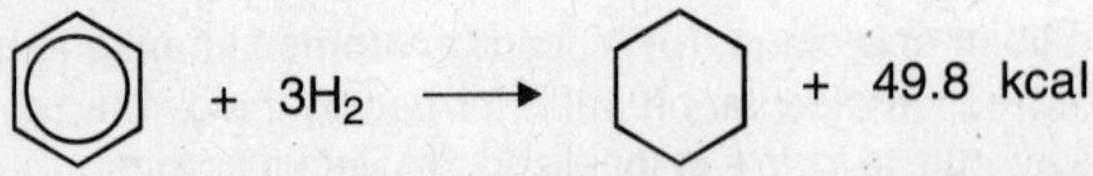

Figure 33.4 makes these energy relationships more evident. The 36 kcal difference between the heat evolved in the hydrogenation of benzene and that estimated for hydrogenation of a compound with three ordinary double bonds is the added stability. This added stability is sometimes called *Resonance Energy*. **Resonance energy is a measure of how much more stable a resonance hybrid structure is than its extreme resonance structures.**

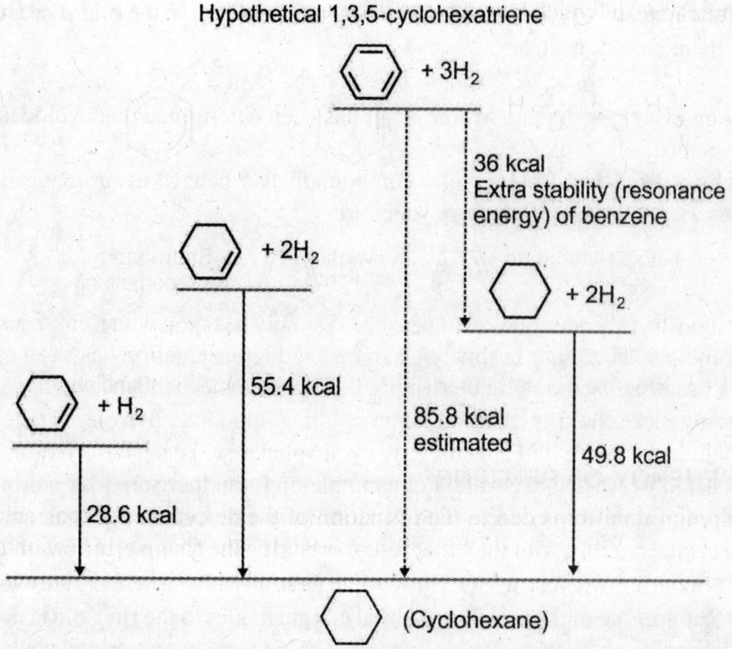

Fig. 33.4. Heats evolved on hydrogenation of one mole of some cyclic compounds.

AROMATICITY (HUCKEL RULE)

The aromatic compounds apparently contain alternate double and single bonds in a cyclic structure, and resemble benzene in chemical behavior. They undergo substitution rather than addition reactions. This characteristic behavior is called **Aromatic Character** or **Aromaticity**. *Aromaticity is, in fact, a property of the sp^2 hybridized planar rings in which the p orbitals (one on each atom) allow cyclic delocalization of π electrons.*

Criteria for Aromaticity

On the basis of the above considerations, can be laid down criteria or rules which help us in knowing whether a particular compound is aromatic or non-aromatic.

Rule 1. An aromatic compound is cyclic and planar.

Rule 2. Each atom in an aromatic ring has a *p* orbital. These *p* orbitals must be parallel so that a continuous overlap is possible around the ring.

Rule 3. The cyclic π molecular orbital (electron cloud) formed by overlap of *p* orbitals must contain $(4n+2)$ π electrons, where n = integer 0, 1, 2, 3 etc. This is known as **Huckel Rule**.

Let us apply these rules to the following examples :

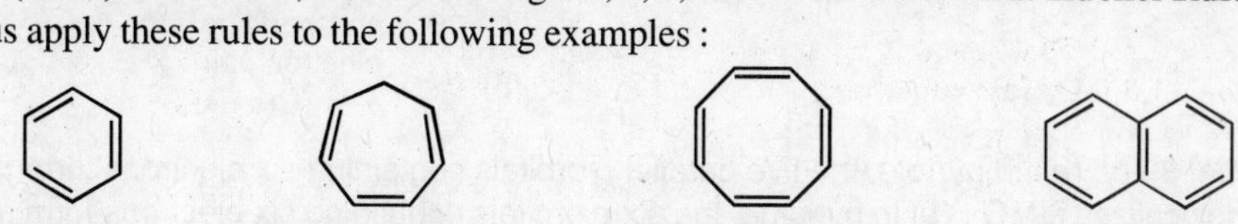

Benzene Cycloheptatriene Cyclooctatetraene Naphthalene

Benzene. It is a cyclic and planar compound. It has a *p* orbital on each carbon of the ring involved in a double bond. It has three double bonds and six π electrons, which is in accordance with Huckel rule.

$$4n+2=6 \quad or \quad 4n=6-2$$
$$\therefore \quad 4n=4 \quad and \quad n=1$$

Cycloheptatriene. It is cyclic and planar. It has three double bonds and six π electrons. But one of the carbons is saturated and does not possess a *p* orbital. Hence a continuous overlap around the ring is not possible. This compound is, therefore, non-aromatic.

Cyclooctatetraene. It is cyclic and has a *p* orbital on each atom of the ring. The Huckel rule is not satisfied, since there are 8π electrons.

$$4n + 2 = 8 \text{ equals } 4n = 6 \qquad \therefore n = 1.5$$

The compound is non-aromatic. Moreover, it has been determined that cyclooctatetraene is not planar but tub-shaped.

Heterocyclic Aromatics. Heterocyclic compounds also behave as aromatic if they obey the aromaticity rules. Examples of such heterocyclics are

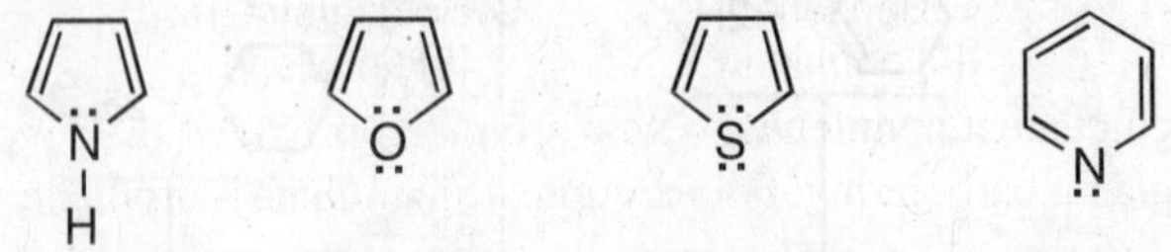

Let us first consider the 5-membered heterocyclic compounds pyrrole, furan, and thiophene. Applying aromaticity rules we find : (1) they are cyclic planar ; (2) each has a *p* orbital on every ring atom ; and (3) each observes Huckel rule. For example, in furan there are four *p* orbitals on the four carbons, containing one electron each. The N atom (also sp^2 hybridized) has an unused pair of electrons in a *p* orbital parallel with the other four *p* orbitals. The four *p* electrons of carbons and the two electrons of N atom form a cyclic π MO. However, in furan and thiophene, there are two unused electron pairs with O or S atom. One of these is in a *p* orbital at right angles to the ring and is not to be counted.

Similarly, in pyridine with three double bonds, each ring atom has a *p* orbital, and the six electrons form the π MO. Therefore, the compound is aromatic. The unshared electron pair on nitrogen is perpendicular to the ring and is not to be counted while applying Huckel rule.

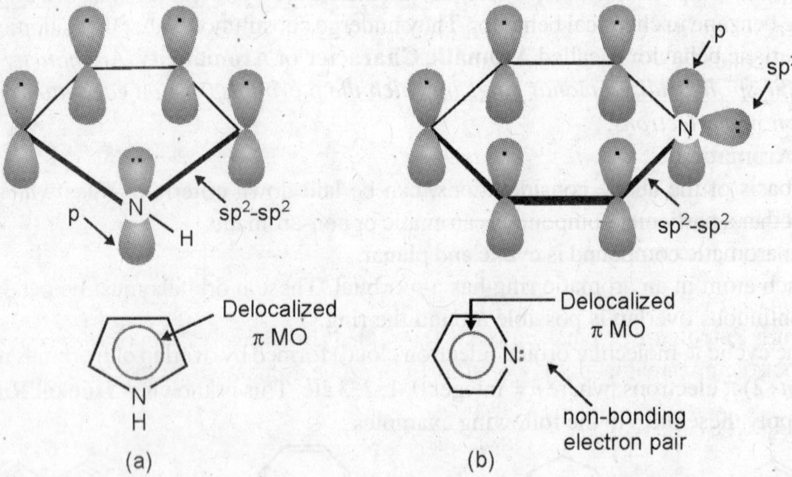

Fig. 33.5. (*a*) In pyrrole, the five parallel *p* orbitals containing six electrons form a delocalized π MO ; (*b*) In pyridine, the six *p* orbitals containing six electrons form a delocalized orbital, the unused sp^2 orbital of nitrogen containing two non-bonding electrons remains as such.

Antiaromatic compounds. Cyclic, planar and completely conjugated compounds that contain 4π electrons are especially unstable and are said to be *antiaromatic*. For example, cyclobutadiene is antiaromatic and very unstable because it contains 4π electrons.

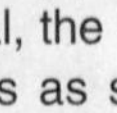

Cyclobutadiene

Cyclobutadiene is planar, cyclic and completely conjugated. It has 4π electrons. Therefore, it is antiaromatic.

WHAT ARE ANNULENES ?

The name annulene has been proposed as a general name for monocyclic compounds having alternating single and double bonds. The ring size of an annulene is indicated by a number in brackets. Thus, benzene is [6]annulene and cyclooctatetraene is [8]annulene.

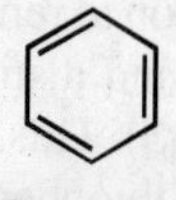
Benzene
[6] annulene

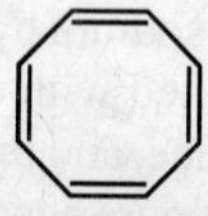
Cyclooctatetraene
[8] annulene

Huckel's rule predicts that annulenes will be aromatic, provided their molecules have $(4n+2)\ \pi$ electrons and have a planar carbon ring. For example, [6]annulene is aromatic whereas [8]annulene is not aromatic.

Points to Remember

Structure and Reactivity of Benzene

1. Benzene has a cyclic, planar, hexagonal structure.
2. All carbon-carbon bond lengths are identical and are intermediate between normal C–C single bonds and C=C double bonds. Benzene does not have alternate single and double bonds.
3. All hydrogens in benzene are equivalent. If a hydrogen is substituted by another atom or group, it does not matter which hydrogen is replaced. Only one product is formed. Always remember that in benzene monosubstitution leads to a single product. Disubstitution yields three isomers.
4. All carbon atoms in benzene are sp^2 hybridized.
5. Typical reactions of benzene are substitution reactions. Addition reactions can be induced, but special conditions are necessary.

BENZENE, C_6H_6

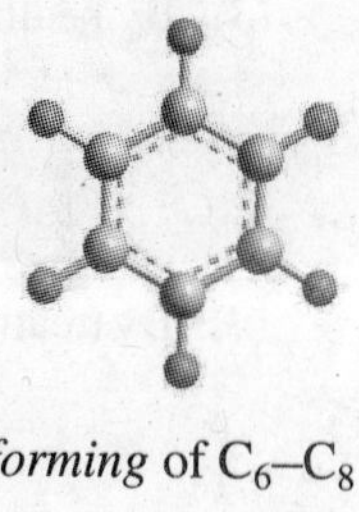

Benzene ring is present in all arenes and their derivatives. Therefore, our study of aromatic compounds must begin with that of benzene. This will lay a sound foundation of the structural features and principles encountered in the study of all classes of aromatic compounds.

LARGE SCALE PREPARATION OF BENZENE

(1) **From Petroleum**. We have already learnt that the *Aromatization* or *Platforming* of C_6–C_8 fraction of petroleum naphtha yields a mixture of benzene, toluene, and xylenes.

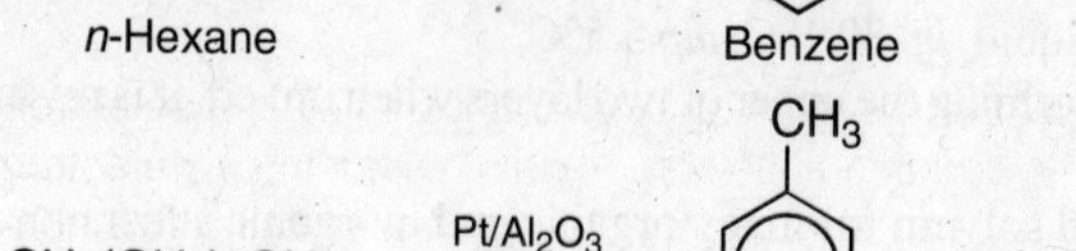

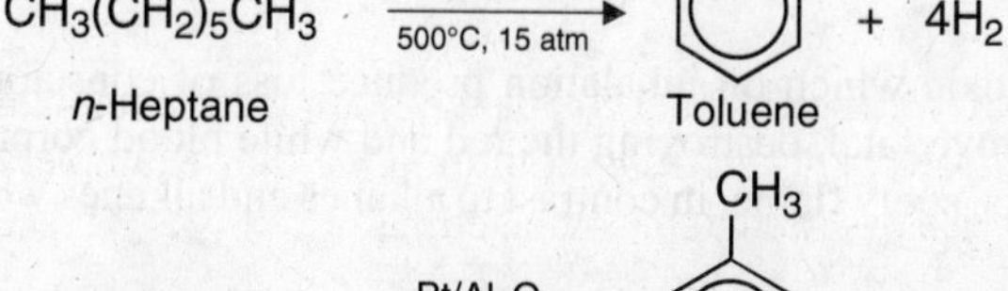

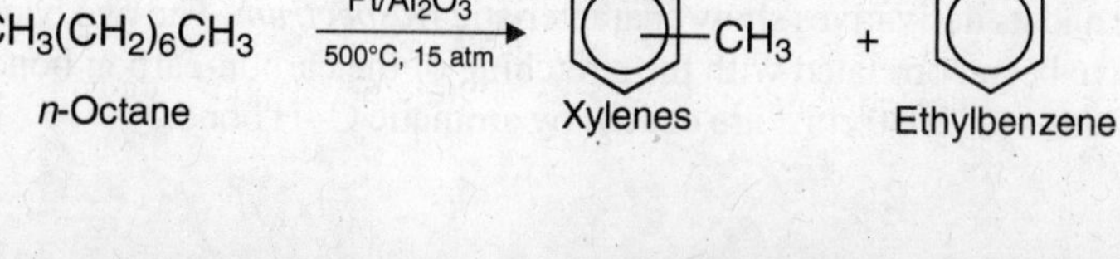

Benzene is recovered from the mixture by solvent extraction and fractional distillation.

(2) **From Light Oil Fraction of Coal-tar**. The high temperature carbonization of bituminous coal is carried by the steel industry. Coal-tar is obtained as a by-product. The distillation of coal-tar gives the *light oil* fraction (up to 200°C). This is washed with concentrated sulfuric acid to remove basic impurities (pyridine), and then with sodium hydroxide solution to remove acidic impurities (phenols). The oil thus purified is finally washed with water ; dried, and distilled. The fraction passing up to 110°C contains benzene, some toluene, and thiophene (C_4H_4S). This is refractionated and the fraction collected between 80-82°C is almost pure benzene. Traces of thiophene present as impurity is removed from the product by heating it with hydrogen under pressure at 400°C in presence of a catalyst and redistillation.

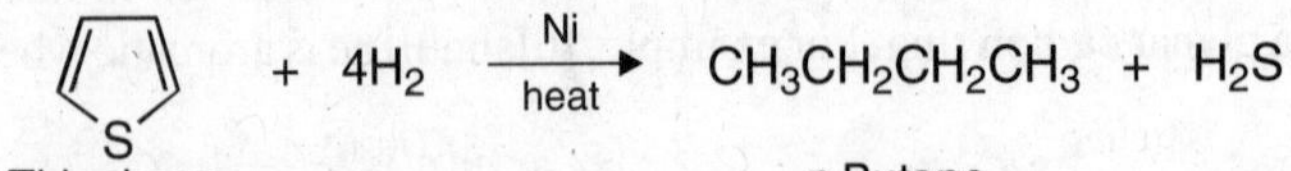

Thiophene *n*-Butane

SMALL SCALE PREPARATION OF BENZENE

Benzene is obtained on large scale from petroleum and coal-tar. The synthetic methods on laboratory scale are of academic interest only. Benzene is obtained :

(*1*) By passing acetylene through red-hot tube at 500°C.

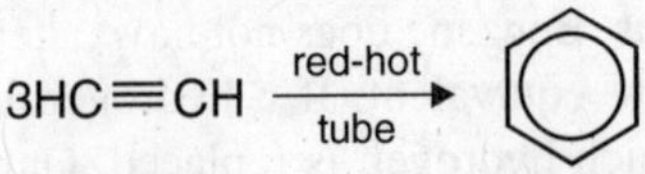

Acetylene Benzene

(2) By heating benzoic acid or its sodium salt with soda-lime (NaOH + CaO).

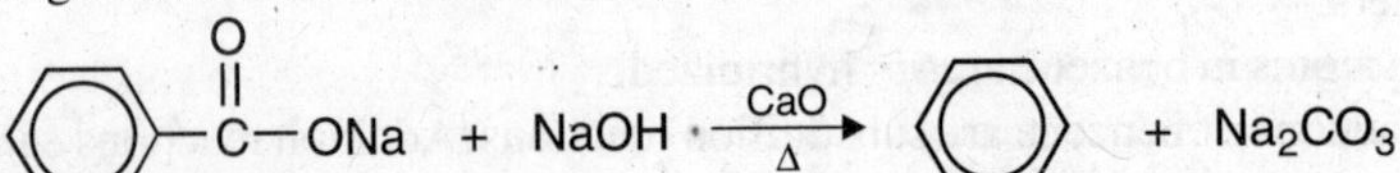

Sodium benzoate Benzene

(*3*) By heating phenol with zinc dust.

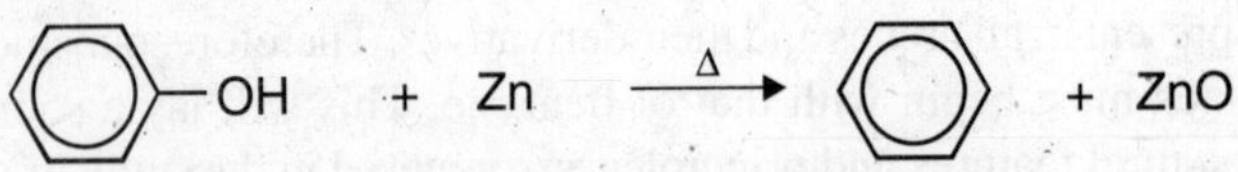

Phenol Benzene

(*4*) By treating chlorobenzene with magnesium followed by treatment with dilute HCl.

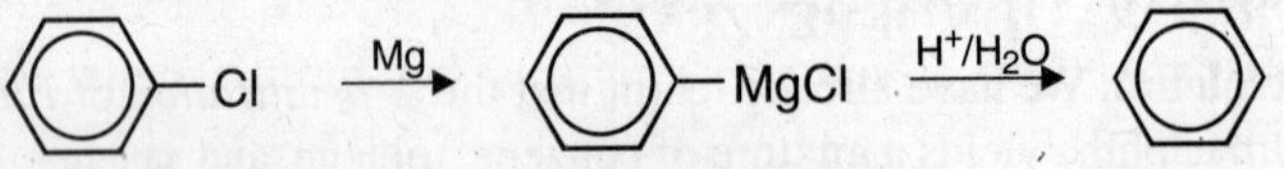

Chlorobenzene Phenylmagnesium chloride Benzene

PHYSICAL PROPERTIES OF BENZENE

(1) Benzene is a colorless liquid, *bp* 80.1°C, *mp* 5.5°C

(2) It is insoluble in water, forming the upper of two layers when mixed. It is miscible with alcohol, ether, and chloroform.

(3) Benzene itself is a good solvent for many organic and inorganic substances *e.g.*, fat, resins, sulfur and iodine.

(4) Its vapors are highly toxic which on inhalation produce loss of consciousness. Benzene poisoning in the long run can prove fatal, destroying the red and white blood corpuscles.

(5) It burns with a luminous, sooty flame, in contrast to alkanes and alkenes which usually burn with a bluish flame.

(6) Benzene and its derivatives show characteristic *IR spectrum*. The two bands near 1600 cm^{-1} and 1500 cm^{-1} have been correlated with the stretching of the carbon-carbon bonds of the aromatic ring. The sharp bands near 3030 cm^{-1} are caused by aromatic C–H bond.

CHEMICAL REACTIONS OF BENZENE

The principal types of reactions of benzene are :

(*a*) Electrophilic Substitution Reactions

(*b*) Addition Reactions

(*c*) Oxidation Reactions

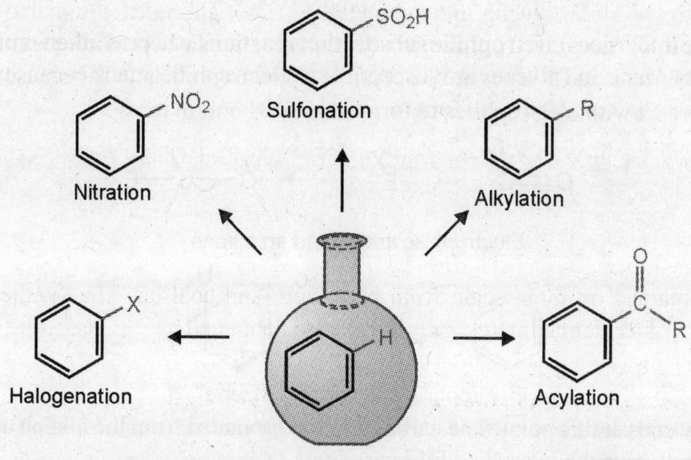

Electrophilic aromatic substitution reactions.

A. ELECTROPHILIC SUBSTITUTION REACTIONS

Benzene undergoes electrophilic substitution reactions. The benzene ring with its delocalized π electrons is an electron-rich system. It is attacked by electrophiles, giving substitution products. These reactions can be represented as :

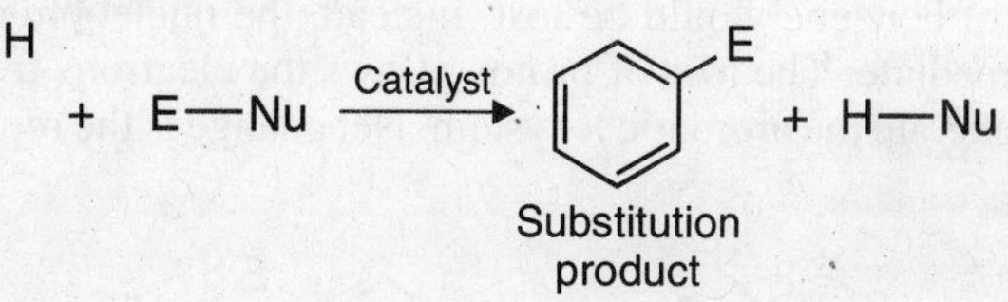

Where E^+ is any electrophile and $Nu:^-$ is a nucleophile. Such reactions in which hydrogen atom of the aromatic ring is replaced by an electrophile are called electrophilic aromatic substitution reactions.

GENERAL MECHANISM. All electrophilic aromatic substitution reactions follow the same three-step mechanism :

Step 1. Formation of an electrophile.

$$E—Nu + \boxed{Catalyst} \longrightarrow E^+ + Nu—\boxed{Catalyst}$$

Step 2. The electrophile attacks the aromatic ring to form a carbonium ion.

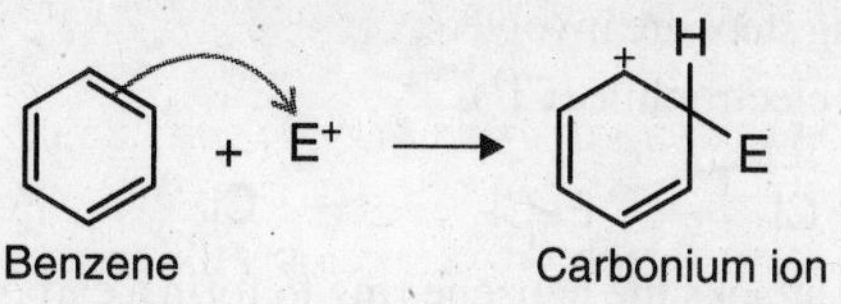

The intermediate ion is resonance-stabilized. It is a hybrid of the following three structures:

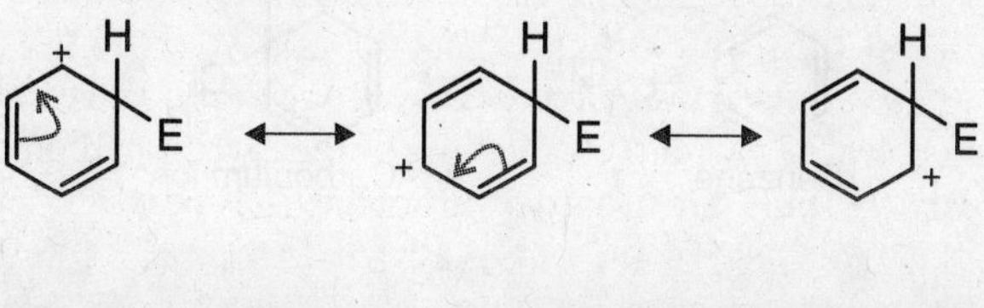

Step 3. Loss of proton gives the substitution product.

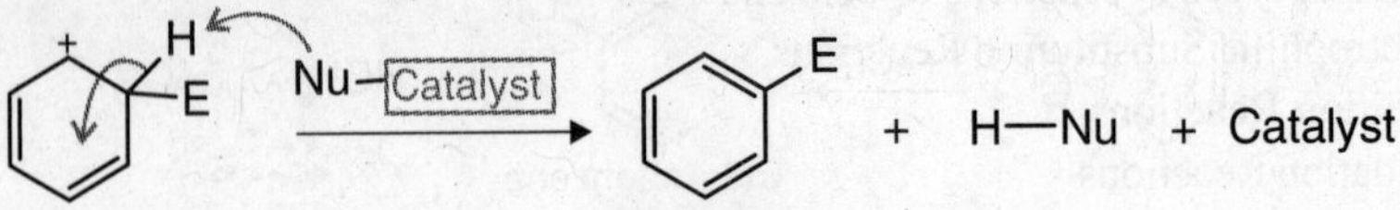

Carbonium ion — Substitution product

Why benzene undergoes electrophilic substitution reactions whereas alkenes undergo addition reactions ? Both benzene and alkenes are susceptible to electrophilic attack because of their exposed π electrons. Both react with electrophiles to form stable carbonium ions.

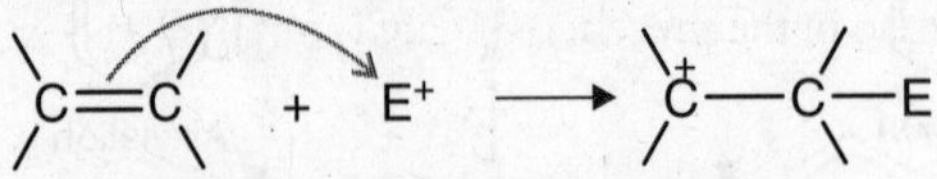

Electrophilic addition to an alkene

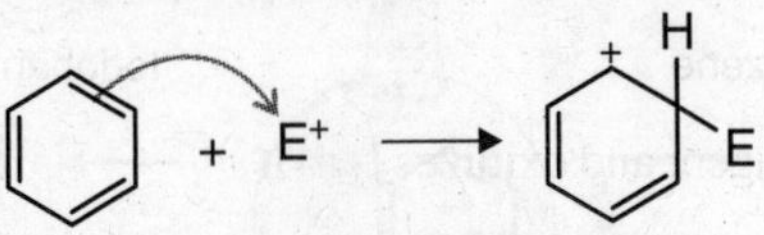

Electrophilic addition to benzene

But similarity ends at this point. The carbonium ion produced from the alkene usually combines with a nucleophile to give the overall addition product :

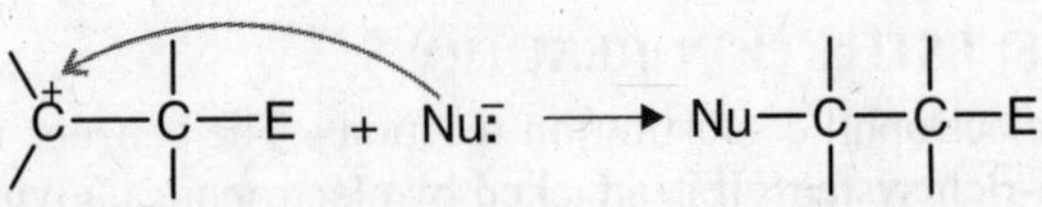

If this happened to benzene, the product would no longer be aromatic. The resonance energy (stabilizing energy) of benzene would be lost. Instead, the nucleophile removes a proton from the carbonium ion intermediate. The loss of proton allows the electrons from the C–H bond to go back into the ring and regenerate the aromatic π system. Net change is the replacement of a hydrogen atom by an electrophile :

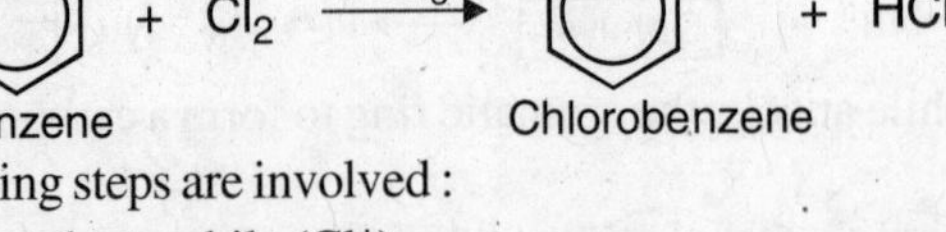

(1) **Halogenation**. Benzene reacts with chlorine in the presence of $FeCl_3$ or $AlCl_3$ at room temperature to form chlorobenzene. Iron powder can be used in place of ferric chloride ($2Fe + 3Cl_2 = 2FeCl_3$).

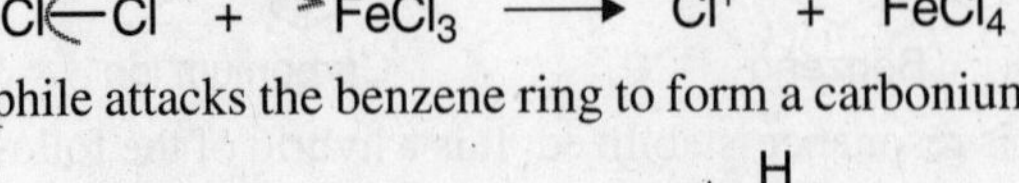

Benzene — Chlorobenzene

MECHANISM. Following steps are involved :

Step 1. Formation of the electrophile (Cl^+).

$$Cl{-}Cl + FeCl_3 \longrightarrow Cl^+ + FeCl_4^-$$

Step 2. The electrophile attacks the benzene ring to form a carbonium ion.

Benzene — Carbonium ion

Step 3. Loss of proton yields chlorobenzene.

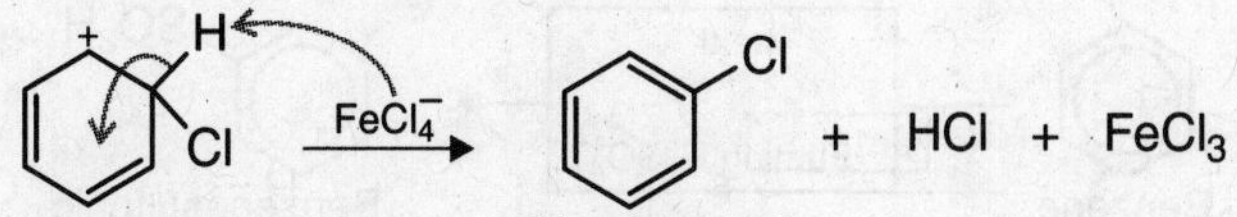

Chlorobenzene

Bromine reacts with benzene similarly.

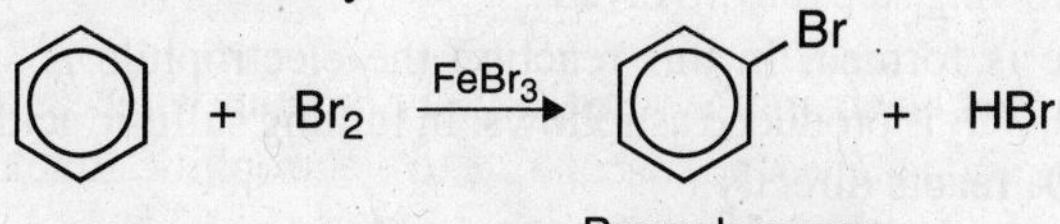

Bromobenzene

Iodine reacts with benzene in the presence of $CuCl_2$ or HNO_3.

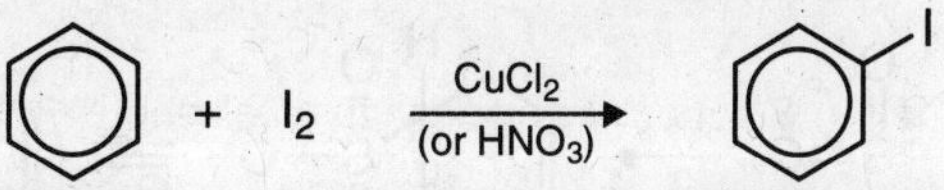

Benzene Iodobenzene

Nitric acid is an oxidizing agent and oxidizes I_2 to I^+.

$$\left.\begin{array}{l} I_2 + CuCl_2 \\ I_2 + HNO_3 \end{array}\right] \longrightarrow \underset{\text{Electrophile}}{I^+}$$

Fluorine is very reactive. Polysubstitution takes place and all benzene hydrogens are replaced by fluorine atoms.

(2) **Nitration**. Benzene reacts with concentrated nitric acid in the presence of concentrated sulfuric acid at 60°C to form nitrobenzene.

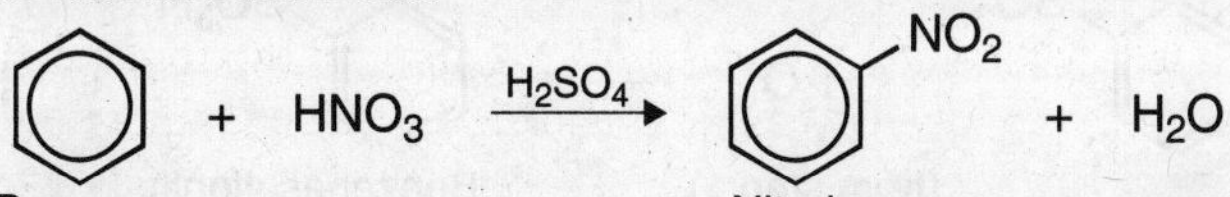

Benzene Nitrobenzene

MECHANISM. Following steps are involved :

Step 1. Formation of the electrophile (NO_2^+).

$$HNO_3 + 2H_2SO_4 \longrightarrow NO_2^+ + 2HSO_4^- + H_3O^+$$

Step 2. The electrophile attacks the benzene ring to form a carbonium ion.

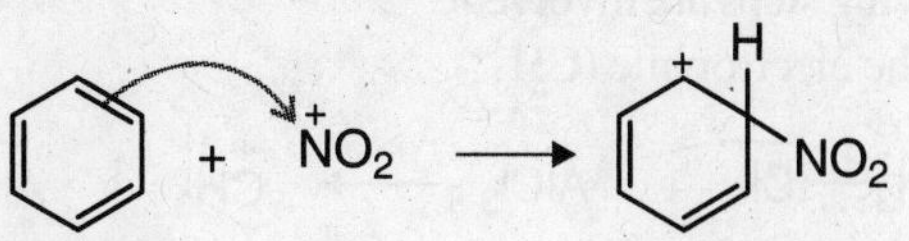

Benzene Carbonium ion

Step 3. Loss of proton yields nitrobenzene.

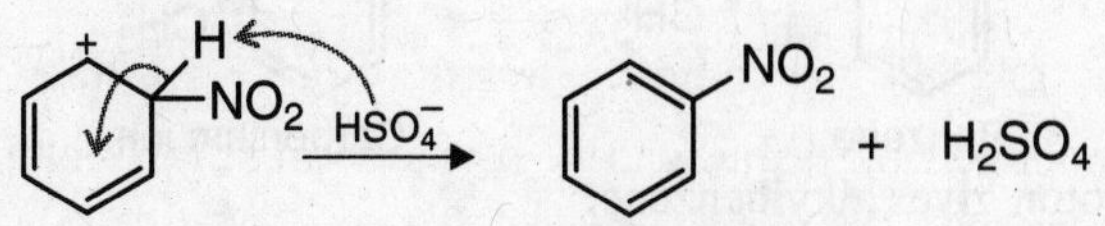

Nitrobenzene

(3) **Sulfonation**. Benzene reacts with concentrated sulfuric acid at 120°C or fuming sulfuric acid at room temperature to give benzenesulfonic acid. *Note* : Fuming sulfuric acid is concentrated sulfuric acid that contains added sulfur trioxide.

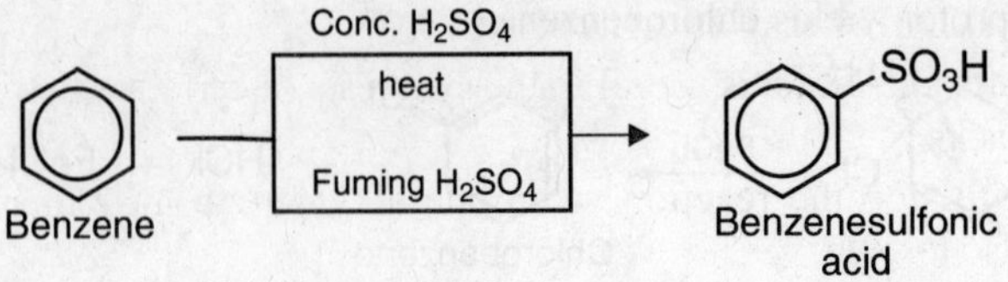

MECHANISM. Following steps are involved :

Step 1. Electrophile is formed. In this reaction the electrophile is sulfur trioxide (SO_3). In concentrated sulfuric acid, SO_3 is produced as follows. In fuming sulfuric acid, this step is unimportant because the dissolved SO_3 reacts directly.

$$2H_2SO_4 \rightleftharpoons SO_3 + H_3O^+ + HSO_4^-$$

Step 2. The electrophile attacks the benzene ring to form a carbonium ion.

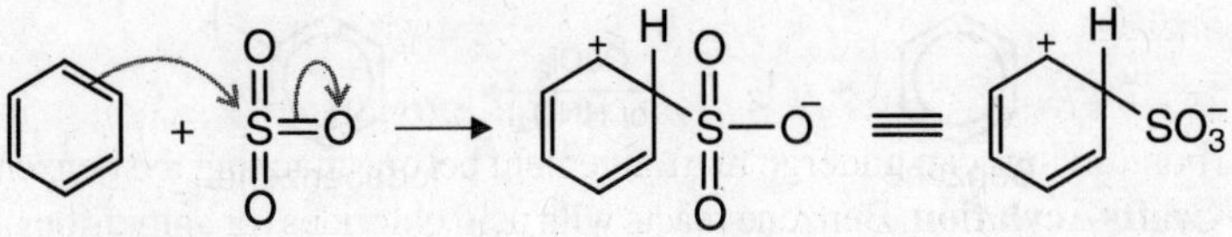

Step 3. Loss of proton.

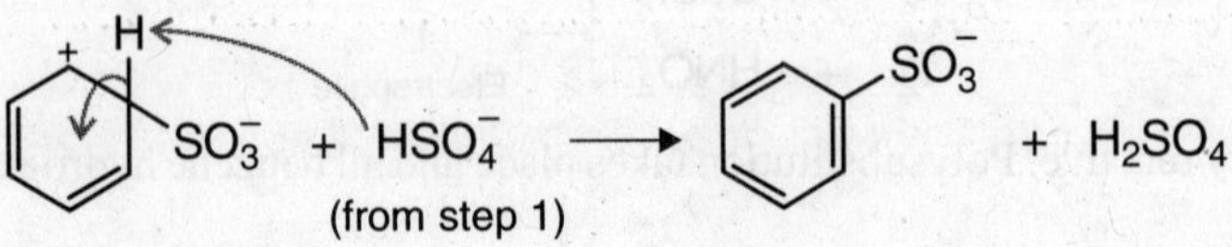

Step 4. Addition of proton gives benzenesulfonic acid.

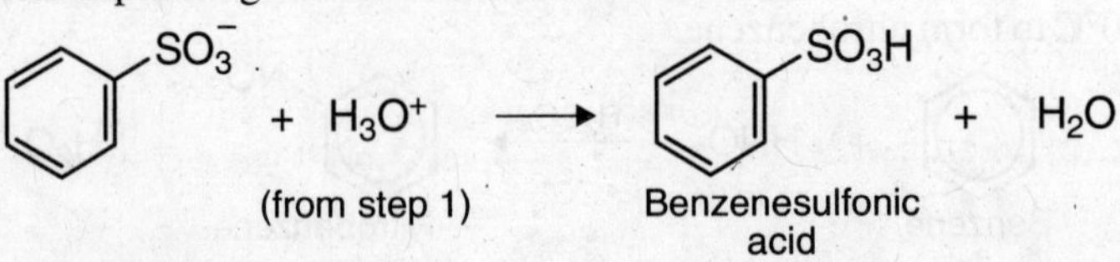

(4) **Friedel-Crafts Alkylation**. Benzene reacts with alkyl halides in the presence of aluminium chloride to form alkylbenzenes. For example,

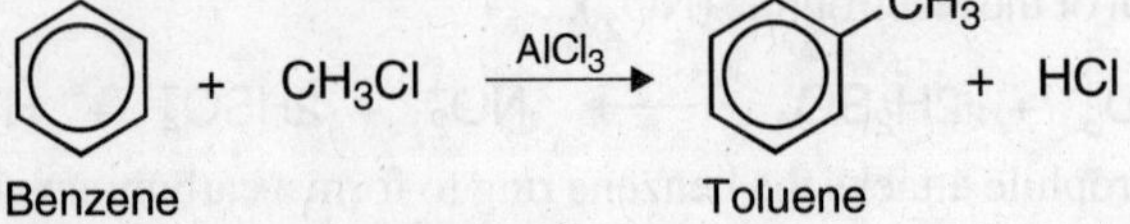

MECHANISM. Following steps are involved :

Step 1. Formation of the electrophile (CH_3^+).

$$CH_3-Cl + AlCl_3 \longrightarrow \overset{+}{C}H_3 + AlCl_4^-$$

Step 2. The electrophile attacks the benzene ring to give a carbonium ion.

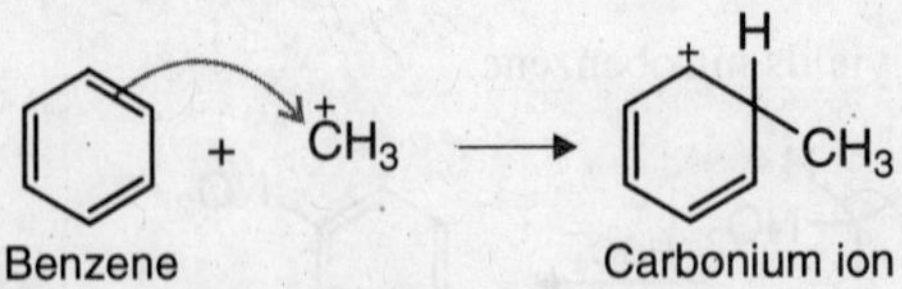

Step 3. Loss of proton gives alkylbenzene.

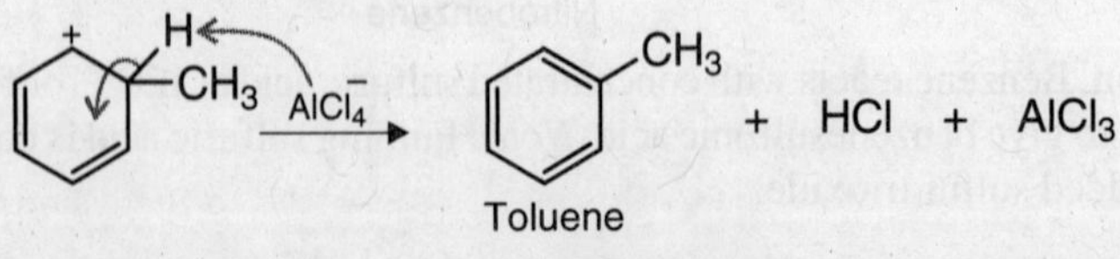

DRAWBACKS OF FRIEDEL-CRAFTS ALKYLATION

While Friedel-Crafts alkylation is useful in the synthesis of certain alkylbenzenes, the reaction has two serious drawbacks :

(1) It is difficult to stop the reaction when one alkyl group has entered the ring. Di- and tri-alkylbenzenes are also formed.

(2) The alkyl groups often tend to rearrange. For example, when benzene is treated with *n*-propyl chloride in the presence of $AlCl_3$, the product is isopropylbenzene (also called cumene) rather than the expected *n*-propylbenzene.

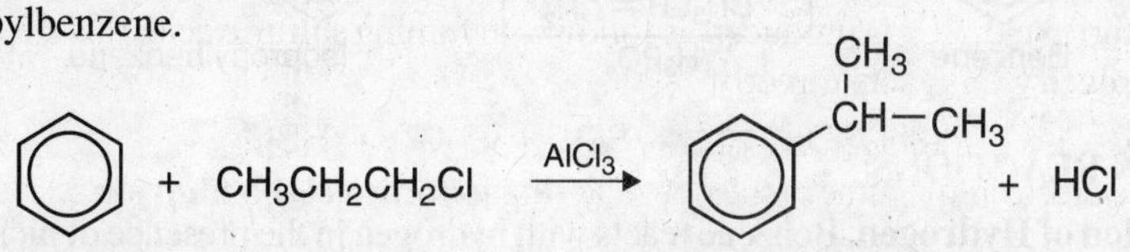

Benzene Isopropylbenzene

This is because the Friedel-Crafts alkylation involves formation of carbonium ion electrophiles (Step 1). These carbonium ions can undergo rearrangement before attacking the benzene ring (Step 2).

(5) **Friedel-Crafts Acylation**. Benzene reacts with acid chlorides (or anhydrides) in the presence of aluminium chloride to give aromatic ketones. For example,

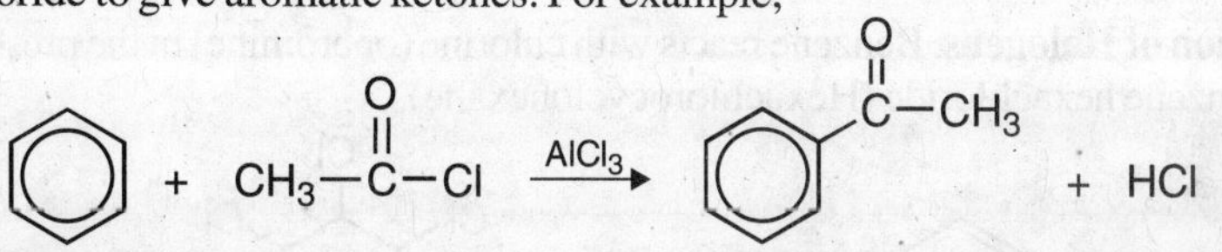

Benzene Acetophenone

MECHANISM. Following steps are involved :

Step 1. Formation of the electrophile $(CH_3-\overset{+}{C}=O)$.

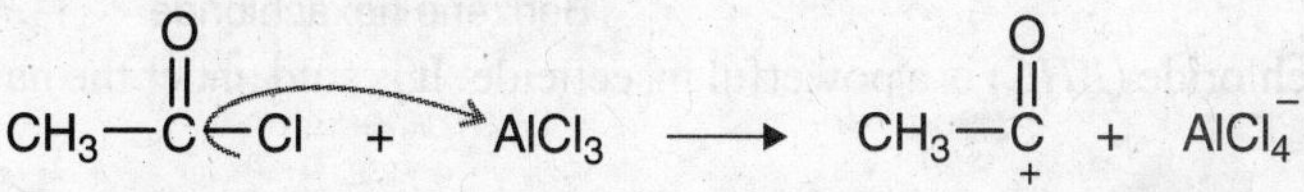

Step 2. The electrophile attacks the benzene ring to form a carbonium ion.

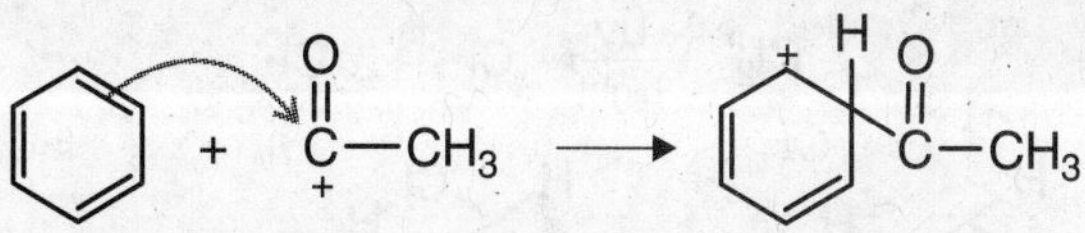

Carbonium ion

Step 3. Loss of proton gives an aromatic ketone.

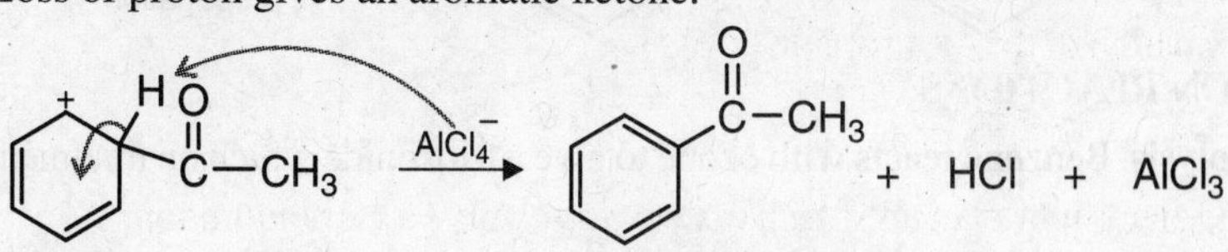

Acetophenone

Since alkylbenzenes other than methyl, ethyl, or *t*-alkyl cannot be prepared from alkyl halides without rearrangement, they are best prepared by *Friedel-Crafts Acylation-Clemmensen Reduction Sequence.* For example,

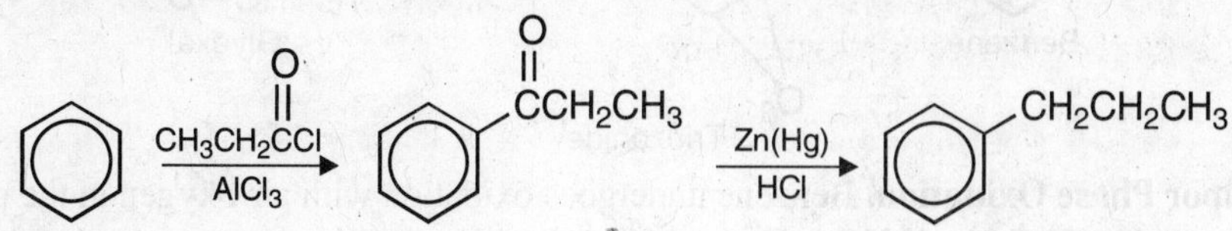

Benzene *n*-Propylbenzene

Alkylbenzenes can also be prepared by treatment of benzene with alcohol or an alkene in the presence of an acid catalyst. In each case the initial step involves the formation of a stable carbonium ion.

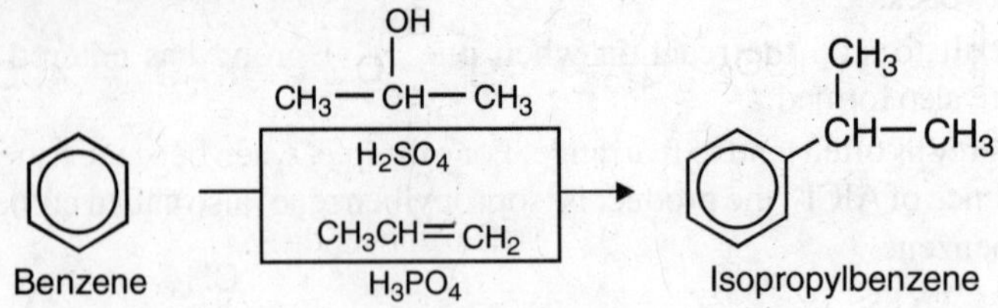

B. ADDITION REACTIONS

(6) **Addition of Hydrogen.** Benzene reacts with hydrogen in the presence of nickel (or platinum) catalyst at 150°C under pressure to form cyclohexane.

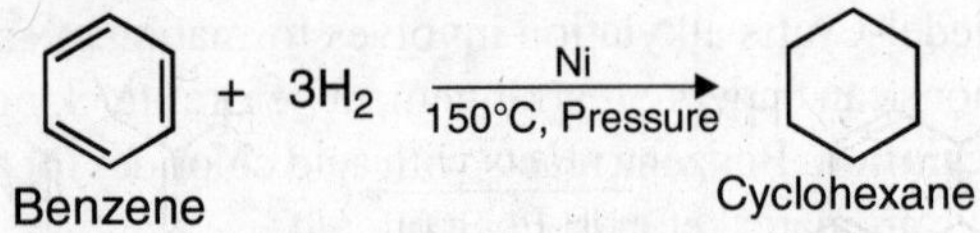

(7) **Addition of Halogens**. Benzene reacts with chlorine (or bromine) in the presence of ultraviolet light to form benzene hexachloride (Hexachlorocyclohexane).

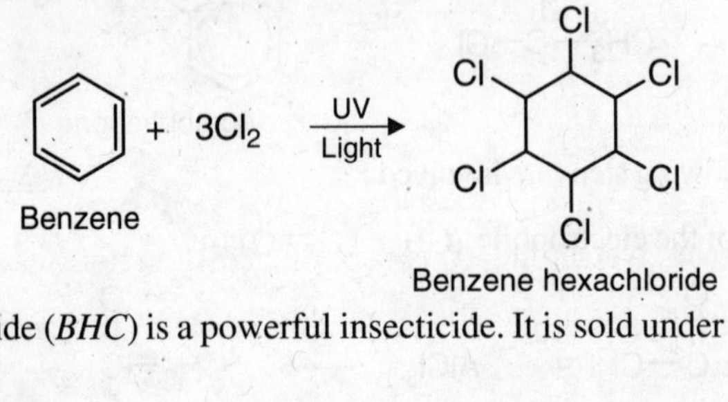

Benzene hexachloride (*BHC*) is a powerful insecticide. It is sold under the name *Gammexane* or *Lindane*.

MECHANISM. The above reaction involves a free-radical mechanism.

$$Cl_2 \xrightarrow[\text{Light}]{\text{UV}} Cl\cdot + Cl\cdot$$

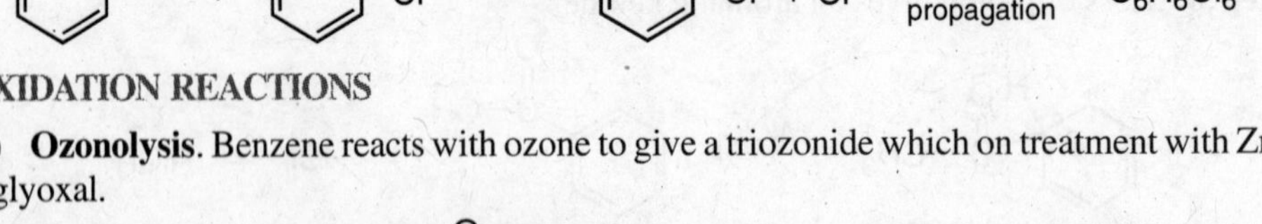

C. OXIDATION REACTIONS

(8) **Ozonolysis**. Benzene reacts with ozone to give a triozonide which on treatment with Zn/H_2O yields glyoxal.

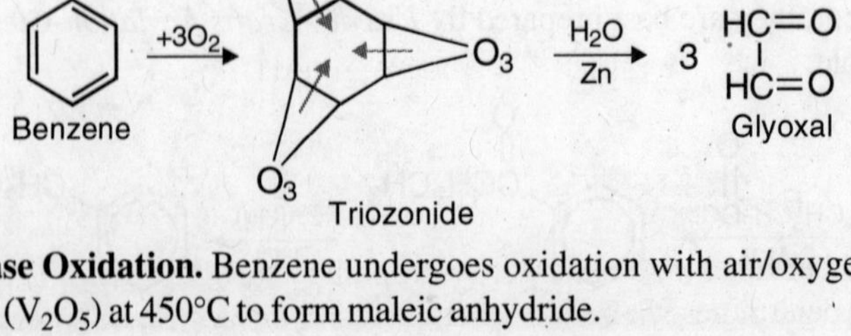

(9) **Vapor Phase Oxidation.** Benzene undergoes oxidation with air/oxygen in the presence of vanadium pentoxide (V_2O_5) at 450°C to form maleic anhydride.

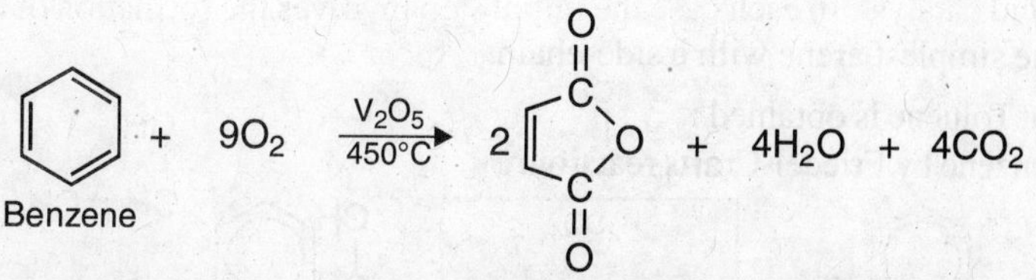

Benzene

Maleic anhydride

D. OTHER REACTIONS

(10) **Birch Reduction**. Benzene and its derivatives can be reduced to nonconjugated 1,4-cyclohexadienes by treatment with sodium or lithium in a mixture of liquid ammonia and an alcohol. The Birch reduction provides a convenient method for making a wide variety of interesting and useful cyclic dienes.

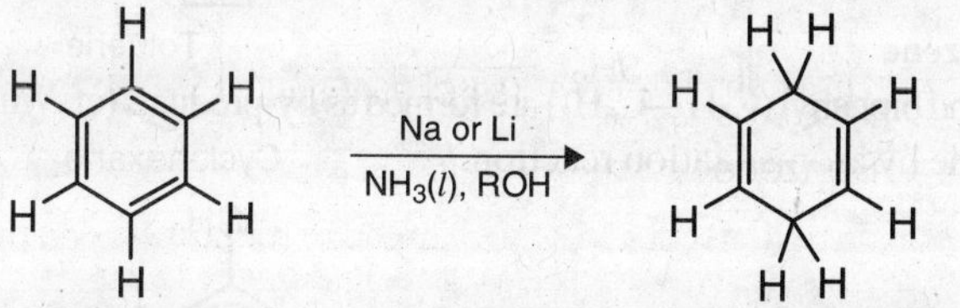

Benzene

1,4-Cyclohexadiene

Points to Remember

Writing Mechanism of Electrophilic Aromatic Substitution Reactions

(1) All electrophilic substitution reactions follow the same 3-step mechanism :

Step 1. Formation of an electrophile.

Step 2. The electrophile attacks the aromatic ring to give a resonance-stabilized carbonium ion (also called arenium ion).

Step 3. Loss of proton gives the substitution product.

(2) Important electrophiles used in aromatic substitution and their sources are :

Electrophile	Source
Cl^+	$Cl_2 + AlCl_3$ or $FeCl_3$
Br^+	$Br_2 + AlBr_3$ or $FeBr_3$
NO_2^+	$HNO_3 + H_2SO_4$
SO_3	Conc. H_2SO_4
	Fuming H_2SO_4
R^+	$RX + AlX_3$ (X = Cl or Br)
	$ROH + H^+$
	An alkene + H^+
$R-\overset{+}{C}=O$	$R-\overset{\overset{O}{\parallel}}{C}-Cl$ + $AlCl_3$

(3) Substituted benzenes also undergo electrophilic substitution. The group already present on the ring determines where further substitution occurs (See Chapter *34*).

TOLUENE, *Methylbenzene*, $C_6H_5CH_3$

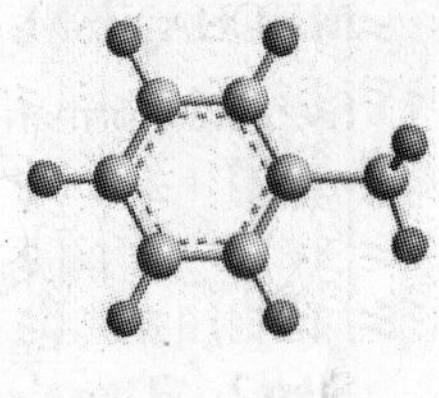

Toluene is the simplest arene with a side-chain.

Preparation. Toluene is obtained :

(*1*) From benzene by Friedel-Crafts reaction.

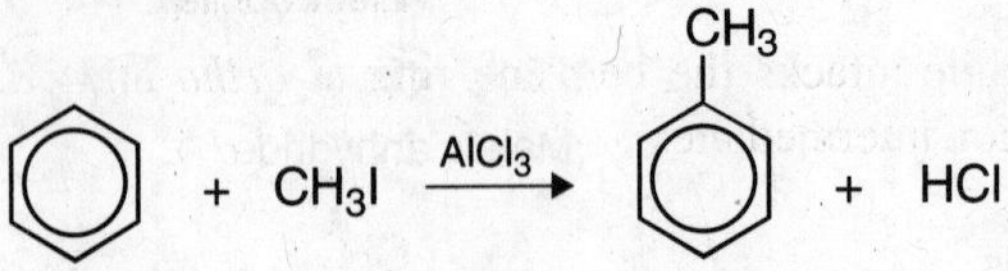

Benzene Toluene

(*2*) From bromobenzene by Wurtz-Fittig reaction.

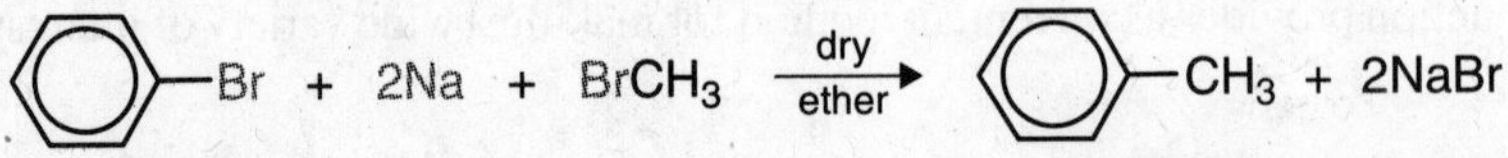

Bromobenzene Toluene

Ethane (CH_3CH_3) and biphenyl (C_6H_5—C_6H_5) are formed as by-products (cf. Wurtz synthesis of alkanes).

(*3*) From *n*-heptane by aromatization reaction.

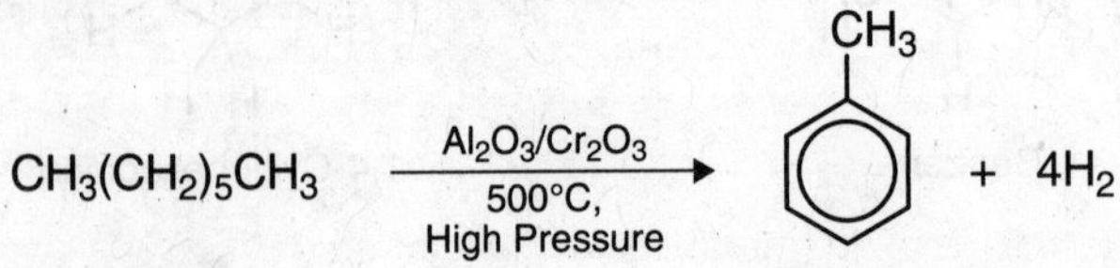

n-Heptane Toluene

This method is used for large-scale production of toluene. *n*-Heptane comes from crude-oil Pt/Al_2O_3 can also be used in place of Al_2O_3/Cr_2O_3.

Properties (*Physical*). Toluene is a colorless liquid, bp 110.6°C, having an odor similar to benzene. It is insoluble in water, but highly soluble in alcohol, ether and benzene. Toluene itself is a good solvent for many organic substances.

(*Chemical*). Toluene has the structural formula :

Toluene undergoes the following reactions :

(a) Electrophilic substitution and Addition reactions of the ring

(b) Substitution and Oxidation reactions of the side-chain

ELECTROPHILIC SUBSTITUTION REACTIONS

The methyl group in toluene is electron-releasing. It activates the ring with regard to electrophilic attack at the *ortho* and *para* positions (See Chapter *34* for explanation). Therefore, toluene gives all the electrophilic substitution reactions undergone by benzene, forming *ortho* and *para* derivatives.

(1) **Halogenation.** Toluene reacts with chlorine or bromine in the presence of $FeCl_3$ (or $AlCl_3$) to give a mixture of *o*- and *p*-derivatives.

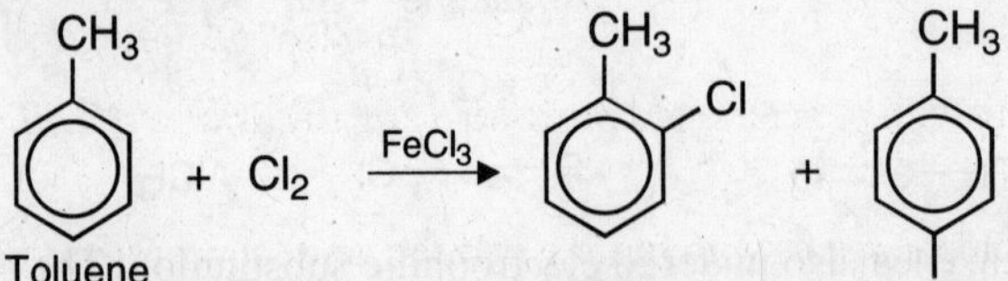

Toluene

o- plus *p*-Chlorotoluene

MECHANISM. Following steps are involved :

Step 1. Formation of the electrophile.

$$Cl{-}Cl + FeCl_3 \longrightarrow \underset{\text{Electrophile}}{Cl^+} + FeCl_4^-$$

Step 2. The electrophile attacks the benzene ring at *ortho* and *para* positions to form the corresponding carbonium ion intermediates.

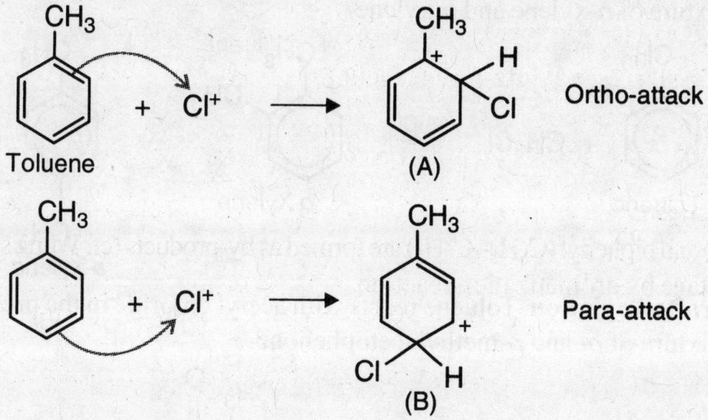

Step 3. Loss of protons from the intermediate carbonium ions yield the products.

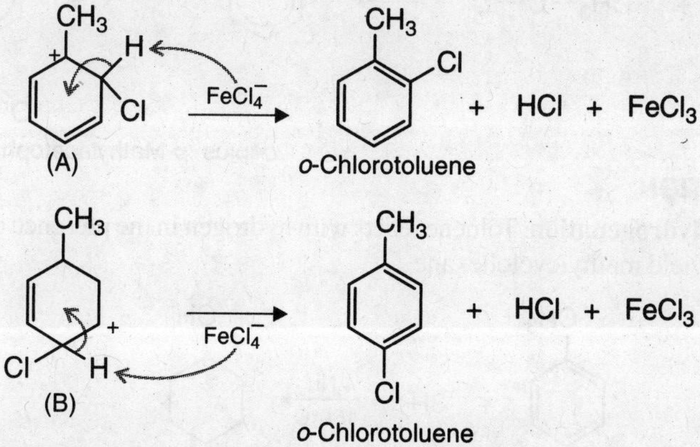

(2) **Nitration**. Toluene reacts with a mixture of concentrated nitric acid and sulfuric acid to give a mixture of *o*- and *p*-nitrotoluene.

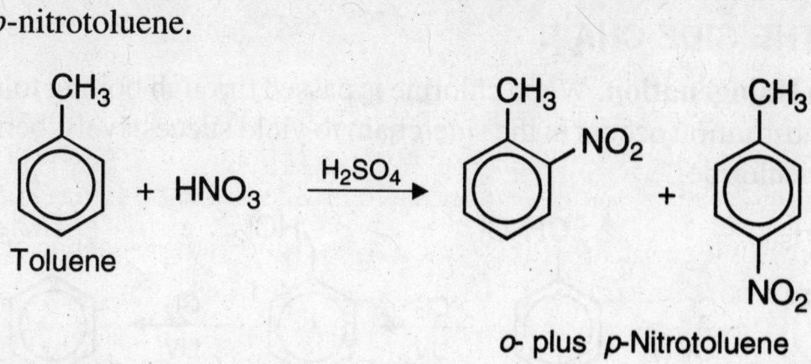

On prolonged heating with concentrated nitric acid and sulfuric acid, *TNT* (2,4,6-trinitrotoluene) is formed.

(3) **Sulfonation**. Toluene reacts with fuming sulfuric acid (concentrated $H_2SO_4 + SO_3$) to form a mixture of *o*- and *p*-toluenesulfonic acid.

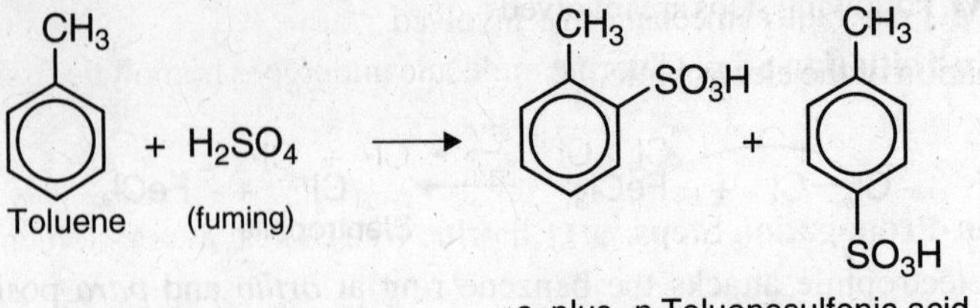

o- plus p-Toluenesulfonic acid

(4) **Friedel-Crafts Alkylation**. Toluene reacts with methyl chloride in the presence of aluminium chloride to give a mixture of *o*-xylene and *p*-xylene.

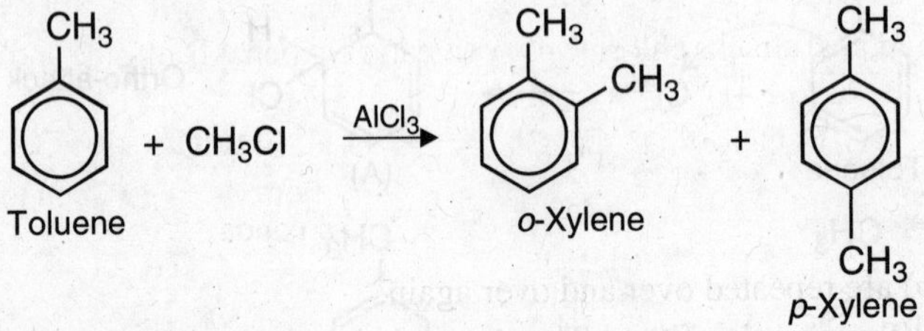

(5) **Friedel-Crafts Acylation**. Toluene reacts with acetyl chloride in the presence of aluminium chloride to give a mixture of *o*- and *p*-methylacetophenone.

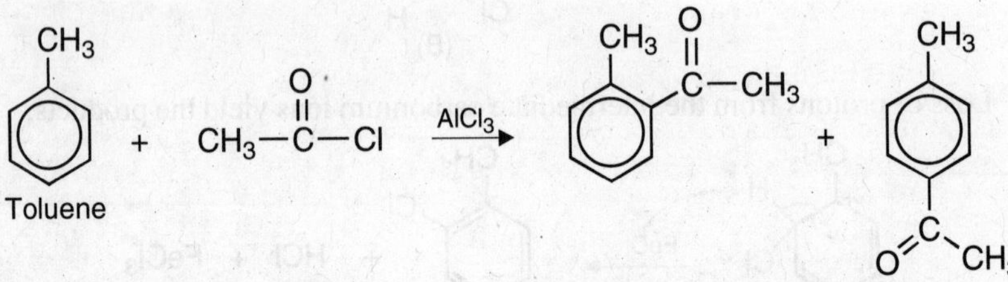

o- plus p-Methylacetophenone

ADDITION REACTION

(6) **Catalytic Hydrogenation**. Toluene reacts with hydrogen in the presence of platinum or nickel catalyst at 150°C to yield methylcyclohexane.

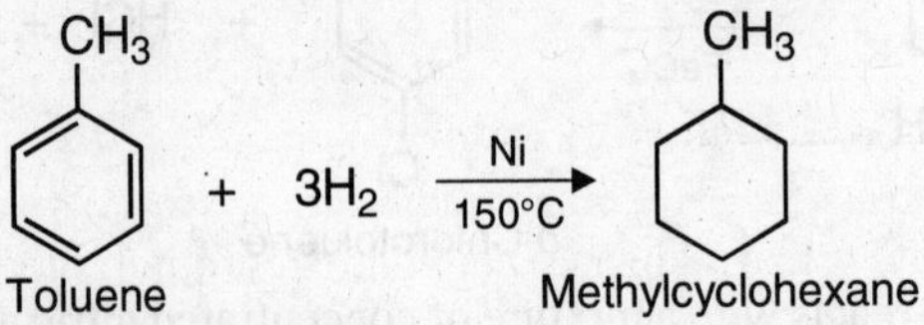

REACTIONS OF THE SIDE-CHAIN

(7) **Side-Chain Halogenation**. When chlorine is passed through boiling toluene in the presence of ultraviolet light, substitution occurs in the side-chain to yield successively, benzyl chloride, benzal chloride, and benzotrichloride.

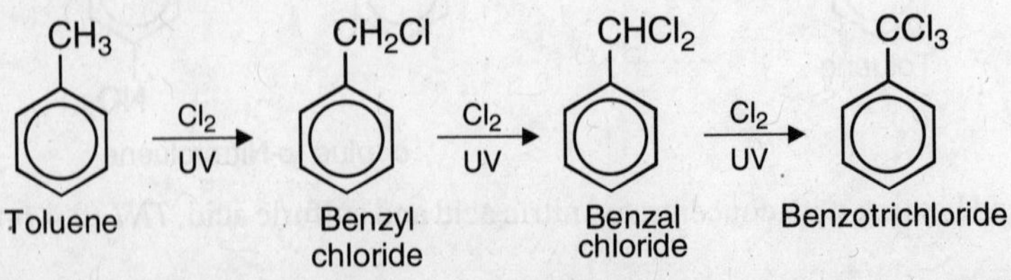

Benzyl chloride on hydrolysis with aqueous NaOH gives benzyl alcohol ($C_6H_5CH_2OH$). Benzal chloride on hydrolysis with aqueous NaOH gives benzaldehyde (C_6H_5CHO). Benzotrichloride on hydrolysis with aqueous NaOH yields benzoic acid (C_6H_5COOH).

MECHANISM. Free- radical mechanism is involved :

Step 1. Chain-Initiation Step. Chlorine molecule undergoes homolytic fission to give chlorine free radicals.

$$Cl : Cl \xrightarrow[\text{light}]{\text{UV}} Cl\cdot + Cl\cdot$$

Step 2. Chain-Propagation Steps. (*a*) Chlorine free radical attacks toluene to form HCl and benzyl free radical.

$$\underset{}{Cl\cdot} + \underset{\text{Toluene}}{H : CH_2C_6H_5} \longrightarrow H:Cl + \cdot CH_2C_6H_5$$

(*b*) Benzyl free radical attacks chlorine molecule to give benzyl chloride and chlorine free radical.

$$C_6H_5CH_2\cdot + Cl : Cl \longrightarrow \underset{\text{Benzyl chloride}}{C_6H_5CH_2 : Cl} + \underset{\text{(Goes to step } a)}{Cl\cdot}$$

Steps (*a*) and (*b*) are repeated over and over again.

Step 3. Chain-Termination Steps. The above chain reaction comes to halt when any two free radicals combine.

$$Cl\cdot + Cl\cdot \longrightarrow Cl—Cl$$

$$C_6H_5CH_2\cdot + Cl\cdot \longrightarrow C_6H_5CH_2—Cl$$

$$C_6H_5CH_2\cdot + \cdot CH_2C_6H_5 \longrightarrow C_6H_5CH_2—CH_2C_6H_5$$

A chlorine free radical can also attack benzyl chloride to give chlorobenzyl free radical. This free radical participates further in the chain reaction to yield benzal chloride.

$$(a)\quad Cl\cdot + \underset{\text{Benzyl chloride}}{H : \overset{Cl}{\overset{|}{C}}HC_6H_5} \longrightarrow H:Cl + \underset{\text{Chlorobenzyl free radical}}{\cdot \overset{Cl}{\overset{|}{C}}HC_6H_5}$$

$$(b)\quad C_6H_5\overset{Cl}{\overset{|}{C}}H\cdot + Cl:Cl \longrightarrow \underset{\text{Benzal chloride}}{C_6H_5\overset{Cl}{\overset{|}{C}}H:Cl} + Cl\cdot$$

Similarly, benzotrichloride is obtained by further chain reaction.

(8) **Side-Chain Oxidation**. (*a*) With hot acidic $KMnO_4$, $Na_2Cr_2O_7$, or concentrated nitric acid, benzoic acid is formed. (*b*) With acidic manganese dioxide or chromyl chloride (CrO_2Cl_2), benzaldehyde is formed.

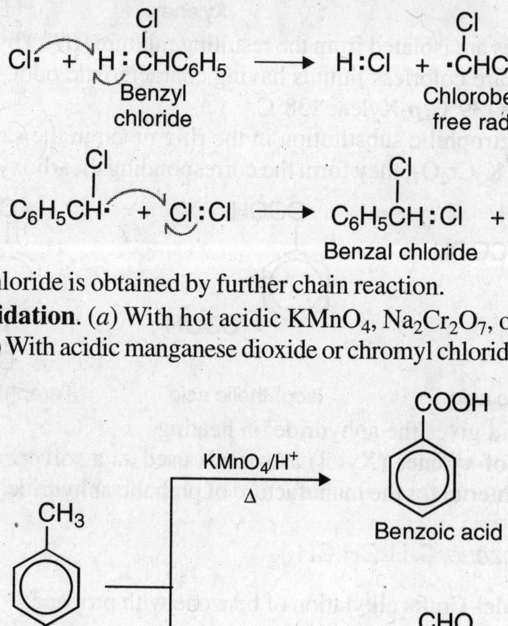

All alkylbenzenes on oxidation with hot acidic $KMnO_4$ or $Na_2Cr_2O_7$ give benzoic acid. The length of the side-chain does not matter. The product is always benzoic acid.

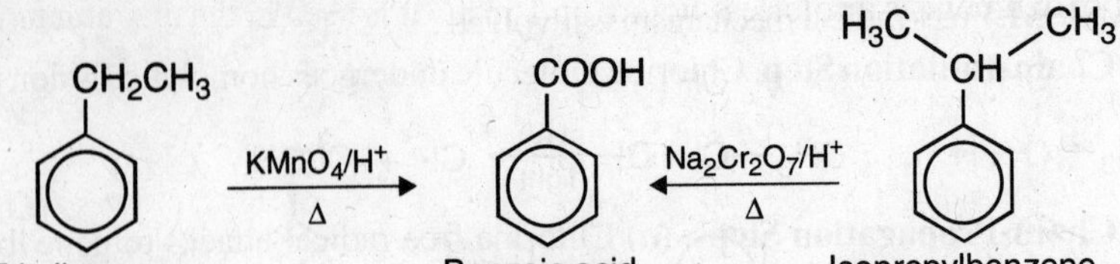

Ethylbenzene Benzoic acid Isopropylbenzene

Uses. Toluene is used : (*1*) for blending petrol ; (*2*) as a solvent for paints, resins, rubber ; (*3*) as a starting material for benzyl derivatives, benzaldehyde, benzoic acid, and *TNT*.

XYLENES, *Dimethylbenzenes*

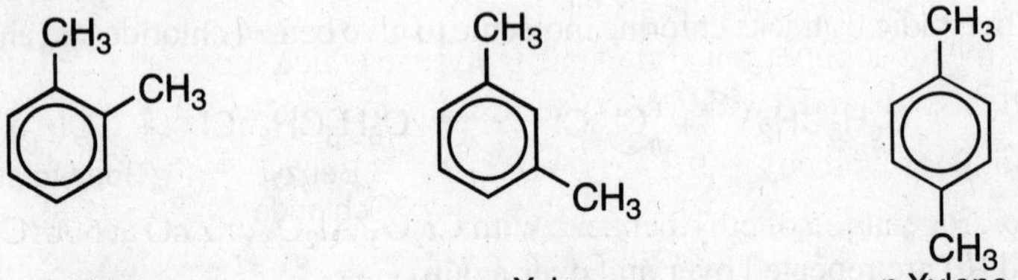

o-Xylene *m*-Xylene *p*-Xylene

Manufacture. Xylenes are produced along with benzene, toluene, and ethylbenzenes, by aromatization of C_6—C_8 fraction of petroleum *naphtha*. For example, *n*-octane gives xylenes as follows :

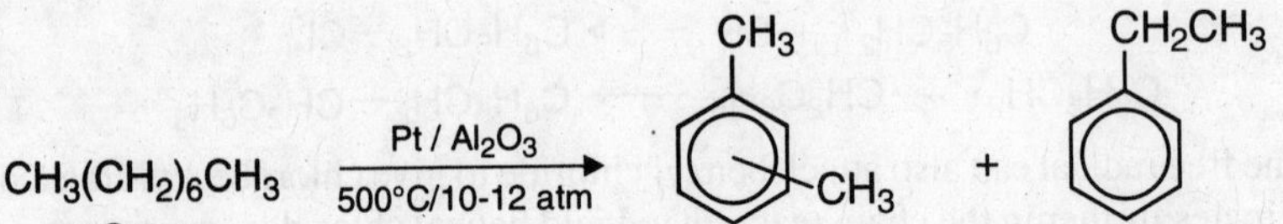

n-Octane Xylenes Ethylbenzene

The individual xylenes are isolated from the resulting mixture (*BTX*) by fractional distillation.

Properties. Xylenes are colorless liquids having characteristic odor. Their boiling points are : *o*-Xylene 144°C, *m*-Xylene 139°C, *p*-Xylene 138°C.

Xylenes undergo electrophilic substitution in the ring much in the same way as toluene. Upon oxidation with $KMnO_4$ or $K_2Cr_2O_7$, they form the corresponding dicarboxylic acids.

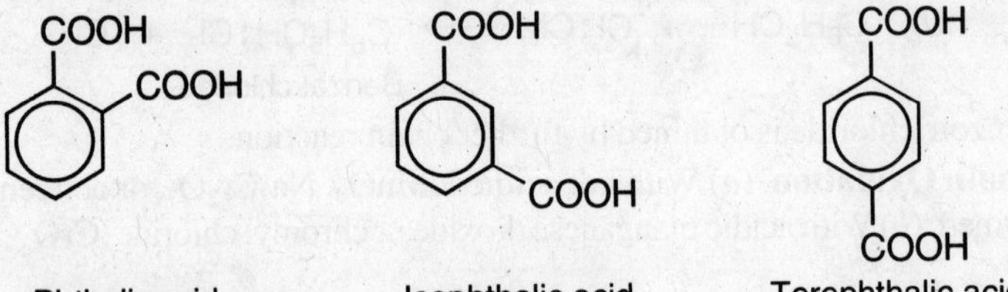

Phthalic acid Isophthalic acid Terephthalic acid

Of these, phthalic acid gives the anhydride on heating.

Uses. (1) A mixture of xylenes (Xylol) as such is used as a solvent and diluent for lacquers ; (2) *o*-Xylene is the raw material for the manufacture of phthalic anhydride.

CUMENE, *Isopropylbenzene*, $C_6H_5CH(CH_3)_2$

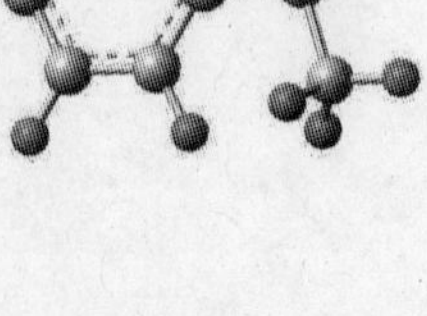

It is obtained by Friedel-Crafts alkylation of benzene with propene.

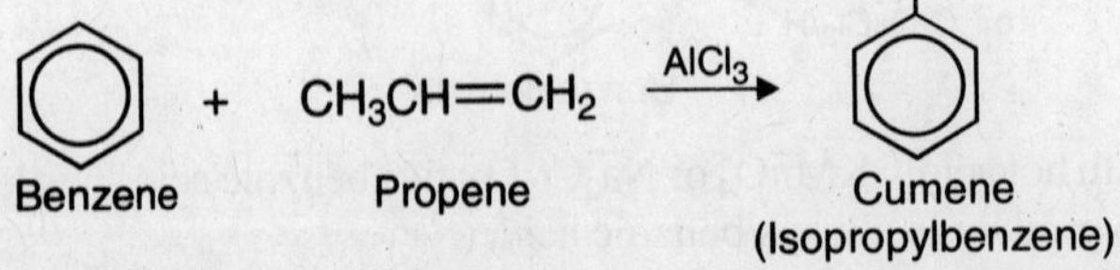

Benzene Propene Cumene (Isopropylbenzene)

Properties. Cumene is a colorless liquid, bp 153°C. It is used in the manufacture of phenol.

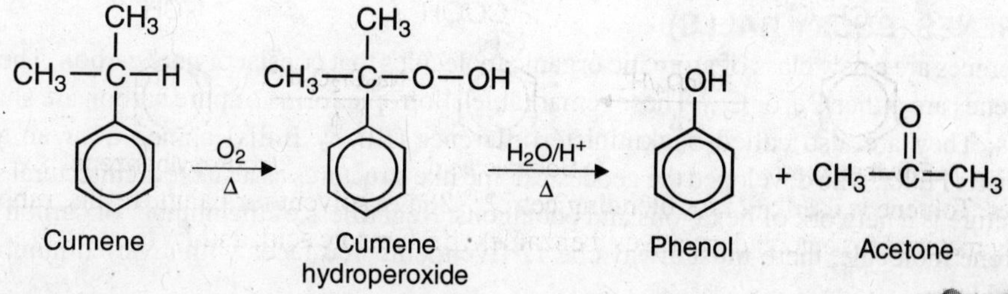

STYRENE, *Vinylbenzene*. $C_6H_5CH=CH_2$

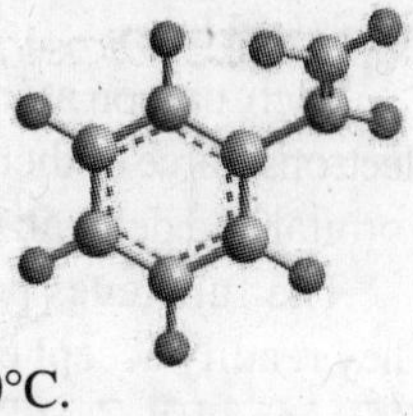

Industrially, the most important arylalkene is styrene which serves as a monomer for polystyrene and a wide variety of useful copolymers.

Preparation. It is obtained :

(1) By dehydrogenation of ethylbenzene with Cr_2O_3/Al_2O_3 or ZnO at 600°C.

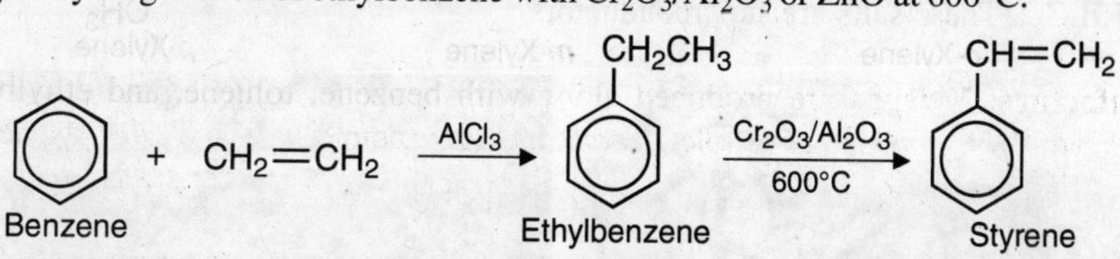

(2) By dehydrohalogenation of 1-chloro-1-phenylethane with alcoholic KOH.

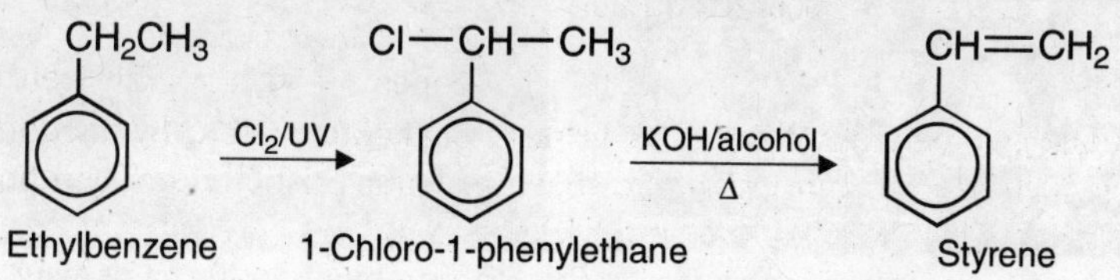

Properties. Styrene is colorless liquid, *bp* 145°C. It gives the following reactions :

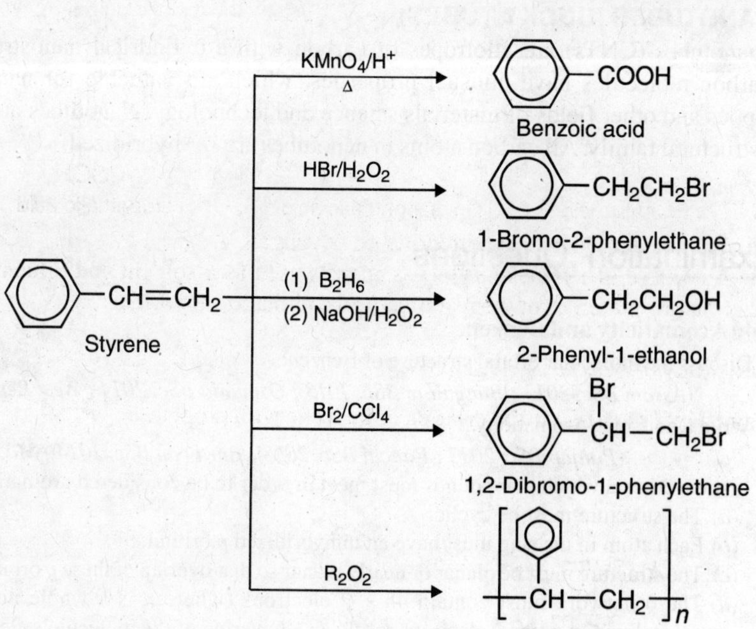

FUTURISTIC MATERIALS

FULLERENES (BUCKY BALLS)

Fullerenes are a new class of aromatic organic molecules that consist of pure carbon. The formula for fullerenes are either C_{60} or C_{70}. These remarkable allotropic forms of pure carbon are shaped like foot balls. They are also called buckministerfullerenes (Bucky Balls) named after an architect, Buckminister Fuller, who developed the geodesic dome like structures as a viable architectural structure. This structure is a network of hexagons and pentagons. Regardless of the number of carbon atoms in the fullerene molecule, there must always be 12 five membered faces with a varied number of six membered faces.

Each carbon atom in the fullerene molecule is sp^2 hybridized. Since each carbon has four valence electrons, three of them are in the three sp^2 hybrid orbitals with the fourth electron is in the unhybridized *p* orbital and delocalized.

The fullerenes possess fascinating properties because of their unusually high electron affinity. They readily accept electrons from metals such as potassium or rubidium to form fulleride salts like K_3C_{60} and Rb_3C_{60}. These salts are superconductors.

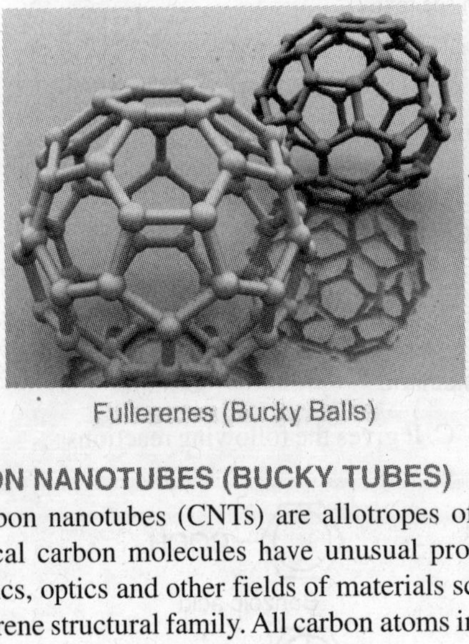

Fullerenes (Bucky Balls)

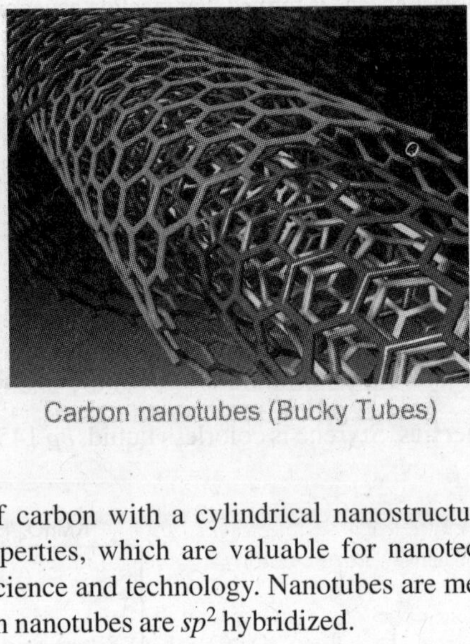

Carbon nanotubes (Bucky Tubes)

CARBON NANOTUBES (BUCKY TUBES)

Carbon nanotubes (CNTs) are allotropes of carbon with a cylindrical nanostructure. These cylindrical carbon molecules have unusual properties, which are valuable for nanotechnology, electronics, optics and other fields of materials science and technology. Nanotubes are members of the fullerene structural family. All carbon atoms in nanotubes are sp^2 hybridized.

Examination Questions

I. Questions on Aromaticity and Benzene

1. Discuss the molecular orbital structure of Benzene.

(*Assam BSc, 2013 ; Bangalore BSc, 2013 ; Osmania BSc, 2014 ; West Bengal BSc, 2015*)

2. Write a note on : Aromaticity.

(*Panjab BSc, 2013 ; Ranchi BSc, 2013 ; Burdwan BSc, 2014 ; Madras BSc, 2015*)

3. List the criteria which compounds must meet in order to be considered aromatic.

Answer. (*a*) The structure must be cyclic.

(*b*) Each atom in the ring must have an unhybridized *p* orbital.

(*c*) The structure must be planar or nearly planar so that overlap of these *p* orbitals is effective.

(*d*) The ð network must contain $4n + 2$ electrons (where *n* is a whole number), so that delocalization of the ð electrons results in a lowering of the molecule's electronic energy.

4. What is Huckel rule ? Write the structure of two compounds that follow this rule.

5. Explain the aromatic character of pyrrole and furan. *(Delhi BSc, 2015)*

6. Explain why pyridine is aromatic ? *(Panjab BSc, 2014)*

7. Which is more stable, cyclobutadiene or 1,3-butadiene ? Explain.

Answer. 1,3-Butadiene is more stable. Cyclobutadiene is antiaromatic since it contains 4 (i.e., 4*n*) π electrons. Antiaromatic systems are less stable than their open-chain counterparts.

8. Classify the following compounds as aromatic, antiaromatic or nonaromatic. Assume polarity of the π network.

Answer. (*a*) is Nonaromatic (*b*) is Nonaromatic
(*c*) is Nonaromatic (*d*) is Antiaromatic

9. How is benzene prepared ? Describe its important reactions.

10. What happens when sodium benzoate is heated with soda-lime ?

Answer. Benzene is formed. For reaction equation see text.

11. What happens when benzene is treated with acetyl chloride in the presence of $AlCl_3$? *(Agra BSc, 2014)*

Answer. Acetophenone is formed. For reaction equation see text.

12. What happens when benzene is treated with methyl chloride in the presence of $AlCl_3$? *(Dibrugarh BSc, 2015)*

Answer. Toluene is formed. For reaction equation see text.

13. How will you distinguish between benzene and cyclohexene ? *(Kerala BSc, 2015)*

Answer. Following tests are used :

(*a*) *Reaction with Br_2 in CCl_4.* Cyclohexene reacts with bromine in carbon tetrachloride solution. Red color of the reagent disappears. Benzene does not react with bromine under these conditions.

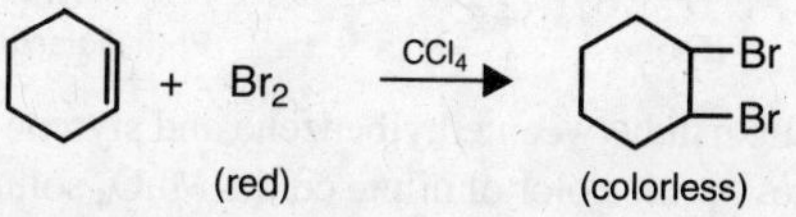

(*b*) *Reaction with dilute cold $KMnO_4$.* Cyclohexene reacts with cold $KMnO_4$ solution to give a brown precipitate of MnO_2. Purple color of the reagent disappears. Benzene does not react with dilute cold $KMnO_4$ solution.

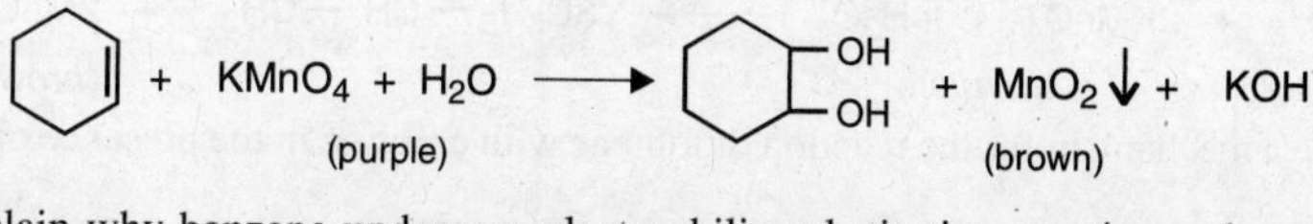

14. Explain why benzene undergoes electrophilic substitution reactions whereas alkenes undergo addition reactions. *(Sagar BSc, 2014 ; Calicut BSc, 2015)*

15. List three limitations usually associated with the Friedel-Crafts alkylation reaction.

Answer. (*a*) Reaction fails when strongly deactivating substituents are present.
(*b*) Rearrangements are possible; The alkyl groups often tend to rearrange.
(*c*) Multiple alkylations frequently occur.

16. Why are Friedel-Crafts acylation reactions not prone to polyacylation ?

Answer. Once the ring is acylated, it is deactivated to further electrophilic attack since the acyl group is electron withdrawing.

17. Give the general mechanism of electrophilic aromatic substitution.

18. Give the mechanism of nitration of benzene.
(North Bengal BSc, 2013 ; Mizoram BSc, 2014 ; Tamil Nadu BSc, 2015)

19. Give the mechanism of chlorination of benzene. *(Utkal BSc, 2015)*

20. Give the mechanism of bromination of benzene. *(Agra BSc, 2014)*

21. Give the mechanism of sulfonation of benzene.

22. Give the mechanism of Friedel-Crafts alkylation of benzene.
(Shivaji BSc, 2013 ; Jiwaji BSc, 2014 ; Indore BSc, 2015)

23. Give the mechanism of Friedel-Crafts acylation of benzene. *(Pondicherry BSc, 2015)*

24. Write a note on : Friedel-Crafts reaction. *(Meerut BSc, 2014)*

25. How will you synthesize the following compounds from benzene :
(*a*) Acetophenone (b) Chlorobenzene
(*c*) Benzophenone (d) Benzenehexachloride

II. Questions on Alkylbenzenes

26. Write the structures of benzene isomers with formula C_8H_{10}.

27. How is toluene prepared ? Describe its important reactions.

28. State the conditions under which chlorination of toluene takes place in the benzene ring and in the side-chain.

29. What happens when ethylbenzene is treated with alkaline $KMnO_4$?

Answer. Benzoic acid is formed. For reaction equation see text.

30. What happens when chlorine is passed through boiling toluene ?
(Madurai BSc, 2014 ; Gulbarga BSc, 2015)

Answer. Substitution takes place in the side-chain to yield successively benzyl chloride, and benzotrichloride. For reaction equation see text.

31. How will you distinguish between ethylbenzene and *o*-xylene ?

Answer. Ethylbenzene gives benzoic acid (mp 122°C) on oxidation with $KMnO_4$. *o*-Xylene gives phthalic acid (mp 184°C).

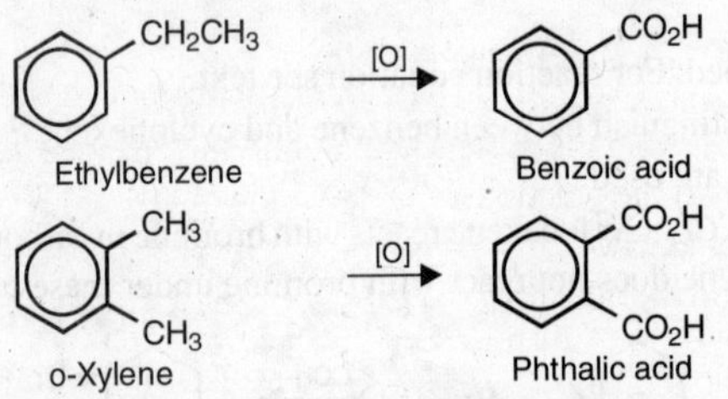

32. How will you distinguish between ethylbenzene and styrene ? *(Delhi BSc, 2015)*

Answer. Styrene decolorizes purple color of dilute cold $KMnO_4$ solution. Brown precipitate of MnO_2 is formed in the reaction. Ethylbenzene does not react with dilute cold $KMnO_4$ solution.

$$\underset{\text{Styrene}}{3C_6H_5CH{=}CH_2} + \underset{(purple)}{2KMnO_4} + 4H_2O \longrightarrow 3C_6H_5-\overset{OH}{\overset{|}{CH}}-\overset{OH}{\overset{|}{CH_2}} + \underset{(brown)}{2MnO_2\downarrow} + 2KOH$$

33. Give a mechanism for the reaction of toluene with chlorine in the presence of UV light.
(Madras BSc, 2014)

34. How will you synthesize ethylbenzene from acetophenone ? *(Anna BSc, 2015)*

Answer. Following step is involved :

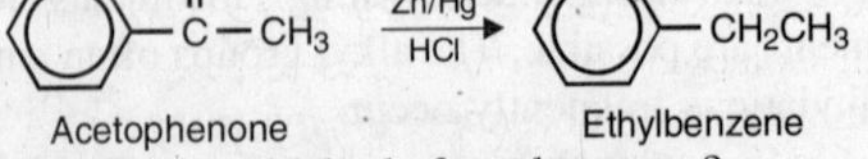

35. How will you synthesize benzaldehyde from benzene ?

36. How will you synthesize benzoic acid from benzene ? *(Jabalpur BSc, 2014)*

Answer. Following steps are involved :

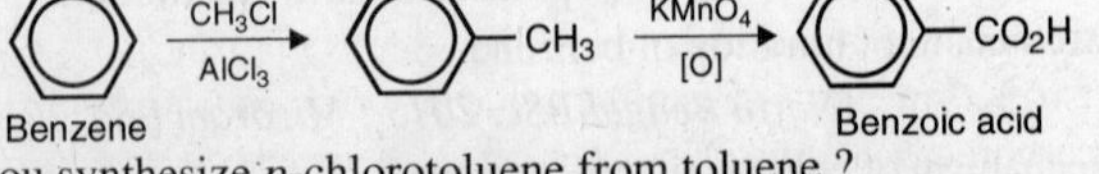

37. How will you synthesize *p*-chlorotoluene from toluene ?

38. How will you synthesize *p*-nitrobenzoic acid from toluene ?

39. How will you synthesize benzyl alcohol from toluene ?

Answer. Following steps are involved :

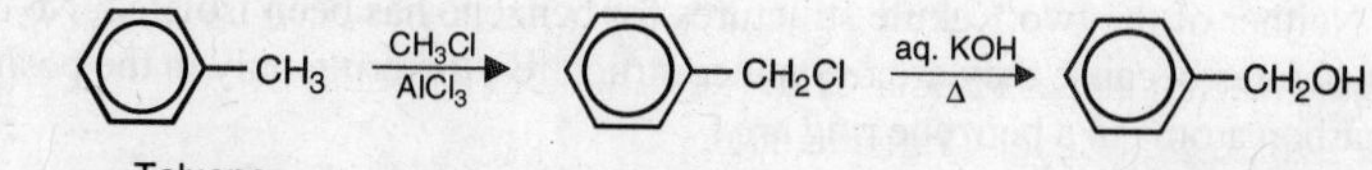

Toluene Benzyl alcohol

40. How will you synthesize styrene from benzene ? *(Mangalore BSc, 2015)*

41. How will you synthesize styrene from bromobenzene ?

Clue. Steps are : *(i)* Mg/ether ; *(ii)* CH_3CHO ; *(iii)* H_2O/H^+ ; and *(iv)* H_2SO_4/heat.

42. How will you synthesize phenylacetylene from styrene ? *(Jabalpur BSc, 2014)*

Answer. Following steps are involved :

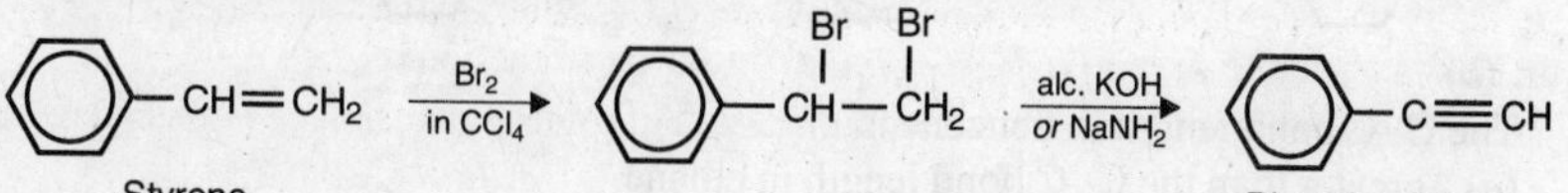

Styrene Phenylacetylene

43. How will you synthesize *n*-propylbenzene from benzene ? *(Kalyani BSc, 2015)*

Answer. Following steps are involved :

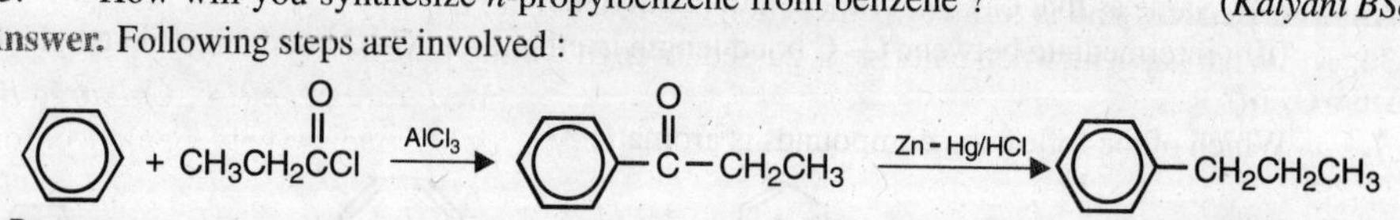

Benzene *n*-Propylbenzene

44. A hydrocarbon, C_9H_{12}, gave *o*-phthalic acid as the only organic product of oxidation. What is the structure and name of the hydrocarbon ?

Answer. The hydrocarbon is *o*-ethyltoluene.

45. A secondary alcohol (*A*), C_3H_8O, reacted with thionyl chloride to give compound (*B*), C_3H_7Cl. Compound (*B*) reacted with benzene in the presence of aluminium chloride to form (*C*), C_9H_{12}. Identify (*A*), (*B*), and (*C*) and write equations for all the reactions.

Answer. (*A*) = Isopropyl alcohol

(*B*) = Isopropyl chloride

(*C*) = Cumene

Multiple Choice Questions

1. Which of the following is an *incorrect* description of benzene ?

(*a*) The CCC bond angles are all equal to 120°.

(*b*) The molecule is planar.

(*c*) The molecule is a 6-membered ring which contains alternating single and double carbon-carbon bonds.

(*d*) The molecule can be drawn as a resonance hybrid of two Kekule structures.

Answer. (*c*)

2. Which of the following statements are *false* about benzene ?

(*a*) It is a planar molecule with bond angles 120°.

(*b*) It is immiscible with water forming the lower layer.

(*c*) It can be converted into cyclohexane by hydrogenation at 200°C in the presence of Ni catalyst.

(*d*) It reacts with ethyl chloride in the presence of aluminium chloride to form ethylbenzene.

Answer. (*b*)

3. Which statement about the structure of benzene is *not* true ?

(*a*) The two Kekule structures of benzene are in equilibrium.

(*b*) The carbon-carbon bond lengths in benzene are greater than the carbon-carbon double bonds in aliphatic compounds.

(c) The molecular geometry of benzene is best described as planar.
(d) The stability of benzene ring is much greater than the stability of 1,3,5-cycloheptatriene.

Answer. (a). Neither of the two Kekule structures for benzene has been isolated. No equilibrium exists between these two structures because they are resonance structures, differing only in the positions of electrons.

4. The carbon atoms in a benzene ring are :
(a) *sp* hybridized (b) sp^3 hybridized
(c) sp^2 hybridized (d) None of these

Answer. (c)

5. Which of the following compounds uses only sp^2 hybridized carbons for bond formation ?

(a) 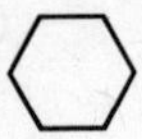 (b)  (c)  (d) 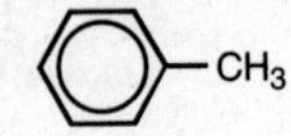

Answer. (b)

6. The C–C bond length in benzene is
(a) greater than the C–C bond length in ethane.
(b) shorter than the C–C bond length in ethylene.
(c) Same as that of C–C bond length in ethylene.
(d) intermediate between C–C bond length in ethane and C–C bond length in ethylene.

Answer. (d)

7. Which of the following compounds is aromatic ?

(a)  (b) 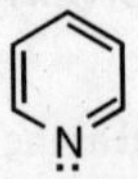

(c)  (d) 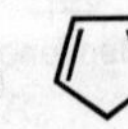

Answer. (b). Use the concept of aromaticity and Huckel rule to arrive at the correct answer.

8. Characteristic reactions of aromatic hydrocarbons are initiated by
(a) Electrophiles (b) Nucleophiles
(c) Free radicals (d) Uncharged molecules

Answer. (a)

9. Phenol on distillation with Zinc dust gives
(a) Phenylzinc (b) Benzene
(c) Cyclohexanone (d) Benzoic acid

Answer. (b)

10. Which of the following statements is *false* about toluene ?
(a) can be prepared by treating benzene with methyl chloride in the presence of $AlCl_3$.
(b) is converted to benzoic acid on refluxing with acidic $KMnO_4$ solution.
(c) on refluxing with concentrated H_2SO_4 gives a mixture of *ortho* and *para* toluenesulfonic acid.
(d) can be nitrated with concentrated nitric acid to give a mixture of *ortho* and *meta* nitrotoluene.

Answer. (d)

11. Which of the following reagents will react with methyl group rather than the benzene ring in methylbenzene ?
(a) Chlorine in the presence of uv light
(b) CH_3Cl in the presence of $AlCl_3$
(c) CH_3COCl in the presence of $AlCl_3$
(d) Hydrogen in the presence of nickel

Answer. (a)

12. Which of the following can be made by the action of CH_3Cl on benzene in the presence of aluminium chloride ?
(a) Ethylbenzene (b) *o*-Xylene
(c) Chlrobenzene (d) *m*-Xylene

Answer. (b)

13. For reactions of ethylbenzene, the ethyl group is considered :
(a) *ortho* director (b) *ortho-para* director
(c) *meta* director (d) *ortho-meta* director

Answer. (b)

14. Toluene reacts with bromine in the presence of uv light to give

(*a*) *m*-Bromotoluene (*b*) Benzyl bromide
(*c*) *o*-Bromotoluene (*d*) Benzoyl bromide

Answer. (*b*)

15. Toluene reacts with chlorine in the presence of $AlCl_3$ to give

(*a*) *o*-Chlorotoluene (*b*) *o*- plus *p*-Chlorotoluene
(*c*) *m*-Chlorotoluene (*d*) *o*- plus *m*-Chlorotoluene

Answer. (*b*)

16. Toluene undergoes oxidation to give

(*a*) Benzyl alcohol (*b*) Quinone
(*c*) Benzaldehyde (*d*) Benzoic acid

Answer. (*d*)

17. Ethylbenzene undergoes oxidation with acidic potassium dichromate to give

(*a*) 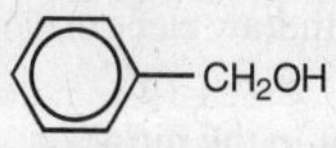

(*b*) 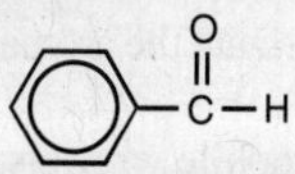

(*c*) 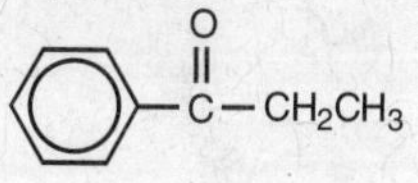

(*d*) 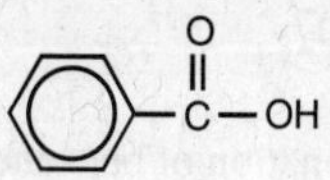

Answer. (*d*)

18. Oxidation of cumene with acidic $K_2Cr_2O_7$ gives

(*a*) Phenylacetic acid (*b*) Benzaldehyde
(*c*) Benzyl alcohol (*d*) Benzoic acid

Answer. (*d*)

19. Oxidation of toluene with chromyl chloride gives benzaldehyde. This reaction is known as

(*a*) Perkin reaction (*b*) Benzoin condensation
(*c*) Étard's reaction (*d*) Ozonolysis

Answer. (*c*)

20. Oxidation of *p*-xylene with acidic potassium dichromate gives

(*a*) 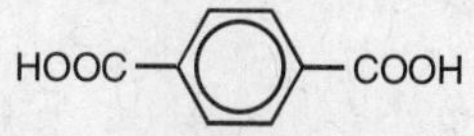

(*b*) 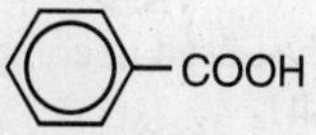

(*c*) 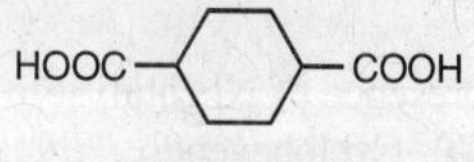

(*d*) 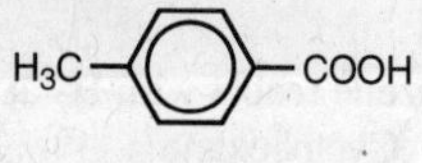

Answer. (*a*)

21. Cresols on distillation with zinc dust gives

(*a*) *o*-Xylene (*b*) Benzene
(*c*) *o*- plus *p*-Xylene (d) Toluene

Answer. (*d*)

22. Benzene undergoes substitution reaction more easily than addition reaction because

(*a*) it has a cyclic structure (*b*) it has three double bonds
(*c*) it has six hydrogen atoms (*d*) there is delocalization of electrons

Answer. (*d*)

23. Benzene reacts with concentrated HNO_3 in the presence of concentrated H_2SO_4 to give nitrobenzene. This reaction is an example of

(*a*) Electrophilic addition (*b*) Nucleophilic addition
(*c*) Electrophilic substitution (*d*) Nucleophilic substitution

Answer. (*c*)

24. Which of the following agents is used in order to make benzene react with concentrated nitric acid to give nitrobenzene ?

(*a*) Concentrated H_2SO_4 (*b*) $FeCl_3$ catalyst
(*c*) Lindlar's catalyst (*d*) Ultraviolet light

Answer. (*a*)

25. Which of the following agents is used in order to make benzene react with acetyl chloride to give acetophenone ?

(*a*) Ultraviolet light (*b*) $AlCl_3$ catalyst
(*c*) Platinum catalyst (*d*) Al_2O_3 catalyst

Answer. (*b*)

26. Which of the following agents is used in order to make benzene react with bromine to give bromobenzene ?

(*a*) Ultraviolet light (*b*) Fe catalyst
(*c*) Nickel catalyst (*d*) Al_2O_3 catalyst

Answer. (*b*)

27. In the Friedel-Craft acetylation of an aromatic ring, the role of the $AlCl_3$ is to

(*a*) Form a $CH_3-\overset{+}{\underset{}{\overset{\overset{O}{\|}}{C}}}$ ion. (*b*) Function as a Lewis base
(*c*) Chlorinate the aromatic ring (*d*) Withdraw electrons from the aromatic ring

Answer. (*a*)

28. The electrophile which is considered to be the active agent in the nitration of benzene is

(*a*) NO_2^- (*b*) NO^+ (*c*) NO_2^+ (*d*) HNO_2^+

Answer. (*c*)

29. In chlorination of benzene, $FeCl_3$ is used to generate

(*a*) Cl^- (*b*) Cl^+ (*c*) Cl_2 (*d*) HCl

Answer. (*b*)

30. In sulfonation of benzene, the attacking species is

(*a*) H^+ (*b*) SO_2 (*c*) SO_3 (*d*) HSO_4^-

Answer. (*c*)

31. Consider the following reaction :

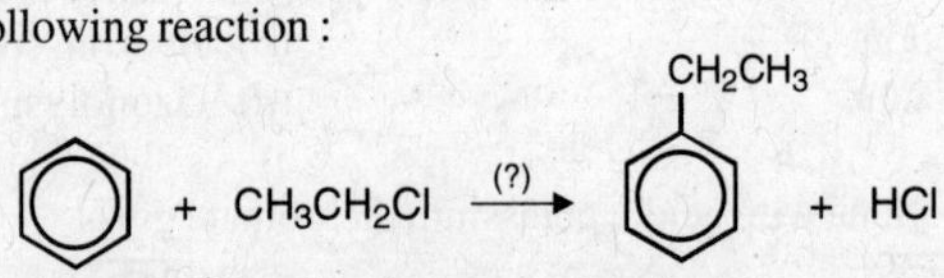

The catalyst used to complete the above reaction is

(*a*) $LiAlH_4$ (*b*) $AlCl_3$ (*c*) Na (*d*) KOH

Answer. (*b*)

32. Benzene reacts with H_2 at 150°C at 30 atm in the presence of Ni catalyst to give

(*a*) Cyclohexane (*b*) Cyclohexene
(*c*) *n*-Hexane (*d*) No reaction occurs

Answer. (*a*)

33. Benzene reacts with chlorine in the presence of $FeCl_3$ catalyst to form

(*a*) Hexachlorobenzene (*b*) Chlorobenzene
(*c*) Hexachlorocyclohexane (*d*) Benzyl chloride

Answer. (*b*)

34. Benzene reacts with acetic anhydride in the presence of $AlCl_3$ to form

(*a*) Acetophenone (*b*) Benzophenone
(*c*) Phenylacetic acid (*d*) Phenyl acetate

Answer. (*a*)

35. Benzene reacts with benzoyl chloride in the presence of anhydrous aluminium chloride to form

(*a*) Benzyl chloride (*b*) Benzaldehyde
(*c*) Benzal chloride (*d*) Benzophenone

Answer. (*d*)

36. Benzene undergoes Friedel-Crafts reaction with isopropyl alcohol in the presence of H_2SO_4 catalyst to give

(*a*) *n*-Propylbenzene (*b*) Benzophenone
(*c*) Isopropylbenzene (*d*) Nothing happens

Answer. (*c*)

37. Benzene reacts with propene in the presence of H_2SO_4 catalyst to give
(*a*) *n*-Propylbenzene (*b*) Benzophenone
(*c*) Cumene (*d*) Nothing happens
Answer. (*c*)

38. Ozonolysis of benzene gives
(*a*) Formic acid (*b*) Glyoxal
(*c*) Formaldehyde (*d*) Glycine
Answer. (*b*)

39. Gammexane is
(*a*) Hexachloroethane (*b*) DDT
(*c*) Hexachlorocyclohexane (*d*) TNT
Answer. (*c*)

40. Under what reaction conditions does the electrophilic chlorination of aromatic compounds usually occur ?
(*a*) $Cl_2, AlCl_3$ (*b*) Cl_2, H_2O (*c*) Cl_2, CCl_4 (*d*) $NaCl, H_2O$
Answer. (*a*)

41. Which of the following is an *incorrect* statement about the bromination of benzene by Br_2 and $FeBr_3$?
(*a*) $FeBr_3$ functions to increase the electrophilicity of Br_2.
(*b*) Formation of the sigma complex is the rate-determining step of the mechanism.
(*c*) The carbanionic intermediate is resonance stabilized.
(*d*) There are two carbon-containing intermediates in the mechanism.
Answer. (*c*)

42. Which of the following is the best choice of reagents to effect the electrophilic iodination of an aromatic ring ?
(*a*) KI, acetone (*b*) I_2, CH_3CN (*c*) KI, HNO_3 (*d*) I_2, HNO_3
Answer. (*d*)

43. Which of the following is another name for cyclobutadiene ?
(*a*) [2]annulene (*b*) [4]annulene (*c*) [6]annulene (*d*) Antibenzene
Answer. (*b*)

44. What is the major difference between an *antiaromatic* and *aromatic* compound ?
(*a*) Antiaromatic compounds have at least one sp^3 hybridized atom in the ring
(*b*) Antiaromatic compounds can assume a chair-like structure while aromatic compounds are nearly flat
(*c*) Aromatic compounds cannot have a charged atom in the structure
(*d*) Only aromatic compounds follow Huckle's rule.
Answer. (*d*)

Finally, we can drink coke with a straw.

34

Directive Effects of Substituents

> **Some students get confused over which group does the directing, the incoming electrophile, E^+, or the initial substituent, —S. Try thinking about it in terms of an aircraft (the E^+) coming into land at an airport (the Ar—S). It is the control tower at the airport on the ground (—S) that does the "directing" of which runway and which ramp the aircraft should go to.**

SUBSTITUTION IN MONOSUBSTITUTED BENZENES

All hydrogen atoms of the benzene ring are equivalent. Therefore, only one monosubstitution product (C_6H_5—S) is possible.

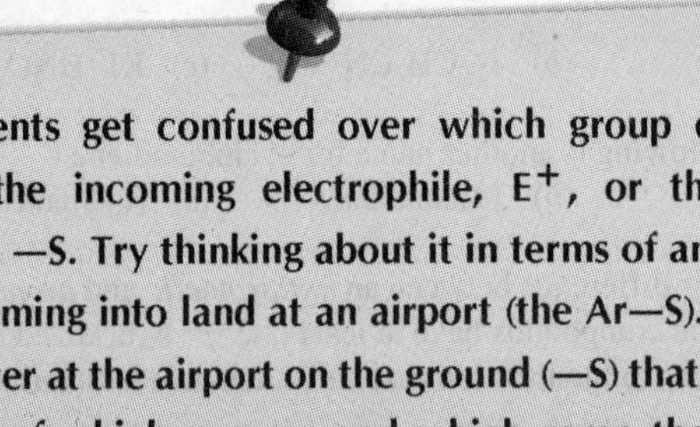

A second substituent, E, can occupy any of the remaining five positions (Fig.34.1)

(1) The positions 2 and 6 are equivalent, and would give the **Ortho Product**.

(2) The positions 3 and 5 are equivalent, and would give the **Meta Product**.

(3) The position 4 is unique and would give the **Para Product**.

TWO TYPES OF INFLUENCE OF SUBSTITUENTS

A substituent (S) already present on the benzene ring exercises two types of influence on further substitution.

(1) **Directive or Orientation Effect**. The first substituent (S) may direct the next incoming substituent (E) to *ortho*, *meta*, or *para* position, depending on the nature of the first substituent. This is called the *Directive* or the *Orientation* (*Orient* = to arrange) *Effect*.

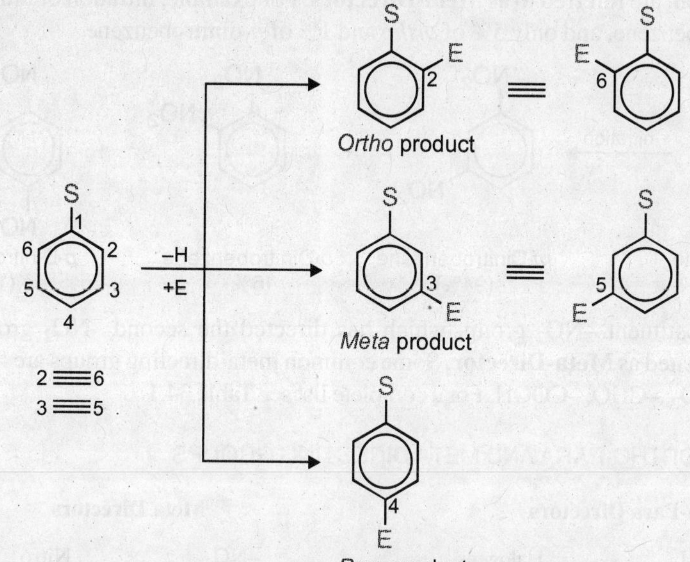

Fig. 34.1. Substitution in monosubstituted benzenes.

(2) **Activity Effects**. The substituent already present may activate or deactivate the benzene ring toward further substitution. These effects are called the *Activity Effects*.

DIRECTIVE EFFECTS OF SUBSTITUENTS

We have seen that in the monosubstituted benzene, C_6H_5–S, there are five available hydrogens. Of these two are *ortho*, two *meta*, and one *para* to S. If substitution was random, on purely statistical basis, the proportion of the disubstituted products would be

Ortho = 2/5 of the total, or 40%
Meta = 2/5 of the total, or 40%
Para = 1/5 of the total, or 20%

However, this distribution is never observed actually. The proportion of the products formed, in fact, is determined by the nature of the first substituent on the ring rather than on any mathematical probability.

Ortho-Para Directing Effect. Certain substituents direct the second substituent to the *ortho* and *para* positions simultaneously. These are called **Ortho-Para Directors.** For example, when phenol is nitrated, the only products obtained are *o*-nitrophenol (53%) and *p*-nitrophenol (47%), and no *meta* isomer is produced.

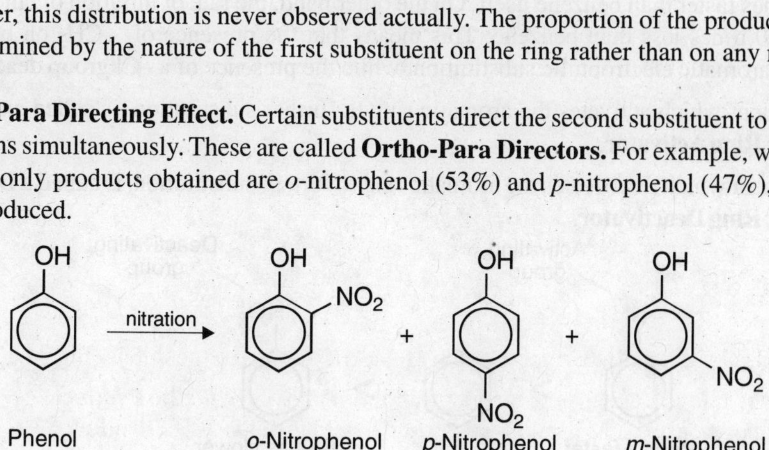

Phenol — *o*-Nitrophenol (53%) — *p*-Nitrophenol (47%) — *m*-Nitrophenol (0%)

The substituent –OH is said to have directed the $-NO_2$ group to *ortho* and *para* positions on the ring. Therefore, –OH is designated as **Ortho-Para Director**.

Some common *ortho-para* directing groups are –Cl, –Br, –I, –OH, $-NH_2$, $-CH_3$, $-C_2H_5$. For a complete list see Table 34.1.

Meta-Directing Effect. The substituents which direct the second incoming substituent primarily to the *meta* position, are referred to as **Meta-Directors**. For example, nitration of nitrobenzene gives 94% of *m*-dinitrobenzene, and only 5% of *ortho* and 1% of *p*-dinitrobenzene.

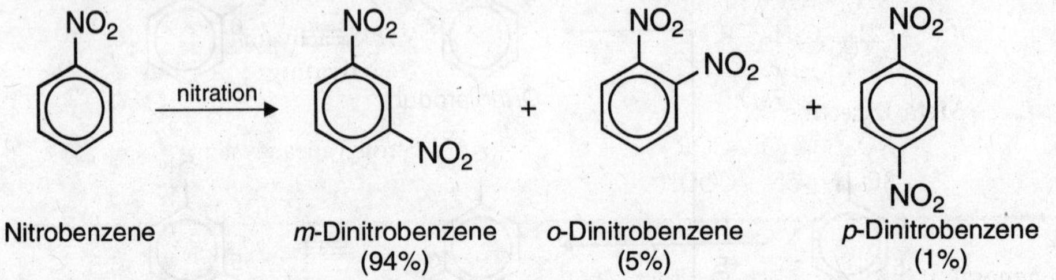

Thus the substituent $-NO_2$ group, which has directed the second $-NO_2$ group to the *meta* position, is designated as **Meta-Director**. Some common meta-directing groups are : $-SO_3H$, –CN, $-NO_2$, –CHO, –COOH. For a complete list see Table 34.1.

Table 34.1 ORTHO-PARA AND META DIRECTING GROUPS

Ortho-Para Directors		Meta Directors	
–F, –Cl, –Br, –I	Halogens	$-NO_2$	Nitro
–OH	Hydroxy	$-SO_3H$	Sulfonic acid
$-OCH_3$, –OR	Alkoxy	–CN	Cyano
$-NH_2$	Amino	–COOH	Carboxylic acid
–NHR, $-NR_2$		–CHO	Aldehyde
$-CH_3$, $-C_2H_5$, –R	Alkyl	–CO–R	Ketone
		$-\overset{+}{N}H_3$	Ammonium ion

EFFECT OF SUBSTITUENTS ON REACTIVITY

As observed earlier, substituent already present on a benzene ring not only directs the position of an incoming group, but also influences the rate of reaction. For example, toluene, $C_6H_5{-}CH_3$, is nitrated 25 times faster than benzene itself. On the other hand, the rate of nitration of chlorobenzene, $C_6H_5{-}Cl$, is 30 times less than benzene. This means that the presence of $-CH_3$ on benzene ring activates it to aromatic electrophilic substitution, while the presence of a –Cl group deactivates it.

A substituent which activates the aromatic ring to further substitution, is called an **Activating Substituent** or **Ring Activator**.

A substituent which deactivates the aromatic ring to further substitution, is called a **Deactivating Substituent** or **Ring Deactivator**.

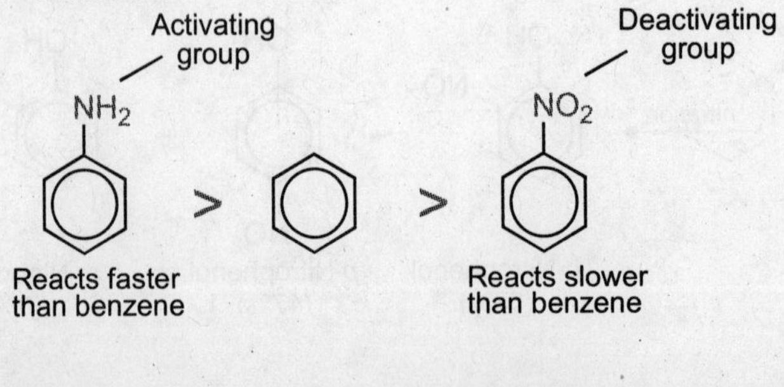

Table 34.2 shows the effects of substituents on reactivity in electrophilic aromatic substitution.

Table 34.2 EFFECT OF SUBSTITUENTS ON REACTIVITY

Substituent	Effect on Reactivity
Ortho-Para Directing	
–OH, –OR	Strongly activating
$–NH_2$, –NHR, $–NR_2$	
$–CH_3$, $–C_2H_5$, –R	Weakly activating
–F, –Cl, –Br, –I	**Deactivating**
Meta Directing	
$–NO_2$, –CHO, –COR	Strongly deactivating
$–SO_3H$, –CN, –COOH	

In general, it has been observed that :

Ortho-para directors activate a ring toward electrophilic substitution, whereas meta directing groups deactivate a ring toward electrophilic substitution. As indicated in the Table 34.2, the only exception to the above generalization occurs in halobenzenes, $C_6H_5–X$.

Although –F, –Cl, –Br, and –I are ortho-para directors, these substituents deactivate an aromatic ring in electrophilic substitution.

THEORY OF DIRECTIVE EFFECTS

The resonance theory clearly explains why certain substituents are *ortho-para* directing, while others are *meta*-directing. Let us examine the various resonance forms of phenol, in which the substituent –OH is strongly *ortho-para* directing.

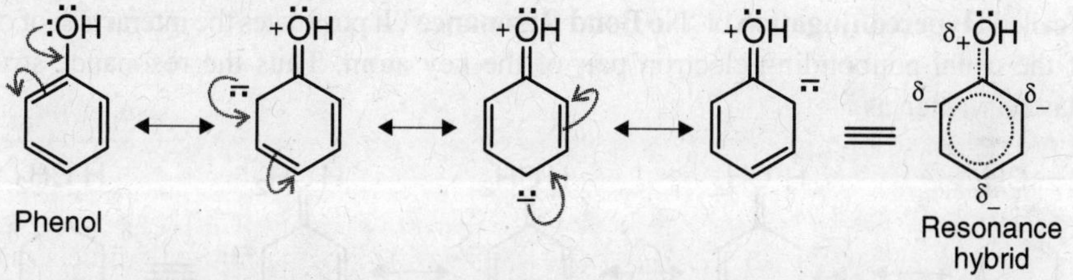

There are two nonbonding electron pairs on the oxygen atom attached to the ring. One of these is distributed into the ring by interaction with the π system as shown above. Notice that in the resonance forms, the *ortho* and *para* positions have a greater electron density than does the *meta* positions. Therefore, the resonance hybrid has negative charges in the *ortho* and *para* positions with electron delocalization as indicated by dots. The electrophile (E^+) would naturally attack at these electron-rich centers, regardless of its nature.

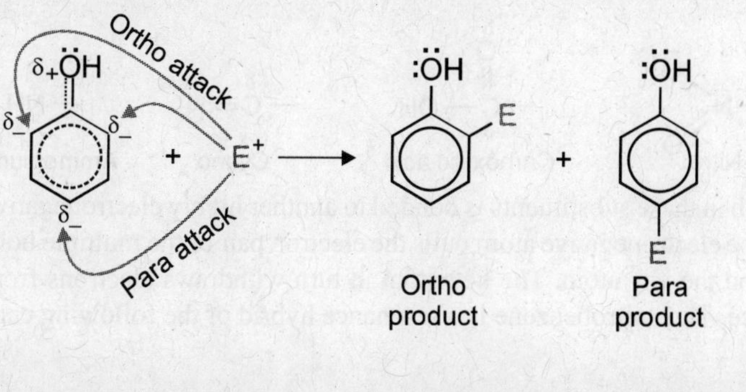

Similarly, the *ortho-para* directive influence of $-NH_2$ and $-Cl$ groups can be explained by the resonance theory. The nitrogen atom of $-NH_2$ group and the $-Cl$ atom possess a nonbonding electron pair, which is distributed into the ring.

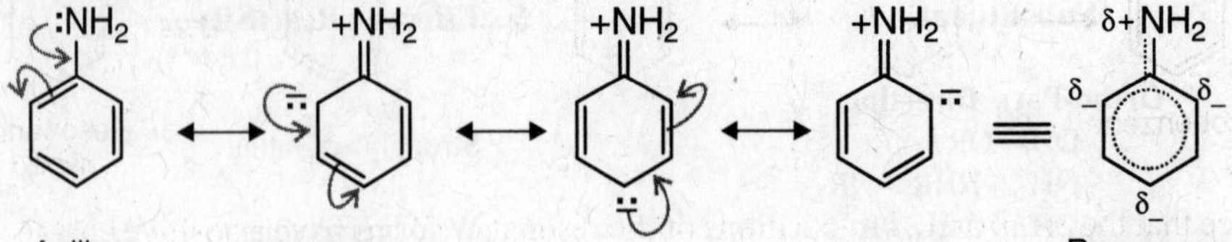

Aniline — Resonance hybrid

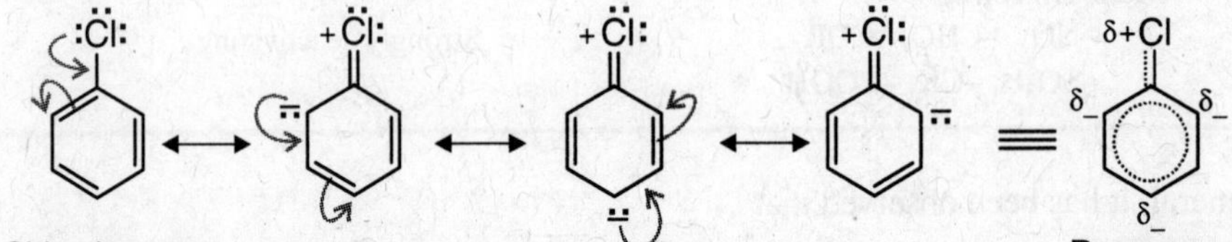

Chlorobenzene — Resonance hybrid

Presence of Nonbonding Electron Pair on Ortho-Para Directing Substituents. It may be noted that all *ortho-para* directing substituents or groups, possess a nonbonding electron pair on the 'key atom'.

$-\ddot{\underset{..}{Cl}}:\quad -\ddot{\underset{..}{Br}}:\quad -\ddot{O}H\quad -\ddot{O}CH_3\quad -\ddot{N}H_2$

Methyl Group is an Exception. The only exception to the above rule is the methyl or alkyl group. It is *ortho-para* directing, although it has no nonbonding electron pair on the key atom, carbon.

The *ortho-para* directive influence of CH_3 group is explained on the basis of special type of resonance called **Hyperconjugation** or **'No Bond Resonance'**. It postulates the interaction of σ electrons instead of the usual nonbonding electron pair of the key atom. Thus the resonance structures of toluene may be written as :

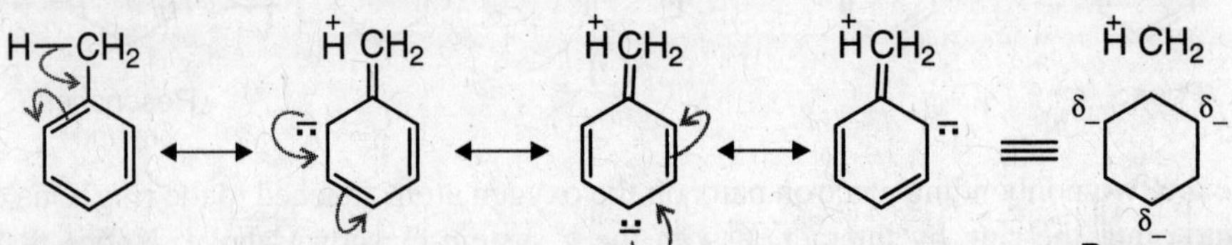

Toluene — Resonance hybrid

Evidently, the *ortho* and *para* positions will be the centers of electrophilic attack.

Resonance Theory and Meta-Directing Effect. Let us first examine the structures of some common *meta*-directing substituents.

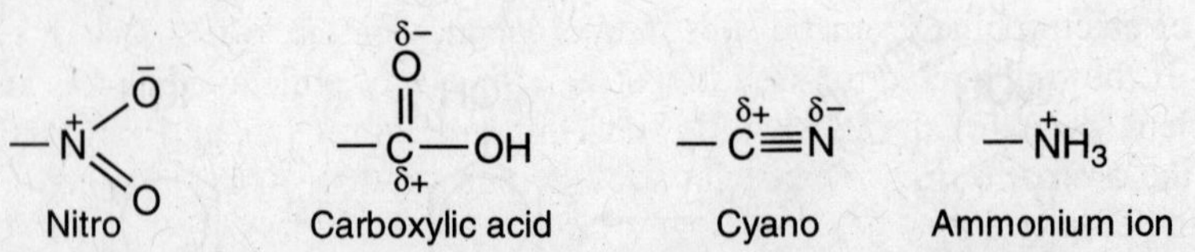

Nitro — Carboxylic acid — Cyano — Ammonium ion

The key atom in these substituents is bonded to another highly electronegative atom by a double or triple bond. The electronegative atom pulls the electron pair of the multiple bond and thus places a positive charge on the key atom. The key atom in turn withdraws electrons from the benzene ring causing resonance. Thus nitrobenzene is a resonance hybrid of the following canonical forms.

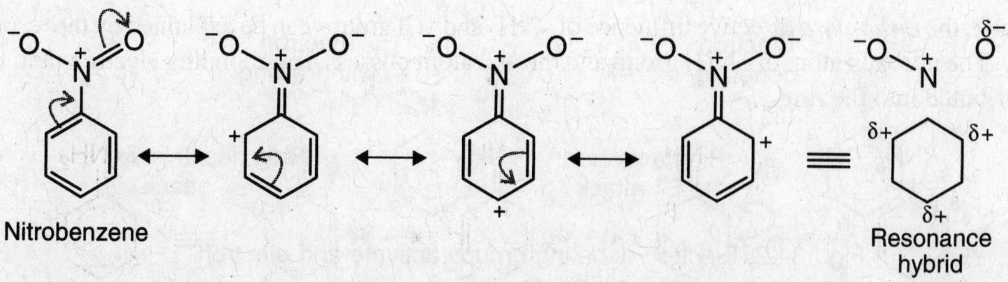

Notice that the *ortho* and *para* positions of the resonance forms have a positive charge and there is no scope of electrophile (E^+) attack at these positions. The electrophile attacks the *meta* positions which are relatively electron-rich.

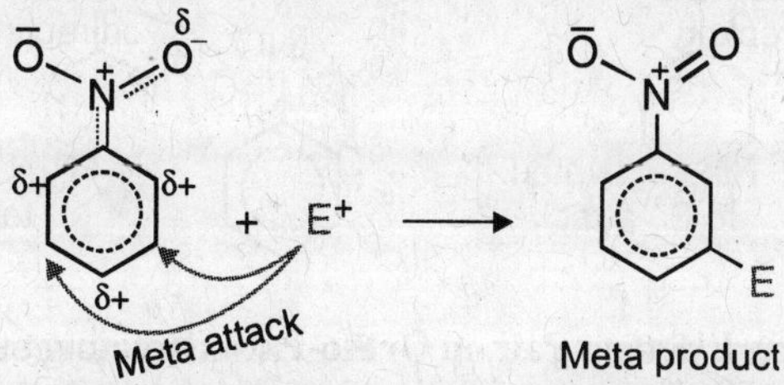

Carboxylic acid and benzonitrile show resonance analogous to nitrobenzene. The resonance hybrid having positive charge on the *ortho* and *para* positions is attacked by the electrophile at the *meta* positions.

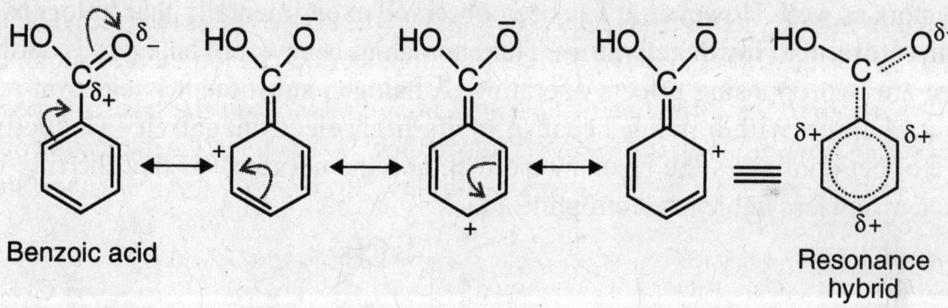

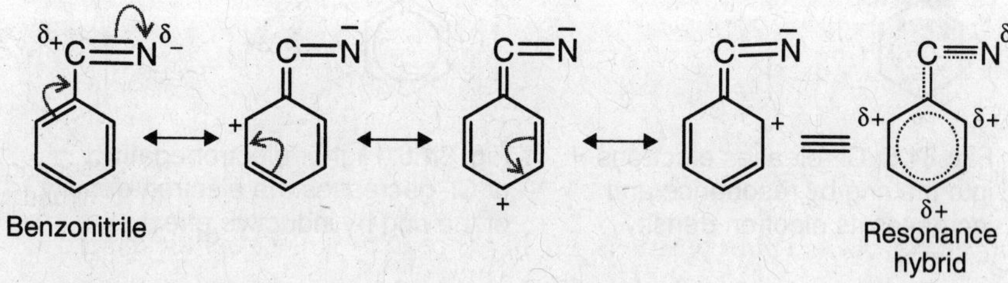

THEORY OF ACTIVATING AND DEACTIVATING EFFECTS

The rate of electrophilic aromatic substitution depends on the availability of electrons in the benzene ring. If the ring is electron-rich (negative), the electrophile attack is faster. If the ring is electron-deficient (positive), the attack is slower. Thus an electron donating or releasing substituent will activate the aromatic ring toward electrophilic substitution, while an electron-withdrawing substituent will deactivate it.

We have already studied that *ortho-para* directing groups (OH, NH_2, CH_3) release electrons into the ring by resonance, making it electron rich, favoring the *ortho, para*-positions in particular. **Thus the ortho-para directing substituents are also ring activators.**

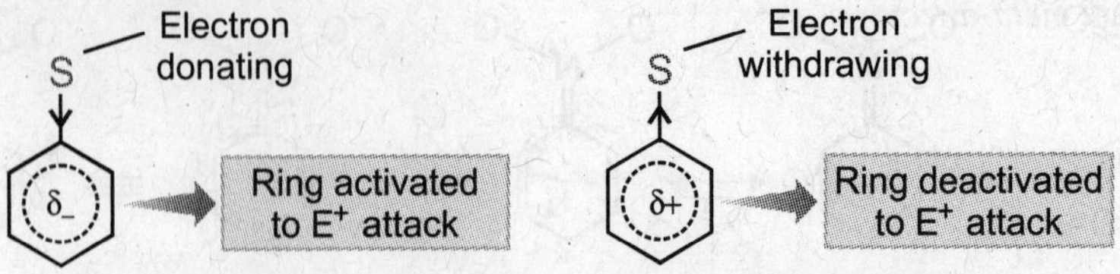

Fig. 34.2. Electron-donating groups activate and electron-withdrawing groups deactivate the aromatic ring.

The *meta*-directing groups (NO_2, SO_3H, COOH) withdraw electrons from the ring by resonance, making it electron-deficient (+). Therefore, **meta-directing groups are ring deactivators.**

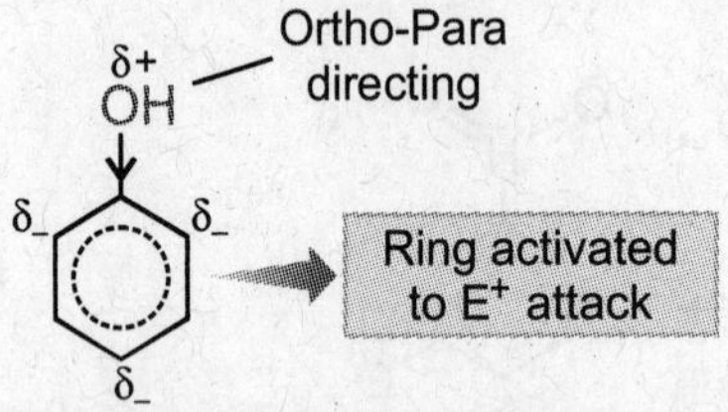

Fig. 34.3. Ortho-para directing substituents being electron-releasing activate the ring to electrophile attack.

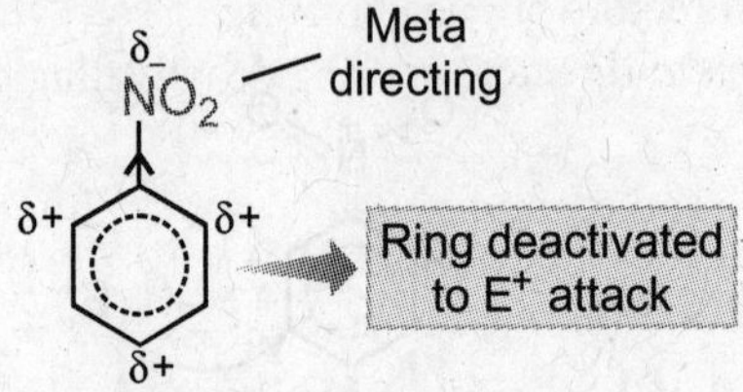

Fig. 34.4. Meta-directing substituents being electron-withdrawing deactivate the ring to electrophile attack.

Anomalous Behavior of Halogens. We have seen above that the *o,p*-directing substituents are, in general, activators as well. However, it has been observed experimentally that **halogens, although ortho-para directors, are ring deactivators**. This anomalous behavior of halogens is attributed to the fact that there are two opposing effects operating. A halogen substituent is electron-releasing by resonance but electron-withdrawing because of its high electronegativity (inductive effect). The inductive effect dominates the resonance effect, and the net result is that the ring is rendered electron-deficient and less liable to electrophile attack.

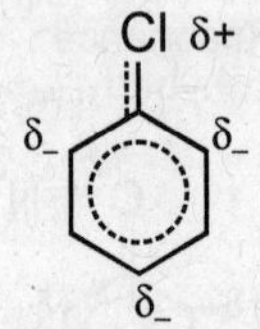

Fig. 34.5. Cl releases electrons into the ring by resonance and increases its electron density.

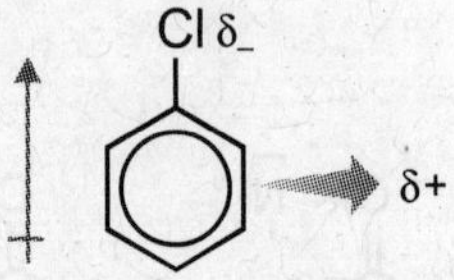

Fig. 34.6. Higher electronegativity of Cl, decreases the electron-density of the ring by inductive effect.

Points to Remember

Determining the directive effect and activating (or deactivating) effect of the various substituents/groups.

(1) Alkyl groups, phenyl group, and substituents which have one or more nonbonding electron pairs (lone pairs) on the atom adjacent to the ring are *ortho*, *para*-directors.

$-CH_3$ $-CH_2CH_3$ 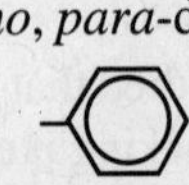

$-\ddot{O}H$ $-\ddot{O}R$ $-\ddot{N}H_2$ $-\ddot{X}:$ $-\ddot{N}HCOCH_3$

(2) Substituents which have a positive (or partial positive) charge on the atom adjacent to the ring are *meta*-directors.

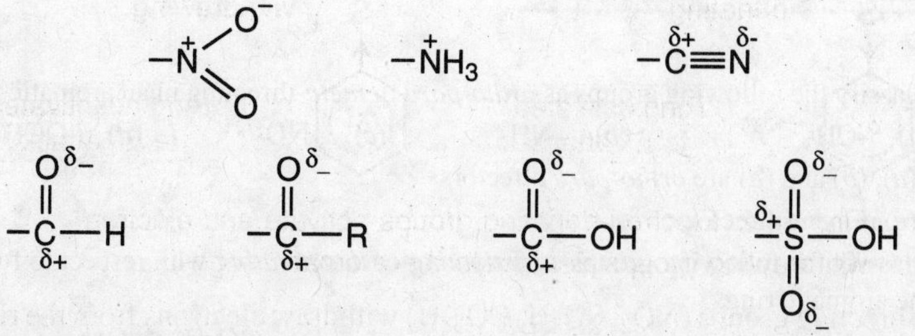

(3) All *ortho, para*-directors are ring activating groups. Halogens are exceptions. They are *o,p*-directors but deactivating.

(4) All *meta*-directors are ring deactivating groups.

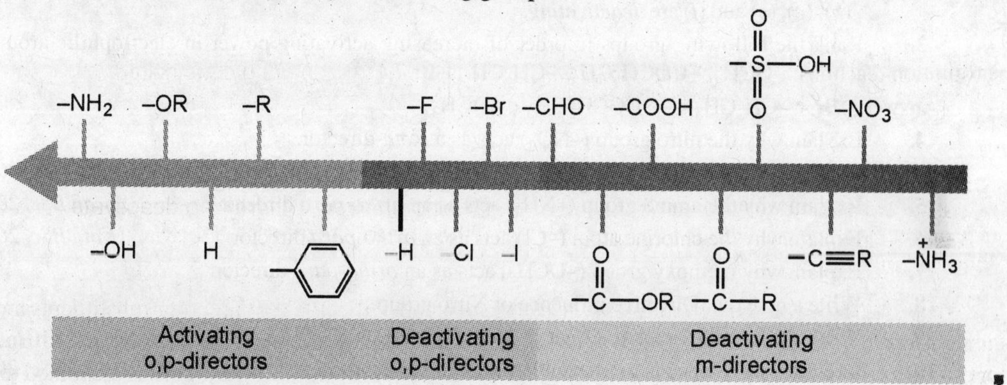

SUBSTITUTION IN DISUBSTITUTED BENZENES

Aromatic substitution reactions of disubstituted benzenes usually give a mixture of products because of competing directive influence of both substituents. However, following generalizations can be made :

(1) **If both substituents are activators, the position of the entering group is primarily governed by the directive influence of the stronger activator**. In the following examples, the arrows are used to indicate the possible positions that may be taken up by an incoming group. Remember that the position between two *meta* substituents is usually the least substituted owing to steric hindrance.

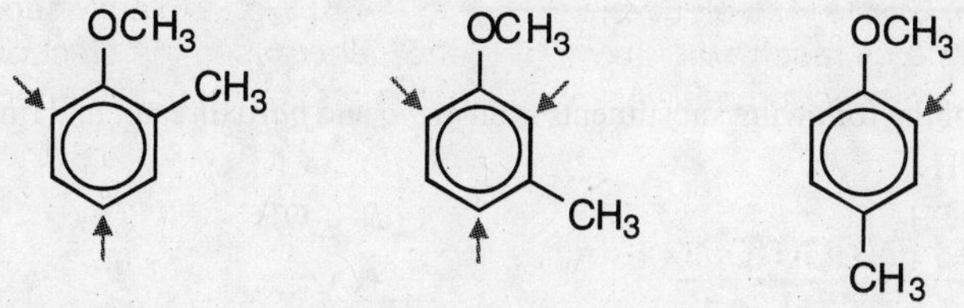

(2) **If one substituent is an activator and the other a deactivator, the position of the entering group will be dictated mainly by the activator.**

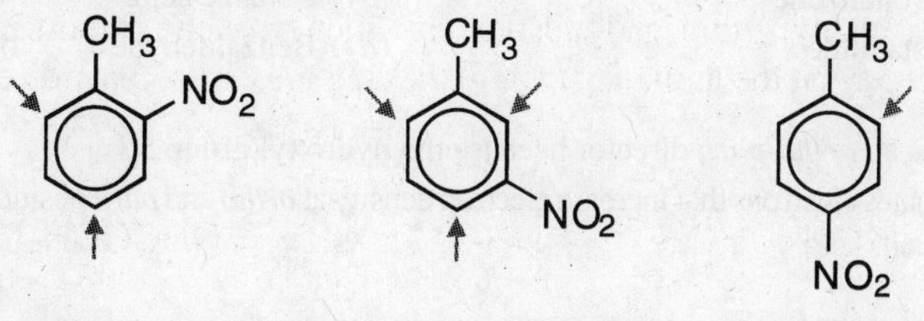

Examination Questions

1. Classify the following groups as *ortho-para* or *meta*-directing in an aromatic nitration reaction.

(*a*) $-CH_3$ (*b*) $-NH_2$ (*c*) $-NO_2$ (*d*) $-OCH_3$

Answer. (*a*), (*b*) and (*d*) are *ortho-para* directors

(*c*) is *meta*-director

2. Classify the following groups as *activating* or *deactivating* with respect to further electrophilic substitution of the aromatic ring.

(*a*) $-NH_2$ (*b*) $-NO_2$ (*c*) $-SO_3H$

(*d*) $-CH_3$ (*e*) $-Cl$ (*f*) $-CN$

Answer. (*a*) and (*d*) are *activating*.

(*b*), (*c*), (*e*) and (*f*) are *deactivating*.

3. Rank the following groups in order of increasing activating power in electrophilic aromatic substitution reactions: $-OCH_3$, $-COCH_2CH_3$, $-CH_2CH_3$, $-Br$.

Answer. $-Br < -CH_2CH_3 < -OCOCH_2CH_3 < -OCH_3$

4. Explain why the nitro group ($-NO_2$) acts as a *meta*-director.

(*Cochin BSc, 2014 ; Lucknow BSc, 2015*)

5. Explain why the amino group ($-NH_2$) acts as an *ortho-para* director. (*Jiwaji BSc, 2014*)

6. Explain why the chlorine atom ($-Cl$) acts as an *ortho-para* director. (*Kota BSc, 2015*)

7. Explain why methoxy group ($-OCH_3$) acts as an *ortho-para* director.

8. Write a note on : Directive influence of Nitro group.

9. Write a note on : Directive effect of Methyl group

10. Explain why toluene is more readily nitrated than benzene. (*West Bengal BSc, 2014*)

11. Explain why benzene is more readily nitrated than nitrobenzene.

(*Hyderabad BSc, 2014 ; Baroda BSc, 2015*)

12. Explain why phenol is nitrated more readily than benzene.

13. Explain why phenol is brominated more readily than benzene.

14. Explain why aniline is more reactive than benzene in Friedel-Crafts reactions.

(*Osmania BSc, 2014*)

Multiple Choice Questions

1. Which of the following substituents is an *ortho* and *para* director and ring deactivating ?

(*a*) $-NH_2$ (*b*) $-Cl$

(*c*) $-OCH_3$ (*d*) $-OH$

Answer. (*b*)

2. Which of the following will undergo substitution in the *ortho* and *para* positions rather than in the *meta* position ?

(*a*) Nitrobenzene (*b*) Benzoic acid

(*c*) Acetanilide (*d*) Benzaldehyde

Answer. (*c*)

3. Phenol is an *ortho-para* director because the hydroxyl group :

(*a*) donates electrons that increase electron density at *ortho* and *para* positions favoring nucleophilic attack.

(*b*) donates electrons that increase electron density at *ortho* and *para* positions favoring electrophilic attack.

(*c*) donates electrons to the *ortho* and *para* positions and attracts electrons away from *meta* positions favoring nucleophilic attack of the ring.

(*d*) donates electrons to the *ortho* and *para* positions and attracts electrons away from *meta* positions favoring electrophilic attack on the ring.

Answer. (*b*)

4. Electron-withdrawing groups are *meta* directors because :

(*a*) the carbonium ion intermediate has a negative charge on the *meta* position.

(*b*) the more stable resonance hybrid occurs with *meta* attachment of the electrophile.

(*c*) the less stable resonance hybrid occurs with *meta* attachment of the electrophile.

(*d*) the carbonium ion intermediate has a positive charge on the *meta* position.

Answer. (*b*). When the electrophile attaches to the benzene ring, a carbonium ion is formed. This carbonium ion is relatively stable because the charge can be delocalized about the ring. *Ortho* or *para* attack results in a resonance form, with the positive charge on the carbon bearing the electron-withdrawing group (highly unstable). This resonance form does not occur with *meta* placement of the incoming electrophile.

5. Which compound undergoes substitution reactions faster than benzene ?

(*a*) 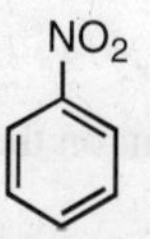

(*b*) 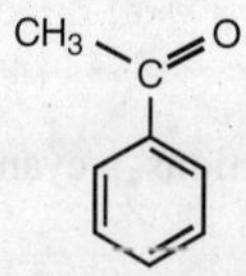

(*c*) 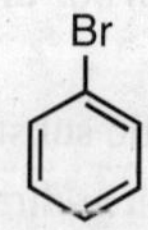

(*d*) 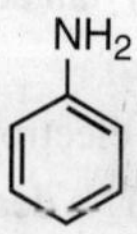

Answer. (*d*). Remember that activating (electron-donating) groups on benzene result in enhanced electrophilic substitution.

6. Which compound undergoes substitution reactions slower than benzene ?

(*a*) 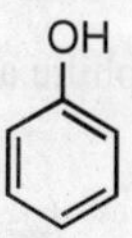

(*b*) 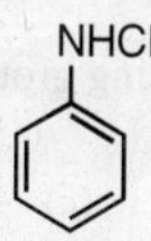

(*c*) 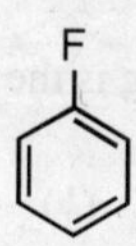

(*d*) 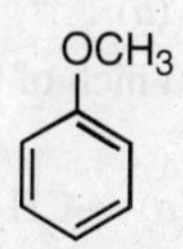

Answer. (*c*). The halogen (–F) is the only deactivating group because of its electronegativity and its low resonance potential.

7. Compared to benzene, nitration of toluene takes place at

(*a*) Same rate (*b*) Faster rate

(*c*) Slower rate (*d*) Can not predict

Answer. (*b*)

8. Which of the following compounds undergoes nitration most readily ?

(*a*) Benzene (*b*) Toluene

(*c*) Nitrobenzene (*d*) Benzoic acid

Answer. (*b*)

9. Which of the following compounds is most readily sulfonated ?

(*a*) Benzene (*b*) Chlorobenzene

(*c*) Toluene (*d*) Nitrobenzene

Answer. (*c*)

10. *p*-Nitrotoluene on treatment with chlorine in the presence of $FeCl_3$ gives

(*a*) *m*-Chlorotoluene (*b*) 2-Chloro-4-nitrotoluene

(*c*) *p*-Chlorotoluene (*d*) 2-Nitro-4-chlorotoluene

Answer. (*b*)

11. In electrophilic aromatic substitution reactions a bromine substituent:

(*a*) is a deactivator and a *m*-director.

(*b*) is a deactivator and an *o,p*-director.

(*c*) is an activator and a *m*-director.

(*d*) is an activator and an *o,p*-director.

Answer. (*b*)

12 In electrophilic aromatic substitution reactions, a $-NHCOCH_3$ substituent on the aromatic ring is:

(*a*) a deactivator and a *m*-director.

(*b*) a deactivator and an *o,p*-director.

(*c*) an activator and a *m*-director.

(*d*) an activator and an *o,p*-director.

Answer. (*d*)

13. In electrophilic aromatic substitution reactions, a $-CO_2H$ substituent on the aromatic ring is:

(*a*) a deactivator and a *m*-director.

(*b*) a deactivator and an *o,p*-director.

(*c*) an activator and a *m*-director.

(*d*) an activator and an *o,p*-director.

Answer. (*a*)

14. In electrophilic aromatic substitution reactions, a cyano substituent on the aromatic ring is:

(*a*) a deactivator and a *m*-director.

(*b*) a deactivator and an *o,p*-director.

(*c*) an activator and a *m*-director.

(*d*) an activator and an *o,p*-director.

Answer. (*a*)

15. Which of the following is the strongest activating group in electrophilic aromatic substitution reactions ?

(*a*) $-CH_2CH_3$ (*b*) $-OCH_3$ (*c*) $-NO_2$ (*d*) $-N(CH_3)_2$

Answer. (*d*)

16. Which of the following compounds will react most rapidly when treated with CH_3CH_2Cl and $AlCl_3$?

(*a*) benzene (*b*) chlorobenzene

(*c*) nitrobenzene (*d*) anisole

Answer. (*d*)

17. Which of the following compounds will undergo bromination most rapidly using Br_2, $FeBr_3$?

(*a*) *p*-methylacetanilide (*b*) bromobenzene

(*c*) acetanilide (*d*) benzenesulfonic acid

Answer. (*a*)

18. Which of the following compounds undergoes reaction with HNO_3/H_2SO_4 at the fastest rate ?

(*a*) ethylbenzene (*b*) benzenesulfonic acid

(*c*) nitrobenzene (*d*) chlorobenzene

Answer. (*a*)

19. Which of the following compounds will react least rapidly when treated with CH_3CH_2Cl and $AlCl_3$?

(*a*) Acetanilide (*b*) Toluene

(*c*) Benzene (*d*) Bromobenzene

Answer. (*d*)

20. Which of the following compounds will undergo bromination least rapidly when treated with Br_2 and $FeBr_3$?

(*a*) *p*-Methylacetanilide (*b*) Bromobenzene
(*c*) Acetanilide (*d*) Benzenesulfonic acid

Answer. (*d*)

21. Rank the following compounds in order of increasing reactivity towards chlorination with $Cl_2/AlCl_3$ (slowest reacting to fastest).

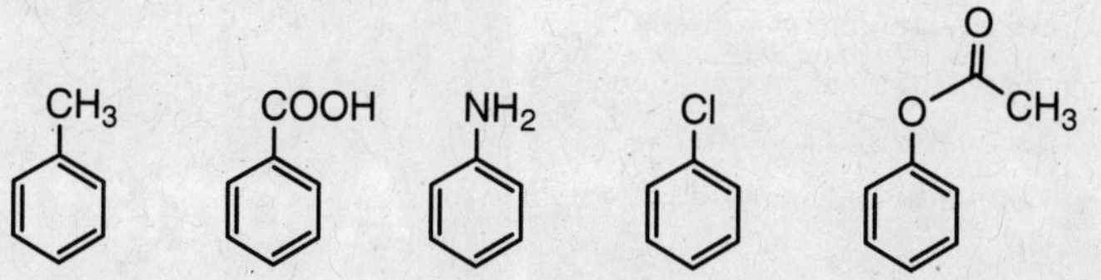

1 2 3 4 5

(*a*) 2 < 4 < 1 < 3 < 5 (*b*) 4 < 2 < 1 < 3 < 5
(*c*) 2 < 4 < 5 < 1 < 3 (*d*) 2 < 4 < 1 < 5 < 3

Answer. (*d*)

22. Which sequence correctly ranks the following aromatic rings in order of increasing rate of reactivity in an electrophilic aromatic substitution reaction ?

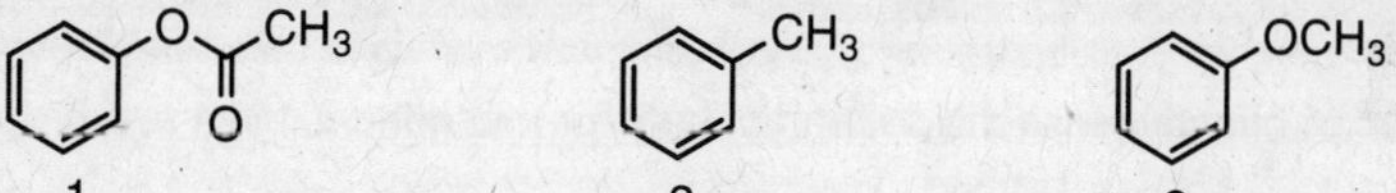

1 2 3

Answer. (*d*)

23. Which sequence correctly ranks the following aromatic rings in order of increasing rate of reactivity with chlorine and aluminum chloride ?

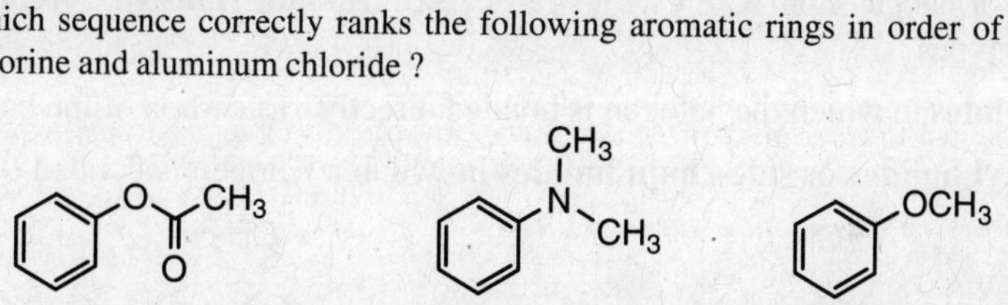

1 2 3

(*a*) 1 < 2 < 3 (*b*) 2 < 3 < 1
(*c*) 3 < 2 < 1 (*d*) 1 < 3 < 2

Answer. (*d*)

35 Aromatic Halogen Compounds

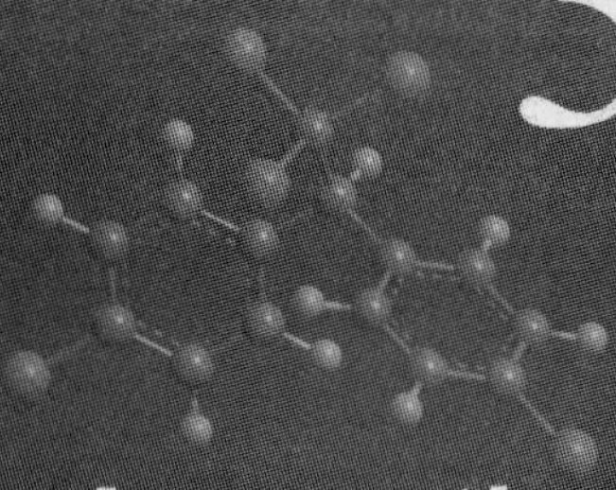

DDT is an effective insecticide, particularly for mosquitoes, flies and crop pests.

The derivatives of arenes in which a hydrogen of the benzene ring or the side-chain has been replaced by a halogen atom X (F, Cl, Br, I), are called **Aromatic Halogen Compounds.** Thus, they are of two types :

(1) **Aryl halides** in which the halogen is bonded directly to a carbon of the benzene ring.

(2) **Arylalkyl halides** or **side-chain halides** in which a halogen is bonded to a carbon of the side-chain.

Chlorobenzene
(Cl bonded to ring)
Aryl halide

Benzyl chloride
(Cl bonded to side-chain)
Arylalkyl halide

In fact, arylalkyl halides are aryl-substituted alkyl halides and show reactions of alkyl halides. Aryl halides, on the other hand, have markedly different chemical behavior and will be discussed first.

ARYL HALIDES

Aryl halides have a halogen atom directly bonded to an aromatic ring. According to IUPAC system, they are named as **Haloarenes**, although common names (as aryl halides) are also in use. If more than one halogen atom is present, their positions on the ring are indicated by numbers or appropriate prefixes : *ortho, meta, para*.

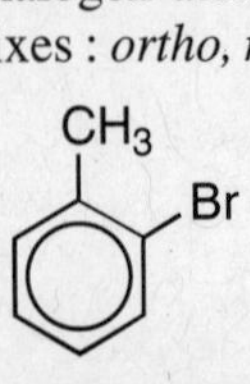

o-Bromotoluene
(2-bromotoluene)

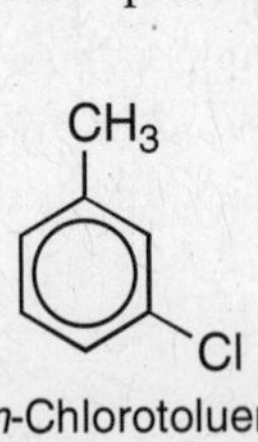

m-Chlorotoluene
(3-chlorotoluene)

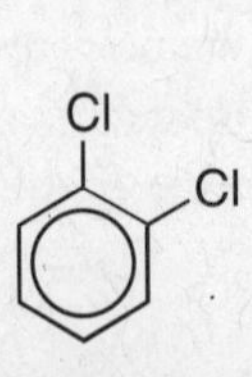

o-Dichlorobenzene
(1,2-dichlorobenzene)

STRUCTURE OF ARYL HALIDES

Structure of aryl halides is explained by taking the example of chlorobenzene.

All ring atoms in benzene and chlorobenzene are sp^2 hybridized.

The structure of chlorobenzene is same as that of benzene, except that the chlorine atom is bonded to a carbon atom of the ring that has a *p* orbital containing a lone pair of electrons. This *p* orbital of chlorine interacts with the *p* orbitals of the ring atoms to form a delocalized π molecular orbital. The delocalized π molecular orbital envelopes the *six* carbons of the ring and the chlorine atom. In other words, the *six* electrons of the ring carbons (one belonging to each sp^2 hybridized carbon) and *two* electrons of the chlorine atom are now associated with all the *seven* atoms. This delocalization of electrons gives some double bond character to the C–Cl bond.

In alkyl halides the molecular orbital is sp^3–*p* for C–X bond. In aryl halides it is sp^2–*p* for the C–X bond. The higher percentage of *s* character in the bond orbitals accounts for the bond shortening.

(1) The delocalization introduces some extra bonding between the carbon and the chlorine, making the bond stronger. This has a major effect on the reactions of compounds like chlorobenzene.

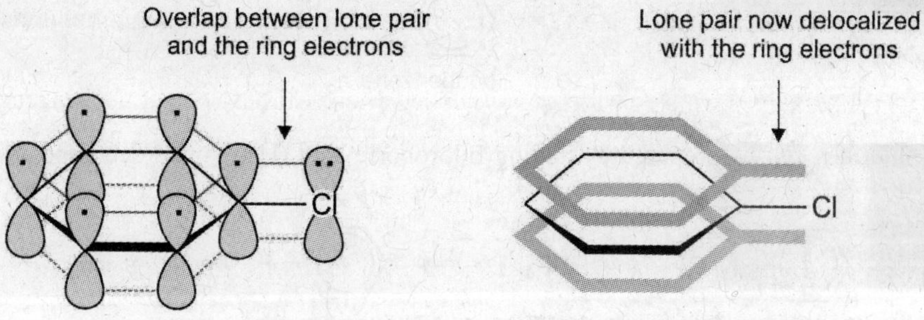

Fig. 35.1. Molecular orbital structure of chlorobenzene. There is interaction between the delocalized electrons in the benzene ring and one of the lone pairs on the chlorine atom.

(2) There is also some movement of electrons away from the chlorine towards the ring. Chlorine is quite electronegative and usually draws electrons in the carbon-chlorine bond towards itself. In this case, this is offset to some extent by the movement of electrons back towards the ring in the delocalization. The molecule is less polar than you would otherwise have expected.

METHODS OF PREPARATION

Aryl halides are prepared by the following methods :

(1) **Direct Halogenation**. Aryl chlorides and aryl bromides are generally prepared from arenes by reaction with chlorine or bromine in the presence of a catalyst such as $FeCl_3$, $FeBr_3$ or $AlCl_3$.

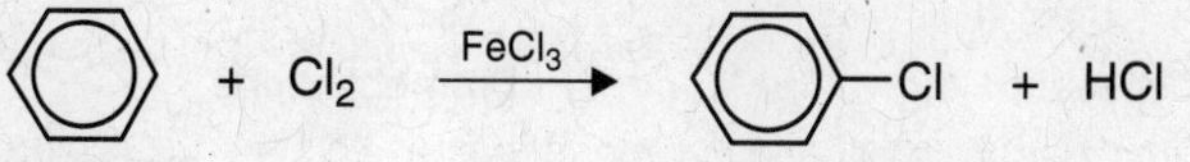

Benzene Chlorobenzene

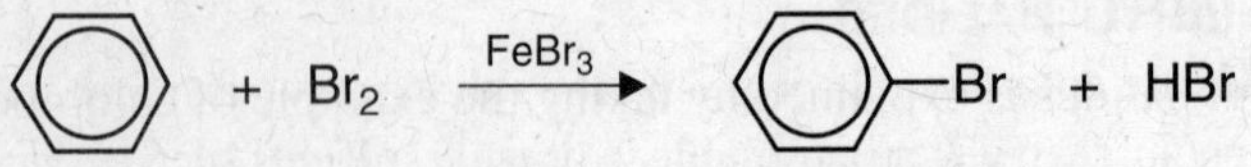

Benzene　　　　Bromobenzene

Aryl iodides are prepared by reaction of arenes with iodine in the presence of reagents such as silver perchlorate, HgO or HNO_3. The reagent, removes HI, as soon as it is produced in the reaction mixture so that it does not reduce aryl iodide back to the arene.

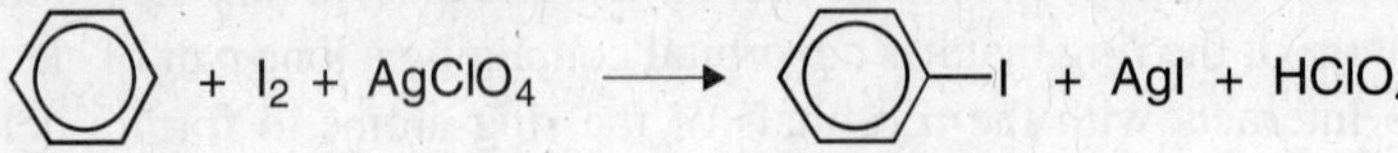

Benzene　　　　Iodobenzene

(2) **Decomposition of Diazonium Salts**. Aryl halides can be prepared most satisfactorily by the decomposition of aryldiazonium salts in the presence of appropriate metal halides.

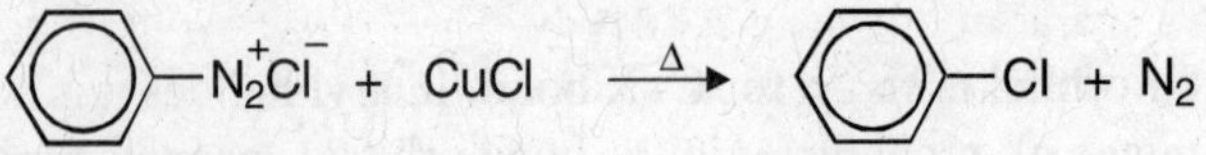 Sandmeyer Reaction

Benzenediazonium chloride　　　　Chlorobenzene

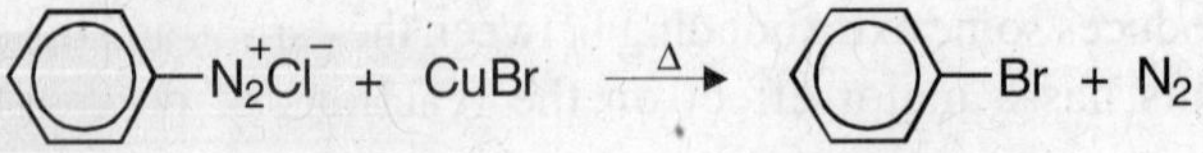 Sandmeyer Reaction

Bromobenzene

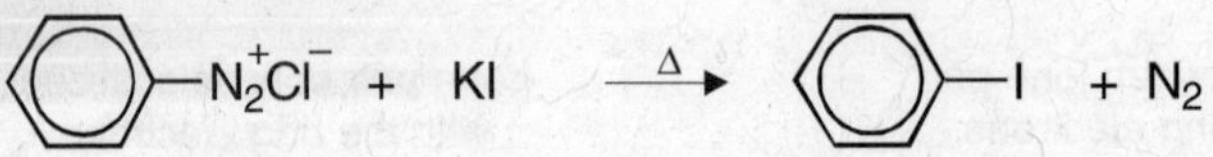

Iodobenzene

Aryl fluorides can be prepared by adding fluoroboric acid (HBF_4) and decomposition of the product.

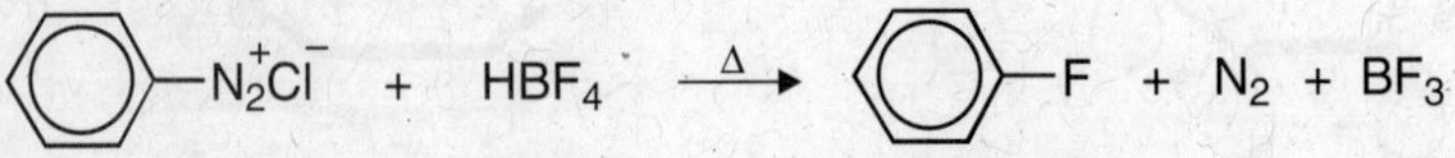

Benzenediazonium chloride　　(*or* $NaBF_4$)　　Fluorobenzene

Benzenediazonium chloride is prepared from benzene as follows :

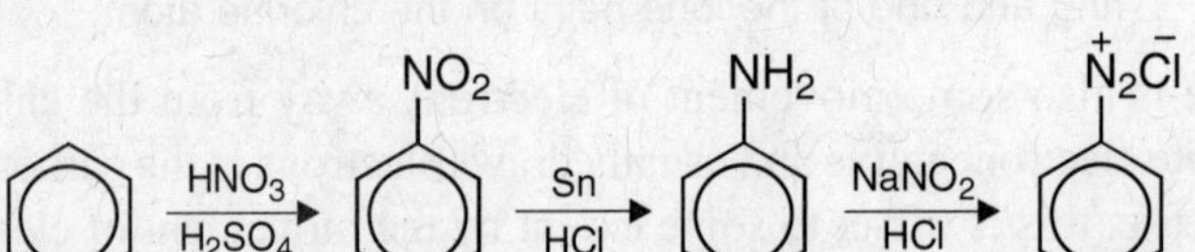

(3) **From Silver Salt of Carboxylic Acids**. The silver salt of carboxylic acids when heated with bromine, readily decarboxylates to form aryl bromide.

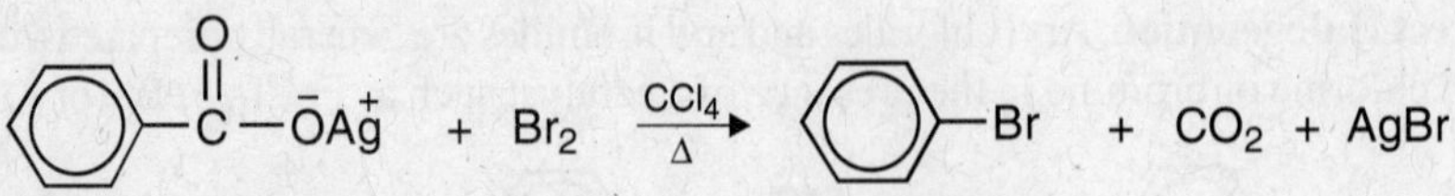

Silver benzoate　　　　Bromobenzene

PHYSICAL PROPERTIES

(1) Aryl halides are colorless oils or crystalline solids having an agreeable odor.

(2) They are insoluble in water but soluble in organic solvents such as ethanol and diethyl ether.

(3) They have density greater than one.

(4) Their boiling points, melting points and densities rise gradually as we pass from fluoro to iodo compounds.

Table 35.1 PHYSICAL CONSTANTS OF HALOBENZENES

Halobenzene	bp °C	mp °C	d (g/ml at 20°C)
Fluorobenzene, C_6H_5F	85	–42	1.031
Chlorobenzene, C_6H_5Cl	132	–45	1.112
Bromobenzene, C_6H_5Br	156	–31	1.495
Iodobenzene, C_6H_5I	188	–30	1.838

(5) Their *UV Spectra* show a characteristic band at 225 mμ. The *IR Spectra* show bands in the region 690-710 and 730-770 cm^{-1}, in addition to the usual bands of the benzene ring.

CHEMICAL PROPERTIES

The molecule of an aryl halide is made of :

(*a*) An aromatic ring

(*b*) One or more halogen atoms directly bonded to the ring.

The reactions of aryl halides are, therefore, due to the aromatic ring and the halogen atoms.

REACTIONS INVOLVING THE HALOGEN ATOM

(1) **Nucleophilic Displacement Reactions**. Unlike alkyl halides, aryl halides do not undergo nucleophilic displacement reactions readily. The halogen bonded to the ring cannot be replaced by OH, NH_2, CN by reaction with aqueous sodium hydroxide, ethanolic ammonia or potassium cyanide. Thus,

$$C_6H_5Cl + NaOH\ (aq) \longrightarrow \text{No Reaction}$$
$$+ NH_3\ (ethanolic) \longrightarrow \text{No Reaction}$$
$$+ KCN \longrightarrow \text{No Reaction}$$

The reasons for this unusual lack of reactivity is that C–X bond in aryl halides is short and strong compared to that in alkyl halides.

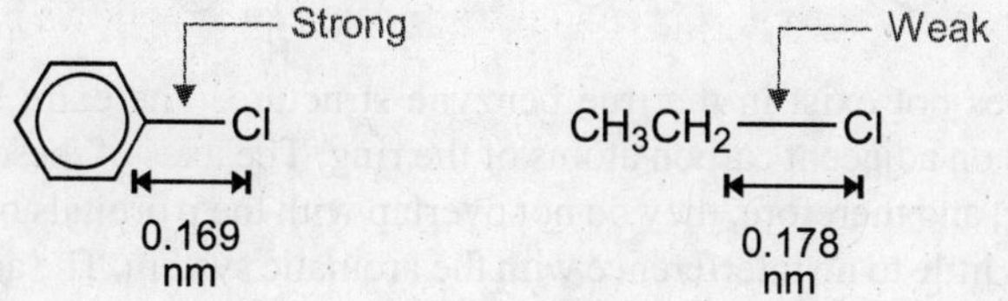

This is so because :

(*i*) The carbon-chlorine bond in chlorobenzene has some double bond character as a result of resonance.

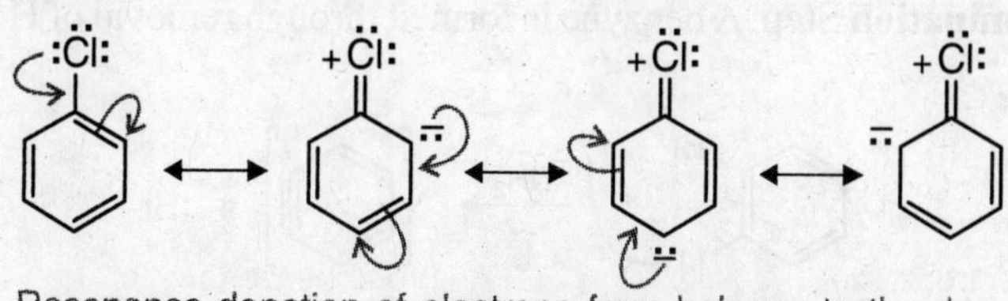

Resonance donation of electrons from halogen to the ring
(C–X bond has double bond character).

(*ii*) The orbitals of the carbon atom in C–Cl bond of aryl chloride are sp^2 hybridized while that in alkyl chloride are sp^3 hybridized. Therefore, the carbon-chlorine bond in aryl chloride has more *s* character and is stronger than the one in alkyl chloride.

Csp^2 —— Clp (more *s* character) *Strong* Csp^3 —— Clp (less *s* character) *Weak*

Another reason for the low reactivity of aryl halides in nucleophilic displacement reactions is that the aromatic ring is a centre of high electron density because of the π electrons. Hence a nucleophile is discouraged from attacking the ring-carbon bearing halogen.

In light of the above facts, nucleophilic displacement of halogen of aryl halides can take place either under drastic conditions or in the presence of electron-withdrawing groups on the aromatic ring.

(*a*) **Under drastic conditions**. The halogen of aryl halides can be displaced by OH, NH_2 or CN with appropriate nucleophilic reagents at high temperature and pressure.

$$\underset{\text{Chlorobenzene}}{C_6H_5Cl} + NaOH \xrightarrow[300\text{ atm}]{350°C} \underset{\text{Phenol}}{C_6H_5OH} + NaCl$$

$$C_6H_5Cl + 2NH_3 \xrightarrow[50\text{ atm}]{250°C} \underset{\text{Aniline}}{C_6H_5NH_2} + NH_4Cl$$

$$C_6H_5Br + CuCN \xrightarrow[\text{pyridine}]{250°C} \underset{\text{Phenyl cyanide}}{C_6H_5CN} + CuBr$$

With strong bases such as $NaNH_2$ or KNH_2. Aryl halides react with sodium amide ($NaNH_2$) or potassium amide (KNH_2) in liquid ammonia to form the respective anilines.

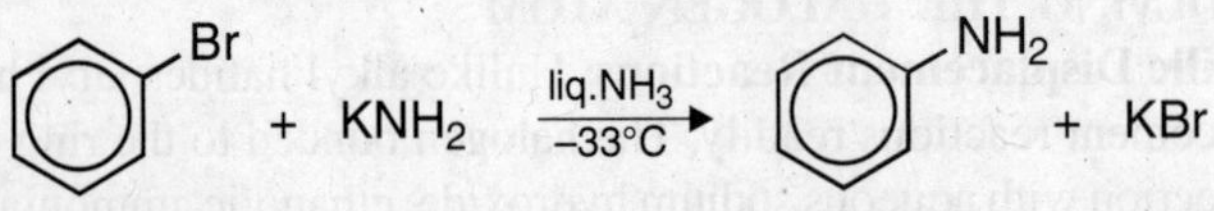

Bromobenzene Aniline

The mechanism for the formation of aniline proceeds *via* a benzyne intermediate. A **benzyne** is a benzene molecule that contains a *theoretical* triple bond. Thus, the following structure represents benzyne:

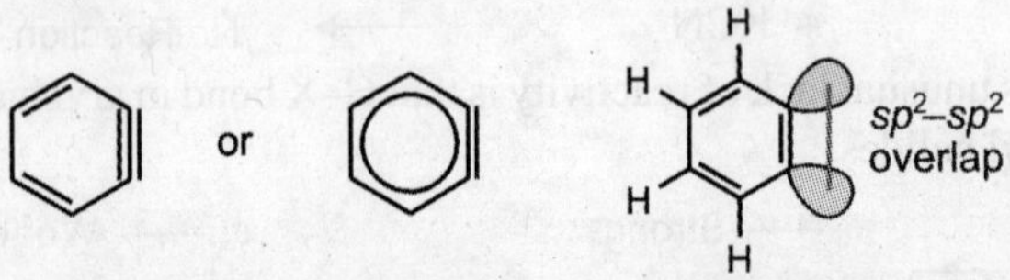

A triple bond does not exist in the true benzyne structure. The extra bond results from the overlap of sp^2 orbitals on adjacent carbon atoms of the ring. The axes of these sp^2 orbitals are in the same plane as the ring, and therefore, they do not overlap with the p orbitals of the aromatic system. Consequently, there is little to no interference with the aromatic system. The additional bond is weak and benzyne is thus highly unstable and highly reactive.

Mechanism. Following elimination-addition mechanism is involved :

Step 1. Elimination Step. A benzyne is formed through removal of H^+ by NH_2^- and loss of Br^-.

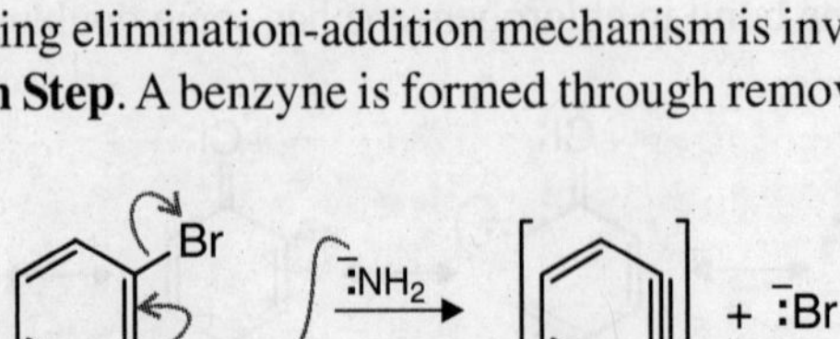

Bromobenzene Benzyne (unstable)

Step 2. Addition Step. A carbanion is formed by nucleophilic addition of NH_2^- to benzyne. The carbanion abstracts a proton from the solvent (NH_3) to give the substitution product.

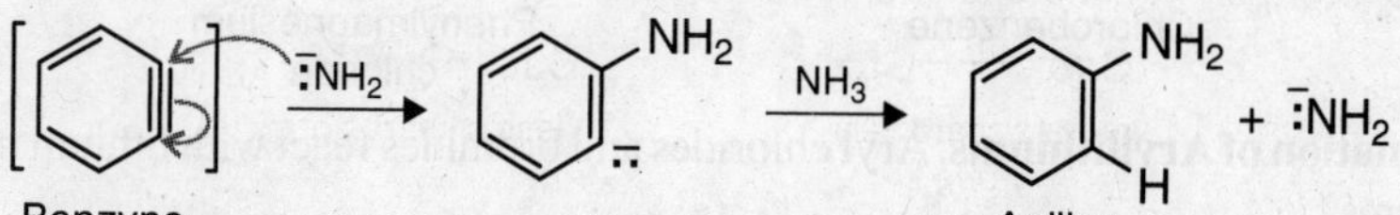

Benzyne Aniline

The above elimination-addition mechanism is known as the **Benzyne Mechanism**.

(*b*) **In the presence of ring activating substituents**. Aryl halides having electron-withdrawing groups such as NO_2 or CN in *ortho* or *para* position to the halogen, undergo nucleophilic substitution reactions relatively easily. For example,

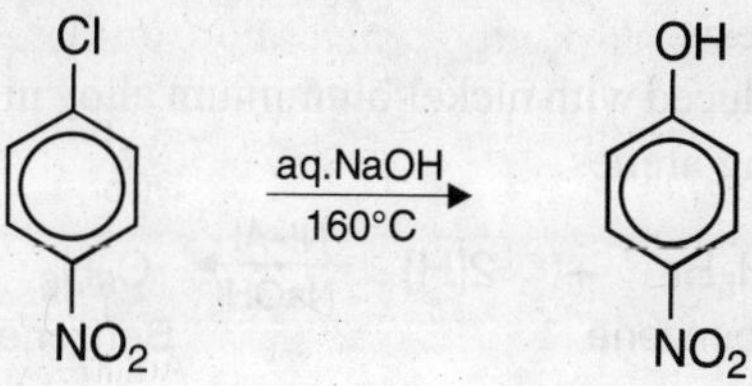

p-Nitrochlorobenzene *p*-Nitrophenol

(2) **Wurtz-Fittig Reaction**. Aryl halides couple with alkyl halides when heated with sodium in ether solution, to form alkylbenzenes.

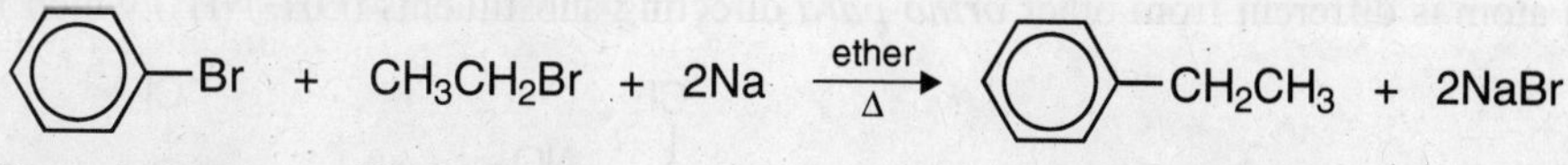

Bromobenzene Ethyl bromide Ethylbenzene

This reaction known as Wurtz-Fittig reaction is used for the synthesis of some alkylbenzenes, although the yield is poor.

(3) **Ullmann Reaction**. Aryl iodides and bromides when heated with copper form biaryl compounds in which two benzene rings are bonded together.

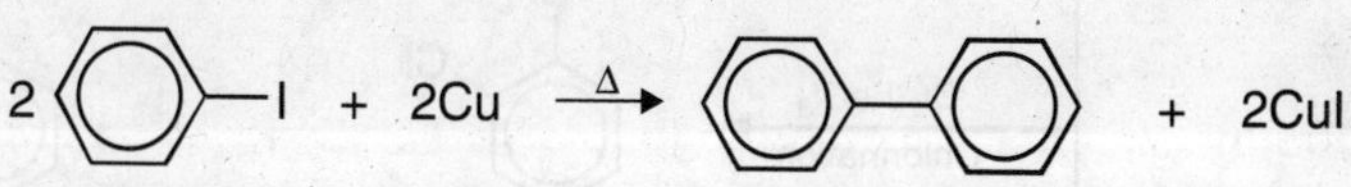

Iodobenzene Biphenyl

Aryl chlorides also participate in Ullmann reaction if the Cl atom is activated by the presence of NO_2 group in the *ortho* and *para* positions.

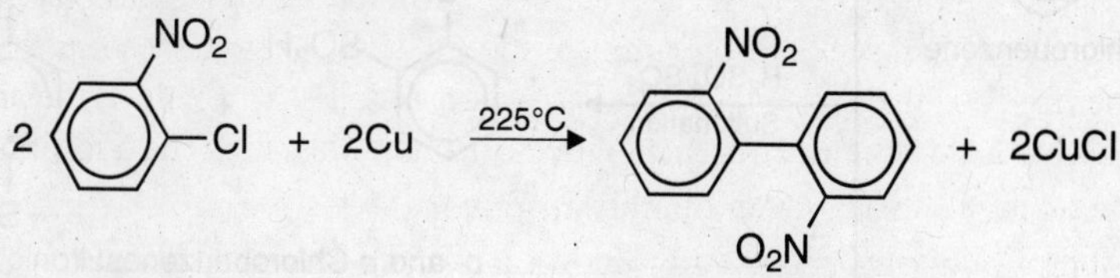

o-Nitrochlorobenzene 2,2′-Dinitrobiphenyl

(4) **Formation of Grignard Reagents**. Aryl iodides and aryl bromides form Grignard reagents in the usual fashion. Aryl chlorides do so by using tetrahydrofuran (THF) instead of ether.

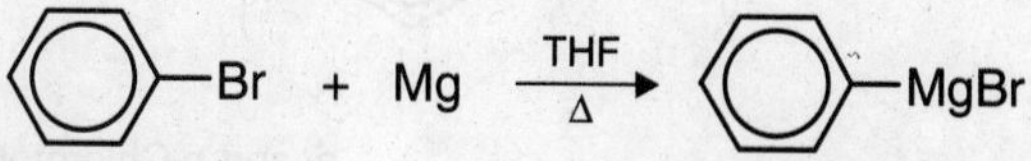

Bromobenzene Phenylmagnesium bromide

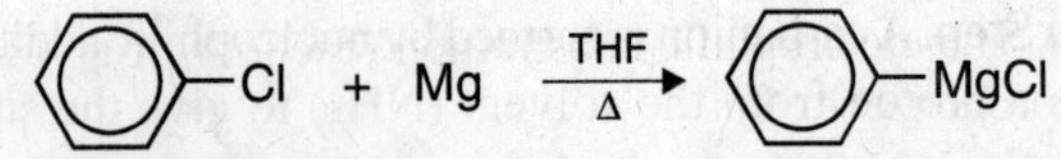

Chlorobenzene Phenylmagnesium chloride

(5) **Formation of Aryllithiums.** Aryl chlorides and bromides react with lithium metal in ether to form aryllithiums.

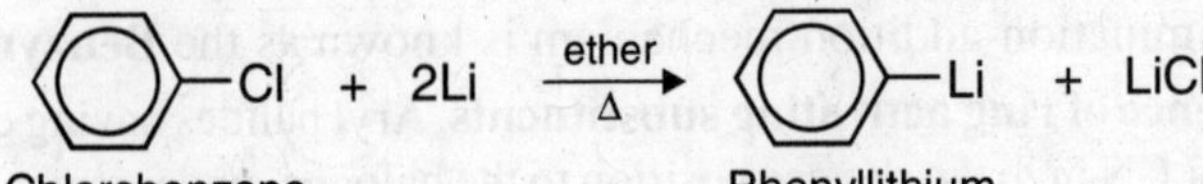

Chlorobenzene Phenyllithium

Aryllithiums give the same reactions as the Grignard reagents and are more reactive. They are now frequently used in synthesis.

(6) **Reduction.** When reduced with nickel-aluminium alloy in the presence of sodium hydroxide, aryl halides give corresponding arenes.

$$\underset{\text{Bromobenzene}}{C_6H_5Br} + 2[H] \xrightarrow[\text{NaOH}]{\text{Ni–Al}} \underset{\text{Benzene}}{C_6H_6} + HBr$$

REACTIONS INVOLVING THE AROMATIC RING

(7) **Electrophilic Aromatic Substitution Reactions.** Aryl halides undergo the typical electrophilic aromatic substitution reactions though less readily than benzene, in *ortho* and *para* positions. The halogen atom is different from other *ortho-para* directing substituents (OH, NH_2) which facilitate

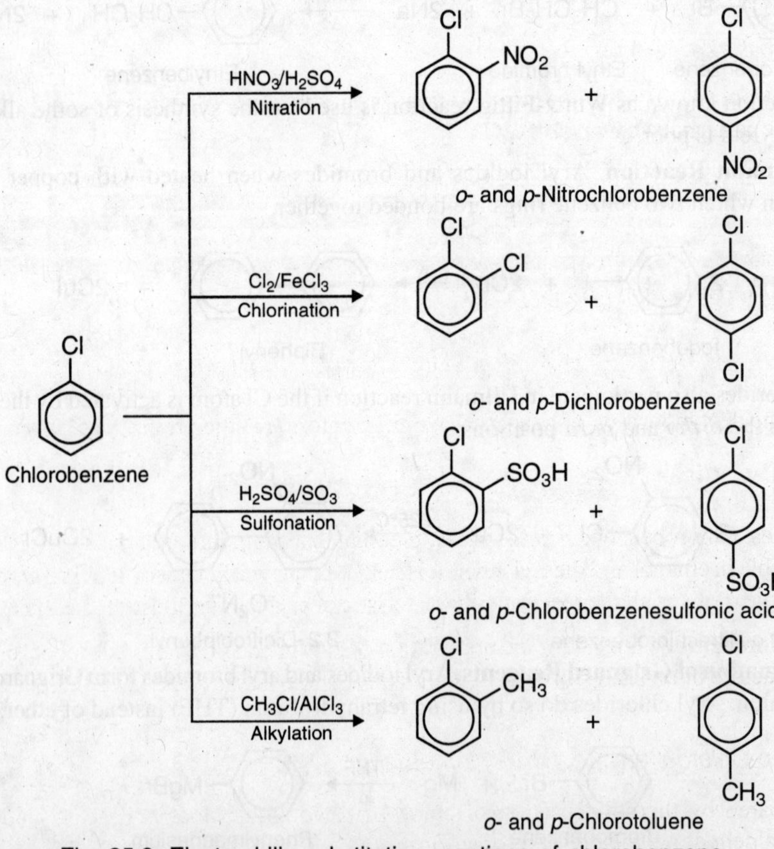

Fig. 35.2. Electrophilic substitution reactions of chlorobenzene.

electrophilic substitution. The electrophilic substitution reactions of aryl halides are illustrated by taking the example of chlorobenzene.

EXPLANATION. The halogen atom in aryl halides is highly electronegative relative to carbons and hence C–Cl bond is polarized. This makes the carbon atom positive and it pulls out electrons from the benzene ring. (*Inductive Effect*)

On the other hand, the resonance effect causes the shift of electrons from the halogen to the ring.

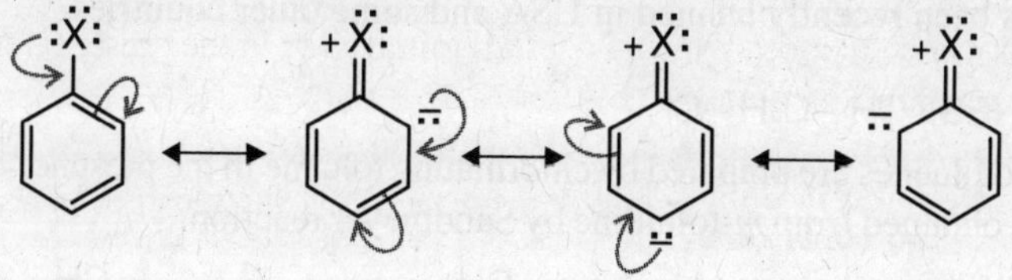

Of the two opposing effects, the inductive effect overweighs the resonance effect. The net result is the decrease of electron density in the benzene ring which is deactivated to electrophilic (E^+) attack. Thus electrophilic substitution in the ring takes place with difficulty than in benzene itself, and at *ortho* and *para* positions which are relatively electron rich.

CHLOROBENZENE, C_6H_5Cl

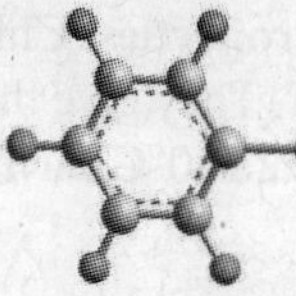

Preparation. (1) By passing dry chlorine through benzene using ferric chloride as a catalyst at about 60°C.

$$C_6H_6 + Cl_2 \xrightarrow[60°C]{FeCl_3} C_6H_5Cl + HCl$$

(2) By passing a mixture of benzene vapor, hydrogen chloride and air over cupric oxide as a catalyst at 250°C. (*Rasching Process*)

$$C_6H_6 + HCl + \frac{1}{2}O_2 \xrightarrow[250°C]{CuO} C_6H_5Cl + H_2O$$

(3) By the decomposition of benzenediazonium chloride in the presence of cuprous chloride.

$$C_6H_5\overset{+}{N_2}\overset{-}{Cl} \xrightarrow{CuCl} C_6H_5Cl + N_2$$

Properties. Chlorobenzene is a colorless, pleasant smelling liquid, *bp* 132°C. It is insoluble in water but soluble in ethanol and diethyl ether. It is heavier than water, density is 1.112 g/ ml at 20°C. It is steam-volatile. It gives all the reactions characteristic of aryl halides illustrated earlier.

Uses. Chlorobenzene is used for the preparation of : (1) phenol; (2) nitrochlorobenzenes required for the manufacture of azo and sulfur dyes, fungicides, preservatives; and (3) DDT.

DDT, *2,2-di(p-Chlorophenyl)-1,1,1-trichloroethane*

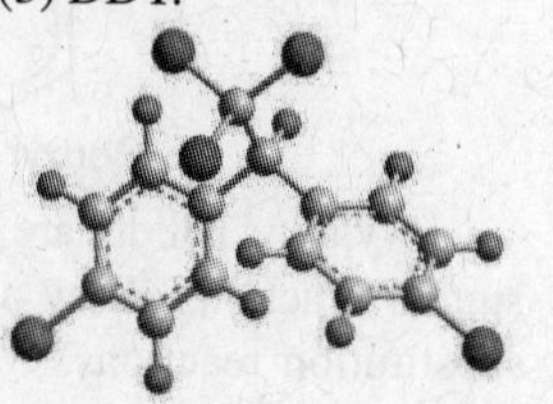

It is prepared by the condensation of chlorobenzene with chloral (trichloroacetaldehyde) in the presence of sulfuric acid.

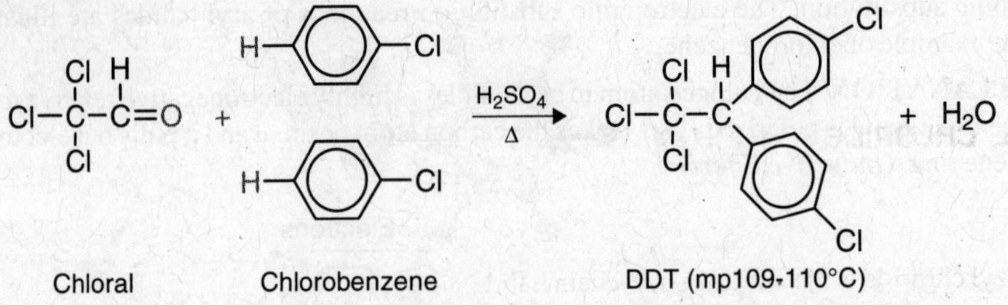

DDT is an effective insecticide, particularly for mosquitoes, flies and crop pests. It is responsible for the almost complete eradication of malaria-carrying mosquitoes and thus making the world free from malaria. Unfortunately it is toxic to humans and plays havoc with several species of useful birds and fish. Its use has been recently banned in USA and some other countries.

CHLOROTOLUENES, $CH_3–C_6H_4–Cl$

o- and *p*-Chlorotoluenes are prepared by chlorinating toluene in the presence of $FeCl_3$ as catalyst. *m*-Chlorotoluene is obtained from *m*-toluidine by Sandmeyer reaction.

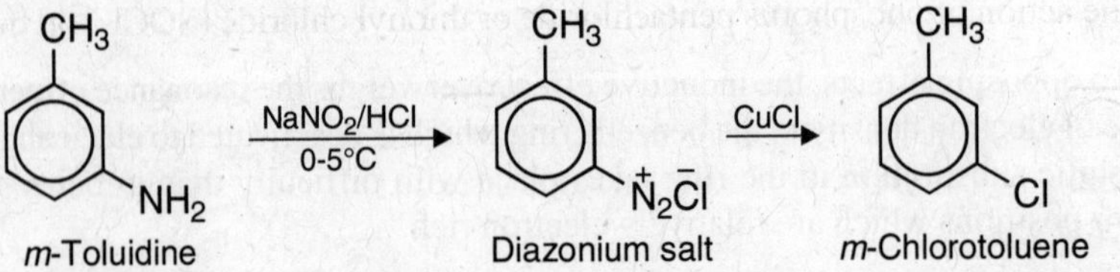

Properties. Chlorotoluenes are heavy, colorless liquids, bp : *ortho* 159°C, *meta* 162°C, *para* 162°C. They give the usual reactions of aryl halides. When *p*-chlorotoluene is treated with aqueous alkali at 340°C, an interesting reaction takes place to yield *p*- and *m*-cresol.

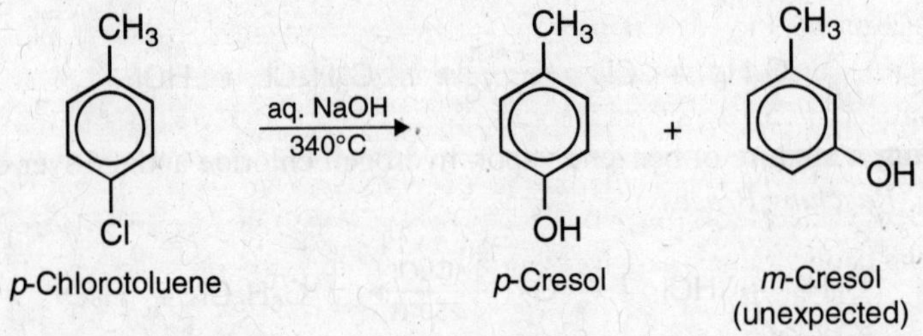

Upon oxidation with alkaline potassium permanganate, chlorotoluenes form the respective chlorobenzoic acids. This reaction is used to distinguish chlorotoluenes from benzyl chloride ($C_6H_5CH_2Cl$) which upon similar oxidation gives benzoic acid.

ARYLALKYL HALIDES

They are aryl-substituted alkyl halides and are also called **Side-Chain Aryl Halides**. Examples are,

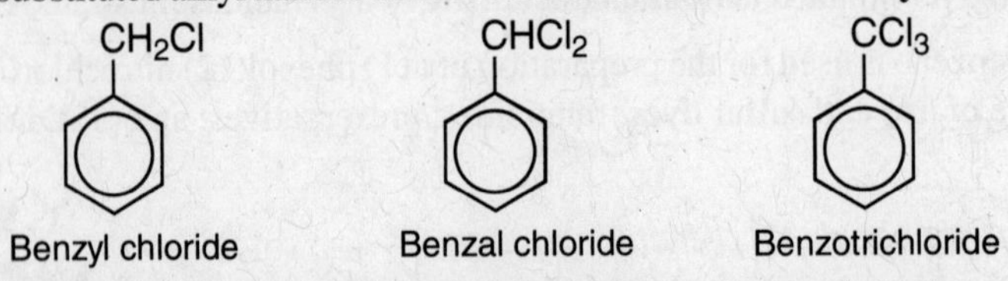

Arylalkyl halides are prepared by methods similar to those of their aliphatic counterparts and give similar reactions. They are more reactive than alkyl halides and also give electrophilic aromatic substitution reactions.

BENZYL CHLORIDE, $C_6H_5CH_2Cl$

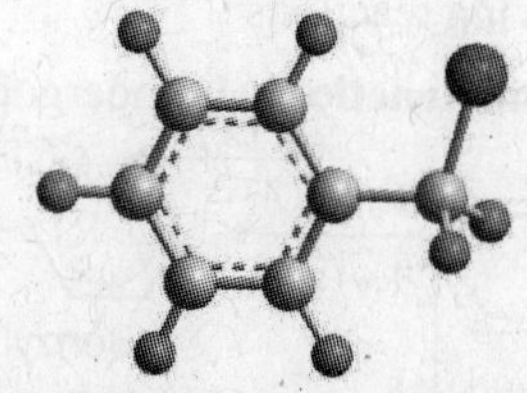

Benzyl chloride is the simplest side-chain halide.

Preparation. Benzyl chloride is prepared :

(1) By passing chlorine into boiling toluene in the presence of UV light till the required increase in weight is reached.

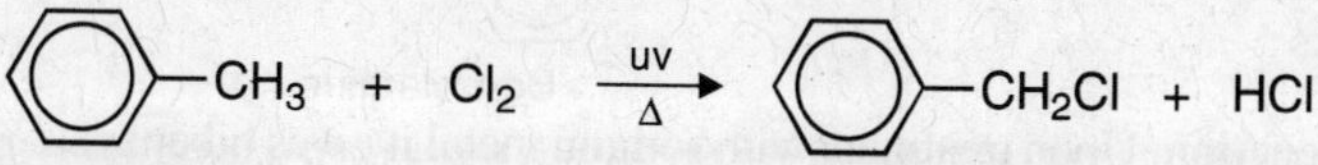

Toluene Benzyl chloride

(2) By the action of phosphorus pentachloride or thionyl chloride ($SOCl_2$) on benzyl alcohol.

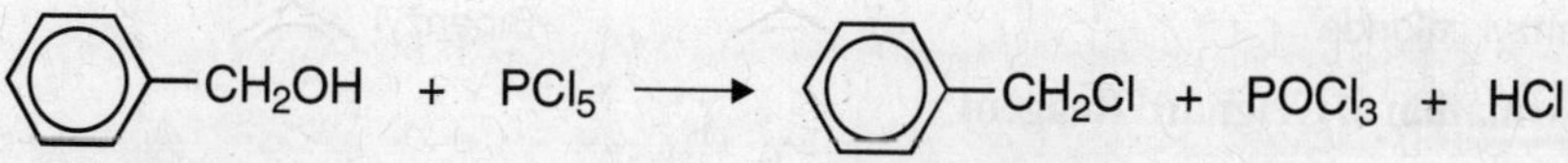

Benzyl alcohol Benzyl chloride

Properties. (*Physical*). Benzyl chloride is a colorless unpleasant smelling liquid, bp 179°C. Its vapors have irritating action on eyes and nose, causing flow of tears. It is insoluble in water but soluble in most organic solvents.

In the molecular orbital model of benzyl ion, the *p* orbitals of the Cl atom overlap with the available *p* orbitals of the benzene ring carbons. The extended molecular orbital has the positive charge of the benzyl ion distributed over all the seven carbons due to delocalization of π electrons. This makes the benzyl ion greatly stable.

(*Chemical*). Benzyl chloride is many times more reactive than alkyl halides. The increased reactivity of the chlorine atom is due to the ready formation of a stable carbonium ion.

Benzyl chloride Benzyl carbonium ion (stable)

The stability of the benzyl carbonium ion ($C_6H_5CH_2^+$) arises from the distribution of the positive charge over the benzene ring due to resonance.

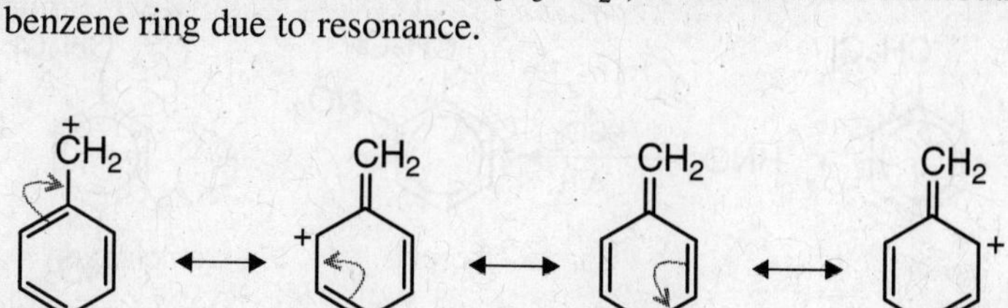

Benzyl carbonium ion

Benzyl chloride gives the following reactions :

(1) **Nucleophilic Displacement Reactions**. It undergoes nucleophilic displacement reactions like the alkyl halides. Thus,

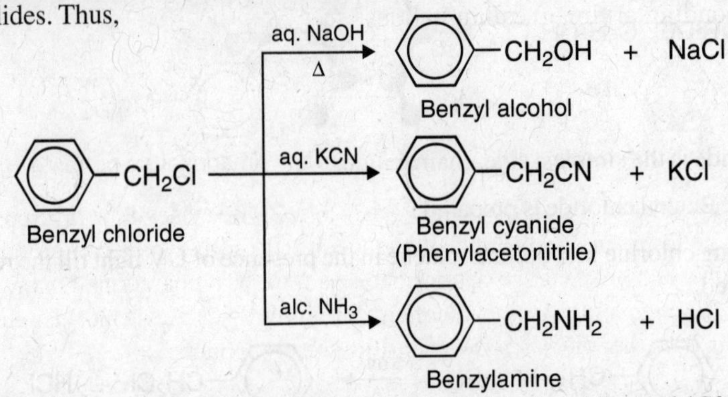

(2) **Wurtz Reaction**. Upon treatment with sodium metal it gives bibenzyl (*Cf*: alkyl halides).

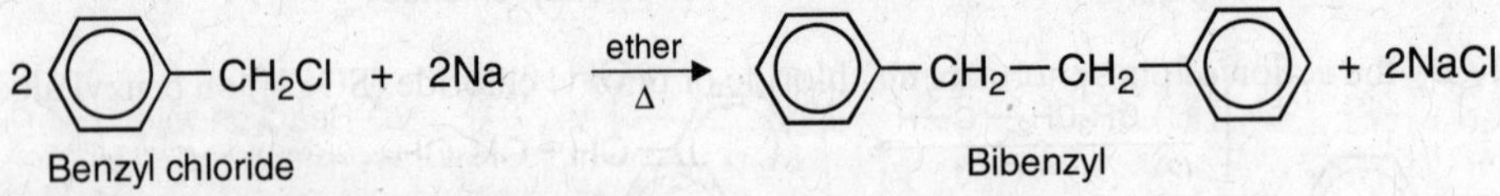

(3) **Formation of Grignard Reagent**.

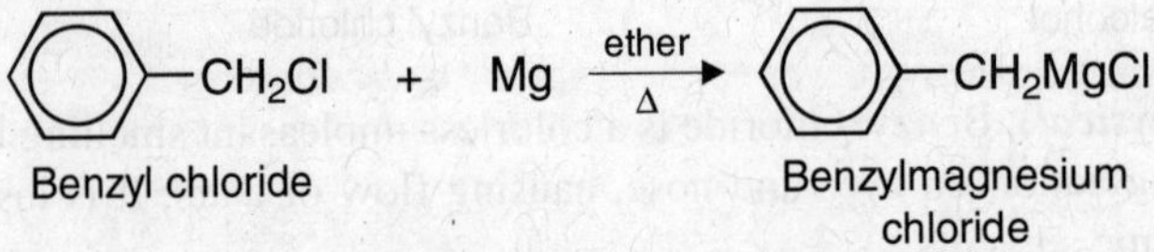

(4) **Oxidation**. When treated with dilute nitric acid or alkaline potassium permanganate, the CH_2Cl group is oxidized to COOH group, forming benzoic acid.

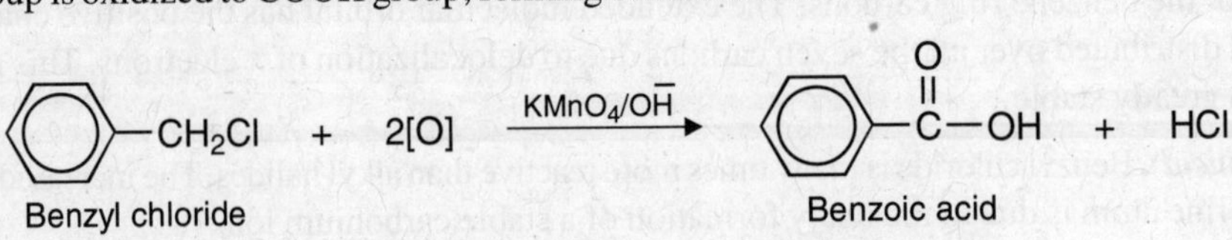

(5) **Reduction**. When reduced with Zn-Cu couple, it yields toluene.

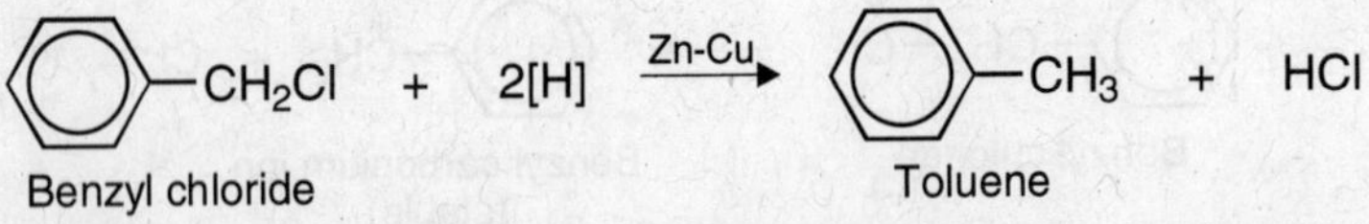

(6) **Electrophilic Aromatic Substitution**. It gives the usual electrophilic substitution reactions of the benzene ring in *ortho* and *para* positions. For example,

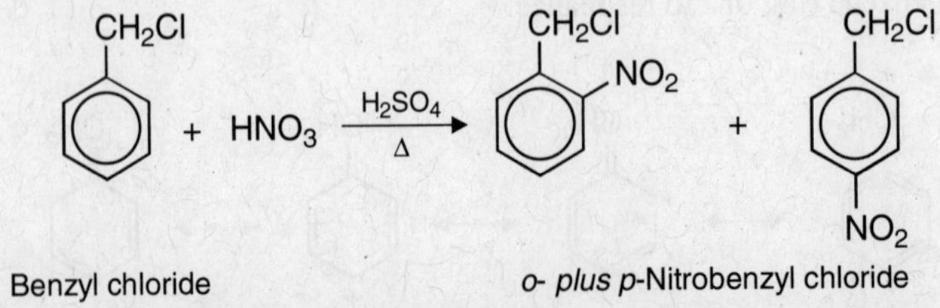

Uses. Benzyl chloride is used : (1) for the manufacture of benzyl alcohol, benzylamine, and benzaldehyde ; (2) for introducing benzyl group in organic molecules.

ARYLMAGNESIUM HALIDES

Preparation. Aryl halides react with magnesium in tetrahydrofuran (in place of ether) to form the corresponding arylmagnesium halides.

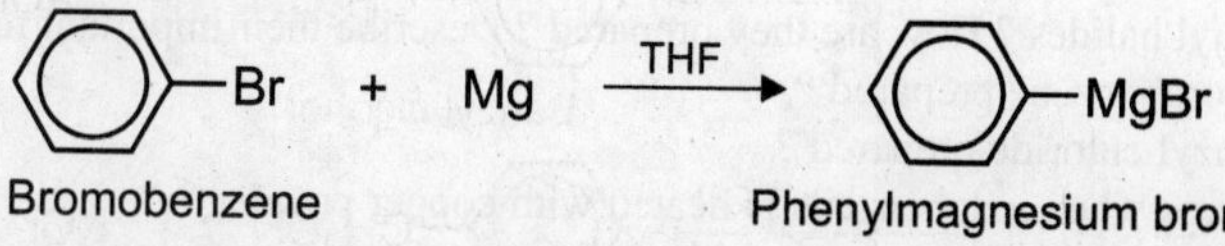

Properties. The order of reactivity is ArI > ArBr > ArCl. Because of the high cost of the iodides and low reactivity of chlorides, aryl bromides are generally taken as a good compromise in laboratory synthesis. Arylmagnesium bromides so obtained are used for the preparation of many useful products including *primary, secondary* and *tertiary* alcohols (See Table 35.2). Aryl chlorides and aryl bromides react with lithium in dry ether or THF to form aryllithium compounds.

Table 35.2 REACTIONS OF ARYL GRIGNARD REAGENTS

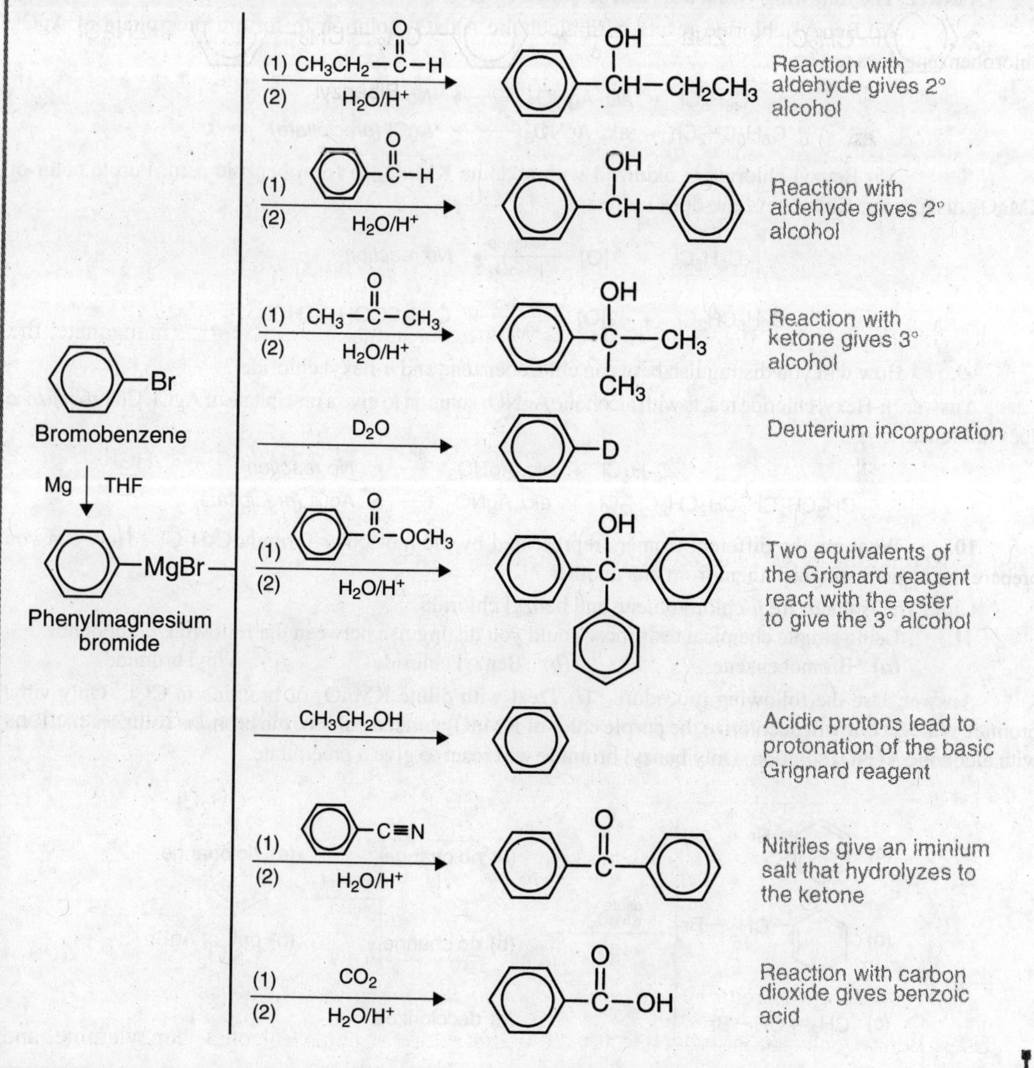

Examination Questions

1. What are aryl halides ? How are they prepared ? Describe their important reactions.
2. How is chlorobenzene prepared ?
3. How is benzyl chloride prepared ?
4. What happens when iodobenzene is heated with copper powder ?

Answer. Diphenyl is formed. For reaction equation see text.

5. What happens when benzyl chloride is treated with aqueous NaOH ?

Answer. Benzyl alcohol is formed. For reaction equation see text.

6. Explain why chlorobenzene is less reactive than benzyl chloride.
(*Delhi BSc, 2014 ; Panjab BSc, 2014 ; Anna BSc, 2015*)
7. Write a note on : Ullmann Reaction.
8. How will you distinguish between chlorobenzene and benzyl chloride ?
(*Bhopal BSc, 2014 ; Karnataka BSc, 2015*)

Answer. The following visual tests can be used :

(*a*) Benzyl chloride reacts with alcoholic $AgNO_3$ solution to form a precipitate of AgCl. Chlorobenzene does not react.

$$C_6H_5Cl + \text{alc. } AgNO_3 \longrightarrow \textit{No reaction}$$
$$C_6H_5CH_2Cl + \text{alc. } AgNO_3 \longrightarrow AgCl \ (\textit{precipitate})$$

(*b*) Benzyl chloride is oxidized with alkaline $KMnO_4$ to form benzoic acid. Purple color of $KMnO_4$ disappears. Chlorobenzene does not react.

$$C_6H_5Cl + [O] \xrightarrow[KMnO_4]{\text{alkaline}} \textit{No reaction}$$
$$C_6H_5CH_2Cl + 2[O] \xrightarrow[KMnO_4]{\text{alkaline}} C_6H_5COOH + HCl$$

9. How will you distinguish between chlorobenzene and *n*-hexyl chloride ?

Answer. *n*-Hexyl chloride reacts with alcoholic $AgNO_3$ solution to give a precipitate of AgCl. Chlorobenzene does not react.

$$C_6H_5Cl + \text{alc. } AgNO_3 \longrightarrow \textit{No reaction}$$
$$CH_3CH_2CH_2CH_2CH_2CH_2Cl + \text{alc. } AgNO_3 \longrightarrow AgCl \ (\textit{precipitate})$$

10. What are the different isomers represented by the molecular formula C_7H_7Cl ? How can you prepare them and distinguish them from one another ?

Clue. The isomers are *p*-chlorotoluene and benzyl chloride.

11. Using simple chemical tests, how could you distinguish between the following compounds ?
(*a*) Bromobenzene (*b*) Benzyl chloride (*c*) Vinyl bromide

Answer. Use the following procedure : (*i*) Treat with dilute $KMnO_4$ or bromine in CCl_4. Only vinyl bromide will react. It will decolorize the purple color of $KMnO_4$ and red color of the bromine solution. (*ii*) Treat with alcoholic $AgNO_3$ solution. Only benzyl bromide will react to give a precipitate.

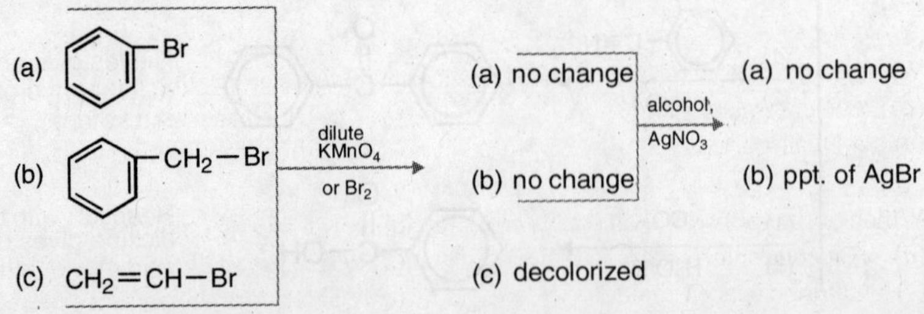

12. How will you synthesize benzyl alcohol from bromobenzene ? *(Lucknow BSc, 2014)*

Answer. Following steps are involved :

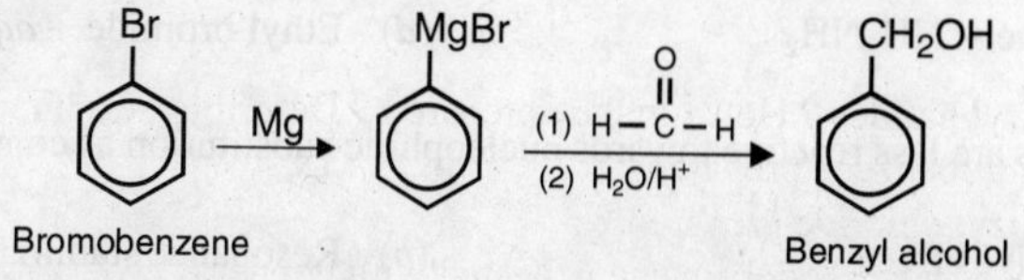

13. How will you synthesize benzoic acid from bromobenzene ? *(Madras BSc, 2015)*

Answer. Following steps are involved :

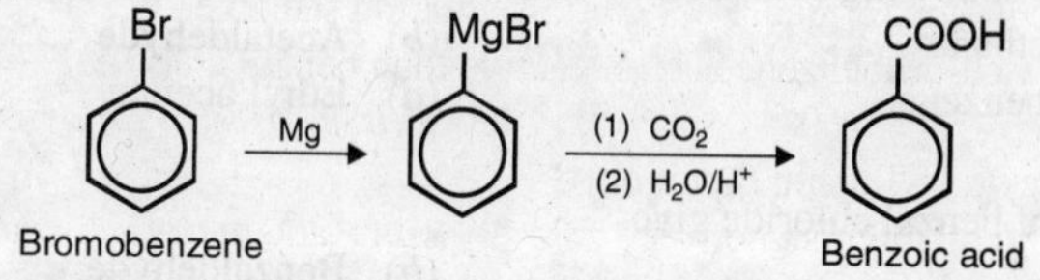

14. How will you synthesize phenylacetic acid from toluene ? *(Gujarat BSc, 2015)*

Answer. Following steps are involved :

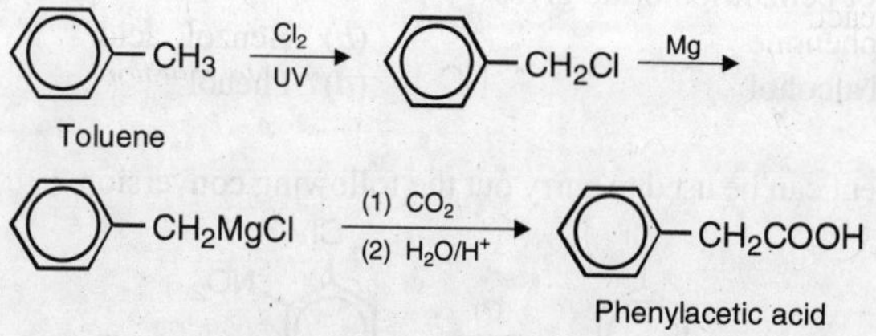

15. How will you synthesize DDT from chlorobenzene ?

(Mysore BSc, 2014 ; Patna BSc, 2015)

16. Three isomers (*A*), (*B*), and (*C*) of formula C_8H_9Br gave the following compounds on oxidation :

(*A*) gives Benzoic acid

(*B*) gives Phthalic acid

(*C*) gives *p*-Chlorobenzoic acid

(*A*) is optically active. Both (*A*) and (*B*), but not (*C*), give a white precipitate when warmed with alcoholic $AgNO_3$ solution. Identify (*A*), (*B*), and (*C*)

Answer. (*A*) = 1-Bromo-1-phenylethane

(*B*) = *o*-Methylbenzyl bromide

(*C*) = *p*-Bromoethylbezene

Multiple Choice Questions

1. How many isomeric dichlorobenzenes are there ?

(*a*) 3 (*b*) 6 (*c*) 9 (*d*) 12

Answer. (*a*)

2. Which of the following compounds has maximum dipole moment ?

(*a*) *p*-Dichlorobenzene (*b*) *m*-Dichlorobenzene

(*c*) *o*-Dichlorobenzene (*d*) Carbon tetrachloride

Answer. (*c*)

3. Which of the following reagents is most suitable for preparing chlorobenzene from benzene ?

(*a*) Aqueous chlorine (*b*) Chlorine in the presence of ultraviolet light

(*c*) Chlorine in acetic acid (*d*) Chlorine in the presence of $FeCl_3$

Answer. (*d*)

4. Which of the reactions between the following pairs of reagents proceeds by a mechanism that may be described as an electrophilic substitution ?

(*a*) Benzene + $Br_2/FeBr_3$ (*b*) Propene + Br_2
(*c*) Acetone + NH_2NH_2 (*d*) Ethyl bromide + *aq* KOH

Answer. (*a*)

5. Aryl halides are less reactive towards nucleophilic substitution as compared to alkyl halides due to

(*a*) Inductive effect (*b*) Resonance stabilization
(*c*) Tautomerism (*d*) Stereoisomerism

Answer. (*b*)

6. Which of the following compounds would be unaffected by boiling aqueous NaOH ?

(*a*) Chloroethane (*b*) Acetaldehyde
(*c*) Chlorobenzene (*d*) Ethyl acetate

Answer. (*c*)

7. Hydrolysis of benzal chloride gives

(*a*) Phenol (*b*) Benzaldehyde
(*c*) Benzyl alcohol (*d*) Benzoyl chloride

Answer. (*b*)

8. Hydrolysis of benzotrichloride gives

(*a*) Benzophenone (*b*) Benzoic acid
(*c*) Benzyl alcohol (*d*) Phenol

Answer. (*b*)

9. Which reagent can be used to carry out the following conversion ?

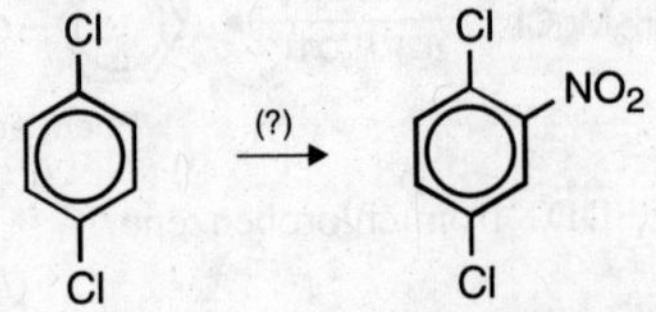

(*a*) $HNO_3 + H_2SO_4$ (*b*) $NaNO_2$
(*c*) $NaNO_2$ - HCl at 5°C (*d*) NO_2

Answer. (*a*)

10. The following compound is called ?

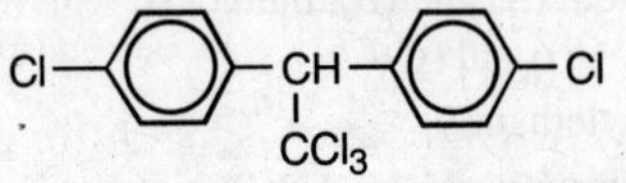

(*a*) Chloral (*b*) DDT
(*c*) Lindane (*d*) BHC

Answer. (*b*)

11. DDT is prepared by the reaction of chlorobenzene with (in the presence of conc. H_2SO_4)

(*a*) Chloral (*b*) Chlorine
(*c*) Chloroform (*d*) Carbon tetrachloride

Answer. (*a*)

12. Which of the following reagents can be used to distinguish between chlorobenzene and allyl chloride ?

(*a*) H_2/Ni (*b*) Br_2 in CCl_4
(*c*) Zn/HCl (*d*) NH_2NH_2

Answer. (*b*)

13. Which of the following reagents can be used to distinguish between chlorobenzene and benzyl chloride ?

(*a*) alcoholic $AgNO_3$ (*b*) Br_2 in CCl_4
(*c*) KCN (*d*) Br_2 in H_2O

Answer. (*a*)

36

Aromatic Sulfonic Acids

Saccharin is 500 times sweeter than table sugar.
It is widely used as calorie-free sugar substitute for diabetics.

They may be regarded as derivatives of sulfuric acid in which one OH group has been replaced by an aryl group, Ar.

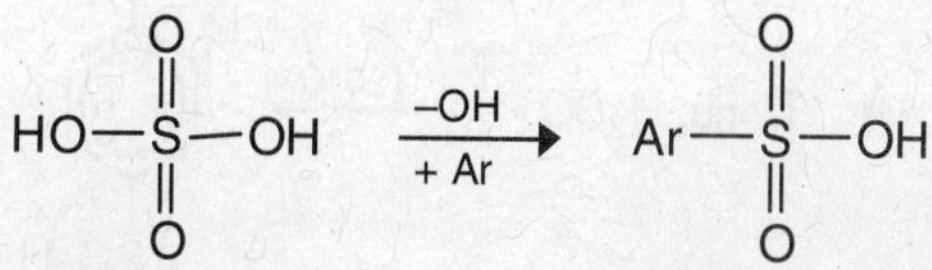

The functional group in this class of compounds is the **Sulfonic acid** or **Sulfo group**,

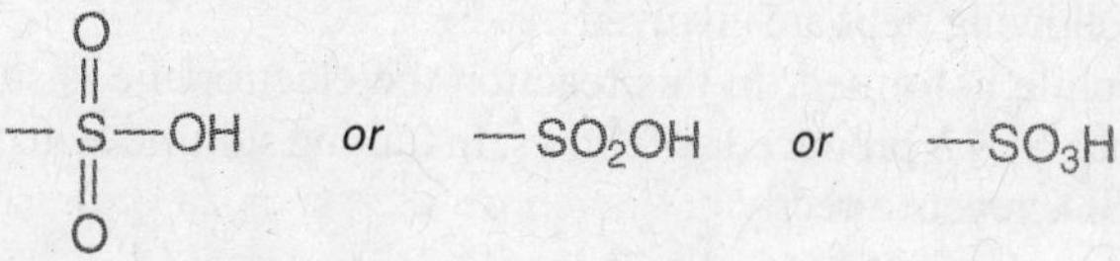

In modern practice, the class name 'Sulfonic Acids' is used to include any compound in which the $-SO_3H$ group is bonded to an aromatic ring.

NOMENCLATURE

Their systematic name is **Arenesulfonic Acids**. The name of an individual member is derived by appending the suffix *-sulfonic acid* to the parent arene.

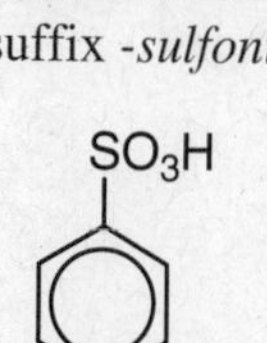

Benzenesulfonic acid

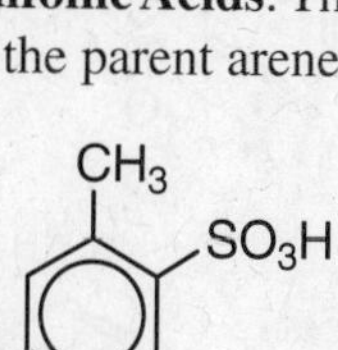

o-Toluenesulfonic acid

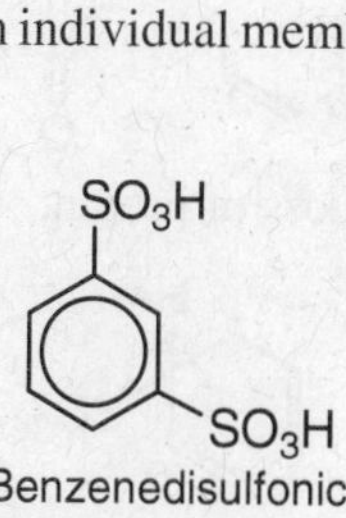

1,3-Benzenedisulfonic acid

The substituted sulfonic acids are named by adding the suffix -sulfonic acid to the name of the parent compound. The –COOH group has priority over $-SO_3H$ in naming.

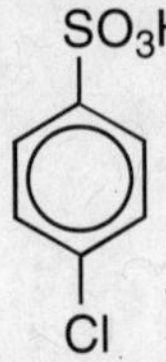
p-Chlorobenzenesulphonic acid

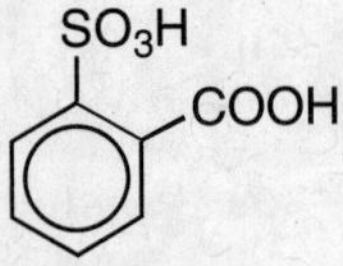
o-Sulphobenzoic acid

The derivatives of sulfonic acids involving the $-SO_3OH$ group are named similar to their carboxylic acid counterparts. Thus we have *Sulfonyl chlorides* ($ArSO_2Cl$), *Sulfonamides* ($ArSO_2NH_2$), *Sulfonic esters* ($ArSO_2OR$).

METHODS OF PREPARATION

Aromatic sulfonic acids are prepared by the following methods :

(1) **Direct Sulfonation**. Arenes react with fuming sulfuric acid at room temperature to give sulfonic acids. Fuming sulfuric acid is concentrated sulfuric acid that contains added sulfur trioxide. Sulfonation can also be carried with concentrated sulfuric acid alone. Such reactions are slow, and heating is generally required. Examples are,

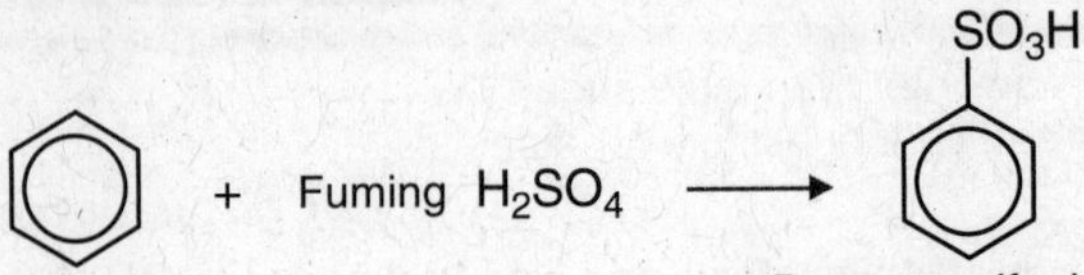

Benzene → Benzenesulfonic acid

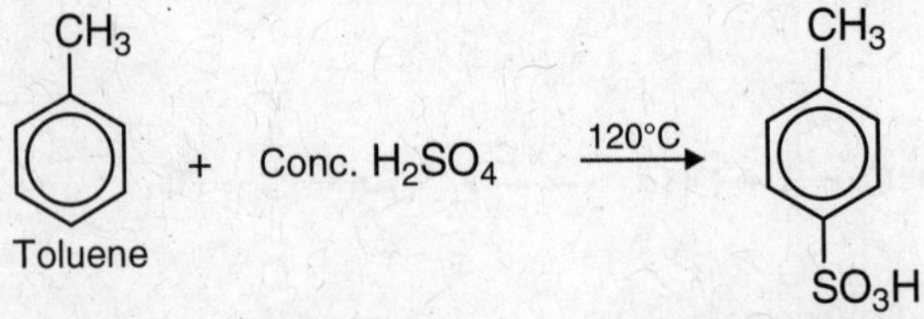

Toluene → *p*-Toluenesulfonic acid (*plus* ortho-isomer)

MECHANISM. Following steps are involved :

Step 1. Electrophile is formed. In this reaction the electrophile is sulfur trioxide (SO_3). In concentrated sulfuric acid, SO_3 is produced as follows. In fuming sulfuric acid, this step is unimportant because the dissolved SO_3 reacts directly.

$$2H_2SO_4 \rightleftharpoons SO_3 + H_3O^+ + HSO_4^-$$

Step 2. The electrophile attacks the benzene ring to form a carbonium ion.

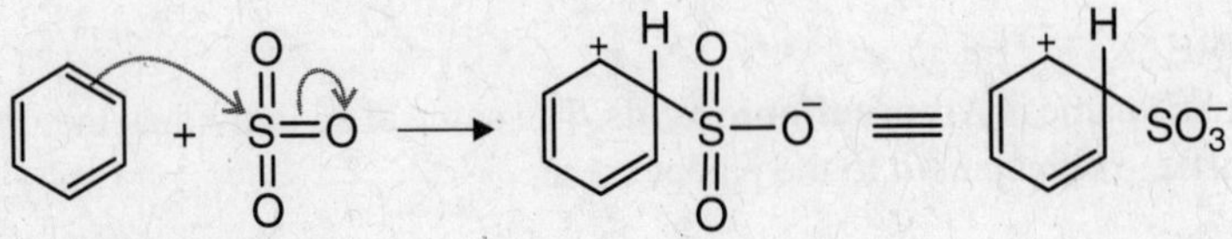

Step 3. Loss of proton.

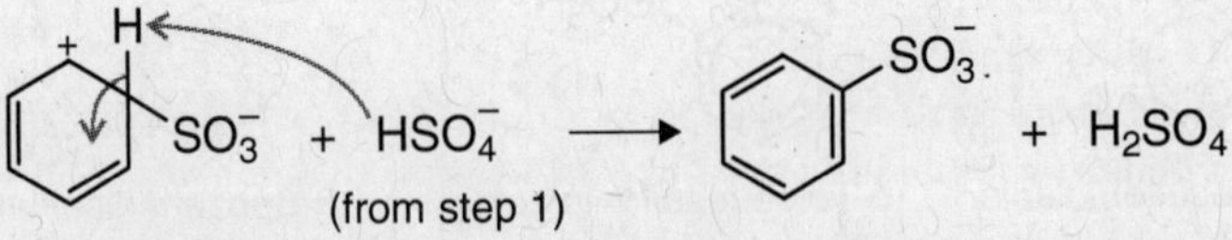

Step 4. Addition of proton gives benzenesulfonic acid.

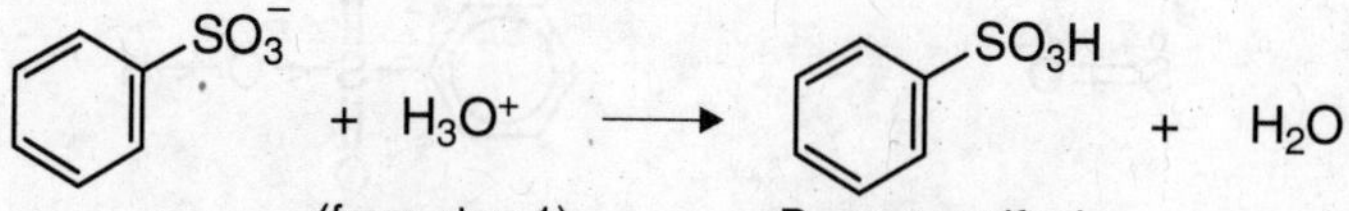

(from step 1) Benzenesulfonic acid

(2) **From Sulfonyl Chlorides.** Arenes react with excess of chlorosulfonic acid to form sulfonyl chlorides. These on acid-hydrolysis yield sulfonic acids.

$$\underset{\text{Arene}}{ArH} \xrightarrow{ClSO_3H} \underset{\text{Sulfonyl chloride}}{ArSO_2Cl} \xrightarrow{H_2O/H^+} \underset{\text{Arenesulfonic acid}}{ArSO_3H}$$

(3) **From Thiophenols.** Thiophenols undergo oxidation with alkaline $KMnO_4$ to give sulfonic acids. For example,

$$\underset{\text{Thiophenol}}{C_6H_5SH} \xrightarrow{KMnO_4} \underset{\text{Benzenesulfonic acid}}{C_6H_5SO_3H}$$

Recovery from Sulfonation Mixtures. It is difficult to isolate sulfonic acids because they are highly soluble in water and are nonvolatile. They are generally recovered from sulfonation mixtures as sodium salts. To the diluted sulfonation mixture ($ArSO_3H + H_2SO_4$), add calcium carbonate and filter. To the filtrate containing dissolved calcium arenesulfonate, then add sodium carbonate and again filter. The filtrate is evaporated to recover solid sodium arenesulfonate.

$$\begin{matrix} H_2SO_4 \\ + \\ ArSO_3H \end{matrix} \xrightarrow{\textit{Add } CaCO_3} \begin{matrix} CaSO_4\downarrow \\ + \\ (ArSO_3)_2Ca \end{matrix} \xrightarrow{\textit{Filter \& add } Na_2CO_3} \begin{matrix} CaCO_3\downarrow \\ + \\ ArSO_3Na \end{matrix} \xrightarrow[\textit{evaporate}]{\textit{Filter \&}} ArSO_3Na$$

Sodium salts of sulfonic acids can be used in place of the free sulfonic acids for many reactions.

PHYSICAL PROPERTIES

(1) Aromatic sulfonic acids are white crystalline solids which are highly hygroscopic and form syrups in moist air. That is why they are invariably used in the form of their salts.

(2) They do not have sharp melting points since they decompose on heating.

(3) They are extremely soluble in water but sparingly soluble in organic solvents.

(4) In the *IR Spectrum*, the O–H bond of sulfonic acids appears at 3100-3500 cm^{-1}, near about the same range as for phenols and alcohols.

CHEMICAL PROPERTIES

Sulfonic acids are strong acids (about as strong as sulfuric acid) and are completely ionized in solution.

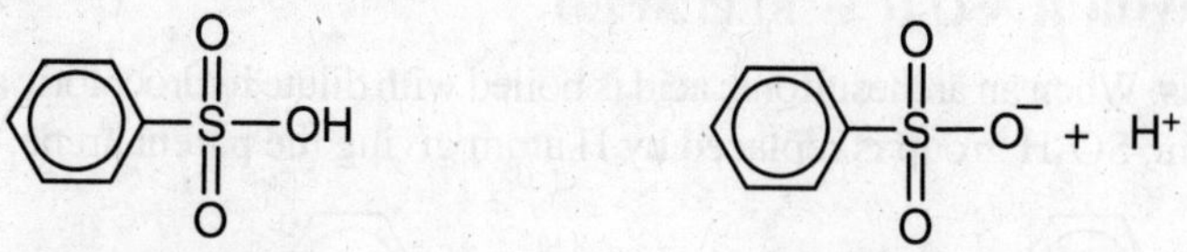

Benzenesulfonic acid Sulfonate ion (stable)

Their high acidity is due to two factors :

(*a*) The two S=O bonds are polarized since sulfur atom has higher electronegativity than oxygen atom. The electron-withdrawing inductive effect so caused facilitates the release of proton.

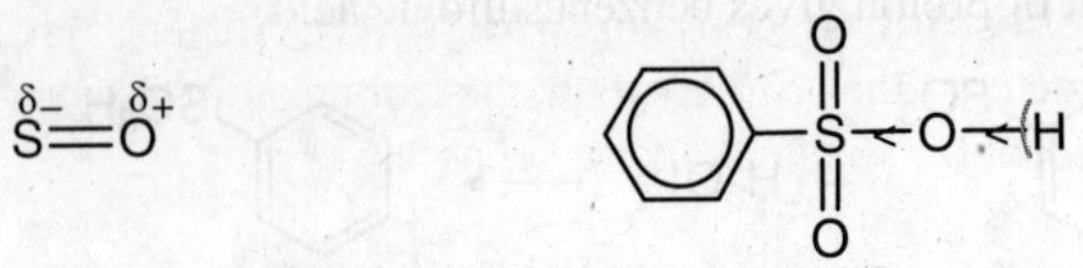

(*b*) The sulfonate anion once formed is resonance stabilized.

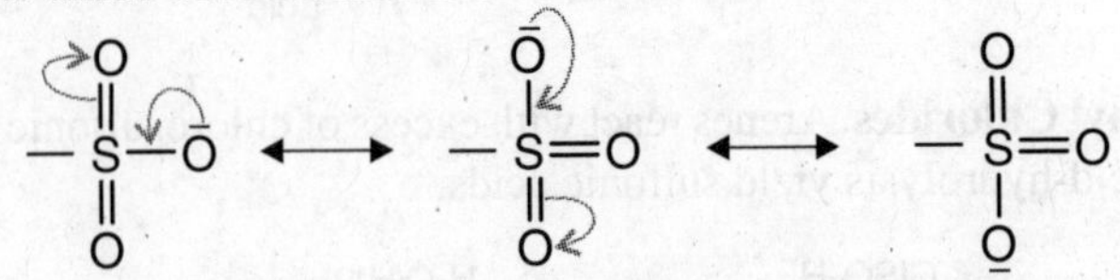

Both these factors are more pronounced in case of sulfonic acids than carboxylic acids. The pK_a of benzenesulfonic acid is 0.70. Electron withdrawing substituents at the *o*- or *p*-positions increase the acidity. For example, 2,4-dinitrobenzenesulfonic acid is a stronger acid than sulfuric acid. Sulfonic acids give the following reactions :

REACTIONS OF –OH OF –SO_3H GROUP

(1) **Reaction with Bases to form Salts**. Aromatic sulfonic acids react with inorganic bases (hydroxides, carbonates, oxides) to form soluble salts called **Sulfonates**.

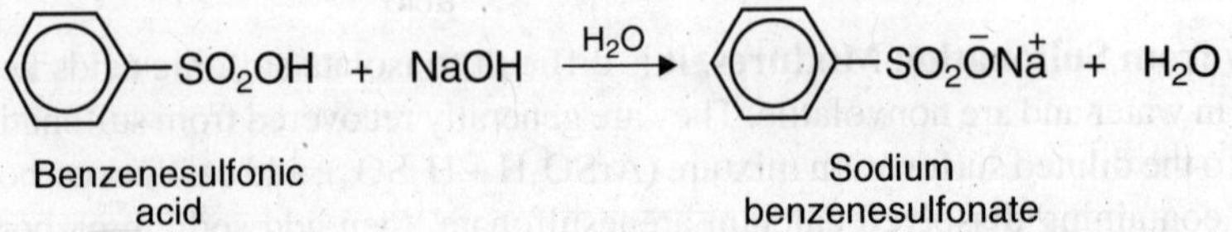

(2) **Reaction with PCl_5 or $SOCl_2$**. When heated, with phosphorus pentachloride or thionyl chloride, sulfonic acids or their salts are converted to the corresponding **Sulfonyl Chlorides**.

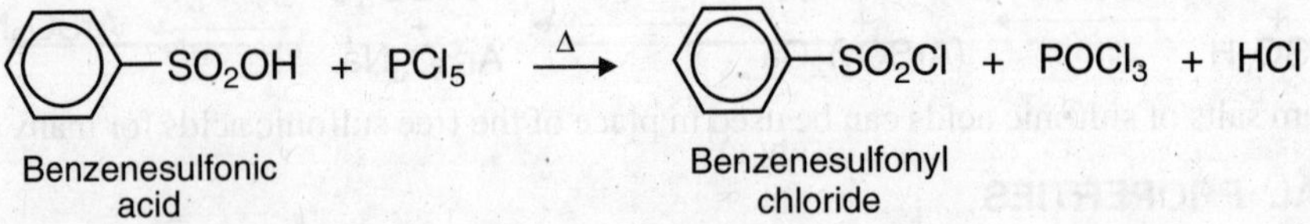

The sulfonyl chlorides behave like the acid chlorides of carboxylic acids. They react with ammonia to give *Sulfonamides* which have definite melting points and are used for identifying sulfonic acids.

(3) **Formation of Esters**. Unlike carboxylic acids, sulfonic acids do not form esters directly by reaction with alcohols. But they do so *via* sulfonyl chlorides.

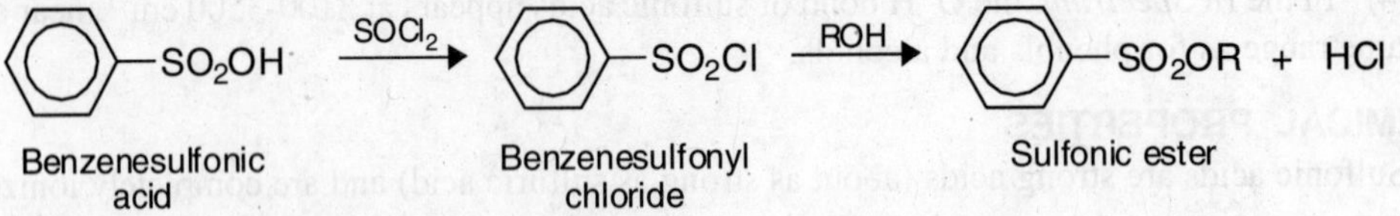

REACTIONS IN WHICH –SO_3H IS REPLACED

(4) **Hydrolysis**. When an arenesulfonic acid is boiled with dilute hydrochloric acid (or dil H_2SO_4), or heated with steam, SO_3H group is replaced by H atom giving the parent arene.

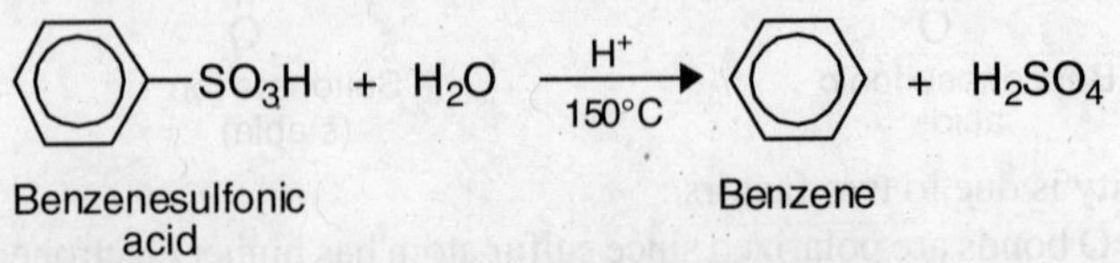

This process which is the reverse of sulfonation is called **Desulfonation**.

(5) **Nucleophilic Substitution Reactions with NaOH, $NaNH_2$, NaCN, NaSH**. Because of the electron-attracting character of the $-SO_3H$ group, the carbon atom to which it is attached, is electron deficient and hence susceptible to attack by nucleophiles.

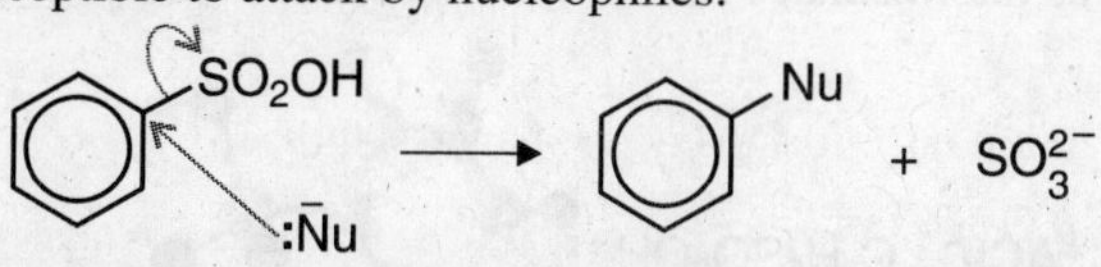

This type of reaction is carried out by fusing together salt of sulfonic acid and the appropriate reagent providing the nucleophile.

(*a*) Fusion with Sodium hydroxide (Nu, $:\bar{O}H$) to form **Phenol**.

$$\underset{\text{Sodium benzenesulfonate}}{C_6H_5SO_2\bar{O}\overset{+}{Na}} + \overset{+}{Na}\bar{O}H \xrightarrow{\text{fuse}} \underset{\text{Sodium phenoxide}}{C_6H_5\bar{O}\overset{+}{Na}} + Na_2SO_3$$

$$C_6H_5ONa \xrightarrow{HCl} \underset{\text{Phenol}}{C_6H_5OH} + NaCl$$

(*b*) Fusion with Sodamide (Nu, $:\bar{N}H_2$) to form **Amine**.

$$C_6H_5SO_2\bar{O}\overset{+}{Na} + \overset{+}{Na}\bar{N}H_2 \xrightarrow{\text{fuse}} \underset{\text{Aniline}}{C_6H_5NH_2} + Na_2SO_3$$

(*c*) Fusion with Sodium cyanide (Nu, $:\bar{C}N$) to form **Nitriles**.

$$C_6H_5SO_2\bar{O}\overset{+}{Na} + \overset{+}{Na}\bar{C}N \xrightarrow{\text{fuse}} \underset{\text{Benzonitrile}}{C_6H_5CN} + Na_2SO_3$$

(*d*) Fusion with Sodium hydrogensulfide (Nu, $:\bar{S}H$) to form **Thiol**.

$$C_6H_5SO_2\bar{O}\overset{+}{Na} + NaSH \xrightarrow{\text{fuse}} \underset{\text{Thiophenol}}{C_6H_5SH} + Na_2SO_3$$

REACTIONS OF BENZENE RING

(6) The $-SO_3H$ group is deactivating and *meta* director. Therefore, the sulfonic acids undergo the usual substitutions such as halogenation, nitration and sulfonation under more difficult conditions than for benzene itself. The incoming substituent occupies the *meta* position.

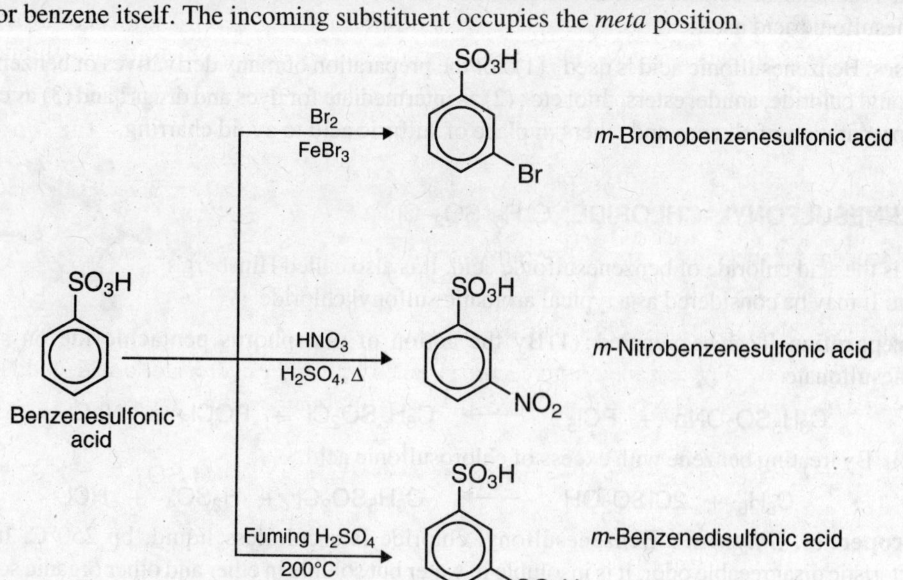

USES OF SULFONIC ACIDS

(1) Sulfonic acids with long alkyl side-chains find considerable use as detergents.

(2) They are used as intermediates for the synthesis of sulfa drugs, saccharin, and certain azo dyes.

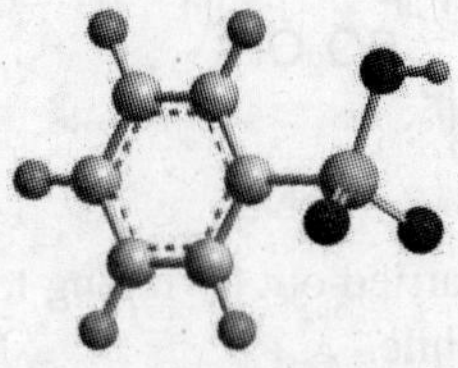

BENZENESULFONIC ACID, $C_6H_5SO_2OH$

It is the parent member of the class in which one $-SO_3H$ group is bonded to benzene ring.

Preparation. Benzene reacts with concentrated sulfuric acid or fuming sulfuric acid (oleum) to form benzenesulfonic acid. Fuming sulfuric acid (oleum) is SO_3 + conc H_2SO_4. If this reaction is carried out at higher temperatures, benzene-1,3-disulfonic acid and benzene- 1,3,5-trisulfonic acid are formed.

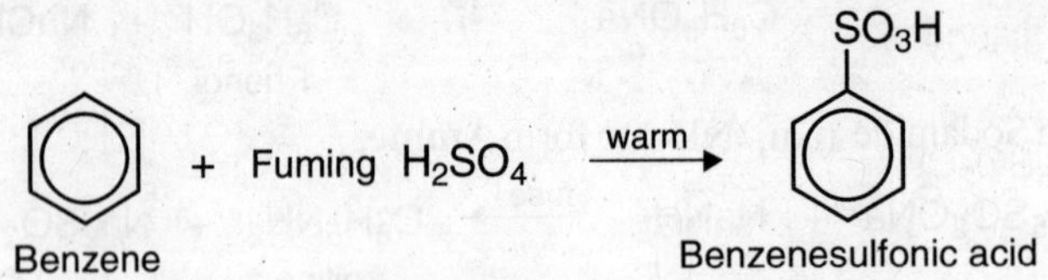

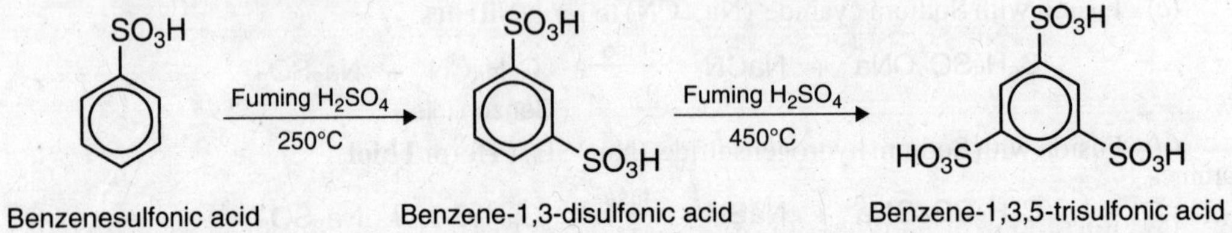

Properties. Benzenesulfonic acid crystallizes from water forming white leaflets containing $1\frac{1}{2}$ H_2O. The anhydrous compound melts at 65–66°C. It is very deliquescent. It is highly soluble in water and its solutions are strongly acidic, nearly as strong as sulfuric acid. Chemically, it gives all the general reactions of sulfonic acids. These have already been discussed by taking example of benzenesulfonic acid itself.

Uses. Benzenesulfonic acid is used : (1) for the preparation of many derivatives of benzene such as sulfonyl chloride, amide, esters, thiol etc ; (2) as intermediate for dyes and drugs ; and (3) as catalyst in the preparation of alkenes and ethers in place of sulfuric acid to avoid charring.

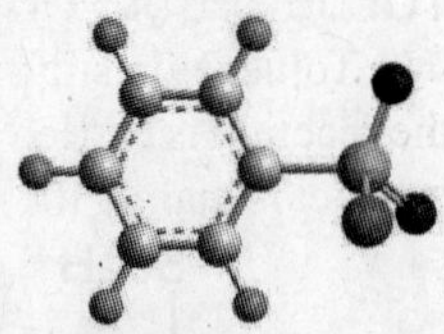

BENZENESULFONYL CHLORIDE, $C_6H_5{-}SO_2{-}Cl$

It is the acid chloride of benzenesulfonic acid. It is also called Hinsberg's reagent. It may be considered as a typical aromatic sulfonyl chloride.

Preparation. It can be prepared : (1) By the action of phosphorus pentachloride on sodium benzenesulfonate.

$$C_6H_5SO_2ONa + PCl_5 \longrightarrow C_6H_5SO_2Cl + POCl_3 + NaCl$$

(2) By treating benzene with excess of chlorosulfonic acid.

$$C_6H_6 + 2ClSO_2OH \longrightarrow C_6H_5SO_2Cl + H_2SO_4 + HCl$$

Properties. (*Physical*). Benzenesulfonyl chloride is a colorless liquid, bp 250°C. It has a characteristic disagreeable odor. It is insoluble in water but soluble in ether and other organic solvents.

(*Chemical*). It gives the reactions of the acid chlorides.

(1) **Hydrolysis**. When heated with water in the presence of alkali, it is hydrolysed to benzenesulfonic acid.

$$C_6H_5SO_2Cl + HOH \xrightarrow[\Delta]{OH^-} C_6H_5SO_2OH + HCl$$

(2) **Formation of Sulfonic Esters**. It reacts with alcohols in the presence of pyridine to form Alkyl benzenesulfonates.

$$C_6H_5SO_2Cl + HOR \xrightarrow{C_5H_5N} C_6H_5SO_2OR + C_5H_5\overset{+}{N}H\overset{-}{Cl}$$

(3) **Formation of Sulfonamide**. It reacts with concentrated ammonia to form benzenesulfonamide.

$$C_6H_5SO_2Cl + 2NH_3 \longrightarrow C_6H_5SO_2NH_2 + NH_4Cl$$

(4) **Formation of *N*-Substituted Sulfonamides**. It reacts with 1° and 2° amines to form *N*-alkyl and *N,N*-dialkylsulfonamides.

$$C_6H_5SO_2Cl + \underset{\text{1° amine}}{H{-}\overset{H}{\overset{|}{N}}{-}R} \longrightarrow \underset{N\text{-Alkylbenzenesulfonamide}}{C_6H_5SO_2{-}\overset{H}{\overset{|}{N}}{-}R} + HCl$$

$$C_6H_5SO_2Cl + \underset{\text{2° amine}}{H{-}\overset{R}{\overset{|}{N}}{-}R} \longrightarrow \underset{N,N\text{-Dialkylbenzenesulfonamide}}{C_6H_5SO_2{-}\overset{R}{\overset{|}{N}}{-}R} + HCl$$

This reaction forms the basis of the classical **Hinsberg Test** for distinguishing 1°, 2° and 3° amines.

(5) **Formation of Sulfones**. It gives Friedel-Crafts reaction with aromatic compounds to form *Sulfones*.

$$C_6H_5SO_2Cl + C_6H_6 \xrightarrow{AlCl_3} \underset{\text{Phenylsulfone}}{C_6H_5{-}SO_2{-}C_6H_5} + HCl$$

Uses. Benzenesulfonyl chloride is used : (1) for preparing benzenesulfonic acid derivatives; (2) for introducing sulfonyl group ($-SO_2-$) in alcohols, amines, benzene ring *i.e.*, for **Sulfonylation** ; (3) as a laboratory reagent for distinguishing 1°, 2°, 3° amines.

TOLUENESULFONIC ACIDS, $CH_3-C_6H_4-SO_3H$

Toluene on sulfonation with concentrated sulfuric acid forms *o*- and *p*-toluenesulfonic acids. If the reaction is carried above 100°C, the *p*-isomer is the main product, while at lower temperatures the product is mainly the *o*-isomer.

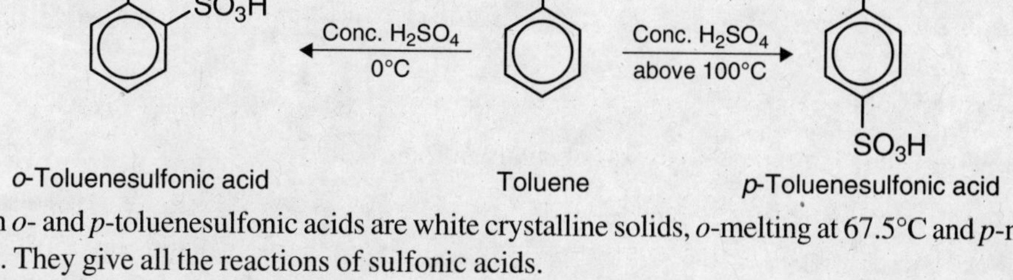

o-Toluenesulfonic acid   Toluene   *p*-Toluenesulfonic acid

Both *o*- and *p*-toluenesulfonic acids are white crystalline solids, *o*-melting at 67.5°C and *p*-melting at 106°C. They give all the reactions of sulfonic acids.

p-Toluenesulfonyl chloride (**Tosyl chloride**) is used to prepare tosyl esters from alcohols. This reaction is called **Tosylation**.

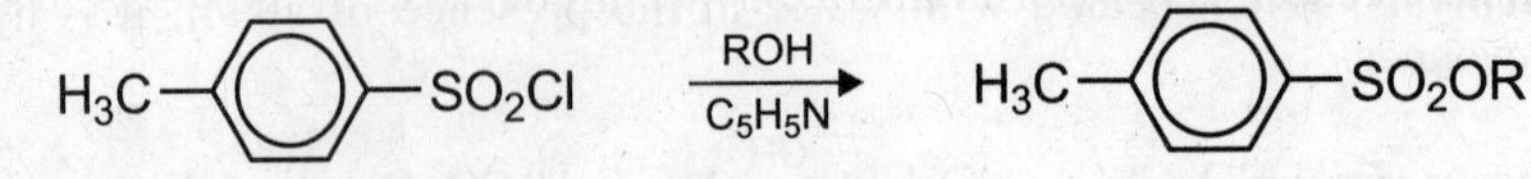

p-Toluenesulfonyl chloride Tosylate

o-Toluenesulfonyl chloride is used in the manufacture of saccharin. *p*-Toluenesulfonyl chloride is used in the preparation of chloramine-T.

SACCHARIN, *o-Benzenesulfonamide*

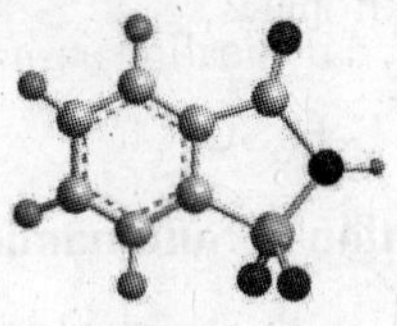

Saccharin is prepared from toluene by the following steps :

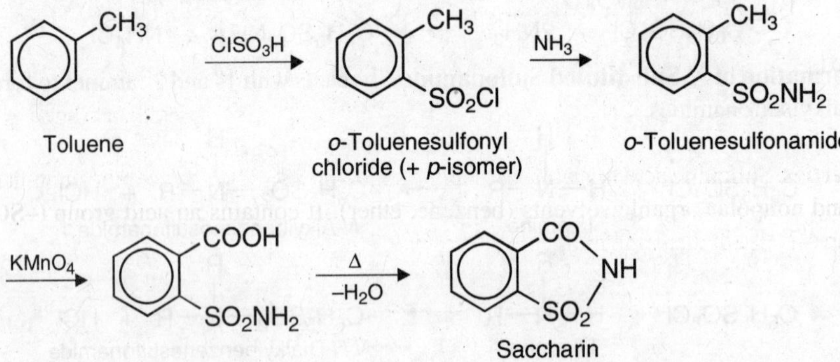

Saccharin is a colorless, crystalline compound, mp 224°C. It is approximately 500 times sweeter than table sugar and is placed in the market as soluble sodium salt (*Sodium Saccharin*). Despite its sweet taste, saccharin leaves a slight bitter after taste. It is widely used as a calorie-free sugar substitute for diabetics and the weight-conscious. It is also used as a sweetening agent in tooth pastes, mouth washes, and aerated drinks.

CHLORAMINE-T

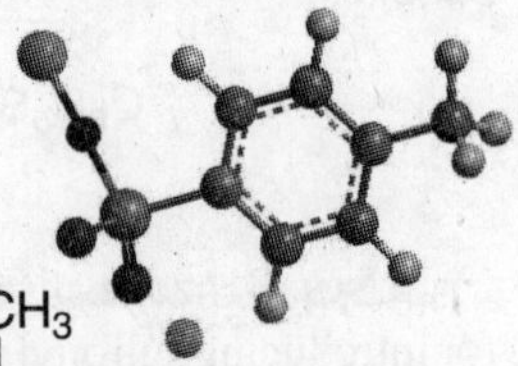

Chloramine-T is prepared from toluene by the following steps :

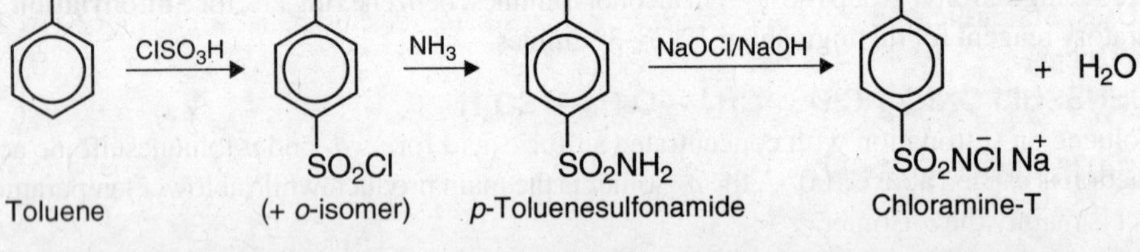

Chloramine-T reacts with water to liberate hypochlorous acid. It is an effective antiseptic for treatment of wounds.

SULFANILIC ACID, *p-Aminobenzenesulfonic acid*

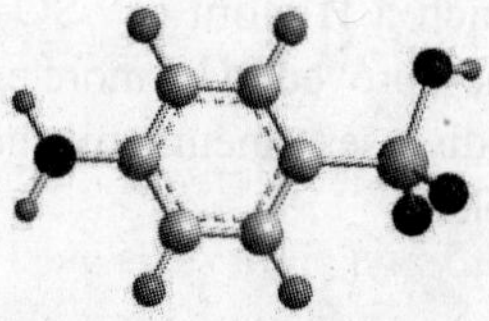

It is prepared by heating aniline with concentrated sulfuric acid at 180°C.

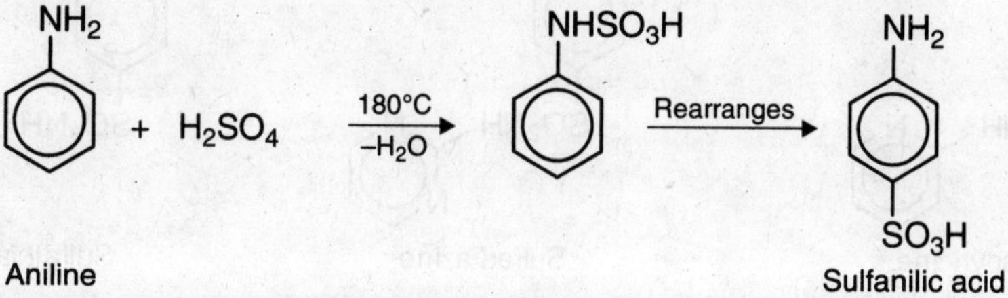

Aniline Sulfanilic acid

Properties. Sulfanilic acid is a white crystalline solid, mp 280°C with decomposition. It is insoluble in water and nonpolar organic solvents (benzene, ether). It contains an acid group ($-SO_3H$) and a basic group ($-NH_2$) in the same molecule. Therefore, it exists as inner-salt *Zwitterion*.

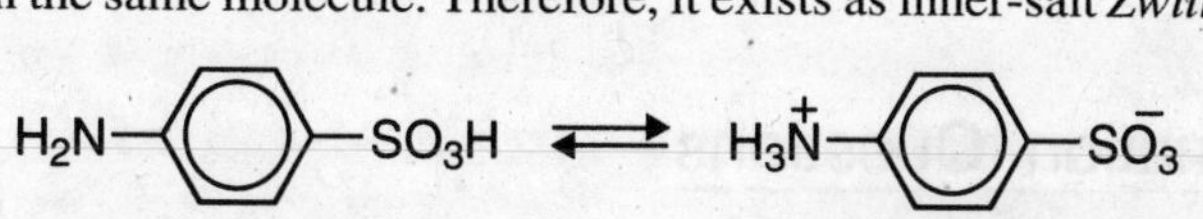

Zwitterion

Sulfanilic acid dissolves in sodium hydroxide to form soluble salt since the alkali can abstract a proton from the $-\overset{+}{N}H_3$ group of the zwitterion.

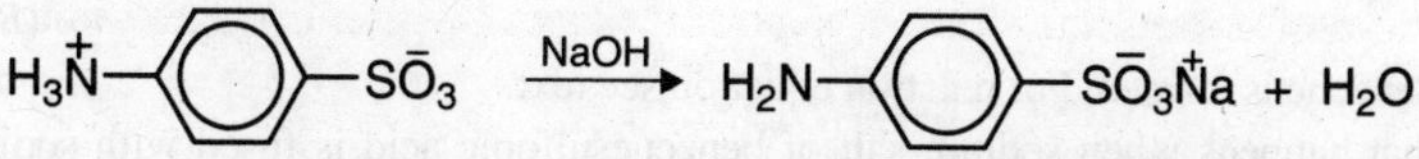

Sodium sulfanilate

The $-SO_3H$ group of sulfanilic acid can be replaced by $-NO_2$ group on treatment with nitric acid. The $-NH_2$ group can be diazotized and coupled with phenols to form dyes.

Uses. Sulfanilic acid is an important dye intermediate. Its substituted amides form the valuable **Sulfa Drugs**.

SULFANILAMIDE

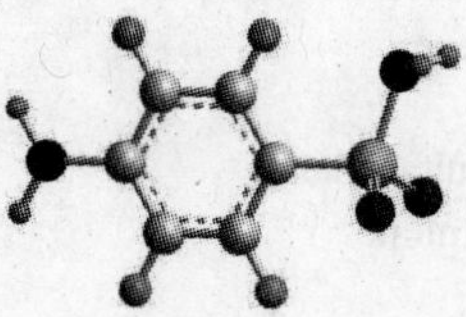

It is prepared from acetanilide by the following steps :

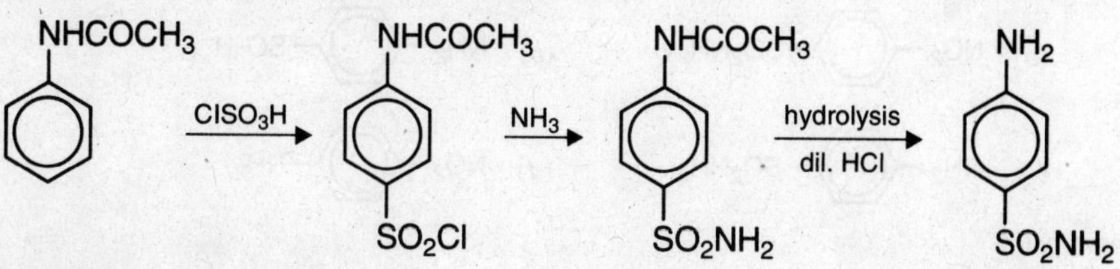

Acetanilide Sulfanilamide

Acetanilide required above is obtained by acetylating aniline with acetyl chloride.

Properties. Sulfanilamide is a white crystalline solid, mp 166.5°C. It is an antimicrobial drug. Its derivatives in which a H atom of $-SO_2NH_2$ has been replaced by some other group (usually a heterocyclic amine), proved to be more effective such drug. Sulfanilamide and its derivatives which were used to cure diseases (pneumonia, gonorrhea, etc) caused by microorganisms were called **Sulfa Drugs**. For example,

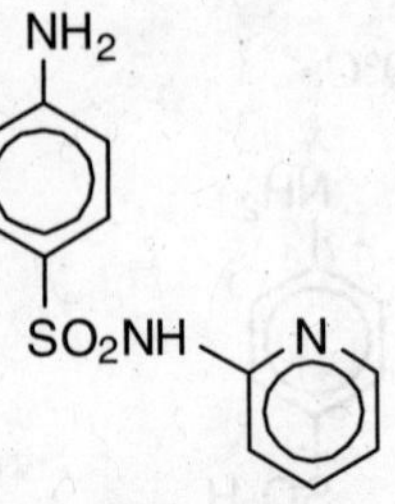
Sulfapyridine

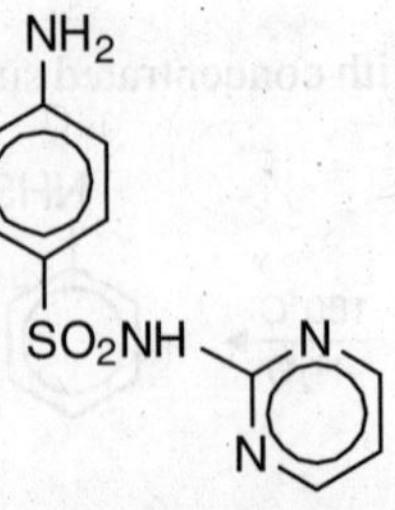
Sulfadiazine

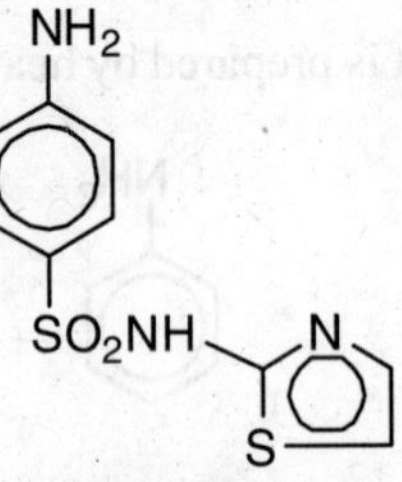
Sulfathiazole

On account of their harmful side effects, the use of sulfa drugs is now limited to veterinary medicine. In human treatment these have been largely replaced by safer and more effective antibiotics such as penicillin and auromycin.

Examination Questions

1. How is benzenesulfonic acid prepared in the laboratory ?
2. Describe the important reactions of benzenesulfonic acid.
3. What happens when benzenesulfonic acid is heated with dilute H_2SO_4 at 150°C ? *(Bhopal BSc, 2015)*

Answer. Benzene is formed. For reaction equation see text.

4. What happens when sodium salt of benzenesulfonic acid is fused with sodium or potassium cyanide ? *(Cuttak BSc, 2014)*

Answer. Benzonitrile is formed. For reaction equation see text.

5. What happens when sodium salt of benzenesulfonic acid is fused with sodium hydroxide ? *(Kakatiya BSc, 2014)*

Answer. Sodium phenoxide is formed. For reaction equation see text.

6. How is benzenesulfonyl chloride prepared ? Describe its important reactions. *(Delhi BSc, 2015)*
7. Discuss the mechanism of sulfonation of benzene.
8. Sulfa drugs are all structurally related to :
   (*a*) Nitrobenzene (*b*) Sulfanilamide
   (*c*) Pyridine (*d*) Aniline

Answer. (*b*)

9. Which of the following compounds is sulfanilamide ?

(*a*) 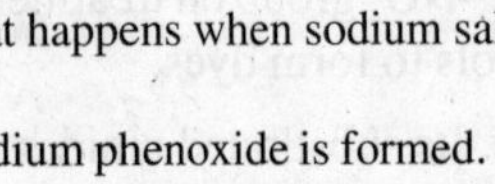
(*b*) 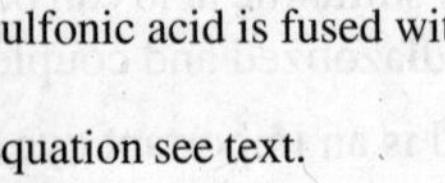
(*c*) 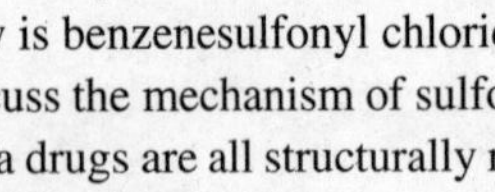
(*d*) 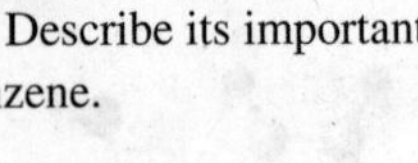

Answer. (*c*)

10. Sulfanilamide
   (*a*) is a sulfur-containing amino acid
   (*b*) is the compound from which other sulfa drugs are derived

(*c*) is an orange-colored aniline dye

(*d*) is commonly administered in aqueous solution

Answer. (*b*)

11. How will you synthesize *p*-toluenesulfonic acid from benzene ?

Answer. Following steps are involved :

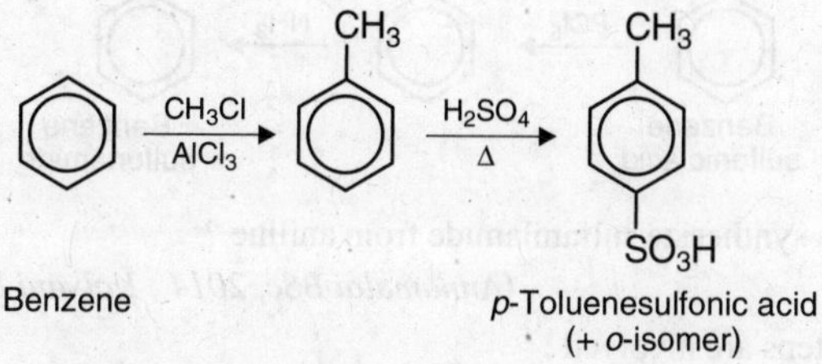

12. How will you synthesize phenol from benzenesulfonic acid ?

Answer. Following steps are involved :

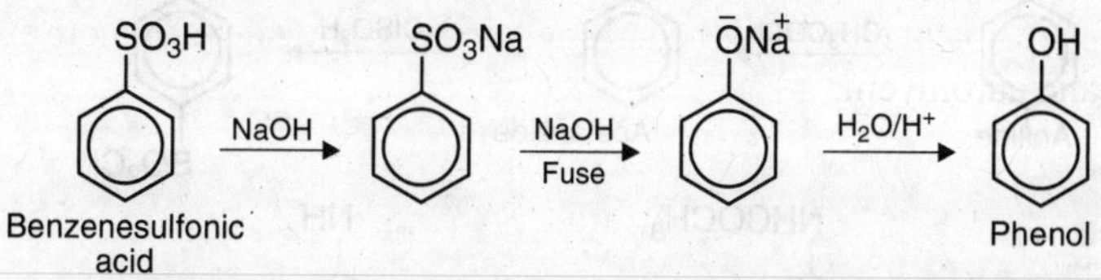

13. How will you synthesize ethylbenzene from benzenesulfonic acid ?

Answer. Following steps are involved :

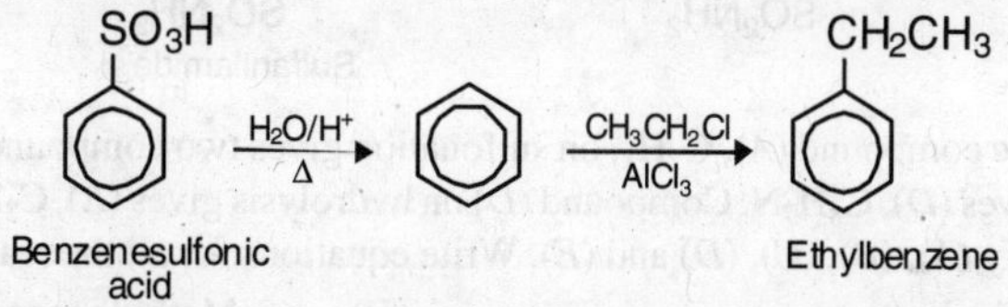

14. How will you synthesize benzoic acid from benzenesulfonic acid ? (*Lucknow BSc, 2014*)

Answer. Following steps are involved :

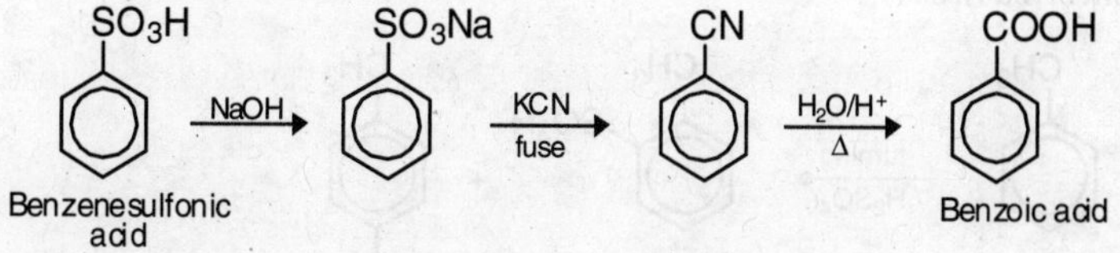

15. How will you synthesize benzylamine from benzenesulfonic acid ? (*Dibrugarh BSc, 2015*)

Answer. Following steps are involved :

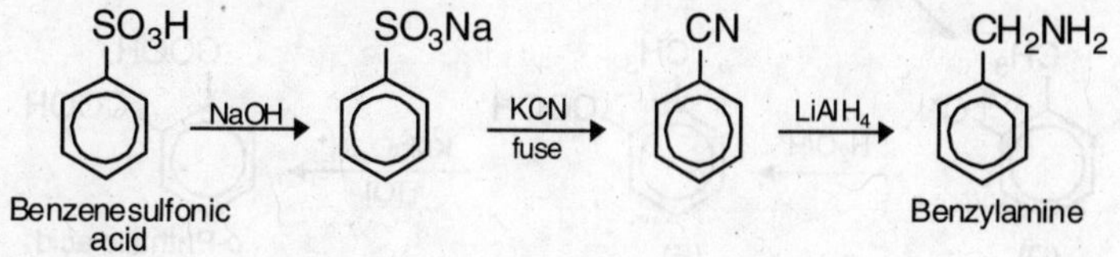

16. How will you synthesize sulfanilic acid from aniline ?

(*Anna BSc, 2014 ; Indore BSc, 2014 ; Nagpur BSc, 2015*)

17. How will you synthesize sulfanilamide from aniline ?

18. How will you synthesize chloramine-T from toluene ?

19. How will you synthesize saccharin from toluene ? *(Mysore BSc, 2014)*

20. How will you synthesize benzenesulfonamide from benzenesulfonic acid ?

Answer. Following steps are involved :

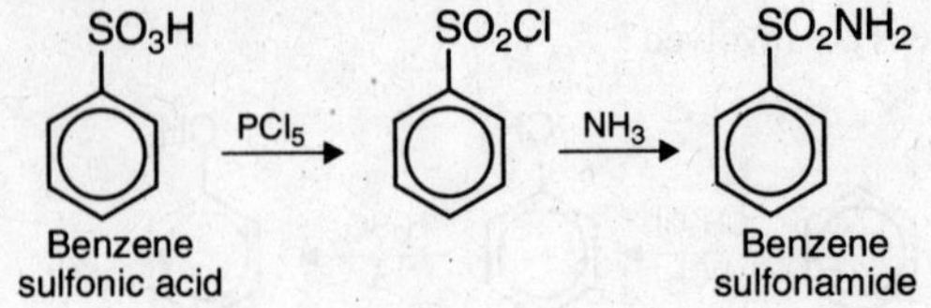

21. How will you synthesize sulfanilamide from aniline ?

(Annamalai BSc, 2014 ; Kalyani BSc, 2014 ; Utkal BSc, 2015)

Answer. Following steps are involved :

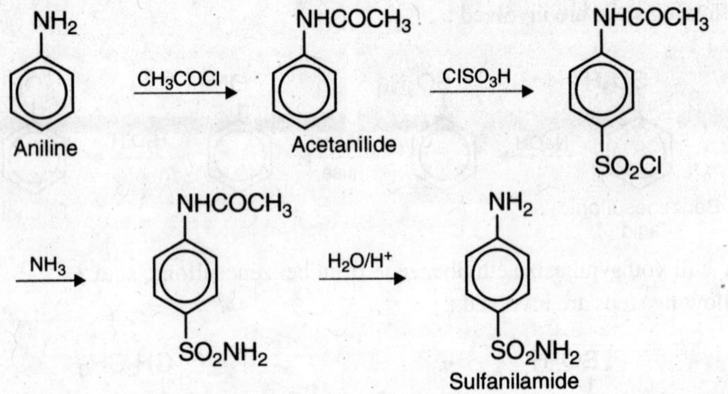

22. An aromatic compound (*A*), C_7H_8, on sulfonation gives two compounds (*B*) and (*C*). Compound (*B*) on fusion with KCN gives (*D*), C_8H_7N. Compound (*D*) on hydrolysis gives (*E*), $C_8H_8O_2$, which on hydrolysis gives phthalic acid. Identify (*A*), (*B*), (*C*), (*D*) and (*E*). Write equations for all the reactions.

Answer. (*A*) = Toluene

(*B*) = *o*-Toluenesulfonic acid

(*C*) = *p*-Toluenesulfonic acid

(*D*) = 2-Methylbenzonitrile

(*E*) = *o*-Toluic acid

Reactions Involved Are :

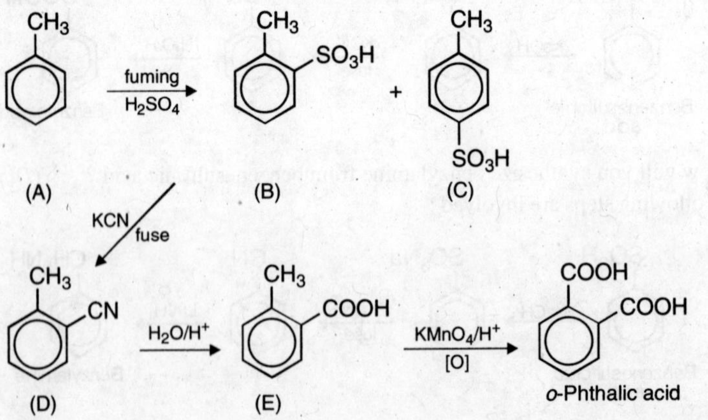

37

Aromatic Nitro Compounds

TNT (2,4,6-Trinitrotoluene) has a long history as military and industrial explosive. In modern applications, particularly mining and building demolition, TNT has replaced nitroglycerine.

They are the derivatives of aromatic hydrocarbons in which one or more ring H atoms have been replaced by $-NO_2$ group.

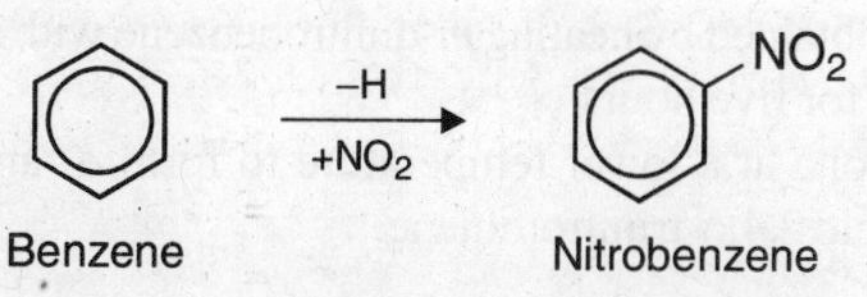

Benzene Nitrobenzene

In modern perspective, aromatic nitro compounds are molecules in which $-NO_2$ group is directly bonded to an aromatic ring. They have the general formula $Ar-NO_2$.

The functional group $-NO_2$ may be represented as

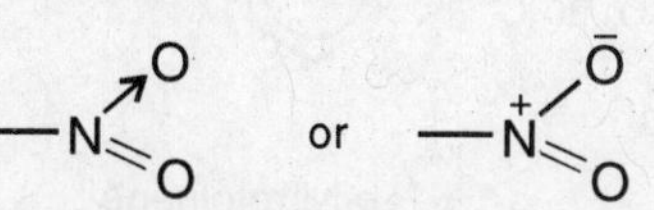

Bond length and dipole measurements show that the two nitrogen-oxygen bond lengths are equal (1·2 Å). Therefore, the structure of $-NO_2$ groups is best regarded as a resonance hybrid of the following two resonance forms (1 and 2).

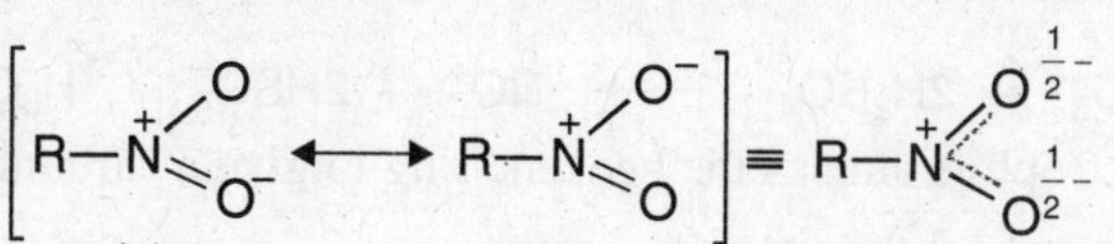

(1) (2) Resonance hybrid

NOMENCLATURE

The aromatic nitro compounds are named by inserting the word 'nitro' in front of the name of the parent hydrocarbon. The position of the $-NO_2$ group is indicated by prefixing *o*-, *m*- or *p*-, or a numeral where necessary.

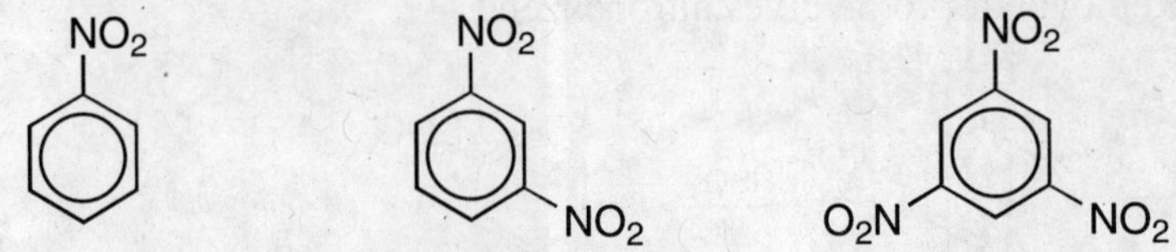

Nitrobenzene — *m*-Dinitrobenzene (1,3-dinitrobenzene) — *s*-Trinitrobenzene (1,3,5-trinitrobenzene)

METHODS OF PREPARATION

(1) **Direct Nitration.** Nitration of an aromatic compound is usually carried out by heating it with a mixture of concentrated nitric acid and sulfuric acid.

$$ArH + HONO_2 \xrightarrow[\Delta]{H_2SO_4} ArNO_2 + H_2O$$

The extent of nitration (introduction of one or more NO_2 groups) is controlled by varying the strength of the 'nitrating mixture' and temperature of the reaction. For unreactive compounds ($C_6H_5NO_2$, $C_6H_5SO_3H$), conc. nitric acid and conc. sulfuric acid may be used at elevated temperatures. For reactive compounds (C_6H_5OH), dilute nitric acid at lower temperatures would suffice. For example, benzene can be nitrated with conc. HNO_3 and conc. H_2SO_4 at 60°C, while the second –NO_2 group can be introduced by using conc. HNO_3 and conc. H_2SO_4 acid at 90°C.

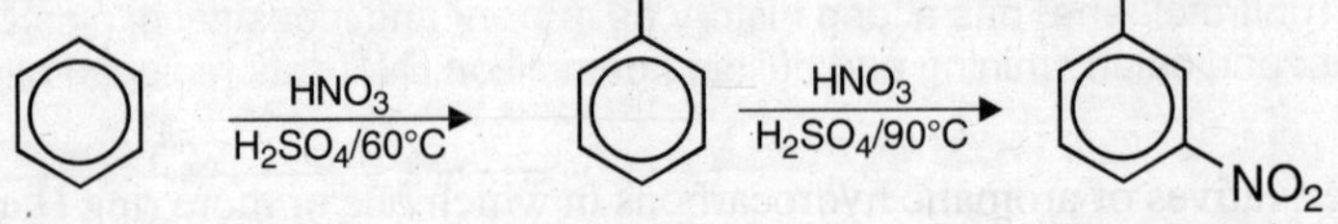

Benzene — Nitrobenzene — *m*-Dinitrobenzene

1,3,5-Trinitrobenzene is obtained by heating *m*-dinitrobenzene with a large excess of fuming nitric acid and fuming sulfuric acid for five hours.

Nitration of toluene is done at a lower temperature to form *o*- and *p*-nitrotoluenes. These are further nitrated at 100°C to yield 2,4,6-trinitrotoluene.

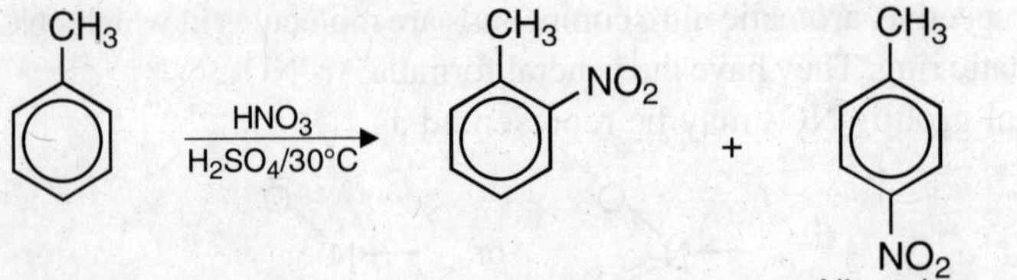

Toluene — *o*-Nitrotoluene — *p*-Nitrotoluene

MECHANISM. Three steps are involved :

Step 1. Formation of the electrophile (NO_2^+).

$$HNO_3 + 2H_2SO_4 \longrightarrow NO_2^+ + 2HSO_4^- + H_3O^+$$

Step 2. The electrophile attacks the benzene ring to give a carbonium ion.

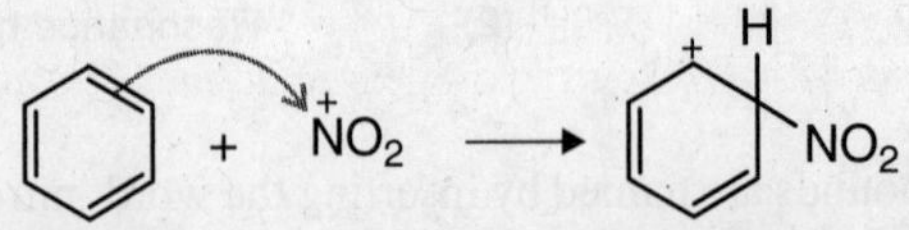

Benzene — Carbonium ion

Step 3. Removal of proton gives nitrobenzene.

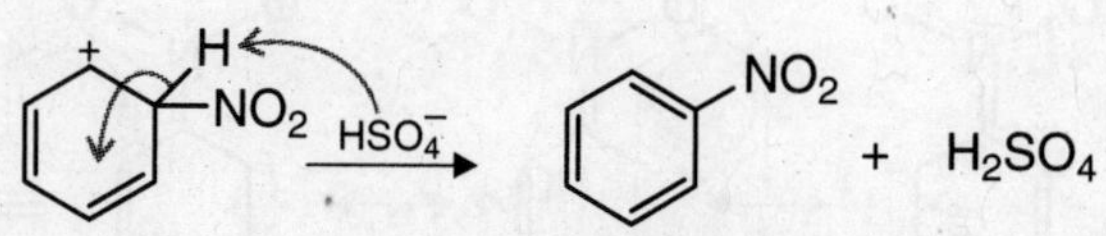

Nitrobenzene

(2) **Oxidation of Arylamines**. Arylamines undergo oxidation with trifluoroperacetic acid (CF_3CO_3H) or hydrogen peroxide in acetic acid (H_2O_2/CH_3COOH) to give the corresponding aromatic nitro compounds.

$$Ar-NH_2 \xrightarrow{CF_3CO_3H} Ar-NO_2$$

$$Ar-NH_2 \xrightarrow{H_2O_2/CH_3COOH} Ar-NO_2$$

(3) **From Diazonium Salts.** Aromatic nitro compounds are also produced by the action of nitrous acid on diazonium salts in the presence of cuprous oxide.

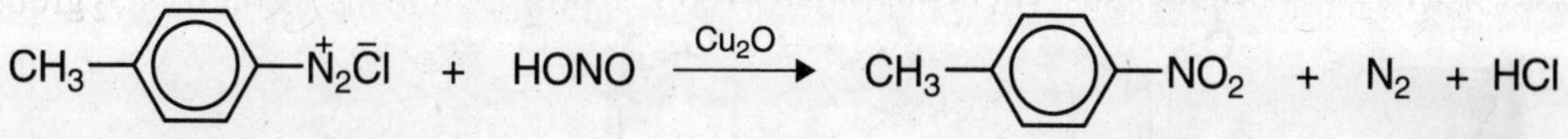

p-Methylbenzenediazonium chloride　　　　*p*-Nitrotoluene

This method is used when direct nitration cannot give the desired nitro compound.

PHYSICAL PROPERTIES

(1) A few lower aromatic nitro compounds (nitrobenzene, nitrotoluenes) are pale yellow liquids having pleasant odor. Others are crystalline solids having no odor.

(2) These are heavier than water and almost insoluble in it, but readily soluble in organic solvents such as benzene, ethanol and ether.

(3) They have high boiling points and melting points which is explained by the highly polar nature of $-NO_2$ group.

Compound	bp°C	mp°C
Nitrobenzene	211	5.7
o-Dinitrobenzene	318	118
m-Dinitrobenzene	302	90
p-Dinitrobenzene	298	173
s-Trinitrobenzene	—	122

Compound	bp°C	mp°C
o-Nitrotoluene	222	3
m-Nitrotoluene	233	16
p-Nitrotoluene	238	52
s-Trinitrotoluene	—	80

(4) They are toxic and are quickly absorbed through the skin.

(5) Many *di*- and *tri*-nitro compounds are highly explosive ; when strongly heated they detonate with great violence.

(6) $-NO_2$ group bonded to benzene ring shows UV absorption maxima at 260 and 330 mμ. In the IR spectrum, bands associated with $-NO_2$ group appear at about 1350 and 1500 cm^{-1}.

CHEMICAL PROPERTIES

The structure of an aromatic nitro compound *e.g.*, nitrobenzene can be represented by its resonance hybrid.

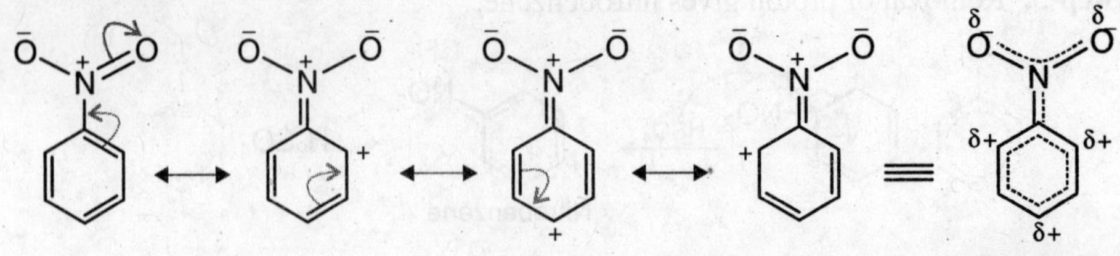

Resonance hybrid

(*a*) The NO_2 group, being made of nitrogen and oxygen only, shows no other reactions excepting those of reduction. C–N bond in hybrid has partial double bond character and hence the NO_2 group is firmly bonded to the ring. For this reason, nitro group cannot ordinarily be replaced by other atoms or groups.

(*b*) The benzene ring is deactivated to electrophiles (E^+) and activated to nucleophiles (Nu^-) due to withdrawl of electrons from it by inductive and resonance effects. As evident from the hybrid structure, the ring will undergo electrophilic substitution with difficulty at relatively electron-rich *meta* positions, and nucleophilic substitution at the *ortho* and *para* positions bearing positive charge.

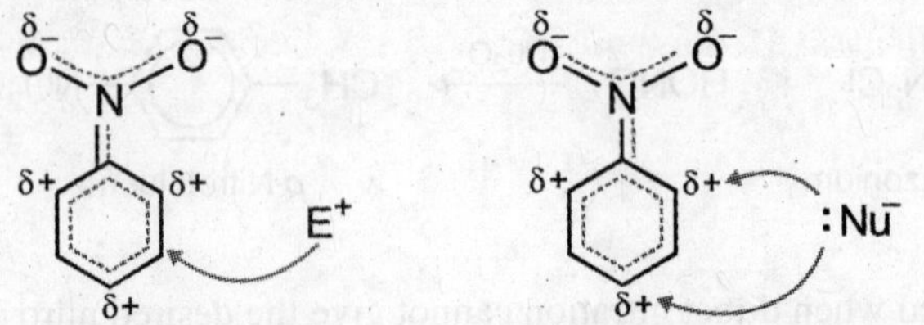

Electrophile attacks *meta* position
Nucleophile attacks *ortho* and *para* positions

REACTIONS INVOLVING –NO_2 GROUP

(1) **Reduction of Nitro Compounds**. Aromatic nitro compounds give a variety of interesting products depending on the reagent and conditions used. The different stages of reduction for nitrobenzene are :

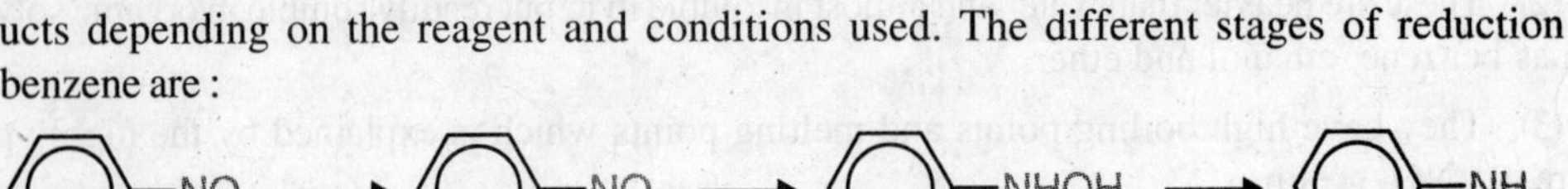

The intermediate substances nitrosobenzene and phenylhydroxylamine may further undergo condensation reactions to form bimolecular reduction products. Which of the products listed above is actually obtained, will be determined by whether the reduction is carried in acid, neutral or alkaline medium.

(*a*) **Reduction in Acid Medium gives Aniline**. When reduced with tin or iron and hydrochloric acid, aromatic nitro compounds are converted to arylamines. Thus,

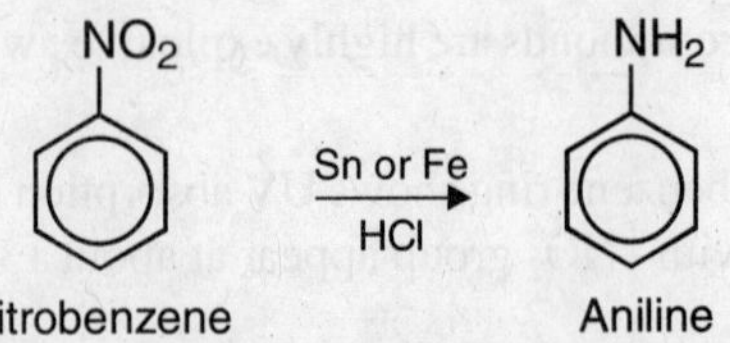

The reduction of –NO_2 to –NH_2 can also be accomplished by (*i*) Catalytic hydrogenation (H_2 in the presence of Ni or Pt) ; (*ii*) Lithium aluminium hydride ; and (*iii*) Ammonium sulfide.

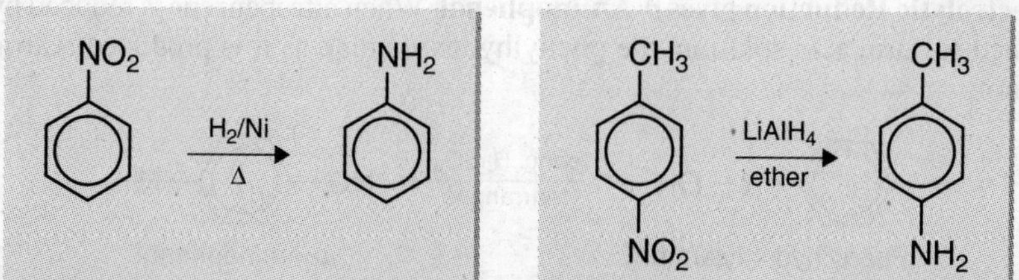

Nitrobenzene Aniline *p*-Nitrotoluene *p*-Toluidine

Selective Reduction. The reduction of *m*-dinitrobenzene with ammonium sulfide or sodium hydrosulfide (NaSH) reduces only one $-NO_2$ group on dinitro aromatic compounds.

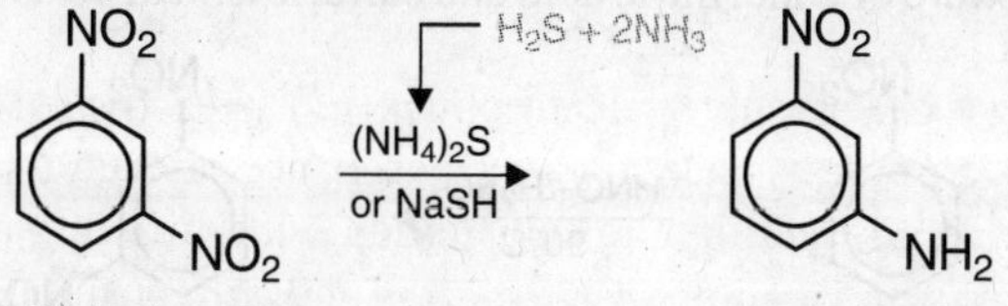

m-Dinitrobenzene *m*-Nitroaniline

(*b*) **Reduction in Neutral Medium gives Phenylhydroxylamine.** When reduced with a neutral reducing agent like zinc dust and aqueous ammonium chloride, aromatic nitro compounds form arylhydroxylamines.

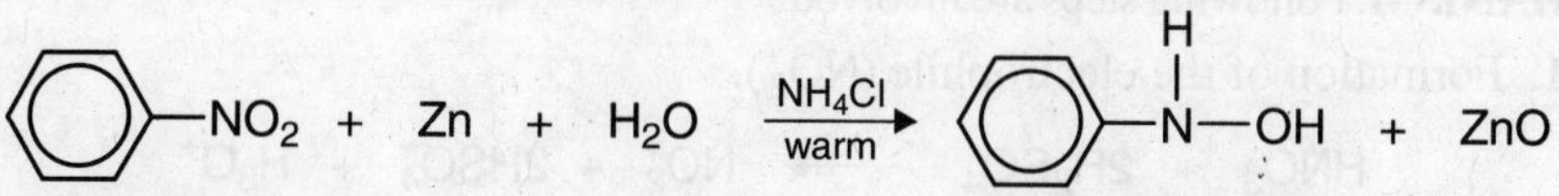

Nitrobenzene Phenylhydroxylamine

(*c*) **Reduction in Alkaline Medium gives Azoxybenzene, Hydrazobenzene, and Azobenzene.** As already mentioned, nitrobenzene on reduction forms the monomolecular intermediate products nitrosobenzene (C_6H_5NO) and phenylhydroxylamine (C_6H_5NHOH). In alkaline medium, these undergo bimolecular condensation reactions which could be stated in the following sequence :

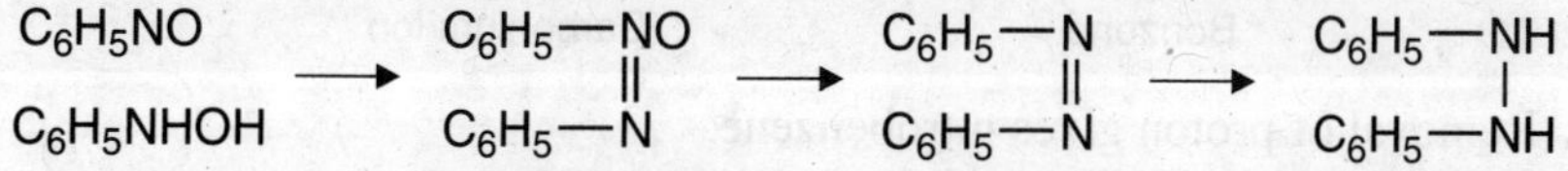

Azoxybenzene Azobenzene Hydrazobenzene

Each of the bimolecular reduction products listed above can be obtained from nitrobenzene by using appropriate reducing agents in alkaline medium.

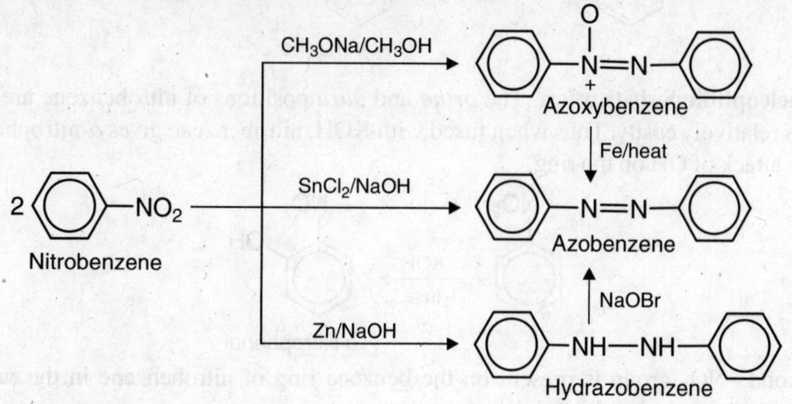

Any of these reduction products can be reduced to aniline with sodium in ethanol.

(*d*) **Electrolytic Reduction gives *p*-Aminophenol.** When nitrobenzene is reduced by electrolysis in concentrated sulfuric acid solution, the phenylhydoxylamine as it is produced rearranges to give *p*-aminophenol.

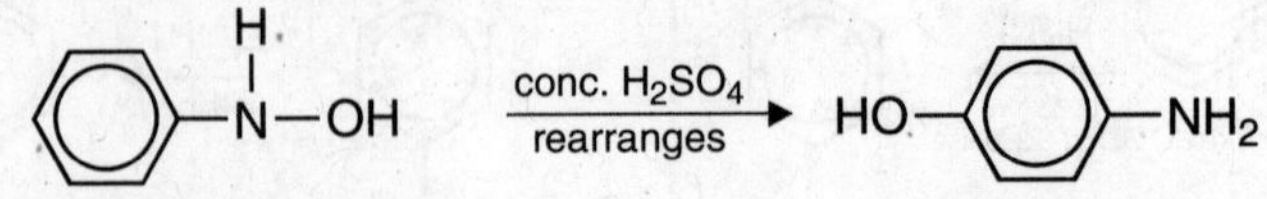

Phenylhydroxylamine p-Aminophenol

REACTIONS OF BENZENE RING

(2) **Electrophilic Substitution.** As already observed, $-NO_2$ group in nitrobenzene exerts *meta* directing and *deactivating* effect on electrophilic substitution in the benzene ring. Thus nitrobenzene can be nitrated with a mixture of conc. nitric acid and sulfuric acid at 90°C to give *m*-dinitrobenzene.

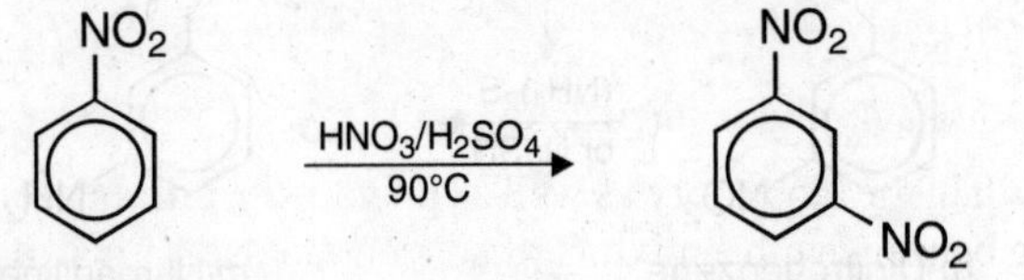

Nitrobenzene m-Dinitrobenzene

Sulfonation of nitrobenzene can be carried with fuming sulfuric acid at 100°C to give *m*-nitrobenzenesulfonic acid. **Aromatic nitro compounds do not undergo Friedel-Crafts reactions.**

MECHANISM. Following steps are involved :

Step 1. Formation of the electrophile (NO_2^+).

$$HNO_3 + 2H_2SO_4 \longrightarrow NO_2^+ + 2HSO_4^- + H_3O^+$$

Step 2. The electrophile attacks the benzene ring to give a carbonium ion.

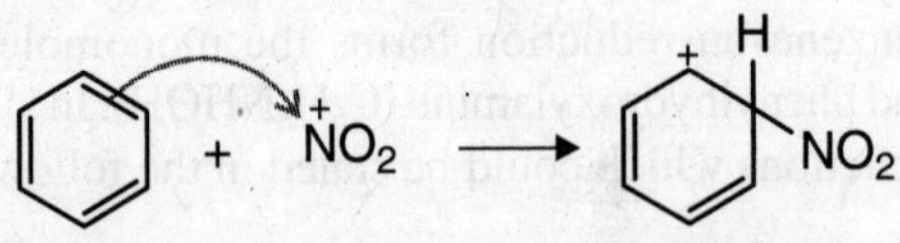

Benzene Carbonium ion

Step 3. Removal of proton gives nitrobenzene.

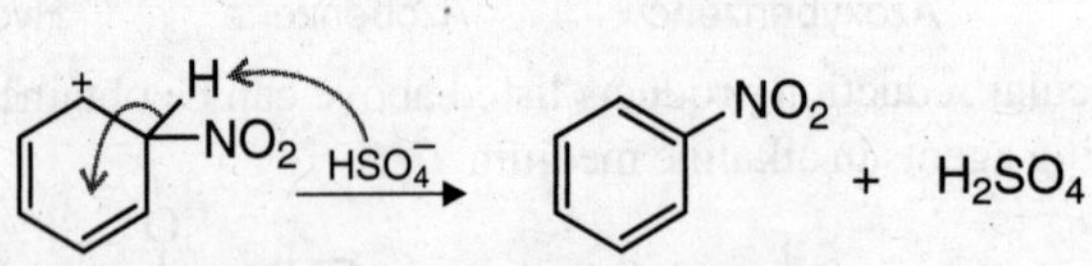

Nitrobenzene

(3) **Nucleophilic Substitution.** The *ortho* and *para* positions of nitrobenzene are attacked by nucleophiles relatively easily. Thus when fused with KOH, nitrobenzene gives *o*-nitrophenol through nucleophilic attack of OH on the ring.

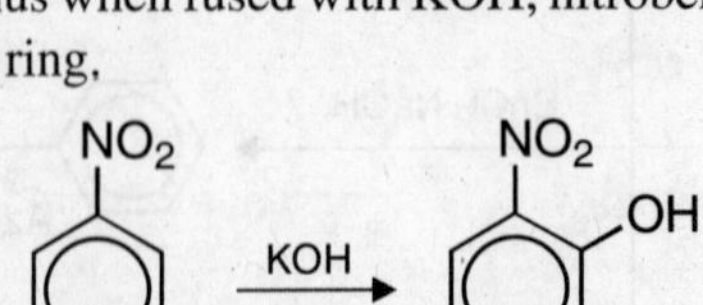

o-Nitrophenol

If a second $-NO_2$ group is present on the benzene ring of nitrobenzene in the *ortho* or *para* position, it undergoes nucleophilic displacement.

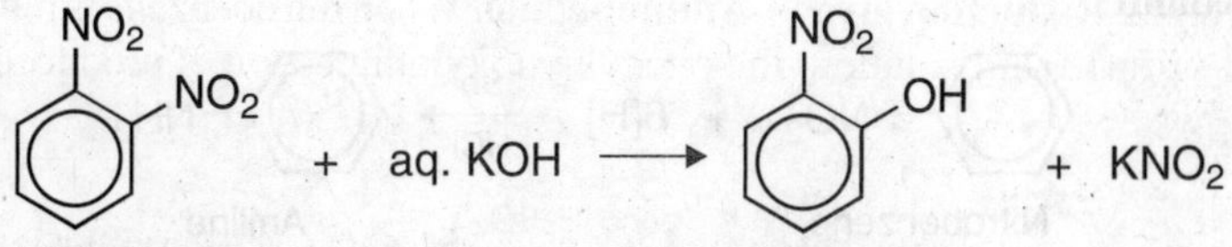

o-Dinitrobenzene　　　　*o*-Nitrophenol

USES OF NITRO COMPOUNDS

Aromatic nitro compounds are used : (1) as solvents on account of their high polarity ; (2) for the synthesis of aromatic amino compounds since they are easily available by direct nitration of benzene and other aromatic compounds ; (3) as explosives, particularly TNB and TNT.

NITROBENZENE, $C_6H_5NO_2$

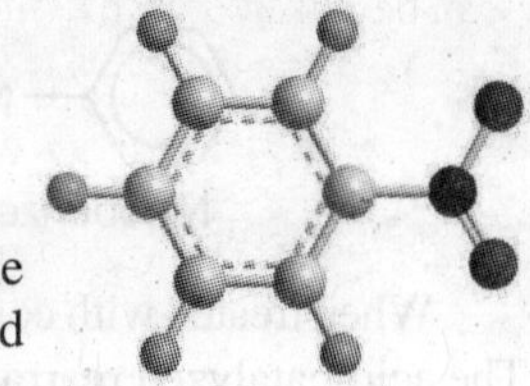

It is the most important aromatic nitro compound.

Preparation. Both in the laboratory as well as in industry, nitrobenzene is prepared by heating benzene with a mixture of concentrated nitric acid and sulfuric acid at 50°C.

"Now then . . . tell this court what happened when you told your teacher that the *nitro* group is an *ortho-para* director"

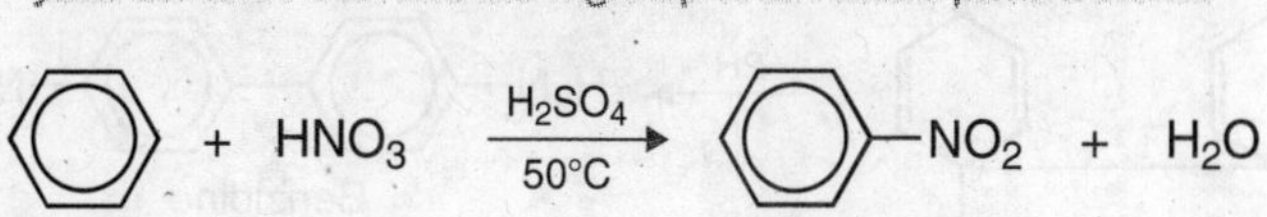

Benzene　　　　Nitrobenzene

The resulting nitrobenzene is separated, washed with sodium carbonate solution to remove any acid, and purified by distillation using an air condenser. The laboratory procedure and the one used on a large scale differ only in working details.

Properties. (***Physical***) Nitrobenzene is a pale yellow liquid (bp 211°C ; sp. gr. 1.087) having odor of bitter almonds. It is insoluble in water but dissolves readily in most organic solvents. It is steam-volatile. Nitrobenzene is highly toxic.

(***Chemical***) The chemical reactions of nitrobenzene are those of the nitro group and the phenyl group. These have already been discussed. The only important reaction of $-NO_2$ group is that of reduction in acid medium, neutral medium or alkaline medium.

(1) **Acid medium**

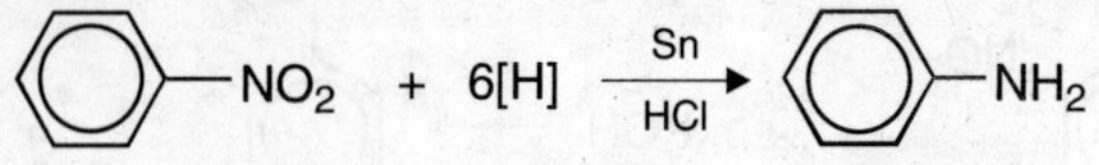

Nitrobenzene　　Aniline (as salt)

(2) **Neutral medium**

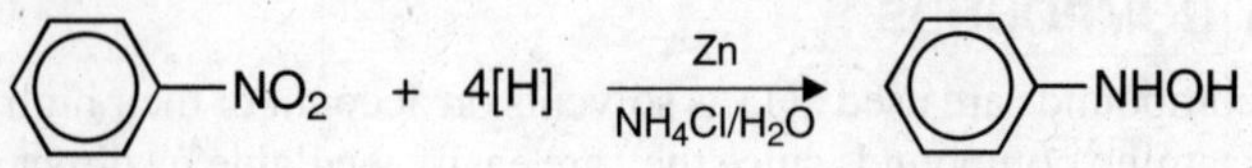

Nitrobenzene　　Phenylhydroxylamine

(3) **Alkaline medium**

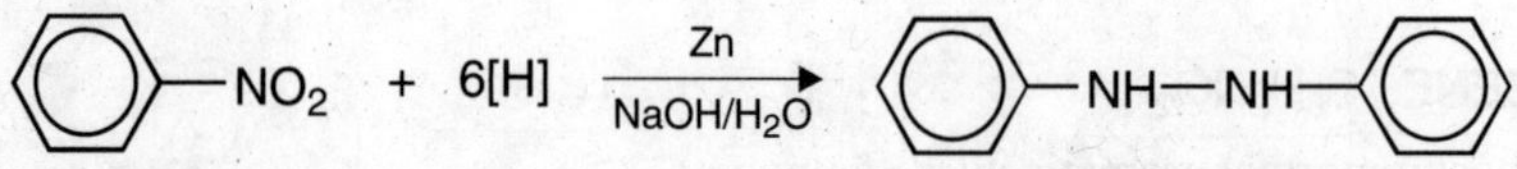

Nitrobenzene　　Hydrazobenzene

When treated with concentrated HCl, hydrazobenzene undergoes rearrangement to give benzidine. The acid-catalyzed rearrangement of hydrazo compounds is called **Benzidine Rearrangement**. For example,

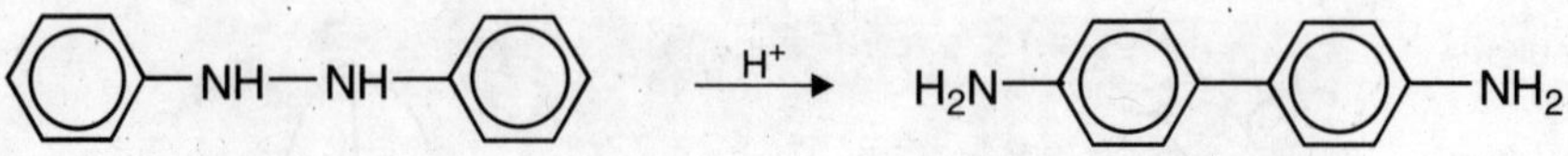

Hydrazobenzene　　Benzidine

MECHANISM. It involves the formation of a bond between the *para* positions of the two rings, and cleavage of the nitrogen-nitrogen bond. Following four steps are involved :

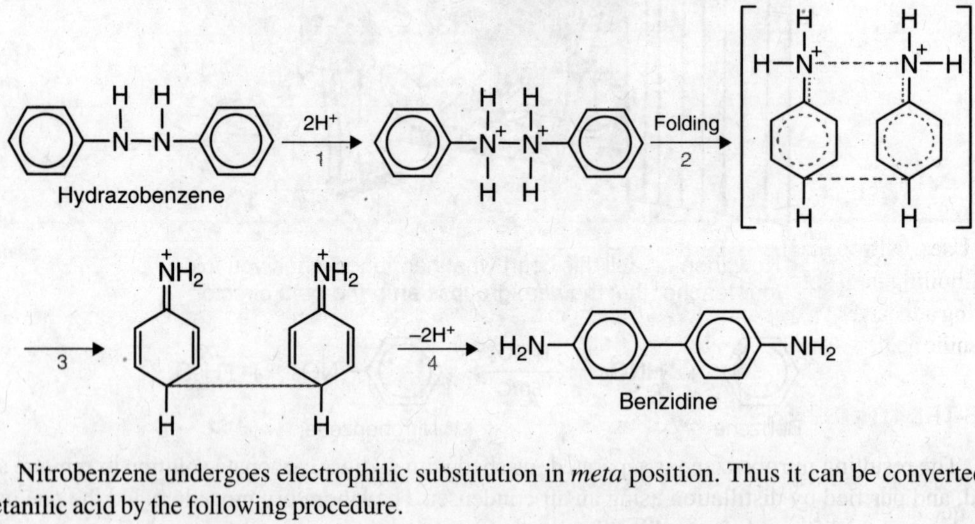

Nitrobenzene undergoes electrophilic substitution in *meta* position. Thus it can be converted to metanilic acid by the following procedure.

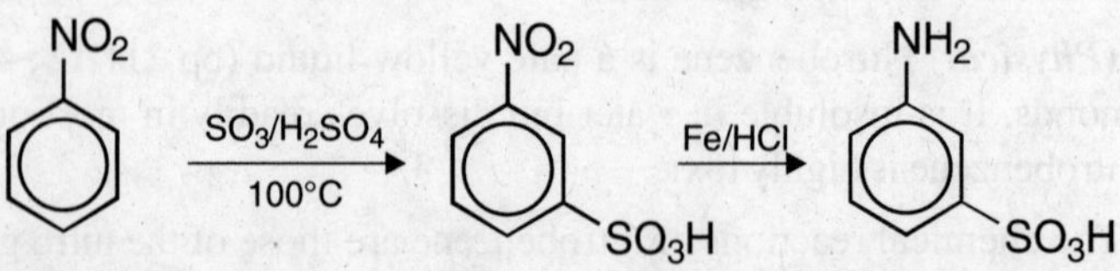

Nitrobenzene　　Metanilic acid

Both benzidine and metanilic acid are dye intermediates.

Table 37.1 REDUCTION PRODUCTS OF NITROBENZENE AND THEIR CONVERSIONS

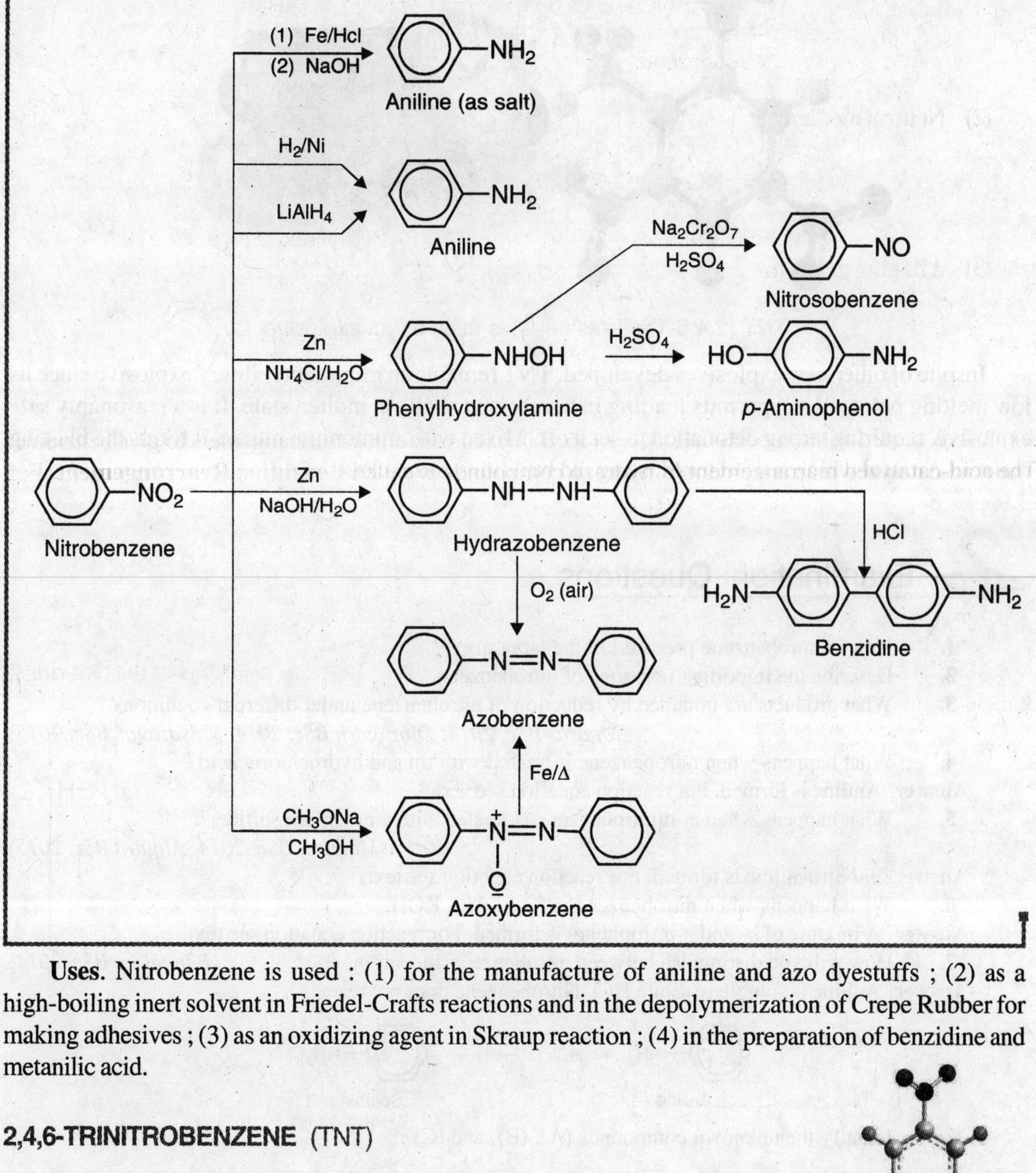

Uses. Nitrobenzene is used : (1) for the manufacture of aniline and azo dyestuffs ; (2) as a high-boiling inert solvent in Friedel-Crafts reactions and in the depolymerization of Crepe Rubber for making adhesives ; (3) as an oxidizing agent in Skraup reaction ; (4) in the preparation of benzidine and metanilic acid.

2,4,6-TRINITROBENZENE (TNT)

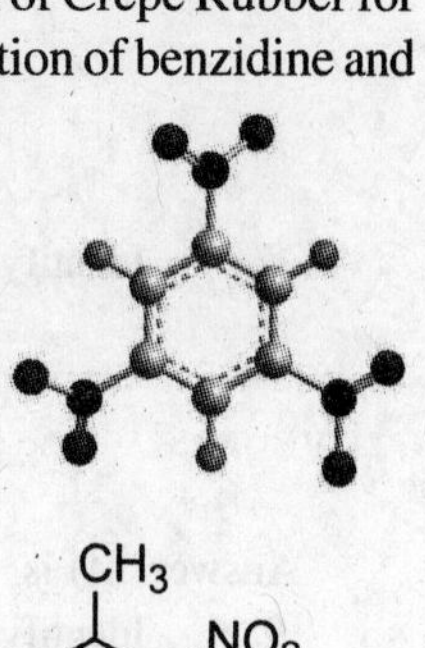

It is made by treating toluene with a mixture of nitric acid and sulfuric acid in a tank at 140°C, then at 180°C, and finally at 230°C.

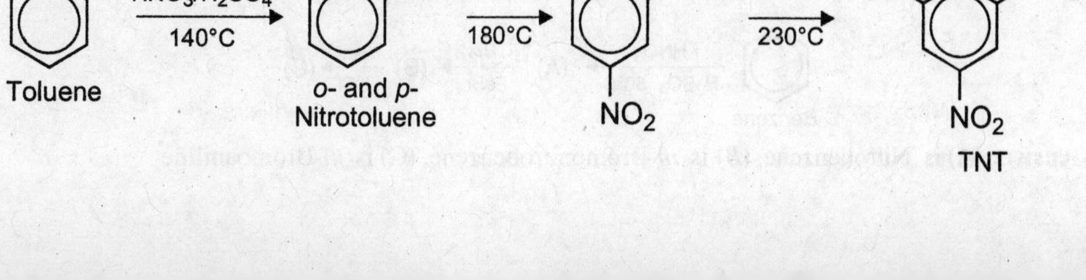

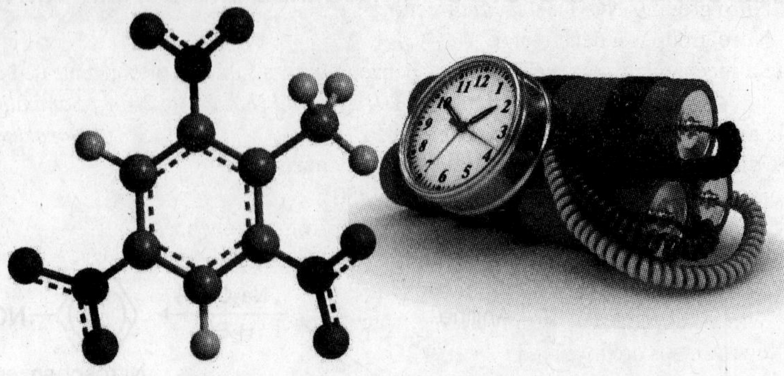

TNT (2,4,6-Trinitrotoluene) is used as an explosive.

Inspite of other new explosives developed, TNT remains an important military explosive since its low melting point (81°C) permits loading in bombs and shells in molten state. It is a reasonably safe explosive, requiring strong detonation to set it off. Mixed with ammonium nitrate, it forms the blasting material **Amatol** which is used for blasting old buildings and rocks.

Examination Questions

1. How is nitrobenzene prepared in the laboratory ?
2. Describe the important reactions of nitrobenzene.
3. What products are obtained by reduction of nitrobenzene under different conditions ?
(Madras BSc, 2014; Dibrugarh BSc, 2014; Vidyasagar BSc, 2015)
4. What happens when nitrobenzene is treated with tin and hydrochloric acid ?

Answer. Aniline is formed. For reaction equation see text.

5. What happens when *m*-dinitrobenzene is treated with ammonium sulfide ?
(Madurai BSc, 2014; Aligarh BSc, 2015)

Answer. *m*-Nitroaniline is formed. For reaction equation see text.

6. What happens when nitrobenzene is fused with KOH.

Answer. A mixture of *o*- and *p*-nitrophenol is formed. For reaction equation see text.

7. How will you distinguish between nitrobenzene and aniline ? *(Bangalore BSc, 2014)*

Answer. Aniline is soluble in dilute HCl. Nitrobenzene does not react.

Aniline Soluble salt

8. Identify the unknown compounds (A), (B), and (C) :

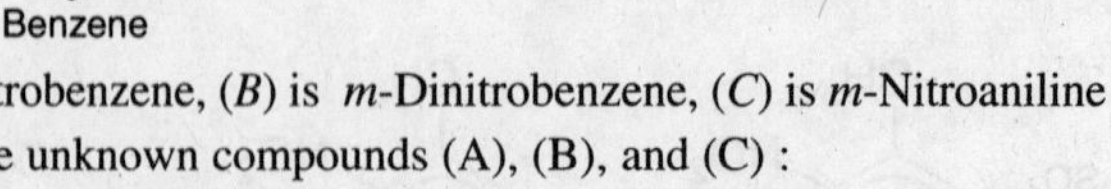

Benzene

Answer. *(A)* is Nitrobenzene, *(B)* is *m*-Dinitrobenzene, *(C)* is *m*-Nitroaniline

9. Identify the unknown compounds (A), (B), and (C) :

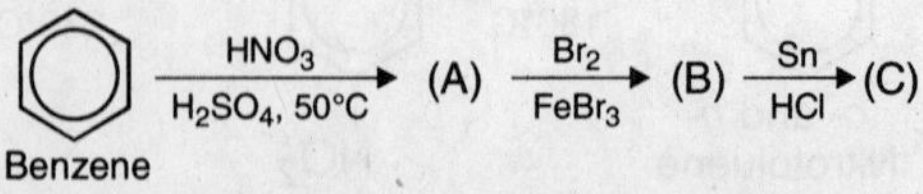

Benzene

Answer.*(A)* is Nitrobenzene, *(B)* is *m*-Bromonitrobenzene, (C) is *m*-Bromoaniline

10. Explain the following facts :
(*a*) Nitro group (–NO_2) is a *meta* director.
(*b*) Nitro group is a deactivator.

11. Write a mechanism for the reaction of benzene with a mixture of concentrated nitric acid and sulfuric acid. (*Mizoram BSc, 2014 ; Tamil Nadu BSc, 2014 ; Sambalpur BSc, 2015*)

12. Write a note on : Benzidine Rearrangement. (*Gorakhpur BSc, 2014*)

13. How will you synthesize *p*-nitrotoluene from benzene ?
Clue. Steps are : (*i*) $Cl_2/AlCl_3$; (*ii*) $HNO_3/H_2SO_4/50°C$.

14. How will you synthesize phenylhydroxylamine from nitrobenzene ?

15. How will you synthesize benzidine from nitrobenzene ?

16. How will you synthesize *TNT* from benzene ? (*Agra BSc, 2015*)

17. How will you synthesize *m*-nitroaniline from nitrobenzene ? (*Annamalai BSc, 2015*)
Answer. Following steps are involved :

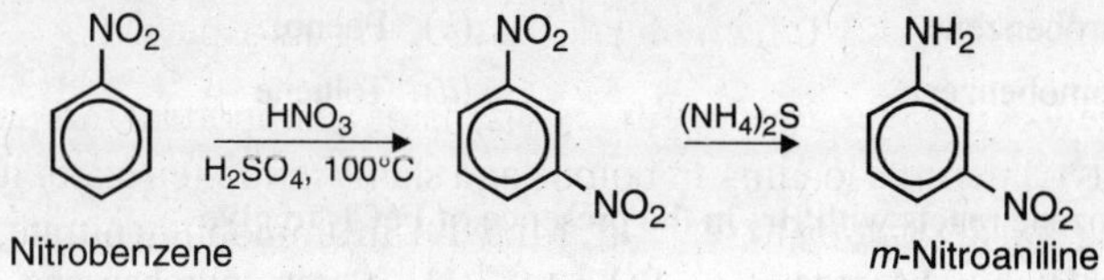

18. How will you synthesize *p*-aminophenol from nitrobenzene ? (*Burdwan BSc, 2014*)
Answer. Following steps are involved :

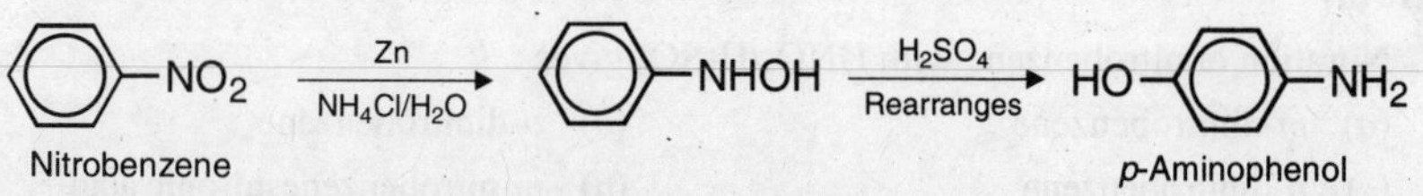

19. How will you synthesize azobenzene from nitrobenzene ? (*Bhopal BSc, 2014*)
Answer. Following steps are involved :

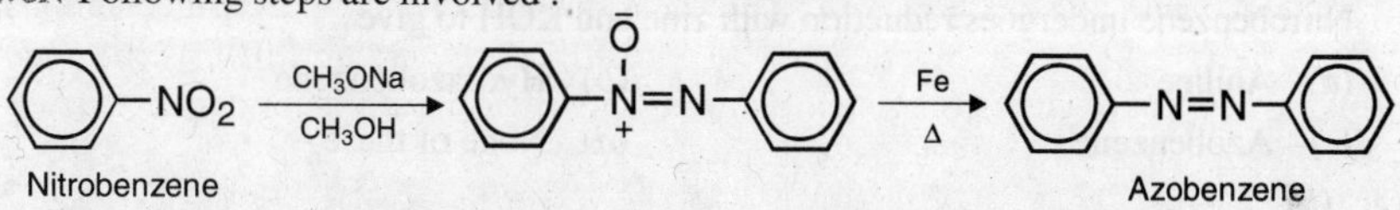

20. How will you synthesize 1,3,5-trinitrobenzene (*TNB*) from toluene ? (*Vikram BSc, 2015*)
Answer. Following steps are involved :

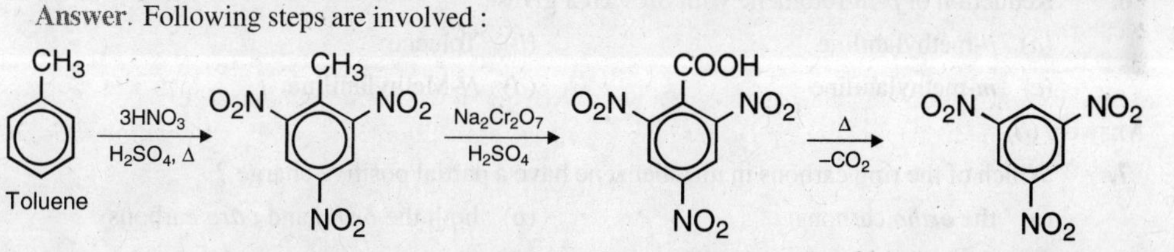

1,3,5-Trinitrobenzene (TNB)

21. A compound $C_7H_5O_6N_3$ (*A*) undergoes oxidation with acidified potassium dichromate to give a mono carboxylic acid (*B*) $C_7H_3O_8N_3$. When (*B*) is heated in acetic acid solution, (*C*) $C_6H_3O_6N_3$ is formed. What are (*A*), (*B*), and (*C*) ? Give equations for the reactions involved.

Answer. (A) = 2,4,6-Trinitrotoluene (*TNT*)
(B) = 2,4,6-Trinitrobenzoic acid
(C) = 1,3,5-Trinitrobenzene (*TNB*)

Reactions involved are :

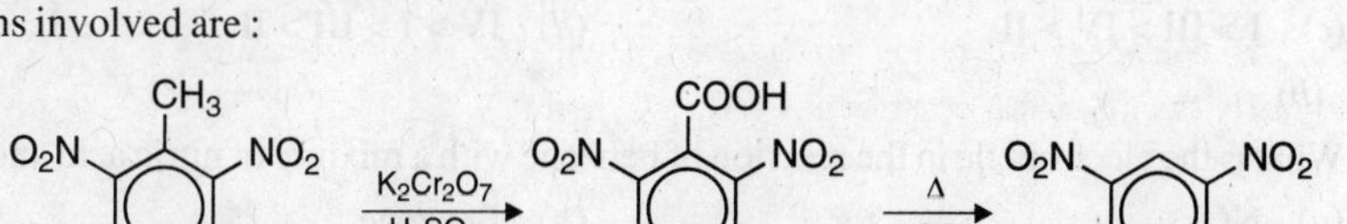

(A) $C_7H_5O_6N_3$ (B) $C_7H_3O_8N_3$ (C) $C_6H_3O_6N_3$

Multiple Choice Questions

1. Which of the following statements is False/Incorrect about nitrobenzene ?

(*a*) is formed by nitration of benzene at temperatures below 60°C.

(*b*) can be further nitrated to give *m*-dinitrobenzene.

(*c*) can be oxidized with $KMnO_4$ to give benzoic acid.

(*d*) can be reduced with Sn/HCl to give aniline.

Answer. (*c*)

2. Which of the following compounds reacts least rapidly in electrophilic substitution reactions ?

(*a*) Nitrobenzene (*b*) Phenol

(*c*) Bromobenzene (*d*) Toluene

Answer. (*a*)

3. Nitrobenzene reacts with Br_2 in the presence of $FeCl_3$ to give

(*a*) *o*-chloronitrobenzene (*b*) *o*-bromonitrobenzene

(*c*) *m*-chloronitrobenzene (*d*) *m*-bromonitrobenzene

Answer. (*d*)

4. Nitration of nitrobenzene with HNO_3/H_2SO_4 gives :

(*a*) *m*-dinitrobenzene (*b*) *o*-dinitrobenzene

(*c*) *p*-dinitrobenzene (*d*) *m*-nitrobenzenesulfonic acid

Answer. (*a*)

5. Nitrobenzene undergoes reduction with zinc and KOH to give

(*a*) Aniline (*b*) Hydrazobenzene

(*c*) Azobenzene (*d*) None of these

Answer. (*b*)

6. Reduction of *p*-nitrotoluene with Sn + HCl gives :

(*a*) *p*-methylaniline (*b*) Toluene

(*c*) *m*-methylaniline (*d*) *N*-Methylaniline

Answer. (*a*)

7. Which of the ring carbons in nitrobenzene have a partial positive charge ?

(*a*) the *ortho* carbons (*b*) both the *ortho* and *para* carbons

(*c*) the *meta* carbons (*d*) the *para* carbons

Answer. (*b*)

8. What is the order of reactivity of the following compounds in electrophilic substitution ? (more reactive > less reactive)

I. Toluene II. Nitrobenzene III. Benzene IV. Phenol

(*a*) I > II > III > IV (*b*) IV > III > II > I

(*c*) I > III > IV > II (*d*) IV > I > III > II

Answer. (*b*)

9. What is the electrophile in the reaction of benzene with a mixture of nitric acid and sulfuric acid ?

(*a*) NO_2^+ (*b*) Benzene

(*c*) NO^+ (*d*) NO_2^-

Answer. (*a*)

10. What product will be obtained from the reaction of nitrobenzene with ethyl chloride and aluminium trichloride ?

(*a*) there will be no reaction (*b*) 3-Ethylnitrobenzene

(*c*) 2-Ethylnitrobenzene (*d*) 4-Ethylnitrobenzene

Answer. (*a*)

11. Predict the major product of the following reaction.

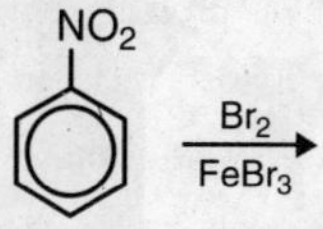

(*a*) *o*-Bromonitrobenzene and *p*-bromonitrobenzene

(*b*) *m*-Bromonitrobenzene

(*c*) *o*-Bromaniline and *p*-bromoaniline

(*d*) *m*-Bromoaniline

Answer. (*b*)

12. Give the product of the following reaction

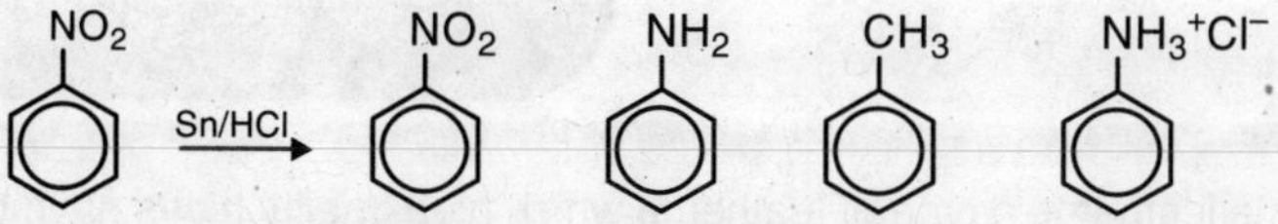

(a) (b) (c) (d)

Answer. (*d*). The nitrobenzene is reduced by Sn/HCl, the product before it is treated with base is the ionic salt, aniline hydrochloride. This compound is a solid.

13. A C_8H_{10} hydrocarbon is nitrated by HNO_3 and sulfuric acid. Two, and only two, $C_8H_9NO_2$ isomers are obtained. Which of the following fits this evidence ?

(*a*) Ethylbenzene (*b*) *ortho*-Xylene

(*c*) *meta*-Xylene (*d*) *para*-Xylene

Answer. (*b*)

14. Which reaction sequence would be best for preparing 3,5-dibromoaniline from nitrobenzene ?

(*a*) *i.* H_2/Pt *ii.* Br_2 in ether

(*b*) *i.* excess $Br_2/FeBr_3$ *ii.* H_2

(*c*) *i.* H_2/Pt *ii.* H_2SO_4, heat *iii.* excess HBr

(*d*) *i.* H_2SO_4, heat *ii.* excess $Br_2/FeBr_3$, heat *iii.* 3 H_2/Pt

Answer. (*b*)

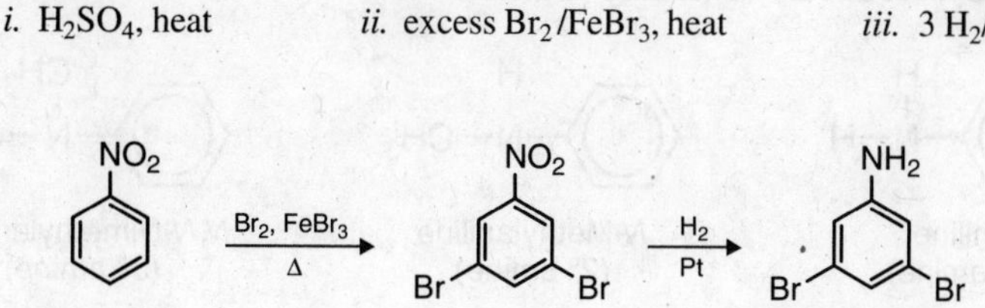

Nitrobenzene 3,5-Dinitroaniline

38 Aromatic Amines

Aniline leather is a type of leather in which high quality hides have been treated with aniline as a dye. This produces a delicate, soft, supple leather.

These are the derivatives of aromatic hydrocarbons in which a hydrogen of the benzene ring has been replaced by an amino group, $-NH_2$.

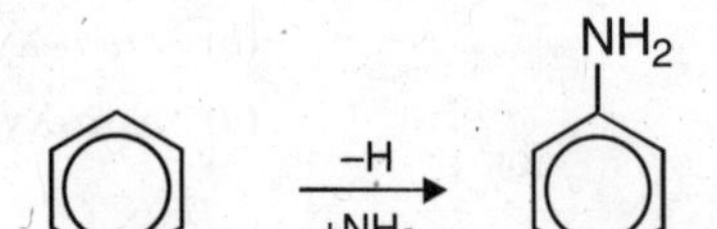

Benzene → Aniline

All such compounds in which an amino or substituted amino group is bonded directly to an aromatic ring are termed **Aromatic Amines**. Like their aliphatic counterparts, aromatic amines could be designated as primary, secondary or tertiary amines.

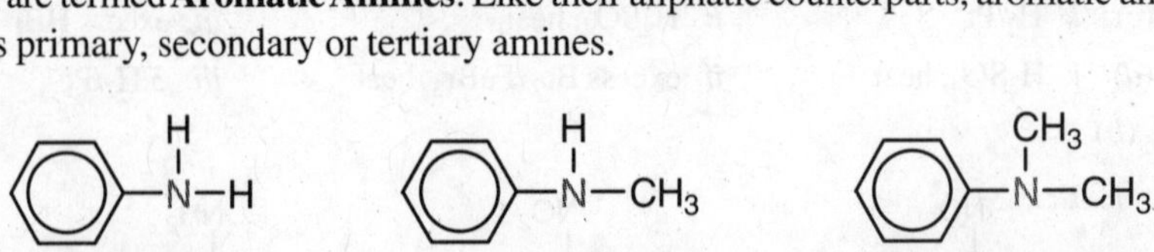

Aniline (1° amine) | *N*-Methylaniline (2° amine) | *N,N*-Dimethylaniline (3° amine)

The derivatives of arenes in which NH_2 is bonded to a side-chain are to be regarded as arylsubstituted aliphatic amines.

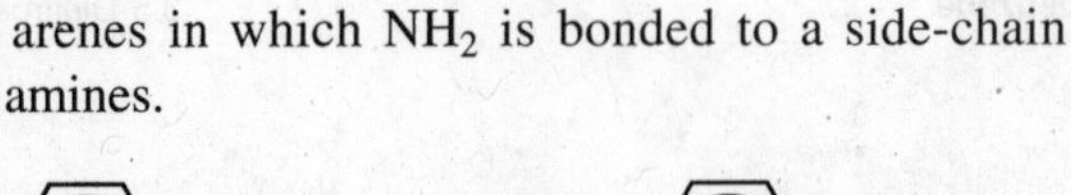

Toluene → Benzylamine

NOMENCLATURE

Simple aromatic amines are named as derivatives of aniline. An aminotoluene is given the special name of *toluidine*. Substituents on the nitrogen atom are named as prefixes preceded by *N*-, or *N,N*-.

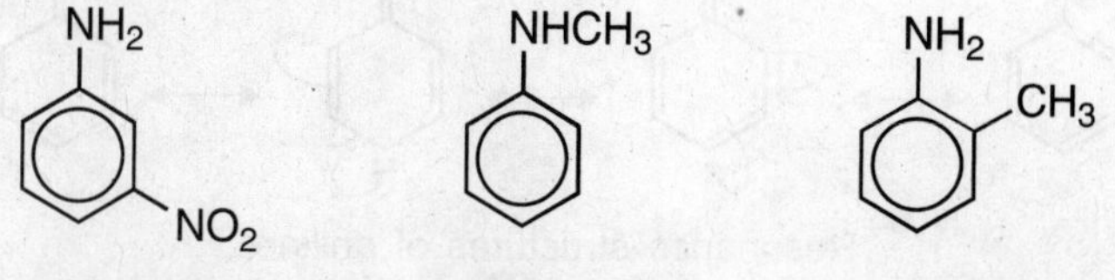

m-Nitroaniline *N*-Methylaniline *o*-Toluidine

STRUCTURE

Consider the example of aniline to descuss the molecular orbital/structure of aromatic amines. In aniline, all ring carbon atoms are sp^2 hybridized and the nitrogen atom of the $-NH_2$ group is sp^3 hybridized.

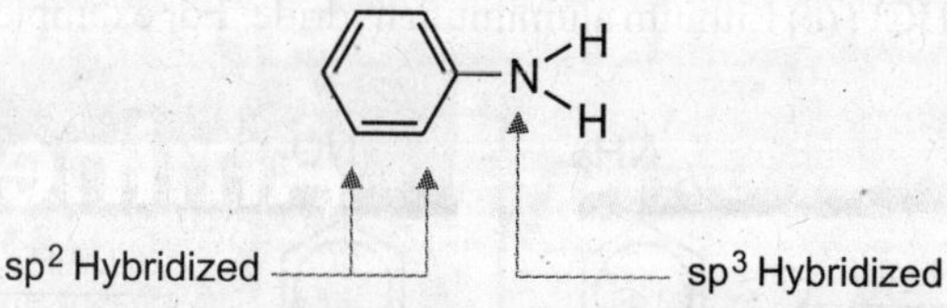

Figure 38.1 shows the formation of sigma (σ) bonds in aniline.

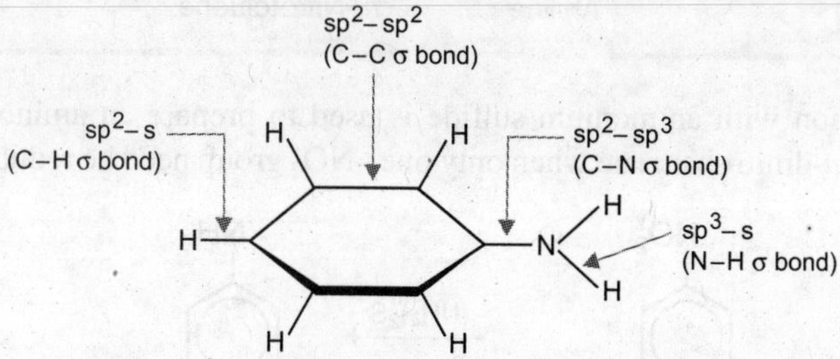

Fig.38.1. Formation of sigma (σ) bonds in aniline.

All ring carbon atoms also have a *p* orbital and these are perpendicular to the plane of σ bonds of the ring. The lateral overlap of these *p* orbitals produces a delocalized π molecular orbital.

The lone-pair electrons on the nitrogen atom resides in the sp^3 hybrid orbital. These lone-pair electrons on the nitrogen atom in the $-NH_2$ group overlap with the delocalized π molecular orbital of the benzene ring.

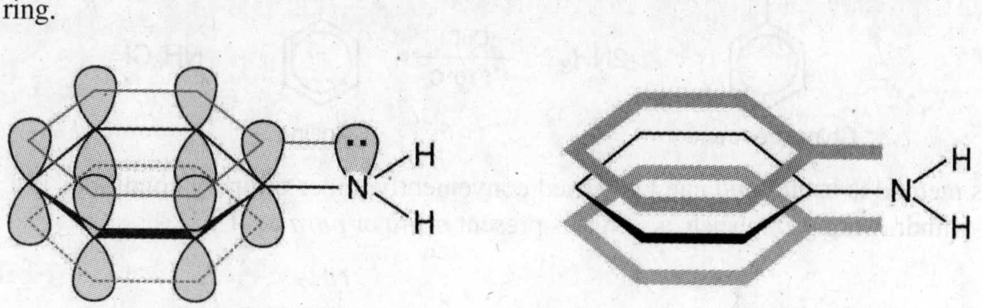

Fig.38.2. Adjacent *p* orbitals overlap to form the delocalized π molecular orbital. The lone-pair electrons on the nitrogen atom reside in sp^3 orbital and interact with the delocalized π molecular orbital to form an extended π molecular orbital containing eight electrons.

According to the resonance theory, aniline is considered to be hybrid of the following resonance structures :

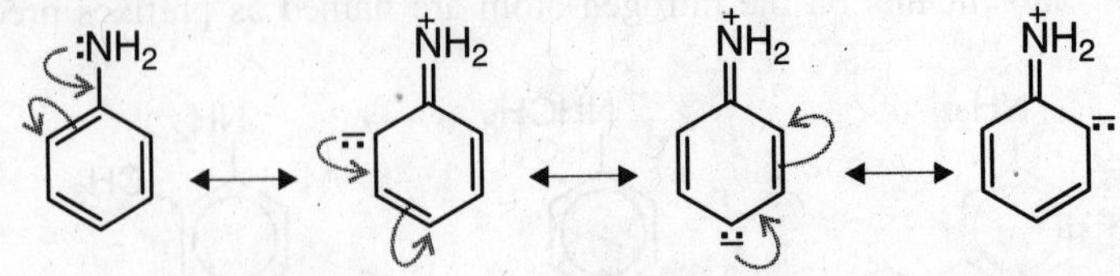

Resonance structures of aniline

(The lone-pair of electrons on nitrogen is delocalized on the benzene ring)

METHODS OF PREPARATION

(1) **Reduction of Nitro Compounds.** This is a very convenient and most widely used method of preparing aromatic amines. The reduction is carried with (*a*) H_2 in the presence of Ni, Pd or Pt as catalyst ; (*b*) Sn or Fe and HCl ; (*c*) Lithium aluminium hydride. For example,

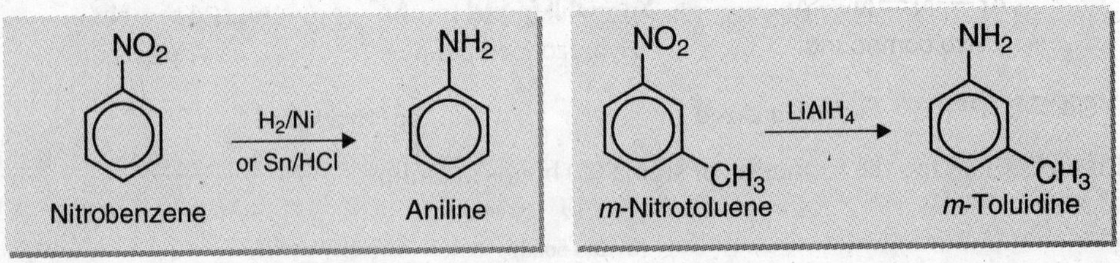

Selective reduction with ammonium sulfide is used to prepare an amino compound such as *m*-nitroaniline from *m*-dinitrobenzene when only one $-NO_2$ group need be reduced.

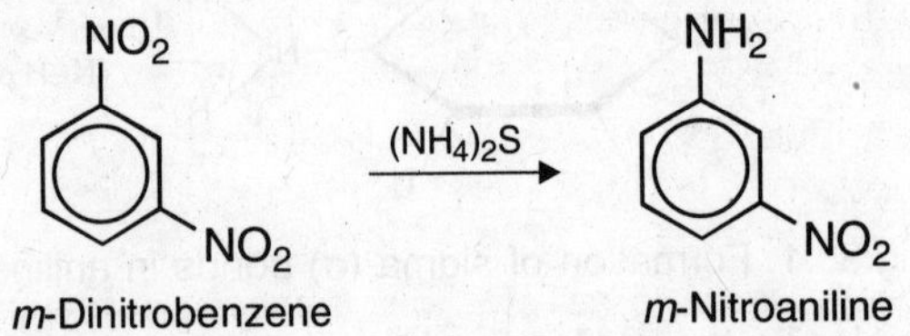

(2) **Ammonolysis of Aryl Chlorides.** Aniline is prepared by treating chlorobenzene with ammonia in the presence of copper salts at high temperature and pressure.

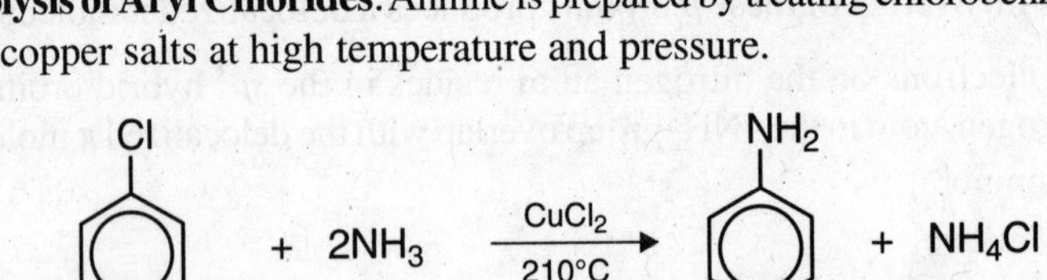

This method is useful and can be carried conveniently under ordinary conditions if a strongly electron-withdrawing group such as $-NO_2$ is present *ortho* or *para* to $-Cl$.

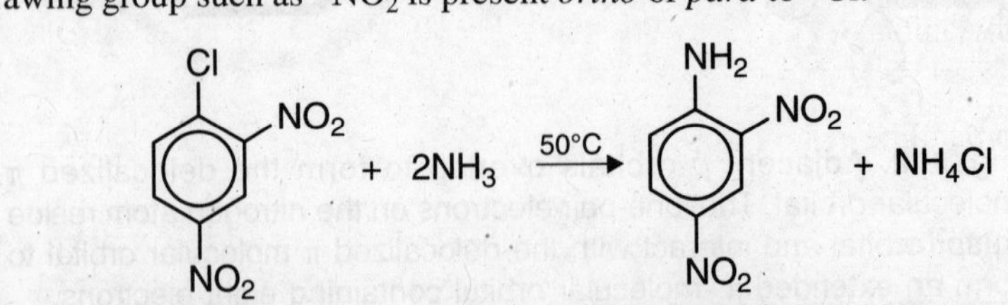

(3) **Hofmann Rearrangement**. When treated with bromine or chlorine in an alkaline solution, an aryl amide undergoes Hofmann rearrangement to yield a primary aromatic amine.

$$\underset{\text{Amide}}{Ar-\overset{O}{\overset{\|}{C}}-NH_2} \xrightarrow[(Br_2+NaOH)]{NaOBr} \underset{1^\circ \text{ amine}}{Ar-NH_2} + CO_2$$

For example,

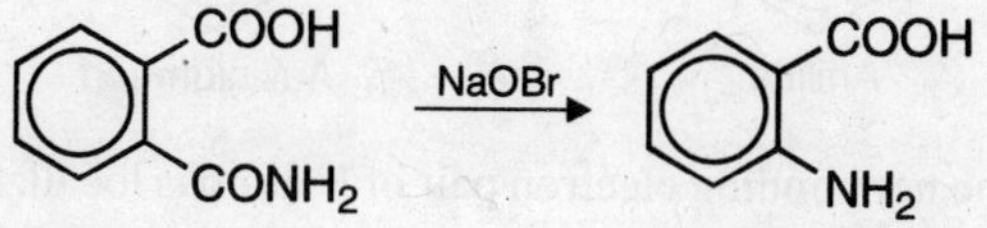

o-Aminobenzoic acid

(4) **Reduction of Azo Compounds**. Azo compounds when reduced with hydrogen in presence of nickel as catalyst, give arylamines.

$$\underset{\text{Azo compound}}{Ar-N=N-Ar} \xrightarrow[Ni,\ \Delta]{H_2} \underset{\text{Hydrazo compound}}{Ar-NH-NH-Ar} \xrightarrow[Ni,\ \Delta]{H_2} \underset{\text{Arylamine}}{2Ar-NH_2}$$

PHYSICAL PROPERTIES

(1) Aromatic amines in general are colorless liquids or solids, having a characteristic odor which is not pleasant. They turn brown in air owing to oxidation.

(2) On account of their high molecular weights and intermolecular hydrogen bonding ability, their boiling points are relatively high.

(3) They are sparingly soluble in water but dissolve in benzene and other organic solvents.

(4) They are steam-volatile and, therefore, can be purified by steam distillation.

(5) They are highly toxic substances. Some of them are carcinogens (cancer causing).

(6) In UV spectra, aniline shows two absorption maxima at 230 and 280 mμ. In IR spectra, primary aromatic amines give bands in the 650–900 and 1500–1640 cm^{-1} region.

BASICITY OF AMINES

Table 38.1 shows that aromatic amines are much less basic in comparison to aliphatic amines.

Table 38.1 pK_b VALUES OF SOME AROMATIC AND ALIPHATIC AMINES. HIGHER pK_b VALUE MEANS LOWER BASICITY.

Amine	Structure	pK_b
Methylamine	CH_3NH_2	3.36
Dimethylamine	$(CH_3)_2NH$	3.27
Trimethylamine	$(CH_3)_3N$	4.19
Aniline	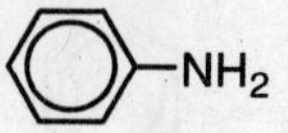	9.37
p-methylaniline		8.92
p-chloroaniline	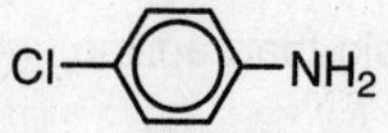	9.85
p-nitroaniline		13.0

The basicity of a compound is a measure of a compound's ability to accept a proton (H^+). Aniline accepts a proton as follows :

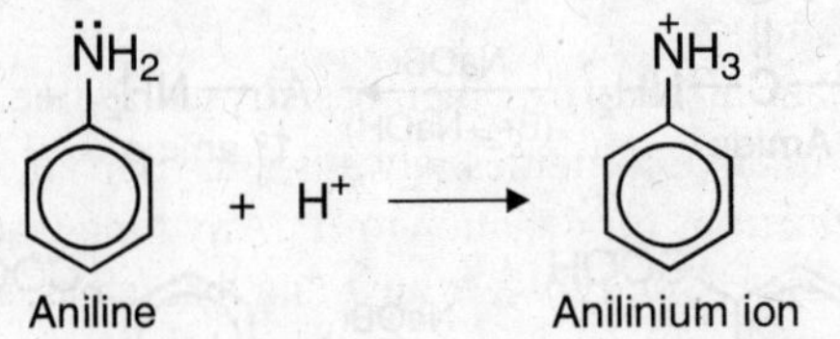

In aliphatic amines, the nonbonding electron pair of N atom is localized and is fully available for coordination with a proton. On the other hand, in aromatic amines, the nonbonding electron pair is delocalized into the benzene ring by resonance.

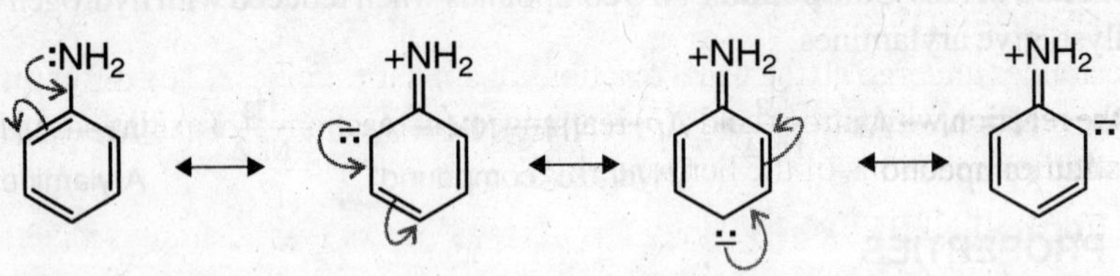

Thus, we have

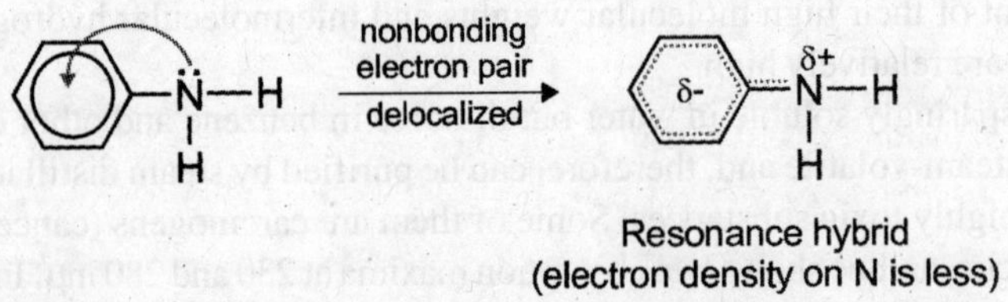

The nonbonding electron pair in the hybrid being dispersed over the benzene ring as also the N atom, is less available for protonation. Hence aromatic amines are less basic than aliphatic amines which show no resonance.

o-Nitroaniline is even more weakly basic than aniline because the electron-withdrawing effect on the ring lowers the electron density on the N atom, making the electron pair even less available for protonation. In *p*-toluidine the CH_3 group is electron-donating and increases the electron density on the N atom, making the electrons available for protonation more than in aniline. Hence toluidine is more basic relative to aniline.

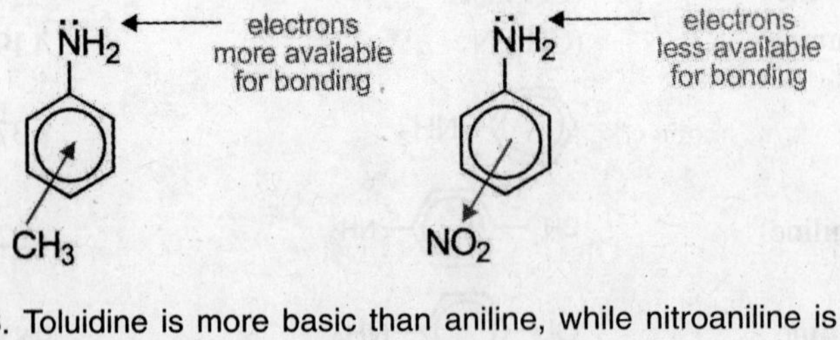

Fig. 38.3. Toluidine is more basic than aniline, while nitroaniline is less so.

Diphenylamine, C_6H_5–NH–C_6H_5, is a weaker base (pK_b = 12) than aniline (pK_b = 9.37) . It is because it contains two phenyl groups instead of one. The nonbonding electron pair of N is still less available as it is distributed over two phenyl groups.

Points to Remember

Basicity of Aromatic Amines

1. Aromatic (and aliphatic amines) owe their basicity to the lone pair of electrons on nitrogen.
2. In determining the relative basicities, always ask yourself about the availability of the nitrogen lone pair electrons for donation to H^+. Anything that withdraws electron density from nitrogen decreases basicity, while anything that releases electron density to nitrogen increases basicity.
3. Electron-releasing groups ($-CH_3$, $-OCH_3$, $-NH_2$) increase the basicity of aromatic amines.
4. Electron-withdrawing groups ($-NO_2$, $-Br$, $-COOH$) decrease the basicity of aromatic amines.

CHEMICAL PROPERTIES

Aromatic amines undergo all the usual reactions of aliphatic amines. They differ from the latter in respect of (*a*) the reaction with nitrous acid ; (*b*) rearrangement reactions ; (*c*) oxidations ; (*d*) electrophilic aromatic substitution reactions of the benzene ring.

(1) **Salt Formation**. Being basic, they form crystalline salts with strong mineral acids such as HCl or H_2SO_4.

$$\mathrm{ArNH_2 + HCl \longrightarrow Ar\overset{+}{N}H_3\overset{-}{Cl}}$$

Arylammonium chloride

(2) **Alkylation**. They react with alkyl halides to form 2° and 3° amines. The amines so formed react with HX (one of the products) to form salts. These salts can be treated with aq NaOH to give free 2° or 3° amines.

$$\underset{1°\text{ amine}}{\mathrm{ArNH_2}} + \mathrm{RX} \longrightarrow \underset{2°\text{ amine}}{\mathrm{Ar{-}NHR}} + \mathrm{HX} \longrightarrow \mathrm{Ar{-}\overset{+}{N}H_2R\,\overset{-}{X}} \xrightarrow{\mathrm{NaOH}} \underset{2°\text{ amine}}{\mathrm{Ar{-}NHR}}$$

$$\underset{2°\text{ amine}}{\mathrm{Ar{-}NHR}} + \mathrm{RX} \longrightarrow \underset{3°\text{ amine}}{\mathrm{Ar{-}NR_2}} + \mathrm{HX} \longrightarrow \mathrm{Ar{-}\overset{+}{N}HR_2\,\overset{-}{X}} \xrightarrow{\mathrm{NaOH}} \underset{3°\text{ amine}}{\mathrm{Ar{-}NR_2}}$$

(3) **Acylation**. With acid halides and anhydrides, they form *N*-Arylamides.

$$\mathrm{ArNH_2} + \underset{\text{Acetyl chloride}}{\mathrm{CH_3{-}\overset{O}{\overset{\|}{C}}{-}Cl}} \longrightarrow \mathrm{ArNH{-}\overset{O}{\overset{\|}{C}}{-}CH_3 + HCl}$$

$$\mathrm{ArNH_2} + \underset{\text{Acetic anhydride}}{\mathrm{(CH_3{-}\overset{O}{\overset{\|}{C}})_2O}} \longrightarrow \mathrm{ArNH{-}\overset{O}{\overset{\|}{C}}{-}CH_3 + CH_3COOH}$$

(4) **Reaction with Aldehydes**. They react with aldehydes to form *imines* (*Schiff's Bases*).

$$\mathrm{ArNH_2 + RCH{=}O} \longrightarrow \underset{\text{Imine}}{\mathrm{Ar{-}N{=}CH{-}R}} + \mathrm{H_2O}$$

(5) **Carbylamine Reaction**. 1° Aromatic amines react with ethanolic potassium hydroxide and chloroform to form *carbylamines* or *isonitriles* which have a disagreeable odor. (*Test*)

$$ArNH_2 + CHCl_3 + 3KOH \xrightarrow{\Delta} \underset{\text{Carbylamine}}{ArNC} + 3KCl + 3H_2O$$

(6) **Reaction with Nitrous acid**. (*a*) 1° Aromatic amines undergo diazotization to give diazonium salts. (*Test*)

$$ArNH_2 + HONO \xrightarrow[0\text{–}5°C]{HCl} Ar\overset{+}{N_2}\overset{-}{Cl} + 2H_2O$$

Aliphatic 1° amines do not form diazonium salts.

(*b*) 2° Amines undergo *N*-nitrosation.

$$Ar{-}N(H){-}R + HONO \xrightarrow{HCl} \underset{\text{Nitrosoamine}}{Ar{-}N(N{=}O){-}R} + H_2O$$

(*c*) 3° Amines undergo *C*-nitrosation.

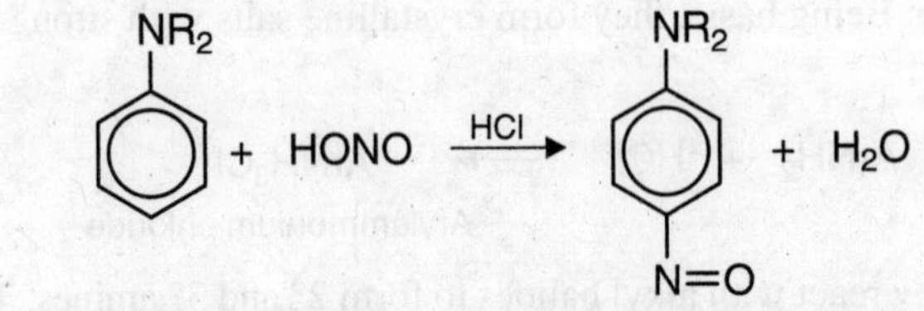

3° amine $\qquad$ *N,N*-Dialkyl-*p*-nitrosoaniline

(7) **Oxidation**. Aromatic amines are readily oxidized to give products depending on conditions. For example, aniline on oxidation with potassium dichromate and sulfuric acid gives *p*-benzoquinone.

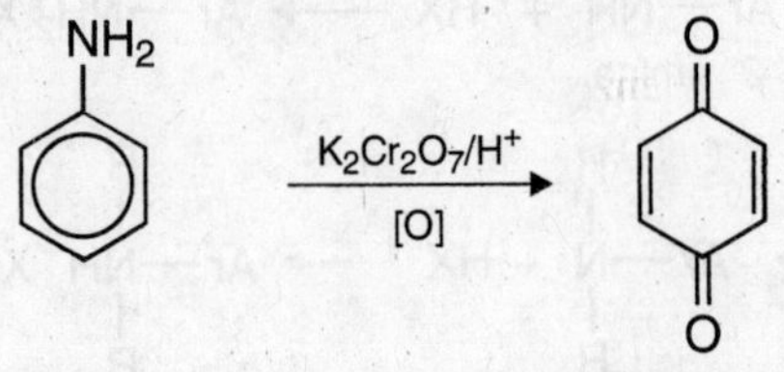

Aniline $\qquad$ *p*-Benzoquinone

Pertrifluoroacetic acid (CF_3CO_3H) oxidizes the $-NH_2$ group to $-NO_2$ group.

$$Ar{-}NH_2 \xrightarrow{CF_3CO_3H} Ar{-}NO_2$$

(8) **Hofmann-Martius Rearrangement**. With *N*, *N*-dialkylanilines in strong acid media at 400°C, intermolecular migration occurs.

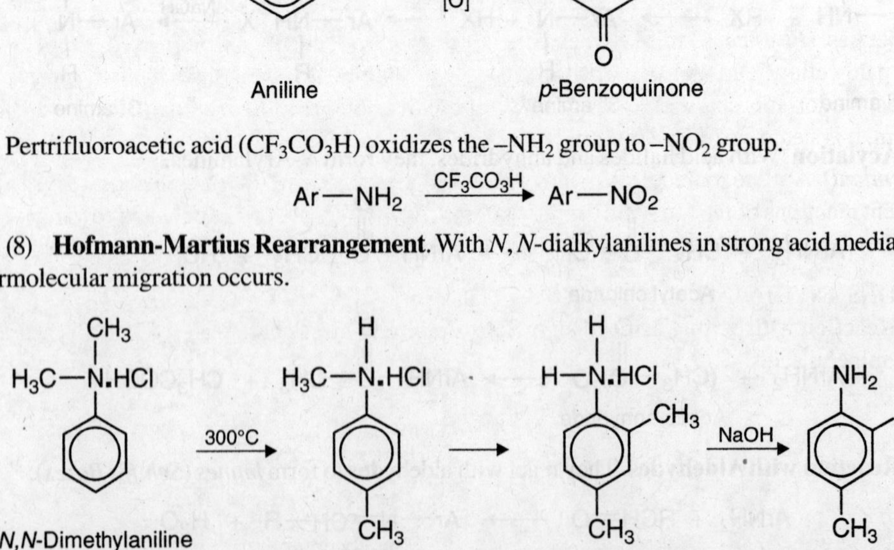

N,N-Dimethylaniline hydrochloride $\qquad$ 2,4-Dimethylaniline

(9) **Electrophilic Aromatic Substitution**. Aromatic amines are highly activated towards substitution in the ring by electrophilic reagents. The $-NH_2$ group directs the new entrant to the *ortho* and *para* positions. (For discussion see Chapter *34*).

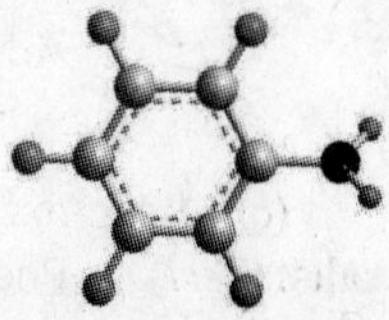

ANILINE, *Phenylamine, Aminobenzene*, $C_6H_5NH_2$

It is the most important aromatic amine and is the best representative of this class of compounds.

Laboratory Preparation. Aniline is prepared in the laboratory by reducing nitrobenzene with tin and concentrated hydrochloric acid.

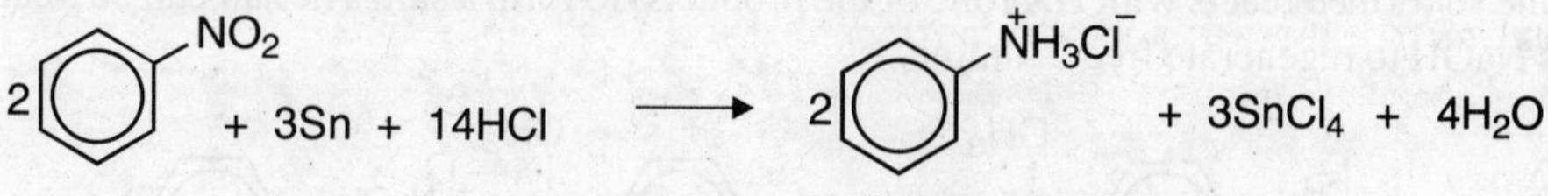

Anilinium chloride

The acid salt of aniline is then treated with sodium hydroxide solution to liberate aniline.

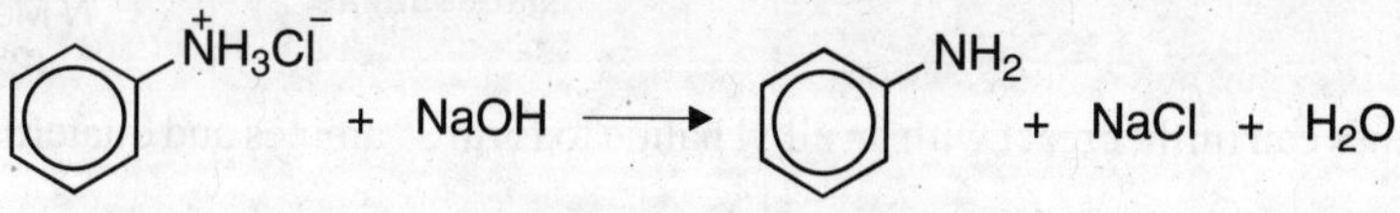

Aniline

Aniline is separated from the mixture by steam distillation.

Manufacture. Aniline is manufactured by three methods :

(1) By vapor-phase catalytic reduction of nitrobenzene.

$$C_6H_5NO_2 + 3H_2 \xrightarrow[270°C]{Cu/SiO_2} C_6H_5NH_2 \text{ (Aniline)} + H_2O$$

(2) By reduction of nitrobenzene with iron and dil. HCl.

$$C_6H_5NO_2 \xrightarrow[\Delta]{Fe/HCl} C_6H_5\overset{+}{N}H_3\overset{-}{Cl} \xrightarrow{Ca(OH)_2} C_6H_5NH_2$$

(3) By ammonolysis of chlorobenzene.

$$C_6H_5Cl + 2NH_3 \xrightarrow[250°C,\ 50\ atm]{Cu_2O} C_6H_5NH_2 + NH_4Cl$$

Properties. (*Physical*). Aniline when pure is a colorless oil, bp 184.4°C, density 1.002 at 20°C. It becomes pale yellow, and then rapidly darkens on exposure to air owing to oxidation. It has a faint characteristic odor. It is sparingly soluble in water but dissolves readily in ethanol, diethyl ether and chloroform. It is steam-volatile. Aniline is toxic.

(*Chemical*). Aniline molecule is made of one NH_2 group bonded to benzene ring. It gives all the replacement reactions of the aliphatic primary amines, and differs only in its behavior towards nitrous acid. The benzene ring gives the usual electrophilic substitution reactions in *ortho* and *para* positions.

REACTIONS INVOLVING NH_2 GROUP

(1) **Reaction with Strong acids to form Salts**. Aniline is a weak base ($pK_b = 4.2\times10^{-10}$) and forms salts with mineral acids.

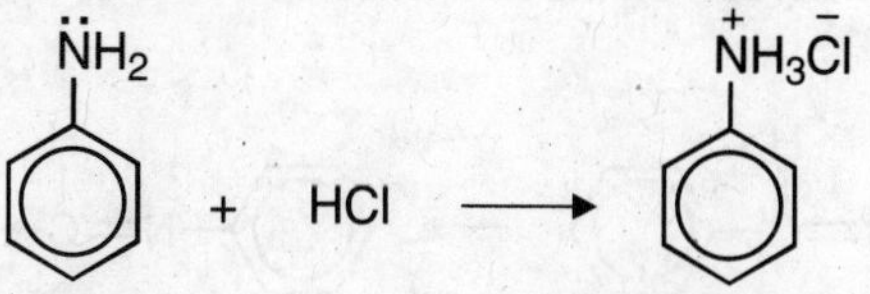

Aniline Phenylammonium chloride

These salts are decomposed by alkalis to regenerate aniline.

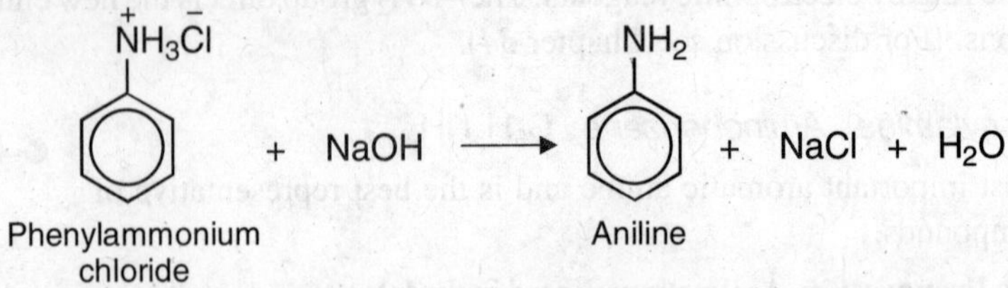

(2) **Reaction with Alkyl halides ; Alkylation.** Aniline reacts with alkyl halides to form 2° amine. The amine so formed reacts with HX (one of the products) to form a salt. The salt can be treated with aqueous NaOH to regenerate free 2° amine.

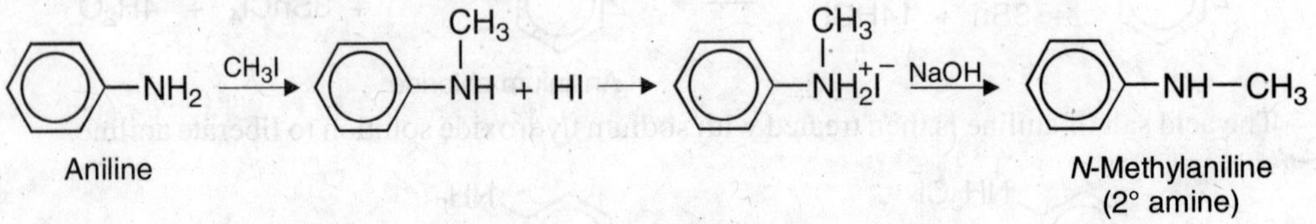

The 2° amines can further react with an alkyl halide to form 3° amines and quaternary ammonium salts.

(3) **Reaction with Acid Chlorides ; Acetylation**. Aniline reacts with acid chlorides (or acid anhydrides) to form the corresponding amides called **Anilides**.

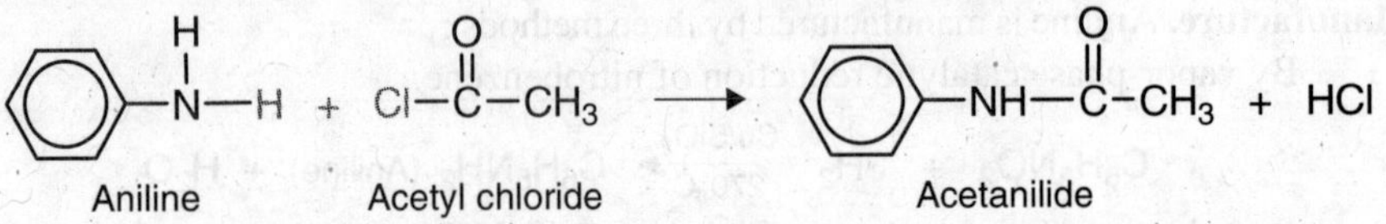

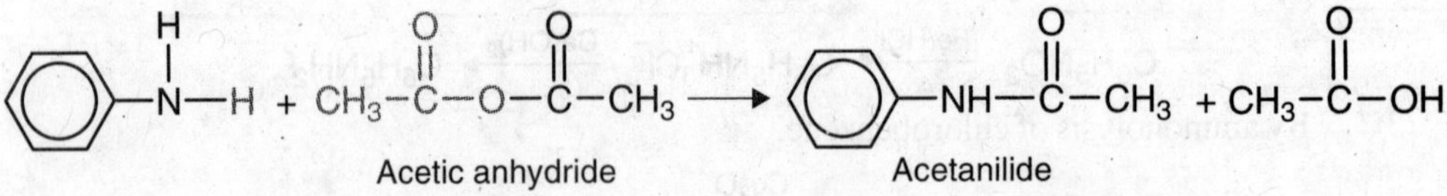

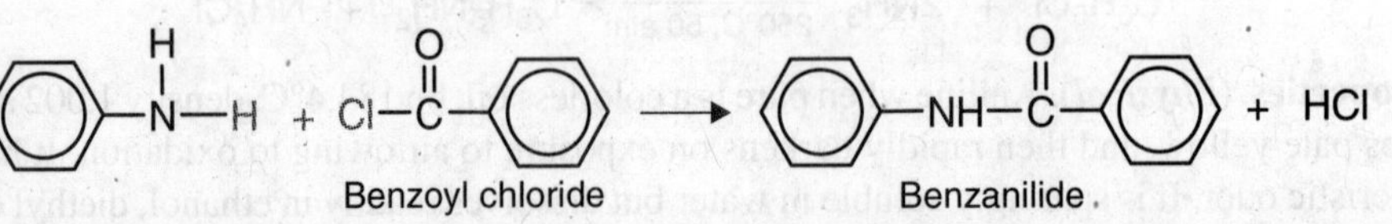

(4) **Reaction with Benzenesulfonyl Chloride**. Aniline reacts with benzenesulfonyl chloride to form *N*-phenylbenzenesulfonamide.

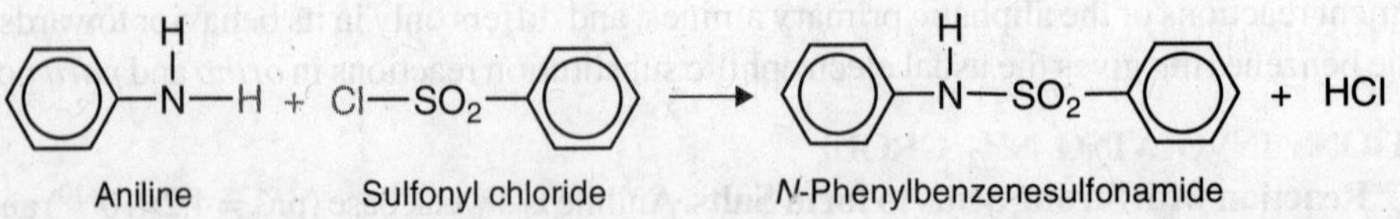

(5) **Reaction with Benzaldehyde**. Aniline condenses with benzaldehyde to give benzylideneaniline (Schiff's base).

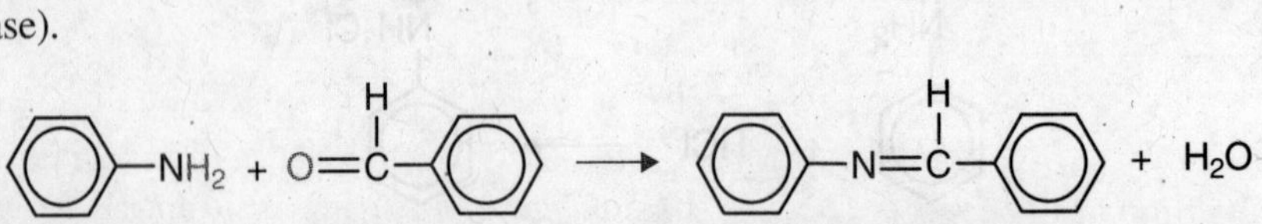

(6) **Carbylamine Reaction**. Aniline reacts with chloroform and alcoholic KOH to give phenyl isocyanide (phenyl carbylamine) which has an offensive smell.

$$\underset{\text{Aniline}}{C_6H_5NH_2} + CHCl_3 + 3KOH \longrightarrow \underset{\text{Phenyl isocyanide}}{C_6H_5NC} + 3KCl + 3H_2O$$

(7) **Reaction with Nitrous Acid**. Aniline reacts with nitrous acid (from $NaNO_2$ + HCl) in the presence of HCl at 0–5°C to give benzenediazonium chloride.

$$NaNO_2 + HCl \longrightarrow HNO_2 + NaCl$$

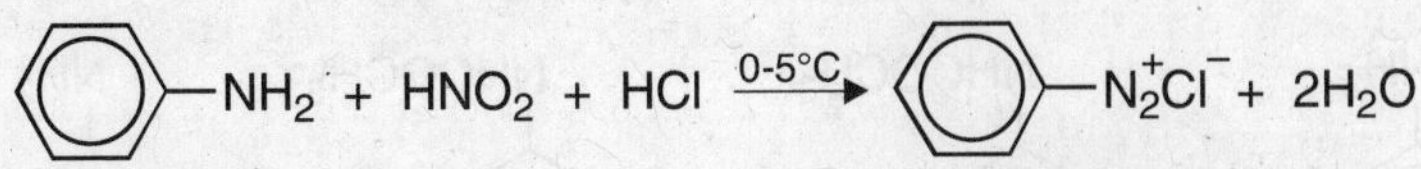

Aniline   Benzenediazonium chloride

(8) **Coupling with Diazonium salts**. Aniline reacts with benzenediazonium chloride to give *p*-aminoazobenzene.

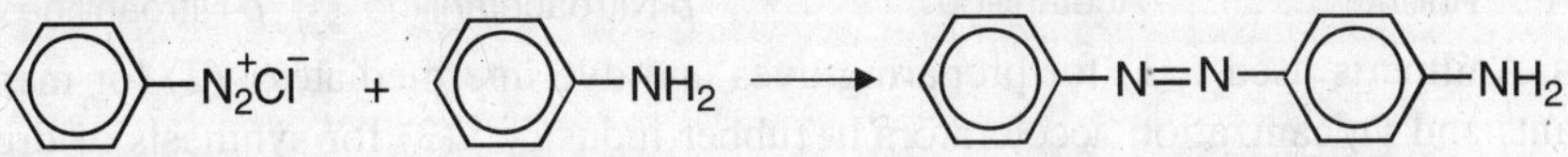

Benzenediazonium chloride   Aniline   *p*-Aminoazobenzene

(9) **Oxidation**. Aniline undergoes oxidation with potassium dichromate to give *p*-benzoquinone.

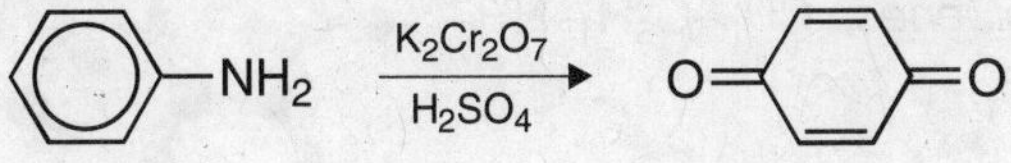

Aniline   *p*-Benzoquinone

SUBSTITUTION REACTIONS OF BENZENE RING

(10) **Halogenation**. Halogenation with bromine or with chlorine is extremely rapid. When bromine water or chlorine water is added to aniline at room temperature, a white precipitate of 2,4,6-tribromo- or 2,4,6-trichloroaniline is immediately formed.

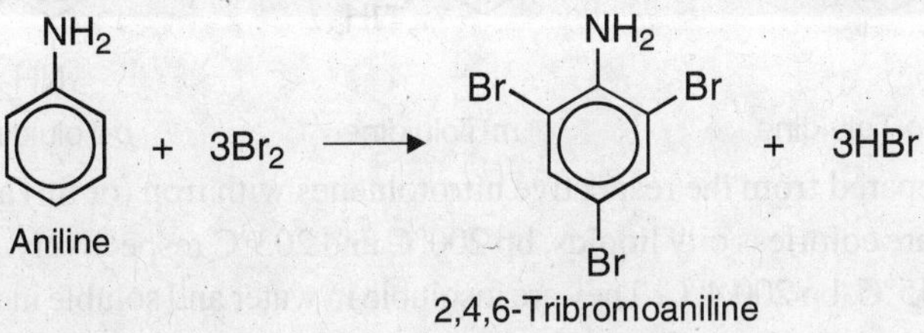

Aniline   2,4,6-Tribromoaniline

(11) **Sulfonation**. On heating with fuming sulfuric acid, aniline is sulfonated to form mainly sulfanilic acid (*p*-aminobenzenesulfonic acid).

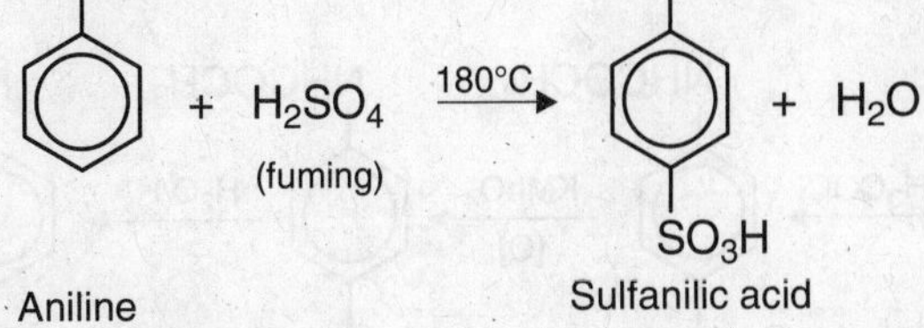

Aniline   Sulfanilic acid

(12) **Nitration**. Aniline reacts with a mixture of concentrated nitric acid and sulfuric acid to give *m*-nitroaniline. The yield is low because of accompanying oxidation.

Why Meta Product ? Nitric acid is a proton acid. It reacts with the amino group to form anilinium ion, $C_6H_5NH_3^+$. Because of the positive charge, the $-NH_3^+$ group is *meta* directing.

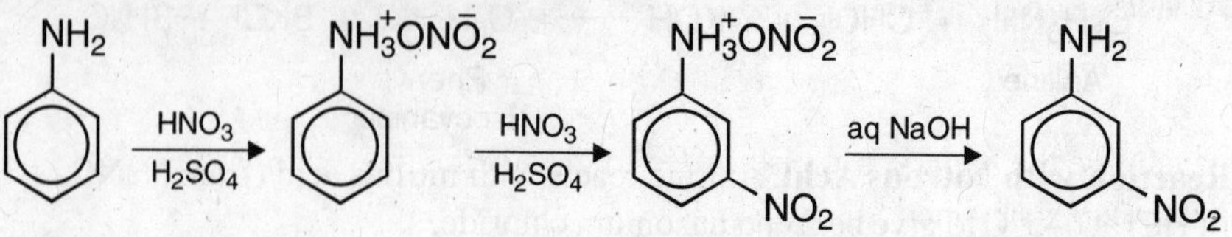

Aniline m-Nitroaniline

p-Nitroaniline can be obtained by first *protecting* the $-NH_2$ group by acylation.

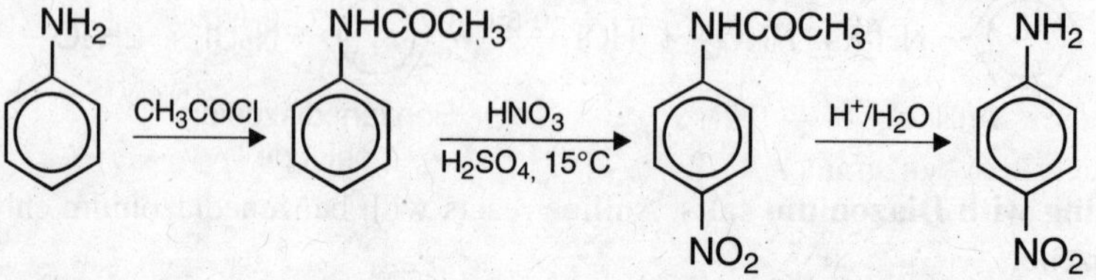

Aniline Acetanilide p-Nitroacetanilide p-Nitroaniline

Uses. Aniline is used : (1) for preparing dyes and dye intermediates ; (2) for manufacture of antioxidants and vulcanization accelerators in rubber industry ; (3) for synthesis of drugs, notably *sulfa drugs* ; (4) for making isocyanates required for polyurethane plastics used for insulation.

TOLUIDINES, *Aminotoluenes*, $CH_3-C_6H_4-NH_2$

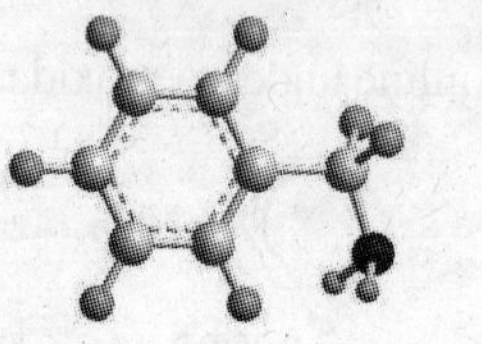

The three aminotoluenes are commonly called *toluidines*.

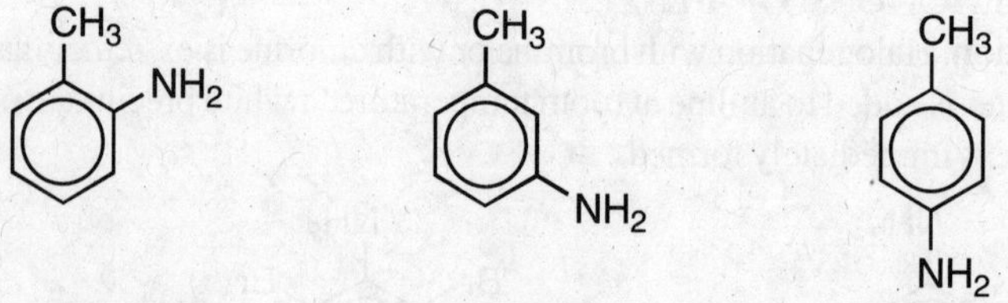

o-Toluidine m-Toluidine p-Toluidine

They may be prepared from the respective nitrotoluenes with iron (or tin) and hydrochloric acid. *o*- and *m*-Toluidines are colorless oily liquids, bp 200°C and 203°C respectively. *p*-Toluidine is a white crystalline solid, mp 45°C, bp 200.4°C. They are insoluble in water and soluble in most organic solvents. At the NH_2 group they give all reactions of aniline.

The oxidation of toluidines after protecting the NH_2 group by acetylation, and subsequent hydrolysis yields the corresponding aminobenzoic acids.

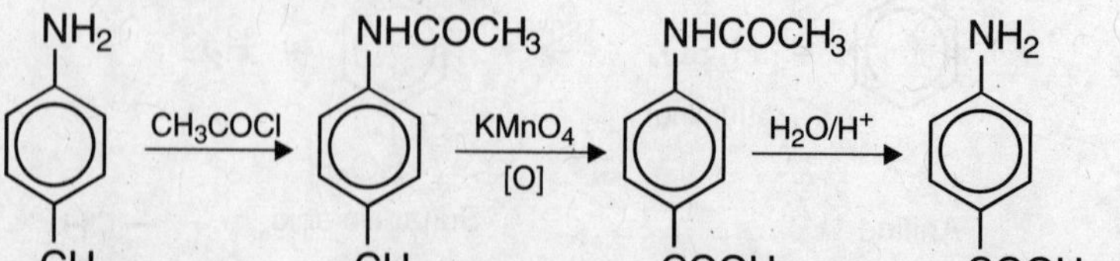

p-Toluidine p-Aminobenzoic acid

BENZYLAMINE, $C_6H_5CH_2NH_2$

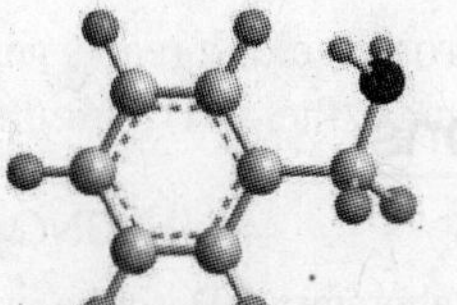

In benzylamine the *amino* group is bonded to the side-chain.

Preparation. It is obtained :

(1) By reduction of benzonitrile.

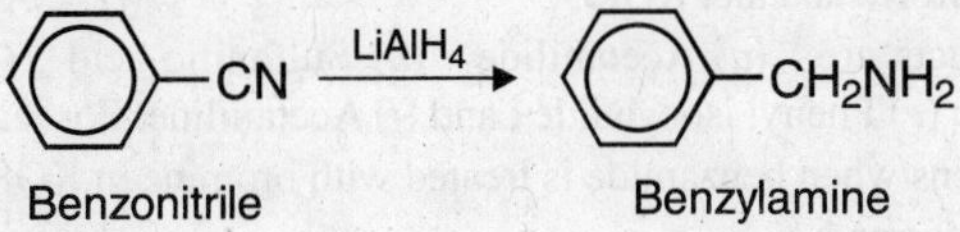

(2) By reduction of benzamide.

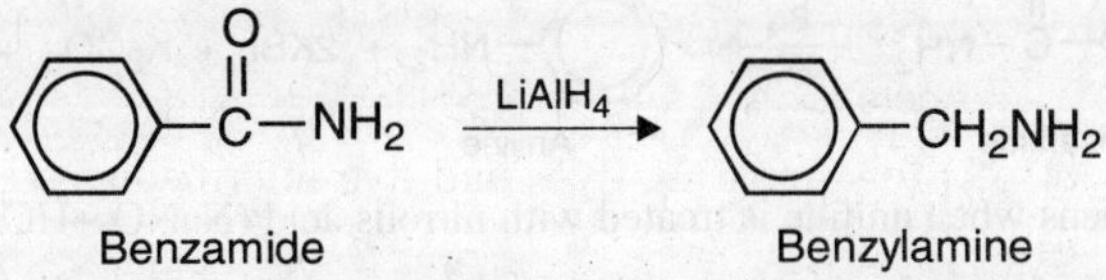

(3) By reduction of benzaldehyde with ammonia and subsequent reduction. (**Reductive Amination**).

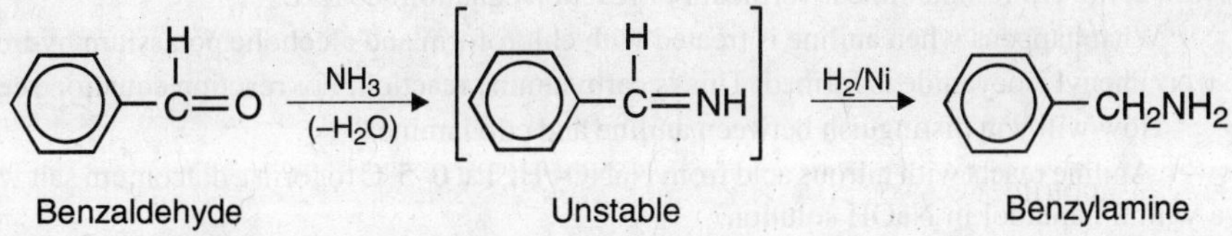

Properties. Benzylamine is a colorless liquid, bp 185°C. It is sparingly soluble in water and is slightly weaker base ($pK_b = 4.67$) than methylamine. It gives the following reactions:

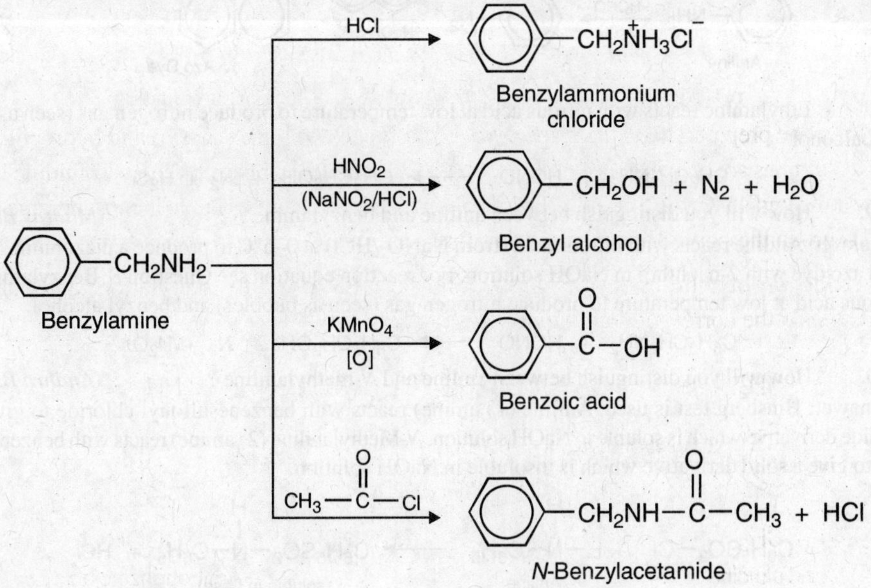

Examination Questions

1. How is aniline prepared in the laboratory ? Describe its important reactions.
2. How is benzylamine prepared ? Describe its important reactions.
3. How does aniline react with

   (*a*) Acetic anhydride (*b*) Conc H_2SO_4 at 180°C

   (*c*) Nitrous acid at 0-5°C (*d*) Bromine water

   (*e*) Chloroform and alc. KOH (*f*) Acetyl chloride

Answer. The products are : (*a*) Acetanilide ; (*b*) Sulfanilic acid ; (*c*) Benzenediazonium chloride ; (*d*) 2,4,6-Tribromoaniline ; (*e*) Phenyl isocyanide ; and (*f*) Acetanilide. For reaction equations see text.

4. What happens when benzamide is treated with bromine in KOH solution ? (*Bihar BSc, 2014*)

Answer. Aniline is formed.

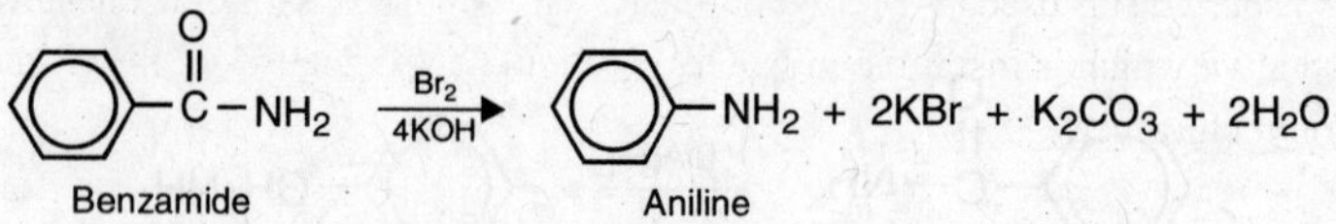

5. What happens when aniline is treated with nitrous acid ($NaNO_2$+HCl) at 0-5°C ? (*Baroda BSc, 2015*)

Answer. Benzenediazonium chloride is formed. For reaction equation see text.

6. What happens when aniline is treated with bromine water ?

Answer. 2,4,6-Tribromoaniline is formed. For reaction equation see text.

7. What happens when aniline is treated with chloroform and alcoholic potassium hydroxide ?

Answer. Phenyl isocyanide is formed. This is carbylamine reaction. For reaction equation see text.

8. How will you distinguish between aniline and ethylamine ?

Answer. Aniline reacts with nitrous acid from $NaNO_2$/HCl at 0–5°C to form a diazonium salt which gives red azo dye with 2-naphthol in NaOH solution.

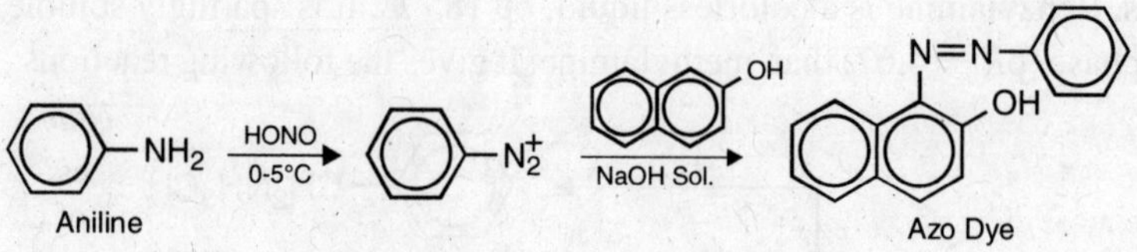

Ethylamine reacts with nitrous acid at low temperature to produce nitrogen gas (seen as bubbles) and ethyl alcohol.

$$CH_3CH_2NH_2 + HONO \longrightarrow CH_3CH_2OH + N_2 + H_2O$$

9. How will you distinguish between aniline and benzylamine ? (*Madras BSc, 2014*)

Answer. Aniline reacts with nitrous acid (from $NaNO_2$/HCl) at 0–5°C to produce a diazonium salt which gives red azo dye with 2-naphthol in NaOH solution. For reaction equation see Question *8*. Benzylamine reacts with nitrous acid at low temperature to produce nitrogen gas (seen as bubbles) and benzyl alcohol.

$$C_6H_5CH_2NH_2 + HONO \longrightarrow C_6H_5CH_2OH + N_2 + H_2O$$

10. How will you distinguish between aniline and *N*-methylaniline ? (*Andhra BSc, 2015*)

Answer. Hinsberg test is used. Aniline (1° amine) reacts with benzenesulfonyl chloride to give a solid sulfonamide derivative which is soluble in NaOH solution. *N*-Methylaniline (2° amine) reacts with benzenesulfonyl chloride to give a solid derivative which is insoluble in NaOH solution.

$$C_6H_5SO_2{-}Cl + \underset{\text{Aniline}}{H{-}\overset{\displaystyle H}{\overset{|}{N}}{-}C_6H_5} \longrightarrow \underset{\text{(soluble in base)}}{C_6H_5SO_2{-}\overset{\displaystyle H}{\overset{|}{N}}{-}C_6H_5} + HCl$$

$$C_6H_5SO_2{-}Cl + H{-}\underset{|}{\overset{CH_3}{N}}{-}C_6H_5 \longrightarrow C_6H_5SO_2{-}\overset{CH_3}{N}{-}C_6H_5 + HCl$$

N-Methylaniline (insoluble in base)

11. How will you distinguish between aniline and *N,N*-dimethylaniline ?

Answer. Hinsberg test is used. Aniline (1° amine) reacts with Benzenesulfonyl chloride to give a solid sulfonamide derivative which is soluble in NaOH solution. *N,N*-Dimethylaniline (3° amine) does not react with Benzenesulfonyl chloride.

$$C_6H_5SO_2{-}Cl + H{-}\overset{H}{N}{-}C_6H_5 \longrightarrow C_6H_5SO_2{-}\overset{H}{N}{-}C_6H_5 + HCl$$

Aniline (soluble in base)

12. How will you distinguish between *N*-methylaniline and *N,N*-dimethylaniline ?

(*Karnataka BSc, 2014 ; Gujarat BSc, 2015*)

Answer. Hinsberg test is used. *N*-Methylaniline (2° amine) reacts with Benzenesulfonyl chloride to give a sulfonamide derivative which is insoluble in NaOH solution. *N,N*-Dimethylaniline (3° amine) does not react with Benzenesulfonyl chloride.

$$C_6H_5SO_2{-}Cl + H{-}\overset{CH_3}{N}{-}C_6H_5 \longrightarrow C_6H_5SO_2{-}\overset{CH_3}{N}{-}C_6H_5 + HCl$$

N-Methylaniline (insoluble in base)

13. Explain why *p*-nitroaniline is less basic than aniline. (*Madurai BSc, 2013*)

Answer. This is because the $-NO_2$ group is electron-withdrawing. It makes the $-NH_2$ group of *p*-nitroaniline more electron-poor than the $-NH_2$ group of aniline. Thus the lone pair of electrons on nitrogen in *p*-nitroaniline is less available for donation to H^+. It is, therefore, less basic.

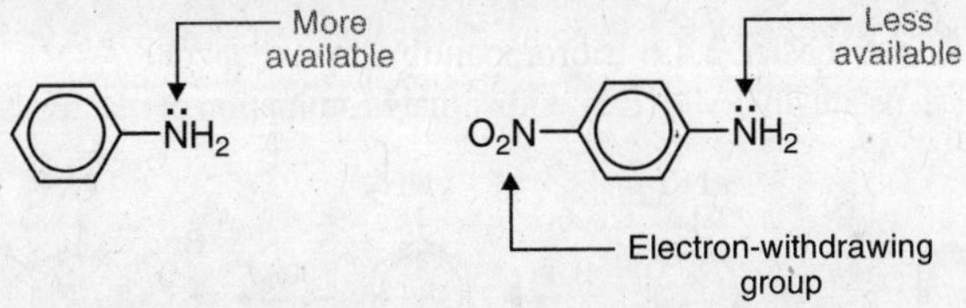

14. Explain why *p*-methylaniline is more basic than aniline. (*Bundelkhand BSc, 2014*)

Answer. This is because the methyl group is electron-donating. It makes the $-NH_2$ group of *p*-methylaniline more electron-rich than the $-NH_2$ group of aniline. Thus the lone pair of electrons on nitrogen in *p*-methylaniline is more available for donation to H^+. It is therefore, more basic.

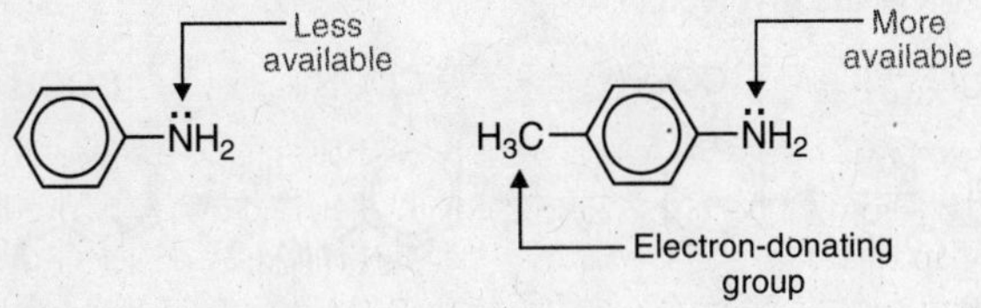

15. Would you expect diphenylamine to be more basic or less basic than aniline ? Explain.

(*North Bengal BSc, 2015*)

Answer. Diphenylamine is less basic than aniline.

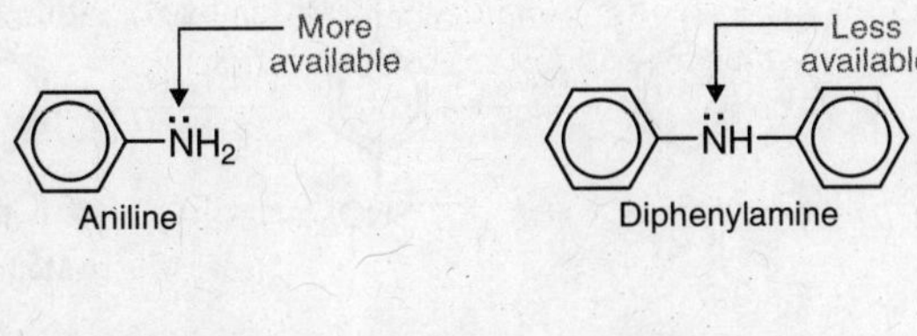

The lone pair of electrons on nitrogen in diphenylamine can overlap with the π electron system of both the aromatic rings. This causes greater delocalization of the lone pair of electrons than that occurs in aniline. Thus the lone pair of electrons on nitrogen in diphenylamine is less available for donation to H^+. It is, therefore, less basic.

16. How will you synthesize aniline from benzene ? *(Aligarh BSc, 2014)*

17. How will you synthesize acetanilide from aniline ?

18. How will you synthesize aniline from benzoic acid ? *(Bhopal BSc, 2015)*

Answer. Following steps are involved :

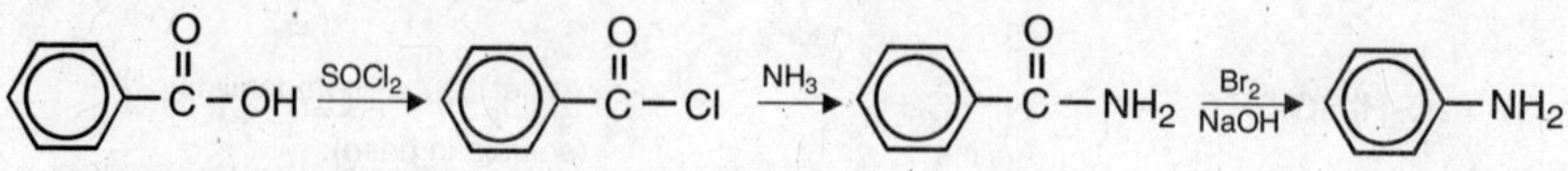

Benzoic acid — Aniline

19. How will you synthesize benzylamine from benzoic acid ?

Answer. Following steps are involved :

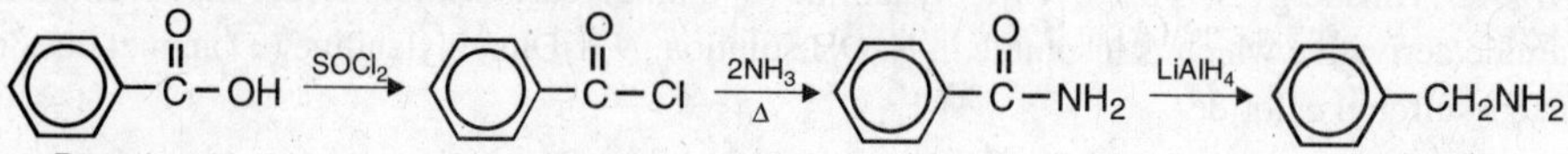

Benzoic acid — Benzylamine

20. How will you synthesize *N,N*-dimethylaniline from benzene ?

Answer. Following steps are involved :

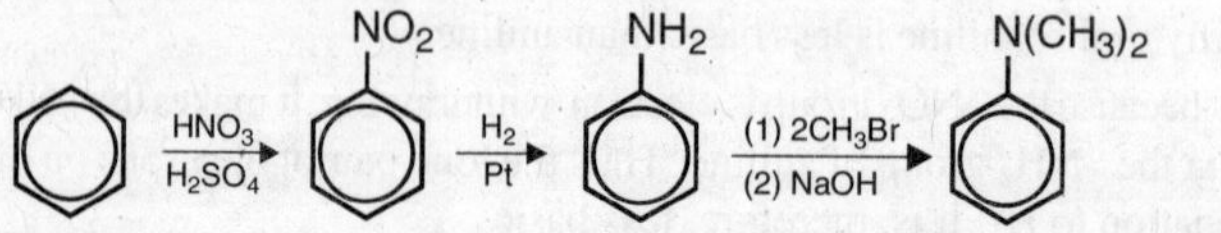

Benzene — *N,N*-Dimethylaniline

21. How will you synthesize 2,4,6-tribromoaniline from benzene ? *(Panjab BSc, 2015)*

Answer. Following steps are involved. Remember that bromination of reactive aniline requires no catalyst.

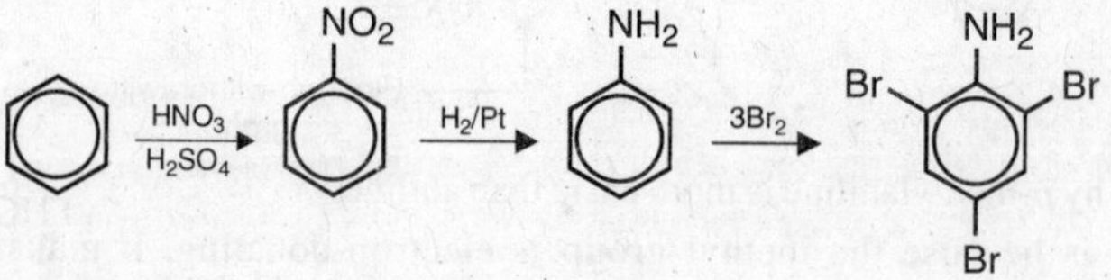

Benzene — 2,4,6-Tribromoaniline

22. How will you synthesize *m*-aminobenzoic acid from toluene ? *(Shivaji BSc, 2015)*

Answer. Following steps are involved :

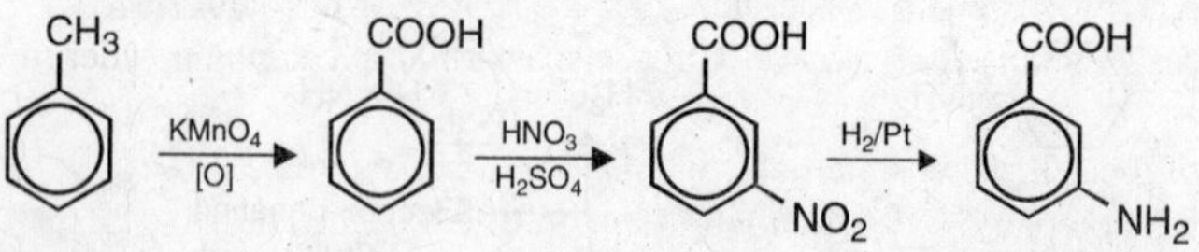

Toluene — *m*-aminobenzoic acid

23. How will you synthesize *p*-methylaniline (*p*-toluidine) from benzene ?

Answer. Following steps are involved :

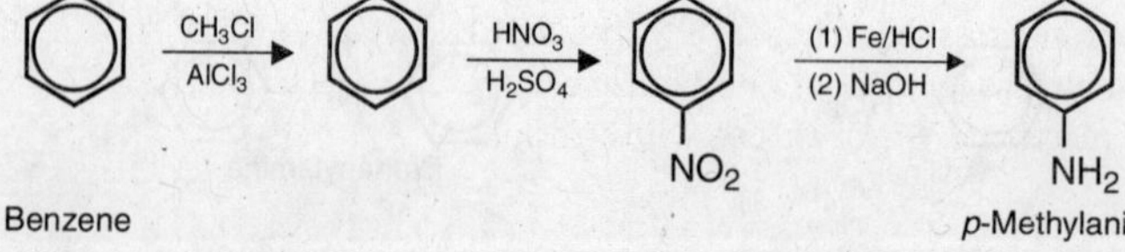

Benzene — *p*-Methylaniline

24. How will you synthesize *m*-nitroaniline from aniline ?

25. How will you synthesize *p*-nitroaniline from aniline ?

26. How will you synthesize *p*-bromoaniline from aniline ? *(Kolkata BSc, 2015)*

Answer. Following steps are involved :

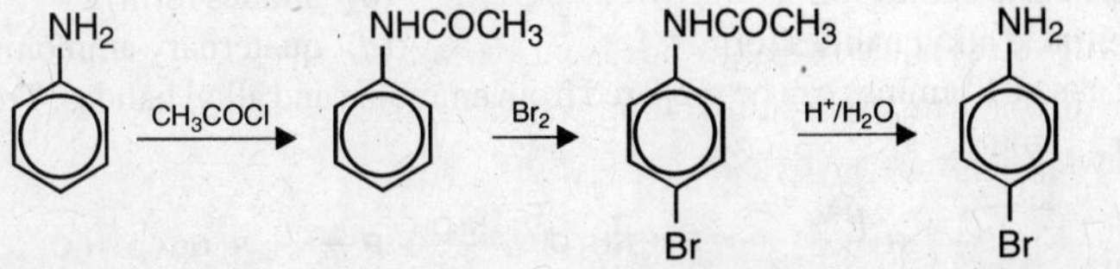

27. An amine (*A*) has the formula C_7H_9N. On treatment with benzenesulfonyl chloride, (*A*) formed a precipitate which was soluble in KOH solution. When the amine (*A*) was treated with nitrous acid at 0°C, a reaction occurred, and evolution of a gas took place immediately (before the solution reached room temperature). Write the structure for (*A*) that is consistent with the data given.

Answer. The compound (*A*) is Benzylamine.

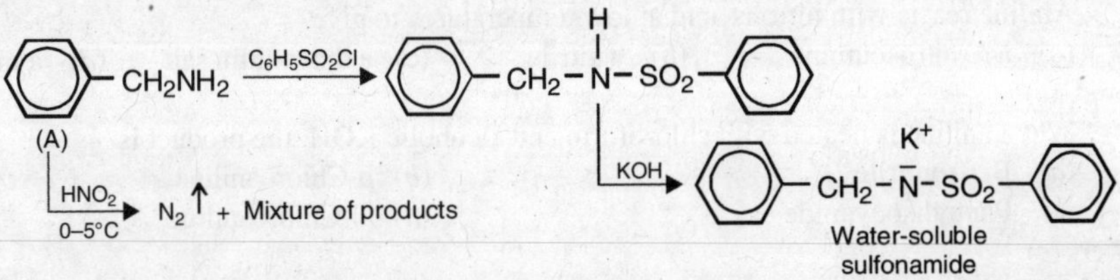

Multiple Choice Questions

1. Which of the following statements is false about aniline ?
   (*a*) It is a primary aromatic amine.
   (*b*) It is a weak base which forms salts with acids.
   (*c*) It liberates nitrogen on treatment with ice-cold nitrous acid.
   (*d*) It can be prepared by the reduction of nitrobenzene with tin and HCl.

Answer. (*c*)

2. Which of the following statements is correct about aniline ?
   (*a*) is weak acid.
   (*b*) diazotises to benzenediazonium chloride when treated with ice-cold nitrous acid.
   (*c*) can be nitrated with concentrated HNO_3 to form *p*-nitroaniline.
   (*d*) gives an intense green color with a suspension of bleaching powder in water.

Answer. (*b*)

3. Which of the following statements is false about benzylamine ?
   (*a*) can be prepared by reduction of benzonitrile with $LiAlH_4$.
   (*b*) cannot be prepared by reduction of benzamide with $LiAlH_4$.
   (*c*) reacts with nitrous acid to give benzyl alcohol.
   (*d*) reacts with $KMnO_4$ to give benzoic acid.

Answer. (*b*)

4. Aniline is prepared by
   (*a*) the reaction of benzene with ammonia
   (*b*) the reduction of nitrobenzene with Sn/HCl
   (*c*) the dehydrogenation of nitrobenzene
   (*d*) the reaction of nitrobenzene with I_2/NaoH

Answer. (*b*)

5. Benzamide reacts with Br_2 and KOH to give

(*a*) Benzene (*b*) Benzylamine (*c*) Aniline (*d*) Benzonitrile

Answer. (*b*)

6. When ammonia is added to an alkyl halide, in the presence of base :

(*a*) primary amines form (*b*) amides form

(*c*) nitrated alkyl halides form (*d*) quaternary ammonium salts form

Answer. (*a*). Salts of 1° amines can be prepared from ammonia and alkyl halides. Treating the ammonium salt with base gives a 1° amine.

$$\ddot{N}H_3 + R-X \longrightarrow R-\overset{+}{N}H_3\ \overset{-}{Cl} \xrightarrow{NaOH} R-NH_2 + NaX + H_2O$$

7. Aniline reacts with bromine water to form

(*a*) Bromobenzene (*b*) *m*-Bromoaniline

(*c*) 2,4,6-Tribromoaniline (*d*) *o*- plus *p*-Bromoaniline

Answer. (*c*)

8. Which of the following will give a precipitate with excess bromine water ?

(*a*) Bromobenzene (*b*) Aniline (*c*) Nitrobenzene (*d*) Toluene

Answer. (*b*)

9. Aniline reacts with nitrous acid at low temperatures to give

(*a*) a *N*-nitrosoamine (*b*) a nitrile (c) a diazonium salt (*d*) a nitrite salt

Answer. (*c*)

10. When aniline is heated with chloroform and alcoholic KOH, the product is

(*a*) Benzonitrile (*b*) *p*-Chloroaniline

(*c*) Phenylisocyanide (*d*) *m*-Chloroaniline

Answer. (*c*)

11. Aniline reacts with acetic anhydride to give :

(*a*) *N*-Methylaniline (*b*) *p*-Aminoacetophenone

(*c*) Acetanilide (*d*) *m*-Aminoacetophenone

Answer. (*c*)

12. Aniline undergoes oxidation with $Na_2Cr_2O_7/H_2SO_4$ to give

(*a*) Schiff's base (*b*) *p*-Benzoquinone

(*c*) Phenol (*d*) Benzoic acid

Answer. (*b*)

13. Which of the following reagents does not react with aniline ?

(*a*) Acetyl chloride (*b*) Acetic anhydride

(*c*) Ammonia (*d*) Nitrous acid

Answer. (c)

14. Rank the following compounds in order of increasing base strength :

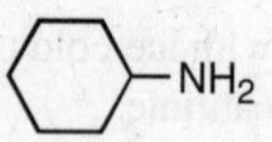
1

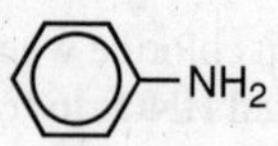
2

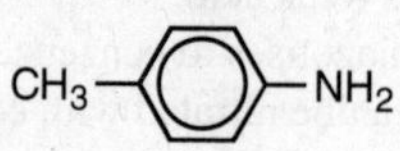
3

(*a*) 1, 2, 3 (*b*) 3, 2, 1

(*c*) 2, 3, 1 (*d*) 2, 1, 3

Answer. (*c*). The aromatic amines are weaker bases than the nonaromatic cyclic amines because the lone-pair electrons on nitrogen are delocalized over the aromatic ring. This delocalization stabilizes the compound but also decreases the ability of the nitrogen to act as a base. Compound (3) is a stronger base than aniline because of the electron-releasing substituent on the ring.

15. Rank the following compounds in order of increasing base strength :

NH_3
1

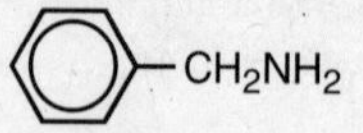
2

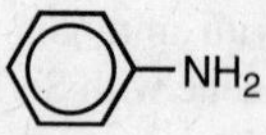
3

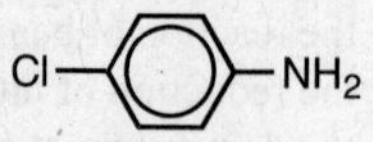
4

(*a*) 1, 2, 3, 4 (*b*) 4, 2, 3, 1

(*c*) 1, 4, 3, 2 (*d*) 4, 3, 1, 2

Answer. (*d*). Of the aromatic amines, the weakest base is the one with the electron-withdrawing substituent: (4) is the weakest followed by aniline (3). Compound (2) is the strongest base because its electrons are not delocalized effectively into the aromatic ring, and it is therefore a primary amine, and its pK_a is slightly lower than that of ammonia.

16. Which of the following compounds is most basic ?

(*a*) Aniline (*b*) Benzylamine (*c*) Acetanilide (*d*) *p*-nitroaniline

Answer. (*b*)

17. Which of the following is most basic ?

(*a*) CH_3NH_2 (*b*) $(CH_3)_3N$ (*c*) $(CH_3)_2NH$ (*d*) 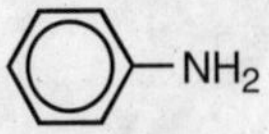

Answer. (*c*)

18. Which of the following is most basic ?

(*a*) $(CH_3)_2NH$ (*b*) 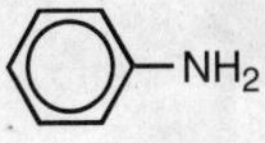

(*c*) $CH_3-\overset{O}{\overset{\|}{C}}-NH_2$ (*d*) NH_3

Answer. (*a*)

19. Which of the following compounds is most basic ?

(*a*) 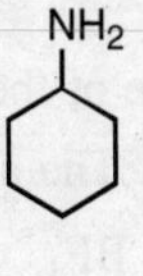 (*b*)  (*c*) 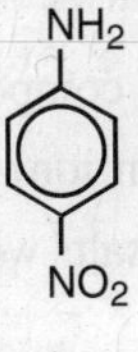 (*d*) 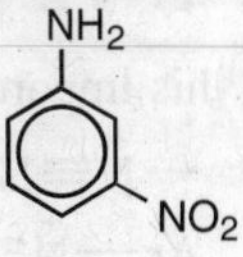

Answer. (*a*)

20. Which of the following is least basic ?

(*a*) Aniline (*b*) *p*-Nitroaniline

(*c*) Ammonia (*d*) Trimethylamine

Answer. (*b*)

21. Which of the following is least basic ?

(*a*) CH_3NH_2 (*b*) $(CH_3)_3N$

(*c*) $(CH_3)_2NH$ (*d*) 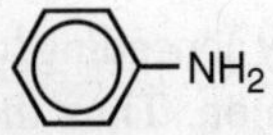

Answer. (*d*)

22. Which of the following compounds is least basic ?

(*a*) 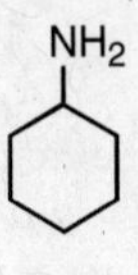 (*b*)  (*c*)  (*d*) 

Answer. (*c*)

23. Which of the following compounds will be most soluble in water at room temperature ?

(*a*) Phenol (*b*) Phenylammonium chloride

(*c*) Benzoic acid (*d*) Benzylamine

Answer. (*b*)

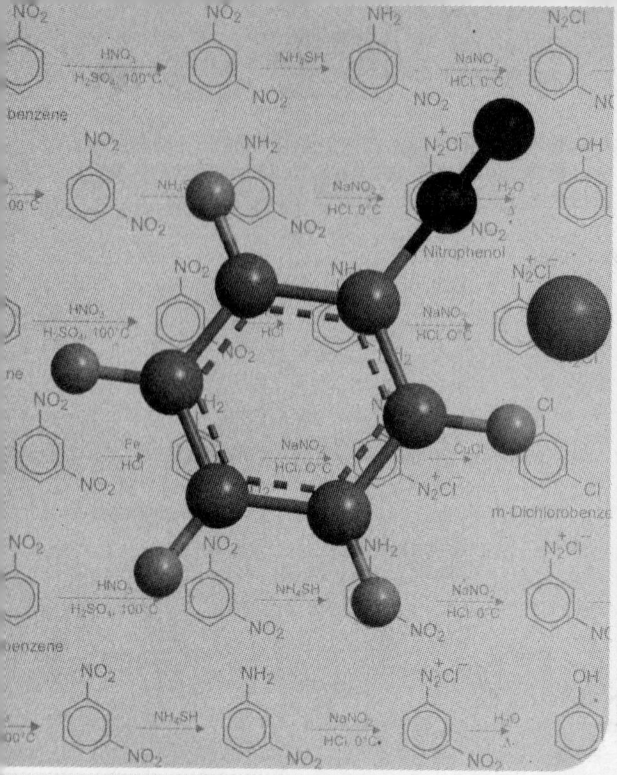

39

Aryldiazonium Salts

This important class of compounds is characterized by the presence of the functional group $-\overset{+}{N}\equiv N$ (diazonium ion) directly bonded to an aryl group, Ar. The aryldiazonium ion, $Ar-\overset{+}{N}\equiv N$, forms salts with anions such as $\overline{Cl}$, $\overline{Br}$, NO_2^-, HSO_4^-, BF_4^-, etc. These salts are called **Aryldiazonium Salts**.

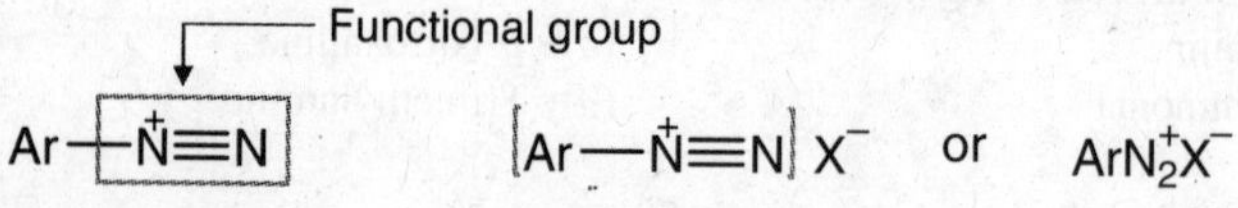

Aryldiazonium ion Aryldiazonium salt

NOMENCLATURE

Diazonium salts are named by appending 'diazonium' to the name of the parent aromatic compound followed by the name of the anion. Thus the diazonium salts derived from arenes are named as *arenediazonium salts*. For example,

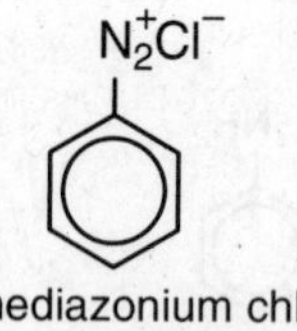

Benzenediazonium chloride

STRUCTURE

The electronic structure of diazonium salts $(Ar\overset{+}{N}_2\overset{-}{X})$ is written as :

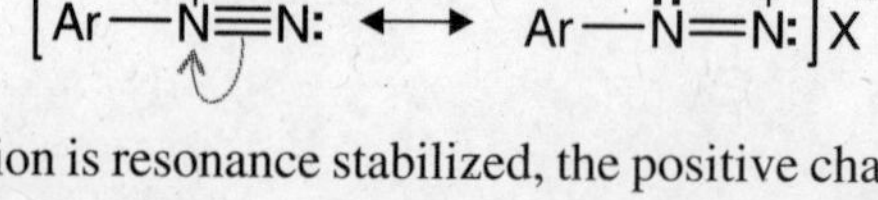

Thus the diazonium cation is resonance stabilized, the positive charge being distributed over the two N atoms.

The stability of aryldiazonium salts relative to alkyldiazonium salts ($R\overset{+}{N_2}\overset{-}{X}$) is also due to resonance involving the benzene ring.

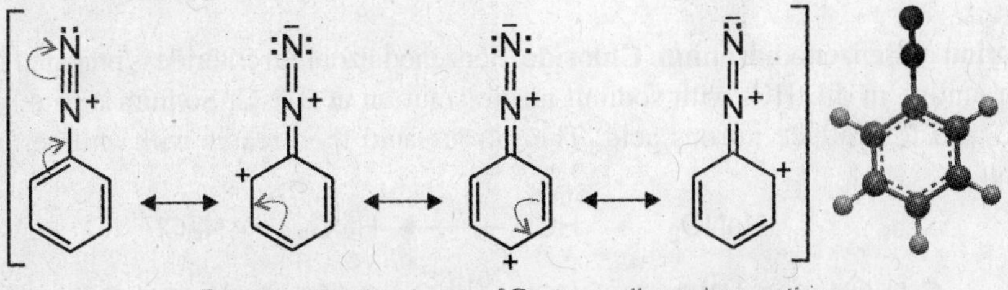

Resonance structures of Benzenediazonium cation.

Thus the C–N bond acquires partial double bond character and is stronger than C–N bond in alkyldiazonium salts. The aliphatic diazonium salts are unstable because R group is incapable of participating in resonance.

Molecular Orbital Structure. In acid solution, arenediazonium salts have diazonium ion structure with a linear C–N–N Bond. The nitrogen atoms are *sp* hybridized and the ring carbon atoms are sp^2 hybridized.

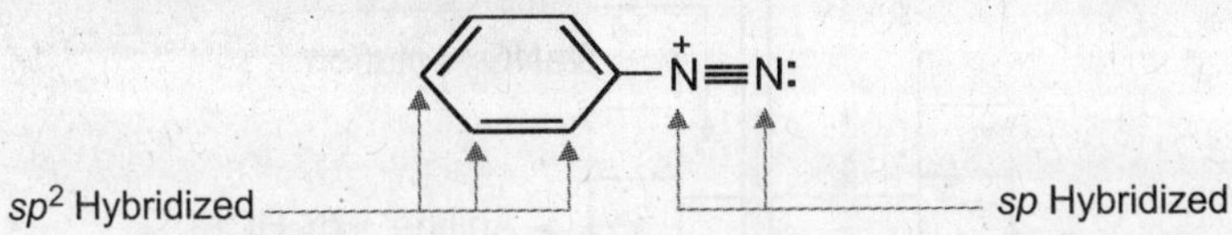

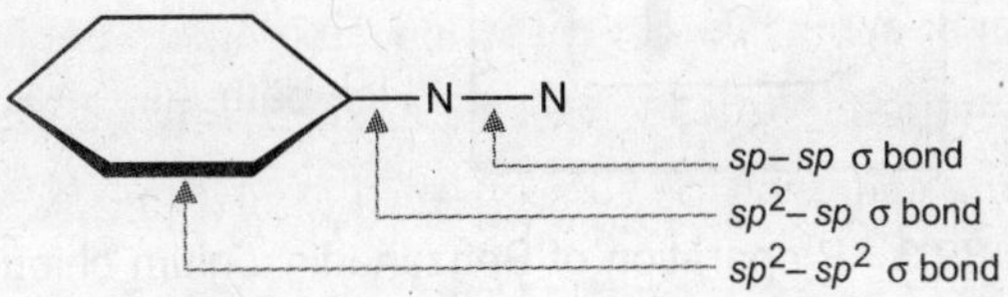

Fig.39.1. Formation of σ bonds in benzenediazonium ion.

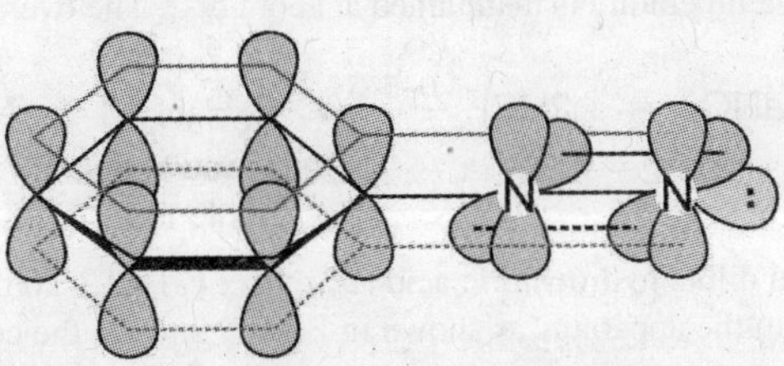

Fig.39.2. Formation of π molecular orbitals in benzenediazonium ion.

The parallel *p* orbitals on nitrogen atoms and the benzene carbon atoms overlap to form a delocalized π molecular orbital. This delocalization is responsible for the relative stability of these compounds. Recall that the aliphatic diazonium ions are very unstable (No delocalization).

PREPARATION

Aryldiazonium salts are prepared by the action of an aromatic primary amine with nitrous acid at about 0–5°C. The nitrous acid is obtained *in situ* by taking sodium nitrite and a mineral acid (HCl or H_2SO_4) in the reaction mixture.

$$NaNO_2 + HCl \longrightarrow HNO_2 + NaCl$$

$$\underset{\text{Aromatic 1° amine}}{ArNH_2} + HNO_2 + HCl \xrightarrow{0\text{–}5°C} \underset{\text{Aryldiazonium chloride}}{Ar\overset{+}{N_2}\overset{-}{Cl}} + 2H_2O$$

The process of converting an amine into the diazonium salt is called diazotization. The procedure of diazotization may be illustrated by taking example of the preparation of benzenediazonium chloride from aniline.

Preparation of Benzenediazonium Chloride. Benzenediazonium chloride is prepared by treating a solution of aniline in dil. HCl with sodium nitrite solution at 0-5°C. Sodium nitrite reacts with hydrochloric acid to produce nitrous acid. The nitrous acid then reacts with aniline to give the diazonium salt.

$$NaNO_2 + HCl \longrightarrow HNO_2 + NaCl$$

$$\underset{\text{Aniline}}{C_6H_5NH_2} + HNO_2 + HCl \xrightarrow{0\text{–}5°C} \underset{\text{Benzenediazonium chloride}}{C_6H_5\overset{+}{N_2}\overset{-}{Cl}} + 2H_2O$$

This method is used for the preparation of benzenediazonium chloride both in the laboratory and on a large scale.

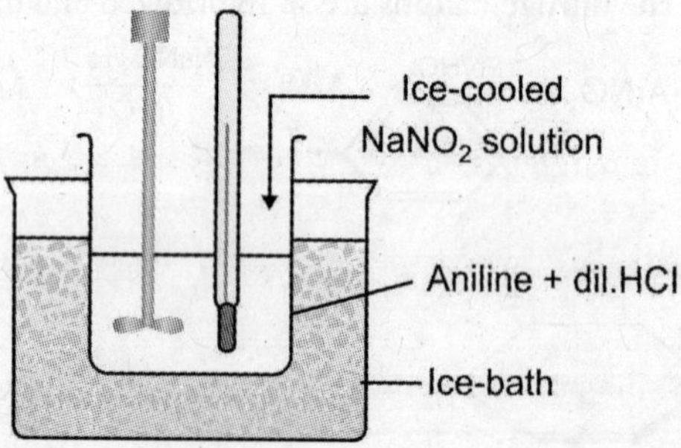

Fig. 39.3. Preparation of Benzenediazonium chloride.

Laboratory Preparation. A cooled solution of sodium nitrite in water is added to a solution of aniline in dil. HCl cooled in an ice bath. The temperature is maintained at about 5°C. The overall reaction is :

$$\underset{\text{Aniline}}{C_6H_5NH_2} + NaNO_2 + 2HCl \xrightarrow{0\text{–}5°C} \underset{\text{Benzenediazonium chloride}}{C_6H_5\overset{+}{N_2}\overset{-}{Cl}} + NaCl + 2H_2O$$

Dissolve 5 ml of aniline in dilute hydrochloric acid (20 ml conc. HCl + 20 ml H_2O) in a beaker. Place the beaker in the ice bath and set up the apparatus as shown in Fig. 39.3. Cool the contents of the beaker to about 5°C. Slowly add with constant stirring sodium nitrite solution (10 g in 50 ml water) previously cooled to about 5°C. Make sure that the temperature does not rise above 10°C. The solution of benzenediazonium chloride so obtained is used immediately for further reaction. **The dry diazonium salt is unstable and explosive, and is seldom separated in the solid state.**

MECHANISM. Following steps are involved : (*1*) Nitrous acid ($NaNO_2$/HCl) furnishes nitrosonium ions $[N{=}O]^+$; (*2*) Nitrosonium ions combine with aniline to form nitrosoaniline ; (*3*) Nitrosoaniline undergoes rearrangement followed by ; (*4*) Dehydration to form diazonium ion which takes up Cl^- to yield the diazonium salt.

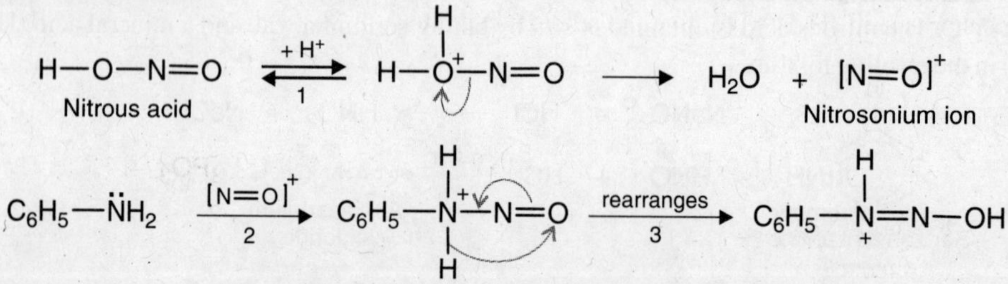

$$\xrightarrow[4]{-H_2O} C_6H_5-\overset{+}{N}\equiv N \xrightarrow{+Cl^-} C_6H_5-\overset{+}{N}\equiv N\overset{-}{Cl}$$

Benzenediazonium chloride

PHYSICAL PROPERTIES

(1) Aryldiazonium salts are colorless, crystalline solids which turn brown on exposure to air.

(2) Most of them are unstable and explode when warmed or subjected to shock, when dry. Diazonium fluoroborates, $Ar\overset{+}{N}_2\overset{-}{B}F_4$, can be isolated and dried without decomposition.

(3) They are soluble in water, sparingly soluble in ethanol and glacial acetic acid, and insoluble in ether.

CHEMICAL PROPERTIES

Aryldiazonium salts are highly reactive compounds and undergo many useful reactions. They have proved to be valuable synthetic intermediates for the preparation of a fairly wide range of aromatic derivatives.

$$ArH \xrightarrow[H_2SO_4]{HNO_3} ArNO_2 \xrightarrow{Fe/HCl} ArNH_2 \xrightarrow[5°C]{NaNO_2/HCl} Ar\overset{+}{N}_2\overset{-}{Cl} \xrightarrow[\text{reactions}]{\text{Synthetic}}$$

As synthetic reagents, diazonium salts compete with Grignard reagents in versatility. The reactions of diazonium salts may be divided into two main types :

(*a*) Those in which the $-\overset{+}{N}_2\overset{-}{X}$ group is replaced by another univalent atom or group, with the liberation of N_2 gas.

(*b*) Those in which the two N atoms are retained in the product.

We will illustrate the **Synthetic Applications of Diazonium Salts** by taking example of benzenediazonium chloride.

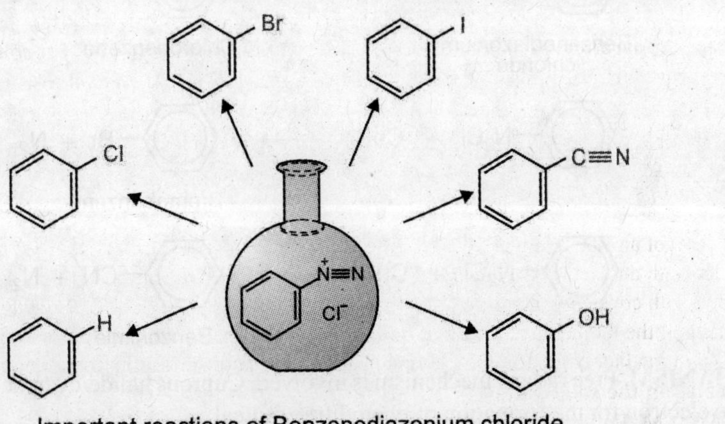

Important reactions of Benzenediazonium chloride.

REACTIONS IN WHICH $-\overset{+}{N}_2\overset{-}{X}$ IS REPLACED

(1) **Reduction to Arenes (*Replacement of N_2X by H*).** A diazonium salt is reduced by hypophosphorus acid (H_3PO_2) to give the respective arene.

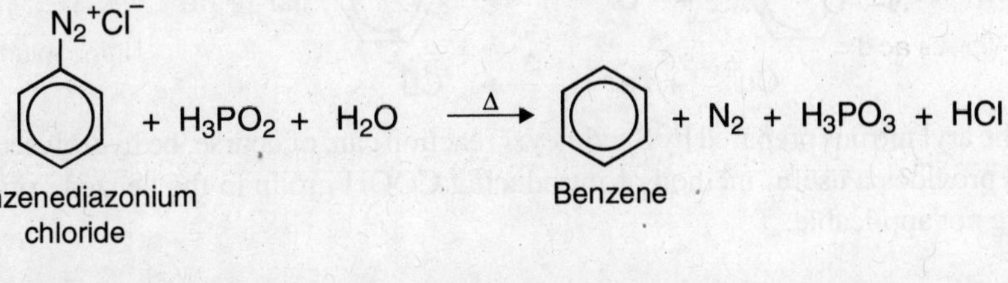

Benzenediazonium chloride　　Benzene

(2) **Synthesis of Phenols (*Replacement of* $-N_2X$ *by OH*).** Aqueous solution of the diazonium salt is acidified and heated to form phenol.

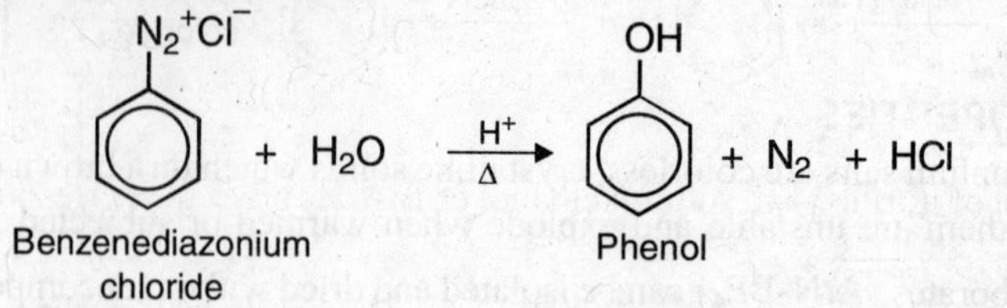

This is a useful reaction for preparing substituted phenols not obtained by electrophilic aromatic substitution of phenol.

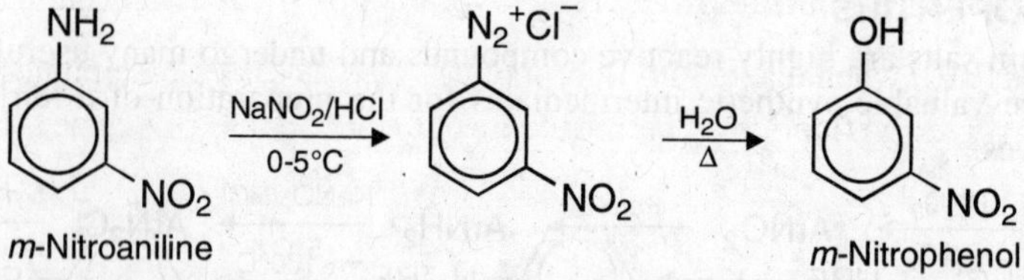

(3) **Synthesis of Aryl chlorides, Aryl bromides, Aryl nitriles (*Replacement of* $-N_2X$ *by Cl, Br, CN*).** When a cold diazonium salt solution is treated with cuprous chloride, cuprous bromide, or cuprous cyanide, the product is aryl chloride, aryl bromide, or aryl nitrile. These reactions are known as **Sandmeyer Reactions.**

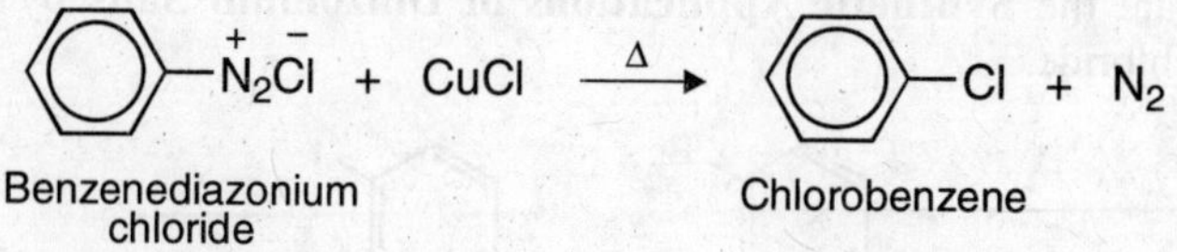

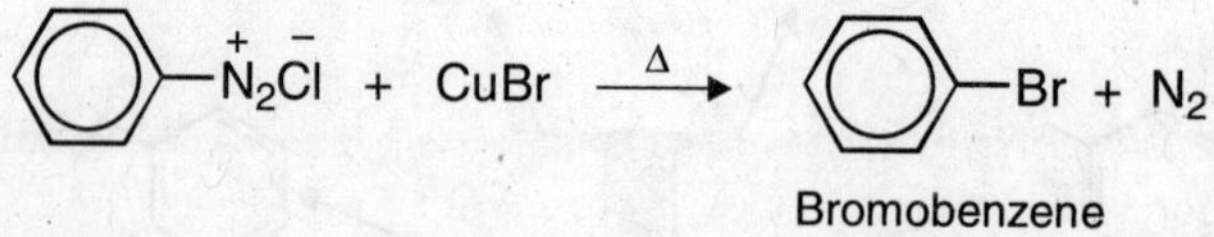

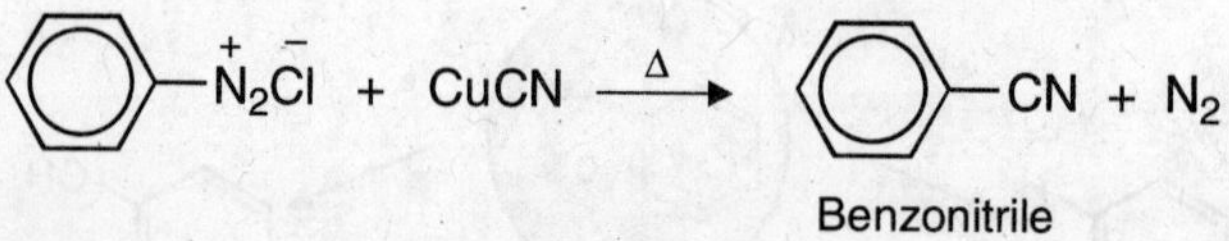

MECHANISM. Free radical mechanism is involved. Cuprous halide/cyanide acts as a catalyst. It provides the electron for the formation of phenyl free radical.

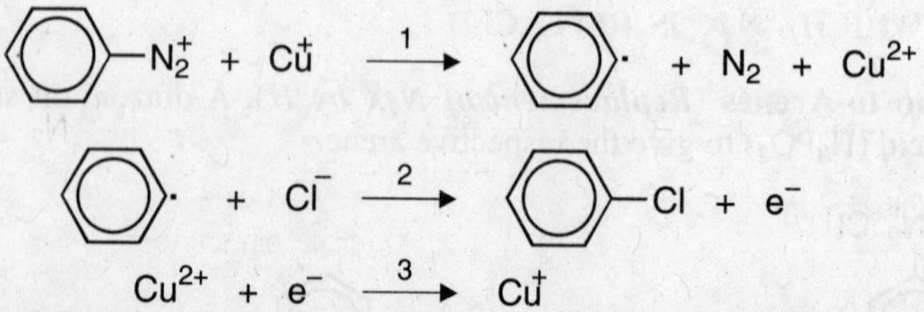

The aryl nitriles prepared by Sandmeyer reaction can, of course, be hydrolysed to carboxylic acids. This provides a useful method of introducing COOH group in the benzene ring where other methods are not applicable.

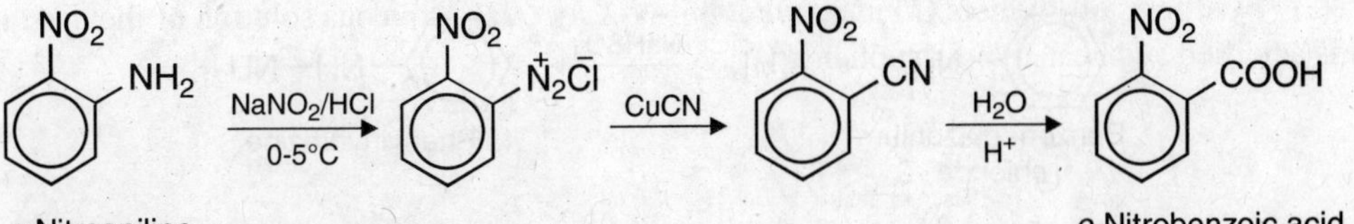

o-Nitroaniline   *o*-Nitrobenzoic acid

The reduction of nitriles with sodium/ethanol or $LiAlH_4$ gives the corresponding amine.

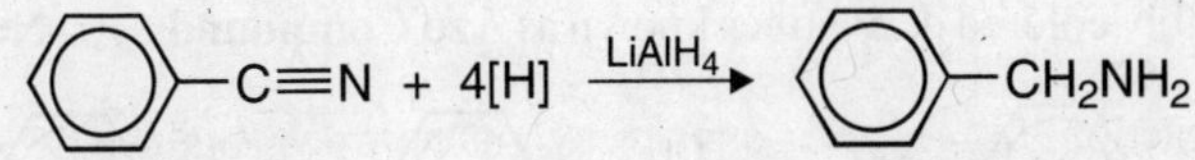

Benzonitrile   Benzylamine

(4) **Synthesis of Aryl fluorides (*Replacement of* $-N_2X$ *by F*).** The diazonium salt is treated with fluoroboric acid (HBF_4) when diazonium fluoroborate ($C_6H_5\overset{+}{N}_2\overset{-}{BF_4}$) precipitates. This is isolated, dried and heated to give aryl fluorides. This procedure of preparing fluorobenzene is called **Schiemann Reaction**.

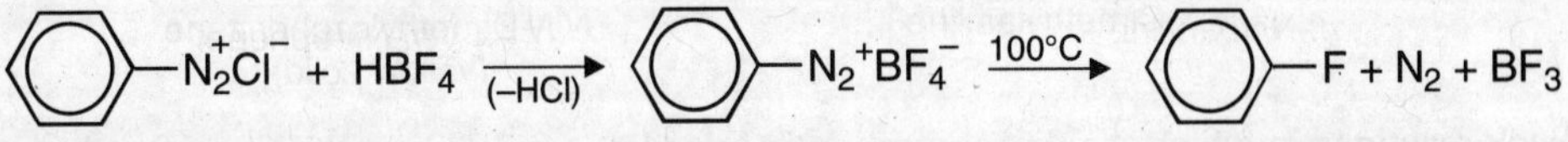

Benzenediazonium chloride   Fluorobenzene

(5) **Synthesis of Aryl iodides (*Replacement of* $-N_2X$ *by I*).** When aqueous solution of a diazonium salt is heated with excess of potassium iodide, aryl iodide is obtained.

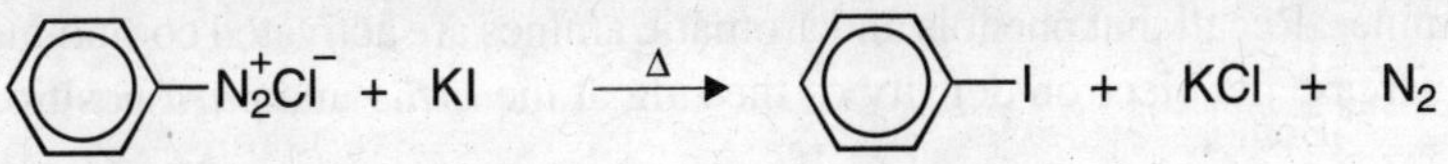

Benzenediazonium chloride   Iodobenzene

(6) **Synthesis of Thiophenols (*Replacement of* $-N_2X$ *by SH*).** Reaction of diazonium salts with potassium hydrosulfide yields thiophenols.

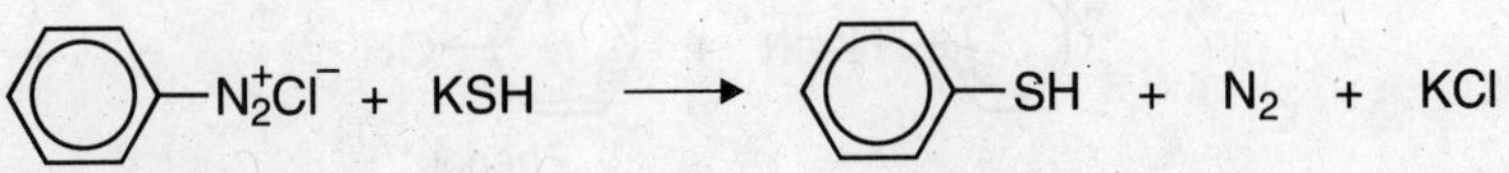

Benzenediazonium chloride   Thiophenol

(7) **Synthesis of Ethers (*Replacement of* $-N_2X$ *by* $-OR$).** Reaction of diazonium salts with alcohols yields alkoxybenzenes.

Benzenediazonium chloride   Anisole (Methoxybenzene)

REACTIONS IN WHICH N-ATOMS ARE RETAINED

(8) **Reduction to Arylhydrazines.** When reduced with sodium bisulfite ($NaHSO_3$), or stannous chloride, aryldiazonium salts yield arylhydrazines.

Phenylhydrazine is used as a reagent in the identification of carbonyl compounds and sugars.

(9) **Diazo Coupling Reactions**. Diazonium salts react readily with phenols and dialkylanilines ($C_6H_5NR_2$) to form highly colored derivatives known as **Azo Compounds**, Ar–N=N–Ar. For example,

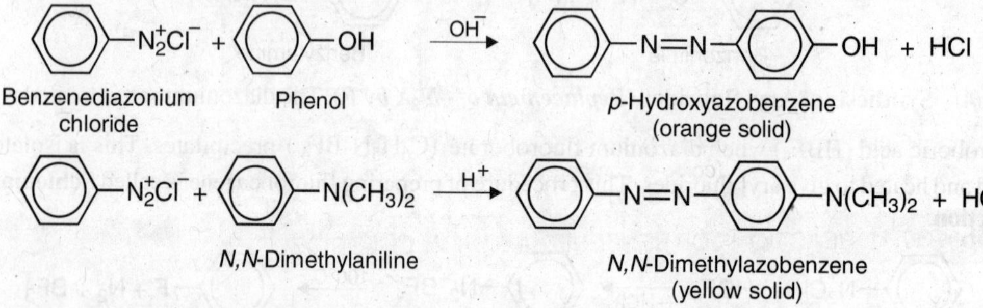

Such reactions in which the diazonium ion is joined or 'coupled' to the ring of a phenol or an aromatic amine are called **Diazo Coupling Reactions**. While the coupling with phenols is carried in slightly alkaline solution, the coupling with aromatic amines is carried in slightly acidic solutions.

MECHANISM. Diazo coupling reactions take place by a typical electrophilic aromatic substitution mechanism. The benzenediazonium ion acts as an electrophile and attacks the *para* position of phenols and aromatic amines. Recall that phenols and aromatic amines are activated compounds. The –OH and –NH_2 groups increase the electron density of the ring at the *ortho* and *para* positions.

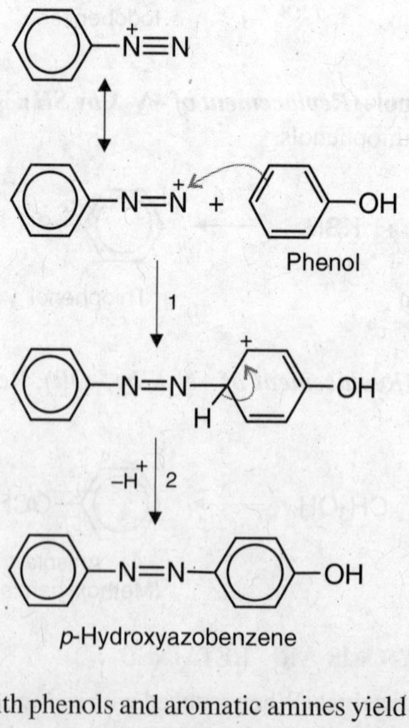

Diazo coupling reactions with phenols and aromatic amines yield a large number of dyes, the **Azo Dyes**.

(10) **Reaction with Amines**. Diazonium salts react with primary and secondary amines in mildly alkaline solutions to form diazoaminobenzenes. Here the coupling takes place at the *N*-atom of the amine and not at the ring. However, the diazoaminobenzene on heating in acidic medium rearranges to give the *C*-azo compound.

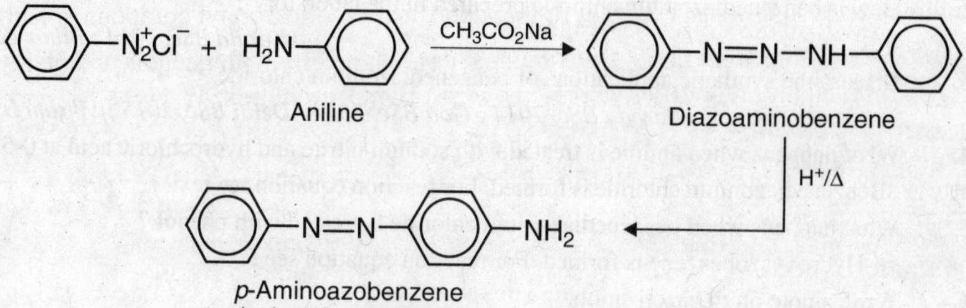

p-Aminoazobenzene (colored compound)

Table 39.1 REACTIONS OF BENZENEDIAZONIUM CHLORIDE

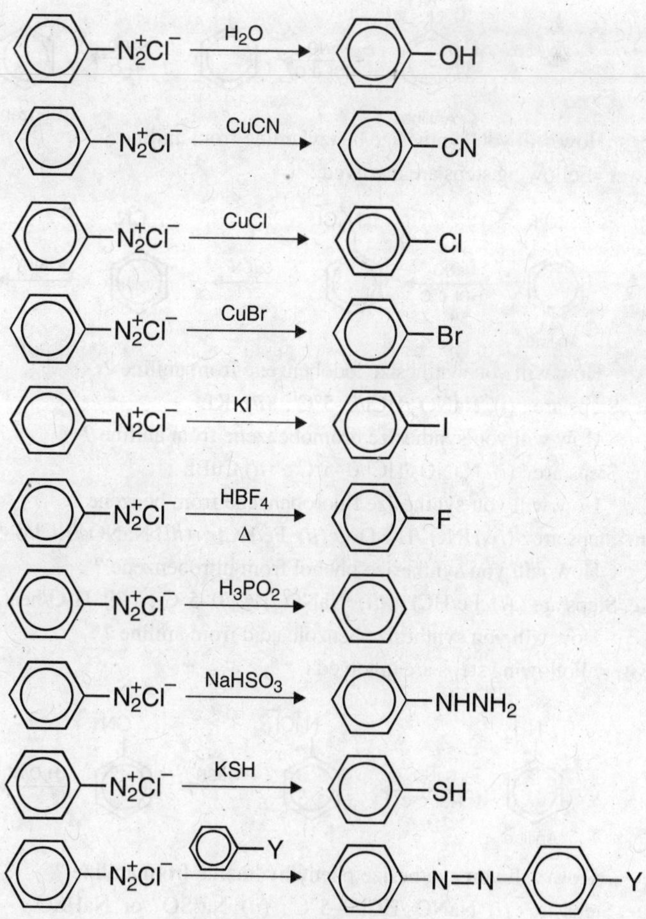

($Y = OH, NH_2, NHR, NR_2$)

Examination Questions

1. Discuss the structure of benzenediazonium chloride.
2. How is benzenediazonium chloride prepared in the laboratory ? *(Kerala BSc, 2014 ; Bihar BSc, 2015)*
3. Discuss the synthetic applications of benzenediazonium chloride. *(Baroda BSc, 2014 ; Goa BSc, 2014 ; Delhi BSc, 2015 ; Vikram BSc, 2015)*
4. What happens when aniline is treated with sodium nitrite and hydrochloric acid at 0-5°C ?

Answer. Benzenediazonium chloride is formed. For reaction equation see text.

5. What happens when benzenediazonium chloride is treated with phenol ?

Answer. *p*-Hydroxyazobenzene is formed. For reaction equation see text.

6. Write a note on : Diazotization.
7. Write a note on : Sandmeyer Reaction.
8. Write a note on : Diazo Coupling Reactions. *(Madras BSc, 2015)*
9. How will you synthesize benzene from aniline ? *(Calicut BSc, 2014)*

Answer. Following steps are involved :

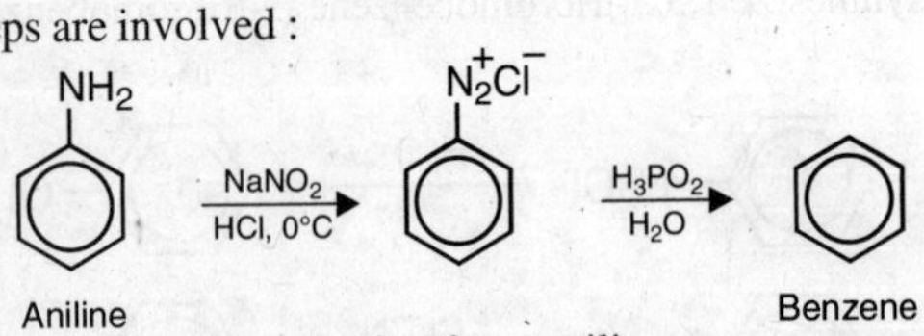

10. How will you synthesize benzylamine from aniline. *(Indore BSc, 2014)*

Answer. Following steps are involved :

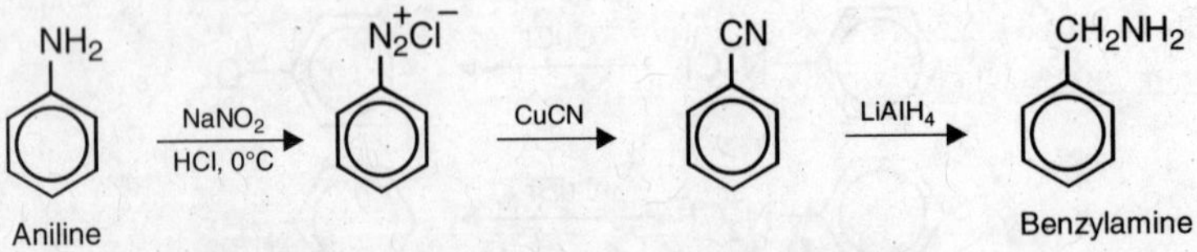

11. How will you synthesize iodobenzene from aniline ? *(Agra BSc, 2015)*

Clue. Steps are : (*i*) $NaNO_2/HCl/0–5°C$; (*ii*) KI.

12. How will you synthesize bromobenzene from aniline ?

Clue. Steps are : (*i*) $NaNO_2/HCl/0–5°C$; (*ii*) CuBr.

13. How will you synthesize fluorobenzene from benzene ? *(Madurai BSc, 2014)*

Clue. Steps are : (*i*) HNO_3/H_2SO_4 ; (*ii*) Fe/HCl ; (*iii*) $NaNO_2/HCl/0–5°C$; (*iv*) HBF_4.

14. How will you synthesize phenol from nitrobenzene ?

Clue. Steps are : (*i*) Fe/HCl ; (*ii*) $NaNO_2/HCl/0–5°C$; (*iii*) H_2O/heat.

15. How will you synthesize benzoic acid from aniline ? *(Rohilkhand BSc, 2014)*

Answer. Following steps are involved :

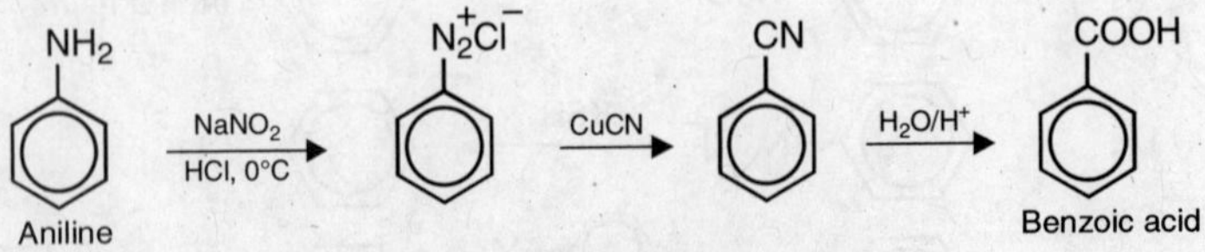

16. How will you synthesize phenylhydrazine from aniline ? *(Andhra BSc, 2014)*

Clue. Steps are : (*i*) $NaNO_2/HCl/0–5°C$; (*ii*) Na_2SO_3 or $NaHSO_3$.

17. How will you synthesize *p*-hydroxyazobenzene from nitrobenzene ?

Clue. Steps are : (*i*) $NaNO_2/HCl/O–5°C$; (*ii*) C_6H_5OH.

18. How will you synthesize *m*-dichlorobenzene from benzene ? *(Delhi BSc, 2008)*
Answer. Following steps are involved :

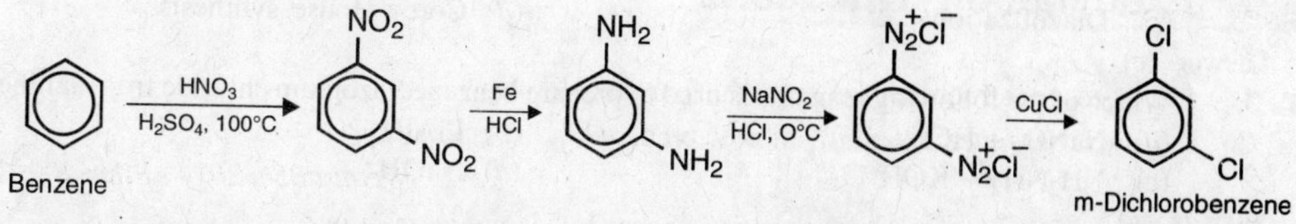

19. How will you synthesize *m*-bromochlorobenzene from nitrobenzene ?
Clue. Steps are : (i) $Br_2/FeBr_3$; *(ii)* Fe/HCl ; *(iii)* $NaNO_2/HCl/0–5°C$; *(iv)* CuCl.

20. How will you synthesize *m*-nitrophenol from nitrobenzene ? *(Calcutta BSc, 2008)*
Answer. Following steps are involved :

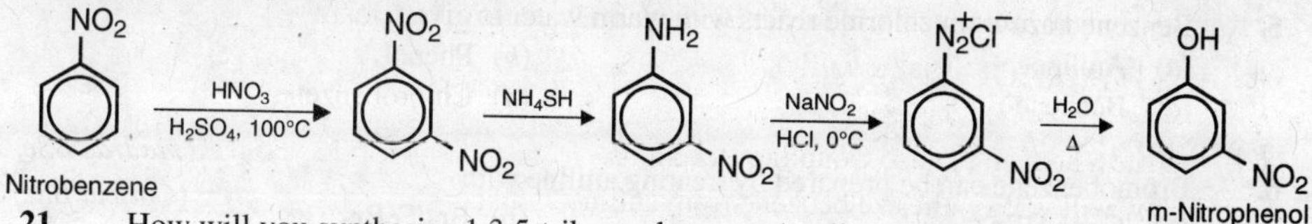

21. How will you synthesize 1,3,5-tribromobenzene (*s*-tribromobenzene) from aniline ?
(Burdwan BSc, 2007 ; Mangalore BSc, 2008)
Answer. Following steps are involved :

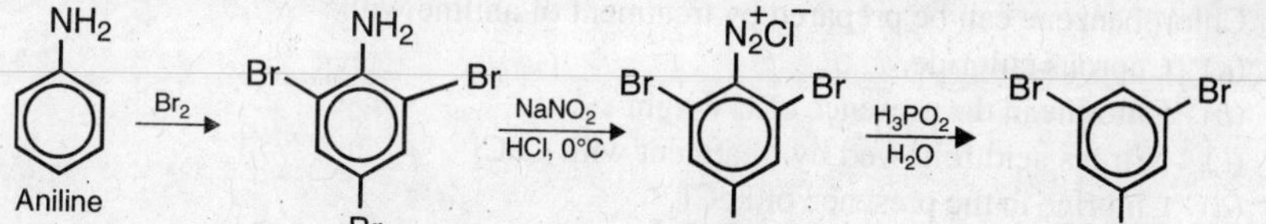

22. How will you synthesize *m*-bromotoluene from toluene.
Answer. Following steps are involved :

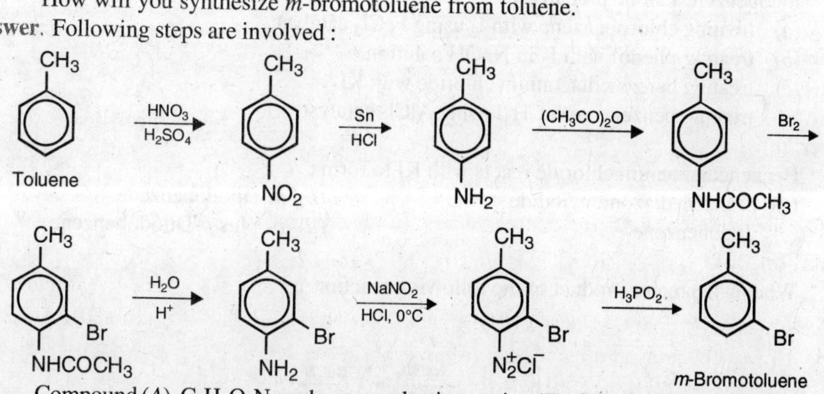

23. Compound *(A)*, $C_7H_7O_2N$, undergoes reduction to give *(B)*, C_7H_9N. Compound *(B)* on diazotization followed by treatment with CuCN gives *(C)*, C_8H_7N. Hydrolysis of *(C)* followed by oxidation gives *(D)*, $C_8H_6O_4$. Compound *(D)* gives only one monosubstitution product. Identify *(A)* to *(D)*.
Answer. *(A)* = *p*-Nitrotoluene *(B)* = *p*-Methylaniline
(C) = *p*-Cyanotoluene *(D)* = Terephthalic acid

Multiple Choice Questions

1. The diazonium salts are the reaction products of the reaction of nitrous acid with
(a) primary aliphatic amines *(b)* primary aromatic amines
(c) secondary aliphatic amines *(d)* secondary aromatic amines
Answer. *(b)*

2. Preparation of a diazonium salt from a primary aromatic amine is known as :
(a) Coupling reaction
(b) Sandmeyer reaction
(c) Diazotization
(d) Corey-House synthesis

Answer. (c)

3. Which of the following reagents is used to prepare benzenediazonium chloride from aniline ?
(a) $NaNO_2 + HCl$
(b) $LiAlH_4$
(c) $NH_2NH_2 + KOH$
(d) NaOH

Answer. (a)

4. Which of the following are optimum temperature conditions for making benzenediazonium chloride from aniline ?
(a) 0°C to 10°C
(b) 20°C to 25°C
(c) 30°C to 40°C
(d) 45°C to 50°C

Answer. (a)

5. Benzenediazonium chloride reacts with warm water to give
(a) Aniline
(b) Phenol
(c) Benzene
(d) Chlorobenzene

Answer. (b)

6. Bromobenzene can be prepared by treating aniline with
(a) Conc HBr
(b) $Br_2/FeBr_3$
(b) CuBr
(d) Nitrous acid *then* CuBr

Answer. (d)

7. Chlorobenzene can be prepared by treatment of aniline with
(a) Cuprous chloride
(b) Chlorine in the presence of UV light
(c) Nitrous acid followed by treatment with CuCl
(d) Chlorine in the presence of $FeCl_3$

Answer. (c)

8. Iodobenzene can be prepared by
(a) treating chlorobenzene with I_2 using $FeCl_3$ catalyst.
(b) treating phenol with I_2 in NaOH solution.
(c) treating benzenediazonium chloride with KI
(d) treating benzene with CH_3I using $AlCl_3$ catalyst.

Answer. (c)

9. Benzenediazonium chloride reacts with KI to form :
(a) Benzenediazonium iodide
(b) *m*-Diiodobenzene
(c) Iodobenzene
(d) *o*- plus *p*-Diiodobenzene

Answer. (c)

10. What is the major product of the following reaction ?

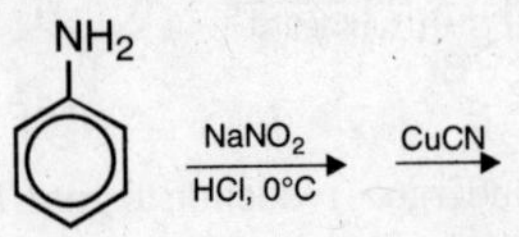

(a) Benzonitrile
(b) Benzoic acid
(c) Nitrobenzene
(d) Benzenediazonium chloride

Answer. (a)

11. Benzenediazonium chloride reacts with hypophosphorus acid to form :
(a) Phenol
(b) Benzaldehyde
(c) Aniline
(d) Benzene

Answer. (d)

12. Benzenediazonium chloride reacts with phenol to form :
(a) *p*-chlorophenol
(b) Chlorobenzene
(c) *p*-Hydroxyazobenzene
(d) DDT

Answer. (c)

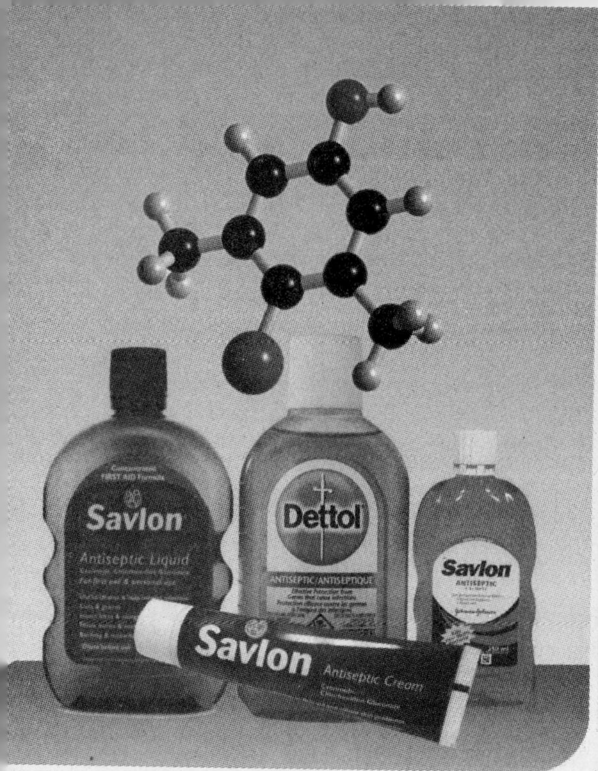

40

Phenols and Aromatic Ethers

Chloroxylenol (4-chloro-3,5-dimethylphenol), isopropyl alcohol, and pine oil are the main constituents of Dettol.

Phenols are compounds containing an –OH group attached directly to an aromatic ring. Like alcohols, they may be monohydric or polyhydric according to the number of –OH groups that they contain. Phenols are usually named by common system or as derivatives of the parent phenol (C_6H_5OH). Examples are :

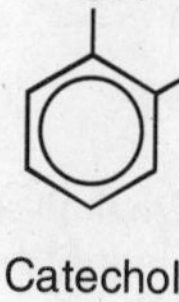

Phenol

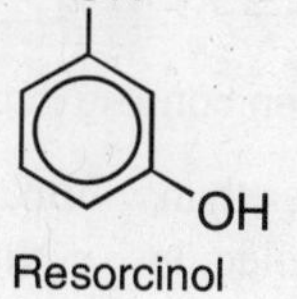

o-Cresol
(2-methylphenol)

p-Nitrophenol
(4-nitrophenol)

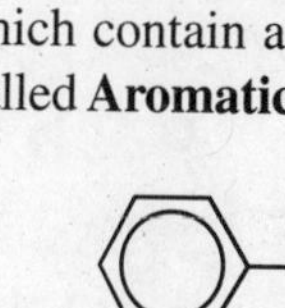

Catechol

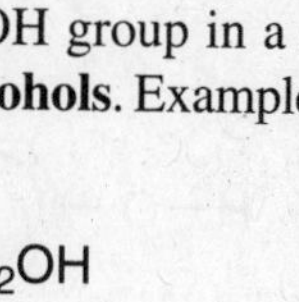

Resorcinol

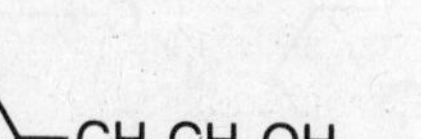

Quinol

Compounds which contain an –OH group in a side-chain attached to an aromatic ring are not phenols. They are called **Aromatic Alcohols**. Examples are :

Benzyl alcohol

2-Phenylethanol

PHYSICAL PROPERTIES OF PHENOLS

(1) Most pure phenols are colourless liquids or solids, although they are often found to contain a red tint owing to the presence of oxidation products.

(2) Phenols have a characteristic '*Carbolic*' odor, which in case of phenol itself is highly toxic.

(3) The boiling points of phenols are higher than aliphatic alcohols of comparable molecular weights. This is due to stronger intermolecular hydrogen bonding in phenols relative to alcohols.

Compound	MW	bp°C
Phenol	94	182
Cyclohexanol	100	162
1-Hexanol	102	157

The introduction of a hydroxy group into an already substituted aromatic ring, especially in the *para* position to methyl, halogen, or nitro group, produces a marked increase in the boiling point. The effect on nitrophenol is shown below :

Compound	mp°C	bp°C
Phenol	42	181
Nitrobenzene	5.5	210
o-Nitrophenol	45	217
m-Nitrophenol	96	—
p-Nitrophenol	114	245

The greater volatility of the *ortho* isomer is due to *intramolecular* hydrogen bonding whereas the higher boiling point of the *para* isomer is due to *intermolecular* hydrogen bonding.

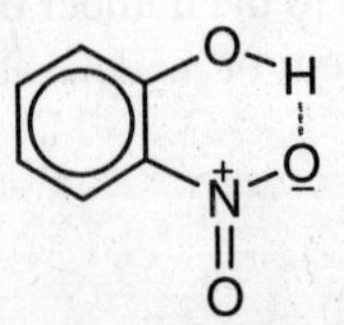

Intramolecular hydrogen bonding in *o*-Nitrophenol

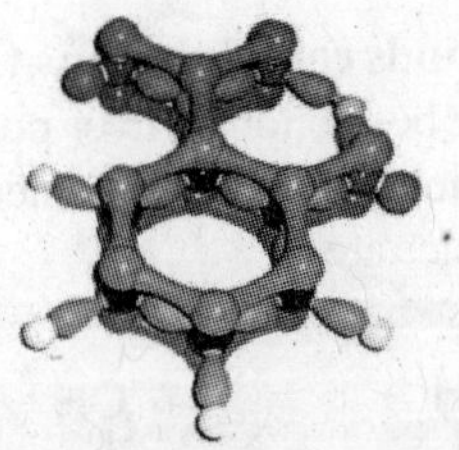

Orbital model of *o*-Nitrophenol showing intramolecular hydrogen bonding.

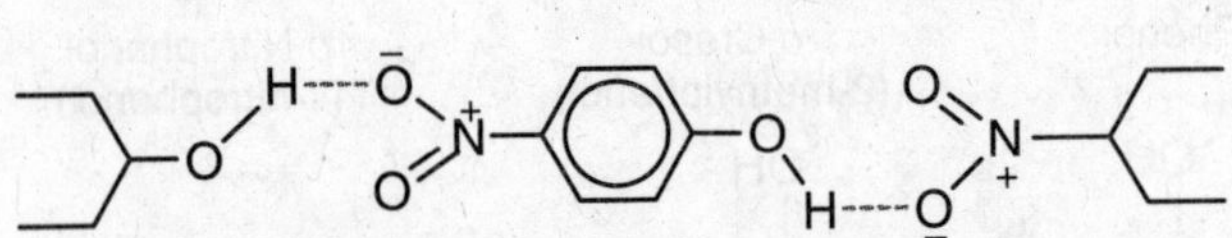

Intermolecular hydrogen bonding in *p*-Nitrophenol

Also, *p*-nitrophenol is more soluble in water than *o*-nitrophenol. This is because *p*-nitrophenol can associate with itself and also with water molecules. *o*-Nitrophenol cannot form intermolecular hydrogen bonds.

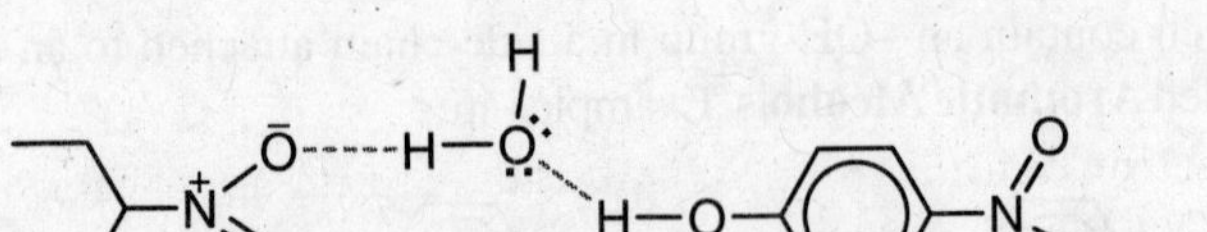

Intermolecular hydrogen bonds between water and *p*-Nitrophenol

The low solubility and higher volatility of *o*-nitrophenol enables it to be easily separated from *p*-nitrophenol by steam distillation.

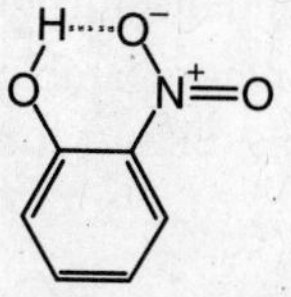

ortho nitrophenol
(Intermolecular hydrogen bonding)

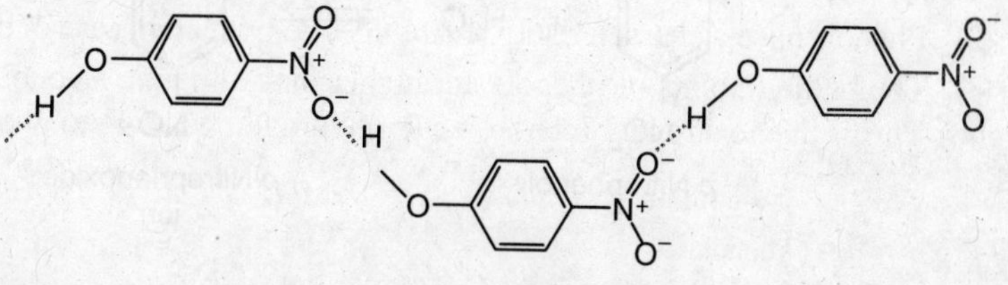

para nitrophenol
(Intermolecular hydrogen bonding)

ACIDITY OF PHENOLS

Phenols are much more acidic than alcohols but less so than carboxylic acids or even carbonic acid. This is indicated in order of acidity constants :

	Formula	**K_a (approx)**
Phenols	Ar–OH	10^{-19}
Alcohols	R–OH	10^{-18}
Carboxylic acids	R–COOH	10^{-5}
Carbonic acid	H_2CO_3	10^{-7}

Phenols are acidic due to the formation of stable phenoxide ions in aqueous solutions. For example, phenol itself gives phenoxide ion on dissociation.

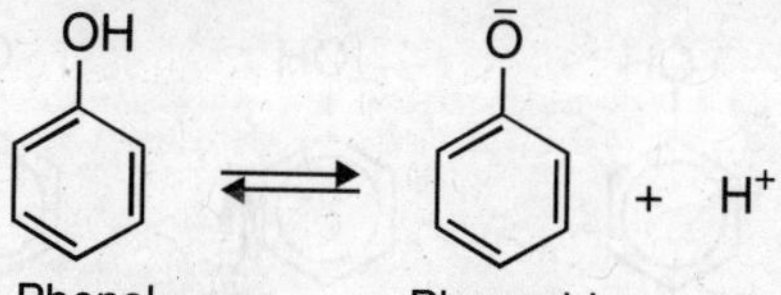

Phenol Phenoxide ion

The phenoxide ion is stable due to resonance :

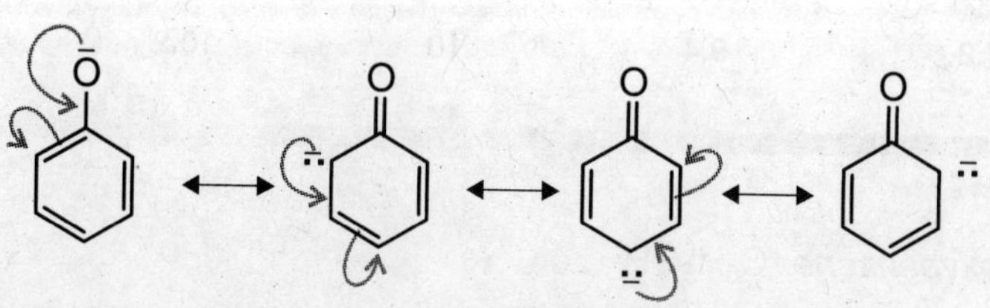

Notice that the negative charge is spread throughout the benzene ring, and thereby effectively dispersed. This charge delocalization is a stabilizing factor in the phenoxide ion. On the other hand, no resonance is possible in alkoxide ions (RO^-) derived from alcohols. The negative charge is concentrated (localized) on a single oxygen atom. Consequently, alcohols are much weaker acids than phenols.

EFFECT OF SUBSTITUENTS ON ACIDITY

(1) **Effect of Electron-Withdrawing Substituents**. An electron-withdrawing group (*e.g.*,–NO_2, –Cl, –CN, –CHO, –COOH) on the aromatic ring is acid-strengthening. It enables the ring to withdraw more electrons from the phenoxy oxygen. This stabilizes the phenoxide ion still further and results in a stronger acid. For example, *p*-nitrophenol is more acidic than phenol.

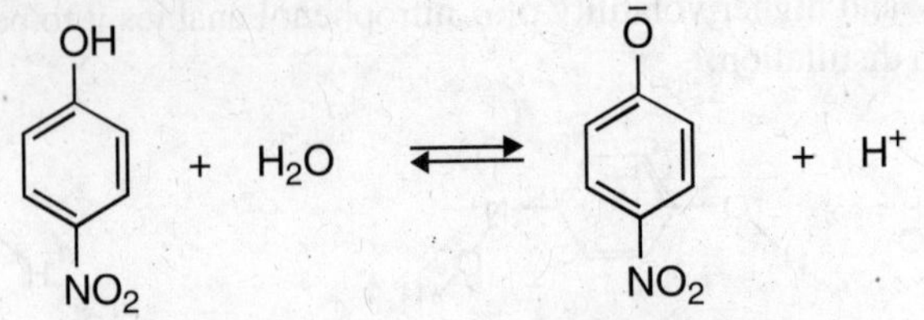

p-Nitrophenol   *p*-Nitrophenoxide ion

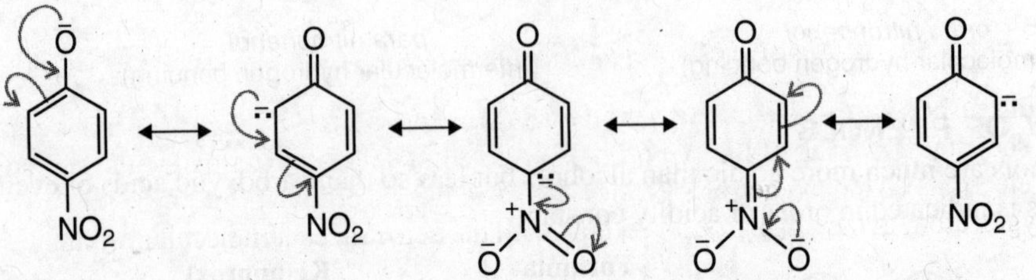

Resonance forms of *p*-Nitrophenoxide ion

(2) **Effect of Electron-Releasing Substituents**. An electron-releasing group (*e.g.*, $-CH_3$, $-OCH_3$, $-NH_2$) on the aromatic ring is acid-weakening. It strengthens the negative charge on phenoxy oxygen and inhibits the charge delocalization due to resonance. This destabilizes the phenoxide ion and results in a weaker acid. For example, *p*-cresol is less acidic than phenol.

The relative acid-strengths of some phenols are as follows :

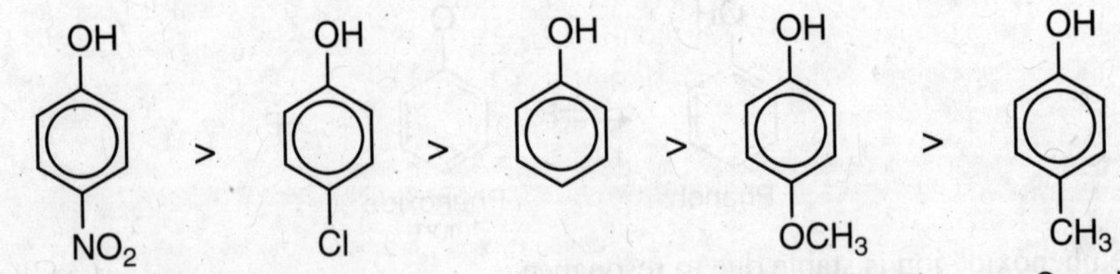

	p-Nitrophenol	*p*-Chlorophenol	Phenol	*p*-Methoxyphenol	*p*-Cresol
pK_a	7.2	9.4	10	10.2	10.3

PHENOL, *Hydroxybenzene*, C_6H_5OH

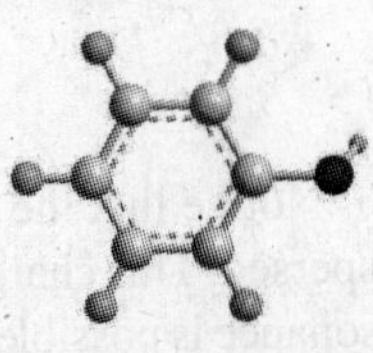

Phenol is the simplest and most important member of the series

Structure. In phenol, all ring carbon atoms are sp^2 hybridized and the oxygen atom of the –OH group is sp^3 hybridized. Fig. 40.1 shows the formation of sigma (σ) bonds in phenol.

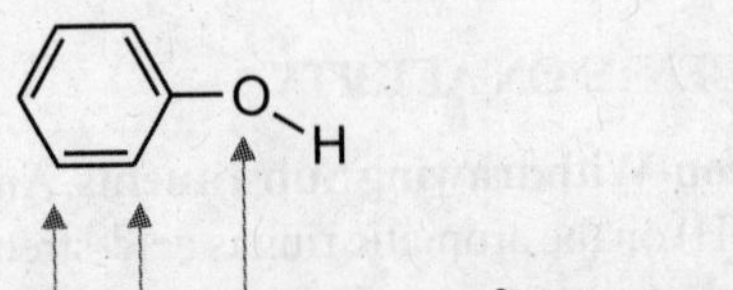

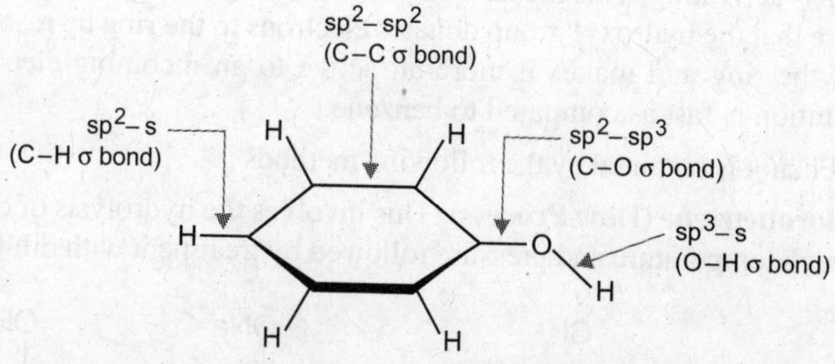

Fig. 40.1. Formation of sigma (σ) bonds in phenol.

All ring carbon atoms also have a *p* orbital and these are perpendicular to the plane of σ bonds of the ring. The lateral overlap of these *p* orbitals produces a delocalized π molecular orbital. The two lone-pair electrons on the oxygen atom reside in the sp^3 hybrid orbitals. One of the lone-pair electrons on the oxygen atom in the –OH group overlaps with the delocalized π molecular orbital.

Fig.40.2. Adjacent *p* orbitals overlap to form the delocalized π molecular orbital. One of the lone-pair electrons on the oxygen atom resides in sp^3 orbital and interacts with the delocalized π molecular orbital to form an extended π molecular orbital containing 8 electrons.

According to the resonance theory, phenol is considered to be hybrid of the following resonance structures.

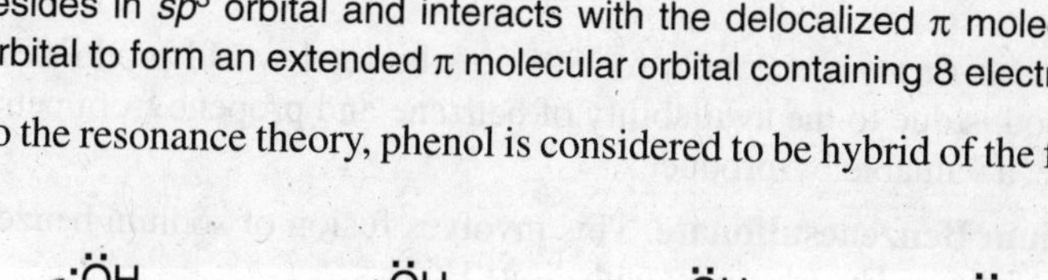

Resonance structures of phenol.

Notice that the *ortho* and *para* positions in the above resonance structures carry a negative charge. An electrophile (E^+) will attack these positions (opposite charges attract). Thus, the hydroxyl group directs all electrophiles to the *ortho* and *para* positions.

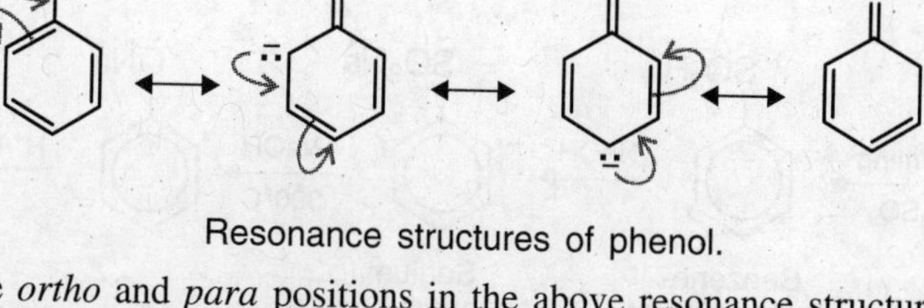

The –OH group is activating. This means that phenol undergoes electrophilic substitution faster than benzene. Notice that the hydroxyl group donates electrons to the ring by resonance. It increases electron density of the ring and makes it more attractive to an incoming electrophile. Thus, the electrophilic substitution is fast as compared to benzene.

Preparation. Phenol is obtained by the following methods :

(1) **From Chlorobenzene (Dow Process).** This involves the hydrolysis of chlorobenzene with aqueous NaOH at high temperature and pressure followed by treatment with dilute HCl.

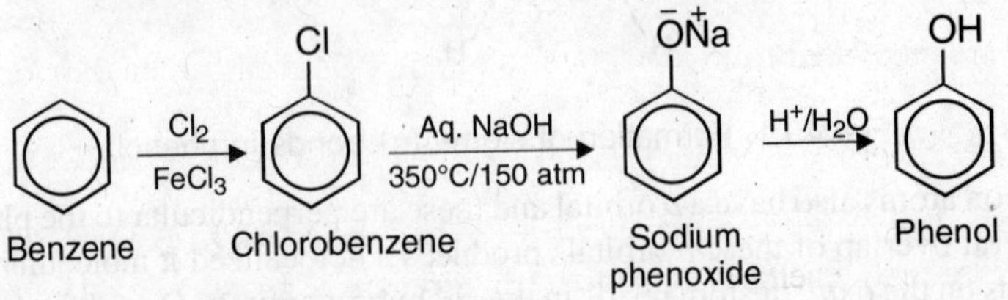

This process was first introduced in 1928 by the Dow Chemical Company of U.S.A.

(2) **From Cumene.** This involves air-oxidation of cumene (Isopropylbenzene) followed by treatment with dilute HCl.

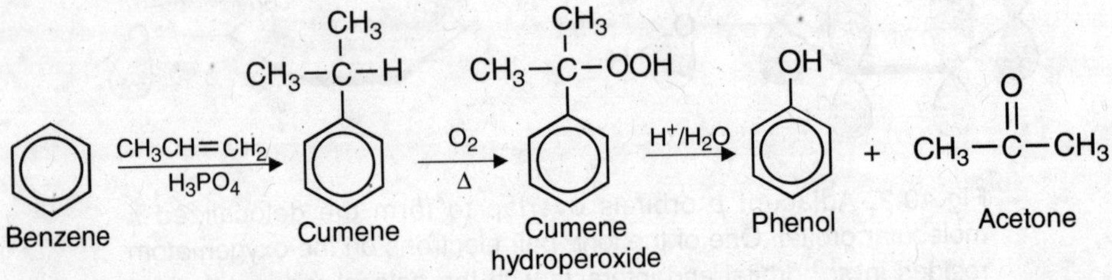

The above *Cumene Process* accounts for 80% of the total world production of phenol. The success of this method is due to the availability of benzene and propene from petroleum and to the formation of acetone, a valuable by-product.

(3) **From Sodium Benzenesulfonate.** This involves fusion of sodium benzenesulfonate with solid NaOH at 300°C followed by treatment with dilute HCl.

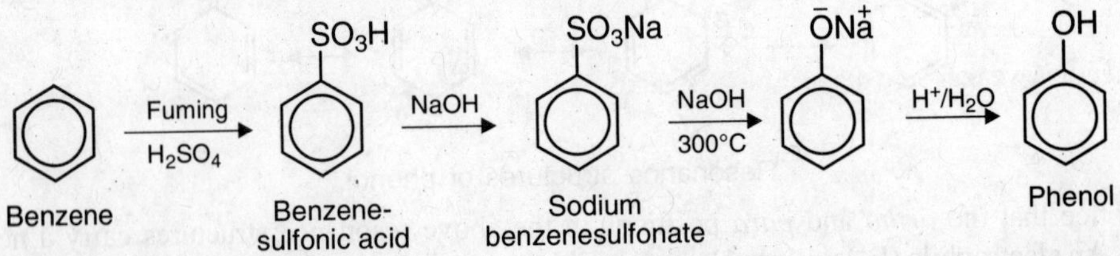

(4) **From Benzenediazonium Salts.** This reaction can be performed easily in the laboratory and simply requires warming a solution of benzenediazonium chloride, prepared from aniline, on a water bath at 50°C.

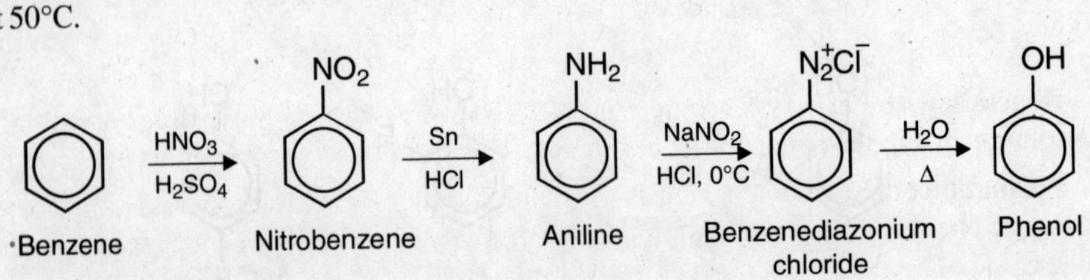

The phenol is recovered by steam distillation and extracted with diethyl ether.

(5) **From Coal Tar**. Coal tar provides a natural source of phenol and cresols, but nowadays provides less than 10% of the total supply. The *middle oil* fraction (170-240°C) of coal tar contains phenols, cresols, and naphthalene. The oil, when cooled, deposits solid naphthalene, which is removed by centrifuging the mixture. The oil left is agitated with NaOH solution when phenol and cresols dissolve as sodium salts.

$$C_6H_5OH + NaOH \longrightarrow C_6H_5\overset{-}{O}\overset{+}{Na} + H_2O$$

The phenols are recovered from the above solution by passing carbon dioxide through it.

$$2C_6H_5\overset{-}{O}\overset{+}{Na} + CO_2 + H_2O \longrightarrow 2C_6H_5OH + Na_2CO_3$$

Phenol is finally isolated from the resulting mixture of phenols by fractional distillation.

Properties. (*Physical*). Phenol is a colourless, hygroscopic, crystalline solid (*mp* 42°C ; *bp* 182°C). The liquid form of phenol containing about 5% water is known as **Carbolic Acid**. Below 65.8°C phenol is only partially miscible with water but above this temperature it is miscible in all proportions. It is readily soluble in ethanol and other organic solvents. It turns pink on exposure to air and light. It has a distinctive odor. Phenol is a poison when taken orally and produces painful blisters in contact with skin. The IR absorption frequencies of phenol are : 3200-3650 cm^{-1} (O–H stretching) ; 1230 cm^{-1} (C–O stretching) ; and 1400-1300 cm^{-1} (O–H bending).

(*Chemical*). The reactions of phenol are the reactions of the –OH group and the benzene ring.

REACTIONS OF –OH GROUP

(1) **Formation of Salts**. Phenol is acidic. It reacts with sodium hydroxide or sodium metal to form salts.

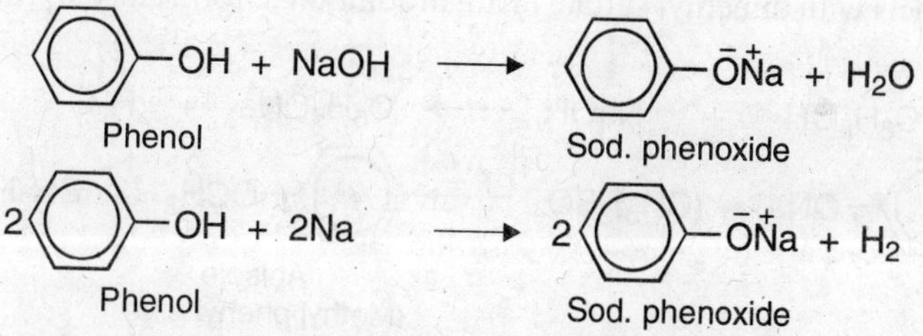

Phenol is weaker acid than carboxylic acids. It does not react with sodium carbonate or bicarbonate.

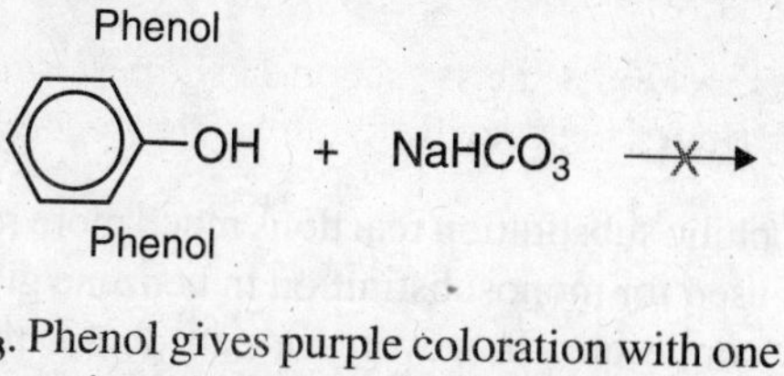

(2) **Reaction with $FeCl_3$**. Phenol gives purple coloration with one or two drops of neutral ferric chloride solution due to the formation of a complex. This reaction is given by most phenols.

(3) **Formation of Esters**. Phenol reacts with acid chlorides (or acid anhydrides) in aqueous alkali solution to give phenyl esters. The alkali first forms the phenoxide ion which then reacts with the acid chloride.

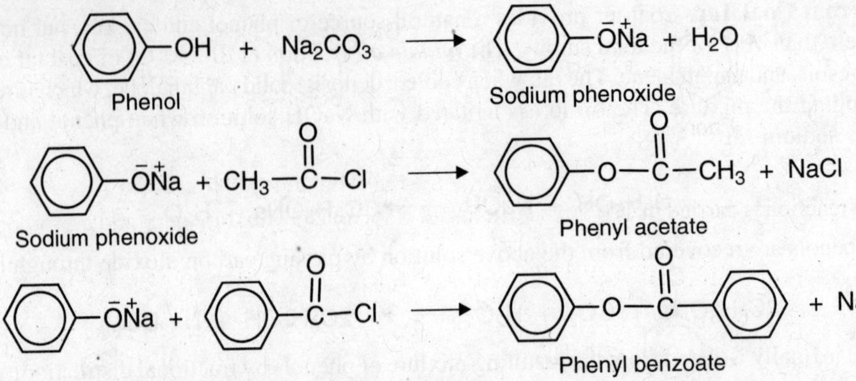

The reaction of phenol with benzoyl chloride is known as *Schotten-Baumann reaction.*

(4) **Formation of Ethers**. (*a*) Phenol reacts with alkyl halides in alkali solution to form phenyl ethers. The alkali first forms the phenoxide ion which then reacts with the alkyl halide.

$$C_6H_5OH + NaOH \longrightarrow C_6H_5\bar{O}\overset{+}{Na} + H_2O$$

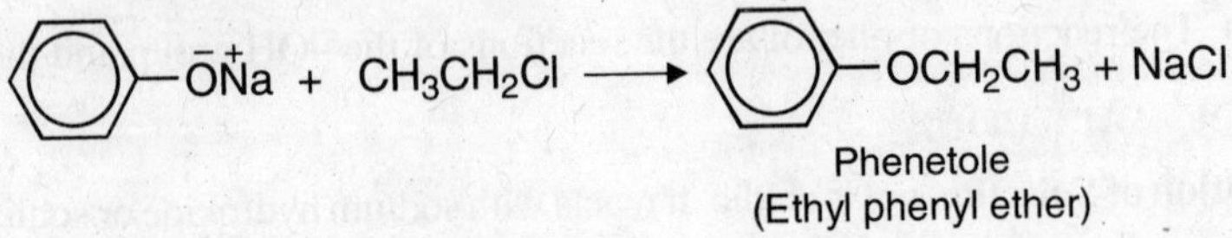

(*b*) Phenol reacts with dimethyl sulfate in alkali solution to form anisole (Methyl phenyl ether).

$$C_6H_5OH + NaOH \longrightarrow C_6H_5\bar{O}\overset{+}{Na} + H_2O$$

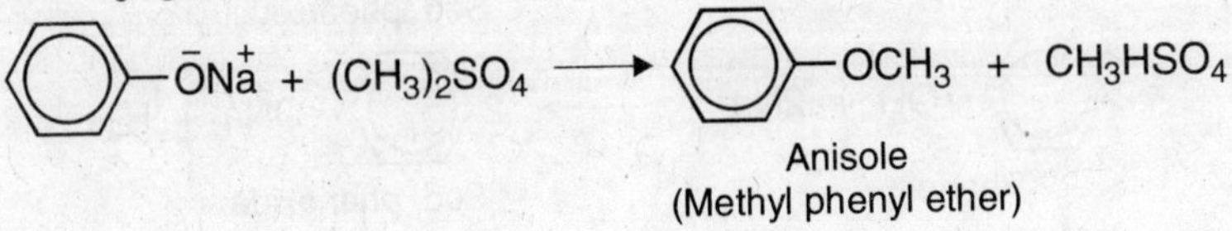

(5) **Reaction with Zinc Dust**. When phenol is distilled with zinc dust, low yield of benzene is obtained.

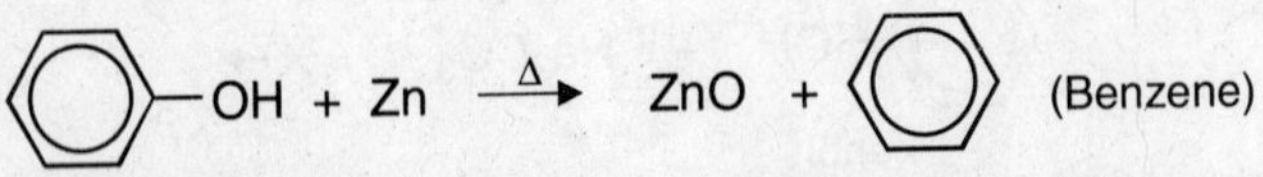

REACTIONS OF BENZENE RING

Phenol undergoes electrophilic substitution reactions much more readily as compared to benzene ring. The reaction conditions used for monosubstitution in benzene give trisubstitution with phenol. By moderating the reaction conditions, however, we can obtain the monosubstitution product. Remember that the –OH group in phenol is an *o,p*-director and activator (See Chapter *34* for discussion).

(6) **Halogenation**. Phenol reacts with bromine water (aqueous bromine) to give precipitate of 2,4,6-tribromophenol. Chlorine reacts in the same way.

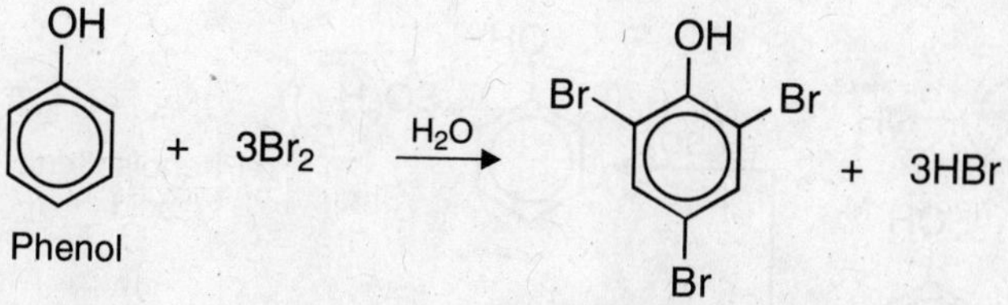

2,4,6-Tribromophenol

If the reaction is carried in CS_2 or CCl_4 (non-polar solvents), a mixture of *o*- and *p*-bromophenol is formed.

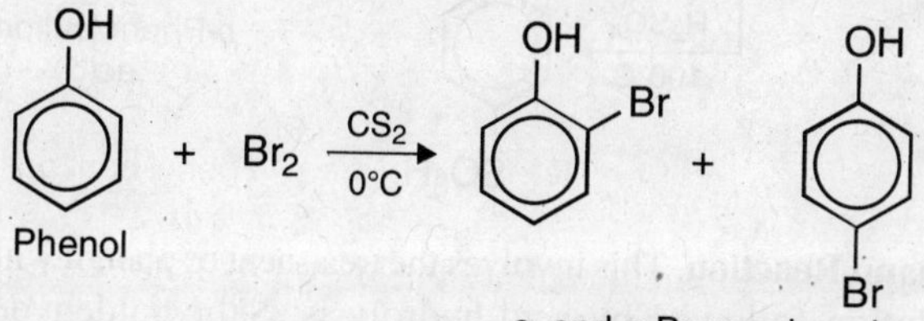

o- and *p*-Bromophenol

(7) **Nitration**. Phenol reacts with dilute nitric acid to give a mixture of *o*- and *p*-nitrophenol.

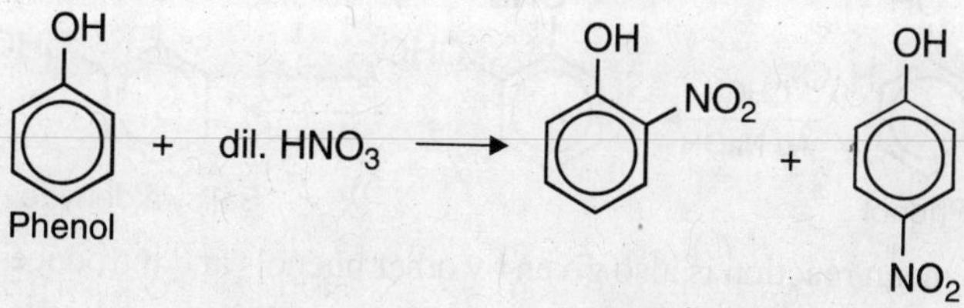

o- and *p*-Nitrophenol

With concentrated nitric acid, picric acid is formed.

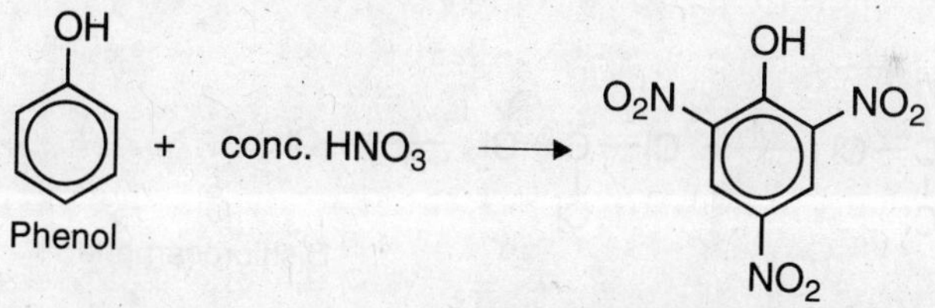

Picric acid (2,4,6-Trinitrophenol)

(8) **Nitrosation**. Phenol reacts with nitrous acid (obtained from $NaNO_2$/HCl at 5°C) to form *p*-nitrosophenol exclusively.

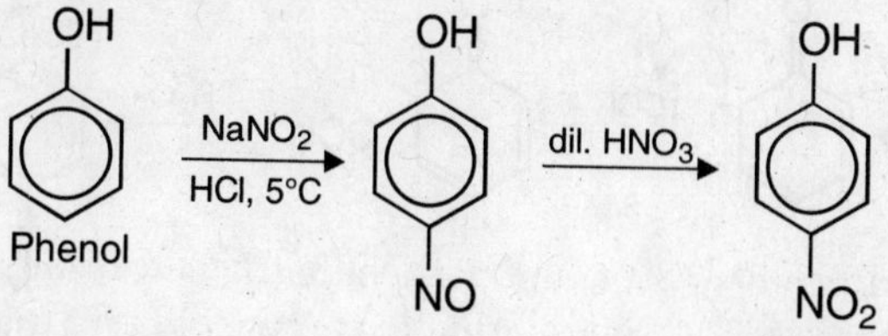

p-Nitrophenol

p-Nitrosophenol on oxidation with dilute nitric acid gives *p*-nitrophenol exclusively.

(9) **Sulfonation**. When phenol is treated with concentrated sulfuric acid at 20°C, *o*-phenolsulfonic acid is the main product. At 100°C, *p*-phenolsulfonic acid is the main product.

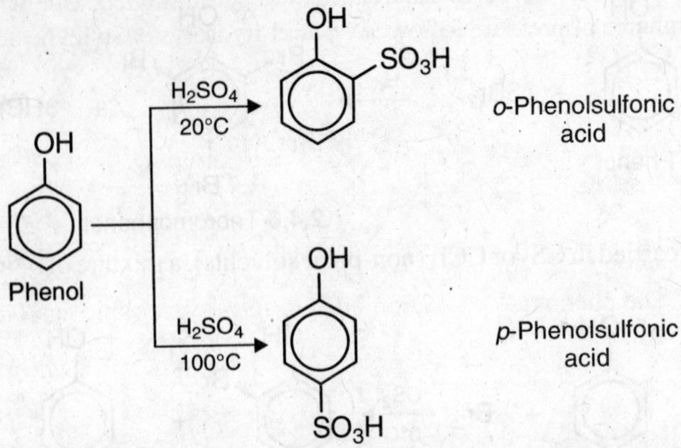

(10) **Reimer-Tiemann Reaction**. This involves the treatment of phenol with chloroform in aqueous sodium hydroxide solution followed by acid-hydrolysis. Salicylaldehyde is formed. If carbon tetrachloride is used in place of chloroform, salicylic acid is formed.

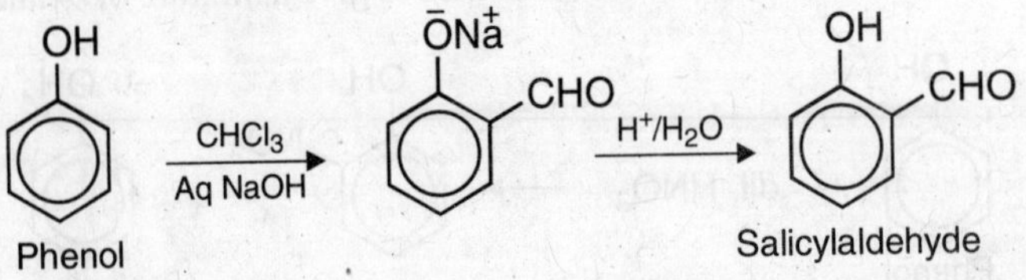

The Reimer-Tiemann reaction is also given by other phenols and introduces –CHO group in the *ortho* position.

MECHANISM. In the above reaction, chloroform first reacts with sodium hydroxide to produce dichlorocarbene.

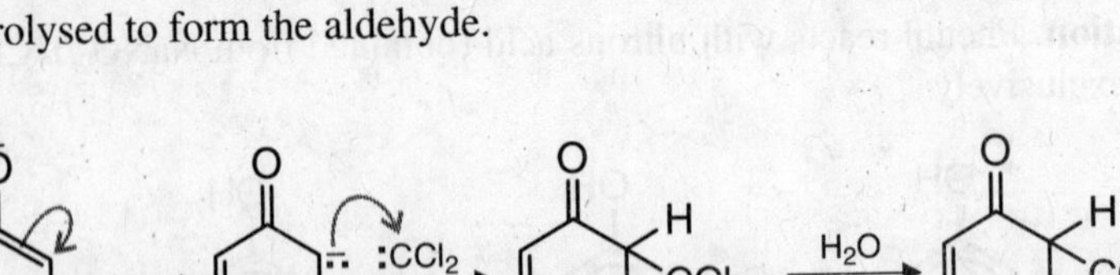

The electron-deficient dichlorocarbene then reacts with sodium phenoxide to form a dihalide. This is then hydrolysed to form the aldehyde.

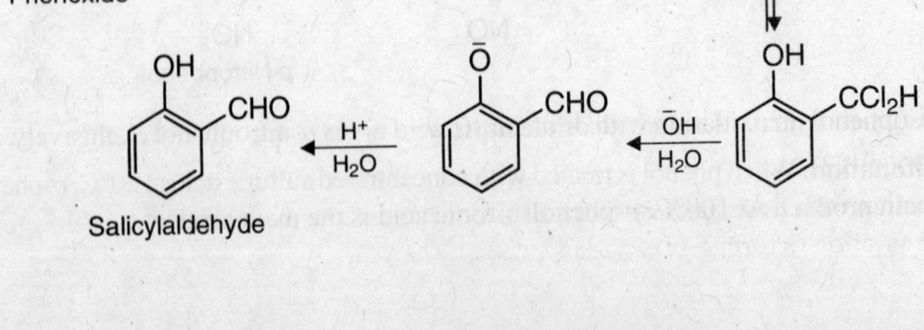

(11) **Kolbe Reaction**. This involves the treatment of sodium phenoxide with carbon dioxide at 125°C under 6 atmospheres of pressure followed by acid-hydrolysis. Salicylic acid is formed.

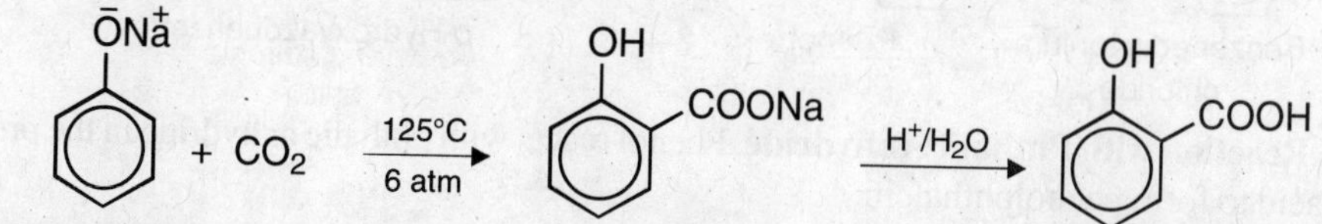

Sod. phenoxide — Sod.salicylate — Salicylic acid

MECHANISM. The phenoxide carbanion adds to the electrophilic carbon of carbon dioxide. Three steps are involved :

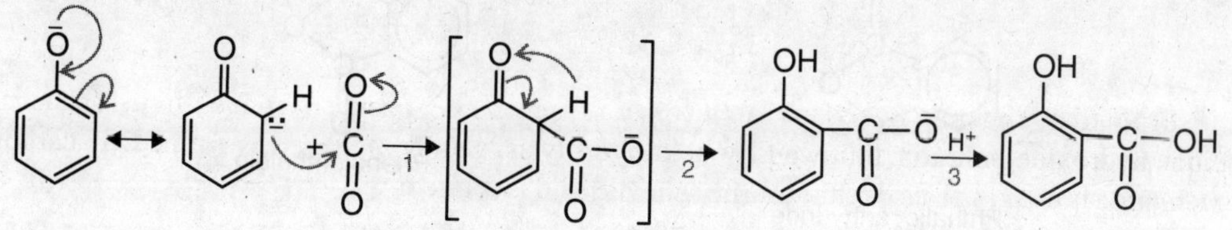

The Kolbe reaction is also given by other phenols and introduces –COOH group in the *ortho* position. This reaction is also called *Kolbe-Schmidt reaction*.

(12) **Gatterman Reaction**. This involves the treatment of phenol with a mixture of hydrogen cyanide and hydrogen chloride in the presence of aluminium chloride catalyst. Salicylaldehyde is formed.

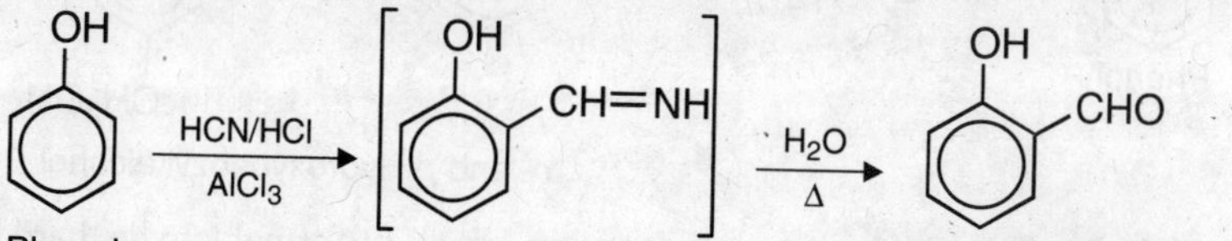

Phenol

Salicylaldehyde
(*o*-Hydroxybenzaldehyde)

(13) **Fries Rearrangement**. The phenol is first treated with acetic anhydride in the presence of aqueous sodium hydroxide to give phenyl acetate. The ester is then heated with aluminium chloride catalyst when the acyl group migrates from the phenolic oxygen to an *ortho* or *para* position of the ring. The product is a mixture of *o*- and *p*-hydroxyacetophenone.

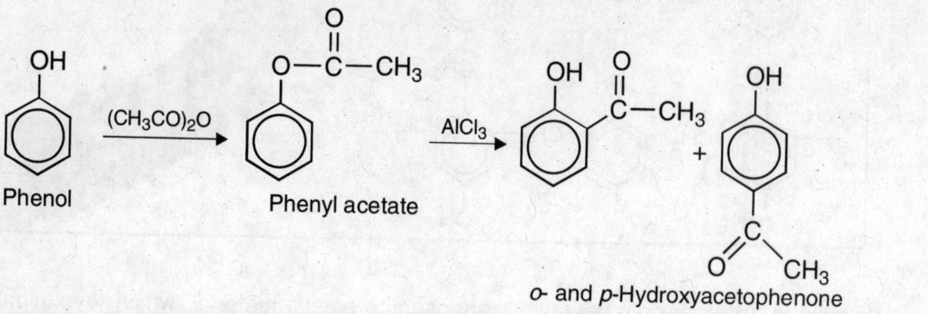

o- and *p*-Hydroxyacetophenone

The Fries rearrangement is also given by other phenols and introduces –COR group in the *ortho* and/or *para* positions.

(14) **Reaction with Benzenediazonium Chloride**. Phenol couples with benzenediazonium chloride in an alkaline solution to form *p*-hydroxyazobenzene (a dye).

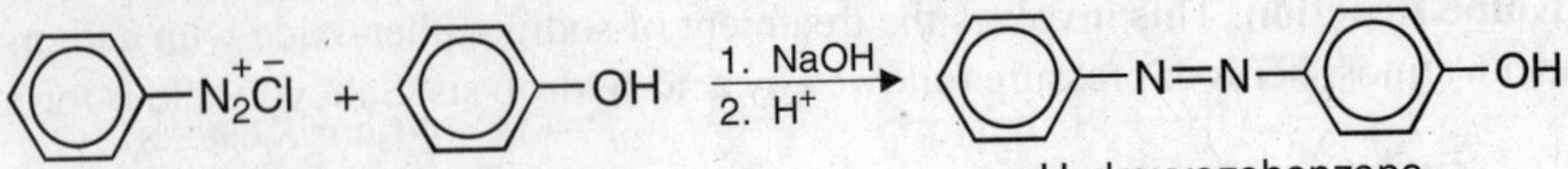

Benzenediazonium chloride — Phenol — *p*-Hydroxyazobenzene

(15) **Reaction with Phthalic Anhydride**. Phenol reacts with phthalic anhydride in the presence of sulfuric acid to form phenolphthalein.

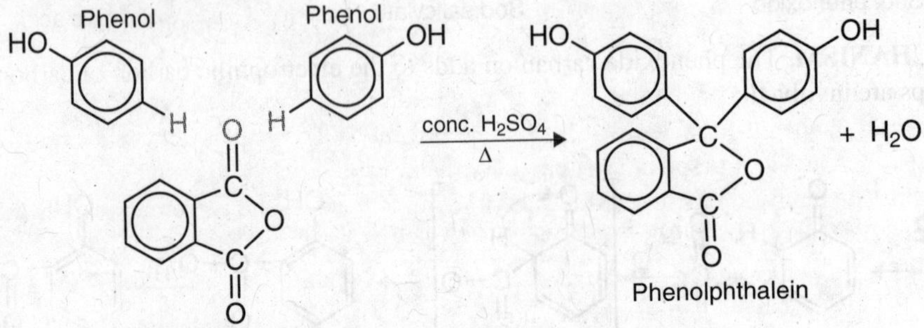

Phthalic anhydride — Phenolphthalein

(16) **Reaction with Formaldehyde**. When phenol is treated with an alkaline solution of formaldehyde, a mixture of *o*- and *p*-hydroxybenzyl alcohol is formed.

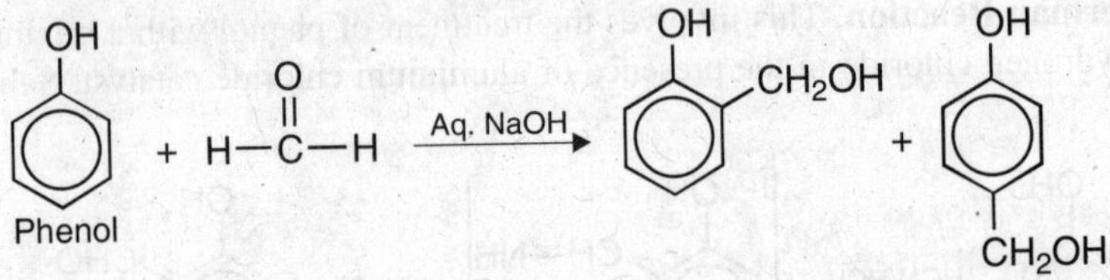

Phenol — *o*- and *p*-Hydroxybenzyl alcohol

If the reaction is carried at high temperature and in excess of formaldehyde, hard thermosetting plastic *Bakelite* is formed.

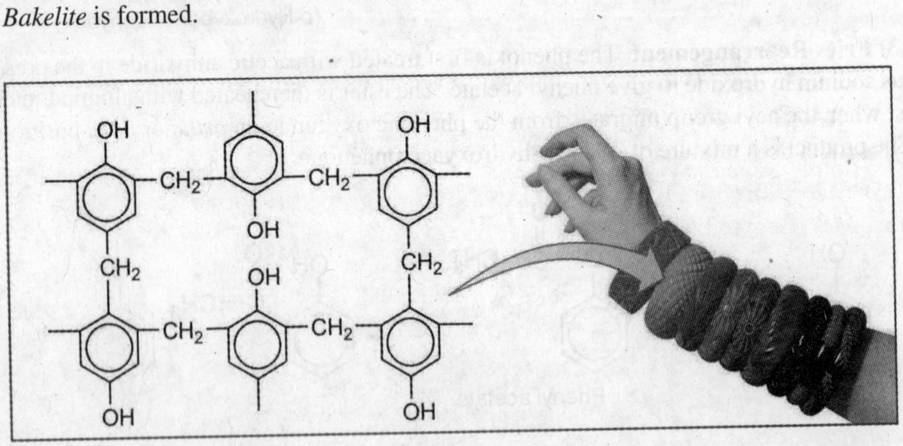

Bakelite is an amazingly versatile thermosetting plastic material, which revolutionized the manufacture of mass produced small consumer goods. Best known applications were the manufacture of telephones, cameras and radios for the home, although it was also widely used in commercial and industrial settings, particularly for electrical fittings appliances, because of the excellent insulating and heat resisting properties.

(17) **Oxidation**. Phenol undergoes oxidation with air or chromic acid to form *p*-benzoquinone.

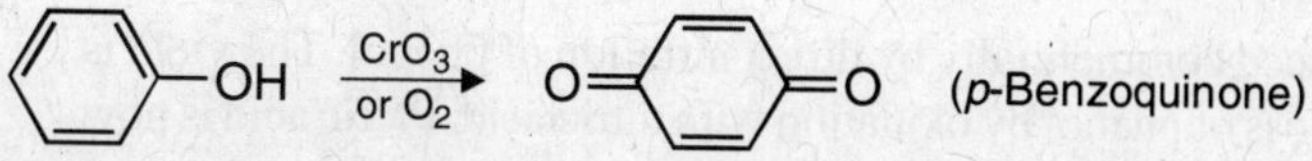

(18) **Catalytic Hydrogenation**. Phenol on catalytic hydrogenation gives cyclohexanol. A mixture of its vapor and hydrogen is passed over nickel catalyst at 160°C.

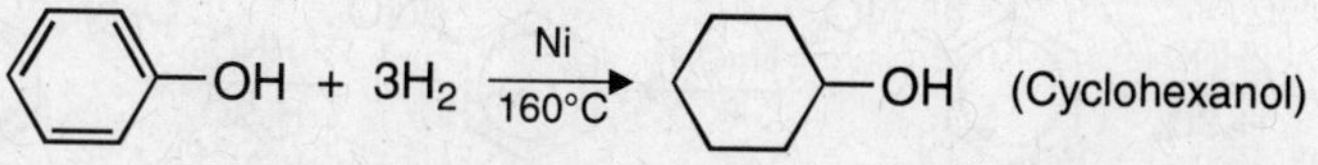

Uses of Phenol. (1) About half of the total world's production of phenol (two billion pounds) is used for making phenol-formaldehyde resins *e.g.*, Bakelite ; (2) Phenol is used as a starting material for ; (*i*) drugs such as Salol, Aspirin ; (*ii*) phenolphthalein and several other dyes : (*iii*) explosives *e.g.*, picric acid ; (*iv*) many other valuable products of common use : wood preservatives, herbicides, nylon, antiseptics (**Dettol**), gasoline additives etc.

COMPARISON OF PHENOL AND ETHYL ALCOHOL

The following six reactions describe the basic differences in the reactivity of phenol and ethyl alcohol :

(1) **Reaction with NaOH**. Phenol is more acidic than ethyl alcohol. It reacts with NaOH to give sodium phenoxide. Ethyl alcohol does not react with NaOH.

$$C_6H_5OH + NaOH \longrightarrow C_6H_5\bar{O}\overset{+}{Na} + H_2O$$

$$CH_3CH_2OH + NaOH \longrightarrow \text{No reaction}$$

(2) **Catalytic Reduction**. Phenol undergoes reduction with H_2/Ni to form cyclohexanol. Ethyl alcohol does not give this reaction.

$$C_6H_5OH + 3H_2(Ni) \longrightarrow C_6H_{11}OH \text{ (Cyclohexanol)}$$

(3) **Reaction with Diazonium Salts**. Phenol reacts with benzenediazonium chloride to form a dye. Ethyl alcohol does not form dyes.

$$C_6H_5N_2^+Cl^- + C_6H_5OH \longrightarrow C_6H_5N{=}NC_6H_4OH \text{ (Azo Dye)}$$

(4) **Reaction with FeCl$_3$**. Phenol gives purple coloration with neutral ferric chloride solution. Ethyl alcohol does not react.

(5) **Reaction with Carboxylic Acids**. Ethyl alcohol reacts with carboxylic acids to form esters. Phenol does not give this reaction.

$$CH_3-\overset{O}{\overset{\|}{C}}-OH + CH_3CH_2OH \longrightarrow CH_3-\overset{O}{\overset{\|}{C}}-OCH_2CH_3 + H_2O$$

$$CH_3-\overset{O}{\overset{\|}{C}}-OH + C_6H_5OH \longrightarrow \text{No reaction}$$

(6) **Reaction with Halogen Acids**. Ethyl alcohol reacts with halogen acids to form alkyl halides. Phenol does not give this reaction.

$$CH_3CH_2OH + HBr \longrightarrow CH_3CH_2Br + H_2O$$

$$C_6H_5OH + HBr \longrightarrow \text{No reaction}$$

PICRIC ACID, *2,4,6-Trinitrophenol*

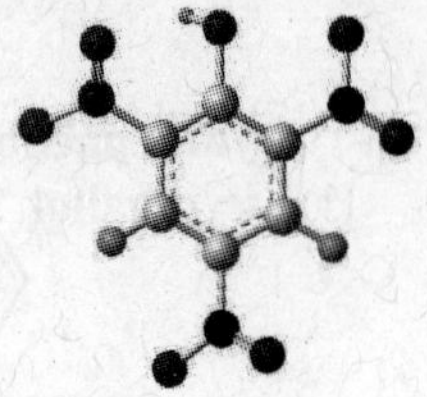

It is not produced commercially by direct nitration of phenol. The yield is poor owing to the loss of phenol by oxidation with nitric acid. Picric acid is now obtained through a synthesis that begins with chlorobenzene.

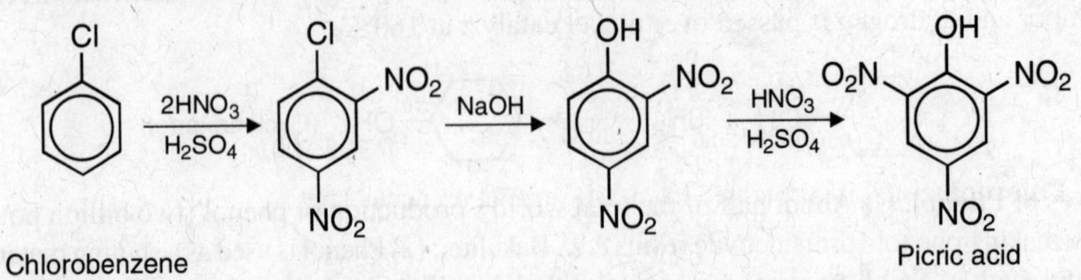

Picric acid is a yellow crystalline solid, mp 122°C. It is sparingly soluble in cold water giving intensely yellow solutions. It is poisonous. Picric acid, owing to the presence of three electron-withdrawing $-NO_2$ groups, is nearly as acidic as mineral acids. Unlike other phenols, it decomposes cabonates. Picric acid is used for making explosives **Lyddite** and **Melinite**. It is a valuable yellow dye for wool.

DI- AND TRIHYDRIC PHENOLS

Dihydric phenols are obtained by methods similar to those used for phenol.

Dihydric phenols are colourless solids having melting points : Catechol 104°C ; Resorcinol 110°C ; Quinol (Hydroquinone) 173°C. Their reactions are similar to phenol.

(1) **Catechol**. It is obtained by hydrolysis of *o*-dichlorobenzene with dilute sodium hydroxide solution at 200°C and in the presence of copper sulfate catalyst.

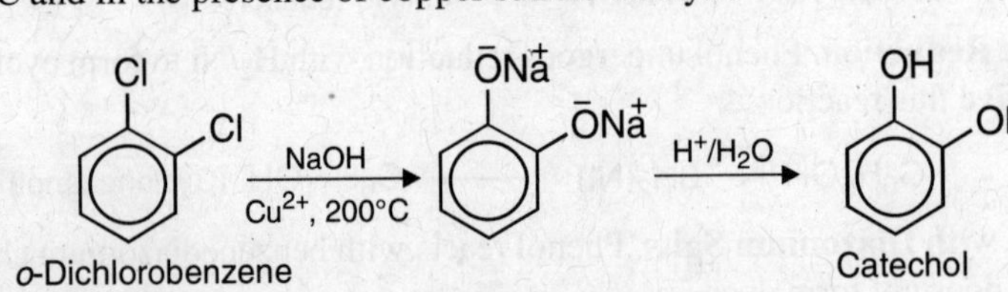

(2) **Resorcinol**. It is obtained from *m*-benzenesulfonic acid.

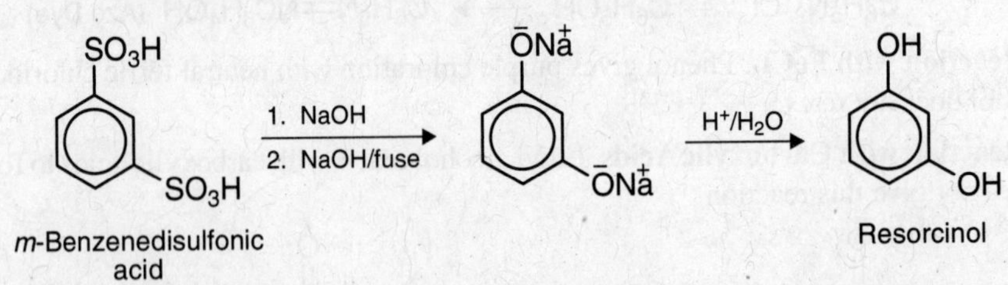

(3) **Quinol** (**Hydroquinone**). It is obtained by reduction of *p*-benzoquinone with iron in the presence of water. *p*-Benzoquinone is prepared by oxidation of aniline.

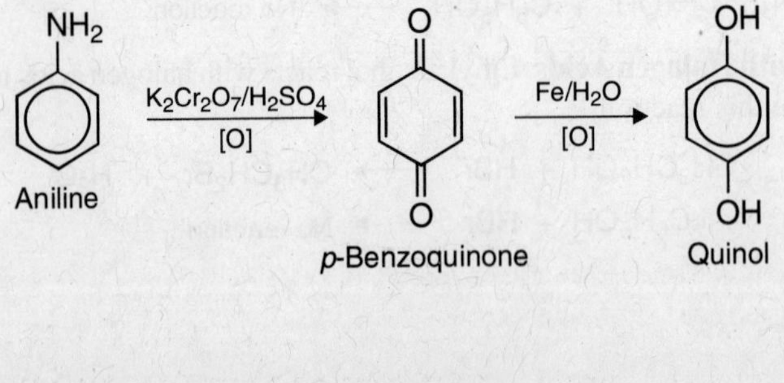

Trihydric phenols. Pyrogallol and phloroglucinol are important trihydric phenols.

(1) **Pyrogallol**. It is obtained by heating gallic acid at 220°C.

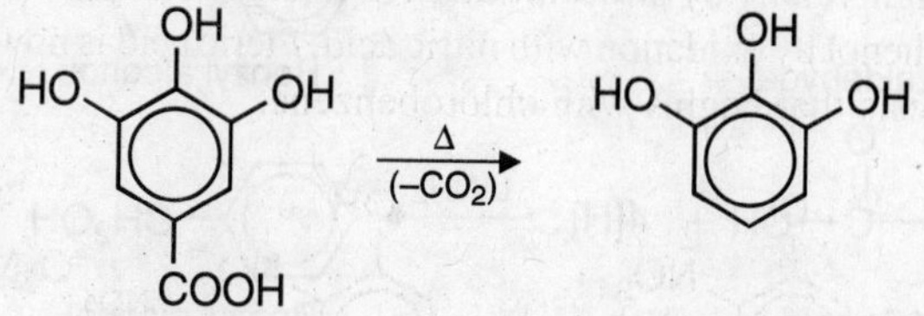

Gallic acid → Pyrogallol

(2) **Phloroglucinol**. It is obtained from 2,4,6-trinitrotoluene (TNT).

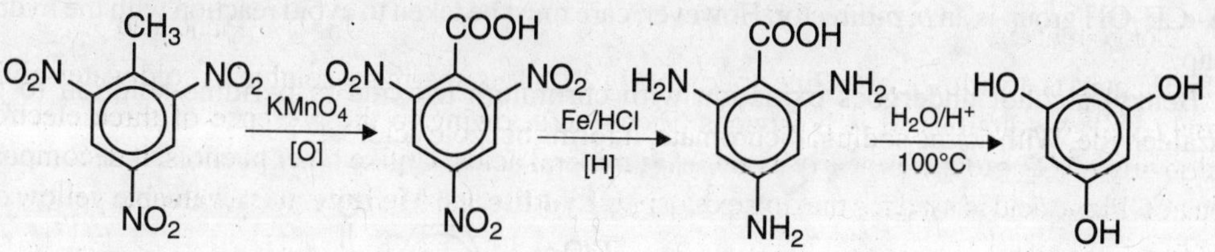

TNT → Phloroglucinol

Houben-Hoesch Reaction. This reaction involves the treatment of a dihydric or a trihydric phenol with an alkyl cyanide in the presence of zinc chloride and hydrogen chloride. The product on hydrolysis yields a phenolic ketone.

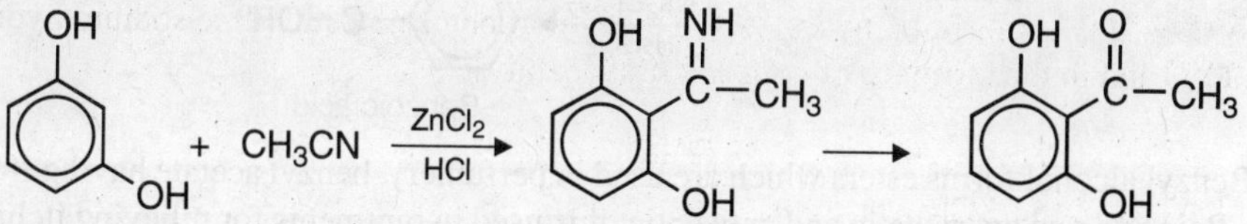

Resorcinol (Intermediate) 2,6-Dihydroxyacetophenone

The Houben-Hoesch reaction provides a method for indirect ring acylation of polyhydric phenols. Phenol does not give this reaction.

AROMATIC ALCOHOLS

Aromatic alcohols are compounds containing a hydroxy group in the side-chain, and may be regarded as aryl derivatives of the aliphatic alcohols.

BENZYL ALCOHOL, $C_6H_5CH_2OH$

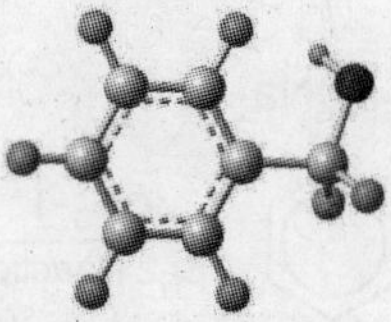

It is the simplest aromatic alcohol and is isomeric with cresols.

Preparation. Benzyl alcohol is obtained :

(1) By hydrolysis of benzyl chloride with aqueous sodium hydroxide.

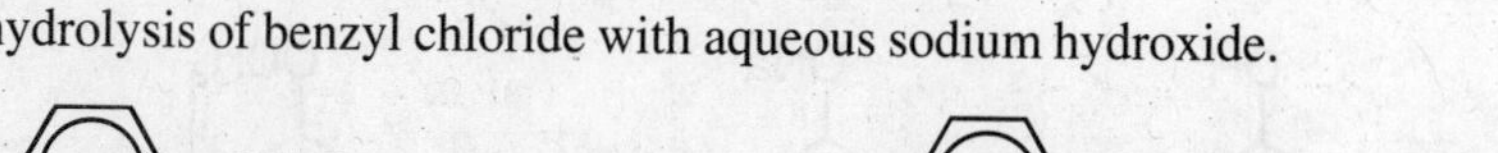

Benzyl chloride → Benzyl alcohol

(2) By reduction of benzaldehyde or benzoic acid.

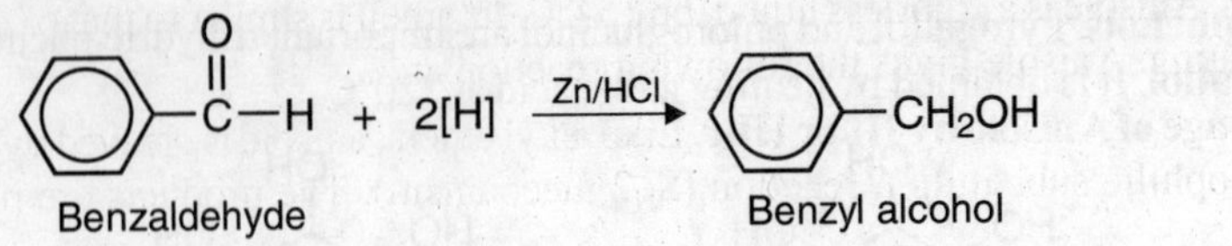

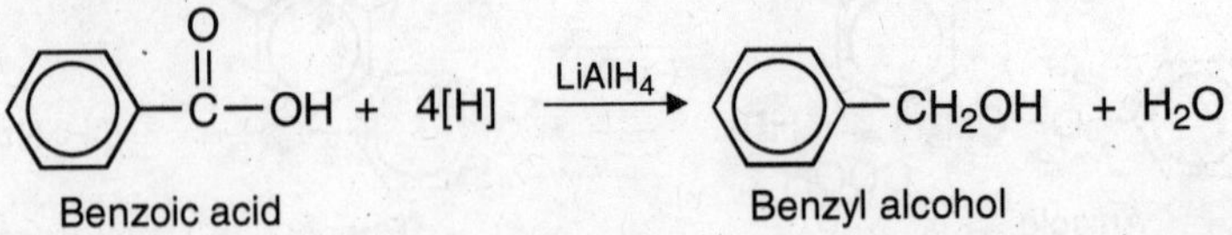

Properties. Benzyl alcohol is a colourless liquid, *bp* 205°C. It is sparingly soluble in water. It behaves like aliphatic primary alcohols and undergoes electrophilic substitution in the benzene ring. The $-CH_2OH$ group is an *o*, *p*-director. However, care must be taken to avoid reaction with the hydroxy group.

Benzyl alcohol undergoes oxidation with chromium trioxide in pyridine solution to give benzaldehyde. With acidic sodium dichromate, it forms benzoic acid.

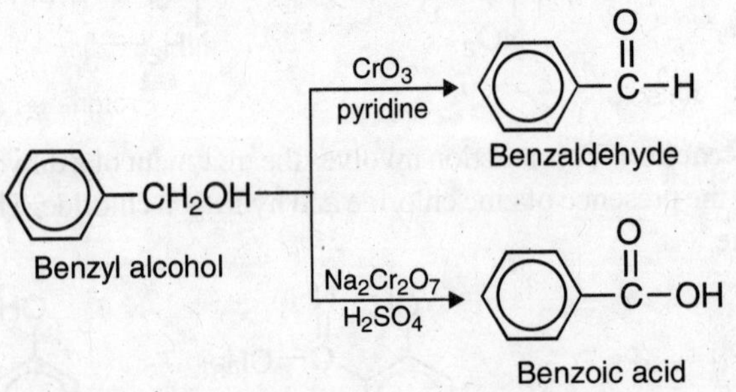

Uses. Benzyl alcohol forms esters which are used in perfumery, benzyl acetate having fragrance of *Jasmine*. Being a local anaesthetic and antiseptic, it is used in ointments for relieving itching.

AROMATIC ETHERS

ANISOLE, *Methoxybenzene*, $C_6H_5OCH_3$

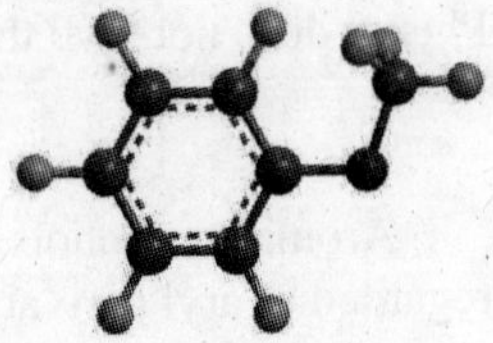

Preparation. Anisole can be prepared as follows :

(1) By treatment of sodium phenoxide with methyl iodide. (*Williamson* synthesis)

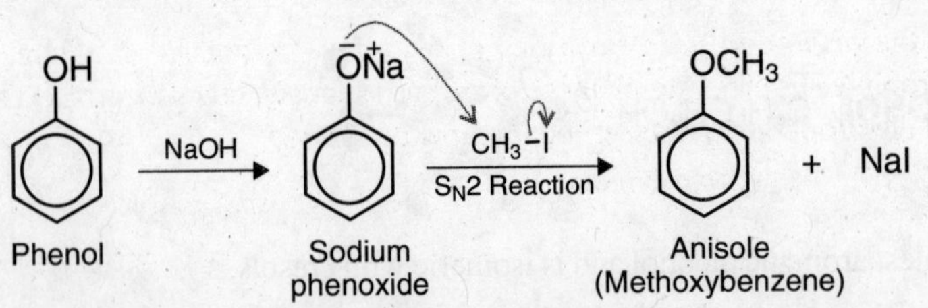

(2) By treatment of sodium phenoxide with dimethyl sulfate.

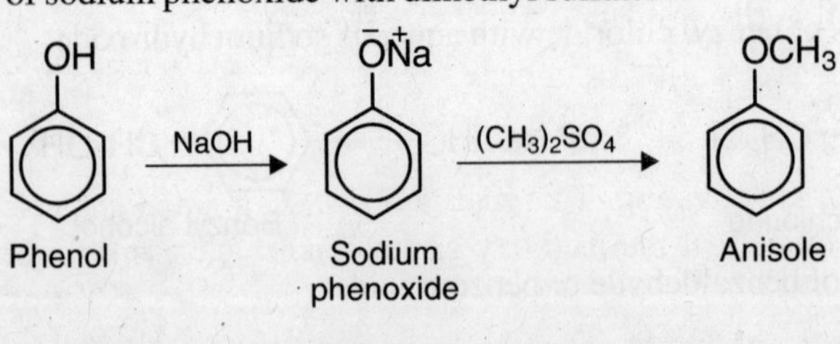

Properties. Anisole is a colorless liquid, bp 154°C. Its smell is similar to that of *anise seed*, and can be quite nauseating. Anisole gives the following reactions :

(1) **Cleavage of Anisole by HI or HBr**. Like alkyl ethers, anisole is cleaved by strong acids HI or HBr in a nucleophilic substitution reaction (S_N2 mechanism). The products are phenol and methyl bromide.

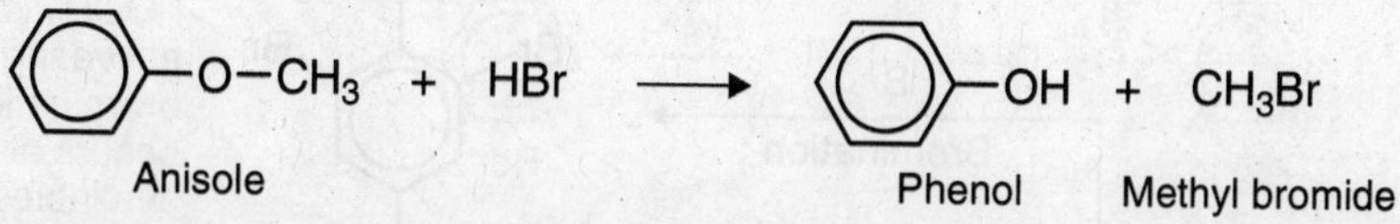

The phenol does not react further to produce bromophenol because the phenol carbon-oxygen bond is very strong because of delocalization and because phenyl cations do not form readily.

MECHANISM. Following steps are involved :

Step 1. Protonation of ether oxygen in anisole.

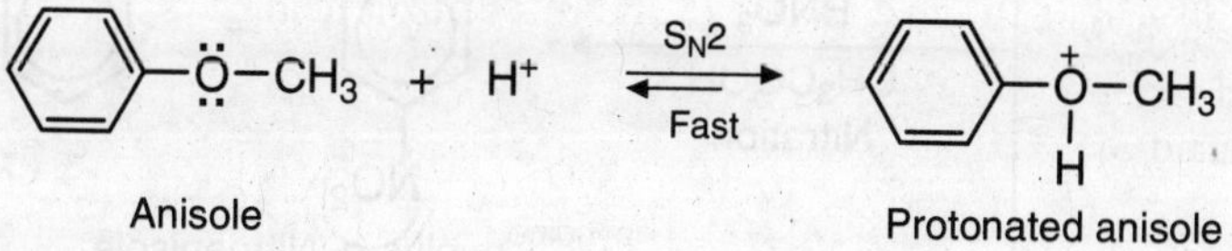

Step 2. The bromide ion attacks protonated anisole to form phenol and methyl bromide.

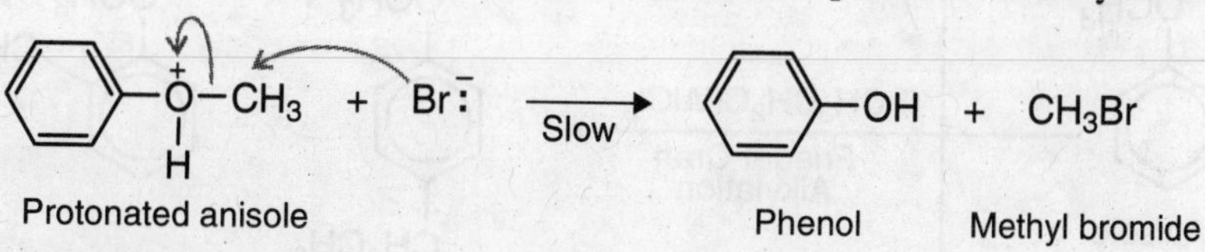

(2) **Electrophilic Substitution**. The methoxy group in anisole is an *ortho*, *para* director and strongly activating.

EXPLANATION. Anisole can be represented as a hybrid of the following resonance structures :

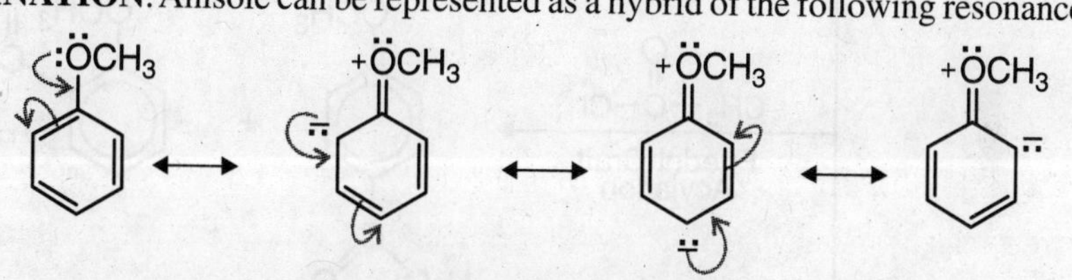

Resonance structures of anisole.

Notice that the *ortho* and *para* positions in the above resonance structures carry a negative charge. An electrophile (E^+) will attack these positions (opposite charges attract). Thus, the methoxy group directs all electrophiles to the *ortho* and *para* positions.

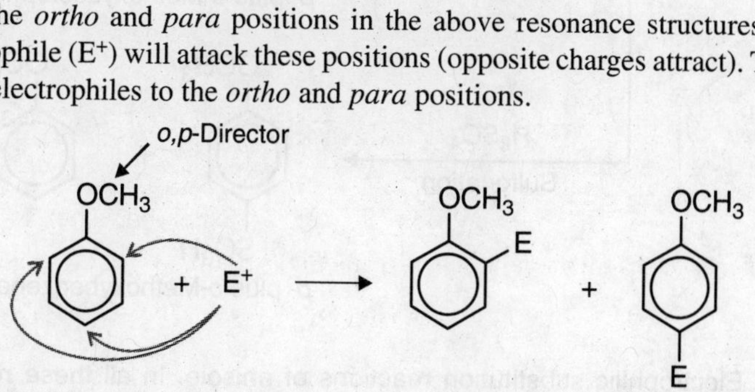

The $-OCH_3$ group is activating. This means that anisole undergoes electrophilic substitution faster than benzene. Notice that the methoxy group donates electrons to the ring by resonance. It increases electron density of the ring and makes it more attractive to an incoming electrophile. Thus, the electrophilic substitution is fast as compared to benzene.

Examples of electrophilic substitution reactions of anisole are given in Figure 40.3.

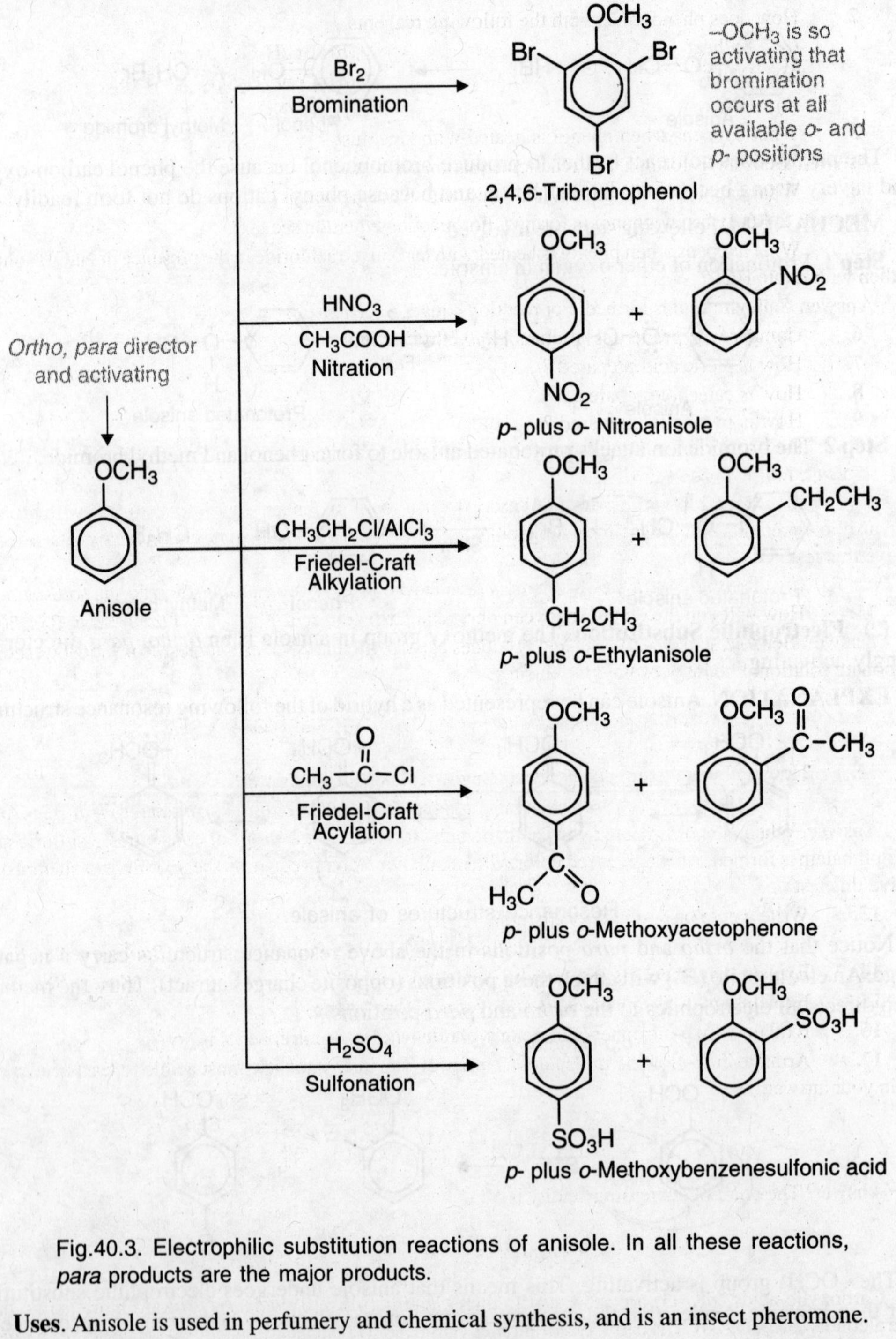

Fig.40.3. Electrophilic substitution reactions of anisole. In all these reactions, *para* products are the major products.

Uses. Anisole is used in perfumery and chemical synthesis, and is an insect pheromone.

Examination Questions

1. How is phenol prepared ? Give four methods.

2. How does phenol react with the following reagents ?

(*a*) Zn/heat (*b*) Br_2/H_2O

(*c*) HNO_3 (*d*) Conc. H_2SO_4

(*e*) HNO_2 (*f*) H_2/Ni

3. What happens when phenol is heated with zinc dust ?

Answer. Benzene is formed. For reaction equation see text.

4. What happens when phenol is treated with bromine water ?

Answer. 2,4,6-Tribromophenol is formed. For reaction equation see text.

5. What happens when phenol is heated with carbon tetrachloride in the presence of NaOH solution and then with dilute HCl ?

Answer. Salicylic acid is formed. For reaction equation see text

6. Compare the properties of phenol and ethyl alcohol.

7. How is picric acid prepared ? (*Kanpur BSc, 2015*)

8. How is catechol prepared ?

9. How is pyrogallol prepared ?

10. How will you distinguish between phenol and ethyl alcohol ? (*Osmania BSc, 2014*)

Answer. Following tests are used :

(*a*) **Reaction with Bromine Water.** When a solution of phenol in water is treated with bromine water, a yellow precipitate of 2,4,6-tribromophenol is formed. For reaction equation see text. Ethyl alcohol does not give this test.

(*b*) **Reaction with $FeCl_3$.** Phenols give purple color with neutral ferric chloride solution.

11. How will you distinguish between phenol and benzoic acid ? (*Tamil Nadu BSc, 2014*)

Answer. Benzoic acid dissolves in and produces carbon dioxide (seen as bubbles) from aqueous sodium bicarbonate solution. Phenol does not give this test.

$$\underset{\text{Benzoic acid}}{C_6H_5COOH} + NaHCO_3 \longrightarrow \underset{\text{Sodium benzoate}}{C_6H_5CO\overset{-}{O}\overset{+}{Na}} + CO_2 + H_2O$$

12. How will you distinguish between phenol and benzyl alcohol ?

(*Aligarh BSc, 2014 ; Madras BSc, 2015 ; North Eastern Hill BSc, 2015*)

Answer. When phenol is heated with phthalic anhydride and a few drops of concentrated sulfuric acid, phenolphthalein is formed, which gives red color with alkali. For reaction equation see text. Benzyl alcohol does not give this test.

13. Write a note on : Acidity of Phenols.

(*Baroda BSc, 2014 ; Patna BSc, 2014 ; Bangalore BSc, 2015*)

14. Explain why phenol is more acidic than ethyl alcohol ? (*Ranchi BSc, 2014*)

15. Explain why *p*-nitrophenol is a stronger acid than phenol ?

16. Explain why *o*-nitrophenol is steam volatile whereas *p*-nitrophenol is not ?

17. Arrange the following molecules in order of decreasing acidity (most acidic to least acidic) and explain your answer.

CH_3CH_2OH CH_3CO_2H 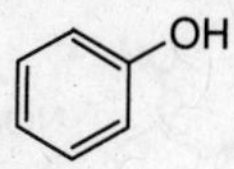

Answer. The order of decreasing acidity is :

CH_3CH_2OH > 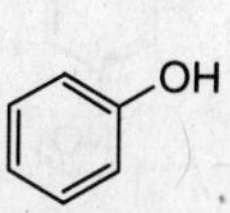

> CH_3CO_2H

Carboxylic acids ($pK_a \approx 5$) are more acidic than phenols ($pK_a \approx 10$) which is turn are more acidic than simple alcohols ($pK_a \approx 16$-20). The carboxylate anion is stabilized by resonance that allows the negative charge to be delocalized onto a second electronegative oxygen atom. The phenoxide ion can also be stabilized by resonance but the charge ends up on C atoms. In contrast, the alkoxide ion has no resonance stabilization since there is no π system.

18. Arrange the following molecules in order of decreasing acidity (most acidic to least acidic) and explain your answer.

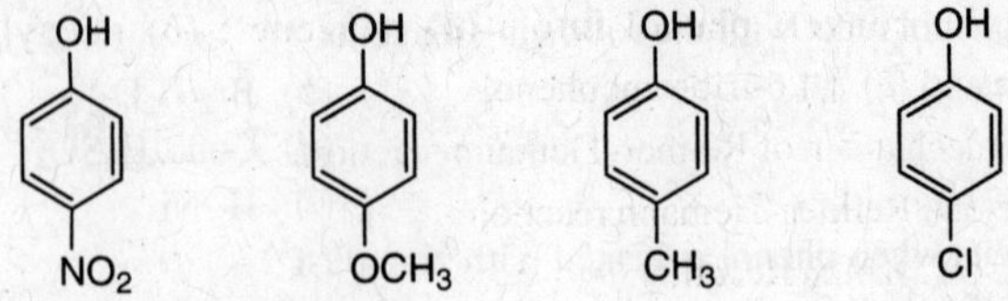

Answer. The order of decreasing acidity is :

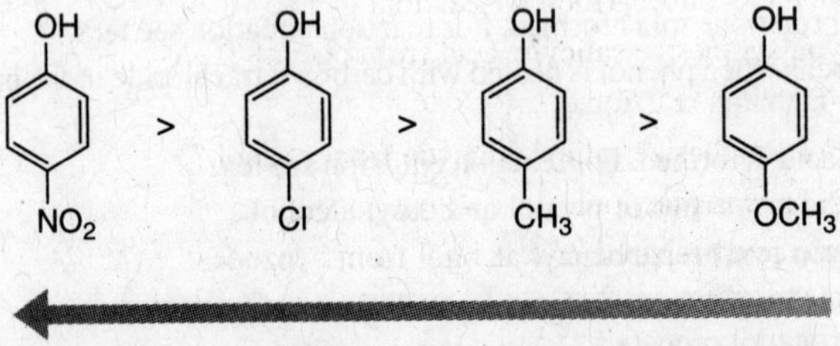

All these structures are *p*-substituted phenols, so the substituent must be the controlling factor. The effect the substituent exerts on the benzene ring will influence the stability of the phenoxide ion. A strong electron withdrawing group will stabilize the phenoxide ion making the phenol *more* acidic whereas a strong electron donor will destabilize the phenoxide ion making the phenol *less* acidic. A nitro group is strongly withdrawing due to resonance, a chloro group is weakly electron withdrawing due to inductive effects, a methyl group is weak electron donor and a methoxy group a strong electron donor due to resonance.

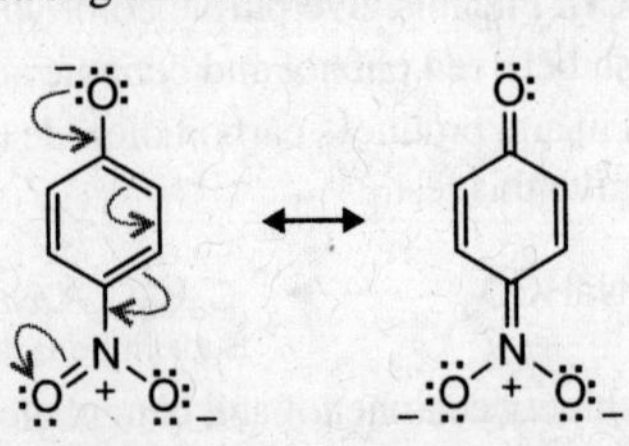

Resonance delocalization of the negative charge from the phenoxide ion to the oxygen of the *p*-nitro group. *p*-Nitrophenol has a similar acidity to the *o*-isomer.

19. Arrange the following molecules in order of decreasing acidity (most acidic to least acidic) and explain your answer.

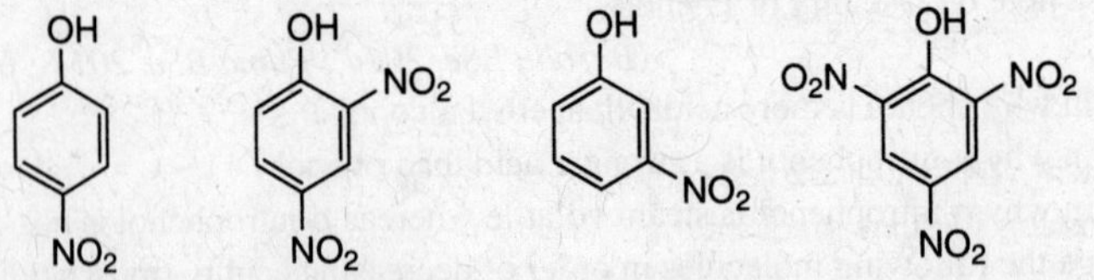

Answer. The order of decreasing acidity is :

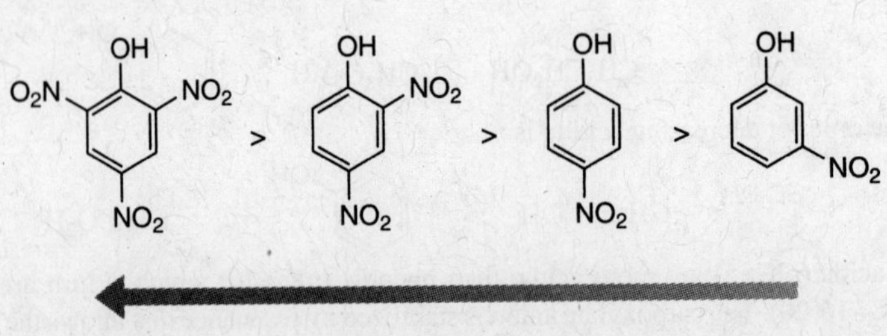

All are nitrophenols. Nitro groups stabilize the phenoxide ion by resonance electron withdrawal that allows the negative charge to be moved to an electronegative oxygen atom in the nitro group when the nitro group is *ortho-* or *para-* to the –OH group. The more nitro groups there in these positions, the greater the stabilization of the phenoxide ion and the more acidic the phenol.

20. How will you convert phenol into : (*a*) benzene ; (*b*) phenyl acetate ; (*c*) anisole ; (*d*) *p*-hydroxyazobenzene ; and (*e*) 2,4,6-tribromophenol.

21. Discuss the mechanism of Reimer-Tiemann reaction. (*Allahabad BSc, 2014*)

22. Write a note on : Reimer-Tiemann reaction.

23. Write a note on : Kolbe Reaction.

24. Write a note on : Fries Rearrangement.

25. Write a note on : Houben-Hoesch Reaction.

26. How will you synthesize salicylic acid from phenol ? (*Bhopal BSc, 2015*)

Clue. Use Reimer-Tiemann reaction.

27. How will you synthesize salicylaldehyde from phenol ? (*Pune BSc, 2014*)

Clue. Use Reimer-Tiemann reaction.

28. How will you synthesize benzyl alcohol from benzene ? (*Jiwaji BSc, 2015*)

Clue. Steps are : (*i*) $Br_2/FeBr_3$; (*ii*) Mg ; (*iii*) HCHO ; (*iv*) H_2O/H^+

29. How will you synthesize benzyl alcohol from toluene ?

Clue. Steps are : (*i*) Br_2/UV ; (*ii*) aq. NaOH/heat

30. How will you synthesize 1-phenylethanol from benzene ?

Answer. Following steps are involved :

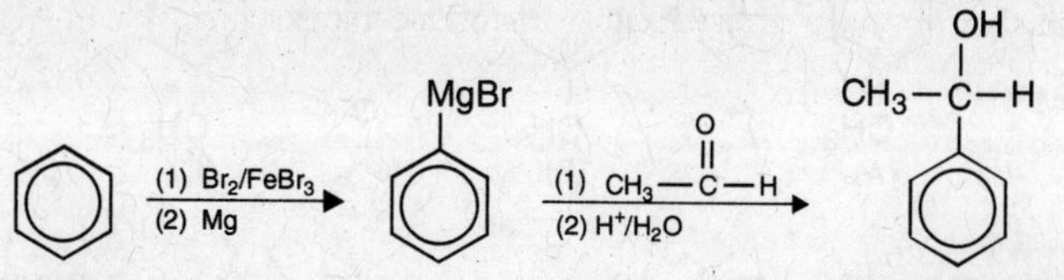

Benzene 1-Phenylethanol

31. How will you synthesize 2-phenylethanol from benzene ? (*Delhi BSc, 2015*)

Answer. Following steps are involved :

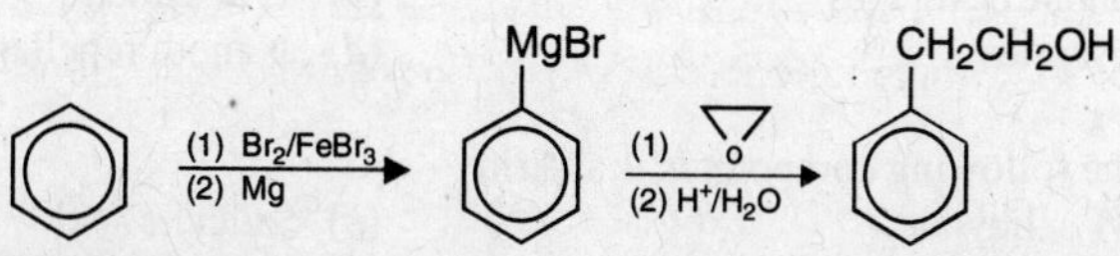

Benzene 2-Phenylethanol

32. How will you synthesize resorcinol from benzene ? (*North Bengal BSc, 2014*)

Answer. Following steps are involved :

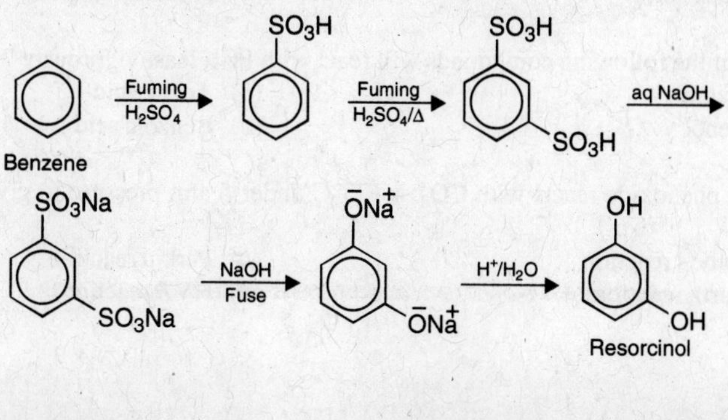

Resorcinol

33. Compound (A), C_9H_{10}, decolorizes both aqueous potassium permanganate solution and bromine in carbon tetrachloride solution. Ozonolysis of (A) yields (B), C_8H_8O, and (C), CH_2O. Treatment of (A) with diborane followed by treatment with alkaline hydrogen peroxide yields (D), $C_9H_{12}O$. Identify (A), (B), and (C) and show the reactions involved.

Answer. (A) = 2-Phenylpropene (B) = Acetophenone
(C) = Formaldehyde (D) = 2-Phenyl-1-propanol

Reactions Involved Are :

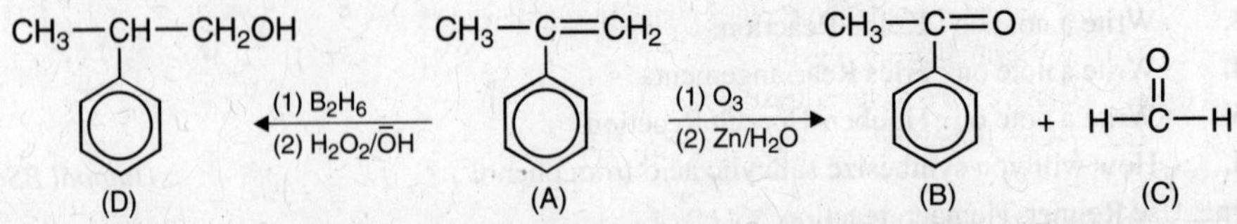

34. Compound (A), C_7H_8O, is soluble in aqueous sodium hydroxide solution. Hydrogenation of (A) with hydrogen and nickel gives (B), $C_7H_{14}O$, which is insoluble in aqueous sodium hydroxide but does react with sodium metal with the liberation of hydrogen. Oxidation of (B) with sodium dichromate in acidic solution gives (C), $C_7H_{12}O$. Identify (A), (B), and (C) and show the reactions involved.

Answer. (A) = *p*-Cresol (B) = 4-Methylcyclohexanol
(C) = 4-Methylcyclohexanone

Reactions Involved Are :

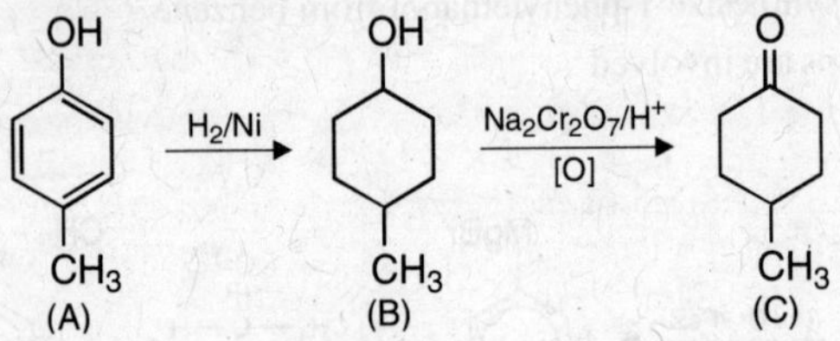

Multiple Choice Questions

1. Phenol is used
(*a*) in alcoholic beverages (*b*) as anesthetic
(*c*) in antiseptics (*d*) as moth repellant

Answer. (*c*)

2. Which of the following compounds is aspirin
(*a*) Methyl salicylate (*b*) Salicylic acid
(*c*) Phenyl salicylate (*d*) Acetylsalicylic acid

Answer. (*d*)

3. When phenol is treated with neutral $FeCl_3$ solution, it develops
(*a*) Violet color (*b*) Yellow color
(*c*) Green color (*d*) Nothing happens

Answer. (*a*)

4. Which of the following compounds will react with PCl_5 least vigorously ?
(*a*) Ethanol (*b*) Acetic acid
(*c*) Phenol (*d*) Benzoic acid

Answer. (*c*)

5. Sodium phenoxide reacts with CO_2 at 125°C under 5 atm pressure to give salicylic acid. This reaction is called
(*a*) Kolbe's reaction (*b*) Perkin reaction
(*c*) Wurtz reaction (*d*) HVZ reaction

Answer. (*a*)

6. Phenol reacts with excess bromine water to give

(*a*) *o*- plus *p*-Bromophenol (*b*) Bromobenzene
(*c*) 2,4,6-Tribromophenol (*d*) *m*-Bromophenol

Answer. (*c*)

7. Resorcinol on distillation with zinc dust gives

(*a*) Benzene (*b*) Cyclohexane
(*c*) Toluene (*d*) *m*-Xylene

Answer. (*a*)

8. Bakelite has the following characteristics

(*a*) A polymer made from phenol and formaldehyde
(*b*) A thermosetting plastic
(*c*) Can be used as an adhesive
(*d*) All of the above

Answer. (*d*)

9. Anisole is formed when phenol is treated with

(*a*) $CH_3I/NaOH$ (*b*) $CH_3CH_2I/NaOH$
(*c*) $CHCl_3/NaOH$ (*d*) Acetic anhydride

Answer. (*a*)

10. Anisole on heating with concentrated HI gives

(*a*) Iodobenzene (*b*) Phenol + CH_3I
(*c*) Iodobenzene + CH_3OH (*d*) Phenol + CH_3OH

Answer. (*b*)

11. Which group forms the strongest H-bonds to water molecules ?

(*a*) Alcohols (*b*) Ethers
(*c*) Phenols (*d*) All equally strong

Answer. (*c*)

12. The compound which is most capable of hydrogen bonding is :

(*a*) $CH_3OCH_2CH_3$ (*b*) $CH_3CH_2CH_2CH_3$

(*c*) 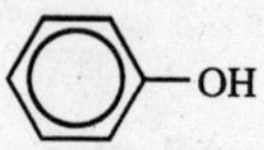 (*d*) $CH_3CH_2—S—S—CH_2CH_3$

Answer. (*c*)

13. Rank the following molecules in decreasing order of boiling points :

 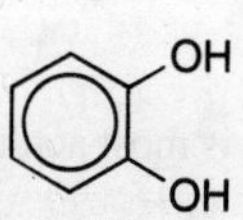 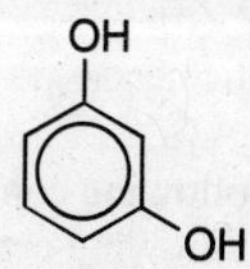

(1) (2) (3)

(*a*) (3) > (2) > (1) (*b*) (2) > (3) > (1)
(*c*) (3) > (1) > (2) (*d*) (2) > (1) > (3)

Answer. (*a*). Hydroxyl groups can more easily form intramolecular H-bonds in the *ortho* position than in the *meta* or *para* positions. These intramolecular H-bonds decrease boiling point. Therefore, compound (3) has a higher boiling point than compound (2). Compound (2), with two hydroxyl groups, has a still higher boiling point than phenol.

14. Rank the following compounds in order of increasing boiling point :

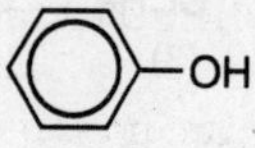 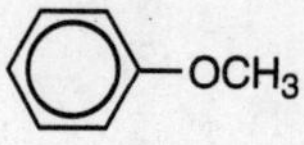 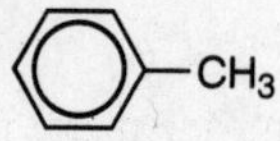

(1) (2) (3)

(*a*) (1) < (2) < (3) (*b*) (3) < (1) < (2)
(*c*) (1) < (3) < (2) (*d*) (3) < (2) < (1)

Answer. (*d*)

15. Phenol is acidic because of
(*a*) Resonance (*b*) Electromeric effect
(*c*) Inductive effect (*d*) Peroxide effect

Answer. (*a*)

16. Which of the following has the most acidic proton ?
(*a*) Thiol (*b*) Alcohol
(*c*) Phenol (*d*) Ether

Answer. (*c*)

17. Phenol is
(*a*) stronger acid than acetic acid (*b*) weaker acid than acetic acid
(*c*) stronger base than methylamine (*d*) weaker base than methylamine

Answer. (*b*)

18. Which of the following compounds is least acidic ?
(*a*) HCl (*b*) Phenol
(*c*) Acetylene (*d*) Picric acid

Answer. (*c*)

19. Which of the following compounds is least acidic ?
(*a*) Formic acid (*b*) Ethanol
(*c*) Acetic acid (*d*) Phenol

Answer. (*b*)

20. Which of the following compounds is most acidic ?
(*a*) Formic acid (*b*) Ethanol
(*c*) Acetic acid (*d*) Phenol

Answer. (*a*)

21. Which of the following compounds is most acidic ?
(*a*) Water (*b*) Cyclohexanol
(*c*) Ethanol (*d*) Phenol

Answer. (*d*)

22. Which one is most acidic ?
(*a*) Hexanol (*b*) Phenol
(*c*) Water (*d*) Diisopropyl ether

Answer. (*b*)

23. Which of the following compounds is most acidic ?
(*a*) *o*-Cresol (*b*) *p*-Nitrophenol
(*c*) *p*-Cresol (*d*) *p*-Chlorophenol

Answer. (*b*)

24. Which of the following compounds is most acidic ?
(*a*) Phenol (*b*) *m*-Nitrophenol
(*c*) *m*-Cresol (*d*) *m*-Bromophenol

Answer. (*b*)

25. Which molecule is the most acidic ?

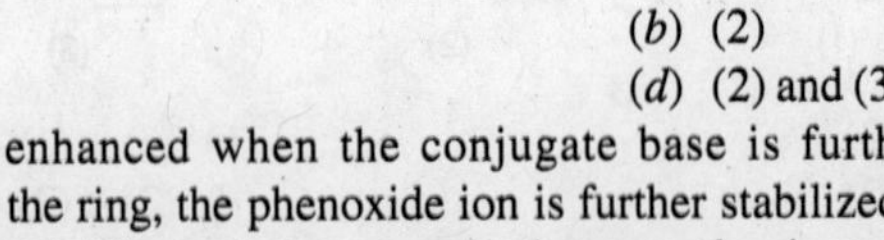

(1) (2) (3)

(*a*) (1) (*b*) (2)
(*c*) (3) (*d*) (2) and (3) have similar acidities

Answer. (*b*). Acidity is enhanced when the conjugate base is further stabilized. By placing an electron-withdrawing group on the ring, the phenoxide ion is further stabilized when the charge spreads over more atoms (greater delocalization). The methoxy group is electron-releasing and destabilizes the phenoxide.

26. Which of the following compounds is most acidic ?

(a)

(b)

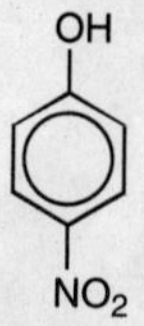
(c)

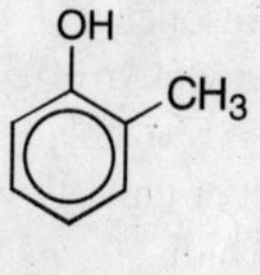
(d)

Answer. (*c*)

27. Which of the following compounds is least acidic ?

(a)

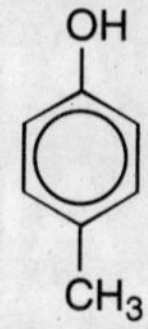
(b)

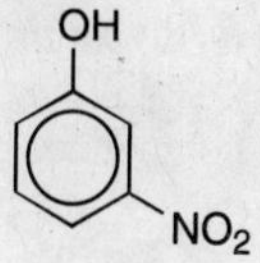
(c)

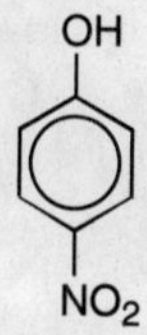
(d)

Answer. (*b*)

28. Arrange the following compounds in order of increasing acidity :

(1)

(2)

(3)

(*a*) (1) < (3) < (2)
(*b*) (1) < (2) < (3)
(*c*) (3) < (2) < (1)
(*d*) (2) < (1) < (3)

Answer. (*a*)

29. Phenols can be easily distinguished from alcohols because

(*a*) phenols are soluble in NaOH, but alcohols are not
(*b*) alcohols are soluble in NaOH, but phenols are not
(*c*) phenols are soluble in $NaHCO_3$, but alcohols are not
(*d*) alcohols are soluble in $NaHCO_3$, but phenols are not

Answer. (*a*)

41

Aromatic Aldehydes and Ketones

Cinnamaldehyde is the organic compound that gives cinnamon its flavor and odor.

AROMATIC ALDEHYDES

Aromatic aldehydes are compounds in which the –CHO group is attached directly to an aromatic ring. They are usually known by the common names or as derivatives of the parent benzaldehyde. Examples are :

Benzaldehyde

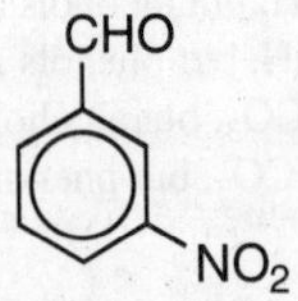

m-Nitrobenzaldehyde

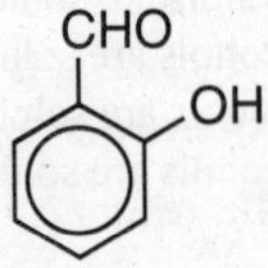

Salicylaldehyde

The compounds in which the –CHO group is not attached directly to the ring are considered as aryl substituted aliphatic aldehydes. Examples are :

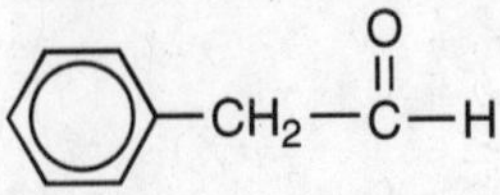

Phenylacetaldehyde

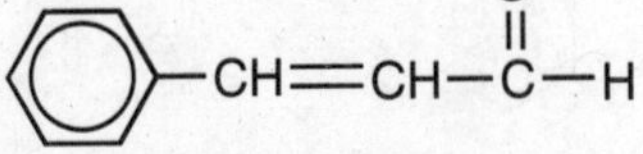

Cinnamaldehyde

BENZALDEHYDE, C_6H_5CHO

It is the simplest aromatic aldehyde.

Structure. In benzaldehyde, all ring carbons, the carbonyl carbon, and the carbonyl oxygen are sp^2 hybridized. It is a planar molecule. Fig. 41.1 shows the formation of sigma (σ) bonds in benzaldehyde. All σ bonds lie in one plane.

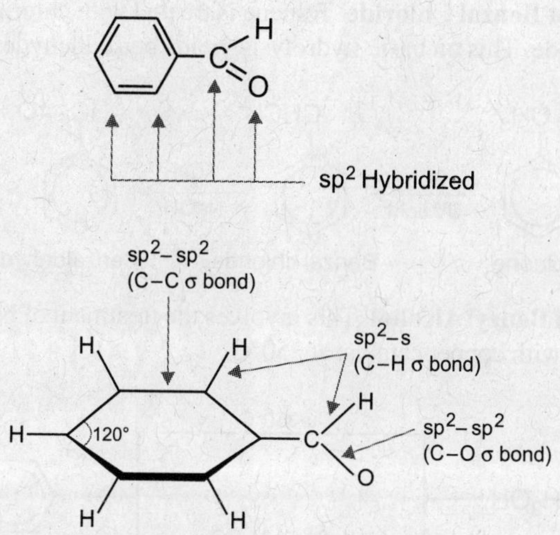

Fig. 41.1. Sigma (σ) bonds in benzaldehyde. All angles are 120°.

All carbon atoms and the oxygen atom in benzaldehyde also have a *p* orbital and these are perpendicular to the plane of σ bonds. The lateral overlap of these *p* orbitals produces a delocalized π molecular orbital. The two lone-pair electrons on the oxygen atom reside in the sp^2 hybrid orbitals.

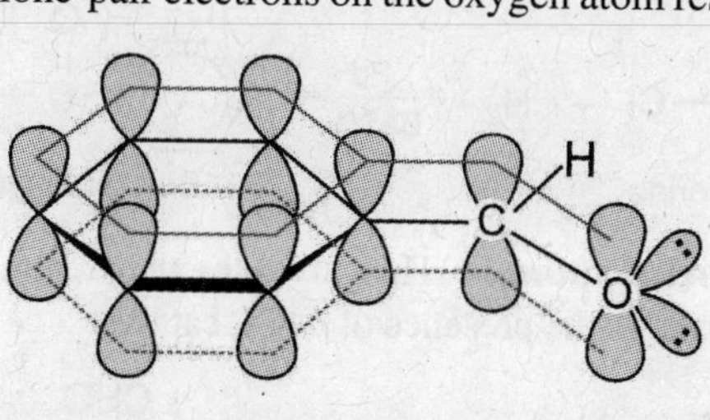

Fig. 41.2. Adjacent *p* orbitals overlap to form the delocalized π molecular orbital of benzaldehyde. Hydrogens bonded to the ring carbons are not shown for clarity.

According to the resonance theory, benzaldehyde is considered to be hybrid of the following resonance structures.

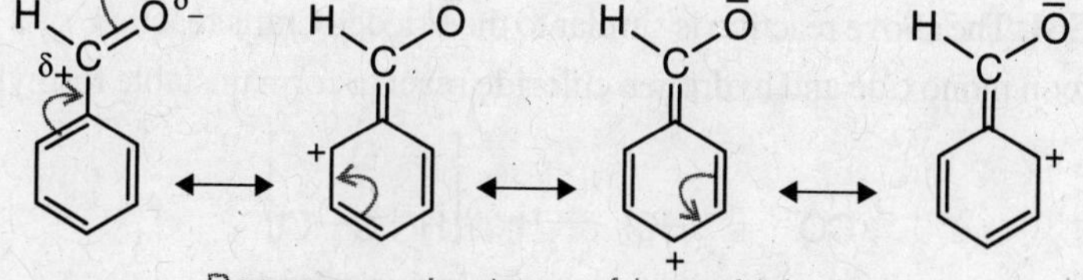

Resonance structures of benzaldehyde.

Preparation. Benzaldehyde is obtained :

(1) **By Oxidation of Toluene**. This involves the treatment of toluene with oxygen/air in the presence of vanadium pentoxide catalyst at 350°C.

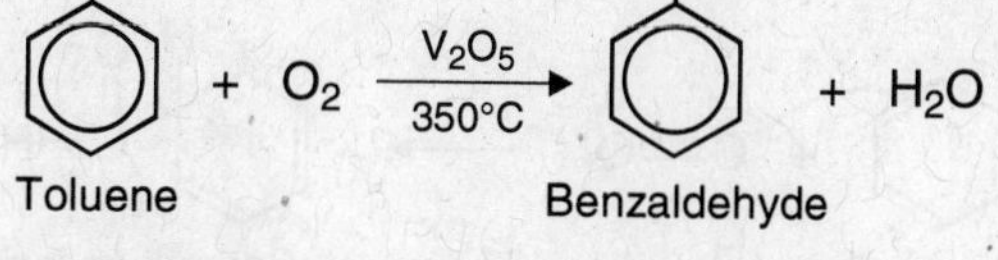

(2) **By Hydrolysis of Benzal Chloride**. Toluene is treated with chlorine in the presence of UV light to give benzal chloride. This on basic-hydrolysis yields benzaldehyde.

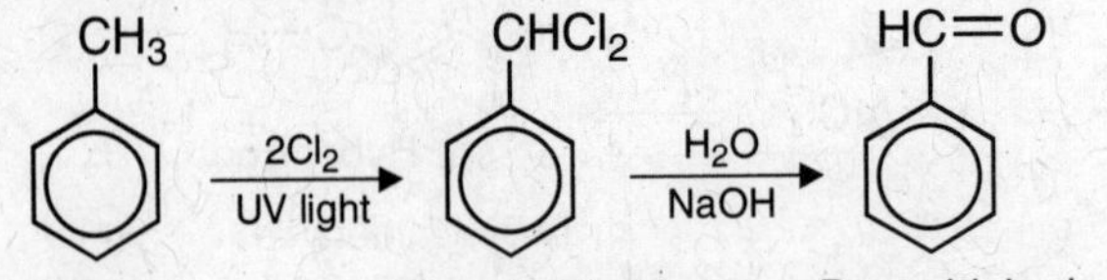

Toluene   Benzal chloride   Benzaldehyde

(3) **By Oxidation of Benzyl Alcohol**. This involves the treatment of benzyl alcohol with acidic potassium dichromate or with copper catalyst at 350°C.

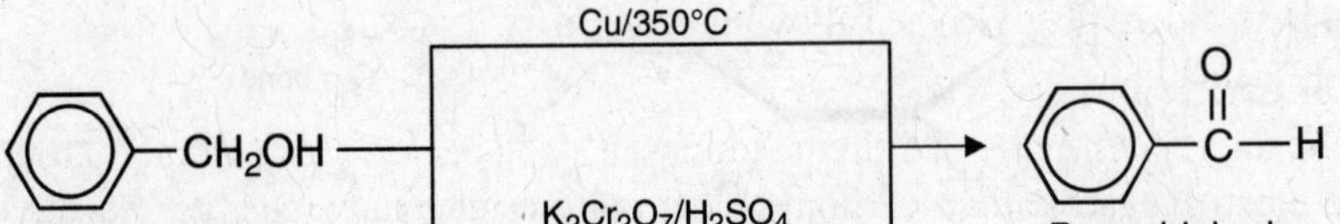

Benzyl alcohol   Benzaldehyde

(4) **By Rosenmund Reduction**. This involves the treatment of benzoyl chloride with hydrogen in the presence of palladium catalyst "Poisoned" with barium sulfate.

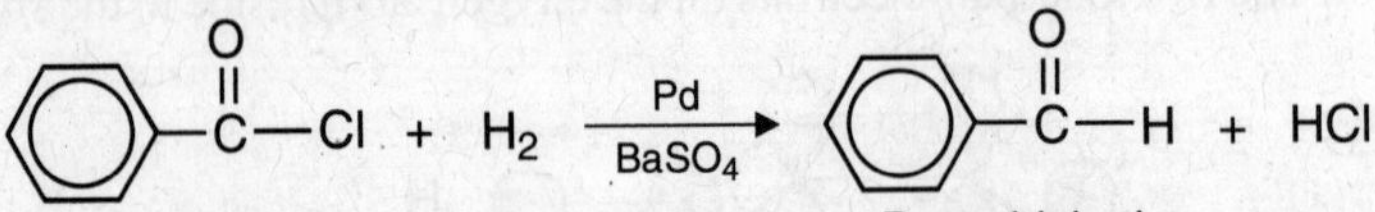

Benzoyl chloride   Benzaldehyde

(5) **By Gattermann-Koch Synthesis**. This involves the treatment of benzene with carbon monoxide and hydrogen chloride in the presence of $AlCl_3$ catalyst.

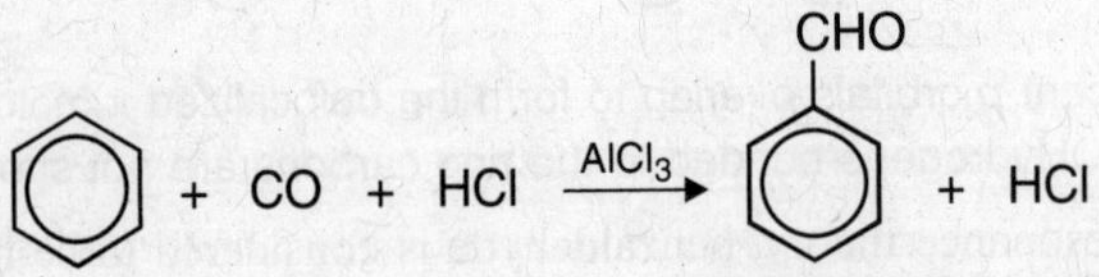

Benzene   Benzaldehyde

MECHANISM. The above reaction is similar to the Friedel-Crafts reaction.

Step 1. Carbon monoxide and hydrogen chloride react to form unstable formyl chloride.

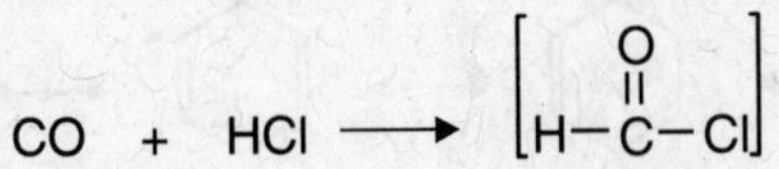

Formyl chloride

Step 2. Formation of the electrophile $H-\overset{+}{C}=O$.

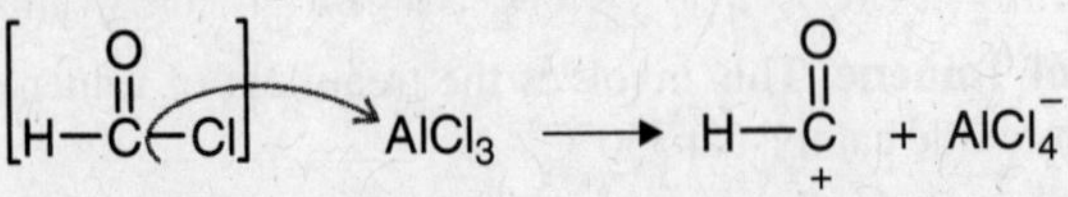

Step 3. The electrophile attacks the benzene ring to give a carbonium ion.

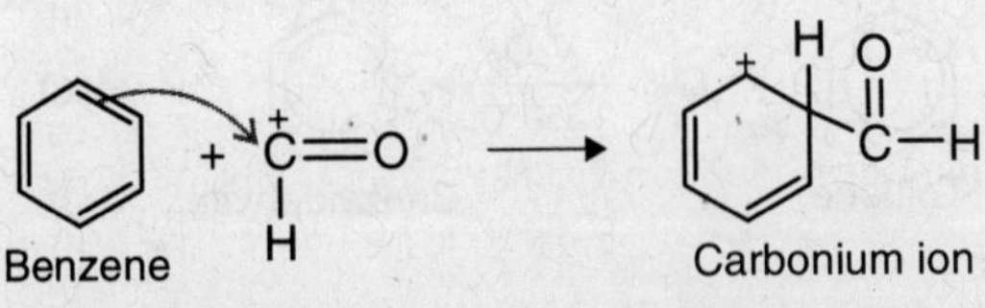

Benzene   Carbonium ion

Step 4. Removal of proton gives benzaldehyde.

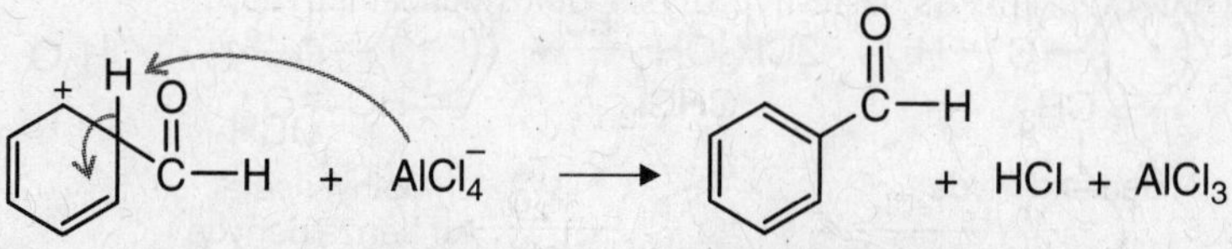

Benzaldehyde

Properties. (*Physical*). Benzaldehyde is a colorless liquid, bp 179°C. It has a pleasant almond-like odor. It is soluble in ethanol and diethyl ether. It is steam-volatile. The IR spectrum of benzaldehyde shows strong absorption bands at 1730-1700 cm^{-1} (due to C=O stretching) and at 2820-2700 cm^{-1} (due to C–H stretching).

(*Chemical*). Benzaldehyde gives the following reactions :

REACTIONS OF CARBONYL GROUP

Benzaldehyde undergoes many addition reactions. This is because the carbonyl groups is polar (oxygen is more electronegative than carbon). Electrophilic reagents attack the negative oxygen, while nucleophiles attack the carbonyl carbon.

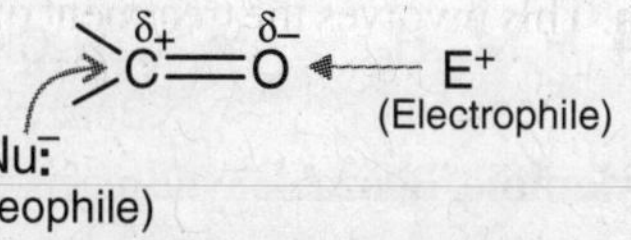

Thus the addition of a polar reagent can be predicted with a fair degree of certainty : **The positive part of the adding reagent always goes to the carbonyl oxygen, while the negative part always goes to the carbonyl carbon.**

(1) **Addition of HCN**. Benzaldehyde reacts with HCN to give a cyanohydrin. This on acid-hydrolysis yields mandelic acid.

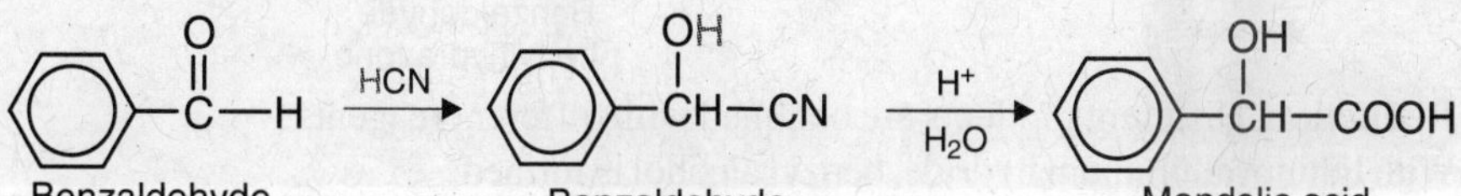

Benzaldehyde — Benzaldehyde cyanohydrin — Mandelic acid

(2) **Addition of Grignard Reagents**. Benzaldehyde reacts with Grignard reagents to form secondary alcohols after hydrolysis of the addition product.

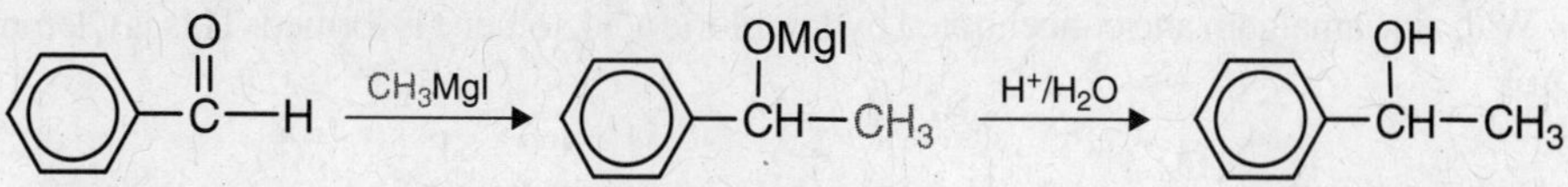

Benzaldehyde — (addition product) — 1-Phenyl-1-ethanol

(3) **Addition of Sodium Bisulfite**. Benzaldehyde reacts with sodium bisulfite to give the bisulfite addition product.

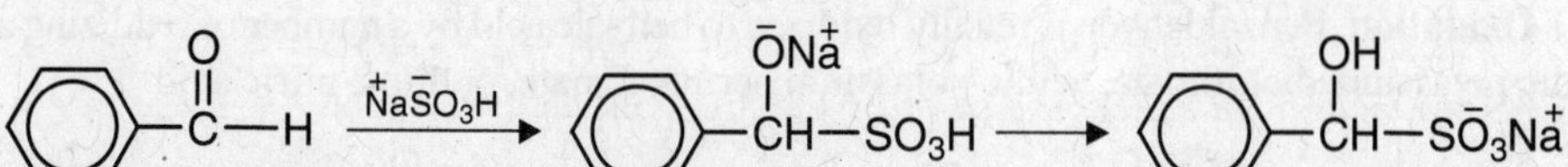

Benzaldehyde — (rearranges) — Benzaldehyde bisulfite

(4) **Addition of Alcohols**. Benzaldehyde reacts with alcohols in the presence of acid catalyst to form an acetal.

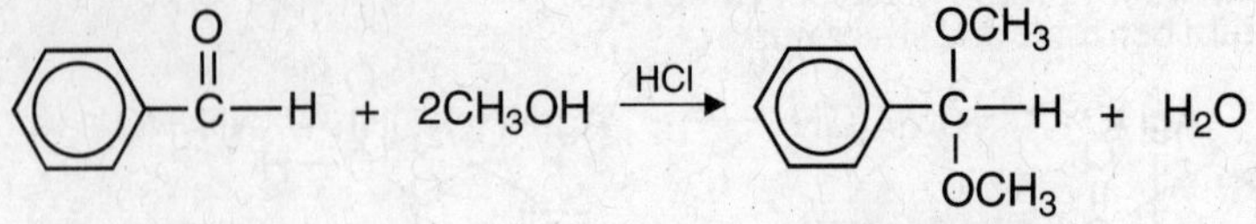

Benzaldehyde — Methyl acetal of benzaldehyde

(5) **Reaction with PCl_5**. Benzaldehyde reacts with phosphorus pentachloride to form benzal chloride.

$$\underset{\text{Benzaldehyde}}{C_6H_5CH{=}O} + PCl_5 \longrightarrow \underset{\text{Benzal chloride}}{C_6H_5CHCl_2} + POCl_3$$

(6) **Reaction with Ammonia**. Benzaldehyde reacts with ammonia to form hydrobenzamide.

$$3C_6H_5CH{=}O + 2NH_3 \longrightarrow \underset{\text{Hydrobenzamide}}{(C_6H_5CH{=}N)_2CHC_6H_5} + 3H_2O$$

(7) **Reaction with Amines**. Benzaldehyde reacts with primary aliphatic or aromatic amines to form Schiff's bases.

$$C_6H_5CH{=}O + \underset{\text{Aniline}}{H_2NC_6H_5} \longrightarrow \underset{\text{Benzylideneanilide}}{C_6H_5CH{=}NC_6H_5} + H_2O$$

(8) **Reaction with Hydroxylamine**. Benzaldehyde reacts with hydroxylamine to form an oxime.

$$C_6H_5CH{=}O + H_2N{-}OH \longrightarrow \underset{\text{Benzaldoxime}}{C_6H_5CH{=}N{-}OH} + H_2O$$

(9) **Reaction with Phenylhydrazine**. Benzaldehyde reacts with phenylhydrazine to give a phenylhydrazone.

$$C_6H_5CH{=}O + H_2N{-}NHC_6H_5 \longrightarrow \underset{\text{Benzaldehyde phenylhydrazone}}{C_6H_5CH{=}N{-}NHC_6H_5} + H_2O$$

(10) **Reduction**. Different products are obtained with different reagents :

(*a*) With lithium aluminium hydride, benzyl alcohol is formed.

$$\underset{\text{Benzaldehyde}}{C_6H_5{-}\overset{O}{\overset{\|}{C}}{-}H} + 2[H] \xrightarrow{LiAlH_4} \underset{\text{Benzyl alcohol}}{C_6H_5{-}CH_2OH}$$

(*b*) With zinc amalgam and concentrated hydrochloric acid, toluene is formed. This is Clemmensen reduction.

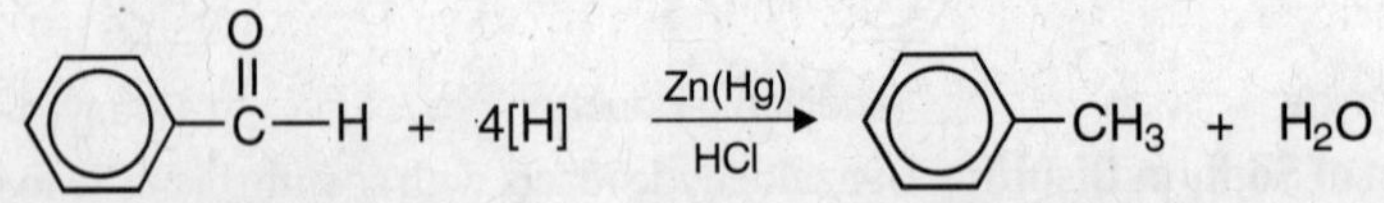

Benzaldehyde — Toluene

(11) **Oxidation**. Benzaldehyde is readily oxidized to benzoic acid by a number of oxidizing agents, e.g. acidic potassium dichromate, acidic potassium permanganate, or dilute nitric acid.

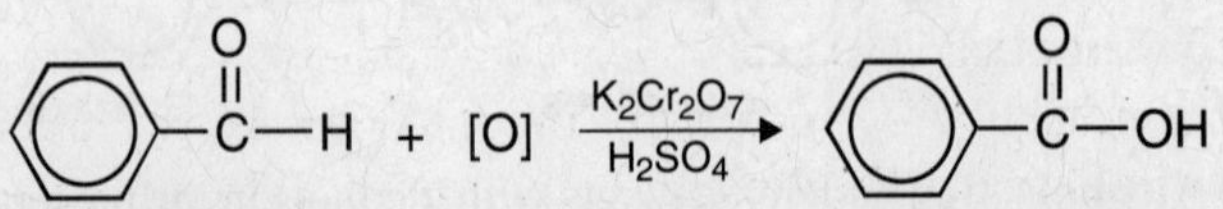

Benzaldehyde — Benzoic acid

Benzaldehyde also undergoes oxidation with Tollens' reagent (ammoniacal silver nitrate solution) to give ammonium benzoate and silver mirror.

$$AgNO_3 + 3NH_4OH \longrightarrow Ag(NH_3)_2OH + NH_4NO_3 + 2H_2O$$

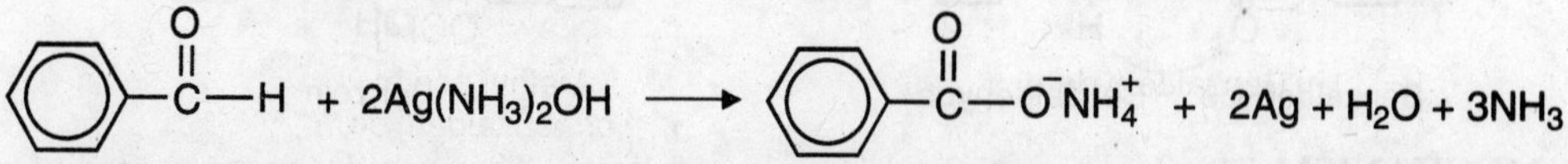

Benzaldehyde does not react with Fehling's solution.

(12) **Cannizzaro Reaction**. The Cannizzaro reaction involves the treatment of an aldehyde (without an α-hydrogen) with concentrated NaOH or KOH. The aldehyde undergoes an oxidation-reduction. For example, when benzaldehyde is heated with concentrated NaOH, a mixture of benzyl alcohol and sodium benzoate is formed.

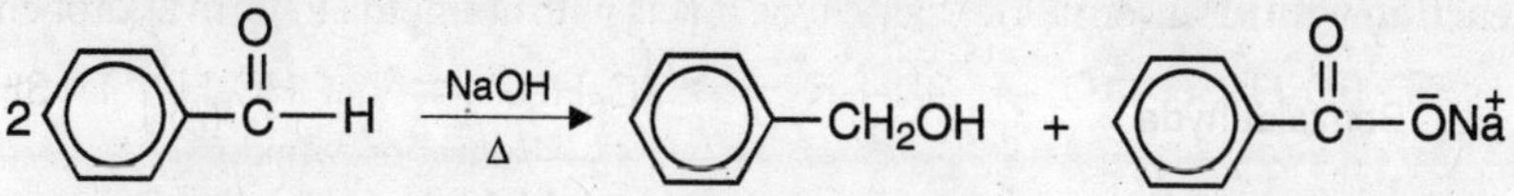

Benzaldehyde Benzyl alcohol Sodium benzoate

MECHANISM. The above reaction involves three steps

Step 1. Attack of OH^- on the carbonyl carbon.

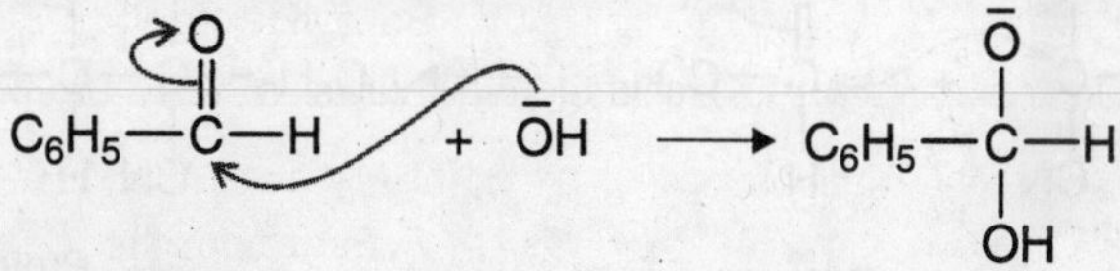

Benzaldehyde (A)

Step 2. Hydride-ion transfer.

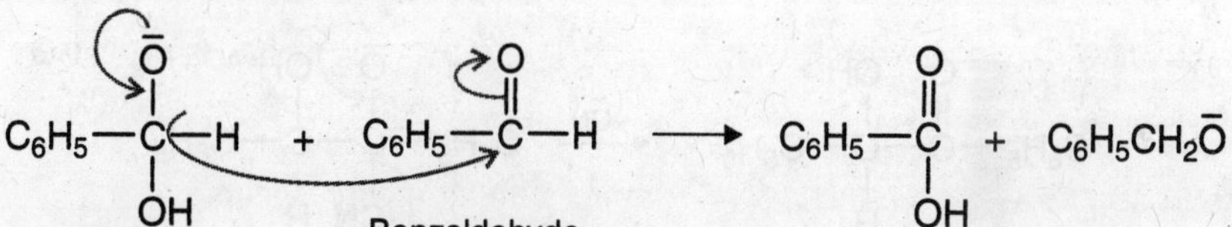

(A) Benzaldehyde

Step 3. Acid-base reaction.

$$C_6H_5-\overset{O}{\overset{\|}{C}}-OH + C_6H_5CH_2\bar{O} + \overset{+}{Na} \longrightarrow \underset{\text{Sodium benzoate}}{C_6H_5-\overset{O}{\overset{\|}{C}}-\bar{O}\overset{+}{Na}} + \underset{\text{Benzyl alcohol}}{C_6H_5CH_2OH}$$

Crossed Cannizzaro Reaction. Two different aldehydes, each lacking α-hydrogen, engage in a crossed Cannizzaro reaction when heated in alkaline solution. For example, when benzaldehyde is heated with formaldehyde in the presence of NaOH, a mixture of benzyl alcohol and sodium formate is formed.

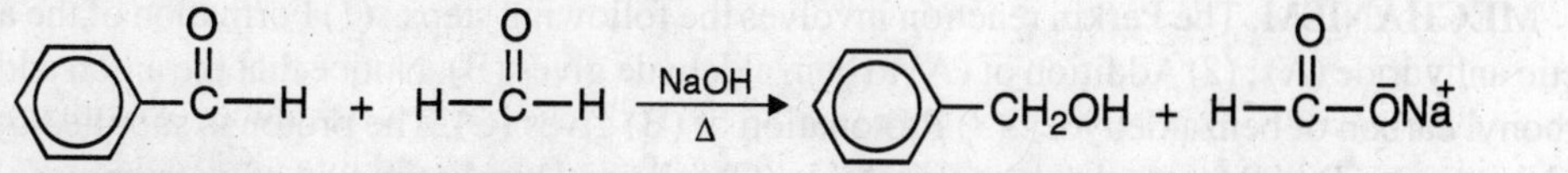

Benzaldehyde Formaldehyde Benzyl alcohol Sodium formate

(13) **Benzoin Condensation.** The benzoin condensation involves the treatment of an aromatic aldehyde with aqueous alcoholic KCN (or NaCN). The products are α-hydroxy ketones. For example, benzaldehyde reacts with alcoholic KCN to form benzoin. Notice that two molecules of benzaldehyde take part in the reaction.

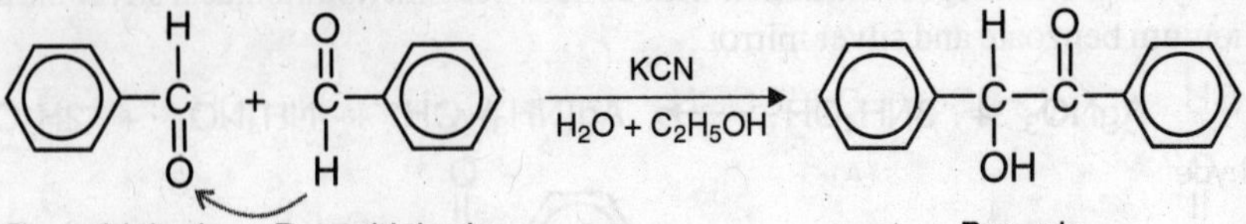

Benzaldehyde Benzaldehyde Benzoin

MECHANISM. The above reaction takes place as follows : The cyanide ion reacts readily with benzaldehyde, and enables the formation of carbanion (A).

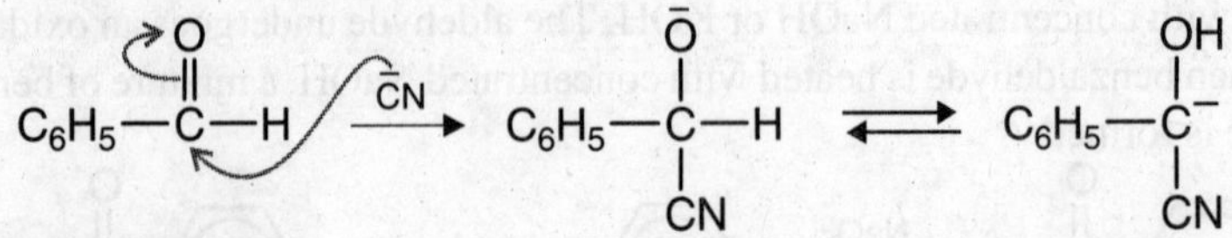

Benzaldehyde (A)

The carbanion (A) can then react with another benzaldehyde molecule. Subsequent proton-transfer and loss of the cyanide ion gives benzoin.

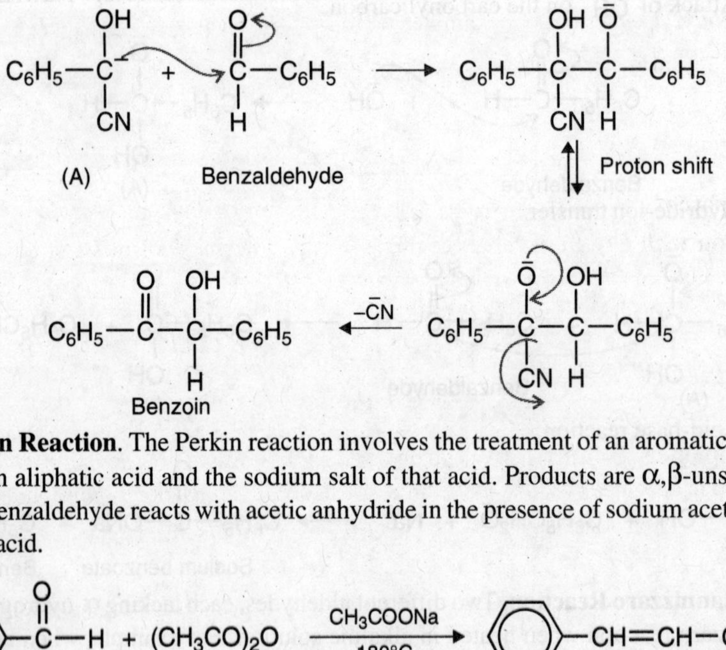

(14) **Perkin Reaction**. The Perkin reaction involves the treatment of an aromatic aldehyde with anhydride of an aliphatic acid and the sodium salt of that acid. Products are α,β-unsaturated acids. For example, benzaldehyde reacts with acetic anhydride in the presence of sodium acetate at 180°C to form cinnamic acid.

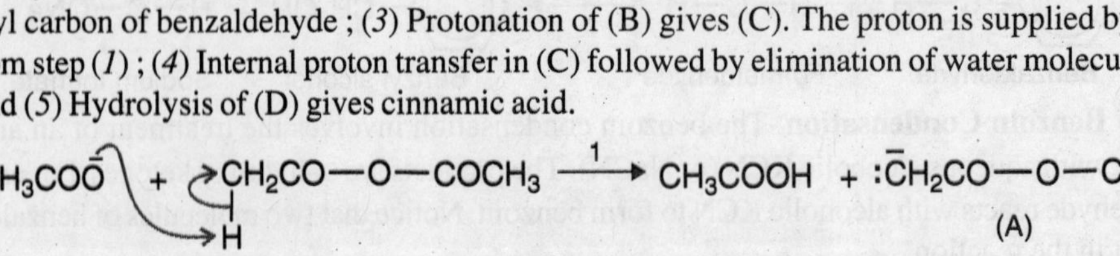

Benzaldehyde Cinnamic acid

MECHANISM. The Perkin reaction involves the following steps : (*1*) Formation of the anion of acetic anhydride (A) ; (*2*) Addition of (A) to benzaldehyde gives (B). Notice that the anion adds to the carbonyl carbon of benzaldehyde ; (*3*) Protonation of (B) gives (C). The proton is supplied by acetic acid from step (*1*) ; (*4*) Internal proton transfer in (C) followed by elimination of water molecule gives (D) ; and (*5*) Hydrolysis of (D) gives cinnamic acid.

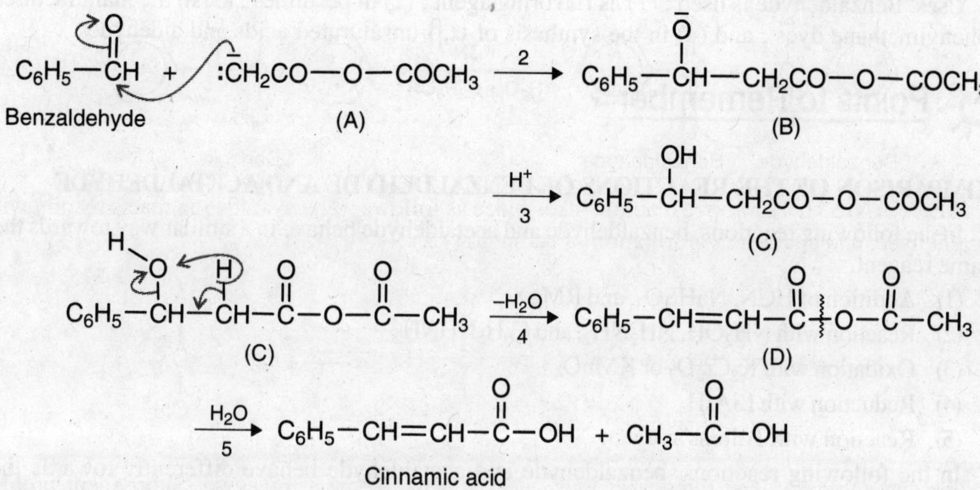

(15) **Mixed Aldol Condensation**. This involves the treatment of benzaldehyde with aliphatic aldehydes or ketones (containing α-hydrogen atoms) in the presence of NaOH. Products are unsaturated carbonyl compounds. For example, benzaldehyde reacts with acetaldehyde in the presence of NaOH to form cinnamaldehyde.

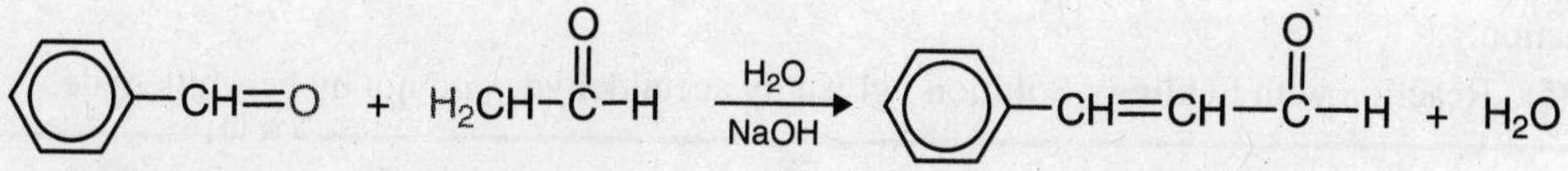

Benzaldehyde Acetaldehyde Cinnamaldehyde

(16) **Reaction with Cl_2**. In the absence of a catalyst, chlorine substitutes in the –CHO group, rather than in the benzene ring, to form benzoyl chloride.

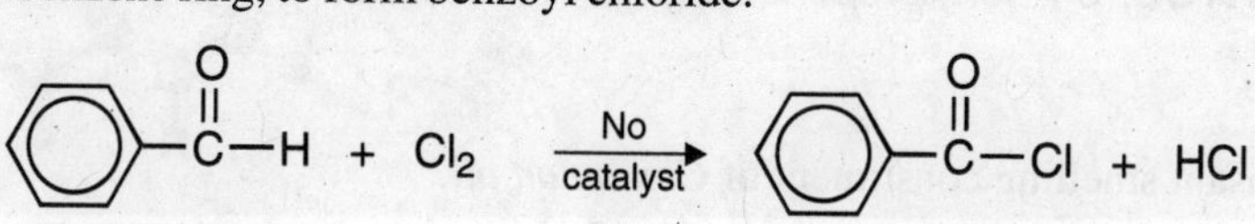

Benzaldehyde Benzoyl chloride

(17) **Electrophilic Substitution**. Benzaldehyde undergoes nitration, sulfonation, and halogenation. The –CHO group in benzaldehyde is a *meta* director and deactivating (For discussion see Chapter *34*).

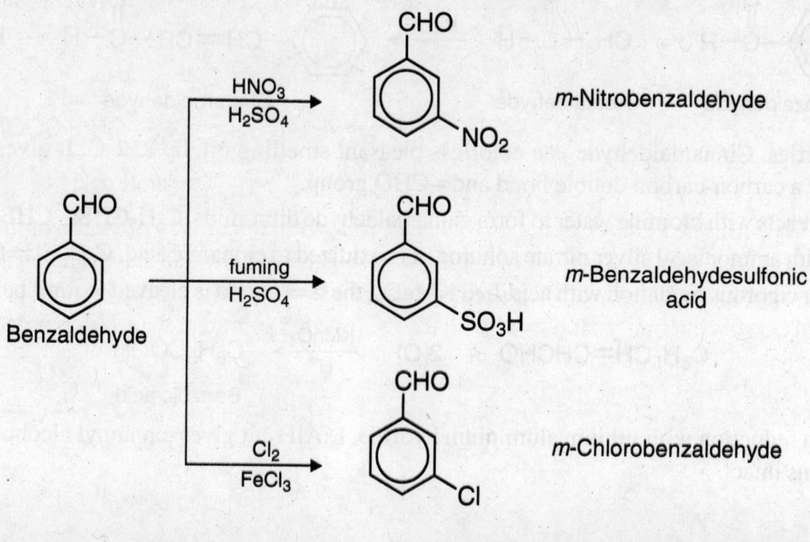

Uses. Benzaldehyde is used : (*1*) as flavoring agent ; (*2*) in perfumes ; (*3*) in the manufacture of triphenylmethane dyes ; and (*4*) in the synthesis of α,β-unsaturated acids and aldehydes.

Points to Remember

COMPARISON OF THE REACTIONS OF BENZALDEHYDE AND ACETALDEHYDE

In the following reactions, benzaldehyde and acetaldehyde behave in a similar way towards the same reagent.

(1) Addition of HCN, $NaHSO_3$, and RMgX
(2) Reaction with NH_2OH, NH_2NH_2, and $C_6H_5NHNH_2$
(3) Oxidation with $K_2Cr_2O_7$ or $KMnO_4$
(4) Reduction with $LiAlH_4$
(5) Reaction with Tollens' reagent

In the following reactions, benzaldehyde and acetaldehyde behave differently towards the same reagent.

(1) Iodoform Reaction : given by acetaldehyde and not by benzaldehyde.
(2) Aldol Condensation : given by acetaldehyde and not by benzaldehyde.
(3) Cannizzaro Reaction : given by benzaldehyde and not by acetaldehyde.
(4) Reaction with NH_3 : benzaldehyde condenses with it whereas acetaldehyde gives addition reaction.
(5) Reaction with Fehling's Solution : given by acetaldehyde and not by benzaldehyde.

CINNAMALDEHYDE, *3-Phenylpropenal*, $C_6H_5CH{=}CHCHO$

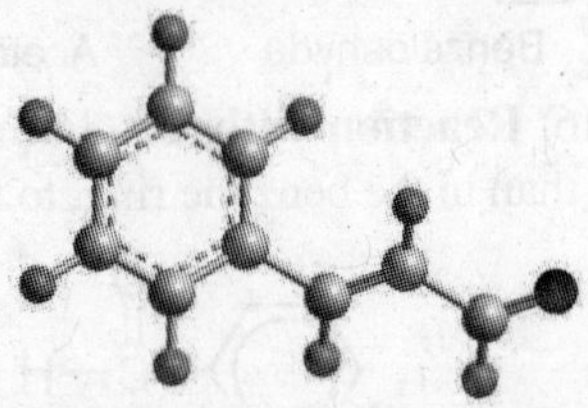

It is the pleasant smelling constituent of *Cinnamon oil*.

Preparation. Cinnamaldehyde is prepared by Claisen condensation. It involves the treatment of benzaldehyde with acetaldehyde in the presence of sodium hydroxide.

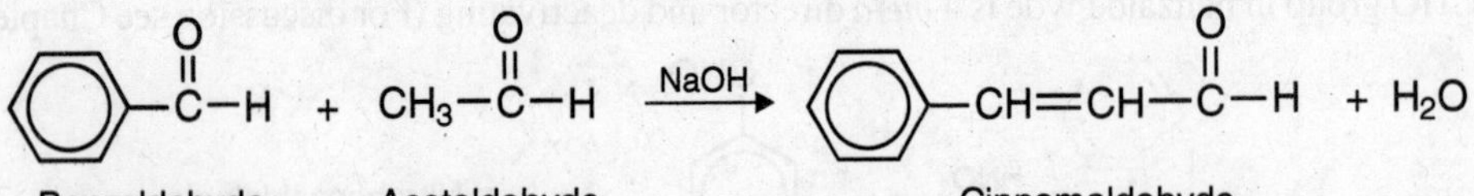

Benzaldehyde Acetaldehyde Cinnamaldehyde

Properties. Cinnamaldehyde is a colorless pleasant smelling oil, bp 252°C. It gives the usual reactions of a carbon-carbon double bond and –CHO group.

(1) It reacts with bromine water to form cinnamaldehyde dibromide, $C_6H_5CHBr{-}CHBrCHO$.
(2) With ammoniacal silver nitrate solution, it is oxidized to cinnamic acid, $C_6H_5CH{=}CHCO_2H$.
(3) On vigorous oxidation with acidified $KMnO_4$, the C=C bond is cleaved to form benzoic acid.

$$C_6H_5CH{=}CHCHO + 2[O] \xrightarrow[H^+]{KMnO_4} \underset{\text{Benzoic acid}}{C_6H_5COOH}$$

(4) On reduction with lithium aluminium hydride, $LiAlH_4$, it gives cinnamyl alcohol, the C=C bond remains intact.

$$C_6H_5CH{=}CHCHO \; + \; 2[H] \xrightarrow{LiAlH_4} \underset{\text{Cinnamyl alcohol}}{C_6H_5CH{=}CHCH_2OH}$$

Cinnamaldehyde is used as *Cinnamon essence*.

SALICYLALDEHYDE, *o-Hydroxybenzaldehyde*

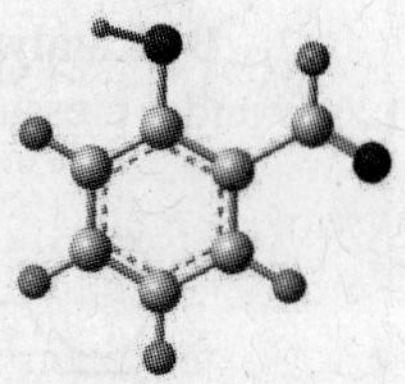

Preparation. Salicylaldehyde is prepared by Reimer-Tiemann reaction. This involves the treatment of a phenol with chloroform and aqueous sodium hydroxide followed by treatment with dilute acid. The –CHO group is introduced *ortho* to the –OH group.

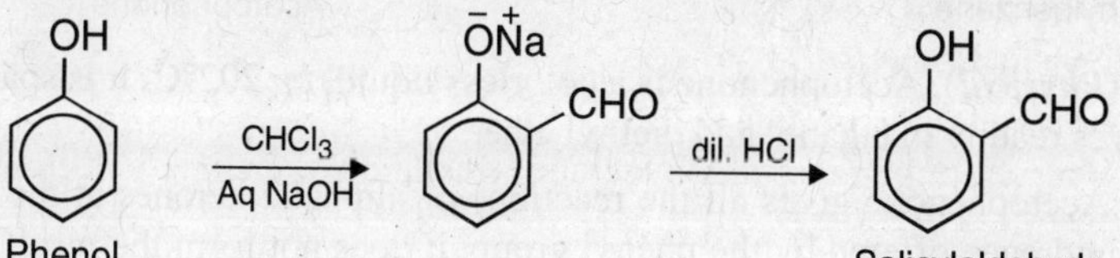

Phenol Salicylaldehyde

Properties. Salicylaldehyde is a colorless oil, bp 197°C, having an odor like bitter almonds. (*1*) Its aqueous solution gives a violet coloration with ferric chloride. (*2*) It can be oxidized to salicylic acid and reduced to *o*-hydroxybenzyl alcohol. (*3*) The OH is tied up by an intramolecular hydrogen bonding. This prevents association with water, thereby decreasing its water-solubility. Salicylaldehyde has also a higher boiling point compared to *m*- and *p*-isomers due to the hydrogen bonding.

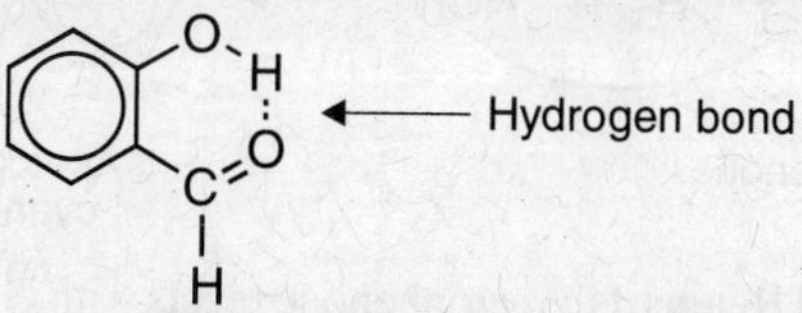

AROMATIC KETONES

Aromatic ketones are compounds in which a carbonyl group is attached to either two aryl groups or one aryl group and one alkyl group. Examples are :

Acetophenone Benzophenone

Aromatic ketones may be prepared by the methods used for aliphatic ketones but they are generally prepared by Friedel-Crafts acylation. They give the general reactions of their aliphatic counterparts but fail to undergo addition reaction with sodium bisulfite. They also give the usual electrophilic reactions of the benzene ring.

ACETOPHENONE, *Methyl phenyl ketone*, $C_6H_5{-}\overset{O}{\overset{\|}{C}}{-}CH_3$

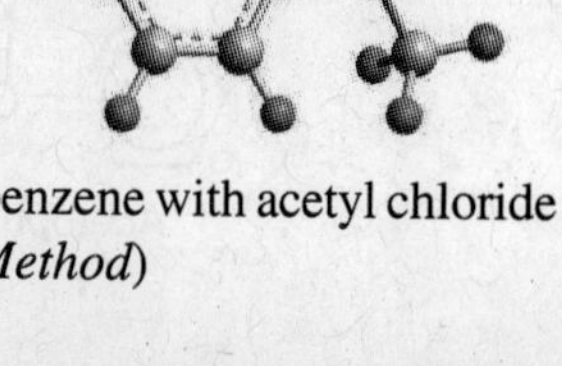

It is the simplest aromatic ketone.

Preparation. Acetophenone is obtained :

(1) **By Friedel-Crafts Acylation**. This involves the treatment of benzene with acetyl chloride or acetic anhydride in the presence of aluminium chloride. (*Laboratory Method*)

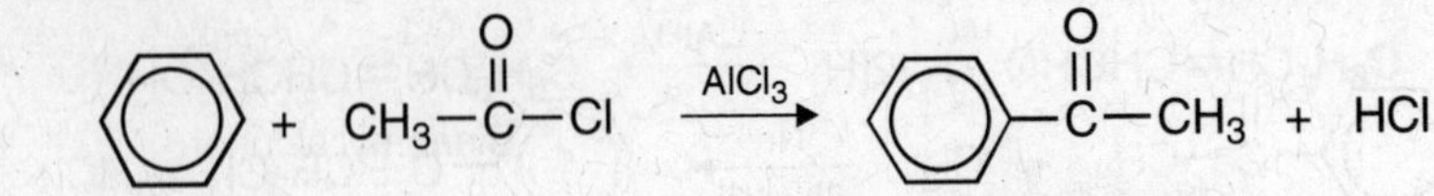

Benzene Acetophenone

(2) **By Catalytic Air-Oxidation**. This involves the treatment of ethylbenzene with oxygen/air at 126°C under pressure in the presence of manganese acetate. (*Commercial Method*)

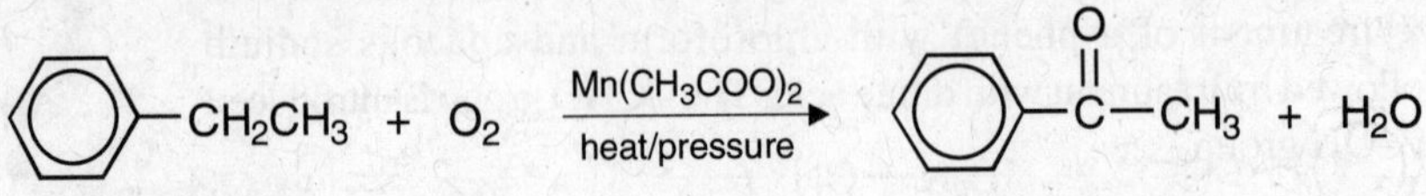

Ethylbenzene Acetophenone

Properties. (*Physical*). Acetophenone is a colorless liquid, *bp* 202°C. It is sparingly soluble in water, but dissolves readily in ethanol and diethyl ether.

(*Chemical*). Acetophenone gives all the reactions of aliphatic ketones at the carbonyl group. Owing to steric hindrance offered by the phenyl group, it does not form the bisulfite compound. It gives the usual electrophilic substitution reactions of the benzene ring in the *meta* position. Its important reactions are given below :

(1) **Addition of HCN**. Acetone reacts with HCN to give acetophenone cyanohydrin. Remember that the positive part of the adding reagent always goes to the negative carbonyl oxygen.

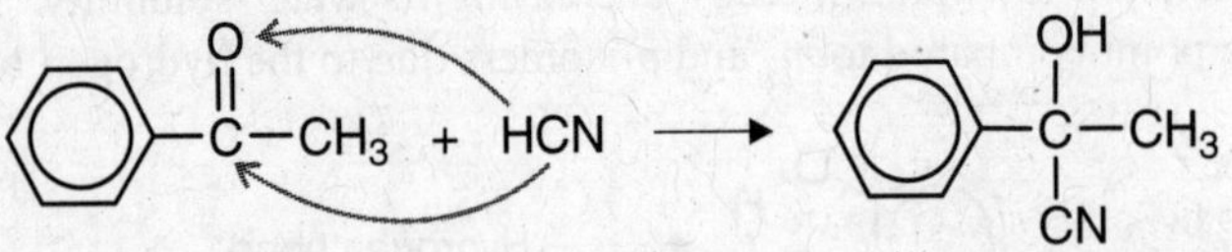

Acetophenone Acetophenone cyanohydrin

(2) **Addition of Grignard Reagents**. Acetophenone reacts with Grignard reagents to form tertiary alcohols after hydrolysis of the addition product.

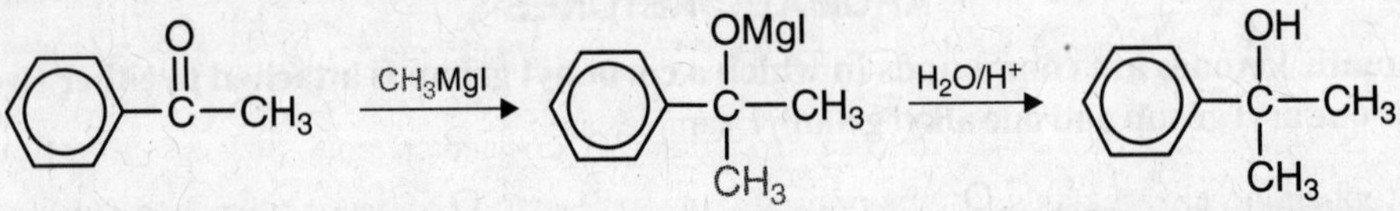

Acetophenone Dimethylphenylcarbinol

(3) **Oxidation**. Acetophenone undergoes oxidation with acidic potassium dichromate or potassium permanganate to give benzoic acid.

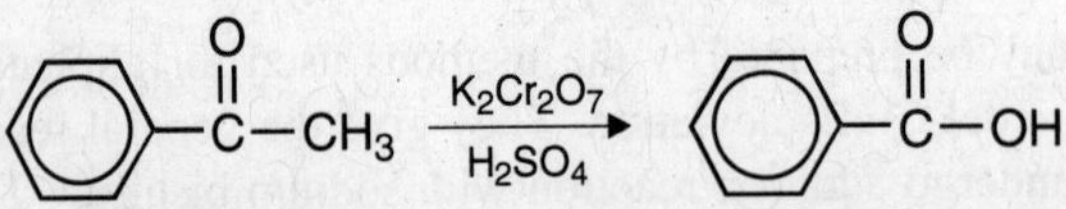

Acetophenone Benzoic acid

(4) **Reduction**. Acetophenone undergoes reduction with sodium amalgam in HCl to give ethylbenzene. (*Clemmensen reduction*)

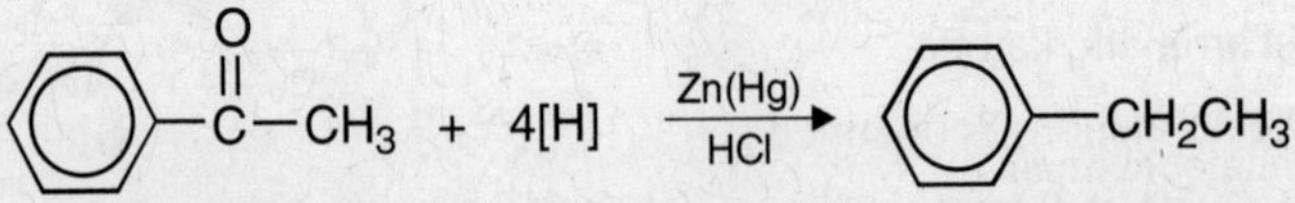

Acetophenone Ethylbenzene

(5) **Chlorination**. Acetophenone reacts with chlorine (in the absence of $AlCl_3$) to form phenacyl chloride.

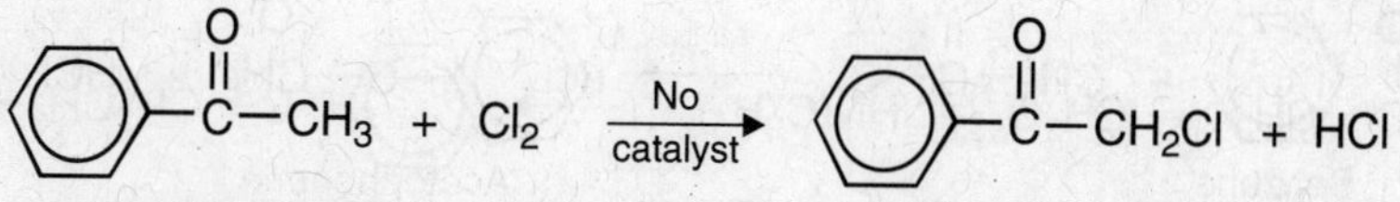

Acetophenone Phenacyl chloride

Phenacyl chloride is a powerful lachrymator and is used by police to disperse mobs.

(6) **Reaction with Hydroxylamine**. Acetophenone reacts with hydroxylamine to form methyl phenyl ketoxime.

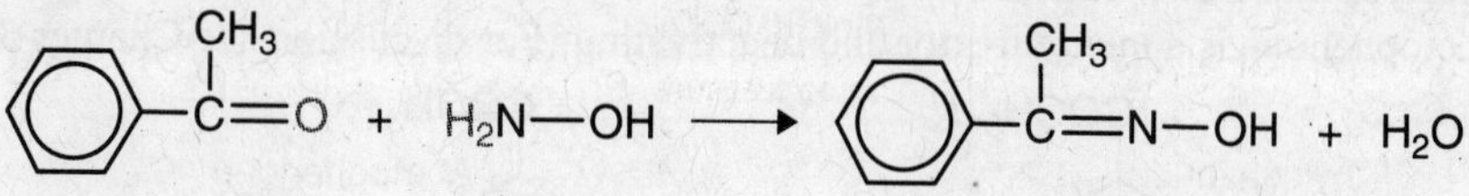

Acetophenone Methyl phenyl ketoxime

The ketoximes produced above undergo rearrangement in the presence of dilute acids or PCl_5 to give *N*-substituted amides. This is called **Beckmann Rearrangement**.

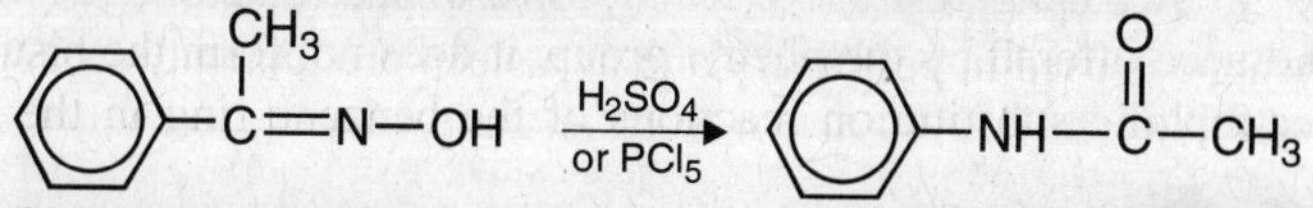

Methyl phenyl ketoxime *N*-Phenylacetamide

MECHANISM. The above reaction takes place by the following steps :

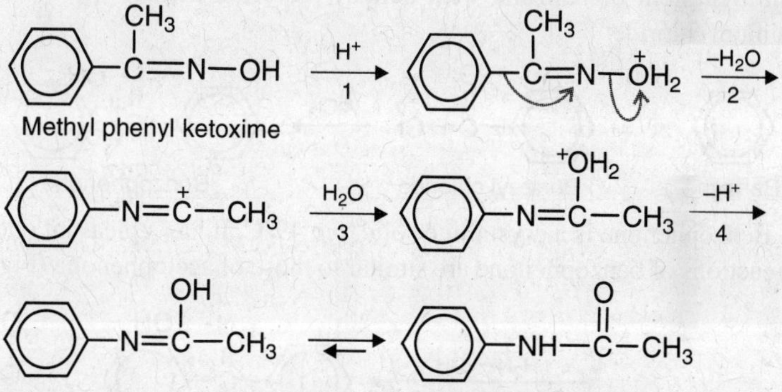

N-Phenylacetamide

Acetophenone also reacts with phenylhydrazine to give the corresponding phenylhydrazone.

(7) **Iodoform Reaction**. Acetophenone is a methyl ketone and reacts with iodine in the presence of sodium hydroxide to form iodoform.

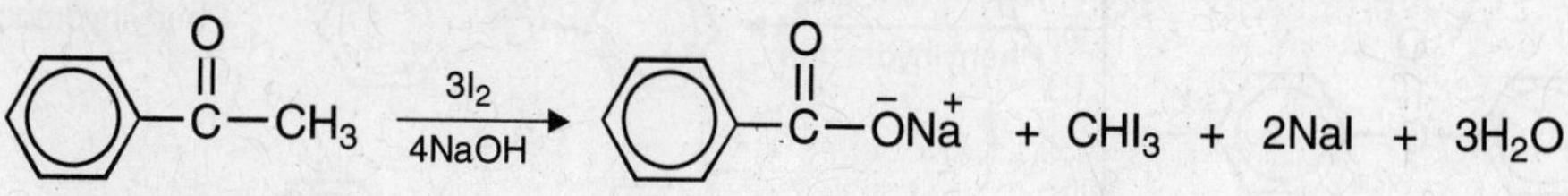

Acetophenone Iodoform

(8) **Aldol-Type Condensation**. Acetophenone undergoes self-addition in the presence of aluminium *t*-butoxide to form **Dypnone**.

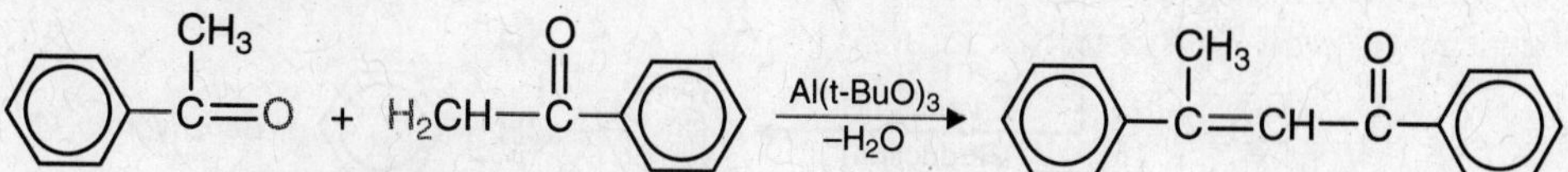

Two molecules of acetophenone Dypnone

(9) **Mannich Reaction**. Acetophenone reacts with formaldehyde and ammonia or a primary amine or a secondary amine to form β-aminoketones. For example,

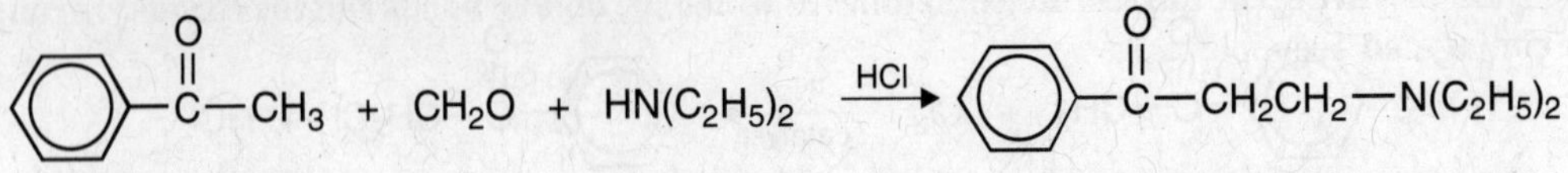

Acetophenone Diethylamine β-Aminoketone

All aromatic ketones containing an α-hydrogen give this reaction. The products of the Mannich reaction are known as **Mannich Bases**.

(10) **Electrophilic Substitution**. Acetophenone undergoes nitration and sulfonation. The $-COCH_3$ group in acetophenone is a *meta* director and deactivating (For discussion see Chapter *34*)

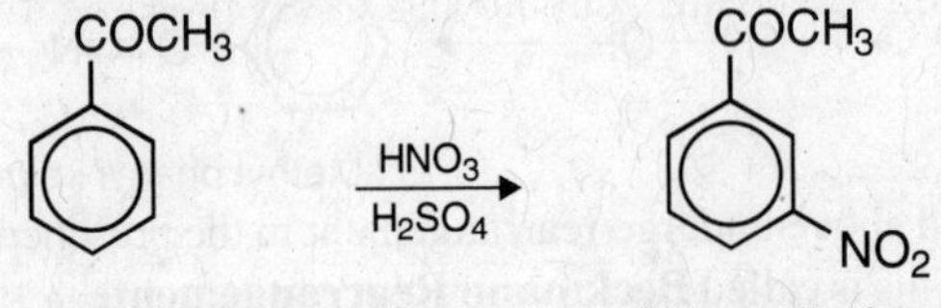

Acetophenone *m*-Nitroacetophenone

Uses. Acetophenone is used in perfumery and as a sleep producing drug, *Hypnone*.

BENZOPHENONE, *Diphenyl ketone*, $C_6H_5-\overset{O}{\overset{\|}{C}}-C_6H_5$

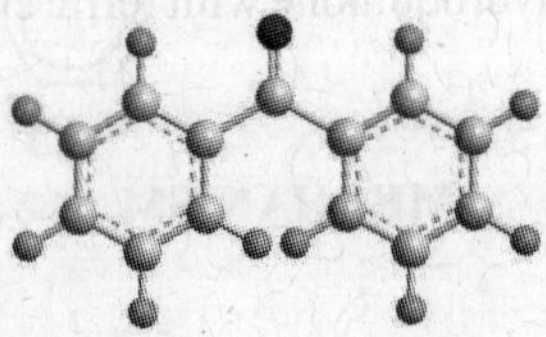

Preparation. Benzophenone is obtained by Friedel-Crafts acylation. This involves the treatment of benzene with benzoyl chloride in the presence of aluminium chloride.

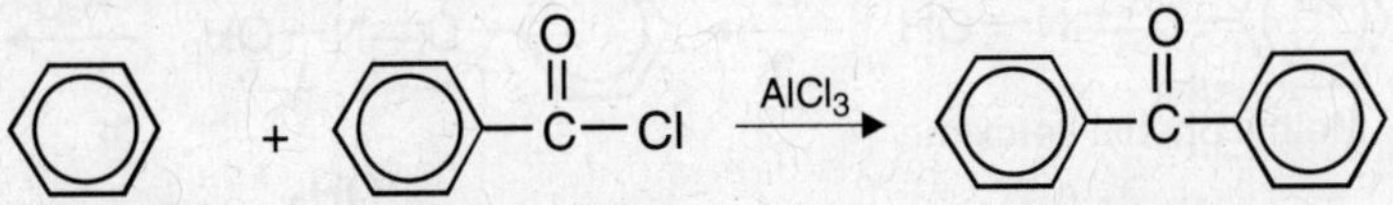

Benzene Benzoyl chloride Benzophenone

Properties. Benzophenone is a crystalline solid, *mp* 49°C. It has a pleasant odor and is used in perfumery. The reactions of benzophenone are similar to those of acetophenone (Fig. 41.1).

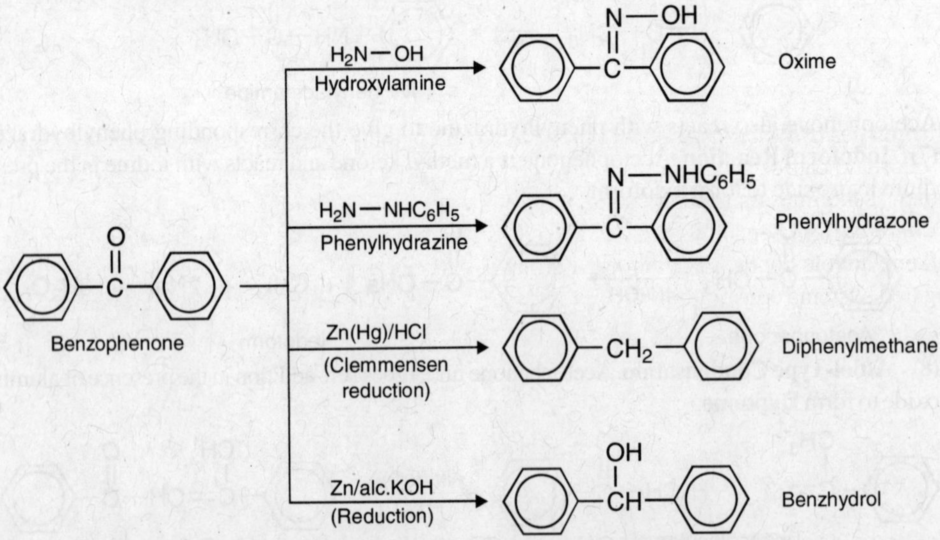

Fig. 41.3. Important reactions of Benzophenone.

QUINONES

Quinones are cyclic unsaturated diketones in which the double bonds and the carbonyl groups are conjugated. Examples are :

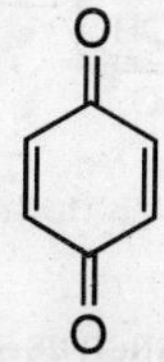

p-Benzoquinone

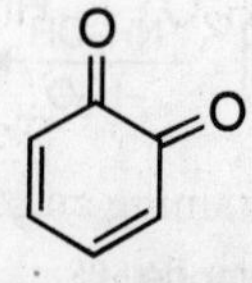

o-Benzoquinone
(*unstable*)

Quinones are not aromatic compounds, but they are easily prepared from, and easily converted into aromatic compounds.

p-BENZOQUINONE, *p-Quinone*

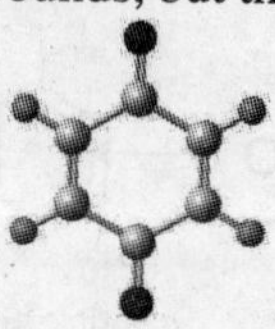

It is the most important quinone.

Preparation. (1) In the laboratory, *p*-benzoquinone is obtained by the mild oxidation of hydroquinone with ferric chloride.

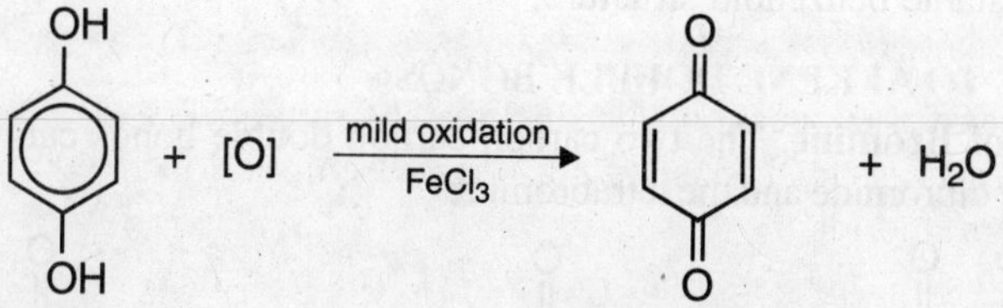

Hydroquinone *p*-Benzoquinone

(2) It is produced industrially by the oxidation of aniline with manganese dioxide and sulfuric acid.

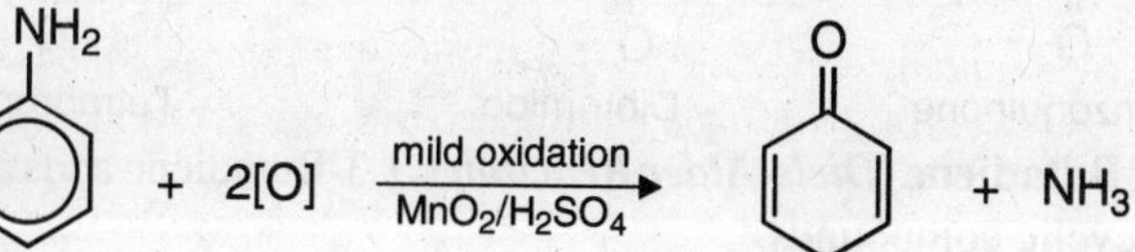

Aniline

p-Benzoquinone

Properties (*Physical*). *p*-Benzoquinone forms deep-yellow prismatic crystals, mp 116°C. It has a sharp odor and causes sneezing. It is only slightly soluble in water but dissolves readily in ethanol and ether. It sublimes on heating and is steam-volatile.

(*Chemical*). *p*-Benzoquinone has a cyclic structure made of two ketonic carbonyl groups and two alkene double bonds. Each carbonyl group is in conjugation with an alkene bond. The two such conjugated systems constituting the molecule are shown below by the dividing dotted line.

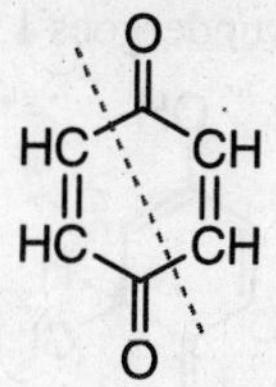

Hence the reactions of *p*-benzoquinone are those due to the carbonyl groups, the alkene bonds, and the conjugated systems –CH=CH–C=O as such.

REACTIONS DUE TO CARBONYL GROUPS

(1) **With NH_2OH**. It reacts with hydroxylamine to form monoxime and dioxime.

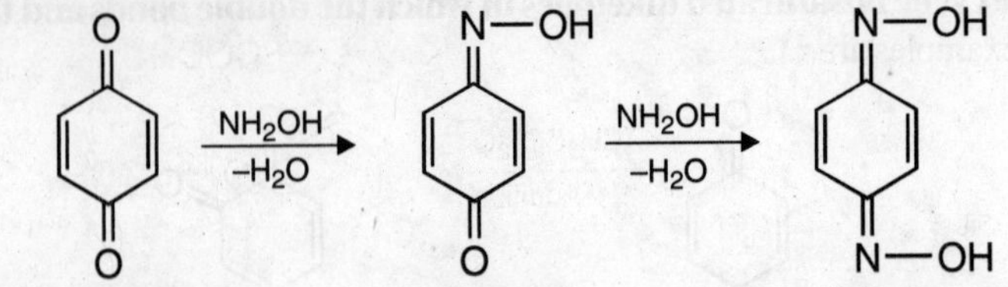

p-Benzoquinone Monoxime Dioxime

(2) **Reduction**. It is reduced with a mild oxidizing agent such as sulfurous acid (H_2SO_3) to form **Hydroquinone**.

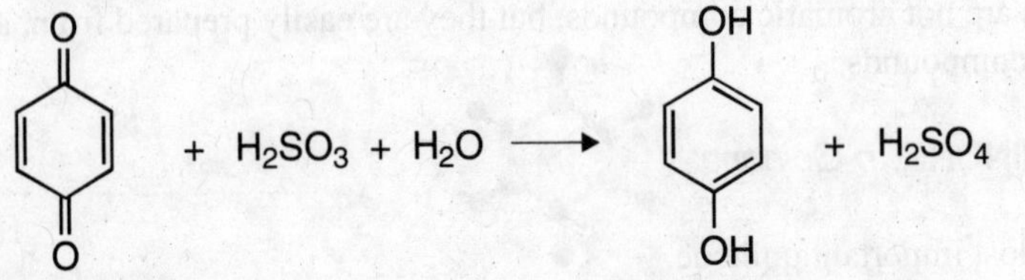

p-Benzoquinone Hydroquinone

Therefore, *p*-benzoquinone itself acts as an oxidizing agent. In this reaction the quinonoid structure passes to the more stable benzenoid structure.

REACTIONS DUE TO ALKENE DOUBLE BONDS

(3) **Addition of Bromine**. The two carbon-carbon double bonds can undergo addition with bromine to form the dibromide and the tetrabromide.

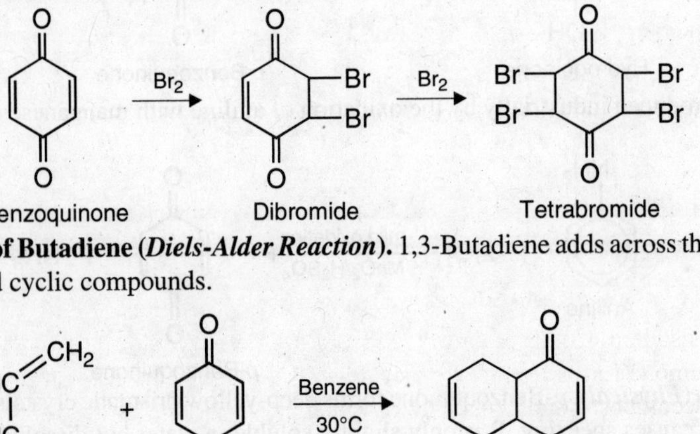

p-Benzoquinone Dibromide Tetrabromide

(4) **Addition of Butadiene (*Diels-Alder Reaction*).** 1,3-Butadiene adds across the C=C bond of the quinone to yield cyclic compounds.

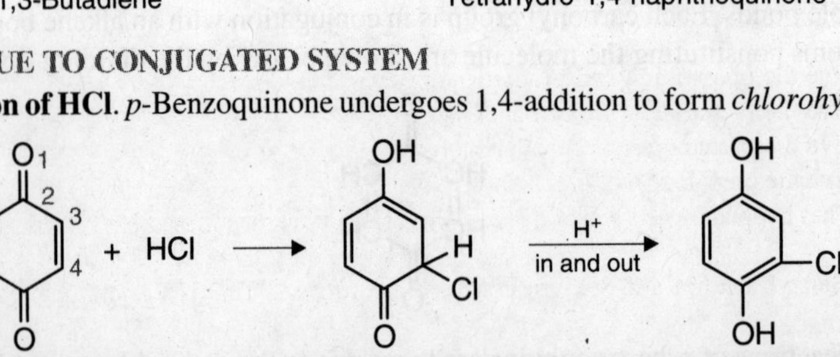

1,3-Butadiene Tetrahydro-1,4-naphthoquinone

REACTIONS DUE TO CONJUGATED SYSTEM

(5) **Addition of HCl**. *p*-Benzoquinone undergoes 1,4-addition to form *chlorohydroquinone*.

p-Benzoquinone Chlorohydroquinone

(6) **Addition of Acetic anhydride.** When quinone is treated with acetic anhydride in the presence of sulfuric acid at 50°C, a molecule of the anhydride is added first.

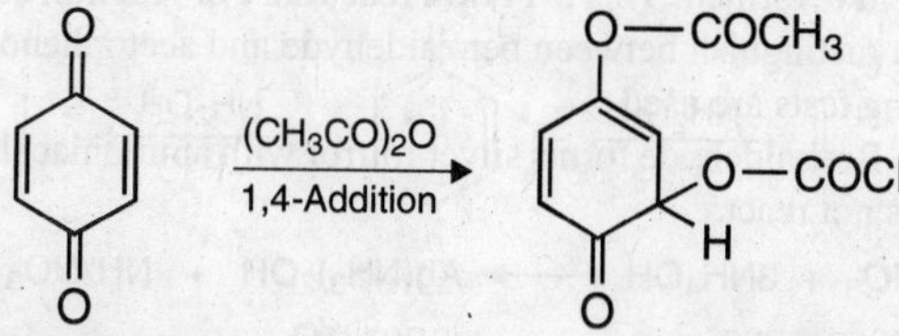

p-Benzoquinone 1,4-Adduct

The 1,4-adduct undergoes migration of H^+ to form a phenolic OH which is acetylated, the final product being 1,2,4-triacetoxybenzene.

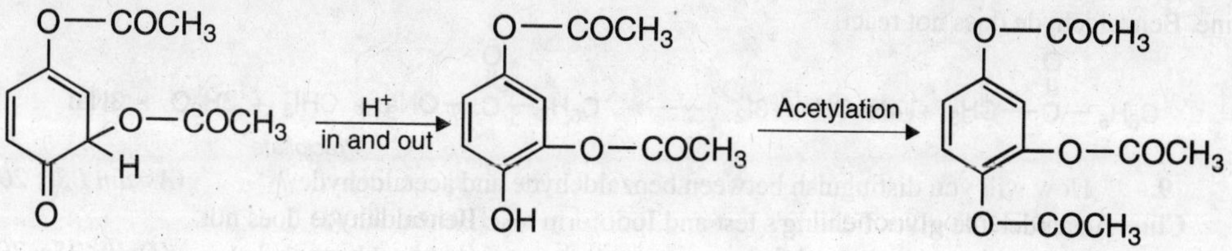

1,4-Adduct 1,2,4-Triacetoxybenzene

Uses. *p*-Benzoquinone is used : (1) as a mild oxidizing agent in organic synthesis ; (2) as *quinhydrone* (equimolar complex of quinone and hydroquinone) for determination of pH value potentiometrically.

o-BENZOQUINONE, *o*-Quinone

It is prepared by the oxidation of catechol with silver oxide.

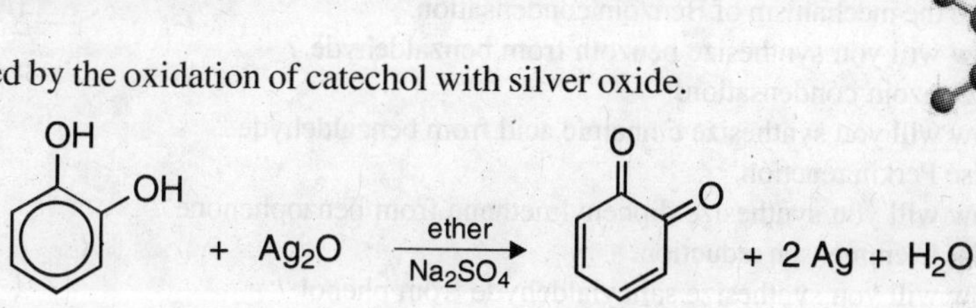

o-Benzoquinone

Anhydrous sodium sulfate is added to absorb the water produced in the reaction.

o-Benzoquinone forms light-red crystals, *mp* 60–70°C, with decomposition. It is exceedingly unstable and decomposes spontaneously on standing.

Examination Questions

1. Give the preparation and properties of benzaldehyde.
2. Give the preparation of acetophenone.
3. Give the preparation of benzophenone.
4. Give the preparation of cinnamaldehyde.
5. What happens when benzaldehyde is heated with concentrated NaOH solution ?

(Bilaspur BSc, 2015)

Answer. Benzyl alcohol and sodium benzoate are formed. This is Cannizzaro reaction. For reaction equation see text.

6. What happens when benzaldehyde is heated with alcoholic KCN solution ?

Answer. Benzoin is formed. This is Benzoin condensation. For reaction equation see text.

7. What happens when benzaldehyde is heated with acetic anhydride in the presence of sodium acetate at 180°C ?

Answer. Cinnamic acid is formed. This is Perkin reaction. For reaction equation see text.

8. How will you distinguish between benzaldehyde and acetophenone ? *(Sambalpur BSc, 2015)*

Answer. The following tests are used :

(*a*) **Tollens' Test**. Benzaldehyde forms silver mirror with ammoniacal silver nitrate solution (Tollens' reagent). Acetophenone does not react.

$$AgNO_3 + 3NH_4OH \longrightarrow Ag(NH_3)_2OH + NH_4NO_3 + 2H_2O$$

$$C_6H_5-\overset{\overset{\large O}{\|}}{C}-H + 2Ag(NH_3)_2OH \longrightarrow C_6H_5-\overset{\overset{\large O}{\|}}{C}-ONH_4 + \underset{mirror}{2Ag} + 3NH_3 + H_2O$$

(*b*) **Iodoform Test**. Acetophenone forms yellow precipitate of iodoform with alkaline solution of iodine. Benzaldehyde does not react.

$$C_6H_5-\overset{\overset{\large O}{\|}}{C}-CH_3 + 4NaOH + 3I_2 \longrightarrow C_6H_5-\overset{\overset{\large O}{\|}}{C}-ONa + \underset{Precipitate}{CHI_3} + 3H_2O + 3NaI$$

9. How will you distinguish between benzaldehyde and acetaldehyde ? *(Assam BSc, 2014)*

Clue. Acetaldehyde gives Fehling's test and Iodoform test. Benzaldehyde does not.

10. How will you distinguish between acetophenone and benzophenone ? *(Delhi BSc, 2015)*

Clue. Acetophenone gives Iodoform test. Benzophenone does not.

11. Write a note on : Perkin Reaction. *(Calicut BSc, 2014)*

12. Write a note on : Cannizzaro Reaction. *(Orissa BSc, 2014)*

13. Write a note on : Beckmann Rearrangement. *(Indore BSc, 2015)*

14. Write a note on : Benzoin Condensation. *(Osmania BSc, 2014)*

15. Give the mechanism of Cannizzaro reaction. *(Shivaji BSc, 2014)*

16. Give the mechanism of Perkin reaction. *(Andhra BSc, 2015)*

17. Give the mechanism of Benzoin condensation. *(Nagpur BSc, 2014)*

18. How will you synthesize benzoin from benzaldehyde ?

Clue. Use Benzoin condensation.

19. How will you synthesize cinnamic acid from benzaldehyde ? *(Anna BSc, 2015)*

Clue. Use Perkin reaction.

20. How will you synthesize diphenylmethane from benzophenone ?

Clue. Use Clemmensen reduction.

21. How will you synthesize salicylaldehyde from phenol ?

Clue. Use Reimer-Tiemann reaction.

22. How will you synthesize ethylbenzene from acetophenone ?

Clue. Use Clemmensen reduction.

23. How will you synthesize benzaldehyde from benzene ? *(Jammu BSc, 2014)*

Answer. Following steps are involved :

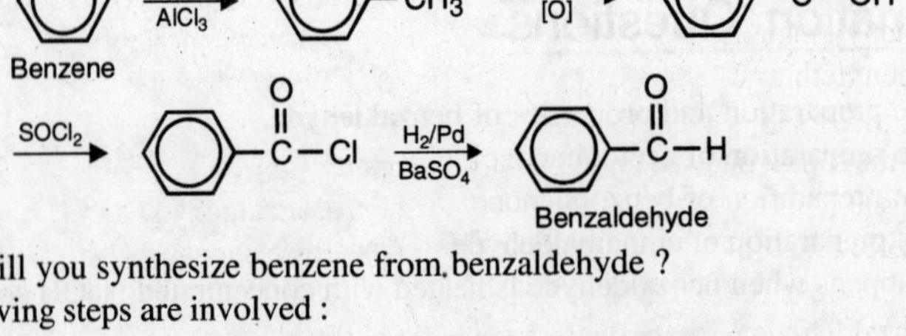

24. How will you synthesize benzene from benzaldehyde ? *(Bihar BSc, 2014)*

Answer. Following steps are involved :

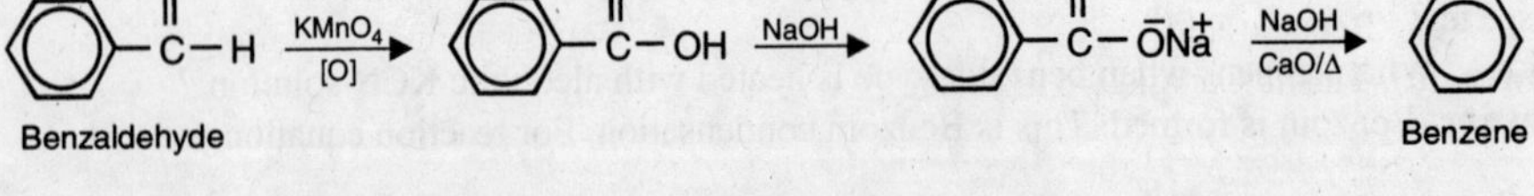

25. How will you synthesize Mandelic acid from benzaldehyde ? (*Aligarh BSc, 2014*)

Clue. Steps are : (*i*) HCN ; (*ii*) H_2O/H^+.

26. How will you synthesize phenylacetic acid from benzaldehyde ? (*Panjab BSc, 2015*)

Answer. Following steps are involved :

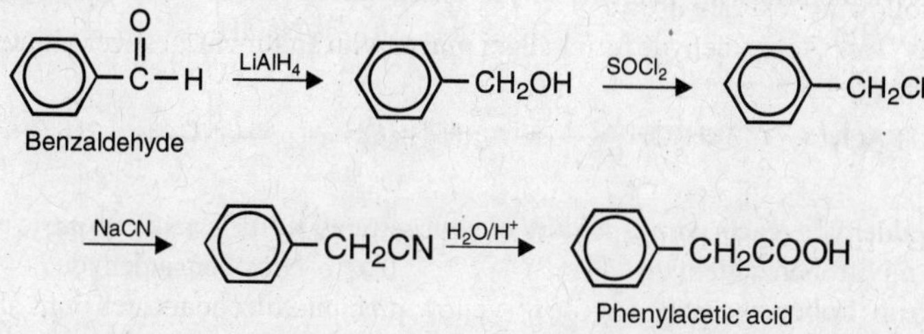

27. A compound of formula C_8H_8O forms an oxime but does not reduce Tollens' reagent. What is the compound ?

Answer. C_8H_8O corresponds to Acetophenone. It forms an oxime but does not reduce Tollens' reagent.

28. Compound (*A*), C_7H_8, is chlorinated in the presence of ultraviolet light to give (*B*). Compound (*B*) on hydrolysis with dilute NaOH gives (*C*). Treatment of (*C*) with acetic anhydride by Perkin reaction produces (*D*) which is an acid (MW=148). Deduce the structural formulas of (*A*), (*B*), (*C*) and (*D*). Write equations for the reactions involved.

Answer. (*A*) is Toluene ; (*B*) is Benzal chloride ; (*C*) is Benzaldehyde ; and (*D*) is Cinnamic acid.

29. Compound (*A*), C_8H_9Cl, is hydrolysed by aqueous NaOH to give (*B*), $C_8H_{10}O$, Compound (*B*) can be oxidized under mild reaction conditions to compound (*C*), C_8H_8O. Compound (*C*) gives positive Iodoform Test. Deduce the structural formulas of (*A*), (*B*), and (*C*). Write equations for the reactions involved.

Answer. (*A*) is 1-Chloro-1-phenylethane ; (*B*) is 1-Phenylethanol ; and (*C*) is Acetophenone.

Multiple Choice Questions

1. Benzoyl chloride reacts with hydrogen in the presence of palladium and barium sulfate to give :

(*a*) Benzyl alcohol (*b*) Benzaldehyde
(*c*) Cinnamic acid (*d*) Benzophenone

Answer. (*b*). This is Rosenmund reduction.

2. Cinnamaldehyde is prepared by the treatment of :

(*a*) benzaldehyde with acetaldehyde in the presence of NaOH
(*b*) benzaldehyde with formaldehyde in the presence of NaOH
(*c*) benzophenone with acetaldehyde in the presence of NaOH
(*d*) benzophenone with formaldehyde in the presence of NaOH

Answer. (*a*)

3. Benzaldehyde reacts with methylmagnesium iodide (followed by hydrolysis) to form

(*a*) 1-Phenylethanol (*b*) Acetophenone
(*c*) 2-Phenylethanol (*d*) Benzophenone

Answer. (*a*)

4. Which of the following will undergo Aldol condensation ?

(*a*) Acetone (*b*) Benzaldehyde
(*c*) Benzoic acid (*d*) Benzophenone

Answer. (*a*)

5. When benzaldehyde is heated with concentrated NaOH, it gives :

(*a*) Benzyl alcohol (*b*) Sodium benzoate
(*c*) Benzoic acid (*d*) Benzyl alcohol + Sodium benzoate

Answer. (*d*). This is Cannizzaro reaction.

6. Benzaldehyde reacts with acetic anhydride in the presence of sodium acetate at 180°C to give :

(*a*) Mandelic acid (*b*) Cinnamic acid

(*c*) Malonic acid (*d*) Benzoic acid

Answer. (*b*). This is Perkin reaction.

7. Which of the following reagents will give an addition reaction with benzaldehyde ?

(*a*) $NaHSO_3$ (*b*) $I_2/NaOH$

(*c*) H_2SO_4 (*d*) $CH_3—\overset{\overset{\displaystyle O}{\|}}{C}—Cl$

Answer. (*a*)

8. Benzaldehyde reacts with a mixture of concentrated sulfuric acid and nitric acid to give

(*a*) *p*-Nitrobenzaldehyde (*b*) *o*-Nitrobenzaldehyde

(*c*) *p*-Nitrobenzoic acid (*d*) *m*-Nitrobenzaldehyde

Answer. (*d*)

9. Acetophenone reacts with Br_2 in the presence of $FeCl_3$ to form :

(*a*) *m*-Bromoacetophenone (*b*) *m*-Dibromobenzene

(*c*) *o*- + *p*-Bromoacetophenone (*d*) *p*-Dibromobenzene

Answer. (*a*)

10. Which of the following statements is NOT correct :

(*a*) Tollens' test is given by both aliphatic and aromatic aldehydes.

(*b*) Fehling's test is given by aliphatic aldehydes. It is not given by aromatic aldehydes.

(*c*) Iodoform test is given by methyl ketones and acetaldehydes.

(*d*) Lucas test is given by both aliphatic and aromatic aldehydes.

Answer. (*d*)

11. Benzaldehyde can be distinguished from acetophenone by using :

(*a*) Tollens' reagent (*b*) Lucas reagent

(*c*) Fehling's reagent (*d*) All of these

Answer. (*a*)

12. Benzaldehyde does not react with

(*a*) Tollens' reagent (*b*) Phenylhydrazine

(*c*) Fehling's solution (*d*) $NaHSO_3$

Answer. (*c*)

13. Which of the following will give a positive Tollens' test ?

(*a*) 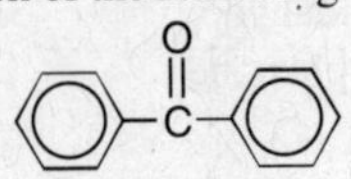

(*b*) 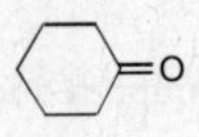

(*c*) $CH_3CH_2—\overset{\overset{\displaystyle O}{\|}}{C}—H$

(*d*) 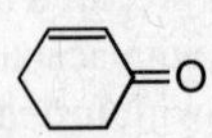

Answer. (*c*)

14. Which of the following compounds yields a yellow precipitate when warmed with a mixture of iodine in aqueous sodium hydroxide ?

(*a*) Acetophenone (*b*) Benzaldehyde

(*c*) Benzophenone (*d*) Propanal

Answer. (*a*)

15. A compound (*A*) gave a positive iodoform reaction and left a solution containing a monocarboxylate ion. Which of the following could (*A*) be ?

(*a*) 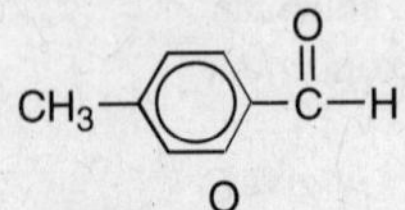

(*b*) 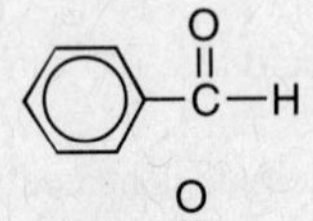

(*c*) 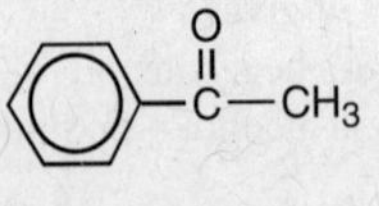

(*d*) 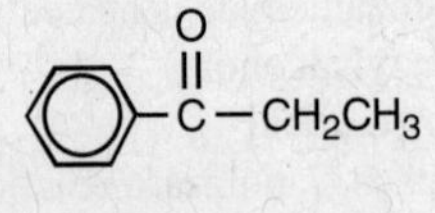

Answer. (*c*)

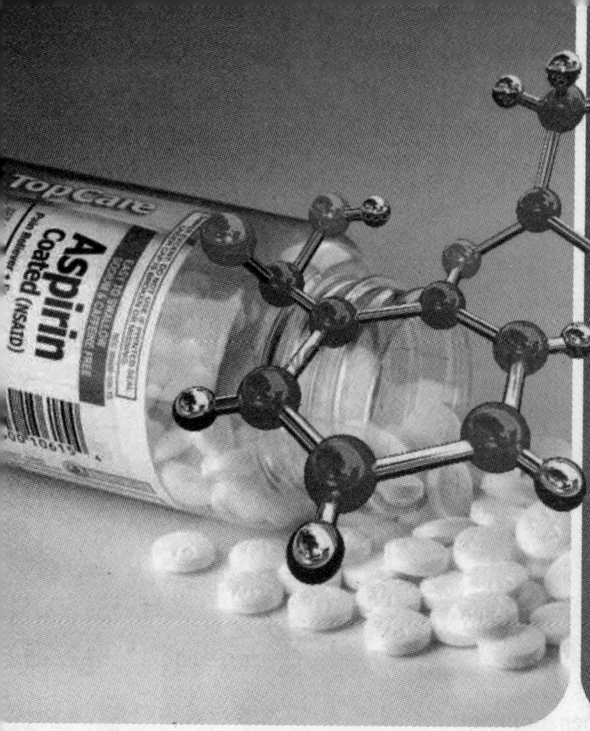

42

Aromatic Carboxylic Acids

Aspirin has completed 100 years of use. Aspirin reduces fever. It is an effective pain reliever. People with history of heart problems often take a precautionary Aspirin per day.

Aromatic acids are compounds in which one or more carboxyl groups (–COOH) are attached directly to the aromatic ring. They are named by the common system or as derivatives of the parent benzoic acid. For example,

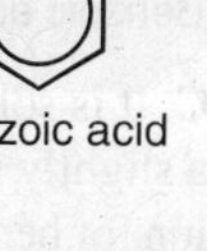

Benzoic acid

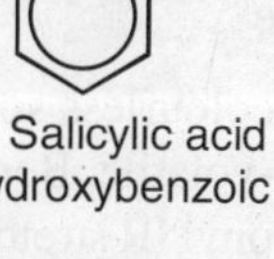

Salicylic acid (*o*-Hydroxybenzoic acid)

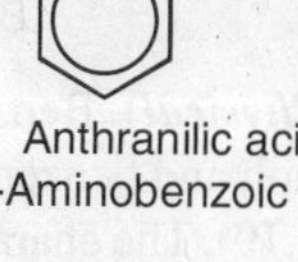

Anthranilic acid (*o*-Aminobenzoic acid)

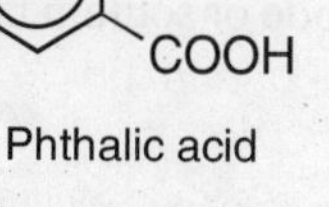

Phthalic acid

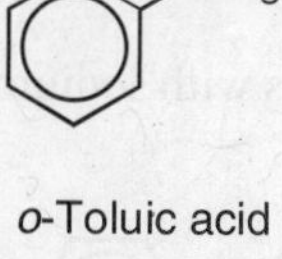

o-Toluic acid

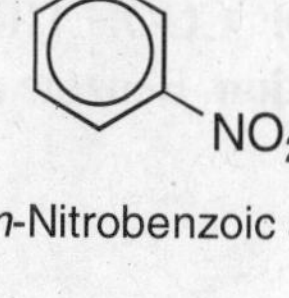

m-Nitrobenzoic acid

Phenylacetic acid ($C_6H_5CH_2COOH$) and other similar compounds in which the carboxyl group is not attached directly to the aromatic ring are called *side-chain aromatic acids*.

Aromatic acids are prepared by the same general methods as used for aliphatic carboxylic acids. In addition, they may be obtained by oxidation of side-chain aromatic hydrocarbons. Aromatic acids are colourless crystalline compounds. They are soluble in hot water and organic solvents. Their reactions are similar to those of aliphatic carboxylic acids. They also give the reactions of the aromatic ring.

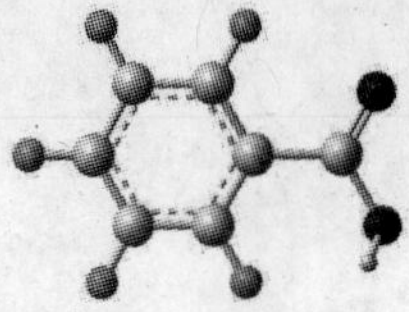

BENZOIC ACID, C_6H_5COOH

It is the simplest aromatic carboxylic acid.

Preparation. Benzoic acid is prepared :

(1) By oxidation of alkylbenzenes with acidic potassium permanganate or sodium dichromate.

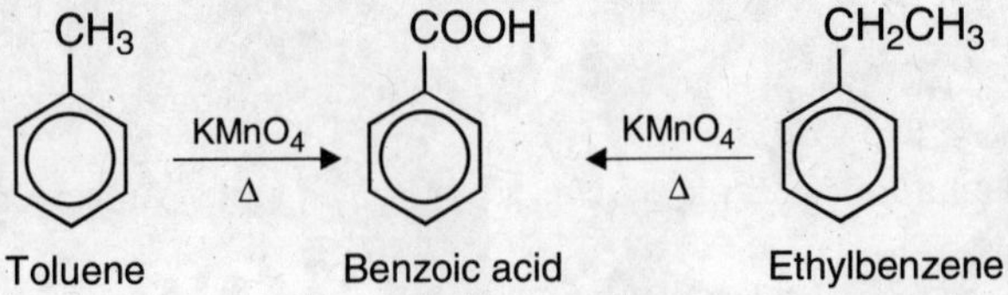

(2) By reaction of phenylmagnesium bromide with carbon dioxide followed by acid-hydrolysis of the addition product.

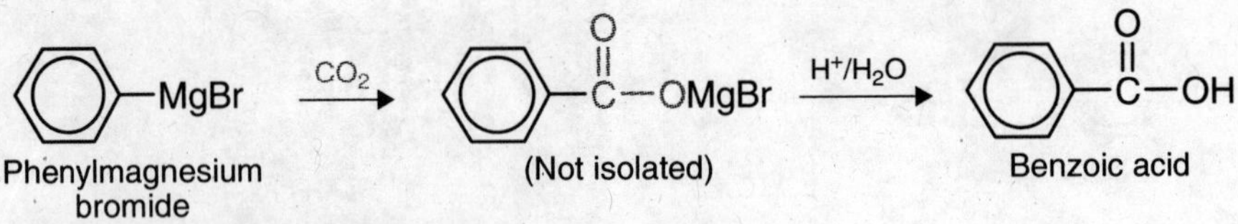

(3) By acid-hydrolysis of phenyl cyanide (benzonitrile).

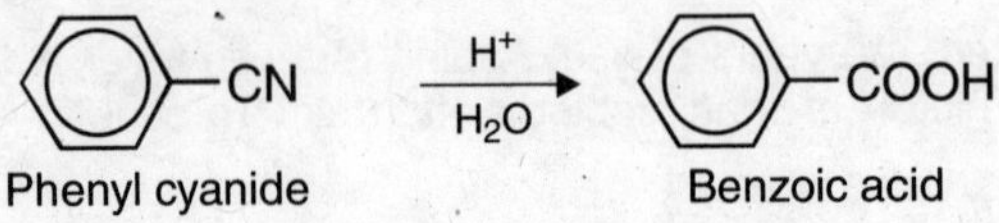

(4) By basic-hydrolysis of benzotrichloride. Benzotrichloride is obtained by chlorination of toluene.

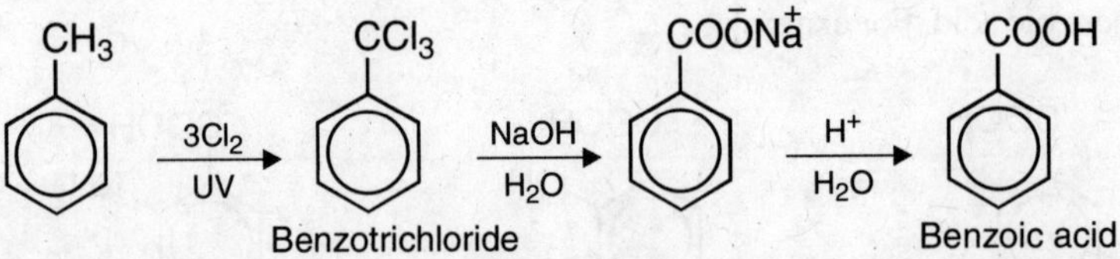

Properties. (*Physical*). Benzoic acid is a colourless solid, *mp* 122°C. It is soluble in hot water, diethyl ether, ethanol, and benzene. It is steam-volatile. Benzoic acid is a slightly stronger acid than acetic acid ($pK_a = 4.19$). The characteristic carbonyl IR stretching absorption for benzoic acid is found at 1710–1720 cm^{-1}

(*Chemical*). The reactions of benzoic acid are those of the carboxyl group and the benzene ring.

A. REACTIONS OF COOH GROUP

(1) **Salt formation.** Benzoic acid reacts with sodium hydroxide or sodium bicarbonate to form sodium benzoate.

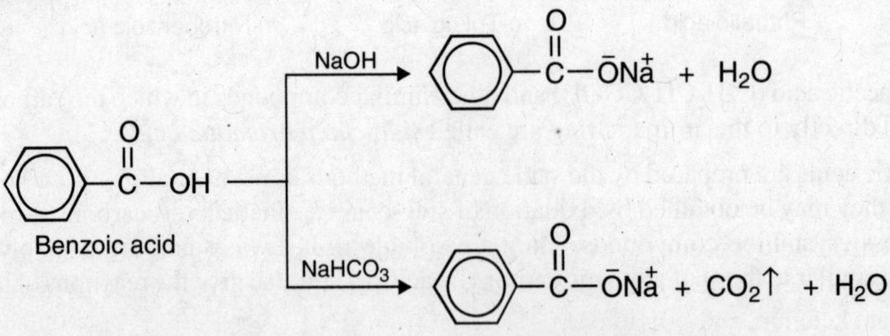

(2) **Ester formation**. It reacts with alcohols in the presence of concentrated sulfuric acid catalyst to form esters.

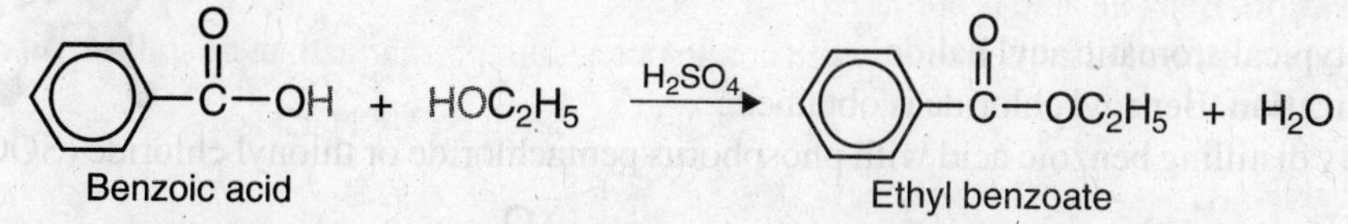

(3) **Acyl halide formation**. Benzoic acid reacts with phosphorus pentachloride or thionyl chloride to form benzoyl chloride.

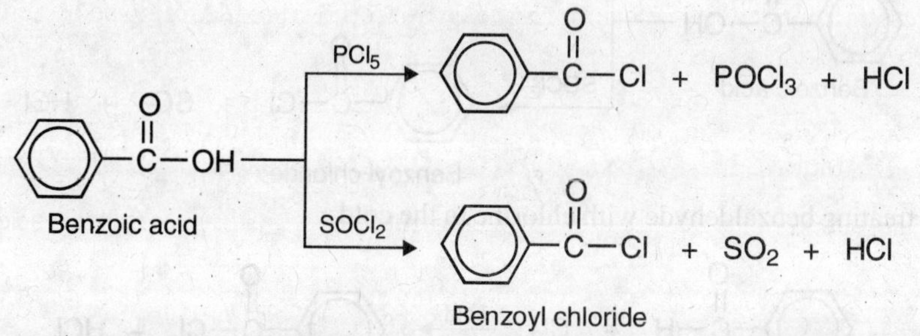

(4) **Reduction to Benzyl alcohol**. Benzoic acid undergoes reduction with lithium aluminium hydride to give benzyl alcohol.

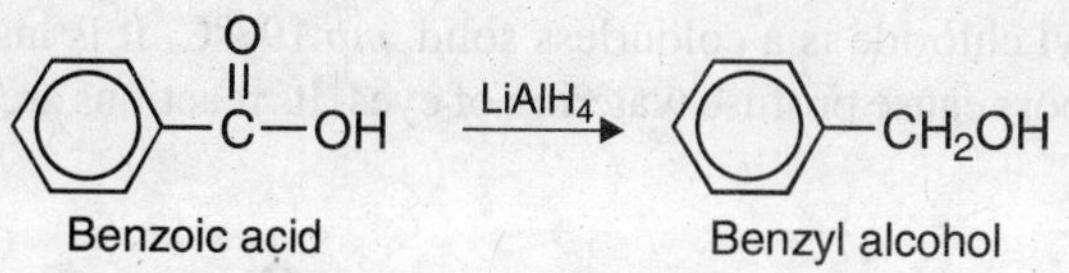

Borane (BH_3) in tetrahydrofuran (THF) is also used for converting benzoic acid into benzyl alcohol. Reaction with borane occurs rapidly at room temperature, and this procedure is often preferred to reduction with $LiAlH_4$ because of the relative ease and safety of using BH_3.

(5) **Decarboxylation**. When heated with soda-lime (CaO/NaOH), benzoic acid loses a molecule of carbon dioxide to yield benzene.

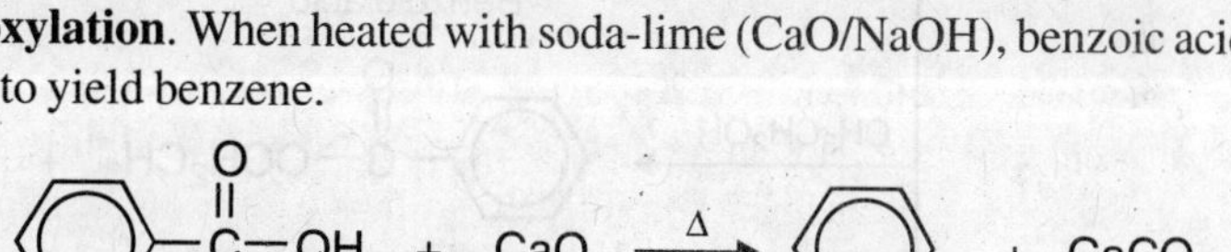

B. REACTIONS OF BENZENE RING

(6) **Electrophilic Substitution**. The benzene ring of benzoic acid undergoes the usual electrophilic substitution reactions. Recall that the –COOH group is a *meta* director and deactivating. For example,

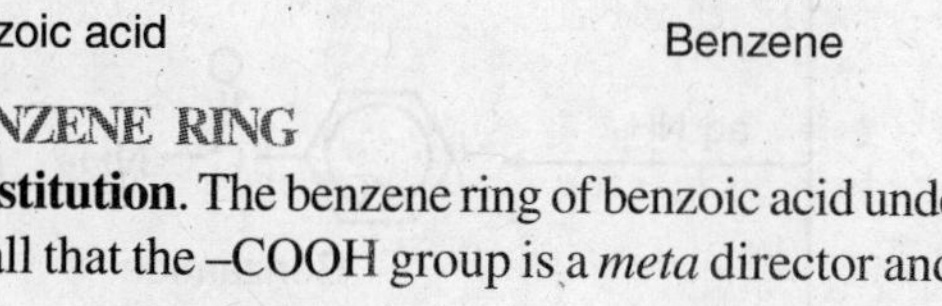

Uses. Benzoic acid is used : (*1*) as *germicide*–in medicine for urinary infection and in vapor form for disinfecting bronchial tubes and ; (*2*) as *food preservative*–sodium benzoate is used for preserving pickles, tomato ketchup, and fruit juices.

BENZOYL CHLORIDE, $C_6H_5-\overset{O}{\overset{\|}{C}}-Cl$

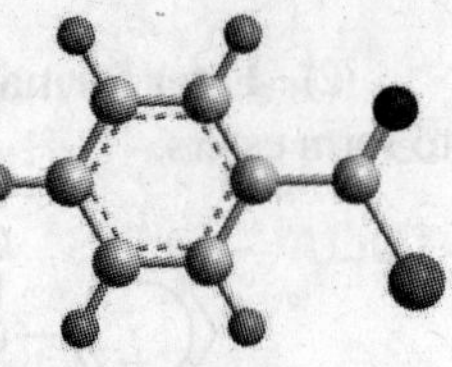

It is a typical aromatic acyl halide.

Preparation. Benzoyl chloride is obtained :

(1) By distilling benzoic acid with phosphorus pentachloride or thionyl chloride ($SOCl_2$).

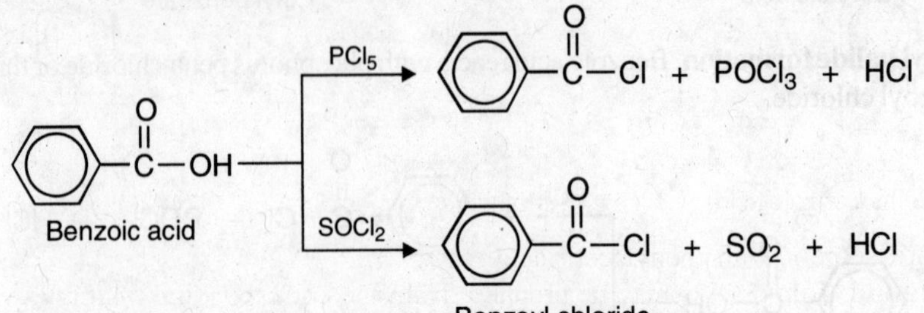

(2) By treating benzaldehyde with chlorine in the cold.

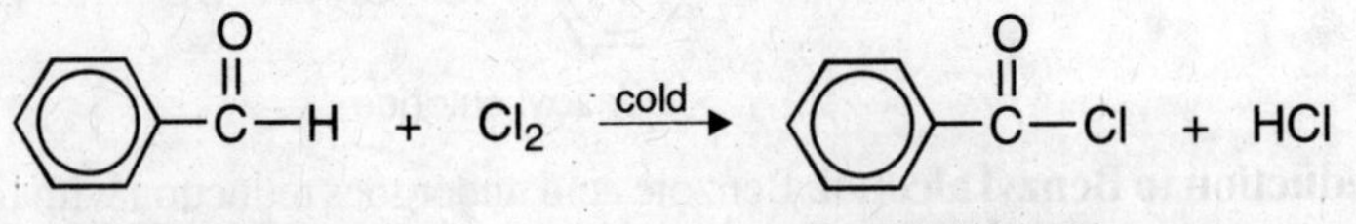

Properties. Benzoyl chloride is a colourless solid, *mp* 197°C. It is insoluble in water. It has a pungent odor and its vapors cause profuse watering of eyes. Its reactions are similar to those of acetyl chloride (Fig. 42.1).

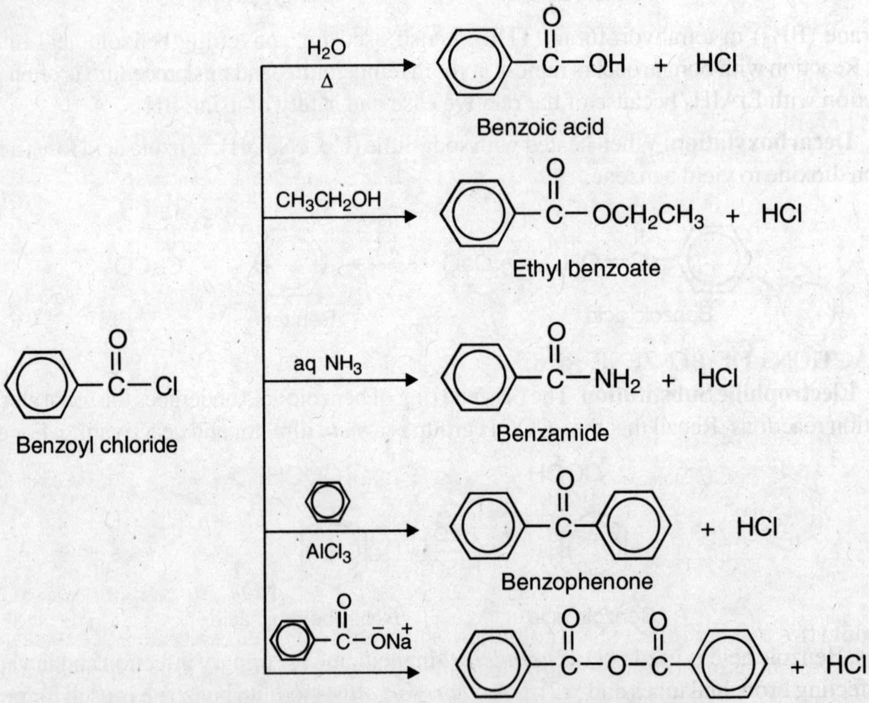

Fig. 42.1. Important reactions of Benzoyl chloride.

Uses. Benzoyl chloride is used : (1) in tear gases to disperse mobs ; and (2) for introducing benzoyl group ($C_6H_5CO–$) in hydroxy and amine compounds.

TOLUIC ACIDS, *Methylbenzoic acids*, $C_6H_4(CH_3)COOH$

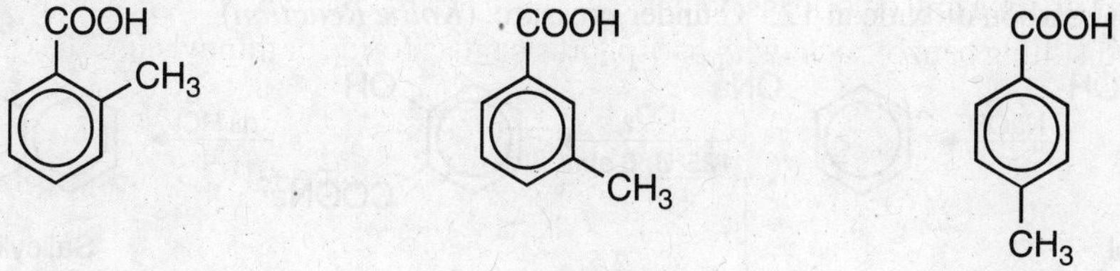

o-Toluic acid (2-Methylbenzoic acid) | *m*-Toluic acid (3-Methylbenzoic acid) | *p*-Toluic acid (4-Methylbenzoic acid)

These are isomeric with phenylacetic acid.

Preparation. Toluic acids may be prepared from the corresponding toluidines *via* diazonium salts. For example,

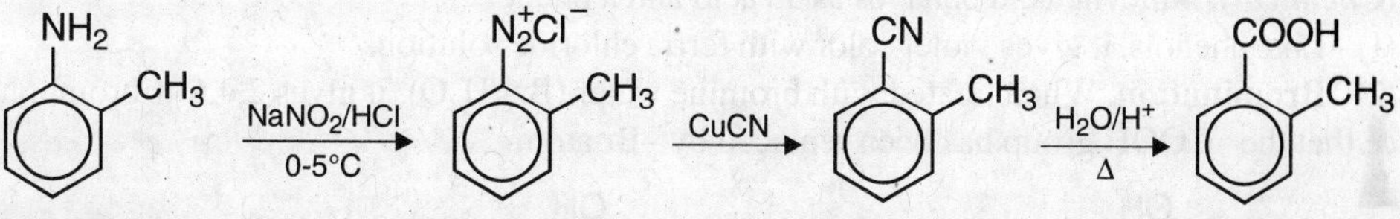

o-Toluidine → *o*-Toluic acid

Properties. The toluic acids are colourless solids, mp : *ortho* 140°C ; *meta* 111°C ; *para* 178°C. The reactions of toluic acids are similar to those of benzoic acid. On oxidation with dilute nitric acid or $KMnO_4$, they give the respective dicarboxylic acids.

ANTHRANILIC ACID, *o-Aminobenzoic acid*

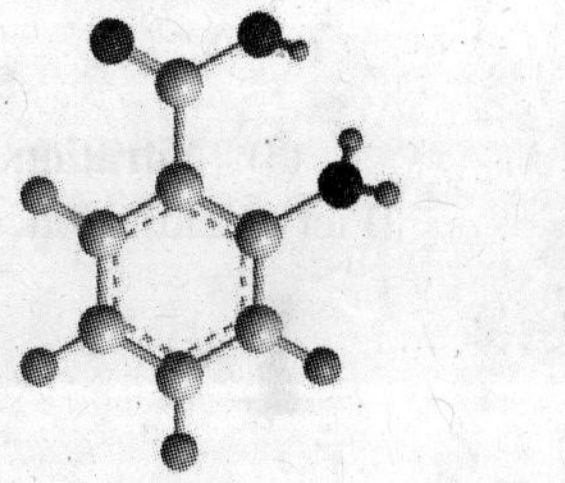

Of the three aminobenzoic acids, only the *ortho* isomer is important.

Preparation. Anthranilic acid may be prepared by reduction of *o*-nitrobenzoic acid with Sn/HCl. Commercially, it is obtained from phthalic acid by the following steps :

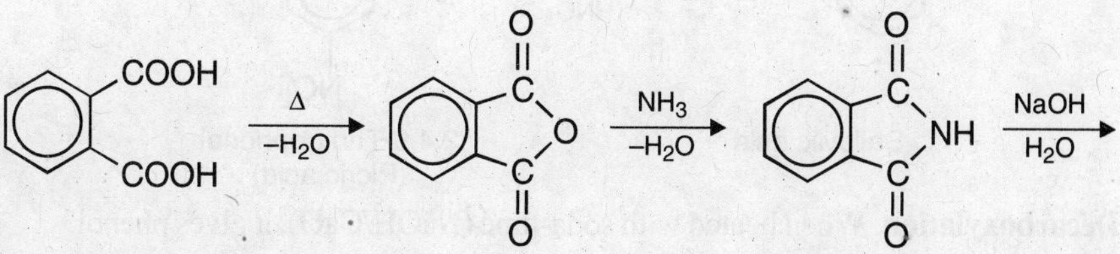

Phthalic acid | Phthalimide

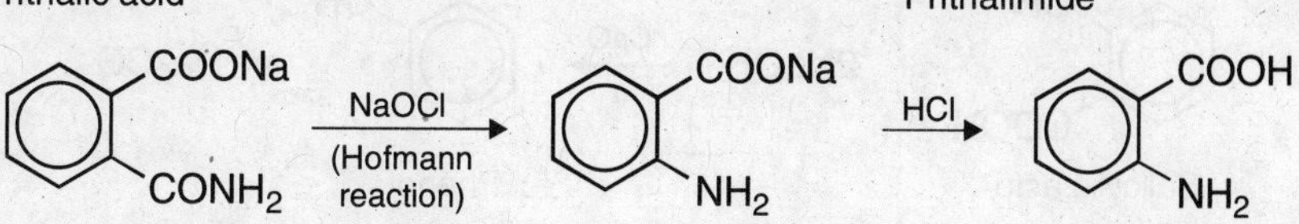

Anthranilic acid

Properties. Anthranilic acid is a colourless solid, *mp* 145°C. It is soluble in water and ethanol. It gives all the reactions of an aromatic amine and carboxylic acid. On heating, it undergoes decarboxylation to give aniline.

Uses. Anthranilic acid is used in the synthesis of dyes.

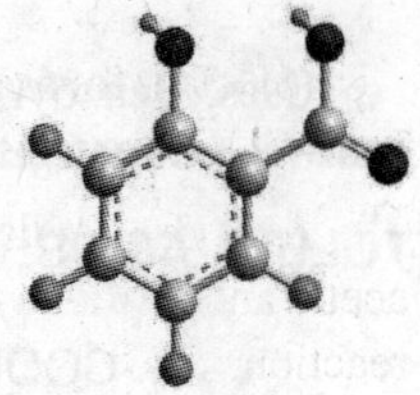

SALICYLIC ACID, *o-Hydroxybenzoic acid*

Of the three hydroxybenzoic acids, only the *ortho* isomer (Salicylic acid) is important.

Preparation. Salicylic acid is obtained commercially by heating sodium phenoxide and carbon dioxide at 125°C under pressure. (*Kolbe Reaction*)

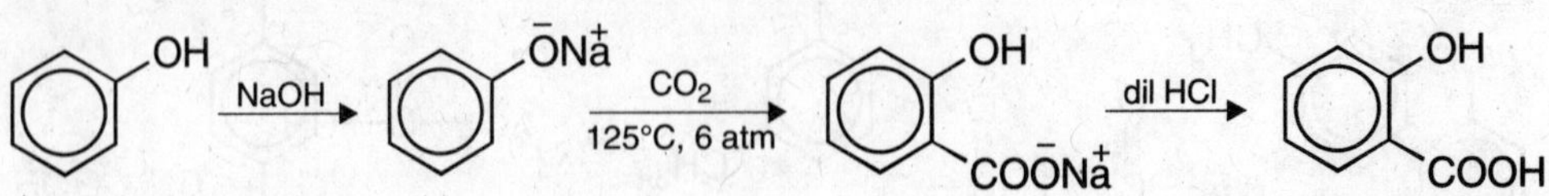

Phenol — Salicylic acid

In fact, the reaction yields a mixture of *o*- and *p*-isomers. Salicylic acid being more volatile is separated by steam-distillation.

Properties. (*Physical*). Salicylic acid is a colourless solid, *mp* 159°C. It is insoluble in cold water but dissolves in hot water, ethanol, and ether. It sublimes below its melting point and is steam-volatile. It is a stronger acid than benzoic acid.

(*Chemical*). Salicylic acid behaves as an acid and a phenol.

(1) Like phenols, it gives violet color with ferric chloride solution.

(2) **Bromination.** When treated with bromine water (Br_2/H_2O), it gives 2,4,6-tribromophenol. Notice that the –COOH group has been replaced by –Br atom.

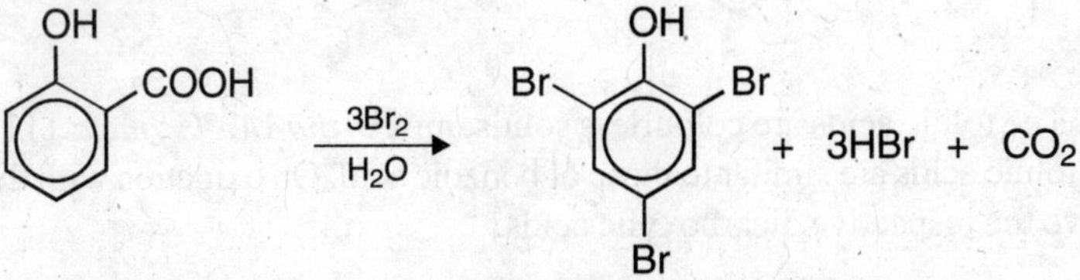

Salicylic acid — 2,4,6-Tribromophenol

(3) **Nitration**. Salicylic acid undergoes nitration with fuming nitric acid to give 2,4,6-trinitrophenol (Picric acid). Notice that –COOH group has been replaced by $–NO_2$ group.

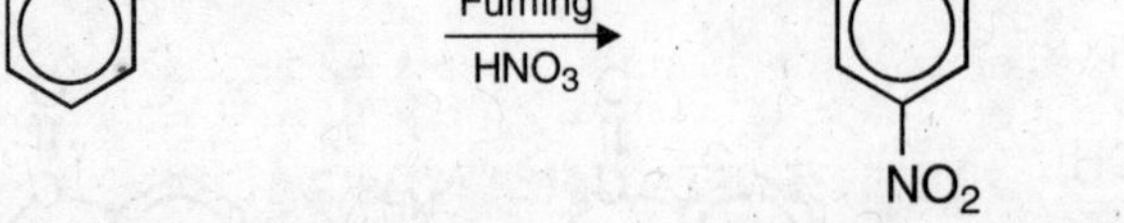

Salicylic acid — 2,4,6-Trinitrophenol (Picric acid)

(4) **Decarboxylation**. When heated with soda-lime (NaOH/CaO), it gives phenol.

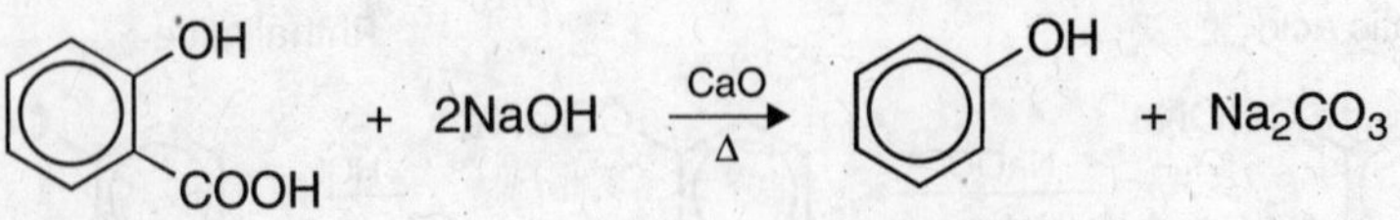

Salicylic acid — Phenol

(5) **Reaction with zinc dust**. When heated with zinc dust, it gives benzoic acid.

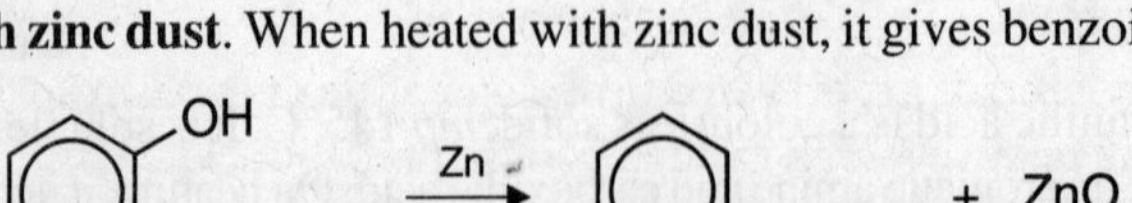

Salicylic acid — Benzoic acid

(6) **Salt formation and Ester formation.** Salicylic acid forms salts (Salicylates) both at the –OH and –COOH groups. Its important derivatives are :

Acetylsalicylic Acid (Aspirin). It is prepared by heating salicylic acid with acetyl chloride or acetic anhydride in the presence of phosphoric acid. Notice that only the OH group is involved in the reaction.

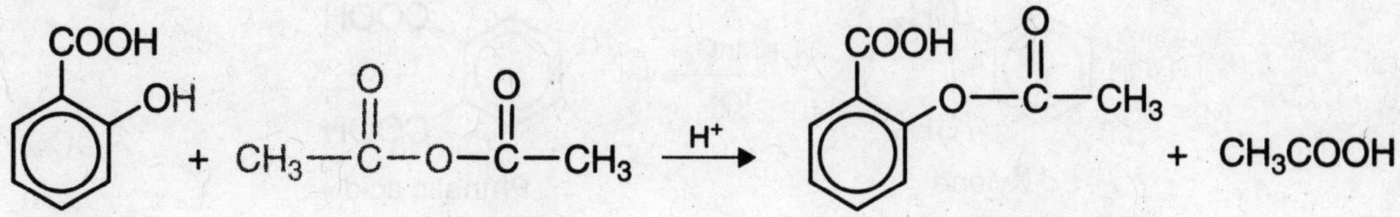

Salicylic acid | Acetic anhydride | Acetylsalicylic acid (Aspirin)

Acetylsalicylic acid is a colourless solid, *mp* 135°C. It reduces fever but does not lower the body temperature. It is also used as a pain killer. It is sold under the trade name **Aspirin**.

Methyl Salicylate (Oil of Wintergreen). It is prepared by esterifying salicylic acid with methanol in the presence of sulfuric acid. Notice that only the –COOH group is involved in the reaction.

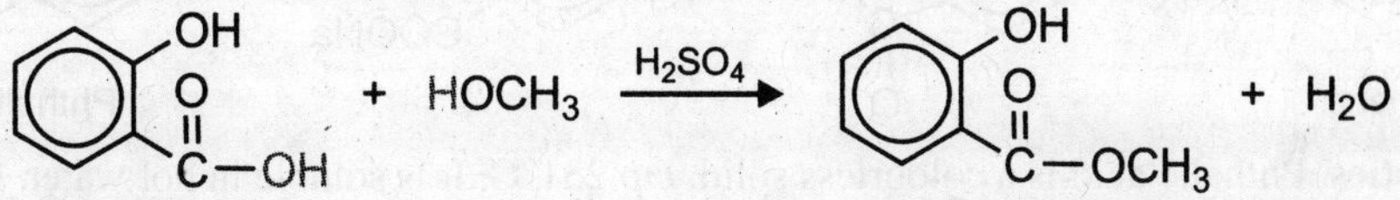

Salicylic acid | Methyl salicylate (Oil of Wintergreen)

Methyl salicylate is colourless, pleasant smelling liquid, *bp* 224°C. It is used in hair tonics and ointments for treating aches, sprains, and bruises.

Phenyl Salicylate (Salol). It is obtained by heating salicylic acid with phenol in the presence of $POCl_3$. Notice that only the –COOH group is involved in the reaction.

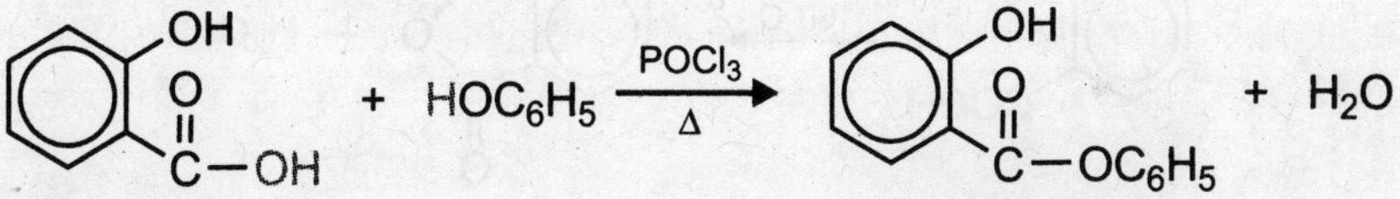

Salicylic acid | Phenyl salicylate (Salol)

Phenyl salicylate is a colourless solid, *mp* 43°C. It is used as an antiseptic.

BENZENEDICARBOXYLIC ACIDS

The *ortho*, *meta*, and *para*-benzenedicarboxylic acids have the common names phthalic, isophthalic, and terephthalic acids respectively.

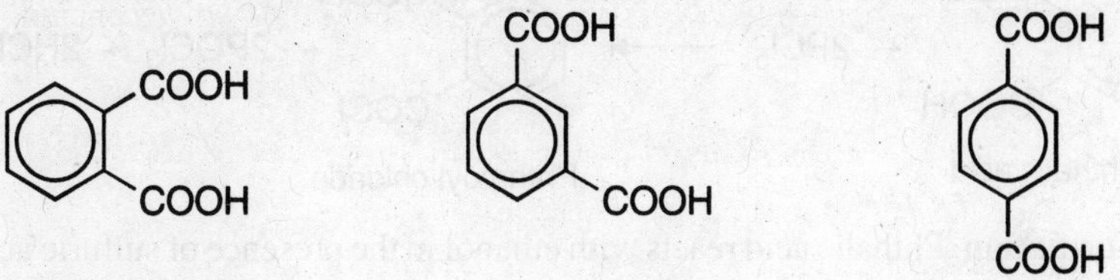

Phthalic acid | Isophthalic acid | Terephthalic acid

Benzenedicarboxylic acids are prepared by the oxidation of the corresponding xylenes. They are colourless solids having relatively high melting points : phthalic acid 231°C ; isophthalic acid 348°C ; terephthalic acid 300°C. They give the usual reactions of carboxylic acids. Phthalic acid is distinguished from the *meta* and *para* isomers as it forms anhydride due to close proximity of the two –COOH groups on the benzene ring.

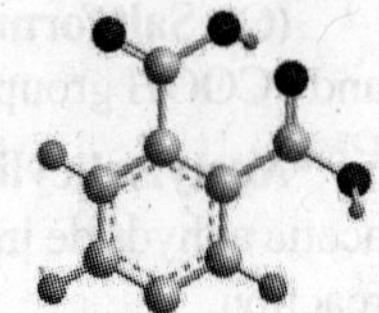

PHTHALIC ACID, 1,2-*Benzenedicarboxylic acid*, $C_6H_4(COOH)_2$

Phthalic acid is the most important of the three benzenedicarboxylic acids.

Preparation. Phthalic acid is obtained :

(1) By oxidation of *o*-xylene with potassium permanganate.

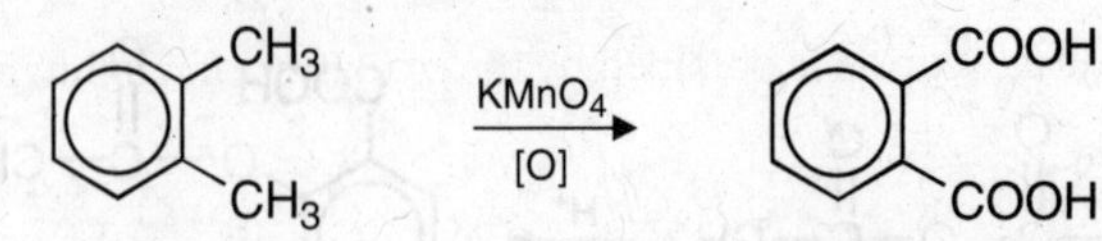

o-Xylene Phthalic acid

(2) By passing a mixture of naphthalene vapor and oxygen/air over vanadium pentoxide catalyst at 350°C. Phthalic anhydride is first formed which on hydrolysis yields phthalic acid.

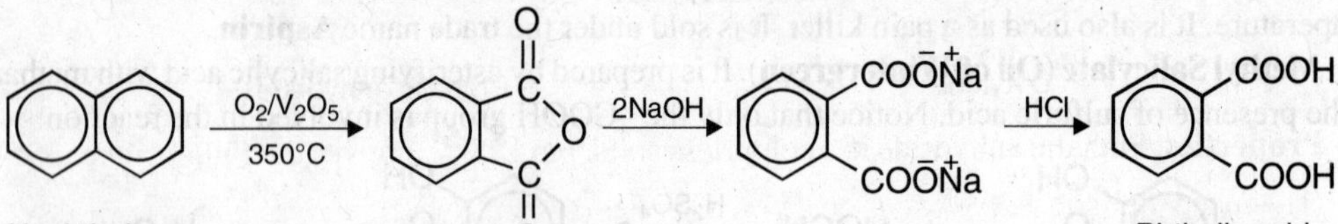

Naphthalene Phthalic acid

Properties. Phthalic acid is a colourless solid, *mp* 231°C. It is soluble in hot water. It gives the following reactions :

(1) **Action of heat**. When heated at 160°C, it loses a molecule of water to form phthalic anhydride. Only phthalic acid forms anhydride, while the *meta* and *para* isomers fail to do so.

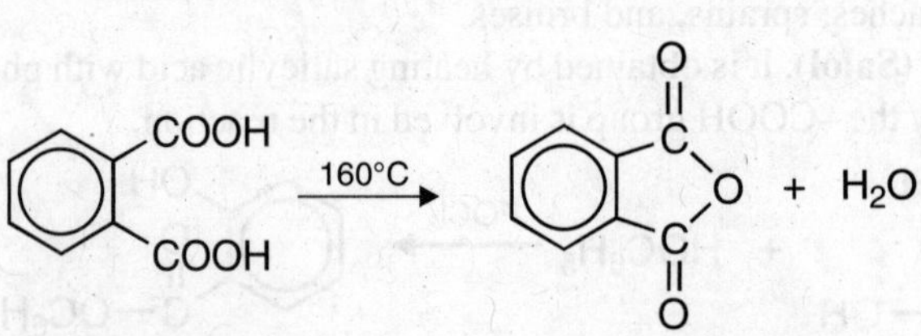

Phthalic acid Phthalic anhydride

When heated in the presence of soda-lime (or KOH), Phthalic acid undergoes decarboxylation to give benzene.

(2) **Reaction with PCl_5**. When heated with phosphorus pentachloride or thionyl chloride ($SOCl_2$), it gives phthaloyl chloride.

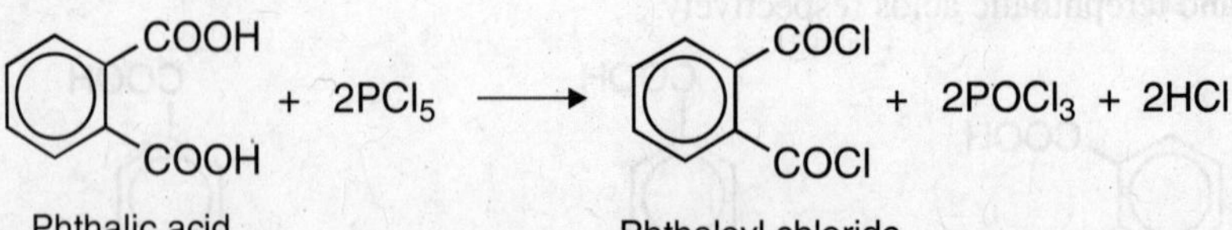

Phthalic acid Phthaloyl chloride

(3) **Ester formation**. Phthalic acid reacts with ethanol in the presence of sulfuric acid catalyst to form diethyl phthalate.

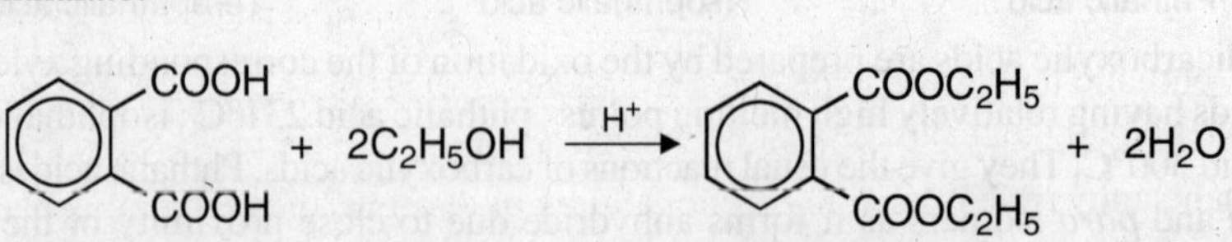

Phthalic acid Diethyl phthalate

PHTHALIC ANHYDRIDE, $C_6H_4(CO)_2O$

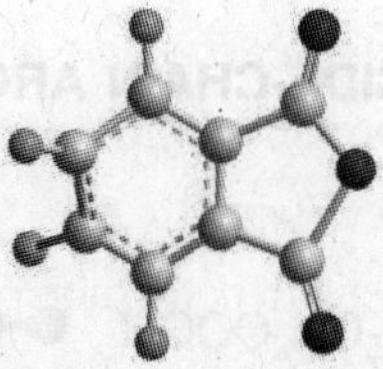

Preparation. Phthalic anhydride is obtained by the oxidation of naphthalene or xylene.

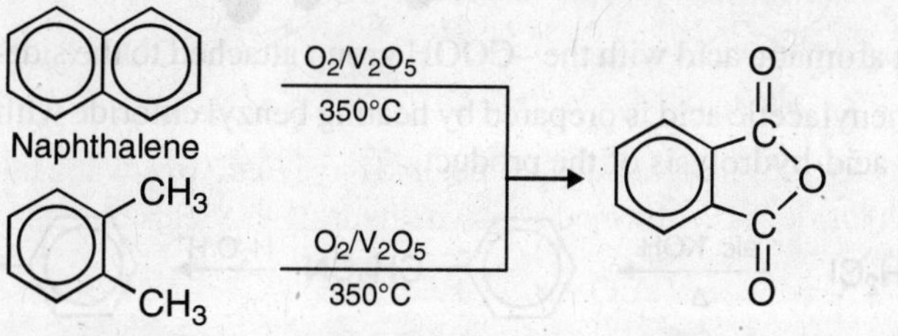

Naphthalene · *o*-Xylene · Phthalic anhydride

Properties. Phthalic anhydride is a colourless solid, *mp* 128°C. It gives the following reactions:

(1) When phthalic anhydride is heated with water (or dilute acid), it gives phthalic acid.

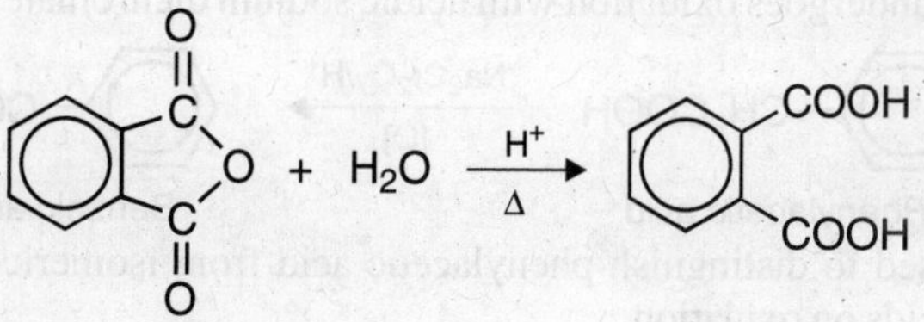

Phthalic anhydride · Phthalic acid

(2) When phthalic anhydride is heated with aqueous ammonia, it gives phthalimide.

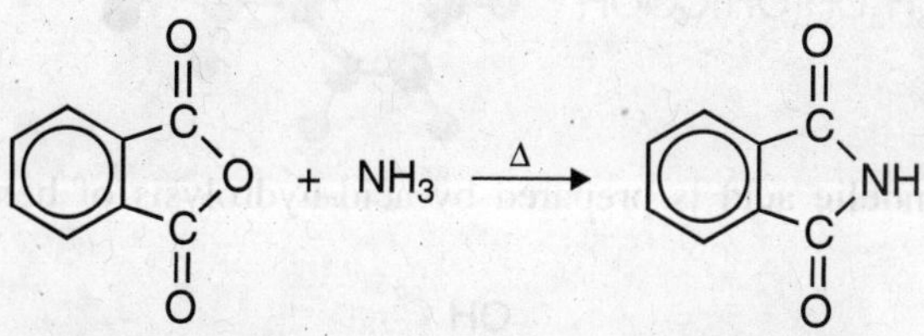

Phthalic anhydride · Phthalimide

(3) When heated with excess of methanol in the presence of concentrated sulfuric acid, it gives dimethyl phthalate.

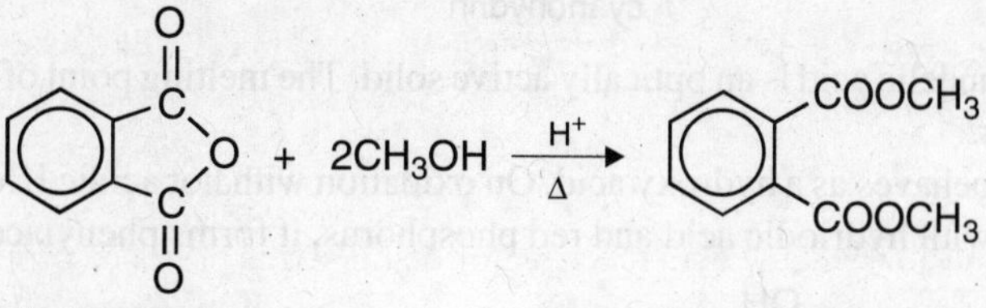

Phthalic anhydride · Dimethyl phthalate

(4) When phthalic anhydride is heated with phenol and resorcinol in the presence of sulfuric acid, it forms phenolphthalein and fluorescein respectively.

Uses. Phthalic anhydride is used in the preparation of dyes, plastics (glyptals), plasticizers, and benzoic acid.

SIDE-CHAIN AROMATIC ACIDS

PHENYLACETIC ACID, $C_6H_5CH_2COOH$

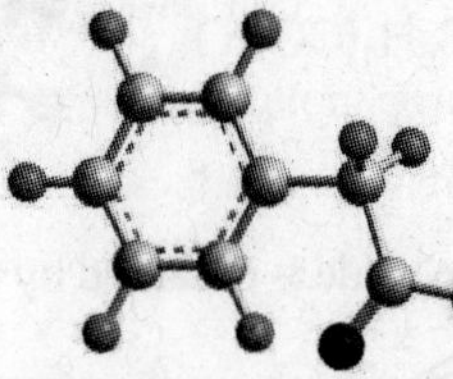

It is the simplest aromatic acid with the –COOH group attached to the side-chain.

Preparation. Phenylacetic acid is prepared by heating benzyl chloride with alcoholic potassium cyanide followed by acid-hydrolysis of the product.

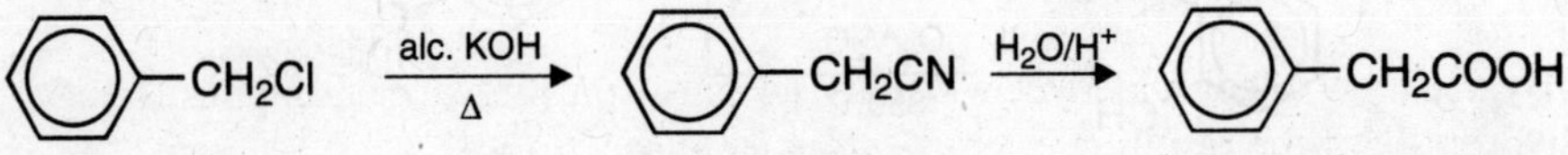

Benzyl chloride   Benzyl cyanide   Phenylacetic acid

Properties. Phenylacetic acid is a colourless solid, *mp* 77°C. It is soluble in hot water. It is a stronger acid than acetic acid.

Phenylacetic acid undergoes oxidation with acidic sodium dichromate to give benzoic acid.

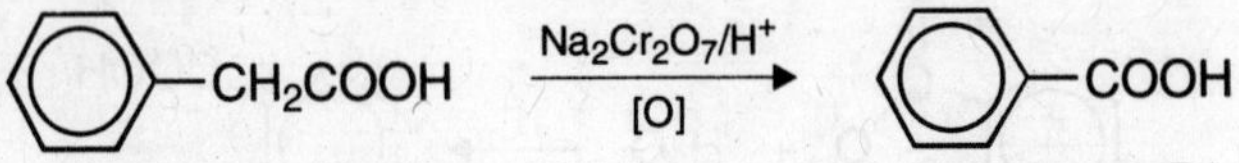

Phenylacetic acid   Benzoic acid

This reaction is used to distinguish phenylacetic acid from isomeric toluic acids which yield benzenedicarboxylic acids on oxidation.

MANDELIC ACID, $C_6H_5CH(OH)COOH$

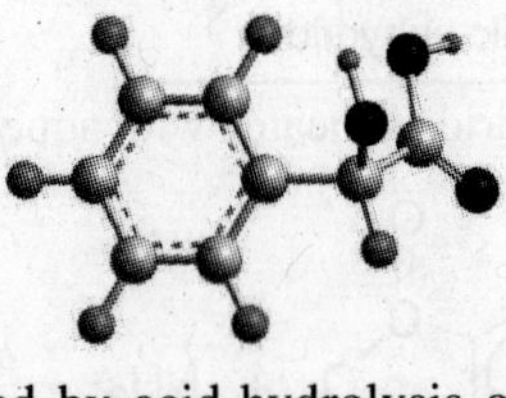

Preparation. Mandelic acid is prepared by acid-hydrolysis of benzaldehyde cyanohydrin (mandelonitrile).

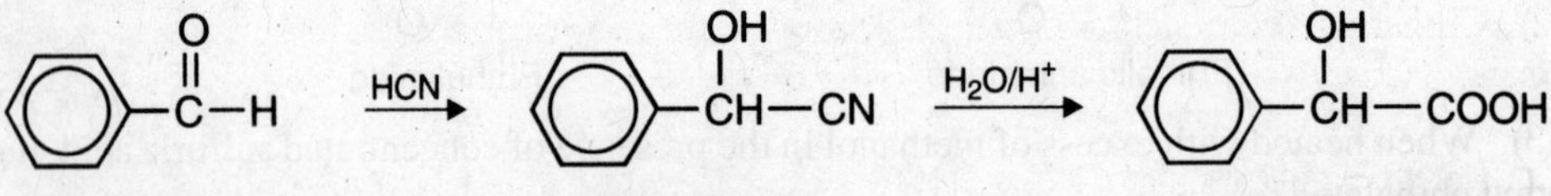

Benzaldehyde   Benzaldehyde cyanohydrin   Mandelic acid

Properties. Mandelic acid is an optically active solid. The melting point of the DL-acid is 118°C. It is soluble in water.

Mandelic acid behaves as a hydroxy acid. On oxidation with hot acidic $KMnO_4$, it gives benzoic acid. On reduction with hydriodic acid and red phosphorus, it forms phenylacetic acid.

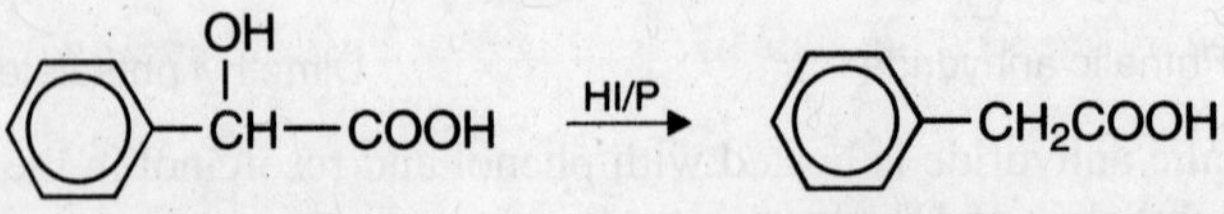

Mandelic acid   Phenylacetic acid

Mandelic acid is used as an antiseptic.

CINNAMIC ACID, $C_6H_5CH{=}CHCOOH$

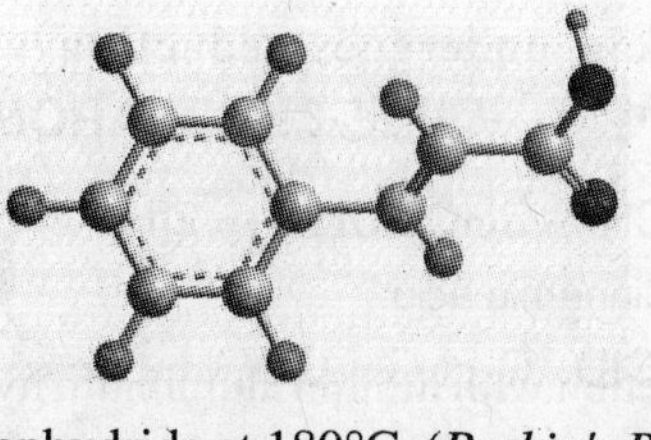

Cinnamic acid is 3-Phenylpropenoic acid.

Preparation. Cinnamic acid is obtained :

(1) By heating benzaldehyde with acetic anhydride at 180°C. (*Perkin's Rection*)

$$\underset{\text{Benzaldehyde}}{C_6H_5CH{=}O} + \underset{\text{Acetic anhydride}}{(CH_3\overset{O}{\overset{\|}{C}})_2O} \xrightarrow[180°C]{CH_3COONa} \underset{\text{Cinnamic acid}}{C_6H_5CH{=}CHCOOH} + CH_3COOH$$

(2) By treatment of benzaldehyde with ethyl acetate in the presence of sodium ethoxide followed by acid-hydrolysis of the resulting ester. (*Claisen Condensation*)

$$\underset{\text{Benzaldehyde}}{C_6H_5CH{=}O} + \underset{\text{Ethyl acetate}}{CH_3COOC_2H_5} \xrightarrow[(-H_2O)]{C_2H_5ONa} \underset{\text{Ester}}{C_6H_5CH{=}CHCOOC_2H_5}$$

$$\xrightarrow{H_2O} \underset{\text{Cinnamic acid}}{C_6H_5CH{=}CHCOOH}$$

(3) By heating benzaldehyde with malonic acid in alcoholic ammonia solution. (*Knoevenagel's Reaction*)

$$C_6H_5CH{=}O + \underset{\text{Malonic acid}}{CH_2(COOH)_2} \xrightarrow[\Delta]{NH_3} \underset{\text{Cinnamic acid}}{C_6H_5CH{=}CHCOOH} + CO_2 + H_2O$$

Properties. Cinnamic acid is a crystalline solid, *mp* 133°C. It is sparingly soluble in water. It behaves as an α,β-unsaturated acid and a benzene derivative.

(1) Cinnamic acid reacts with sodium hydroxide to form sodium cinnamate.

$$\underset{\text{Cinnamic acid}}{C_6H_5CH{=}CHCOOH} + NaOH \longrightarrow \underset{\text{Sod. cinnamate}}{C_6H_5CH{=}CHCOONa} + H_2O$$

(2) Cinnamic acid reacts with ethyl alcohol in the presence of an acid catalyst to give ethyl cinnamate.

$$\underset{\text{Cinnamic acid}}{C_6H_5CH{=}CHCOOH} + C_2H_5OH \xrightarrow{H^+} \underset{\text{Ethyl cinnamate}}{C_6H_5CH{=}CHCOOC_2H_5} + H_2O$$

(3) Cinnamic acid reacts with PCl_5 or $SOCl_2$ to form cinnamoyl chloride.

$$\underset{\text{Cinnamic acid}}{C_6H_5CH{=}CHCOOH} \xrightarrow{PCl_5} \underset{\text{Cinnamoyl chloride}}{C_6H_5CH{=}CHCOCl} + POCl_3 + HCl$$

(4) Cinnamic acid reacts with bromine in carbon tetrachloride to give dibromocinnamic acid.

$$\underset{\text{Cinnamic acid}}{C_6H_5CH{=}CHCOOH} + Br_2 \xrightarrow{CCl_4} \underset{\text{Dibromocinnamic acid}}{C_6H_5{-}\overset{Br}{\overset{|}{C}H}{-}\overset{Br}{\overset{|}{C}H}{-}COOH}$$

(5) Cinnamic acid undergoes reduction with sodium amalgam and water to give 3-phenylpropanoic acid.

$$\underset{\text{Cinnamic acid}}{C_6H_5CH{=}CHCOOH} + 2[H] \xrightarrow[H_2O]{Na/Hg} \underset{\text{3-Phenylpropanoic acid}}{C_6H_5CH_2CH_2COOH}$$

It undergoes reduction with lithium aluminium hydride to form 3-phenyl-2-propenol.

$$\underset{\text{Cinnamic acid}}{C_6H_5CH{=}CHCOOH} + [H] \xrightarrow{LiAlH_4} \underset{\text{3-Phenyl-2-propenol}}{C_6H_5CH{=}CHCH_2OH}$$

(6) Cinnamic acid undergoes oxidation with hot acidic $KMnO_4$ to give benzoic acid.

$$\underset{\text{Cinnamic acid}}{C_6H_5CH{=}CHCOOH} + [O] \xrightarrow[\Delta]{KMnO_4} \underset{\text{Benzoic acid}}{C_6H_5COOH}$$

(7) Cinnamic acid undergoes decarboxylation with soda-lime to yield styrene.

$$\underset{\text{Cinnamic acid}}{C_6H_5CH{=}CHCOOH} + NaOH \xrightarrow[\Delta]{CaO} \underset{\text{Styrene}}{C_6H_5CH{=}CH_2} + Na_2CO_3 + H_2O$$

(8) It undergoes electrophilic substitution reactions to form *o,p*-isomers. For example,

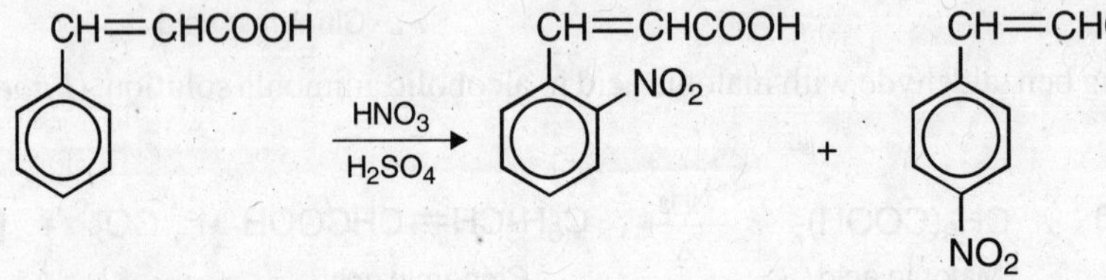

Cinnamic acid — *o*- and *p*-Nitrocinnamic acid

ACIDITY OF SUBSTITUTED BENZOIC ACIDS

It might be expected that resonance-stabilization by the aromatic π molecular orbital would play a large role in the relative acidity of benzoic acid and substituted benzoic acids. However, this is not so. The negative charge of the carboxylate ion is shared by the two carboxylate oxygen atoms but cannot be effectively delocalized by the aromatic ring. This is because, the oxygens of the carboxylate anion are not directly adjacent to the aromatic ring ; resonance structures cannot be drawn.

Because the benzene ring does not participate in resonance-stabilization of the carboxylate

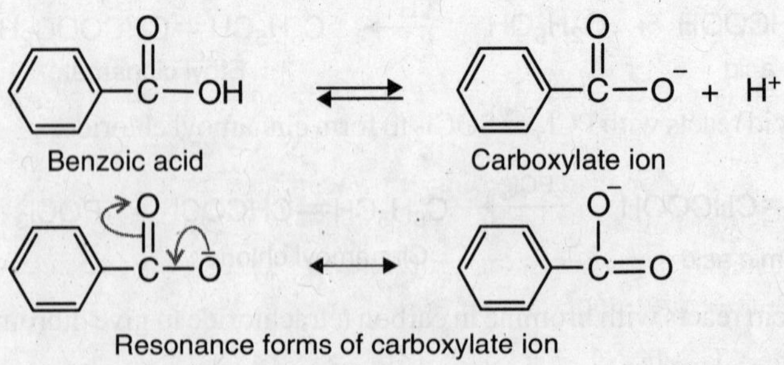

Resonance forms of carboxylate ion

group, substituents on a benzene ring influence acidity primarily by the *inductive effect.* Regardless of the position of substitution, an *electron-withdrawing* group usually increases the acidity of a benzoic acid (Table 42.1).

Table 42.1 ACIDITY CONSTANTS OF SUBSTITUTED BENZOIC ACIDS

Acid	pK_a
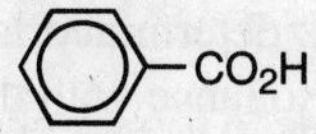	4.19
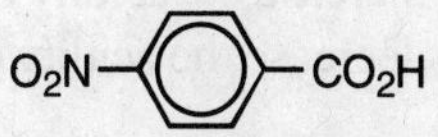	3.43
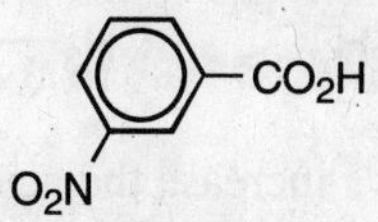	3.49
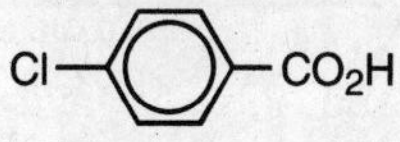	3.98
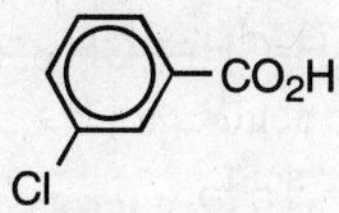	3.83
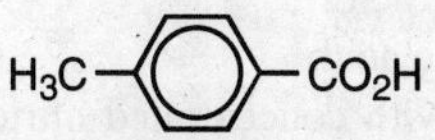	4.37
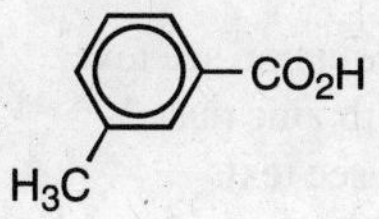	4.27

The reason that an electronegative substituent increases the acid strength are again destabilization of the acid and stabilization of the anion.

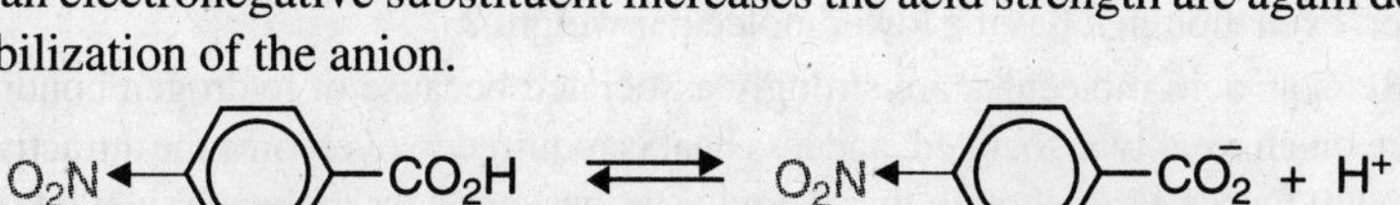

Destabilized by electron-withdrawal — Stabilized by electron-withdrawal

An *electron-releasing* alkyl substituent that is *p*- or *m*- to the carboxyl group decreases the acidity of a benzoic acid.

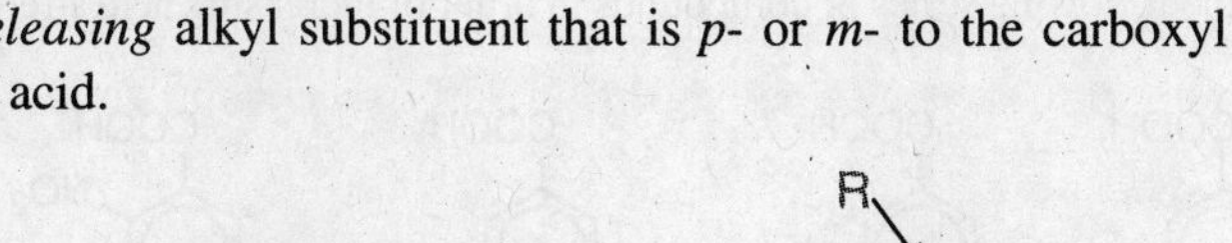

p- or *m*-alkyl group decreases acid strength

All *ortho*-substituents (whether electron-withdrawing *or* electron-releasing) increase acidity of a benzoic acid. The reasons for this are not entirely known, although there may be *steric effect* from having two substituents *ortho* to one another.

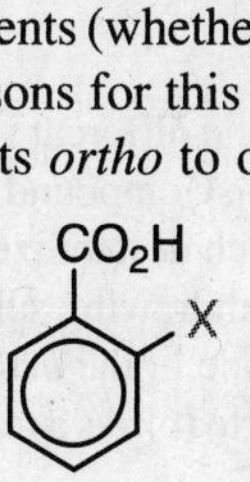

Ortho-substituents increase acid strength

Points to Remember

Predicting Acid Strengths of Substituted Benzoic Acids

(1) The strength of the substituted benzoic acids is determined largely by *inductive effects* because the $-CO_2^-$ group does not enter into resonance with the aromatic ring.

(2) Electron-withdrawing substituents ($-NO_2$, $-Cl$) increase the acidity of a benzoic acid.

(3) Electron-releasing substituents ($-CH_3$, $-OCH_3$) decrease the acidity of a benzoic acid.

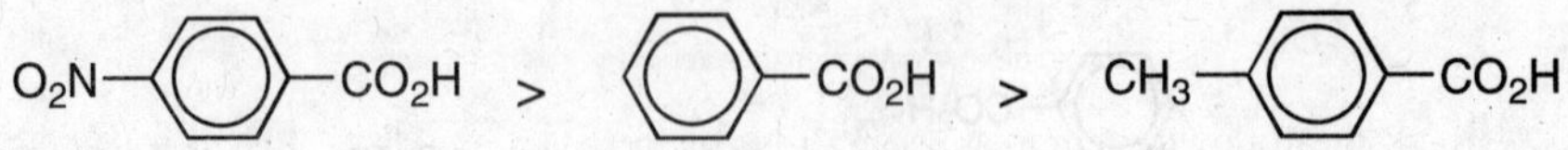

(4) Substituents at the *ortho* position almost always increase the acidity of a benzoic acid.

Examination Questions

1. Give the preparation and properties of benzoic acid.

2. Give the preparation and properties of phthalic acid.

3. Give the preparation and properties of salicylic acid.

4. Give the preparation and properties of cinnamic acid.

5. What happens when benzoic acid is treated with concentrated nitric acid in the presence of concentrated sulfuric acid ? *(Bihar BSc, 2014)*

Answer. *m*-Nitrobenzoic acid is formed. For reaction equation see text.

6. What happens when salicylic acid is heated with zinc dust ?

Answer. Benzoic acid is formed. For reaction equation see text.

7. Explain why nitration of benzoic acid gives *m*-nitrobenzoic acid.

8. Benzoic acid boils at 249°C, but its methyl ester boils at 199°C. Why is the boiling point of the acid so much higher, even though it has the lower molecular weight ?

Answer. Benzoic acid molecules are strongly associated because of hydrogen bonding. Molecules of methyl benzoate are much weakly associated, and less heat is required to overcome the attractive forces between molecules of the methyl ester. Even though the methyl ester has a greater molecular weight, it boils at a lower temperature than the acid.

9. Explain why benzoic acid is stronger than acetic acid. *(Panjab BSc, 2015)*

10. Arrange the following series of compounds in order of increasing acid strength. Give reasons for the order you select.

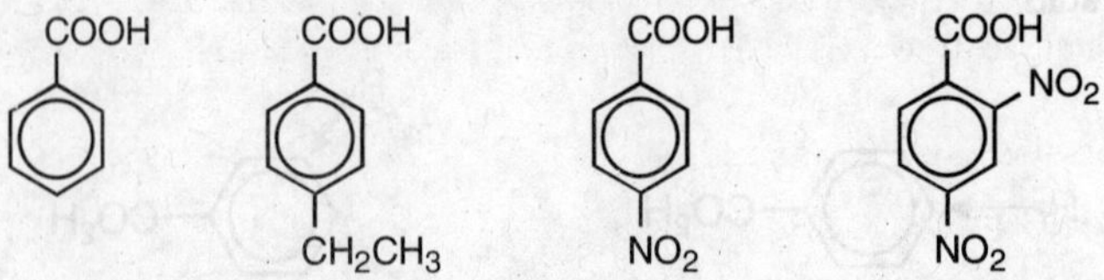

(*a*) (*b*) (*c*) (*d*)

Answer. The order is :

(b) < (a) < (c) < (d)

Notice that we are comparing the acid strength of benzoic acid (*a*) and different substituted benzoic acids. Consider the effect of substituents on the stability of the carboxylate anion. Compound (*b*) is a weaker acid than (*a*) because the ethyl group in (*b*) is an electron-releasing substituent which destabilizes the carboxylate anion. Compounds (*c*) and (*d*) contain nitro substituents, which are electron-withdrawing substituents. In (*c*) there is one NO_2 group in the *para* position ; in (*d*) there are two NO_2 groups, one in the *ortho*, the other in the *para* position. The anion resulting from (*d*) is much more stable than the one due to (*c*). As a result, (*d*) is a stronger acid than (*c*).

11. Rank the following compounds in order of increasing acidity : *p*-Methylbenzoic acid, *p*-Bromobenzoic acid, and *p*-Nitrobenzoic acid. (*Banaras BSc, 2013*)

Answer. The acidity order is :

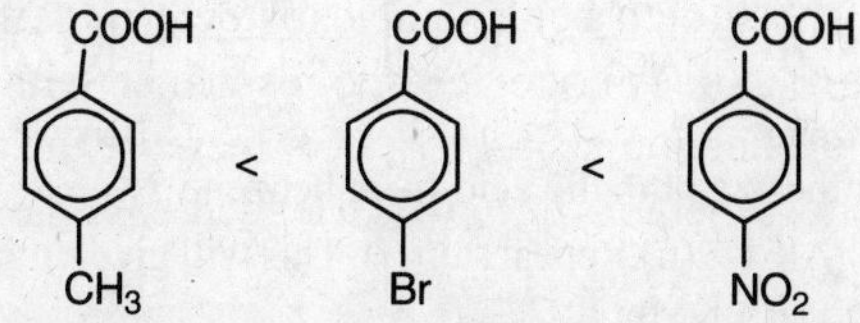

Least acidic → Most acidic

12. Rank the following compounds in order of increasing acidity : Benzoic acid, Ethanol, and *p*-Methylbenzoic acid. (*Jammu BSc, 2014 ; Mysore BSc, 2015*)

Answer. The acidity order is :

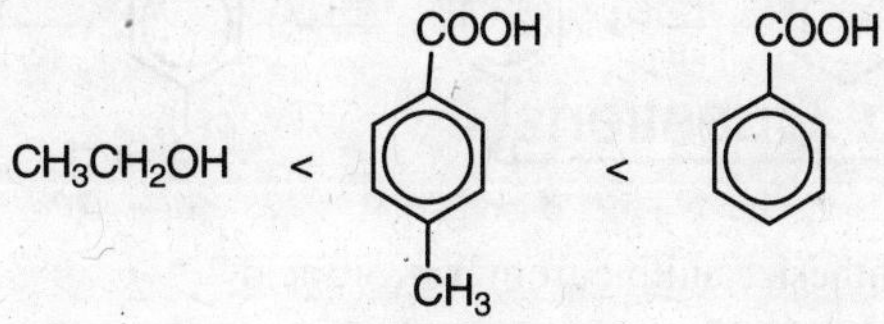

Least acidic → Most acidic

13. Rank the following compounds in order of increasing acidity : Sulfuric acid, Methanol, Phenol, *p*-Nitrophenol, and Acetic acid.

Answer. The acidity order is :

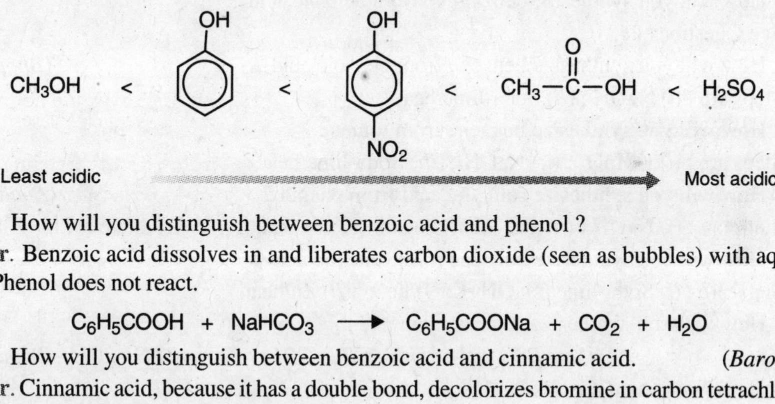

Least acidic → Most acidic

14. How will you distinguish between benzoic acid and phenol ?

Answer. Benzoic acid dissolves in and liberates carbon dioxide (seen as bubbles) with aqueous sodium bicarbonate. Phenol does not react.

$$C_6H_5COOH + NaHCO_3 \longrightarrow C_6H_5COONa + CO_2 + H_2O$$

15. How will you distinguish between benzoic acid and cinnamic acid. (*Baroda BSc, 2015*)

Answer. Cinnamic acid, because it has a double bond, decolorizes bromine in carbon tetrachloride. Benzoic acid does not react.

$$\underset{}{C_6H_5CH{=}CHCOOH} + \underset{\text{red}}{Br_2\ (\text{in } CCl_4)} \longrightarrow \underset{\text{colourless}}{C_6H_5-\overset{\overset{\displaystyle Br}{|}}{CH}-\overset{\overset{\displaystyle Br}{|}}{CH}-COOH}$$

16. How will you distinguish between phenol and salicylic acid ? (*Anna BSc, 2014*)

Clue. When phenol is heated with phthalic anhydride in the presence of concentrated sulfuric acid, phenolphthalein is formed. Red color is produced when aqueous NaOH solution is added to the reaction mixture. For reaction equation see text. Salicylic acid does not give this test.

17. How will you distinguish between salicylic acid and benzoic acid ?

Clue. Salicylic acid reacts with bromine-water to give a white precipitate of 2,4,6-tribromophenol. The red color of bromine disappears. For reaction equation see text. Benzoic acid does not give this test. Also, salicylic acid gives a violet color with $FeCl_3$ solution. This color is due to the formation of an iron complex. Benzoic acid does not give this test.

18. How will you synthesize benzoic acid from benzene ? (*North Bengal BSc, 2015*)

Answer. Following steps are involved :

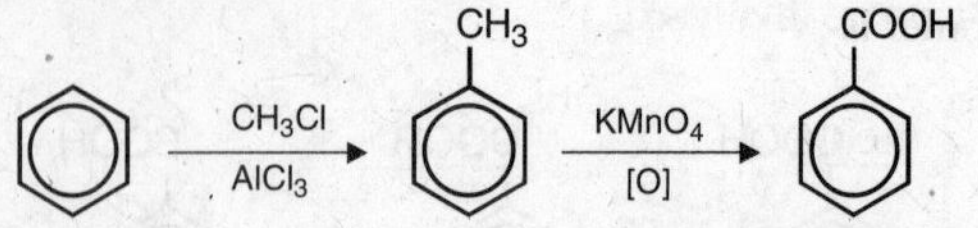

Benzene → Benzoic acid

19. How will you synthesize phthalic acid from benzene ?

Clue. Steps are : (*i*) $CH_3Cl/AlCl_3$; (*ii*) Repeat step (*i*). This will give a mixture of *o*- and *p*-xylene. Separate and use *o*-xylene in the next step ; (*iii*) $KMnO_4$.

20. How will you synthesize *p*-toluic acid from benzene ?

Answer. Following steps are involved :

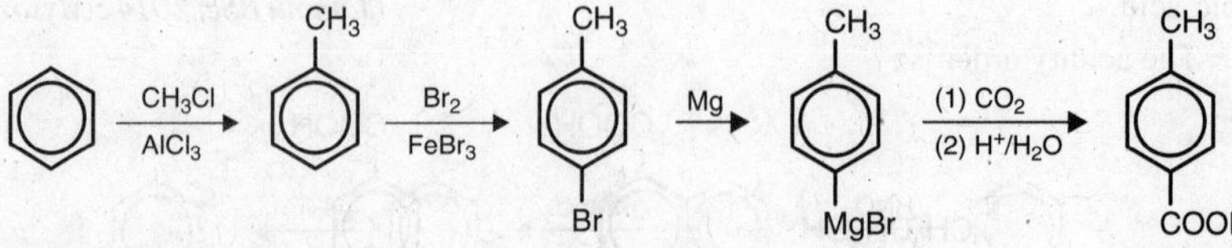

Benzene (+ o-isomer) *p*-Toluic acid

21. How will you synthesize aniline from benzoic acid ?

Answer. Following steps are involved :

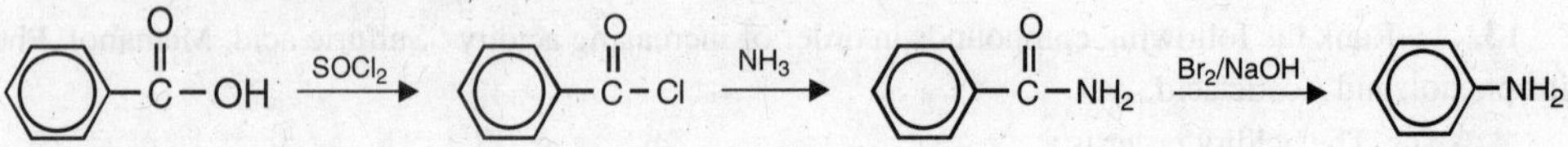

Benzoic acid Aniline

22. How will you synthesize benzamide from benzoic acid ?

Clue. See Questions *21*.

23. How will you synthesize benzene from benzoic acid ? (*Bhopal BSc, 2014*)

Clue. Steps are : (*i*) NaOH ; (*ii*) Soda-lime/heat.

24. How will you synthesize benzene from toluene ?

Clue. Steps are : (*i*) $KMnO_4$; (*ii*) NaOH ; (*iii*) Soda-lime/heat.

25. How will you synthesize salicylic acid from phenol ? (*Andhra BSc, 2015*)

Clue. Steps are : (*i*) NaOH ; (*ii*) CO_2/125°C/6 atm (Kolbe reaction) ; (*iii*) dilute HCl.

26. How will you synthesize benzene from salicylic acid ?

Clue. Steps are : (*i*) Soda-lime (NaOH+CaO)/heat ; (*ii*) Zn/heat.

27. How will you synthesize Aspirin from phenol ?

(*Pune BSc, 2014 ; Shivaji BSc, 2014 ; Kalyani BSc, 2015 ; Tripura BSc, 2015*)

Answer. Following steps are involved :

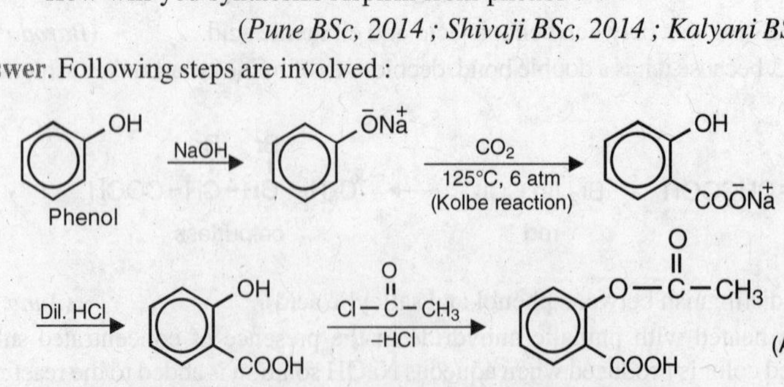

Phenol

Aspirin (Acetylsalicylic acid)

28. How will you synthesize phenylacetic acid from toluene ? (*Delhi BSc, 2015*)

Answer. Following steps are involved :

Toluene Phenylacetic acid

29. How will you synthesize Oil of Wintergreen (Methyl salicylate) from phenol ?

(Kanpur BSc, 2014 ; Periyar BSc, 2015)

Answer. Following steps are involved :

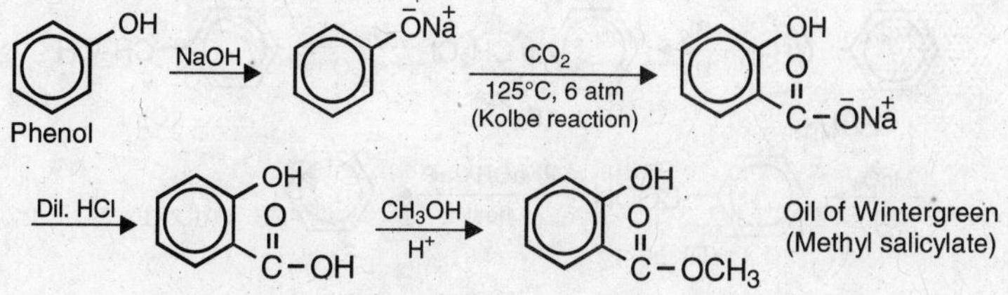

30. How will you synthesize *p*-aminobenzoic acid (PABA) from toluene ?

Answer. Following steps are involved :

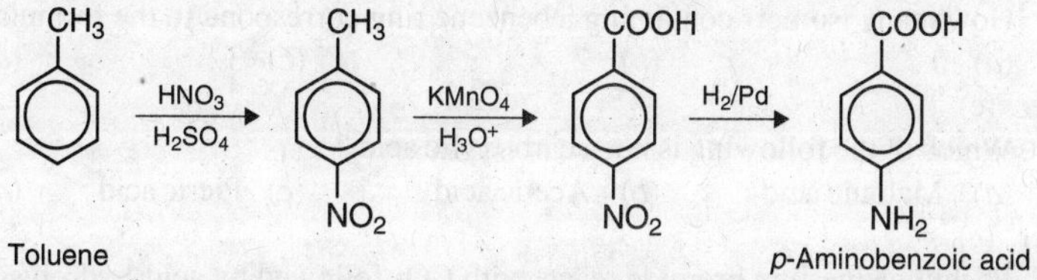

31. How will you synthesize cinnamic acid from benzene ? *(Annamalai BSc, 2015)*

Answer. Following steps are involved :

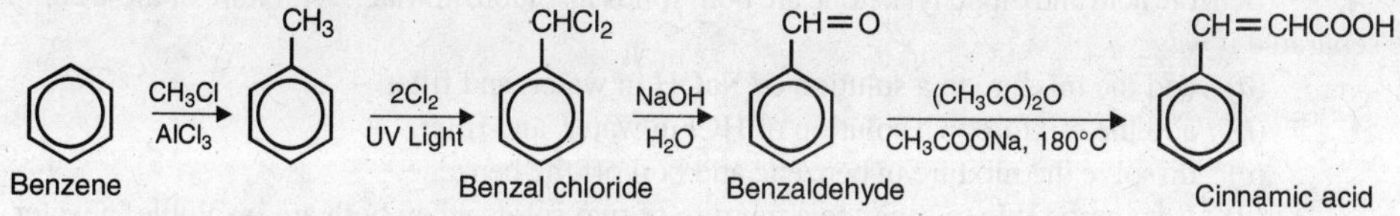

32. How will you synthesize *m*-nitrobenzoic acid from benzene ?

Clue. Steps are : *(i)* $CH_3Cl/AlCl_3$; *(ii)* $KMnO_4$; *(iii)* HNO_3/H_2SO_4.

33. Compound (A) and (B) are isomers having the molecular formula C_8H_{10}. Oxidation of (A) gives benzoic acid, while oxidation of (B) gives phthalic acid which forms an anhydride on heating. Give the structures of (A) and (B) and write equations for the reactions involved.

Answer. *(A)* is ethylbenzene ; *(B)* is *o*-xylene.

Reactions Involved Are :

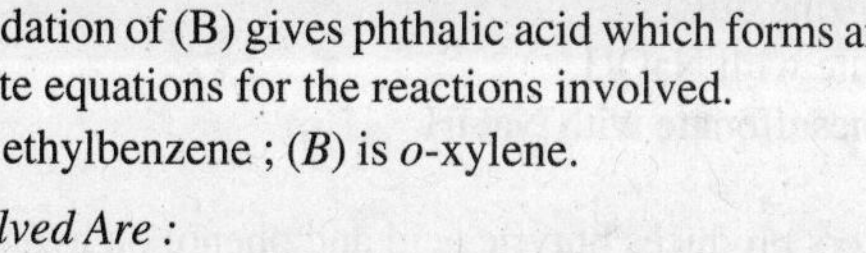

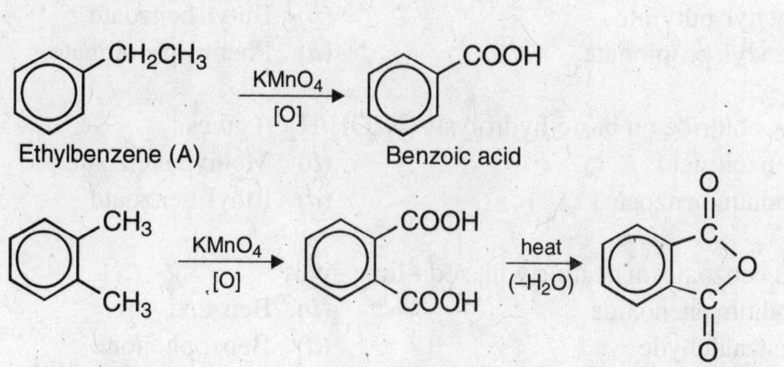

34. An organic compound (A), C_7H_8, on treatment with chlorine in the presence of ultraviolet light at its boiling point gives (B), C_7H_7Cl. The compound (B) on hydrolysis with aqueous NaOH gives (C), C_7H_8O, which on oxidation with $KMnO_4$ forms an acid (D), $C_7H_6O_2$. The acid (D) on distillation with sodalime gives benzene. Deduce the structural formulas of (A), (B), (C), and (D). Write equations for the reactions involved.

Answer. (*A*) is toluene ; (*B*) is benzyl chloride ; (*C*) is benzyl alcohol ; and (*D*) is benzoic acid.

Reactions Involved Are :

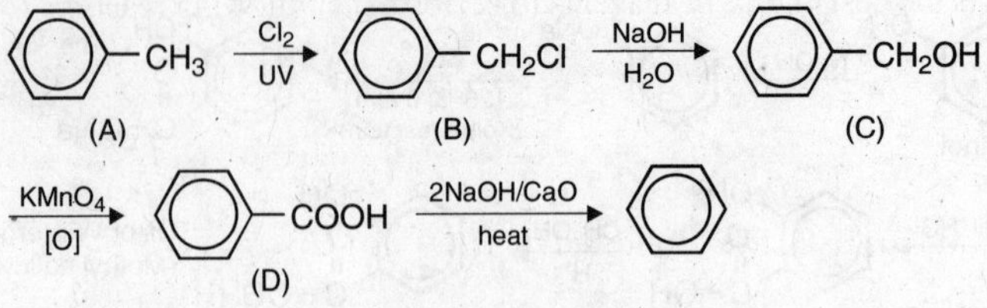

Multiple Choice Questions

1. How many isomers containing a benzene ring correspond to the formula $C_6H_4(COOH)_2$?
(*a*) 1 (*b*) 2 (*c*) 3 (*d*) 4

Answer. (c)

2. Which of the following is not a carboxylic acid ?
(*a*) Malonic acid (*b*) Acetic acid (*c*) Picric acid (*d*) Adipic acid

Answer. (*c*)

3. Phenylmagnesium bromide reacts with CO_2 followed by acid-hydrolysis to form :
(*a*) Phenol (*b*) Benzoic acid (*c*) Bromobenzene (*d*) Acetophenone

Answer. (*b*)

4. Benzoic acid and diphenyl ketone are both solids insoluble in water. A mixture of these compounds can be separated if we
(*a*) add the mixture to a solution of NaOH in water, and filter.
(*b*) add the mixture to a solution of HCl in water, and filter.
(*c*) dissolve the mixture in benzene and boil off the benzene.
(*d*) It is impossible to separate a mixture of two solids when both are insoluble in water.

Answer. (*a*)

5. Which of the following reactions does not produce benzene as one of the products ?
(*a*) Treating benzenediazonium chloride with H_3PO_2
(*b*) Distilling phenol with zinc dust
(*c*) Fusing sodium benzoate with NaOH
(*d*) Fusing sodium benzenesulfonate with NaOH

Answer. (*d*)

6. Which of the following esters produces butyric acid and phenol on hydrolysis ?
(*a*) Phenyl butyrate (*b*) Butyl benzoate
(*c*) Benzyl propionate (*d*) Phenyl propionate

Answer. (*a*)

7. Benzoyl chloride on basic-hydrolysis ($NaOH/H_2O$) gives :
(*a*) Benzoic acid (*b*) Methyl benzoate
(*c*) Sodium benzoate (*d*) Ethyl benzoate

Answer. (*c*)

8. Sodium benzoate on heating with soda-lime gives :
(*a*) Sodium phenoxide (*b*) Benzene
(*c*) Benzaldehyde (*d*) Benzophenone

Answer. (*b*)

9. What is the best choice of reagents to perform the following reaction ?

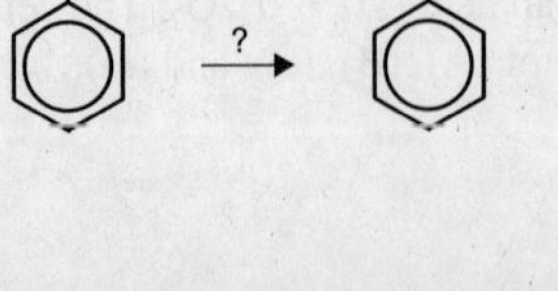

(*a*) NaCN followed by H_3O^+ (*b*) CH_3OH, H_2SO_4
(*c*) Mg, followed by CO_2, then H_3O^+ (*d*) HCOOH, H_2O, heat

Answer. (*c*)

10. What is the best choice of reagents to perform the following reaction ?

(*a*) $KMnO_4$, H_2O, heat, followed by H_3O^+ (*b*) CO_2
(*c*) H_2O (*d*) NaOH, H_2O, heat

Answer. (*a*)

11. Methyl benzoate on hydrolysis gives
(*a*) Acetic acid (*b*) Benzoic acid
(*c*) Picric acid (*d*) Phenylacetic acid

Answer. (*b*)

12. Aspirin is produced by heating salicylic acid with
(*a*) acetic anhydride in the presence of phosphoric acid
(*b*) benzoic anhydride in the presence of phosphoric acid
(*c*) methyl alcohol in the presence of sulfuric acid
(*d*) phenol in the presence of sulfuric acid

Answer. (*a*)

13. Oil of Wintergreen (Methyl salicylate) is obtained by treating salicylic acid with
(*a*) Methane at 150°C (*b*) CH_3OH and conc. H_2SO_4
(*c*) Methyl chloride (*d*) NaOH and then CH_3OH

Answer. (*b*)

14. Which of the following statements is true :
(*a*) *o*-Nitrobenzoic acid is more acidic than *p*-nitrophenol
(*b*) *o*-Nitrobenzoic acid is less acidic than *p*-nitrophenol
(*c*) *o*-Nitrobenzoic acid and *p*-nitrobenzoic acid have the same acid strength
(*d*) None of these statements is true

Answer. (*a*)

15. The order of increasing acidity is :
(*a*) chloroacetic acid > formic acid > benzoic acid
(*b*) benzoic acid > formic acid > chloroacetic acid
(*c*) all are the same
(*d*) not determinable from available data

Answer. (*a*)

16. Which of the following is the strongest acid ?
(*a*) Trichloroacetic acid (*b*) Phenol
(*c*) Acetic acid (*d*) Benzoic acid

Answer. (*a*)

17. Which of the following is the strongest acid ?
(*a*) Ethanol (*b*) Phenol
(*c*) Anisole (*d*) Benzoic acid

Answer. (*d*)

18. Which acid is weaker than benzoic acid ?
(*a*) *p*-Methylbenzoic acid (*b*) *p*-Chlorobenzoic acid
(*c*) *p*-Nitrobenzoic acid (*d*) *o*-Chlorobenzoic acid

Answer. (*a*)

19. Compound (*A*) when reacted with PCl_5 and then with ammonia gave (*B*). Compound (*B*) when treated with bromine and NaOH, produced (*C*). Compound (*C*) on treatment with $NaNO_2$/HCl at 0°C and then boiling with H_2O produced o-cresol. Compound (*A*) is
(*a*) *o*-Toluic acid (*b*) *o*-Chlorotoluene
(*c*) *m*-Toluic acid (*d*) *o*-Bromotoluene

Answer. (*a*)

43

Dyes: Color and Constitution

WHY DO WE SEE COLOR ?

Visible light consists of electromagnetic radiations of wavelengths 400-750 nm. Each wavelength is associated with a definite energy and produces a characteristic color sensation as it strikes the retina of the human eye. Light possessing all wavelengths in the visible range appears white. The complete absence of these wavelengths causes darkness, black.

When a beam of visible light strikes a colored substance, certain wavelengths are absorbed, and others are reflected. Those reflected compose the color of the substance. For example, if a substance absorbs wavelengths in the blue-green region of the spectrum, it appears red which is the color of the remaining wavelengths. On the contrary, if red light is absorbed, the substance would look blue-green. The table below correlates the wavelengths absorbed with complementary colors observed.

Table 43.1 RELATIONSHIP OF COLOR ABSORBED AND COMPLEMENTARY COLOR OBSERVED

Wavelength absorbed (nm)	Color absorbed	Complementary color
400-435	violet	yellow-green
435-480	blue	yellow
480-490	blue-green	orange
490-500	green-blue	red
500-560	green	purple
560-580	yellow-green	violet
580-595	yellow	blue
595-605	orange	green-blue
605-750	red	blue-green

Note : 1 nm = 1 mμ = 10 Å

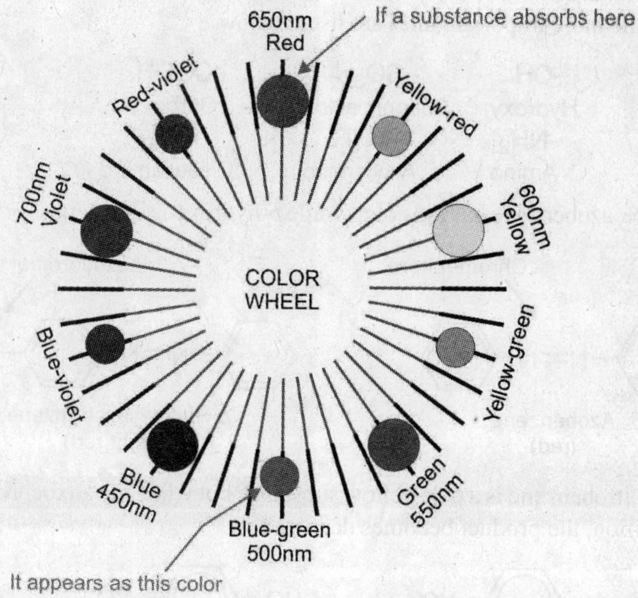

COLOR AND CONSTITUTION

There are two theories of color and its relation to chemical structure :

(1) **Chromophore-Auxochrome Theory.** In an attempt to correlate color with molecular structure, Otto Witt put forth his theory of 'Color and Constitution' as early as 1876. This theory which was originally named after him is often referred to as '*The Chromophore-Auxochrome Theory*'. It postulates that :

(*i*) **The color of organic compounds is due to the presence of certain multiple bonded groups called chromophores** (*Gr., Chroma* = color ; *Phorein* = to bear). A few important chromophore groups are :

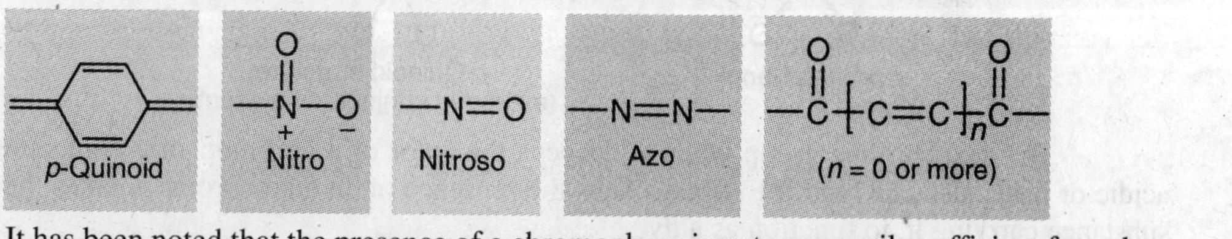

It has been noted that the presence of a chromophore is not necessarily sufficient for color. To make a substance colored, the chromophore has to be conjugated with an extensive system of alternate single and double bonds as exists in aromatic rings. Thus, nitromethane is colorless, while 1-nitronaphthalene is yellow.

CH_3NO_2 Nitromethane (colorless)

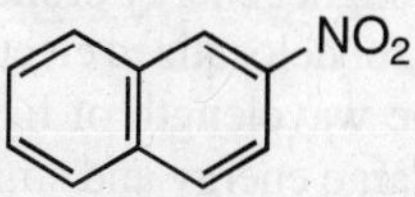

2-Nitronaphthalene (yellow)

(*ii*) **Certain groups, while not producing color themselves, when present along with chromophores in an organic substance, intensify the color. Such color assisting groups are called**

auxochromes (Gr., *Auxanein* = to increase ; *Chroma* = color). The auxochromes are acidic or basic functions, of which the more important ones are listed below.

Acidic :	–OH Hydroxy	$-SO_3H$ Sulfonic acid	–COOH Carboxyl
Basic :	$-NH_2$ Amino	–NHR Alkylamino	$-NR_2$ Dialkylamino

For example, azobenzene has red color, while *p*-hydroxyazobenzene is brilliant red.

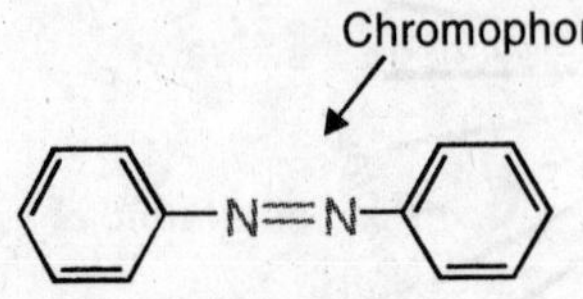

Azobenzene
(red)

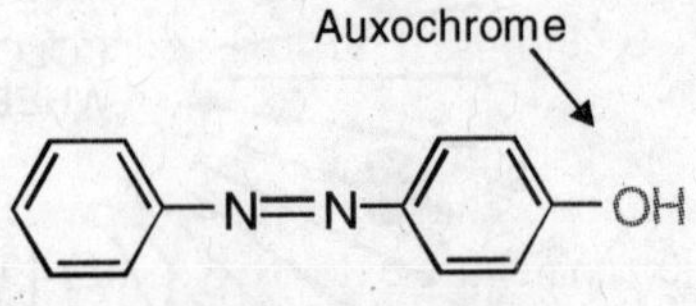

p-Hydroxyazobenzene
(bright red)

Similarly, nitrobenzene is a pale yellow substance but when the auxochrome –OH is present in ortho or para position, the product becomes deep yellow.

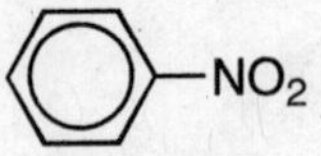
Nitrobenzene
(pale yellow)

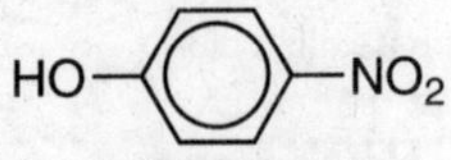
p-Nitrophenol
(deep yellow)

The auxochrome –OH deepens the color by extending the conjugated system between the chromophore and the auxochrome due to resonance.

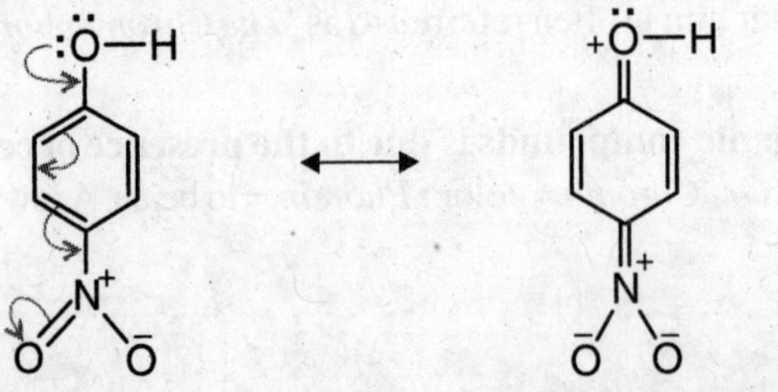
p-Nitrophenol

p-Quinoid structure
(extended conjugated system)

The auxochromic group while it deepens the color of a chromophoric compound, being acidic or basic also can bind it to fibers. Thus the presence of an auxochrome enables the colored substance carrying it, to function as a dye.

(2) **Modern Theory of Color**. When a molecule absorbs light, its bonding electrons are excited. In other words the bonding electrons absorb energy and are promoted to orbitals of higher energy. The energy absorbed by an electron in such a transition is equal to difference of the energy of MO in the ground state and the energy of MO to which the electron is promoted. The relative energies of the MOs in the ground state are :

$$\pi \textbf{ delocalized} > \pi > \sigma$$

Therefore, to promote an electron to the same higher energy orbital, maximum energy will be required for σ electron, less for π electron, and still less for delocalized electron. Further, the smaller the energy required for promotion, the longer will be the wavelength of light absorbed. For example, methane having σ bonds only, on excitation absorbs large energy and smaller wavelengths in the far ultraviolet region. Hence it is colorless. Ethylene has a π bond and absorbs less energy and longer wavelengths still fall in the near ultraviolet region, and ethylene is colorless.

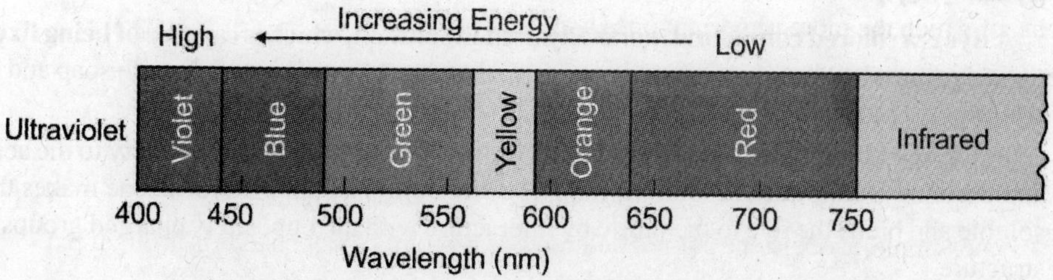

Fig. 43.1 Visible spectrum showing relationships between energy, wavelengths and colors.

In a compound having a highly conjugated system of single and double bonds, the carbon *p*-orbitals overlap and cause extensive delocalization of the π electrons.

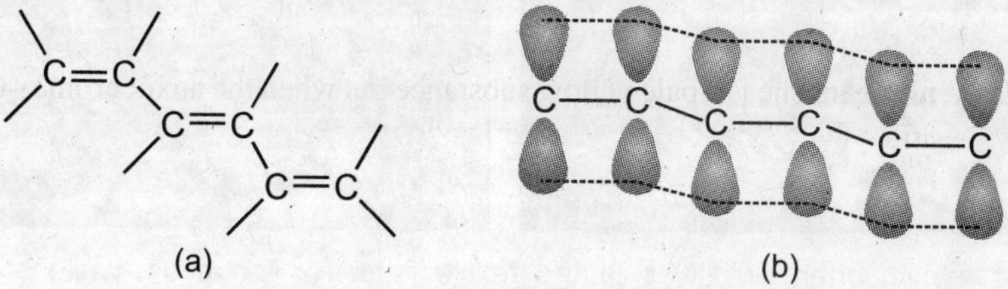

Fig. 43.2. (*a*) Conjugated double bonds ; (*b*) Overlapping *p*-orbitals cause π electron delocalization.

The delocalized electrons in such a system being in a higher energy orbital than σ and π orbital, require less energy for excitation. They absorb longer wavelengths which fall in the range of the visible spectrum and hence the compound shows color. Thus, β-carotene, the orange pigment of carrots, with eleven conjugated double bonds, has a highly π-delocalized electron system covering 22 carbon atoms.

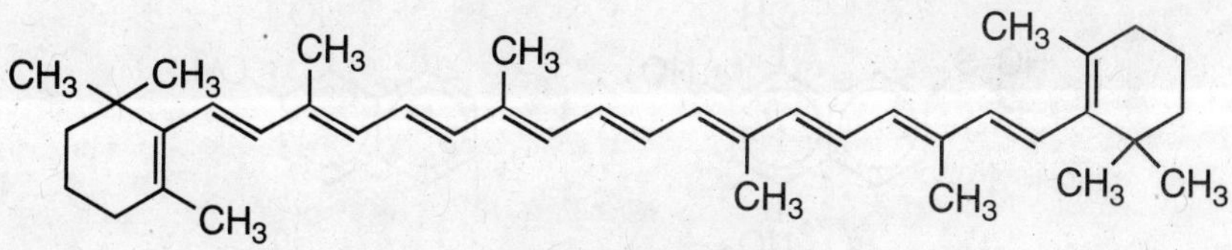

β-Carotene (orange)

Benzene has a ring structure made of three conjugated double bonds. The carbon *p*-orbitals are involved in forming 6π delocalized electron system. It absorbs wavelengths in the ultraviolet region and is colorless. When it is conjugated with chromophoric groups which contain double bonds, electron delocalization is increased and the compound is colored. Electron delocalization is further enhanced by groups such as $-OH$ and $-NH_2$ which can donate unshared electron pair for still greater extension of the conjugated system. Thereby the absorption is shifted to a longer wavelength (toward red), making the color more intense. The extended chromophoric systems delocalized by resonance are responsible for the color of many dyes and indicators.

In context of the above theory, chromophore is now defined as any isolated functional group which shows absorption in ultraviolet or visible region (200-800 nm). Any groups that cause deepening of the color (absorption shifted toward red) are known as **Bathochromic Groups** and the effect **as Bathochromic Shift**. The groups that cause the opposite effect (*Shift toward violet*) are termed **Hypsochromic Groups** and the effect is referred to as **Hypsochromic Shift**.

WHAT IS A DYE ?

A dye is a colored compound, normally used in solution, which is capable of being fixed to a fabric. The dye must be 'fast' or chemically stable so that the color will not wash with soap and water, or fade on exposure to sunlight (ultraviolet light).

A dye owes its color to the presence of a chromophore and its fixing property to the acidic or basic auxochromic groups such as OH, SO_3H, NH_2, NR_2, etc. The polar auxochrome makes the dye water-soluble and binds the dye to the fabric by interaction with the oppositely charged groups of the fabric structure.

Dyes are important additives in the textile industry. Azo dyes, while still in widespread use, are becoming less attractive for this purpose because some have been found to degrade to carcinogenic benzenamines.

CLASSIFICATION OF DYES BY STRUCTURE

Dyes may be classified according to the type of *chromophores* present in their structures. This method of classification includes the following main types :

(1) **Nitro and Nitroso Dyes**. The NO_2 and NO groups are chromophores in this class of dyes. Examples are,

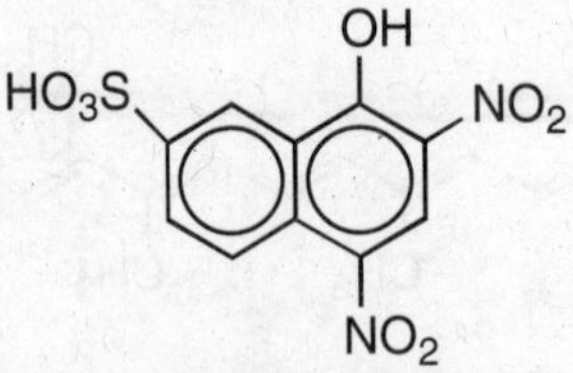

Naphthol yellow S

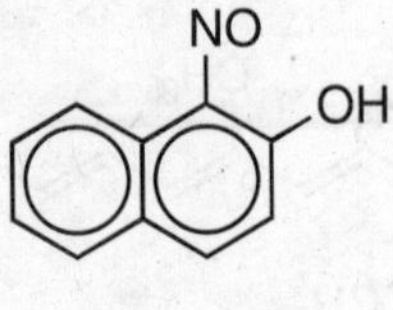

Mordant green 4

(2) **Azo Dyes**. The azo dyes contain one or more azo groups, –N=N–, as the primary chromophore. The common auxochromes are NH_2, NR_2, OH, SO_3H, etc. Azo dyes form the largest and most important group of synthetic dyes. They are highly colored and can prepared by diazotizing an aromatic amine and subsequent coupling with a suitable aromatic phenol or amine. By varying the substituents present in both the diazonium salt and the coupling compound a series of azo dyes can be produced with almost any color. Examples are,

(*a*) **Para Red**

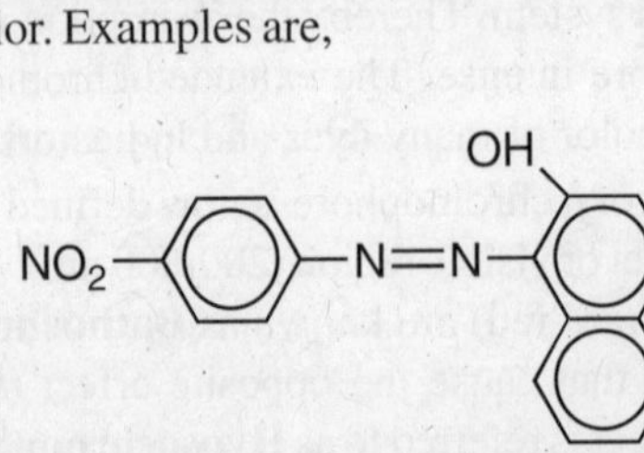

It was the first *azo dye* to be prepared. Para red is obtained by the reaction of diazotized *p*-nitroaniline with β-naphthol on fabric itself.

(*b*) **Methyl Orange.** Methyl orange is obtained from sulfanilic acid by the following steps :

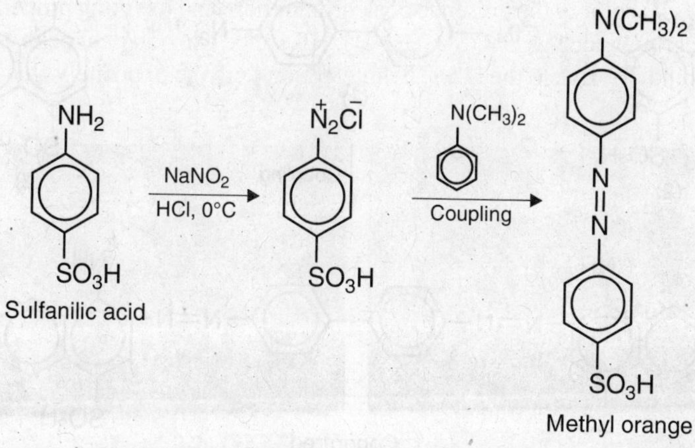

Methyl orange imparts orange color to wool and silk but the color is not fast to sunlight or washing. It is a valuable indicator for acid-base titrations because it gives yellow color in basic solution and red color in acid solution. The change in color is due to the change in the structure of the ion.

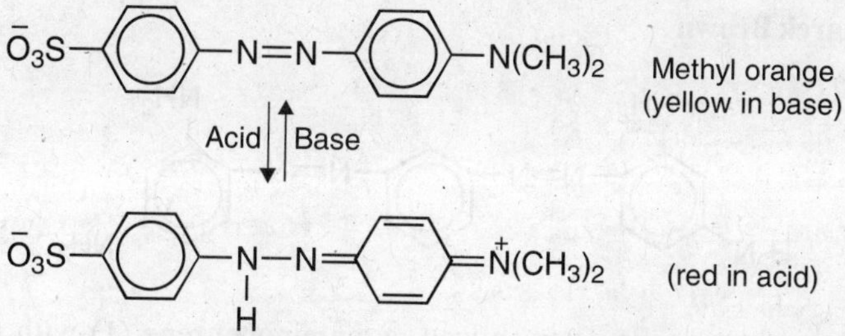

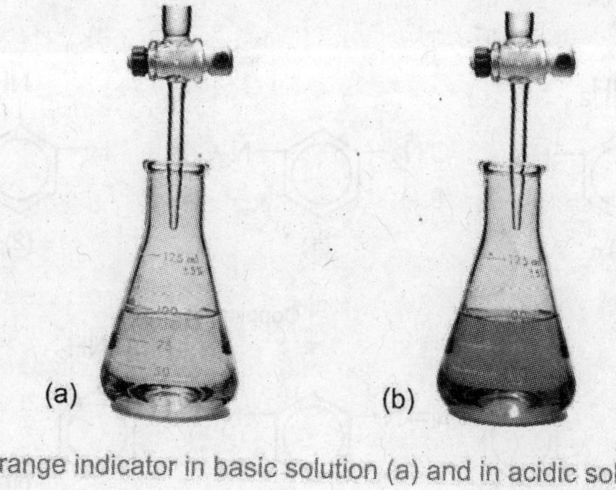

Methyl orange indicator in basic solution (a) and in acidic solution (b).

(*c*) **Congo Red**

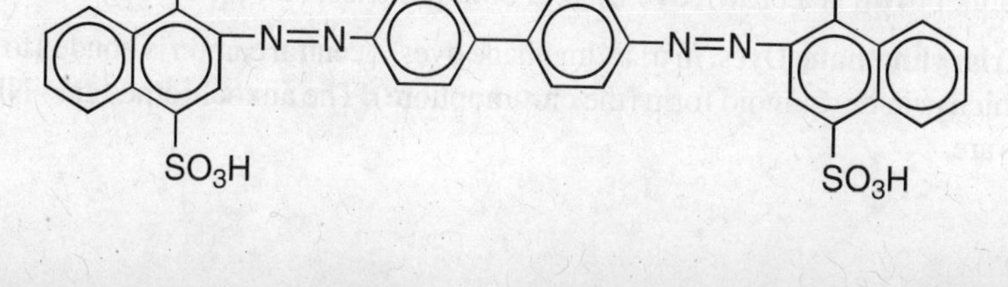

Congo red contains two *azo* groups. It is obtained by coupling tetrazotized benzidine (1) with two molecules of naphthionic acid (2).

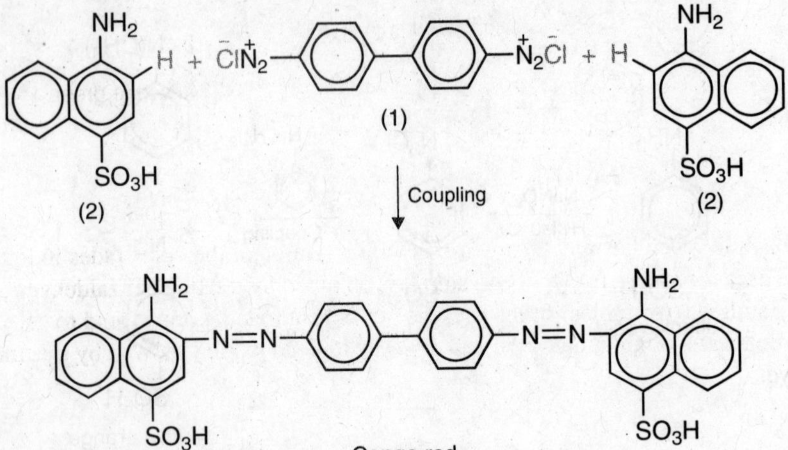

Congo red

It is a direct dye and its sodium salt is used for dyeing cotton red from aqueous solutions. Congo red is also used as an indicator, being red in alkali and blue in acid solution.

(*d*) **Bismarck Brown**

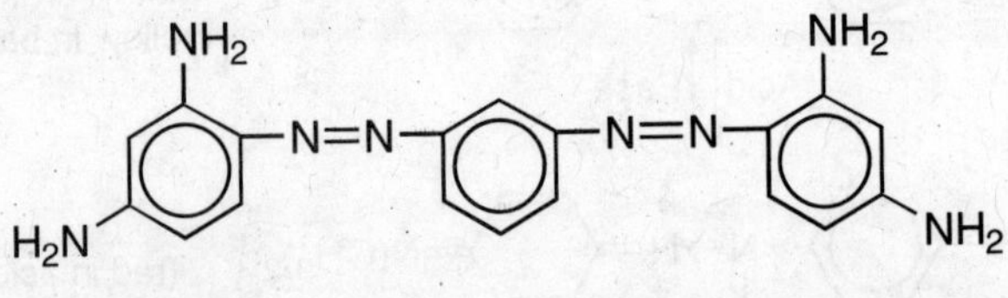

It is obtained by coupling tetrazotized *m*-diaminobenzene (1) with two molecules of *m*-diaminobenzene (2).

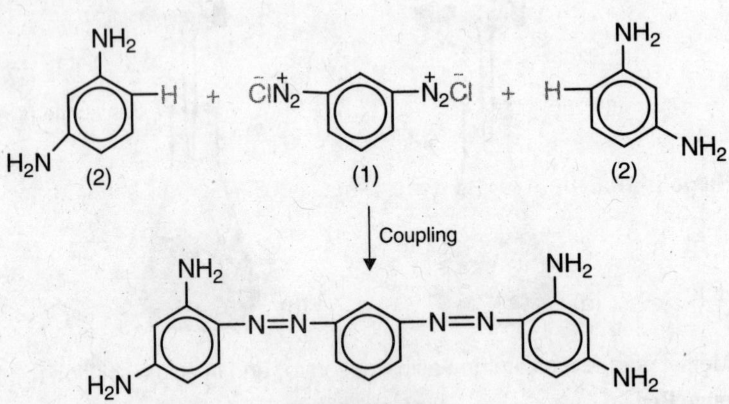

Bismarck brown

Bismarck brown is a brown dye used in boot polishes, and for dyeing wool and cotton.

(3) **Triarylmethane Dyes.** In triarylmethane dyes, a central carbon is bonded to three aromatic rings, one of which is in the quinoid form (the chromophore). The auxochromes are $-NH_2$, $-NR_2$, and $-OH$. Examples are,

(*a*) **Malachite Green**

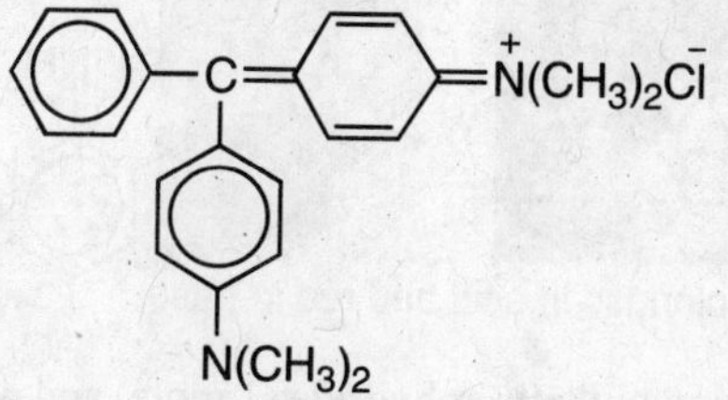

Malachite green

Malachite green has a deep green-blue color. Although the color fades in light, Malachite green is used as a *direct dye* for wool and silk. It is obtained by treating benzaldehyde (1 mole) with *N,N*-dimethylaniline (two moles) in the presence of concentrated sulfuric acid to give **Leuco Base** (Gr. *leuco* = colorless). Oxidation of the *leuco base* with lead oxide followed by treatment with HCl yields the dye.

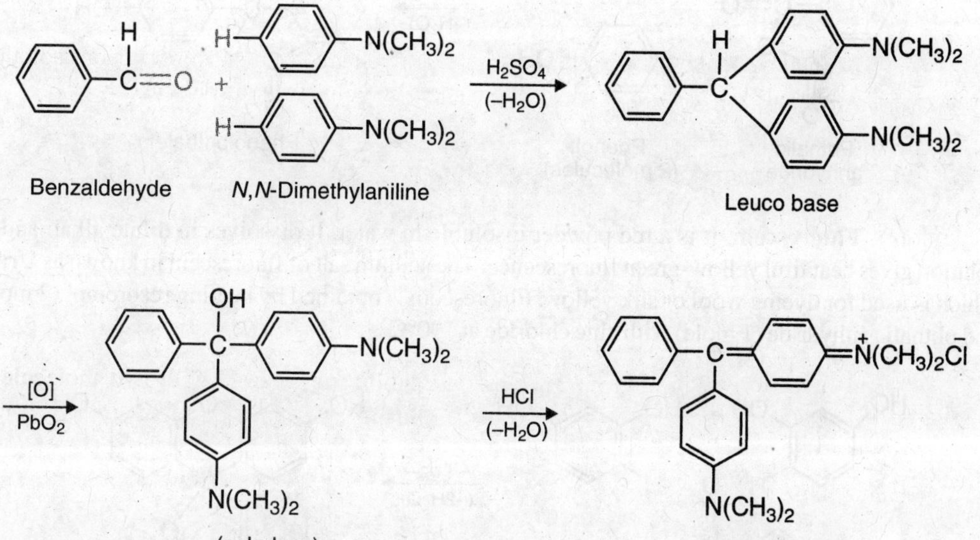

(*b*) **Phenolphthalein**. It is also a triarylmethane dye but it is better known as an acid-base indicator.

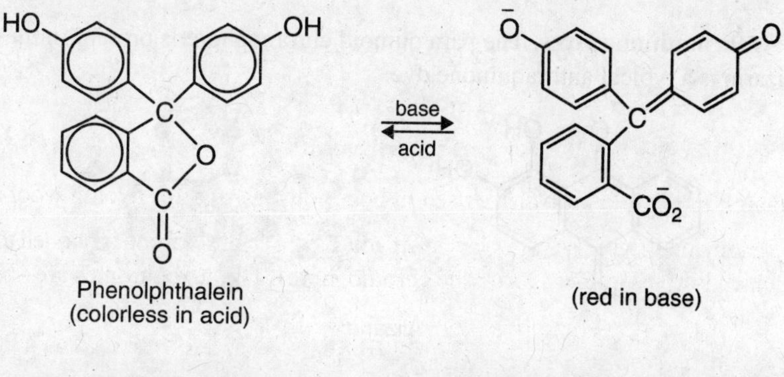

Phenolphthalein is colorless in acid and red in base.

Phenolphthalein is prepared by heating phthalic anhydride (1 mole) and phenol (two moles) in the presence of anhydrous zinc chloride at 120°C.

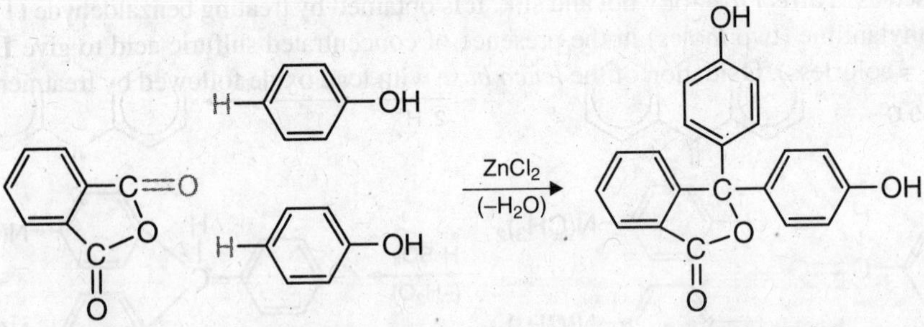

Phthalic anhydride   Phenol (2 molecules)   Phenolphthalein

(*c*) **Fluorescein**. It is a red powder insoluble in water. It dissolves in dilute alkali and the solution gives beautiful yellow-green fluorescence. The sodium salt of fluorescein in known as **Uraine** which is used for dyeing wool or silk yellow. Fluorescein is obtained by heating resorcinol (2 moles) and phthalic anhydride (1 mole) with zinc chloride at 190°C.

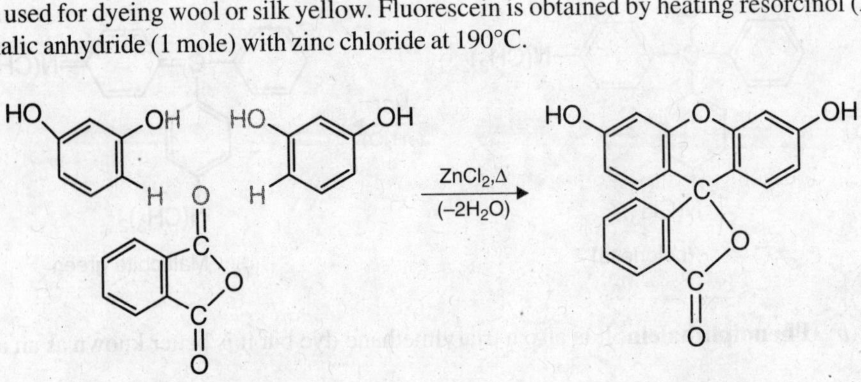

Phthalic anhydride   Fluorescein

(4) **Anthraquinone Dyes.** The para quinoid chromophore is present in these anthracene-type dyes. Alizarin is a typical anthraquinone dye.

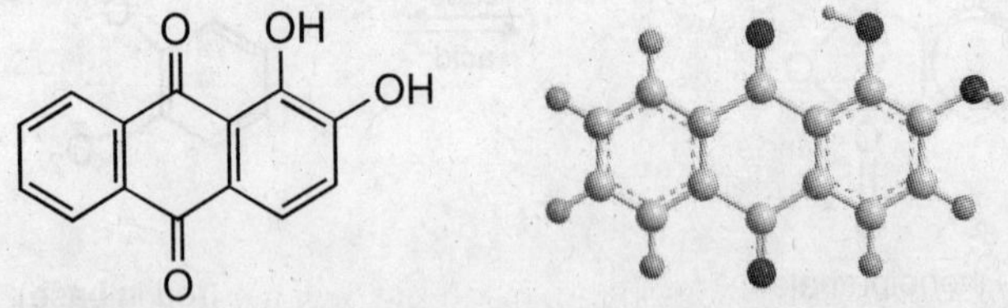

Alizarin

Alizarin is obtained from phthalic anhydride and benzene by the following steps.

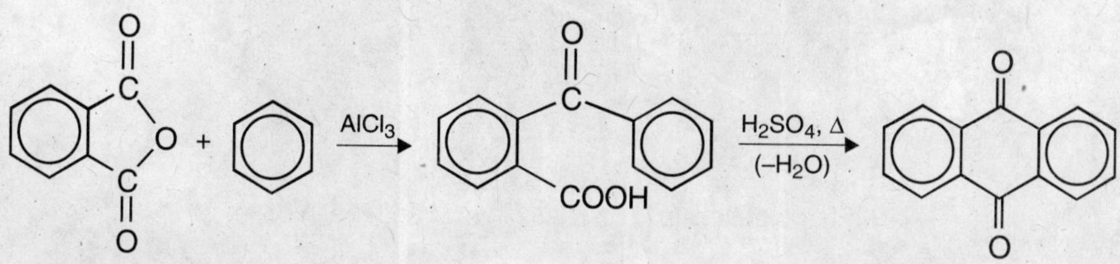

Phthalic anhydride — o-Benzoylbenzoic acid — Anthraquinone

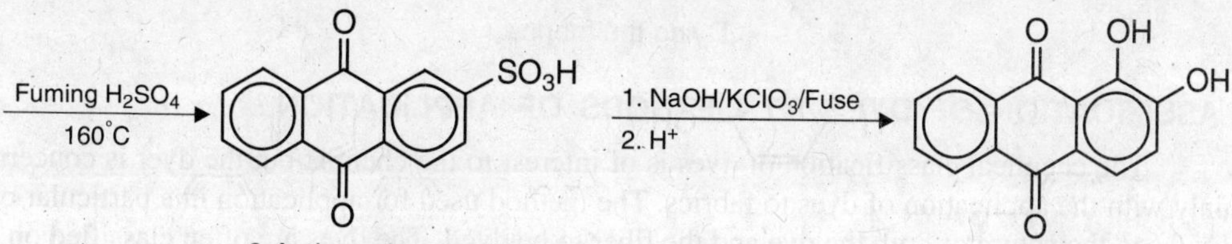

2-Anthraquinonesulfonic acid — Alizarin

Alizarin forms ruby red crystals which dissolve in alkali to give purple solutions. It is used to dye wool and cotton.

(5) **Indigo Dyes**. Indigo is an example of the type of dyes which contain carbonyl chromophore.

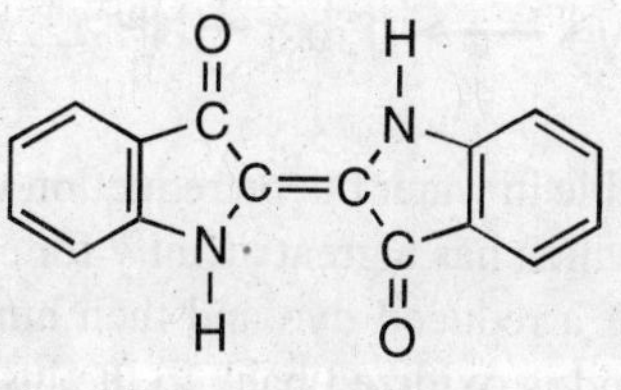

Indigo

Indigo is obtained from anthranilic acid and chloroacetic acid by the following steps.

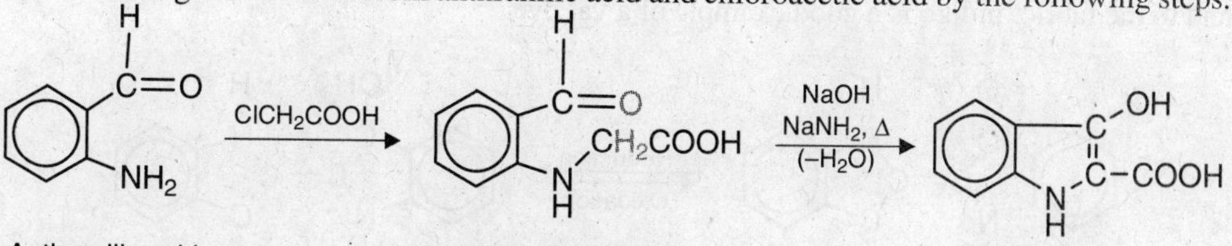

Anthranilic acid — N-Phenylglycine-o-carboxylic acid — Indoxylic acid (unstable)

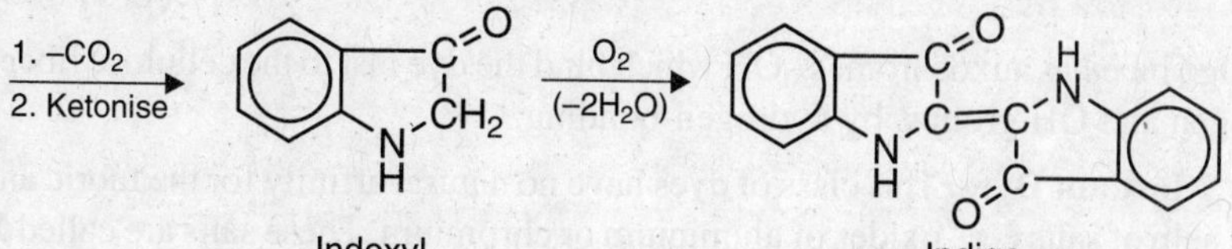

Indoxyl (2 moles) — Indigo

It is a dark-blue crystalline compound, insoluble in water. It is used for dyeing cotton by the *Vat Process*.

Dying the fabrics.

CLASSIFICATION OF DYES BY METHODS OF APPLICATION

The chemical classification of dyes is of interest to the chemist but the dyer is concerned mainly with the application of dyes to fabrics. The method used for application in a particular case depends on the nature of both the dye and the fiber to be dyed. The dyes are often classified on the basis of technique employed for their application.

(1) **Direct Dyes**. These can be applied to a fabric by direct immersion in a water solution of the dye. A direct dye contains acidic or basic auxochrome which combines with the opposite polar group present in the chemical structure of the fiber. Wool and silk are readily dyed by this method. **Martius Yellow**, a typical direct dye, has the acidic auxochrome –OH which interacts with the basic $-NH_2$ group of wool or silk.

$$\underset{\text{(Wool or silk)}}{\text{Fiber}-NH_2} + \text{HO}-\text{Dye} \longrightarrow \text{Fiber}-\overset{+}{N}H_3 \cdots\cdots \bar{O}-\text{Dye}$$

(2) **Vat Dyes**. These dyes are insoluble in water but on reduction with sodium hydrosulfide in a *vat* form a colorless soluble compound which has a great affinity for cotton and other cellulose fibers. The cloth is soaked in the solution of a reduced dye and then hung in air, or treated with oxidants. As a result, the colourless compound is oxidized back to the insoluble dye which is now bound to the fabric. Indigo is a good example of a vat dye.

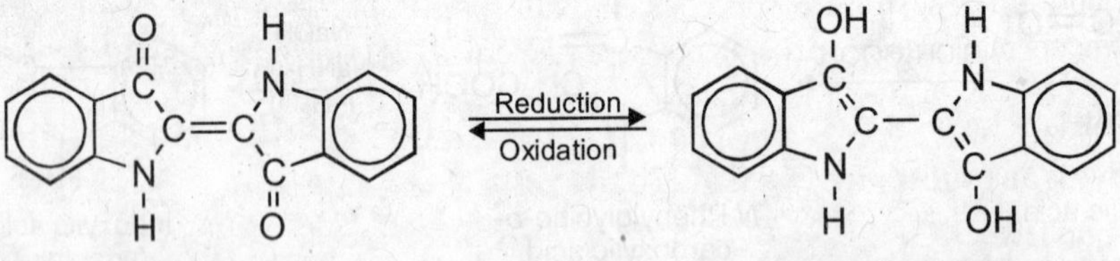

Indigo (blue) Indigo white (colorless)

Indigo blue has auxochromes –OH which bind the dye fast to the cellulose fiber that contains ethereal oxygen and OH groups, by hydrogen-bonding.

(3) **Mordant Dyes.** This class of dyes have no natural affinity for the fabric and are applied to it with the help of salts e.g., oxides of aluminium or chromium. These salts are called **Mordants**. A fiber such as cotton is first treated with a mordant and then with the dye solution. The mordant forms an insoluble coordination complex between the fiber and the dye, and binds the two. The insoluble complex compound appears in the form of 'lakes' that are fast to light and washing. Alizarin is an example of a mordant dye.

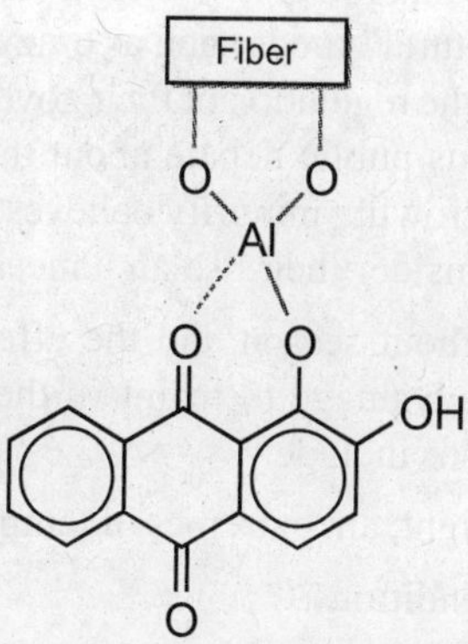

Fig. 43.3 Mordant action of Alizarin.

The mordant dyeing is most suitable for wool and nylon.

(4) **Azoic Dyes (Ingrain Dyes)**. In this method of dyeing, the water insoluble azo dye is produced in the fabric itself. The cloth is first soaked in the solution of a coupling reagent-usually a phenol or naphthol. Then it is immersed in the solution of an appropriate diazonium salt. The dye that is developed in the fabric binds fast to it *via* auxochromes. The azoic dyeing is particularly suitable for cotton and other cellulosic fibers but may also be used for *nylon*.

(5) **Disperse Dyes**. These dyes are insoluble in water but can be dispersed in a colloidal form in water. The fabric is immersed in the colloidal dispersion of the dye. The fine dye particles are absorbed into the 'crystal structure' of the fabric. Disperse dyes are used with modern synthetic fabrics such as nylon, orlon, polyesters and cellulose acetate.

ORGANIC PIGMENTS

The difference between a dye and a pigment is that a dye is actually absorbed by the material to be colored, whereas a pigment is applied to the surface. Pigments are highly insoluble powders of very fine particle size. They are applied in the form of dispersion in water or some other liquid. Pigments are used in paints, plastics, rubber, printing inks, and other materials for imparting color, opaqueness, and other desirable properties. In paints, for example, pigments may also impart anticorrosive properties.

FOOD DYES

A food dye is any substance that is added to food to change its color.

Purpose of Food Dyes

People learn to associate certain colors with certain flavors, and this causes the **color of food** to influence the perceived flavor, in anything from fruit gums to wine. For this reason, food manufacturers add dyes to their products. Sometimes the aim is to simulate a natural color as perceived by the consumer, such as adding red coloring to grace cherries (which would otherwise be beige), but sometimes it is for effect, like the green ketchup that Heinz launched in 2000.

While most consumers are aware that foods with bright, unnatural colors are artificially colored, few people know that apparently "natural" foods such as oranges are sometimes also dyed to mask natural variations in color. Most trust the regulation of their governments to keep any unsafe additives off the market, but there is a vigorous public debate about the safety of many food colorings - a notable example being tartrazine. A growing minority believes that the effects of colorings have not been well enough researched and consider their use an unnecessary risk.

Color variation in foods throughout season and the effects of processing and storing often make color addition a commercial advantage to maintain the color expected or preferred by the consumer. Some of the primary reasons include:

- Offsetting color loss due to light, air, extremes of temperature, moisture, and the storage conditions.
- Masking natural variations in color.
- Enhancing naturally occurring colors.
- Providing identity to foods.
- Protecting flavors and vitamins from damage by light.

Safety Concerns

Food dyes are tested for safety by various bodies around the world. In India, CI numbers are given to the approved dyes, in the United States, FD & C numbers are used, while in the European Union E numbers are used.

Points to Remember

1. For thousands of years, dyes were obtained from natural sources, such as plants and animals. Inspite of the fact that synthetic dyes have replaced many natural dyes for commercial use, natural dyes still hold a fascination and are used extensively by artisans around the world.
2. Today the number of synthetic dyes and colorants exceeds 7,000.
3. **A dye is a colored compound that is used to impart color to a fabric. All colored compounds are not dyes. To be useful as a dye, a compound must show fastness to light, washing, heat, and bleaching.**
4. Compounds with single and isolated double bonds do not show color.
5. Compounds with large number of conjugated double bonds show color.
6. Pigment is a coloring matter, usually in the form of an insoluble powder, mixed with oil, water, etc. to make paints.
7. **Indicators are water soluble organic dyes that change color when a particular pH is reached. These are weak acids in which the conjugate base is a different color.**
8. Dyes are not just for fabrics. They have been added to food for centuries to enhance its appearance. Color is a big factor in what foods we choose and how much we enjoy a food. Many food dyes have been banned through the years because they were shown to cause cancer. While some food colorings are natural, 95% of those used are artificial.

Examination Questions

1. Discuss the theories of color and constitution. (*Panjab BSc, 2003 ; Gujarat BSc, 2014 ; Delhi BSc, 2015*)
2. What is a dye ? How are dyes classified on the basis of structure ? (*Merrut BSc, 2014 ; Ambedkar BSc, 2015*)
3. What is a dye ? How are dyes classified on the basis of application ?
4. Write notes on :
   (*a*) Azo dyes (*Andhra BSc, 2015*)
   (*b*) Direct dyes (*Nagpur BSc, 2013 ; Dibrugarh BSc, 2014*)
   (*c*) Vat dyes
   (*d*) Mordant dyes
5. Give a method of preparation and use of Methyl orange. (*Madras BSc, 2014*)
6. Give a method of preparation and use of Congo red. (*Anna BSc, 2014 ; Kerala BSc, 2015*)
7. Give a method of preparation and use of Bismarck brown.
8. Give a method of preparation and use of Malachite green.
9. Give a method of preparation and use of Phenolphthalein. (*Burdwan BSc, 2014 ; Kanpur BSc, 2015*)
10. Give a method of preparation and use of Fluorescein.
11. Give a method of preparation and use of Alizarin.
12. Give a method of preparation and use of Indigo. (*Manipal BSc, 2014*)
13. How will you synthesize *N,N*-Dimethyl-*p*-hydroxyazobenzene from benzene?

Answer. Following steps are involved :

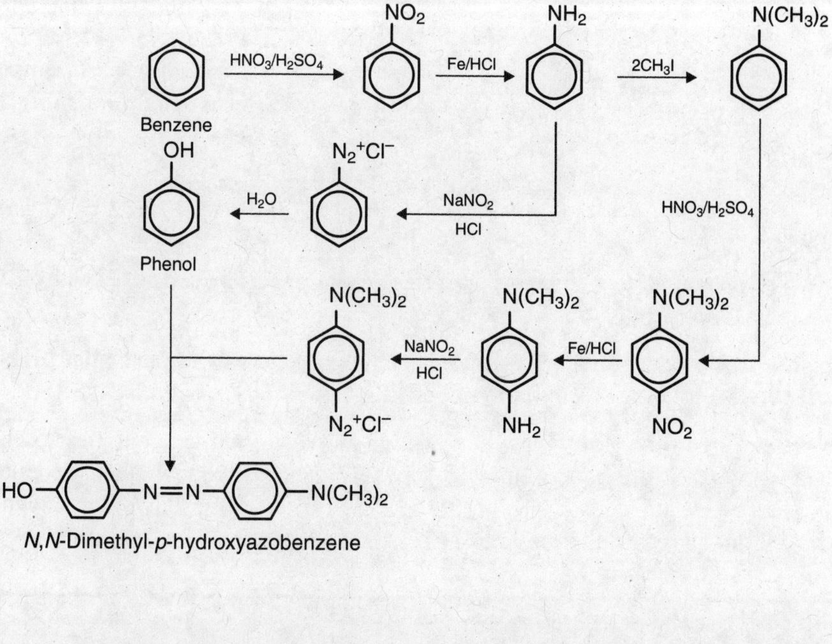

N,N-Dimethyl-*p*-hydroxyazobenzene

Multiple Choice Questions

1. The reason why materials appear colored is

(*a*) The selective absorption of spectral colors

(*b*) The interaction between the light and the electrons of the dye molecules

(*c*) The composition of the white light from various spectral colors

(*d*) All of these

Answer. (*d*)

2. Which of the following is a chromophore ?

(*a*) $-NO_2$ (*b*) $-SO_3H$ (*c*) $-OH$ (*d*) $-COOH$

Answer. (*a*)

3. Which of the following is an auxochrome ?

(*a*) $-N{=}O$ (*b*) $-NO_2$ (*c*) $-N{=}N-$ (*d*) $-OH$

Answer. (*d*)

4. The water-solubility of dyes can be increased by introducing

(*a*) SO_3Na groups (*b*) COOH groups (*c*) OH groups (*d*) All of these

Answer. (*d*)

5. Which dyes become linked to the fiber by chemical reaction ?

(*a*) Acid dyes (*b*) Direct dyes (*c*) Disperse dyes (*d*) None of these

Answer. (*a*)

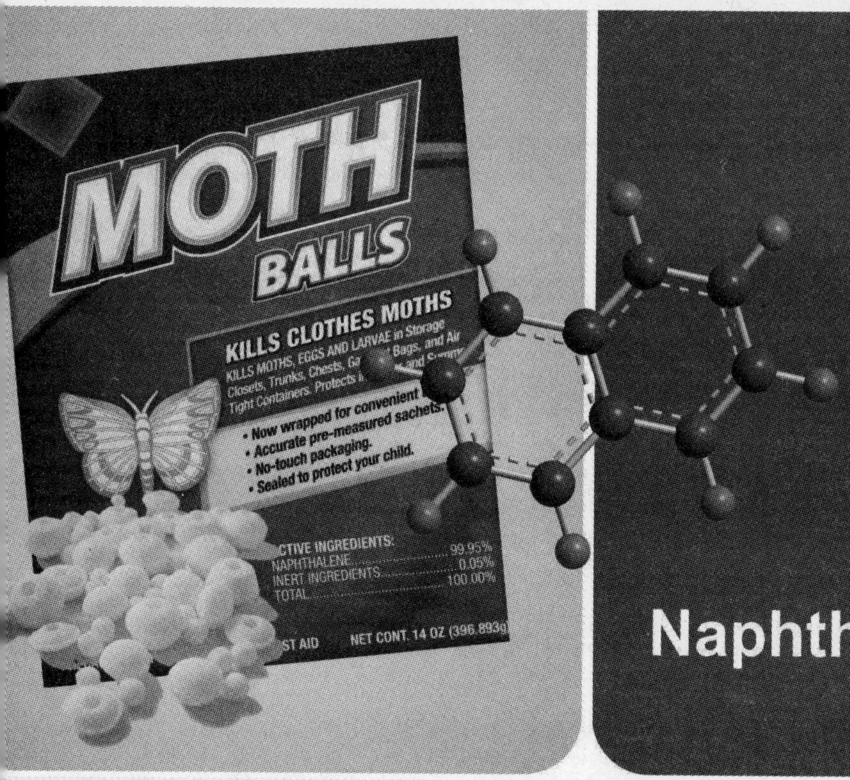

44

Naphthalene

Fused or condensed aromatic hydrocarbons contain more than one ring and have two carbons shared by two (or more) aromatic rings. The most important members of this class are naphthalene, anthracene and phenanthrene.

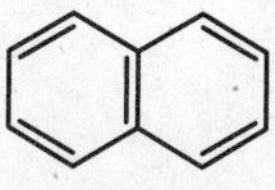
Naphthalene

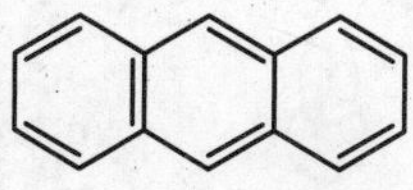
Anthracene

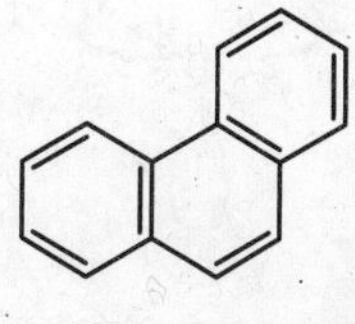
Phenanthrene

Naphthalene has two carbon atoms shared by two rings. Anthracene and phenanthrene have two pairs of carbon atoms shared by two rings, each pair being shared by a separate pair of rings.

Naphthalene is the largest single constituent (6 to 10 per cent) of coal-tar. It is obtained industrially by chilling the middle oil fraction (bp 170-230°C), when crude naphthalene crystallizes out. It is purified by washing it successively with dilute sulfuric acid (to remove basic impurities), sodium hydroxide (to remove acidic impurities), and water. Finally the solid is sublimed or steam-distilled to give pure naphthalene.

NOMENCLATURE OF NAPHTHALENE DERIVATIVES

In naming derivatives of naphthalene, the numbering system shown below is used. The numbers selected to denote the position of a substituent on the naphthalene rings should be as small as possible.

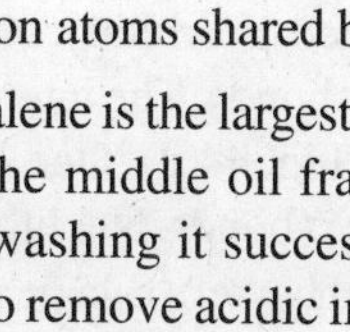

Naphthalene

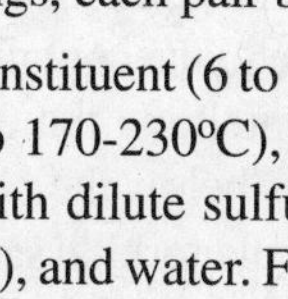

1-Bromonaphthalene

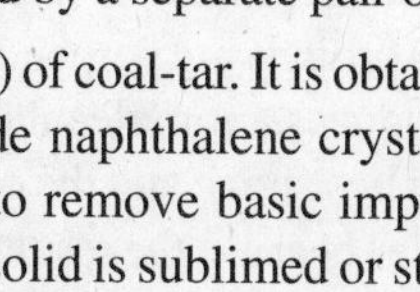

1,6-Dimethylnaphthalene

An alternative nomenclature system is occasionally used when there is only one substituent on the naphthalene skeleton. This system uses the Greek letters α (alpha) and β (beta) to designate the two possible orientations of the single substituent.

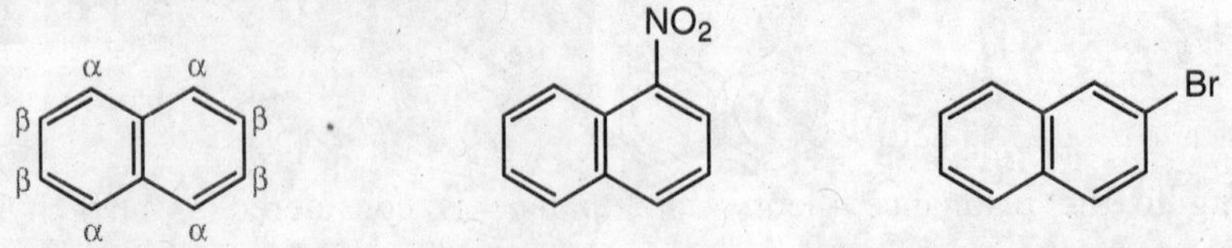

Naphthalene α-Nitronaphthalene β-Bromonaphthalene

STRUCTURE OF NAPHTHALENE

All ten carbon atoms in naphthalene are sp^2 hybridized. The sp^2 hybrid orbitals overlap with each other and with *s* orbitals of the eight hydrogen atoms forming C—C and C—H σ bonds (Fig.44.1). Since the σ bonds result from the overlap of trigonal sp^2 orbitals, all carbon and hydrogen atoms in naphthalene lie in one plane. This has been confirmed by X-ray diffraction studies.

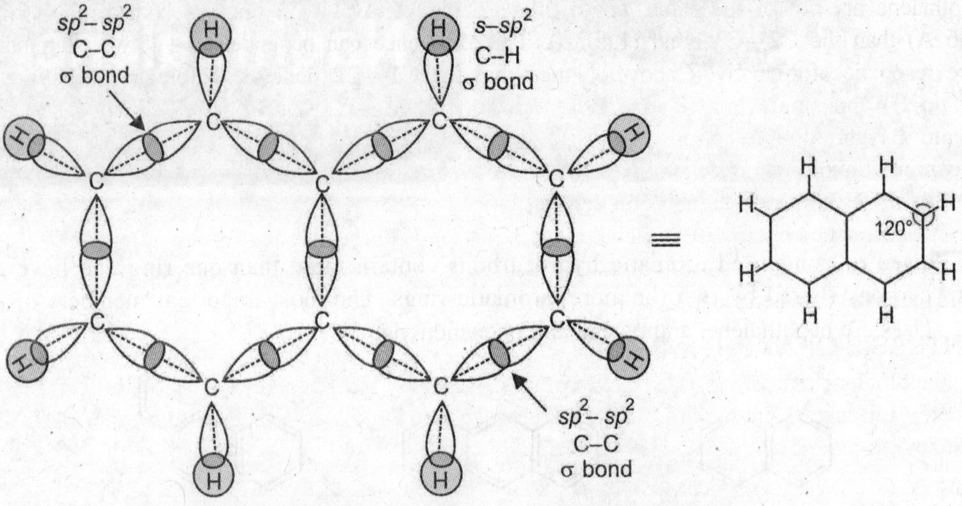

Fig. 44.1. Formation of the σ bonds in naphthalene.
All σ bonds in naphthalene lie in one plane. All bond angles are 120°.

Also each carbon atom in naphthalene possesses an unhybridized *p* orbital containing one electron. These *p* orbitals are perpendicular to the plane containing the σ bonds. The lateral overlap of the *p* orbitals produces a π molecular orbital containing ten electrons (Fig. 44.2). One half of this π molecular orbital lies above and the other half lies below the plane of the σ bonds. Naphthalene shows aromatic properties because the resulting π molecular orbital satisfies the Huckel's rule (n =2 in $4n + 2$).

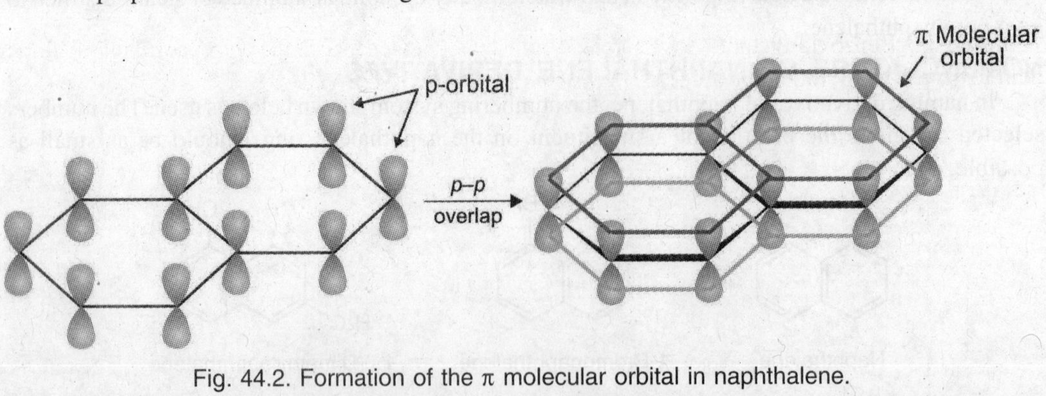

Fig. 44.2. Formation of the π molecular orbital in naphthalene.

A common shorthand representation of naphthalene is simply two fused hexagons with a circle inside each hexagon. The circle represent the π molecular orbital.

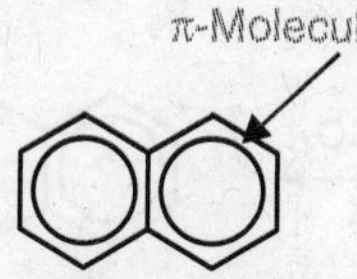

According to the resonance theory, naphthalene is considered to be a hybrid of the following *three* forms.

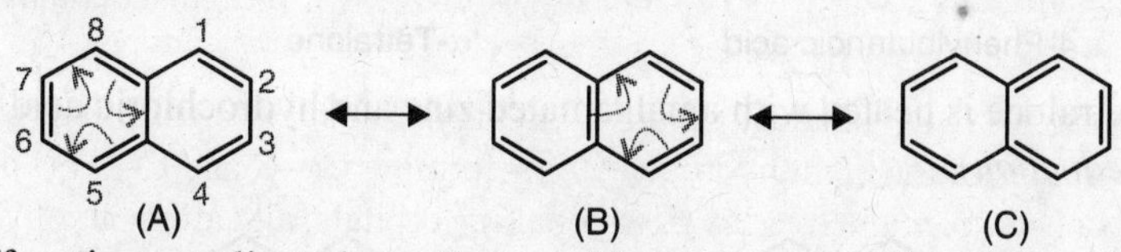

X-Ray diffraction studies show that, unlike benzene, all carbon-carbon bonds in naphthalene are not of the same length. In particular, the C1—C2 bond is considerably shorter (1.36 Å) than the C2—C3 bond (1.40 Å). This difference can be understood if we examine the three resonance forms given above. Notice that the C1—C2 bond is double in two structures (A and B) and single in only one (C); whereas the C2—C3 bond is single in two structures (A and B) and double in only one (C). We would, therefore, expect the C1—C2 bond to have more double-bond character (shorter bond length), and the C2—C3 bond to have more single-bond character (longer bond length).

The resonance energy of naphthalene is about 61 kcal/mole. This value is less than twice the amount of a single benzene ring (36 kcal/mole). As a result, naphthalene is somewhat less aromatic (more reactive) than benzene.

SYNTHESIS OF NAPHTHALENE

Naphthalene may be obtained by **Haworth Synthesis**. This involves the following 5 steps:

Step 1. Benzene and succinic anhydride are heated in the presence of aluminium chloride to form 3-benzoylpropanoic acid. (*Friedel-Crafts Acylation*)

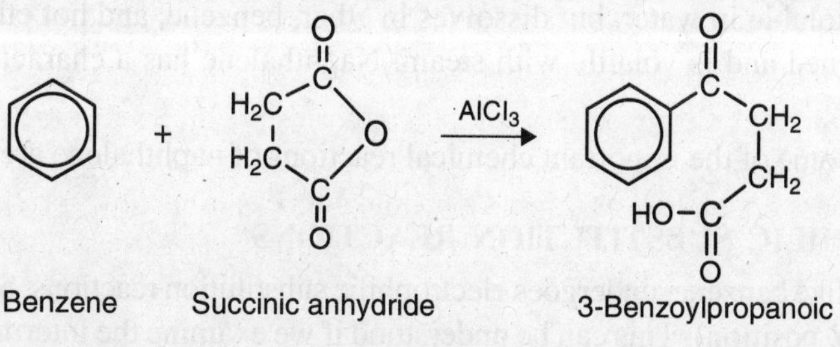

Step 2. 3-Benzoylpropanoic acid is treated with amalgamated zinc in the presence of hydrochloric acid to give 4-phenylbutanoic acid. (*Clemmensen Reduction*)

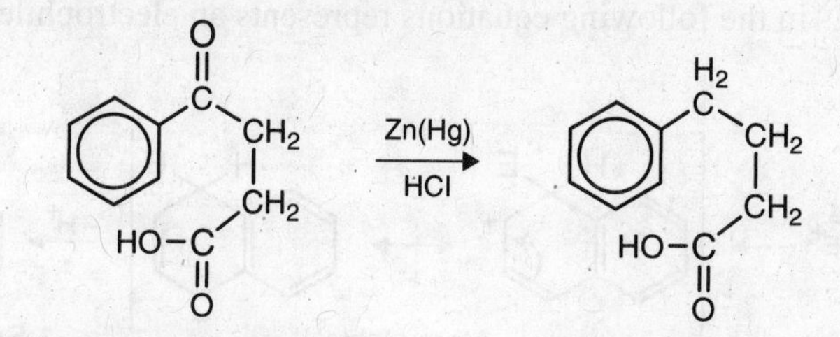

Step 3. 4-Phenylbutanoic acid is heated with concentrated sulfuric acid or polyphosphoric acid to form α-tetralone. (*Ring-closure Reaction*)

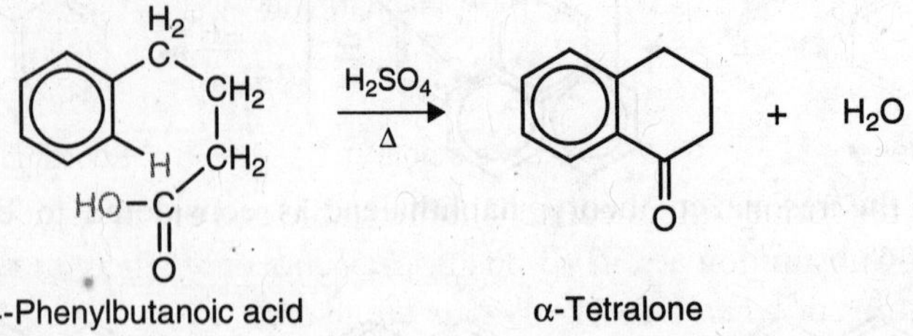

Step 4. α-Tetralone is heated with amalgamated zinc and hydrochloric acid to give tetralin. (*Clemmensen Reduction*)

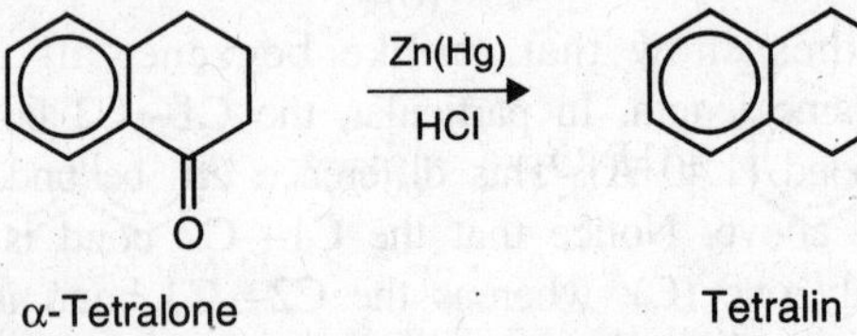

Step 5. Tetralin is heated with palladium to yield naphthalene. (*Aromatization Reaction*)

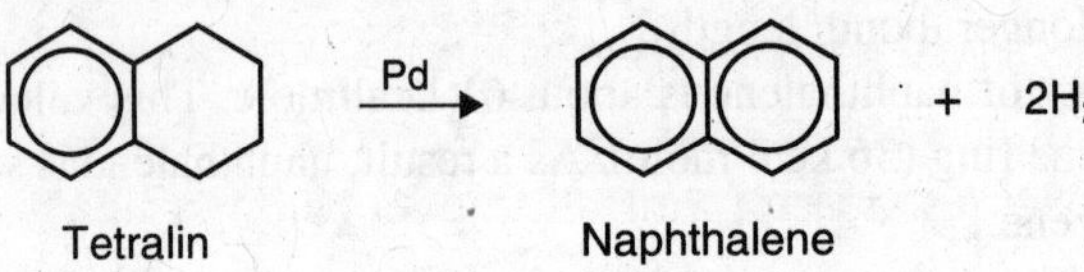

Substituted naphthalenes can be obtained by using toluene, bromobenzene, or anisole instead of benzene in *Step (1)*.

PROPERTIES OF NAPHTHALENE

(Physical). Naphthalene is a colorless crystalline solid. It melts at 82°C and boils at 218°C. Naphthalene is insoluble in water, but dissolves in ether, benzene, and hot ethanol. It sublimes readily when warmed and is volatile with steam. Naphthalene has a characteristic 'moth ball' odor.

(Chemical). Some of the important chemical reactions of naphthalene are described below:

A. ELECTROPHILIC SUBSTITUTION REACTIONS

Naphthalene, like benzene, undergoes electrophilic substitution reactions. Substitution occurs primarily at C-1 (α-position). This can be understood if we examine the intermediate carbonium ion. Two resonance forms can be written for the intermediate carbonium ion obtained from the attack at C-1 (without involving the other ring), whereas only one such form is possible for substitution at C-2. E^+ in the following equations represents an electrophile.

Attack at C-1

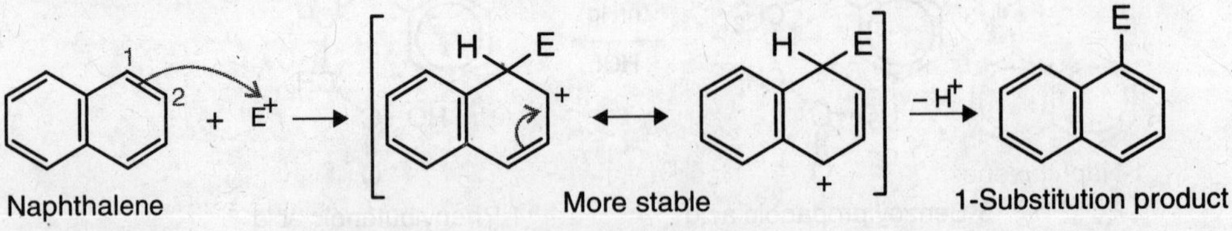

Attack at C-2

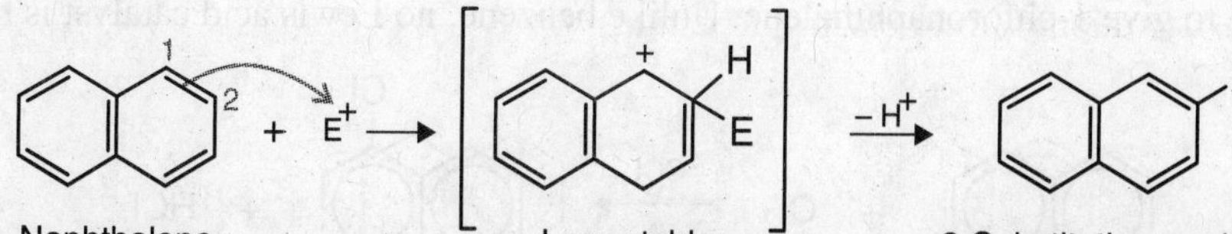

Naphthalene　　　Less stable　　　2-Substitution product

Consequently the former intermediate is more stable and the product with a substituent at C-1 predominates. Substitution at C-2 (β-position) occurs only when the reactions are carried at higher temperatures or when bulkier solvents are used.

(1) **Nitration.** Naphthalene undergoes nitration with concentrated nitric acid in the presence of sulfuric acid at 60°C to produce 1-nitronaphthalene.

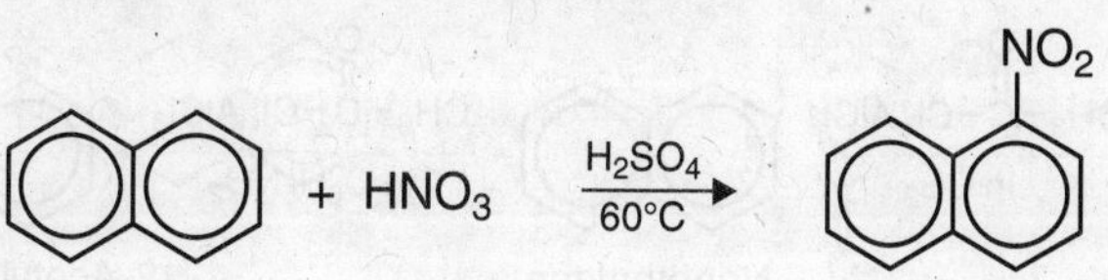

Naphthalene　　　1-Nitronaphthalene

MECHANISM. The mechanism of this reaction involves the following three steps :

Step 1. The electrophile NO_2^+ is generated.

$$2H_2SO_4 + HONO_2 \longrightarrow \overset{+}{NO_2} + 2H\overset{-}{SO_4} + H_3\overset{+}{O}$$

Step 2. The electrophile attacks naphthalene at C-1 to form the resonance-stabilized carbonium ion (only those resonance forms are shown below which do not involve the other ring).

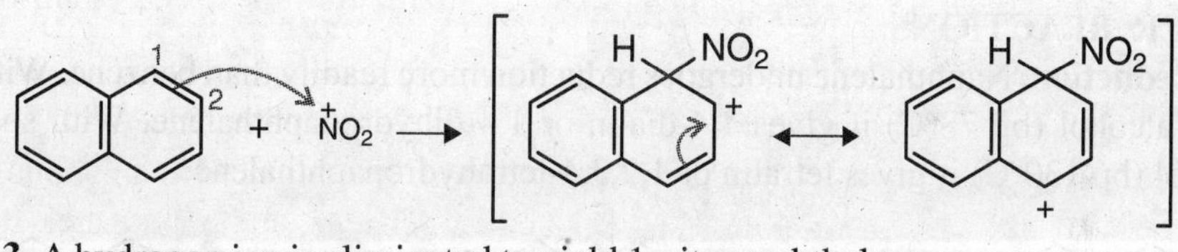

Step 3. A hydrogen ion is eliminated to yield 1-nitronaphthalene.

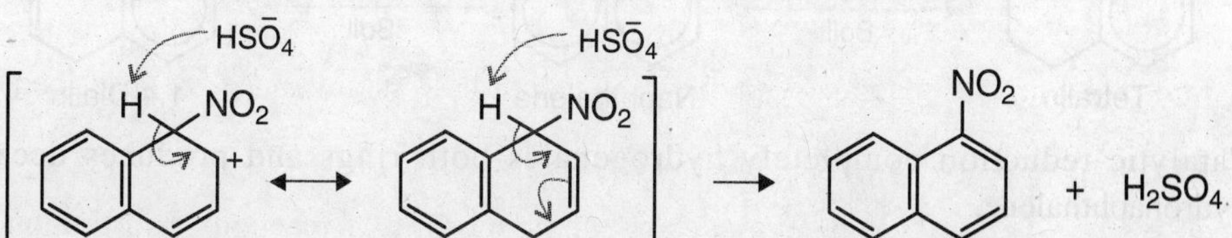

1-Nitronaphthalene

Similar steps may be written for other electrophilic substitution reactions of naphthalene.

(2) **Sulfonation.** Naphthalene undergoes sulfonation with concentrated sulfuric acid at 60°C to form 1-naphthalenesulfonic acid. If the reaction is carried at 165°C, 2-naphthalenesulfonic acid is obtained.

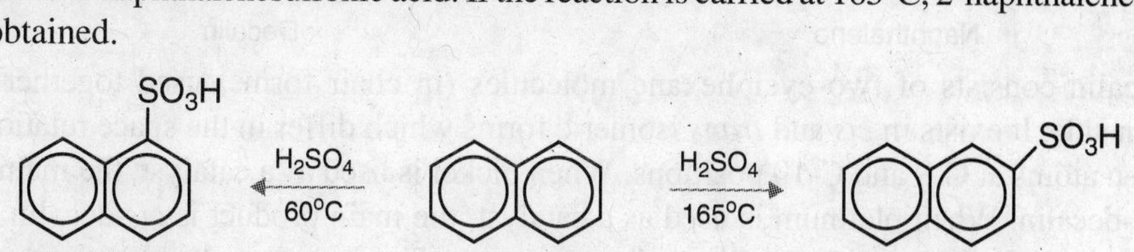

1-Naphthalene-sulfonic acid　　　Naphthalene　　　2-Naphthalene-sulfonic acid

(3) **Halogenation.** Naphthalene undergoes chlorination or bromination in boiling carbon tetrachloride to give 1-chloronaphthalene. Unlike benzene, no Lewis acid catalyst is required.

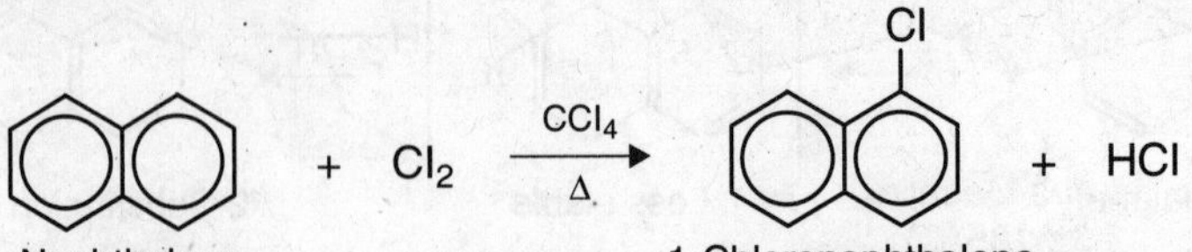

Naphthalene 1-Chloronaphthalene

(4) **Friedel-Crafts Acylation.** Naphthalene undergoes acylation with acetyl chloride and aluminium chloride in carbon disulfide to give 1-acetylnaphthalene. When nitrobenzene is used as a solvent, 2-acetylnaphthalene is obtained.

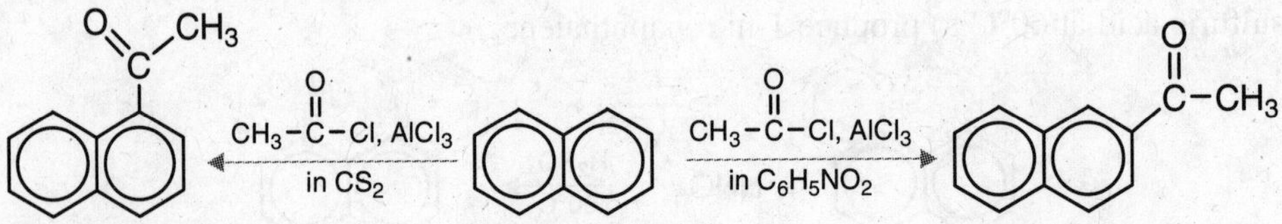

1-Acetylnaphthalene Naphthalene 2-Acetylnaphthalene

(5) **Friedel-Crafts Alkylation.** Naphthalene undergoes alkylation with alkyl halides (methyl halides do not react) in the presence of aluminium chloride to give 2-alkylnaphthalene.

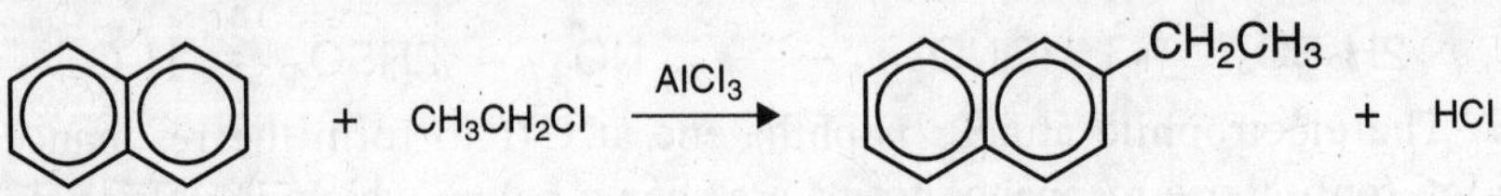

Naphthalene 2-Ethylnaphthalene

B. OTHER REACTIONS

(6) **Reduction.** Naphthalene undergoes reduction more readily than benzene. With sodium and ethyl alcohol (bp 78°C) it gives 1,4-dialin or 1,4-dihydronaphthalene. With sodium and isopentanol (bp 130°C) it gives tetralin or 1,2,3,4-tetrahydronaphthalene.

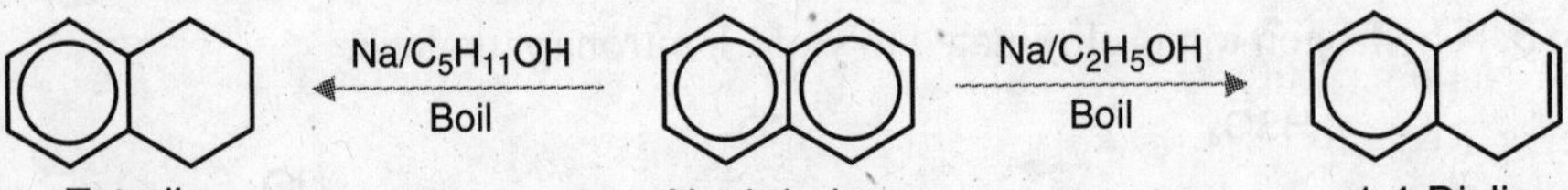

Tetralin Naphthalene 1,4-Dialin

Catalytic reduction completely hydrogenates both rings and produces decalin or decahydronaphthalene.

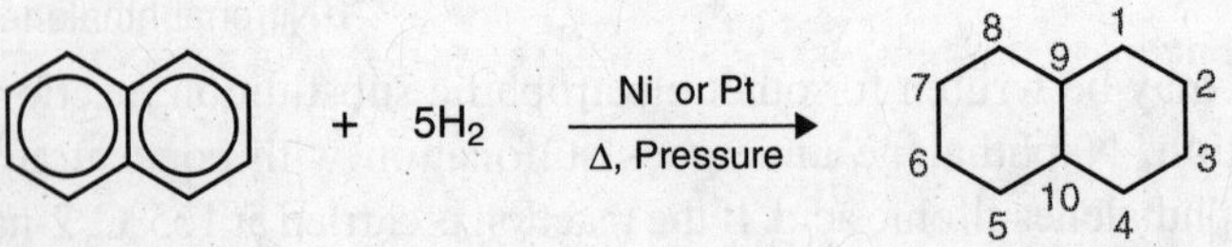

Naphthalene Decalin

Decalin consists of two cyclohexane molecules (in chair forms) used together along a common side. It exists in *cis* and *trans* isomeric forms which differ in the space relationship of hydrogen atoms at C-9 and C-10 positions. When nickel is used as a catalyst, the main product is *trans*-decalin. When platinum is used as a catalyst, the main product is *cis*-decalin. Decalin sold commercially is the mixture of both these forms, and is used as a solvent in the varnish and lacquer trade.

trans-Decalin (bp185°C) *cis*-Decalin (bp195°C)

(7) **Oxidation.** Naphthalene is much more easily oxidized than benzene. With chromium trioxide in acetic acid at room temperature, it gives 1,4-naphthaquinone.

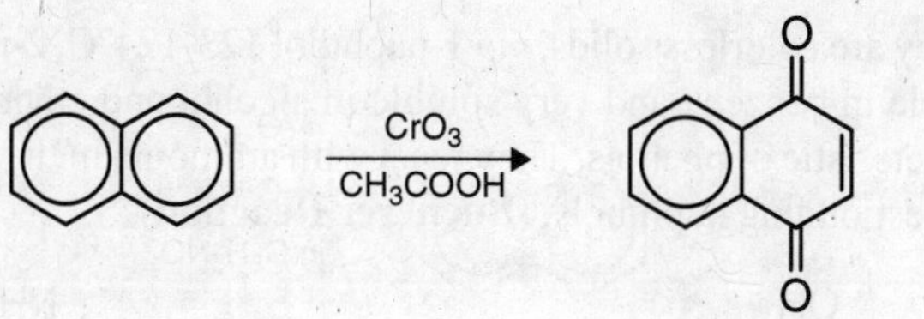

Naphthalene 1,4-Naphthaquinone

Oxidation of naphthalene with oxygen and vanadium pentoxide at 475°C yields phthalic anhydride. This method is used industrially.

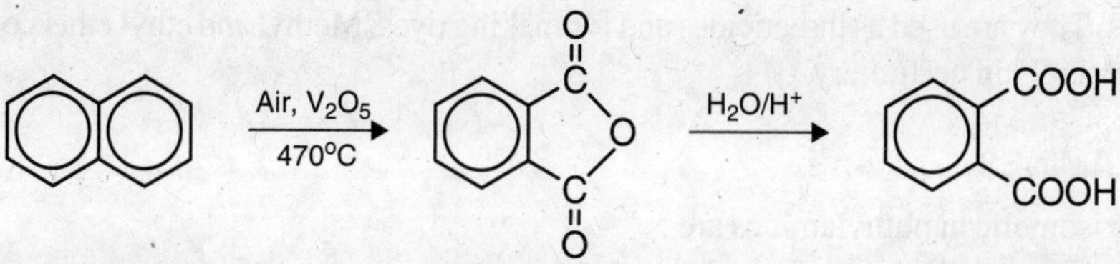

Naphthalene Phthalic anhydride Phthalic acid

Phthalic acid is an important industrial material. It is used in the manufacture of resins paints, dyes and plastics.

Uses. Naphthalene as 'moth balls' has been used to protect woolen goods from moths for many years. Recently, *p*-dichlorobenzene has replaced naphthalene in the manufacture of moth balls, as it has a less obnoxious odor. It is used for increasing the illuminating power of coal gas. Naphthalene is used in the manufacture of phthalic anhydride, carbaryl for insecticides, 2-naphthol for dyes and several medicinal products.

NAPHTHOLS

The monohydroxy derivatives of naphthalene are called naphthols.

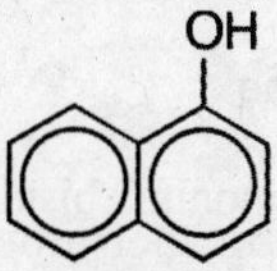

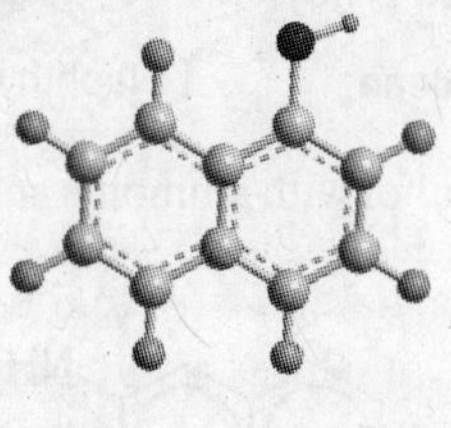

1-Naphthol (α-Naphthol)

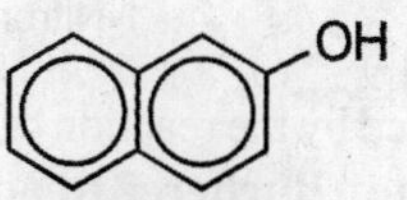

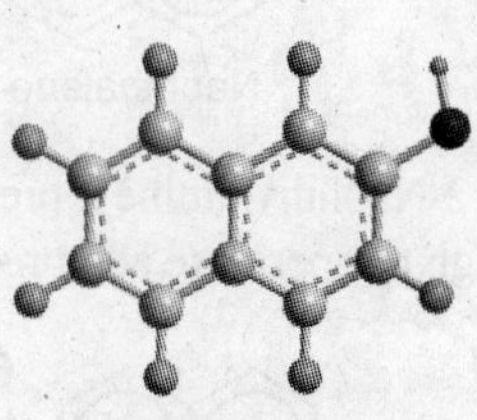

2-Naphthol (β-Naphthol)

Preparation. Both 1- and 2-naphthols are prepared from the corresponding naphthalenesulfonic acids by fusion with sodium hydroxide (300°C) followed by acidification.

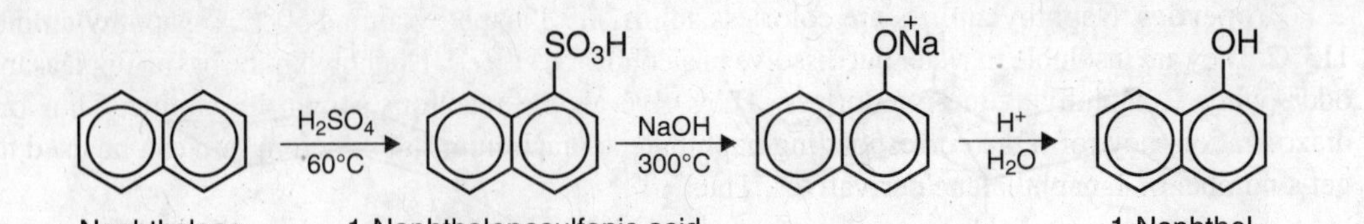

Naphthalene — 1-Naphthalenesulfonic acid — 1-Naphthol

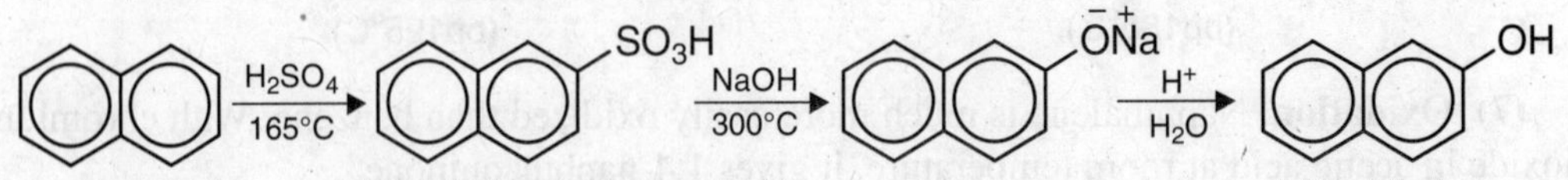

Naphthalene — 2-Naphthalenesulfonic acid — 2-Naphthol

Properties. They are colorless solids, *mp* 1-naphthol 123-124°C, 2-naphthol 96°C. They are insoluble in water, soluble in benzene and very soluble in alcohol and ether. Naphthols give all the chemical reactions characteristic of phenols. They react with ammonia in the presence of ammonium bisulfite to yield the corresponding naphthols. (**Bucherer Reaction**)

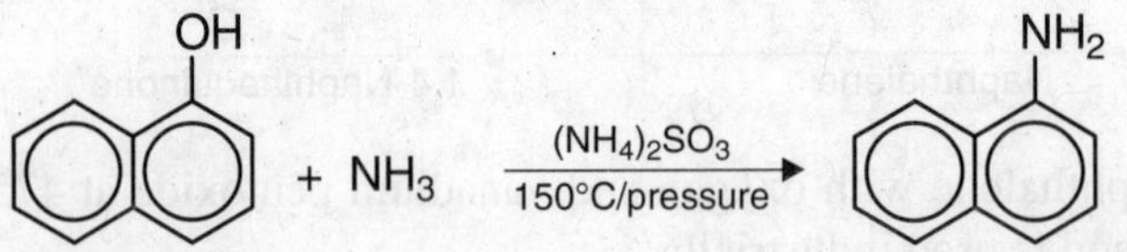

1-Naphthol — 2-Naphthylamine

Uses. They are used as insecticides and for making dyes. Methyl and ethyl ethers of β-naphthol are used extensively in perfumery.

NAPHTHYLAMINES

Two isomeric naphthylamines are :

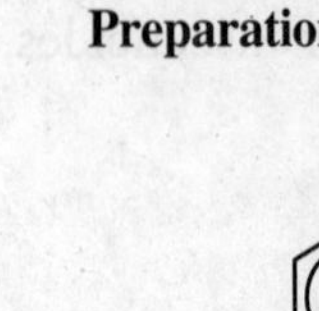

1-Naphthylamine
(α-Naphthylamine)

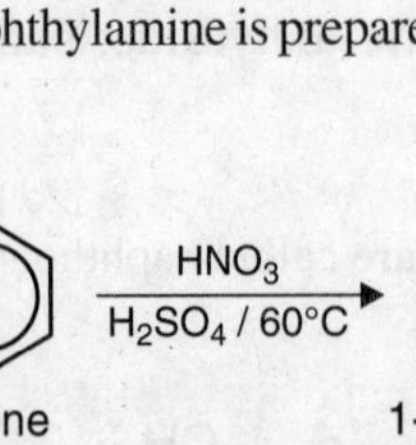

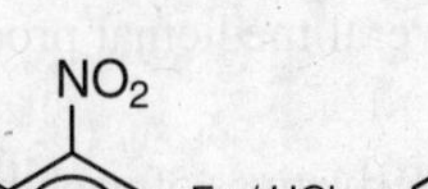

2-Naphthylamine
(β-Naphthylamine)

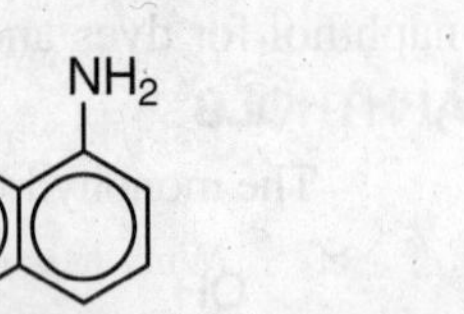

Preparation. α-Naphthylamine is prepared by the reduction of the corresponding nitronaphthalene.

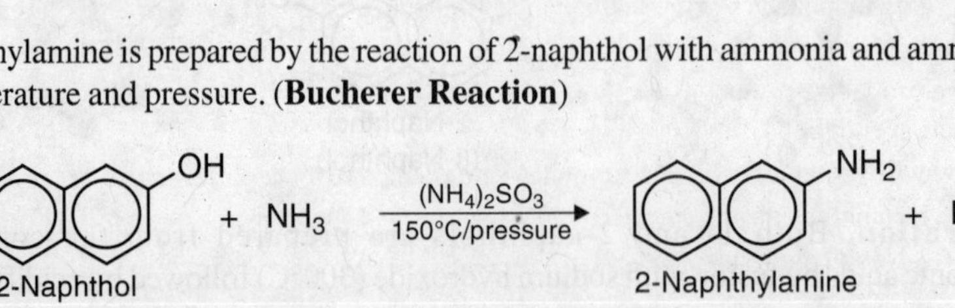

Naphthalene — 1-Nitronaphthalene — 1-Naphthylamine

2-Naphthylamine is prepared by the reaction of 2-naphthol with ammonia and ammonium bisulfite at high temperature and pressure. (**Bucherer Reaction**)

2-Naphthol — 2-Naphthylamine

Properties. Naphthylamines are colorless solids, *mp* 1-naphthylamine 50°C, 2-naphthylamine 113°C. They are insoluble in water but dissolve in alcohol and ether. 1-Naphthylamine has an unpleasant odor, while 2-naphthylamine is odorless. They give all the reactions shown by aniline. Thus on diazotization they form the corresponding naphthalene diazonium salt which in turn can be used to get a number of 1-naphthalene derivatives. Thus,

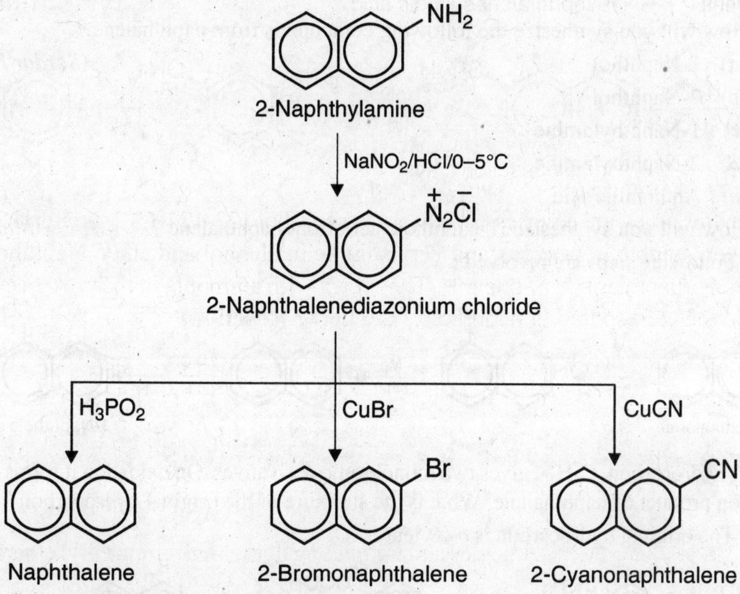

Uses. They are used in the manufacture of dyes.

Examination Questions

1. Write the structural formulas of (*a*) Naphthalene ; (*b*) Anthracene ; (*c*) Phenanthrene.

2. Discuss the molecular orbital structure of naphthalene ? (Baroda *BSc, 2015*)

3. How is naphthalene synthesized ? Describe its important reactions and uses. (*Kakatiya BSc, 2014 ; Manipal BSc, 2015*)

4. Write notes on :

(*a*) Molecular orbital structure of naphthalene. (*Burdwan BSc, 2014*)

(*b*) Electrophilic substitution of naphthalene. (*Orissa BSc, 2015*)

(*c*) Haworth's synthesis of naphthalene.

5. Explain why the electrophilic substitution in naphthalene takes place mainly at the 1-position. (*Baroda BSc, 2014 ; Andhra BSc, 2015*)

6. What happens when

(*a*) Naphthalene is treated with concentrated sulfuric acid at 165°C.

(*b*) Naphthalene is warmed with concentrated nitric acid in the presence of sulfuric acid.

(*c*) Naphthalene is heated with oxygen in the presence of vanadium pentoxide.

Answer. The products are : (*a*) 2-Naphthalenesulfonic acid ; (*b*) 1-Nitronaphthalene ; and (*c*) Phthalic anhydride. For reaction equations see text.

7. How will you synthesize the following compounds from naphthalene ?

(*a*) 1-Naphthalenesulfonic acid

(*b*) 2-Naphthalenesulfonic acid

(*c*) 2-Acetylnaphthalene
(*d*) Phthalic anhydride
(*e*) Tetralin

Answer. These are all single-step conversions. Following reagents are used : (*a*) conc H_2SO_4 at 40°C ; (*b*) conc H_2SO_4 at 165°C ; (*c*) $CH_3COCl/AlCl_3$ in nitrobenzene ; (*d*) O_2/V_2O_5 at 500°C; and (*e*) $Na/C_5H_{11}OH$, heat. For reaction equations see text.

8. How will you synthesize the following compounds from naphthalene ?
(*a*) 1-Naphthol (*Sardar Patel BSc, 2014*)
(*b*) 2-Naphthol
(*c*) 1-Naphthylamine (*Agra BSc, 2015*)
(*d*) 2-Naphthylamine
(*e*) Anthranilic acid (*Goa BSc, 2014*)

9. How will you synthesize 1-naphthoic acid from naphthalene ? (*Madras BSc, 2015*)

Answer. Following steps are involved :

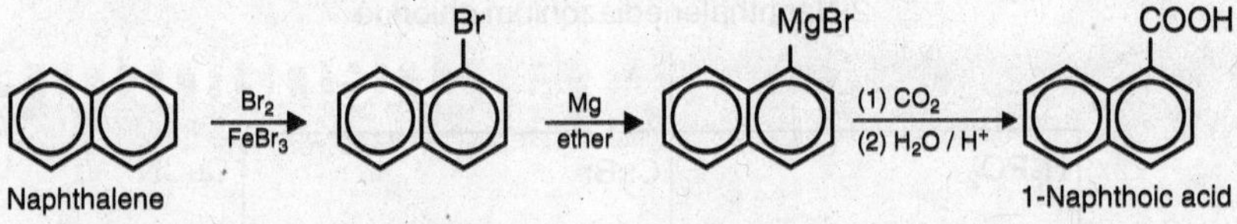

10. A hydrocarbon, C_8H_{10}, gives two monobromo derivatives. On oxidation it forms an acid isomeric with the oxidation product of naphthalene. What is the structure of the original hydrocarbon ?

Answer. The original hydrocarbon is *o*-Xylene.

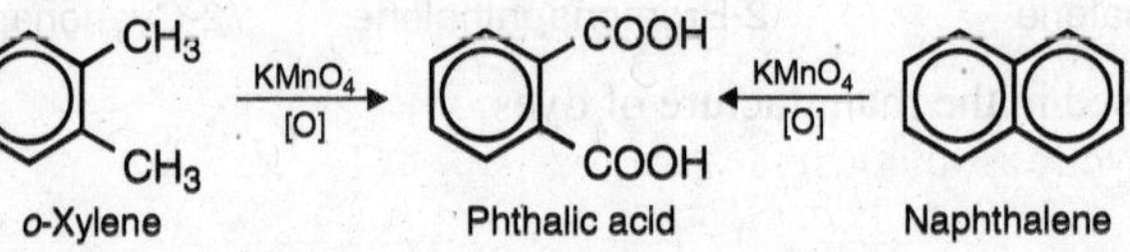

Multiple Choice Questions

1. Unlike ________, the carbon-carbon bonds in naphthalene are not of the same length.
(*a*) Pyridine (*b*) Benzene (*c*) Ethylene (*d*) Alkane

Answer. (*b*)

2. All carbon atoms in naphthalene are
(*a*) *sp* hybridized (*b*) sp^2 hybridized (*c*) sp^3 hybridized (*d*) None of these

Answer. (*b*)

3. Naphthalene undergoes oxidation with $Na_2Cr_2O_7 / H_2SO_4$ to form
(*a*) Phthalic acid (*b*) Benzoic acid (*c*) Tetralin (*d*) Phenylacetic acid

Answer. (*a*)

4. Naphthalene undergoes reduction with H_2 in the presence of Ni catalyst at high temperature and pressure to give
(*a*) Phthalic acid (*b*) Decalin (*c*) Benzoic acid (*d*) Tetralin

Answer. (*b*)

5. Naphthalene undergoes nitration with HNO_3/H_2SO_4 at 60°C to give mainly
(*a*) 1-Nitronaphthalene (*b*) 1,2-Dinitronaphthalene
(*c*) 2-Nitronaphthalene (*d*) 1,5-Dinitronaphthalene

Answer. (*a*)

45

Anthracene and Phenanthrene

The black material in Diesel exhaust consists of small particles that are rich in polynuclear aromatic hydrocarbons.

There are two tricyclic aromatic hydrocarbons :

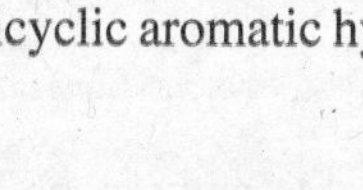

Anthracene

Phenanthrene

Anthracene occurs in coal-tar (less than 1 per cent) and is obtained from the green oil fraction (bp 270-360°C). On cooling this fraction crude anthracene crystallizes out. Crude anthracene contains phenanthrene and carbazole as impurities. It is purified by washing it successively with solvent naphtha (to remove phenanthrene) and pyridine (to remove carbazole). Finally the solid is sublimed to give pure anthracene.

In naming derivatives of anthracene, the numbering system shown below is used. The numbers selected to denote the position of a substituent on the anthracene rings should be as small as possible.

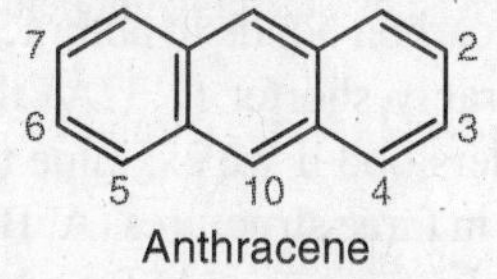

Anthracene

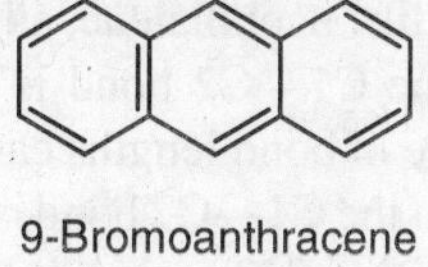

9-Bromoanthracene

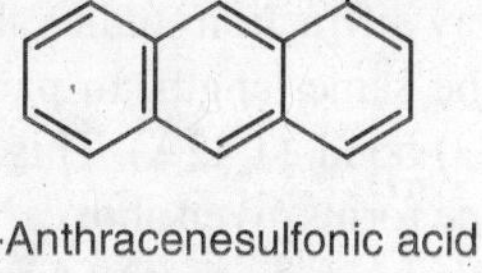

1-Anthracenesulfonic acid

STRUCTURE OF ANTHRACENE

All fourteen carbon atoms in anthracene are sp^2 hybridized. The sp^2 hybrid orbitals overlap with each other and with *s* orbitals of the ten hydrogen atoms forming C—C and C—H σ bonds. Since the bonds result from the overlap of trigonal sp^2 orbitals, all carbon and hydrogen atoms in anthracene lie in the same plane (Fig. 45.1). This has been confirmed by X-ray diffraction studies.

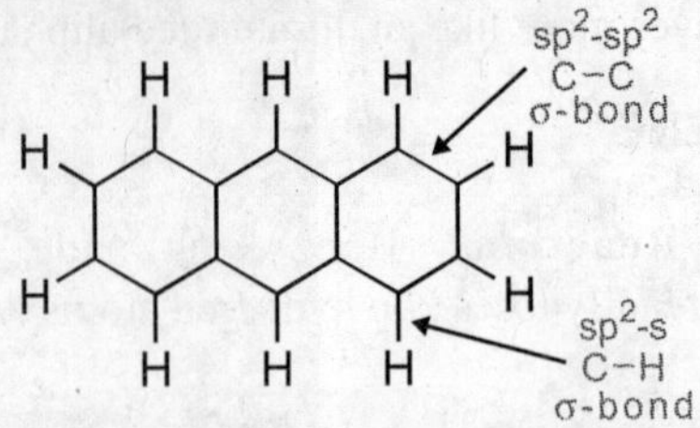

Fig. 45.1. All carbon atoms in anthracene are sp^2 hybridized. All atoms in anthracene lie in the same plane.

Also each carbon atom in anthracene possesses an unhybridized *p* orbital containing one electron. These *p* orbitals are perpendicular to the plane of the σ bonds. The lateral overlap of these *p* orbitals produces a π molecular orbital containing fourteen electrons (Fig. 45.2). One half of this π molecular orbital lies above and the other half lies below the plane of the σ bonds. Anthracene shows aromatic properties because the resulting π molecular orbital satisfies the Huckel's rule ($n = 3$ in $4n + 2$).

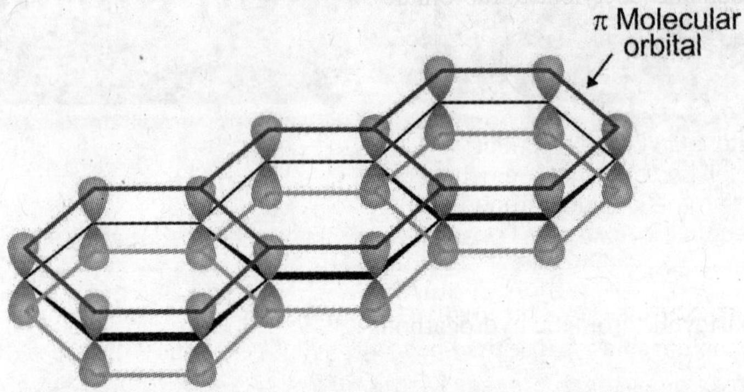

Fig. 45.2. Formation of π molecular orbital in anthracene.

A common shorthand representation of anthracene is simply three linearly fused hexagons with a circle inside each hexagon. The circles represent the π molecular orbital.

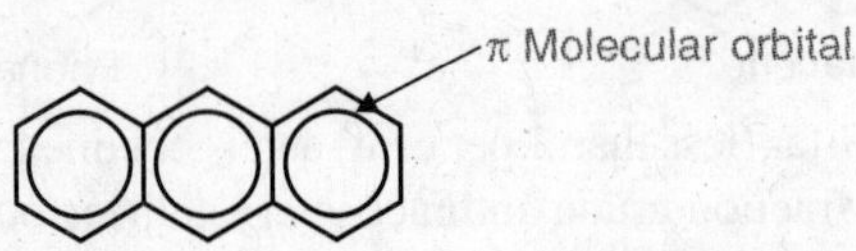

According to the resonance theory, anthracene is considered to be a hybrid of the following *four* canonical forms.

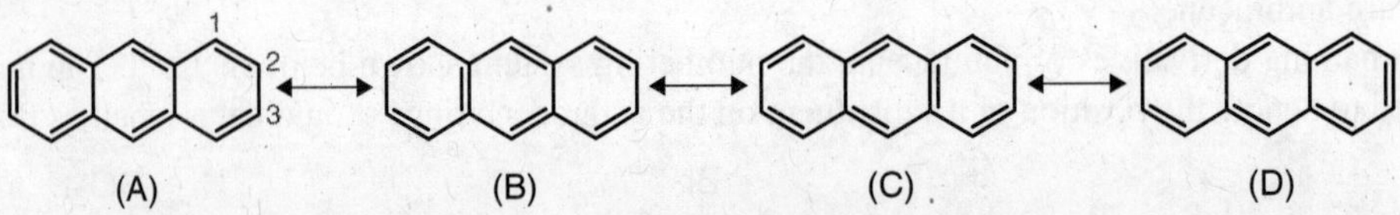

X-ray diffraction studies show that, like naphthalene, all carbon-carbon bonds in anthracene are not of the same length. In particular, the C1—C2 bond is considerably shorter (1.37Å) than the (C2—C3) bond. (1.42Å). This difference in bond lengths can be understood if we examine the four resonance forms given above. Notice that the C1—C2 bond is double in three structures (A, B and C) and single in only one (D); whereas the C2—C3 bond is single in three structures (A, B and C), and double in only one (D). We would, therefore, expect the C1—C2 bond to have more double-bond character (shorter bond length), and the C2—C3 bond to have more single-bond character (longer bond length).

The resonance energy of anthracene is 84 kcal/mole. This averages to 28 kcal/mole per ring, which is substantially lower than that of benzene (36 kcal/mole). As a result, anthracene is much less aromatic than benzene and behaves more like an unsaturated aliphatic hydrocarbon.

SYNTHESIS OF ANTHRACENE

Anthracene may be obtained :

(1) **By Friedel-Crafts Reactions.** Benzyl chloride reacts with itself to form 9,10-dihydroanthracene, which readily loses two hydrogen atoms to yield anthracene.

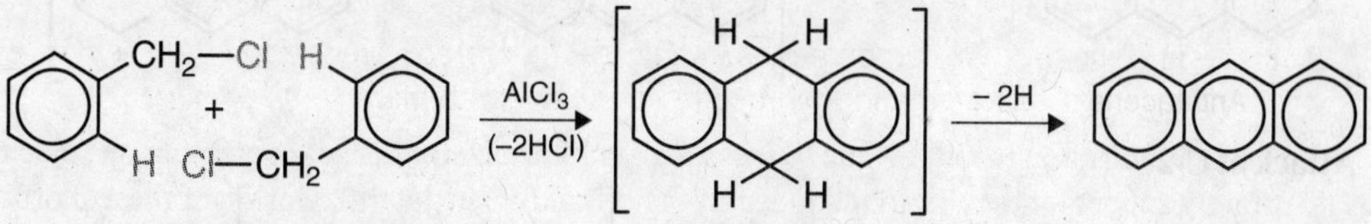

Benzyl chloride (2 molecules) 9,10-Dihydroanthracene Anthracene

Anthracene may also be prepared by the Friedel-Crafts reaction between benzene and 1,1,2,2-tetrabromoethane (acetylene tetrabromide).

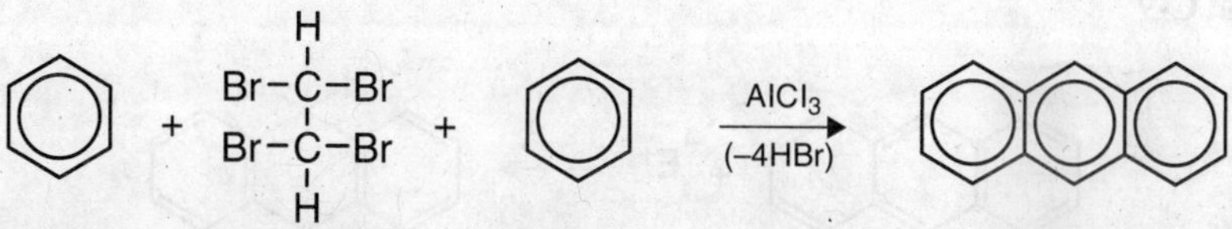

Benzene Acetylene tetrabromide Benzene Anthracene

(2) **By Haworth Synthesis.** This involves the treatment of benzene with phthalic anhydride in the presence of aluminium chloride to form *o*-benzoylbenzoic acid. This is then heated with concentrated sulfuric acid to give 9,10-anthraquinone. Distillation of the anthraquinone with zinc dust yields anthracene.

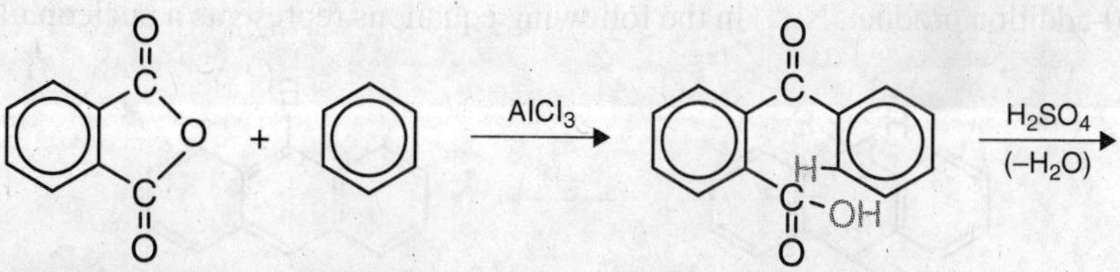

Phthalic anhydride Benzene *o*-Benzoylbenzoic acid

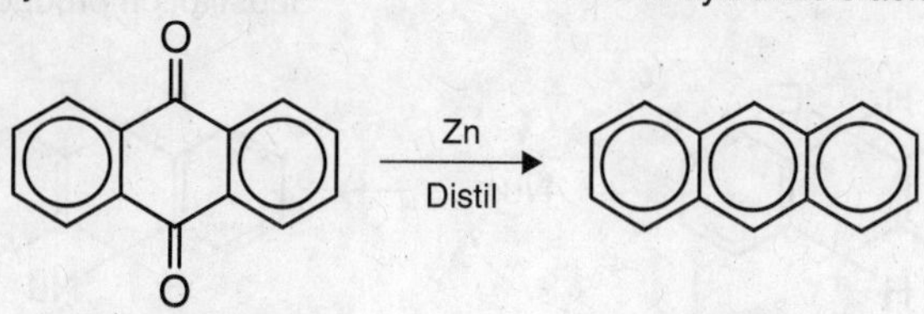

9,10-Anthraquinone Anthracene

PROPERTIES OF ANTHRACENE

(Physical). Anthracene is a colorless solid. It melts at 218°C and boils at 340°C. Anthracene is insoluble in water, but dissolves in benzene. It shows a strong blue fluorescence when exposed to ultraviolet light. This fluorescent property of anthracene is used in criminal detection work, since a small amount of finely powdered anthracene on clothing, skin, money, etc., is not detected under ordinary light but easily noticed when exposed to ultraviolet light.

(Chemical). Anthracene undergoes addition and electrophilic substitution reactions. These reactions preferentially occur at the C-9 and C-10 positions. This can be understood if we examine the intermediate carbonium ions obtained from attack at C-1, C-2, and C-9 (all other positions are equivalent to either 1 or 2 or 9 by symmetry). E^+ in the following equations represents an electrophile.

Attack at C-1

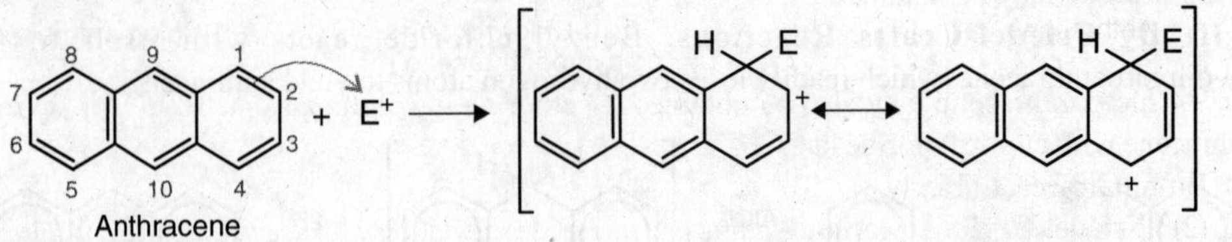

Anthracene

Attack at C-2

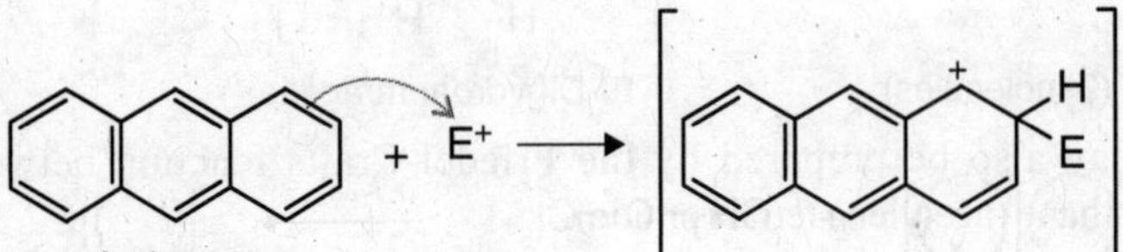

Anthracene

Attack at C-9

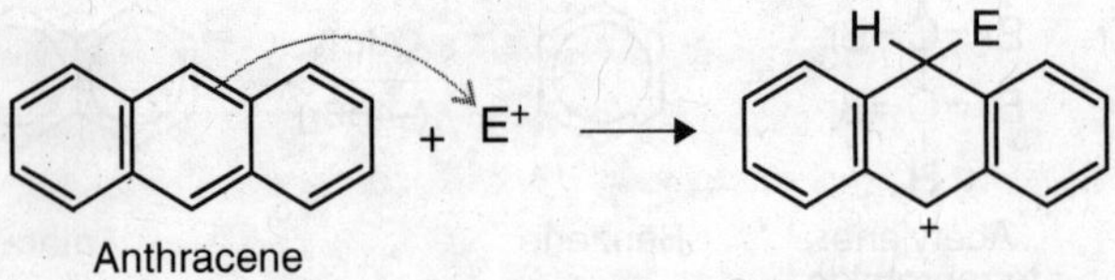

Anthracene

Attack at C-9 yields a carbonium ion intermediate in which two benzene rings are retained; whereas attack at C-1 or C-2 yields an intermediate in which a naphthalene system is retained. The former intermediate is more stable and its formation favored because the resonance energy of two benzene rings (2 x 36 = 72 kcal) exceeds that of napthalene (61 kcal). This carbonium ion intermediate can lose a proton to give the corresponding substitution product, or it can react with a nucleophile to form the 9,10-addition product. Nu:⁻ in the following equations represents a nucleophile.

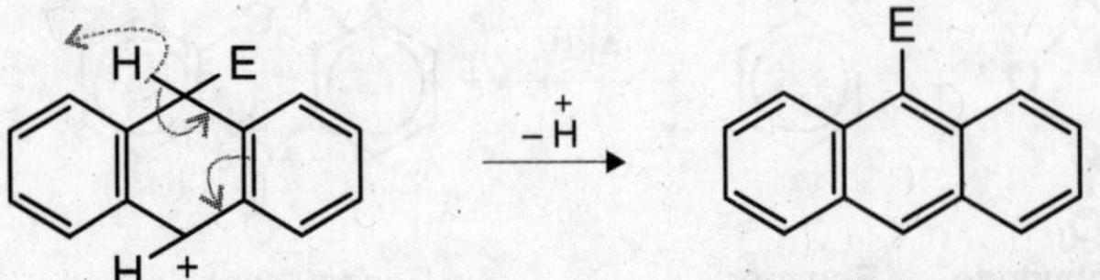

Substitution product

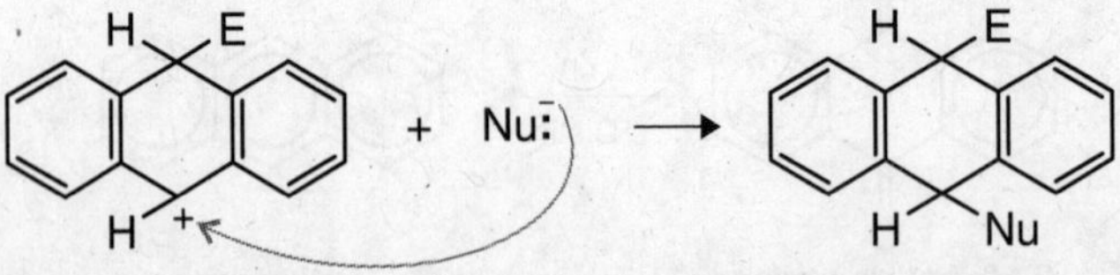

Addtion product

Substitution at C-1 or C-2 occurs only when the reaction is reversible (as in the case of sulfonation). The main chemical reactions of anthracene are described below :

(1) **Reaction with Halogens.** Anthracene reacts with chlorine in carbon tetrachloride at room temperature to give 9,10-dichloro-9,10-dihydroanthracene (an addition product). On heating, this addition product loses a molecule of hydrogen chloride by 1,4-elimination to form 9-chloroanthracene (a substitution product).

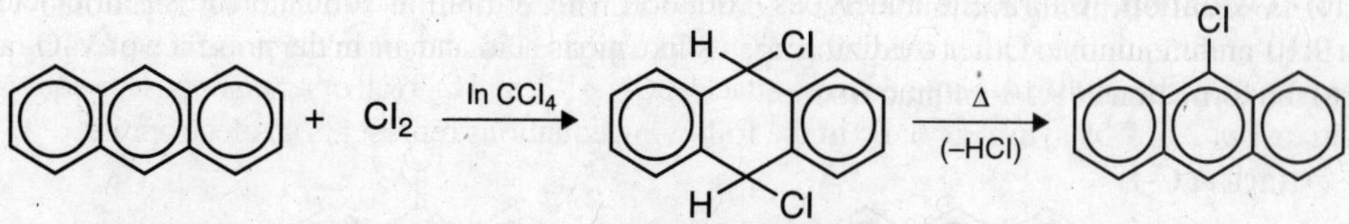

Anthracene 9,10-Chloro-9,10-dihydroanthracene 9-Chloroanthracene

9-Chloroanthracene may also be obtained by chlorinating anthracene at 100°C, or by treating anthracene with cupric chloride in CCl_4.

Bromine reacts similarly.

(2) **Friedel-Crafts Acylation.** Anthracene undergoes acylation with acetyl chloride and aluminium chloride to form 9-acetylanthracene.

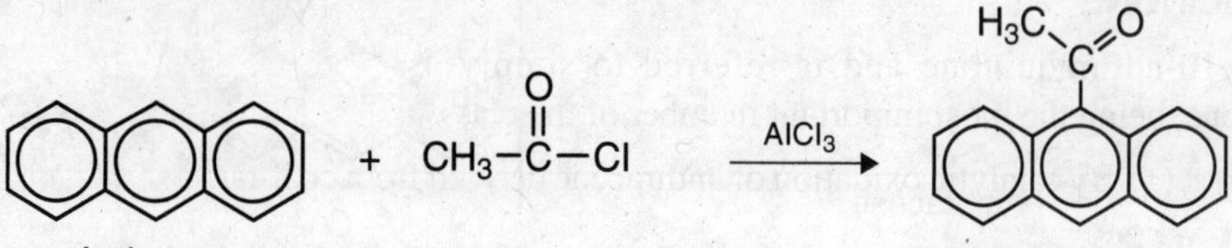

Anthracene 9-Acetylanthracene

(3) **Nitration.** Anthracene undergoes nitration with concentrated nitric acid in acetic anhydride at room temperature to yield a mixture of 9-nitroanthracene and 9,10-dinitroanthracene. The usual nitrating mixture ($HNO_3 + H_2SO_4$) is not used because it leads to the formation of 9,10-anthraquinone by oxidation.

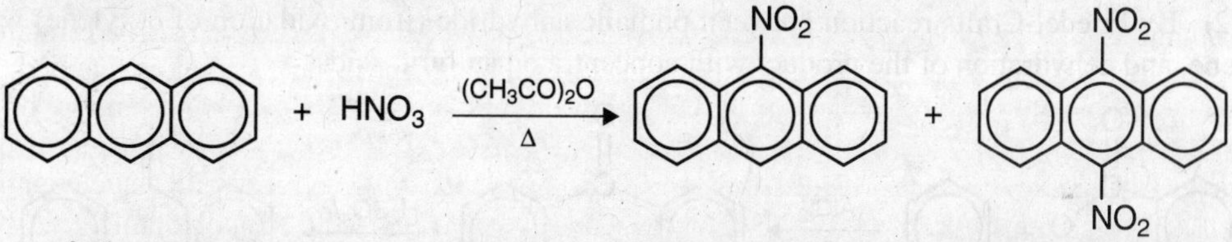

Anthracene 9-Nitroanthracene 9,10-Dinitroanthracene

(4) **Sulfonation.** Anthracene undergoes sulfonation with concentrated sulfuric acid to yield a mixture of 1-anthracenesulfonic acid and 2-anthracenesulfonic acid. At lower temperatures 1-anthracenesulfonic acid is the major product; where as at higher temperatures 2-anthracenesulfonic acid is the major product.

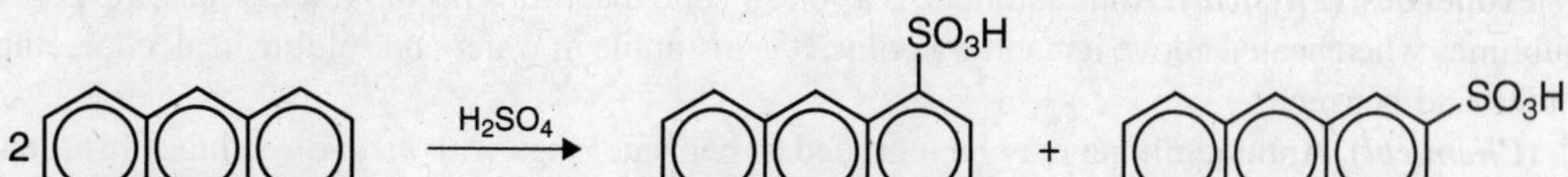

Anthracene 1-Anthracenesulfonic acid 2-Anthracenesulfonic acid

(5) **Reduction.** Anthracene undergoes reduction with sodium and ethyl alcohol to form 9,10-dihydroanthracene.

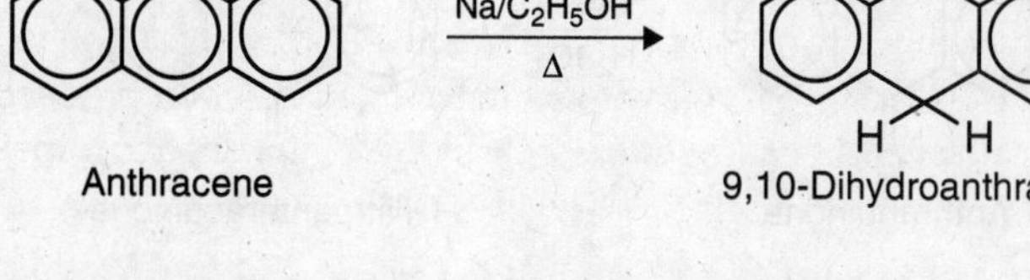

Anthracene 9,10-Dihydroanthracene

(6) **Oxidation.** Anthracene undergoes oxidation with sodium dichromate and sulfuric acid to form 9,10-anthraquinone. Other oxidizing agents like nitric acid and air in the presence of V_2O_5 also lead to the formation of 9,10-anthraquinone.

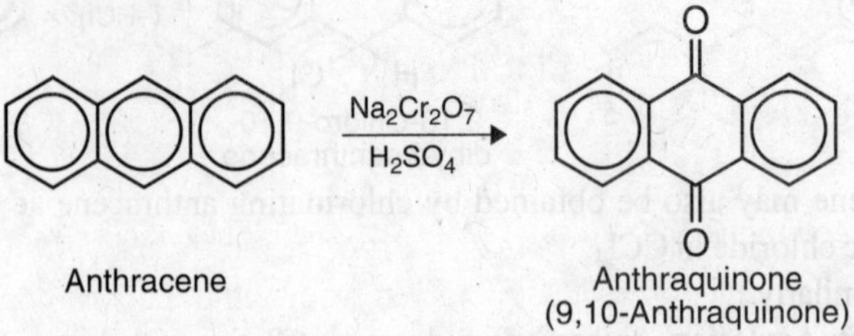

Use. Anthracene is used in the manufacture of anthraquinone.

ANTHRAQUINONE

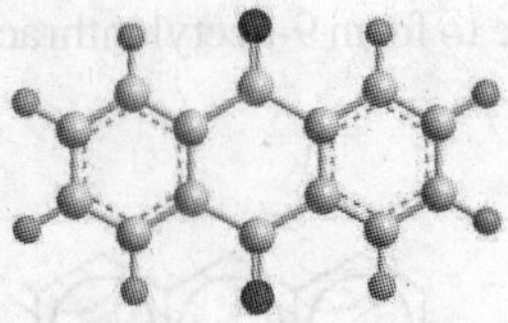

It is 9,10-anthraquinone and is referred to simply as anthraquinone, being the most important member of the class.

Manufacture. (1) By catalytic oxidation of anthracene derived from coal-tar.

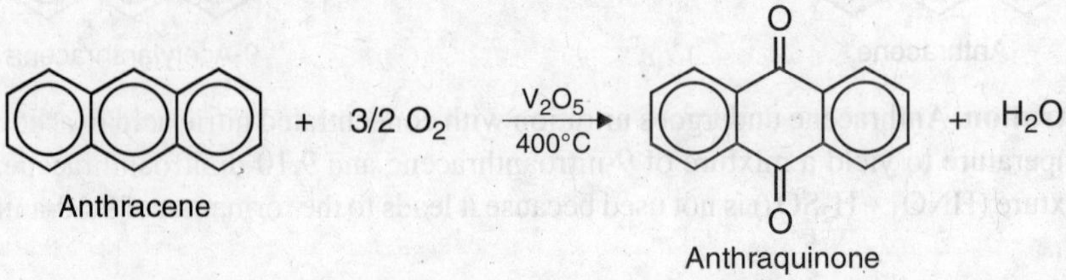

(2) By Friedel-Crafts reaction between phthalic anhydride (from oxidation of *o*-xylene) with benzene, and dehydration of the product with concentrated sulfuric acid.

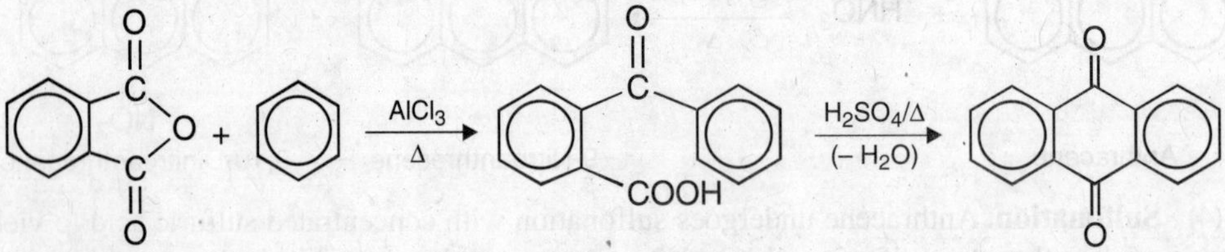

Properties. (*Physical*). Anthraquinone is a yellow solid that forms needle-like crystals, mp 286°C. It sublimes when heated above its melting point. It is insoluble in water and soluble in alcohol, ether, acetone and benzene.

(*Chemical*). Anthraquinone may be regarded as benzene rings with *meta* directing substituents (–CO–) in *ortho* positions. As a result, all the four positions in each ring are deactivated. Anthraquinone, therefore, gives electrophilic substitution reactions only under drastic conditions.

(1) **Nitration.** When nitrated with concentrated nitric acid and sulfuric acid, anthraquinone yields 1-nitroanthraquinone.

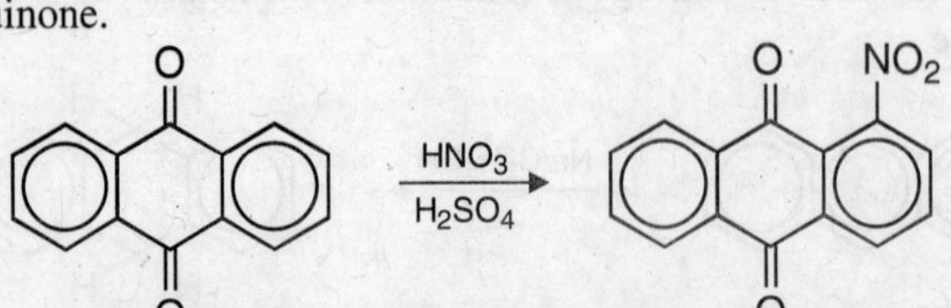

(2) **Sulfonation.** When sulfonated with fuming sulfuric acid at 160°C, it forms 2-anthraquinonesulfonic acid. This is a useful intermediate in the synthesis of alizarin.

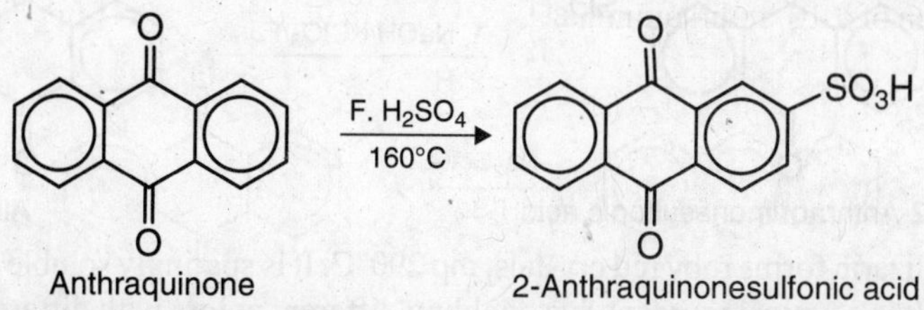

(3) **Reduction.** The reduction product obtained from anthraquinone depends on the nature of the reducing agent used.

(a) When reduced with tin and hydrochloric acid in acetic acid, anthraquinone yields *anthrone*.

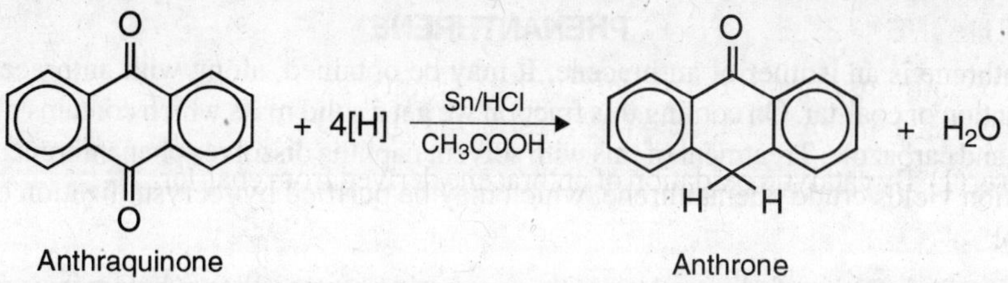

(*b*) When reduced with sodium hydrosulfite ($Na_2S_2O_4$) and alkali, it forms 9,10-dihydroxyanthracene.

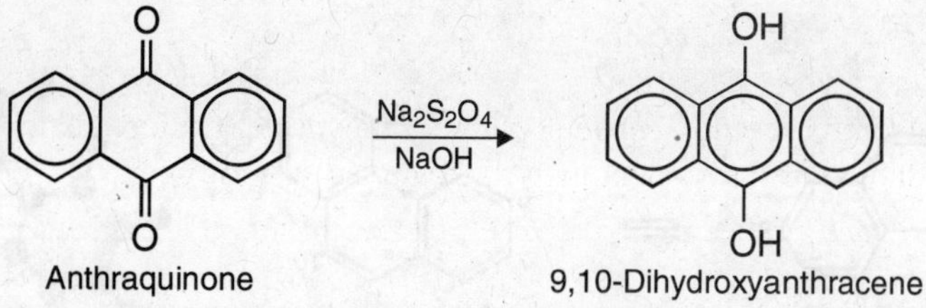

Uses. Anthraquinone is used in the manufacture of alizarin and several other dyes.

ALIZARIN, *1,2-Dihydroxyanthraquinone*

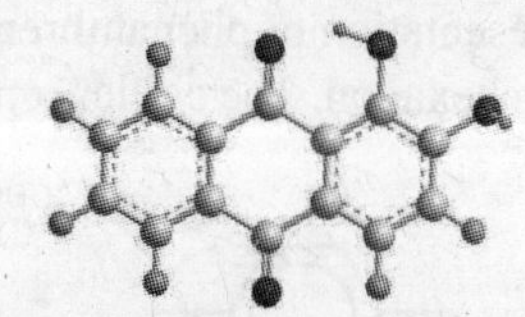

Alizarin was once used as a dye of splendid red color. It was first obtained from the roots of the madder plant (Fr. *alizarin* = madder). Later it was produced commercially as a synthetic product.

Synthetic Preparation. Anthraquinone is first prepared by Friedel-Crafts reaction between phthalic anhydride and benzene, followed by dehydration with concentrated sulfuric acid ; (2) Anthraquinone is then sulfonated with fuming sulfuric acid to form 2-anthraquinonesulfonic acid ; (3) This is finally fused with sodium hydroxide and potassium chlorate, and acidified ; the $-SO_3H$ is replaced by –OH and another –OH is introduced in 2-position by oxidation.

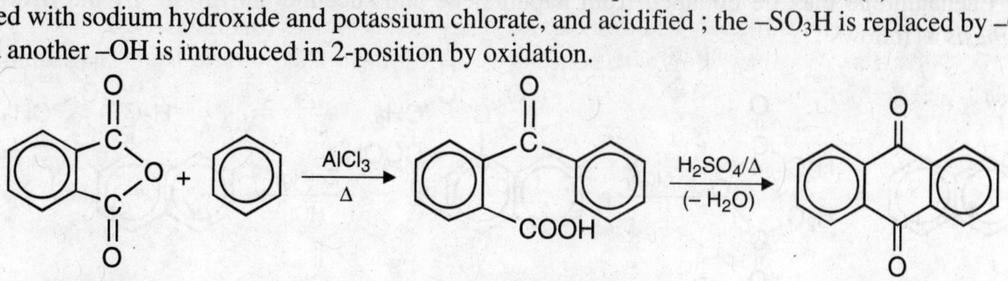

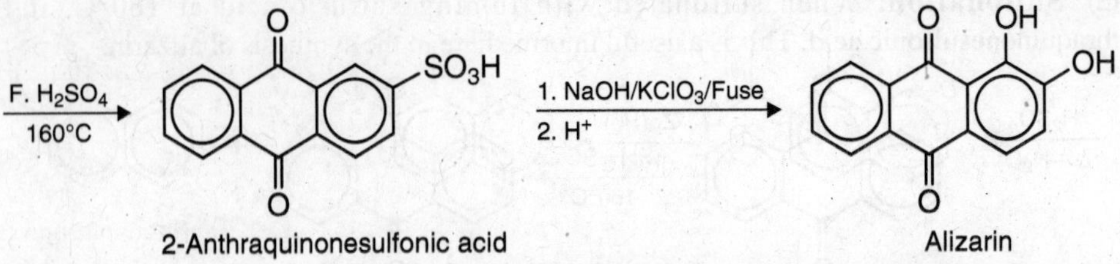

Properties. Alizarin forms ruby red crystals, mp 290°C. It is sparingly soluble in water, soluble in alcohol and ether. It is a typical mordant dye, yielding different colors with different mordants. Thus an aluminium mordant gives a bright-red color (Turkey red). The use of alizarin as a dye is now obsolete. However, it is the parent of many other dyes and pigments in current use.

PHENANTHRENE

Phenanthrene is an isomer of anthracene. It may be obtained, along with anthracene, from the green oil fraction of coal-tar. On cooling this fraction we get a solid mass which contains phenanthrene, anthracene, and carbazole. Treatment of this with solvent naphtha dissolves phenanthrene. Evaporation of this solution yields crude phenanthrene, which may be purified by recrystallization of the picrate from ethanol.

In naming derivatives of phenanthrene, the numbering system shown below is used. As before, the numbers selected to denote the position of a substituent on phenanthrene rings should be as small as possible.

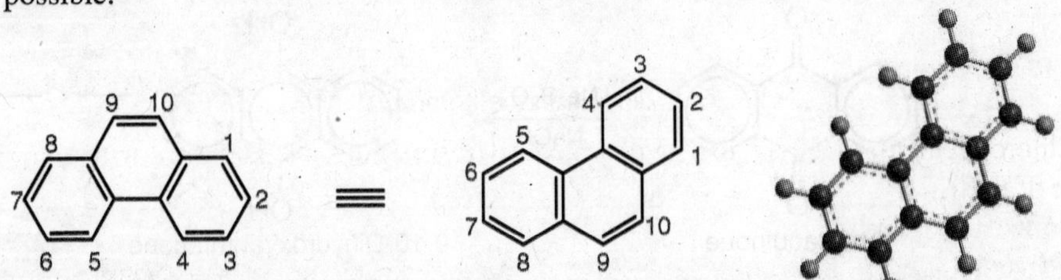

A common shorthand representation of phenanthrene is simply three hexagons fused in angular fashion with a circle inside each hexagon. The circles represent the π molecular orbital.

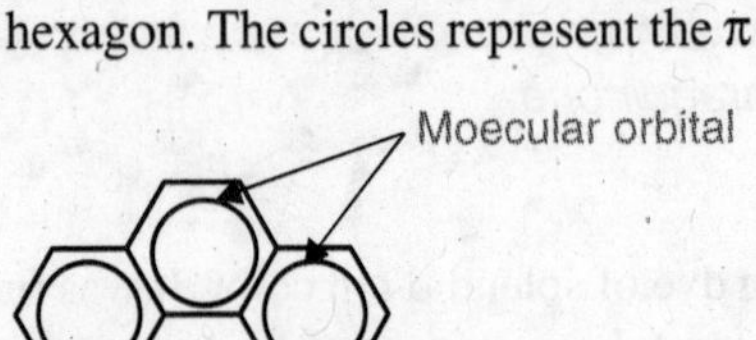

SYNTHESIS OF PHENANTHRENE

Phenanthrene may be obtained from naphthalene and succinic anhydride by the *Haworth synthesis* as follows :

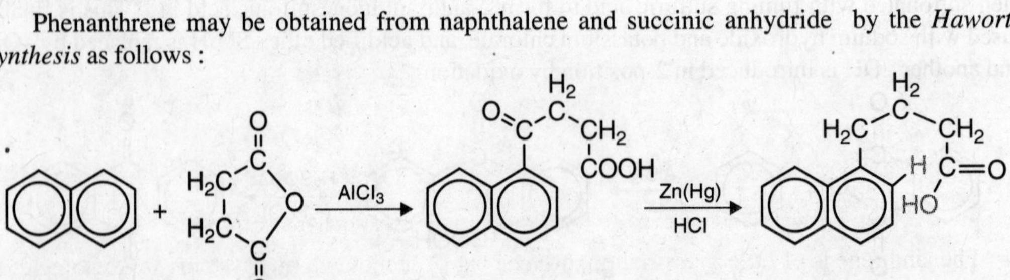

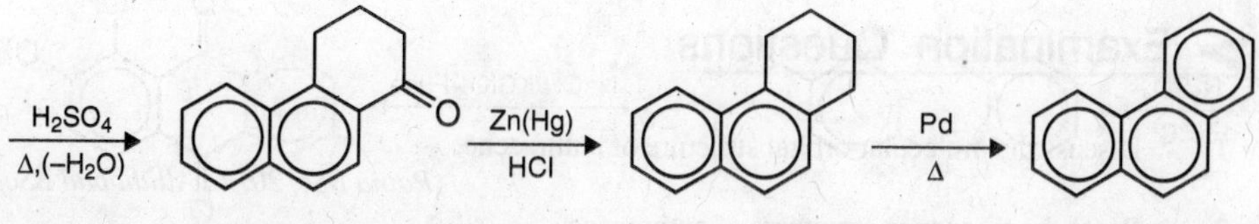

Phenanthrene

PROPERTIES OF PHENANTHRENE

(*Physical*). Phenanthrene is a colorless solid, mp 100°C. It is insoluble in water, but dissolves readily in ethanol, benzene, and ether. Phenanthrene gives blue fluoroscence in benzene solution.

(*Chemical*). Phenanthrene undergoes oxidation, addition, and electrophilic substitution reactions. As with anthracene, these reactions preferentially occur at the C-9 and C-10 positions.

Some of the important reactions of phenanthrene are described below :

(1) **Reaction with Halogens.** Phenanthrene reacts with chlorine in carbon tetrachloride at room temperature to give 9,10-dichloro-9,10-dihydrophenanthrene (an addition product). On heating, this loses a molecule of hydrogen chloride to yield 9-chlorophenanthrene (a substitution product).

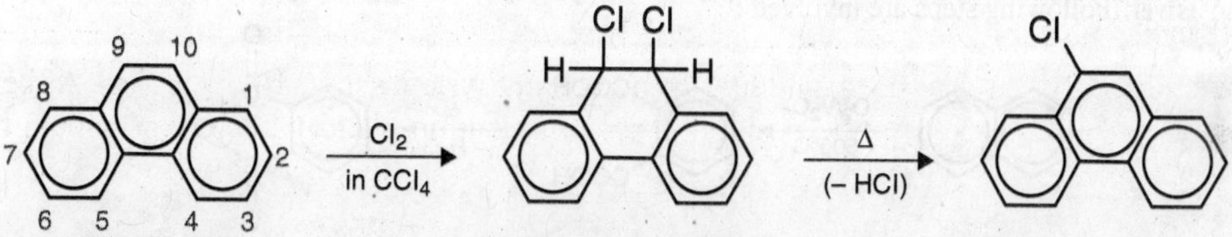

Phenanthrene | 9,10-Dichloro-9,10-dihydrophenanthrene | 9-Chlorophenanthrene

9-Chlorophenanthrene may also be obtained by treatment of phenanthrene with chlorine in the presence of $FeCl_3$.

Bromine reacts similarly.

(2) **Friedel-Crafts Acylation.** Phenanthrene undergoes acylation with acetyl chloride in the presence of aluminium chloride at 0°C to give 9-acetylphenanthrene.

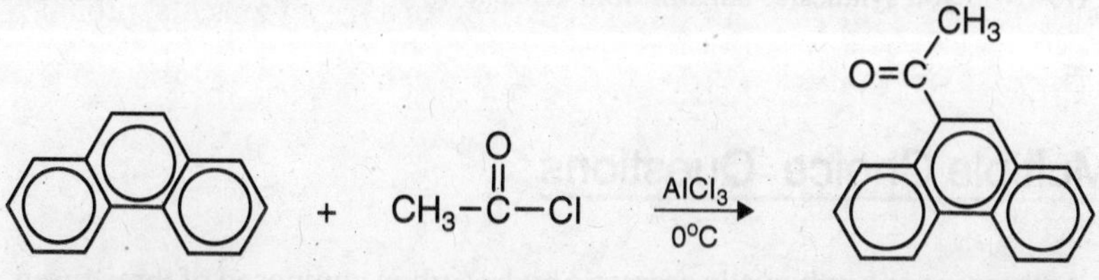

Phenanthrene | 9-Acetylphenanthrene

(3) **Nitration.** Phenanthrene undergoes nitration with concentrated nitric acid and sulfuric acid to yield 9-nitrophenanthrene.

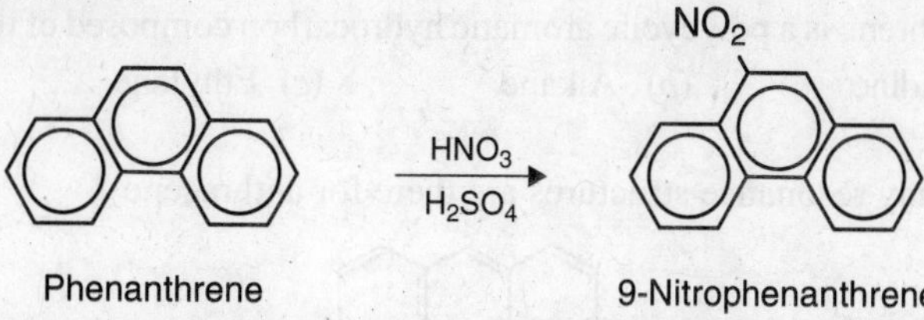

Phenanthrene | 9-Nitrophenanthrene

Phenanthrene is of little industrial importance but phenanthrene ring systems are found widely distributed in natural compounds such as carcinogenic (meaning cancer-producing) hydrocarbons, bile acids, sex hormones, morphine alkaloids, and cholesterol.

Examination Questions

1. Discuss the molecular orbital structure of anthracene.
(Patna BSc, 2014 ; Allahabad BSc, 2015)
2. Write the resonance structures of anthracene.
3. How is anthracene synthesized ? Describe its important reactions. *(Kalyani BSc, 2014)*
4. What happens when anthracene is treated with :
(*a*) Bromine in CCl_4 *(Burdwan BSc, 2015)*
(*b*) Nitric acid and acetic acid *(North Bengal BSc, 2014)*
(*c*) Sodium and ethyl alcohol
(*d*) Sodium dichromate and sulfuric acid
(*e*) Oxygen in the presence of V_2O_5 at 500°C *(Mysore BSc, 2014)*

Answer. The products are : (*a*) 9-Bromoanthracene; (*b*) 9-Nitroanthracene ; (*c*) 9,10-Dihydroanthracene ; (*d*) Anthraquinone ; and (*e*) Anthraquinone. For reaction equations, see text.

5. How will you synthesize anthracene from benzene ?
6. How will you synthesize anthraquinone from naphthalene ? *(Osmania BSc, 2015)*

Answer. Following steps are involved :

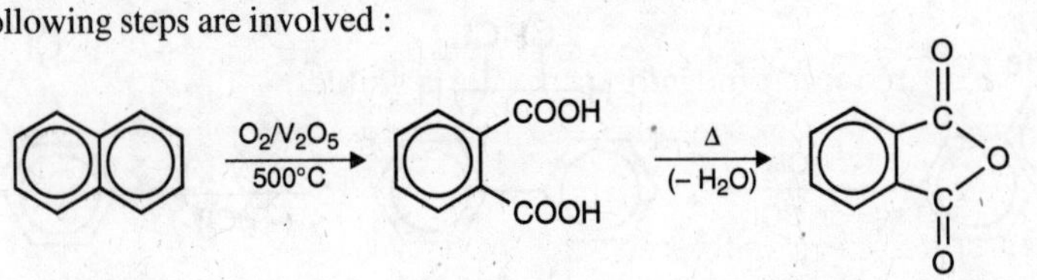

Naphthalene Phthalic acid Phthalic anhydride

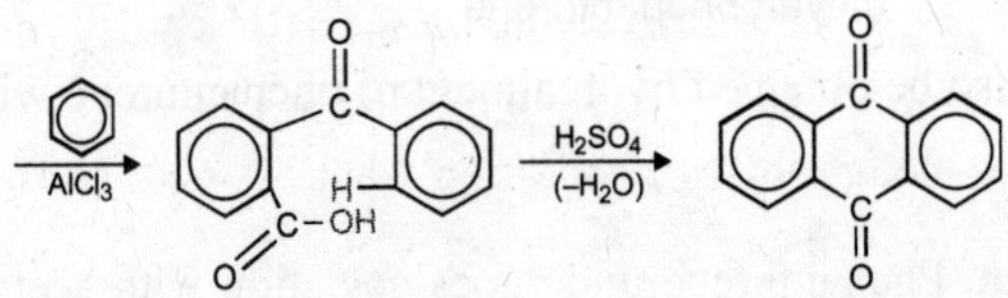

Anthraquinone

7. How will you synthesize alizarin from benzene ? *(Jabalpur BSc, 2014)*

Multiple Choice Questions

1. Anthracene is a polycyclic aromatic hydrocarbon composed of three fused_____ rings.
(*a*) Ethylene (*b*) Alkane (*c*) Pyridine (*d*) Benzene

Answer. (*d*)

2. Phenanthrene is a polycyclic aromatic hydrocarbon composed of three fused_____ rings.
(*a*) Pyridine (*b*) Alkane (*c*) Ethylene (*d*) Benzene

Answer. (*d*)

3. How many resonance structures are there for anthracene ?

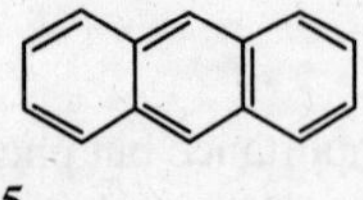

(*a*) 6 (*b*) 5 (*c*) 4 (*d*) 2

Answer. (*c*)

4. Which of the following compounds is phenanthrene ?

(*a*) 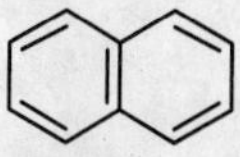 (*b*) 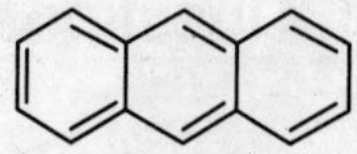

(*c*) 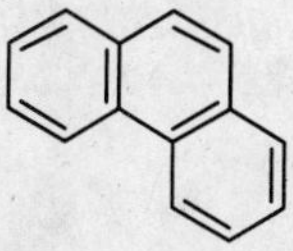 (*d*) 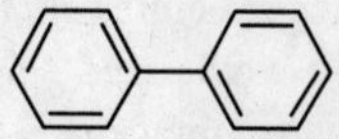

Answer. (*c*)

5. How many resonance structures are there for phenanthrene ?

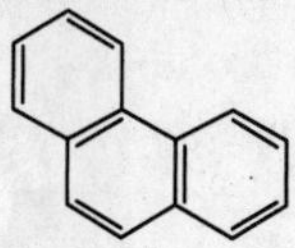

(*a*) 6 (*b*) 5 (*c*) 4 (*d*) 3

Answer. (*b*)

6. How many mono-bromoanthracenes are possible ?

(*a*) 1 (*b*) 2 (*c*) 3 (*d*) 4

Answer. (*c*)

7. All carbon atoms in anthracene are

(*a*) *sp* hybridized (*b*) sp^3 hybridized

(*c*) sp^2 hybridized (*d*) None of these

Answer. (*c*)

8. Anthracene undergoes electrophilic substitution reactions mainly at

(*a*) C-1 (*b*) C-2 (*c*) C-9 (*d*) C-1 and C-2

Answer. (*c*)

9. Anthracene undergoes oxidation with O_2/V_2O_5 at 500°C to give

(*a*) Benzoic acid (*b*) Anthraquinone

(*c*) Phthalic acid (*d*) Benzophenone

Answer. (*b*)

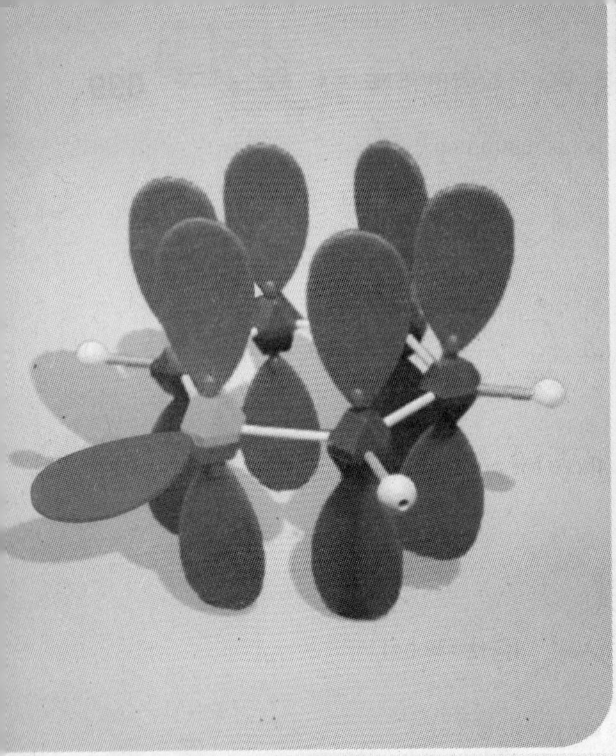

46

Heterocyclic Compounds

Heterocyclic compounds are cyclic compounds in which the ring atoms are of carbon and some other element. The atom of the other element (For example, N, S, or O) is called the **hetero atom**. Some of the important heterocyclic compounds are :

5-Membered Heterocyclic Compounds :

Pyrrole

Furan

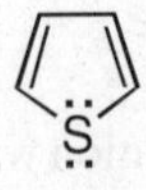

Thiophene

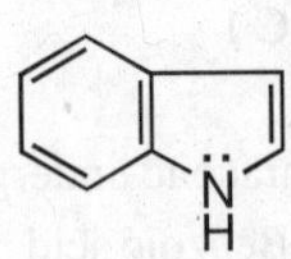

Indole

6-Membered Heterocyclic Compounds :

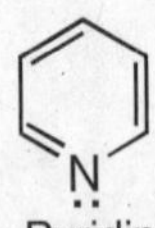

Pyridine

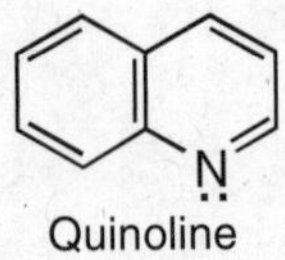

Quinoline

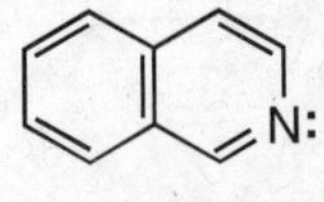

Isoquinoline

Most of the heterocyclic compounds are known by their common names. For naming their derivatives, the hetero atom is always numbered as 1. The carbon atoms next to the hetero atom are sometimes referred to as the α-carbon atoms and those further away as β- and γ-carbon atoms.

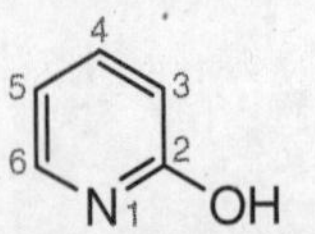

2-Hydroxypyridine

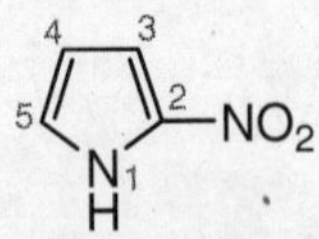

2-Nitropyrrole

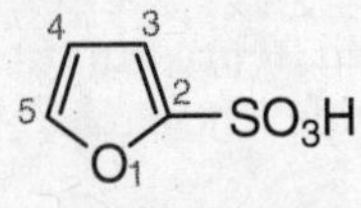

2-Furansulfonic acid

SINGLE-RING HETEROCYCLES

PYRROLE, *Azole*, C_4H_5N

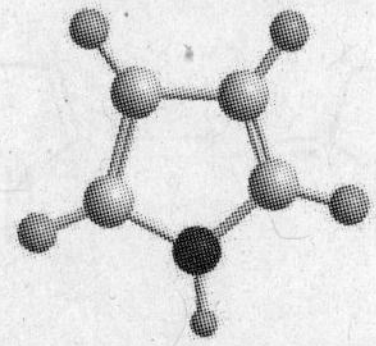

The pyrrole ring system is important as it is found in many natural products including hemoglobin, chlorophyll, and alkaloids.

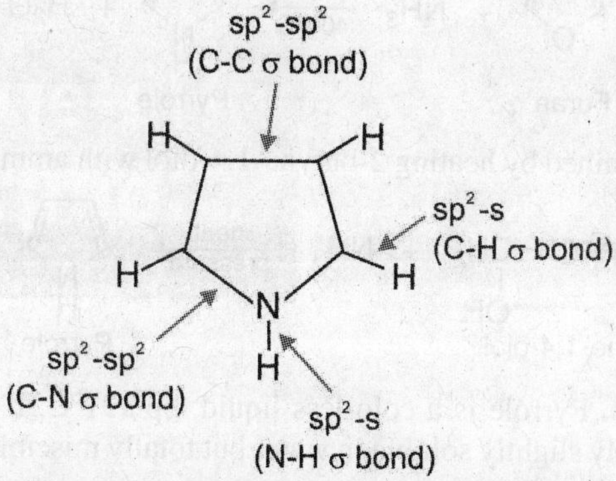

Fig. 46.1. Sigma (σ) bonds in pyrrole.

Structure. All ring atoms in pyrrole (*4* carbons + *1* nitrogen) are sp^2 hybridized. The sp^2 hybrid orbitals overlap with each other and with *s* atomic orbitals of the five hydrogens forming C–C, C–N, C–H, and N–H σ bonds (See Fig. 46.1). All these σ bonds lie in one plane.

Each ring atom in pyrrole also has an unhybridized *p* orbital and these are perpendicular to the plane of σ bonds. The *p* orbitals on carbons contain one electron each and the *p* orbital on nitrogen contains two electrons (the lone pair). The lateral overlap of these *p* orbitals produces a molecular orbital containing six electrons (Fig. 46.2). Pyrrole shows aromatic properties or aromaticity because the resulting π molecular orbital satisfies the Huckel's rule ($n = 1$ in $4n + 2$).

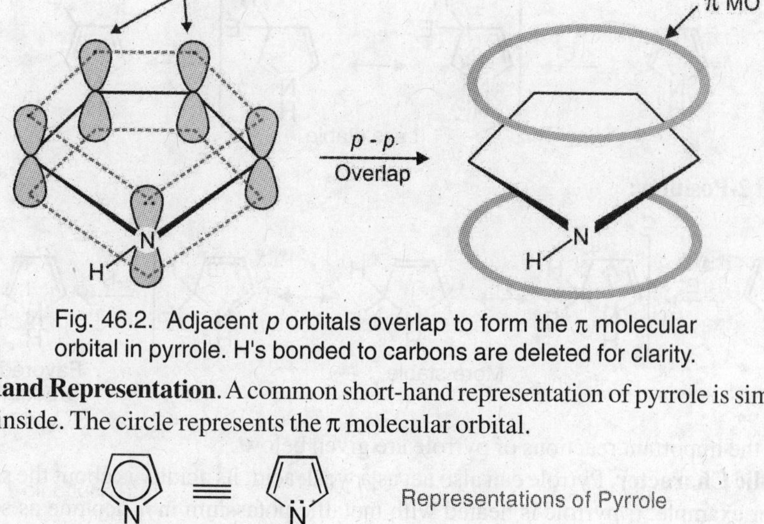

Fig. 46.2. Adjacent *p* orbitals overlap to form the π molecular orbital in pyrrole. H's bonded to carbons are deleted for clarity.

Short-Hand Representation. A common short-hand representation of pyrrole is simply pentagon with a circle inside. The circle represents the π molecular orbital.

Representations of Pyrrole

Resonance Structures. According to the resonance theory, pyrrole is considered to be hybrid of the following five resonance structures.

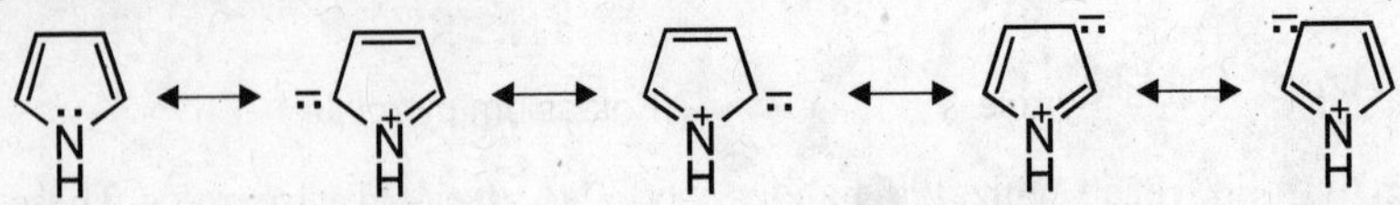

Preparation. Pyrrole is obtained :

(1) Industrially pyrrole is obtained by passing a mixture of furan and ammonia over alumina at 400°C.

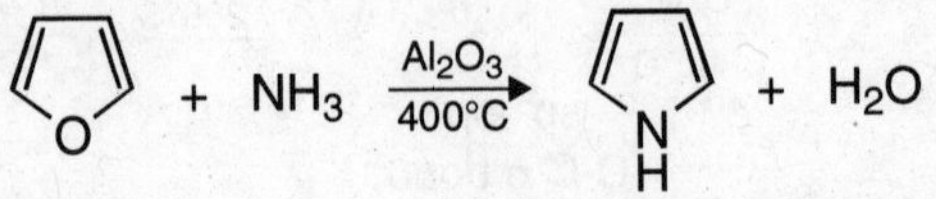

Furan Pyrrole

(2) Pyrrole can be obtained by heating 2-butyne-1,4-diol with ammonia under pressure.

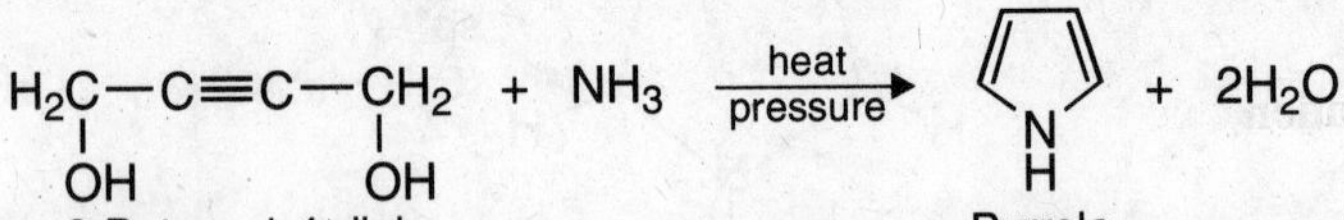

2-Butyne-1,4-diol Pyrrole

Properties. (*Physical*). Pyrrole is a colorless liquid, bp 131°C. It turns brown in the air and gradually resinifies. It is only slightly soluble in water but totally miscible with ether or ethanol.

(*Chemical*). Pyrrole is aromatic and more reactive than benzene. It gives electrophilic substitution reactions such as halogenation, nitration, and sulfonation. It also undergoes diazotization and Reimer-Tiemann reactions, while benzene does not.

Pyrrole undergoes electrophilic substitution at the 2-position. Attack of the electrophile at 3-position in pyrrole leads to an intermediate with only two resonance structures. Three resonance structures are possible for the intermediate produced by attack at 2-position. That is, the intermediate produced by attack at 2-position is more stable. This is the reason that electrophilic attack occurs at 2-position rather than at 3-position. E^+ in the following equations represents an electrophile.

Attack at 3-Position :

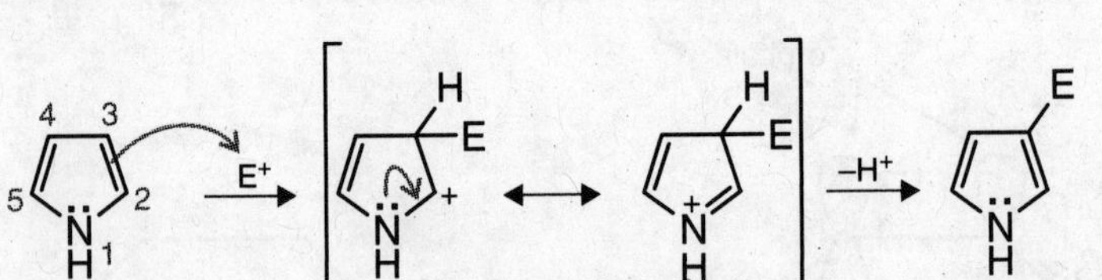

Less stable

Attack at 2-Position :

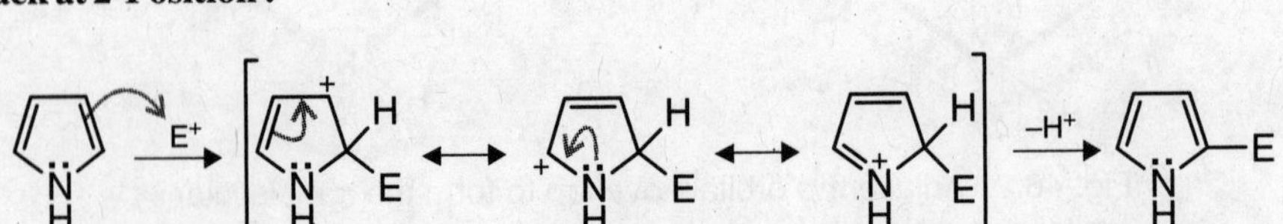

More stable Favored product

Some of the important reactions of pyrrole are given below :

(1) **Acidic Character**. Pyrrole can also act as a weak acid; its acidity is about the same as that of acetylene. For example, if pyrrole is heated with metallic potassium in *n*-heptane as solvent, stable potassium pyrrolide is formed

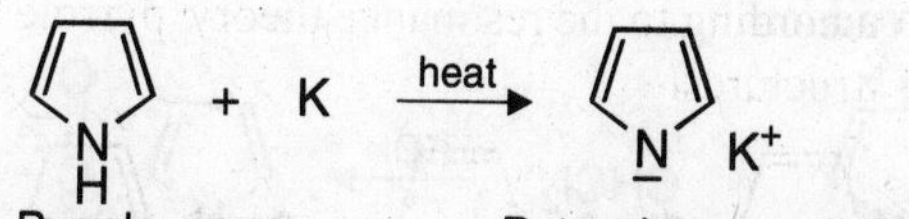

Pyrrole   Potassium pyrrolide

Potassium pyrrolide reacts with alkyl halides at 60°C to give *N*-alkylpyrroles. These on heating to 200°C readily rearrange to *C*-alkylpyrroles.

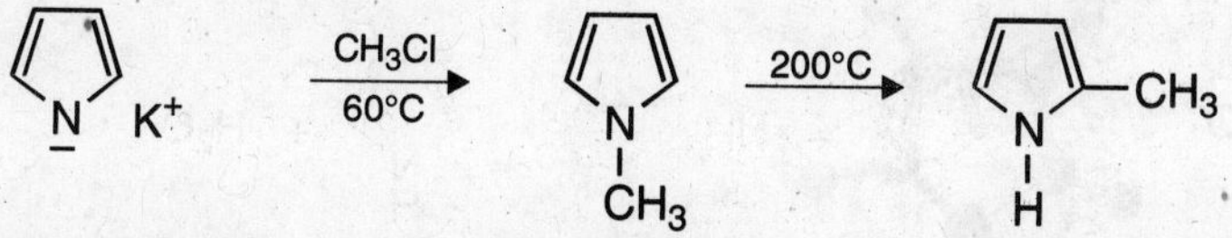

Potassium pyrrolide   1-Methylpyrrole   2-Methylpyrrole

(2) **Electrophilic Substitution Reactions.** Pyrrole undergoes electrophilic substitution reactions at 2-position.

(*a*) **Halogenation**

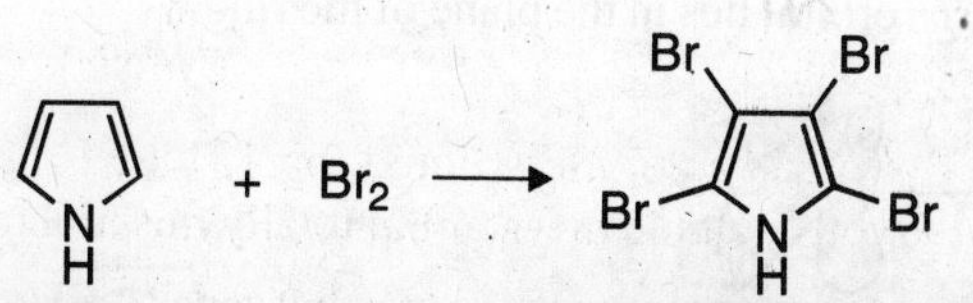

Pyrrole   2,3,4,5-Tetrabromopyrrole

(*b*) **Nitration**

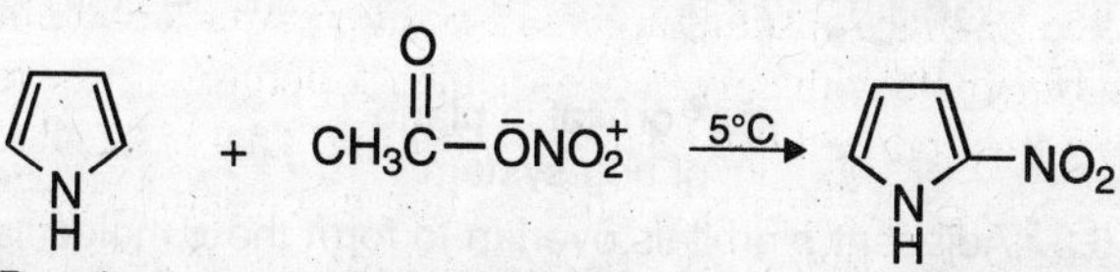

Pyrrole   Acetyl nitrate   2-Nitropyrrole

(*c*) **Sulfonation**

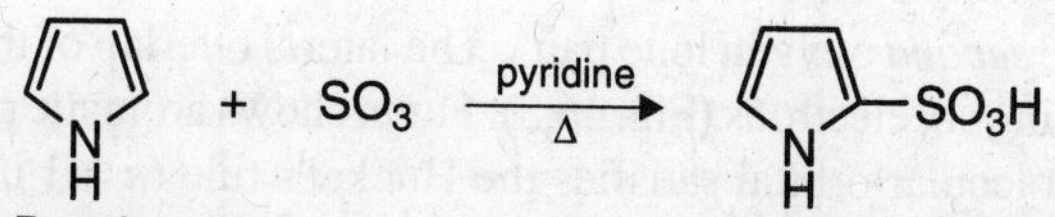

Pyrrole   2-Pyrrolesulfonic acid

(*d*) **Friedel-Crafts Acylation**

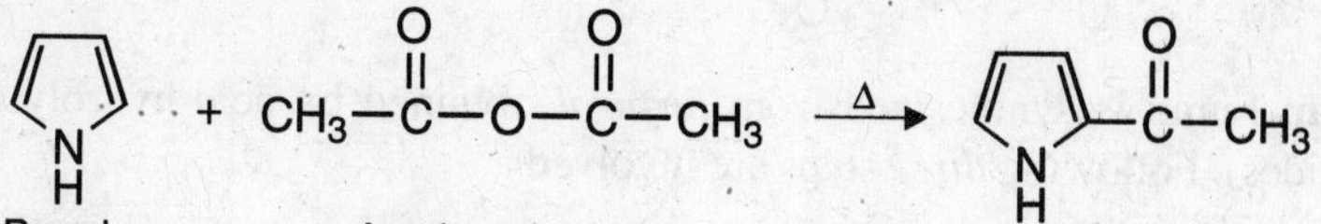

Pyrrole   Acetic anhydride   2-Acetylpyrrole

(*e*) **Diazotization**

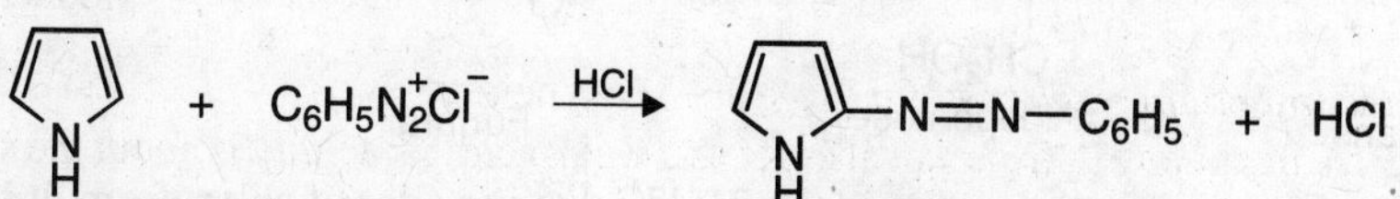

Pyrrole   2-Phenylazopyrrole

(*f*) **Reimer-Tiemann Reaction**

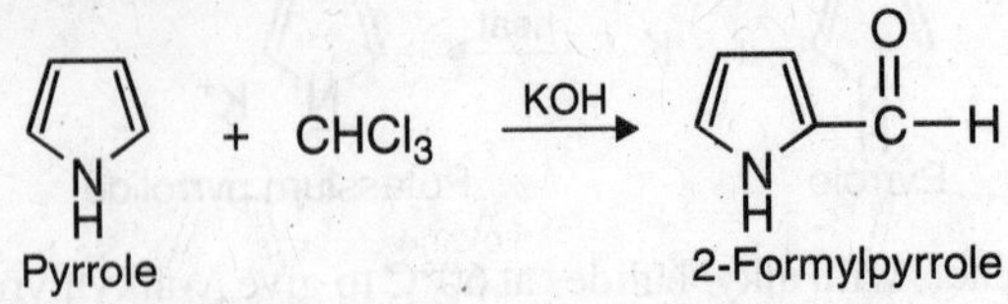

Uses. Pyrrole is used : (1) as a commercial solvent ; (2) for pharmaceuticals.

FURAN, C_4H_4O

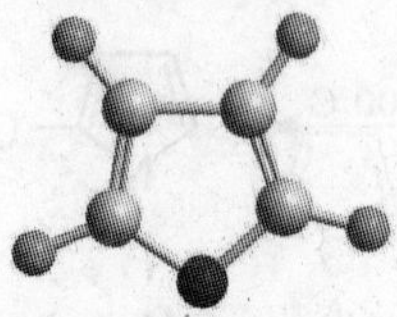

Structure. All ring atoms in furan (*4* Carbons + *1* Oxygen) are sp^2 hybridized. The sp^2 hybrid orbitals overlap with each other and with the s atomic orbitals of the four hydrogens forming C–C, C–O, and C–H σ bonds. All these σ bonds lie in one plane. The third sp^2 orbital on oxygen is completely filled. The unshared pair of electrons in this sp^2 orbital accounts for one of the oxygen lone pairs. This completely filled sp^2 orbital lies in the plane of the ring.

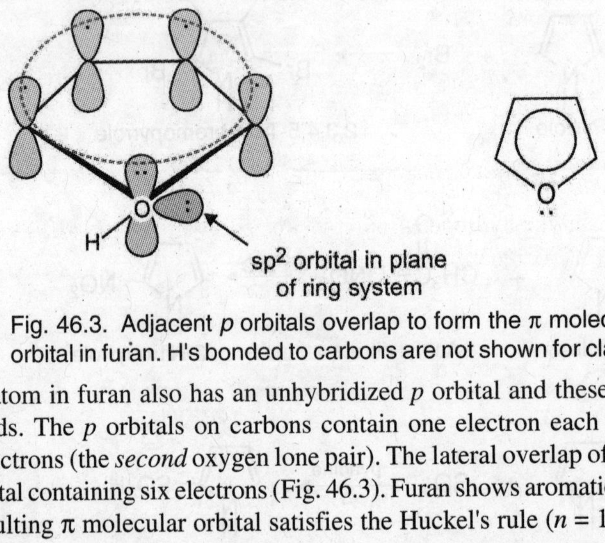

Fig. 46.3. Adjacent *p* orbitals overlap to form the π molecular orbital in furan. H's bonded to carbons are not shown for clarity.

Each ring atom in furan also has an unhybridized *p* orbital and these are perpendicular to the plane of σ bonds. The *p* orbitals on carbons contain one electron each and *p* orbital on oxygen contains two electrons (the *second* oxygen lone pair). The lateral overlap of these *p* orbitals produces a molecular orbital containing six electrons (Fig. 46.3). Furan shows aromatic properties or aromaticity because the resulting π molecular orbital satisfies the Huckel's rule ($n = 1$ in $4n + 2$). Furan can be represented as :

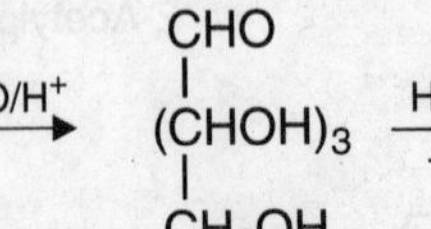

Representations of Furan

Preparation. Furan is synthesized from *furfural* obtained by acid-hydrolysis of pentosans (C_5-polysaccharides). Following *three* steps are involved :

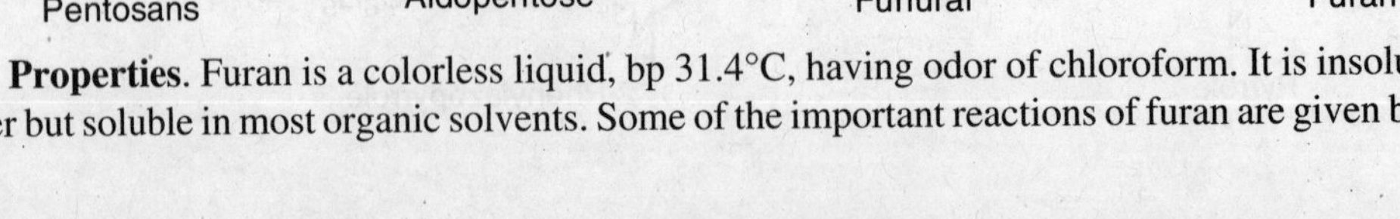

Properties. Furan is a colorless liquid, bp 31.4°C, having odor of chloroform. It is insoluble in ether but soluble in most organic solvents. Some of the important reactions of furan are given below :

(1) **Electrophilic Substitution Reactions**. Furan, like pyrrole, undergoes electrophilic substitution reactions at 2-position.

(*a*) **Halogenation**

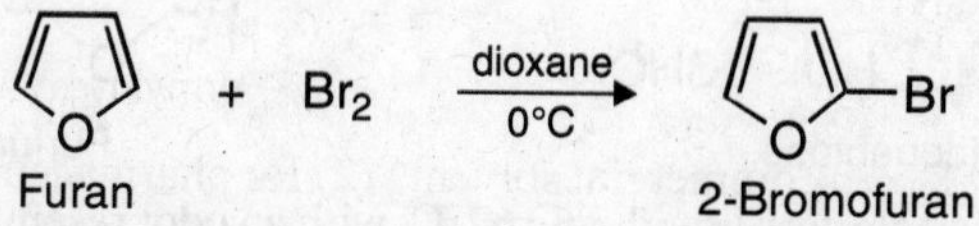

(*b*) **Nitration**

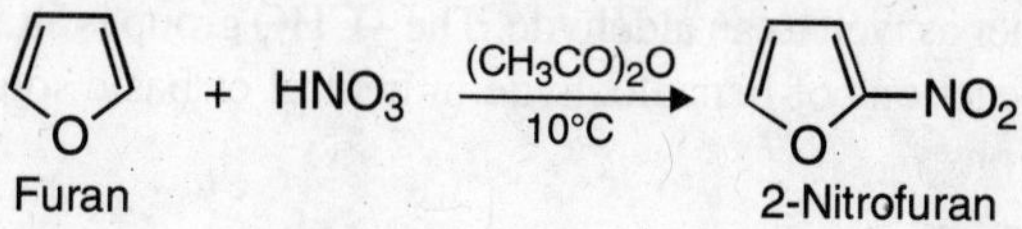

(*c*) **Sulfonation**

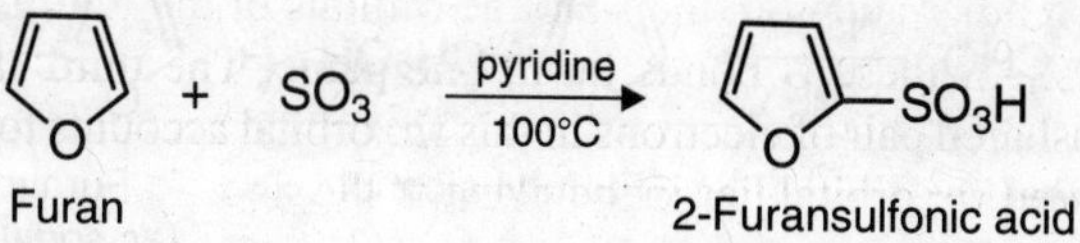

(*d*) **Friedel-Crafts Acylation**

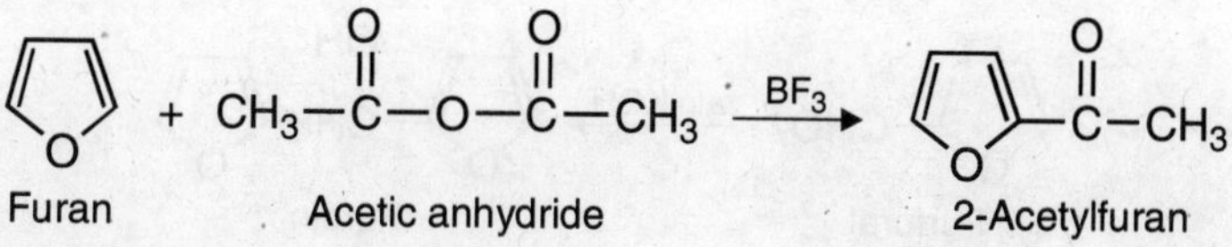

(2) **Reduction**. On catalytic hydrogenation, it yields tetrahydrofuran (THF) which is used as a solvent in place of ether in the Grignard reactions.

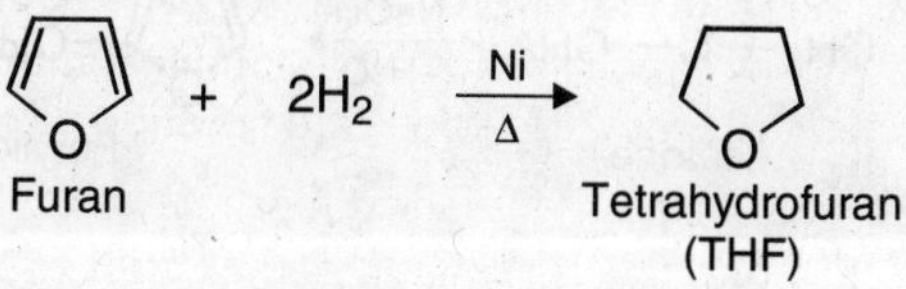

FURFURAL, $C_4H_3O.CHO$

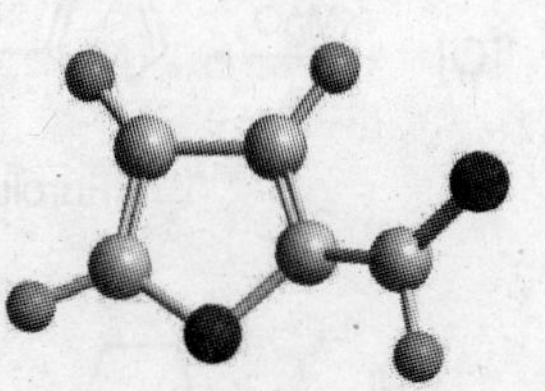

Preparation. Furfural is obtained from agricultural waste materials such as corncobs, oat hulls, rice hulls, and bagasse, which are rich in pentosans (C_5-polysaccharides). The pentosans on acid-hydrolysis give pentoses (mainly xylose) that dehydrate to yield furfural on distillation.

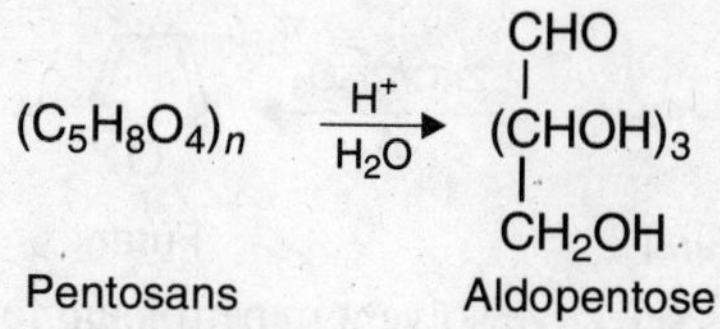

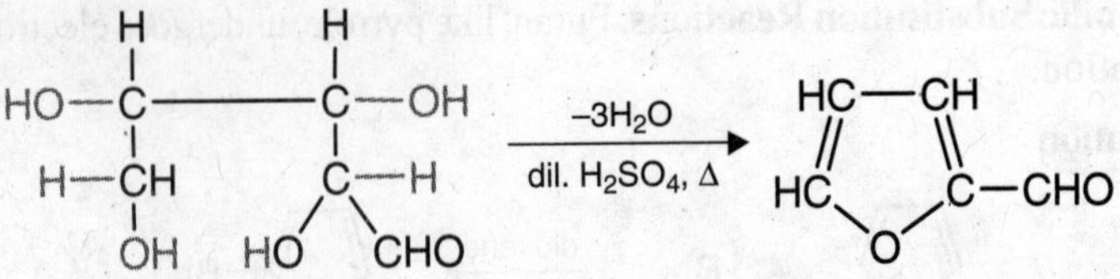

Aldopentose Furfural

Properties. Furfural is a colorless liquid, bp 162°C, with an odor resembling that of benzaldehyde. It is soluble in water and is miscible with most organic solvents.

Furfural is a cyclic ether as well as an aldehyde. The –CHO group is attached to an aromatic furan ring, and it gives all the reactions of benzaldehyde in neutral or basic solutions. Furfural gives the following important reactions :

(*a*) **Cannizzaro Reaction**

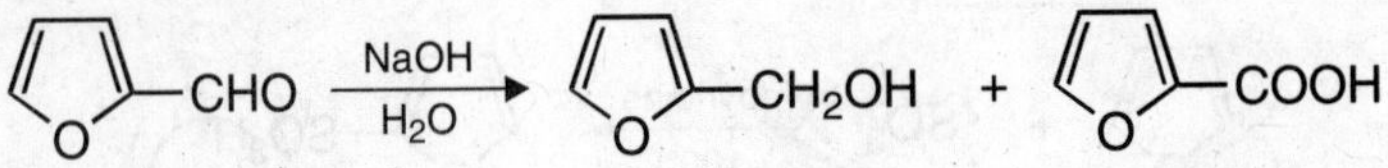

Furfural Furfuryl alcohol Furoic acid (as sodium salt)

(*b*) **Benzoin-type Condensation**

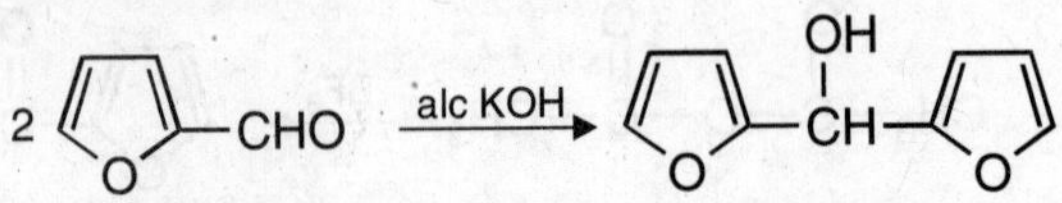

Furfural Furoin

(*c*) **Claisen-Schmidt Condensation**

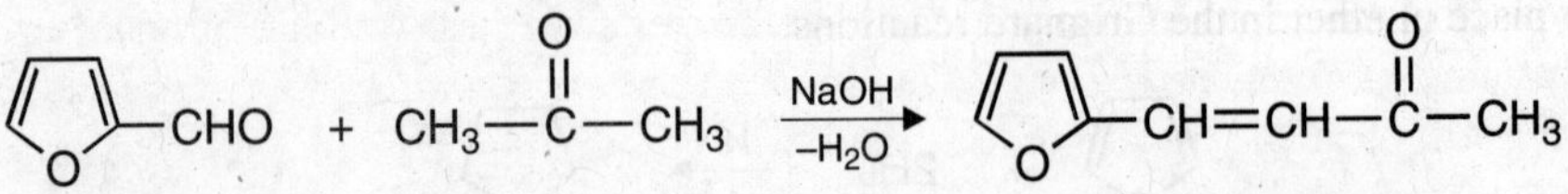

Furfural Acetone Furylidene acetone

(*d*) **Oxidation**

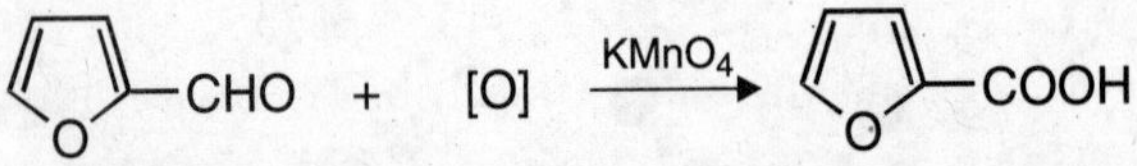

Furfural Furoic acid

(*e*) **Reduction**

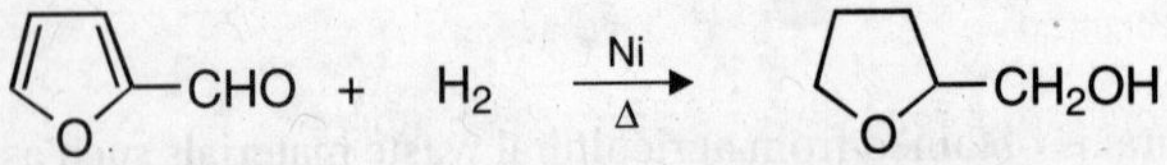

Furfural Tetrahydrofurfuryl alcohol

(*f*) **Decarboxylation**

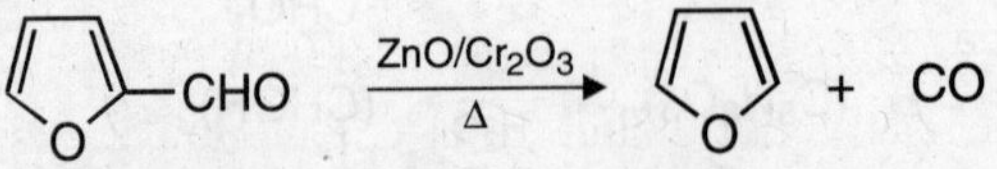

Furfural Furan

Uses. Furfural is used : (1) as selective solvent in petroleum refining ; (2) in the preparation of furan and furan derivatives.

THIOPHENE, C_4H_4S

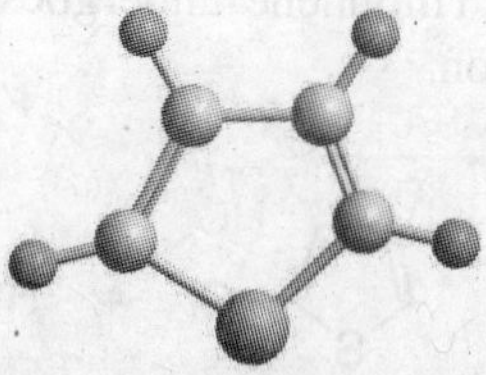

Structure. All ring atoms in thiophene (*4* carbons + *1* sulfur) are sp^2 hybridized. The sp^2 hybrid orbitals overlap with each other and with the *s* atomic orbitals of the four hydrogens forming C–C, C–S, and C–H σ bonds. All these σ bonds lie in one plane. The third sp^2 orbital on sulfur is completely filled. The unshared pair of electrons in this sp^2 orbital accounts for one of the sulfur lone pairs. This completely filled sp^2 orbital lies in the plane of the ring.

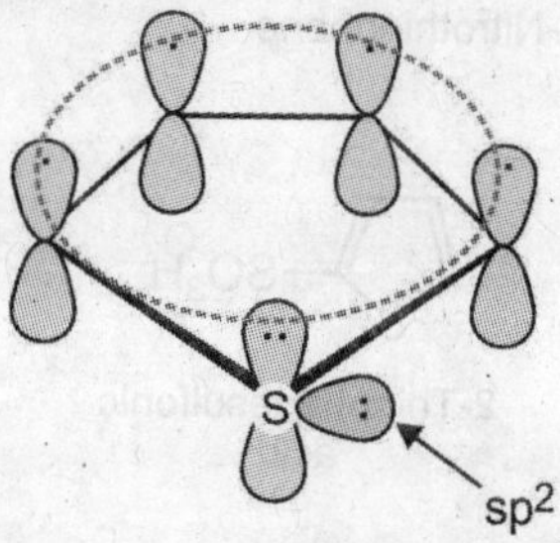

Fig. 46.4. Adjacent *p* orbitals overlap to form the π molecular orbital in thiophene. H's bonded to carbons are not shown for clarity.

Each ring atom in thiophene also has an unhybridized *p* orbital and these are perpendicular to the plane of σ bonds. The *p* orbitals on carbons contain one electron each and *p* orbital on sulfur contains two electrons (the *second* sulfur lone pair). The lateral overlap of these *p* orbitals produces a molecular orbital containing six electrons (Fig. 46.4). Thiophene shows aromatic properties or aromaticity because the resulting π molecular orbital satisfies the Huckel's rule ($n = 1$ in $4n + 2$). Thiophene can be represented as :

Representations of Thiophene

Preparation. (1) By heating sodium succinate with phosphorus trisulfide.

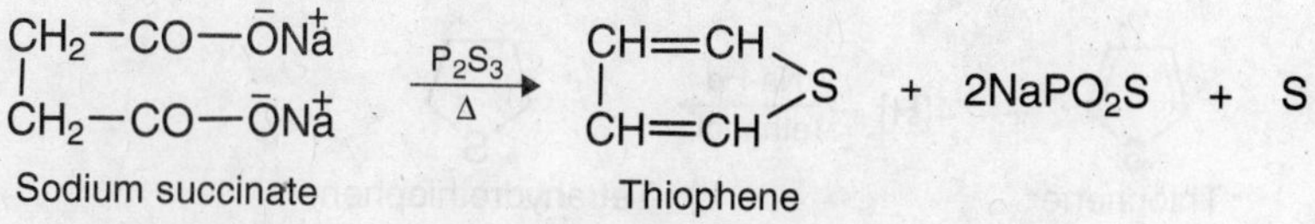

(2) By the high-temperature reaction of sulfur with butane.

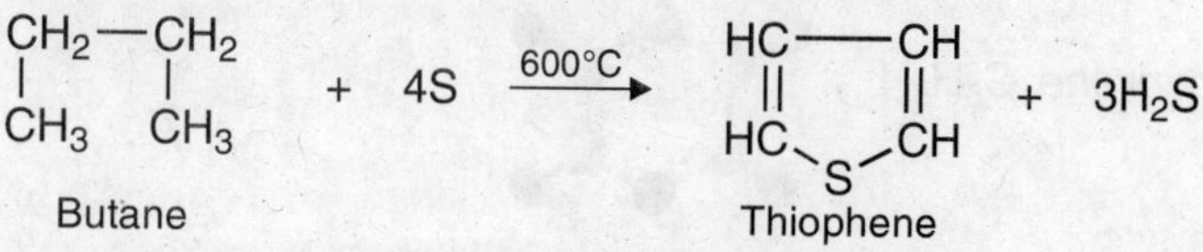

Properties. Thiophene is a colorless liquid, bp 84°C, having odor very similar to that of benzene. It is insoluble in water but freely soluble in ethanol, ether, and acetone.

Thiophene is much more reactive than benzene. Thus thiophene undergoes the electrophilic substitution reactions like benzene, under moderate conditions.

(1) **Electrophilic Substitution Reactions**. Thiophene undergoes chlorination, nitration, sulfonation, and Friedel-Crafts acylation at 2-position.

(*a*) **Chlorination**

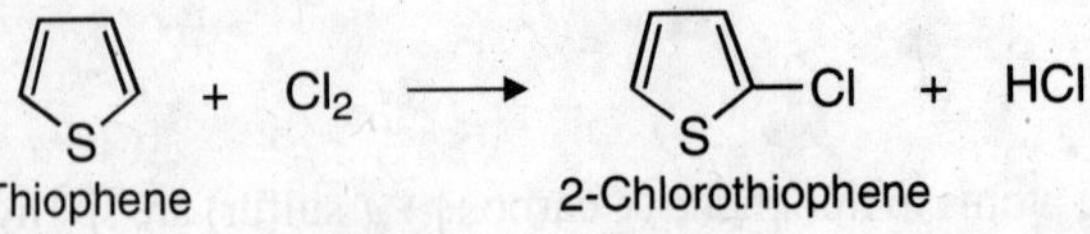

(*b*) **Nitration**

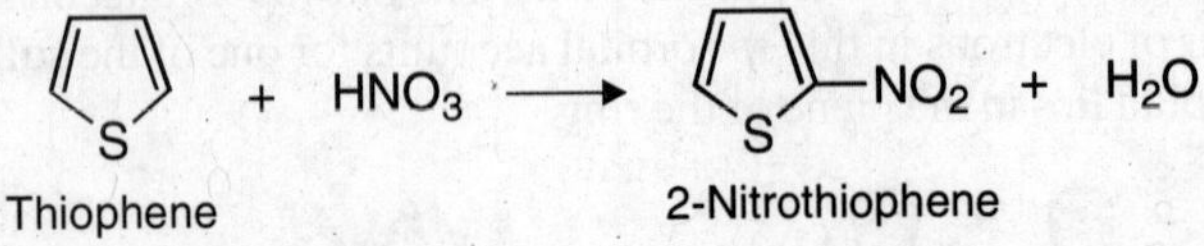

(*c*) **Sulfonation**

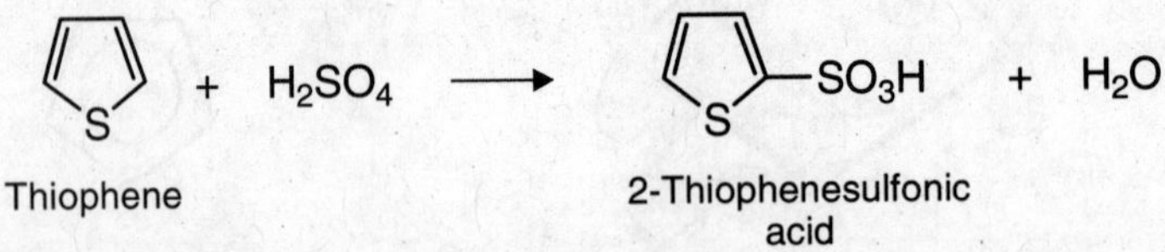

(*d*) **Friedel-Crafts Acylation**

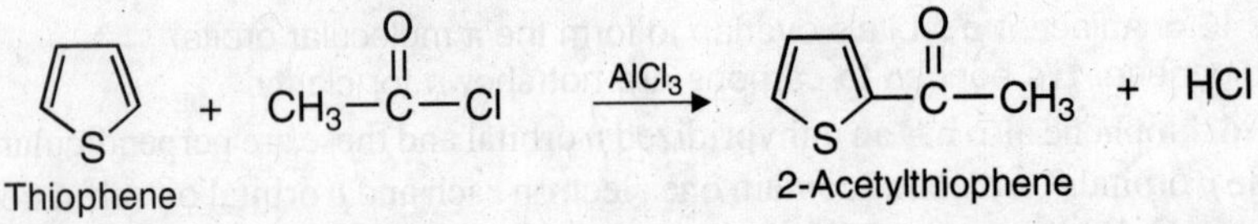

(2) **Desulfurization**. Catalytic reduction of thiophene with nickel catalyst results in the removal of sulfur to form *n*-butane.

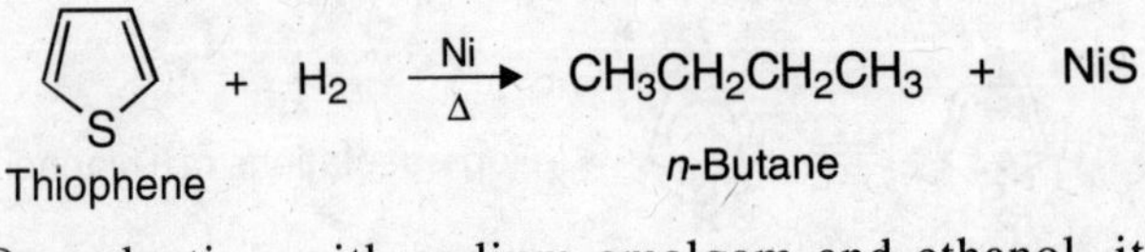

(3) **Reduction**. On reduction with sodium amalgam and ethanol, it is converted to tetrahydrothiophene.

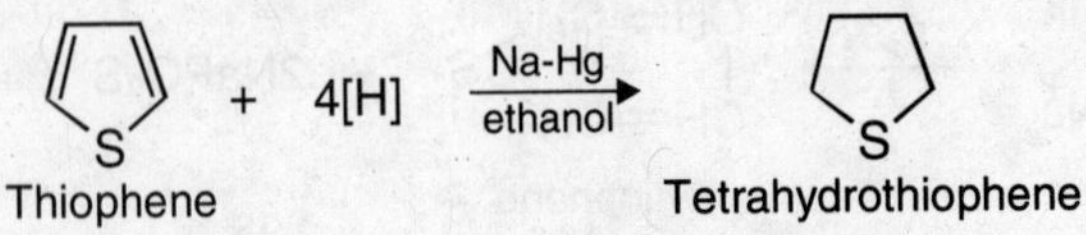

PYRIDINE, *Azabenzene*, C_5H_5N

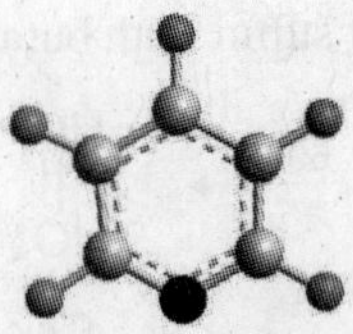

Pyridine occurs in coal tar (0.1%) and in the distillate from bones (bone oil) and has been produced industrially from these sources.

Structure. All ring atoms in pyridine (*5* carbons + *1* nitrogen) are sp^2 hybridized. Two of the sp^2 orbitals on each atom overlap with each other to form the C–C and C–N σ bonds. The third sp^2 orbital

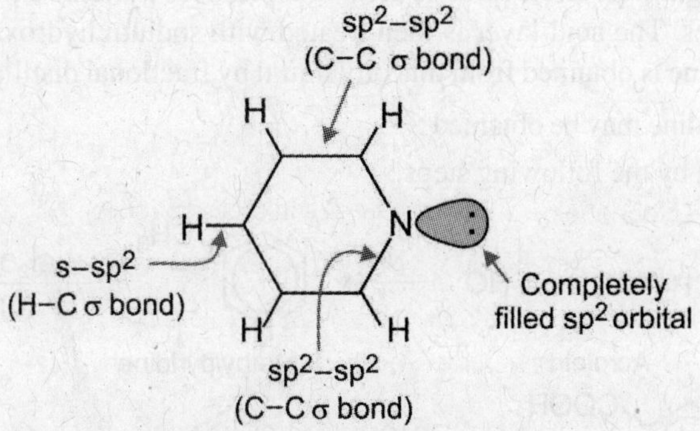

Fig. 46.5 Sigma (σ) bonds in pyridine.

on each carbon overlaps with an *s* orbital from hydrogen to form the C–H σ bonds. The third sp^2 orbital on nitrogen is completely filled. The unshared pair of electrons in this orbital is referred to as the nitrogen lone pair. All bonds in pyridine lie in one plane and all bond angles are 120°.

Each ring atom in pyridine also has an unhybridized *p* orbital containing an electron. These *p* orbitals are perpendicular to the plane containing (σ) bonds. The lateral overlap of the *p* orbitals produces a delocalized π molecular orbital containing six electrons. One half of this π MO lies above and the other half lies below the plane of the σ bonds. Pyridine shows aromatic properties or aromaticity because the resulting π molecular orbital satisfies the Huckel's rule ($n = 1$ in $4n + 2$).

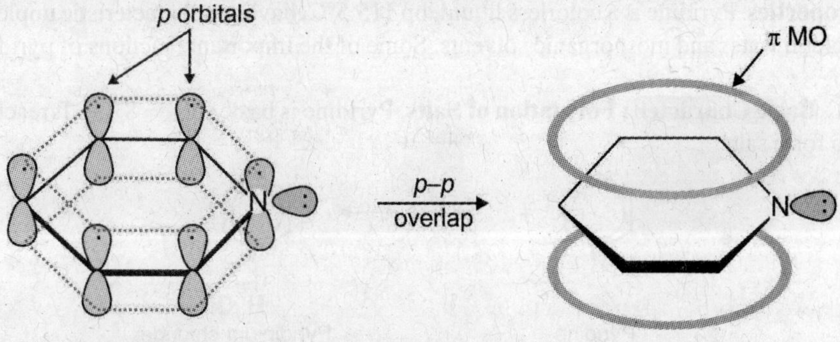

Fig. 46.6. Adjacent *p* orbitals overlap to form the π molecular orbital in pyridine. H's bonded to carbon atoms are not shown for clarity.

Short-Hand Representation. A common short-hand representation of pyridine is simply a hexagon with a circle inside. The circle represents the delocalized molecular orbital.

Representations of Pyridine

Resonance Structures. According to the resonance theory, pyridine is considered to be hybrid of the following resonance structures.

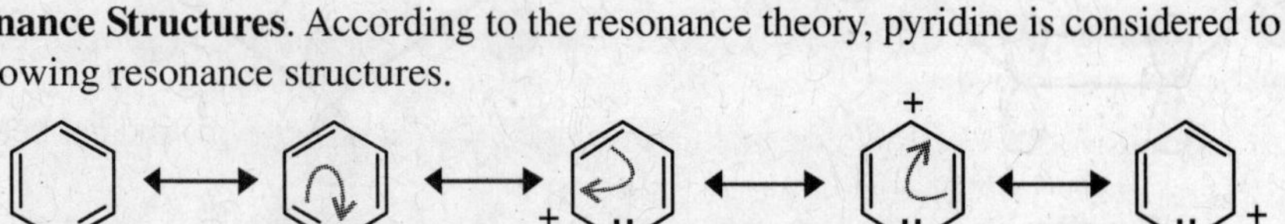

Isolation from Coal-tar. Pyridine occurs in the *light oil* fraction of coal-tar (bp upto 170°C). It can be isolated by extracting it with dilute sulfuric acid. This removes pyridine and other bases in the acid-layer as soluble sulfates. The acid-layer is then treated with sodium hydroxide when a dark brown liquid separates. Pyridine is obtained from this oily liquid by fractional distillation.

Preparation. Pyridine may be obtained :

(1) From acrolein by the following steps :

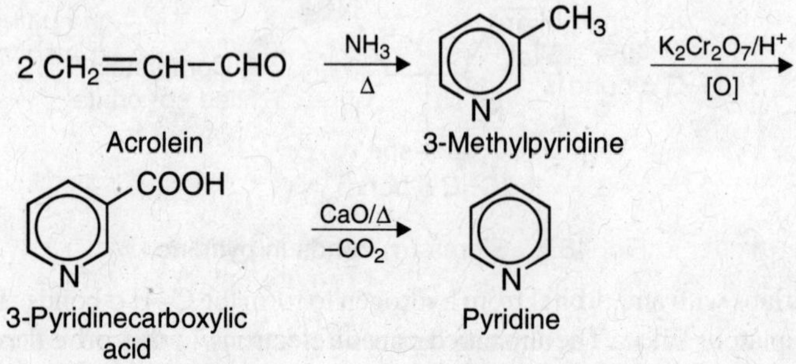

(2) By heating a mixture of acetylene, ammonia, and formaldehyde dimethylacetal in the presence of alumina at 500°C. (*Industrial Method*)

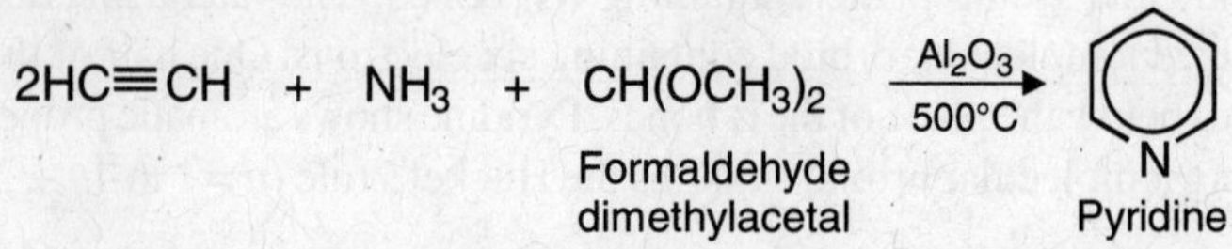

Properties. Pyridine is a colorless liquid, bp 115.5°C, having a characteristic unpleasant odor. It is soluble in water and most organic solvents. Some of the important reactions of pyridine are given below :

(1) **Basic Character ; Formation of Salts**. Pyridine is basic ($pK_b = 8.75$). It reacts with strong acids to form salts.

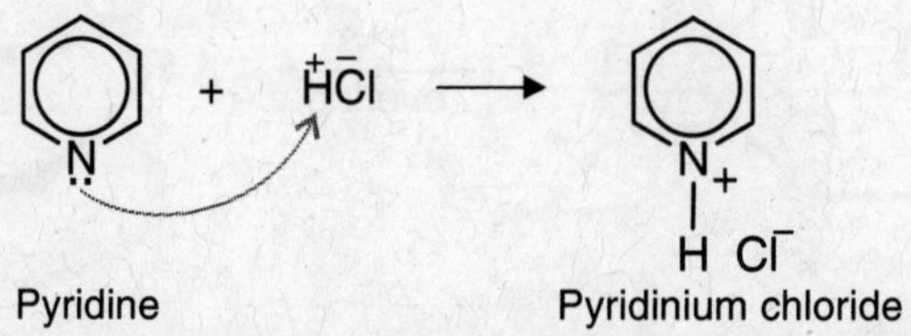

EXPLANATION. The reason for the basic character of pyridine is that the nitrogen lone pair electrons are in sp^2 hybrid orbital and are not involved in the formation of the delocalized π molecular orbital. It is readily available for the formation of a new N–H bond with proton.

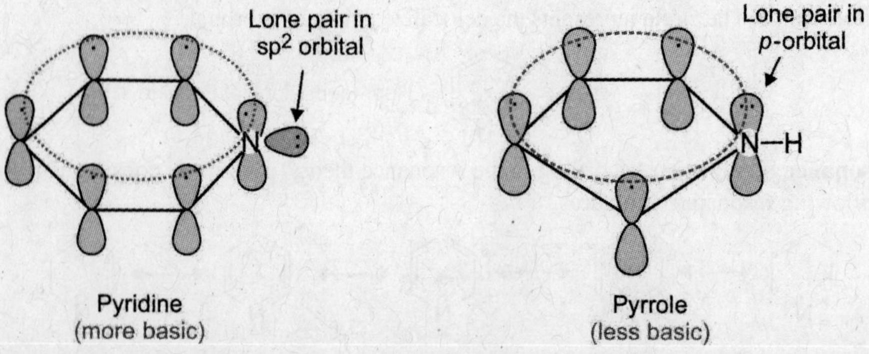

Pyridine is more basic than pyrrole. This is because the nitrogen lone pair electrons in pyrrole are in *p* orbital and form part of the delocalized π molecular orbital. They are not available for the formation of a new N–H bond with proton.

Pyridine is less basic than aliphatic amines. Pyridine is less basic ($pK_b = 8.75$) than aliphatic amines ($pK_b = 4$). This is due to the difference in the nature of hybrid orbitals containing the nitrogen lone pair electrons.

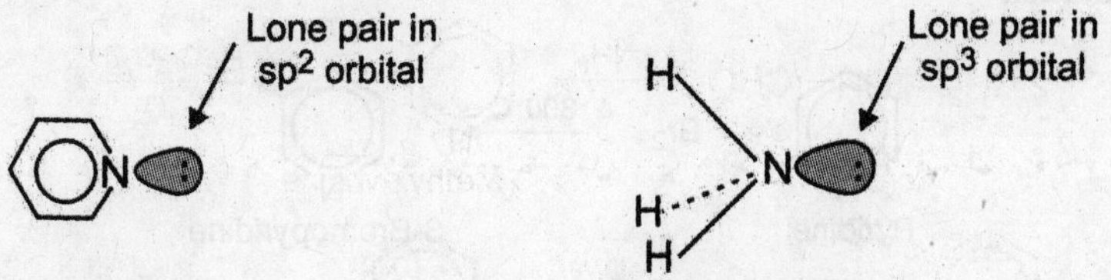

Pyridine (less basic) Aliphatic amines (more basic)

In pyridine it is in an sp^2 orbital ; in aliphatic amines it is in an sp^3 orbital. We know that sp^2 orbitals are smaller (due to more *s* character) than sp^3 orbitals. This means that the lone pair of electrons on nitrogen in pyridine is more closely associated with the nitrogen nucleus. It is, therefore, less available for the formation of bond with proton and consequently the relative basicity is reduced.

(2) **Electrophilic Substitution Reactions** Pyridine undergoes electrophilic substitution reactions at 3-position only under vigorous conditions.

EXPLANATION. Attack of the electrophile at 2-position (*or* 4-position) in pyridine leads to an intermediate with only two important resonance contributing structures. Three resonance structures are possible for the intermediate produced by attack at 3-position. That is, the intermediate produced by the attack at 3-position is more stable. This is the reason that electrophilic attack occurs at 3-position. E^+ in the following equations represents an electrophile.

Attack at 2-Position :

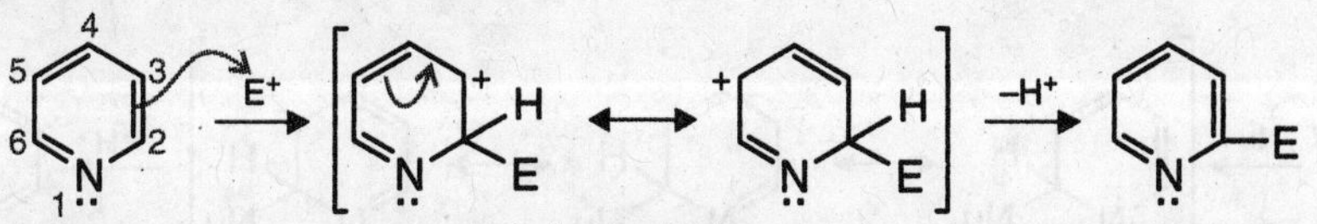

Less stable

Attack at 3-Position :

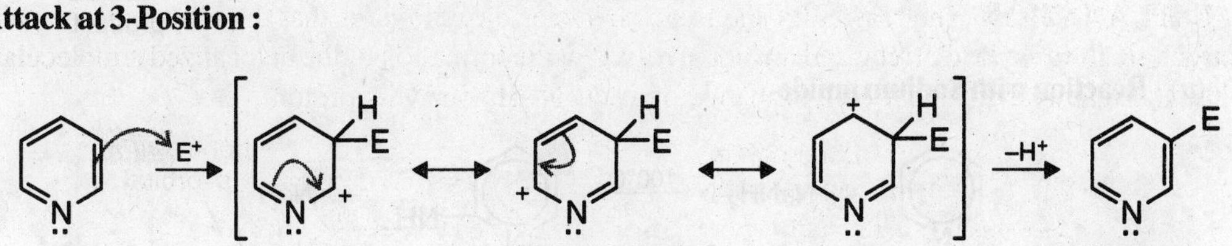

More stable Favored product

(*a*) **Nitration**

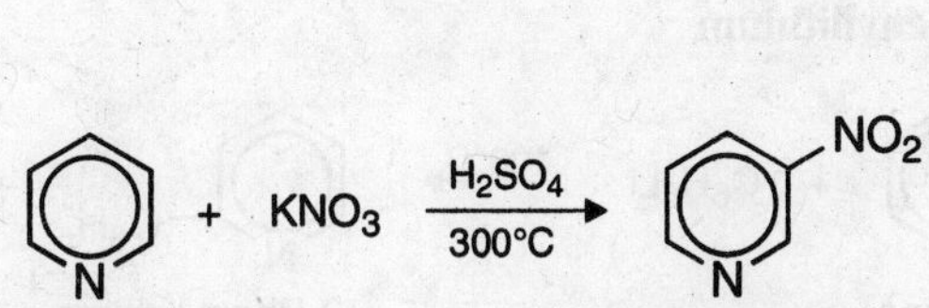

Pyridine 3-Nitropyridine

(*b*) **Sulfonation**

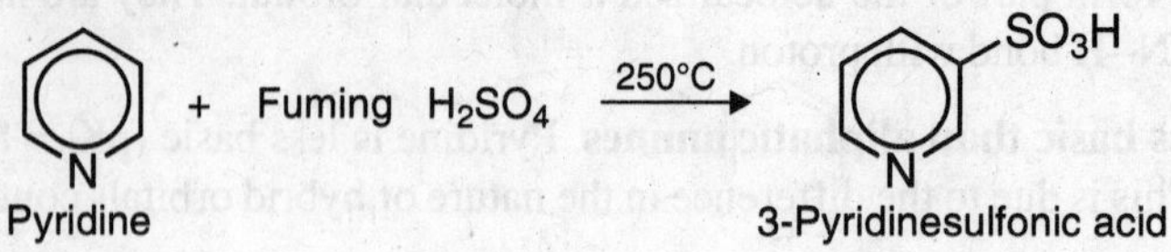

(*c*) **Bromination**

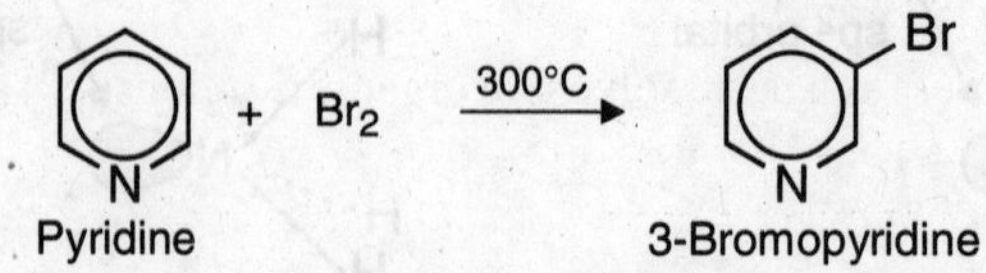

(3) **Nucleophilic Substitution Reactions**. Pyridine undergoes nucleophilic substitution reactions at 2-position.

EXPLANATION. Attack of the nucleophile at 3-position in pyridine leads to an intermediate with three resonance contributing structures. Attack of the nucleophile at 2-position (or 4-position) also gives an intermediate with three resonance structures. (Attack at 4-position resembles attack at 2-position). This position is attacked only when 2-position is blocked). $Nu^{:-}$ in the following equations represents a nucleophile.

Attack at 3-Position :

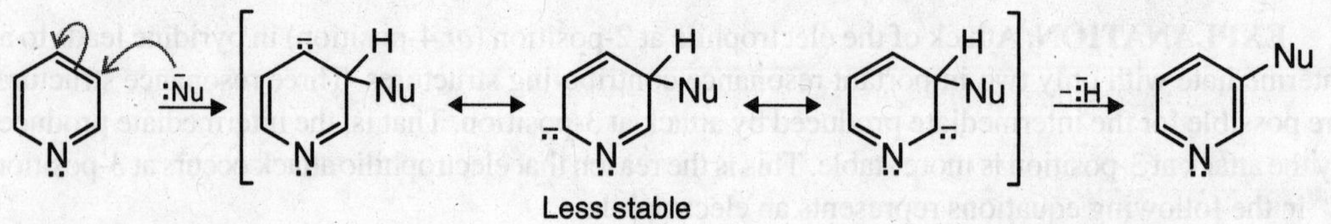

Attack at 2-Position :

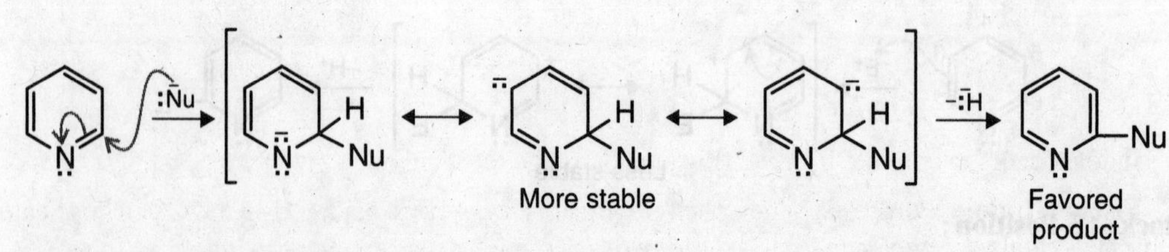

(*a*) **Reaction with Sodium amide**

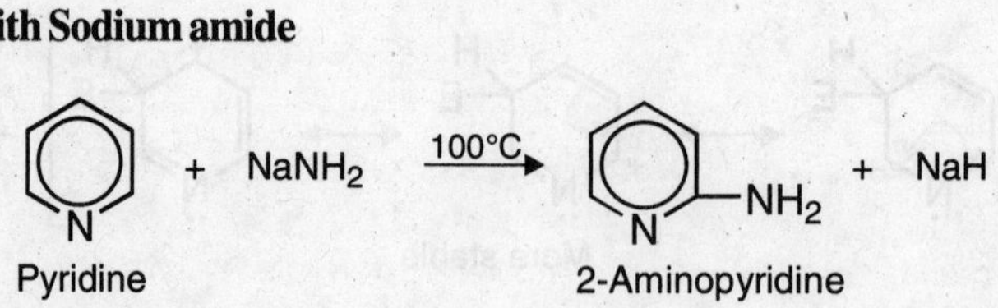

(*b*) **Reaction with Phenyllithium**

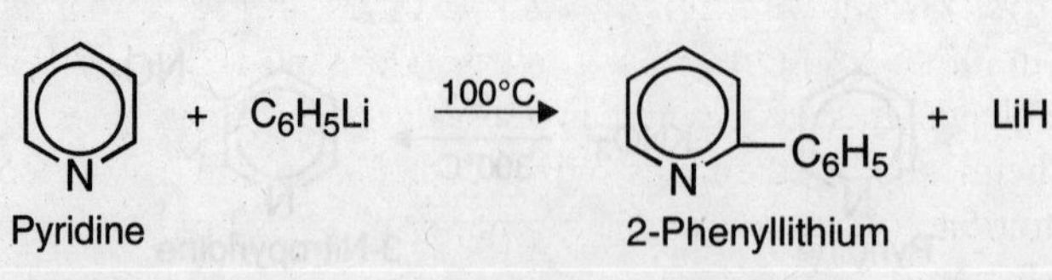

(4) **Reduction**. Pyridine undergoes reduction with H_2/Ni or Na/C_2H_5OH to form piperidine.

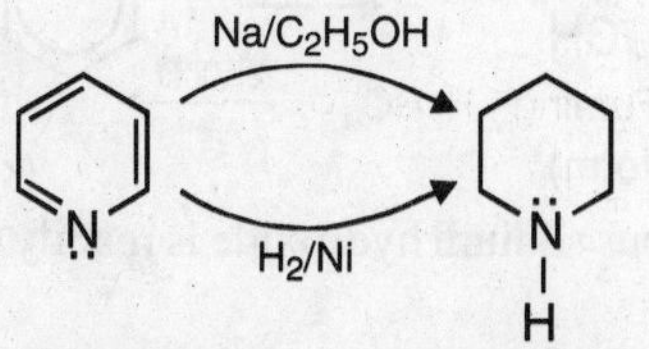

Pyridine Piperidine

Uses. Pyridine is used : (*1*) as a basic solvent in organic reactions ; (*2*) to denature alcohol ; (*3*) for preparing sulfapyridine.

Several mouthwashes contain a pyridinium chloride salt as an antibacterial agent

INDOLE, *2,3-Benzopyrrole*, C_8H_7N

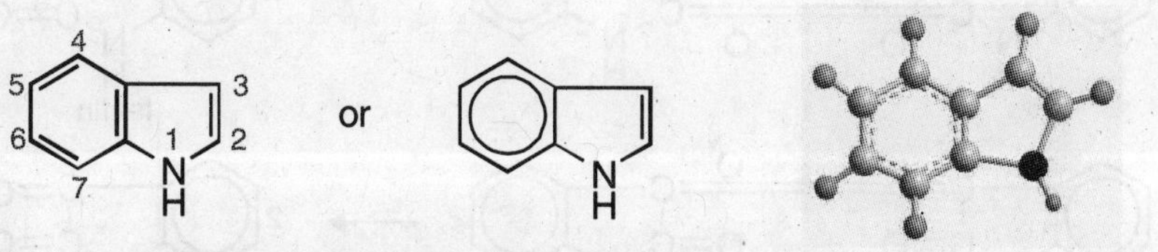

Indole occurs in coal-tar, jasmine flower and orange blossom oil.

Fischer-Indole Synthesis. When phenylhydrazone of pyruvic acid, CH_3–CO–COOH, is heated with $ZnCl_2$ catalyst, it yields indole-2-carboxylic acid which decarboxylates to indole.

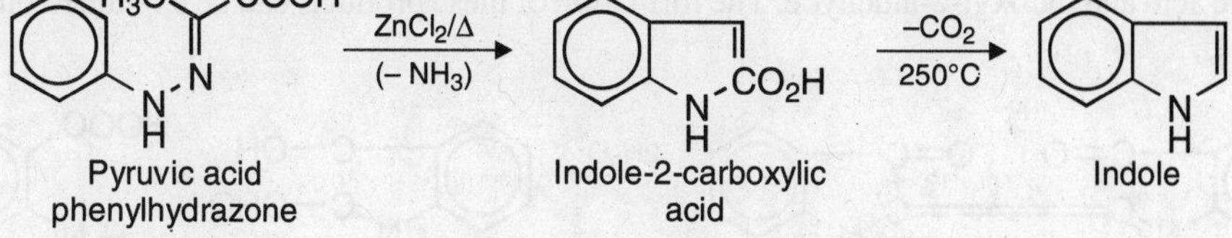

Pyruvic acid phenylhydrazone Indole-2-carboxylic acid Indole

Properties. (*Physical*). Indole forms colorless crystals, mp 52.5°C. It has a strong fecal odor. Curiously enough, in dilute solutions it has flowery odor, and is used in perfumery for preparing jasmine and orange blends. Indole is soluble in hot water, alcohol and ether.

(*Chemical*). Chemically, it resembles pyrrole. It is basic and gives electrophilic substitution reactions such as nitration, sulfonation, and halogenation at 3-position. Its derivative **Indoxyl** exhibits tautomerism and behaves as a phenol.

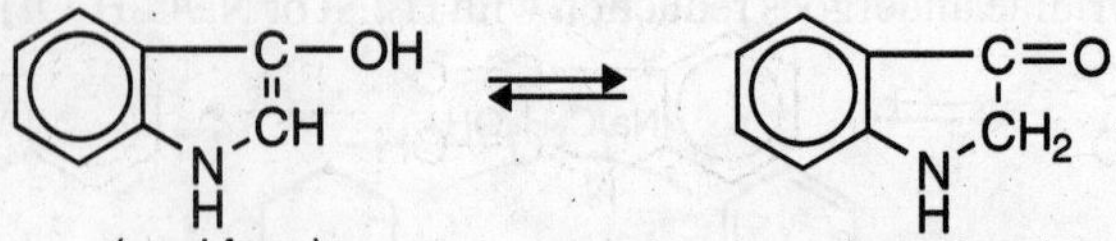

(enol form) (keto form)

A solution of indoxyl in aqueous sodium hydroxide is readily oxidized by air to the insoluble blue dye, *indigo*.

INDIGO, *Indigotin*, $C_{16}H_{10}N_2O_2$

Indigo is one of the oldest blue dyes used by man. At one time it was obtained in large quantities in India from the indigo plant (*Indigofera tinctoria*). The dye takes its name from the Greek word *Indikos*, meaning 'of India'. The preparation of indigo has now been entirely superseded by synthetic processes.

Structure of Indigo

(1) Indigo has the molecular formula $C_{16}H_{10}O_2N_2$.

(2) **Presence of Two Isatin Units**. Oxidation of indigo with nitric acid yields two molecules of isatin.

$$\underset{\text{Indigo}}{C_{16}H_{10}O_2N_2} + 2[O] \xrightarrow{HNO_3} \underset{\text{Isatin}}{2C_8H_5O_2N} \text{ or } 2$$

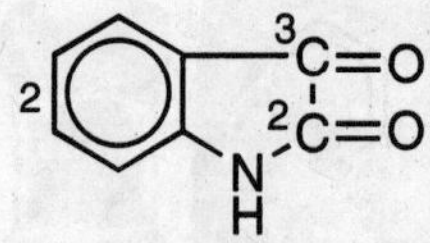

This indicates that indigo molecule is made of two isatin units joined through 2 or 3 carbons by a double bond. The two possible structures of indigo would be I or II, each of which permits the formation of isatin by oxidative cleavage.

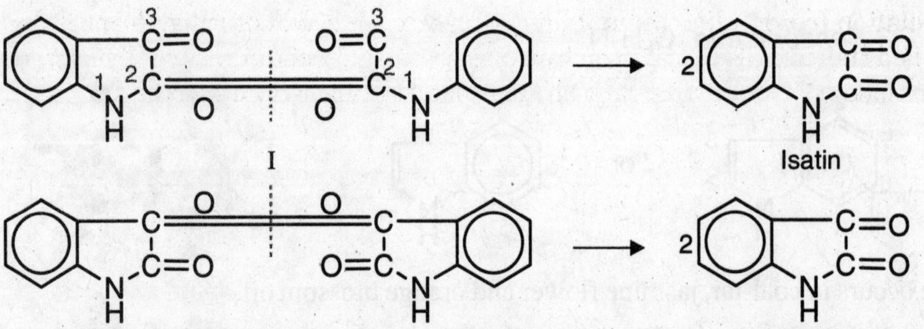

I Isatin

II Isatin

(3) **Isatin Units joined through C-2**. On treatment with dilute sodium hydroxide, indigo forms anthranilic acid and indoxyl-2-aldehyde. The formation of these products can be explained only by the structure I.

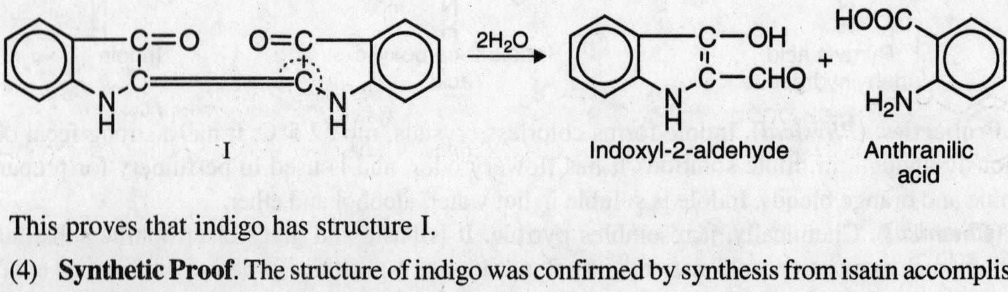

I Indoxyl-2-aldehyde Anthranilic acid

This proves that indigo has structure I.

(4) **Synthetic Proof**. The structure of indigo was confirmed by synthesis from isatin accomplished by Baeyer (1872).

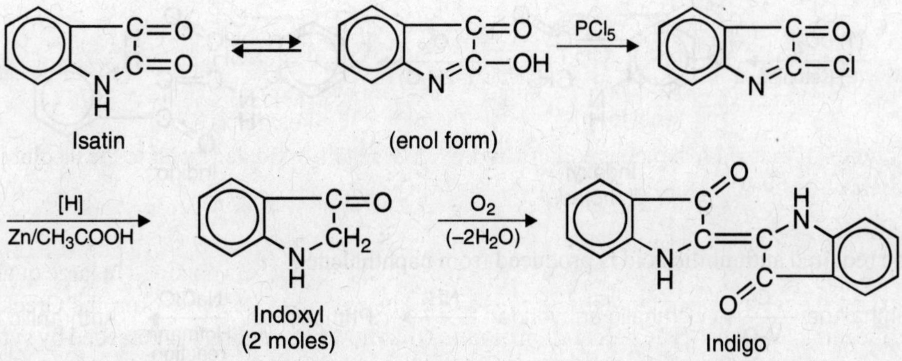

(5) **Indigo is Trans Form**. Indigo, in which the two isatin units are joined by a carbon-carbon double bond, is capable of exhibiting geometrical isomerism. The cis- and trans forms may be written as :

X-ray analysis shows that ordinary indigo is the more stable transform.

Manufacture of Indigo

(1) **Isolation from Indigo Plant**. Indigo occurs in the leaves of indigo plant as β-glucoside of indoxyl, called **Indican**. The indican on hydrolysis with hydrochloric acid or by *enzymes* present in plants, forms indoxyl. Upon oxidation with atmospheric oxygen, it yields indigo.

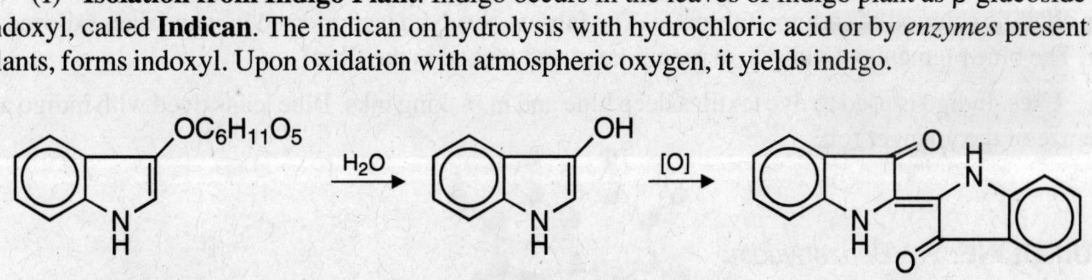

(2) **Commercial Synthesis**. This involves the reaction of anthranilic acid with chloroacetic acid to form *N*-phenylglycine-*o*-carboxylic acid. It is followed by fusion with a mixture of sodium hydroxide and sodium amide to get unstable indoxylic acid which decarboxylates to indoxyl. Indoxyl on oxidation with atmospheric oxygen is converted to indigo.

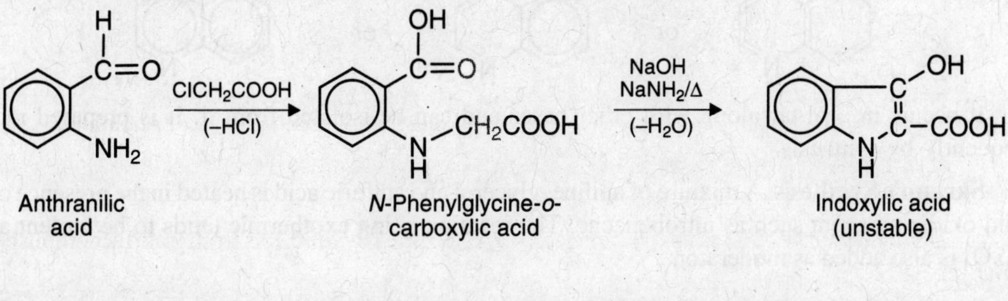

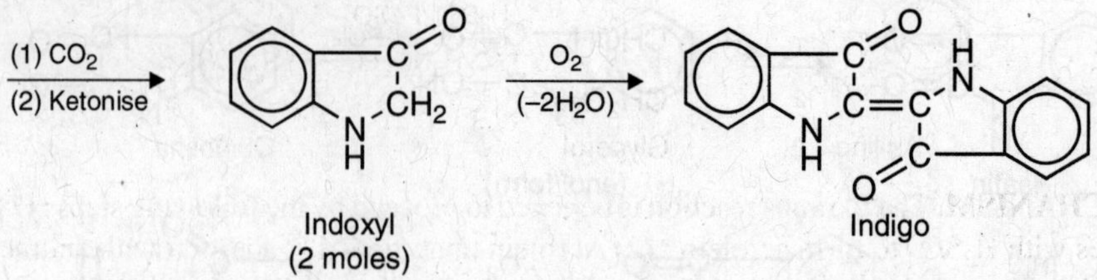

The required anthranilic acid is produced from naphthalene :

$$\text{Naphthalene} \xrightarrow[V_2O_5]{O_2} \text{Phthalic anhydride} \xrightarrow{NH_3} \text{Phthalimide} \xrightarrow[\text{Hofmann reaction}]{NaClO} \text{Anthranilic acid}$$

Properties. Indigo is a dark blue powder with a copper lustre, mp 390°C. It is insoluble in water, ether and alcohol. When reduced with sodium dithionite ($Na_2S_2O_4$) in the presence of sodium hydroxide, indigo is transformed into **Leucoindigo** (or *Indigo White*). The blue color is regenerated when this solution is exposed to air, or more quickly by using an oxidizing agent such as sodium perborate.

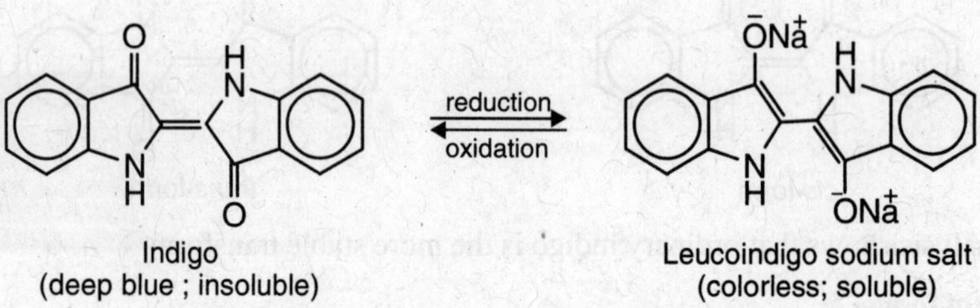

This reaction forms the basis of the technique of **Vat Dyeing** with Indigo. Indigo is suspended in the alkaline solution of sodium dithionite. The fabric to be dyed is immersed in it and then exposed to air. The blue pigments of indigo are precipitated and locked in the fibre.

Uses. Indigo is used to dye textiles deep blue and in making inks. Blue jeans dyed with indigo are a craze of the younger folk.

QUINOLINE, *2,3-Benzopyridine*

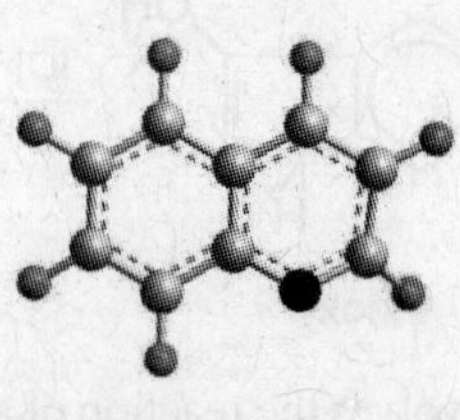

It is a bicyclic heterocycle having a benzene ring fused with a pyridine ring in 2,3-positions. It could be considered as naphthalene in which one CH in 2-position has been replaced by N. The official name of quinoline is **1-Azanaphthalene**.

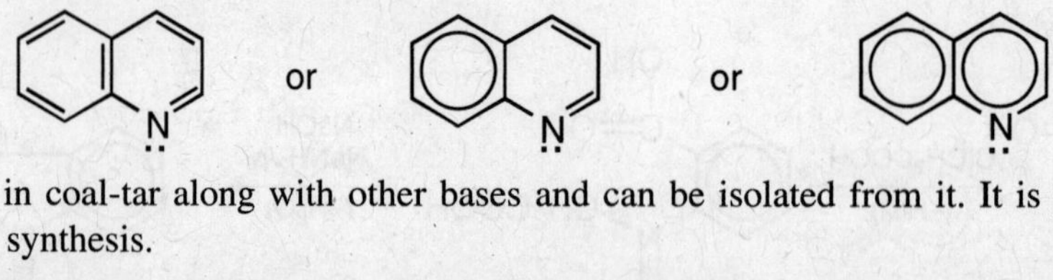

It occurs in coal-tar along with other bases and can be isolated from it. It is prepared more frequently by synthesis.

Skraup's Synthesis. A mixture of aniline, glycerol and sulfuric acid is heated in the presence of a mild oxidizing agent such as nitrobenzene. The reaction being exothermic tends to be violent and $FeSO_4$ is also added as moderator.

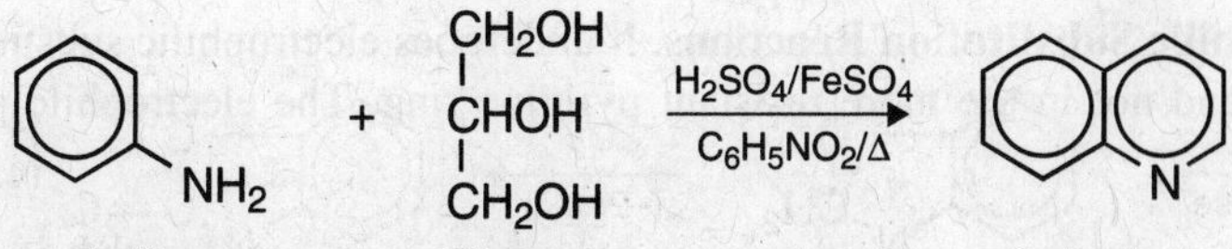

Aniline Glycerol Quinoline

MECHANISM. The Skraups reaction is believed to proceed by the following steps : (1) Glycerol dehydrates with H_2SO_4 to form acrolein ; (2) Acrolein undergoes 1,4-addition with aniline ; (3) The resulting β-anilinopropionaldehyde cyclizes to dihydroquinoline ; (4) Dihydroquinoline is oxidized to quinoline, while nitrobenzene is reduced to aniline which is reused.

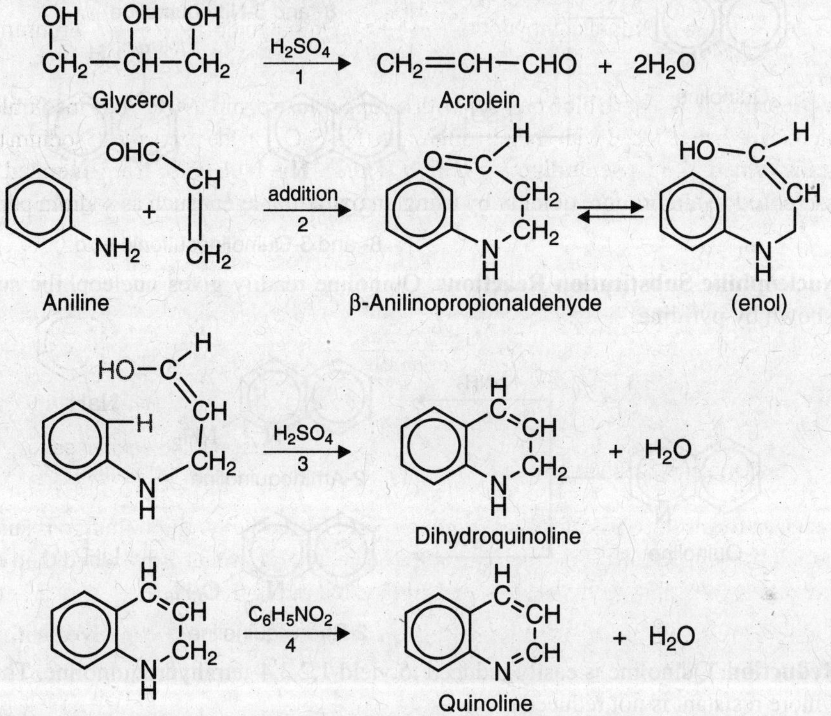

Properties. Quinoline is a colorless liquid, bp 238°C, having a disagreeable, pyridine-like odor. It is soluble in water and completely miscible with alcohol, ether and acetone. Chemically, it gives all the reactions of pyridine and electrophilic substitution reactions of the benzene ring.

(1) **Formation of Salts**. Quinoline is a slightly weaker base ($K_b = 6.3 \times 10^{-10}$) than pyridine. Like other tertiary bases it forms salts with acids and reacts with alkyl halides yielding quaternary salts.

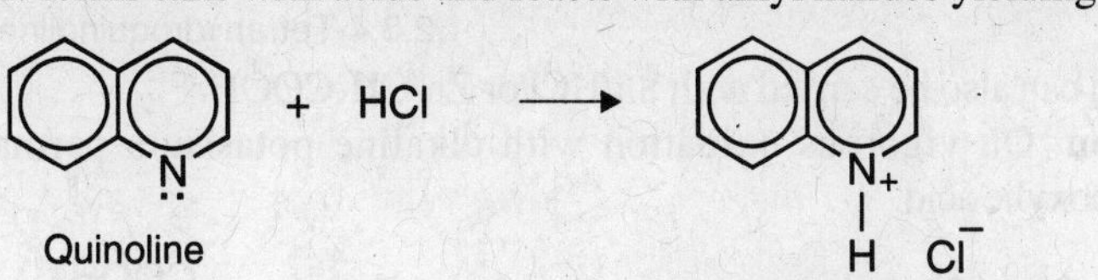

Quinoline Quinoline hydrochloride

Quinoline *N*-Methylquinolinium iodide

(2) **Electrophilic Substitution Reactions**. It undergoes electrophilic substitution reactions in the benzene ring and not in the more resistant pyridine ring. The electrophile preferably attacks positions 8 and 5.

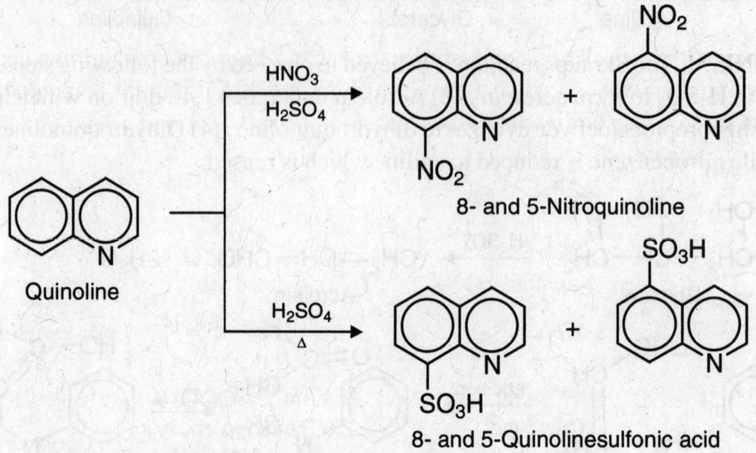

(3) **Nucleophilic Substitution Reactions**. Quinoline readily gives nucleophilic substitution reactions shown by pyridine.

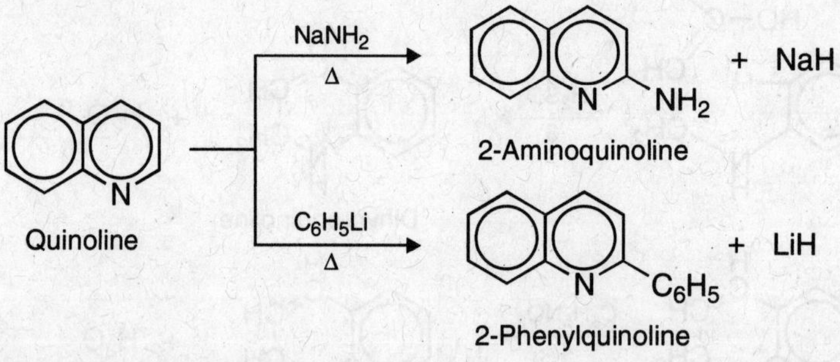

(4) **Reduction**. Quinoline is easily reduced to yield 1,2,3,4-tetrahydroquinoline. The benzene ring being more resistant is not reduced.

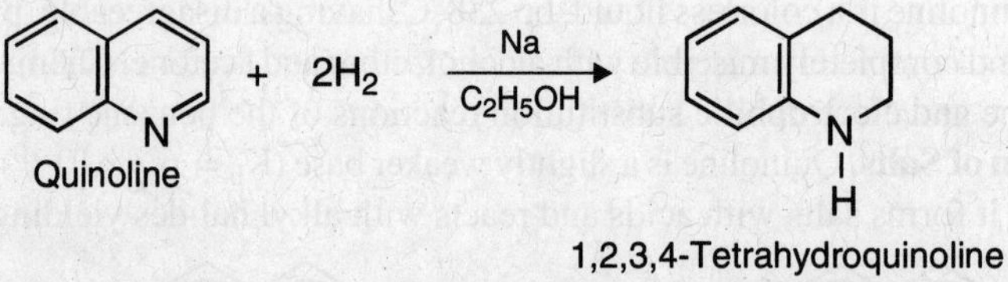

This reduction can also be carried with Sn/HCl or Zn/CH_3COOH.

(5) **Oxidation**. On vigorous oxidation with alkaline potassium permanganate, it yields pyridine-2,3-dicarboxylic acid.

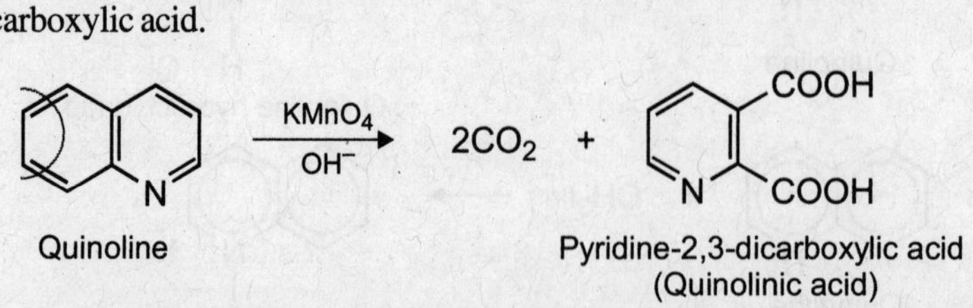

On heating, quinolinic acid decarboxylates to yield pyridine-3-carboxylic acid (nicotinic acid).

Uses. Quinoline is used : (1) as a high-boiling basic solvent in organic reactions : (2) in the manufacture of pharmaceuticals, dyes, and insecticides.

Structure. (1) It has the molecular formula C_9H_7N.

(2) It gives the electrophilic substitution reactions like benzene and forms salts with acids just like pyridine.

(3) On oxidation with alkaline potassium permanganate it forms 2,3-pyridinedicarboxylic acid. From analogy of oxidation of naphthalene to form phthalic acid, it is argued that quinoline is made of a benzene ring fused with pyridine in 2,3-positions.

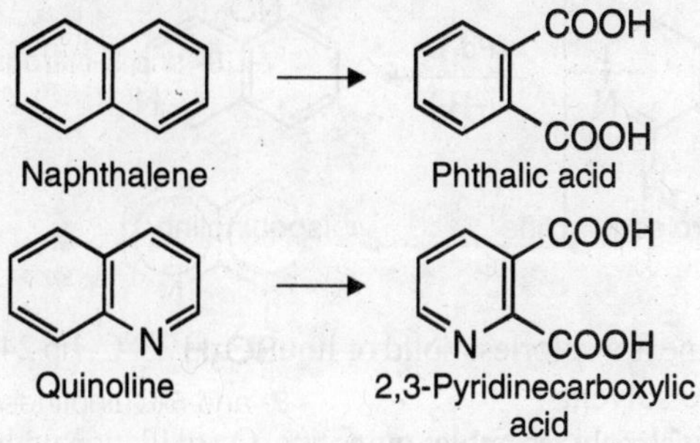

(4) The structure of quinoline is confirmed by **Friedlander Synthesis** which involves the condensation of *o*-aminobenzaldehyde with acetaldehyde in the presence of an alkali.

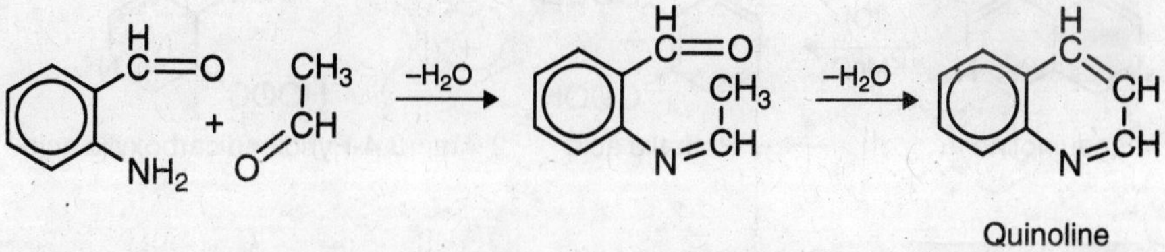

(5) Quinoline is considered as a resonance hybrid of three canonical forms.

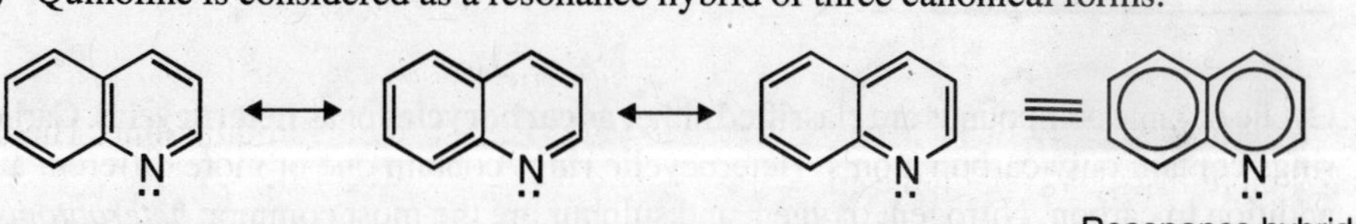

(6) Quinoline is a 10π electron system like naphthalene. It obeys Huckel rule ($n = 2$ in $4n + 2$) which proves its aromatic character.

ISOQUINOLINE, 4,5-*Benzopyridine*, 2-*Azanaphthalene*

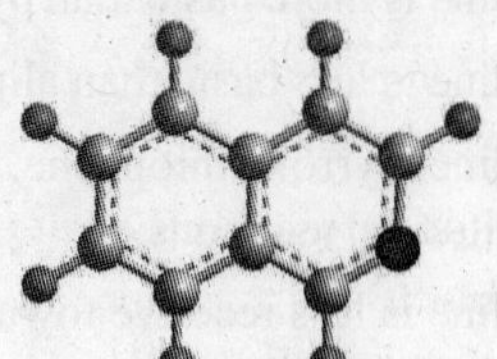

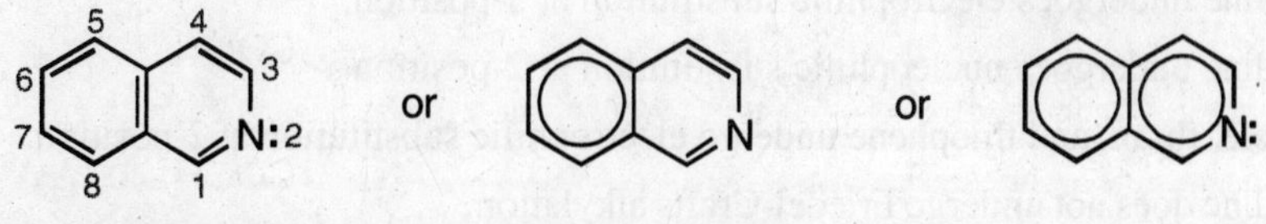

Isoquinoline is found as a part of the total structure of many alkaloids.

Bischler-Napieralski Synthesis. Isoquinoline is easily prepared from 2-phenylethylamine by the following steps.

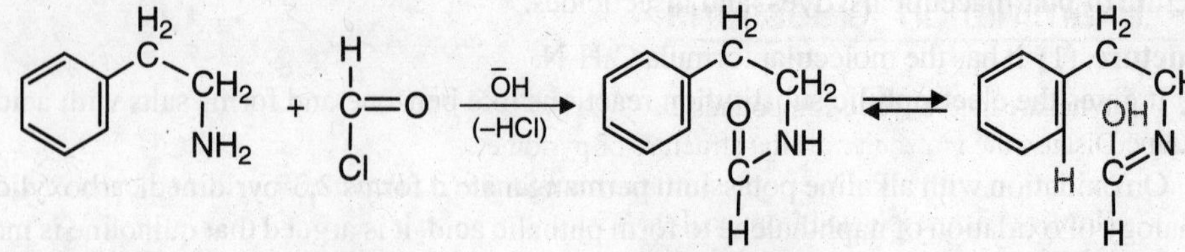

2-Phenylethylamine Formyl chloride *N*-Formyl-2-phenylethylamine (enol form)

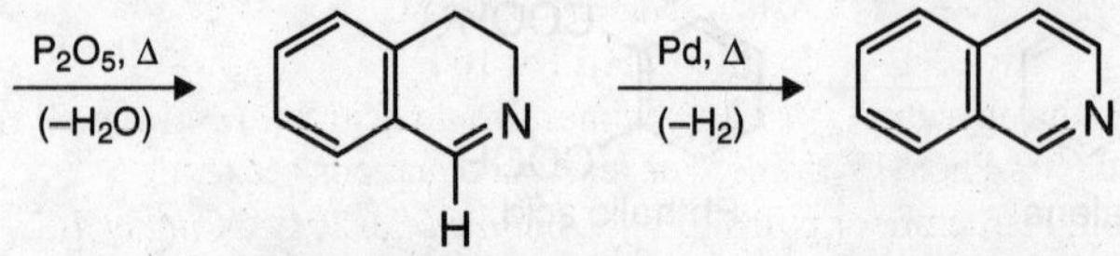

3,4-Dihydroisoquinoline Isoquinoline

Properties. Isoquinoline is a colorless solid or liquid, mp 23°C, bp 243°C. It is insoluble in water, and soluble in most organic solvents.

In chemical properties, it closely resembles quinoline. On oxidation with permanganate, isoquinoline forms phthalic acid and 3,4-pyridinedicarboxylic acid.

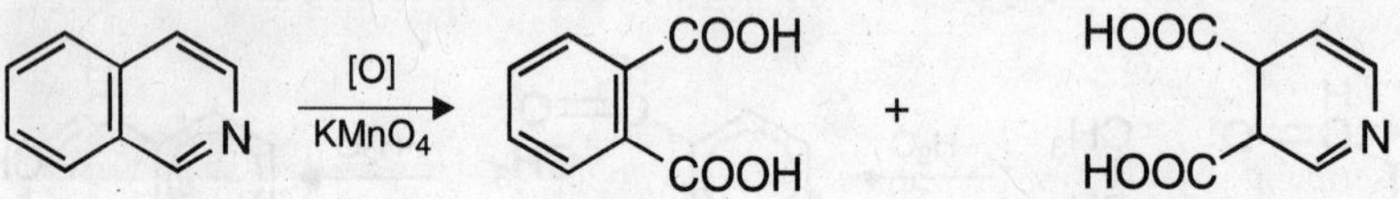

Isoquinoline Phthalic acid 3,4-Pyridinedicarboxylic acid

Points to Remember

1. Cyclic organic compounds are classified either as **carbocycles** or as **heterocycles**. Carbocyclic rings contain only carbon atoms. Heterocyclic rings contain one or more different atoms in addition to carbon. Nitrogen, oxygen, and sulphur are the most common *heteroatoms*.
2. All ring atoms in pyridine, pyrrole, thiophene, and furan are sp^2 hybridized.
3. Pyridine is more basic than pyrrole.
4. Pyridine is less basic than aliphatic amines.
5. Pyridine, pyrrole, thiophene, and furan are aromatic compounds. They undergo electrophilic substitution reactions.
6. Pyridine is less reactive towards electrophilic substitution than benzene.
7. Pyridine undergoes electrophilic substitution at 3-position.
8. Pyridine undergoes nucleophilic substitution at 2-position.
9. Pyrrole, furan, and thiophene undergo electrophilic substitution at 2-position.
10. Pyridine does not undergo Friedel-Crafts alkylation.
11. Quinoline undergoes electrophilic substitution at 8-position and 5-position.
12. Quinoline undergoes nucleophilic substitution at 2-position.

Examination Questions

1. What are heterocyclic compounds ?
2. Discuss the molecular orbital structure of pyridine.

(Madras BSc, 2013 ; Banaras BSc, 2014 ; Delhi BSc, 2015)

3. How is pyridine isolated from coal-tar ?
4. How is pyridine synthesized ? How does it react with the following reagents ?

   (*a*) KNO_3/H_2SO_4 at 300°C (*b*) fuming H_2SO_4 at 250°C
   (*c*) Br_2 at 300°C (*d*) $NaNH_2$ at 100°C
   (*e*) H_2/Ni (*f*) C_6H_5Li at 100°C

Answer. The products are : (*a*) 3-Nitropyridine ; (*b*) 3-Pyridinesulfonic acid ; (*c*) 3-Bromopyridine ; (*d*) 2-Aminopyridine; (*e*) Piperidine ; and (*f*) 2-Phenylpyridine. For reaction equations see text.

5. Discuss the molecular orbital structure of pyrrole. *(Andhra BSc, 2014 ; Cuttack BSc, 2015)*
6. How is pyrrole synthesized ? What happens when pyrrole is treated with the following reagents:

   (*a*) Nitric acid in acetic anhydride at –10°C
   (*b*) Sulfur trioxide in pyridine
   (*c*) Benzenediazonium chloride
   (*d*) Bromine in alcohol

Answer. The products are : (*a*) 2-Nitropyrrole ; (*b*) 2-Pyrrolesulfonic acid ; (*c*) 2-Phenylazopyrrole ; (*d*) Tetrabromopyrrole. For reaction equations see text.

7. How is pyridine synthesized ? Describe its important reactions.
8. How is pyrrole synthesized ? Describe its important reactions.
9. How is furan synthesized ? Describe its important reactions.
10. How is thiophene synthesized ? Describe its important reactions.
11. Describe the preparation and properties of furfural.
12. How is quinoline synthesized ? Describe its important reactions.
13. How is the structure of quinoline established ?
14. How is the structure of indigo established ?
15. Write notes on :

   (*a*) Molecular orbital structure of Pyridine *(Madurai BSc, 2014)*
   (*b*) Aromaticity of Pyridine *(Rani Durgavati BSc, 2015)*
   (*c*) Skraup Quinoline Synthesis
   (*d*) Bischler-Napieralski Synthesis *(Magadh BSc, 2015)*

16. Explain why pyridine and pyrrole are aromatic.

(Mysore BSc, 2014 ; North Eastern Hills BSc, 2015)

Answer. There are three requirements (**Aromaticity Rules**) for a molecule to be aromatic :

(1) An aromatic molecule is cyclic and planar.

(2) Each atom in an aromatic ring has a *p* orbital. These *p* orbitals must be parallel so that a continuous overlap is possible around the ring.

(3) The cyclic π molecular orbital formed by overlap of *p* orbitals must contain $(4n + 2)pi$ electrons, where n = integer 0, 1, 2, 3, etc. This is known as **Huckel Rule**.

Pyridine and pyrrole are both aromatic. Both obey the above aromaticity rules : they are cyclic and planar; each has a *p* orbital on every ring atom ; and each obeys the Huckel Rule (both have 6π electrons ; $n = 1$).

17. Explain why pyridine does not undergo Friedel-Crafts reactions.

Answer. This is because the Lewis acids ($FeCl_3$ or $AlCl_3$) which are used as catalysts in these reactions coordinate with the lone pair of electrons on nitrogen to form a complex.

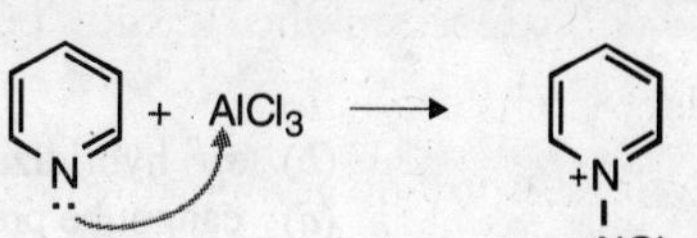

Pyridine-Aluminium complex

18. Explain why

(*a*) Pyridine is more basic than pyrrole. (*Vikram BSc, 2014*)

(*b*) Pyridine is less basic than aliphatic amines. (*Allahabad BSc, 2015*)

(*c*) Pyridine is more basic than aniline.

Answer. (*b*) Pyridine is less basic ($pK_b = 8.75$) than aliphatic amines ($pK_b = 4$). This is due to the difference in the nature of hybrid orbitals containing the nitrogen lone pair electrons.

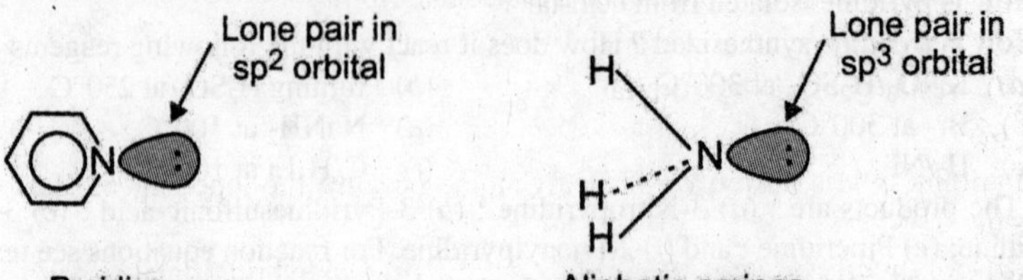

Pyridine (less basic) Aliphatic amines (more basic)

In pyridine it is in an sp^2 orbital ; in aliphatic amines it is in an sp^3 orbital. We know that sp^2 orbitals are smaller (due to more *s* character) than sp^3 orbitals. This means that the lone pair of electrons on nitrogen in pyridine is more closely associated with the nitrogen nucleus. It is , therefore, less available for the formation of bond with proton and consequently the relative basicity is reduced.

19. Explain why :

(*a*) Pyridine undergoes electrophilic substitution at 3-position

(*b*) Pyridine undergoes nucleophilic substitution at 2-position

(*c*) Pyrrole undergoes electrophilic substitution at 2-position

20. How will you synthesize 2-methylquinoline from aniline ?

Answer. Following steps are involved :

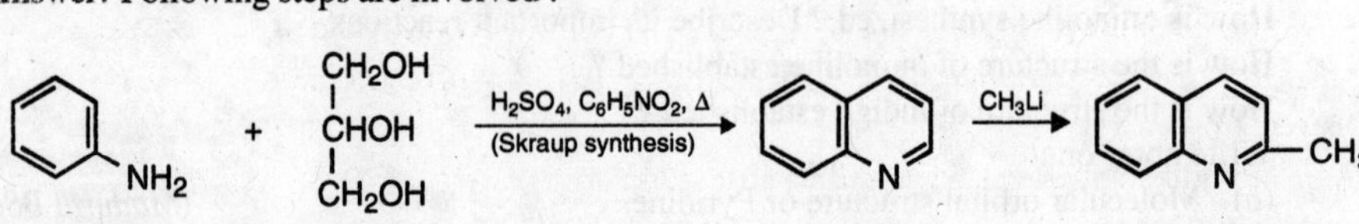

Aniline Glycerol Quinoline 2-Methylquinoline

21. How will you synthesize indigo from anthranilic acid ?

Multiple Choice Questions

1. Which of the following is not a heterocyclic compound ?

(*a*) 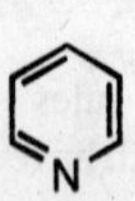 (*b*) 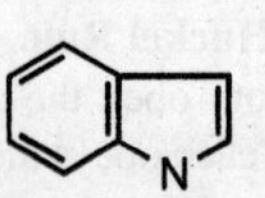 (*c*) 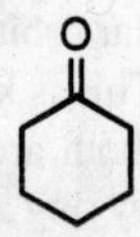 (*d*) 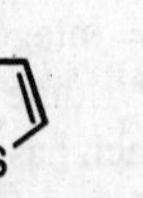

Answer. (*c*)

2. Which of the following heterocyclic compounds is not aromatic ?

(*a*) Pyridine (*b*) Pyrrole (*c*) Furan (*d*) Piperidine

Answer. (*d*)

3. The 'N' atom in pyridine is

(*a*) sp^3 hybridized (*b*) sp^2 hybridized

(*c*) *sp* hybridized (*d*) cannot be predicted

Answer. (*b*)

4. Pyridine has a delocalized π molecular orbital containing

(*a*) 4 electrons (*b*) 6 electrons
(*c*) 8 electrons (*d*) 12 electrons

Answer. (*b*)

5. Pyrrole is less basic than pyridine because the lone-pair of electrons on N-atom in pyrrole

(*a*) is part of the delocalized π molecular orbital.
(*b*) is not part of the delocalized π molecular orbital.
(*c*) resides in sp^2 hybrid orbital
(*d*) resides in *sp* hybrid orbital

Answer. (*a*)

6. Pyridine is less basic than trimethylamine because the lone-pair of electrons on N-atom in pyridine resides in

(*a*) sp^2 hybrid orbital (*b*) *sp* hybrid orbital
(*c*) sp^3 hybrid orbital (*d*) *p*-orbital

Answer. (*a*)

7. Pyridine reacts with HCl to form

(*a*) Pyridinium chloride (*b*) 2-Chloropyridine
(*c*) 3-Chloropyridine (*d*) All of these

Answer. (*a*)

8. Pyridine undergoes electrophilic substitution with fuming H_2SO_4 at 350°C to give

(*a*) 2-Pyridinesulfonic acid (*b*) 4-Pyridinesulfonic acid
(*c*) 3-Pyridinesulfonic acid (*d*) None of these

Answer. (*c*)

9. Pyridine reacts with a mixture of KNO_3 and H_2SO_4 at 300°C to give

(*a*) 1-Nitropyridine (*b*) 2-Nitropyridine
(*c*) 3-Nitropyridine (*d*) 4-Nitropyridine

Answer. (*c*)

10. Pyridine undergoes nuclophilic substitution with $NaNH_2$ at 100°C to form

(*a*) 2-Aminopyridine (*b*) 3-Aminopyridine
(*c*) 4-Aminopyridine (*d*) None of these

Answer. (*a*)

11. Which of the following reagents will react with pyrrole to form 2-formylpyrrole ?

(*a*) HCOOH (*b*) $CHCl_3$/KOH
(*c*) H_2O_2 (*d*) $(CH_3CO)_2O/SnCl_4$

Answer. (*b*)

12. Furan reacts with ammonia in the presence of alumina at 400°C to give

(*a*) Pyridine (*b*) Furfural
(*c*) Pyrrole (*d*) Furoic acid

Answer. (*c*)

13. Which of the following reagents will react with furan to form 2-furansulfonic acid ?

(*a*) SO_3 in pyridine at 100°C (*b*) Dilute H_2SO_4 at 200°C
(*c*) SO_2 at 100°C (*d*) Dilute H_2SO_4 at 100°C

Answer. (*a*)

14. When aniline is heated with glycerol in the presence of sulfuric acid and nitrobenzene, it gives quinoline. This reaction in called

(*a*) Fischer synthesis (*b*) Skraup synthesis
(*c*) Diazotization (*d*) Corey-House synthesis

Answer. (*b*)

CLASSIFICATION OF ALKALOIDS

The individual alkaloids are frequently called by *common names*. These end in *-ine*, indicating that all alkaloids are *amines*. Alkaloids are divided into classes depending on the prominent heterocyclic ring system present in the molecule. Thus we have :

(1) **Pyridine alkaloids** *e.g.*, Nicotine

(2) **Piperidine alkaloids** *e.g.*, Coniine, piperine

(3) **Quinoline alkaloids** *e.g.*, Quinine

(4) **Isoquinoline alkaloids** *e.g.*, Papaverine, morphine

(5) **Indole alkaloids** *e.g.*, Gramine, strychnine

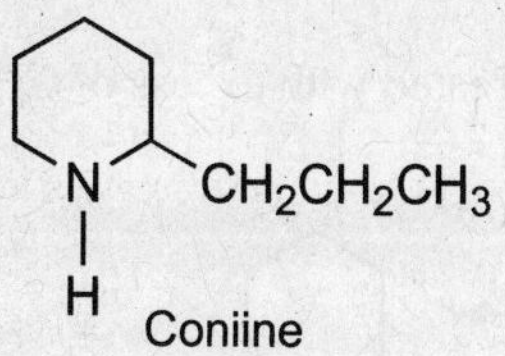

Coniine
(a piperidine alkaloid)

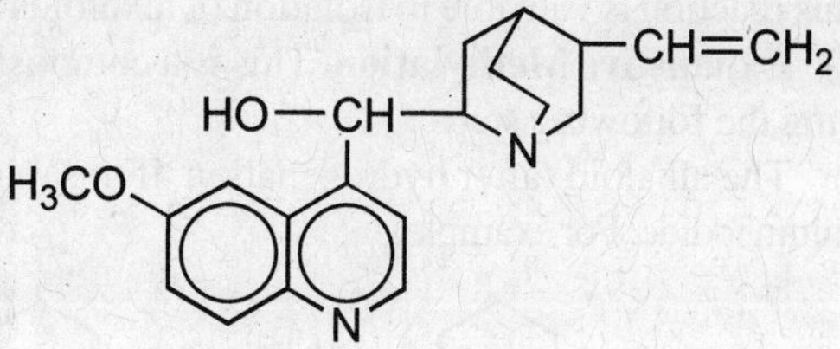

Quinine
(a quinoline alkaloid)

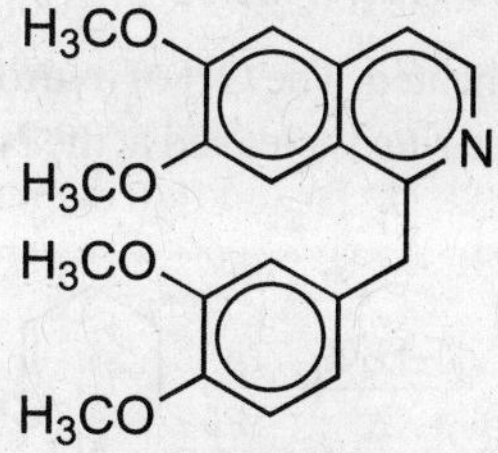

Papaverine
(an isoquinoline alkaloid)

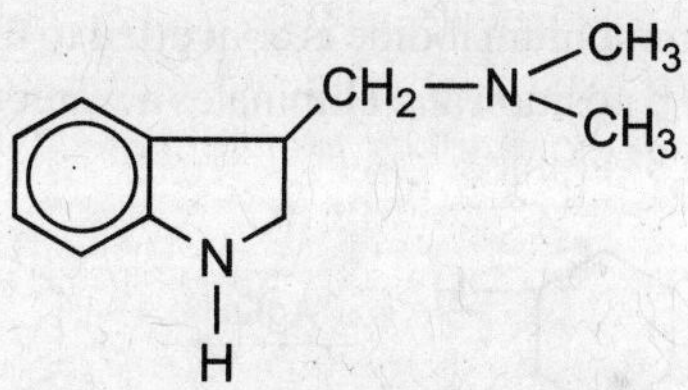

Gramine
(an indole alkaloid)

Note : In the *line formulas* corners or junctions, or end of lines represent carbons with appropriate number of H atoms.

OCCURRENCE AND EXTRACTION

Alkaloids occur chiefly in plants of the *dicotyledonous* family and are localized in seeds, leaves, bark or root of the plant. They are generally found as salts of plant acids such as oxalic, malic, tartaric or citric acid. An alkaloid often occurs in the plant source along with several other closely related alkaloids.

The powdered plant material is digested with water acidulated with HCl (or H_2SO_4). The mineral acid liberates the alkaloid from their salts with plant acids and converts them into hydrochlorides. The aqueous layer containing the hydrochlorides of alkaloids and soluble plant impurities (coloring matter, carbohydrates) is made basic, say with NaOH solution. The insoluble alkaloids that are set free, precipitate out. The solid mass so obtained is then extracted with ether when alkaloids pass into solution and impurities are left behind. The ether extract is evaporated to give a mixture of pure alkaloids. The required alkaloid is then separated from the mixture by fractional crystallization, chromatography and ion-exchange method.

GENERAL PROPERTIES

(Physical). (1) Most alkaloids are colorless, crystalline solids, while a few (*coniine, nicotine*) are liquids.

(2) In general they are insoluble in water but the liquid alkaloids are soluble. They dissolve readily in organic solvents such as ethanol, ether or chloroform.

(3) They induce marked physiological activity in humans. Most of them have an intensely bitter taste and are poisons.

(4) They are all optically active, the majority being levorotatory.

(*Chemical*). Alkaloids give the reactions of cyclic amines as also of any other functional group (OH, CO_2H, OCH_3, >CO, CO_2R, $CONH_2$) present.

(1) Formation of Salts. Like amines they form salts both with inorganic and organic acids. These salts when treated with a base (NaOH, Na_2CO_3) are decomposed to liberate free alkaloid.

$$\text{(Alk)NH} + \text{HCl} \longrightarrow \text{(Alk)}\overset{+}{\text{N}}\text{H}_2\overset{-}{\text{Cl}} \xrightarrow{\text{NaOH}} \text{(Alk)NH}$$

This reaction is valuable in isolation of alkaloids from plant materials.

(2) Exhaustive Methylation. This is a composite reaction of an alkaloid (heterocyclic amine), involving the following steps :

(*i*) The alkaloid (after hydrogenation, if unsaturated) is treated with excess of CH_3I to form 4° ammonium iodide. For example,

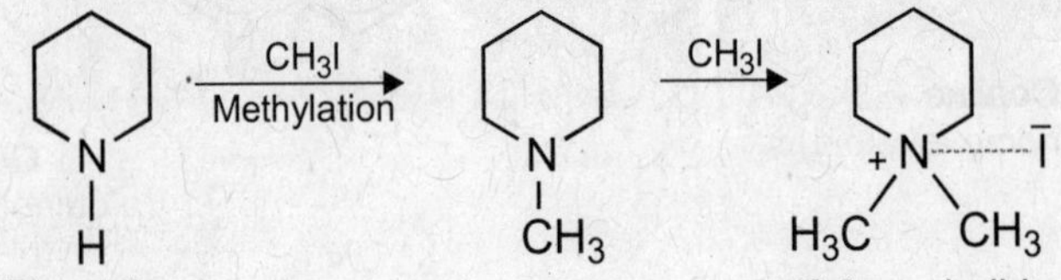

Piperidine   4° Amm iodide

(*ii*) 4° Ammonium iodide is converted to the hydroxide and heated. The OH of hydroxide extracts a H atom from β position and eliminates a water molecule. Also, the ring is cleaved at the N atom to give an open-chain 3° amine.

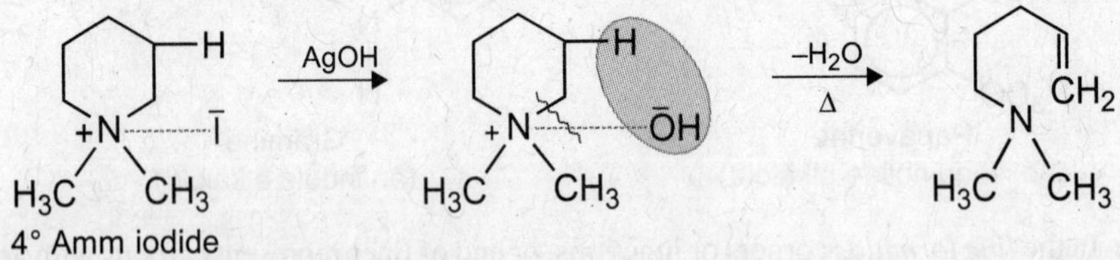

4° Amm iodide

(*iii*) The steps (*i*) and (*ii*) are repeated when a second cleavage at the N atom gives an unsaturated hydrocarbon which isomerizes to a conjugated diene.

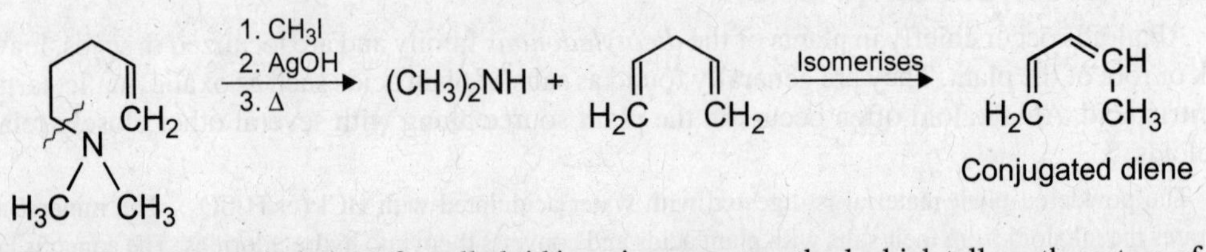

Conjugated diene

The exhaustive methylation of an alkaloid is an important method which tells us the nature of the carbon skeleton in the heterocyclic system.

IMPORTANT ALKALOIDS

CONIINE, $C_8H_{17}N$

As already shown, coniine is one of the simplest alkaloids belonging to 'piperidine group'. Its structural formula can be written as

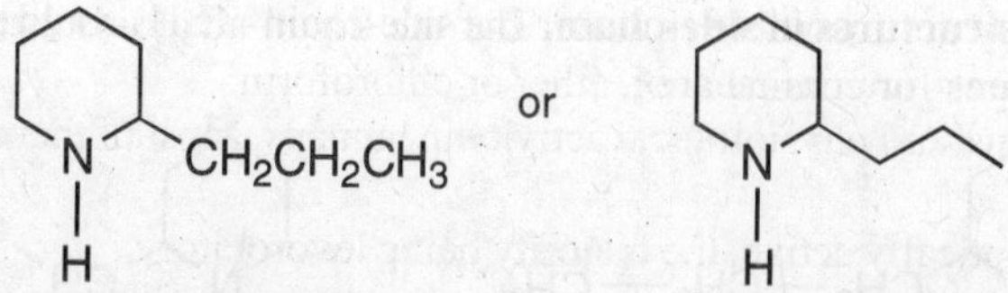

Coniine
(2-*n*-Propylpiperidine)

Line formula

Coniine occurs in seeds and other parts of the hemlock herb (*conium maculatum*). It is one of history's most poisonous alkaloids. Socrates, the great Greek philosopher, was executed in 400 BC by drinking a cup of hemlock.

The poison given to Socrates, coniine, comes from the Hemlock plant (Conium maculatum).

Isolation. The powdered hemlock seeds are distilled with sodium hydroxide solution. The distillate is extracted with ether. The ethereal extract is evaporated and the oily residue is carefully fractionated. The fraction distilling at 167°C is coniine.

Properties. Coniine is a colorless unpleasant smelling liquid, bp 167°C. It is sparingly soluble in water but fairly soluble in organic solvents, readily soluble in ethanol. The natural alkaloid is dextrorotatory, $[\alpha]_D = +15.6$. It is strongly basic and forms salts with acids such as hydrochloric acid and hydrobromic acid. Both coniine and its salts are highly toxic to humans causing paralysis, convulsions and death.

Structure. (1) Molecular formula of coniine is found to be $C_8H_{17}N$.

(2) **Coniine contains piperidine ring with 2-side chain.** When distilled with zinc dust, coniine gives conyrine ($C_8H_{11}N$) which on oxidation with $KMnO_4$ gives pyridine-2-carboxylic acid.

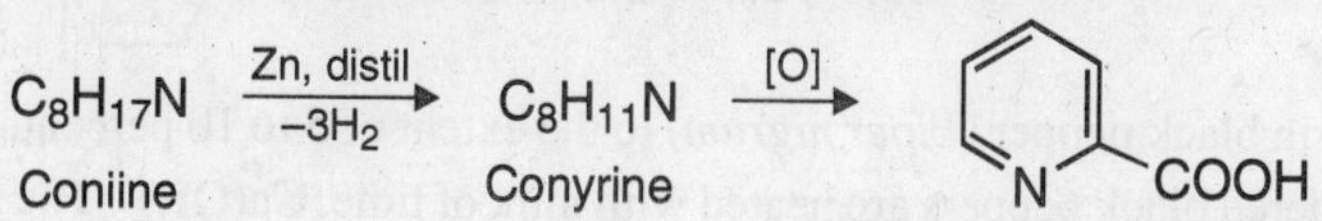

Pyridine-2-carboxylic acid

This shows that : (*i*) Conyrine is made of a pyridine ring, C_5H_4N–, with a side-chain in 2-position ; (*ii*) The side-chain is C_3H_7 ($C_8H_{11}N - C_5H_4N$) ; (*iii*) Coniine is hexahydroconyrine. Thus the structural formulas of conyrine and coniine may be written as :

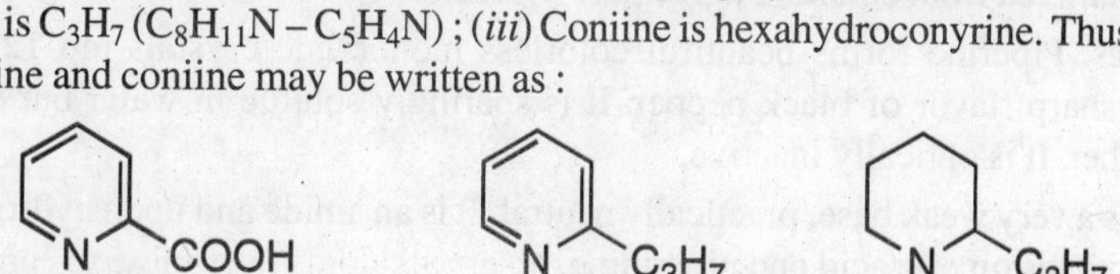

Pyridine-2-carboxylic acid　　Conyrine　　Coniine

(3) **Two possible structures of side-chain.** The side-chain $-C_3H_7$ could be *normal* or *iso*, so that the two possible structures for coniine are :

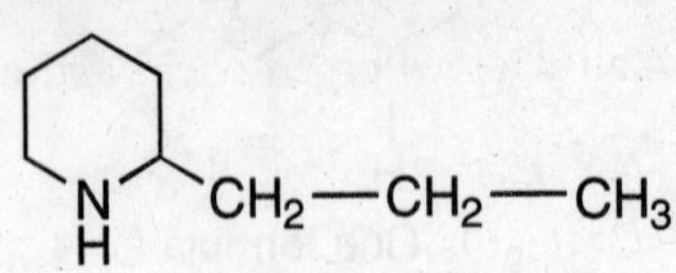

I

(*normal* side-chain)

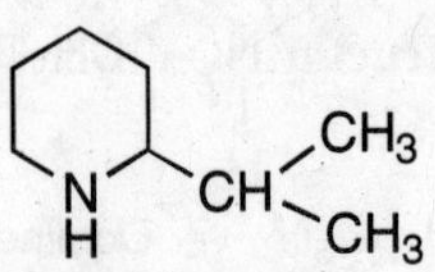

II

(*iso* side-chain)

(4) **HI reduces coniine to *n*-octane.** The structure I is actually shown to be correct as the natural alkaloid when heated with hydriodic acid at 300°C is reduced to give *n*-octane.

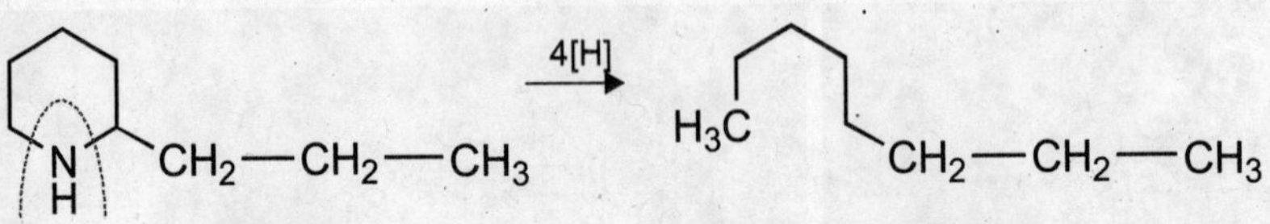

Coniine → *n*-Octane

(5) **Synthesis.** The structure of coniine has been confirmed by the following synthesis :

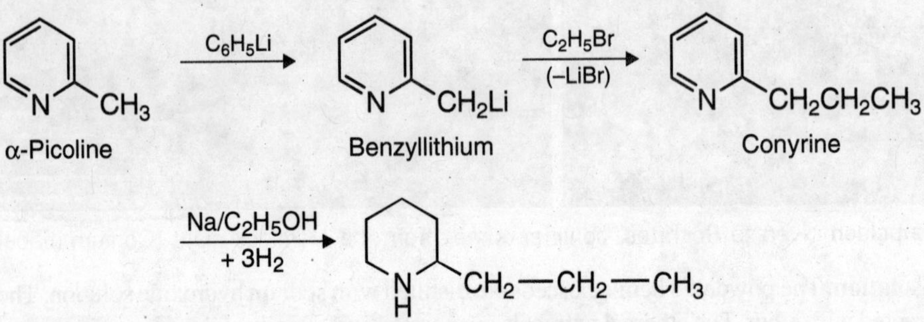

α-Picoline → Benzyllithium → Conyrine → (±)-Coniine

The racemic coniine is resolved by means of (+)-tartaric acid. The (+)-coniine so obtained is identical with the natural alkaloid.

PIPERINE, $C_{17}H_{19}O_3N$

It is a piperidine alkaloid having the structural formula

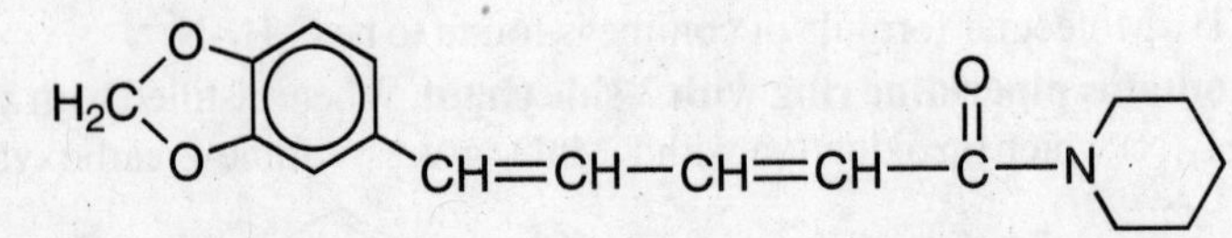

Piperine occurs in black pepper (*Piper nigrum*) to the extent of 5 to 10 percent.

Isolation. Powdered black peppers are heated with milk of lime, $Ca(OH)_2$. The resulting mass is evaporated to dryness. The residue is extracted with ether. The ether solution is evaporated and the residue is crystallized from ethanol to give pure piperine.

Properties. Piperine forms beautiful colorless monoclinic crystals mp 129.5°C, having the characteristic sharp flavor of black pepper. It is sparingly soluble in water but dissolves easily in ethanol and ether. It is optically inactive.

Piperine is a very weak base, practically neutral. It is an amide and upon hydrolysis with aqueous base (or acid) yields piperic acid and piperidine.

Piperine is used occasionally in favoring and to relieve colic pain.

Structure. (1) As shown by elemental analysis and molecular weight determination, the molecular formula of piperine is $C_{17}H_{19}O_3N$.

(2) **Amide Structure.** On hydrolysis with alkali it gives piperic acid and a secondary base, piperidine.

$$\underset{\text{Piperine}}{C_{17}H_{19}O_3N} + H_2O \xrightarrow[\Delta]{KOH} \underset{\text{Piperic acid}}{C_{11}H_9O_2.COOH} + \underset{\text{Piperidine}}{C_5H_{10}NH}$$

This reaction indicates that piperine is the piperidine-amide of piperic acid and its structure could be derived as

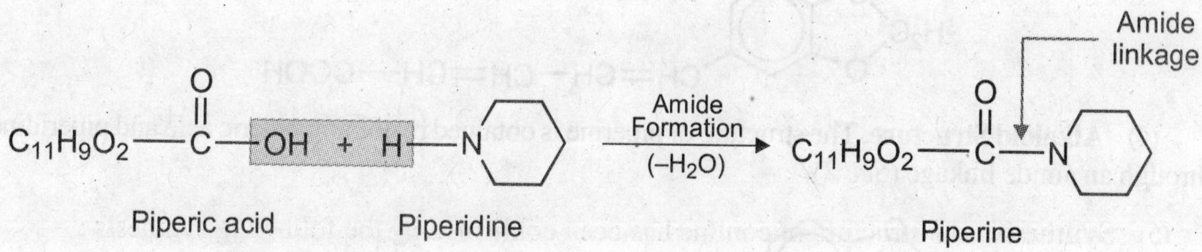

If we can now find the structure of piperic acid, complete molecular structure of the alkaloid could be written.

(3) **Structure of Piperic Acid ($C_{12}H_{10}O_4$) :**

(*i*) As shown by routine chemical tests, it contains one COOH group and two ethylene double bonds.

(*ii*) **Structure of Side-chain.** When oxidized with potassium permanganate, piperic acid first forms an aldehyde, piperonal, and then piperonylic acid.

$$\underset{\text{Piperic acid}}{C_{12}H_{10}O_4} \xrightarrow{[O]} \underset{\text{Piperonal}}{C_8H_6O_3} \xrightarrow{[O]} \underset{\text{Piperonylic acid}}{C_8H_6O_4}$$

Piperonylic acid is C_4H_4 less than piperic acid and so is piperonal. Since piperonal shows negative test for COOH, the (C_4H_4)COOH must be present as side-chain of the aromatic moiety in piperic acid. To explain the above oxidation reactions we construct the side-chain having two double bonds and a COOH group as below.

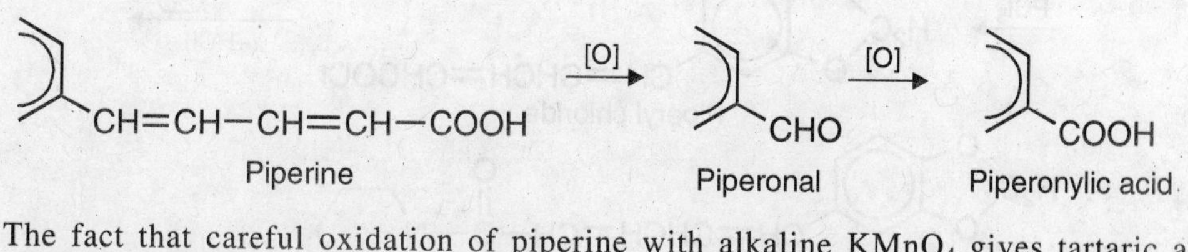

The fact that careful oxidation of piperine with alkaline $KMnO_4$ gives tartaric acid, $HO_2C.CHOH.CHOH.CO_2H$, confirms that the side-chain in piperic acid is a straight chain with carbon-carbon double bonds in alternate positions.

(*iii*) **Structure of Piperonal.** When boiled with concentrated hydriodic acid, piperonylic acid gives methylene iodide (CH_2I_2) and protocatechuic acid (3,4-dihydroxybenzoic acid). This shows that piperonylic acid is methylene ether of protocatechuic acid. Thus,

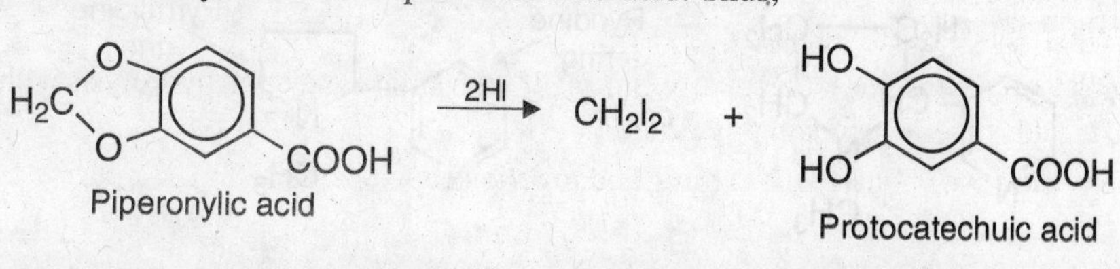

Piperonal is the aldehyde corresponding to piperonylic acid and its structure can be written as :

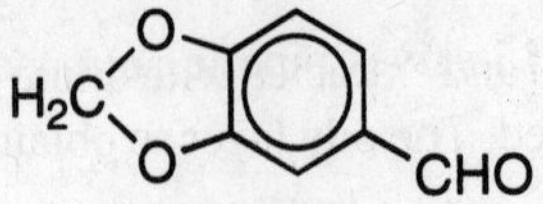

Piperonal

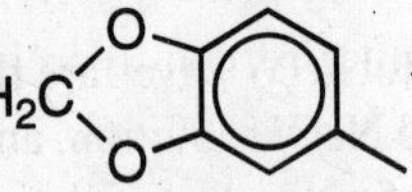

Aromatic moiety in piperic acid

(*iv*) Knowing the structure of the side-chain from (*ii*), we can construct the possible structural formula of piperic acid.

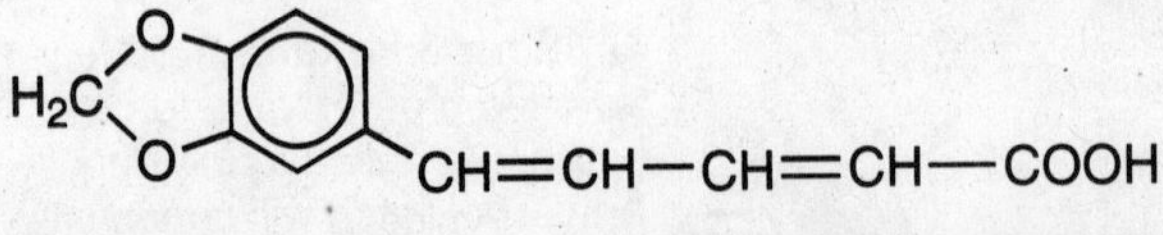

(4) **Alkaloid Structure.** The structure of piperine is obtained by joining piperic acid and piperidine through an amide linkage (*See* 2).

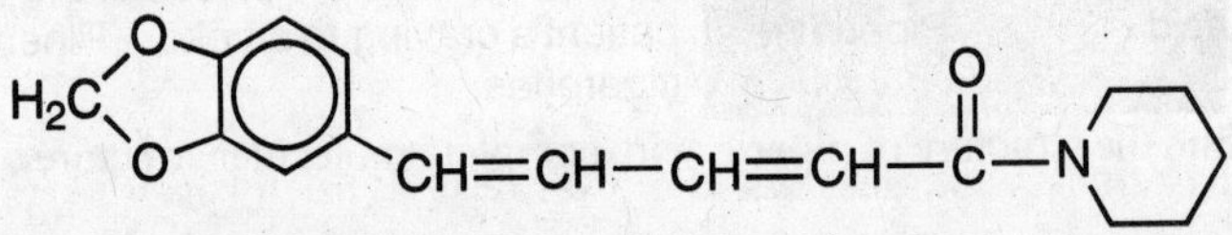

(5) **Synthesis.** The structure of piperine has been confirmed by the following synthesis :

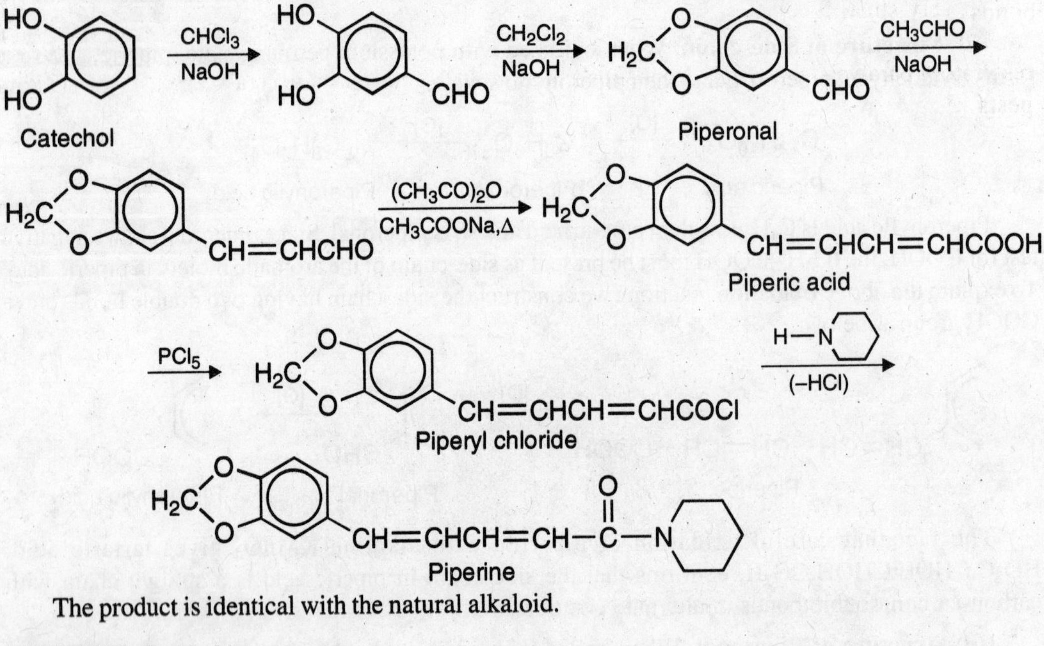

The product is identical with the natural alkaloid.

NICOTINE, $C_{10}H_{14}N_2$

It is a pyridine (or pyridine-pyrrolidine) alkaloid having the structural formula

Nicotine occurs to the extent of 2-8 per cent in dried leaves of the tobacco plant (*Nicotiana tabacum*).

Isolation. Dry tobacco leaves are finely powdered and extracted with dilute acid solution. To the acid extract is added NaOH solution, and steam distilled. The oily layer so obtained is crude nicotine which is purified by fractional distillation in vacuum.

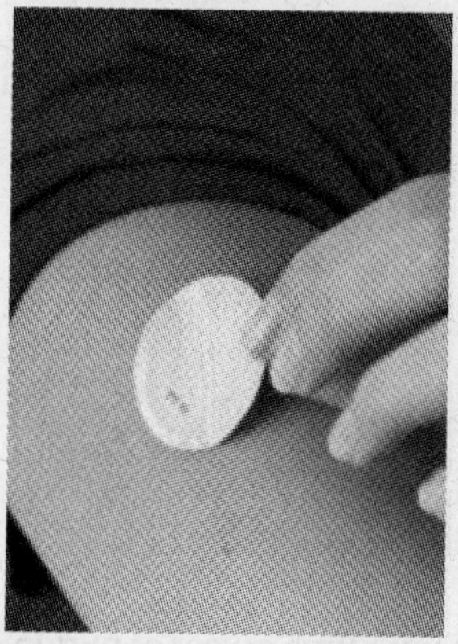

Quitting Smoking. A nicotine patch, a type of transdermal patch, is applied to the upper arm. The patch will continually release a small amount of nicotine into the body through the bloodstream. This will reduce the patient's craving to smoke cigarettes.

Properties. Nicotine is a colorless liquid, bp 245.5°C. It has tobacco-like smell and a burning 'alkali taste'. It is soluble in water and also in organic solvents. The natural alkaloid is dextrorotatory, $[\alpha]_D = +163.2°$.

In very small doses nicotine stimulates the central nervous system for a while. That is why smokers crave for it *via* cigarettes, cigars or pipes. In large doses, the alkaloid causes depression, respiratory paralysis, and death. It kills insects on contact and is used in sprays against leaf-sucking pests.

He had tried many ways to quit smoking but only **one** worked.

Structure. (1) The molecular formula of nicotine has been found to be $C_{10}H_{14}N_2$.

(2) **Presence of Two 3° N-atoms.** When treated with methyl iodide, it forms dimethiodide showing that both the N-atoms of nicotine are tertiary in character.

$$\underset{\text{Nicotine}}{C_{10}H_{14}N_2} + 2CH_3I \longrightarrow \underset{\text{Dimethiodide}}{C_{10}H_{14}\left(-\overset{+}{N}\overset{-}{I}(CH_3)\right)_2}$$

47 Alkaloids

Alkaloids (meaning *alkali-like*) are an important class of basic organic compounds that occur in higher plants. They are known to contain one or more nitrogen heterocyclic rings as an integral part of their structure. They produce striking physiological effects when administered to humans. Thus, some alkaloids stimulate the central nervous system, while others cause paralysis ; some alkaloids act as pain-relievers, others as local anesthetics ; and still others act against infectious microorganisms. Most alkaloids are toxic and yet many find use in medicine.

WHAT ARE ALKALOIDS ?

A precise definition of the term *alkaloid* covering such a wide variety of compounds of this class is difficult. However, a broad definition may be stated as : **Alkaloids are plant-produced basic (alkaline) compounds which contain one or more nitrogen heterocyclic rings and can have profound physiological effect an humans.**

The above definition covers a large majority of alkaloids but it is by no means perfect. Its inexactness is evident from facts such as : (1) Piperine, the alkaloid of pepper, is neither basic nor does it produce any physiological effect ; (2) Caffeine, a basic nitrogen heterocycle, a heart stimulant and a diuretic, is usually not considered an alkaloid ; (3) The alkaloid Marijuana contains no nitrogen ; and (4) Ephiderene is an open chain alkaloid.

Points to Remember

1. Alkaloids occur in plants.
2. Alkaloids contain nitrogen heterocyclic rings.
3. Alkaloids are basic compounds.
4. Alkaloids have a profound physiological effect on humans and act as poisons, hallucinogens, stimulants, and depressants.
5. In the line formulas (as shown above for coniine and nicotine) corners or junctions represent carbons with appropriate number of hydrogens.

(3) **Presence of a Pyridine ring.** On oxidation with chromic acid or nitric acid, nicotine gives nicotinic acid (pyridine-3-carboxylic acid).

$$\underset{\text{Nicotine}}{C_{10}H_{14}N_2} \xrightarrow{[O]} \underset{\text{Nicotinic acid}}{(C_5H_4N)COOH} \quad \text{or}$$

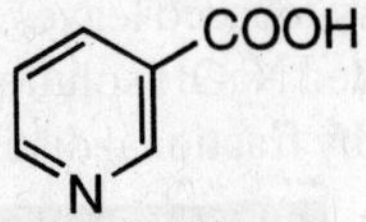

This shows that the alkaloid contains a pyridine ring (C_5H_4N) with a side-chain.

$$\underset{\text{Nicotine}}{C_{10}H_{14}N_2} - C_5H_4N = \underset{\text{side-chain}}{C_5H_{10}N}$$

The formula of nicotine may be written as

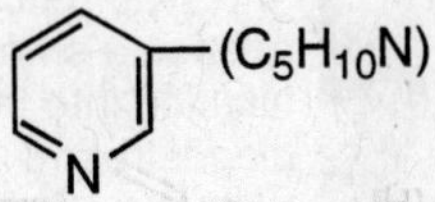

(4) **Structure of the Side-chain :**

(*i*) Nicotine zinc chloride complex on distillation with lime gives (*a*) pyridine ; (*b*) pyrrole ; and (*c*) methylamine.

$$C_{10}H_{14}N_2.ZnCl_2 \xrightarrow[\Delta]{\text{lime}}$$

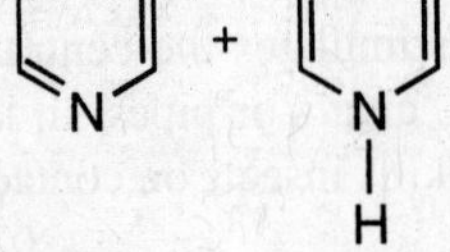

$$+ \underset{\text{Methylamine}}{CH_3NH_2}$$

Pyridine Pyrrole

This indicates that the side-chain is a pyrrole derivative.

(*ii*) When heated with concentrated hydriodic acid at 200-300°C, nicotine forms methyl iodide. This proves that CH_3 group in the side-chain is attached to N-atom.

$$\underset{\text{3°nitrogen}}{>N-CH_3} + HI \xrightarrow{\Delta} >N-H + CH_3I$$

(*iii*) To account for the side-chain, $(C_4H_7)N–CH_3$, it may be written as *N*-methylpyrrolidine group.

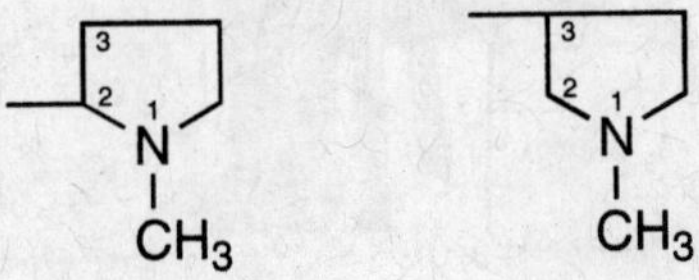

(*iv*) Nicotine hydriodide when treated with methyl iodide, forms nicotine isomethiodide which on oxidation with potassium ferricyanide yields *nicotone*. Nicotone on oxidation with chromium trioxide produces *hygrinic acid*.

$$\underset{\substack{\text{Nicotine}\\\text{isomethiodide}}}{C_{10}H_{14}N_2.CH_3I} \xrightarrow{[O]} \underset{\text{Nicotone}}{C_{10}H_{13}N_2O.CH_3} \xrightarrow{[O]}$$

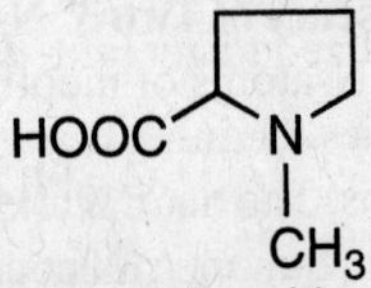

Hygrinic acid

Therefore, the pyrrolidine ring is bonded to pyridine ring through 2-position.

(5) **Structural formula.** From (3) and (4), the structural formula of nicotine may be constructed as

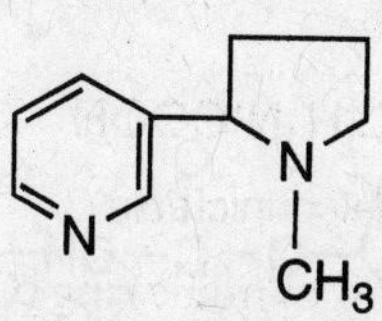

(6) **Synthesis.** The structure of nicotine has been confirmed by the following synthesis :

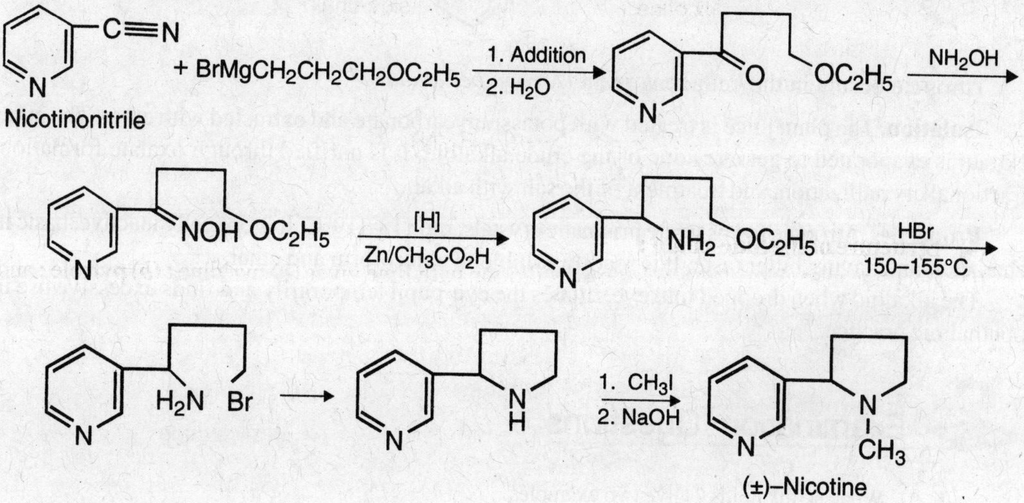

The product was resolved by means of (+)-tartaric acid. The synthetic (–)-nicotine was found to be identical with the natural alkaloid.

QUININE, $C_{20}H_{24}N_2O_2$

It is a quinoline alkaloid and has the structural formula

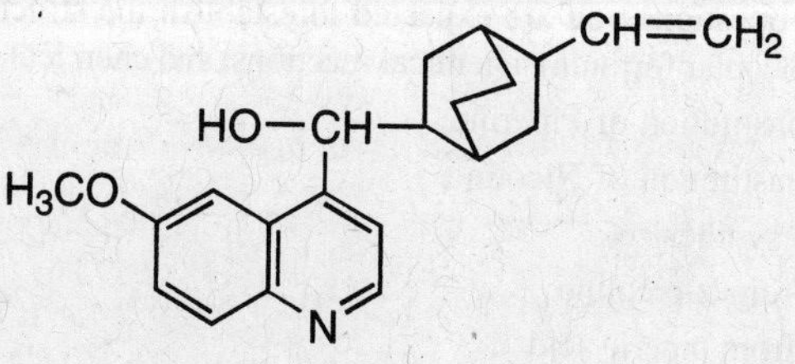

Quinine is present along with other alkaloids in the bark of *cinchona tree* which is cultivated in Java, Sri Lanka and India.

Isolation. The cinchona bark is finely powdered and treated with alkali. The alkaline mass is then extracted with petroleum ether. The extract is washed with dilute H_2SO_4 when the alkaloids dissolve as sulfates. The aqueous layer is concentrated when quinine sulfate being relatively less soluble, separates. It is recrystallized to get pure quinine sulfate.

Properties. Quinine is a white solid, mp 177°C. It is sparingly soluble in water to give intensely bitter solutions. The natural alkaloid is levorotatory, $[\alpha]_D = -148.2°$. It is weakly basic and forms salts with acids *e.g.*, quinine sulfate, $(C_{20}H_{24}O_2N_2)H_2SO_4$.

Quinine has been used for centuries as specific for treatment of malaria. Today, it has been largely replaced by other synthetic antimalarials, such as chloroquine.

ATROPINE, $C_{17}H_{23}NO_3$

The alkaloid contains a pyridine ring and a pyrrole ring with the N-atom common to both. It has the structural formula

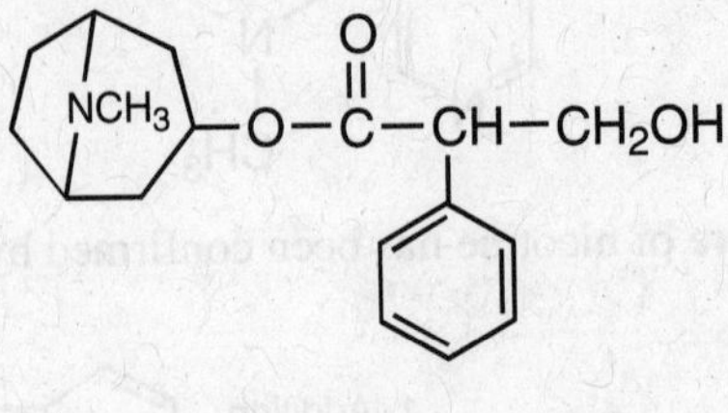

Atropine occurs in the belladona plant (*Atropa belladona*).

Isolation. The plant juice is treated with potassium carbonate and extracted with ether. The ether extract is evaporated to get a residue of the crude alkaloid. It is purified through oxalate formation, fractional crystallization, and treatment of the salt with alkali.

Properties. Atropine forms white prismatic crystals, mp 118-119°C. It is optically inactive, basic in character, and having bitter taste. It is readily soluble in chloroform and ether.

The alkaloid when dropped into eye, dilates the eye-pupil temporarily and finds excessive use in ophthalmic practice.

Examination Questions

1. What are alkaloids ? Give two examples.
2. How are alkaloids classified ?
3. What are alkaloids ? How are they isolated from plants ? Describe their general properties. *(Andhra BSc, 2014)*
4. How is nicotine isolated from tobacco leaves ?
5. Discuss the constitution (structure) of Coniine.

   Note. In such questions you are expected to establish the structure on the basis of analytical properties (molecular formula), chemical reactions, and chemical synthesis.
6. Discuss the constitution of Piperine.
7. Discuss the constitution of Nicotine.
8. How will you synthesize
   (*a*) Coniine from α-picoline
   (*b*) Piperine from piperic acid
   (*c*) Piperic acid from catechol
   (*d*) Nicotine from nicotinonitrile
   (*e*) Pyridine-2-carboxylic acid from coniine
9. Write a note on : Hofmann exhaustive methylation. *(Rajasthan BSc, 2015)*
10. What is the structural formula of Quinine ? What is its use.
11. What is the structural formula of Atropine ? What is its use.
12. (*True/False*) Alkaloids are natural products that contain mostly basic nitrogen atoms.

Answer. *True*

13. (*True/False*) India is the leading country for incorporating traditional herbal medicine, acupuncture, and modern medicine in its health care system.

Answer. *False*

14. (*True/False*) Herbal remedies are regulated by the Indian Food and Drug Administration and are required to undergo the same rigorous testing and approval as prescription drugs.

Answer. *False*

15. (*True/False*)The most effective treatment for malaria is quinine, the active alkaloid in the bark of the *Cinchona* tree.

Answer. *True*

Multiple Choice Questions

1. Alkaloids are :

(*a*) natural products, found in animals, that contain one or more alcohol groups
(*b*) natural products, found in animals, that contain one or more nitrogen heteroatoms
(*c*) natural products, found in plants, that contain one or more nitrogen hetero atoms
(*d*) alcohol-like compounds

Answer. (*c*)

2. Alkaloids

(*a*) contain nitrogen. (*b*) have a bitter taste.
(*c*) are usually acidic. (*d*) only (*a*) and (*b*) are correct.

Answer. (*d*)

3. Quinine is obtained from the bark of which tree ?

(*a*) Cinchona (*b*) Redwood
(*c*) Bunyan (*d*) Eucalyptus

Answer. (*a*)

4. Nicotine causes which of these changes in the body ?

(*a*) lowers body temperature (*b*) increases heart rate
(*c*) increases blood pressure (*d*) (*b*) and (*c*)

Answer. (*d*)

5. What happens when nicotine comes into contact with the brain ?

(*a*) tissues swell (*b*) dopamine is released
(*c*) nerve centers shut down (*d*) adrenaline increases

Answer. (*b*)

6. Which of these is the most significant health risk for smokers ?

(*a*) cancer (*b*) high blood pressure
(*c*) heart attacks (*d*) arthritis

Answer. (*a*)

7. Smokers crave cigarettes because they ....

(*a*) like the smell of cigarette smoke
(*b*) are addicted to nicotine
(*c*) like the way cigarettes make them look

Answer. (*b*)

48

Terpenes

Terpenes are a class of compounds that give plants their odor, flavor, and in some cases color. They occur widely in the leaves and fruits of higher plants as conifers, citrus, and eucalyptus. *When the plant source is distilled with steam, the oily materials so obtained are called* **Essential Oils** (essential parts of plants).

The essential oils are generally composed of mixture of either the hydrocarbons (*polyenes*) having general formula $(C_5H_8)_n$, or their oxygen derivatives (alcohols, aldehydes, ketones). This class of compounds are designated as **Terpenes** or **Terpenoids**.

Points to Remember

1. Terpenes are isolated from plants.
2. An isoprene unit is a chain of four carbons with one carbon attached to the second carbon of the four-carbon chain.
3. All terpenes contain 2,3,4,5,6 or 8 isoprene units.
4. The structure of terpenes are frequently represented with line formulas. A corner is meant to represent a carbon atom with appropriate number of hydrogens. A line branching off the structure represents a methyl branch, not a hydrogen. For example,

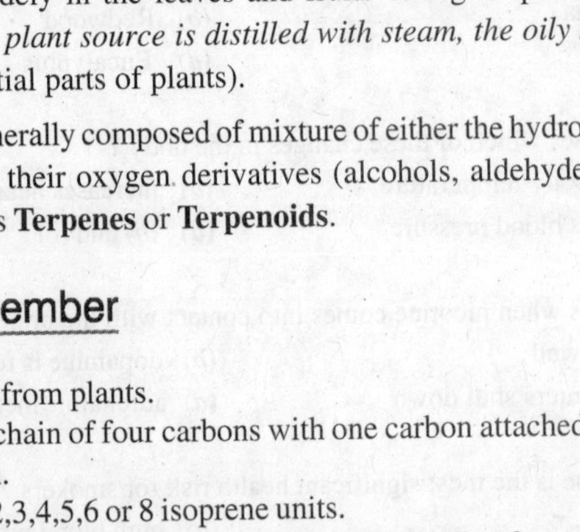

Myrcene Line formula

ISOPRENE RULE

From a study of the molecular structure of a large number of the then known terpenes, Otto Wallach (Noble Prize 1910) gave the so-called **Isoprene Rule**. It states that : **The molecules of all terpenes are constructed of two or more isoprene (iso-C_5) units, usually joined in a head-to-tail fashion.** Isoprene is 2-methyl-1,3-butadiene and may be represented as :

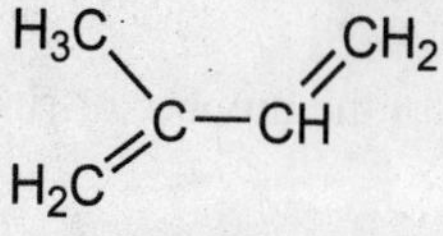

Isoprene

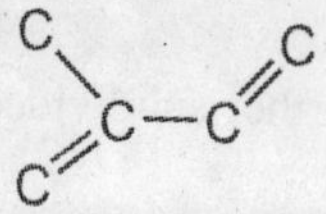

Condensed formula

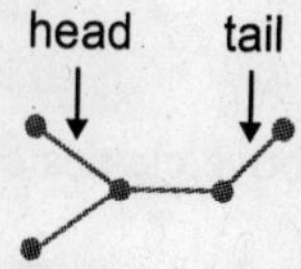

Isoprene skeleton

In applying Isoprene rule, we look only for the skeletal unit iso-C_5, neglecting the number and position of double bonds. Thus the terpene known as *myrcene* could be dissected by the dashed line into two isoprene units that are joined 'head-to-tail'.

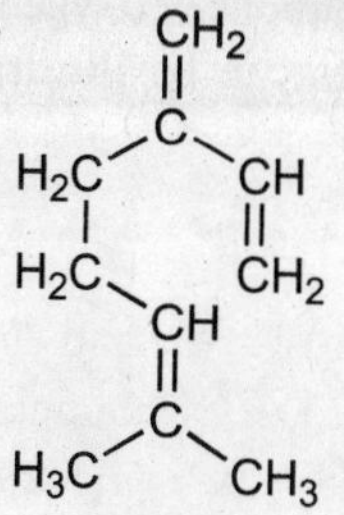

Myrcene

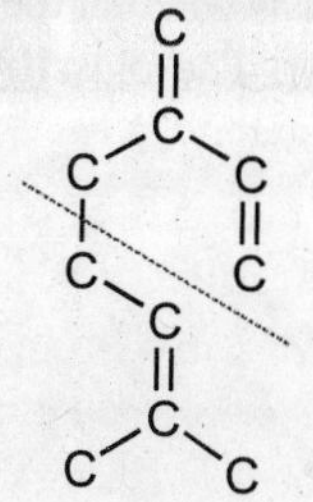

Condensed formula

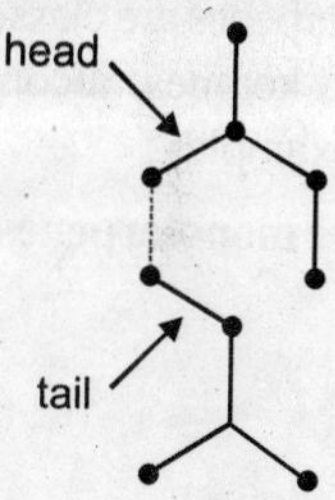

Carbon skeleton (dotted bond shows head-to-tail arrangement)

Isoprene rule although generally valid, is not universal. However, it has proved of great help in deriving the structure of terpenes.

LINE FORMULAS OF TERPENES

The structures are frequently written with line formulas. The carbon-carbon bonds are represented by lines. The carbon atoms with appropriate number of hydrogens are understood to be present at corners or junctions, and the ends of lines. Thus the line formulas of *myrcene* and *menthol* may be written as

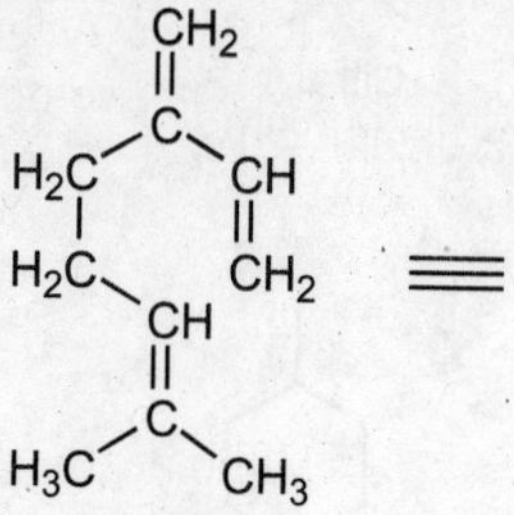

Myrcene

Line formula

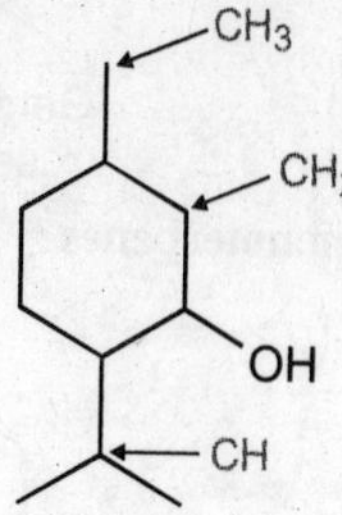

Menthol

CLASSIFICATION OF TERPENES

Terpenes are classified according to the number of isoprene units (C_5) in the molecule. The simplest terpenes have two isoprene units (ten carbons) and are called **Monoterpenes**. In fact, the designation **Terpenes** is by custom specifically reserved for the C_{10} compounds. Other classes are listed below :

Monoterpenes	Two isoprene units	C_{10}
Sesquiterpenes	Three isoprene units	C_{15}
Diterpenes	Four isoprene units	C_{20}
Triterpenes	Six isoprene units	C_{30}
Tetraterpenes	Eight isoprene units	C_{40}

The above classes of terpenes are further subdivided according to the number of rings in the molecule :

Acyclic	No rings (Straight chain)
Monocyclic	One ring
Bicyclic	Two rings
Tricyclic	Three rings, and so on.

Then terpenes are classified on the basis whether they are pure hydrocarbons or oxygen derivatives (aldehydes, ketones, alcohols or ethers). The classification of terpenes can be illustrated by the following examples :

Acyclic monoterpenes :

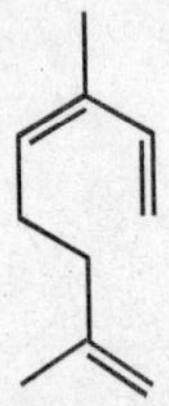

Ocimene
(basil oil)

Myrcene
(bayberry oil)

Acyclic oxygenated monoterpenes :

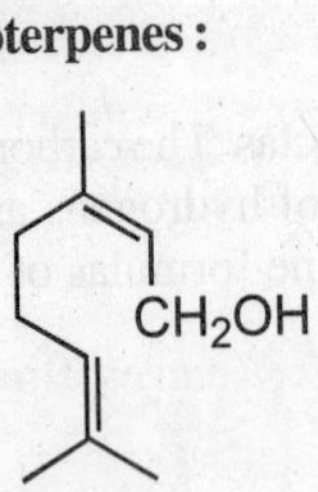

Geraniol
(rose oil)

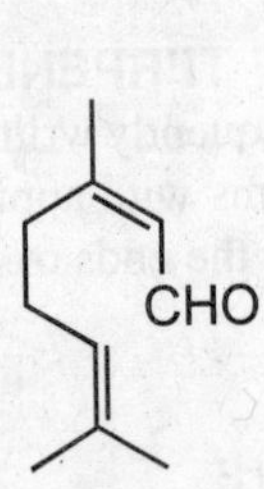

Citral
(lemon oil)

Monocyclic monoterpenes :

Limonene
(lemon and orange oil)

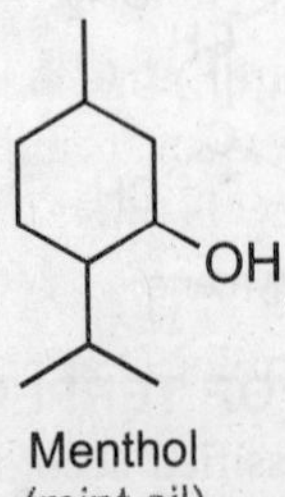

Menthol
(mint oil)

Bicyclic monoterpenes :

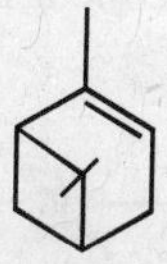

α-Pinene
(turpentine oil)

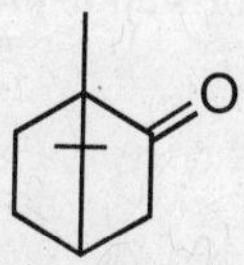

Camphor
(camphor oil)

The discussion of *sesqui-*, *di-*, *tri-*, and *tetraterpenes* is beyond the scope of this book.

Sesquiterpenes

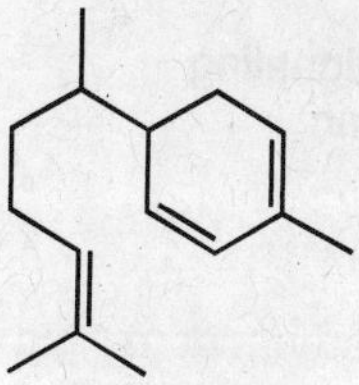

Zingiberene
(ginger oil)

Diterpene

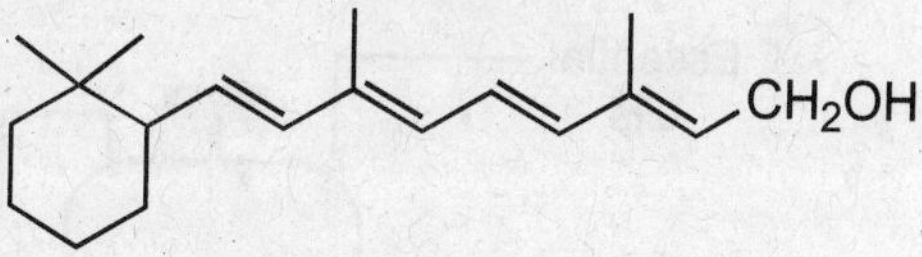

Vitamin A

Myrcene is found in hops. It is a muscle relaxant and sleep aid.

α-Pinene is found in turpentine oil and pine needles. It is anti-inflammatory and anti-bacterial.

ISOLATION OF TERPENES

Essential oils are first extracted from the plant source (leaves, flowers, stem or root) mainly by three methods :

(1) **Steam Distillation.** The mascerated plant material is steam distilled. The oil, if any, is collected separately. The aqueous steam distillate is saturated with salt and extracted with a purified solvent as light petroleum or benzene. The combined oil and the solvent extract are dried. The solvent is then removed by evaporation under reduced pressure to give the essential oil.

(2) **Direct Solvent Extraction.** If a particular terpene is decomposed under the conditions of steam distillation, the plant material is directly extracted with *light petroleum* or *ether* at room temperature. The extract is filtered and the solvent removed by evaporation under *vacuo* to recover the essential oil.

(3) **By Fat Adsorption (*Enfleurage Process*).** The flower petals are spread over a molten layer of fat (tallow and lard) for several days. The fat enriched with adsorbed essential oil from petals, is stirred with pure ethanol. The ethanol extract is then evaporated at 0°C in vacuum to give the essential oil.

The essential oils obtained as above are usually mixture of many terpenes. These are separated by fractional distillation or vapor-phase chromatography. Chemical methods are also used where possible.

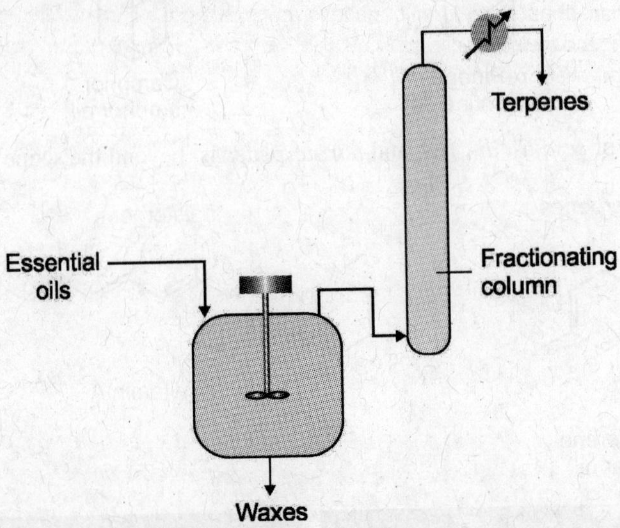

Fig. 48.1. Separation of terpenes from essential oils.

GENERAL PROPERTIES OF TERPENES

(Physical). (1) Most terpenes are colourless, fragrant liquids having high refractive indices. A few of them e.g., camphor, are solids.

(2) They are lighter than water and readily volatile in steam.

(3) They are soluble in organic solvents (light petroleum, ether, benzene) but usually not in water.

(4) They are optically active and a number of them possess antiseptic properties.

(Chemical). (1) They give the usual addition reactions of the carbon-carbon double bonds (*e.g.*, with HBr, Br_2, H_2, HOCl, O_3) and the aliphatic rings.

(2) They also show reactions characteristic of the functional groups such as –CHO, >CO, and –OH, when present in the molecule.

IMPORTANT TERPENES

CITRAL, $C_{10}H_{16}O$

It is the most important acyclic (open chain) monoterpenoid and has the structural formula.

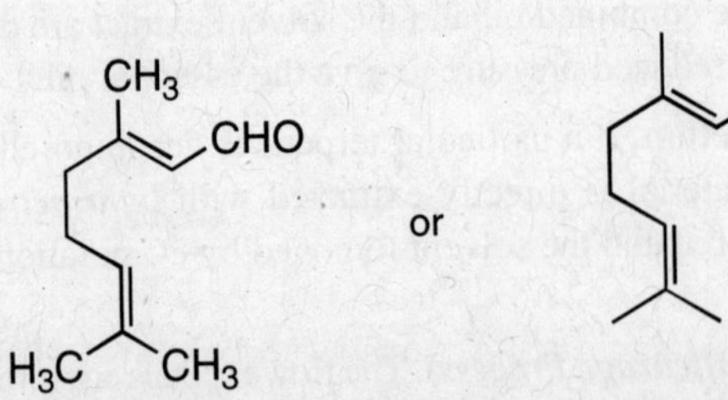

Citral occurs to the extent of 80 per cent in lemongrass oil. It is also present in the oil of citrus fruits (citron, orange, lemon)

Preparation. It is obtained from lemongrass oil by treatment with sodium bisulfite solution. Crystalline citral bisulfite derivative separates. This upon hydrolysis with sodium carbonate gives pure citral.

Properties. Citral is a colourless liquid, bp 224-228°C. It has a strong lemon-like odor. It exhibits geometrical isomerism about the double bonded carbons carrying CH_3 and CHO groups. The *cis*-isomer is known as Citral-*a* and the *trans*-isomer Citral-*b*. Ordinary citral obtained from lemongrass oil is, in fact, mixture of Citral-*a* (90%) and Citral-*b* (10%)

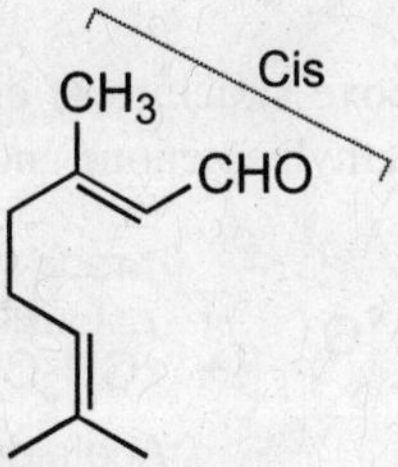

Citral-*a*

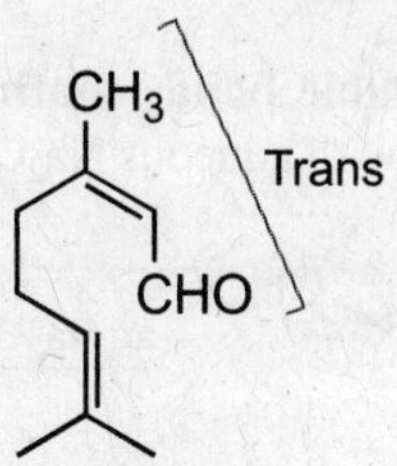

Citral-*b*

Uses. Citral is used extensively in the perfume and flavor industry to stimulate lemon-like odor, and for the manufacture of vitamin A. Recently citral has become important as a drug for reducing blood pressure.

Structure. (1) Elemental analysis and molecular weight determinations show that the molecular formula of citral is $C_{10}H_{16}O$.

(2) **Presence of two C=C.** It adds two molecules of bromine to form a tetrabromide, indicating the presence of two carbon-carbon double bonds.

$$\underset{\text{Citral}}{C_{10}H_{16}O} + 2Br_2 \longrightarrow \underset{\text{Tetrabromide}}{C_{10}H_{16}OBr_4}$$

(3) **Presence of CHO.** It forms an oxime and on oxidation gives geranic acid without loss of any carbon atom.

$$\underset{\text{Citral}}{(C_9H_{15})-\overset{\overset{\displaystyle O}{\|}}{C}-H} + [O] \xrightarrow{Ag_2O} \underset{\text{Geranic acid}}{(C_9H_{15})-\overset{\overset{\displaystyle O}{\|}}{C}-OH}$$

(4) **Carbon skeleton.** Simmler (1891) suggested that two isoprene units are joined head-to-tail in citral (*Isoprene rule*), and its carbon skeleton was

(5) **Position of C=C bonds indicated.** The oxidation of citral with alkaline permanganate, followed by chromic acid, gives acetone, levulinic acid, and oxalic acid. This is accountable only if the position of the double bonds is as shown in the formula below.

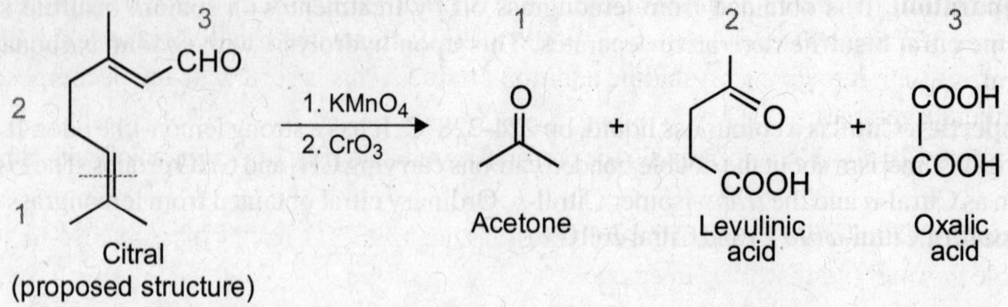

(6) **Position of double bonds confirmed.** The above structure of citral is supported by the decomposition of citral with aqueous Na_2CO_3 to give methylheptenone and acetaldehyde.

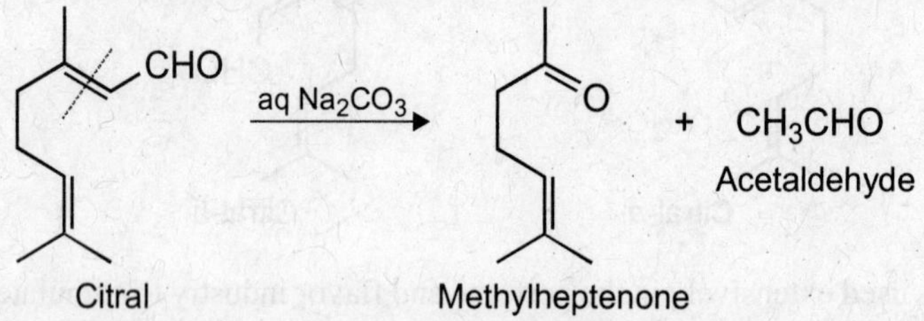

(7) **Synthesis.** The structure of citral was confirmed by the following synthesis from methylheptenone. The various steps are (*i*) Reformatsky reaction ; (*ii*) Hydrolysis with dilute acid; (*iii*) Dehydration with acetic anhydride ; (*iv*) Distilling calcium geranate with calcium formate.

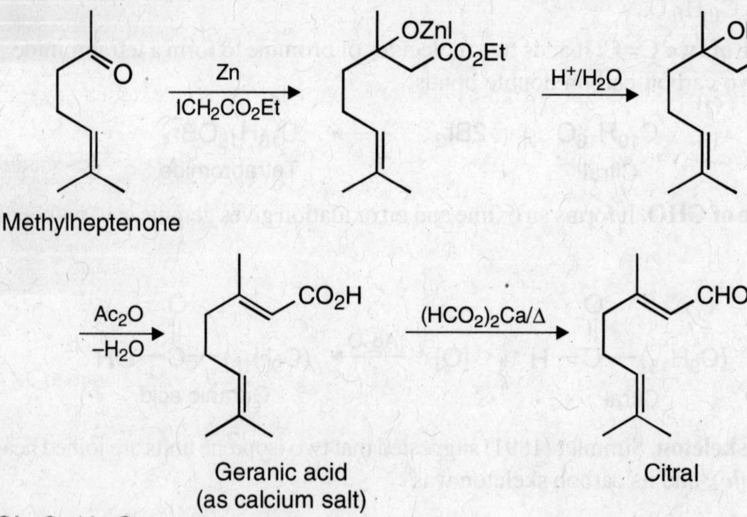

GERANIOL, $C_{10}H_{18}O$

It is an acyclic terpene alcohol and has the structure :

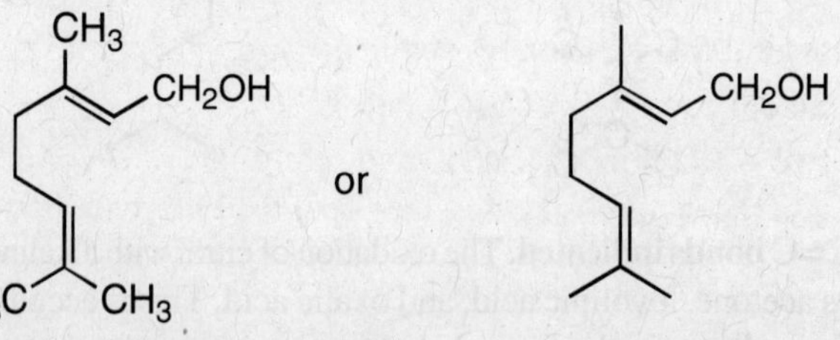

Geraniol occurs in oils of rose, palmrosa, geranium, citronella, lemongrass, and lavender.

Preparation. (1) **From Palmrosa Oil.** This is obtained cheap from a wild-growing grass, *Cymbopogon Martini*, by steam distillation. When the oil is treated with anhydrous calcium chloride, geraniol reacts with it to give a crystalline addition product. This is separated and decomposed with water to liberate geraniol.

(2) **From Citral-a**. Geraniol may also be prepared by reduction of Citral-*a* (geraniol) with aluminium amalgam.

Properties. (*Physical*). Geraniol is a colourless liquid, bp 230°C, having pleasant rose-like odor. It is insoluble in water but dissolves in ethanol.

Geraniol exhibits geometrical isomerism and it is *cis*-form. The *trans*-isomer is another terpene **Nerol** (from *neroli oil*), colourless liquid, bp 225°C, having rose-like odor.

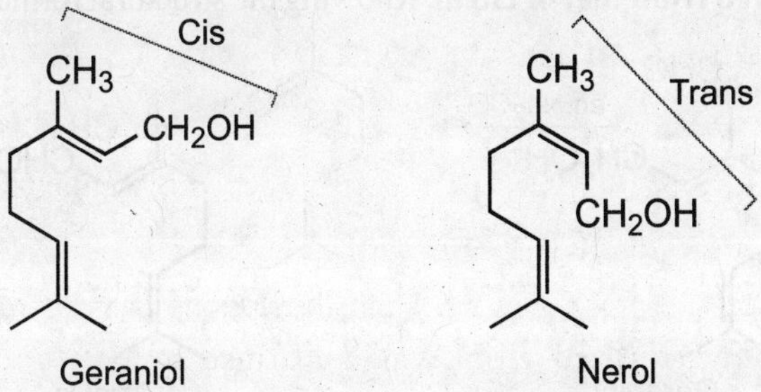

Geraniol (*cis*-form)   Nerol (*trans*-form)

(*Chemical*). The structural formula of geraniol has two carbon-carbon double bonds and a primary alcohol group. Thus it gives the reactions of dienes as also of primary alcohols. Two of its important reactions are listed below.

(1) **Oxidation.** Upon oxidation, geraniol is first converted to citral-*a* and then to geranic acid.

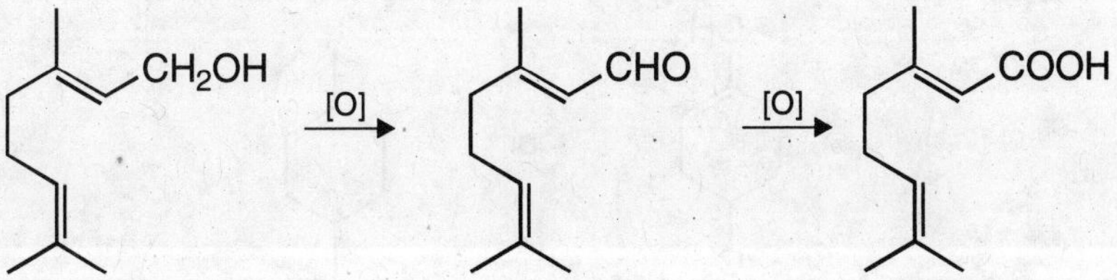

Geraniol   Citral-a   Geranic acid

(2) **Cyclization.** When treated with dilute sulfuric acid, geraniol forms α-terpineol.

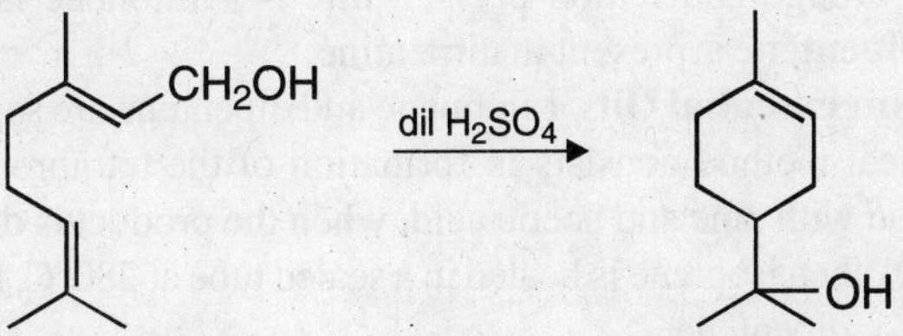

Geraniol   α-Terpineol

Uses. Geraniol and nerol are extensively used in perfumery to stimulate the odor of roses.

Structure. (1) Geraniol has the molecular formula $C_{10}H_{18}O$.

(2) **Presence of two C=C bonds**. It adds two molecules of bromine, and two molecules of hydrogen on catalytic hydrogenation, showing the presence of two carbon-carbon double bonds.

$$\underset{\text{Geraniol}}{C_{10}H_{18}O} + 2Br_2 \longrightarrow \underset{\text{Tetrabromide}}{C_{10}H_{18}OBr_4}$$

$$C_{10}H_{18}O \quad + \quad 2H_2 \xrightarrow[\Delta]{Ni} C_{10}H_{22}O$$

Saturated alcohol

(3) **Presence of CH_2OH.** On oxidation it gives an aldehyde (citral-*a*) and then an acid (geranic acid) containing the same number of carbon atoms as geraniol.

$$\underset{\text{Geraniol}}{(C_9H_{15})CH_2OH} \xrightarrow{[O]} \underset{\text{Citral-}a}{(C_9H_{15})CHO} \xrightarrow{[O]} \underset{\text{Geranic acid}}{(C_9H_{15})COOH}$$

This proves that geraniol is a primary alcohol and the arrangement of C atoms as also the positions of the two double bonds are the same as in Citral-*a*.

(4) **Structure derived from that of citral.** Knowing the structural formula of citral-*a*, geraniol is assigned the structure.

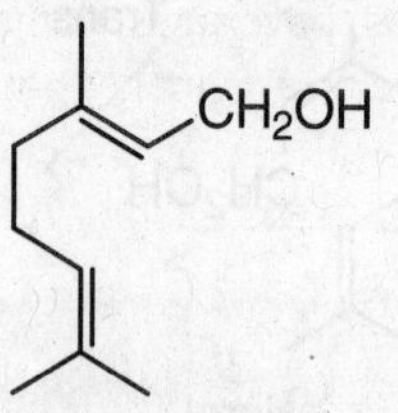

Geraniol

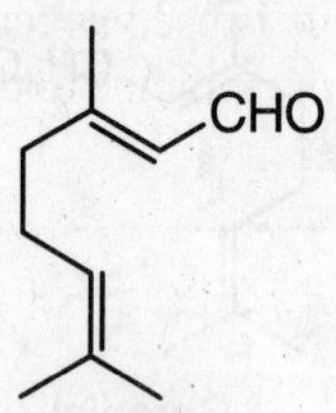

Citral-*a*
(*cis*-form)

Therefore, geraniol is also the *cis*-form which is confirmed by NMR studies.

LIMONENE, $C_{10}H_{16}$

It is the most widely distributed terpene and has the structure :

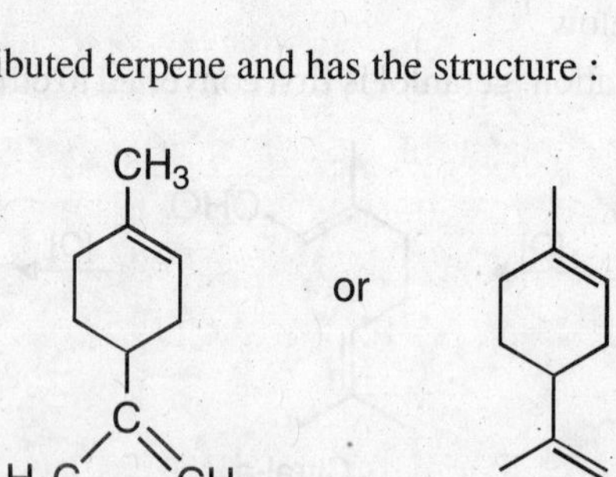

(+)-Limonene occurs in orange and lemon peels, while (–)-limonene is found in pine needles. (±)-limonene known as **Dipentene** is present in turpentine.

Preparation. (1) **From Essential Oils.** Limonene and dipentene are separated from the relevant essential oils. The chemical method consists in formation of the tetrabromide by addition of Br_2, followed by debromination with zinc and acetic acid, when the product is dipentene.

(2) **From Isoprene.** When isoprene is heated in a sealed tube at 280°C, Diels-Alder reaction with itself yields (±)-limonene or dipentene.

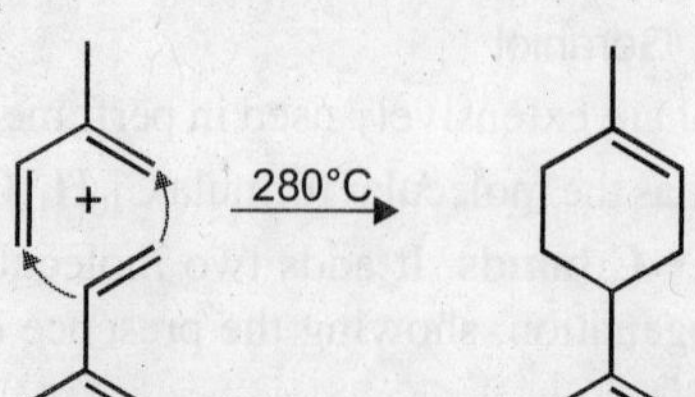

Isoprene (2 molecules) Dipentene

Properties. (***Physical***). Limonene is a liquid, bp 178°C. It has lemon-like odor. It is insoluble in water, but dissolves in ethanol and diethyl ether.

(***Chemical***). It gives the reactions of an alkadiene. On catalytic hydrogenation (S, Δ) it gives *p*-menthane, while with bromine it yields a tetrabromide. When heated in the presence of palladium (catalyst), it disproportionates to *p*-cymene and *p*-menthane.

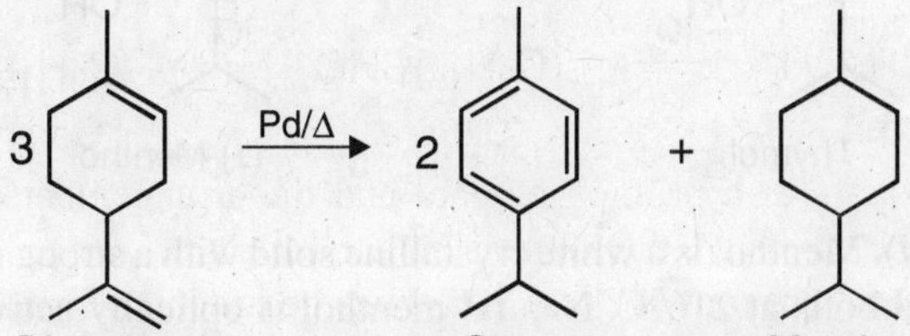

Dipentene *p*-Cymene *p*-Menthane

Uses. (1) Limonene is used as favoring agent in foods, beverages, dental and shaving creams ; (2) Dipentene is used in medicine and for the manufacture of *p*-cymene, *p*-menthane, and synthetic resins.

(+)-Limonene

"Being a solvent of cholesterol, (+)-limonene has been used clinically to dissolve cholesterol containing gallstones. (+)-Limonene has well-established chemopreventive activity against many types of cancer

(+)-Limonene is found in citrus essential oils.

Orange 90%	Grapefruit 88%	Lemon 65%	Lime 55%

MENTHOL, $C_{10}H_{20}O$

It is a monocyclic monoterpene secondary alcohol, having the structure :

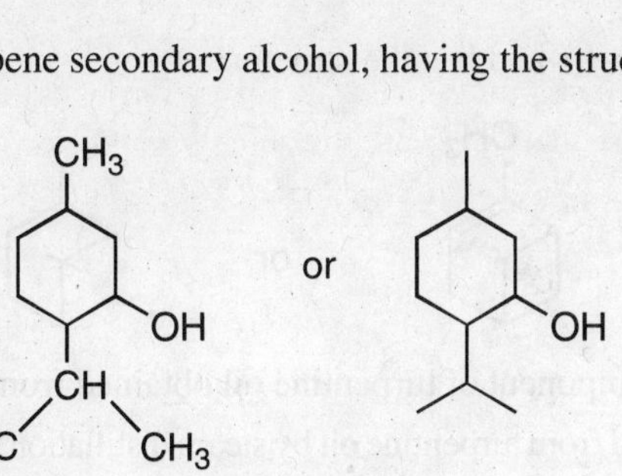

Menthol occurs in peppermint oil and Japanese mint oil.

Preparation. (1) **From Peppermint or Mint Oil**. The oil is chilled when crystals of menthol separate. These are purified by recrystallization.

(2) **From Thymol.** Menthol is also obtained by catalytic hydrogenation of thymol.

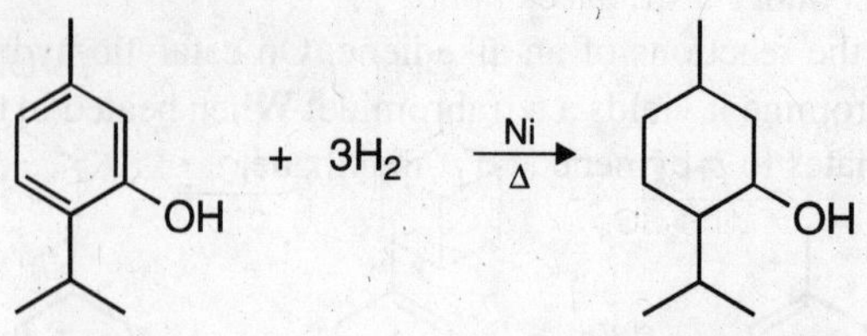

Thymol (±)-Menthol

Properties. (*Physical*). Menthol is a white crystalline solid with a strong minty odor and cooling taste. It melts at 42°C and boils at 216°C. Natural menthol is optically active, the (–)-form. It has anaesthetic and antiseptic action.

(*Chemical*). It gives the reactions of an alicyclic secondary alcohol.

(1) **Oxidation.** Upon oxidation with chromic acid, it gives the corresponding ketone, menthone.

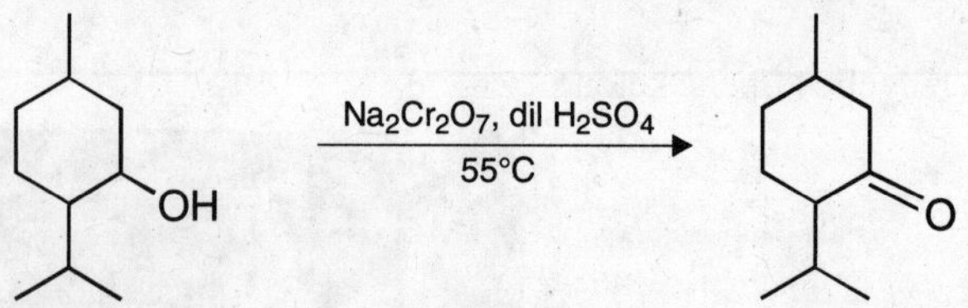

Menthol Menthone

(2) **Dehydration**. When heated with potassium hydrogen sulfate, it gives menthene.

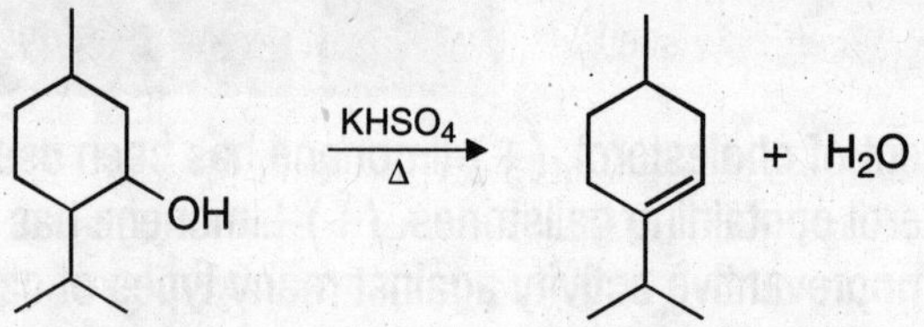

Menthol Menthene

Uses. Menthol, on account of its anaesthetic and antiseptic action, pleasant odor and cooling taste, is used : (1) in pharmaceuticals as local anaesthetic and for relief of headache ; (2) in mouth washes, nasal sprays and inhalers (**Vick's Inhaler**), cough drops, chewing-gums, and mentholated cigarettes ; (3) in cosmetics – toothpastes, shaving creams, shaving lotions, face creams and powders.

α-PINENE, $C_{10}H_{16}$

It is a *bicyclic monoterpene* and has the structure :

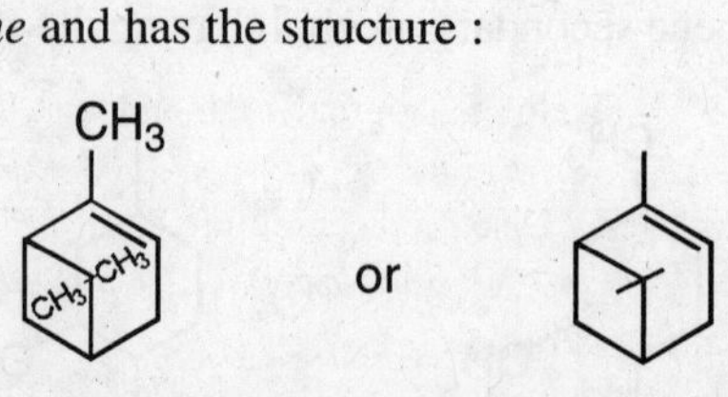

α-Pinene is the chief component of turpentine oil obtained from the pine tree.

Preparation. It is isolated from turpentine oil by steam distillation followed by fractional distillation. It is purified by conversion into nitrosyl chloride which upon treatment with aniline liberates α-pinene.

Properties. α-Pinene is a colourless liquid, bp 156°C. It has a characteristic odor and is optically active, $[\alpha]_D = 48.8°$.

(1) **Reaction with HCl**. It reacts with dry hydrogen chloride at 0°C to form bornyl chloride, mp 131°C, with a faint camphor odor ('artificial camphor').

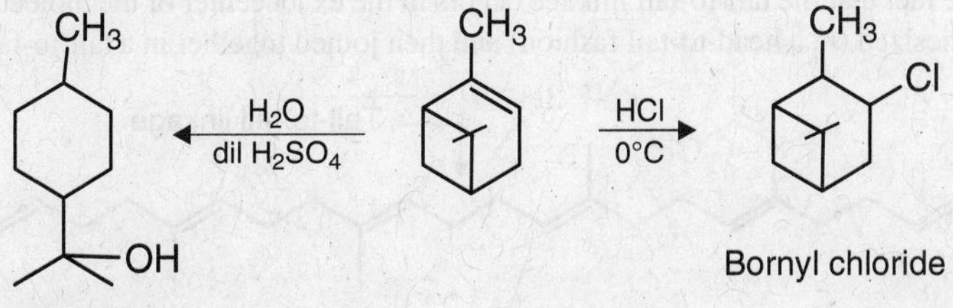

(2) **Reaction with H_2SO_4**. When treated with dilute H_2SO_4, the 4-carbon ring in terpene cleaves to form α-Terpineol.

Uses. Turpentine oil which is mainly α-pinene is used as a thinner for paints and for the commercial synthesis of camphor.

CAMPHOR, $C_{10}H_{16}O$

It is a well-known bicyclic terpene ketone

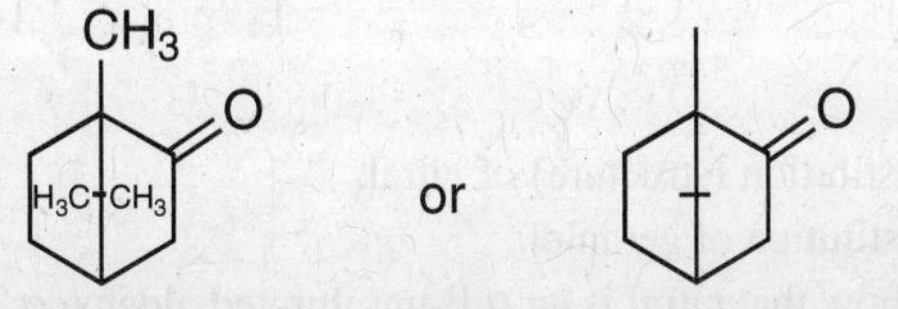

Preparation. It is prepared by chilling *Camphor oil* obtained by steam distillation of wood and leaves of the camphor tree which grows in Formosa. Now it is mostly produced by a synthetic method starting from α-pinene.

Properties. Camphor is a colourless crystalline solid, sp gr 0.999, mp 179°C, bp 209.1°C. It sublimes at room temperature and its vapor have a strong characteristic smell. Natural camphor is dextrorotatory, $[\alpha]_D = +44°$. Chemically it behaves as a ketone. It forms an oxime and on reduction forms a secondary alcohol, *Borneol*.

Uses. Camphor is used as : (*i*) a medicinal, incense ; (*ii*) plasticizer ; (*iii*) moth repeller ; and (*iv*) in embalming fluids.

Examination Questions

1. What are terpenes ? How are they classified ?

2. Write a note on : Isoprene Rule (*Agra BSc, 2013 ; Baroda BSc, 2015*)

3. What is Isoprene rule ? Indicate the isoprene units in the structure of citral and α-pinene. (*Delhi BSc, 2014 ; Dibrugarh BSc, 2015*)

4. How are terpenes classified ? Discuss the structure of any one terpene, along with its synthesis. (*Burdwan BSc, 2013 ; Madras BSc, 2014 ; Delhi BSc, 2015*)

5. Mark the isoprene units in geraniol.

Answer.

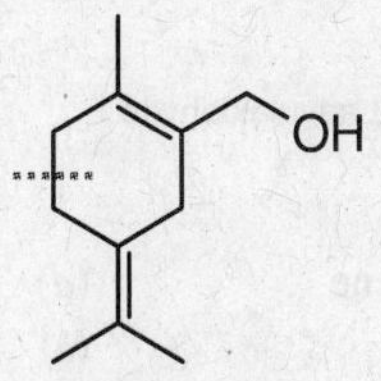

Geraniol

6. Mark the isoprene units in squalene. One of the linkages in squalene is tail-to-tail. What does this suggest about how squalene is synthesized in nature ?

Answer. The fact that the tail-to-tail linkage occurs in the exact center of the molecule indicates that the two halves are synthesized (in a head-to-tail fashion) and then joined together in a tail-to-tail linkage.

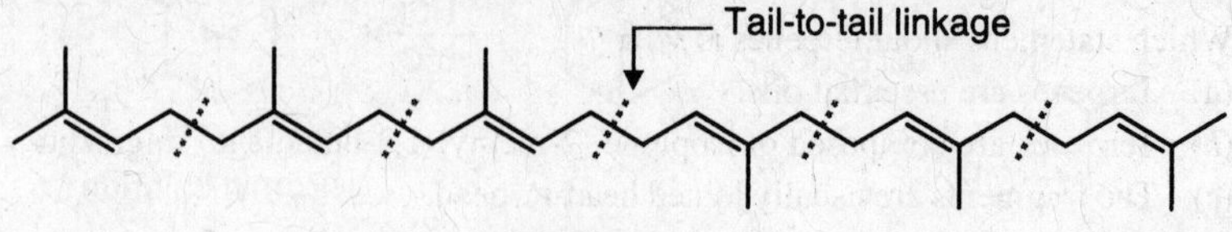

Squalene

7. Mark the isoprene units in β-carotene.

Answer. Squalene, β-carotene is synthesized in the same way. Two halves are synthesized (in a head-to-tail fashion) and then joined together in a tail-to-tail linkage.

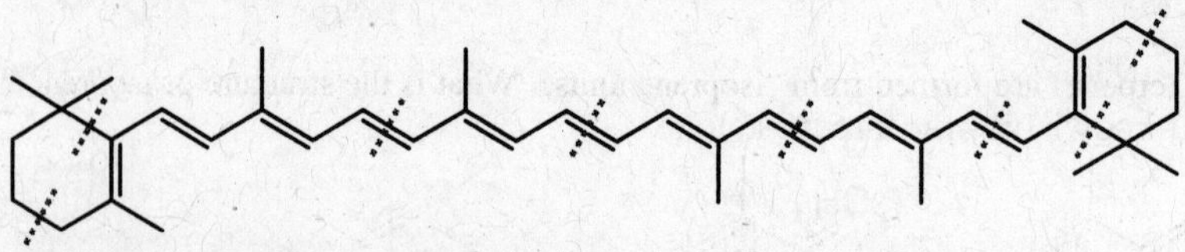

8. Discuss the constitution (structure) of citral.

9. Discuss the constitution of geraniol.

10. How will you show that citral is an α,β-unsaturated aldehyde ?

11. How can position of double bonds be established in citral ?

12. What are the structural formulas and uses of :

(*a*) Menthol (*b*) α-Pinene (*c*) Camphor

13. What happens when :

(*a*) Citral is treated with aqueous Na_2CO_3

(*b*) Geraniol is oxidized with $Na_2Cr_2O_7/H_2SO_4$ (*Goa BSc, 2015*)

(*c*) Isoprene is heated in a sealed tube at 280°C

(*d*) Thymol is heated with hydrogen in the presence of nickel catalyst

(*e*) Menthol is heated with $KHSO_4$ (*Kalyani BSc, 2014*)

(*f*) α-Pinene is heated with dilute H_2SO_4

Answer. The products are : (*a*) Methylheptenone and acetaldehyde ; (*b*) Geranic acid ; (*c*) Dipentene ; (*d*) Menthol ; (*e*) Menthene ; and (*f*) α-Terpineol. Reactions are given in the text.

14. What happens when citral is subjected to ozonolysis ? (*Madurai BSc, 2015*)

Answer. Mixture of laevualdehyde, glyoxal and acetone is formed.

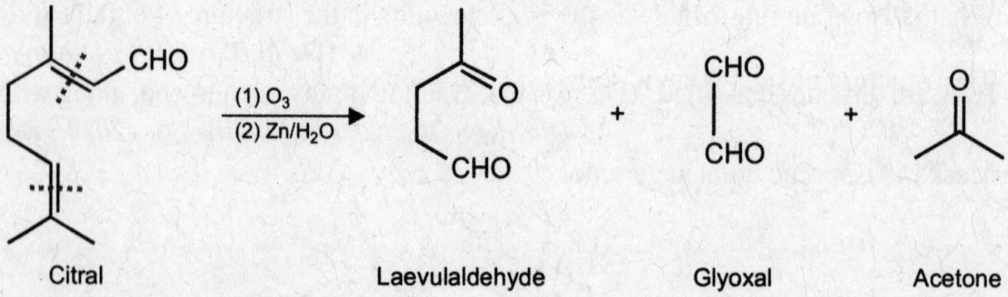

Citral Laevulaldehyde Glyoxal Acetone

15. How will you synthesize :

(*a*) α-Terpineol from α-pinene

(*b*) Geraniol from citral

(*c*) *p*-Cymene from citral

(*d*) Menthol from thymol

(*e*) Citral from methylheptanone

Multiple Choice Questions

1. Which statement about terpenes is *false* ?

(*a*) Terpenes are essential oils.

(*b*) Terpenes are composed of isoprene (2-methyl-1,3-butadiene) fragments.

(*c*) The fragments are usually joined head-to-head.

(*d*) Terpenes are also known as isoprenoids.

Answer. (*c*)

2. Which structure is the of isoprene ?

(*a*) $CH_3–CH(CH_3)–CH=CH_2$ (*b*) $H_2C=C(CH_3)–CH=CH_2$

(*c*) $CH_3–CH_2–CH_2–CH=CH_2$ (*d*) $CH_3–CH=CH–CH=CH_2$

Answer. (*b*)

3. Terpenes are formed from "isoprene units." What is the structure of isoprene ?

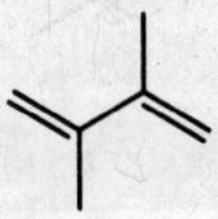
I

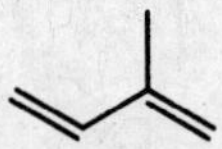
II

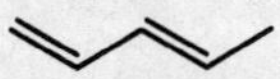
III

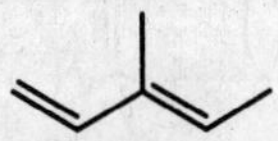
IV

(*a*) I (*b*) II (*c*) III (*d*) IV

Answer. (*b*)

4. What is the "isoprene rule" ?

(*a*) Terpenes contain "head-to-head" connected isoprenes.

(*b*) Terpenes contain "head-to-tail" connected isoprenes.

(*c*) Terpenes contain "tail-to-tail" connected isoprenes.

(*d*) Terpenes contain 5_n atoms.

Answer. (*b*)

5. How many isoprene units are there in monoterpene ?

(*a*) 1 (*b*) 2 (*c*) 3 (*d*) 4

Answer. (*b*)

6. How many C_5 units are there in a diterpene ?

(*a*) 1 (*b*) 2 (*c*) 3 (*d*) 4

Answer. (*d*)

7. How many double bonds occur in the triterpene squalene ?

(*a*) 1 (*b*) 2 (*c*) 3 (*d*) 6

Answer. (*d*)

8. The structure of menthone is shown below. Methone is a

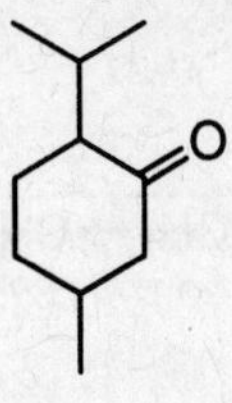

(*a*) monoterpene (*b*) sesquiterpene (*c*) diterpene (*d*) triterpene

Answer. (*a*)

49 Synthetic Polymers

The Miracle of Teflon. Teflon is heat resistant and chemically inert.

More chemists work in the field of *polymer chemistry* than in any other area of chemistry. A quick rundown of the familiar names of some typical man-made synthetic polymers will show how closely these products touch the lives of everyone : Nylon, PVC, Teflon, Polyethylene, Dacron, Orlon, Rubbers. When we also consider some natural polymers such as cellulose, starch, proteins, and natural rubber, the magnitude of polymer chemistry becomes apparent. In this chapter we will study only the chemistry of synthetic polymers.

Polymers are high molecular weight compounds whose structures are made up of a large number of simple repeating units. The repeating units are usually obtained from low molecular weight simple compounds referred to as monomers. The reaction by which monomers are converted into polymers is known as polymerization. The formation of polyethylene from ethylene is an example of polymerization reaction.

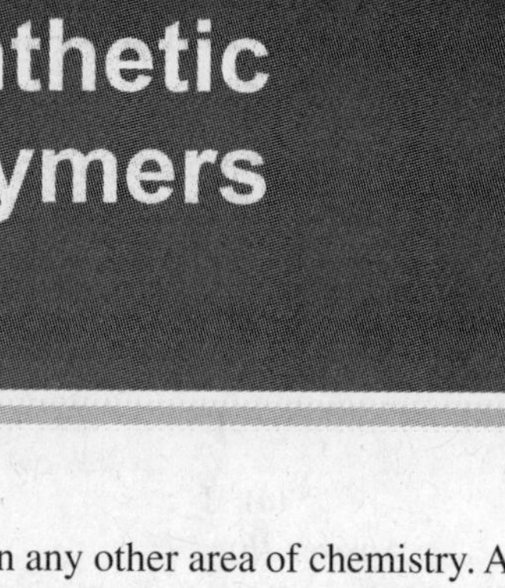

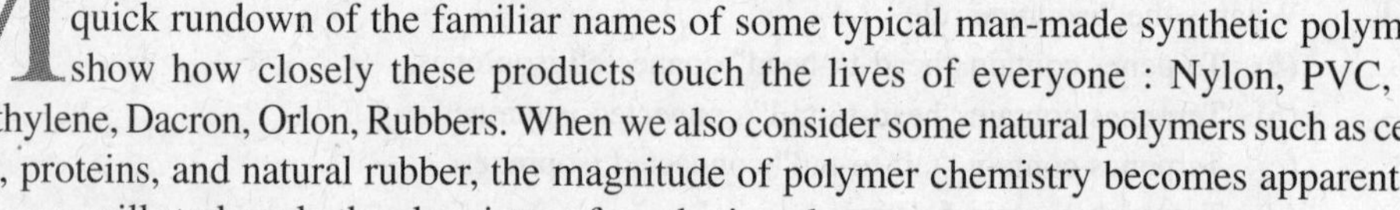

Polymers which are synthesized from only one kind of monomer are called **Homopolymers**. Polymers which are prepared from more than one kind of monomer are called **Copolymers**.

SHORTHAND FOR WRITING POLYMER REACTIONS

The polymerization reactions are time-consuming to write. Chemists have a shorthand way of writing the structural formula for a polymer molecule and the equation for a polymerization reaction. The monomer is shown on the *left-hand* (reactants) side of the equation. On the right-hand (products) side, the repeating polymer unit is enclosed in brackets, and lines pass through the brackets to show that the chain continues in both directions. The subscript *n* shows how many times the repeating unit appears in the full polymer ; generally the value of *n* is in the thousands.

$$\underset{\text{Ethylene}}{n\,CH_2{=}CH_2} \xrightarrow{\text{Polymerization}} \underset{\text{Polyethylene}}{-\!\!\left[CH_2{-}CH_2\right]_n\!\!-}$$

Frequently the end groups of polymers are unknown-they may arise from impurities in the reaction mixture. In some cases, the end groups can be controlled. The properties of a polymer are governed almost entirely by the bulk of the polymer molecule rather than the end groups. To emphasize the basic structure of the polymer, it is customary not to include the end groups in the formula.

Polymers are said to be **linear** if the repeating units are joined together like links in a chain. The chains may be **branched**, or joined together by cross-links or both. More extensive cross-linking may lead to the formation of **3D cross-linked** polymer (Fig. 49.1).

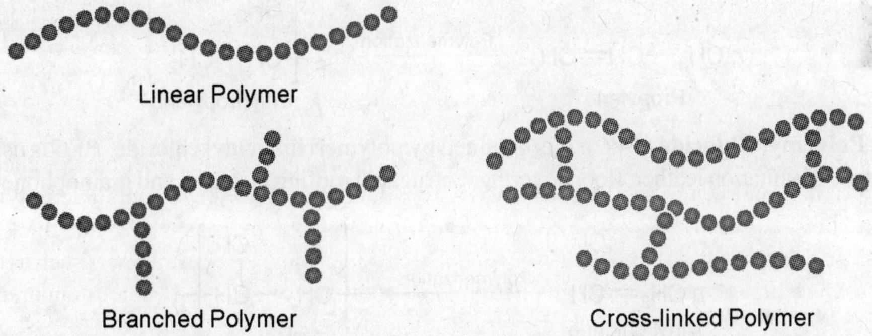

Fig. 49.1. Polymer structure. Each circle represents a single repeating unit.

CLASSIFICATION OF POLYMERS

There are two main types of polymers : Addition polymers and Condensation polymers.

ADDITION POLYMERS (*Chain-Growth Polymers*)

Addition polymers are formed by combination of alkene monomers to produce a single huge molecule only. These reactions are catalyzed by peroxides or acids. The reactions require pressures of 1000 atmospheres at 200°C. Much lower temperatures and pressures can be used with so called **Ziegler Catalysts** which consist of a trialkyl aluminium and titanium tetrachloride in an inert solvent.

(1) **Polyethylene (Polythene).** It is obtained by polymerizing ethylene. Polyethylene has been produced commercially since 1943. It is used in the manufacture of houseware such as buckets and dustbins, carpet backing, packaging materials, and cable insulation.

$$\underset{\text{Ethylene}}{n\,CH_2{=}CH_2} \xrightarrow{\text{Polymerization}} \underset{\text{Polyethylene}}{-\!\!\left[CH_2{-}CH_2\right]_n\!\!-}$$

Notice that the monomer (ethylene) contains a double bond and the polymer does not. The electrons of the monomer π bond have moved and are used to link one monomer unit to another by σ bonds as indicated by extended lines in abbreviated polymer formula. The backbone of the polymer consists of the carbon atoms that originally formed the double bonds. Nothing is lost. The monomers simply add to each other.

Polyethylene bottles.

Polypropylene chairs.

(2) **Polypropylene.** It is obtained by polymerizing propylene. Polypropylene is used in the manufacture of houseware (pans, drinking glasses), medical equipment (can be sterilized), toys, electronic components, tubes, pipes, and fibres.

$$\underset{\text{Propylene}}{n\,CH_3-CH=CH_2} \xrightarrow{\text{Polymerization}} \underset{\text{Polypropylene}}{\left[-\underset{|}{\overset{CH_3}{CH}}-CH_2-\right]_n}$$

(3) **Polyvinyl Chloride** (*PVC*). It is obtained by polymerizing vinyl chloride. *PVC* is used in the manufacture of imitation leather, floor covering, corrugated roofing material, and gramophone records.

$$\underset{\text{Vinyl chloride}}{n\,CH_2=\overset{Cl}{\overset{|}{CH}}} \xrightarrow{\text{Polymerization}} \underset{\text{PVC}}{\left[-CH_2-\overset{Cl}{\overset{|}{CH}}-\right]_n}$$

Vinyl chloride is obtained from acetylene by treatment with HCl in the presence of $HgCl_2$.

$$\underset{\text{Acetylene}}{HC\equiv CH} + HCl \xrightarrow{HgCl_2} \underset{\text{Vinyl chloride}}{CH_2=CH-Cl}$$

(4) **Orlon.** It is obtained by polymerizing vinyl cyanide (acrylonitrile). Orlon is used as a synthetic textile fiber in clothing and carpeting.

$$\underset{\text{Vinyl cyanide}}{n\,CH_2=\overset{CN}{\overset{|}{CH}}} \xrightarrow{\text{Polymerization}} \underset{\text{Orlon}}{\left[-CH_2-\overset{CN}{\overset{|}{CH}}-\right]_n}$$

Vinyl cyanide can be prepared : (*a*) By treating acetylene with HCN in the presence of $Ba(CN)_2$; or (*b*) By catalytic oxidation of propylene in the presence of ammonia.

$$\underset{\text{Acetylene}}{HC\equiv CH} + HCN \xrightarrow{Ba(CN)_2} \underset{\text{Vinyl cyanide}}{CH_2=CH-CN}$$

$$\underset{\text{Propylene}}{CH_3CH=CH_2} + 3O_2 \xrightarrow[450°C]{2NH_3} \underset{\text{Vinyl cyanide}}{CH_2=CH-CN} + 6H_2O$$

PVC (Polyvinyl chloride) doors.

PVC tubes.

(5) **Teflon** (*PTFE*). It is obtained by polymerizing tetrafluoroethylene. Teflon is familiar because of its use as nonstick coating particularly for cooking utensils. *Nonstick frying pans have teflon coating.* Because of its low chemical reactivity, excellent toughness, electrical and heat resistance, teflon is used as insulation for electrical items and in the manufacture of gaskets and valves.

$$\underset{\text{Tetrafluoroethylene}}{n\,CF_2{=}CF_2} \xrightarrow{\text{Polymerization}} \underset{\text{Teflon}}{\left[-\underset{F}{\overset{F}{\underset{|}{\overset{|}{C}}}}-\underset{F}{\overset{F}{\underset{|}{\overset{|}{C}}}}- \right]_n}$$

Tetrafluoroethylene is obtained from chloroform as follows :

$$\underset{\text{Chloroform}}{CHCl_3} + 2HF \longrightarrow \underset{\text{Chlorodifluoromethane}}{CHClF_2 + 2HCl}$$

$$CHClF_2 \xrightarrow{900°C} \underset{\text{Tetrafluoroethylene}}{CF_2{=}CF_2} + 2HCl$$

Teflon coated non-stick pans.

Teflon is used as roof material for sports stadiums.

Discovery of Teflon. Teflon was discovered accidentally by a polymer chemist working at DuPont, USA. He was working with tetrafluoroethylene, a gaseous monomer which was being maintained under pressure in gas cylinders. One day the chemist turned the valve to obtain some of the gaseous monomer, but found none coming out though the tank should have been nearly full. He sawed open the steel tank and found that the monomer had polymerized inside the tank to a hard, inert material. With the knowledge that tetrafluoroethylene could indeed be polymerized, chemists at DuPont were soon able to find reaction conditions of initiating the process in the laboratory.

He is doing that ever since he read that many great discoveries are made by accident.

(6) **Polystyrene.** It is obtained by polymerizing styrene. Polystyrene is used in the manufacture of food containers, cosmetic bottles, television cabinets, plastic cups, packaging, and toys.

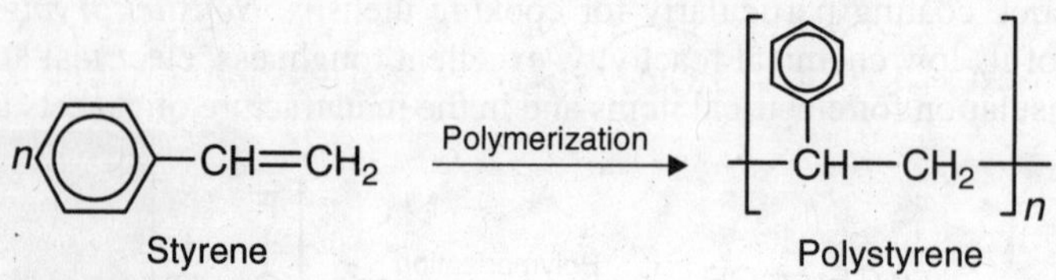

Styrene is prepared from benzene as follows :

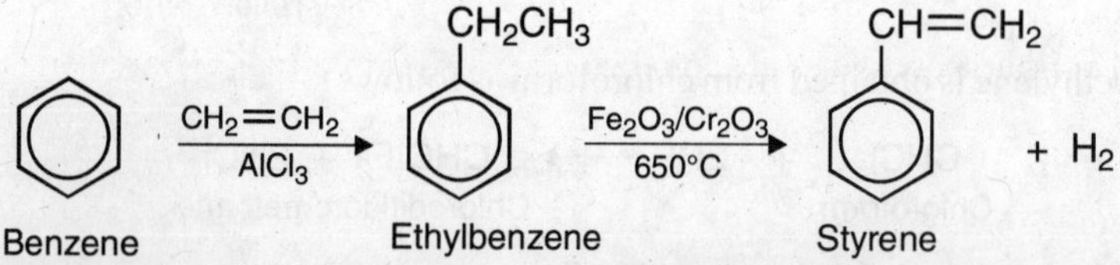

Polystyrene products.

Polystyrene art work.

MECHANISM OF ADDITION POLYMERIZATION

Addition polymerization takes place either by a free radical mechanism *or* ionic mechanism depending upon the reagents used.

FREE-RADICAL POLYMERIZATION

Free-radical polymerization is catalyzed by organic peroxides or other reagents which decompose to give free radicals. Following steps are involved :

(1) **Chain Initiation.** Organic peroxides undergo homolytic fission to form free radicals.

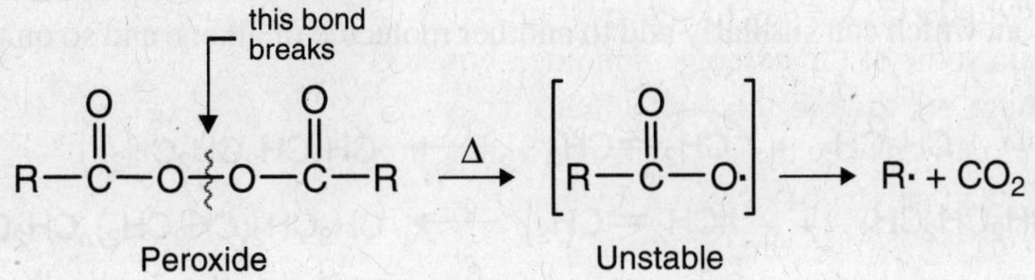

(2) **Chain Propagation.** Free radical produced in the above step adds to an alkene molecule to form a new free radical.

$$R\cdot + CH_2{=}CH_2 \longrightarrow R{-}CH_2{-}\dot{C}H_2$$

This free radical can attack another alkene molecule and so on.

$$RCH_2\dot{C}H_2 + CH_2{=}CH_2 \longrightarrow RCH_2CH_2CH_2\dot{C}H_2$$

$$RCH_2\dot{C}H_2 + nCH_2{=}CH_2 \longrightarrow R(CH_2CH_2)_nCH_2\dot{C}H_2$$

(3) **Chain Termination.** The above chain reaction comes to halt when two free radical chains combine.

$$2R(CH_2CH_2)_nCH_2\dot{C}H_2 \longrightarrow \underset{\text{Polyethylene (Polymer)}}{R(CH_2CH_2)_nCH_2CH_2{:}CH_2CH_2(CH_2CH_2)_nR}$$

Benzoyl or *t*-butyl peroxides are common reagents for free-radical polymerization.

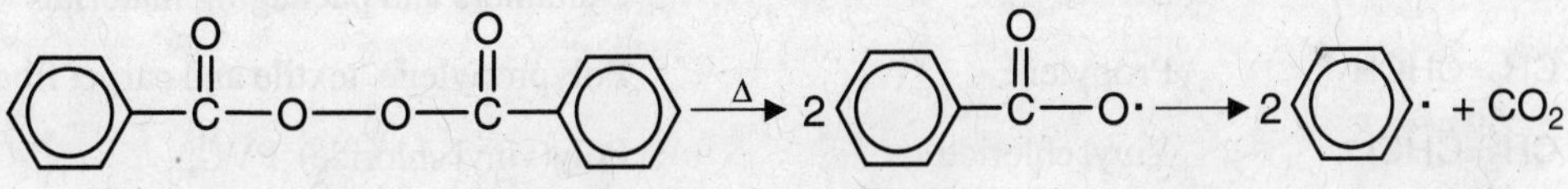

$$(CH_3)_3C{-}O{-}O{-}C(CH_3)_3 \xrightarrow{\Delta} 2(CH_3)_3CO\cdot \longrightarrow (CH_3)_2C{=}O + \dot{C}H_3$$

Phenyl or methyl free radicals will then initiate free-radical polymerization.

Free-radical polymerization can also be promoted by a mixture of ferrous sulfate and hydrogen peroxide ($FeSO_4 + H_2O_2$). These two compounds react to produce hydroxyl radicals ($HO^{\cdot}$) which act as chain initiators. Ferrous sulfate and hydrogen peroxide mixture is used as catalyst in the manufacture of Orlon and Teflon.

IONIC POLYMERIZATION

Ionic polymerization is initiated by acids such as H_2SO_4, HF, or BF_3 in H_2O. Following steps are involved :

(1) **Chain Initiation.** The acid catalyst gives proton. *Remember : Only one of the following acids is used as a catalyst. Not all of them.*

$$H_2SO_4 \longrightarrow H^+ + HSO_4^-$$

$$HF \longrightarrow H^+ + F^-$$

$$BF_3 + H_2O \longrightarrow H^+ + BF_3(OH)^-$$

The proton adds to the carbon-carbon double bond of alkene to give a stable carbonium ion.

$$H^+ + CH_2{=}CH_2 \longrightarrow CH_3\overset{+}{C}H_2$$

(2) **Chain Propagation.** The carbonium ion adds to another molecule of alkene to produce a new carbonium ion, which can similarly add to another molecule of alkene and so on.

$$CH_3\overset{+}{C}H_2 + CH_2{=}CH_2 \longrightarrow CH_3CH_2CH_2\overset{+}{C}H_2$$

$$CH_3CH_2CH_2\overset{+}{C}H_2 + nCH_2{=}CH_2 \longrightarrow CH_3CH_2(CH_2CH_2)_nCH_2\overset{+}{C}H_2$$

(3) **Chain Termination.** The chain reaction can be halted by combination with an anion (negative ions) or by loss of a proton.

$$CH_3CH_2(CH_2CH_2)_n\underset{}{\overset{H\ \ HSO_4^-}{CH}}{-}\overset{+}{C}H_2 \longrightarrow \underset{\text{Polyethylene (Polymer)}}{CH_3CH_2(CH_2CH_2)_nCH{=}CH_2} + H_2SO_4$$

Points to Remember

Important addition polymers derived from substituted ethylenes and their common uses are described below:

Monomer Formula	Common Name	Polymer Name(s) and Common Uses
$CH_2{=}CH_2$	Ethylene	Polyethylene, Polythene; break-resistant containers and packaging materials
$CH_2{=}CHCH_3$	Propylene	Polypropylene, textile and carpet fibers
$CH_2{=}CHCl$	Vinyl chloride	Poly(vinyl chloride), PVC; construction tubing
$CH_2{=}CCl_2$	1,1-Dichloroethylene	Poly(1,1-dichloroethylene), Saran; food packaging
$CH_2{=}CHCN$	Acrylonitrile	Polyacrylonitrile, Orlon; acrylics and acrylates
$CF_2{=}CF_2$	Tetrafluoroethylene	Polytetrafluoroethylene, Teflon; nonstick coatings
$CH_2{=}CHC_6H_5$	Styrene	Polystyrene, Styrofoam; insulating materials
$CH_2{=}CHCO_2CH_2CH_3$	Ethyl acrylate	Poly(ethyl acrylate); latex paints
$CH_2{=}\underset{\displaystyle CH_3}{\underset{\vert}{C}}CO_2CH_3$	Methyl methacrylate	Poly(methyl methacrylate), Lucite, Plexiglas; glass substitutes

CONDENSATION POLYMERS *(Step-Growth Polymers)*

Condensation polymers are formed by combination of monomers with the elimination of simple molecules such as H_2O or CH_3OH. There are two main types of condensation polymers : Polyesters or Polyamides. Examples are :

(1) **Terylene (Dacron)** is the most important polyester. It is obtained by heating ethylene glycol with dimethyl terephthalate at 300°C in the presence of metal oxides. The polymer melt can be spun into fibres and combined with such naturally occurring fibres as cotton. The combination of cotton and polyester yields a fabric (e.g. 60/40 cloth) that dries quickly without wrinkling, yet still retains the cool and comfort of 100% cotton. Polyesters have truly revolutionized the cloth industry.

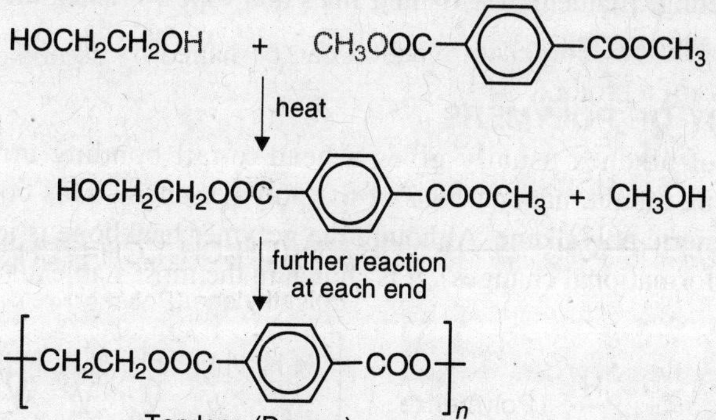

Terylene (Dacron)

Ethylene glycol is manufactured by catalytic oxidation of ethylene to ethylene oxide which is then treated with dilute acid.

$$\underset{\text{Ethylene}}{CH_2{=}CH_2} \xrightarrow[300°C]{O_2/Ag} H_2C\overset{O}{—}CH_2 \xrightarrow{H^+/H_2O} \underset{\text{Ethylene glycol}}{HOCH_2CH_2OH}$$

Dimethyl terephthalate is obtained by catalytic oxidation of *p*-xylene followed by esterification.

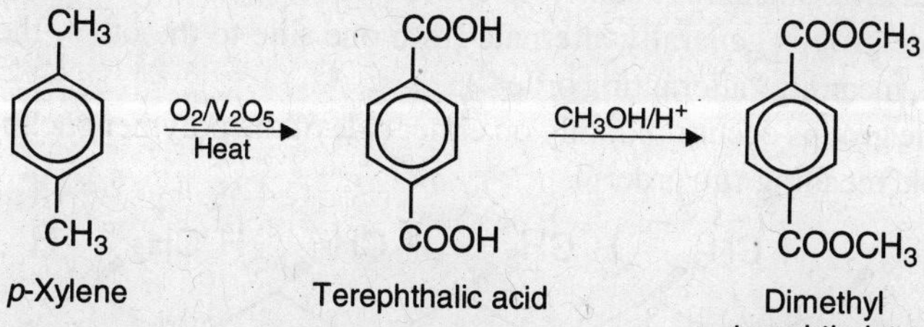

p-Xylene Terephthalic acid Dimethyl terephthalate

(2) **Nylon-6,6** is the most important polyamide. It is obtained by heating adipic acid with hexamethylene diamine under nitrogen at 200°C. Nylon-6,6 derives its name from its starting materials, adipic acid and hexamethylene diamine, both of which have *six* carbons.

$$HOOC(CH_2)_4COOH + HNH(CH_2)_6NH_2 \xrightarrow[-H_2O]{\text{heat}} HOOC(CH_2)_4CONH(CH_2)_6NH_2$$

$$\xrightarrow[\text{at each end}]{\text{further reaction}} \underset{\text{Nylon-6,6}}{\left[CO(CH_2)_4CONH(CH_2)_6NH\right]_n}$$

Both starting materials can be prepared from 1,3-butadiene as shown below :

$$\underset{\text{1,3-Butadiene}}{CH_2{=}CHCH{=}CH_2} \xrightarrow[\text{(1,4-Addition)}]{Cl_2} ClCH_2CH{=}CHCH_2Cl \xrightarrow{NaCN}$$

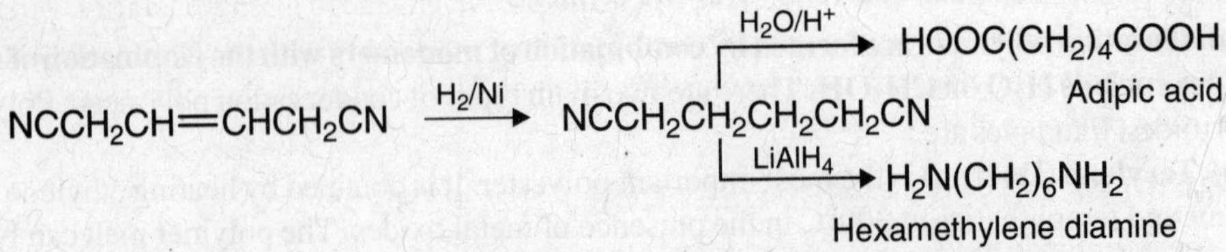

Nylon-6,6 was developed as a synthetic fiber for the production of stockings and other wearing apparel. It was introduced to the public at the New York World's Fair in 1939. It is used to make fibers for clothing and carpeting, filaments for fishing lines and ropes, bristles for brushes, and molded objects such as gears and bearings.

STEREOCHEMISTRY OF POLYMERS

Polymerization of alkenes usually gives a head-to-tail bonding arrangement, with any substituent(s) appearing on alternate carbons of the polymer chain. This bonding arrangement is shown below for a generic polyalkene. Although the polymer backbone is joined by single bonds (and can undergo conformational changes), it is shown in the most stable all-anti conformation.

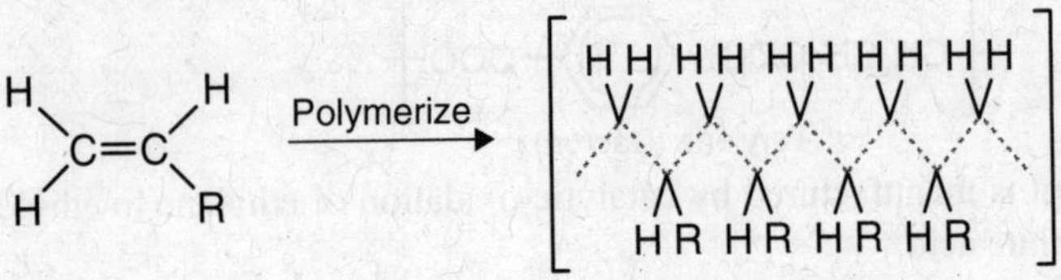

The stereochemistry of the side groups (R) in the polymer has a major effect on the polymer's properties. The polymer has many chirality centers, raising the possibility of millions of stereoisomers. Polymers are grouped into three classes, according to their predominant stereochemistry.

(*a*) If the side groups are generally on the same side of the polymer backbone, the polymer is called *isotactic* (Greek, iso, meaning "same," and tactic, meaning "order").

(*b*) If the side groups generally alternate from one side to the other, the polymer is called *syndiotactic* (Greek, meaning "alternating order").

(*c*) If the side groups occur randomly on either side of the polymer backbone, the polymer is called *atactic* (Greek, meaning "no order").

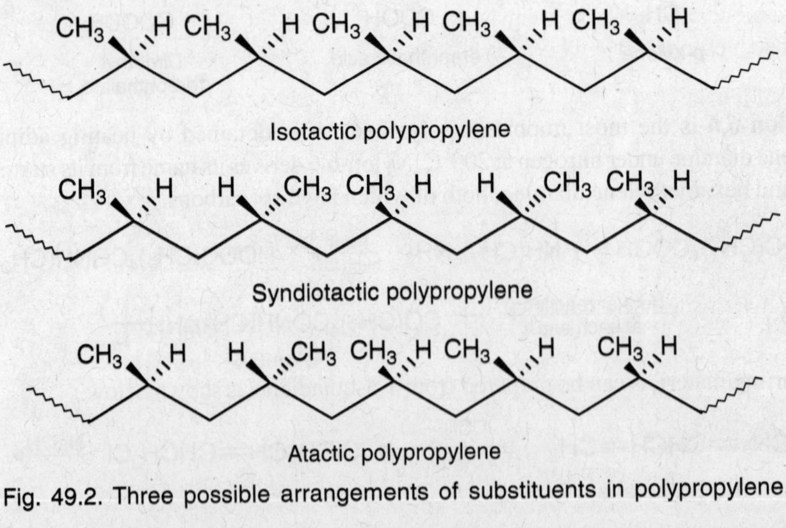

Fig. 49.2. Three possible arrangements of substituents in polypropylene.

In most cases, isotactic and syndiotactic polymers have enhanced strength, clarity, and thermal properties over the atactic form of the polymer. The three possible arrangements of substituents are shown using polypropylene as an example. The three possible arrangements of substituents can be shown using polypropylene (Fig.49.2).

THERMOPLASTIC AND THERMOSETTING POLYMERS

A thermoplastic polymer is one which softens on heating and becomes hard on cooling. Most addition polymers and some condensation polymers are thermoplastic. Examples are : Polyethylene, Polypropylene, Polystyrene, and Nylone-6,6.

A thermosetting polymer is one which becomes hard on heating. It can not be softened by heating. Such polymers are prepared in two stages. The first step is the formation of long-chain molecules which are capable of further reaction with each other. The second stage is the application of heat which causes a reaction to occur between the chains, thus producing a complex cross-linked polymer. Example is *Bakelite* which is made by reaction between phenol and formaldehyde.

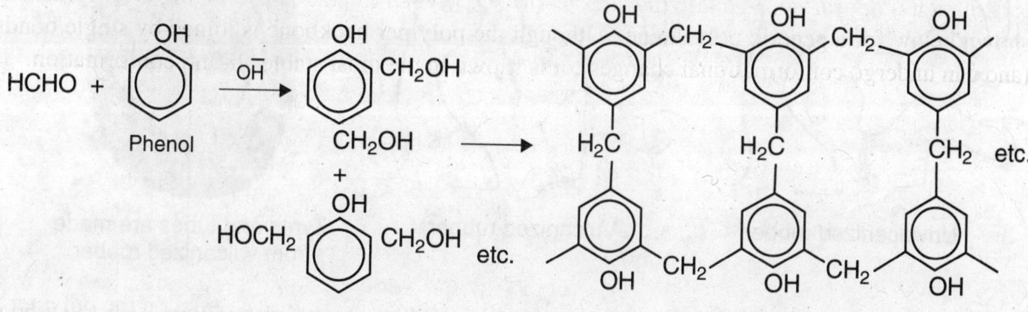

Bakelite

Bakelite is widely used for making moulded products such as handles on electrical and cooking utensils, electrical plates, switches, and appliances.

NATURAL AND SYNTHETIC RUBBERS

NATURAL RUBBER

When Christopher Columbus arrived in the New World, he saw native Americans play a game with balls made from the gum of rubber trees. The natives also coated outer garments with the material for protection against rain. They also knew how to prepare footwear and bottles by coating clay moulds with rubber and allowing them to dry. The name *rubber* was given to the substance by Joseph Priestley (discoverer of oxygen), who used it to rub out pencil marks.

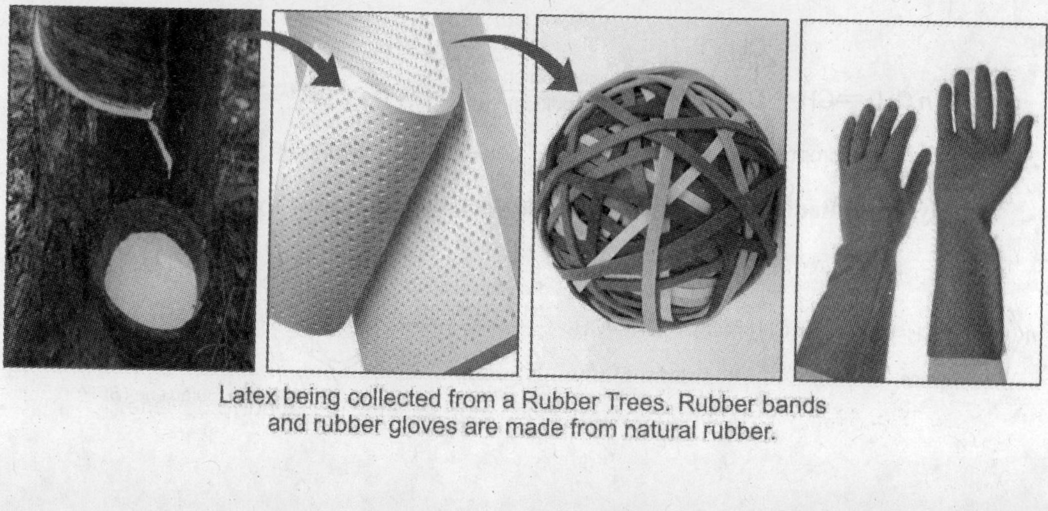

Latex being collected from a Rubber Trees. Rubber bands and rubber gloves are made from natural rubber.

Natural rubber is a hydrocarbon polymer built up from the monomer isoprene.

$$n\,H_2C{=}\underset{\displaystyle\text{Isoprene}}{\overset{\displaystyle CH_3}{|}{C}}{-}CH{=}CH_2 \xrightarrow[\text{(1,4-Addition)}]{\text{Polymerization}} \underset{\text{Natural Rubber}}{\left[-CH_2-\overset{CH_3}{\overset{|}{C}}{=}CH-CH_2-\right]_n}$$

Raw rubber obtained from milky sap (*latex*) of the rubber tree does not possess the characteristics of the rubber with which we are familiar. In order to give it strength and elasticity it is **vulcanized.** In the vulcanization process, raw rubber is mixed with small amount of sulfur and heated. The sulfur reacts with the polymer molecules forming a cross-linked net work.

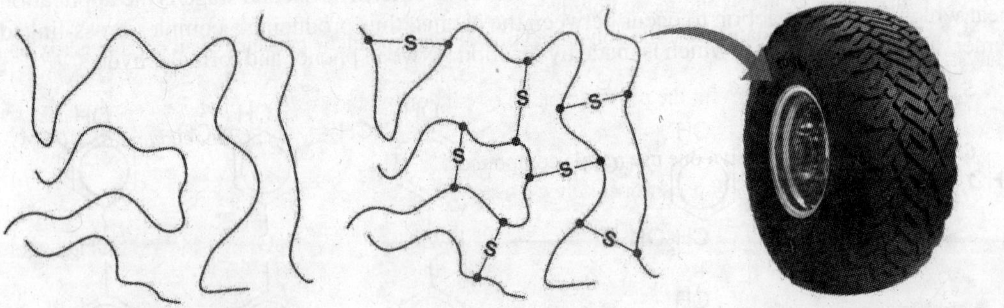

Unvulcanized rubber　　Vulcanized rubber　　Tyres and tubes are made from vulcanized rubber.

This cross-linking gives mechanical strength to the rubber. In addition, fillers such as carbon black and zinc oxide are usually added to the crude rubber before vulcanization in order to improve its wearing characteristics.

SYNTHETIC RUBBERS

(*a*) **Neoprene Rubber.** It was the first synthetic rubber manufactured on large scale. The monomer, chloroprene, is prepared from acetylene. Neoprene is particularly resistant to chemical action and is therefore used in making hoses for petrol and oil, and containers for corrosive chemicals.

$$2HC{\equiv}CH \xrightarrow[NH_4Cl]{Cu_2Cl_2} \underset{\text{Vinylacetylene}}{CH_2{=}CH{-}C{\equiv}CH} \xrightarrow{HCl}$$

$$n\,\underset{\text{Chloroprene}}{CH_2{=}CH{-}\overset{Cl}{\overset{|}{C}}{=}CH_2} \xrightarrow[\text{(1,4-Addition)}]{\text{Polymerization}} \underset{\text{Neoprene Rubber}}{\left[-CH_2-CH{=}\overset{Cl}{\overset{|}{C}}-CH_2-\right]_n}$$

(b) **Styrene-Butadiene Rubber (*SBR*).** It is a copolymer made from styrene and 1,3-butadiene.

Examination Questions

1. What is polymerization ? Illustrate with two suitable examples.
2. Describe the difference between thermosetting and thermoplastic polymers. *(Kerala BSc, 2014 ; Panjab BSc, 2015)*
3. Write short notes on :
   (*a*) Addition Polymerization *(Dibrugarh BSc, 2014)*
   (*b*) Condensation Polymerization *(Madras BSc, 2015)*
   (*c*) Synthetic Rubbers *(North Eastern Hill BSc, 2014)*
   (*d*) Atactic and Syndrotactic Polymers
4. Write a mechanism for the polymerization of ethylene in the presence of an organic peroxide as catalyst. *(Delhi BSc, 2015)*
5. Write a mechanism for the polymerization of ethylene in the presence of an acid catalyst. *(Bundelkhand BSc, 2014)*
6. What is teflon ? Give one use of this compound.
7. Draw a shorthand representation of the polymer produced from each of the following monomers:
   (*a*) $CH_2{=}CHCN$ (*b*) $CH_3CH{=}CH_2$ (*c*) $CH_2{=}CF_2$ (*d*) $C_6H_5CH{=}CH_2$

Answer.

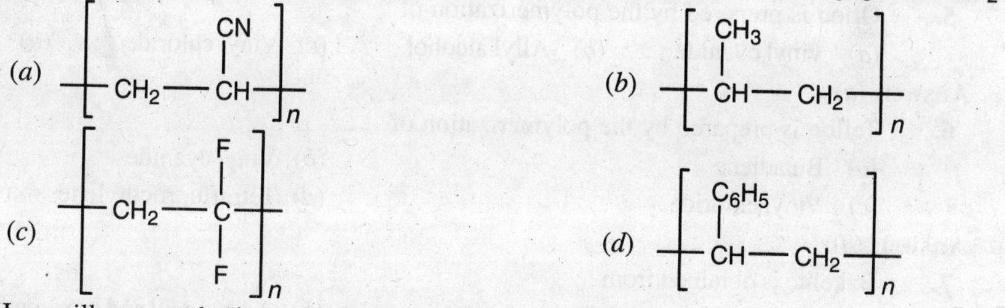

8. How will you synthesize polyvinyl chloride from acetylene ?

Answer. Following steps are involved :

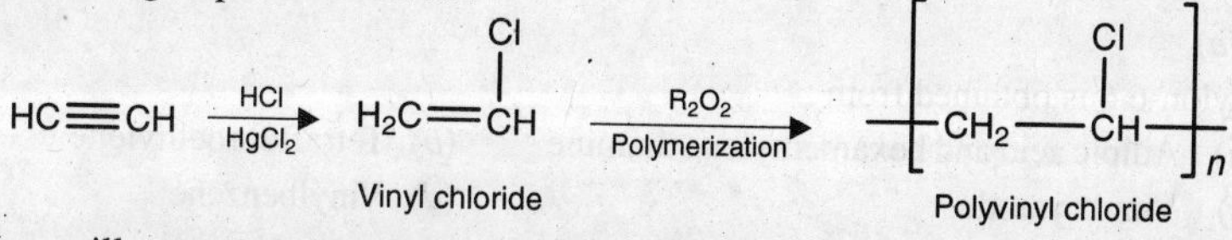

9. How will you synthesize Orlon from acetylene ? *(Agra BSc, 2014)*

Answer. Following steps are involved :

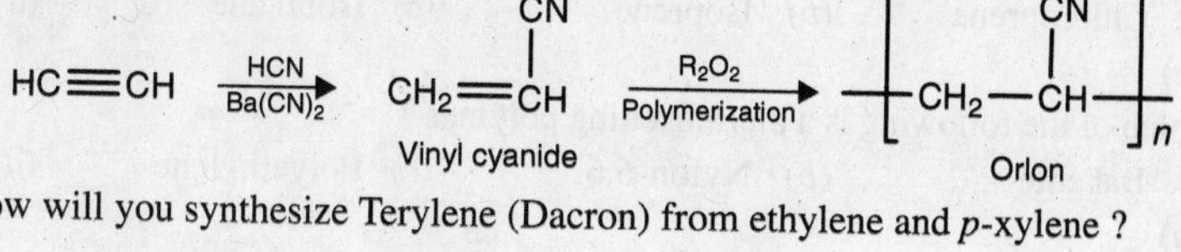

10. How will you synthesize Terylene (Dacron) from ethylene and *p*-xylene ? *(Jiwaji BSc, 2015)*
11. How will you synthesize Nylon-6,6 from 1,3-butadiene ? *(Calicut BSc, 2014)*
12. How will you synthesize Neoprene from acetylene ? *(Kalyani BSc, 2015)*
13. How will you synthesize Polystyrene from benzene ? *(Aligarh BSc, 2014)*

Answer. Following steps are involved :

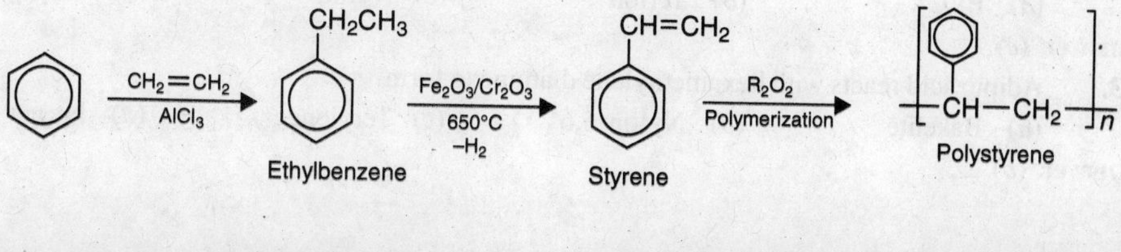

Multiple Choice Questions

1. What term is used for the repeating units of a polymer ?
(*a*) monomer (*b*) unit residue (*c*) condensate (*d*) unit structure

Answer. (*a*)

2. What term identifies polymers composed of identical monomer units ?
(*a*) homopolymers (*b*) isopolymers
(*c*) peripolymers (*d*) unipolymers

Answer. (*a*)

3. Which of the following is a copolymer ?
(*a*) Teflon (*b*) Polystrene (*c*) Orlon (*d*) Nylon 6,6

Answer. (*d*)

4. What monomers are most commonly used in addition polymerization ?
(*a*) esters (*b*) carboxylic acids
(*c*) alkenes (*d*) amides

Answer. (*c*)

5. Orlon is prepared by the polymerization of
(*a*) Vinyl cyanide (*b*) Allyl alcohol (*c*) Vinyl chloride (*d*) Allyl chloride

Answer. (*a*)

6. Teflon is prepared by the polymerization of
(*a*) Butadiene (*b*) Vinyl cyanide
(*c*) Vinyl chloride (*d*) Tetrafluoroethylene

Answer. (*d*)

7. Bakelite is obtained from :
(*a*) Phenol and formaldehyde (*b*) Adipic acid and hexamethylene diamine
(*c*) Neoprene (*d*) Dimethyl terephthalate and ethylene glycol

Answer. (*a*)

8. Nylon-6,6 is obtained from :
(*a*) Adipic acid and hexamethylene diamine (*b*) Tetrafluoroethylene
(*c*) Vinyl cyanide (*d*) Vinylbenzene

Answer. (*a*)

9. Neoprene is a polymer of the following monomer
(*a*) Chloroprene (*b*) Isoprene (*c*) Isobutane (*d*) Isopentene

Answer. (*a*)

10. Which of the following is a thermosetting polymer ?
(*a*) Bakelite (*b*) Nylon-6,6 (*c*) Polyethylene (*d*) Teflon

Answer. (*a*)

11. Which of the following is an example of a condensation polymer ?
(*a*) Nylon-6,6 (*b*) Teflon (*c*) Polypropylene (*d*) Orlon

Answer. (*a*)

12. Which of the following polymers contain nitrogen
(*a*) PVC (*b*) Teflon (*c*) Nylon (*d*) Terylene

Answer. (*c*)

13. Adipic acid reacts with hexamethylene diamine to form
(*a*) Bakelite (*b*) Nylon-6,6 (*c*) Terylene (*d*) Nylon-6,8

Answer. (*b*)

In 1912, Casimir Funk, a Polish biochemist, coined the word *vitamine* (from the Latin *vita*, life) for these missing factors. Funk thought that they all contained the amino group. The final '*e*' was dropped after it was found that not all the factors were amines. The generic term became *vitamin*.

Since organisms differ in their synthetic abilities, a substance that is a vitamin for one species may not be so for another. Over the years, scientists have isolated 13 vitamins needed by humans. Researchers today are in general agreement that no more vitamins are likely to be found. One reason is that about 50 years has elapsed since the last one (vitamin B_{12}) was discovered, despite an intensive search for others. Another reason is that many people have lived for years on intravenous solutions containing only the known vitamins and other nutrients without exhibiting signs of vitamin deficiency.

The vitamins are divided into two broad categories, the **fat-soluble vitamins**, including A, D, E, and K, and the **water-soluble vitamins**, made up of the B complex and vitamin C. All the fat-soluble vitamins incorporate a high proportion of hydrocarbon structural elements. There are one or two oxygen atoms present, but the compounds as a whole are nonpolar. In contrast, a water-soluble vitamin contains a high proportion of the electronegative atoms oxygen and nitrogen, which can form hydrogen bonds to water, therefore, the molecule as a whole is soluble in water.

FAT-SOLUBLE VITAMINS

The fat-soluble vitamins are those soluble in fats and in fat solvents. Included in this classification are vitamins A, D, E, and K. The role of vitamins as food accessories usually is considered in relation to nutritional deficiencies. It should be pointed out that doses of the fat-soluble vitamins, when given far in excess of normal requirements, also can have toxic effects. Vitamin poisoning occurred in a number of arctic explorers, who became seriously ill after eating polar bear liver. There also have been numerous cases of vitamin poisoning in infants. Young mothers, eager to fulfill all vitamin requirements of their first offspring, sometimes give overdoses of fat-soluble vitamins. Poisoning by water-soluble vitamins is not possible because any amounts not required are voided from the body in urine.

Each of the fat-soluble vitamins is discussed briefly in the following sections.

(*a*) **Vitamin A** may be obtained from the coloring matter of many green and yellow vegetables. Vitamin A, as such, is not found in plants, only β-carotene, its precursor or *provitamin*. The β-carotene molecule when cleaved in the center of the linear chain and converted at each end to an alcohol function, yields two molecules of vitamin A. Other sources of vitamin A are fish-liver oil, the livers of other animals, eggs, butter, and cheese.

Vitamin A_1, or *retinol*, is oxidized in the body to the aldehyde 11-*cis-retinal*, which combines with a protein called *opsin* to produce a light-sensitive substance *rhodopsin*. Rhodopsin is located in the retina and, upon absorption of radiation within the visible range, causes an isomerization of *cis*-retinal to the *trans* form.

There are two vitamins A. One is known as a vitamin A_1, or retinol, and the other as vitamin A_2, or 3-dehydroretinol. Vitamin A_2 is found in the liver oils of fresh water fish and differs structurally from vitamin A_1, found in the livers of cod and other salt-water fish, by having a second double bond between carbons 3 and 4. Physiologically, the two vitamins have the same activity, and both are called vitamin A.

A deficiency of vitamin A causes night blindness - an inability to see in dim light. Another disease of the eye known as *xerophthalmia*, in which the tear glands cease to function, results from a lack of vitamin A.

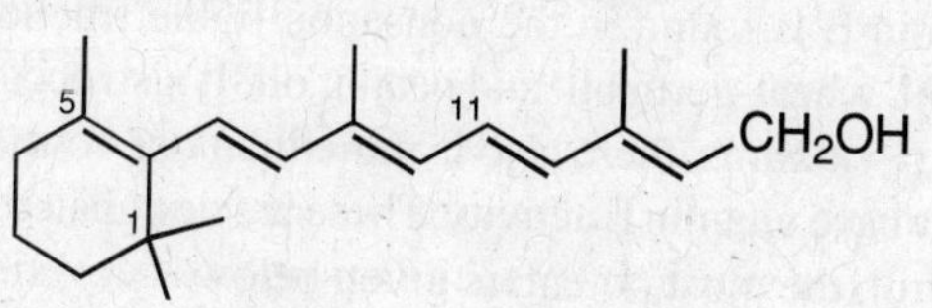

Vitamin A (*trans*-retinol)

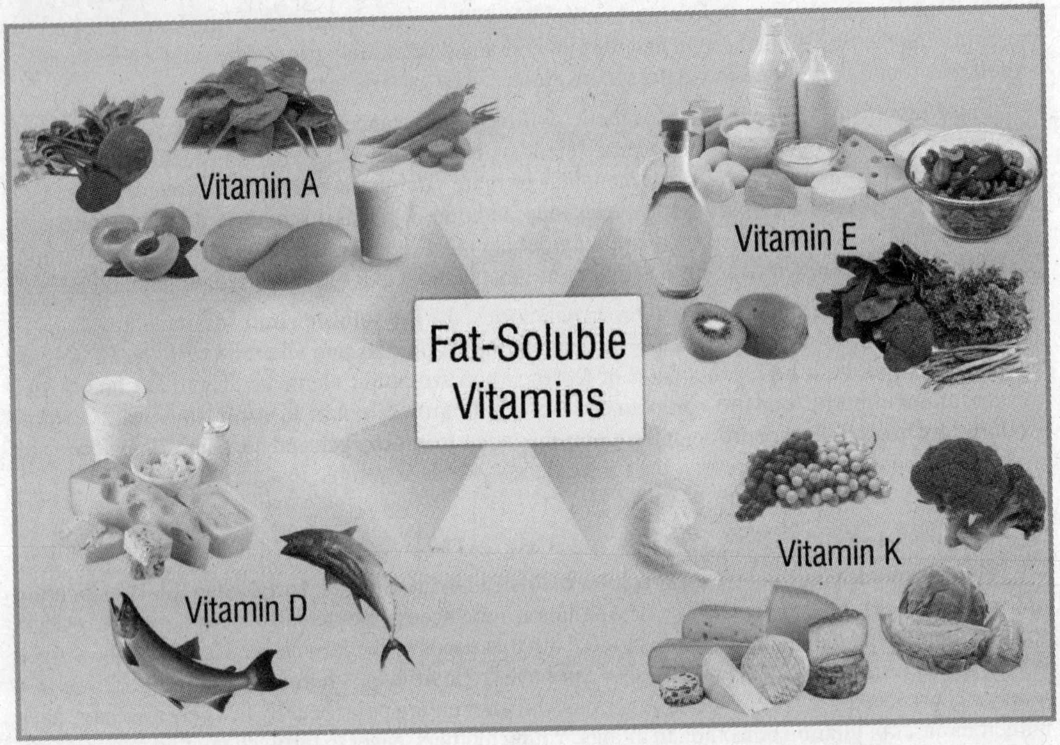

(*b*) **Vitamin D** is sometimes referred to as the "antirachitic" vitamin. It is related to the proper deposition of calcium phosphate and controls the normal development of the teeth and bones. There are ten or more compounds that have antirachitic properties and are designated D_1, D_2, D_3 etc. Vitamin D from fish oils is D_3, while that produced by irradiation of the skin with ultraviolet or sunlight is D_2. Vitamin D_2 is known as *calciferol* and is derived from ergosterol, a plant sterol. Milk, by irradiation, is fortified with vitamin D. Vitamin D is produced in or on the skin by irradiation of the sterols present there. Sunlight supplies *not* the vitamin, only the necessary radiation. Vitamin D_2 has the following structure.

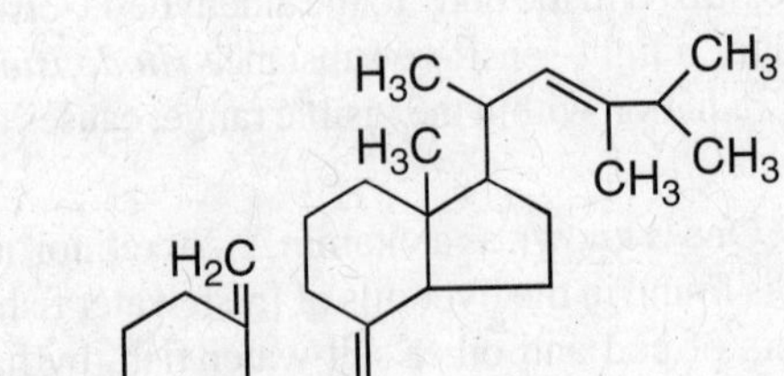

Calciferol (Vitamin D_2)

(*c*) **Vitamin E,** sometimes called the fertility factor, is related to the proper functioning of the reproductive system. Vitamin E is found in the nonsaponifiable fraction of vegetable oils such as corn-germ oil, cottonseed oil, wheat-germ oil, and peanut oil. It also occurs in green leafy vegetables. As in the case of the A and D vitamins, there also is more than one form of vitamin E. Four different structures called *tocopherols* have vitamin E activity. These are designated α-, β-, γ-, and δ-tocopherols. The structure of α-tocopherol, the most potent, is given below.

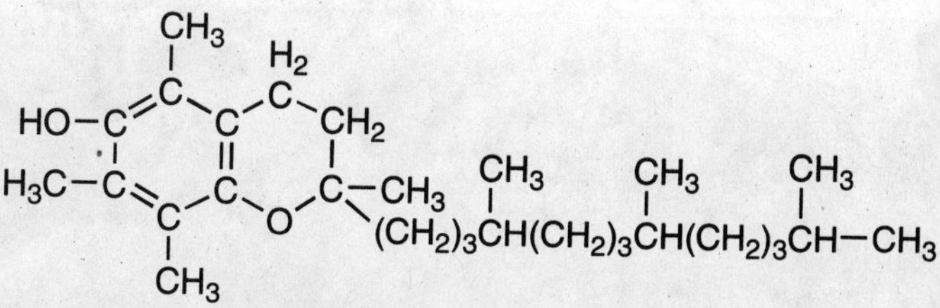

α-Tocopherol

Vitamin E is also used as an antioxidant for the prevention of oxidative rancidity in vegetable oils.

(*d*) **Vitamin K** is the antihemorrhagic factor related to the blood clotting mechanism. This vitamin is important especially from a surgical standpoint.

There are at least two K vitamins. Vitamin K_1 is obtained from the alfalfa leaf ; vitamin K_2 is produced by bacterial action in the intestinal canal. The structure of vitamin K_1 is given below.

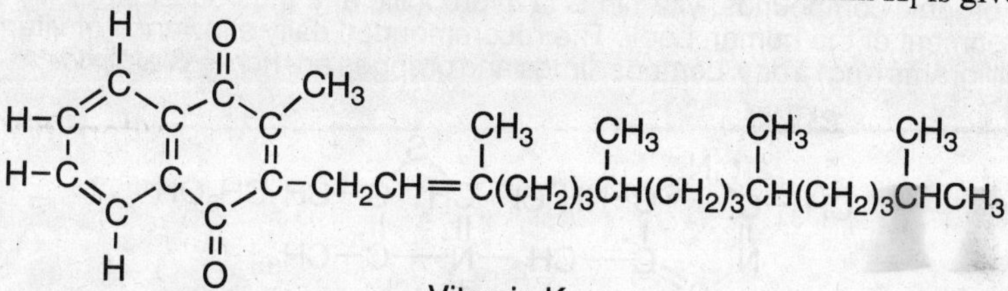

Vitamin K_1

WATER-SOLUBLE VITAMINS

The water-soluble vitamins include vitamin C and all other vitamins designated B. The latter are collectively referred to as the vitamin B complex. The water-soluble vitamins bear less resemblance to each other than do the fat-soluble vitamins. Whereas the fat-soluble vitamins are isoprenoid structures and largely of hydrocarbon composition, the water-soluble vitamins all possess polar groupings to render them water-soluble.

(*a*) **Vitamin C,** now called **ascorbic acid,** is the vitamin that prevents scurvy. A person with scurvy suffers pain in the joints and hemorrhages from the mucous membranes of the mouth. The gums especially are affected and become red, ulcerated, and even gangrenous. Ascorbic acid is abundantly found in citrus fruits, tomatoes, green peppers, and parsley. It should be obtained from fresh sources for it loses its potency when heated or when exposed to air for any length of time. Ascorbic acid is a hexose derivative with the following structure.

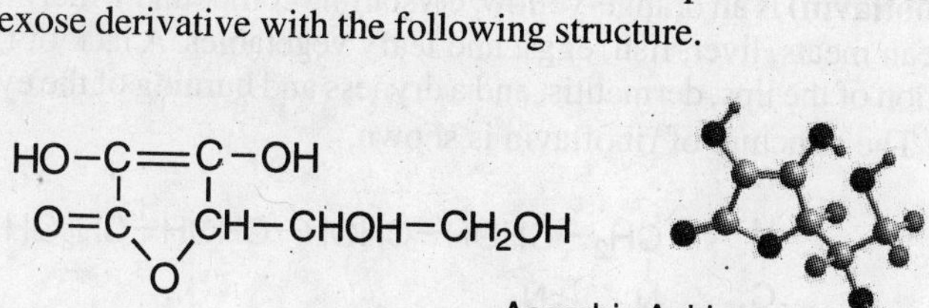

Ascorbic Acid

(*b*) **The B Vitamin Complex.** Vitamins designated as B vitamins were at one time all thought to be the same. The isolation of each new factor from vitamin B preparations has led to a designation of B_1, B_2, B_3, etc, for each new factor. The B vitamins appear to play an important role in energy metabolism. A brief discussion of the principal B vitamins is given in the following sections.

Vitamin B_1 (Thiamine), a deficiency of which causes *beriberi* in humans, is present in whole cereal grains, legumes, lean meat, nuts, and yeast. The vitamin has been found to contain both the pyrimidine and thiazole heterocyclic systems. The structure of thiamine has been determined, and the vitamin has been prepared. Thiamine usually is prepared in the form of an acid salt.

Unlike most organic compounds, vitamin C is hydrophilic and thus passes quickly through the watery environment of the human body. The recommended daily allowance of vitamin C for an adult is 60 milligrams (mg) a day. Lemons, limes and oranges are rich in vitamin C.

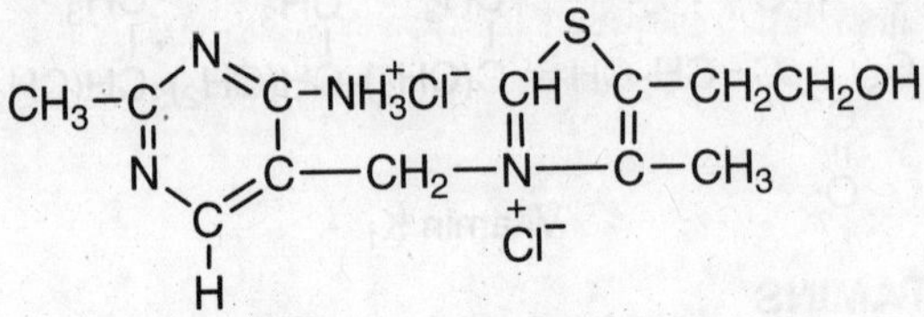

Thiamine hydrochloride (Vitamin B_1)

Thiamine occurs in nature either as the pyrophosphoric acid ester or as the free vitamin.

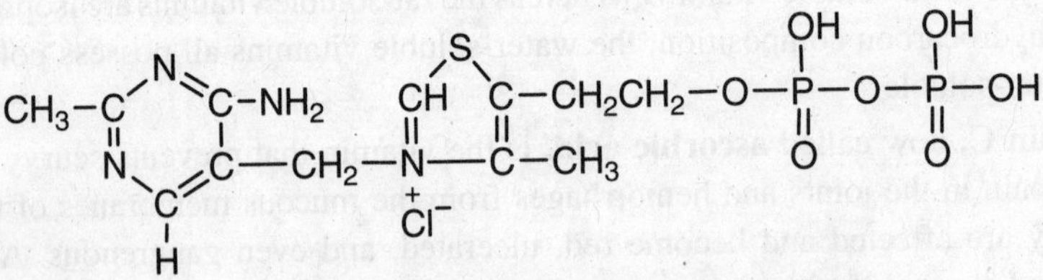

Thiamine hydrochloride pyrophosphate

Vitamin B_2 (Riboflavin) is an orange-yellow, crystalline compound widely distributed in nature. It is found in milk, lean meats, liver, fish, eggs, and leafy vegetables. A lack of riboflavin in the diet causes an inflammation of the lips, dermatitis, and a dryness and burning of the eyes, accompanied by a sensitivity to light. The structure of riboflavin is shown.

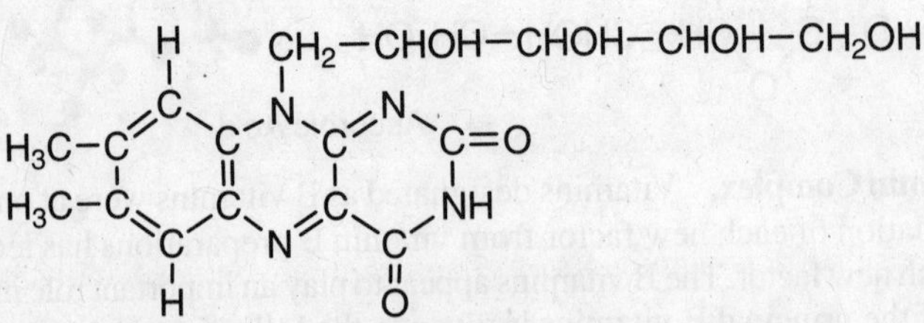

Riboflavin

Niacin (Nicotinic acid), another of the B vitamins, is the antipellagra factor. Pellagra is a disease characterized by dermatitis, a pigmentation and thickening of the skin, and soreness and inflammation of the tongue and mouth.

Niacin may be found in most of the same foods that supply riboflavin. Especially rich sources of niacin are lean meats and liver. Although present in whole cereal grains, niacin is lost in the milling process. The niacin structure is part of the enzyme *nicotinamide-adenine-dinucleotide* (NAD). The structural formula of niacin, shown below, is relatively simple when compared to those of other vitamins. The oxidation of nicotine or β-picoline (3-methylpyridine) produces nicotinic acid, or niacin.

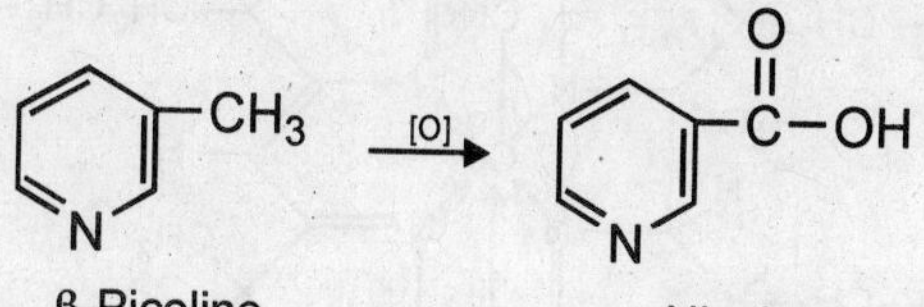

Vitamin B_6 (Pyridoxine) is a vitamin whose function appears to be intimately related to the proper metabolism of fats and amino acids. Meat, fish, egg yolk, and whole cereal grains are rich sources of vitamin B_6. The exact requirements of vitamin B_6 for adult humans has not been established, but a lack of pyridoxine of vitamin B_6 for adult humans has not been established, but a lack of pyridoxine in the diet of experimental animals leads to dermatitis, anemia, and epileptic seizures. The structures of pyridoxine and two of its derivatives which also show vitamin B_6 activity are given below.

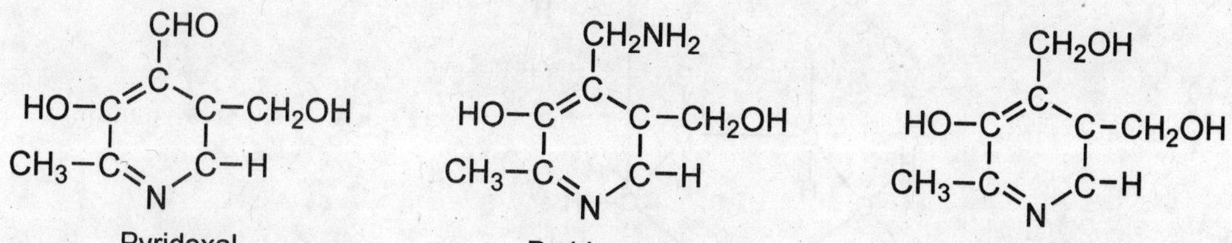

Vitamin B_{12} is the vitamin that prevents pernicious anemia. It is a dark red crystalline compound with a complex structure. Like heme and chlorophyll-a, it also contains a porphyrin nucleus, but a unique feature of its structure is the presence of trivalent cobalt. Vitamin B_{12} was the first cobalt-containing organic compound to be found. It sometimes is referred to as *cobalamin*. It is found most abundantly in liver but also occurs in meat, eggs, and seafood. The structure of vitamin B_{12} is shown in Fig. 50.1.

Points to Remember

1. The vitamins are trace nutrients required in the diet of animals for proper growth, development, and good health.
2. The absence of sufficient amounts of vitamins in the diet results in deficiency diseases.
3. Vitamins are generally classified as fat-soluble or water-soluble.
4. The fat-soluble vitamins are vitamins A, D, E, and K ; the water-soluble vitamins are vitamin C and those usually designated B vitamins.

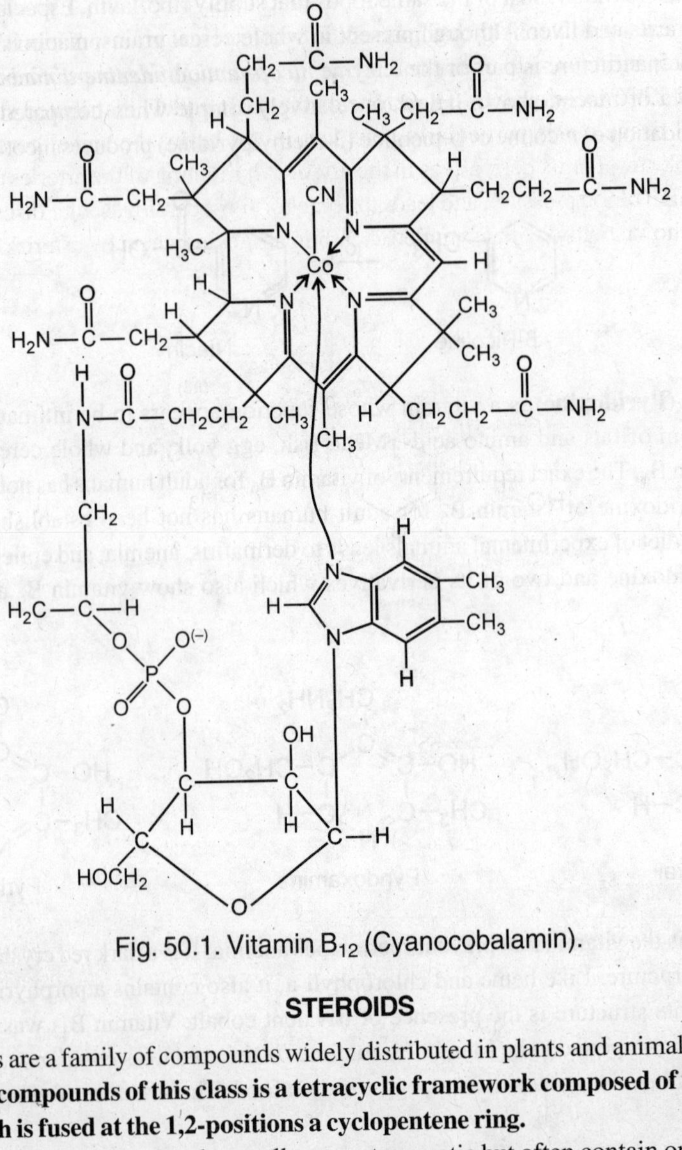

Fig. 50.1. Vitamin B_{12} (Cyanocobalamin).

STEROIDS

The steroids are a family of compounds widely distributed in plants and animals. **Common to the structure of all compounds of this class is a tetracyclic framework composed of the phenanthrene nucleus to which is fused at the 1,2-positions a cyclopentene ring.**

The rings in the steroid molecule usually are not aromatic but often contain one or more isolated double bonds. The total structure of one steroid differs from that of another, usually by a variation in the side chain or by a variation in the number and type of functional groups. To the family of steroids with this common ring system belong the sterols, the sex hormones, the bile acids, and other biologically important materials.

For purposes of nomenclature the steroid ring skeleton is numbered as given below.

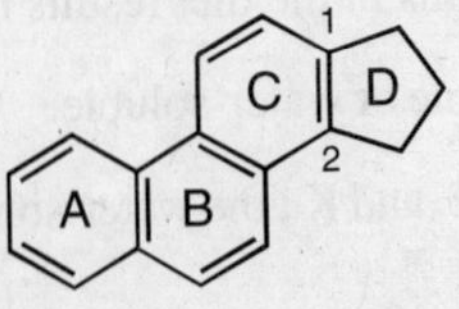

1,2-Cyclopentenophenanthrene

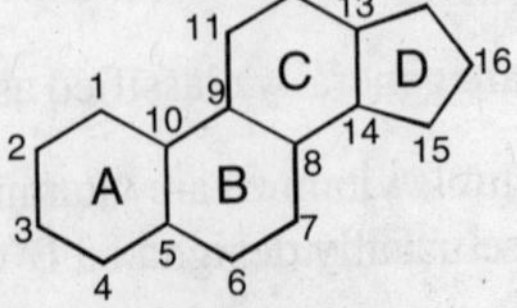

Steroid ring system

CHOLESTEROL

The **sterols** are solid alcohols that possess a hydroxyl group at position 3, a double bond between carbons 5 and 6, a side chain on carbon 17, and methyl groups joined to ring carbons numbered 10 and 13. *Cholesterol,* $C_{27}H_{46}O$, one of the most widely distributed sterols, is found in almost all animal tissue but is particularly abundant in the brain, the spinal cord, and gallstones. Deposition of cholesterol or its derivatives in the arteries (hardening of the arteries) restricts the flow of blood, causes high blood pressure, and leads to some forms of cardiovascular disease. The structure of cholesterol is shown below with asymmetric carbon atoms indicated by asterisks.

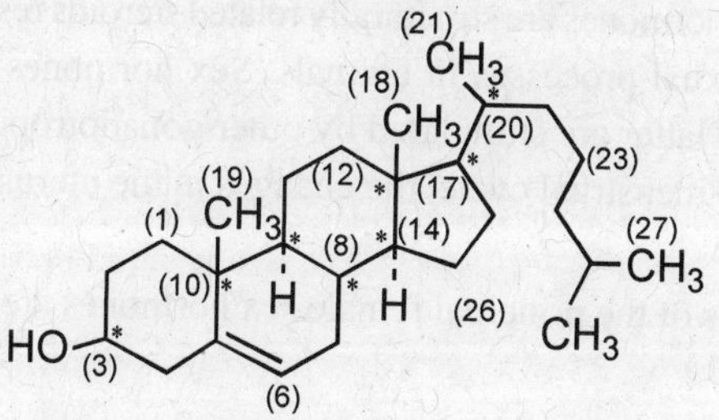

Structure and numbering system of Cholesterol

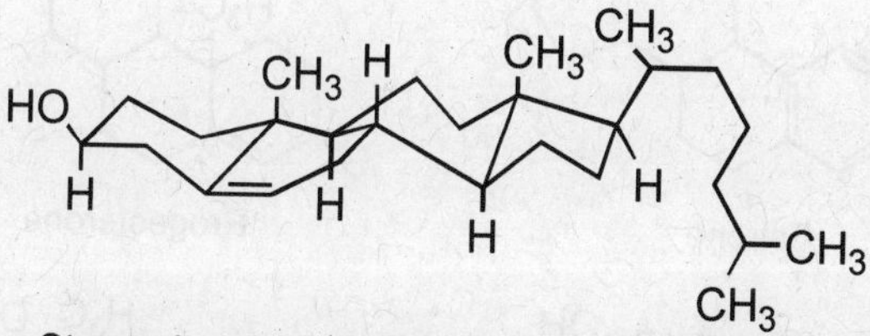

Stereochemical configuration of Cholesterol

Green and yellow vegetables are primary dietary source of vitamin A precursor compounds.

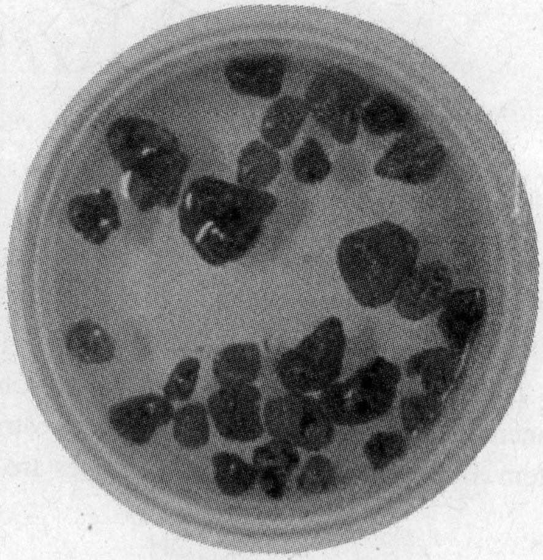

Human gallstones are almost pure cholesterol.

ERGOSTEROL

Although cholesterol is found only in animals, a large number of closely related compounds known as **phytosterols** are found in plants. One of these, *ergosterol,* $C_{28}H_{44}O$, is produced by yeast. Ergosterol is of particular interest because, when irradiated, it yield calciferol, vitamin D_2.

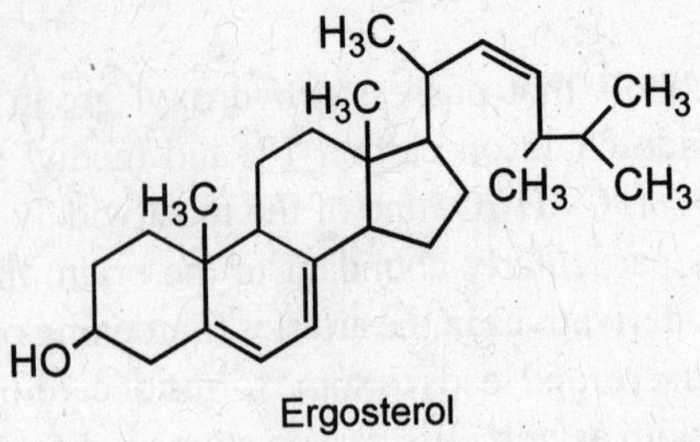

Ergosterol

SEX HORMONES

The male and female sex hormones are structurally related steroids responsible for the development of sex characteristics and sexual processes in animals. Sex hormones are produced in the gonads (ovaries and testes) when the latter are stimulated by other gonadotropic hormones. The female sex hormones are involved in the menstrual cycle, the changes in the uterus, and the preparation for and maintenance of pregnancy.

The structures and names of the principal female sex hormones are given below.

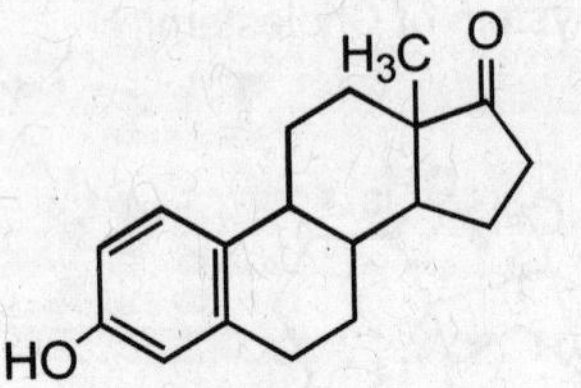

Estrone

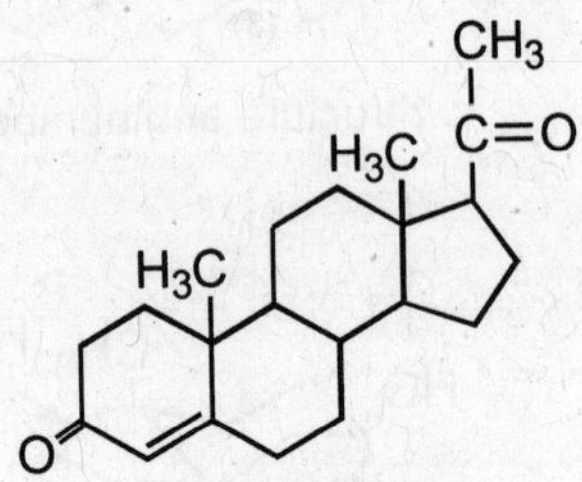

Progesterone

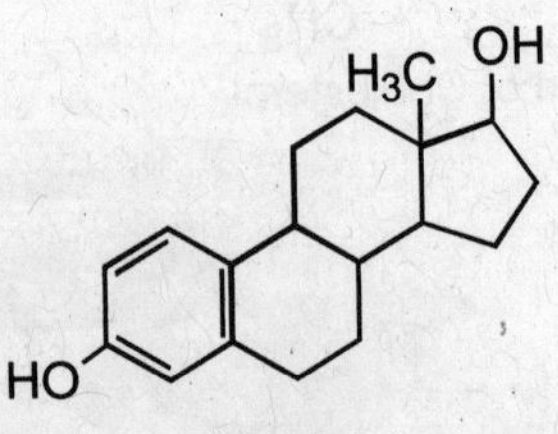

Estradiol

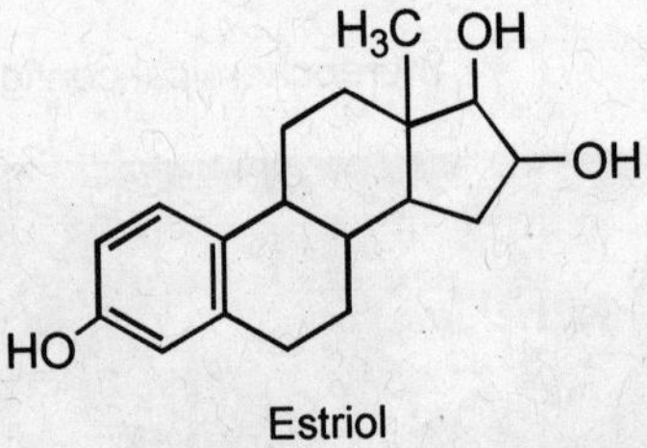

Estriol

Oral contraceptives have been developed that contain synthetic compounds structurally similar to progesterone and estriol but modified chemically to permit easier assimilation into the bloodstream. These synthetic agents, when taken orally, suppress ovulation and mimic pregnancy.

The structures of two synthetic progesterones and a synthetic estrogen are shown. Estrogen is a generic term for a substance that induces estrus - the cyclic phenomenon of the female reproductive system. Both types of synthetic hormones are used either in combined form or sequentially.

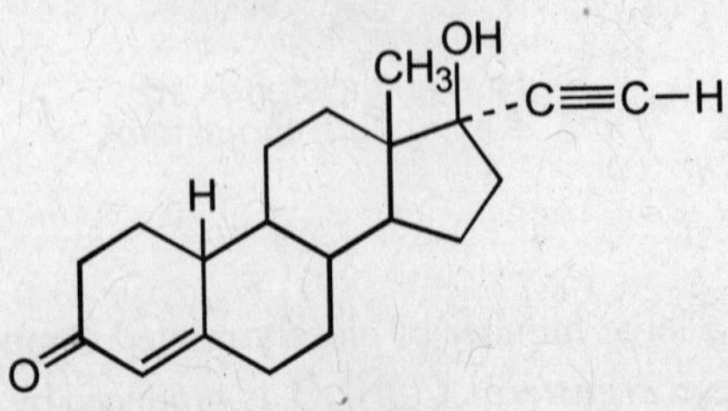

(a)
Northindrone

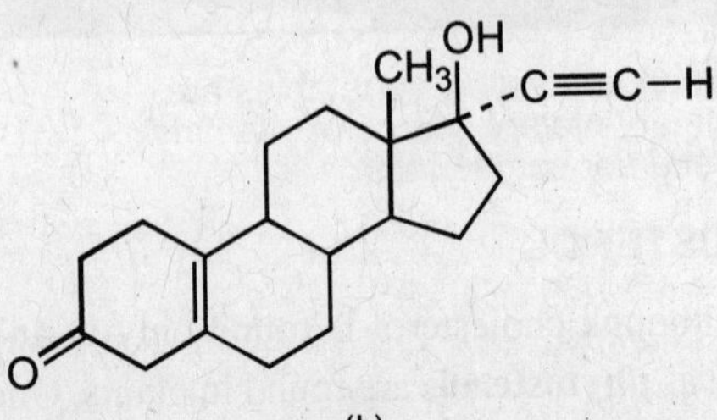

(b)
Norethynodrel

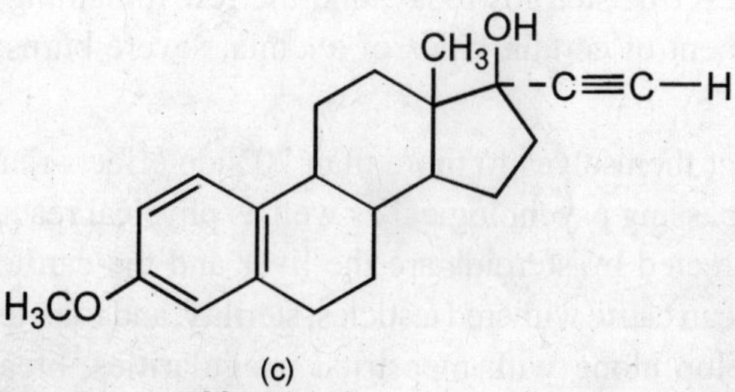

(c)
Mestranol

Orthonovum = (a) + (c)
Enovide = (b) + (c)

The male sex hormones, called **androgens,** except for the absence of an aromatic ring, are very similar in structure to the female hormones. The male sex hormones control the development of the male genital tract and the secondary male characteristics - e.g., beard, voice. The structures of the principal male sex hormones are shown below.

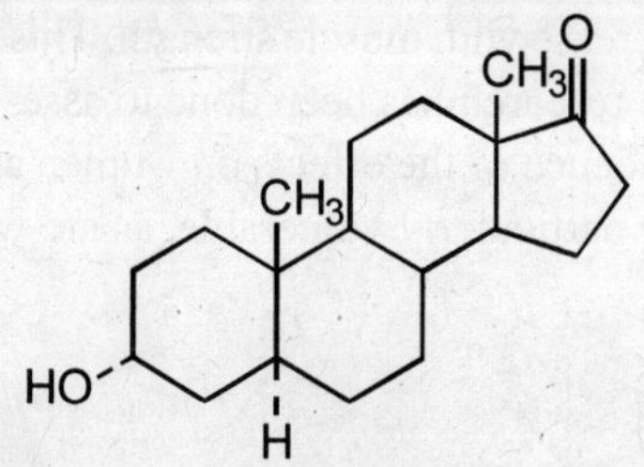

Androsterone

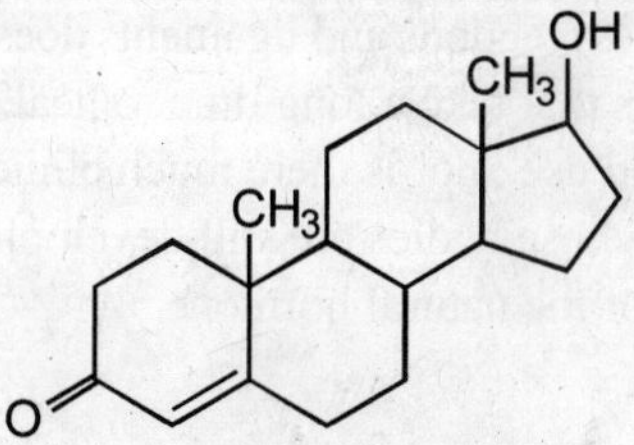

Testosterone

ADRENAL STEROIDS

Cortisone, and its derivative, *17-hydroxycorticosterone,* are two steroids produced by the adrenal cortex. These compounds have been used with beneficial results in the treatment of inflammatory and allergic diseases. The structures of cortisone and its reduced form, 17-hydroxycorticosterone, are given below.

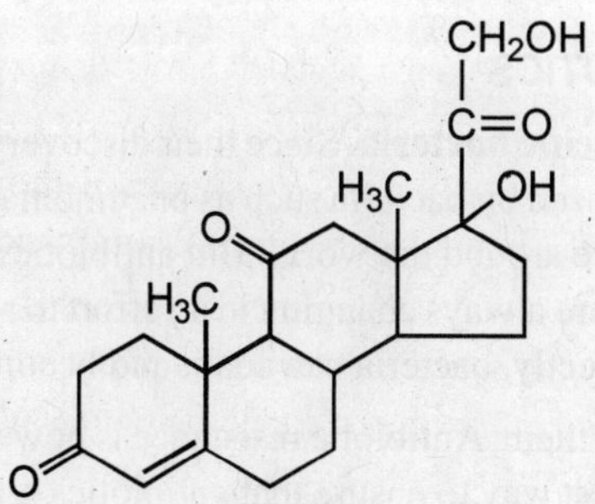

Cortisone

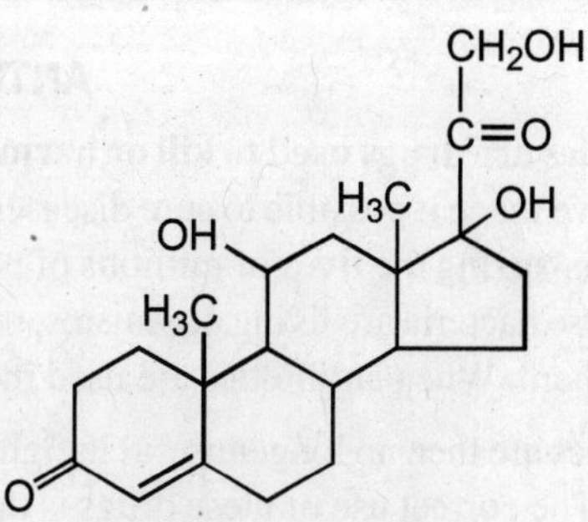

17-Hydroxycorticosterone

ANABOLIC STEROIDS

Anabolic steroids are a group of powerful compounds that are closely related chemically to the male sex hormone testosterone. These artificial substances were developed in the 1930's originally to help men whose bodies produced inadequate amounts of the natural hormone that is responsible for the development of masculine characteristics occurring at puberty, such as lowering of voice and growth of body hair.

Physicians seldom prescribe steroids today, and the few remaining medical uses for them are generally limited to treatment of certain kinds of anemia, severe burns, and some types of breast cancer.

Steroid abusers subject themselves to more than 70 side effects ranging in severity from liver cancer to acne and encompassing psychological as well as physical reactions. The parts of the body that are most seriously affected by steroids are the liver and the cardiovascular and reproductive systems. In males, steroids can cause withered testicles, sterility, and impotence. In females, irreversible masculine traits can develop along with menstrual irregularities, breast reduction, and sterility. Psychological effects in both sexes include aggressive, combative behavior known as "roid rage" and depression. Some side effects may not show up for years, such as heart attacks and strokes, and some might not even be recognized as side effects, such as failure to achieve full height potential because of arrested bone development during adolescence.

Most steroids used illegally are obtained through the black market from underground laboratories and foreign sources. The quality and purity of such drugs are questionable at best. Furthermore, while steroids may build muscles, their intended effect – increased strength – may be offset by the fact that the strength of tendons and ligaments doesn't increase with muscle strength. This imbalance may result in injuries that take a long time to heal. Little research has been done to assess the long-term effects of steroid use ; nor is there much clinical evidence of the effects on women and adolescents. Young people whose bodies are still developing are particularly vulnerable, as are women, because they have less of the natural hormone.

Points to Remember

1. The steroids are natural products that have in common the 4-cycle, skeletal structure of 1,2-pentenophenanthrene.
2. Steroids with one or more hydroxyl groups may be called *sterols.*
3. The most abundant steroids of animal origin are cholesterol, ergosterol, the sex hormones, and cortisone.

ANTIBIOTICS

Antibiotics are drugs used to kill or harm specific bacteria. Since their discovery in the 1930s, antibiotics have made it possible to cure diseases caused by bacteria such as pneumonia, tuberculosis, and meningitis-saving the lives of millions of people around the world. But antibiotics must be used wisely. Because bacteria are living organisms, they are always changing in an effort to resist the drugs that can kill them. When antibiotics are used incorrectly, bacteria can adapt and become resistant.

Antibiotics are then no longer useful in fighting them. Antibiotic resistance is now a major public health issue. The correct use of these drugs is the best way to ensure that antibiotics remain useful in treating infections.

PENICILLINS

Penicillin, the first antibiotic to be isolated from a mold, was introduced into clinical practice in 1941 with remarkable results. It was found to be active against a number of *gram-positive* microorganisms of the cocci type and against spirochetes. *Gram-positive* refers to organisms that stain "positive" when treated with gram stain : *gram-negative* refers to those that stain "negative" to the same reagent. A *broad spectrum antibiotic* is one that may be effective against both gram-negative and gram-positive organisms. A number of penicillins have been prepared. Their skeletal structures are the same, but they differ from one another in the character of the side chain R (see formula for penicillin). Commercial preparations of penicillin are mainly penicillin G ($R = C_6H_5CH_2–$, benzyl).

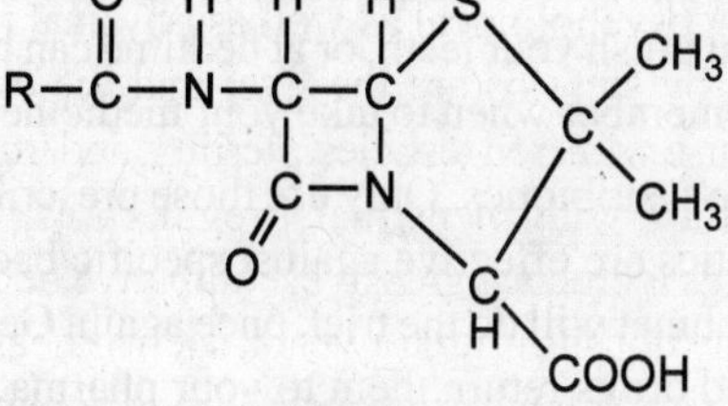

Penicillin

STREPTOMYCIN

A more recent antibiotic is **streptomycin**, whose structure is that of a trisaccharide. Streptomycin is active against *gram-negative* bacteria and is used in tuberculosis therapy. The structure of streptomycin is indicated below.

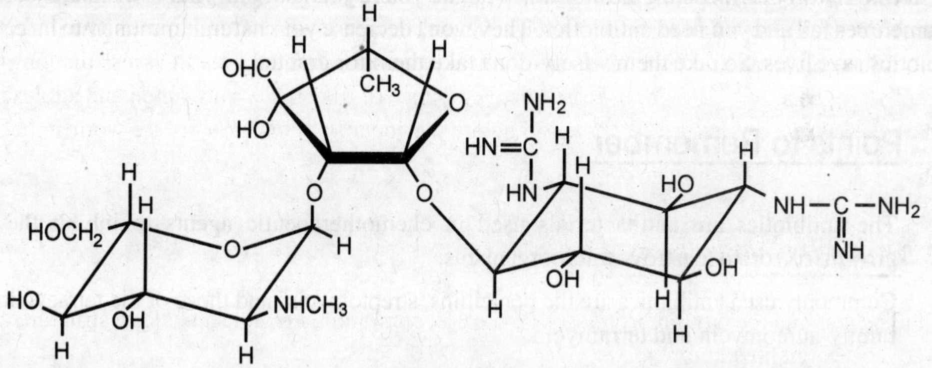

Streptomycin

TETRACYCLINES

The **tetracyclines** comprise a family of compounds, each member of which has a four-fused-ring structure. Each member differs from the other only in minor detail. Tetracycline, its 7-chloro derivative *aureomycin,* and its 5-hydroxy derivative, *terramycin,* are *broad spectrum* antibiotics widely used against a number of bacterial and viral diseases.

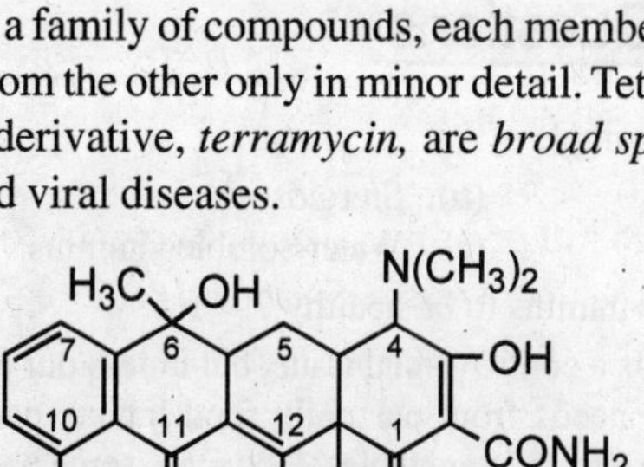

Tetracycline

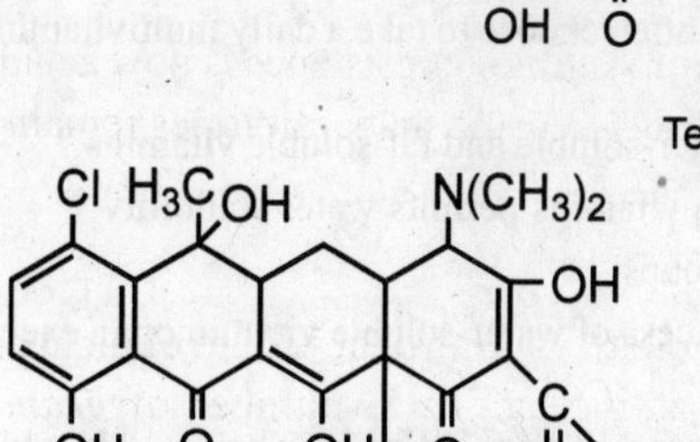

Aureomycin

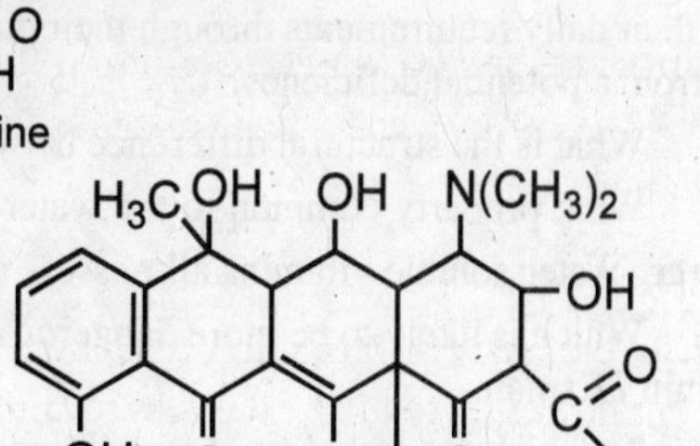

Terramycin

THE DOs AND DON'Ts OF ANTIBIOTICS

DON'T insist that your doctor prescribe an antibiotics. Antibiotics can only cure bacterial infections, they are useless against a cold, flu, or other viral infections.

DON'T forget to take your antibiotic. Simple reminders such as keeping them in view or taking them around mealtime, when you brush your teeth, or at bedtime can be very helpful. Your pharmacist can help you with a system to remember when to take your medicine.

DON'T share other people's antibiotics. Only use those prescribed specifically for your current illness. Since individual antibiotics are effective against specific bacteria, don't assume the leftover ones you have in the medicine cabinet will do the trick once again. Get rid of any leftover antibiotic as you would any other old, unused drugs return them to your pharmacist for safe disposal.

DON'T ignore unexpected side effects ; tell your doctor or pharmacist about severe or unusual reactions. Sometimes, your doctor may prescribe a different antibiotic, or a change in the schedule or dosage, that may make it easier to take.

DON'T worry about taking antibiotics. Whether you're young or old, your own natural defenses can sometimes fail and you need antibiotics. They won't decrease your natural immunity to infections. Antibiotics save lives. So take them wisely-don't take them for granted.

Points to Remember

1. The antibiotics are antibacterials used in chemotherapeutic agents to inhibit the growth of, or to destroy, microorganisms.
2. Commonly used antibiotics are the penicillins, streptomycin, and those of the tetracycline family-aureomycin and terramycin.

Examination Questions

1. Define and give an example of

(*a*) Vitamins (*b*) Steroids (*c*) Fat-soluble vitamins

(*d*) Antibiotic (*e*) Water-soluble vitamins

2. Do we need to take vitamins to be healthy ?

Answer. Probably not. This is a controversial issue, but unless our diet is severely limited, we are more than likely meeting our nutritional needs from our daily food intake, particularly if we eat a balanced diet containing at least five portions of fruits and vegetables. However, some people don't know whether or not they are meeting their daily requirements through their diet, they often choose to take a daily multivitamin to protect themselves from a potential deficiency.

3. What is the structural difference between water-soluble and fat-soluble vitamins ?

4. What property common to the water-soluble vitamins permits water solubility ?

Answer. Water-soluble vitamins all possess polar groups.

5. Which is likely to be more dangerous – an excess of water-soluble vitamin or an excess of a fat-soluble vitamin ? Explain.

6. To which general class of compounds does ascorbic acid (Vitamin C) belong ? In which general class of compounds could we classify all other water-soluble vitamins ?

Answer. Ascorbic acid (Vitamin C) is a carbohydrate derivative and an oxygen-containing heterocycle. All other water-soluble vitamins possess nitrogen-containing heterocycles.

7. Can vitamin E make one look or feel younger ?

Answer. Nothing can slow the natural ageing process, but this vitamin will help decrease the severity of diseases associated with older age, including coronary heart disease, certain forms of cancer and cataracts. Certainly there is no reason to discourage older people from taking additional vitamin E even though it won't actually delay the ageing process. Good sources of vitamin E include wholegrain cereals, nuts and seeds.

8. Identify each vitamin as water-soluble or fat-soluble.

(*a*) Vitamin A (*b*) Vitamin B_6
(*c*) Vitamin B_{12} (*d*) Vitamin C
(*e*) Vitamin K

Answer. (*a*) and (*e*) are fat-soluble.
(*b*), (*c*) and (*d*) are water-soluble.

9. Identify the vitamin deficiency associated with

(*a*) Scurvy (*b*) Night blindness
(*c*) Beriberi

Answer. (*a*) Vitamin C, (*b*) Vitamin A, and (*c*) Vitamin B_1.

Multiple Choice Questions

1. Vitamins are

(*a*) only necessary when one is ill
(*b*) only needed by athletes involved in heavy endurance training
(*c*) essential organic substances needed in minute amounts by the body to perform, a highly specific metabolic function
(*d*) metallic elements found in enzymes and hormones

Answer. (*c*)

2. When it comes to vitamins, people :

(*a*) Can generally get all that they need by eating a healthy, varied diet
(*b*) Must take vitamin supplements to get all the vitamins they need
(*c*) Needn't worry about eating a balanced diet if they take a multivitamin
(*d*) Should cut back because vitamins are high in calories

Answer. (*a*)

3. How many known vitamins are there ?

(*a*) 6
(*b*) 13
(*c*) 19
(*d*) 26, one for each letter in the alphabet

Answer. (*b*)

4. Vitamins can be grouped into two broad categories

(*a*) fat-soluble and water-soluble (*b*) acids and bases
(*c*) acids and esters (*d*) esters and amines

Answer. (*a*)

5. The fat-soluble vitamins are

(*a*) A, C and E (*b*) A, D, E and K
(*c*) A, B, C and D (*d*) folacin, niacin, thiamine and iron

Answer. (*b*)

6. Fat-soluble vitamins are
   (*a*) stored in the body
   (*b*) excreted from the body
   (*c*) rapidly decomposed in the body
   (*d*) less important than water-soluble

Answer. (*a*)

7. Which of the following is NOT a water-soluble vitamin ?
   (*a*) Vitamin A (*b*) Vitamin C
   (*c*) Vitamin B_6 (*d*) Vitamin B_{12}

Answer. (*a*)

8. Which of the following is NOT a vitamin
   (*a*) Thiamin (*b*) Riboflavin
   (*c*) Calcium (*d*) Vitamin H (biotin)

Answer. (*c*)

9. Which vitamin is best known to be good for cell growth, your immune system and maintaining good eyesight ?
   (*a*) Vitamin D (*b*) Vitamin A
   (*c*) Vitamin C (*d*) Vitamin 20-20

Answer. (*b*)

10. The chemical name for vitamin C is
   (*a*) linoleic acid (*b*) hydrochloric acid
   (*c*) protease inhibitor (*d*) ascorbic acid

Answer. (*d*)

11. A deficiency of vitamin C causes
   (*a*) Ascorbic acid tingling (*b*) Beriberi
   (*c*) Pellagra (*d*) Scurvy

Answer. (*d*)

12. In addition to citrus fruits, which of the following are good sources of vitamin C ?
   (*a*) Milk and yogurt (*b*) Broccoli and green peppers
   (*c*) Whole wheat bread and cereal (*d*) All meats except red meat

Answer. (*b*)

13. Under the influence of sunlight, _______ is synthesized in our skin from _______.
   (*a*) Vitamin C, glucose (*b*) Vitamin E, starch
   (*c*) Vitamin D, cholesterol (*d*) Vitamin B1, protein

Answer. (*c*)

14. For strong bones ________ is also required because it helps absorb calcium.
   (*a*) Vitamin D (*b*) Vitamin K
   (*c*) Beta-carotene (*d*) Vitamin E

Answer. (*a*)

15. Some food sources of Vitamin D are:
   (*a*) Fruits and vegetables (*b*) Salmon and egg yolks
   (*c*) Butter and fortified milk (*d*) Both (*b*) and (*c*)

Answer. (*d*)

16. A good food source of ascorbic acid is
   (*a*) meat (*b*) citrus fruits
   (*c*) lettuce (*d*) coffee

Answer. (*b*)

17. Vitamin C is required for the production and maintenance of:

(*a*) Collagen (*b*) Hormone

(*c*) Ascorbic Acid (*d*) Red Blood Cells

Answer. (*a*)

18. The deficiency disorder associated with low levels of vitamin B1 is

(*a*) dryness of the skin. (*b*) beri beri.

(*c*) night blindness. (*d*) scurvy.

Answer. (*b*)

19. Common food sources of Vitamin A are:

(*a*) Milk, eggs, butter, cheese, cream, and liver

(*b*) White sugar, honey, and sugar cane

(*c*) Broccoli, apricots, cantaloupe, carrots, sweet potato, spinach

(*d*) Both (*a*) and (*c*)

Answer. (*d*)

20. A deficiency of thiamin is known to cause what condition ?

(*a*) Scurvy (*b*) Angular chelitis

(*c*) Beriberi (*d*) Pellagra

Answer. (*c*)

21. The vitamin known as thiamin, is also known as

(*a*) Vitamin B_1 (*b*) Vitamin B_2

(*c*) Vitamin B_6 (*d*) Vitamin B_{12}

Answer. (*a*)

22. The vitamin, riboflavin, is also known as

(a) Vitamin B_1 (b) Vitamin B_2

(c) Vitamin B_6 (d) Vitamin C

Answer. (*b*)

23. Cholesterol that is present in the blood serum is closely associated with

(*a*) hardening of the arteries (*b*) overactive kidneys

(*c*) diabetes (*d*) osteoporosis

Answer. (*a*)

You must learn to predict trends in acidity, basicity, boiling points and water-solubility. It is an essential skill and requires thorough understanding of the principles of organic chemistry.

51

Predicting Trends

Predicting trends in acidity, basicity, boiling points and water-solubility is an essential skill. It requires thorough understanding of the principles of organic chemistry and extensive practice. Many problems have already been discussed in this book. More are given below to help you solve such problems more effectively.

Only idiots use a coin to predict trends in acidities, basicities etc. Such questions require reasoning and practice. Follow the guidelines provided in this chapter.

Points to Remember

PREDICTING TRENDS IN ACIDITIES

- Acids donate a proton. That is, they lose a proton.
- Acidity (acid-strength) depends on the ease with which they lose a proton.
- Aliphatic carboxylic acids are acidic and lose a proton readily. This is because the carboxylate ion formed by ionization or reaction with a base is stabilized by resonance.
- Acidity of substituted aliphatic carboxylic acids depends on the inductive effect and position of the substituents. Electron-releasing alkyl groups decrease the acidity. Electron-withdrawing substitutents (F, Cl, Br, NO_2, CN, SO_3H, COOH) increase the acidity.
- Phenol is acidic. This is because the phenoxide ion formed by ionization or reaction with a base is stabilized by resonance.
- Acidity of phenols depends on the nature and position of the substituent. Electron-releasing alkyl groups decrease the acidity of phenols. Electron-withdrawing substituents (NO_2, Cl, CN) increase the acidity of phenols.
- Phenols are much more acidic than alcohols. This is because alcohols give alkoxide ions on loss of a proton. The alkoxide ions so formed are not resonance-stabilized.
- *p*-Nitrophenol is almost 1000 times more acidic than phenol.
- Picric acid (2,4,6-Trinitrophenol) is as strong as HCl.
- Alcohols are weaker acids than water.
- Thiols are more acidic than alcohols.
- Aromatic carboxylic acids are acidic and lose a proton readily. This is because the carboxylate ion formed by ionization or reaction with a base is stabilized by resonance.
- Acidity of substituted benzoic acids depends on the nature and position of the substituent. Electron-releasing substituents (CH_3, OCH_3) decrease the acidity of a benzoic acid. Electron-withdrawing substituents increase the acidity of a benzoic acid.
- Almost all ortho substituents increase the acidity of a benzoic acid.
- The acidity of a compound may be expressed quantitatively in terms of its acidity constant, pKa. Stronger acids will have smaller pK_a. Weaker acids will have larger pK_a.

Examination Questions

1. Explain why acetic acid is a stronger acid than ethyl alcohol.

Answer. Acetic acid is a stronger acid than ethyl alcohol because the acetate ion formed by ionization is stabilized by resonance. The negative charge is delocalized over two oxygen atoms. In the ethoxide ion, on the contrary, the negative charge is localized on the single oxygen atom. There is no resonance. Ethyl alcohol is, therefore, much weaker than acetic acid.

Acetate ion
(Resonance structures)

Ethoxide ion
(No resonance structures are possible)

2. Explain why trichloroacetic acid is a stronger acid than acetic acid. *(Aligarh BSc, 2014)*

Answer. Electron-withdrawing substituents increase the acidity. This is because the electron-withdrawing substituents decrease the negative charge on the carboxylate ion and stabilize it. The loss of proton becomes relatively easy. Thus trichloroacetic acid is stronger than acetic acid as it contains three electron-withdrawing chlorine atoms.

3. Explain why ethanethiol is more acidic than ethanol.

Answer. The SH bond in ethanethiol (CH_3CH_2SH) is much weaker than the OH bond in ethanol (CH_3CH_2OH). This is because of poor overlap of 1*s* orbital on hydrogen with the large 3*p* orbital on sulfur. As a result ethanethiol loses a proton more readily than ethanol.

4. Explain why benzoic acid is a stronger acid than acetic acid. *(Jiwaji BSc, 2014)*

Answer. Benzoic acid is a stronger acid than acetic acid because of the greater stability of the benzoate ion due to resonance.

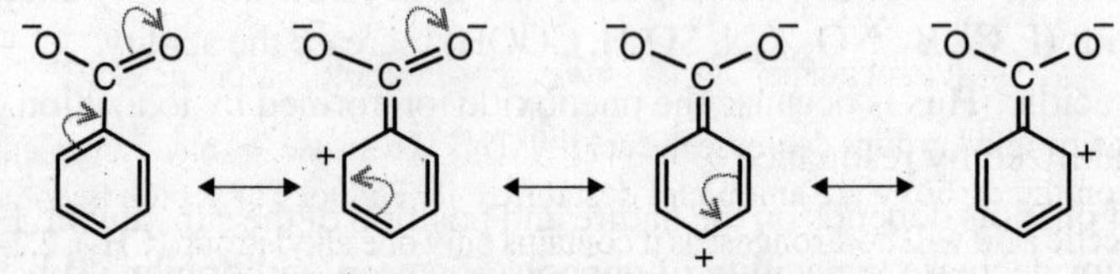

Resonance structures of benzoate ion

5. Explain why phenol is a stronger acid than ethyl alcohol. *(Saurashtra BSc, 2015)*

Answer. Phenol is a stronger acid than ethyl alcohol because the phenoxide ion formed by ionization is stabilized by resonance. In ethyl alcohol, no resonance is possible and the negative charge is localized on the single oxygen atom.

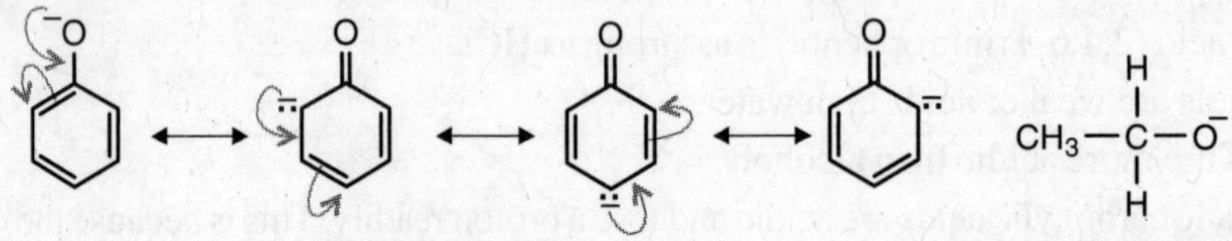

Phenoxide ion (Resonance structures) — Ethoxide ion (No resonance structures are possible)

6. Explain why *p*-nitrophenol is a much stronger acid than phenol.

Answer. Phenols are acidic compounds due to the dissociation shown below :

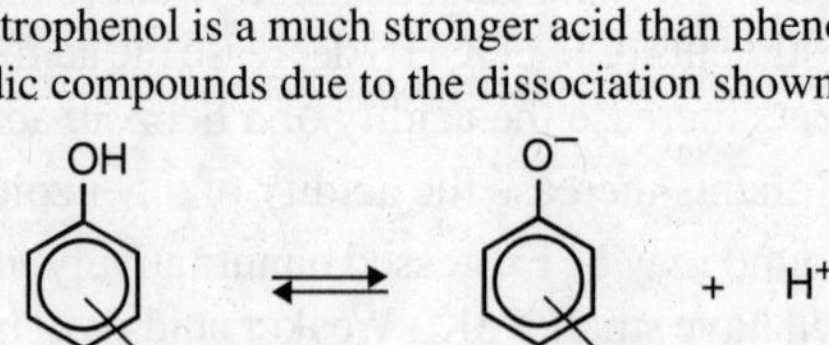

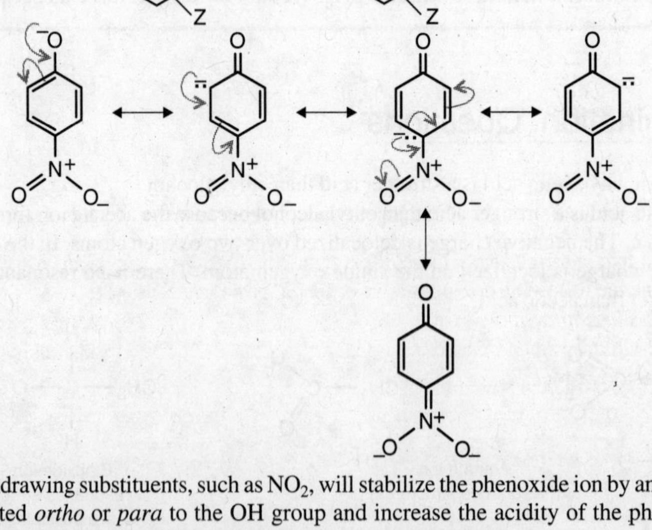

Electron withdrawing substituents, such as NO_2, will stabilize the phenoxide ion by an additional resonance structure when located *ortho* or *para* to the OH group and increase the acidity of the phenol. This additional structure has the charge spread over to the attached group as shown above.

The phenoxide ion obtained from phenol does not have the additional resonance structure in which the charge is spread to the attached group. Therefore, it is less stable and less easily formed. Thus *p*-nitrophenol is much more acidic than phenol.

7. Arrange the following compounds in order of increasing acidity and explain your answer.
(*a*) Acetic acid
(*b*) 2-Methylpropanoic acid
(*c*) 2,2-Dimethylpropanoic acid

Answer. The acidity order is:

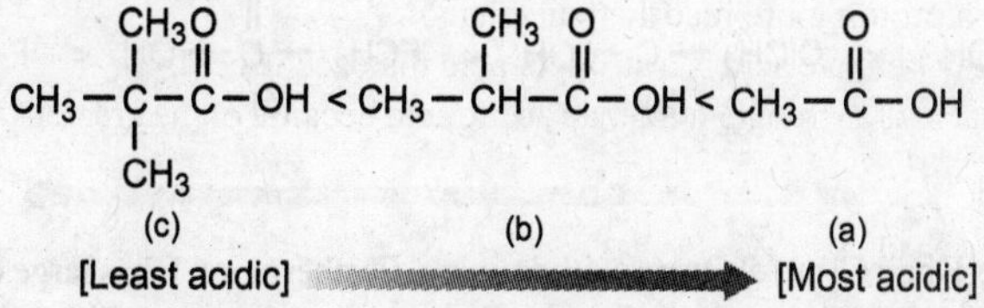

Electron-releasing alkyl groups decrease the acidity. This is because the electron-releasing groups increase the negative charge on the carboxylate anion and destabilize it. The loss of proton becomes easy. Thus in the above compounds, acetic acid will be strongest as it contains only one alkyl group (CH_3). 2,2-Dimethylpropanoic acid will be weakest as it contains three alkyl groups ($3 \times CH_3$).

8. Arrange the following compounds in order of increasing acidity and explain your answer:
(*a*) Butanoic acid
(*b*) 2-Chlorobutanoic acid
(*c*) 3-Chlorobutanoic acid

Answer. The acidity order is:

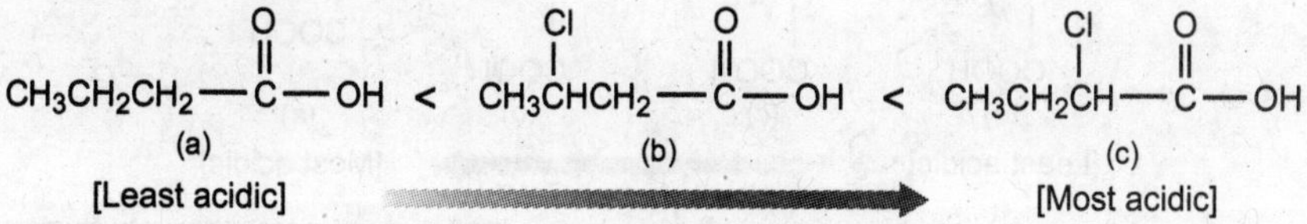

The 2-chloro substituent increases the acidity of butanoic acid substantially, due to inductive effect. The 3-chloro substituent exerts a similar but much smaller effect, because C–Cl bond is now further away from the carboxyl group. Inductive effect falls off rapidly with distance.

9. Arrange the following compounds in order of increasing acidity:
(*a*) 3-Chloropropanoic acid
(*b*) 2,2-Dichloropropanoic acid
(*c*) 2-Chloropropanoic acid (*Ranchi BSc, 2015*)

Answer. The acidity order is :

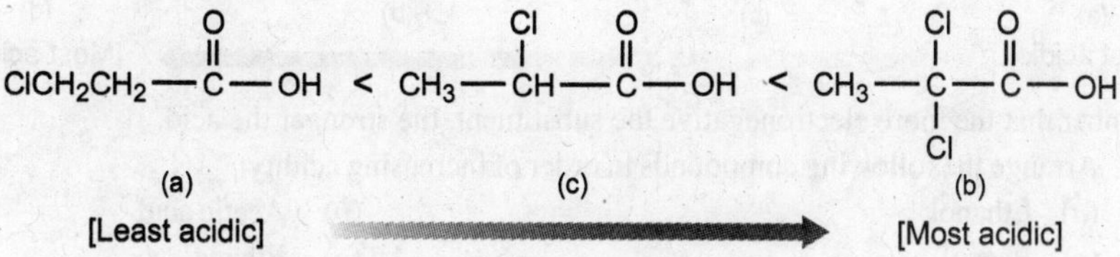

Remember that two chlorines have more effect than one. Inductive effect falls off with distance.

10. Arrange the following compounds in order of increasing acidity:
(*a*) 2-Fluoropropanoic acid
(*b*) 2-Bromopropanoic acid
(*c*) 2-Chloropropanoic acid
(*d*) Propanoic acid

Answer. The acidity order is:

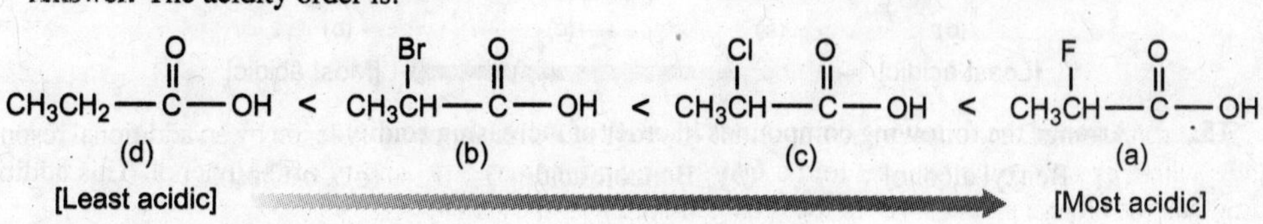

Remember that the more electronegative the substituent, the stronger the acid.

11. Arrange the following compounds in order of increasing acidity:

(*a*) Trifluoroacetic acid (*b*) Chloroacetic acid
(*c*) Fluoroacetic acid (*d*) Acetic acid

(*Awadh BSc, 2014*)

Answer. The acidity order is:

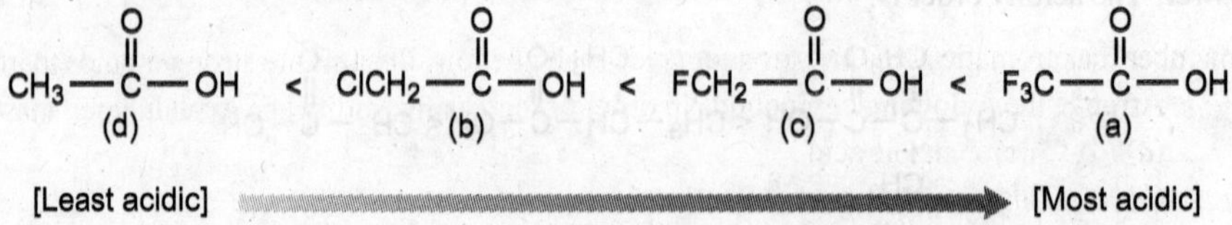

The halogen atoms F and Cl are electron-withdrawing. Furthermore, F has large electronegativity than that of Cl. Electron-withdrawing substituents can disperse the negative charge, stabilize the carboxylate anion, and thus increase acidity.

12. Arrange the following compounds in order of increasing acidity:

(*a*) Oxalic acid (*b*) Malonic acid
(*c*) Succinic acid (*d*) Glutaric acid

Answer. The acidity order is:

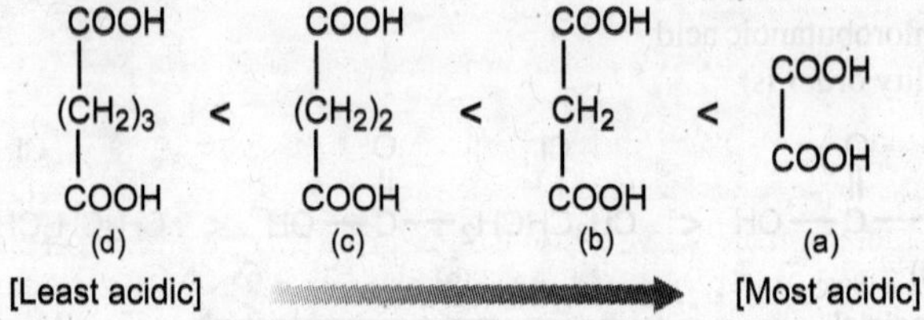

Remember that COOH group is electron-withdrawing group. The inductive effect falls off with distance.

13. Arrange the following compounds in order of increasing acidity:

(*a*) CH_3COOH (*b*) $ClCH_2COOH$
(*c*) $HSCH_2COOH$ (*d*) FCH_2COOH

Answer. The acidity order is:

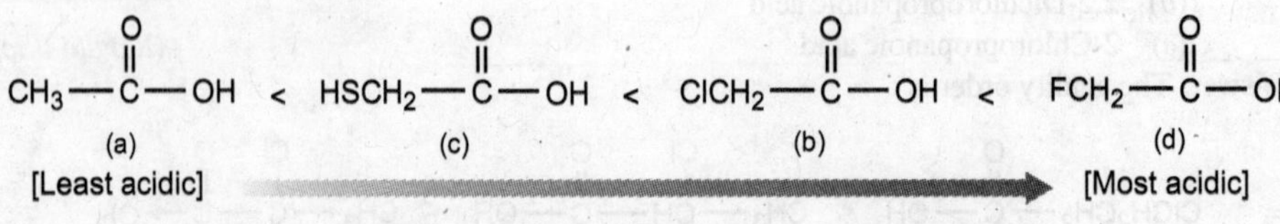

Remember that the more electronegative the substituent, the stronger the acid.

14. Arrange the following compounds in order of increasing acidity:

(*a*) Ethanol (*b*) Acetic acid
(*c*) Phenol (*d*) Ethane

Answer. The acidity order is:

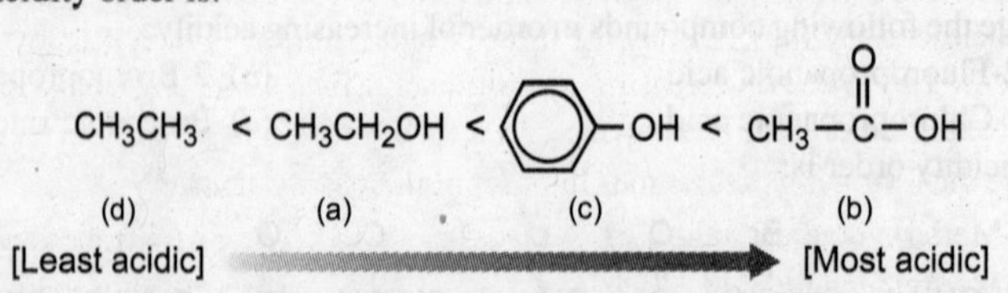

15. Arrange the following compounds in order of increasing acidity:

(*a*) Benzyl alcohol (*b*) Benzoic acid (*c*) *o*-Cresol

Answer. The acidity order is:

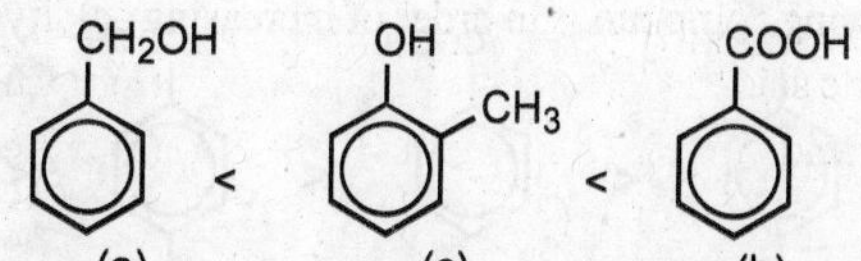

(a) < (c) < (b)

[Least acidic] → [Most acidic]

Remember that aromatic acids are stronger acids than phenols. Phenols are stronger acids than alcohols.

16. **Arrange the following compounds in order of increasing acidity and explain your answer.**

(*a*) *o*-Chlorobenzoic acid

(*b*) *m*-Chlorobenzoic acid

(*c*) *p*-Chlorobenzoic acid

Answer. The acidity order is:

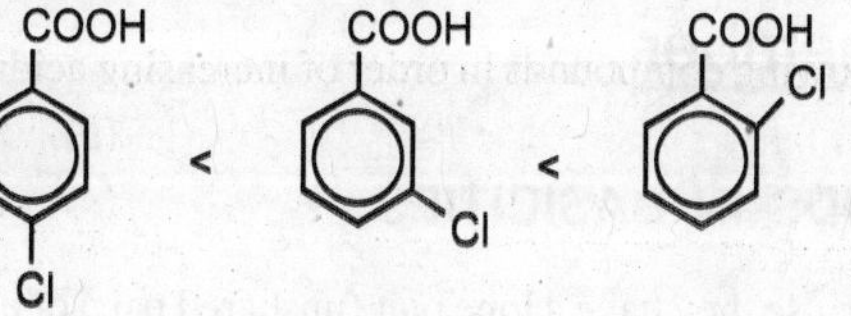

(c) < (b) < (a)

[Least acidic] → [Most acidic]

All three chloro acids are stronger than benzoic acid. This is because Cl is an electron-withdrawing substituent. It stabilizes the carboxylate anion.

Among the three chlorobenzoic acids, ortho isomer is the strongest acid, since in this isomer the chloro substituent is closest to the carboxyl group and exerts maximum electron-withdrawing inductive effect. The effect falls off with distance. The para isomer is the weakest acid because the electron-withdrawing Cl is the greatest distance from the carboxyl group, stabilizing the anion the least.

17. **Arrange the following compounds in order of increasing acidity:**

(*a*) Phenol (*b*) *p*-Chlorophenol

(*c*) *p*-Cresol (*d*) Cyclohexanol

(*Madras BSc, 2014*)

Answer. The acidity order is:

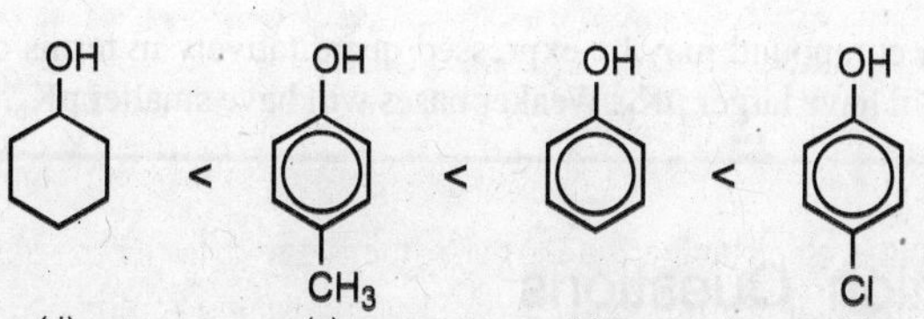

(d) < (c) < (a) < (b)

[Least acidic] → [Most acidic]

Phenols are much more acidic than alcohols. Therefore cyclohexanol, which is the only alcohol in the group, would be least acidic of the four. Among the three phenols, acidity is increased by electron-withdrawing substituents and decreased by electron-releasing substituents. Since Cl is electron-withdrawing and CH_3 is electron-releasing, the expected order is an shown.

18. Arrange the following compounds in order of increasing acidity:

(*a*) *p*-Methoxybenozic acid (*b*) *p*-Nitrobenzoic acid

(*c*) *p*-Methylbenzoic acid (*d*) *p*-Chlorobenzoic acid

(*e*) Benzoic acid

Answer. The acidity order is:

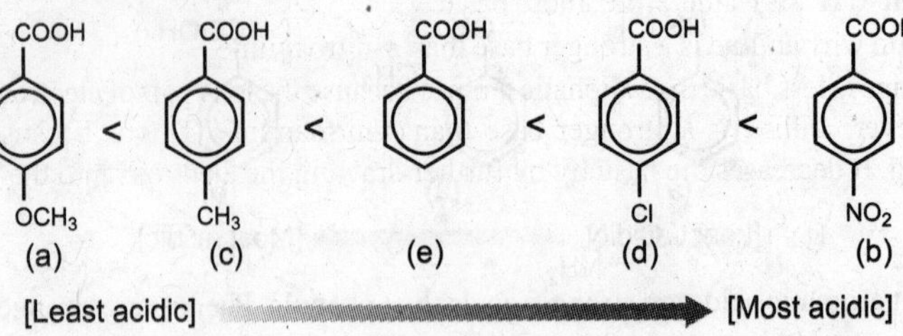

The OCH_3 and CH_3 groups are electron-releasing substituents which can destabilize the carboxylate anion, and thus decrease acidity. The Cl and NO_2 groups are electron-withdrawing substituents which stabilize the anion and increase acidity. Thus in the given set *p*-methoxybenzoic acid is least acidic, and *p*-nitrobenzoic acid is most acidic.

Points to Remember

PREDICTING TRENDS IN BASICITIES

- Amines are basic because they have a lone pair (unshared pair) of electrons on nitrogen. This lone pair of electrons is available for formation of new bond with a proton or Lewis acids.
- Basicity (base-strength) depends on the degree of availability of the lone pair of electrons.
- Any factor which increases the availability of the lone pair of electrons on nitrogen for protonation, increases the basicity, and *vice versa*.
- Electron-releasing groups on the nitrogen atom of amines increase the basicity and electron-withdrawing groups decrease the basicity.
- Aliphatic amines are stronger bases than ammonia.
- Aromatic amines are less basic than aliphatic amines. This is because the lone pair of electrons on nitrogen in aromatic amines is less available due to delocalization.
- Electron-releasing groups increase the basicity of aromatic amines. Electron-withdrawing groups on the aromatic ring decrease the basicity.
- The basicity of a compound may be expressed quantitatively in terms of its basicity constant, pk_b. Stronger bases will have larger pK_b. Weaker bases will have smaller pK_b.

Examination Questions

1. Explain why aniline is less basic than methylamine. *(Bundelkhand BSc, 2015)*

Answer. Aniline is less basic because the lone pair of electrons on nitrogen is less available for protonation. It is delocalized into the benzene ring by resonance.

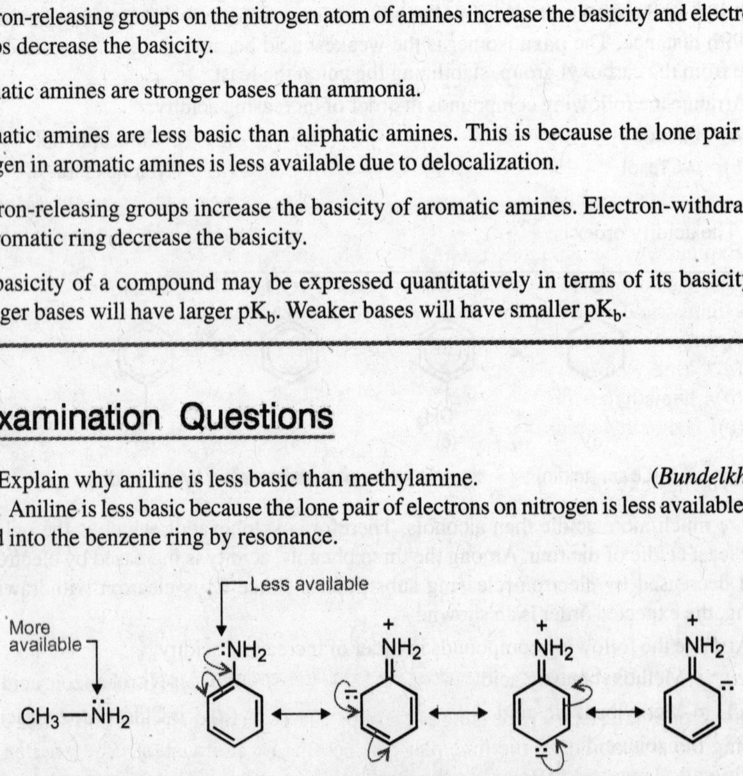

Methylamine

Resonance structures of aniline

In methylamine, no such delocalization is possible. The lone pair of electrons on nitrogen are more available for donation to H^+. It is, therefore, more basic.

2. Explain why aniline is a stronger base than *p*-nitroaniline.

Answer. Aniline is less basic than aliphatic amines because the lone pair of electrons is drawn into the ring by resonance. However, aniline is a stronger base than *p*-nitroaniline. This is because the *p*-nitro group is electron-withdrawing. It decreases the basicity by further drawing the electrons into the benzene ring.

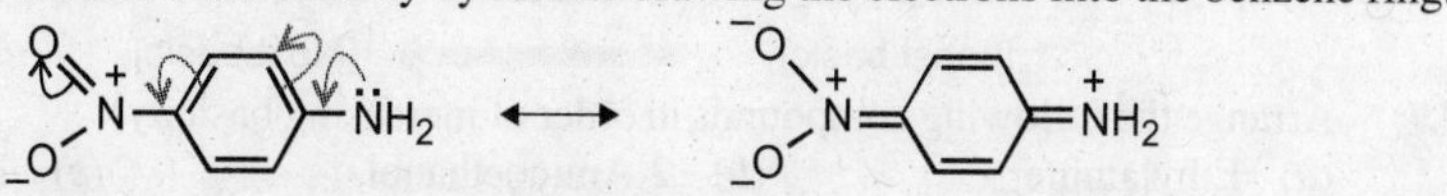

p-Nitroaniline

Notice that the possibilities for delocalization of the lone pair of electrons are greater in *p*-nitroaniline than in aniline.

3. Explain why benzylamine is more basic than *N*-methylamine.

Answer. *N*-Methylaniline is less basic than benzylamine because the lone pair of electrons in *N*-Methylaniline is delocalized into the benzene ring. It is less available for protonation.

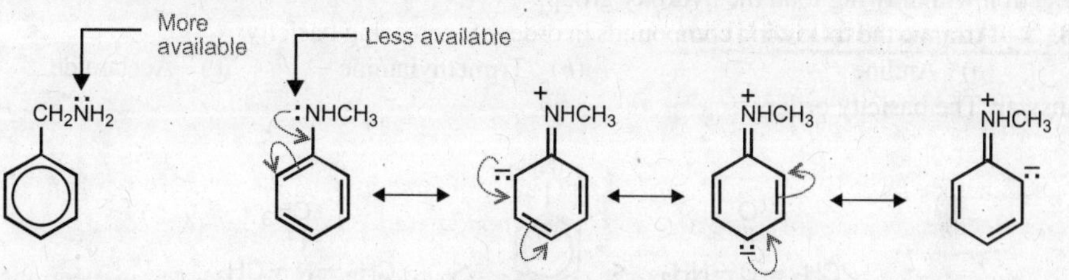

Benzylamine — Resonance forms of *N*-methylaniline

4. Explain why amines are more basic than amides. (*Tezpur BSc, 2015*)

Answer. In simple amines, the lone pair of electrons is usually localized on nitrogen, and hence available for protonation. In amides, on the other hand, the lone pair on nitrogen is delocalized to the carbonyl oxygen through resonance.

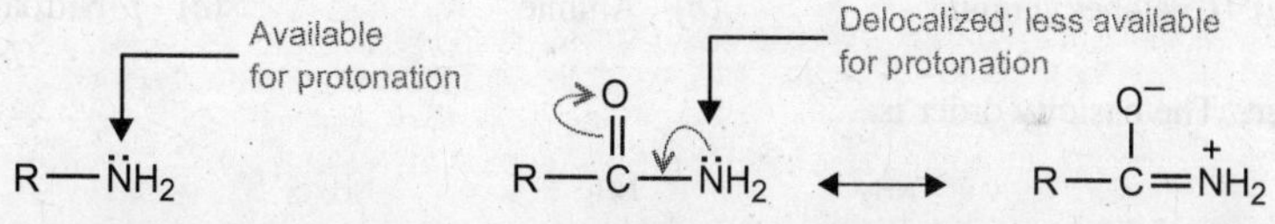

Amine — Amide

5. Explain why methylamine is more basic than methanol.

Answer. Methylamine is a stronger base because N is less electronegative than O. Its lone pair of electrons is less tightly held, so it is more readily available for protonation.

6. Arrange the following compounds in order of increasing basicity and explain your answer:

(*a*) Methylamine
(*b*) Dimethylamine
(*c*) Trimethylamine

(*Punjabi BSc, 2014*)

Answer. The basicity order is:

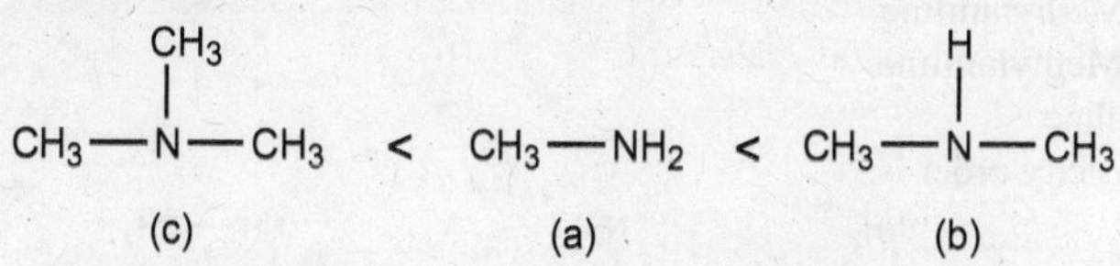

(c) < (a) < (b)

[Least basic] → [Most basic]

The methyl groups are electron releasing (+I groups). They increase the electron density around nitrogen thereby increasing the availability of the lone pair of electrons for protonation. **The greater the number of electron releasing alkyl groups, the greater the availability of nitrogen's lone pair and stronger the acid.**

This explains why dimethylamine is a stronger base than methylamine. However, trimethylamine is a weaker base than both dimethylamine and methylamine. Although in trimethylamine the electron density is further increased, the steric crowding of these methyl groups makes the approach and bonding by a proton relatively difficult. The electrons are there but the path is blocked.

Remember that for alkylamines, the basicity order is:

3° Amines < 1° Amines < 2° Amines

[Least basic] ⟶ [Most basic]

7. Arrange the following compounds in order of increasing basicity:
(*a*) Ethylamine (*b*) 2-Aminoethanol (*c*) 2-Fluoroethylamine

Answer. The basicity order is:

$FCH_2CH_2NH_2$ (c) < $NH_2CH_2CH_2OH$ (b) < $CH_3CH_2NH_2$ (a)

[Least basic] ⟶ [Most basic]

The hydroxy and fluoro substituents are electron withdrawing (Inductive effect). The fluoro substituent is more electron withdrawing than the hydroxy group.

8. Arrange the following compounds in order of increasing basicity:
(*a*) Aniline (*b*) Trimethylamine (*c*) Acetamide

Answer. The basicity order is:

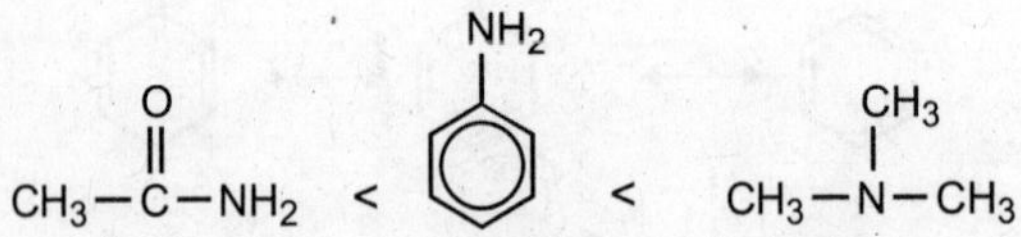

(c) (a) (b)

[Least basic] ⟶ [Trimethylamine]

Remember that amides are less basic than amines, and aromatic amines are less basic than aliphatic amines.

9. Arrange the following compounds in order of increasing acidity:
(*a*) Cyclohexylamine (*b*) Aniline (*c*) *p*-Nitroaniline

(*Indore BSc, 2014*)

Answer. The basicity order is:

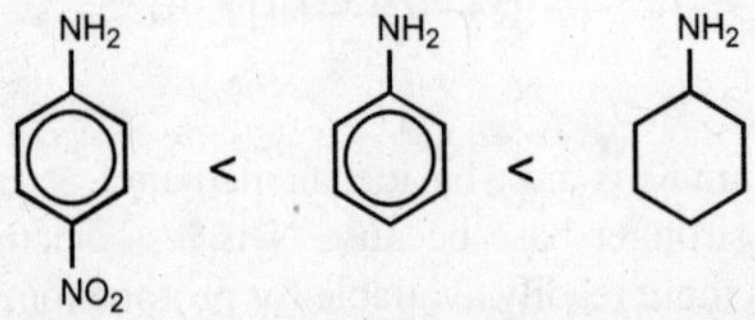

(c) (b) (a)

[Least basic] ⟶ [Most basic]

Aliphatic amines are more basic than aromatic amines. Nitro group in the *p*-position reduces basicity because it withdraws electron-density by its inductive and resonance effects.

10. Arrange the following compounds in order of increasing basicity:
(*a*) *N*-Methylaniline
(*b*) *m*-Methylaniline
(*c*) Aniline

Answer. The basicity order is:

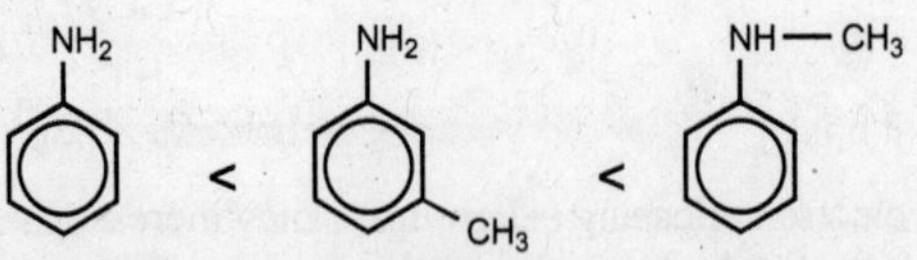

(c) (b) (a)

[Least basic] ⟶ [Most basic]

Methyl group increases basicity more when directly on the N atom than when farther away.

11. Arrange the following compounds in order of increasing basicity:

(*a*) Aniline (*b*) Benzylamine (*c*) 2-Phenylethylamine

Answer. The basicity order is:

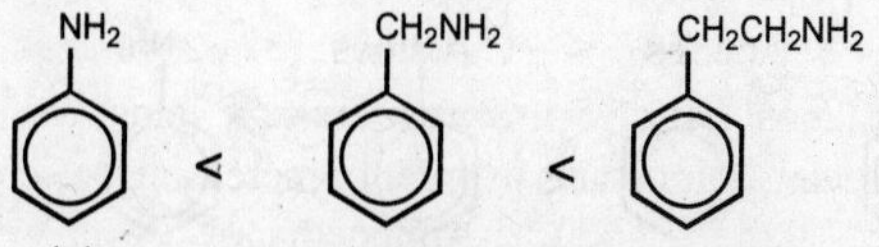

(a) (b) (c)

[Least basic] → [Most basic]

Benzylamine and 2-phenylethylamine are aliphatic amines. Benzylamine is a weaker base because the phenyl group is electron withdrawing (–I Effect). Aniline is aromatic. The lone pair of electrons on nitrogen in aniline is less available due to delocalization.

12. Arrange the following compounds in order of increasing basicity :

(*a*) Ammonia (*b*) Methylamine
(*c*) Aniline (*d*) *p*-Nitroaniline
(*e*) *p*-Methylaniline

Answer. The basicity order is:

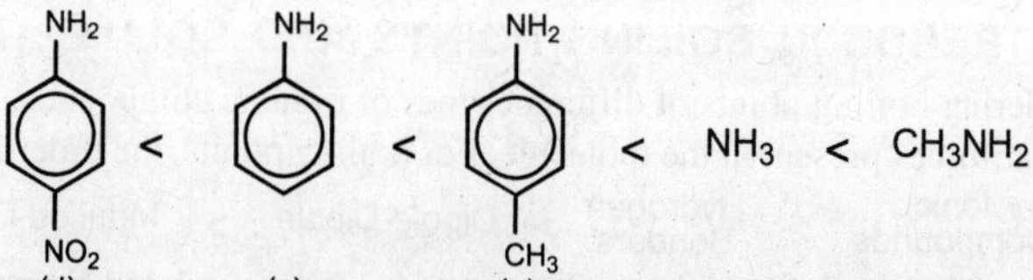

(d) (c) (e) (a) (b)

[Least basic] → [Most basic]

In general, aliphatic amines are more basic than aromatic amines. This is because the lone pair of electrons on nitrogen in aromatic amines is less available due to delocalization.

Methylamine is more basic than ammonia. This is because the methyl group is electron releasing group. It increases the electron density around nitrogen, thereby increasing the availability of the lone pair of electrons for protonation.

Of the aromatic amines, the *p*-methylaniline is a stronger base because of the electron releasing effect of the methyl group. The *p*-nitroaniline is the weaker of the aromatic bases due to the electron withdrawing effect of the nitro group.

13. Arrange the following compounds in order of increasing basicity:

(*a*) Acetanilide (*b*) Aniline
(*c*) Cyclohexylamine (*d*) *N*-Ethylaniline

Answer. The order of basicity is:

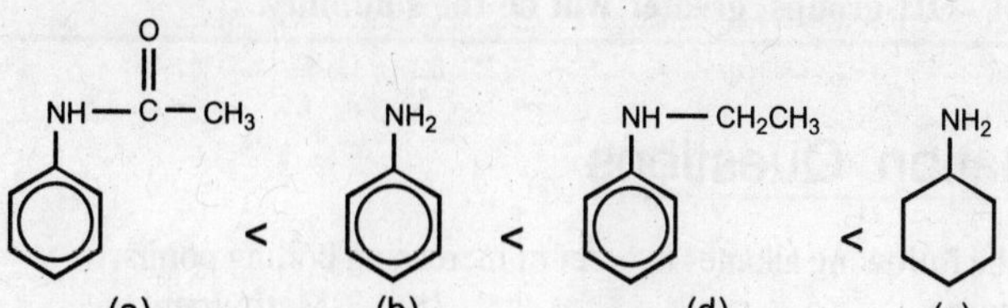

(a) (b) (d) (c)

[Least basic] → [Most basic]

Remember that amides are less basic than amines, and aromatic amines are less basic than aliphatic amines. The two aromatic amines are placed in the order shown because the ethyl group is electron-releasing compared to hydrogen.

14. Arrange the following aromatic amines in order of increasing basicity:

(*a*) *p*-Chloroaniline (*b*) *p*-Nitroaniline
(*c*) *p*-Methylaniline (*d*) *p*-Methoxyaniline
(*e*) Aniline

Answer. The basicity order is:

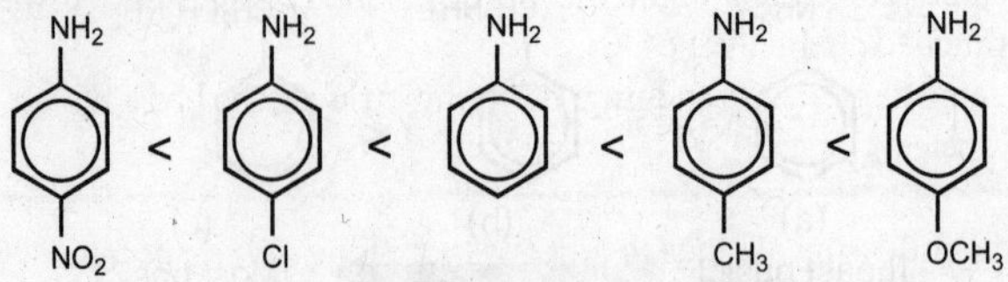

(b) (a) (e) (c) (d)

[Least basic] → [Most basic]

The CH_3 and OCH_3 substituents are electron releasing substituents. They increase the electron density around nitrogen and increase the basicity.

The NO_2 and Cl substituents are electron withdrawing substituents. They decrease the electron density around nitrogen and decrease the basicity.

Points to Remember

PREDICTING TRENDS IN BOILING POINTS AND SOLUBILITIES

- When considering boiling points of different types of organic compounds, first consider the type of intermolecular forces present in the molecule. For boiling points, the order is :

  Ionic Compounds > Hydrogen Bonders > Dipole-Dipole > Induced Dipole

  Branching, etc, need not be considered. The nonpolar compound will have the lowest boiling point.
- In the absence of hydrogen bonding, boiling points increase with molecular weight. The higher the molecular weight, the higher is the boiling point.
- **For compounds of similar molecular weights, boiling points increase with the degree of hydrogen bonding. The greater the degree of hydrogen bonding, the higher will be the boiling point.**
- For alcohols of same molecular weights, branched-chain alcohols will have lower boiling points. This is because the decreased surface area leads to decreased van der Waals forces.
- For alcohols of different molecular weight, the straight-chain alcohol containing the greater number of carbon atoms will have higher boiling point. This is because of increased van der Waals forces.
- **Compounds capable of forming hydrogen bonds with water will be soluble in water.**
- Degree of hydrogen bonding, chain-length (molecular weight), and branching will determine the extent of solubility.
- Compounds with shorter nonpolar chain are most soluble in water.
- **Among alcohols, the solubility in water depends mainly on the number of -OH groups. Greater the number of –OH groups, greater will be the solubility.**

Examination Questions

1. Arrange the following alkanes in order of increasing boiling points:

(*a*) *n*-Hexane (*b*) 3-Methylpentane
(*c*) 2,2-Dimethylbutane (*d*) *n*-Heptane
(*e*) 2-Methylpentane *(Delhi BSc, 2015)*

Answer. The boiling point order is:

(c) < (e) < (b) < (a) < (d)

[Lowest bp] → [Highest bp]

Remember the following facts for the prediction of the boiling point of alkanes:

(1) The larger its molecular weight, the higher its boiling point.

(2) With the same molecular weight, the straight-chain alkane has a higher boiling point than that of the branched-chain alkane.

(3) With the same molecular weight, the more branched alkanes have lower boiling point than that of the less branched chain alkanes.

(4) With the same degree of branching, the symmetric alkane has a higher boiling point than the less symmetric one.

2. Arrange the following alkanes in order of increasing boiling points:

(*a*) *n*-Pentane (*b*) *n*-Hexane
(*c*) *n*-Heptane (*d*) *n*-Butane

Answer. The boiling point order is:

(d) < (a) < (b) < (c)

[Lowest bp] → [Highest bp]

3. Arrange the following alkanes in order of increasing boiling points:

(*a*) *n*-Hexane (*b*) 2-Methylpentane
(*c*) *n*-Pentane (*d*) *n*-Butane

Answer. The boiling point order is:

(d) < (c) < (b) < (a)

[Lowest bp] → [Highest bp]

4. Arrange the following alkanes in order of increasing boiling points :

(*a*) *n*-Octane
(*b*) 2,2,3-Trimethylpentane
(*c*) 2-Methylheptane

Answer. The boiling point order is :

(b) < (c) < (a)

[Lowest bp] → [Highest bp]

5. Arrange the following compounds in order of increasing boiling points and explain your answer :

(*a*) Ethyl fluoride (*b*) Ethylamine
(*c*) Dimethyl ether (*d*) Ethyl alcohol

Answer. The boiling point order is:

$$\underset{(a)}{CH_3CH_2F} < \underset{(c)}{CH_3—O—CH_3} < \underset{(b)}{CH_3CH_2NH_2} < \underset{(d)}{CH_3CH_2OH}$$

[Lowest bp] → [Highest bp]

Since the molecular weight of these compounds are about the same, van der Waals forces are equal and are not a factor here.

The factors which control the boiling points are hydrogen bonding and polarity. Ethyl fluoride and dimethyl ether can not hydrogen bond and thus boil at a low temperature. But since the ether is more polar than the alkyl halides, the ether will show the higher boiling point. Ethylamine and ethyl alcohol can hydrogen bond. Since the alcohol forms a stronger hydrogen bond than the amine, the alcohol will boil at a higher temperature than the amine.

6. Arrange the following compounds, which are nearly of identical molecular weights, in order of increasing boiling points. Explain your answer.

(*a*) 1-Aminobutane (*b*) 1-Butanol
(*c*) Methyl *n*-propyl ether (*d*) Pentane

Answer. The boiling point order is :

$$\underset{(d)}{CH_3CH_2CH_2CH_2CH_3} < \underset{(c)}{CH_3OCH_2CH_2CH_3} < \underset{(a)}{CH_3CH_2CH_2CH_2NH_2} < \underset{(b)}{CH_3CH_2CH_2CH_2OH}$$

[Lowest bp] → [Highest bp]

Hydrogen bonding is not possible in the first two compounds, but C–O bonds are polar, giving the ether a higher boiling point than the alkane. The hydrogen bonds in alcohols (O–H- -O) are more effective than those in amines (N–H- -O), explaining the order of the last two compounds.

7. Arrange the following compounds in order of increasing boiling points :

(*a*) *n*-Hexane (*b*) 1-Hexanol
(*c*) 2-Methylpentanol (*d*) 2,2-Dimethylbutanol
(*e*) 1-Heptanol

Answer. The boiling point order is :

(a) < (d) < (c) < (b) < (e)

[Lowest bp] ⟶ [Highest bp]

Notice that :

(1) *n*-Hexane (MW 86) is a nonpolar molecule.
(2) 1-Hexanol (MW 102) is a polar straight-chain alcohol.
(3) 2-Methylpentanol (MW 102) is a polar branched-chain alcohol. 2-Methylpetanol refers to 2-Methyl-1-pentanol.
(4) 2,2-Dimethylbutanol (MW 102) is a polar and more branched-chain alcohol.
(5) 1-Heptanol (MW 134) is a polar straight-chain alcohol.

8. Arrange the following compounds in order of increasing boiling points and explain your answer :

(*a*) Acetic acid (*b*) 2-Hydroxyethanal (*c*) Methyl acetate

Answer. The boiling point order is :

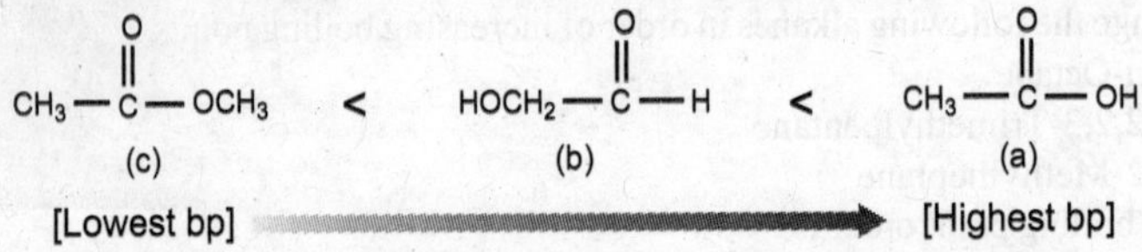

The lowest boiling compound has no O–H bonds and can not hydrogen bond. The next has one O–H bond and can hydrogen bond. In the compound with the highest boiling point, the O–H is polarized by the C=O group making the hydrogen bonding even stronger. Also in acetic acid, two molecules can orient so that hydrogen bonding can occur in two positions.

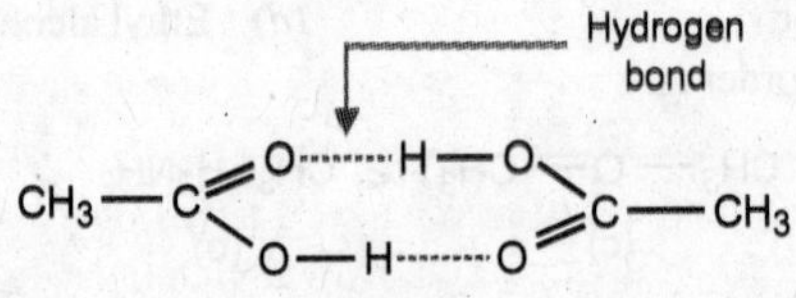

Acetic acid

9. Which will have the higher boiling point, ethyl alcohol (MW 46) or ethanethiol (MW 62) ? Explain your answer. (*Sambalpur BSc, 2014*)

Answer. Ethyl alcohol has higher boiling point than ethanethiol.

$$\underset{\text{bp } 79°C}{CH_3CH_2OH} > \underset{\text{bp } 36°C}{CH_3CH_2SH}$$

Boiling point normally increases with increased molecular weight. This is because hydrogen bonding is possible in ethyl alcohol. No hydrogen bonding is possible in ethanethiol.

10. Explain the fact that carbonyl compounds boil at higher temperature than hydrocarbons but at lower temperatures than alcohols of comparable molecular weight.

Answer. Being polar, carbonyl compounds, tend to associate : the positive part of the molecule is attracted to the negative part of another molecule. This attractive force, not significant in hydrocarbons, requires heat energy to overcome when the substance is converted from liquid to vapor. Thus carbonyl compounds have higher boiling points than hydrocarbons.

Carbonyl compounds have no O–H bonds and cannot form hydrogen bonds with one another, as alcohols can. Thus carbonyl compounds have lower boiling point than alcohols.

11. **Which will have the higher boiling point, pentanoic acid or 2,2-dimethylpropanoic acid ? Explain**

Answer. Pentanoic acid has higher boiling point than 2,2-dimethylpropanoic acid.

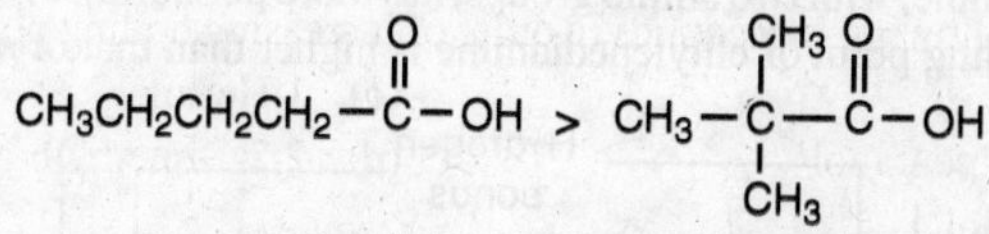

bp 187°C   bp 164°C

The molecular weight of these compounds are identical. But 2,2-dimethylpropanoic acid is a branched-chain carboxylic acid. Remember chain-branching lowers the boiling point.

12. **Which will have the higher boiling point, propanoic acid or 1-butanol ? Explain.**

(Patna BSc, 2013)

Answer. Propanoic acid has higher boiling point than 1-butanol. The molecular weights of these compounds are identical (74). But carboxylic acids hydrogen-bond more effectively than alcohols.

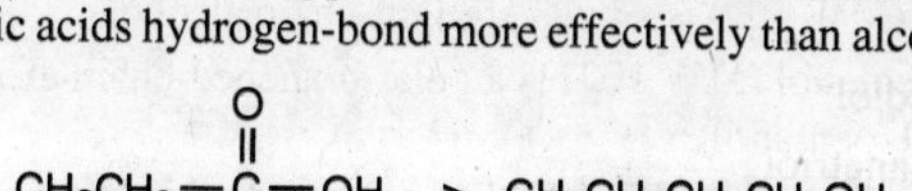

bp 141°C   bp 118°C

13. **Which has the higher boiling point, butanoic acid or ethyl acetate ? Explain**

(Kalyani BSc, 2014)

Answer. The two compounds are isomers and have the same molecular weight. Butanoic acid has an OH group and is thus capable of hydrogen bonding. Ethyl acetate can not hydrogen bond. Butanoic acid is higher boiling.

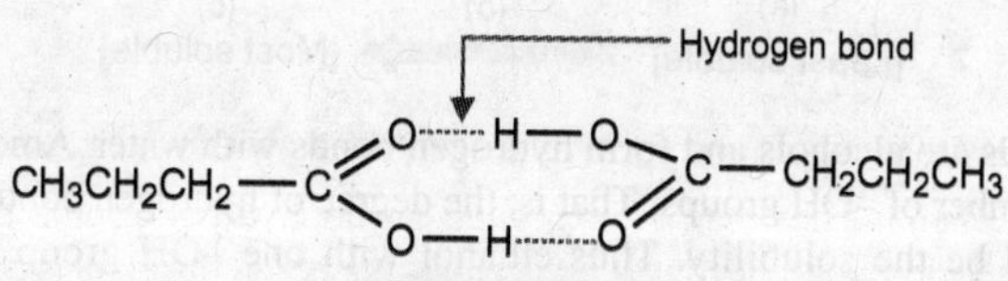

Butanoic acid

14. **Explain why trimethylamine, $(CH_3)_3N$, has considerably lower boiling point (3°C) than *n*-propylamine, $CH_3CH_2CH_2NH_2$ (49°C) even though these two compounds have the same molecular weight.**

(Dibrugarh BSc, 2013)

Answer. Molecules of *n*-propylamine can form hydrogen bonds to each other.

The molecules of trimethylamine, because they have no hydrogens attached to nitrogen, cannot form hydrogen bonds to each other. Thus the boiling point of trimethylamine is lower than that of *n*-propylamine.

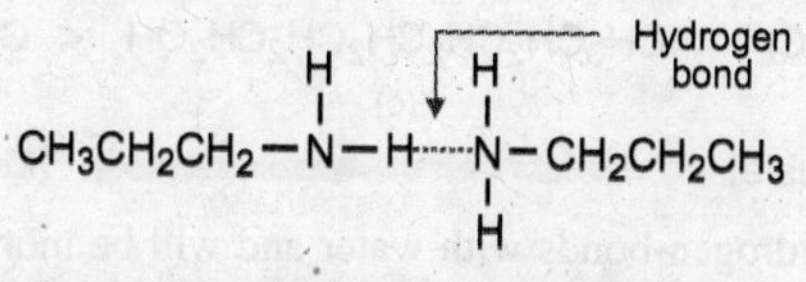

15. **Explain why dimethylamine has a higher boiling point than trimethylamine, even though the latter has higher molecular weight.**

Answer. Hydrogen bonding is possible in dimethylamine, but not in trimethylamine:

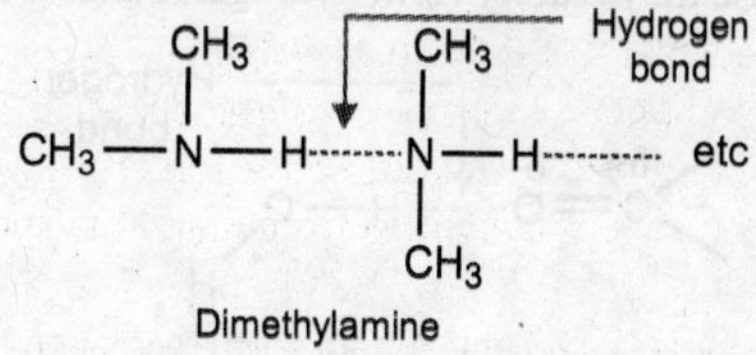

Dimethylamine

Energy is required to break these bonds to vaporize dimethylamine, thus raising its boiling point.

16. Explain why ethylenediamine, $H_2NCH_2CH_2NH_2$, has considerably higher boiling point than *n*-propylamine, $CH_3CH_2CH_2NH_2$. *(Kerala BSc, 2014)*

Answer. Ethylenediamine, with two amino groups, has more possibilities for hydrogen bonding than does *n*-propylamine. Thus the boiling point of ethylenediamine is higher than that of *n*-propylamine.

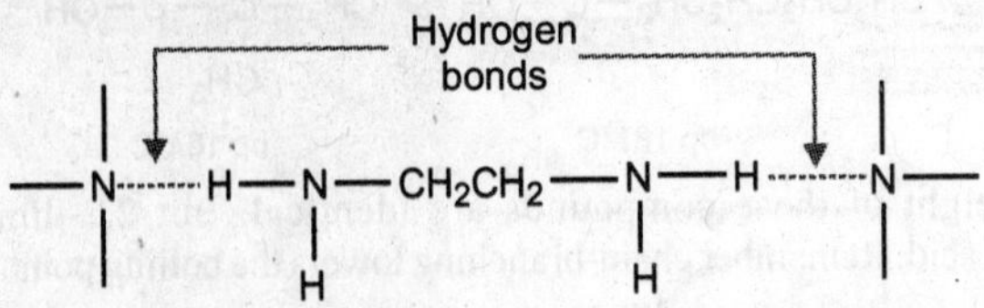

17. Arrange the following compounds in order of increasing solubility in water, and explain your answer.

(*a*) Ethanol

(*b*) 1,2-Ethanediol

(*c*) 1,2,3-Propanetriol

Answer. The solubility order is:

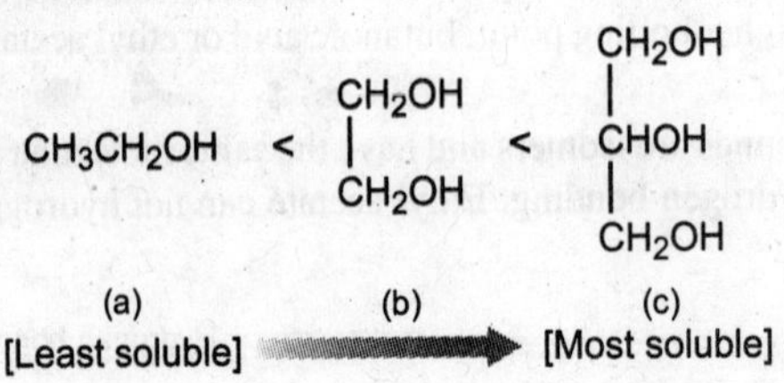

All three compounds are alcohols and form hydrogen bonds with water. Among alcohols, the solubility depends mainly on the number of –OH groups. That is, the degree of hydrogen bonding. Greater the number of –OH groups, greater will be the solubility. Thus ethanol with one –OH group will be least soluble, and 1,2,3-propanetriol with three –OH groups will be most soluble.

18. Arrange the following compounds in order of increasing solubility in water, and explain your answer.

(*a*) Ethanol

(*b*) Ethyl chloride

(*c*) I-Hexanol

Answer. The solubility order is:

$$\underset{(b)}{CH_3CH_2Cl} < \underset{(c)}{CH_3CH_2CH_2CH_2CH_2CH_2OH} < \underset{(a)}{CH_3CH_2OH}$$

[Least soluble] → [Most soluble]

Both alcohols can form hydrogen-bonds with water and will be more soluble than ethyl chloride. The lower molecular weight alcohol (ethanol) will be more soluble (it has a shorter hydrophobic carbon chain).

19. Explain why carbonyl compounds of lower molecular weight are water-soluble.

Answer. Carbonyl compounds cannot form hydrogen bonds with themselves. However, carbonyl compounds are water-soluble because they readily form hydrogen bonds with water molecules.

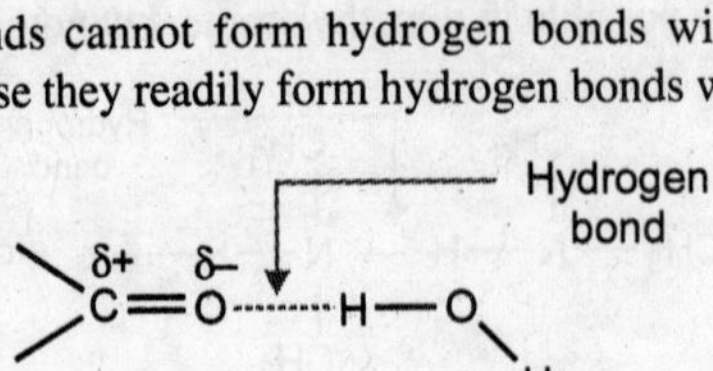

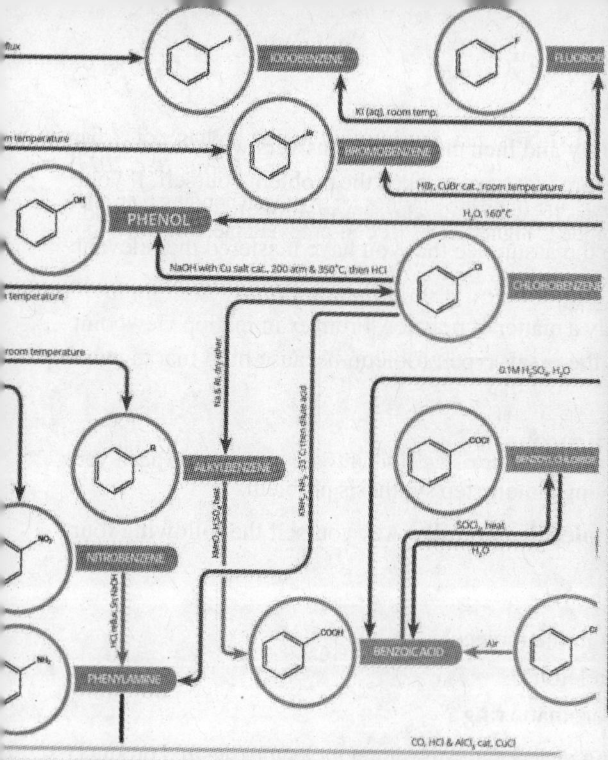

52

Multistep Synthesis

Many students complain that there are not enough 'Multistep Synthesis' problems in their books to prepare them for Entrance or University Examinations. The aim of this chapter is to add to the student's collective knowledge of organic chemistry and his self-confidence by providing exam-type problems.

Most of the *Multistep Synthesis* problems given in this chapter have been carefully selected from recent examinations. They cover a wide range of reactions included in this textbook. It is hoped that these additional problems will provide plenty of practice and improve your problem solving skills.

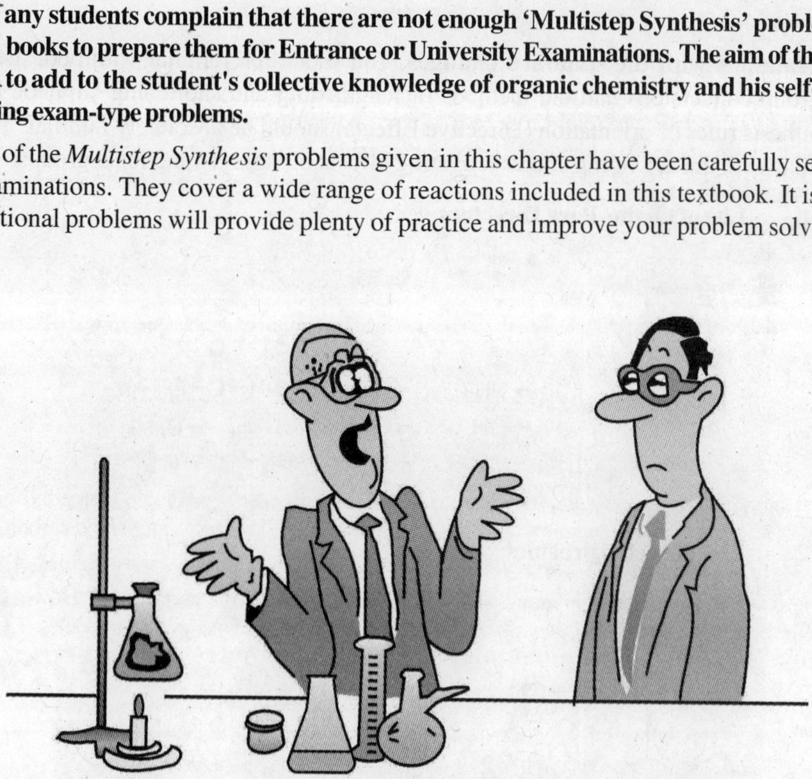

It was a brilliant experiment. Unfortunately, I forgot to note down what I did.

You should work these problems independently and then check your answers with those given here. Resist the temptation to check the answer before you have worked the problem yourself. If your answer is not correct, you know what portions of the book you need to study more thoroughly. And, if you have solved a problem correctly, you have the assurance that you have mastered the relevant portions of the book.

Solving multistep synthesis problems is simply a matter of practice. From examination viewpoint the synthesis problem is a far better choice than the essay-type problem, because high marks in an essay-type problem are hard to obtain.

MULTISTEP SYNTHESIS : TIPS AND TRICKS

Keep the following points in mind while solving a multistep synthesis problem :

(1) **Examine the structure of the target molecule carefully.** Ask yourself the following four questions :

(a) What is the type of the compound ?

(b) What functional groups are present in the molecule ?

(c) What is the nature of the carbon skeleton ?

(d) If the target molecule contains any aromatic ring ?

(2) **Work backwards.** When both the starting material and the target molecule (desired product) are given, the temptation is to begin automatically with the starting material. However, it is better to work backwards from the target molecule. This procedure only limits the possible answers, and not the general solution.

(3) **Be familiar with the standard changes.** You should be familiar with both the standard functional group conversions and the methods of lengthening and shortening aliphatic chains. In aromatic synthesis rules of orientation (Directive Effects) should be absolutely familiar. You should also be familiar with the activating and deactivating effects of various functional groups.

Remember 1. **List of Ortho-Para Directors**

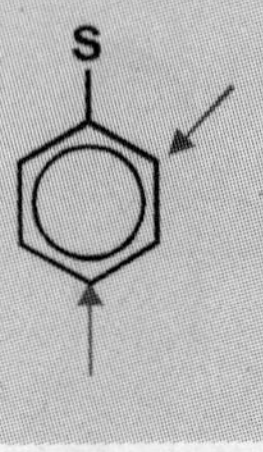

S			
$-NH_2$,	$-NR_2$,	$-NH-\overset{O}{\overset{\|\|}{C}}-CH_3$	
$-OH$,	$-OCH_3$,	$-CH_3$,	$-C_6H_5$
$-Cl$,	$-Br$,	$-I$,	$-F$

Remember 2. **List of Meta Directors**

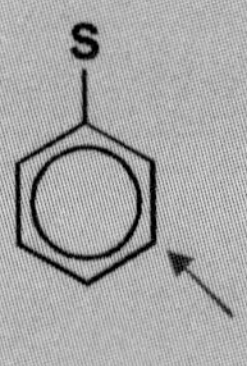

S			
$-CN$,	$-SO_3H$,	$-NO_2$,	$-\overset{+}{N}H_3$
$-\overset{O}{\overset{\|\|}{C}}-OH$,	$-\overset{O}{\overset{\|\|}{C}}-CH_3$,	$-\overset{O}{\overset{\|\|}{C}}-H$,	$-\overset{O}{\overset{\|\|}{C}}-OCH_3$

Remember 3. **List of compounds which do not give Friedel-Crafts reactions**

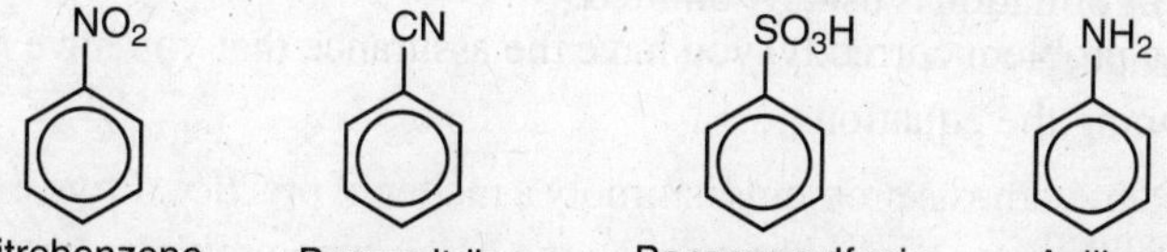

Nitrobenzene Benzonitrile Benzenesulfonic acid Aniline

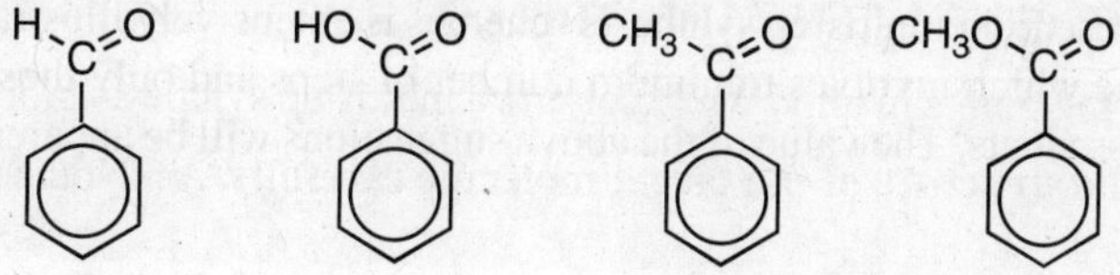

Benzaldehyde Benzoic acid Acetophenone Methyl benzoate

(4) **Never make up a reaction** that seems to bring together the right structural parts. The danger here is that the chemistry will be incorrect because of the energy changes or the reaction violates a principle of mechanism. To illustrate, consider the following reaction, which seems to bring the right parts together.

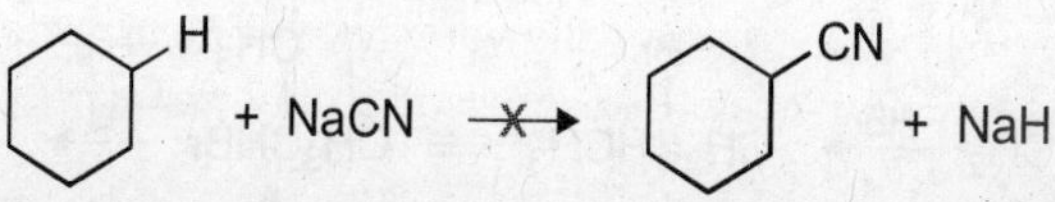

The only merit this *proposed synthesis* has is that the reagents are familiar and available. It has no chance of working in the laboratory. In this case, the reaction is both energetically unfavorable and mechanistic nonsense. Cyanide ion, an excellent nucleophile, can not attack cyclohexane because no qualified leaving group is present. The hydride ion, $H:^-$, is not a leaving group.

Sometimes the simple error of proposing a made-up reaction is compounded by adding exotic conditions in an attempt to make the process work. This is usually hopeless. The non-reaction above cannot work by adding an acid, base, sunlight or high temperature. You need to learn and apply the standard types of synthesis reactions given in this textbook.

(5) **Other common errors.** Once a final solution has been obtained, the steps must be checked for pitfalls. Common errors are:

(a) Failure to use Markovnikov rule correctly.

(b) Failure to use Saytzeff rule correctly.

(c) Failure to show reagents and reaction conditions.

(d) Failure to use directive effects correctly.

(e) Use of NH_3 to convert an alkyl halide into an amine. This is a poor reaction. It gives a mixture of products. Use an indirect method.

(f) Use of Wurtz reaction to prepare odd-numbered carbon-chains. Use Corey-House synthesis instead.

(6) **Representing multistep synthesis.** The language of organic chemistry includes structures, mechanisms, and reactions communicated by visual symbols. Multistep synthesis problems, which have a different emphasis, have a style of their own. The objective is to bring the sequence of functional group conversions and skeletal changes into clear focus. The three essentials of this style are:

(a) Structural formula of the starting material.

(b) Reagents and reaction conditions on reaction arrows to represent the reaction type in the most concise form possible.

(c) Structural formula of the product (Target Molecule).

The following information is usually omitted:

(a) By-products
(b) Balancing the equations
(c) Yields
(d) Mechanism

Try to avoid the temptation to show a mechanism just because you know it. In general, in a multistep synthesis problem, it will earn you no credit, and you may lose if you get it wrong. The correct style for representing multistep synthesis schemes is extensively illustrated in this chapter. A good synthesis is one which involves minimum number of steps and only those reactions which do not give mixture of products. The value of the above suggestions will be apparent from the fõllowing examples:

Examination Questions

(1) 2-Methylhexane from propene *(Anna BSc, 2014)*

Key step is Corey-House synthesis.

$$\underset{\text{Propene}}{CH_3CH{=}CH_2} \xrightarrow{HBr} CH_3\overset{\overset{Br}{|}}{C}HCH_3 \equiv CH_3\overset{\overset{CH_3}{|}}{C}HBr \xrightarrow{Li} \underset{\text{(2 molecules)}}{CH_3\overset{\overset{CH_3}{|}}{C}HLi}$$

$$\xrightarrow{CuBr} (CH_3\overset{\overset{CH_3}{|}}{C}H)_2CuLi \xrightarrow{CH_3CH_2CH_2CH_2Br} \underset{\text{2-Methylhexane}}{CH_3\overset{\overset{CH_3}{|}}{C}HCH_2CH_2CH_2CH_3}$$

(2) Pentane from propene *(Utkal BSc, 2015)*

Key step is Corey-House synthesis.

$$\underset{\text{Propene}}{CH_3CH{=}CH_2} \xrightarrow[R_2O_2]{HBr} CH_3CH_2CH_2Br \xrightarrow{2Li} CH_3CH_2CH_2Li$$

$$\xrightarrow{CuBr} (CH_3CH_2CH_2)_2CuLi \xrightarrow{CH_3CH_2Br} \underset{\text{Pentane}}{CH_3CH_2CH_2CH_2CH_3}$$

(3) 1,2-Dichloroethane from ethyl alcohol

Key step is reaction of ethyl alcohol with concentrated H_2SO_4 at 170°C to form ethylene (Dehydration of alcohols).

$$\underset{\text{Ethyl alcohol}}{CH_3CH_2OH} \xrightarrow[\Delta]{H_2SO_4} CH_2{=}CH_2 \xrightarrow{Cl_2} \underset{\text{1,2-Dichloroethane}}{ClCH_2CH_2Cl}$$

(4) 1-Bromopropane from 2-bromopropane

Key step is reaction of propene with HBr in the presence of peroxide to form 1-bromopropane (Anti-Markovnikov addition).

$$\underset{\text{2-Bromopropane}}{CH_3{-}\overset{\overset{Br}{|}}{C}H{-}CH_3} \xrightarrow[\text{alcohol}/\Delta]{KOH} CH_3{-}CH{=}CH_2 \xrightarrow[\text{peroxide}]{HBr} \underset{\text{1-Bromopropane}}{CH_3CH_2CH_2Br}$$

(5) 1-Propanol from propane *(Madurai BSc, 2014)*

Key step is the reaction of a mixture of 1-bromopropane and 2-bromopropane with alcoholic KOH to form propene.

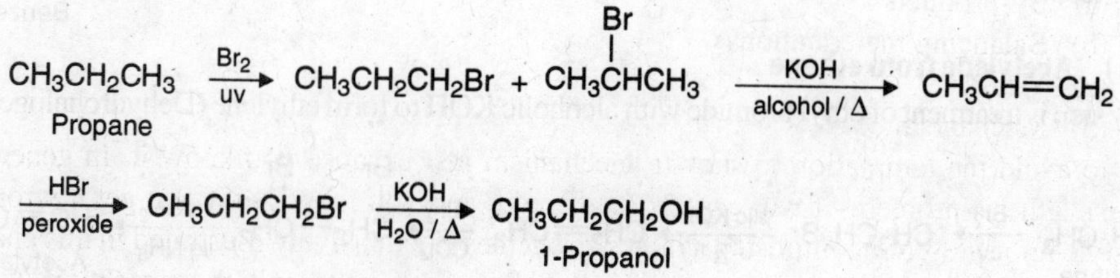

(6) 2-Methyl-1-propanol from 2-methyl-2-propanol

Key step is hydroboration-oxidation reaction. Net result is the anti-Markovnikov addition of elements of water.

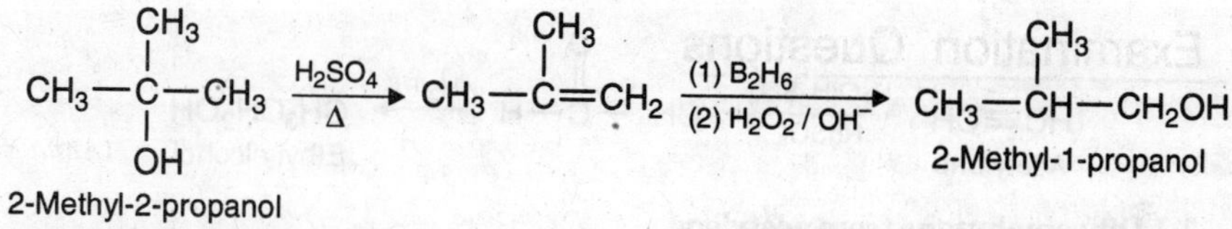

(7) Allyl chloride from propane *(Vishwa Bharati BSc, 2015)*

Key step is the treatment of propane with Cl_2 at 500°C (Allylic substitution).

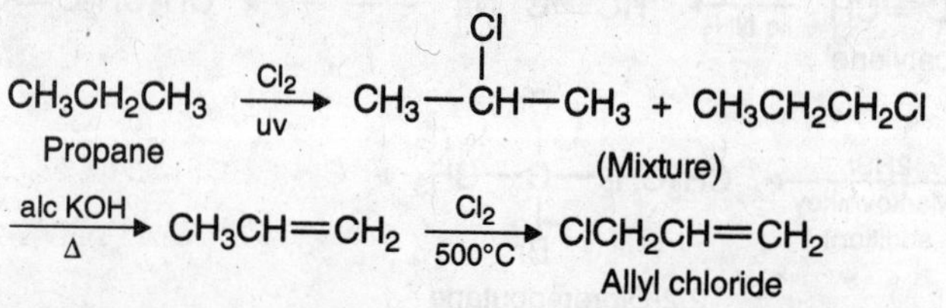

(8) Allyl iodide from isopropyl chloride *(Mysore BSc, 2014)*

Key reaction is treatment of allyl chloride with NaI/acetone (Halogen exchange) to form allyl iodide.

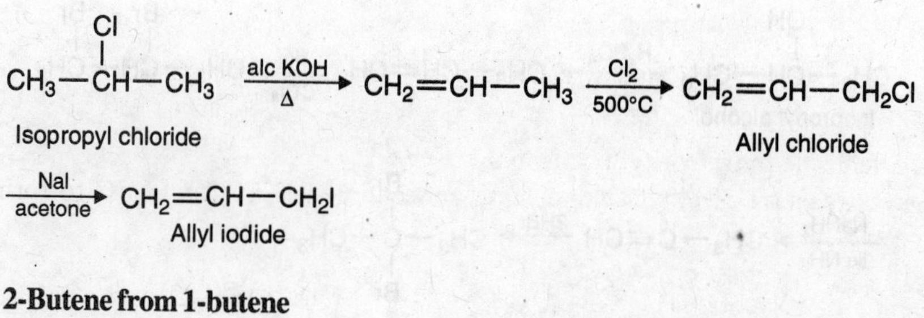

(9) 2-Butene from 1-butene

Key step is addition of HBr to 1-butene to form 2-bromobutane (Markovnikov addition).

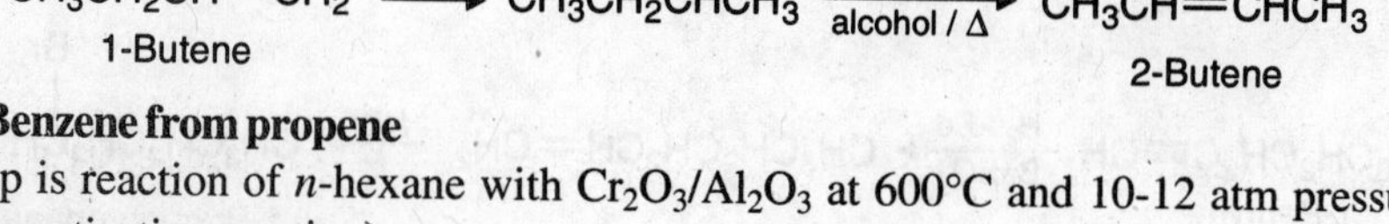

(10) Benzene from propene

Key step is reaction of *n*-hexane with Cr_2O_3/Al_2O_3 at 600°C and 10-12 atm pressure to form benzene (Aromatization reaction).

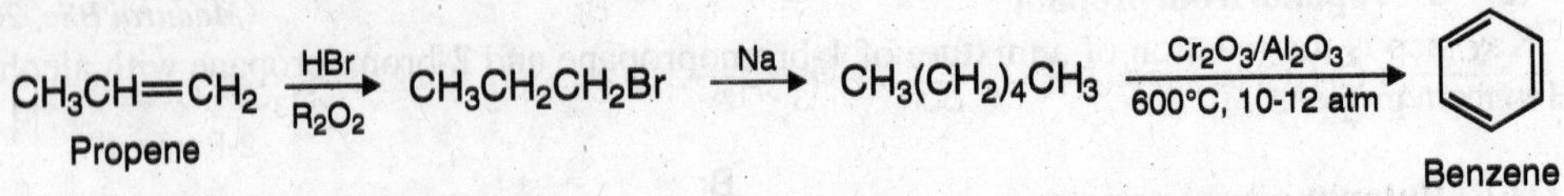

(11) Acetylene from ethane

Key step is treatment of ethyl bromide with alcoholic KOH to form ethylene (Dehydrohalogenation).

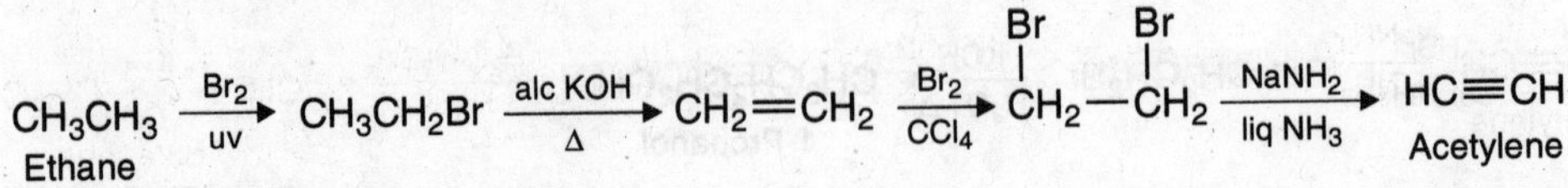

(12) Ethyl alcohol from acetylene

Key step is treatment of acetylene with dilute H_2SO_4 in the presence of mercuric sulfate to give acetaldehyde.

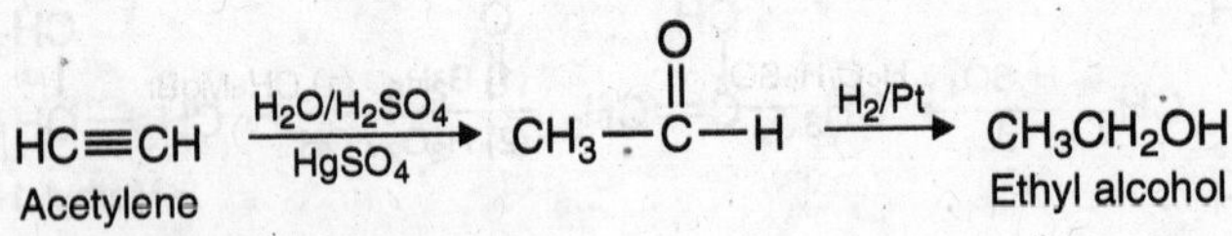

(13) 2,2-Dibromobutane from acetylene

Key step is the reaction of sodium acetylide with ethyl bromide to form 1-butyne.

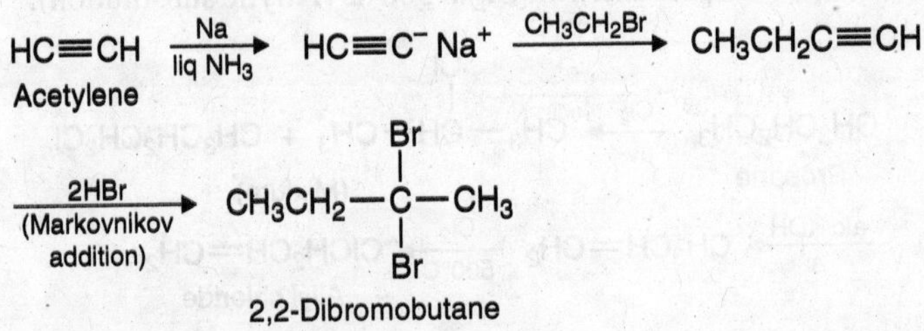

(14) 2,2-Dibromopropane from isopropyl alcohol

Key step is the treatment of 1,2-dibromopropane with $NaNH_2$ to give propyne.

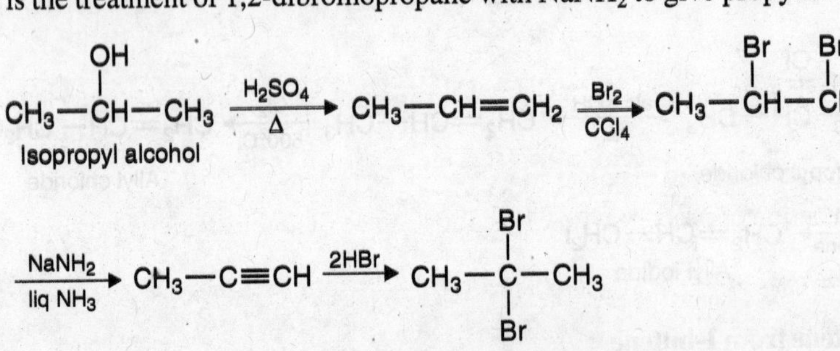

(15) 2-Pentyne from 1-pentyne

Key step is treatment of 1-pentyne with Lindlar's catalyst (H_2/Pd-$BaSO_4$) to form 1-pentene.

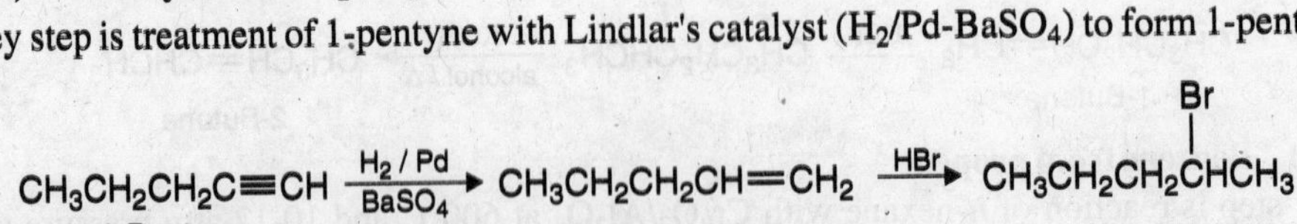

$$\xrightarrow[\text{alcohol}/\Delta]{\text{KOH}} CH_3CH_2CH{=}CHCH_3 \xrightarrow[CCl_4]{Br_2} CH_3CH_2\underset{}{CH(Br)}{-}CH(Br)CH_3 \xrightarrow{NaNH_2} CH_3CH_2C{\equiv}CCH_3$$

2-Pentyne

(16) Butanone from acetylene *(Madras BSc, 2014)*

Key step is hydration of 1-butyne with H_2O in the presence of H_2SO_4 and $HgSO_4$.

$$\underset{\text{Acetylene}}{HC{\equiv}CH} \xrightarrow[\text{liq } NH_3]{NaNH_2} HC{\equiv}C\bar{:}\overset{+}{Na} \xrightarrow{CH_3CH_2Br} HC{\equiv}CCH_2CH_3 \xrightarrow[HgSO_4]{H_2O/H_2SO_4} CH_3{-}\overset{O}{\overset{\|}{C}}{-}CH_2CH_3$$

Butanone (Ethyl methyl ketone)

(17) Acetone from acetylene *(Lucknow BSc, 2015)*

Method 1. Key step is Grignard synthesis.

$$\underset{\text{Acetylene}}{HC{\equiv}CH} \xrightarrow[HgSO_4]{H_2O/H_2SO_4} CH_3{-}\overset{O}{\overset{\|}{C}}{-}H \xrightarrow[(2)\ H_2O/H^+]{(1)\ CH_3MgBr}$$

$$CH_3{-}\overset{OH}{\overset{|}{C}H}{-}CH_3 \xrightarrow[{[O]}]{Na_2Cr_2O_7/H^+} \underset{\text{Acetone}}{CH_3{-}\overset{O}{\overset{\|}{C}}{-}CH_3}$$

Method 2. Key step is hydration of propyne.

$$\underset{\text{Acetylene}}{HC{\equiv}CH} \xrightarrow[\text{or Na/liq } NH_3]{NaNH_2} HC{\equiv}C:^{-}\overset{+}{Na} \xrightarrow{CH_3I}$$

$$CH_3C{\equiv}CH \xrightarrow[HgSO_4]{H_2O/H_2SO_4} \underset{\text{Acetone}}{CH_3{-}\overset{O}{\overset{\|}{C}}{-}CH_3}$$

(18) 1,3-Butadiene from 1-butene *(Jiwaji BSc, 2014)*

Key step is the reaction of 1-butene with Cl_2 at 600°C (Allylic substitution) to give 3-chloro-1-butene.

$$\underset{\text{1-Butene}}{CH_3{-}CH_2{-}CH{=}CH_2} \xrightarrow[600°C]{Cl_2} CH_3{-}\overset{Cl}{\overset{|}{C}H}{-}CH{=}CH_2 \xrightarrow[\Delta]{\text{alc KOH}} \underset{\text{1,3-Butadiene}}{CH_2{=}CH{-}CH{=}CH_2}$$

(19) 1,5-Hexadiene from propene

Key step is reaction of propene with Cl_2 at 500°C to form 3-chloropropene (Allylic substitution).

$$\underset{\text{Propene}}{CH_3CH{=}CH_2} \xrightarrow[500°C]{Cl_2} \underset{\text{(2-molecules)}}{ClCH_2CH{=}CH_2} \xrightarrow[\text{(Wurtz)}]{Na} \underset{\text{1,5-Hexadiene}}{CH_2{=}CHCH_2CH_2CH{=}CH_2}$$

(20) 1,3-Cyclohexadiene from cyclohexane

Key step is treatment of cyclohexene with Cl_2 at 600°C to give 3-chlorocyclohexene.

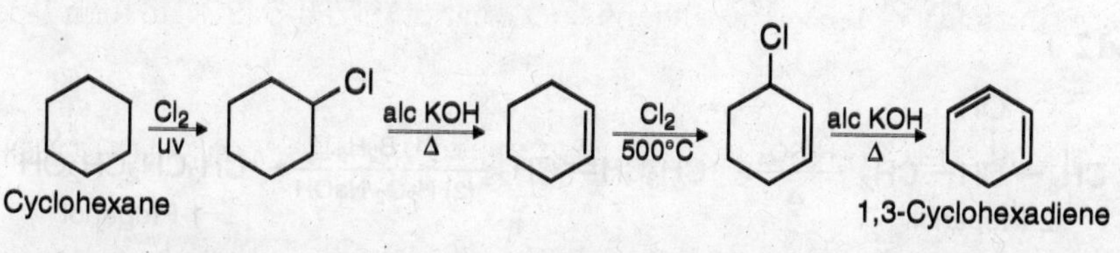

Cyclohexane 1,3-Cyclohexadiene

(21) 1,2-Dibromopropane from 1-bromopropane

Key step is the reaction of 1-bromopropane with alcoholic KOH to form propene.

$$\underset{\text{1-Bromopropane}}{CH_3CH_2CH_2Br} \xrightarrow[\text{alcohol / }\Delta]{\text{KOH}} CH_3CH{=}CH_2 \xrightarrow{Br_2} \underset{\text{1,2-Dibromopropane}}{CH_3{-}\overset{\overset{\displaystyle Br}{|}}{CH}{-}\overset{\overset{\displaystyle Br}{|}}{CH_2}}$$

(22) Allyl iodide from isopropyl chloride *(Cochin BSc, 2015)*

Key step is the reaction of allyl chloride with NaI/acetone to form allyl iodide (Halogen Exchange reaction).

$$\underset{\text{Isopropyl chloride}}{CH_3{-}\overset{\overset{\displaystyle Cl}{|}}{CH}{-}CH_3} \xrightarrow[\text{alcohol / }\Delta]{\text{KOH}} CH_2{=}CH{-}CH_3 \xrightarrow[500°C]{Cl_2} CH_2{=}CH{-}CH_2Cl$$

$$\xrightarrow[\text{acetone / }\Delta]{\text{NaI}} \underset{\text{Allyl iodide}}{CH_2{=}CH{-}CH_2I}$$

(23) 2-Propanol from 1-propanol *(Calicut BSc, 2014)*

Method 1

$$\underset{\text{1-Propanol}}{CH_3CH_2CH_2OH} \xrightarrow[\Delta]{H_2SO_4} CH_3{-}CH{=}CH_2 \xrightarrow{\text{HBr}} CH_3{-}\overset{\overset{\displaystyle Br}{|}}{CH}{-}CH_3$$

$$\xrightarrow{\text{aq KOH}} \underset{\text{2-Propanol}}{CH_3{-}\overset{\overset{\displaystyle OH}{|}}{CH}{-}CH_3}$$

Method 2

$$\underset{\text{1-Propanol}}{CH_3CH_2CH_2OH} \xrightarrow[\Delta]{H_2SO_4} CH_3CH{=}CH_2 \xrightarrow[\Delta]{H_2O/H^+} \underset{\text{2-Propanol}}{CH_3{-}\overset{\overset{\displaystyle OH}{|}}{CH}{-}CH_3}$$

(24) 1-Propanol from 2-propanol *(Calicut BSc, 2014)*

Method 1

$$\underset{\text{2-Propanol}}{CH_3{-}\overset{\overset{\displaystyle OH}{|}}{CH}{-}CH_3} \xrightarrow[\Delta]{H_2SO_4} CH_3CH{=}CH_2 \xrightarrow[\text{peroxide}]{\text{HBr}} CH_3CH_2CH_2Br$$

$$\xrightarrow{\text{aq KOH}} \underset{\text{1-Propanol}}{CH_3CH_2CH_2OH}$$

Method 2

$$\underset{\text{2-Propanol}}{CH_3{-}\overset{\overset{\displaystyle OH}{|}}{CH}{-}CH_3} \xrightarrow[\Delta]{H_2SO_4} CH_3CH{=}CH_2 \xrightarrow[\text{(2) }H_2O_2/NaOH]{\text{(1) }B_2H_6} \underset{\text{1-Propanol}}{CH_3CH_2CH_2OH}$$

(25) 1-Butanol from methyl alcohol

Key step is the reaction of ethylmagnesium bromide with ethylene oxide followed by acid-hydrolysis.

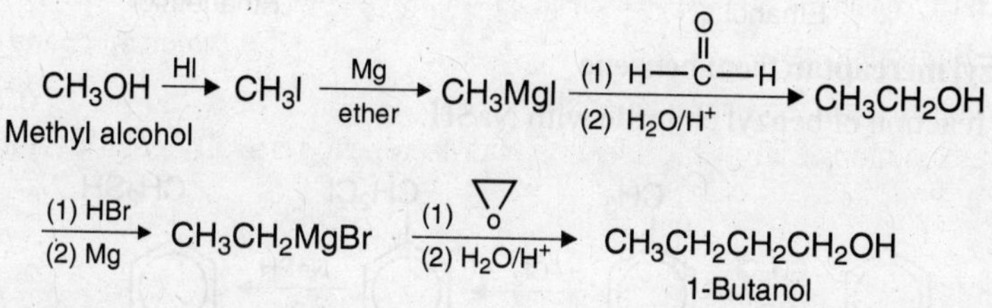

(26) 2-Butanol from acetylene *(Vikram BSc, 2014)*

Key step is Grignard synthesis.

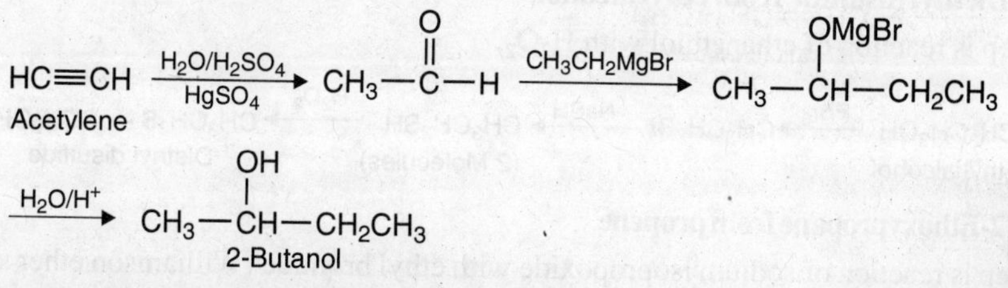

(27) *tert*-Butyl alcohol from isobutyl alcohol

Key step is acid-catalyzed hydration of 2-methylpropene.

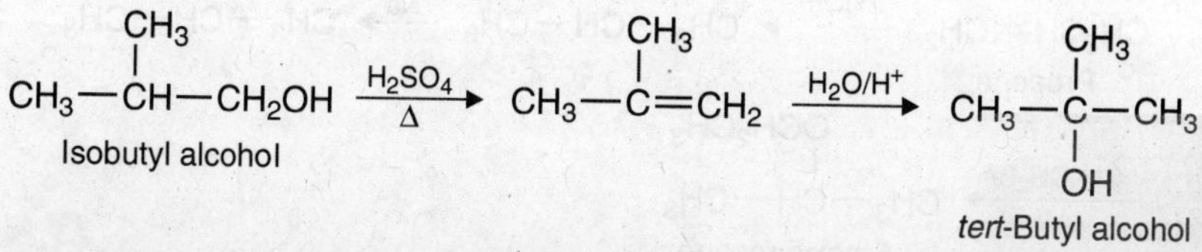

(28) Isobutyl alcohol from tert-butyl alcohol

Key reaction is hydroboration-oxidation of 2-methylpropene.

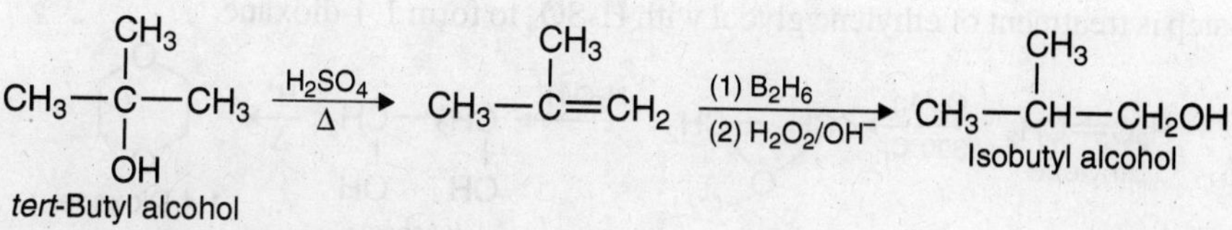

(29) Isobutyl alcohol from n-propyl alcohol *(Gulbarga BSc, 2015)*

Key step is Grignard synthesis.

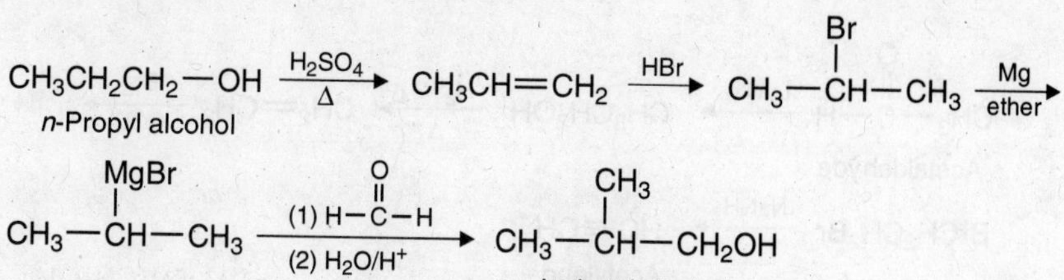

(30) Ethanethiol from ethanol

Key step is reaction of ethyl bromide with NaSH.

$$\underset{\text{Ethanol}}{CH_3CH_2OH} \xrightarrow{PBr_3} CH_3CH_2Br \xrightarrow{NaSH} \underset{\text{Ethanethiol}}{CH_3CH_2SH}$$

(31) Benzyl mercaptan from benzene

Key step is reaction of benzyl chloride with NaSH.

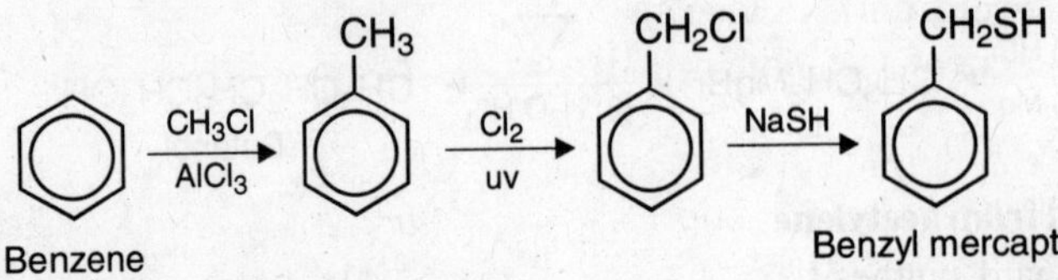

(32) Diethyl disulfide from ethyl alcohol

Key step is reaction of ethanethiol with H_2O_2.

$$\underset{\text{Ethyl alcohol}}{CH_3CH_2OH} \xrightarrow{PBr_3} CH_3CH_2Br \xrightarrow{NaSH} \underset{\text{(2 Molecules)}}{CH_3CH_2SH} \xrightarrow{H_2O_2} \underset{\text{Diethyl disulfide}}{CH_3CH_2S-SCH_2CH_3}$$

(33) 2-Ethoxypropane from propene

Key step is reaction of sodium isopropoxide with ethyl bromide (Williamson ether synthesis).

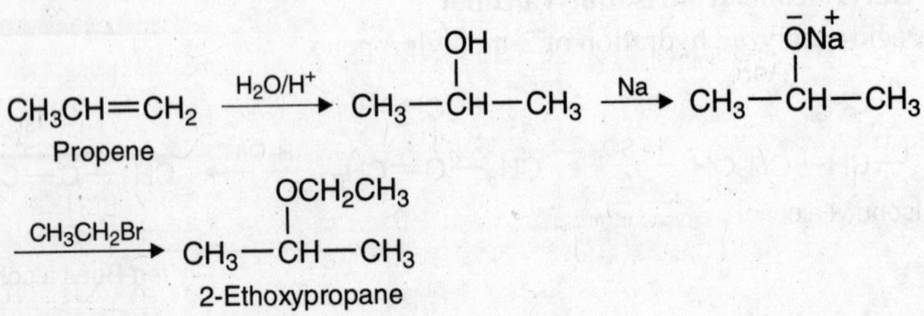

(34) 1,4-Dioxane from ethylene

Key step is treatment of ethylene glycol with H_2SO_4 to form 1,4-dioxane.

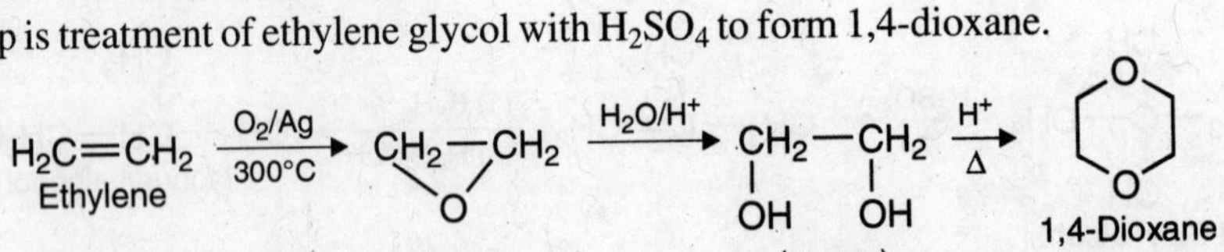

(35) Acetylene from acetaldehyde

Key step is reduction of acetaldehyde with $LiAlH_4$ to form ethyl alcohol.

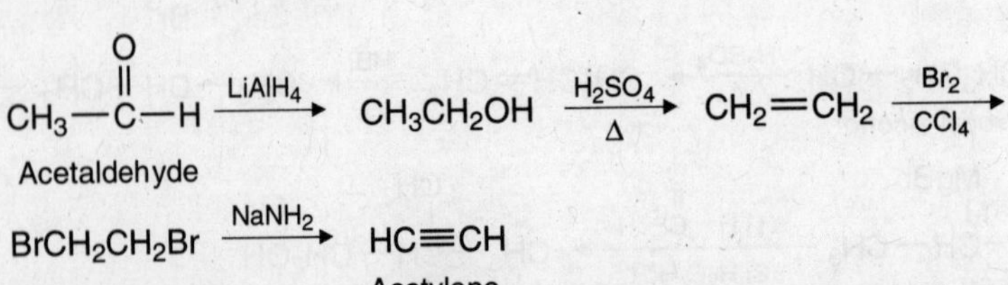

(36) Acetaldehyde from formaldehyde

Key reaction is oxidation of ethyl alcohol to acetaldehyde. Remember : Primary alcohols can be oxidized to aldehydes by heating with copper or by treatment with pyridinium chlorochromate ($C_5H_5NH^+CrO_3Cl^-$ in CH_2Cl_2). Pyridinium chlorochromate is abbreviated as PCC.

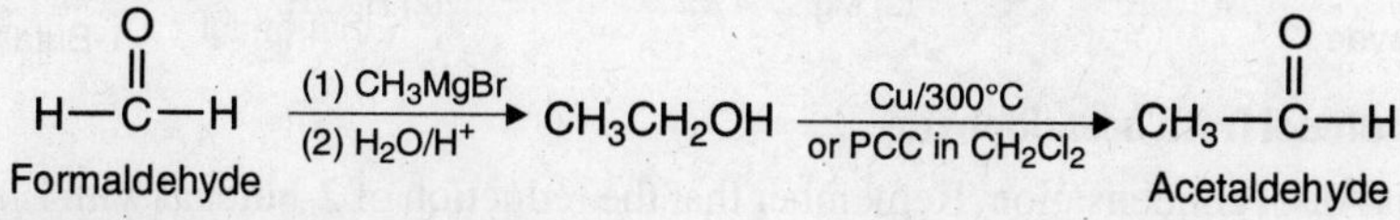

(37) Formaldehyde from acetaldehyde

Key step is treatment of acetamide with Br_2/KOH to form methylamine (Hofmann degradation reaction).

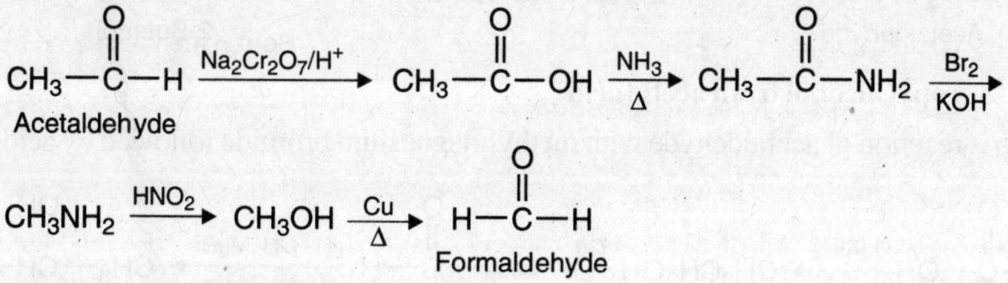

(38) Acetone from acetaldehyde

Key step is oxidation of acetaldehyde with $KMnO_4$ to give acetic acid.

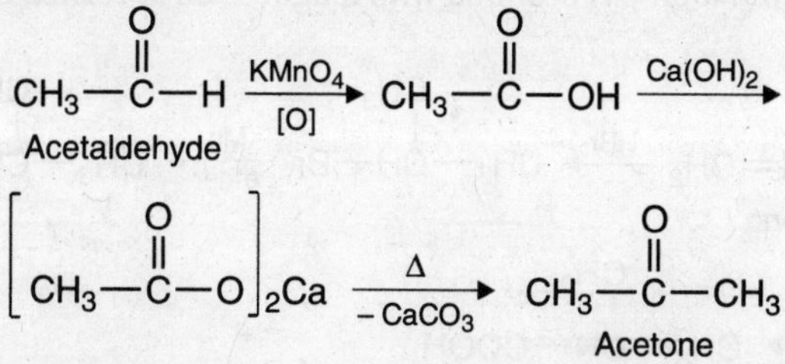

(39) Isohexane from acetone

Key step is Aldol condensation of acetone.

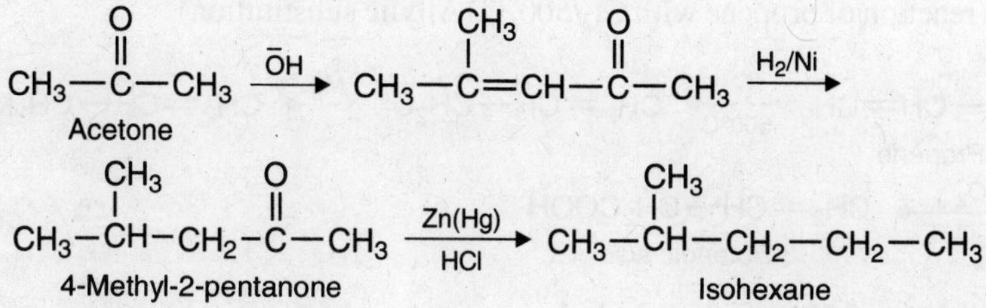

(40) 1-Butanol from acetaldehyde *(North Bengal BSc, 2015)*

Method 1. Key step is Aldol condensation. Remember that the reduction of 2-butenal with H_2 in the presence of nickel catalyst reduces both the carbon-carbon double bond and the aldehyde group.

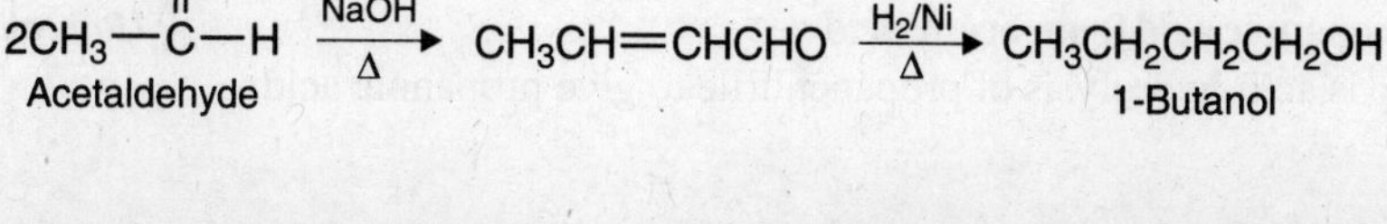

Method 2. Key step is Grignard reaction.

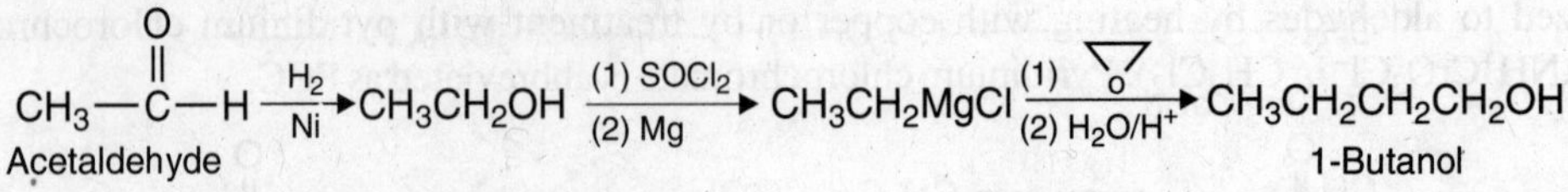

(41) 2-Butenol from acetaldehyde

Key step is Aldol condensation. Remember that the reduction of 2-butenal with $LiAlH_4$ does not effect the carbon-carbon double bond.

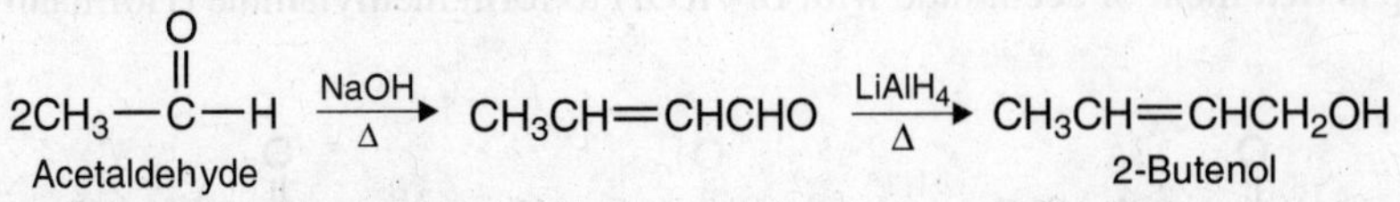

(42) Isopropyl alcohol from acetic acid

Key step is reaction of acetaldehyde with methylmagnesium bromide followed by acid-hydrolysis.

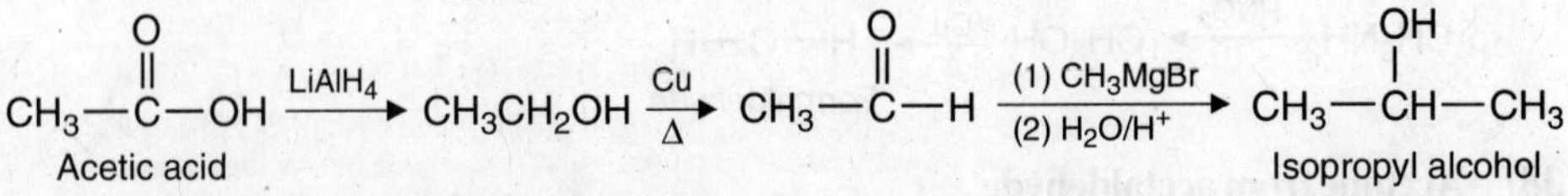

(43) 2-Methylpropanoic acid from propene

Key step is treatment of isopropyl bromide with magnesium to form a Grignard reagent.

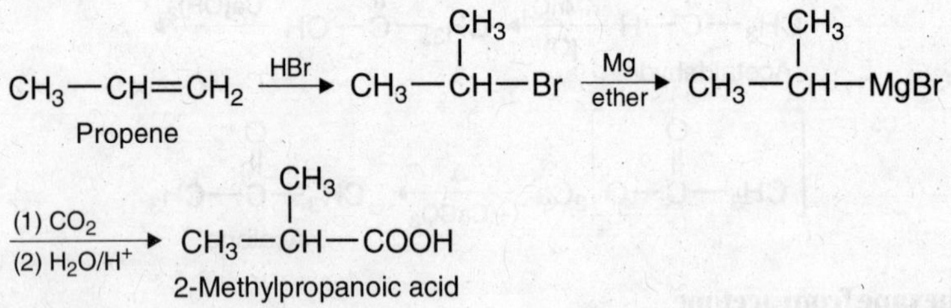

(44) 3-Butenoic acid from propene *(West Bengal BSc, 2015)*

Key step is reaction of propene with Cl_2/500°C (Allylic substitution).

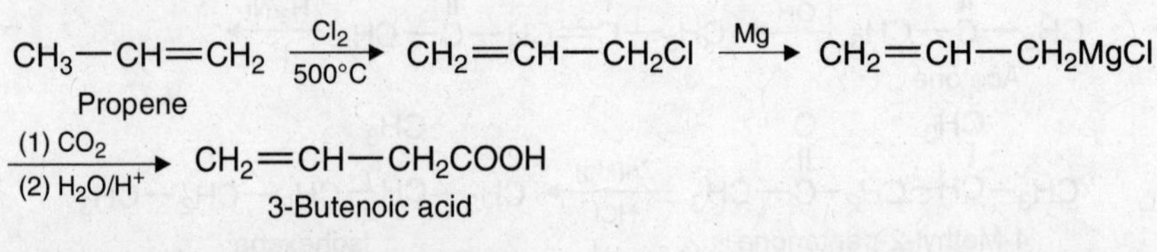

(45) Propanoic acid from ethyl alcohol

Key step is carbonation of ethylmagnesium bromide followed by acid-hydrolysis.

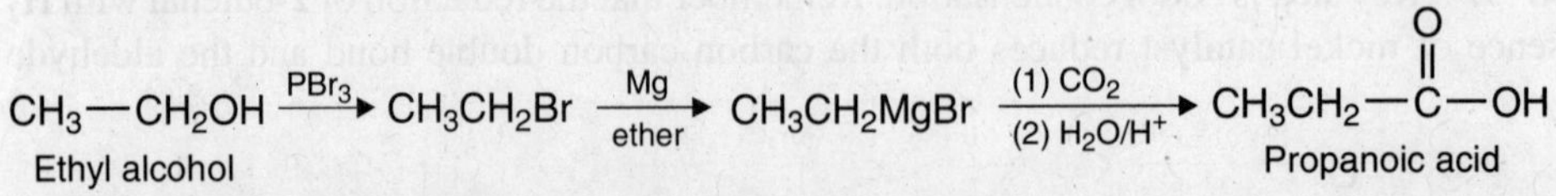

(46) Propanoic acid from acetic acid *(Rajasthan BSc, 2014)*

Key step is acid-hydrolysis of propanenitrile to give propanoic acid.

$$CH_3-\overset{\overset{O}{\|}}{C}-OH \xrightarrow{LiAlH_4} CH_3CH_2OH \xrightarrow{SOCl_2} CH_3CH_2Cl \xrightarrow{NaCN}$$

Acetic acid

$$CH_3CH_2CN \xrightarrow{H_2O/H^+} CH_3CH_2COOH$$

Propanoic acid

(47) Acetic acid from propanoic acid

Key step is reaction of propanamide with Br_2/NaOH to form ethylamine (Hofmann degradation reaction).

$$CH_3CH_2-\overset{\overset{O}{\|}}{C}-OH \xrightarrow[\Delta]{NH_3} CH_3CH_2-\overset{\overset{O}{\|}}{C}-NH_2 \xrightarrow[NaOH]{Br_2} CH_3CH_2NH_2$$

Propanoic acid

$$\xrightarrow{HNO_2} CH_3CH_2OH \xrightarrow{Na_2Cr_2O_7/H^+} CH_3-\overset{\overset{O}{\|}}{C}-OH$$

Acetic acid

(48) Malonic acid from acetic acid *(Jharkhand BSc, 2014)*

Key step is reaction of acetic acid with Cl_2/P to from α-chloroacetic acid (Hell-Volhard-Zelinsky reaction).

$$CH_3-\overset{\overset{O}{\|}}{C}-OH \xrightarrow[P]{Cl_2} ClCH_2-\overset{\overset{O}{\|}}{C}-OH \xrightarrow{NaCN} CNCH_2-\overset{\overset{O}{\|}}{C}-OH$$

Acetic acid

$$\xrightarrow{H_2O/H^+} HOOC-CH_2-COOH$$

Malonic acid

(49) Ethyl acetate from ethyl alcohol *(Orissa BSc, 2015)*

Method 1. Key step is treatment of acetyl chloride with sodium ethoxide.

$$CH_3CH_2OH \xrightarrow{KMnO_4/H^+} CH_3-\overset{\overset{O}{\|}}{C}-OH \xrightarrow{SOCl_2} CH_3-\overset{\overset{O}{\|}}{C}-Cl$$

Ethyl alcohol

$$\xrightarrow{CH_3CH_2O^-Na^+} CH_3-\overset{\overset{O}{\|}}{C}-OCH_2CH_3$$

Ethyl acetate

Method 2. Key step is treatment of acetic acid with ethyl alcohol in the presence of acid catalyst.

$$CH_3CH_2OH \xrightarrow{KMnO_4/H^+} CH_3-\overset{\overset{O}{\|}}{C}-OH \xrightarrow[H^+]{CH_3CH_2OH} CH_3-\overset{\overset{O}{\|}}{C}-OCH_2CH_3$$

Ethyl alcohol | Ethyl acetate

(50) Ethylamine from ethyl alcohol

Key step is treatment of acetyl chloride with ammonia followed by reduction with $LiAlH_4$.

$$\underset{\text{Ethyl alcohol}}{CH_3CH_2OH} \xrightarrow[{[O]}]{KMnO_4} CH_3-\overset{O}{\overset{\|}{C}}-OH \xrightarrow{SOCl_2} CH_3-\overset{O}{\overset{\|}{C}}-Cl \xrightarrow{NH_3}$$

$$CH_3-\overset{O}{\overset{\|}{C}}-NH_2 \xrightarrow{LiAlH_4} \underset{\text{Ethylamine}}{CH_3CH_2NH_2}$$

(51) Ethylamine from n-propylamine *(Patna BSc, 2014)*

Key step is treatment of propanamide with $Br_2/NaOH$ to form ethylamine (Hofmann degradation reaction).

$$\underset{n\text{-Propylamine}}{CH_3CH_2CH_2NH_2} \xrightarrow{HNO_2} CH_3CH_2CH_2OH \xrightarrow[{[O]}]{Na_2Cr_2O_7/H^+} CH_3CH_2-\overset{O}{\overset{\|}{C}}-OH$$

$$\xrightarrow[\Delta]{2NH_3} CH_3CH_2-\overset{O}{\overset{\|}{C}}-NH_2 \xrightarrow[NaOH]{Br_2} \underset{\text{Ethylamine}}{CH_3CH_2NH_2}$$

(52) n-Propylamine from ethylamine *(Sambalpur BSc, 2014)*

Key step is reduction of propanenitrile with $LiAlH_4$ to give *n*-propylamine.

$$\underset{\text{Ethylamine}}{CH_3CH_2NH_2} \xrightarrow{HNO_2} CH_3CH_2OH \xrightarrow{SOCl_2} CH_3CH_2Cl \xrightarrow{NaCN}$$

$$CH_3CH_2CN \xrightarrow[\text{(or } H_2/Ni)]{LiAlH_4} \underset{n\text{-Propylamine}}{CH_3CH_2CH_2NH_2}$$

(53) n-Propylamine from 1-propanol *(Sambalpur BSc, 2014)*

Key step is reduction of an amide with $LiAlH_4$ to give an amine.

$$\underset{\text{1-Propanol}}{CH_3CH_2CH_2OH} \xrightarrow[\Delta]{Na_2Cr_2O_7/H^+} CH_3CH_2-\overset{O}{\overset{\|}{C}}-OH \xrightarrow{SOCl_2} CH_3CH_2-\overset{O}{\overset{\|}{C}}-Cl$$

$$\xrightarrow[\Delta]{2NH_3} CH_3CH_2-\overset{O}{\overset{\|}{C}}-NH_2 \xrightarrow{LiAlH_4} \underset{n\text{-Propylamine}}{CH_3CH_2CH_2NH_2}$$

(54) n-Butylamine from 1-Propanol

Key step is reduction of nitrile with $LiAlH_4$ or H_2/Ni.

$$\underset{\text{1-Propanol}}{CH_3CH_2CH_2OH} \xrightarrow{PBr_3} CH_3CH_2CH_2Br \xrightarrow{NaCN} CH_3CH_2CH_2CN$$

$$\xrightarrow{LiAlH_4} \underset{n\text{-Butylamine}}{CH_3CH_2CH_2CH_2NH_2}$$

(55) Ethyl methyl amine from ethanol

Key step is reaction of acetaldehyde with methylamine followed by reaction with H_2/Ni (Reductive amination).

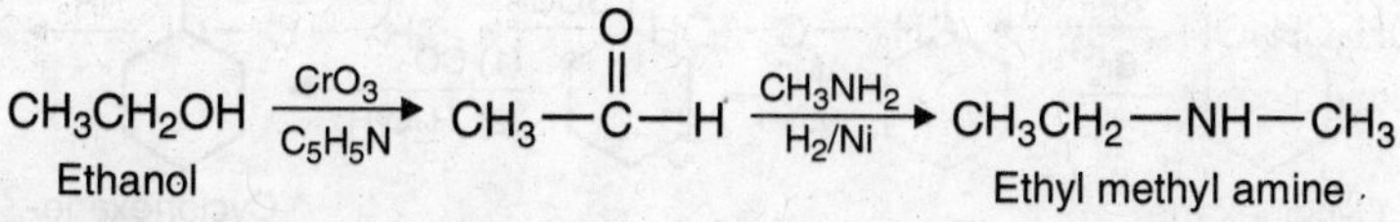

(56) Diethylamine from ethyl alcohol

Key step is reaction of acetyl chloride with ethylamine followed by reduction with $LiAlH_4$.

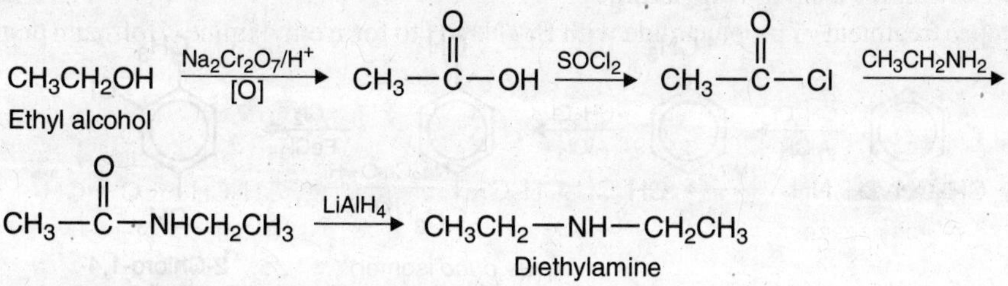

(57) Triethylamine with ethyl alcohol

Key step is reaction of acetyl chloride with diethylamine followed by reduction with $LiAlH_4$.

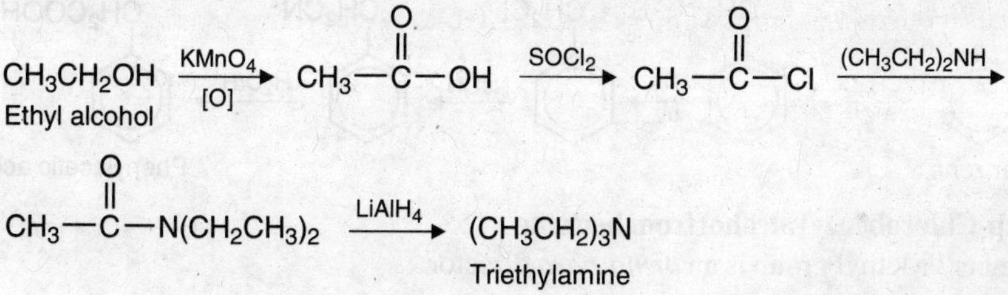

(58) 1,6-Hexanediamine from 1,3-butadiene

Key reaction is treatment of 1,3-butadiene with Cl_2 to give a 1,4-addition product.

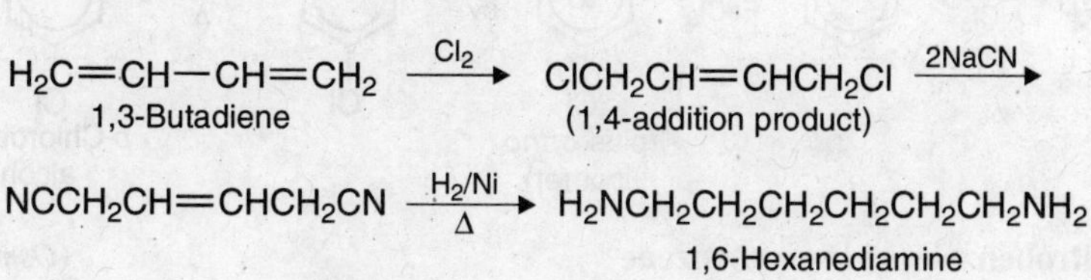

(59) Cyclohexanone from cyclohexane *(Delhi BSc, 2015)*

Key step is the oxidation with of cyclohexanol with $Na_2Cr_2O_7/H^+$ to give cyclohexanone.

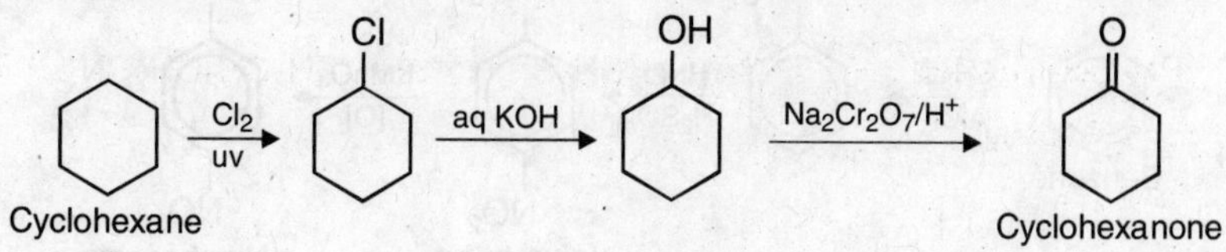

(60) Cyclohexanecarboxylic acid from cyclohexane

Key reaction is treatment of Grignard reagent with CO_2 followed by hydrolysis.

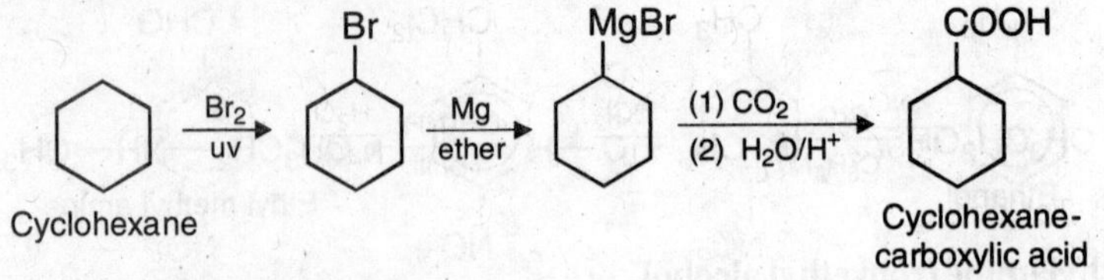

(61) 2-Chloro-1,4-dimethylbenzene from benzene

All four positions in *p*-xylene are identical.

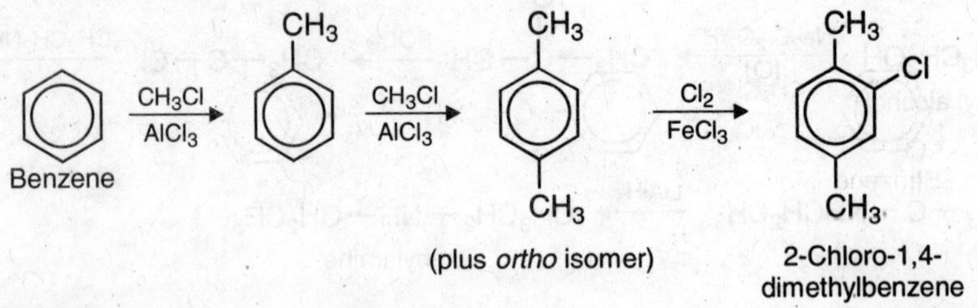

(62) Phenylacetic acid from benzene *(Annamalai BSc, 2014)*

Key step is acid-hydrolysis of benzonitrile.

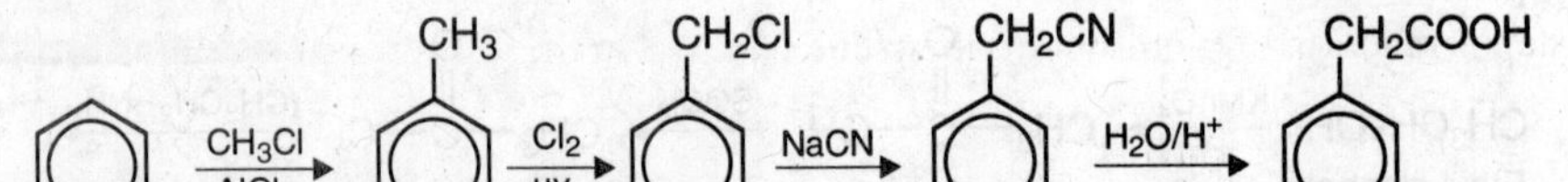

(63) p-Chlorobenzyl alcohol from benzene

Remember : Methyl group is an *ortho-para* director.

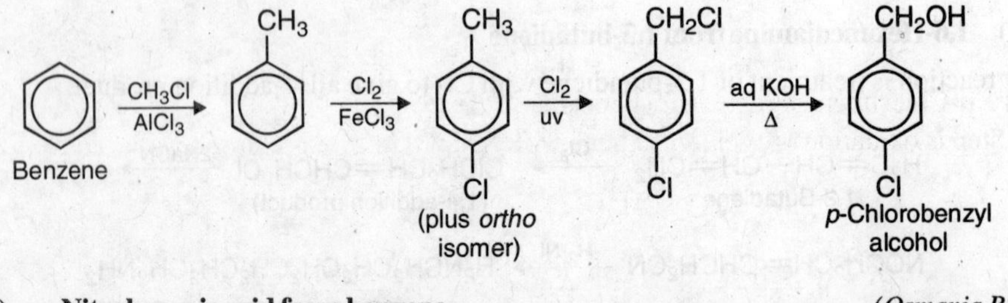

(64) p-Nitrobenzoic acid from benzene *(Osmania BSc, 2015)*

Key step is conversion of toluene into *p*-nitrotoluene : Remember : $-CH_3$ group is *ortho-para* director.

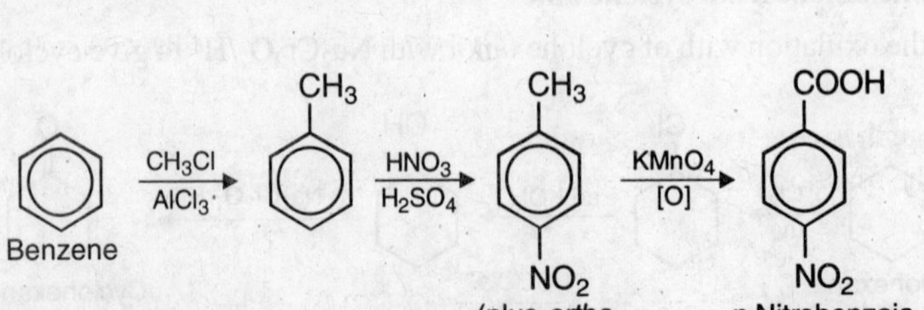

(65) p-Nitrobenzaldehyde from toluene

Key step is hydrolysis of *p*-nitrobenzal chloride to give *p*-nitrobenzaldehyde.

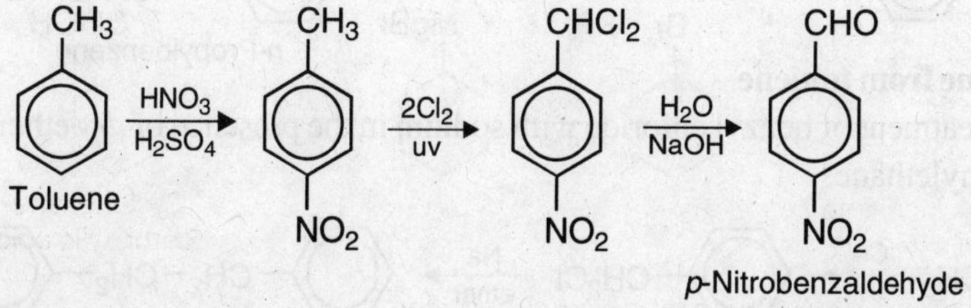

(66) Styrene from benzene *(Panjab BSc, 2014)*

Key step is reaction of benzene with ethyl chloride in the presence of $AlCl_3$ (Friedel-Crafts alkylation).

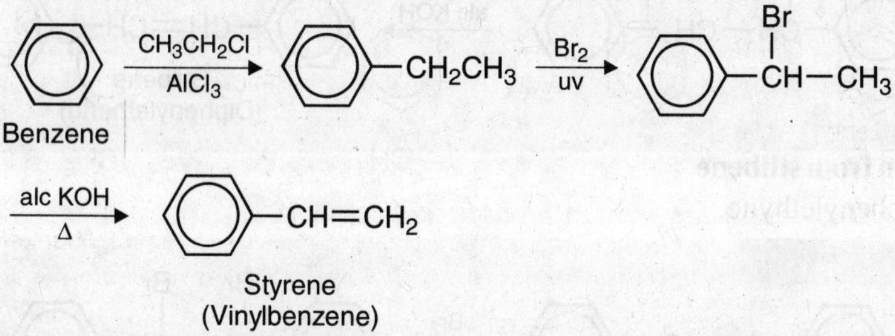

(67) p-Chlorostyrene from benzene

Key step is reaction of *p*-chloroethylbenzene with Cl_2 in the presence of UV light.

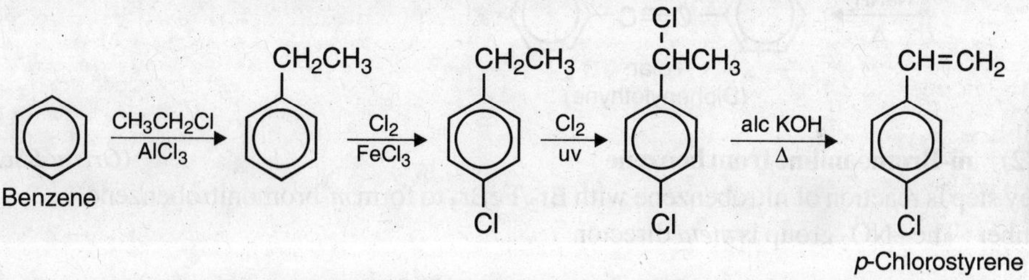

(68) p-Chlorobenzoic acid from benzene

Key step is oxidation of *p*-chlorotoluene with $KMnO_4$ to give *p*-chlorobenzoic acid.

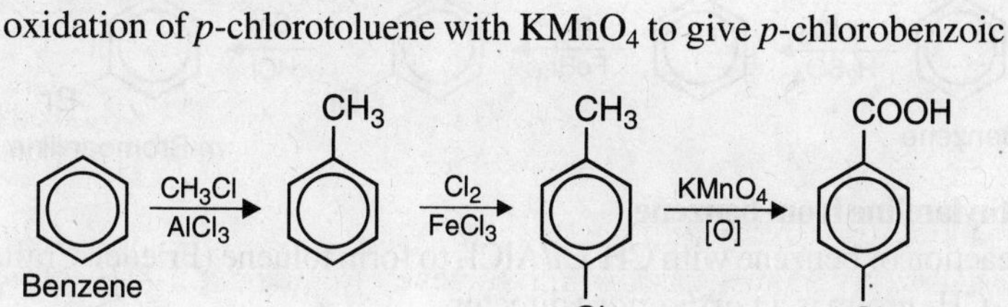

(69) n-Propylbenzene from propanoic acid

Key step is the reaction of ethyl phenyl ketone with NH_2NH_2/KOH (Wolf-Kishner reduction).

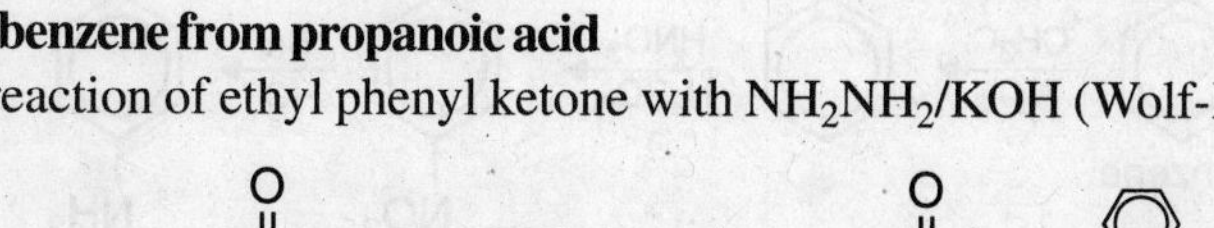

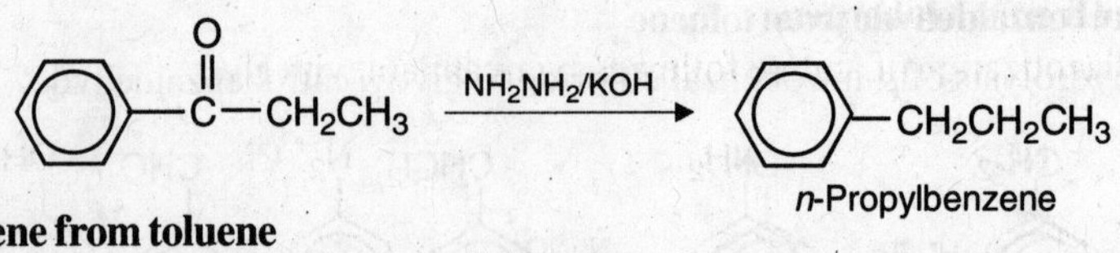

(70) Stilbene from toluene

Key step is treatment of benzyl chloride with sodium in the presence of *dry* ether (Wurtz reaction) to give 1,2-diphenylethane.

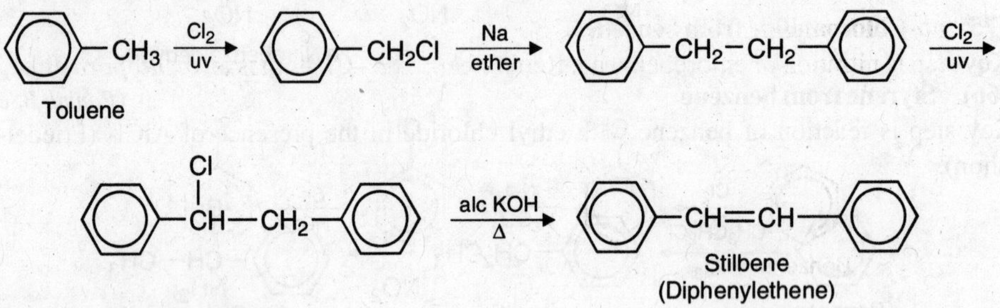

(71) Tolan from stilbene

Tolan is diphenylethyne.

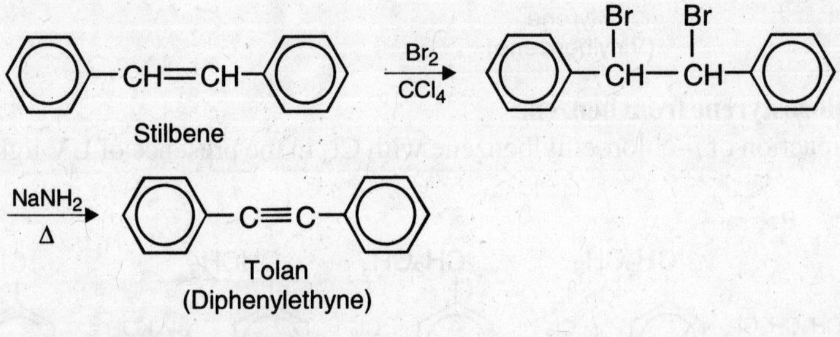

(72) m-Bromoaniline from benzene *(Orissa BSc, 2014)*

Key step is reaction of nitrobenzene with $Br_2/FeBr_3$ to form *m*-bromonitrobenzene.
Remember : The $-NO_2$ group is *meta* director.

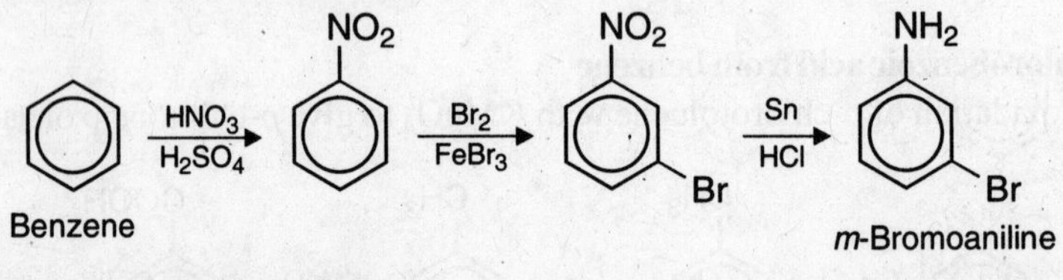

(73) p-Methylaniline from benzene

Key step is reaction of benzene with $CH_3Cl/AlCl_3$ to form toluene (Friedel-Crafts reaction).
Remember : The $-CH_3$ group is an *ortho-para* director.

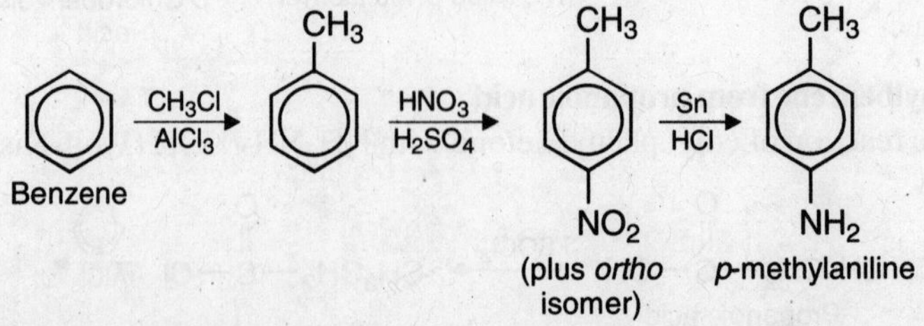

(74) Phenol from nitrobenzene

Key step is diazotization of aniline followed by treatment with H_2O.

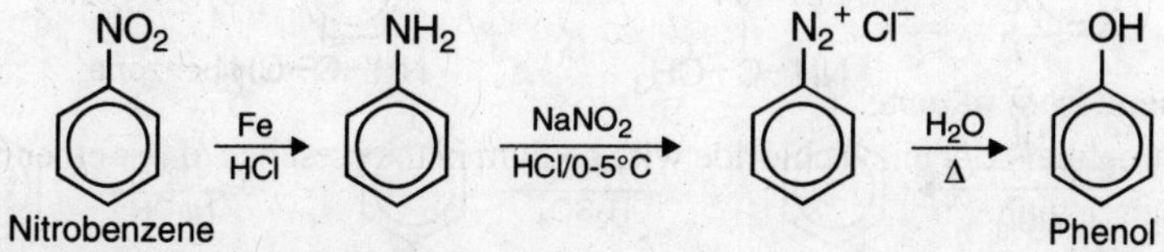

(75) p-Chloroaniline from benzene

Key step is nitration of chlorobenzene. Remember : The –Cl atom is an *ortho-para* director.

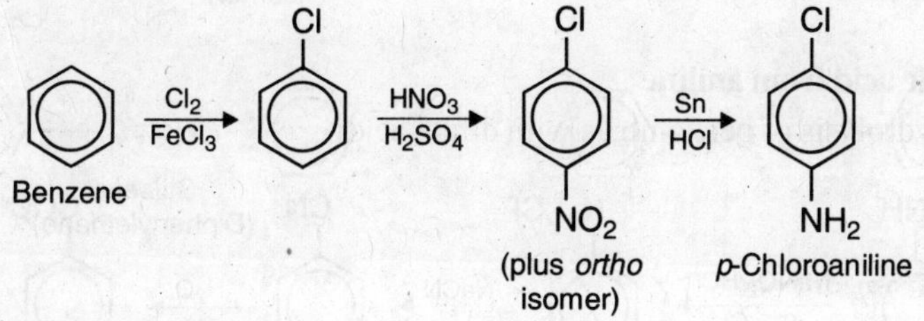

(76) p-Toluidine from benzene (*Periyar BSc, 2015*)

Key step is nitration of toluene.

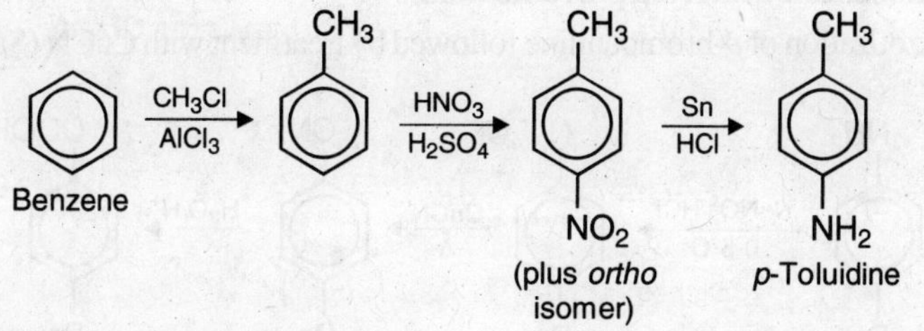

(77) p-Nitroaniline from p-nitrotoluene

Key step is reaction of *p*-Nitrobenzamide with Br_2/NaOH (Hofmann degradation).

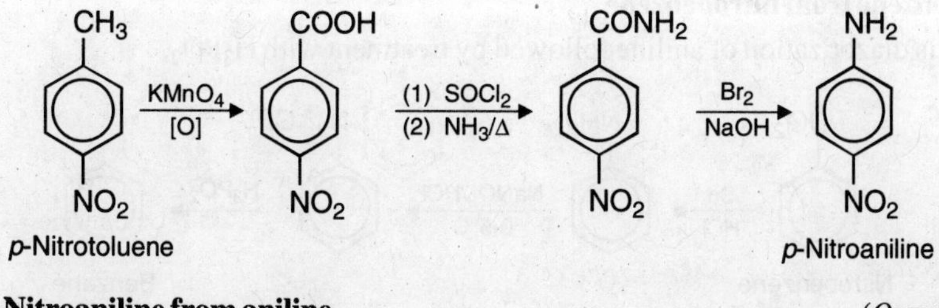

(78) p-Nitroaniline from aniline (*Osmania BSc, 2014*)

Remember : $-NH_2$ is an *ortho-para* director. However, direct nitration of aniline with HNO_3/H_2SO_4 gives a *meta* product. This is because, aniline is basic and reacts with acid to form anilinium ion which gives a *meta* product.

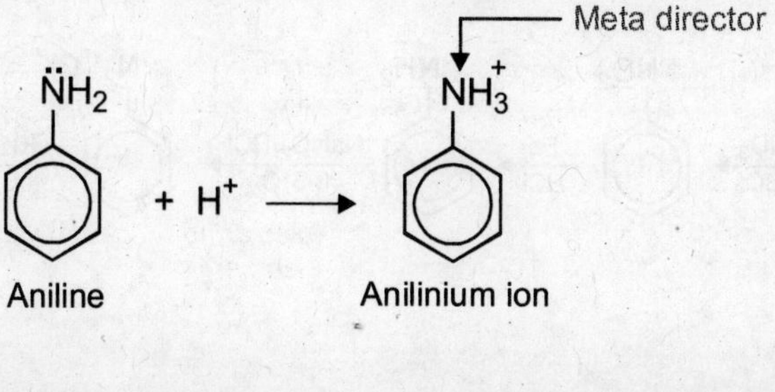

This synthesis requires protection of the $-NH_2$ group.

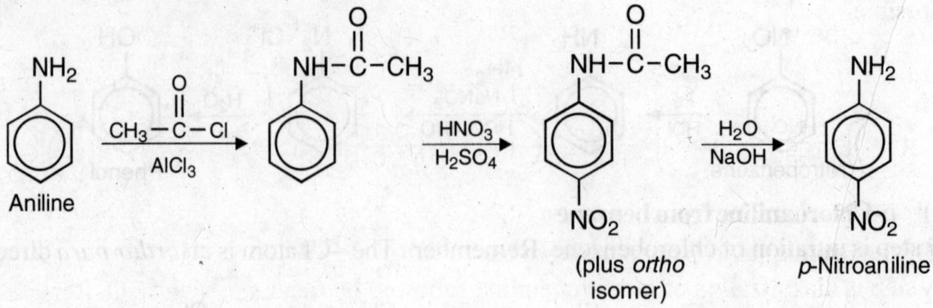

(79) **Benzoic acid from aniline**

Key step is hydrolysis of benzonitrile with dilute acid.

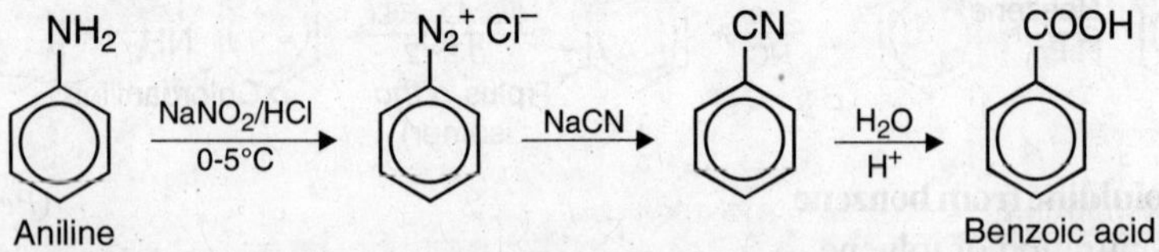

(80) **p-Bromobenzoic acid from p-bromoaniline**

Key step is diazotization of *p*-bromoaniline followed by treatment with CuCN (Sandmeyer reaction).

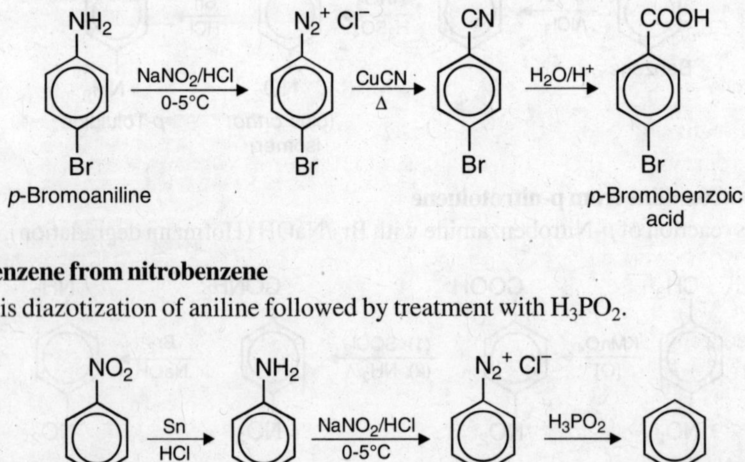

(81) **Benzene from nitrobenzene**

Key step is diazotization of aniline followed by treatment with H_3PO_2.

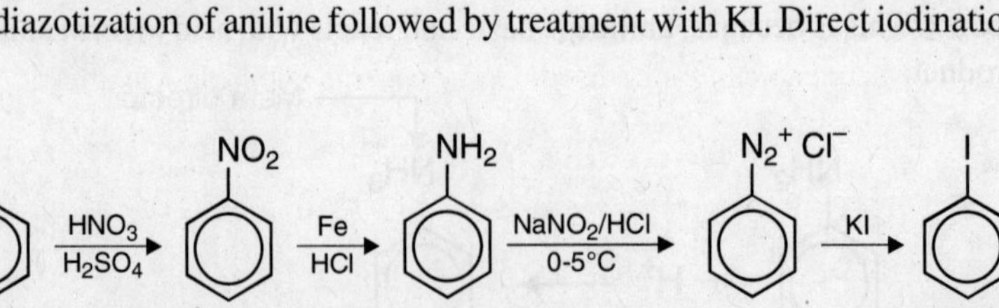

(82) **Iodobenzene from benzene**

Key step is diazotization of aniline followed by treatment with KI. Direct iodination of benzene is not possible.

(83) Fluorobenzene from benzene

Key step is diazotization of aniline followed by treatment with HBF_4. Direct fluorination of benzene is not possible.

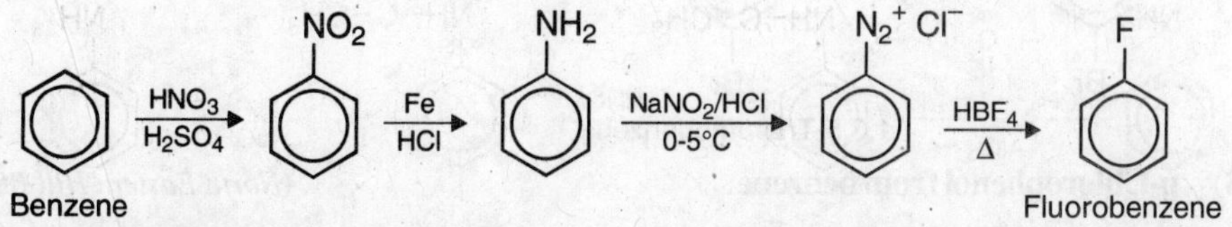

(84) m-Bromophenol from nitrobenzene *(Anna BSc, 2015)*

Key step is diazotization of *m*-bromoaniline followed by treatment with H_2O. Remember : The $-NO_2$ group is a *meta* director.

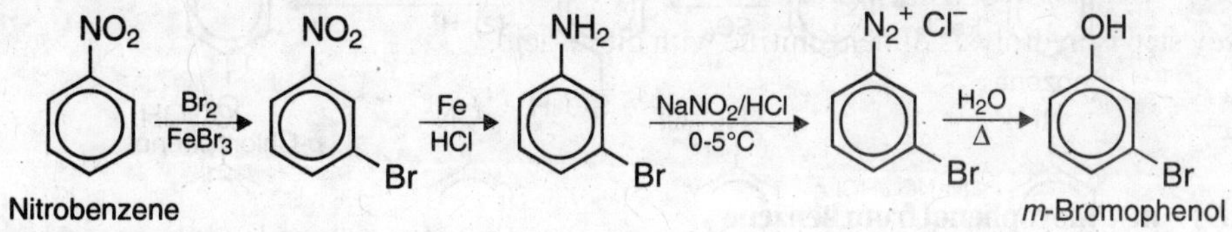

(85) m-Dibromobenzene from nitrobenzene

Key step is diazotization of *m*-bromoaniline followed by treatment with CuBr (Sandmeyer reaction).

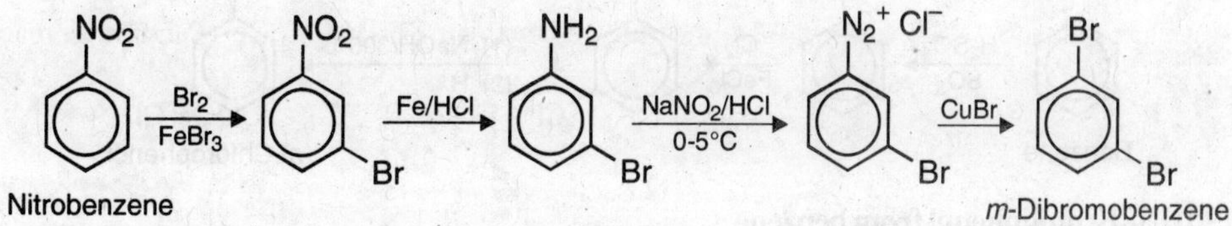

(86) m-Bromochlorobenzene from benzene

Key step is diazotization of *m*-chloroaniline followed by treatment with CuBr (Sandmeyer reaction).

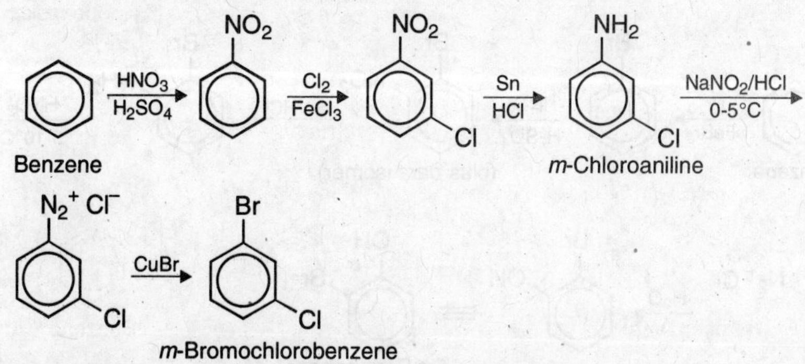

(87) 1,3,5-Tribromobenzene from benzene *(Delhi BSc, 2015)*

Notice that all bromine atoms on benzene in the target molecule are *meta* to each other. Therefore, direct bromination of benzene cannot be used. Carry out this synthesis *via* aniline.

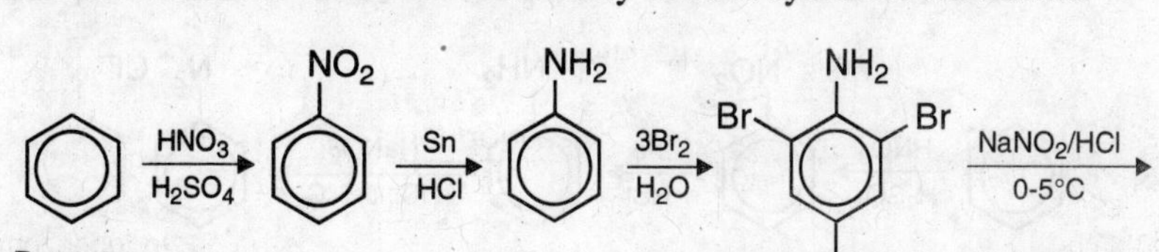

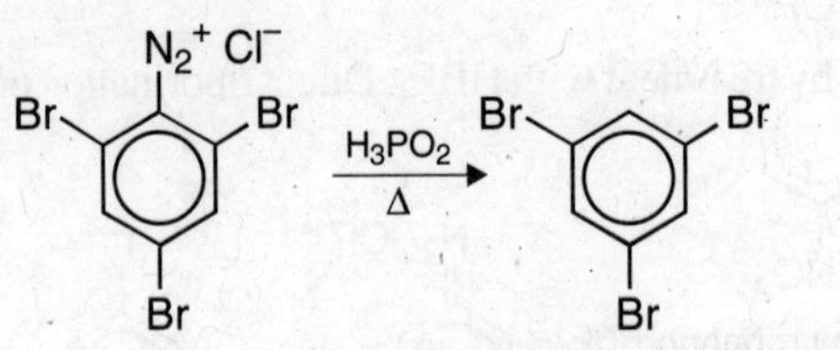

1,3,5-Tribromobenzene

(88) p-Chlorophenol from benzene *(North Eastern Hill BSc, 2014)*

Remember : –Cl is an *ortho-para* director.

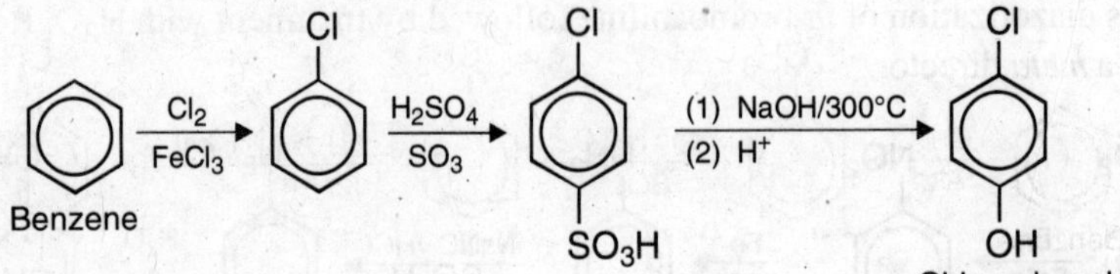

Benzene   *p*-Chlorophenol

(89) m-Chlorophenol from benzene

Remember : $-SO_3H$ is a *meta*-director.

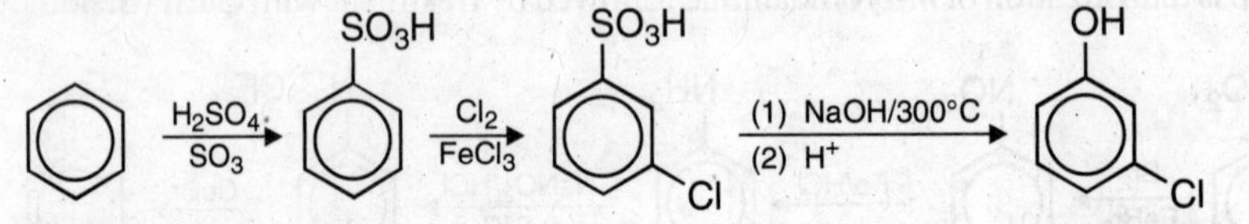

Benzene   *m*-Chlorophenol

(90) o-Bromophenol from benzene

Key steps are diazotization of *o*-bromoaniline followed by heating with H_2O.

Remember : – Br is an *ortho-para* director.

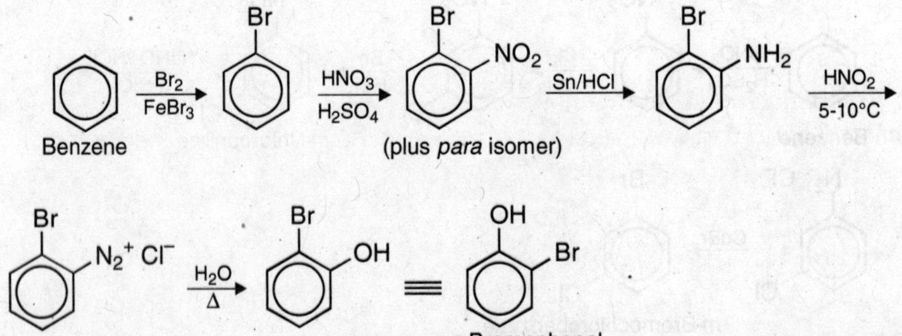

Benzene   (plus *para* isomer)   *o*-Bromophenol

(91) 2,4,6-Trinitrophenol from benzene *(Madras BSc, 2014)*

Key step is nitration of phenol with concentrated HNO_3+H_2SO_4 to form 2,4,6-trinitrophenol.

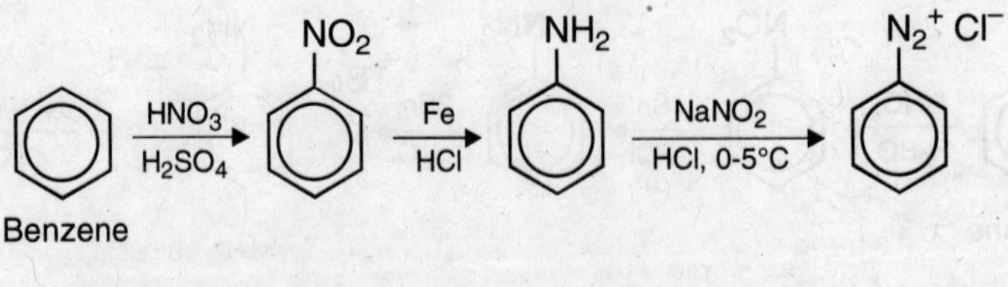

Benzene

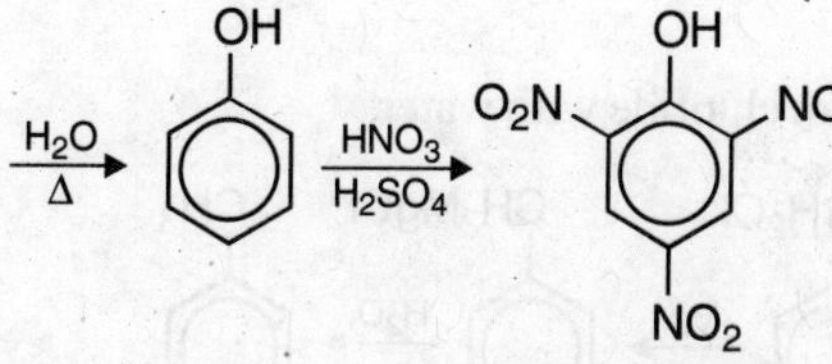

2,4,6-Trinitrophenol

(92) p-Cresol from benzene

Key step is sulfonation of toluene with fuming sulfuric acid (concentrated $H_2SO_4 + SO_3$). Remember : $-CH_3$ group is an *ortho-para* director.

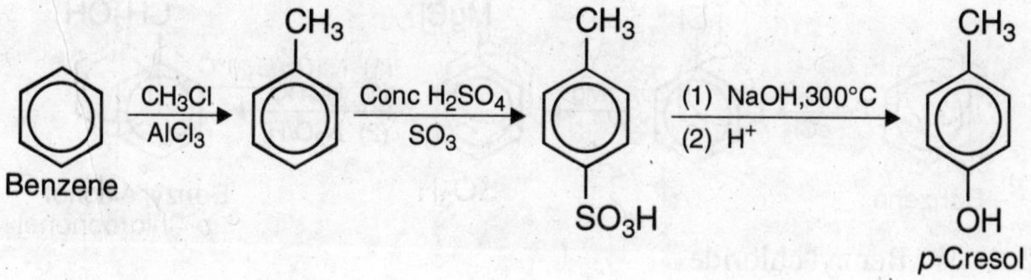

(93) o-Cresol from toluene

Key step is diazotization of *o*-toluidine followed by treatment with H_2O.

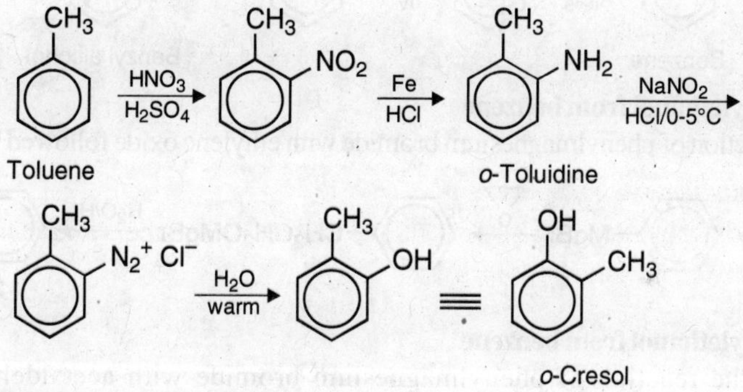

(94) Resorcinol from Benzenesulfonic acid (*Jadavpur BSc, 2015*)

Remember : $-SO_3H$ is a *meta*-director.

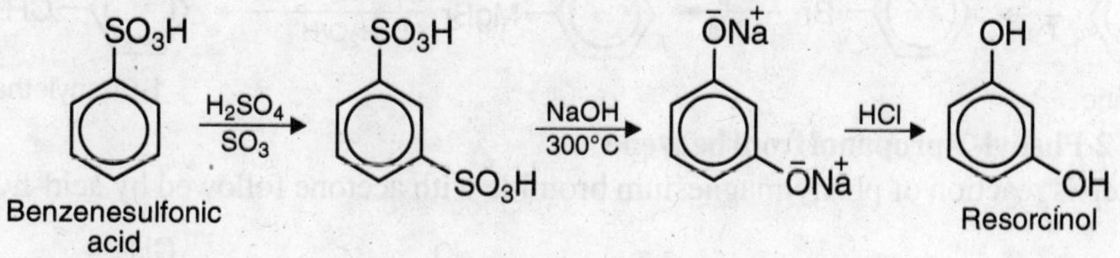

(95) Benzyl alcohol from toluene

Key step is reduction of benzoic acid with $LiAlH_4$.

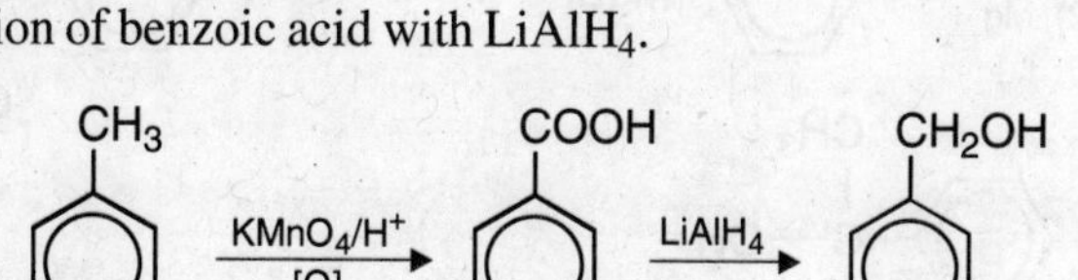

(96) Toluene from benzyl alcohol

Key step is treatment of benzylmagnesium chloride with water.

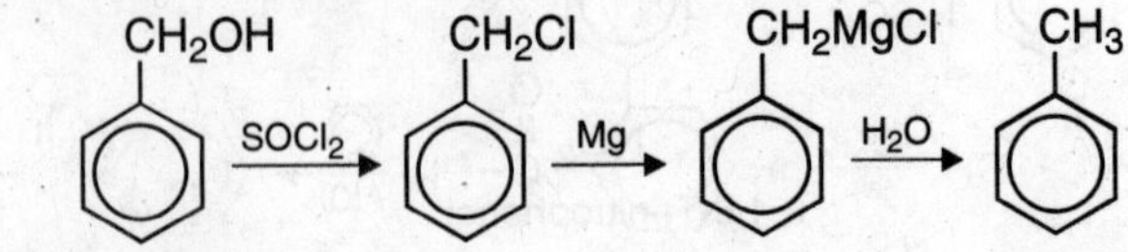

Benzyl alcohol Toluene

(97) Benzyl alcohol from benzene

Method 1. *Via* Phenylmagnesium chloride

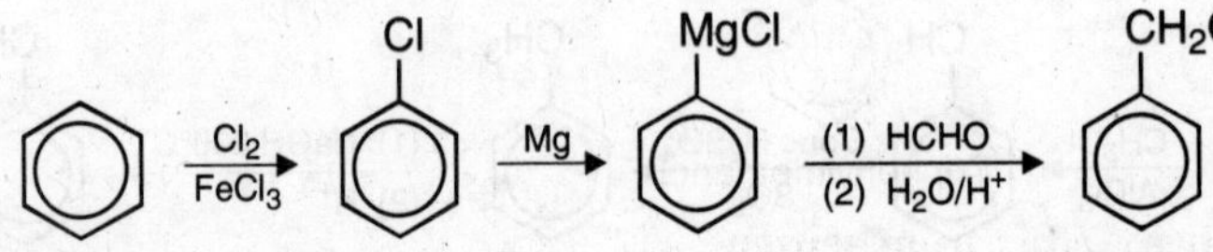

Benzene Benzyl alcohol

Method 2. *Via* Benzyl chloride

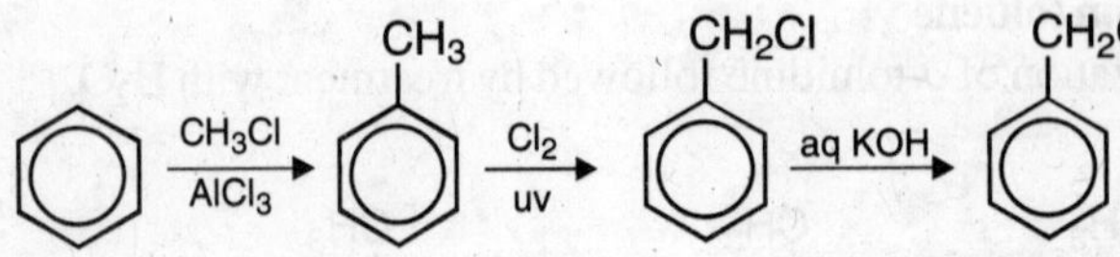

Benzene Benzyl alcohol

(98) 2-Phenylethanol from benzene *(Gujarat BSc, 2014)*

Key step is reaction of phenylmagnesium bromide with ethylene oxide followed by acid-hydrolysis.

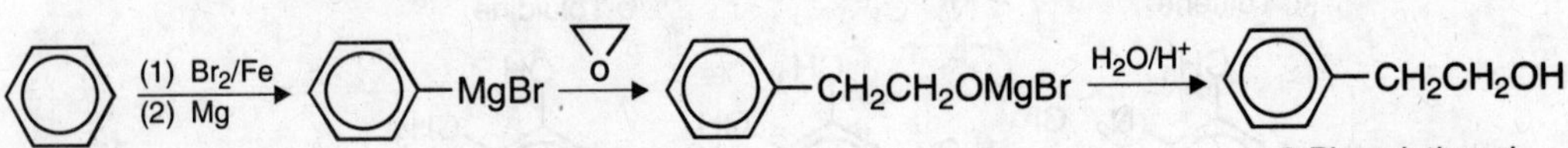

Benzene 2-Phenylethanol

(99) 1-Phenylethanol from benzene *(Mysore BSc, 2015)*

Key step is the reaction of phenylmagnesium bromide with acetaldehyde followed by acid-hydrolysis.

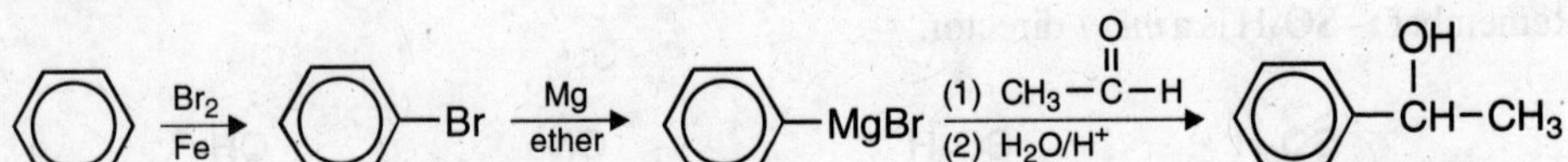

Benzene 1-Phenylethanol

(100) 2-Phenyl-2-propanol from benzene

Key step is reaction of phenylmagnesium bromide with acetone followed by acid-hydrolysis.

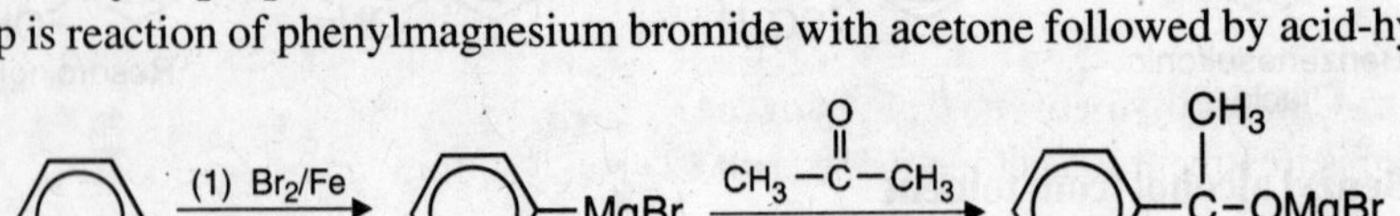

Benzene

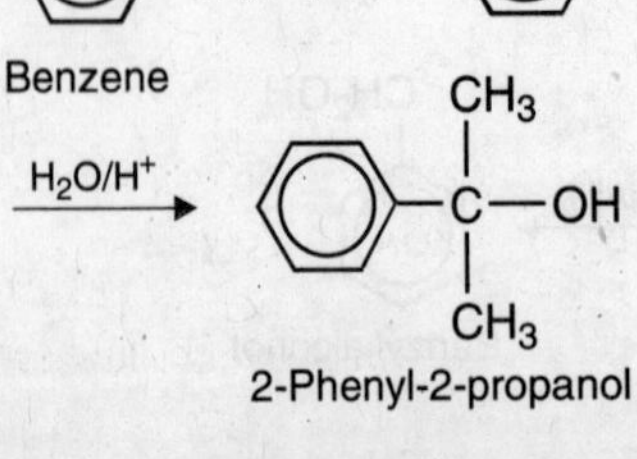

2-Phenyl-2-propanol

(101) Triphenylmethanol from benzoic acid

Key step is treatment of benzophenone with phenylmagnesium chloride followed by acid-hydrolysis.

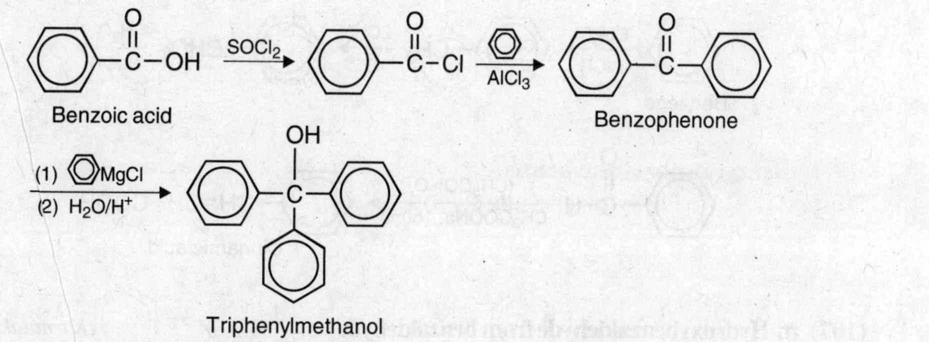

(102) Benzyl phenyl ether from benzene

Key step is treatment of sodium phenoxide with benzyl bromide (Williamson ether synthesis).

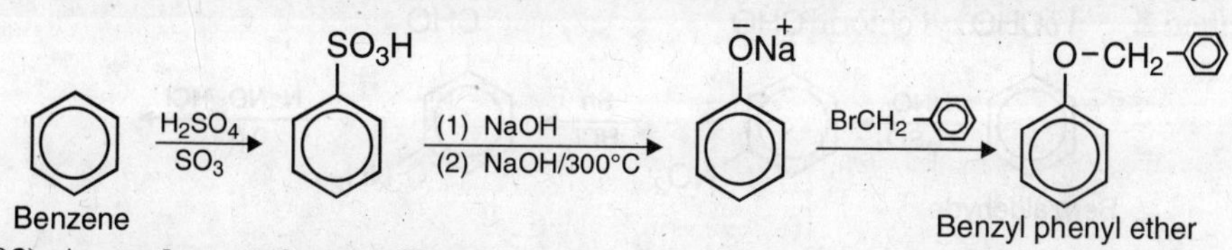

(103) Acetophenone from aniline

Key step is treatment of benzonitrile with methylmagnesium bromide followed by hydrolysis.

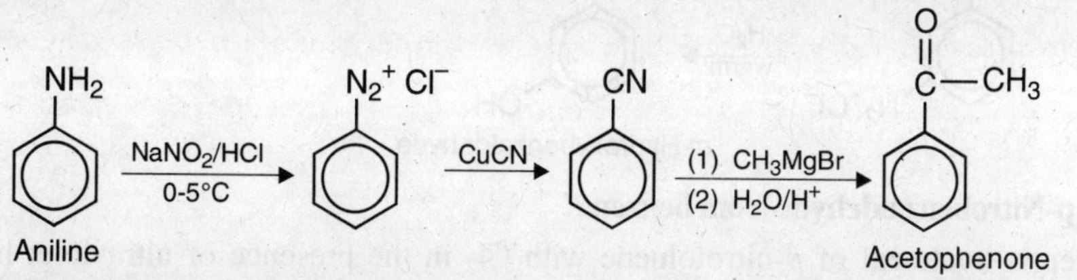

(104) p-Chloroethylbenzene from benzene

Key step is treatment of acetophenone with NH_2NH_2/KOH (Wolf-Kishner reduction).

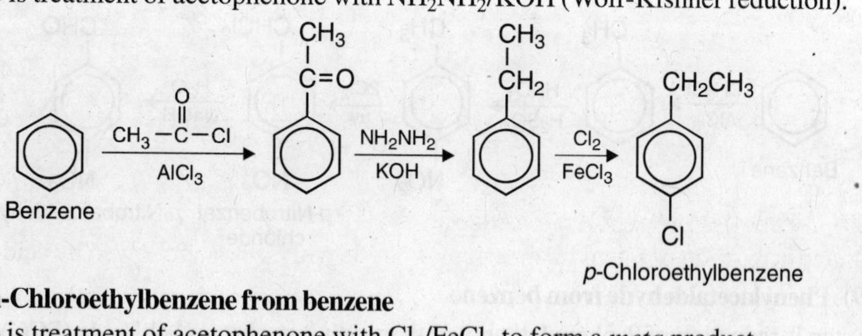

(105) m-Chloroethylbenzene from benzene

Key step is treatment of acetophenone with $Cl_2/FeCl_3$ to form a *meta* product.

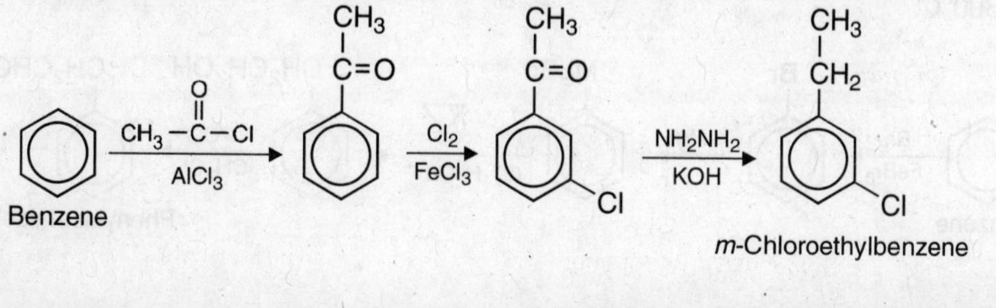

(106) Cinnamic acid from benzene (*Magadh BSc, 2014*)

Key step is treatment of benzaldehyde with acetic anhydride and sodium acetate at 180°C (Perkin reaction).

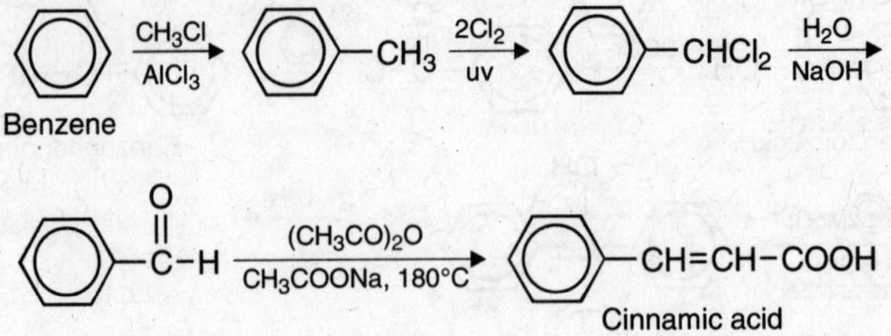

(107) m-Hydroxybenzaldehyde from benzaldehyde (*Karnataka BSc, 2015*)

Remember : –CHO is a *meta* director.

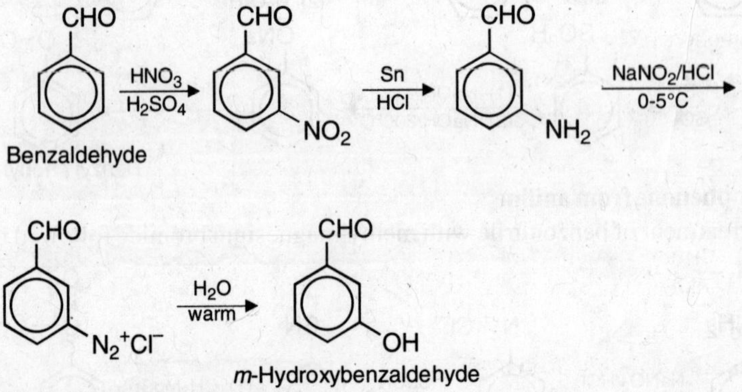

(108) p-Nitrobenzaldehyde from benzene

Key step is treatment of *p*-nitrotoluene with Cl_2 in the presence of ultraviolet light to form *p*-nitrobenzal chloride followed by hydrolysis.

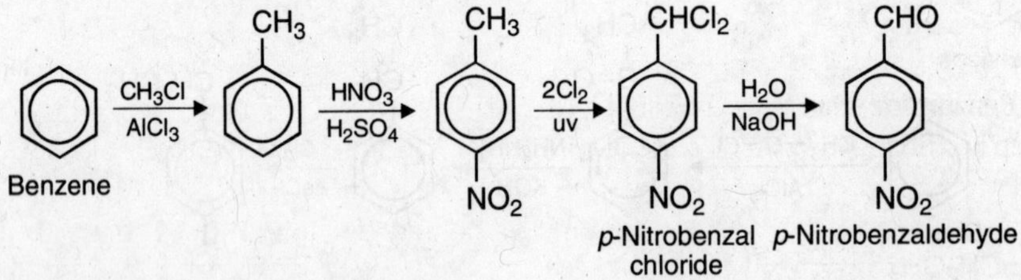

(109) Phenylacetaldehyde from benzene

Key step is reduction of 2-phenylethanol with pyridinium chromyl chloride (PCC) in CH_2Cl_2 or copper at 300°C.

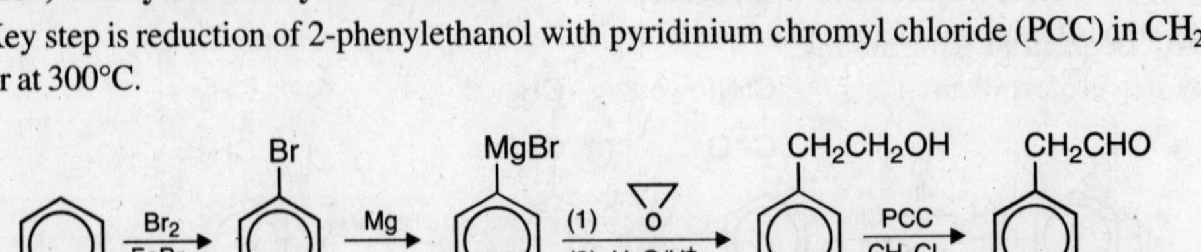

(110) Phenylacetic acid from benzyl chloride *(Indore BSc, 2015)*

Method 1. *Via* Grignard synthesis

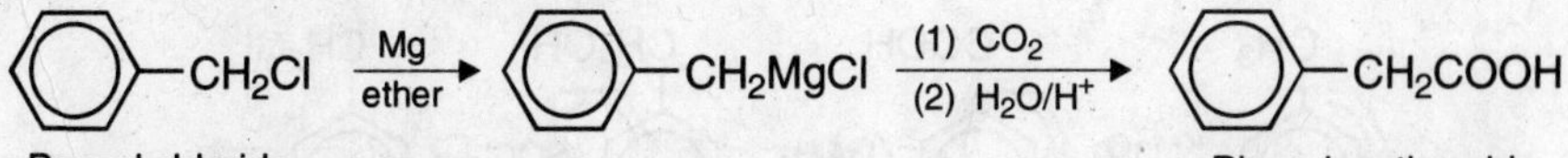

Benzyl chloride — Phenylacetic acid

Method 2. *Via* Nitrile

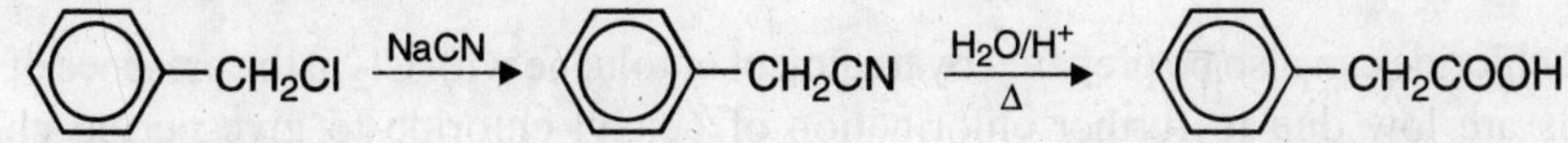

Benzyl chloride — Phenylacetic acid

(111) Benzoic acid from bromobenzene

Key reaction is treatment of phenylmagnesium bromide with carbon dioxide followed by acid-hydrolysis.

Bromobenzene — Benzoic acid

(112) Benzamide from benzene *(Sambalpur BSc, 2015)*

Key step is treatment of benzoyl chloride with ammonia to give benzamide.

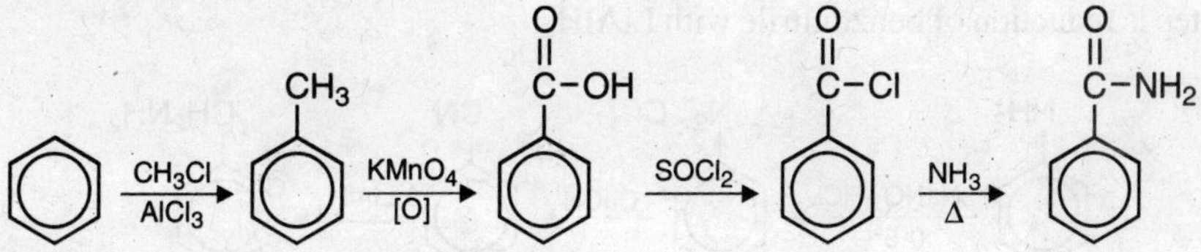

Benzene — Benzamide

(113) m-Bromobenzamide from benzene

Remember : –$CONH_2$ group is a *meta*-director.

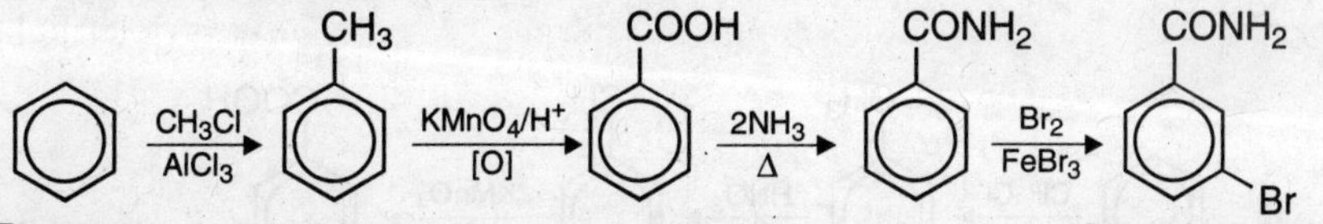

Benzene — *m*-Bromobenzamide

(114) Benzamide from benzyl alcohol *(Agra BSc, 2014)*

Key step is oxidation of benzyl alcohol with acidic $K_2Cr_2O_7$ to give benzoic acid.

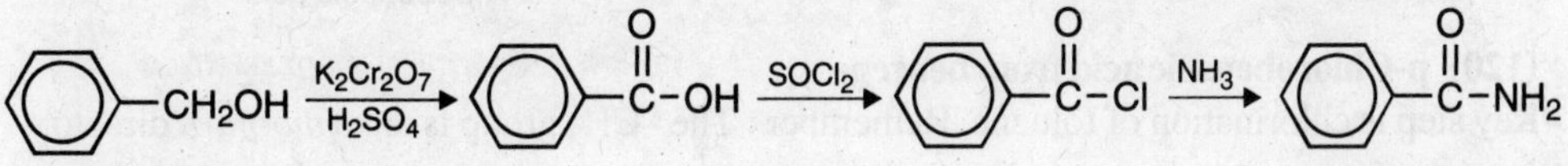

Benzyl alcohol — Benzamide

(115) Benzamide from aniline

Key step is diazotization of aniline followed by treatment with CuCN.

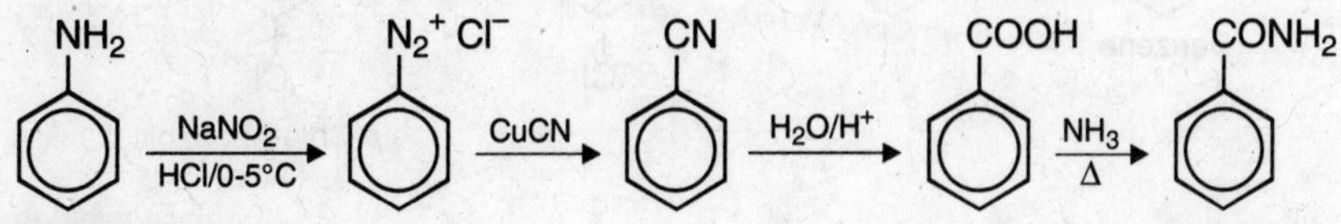

Aniline — Benzamide

(116) Benzyl chloride from toluene

Key step is treatment of benzyl alcohol with $SOCl_2$.

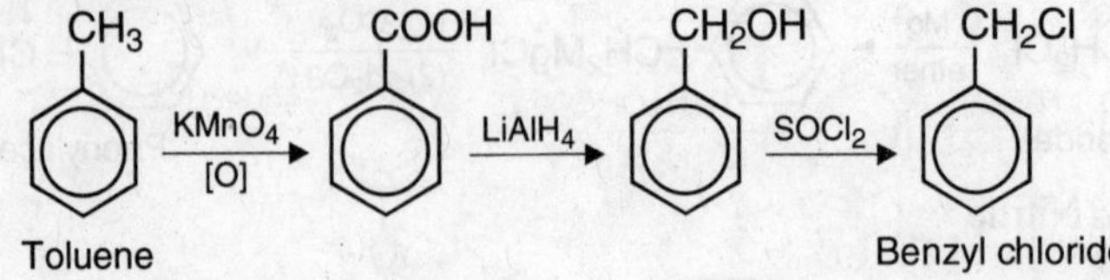

Benzyl chloride can also be prepared by treatment of toluene with Cl_2 in the presence of ultraviolet light. Yields are low due to further chlorination of benzyl chloride to give benzal chloride and benzotrichloride.

(117) Benzylamine from toluene

Key step is reduction of benzamide with $LiAlH_4$.

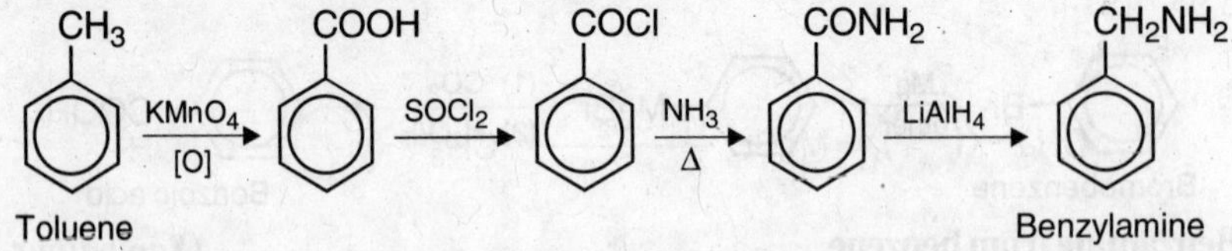

(118) Benzylamine from aniline *(Panjab BSc, 2015)*

Key step is reduction of benzonitrile with $LiAlH_4$.

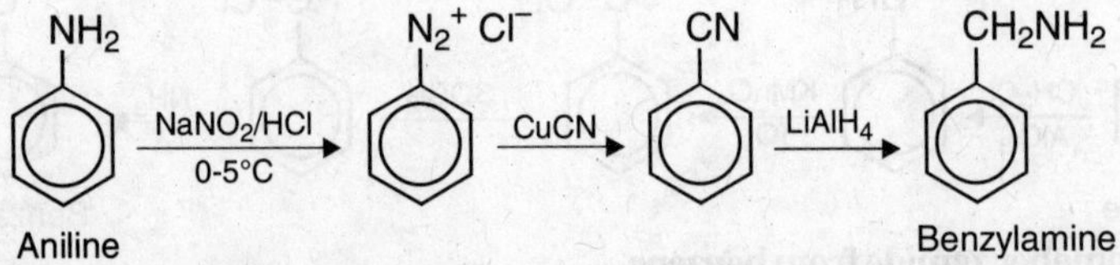

(119) p-Nitrobenzoic acid from benzene

Key step is nitration of toluene. Remember : The $-CH_3$ group is an *ortho-para* director.

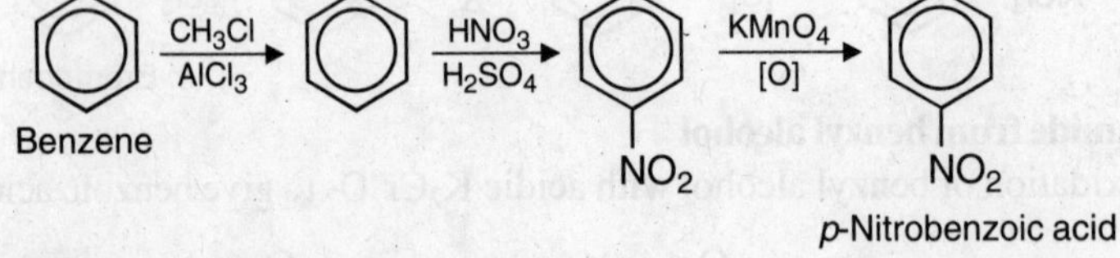

(120) p-Chlorobenzoic acid from benzene

Key step is chlorination of toluene. Remember : The $-CH_3$ group is an *ortho-para* director.

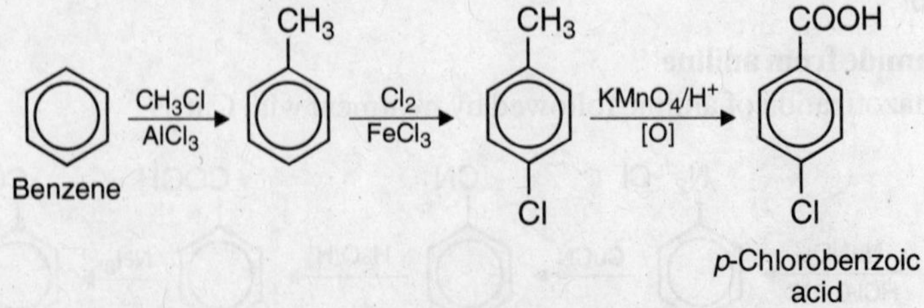

(121) p-Hydroxybenzoic acid from toluene

Key step is sulfonation of toluene with fuming sulfuric acid (Conc. $H_2SO_4 + SO_3$).

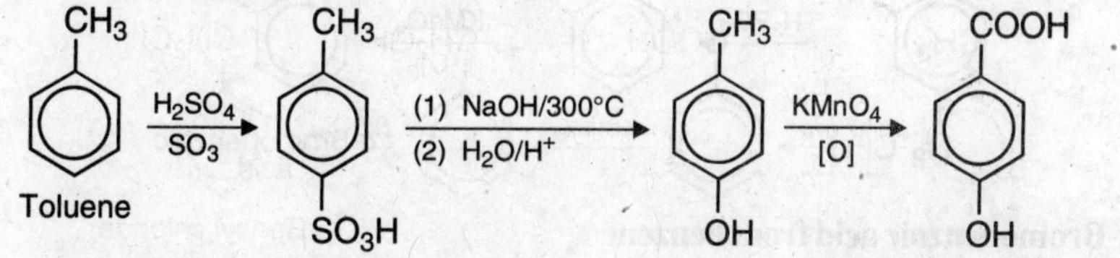

p-Hydroxybenzoic acid

(122) p-Toluic acid from benzene *(Bihar BSc, 2015)*

Key step is treatment of toluene with $Br_2/FeBr_3$ to form *p*-bromotoluene.

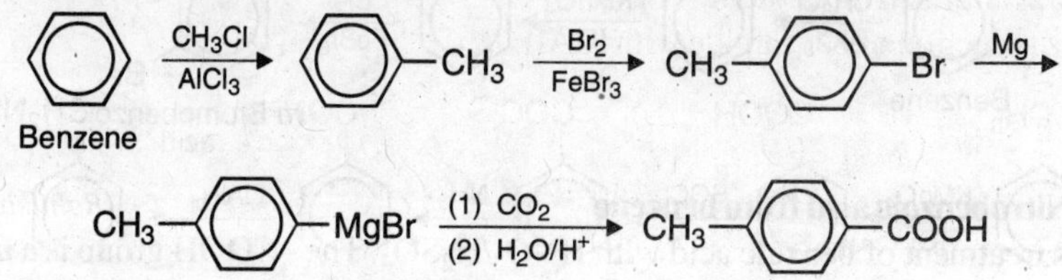

p-Toluic acid

(123) p-Methylbenzoic acid from benzene *(Bundelkhand BSc, 2014)*

Method 1. *Via* Grignard reaction

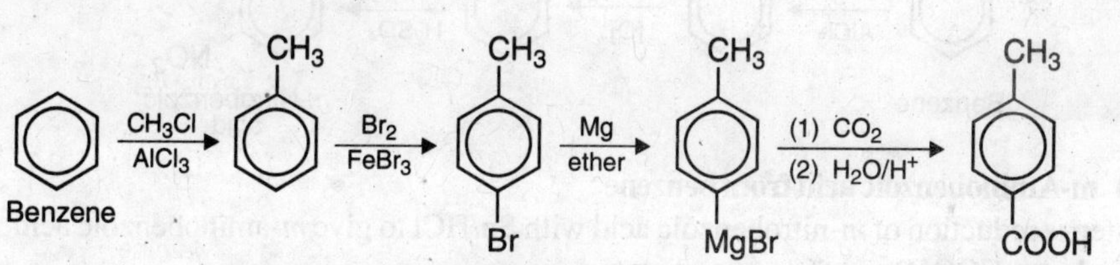

(plus *ortho* isomer)

p-Methylbenzoic acid

Method 2. *Via* Iodoform reaction

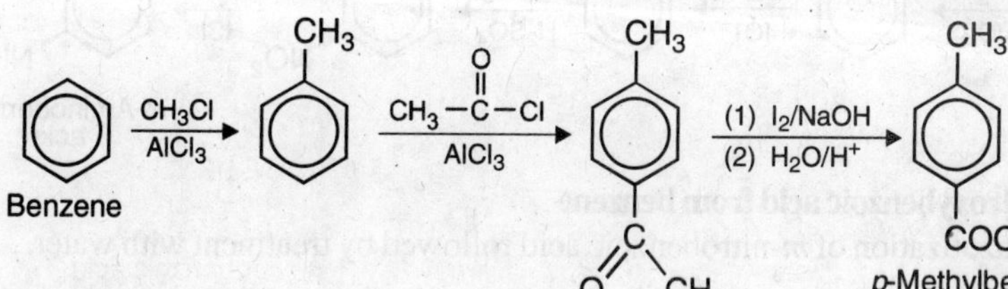

p-Methylbenzoic acid

(124) o-Bromobenzoic acid from p-nitrotoluene

Key step is bromination of *p*-nitrotoluene. Remember : The $-CH_3$ group is an *ortho-para* director and $-NO_2$ group is a *meta* director.

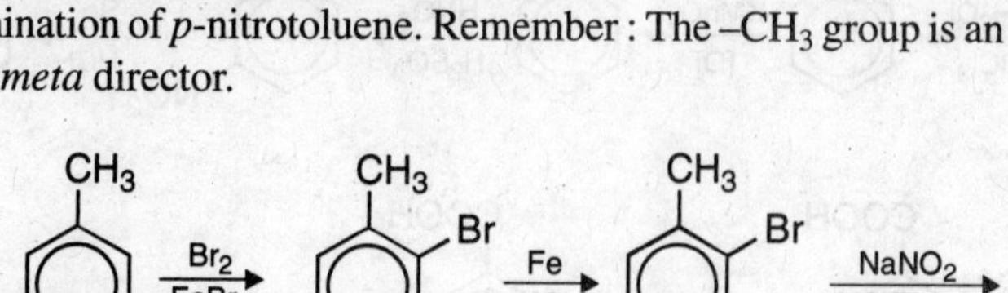

p-Nitrotoluene

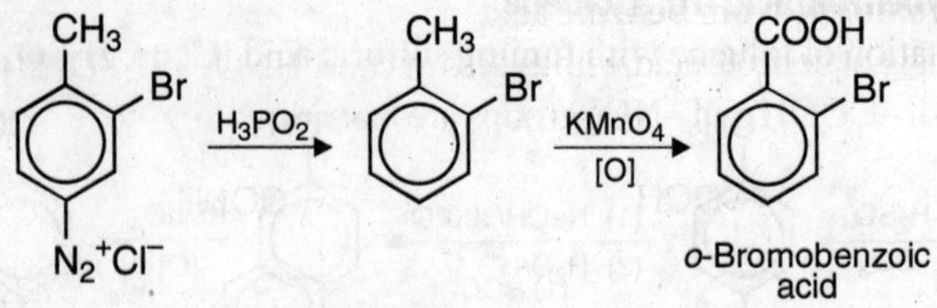

(125) m-Bromobenzoic acid from benzene

Key step is treatment of benzoic acid with $Br_2/FeBr_3$ to form *m*-bromobenzoic acid.

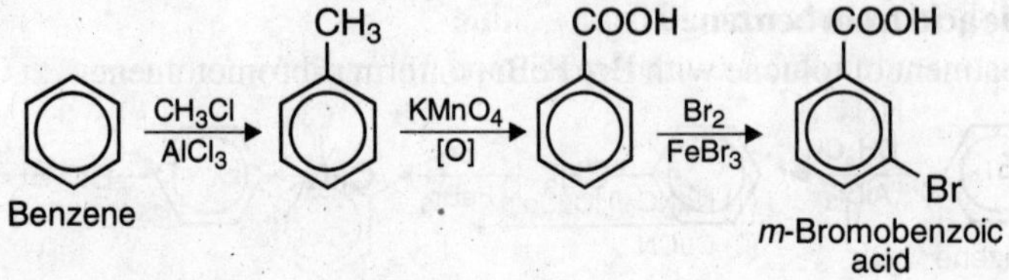

(126) m-Nitrobenzoic acid from benzene

(*Rohilkhand BSc, 2015*)

Key step is treatment of benzoic acid with HNO_3/H_2SO_4. The –COOH group is a *meta* director.

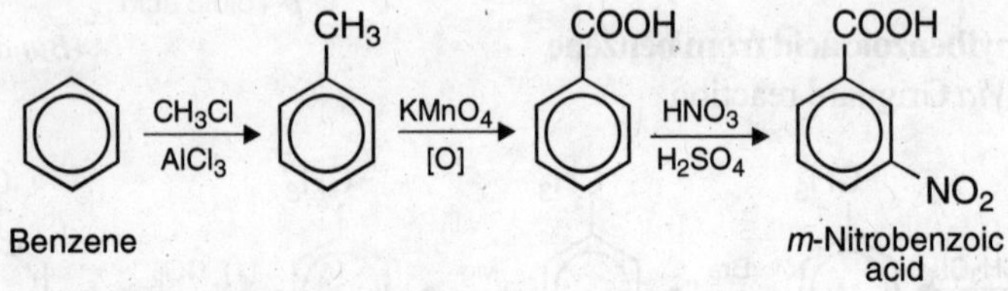

(127) m-Aminobenzoic acid from benzene

Key step is reduction of *m*-nitrobenzoic acid with Sn/HCl to give *m*-aminobenzoic acid.
Remember : –COOH group is a *meta* director.

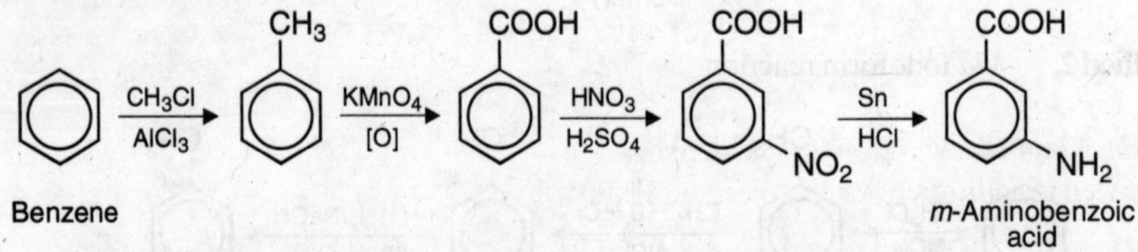

(128) m-Hydroxybenzoic acid from benzene

Key step is diazotization of *m*-nitrobenzoic acid followed by treatment with water.

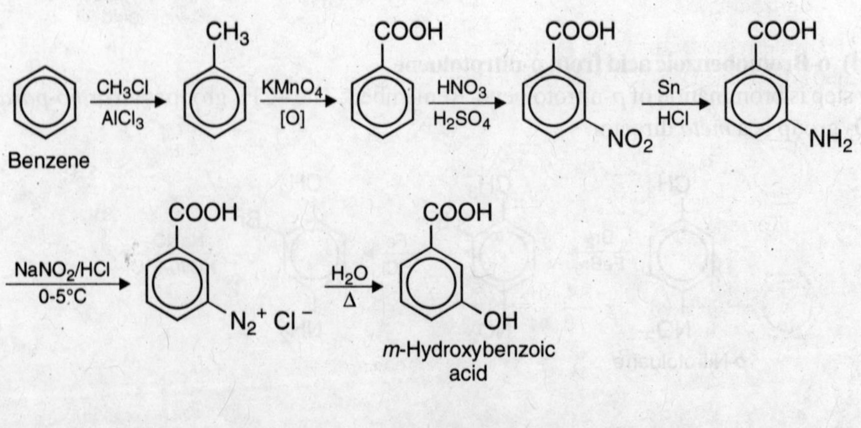

(129) 3,5-Dinitroaniline from benzoic acid *(Manipal BSc, 2014)*

Key step is treatment of benzoic acid with excess of conc. $HNO_3 + H_2SO_4$ to form 3,5-dinitrobenzoic acid. Remember : Both $-COOH$ and $-NO_2$ groups are *meta* directors.

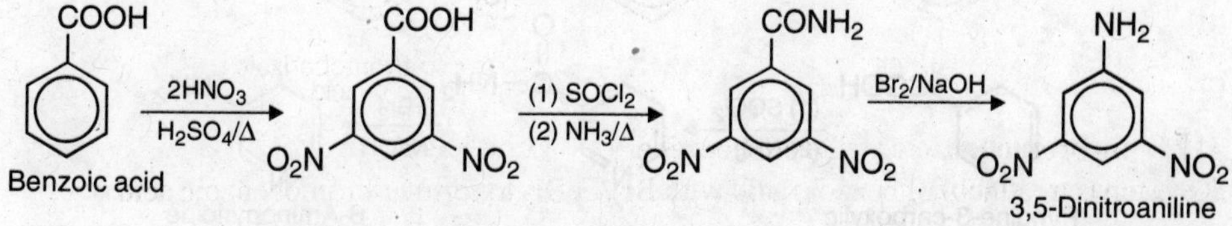

(130) 2,4,6-Tribromobenzoic acid from aniline

Key step is diazotization of 2,4,6-tribromoaniline followed by treatment with CuCN.

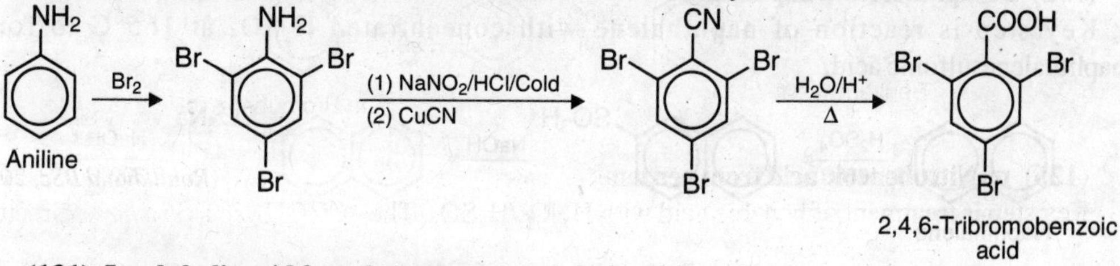

(131) Isophthalic acid from benzene

Key reaction is diazotization of *m*-dinitrobenzene followed by treatment with CuCN.

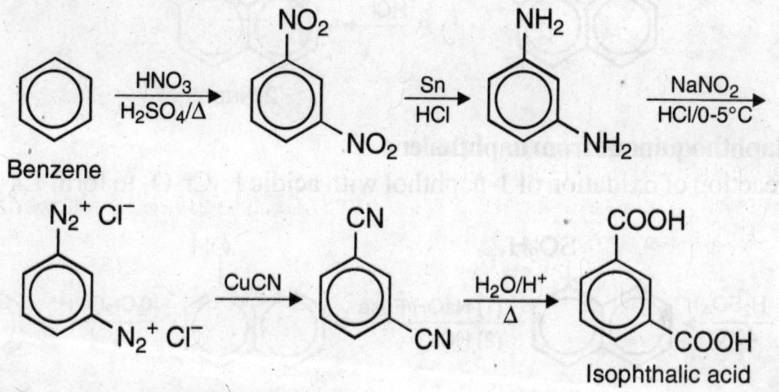

(132) Mandelic acid from benzaldehyde *(Jammu BSc, 2014)*

Key step is addition of HCN to benzaldehyde.

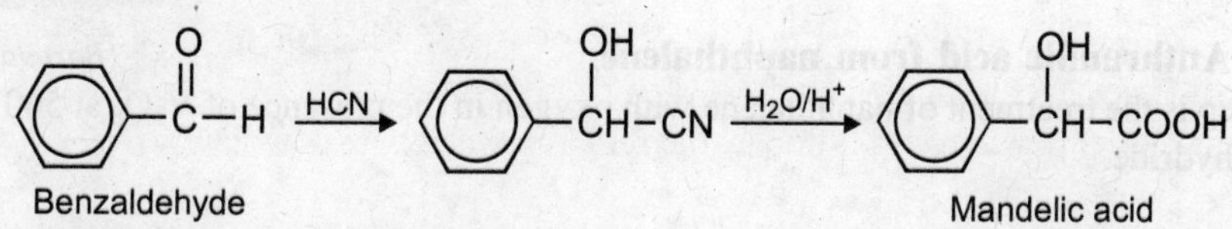

(133) Diphenylmethane from toluene

Key step is Clemmensen (NH_2NH_2/NaOH) or Wolf-Kishner reduction (Zn-Hg/HCl) of benzophenone.

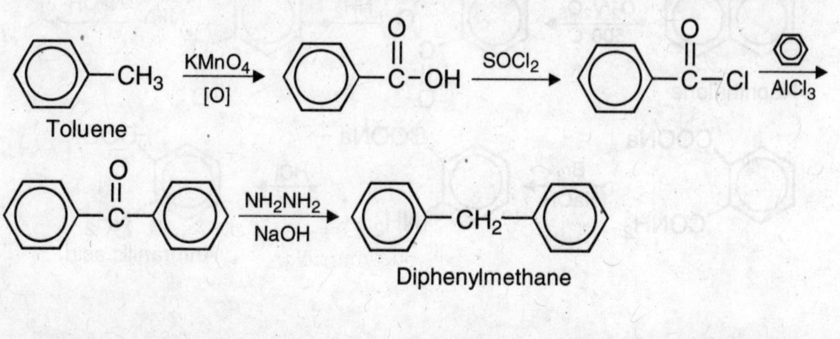

(134) 3-Aminopyridine from pyridine-3-carboxylic acid

Key step is the conversion of the $-CONH_2$ into $-NH_2$ group by treatment with Br_2/NaOH (Hofmann degradation).

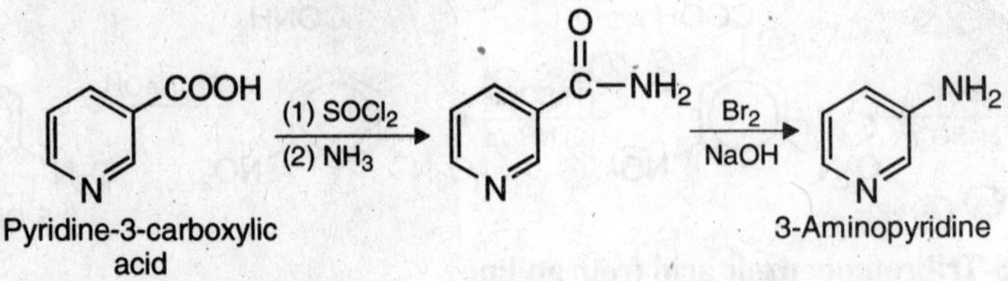

(135) 2-Naphthol from naphthalene (*Rani Durgawati BSc, 2014*)

Key step is reaction of naphthalene with concentrated H_2SO_4 at 165°C to form 2-naphthalenesulfonic acid.

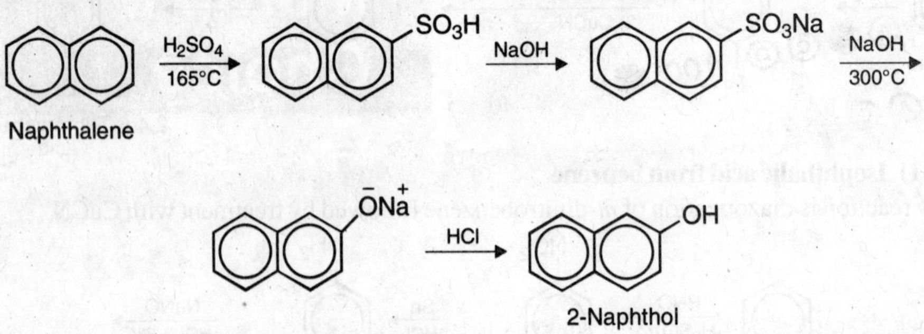

(136) 1,4-Naphthoquinone from naphthalene

Key step is reaction of oxidation of 1-naphthol with acidic $K_2Cr_2O_7$ to form 1,4-naphthoquinone.

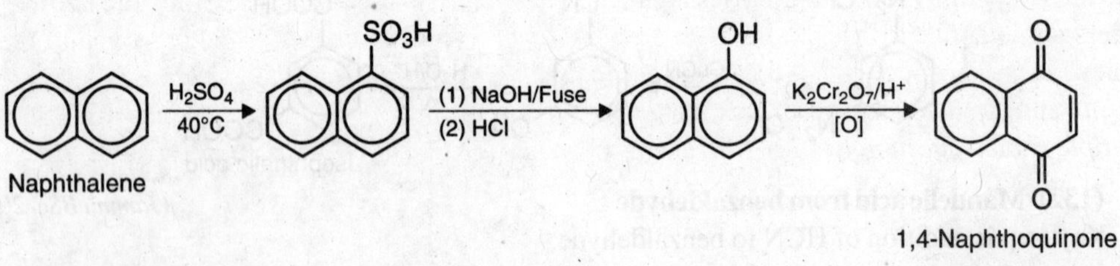

(137) Anthranilic acid from naphthalene (*Baroda BSc, 2015*)

Key step is the treatment of naphthalene with oxygen in the presence of V_2O_5 at 500°C to form phthalic anhydride.

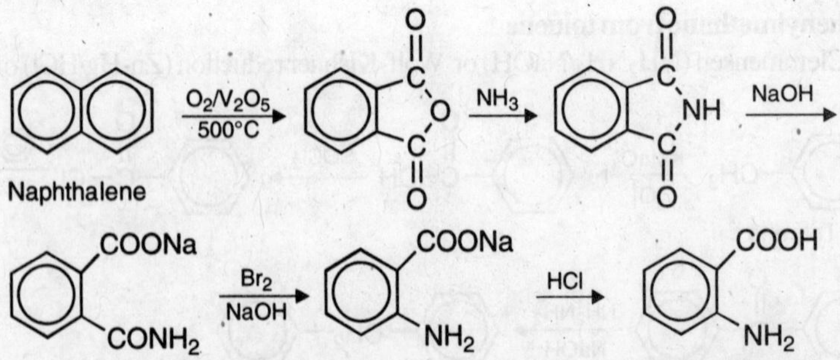

53

500 New Multiple Choice Questions

Each year, more than a million students spend three grueling hours taking the Entrance Tests that engineering colleges and universities use to help them decide which students to admit. More and more students and teachers have became frustrated because they feel that these tests are not entirely fair. Some critics argue that these tests do not actually test any true knowledge. Others argue about whether these tests are predictive of future success or simply a gauge of past achievement. The fact remains that these entrance tests are there ! To do well, you need to know both the *ins* and *outs* of these tests and the material they cover. Most of these entrance exams have *multiple choice questions*. Many BSc exams now also have multiple choice questions.

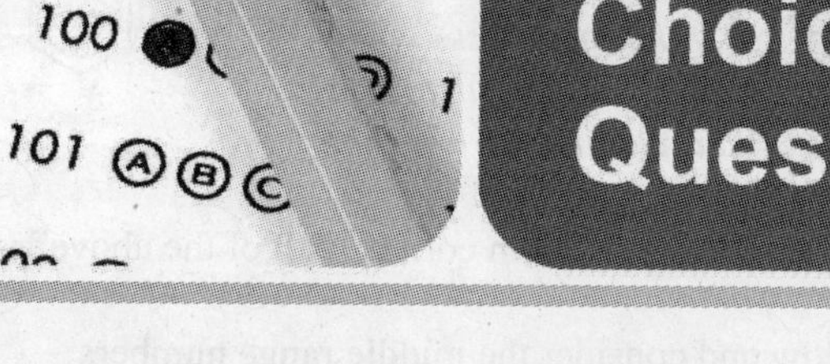

Success on a multiple choice exam
depends equally on How Fast and How Right You are!

STRATEGIES FOR TAKING MULTIPLE CHOICE TESTS

Multiple choice questions usually include a phrase or stem followed by three to five options/choices.

Test strategies :

- Read the directions carefully
- Know if you are penalized for guessing
- Answer easy questions first

Answering options :

- *Improve your odds, think critically*
- *Hide the options, read the stem, and try to answer*
  Select the option that most closely matches your answer
- *Read the stem with each option*
  Treat each option as a true-false question, and choose the "most true"

Strategies to answer difficult questions :

- *Eliminate options you know to be incorrect*
- *Question options that are totally unfamiliar to you*
- *Question options that contain negative or absolute words*
- *"All of the above"*
  If you know two of the three options seem correct, "all of the above" is a strong possibility
- *Number answers :*
  Toss out the high and low and consider the middle range numbers
- *"Look alike options"*
  Probably one is correct; choose the best but eliminate choices that mean basically the same thing, and thus cancel each other out
- *Echo options :*
  If two options are opposite each other, chances are one of them is correct
- *If two alternatives seem correct,*
  compare them for differences,
  then refer to the stem to find your best answer

Guessing :

- *Always guess when there is no penalty*
  for guessing or you can eliminate options
- *Don't guess if you are penalized*
  for guessing and if you have no basis for your choice
- *Don't change your answer*
  unless you are sure of the correction

When taking a test you are demonstrating your ability to understand your course material. Successful test taking avoids carelessness. Stay relaxed and confident. Remind yourself that you are well-prepared and going to do well. If you find yourself anxious, take several slow, deep breaths to relax. Don't talk about the tests to other students just before it; anxiety is contagious.

This chapter includes 500 new multiple choice questions. The purpose is to assess the breadth of student's knowledge and understanding of the basic concepts of organic chemistry. Remember that the multiple choice questions :

1. are not simply definitions.
2. require application of recalled information, not simple recall.
3. require fine distinctions between correct and nearly-correct statements.
4. emphasize conceptual understanding

500 Practice Problems

1. The usual number of bonds found around an uncharged carbon is

(*a*) 1 (*b*) 2 (*c*) 3 (*d*) 4

Answer. (*d*)

2. The normal number of bonds found around an uncharged oxygen in an organic compound is

(*a*) 1 (*b*) 2 (*c*) 3 (*d*) 4

Answer. (*b*)

3. Which is not a pair of homologs ?

(*a*) methylamine and ethylamine
(*b*) propane and butane
(*c*) methylamine and methanol
(*d*) methanol and ethanol

Answer. (*c*)

4. Which does not represent a pair of homologs ?

(*a*) methylamine and ethylamine
(*b*) cyclopropane and cyclobutane
(*c*) pentane and pentene
(*d*) pentane and hexane

Answer. (*c*)

5. What is the empirical formula of a compound that contains 72 g carbon, 18 g hydrogen, and 192 g oxygen ? (The masses of H, C and O are 1, 12 and 16).

(*a*) CH_4O_4 (*b*) CHO
(*c*) CH_3O_2 (*d*) CH_4O_2

Answer. (*c*)

6. What is the total number of bonding electron pairs in ethane ?

(*a*) 7 (*b*) 8 (*c*) 5 (*d*) 6

Answer. (*a*)

7. What is the value of the H–C–H angle in methane ?

(*a*) 180° (*b*) 110° (*c*) 120° (*d*) 109.5°

Answer. (*d*)

8. What orbitals overlap to form the C–C bond in ethane ?

(*a*) two 2*s* orbitals
(*b*) one *s* orbital and three *p* orbitals
(*c*) two 2*p* orbitals
(*d*) two $2sp^3$ orbitals

Answer. (*d*)

9. Which best describes the C–C bonds in $H_3C–CH_2–CH_3$?

(*a*) $sp^2–sp^2$ (*b*) $sp^2–sp^3$
(*c*) $sp^3–sp^3$ (*d*) $sp^3–sp$

Answer. (*c*)

10. What orbitals are used for the formation of the C–H bonds in ethene ?

(*a*) C(*sp*) H(*s*) (*b*) C(sp^2) H(*s*)
(*c*) C(sp^3) H(*s*) (*d*) C(sp^2) H(*p*)

Answer. (*b*)

11. Which hybridization is associated with bond angles of about 120 degrees ?

(*a*) *sp* (*b*) sp^2 (*c*) sp^3 (*d*) sp^4

Answer. (*b*)

12. In ethyne, which of carbon's orbitals overlap to form the sigma bond component of the carbon-carbon triple bond ?

(*a*) two carbon 2*p* orbitals

(b) two carbon 2*s* orbitals
(c) two carbon 2*sp* orbitals
(d) two carbon $2sp^2$ orbitals

Answer. (c)

13. Alkynes contain two π-bonds. Which statement best describes how these π-bonds are formed ?
(a) Each π-bond is formed by side-by-side C(*p*)-C(*p*) overlap.
(b) Each π-bond is formed by end-on C(*sp*)-C(*sp*) overlap.
(c) Each π-bond is formed by end-on C(*p*)-C(*p*) overlap.
(d) Each p-bond is formed by side-by-side C(*sp*)-C(*sp*) overlap.

Answer. (a)

14. How long is a C–C triple bond ?
(a) 1.20 Å (b) 0.1 Å (c) 1.00 Å (d) 0.10 Å

Answer. (a)

15. Which of the following contain *pi* bonds ?
(a) Alkynes (b) Aldehydes
(c) ketones (d) All of these

Answer. (d)

16. Which of the following contains an *sp*-hybridized carbon ?
(a) Propyne, $CH_3-C{\equiv}C-H$
(b) Ethanenitrile, $CH_3-C{\equiv}N$
(c) Carbon dioxide, O=C=O
(d) All of the above

Answer. (d)

17. What is the hybridization of the carbon atom in carbon dioxide ?
(a) *p* (b) *sp* (c) sp^2 (d) sp^3

Answer. (b)

18. In the molecule allene, what is the hybridization of carbon #2 ?

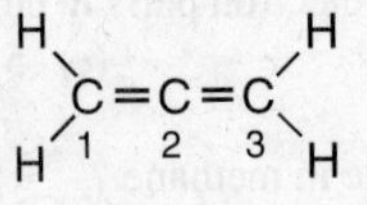

Allene

(a) *sp* (b) sp^2
(c) sp^3 (d) unhybridized

Answer. (a)

19. What kind of orbitals do the nonbonding electrons of the oxygen in an alcohol occupy ?
(a) They occupy *s* orbitals (b) They occupy sp^2 orbitals
(c) They occupy sp orbitals (d) They occupy sp^3 orbitals

Answer. (d)

20. What atomic orbitals overlap to form the sigma bond component of the CN triple bond ?
(a) C, *sp*; N, sp^2 (b) C, sp^2, N, *sp*
(c) C, *sp*; N, *sp* (d) C, sp^2; N sp^2

Answer. (c)

21. Which of the following contains an *sp*-hybridized nitrogen ?
(a) NH_3 (b) $CH_3CH_2NH_2$ (c) CH_3NH_2 (d) HCN

Answer. (d)

22. What is the hybridization of the orbital occupied by the lone pair of electrons in ammonia ?
(a) sp^3 (b) sp^4 (c) *sp* (d) sp^2

Answer. (a)

23. Which of the following statements is generally true ?
(a) The more *s*-character, the longer the bond
(b) The more *p*-character, the shorter the bond
(c) The more *s*-character, the stronger the bond
(d) The more *p*-character, the stronger the bond

Answer. (c)

24. Which is the correct order of C–H bond lengths in ethane, ethene, and ethyne ?
(*a*) CH(ethane) < CH(ethene) < CH(ethyne)
(*b*) CH(ethane) > CH(ethene) > CH(ethyne)
(*c*) CH(ethane) = CH(ethene) = CH(ethyne)
(*d*) CH(ethane) < CH(ethene) = CH(ethyne)

Answer. (*a*)

25. Which of the following is true about resonance forms ?
(*a*) They are not always equivalent.
(*b*) They differ only in the position of electrons.
(*c*) Together they represent a more stable species than any of them individually.
(*d*) All of the above

Answer. (*d*)

26. Which of the following statements is NOT true ?
(*a*) Resonance forms must obey normal valency rules
(*b*) Resonance forms are in equilibrium
(*c*) The hybrid is more stable than the individual resonance forms
(*d*) Resonance forms differ only in the distribution of electrons

Answer. (*b*)

27. The electronegativity of an atom refers to its ability to ....
(*a*) form ionic bonds
(*b*) attract electrons to itself
(*c*) push electrons away from it
(*d*) be a strong base

Answer. (*b*)

28. What is the electronegativity difference between lithium and carbon ?
(*a*) 1.0 (*b*) 1.5 (*c*) 2.0 (*d*) 0.5

Answer. (*b*)

29. Which of the following has the largest electronegativity ?
(*a*) C (*b*) F (*c*) N (*d*) O

Answer. (*b*)

30. Which of the following has a molecular dipole ?
(*a*) H_2S (*b*) CH_2Cl_2 (*c*) FCH_3 (*d*) All of these

Answer. (*d*)

31. Which is the correct order of bond polarity ?
(*a*) C-F > C-O = C-N < C-C
(*b*) C-F > C-O > C-N > C-C
(*c*) C-F < C-O < C-N < C-C
(*d*) C-F = C-O < C-N < C-C

Answer. (*b*)

32. Which of the following molecules has the largest dipole moment ?
(*a*) CF_4 (*b*) CH_4 (*c*) CCl_4 (*d*) H_2O

Answer. (*d*)

33. Among the following molecules, which one has the largest dipole moment ?
(*a*) chloroethane
(*b*) 1,1,2,2-tetrachloroethene
(*c*) 1,1-dichloroethene
(*d*) 1,1,1-trichloroethene

Answer. (*d*)

34. What is the source of van der Waals attractive forces ?
(*a*) charge-induced dipole interactions
(*b*) charge-dipole interactions
(*c*) dipole-dipole interactions
(*d*) induced dipole-induced dipole interactions

Answer. (*d*)

35. Hydrogen bonding of water molecules involves which kind of interaction ?
(*a*) attractive interaction between the O of one water molecule and the O of another
(*b*) bond formation between both H's of one water molecule with the O of another
(*c*) coordination of an H of one water molecule with an H of another
(*d*) coordination of an H of one water molecules with the O of another

Answer. (*d*)

36. Which of the following represents a pair of constitutional isomers ?
(*a*) isopentane and 2-methylbutane
(*b*) neopentane and 2,2-dimethylpropane
(*c*) isohexane and 2-methylpentane
(*d*) neopentane and isopentane

Answer. (*d*)

37. Which pair of compounds below are constitutional isomers ?
(*a*) Cyclohexane and *trans*-3-hexene
(*b*) 1,3-Butadiene and 2-butyne
(*c*) Methylcyclobutane and 1-pentene
(*d*) All of the above

Answer. (*d*)

38. Which of the following is an isomer of octane ?
(*a*) 2,3,4-Trimethyloctane
(*b*) 2,3-Dimethyloctane
(*c*) 2,2,3,3-Tetramethylbutane
(*d*) All of the above

Answer. (*c*)

39. How many structural isomers can be formed by monochlorination of methylcyclohexane ?
(*a*) 5 (*b*) 4 (*c*) 1 (*d*) 2

Answer. (*a*)

40. The prefix *trans* in the name of a cycloalkane indicates
(*a*) that two substituents point toward opposite faces of the ring.
(*b*) that two substituents point toward the same face of the ring.
(*c*) that two substituents are on the same ring carbon.
(*d*) that two substituents are on adjacent ring carbons.

Answer. (*a*)

41. Which of the following can exist as *cis* or *trans* ?
(*a*) Methylcyclobutane
(*b*) 1,1-Dimethylcyclobutane
(*c*) 1,2-Dimethylcyclobutane
(*d*) 1-Butene

Answer. (*c*)

42. Which of the following has *cis* and *trans* isomers ?
(*a*) 2-Methyl-2-butene (*b*) 1-Heptene
(*c*) Cyclobutene (*d*) None of these

Answer. (*d*)

43. Which pair of compounds below are stereoisomers ?
(*a*) 1-pentene and *trans* 2-pentene (*b*) *cis*-2-pentene and cyclopentane
(*c*) *cis*-2-pentene and *trans*-2-pentene (*d*) none of these

Answer. (*c*)

44. Which most correctly describes an optically active molecule ?
(*a*) a molecule that changes the direction of the light beam
(*b*) a molecule that changes the frequency of the light
(*c*) a molecule that rotates the plane of plane-polarized light
(*d*) a molecule that absorbs plane-polarized light

Answer. (*c*)

45. Which is the sequence of devices in a polarimeter ?
(*a*) light source - polarizer - sample - analyzer - viewer
(*b*) light source - analyzer - sample - polarizer - viewer
(*c*) light source - sample - polarizer - analyzer - viewer
(*d*) light source - analyzer - polarizer - sample - viewer

Answer. (*a*)

46. Which equation is used to calculate the specific rotation. (*Alpha* represents the observed rotation)
(*a*) alpha × cell length
(*b*) alpha/(cell length × concentration)
(*c*) (alpha × cell length)/concentration
(*d*) alpha × cell length × concentration

Answer. (*b*)

47. A compound *A* shows an optical rotation of 10 degrees when measured in a cell that is 5 cm long and with a concentration of 1 g/ml. What is the specific rotation of compound *A* ?
(*a*) 20 (*b*) 5 (*c*) 10 (*d*) 2

Answer. (*a*)

48. What is the maximum number of stereoisomers for a compound with *n* chiral centers ?
(*a*) 2^{2n} (*b*) n^2 (*c*) $2n$ (*d*) 2^n

Answer. (*d*)

49. Which of these methanes is chiral ?
(*a*) difluoromethane (*b*) dibromodichloromethane
(*c*) fluorochloromethane (*d*) deuterofluorochloromethane

Answer. (*d*)

50. Which has one chiral center ?
(*a*) 2-chloropentane (*b*) 1-chloropropane
(*c*) 3-chloropentane (*d*) 2-chloropropane

Answer. (*a*)

51. How many chiral centres are there in tartaric acid and how many different stereoisomers exist for this acid ?
(*a*) 2 chiral centers ; 2 isomers
(*b*) 2 chiral centers ; 1 isomer
(*c*) 1 chiral center ; 1 isomer
(*d*) 2 chiral centers ; 3 isomers

Answer. (*d*)

52. How many chiral centers are contained in 3-methylhexane ?
(*a*) one (*b*) two
(*c*) three (*d*) four

Answer. (*a*)

53. How many stereoisomers of 1-bromo-2-chloroheptane can exist ?
(*a*) 2 : a pair of enantiomers
(*b*) 4 : two chiral pairs of configurational isomers
(*c*) 3 : one pair of enantiomers and a meso form
(*d*) 5 : two pairs of enantiomers and meso form

Answer. (*a*)

54. Which of the following is optically active ?
(*a*) A racemic mixture of 2-bromobutane
(*b*) 2-Bromopropane
(*c*) (S)-2-Bromobutane
(*d*) None of the above

Answer. (*c*)

55. How many stereoisomers exist for 2,3-dibromobutane ?
(*a*) four stereoisomers : two pairs of diastereoisomers
(*b*) two stereoisomers : the meso form and its enantiomer

(*c*) only one stereoisomer : the meso form
(*d*) three stereoisomers : the meso form and a pair of enantiomers

Answer. (*d*)

56. The term that best describes the relationship represented *by* the Fischer projections below is

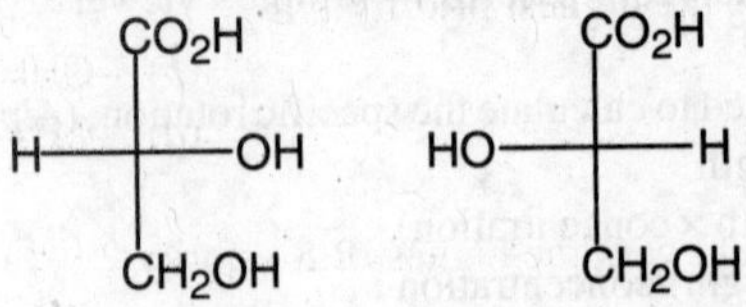

(*a*) same (*b*) enantiomers
(*c*) diastereomers (*d*) constitutional isomers

Answer. (*b*)

57. Which is true about enantiomers ?
(*a*) They have identical boiling points.
(*b*) They have identical melting points.
(*c*) They have identical solubility in achiral solvents.
(*d*) All of the above

Answer. (*d*)

58. What is the relation between an enantiomer's configuration and its specific rotation ?
(*a*) R compounds always are levorotatory.
(*b*) R compounds usually are dextrorotatory with few exceptions.
(*c*) There is no relation.
(*d*) R compounds always are dextrorotatory.

Answer. (*c*)

59. Which is true about diastereomers ?
(*a*) They have identical boiling points.
(*b*) They have identical melting points.
(*c*) They have identical solubility in achiral solvents.
(*d*) None of the above

Answer. (*d*)

60. Which best defines a meso compound ?
(*a*) a molecule with a plane of symmetry and no chiral center
(*b*) a molecule that cannot be superposed on its mirror image even though it has no chiral centers
(*c*) a molecule that can be superposed on its mirror image and that does contain chiral centers
(*d*) a molecule that has no mirror image

Answer. (*c*)

61. Which of the following compounds is a meso compound ?
(*a*) *cis*-1,2-dichlorocyclopropane
(*b*) 1,1-dichlorocyclopropane
(*c*) *cis*-1-chloro-2-bromocyclopropane
(*d*) *trans*-1,2-dichlorocyclopropane

Answer. (*a*)

62. In the Cahn-Ingold-Prelog R,S nomenclature, what does the S represent ?
(*a*) sonorous (*b*) sinister
(*c*) spatial (*d*) solid

Answer. (*b*)

63. What is the correct R,S ranking of the substituents ?
(*a*) hydrogen > methyl > ethyl > bromine
(*b*) ethyl > methyl > hydrogen > bromine
(*c*) bromine > ethyl > methyl > hydrogen
(*d*) bromine > ethyl > hydrogen > methyl

Answer. (*c*)

64. Which substituent has the highest priority ?

(*a*) $-CBr_2-H$ (*b*) $-CH_2-Br$
(*c*) $-CH_2-Cl$ (*d*) $-CH_2-I$

Answer. (*d*)

65. Which substituent has the highest priority ?

(*a*) $-CH_2-OH$ (*b*) $-CH=O$
(*c*) $-CH_2-CH_2-OH$ (*d*) $-CH_2-OCH_3$

Answer. (*b*)

66. Which is the substituent with the highest R,S priority ?

(*a*) $-CH_2-OH$ (*b*) $-CH=S$
(*c*) $-CH=O$ (*d*) $-CH_2-CH_2-SH$

Answer. (*b*)

67. Which substituent has the highest (R,S) priority ?

(*a*) $-CH_2-NH_2$ (*b*) $-CH_2-F$
(*c*) $-CH_2-OH$ (*d*) $-CH_2-CH_3$

Answer. (*b*)

68. Which of the following has the highest CIP priority ?

(*a*) $-CH(CH_3)_2$ (*b*) $-CH_2CH_2CH_3$
(*c*) $-CH_2CH=CH_2$ (*d*) $-CH_2CH_2CH_2CH_3$

Answer. (*a*)

69. Which group below has the highest CIP priority ?

(*a*) $-C\equiv N$ (*b*) $-CH_2NH_2$
(*c*) $-C(CH_3)_3$ (*d*) $-CH_2CH_2CH_2CH_3$

Answer. (*a*)

70. Carbon number one of 1-bromopropane

(*a*) bears a partial negative charge.
(*b*) bears a partial positive charge.
(*c*) acts as a nucleophile.
(*d*) acts as a leaving group.

Answer. (*b*)

71. Which statement best characterizes the polarity of a C–Cl bond ?

(*a*) The bond is ionic and polar with chlorine carrying essentially a full negative charge.
(*b*) The bond is covalent and polar with chlorine carrying a partial negative charge.
(*c*) The bond is covalent and essentially nonpolar.
(*d*) The bond is covalent and polar with chlorine carrying a partial positive charge.

Answer. (*b*)

72. When a bond between two atoms breaks to give a positive ion and a negative ion, this is referred to as.......

(*a*) Homolytic Bond Cleavage
(*b*) Nucleophilic Bond Cleavage
(*c*) Heterolytic Bond Cleavage
(*d*) Equivalent Bond Cleavage

Answer. (*c*)

73. The notation used for representing a reaction mechanism requires that curved arrows

(*a*) indicate the direction of atomic motion
(*b*) show the direction of electron flow (i.e., nucleophile attacking electrophile)
(*c*) show the direction of positive charge flow (i.e. electrophile attacking nucleophile)
(*d*) All of the above

Answer. (*b*)

74. What is the hybridization of the carbon of the methyl cation, CH_3^+ ?

(*a*) C-atom is sp^2-hybridized.

(*b*) C-atom is not hybridized.
(*c*) C-atom is sp^3-hybridized.
(*d*) C-atom is *sp*-hybridized.

Answer. (*a*)

75. Which best describes the geometry and the C-hybridization of a free radical center ?
(*a*) planar, trigonal *sp* C-atom
(*b*) pyramidal, tris-coordinated sp^3 C-atom
(*c*) planar, trigonal sp^2 C-atom
(*d*) pyramidal, *tetra*-coordinated sp^3 C-atom

Answer. (*c*)

76. Which of the following carbanions is the most stable ?
(*a*) a methyl carbanion
(*b*) a *tert*-butyl carbanion
(*c*) an ethyl carbanion
(*d*) a *sec*-butyl carbanion

Answer. (*a*)

77. Which kind of free radical is most stable ?
(*a*) secondary radical
(*b*) primary radical
(*c*) methyl radical
(*d*) tertiary radical

Answer. (*d*)

78. Which correctly describes the order of stabilities of the various types of free radicals ?
(*a*) benzyl > allyl > tert > sec > prim
(*b*) tert > sec > prim > allyl > benzyl
(*c*) allyl > benzyl > tert > sec > prim
(*d*) tert > sec > prim > benzyl > allyl

Answer. (*a*)

79. What reagent is the best electrophile ?
(*a*) bromine, Br_2 (*b*) bromine cation, Br^+
(*c*) bromide, Br^- (*d*) bromine radical, $Br^\cdot$

Answer. (*b*)

80. Which of the following is not true about electromagnetic radiation ?
(*a*) Energy is directly proportional to wavenumber.
(*b*) Frequency is directly proportional to wavelength.
(*c*) Energy is inversely proportional to wavelength.
(*d*) Frequency is directly proportional to wavenumber.

Answer. (*b*)

81. Which of the following represents the correct relative energies ?
(*a*) ultraviolet light > visible light > infrared radiation > X-rays > microwaves
(*b*) X-rays > infrared radiation > ultraviolet light > visible light > microwaves
(*c*) X-rays > ultraviolet light > visible light > infrared radiation > microwaves
(*d*) X-rays > visible light > ultraviolet light > infrared radiation > microwaves

Answer. (*c*)

82. What is the definition of "wavenumber" ?
(*a*) 1/wavelength
(*b*) wavelength/light speed
(*c*) light speed/wavelength
(*d*) wavelength × light speed

Answer. (*a*)

83. Which of the following compounds absorbs light with the longest wavelength ?
(*a*) 1,3,5,7-octatetraene
(*b*) 1,3,5-octatetriene
(*c*) 1,3-butadiene
(*d*) 1,3,5-hextriene

Answer. (*a*)

84. Which compound below will have the longest λ_{max} ?
(*a*) Cyclohexene
(*b*) Benzene
(*c*) Toluene
(*d*) Nitrobenzene

Answer. (*d*)

85. Which compound shows the longest λ_{max} in the UV-vis spectrum ?
(*a*) Toluene
(*b*) Benzoic acid
(*c*) *p*-Vinylbenzoic acid
(*d*) Styrene

Answer. (*c*)

86. What does IR spectroscopy allows us to determine ?
(*a*) the number of carbons in a compound
(*b*) the kinds of bonds in a compound
(*c*) the molecular formula of a compound
(*d*) the carbon-hydrogen framework of a compound

Answer. (*b*)

87. Which of the following solvents is the best to use when taking an IR spectrum ?
(*a*) CCl_4 (*b*) methanol
(*c*) water (*d*) ethanol

Answer. (*a*)

88. Which area of the IR spectrum is called the "functional group region" ?
(*a*) 4,000-3,000 cm^{-1} (*b*) 2,000-400 cm^{-1}
(*c*) 4,000-2,000 cm^{-1} (*d*) 4,000-1,000 cm^{-1}

Answer. (*d*)

89. Which area of the IR spectrum is called the "fingerprint region" ?
(*a*) 2000-1000 cm^{-1} (*b*) 2000-400 cm^{-1}
(*c*) 3000-1000 cm^{-1} (*d*) 1400-600 cm^{-1}

Answer. (*d*)

90. Where in an IR spectrum does a carbon-carbon triple bond shows a stretching vibration ?
(*a*) 3300 cm^{-1} (*b*) 2700 cm^{-1}
(*c*) 2100 cm^{-1} (*d*) 3100 cm^{-1}

Answer. (*c*)

91. An absorption band at 1600 cm^{-1} indicated that a compound contains which of the following groups ?
(*a*) O–H (*b*) C=C
(*c*) C–H (*d*) C=O

Answer. (*b*)

92. Which of the following absorption bands could be used to distinguish between dimethylamine and trimethylamine ?
(*a*) 2950 cm^{-1} (*b*) 1750-1700 cm^{-1}
(*c*) 2100 cm^{-1} (*d*) 3500-3300 cm^{-1}

Answer. (*d*)

93. Where is an IR spectrum would you find an absorption band due to carbon-oxygen double bond ?
(*a*) 3300-3000 cm^{-1} (*b*) 1250-1050 cm^{-1}
(*c*) 1600-1500 cm^{-1} (*d*) 1780-1650 cm^{-1}

Answer. (*d*)

94. A strong peak between 1700 and 1760 cm^{-1} in an infrared spectrum most likely indicates
(*a*) the presence of an alkene.
(*b*) the presence of a saturated compound.
(*c*) the presence of an alkene.
(*d*) the presence of a carbonyl compound.
Answer. (*d*)

95. Which class of compounds below shows a strong IR absorption near 1700 cm^{-1}.
(*a*) Alcohol (*b*) Ether
(*c*) Alkyne (*d*) Aldehyde
Answer. (*d*)

96. Where in the infrared spectrum would you find a strong absorption band representing the stretching frequency of the C=O group of simple aldehydes and ketones ?
(*a*) near 3450 cm^{-1} (*b*) near 1490 cm^{-1}
(*c*) near 1710 cm^{-1} (*d*) near 810 cm^{-1}
Answer. (*c*)

97. What carboxylic acid derivative often shows two strong absorptions in the carbonyl region of the IR ?
(*a*) lactone (*b*) acid halide
(*c*) nitrile (*d*) anhydride
Answer. (*d*)

98. What type of radiation is used in nuclear magnetic resonance spectroscopy ?
(*a*) Visible light (*b*) Ultraviolet light
(*c*) Radio waves (*d*) Microwaves
Answer. (*c*)

99. Which molecule is used as the internal reference compound in proton NMR spectroscopy ?
(*a*) TNS (*b*) DMS
(*c*) TMS (*d*) DMF
Answer. (*c*)

100. In what region of the NMR spectrum would you find the signal for the acidic hydrogen of a carboxylic acid ?
(*a*) 10-12 ppm (*b*) 7.5-10 ppm
(*c*) 2.5-5 ppm (*d*) 0-2.5 ppm
Answer. (*a*)

101. How many signals would you expect to see in the proton NMR spectrum of butane ?
(*a*) 2 (*b*) 3 (*c*) 1 (*d*) 4
Answer. (*a*)

102. Which of the following compounds would have the signal for its methylene hydrogens farthest downfield ?
(*a*) ethyl fluoride (*b*) ethyl iodide
(*c*) ethyl bromide (*d*) ethyl chloride
Answer. (*a*)

103. How many signals will vinyl chloride have in its proton NMR spectrum ?
(*a*) 4 (*b*) 1 (*c*) 3 (*d*) 2
Answer. (*c*)

104. How many signals does 2,2,4-trimethylpentane have in its proton NMR spectrum ?
(*a*) 3 (*b*) 4 (*c*) 2 (*d*) 5
Answer. (*b*)

105. The proton NMR spectrum of methyl propyl ketone has :
(*a*) 4 signals ; a singlet, 2 triplets, and a quartet
(*b*) 3 signals ; a singlet, a triplet, and a quartet
(*c*) 3 signals ; 2 triplets, and a multiplet
(*d*) 4 signals ; a singlet, 2 triplets, and a multiplet
Answer. (*d*)

106. In which order are the various parts of a mass spectrometer found ?

(*a*) sample - ionization chamber - collector - magnet - recorder
(*b*) sample - ionization chamber - magnet - collector - recorder
(*c*) ionization chamber - sample - magnet - collector - recorder
(*d*) ionization chamber - sample - collector - magnet - recorder

Answer. (*b*)

107. Which of the following is required in order to obtain a mass spectrum ?

(*a*) The compound must lose a pair of electrons.
(*b*) The compound must lose an electron.
(*c*) The compound must be water soluble.
(*d*) The compound must gain an electron.

Answer. (*b*)

108. What is plotted on the x and y axes in a mass spectrum ?

(*a*) y-axis = abundance of ion ; x-axis = charge
(*b*) y-axis = charge on ion ; x-axis = mass
(*c*) y-axis = abundance of ion ; x-axis = mass
(*d*) y-axis = abundance of ion ; x-axis = mass/charge

Answer. (*d*)

109. What is a molecular ion ?

(*a*) A compound that has gained a pair of electrons.
(*b*) A compound that has gained an electron.
(*c*) A compound that has lost an electron.
(*d*) A compound that has lost a pair of electrons.

Answer. (*c*)

110. What is the "base peak" in a mass spectrum ?

(*a*) The peak with the greatest intensity.
(*b*) The peak with the highest m/z value.
(*c*) The peak with the lowest m/z value.
(*d*) The peak with the lowest intensity.

Answer. (*a*)

111. What is the mass of the molecular ion generated from ethane ?

(*a*) 30 (*b*) 27 (*c*) 28 (*d*) 15

Answer. (*a*)

112. The spectrum of pentane shows a peak at m/z = 29. What species causes this peak ?

(*a*) ethyl cation (*b*) ethyl radical cation
(*c*) ethyl radical (*d*) none of these

Answer. (*a*)

113. The IUPAC recommendation for naming branches alkanes specifies that

(*a*) the longest continuous chain is the parent.
(*b*) alkyl branches are listed in alphabetical order.
(*c*) numbers are used to specify the location of substituents along the chain.
(*d*) All of the above

Answer. (*d*)

114. The abbreviation "R" is often used in organic chemistry structural formulas to indicate

(*a*) a functional group
(*b*) a reactive site on the molecule
(*c*) a relatively unreactive alkyl group
(*d*) All of the above

Answer. (*c*)

115. How many hydrogen atoms will an alkane molecule with 22 carbons have ?
(*a*) 10 (*b*) 22 (*c*) 44 (*d*) 46

Answer. (*d*)

116. The correct name for an unbranched 10-carbon alkane is
(*a*) octane (*b*) cyclodecane
(*c*) decane (*d*) cyclohexane

Answer. (*c*)

117. What is the IUPAC name of isopentane ?
(*a*) 2,2-dimethylpropane (*b*) 2-methylpentane
(*c*) 1-methylbutane (*d*) 2-methylbutane

Answer. (*d*)

118. What is the IUPAC name of neopentane ?
(*a*) 1-methylbutane (*b*) 1,1-dimethylpropane
(*c*) 2-methylbutane (*d*) 2,2-dimethylpropane

Answer. (*d*)

119. Alkanes are
(*a*) water soluble
(*b*) polar
(*c*) relatively unreactive compared with many other types of organic compounds
(*d*) All of the above

Answer. (*c*)

120. How many of the following five 1-alkenes - propene, butene, pentene, hexene, and heptene - are liquid at room temperature ?
(*a*) 2 (*b*) 3 (*c*) 4 (*d*) 5

Answer. (*b*)

121. Generally, as the amount of branching among alkane isomers increases, the
(*a*) BP increases
(*b*) BP decreases
(*c*) number of tertiary and quaternary carbon decreases
(*d*) number of primary carbons decreases

Answer. (*b*)

122. Which of the following has the highest boiling point ?
(*a*) Hexane (*b*) 2-Methylpentane
(*c*) 2,2-Dimethylbutane (*d*) 2,3-Dimethylbutane

Answer. (*a*)

123. Which one of the following hexanes has the lowest boiling point ?
(*a*) 3-methylpentane
(*b*) hexane
(*c*) 2,2-dimethylbutane
(*d*) 2,3-dimethylbutane

Answer. (*c*)

124. Which is true about Gilman reagents ?
(*a*) They are prepared from Grignard reagents.
(*b*) The are prepared from alkyllithiums.
(*c*) They are used to prepare alkyl halides.
(*d*) All of the above

Answer. (*b*)

125. Which of the following structures is a Gilman reagent ?
(*a*) $(CH_3)_2ZnLi$ (*b*) $(CH_3)_2Zn_2Li$
(*c*) $(CH_3)_2CuLi$ (*d*) $(CH_3)_3CuLi$

Answer. (*c*)

126. Which compound could be most cleanly made by free-radical chlorination of an alkane ?
(*a*) 1-Chloropentane

(*b*) 1-Chloro-3-methylbutane
(*c*) 1-Chloro-2-methylbutane
(*d*) 1-Chloro-2,2-dimethylpropane

Answer. (*d*)

127. The chlorination of methane may yield chloromethane, dichloromethane, trichloromethane, and tetrachloromethane. What measure ensures that chloromethane is the principal product formed ?
(*a*) Use excess chlorine
(*b*) Use high pressure
(*c*) Use low temperature and low pressure
(*d*) Use excess methane

Answer. (*d*)

128. Which occurs when methane is combined with fluorine in the presence of light ?
(*a*) explosive exothermic reaction
(*b*) slow endothermic reaction
(*c*) fast exothermic reaction
(*d*) no reaction

Answer. (*a*)

129. Which of the following is produced by complete combustion of alkanes ?
(*a*) Alcohols
(*b*) Alkyl halides
(*c*) Alkenes
(*d*) None of the above

Answer. (*d*)

130. Which is the balanced equation for the complete combustion of methane ?
(*a*) $CH_4 + 2O_2 \longrightarrow CO_2 + 2H_2O$
(*b*) $4CH_4 + 7O_2 \longrightarrow 2CO + 2CO_2 + 8H_2O$
(*c*) $CH_4 + 3O_2 \longrightarrow CO_2 + 2H_2O_2$
(*d*) $2CH_4 + 3O_2 \longrightarrow 2CO + 4H_2O$

Answer. (*a*)

131. Which sequence correctly describes the steps involved in a free radical chain reaction ?
(*a*) initiation, propagation, termination
(*b*) propagation, initiation, termination
(*c*) initiation, termination, propagation
(*d*) propagation, termination, initiation

Answer. (*a*)

132. Which pair of structures represents *conformational* isomers ?
(*a*) isopentane and neopentane
(*b*) hexane and cyclohexane
(*c*) eclipsed and staggered ethane
(*d*) hexene and cyclohexane

Answer. (*c*)

133. Name the hydrocarbon $CH_3–CH=CH–CH_2CH_2CH_3$ using IUPAC nomenclature.
(*a*) 1-hexene (*b*) 2-hexane
(*c*) 2-pentene (*d*) 2-hexene

Answer. (*b*)

134. What is the hybridization of the C-atom in ethene ?
(*a*) *sp* (*b*) pure *p*
(*c*) sp^3 (*d*) sp^2

Answer. (*d*)

135. How long is the C=C double bond in ethene ?
(*a*) 1.34 Å (*b*) 1.20 Å
(*c*) 1.44 Å (*d*) 1.54 Å

Answer. (*a*)

136. What is the approximate energy of a pi-bond in ethene ?
(*a*) 66 kcal/mol (*b*) 69 kcal/mol
(*c*) 63 kcal/mol (*d*) 60 kcal/mol

Answer. (*c*)

137. Which of the following alkenes is the most stable ?
(*a*) 2,4-dimethyl-2-pentene
(*b*) 2-methyl-2-pentene
(*c*) 3-methyl-2-pentene
(*d*) 2,3-dimethyl-2-pentene

Answer. (*d*)

138. An alkene hydrogenation reaction is normally catalyzed by
(*a*) acid (*b*) hydrogen
(*c*) metal (*d*) solvent

Answer. (*c*)

139. In the free radical addition of HBr to alkenes, peroxide serves as a
(*a*) catalyst (*b*) intermediate
(*c*) initiator (*d*) All of these

Answer. (*c*)

140. Which reagent below is often used to convert an alkene into an alkyl halide ?
(*a*) Thionyl chloride, $SOCl_2$
(*b*) Phosphorus tribromide, PBr_3
(*c*) Cuprous iodide (CuI) in ether
(*d*) HBr

Answer. (*d*)

141. Which reagents/conditions are *not* used for dehydration reactions ?
(*a*) Concentrated sulfuric acid
(*b*) Concentrated phosphoric acid
(*c*) Phosphorus oxychloride
(*d*) KOH/ethanol

Answer. (*d*)

142. The reagent NBS is used for
(*a*) preparing Gilman reagents.
(*b*) preparing Grignard reagents.
(*c*) preparing alkenes out of alkyl halides.
(*d*) bromination of allylic positions.

Answer. (*d*)

143. Which reagent below can be used to convert cyclohexanol into cyclohexene ?
(*a*) PBr_3 (*b*) PCC
(*c*) H_2SO_4 (*d*) $KMnO_4$

Answer. (*c*)

144. Which substrates cannot be reduced with lithium aluminium hydride, LAH ?
(*a*) alkenes (*b*) ketones
(*c*) esters (*d*) aldehydes

Answer. (*a*)

145. What is Markovnikov's rule ?
(*a*) Addition of a proton to an alkene or alkyne will occur at the least-substituted carbon.
(*b*) Addition of a hydrogen radical to an alkene or alkyne will occur at the least-substituted carbon.
(*c*) Addition of a proton to an alkene or alkyne will occur at the most-substituted carbon.
(*d*) In the addition of HX, the X will be added to the least-substituted carbon.

Answer. (*a*)

146. Which one of the following statements about the Markovnikov rule is *not* true ?
(*a*) When HX adds to an asymmetrical alkene, the halogen attaches to the most substituted

alkene C-atom.

(*b*) When HX adds to an asymmetrical alkene, the protonation leads to the most stable carbocation intermediate.

(*c*) In Markovnikov additions, the alkene C-atom with the most H atoms will be protonated.

(*d*) When HX adds to an asymmetrical alkene, the halogen attaches to the least substituted alkene C-atom.

Answer. (*d*)

147. What product is formed by bromination of 2-methyl-1-pentene ?

(*a*) 2-bromo-2-methyl-1-pentene
(*b*) 1,2-dibromo-2-methylpentane
(*c*) 1-bromo-2-methyl-1-pentene
(*d*) 2,2-dibromo-2-methylpentane

Answer. (*b*)

148. Which product is formed in the peroxide-catalyzed hydrobromination of 1-methylcyclohexene ?

(*a*) 4-bromo-1-methylcyclohexane
(*b*) 1-bromo-1-methylcyclohexane
(*c*) 3-bromo-1-methylcyclohexane
(*d*) 2-bromo-1-methylcyclohexane

Answer. (*d*)

149. What is the major product obtained from the acid-catalyzed dehydration of $(CH_3)_2C(OH)\text{-}CH_2CH_3$?

(*a*) $(CH_3)_2C{=}CH{-}CH_3$
(*b*) $(CH_3)_2CH{-}CH{=}CH_2$
(*c*) $CH_3{-}CH{=}CH(CH_3){-}CH_3$
(*d*) $CH_2{=}CH(CH_3){-}CH_2CH_3$

Answer. (*a*)

150. What product is formed in the peroxide-catalyzed hydrobromination of 1-pentene ?

(*a*) $CH_3{-}CH_2{-}CH_2{-}CHBr{-}CH_3$
(*b*) $(CH_3)_2CH{-}CH_2{-}CH_2{-}Br$
(*c*) $CH_3{-}CH_2{-}CH_2{-}CH_2{-}Br$
(*d*) $CH_3{-}CH_2{-}CH_2{-}CH_2{-}CH_2{-}Br$

Answer. (*d*)

151. What product is formed by the reaction of 1-octene with 1. BH_3/THF and 2. HO^-, H_2O_2, H_2O ?

(*a*) $CH_3{-}CH_2{-}CH_2{-}CH_2{-}CH_2{-}CH_2{-}CH(OH){-}CH_3$
(*b*) $CH_3{-}CH_2{-}CH_2{-}CH_2{-}CH(CH_3){-}CH(OH){-}CH_3$
(*c*) $[CH_3{-}CH_2{-}CH_2{-}CH_2{-}CH_2{-}CH_2{-}CH_2{-}CH_2]_3B$
(*d*) $CH_3{-}CH_2{-}CH_2{-}CH_2{-}CH_2{-}CH_2{-}CH_2{-}CH_2{-}OH$

Answer. (*d*)

152. Which alkene below would give different products for HBr addition depending on whether or not peroxide was present ?

(*a*) cyclohexene
(*b*) *trans*-3-hexene
(*c*) 2,3-dimethyl-2-butene
(*d*) None of the above

Answer. (*c*)

153. The addition of HBr to 3-methyl-1-butene leads to several products. One of these products, the major product in fact, is formed *via* a rearranged carbocation. Identify that product.

(*a*) $CH_3{-}CH_2{-}CH_2{-}CH_2{-}CH_2Br$
(*b*) $(CH_3)_2CH{-}CHBr{-}CH_3$
(*c*) $CH_3{-}CH_2{-}CBrH{-}CH_2{-}CH_3$
(*d*) $(CH_3)_2CBr{-}CH_2{-}CH_3$

Answer. (*d*)

154. Which steps below will convert an alkene into an alcohol ?

(*a*) Water and acid catalyst

(*b*) Water and mercury acetate followed by sodium borohydride
(*c*) Borane followed by alkaline hydrogen peroxide
(*d*) All of the above

Answer. (*d*)

155. The first step in the mechanism for electrophilic addition of HBr to an alkene is
(*a*) attachment of the bromide to the most substituted carbon
(*b*) a carbocation rearrangement
(*c*) protonation of an alkene carbon to yield a carbocation
(*d*) None of the above

Answer. (*c*)

156. Which is true about the mechanism for acid-catalyzed alkene hydration ?
(*a*) The catalyst is regenerated in the last step
(*b*) Water is the nucleophile
(*c*) There is a carbocation intermediate
(*d*) All of the above

Answer. (*d*)

157. Which is true about alkynes ?
(*a*) They have at least four atoms on the same axis.
(*b*) They have at least two pi bonds.
(*c*) They normally react with bromine to yield addition products.
(*d*) All of the above

Answer. (*d*)

158. Which statement is true about alkynes ?
(*a*) Terminal alkynes have acidic hydrogens.
(*b*) They can be prepared *via* elimination reactions.
(*c*) Their systematic IUPAC names end in "yne."
(*d*) All of the above

Answer. (*d*)

159. Indicate the correct order for the acidities of the hydrocarbons.
(*a*) ethane < ethene = ethyne
(*b*) ethane < ethene < ethyne
(*c*) ethane > ethene > ethyne
(*d*) ethane > ethene < ethyne

Answer. (*b*)

160. Which of the following is the most acidic ?
(*a*) Ethane (*b*) Ethene
(*c*) Ethyne (*d*) Ammonia

Answer. (*c*)

161. Acetylene is used in torches. Which reaction is occurring in the flame of a torch ?
(*a*) oxidation of acetylene to ethene and water
(*b*) oxidation of acetylene to carbon monoxide and water
(*c*) oxidation of acetylene to carbon (soot) and water
(*d*) oxidation of acetylene to carbon dioxide and water

Answer. (*d*)

162. The hydration of acetylene requires strong acid catalysis. Which catalyst gives the best results ?
(*a*) H_2SO_4
(*b*) $HgSO_4$
(*c*) H_2SO_4 and $MgSO_4$
(*d*) H_2SO_4 and $HgSO_4$

Answer. (*d*)

163. Which reagent reacts with calcium carbide to produce acetylene gas ?
(*a*) O_2 (*b*) H_2O
(*c*) H_2 (*d*) Ca

Answer. (*b*)

164. Which one of the following salts yields brick-red precipitate when it reacts with terminal alkynes ?
(*a*) $Mg(NH_3)_2(NO_3)_2$
(*b*) $Na(NH_3)_2NO_3$
(*c*) $Ag(NH_3)_2NO_3$
(*d*) $Cu(NH_3)_2NO_3$

Answer. (*d*)

165. The reaction of an alkyne with hydrogen gas over palladium metal
(*a*) produces alkanes
(*b*) is a reduction reaction
(*c*) is called hydrogenation
(*d*) All of the above

Answer. (*d*)

166. Which compound below will yield carbon dioxide as one of its cleavage products when treated with ozone and in aqueous acid ?
(*a*) 1-Hexyne
(*b*) 2-Hexyne
(*c*) 3-Hexyne
(*d*) All of the above

Answer. (*a*)

167. Which of the following conditions will not cleave alkynes ?
(*a*) H_2O, $HgSO_4$, H_2SO_4
(*b*) O_3, H_3O^+
(*c*) $KMnO_4$, H_3O^+
(*d*) All of the above

Answer. (*a*)

168. When 1-heptyne is treated with $KMnO_4/H^+$, the resulting product is
(*a*) 1,3-heptadiene
(*b*) $CH_3(CH_2)_4CO_2H + CO_2$
(*c*) heptane
(*d*) $HO(CH_2)_7OH$

Answer. (*b*)

169. When acetylene is treated with excess of hydrogen bromide, the resulting product is
(*a*) 1,1-dibromoethane (*b*) 1-bromoethane
(*c*) 1,2-dibromoethane (*d*) 1,3-dibromoethane

Answer. (*a*)

170. In the presence of excess hydrogen bromide, what is the major product of the addition of HBr to 2-pentyne ?
(*a*) 1-bromopentene (*b*) 1,1-dibromopentane
(*c*) 2-bromopentene (*d*) 2,2-dibromopentane

Answer. (*d*)

171. What alkyne is the best starting material for the synthesis of the acetone $H_3C–CO–CH_3$, *via* acid-catalyzed hydration ?
(*a*) ethyne (*b*) propyne
(*c*) 1-butyne (*d*) 2-butyne

Answer. (*b*)

172. What alkyne is the best starting material for the synthesis of the 2-butanone, $H_3C–CO–CH_2–CH_3$, *via* acid-catalyzed hydration ?
(*a*) ethyne (*b*) 2-butyne
(*c*) 1-butyne (*d*) propyne

Answer. (*b*)

173. What conditions will convert propyne into 2-heptyne ?
(*a*) $NaNH_2$ then CH_3CH_2Cl

(*b*) $NaNH_2$ then $CH_3CH_2CH_2Cl$
(*c*) $NaNH_2$ then $CH_3CH_2CH_2CH_2Cl$
(*d*) $NaNH_2$ then $CH_3CH_2CH_2CH_2CH_2Cl$

Answer. (*c*)

174. What product would be obtained from the reaction of 2,3-dibromobutane with excess $NaNH_2$?
(*a*) 1-butyne
(*b*) 2-butyne
(*c*) 2-bromo-2-butene
(*d*) 1,3-butadiene

Answer. (*b*)

175. Which of the following reaction sequences can be used to convert 2-butene into 2-butyne ?
(*a*) Br_2/CCl_4, excess $NaNH_2$
(*b*) Br_2/CCl_4, excess NaOH
(*c*) Br_2/H_2O, excess NaOH
(*d*) Br_2/H_2O, excess $NaNH_2$

Answer. (*a*)

176. Which of the following dienes below is the most stable ?
(*a*) $H_2C{=}CH{-}CH{=}CH{-}CH_2CH_3$
(*b*) $H_2C{=}CHCH_2CH_2CH{=}CH_2$
(*c*) $H_2C{=}C{=}CHCH_2CH_2CH_3$
(*d*) All of the above are isomers and therefore have the same stability

Answer. (*a*)

177. Which of the following dienes would you expect to be most stable ?

(*a*) 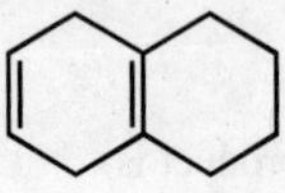
(*b*) 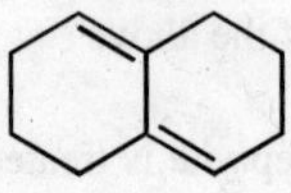

(*c*) 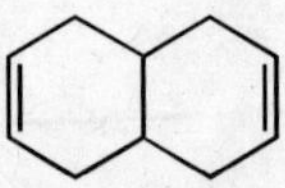
(*d*) 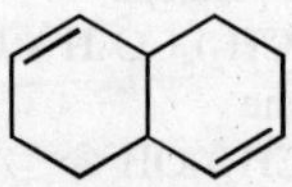

Answer. (*b*)

178. 1,4-Addition of one equivalent of HBr to 2,4-hexadiene yields
(*a*) 1-bromo-2-hexene
(*b*) 1-bromo-3-hexene
(*c*) 4-bromo-2-hexene
(*d*) 2-bromo-3-hexene

Answer. (*d*)

179. Which of the following statements is true about a Diels-Alder reaction ?
(*a*) They are concerted
(*b*) They are stereospecific
(*c*) They are also called [4+2] cycloadditions
(*d*) All of the above

Answer. (*d*)

180. The systematic IUPAC name for *tert*-butyl bromide is
(*a*) tribromobutane
(*b*) 1-bromobutane
(*c*) 1-bromo-1,1-dimethylethane
(*d*) 2-bromo-2-methylpropane

Answer. (*d*)

181. The systematic IUPAC name for chloroform is
(*a*) carbon tetrachloride
(*b*) methyl chloride
(*c*) chloromethane

(*d*) trichloromethane

Answer. (*d*)

182. Which of the following is a method for synthesizing alkyl halides ?

(*a*) HX + alkane
(*b*) HX + alkene
(*c*) Alcohol + X_2
(*d*) All of the above

Answer. (*b*)

183. What is the best set of conditions for the following conversion ?

$$CH_3OH \longrightarrow CH_3Cl$$

(*a*) HCl (*b*) $SOCl_2$
(*c*) Cl_2 and *hv* (*d*) NBS

Answer. (*b*)

184. What reagent best converts an alcohol into the corresponding alkyl bromide ?

(*a*) NaBr, acetic acid (*b*) NaBr, H_2SO_4
(*c*) LiBr (*d*) NaBr

Answer. (*b*)

185. Which reaction below is an elimination ?

(*a*) Hydroboration (*b*) Oxymercuration
(*c*) Hydrogenation (*d*) Dehydrohalogenation

Answer. (*d*)

186. In a nucleophilic substitution on an alkyl halide, the halogen will usually be the

(*a*) nucleophile (*b*) substrate
(*c*) leaving group (*d*) solvent

Answer. (*c*)

187. Which one of the following items does not affect the mechanism of nucleophilic substitution in a significant fashion ?

(*a*) intensity of light present
(*b*) reactivity of the nucleophile
(*c*) structure of the alkyl halide
(*d*) concentration of the nucleophile

Answer. (*a*)

188. Which rate law describes unimolecular nucleophilic substitution reactions ?

(*a*) Rate = *k*[nucleophile]
(*b*) Rate = *k*[alkyl halide]^2
(*c*) Rate = *k*[alkyl halide][nucleophile]
(*d*) Rate = *k*[alkyl halide]

Answer. (*d*)

189. Which one of the following statements describe the order of reaction of the bimolecular nucleophilic substitution ?

(*a*) The S_N2 reaction is second-order in alkyl halide
(*b*) The S_N2 reaction is zero-order in nucleophile but second-order in alkyl halide
(*c*) The S_N2 reaction is first-order in nucleophile and first-order in alkyl halide
(*d*) The S_N2 reaction is second-order in nucleophile

Answer. (*c*)

190. Which statement best describes the mechanism of the S_N2 reaction ?

(*a*) front-side attack with inversion of configuration
(*b*) back-side attack with inversion of configuration
(*c*) back-side attack with retention of configuration
(*d*) front-side attack with retention of configuration

Answer. (*b*)

191. Which rate equation describes an E2 reaction ?

(*a*) Rate = [alkyl halide] [base]

(*b*) Rate = [alkyl halide]^2 [base]
(*c*) Rate = [alkyl halide] [base]^2
(*d*) Rate = *k* [alkyl halide]

Answer. (*a*)

192. Which statement describes the kinetics of an E2 reaction ?
(*a*) First-order in alkyl halide, first-order in base, and first-order overall.
(*b*) Second-order in alkyl halide, and second-order overall.
(*c*) First-order in alkyl halide, and first-order overall.
(*d*) First-order in alkyl halide, first-order in base, and second-order overall.

Answer. (*d*)

193. An E2 reaction of an alkyl halide
(*a*) occurs in two steps.
(*b*) occurs *via* a carbocation.
(*c*) will yield an alkene.
(*d*) All of the above

Answer. (*c*)

194. The first step in the E1 mechanism
(*a*) is solvolysis.
(*b*) is the same as that of the S_N1 mechanism.
(*c*) yields a carbocation.
(*d*) All of the above

Answer. (*d*)

195. Which of the following is the rate equation for an E1 reaction ?
(*a*) Rate = [alkyl halide]^2[base]
(*b*) Rate = *k*[alkyl halide]^2
(*c*) Rate = *k*[alkyl halide]
(*d*) Rate = [alkyl halide][base]

Answer. (*c*)

196. Which metal is employed in the preparation of a "Grignard" reagent ?
(*a*) Mg (*b*) Zn (*c*) Li (*d*) Cu

Answer. (*a*)

197. Which of the following solvents can be used for the preparation of a Grignard reagent ?
(*a*) formic acid (*b*) water
(*c*) methanol (*d*) diethyl ether

Answer. (*d*)

198. Which of the following best describes the carbon-lithium bond ?
(*a*) The carbon-lithium bond is covalent and somewhat polar.
(*b*) The carbon-lithium bond is almost ionic with carbon negative and lithium positive.
(*c*) The carbon-lithium bond is covalent and nonpolar.
(*d*) The carbon-lithium bond is almost ionic with carbon positive and lithium negative.

Answer. (*b*)

199. If CH_3CH_2Li were treated with water, the result would be
(*a*) $CH_3CH_2OH + LiH$
(*b*) $CH_3CH_3 + LiOH$
(*c*) $CH_3CH_2CH_2CH_3 + Li$
(*d*) No reaction

Answer. (*b*)

200. In the reaction of ethyllithium with water, what is the fate of the lithium ?
(*a*) Li (*b*) LiOEt
(*c*) LiOH (*d*) Li_2O

Answer. (*c*)

201. Which of the following reagents would you expect to yield a normal addition product in a reaction with 2-cyclohexenone ?
(*a*) $[(CH_3CH_2)_2Cu]Li$

(*b*) $(CH_3CH_2)_2NH$
(*c*) CH_3CH_2MgBr
(*d*) All of the above

Answer. (*c*)

202. What intermediate is formed in the reaction of a nitrile with a Grignard reagent to form a ketone ?
(*a*) the magnesium salt of a carboxylic acid
(*b*) an amide
(*c*) the magnesium salt of an imine
(*d*) an amine

Answer. (*c*)

203. What type of alcohol is generated in the reaction between formaldehyde and a Grignard reagent ?
(*a*) no alcohol is generated
(*b*) a secondary alcohol
(*c*) a primary alcohol
(*d*) a tertiary alcohol

Answer. (*c*)

204. Which of the following could be prepared from a Grignard reagent and an ester ?
(*a*) 1-Butanol
(*b*) 2-Butanol
(*c*) *tert*-Butyl alcohol
(*d*) All of the above

Answer. (*c*)

205. Which of the following could be prepared from a Grignard reagent and an epoxide ?
(*a*) 1-Butanol
(*b*) 2-Butanol
(*c*) *tert*-Butyl alcohol
(*d*) All of the above

Answer. (*a*)

206. What product is formed in the reaction of methylmagnesium bromide with ethylene oxide ?
(*a*) $CH_3–CH_2–CH_2–OH$
(*b*) $CH_3–CH(OH)–CH_2–CH_3$
(*c*) $(CH_3)_3C–CH_2–OH$
(*d*) $(CH_3)_2CH–CH_2–OH$

Answer. (*a*)

207. Which reagents below would convert 1-bromocyclopentene to 1-cyclopentenecarboxylic acid ?
(*a*) $LiAlH_4$, then H_3O^+
(*b*) NaCN, then H_3O^+
(*c*) Mg, ether, then CO_2, then H_3O^+
(*d*) $KMnO_4$, H_3O^+

Answer. (*c*)

208. Which Grignard reagent is best suited for the synthesis of 3-methyl-1-butanol, $(CH_3)_2CH–CH_2–CH_2–OH$, from an aldehyde or a ketone ?
(*a*) $CH_3–CH_2–CH_2–MgBr$
(*b*) $(CH_3)_2CH–CH_2–CH_2MgBr$
(*c*) $(CH_3)_2CH–CH_2–MgBr$
(*d*) $(CH_3)_2CH–MgBr$

Answer. (*c*)

209. Which product results from the reaction of propanoyl chloride with two equivalents of CH_3CH_2MgBr followed by aqueous acid ?
(*a*) 2-Pentanone
(*b*) 3-Ethyl-3-pentanol
(*c*) Propanoic acid + acetic acid
(*d*) Acetic anhydride

Answer. (*b*)

210. What is the product of the reaction of ethyl formate with CH_3Li, then H_3O^+ ?
(*a*) ethane (*b*) ethyl acetate
(*c*) acetaldehyde (*d*) isopropyl alcohol

Answer. (*d*)

211. Identify the primary alcohol among the following alcohols.
(*a*) 2-methyl-2-propanol
(*b*) benzyl alcohol
(*c*) cyclohexanol
(*d*) phenol

Answer. (*b*)

212. Which structure is that of allyl alcohol ?
(*a*) CH_2=CH–CH_2–OH
(*b*) CH_3–CH_2–CH=O
(*c*) CH_2=C=CH–OH
(*d*) CH_3–CH=CH–OH

Answer. (*a*)

213. Identify the secondary alcohol among the following :
(*a*) cyclohexanol (*b*) benzyl alcohol
(*c*) phenol (*d*) hydroquinone

Answer. (*a*)

214. Which is a *tertiary* alcohol with the formula C_4H_7OH ?
(*a*) 2-methylcyclopropanol
(*b*) 1-methylcyclopropanol
(*c*) cyclopropanol
(*d*) cyclobutanol

Answer. (*b*)

215. What types of orbitals are occupied by the oxygen lone pairs in the alcohol group ?
(*a*) *sp*-hybrids (*b*) sp^3-hybrids
(*c*) sp^2-hybrids (*d*) pure *p*-orbitals

Answer. (*d*)

216. What type of orbitals are involved in the C–O bond in alcohols ?
(*a*) (C)sp^3–(O)sp^3 (*b*) (C)sp^2–(O)sp^2
(*c*) (C)*sp*–(O)sp^2 (*d*) (C)sp^2–(O)*sp*

Answer. (*a*)

217. What are the main components of "synthesis gas" ?
(*a*) mixture of CO and H_2 (*b*) mixture of CO_2 and H_2
(*c*) mixture of CO and H_2O (*d*) mixture of CO_2 and H_2O

Answer. (*a*)

218. Which of the following plants cannot be used to produce ethanol ?
(*a*) wood (*b*) grapes (*c*) barley (*d*) corn

Answer. (*a*)

219. What quantity of ethanol is fatal for an adult ?
(*a*) 100 ml (*b*) 200 ml
(*c*) 300 ml (*d*) 50 ml

Answer. (*c*)

220. Which alcohol is the most acidic ?
(*a*) 1,1-dichloroethanol (*b*) methanol
(*c*) ethanol (*d*) 1-chloroethanol

Answer. (*a*)

221. Which alcohol has the lowest solubility in water ?
(*a*) propanol (*b*) butanol
(*c*) ethanol (*d*) pentanol

Answer. (*d*)

222. Which is the boiling point of ethanol ?
(*a*) 98 °C (*b*) 88°C (*c*) 78°C (*d*) 68°C

Answer. (*c*)

223. Which of the following applies to secondary alcohols ?
(*a*) They are more acidic than phenols
(*b*) They react with $NaBH_4$ and then aqueous acid to yield aldehydes or ketones
(*c*) Their boiling points are higher than those of similar alkanes or alkyl halides
(*d*) All of the above

Answer. (*c*)

224. Which of the following molecules has the highest boiling point ?
(*a*) methane (*b*) methyl ethyl ether
(*c*) butane (*d*) butanol

Answer. (*d*)

225. Which of the following has the highest boiling point ?
(*a*) 1-Propanol
(*b*) Isopropyl alcohol
(*c*) Glycerol (1,2,3-propanetriol)
(*d*) Ethanol

Answer. (*c*)

226. Which combination of chemicals is known as the "Lucas reagent" ?
(*a*) NaCl, $SnCl_2$ (*b*) HCl, $ZnCl_2$
(*c*) HCl, $CuCl_2$ (*d*) HCl, $SnCl_2$

Answer. (*b*)

227. How quickly would a separate layer form if *tert*-butanol was submitted to the Lucas test ?
(*a*) about 2 minutes
(*b*) less than 1 minute
(*c*) about 1 hour
(*d*) about 5 minutes

Answer. (*b*)

228. The reaction of alcohols with hydrogen halides can be accelerated with $ZnCl_2$. Why ?
(*a*) $ZnCl_2$ acts as a Bronsted base
(*b*) $ZnCl_2$ acts as a Bronsted acid
(*c*) $ZnCl_2$ acts as a Lewis acid
(*d*) $ZnCl_2$ acts as a Lewis base

Answer. (*c*)

229. Aside from conc. sulfuric acid, what other strong acid is employed in the dehydration of alcohols ?
(*a*) conc. acetic acid (*b*) conc. phosphoric acid
(*c*) conc. acetone (*d*) conc. chromic acid

Answer. (*b*)

230. Why is the oxidation of a primary alcohol with a mixture of sodium dichromate and sulfuric acid not a good method for the preparation of the corresponding aldehyde ?
(*a*) The product will be the corresponding alkane.
(*b*) A mixture of sodium dichromate and sulfuric acid will not oxidize a primary alcohol.
(*c*) Any aldehyde produced will be oxidized further.
(*d*) The product will be the corresponding ketone.

Answer. (*c*)

231. Which one of the following alcohols *cannot* be oxidized to a carbonyl compound ?
(*a*) 2-pentanol
(*b*) 2-butanol
(*c*) 3-methyl-3-pentanol
(*d*) pentanol

Answer. (*c*)

232. What product is formed by the reaction of 3-methyl-1-butene with 1. BH_3/THF and 2. HO^-, H_2O_2, H_2O ?
(*a*) $(CH_3)_2CH{-}CH_2{-}CH_2{-}OH$
(*b*) $[(CH_3)_2CH{-}CH_2{-}CH_2]_3B$
(*c*) $CH_3{-}CH_2{-}CH_2{-}CH_2{-}CH_2{-}OH$
(*d*) $(CH_3)_2CH{-}C(OH)H{-}CH_3$

Answer. (*a*)

233. What is the major product obtained from the acid-catalyzed dehydration of 1-methylcyclohexanol ?
(*a*) methylcyclohexane (*b*) 2-methylcyclohexene
(*c*) 3-methylcyclohexene (*d*) 1-methylcyclohexene

Answer. (*d*)

234. Treating a ketone with $NaBH_4$ or $LiAlH_4$ followed by H_3O^+ will usually produce
(*a*) a phenol (*b*) a primary alcohol
(*c*) a secondary alcohol (*d*) a tertiary alcohol

Answer. (*c*)

235. Which conditions below would convert 1-butene into $CH_3CH_2CH_2CH_2OH$?
(*a*) H_2O/H_3O^+
(*b*) $Hg(O_2CCH_3)_2$ followed by $NaBH_4$
(*c*) BH_3 followed by H_2O_2/OH^-
(*d*) All of the above

Answer. (*c*)

236. Which of the following propane derivatives has the highest boiling point ?
(*a*) 2-propanol (*b*) glycerol
(*c*) propylene glycol (*d*) 1-propanol

Answer. (*b*)

237. Treatment of a 1,2 diol (vicinal diol) with periodic acid results in
(*a*) cleavage to alkenes and alkanes
(*b*) cleavage to carbon dioxide and water
(*c*) cleavage to carboxylic acids and alcohols
(*d*) none of the above

Answer. (*d*)

238. What is nitroglycerine ? Glycerine is 1,2,3-propanetriol.
(*a*) the mononitrate of glycerine, ester at C2
(*b*) the dinitrate of glycerine, esters at C1 and C3
(*c*) the mononitrate of glycerine, ester at C1
(*d*) the trinitrate of glycerine

Answer. (*d*)

239. What is the key step of the generic pinacol rearrangement ?
(*a*) methyl shift in a carbocation
(*b*) hydrogen shift in a carbocation
(*c*) methyl shift in a carbanion
(*d*) hydrogen shift in a carbanion

Answer. (*a*)

240. Another name for a thiol is a
(*a*) hemiacetal (*b*) mercaptan
(*c*) disulfide (*d*) ketone

Answer. (*b*)

241. Which of the following applies to thiols ?
(*a*) They can be made from alkyl halides.
(*b*) They often stink.
(*c*) They can be interconverted with disulfides.
(*d*) All of the above.

Answer. (*d*)

242. Reaction of isopropyl mercaptan with I_2 will most likely yield
(*a*) isopropyl alcohol
(*b*) 2-iodopropane
(*c*) a ketone
(*d*) $(CH_3)_2CHSSCH(CH_3)_2$

Answer. (*d*)

243. The structure R–S–R', where R and R' are alkyl groups is a
(*a*) thiol (*b*) mercaptan
(*c*) sulfide (*d*) disulfide

Answer. (*c*)

244. Which reagent (s) below will convert thiols into disulfides ?
(*a*) Thiourea followed by aqueous alkali
(*b*) Zinc metal in aqueous acid
(*c*) Molecular bromine, Br_2
(*d*) All of the above

Answer. (*c*)

245. Which of the following reactions is used to prepare a thioether ?
(*a*) Reaction of a sulfonate with an alkyl bromide
(*b*) Reaction of a sulfite with an alkyl bromide
(*c*) Reaction of a sulfate with an alkyl bromide
(*d*) Reaction of a sulfide with an alkyl bromide

Answer. (*d*)

246. What is the total number of nonbonding electron pairs in dimethyl ether ?
(*a*) 2 (*b*) 1 (*c*) 3 (*d*) 0

Answer. (*a*)

247. Which of the following has the highest boiling point ?
(*a*) $HOCH_2CH_2CH_2OH$
(*b*) $CH_3OCH_2OCH_3$
(*c*) $CH_3OCH_2CH_2OH$
(*d*) $HOCH_2CH_2CH_3$

Answer. (*a*)

248. Which of the following would have the lowest boiling point ?
(*a*) $HOCH_2CH_2CH_2CH_2OH$
(*b*) $HOCH_2CH_2CH_2OCH_3$
(*c*) $HOCH_2CH_2OCH_2CH_3$
(*d*) $CH_3OCH_2CH_2OCH_3$

Answer. (*d*)

249. Consider the Williamson ether synthesis. Which statement does not apply ?
(*a*) The R group of the alkoxide must be primary.
(*b*) Alkyl iodides and alkyl bromides work well.
(*c*) Alkoxide ion reacts with an alkyl halide in a nucleophilic substitution.
(*d*) The R group of the alkyl halide must be primary

Answer. (*a*)

250. Which of the following ethers would be obtained in greatest yield directly from alcohols ?
(*a*) ethyl methyl ether
(*b*) ethyl propyl ether
(*c*) methyl propyl ether
(*d*) dipropyl ether

Answer. (*d*)

251. What reagents should be used to synthesize methyl isopropyl ether ?
(*a*) isopropyl chloride and methyl chloride
(*b*) methyl chloride and methoxide ion

(*c*) methyl chloride and isopropoxide ion
(*d*) methyl alcohol and isopropoxide ion

Answer. (*c*)

252. Which of the following reagents should be used to prepare *tert*-butyl ethyl ether *via* a Williamson ether synthesis ?
(*a*) ethyl alcohol and *tert*-butoxide ion
(*b*) ethyl bromide and *tert*-butoxide ion
(*c*) *tert*-butyl alcohol and ethoxide ion
(*d*) *tert*-butyl iodide and ethyl bromide

Answer. (*b*)

253. Which ether below could be prepared cleanly by alcohol dehydration ?
(*a*) 1-Methoxybutane
(*b*) 2-Methoxybutane
(*c*) Dibutyl ether
(*d*) All of the above

Answer. (*c*)

254. Cleavage of the cyclic ether tetrahydrofuran with HI will likely produce
(*a*) ethyl iodide and ethanol
(*b*) 4-iodo-1-butanol
(*c*) 1,4-diiodobutane
(*d*) 1,4-butanediol

Answer. (*b*)

255. What is a crown ether ?
(*a*) a cyclic compound with several ether linkages
(*b*) an acylic symmetrical ether
(*c*) an ether that incorporates a crown
(*d*) a royal ether

Answer. (*a*)

256. What is the product of the acid-catalyzed opening of an epoxide ?
(*a*) 1,2-endiol (*b*) 1,2-diol
(*c*) 1,3-diol (*d*) 1,1-diol

Answer. (*b*)

257. In what way do aldehydes and ketones differ from each other ?
(*a*) Ketones contain the carbonyl group, C=O ; aldehydes contain the hydroxyl group, C–OH.
(*b*) In aldehydes, the carbonyl group must be bonded to at least one hydrogen ; in ketones the carbonyl group must be bonded to two carbons.
(*c*) Aldehydes contain the elements C and H ; ketones contain C, H, and O.
(*d*) The carbonyl carbon is sp^2 hybridized in aldehydes but is sp^3 hybridized in ketones.

Answer. (*b*)

258. What is the hybridization of the carbon of a carbonyl group ?
(*a*) sp^3 (*b*) sp^2 (*c*) sp (*d*) s^3p^2

Answer. (*b*)

259. Which is the true statement ?
(*a*) The double bond of a carbonyl group is longer and stronger than the double bond of an alkene.
(*b*) The double bond of a carbonyl group is shorter but weaker than the double bond of an alkene.
(*c*) The double bond of a carbonyl group is longer but weaker than the double bond of an alkene.
(*d*) The double bond of a typical carbonyl group is shorter and stronger than the double bond of a typical alkene.

Answer. (*d*)

260. Which one of the following has the highest boiling point ?
(*a*) acetone (*b*) methyl ethyl ketone
(*c*) acetaldehyde (*d*) propionaldehyde

Answer. (*b*)

261. In cyanohydrin formation, the hybridization of the carbonyl carbon :
(*a*) changes from sp^3 to sp^2
(*b*) changes from sp^2 to sp^3
(*c*) changes from sp^2 to sp
(*d*) changes from sp^3 to sp
Answer. (*b*)

262. Which of the following would react fastest with a given nucleophile ?
(*a*) Methanal (formaldehyde)
(*b*) Ethanal (acetaldehyde)
(*c*) Propanone (acetone)
(*d*) 2-Butanone
Answer. (*a*)

263. Which compound below reacts with Tollens' reagent ?
(*a*) 1-Pentanol (*b*) Pentanal
(*c*) Pentanoic acid (*d*) None of the these
Answer. (*b*)

264. Which functional group below does Tollens' reagent selectively oxidize ?
(*a*) Carboxylic acid (*b*) Primary alcohol
(*c*) Aldehyde (*d*) None of the above
Answer. (*c*)

265. What transformation does Tollens' reagent cause ?
(*a*) conversion of aldehydes to ketones
(*b*) reduction of carboxylic acids
(*c*) oxidation of ketones
(*d*) oxidation of aldehydes
Answer. (*d*)

266. Which of the following is a Wittig reagent ?
(*a*) Ph_3P
(*b*) $[Ph_3PCH_3]^+I^-$
(*c*) $Ph_3P{=}CH_2$
(*d*) None of these
Answer. (*c*)

267. What is the Wittig reaction most useful for ?
(*a*) synthesis of ketones
(*b*) preparation of aldehydes
(*c*) synthesis of alkenes
(*d*) preparation of alkylbenzenes
Answer. (*c*)

268. For which of the following oxidation reactions does one need to use pyridinium chlorochromate?
(*a*) acetaldehyde to acetic acid
(*b*) ethanol to acetaldehyde
(*c*) methanol to formic acid
(*d*) ethanol to acetic acid
Answer. (*b*)

269. What derivative of aldehydes and ketones is known as a *Schiff base* ?
(*a*) acetal (*b*) imine
(*c*) oxime (*d*) hydrazone
Answer. (*b*)

270. Which reaction below will yield an aldehyde ?
(*a*) Ozonolysis of 2,3-dimethyl-2-butene
(*b*) Treating 1-propanol with PCC
(*c*) Treating 1-propanol with Jones' reagent
(*d*) All of the above
Answer. (*b*)

271. Which one *cannot* be used to produce an aldehyde or ketone ?
(*a*) hydroboration-oxidation of 1-butyne
(*b*) Friedel-Crafts acylation of nitrobenzene
(*c*) oxidation of 2-butanol with sodium dichromate in sulfuric acid
(*d*) hydration of 2-butyne with Hg^{2+}, H_2SO_4, H_2O

Answer. (*b*)

272. What reacts with an aldehyde or ketone to produce an oxime ?
(*a*) an amine (*b*) hydrazine
(*c*) hydroxylamine (*d*) semicarbazide

Answer. (*c*)

273. Which is correct concerning an Aldol addition reaction ?
(*a*) The product is a beta-hydroxy aldehyde or ketone
(*b*) The product will dehydrate at higher temperature if possible
(*c*) At least one reactant had alpha hydrogen
(*d*) All of the above

Answer. (*d*)

274. What is the product of the Aldol condensation ?
(*a*) an alpha, beta-unsaturated ester
(*b*) a beta-hydroxy aldehyde or ketone
(*c*) a beta-hydroxy amide
(*d*) an alpha-hydroxy anhydride

Answer. (*b*)

275. What is the total number of Aldol products produced on all combinations of self-condensation and crossed-condensation of a mixture of propanal and butanal ?
(*a*) 4 (*b*) 2 (*c*) 5 (*d*) 3

Answer. (*a*)

276. What step occurs in an Aldol condensation ?
(*a*) alkylation of an enolate with an alkyl bromide
(*b*) attack by a hydroxide anion on a carbonyl carbon
(*c*) a nucleophilic attack on a C=C double bond by a carbonyl carbon
(*d*) nucleophilic attack by an enol or enolate carbon on a carbonyl carbon

Answer. (*d*)

277. The product of a Claisen condensation between two esters is usually
(*a*) an alpha-hydroxy ester
(*b*) a beta-hydroxy ester
(*c*) an alpha-keto ester
(*d*) a beta-keto ester

Answer. (*d*)

278. What is the defining functional group of a carboxylic acid ?
(*a*) an –OH attached to a carbonyl group
(*b*) an –OH group attached to an aromatic ring
(*c*) an –OH group attached to a C=C double bond
(*d*) an –H attached to a carbonyl group

Answer. (*a*)

279. An unknown, crystalline substance is probably a carboxylic acid if it :
(*a*) is insoluble in both dilute HCl and dilute $NaHCO_3$, but dissolves in benzene with evolution of bubbles of carbon dioxide.
(*b*) is insoluble in water but dissolves in sodium bicarbonate solution with evolution of bubbles of carbon dioxide.
(*c*) dissolves in dilute HCl and crystallizes out on addition of dilute NaOH.
(*d*) dissolves in water and crystallizes out on addition of ethyl alcohol.

Answer. (*b*)

280. What is the major carboxylic acid found in vinegar ?
(*a*) formic acid (*b*) acetic acid
(*c*) propionic acid (*d*) malonic acid

Answer. (*b*)

281. What is the common name of the 1-carbon carboxylic acid that bears the IUPAC name methanoic acid ?
(*a*) formic acid (*b*) acetic acid
(*c*) methanic acid (*d*) monaic acid

Answer. (*a*)

282. Consider the molecule $CH_3–CH(OH)–CH_2–COOH$. Which name best describes this molecule ?
(*a*) 3-hydroxybutanoic acid
(*b*) 3-hydroxybutane carboxylic acid
(*c*) 2-hydroxybutane carboxylic acid
(*d*) 2-hydroxybutanoic acid

Answer. (*a*)

283. Which of the following compounds is peroxyformic acid ?
(*a*) H_3CCO_3H (*b*) HCO_3H
(*c*) HCO_2H (*d*) $PhCO_3H$

Answer. (*b*)

284. Which compound below has the highest boiling point ?
(*a*) 1-Hexanol
(*b*) Dipropyl ether
(*c*) Hexanoic acid
(*d*) Hexanedioic acid

Answer. (*d*)

285. Which acid below is the strongest ?
(*a*) Butanoic acid
(*b*) 2-Chlorobutanoic acid
(*c*) 3-Chlorobutanoic acid
(*d*) 4-Chlorobutanoic acid

Answer. (*b*)

286. Which of the following compounds is the weakest acid ?
(*a*) acetic acid, CH_3CO_2H (*b*) methanol, CH_3OH
(*c*) formic acid, HCO_2H (*d*) ammonium ion, NH_4^+

Answer. (*b*)

287. Which acid exhibits the lowest value for its pKa ?
(*a*) acetic acid (*b*) 2-chlorobutanoic acid
(*c*) 3-chlorobutanoic acid (*d*) trichloroacetic acid

Answer. (*d*)

288. What is the order of acidity (most acidic to least acidic) ?

CH_3CO_2H	$CH_3OCH_2CO_2H$	CH_3CH_2OH
I	II	III

(*a*) I, II, III
(*b*) III, II, I
(*c*) III, I, II
(*d*) II, I, III

Answer. (*d*)

289. Which reaction below will yield acetic acid ?
(*a*) Treating 2-butene with acidic aqueous potassium permanganate.
(*b*) Treating 2-butyne with ozone and aqueous acid.
(*c*) Treating ethanal with Tollens' reagent then aqueous acid.
(*d*) All of the above.

Answer. (*d*)

290. What will *not* yield acetic acid ?
(*a*) CH_3–Mg–Br and CO_2
(*b*) ethane, $Na_2Cr_2O_7$, H_2SO_4, heat
(*c*) acetonitrile, HCl, and water
(*d*) 2-butene and hot conc. $KMnO_4$

Answer. (*b*)

291. Which compound below will not undergo hydrolysis to yield acetic acid ?
(*a*) Acetonitrile
(*b*) Ethanoic anhydride
(*c*) Acetaldehyde
(*d*) Acetyl chloride

Answer. (*c*)

292. Which reaction yields butanoic acid ?
(*a*) Ozonolysis of 4-octyne
(*b*) Heat $CH_3CH_2CH_2CN$ with aqueous acid
(*c*) Add $CH_3CH_2CH_2MgBr$ to CO_2, then add aqueous acid
(*d*) All of the above

Answer. (*d*)

293. Which reaction is possible for pentanoic acid ?
(*a*) Reaction with base to yield pentanoate salt
(*b*) Alpha substitution to yield pentanoic acid with a substituent on carbon number 2
(*c*) Reduction to 1-pentanol
(*d*) All of the above

Answer. (*d*)

294. The name of the organic product that results from reaction of propanoic acid with potassium hydroxide is
(*a*) 2-hydroxypropanoic acid
(*b*) potassium propanide
(*c*) propanoic potassiate
(*d*) potassium propanoate

Answer. (*d*)

295. Which carboxylic acids are formed by ozonolysis of 2-pentyne ?
(*a*) one molecule of acetic acid and one molecule of formic acid
(*b*) two molecules of acetic acid
(*c*) one molecule of acetic acid and one molecule of propionic acid
(*d*) two molecules of propionic acid

Answer. (*c*)

296. What product is isolated from the reaction of a carboxylic acid with $LiAlH_4$?
(*a*) the lithium salt of the acid
(*b*) a secondary alcohol
(*c*) an aldehyde
(*d*) a primary alcohol

Answer. (*d*)

297. The HVZ reaction
(*a*) works only on methyl ketones.
(*b*) involves an enamine intermediate.
(*c*) produces an *alpha*-bromocarboxylic acid.
(*d*) All of the above

Answer. (*c*)

298. Which of the following is not classified as a derivative of a carboxylic acid ?
(*a*) ester
(*b*) anhydride
(*c*) amide
(*d*) aldehyde

Answer. (*d*)

299. What is the common name of the acid derivative known by the IUPAC name methyl methanoate ?
(*a*) ethyl acetate (*b*) ethyl formate
(*c*) methyl acetate (*d*) methyl formate

Answer. (*d*)

300. What is the common name of the carboxylic acid that bears the IUPAC name propanedioic acid ?
(*a*) succinic acid (*b*) phthalic acid
(*c*) malonic acid (*d*) oxalic acid

Answer. (*c*)

301. What acid derivative is often responsible for the fragrance or flavor of fruits and plants ?
(*a*) anhydride (*b*) acid chloride
(*c*) nitrile (*d*) ester

Answer. (*d*)

302. In a nucleophilic acyl substitution reaction
(*a*) the nucleophile attacks the carbonyl as the leaving group departs simultaneously.
(*b*) the nucleophile attacks the carbonyl and then the leaving group departs.
(*c*) the nucleophile attacks the carbonyl after the leaving group departs.
(*d*) the nucleophile attacks the leaving group, causing it to depart.

Answer. (*b*)

303. Which compound below will react fastest in a nucleophilic acyl substitution with water ?
(*a*) Acetyl chloride (*b*) Acetic anhydride
(*c*) Ethyl acetate (*d*) Acetamide

Answer. (*a*)

304. Why are acid halides particularly effective reagents for the synthesis of amides, esters, and other derivatives of carboxylic acids ?
(*a*) The halide anion is an excellent leaving group.
(*b*) Halide substituents are highly susceptible to nucleophilic attack.
(*c*) Halide substituents are highly susceptible to electrophilic attack.
(*d*) The carbonyl carbon is sp^2 hybridized and is trigonal.

Answer. (*a*)

305. Which reagent converts a carboxylic acid to an acid chloride ?
(*a*) $SOCl_2$ (*b*) formyl chloride
(*c*) Cl_2 (*d*) HCl

Answer. (*a*)

306. What is the product of the reaction of acetyl chloride with ammonia ?
(*a*) methylamine (*b*) ethylamine
(*c*) acetamide (*d*) acetonitrile

Answer. (*c*)

307. For a given carbon content, which of the following have the highest boiling points ?
(*a*) carboxylic acids (*b*) nitriles
(*c*) methyl esters (*d*) primary amides

Answer. (*d*)

308. What is the product of the dehydration of an amide with P_2O_5 ?
(*a*) an alkene (*b*) an imide
(*c*) a nitrile (*d*) an ether

Answer. (*c*)

309. What is produced when propanamide is heated with aqueous acid ?
(*a*) CH_3CH_2COOH *and* NH_4^+
(*b*) $CH_3CH_2CH_2NH_2$
(*c*) $CH_3CH_2C{\equiv}N$
(*d*) $CH_3CH_2CONH_3^+$

Answer. (*a*)

310. Which reagents are used in the most general preparation of acid anhydrides ?
(*a*) carboxylic acid and an ester
(*b*) an acid chloride and the salt of a carboxylic acid
(*c*) an acid chloride and an ester
(*d*) the salt of a carboxylic acid and an ester

Answer. (*b*)

311. Which pair of reactants below will yield butanoic anhydride ?
(*a*) Butanoyl chloride and 1-butanol
(*b*) Butanoic acid and water
(*c*) Butanoyl chloride and sodium butanoate
(*d*) All of the above

Answer. (*c*)

312. Which reagents will convert propanoic acid to propanoic anhydride ?
(*a*) H_2O, H_3O^+
(*b*) $SOCl_2$, then CH_3CH_2COONa
(*c*) $CH_3CH_2CH_2MgBr$, then H_3O^+
(*d*) All of the above

Answer. (*b*)

313. Which reaction below will lead to an ester ?
(*a*) Reaction of pentanoic acid with ethanol in the presence of dry acid catalyst
(*b*) Reaction of pentanoyl chloride with ethanol
(*c*) Reaction of pentanoic anhydride with ethanol
(*d*) All of the above

Answer. (*d*)

314. To make an ester by Fischer esterification, one would begin with
(*a*) an alcohol and an acyl chloride
(*b*) an alcohol and an anhydride
(*c*) an alcohol and an amide
(*d*) an alcohol and a carboxylic acid

Answer. (*d*)

315. Consider the Fischer esterification of methanol with formic acid. Which statement does not apply ?
(*a*) The ester formed is methyl formate.
(*b*) This is an equilibrium reaction.
(*c*) Esterification releases water.
(*d*) Adding more mineral acid shifts the equilibrium towards more ester formation.

Answer. (*d*)

316. What is the function of the acid catalyst in the first step of the Fischer esterification of a carboxylic acid ?
(*a*) to protonate the –OH oxygen of the carboxyl group
(*b*) to protonate the carbonyl oxygen
(*c*) to protonate the carbonyl carbon
(*d*) to protonate the –OH group of the alcohol

Answer. (*b*)

317. What is the product of the reaction of benzyl acetate and $CH_3O^-Na^+$?
(*a*) phenyl benzoate
(*b*) methyl benzoate
(*c*) methyl acetate
(*d*) ethyl acetate

Answer. (*c*)

318. Which products result from the reaction $CH_3CH_2COOCH_3$ with H_2O, NaOH ?
(*a*) $CH_3CH_2COOH + HOCH_3$
(*b*) $CH_3CH_2COO^-Na^+ + HOCH_3$
(*c*) $CH_3CH_2CH_2OH + HOCH_3$
(*d*) Propanoic anhydride + $NaOCH_3$

Answer. (*b*)

319. What is always a product of the acetoacetic ester synthesis ?

(*a*) CO (*b*) H_2O

(*c*) CO_2 (*d*) $R–CH_2–CO_2H$

Answer. (*c*)

320. What is the final product of the acetoacetic ester synthesis ?

(*a*) substituted derivatives of acetic acid

(*b*) substituted derivatives of acetone

(*c*) substituted derivatives of acetaldehyde

(*d*) substituted derivatives of a malonic acid

Answer. (*b*)

321. Methyl ethyl ketone (2-butanone) can be produced easily from which of the following sets of reactants ?

(*a*) methyl bromide and propanal

(*b*) ethyl bromide and acetaldehyde

(*c*) methyl bromide and malonic ester

(*d*) methyl bromide and acetoacetic ester

Answer. (*d*)

322. The acetoacetic ester synthesis is used to prepare

(*a*) α,β-unsaturated carbonyl compounds

(*b*) methyl ketones

(*c*) enamines

(*d*) benzoic acid

Answer. (*b*)

323. Treating acetoacetic ester with ethoxide/ethanol, then benzyl bromide, then hot aqueous acid would yield

(*a*) 3-phenyl-2-propanone

(*b*) 4-phenyl-2-butanone

(*c*) 3-phenyl-2-butanone

(*d*) 1,3-diphenyl-2-propanone

Answer. (*b*)

324. Which reagent below would be useful if preparing pentanoic acid from malonic ester ?

(*a*) 1-Bromopropane (*b*) 2-Bromopropane

(*c*) 1-Bromopentane (*d*) 2-Bromopentane

Answer. (*a*)

325. What term applies to an equilibrium between an enol and its corresponding carbonyl compound ?

(*a*) electrophilic alpha substitution

(*b*) alpha rearrangement

(*c*) keto-enol tautomerism

(*d*) ketonic equilibrium

Answer. (*c*)

326. Which is the *tautomer* of a ketone ?

(*a*) ketene (*b*) ketal

(*c*) *al*cohol (*d*) enol

Answer. (*d*)

327. What functional group is present in all *triglycerides* ?

(*a*) ester (*b*) carboxylic acid

(*c*) alcoholic –OH (*d*) the –C=C– double bond

Answer. (*a*)

328. What would be *least* useful for extracting a lipid from plant cells ?

(*a*) water (*b*) benzene

(*c*) chloroform (*d*) diethyl ether

Answer. (*a*)

329. What commercial product has the sodium salt of long-chain carboxylic acids as its major component ?
(*a*) aspirin (*b*) baking soda
(*c*) artificial sweeteners (*d*) soap

Answer. (*d*)

330. What reaction produces soap as a product ?
(*a*) hydrolysis of large-ring lactones in ethylene glycol
(*b*) acid anhydride of a dicarboxylic acid and a heterocyclic amine
(*c*) triester of glycerol with long-chain acids and NaOH
(*d*) acid chloride of an unsaturated acid and an alcohol

Answer. (*c*)

331. Soap is manufactured by the reaction of triglycerides with which reagent ?
(*a*) H_2SO_4 and water (*b*) H_2O_2
(*c*) H_2 and a catalyst (*d*) NaOH and water

Answer. (*d*)

332. What acts as the hydrophobic portion of a soap molecule ?
(*a*) the Na^+ cation
(*b*) the glycerol portion of the triglyceride
(*c*) glycerol
(*d*) the long hydrocarbon chain

Answer. (*d*)

333. The dispersal of micelles of grease in soapy water results in the formation of which of the following ?
(*a*) wax (*b*) emulsion
(*c*) solution (*d*) alloy

Answer. (*b*)

334. What process is used to convert vegetable oils into margarine and other solid or semisolid vegetable shortenings ?
(*a*) bromination (*b*) hydrolysis
(*c*) catalytic hydrogenation (*d*) oxidation

Answer. (*c*)

335. What is the composition of waxy, hard-water curd or scum formed when soap is used in hard water ?
(*a*) hydrates of carboxylic acids and their salts
(*b*) calcium, magnesium, or iron salts of long-chain carboxylic acids
(*c*) products of the loss of CO_2 from long-chain carboxylic acids
(*d*) sodium salts of long-chain carboxylic acids

Answer. (*b*)

336. What is the most significant difference between the molecular structures of soaps and the most widely used classes of synthetic detergents ?
(*a*) Synthetic detergents contain nitrogen atoms ; soaps do not.
(*b*) Soaps are sodium salts ; synthetic detergents are covalent bases.
(*c*) Synthetic detergents contain the sulfonate functional group ; soaps contain the carboxylate functional group.
(*d*) Soaps contain aromatic rings ; synthetic detergents do not.

Answer. (*c*)

337. What is the hybridization of the orbital containing nitrogen's nonbonding electrons in $CH_3–NH–CH_2–CH_3$?
(*a*) The orbital containing the nonbonding electrons is not hybridized
(*b*) sp^2
(*c*) sp^3
(*d*) sp

Answer. (*a*)

338. Which one of the following is not affected (or is least affected) by the lone pair of electrons on an amine's nitrogen ?
(*a*) solubility in alcohols and in water

(*b*) hydrogen-bond formation
(*c*) melting point
(*d*) basicity

Answer. (*c*)

339. Which offers the best method for the conversion of an amide to an amine ?
(*a*) Na, NH_3 (*b*) $NaNO_2$, HCl
(*c*) $LiAlH_4$ (*d*) H_2, Pt, heat, pressure

Answer. (*c*)

340. What is the product of the Hofmann rearrangement of an amide ?
(*a*) an amine (*b*) an *N,N*-dialkylamide
(*c*) an alcohol (*d*) a nitrile

Answer. (*a*)

341. Which one of the following is a reagent in the Gabriel amine synthesis ?
(*a*) an acyl or aryl halide (*b*) hydroxylamine
(*c*) a Schiff base (*d*) phthalimide

Answer. (*d*)

342. The Hofmann rearrangement of which of the following produces ethylamine as a product ?
(*a*) $CH_3–CO–CH_3$
(*b*) $CH_3–CH_2–CO–CH_2–CH_3$
(*c*) $CH_3–CO–Cl$
(*d*) $CH_3–CH_2–CO–NH_2$

Answer. (*d*)

343. Which one of the following *cannot* be prepared by reductive amination ?
(*a*) triethylamine (*b*) aniline
(*c*) benzylamine (*d*) methylamine

Answer. (*b*)

344. What set of reagents is used to convert a secondary amine to an *N*-nitrosamine ?
(*a*) $CH_3–NO_2$ and HCl
(*b*) $CH_3–NH_2$ and H_2O_2
(*c*) N_2 and H_2SO_4
(*d*) $NaNO_2$ and HCl

Answer. (*d*)

345. Which method below can be used to prepare a tertiary amine ?
(*a*) Reduction of an amide with $LiAlH_4$ then water
(*b*) The Gabriel amine synthesis
(*c*) The Hofmann rearrangement
(*d*) All of the above

Answer. (*a*)

346. Which converts methyl bromide to ethylamine ?
(*a*) a large excess of NH_3
(*b*) –CN, then $LiAlH_4$, then H_2O
(*c*) acetyl chloride, then $LiAlH_4$
(*d*) $NaNO_2$ and HCl

Answer. (*b*)

347. What is the total number of bonding electron pairs in acetonitrile ?
(*a*) 6 (*b*) 5 (*c*) 4 (*d*) 7

Answer. (*d*)

348. Which product below results from treatment of acetonitrile with lithium aluminum hydride then water ?
(*a*) Acetaldehyde, $CH_3CH=O$
(*b*) Ethanol, CH_3CH_2OH
(*c*) Ethylamine, $CH_3CH_2NH_2$
(*d*) None of the above

Answer. (*c*)

349. What is the molecular formula for the homologous series of cycloalkanes ?
(*a*) C_nH_{2n} (*b*) C_nH_{2n-2} (*c*) C_nH_{2n+2} (*d*) C_nH_{2n+1}
Answer. (*a*)

350. What is the general molecular formula for cyclic alkenes ?
(*a*) C_nH_{2n+2} (*b*) C_nH_{2n} (*c*) C_nH_{2n+4} (*d*) C_nH_{2n-2}
Answer. (*d*)

351. What is the product of the reaction of 2-butene and diazomethane ?
(*a*) methylcyclopropane (*b*) 1,1-dimethylcyclopropane
(*c*) 1,3-dimethylcyclopropane (*d*) 1,2-dimethylcyclopropane
Answer. (*d*)

352. What is the product of the reaction between 1,2-dichloro-1-propene and diazomethane ?
(*a*) 1,2-dichloro-1-methylcyclopropane
(*b*) 2,2-dichloro-1-methylcyclopropane
(*c*) 1,1-dichloro-2-methylcyclopropane
(*d*) 1,2-dichlorocyclobutane
Answer. (*a*)

353. Three membered rings readily undergo ring opening reactions because :
(*a*) they are strongly positive
(*b*) energy is released by relieving angle and torsional strain
(*c*) they are highly polarized
(*d*) they are strongly negative
Answer. (*b*)

354. Which of the following is a balanced equation for the complete combustion of cyclopropane ?
(*a*) $C_3H_6 + 3O_2 \longrightarrow 3CO + 3H_2O$
(*b*) $2C_3H_6 + 9O_2 \longrightarrow 6CO_2 + 6H_2O$
(*c*) $2C_3H_6 + 9O_2 \longrightarrow 4CO_2 + 6H_2CO + 4H_2O$
(*d*) $C_3H_8 + 5O_2 \longrightarrow 3CO_2 + 4H_2O$
Answer. (*b*)

355. Compound A has the formula $C_{10}H_{16}$ and reacts with excess hydrogen over palladium to yield compound B with the formula $C_{10}H_{18}$. How many rings are in A ?
(*a*) 1 (*b*) 2 (*c*) 3 (*d*) 4
Answer. (*b*)

356. Which of the following statements concerning the ring strain in cyclopropene is correct ?
(*a*) The C–C=C angle is reduced by 30 degrees compared to its normal value in propene.
(*b*) The C–C=C angle is reduced by 40 degrees compared to its normal value in propene.
(*c*) The C–C=C angle is reduced by 50 degrees compared to its normal value in propene.
(*d*) The C–C=C angle is reduced by 60 degrees compared to its normal value in propene.
Answer. (*d*)

357. Which of the following statements concerning the ring strain in cyclobutene is correct ?
(*a*) The C–C=C angle is reduced by 20 degrees compared to its normal value in butene.
(*b*) The C–C=C angle is reduced by 30 degrees compared to its normal value in butene.
(*c*) The C–C=C angle is reduced by 40 degrees compared to its normal value in butene.
(*d*) The C–C=C angle is reduced by 60 degrees compared to its normal value in butene.
Answer. (*b*)

358. Which statement concerning ring strain of cycloalkanes is *not* correct ?
(*a*) Cyclopentane and cycloheptane have about the same ring strain.
(*b*) Cyclopropane has the greatest amount of ring strain.
(*c*) Cyclohexene is free of ring strain.
(*d*) Strained molecules have lower heats of formation than unstrained molecules.
Answer. (*d*)

359. What is the difference between C–C–C angles of a planar, hexagonal cyclohexane ring and the chair structure of the actual cyclohexane molecule ?
(*a*) They are the same in both : 120.0° in a planar hexagon : 120.0° in cyclohexane
(*b*) 109.5° in a planar hexagon : 120.0° in cyclohexane

(*c*) 120.0° in a planar hexagon : 109.5° in cyclohexane
(*d*) They are the same in both : 109.5° in a planar hexagon : 109.5° in cyclohexane

Answer. (*c*)

360. In general, substituents on chair cyclohexane ring prefer to be
(*a*) *syn* (*b*) eclipsed
(*c*) axial (*d*) equatorial

Answer. (*d*)

361. In the most stable conformer of *cis*-1,3-dimethylcyclohexane, the methyl groups are
(*a*) *trans*
(*b*) both axial
(*c*) both equatorial
(*d*) on the same carbon

Answer. (*c*)

362. Which class of compounds is referred to as aldoses ?
(*a*) polyhydroxy aldehydes
(*b*) compounds containing both aldehyde and ketonic carbonyl groups
(*c*) polycyclic aldehydes
(*d*) polyhydroxy ketones

Answer. (*a*)

363. Which of the following is a ketose ?
(*a*) Glucose (*b*) Fructose
(*c*) Ribose (*d*) All of these

Answer. (*b*)

364. To what class of monosaccharides does glyceraldehyde belong ?
(*a*) ketotriose (*b*) aldotetrose
(*c*) aldotriose (*d*) ketotetrose

Answer. (*c*)

365. In solution, glucose exists almost entirely as :
(*a*) a 6-membered hemiacetal ring.
(*b*) a 5-membered hemiacetal ring.
(*c*) a disaccharide consisting of two 6-membered rings.
(*d*) an uncyclized, straight-chain aldohexose.

Answer. (*a*)

366. In solution, fructose exists as a
(*a*) disaccharide consisting of two furanose rings.
(*b*) single pyranose ring.
(*c*) single furanose ring.
(*d*) disaccharide consisting of one furanose ring and one pyranose ring.

Answer. (*c*)

367. The α and β forms of glucopyranose are
(*a*) stereoisomers (*b*) diastereomers
(*c*) anomers (*d*) All of these

Answer. (*d*)

368. What molecular characteristic defines the D series of sugars ?
(*a*) the direction of rotation of polarized light
(*b*) the chirality of the chiral carbon closest to the carbonyl group
(*c*) the chirality of the chiral carbon nearest the bottom in the Fischer projection
(*d*) the location of the carbonyl group on the carbohydrate chain

Answer. (*c*)

369. Which of the following are anomers ?
(*a*) β-D-glucopyranose and β-L-glucopyranose
(*b*) β-D-glucopyranose and β-D-glucofuranose
(*c*) β-D-glucopyranose and α-D-glucopyranose
(*d*) All of the above

Answer. (*c*)

370. D-glucose and L-glucose are
(*a*) anomers (*b*) disaccharides
(*c*) enatiomers (*d*) Ketoses

Answer. (*c*)

371. In terms of the sugars present, what is the composition of honey ?
(*a*) a mixture of glucose and fructose
(*b*) a mixture of glucose and galactose
(*c*) a mixture of sucrose and galactose
(*d*) a mixture of fructose and galactose

Answer. (*a*)

372. What conclusion can be drawn from the observation that two aldopentoses give the same osazone ?
(*a*) One aldopentose exists as a glycoside, the other as a hemiacetal.
(*b*) The two aldopentoses are epimers.
(*c*) One aldopentose is D, the other is L.
(*d*) The two aldopentoses are enantiomers.

Answer. (*b*)

373. What does a positive Tollens' test indicates ?
(*a*) A carbohydrate can exist as an open-chain aldose or ketose.
(*b*) A carbohydrate contains an oxidizable $-CH_2-OH$ group.
(*c*) A compound is a carbohydrate.
(*d*) A carbohydrate is a glycoside.

Answer. (*a*)

374. What oxidation can be carried out with bromine water ?
(*a*) oxidation of the terminal $-CH_2-OH$ to $-CO_2H$
(*b*) oxidation of the terminal $-CH_2-OH$ to an aldehyde carbonyl
(*c*) oxidation of a ketose to an aldonic acid
(*d*) oxidation of an aldose to an aldonic acid

Answer. (*d*)

375. What reagent converts sugars to osazones ?
(*a*) NO_2 (*b*) $HO-NH_2$
(*c*) $C_6H_5-NH-NH_2$ (*d*) NH_3

Answer. (*c*)

376. Which conditions below will oxidize an aldose to a carboxylic acid ?
(*a*) $Ag(NH_3)_2^+$, then H_3O^+
(*b*) Cu^{2+}, then H_3O^+
(*c*) HNO_3, warm
(*d*) All of the above

Answer. (*d*)

377. In the chain-lengthening Killiani-Fischer synthesis, what is the source of the new carbon atom ?
(*a*) HCN (*b*) CH_3-Br
(*c*) $(CH_3)_2CuLi$ (*d*) formaldehyde

Answer. (*a*)

378. What is the product obtained when the Killani-Fischer synthesis is carried out on D-glyceraldehyde ?
(*a*) D-ribose
(*b*) L-glyceraldehyde
(*c*) D- and L-erythrose
(*d*) D-threose and D-erythrose

Answer. (*d*)

379. Which of the following is a polysaccharide ?
(*a*) Glucose (*b*) Ribose
(*c*) Sucrose (*d*) Cellulose

Answer. (*d*)

380. In which of the following disaccharides does the glycosidic link joins two anomeric carbons, one from each ring ?

(*a*) Cellobiose (*b*) maltose
(*c*) sucrose (*d*) lactose

Answer. (*c*)

381. What monosaccharide is obtained on hydrolysis of cellulose ?

(*a*) galactose (*b*) sucrose
(*c*) glucose (*d*) fructose

Answer. (*c*)

382. What is the principal difference between cellulose and amylose of starch.

(*a*) The glycosidic linkages of amylose are alpha-1,4' ; those of cellulose are beta-1,4'.
(*b*) Tollens' test is positive for amylose ; negative for cellulose.
(*c*) Amylose is a polymer of L-glucose ; cellulose is a polymer of D-glucose.
(*d*) Amylose is a polymer of fructose ; cellulose is a polymer of glucose.

Answer. (*a*)

383. How many standard α-amino acids – those found in nearly all animal proteins – are there ?

(*a*) 20 (*b*) 5 (*c*) 55 (*d*) 10

Answer. (*a*)

384. What is the name of the simplest amino acid, $H_2N–CH_2–CO_2H$?

(*a*) leucine
(*b*) glycine
(*c*) alanine
(*d*) valine

Answer. (*b*)

385. What substituents are found on the *alpha* carbon of the amino acid *alanine* ?

(*a*) 2 hydrogens ; 1 NH_2 group ; 1 CO_2H group
(*b*) 1 hydrogen ; 2 CH_3 groups ; 1 CO_2H group
(*c*) 2 hydrogens ; 1 CH_3 group ; 1 CO_2H group
(*d*) 1 hydrogen ; 1 NH_2 group ; 1 CH_3 group ; 1 CO_2H group

Answer. (*d*)

386. What organic functional group forms the *peptide* bond of proteins ?

(*a*) ether (*b*) amide
(*c*) ester (*d*) acetal

Answer. (*b*)

387. What represents the tripeptide *glycylarginylproline* ?

(*a*) LRN (*b*) LAP
(*c*) PAP (*d*) GAP

Answer. (*d*)

388. The isoelectric point of an amino acid is the

(*a*) pH at which all the functional groups are protonated.
(*b*) same as the pKa for that acid.
(*c*) pH at which no net migration occurs during electrophoresis.
(*d*) All of the above

Answer. (*c*)

389. Electrophoresis is useful for separating a mixture of amino acids that have substantially different:

(*a*) dipole moments (*b*) isoelectric points
(*c*) boiling points (*d*) melting points

Answer. (*b*)

390. Which reaction scheme will lead to alanine ?

(*a*) CH_3CH_2COOH treated with excess ammonia, then DCC, then CF_3COOH

(*b*) CH_3CH_2COOH treated with PBr_3 and Br_2, then water, then excess ammonia
(*c*) CH_3COOH treated with PBr_3 and Br_2, then water, then excess ammonia
(*d*) None of the above

Answer. (*b*)

391. Which of the following is the best starting material for a Strecker synthesis of serine ?
(*a*) $HOCH_2CH_2COOH$
(*b*) $HOCH_2COOH$
(*c*) $HOCH_2CHO$
(*d*) $HOCH_2CH_2CH_2OH$

Answer. (*c*)

392. What maintains the primary structure of a protein ?
(*a*) dipole-dipole interactions
(*b*) covalent bonds
(*c*) van der Waals forces
(*d*) hydrogen bonds

Answer. (*b*)

393. What is the *primary structure* of a protein ?
(*a*) the combination of the proteins and any prosthetic groups it may be bonded to
(*b*) the sequence of the amino acids in the protein chain and any disulfide bridges
(*c*) the first of several structural features to be discovered
(*d*) the most stable of all possible conformations of the protein

Answer. (*b*)

394. What forces usually maintain the secondary structure of a protein ?
(*a*) -S-S- links (*b*) van der Waals forces
(*c*) hydrogen bonds (*d*) ionic bonds

Answer. (*c*)

395. The α-helix is an example of what level of protein structure ?
(*a*) Primary (*b*) Secondary
(*c*) Tertiary (*d*) Quaternary

Answer. (*b*)

396. Which of the following forces helps to maintain tertiary protein structure ?
(*a*) salt bridges (*b*) side chain solubility
(*c*) disulfide links (*d*) All of these

Answer. (*d*)

397. What holds the two strands of DNA together ?
(*a*) dipole-dipole interactions
(*b*) hydrogen bonds
(*c*) van der Waals forces
(*d*) ionic bonds

Answer. (*b*)

398. What term refers to a change in a secondary or higher structure of a protein, usually causing a change in properties ?
(*a*) deproteination (*b*) depeptidation
(*c*) destabilization (*d*) denaturation

Answer. (*d*)

399. What process is used to cleave a disulfide bridge in a polypeptide or protein molecule ?
(*a*) mild reduction (*b*) mild oxidation
(*c*) treatment with weak acid (*d*) treatment with weak base

Answer. (*b*)

400. Which term most accurately describes what happens when milk curdles ?
(*a*) protein conjugation (*b*) protein denaturation
(*c*) peptide bond cleavage (*d*) protein synthesis

Answer. (*b*)

401. Of what value is *ninhydrin* in amino acid chemistry ?
(*a*) detecting the presence of amino acids
(*b*) distinguishing among acidic, neutral, and basic amino acids
(*c*) distinguishing between D and L stereoisomers
(*d*) determining the isoelectric point of an amino acid

Answer. (*a*)

402. Which of the following is a requirement for a compound to be aromatic ?
(*a*) A strong odor
(*b*) Complete conjugation around a ring
(*c*) Must be benzene or a substituted benzene
(*d*) All of the above

Answer. (*b*)

403. Which of the following heterocyclic compounds is aromatic ?

(*a*) 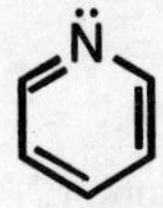
(*b*) 
(*c*) 
(*d*) All of the above

Answer. (*d*)

404. Which of the following ions is aromatic ?

(*a*) 
(*b*) 
(*c*) 
(*d*) All of the above

Answer. (*b*)

405. What kind of reactions do aromatic compounds such as benzene undergo ?
(*a*) electrophilic aromatic substitution reactions
(*b*) electrophilic addition reactions
(*c*) nucleophilic addition reactions
(*d*) nucleophilic aromatic substitution reactions

Answer. (*a*)

406. How many resonance contributors does the carbocation have that is formed as an intermediate when benzene undergoes an electrophilic aromatic substitution reaction ?

(*a*) 1 (*b*) 2 (*c*) 3 (*d*) 4

Answer. (*c*)

407. What product results from the reaction of benzene, $AlCl_3$, and *t*-butyl chloride ?
(*a*) chlorobenzene
(*b*) *o*-xylene
(*c*) toluene
(*d*) *t*-butylbenzene

Answer. (*d*)

408. What is produced when benzene reacts with a mixture of nitric acid and sulfuric acid ?
(*a*) C_6H_5–OH
(*b*) C_6H_5–SO_2
(*c*) C_6H_5–NO_2
(*d*) C_6H_5–SO_3H

Answer. (*c*)

409. Benzene is converted to nitrobenzene by a mixture of sulfuric acid and nitric acid. What is the catalyst for this reaction ?
(*a*) the nitronium ion
(*b*) nitric acid
(*c*) the sulfonium ion
(*d*) sulfuric acid

Answer. (*d*)

410. In the reaction of benzene with fuming sulfuric acid to produce benzenesulfonic acid, what is the effective electrophile ?

(*a*) SO_2^+ (*b*) H_2SO_4
(*c*) SO_3 (*d*) HSO_3^+

Answer. (*c*)

411. Which of the following is the most effective catalyst for the reaction of benzene and Br_2 to produce bromobenzene ?

(*a*) $FeBr_3$ (*b*) KBr
(*c*) HBr (*d*) HNO_3

Answer. (*a*)

412. What is the best method for converting benzene to *n*-propylbenzene ?

(*a*) benzene, $CH_3–CH_2–CO–Cl$, $AlCl_3$; then Zn(Hg) and aq. HCl
(*b*) benzene, $Cl–CH_2–CH_2–CH_3$, $AlCl_3$
(*c*) toluene, ethyl chloride, $AlCl_3$
(*d*) benzene, $Cl–CO–CH_3$, $AlCl_3$

Answer. (*a*)

413. Which reaction below is an example of electrophilic aromatic substitution ?

(*a*) Bromobenzene reacted with Mg, ether
(*b*) Toluene reacted with excess H_2 over Ni catalyst
(*c*) *p*-Bromotoluene reacted with hot aqueous acid and $KMnO_4$
(*d*) Toluene reacted with Br_2, $FeBr_3$

Answer. (*d*)

414. What starting material and conditions would produce benzene hexachloride ?

(*a*) cyclohexane, excess HCl, heat, pressure
(*b*) benzene, excess Cl_2, $FeCl_3$
(*c*) benzene, excess Cl_2, and either heat and pressure or ultraviolet light
(*d*) benzene, excess HCl, heat, pressure

Answer. (*c*)

415. The most likely product from the reaction of toluene with NBS and peroxide is

(*a*) ortho-boromotoluene (*b*) meta-bromotoluene
(*c*) para-bromotoluene (*d*) benzyl bromide

Answer. (*d*)

416. What is produced by the reaction of toluene with molecular hydrogen at high pressure and in the presence of a catalyst such as Pt, Pd, or Ni ?

(*a*) cyclohexane
(*b*) benzene
(*c*) methylcyclohexane
(*d*) 1-methyl-1,3-cyclohexadiene

Answer. (*c*)

417. Which ring position(s) of toluene are most reactive toward electrophilic aromatic substitution ?

(*a*) ortho and para (*b*) meta and para
(*c*) ortho and meta (*d*) meta

Answer. (*a*)

418. What is the major product of Friedel-Crafts acylation of toluene with acetyl chloride ?

(*a*) ortho $CH_3–C_6H_5–CO–CH_3$
(*b*) $C_6H_5–CO–CH_3$
(*c*) para $CH_3–C_6H_5–CO–CH_3$
(*d*) meta $CH_3–C_6H_5–CO–CH_3$

Answer. (*c*)

419. Which one of the following compounds will produce one and only one monobromination product ?

(*a*) *m*-nitroanisole (*b*) *m*-dinitrobenzene
(*c*) *o*-chlorotoluene (*d*) *o*-xylene

Answer. (*b*)

420. Which of the following is an ortho/para director ?
(*a*) –COOH (*b*) –Br
(*c*) –CN (*d*) –CHO
Answer. (*b*)

421. Which compound below is likely to nitrate at the meta position ?

(*a*) 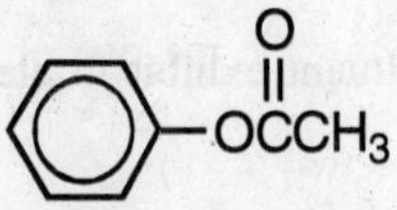

(*b*) 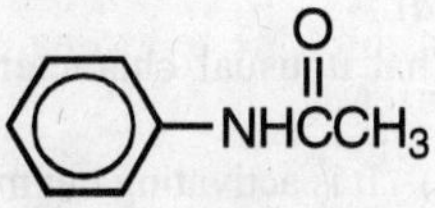

(*c*) 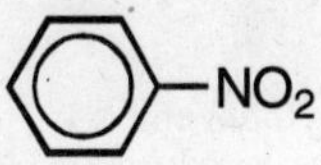

(*d*) All of these

Answer. (*c*)

422. The $-NO_2$ group deactivates the benzene ring toward electrophilic aromatic substitution. At which ring position(s) is this deactivation least effective ?
(*a*) meta (*b*) ortho and para
(*c*) para (*d*) ortho
Answer. (*a*)

423. Which one of the following is least reactive toward electrophilic aromatic substitution ?
(*a*) nitrobenzene (*b*) phenol
(*c*) ethylbenzene (*d*) benzene
Answer. (*a*)

424. Which compound below will likely undergo acylation the fastest ?

(*a*) 
(*b*) 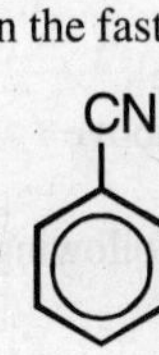

(*c*) 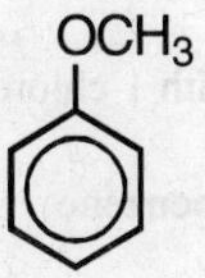

(*d*) 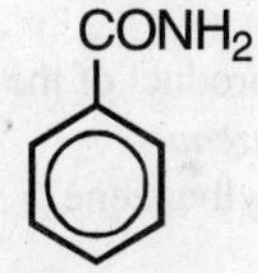

Answer. (*c*)

425. Which of the following is not true about the relative reactivity of the two compounds toward electrophilic aromatic substitution ?
(*a*) Phenol is more reactive than toluene.
(*b*) Bromobenzene is more reactive than benzaldehyde.
(*c*) Chlorobenzene is more reactive than benzene.
(*d*) Benzene is more reactive than nitrobenzene.
Answer. (*c*)

426. Which of the following is not a true statement ?
(*a*) A methyl substituent is an ortho/para director.
(*b*) A chloro substituent is an ortho/para director.
(*c*) A protonated amino substituent is an ortho/para director.
(*d*) A nitro substituent is a meta director.
Answer. (*c*)

427. What effect does a $-NH-CO-CH_3$ substituent have on electrophilic aromatic substitution ?
(*a*) It is deactivating and meta directing.
(*b*) It is activating and ortho/para directing.
(*c*) It is deactivating and ortho/para directing.
(*d*) It is activating and meta directing.
Answer. (*b*)

428. Which of the following combinations would you expect in the nitration of bromobenzene ?
(*a*) ortho/para product and a slower reaction than benzene
(*b*) ortho/para product and a faster reaction than benzene
(*c*) meta product and a slower reaction than benzene
(*d*) meta product and a faster reaction than benzene

Answer. (*a*)

429. What unusual characteristic does a halogen substituent exhibit in electrophilic aromatic substitutions ?
(*a*) It is activating but meta directing.
(*b*) It is activating but exclusively ortho directing.
(*c*) It is deactivating but ortho/para directing.
(*d*) It is deactivating and directs to all positions equally.

Answer. (*c*)

430. Halogens are deactivating yet ortho/para directing in electrophilic aromatic substitution. How do we explain this ?
(*a*) a combination of inductive electron release and resonance electron release
(*b*) inductive electron withdrawal with no resonance effect at all
(*c*) a combination of inductive electron withdrawal and resonance electron release
(*d*) resonance electron release with no inductive effect at all

Answer. (*a*)

431. Which of the following compounds will undergo a Friedel-Crafts acylation reaction ?
(*a*) nitrobenzene
(*b*) benzaldehyde
(*c*) acetophenone
(*d*) None of the above

Answer. (*d*)

432. Which one of the following will not undergo a Friedel-Crafts alkylation ?
(*a*) benzene (*b*) toluene
(*c*) benzoic acid (*d*) *p*-chlorotoluene

Answer. (*c*)

433. What is the product of the Friedel-Crafts reaction of benzene with 1-chloropropane ?
(*a*) *ethylbenzene* (*b*) *o*-xylene
(*c*) isopropylbenzene (*d*) *n*-propylbenzene

Answer. (*c*)

434. Which of the following will likely react with KOH/H_2O at elevated temperature and pressure to yield a single organic product *via* the benzyne pathway ?
(*a*) Chlorobenzene (*b*) *m*-chlorotoluene
(*c*) *p*-Chlorotoluene (*d*) *o*-Chlorotoluene

Answer. (*a*)

435. What is the common name for a solution of 7% of SO_3 in sulfuric acid ?
(*a*) supersulfuric acid (*b*) hypersulfuric acid
(*c*) sulfurous sulfuric acid (*d*) fuming sulfuric acid

Answer. (*d*)

436. What step(s) below will convert benzene to *p*-bromobenzenesulfonic acid ?
(*a*) H_2SO_4/HNO_3 and then $Br_2/FeBr_3$
(*b*) $Br_2/FeBr_3$ and then H_2SO_4/HNO_3
(*c*) H_2SO_4/SO_3 and then $Br_2/FeBr_3$
(*d*) $Br_2/FeBr_3$ and then H_2SO_4/SO_3

Answer. (*d*)

437. The aromatic amine C_6H_5–NH_2 is commonly known as
(*a*) aniline (*b*) aminobenzene
(*c*) benzylamine (*d*) phenylamine

Answer. (*a*)

438. Aromatic amines, such as aniline, are weaker bases than the simple aliphatic amines largely because of

(*a*) solvation effects
(*b*) resonance effects
(*c*) the steric size of alkyl groups in the aliphatic amines
(*d*) inductive effects

Answer. (*b*)

439. What is the order of basicity (most basic to least) ?

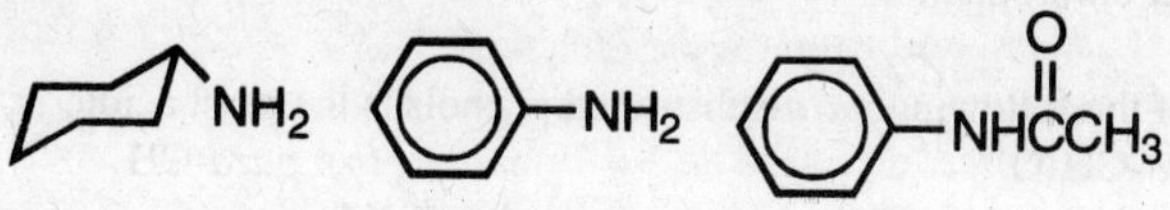

I II III

(*a*) I, II, III
(*b*) I, III, II
(*c*) III, II, I
(*d*) II, I, II

Answer. (*a*)

440. Aniline could be converted to benzene through which reaction sequence ?

(*a*) H_2SO_4, HNO_3, followed by H_2, Ni
(*b*) Excess CH_3I followed by aqueous AgO_2 and heat
(*c*) HONO, H_2SO_4 followed H_3PO_2
(*d*) None of the above

Answer. (*c*)

441. How can aniline be converted into benzene ?

(*a*) (1) diazotization (2) H_3PO_4
(*b*) Wolff-Kishner reduction
(*c*) (1) diazotization (2) H_3PO_2
(*d*) Clemmenson reduction

Answer. (*a*)

442. Which produces *m*-bromaniline as a product ?

(*a*) nitrobenzene, Br_2, $FeBr_3$, then Zn, HCl, then HO^-
(*b*) benzene, Br_2, $FeBr_3$, then HNO_3, H_2SO_4, then Zn, HCl, then HO^-
(*c*) aniline, Br_2, $FeBr_3$
(*d*) bromobenzene, NH_3, formaldehyde, Sn, H_2SO_4, then HO^-

Answer. (*a*)

443. Why is the diazotization of primary alkylamines not a useful procedure in organic syntheses ?

(*a*) Primary alkylamines do not undergo diazotization.
(*b*) The reaction is too slow to be practical.
(*c*) The diazonium salt of primary alkylamines decomposes rapidly to a complex mixture of product.
(*d*) The diazonium salt obtained as a product is too stable to react further.

Answer. (*c*)

444. Which offers the best preparation of fluorobenzene ?

(*a*) aniline + HF
(*b*) aniline + SO_2F_2, then $NaNO_2$ and HF
(*c*) aniline + F_2, in the presence of FeF_3
(*d*) benzenediazonium chloride and HBF_4, heat

Answer. (*d*)

445. Which of the following is most acidic ?

(*a*) Phenol (*b*) Cyclohexanol
(*c*) Benzoic acid (*d*) *p*-Ethylbenzoic acid

Answer. (*c*)

446. Which compound is most acidic ?

(*a*) Phenol (*b*) *p*-Nitrophenol

(*c*) *p*-Methoxyphenol (*d*) *p*-Ethylphenol

Answer. (*b*)

447. Which phenol is the most acidic ?

(*a*) *meta*-chlorophenol

(*b*) *para*-nitrophenol

(*c*) *meta*-nitrophenol

(*d*) *para*-chlorophenol

Answer. (*b*)

448. Which of the following *para*-substituted phenols is the most acidic ?

(*a*) *para*-CH_3O (*b*) *para*-CH_3

(*c*) *para*-NO_2 (*d*) *para*-Cl

Answer. (*c*)

449. Which one of the following molecules is most likely to form an intramolecular hydrogen bond ?

(*a*) catechol (*b*) resorcinol

(*c*) hydroquinone (*d*) phenol

Answer. (*a*)

450. Which of the following reactions will likely yield a phenol ?

(*a*) An aryldiazonium (ArN_2^+) treated with aqueous acid

(*b*) Nucleophilic aromatic substitution by hydroxide on ArCl

(*c*) Heating $ArCH(CH_3)_2$ in air and then aqueous acid

(*d*) All of the above

Answer. (*d*)

451. What represents the best method for producing salicylic acid ?

(*a*) phenol, base, carbon dioxide, then acid

(*b*) benzoic acid, strong base, high temperature

(*c*) toluene, strong base, water, high temperature

(*d*) bromobenzene, carbonic acid, high temperature

Answer. (*a*)

452. What characteristics of the methoxy group causes it to direct electrophiles to the *ortho* and *para* positions of anisole ?

(*a*) removal of ring electrons through resonance

(*b*) removal of ring electrons because of oxygen's electronegativity

(*c*) donation of the oxygen's nonbonding electrons to the ring through resonance

(*d*) interaction of the hydrogens of the methyl group

Answer. (*c*)

453. Which reagents and/or conditions will convert 1-phenyl-3-pentanone to pentylbenzene ?

(*a*) H_2, Pd, ethanol

(*b*) $NaBH_4$, then H_3O^+

(*c*) Zn(Hg), H_3O^+

(*d*) All of the above

Answer. (*c*)

454. Which reaction below will yield a ketone ?

(*a*) Treating a secondary alcohol with PCC

(*b*) Friedel-Crafts acylation of benzene

(*c*) Treating an alkyne with H_2O, $HgSO_4$, H_2SO_4

(*d*) All of the above

Answer. (*d*)

455. What is the product of the following reaction ?

$$C_6H_5-\overset{\overset{\large O}{\|}}{C}-CH_3 \xrightarrow[HCl]{Zn/Hg}$$

(*a*) C_6H_5–CHOH–CH_3

(*b*) $C_6H_5–CH_3$
(*c*) $C_6H_5–CO_2H$
(*d*) $C_6H_5–CH_2–CH_3$

Answer. (*d*)

456. Which acid below is the strongest ?
(*a*) Benzoic acid
(*b*) *p*-Methylbenzoic acid
(*c*) *p*-(Trichloromethyl)benzoic acid
(*d*) *p*-(Trifluoromethyl)benzoic acid

Answer. (*d*)

457. Which of the following compounds will form benzoic acid when heated in an acidic solution of potassium permanganate ?
(*a*) propylbenzene (*b*) toluene
(*c*) *sec*-butylbenzene (*d*) all of these

Answer. (*d*)

458. Which represents the product of the reaction of para $O_2N–(C_6H_4)–CH_2–CH_3$ with hot $Na_2Cr_2O_7$?
(*a*) *p*-nitrobenzoic acid (*b*) benzoic acid
(*c*) nitrobenzene (*d*) *p*-aminobenzoic acid

Answer. (*a*)

459. What is the product of the reaction to benzoic acid and ethyl alcohol in the presence of a catalytic amount of acid ?
(*a*) ethylbenzene (*b*) benzyl methyl ether
(*c*) acetophenone (*d*) ethyl benzoate

Answer. (*d*)

460. What reactant below would yield benzoic acid when treated with hot aqueous potassium permanganate ?
(*a*) *p*-Xylene (*b*) $PhCH_2CH_2Ph$
(*c*) Benzene (*d*) All of these

Answer. (*b*)

461. Which step(s) below will convert toluene to *m*-chlorobenzoic acid ?
(*a*) Water and Cl_2
(*b*) $Cl_2/FeCl_3$ followed by hot aqueous permanganate
(*c*) Hot aqueous permanganate and then $Cl_2/FeCl_3$
(*d*) None of the above

Answer. (*d*)

462. The characteristic of all the *phthalic acids* is that they :
(*a*) are derived from animal fat.
(*b*) contain aromatic ring bearing two carboxyl groups.
(*c*) are acids with triple bond conjugated to the carbonyl group.
(*d*) are isolated from insects.

Answer. (*b*)

463. Which of the following pairs of reagents reacts to form aspirin ?
(*a*) phenylacetic acid and benzoic anhydride
(*b*) benzoic acid and ethyl acetate
(*c*) salicylic acid and acetic anhydride
(*d*) ethyl benzoate and methyl salicylate

Answer. (*c*)

464. Heterocyclic compounds, or heterocycles, are :
(*a*) compounds in which one or more of the atoms has an unusual bonding pattern
(*b*) compounds containing unusual rings
(*c*) cyclic compounds in which one or more of the atoms of the ring are hetero atoms
(*d*) compounds containing at least two different rings

Answer. (*c*)

465. A hetero atom is
(*a*) any atom other than carbon
(*b*) any atom with an odd atomic number
(*c*) any atom other than hydrogen
(*d*) any atom with an even atomic number

Answer. (*a*)

466. Pyridine can be described as :
(*a*) a benzene ring with one of the carbons replaced by a nitrogen
(*b*) a five-membered ring containing oxygen
(*c*) an organosulfur compound
(*d*) a six-membered ring containing two nitrogen

Answer. (*a*)

467. Pyrrole is
(*a*) a five-membered-ring that contains oxygen
(*b*) a six-membered-ring that contains oxygen
(*c*) a five-membered-ring that contains nitrogen
(*d*) a six-membered-ring that contains nitrogen

Answer. (*c*)

468. Furan is
(*a*) a five-membered-ring containing oxygen
(*b*) a six-membered-ring containing nitrogen
(*c*) a six-membered-ring containing oxygen
(*d*) a five-membered-ring containing nitrogen

Answer. (*a*)

469. Name three five-membered-ring heterocycles.
(*a*) pyrimidine, furan, and thiophene
(*b*) pyrrole, furan, and thiophene
(*c*) pyrrole, pyrimidine, and quinoline
(*d*) pyrrole, quinoline, and thiophene

Answer. (*b*)

470. Pyrrole, furan, and thiophene have what in common ?
(*a*) they are six-membered-ring heterocycles
(*b*) they react exothermically with water
(*c*) they are well-known by their IUPAC names
(*d*) they are five-membered-ring heterocycles

Answer. (*d*)

471. Saturated heterocycles contain :
(*a*) only sp^2 atoms
(*b*) only double bonds
(*c*) as much water as they can absorb
(*d*) no double bonds

Answer. (*d*)

472. A six-membered ring containing oxygen atoms in the one and four positions is called :
(*a*) 1,4-oxane
(*b*) 1,4-oxoniumcyclohexane
(*c*) 1,4-dioxane
(*d*) 1,4-dioxocyclohexane

Answer. (*c*)

473. Which one of the following is the strongest base ?
(*a*) pyrrole (*b*) methylamine
(*c*) ammonia (*d*) aniline

Answer. (*b*)

474. Which of the following is the weakest base ?

(*a*) methylamine (*b*) aniline

(*c*) piperidine (*d*) acetonitrile

Answer. (*d*)

475. Which of the following heterocyclic compounds is aromatic ?

(*a*) 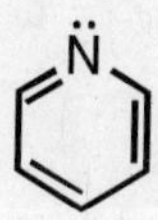

(*b*) 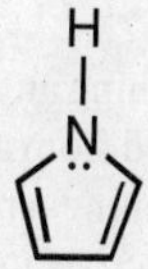

(*c*) 

(*d*) All of these

Answer. (*d*)

476. Pyridine is

(*a*) a tertiary amine .

(*b*) more reactive than benzene towards nucleophilic aromatic substitution

(*c*) less reactive than benzene towards electrophilic aromatic substitution

(*d*) all of the above

Answer. (*a*)

477. Pyridine reacts with electrophiles at :

(*a*) the nitrogen atom and C–3

(*b*) It does not react with electrophiles

(*c*) the nitrogen atom only

(*d*) the nitrogen atom and C–2

Answer. (*a*)

478. The reaction of $NaNH_2$ and pyridine to produce 2-aminopyridine is called the :

(*a*) Steffenson reaction

(*b*) Chichiban reaction

(*c*) Gaswick protocol

(*d*) Amination procedure

Answer. (*b*)

479. Electrophilic aromatic substitution in pyrrole occurs preferentially at

(*a*) C–3 (*b*) the nitrogen atom

(*c*) C–2 (*d*) it does not occur

Answer. (*c*)

480. Which of the following statements is true ?

(*a*) Thiophene is an aromatic compound that undergoes electrophilic aromatic substitution reactions.

(*b*) Thiophene is an aromatic compound that undergoes nucleophilic aromatic substitution reactions.

(*c*) Thiophene is an aromatic compound that undergoes electrophilic addition reactions.

(*d*) Thiophene is more reactive than furan towards electrophilic substitution.

Answer. (*a*)

481. Electrophilic aromatic substitution in thiophene occurs preferentially at :

(*a*) the nitrogen atom (*b*) C–2

(*c*) C–3 (*d*) does not occur

Answer. (*b*)

482. Which of the following is not true about isoquinoline :

(*a*) it is aromatic (*b*) it is benzopyridine

(*c*) it has an isopropyl substituent (*d*) it has a nitrogen hetero atom

Answer. (*c*)

483. Imidazole is an aromatic compound because
(*a*) it contains sp^2 hybridized atoms
(*b*) it smells like springtime in the everglades
(*c*) it is cyclic, planar, and has three pairs of delocalized *pi* electrons
(*d*) it is planar, conjugated compound

Answer. (*c*)

484. What are alkaloids ?
(*a*) amines containing an aromatic ring
(*b*) amines derived from alkenes
(*c*) amines containing a nitrogen atom as part of a ring structure
(*d*) biologically active amines usually found in plants ?

Answer. (*d*)

485. Alkaloids are :
(*a*) natural products, found in animals, that contain one or more alcohol groups
(*b*) natural products, found in animals, that contain one or more nitrogen heteroatoms
(*c*) natural products, found in plants, that contain one or more nitrogen hetero atoms
(*d*) alcohol-like compounds

Answer. (*c*)

486. What is the major source of *terpenes* ?
(*a*) petroleum (*b*) sea water
(*c*) animal fats (*d*) plant oils

Answer. (*d*)

487. What is the major characteristic of terpenes ?
(*a*) analgesic properties (*b*) density
(*c*) color (*d*) odor

Answer. (*d*)

488. What is the fundamental structural unit of *terpene* molecules ?
(*a*) styrene (*b*) cyclohexene
(*c*) isoprene (*d*) isobutylene

Answer. (*c*)

489. What is the minimum number of carbon atoms present in a terpene ?
(*a*) 15 (*b*) 12 (*c*) 10 (*d*) 5

Answer. (*c*)

490. How many carbon atoms are in a *diterpene* ?
(*a*) 5 (*b*) 15 (*c*) 20 (*d*) 10

Answer. (*c*)

491. What constitutes a terpene's "head" ?
(*a*) a C=C double bond
(*b*) a carboxylate group
(*c*) a branched alkyl group
(*d*) an aldehyde functional group

Answer. (*c*)

492. What term is used for the repeating units of a polymer ?
(*a*) unit structure (*b*) condensate (*c*) unit residue (*d*) monomer

Answer. (*d*)

493. What monomers are most commonly used in addition polymerization ?
(*a*) amides (*b*) esters (*c*) cycloalkanes (*d*) alkenes

Answer. (*d*)

494. What is the characteristic of all condensation polymers ?
(*a*) Each polymerizing molecule adds to the end of a growing polymeric chain.
(*b*) The polymeric product is an ester.
(*c*) Condensation is observed on the walls of the reaction vessel.
(*d*) A small molecule is lost as bonds form between the polymerizing molecules.

Answer. (*d*)

495. Which of the following is a copolymer ?

(*a*) Natural rubber (*b*) Nylon-6,6

(*c*) Orlon (*d*) Teflon

Answer. (*b*)

496. The "n" subscript outside the parentheses in a structural formula representation of a polymer denotes

(*a*) one or two

(*b*) no more than three

(*c*) at least 10,000

(*d*) a large but imprecise number of repeating units

Answer. (*d*)

497. What is used to initiate a free-radical polymerization ?

(*a*) benzoyl peroxide (*b*) benzoic acid

(*c*) styrene (*d*) phthalic acid

Answer. (*a*)

498. What polymer is responsible for most of the wrinkle-free character of modern, permanent press fabrics ?

(*a*) polyester (*b*) polystyrene

(*c*) polyethylene (*d*) polyurethane

Answer. (*a*)

499. In which polymer is $(-CF_2-CF_2-)_n-$ the repeating unit ?

(*a*) Teflon (*b*) Orlon (*c*) Plexiglas (*d*) Lucite

Answer. (*a*)

500. *Latex* is the source of :

(*a*) isoprene (*b*) nylon

(*c*) natural rubber (*d*) collagen

Answer. (*c*)

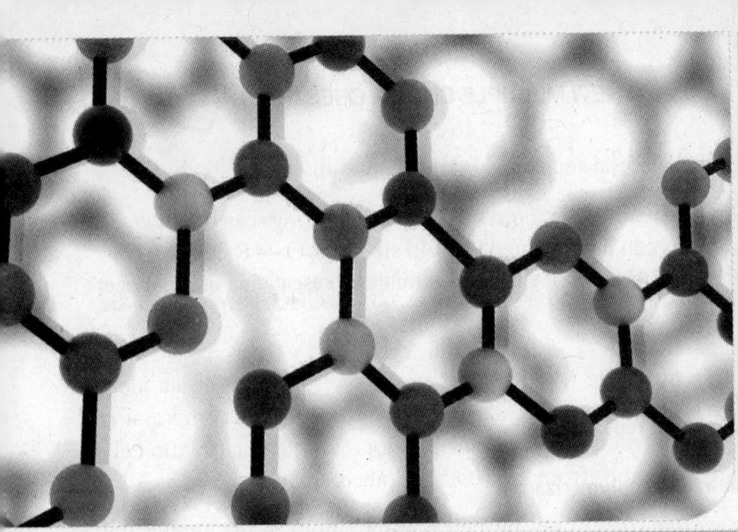

Glossary

Absolute configuration. The actual three-dimensional structure of a chiral molecule. Absolute configurations are specified verbally by the R,S convention and are represented on paper by Fischer projections.

Acetal. A functional group consisting of two ether-type oxygen atoms bound to the same carbon, $R_2C(OR_2)$.

Achiral. Having a lack of handedness. A molecule is achiral if it has a plane of symmetry and is thus superimposable on its mirror image.

Activating group. An electron-donating group such as hydroxyl (–OH) or amino ($-NH_2$) that increases the reactivity of an aromatic ring toward electrophilic aromatic substitution. All activating groups are *ortho/para* directing.

Activation energy. The difference in energy levels between ground state and transition state. The amount of activation energy required by a reaction determines the rate at which the reaction proceeds. The majority of organic reactions have activation energies of 10-25 kcal/mol.

Acylation. The introduction of an acyl group, –COR, onto a molecule. For example, acylation of an alcohol yields an ester ($R'OH \rightarrow R'OCOR$), acylation of an amine yields an amide ($R'NH_2 \rightarrow R'NHCOR$), and acylation of an aromatic ring yields an alkyl aryl ketone ($ArH \rightarrow ArCOR$).

Acylium ion. A resonance-stabilized carbocation in which the positive charge is located at a carbonyl-group carbon, $R{-}\overset{+}{C}{=}O$. Acylium ions are strongly electrophilic and are involved as intermediates in Friedel-Crafts acylation reactions.

1,4-Addition. Addition of a reagent to the ends of a conjugated π system. Conjugated dienes yield 1,4 adducts when treated with electrophiles such as HCl.

Addition polymer. A polymer produced by a chain reaction procedure in which an initiator adds to a carbon-carbon double bond to yield a reactive intermediate. The chain is then built as monomers add successively to the reactive end of the growing chain. Addition polymers are formed by combination of alkene monomers to produce a single huge molecule only.

Aldaric acid. The dicarboxylic acid resulting from oxidation of an aldose.

Aldonic acid. The monocarboxylic acid resulting from mild oxidation of an aldose.

Alicyclic. Referring to an aliphatic cyclic hydrocarbon such as a cycloalkane or cycloalkene.

Aliphatic. Referring to a non-aromatic hydrocarbon such as a simple alkane, alkene, or alkyne.

Alkylation. Introduction of an alkyl group onto a molecule. For example, aromatic rings can be alkylated to yield arenes (ArH $\rightarrow$ ArR), alkoxide anions can be alkylated to yield ethers (R'O $\rightarrow$ R'OR).

Allylic. Used to refer to the position next to a double bond. For example, $CH_2=CHCH_2Br$ is an allylic bromide.

Angle strain. The strain introduced into a molecule when a bond angle is deformed from its ideal value. Angle strain is particularly important in small-ring cycloalkanes where it results from compression of bond angles to less than their ideal tetrahedral values. For example, cyclopropane has approximately 22 kcal/mol angle strain owing to bond deformations from the 109.5° tetrahedral angle to 60°.

Anomers. Cyclic stereoisomers of sugars that differ only in their configurations at the hemiacetal (anomeric) carbon.

Anti conformation. The geometric arrangement around a carbon-carbon single bond, in which the two largest substituents are 180° apart, as viewed in a Newman projection.

Aromaticity. The special characteristics of cyclic conjugated π electron systems that result from their electronic structures. These characteristics include unusual stability and a tendency to undergo substitution reactions rather than addition reactions on treatment with electrophiles. Aromatic molecules must be planar, cyclic, conjugated species that have $4n + 2\pi$ electrons.

Asymmetric center. See chiral center.

Atactic polymers. Addition polymers that have a random stereochemical arrangement of substituents on the polymer backbone. These polymers result from high temperature radical initiated polymerization of alkene monomers.

Aufbau principle. A guide for determining the ground-state electronic configuration of elements by filling the lowest energy orbitals first.

Axial bond. A bond to chair cyclohexane that lies along the ring axis perpendicular to the rough plane of the ring.

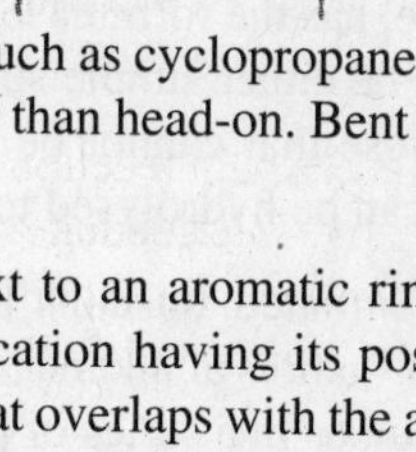

Bent bonds. The bonds in small rings such as cyclopropane that bend away from the internuclear line and overlap at a slight angle, rather than head-on. Bent bonds are highly strained and highly reactive.

Benzylic. Referring to the position next to an aromatic ring. For example, a benzylic cation is a resonance stabilized, conjugated carbocation having its positive charge located on a carbon atom next to the benzene ring in a π orbital that overlaps with the aromatic π system.

Benzyne. An unstable intermediate having a triple bond in a benzene ring.

Bimolecular reaction. A reaction that occurs between two reagents.

Block co-polymer. A polymer consisting of alternating homopolymer blocks. Block co-polymers are usually prepared by initiating addition polymerization of one monomer, followed by addition of an excess of a second monomer.

Boat cyclohexane. A three-dimensional conformation of cyclohexane that bears a slight resemblance to a boat. Boat cyclohexane has no angle strain, but has a large number of eclipsing interactions that make it less stable than chair cyclohexane.

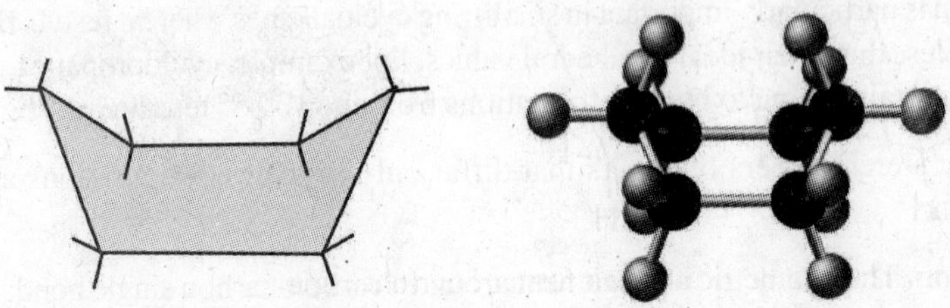

Boat form of cyclohexane

Bond angle. The angle formed between two adjacent bonds.

Bond-dissociation energy. The amount of energy needed to homolytically break a bond to produce two radical fragments.

Bond length. The distance between the nuclei of two atoms that are bonded to each other.

Bond strength. See Bond-dissociation energy.

Bronsted acid. A substance that donates a hydrogen ion (proton) to a base.

Carbanion. A carbon-anion, or substance that contains a trivalent, negatively charged carbon atom ($R_3C{:}^-$). Carbanions are sp^3 hybridized and have eight electrons in the outer shell of the negatively charged carbon.

Carbene. A neutral substance that contains a divalent carbon atom having only six electrons in its outer shell ($R_2C{:}$).

Carbocation. A carbon-cation, or substance that contains a trivalent, positively charged carbon atom having six electrons in its outer shell (R_3C^+). Carbocations are planar and sp^2 hybridized. They are also called carbonium ions.

Carbocycle. A cyclic molecule that has only carbon atoms in ring.

Carbohydrate. A polyhydroxy aldehyde or polyhydroxy ketone. The name derives from the fact that glucose, the most abundant carbohydrate, has the formula $C_6H_{12}O_6$ and was originally thought to be a hydrate of carbon. Carbohydrates can be either simple sugars such as glucose or complex sugars such as cellulose. Simple sugars are those that cannot be hydrolysed to yield smaller molecules, whereas complex sugars are those that can be hydrolysed to yield simpler sugars.

Chain reaction. A reaction that, once initiated, sustains itself in an endlessly repeating cycle of propagation steps. The free radical chlorination of alkanes is an example of a chain reaction that is initiated by UV light and that then continues in a series of propagation steps.

Step 1	:	Initiation	:	$Cl_2 \rightarrow 2\,Cl\cdot$
Steps 2	:	Propagation	:	$Cl\cdot + CH_4 \rightarrow HCl + \cdot CH_3$
and 3				$\cdot CH_3 + Cl_2 \rightarrow CH_3Cl + Cl\cdot$

Step 4 : Termination : R· + R· → R–R

Chair cyclohexane. A three dimensional conformation of cyclohexane that resembles the shape of a chair. The chair form of cyclohexane has no angle strain. It represents the lowest energy conformation of the molecule.

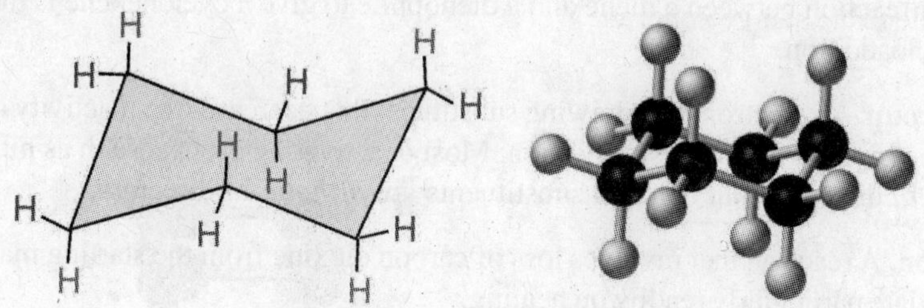

Chair form of cyclohexane

Chiral. Having handedness. Chiral molecules are those that do not have a plane of symmetry and are therefore not superimposable on their mirror image. A chiral molecule thus exists in two forms, one right-handed and one left-handed. The most common (though not the only) cause of chirality in a molecule is the presence of a carbon atom that is bonded to four different substituents.

Chiral center. An atom (usually carbon) that is bonded to four different groups and is therefore chiral.

Chromatography. A technique for separating a mixture of compounds into pure components. Chromatography operates on the principle of differential adsorption whereby different compounds absorb on a stationary support phase and are then carried along at different rates by a mobile phase.

Cis-trans isomers. Special kind of stereoisomers that differ in their stereochemistry about a double bond or on a ring. Cis-trans isomers are also called geometric isomers.

Concerted. Referring to a reaction that takes place in a single step without intermediates. For example, the Diels-Alder cycloaddition reaction is a concerted process.

Condensation polymer. A polymer produced by a series of polar reactions between two difunctional monomers. The polymer normally has the two monomer units in alternating order and usually has other atoms in addition to carbon in the polymer backbone. Nylon-6,6, a polyamide produced by reaction between a diacid and a diamine, is an example of such a polymer.

Configuration. The three-dimensional arrangement of atoms bonded to a chiral center relative to the stereochemistry of other chiral centers in the same molecule.

Conformation. The exact three-dimensional shape of a molecule at any given time, assuming that rotation around single bonds is frozen.

Conjugate base. The anion that results from dissociation of a Bronsted acid.

Conjugation. A series of alternating single and multiple bonds with overlapping *p* orbitals. For example, 1,3-butadiene is a conjugated diene, 3-buten-2-one is a conjugated enone, and benzene is a cyclic conjugated triene.

Constitutional isomers. Isomers that have their atoms connected in a different order. For example, butane and 2-methylpropane are constitutional isomers.

Copolymer. A polymer formed by addition polymerization of a mixture of two or more different monomer units.

Covalent bond. A bond formed by sharing electrons between two nuclei.

Cracking. A process used in petroleum refining in which large alkanes are thermally cracked into smaller fragments.

Cycloaddition. A reaction in which two reactants add together in a single step to yield a cyclic product. The Diels-Alder reaction between a diene and a dienophile to give a cyclohexene is the best-known example of a cycloaddition.

Deactivating group. An electron-withdrawing substituent that decreases the reactivity of an aromatic ring towards electrophilic aromatic substitution. Most deactivating groups, such as nitro, cyano and carbonyl are *meta* directors, but halogen substituents are *ortho/para* directors.

Decarboxylation. A reaction that involves loss of carbon dioxide from the starting material. β-Keto acids decarboxylate particularly readily on heating.

Dehydration. A reaction that involves loss of water from the starting material. Most alcohols can be dehydrated to yield alkenes.

Delocalization. A spreading out of electron density over a conjugated π-electron system. For example, allylic cations and allylic anions are delocalized because their charges are spread out by resonance stabilization over the entire π electron system.

Denaturation. The physical changes that occur in proteins when secondary and tertiary structures are disrupted. Denaturation is usually brought about by heat treatment or by a change in pH and is accompanied by a loss of biological activity.

Dextrorotatory. A word used to describe an optically active substance that rotates the plane of polarization of plane-polarized light in a right-handed (clockwise) direction. The direction of rotation is not, however, related to the absolute configuration of the molecule.

Diastereomer. A term that indicates the relationship between non-mirror-image stereoisomers. Diastereomers are stereoisomers that have the same configuration at one or more chiral centers, but differ at other chiral centers.

Diazotization. The conversion of a primary amine, RNH_2 into diazonium salt, RN_2^+ by treatment with nitrous acid. Aryl diazonium salts are stable, but alkyl diazonium salts are extremely reactive and are rarely isolable.

Dielectric constant. A measure of the ability of a solvent to act as an insulator of electric charge. Solvents that have high dielectric constants are highly polar.

Dienophile. A compound containing a double bond that can take part in the Diels-Alder cycloaddition reaction. The most reactive dienophiles are those that have electron-withdrawing groups such as nitro, cyano, or carbonyl on the double bond.

Dipolar aprotic solvent. A dipolar solvent that cannot function as a hydrogen ion donor.

Dipole moment. A measure of the polarity of a molecule. A dipole moment arises when the centers of gravity of positive and negative charges within a molecule do not coincide.

Dissociation constant. A measure of the extent to which a molecule dissociates into ions. The dissociation constant for the reaction $AB \rightarrow A^+ + B^-$ is given by the expression :

$$K_{diss} = \frac{[A^+][B^-]}{[A][B]}$$

where the quantities in brackets represent the molar concentrations of the reactants and products.

DNA. Deoxyribonucleic acid, the biopolymer consisting of deoxyribonucleotide units linked together through phosphate-sugar bonds. DNA is found in the nucleus of cells and contains an organism's genetic information.

Eclipsed conformation. The arrangement around a carbon-carbon single bond in which the bonds to substituents on one carbon are parallel to the bonds to substituents on the neighboring carbon as viewed in a Newman projection. For example, the eclipsed conformation of ethane has the C–H bonds on one carbon lined up with the C–H bonds on the neighboring carbon.

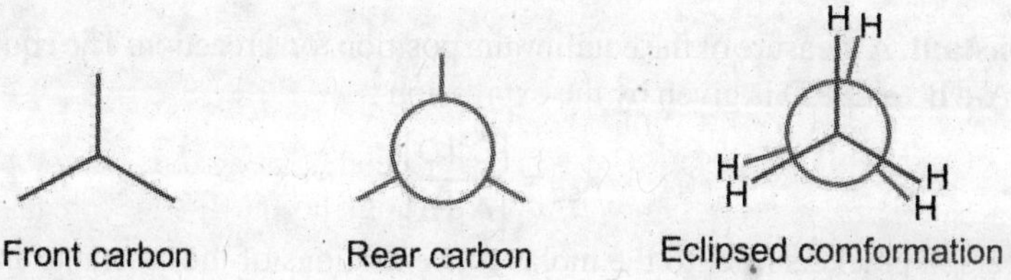

Elastomers. Amorphous polymers that have the ability to stretch out and then return to their previous shape. These polymers have irregular shapes that prevent crystalline formation and have little cross-linking between chains.

Electron affinity. The measure of an atom's tendency to gain an electron and form an anion. Elements on the right side of the periodic table such as the halogens have higher electron affinities than elements on the left side.

Electronegativity. The ability of an atom to attract electrons and thereby polarize a bond. As a general rule, electronegativity increases in going across the periodic table from right to left and in going from bottom to top.

Electrophile. An electron-lover, or substance that accepts an electron pair from a nucleophile in a polar bond forming reaction.

Electrophoresis. A technique used for separating charged organic molecules, particularly proteins and amino acids. The mixture to be separated is placed on a buffered gel or paper and an electric potential is applied across the ends of the apparatus. Negatively charged molecules migrate towards the positive electrode and positively charged molecules migrate towards the negative electrode.

Empirical formula. A formula that gives the relative proportions of elements in a compound in smallest whole numbers.

Enantiomers. Stereoisomers of a chiral substance that have a mirror image relationship. Enantiomers must have opposite configurations at all chiral centers in the molecule.

Endothermic. A term used to describe reactions that absorb energy and that therefore have positive enthalpy changes. In reaction energy diagrams, the products of endothermic reactions have higher energy levels than the starting materials.

Entgegen (E). A term used to describe the stereochemistry of a carbon-carbon double bond. The two groups on each carbon are first assigned priorities according to the Cahn-Ingold-Prelog sequence rules and the two carbons are then compared. If the high priority groups on each carbon are on opposite sides of the double bond, the bond has E geometry.

Enzyme. A biological catalyst. Enzymes are large proteins that catalyze specific biochemical reactions.

Epoxide. A three-membered cyclic ether functional group.

Equatorial bond. A bond to cyclohexane that lies along the rough equator of the ring.

Equational bonds

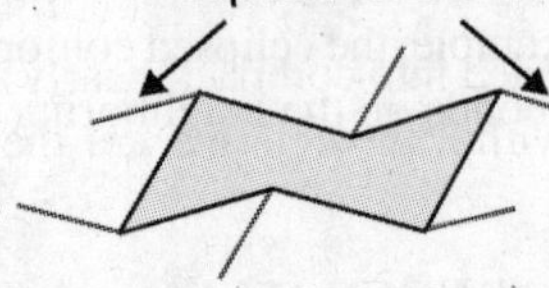

Equilibrium constant. A measure of the equilibrium position for a reaction. The equilibrium constant for the reaction A + B → C + D is given by the expression :

$$K_{eq} = \frac{[C][D]}{[A][B]}$$

where the numbers in brackets refer to the molar concentrations of the reactants and products.

Essential oil. The volatile oil that is obtained by steam distillation of a plant extract.

Excited-state configuration. An electronic configuration having a higher energy level than the ground state.

Exothermic. A term used to describe reactions that release energy and that therefore have negative enthalpy changes. On reaction energy diagrams, the products of exothermic reactions have energy levels lower than those of starting materials.

Fat. A solid triacylglycerol derived from animal sources.

Fibres. Thin threads produced by extruding a molten polymer through small holes in a die.

Fibrous protein. Proteins that consist of polypeptide chains arranged side by side in long threads. These proteins are tough, insoluble in water and are used in nature for structural materials such as hair, hooves, and fingernails.

Fischer projection. A means of depicting the absolute configuration of chiral molecules on a flat page. A Fischer projection employs a cross to represent the chiral center ; the horizontal arms of the cross represent bonds coming out of the plane of the page, whereas the vertical arms of the cross represent bonds going back into the plane of the page.

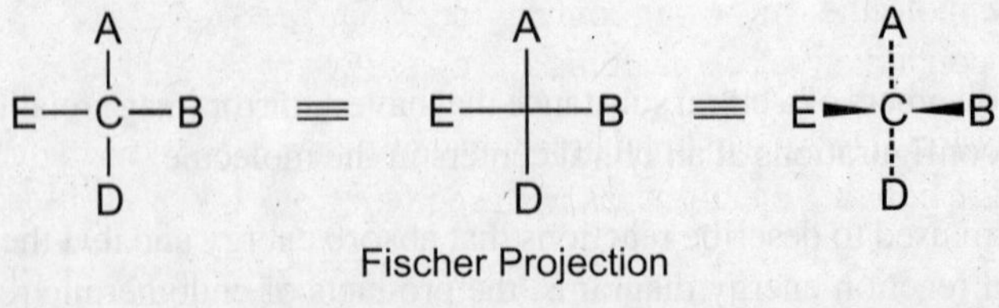

Fischer Projection

Functional group. An atom or group of atoms that is part of a larger molecule and that has a characteristic chemical reactivity. Functional groups display the same chemistry in all molecules of which they are a part.

Gauche conformation. The conformation of butane in which the two methyl groups lie 60° apart, as viewed in a Newman projection. This conformation has 0.9 kcal/mol steric strain.

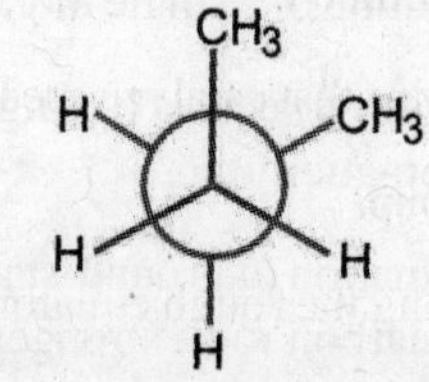

Gauche conformation

Geometric isomers. An old term for *cis-trans* isomers.

Globular protein. Proteins that are coiled into compact, nearly spherical shapes. These proteins are generally water soluble, are mobile within the cell and are the structural class to which enzymes belong.

Glycol. A 1,2-diol such as ethylene glycol, $HOCH_2CH_2OH$.

Glycoside. A cyclic acetal formed by reaction of a sugar with another alcohol.

Ground state. The most stable, lowest energy electronic configuration of a molecule.

Halohydrin. A 1,2-disubstituted haloalcohol such as is obtained on addition of HOBr to an alkene.

Halonium ion. A species containing a positively charged, divalent halogen. Three-membered ring bromonium ions are implicated as intermediates in the electrophilic addition of bromine to alkenes.

Haworth projection. A means of viewing stereochemistry in cyclic hemiacetal forms of sugar. Haworth projections are drawn so that the ring is flat and is viewed from an oblique angle with the hemiacetal oxygen at the upper right.

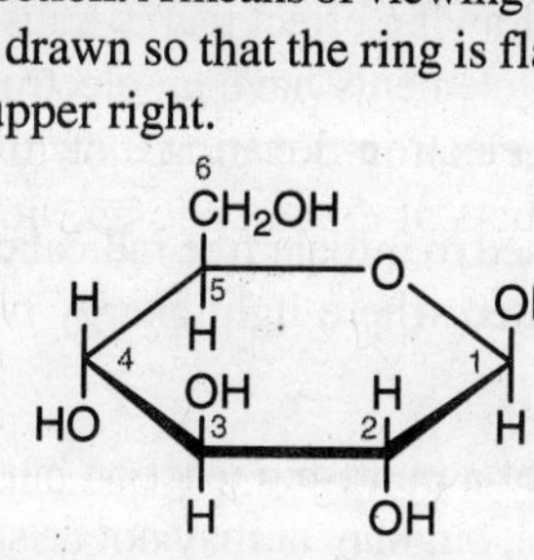

Haworth projection of gulcose

Heat of combustion. The amount of heat released when a compound is burned in calorimeter according to the equation :

$$C_nH_m + O_2 \longrightarrow n\,CO_2 + \frac{m}{2}H_2O$$

Heat of hydrogenation. The amount of heat released when a carbon-carbon double bond is hydrogenated. Comparison of heats of hydrogenation for different alkenes allows one to determine the stability of the different double bonds.

Heterocycle. A cyclic molecule whose ring contains more than one kind of atom. For example, pyridine is a heterocycle that contains five cabon atoms and one nitrogen atom in its ring.

Heterolytic bond fission. The kind of bond breaking that occurs in polar reactions when one fragment leaves with both of the bonding electrons, as in the equation :

$$A:B \longrightarrow A^+ + B^{:-}$$

Homolytic bond fission. The kind of bond breaking that occurs in free radical reactions when each fragment leaves with one bonding electron according to the equation :

$$A:B \longrightarrow A\cdot + B\cdot$$

Homopolymer. A polymer made by addition polymerization of a single monomer unit.

Huckel's rule. A rule stating that monocyclic conjugated molecules having $4n + 2\pi$ electrons (n = an integer) show the unusual stability associated with aromaticity.

Hybrid orbital. An orbital that is derived from a combination of ground-state (s, p, d) atomic orbitals. Hybrid orbitals, such as the sp^3, sp^2 and sp hybrids of carbon form stronger bonds than ground-state atomic orbitals.

Hydration. Addition of water to a molecule, such as occurs when alkenes are treated with strong sulfuric acid.

Hydroboration. Addition of borane (BH_3) or an alkyl borane to an alkene. The resultant trialkylborane products are useful synthetic intermediates that can be oxidized to yield alcohols.

Hydrogenation. Addition of hydrogen to a double or triple bond to yield the saturated product.

Hydrogen bond. A weak (5 kcal/mol) attraction between a hydrogen atom bonded to an electronegative element and an electron lone pair on another atom.

Hyperconjugation. A weak stabilizing interaction that results from overlap of a p orbital with a neighboring σ bond. Hyperconjugation is important in stabilizing carbocations and in stabilizing substituted alkenes.

Inductive effect. The electron attracting or electron withdrawing effect that is transmitted through σ bonds as the result of a nearby dipole. Electronegative elements have an electron-withdrawing inductive effect, whereas electropositive elements have an electron-donating inductive effect.

Initiator. A substance with an easily broken bond that is used to initiate free radical chain reactions. For example, free radical chlorination of alkanes is initiated where light energy breaks the weak chlorine-chlorine bond to form chlorine free radicals.

Intermediate. A species that is formed during the course of a multistep reaction but is not the final product. Intermediates are more stable than transition states, but may or may not be stable enough to be isolated.

Intramolecular, Intermolecular. Reactions that occur within the same molecule are intramolecular, whereas reactions that occur between two molecules are intermolecular.

Ionic bond. A bond between two ions due to the electrical attraction of unlike charges. Ionic bonds are formed between strongly electronegative elements (such as the halogens) and strongly electropositive elements (such as the alkali metals).

Ionization energy. The amount of energy required to remove an electron from an atom. Elements on the far right of the periodic table have high ionization energies, and elements on the far left of the periodic table have low ionization energies.

Isoelectric point. The pH at which the number of positive charges and the number of negative charges on a protein or amino acid are exactly balanced.

Isomers. Compounds that have the same molecular formula but have different structures.

Isoprene rule. An observation to the effect that terpenes appear to be made up of isoprene (2-methyl-1,3-butadiene) units connected in a head-to-tail fashion. Monoterpenes have two isoprene units, sesquiterpenes have three isoprene units, diterpenes have four isoprene units and so on.

Isotactic polymer. An addition polymer in which all substituents on the polymer backbone have the same three- dimensional orientation.

Kinetics. Referring to rates of reactions. Kinetics measurements can be extremely important in helping to determine reaction mechanisms.

Leaving group. The group that is replaced in a substitution reaction. The best leaving groups in nucleophilic substitution reactions are those that form the most stable, least basic, anions.

Levorotatory. Used to describe an optically active substance that rotates the plane of polarization of plane-polarized light in a left-handed (counterclockwise) direction.

Lewis acid. A substance having a vacant low-energy orbital that can accept an electron pair from a base. All electrophiles are Lewis acids, but transition metal salts such as $AlCl_3$, and $ZnCl_2$ are particularly good ones.

Lewis base. A substance that donates an electron lone pair to an acid. All nucleophiles are Lewis bases.

Lewis structure. A representation of a molecule showing covalent bonds as a pair of electron dots between atoms.

Lipid. A naturally occurring substance isolated from cells and tissues by extraction with nonpolar solvents. Lipids belong to many different structural classes, including fats, terpenes, prostaglandins, and steroids.

Lipophilic. Fat-loving. Long non-polar hydrocarbon chains tend to cluster together in polar solvents because of their lipophilic properties.

Lone-pair electrons. Non-bonding electron pairs that occupy valence orbitals. It is the lone-pair electrons that are used by nucleophiles in their reactions with electrophiles.

Markovnikov's rule. A guide for determining the orientation of electrophilic addition reactions. In the addition of HX to an alkene, the hydrogen atom becomes bonded to the alkene carbon that has fewer alkyl substituents. A modern statement of this same rule is that electrophilic addition reactions proceed *via* the most stable carbocation intermediate.

Mechanism. A complete description of how a reaction occurs. A mechanism must account for all starting materials and all products and must describe the details of each individual step in the overall reaction process.

Meso. A *meso* compound is one that contains chiral centers but is nevertheless achiral by virtue of a symmetry plane.

Micelle. A spherical cluster of soap-like molecules that aggregate in aqueous solution. The ionic *heads* of the molecules lie on the outside where they are solvated by water and the organic *tails* bunch together on the inside of the micelle.

Mobile phase. The solvent (either gas or liquid) used in chromatography to move material along the solid adsorbent phase.

Molecular formula. An expression of the total numbers of each kind of atom present in a molecule. The molecular formula must be a whole-number multiple of the empirical formula.

Molecular orbital. An orbital that is the property of the entire molecule rather than of an individual atom. Molecular orbitals result from overlap of two or more atomic orbitals when bonds are formed.

Monomer. The simple starting units from which polymers are made.

Mutarotation. The spontaneous change in optical rotation observed when a pure anomer of a sugar is dissolved in water. Mutarotation is caused by the reversible opening and closing of the acetal linkage, which yields an equilibrium mixture of anomers.

Newman projection. A means of indicating stereochemical relationships between substituent groups on neighboring carbons. The carbon-carbon bond is viewed end-on and the carbons are indicated by a circle. Bonds radiating from the center of the circle are attached to the front carbon, and bonds radiating from the edge of the circle are attached to the rear carbon.

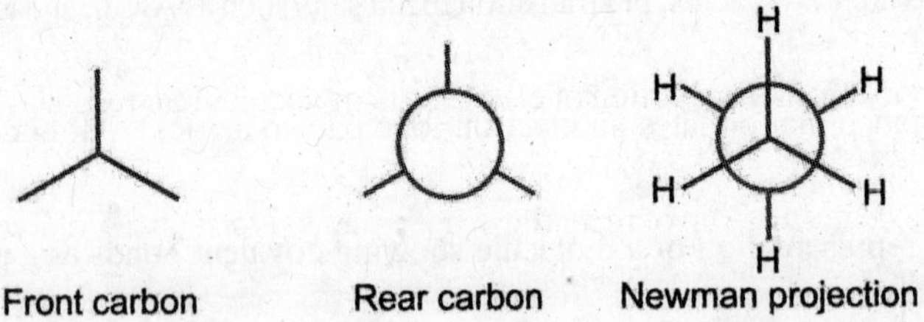

Normal alkane. A straight-chain alkane, as opposed to a branched alkane. Normal alkanes are denoted by prefix *n*, as in *n*-C_4H_{10} (*n*-butane).

Nucleophile. A nucleus-lover, or species that donates an electron pair to an electrophile in a polar bond-forming reaction. Nucleophiles are also Lewis bases.

Olefin. An alternative name for an alkene.

Optical isomers. Enantiomers. Optical isomers are isomers that have a mirror image relationship.

Optically active. A substance that rotates the plane of polarization of plane-polarized light. Note that an optically active sample must contain chiral molecules, but that all samples with chiral molecules are not optically active. Thus, a racemic sample is optically inactive even though the individual molecules are chiral.

Orbital. The volume of space in which an electron is most likely to be found.

Ozonide. The product formed by addition of ozone to a carbon-carbon double bond. Ozonides are usually treated with a reducing agent, such as zinc in water, to produce carbonyl compounds.

Paraffins. A trivial name for alkanes.

Pauli exclusion principle. A statement of the fact that not more than two electrons can occupy the same orbital, and that these two must have spins of opposite sign.

Peptides. Amino acid polymers in which the individual amino acid residues are linked by amide bonds.

Peroxide. A molecule containing an oxygen-oxygen bond functional group, R-O-O-R' or R-O-O-H. The peroxides present as explosive impurities in ether solvents are usually of the latter type. Since the oxygen-oxygen bond is weak and easily broken, peroxides are often used to initiate free radical chain reactions.

Phospholipids. Lipids that contain a phosphate residue. For example, phosphoglycerides contain a glycerol backbone linked to two fatty acids and a phosphoric acid.

Pi bond. The covalent bond formed by sideways overlap of atomic orbitals. For example, carbon-carbon double bonds contain a π bond formed by sideways overlap of two p orbitals.

Plane of symmetry. An imaginary plane that bisects a molecule such that one half of the molecule is the mirror image of the other half. Molecules containing a plane of symmetry are *achiral*.

Plane-polarized light. Ordinary light that has its electric vectors in a single plane rather than in random planes. The plane of polarization is rotated when the light is passed through a solution of a *chiral* substance.

Polar reaction. A reaction in which bonds are made when a nucleophile donates two electrons to an electrophile and in which bonds are broken when one fragment leaves with both electrons from the bond.

Polarity. The unsymmetrical distribution of electrons in molecules that results when one atom attracts electrons more strongly than another.

Polymer. A large molecule made up of repeating smaller units. For example, polyethylene is a synthetic polymer made from repeating ethylene units.

Primary, secondary, tertiary, quaternary. Terms used to describe the substitution pattern at a specific site. A primary site has one organic substituent attached to it, a secondary site has two organic substituents, a tertiary site has three and a quaternary site has four.

	Carbon	Hydrogen	Alcohol	Amine
Primary	RCH_3	RCH_3	RCH_2OH	RNH_2
Secondary	R_2CH_2	R_2CH_2	R_2CHOH	R_2NH
Tertiary	R_3CH	R_3CH	R_3COH	R_3N
Quaternary	R_4C			

Primary structure. The amino acid sequence in a protein.

Principle of maximum overlap. The strongest bonds are formed when overlap between orbitals is greatest.

Propagation step. The step or series of steps in a free radical chain reaction that carry on the chain. The propagation steps must yield both product and a reactive intermediate to carry on the chain.

Protecting group. A group that is introduced to protect a sensitive functional group towards reaction elsewhere in the molecule. After serving its protective function, the group is then removed. For example, ketones and aldehydes are often protected as acetals by reaction with ethylene glycol.

Protein. A large peptide containing fifty or more amino acid residues. Proteins serve both as structural materials (hairs, horns, fingernails) and as enzymes that control an organism's chemistry.

Protic solvent. A solvent such as water or alcohol that can serve as a proton donor. Protic solvents are particularly good at stabilizing anions by hydrogen bonding, thereby lowering their reactivity.

Quaternary. (see primary)

Quaternary structure. The highest level of protein structure, involving a specific aggregation of individual proteins into a large cluster.

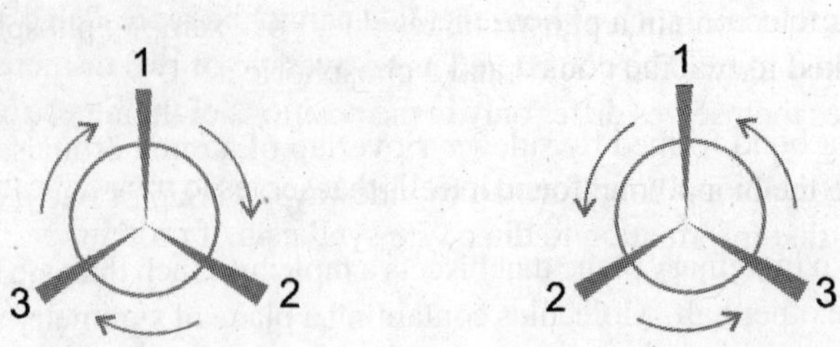

R configuration S configuration

R, S convention. A method for defining the absolute configuration around chiral centers. The Cahn-Ingold-Prelog sequence rules are used to assign relative priorities to the four substituents on the chiral center and the center is oriented such that the group of lowest (fourth priority) faces directly away from the viewer. If the three remaining substituents have a right-handed or clockwise relationship in going from first to second to third priority, then the chiral center is denoted R (*rectus,* right). If the three remaining substituents have a left-handed or counterclockwise relationship, the chiral center is denoted S (*sinister,* left).

Racemic mixture. A mixture consisting of equal parts (+) and (-) enantiomers of a chiral substance. Even though the individual molecules are chiral, racemic mixtures are optically inactive.

Racemization. The process whereby one enantiomer of a chiral molecule becomes converted into a 50:50 mixture of enantiomers, thus losing its optical activity. For example, this might happen during an S_N1 reaction of a chiral alkyl halide.

Radical. When used in organic nomenclature, the word radical refers to a part of a molecule that appears in its name for example, the phenyl in phenyl acetate. Chemically, however, a radical is a species that has an odd number of electrons, such as the chlorine free radical (Cl·).

Radical reaction. A reaction in which bonds are made by donation of one electron from each of two reagents and in which bonds are broken when each fragment leaves with one electron.

Reaction energy diagram. A pictorial representation of the course of a reaction, in which potential energy is plotted as a function of reaction progress: Starting materials, transition states, intermediates and final products are all represented and their appropriate energy levels are indicated.

Reducing sugar. Any sugar that reduces silver ion in the Tollens' test or cupric ion in the Fehling's or Benedict's tests. All sugars that are aldehydes or that can be readily converted into aldehydes are reducing. Glycosides, however, are not reducing sugars.

Refining. The process by which petroleum is converted into gasoline and other useful products.

Resolution. The process by which a racemic mixture is separated into its two pure enantiomers. For example, a racemic carboxylic acid might be converted by reaction with a chiral amine base into a diastereomeric mixture of salts, which could be separated by fractional crystallization. Preparation of the free acids would then yield the two pure enantiomeric acids.

Resonance effect. The effect by which substituents donate or withdraw electron density through orbitals overlap with neighboring π bonds. For example, an oxygen or nitrogen substituent donates electron density to an aromatic ring by overlap of the O or N orbital with the aromatic ring by *p*-orbitals. A carbonyl substituent, however, withdraws electron density from an aromatic ring by *p*-orbital overlap. These effects are particularly important in determining whether a given group is *meta* directing or *ortho*, *para*-directing in electrophilic aromatic substitution reactions.

Resonance hybrid. A molecule, such as benzene, that cannot be represented adequately by a single Kekule structure but must instead be considered as an average of two or more resonance structures The resonance structures themselves differ only in the positions of their electrons, not of their nuclei.

RNA. Ribonucleic acid, the biopolymer found in cells that serves to transcribe the genetic information found in DNA and use that information to direct the synthesis of proteins.

Saccharide. A sugar.

Saponification. An old term for the base-induced hydrolysis of an ester to yield a carboxylic acid salt.

Saturated. A saturated molecule is one that has only single bonds and thus cannot undergo addition reactions. Alkanes, for example, are saturated, but alkenes are unsaturated.

Second-order reaction. A reaction whose rate-limiting step is bimolecular and whose kinetics are therefore dependent on the concentration of two reagents.

Secondary. (See primary)

Secondary structure. The level of protein substructure that involves organization of chain sections into ordered arrangements such as β-pleated sheets or α-helices.

Sequence rules. A series or rules devised by Cahn, Ingold and Prelog for assigning relative priorities to substituent groups on a double-bond carbon atom or on a chiral center. Once priorities have been established, E, Z-double bond geometry and R,S-configurational assignments can be made.

Sigma bond. A covalent bond formed by head-on overlap of atomic orbitals.

Skew conformation. Any conformation about a single bond that is intermediate between staggered and eclipsed. (See staggered conformation ; eclipsed conformation)

Soap. The mixture of long-chain fatty acid salts obtained on basic hydrolysis of animal fat.

sp Orbital. A hybrid orbital derived from the combination of an *s* and *p* atomic orbital. The two *sp* orbitals that result from hybridization are oriented at an angle of 180° to each other.

sp^2 Orbital. A hybrid orbital derived by combination of an *s* atomic orbital with two *p* atomic orbitals. The three sp^2 hybrid orbitals that result lie in a plane at angles of 120° to each other.

sp^3 Orbital. A hybrid orbital derived by combination of an *s* atomic orbital with three *p* atomic orbitals. The four sp^3 hybrid orbitals that result are directed towards the corners of a tetrahedron at angles of 109.5° to each other.

Staggered conformation. The three-dimensional arrangement of atoms around a carbon-carbon single bond in which the bonds on one carbon exactly bisect the bond angles on the second carbon as viewed end-on.

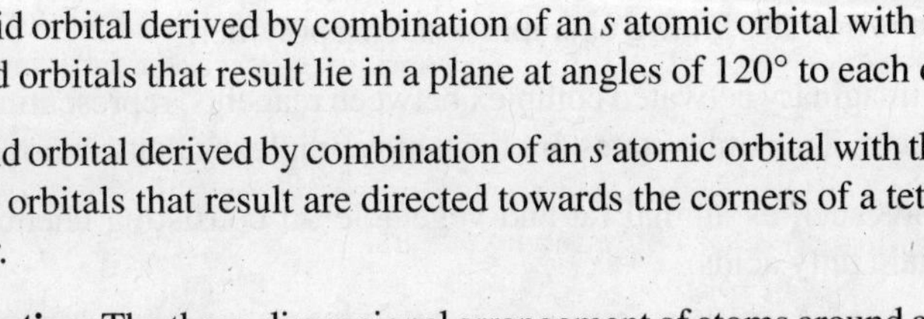

Front carbon   Rear carbon   Staggered conformation

Stationary phase. The solid support used in chromatography. The molecules to be chromatographically separated adsorb on the stationary phase and are moved along by the mobile phase. Silica gel (hydrated SiO_2) and alumina (Al_2O_3) are often used as stationary phases in column chromatography of organic mixtures.

Stereochemistry. The branch of chemistry concerned with the three-dimensional arrangement of atoms in molecules.

Stereoisomers. Isomers that have their atoms connected in the same order but that have different three-dimensional arrangements. The term stereoisomer includes both enantiomers and diastereomers but does not include constitutional isomers.

Stereospecific. A term indicating that only a single stereoisomer is produced in a given reaction, rather than a mixture.

Steric strain. The strain imposed on a molecule when two groups are too close together and try to occupy the same space. Steric strain is responsible both for the greater stability of *trans* versus *cis* alkenes and for the greater stability of *equatorially* substituted versus *axially* substituted cyclohexanes.

Syn stereochemistry. A syn addition reaction is one in which the two ends of the double bond are attacked from the same side.

Syndiotactic polymer. An addition polymer in which the substituents on the polymer backbone have a regular alternating stereochemistry.

Terpenes. Lipids that are formally derived by head-to-tail polymerization of isoprene units.

Tertiary. (see primary)

Tertiary structure. The level of protein structure that involves the manner in which the entire protein chain is folded into a specific three-dimensional arrangement.

Thermoplastic. A polymer that is hard at room temperature but that becomes soft and pliable when heated. Thermoplastics are used for the manufacture of a variety of moulded objects.

Thermosetting resin. A polymer that is highly cross-linked and that sets into a hard, insoluble mass when heated. Bakelite is the best-known example of such a polymer.

Transition state. An imaginary activated complex between reagents, representing the highest energy point on a reaction curve. Transition states are unstable complexes that cannot be isolated.

Triacylglycerol. Lipids such as animal fat and vegetable oil consisting chemically of triesters of glycerol with long-chain fatty acids.

Unimolecular reaction. A reaction that occurs by spontaneous transformation of the starting material without the intervention of other reagents. For example, the dissociation of a tertiary alkyl halide in the SN1 reaction is a unimolecular process.

Unsaturated. An unsaturated molecule is one that has multiple bonds and can undergo addition reactions. Alkenes and alkynes, for example, are unsaturated.

Van der Waals forces. The attractive forces between molecules that are caused by dipole-dipole interactions. Van der Waals forces are one of primary forces responsible for holding molecules together in the liquid state.

Vicinal. A term that refers to a 1,2-disubstitution pattern. For example, 1,2-dibromoethane is a vicinal dibromide.

Vinylic. A term that refers to a substituent at a double-bond carbon atom. For example, chloroethylene is a vinylic chloride and enols are vinylic alcohols.

Vulcanization. A process for hardening rubber by heating in presence of elemental sulfur. The sulfur functions by forming cross links between polymer chains.

Ylides. A neutral dipolar molecule in which the positive and negative charges are adjacent. For example, the phosphoranes used in Wittig reactions are ylides.

Zusammen (Z). A term used to describe the stereochemistry of a carbon-carbon double bond. The two groups on each carbon are assigned priorities according to the Cahn-Ingold-Prelog sequence rules and the two carbons are compared. If the high priority groups on each carbon are on the same side of the double bond, the bond has Z geometry.

Zwitterion. A neutral dipolar molecule in which the positive and negative charges are not adjacent. For example, amino acids exist as zwitterions. (Zwitterions are also called betains).

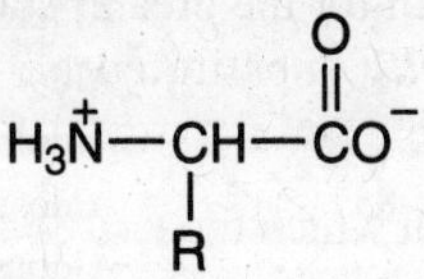

An amino acid zwitterion

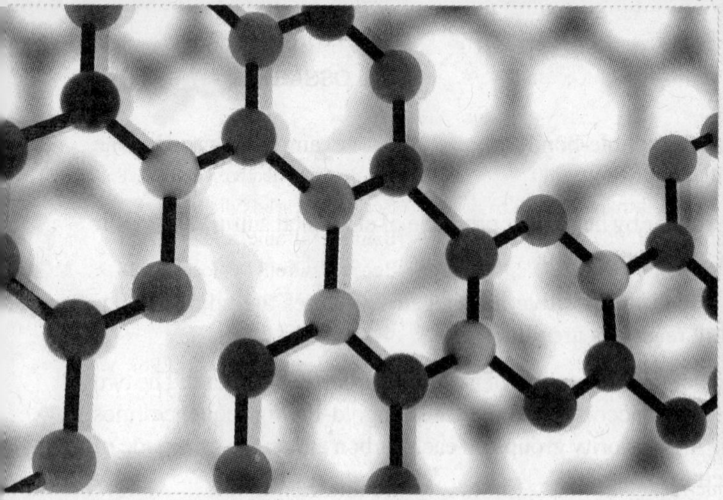

Index

B

D

N

Q

R

S

T

U

V

W

X

Z